Sullivan's Guide to Putting It Together

Putting It Together Sections	Objective	Page(s)
5.6 Putting It Together: Which Method Do I Use?	① Determine the appropriate probability rule to use ② Determine the appropriate counting technique to use	302–303 303–305
9.5 Putting It Together: Which Method Do I Use?	① Determine the appropriate confidence interval to construct	455–456
10.6 Putting It Together: Which Method Do I Use?	① Determine the appropriate hypothesis test to perform (one sample)	510
11.5 Putting It Together: Which Method Do I Use?	① Determine the appropriate hypothesis test to perform (two samples)	567–568

Putting It Together Exercises	Skills Utilized	Section(s) Covered	Page(s)
1.2.24 Passive Smoke	Variables, observational studies, designed experiments	1.1, 1.2	21
1.4.37 Comparing Sampling Methods	Simple random sampling and other sampling techniques	1.3, 1.4	36
1.4.38 Thinking about Randomness	Random sampling	1.3, 1.4	36
2.1.29 Online Homework	Variables, designed experiments, bar graphs	1.1, 1.2, 1.6, 2.1	75
2.2.47 Time Viewing a Webpage	Graphing data	2.2	93
2.2.48 Which Graphical Summary?	Choosing the best graphical summary	2.1, 2.2	93
2.3.19 Rates of Return on Stocks	Relative frequency distributions, relative frequency histograms, relative frequency polygons, ogives	2.2, 2.3	100
2.3.20 Shark!	Graphing data	2.3	100
3.1.41 Shape, Mean, and Median	Discrete vs. continuous data, histograms, shape of a distribution, mean, median, mode, bias	1.1, 1.4, 2.2, 3.1	130
3.5.17 Earthquakes	Mean, median, range, standard deviation, relative frequency histogram, boxplots, outliers	2.2, 3.1, 3.2, 3.4, 3.5	171
3.5.18 Paternal Smoking	Observational studies, designed experiments, lurking variables, mean, median, standard deviation, quartiles, boxplots	1.2, 1.6, 3.1, 3.2, 3.4, 3.5	171–172
4.2.29 Housing Prices	Scatter diagrams, correlation, linear regression	4.1, 4.2	209
4.2.30 Smoking and Birth Weight	Observational study vs. designed experiment, prospective studies, scatter diagrams, linear regression, correlation vs. causation, lurking variables	1.2, 4.1, 4.2	209–210
4.3.31 A Tornado Model	Explanatory and response variables, scatter diagrams, correlation, least-square regression, interpret slope, coefficient of determination, residual plots, residual analysis	4.1, 4.2, 4.3	224
4.3.32 Exam Scores	Building a linear model	4.1, 4.2, 4.3	224
5.1.54 Drug Side Effects	Variables, graphical summaries of data, experiments, probability	1.1, 1.6, 2.1, 5.1	261
5.2.44 Speeding Tickets	Contingency tables, marginal distributions, empirical probabilities	4.4, 5.1	272
5.2.45 Red Light Cameras	Variables, relative frequency distributions, bar graphs, mean, standard deviation, probability, Simpson's Paradox	1.1, 2.1, 3.1, 3.2, 4.4, 5.1, 5.2	272–273
6.1.35 Sullivan Statistics Survey I	Mean, standard deviation, probability, probability distributions	3.1, 3.2, 5.1, 6.1	327
6.2.55 Beating the Stock Market	Expected value, binomial probabilities	6.1, 6.2	342
7.2.52 Birth Weights	Relative frequency distribution, histograms, mean and standard deviation from grouped data, normal probabilities	2.1, 2.2, 3.3, 7.2	377
7.3.13 Demon Roller Coaster	Histograms, distribution shape, normal probability plots	2.2, 7.3	382
8.1.33 Playing Roulette	Probability distributions, mean and standard deviation of a random variable, sampling distributions	6.1, 8.1	406–407
9.1.47 Hand Washing	Observational studies, bias, confidence intervals	1.2, 1.5, 9.1	434
9.2.49 Smoking Cessation Study	Experimental design, confidence intervals	1.6, 9.1, 9.2	448
10.2.38 Lupus	Observational studies, retrospective vs. prospective studies, bar graphs, confidence intervals, hypothesis testing	1.2, 2.1, 9.1, 10.2	493
10.2.39 Naughty or Nice?	Experimental design, determining null and alternative hypotheses, binomial probabilities, interpreting P-values	1.6, 6.2, 10.1, 10.2	493

(continued)

Putting It Together Exercises	Skills Utilized	Section(s) Covered	Page(s)
11.1.36 Salk Vaccine	Completely randomized design, hypothesis testing	1.6, 11.1	536
11.2.18 Glide Testing	Matched pairs design, hypothesis testing	1.6, 11.2	546
11.3.23 Online Homework	Completely randomized design, confounding, hypothesis testing	1.6, 11.3	557–558
12.1.27 The V-2 Rocket in London	Mean of discrete data, expected value, Poisson probability distribution, goodness-of-fit	6.1, 6.3, 12.1	591
12.1.28 Weldon's Dice	Addition Rule for Disjoint Events, classical probability, goodness-of-Fit	5.1, 5.2, 12.1	591
12.2.21 Women, Aspirin, and Heart Attacks	Population, sample, variables, observational study vs. designed experiment, experimental design, compare two proportions, chi-square test of homogeneity	1.1, 1.2, 1.6, 11.1, 12.2	606
13.1.27 Psychological Profiles	Standard deviation, sampling methods, two-sample t-test, Central Limit Theorem, one-way Analysis of Variance	1.4, 3.2, 8.1, 11.2, 13.1	634
13.2.17 Time to Complete a Degree	Observational studies; sample mean, sample standard deviation, confidence intervals for a mean, one-way Analysis of Variance, Tukey's test	1.2, 3.1, 3.2, 9.2, 13.1, 13.2	643
13.4.22 Students at Ease	Population, designed experiments versus observational studies, sample means, sample standard deviation, two sample t-tests, one-way ANOVA, interaction effects, non-sampling error	1.1, 1.2, 3.1, 3.2, 11.3, 13.1, 13.4	665–666
14.6.8 Purchasing Diamonds	Level of measurement, correlation matrix, multiple regression, confidence and prediction intervals	1.1, 14.3, 14.4, 14.6	735

Updated for this edition is the Student Activity Workbook. The Activity Workbook includes many tactile activities for the classroom. In addition, the workbook includes activities based on statistical applets. Below is a list of the applet activities.

Applet	Section	Activity Name
Mean versus Median	3.1	Understanding Measures of Center
Standard Deviation	3.2	Exploring Standard Deviation
Correlation by Eye	4.1	Exploring Properties of the Linear Correlation Coefficient
Regression by Eye	4.2	Minimizing the Sum of the Squared Residuals
Regression Influence	4.3	Understanding Influential Observations
Rolling a Single Die	5.1	Demonstrating the Law of Large Numbers
Binomial Distribution	6.2	Exploring a Binomial Distribution from Multiple Perspectives
Baseball Applet	6.2	Using Binomial Probabilities in Baseball
Sampling Distributions	8.1	Sampling from Normal and Non-normal Populations
Sampling Distributions Binary	8.2	Describing the Distribution of the Sample Proportion
Confidence Intervals for a Proportion	9.1	Exploring the Effects of Confidence Level, Sample Size, and Shape I
Confidence Intervals for a Mean	9.2	Exploring the Effects of Confidence Level, Sample Size, and Shape II
Political Poll Applet	10.2	The Logic of Hypothesis Testing
Hypothesis Tests for a Proportion	10.2	Understanding Type I Error Rates
Cola Applet	10.2	Testing Cola Preferences
Hypothesis Tests for a Mean	10.3	Understanding Type I Error Rates
Randomization Test Warts	11.1	Making an Inference about Two Proportions
Randomization Test Basketball	11.2	Predicting Basketball Game Outcomes
Randomization Test Sentence	11.2	Considering the Effects of Grammar
Randomization Test Kiss	11.2	Analyzing Kiss Data
Randomization Test Algebra	11.3	Using Randomization Test for Independent Means
Randomization Test Market	11.3	Comparing Bull and Bear Markets
Randomization Test Zillow	14.1	Using a Randomization Test for Correlation
Randomization Test Brain Size	14.1	Using a Randomization Test for Correlation

STATISTICS

INFORMED DECISIONS USING DATA 5e

Michael Sullivan, III

Joliet Junior College

PEARSON

Boston   Columbus   Hoboken   Indianapolis   New York   San Francisco
Amsterdam   Cape Town   Dubai   London   Madrid   Milan   Munich   New York   Paris   Montréal   Toronto
Delhi   Mexico City   São Paulo   Sydney   Hong Kong   Seoul   Singapore   Taipei   Tokyo

Editorial Director: Chris Hoag
Editor in Chief: Deirdre Lynch
Acquisitions Editor: Patrick Barbera
Editorial Assistant: Justin Billing
Program Team Lead: Karen Wernholm
Program Manager: Danielle Simbajon
Project Team Lead: Peter Silvia
Project Manager: Tamela Ambush
Senior Media Producer: Vicki Dreyfus
Media Producer: Jean Choe
QA Manager, Assessment Content: Marty Wright
Senior Content Developer: John Flanagan
MathXL Senior Project Manager: Bob Carroll
Field Marketing Manager: Andrew Noble
Product Marketing Manager: Tiffany Bitzel
Marketing Assistant: Jennifer Myers
Senior Technical Art Specialist: Joe Vetere
Manager Rights and Permissions: Gina M. Cheselka
Procurement Specialist: Carol Melville
Associate Director Art/Design: Andrea Nix
Senior Design Specialist: Heather Scott
Composition: Lumina Datamatics, Inc.
Cover Design: Tamara Newnam
Cover Image: Lucky Dragon USA/Fotolia

Library of Congress Cataloging-in-Publication Data

Sullivan, Michael, III, 1967—
   Statistics : informed decisions using data / Michael Sullivan, III, Joliet Junior College.
—5th edition.
      pages cm
   Includes index.
   ISBN 0-13-413353-6 (hardcover) — ISBN 0-13-413542-3 (instructor hardcover)
   1. Statistics—Textbooks. I. Title.
   QA276.12.S85 2017
519.5—dc23

                                                2015011663

MICROSOFT AND/OR ITS RESPECTIVE SUPPLIERS MAKE NO REPRESENTATIONS ABOUT THE SUITABILITY OF THE INFORMATION CONTAINED IN THE DOCUMENTS AND RELATED GRAPHICS PUBLISHED AS PART OF THE SERVICES FOR ANY PURPOSE. ALL SUCH DOCUMENTS AND RELATED GRAPHICS ARE PROVIDED "AS IS" WITHOUT WARRANTY OF ANY KIND. MICROSOFT AND/OR ITS RESPECTIVE SUPPLIERS HEREBY DISCLAIM ALL WARRANTIES AND CONDITIONS WITH REGARD TO THIS INFORMATION, INCLUDING ALL WARRANTIES AND CONDITIONS OF MERCHANTABILITY, WHETHER EXPRESS, IMPLIED OR STATUTORY, FITNESS FOR A PARTICULAR PURPOSE, TITLE AND NON-INFRINGEMENT. IN NO EVENT SHALL MICROSOFT AND/OR ITS RESPECTIVE SUPPLIERS BE LIABLE FOR ANY SPECIAL, INDIRECT OR CONSEQUENTIAL DAMAGES OR ANY DAMAGES WHATSOEVER RESULTING FROM LOSS OF USE, DATA OR PROFITS, WHETHER IN AN ACTION OF CONTRACT, NEGLIGENCE OR OTHER TORTIOUS ACTION, ARISING OUT OF OR IN CONNECTION WITH THE USE OR PERFORMANCE OF INFORMATION AVAILABLE FROM THE SERVICES.

THE DOCUMENTS AND RELATED GRAPHICS CONTAINED HEREIN COULD INCLUDE TECHNICAL INACCURACIES OR TYPOGRAPHICAL ERRORS. CHANGES ARE PERIODICALLY ADDED TO THE INFORMATION HEREIN. MICROSOFT AND/OR ITS RESPECTIVE SUPPLIERS MAY MAKE IMPROVEMENTS AND/OR CHANGES IN THE PRODUCT(S) AND/OR THE PROGRAM(S) DESCRIBED HEREIN AT ANY TIME. PARTIAL SCREEN SHOTS MAY BE VIEWED IN FULL WITHIN THE SOFTWARE VERSION SPECIFIED.

MICROSOFT® WINDOWS®, AND MICROSOFT OFFICE® ARE REGISTERED TRADEMARKS OF THE MICROSOFT CORPORATION IN THE U.S.A. AND OTHER COUNTRIES. THIS BOOK IS NOT SPONSORED OR ENDORSED BY OR AFFILIATED WITH THE MICROSOFT CORPORATION.

Copyright © 2017, 2013, 2010 by Pearson Education, Inc. or its affiliates. All Rights Reserved. Printed in the United States of America. This publication is protected by copyright, and permission should be obtained from the publisher prior to any prohibited reproduction, storage in a retrieval system, or transmission in any form or by any means, electronic, mechanical, photocopying, recording, or otherwise. For information regarding permissions, request forms and the appropriate contacts within the Pearson Education Global Rights & Permissions Department, please visit www.pearsoned.com/permissions/.

Acknowledgements of third party content appear on page PC-1, which constitutes an extension of this copyright page.

PEARSON, ALWAYS LEARNING, MYSTATLAB are exclusive trademarks owned by Pearson Education, Inc. or its affiliates in the United States and/or other countries.

Unless otherwise indicated herein, any third-party trademarks that may appear in this work are the property of their respective owners and any references to third-party trademarks, logos or other trade dress are for demonstrative or descriptive purposes only. Such references are not intended to imply any sponsorship, endorsement, authorization, or promotion of Pearson's products by the owners of such marks, or any relationship between the owner and Pearson Education, Inc. or its affiliates, authors, licensees or distributors.

1 2 3 4 5 6 7 8 9 10—RRD—19 18 17 16 15

www.pearsonhighered.com

ISBN 10:   0-13-413353-6
ISBN 13: 978-0-13-413353-9

To My Wife Yolanda
and My Children
Michael, Kevin, and Marissa

Contents

Describing the Relation between Two Variables 180

PART 3   Probability and Probability Distributions 245

Probability 246

Discrete Probability Distributions 315

The Normal Probability Distribution 355

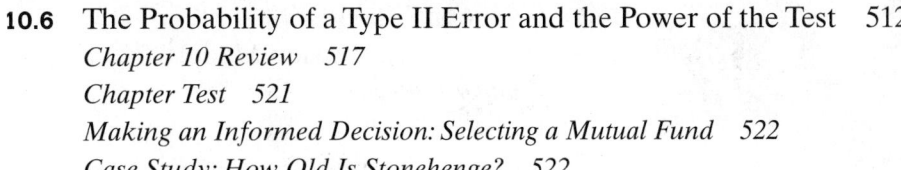

Preface to the Instructor

Capturing a Powerful and Exciting Discipline in a Textbook

Statistics is a powerful subject, and it is one of my passions. Bringing my passion for the subject together with my desire to create a text that would work for me, my students, and my school led me to write the first edition of this textbook. It continues to motivate me as I reflect on changes in students, in the statistics community, and in the world around us.

When I started writing, I used the manuscript of this text in class. My students provided valuable, insightful feedback, and I made adjustments based on their comments. In many respects, this text was written by students and for students. I also received constructive feedback from a wide range of statistics faculty, which has refined ideas in the book and in my teaching. I continue to receive valuable feedback from both faculty and students, and this text continues to evolve with the goal of providing clear, concise, and readable explanations, while challenging students to think statistically.

In writing this edition, I continue to make a special effort to abide by the Guidelines for Assessment and Instruction in Statistics Education (GAISE) for the college introductory course endorsed by the American Statistical Association (ASA). The GAISE Report gives six recommendations for the course:

1. Emphasize statistical literacy and develop statistical thinking
2. Use real data in teaching statistics
3. Stress conceptual understanding
4. Foster active learning
5. Use technology for developing conceptual understanding
6. Use assessments to improve and evaluate student learning

Changes to this edition and the hallmark features of the text reflect a strong adherence to these important GAISE guidelines.

Putting It Together

When students are learning statistics, often they struggle with seeing the big picture of how it all fits together. One of my goals is to help students learn not just the important concepts and methods of statistics but also how to put them together.

On the inside front cover, you'll see a pathway that provides a guide for students as they navigate through the process of learning statistics. The features and chapter organization in the fifth edition reinforce this important process.

New to This Edition

- **Over 350 New and Updated Exercises** The fifth edition makes a concerted effort to require students to write a few sentences that explain the results of their statistical analysis. To reflect this effort, the answers in the back of the text provide recommended explanations of the statistical results. In addition, exercises have been written to require students to understand pitfalls in faulty statistical analysis.

- **Over 100 New and Updated Examples** The examples continue to engage and provide clear, concise explanations for the students while following the Problem, Approach, Solution presentation. Problem lays out the scenario of the example, Approach provides insight into the thought process behind the methodology used to solve the problem, and Solution goes through the solution utilizing the methodology suggested in the approach.

- **Videos** The suite of videos available with this edition has been extensively updated. Featuring the author and George Woodbury, there are both instructional videos that develop statistical concepts and example videos. Most example videos have both by-hand solutions and technology solutions (where applicable). In addition, each Chapter Test problem has video solutions available.

- **Retain Your Knowledge** A new problem type. The Retain Your Knowledge problems occur periodically at the end of section exercises. These problems are meant to assist students in retaining skills learned earlier in the course so that the material is fresh for the final exam.

- **Big Data Problems** Data is ubiquitous today. The ability to collect data from a variety of sources has resulted in very large data sets. While analysis of data sets with tens of thousands of observations with thousands of variables is not practical at the introductory level, it is important for students to analyze data sets with more than fifty observations. These problems are marked with a ⛰ icon and the data is available at www.pearsonhighered.com/sullivanstats.

- **Technology Help in MyStatLab** Problems in MyStatLab that may be analyzed using statistical packages now have an updated technology help feature. Marked with a 🖥 icon, this features provides step-by-step instructions on how to obtain results using StatCrunch, TI-84 Plus/TI-84 Plus C, and Excel.

- **Instructor Resource Guide** The Instructor Resource Guide provides an overview of the chapter. It also details points to emphasize within each section and suggestions for presenting the material. In addition, the guide provides examples that may be used in the classroom.

Hallmark Features

- **Student Activity Workbook** The updated activity workbook contains many in-class activities that may be used to enhance your students' conceptual understanding of statistical concepts. The activities involve many tactile and applet-based simulations. Applets for the activities may be found at www.pearsonhighered.com/sullivanstats. In addition, the activity workbook includes many exercises that introduce **simulation** and **randomization methods** for statistical inference.

- Chapter 10 has simulation techniques that are powerful introductions to the logic of hypothesis testing. There are two activities that utilize simulation techniques. It also contains an activity on using Bootstrapping to test hypotheses for a single mean.
- Chapter 11 has randomization techniques for analyzing the difference of two proportions and the difference of two means. There are four activities for analyzing the difference of two proportions and two activities for analyzing the difference of two means.
- Chapter 14 has randomization techniques for analyzing the strength of association between two quantitative variables. There are two activities for a randomization test for correlation.

The workbook is accompanied by an instructor resource guide with suggestions for incorporating the activities into class.

- Because the use of **Real Data** piques student interest and helps show the relevance of statistics, great efforts have been made to extensively incorporate real data in the exercises and examples.
- **Putting It Together** sections appear in Chapters 5, 9, 10, and 11. The problems in these sections are meant to help students identify the correct approach to solving a problem. Many new exercises have been added to these sections that mix in inferential techniques from previous sections. Plus, there are new problems that require students to identify the inferential technique that may be used to answer the research objective (but no analysis is required). For example, see Problems 23 to 29 in Section 10.5.
- **Step-by-Step Annotated Examples** guide a student from problem to solution in three easy-to-follow steps.
- **"Now Work"** problems follow most examples so students can practice the concepts shown.
- Multiple types of **Exercises** are used at the end of sections and chapters to test varying skills with progressive levels of difficulty. These exercises include **Vocabulary and Skill Building**, **Applying the Concepts**, and **Explaining the Concepts**.
- **Chapter Review** sections include:
  - **Chapter Summary**.
  - A list of key chapter **Vocabulary**.
  - A list of **Formulas** used in the chapter.
  - **Chapter Objectives** listed with corresponding review exercises.
  - **Review Exercises** with all answers available in the back of the book.
  - **Chapter Test** with all answers available in the back of the book. In addition, the Chapter Test problems have **video solutions** available.
- Each chapter concludes with **Case Studies** that help students apply their knowledge and promote active learning.

Integration of Technology

This book can be used with or without technology. Should you choose to integrate technology in the course, the following resources are available for your students:

- Technology Step-by-Step guides are included in applicable sections that show how to use Minitab®, Excel®, the TI-83/84, and StatCrunch to complete statistics processes.
- Any problem that has 12 or more observations in the data set has a ⊿ icon indicating that data set is included on the companion website (http://www.pearsonhighered.com/sullivanstats) in various formats. Any problem that has a very large data set that is not printed in the text has a ⊿ icon, which also indicates that the data set is included on the companion website. These data sets have many observations and often many variables.
- Where applicable, exercises and examples incorporate output screens from various software including Minitab, the TI-83/84 Plus C, Excel, and StatCrunch.
- Twenty new Applets are included on the companion website and connected with certain activities from the Student Activity Workbook, allowing students to manipulate data and interact with animations. See the front inside cover for a list of applets.
- Accompanying Technology Manuals are available that contain detailed tutorial instructions and worked out examples and exercises for the TI-83/84 and 89 and Excel.

Companion Website Contents

- Data Sets
- Twenty new Applets (See description on the insert in front of the text.)
- Formula Cards and Tables in PDF format
- Additional Topics Folder including:
  - Sections 4.5, 5.7, and 6.4
  - Appendix A and Appendix B
- A copy of the questions asked on the Sullivan Statistics Survey I and Survey II
- Consumer Reports projects that were formerly in the text

Key Chapter Content Changes

Chapter 1 Data Collection

The chapter now includes an expanded discussion of confounding, including a distinction between lurking variables and confounding variables.

Chapter 4 Describing the Relation between Two Variables

Section 4.3 now includes a brief discussion of the concept of leverage in the material on identifying influential observations. The conditional bar graphs in Section 4.4 have been drawn so that each category of the explanatory

PREFACE TO THE INSTRUCTOR

variable is grouped. This allows the student to see the complete distribution of each category of the explanatory variable. In addition, the material now includes stacked (or segmented) conditional bar graphs.

Chapter 6 Discrete Probability Distributions

The graphical representation of discrete probability distributions no longer is presented as a probability histogram. Instead, the graph of a discrete probability distribution is presented to emphasize that the data is discrete. Therefore, the graph of discrete probability distributions is drawn using vertical lines above each value of the random variable to a height that is the probability of the random variable.

Chapter 7 The Normal Probability Distribution

The assessment of normality of a random variable using normal probability plots has changed. We no longer rely on normal probability plots drawn using Minitab. Instead, we utilize the correlation between the observed data and normal scores. This approach is based upon the research of S.W. Looney and T. R. Gulledge in their paper, "Use of the Correlation Coefficient with Normal Probability Plots," published in the *American Statistician*. This material may be skipped without loss of continuity (especially for those who postponed the material in Chapter 4). Some problems from Chapter 9 through 13 may need to be skipped or edited, however.

Chapter 9 Estimating the Value of a Parameter

The Putting It Together section went through an extensive renovation of the exercises. Emphasis is placed on identifying the variable of interest in the study (in particular, whether the variable is qualitative or quantitative). In addition, there are problems that simply require the student to identify the type of interval that could be constructed to address the research concerns.

Chapter 10 Hypothesis Testing Regarding a Parameter

The Putting It Together section went through an extensive revision. Again, emphasis is placed on identifying the variable of interest in the study. The exercises include a mix of hypothesis tests and confidence intervals. Plus, there are problems that require the student to identify the type of inference that could be constructed to address the research.

Chapter 11 Inference on Two Samples

The material on inference for two dependent population proportions is now covered in Section 12.3 utilizing the chi-square distribution. As in Chapter 9 and Chapter 10, the Putting It Together section's exercises were revised extensively. There is a healthy mix of two-sample and single-sample analysis (both hypothesis tests and confidence intervals). This will help students to develop the ability to determine the type of analysis required for a given research objective.

Chapter 12 Comparing Three or More Means

In Section 12.2, we now emphasize how to distinguish between the chi-square test for independence and the chi-square test for homogeneity of proportions. The material on inference for two dependent proportions formerly in Section 11.1 is now a stand-alone Section 12.3 so that we might use chi-square methods to analyze the data.

Chapter 13 Comparing Three or More Means

The Analysis of Variance procedures now include construction of normal probability plots of the residuals to verify the normality requirement.

Chapter 14 Inference on the Least-Squares Regression Model and Multiple Regression

Section 14.3 Multiple Regression from the fourth edition has been expanded to four sections. The discussion now includes increased emphasis on interaction, dummy variables, and polynomial regression. Building regression models is now its own section and includes stepwise, forward, and backward regression model building.

Flexible to Work with Your Syllabus

To meet the varied needs of diverse syllabi, this book has been organized to be flexible.

You will notice the "Preparing for This Section" material at the beginning of each section, which will tip you off to dependencies within the course. The two most common variations within an introductory statistics course are the treatment of regression analysis and the treatment of probability.

- **Coverage of Correlation and Regression** The text was written with the descriptive portion of bivariate data (Chapter 4) presented after the descriptive portion of univariate data (Chapter 3). Instructors who prefer to postpone the discussion of bivariate data can skip Chapter 4 and return to it before covering Chapter 14. (Because Section 4.5 on nonlinear regression is covered by a select few instructors, it is located on the website that accompanies the text in Adobe PDF form, so that it can be easily printed.)
- **Coverage of Probability** The text allows for light to extensive coverage of probability. Instructors wishing to minimize probability may cover Section 5.1 and skip the remaining sections. A mid-level treatment of probability can be accomplished by covering Sections 5.1 through 5.3. Instructors who will cover the chi-square test for independence will want to cover Sections 5.1 through 5.3. In addition, an instructor who will cover binomial probabilities will want to cover independence in Section 5.3 and combinations in Section 5.5.

Acknowledgments

Textbooks evolve into their final form through the efforts and contributions of many people. First and foremost, I would like to thank my family, whose dedication to this project was just as much as mine:  my wife, Yolanda, whose words of encouragement and support were unabashed, and my children, Michael, Kevin, and Marissa, who have been supportive throughout their childhood and now into adulthood (my how time flies). I owe each of them my sincerest gratitude. I would also like to thank the entire Mathematics Department at Joliet Junior College and my colleagues who provided support, ideas, and encouragement to help me complete this project. From Pearson Education: I thank Patrick Barbera, whose editorial expertise has been an invaluable asset; Deirdre Lynch, who has provided many suggestions that clearly demonstrate her expertise; Tamela Ambush, who provided organizational skills that made this project go smoothly; Tiffany Bitzel and Andrew Noble, for their marketing savvy and dedication to getting the word out; Vicki Dreyfus, for her dedication in organizing all the media; Jenna Vittorioso, for her ability to control the production process; Dana Bettez for her editorial skill with the Instructor's Resource Guide; and the Pearson sales team, for their confidence and support of this book.

I also want to thank Ryan Cromar, Susan Herring, Craig Johnson, Kathleen McLaughlin, Alana Tuckey, and Dorothy Wakefield for their help in creating supplements. A big thank-you goes to Brad Davis and Jared Burch, who assisted in verifying answers for the back of the text and helped in proofreading. I would also like to acknowledge Kathleen Almy and Heather Foes for their help and expertise in developing the Student Activity Workbook. Finally, I would like to thank George Woodbury for helping me with the incredible suite of videos that accompanies the text. Many thanks to all the reviewers, whose insights and ideas form the backbone of this text. I apologize for any omissions.

CALIFORNIA *Charles Biles, Humboldt State University • Carol Curtis, Fresno City College • Jacqueline Faris, Modesto Junior College • Freida Ganter, California State University–Fresno • Sherry Lohse, Napa Valley College • Craig Nance, Santiago Canyon College • Diane Van Deusen, Napa Valley College* **COLORADO** *Roxanne Byrne, University of Colorado–Denver* **CONNECTICUT** *Kathleen McLaughlin, Manchester Community College • Dorothy Wakefield, University of Connecticut • Cathleen M. Zucco Teveloff, Trinity College* **DISTRICT OF COLUMBIA** *Monica Jackson, American University • Jill McGowan, Howard University* **FLORIDA** *Randall Allbritton, Daytona Beach Community College • Greg Bloxom, Pensacola State College • Anthony DePass, St. Petersburgh College Clearwater • Kelcey Ellis, University of Central Florida • Franco Fedele, University of West Florida • Laura Heath, Palm Beach Community College • Perrian Herring, Okaloosa Walton College • Marilyn Hixson, Brevard Community College • Daniel Inghram, University of Central Florida • Philip Pina, Florida Atlantic University • Mike Rosenthal, Florida International University • James Smart, Tallahassee Community College* **GEORGIA** *Virginia Parks, Georgia Perimeter College • Chandler Pike, University of Georgia • Jill Smith, University of Georgia • John Weber, Georgia Perimeter College* **HAWAII** *Eric Matsuoka at Leeward Community College* **IDAHO** *K. Shane Goodwin, Brigham Young University • Craig Johnson, Brigham Young University • Brent Timothy, Brigham Young University • Kirk Trigsted, University of Idaho* **ILLINOIS** *Grant Alexander, Joliet Junior College • Kathleen Almy, Rock Valley College • John Bialas, Joliet Junior College • Linda Blanco, Joliet Junior College • Kevin Bodden, Lewis & Clark Community College • Rebecca Bonk, Joliet Junior College • Joanne Brunner, Joliet Junior College • James Butterbach, Joliet Junior College • Robert Capetta, College of DuPage • Elena Catoiu, Joliet Junior College • Faye Dang, Joliet Junior College • Laura Egner, Joliet Junior College • Jason Eltrevoog, Joliet Junior College • Erica Egizio, Lewis University • Heather Foes, Rock Valley College • Randy Gallaher, Lewis & Clark Community College • Melissa Gaddini, Robert Morris University • Iraj Kalantari, Western Illinois University • Donna Katula, Joliet Junior College • Diane Long, College of DuPage • Heidi Lyne, Joliet Junior College • Jean McArthur, Joliet Junior College • Patricia McCarthy, Robert Morris University • David McGuire, Joliet Junior College • Angela McNulty, Joliet Junior College • Andrew Neath, Southern Illinois University-Edwardsville • Linda Padilla, Joliet Junior College • David Ruffato, Joliet Junior College • Patrick Stevens, Joliet Junior College • Robert Tuskey, Joliet Junior College • Stephen Zuro, Joliet Junior College* **INDIANA** *Susitha Karunaratne, Purdue University North Central • Jason Parcon, Indiana University–Purdue University Ft. Wayne • Henry Wakhungu, Indiana University* **KANSAS** *Donna Gorton, Butler Community College • Ingrid Peterson, University of Kansas* **LOUISIANA** *Melissa Myers, University of Louisiana at Lafayette* **MARYLAND** *Nancy Chell, Anne Arundel Community College • John Climent, Cecil Community College • Rita Kolb, The Community College of Baltimore County • Jignasa Rami, Community College of Baltimore County • Mary Lou Townsend, Wor-Wic Community College* **MASSACHUSETTS** *Susan McCourt, Bristol Community College • Daniel Weiner, Boston University • Pradipta Seal, Boston University of Public Health* **MICHIGAN** *Margaret M. Balachowski, Michigan Technological University • Diane Krasnewich, Muskegon Community College • Susan Lenker, Central Michigan University • Timothy D. Stebbins, Kalamazoo Valley Community College • Sharon Stokero, Michigan Technological University • Alana Tuckey, Jackson Community College* **MINNESOTA** *Mezbhur Rahman, Minnesota State University* **MISSOURI** *Farroll Tim Wright, University of Missouri–Columbia* **NEBRASKA** *Jane Keller, Metropolitan Community College* **NEW YORK** *Jacob Amidon, Finger Lakes Community College •*

Stella Aminova, Hunter College • Jennifer Bergamo, Onondaga Community College • Kathleen Cantone, Onondaga Community College • Pinyuen Chen, Syracuse University • Sandra Clarkson, Hunter College of CUNY • Rebecca Daggar, Rochester Institute of Technology • Bryan Ingham, Finger Lakes Community College • Anne M. Jowsey, Niagara County Community College • Maryann E. Justinger, Erie Community College–South Campus • Bernadette Lanciaux, Rochester Institute of Technology • Kathleen Miranda, SUNY at Old Westbury • Robert Sackett, Erie Community College–North Campus • Sean Simpson, Westchester Community College • Bill Williams, Hunter College of CUNY **NORTH CAROLINA** *Fusan Akman, Coastal Carolina Community College • Mohammad Kazemi, University of North Carolina–Charlotte • Janet Mays, Elon University • Marilyn McCollum, North Carolina State University • Claudia McKenzie, Central Piedmont Community College • Said E. Said, East Carolina University • Karen Spike, University of North Carolina–Wilmington • Jeanette Szwec, Cape Fear Community College* **NORTH DAKOTA** *Myron Berg, Dickinson State University • Ronald Degges, North Dakota State University* **OHIO** *Richard Einsporn, The University of Akron • Michael McCraith, Cuyaghoga Community College* **OREGON** *Daniel Kim, Southern Oregon University • Jong Sung Kin, Portland State University* **SOUTH CAROLINA** *Diana Asmus, Greenville Technical College • Dr. William P. Fox, Francis Marion University • Cheryl Hawkins, Greenville Technical College • Rose Jenkins, Midlands Technical College • Lindsay Packer, College of Charleston • Laura Shick, Clemson University* **TENNESSEE** *Tim Britt, Jackson State Community College • Nancy Pevey, Pellissippi State Technical Community College • David Ray, University of Tennessee–Martin* **TEXAS** *Edith Aguirre, El Paso Community College • Ivette Chuca, El Paso Community College • Aaron Gutknecht, Tarrant County College • Jada Hill, Richland College • David Lane, Rice University • Alma F. Lopez, South Plains College • Shanna Moody, University of Texas at Arlington* **UTAH** *Joe Gallegos, Salt Lake City Community College • Alia Maw, Salt Lake City Community College* **VIRGINIA** *Kim Jones, Virginia Commonwealth University • Vasanth Solomon, Old Dominion University* **WEST VIRGINIA** *Mike Mays, West Virginia University* **WISCONSIN** *William Applebaugh, University of Wisconsin–Eau Claire • Carolyn Chapel, Western Wisconsin Technical College • Beverly Dretzke, University of Wisconsin–Eau Claire • Jolene Hartwick, Western Wisconsin Technical College • Thomas Pomykalski, Madison Area Technical College • Walter Reid, University of Wisconsin-Eau Claire*

Michael Sullivan, III
Joliet Junior College

BREAK THROUGH
To improving results

Resources for Success

MyStatLab™ Online Course for
Statistics: Informed Decisions Using Data 5e by Michael Sullivan, III
(access code required)

MyStatLab is available to accompany Pearson's market leading text offerings. To give students a consistent tone, voice, and teaching method each text's flavor and approach is tightly integrated throughout the accompanying MyStatLab course, making learning the material as seamless as possible.

New! Technology Support Videos

In these videos, the author demonstrates the easy-to-follow steps needed to solve a problem in several different formats—by-hand, TI-84 Plus C, and StatCrunch.

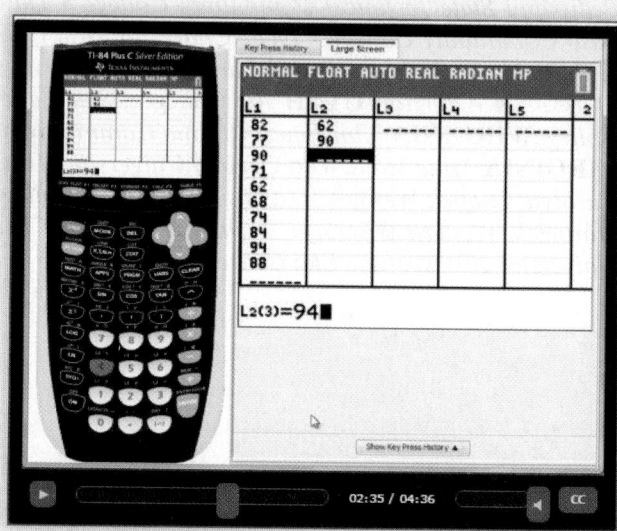

Technology Step-by-Step - StatCrunch

Approximating the Mean and Standard Deviation from Grouped Data

1. If necessary, enter the summarized data into the spreadsheet. Name the columns.
2. Select **Stat**, highlight **Summary Stats**, and select **Grouped/Binned data**.
3. Choose the column that contains the class under the "Bins in:" drop-down menu. Choose the column that contains the frequencies in the "Counts in:" drop-down menu. Select the "Consecutive lower limits" radio button for defining the midpoints. Click Compute!.

Technology Step-by-Step

Technology Step-by-Step guides show how to use StatCrunch®, Excel®, and the TI-84 graphing calculators to complete statistics processes.

Interactive Applets

Applets are a powerful tool for developing statistical concepts and enhancing understanding. There are twenty new applets that accompany the text and many activities in the Student Activity Workbook that utilize these applets.

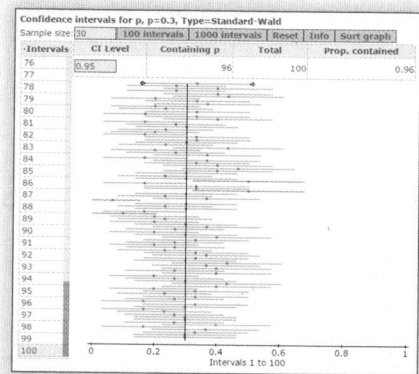

Resources for Success

BREAKTHROUGH
To improving results

MyStatLab™ Online Course
(access code required)

MyStatLab from Pearson is the world's leading online resource for teaching and learning statistics; integrating interactive homework, assessment, and media in a flexible, easy-to-use format. MyStatLab is a course management system that helps individual students succeed.

- The author analyzed aggregated student usage and performance data from MySTatLab for the previous edition of this text. The results of this analysis helped improve the quality and quantity of exercises that matter the most to instructors and students.

- MyStatLab can be implemented successfully in any environment—lab-based, traditional, fully online, or hybrid—and demonstrates the quantifiable difference that integrated usage has on student retention, subsequent success, and overall achievement.

- MyStatLab's comprehensive gradebook automatically tracks students' results on tests, quizzes, homework, and in the study plan. Instructors can use the gradebook to provide positive feedback or intervene if students have trouble. Gradebook data can be easily exported to a variety of spreadsheet programs, such as Microsoft Excel.

MyStatLab provides **engaging experiences** that personalize, stimulate, and measure learning for each student. In addition to the resources below, each course includes a full interactive online version of the accompanying textbook.

- **Personalized Learning:** MyStatLab's personalized homework, and adaptive and companion study plan features allow your students to work more efficiently spending time where they really need to.

- **Tutorial Exercises with Multimedia Learning Aids:** The homework and practice exercises in MyStatLab align with the exercises in the textbook, and most regenerate algorithmically to give students unlimited opportunity for practice and mastery. Exercises offer immediate helpful feedback, guided solutions, sample problems, animations, videos, statistical software tutorial videos and eText clips for extra help at point-of-use.

- **Learning Catalytics™:** MyStatLab now provides Learning Catalytics—an interactive student response tool that uses students' smartphones, tablets, or laptops to engage them in more sophisticated tasks and thinking.

- **Videos** tie statistics to the real world.

  - **StatTalk Videos:** Fun-loving statistician Andrew Vickers takes to the streets of Brooklyn, NY, to demonstrate important statistical concepts through interesting stories and real-life events. This series of 24 fun and engaging videos will help students actually understand statistical concepts. Available with an instructor's user guide and assessment questions.

- **Additional Question Libraries:** In addition to algorithmically regenerated questions that are aligned with your textbook, MyStatLab courses come with two additional question libraries:

  - **450 exercises in Getting Ready for Statistics** cover the developmental math topics students need for the course. These can be assigned as a prerequisite to other assignments, if desired.

  - **1000 exercises** in the **Conceptual Question Library** require students to apply their statistical understanding.

- **StatCrunch™:** MyStatLab integrates the web-based statistical software, StatCrunch, within the online assessment platform so that students can easily analyze data sets from exercises and the text. In addition, MyStatLab includes access to www.statcrunch.com, a vibrant online community where users can access tens of thousands of shared data sets, create and conduct online surveys, perform complex analyses using the powerful statistical software, and generate compelling reports.

- **Statistical Software, Support and Integration:** We make it easy to copy our data sets, from both the eText and the MyStatLab questions, into software such as StatCrunch, Minitab®, Excel®, and more. Students have access to a variety of support tools—Technology Tutorial Videos, Technology Study Cards, and Technology Manuals for select titles—to learn how to effectively use statistical software.

MyStatLab Accessibility:

- MyStatLab is compatible with the JAWS screen reader, and enables multiple-choice, fill-in-the-blank and free-response problem-types to be read, and interacted with via keyboard controls and math notation input. MyStatLab also works with screen enlargers, including ZoomText, MAGic®, and SuperNova. And all MyStatLab videos accompanying texts with copyright 2009 and later have closed captioning.

- More information on this functionality is available at http://mystatlab.com/accessibility.

And, MyStatLab comes from an **experienced partner** with educational expertise and an eye on the future.

- Knowing that you are using a Pearson product means knowing that you are using quality content. That means that our eTexts are accurate and our assessment tools work. It means we are committed to making MyStatLab as accessible as possible.

- Whether you are just getting started with MyStatLab, or have a question along the way, we're here to help you learn about our technologies and how to incorporate them into your course.

To learn more about how MyStatLab combines proven learning applications with powerful assessment, visit **www.mystatlab.com** or contact your Pearson representative.

MathXL® for Statistics Online Course (access code required)

MathXL® is the homework and assessment engine that runs MyStatLab. (MyStatLab is MathXL plus a learning management system.)

With MathXL for Statistics, instructors can:

- Create, edit, and assign online homework and tests using algorithmically generated exercises correlated at the objective level to the textbook.
- Create and assign their own online exercises and import TestGen® tests for added flexibility.
- Maintain records of all student work, tracked in MathXL's online gradebook.

With MathXL for Statistics, students can:

- Take chapter tests in MathXL and receive personalized study plans and/or personalized homework assignments based on their test results.
- Use the study plan and/or the homework to link directly to tutorial exercises for the objectives they need to study.
- Students can also access supplemental animations and video clips directly from selected exercises.
- Knowing that students often use external statistical software, we make it easy to copy our data sets, both from the eText and the MyStatLab questions, into software like StatCrunch™, Minitab, Excel and more.

MathXL for Statistics is available to qualified adopters. For more information, visit our website at www.mathxl.com, or contact your Pearson representative.

StatCrunch™

StatCrunch is powerful web-based statistical software that allows users to perform complex analyses, share data sets, and generate compelling reports of their data. The vibrant online community offers tens of thousand shared data sets for students to analyze.

- **Collect.** Users can upload their own data to StatCrunch or search a large library of publicly shared data sets, spanning almost any topic of interest. Also, an online survey tool allows users to quickly collect data via web-based surveys.
- **Crunch.** A full range of numerical and graphical methods allow users to analyze and gain insights from any data

set. Interactive graphics help users understand statistical concepts, and are available for export to enrich reports with visual representations of data.

- **Communicate.** Reporting options help users create a wide variety of visually-appealing representations of their data.

Full access to StatCrunch is available with a MyStatLab kit, and StatCrunch is available by itself to qualified adopters. StatCrunch Mobile is also now available; just visit www.statcrunch.com from the browser on your smart phone or tablet. For more information, visit our website at www.statcrunch.com, or contact your Pearson representative.

TestGen®

TestGen® (www.pearsoned.com/testgen) enables instructors to build, edit, print, and administer tests using a computerized bank of questions developed to cover all the objectives of the text. TestGen is algorithmically based, allowing instructors to create multiple but equivalent versions of the same question or test with the click of a button. Instructors can also modify test bank questions or add new questions. The software and testbank are available for download from Pearson's Instructor Resource Center.

Learning Catalytics™
Foster student engagement and peer-to-peer learning

Generate class discussion, guide your lecture, and promote peer-to-peer learning with real-time analytics. MyMathLab and MyStatLab now provide Learning Catalytics—an interactive student response tool that uses students' smartphones, tablets, or laptops to engage them in more sophisticated tasks and thinking.

Instructors, you can:

- Pose a variety of open-ended questions that help your students develop critical thinking skills
- Monitor responses to find out where students are struggling
- Use real-time data to adjust your instructional strategy and try other ways of engaging your students during class
- Manage student interactions by automatically grouping students for discussion, teamwork, and peer-to-peer learning

Resources for Success

BREAKTHROUGH
To improving results

Instructor Resources

Annotated Instructor's Edition
(ISBN 978-0-13-413542-7; ISBN 10: 0-13-413542-3)
Margin notes include helpful teaching tips, suggest alternative presentations, and point out common student errors. Exercise answers are included in the exercise sets when space allows. The complete answer section is located in the back of the text.

Instructor's Resource Center
All instructor resources can be downloaded from www.pearsonhighered.com/irc. This is a password-protected site that requires instructors to set up an account or, alternatively, instructor resources can be ordered from your Pearson Higher Education sales representative.

Instructor's Solutions Manual (Download only)
by GEX Publishing Services.
Fully worked solutions to every textbook exercise, including the chapter review and chapter tests. Case Study Answers are also provided. Available from the Instructor's Resource Center and MyStatLab.

Online Test Bank
A test bank derived from TestGen is available on the Instructor's Resource Center. There is also a link to the TestGen website within the Instructor Resource area of MyStatLab.

PowerPoint® Lecture Slides
Free to qualified adopters, this classroom lecture presentation software is geared specifically to the sequence and philosophy of *Statistics: Informed Decisions Using Data*. Key graphics from the book are included to help bring the statistical concepts alive in the classroom. Slides are available for download from the Instructor's Resource Center and MyStatLab.

New! Instructor's Guide for Student Activity Workbook (Download only)
by Heather Foes and Kathleen Almy, *Rock Valley College* and Michael Sullivan, III, *Joliet Junior College*. Accompanies the activity workbook with suggestions for incorporating the activities into class. The Guide is available from the Instructor's Resource Center and MyStatLab.

New! Instructor's Resource Guide (Download only)
by Michael Sullivan, III, *Joliet Junior College*.
This guide presents an overview of each chapter along with details about concepts that should be emphasized for each section. The resource guide also provides additional examples (complete with solutions) for each section that may be used for classroom presentations. This Guide is available from the Instructor's Resource Center and MyStataLab.

Student Resources

New! Author in the Classroom Videos
by Michael Sullivan, III, *Joliet Junior College* and George Woodbury, *College of the Sequoias*

The suite of videos available with this edition has been extensively updated. Featuring the author and George Woodbury, there are both instructional videos that develop statistical concepts and example videos. Most example videos have both by-hand solutions and technology solutions (where applicable). In addition, each Chapter Test problem has video solutions available.

Student's Solutions Manual
by GEX Publishing Services
(ISBN 13: 978-0-13-413540-3; ISBN 10: 0-13-413540-7)
Fully worked solutions to odd-numbered exercises with all solutions to the chapter reviews and chapter tests.

Technology Manuals
The following technology manuals contain detailed tutorial instructions and worked-out examples and exercises. Available on the Pearson site: www.pearsonhighered.com/mathstatsresources

- **Excel Manual (including XLSTAT)** by Alana Tuckey, *Jackson Community College*
- **Graphing Calculator Manual for the TI-83/84 Plus and TI-89** by Kathleen McLaughlin and Dorothy Wakefield

New! Student Activity Workbook
by Heather Foes and Kathleen Almy, *Rock Valley College*, and Michael Sullivan, III, *Joliet Junior College*
(ISBN 13: 978-0-13-411610-5; ISBN 10: 0-13-411610-0)
Includes classroom and applet activities that allow students to experience statistics firsthand in an active learning environment. Also introduces resampling methods that help develop conceptual understanding of hypothesis testing.

Applications Index

Getting the Information You Need

Statistics is a process—a series of steps that leads to a goal. This text is divided into four parts to help the reader see the process of statistics.

Part 1 focuses on the first step in the process, which is to determine the research objective or question to be answered. Then information is obtained to answer the questions stated in the research objective.

1

Data Collection

Outline

Making an Informed Decision

 It is your senior year of high school. You will have a lot of exciting experiences in the upcoming year, plus a major decision to make—which college should I attend? The choice you make may affect many aspects of your life—your career, where you live, your significant other, and so on, so you don't want to simply choose the college that everyone else picks. You need to design a questionnaire to help you make an informed decision about college. In addition, you want to know how well the college you are considering educates its students. See Making an Informed Decision on page 59.

PUTTING IT TOGETHER

Statistics plays a major role in many aspects of our lives. It is used in sports, for example, to help a general manager decide which player might be the best fit for a team. It is used in politics to help candidates understand how the public feels about various policies. And statistics is used in medicine to help determine the effectiveness of new drugs.

Used appropriately, statistics can enhance our understanding of the world around us. Used inappropriately, it can lend support to inaccurate beliefs. Understanding statistical methods will provide you with the ability to analyze and critique studies and the opportunity to become an informed consumer of information. Understanding statistical methods will also enable you to distinguish solid analysis from bogus "facts."

To help you understand the features of this text and for hints to help you study, read the *Pathway to Success* on the front inside cover of the text.

1.1   Introduction to the Practice of Statistics

Objectives

① Define statistics and statistical thinking

② Explain the process of statistics

③ Distinguish between qualitative and quantitative variables

④ Distinguish between discrete and continuous variables

⑤ Determine the level of measurement of a variable

① Define Statistics and Statistical Thinking

What is statistics? Many people say that statistics is numbers. After all, we are bombarded by numbers that supposedly represent how we feel and who we are. For example, we hear on the radio that 50% of first marriages, 67% of second marriages, and 74% of third marriages end in divorce (Forest Institute of Professional Psychology, Springfield, MO).

Another interesting consideration about the "facts" we hear or read is that two different sources can report two different results. For example, an October 23, 2014 poll by ABC News and the *Washington Post* indicated that 70% of Americans believed the country was on the wrong track. However, an October 30, 2014 poll by NBC News and the *Wall Street Journal* indicated that 63% of Americans believed the country was on the wrong track. Is it possible that the percent of Americans who believe the country is on the wrong track could decrease by 7% in one week, or is something else going on? Statistics helps to provide the answer.

Certainly, statistics has a lot to do with numbers, but this definition is only partially correct. Statistics is also about where the numbers come from (that is, how they were obtained) and how closely the numbers reflect reality.

Definition   **Statistics** is the science of collecting, organizing, summarizing, and analyzing information to draw conclusions or answer questions. In addition, statistics is about providing a measure of confidence in any conclusions.

Let's break this definition into four parts. The first part states that statistics involves the collection of information. The second refers to the organization and summarization of information. The third states that the information is analyzed to draw conclusions or answer specific questions. The fourth part states that results should be reported using some measure that represents how convinced we are that our conclusions reflect reality.

What is the information referred to in the definition? The information is **data**, which the *American Heritage Dictionary* defines as "a fact or proposition used to draw a conclusion or make a decision." Data can be numerical, as in height, or nonnumerical, as in gender. In either case, data describe characteristics of an individual.

In Other Words
Anecdotal means that the information being conveyed is based on casual observation, not scientific research.

Analysis of data can lead to powerful results. Data can be used to offset anecdotal claims, such as the suggestion that cellular telephones cause brain cancer. After carefully collecting, summarizing, and analyzing data regarding this phenomenon, it was determined that there is no link between cell phone usage and brain cancer. See Examples 1 and 2 in Section 1.2.

Because data are powerful, they can be dangerous when misused. The misuse of data usually occurs when data are incorrectly obtained or analyzed. For example, radio or television talk shows regularly ask poll questions for which respondents must call in or use the Internet to supply their vote. Most likely, the individuals who are going to call in are those who have a strong opinion about the topic. This group is not likely to be representative of people in general, so the results of the poll are not meaningful. Whenever we look at data, we should be mindful of where the data come from.

Even when data tell us that a relation exists, we need to investigate. For example, a study showed that breast-fed children have higher IQs than those who were not breast-fed. Does this study mean that a mother who breast-feeds her child will increase the child's IQ? Not necessarily. It may be that some factor other than breast-feeding contributes to the IQ of the children. In this case, it turns out that mothers who breast-feed generally have higher IQs than those who do not. Therefore, it may be genetics that leads to the higher IQ, not breast-feeding.* This illustrates an idea in statistics known as the *lurking variable*. A good statistical study will have a way of dealing with lurking variables.

A key aspect of data is that they vary. Consider the students in your classroom. Is everyone the same height? No. Does everyone have the same color hair? No. So, within groups there is variation. Now consider yourself. Do you eat the same amount of food each day? No. Do you sleep the same number of hours each day? No. So even considering an individual there is variation. Data vary. One goal of statistics is to describe and understand the sources of variation. Variability in data may help to explain the different results obtained by the ABC News/*Washington Post* and NBC News/*Wall Street Journal* polls described at the beginning of this section.

Because of this variability, the results that we obtain using data can vary. In a mathematics class, if Bob and Jane are asked to solve $3x + 5 = 11$, they will both obtain $x = 2$ as the solution when they use the correct procedures. In a statistics class, if Bob and Jane are asked to estimate the average commute time for workers in Dallas, Texas, they will likely get different answers, even though both use the correct procedure. The different answers occur because they likely surveyed different individuals, and these individuals have different commute times. Bob and Jane would get the same result if they both asked *all* commuters or the same commuters about their commutes, but how likely is this?

So, in mathematics when a problem is solved correctly, the results can be reported with 100% certainty. In statistics, when a problem is solved, the results do not have 100% certainty. In statistics, we might say that we are 95% confident that the average commute time in Dallas, Texas, is between 20 and 23 minutes. Uncertain results may seem disturbing now but will feel more comfortable as we proceed through the course.

Without certainty, how can statistics be useful? Statistics can provide an understanding of the world around us because recognizing where variability in data comes from can help us to control it. Understanding the techniques presented in this text will provide you with powerful tools that will give you the ability to analyze and critique media reports, make investment decisions, or conduct research on major purchases. This will help to make you an informed citizen, consumer of information, and critical and statistical thinker.

② Explain the Process of Statistics

Consider the following scenario.

NOTE

Obtaining a truthful response to a question such as this is challenging. In Section 1.5, we present some techniques for obtaining truthful responses to sensitive questions. •

You are walking down the street and notice that a person walking in front of you drops $100. Nobody seems to notice the $100 except you. Since you could keep the money without anyone knowing, would you keep the money or return it to the owner?

Suppose you wanted to use this scenario as a gauge of the morality of students at your school by determining the percent of students who would return the money. How might you do this? You could attempt to present the scenario to every student at the school, but this would be difficult or impossible if the student body is large. A second possibility is to present the scenario to 50 students and use the results to make a statement about all the students at the school.

*In fact, a study found that a gene called FADS2 is responsible for higher IQ scores in breast-fed babies.
Source: Duke University, "Breastfeeding Boosts IQ in Infants with 'Helpful' Genetic Variant," *Science Daily* 6 November 2007.

Figure 1

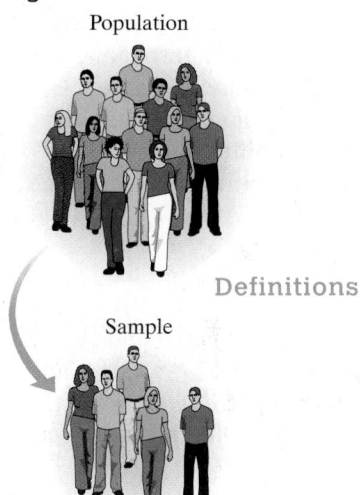

Population

Sample

Individual

Definitions   The entire group to be studied is called the **population**. An **individual** is a person or object that is a member of the population being studied. A **sample** is a subset of the population that is being studied. See Figure 1.

In the $100 study presented, the population is all the students at the school. Each student is an individual. The sample is the 50 students selected to participate in the study.

Suppose 39 of the 50 students stated that they would return the money to the owner. We could present this result by saying that the percent of students in the survey who would return the money to the owner is 78%. This is an example of a *descriptive statistic* because it describes the results of the sample without making any general conclusions about the population.

Definitions   A **statistic** is a numerical summary of a sample. **Descriptive statistics** consist of organizing and summarizing data. Descriptive statistics describe data through numerical summaries, tables, and graphs.

So 78% is a statistic because it is a numerical summary based on a sample. Descriptive statistics make it easier to get an overview of what the data are telling us.

If we extend the results of our sample to the population, we are performing *inferential statistics*.

Definition   **Inferential statistics** uses methods that take a result from a sample, extend it to the population, and measure the reliability of the result.

The generalization contains uncertainty because a sample cannot tell us everything about a population. Therefore, inferential statistics includes a level of confidence in the results. So rather than saying that 78% of all students would return the money, we might say that we are 95% confident that between 74% and 82% of all students would return the money. Notice how this inferential statement includes a *level of confidence* (measure of reliability) in our results. It also includes a range of values to account for the variability in our results.

One goal of inferential statistics is to use statistics to estimate *parameters*.

Definition   A **parameter** is a numerical summary of a population.

EXAMPLE 1   Parameter versus Statistic

Suppose 48.2% of all students on your campus own a car. This value represents a parameter because it is a numerical summary of a population. Suppose a sample of 100 students is obtained, and from this sample we find that 46% own a car. This value represents a statistic because it is a numerical summary of a sample.

● **Now Work Problem 7**

The methods of statistics follow a process.

┌─ **CAUTION!** ─────────────┐
Many nonscientific studies are
based on *convenience samples*,
such as Internet surveys or
phone-in polls. The results of any
study performed using this type of
sampling method are not reliable.
└──────────────────────────┘

The Process of Statistics

1. *Identify the research objective.* A researcher must determine the question(s) he or she wants answered. The question(s) must clearly identify the population that is to be studied.
2. *Collect the data needed to answer the question(s) posed in (1).* Conducting research on an entire population is often difficult and expensive, so we typically look at a sample. This step is vital to the statistical process, because

(continued)

if the data are not collected correctly, the conclusions drawn are meaningless. Do not overlook the importance of appropriate data collection. We discuss this step in detail in Sections 1.2 through 1.6.

3. *Describe the data.* Descriptive statistics allow the researcher to obtain an overview of the data and can help determine the type of statistical methods the researcher should use. We discuss this step in detail in Chapters 2 through 4.

4. *Perform inference.* Apply the appropriate techniques to extend the results obtained from the sample to the population and report a level of reliability of the results. We discuss techniques for measuring reliability in Chapters 5 through 8 and inferential techniques in Chapters 9 through 15.

EXAMPLE 2   The Process of Statistics: Minimum Wage

CBS News and the *New York Times* conducted a poll September 12–15, 2014, and asked, "As you may know, the federal minimum wage is currently $7.25 an hour. Do you favor or oppose raising the minimum wage to $10.10?" The following statistical process allowed the researchers to conduct their study.

1. *Identify the research objective.* The researchers wanted to determine the percentage of adult Americans who favor raising the minimum wage. Therefore, the population being studied was adult Americans.

2. *Collect the data needed to answer the question posed in (1).* It is unreasonable to expect to survey the more than 200 million adult Americans to determine how they feel about the minimum wage. So the researchers surveyed a sample of 1009 adult Americans. Of those surveyed, 706 stated they favor an increase in the minimum wage to $10.10 an hour.

3. *Describe the data.* Of the 1009 individuals in the survey, 70% (= 706/1009) believe the minimum wage should be raised to $10.10 an hour. This is a descriptive statistic because it is a numerical summary of the data.

4. *Perform inference.* CBS News and the *New York Times* wanted to extend the results of the survey to all adult Americans. Remember, when generalizing results from a sample to a population, the results are uncertain. To account for this uncertainty, researchers reported a 3% *margin of error.* This means that CBS News and the *New York Times* feel fairly certain (in fact, 95% certain) that the percentage of *all* adult Americans who favor an increase in the minimum wage to $10.10 an hour is somewhere between 67% (70% – 3%) and 73% (70% + 3%).

● Now Work Problem 49

---

③ **Distinguish between Qualitative and Quantitative Variables**

Once a research objective is stated, a list of the information we want to learn about the individuals must be created. **Variables** are the characteristics of the individuals within the population. For example, recently my son and I planted a tomato plant in our backyard. We collected information about the tomatoes harvested from the plant. The individuals we studied were the tomatoes. The variable that interested us was the weight of a tomato. My son noted that the tomatoes had different weights even though they came from the same plant. He discovered that variables such as weight may vary.

If variables did not vary, they would be constants, and statistical inference would not be necessary. Think about it this way: If each tomato had the same weight, then knowing the weight of one tomato would allow us to determine the weights of all tomatoes. However, the weights of the tomatoes vary. One goal of research is to learn the causes of the variability so that we can learn to grow plants that yield the best tomatoes.

Variables can be classified into two groups: *qualitative* or *quantitative*.

Definitions

Qualitative, or categorical, variables allow for classification of individuals based on some attribute or characteristic.

Quantitative variables provide numerical measures of individuals. The values of a quantitative variable can be added or subtracted and provide meaningful results.

Many examples in this text will include a suggested **approach**, or a way to look at and organize a problem so that it can be solved. The approach will be a suggested method of *attack* toward solving the problem. This does not mean that the approach given is the only way to solve the problem, because many problems have more than one approach leading to a correct solution.

EXAMPLE 3   Distinguishing between Qualitative and Quantitative Variables

Problem Determine whether the following variables are qualitative or quantitative.
(a) Gender
(b) Temperature
(c) Number of days during the past week that a college student studied
(d) Zip code

Approach Quantitative variables are numerical measures such that meaningful arithmetic operations can be performed on the values of the variable. Qualitative variables describe an attribute or characteristic of the individual that allows researchers to categorize the individual.

Solution
(a) Gender is a qualitative variable because it allows a researcher to categorize the individual as male or female. Notice that arithmetic operations cannot be performed on these attributes.
(b) Temperature is a quantitative variable because it is numeric, and operations such as addition and subtraction provide meaningful results. For example, 70°F is 10°F warmer than 60°F.
(c) Number of days during the past week that a college student studied is a quantitative variable because it is numeric, and operations such as addition and subtraction provide meaningful results.
(d) Zip code is a qualitative variable because it categorizes a location. Notice that, even though zip codes are numeric, adding or subtracting zip codes does not provide meaningful results.

● **Now Work Problem 15**

Example 3(d) shows us that a variable may be qualitative while having numeric values. Just because the value of a variable is numeric does not mean that the variable is quantitative.

④ **Distinguish between Discrete and Continuous Variables**

We can further classify quantitative variables into two types: *discrete* or *continuous*.

Definitions

In Other Words
If you count to get the value of a quantitative variable, it is discrete. If you measure to get the value of a quantitative variable, it is continuous.

A **discrete variable** is a quantitative variable that has either a finite number of possible values or a countable number of possible values. The term *countable* means that the values result from counting, such as 0, 1, 2, 3, and so on. A discrete variable cannot take on every possible value between any two possible values.

A **continuous variable** is a quantitative variable that has an infinite number of possible values that are not countable. A continuous variable may take on every possible value between any two values.

Figure 2 illustrates the relationship among qualitative, quantitative, discrete, and continuous variables.

Figure 2

Qualitative variables

Quantitative variables

Discrete variables

Continuous variables

EXAMPLE 4 Distinguishing between Discrete and Continuous Variables

Problem Determine whether the quantitative variables are discrete or continuous.
(a) The number of heads obtained after flipping a coin five times.
(b) The number of cars that arrive at a McDonald's drive-thru between 12:00 P.M. and 1:00 P.M.
(c) The distance a 2014 Toyota Prius can travel in city driving conditions with a full tank of gas.

Approach A variable is discrete if its value results from counting. A variable is continuous if its value is measured.

Solution
(a) The number of heads obtained by flipping a coin five times is a discrete variable because we can count the number of heads obtained. The possible values of this discrete variable are 0, 1, 2, 3, 4, 5.
(b) The number of cars that arrive at a McDonald's drive-thru between 12:00 P.M. and 1:00 P.M. is a discrete variable because we find its value by counting the cars. The possible values of this discrete variable are 0, 1, 2, 3, 4, and so on. Notice that this number has no upper limit.
(c) The distance traveled is a continuous variable because we measure the distance (miles, feet, inches, and so on).

• **Now Work Problem 23**

Continuous variables are often rounded. For example, if a certain make of car gets 24 miles per gallon (mpg) of gasoline, its miles per gallon must be greater than or equal to 23.5 and less than 24.5, or $23.5 \leq mpg < 24.5$.

The type of variable (qualitative, discrete, or continuous) dictates the methods that can be used to analyze the data.

The list of observed values for a variable is **data**. Gender is a variable; the observations male and female are data. **Qualitative data** are observations corresponding to a qualitative variable. **Quantitative data** are observations corresponding to a quantitative variable. **Discrete data** are observations corresponding to a discrete variable. **Continuous data** are observations corresponding to a continuous variable.

EXAMPLE 5 Distinguishing between Variables and Data

Problem Table 1 presents a group of selected countries and information regarding these countries as of July, 2014. Identify the individuals, variables, and data in Table 1.

Table 1			
Country	**Government Type**	**Life Expectancy (years)**	**Population (in millions)**
Australia	Federal parliamentary democracy	82.07	22.5
Canada	Constitutional monarchy	81.67	34.8
France	Republic	81.66	66.3
Morocco	Constitutional monarchy	76.51	33.0
Poland	Republic	76.65	38.3
Sri Lanka	Republic	76.35	21.9
United States	Federal republic	79.56	318.9

Source: CIA *World Factbook*

Approach An individual is an object or person for whom we wish to obtain data. The variables are the characteristics of the individuals, and the data are the specific values of the variables.

Solution The individuals in the study are the countries: Australia, Canada, and so on. The variables measured for each country are *government type*, *life expectancy*, and *population*. The variable *government type* is qualitative because it categorizes the individual. The variables *life expectancy* and *population* are quantitative.

The quantitative variable *life expectancy* is continuous because it is measured. The quantitative variable *population* is discrete because we count people. The observations are the data. For example, the data corresponding to the variable *life expectancy* are 82.07, 81.67, 81.66, 76.51, 76.65, 76.35, and 79.56. The following data correspond to the individual Poland: a republic government with residents whose life expectancy is 76.65 years and population is 38.3 million people. Republic is an instance of qualitative data that results from observing the value of the qualitative variable *government type*. The life expectancy of 76.65 years is an instance of quantitative data that results from observing the value of the quantitative variable *life expectancy*.

● **Now Work Problem 45**

⑤ Determine the Level of Measurement of a Variable

Rather than classify a variable as qualitative or quantitative, we can assign a level of measurement to the variable.

Definitions

A variable is at the **nominal level of measurement** if the values of the variable name, label, or categorize. In addition, the naming scheme does not allow for the values of the variable to be arranged in a ranked or specific order.

In Other Words
The word *nominal* comes from the Latin word **nomen**, which means to name. When you see the word **ordinal**, think order.

A variable is at the **ordinal level of measurement** if it has the properties of the nominal level of measurement, however the naming scheme allows for the values of the variable to be arranged in a ranked or specific order.

A variable is at the **interval level of measurement** if it has the properties of the ordinal level of measurement and the differences in the values of the variable have meaning. A value of zero does not mean the absence of the quantity. Arithmetic operations such as addition and subtraction can be performed on values of the variable.

A variable is at the **ratio level of measurement** if it has the properties of the interval level of measurement and the ratios of the values of the variable have meaning. A value of zero means the absence of the quantity. Arithmetic operations such as multiplication and division can be performed on the values of the variable.

Nominal or ordinal variables are also qualitative variables. Interval or ratio variables are also quantitative variables.

EXAMPLE 6   Determining the Level of Measurement of a Variable

Problem For each of the following variables, determine the level of measurement.
(a) Gender
(b) Temperature
(c) Number of days during the past week that a college student studied
(d) Letter grade earned in your statistics class

Approach For each variable, we ask the following: Does the variable simply categorize each individual? If so, the variable is nominal. Does the variable categorize *and* allow ranking of each value of the variable? If so, the variable is ordinal. Do differences in values of the variable have meaning, but a value of zero does not mean the absence of the quantity? If so, the variable is interval. Do ratios of values of the variable have meaning *and* there is a natural zero starting point? If so, the variable is ratio.

Solution
(a) Gender is a variable measured at the nominal level because it only allows for categorization of male or female. Plus, it is not possible to rank gender classifications.
(b) Temperature is a variable measured at the interval level because differences in the value of the variable make sense. For example, 70°F is 10°F warmer than 60°F. Notice that the ratio of temperatures does not represent a meaningful result. For example, 60°F is not twice as warm as 30°F. In addition, 0°F does not represent the absence of heat.
(c) Number of days during the past week that a college student studied is measured at the ratio level, because the ratio of two values makes sense and a value of zero has meaning. For example, a student who studies four days studies twice as many days as a student who studies two days.
(d) Letter grade is a variable measured at the ordinal level because the values of the variable can be ranked, but differences in values have no meaning. For example, an A is better than a B, but A − B has no meaning.

• **Now Work Problem 31**

When classifying variables according to their level of measurement, it is extremely important that we recognize what the variable is intended to measure. For example, suppose we want to know whether cars with 4-cylinder engines get better gas mileage than cars with 6-cylinder engines. Here, engine size represents a category of data and so the variable is nominal. On the other hand, if we want to know the average number of cylinders in cars in the United States, the variable is classified as ratio (an 8-cylinder engine has twice as many cylinders as a 4-cylinder engine).

1.1 Assess Your Understanding

Vocabulary and Skill Building

1. Define statistics.

2. Explain the difference between a population and a sample.

3. A(n) _____ is a person or object that is a member of the population being studied.

4. _____ statistics consists of organizing and summarizing information collected, while _____ statistics uses methods that generalize results obtained from a sample to the population and measure the reliability of the results.

5. A(n) _____ is a numerical summary of a sample.
A(n) _____ is a numerical summary of a population.

6. _____ are the characteristics of the individuals of the population being studied.

In Problems 7–14, determine whether the underlined value is a parameter or a statistic.

NW **7. State Government** Following the 2014 national midterm election, *18%* of the governors of the 50 United States were female. *Source:* National Governors Association

8. Calculus Exam The average score for a class of 28 students taking a calculus midterm exam was 72%.

9. School Bullies In a national survey of 1300 high school students (grades 9 to 12), *32%* of respondents reported that someone had bullied them at school. *Source:* Bureau of Justice Statistics

10. Drug Use In a national survey on substance abuse, 9.6% of respondents aged 12 to 17 reported using illicit drugs within the past month. *Source:* Substance Abuse and Mental Health Services Administration, *Results from the 2012 National Survey on Drug Use and Health: National Findings*

11. Batting Average Ty Cobb is one of Major League Baseball's greatest hitters of all time, with a career batting average of 0.366. *Source:* baseball-almanac.com

12. Moonwalkers Only 12 men have walked on the moon. The average time these men spent on the moon was 43.92 hours. *Source:* www.theguardian.com

13. Hygiene Habits A study of 6076 adults in public rest rooms (in Atlanta, Chicago, New York City, and San Francisco) found that 23% did not wash their hands before exiting.
Source: American Society of Microbiology and the Soap and Detergent Association

14. Public Knowledge Interviews of 100 adults 18 years of age or older, conducted nationwide, found that 44% could state the minimum age required for the office of U.S. president.
Source: Newsweek Magazine

In Problems 15–22, classify the variable as qualitative or quantitative.

NW **15.** Nation of origin

16. Number of siblings

17. Grams of carbohydrates in a doughnut

18. Number on a football player's jersey

19. Number of unpopped kernels in a bag of microwave popcorn

20. Assessed value of a house

21. Phone number

22. Student ID number

In Problems 23–30, determine whether the quantitative variable is discrete or continuous.

NW **23.** Goals scored in a season by a soccer player

24. Volume of water lost each day through a leaky faucet

25. Length (in minutes) of a country song

26. Number of Sequoia trees in a randomly selected acre of Yosemite National Park

27. High temperature on a randomly selected day in Memphis, Tennessee

28. Internet connection speed in kilobytes per second

29. Points scored in an NCAA basketball game

30. Air pressure in pounds per square inch in an automobile tire

In Problems 31–38, determine the level of measurement of each variable.

NW **31.** Nation of origin

32. Movie ratings of one star through five stars

33. Volume of water used by a household in a day

34. Year of birth of college students

35. Highest degree conferred (high school, bachelor's, and so on)

36. Eye color

37. Assessed value of a house

38. Time of day measured in military time

In Problems 39–44, a research objective is presented. For each, identify the population and sample in the study.

39. The Gallup Organization contacts 1028 teenagers who are 13 to 17 years of age and live in the United States and asks whether or not they had been prescribed medications for any mental disorders, such as depression or anxiety.

40. A quality-control manager randomly selects 50 bottles of Coca-Cola that were filled on October 15 to assess the calibration of the filling machine.

41. A farmer interested in the weight of his soybean crop randomly samples 100 plants and weighs the soybeans on each plant.

42. Every year the U.S. Census Bureau releases the *Current Population Report* based on a survey of 50,000 households. The goal of this report is to learn the demographic characteristics, such as income, of all households within the United States.

43. Folate and Hypertension Researchers want to determine whether or not higher folate intake is associated with a lower risk of hypertension (high blood pressure) in women (27 to 44 years of age). To make this determination, they look at 7373 cases of hypertension in these women and find that those who consume at least 1000 micrograms per day (μg/d) of total folate had a decreased risk of hypertension compared with those who consume less than 200 μg/d. *Source:* John P. Forman, MD; Eric B. Rimm, ScD; Meir J. Stampfer, MD; Gary C. Curhan, MD, ScD, "Folate Intake and the Risk of Incident Hypertension among US Women," *Journal of the American Medical Association* 293:320–329, 2005

44. A community college notices that an increasing number of full-time students are working while attending the school. The administration randomly selects 128 students and asks how many hours per week each works.

In Problems 45 and 46, identify the individuals, variables, and data corresponding to the variables. Determine whether each variable is qualitative, continuous, or discrete.

NW **45. Driver's License Laws** The following data represent driver's license laws for various states.

State	Minimum Age for Driver's License (unrestricted)	Mandatory Belt Use Seating Positions	Maximum Allowable Speed Limit (cars on rural interstate), mph
Alabama	17	Front	70
Colorado	17	Front	75
Indiana	18	All	70
North Carolina	16	All	70
Wisconsin	18	All	65

Source: Governors Highway Safety Association

46. BMW Cars The following information relates to the 2011 model year product line of BMW automobiles.

Model	Body Style	Weight (lb)	Number of Seats
3 Series	Coupe	3362	4
5 Series	Sedan	4056	5
6 Series	Convertible	4277	4
7 Series	Sedan	4564	5
X3	Sport utility	4012	5
Z4	Roadster Coupe	3505	2

Source: www.motortrend.com

Applying the Concepts

47. Smoker's IQ A study was conducted in which 20,211 18-year-old Israeli male military recruits were given an exam to measure IQ. In addition, the recruits were asked to disclose their smoking status. An individual was considered a smoker if he smoked at least one cigarette per day. The goal of the study was to determine whether adolescents aged 18 to 21 who smoke have a lower IQ than nonsmokers. It was found that the average IQ of the smokers was 94, while the average IQ of the nonsmokers was 101. The researchers concluded that lower IQ individuals are more likely to choose to smoke, not that smoking makes people less intelligent.

Source: Weiser, M., Zarka, S., Werbeloff, N., Kravitz, E. and Lubin, G. (2010). "Cognitive Test Scores in Male Adolescent Cigarette Smokers Compared to Non-smokers: A Population-Based Study." *Addiction.* 105:358–363. doi: 10.1111/j.1360-0443.2009.02740.x)

(a) What is the research objective?
(b) What is the population being studied? What is the sample?
(c) What are the descriptive statistics?
(d) What are the conclusions of the study?

48. A Cure for the Common Wart A study conducted by researchers was designed "to determine if application of duct tape is as effective as cryotherapy (liquid nitrogen applied to the wart for 10 seconds every 2 to 3 weeks) in the treatment of common warts." The researchers randomly divided 51 patients into two groups. The 26 patients in group 1 had their warts treated by applying duct tape to the wart for 6.5 days and then removing the tape for 12 hours, at which point the cycle was repeated for a maximum of 2 months. The 25 patients in group 2 had their warts treated by cryotherapy for a maximum of six treatments. Once the treatments were complete, it was determined that 85% of the patients in group 1 and 60% of the patients in group 2 had complete resolution of their warts. The researchers concluded that duct tape is significantly more effective in treating warts than cryotherapy.

Source: Dean R. Focht III, Carole Spicer, Mary P. Fairchok. "The Efficacy of Duct Tape vs. Cryotherapy in the Treatment of Verruca Vulgaris (The Common Wart)," *Archives of Pediatrics and Adolescent Medicine,* 156(10), 2002

(a) What is the research objective?
(b) What is the population being studied? What is the sample?
(c) What are the descriptive statistics?
(d) What are the conclusions of the study?

NW **49. Government Waste** Gallup News Service conducted a survey of 1017 American adults aged 18 years or older, September 4–7, 2014. The respondents were asked, "Of every tax dollar that goes to the federal government in Washington, D.C., how many cents of each dollar would you say are wasted?" Of the 1017 individuals surveyed, 35% indicated that 51 cents or more is wasted. Gallup reported that 35% of all adult Americans 18 years or older believe the federal government wastes at least 51 cents of each dollar spent, with a margin of error of 4% and a 95% level of confidence.

(a) What is the research objective?
(b) What is the population?
(c) What is the sample?
(d) List the descriptive statistics.
(e) What can be inferred from this survey?

50. Investment Decision The Gallup Organization conducted a survey of 1018 adults, aged 18 and older, living in the United States and asked, "If you had a thousand dollars to spend, do you think investing it in the stock market would be a good or bad idea?" Of the 1018 adults, 46% said it would be a bad idea. The Gallup Organization reported that 46% of all adults, aged 18 and older, living in the United States thought it was a bad idea to invest $1000 in the stock market with a 4% margin of error with 95% confidence.

(a) What is the research objective?
(b) What is the population?
(c) What is the sample?
(d) List the descriptive statistics.
(e) What can be inferred from this survey?

51. What Level of Measurement? It is extremely important for a researcher to clearly define the variables in a study because this helps to determine the type of analysis that can be performed on the data. For example, if a researcher wanted to describe baseball players based on jersey number, what level of measurement would the variable *jersey number* be? Now suppose the researcher felt that certain players who were of lower caliber received higher numbers. Does the level of measurement of the variable change? If so, how?

52. Interpreting the Variable Suppose a fundraiser holds a raffle for which each person who enters the room receives a ticket numbered 1 to N, where N is the number of people at the fundraiser. The first person to arrive receives ticket number 1, the second person receives ticket number 2, and so on. Determine the level of measurement for each of the following interpretations of the variable *ticket number*.

(a) The winning ticket number.

(b) The winning ticket number was announced as 329. An attendee noted his ticket number was 294 and stated, "I guess I arrived too early."

(c) The winning ticket number was announced as 329. An attendee looked around the room and commented, "It doesn't look like there are 329 people in attendance."

53. Analyze the Article Read the newspaper article and answer the following questions:

(a) What is the research question the study addresses?

(b) What is the sample?

(c) What type of variable is season in which you were born?

(d) What can be said (in general) about individuals born in summer? Winter?

(e) What conclusion was drawn from the study?

Season of Birth Affects Your Mood Later In Life by Nicola Fifield

Babies born in the summer are much more likely to suffer from mood swings when they grow up while those born in the winter are less likely to become irritable adults, scientists claim.

Researchers studied 400 people and matched their personality type to when in the year they were born.

They claim that people born at certain times of the year have a far greater chance of developing certain types of temperaments, which can lead to mood disorders.

The scientists, from Budapest, said this was because the seasons had an influence on certain monoamine neurotransmitters, such as dopamine and serotonin, which control mood, however more research was needed to find out why.

They discovered that the number of people with a "cyclothymic" temperament, characterized by rapid, frequent swings between sad and cheerful moods, was significantly higher in those born in the summer.

Those with a hyperthymic temperament, a tendency to be excessively positive, was significantly higher among those born in the spring and summer.

The study also found that those born in the autumn were less likely to be depressive, while those born in winter were less likely to be irritable.

Lead researcher, assistant professor Xenia Gonda, said: "Biochemical studies have shown that the season in which you are born has an influence on certain monoamine neurotransmitters, such as dopamine and serotonin, which

is detectable even in adult life. This led us to believe that birth season may have a longer-lasting effect.

"Our work looked at 400 subjects and matched their birth season to personality types in later life.

"Basically, it seems that when you are born may increase or decrease your chance of developing certain mood disorders.

Professor Gonda added: "We can't yet say anything about the mechanisms involved.

What we are now looking at is to see if there are genetic markers which are related to season of birth and mood disorder".

The study may well provide a clue as to why some of the nation's best known personalities are good natured, while others are slightly grumpier.

The Duchess of Cambridge was born in winter, on January 9, which according to the study, means she is less likely to be irritable while Roy Keane, the famously hot-headed former Manchester United footballer, was born in August, when the scientists say people are more likely to have mood swings.

Mary Berry, the ever-cheerful presenter of the Great British Bake Off, was born in the Spring, when, according to the study, people are more likely to be excessively positive.

The study is being presented at the annual conference of the European College of Neuropsychopharmacology (ECNP) in Berlin, Germany, on Sunday.

Professor Eduard Vieta, from the ECNP, said: "Although both genetic and environmental factors are involved in one's temperament, now we know that the season at birth plays a role too.

"And the finding of "high mood" tendency (hyperthymic temperament) for those born in summer is quite intriguing."
The Telegraph, October 19, 2014

Source: Season of Birth Affects Your Mood Later In Life by Nicola Fifield from The Telegraph. Copyright © 2014 by Telegraph Media Group Limited.

Explaining the Concepts

54. Contrast the differences between qualitative and quantitative variables.

55. Discuss the differences between discrete and continuous variables.

56. In your own words, define the four levels of measurement of a variable. Give an example of each.

57. Explain what is meant when we say "data vary." How does this variability affect the results of statistical analysis?

58. Explain the process of statistics.

59. The age of a person is commonly considered to be a continuous random variable. Could it be considered a discrete random variable instead? Explain.

1.2 Observational Studies versus Designed Experiments

Objectives   ① Distinguish between an observational study and an experiment

② Explain the various types of observational studies

① Distinguish between an Observational Study and an Experiment

Once our research question is developed, we must develop methods for obtaining the data that can be used to answer the questions posed in our research objective. There are two methods for collecting data, *observational studies* and *designed experiments*. To see the difference between these two methods, read the following two studies.

EXAMPLE 1   Cellular Phones and Brain Tumors

Researchers wanted to determine whether there is an association between mobile phone use and brain tumors. To do so, 791,710 middle-aged women in the United Kingdom were followed over a period of 7 years. During this time, there were 1261 incidences of brain tumors. The researchers compared the women who never used a mobile phone to those who used mobile phones and found no significant difference in the incidence rate of brain tumors between the two groups.

Source: Benson, V. S. et al. "Mobile Phone Use and Risk of Brain Neoplasms and Other Cancers: Prospective Study," *International Journal of Epidemiology* 2013 Jun; 42(3); 792–802   •

EXAMPLE 2   Cellular Phones and Brain Tumors

Researchers examined "whether chronic exposure to radio frequency (RF) radiation at two common cell phone signals—835.62 megahertz, a frequency used by analogue cell phones, and 847.74 megahertz, a frequency used by digital cell phones—caused brain tumors in rats." To do so, the researchers randomly divided 480 rats into three groups. The rats in group 1 were exposed to the analogue cell phone frequency; the rats in group 2 were exposed to the digital frequency; the rats in group 3 received no radiation. The exposure was done for 4 hours a day, 5 days a week for 2 years. The rats in all three groups were treated the same, except for the RF exposure.

   After 505 days of exposure, the researchers ". . . found no statistically significant increases in any tumor type, including brain, liver, lung or kidney, compared to the control group."

Source: M. La Regina, E. Moros, W. Pickard, W. Straube, J. L. Roti Roti, "The Effect of Chronic Exposure to 835.62 MHz FMCW or 847.74 MHz CDMA on the Incidence of Spontaneous Tumors in Rats," *Bioelectromagnetic Society Conference,* June 25, 2002   •

   In both studies, the goal was to determine if radio frequencies from cell phones increase the risk of contracting brain tumors. Whether or not brain cancer was contracted is the *response variable*. The level of cell phone usage is the *explanatory variable*. In research, we wish to determine how varying the amount of an **explanatory variable** affects the value of a **response variable**.

   What are the differences between the studies in Examples 1 and 2? Obviously, in Example 1 the study was conducted on humans, while the study in Example 2 was conducted on rats. However, there is a bigger difference. In Example 1, no attempt was made to influence the individuals in the study. The researchers simply followed the women over time to determine their use of cell phones. In other words, no attempt was made to influence the value of the explanatory variable, radio-frequency exposure (cell phone use). Because the researchers simply recorded the behavior of the participants, the study in Example 1 is an *observational study*.

Definition | An **observational study** measures the value of the response variable without attempting to influence the value of either the response or explanatory variables. That is, in an observational study, the researcher observes the behavior of the individuals without trying to influence the outcome of the study.

In the study in Example 2, the researchers obtained 480 rats and divided the rats into three groups. Each group was *intentionally* exposed to various levels of radiation. The researchers then compared the number of rats that had brain tumors. Clearly, there was an attempt to influence the individuals in this study because the value of the explanatory variable (exposure to radio frequency) was influenced. Because the researchers controlled the value of the explanatory variable, we call the study in Example 2 a *designed experiment*.

Definition

• Now Work Problem 9

If a researcher assigns the individuals in a study to a certain group, intentionally changes the value of an explanatory variable, and then records the value of the response variable for each group, the study is a **designed experiment**.

Which Is Better? A Designed Experiment or an Observational Study?

To answer this question, let's consider another study.

EXAMPLE 3   Do Flu Shots Benefit Seniors?

Researchers wanted to determine the long-term benefits of the influenza vaccine on seniors aged 65 years and older by looking at records of over 36,000 seniors for 10 years. The seniors were divided into two groups. Group 1 were seniors who chose to get a flu vaccination shot, and group 2 were seniors who chose not to get a flu vaccination shot. After observing the seniors for 10 years, it was determined that seniors who get flu shots are 27% less likely to be hospitalized for pneumonia or influenza and 48% less likely to die from pneumonia or influenza.

Source: Kristin L. Nichol, MD, MPH, MBA, James D. Nordin, MD, MPH, David B. Nelson, PhD, John P. Mullooly, PhD, Eelko Hak, PhD. "Effectiveness of Influenza Vaccine in the Community-Dwelling Elderly," *New England Journal of Medicine* 357:1373–1381, 2007

Wow! The results of this study sound great! All seniors should go out and get a flu shot. Right? Not necessarily. The authors were concerned about *confounding*. They were concerned that lower hospitalization and death rates may have been due to something other than the flu shot. Could it be that seniors who get flu shots are more health conscious or are able to get to the clinic more easily? Does race, income, or gender play a role in whether one might contract (and possibly die from) influenza?

Definition | **Confounding** in a study occurs when the effects of two or more explanatory variables are not separated. Therefore, any relation that may exist between an explanatory variable and the response variable may be due to some other variable or variables not accounted for in the study.

Confounding is potentially a major problem with observational studies. Often, the cause of confounding is a *lurking variable*.

Definition A **lurking variable** is an explanatory variable that was not considered in a study, but that affects the value of the response variable in the study. In addition, lurking variables are typically related to explanatory variables considered in the study.

In the influenza study, possible lurking variables might be age, health status, or mobility of the senior. How can we manage the effect of lurking variables? One possibility is to look at the individuals in the study to determine if they differ in any significant way. For example, it turns out in the influenza study that the seniors who elected to get a flu shot were actually *less* healthy than those who did not. The researchers also accounted for race and income. The authors identified another potential lurking variable, *functional status*, meaning the ability of the seniors to conduct day-to-day activities on their own. The authors were able to adjust their results for this variable as well.

Even after accounting for all the potential lurking variables in the study, the authors were still careful to conclude that getting an influenza shot is *associated* with a lower risk of being hospitalized or dying from influenza. The authors used the term *associated*, instead of saying the influenza shots *caused* a lower risk of death, because the study was observational.

 Observational studies do not allow a researcher to claim causation, only association.

Designed experiments, on the other hand, are used whenever control of certain variables is possible and desirable. This type of research allows the researcher to identify certain cause and effect relationships among the variables in the study.

So why ever conduct an observational study if we can't claim causation? Often, it is unethical to conduct an experiment. Consider the link between smoking and lung cancer. In a designed experiment to determine if smoking causes lung cancer in humans, a researcher would divide a group of volunteers into group 1 who would smoke a pack of cigarettes every day for the next 10 years, and group 2 who would not smoke. In addition, eating habits, sleeping habits, and exercise would be controlled so that the only difference between the two groups was smoking. After 10 years the experiment's researcher would compare the proportion of participants in the study who contract lung cancer in the smoking group to the nonsmoking group. If the two proportions differ significantly, it could be said that smoking causes cancer. This designed experiment is able to control many of the factors that might affect whether one contracts lung cancer that would not be controlled in an observational study, however, it is a very unethical study.

Other reasons exist for conducting observational studies over designed experiments. An article in support of observational studies states, "observational studies have several advantages over designed experiments, including lower cost, greater timeliness, and a broader range of patients." (*Source:* Kjell Benson, BA, and Arthur J. Hartz, MD, PhD. "A Comparison of Observational Studies and Randomized, Controlled Trials," *New England Journal of Medicine* 342:1878–1886, 2000).

One final thought regarding confounding. In designed experiments, it is possible to have two explanatory variables in a study that are related to each other and related to the response variable. For example, suppose Professor Egner wanted to conduct an experiment in which she compared student success using online homework versus traditional textbook homework. To do the study, she taught her morning statistics class using the online homework and her afternoon class using traditional textbook homework. At the end of the semester, she compared the final exam scores for the online section to the textbook section. If the morning section had higher scores, could Professor Egner conclude that online homework is the cause of higher exam scores? Not necessarily. It is possible that the morning class had students who were more motivated. It is impossible to know whether the outcome was due to the online homework or to the time at which the class was taught. In this sense, we say that the time of day the class is taught is a *confounding variable*.

Definition

> A **confounding variable** is an explanatory variable that was considered in a study whose effect cannot be distinguished from a second explanatory variable in the study.

The big difference between lurking variables and confounding variables is that lurking variables are not considered in the study (for example, we did not consider lifestyle in the pneumonia study) whereas confounding variables are measured in the study (for example, we measured morning versus afternoon classes).

So lurking variables are related to both the explanatory and response variables, and this relation is what creates the apparent association between the explanatory and response variable in the study. For example, lifestyle (healthy or not) is associated with the likelihood of getting an influenza shot as well as the likelihood of contracting pneumonia or influenza.

A confounding variable is a variable in a study that does not necessarily have any association with the other explanatory variable, but does have an effect on the response variable. Perhaps morning students are more motivated, and this is what led to the higher final exam scores, not the homework delivery system.

The bottom line is that both lurking variables and confounding variables can confound the results of a study, so a researcher should be mindful of their potential existence.

We will continue to look at obtaining data through various types of observational studies until Section 1.6, when we will look at designed experiments.

② Explain the Various Types of Observational Studies

There are three major categories of observational studies: (1) cross-sectional studies, (2) case-control studies, and (3) cohort studies.

Cross-sectional Studies   These observational studies collect information about individuals at a specific point in time or over a very short period of time.

For example, a researcher might want to assess the risk associated with smoking by looking at a group of people, determining how many are smokers, and comparing the rate of lung cancer of the smokers to the nonsmokers.

An advantage of cross-sectional studies is that they are cheap and quick to do. However, they have limitations. For our lung cancer study, individuals might develop cancer after the data are collected, so our study will not give the full picture.

Case-control Studies   These studies are **retrospective**, meaning that they require individuals to look back in time or require the researcher to look at existing records. In case-control studies, individuals who have a certain characteristic may be matched with those who do not.

For example, we might match individuals who smoke with those who do not. When we say "match" individuals, we mean that we would like the individuals in the study to be as similar (homogeneous) as possible in terms of demographics and other variables that may affect the response variable. Once homogeneous groups are established, we would ask the individuals in each group how much they smoked over the past 25 years. The rate of lung cancer between the two groups would then be compared.

A disadvantage to this type of study is that it requires individuals to recall information from the past. It also requires the individuals to be truthful in their responses. An advantage of case-control studies is that they can be done relatively quickly and inexpensively.

Cohort Studies   A cohort study first identifies a group of individuals to participate in the study (the cohort). The cohort is then observed over a long period of time. During this period, characteristics about the individuals are recorded and some individuals will be exposed to certain factors (not intentionally) and others will not. At the end of the study the value of the response variable is recorded for the individuals.

Typically, cohort studies require many individuals to participate over long periods of time. Because the data are collected over time, cohort studies are **prospective**. Another problem with cohort studies is that individuals tend to drop out due to the long time frame. This could lead to misleading results. That said, cohort studies are the most powerful of the observational studies.

One of the largest cohort studies is the Framingham Heart Study. In this study, more than 10,000 individuals have been monitored since 1948. The study continues to this day, with the grandchildren of the original participants taking part in the study. This cohort study is responsible for many of the breakthroughs in understanding heart disease. Its cost is in excess of $10 million.

Some Concluding Remarks about Observational Studies versus Designed Experiments

Is a designed experiment always superior to an observational study? Not necessarily. Plus, observational studies play a role in the research process. For example, because cross-sectional and case-control observational studies are relatively inexpensive, they allow researchers to explore possible associations prior to undertaking large cohort studies or designing experiments.

Also, it is not always possible to conduct an experiment. For example, we could not conduct an experiment to investigate the perceived link between high tension wires and leukemia (on humans). Do you see why?

• Now Work Problem 19

Existing Sources of Data and Census Data

The saying "*There is no point in reinventing the wheel*" applies to spending energy obtaining data that already exist. If a researcher wishes to conduct a study and an appropriate data set exists, it would be silly to collect the data from scratch. For example, various federal agencies regularly collect data that are available to the public. Some of these agencies include the Centers for Disease Control and Prevention (www.cdc.gov), the Internal Revenue Service (www.irs.gov), and the Department of Justice (www.bjs.gov). A great website that lists virtually all the sources of federal data is fedstats.sites.usa.gov. Another great source of data is the General Social Survey (GSS) administered by the University of Chicago. This survey regularly asks "demographic and attitudinal questions" of individuals around the country. The website is https://gssdataexplorer.norc.org.

Another source of data is a *census*.

Definition

> A **census** is a list of all individuals in a population along with certain characteristics of each individual.

The United States conducts a census every 10 years to learn the demographic makeup of the United States. Everyone whose usual residence is within the borders of the United States must fill out a questionnaire packet. The cost of obtaining the census in 2010 was approximately $5.4 billion; about 635,000 temporary workers were hired to assist in collecting the data.

Why is the U.S. Census so important? The results of the census are used to determine the number of representatives in the House of Representatives in each state, congressional districts, distribution of funds for government programs (such as Medicaid), and planning for the construction of schools and roads. The first census of the United States was obtained in 1790 under the direction of Thomas Jefferson. It is a constitutional mandate that a census be conducted every 10 years.

Is the United States successful in obtaining a census? Not entirely. Some individuals go uncounted due to illiteracy, language issues, and homelessness. Given the political stakes that are based on the census, politicians often debate how to count these individuals. Statisticians have offered solutions to the counting problem. If you wish, go to www.census.gov and in the search box type *count homeless*. You will find many articles on the Census Bureau's attempt to count the homeless. The bottom line is that even census data can have flaws.

1.2 Assess Your Understanding

Vocabulary and Skill Building

1. In your own words, define explanatory variable and response variable.

2. What is an observational study? What is a designed experiment? Which allows the researcher to claim causation between an explanatory variable and a response variable?

3. Explain what is meant by confounding. What is a lurking variable? What is a confounding variable?

4. Given a choice, would you conduct a study using an observational study or a designed experiment? Why?

5. What is a cross-sectional study? What is a case-control study? Which is the superior observational study? Why?

6. The data used in the influenza study presented in Example 3 were obtained from a cohort study. What does this mean? Why is a cohort study superior to a case-control study?

7. Explain why it would be unlikely to use a designed experiment to answer the research question posed in Example 3.

8. What does it mean when an observational study is retrospective? What does it mean when an observational study is prospective?

In Problems 9–16, determine whether the study depicts an observational study or an experiment.

NW 9. Cancer Study The American Cancer Society is beginning a study to learn why some people never get cancer. To take part in the study, a person must be 30–65 years of age and never had cancer. The study requires that the participants fill out surveys about their health and habits and give blood samples and waist measurements. These surveys must be filled out every two years. The study is expected to last for the next 20 years.

10. Rats with cancer are divided into two groups. One group receives 5 milligrams (mg) of a medication that is thought to fight cancer, and the other receives 10 mg. After 2 years, the spread of the cancer is measured.

11. Seventh-grade students are randomly divided into two groups. One group is taught math using traditional techniques; the other is taught math using a reform method. After 1 year, each group is given an achievement test to compare proficiency.

12. Hair and Heart Disease A study in which balding men were compared with non-balding men at one point in time found that balding men were 70% more likely to have heart disease.
Source: USA Today, April 4, 2013

13. A survey is conducted asking 400 people, "Do you prefer Coke or Pepsi?"

14. Two hundred people are asked to perform a taste test in which they drink from two randomly placed, unmarked cups and asked which drink they prefer.

15. Sixty patients with carpal tunnel syndrome are randomly divided into two groups. One group is treated weekly with both acupuncture and an exercise regimen. The other is treated weekly with the exact same exercise regimen, but no acupuncture. After 1 year, both groups are questioned about their level of pain due to carpal tunnel syndrome.

16. Conservation agents netted 250 large-mouth bass in a lake and determined how many were carrying parasites.

Applying the Concepts

17. Happiness and Your Heart Is there an association between level of happiness and the risk of heart disease? Researchers studied 1739 people over a 10 year period and asked questions about their daily lives and the hassles they face. The researchers also determined which individuals in the study experienced any type of heart disease. After their analysis, they concluded that happy individuals are less likely to experience heart disease.
Source: European Heart Journal 31 (9):1065–1070, February 2010.

(a) What type of observational study is this? Explain.

(b) What is the response variable? What is the explanatory variable?

(c) In the report, the researchers stated that "the research team also hasn't ruled out that a common factor like genetics could be causing both the emotions and the heart disease." Use the language introduced in this section to explain what this sentence means.

18. Daily Coffee Consumption Is there an association between daily coffee consumption and the occurrence of skin cancer? Researchers asked 93,676 women to disclose their coffee-drinking habits and also determined which of the women had nonmelanoma skin cancer. The researchers concluded that consumption of six or more cups of caffeinated coffee per day was associated with a reduction in nonmelanoma skin cancer.
Source: European Journal of Cancer Prevention, 16(5): 446–452, October 2007.

(a) What type of observational study was this? Explain.

(b) What is the response variable in the study? What is the explanatory variable?

(c) In their report, the researchers stated that "After adjusting for various demographic and lifestyle variables, daily consumption of six or more cups was associated with a 30% reduced prevalence of nonmelanoma skin cancer." Why was it important to adjust for these variables?

NW 19. Television in the Bedroom Is a television (TV) in the bedroom associated with obesity? Researchers questioned 379 twelve-year old adolescents and concluded that the body mass index (BMI) of the adolescents who had a TV in their bedroom was significantly higher than the BMI of those who did not have a TV in their bedroom.
Source: Christelle Delmas, Carine Platat, Brigette Schweitzer, Aline Wagner, Mohamed Oujaa, and Chantal Simon. "Association Between Television in Bedroom and Adiposity Throughout Adolescence," *Obesity*, 15:2495–2503, 2007.

(a) Why is this an observational study? What type of observational study is this?

(b) What is the response variable in the study? What is the explanatory variable?

(c) Can you think of any lurking variables that may affect the results of the study?

(d) In the report, the researchers stated, "These results remain significant after adjustment for socioeconomic status." What does this mean?

(e) Can we conclude that a television in the bedroom causes a higher body mass index? Explain.

20. Get Married, Gain Weight Are young couples who marry or cohabitate more likely to gain weight than those who stay single? Researchers followed 8000 men and women for 7 years.

At the start of the study, none of the participants were married or living with a romantic partner. The researchers found that women who married or cohabitated during the study gained 9 pounds more than single women, and married or cohabitating men gained, on average, 6 pounds more than single men.

(a) Why is this an observational study? What type of observational study is this?

(b) What is the response variable in the study? What is the explanatory variable?

(c) Identify some potential lurking variables in this study.

(d) Can we conclude that getting married or cohabiting causes one to gain weight? Explain.

21. **Midwives** Researchers Sally Tracy and associates undertook a cross-sectional study looking at the method of delivery and cost of delivery for first-time "low risk" mothers under three delivery scenarios:

(1) Caseload midwifery

(2) Standard hospital care

(3) Private obstetric care

The results of the study revealed that 58.5% of all births with midwifery were vaginal deliveries compared with 48.2% of standard hospital births and 30.8% of private obstetric care. In addition, the costs of delivery from midwifery was $3903.78 compared with $5279.23 for standard hospital care and $5413.69 for private obstetric care.

Source: Sally K Tracy, Alec Welsh, Donna Hartz, Anne Lainchbury, Andrew Bisits, Jan White, and Mark Tracy "Caseload midwifery compared to standard or private obstetric care for first time mothers in a public teaching hospital in Australia: a cross sectional study of cost and birth outcomes" *BMC Pregnancy and Childbirth* 2014, 14:46

(a) Why is this a cross-sectional observational study?

(b) Name the explanatory variable in the study.

(c) Name the two response variables in the study and determine whether each is qualitative or quantitative.

22. **Web Page Design** Magnum, LLC, is a web page design firm that has two designs for an online hardware store. To determine which is the more effective design, Magnum uses one page in the Denver area and a second page in the Miami area. For each visit, Magnum records the amount of time visiting the site and the amount spent by the visitor.

(a) What is the explanatory variable in this study? Is it qualitative or quantitative?

(b) What are the two response variables? For each response variable, state whether it is qualitative or quantitative.

(c) Explain how confounding might be an issue with this study.

23. **Analyze the Article** Write a summary of the following opinion. The opinion was posted at abcnews.com. Include the type of study conducted, possible lurking variables, and conclusions. What is the message of the author of the article?

Power Lines and Cancer—To Move or Not to Move

New Research May Cause More Fear Than Warranted, One Physician Explains

OPINION by JOSEPH MOORE, M.D.

A recent study out of Switzerland indicates there might be an increased risk of certain blood cancers in people with prolonged exposure to electromagnetic fields, like those generated from high-voltage power lines.

If you live in a house near one of these high-voltage power lines, a study like this one might make you wonder whether you should move.

But based on what we know now, I don't think that's necessary. We can never say there is no risk, but we can say that the risk appears to be extremely small.

"Scare Science"

The results of studies like this add a bit more to our knowledge of potential harmful environmental exposures, but they should also be seen in conjunction with the results of hundreds of studies that have gone before. It cannot be seen as a definitive call to action in and of itself.

The current study followed more than 20,000 Swiss railway workers over a period of 30 years. True, that represents a lot of people over a long period of time.

However, the problem with many epidemiological studies, like this one, is that it is difficult to have an absolute control group of people to compare results with. The researchers compared the incidence of different cancers of workers with a high amount of electromagnetic field exposure to those workers with lower exposures.

These studies aren't like those that have identified definitive links between an exposure and a disease— like those involving smoking and lung cancer. In those studies, we can actually measure the damage done to lung tissue as a direct result of smoking. But usually it's very difficult for the conclusions of an epidemiological study to rise to the level of controlled studies in determining public policy.

Remember the recent scare about coffee and increased risk of pancreatic cancer? Or the always-simmering issue of cell phone use and brain tumors?

As far as I can tell, none of us have turned in our cell phones. In our own minds, we've decided that any links to cell phone use and brain cancer have not been proven definitively. While we can't say that there is absolutely no risk in using cell phones, individuals have determined on their own that the potential risks appear to be quite small and are outweighed by the benefits.

Findings Shouldn't Lead to Fear

As a society, we should continue to investigate these and other related exposures to try to prove one way or another whether they are disease-causing. If we don't continue to study, we won't find out. It's that simple.

When findings like these come out, and I'm sure there will be more in the future, I would advise people not to lose their heads. Remain calm. You should take the results as we scientists do—as intriguing pieces of data about a problem we will eventually learn more about, either positively or negatively, in the future. It should not necessarily alter what we do right now.

What we can do is take actions that we know will reduce our chances of developing cancer.

Stop smoking and avoid passive smoke. It is the leading cause of cancer that individuals have control over.

Whenever you go outside, put on sunscreen or cover up.

Eat a healthy diet and stay physically active.

Make sure you get tested or screened. Procedures like colonoscopies, mammograms, pap smears and prostate exams can catch the early signs of cancer, when the chances of successfully treating them are the best.

Taking the actions above will go much farther in reducing your risks for cancer than moving away from power lines or throwing away your cell phone.

Dr. Joseph Moore is a medical oncologist at Duke University Comprehensive Cancer Center.

Source: Copyright © by Joseph Moore.

24. **Putting It Together: Passive Smoke?** The following abstract appears in *The New England Journal of Medicine:*

BACKGROUND. The relation between passive smoking and lung cancer is of great public health importance. Some previous studies have suggested that exposure to environmental tobacco smoke in the household can cause lung cancer, but others have found no effect. Smoking by the spouse has been the most commonly used measure of this exposure.

METHODS. In order to determine whether lung cancer is associated with exposure to tobacco smoke within the household, we conducted a case-control study of 191 patients with lung cancer who had never smoked and an equal number of persons without lung cancer who had never smoked. Lifetime residential histories including information on exposure to environmental tobacco smoke were compiled and analyzed. Exposure was measured in terms of "smoker-years," determined by multiplying the number of years in each residence by the number of smokers in the household.

RESULTS. Household exposure to 25 or more smoker-years during childhood and adolescence doubled the risk of lung cancer. Approximately 15 percent of the control subjects who had never smoked reported this level of exposure. Household exposure of less than 25 smoker-years during childhood and adolescence did not increase the risk of lung cancer. Exposure to a spouse's smoking, which constituted less than one third of total household exposure on average, was not associated with an increase in risk.

CONCLUSIONS. The possibility of recall bias and other methodologic problems may influence the results of case-control studies of environmental tobacco smoke. Nonetheless, our findings regarding exposure during early life suggest that approximately 17 percent of lung cancers among nonsmokers can be attributed to high levels of exposure to cigarette smoke during childhood and adolescence.

(a) What is the research objective?
(b) What makes this study a case-control study? Why is this a retrospective study?
(c) What is the response variable in the study? Is it qualitative or quantitative?
(d) What is the explanatory variable in the study? Is it qualitative or quantitative?
(e) Can you identify any lurking variables that may have affected this study?
(f) What is the conclusion of the study? Can we conclude that exposure to smoke in the household causes lung cancer?
(g) Would it be possible to design an experiment to answer the research question in part (a)? Explain.

1.3   Simple Random Sampling

Objective   ① Obtain a simple random sample

Sampling

Besides the observational studies that we looked at in Section 1.2, observational studies can also be conducted by administering a survey. When administering a survey, the researcher must first identify the population that is to be targeted. For example, the Gallup Organization regularly surveys Americans about various pop-culture and political issues. Often, the population of interest is Americans aged 18 years or older. Of course, the Gallup Organization cannot survey *all* adult Americans (there are over 200 million), so instead the group typically surveys a *random sample* of about 1000 adult Americans.

Definition   **Random sampling** is the process of using chance to select individuals from a population to be included in the sample.

For the results of a survey to be reliable, the characteristics of the individuals in the sample must be representative of the characteristics of the individuals in the population. The key to obtaining a sample representative of a population is to let *chance* or *randomness* play a role in dictating which individuals are in the sample, rather than convenience. **If convenience is used to obtain a sample, the results of the survey are meaningless.**

Suppose that Gallup wants to know the proportion of adult Americans who consider themselves to be baseball fans. If Gallup obtained a sample by standing outside of Fenway Park (home of the Boston Red Sox professional baseball team), the survey results are not likely to be reliable. Why? Clearly, the individuals in the sample do not accurately reflect the makeup of the entire population. As another example, suppose you wanted to learn the proportion of students on your campus who work. It might be convenient to survey the students in your statistics class, but do these students represent the overall student body? Does the proportion of freshmen, sophomores, juniors, and seniors in your class mirror the proportion of freshmen, sophomores, juniors, and seniors on campus? Does the proportion of males and females in your class resemble the proportion of males and females across campus? Probably not. For this reason, the convenient sample is not representative of the population, which means any results reported from your survey are misleading.

We will discuss four basic sampling techniques: *simple random sampling*, *stratified sampling*, *systematic sampling*, and *cluster sampling*. These sampling methods are designed so that any selection biases introduced (knowingly or unknowingly) by the surveyor during the selection process are eliminated. In other words, the surveyor does not have a choice as to which individuals are in the study. We will discuss simple random sampling now and the remaining three types of sampling in Section 1.4.

❶ Obtain a Simple Random Sample

The most basic sample survey design is *simple random sampling*.

Definition

A sample of size n from a population of size N is obtained through **simple random sampling** if every possible sample of size n has an equally likely chance of occurring. The sample is then called a **simple random sample**.

In Other Words
Simple random sampling is like selecting names from a hat.

The number of individuals in the sample is always less than the number of individuals in the population.

EXAMPLE 1   Illustrating Simple Random Sampling

Problem Sophia has four tickets to a concert. Six of her friends, Yolanda, Michael, Kevin, Marissa, Annie, and Katie, have all expressed an interest in going to the concert. Sophia decides to randomly select three of her six friends to attend the concert.
(a) List all possible samples of size $n = 3$ from the population of size $N = 6$. Once an individual is chosen, he or she cannot be chosen again.
(b) Comment on the likelihood of the sample containing Michael, Kevin, and Marissa.

Approach List all possible combinations of three people chosen from the six. Remember, in simple random sampling, each sample of size 3 is equally likely to occur.

Solution
(a) The possible samples of size 3 are listed in Table 2.

Table 2			
Yolanda, Michael, Kevin	Yolanda, Michael, Marissa	Yolanda, Michael, Annie	Yolanda, Michael, Katie
Yolanda, Kevin, Marissa	Yolanda, Kevin, Annie	Yolanda, Kevin, Katie	Yolanda, Marissa, Annie
Yolanda, Marissa, Katie	Yolanda, Annie, Katie	Michael, Kevin, Marissa	Michael, Kevin, Annie
Michael, Kevin, Katie	Michael, Marissa, Annie	Michael, Marissa, Katie	Michael, Annie, Katie
Kevin, Marissa, Annie	Kevin, Marissa, Katie	Kevin, Annie, Katie	Marissa, Annie, Katie

From Table 2, we see that there are 20 possible samples of size 3 from the population of size 6. We use the term *sample* to mean the individuals in the sample.
(b) Only 1 of the 20 possible samples contains Michael, Kevin, and Marissa, so there is a 1 in 20 chance that the simple random sample will contain these three. In fact, all the samples of size 3 have a 1 in 20 chance of occurring.

● **Now Work Problem 7**

Obtaining a Simple Random Sample

The results of Example 1 leave one question unanswered: How do we select the individuals in a simple random sample? We could write the names of the individuals in the population on different sheets of paper and then select names from a hat. Often, however, the size of the population is so large that performing simple random sampling in this fashion is not practical. Instead, each individual in the population is assigned a unique number between 1 and N, where N is the size of the population. Then n distinct random numbers from this list are selected, where n represents the size of the sample. To number the individuals in the population, we need a **frame**—a list of all the individuals within the population.

In Other Words
A frame lists all the individuals in a population. For example, a list of all registered voters in a particular precinct might be a frame.

EXAMPLE 2 Obtaining a Simple Random Sample Using a Table of Random Numbers

Problem The accounting firm of Senese and Associates has grown. To make sure their clients are still satisfied with the services they are receiving, the company decides to send a survey out to a simple random sample of 5 of its 30 clients.

Approach

Step 1 The clients must be listed (the frame) and numbered from 01 to 30.

Step 2 Five unique numbers will be randomly selected. The clients corresponding to the numbers are sent a survey. This process is called *sampling without replacement*. In a **sample without replacement**, an individual who is selected is removed from the population and cannot be chosen again. In a **sample with replacement**, a selected individual is placed back into the population and could be chosen a second time. We use sampling without replacement so that we don't select the same client twice.

Solution

Step 1 Table 3 shows the list of clients. We arrange them in alphabetical order, although this is not necessary, and number them from 01 to 30.

Table 3		
01. ABC Electric	11. Fox Studios	21. R&Q Realty
02. Brassil Construction	12. Haynes Hauling	22. Ritter Engineering
03. Bridal Zone	13. House of Hair	23. Simplex Forms
04. Casey's Glass House	14. John's Bakery	24. Spruce Landscaping
05. Chicago Locksmith	15. Logistics Management, Inc.	25. Thors, Robert DDS
06. DeSoto Painting	16. Lucky Larry's Bistro	26. Travel Zone
07. Dino Jump	17. Moe's Exterminating	27. Ultimate Electric
08. Euro Car Care	18. Nick's Tavern	28. Venetian Gardens Restaurant
09. Farrell's Antiques	19. Orion Bowling	29. Walker Insurance
10. First Fifth Bank	20. Precise Plumbing	30. Worldwide Wireless

Step 2 A table of random numbers can be used to select the individuals to be in the sample. See Table 4 on the next page.* We pick a starting place in the table by closing our eyes and placing a finger on it. This method accomplishes the goal of being random. Suppose we start in column 4, row 13. Because our data have two digits, we select two-digit numbers from the table using columns 4 and 5. We select numbers between 01 and 30, inclusive, and skip 00, numbers greater than 30, and numbers already selected.

---

*Each digit is in its own column. The digits are displayed in groups of five for ease of reading. The digits in row 1 are 893922321274483, and so on. The first digit, 8, is in column 1; the second digit, 9, is in column 2; the ninth digit, 1, is in column 9.

(continued)

Table 4

Column 4

Row 13

Row Number	Column Number 01–05	06–10	11–15	16–20	21–25	26–30	31–35	36–40	41–45	46–50
01	89392	23212	74483	36590	25956	36544	68518	40805	09980	00467
02	61458	17639	96252	95649	73727	33912	72896	66218	52341	97141
03	11452	74197	81962	48433	90360	26480	73231	37740	26628	44690
04	27575	04429	31308	02241	01698	19191	18948	78871	36030	23980
05	36829	59109	88976	46845	28329	47460	88944	08264	00843	84592
06	81902	93458	42161	26099	09419	89073	82849	09160	61845	40906
07	59761	55212	33360	68751	86737	79743	85262	31887	37879	17525
08	46827	25906	64708	20307	78423	15910	86548	08763	47050	18513
09	24040	66449	32353	83668	13874	86741	81312	54185	78824	00718
10	98144	96372	50277	15571	82261	66628	31457	00377	63423	55141
11	14228	17930	30118	00438	49666	65189	62869	31304	17117	71489
12	55366	51057	90065	14791	62426	02957	85518	28822	30588	32798
13	96101	30646	35526	90389	73634	79304	96635	06626	94683	16696
14	38152	55474	30153	26525	83647	31988	82182	98377	33802	80471
15	85007	18416	24661	95581	45868	15662	28906	36392	07617	50248
16	85544	15890	80011	18160	33468	84106	40603	01315	74664	20553
17	10446	20699	98370	17684	16932	80449	92654	02084	19985	59321
18	67237	45509	17638	65115	29757	80705	82686	48565	72612	61760
19	23026	89817	05403	82209	30573	47501	00135	33955	50250	72592
20	67411	58542	18678	46491	13219	84084	27783	34508	55158	78742

We skip 52 because it is larger than 30.

The first number in the list is 01, so the client corresponding to 01 will receive a survey. Reading down, the next number in the list is 52, which is greater than 30, so we skip it. Continuing down the list, the following numbers are selected from the list:

$$01, 07, 26, 11, 23$$

We display each of the random numbers used to select the individuals in the sample in boldface type in Table 4 to help you to understand where they came from. The clients corresponding to these numbers are

ABC Electric, Dino Jump, Travel Zone, Fox Studios, Simplex Forms •

EXAMPLE 3   Obtaining a Simple Random Sample Using Technology

Problem Find a simple random sample of five clients for the problem presented in Example 2.

Approach The approach is similar to that given in Example 2.

Step 1   Obtain the frame and assign the clients numbers from 01 to 30.

Step 2   Randomly select five numbers using a random number generator. To do this, we must first set the *seed*. The **seed** is an initial point for the generator to start creating random numbers—like selecting the initial point in the table of random numbers. The seed can be any nonzero number. Statistical software such as StatCrunch, Minitab, or Excel can be used to generate random numbers, but we will use a TI-84 Plus C graphing calculator. The steps for obtaining random numbers using StatCrunch, Minitab, Excel, and the TI-83/84 Plus/84 Plus C graphing calculator can be found in the Technology Step-by-Step beginning on page 25.

Solution

Step 1   Table 3 on page 23 shows the list of clients and numbers corresponding to the clients.

Step 2   Figure 3(a) shows the seed set at 34 on a TI-84 Plus C graphing calculator. Now we can generate a list of random numbers, which are shown in Figure 3(b).

Figure 3

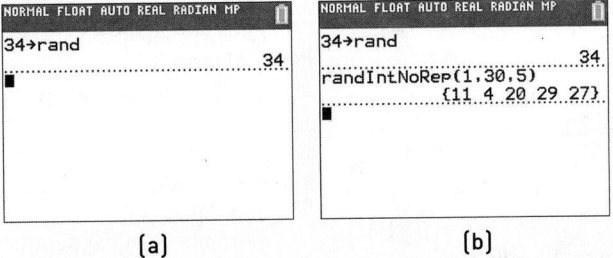

| (a) | (b) |

The following numbers are generated by the calculator:

$$11, 4, 20, 29, 27$$

The clients corresponding to these numbers are the clients to be surveyed: Fox Studios, Casey's Glass House, Precise Plumbing, Walker Insurance, and Ultimate Electric.   ●

● Now Work Problem 11

Using Technology

If you are using a different statistical package or type of calculator, the random numbers generated will likely be different. This does not mean you are wrong. There is no such thing as a wrong random sample as long as the correct procedures are followed.

CAUTION!
Random-number generators are not truly random, because they are programs, and programs do not act "randomly." The seed dictates the random numbers that are generated.

Notice an important difference in the solutions of Examples 2 and 3. Because both samples were obtained randomly, they resulted in different individuals in the sample! For this reason, each sample will likely result in different descriptive statistics. Any inference based on each sample *may* result in different conclusions regarding the population. This is the nature of statistics. Inferences based on samples will vary because the individuals in different samples vary.

Technology Step-by-Step   Obtaining a Simple Random Sample

TI-83/84 Plus

1. Enter any nonzero number (the seed) on the HOME screen.
2. Press the STO ▶ button.
3. Press the MATH button.
4. Highlight the PRB menu and select 1: rand.
5. From the HOME screen press ENTER.
6. Press the MATH button. Highlight the PRB menu and select 5: randInt(.
7. With randInt(on the HOME screen, enter 1, N), where N is the population size. For example, if $N = 500$, enter the following:

       randInt(1,500)

Press ENTER to obtain the first individual in the sample. Continue pressing ENTER until the desired sample size is obtained.

TI-84 Plus C

1. Enter any nonzero number (the seed) on the HOME screen.
2. Press the STO ▶ button.
3. Press the MATH button.
4. Highlight the PROB menu and select 1: rand.
5. From the HOME screen press ENTER.
6. Press the MATH button.
7. Highlight the PROB menu and select 8: randIntNoRep(.
8. Type in the values for lower, upper, and n.
9. Highlight Paste. Press ENTER. Press ENTER a second time from the HOME screen.

Minitab

1. Select the **Calc** menu and highlight **Set Base . . . .**
2. Enter any seed number you desire. Note that it is not necessary to set the seed, because Minitab uses the time of day in seconds to set the seed.
3. Select the **Calc** menu, highlight **Random Data**, and select **Integer . . . .**

4. Fill in the following window with the appropriate values. To obtain a simple random sample for the situation in Example 2, we would enter the following:

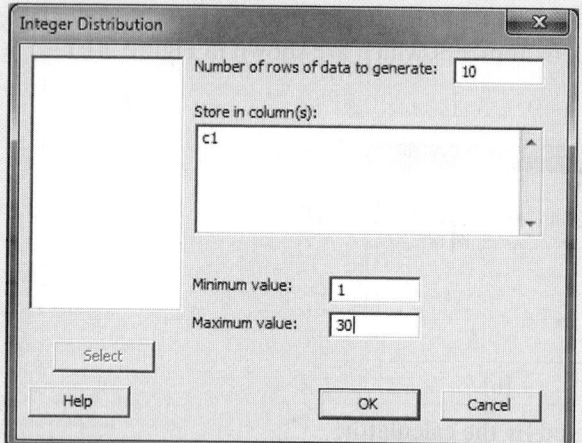

Generate 10 rows of data (instead of 5) in case any of the random numbers repeat. Select OK, and the random numbers will appear in column 1 (C1) in the spreadsheet.

Excel

1. Be sure the Data Analysis ToolPak is activated. This is done by selecting **File** and then **Options**. Select **Add-Ins**. Under **Manage:**, select **Excel Add-Ins**. Click **Go**. . . . Check the **Analysis ToolPak** box. Click **OK**.
2. Select **Data** and then **Data Analysis**. Highlight **Random Number Generation** and select **OK**.
3. Fill in the window with the appropriate values. To obtain a simple random sample for the situation in Example 2, fill in the following:

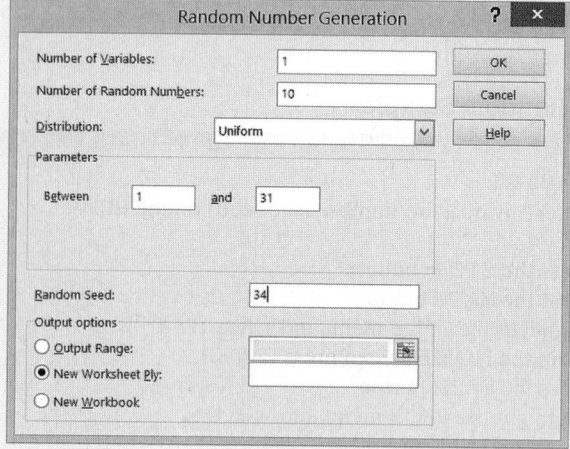

Generate 10 rows of data (instead of 5) in case any of the random numbers repeat. Notice also that the parameter is between 1 and 31, so any value greater than or equal to 1 and less than or equal to 31 is possible. In the unlikely event that 31 appears, simply ignore it. Select **OK** and the random numbers will appear in column 1 (A1) in the spreadsheet. Ignore any values to the right of the decimal place.

StatCrunch

1. Select **Data**, highlight **Simulate**, then select **Discrete Uniform**.
2. Fill in the window with the appropriate values. To obtain a simple random sample for the situation in Example 2, enter the values shown in the figure. Generate 10 rows of data (instead of 5) is in case any of the random numbers repeat. Click Compute!, and the random numbers will appear in the spreadsheet. *Note:* You could also select the dynamic seed radio button, if you like, to set the seed.

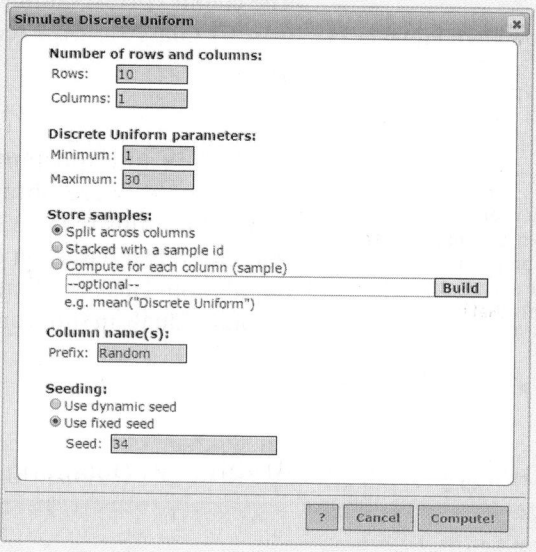

1.3 Assess Your Understanding

Vocabulary and Skill Building

1. What is a frame?
2. Define simple random sampling.
3. What does it mean when sampling is done without replacement?
4. What is random sampling? Why is it used and how does it compare with convenience sampling?
5. **Literature** As part of a college literature course, students must read three classic works of literature from the provided list. Obtain a simple random sample of size 3 from this list. Write a short description of the process you used to generate your sample.

Pride and Prejudice	The Scarlet Letter
As I Lay Dying	The Jungle
Death of a Salesman	Huckleberry Finn
The Sun Also Rises	Crime and Punishment
A Tale of Two Cities	

6. Team Captains A coach must select two players to serve as captains. He wants to randomly select two players to be the captains. Obtain a simple random sample of size 2 from the following list: Mady, Breanne, Evin, Tori, Emily, Clair, Caty, Jory, Payton, Jordyn. Write a short description of the process you used to generate your sample.

NW 7. Course Selection A student entering a doctoral program in educational psychology is required to select two courses from the list of courses provided as part of his or her program.

   EPR 616, *Research in Child Development*
   EPR 630, *Educational Research Planning and Interpretation*
   EPR 631, *Nonparametric Statistics*
   EPR 632, *Methods of Multivariate Analysis*
   EPR 645, *Theory of Measurement*
   EPR 649, *Fieldwork Methods in Educational Research*
   EPR 650, *Interpretive Methods in Educational Research*

   (a) List all possible two-course selections.
   (b) Comment on the likelihood that the pair of courses EPR 630 and EPR 645 will be selected.

8. Merit Badge Requirements To complete the Citizenship in the World merit badge, one must select two of the following seven organizations and describe their role in the world.

Source: Boy Scouts of America

   1. The United Nations
   2. The World Court
   3. World Organization of the Scout Movement
   4. The World Health Organization
   5. Amnesty International
   6. The International Committee of the Red Cross
   7. CARE

   (a) List all possible pairs of organizations.
   (b) Comment on the likelihood that the pair "The United Nations" and "Amnesty International" will be selected.

Applying the Concepts

9. Sampling the Faculty A community college employs 87 full-time faculty members. To gain the faculty's opinions about an upcoming building project, the college president wishes to obtain a simple random sample that will consist of 9 faculty members. He numbers the faculty from 1 to 87.

   (a) Using Table I from Appendix A, the president closes his eyes and drops his ink pen on the table. It points to the digit in row 5, column 22. Using this position as the starting point and proceeding downward, determine the numbers for the 9 faculty members who will be included in the sample.
   (b) If the president uses technology, determine the numbers for the 9 faculty members who will be included in the sample.

10. Sampling the Students The same community college from Problem 9 has 7656 students currently enrolled in classes. To gain the students' opinions about an upcoming building project, the college president wishes to obtain a simple random sample of 20 students. He numbers the students from 1 to 7656.

   (a) Using Table I from Appendix A, the president closes his eyes and drops his ink pen on the table. It points to the digit in row 11, column 32. Using this position as the starting point and proceeding downward, determine the numbers for the 20 students who will be included in the sample.
   (b) If the president uses technology, determine the numbers for the 20 students who will be included in the sample.

NW 11. Obtaining a Simple Random Sample The following table lists the 50 states.

1. Alabama	19. Maine	35. Ohio
2. Alaska	20. Maryland	36. Oklahoma
3. Arizona	21. Massachusetts	37. Oregon
4. Arkansas	22. Michigan	38. Pennsylvania
5. California	23. Minnesota	39. Rhode Island
6. Colorado	24. Mississippi	40. South Carolina
7. Connecticut	25. Missouri	41. South Dakota
8. Delaware	26. Montana	42. Tennessee
9. Florida	27. Nebraska	43. Texas
10. Georgia	28. Nevada	44. Utah
11. Hawaii	29. New Hampshire	45. Vermont
12. Idaho	30. New Jersey	46. Virginia
13. Illinois	31. New Mexico	47. Washington
14. Indiana	32. New York	48. West Virginia
15. Iowa	33. North Carolina	49. Wisconsin
16. Kansas	34. North Dakota	50. Wyoming
17. Kentucky	33. North Carolina	
18. Louisiana	34. North Dakota	

   (a) Obtain a simple random sample of size 10 using Table I in Appendix A, a graphing calculator, or computer software.
   (b) Obtain a second simple random sample of size 10 using Table I in Appendix A, a graphing calculator, or computer software.

12. Obtaining a Simple Random Sample The following table lists the 44 presidents of the United States.

1. Washington	16. Lincoln	31. Hoover
2. J. Adams	17. A. Johnson	32. F. D. Roosevelt
3. Jefferson	18. Grant	33. Truman
4. Madison	19. Hayes	34. Eisenhower
5. Monroe	20. Garfield	35. Kennedy
6. J. Q. Adams	21. Arthur	36. L. B. Johnson
7. Jackson	22. Cleveland	37. Nixon
8. Van Buren	23. B. Harrison	38. Ford
9. W. H. Harrison	24. Cleveland	39. Carter
10. Tyler	25. McKinley	40. Reagan
11. Polk	26. T. Roosevelt	41. G. H. Bush
12. Taylor	27. Taft	42. Clinton
13. Fillmore	28. Wilson	43. G. W. Bush
14. Pierce	29. Harding	44. Obama
15. Buchanan	30. Coolidge	

   (a) Obtain a simple random sample of size 8 using Table I in Appendix A, a graphing calculator, or computer software.
   (b) Obtain a second simple random sample of size 8 using Table I in Appendix A, a graphing calculator, or computer software.

13. Obtaining a Simple Random Sample The president of the student government wants to conduct a survey to determine the student body's opinion regarding student services. The administration provides you with a list of the names and phone numbers of the 19,935 registered students.

(a) Discuss the procedure you would follow to obtain a simple random sample of 25 students.

(b) Obtain this sample.

14. Obtaining a Simple Random Sample The mayor of Justice, Illinois, asks you to poll the residents of the village and provides you with a list of the names and phone numbers of the 5832 residents of the village.

(a) Discuss the procedure you would follow to obtain a simple random sample of 20 residents.

(b) Obtain this sample.

15. Future Government Club The Future Government Club wants to sponsor a panel discussion on the upcoming national election. The club wants four of its members to lead the panel discussion. Obtain a simple random sample of size 4 from the below table. Write a short description of the process you used to generate your sample.

Blouin	Fallenbuchel	Niemeyer	Rice
Bolden	Grajewski	Nolan	Salihar
Bolt	Haydra	Ochs	Tate
Carter	Keating	Opacian	Thompson
Cooper	Khouri	Pawlak	Trudeau
Debold	Lukens	Pechtold	Washington
De Young	May	Ramirez	Wright
Engler	Motola	Redmond	Zenkel

16. Worker Morale The owner of a private food store is concerned about employee morale. She decides to survey the employees to learn about work environment and job satisfaction. Obtain a simple random sample of size 5 from the names in the given table. Write a short description of the process you used to generate your sample.

Archer	Foushi	Kemp	Oliver
Bolcerek	Gow	Lathus	Orsini
Bryant	Grove	Lindsey	Salazar
Carlisle	Hall	Massie	Ullrich
Cole	Hills	McGuffin	Vaneck
Dimas	Houston	Musa	Weber
Ellison	Kats	Nickas	Zavodny
Everhart			

1.4 Other Effective Sampling Methods

Objectives

① Obtain a stratified sample

② Obtain a systematic sample

③ Obtain a cluster sample

The goal of sampling is to obtain as much information as possible about the population at the least cost. Remember, we are using the word *cost* in a general sense. Cost includes monetary outlays, time, and other resources. With this goal in mind, we may find it advantageous to use sampling techniques other than simple random sampling.

① Obtain a Stratified Sample

Under certain circumstances, *stratified sampling* provides more information about the population for less cost than simple random sampling.

Definition A **stratified sample** is obtained by separating the population into nonoverlapping groups called *strata* and then obtaining a simple random sample from each stratum. The individuals within each stratum should be homogeneous (or similar) in some way.

In Other Words
Stratum is singular, while **strata** is plural. The word **strata** means divisions. So a stratified sample is a simple random sample of different divisions of the population.

For example, suppose Congress was considering a bill that abolishes estate taxes. In an effort to determine the opinion of her constituency, a senator asks a pollster to conduct a survey within her state. The pollster may divide the population of registered voters within the state into three strata: Republican, Democrat, and Independent. This grouping makes sense because the members within each of the three parties may have the same opinion regarding estate taxes, but opinions among parties may differ. The main criterion in performing a stratified sample is that each group (stratum) must have a common attribute that results in the individuals being similar within the stratum.

An advantage of stratified sampling over simple random sampling is that it may allow fewer individuals to be surveyed while obtaining the same or more information. This result occurs because individuals within each subgroup have similar characteristics, so opinions within the group are not as likely to vary much from one individual to the next. In addition, a stratified sample guarantees that each stratum is represented in the sample.

EXAMPLE 1 Obtaining a Stratified Sample

Problem The president of DePaul University wants to conduct a survey to determine the community's opinion regarding campus safety. The president divides the DePaul community into three groups: resident students, nonresident (commuting) students, and staff (including faculty) so that he can obtain a stratified sample. Suppose there are 6204 resident students, 13,304 nonresident students, and 2401 staff, for a total of 21,909 individuals in the population. The president wants to obtain a sample of size 100, with the number of individuals selected from each stratum weighted by the population size. So resident students make up $6204/21{,}909 = 28\%$ of the sample, nonresident students account for 61% of the sample, and staff constitute 11% of the sample. A sample of size 100 requires a stratified sample of $0.28(100) = 28$ resident students, $0.61(100) = 61$ nonresident students, and $0.11(100) = 11$ staff.

Approach To obtain the stratified sample, conduct a simple random sample within each group. That is, obtain a simple random sample of 28 resident students (from the 6204 resident students), a simple random sample of 61 nonresident students, and a simple random sample of 11 staff. Be sure to use a different seed for each stratum.

Solution Using Minitab, with the seed set to 4032 and the values shown in Figure 4, we obtain the following sample of staff:

$$240, 630, 847, 190, 2096, 705, 2320, 323, 701, 471, 744$$

Figure 4

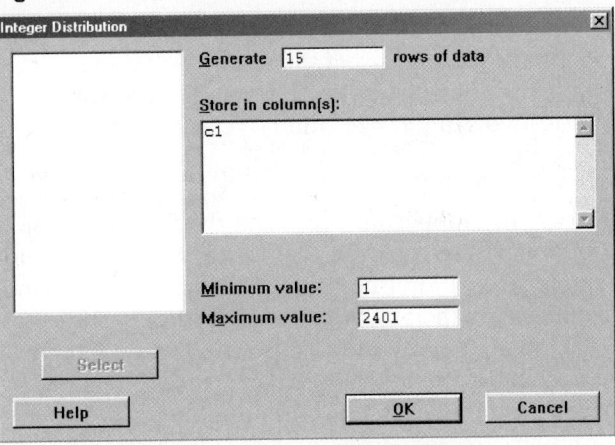

·-CAUTION!-------------------
Do not use the same seed (or starting point in Table I) for each stratum in a stratified sample, because we want the simple random samples within each stratum to be independent of each other.

Repeat this procedure for the resident and nonresident students using a different

● Now Work Problem 25 seed.

An advantage of stratified sampling over simple random sampling is that the researcher can determine characteristics within each stratum. This allows an analysis to be performed on each stratum to see if any significant differences among them exist. For example, we could analyze the data obtained in Example 1 to see if there is a difference in the opinions of students versus staff.

② Obtain a Systematic Sample

In both simple random sampling and stratified sampling, a frame—a list of the individuals in the population being studied—must exist. Therefore, these sampling techniques require some preliminary work before obtaining the sample. A sampling technique that does not require a frame is *systematic sampling*.

Definition   A **systematic sample** is obtained by selecting every kth individual from the population. The first individual selected corresponds to a random number between 1 and k.

Because systematic sampling does not require a frame, it is a useful technique when you cannot obtain a list of the individuals in the population.

To obtain a systematic sample, select a number k, randomly select a number between 1 and k and survey that individual, then survey every kth individual there after. For example, we might decide to survey every $k = $ 8th individual. We randomly select a number between 1 and 8, such as 5. This means we survey the 5th, $5 + 8 = $ 13th, $13 + 8 = $ 21st, $21 + 8 = $ 29th, and so on, individuals until we reach the desired sample size.

EXAMPLE 2   Obtaining a Systematic Sample without a Frame

Problem The manager of Kroger Food Stores wants to measure the satisfaction of the store's customers. Design a sampling technique that can be used to obtain a sample of 40 customers.

Approach A frame of Kroger customers would be difficult, if not impossible, to obtain. Therefore, it is reasonable to use systematic sampling by surveying every kth customer who leaves the store.

Solution The manager decides to obtain a systematic sample by surveying every 7th customer. He randomly determines a number between 1 and 7, say 5. He then surveys the 5th customer exiting the store and every 7th customer thereafter, until he has a sample of 40 customers. The survey will include customers 5, 12, 19, . . . , 278.*   ●

But how do we select the value of k? If the size of the population is unknown, there is no mathematical way to determine k. The value of k must be small enough to achieve our desired sample size, and large enough to obtain a sample that is representative of the population.

To clarify this point, let's revisit Example 2. If k is too large, say 30, we will survey every 30th shopper, starting with the 5th. A sample of size 40 would require that 1175 shoppers visit Kroger on that day. If Kroger does not have 1175 shoppers, the desired sample size will not be achieved. On the other hand, if k is too small, say 4, the store would survey the 5th, 9th, . . . , 161st shopper. The 161st shopper might exit the store at 3 P.M., so our survey would not include any of the evening shoppers. This sample is not representative of *all* Kroger patrons! An estimate of the size of the population would help to determine an appropriate value for k.

*Because we are surveying 40 customers, the first individual surveyed is the 5th, the second is the $5 + 7 = 12$th, the third is the $5 + (2) 7 = 19$th, and so on, until we reach the 40th, which is the $5 + (39) 7 = 278$th shopper.

To determine the value of k when the size of the population, N, is known is relatively straightforward. Suppose the population size is $N = 20,325$ and we desire a sample of size $n = 100$. To guarantee that individuals are selected evenly from both the beginning and the end of the population (such as early and late shoppers), we compute N/n and round down to the nearest integer. For example, $20,325/100 = 203.25$, so $k = 203$. Then we randomly select a number between 1 and 203 and select every 203rd individual thereafter. So, if we randomly selected 90 as our starting point, we would survey the 90th, 293rd, 496th, . . . , 20,187th individuals.

We summarize the procedure as follows:

Steps in Systematic Sampling

1. If possible, approximate the population size, N.
2. Determine the sample size desired, n.
3. Compute $\dfrac{N}{n}$ and round down to the nearest integer. This value is k.
4. Randomly select a number between 1 and k. Call this number p.
5. The sample will consist of the following individuals:

$$p, p + k, p + 2k, \ldots, p + (n - 1)k$$

• Now Work Problem 27

Because systematic sampling does not require a frame, it typically provides more information for a given cost than does simple random sampling. In addition, systematic sampling is easier to employ, so there is less likelihood of interviewer error occurring, such as selecting the wrong individual to be surveyed.

③ Obtain a Cluster Sample

A fourth sampling method is called *cluster sampling*. Like the previous three sampling methods, this method has benefits under certain circumstances.

Definition

A **cluster sample** is obtained by selecting all individuals within a randomly selected collection or group of individuals.

In Other Words
Imagine a mall parking lot. Each subsection of the lot could be a cluster (Section F-4, for example).

Suppose a school administrator wants to learn the characteristics of students enrolled in online classes. Rather than obtaining a simple random sample based on the frame of all students enrolled in online classes, the administrator could treat each online class as a cluster and then obtain a simple random sample of these clusters. The administrator would then survey *all* the students in the selected clusters. This reduces the number of classes that get surveyed.

EXAMPLE 3   Obtaining a Cluster Sample

Problem A sociologist wants to gather data regarding household income within the city of Boston. Obtain a sample using cluster sampling.

Approach The city of Boston can be set up so that each city block is a cluster. Once the city blocks have been identified, obtain a simple random sample of the city blocks and survey all households on the blocks selected.

Solution Suppose there are 10,493 city blocks in Boston. First, the sociologist must number the blocks from 1 to 10,493. Suppose the sociologist has enough time and money to survey 20 clusters (city blocks). The sociologist should obtain a simple

(continued)

·CAUTION!--------------------------
Stratified and cluster samples are different. In a stratified sample, we divide the population into two or more homogeneous groups. Then we obtain a simple random sample from each group. In a cluster sample, we divide the population into groups, obtain a simple random sample of some of the groups, and survey *all* individuals in the selected groups.

random sample of 20 numbers between 1 and 10,493 and survey all households from the clusters selected. Cluster sampling is a good choice in this example because it reduces the travel time to households that is likely to occur with both simple random sampling and stratified sampling. In addition, there is no need to obtain a frame of all the households with cluster sampling. The only frame needed is one that provides information regarding city blocks. •

The following are a few of the questions that arise in cluster sampling:

- How do I cluster the population?
- How many clusters do I sample?
- How many individuals should be in each cluster?

First, we must determine whether the individuals within the proposed cluster are homogeneous (similar individuals) or heterogeneous (dissimilar individuals). In Example 3, city blocks tend to have similar households. Survey responses from houses on the same city block are likely to be similar. This results in duplicate information. We conclude that if the clusters have homogeneous individuals it is better to have more clusters with fewer individuals in each cluster.

What if the cluster is heterogeneous? Under this circumstance, the heterogeneity of the cluster likely resembles the heterogeneity of the population. In other words, each cluster is a scaled-down representation of the overall population. For example, a quality-control manager might use shipping boxes that contain 100 light bulbs as a cluster, since the rate of defects within the cluster would resemble the rate of defects in the population, assuming the bulbs are randomly placed in the box. Thus, when each cluster is heterogeneous, fewer clusters with more individuals in each cluster are appropriate.

● **Now Work Problem 13**

Convenience Sampling

In the four sampling techniques just presented, the individuals are selected randomly. Inappropriate sampling methods are those in which the individuals are not randomly selected.

Have you ever watched a talk show where the host asks listeners to reply to a poll through Twitter? This is a non-scientific data collection method, so the results of any analysis are suspect because the data was obtained using a *convenience sample*.

Definition

A **convenience sample** is a sample in which the individuals are easily obtained and not based on randomness.

·CAUTION!--------------------
Studies that use convenience sampling generally have results that are suspect. The results should be looked on with extreme skepticism.

The most popular of the many types of convenience samples are those in which the individuals in the sample are **self-selected** (the individuals themselves decide to participate in a survey). These are also called **voluntary response** samples. One example of self-selected sampling is phone-in polling; a radio personality will ask his or her listeners to phone the station to submit their opinions. Another example is the use of the Internet to conduct surveys. For example, a television news show will present a story regarding a certain topic and ask its viewers to "tell us what you think" by completing a questionnaire online or phoning in an opinion. Both of these samples are poor designs because the individuals who decide to be in the sample generally have strong opinions about the topic. A more typical individual in the population will not bother phoning or logging on to a computer to complete a survey. Any inference made regarding the population from this type of sample should be made with extreme caution.

Convenience samples yield unreliable results because the individuals participating in the survey are not chosen using random sampling. Instead, the interviewer or participant selects who is in the survey. Would an interviewer select an ornery individual? Of course not! Therefore, the sample is likely not to be representative of the population.

Multistage Sampling

In practice, most large-scale surveys obtain samples using a combination of the techniques just presented.

As an example of multistage sampling, consider Nielsen Media Research. Nielsen randomly selects households and monitors the television programs these households are watching through a People Meter. The meter is an electronic box placed on each TV within the household. The People Meter measures what program is being watched and who is watching it. Nielsen selects the households with the use of a two-stage sampling process.

Stage 1 Using U.S. Census data, Nielsen divides the country into geographic areas (strata). The strata are typically city blocks in urban areas and geographic regions in rural areas. About 6000 strata are randomly selected.

Stage 2 Nielsen sends representatives to the selected strata and lists the households within the strata. The households are then randomly selected through a simple random sample.

Nielsen sells the information obtained to television stations and companies. These results are used to help determine prices for commercials.

As another example of multistage sampling, consider the sample used by the Census Bureau for the Current Population Survey. This survey requires five stages of sampling:

Stage 1 Stratified sample

Stage 2 Cluster sample

Stage 3 Stratified sample

Stage 4 Cluster sample

Stage 5 Systematic sample

This survey is very important because it is used to obtain demographic estimates of the United States in noncensus years. Details about the Census Bureau's sampling method can be found in *The Current Population Survey: Design and Methodology*, Technical Paper No. 40.

Sample Size Considerations

Throughout our discussion of sampling, we did not mention how to determine the sample size. Researchers need to know how many individuals they must survey to draw conclusions about the population within some predetermined margin of error. They must find a balance between the reliability of the results and the cost of obtaining these results. Time and money determine the level of confidence researchers will place on the conclusions drawn from the sample data. The more time and money researchers have available, the more accurate the results of the statistical inference.

In Sections 9.1 and 9.2, we discuss techniques for determining the sample size required to estimate characteristics regarding the population within some margin of error. (For a detailed discussion of sample size considerations, consult a text on sampling techniques such as *Elements of Sampling Theory and Methods* by Z. Govindarajulu, Pearson, 1999.)

Summary

Figure 5 on the next page illustrates the four sampling techniques presented.

Figure 5

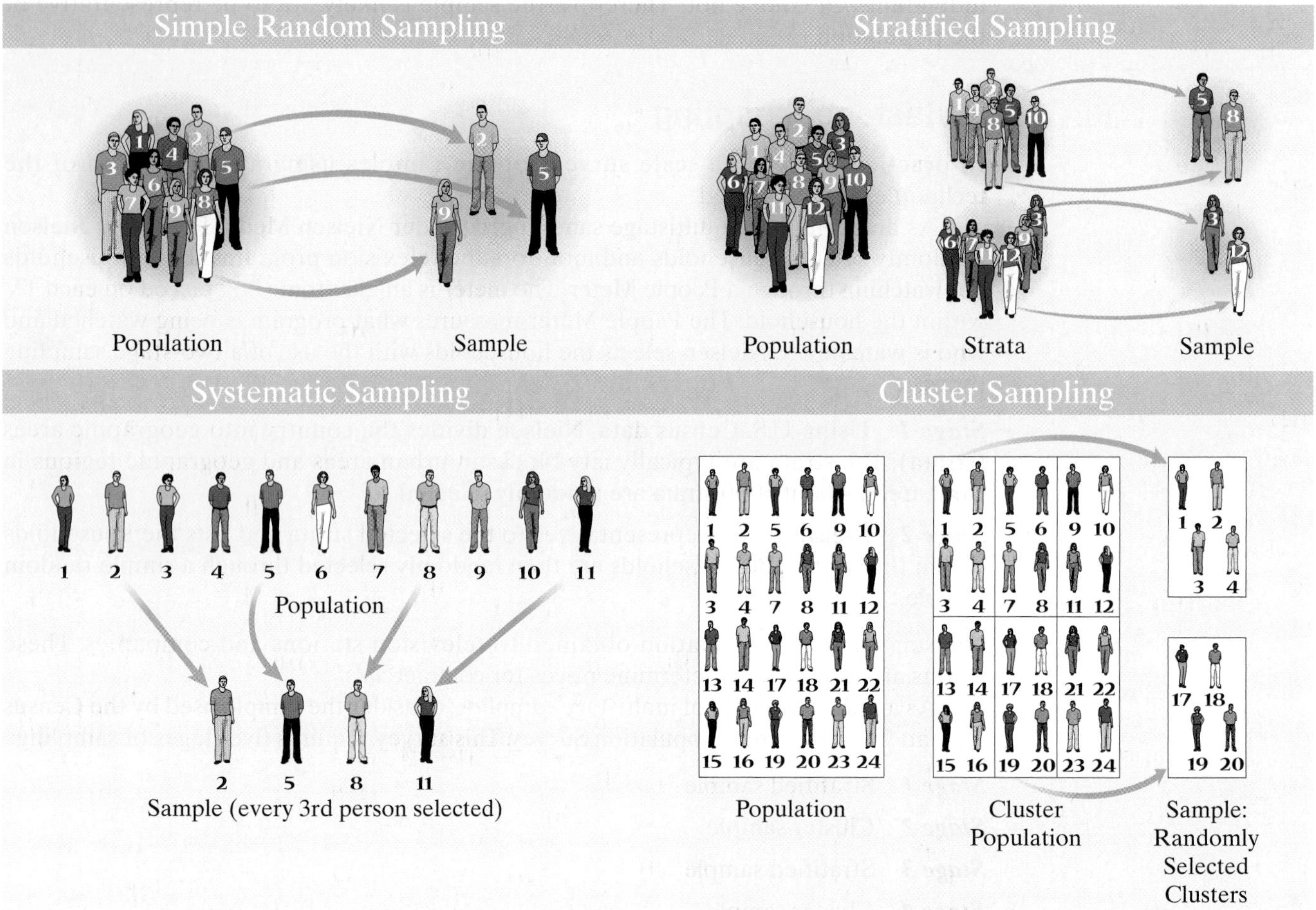

Simple Random Sampling

Population → Sample

Stratified Sampling

Population → Strata → Sample

Systematic Sampling

Population

Sample (every 3rd person selected)

Cluster Sampling

Population

Cluster Population

Sample: Randomly Selected Clusters

1.4 Assess Your Understanding

Vocabulary and Skill Building

1. Describe a circumstance in which stratified sampling would be an appropriate sampling method.

2. Which sampling method does not require a frame?

3. Why are convenience samples ill advised?

4. A(n) _____ sample is obtained by dividing the population into groups and selecting all individuals from within a random sample of the groups.

5. A(n) _____ sample is obtained by dividing the population into homogeneous groups and randomly selecting individuals from each group.

6. True or False: When taking a systematic random sample of size n, every group of size n from the population has the same chance of being selected.

7. True or False: A simple random sample is always preferred because it obtains the same information as other sampling plans but requires a smaller sample size.

8. True or False: When conducting a cluster sample, it is better to have fewer clusters with more individuals when the clusters are heterogeneous.

9. True or False: Inferences based on voluntary response samples are generally not reliable.

10. True or False: When obtaining a stratified sample, the number of individuals included within each stratum must be equal.

In Problems 11–22, identify the type of sampling used.

11. To estimate the percentage of defects in a recent manufacturing batch, a quality-control manager at Intel selects every 8th chip that comes off the assembly line starting with the 3rd until she obtains a sample of 140 chips.

12. To determine the prevalence of human growth hormone (HGH) use among high school varsity baseball players, the State Athletic Commission randomly selects 50 high schools. All members of the selected high schools' varsity baseball teams are tested for HGH.

NW 13. To determine customer opinion of its boarding policy, Southwest Airlines randomly selects 60 flights during a certain week and surveys all passengers on the flights.

14. A member of Congress wishes to determine her constituency's opinion regarding estate taxes. She divides her constituency into three income classes: low-income households, middle-income households, and upper-income households. She then takes a simple random sample of households from each income class.

15. In an effort to identify whether an advertising campaign has been effective, a marketing firm conducts a nationwide poll by randomly selecting individuals from a list of known users of the product.

16. A radio station asks its listeners to call in their opinion regarding the use of U.S. forces in peacekeeping missions.

17. A farmer divides his orchard into 50 subsections, randomly selects 4, and samples all the trees within the 4 subsections to approximate the yield of his orchard.

18. A college official divides the student population into five classes: freshman, sophomore, junior, senior, and graduate student. The official takes a simple random sample from each class and asks the members' opinions regarding student services.

19. A survey regarding download time on a certain website is administered on the Internet by a market research firm to anyone who would like to take it.

20. The presider of a guest-lecture series at a university stands outside the auditorium before a lecture begins and hands every fifth person who arrives, beginning with the third, a speaker evaluation survey to be completed and returned at the end of the program.

21. To determine his DSL Internet connection speed, Shawn divides up the day into four parts: morning, midday, evening, and late night. He then measures his Internet connection speed at 5 randomly selected times during each part of the day.

22. 24 Hour Fitness wants to administer a satisfaction survey to its current members. Using its membership roster, the club randomly selects 40 club members and asks them about their level of satisfaction with the club.

23. A salesperson obtained a systematic sample of size 20 from a list of 500 clients. To do so, he randomly selected a number from 1 to 25, obtaining the number 16. He included in the sample the 16th client on the list and every 25th client thereafter. List the numbers that correspond to the 20 clients selected.

24. A quality-control expert wishes to obtain a cluster sample by selecting 10 of 795 clusters. She numbers the clusters from 1 to 795. Using Table I from Appendix A, she closes her eyes and drops a pencil on the table. It points to the digit in row 8, column 38. Using this position as the starting point and proceeding downward, determine the numbers for the 10 clusters selected.

Applying the Concepts

NW **25. Stratified Sampling** The Future Government Club wants to sponsor a panel discussion on the upcoming national election. The club wants to have four of its members lead the panel discussion. To be fair, however, the panel should consist of two Democrats and two Republicans. From the list of current members of the club, obtain a stratified sample of two Democrats and two Republicans to serve on the panel.

Democrats			
Bolden	Fallenbuchel	Motola	Ramirez
Bolt	Haydra	Nolan	Tate
Carter	Khouri	Opacian	Washington
Debold	Lukens	Pawlak	Wright

Republicans			
Blouin	Grajewski	Ochs	Salihar
Cooper	Keating	Pechtold	Thompson
De Young	May	Redmond	Trudeau
Engler	Niemeyer	Rice	Zenkel

26. Stratified Sampling The owner of a private food store is concerned about employee morale. She decides to survey the managers and hourly employees to see if she can learn about work environment and job satisfaction. From the list of workers at the store, obtain a stratified sample of two managers and four hourly employees to survey.

Managers		Hourly Employees		
Carlisle	Oliver	Archer	Foushi	Massie
Hills	Orsini	Bolcerek	Gow	Musa
Kats	Ullrich	Bryant	Grove	Nickas
Lindsey	McGuffin	Cole	Hall	Salazar
		Dimas	Houston	Vaneck
		Ellison	Kemp	Weber
		Everhart	Lathus	Zavodny

NW **27. Systematic Sample** The human resource department at a certain company wants to conduct a survey regarding worker morale. The department has an alphabetical list of all 4502 employees at the company and wants to conduct a systematic sample.

(a) Determine k if the sample size is 50.

(b) Determine the individuals who will be administered the survey. More than one answer is possible.

28. Systematic Sample To predict the outcome of a county election, a newspaper obtains a list of all 945,035 registered voters in the county and wants to conduct a systematic sample.

(a) Determine k if the sample size is 130.

(b) Determine the individuals who will be administered the survey. More than one answer is possible.

29. Which Method? The mathematics department at a university wishes to administer a survey to a sample of students taking college algebra. The department is offering 32 sections of college algebra, similar in class size and makeup, with a total of 1280 students. They would like the sample size to be roughly 10% of the population of college algebra students this semester. How might the department obtain a simple random sample? A stratified sample? A cluster sample? Which method do you think is best in this situation?

30. Good Sampling Method? To obtain students' opinions about proposed changes to course registration procedures, the administration of a small college asked for faculty volunteers who were willing to administer a survey in one of their classes. Twenty-three faculty members volunteered. Each faculty member gave the survey to all the students in one course of their choosing. Would this sampling method be considered a cluster sample? Why or why not?

31. Sample Design The city of Naperville is considering the construction of a new commuter rail station. The city wishes to survey the residents of the city to obtain their opinion regarding the use of tax dollars for this purpose. Design a sampling method to obtain the individuals in the sample. Be sure to support your choice.

32. Sample Design A school board at a local community college is considering raising the student services fees. The board wants to obtain the opinion of the student body before proceeding. Design a sampling method to obtain the individuals in the sample. Be sure to support your choice.

33. Sample Design Target wants to open a new store in the village of Lockport. Before construction, Target's marketers want to obtain some demographic information regarding the area under consideration. Design a sampling method to obtain the individuals in the sample. Be sure to support your choice.

34. Sample Design The county sheriff wants to determine if a certain highway has a high proportion of speeders traveling on it. Design a sampling method to obtain the individuals in the sample. Be sure to support your choice.

35. Sample Design A pharmaceutical company wants to conduct a survey of 30 individuals who have high cholesterol. The company has obtained a list from doctors throughout the country of 6600 individuals who are known to have high cholesterol. Design a sampling method to obtain the individuals in the sample. Be sure to support your choice.

36. Sample Design A marketing executive for Coca-Cola, Inc., wants to identify television shows that people in the Boston area who typically drink Coke are watching. The executive has a list of all households in the Boston area. Design a sampling method to obtain the individuals in the sample. Be sure to support your choice.

37. Putting It Together: Comparing Sampling Methods Suppose a political strategist wants to get a sense of how American adults aged 18 years or older feel about health care and health insurance.

(a) In a political poll, what would be a good frame to use for obtaining a sample?

(b) Explain why simple random sampling may not guarantee that the sample has an accurate representation of registered Democrats, registered Republicans, and registered Independents.

(c) How can stratified sampling guarantee this representation?

38. Putting It Together: Thinking about Randomness What is random sampling? Why is it necessary for a sample to be obtained randomly rather than conveniently? Will randomness guarantee that a sample will provide accurate information about the population? Explain.

39. Research the origins of the Gallup Poll and the current sampling method the organization uses. Report your findings to the class.

40. Research the sampling methods used by a market research firm in your neighborhood. Report your findings to the class. The report should include the types of sampling methods used, number of stages, and sample size.

1.5 Bias in Sampling

Objective ① Explain the sources of bias in sampling

① Explain the Sources of Bias in Sampling

So far we have looked at *how* to obtain samples, but not at some of the problems that inevitably arise in sampling. Remember, the goal of sampling is to obtain information about a population through a sample.

Definition

If the results of the sample are not representative of the population, then the sample has **bias**.

In Other Words
The word *bias* could mean to give preference to selecting some individuals over others; it could also mean that certain responses are more likely to occur in the sample than in the population.

There are three sources of bias in sampling:

1. Sampling bias
2. Nonresponse bias
3. Response bias

Sampling Bias
Sampling bias means that the technique used to obtain the sample's individuals tends to favor one part of the population over another. Any convenience sample has sampling bias because the individuals are not chosen through a random sample.

Sampling bias also results due to **undercoverage**, which occurs when the proportion of one segment of the population is lower in a sample than it is in the population. Undercoverage can result if the frame used to obtain the sample is incomplete or not representative of the population. Some frames, such as the list of all registered voters, may seem easy to obtain; but even this frame may be incomplete since people who recently registered to vote may not be on the published list of registered voters.

Sampling bias can lead to incorrect predictions. For example, the magazine *Literary Digest* predicted that Alfred M. Landon would defeat Franklin D. Roosevelt in the 1936 presidential election. The *Literary Digest* conducted a poll based on a list of its subscribers, telephone directories, and automobile owners. On the basis of the results, the *Literary Digest* predicted that Landon would win with 57% of the popular vote. However, Roosevelt won the election with about 62% of the popular vote. This election took place during the height of the Great Depression. In 1936, most subscribers to the magazine, households with telephones, and automobile owners were Republican, the party of Landon. Therefore, the choice of the frame used to conduct the survey led to an incorrect prediction due to sampling bias. Essentially, there was undercoverage of Democrats.

It is difficult to gain access to a *complete* list of individuals in a population. For example, public-opinion polls often use random digit dialing (RDD) telephone surveys, which implies that the frame is all households with telephones (landlines or cell phones). This method of sampling excludes households without telephones, as well as homeless people. If these people differ in some way from people with telephones or homes, the results of the sample may not be valid.

Nonresponse Bias

Nonresponse bias exists when individuals selected to be in the sample who do not respond to the survey have different opinions from those who do. Nonresponse can occur because individuals selected for the sample do not wish to respond or the interviewer was unable to contact them.

All surveys will suffer from nonresponse. The federal government's Current Population Survey has a response rate of about 92%, but it varies depending on the age of the individual. For example, the response rate for 20- to 29-year-olds is 85%, and for individuals 70 and older, it is 99%. Response rates in random digit dialing telephone surveys are typically around 70%, e-mail survey response rates hover around 40%, and mail surveys can have response rates as high as 60%.

Nonresponse bias can be controlled using callbacks. For example, if a mailed questionnaire was not returned, a callback might mean phoning the individual to conduct the survey. If an individual was not at home, a callback might mean returning to the home at other times in the day.

Another method to improve nonresponse is using rewards, such as cash payments for completing a questionnaire, or incentives such as a cover letter that states that the responses to the questionnaire will determine future policy. For example, I received $1 with a survey regarding my satisfaction with a recent purchase. The $1 "payment" was meant to make me feel guilty enough to fill out the questionnaire. As another example, a city may send out questionnaires to households and state in a cover letter that the responses to the questionnaire will be used to decide pending issues within the city.

Let's consider the *Literary Digest* poll again. The *Literary Digest* mailed out more than 10 million questionnaires and 2.3 million people responded. The rather low response rate (23%) contributed to the incorrect prediction. After all, Roosevelt was the incumbent president and only those who were unhappy with his administration were likely to respond. In the same election, the 35-year-old George Gallup predicted that Roosevelt would win the election in his survey involving 50,000 people.

Response Bias

Response bias exists when the answers on a survey do not reflect the true feelings of the respondent. Response bias can occur in a number of ways.

Interviewer Error   A trained interviewer is essential to obtain accurate information from a survey. A skilled interviewer can elicit responses from individuals and make the interviewee feel comfortable enough to give truthful responses. For example, a good interviewer can obtain truthful answers to questions as sensitive as "Have you ever cheated on your taxes?" Do not be quick to trust surveys conducted by poorly trained interviewers. Do not trust survey results if the sponsor has a vested interest in the results of the survey. Would you trust a survey conducted by a car dealer that reports 90% of customers say they would buy another car from the dealer?

Misrepresented Answers   Some survey questions result in responses that misrepresent facts or are flat-out lies. For example, a survey of recent college graduates may find that self-reported salaries are inflated. Also, people may overestimate their abilities. For example, ask people how many push-ups they can do in 1 minute, and then ask them to do the push-ups. How accurate were they?

> **CAUTION!**
> The wording of questions can significantly affect the responses and, therefore, the validity of a study. A great source for phrasing survey questions is Saris, W. E. and Gallhofer, I. N. (2007) *Design, Evaluation, and Analysis of Questionnaires for Survey Research.* Hoboken, NJ: Wiley.

Wording of Questions   The way a question is worded can lead to response bias in a survey, so questions must always be asked in balanced form. For example, the "yes/no" question

Do you oppose the reduction of estate taxes?

should be written

Do you favor or oppose the reduction of estate taxes?

The second question is balanced. Do you see the difference? Consider the following report based on studies from Schuman and Presser (*Questions and Answers in Attitude Surveys*, 1981, p. 277), who asked the following two questions:

(A)   Do you think the United States should forbid public speeches against democracy?
(B)   Do you think the United States should allow public speeches against democracy?

For those respondents presented with question A, 21.4% gave "yes" responses, while for those given question B, 47.8% gave "no" responses. The conclusion you may arrive at is that most people are not willing to forbid something, but more people are willing not to allow something. These results illustrate how wording a question can alter a survey's outcome.

Another consideration in wording a question is not to be vague. The question "How much do you study?" is too vague. Does the researcher mean how much do I study for all my classes or just for statistics? Does the researcher mean per day or per week? The question should be written "How many hours do you study statistics each week?"

Ordering of Questions or Words   Many surveys will rearrange the order of the questions within a questionnaire so that responses are not affected by prior questions. Consider an example from Schuman and Presser in which the following two questions were asked:

(A)   Do you think the United States should let Communist newspaper reporters from other countries come in here and send back to their papers the news as they see it?
(B)   Do you think a Communist country such as Russia should let American newspaper reporters come in and send back to America the news as they see it?

For surveys conducted in 1980 in which the questions appeared in the order (A, B), 54.7% of respondents answered "yes" to A and 63.7% answered "yes" to B. If the questions were ordered (B, A), then 74.6% answered "yes" to A and 81.9% answered "yes" to B. When Americans are first asked if U.S. reporters should be allowed to report Communist news, they are more likely to agree that Communists should be allowed to report American news. Questions should be rearranged as much as possible to help reduce effects of this type.

Pollsters will also rearrange words within a question. For example, the Gallup Organization routinely asks the following question of adults aged 18 years or older:

Do you [rotated: approve (or) disapprove] of the job "the current president" is doing as president?

The words *approve* and *disapprove* are rotated to remove the effect that may occur by writing the word *approve* first in the question.

Type of Question   One of the first considerations in designing a question is determining whether the question should be *open* or *closed*.

An **open question** allows the respondent to choose his or her response:

What is the most important problem facing America's youth today?

A **closed question** requires the respondent to choose from a list of predetermined responses:

What is the most important problem facing America's youth today?

(a) Drugs
(b) Violence
(c) Single-parent homes
(d) Promiscuity
(e) Peer pressure

In closed questions, the possible responses should be rearranged because respondents are likely to choose early choices in a list rather than later choices.

An open question should be phrased so that the responses are similar. (You don't want a wide variety of responses.) This allows for easy analysis of the responses. Closed questions limit the number of respondent choices and, therefore, the results are much easier to analyze. The limited choices, however, do not always include a respondent's desired choice. In that case, the respondent will have to choose a secondary answer or skip the question.

Survey designers recommend conducting pretest surveys with open questions and then using the most popular answers as the choices on closed-question surveys. Another issue to consider in the closed-question design is the number of possible responses. The option "no opinion" should be omitted, because this option does not allow for meaningful analysis. The goal is to limit the number of choices in a closed question without forcing respondents to choose an option they do not prefer, which would make the survey have response bias.

Data-entry Error   Although not technically a result of response bias, data-entry error will lead to results that are not representative of the population. Once data are collected, the results may need to be entered into a computer, which could result in input errors. Or, a respondent may make a data entry error. For example, 39 may be entered as 93. It is imperative that data be checked for accuracy. In this text, we present some suggestions for checking for data error.

Can a Census Have Bias?

The discussion so far has focused on bias in samples, but bias can also occur when conducting a census. A question on a census form could be misunderstood, thereby leading to response bias in the results. We also mentioned that it is often difficult to contact each individual in a population. For example, the U.S. Census Bureau is challenged to count each homeless person in the country, so the census data published by the U.S. government likely suffers from nonresponse bias.

Sampling Error versus Nonsampling Error   Nonresponse bias, response bias, and data-entry errors are types of *nonsampling error*. However, whenever a sample is used to learn information about a population, there will inevitably also be *sampling error*.

Definitions

Nonsampling errors result from undercoverage, nonresponse bias, response bias, or data-entry error. Such errors could also be present in a complete census of the population. **Sampling error** results from using a sample to estimate information about a population. This type of error occurs because a sample gives incomplete information about a population.

In Other Words
We can think of sampling error as error that results from using a subset of the population to describe characteristics of the population. Nonsampling error is error that results from obtaining and recording the information collected.

By incomplete information, we mean that the individuals in the sample cannot reveal all the information about the population. Suppose we wanted to determine the average age of the students enrolled in an introductory statistics course. To do this, we obtain a simple random sample of four students and ask them to write their age on a sheet of paper and turn it in. The average age of these four students is 23.25 years. Assume that no students lied about their age or misunderstood the question, and the sampling was done appropriately. If the actual average age of all 30 students in the class (the population) is 22.91 years, then the sampling error is 23.25 − 22.91 = 0.34 year. Now suppose the same survey is conducted, but this time one student lies about his age. Then the results of the survey will also have nonsampling error.

 1.5 Assess Your Understanding

Vocabulary and Skill Building

1. What is a closed question? What is an open question? Discuss the advantages and disadvantages of each type of question.

2. What does it mean when a part of the population is under-represented?

3. What is bias? Name the three sources of bias and provide an example of each. How can a census have bias?

4. Distinguish between nonsampling error and sampling error.

In Problems 5–16, the survey has bias. (a) Determine the type of bias. (b) Suggest a remedy.

5. A retail store manager wants to conduct a study regarding the shopping habits of his customers. He selects the first 60 customers who enter his store on a Saturday morning.

6. The village of Oak Lawn wishes to conduct a study regarding the income level of households within the village. The village manager selects 10 homes in the southwest corner of the village and sends an interviewer to the homes to determine household income.

7. An antigun advocate wants to estimate the percentage of people who favor stricter gun laws. He conducts a nationwide survey of 1203 randomly selected adults 18 years old and older. The interviewer asks the respondents, "Do you favor harsher penalties for individuals who sell guns illegally?"

8. Suppose you are conducting a survey regarding the sleeping habits of students. From a list of registered students, you obtain a simple random sample of 150 students. One survey question is "How much sleep do you get?"

9. A polling organization conducts a study to estimate the percentage of households that speaks a foreign language as the primary language. It mails a questionnaire to 1023 randomly selected households throughout the United States and asks the head of household if a foreign language is the primary language spoken in the home. Of the 1023 households selected, 12 responded.

10. Cold Stone Creamery is considering opening a new store in O'Fallon. Before opening, the company wants to know the percentage of households in O'Fallon that regularly visit an ice cream shop. The market researcher obtains a list of households in O'Fallon, randomly selects 150, and mails a questionnaire that asks about ice cream eating habits and flavor preferences. Of the 150 questionnaires mailed, 4 are returned.

11. A newspaper article reported, "The *Cosmopolitan* magazine survey of more than 5000 Australian women aged 18–34 found about 42 percent considered themselves overweight or obese."
Source: Herald Sun, September 9, 2007

12. A health teacher wants to research the weight of college students. She obtains the weights for all the students in her 9 A.M. class by looking at their driver's licenses or state IDs.

13. A magazine is conducting a study on the effects of infidelity in a marriage. The editors randomly select 400 women whose husbands were unfaithful and ask, "Do you believe a marriage can survive when the husband destroys the trust that must exist between husband and wife?"

14. A textbook publisher wants to determine what percentage of college professors either require or recommend that their students purchase textbook packages with supplemental materials. The publisher sends surveys by e-mail to a random sample of 320 faculty members who have registered with its website. The publisher reports that 80% of college professors require or recommend that their students purchase some type of textbook package.

15. Suppose you are conducting a survey regarding illicit drug use among teenagers in the Baltimore school district. You obtain a cluster sample of 12 schools within the district and sample all sophomore students in the randomly selected schools. The survey is administered by the teachers.

16. To determine the public's opinion of the police department, the police chief obtains a cluster sample of 15 census tracts within his jurisdiction and samples all households in the randomly selected tracts. Uniformed police officers go door to door to conduct the survey.

Applying the Concepts

17. Response Rates Surveys tend to suffer from low response rates. Based on past experience, a researcher determines that the typical response rate for an e-mail survey is 40%. She wishes to obtain a sample of 300 respondents, so she e-mails the survey to 1500 randomly selected e-mail addresses. Assuming the response rate for her survey is 40%, will the respondents form an unbiased sample? Explain.

18. Delivery Format The General Social Survey asked, "About how often did you have sex in the past 12 months?" About 47% of respondents indicated they had sex at least once a week. In an internet survey for a marriage and family wellness center, respondents were asked, "How often do you and your partner have sex (on average)?" About 31% of respondents indicated they had sex with their partner at least once a week. Explain how the delivery method for such a question could result in biased responses.

19. Order of the Questions Consider the following two questions:

(a) Suppose that a rape is committed in which the woman becomes pregnant. Do you think the criminal should or should not face additional charges if the woman becomes pregnant?

(b) Do you think abortions should be legal under any circumstances, legal under certain circumstances, or illegal in all circumstances?

Do you think the order in which the questions are asked will affect the survey results? If so, what can the pollster do to alleviate this response bias?

20. Order of the Questions Consider the following two questions:

(a) Do you believe that the government should or should not be allowed to prohibit individuals from expressing their religious beliefs at their place of employment?

(b) Do you believe that the government should or should not be allowed to prohibit teachers from expressing their religious beliefs in public school classrooms?

Do you think the order in which the questions are asked will affect the survey results? If so, what can the pollster do to alleviate this response bias? Discuss the choice of the word *prohibit* in the survey questions.

21. Improving Response Rates Suppose you are reading an article at psychcentral.com and the following text appears in a pop-up window:

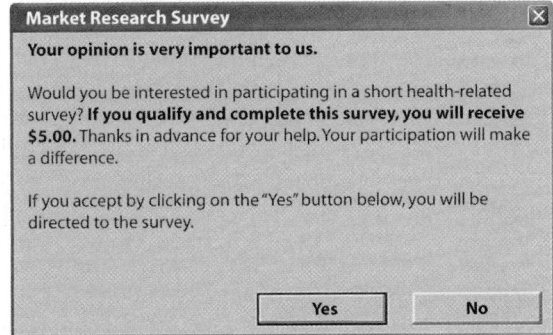

What tactic is the company using to increase the response rate for its survey?

22. Rotating Choices Consider this question from a recent Gallup poll:

Which of the following approaches to solving the nation's energy problems do you think the U.S. should follow right now—[ROTATED: emphasize production of more oil, gas and coal supplies (or) emphasize more conservation by consumers of existing energy supplies]?

Why is it important to rotate the two choices presented in the question?

23. Random Digit Dialing Many polls use random digit dialing (RDD) to obtain a sample, which means a computer randomly generates phone numbers of landlines and cell phones. What is the frame for this type of sampling? Who would be excluded from the survey and how might this affect the results of the survey?

24. Caller ID How do you think caller ID has affected phone surveys?

25. Don't Call Me! The Telephone Consumer Protection Act (TCPA) allows consumers to put themselves on a do-not-call registry. If a number is on the registry, commercial telemarketers are not allowed to call you. Do you believe this has affected the ability of surveyors to obtain accurate polling results? If so, how?

26. Current Population Survey In the federal government's Current Population Survey, the response rate for 20- to 29-year-olds is 85%; for individuals at least 70 years of age it is 99%. Why do you think this is?

27. Exit Polling During every election, pollsters conduct exit polls to help determine which candidate people voted for. During the 2004 presidential election, pollsters incorrectly predicted John Kerry the winner over George Bush. When asked how this error could have happened, the pollsters cited interviewer error due to the fact that many of the interviewers were young, only young voters agreed to be interviewed, and young voters tended to favor Kerry. Plus, young interviewers tend to make more data-entry mistakes. In addition, the method of selecting individuals to be interviewed led to selecting a higher proportion of female voters, and Kerry was favored by females. In some precincts, interviewers were denied access to voters. Research the 2004 exit polling fiasco. Explain which nonsampling errors led to the incorrect conclusion regarding the 2004 election.

28. Write Your Own Survey Develop a survey that you could administer using online survey tools such as StatCrunch, surveymonkey.com, or polldaddy.com. Administer the survey. Did the responses accurately reflect the goals of each question? What types of nonsampling error did you encounter in the survey? If you invited individuals to take the survey via an e-mail, what type of response rate did you obtain? What approach did you take to increase response rate?

29. Wording Survey Questions In the early 1990s, Gallup asked Americans whether they supported the United States bombing Serbian forces in Bosnia. In this survey, 35% of respondents supported the idea. The very same day, ABC News asked whether Americans would support the United States, along with its allies in Europe, bombing Serbian forces in Bosnia. In this survey, 65% supported the idea. Explain the difference in the wording of the question. What does this suggest?

30. Wording Survey Questions Write a survey question that contains strong wording and one that contains tempered wording. Post each question in an online survey site such as StatCrunch, surveymonkey.com, or polldaddy.com. Administer the survey to at least 25 different people for each question. How does the wording affect the response?

31. Order in Survey Questions Write two questions that could have different responses, depending on the order in which the questions are presented. Or write a single question such that the order in which words are presented could affect the response. Administer the survey to at least 25 different people for each question. Did the results differ?

32. Informed Opinions People often respond to survey questions without any knowledge of the subject matter. A common example of this is the discussion on banning dihydrogen monoxide. The Centers for Disease Control (CDC) reports that there were 1423 deaths due to asbestos in 2005, but over 3443 deaths were attributed to dihydrogen monoxide in 2007. Articles and websites such as www.dhmo.org tell how this substance is widely used despite the dangers associated with it. Many people have joined the cause to ban this substance without realizing that dihydrogen monoxide is simply water (H_2O). Their eagerness to protect the environment or their fear of seeming uninformed may be part of the problem. Put together a survey that asks individuals whether dihydrogen monoxide should or should not be banned. Give the survey to 20 randomly selected students around campus and report your results to the class. An example survey might look like the following:

Dihydrogen monoxide is colorless, odorless, and kills thousands of people every year. Most of these deaths are caused by accidental inhalation, but the dangers of dihydrogen monoxide do not stop there. Prolonged exposure to its solid form can severely damage skin tissue. Symptoms of ingestion can include excessive sweating and urination and possibly a bloated feeling, nausea, vomiting, and body electrolyte imbalance. Dihydrogen monoxide is a major component of acid rain and can cause corrosion after coming in contact with certain metals.

Do you believe that the government should or should not ban the use of dihydrogen monoxide?

33. Name two biases that led to the *Literary Digest* making an incorrect prediction in the presidential election of 1936.

34. Research on George Gallup Research the polling done by George Gallup in the 1936 presidential election. Write a report on your findings and include information about the sampling technique and sample size. Next, research the polling done by Gallup for the 1948 presidential election. Did Gallup accurately predict the outcome of the election? What lessons were learned by Gallup?

35. The Challenge in Polling One of the challenges in polling for elections is deciding who to include in your frame and who might actually turn out to vote.

(a) Suppose you were asked to conduct a poll for a senatorial election. Explain how you might design your sample. In your explanation include a discussion of the difference between "registered voters" and "likely voters." What role would stratification play in your sampling?

(b) Voter turnout is different for presidential election cycles (2012, 2016, 2020, and so on) versus non-presidential election cycles (2014, 2018, 2022, and so on). Explain the role election cycle plays in voter turnout and explain how this may affect your sampling methodology.

(c) During the 2014 election, Nate Silver of FiveThirtyEight said "the pre-election polling averages (not the FiveThirtyEight forecasts, which also account for other factors) in the 10 most competitive Senate races had a 6-percentage point Democratic bias as compared to the votes counted in each state so far." Explain what this means and explain how this would have impacted polling results compared with actual results.

Explaining the Concepts

36. Why is it rare for frames to be completely accurate?

37. What are some solutions to nonresponse?

38. Discuss the benefits of having trained interviewers.

39. What are the advantages of having a presurvey when constructing a questionnaire that has closed questions?

40. Discuss the pros and cons of telephone interviews that take place during dinner time in the early evening.

41. Why is a high response rate desired? How would a low response rate affect survey results?

42. Discuss why the order of questions or choices within a questionnaire are important in sample surveys.

43. Suppose a survey asks, "Do you own any CDs?" Explain how this could be interpreted in more than one way. Suggest a way in which the question could be improved.

44. Discuss a possible advantage of offering rewards or incentives to increase response rates. Are there any disadvantages?

1.6 The Design of Experiments

Objectives
① Describe the characteristics of an experiment
② Explain the steps in designing an experiment
③ Explain the completely randomized design
④ Explain the matched-pairs design
⑤ Explain the randomized block design

The major theme of this chapter has been data collection. Section 1.2 briefly discussed the idea of an experiment, but the main focus was on observational studies. Sections 1.3 through 1.5 focused on sampling and surveys. In this section, we further develop the idea of collecting data through an experiment.

① ## Describe the Characteristics of an Experiment

Remember, in an observational study, if an association exists between an explanatory variable and response variable the researcher cannot claim causality. To demonstrate how changes in the explanatory variable *cause* changes in the response variable, the researcher needs to conduct an *experiment*.

Definitions
An **experiment** is a controlled study conducted to determine the effect varying one or more explanatory variables or **factors** has on a response variable. Any combination of the values of the factors is called a **treatment**.

Historical Note

Sir Ronald Fisher, often called the Father of Modern Statistics, was born in England on February 17, 1890. He received a BA in astronomy from Cambridge University in 1912. In 1914, he took a position teaching mathematics and physics at a high school. He did this to help serve his country during World War I. (He was rejected by the army because of his poor eyesight.) In 1919, Fisher took a job as a statistician at Rothamsted Experimental Station, where he was involved in agricultural research. In 1933, Fisher became Galton Professor of Eugenics at Cambridge University, where he studied Rh blood groups. In 1943 he was appointed to the Balfour Chair of Genetics at Cambridge. He was knighted by Queen Elizabeth in 1952. Fisher retired in 1957 and died in Adelaide, Australia, on July 29, 1962. One of his famous quotations is "To call in the statistician after the experiment is done may be no more than asking him to perform a postmortem examination: he may be able to say what the experiment died of."

In an experiment, the **experimental unit** is a person, object, or some other well-defined item upon which a treatment is applied. We often refer to the experimental unit as a **subject** when he or she is a person. The subject is analogous to the individual in a survey.

The goal in an experiment is to determine the effect various treatments have on the response variable. For example, we might want to determine whether a new treatment is superior to an existing treatment (or no treatment at all). To make this determination, experiments require a *control group*. A **control group** serves as a baseline treatment that can be used to compare it to other treatments. For example, a researcher in education might want to determine if students who do their homework using an online homework system do better on an exam than those who do their homework from the text. The students doing the text homework might serve as the control group (since this is the currently accepted practice). The factor is the type of homework. There are two treatments: online homework and text homework.

A second method for defining the control group is through the use of a *placebo*. A **placebo** is an innocuous medication, such as a sugar tablet, that looks, tastes, and smells like the experimental medication.

In an experiment, it is important that each group be treated the same way. It is also important that individuals do not adjust their behavior because of the treatment they are receiving. For this reason, many experiments use a technique called *blinding*. **Blinding** refers to nondisclosure of the treatment an experimental unit is receiving. There are two types of blinding: *single blinding* and *double blinding*.

Definitions

In **single-blind** experiments, the experimental unit (or subject) does not know which treatment he or she is receiving. In **double-blind** experiments, neither the experimental unit nor the researcher in contact with the experimental unit knows which treatment the experimental unit is receiving.

EXAMPLE 1   The Characteristics of an Experiment

Problem Lipitor is a cholesterol-lowering drug made by Pfizer. In the Collaborative Atorvastatin Diabetes Study (CARDS), the effect of Lipitor on cardiovascular disease was assessed in 2838 subjects, ages 40 to 75, with type 2 diabetes, without prior history of cardiovascular disease. In this placebo-controlled, double-blind experiment, subjects were randomly allocated to either Lipitor 10 mg daily (1428) or placebo (1410) and were followed for 4 years. The response variable was the occurrence of any major cardiovascular event.

Lipitor significantly reduced the rate of major cardiovascular events (83 events in the Lipitor group versus 127 events in the placebo group). There were 61 deaths in the Lipitor group versus 82 deaths in the placebo group.

(a) What does it mean for the experiment to be placebo-controlled?
(b) What does it mean for the experiment to be double-blind?
(c) What is the population for which this study applies? What is the sample?
(d) What are the treatments?
(e) What is the response variable? Is it qualitative or quantitative?

Approach Apply the definitions just presented.

Solution
(a) The placebo is a medication that looks, smells, and tastes like Lipitor. The placebo control group serves as a baseline against which to compare the results from the group receiving Lipitor. The placebo is also used because people tend to behave differently when they are in a study. By having a placebo control group, the effect of this is neutralized.
(b) Since the experiment is double-blind, the subjects, as well as the individual monitoring the subjects, do not know whether the subjects are receiving Lipitor

(continued)

or the placebo. The experiment is double-blind so that the subjects receiving the medication do not behave differently from those receiving the placebo and so the individual monitoring the subjects does not treat those in the Lipitor group differently from those in the placebo group.

(c) The population is individuals from 40 to 75 years of age with type 2 diabetes without a prior history of cardiovascular disease. The sample is the 2838 subjects in the study.

(d) The treatments are 10 mg of Lipitor or a placebo daily.

(e) The response variable is whether the subject had any major cardiovascular event, such as a stroke, or not. It is a qualitative variable.

● Now Work Problem 7

●

❷ Explain the Steps in Designing an Experiment

To **design** an experiment means to describe the overall plan in conducting the experiment. Conducting an experiment requires a series of steps.

Step 1 **Identify the Problem to Be Solved.** The statement of the problem should be as explicit as possible and should provide the experimenter with direction. The statement must also identify the response variable and the population to be studied. Often, the statement is referred to as the *claim*.

Step 2 **Determine the Factors That Affect the Response Variable.** The factors are usually identified by an expert in the field of study. In identifying the factors, ask, "What things affect the value of the response variable?" After the factors are identified, determine which factors to fix at some predetermined level, which to manipulate, and which to leave uncontrolled.

Step 3 **Determine the Number of Experimental Units.** As a general rule, choose as many experimental units as time and money allow. Techniques (such as those in Sections 9.1 and 9.2) exist for determining sample size, provided certain information is available.

Step 4 **Determine the Level of Each Factor.** There are two ways to deal with the factors: control or randomize.

1. Control: There are two ways to control the factors.
   (a) Set the level of a factor at one value throughout the experiment (if you are *not* interested in its effect on the response variable).
   (b) Set the level of a factor at various levels (if you are interested in its effect on the response variable). The combinations of the levels of all varied factors constitute the treatments in the experiment.

2. Randomize: Randomly assign the experimental units to treatment groups. Because it is difficult, if not impossible, to identify all factors in an experiment, randomly assigning experimental units to treatment groups mutes the effect of variation attributable to factors (explanatory variables) not controlled.

Step 5 **Conduct the Experiment.**

(a) **Replication** occurs when each treatment is applied to more than one experimental unit. Using more than one experimental unit for each treatment ensures the effect of a treatment is not due to some characteristic of a single experimental unit. It is a good idea to assign an equal number of experimental units to each treatment.

(b) Collect and process the data. Measure the value of the response variable for each replication. Then organize the results. The idea is that the value of the response variable for each treatment group is the *same* before the experiment because of randomization. Then any *difference* in the value of the response variable among the different treatment groups is a result of differences in the level of the treatment.

Step 6 **Test the Claim.** This is the subject of inferential statistics. **Inferential statistics** is a process in which generalizations about a population are made on the basis of results obtained from a sample. Provide a statement regarding the level of confidence in the generalization. Methods of inferential statistics are presented in Chapters 9 through 15.

❸ Explain the Completely Randomized Design

The steps just given apply to any type of designed experiment. We now concentrate on the simplest type of experiment.

Definition   A **completely randomized design** is one in which each experimental unit is randomly assigned to a treatment.

EXAMPLE 2   A Completely Randomized Design

Problem A farmer wishes to determine the optimal level of a new fertilizer on his soybean crop. Design an experiment that will assist him.

Approach Follow the steps for designing an experiment.

Solution

Step 1   The farmer wants to identify the optimal level of fertilizer for growing soybeans. We define *optimal* as the level that maximizes yield. So the response variable will be crop yield.

Step 2   Some factors that affect crop yield are fertilizer, precipitation, sunlight, method of tilling the soil, type of soil, plant, and temperature.

Step 3   In this experiment, we will plant 60 soybean plants (experimental units).

Step 4   List the factors and their levels.

In Other Words
The various levels of the factor are the treatments in a completely randomized design.

- **Fertilizer.** This factor will be controlled and set at three levels. We wish to measure the effect of varying the level of this variable on the response variable, yield. We will set the treatments (level of fertilizer) as follows:

    Treatment A: 20 soybean plants receive no fertilizer.
    Treatment B: 20 soybean plants receive 2 teaspoons of fertilizer per gallon of
       water every 2 weeks.
    Treatment C: 20 soybean plants receive 4 teaspoons of fertilizer per gallon of
       water every 2 weeks.

See Figure 6.

Figure 6

- **Precipitation.** The amount of rainfall cannot be controlled, but the amount of watering done can be controlled. Each plant will receive the same amount of precipitation.
- **Sunlight.** This uncontrollable factor will be roughly the same for each plant.
- **Method of tilling.** Control this factor by using round-up ready method of tilling for each plant.
- **Type of soil.** Certain aspects of the soil, such as level of acidity, can be controlled. In addition, each plant will be planted within a 1-acre area, so it is reasonable to assume that the soil conditions for each plant are equivalent.
- **Plant.** There may be variation from plant to plant. To account for this, randomly assign the plants to a treatment.
- **Temperature.** This factor is uncontrollable, but will be the same for each plant.

Step 5

(a) Randomly assign each plant to a treatment group. First, number the plants from 1 to 60 and randomly generate 20 numbers. The plants corresponding to these numbers get treatment A. Next number the remaining plants 1 to 40 and randomly generate 20 numbers. The plants corresponding to these numbers get treatment B. The remaining plants get treatment C. Now till the soil, plant the soybean plants, and fertilize according to the schedule prescribed.

(b) At the end of the growing season, determine the crop yield for each plant.

Step 6   Determine any differences in yield among the three treatment groups. Figure 7 on the following page illustrates the experimental design.

(continued)

Figure 7

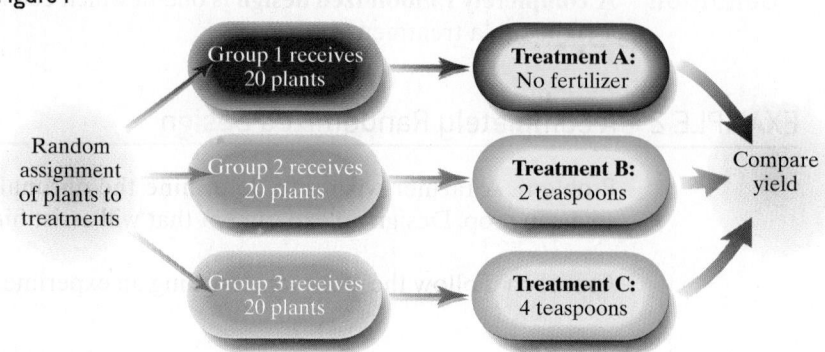

Example 2 is a completely randomized design because the experimental units (the plants) were randomly assigned to the treatments. It is the most popular experimental design because of its simplicity, but it is not always the best. We discuss inferential procedures for the completely randomized design with two treatments and quantitative response variable in Section 11.3 and with three or more treatments in Section 13.1. We discuss inferential procedures for the completely randomized design for a qualitative response variable in Sections 11.1 and 12.2.

• **Now Work Problem 9**

④ **Explain the Matched-Pairs Design**

Another type of experimental design is called a *matched-pairs design*.

Definition

> A **matched-pairs design** is an experimental design in which the experimental units are paired up. The pairs are selected so that they are related in some way (that is, the same person before and after a treatment, twins, husband and wife, same geographical location, and so on). There are only two levels of treatment in a matched-pairs design.

In matched-pairs design, one matched individual will receive one treatment and the other receives a different treatment. The matched pair is randomly assigned to the treatment using a coin flip or a random-number generator. We then look at the difference in the results of each matched pair. One common type of matched-pairs design is to measure a response variable on an experimental unit before and after a treatment is applied. In this case, the individual is matched against itself. These experiments are sometimes called before–after or pretest–posttest experiments.

EXAMPLE 3 A Matched-Pairs Design

Problem An educational psychologist wants to determine whether listening to music has an effect on a student's ability to learn. Design an experiment to help the psychologist answer the question.

Approach We will use a matched-pairs design by matching students according to IQ and gender (just in case gender plays a role in learning with music).

Solution Match students according to IQ and gender. For example, match two females with IQs in the 110 to 115 range.

For each pair of students, flip a coin to determine which student is assigned the treatment of a quiet room or a room with music playing in the background.

Each student will be given a statistics textbook and asked to study Section 1.1. After 2 hours, the students will enter a testing center and take a short quiz on the material in the section. Compute the difference in the scores of each matched pair. Any differences in scores will be attributed to the treatment. Figure 8 illustrates the design.

Figure 8

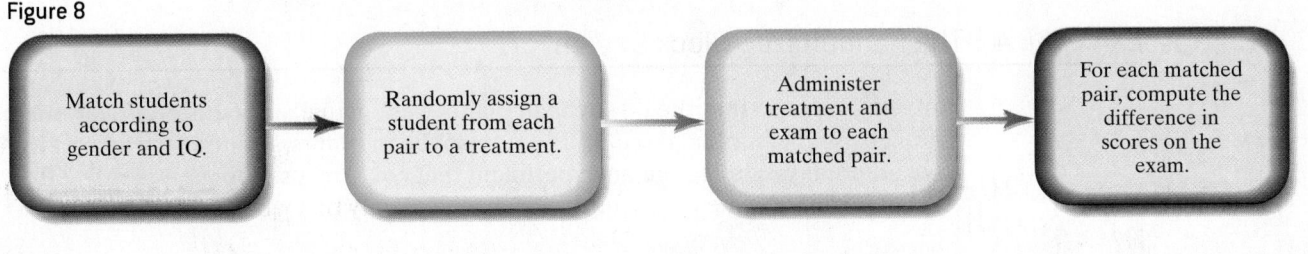

Match students according to gender and IQ. → Randomly assign a student from each pair to a treatment. → Administer treatment and exam to each matched pair. → For each matched pair, compute the difference in scores on the exam.

● **Now Work Problem 11**

We discuss statistical inference for the matched-pairs design for a qualitative response variable in Section 12.3 and quantitative response variable in Section 11.2.

⑤ **Explain the Randomized Block Design**

The completely randomized design is the simplest experimental design. However, its simplicity can lead to flaws. Before we introduce a slightly more complicated experimental design, consider the following story.

I coach my son's soccer team, which is comprised of four 10-year-olds, six 9-year-olds, and four 8-year-olds. After each practice, I like to have a 15-minute scrimmage where I randomly assign seven players to each team. I quickly learned that randomly assigning players sometimes resulted in very unequal teams (such as a team of four 10-year-olds and three 9-year-olds). So I learned that I should randomly assign two 10-year-olds to one team and the other two 10-year-olds to the other team. I then assigned three 9-year-olds to each team and two 8-year-olds to each team. Using the language of statistics, I was *blocking* by age.

Definitions   Grouping together similar (homogeneous) experimental units and then randomly assigning the experimental units within each group to a treatment is called **blocking**. Each group of homogeneous individuals is called a **block**.

With the soccer story in mind, let's revisit our crop-yield experiment from Example 2, where we assumed that the soybean plants were the same variety. Suppose we in fact had two varieties of soybeans: Chemgro and Pioneer. Using the completely randomized design, it is impossible to know whether any differences in yield were due to the differences in fertilizer or to the differences in variety.

Perhaps the Chemgro variety naturally has higher yield than Pioneer. If a majority of the soybean plants randomly assigned to treatment C was Chemgro, we would not know whether to attribute the different crop yields to the level of fertilizer or to the plant variety. So, plant variety is a confounding variable. Recall, a confounding variable exists when the effect of two factors (explanatory variables) on the response variable cannot be distinguished.

To resolve issues of this type, we introduce a third experimental design.

Definition   A **randomized block design** is used when the experimental units are divided into homogeneous groups called blocks. Within each block, the experimental units are randomly assigned to treatments.

In a randomized block design, we do not wish to determine whether the differences between blocks result in any difference in the value of the response variable. Our goal is to remove any variability in the response variable that may be attributable to the block.

EXAMPLE 4   The Randomized Block Design

Problem Suppose that the 60 soybean plants were actually two different varieties: 30 Chemgro soybean plants and 30 Pioneer soybean plants and that the two varieties have different yields. Design an experiment that could be used to measure the effect of level of fertilizer while taking into account the variety of soybean.

Approach Because the yields of the two varieties may differ, use a randomized block design, blocking by variety, to determine which level of fertilizer results in the highest yield.

Figure 9

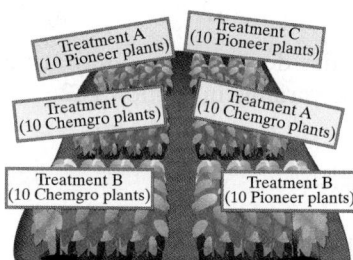

Solution In a randomized block design, divide the 60 plants into two blocks: 30 Chemgro plants in block 1 and 30 Pioneer plants in block 2. Within each block, randomly assign 10 plants to each treatment (A, B, or C). The other factors that may affect yield are controlled as they were in Example 2. See Figure 9.

The logic is each block contains plants of the same variety, so plant variety does not affect the value of the response variable. Figure 10 illustrates the design. Notice that we do not compare yields across plant variety because this was not our goal—we already know which variety yields more. The reason for blocking is to reduce variability due to plant variety by comparing yields within each variety.

Figure 10

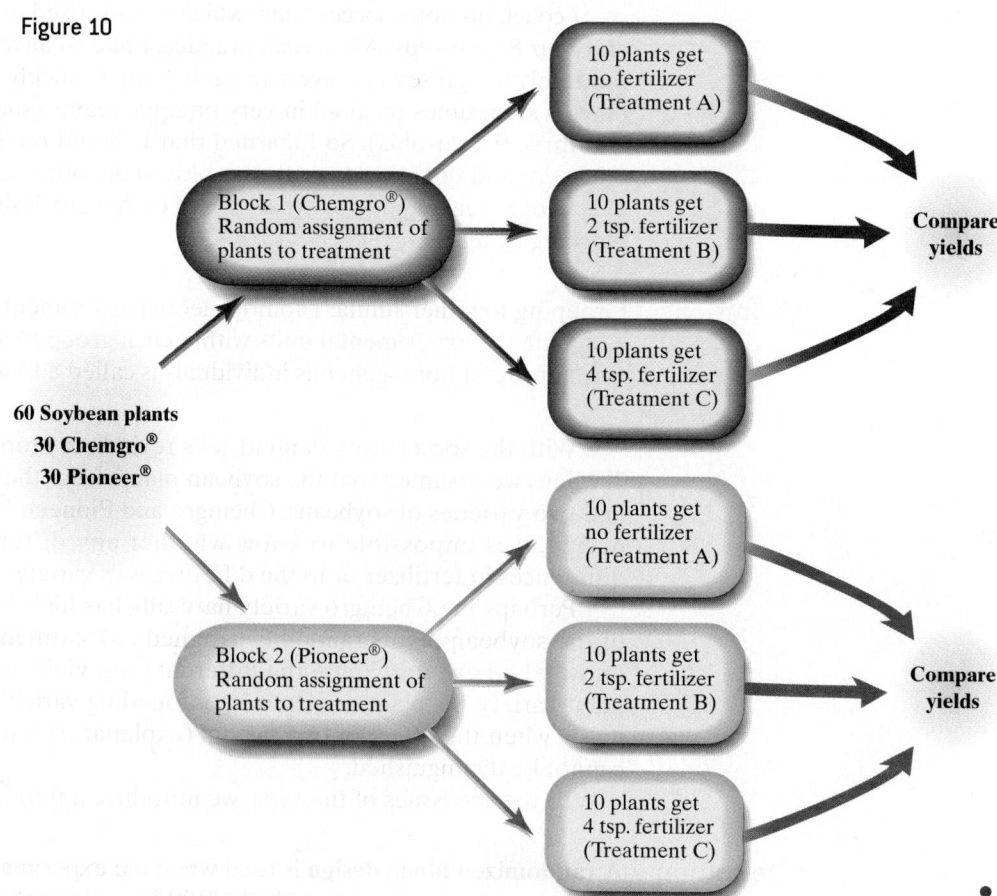

● **Now Work Problem 19**

One note about the relation between a designed experiment and simple random sampling: It is often the case that the experimental units selected to participate in a study are not randomly selected. This is because we often need the experimental units to have some common trait, such as high blood pressure. For this reason, participants in experiments are recruited or volunteer to be in a study. However, once we have the experimental units, we use simple random sampling to assign them to treatment groups. With random assignment we assume that the participants are similar at the start of the experiment. Because the treatment is the only difference between the groups, we can say the treatment *caused* the difference observed in the response variable.

 ## 1.6 Assess Your Understanding

Vocabulary

1. Define the following:
(a) Experimental unit
(b) Treatment
(c) Response variable
(d) Factor
(e) Placebo
(f) Confounding

2. What is replication in an experiment?

3. Explain the difference between a single-blind and a double- blind experiment.

4. A(n) _____ _____ design is one in which each experimental unit is randomly assigned to a treatment. A(n) _____ _____ design is one in which the experimental units are paired up.

5. Grouping together similar experimental units and then randomly assigning the experimental units within each group to a treatment is called _____.

6. True or False: Generally, the goal of an experiment is to determine the effect that treatments will have on the response variable.

Applying the Concepts

NW 7. Chew Your Food Researchers wanted to determine the association between number of times one chews food and food consumption. They identified 45 individuals who were 18 to 45 years of age. First, the researchers determined a baseline for number of chews before swallowing food. Next, each participant attended three sessions to eat pizza for lunch until comfortably full by chewing each portion of food 100%, 150%, and 200% of their baseline number of chews before swallowing. Food intake for each of the three chewing treatments was then measured. It was found that food consumption was reduced significantly, by 9.5% and 14.8%, respectively, for the 150% and 200% number of chews compared to the baseline.

Source: Yong Zhu and James H. Hollis. "Increasing the Number of Chews before Swallowing Reduces Meal Size in Normal-Weight, Overweight, and Obese Adults," *Journal of the Academy of Nutrition and Dietetics*, 11 November 2013.

(a) What is the research objective of the study?
(b) What is the response variable in this study? Is it quantitative or qualitative?
(c) What is the explanatory variable in this study? Is it quantitative or qualitative?
(d) Who are the experimental units?
(e) How is control used in this study?
(f) Each individual chewed 100%, 150%, and 200% of their baseline number of chews before swallowing. This is referred to as a *repeated-measures study* since the same participants were exposed to each treatment. The order in which chewing took place (100% versus 150% versus 200%) was determined randomly. Explain why this is important.

8. Alcohol Dependence To determine if topiramate is a safe and effective treatment for alcohol dependence, researchers conducted a 14-week trial of 371 men and women aged 18 to 65 years diagnosed with alcohol dependence. In this double-blind, randomized, placebo-controlled experiment, subjects were randomly given either 300 milligrams (mg) of topiramate (183 subjects) or a placebo (188 subjects) daily, along with a weekly compliance enhancement intervention. The variable used to determine the effectiveness of the treatment was self-reported percentage of heavy drinking days. Results indicated that topiramate was more effective than placebo at reducing the percentage of heavy drinking days. The researchers concluded that topiramate is a promising treatment for alcohol dependence.

Source: Bankole A. Johnson, Norman Rosenthal, et al. "Topiramate for Treating Alcohol Dependence: A Randomized Controlled Trial," *Journal of the American Medical Association*, 298(14):1641–1651, 2007

(a) What does it mean for the experiment to be placebo-controlled?

(b) What does it mean for the experiment to be double-blind? Why do you think it is necessary for the experiment to be double-blind?

(c) What does it mean for the experiment to be randomized?

(d) What is the population for which this study applies? What is the sample?

(e) What are the treatments?

(f) What is the response variable?

NW 9. School Psychology A school psychologist wants to test the effectiveness of a new method for teaching reading. She recruits 500 first-grade students in District 203 and randomly divides them into two groups. Group 1 is taught by means of the new method, while group 2 is taught by traditional methods. The same teacher is assigned to teach both groups. At the end of the year, an achievement test is administered and the results of the two groups are compared.

(a) What is the response variable in this experiment?

(b) Think of some of the factors in the study. How are they controlled?

(c) What are the treatments? How many treatments are there?

(d) How are the factors that are not controlled dealt with?

(e) Which group serves as the control group?

(f) What type of experimental design is this?

(g) Identify the subjects.

(h) Draw a diagram similar to Figure 7, 8, or 10 to illustrate the design.

10. Pharmacy A pharmaceutical company has developed an experimental drug meant to relieve symptoms associated with the common cold. The company identifies 300 adult males 25 to 29 years old who have a common cold and randomly divides them into two groups. Group 1 is given the experimental drug, while group 2 is given a placebo. After 1 week of treatment, the proportion that still have cold symptoms in each group are compared.

(a) What is the response variable in this experiment?

(b) Think of some of the factors in the study. How are they controlled?

(c) What are the treatments? How many treatments are there?

(d) How are the factors that are not controlled dealt with?

(e) What type of experimental design is this?

(f) Identify the subjects.

(g) Draw a diagram similar to Figure 7, 8, or 10 to illustrate the design.

NW 11. Whiter Teeth An ad for Crest Whitestrips Premium claims that the strips will whiten teeth in 7 days and the results will last for 12 months. A researcher who wishes to test this claim studies 20 sets of identical twins. Within each set of twins, one is randomly selected to use Crest Whitestrips Premium in addition to regular brushing and flossing, while the other just brushes and flosses. Whiteness of teeth is measured at the beginning of the study, after 7 days, and every month thereafter for 12 months.

(a) What type of experimental design is this?

(b) What is the response variable in this experiment?

(c) What are the treatments?

(d) What are other factors (controlled or uncontrolled) that could affect the response variable?

(e) What might be an advantage of using identical twins as subjects in this experiment?

12. Assessment To help assess student learning in her developmental math courses, a mathematics professor at a community college implemented pre- and posttests for her students. A knowledge-gained score was obtained by taking the difference of the two test scores.

(a) What type of experimental design is this?

(b) What is the response variable in this experiment?

(c) What is the treatment?

13. Insomnia Researchers wanted to test the effectiveness of a new cognitive behavioral therapy (CBT) compared with both an older behavioral treatment and a placebo therapy for treating insomnia. They identified 75 adults with insomnia. Patients were randomly assigned to one of three treatment groups. Twenty-five patients were randomly assigned to receive CBT (sleep education, stimulus control, and time-in-bed restrictions), another 25 received muscle relaxation training (RT), and the final 25 received a placebo treatment. Treatment lasted 6 weeks, with follow-up conducted at 6 months. To measure the effectiveness of the treatment, researchers used wake time after sleep onset (WASO). CBT produced larger improvements than did RT or placebo treatment. For example, the CBT-treated patients achieved an average 54% reduction in their WASO, whereas RT-treated and placebo-treated patients, respectively, achieved only 16% and 12% reductions in this measure. Results suggest that CBT treatment leads to significant sleep improvements within 6 weeks, and these improvements appear to endure through 6 months of follow-up.

Source: Jack D. Edinger, PhD; William K. Wohlgemuth, PhD; Rodney A. Radtke, MD; Gail R. Marsh, PhD; Ruth E. Quillian, PhD. "Cognitive Behavioral Therapy for Treatment of Chronic Primary Insomnia," *Journal of the American Medical Association* 285: 1856–1864, 2001

(a) What type of experimental design is this?

(b) What is the population being studied?

(c) What is the response variable in this study?

(d) What are the treatments?

(e) Identify the experimental units.

(f) Draw a diagram similar to Figure 7, 8, or 10 to illustrate the design.

14. Depression Researchers wanted to compare the effectiveness of an extract of St. John's wort with placebo in outpatients with major depression. To do this, they recruited 200 adult outpatients diagnosed as having major depression and having a baseline Hamilton Rating Scale for Depression (HAM-D) score of at least 20. Participants were randomly assigned to receive either St. John's wort extract, 900 milligrams per day (mg/d) for 4 weeks, increased to 1200 mg/d in the absence of an adequate response thereafter, or a placebo for 8 weeks. The response variable was the change on the HAM-D over the treatment period. After analysis of the data, it was concluded that St. John's wort was not effective for treatment of major depression.

Source: Richard C. Shelton, MD, et al. "Effectiveness of St. John's Wort in Major Depression," *Journal of the American Medical Association* 285:1978–1986, 2001

(a) What type of experimental design is this?

(b) What is the population that is being studied?

(c) What is the response variable in this study?

(d) What are the treatments?

(e) Identify the experimental units.

(f) What is the control group in this study?

(g) Draw a diagram similar to Figure 7, 8, or 10 to illustrate the design.

15. The Memory Drug? Researchers wanted to evaluate whether ginkgo, an over-the-counter herb marketed as enhancing memory, improves memory in elderly adults as measured by objective tests. To do this, they recruited 98 men and 132 women older than 60 years and in good health. Participants were randomly assigned to receive ginkgo, 40 milligrams (mg) 3 times per day, or a matching placebo. The measure of memory improvement was determined by a standardized test of learning and memory. After 6 weeks of treatment, the data indicated that ginkgo did not increase performance on standard tests of learning, memory, attention, and concentration. These data suggest that, when taken following the manufacturer's instructions, ginkgo provides no measurable increase in memory or related cognitive function to adults with healthy cognitive function.

Source: Paul R. Solomon et al. "Ginkgo for Memory Enhancement," *Journal of the American Medical Association* 288:835–840, 2002.

(a) What type of experimental design is this?

(b) What is the population being studied?

(c) What is the response variable in this study?

(d) What is the factor that is set to predetermined levels? What are the treatments?

(e) Identify the experimental units.

(f) What is the control group in this study?

(g) Draw a diagram similar to Figure 7, 8, or 10 to illustrate the design.

16. Shrinking Stomach? Researchers wanted to determine whether the stomach shrinks as a result of dieting. To do this, they randomly divided 23 obese patients into two groups. The 14 individuals in the experimental group were placed on a diet that allowed them to consume 2508 kilojoules (kJ) per day for 4 weeks. The 9 subjects in the control group ate as they normally would. To assess the size of the stomach, a latex gastric balloon was inserted into each subject's stomach and filled with the maximum amount of water that could be tolerated by the patient. The volume of water was compared to the volume that could be tolerated at the beginning of the study. The experimental subjects experienced a 27% reduction in gastric capacity, while the subjects in the control group experienced no change in gastric capacity. It was concluded that a reduction in gastric capacity occurs after a restricted diet.

Source: A. Geliebter et al. "Reduced Stomach Capacity in Obese Subjects after Dieting," *American Journal of Clinical Nutrition* 63(2):170–173, 1996

(a) What type of experimental design is this?

(b) What is the population that the results of this experiment apply to?

(c) What is the response variable in this study? Is it qualitative or quantitative?

(d) What are the treatments?

(e) Identify the experimental units.

(f) Draw a diagram similar to Figure 7, 8, or 10 to illustrate the design.

17. Dominant Hand Professor Andy Neill wanted to determine if the reaction time of people differs in their dominant hand versus their nondominant hand. To do this, he recruited 15 students. Each student was asked to hold a yardstick between the index finger and thumb. The student was asked to open the hand, release the yardstick, and then catch the yardstick between the index finger and thumb. The distance that the yardstick fell served as a measure of reaction time. A coin flip was used to determine whether the student would use their dominant hand first or the nondominant hand. Results indicated that the reaction time in the dominant hand exceeded that of the nondominant hand.

(a) What type of experimental design is this?

(b) What is the response variable in this study?

(c) What is the treatment?

(d) Identify the experimental units.

(e) Why did Professor Neill use a coin flip to determine whether the student should begin with the dominant hand or the nondominant hand?

(f) Draw a diagram similar to Figure 7, 8, or 10 to illustrate the design.

18. Golf Anyone? A local golf pro wanted to compare two styles of golf club. One golf club had a graphite shaft and the other had a steel shaft. It is believed that graphite shafts allow a player to hit the ball farther, but the manufacturer of the new steel shaft said the ball travels just as far with its new technology. To test this belief, the pro recruited 10 golfers from the driving range. Each player was asked to hit one ball with the graphite-shafted club and one ball with the new steel-shafted club. The distance that the ball traveled was determined using a range finder. A coin flip was used to determine whether the player hit with the graphite club or the steel club first. Results indicated that the distance the ball was hit with the graphite club was no different than the distance when using the steel club.

(a) What type of experimental design is this?

(b) What is the response variable in this study?

(c) What is the factor that is set to predetermined levels? What is the treatment?

(d) Identify the experimental units.

(e) Why did the golf pro use a coin flip to determine whether the golfer should hit with the graphite first or the steel first?

(f) Draw a diagram similar to Figure 7, 8, or 10 to illustrate the design.

NW **19. Marketing** A marketing research firm wants to determine the most effective method of advertising: print, radio, or television. They recruit 300 volunteers to participate in the study. The chief researcher believes that level of education plays a role in the effectiveness of advertising, so she segments the volunteers by level of education. Of the 300 volunteers, 120 have a high school education, 120 have a college diploma, and 60 have advanced degrees. The 120 volunteers with a high school diploma are randomly assigned to either the print advertising group, the radio group, or the television group. The same procedure is followed for the college graduate and advanced degree volunteers. Each group is exposed to the advertising. After 1 hour, a recall exam is given and the number of correct answers recorded.

(a) What type of experimental design is this?

(b) What is the response variable in this experiment?

(c) What are the treatments?

(d) What variable serves as the block?

(e) Draw a diagram similar to Figure 7, 8, 10 to illustrate the design.

20. Social Work A social worker wants to examine methods that can be used to deter truancy. Three hundred chronically truant students volunteer for the study. Because the social worker believes that socioeconomic class plays a role in truancy, she divides the 300 volunteers according to household income. Of the 300 students, 120 fall in the low-income category, 132 fall in the middle-income category, and the remaining 48 fall in the upper-income category. The students within each income category are randomly divided into three groups. Students in group 1 receive no intervention. Students in group 2 are treated with positive reinforcement in which, for each day the student is not truant, he or she receives a star that can be traded in for rewards. Students in group 3 are treated with negative reinforcement such that each truancy results in a 1-hour detention. However, the hours of detention are cumulative, meaning that the first truancy results in 1 hour of detention, the second truancy results in 2 hours, and so on. After a full school year, the total number of truancies are compared.

(a) What type of experimental design is this?

(b) What is the response variable in this experiment?

(c) What are the treatments?

(d) What variable serves as the block?

(e) Draw a diagram similar to Figure 7, 8, or 10 to illustrate the design.

21. Drug Effectiveness A pharmaceutical company wants to test the effectiveness of an experimental drug meant to reduce high cholesterol. The researcher at the pharmaceutical company has decided to test the effectiveness of the drug through a completely randomized design. She has obtained 20 volunteers with high cholesterol: Ann, John, Michael, Kevin, Marissa, Christina, Eddie, Shannon, Julia, Randy, Sue, Tom, Wanda, Roger, Laurie, Rick, Kim, Joe, Colleen, and Bill. Number the volunteers from 1 to 20. Use a random-number generator to randomly assign 10 of the volunteers to the experimental group. The remaining volunteers will go into the control group. List the individuals in each group.

22. Effects of Alcohol A researcher has recruited 20 volunteers to participate in a study. The researcher wishes to measure the effect of alcohol on an individual's reaction time. The 20 volunteers are randomly divided into two groups. Group 1 serves as a control group in which participants drink four 1-ounce cups of a liquid that looks, smells, and tastes like alcohol in 15-minute increments. Group 2 serves as an experimental group in which participants drink four 1-ounce cups of 80-proof alcohol in 15-minute increments. After drinking the last 1-ounce cup, the participants sit for 20 minutes. After the 20-minute resting period, the reaction time to a stimulus is measured.

(a) What type of experimental design is this?

(b) Use Table I in Appendix A or a random-number generator to divide the 20 volunteers into groups 1 and 2 by assigning the volunteers a number between 1 and 20. Then randomly select 10 numbers between 1 and 20. The individuals corresponding to these numbers will go into group 1.

23. Green Tea You wonder whether green tea lowers cholesterol.

(a) To research the claim that green tea lowers LDL (so-called bad) cholesterol, you ask a random sample of individuals to divulge whether they are regular green tea users or not. You also obtain their LDL cholesterol levels. Finally, you compare the LDL cholesterol levels of the green tea drinkers to those of the non-green tea drinkers. Explain why this is an observational study.

(b) Name some lurking variables that might exist in the study.

(c) Suppose, instead of surveying individuals regarding their tea-drinking habits, you decide to conduct a designed experiment. You identify 120 volunteers to participate in the study and decided on three levels of the treatment: a placebo, one cup of green tea daily, two cups of green tea daily. The experiment is to run for one year. The response variable will be the change in LDL cholesterol for each subject from the beginning of the study to the end. What type of experimental design is this?

(d) Explain how you would use blinding in this experiment.

(e) What is the factor? Is it qualitative or quantitative?

(f) What factors might you attempt to control in this experiment.

(g) Explain how to use randomization in this experiment. How does randomization neutralize those variables that are not controlled?

(h) Suppose you assigned 40 subjects to each of the three treatment groups. In addition, you decided to control the variable exercise by having each subject perform 150 minutes of cardiovascular exercise each week by walking on a treadmill. However, the 40 subjects in the placebo group decided they did not want to walk on the treadmill and skipped the weekly exercise. Explain how exercise is now a confounding variable.

24. Priming for Healthy Food Does alerting shoppers at a grocery store regarding the healthiness (or lack thereof) of energy-dense snack foods change the shopping habits of overweight individuals? To answer this question, researchers randomly gave 42 overweight shoppers a recipe flyer that either contained health information or did not contain the health information. This type of intervention is referred to as *priming*. To determine purchases, the receipts of the participants were reviewed. Results of the study found that shoppers primed with the health- and diet-related words on the recipe bought significantly (almost 75%) fewer unhealthy snacks than those without the primes.

Source: E. K. Papies and associates, "Using Health Primes to Reduce Unhealthy Snack Purchases Among Overweight Consumers in a Grocery Store" *International Journal of Obesity* (2013), 1–6.

(a) What is the research objective?

(b) Who are the subjects?

(c) Explain why blinding is not possible in this study.

(d) What is the explanatory variable in the study.

(e) The response variable was number of unhealthy snacks purchased. Is this quantitative or qualitative?

(f) Another factor in the study was weight status (normal weight vs overweight). Suppose all the normal weight subjects were given the flyer with the prime and overweight subjects were given the flyer without the prime. Explain how confounding would play a role in the study.

25. Batteries An engineer wants to determine the effect of temperature on battery voltage. In particular, he is interested in determining if there is a significant difference in the voltage of the batteries when exposed to temperatures of 90°F, 70°F, and 50°F. Help the engineer design the experiment. Include a diagram to illustrate your design.

26. Tire Design An engineer has just developed a new tire design. However, before going into production, the tire company wants to determine if the new tire reduces braking distance on a car traveling 60 miles per hour compared with radial tires. Design an experiment to help the engineer determine if the new tire reduces braking distance.

27. Octane Does the level of octane in gasoline affect gas mileage? To answer this question, an automotive engineer obtains 60 cars. Twenty of the cars are compact, 20 are full size, and 20 are sport utility vehicles (SUVs). Design an experiment for the engineer. Include a diagram to illustrate your design.

28. School Psychology The school psychologist first presented in Problem 9 worries that girls and boys may react differently to the two methods of instruction. Design an experiment that will eliminate the variability due to gender on the response variable, the score on the achievement test. Include a diagram to illustrate your design.

29. Designing an Experiment Researchers want to know if there is a link between hypertension (high blood pressure) and consumption of salt. Past studies have indicated that the consumption of fruits and vegetables offsets the negative impact of salt consumption. It is also known that there is quite a bit of person-to-person variability in the ability of the body to process and eliminate salt. However, no method exists for identifying individuals who have a higher ability to process salt. The U.S. Department of Agriculture recommends that daily intake of salt should not exceed 2400 milligrams (mg). The researchers want to keep the design simple, so they choose to conduct their study using a completely randomized design.

(a) What is the response variable in the study?

(b) Name three factors that have been identified.

(c) For each factor identified, determine whether the variable can be controlled or cannot be controlled. If a factor cannot be controlled, what should be done to reduce variability in the response variable?

(d) How many treatments would you recommend? Why?

30. Search a newspaper, magazine, or other periodical that describes an experiment. Identify the population, experimental unit, response variable, treatment, factors, and their levels.

31. Research the *placebo effect* and the *Hawthorne effect*. Write a paragraph that describes how each affects the outcome of an experiment.

32. Coke or Pepsi You want to perform an experiment with a goal to determine whether people prefer Coke or Pepsi. Design an experiment that utilizes the completely randomized design. Design an experiment that utilizes the matched-pairs design. In both designs, be sure to identify the response variable, the role of blinding, and randomization. Which design do you prefer? Why?

Explaining the Concepts

33. The Dictator Game In their book *SuperFreakonomics*, authors Steven Levitt and Stephen Dubner describe the research of behavioral economist John List. List recruited customers and dealers at a baseball-card show to participate in an experiment in which the customer would state how much he was willing to pay for a single baseball card. The prices ranged from $4 (lowball) to $50 (premium card). The dealer would then give the customer a card that was supposed to correspond to the offer price. In this setting, the dealer could certainly give the buyer a card worth less than the offer price, but this rarely happened. The card received by the buyer was close in value to the price offered. Next, List went to the trading floor at the show and again recruited customers. But this time the customers approached dealers at their booth. The dealers did not know they were being watched. The scenario went something like this: as the customer approached the dealer's booth, he would say, "Please give me the best Derek Jeter card you can for $20." In this scenario, the dealers consistently ripped off the customers by giving them cards worth much less than the offer price. In fact, the dealers who were the worst offenders were the same dealers who refused to participate in List's study. Do you believe that individuals who volunteer for experiments are scientific do-gooders? That is, do you believe that in designed experiments subjects strive to meet the expectations of the researcher? In addition, do you believe that results of experiments may suffer because many experiments require individuals to volunteer, and individuals who are not do-gooders do not volunteer for studies? Now, explain why control groups are needed in designed experiments and the role they can play in neutralizing the impact of scientific do-gooders.

34. Ebola In October 2014, there was an Ebola breakout in West Africa. At the time, there was no vaccine for the virus, however, there were some experimental drugs that had not yet been approved for humans. Because the spread of the disease was reaching an epidemic, there were calls to initiate randomized trials of an experimental drug on human subjects right away.

(a) Discuss how you would go about designing a randomized trial to assess the efficacy of an experimental Ebola vaccine.

(b) Doctors Without Borders was on the record, prior to any randomized trial, as saying that trials in which subjects are assigned to a control group are unethical. Discuss the ethics behind a randomized trial of a potential life-saving vaccine to test its efficacy while an epidemic is raging.

35. Discuss how a randomized block design is similar to a stratified random sample. What is the purpose of blocking?

36. What is the role of randomization in a designed experiment? If you were conducting a completely randomized design with three treatments and 90 experimental units, describe how you would randomly assign the experimental units to the treatments.

 **Chapter 1 Review**

Summary

We defined statistics as a science in which data are collected, organized, summarized, and analyzed to infer characteristics regarding a population. Statistics also provides a measure of confidence in the conclusions that are drawn. Descriptive statistics consists of organizing and summarizing information, while inferential statistics consists of drawing conclusions about a population based on results obtained from a sample. The population is a collection of individuals about which information is desired and the sample is a subset of the population.

Data are the observations of a variable. Data can be either qualitative or quantitative. Quantitative data are either discrete or continuous.

Data can be obtained from four sources: a census, an existing source, an observational study, or a designed experiment. A census will list all the individuals in the population, along with certain characteristics. Due to the cost of obtaining a census, most researchers opt for obtaining a sample. In observational studies, the response variable is measured without attempting to influence its value. In addition, the explanatory variable is not manipulated. Designed experiments are used when control of the individuals in the study is desired to isolate the effect of a certain treatment on a response variable.

We introduced five sampling methods: simple random sampling, stratified sampling, systematic sampling, cluster sampling, and convenience sampling. All the sampling methods, except for convenience sampling, allow for unbiased statistical inference to be made. Convenience sampling typically leads to an unrepresentative sample and biased results.

Vocabulary

Statistics (p. 3)
Data (pp. 3, 8)
Population (p. 5)
Individual (p. 5)
Sample (p. 5)
Statistic (p. 5)
Descriptive statistics (p. 5)
Inferential statistics (pp. 5, 44)
Parameter (p. 5)
Variable (p. 6)
Qualitative or categorical
    variable (p. 7)
Quantitative variable (p. 7)
Discrete variable (p. 7)
Continuous variable (p. 7)
Qualitative data (p. 8)
Quantitative data (p. 8)
Discrete data (p. 8)
Continuous data (p. 8)
Nominal level of measurement (p. 9)
Ordinal level of measurement (p. 9)
Interval level of measurement (p. 9)
Ratio level of measurement (p. 9)
Explanatory variable (p. 14)

Response variable (p. 14)
Observational study (p. 15)
Designed experiment (p. 15)
Confounding (p. 15)
Lurking variable (p. 16)
Confounding variable (p. 17)
Retrospective (p. 17)
Prospective (p. 18)
Census (p. 18)
Random sampling (p. 21)
Simple random sampling (p. 22)
Simple random sample (p. 22)
Frame (p. 23)
Sampling without replacement (p. 23)
Sampling with replacement (p. 23)
Seed (p. 24)
Stratified sample (p. 28)
Systematic sample (p. 30)
Cluster sample (p. 31)
Convenience sample (p. 32)
Self-selected (p. 32)
Voluntary response (p. 32)
Bias (p. 36)
Sampling bias (p. 36)

Undercoverage (p. 36)
Nonresponse bias (p. 37)
Response bias (p. 37)
Open question (p. 39)
Closed question (p. 39)
Nonsampling error (p. 39)
Sampling error (p. 39)
Experiment (p. 42)
Factors (p. 42)
Treatment (p. 42)
Experimental unit (p. 43)
Subject (p. 43)
Control group (p. 43)
Placebo (p. 43)
Blinding (p. 43)
Single-blind (p. 43)
Double-blind (p. 43)
Design (p. 44)
Replication (p. 44)
Completely randomized design (p. 45)
Matched-pairs design (p. 46)
Blocking (p. 47)
Block (p. 47)
Randomized block design (p. 47)

OBJECTIVES

Section	You should be able to . . .	Example(s)	Review Exercises
1.1	**1** Define statistics and statistical thinking (p. 3)	pp. 3–4	1
	2 Explain the process of statistics (p. 4)	2	7, 14, 15
	3 Distinguish between qualitative and quantitative variables (p. 6)	3	11–13
	4 Distinguish between discrete and continuous variables (p. 7)	4, 5	11–12
	5 Determine the level of measurement of a variable (p. 9)	6	16–19

Section	You should be able to . . .	Example(s)	Review Exercises
1.2	**1** Distinguish between an observational study and an experiment (p. 14)	1–3	20–21, 31(b)
	2 Explain the various types of observational studies (p. 17)	pp. 17–18	6, 22
1.3	**1** Obtain a simple random sample (p. 22)	1–3	28, 30
1.4	**1** Obtain a stratified sample (p. 28)	1	25
	2 Obtain a systematic sample (p. 30)	2	26, 29
	3 Obtain a cluster sample (p. 31)	3	24
1.5	**1** Explain the sources of bias in sampling (p. 36)	pp. 36–40	8, 9, 27
1.6	**1** Describe the characteristics of an experiment (p. 42)	1	5
	2 Explain the steps in designing an experiment (p. 44)	p. 44	10
	3 Explain the completely randomized design (p. 44)	2	31, 35(a), 36
	4 Explain the matched-pairs design (p. 46)	3	32, 36
	5 Explain the randomized block design (p. 47)	4	33, 35(b)

Review Exercises

In Problems 1–5, provide a definition using your own words.

1. Statistics
2. Population
3. Sample
4. Observational study
5. Designed experiment
6. List and describe the three major types of observational studies.
7. What is meant by the *process of statistics*?
8. List and explain the three sources of bias in sampling. Provide some methods that might be used to minimize bias in sampling.
9. Distinguish between sampling and nonsampling error.
10. Explain the steps in designing an experiment.

In Problems 11–13, classify the variable as qualitative or quantitative. If the variable is quantitative, state whether it is discrete or continuous.

11. Number of new automobiles sold at a dealership on a given day
12. Weight in carats of an uncut diamond
13. Brand name of a pair of running shoes

In Problems 14 and 15, determine whether the underlined value is a parameter or a statistic.

14. In a survey of 1011 people age 50 or older, <u>73%</u> agreed with the statement "I believe in life after death."
Source: Bill Newcott. "Life after Death," *AARP Magazine*, Sept./Oct. 2007.

15. **Completion Rate** In the 2015 NCAA Football Championship Game, quarterback Cardale Jones completed <u>70%</u> of his passes for a total of 242 yards.

In Problems 16–19, determine the level of measurement of each variable.

16. Birth year
17. Marital status
18. Stock rating (strong buy, buy, hold, sell, strong sell)
19. Number of siblings

In Problems 20 and 21, determine whether the study depicts an observational study or a designed experiment.

20. A parent group examines 25 randomly selected PG-13 movies and 25 randomly selected PG movies, records the number of sexual innuendos and curse words that occur in each, and then compares the number of sexual innuendos and curse words between the two movie ratings.

21. A sample of 504 patients in early stages of Alzheimer's disease is divided into two groups. One group receives an experimental drug; the other receives a placebo. The advance of the disease in the patients from the two groups is tracked at 1-month intervals over the next year.

22. Read the following description of an observational study and determine whether it is a cross-sectional, a case-control, or a cohort study. Explain your choice.

> The Cancer Prevention Study II (CPS-II) examines the relationship among environmental and lifestyle factors of cancer cases by tracking approximately 1.2 million men and women. Study participants completed an initial study questionnaire in 1982 providing information on a range of lifestyle factors, such as diet, alcohol and tobacco use, occupation, medical history, and family cancer history. These data have been examined extensively in relation to cancer mortality. The vital status of study participants is updated biennially.
>
> *Source:* American Cancer Society.

In Problems 23–26, determine the type of sampling used.

23. On election day, a pollster for Fox News positions herself outside a polling place near her home and asks the first 50 voters leaving the facility to complete a survey.

24. An Internet service provider randomly selects 15 residential blocks from a large city and surveys every household in these 15 blocks to determine the number that would use a high-speed Internet service.

25. Thirty-five sophomores, 22 juniors, and 35 seniors are randomly selected to participate in a study from 574 sophomores, 462 juniors, and 532 seniors at a certain high school.

26. Officers for the Department of Motor Vehicles pull aside every 40th tractor trailer passing through a weigh station, starting with the 12th, for an emissions test.

27. Each of the following surveys has bias. Determine the type of bias and suggest a remedy.

(a) A politician sends a survey about tax issues to a random sample of subscribers to a literary magazine.

(b) An interviewer with little foreign language knowledge is sent to an area where her language is not commonly spoken.

(c) A data-entry clerk mistypes survey results into his computer.

28. Obtaining a Simple Random Sample The mayor of a town wants to conduct personal interviews with small business owners to determine if there is anything he could do to help improve business conditions. The following list gives the names of the companies in the town. Obtain a simple random sample of size 5 from the companies in the town.

Allied Tube and Conduit	Lighthouse Financial	Senese's Winery
Bechstien Construction Co.	Mill Creek Animal Clinic	Skyline Laboratory
Cizer Trucking Co.	Nancy's Flowers	Solus, Maria, DDS
D & M Welding	Norm's Jewelry	Trust Lock and Key
Grace Cleaning Service	Papoose Children's Center	Ultimate Carpet
Jiffy Lube	Plaza Inn Motel	Waterfront Tavern
Levin, Thomas, MD	Risky Business Security	WPA Pharmacy

29. Obtaining a Systematic Sample A quality-control engineer wants to be sure that bolts coming off an assembly line are within prescribed tolerances. He wants to conduct a systematic sample by selecting every 9th bolt to come off the assembly line. The machine produces 30,000 bolts per day, and the engineer wants a sample of 32 bolts. Which bolts will be sampled?

30. Obtaining a Simple Random Sample Based on the Military Standard 105E (ANS1/ASQC Z1.4, ISO 2859) Tables, a lot of 91 to 150 items with an acceptable quality level (AQL) of 1% and a normal inspection plan would require a sample of size 13 to be inspected for defects. If the sample contains no defects, the entire lot is accepted. Otherwise, the entire lot is rejected. A shipment of 100 night-vision goggles is received and must be inspected. Discuss the procedure you would follow to obtain a simple random sample of 13 goggles to inspect.

31. Tooth-Whitening Gum Smoking and drinking coffee have a tendency to stain teeth. In an effort to determine the ability of chewing gum to remove stains on teeth, researchers conducted an experiment in which 64 bovine incisors (teeth) were stained with natural pigments such as coffee for 10 days. Each tooth was randomly assigned to one of four treatments: gum A, gum B, gum C, or saliva. Each tooth group was placed into a device that simulated a human chewing gum. The temperature of the device was maintained at body temperature and the tooth was in the device for 20 minutes. The process was repeated six times (for a total of 120 minutes of chewing). The researcher conducting the experiment did not know which treatment was being applied to each experimental unit. Upon removing a tooth from the chewing apparatus, the color change was measured using a spectrophotometer. The percentage of stain removed by each treatment after 120 minutes is as follows: Gum A, 47.6%; Gum B, 45.2%, Gum C, 21.4%, Saliva, 2.1%.

The researchers concluded that gums A and B removed significantly more stain than gum C or saliva. In addition, gum C removed significantly more stain than saliva.

Source: Michael Moore et al. "In Vitro Tooth Whitening Effect of Two Medicated Chewing Gums Compared to a Whitening Gum and Saliva," *BioMed Central Oral Health* 8:23, 2008.

(a) Identify the research objective.

(b) Is this an observational study or designed experiment? Why?

(c) If observational, what type of observational study is this? If an experiment, what type of experimental design is this?

(d) What is the response variable?

(e) What is the explanatory variable? Is it qualitative or quantitative?

(f) Identify the experimental units.

(g) State a factor that could affect the value of the response variable that is fixed at a set level.

(h) What is the conclusion of the study?

32. Reaction Time Researchers wanted to assess the effect of low alcohol consumption on reaction time in seniors, believing that even low levels of alcohol consumption can impair the ability to walk, thereby increasing the likelihood of falling. They identified 13 healthy seniors who were not heavy consumers of alcohol. The experiment took place in late afternoon. Each subject was instructed to have a light lunch and not to drink any caffeinated drinks in the 4 hours prior to arriving at the lab. The seniors were asked to walk on a treadmill on which an obstacle would appear randomly. The reaction time was measured by determining the time it took the senior to lift his or her foot upon the appearance of the obstacle. First, each senior walked the treadmill after consuming a drink consisting of water mixed with orange juice with the scent and taste of vodka. The senior was then asked to drink two additional drinks (40% vodka mixed with orange juice). The senior then walked on the treadmill again. The average response time increased by 19 milliseconds after the alcohol treatment. The researchers concluded that response times are significantly delayed even for low levels of alcohol consumption.

Source: Judith Hegeman et al. "Even Low Alcohol Concentrations Affect Obstacle Avoidance Reactions in Healthy Senior Individuals," *BMC Research Notes* 3:243, 2010.

(a) What type of experimental design is this?

(b) What is the response variable in this experiment? Is it quantitative or qualitative?

(c) What is the treatment?

(d) What factors were controlled and set at a fixed level in this experiment?

(e) Can you think of any factors that may affect reaction to alcohol that were not controlled?

(f) Why do you think the researchers used a drink that had the scent and taste of vodka to serve as the treatment for a baseline measure?

(g) What was the conclusion of the study? To whom does this conclusion apply?

33. Exam Grades A statistics instructor wants to see if allowing students to use a notecard on exams will affect their overall exam scores. The instructor thinks that the results could be affected by the type of class: traditional, Web-enhanced, or online. He randomly selects half of his students from each type of class to use notecards. At the end of the semester, the average exam grades for those using notecards is compared to the average for those without notecards.

(a) What type of experimental design is this?

(b) What is the response variable in this experiment?

(c) What is the factor that is set to predetermined levels? What are the treatments?

(d) Identify the experimental units.

(e) Draw a diagram similar to Figure 7, 8, or 10 to illustrate the design.

34. Multiple Choice A common tip for taking multiple-choice tests is to always pick (b) or (c) if you are unsure. The idea is that instructors tend to feel the answer is more hidden if it is surrounded by distractor answers. An astute statistics instructor is aware of this and decides to use a table of random digits to select which choice will be the correct answer. If each question has five choices, use Table I in Appendix A or a random-number generator to determine the correct answers for a 20-question multiple-choice exam.

35. Humor in Advertising A marketing research firm wants to know whether information presented in a commercial is better recalled when presented using humor or serious commentary by adults between 18 and 35 years of age. They will use an exam that asks questions of 50 subjects about information presented in the ad. The response variable will be percentage of information recalled.

(a) Create a completely randomized design to answer the question. Be sure to include a diagram to illustrate your design.

(b) The chief of advertising criticizes your design because she knows that women react differently from men to advertising. Redesign the experiment as a randomized block design assuming that 30 of the subjects in your study are female and the remaining 20 are male. Be sure to include a diagram to illustrate your design.

36. Describe what is meant by a matched-pairs design. Contrast this experimental design with a completely randomized design.

37. Internet Search Go to an online science magazine such as *Science Daily* (www.sciencedaily.com) or an open source online medical journal such as BioMed Central (www.biomedcentral.com) and identify an article that includes statistical research.

(a) Was the study you selected a designed experiment or an observational study?

(b) What was the research objective?

(c) What was the response variable in the study?

(d) Summarize the conclusions of the study.

38. Cell Phones Many newspaper articles discuss the dangers of teens texting while driving. Suppose you are a journalist and want to chime in on the discussion. However, you want your article to be more compelling, so you decide to conduct an experiment with one hundred 16- to 19-year-old volunteers. Design an experiment that will assess the dangers of texting while driving. Decide on the type of experiment (completely randomized, matched-pair, or other), a response variable, the explanatory variables, and any controls that will be imposed. Also, explain how you are going to obtain the data without potentially harming your subjects. Write an article that presents the experiment to your readers so that they know what to anticipate in your follow-up article. Remember, your article is for the general public, so be sure to clearly explain the various facets of your experiment.

39. What is the role of randomization in a designed experiment? If you were conducting a completely randomized design with four treatments and 100 experimental units, describe how you would randomly assign the experimental units to the treatments.

 **Chapter Test**

1. List the four components that comprise the definition of statistics.

2. What is meant by the *process of statistics*?

In Problems 3–5, determine if the variable is qualitative or quantitative. If the variable is quantitative, determine if it is discrete or continuous. State the level of measurement of the variable.

3. Time to complete the 500-meter race in speed skating.

4. Video game rating system by the Entertainment Software Rating Board (EC, E, E10+ , T, M, AO, RP)

5. The number of surface imperfections on a camera lens.

In Problems 6 and 7, determine whether the study depicts an observational study or a designed experiment. Identify the response variable in each case.

6. A random sample of 30 digital cameras is selected and divided into two groups. One group uses a brand-name battery, while the other uses a generic plain-label battery. All variables besides battery type are controlled. Pictures are taken under identical conditions and the battery life of the two groups is compared.

7. A sports reporter asks 100 baseball fans if Barry Bonds's 756th homerun ball should be marked with an asterisk when sent to the Baseball Hall of Fame.

8. Contrast the three major types of observational studies in terms of the time frame when the data are collected.

9. Compare and contrast observational studies and designed experiments. Which study allows a researcher to claim causality?

10. Explain why it is important to use a control group and blinding in an experiment.

11. List the steps required to conduct an experiment.

12. A cellular phone company is looking for ways to improve customer satisfaction. They want to select a simple random sample of four stores from their 15 franchises in which to conduct customer satisfaction surveys. Discuss the procedure you would use, and then use the procedure to select a simple random sample of size $n = 4$. The locations are as follows:

Afton	Ballwin	Chesterfield	Clayton	Deer Creek
Ellisville	Farmington	Fenton	Ladue	Lake St. Louis
O'Fallon	Pevely	Shrewsbury	Troy	Warrenton

13. A congresswoman wants to survey her constituency regarding public policy. She asks one of her staff members to obtain a sample of residents of the district. The frame she has available lists 9012 Democrats, 8302 Republicans, and 3012 Independents. Obtain a stratified random sample of

8 Democrats, 7 Republicans, and 3 Independents. Be sure to discuss the procedure used.

14. A farmer has a 500-acre orchard in Florida. Each acre is subdivided into blocks of 5. Altogether, there are 2500 blocks of trees on the farm. After a frost, he wants to get an idea of the extent of the damage. Obtain a sample of 10 blocks of trees using a cluster sample. Be sure to discuss the procedure used.

15. A casino manager wants to inspect a sample of 14 slot machines in his casino for quality-control purposes. There are 600 sequentially numbered slot machines operating in the casino. Obtain a systematic sample of 14 slot machines. Be sure to discuss how you obtained the sample.

16. Describe what is meant by an experiment that has a completely randomized design. Contrast this experimental design with a randomized block design.

17. Each of the following surveys has bias. Identify the type of bias.

(a) A television survey that gives 900 phone numbers for viewers to call with their vote. Each call costs $2.00.

(b) An employer distributes a survey to her 450 employees asking them how many hours each week, on average, they surf the Internet during business hours. Three of the employees complete the survey.

(c) A question on a survey asks, "Do you favor or oppose a minor increase in property tax to ensure fair salaries for teachers and properly equipped school buildings?"

(d) A researcher conducting a poll about national politics sends a survey to a random sample of subscribers to *Time* magazine.

18. Shapely Glasses Does the shape of a glass play a role in determining the amount of time it takes to finish the drink? Researchers identified 159 male and female self-professed social drinkers. One week the subjects were given a 12 ounce beer with either a straight glass or a curved glass. A week later the subjects were given a 12 ounce beer in the other glass. The first week, subjects were given a glass and asked to consume the drink at their own pace while watching television. The time to complete the drink was measured. During the second week, the subjects were given the other shaped glass and asked to complete a computer task. Again, the time to complete the drink was measured. The type of glass given in the first week was determined randomly. The researchers found that the time to complete the drink was significantly faster for the curved glass.

Source: A. S. Attwood, N. E. Scott-Samuel, G. Stothart, M. R. Munafò (2012) "Glass Shape Influences Consumption Rate for Alcoholic Beverages." *PLoS ONE* 7(8): e43007. doi:10.1371/journal.pone.0043007.

(a) What type of experimental design is this?

(b) Who are the subjects?

(c) What is the treatment?

(d) What is the response variable? Is it qualitative or quantitative?

(e) Explain the role randomization plays in this experiment.

(f) Draw a figure that illustrates the design.

19. Nucryst Pharmaceuticals, Inc., announced the results of its first human trial of NPI 32101, a topical form of its skin ointment. A total of 225 patients diagnosed with skin irritations were randomly divided into three groups as part of a double-blind, placebo-controlled study to test the effectiveness of the new topical cream. The first group received a 0.5% cream, the second group received a 1.0% cream, and the third group received a placebo. Groups were treated twice daily for a 6-week period.

Source: www.nucryst.com.

(a) What type of experimental design is this?

(b) What is the factor that is set to predetermined levels? What are the treatments?

(c) What does it mean for this study to be double-blind?

(d) What is the control group for this study?

(e) Identify the experimental units.

(f) Draw a diagram similar to Figure 7, 8, or 10 to illustrate the design.

20. Researchers Katherine Tucker and associates wanted to determine whether consumption of cola is associated with lower bone mineral density. They looked at 1125 men and 1413 women in the Framingham Osteoporosis Study, which is a cohort that began in 1971. The first examination in this study began between 1971 and 1975, with participants returning for an examination every 4 years. Based on results of questionnaires, the researchers were able to determine cola consumption on a weekly basis. Analysis of the results indicated that women who consumed at least one cola per day (on average) had a bone mineral density that was significantly lower at the femoral neck than those who consumed less than one cola per day. The researchers did not find this relation in men.

Source: "Colas, but not other carbonated beverages, are associated with low bone mineral density in older women: The Framingham Osteoporosis Study," *American Journal of Clinical Nutrition* 84: 936–942, 2006.

(a) Why is this a cohort study?

(b) What is the response variable in this study? What is the explanatory variable?

(c) Is the response variable qualitative or quantitative?

(d) The following appears in the article: "Variables that could potentially confound the relation between carbonated beverage consumption and bone mineral density were obtained from information collected (in the questionnaire)." What does this mean?

(e) Can you think of any lurking variables that should be accounted for?

(f) What are the conclusions of the study? Does increased cola consumption cause a lower bone mineral density?

21. Explain the difference between a lurking variable and a confounding variable.

Making an Informed Decision

What College Should I Attend?

One of the most difficult tasks of surveying is phrasing questions so that they are not misunderstood. In addition, questions must be phrased so that the researcher obtains answers that allow for meaningful analysis. We wish to create a questionnaire that can be used to make an informed decision about what college to attend. In addition, we want to develop a plan for determining how well a particular college educates its students.

1. Using the school you are currently attending, determine a sampling plan to obtain a random sample of 20 students.

2. Develop a questionnaire that can be administered to the students. The questions in the survey should obtain demographic information about the student (age, gender) as well as questions that pertain to why they chose the school. Also, develop questions that address the students' satisfaction with their school choice. Finally, develop questions that relate to things you consider important

in a college environment. Administer the survey either by paper or use an online survey program such as Poll Monkey, Google, or StatCrunch.

3. Summarize your findings. Based on your findings, would you choose the school for yourself?

4. A second gauge in determining college choice is how well the school educates its students. Certainly, some schools have higher-caliber students as incoming freshmen than others, but you do not want academic ability upon entering the school to factor in your decision. Rather, you want to measure how much the college has increased its students' skill set. Design an experiment that would allow you to measure the response variable "increase in academic skill set" for a particular college. Provide detail for controls in obtaining this information.

 **CASE STUDY** Chrysalises for Cash

The butterfly symbolizes the notion of personal change. Increasingly, people are turning to butterflies to consecrate meaningful events, such as birthdays, weddings, and funerals. To fill this need, a new industry has hatched.

Due to the possibility of introducing an invasive species, butterfly suppliers are monitored by governmental agencies. Along with following regulations, butterfly suppliers must ensure quality and quantity of their product while maintaining a profit. To this end, an individual supplier may hire independent contractors to hatch the varieties needed. These entrepreneurs are paid a small fee for each chrysalis delivered, with a 50% bonus added for each hatched healthy butterfly. This fee structure provides little room for profit. Therefore, it is important that these contractors deliver a high proportion of healthy butterflies that emerge at a fairly predictable rate.

In Florida, one such entrepreneur specializes in harvesting the black swallowtail butterfly. In nature, the female butterfly seeks plants, such as carrot and parsley, to harvest and lay eggs on. A newly hatched caterpillar consumes the host plant, then secures itself and sheds its skin, revealing a chrysalis. During this resting phase, environmental factors such as temperature and humidity

may affect the transformation process. Typically, the black swallowtail takes about 1 week to complete its metamorphosis and emerge from its chrysalis. The transformation occasionally results in deformities ranging from wings that will not fully open to missing limbs.

The Florida contractor believes that there are differences in quality and emergence time among his broods. Not having taken a scientific approach to the problem, he relies on his memory of seasons past. It seems to him that late-season butterflies emerge sooner and with a greater number of deformities than their early-season counterparts. He also speculates that the type and nutritional value of the food consumed by the caterpillar might contribute to any observed differences. This year he is committed to a more formal approach to his butterfly harvest.

Since it takes 2 days to deliver the chrysalises from the contractor to the supplier, it is important that the butterflies do not emerge prematurely. It is equally important that the number of defective butterflies be minimized. With these two goals in mind, the contractor seeks the best combination of food source, fertilizer, and brood season to maximize his profits. To examine the effects of these variables on emergence time and number of deformed butterflies, the entrepreneur designed the following experiment.

Eight identical pots were filled with equal amounts of a soil and watered carefully to ensure consistency. Two pots of carrot plants and two of parsley were set outside during the early part of the brood season. For the carrot pair, one pot was fed a fixed amount of liquid fertilizer, while the other was fed a nutritionally similar amount of solid fertilizer. The two pots of parsley were similarly fertilized. All four pots were placed next to each other to ensure similar exposures to environmental conditions such as temperature and solar radiation. Five black swallowtail caterpillars of similar age were placed into each container, each allowed to mature and form a chrysalis. The time from chrysalis formation until emergence was reported to the nearest day, along with any defects. The same procedure was followed with the four pots that were placed outdoors during the late brood season.

Write a report describing the experimental goals and design for the entrepreneur's experiment. Follow the procedure outlined in the box on steps in designing and conducting an experiment (p. 44). Step 5(b), of this procedure is provided in the following table and should be included in your report.

Florida Black Swallowtail Chrysalis Experiment Data				
Season	Food	Fertilizer	Number Deformed	Emergence Time (Days)
Early	Parsley	Solid	0	6, 6, 7, 7, 7
Early	Parsley	Liquid	0	6, 7, 7, 8, 8
Early	Carrot	Solid	1	3, 6, 6, 7, 8
Early	Carrot	Liquid	0	6, 6, 7, 8, 8
Late	Parsley	Solid	2	2, 3, 4, 4, 5
Late	Parsley	Liquid	1	2, 3, 4, 5, 5
Late	Carrot	Solid	2	3, 3, 3, 4, 5
Late	Carrot	Liquid	0	2, 4, 4, 4, 5

In your report, provide a general descriptive analysis of these data. Be sure to include recommendations for the combination of season, food source, and type of fertilizer that result in the fewest deformed butterflies while achieving a long emergence time. Conclude your report with recommendations for further experiments. For each proposed experiment, be sure to do the following:

1. State the problem to be solved and define the response variables.

2. Define the factors that affect the response variables.

3. State the number of experimental units.

4. State the treatment.

Remember, statistics is a process. The first chapter (Part 1) dealt with the first two steps in the statistical process: (1) identify the research objective and (2) collect the data needed to answer the questions in the research objective. The next three chapters deal with organizing, summarizing, and presenting the data collected. This step in the process is called *descriptive statistics*.

2

Organizing and Summarizing Data

Making an Informed Decision

Suppose you work for the school newspaper. Your editor approaches you with a special reporting assignment. Your task is to write an article that describes the "typical" student at your school, complete with supporting information. How are you going to do this assignment? See the Decisions project on page 115.

PUTTING IT TOGETHER

Chapter 1 discussed how to identify the research objective and collect data. We learned that data can be obtained from either observational studies or designed experiments. When data are obtained, they are referred to as **raw data**.

The purpose of this chapter is to learn how to organize raw data into a meaningful form so that we can understand what the data are telling us. The first step in determining how to organize raw data is to determine whether the data is qualitative or quantitative.

2.1   Organizing Qualitative Data

Preparing for This Section   Before getting started, review the following:

- Qualitative data (Section 1.1, p. 8)
- Level of measurement (Section 1.1 pp. 9–10)

Objectives   ❶ Organize qualitative data in tables

   ❷ Construct bar graphs

   ❸ Construct pie charts

In this section we will concentrate on tabular and graphical summaries of qualitative data. Sections 2.2 and 2.3 discuss methods for summarizing quantitative data.

❶ Organize Qualitative Data in Tables

Recall that qualitative (or categorical) data provide measures that categorize or classify an individual. When raw qualitative data are collected, we often first determine the number of individuals within each category.

Definition   A **frequency distribution** lists each category of data and the number of occurrences for each category of data.

EXAMPLE 1   Organizing Qualitative Data into a Frequency Distribution

Problem   A physical therapist wants to determine types of rehabilitation required by her patients. To do so, she obtains a simple random sample of 30 of her patients and records the body part requiring rehabilitation. See Table 1. Construct a frequency distribution of location of injury.

Approach   To construct a frequency distribution, create a list of the body parts (categories) and tally each occurrence. Finally, add up the number of tallies to determine the frequency.

Solution   Table 2 shows that the back is the most common body part requiring rehabilitation, with a total of 12.

CAUTION!
The data in Table 2 are still qualitative. The frequency simply represents the count of each category.

Table 1

Back	Back	Hand
Wrist	Back	Groin
Elbow	Back	Back
Back	Shoulder	Shoulder
Hip	Knee	Hip
Neck	Knee	Knee
Shoulder	Shoulder	Back
Back	Back	Back
Knee	Knee	Back
Hand	Back	Wrist

Source: Krystal Catton, student at Joliet Junior College

Table 2

Body Part	Tally	Frequency
Back	┼┼┼┼ ┼┼┼┼ ‖	12
Wrist	‖	2
Elbow			1
Hip	‖	2
Shoulder	‖‖	4
Knee	┼┼┼┼	5
Hand	‖	2
Groin			1
Neck			1

In any frequency distribution, it is a good idea to add up the frequency column to make sure that it equals the number of observations. In Example 1, the frequency column adds up to 30, as it should.

Often, we want to know the *relative frequency* of the categories, rather than the frequency.

Definition

The **relative frequency** is the proportion (or percent) of observations within a category and is found using the formula

$$\text{Relative frequency} = \frac{\text{frequency}}{\text{sum of all frequencies}} \qquad (1)$$

A **relative frequency distribution** lists each category of data together with the relative frequency.

In Other Words
A frequency distribution shows the number of observations that belong in each category. A relative frequency distribution shows the proportion of observations that belong in each category.

EXAMPLE 2 Constructing a Relative Frequency Distribution of Qualitative Data

Problem Using the summarized data in Table 2, construct a relative frequency distribution.

Approach Add all the frequencies, and then use Formula (1) to compute the relative frequency of each category of data.

Solution The sum of all the values in the frequency column in Table 2 is 30.

Now compute the relative frequency of each category. For example, the relative frequency of the category *Back* is 12/30 = 0.4. The relative frequency distribution is shown in Table 3.

Table 3		
Body Part	**Frequency**	**Relative Frequency**
Back	12	$\frac{12}{30} = 0.4$
Wrist	2	$\frac{2}{30} \approx 0.0667$
Elbow	1	0.0333
Hip	2	0.0667
Shoulder	4	0.1333
Knee	5	0.1667
Hand	2	0.0667
Groin	1	0.0333
Neck	1	0.0333
Total	**30**	**1**

From the distribution, the most common body part for rehabilitation is the back.

⚏ Using Technology

Some statistical spreadsheets such as Minitab and StatCrunch have a command that will construct a frequency and relative frequency distribution of raw qualitative data.

• Now Work Problems 21(a)–(b)

It is a good idea to add up the relative frequencies to be sure they sum to 1. In fraction form, the sum should be exactly 1. In decimal form, the sum may differ slightly from 1 due to rounding.

② Construct Bar Graphs

Once raw data are organized in a table, we can create graphs. Just as "a picture is worth a thousand words," pictures of data result in a more powerful message than tables. Try the following exercise: Open a newspaper and look at a table and a graph. Study each. Now put the paper away and close your eyes. What do you see in your mind's eye? Can you recall information more easily from the table or the graph? In general, people are more likely to recall information obtained from a graph than they are from a table.

A common device for graphically representing qualitative data is a bar graph.

Definition   A **bar graph** is constructed by labeling each category of data on either the horizontal or vertical axis and the frequency or relative frequency of the category on the other axis. Rectangles of equal width are drawn for each category. The height of each rectangle represents the category's frequency or relative frequency.

EXAMPLE 3   Constructing a Frequency and a Relative Frequency Bar Graph

Problem   Use the data summarized in Table 3 to construct a frequency bar graph and a relative frequency bar graph.

Approach   Use a horizontal axis to indicate the categories of the data (body parts) and a vertical axis to represent the frequency or relative frequency. Draw rectangles of equal width to the height that is the frequency or relative frequency for each category. The bars do not touch each other.

Solution   Figure 1(a) shows the frequency bar graph, and Figure 1(b) shows the relative frequency bar graph.

Figure 1

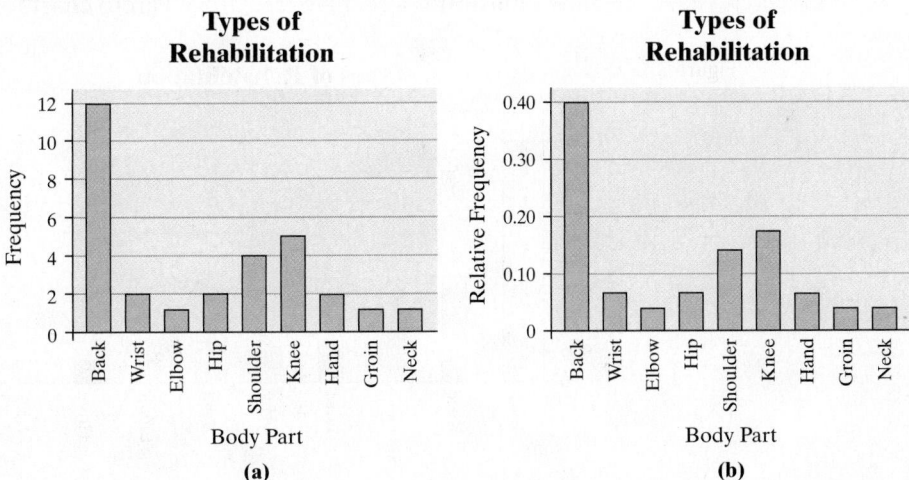

CAUTION!
Graphs that start the scale at some value other than 0 or have bars with unequal widths, bars with different colors, or three-dimensional bars can misrepresent the data.

EXAMPLE 4   Constructing a Frequency or Relative Frequency Bar Graph Using Technology

Problem   Use a statistical spreadsheet to construct a frequency or relative frequency bar graph for the data in Example 1.

Using Technology

The graphs obtained from a different statistical package may differ from those in Figure 2. Some packages use the word *count* in place of *frequency* or *percent* in place of *relative frequency*.

Approach   We will use Excel to construct the frequency and relative frequency bar graphs. The steps for constructing the graphs using Minitab, Excel, and StatCrunch are given in the Technology Step-by-Step on pages 69–70. **Note:** TI-graphing calculators cannot draw frequency or relative frequency bar graphs.

Solution   Figure 2 on the following page shows the frequency and relative frequency bar graphs obtained from Excel.

(continued)

Figure 2

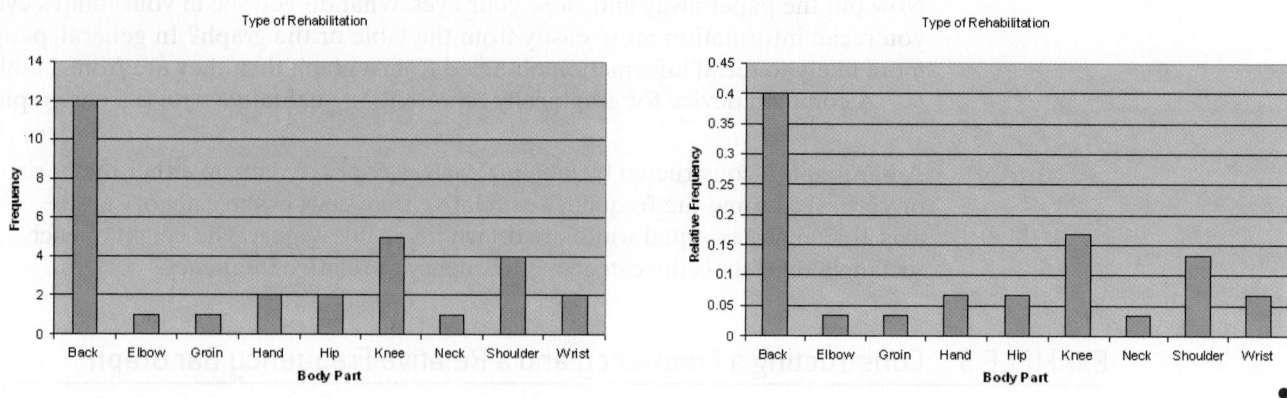

● **Now Work Problems 21(c)–(d)**

In bar graphs, the order of the categories does not usually matter. However, bar graphs that have categories arranged in decreasing order of frequency help prioritize categories for decision-making purposes in areas such as quality control, human resources, and marketing.

Definition

A **Pareto chart** is a bar graph whose bars are drawn in decreasing order of frequency or relative frequency.

Figure 3 illustrates a relative frequency Pareto chart for the data in Table 3.

Figure 3

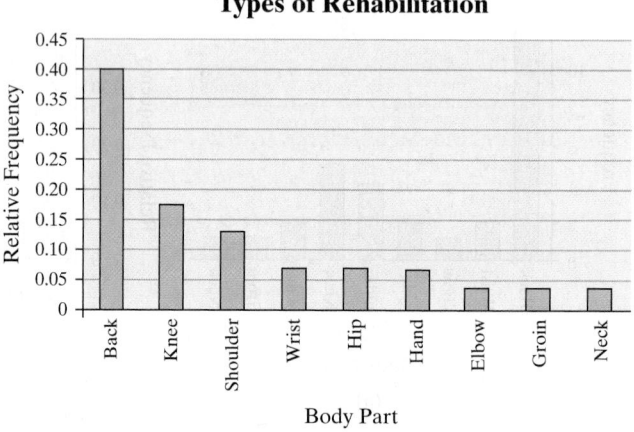

Types of Rehabilitation

Side-by-Side Bar Graphs

Suppose we want to know whether more people are finishing college today than in 1990. We could draw a **side-by-side bar graph** to compare the data for the two different years. Data sets should be compared by using relative frequencies, because different sample or population sizes make comparisons using frequencies difficult or misleading.

EXAMPLE 5   Comparing Two Data Sets

Problem The data in Table 4 represent the educational attainment in 1990 and 2013 of adults 25 years and older who are residents of the United States. The data are in thousands. So 39,344 represents 39,344,000.

(a) Draw a side-by-side relative frequency bar graph of the data.

(b) Make some general conclusions based on the graph.

Table 4		
Educational Attainment	**1990**	**2013**
Not a high school graduate	39,344	24,517
High school diploma	47,643	61,704
Some college, no degree	29,780	34,805
Associate's degree	9,792	20,367
Bachelor's degree	20,833	41,575
Graduate or professional degree	11,478	23,931
Totals	**158,870**	**206,899**

Source: U.S. Census Bureau

Approach First, determine the relative frequencies of each category for each year. To construct side-by-side bar graphs, draw two bars for each category of data, one for 1990, the other for 2013.

Solution Table 5 shows the relative frequency for each category.
(a) The side-by-side bar graph is shown in Figure 4.

Table 5		
Educational Attainment	**1990**	**2013**
Not a high school graduate	0.2476	0.1185
High school diploma	0.2999	0.2982
Some college, no degree	0.1874	0.1682
Associate's degree	0.0616	0.0984
Bachelor's degree	0.1311	0.2009
Graduate or professional degree	0.0722	0.1157

Figure 4

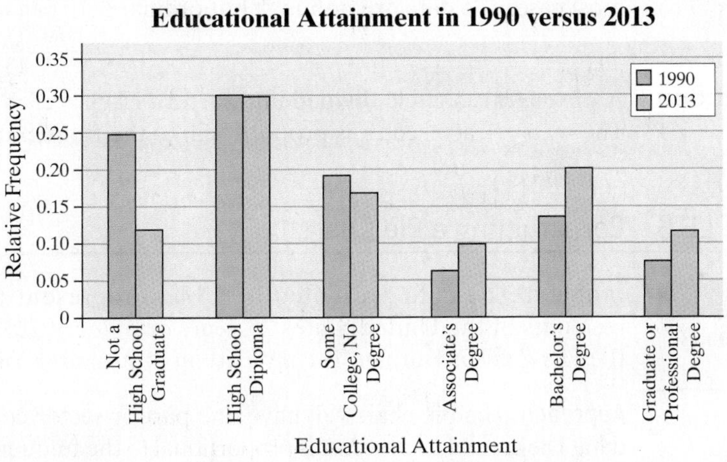

Educational Attainment in 1990 versus 2013

(b) The relative frequency of adults who are not high school graduates in 2013 is less than half that of 1990. In 2013, a much higher percentage of the adult population has at least a bachelor's degree. However, the percentage of the population with a bachelor's degree has not doubled (as the frequencies in Table 4 might suggest). An overall conclusion is that adult Americans are more educated in 2013 than they were in 1990.

• **Now Work Problem 17**

Horizontal Bars

Bar graphs may also be drawn with horizontal bars. Horizontal bars are preferable when category names are lengthy. For example, Figure 5 on the next page uses horizontal bars to display the same data as in Figure 4.

Figure 5

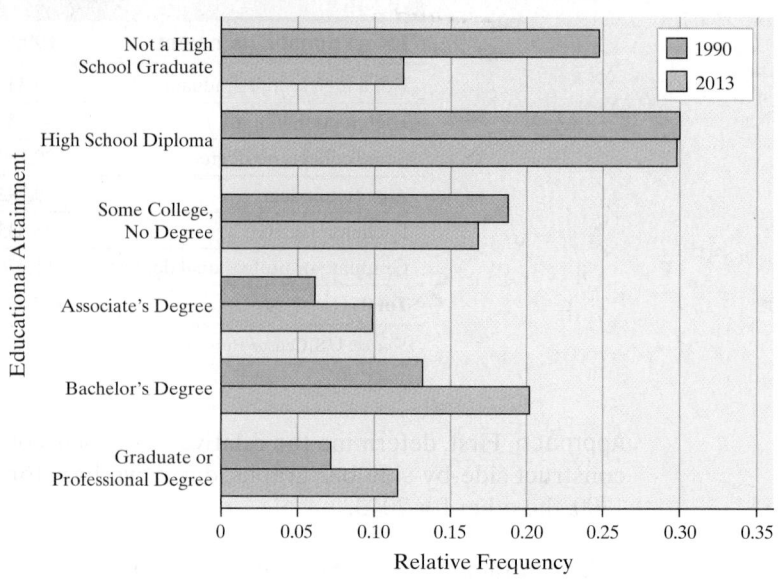

Educational Attainment in 1990 versus 2013

③ Construct Pie Charts

Pie charts are typically used to present the relative frequency of qualitative data. In most cases the data are nominal, but ordinal data can also be displayed in a pie chart.

Definition | A **pie chart** is a circle divided into sectors. Each sector represents a category of data. The area of each sector is proportional to the frequency of the category.

EXAMPLE 6 Constructing a Pie Chart

Problem The data presented in Table 6 represent the educational attainment of residents of the United States 25 years or older in 2013, based on data obtained from the U.S. Census Bureau. The data are in thousands. Construct a pie chart of the data.

Approach The pie chart will have one part or sector corresponding to each category of data. The area of each sector is proportional to the frequency of each category. For example, from Table 5, the proportion of all U.S. residents 25 years or older who are not high school graduates is 0.1185. The category "not a high school graduate" will make up 11.85% of the area of the pie chart. Since a circle has 360 degrees, the degree measure of the sector for this category will be $(0.1185)360° \approx 43°$. Use a protractor to measure each angle.

Solution Use the same approach for the remaining categories to obtain Table 7.

Table 6

Educational Attainment	Frequency
Not a high school graduate	24,517
High school diploma	61,704
Some college, no degree	34,805
Associate's degree	20,367
Bachelor's degree	41,575
Graduate or professional degree	23,931
Totals	**206,899**

Table 7

Educational Attainment	Frequency	Relative Frequency	Degree Measure of Each Sector
Not a high school graduate	24,517	0.1185	43
High school diploma	61,704	0.2982	107
Some college, no degree	34,805	0.1682	61
Associate's degree	20,367	0.0984	35
Bachelor's degree	41,575	0.2009	72
Graduate or professional degree	23,931	0.1157	42

▣ **Using Technology**

Most statistical spreadsheets are capable of drawing pie charts. See the Technology Step-by-Step on pages 69–70 for instructions on drawing pie charts using Minitab, Excel, and StatCrunch. The TI-83 and TI-84 Plus graphing calculators do not draw pie charts.

To construct a pie chart by hand, use a protractor to approximate the angles for each sector. See Figure 6.

●

Figure 6

Educational Attainment, 2013

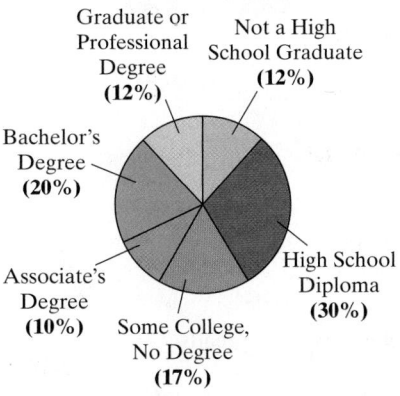

● Now Work Problem 21(e)

To make a pie chart, we need all the categories of the variable under consideration. For example, using Example 1, we could create a bar graph that lists the proportion of patients requiring back, shoulder, or knee rehabilitation, but it would not make sense to construct a pie chart for this situation. Do you see why? Only 70% of the data would be represented.

When should a bar graph or a pie chart be used? Pie charts are useful for showing the division of all possible values of a qualitative variable into its parts. However, because angles are often hard to judge in pie charts, they are not as useful in comparing two specific values of the qualitative variable. Instead the emphasis is on comparing the part to the whole. Bar graphs are useful when we want to compare the different parts, not necessarily the parts to the whole. For example, to get the "big picture" regarding educational attainment in 2013, a pie chart is a good visual summary. However, to compare bachelor's degrees to high school diplomas, a bar graph is a better visual summary. Since bars are easier to draw and compare, some practitioners forgo pie charts in favor of Pareto charts when comparing parts to the whole.

Technology Step-by-Step Drawing Bar Graphs and Pie Charts

TI-83/84 Plus
The TI-83 or TI-84 Plus does not have the ability to draw bar graphs or pie charts.

Minitab
Frequency or Relative Frequency Distributions from Raw Data
1. Enter the raw data in C1.
2. Select **Stat** and highlight **Tables** and select **Tally Individual Variables** . . .
3. Fill in the window with appropriate values. In the "Variables" box, enter C1. Check "counts" for a frequency distribution and/or "percents" for a relative frequency distribution. Click OK.

Bar Graphs from Summarized Data
1. Enter the categories in C1 and the frequency or relative frequency in C2.
2. Select **Graph** and then **Bar Chart**.
3. In the "Bars represent" pull-down menu, select "Values from a table" and highlight "Simple." Press OK.
4. Fill in the window with the appropriate values. In the "Graph variables" box, enter C2. In the "Categorical variable" box, enter C1. By pressing Labels, you can add a title to the graph. Click OK to obtain the bar graph.

Bar Graphs from Raw Data
1. Enter the raw data in C1.
2. Select **Graph** and then **Bar Chart**.
3. In the "Bars represent" pull-down menu, select "Counts of unique values" and highlight "Simple." Press OK.
4. Fill in the window with the appropriate values. In the "Categorical variable" box, enter C1. By pressing Labels, you can add a title to the graph. Click OK to obtain the bar graph.

Pie Chart from Raw or Summarized Data
1. If the data are summarized in a table, enter the categories in C1 and the frequency or relative frequency in C2. If the data are raw, enter the data in C1.
2. Select **Graph** and then **Pie Chart**.
3. Fill in the window with the appropriate values. If the data are summarized, click the "Chart values from a table"

radio button; if the data are raw, click the "Chart counts of unique values" radio button. For summarized data, enter C1 in the "Categorical variable" box and C2 in the "Summary variable" box. If the data are raw, enter C1 in the "Categorical variable" box. By pressing Labels, you can add a title to the graph. Click OK to obtain the pie chart.

Excel
Bar Graphs from Summarized Data
1. Enter the categories in column A and the frequency or relative frequency in column B.
2. Highlight the data to be graphed.
3. Select the Insert menu. Click the "column" or "bar" chart type. Select the chart type in the upper-left corner.
4. Click the "+" to enter axes labels and chart title.

Pie Charts from Summarized Data
1. Enter the categories in column A and the frequencies in column B.
2. Highlight the data to be graphed.
3. Select the Insert menu and click the "pie" chart type. Select the pie chart in the upper-left corner.
4. Click the "+" to enter labels, chart title, and legend.

StatCrunch
Frequency or Relative Frequency Distributions from Raw Data
1. If necessary, enter the raw data into the spreadsheet. Name the column.
2. Select **Stat**, highlight **Tables**, and select **Frequency**.
3. Click on the variable you wish to summarize. Click the type of table you want. If you want both Frequency and Relative Frequency, highlight Frequency, then press Ctrl (or Command on an Apple) and select Relative frequency. Click Compute!

Bar Graphs from Summarized Data
1. If necessary, enter the summarized data into the spreadsheet. Name the variable and frequency (or relative frequency) columns.
2. Select **Graph**, highlight **Bar Plot**, then select **With Summary**.

3. Select the "Categories in:" variable and "Counts in:" variable. Choose the type of bar graph (frequency or relative frequency). Enter labels for the X-axis and Y-axis. Enter a title for the graph. Click Compute!

Bar Graphs from Raw Data

1. If necessary, enter the raw data into the spreadsheet. Name the column variable.

2. Select **Graph**, highlight **Bar Plot**, then highlight **With Data**.

3. Click on the column name of the variable you wish to summarize. Leave the grouping option as "Split bars." Choose the type of bar graph (frequency or relative frequency). Enter labels for the X-axis and Y-axis. Enter a title for the graph. Click Compute!

Side-by-Side Bar Graphs from Summarized Data

1. If necessary, enter the summarized data into the spreadsheet. Name the columns.

2. Select **Graph**, highlight **Chart**, then select **Columns**.

3. Select the column variables that contain the frequency or relative frequency of each category. Select the

column of the variable that has the row labels. Choose the display you would like (vertical or horizontal split bars). Click Compute!

Pie Chart from Summarized Data

1. If necessary, enter the summarized data into the spreadsheet. Name the column.

2. Select **Graph**, highlight **Pie Chart**, then select **With Summary**.

3. Select the "Categories in:" variable and "Counts in:" variable. Choose the display you would like. Enter a title for the graph. Click Compute!

Pie Chart from Raw Data

1. If necessary, enter the raw data into the spreadsheet. Name the column variable.

2. Select **Graph**, highlight **Pie Chart**, then select **With Data**.

3. Click on the column name of the variable you wish to summarize. Choose the display you would like. Enter a title for the graph. Click Compute!

 ## 2.1 Assess Your Understanding

Vocabulary and Skill Building

1. Define raw data in your own words.

2. A frequency distribution lists the _____ of occurrences of each category of data, while a relative frequency distribution lists the _____ of occurrences of each category of data.

3. In a relative frequency distribution, what should the relative frequencies add up to?

4. What is a bar graph? What is a Pareto chart?

5. Flu Season The pie chart shown, the type we see in *USA Today*, depicts the approaches people use to avoid getting the flu.

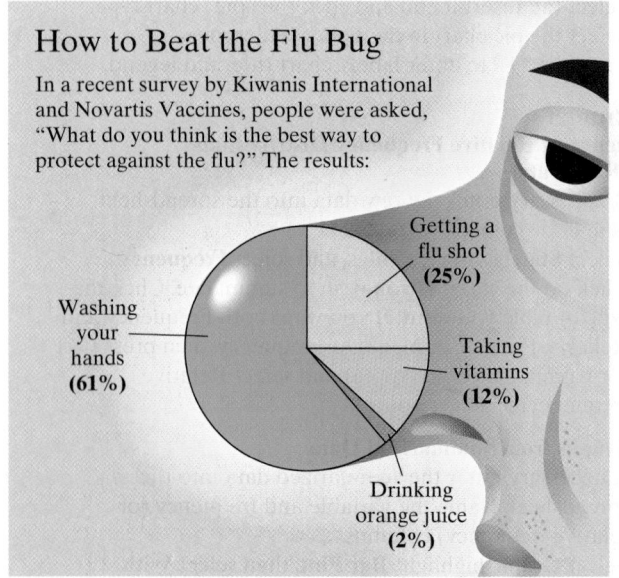

How to Beat the Flu Bug

In a recent survey by Kiwanis International and Novartis Vaccines, people were asked, "What do you think is the best way to protect against the flu?" The results:

- Getting a flu shot (25%)
- Washing your hands (61%)
- Taking vitamins (12%)
- Drinking orange juice (2%)

Source: Kiwanis International and Novartis Vaccines

(a) What is the most common approach? What percentage of the population chooses this method?

(b) What is the least used approach? What percentage of the population chooses this method?

(c) What percentage of the population thinks flu shots are the best way to beat the flu?

6. Cosmetic Surgery This *USA Today*–type chart shows the most frequent cosmetic surgeries for women in 2009.

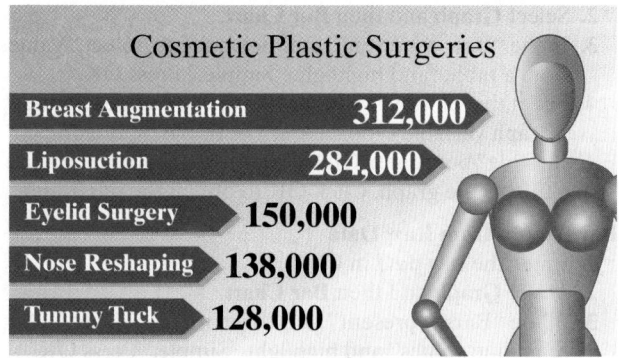

Cosmetic Plastic Surgeries

Breast Augmentation	312,000
Liposuction	284,000
Eyelid Surgery	150,000
Nose Reshaping	138,000
Tummy Tuck	128,000

By Anne R. Carey and Suzy Parker, USA Today
Source: American Society of Plastic Surgeons (plasticsurgery.org)

(a) If women had 1.35 million cosmetic surgeries in 2009, what percent were for tummy tucks?

(b) What percent were for nose reshaping?

(c) How many surgeries are not accounted for in the graph?

7. Most Valuable Player The following Pareto chart shows the position played by the most valuable player (MVP) in the National League since 1931. *Source:* http://www.baseball-almanac.com/

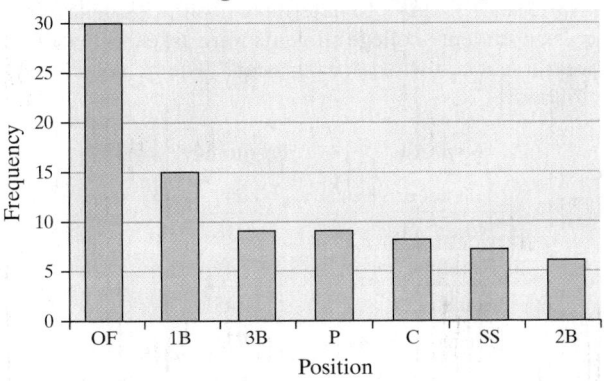

National League Most Valuable Player Award

(a) Which position had the most MVPs?
(b) How many MVPs played first base (1B)?
(c) How many more MVPs played outfield (OF) than first base?
(d) There are three outfield positions (left field, center field, right field). Given this, how might the graph be misleading?

8. Poverty The U.S. Census Bureau uses money income thresholds to define poverty. For example, in 2013 the poverty threshold for a family of four with two children was $23,264. The bar graph represents the number of people living in poverty in the United States in 2013, by ethnicity.

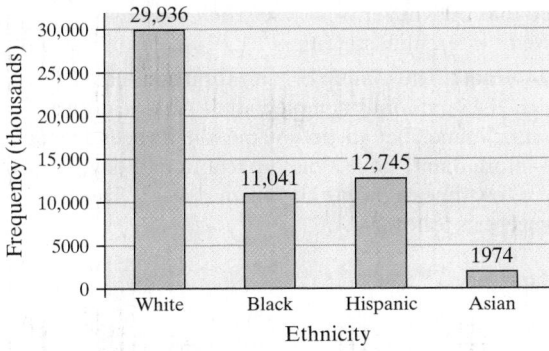

Number in Poverty

Source: U.S. Census Bureau

(a) How many whites were living in poverty in 2013?
(b) Of the impoverished, what percent were Hispanic?
(c) How might this graph be misleading?

9. Divorce The following graph represents the results of a survey, in which a random sample of adult Americans was asked, *"Please tell me whether you personally believe that in general divorce is morally acceptable or morally wrong."*

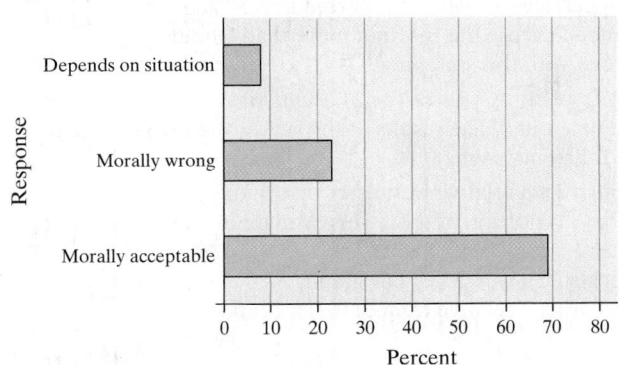

Opinion Regarding Divorce

Source: Gallup

(a) What percent of the respondents believe divorce is morally acceptable?
(b) If there are 240 million adults in America, how many believe that divorce is morally wrong?
(c) If Gallup claimed that the results of the survey indicate that 8% of adult Americans believe that divorce is acceptable in certain situations, would you say this statement is descriptive or inferential? Why?

10. Identity Theft Identity fraud occurs when someone else's personal information is used to open credit card accounts, apply for a job, receive benefits, and so on. The following relative frequency bar graph represents the various types of identity theft based on a study conducted by the Federal Trade Commission.

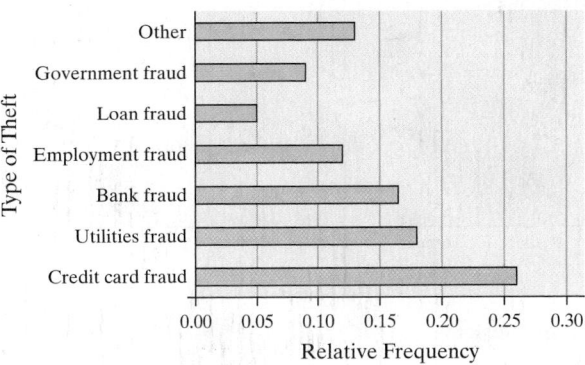

Identity Theft

Source: Federal Trade Commission

(a) Approximately what percentage of identity theft was loan fraud (such as applying for a loan in someone else's name)?
(b) If there were 10 million cases of identity fraud in a recent year, how many were credit card fraud (someone uses someone else's credit card to make a purchase)?

11. Made in America A random sample of 2163 adults (aged 18 and over) was asked, "When you see an ad emphasizing that a product is 'Made in America', are you more likely to buy it, less likely to buy it, or neither more nor less likely to buy it?" The results of the survey are presented in the side-by-side bar graph.

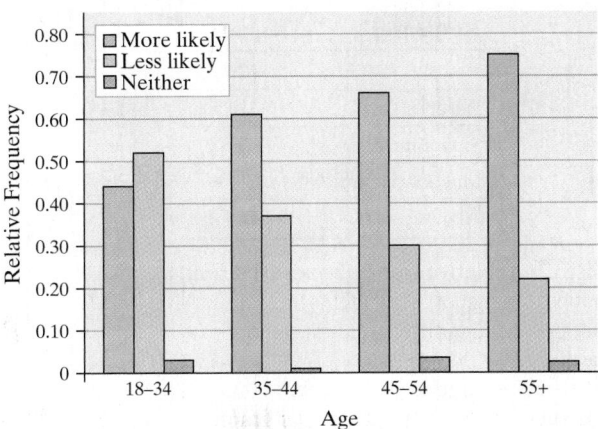

Likelihood to Buy When Made in America

Source: Harris Interactive

(a) What proportion of 18- to 34-year-old respondents are more likely to buy when made in America? What proportion of 35- to 44-year-old respondents are more likely to buy when made America?
(b) Which age group has the greatest proportion who are more likely to buy when made in America?

(c) Which age group has a majority of respondents who are less likely to buy when made in America?

(d) What is the apparent association between age and likelihood to buy when made in America?

12. Desirability Attributes A random sample of 2163 adults (aged 18 and over) was asked, "Given a choice of the following, which one would you most want to be?" The results of the survey are presented in the side-by-side bar graph.

Desirability Attributes

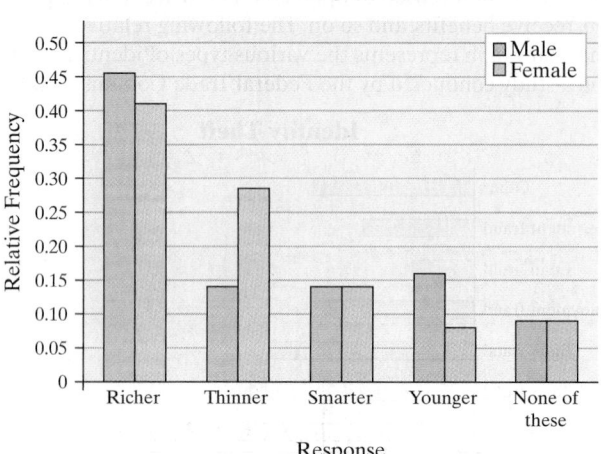

Source: Harris Interactive

(a) What proportion of males would like to be richer? What proportion of females would like to be richer?

(b) Which attribute do females desire more than males?

(c) Which attribute do males prefer over females two-to-one?

(d) Which attribute do males and females desire in equal proportion?

Applying the Concepts

13. College Survey In a national survey conducted by the Centers for Disease Control to determine health-risk behaviors among college students, college students were asked, "How often do you wear a seat belt when riding in a car driven by someone else?" The frequencies were as follows:

Response	Frequency
Never	125
Rarely	324
Sometimes	552
Most of the time	1257
Always	2518

(a) Construct a relative frequency distribution.

(b) What percentage of respondents answered "Always"?

(c) What percentage of respondents answered "Never" or "Rarely"?

(d) Construct a frequency bar graph.

(e) Construct a relative frequency bar graph.

(f) Construct a pie chart.

(g) Suppose that a representative from the Centers for Disease Control says, "52.7% of college students surveyed always wear a seat belt." Is this a descriptive or inferential statement?

14. College Survey In a national survey conducted by the Centers for Disease Control to determine health-risk behaviors among college students, college students were asked, "How often do you wear a seat belt when driving a car?" The frequencies were as follows:

Response	Frequency
I do not drive a car	249
Never	118
Rarely	249
Sometimes	345
Most of the time	716
Always	3093

(a) Construct a relative frequency distribution.

(b) What percentage of respondents answered "Always"?

(c) What percentage of respondents answered "Never" or "Rarely"?

(d) Construct a frequency bar graph.

(e) Construct a relative frequency bar graph.

(f) Construct a pie chart.

(g) Compute the relative frequencies of "Never," "Rarely," "Sometimes," "Most of the time," and "Always," excluding those that do not drive. Compare with those in Problem 13. What might you conclude?

(h) Suppose that a representative from the Centers for Disease Control says, "2.5% of the college students in this survey responded that they never wear a seat belt." Is this a descriptive or inferential statement?

15. Use the Internet? The Gallup organization conducted a survey in which 1025 randomly sampled adult Americans were asked, "How much time, if at all, do you personally spend using the Internet—more than 1 hour a day, up to 1 hour a day, a few times a week, a few times a month or less, or never?" The results of the survey were as follows:

Response	Frequency
More than 1 hour a day	377
Up to 1 hour a day	192
A few times a week	132
A few times a month or less	81
Never	243

(a) Construct a relative frequency distribution.

(b) What proportion of those surveyed never use the Internet?

(c) Construct a frequency bar graph.

(d) Construct a relative frequency bar graph.

(e) Construct a pie chart.

(f) A local news broadcast reported that 37% of adult Americans use the Internet more than 1 hour a day. What is wrong with this statement?

16. Dining Out A sample of 521 adults was asked, "How often do you dine out?" The results of the survey are given in the table on the following page.

(a) Construct a relative frequency distribution.

(b) What proportion of those surveyed dine out once or twice a week?

(c) Construct a frequency bar graph.

(d) Construct a relative frequency bar graph.

Response	Frequency
Several times a week	103
Once or twice a week	204
A few times a month	130
Very rarely	79
Never	5

Source: www.amylovesit.com

NW 17. Texting A survey of U.S. adults and teens (ages 12–17) was administered by Pew Research, to determine the number of texts sent in a single day.

Number of Texts	Adults	Teens
None	173	13
1–10	978	138
11–20	249	69
21–50	249	113
51–100	134	113
101+	153	181

Source: Pew Internet

(a) Construct a relative frequency distribution for adults.
(b) Construct a relative frequency distribution for teens.
(c) Construct a side-by-side relative frequency bar graph.
(d) Compare the texting habits of adults and teens.

18. Educational Attainment The educational attainment of males and females in 2013 was as follows:

Educational Attainment	Males (in millions)	Females (in millions)
Not a high school graduate	12.3	12.2
High school graduate	30.0	31.7
Some college, but no degree	16.5	18.3
Associate's degree	8.8	11.6
Bachelor's degree	19.9	21.7
Advanced degree	11.9	12.1

Source: U.S. Census Bureau

(a) Construct a relative frequency distribution for males.
(b) Construct a relative frequency distribution for females.
(c) Construct a side-by-side relative frequency bar graph.
(d) Compare each gender's educational attainment. Make a conjecture about the reasons for the differences.

19. Dream Job A survey of adult men and women asked, "Which one of the following jobs would you most like to have?" The results of the survey are shown in the table.

(a) Construct a relative frequency distribution for men and women.
(b) Construct a side-by-side relative frequency bar graph.
(c) What are the apparent differences in gender as it pertains to this question?

Response	Men	Women
Professional athlete	40	18
Actor/actress	26	37
President of the United States	13	13
Rock star	13	13
Not sure	7	19

Source: Marist Poll

20. Car Color A survey of 100 randomly selected autos in the luxury car segment and 100 randomly selected autos in the sports car segment that were recently purchased yielded the following colors.

Color	Number of Luxury Cars	Number of Sports Cars
White	25	10
Black	22	15
Silver	16	18
Gray	12	15
Blue	7	13
Red	7	15
Gold	6	5
Green	3	2
Brown	2	7

Source: Based on results from www.infoplease.com

(a) Construct a relative frequency distribution for each car type.
(b) Draw a side-by-side relative frequency bar graph.
(c) Compare the colors for the two car types. Make a conjecture about the reasons for the differences.

NW 21. Walt Disney Stock The table shows the movement of Walt Disney stock for 30 randomly selected trading days. "Up" means the stock price increased in value for the day, "Down" means the stock price decreased in value for the day, and "No Change" means the stock price closed at the same price it closed for the previous day.

Down	Up	Up	Down	Down	Up
Down	Up	Down	Up	Down	Up
Down	Down	Up	Up	Up	Up
Down	Down	Down	Up	Down	Up
No Change	Up	Down	Down	No Change	Down

Source: Yahoo!Finance

(a) Construct a frequency distribution.
(b) Construct a relative frequency distribution.
(c) Construct a frequency bar graph.
(d) Construct a relative frequency bar graph.
(e) Construct a pie chart.

22. Bachelor Party In a survey conducted by Opinion Research, participants were asked, "If you had an 'X-rated' bachelor party, would you tell your fiancé all, edit details, or say nothing?" The data on the following page are based on their results.

(a) Construct a frequency distribution.
(b) Construct a relative frequency distribution.
(c) Construct a frequency bar graph.
(d) Construct a relative frequency bar graph.
(e) Construct a pie chart.

Tell all	Edit details	Tell all	Edit details	Tell all
Edit details	Say nothing	Edit details	Tell all	Tell all
Tell all	Tell all	Edit details	Tell all	Say nothing
Edit details	Tell all	Tell all	Say nothing	Tell all
Tell all	Edit details	Tell all	Tell all	Say nothing

Source: Based on results from Opinion Research

✂ 🛠 **23. Favorite Day to Eat Out** A survey was conducted by Wakefield Research in which participants were asked to disclose their favorite night to order takeout for dinner. The following data are based on their results.

Thursday	Saturday	Friday	Friday	Sunday
Wednesday	Saturday	Friday	Tuesday	Friday
Saturday	Monday	Friday	Friday	Sunday
Friday	Tuesday	Wednesday	Saturday	Friday
Wednesday	Monday	Wednesday	Wednesday	Friday
Friday	Wednesday	Thursday	Tuesday	Friday
Tuesday	Saturday	Friday	Tuesday	Friday
Saturday	Saturday	Saturday	Sunday	Friday

Source: Based on results from Wakefield Research

(a) Construct a frequency distribution.
(b) Construct a relative frequency distribution.
(c) If you own a restaurant, which day would you purchase an advertisement in the local newspaper? Are there any days you would avoid purchasing advertising space?
(d) Construct a frequency bar graph.
(e) Construct a relative frequency bar graph.
(f) Construct a pie chart.

🛠 **24. Blood Type** A phlebotomist draws the blood of a random sample of 50 patients and determines their blood types as shown:

O	O	A	A	O
B	O	B	A	O
AB	B	A	B	AB
O	O	A	A	O
AB	O	A	B	A
O	A	A	O	A
O	A	O	AB	A
O	B	A	A	O
O	O	O	A	O
O	A	O	A	O

(a) Construct a frequency distribution.
(b) Construct a relative frequency distribution.
(c) According to the data, which blood type is most common?
(d) According to the data, which blood type is least common?
(e) Use the results of the sample to conjecture the percentage of the population that has type O blood. Is this an example of descriptive or inferential statistics?
(f) Contact a local hospital and ask them the percentage of the population that is blood type O. Why might the results differ?
(g) Draw a frequency bar graph.
(h) Draw a relative frequency bar graph.
(i) Draw a pie chart.

🛠 **25. President's State of Birth** The following table lists the presidents of the United States (as of May, 2015) and their state of birth.

Birthplace of U.S. President					
President	**State of Birth**	**President**	**State of Birth**	**President**	**State of Birth**
Washington	Virginia	Lincoln	Kentucky	Hoover	Iowa
J. Adams	Massachusetts	A. Johnson	North Carolina	F. D. Roosevelt	New York
Jefferson	Virginia	Grant	Ohio	Truman	Missouri
Madison	Virginia	Hayes	Ohio	Eisenhower	Texas
Monroe	Virginia	Garfield	Ohio	Kennedy	Massachusetts
J. Q. Adams	Massachusetts	Arthur	Vermont	L. B. Johnson	Texas
Jackson	South Carolina	Cleveland	New Jersey	Nixon	California
Van Buren	New York	B. Harrison	Ohio	Ford	Nebraska
W. H. Harrison	Virginia	Cleveland	New Jersey	Carter	Georgia
Tyler	Virginia	McKinley	Ohio	Reagan	Illinois
Polk	North Carolina	T. Roosevelt	New York	G. H. Bush	Massachusetts
Taylor	Virginia	Taft	Ohio	Clinton	Arkansas
Fillmore	New York	Wilson	Virginia	G. W. Bush	Connecticut
Pierce	New Hampshire	Harding	Ohio	Obama	Hawaii
Buchanan	Pennsylvania	Coolidge	Vermont		

(a) Construct a frequency bar graph for state of birth.
(b) Which state has yielded the most presidents?
(c) Explain why the answer obtained in part (b) may be misleading.

26. Highest Elevation The following data represent the land area and highest elevation for each of the seven continents.

Continent	Land Area (square miles)	Highest Elevation (feet)
Africa	11,608,000	19,340
Antarctica	5,100,000	16,066
Asia	17,212,000	29,035
Australia	3,132,000	7,310
Europe	3,837,000	18,510
North America	9,449,000	20,320
South America	6,879,000	22,834

Source: www.infoplease.com

(a) Would it make sense to draw a pie chart for land area? Why? If so, draw a pie chart.
(b) Would it make sense to draw a pie chart for the highest elevation? Why? If so, draw a pie chart.

27. StatCrunch Survey Choose a qualitative variable from the Sullivan StatCrunch Survey I data set at www.pearsonhighered.com/sullivanstats and summarize the variable.

28. StatCrunch Survey Choose a qualitative variable from the Sullivan StatCrunch Survey I data set at www.pearsonhighered.com/sullivanstats and summarize the variable by gender.

29. Putting It Together: Online Homework Keeping students engaged in the learning process greatly increases their chance of success in a course. Traditional lecture-based math instruction has given way to a more student-engaged approach where students interact with the teacher in class and receive immediate feedback to their responses. The teacher presence allows students, when incorrect in a response, to be guided through a solution and then immediately be given a similar problem to attempt.

A researcher conducted a study to investigate whether an online homework system using an attempt–feedback–reattempt approach improved student learning over traditional pencil-and-paper homework. The online homework system was designed to increase student engagement outside class, something commonly missing in traditional pencil-and-paper assignments, ultimately leading to increased learning.

The study was conducted using two first-semester calculus classes taught by the researcher in a single semester. One class was assigned traditional homework and the other was assigned online homework that used the attempt–feedback–reattempt approach. The summaries are based on data from the study.

(a) What is the research objective?
(b) Is this study an observational study or experiment?
(c) Give an example of how the researcher attempted to control variables in the study.
(d) Explain why assigning homework type to entirely separate classes can confound the conclusions of the study.
(e) For the data in the table, (i) identify the variables, (ii) indicate whether the variables are qualitative or quantitative, and (iii) for each quantitative variable, indicate whether the variable is discrete or continuous.

	Prior College Experience		No Prior College Experience	
	Traditional	Online	Traditional	Online
Number of students	10	9	23	18
Average age	22.8	19.4	18.1	18.1
Average exam score	84.5	68.9	79.4	80.6

Grades Earned on Exams (no prior college experience)

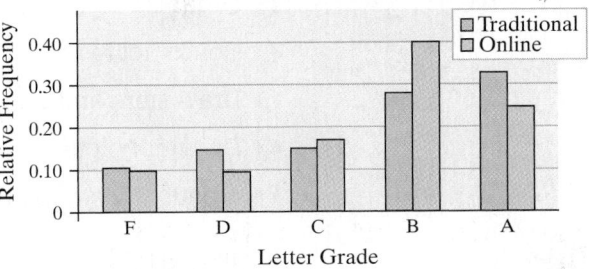

Source: Journal of Computers in Mathematics and Science Teaching 26(1):55–73, 2007

(f) What type of variable is letter grade? What level of measurement is letter grade? Do you think presenting the data in a table from A to F would be a better representation of the data than presenting it in a graph?
(g) What type of graph is displayed?
(h) Could the data in the graph be presented in a pie chart? If so, what is the "whole"? If not, why not?
(i) Considering the students with no prior college experience, how might the table and the graph generate conflicting conclusions?

Explaining the Concepts

30. When should relative frequencies be used when comparing two data sets? Why?

31. Suppose you need to summarize ordinal data in a bar graph. How would you arrange the categories of data on the horizontal axis? Is it possible to make the order of the data apparent in a pie chart?

32. Describe the circumstances in which a bar graph is preferable to a pie chart. When is a pie chart preferred over a bar graph? Are there circumstances in which a pie chart cannot be drawn, but a bar graph could be drawn? What are these circumstances?

33. Consider the information in the chart shown below, which is in the *USA Today* style of graph. Could the information provided be organized into a pie chart? Why or why not?

2.2 Organizing Quantitative Data: The Popular Displays

Preparing for This Section Before getting started, review the following:

- Quantitative variable (Section 1.1, p. 7)
- Discrete variable (Section 1.1, pp. 7–8)
- Continuous variable (Section 1.1, pp. 7–8)

Objectives
1. Organize discrete data in tables
2. Construct histograms of discrete data
3. Organize continuous data in tables
4. Construct histograms of continuous data
5. Draw stem-and-leaf plots
6. Draw dot plots
7. Identify the shape of a distribution

In summarizing quantitative data, we first determine whether the data are discrete or continuous. If the data are discrete with relatively few different values of the variable, then the categories of data (called **classes**) will be the observations (as in qualitative data). If the data are discrete, but with many different values of the variable or if the data are continuous, then the categories of data (the *classes*) must be created using intervals of numbers. We will first present the techniques for organizing discrete quantitative data when there are relatively few different values and then proceed to organizing continuous quantitative data.

1 Organize Discrete Data in Tables

Use the values of the discrete variable to create the classes when the number of distinct data values is small.

EXAMPLE 1   Constructing Frequency and Relative Frequency Distributions from Discrete Data

Problem The manager of a Wendy's fast-food restaurant wants to know the typical number of customers who arrive during the lunch hour. The data in Table 8 represent the number of customers who arrive at Wendy's for 40 randomly selected 15-minute intervals of time during lunch. For example, during one 15-minute interval, seven customers arrived. Construct a frequency and relative frequency distribution of the data.

Table 8							
Number of Arrivals at Wendy's							
7	6	6	6	4	6	2	6
5	6	6	11	4	5	7	6
2	7	1	2	4	8	2	6
6	5	5	3	7	5	4	6
2	2	9	7	5	9	8	5

Approach The number of people arriving could be 0, 1, 2, 3, ... Table 8 shows there are 11 categories of data from this study: 1, 2, 3, ..., 11. Tally the number of observations for each category, count each tally, and create the frequency and relative frequency distributions.

Solution The two distributions are shown in Table 9.

Table 9			
Number of Customers	**Tally**	**Frequency**	**Relative Frequency**
1	|	1	$\frac{1}{40} = 0.025$
2	⫴⫴ |	6	0.15
3	|	1	0.025
4	||||	4	0.1
5	⫴⫴ ||	7	0.175
6	⫴⫴ ⫴⫴ |	11	0.275
7	⫴⫴	5	0.125
8	||	2	0.05
9	||	2	0.05
10		0	0.0
11	|	1	0.025

• Now Work Problems 25(a)–(e)

On the basis of the relative frequencies, 27.5% of the 15-minute intervals had 6 customers arrive at Wendy's during the lunch hour. •

② Construct Histograms of Discrete Data

The *histogram*, a graph used to present quantitative data, is similar to the bar graph.

Definition A **histogram** is constructed by drawing rectangles for each class of data. The height of each rectangle is the frequency or relative frequency of the class. The width of each rectangle is the same and the rectangles touch each other.

EXAMPLE 2 Drawing a Histogram for Discrete Data

Problem Construct a frequency histogram and a relative frequency histogram using the data in Table 9.

--- CAUTION! --------------------
The rectangles in histograms touch, but the rectangles in bar graphs do not touch.

Approach The value of each category of data (number of customers) is on the horizontal axis and the frequency or relative frequency is on the vertical axis. Draw rectangles of equal width centered at the value of each category. For example, the first rectangle is centered at 1. For the frequency histogram, the height of the rectangle is

(continued)

the frequency of the category; for the relative frequency histogram, the height is the relative frequency of the category. Remember, the rectangles touch.

Solution Figures 7(a) and (b) show the frequency and relative frequency histograms, respectively.

Figure 7

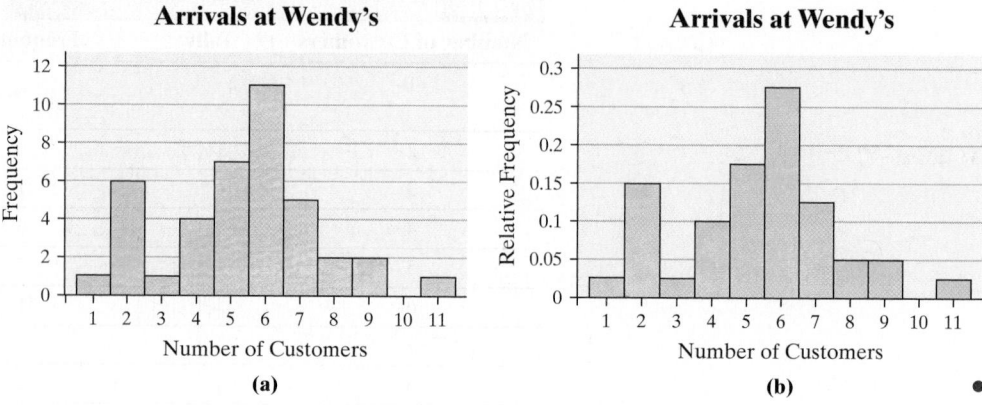

Now Work Problems 25(f)–(g)

③ Organize Continuous Data in Tables

Classes are categories into which data are grouped. When a data set consists of a large number of different discrete data values or when a data set consists of continuous data, we create classes by using intervals of numbers.

Table 10 is a typical frequency distribution created from continuous data. The data represent the number of U.S. residents, ages 25–74, who had earned a bachelor's degree or higher in 2013.

Notice that the data are categorized, or grouped, by intervals of numbers. Each interval represents a class. For example, the first class is 25- to 34-year-old U.S. residents who had a bachelor's degree or higher. We read this interval as follows: "The number of U.S. residents, ages 25–34, with a bachelor's degree or higher was 14,481,000 in 2013." There are five classes in the table, each with a **lower class limit** (the smallest value within the class) and an **upper class limit** (the largest value within the class). The lower class limit for the first class in Table 10 is 25; the upper class limit is 34. The **class width** is the difference between consecutive lower class limits. In Table 10 the class width is $35 - 25 = 10$.

The data in Table 10 are continuous. So the class 25–34 actually represents 25–34.999 . . . , or 25 up to every value less than 35.

Notice that the classes in Table 10 do not overlap. This is necessary to avoid confusion as to which class a data value belongs. Notice also that the class widths are equal for all classes.

One exception to the requirement of equal class widths occurs in open-ended tables. A table is **open ended** if the first class has no lower class limit or the last class has no upper class limit. The data in Table 11 represent the number of births to unmarried mothers in 2012 in the United States. The last class in the table, "40 and over," is open-ended.

Table 10

Age	Number (in thousands)
25–34	14,481
35–44	14,155
45–54	13,802
55–64	12,121
65–74	7009

Source: U.S. Census Bureau

Table 11

Age	Number
10–19	246,561
20–29	994,828
30–39	336,179
40 and over	28,075

Source: National Vital Statisitics Reports, Vol. 63, No. 2

EXAMPLE 3   Organizing Continuous Data into a Frequency and Relative Frequency Distribution

Problem Suppose you are considering investing in a Roth IRA. You collect the data in Table 12, which represent the five-year rate of return (in percent, adjusted for sales charges) for a simple random sample of 40 large-blended mutual funds. Construct a frequency and relative frequency distribution of the data.

In Other Words
For qualitative and many
discrete data, the classes are
formed by using the data. For
continuous data, the classes
are formed by using intervals of
numbers, such as 30–39.

Table 12							
Five-Year Rate of Return of Mutual Funds (as of 11/18/2014)							
10.94	14.60	12.80	16.00	11.93	15.68	9.03	13.40
10.53	13.98	13.86	12.36	13.54	9.94	13.93	13.63
14.12	14.88	14.77	13.13	8.28	19.43	12.98	13.16
12.26	14.20	14.80	13.26	13.67	10.08	14.86	8.71
12.17	10.26	15.22	13.36	13.55	13.90	15.64	12.80

Source: Morningstar.com

┌ CAUTION! ─────────────┐
¦ Watch out for tables with classes ¦
¦ that overlap, such as a first class of ¦
¦ 20–30 and a second class of 30–40. ¦
└────────────────────────┘

Approach To construct a frequency distribution, first create classes of equal width. Table 12 has 40 observations that range from 8.28 to 19.43, so we decide to create the classes such that the lower class limit of the first class is 8 (a little smaller than the smallest data value) and the class width is 1. There is nothing magical about the choice of 1 as a class width. We could have selected a class width of 3 or any other class width. We choose a class width that we think will nicely summarize the data. If our choice doesn't accomplish this, we can always try another. The second class has a lower class limit $8 + 1 = 9$. The classes cannot overlap, so the upper class limit of the first class is 8.99. Continuing in this fashion, we obtain the following classes:

$$8–8.99$$
$$9–9.99$$
$$\vdots$$
$$19–19.99$$

This gives us 12 classes. Tally the number of observations in each class, count the tallies, and create the frequency distribution. Divide the frequency of each class by 40, the number of observations, to obtain the relative frequency.

Solution Tally the data as shown in the second column of Table 13. The third column shows the frequency of each class. From the frequency distribution, we conclude that a five-year rate of return between 13% and 13.99% occurs with the most frequency. The fourth column shows the relative frequency of each class. So, 32.5% of the large-blended mutual funds had a five-year rate of return between 13% and 13.99%.

🖳 Using Technology

Many technologies have a "sort" feature that makes tallying data by hand much easier.

Table 13			
Class (Five-year rate of return)	**Tally**	**Frequency**	**Relative Frequency**
8–8.99	||	2	2/40 = 0.05
9–9.99	||	2	2/40 = 0.05
10–10.99	||||	4	4/40 = 0.10
11–11.99	|	1	1/40 = 0.025
12–12.99	ⅡⅢ |	6	6/40 = 0.15
13–13.99	ⅢⅢ ⅢⅢ |||	13	13/40 = 0.325
14–14.99	ⅢⅢ ||	7	7/40 = 0.175
15–15.99	|||	3	3/40 = 0.075
16–16.99	|	1	1/40 = 0.025
17–17.99		0	0/40 = 0
18–18.99		0	0/40 = 0
19–19.99	|	1	1/40 = 0.025

(continued)

• **Now Work Problems 27(a)–(b)**

One mutual fund had a five-year rate of return between 19% and 19.99%. We might consider this mutual fund worthy of our investment. This type of information would be more difficult to obtain from the raw data. •

Historical Note

Florence Nightingale was born in Italy on May 12, 1820. She was named after the city of her birth. Nightingale was educated by her father, who attended Cambridge University. Between 1849 and 1851, she studied nursing throughout Europe. In 1854, she was asked to oversee the introduction of female nurses into the military hospitals in Turkey. While there, she greatly improved the mortality rate of wounded soldiers. She collected data and invented graphs (the polar area diagram), tables, and charts to show that improving sanitary conditions would lead to decreased mortality rates. In 1869, Nightingale founded the Nightingale School Home for Nurses. After a long and eventful life as a reformer of health care and contributor to graphics in statistics, Florence Nightingale died on August 13, 1910.

Though formulas and procedures exist for creating frequency distributions from raw data, they do not necessarily provide better summaries. There is no one correct frequency distribution for a particular set of data. However, some frequency distributions can better illustrate patterns within the data than others. So constructing frequency distributions is somewhat of an art form. Use the distribution that seems to provide the best overall summary of the data.

The goal in constructing a frequency distribution is to reveal interesting features of the data, but we also typically want the number of classes to be between 5 and 20. When the data set is small, we usually want fewer classes. When the data set is large, we usually want more classes. Why do you think this is reasonable?

Although there is no "right" frequency distribution, there are bad ones. Use the following guidelines to help determine an appropriate lower class limit of the first class and class width.

Guidelines for Determining the Lower Class Limit of the First Class and Class Width

Choosing the Lower Class Limit of the First Class

Choose the smallest observation in the data set or a convenient number slightly lower than the smallest observation in the data set. For example, in Table 12, the smallest observation is 8.28. A convenient lower class limit of the first class is 8.

Determining the Class Width

- Decide on the number of classes. Generally, there should be between 5 and 20 classes. The smaller the data set, the fewer classes you should have. For example, we might choose 12 classes for the data in Table 12.
- Determine the class width by computing

$$\text{Class width} \approx \frac{\text{largest data value} - \text{smallest data value}}{\text{number of classes}}$$

Round this value *up* to a convenient number. For example, using the data in Table 12, we obtain class width $\approx \dfrac{19.43 - 8.28}{12} = 0.929$. We round this up to 1 because this is an easy number to work with. Rounding up may result in fewer classes than were originally intended.

In Other Words
Rounding *up* is different from rounding *off*. For example, 6.2 rounded *up* would be 7, while 6.2 rounded *off* would be 6.

• **Now Work Problems 31(a)–(c)**

Applying these guidelines, to the five-year rate of return data, we would end up with the frequency distribution shown in Table 13.

④ Construct Histograms of Continuous Data

We are now ready to draw histograms of continuous data.

EXAMPLE 4   Drawing a Histogram of Continuous Data

Problem Construct a frequency and relative frequency histogram of the five-year rate-of-return data discussed in Example 3.

Approach To draw the frequency histogram, use the frequency distribution in Table 13. First, label the lower class limits of each class on the horizontal axis. Then,

In Other Words
Creating the classes for summarizing continuous data is an art form. There is no such thing as *the* correct frequency distribution. However, there can be less desirable frequency distributions. The larger the class width, the fewer classes a frequency distribution will have.

for each class, draw a rectangle whose width is the class width and whose height is the frequency. For the relative frequency histogram, the height of the rectangle is the relative frequency.

Solution Figures 8(a) and (b) show the frequency and relative frequency histograms, respectively.

Figure 8

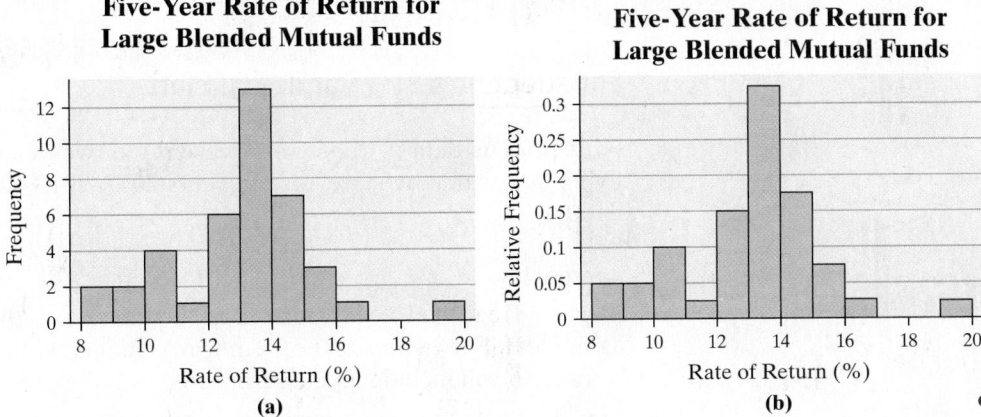

(a)   **(b)**

EXAMPLE 5   Drawing a Histogram for Continuous Data Using Technology

Problem Construct a frequency and relative frequency histogram of the five-year rate-of-return data discussed in Example 3.

Approach We will use StatCrunch to construct the frequency and relative frequency histograms. The steps for constructing the graphs using the TI-83/84 Plus graphing calculators, Minitab, Excel, and StatCrunch, are given in the Technology Step-by-Step on pages 86–87.

Solution Figures 9(a) and (b) show the frequency and relative frequency histograms, respectively, obtained from StatCrunch.

Figure 9

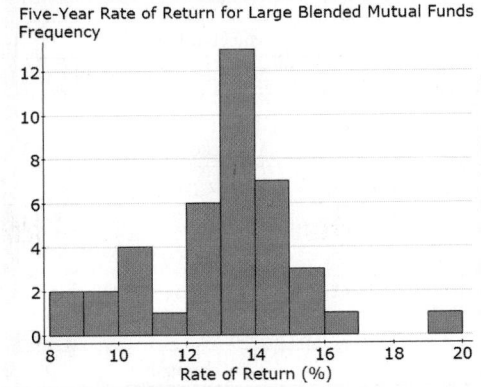
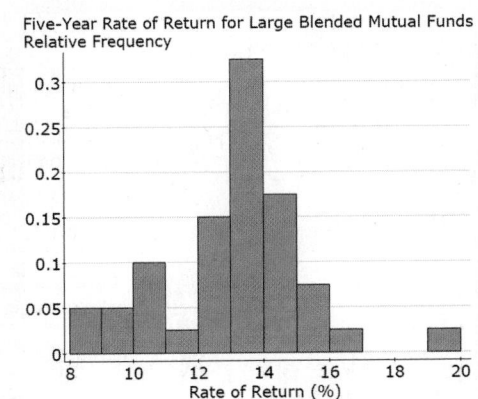

● **Now Work Problems 27(c)–(d)**

Using technology to construct histograms is a convenient and efficient way to explore patterns in data using different class widths.

⑤ Draw Stem-and-Leaf Plots

A **stem-and-leaf plot** is another way to represent quantitative data graphically. In a stem-and-leaf plot (or *stem plot*), use the digits to the left of the rightmost digit to form the **stem**. Each rightmost digit forms a **leaf**. For example, a data value of 147 would have 14 as the stem and 7 as the leaf.

EXAMPLE 6   Constructing a Stem-and-Leaf Plot

Problem The data in Table 14 represent the two-year average percentage of persons living in poverty, by state, for the years 2012–2013. Draw a stem-and-leaf plot of the data.

Approach

Step 1   Treat the integer portion of the number as the stem and the decimal portion as the leaf. For example, the stem for Alabama will be 16 and the leaf will be 4. The stem of 16 will include all data from 16.0 to 16.9.

Step 2   Write the stems vertically in ascending order, and then draw a vertical line to the right of the stems.

Step 3   Write the leaves corresponding to the stem.

Step 4   Within each stem, rearrange the leaves in ascending order. Title the plot and include a legend to indicate what the values represent.

Solution

Step 1   The stem for Alabama is 16 and the corresponding leaf is 4. The stem for Alaska is 10 and its leaf is 5, and so on.

Table 14					
Two-Year Average Percentage of Persons Living in Poverty (2012–2013)					
State	**Percent**	**State**	**Percent**	**State**	**Percent**
Alabama	16.4	Kentucky	18.9	North Dakota	10.6
Alaska	10.5	Louisiana	20.2	Ohio	14.5
Arizona	19.6	Maine	12.5	Oklahoma	16.0
Arkansas	18.6	Maryland	10.1	Oregon	14.3
California	15.4	Massachusetts	11.6	Pennsylvania	13.1
Colorado	11.2	Michigan	14.1	Rhode Island	13.6
Connecticut	10.8	Minnesota	11.0	South Carolina	16.3
Delaware	13.7	Mississippi	22.2	South Dakota	11.5
D.C.	19.9	Missouri	14.5	Tennessee	18.4
Florida	15.1	Montana	14.0	Texas	16.9
Georgia	17.2	Nebraska	11.6	Utah	9.6
Hawaii	12.5	Nevada	16.6	Vermont	10.0
Idaho	13.7	New Hampshire	8.6	Virginia	10.5
Illinois	12.9	New Jersey	10.2	Washington	11.8
Indiana	13.4	New Mexico	21.0	West Virginia	17.0
Iowa	10.5	New York	15.9	Wisconsin	11.2
Kansas	13.6	North Carolina	17.9	Wyoming	10.7

Source: U.S. Census Bureau, *Income and Poverty in the United States, 2013*

Step 2   Since the lowest data value is 8.6 and the highest data value is 22.2, let the stems range from 8 to 22. Write the stems vertically in Figure 10(a), along with a vertical line to the right of the stem.

Step 3   Write the leaves corresponding to each stem. See Figure 10(b).

Step 4   Rearrange the leaves in ascending order, give the plot a title, and add a legend. See Figure 10(c).

Figure 10

In Other Words
The choice of the stem in the construction of a stem-and-leaf plot is also an art form. It acts just like the class width. For example, the stem of 8 in Figure 10 represents the class 8.0–8.9. The stem of 9 represents the class 9.0–9.9. Notice that the class width is 1.0. The number of leaves is the frequency of each category.

Percentage of Persons Living in Poverty

8	8 | 6	8 | 6
9	9 | 6	9 | 6
10	10 | 5 8 5 1 2 6 0 5 7	10 | 0 1 2 5 5 5 6 7 8
11	11 | 2 6 0 6 5 8 2	11 | 0 2 2 5 6 6 8
12	12 | 5 9 5	12 | 5 5 9
13	13 | 7 7 4 6 1 6	13 | 1 4 6 6 7 7
14	14 | 1 5 0 5 3	14 | 0 1 3 5 5
15	15 | 4 1 9	15 | 1 4 9
16	16 | 4 6 0 3 9	16 | 0 3 4 6 9
17	17 | 2 9 0	17 | 0 2 9
18	18 | 6 9 4	18 | 4 6 9
19	19 | 6 9	19 | 6 9
20	20 | 2	20 | 2
21	21 | 0	21 | 0
22	22 | 2	22 | 2

Legend: 8\|6 represents 8.6%

(a)   (b)   (c)

The following summarizes the method for constructing a stem-and-leaf plot.

Construction of a Stem-and-Leaf Plot

Step 1   The stem of a data value will consist of the digits to the left of the rightmost digit. The leaf of a data value will be the rightmost digit.

Step 2   Write the stems in a vertical column in increasing order. Draw a vertical line to the right of the stems.

Step 3   Write each leaf corresponding to the stems to the right of the vertical line.

Step 4   Within each stem, rearrange the leaves in ascending order, title the plot, and include a legend to indicate what the values represent.

EXAMPLE 7   Constructing a Stem-and-Leaf Plot Using Technology

Problem   Construct a stem-and-leaf plot of the poverty data discussed in Example 6.

Approach   We will use Minitab. The steps for constructing the graphs using Minitab or StatCrunch are given in the Technology Step-by-Step on pages 86–87. **Note:** The TI graphing calculators and Excel are not capable of drawing stem-and-leaf plots.

Solution   Figure 11 on the following page shows the stem-and-leaf plot obtained from Minitab.

(continued)

■■ **Using Technology**

In Minitab, there is a column of numbers left of the stem. The (6) indicates that there are 6 observations in the class containing the middle value (called the *median*). The values above the (6) represent the number of observations less than or equal to the upper class limit of the class. For example, 11 states have percentages in poverty less than or equal to 10.9. The values in the left column below the (6) indicate the number of observations greater than or equal to the lower class limit of the class. For example, 19 states have percentages in poverty greater than or equal to 15.0.

• **Now Work Problem 33(a)**

Figure 11

```
Stem-and-leaf of Poverty Rates   N = 51
Leaf Unit = 0.10

    1    8   6
    2    9   6
   11   10   012555678
   18   11   0225668
   21   12   559
  (6)   13   146677
   24   14   01355
   19   15   149
   16   16   03469
   11   17   029
    8   18   469
    5   19   69
    3   20   2
    2   21   0
    1   22   2
```

Notice that the stem-and-leaf plot looks much like a histogram turned on its side. The stem serves as the class. For example, the stem 10 contains all data from 10.0 to 10.9. The leaves represent the frequency (height of the rectangle). Therefore, it is important to space the leaves evenly.

One advantage of the stem-and-leaf plot over frequency distributions and histograms is that the raw data can be retrieved from the stem-and-leaf plot. So, in a stem-and-leaf plot, we could determine the maximum observation. We cannot learn this information from a histogram.

On the other hand, stem-and-leaf plots lose their usefulness when data sets are large or consist of a large range of values. In addition, the steps listed for creating stem-and-leaf plots sometimes must be modified to meet the needs of the data. See Problem 37.

A limitation of the stem-and-leaf plots is that we are effectively forced to use a "class width" of 1.0 even though a larger "width" may be more desirable. This illustrates that we must weigh the advantages against the disadvantages when choosing the type of graph to summarize data.

Split Stems

The data in Table 15 range from 11 to 48. Figure 12 shows a stem-and-leaf plot using the tens digit as the stem and the ones digit as the leaf. The data appear rather bunched. To resolve this problem, we can use **split stems.** For example, rather than using one stem for the class of data 10–19, we could use two stems, one for the 10–14 interval and the second for the 15–19 interval. We do this in Figure 13.

In Other Words
Stem-and-leaf plots are best used when the data set is small and the range of values is not too wide.

In Other Words
Using split stems is like adding more classes to a frequency distribution.

Table 15				
27	17	11	24	36
13	29	22	18	17
23	30	12	46	17
32	48	11	18	23
18	32	26	24	38
24	15	13	31	22
18	21	27	20	16
15	37	19	19	29

Figure 12

```
1 | 1 1 2 3 3 5 5 6 7 7 7 8 8 8 8 9 9
2 | 0 1 2 2 3 3 4 4 4 6 7 7 9 9
3 | 0 1 2 2 6 7 8
4 | 6 8
```

Legend: 1|1 represents 11

Figure 13

```
1 | 1 1 2 3 3
1 | 5 5 6 7 7 7 8 8 8 8 9 9
2 | 0 1 2 2 3 3 4 4 4
2 | 6 7 7 9 9
3 | 0 1 2 2
3 | 6 7 8
4 |
4 | 6 8
```

Legend: 1|1 represents 11

The stem-and-leaf plot shown in Figure 13 reveals the distribution of the data better. As with the determination of class intervals in the creation of frequency histograms, judgment plays a major role. There is no such thing as the correct stem-and-leaf plot. However, a quick comparison of Figures 12 and 13 shows that some plots are better than others.

● Now Work Problem 39

⑥ Draw Dot Plots

One more graph! We draw a **dot plot** by placing each observation horizontally in increasing order and placing a dot above the observation each time it is observed.

EXAMPLE 8   Drawing a Dot Plot

Problem Draw a dot plot for the number of arrivals at Wendy's data from Table 8 on page 76.

Approach The smallest observation in the data set is 1 and the largest is 11. Write the numbers 1 through 11 horizontally. For each observation, place a dot above the value of the observation.

Solution Figure 14 shows the dot plot.

Figure 14

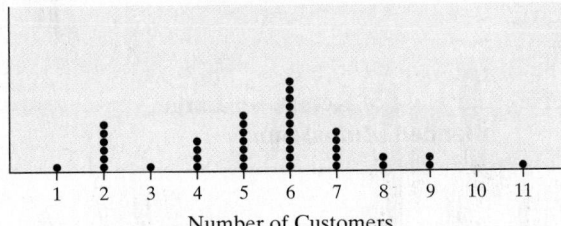

Arrivals at Wendy's

Number of Customers

● Now Work Problem 45

⑦ Identify the Shape of a Distribution

One way that a variable is described is through the shape of its distribution. Distribution shapes are typically classified as symmetric, skewed left, or skewed right. Figure 15 on the following page displays various histograms and the shape of the distribution.

Figures 15(a) and (b) show symmetric distributions. They are symmetric because, if we split the histogram down the middle, the right and left sides are mirror images. Figure 15(a) is a **uniform distribution** because the frequency of each value of the variable is evenly spread out across the values of the variable. Figure 15(b) displays a **bell-shaped distribution** because the highest frequency occurs in the middle and frequencies tail off to the left and right of the middle. That is, the graph looks like the profile of all bell. The distribution in Figure 15(c) is **skewed right**. Notice that the tail to the right of the peak is longer than the tail to the left of the peak. Finally, Figure 15(d) illustrates a distribution that is **skewed left**, because the tail to the left of the peak is longer than the tail to the right of the peak.

Figure 15

CAUTION!
We do not describe qualitative data as skewed left, skewed right, or uniform.

CAUTION!
It is important to recognize that data will not always exhibit behavior that perfectly matches any of the shapes given in Figure 15. To identify the shape of a distribution, some flexibility is required. In addition, people may disagree on the shape, since identifying shape is subjective.

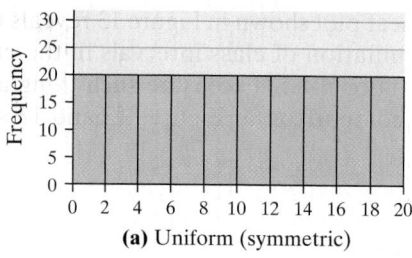

(a) Uniform (symmetric)

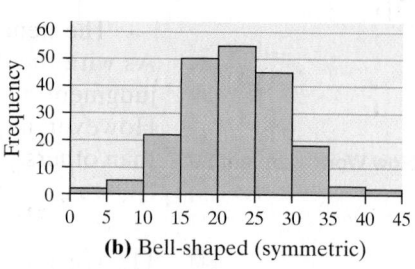

(b) Bell-shaped (symmetric)

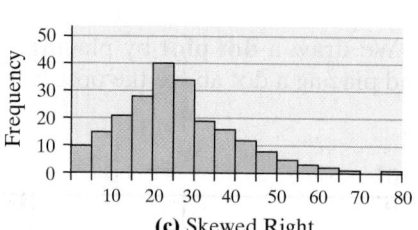

(c) Skewed Right

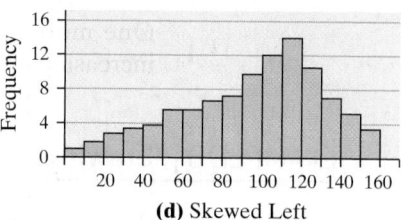

(d) Skewed Left

EXAMPLE 9   Identifying the Shape of a Distribution

Problem Figure 16 displays the histogram obtained in Example 4 for the five-year rate of return for large-blended mutual funds. Describe the shape of the distribution.

Approach We compare the shape of the distribution displayed in Figure 16 with those in Figure 15.

Solution Since the histogram looks most like Figure 15(b), the distribution is fairly symmetric.

Figure 16

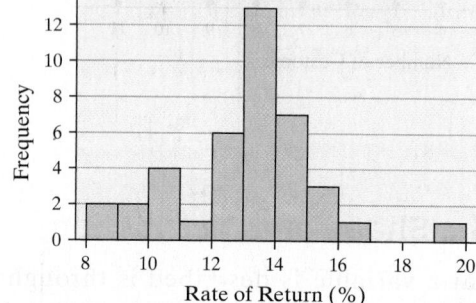

Five-Year Rate of Return for Large Blended Mutual Funds

● Now Work Problem 27(e)

Technology Step-by-Step   Drawing Histograms, Stem-and-Leaf Plots, and Dot Plots

TI-83/84 Plus
Histograms
1. Enter the raw data in L1 by pressing STAT and selecting 1: Edit.
2. Press 2nd Y = to access the StatPlot menu. Select 1: Plot1.
3. Place the cursor on "ON" and press ENTER.
4. Place the cursor on the histogram icon (see the figure) and press ENTER. Press 2nd QUIT to exit Plot 1 menu.
5. Press WINDOW. Set Xmin to the lower class limit of the first class. Set Xmax to the lower class limit of the class

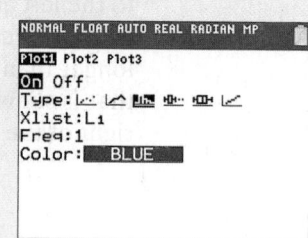

following the class containing the largest value. For example, if the first class is 0–9, set Xmin to 0. If the class width is 10 and the last class is 90–99, set Xmax to 100. Set Xscl to the class width. Set Ymin to 0. Set Ymax to a value larger than the frequency of the class with the highest frequency.

6. Press GRAPH.

Helpful Hints: To determine each class frequency, press TRACE and use the arrow keys to scroll through each class. If you decrease the value of Ymin to a value such as −5, you can see the values displayed on the screen easier. The TI graphing calculators do not draw stem-and-leaf plots or dot plots.

Minitab
Histograms
1. Enter the raw data in C1.
2. Select the **Graph** menu and then **Histogram** . . .
3. Highlight the "simple" icon and press OK.
4. Put the cursor in the "Graph variables" box. Highlight C1 and press Select. Click SCALE and select the Y-Scale Type tab. For a frequency histogram, click the frequency radio button. For a relative frequency histogram, click the percent radio button. Click OK twice.

Note: To adjust the class width and to change the labels on the horizontal axis to the lower class limit, double-click inside one of the bars in the histogram. Select the "binning" tab in the window that opens. Click the cutpoint button and the midpoint/cutpoint positions radio button. In the midpoint/cutpoint box, enter the lower class limits of each class. Click OK.

Stem-and-Leaf Plots
1. With the raw data entered in C1, select the **Graph** menu and then **Stem-and-Leaf**.
2. Select the data in C1 and press OK.

Dot Plots
1. Enter the raw data in C1.
2. Select the **Graph** menu and then **Dotplot**.
3. Highlight the "simple" icon and press OK.
4. Put the cursor in the "Graph variables" box. Highlight C1 and press Select. Click OK.

Excel
Histograms
1. Load the XLSTAT Add-in.
2. Enter the raw data in column A.
3. Select XLSTAT. Click Describing data, then select Histograms.
4. With the cursor in the Data cell, highlight the data in Column A.
5. Click either the Continuous or Discrete radio button.
6. Click the Options tab. Decide on either a certain number of intervals or enter your own lower class limits. To enter your own intervals, enter the lower class limits in Column B.
7. Click the Charts tab. Choose either Frequency or Relative Frequency. Click OK.

StatCrunch
Frequency and Relative Frequency Distributions of Discrete Data
1. Enter the raw data into the spreadsheet. Name the column variable.
2. Select **Stat**, highlight **Tables**, and select **Frequency**.
3. Click on the variable you wish to summarize. Click the Type of table you want. If you want both Frequency and Relative Frequency, highlight Frequency, then press Ctrl (or Command on an Apple) and select Relative frequency. Click Compute!.

Frequency and Relative Frequency Distributions of Continuous Data
1. If necessary, enter the raw data into the spreadsheet. Name the column variable.
2. Select **Data** and then **Bin**.
3. Click the variable you wish to summarize. Click the "Use fixed width bins" radio button. Enter the lower class limit of the first class in the "Start at:" cell. Enter the class width in the "Bin width:" cell. Leave the "Include left endpoint" radio button selected. Click Compute!.
4. Select **Stat** and highlight **Tables**, then **Frequency**.
5. Choose the Bin(column name) variable. Under Type:, select Frequency and Relative Frequency. Click Compute!

Histograms
1. If necessary, enter the raw data into the spreadsheet. Name the column variable.
2. Select **Graph** and then **Histogram**.
3. Click on the variable you wish to summarize. Choose the type of histogram (frequency or relative frequency). You have the option of choosing a lower class limit for the first class by entering a value in the cell marked "Bins: Start at:" You have the option of choosing a class width by entering a value in the cell marked "Bins: Width:" Enter labels for the X-axis and Y-axis. Enter a title for the graph. Click Compute!.

Stem-and-Leaf Plots
1. If necessary, enter the raw data into the spreadsheet. Name the column variable.
2. Select **Graph** and then **Stem and Leaf**.
3. Click on the variable you wish to summarize. Select None for outlier trimming. You also have the option of selecting the leaf unit from the drop-down menu. Click Compute!.

Dot Plots
1. If necessary, enter the raw data into the spreadsheet. Name the column variable.
2. Select **Graph** and then **Dotplot**.
3. Click on the variable you wish to summarize. Enter labels for the X-axis and Y-axis. Enter a title for the graph. Click Compute!.

2.2 Assess Your Understanding

Vocabulary and Skill Building

1. The categories by which data are grouped are called _____.

2. The _____ class limit is the smallest value within the class and the _____ class limit is the largest value within the class.

3. The _____ _____ is the difference between consecutive lower class limits.

4. What does it mean if a distribution is said to be "skewed left"?

5. *True or False*: There is not one particular frequency distribution that is correct, but there are frequency distributions that are less desirable than others.

6. *True or False*: Stem-and-leaf plots are particularly useful for large sets of data.

7. *True or False*: The shape of the distribution shown is best classified as skewed left.

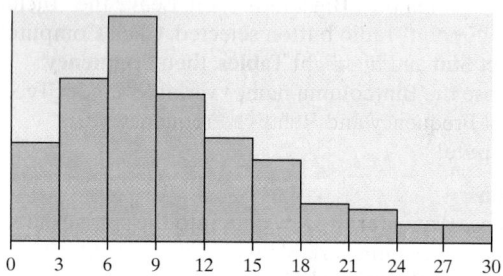

8. *True or False*: The shape of the distribution shown is best classified as uniform.

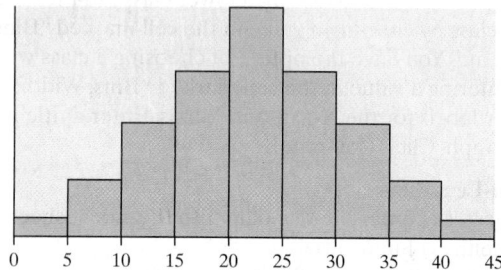

9. Rolling the Dice An experiment was conducted in which two fair dice were thrown 100 times. The sum of the pips showing on the dice was then recorded. The following frequency histogram gives the results.

Sum of Two Dice

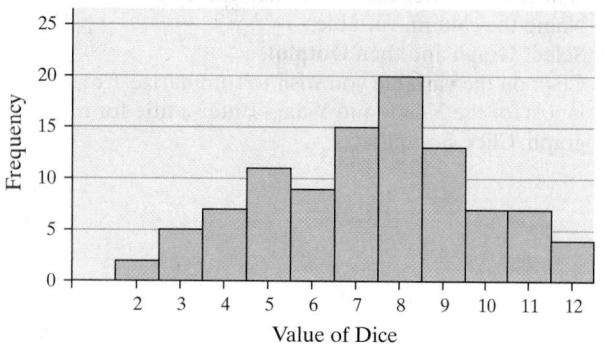

(a) What was the most frequent outcome of the experiment?
(b) What was the least frequent?
(c) How many times did we observe a 7?
(d) How many more 5's were observed than 4's?
(e) Determine the percentage of time a 7 was observed.
(f) Describe the shape of the distribution.

10. Car Sales A car salesman records the number of cars he sold each week for the past year. The following frequency histogram shows the results.

Cars Sold per Week

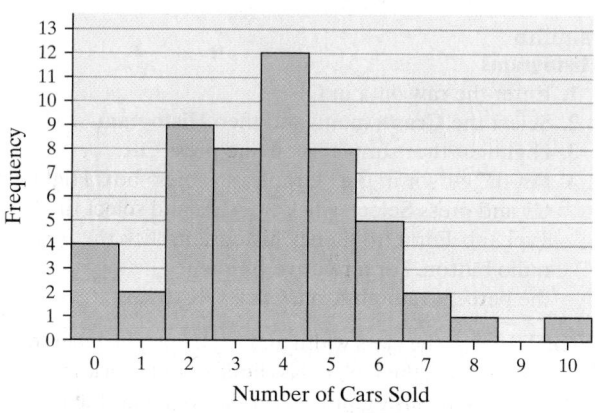

(a) What is the most frequent number of cars sold in a week?
(b) For how many weeks were two cars sold?
(c) Determine the percentage of time two cars were sold.
(d) Describe the shape of the distribution.

11. IQ Scores The following frequency histogram represents the IQ scores of a random sample of seventh-grade students. IQs are measured to the nearest whole number. The frequency of each class is labeled above each rectangle.

IQs of 7th-Grade Students

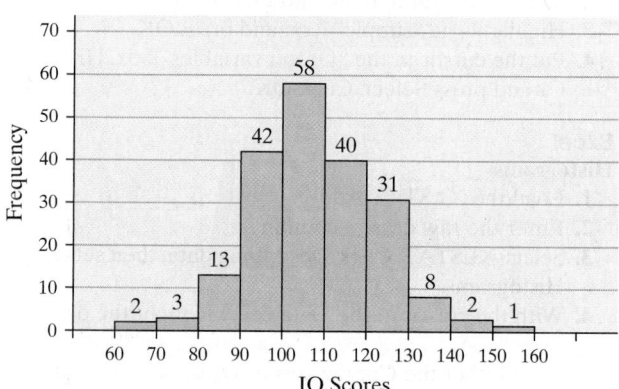

(a) How many students were sampled?
(b) Determine the class width.
(c) Identify the classes and their frequencies.
(d) Which class has the highest frequency?
(e) Which class has the lowest frequency?
(f) What percent of students had an IQ of at least 130?
(g) Did any students have an IQ of 165?

12. Alcohol-Related Traffic Fatalities The frequency histogram represents the number of alcohol-related traffic fatalities by state (including Washington, D.C.) in 2012 according to Mothers Against Drunk Driving.

Alcohol-Related Traffic Fatalities, 2012

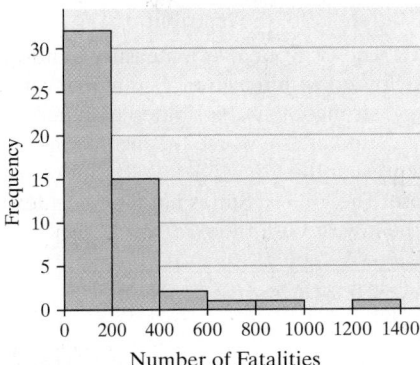

(a) Determine the class width.
(b) Identify the classes.
(c) Which class has the highest frequency?
(d) Describe the shape of the distribution.
(e) A reporter writes the following statement: "According to the data, Texas had 1296 alcohol-related deaths, while Vermont had only 23. So the roads in Vermont are much safer." Explain what is wrong with this statement and how a fair comparison can be made between alcohol-related traffic fatalities in Texas versus Vermont.

In Problems 13 and 14, for each variable presented, state whether you would expect a histogram of the data to be bell-shaped, uniform, skewed left, or skewed right. Justify your reasoning.

13. (a) Annual household incomes in the United States
   (b) Scores on a standardized exam such as the SAT
   (c) Number of people living in a household
   (d) Ages of patients diagnosed with Alzheimer's disease

14. (a) Number of alcoholic drinks consumed per week
   (b) Ages of students in a public school district
   (c) Ages of hearing-aid patients
   (d) Heights of full-grown men

Applying the Concepts

15. Predicting School Enrollment To predict future enrollment in a school district, fifty households within the district were sampled, and asked to disclose the number of children under the age of five living in the household. The results of the survey are presented in the following table.

Number of Children under 5	Number of Households
0	16
1	18
2	12
3	3
4	1

(a) Construct a relative frequency distribution of the data.
(b) What percentage of households has two children under the age of 5?
(c) What percentage of households has one or two children under the age of 5?

16. Free Throws In an experiment, a researcher asks a basketball player to record the number of free throws she shoots until she misses. The experiment is repeated 50 times. The following table lists the distribution of the number of free throws attempted until a miss is recorded.

Number of Free Throws until a Miss	Frequency
1	16
2	11
3	9
4	7
5	2
6	3
7	0
8	1
9	0
10	1

(a) Construct a relative frequency distribution of the data.
(b) What percentage of the time did she first miss on her fourth free throw?
(c) What percentage of the time did she first miss the tenth free throw?
(d) What percentage of the time did she make at least five in a row?

In Problems 17–20, determine the original set of data.

17.
```
1 | 0 1 4
2 | 1 4 4 7 9
3 | 3 5 5 5 7 7 8
4 | 0 0 1 2 6 6 8 9 9
5 | 3 3 5 8
6 | 1 2
```
Legend: 1|0 represents 10

18.
```
24 | 0 4 7
25 | 2 2 3 9 9
26 | 3 4 5 8 8 9
27 | 0 1 1 3 6 6
28 | 2 3 8
```
Legend: 24|0 represents 240

19.
```
1 | 2 4 6
2 | 1 4 7 7 9
3 | 3 3 5 7 7 8
4 | 0 1 1 3 6 6 8 8 9
5 | 3 4 5 8
6 | 2 4
```
Legend: 1|2 represents 1.2

20.
```
12 | 3 7 9 9
13 | 0 4 5 7 8 9 9
14 | 2 4 4 7 7 8 9
15 | 1 2 2 5 6
16 | 0 3
```
Legend: 12|3 represents 12.3

In Problems 21 and 22, find (a) the number of classes, (b) the class limits, and (c) the class width.

21. World Cup The following data represent the top speed (in kilometers per hour) of all the players (except goaltenders) in the 2014 World Cup Soccer Tournament.

Speed (km/hr)	Number of Players
10–13.9	4
14–17.9	7
18–21.9	17
22–25.9	91
26–29.9	282
30–33.9	206

Source: www.fifa.com

22. Earthquakes The following data represent the magnitude of earthquakes worldwide in October 2014.

Magnitude	Number
0–0.9	3371
1.0–1.9	3400
2.0–2.9	1237
3.0–3.9	286
4.0–4.9	1045
5.0–5.9	121
6.0–6.9	7
7.0–7.9	2

Source: U.S. Geological Survey, Earthquake Hazards Program

In Problems 23 and 24, construct (a) a relative frequency distribution, (b) a frequency histogram, and (c) a relative frequency histogram for the given data. Then answer the questions that follow.

23. Use the data in Problem 21. What percentage of players had a top speed between 30 and 33.9 km/h? What percentage of players had a top speed less than 13.9 km/h?

24. Use the data in Problem 22. What percentage of earthquakes registered 4.0 to 4.9? What percentage of earthquakes registered 4.9 or less?

NW 25. Televisions in the Household A researcher with A. C. Nielsen wanted to determine the number of televisions in households. He conducts a survey of 40 randomly selected households and obtains the following data.

1	1	4	2	3	3	5	1
1	2	2	4	1	1	0	3
1	2	2	1	3	1	1	3
2	3	2	2	1	2	3	2
1	2	2	2	2	1	3	1

Source: Based on data from the U.S. Department of Energy

(a) Are these data discrete or continuous? Explain.
(b) Construct a frequency distribution of the data.
(c) Construct a relative frequency distribution of the data.
(d) What percentage of households in the survey have three televisions?
(e) What percentage of households in the survey have four or more televisions?
(f) Construct a frequency histogram of the data.
(g) Construct a relative frequency histogram of the data.
(h) Describe the shape of the distribution.

26. Waiting The data below represent the number of customers waiting for a table at 6:00 P.M. for 40 consecutive Saturdays at Bobak's Restaurant.

11	5	11	3	6	8	6	7
4	5	13	9	6	4	14	11
13	10	9	6	8	10	9	5
10	8	7	3	8	8	7	8
7	9	10	4	8	6	11	8

(a) Are these data discrete or continuous? Explain.
(b) Construct a frequency distribution of the data.
(c) Construct a relative frequency distribution of the data.
(d) What percentage of the Saturdays had 10 or more customers waiting for a table at 6:00 P.M.?
(e) What percentage of the Saturdays had five or fewer customers waiting for a table at 6:00 P.M.?
(f) Construct a frequency histogram of the data.
(g) Construct a relative frequency histogram of the data.
(h) Describe the shape of the distribution.

NW 27. Gini Index The Gini Index is a measure of how evenly income is distributed within a country, ranging from 0 to 100. An index of 0 suggests income is distributed with perfect equality. The higher the number, the worse the income inequality. The data below represent the Gini Index for a random sample of countries. **Note:** The United States has a Gini Index of 45 and Sweden has the lowest Gini Index.

23.0	27.0	30.0	32.0	34.0	36.5	38.5	40.0	41.9	45.0	47.2	50.4	55.1
23.8	27.2	30.3	32.1	34.1	36.7	39.0	40.0	42.4	45.3	47.4	50.8	57.7
24.3	27.4	30.6	32.6	34.2	36.8	39.0	40.1	42.5	45.3	47.5	50.9	58.5
24.7	28.0	30.7	32.7	34.4	36.8	39.0	40.2	43.2	45.5	47.7	51.9	59.2
24.8	28.0	30.9	32.8	34.5	36.8	39.0	40.5	43.7	45.6	47.8	51.9	59.7
25.0	28.2	30.9	33.0	35.2	37.6	39.2	40.8	44.3	45.8	48.3	52.1	61.3
26.0	28.2	31.0	33.2	35.3	37.6	39.4	40.9	44.5	46.0	49.0	53.0	62.9
26.0	28.9	31.3	33.2	35.5	37.6	39.4	41.3	44.6	46.2	50.1	53.2	63.0
26.0	29.0	31.9	33.4	36.2	37.7	39.5	41.5	44.6	46.8	50.2	53.6	63.1
26.3	29.6	31.9	33.7	36.2	37.9	39.7	41.7	44.8	46.9	50.3	53.7	63.2
26.8	30.0	32.0	33.9	36.5	38.0						

Source: CIA World Factbook

With a first class having a lower class limit of 20 and a class width of 5:

(a) Construct a frequency distribution.
(b) Construct a relative frequency distribution.
(c) Construct a frequency histogram of the data.
(d) Construct a relative frequency histogram of the data.
(e) Describe the shape of the distribution.
(f) Repeat parts (a)–(e) using a class width of 10.
(g) Does one frequency distribution provide a better summary of the data than the other? Explain.

28. Average Income The following data represent the median household income (in dollars) for the 50 states and the District of Columbia in 2013.

39,622	42,499	48,801	52,888	55,258	60,907	67,620
39,919	43,749	50,121	53,027	55,700	61,137	67,781
40,241	43,777	50,311	53,774	56,307	61,408	71,322
40,850	44,132	50,553	53,843	57,196	61,782	
41,208	45,369	50,602	53,952	57,528	62,963	
41,381	46,398	51,485	54,453	57,812	62,967	
42,127	47,439	51,767	54,842	60,106	63,371	
42,158	47,886	52,219	54,855	60,675	65,262	

Source: U.S. Census Bureau

With the first class having a lower class limit of 35,000 and a class width of 5000:

(a) Construct a frequency distribution.
(b) Construct a relative frequency distribution.
(c) Construct a frequency histogram of the data.
(d) Construct a relative frequency histogram of the data.

(e) Describe the shape of the distribution.
(f) Repeat parts (a)–(e) using a class width of 10,000.
(g) Does one frequency distribution provide a better summary of the data than the other? Explain.

29. Cigarette Tax Rates The table shows the tax, in dollars, on a pack of cigarettes in each of the 50 states and Washington, DC, as of January 2014. **Note:** The state with the lowest tax is Missouri and the state with the highest tax is New York.

0.425	1.339	0.60	0.17	0.45	1.53	2.52
2.00	0.37	0.36	1.70	0.44	0.62	0.60
2.00	3.20	2.00	0.64	1.25	1.41	2.50
1.15	0.57	2.00	0.80	1.03	1.70	
0.87	1.98	3.51	1.68	1.31	2.62	
0.84	0.995	2.00	2.70	1.60	0.30	
3.40	1.36	2.83	1.66	3.50	3.025	
1.60	0.79	0.68	4.35	0.57	0.55	

Source: Tax Foundation

With a first class having a lower class limit of 0 and a class width of 0.50:

(a) Construct a frequency distribution.
(b) Construct a relative frequency distribution.
(c) Construct a frequency histogram of the data.
(d) Construct a relative frequency histogram of the data.
(e) Describe the shape of the distribution.
(f) Repeat parts (a)–(e) using a class width of 1.
(g) Does one frequency distribution provide a better summary of the data than the other? Explain.

30. Dividend Yield A dividend is a payment from a publicly traded company to its shareholders. The dividend yield of a stock is determined by dividing the annual dividend of a stock by its price. The following data represent the dividend yields (in percent) of a random sample of 28 publicly traded stocks of companies with a value of at least $5 billion.

1.7	0	1.15	0.62	1.06	2.45	2.38
2.83	2.16	1.05	1.22	1.68	0.89	0
2.59	0	1.7	0.64	0.67	2.07	0.94
2.04	0	0	1.35	0	0	0.41

Source: Yahoo! Finance

With the first class having a lower class limit of 0 and a class width of 0.40:

(a) Construct a frequency distribution.
(b) Construct a relative frequency distribution.
(c) Construct a frequency histogram of the data.
(d) Construct a relative frequency histogram of the data.
(e) Describe the shape of the distribution.
(f) Repeat parts (a)–(e) using a class width of 0.8.
(g) Which frequency distribution seems to provide a better summary of the data?

NW 31. Violent Crimes Violent crimes include murder, forcible rape, robbery, and aggravated assault. The data in the next column represent the violent-crime rate (crimes per 100,000 population) by state plus the District of Columbia in 2013.

449.9	1243.7	354.6	260.8	406.8	558.8	316.3
603.2	487.1	222.6	450.9	353.4	321.8	280.5
428.9	378.9	496.9	272.2	244.7	643.6	201.4
469.1	239.2	122.7	259.4	299.7	408.6	
423.1	207.9	476.8	607.6	469.3	205.8	
308.9	414.8	405.5	187.9	247.6	142.6	
283.0	345.7	454.5	290.2	348.7	190.1	
547.4	263.9	230.9	559.1	252.4	295.6	

Source: Federal Bureau of Investigation

(a) If seven classes are to be formed, choose an appropriate lower class limit for the first class and a class width.
(b) Construct a frequency distribution.
(c) Construct a relative frequency distribution.
(d) Construct a frequency histogram of the data.
(e) Construct a relative frequency histogram of the data.
(f) Describe the shape of the distribution.

32. Volume of Altria Group Stock The volume of a stock is the number of shares traded on a given day. The following data, in millions, so that 6.42 represents 6,420,000 shares traded, represent the volume of Altria Group stock traded for a random sample of 35 trading days in 2014.

6.42	23.59	18.91	7.85	7.76
8.51	9.05	14.83	14.43	8.55
6.37	10.30	10.16	10.90	11.20
13.57	9.13	7.83	15.32	14.05
7.84	7.88	17.10	16.58	7.68
7.69	10.22	10.49	8.41	7.85
10.94	20.15	8.97	15.39	8.32

Source: TD Ameritrade

(a) If six classes are to be formed, choose an appropriate lower class limit for the first class and a class width.
(b) Construct a frequency distribution.
(c) Construct a relative frequency distribution.
(d) Construct a frequency histogram of the data.
(e) Construct a relative frequency histogram of the data.
(f) Describe the shape of the distribution.

In Problems 33–36, (a) construct a stem-and-leaf plot and (b) describe the shape of the distribution.

NW 33. Age at Inauguration The following data represent the ages of the presidents of the United States (from George Washington through Barack Obama) on their first days in office.

Note: President Cleveland's age is listed twice, 47 and 55, because he is historically counted as two different presidents, numbers 22 and 24, since his terms were not consecutive.

42	47	50	52	54	55	57	61	64
43	48	51	52	54	56	57	61	65
46	49	51	54	55	56	57	61	68
46	49	51	54	55	56	58	62	69
47	50	51	54	55	57	60	64	

Source: factmonster.com

34. Divorce Rates The data on the next page represent the divorce rate (per 1000 population) for most states in the United States in the year 2011.

Note: The list includes the District of Columbia, but excludes California, Georgia, Hawaii, Indiana, Louisiana, and Minnesota because of failure to report.

4.3	4.1	3.2	3.3	3.2
2.6	3.9	2.9	3.3	4.5
4.0	3.9	4.4	3.6	4.0
5.2	5.6	4.2	2.7	2.7
3.8	2.8	2.9	2.9	3.7
4.8	5.2	3.2	4.3	4.9
2.4	5.3	4.8	2.9	3.9
3.5	4.4	3.1	3.4	3.4
3.8	3.8	2.9	3.7	3.6

Source: U.S. Census Bureau

35. Grams of Fat in a McDonald's Breakfast The following data represent the number of grams of fat in breakfast meals offered at McDonald's.

12	22	27	3	25	30	32	37
27	31	11	16	21	32	22	46
51	55	59	16	36	30	9	24

Source: McDonald's Corporation, *McDonald's USA Nutrition Facts*

36. Gasoline Mileages The following data represent the number of miles per gallon achieved on the highway for small cars for the model year 2014.

35	34	35	34	36	35	38	37	40
35	34	34	36	29	36	43	36	34
36	34	36	35	33	33	29	28	25
35	36	37	36	35	30	36	34	36
34	32	32	31	31	33	33	33	33
31	30	25	23	22	34	31	30	28
34	34	31	33	32	31	31	33	33
27	26	25	23	35	34	32	33	32
30	36	35	33	31	42	42	34	31
33	29	29	30					

Source: fueleconomy.gov

37. Five-Year Rate of Return Sometimes, data must be modified before a stem-and-leaf plot may be constructed. For example, the five-year rate of return data in Table 12 on page 79 is reported to the hundredth decimal place. So, if we used the integer portion of the data as the stem and the decimals as the leaves, then the stems would be 8, 9, 10, ..., 19; but the leaves would be two digits (such as 94, 53, and so on). This is not acceptable since each leaf must be a single digit. To resolve this problem, we round the data to the nearest tenth.

(a) Round the data to the nearest tenth.
(b) Draw a stem-and-leaf plot of the modified data.
(c) Describe the shape of the distribution.

38. Home Appreciation The following data represent the price appreciation in home values between the first quarter of 1991 and the second quarter of 2014 for homes in each of the 50 states plus the District of Columbia. **Note:** The best price appreciation was in the District of Columbia and the worst was in Ohio.

68.95	75.41	104.80	112.66	93.69	87.21	133.24
112.60	207.11	136.09	70.58	117.37	109.16	66.58
349.76	148.61	104.67	209.52	87.45	101.01	80.48
185.08	129.62	93.28	127.05	96.44	107.03	
135.91	121.85	94.46	104.26	109.17	113.11	
115.40	150.33	65.10	80.09	110.42	90.96	
85.00	134.50	104.01	118.06	97.26	125.32	
219.86	197.44	224.92	112.91	145.34	86.68	

Source: Federal Housing Finance Agency

(a) Round each observation to the nearest whole number and draw a stem-and-leaf plot.
(b) Describe the shape of the distribution.
(c) Do you think the stem-and-leaf plot does a good job of visually summarizing the data, or would you recommend a different graphical representation of the data? If you recommend a different graph, which would you recommend? Why?

NW 39. Violent Crimes Use the violent crime rate data from Problem 31 to answer each of the following:

(a) Round the data to the nearest ten (for example, round 603.2 as 600).
(b) Draw a stem-and-leaf plot, treating the hundreds position as the stem and the tens position as the leaf (so that 4|5 represents 450).
(c) Redraw the stem-and-leaf plot using split stems.
(d) In your opinion, which of these plots better summarizes the data? Why?

In Problems 40 and 41, we compare data sets. A great way to compare two data sets is through back-to-back stem-and-leaf plots. The figure represents the number of grams of fat in 20 sandwiches served at McDonald's and 20 sandwiches served at Burger King.

Fat (g) in Fast Food Sandwiches

McDonald's		Burger King
98	0	7
9 8 7 6 6 4 2 0	1	2 2 3 6 6 7
9 8 8 6 6 4 3 3 1	2	1 2 9
	3	0 3 9 9
2	4	4 7
	5	4 7
	6	5 8

Legend: 8|0|7 represents 8 g of fat for McDonald's and 7 g of fat for Burger King

Source: McDonald's Corporation, *McDonald's USA Nutrition Facts*, Burger King Corporation, *Nutritional Information*

40. Academy Award Winners The following data represent the ages on the ceremony date of the Academy Award winners for Best Actor and Best Actress in a leading role for the 37 years from 1978 to 2014.

Best Actor Ages									
30	40	42	37	76	39	50	55	53	45
36	62	43	51	48	44	32	42	54	52
37	38	60	32	45	60	46	40	36	50
47	29	43	37	38	45	39			

Best Actress Ages									
32	41	33	31	74	33	32	22	49	38
61	21	41	26	33	44	80	42	29	33
36	45	45	49	39	34	26	25	33	29
35	35	28	30	29	61	62			

(a) Construct a back-to-back stem-and-leaf plot.
(b) Compare the two populations. What can you conclude from the back-to-back stem-and-leaf plot?

41. Home Run Distances In 1998, Mark McGwire of the St. Louis Cardinals set the record for the most home runs hit in a season with 70 home runs. In 2001, Barry Bonds of the San Francisco Giants broke McGwire's record by hitting 73 home runs. The following data represent the distances, in feet, of each player's home runs in his record-setting season.

Mark McGwire						
360	370	370	430	420	340	460
410	440	410	380	360	350	527
380	550	478	420	390	420	425
370	480	390	430	388	423	410
360	410	450	350	450	430	461
430	470	440	400	390	510	430
450	452	420	380	470	398	409
385	369	460	390	510	500	450
470	430	458	380	430	341	385
410	420	380	400	440	377	370

Barry Bonds						
420	417	440	410	390	417	420
410	380	430	370	420	400	360
410	420	391	416	440	410	415
436	430	410	400	390	420	410
420	410	410	450	320	430	380
375	375	347	380	429	320	360
375	370	440	400	405	430	350
396	410	380	430	415	380	375
400	435	420	420	488	361	394
410	411	365	360	440	435	454
442	404	385				

(a) Construct a back-to-back stem-and-leaf plot.
(b) Compare the two populations. What can you conclude from the back-to-back stem-and-leaf plot?

42. Sullivan Survey Choose a continuous quantitative variable from the Sullivan Survey I data set at www.pearsonhighered.com/sullivanstats and obtain an appropriate graphical summary of the variable.

43. Return on Investment Payscale.com tracks the graduates of all institutions of higher education and determines the return on investment (ROI) of the school's graduates. ROI can be thought of as the investment return on the expenses associated with attending college. Go to www.pearsonhighered.com/sullivanstats and download the file 2_2_43. Construct a relative frequency histogram of the data. Comment on what you notice based on the graphical summary.

44. Sullivan Survey Draw a dot plot of the variable "ideal number of children" from the Sullivan Survey I data set at www.pearsonhighered.com/sullivanstats. Now draw a dot plot

of the variable "ideal number of children" from the survey data set by "gender." Do there appear to be any differences in the ideal number of children for males and females? Is there a better graph that could be drawn to make the comparison easier?

NW 45. Televisions in the Household Draw a dot plot of the televisions per household data from Problem 25.

46. Waiting Draw a dot plot of the waiting data from Problem 26.

47. Putting It Together: Time Viewing a Web Page Nielsen/NetRatings measures the amount of time an individual spends viewing a specific web page. The following data represent the amount of time, in seconds, a random sample of 40 surfers spent viewing a web page. Decide on an appropriate graphical summary and create the graphical summary. Write a few sentences that describe the data. Be sure to include in your description any interesting features the data may exhibit.

19	185	4	104	23
27	73	27	12	16
15	27	51	10	40
111	83	27	31	5
9	75	11	48	65
86	51	69	257	14
20	45	19	13	81
26	42	156	12	114

Source: Based on information provided by Nielsen/NetRatings

48. Putting It Together: Which Graphical Summary? Suppose you just obtained data from a survey in which you learned the following information about 50 individuals: age, income, marital status, number of vehicles in household. For each variable, explain the type of graphical summary you might be able to draw to provide a visual summary of the data.

Explaining the Concepts

49. Why shouldn't classes overlap when summarizing continuous data in a frequency or relative frequency distribution?

50. Discuss the advantages and disadvantages of histograms versus stem-and-leaf plots.

51. Is there such a thing as the correct choice for a class width? Is there such a thing as a poor choice for a class width? Explain your reasoning.

52. Describe the situations in which it is preferable to use relative frequencies over frequencies when summarizing quantitative data.

53. StatCrunch Choose any data set that has at least 50 observations of a quantitative variable. In StatCrunch, open the "Histogram with Sliders" applet. Adjust the bin width and starting point of the histogram. Can the choice of bin width (class width) affect the shape of the histogram? Explain.

54. Sketch four histograms—one skewed right, one skewed left, one bell-shaped, and one uniform. Label each histogram according to its shape. What makes a histogram skewed left? Skewed right? Symmetric?

2.3 Additional Displays of Quantitative Data

Objectives

1. Construct frequency polygons
2. Create cumulative frequency and relative frequency tables
3. Construct frequency and relative frequency ogives
4. Draw time-series graphs

In this section, we continue to organize and summarize quantitative data by presenting additional displays commonly used in statistical analysis: *frequency polygons*, *cumulative distributions*, *ogives*, and *time-series graphs*.

1 Construct Frequency Polygons

Another way to graphically represent quantitative data sets is through *frequency polygons*. They provide the same information as histograms. Before we can provide a method for constructing frequency polygons, we must learn how to obtain the *class midpoint* of a class.

Definition

A **class midpoint** is the sum of consecutive lower class limits divided by 2.

Definition

A **frequency polygon** is a graph that uses points, connected by line segments, to represent the frequencies for the classes. It is constructed by plotting a point above each class midpoint on a horizontal axis at a height equal to the frequency of the class. Next, line segments are drawn connecting consecutive points. Two additional line segments are drawn connecting each end of the graph with the horizontal axis.

To construct a frequency polygon of the data in Table 13 on page 79 (Section 2.2), calculate the class midpoints of each class, as shown in Table 16.

Table 16			
Class (Five-year rate of return)	**Class Midpoint**	**Frequency**	**Relative Frequency**
8–8.99	$\dfrac{8 + 9}{2} = 8.5$	2	0.05
9–9.99	$\dfrac{9 + 10}{2} = 9.5$	2	0.05
10–10.99	10.5	4	0.1
11–11.99	11.5	1	0.025
12–12.99	12.5	6	0.15
13–13.99	13.5	13	0.325
14–14.99	14.5	7	0.175
15–15.99	15.5	3	0.075
16–16.99	16.5	1	0.025
17–17.99	17.5	0	0
18–18.99	18.5	0	0
19–19.99	19.5	1	0.025

Using Technology
Statistical spreadsheets and certain graphing calculators have the ability to create frequency polygons.

Then plot points whose *x*-coordinates are the class midpoints and *y*-coordinates are the frequencies; connect them with line segments. Next determine the midpoint of the class preceding the first class to be 7.5 and the midpoint of the class following the last class to be 20.5. (Do you see why?) Finally, connect each end of the graph with the horizontal axis at $(7.5, 0)$ and $(20.5, 0)$, respectively, and obtain Figure 17.

Figure 17

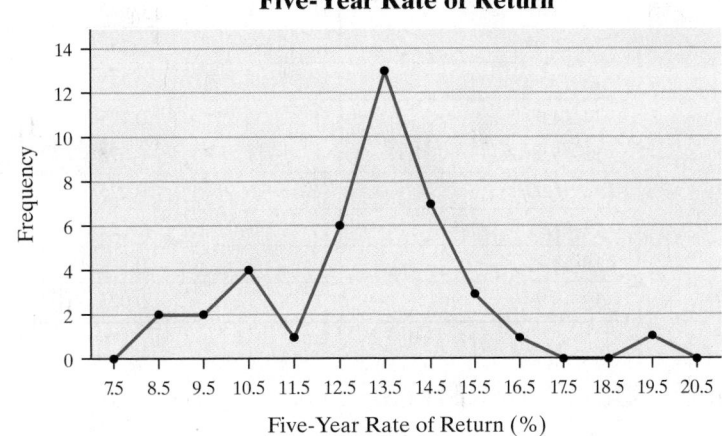

Five-Year Rate of Return

• **Now Work Problem 13(c)**

② ## Create Cumulative Frequency and Relative Frequency Tables

Since quantitative data can be ordered (that is, written in ascending or descending order), they can be summarized in a *cumulative frequency distribution* and a *cumulative relative frequency distribution*.

Definitions

A **cumulative frequency distribution** displays the aggregate frequency of the category. In other words, for discrete data, it displays the total number of observations less than or equal to the category. For continuous data, it displays the total number of observations less than or equal to the upper class limit of a class.

A **cumulative relative frequency distribution** displays the proportion (or percentage) of observations less than or equal to the category for discrete data and the proportion (or percentage) of observations less than or equal to the upper class limit for continuous data.

So the cumulative frequency for the second class is the sum of the frequencies of classes 1 and 2, the cumulative frequency for the third class is the sum of the frequencies of classes 1, 2, and 3, and so on.

Table 17 on the following page displays the cumulative frequency and cumulative relative frequency of the data summarized in Table 13 in Section 2.2. Table 17 shows that 38 of the 40 mutual funds had five-year rates of return of 15.99% or less. The cumulative relative frequency distribution is shown in the fifth column. We see that 95% of the mutual funds had a five-year rate of return of 15.99% or less. Also, a mutual fund with a five-year rate of return of 16% or higher outperformed 95% of its peers.

• **Now Work Problems 13(a)–(b)**

Table 17				
Class (Five-year rate of return)	Frequency	Relative Frequency	Cumulative Frequency	Cumulative Relative Frequency
8–8.99	2	0.05	2	0.05
9–9.99	2	0.05	4	0.1
10–10.99	4	0.1	8	0.2
11–11.99	1	0.025	9	0.225
12–12.99	6	0.15	15	0.375
13–13.99	13	0.325	28	0.7
14–14.99	7	0.175	35	0.875
15–15.99	3	0.075	38	0.95
16–16.99	1	0.025	39	0.975
17–17.99	0	0	39	0.975
18–18.99	0	0	39	0.975
19–19.99	1	0.025	40	1

③ **Construct Frequency and Relative Frequency Ogives**

Recall that the cumulative frequency of a class is the aggregate frequency less than or equal to the upper class limit.

Definition

> An **ogive** (read as "oh jīve") is a graph that represents the cumulative frequency or cumulative relative frequency for the class. It is constructed by plotting points whose x-coordinates are the upper class limits and whose y-coordinates are the cumulative frequencies or cumulative relative frequencies of the class. Then line segments are drawn connecting consecutive points. An additional line segment is drawn connecting the first point to the horizontal axis at a location representing the upper limit of the class that would precede the first class (if it existed).

Construct a relative frequency ogive using the data in Table 17 by plotting points whose x-coordinates are the upper class limits and whose y-coordinates are the cumulative relative frequencies of the classes. Then connect the points with line segments. See Figure 18.

Figure 18

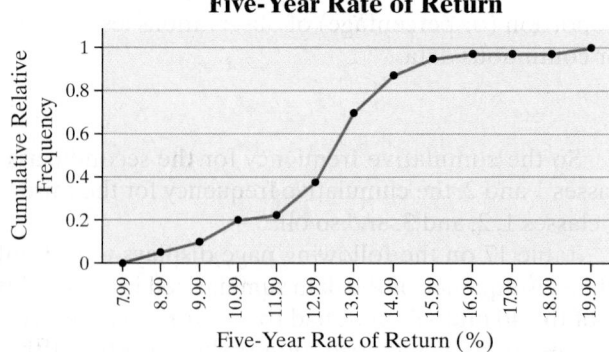

Five-Year Rate of Return

Using Technology

Statistical spreadsheets and certain graphing calculators can draw ogives.

● Now Work Problem 13(d)

Figure 18 shows that 95% of the mutual funds had a five-year rate of return less than or equal to 15.99%.

4 Draw Time-Series Graphs

If the value of a variable is measured at different points in time, the data are referred to as **time-series data**. The closing price of Cisco Systems stock at the end of each month for the past 12 years is an example of time-series data.

Definition | A **time-series plot** is obtained by plotting the time in which a variable is measured on the horizontal axis and the corresponding value of the variable on the vertical axis. Line segments are then drawn connecting the points.

Time-series plots are very useful in identifying trends in the data over time.

EXAMPLE 1   Drawing a Time-Series Plot

Problem The Partisan Conflict Index (PCI) tracks the degree of political disagreement among U.S. politicians in the federal government. It is found by measuring the frequency of newspaper articles reporting disagreement in a given month. Higher values of the index suggest greater conflict among political parties, Congress, and the President. The data in Table 18 represents the PCI in October from 2000 to 2014. Construct a time-series plot of the data. In what year was the index highest?

Approach
Step 1   Plot points for each year, with the date on the horizontal axis and the Partisan Conflict Index on the vertical axis.

Step 2   Connect the points with line segments.

Solution Figure 19 shows the time-series plot. Notice the jump in the PCI following the 2010 midterm elections and the huge spike in October, 2013 the year the PCI was highest. During October, 2013, the federal government shut down because Congress could not agree on a budget, largely due to a standoff over the Affordable Care Act (aka Obamacare).

Table 18

Year	Partisan Conflict Index (PCI)
2000	85.12
2001	70.46
2002	96.87
2003	94.18
2004	123.41
2005	74.39
2006	80.16
2007	88.88
2008	93.00
2009	90.54
2010	147.52
2011	129.75
2012	147.55
2013	252.10
2014	137.58

Source: Federal Reserve Bank of Philadelphia

⚏ Using Technology

Statistical spreadsheets, such as StatCrunch, Excel, or Minitab, and certain graphing calculators, such as the TI-83 or TI-84 Plus, can create time-series graphs.

Figure 19

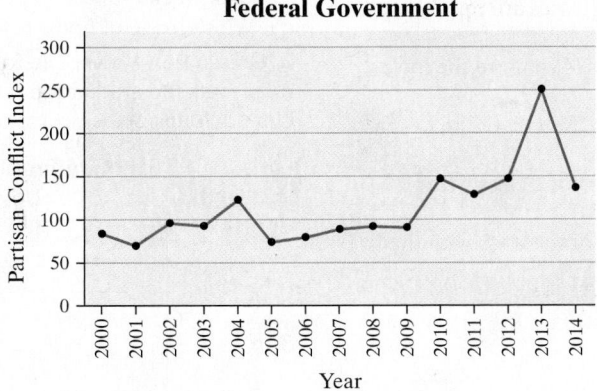

Partisan Conflict Index in the United States Federal Government

● Now Work Problem 15

Technology Step-by-Step Drawing Frequency Polygons, Ogives, and Time-Series Plots

TI-83/84 Plus
The TI-83 and TI-84 Plus can draw all three of these graphs.

Minitab
Minitab can draw all three of these graphs.

Excel
Excel can draw all three of these graphs.

StatCrunch
StatCrunch can draw all three of these graphs.

2.3 Assess Your Understanding

Vocabulary and Skill Building

1. What is an ogive?

2. What are time-series data?

3. *True or False*: When plotting an ogive, the plotted points have *x*-coordinates that are equal to the upper limits of each class.

4. *True or False*: When plotting a frequency polygon, we plot the percentages for each class above the midpoint and connect the points with line segments.

5. Grade Point Average The following frequency polygon represents the cumulative grade point average of a random sample of students at an engineering school.

What Is Your Cumulative Grade Point Average?

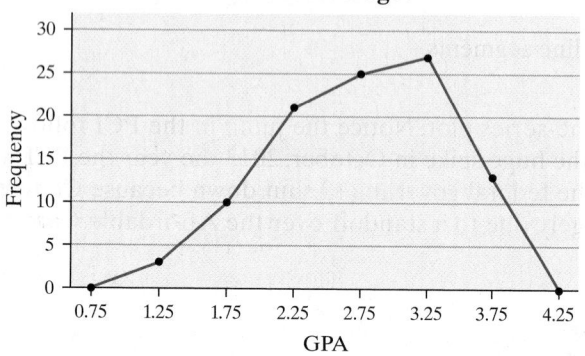

(a) What is the class width? How many classes are represented in the graph?

(b) What is the midpoint of the first class? What are the lower and upper limits of the first class?

(c) What is the midpoint of the last class? What are the lower and upper limits of the last class?

(d) What are the lower and upper limits of the class with 25 students?

(e) What are the lower and upper limits of the class with the fewest students?

6. Exercise The following frequency polygon represents the number of hours each week a random sample of adult Americans exercises.

How Many Hours Do You Exercise Each Week?

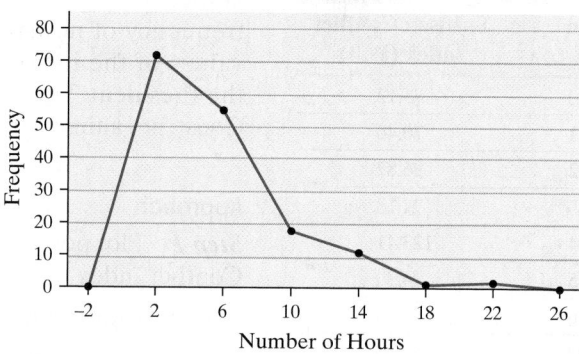

(a) What is the class width? How many classes are represented in the graph?

(b) What is the midpoint of the first class? What are the lower and upper limits of the first class?

(c) What is the midpoint of the last class? What are the lower and upper limits of the last class?

(d) What is the most popular number of hours spent exercising each week?

(e) What is the least popular number of hours spent exercising each week?

(f) Which range of hours had 55 people exercise each week?

7. Graduation Rates The following relative frequency ogive represents the graduation rate of all four-year universities in the United States. *Source:* payscale.com

Graduation Rate of United States Universities

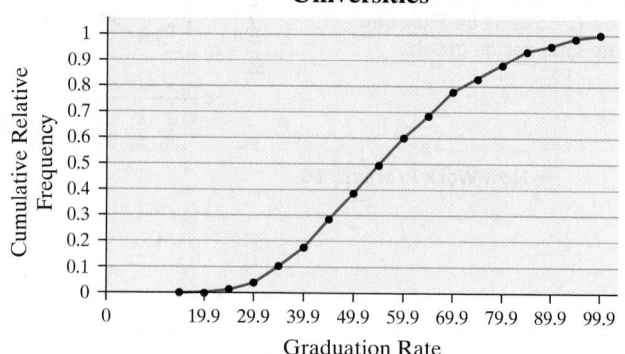

(a) What is the class width?
(b) Approximately 10% of all four-year universities have a graduation rate below what level?
(c) What percent of all four-year universities have a graduation rate less than 59.9%?
(d) Five percent of all four-year universities have a graduation rate above what level?

8. Tornadoes The following relative frequency ogive represents the lengths of a random sample of tornadoes in the United States.

Lengths of Tornadoes in the United States

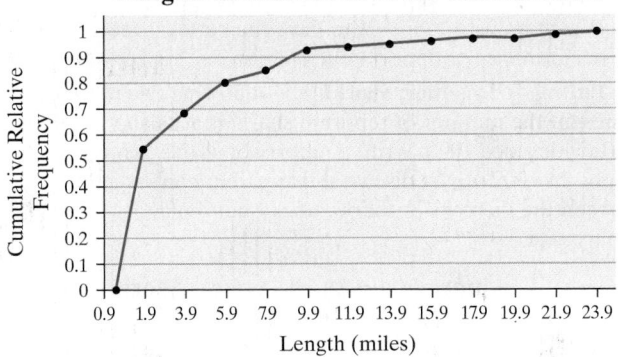

(a) What is the class width?
(b) Approximately 92% of all tornadoes are less than what length?
(c) What percent of all tornadoes have a length less than 5.9 miles?
(d) More than 50% of all tornadoes are less than what length?

9. Misery Index The following time-series plot shows the annual unemployment and inflation rates for the years 1992 through 2014. *Source:* www.miseryindex.us

Unemployment & Inflation Rates

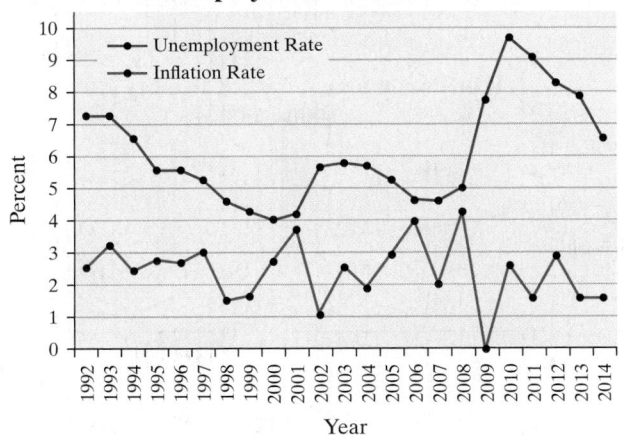

(a) Estimate the unemployment rate in 2011.
(b) In what year was the unemployment rate highest?
(c) In what year was the inflation rate highest?
(d) In what year were the unemployment rate and inflation rate closest? furthest?
(e) The misery index is defined as the sum of the unemployment rate and the inflation rate. According to the misery index, which year was more "miserable", 1999 or 2014?
(f) Comment on the trend in the misery index since 2010.

10. Age at First Marriage The following time-series plot shows the average age at which individuals first marry by gender for each year of the census since 1890.

Average Age at First Marriage by Gender

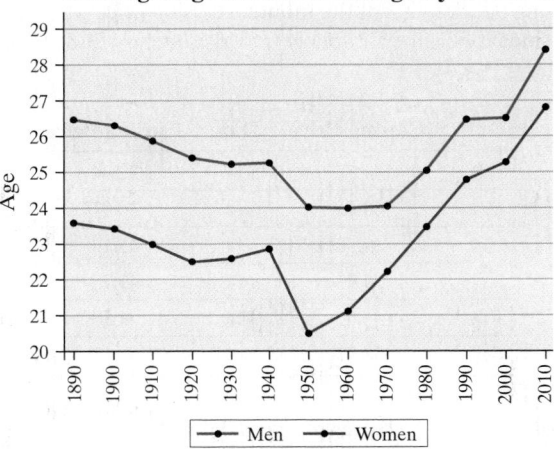

(a) To the nearest year, what was the average age of a man who first married in 1980?
(b) To the nearest year, what was the average age of a woman who first married in 1960?
(c) In which year was the difference in the average age of men and women at which they first married the largest? Approximately, what is the age difference?
(d) In which year was the difference in the average age of men and women at which they first married the least?

Applying the Concepts

In Problems 13–16, use the frequency distributions in the problem indicated from Section 2.2 to do the following:
(a) *Construct a cumulative frequency distribution.*
(b) *Construct a cumulative relative frequency distribution.*
(c) *Draw a frequency polygon.*
(d) *Draw a relative frequency ogive.*

11. Problem 21   **12.** Problem 22
NW 13. Problem 27   **14.** Problem 28

NW 15. BP Stock and an Oil Spill In 2010, there was an explosion on a BP oil rig in the Gulf of Mexico. The following data represent the closing price of BP stock at the end of each month in 2010.

Date	Closing Price	Date	Closing Price
1/10	45.22	7/10	32
2/10	43.53	8/10	28.97
3/10	46.69	9/10	34.24
4/10	42.66	10/10	33.93
5/10	35.72	11/10	33.27
6/10	24.02	12/10	36.74

Source: Yahoo! Finance

(a) Construct a time-series plot of the data.
(b) Percentage change may be found using the formula

Percentage change $= \dfrac{P_2 - P_1}{P_1}$. For example, the percentage change in price from 1/10 to 2/10 was Percentage change $=$

$\dfrac{43.53 - 45.22}{45.22} = -0.037 = -3.7\%$.

That is, BP stock declined 3.7% from January to February in 2010. The extent of the damage of the explosion was determined in June, 2010. What was the percentage change in the BP stock price from May to June, 2010?

16. Twitter Stock Twitter went public in November, 2013. The following data represent the monthly closing price of Twitter stock since it became a publicly traded company for its first year of trading.

Date	Closing Price	Date	Closing Price
11/13	41.57	5/14	32.44
12/13	63.65	6/14	40.97
1/14	64.50	7/14	45.19
2/14	54.91	8/14	49.75
3/14	46.67	9/14	51.58
4/14	38.97	10/14	41.47

Source: Yahoo! Finance

(a) Construct a time-series plot of the data.

(b) See Problem 15(b). What was the percentage change in the Twitter stock price from November to December, 2013? What was the percentage change in the Twitter stock price from November, 2013 to October, 2014? What does this suggest?

17. Federal Debt The following data represent the percentage of total federal debt held by the public as a percentage of gross domestic product (GDP). The GDP of a country is the total value of all goods and services produced within the country in a given year. It can be thought of as the total income of the country for the year. Construct a time-series plot and comment on any trends.

Year	Debt as % GDP	Year	Debt as % GDP
1996	63.97	2006	60.99
1997	62.37	2007	61.81
1998	60.27	2008	67.84
1999	57.99	2009	82.37
2000	54.70	2010	90.44
2001	54.30	2011	95.05
2002	56.45	2012	98.81
2003	58.72	2013	99.54
2004	59.91	2014	102.84
2005	60.37		

Source: http://www.usgovernmentspending.com/federal_debt_chart.html

18. College Enrollment The following data represent the percentage of 18- to 24-year-olds enrolled in college. Construct a time-series plot and comment on any trends.

Year	Percent Enrolled	Year	Percent Enrolled
1998	36.5	2006	37.3
1999	35.6	2007	38.8
2000	35.5	2008	39.6
2001	36.3	2009	41.3
2002	36.7	2010	41.2
2003	37.8	2011	42.0
2004	38.0	2012	41.0
2005	38.9		

Source: National Center for Education Statistics

19. Putting It Together: Rates of Return of Stocks Stocks may be categorized by industry. Go to www.pearsonhighered.com/sullivanstats and download

the file 2_3_19. The data represent the three-year rate of return of stocks categorized as consumer defensive and industrial (as of November 25, 2014).

(a) Construct a relative frequency distribution for each industry. To make an easy comparison, create each frequency distribution so that the lower class limit of the first class is −20 and the class width is 10.

(b) Draw relative frequency histograms for each industry.

(c) On the same graph, construct a relative frequency polygon for each of the two industries.

(d) On the same graph, construct a relative frequency ogive for each of the two industries.

(e) Which industry appears to have the better three-year performance? Support your opinion.

20. Putting It Together: Shark! The following two graphics represent the number of reported shark attacks and fatalities worldwide since 1900. Write a report about the trends in the graphs. In your report discuss the apparent contradiction between the increase in shark attacks, but the decrease in fatality rate.

Unprovoked Shark Attacks Worldwide

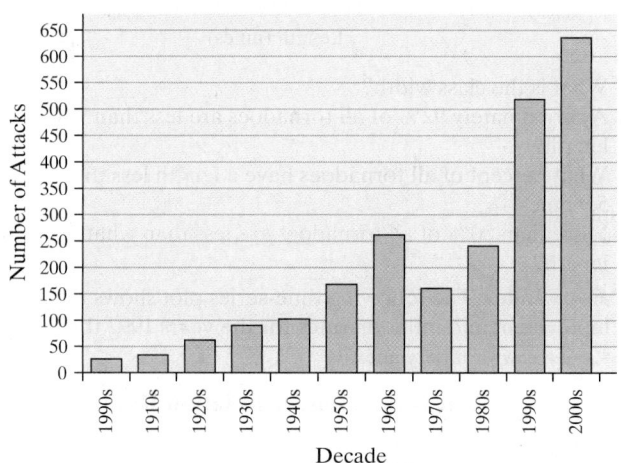

Worldwide Shark Attack Fatality Rate, 1900–2009

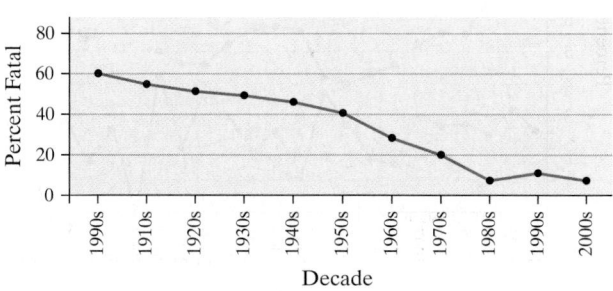

Source: Florida Museum of Natural History

Explaining the Concepts

21. Which do you prefer: histograms, stem-and-leaf plots, or frequency polygons? Be sure to support your opinion. Are there circumstances in which one might be preferred over another?

22. The cumulative relative frequency for the last class must always be 1. Why?

23. What type of variable is required when drawing a time-series plot? Why do we draw time-series plots?

2.4 Graphical Misrepresentations of Data

Objective ❶ Describe what can make a graph misleading or deceptive

❶ Describe What Can Make a Graph Misleading or Deceptive

Statistics: The only science that enables different experts using the same figures to draw different conclusions.—Evan Esar

Statistics often gets a bad rap for having the ability to manipulate data to support any position. One method of distorting the truth is through graphics. We mentioned in Section 2.1 how visual displays send more powerful messages than raw data or even tables of data. Since graphics are so powerful, care must be taken in constructing graphics and in interpreting their messages. Graphics may *mislead* or *deceive*. We will call graphs misleading if they unintentionally create an incorrect impression. We consider graphs deceptive if they purposely create an incorrect impression. In either case, a reader's incorrect impression can have serious consequences. Therefore, it is important to be able to recognize misleading and deceptive graphs.

The most common graphical misrepresentations of data involve the scale of the graph, an inconsistent scale, or a misplaced origin. Increments between tick marks should be constant, and scales for comparative graphs should be the same. Also, because readers usually assume that the baseline, or zero point, is at the bottom of the graph, a graph that begins at a higher or lower value can be misleading.

EXAMPLE 1 Misrepresentation of Data

Problem A home security company located in Minneapolis, Minnesota, develops a summer ad campaign with the slogan "When you leave for vacation, burglars leave for work." According to the city of Minneapolis, roughly 20% of home burglaries occur during the peak vacation months of July and August. The advertisement contains the graphic shown in Figure 20. Explain what is wrong with the graphic.

Figure 20

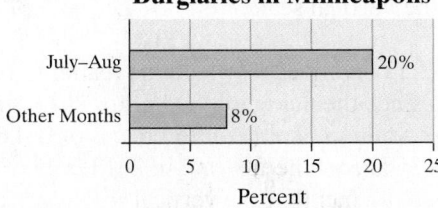

Burglaries in Minneapolis

Approach Look for any characteristics that may mislead a reader, such as inconsistent scales or poorly defined categories.

Solution Consider how the categories of data are defined. The sum of the percentages (the relative frequencies) over all 12 months should be 1. Because $10(0.08) + 0.20 = 1$, it is clear that the bar for Other Months represents an average percent for each month, while the bar for July–August represents the average percent for the months July and August combined. The unsuspecting reader is mislead into thinking that July and August each have a burglary rate of 20%.

Figure 21 on the following page gives a better picture of the burglary distribution. The increase during the month of July is not as dramatic as the bar graph in Figure 20 implies and August actually has fewer burglaries than September or October. In fact, Figure 20 would be considered deceitful because the security company is intentionally trying to convince consumers that July and August are much higher burglary months.

(continued)

Figure 21

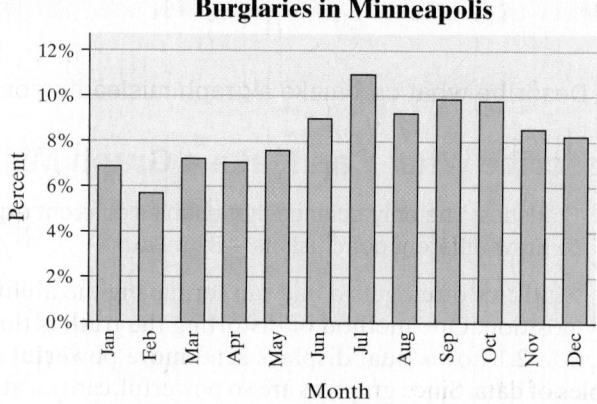

Burglaries in Minneapolis

● Now Work Problem 5

EXAMPLE 2   Misrepresentation of Data by Manipulating the Vertical Scale

Problem A national news organization developed the graphic shown in Figure 22 to illustrate the change in the highest marginal tax rate effective January 1, 2013. Why might this graph be considered misleading?

Figure 22

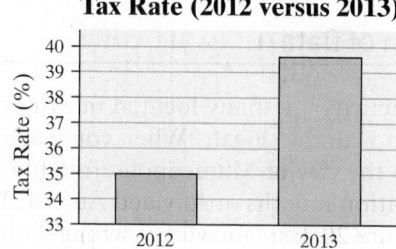

Highest Marginal Income Tax Rate (2012 versus 2013)

Approach We need to find any characteristics that may mislead a reader, such as manipulation of the vertical scale.

Solution The graph in Figure 22 may lead the reader to believe that marginal tax rates are more than tripling since the height of the bar for 2013 is more than three times the length of the bar for 2012 when in fact the difference is only 4.6 percentage points (because the tax rate in 2012 is 35% and the tax rate in 2013 is 39.6%). The reason for this incorrect conclusion is due to the fact that the vertical scale does not begin at 0. A graph that does not distort the difference in tax rates is given in Figure 23. Notice that the increase in the tax rate is still apparent in the graph without the distortion as to the size of the increase.

Figure 23

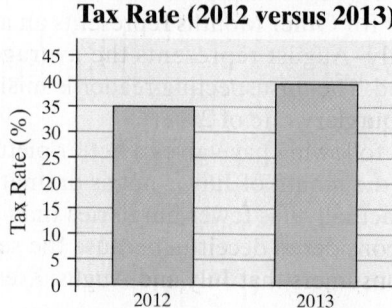

Highest Marginal Income Tax Rate (2012 versus 2013)

EXAMPLE 3   Misrepresentation of Data by Manipulating the Vertical Scale

Problem The time-series graph in Figure 24 depicts the number of residents in the United States living in poverty. Why might this graph be considered misrepresentative?

Figure 24

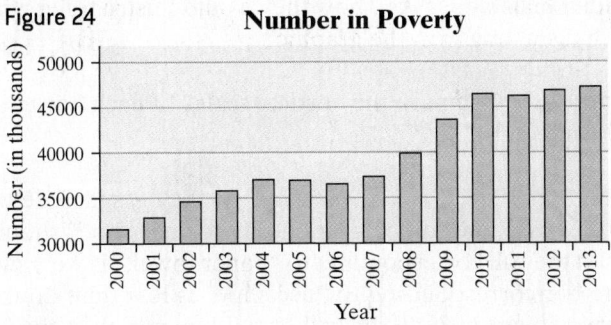

Approach Look for any characteristics that may mislead a reader, such as manipulation of the vertical scale.

Solution The graph may mislead readers to believe that the number in poverty has more than doubled since 2007 because the bar for 2013 is more than twice the length of the bar from 2007. Notice that the vertical axis begins at 30,000 instead of 0. This type of scaling is common when the smallest observed data value is a rather large number. It is not necessarily done intentionally to confuse or mislead the reader. Often, the main purpose in graphs (particularly time-series graphs) is to discover a trend, rather than the actual differences in the data. However, the author of the graph should clearly indicate that the graph does not begin at 0 by including the symbol ⌇ in the vertical scale. This symbol indicates that the scale has been truncated and the graph has a gap. See Figure 25.

Figure 25

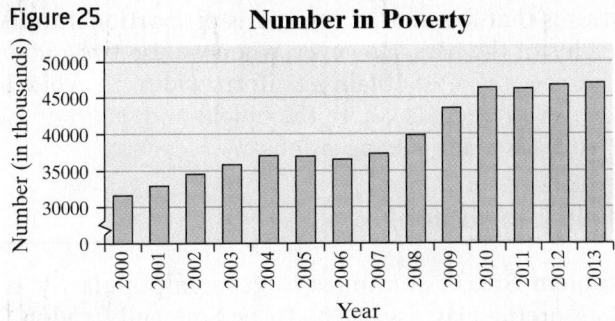

Although the data does stand out in Figure 25, it is better to use a time-series plot when displaying time series data (rather than a bar plot). In addition, it is better to use the percent of the population in poverty rather than the number living in poverty. This is due to the fact that increases in poverty may be due to increases in the population as well as a deterioration of the economy. Figure 26 shows a time-series plot of the percentage of U.S. residents living in poverty. The lack of bars allows us to focus on the trend in the data, rather than the relative size (or area) of the bars. Also, it is clear that the percent in poverty was highest in 2010 and declined in 2013—this is not clear from the bar graphs.

Figure 26

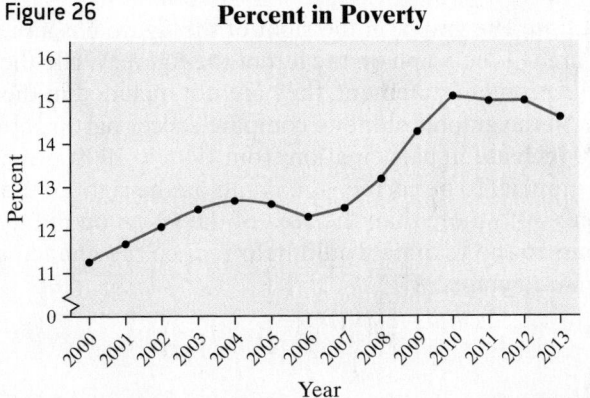

● **Now Work Problem 3**

EXAMPLE 4 Misrepresentation of Data

Problem The bar graph illustrated in Figure 27 is a *USA Today*-type graph. A survey was conducted by Impulse Research for Quilted Northern Confidential in which individuals were asked how they would flush a toilet when the facilities are not sanitary. What's wrong with the graphic?

Figure 27

Approach Compare the vertical scales of each bar to see if they accurately depict the given percentages.

Solution First, it is unclear whether the bars include the roll of toilet paper or not. In either case, the roll corresponding to "use shoe" should be $2.4(=41/17)$ times taller than the roll corresponding to "paper towel." If we include the roll of toilet paper, then the bar corresponding to "use shoe" is less than double the height of "paper towel." If we do not include the roll of toilet paper, then the bar corresponding to "use shoe" is almost exactly double the height of the bar corresponding to "paper towel." The vertical scaling is incorrect.

● Now Work Problem 9

Newspapers, magazines, and websites often go for a "wow" factor when displaying graphs. The graph designer may be more interested in catching the reader's eye than making the data stand out. The two most commonly used tactics are 3-D graphs and pictograms (graphs that use pictures to represent the data). The use of 3-D effects is strongly discouraged, because such graphs are often difficult to read, add little value to the graph, and distract the reader from the data.

When comparing bars, our eyes are really comparing the *areas* of the bars. That is why we emphasized that the bars or classes should be of the same width. Uniform width ensures that the area of the bar is proportional to its height, so we can compare the heights of the bars. However, when we use two-dimensional pictures in place of bars, it is not possible to obtain a uniform width. To avoid distorting the picture when values increase or decrease, both the height and width of the picture must be adjusted. This often leads to misleading graphs.

EXAMPLE 5 Misleading Graphs

Problem Soccer continues to grow in popularity as a sport in the United States. High-profile players such as Hope Solo and Landon Donovan have helped generate renewed interest in the sport at various age levels. In 1991, there were approximately 10 million participants in the United States aged 7 years or older. By 2009 this number had climbed to 14 million. To illustrate this increase, we could create a graphic like the one shown in Figure 28. Describe how the graph may be misleading.
Source: U.S. Census Bureau; National Sporting Goods Association.

Figure 28
Soccer Participation

1991   2009

Approach Look for characteristics of the graph that seem to manipulate the facts, such as an incorrect depiction of the size of the graphics.

Solution The graph on the right of the figure has area that is more than four times the area of the graph on the left of the figure. While the number of participants is given in the problem statement, they are not included in the graph, which makes the reader rely on the graphic alone to compare soccer participation in the two years. There was a 40% increase in participation from 1991 to 2009, not more than 300% as indicated by the graphic. To be correct, the graph on the right of the figure should have an area that is only 40% more than the area of the graph on the left of the figure. Adding the data values to the graphic would help reduce the chance of misinterpretation due to the oversized graph.

● Now Work Problem 13

Figure 29

Soccer Participation

1991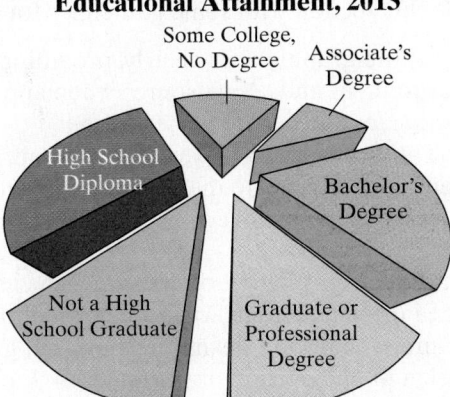

2009

A variation on pictograms is to use a smaller picture repeatedly, with each picture representing a certain quantity. For example, we could present the data from Figure 28 by using a smaller soccer ball to represent 1 million participants. The resulting graphic is displayed in Figure 29. Note how the uniform size of the graphic allows us to make a more accurate comparison of the two quantities.

= 1 million participants

EXAMPLE 6   Misrepresentations of Data: Three-Dimensional Scale

Problem Figure 30 represents the educational attainment (level of education) in 2013 of adults 25 years and older who are U.S. residents. Why might this graph be considered misrepresentative.

Figure 30

Educational Attainment, 2013

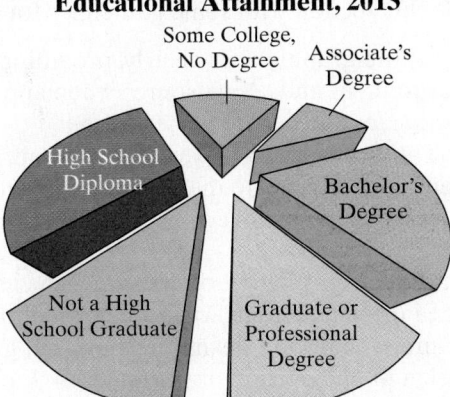

Approach Find any characteristics that may mislead a reader, such as overemphasis on one category of data.

Solution Three-dimensional pie charts tend to overstate the significance of categories closer to the reader. For example, in this graph the category "Graduate or Professional Degree" looks well over twice the size of the category "Associate's Degree." Yet, in the raw data shown in Table 19 there are 23,931,000 individuals who have a Graduate or Professional Degree and 20,367,000 individuals who have an Associate's Degree!

Table 19	
Educational Attainment	**Frequency (000s)**
Not a High School Graduate	24,517
High School Diploma	61,704
Some College, No Degree	34,805
Associate's Degree	20,367
Bachelor's Degree	41,575
Graduate or Professional Degree	23,931

Source: U.S. Census Bureau

Figure 31

Educational Attainment, 2013

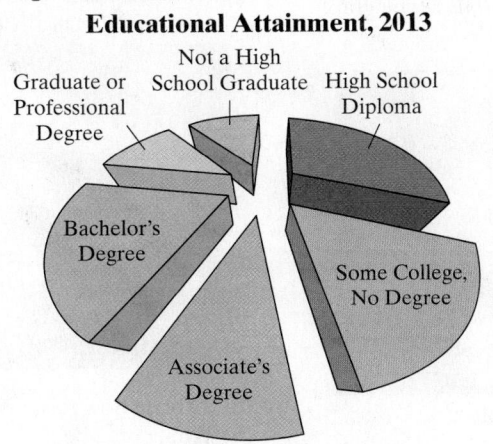

Figure 32

Educational Attainment, 2013

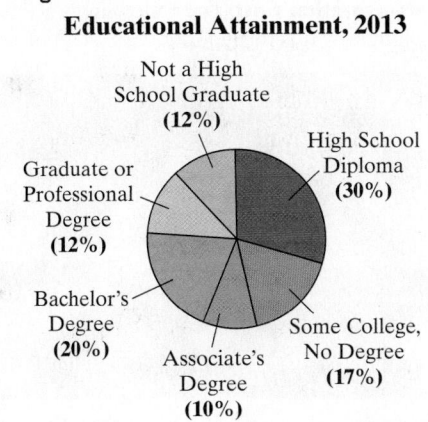

(continued)

In fact, if we rotate the pie chart to bring the category "Associate's Degree" to the front, we obtain the graph shown in Figure 31. Now, "Associate's Degree" appears to be the larger piece of the pie. Clearly, the three-dimensional representation distorts the number of individuals in each category, so these types of graphs should be avoided. Instead, stick to two-dimensional graphs with clear labels as shown in Figure 32. •

The material presented in this section is by no means all-inclusive. There are many ways to create graphs that mislead. Two popular texts written about ways that graphs mislead or deceive are *How to Lie with Statistics* (W. W. Norton & Company, Inc., 1982) by Darrell Huff and *The Visual Display of Quantitative Information* (Graphics Press, 2001) by Edward Tufte.

We conclude this section with some guidelines for constructing good graphics.

- Title and label the graphic axes clearly, providing explanations if needed. Include units of measurement and a data source when appropriate.
- Avoid distortion. Never lie about the data.
- Minimize the amount of white space in the graph. Use the available space to let the data stand out. If you truncate the scales, clearly indicate this to the reader.
- Avoid clutter, such as excessive gridlines and unnecessary backgrounds or pictures. Don't distract the reader.
- Avoid three dimensions. Three-dimensional charts may look nice, but they distract the reader and often lead to misinterpretation of the graphic.
- Do not use more than one design in the same graphic. Sometimes graphs use a different design in one portion of the graph to draw attention to that area. Don't try to force the reader to any specific part of the graph. Let the data speak for themselves.
- Avoid relative graphs that are devoid of data or scales.

 2.4 Assess Your Understanding

Applying the Concepts

1. Inauguration Cost The following is a *USA Today*-type graph. Explain how it is misleading.

2. Burning Calories The following is a *USA Today*-type graph.

(a) Explain how it is misleading.
(b) What could be done to improve the graphic?

NW 3. Median Earnings The graph shows the median income for females from 2007 to 2013 in constant 2013 dollars.

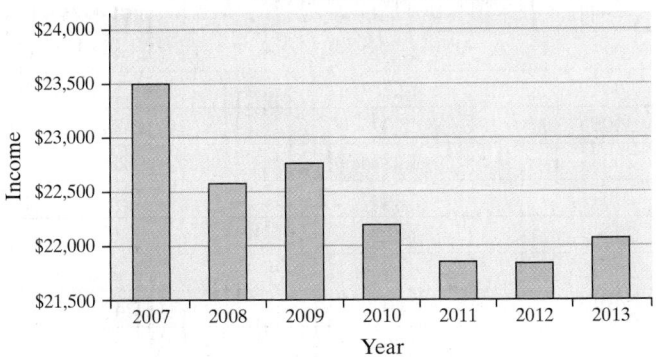

Median Income for Females in Constant 2013 Dollars

Source: U.S. Census Bureau

(a) How is the graph misleading? What does the graph seem to convey?

(b) Redraw the graph so that it is not misleading. What does the new graph seem to convey?

4. Union Membership The following relative frequency histogram represents the proportion of employed people aged 25–64 years old who are members of a union.

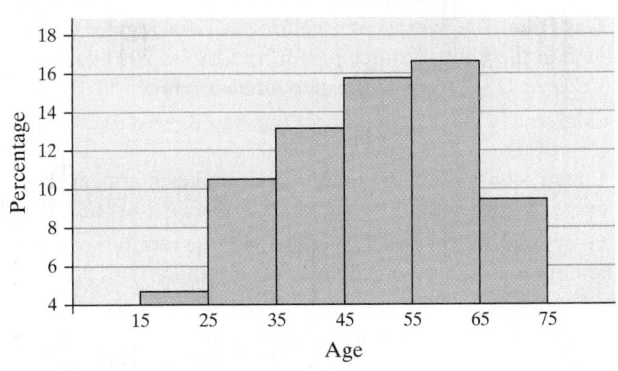

Union Membership

Source: U.S. Bureau of Labor Statistics

(a) Describe how this graph is misleading. What might a reader conclude from the graph?

(b) Redraw the histogram so that it is not misleading.

NW 5. Robberies A newspaper article claimed that the afternoon hours were the worst in terms of robberies and provided the following graph in support of this claim. Explain how this graph is misleading.

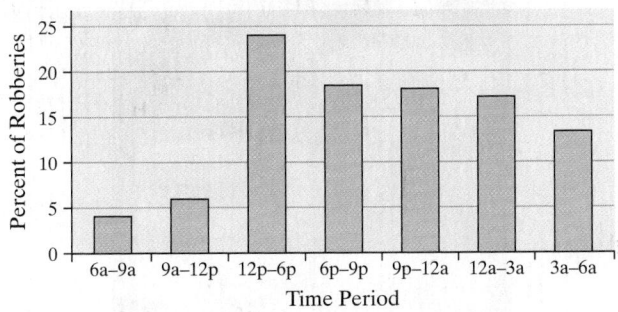

Hourly Crime Distribution (Robbery)

Source: U.S. Statistical Abstract

6. Car Accidents An article in a student newspaper claims that younger drivers are safer than older drivers and provides the following graph to support the claim. Explain how this graph is misleading.

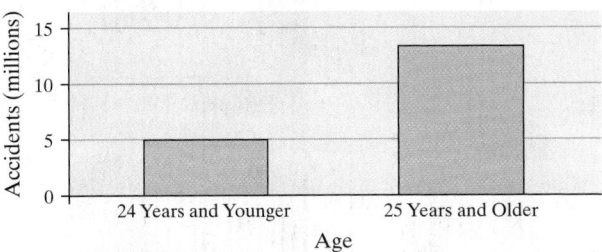

Number of Motor Vehicle Accidents

Source: U.S. Statistical Abstract

7. Tax Revenue The following histogram drawn in StatCrunch represents the total tax collected by the Internal Revenue Service for each state plus Washington, DC. Explain why the graph is misleading.

Note: Vermont paid the least in tax ($4.046 million), while California paid the most ($334.425 million).

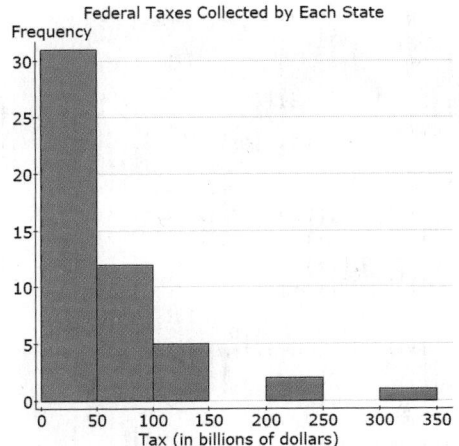

8. You Explain It! Oil Reserves The U.S. Strategic Oil Reserve is a government-owned stockpile of crude oil. It was established after the oil embargo in the mid-1970s and is meant to serve as a national defense fuel reserve, as well as to offset reductions in commercial oil supplies that would threaten the U.S. economy. The graphic on the following page depicts oil reserves in 1977 and 2014.

(a) How many times larger should the graphic for 2014 be than the 1977 graphic (to the nearest whole number)?

(b) The United States imported approximately 7.7 million barrels of oil per day in 2014. At that rate, assuming no change in U.S. oil production, how long would the U.S. strategic oil reserve last if no oil were imported?

U.S. Strategic Oil Reserves (millions of barrels)

1977   2014

Source: U.S. Energy Information Administration

NW 9. Cost of Kids The following is a *USA Today*-type graph based on data from the Department of Agriculture. It represents the percentage of income a middle-income family will spend on their children.

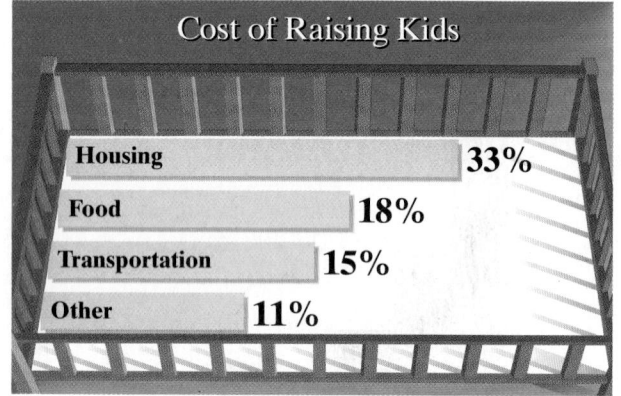

(a) How is the graphic misleading?
(b) What could be done to improve the graphic?

10. Worker Injury The safety manager at Klutz Enterprises provides the following graph to the plant manager and claims that the rate of worker injuries has been reduced by 67% over a 12-year period. Does the graph support his claim? Explain.

Proportion of Workers Injured

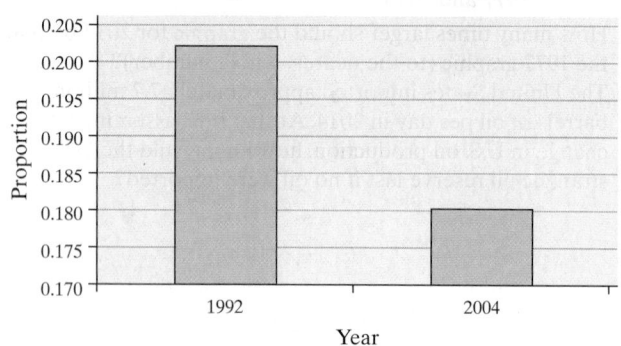

Year

11. Health Care Expenditures The following data represent health care expenditures per capita (per person) and as a percentage of the U.S. gross domestic product (GDP) from 2003

to 2012. Gross domestic product is the total value of all goods and services created during the course of the year.

| Year | Health Care | |
	per Capita	as a percent of GDP
2003	11,512	15.4
2004	12,277	15.5
2005	13,095	15.5
2006	13,858	15.6
2007	14,480	15.9
2008	14,720	16.4
2009	14,418	17.4
2010	14,958	17.4
2011	15,534	17.3
2012	16,245	17.2

Source: Center for Medicare and Medicaid Services, Office of the Actuary

(a) Construct a time-series plot that a politician would create to support the position that health care expenditures are increasing and must be slowed.
(b) Construct a time-series plot that the health care industry would create to refute the opinion of the politician. Is your graph convincing?
(c) Explain how different measures may be used to support two completely different positions.

12. Gas Hike The average per gallon price for regular unleaded gasoline in the United States rose from $1.46 in 2001 to $2.77 in 2015. *Source:* U.S. Energy Information Administration

(a) Construct a graphic that is not misleading to depict this situation.
(b) Construct a misleading graphic that makes it appear the average price roughly quadrupled between 2001 and 2015.

NW 13. Overweight Between 1980 and 2012, the number of adults in the United States who were overweight more than doubled from 15% to 35%.

Source: Centers for Disease Control and Prevention

(a) Construct a graphic that is not misleading to depict this situation.
(b) Construct a misleading graphic that makes it appear that the percent of overweight adults has more than quadrupled between 1980 and 2012.

14. Ideal Family Size The following *USA Today*-type graphic illustrates the ideal family size (total children) based on a survey of adult Americans.

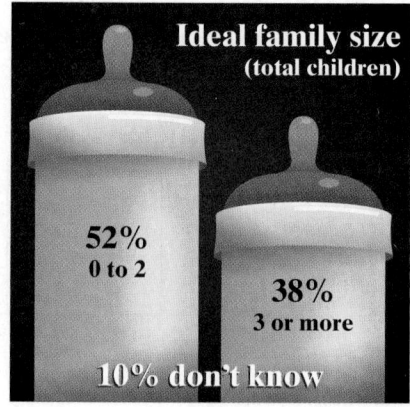

(a) What type of graphic is being displayed?
(b) Describe any problems with the graphic.
(c) Construct a graphic that is not misleading and makes the data stand out.

15. National League Baseball MVP The following pie chart displays the position played by the most valuable player (MVP) in the National League of Major League Baseball from 1931 thru 2013. Explain how the graphic is misleading. What should be done to improve the graphic?

National League MVP [1931–2013]

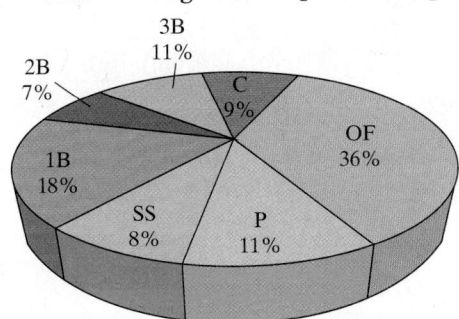

16. Inauguration Day The following graph appears on the Wikipedia website. The intention of the graph is to display the ages of U.S. presidents on Inauguration Day. Explain any problems you see with the graph.

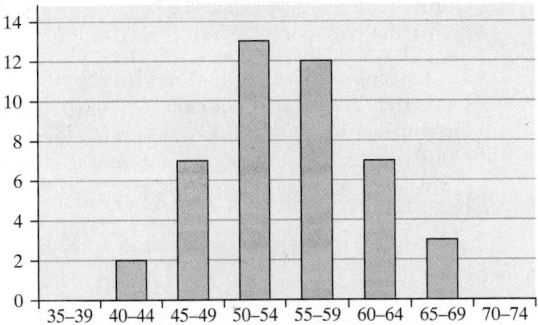

Source: http://en.wikipedia.org/wiki/List_of_Presidents_ of_the_United_States_by_age. Published by Wikimedia Foundation, © 2010.

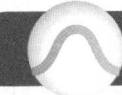 **Chapter 2 Review**

Summary

In this chapter, we learned how to graphically display raw data. Raw data is data that has not yet been organized in any form. It is the data we typically obtain from an observational study or designed experiment. A simple example of raw data would be Male, Female, Female, Male, Female, and so on.

Raw data are first organized into tables. The first step in organizing any data is to determine whether the data are qualitative or quantitative.

Qualitative data may be organized into tables by using the data itself to form categories. For example, if the raw data is for the variable gender, then the categories would be Female and Male. The tables formed are either frequency tables or relative frequency tables. The types of graphs drawn from qualitative data include bar graphs, side-by-side bar graphs, and pie charts.

Quantitative data may also be organized into tables. For discrete data with only a few outcomes, the outcomes form the categories of data. However, for discrete data with a variety of outcomes, or for continuous data, the categories of data are formed using classes. Quantitative data can be organized into frequency, relative frequency, cumulative frequency, and cumulative relative frequency tables. Quantitative data may also be represented graphically in histograms, stem-and-leaf plots, dot plots, frequency polygons, ogives, and time series plots.

Finally, in creating graphs, care must be taken not to draw a graph that misleads or deceives the reader.

Vocabulary

Raw data (p. 62)
Frequency distribution (p. 63)
Relative frequency (p. 64)
Relative frequency distribution (p. 64)
Bar graph (p. 65)
Pareto chart (p. 66)
Side-by-side bar graph (p. 66)
Pie chart (p. 68)
Classes (p. 76)
Histogram (p. 77)

Lower and upper class limits (p. 78)
Class width (p. 78)
Open ended (p. 78)
Stem-and-leaf plot (p. 82)
Stem (p. 82)
Leaf (p. 82)
Split stems (p. 84)
Dot plot (p. 85)
Uniform distribution (p. 85)
Bell-shaped distribution (p. 85)
Skewed right (p. 85)

Skewed left (p. 85)
Class midpoint (p. 94)
Frequency polygon (p. 94)
Cumulative frequency distribution (p. 95)
Cumulative relative frequency distribution (p. 95)
Ogive (p. 96)
Time-series data (p. 97)
Time series plot (p. 97)

OBJECTIVES

Section	You should be able to . . .	Example(s)	Review Exercises
2.1	1 Organize qualitative data in tables (p. 63)	1, 2	2(a), 4(a) and (b)
	2 Construct bar graphs (p. 64)	3 through 5	2(c) and (d), 4(c)
	3 Construct pie charts (p. 68)	6	2(e), 4(d)
2.2	1 Organize discrete data in tables (p. 76)	1	5(a) and (b)
	2 Construct histograms of discrete data (p. 77)	2	5(e) and (f)
	3 Organize continuous data in tables (p. 78)	3	3(a), 6(a) and (b), 7(a) and (b)
	4 Construct histograms of continuous data (p. 80)	4, 5	3(d) and (e), 6(c) and (d), 7(e) and (f)
	5 Draw stem-and-leaf plots (p. 82)	6, 7	8
	6 Draw dot plots (p. 85)	8	5(i)
	7 Identify the shape of a distribution (p. 85)	9	3(d), 5(e), 6(e), 7(e), 8
2.3	1 Construct frequency polygons (p. 94)	pp. 94–95	3(f)
	2 Create cumulative frequency and relative frequency tables (p. 95)	pp. 95–96	3(b) and (c), 5(c) and (d)
	3 Construct frequency and relative frequency ogives (p. 96)	p. 96	3(h) and (i), 7(c) and (d)
	4 Draw time-series graphs (p. 97)	1	10
2.4	1 Describe what can make a graph misleading or deceptive (p. 101)	1 through 6	9(c), 10, 11, 12

Review Exercises

1. Effective Commercial Harris Interactive conducted a poll of U.S. adults and asked, "When there is a voiceover in a commercial, which type of voice is more likely to sell me a car?" Results of the survey are in the bar graph.

Convincing Voice in Purchasing a Car

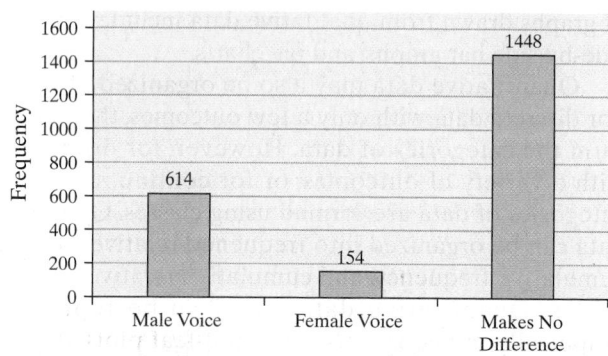

(a) How many participants were in the survey?
(b) What is the relative frequency of the respondents who indicated that it made no difference which voice they heard?
(c) Redraw the graph as a Pareto chart.
(d) Research automotive commercials. Do you believe that auto manufacturers use the results of this survey when developing their commercials?

2. Weapons Used in Homicide The following frequency distribution represents the cause of death in homicides for the year 2013.

Type of Weapon	Frequency
Firearms	8438
Knives or cutting instruments	1486
Personal weapons (hands, fists, etc.)	685
Other weapon or not stated	1621

Source: Crime in the United States, 2013, FBI, Uniform Crime Reports

(a) Construct a relative frequency distribution.
(b) What percentage of homicides was committed using a firearm?
(c) Construct a frequency bar graph.
(d) Construct a relative frequency bar graph.
(e) Construct a pie chart.

3. Live Births The following frequency distribution represents the number of live births (in thousands) in the United States in 2012 by age of mother.

Age of Mother (years)	Births (thousands)
10–14	3
15–19	275
20–24	902
25–29	1128
30–34	1044
35–39	487
40–44	110
45–49	7
50–54	1

Source: National Center for Health Statistics

(a) Construct a relative frequency distribution.
(b) Construct a cumulative frequency distribution.
(c) Construct a cumulative relative frequency distribution.
(d) Construct a frequency histogram. Describe the shape of the distribution.
(e) Construct a relative frequency histogram.
(f) Construct a frequency polygon.
(g) Construct a relative frequency polygon.
(h) Construct a frequency ogive.
(i) Construct a relative frequency ogive.
(j) What percentage of live births were to mothers aged 20 to 24?
(k) What percentage of live births were to mothers of age 30 or older?

4. Political Affiliation One hundred randomly selected registered voters in the city of Naperville was asked their political affiliation: Democrat (D), Republican (R), or Independent (I). The results of the survey are shown below.

D	R	D	R	D	R	D	D	R	D
R	D	D	D	R	R	D	D	D	D
R	R	I	I	D	R	D	R	R	R
I	D	D	R	I	I	R	D	R	R
D	I	R	D	D	D	D	I	I	R
R	I	R	R	I	D	D	D	D	R
D	I	I	D	D	R	R	R	R	D
D	R	R	R	D	D	I	I	D	D
D	D	I	D	R	I	D	D	D	D
R	R	R	R	R	D	R	D	R	D

(a) Construct a frequency distribution of the data.
(b) Construct a relative frequency distribution of the data.
(c) Construct a relative frequency bar graph of the data.
(d) Construct a pie chart of the data.
(e) What appears to be the most common political affiliation in Naperville?

5. Family Size A random sample of 60 couples married for seven years were asked to give the number of children they have. The results of the survey are as follows:

0	2	3	1	3	2
3	2	3	5	3	2
1	1	3	2	3	2
4	3	3	0	4	2
2	0	1	0	3	4
0	3	3	2	1	2
2	3	3	1	3	2
4	3	2	2	0	2
0	1	4	2	2	3
4	3	3	2	4	3

(a) Construct a frequency distribution of the data.
(b) Construct a relative frequency distribution of the data.
(c) Construct a cumulative frequency distribution of the data.
(d) Construct a cumulative relative frequency distribution of the data.
(e) Construct a frequency histogram of the data. Describe the shape of the distribution.
(f) Construct a relative frequency histogram of the data.
(g) What percentage of couples married seven years has two children?
(h) What percentage of couples married seven years has at least two children?
(i) Draw a dot plot of the data.

6. Home Ownership Rates The table shows the home ownership rate in each of the 50 states and Washington, DC, in 2013.

73.7	74.2	70.2	62.7	53.6	68.4	68.3	68.4
64.4	45.0	69.6	73.7	73.3	72.7	63.2	71.1
64.9	66.5	63.4	74.1	64.9	63.0	70.7	
64.5	63.8	66.4	76.5	67.0	72.9	74.6	
54.6	58.2	67.3	70.9	52.5	61.7	70.0	
64.4	69.3	73.3	67.8	66.1	72.9	62.3	
68.5	68.1	67.6	68.6	67.6	66.6	79.2	

Source: U.S. Census Bureau

Note: The state with the highest home ownership rate is West Virginia and the lowest is Washington, DC.
With a lower class limit of the first class of 45 and a class width of 5:

(a) Construct a frequency distribution.
(b) Construct a relative frequency distribution.
(c) Construct a frequency histogram of the data.
(d) Construct a relative frequency histogram of the data.
(e) Describe the shape of the distribution.
(f) Repeat parts (a)–(e) using a lower class limit for the first class of 40 and a class width of 10.
(g) Does one frequency distribution provide a better summary of the data than the other? Explain.

7. Diameter of a Cookie The following data represent the diameter (in inches) of a random sample of 34 Keebler Chips Deluxe™ Chocolate Chip Cookies.

2.3414	2.3010	2.2850	2.3015	2.2850	2.3019	2.2400
2.3005	2.2630	2.2853	2.3360	2.3696	2.3300	2.3290
2.2303	2.2600	2.2409	2.2020	2.3223	2.2851	2.2382
2.2438	2.3255	2.2597	2.3020	2.2658	2.2752	2.2256
2.2611	2.3006	2.2011	2.2790	2.2425	2.3003	

Source: Trina S. McNamara, student at Joliet Junior College

(a) Construct a frequency distribution.
(b) Construct a relative frequency distribution.
(c) Construct a cumulative frequency distribution.
(d) Construct a cumulative relative frequency distribution.
(e) Construct a frequency histogram. Describe the shape of the distribution.
(f) Construct a relative frequency histogram.

8. Time Online The following data represent the average number of hours per week that a random sample of 40 college students spend online. The data are based on the ECAR Study of Undergraduate Students and Information Technology, 2007. Construct a stem-and-leaf diagram of the data, and comment on the shape of the distribution.

18.9	14.0	24.4	17.4	13.7	16.5	14.8	20.8
22.9	22.2	13.4	18.8	15.1	21.9	21.1	14.7
18.6	18.0	21.1	15.6	16.6	20.6	17.3	17.9
15.2	16.4	14.5	17.1	25.7	17.4	18.8	17.1
13.6	20.1	15.3	19.2	23.4	14.5	18.6	23.8

9. Grade Inflation The side-by-side bar graph to the right shows the average grade point average for the years 1991–1992, 1996–1997, 2001–2002, and 2006–2007 for colleges and universities.

(a) Does the graph suggest that grade inflation is a problem in colleges?

(b) Determine the percentage increase in GPAs for public schools from 1991 to 2006. Determine the percentage increase in GPAs for private schools from 1991 to 2006. Which type of institution appears to have the higher inflation?

(c) Do you believe the graph is misleading? Explain.

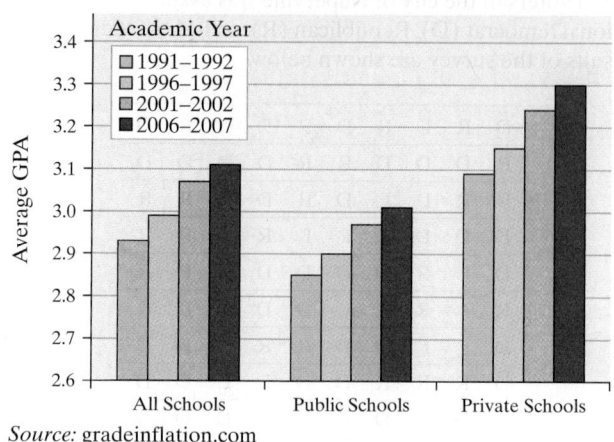

Recent GPA Trends Nationwide

Source: gradeinflation.com

10. Income Distribution The following data represent the percentage of total adjusted gross income (AGI) earned and percentage of tax paid by various income classes. The top 1% represents the percentage of total AGI earned and tax paid by those whose income is higher than 99% of all earners. The bottom 50% represents the percentage of total AGI earned and tax paid by those whose income is in the bottom 50% of all income earners. For example, in 2001, 17.53% of all income earned in the United States was earned by those in the top 1% of all income earners, while 13.81% of all income earned in the United States was earned by those in the bottom 50% of income earners.

Year	Income Share of Top 1% of Earners	Income Share of Bottom 50% of Earners	Income Tax Share of Top 1% of Earners	Income Tax Share of Bottom 50% of Earners
2001	17.53	13.81	33.89	3.97
2002	16.12	14.23	33.71	3.50
2003	16.77	13.99	34.27	3.46
2004	19.00	13.42	36.89	3.30
2005	21.20	12.83	39.38	3.07
2006	22.06	12.51	39.89	2.99
2007	22.83	12.26	40.41	2.89
2008	20.00	12.75	38.02	2.70
2009	17.21	11.88	36.34	2.46
2010	18.87	11.74	37.38	2.36
2011	18.70	11.55	35.06	2.89

Source: The Tax Foundation

(a) Use the data to make a strong argument that adjusted gross incomes are diverging among Americans.

(b) Use the data to make an argument that, while incomes are diverging between the top 1% and bottom 50%, the total taxes paid as a share of income are also diverging.

11. Misleading Graphs In 2013, the average earnings of a high school graduate were $30,286. At $50,738, the average earnings of a recipient of a bachelor's degree were about 66% higher.
Source: U.S. Census Bureau

(a) Construct a misleading graph that a college recruiter might create to convince high school students that they should attend college.

(b) Construct a graph that does not mislead.

12. High Heels The graphic to the right is a *USA Today*–type graph displaying women's preference for shoes.

(a) Which type of shoe is preferred the most? The least?

(b) How is the graph misleading?

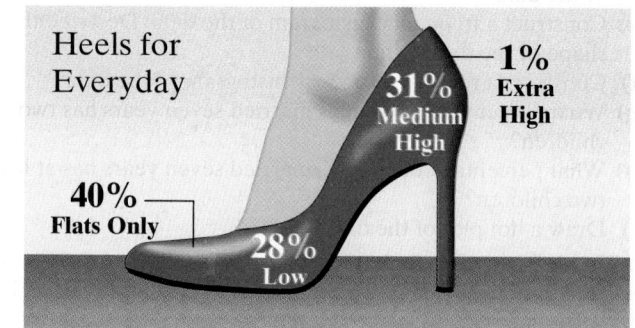

Chapter Test

1. The graph shows the ratings on Yelp for Hot Doug's Restaurant.

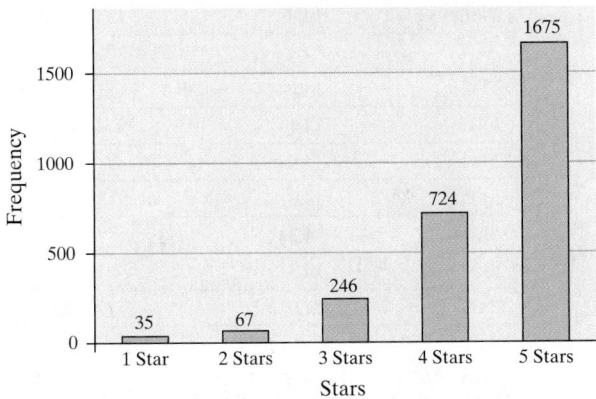

(a) Which was the most popular rating for Hot Doug's?
(b) How many ratings were posted on Hot Doug's?
(c) How many more 5 Star ratings are there than 4 Star ratings?
(d) What percentage of ratings are 5 Star ratings?
(e) Is it appropriate to describe the shape of the distribution as skewed left? Why or why not?

2. A random sample of 1005 adult Americans was asked, "How would you prefer to pay for new road construction?" Results of the survey are below.

Response	Frequency
New tolls	412
Increase gas tax	181
No new roads	412

Source: HNTB Corporation

(a) Construct a relative frequency distribution.
(b) What percent of the respondents indicated they would like to see an increase in gas taxes?
(c) Construct a frequency bar graph.
(d) Construct a relative frequency bar graph.
(e) Construct a pie chart.

3. Interested in knowing the educational background of its customers the Metra Train Company contracted a marketing firm to conduct a survey asking 50 randomly selected commuters at the train station to disclose their educational attainment. The results shown were obtained.

(a) Construct a frequency distribution of the data.
(b) Construct a relative frequency distribution of the data.
(c) Construct a relative frequency bar graph of the data.
(d) Construct a pie chart of the data.
(e) What is the most common educational level of a commuter?

No high school diploma	Some college
High school graduate	No high school diploma
Some college	No high school diploma
No high school diploma	High school graduate
Associate's degree	High school graduate
Bachelor's degree	High school graduate
Bachelor's degree	High school graduate
High school graduate	Associate's degree
Associate's degree	High school graduate
Some college	No high school diploma
Some college	High school graduate
High school graduate	Advanced degree
High school graduate	High school graduate
Bachelor's degree	Bachelor's degree
High school graduate	Some college
Bachelor's degree	Advanced degree
Advanced degree	No high school diploma
Bachelor's degree	High school graduate
Bachelor's degree	No high school diploma
Associate's degree	Some college
Advanced degree	Some college
High school graduate	No high school diploma
Bachelor's degree	No high school diploma
Some college	Some college
High school graduate	High school graduate

4. The following data represent the number of cars that arrived at a McDonald's drive-through between 11:50 A.M. and 12:00 noon each Wednesday for the past 50 weeks:

1	7	3	8	2	3	8	2	6	3
6	5	6	4	3	4	3	8	1	2
5	3	6	3	3	4	3	2	1	2
4	4	9	3	5	2	3	5	5	5
2	5	6	1	7	1	5	3	8	4

(a) Construct a frequency distribution of the data.
(b) Construct a relative frequency distribution of the data.
(c) Construct a cumulative frequency distribution of the data.
(d) Construct a cumulative relative frequency distribution of the data.
(e) Construct a frequency histogram of the data. Describe the shape of the distribution.
(f) Construct a relative frequency histogram of the data.
(g) What percentage of weeks did exactly three cars arrive between 11:50 A.M. and 12:00 noon?
(h) What percentage of weeks did three or more cars arrive between 11:50 A.M. and 12:00 noon?
(i) Draw a dot plot of the data.

5. Dr. Paul Oswiecmiski randomly selects 40 of his 20- to 29-year-old patients and obtains the following data regarding their serum HDL cholesterol:

70	56	48	48	53	52	66	48
36	49	28	35	58	62	45	60
38	73	45	51	56	51	46	39
56	32	44	60	51	44	63	50
46	69	53	70	33	54	55	52

Source: Paul Oswiecmiski

(a) Construct a frequency distribution.
(b) Construct a relative frequency distribution.
(c) Construct a frequency histogram of the data.
(d) Construct a relative frequency histogram of the data.
(e) Describe the shape of the distribution.

6. The following frequency distribution represents the closing prices (in dollars) of homes sold in a Midwest city.

Closing Price ($)	Number of Houses
50,000–99,999	4
100,000–149,999	13
150,000–199,999	19
200,000–249,999	7
250,000–299,999	3
300,000–349,999	2
350,000–399,999	1
400,000–449,999	0
450,000–499,999	1

(a) Construct a cumulative frequency distribution.
(b) Construct a cumulative relative frequency distribution.
(c) What percent of homes sold for less than $200,000?
(d) Construct a frequency polygon. Describe the shape of the distribution.
(e) Construct a frequency ogive.
(f) Construct a relative frequency ogive.

7. The following data represent the time (in minutes) students spent working their Section 1.1 homework from Sullivan's College Algebra course (based on time logged into MyMathLab). Draw a stem-and-leaf diagram of the data and comment on the shape of the distribution.

46	47	110	56	71	109
63	91	111	93	125	78
85	108	73	118	70	89
99	45	73	125	96	109
110	61	40	52	103	126

8. The data below shows birth rate and per capita income (in thousands of 2011 dollars) from 2000 through 2012. Draw a time-series plot for both birth rate and per capita income. Comment on any similarities in the trends.

Year	Birth Rate (births per 1000 women age 15–44)	Per Capita Income (thousands of 2011 dollars)
2000	14.4	29.8
2001	14.1	29.6
2002	14.0	29.1
2003	14.1	29.1
2004	14.0	29.0
2005	14.0	29.4
2006	14.3	30.0
2007	14.3	29.7
2008	14.0	28.8
2009	13.5	28.4
2010	13.0	28.0
2011	12.7	28.1
2012	12.6	28.2

Source: Centers for Disease Control and U.S. Census Bureau

9. The following is a *USA Today*-type graph. Do you think the graph is misleading? Why? If you think it is misleading, what might be done to improve the graph?

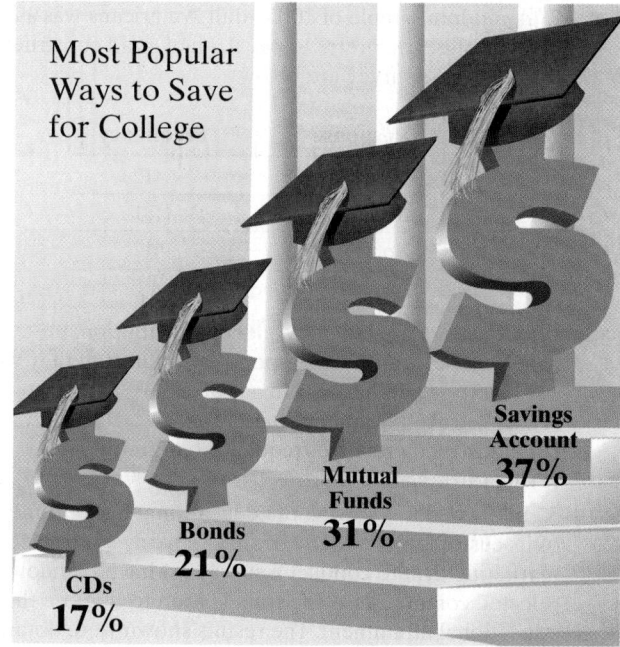

Most Popular Ways to Save for College

Savings Account **37%**

Mutual Funds **31%**

Bonds **21%**

CDs **17%**

Making an Informed Decision

Tables or Graphs?

You work for the school newspaper. Your editor approaches you with a special reporting assignment. Your task is to write an article that describes the "typical" student at your school, complete with supporting information. To write this article, you have to survey at least 40 students and ask them to respond to a questionnaire. The editor would like to have at least two qualitative and two quantitative variables that describe the typical student. The results of the survey will be presented in your article, but you are unsure whether you should present tabular or graphical summaries, so you decide to perform the following experiment.

1. Develop a questionnaire that results in obtaining the values of two qualitative and two quantitative variables. Administer the questionnaire to at least 40 students on your campus.
2. Summarize the data in both tabular and graphical form.
3. Select 20 individuals. (They don't have to be students at your school.) Give the tabular summaries to 10 individuals and the graphical summaries to the other 10. Ask each individual to study the table or graph for 5 seconds. After 1 minute, give a questionnaire that asks various questions regarding the information contained in

the table or graph. For example, if you summarize age data, ask the individual which age group has the highest frequency. Record the number of correct answers for each individual. Which summary results in a higher percentage of correct answers, the tables or the graphs? Write a report that discusses your findings.

4. Now use the data collected from the questionnaire to create some misleading graphs. Again, select 20 individuals. Give 10 individuals the misleading graphs and 10 individuals the correct graphs. Ask each individual to study each graph for 5 seconds. After 1 minute has elapsed, give a questionnaire that asks various questions regarding the information contained in the graphs. Record the number of correct answers for each individual. Did the misleading graphs mislead? Write a report that discusses your findings.

Note: Be sure to check with your school's administration regarding privacy laws and policies regarding studies involving human subjects.

 **CASE STUDY** The Day the Sky Roared

Shortly after daybreak on April 3, 1974, thunder rumbled through the dark skies that covered much of the midwestern United States. Lightning struck areas from the Gulf Coast states to the Canadian border. By the predawn hours of the next day, the affected region of around 490,000 acres was devastated by more than 100 tornadoes. This "super outbreak" was responsible for the deaths of more than 300 people in 11 states. More than 6100 people were injured by the storms, with approximately 27,500 families suffering some kind of loss. The total cost attributed to the disaster was more than $600 million. Amazingly, the storm resulted in six category five tornadoes with wind speeds exceeding 260 miles per hour. To put this figure in perspective, the region endured about one decade's worth of category five tornadoes in a single 24-hour period!

Fujita Wind Damage Scale		
F-scale	**Wind Speed (mph)**	**Damage**
F-0	Up to 72	Light
F-1	73 to 112	Moderate
F-2	113 to 157	Considerable
F-3	158 to 206	Severe
F-4	207 to 260	Devastating
F-5	Above 260	Incredible

Structural engineers and meteorologists are interested in understanding catastrophic events such as this tornado outbreak. Variables such as tornado intensity (as described by the F-scale), tornado duration (time spent by the tornado in contact with the ground), and death demographics can provide insights into these events and their impact on the human population. The following table lists the duration time and F-scale for each tornado in the April 1974 super outbreak.

Tornado Duration Times	
F-scale	**Tornado Duration (minutes)**
F-0	1, 1, 5, 1, 1, 6, 4, 10, 5, 4, 1, 1, 1, 1, 1, 1, 1, 1, 1, 30, 1, 9
F-1	16, 13, 9, 8, 13, 10, 15, 1, 17, 23, 10, 8, 12, 5, 20, 31, 12, 5, 30, 13, 7, 1, 5, 13, 1, 2, 5, 10, 1, 20, 5
F-2	7, 15, 2, 10, 23, 10, 7, 12, 8, 1, 8, 19, 5, 10, 15, 20, 10, 13, 20, 15, 13, 14, 1, 4, 2, 15, 30, 91, 11, 5
F-3	9, 20, 8, 16, 26, 36, 10, 20, 50, 17, 26, 31, 21, 30, 23, 28, 23, 18, 35, 35, 15, 25, 30, 15, 22, 18, 58, 19, 23, 31, 13, 26, 40, 14, 11
F-4	120, 23, 23, 42, 47, 25, 22, 22, 34, 50, 38, 28, 39, 29, 28, 25, 34, 16, 40, 55, 124, 30, 30, 31
F-5	37, 69, 23, 52, 61, 122

The following table presents the number of deaths as a function of F-scale and community size:

Deaths as a Function of F-scale and Community Size			
F-Scale	**Deaths**	**Community Size**	**Deaths**
F-0	0	Rural areas	99
F-1	0	Small communities	77
F-2	14	Small cities	63
F-3	32	Medium cities	56
F-4	129	Large cities	10
F-5	130		

Create a report that graphically displays and discusses the tornado-related data. Your report should include the following:

1. A bar graph or pie chart (or both) that depicts the number of tornadoes by F-scale. Generally, only a little more than 1% of all tornadoes exceed F-3 on the Fujita Wind Damage Scale. How does the frequency of the most severe tornadoes of the April 3–4, 1974, outbreak compare with normal tornado formation?

2. A histogram that displays the distribution of tornado duration for all the tornadoes.

3. Six histograms displaying tornado duration for each of the F-scale categories. Does there appear to be a relationship between duration and intensity? If so, describe this relationship.

4. A bar chart that shows the relationship between the number of deaths and tornado intensity. Ordinarily, the most severe tornadoes (F-4 and F-5) account for more than 70% of deaths. Is the death distribution of this outbreak consistent with this observation?

5. A bar chart that shows the relationship between the number of deaths and community size. Are tornadoes more likely to strike rural areas? Include a discussion describing the number of deaths as a function of community size.

6. A general summary of your findings and conclusions.

Data Source: Abbey, Robert F., and T. Theodore Fujita. "Tornadoes: The Tornado Outbreak of 3–4 April 1974." In *The Thunderstorm in Human Affairs,* 2nd ed., edited by Edwin Kessler, 37–66. Norman, OK: University of Oklahoma Press, 1983. The death figures presented in this case study are based on approximations made from charts by Abbey and Fujita. Additional descriptions of events and normal tornado statistics are derived from Jack Williams's *The Weather Book* (New York: Vintage Books, 1992).

3

Numerically Summarizing Data

Making an Informed Decision

Suppose that you are in the market for a used car. To make an informed decision regarding your purchase, you decide to collect as much information as possible. What information is important in helping you make this decision? See the Decisions project on page 178.

PUTTING IT TOGETHER

When we look at a distribution of data, we should consider three characteristics of the distribution: shape, center, and spread. In the last chapter, we discussed methods for organizing raw data into tables and graphs. These graphs (such as the histogram) allow us to identify the shape of the distribution: symmetric (in particular, bell shaped or uniform), skewed right, or skewed left.

The center and spread are numerical summaries of the data. The center of a data set is commonly called the *average*. There are many ways to describe the *average* value of a distribution. In addition, there are many ways to measure the spread of a distribution. The most appropriate measure of center and spread depends on the distribution's shape.

Once these three characteristics of the distribution are known, we can analyze the data for interesting features, including unusual data values, called *outliers*.

3.1 Measures of Central Tendency

Preparing for This Section Before getting started, review the following:

- Population versus sample (Section 1.1, p. 5)
- Parameter versus statistic (Section 1.1, p. 5)
- Quantitative data (Section 1.1, p. 8)
- Qualitative data (Section 1.1, p. 8)
- Simple random sampling (Section 1.3, pp. 21–25)

Objectives

1. Determine the arithmetic mean of a variable from raw data
2. Determine the median of a variable from raw data
3. Explain what it means for a statistic to be resistant
4. Determine the mode of a variable from raw data

A measure of central tendency numerically describes the average or typical data value. We hear the word *average* in the news all the time:

- The average miles per gallon of gasoline of the 2015 Chevrolet Corvette Z06 in highway driving is 29.
- According to the U.S. Census Bureau, the national average commute time to work in 2013 was 25.4 minutes.
- According to the U.S. Census Bureau, the average household income in 2013 was $51,939.
- The average American woman is 5'4" tall and weighs 142 pounds.

> **CAUTION!**
> Whenever you hear the word *average*, be aware that the word may not always be referring to the mean. One average could be used to support one position, while another average could be used to support a different position.

In this chapter, we discuss the three most widely-used measures of central tendency: the *mean*, the *median*, and the *mode*. In the media (newspapers, blogs, and so on), *average* usually refers to the mean. But beware: some reporters use *average* to refer to the median or mode. As we shall see, these three measures of central tendency can give very different results!

1 Determine the Arithmetic Mean of a Variable from Raw Data

In everyday language, the word *average* often represents the arithmetic mean. To compute the arithmetic mean of a set of data, the data must be quantitative.

Definitions

The **arithmetic mean** of a variable is computed by adding all the values of the variable in the data set and dividing by the number of observations. The **population arithmetic mean**, μ (pronounced "mew"), is computed using all the individuals in a population. The population mean is a parameter.

The **sample arithmetic mean**, $\bar{x}$ (pronounced "x-bar"), is computed using sample data. The sample mean is a statistic.

While other types of means exist (see Problems 39 and 40), the arithmetic mean is generally referred to as the **mean**. We will follow this practice for the remainder of the text.

We usually use Greek letters to represent parameters and Roman letters (such as x or s) to represent statistics. The formulas for computing population and sample means follow:

In Other Words
To find the mean of a set of data, add up all the observations and divide by the number of observations.

If $x_1, x_2, \ldots, x_N$ are the N observations of a variable from a population, then the population mean, μ, is

$$\mu = \frac{x_1 + x_2 + \cdots + x_N}{N} = \frac{\sum x_i}{N} \qquad \textbf{(1)}$$

If $x_1, x_2, \ldots, x_n$ are n observations of a variable from a sample, then the sample mean, $\bar{x}$, is

$$\bar{x} = \frac{x_1 + x_2 + \cdots + x_n}{n} = \frac{\sum x_i}{n} \qquad (2)$$

Note that N represents the size of the population, and n represents the size of the sample. The symbol Σ (the Greek letter capital sigma) tells us to add the terms. The subscript i shows that the various values are distinct and does not serve as a mathematical operation. For example, x_1 is the first data value, x_2 is the second, and so on.

EXAMPLE 1   Computing a Population Mean and a Sample Mean

Problem The data in Table 1 represent the first exam score of 10 students enrolled in Introductory Statistics. Treat the 10 students as a population.
(a) Compute the population mean.
(b) Find a simple random sample of size $n = 4$ students.
(c) Compute the sample mean of the sample found in part (b).

Approach
(a) To compute the population mean, add all the data values (test scores) and divide by the number of individuals in the population.
(b) Recall from Section 1.3 that we can use Table I in Appendix A, a calculator with a random-number generator, or computer software to obtain simple random samples. We will use a TI-84 Plus C graphing calculator.
(c) Find the sample mean by adding the data values corresponding to the individuals in the sample and then dividing by $n = 4$, the sample size.

Solution
(a) Compute the population mean by adding the scores of all 10 students:

$$\sum x_i = x_1 + x_2 + x_3 + \cdots + x_{10}$$
$$= 82 + 77 + 90 + 71 + 62 + 68 + 74 + 84 + 94 + 88$$
$$= 790$$

Divide this result by 10, the number of students in the class.

$$\mu = \frac{\sum x_i}{N} = \frac{790}{10} = 79$$

Although it was not necessary in this problem, we will agree to round the mean to one more decimal place than that in the raw data.

(b) To find a simple random sample of size $n = 4$ from a population of size $N = 10$, we will use the TI-84 Plus C random-number generator with a seed of 14. (Recall that this gives the calculator its starting point to generate the list of random numbers.) Figure 1 shows the students in the sample: Jennifer (62), Juan (88), Ryanne (77), and Dave (68).

(c) Compute the sample mean by adding the scores of the four students:

$$\sum x_i = x_1 + x_2 + x_3 + x_4$$
$$= 62 + 88 + 77 + 68$$
$$= 295$$

Divide this result by 4, the number of individuals in the sample.

$$\bar{x} = \frac{\sum x_i}{n} = \frac{295}{4} = 73.8 \qquad \text{Round to the nearest tenth.}$$

Table 1

Student	Score
1. Michelle	82
2. Ryanne	77
3. Bilal	90
4. Pam	71
5. Jennifer	62
6. Dave	68
7. Joel	74
8. Sam	84
9. Justine	94
10. Juan	88

Figure 1

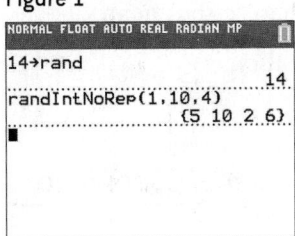

• **Now Work Problem 21**

It helps to think of the mean of a data set as the center of gravity. In other words, the mean is the value such that a histogram of the data is perfectly balanced, with equal weight on each side of the mean. Figure 2 shows a histogram of the data in Table 1 with the mean labeled. The histogram balances at $\mu = 79$.

Figure 2

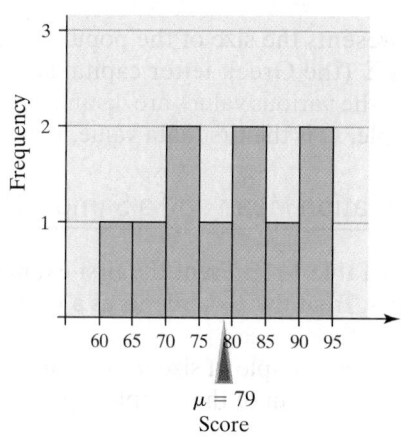

Scores on First Exam

$\mu = 79$
Score

② Determine the Median of a Variable from Raw Data

A second measure of central tendency is the median. To compute the median of a set of data, the data must be quantitative.

Definition

The **median** of a variable is the value that lies in the middle of the data when arranged in ascending order. We use M to represent the median.

In Other Words
To help remember the idea behind the median, think of the median of a highway; it divides the highway in half. So the median divides the data in half, with at most half the data below the median and at most half above it.

Steps in Finding the Median of a Data Set

Step 1 Arrange the data in ascending order.

Step 2 Determine the number of observations, n.

Step 3 Determine the observation in the middle of the data set.

- If the number of observations is odd, then the median is the data value exactly in the middle of the data set. That is, the median is the observation that lies in the $\dfrac{n+1}{2}$ position.

- If the number of observations is even, then the median is the mean of the two middle observations in the data set. That is, the median is the mean of the observations that lie in the $\dfrac{n}{2}$ position and the $\dfrac{n}{2} + 1$ position.

EXAMPLE 2 Determining the Median of a Data Set with an Odd Number of Observations

Problem The data in Table 2 represent the length (in seconds) of a random sample of songs released in the 1970s. Find the median length of the songs.

Approach Follow the steps listed above.

Table 2	
Song Name	**Length**
"Sister Golden Hair"	201
"Black Water"	257
"Free Bird"	284
"The Hustle"	208
"Southern Nights"	179
"Stayin' Alive"	222
"We Are Family"	217
"Heart of Glass"	206
"My Sharona"	240

Solution

Step 1   Arrange the data in ascending order:

$$179, 201, 206, 208, 217, 222, 240, 257, 284$$

Step 2   There are $n = 9$ observations.

Step 3   Since n is odd, the median, M, is the observation exactly in the middle of the data set, 217 seconds (the $\dfrac{n+1}{2} = \dfrac{9+1}{2} = 5$th data value). We list the data in ascending order and show the median in blue.

$$179, 201, 206, 208, 217, 222, 240, 257, 284$$

Notice there are four observations on each side of the median.

EXAMPLE 3   **Determining the Median of a Data Set with an Even Number of Observations**

Problem   Find the median score of the data in Table 1 on page 119.

Approach   Follow the steps listed on the previous page.

Solution

Step 1   Arrange the data in ascending order:

$$62, 68, 71, 74, 77, 82, 84, 88, 90, 94$$

Step 2   There are $n = 10$ observations.

Step 3   Because n is even, the median is the mean of the two middle observations, the fifth $\left(\dfrac{n}{2} = \dfrac{10}{2} = 5\right)$ and sixth $\left(\dfrac{n}{2} + 1 = \dfrac{10}{2} + 1 = 6\right)$ observations with the data written in ascending order. So the median is the mean of 77 and 82:

$$M = \frac{77 + 82}{2} = 79.5$$

Notice that there are five observations on each side of the median.

$$62, 68, 71, 74, 77, 82, 84, 88, 90, 94$$
$$\uparrow$$
$$M = 79.5$$

• Now compute the median of the data in Problem 15 by hand

Notice that 50% (or half) of the students scored less than 79.5 and 50% (or half) of the students scored above 79.5.

EXAMPLE 4   **Finding the Mean and Median Using Technology**

Problem   Use statistical software or a calculator to determine the population mean and median of the student test score data in Table 1 on page 119.

Approach   We will use StatCrunch to obtain the mean and median. The steps for obtaining measures of central tendency using the TI-83/84 Plus graphing calculator, Minitab, Excel, and StatCrunch are given in the Technology Step-by-Step on page 126.

Figure 3

Summary statistics:

Column	n	Mean	Median
Scores	10	79	79.5

Solution   Figure 3 shows the output obtained from StatCrunch. The results agree with the "by hand" solution from Examples 1 and 3.

3 Explain What It Means for a Statistic to Be Resistant

Which measure of central tendency is better to use—the mean or the median? It depends.

EXAMPLE 5   Comparing the Mean and the Median

Problem Yolanda wants to know how much time she typically spends on her cell phone. She goes to her phone's website and records the call length for a random sample of 12 calls, shown in Table 3. Find the mean and median length of a cell phone call. Which measure of central tendency better describes the length of a typical phone call?

Table 3

1	7	4	1
2	4	3	48
3	5	3	6

Source: Yolanda Sullivan's cell phone records

Approach We will find the mean and median using Minitab. To help judge which is the better measure of central tendency, we will also draw a dot plot of the data.

Solution Figure 4 indicates that the mean call length is $\bar{x} = 7.3$ minutes and the median call length is 3.5 minutes. Figure 5 shows a dot plot of the data using Minitab.

Figure 4

Descriptive Statistics: TalkTime

Variable	N	N*	Mean	SE Mean	StDev	Minimum	Q1	Median	Q3	Maximum
TalkTime	12	0	7.25	3.74	12.96	1.00	2.25	3.50	5.75	48.00

Figure 5

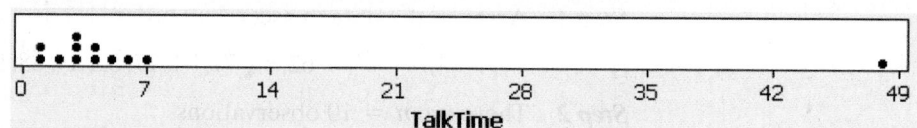

Which measure of central tendency do you think better describes the typical call length? Since only one phone call is longer than the mean, the mean is not representative of the typical call length. So the median is the better measure of central tendency for this data set. •

Look back at the data in Table 3. Suppose Yolanda's 48-minute call was actually a 5-minute call. Then the mean call length would be 3.7 minutes and the median call length would still be 3.5 minutes. So one extreme observation (48 minutes) can cause the mean to increase substantially, but have no effect on the median. In other words, the mean is sensitive to extreme values while the median is not. In fact, if Yolanda's 48-minute call had actually been 148 minutes long, the median would still be 3.5 minutes, but the mean would increase to 15.6 minutes. The median is based on the value of the middle observation, so the value of the largest observation does not affect its computation.

Definition   A numerical summary of data is said to be **resistant** if extreme values (very large or small) relative to the data do not affect its value substantially.

So the median is resistant, while the mean is not resistant.

When data are skewed, there are extreme values in the tail, which tend to pull the mean in the direction of the tail. For example, in skewed-right distributions, there are large observations in the right tail. These observations increase the value of the mean, but have little effect on the median. Similarly, if a distribution is skewed left, the mean tends to be smaller than the median. In symmetric distributions, the mean and the median are close in value. We summarize these ideas in Table 4 and Figure 6.

Table 4

Relation between the Mean, Median, and Distribution Shape

Distribution Shape	Mean versus Median
Skewed left	Mean substantially smaller than median
Symmetric	Mean roughly equal to median
Skewed right	Mean substantially larger than median

Figure 6

Mean or median versus skewness

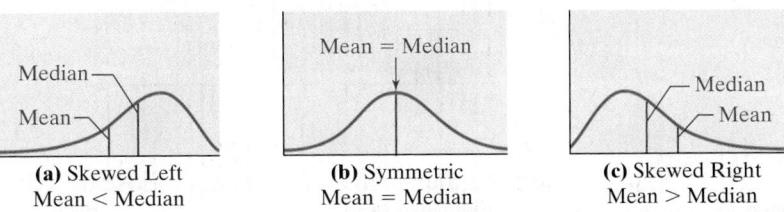

(a) Skewed Left
Mean < Median

(b) Symmetric
Mean = Median

(c) Skewed Right
Mean > Median

A word of caution is in order. The relation between the mean, median, and skewness are guidelines. The guidelines tend to hold up well for continuous data, but when the data are discrete, the rules can be easily violated. See Problem 41.*

You may be asking yourself, "Why would I ever compute the mean?" After all, the mean and median are close in value for symmetric data, and the median is the better measure of central tendency for skewed data. The reason we compute the mean is that much of the statistical inference that we perform is based on the mean. We will have more to say about this in Chapter 8. Plus, the mean uses all the data, while the median relies only on the position of the data.

EXAMPLE 6   Describing the Shape of a Distribution

Table 5

5.8	7.4	9.2	7.0	8.5	7.6
7.9	7.8	7.9	7.7	9.0	7.1
8.7	7.2	6.1	7.2	7.1	7.2
7.9	5.9	7.0	7.8	7.2	7.5
7.3	6.4	7.4	8.2	9.1	7.3
9.4	6.8	7.0	8.1	8.0	7.5
7.3	6.9	6.9	6.4	7.8	8.7
7.1	7.0	7.0	7.4	8.2	7.2
7.6	6.7				

Problem The data in Table 5 represent the birth weights (in pounds) of 50 randomly sampled babies.
(a) Find the mean and the median birth weight.
(b) Describe the shape of the distribution.
(c) Which measure of central tendency better describes the average birth weight?

Approach
(a) This can be done either by hand or technology. We will use a TI-84 Plus C.
(b) Draw a histogram to identify the shape of the distribution.
(c) If the data are roughly symmetric, the mean is the better measure of central tendency. If the data are skewed, the median is the better measure.

Solution
(a) Using a TI-84 Plus C, we find $\bar{x} = 7.49$ and $M = 7.35$. See Figure 7.

Figure 7

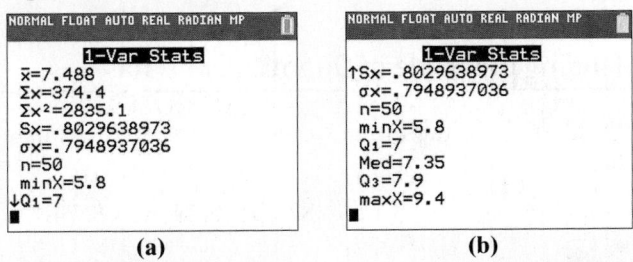

(a)

(b)

(b) Figure 8 on the following page shows the frequency histogram with the mean and median labeled. The distribution is bell shaped. We have further evidence of the shape because the mean and median are close to each other.

(continued)

*This idea is discussed in "Mean, Median, and Skew: Correcting a Textbook Rule" by Paul T. von Hippel. *Journal of Statistics Education*, Volume 13, Number 2 (2005).

Figure 8 Birth weights of 50 randomly selected babies

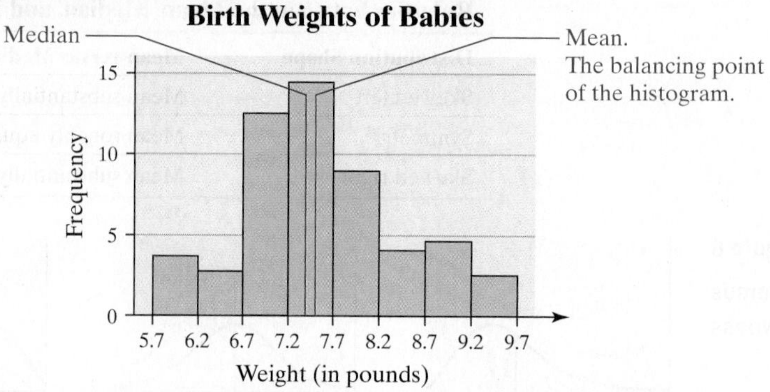

Birth Weights of Babies

(c) Because the mean and median are close in value, we use the mean as the measure of central tendency.

• Now Work Problem 23

---

④ **Determine the Mode of a Variable from Raw Data**

A third measure of central tendency is the mode, which can be computed for either quantitative or qualitative data.

Definition   The **mode** of a variable is the most frequent observation of the variable that occurs in the data set.

To compute the mode, tally the number of observations that occur for each data value. The data value that occurs most often is the mode. A set of data can have no mode, one mode, or more than one mode. If no observation occurs more than once, we say the data have **no mode**.

EXAMPLE 7   Finding the Mode of Quantitative Data

Problem The following data represent the number of O-ring failures on the shuttle *Columbia* for its 17 flights prior to its fatal flight:

$$0, 0, 0, 0, 0, 0, 0, 0, 0, 0, 0, 1, 1, 1, 1, 2, 3$$

Find the mode number of O-ring failures.

Approach Tally the number of times we observe each data value. The data value with the highest frequency is the mode.

Solution The mode is 0 because it occurs most frequently (11 times).

EXAMPLE 8   Finding the Mode of Quantitative Data

Problem Find the mode of the exam score data listed in Table 1, which is repeated here:

$$82, 77, 90, 71, 62, 68, 74, 84, 94, 88$$

Approach Tally the number of times we observe each data value. The data value with the highest frequency is the mode.

• Now compute the mode of the data in Problem 15

Solution Since each data value occurs only once, there is no mode.

A data set can have more than one mode. For example, if the data set in Table 1 had two scores of 77 and 88, the data set would have two modes: 77 and 88. In this case, we say the data are **bimodal.** If a data set has three or more data values that occur with the highest frequency, the data set is **multimodal.** The mode is usually not reported for multimodal data because it is not representative of a typical value. Figure 9(a) shows a distribution with one mode. Figure 9(b) shows a distribution that is bimodal.

Figure 9

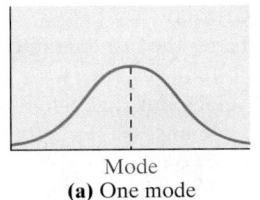

(a) One mode

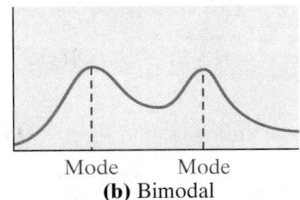

(b) Bimodal

In Other Words
Remember, nominal data are qualitative data that cannot be written in any meaningful order.

We cannot determine the value of the mean or median of data that are nominal. The only measure of central tendency that can be determined for nominal data is the mode.

EXAMPLE 9   Determining the Mode of Qualitative Data

Problem The data in Table 6 represent the location of injuries that required rehabilitation by a physical therapist. Determine the mode location of injury.

Table 6

Hip	Back	Back	Back	Hand	Neck	Knee	Knee	Knee	Hand
Knee	Shoulder	Wrist	Back	Groin	Shoulder	Shoulder	Back	Knee	Back
Hip	Shoulder	Elbow	Back	Back	Back	Back	Back	Back	Wrist

Source: Krystal Catton, student at Joliet Junior College

Approach Determine the location of injury that occurs with the highest frequency.

● **Now Work Problem 29**   **Solution** The mode location of injury is the back, with 12 instances.   ●

Summary: Measures of Central Tendency			
Measure of Central Tendency	**Computation**	**Interpretation**	**When to Use**
Mean	Population mean: $$\mu = \frac{\Sigma x_i}{N}$$ Sample mean: $$\bar{x} = \frac{\Sigma x_i}{n}$$	Center of gravity	When data are quantitative and the frequency distribution is roughly symmetric
Median	Arrange data in ascending order and divide the data set in half	Divides the bottom 50% of the data from the top 50%	When the data are quantitative and the frequency distribution is skewed left or right
Mode	Tally data to determine most frequent observation	Most frequent observation	When the most frequent observation is the desired measure of central tendency or the data are qualitative

Technology Step-by-Step Determining the Mean and Median

TI-83/84 Plus
1. Enter raw data in L1 by pressing STAT and then selecting 1 : Edit.
2. Press STAT, highlight the CALC menu, and select 1 : 1-Var Stats. In the menu, select L1 for List. Clear the entry in FreqList:. Highlight Calculate and press ENTER.

Minitab
1. Enter the data in C1.
2. Select the **Stat** menu, highlight **Basic Statistics,** and then select **Display Descriptive Statistics.**
3. In the **Variables** window, enter C1. Click OK.

Excel
1. Enter the data in column A.
2. Select the **Data** menu and click **Data Analysis.**

3. In the Data Analysis window, highlight **Descriptive Statistics** and click OK.
4. With the cursor in the **Input Range** window, use the mouse to highlight the data in column A.
5. Select the **Summary statistics** option and click OK.

StatCrunch
1. Enter the raw data into the spreadsheet. Name the column variable.
2. Select **Stat**, highlight **Summary Stats**, and select **Columns**.
3. Click on the variable you want to summarize. If you wish to compute certain statistics, hold down the Control (Ctrl) key when selecting the statistic (or Command on an Apple). Click Compute!.

3.1 Assess Your Understanding

Vocabulary and Skill Building

1. What does it mean if a statistic is resistant?

2. In the 2013 Current Population Survey conducted by the U.S. Census Bureau, two average household incomes were reported: $51,939 and $72,641. One of these averages is the mean and the other is the median. Which is the mean? Support your answer.

3. The U.S. Department of Housing and Urban Development (HUD) uses the median to report the average price of a home in the United States. Why do you think HUD uses the median?

4. A histogram of a set of data indicates that the distribution of the data is skewed right. Which measure of central tendency will likely be larger, the mean or the median? Why?

5. If a data set contains 10,000 values arranged in increasing order, where is the median located?

6. *True or False:* A data set will always have exactly one mode.

In Problems 7–10, find the population mean or sample mean as indicated.

7. Sample: 20, 13, 4, 8, 10

8. Sample: 83, 65, 91, 87, 84

9. Population: 3, 6, 10, 12, 14

10. Population: 1, 19, 25, 15, 12, 16, 28, 13, 6

11. For Super Bowl XLVII, there were 82,566 tickets sold for a total value of $218,469,636. What was the mean price per ticket?

12. The median for the given set of six ordered data values is 26.5. What is the missing value? 7 12 21 _____ 41 50

13. **Miles per Gallon** The following data represent the miles per gallon for a 2013 Ford Fusion for six randomly selected vehicles. Compute the mean, median, and mode miles per gallon.
Source: www.fueleconomy.gov

34.0, 33.2, 37.0, 29.4, 23.6, 25.9

14. **Exam Time** The following data represent the amount of time (in minutes) a random sample of eight students took to complete the online portion of an exam in Sullivan's Statistics course. Compute the mean, median, and mode time.

60.5, 128.0, 84.6, 122.3, 78.9, 94.7, 85.9, 89.9

NW 15. **Concrete Mix** A certain type of concrete mix is designed to withstand 3000 pounds per square inch (psi) of pressure. The concrete is poured into casting cylinders and allowed to set for 28 days. The concrete's strength is then measured. The following data represent the strength of nine randomly selected casts (in psi). Compute the mean, median, and mode strength of the concrete (in psi).

3960, 4090, 3200, 3100, 2940, 3830, 4090, 4040, 3780

16. **Flight Time** The following data represent the flight time (in minutes) of a random sample of seven flights from Las Vegas, Nevada, to Newark, New Jersey, on United Airlines. Compute the mean, median, and mode flight time.

282, 270, 260, 266, 257, 260, 267

17. For each of the three histograms shown, determine whether the mean is greater than, less than, or approximately equal to the median. Justify your answer.

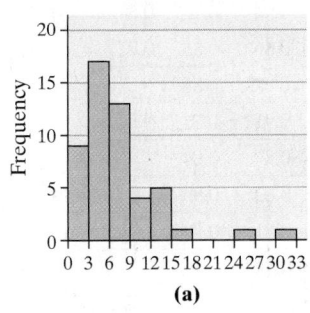

(a)

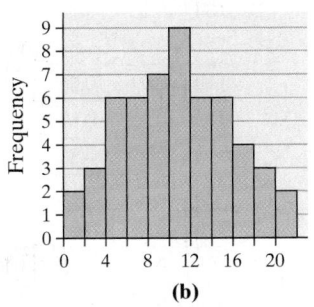

(b)

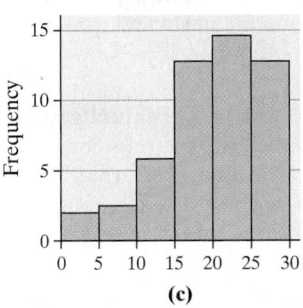

(c)

18. Match the histograms shown to the summary statistics:

	Mean	Median
I	42	42
II	31	36
III	31	26
IV	31	32

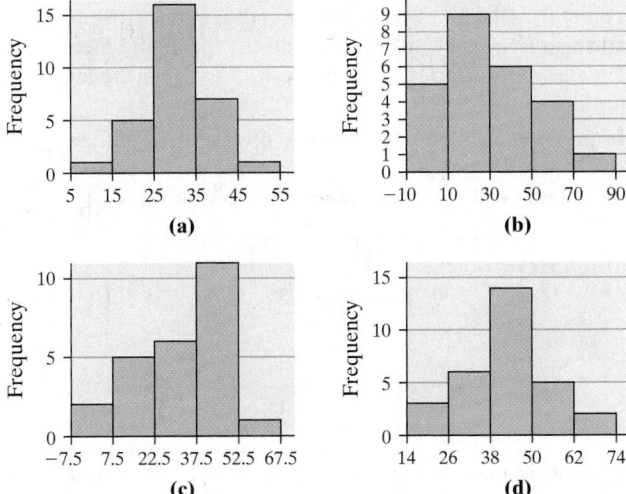

(a)   (b)   (c)   (d)

Applying the Concepts

19. Exam Scores The data in the next column represent exam scores in a statistics class taught using traditional lecture and a class taught using a "flipped" classroom. The "flipped" classroom is one where the content is delivered via video and watched at home, while class time is used for homework and activities.

Traditional	70.8	69.1	79.4	67.6	85.3	78.2	56.2
	81.3	80.9	71.5	63.7	69.8	59.8	
Flipped	76.4	71.6	63.4	72.4	77.9	91.8	78.9
	76.8	82.1	70.2	91.5	77.8	76.5	

Source: Michael Sullivan

(a) Determine the mean and median score for each class. Comment on any differences.
(b) Suppose the score of 59.8 in the traditional course was incorrectly recorded as 598. How does this affect the mean? the median? What property does this illustrate?

20. pH in Water The acidity or alkalinity of a solution is measured using pH. A pH less than 7 is acidic; a pH greater than 7 is alkaline. The following data represent the pH in samples of bottled water and tap water.

Tap	7.64	7.45	7.47	7.50	7.68	7.69
	7.45	7.10	7.56	7.47	7.52	7.47
Bottled	5.15	5.09	5.26	5.20	5.02	5.23
	5.28	5.26	5.13	5.26	5.21	5.24

Source: Emily McCarney, student at Joliet Junior College

(a) Determine the mean, median, and mode pH for each type of water. Comment on the differences between the two water types.
(b) Suppose the pH of 7.10 in tap water was incorrectly recorded as 1.70. How does this affect the mean? the median? What property of the median does this illustrate?

NW 21. Pulse Rates The following data represent the pulse rates (beats per minute) of nine students enrolled in a section of Sullivan's Introductory Statistics course. Treat the nine students as a population.

Student	Pulse
Perpectual Bempah	76
Megan Brooks	60
Jeff Honeycutt	60
Clarice Jefferson	81
Crystal Kurtenbach	72
Janette Lantka	80
Kevin McCarthy	80
Tammy Ohm	68
Kathy Wojdyla	73

(a) Determine the population mean pulse.
(b) Find three simple random samples of size 3 and determine the sample mean pulse of each sample.
(c) Which samples result in a sample mean that overestimates the population mean? Which samples result in a sample mean that underestimates the population mean? Do any samples lead to a sample mean that equals the population mean?

22. Travel Time The data on the next page represent the travel time (in minutes) to school for nine students enrolled in Sullivan's College Algebra course. Treat the nine students as a population.

Student	Travel Time	Student	Travel Time
Amanda	39	Scot	45
Amber	21	Erica	11
Tim	9	Tiffany	12
Mike	32	Glenn	39
Nicole	30		

(a) Determine the population mean for travel time.
(b) Find three simple random samples of size 4 and determine the sample mean for travel time of each sample.
(c) Which samples result in a sample mean that overestimates the population mean? Which samples result in a sample mean that underestimates the population mean? Do any samples lead to a sample mean that equals the population mean?

NW **23. Connection Time** A histogram of the connection time, in seconds, to an Internet service provider for 30 randomly selected connections is shown. The mean connection time is 39.007 seconds and the median connection time is 39.065 seconds. Identify the shape of the distribution. Which measure of central tendency better describes the "center" of the distribution?

Connection Time

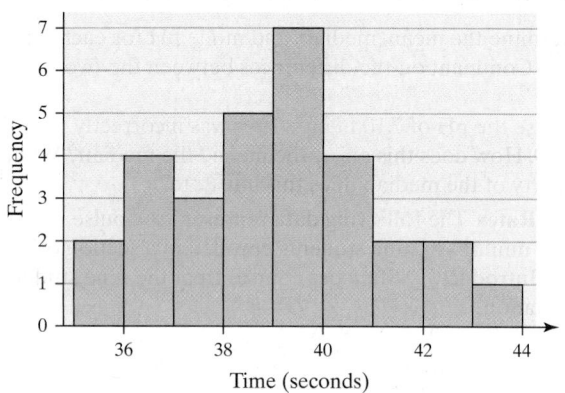

Source: Nicole Spreitzer, student at Joliet Junior College

24. Journal Costs A histogram of the annual subscription cost (in dollars) for 26 biology journals is shown. The mean subscription cost is $1846 and the median subscription cost is $1142. Identify the shape of the distribution. Which measure of central tendency better describes the "center" of the distribution?

Cost of Biology Journals

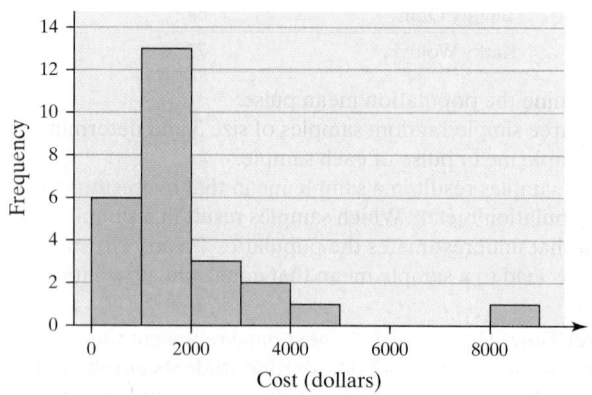

Source: Carol Wesolowski, student at Joliet Junior College

25. M&Ms The following data represent the weights (in grams) of a simple random sample of 50 M&M plain candies.

0.87	0.88	0.82	0.90	0.90	0.84	0.84
0.91	0.94	0.86	0.86	0.86	0.88	0.87
0.89	0.91	0.86	0.87	0.93	0.88	
0.83	0.95	0.87	0.93	0.91	0.85	
0.91	0.91	0.86	0.89	0.87	0.84	
0.88	0.88	0.89	0.79	0.82	0.83	
0.90	0.88	0.84	0.93	0.81	0.90	
0.88	0.92	0.85	0.84	0.84	0.86	

Source: Michael Sullivan

Determine the shape of the distribution of weights of M&Ms by drawing a frequency histogram. Find the mean and median. Which measure of central tendency better describes the weight of a plain M&M?

26. Old Faithful We have all heard of the Old Faithful geyser in Yellowstone National Park. However, there is another, less famous, Old Faithful geyser in Calistoga, California. The following data represent the length of eruption (in seconds) for a random sample of eruptions of the California Old Faithful.

108	108	99	105	103	103	94
102	99	106	90	104	110	110
103	109	109	111	101	101	
110	102	105	110	106	104	
104	100	103	102	120	90	
113	116	95	105	103	101	
100	101	107	110	92	108	

Source: Ladonna Hansen, Park Curator

Determine the shape of the distribution of length of eruption by drawing a frequency histogram. Find the mean and median. Which measure of central tendency better describes the length of eruption?

27. Hours Working A random sample of 25 college students was asked, "How many hours per week typically do you work outside the home?" Their responses were as follows:

0	0	15	20	30
40	30	20	35	35
28	15	20	25	25
30	5	0	30	24
28	30	35	15	15

Determine the shape of the distribution of hours worked by drawing a frequency histogram. Find the mean and median. Which measure of central tendency better describes hours worked?

28. A Dealer's Profit The following data represent the profit (in dollars) of a new-car dealer for a random sample of 40 sales.

781	1038	453	1446	3082
501	451	1826	1348	3001
1342	1889	580	0	2909
2883	480	1664	1064	2978
149	1291	507	261	540
543	87	798	673	2862
1692	1783	2186	398	526
730	2324	2823	1676	4148

Source: Ashley Hudson, student at Joliet Junior College

Determine the shape of the distribution of new-car profit by drawing a frequency histogram. Find the mean and median. Which measure of central tendency better describes the profit?

NW 29. Political Views A sample of 30 registered voters was surveyed in which the respondents were asked, "Do you consider your political views to be conservative, moderate, or liberal?" The results of the survey are shown in the table.

Liberal	Conservative	Moderate
Moderate	Liberal	Moderate
Liberal	Moderate	Conservative
Moderate	Conservative	Moderate
Moderate	Moderate	Liberal
Liberal	Moderate	Liberal
Conservative	Moderate	Moderate
Liberal	Conservative	Liberal
Liberal	Conservative	Liberal
Conservative	Moderate	Conservative

Source: Based on data from the General Social Survey

(a) Determine the mode political view.
(b) Do you think it would be a good idea to rotate the choices conservative, moderate, or liberal in the question? Why?

30. Hospital Admissions The following data represent the diagnosis of a random sample of 20 patients admitted to a hospital. Determine the mode diagnosis.

Cancer	Motor vehicle accident	Congestive heart failure
Gunshot wound	Fall	Gunshot wound
Gunshot wound	Motor vehicle accident	Gunshot wound
Assault	Motor vehicle accident	Gunshot wound
Motor vehicle accident	Motor vehicle accident	Gunshot wound
Motor vehicle accident	Gunshot wound	Motor vehicle accident
Fall	Gunshot wound	

Source: Tamela Ohm, student at Joliet Junior College

31. Resistance and Sample Size Each of the following three data sets in the next column represents the IQ scores of a random sample of adults. IQ scores are known to have a mean

and median of 100. For each data set, determine the mean and median. For each data set recalculate the mean and median, assuming that the individual whose IQ is 106 is accidentally recorded as 160. For each sample size, state what happens to the mean and the median. Comment on the role the number of observations plays in resistance.

Sample of Size 5

106	92	98	103	100

Sample of Size 12

106	92	98	103	100	102
98	124	83	70	108	121

Sample of Size 30

106	92	98	103	100	102
98	124	83	70	108	121
102	87	121	107	97	114
140	93	130	72	81	90
103	97	89	98	88	103

32. Mr. Zuro finds the mean height of all 14 students in his statistics class to be 68.0 inches. Just as Mr. Zuro finishes explaining how to get the mean, Danielle walks in late. Danielle is 65 inches tall. What is the mean height of the 15 students in the class?

33. Missing Exam Grade A professor has recorded exam grades for 20 students in his class, but one of the grades is no longer readable. If the mean score on the exam was 82 and the mean of the 19 readable scores is 84, what is the value of the unreadable score?

34. Survival Rates Unfortunately, a friend of yours has been diagnosed with cancer. A histogram of the survival time (in months) of patients diagnosed with this form of cancer is shown in the figure; the median survival time is 11 months, while the mean survival time is 69 months. What words of encouragement should you share with your friend from a statistical point of view?

Survival Time of Patients

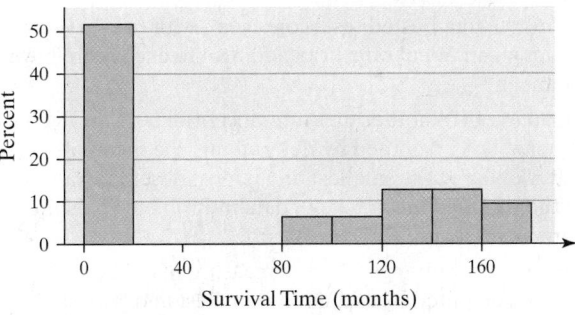

35. Blood Alcohol Concentration Go to http://www.pearsonhighered.com/sullivanstats to obtain the data 3_1_35. The data represent the blood alcohol concentration (BAC), in percent, of a random sample of drivers involved in fatal car accidents. A BAC of 0 indicates that no alcohol was present. Draw a histogram of the data, describe the shape, and determine the mean and median BAC of drivers in fatal accidents. Which measure of central tendency better describes the typical BAC of drivers in fatal accidents? Explain.

36. Sullivan Survey Go to http://www.pearsonhighered.com/ sullivanstats and download the SullivanStatsSurveyI data. The data represent the results of a survey conducted by the author. The column "Texts" represents the number of texts the individual sent for the month prior to the survey. Draw a relative frequency histogram and determine the mean and median number of texts for each gender. Make some general comments about the results.

37. Linear Transformations Benjamin owns a small Internet business. Besides himself, he employs nine other people. The salaries earned by the employees are given next in thousands of dollars (Benjamin's salary is the largest, of course):

$$30, 30, 45, 50, 50, 50, 55, 55, 60, 75$$

(a) Determine the mean, median, and mode for salary.

(b) Business has been good! As a result, Benjamin has a total of $25,000 in bonus pay to distribute to his employees. One option for distributing bonuses is to give each employee (including himself) $2500. Add the bonuses under this plan to the original salaries to create a new data set. Recalculate the mean, median, and mode. How do they compare to the originals?

(c) As a second option, Benjamin can give each employee a bonus of 5% of his or her original salary. Add the bonuses under this second plan to the original salaries to create a new data set. Recalculate the mean, median, and mode. How do they compare to the originals?

(d) As a third option, Benjamin decides not to give his employees a bonus at all. Instead, he keeps the $25,000 for himself. Use this plan to create a new data set. Recalculate the mean, median, and mode. How do they compare to the originals?

38. Linear Transformations Use the five test scores of 65, 70, 71, 75, and 95 to answer the following questions:

(a) Find the sample mean.

(b) Find the median.

(c) Which measure of central tendency better describes the typical test score?

(d) Suppose the professor decides to curve the exam by adding 4 points to each test score. Compute the sample mean based on the adjusted scores.

(e) Compare the unadjusted test score mean with the curved test score mean. What effect did adding 4 to each score have on the mean?

39. Trimmed Mean Another measure of central tendency is the trimmed mean. It is computed by determining the mean of a data set after deleting the smallest and largest observed values. Compute the trimmed mean for the data in Problem 25. Is the trimmed mean resistant? Explain.

40. Midrange The midrange is also a measure of central tendency. It is computed by adding the smallest and largest observed values of a data set and dividing the result by 2; that is,

$$\text{Midrange} = \frac{\text{largest data value} + \text{smallest data value}}{2}$$

Compute the midrange for the data in Problem 25. Is the midrange resistant? Explain.

41. Putting It Together: Shape, Mean and Median As part of a semester project in a statistics course, Carlos surveyed a sample of 50 high school students and asked, "How many days in the past week have you consumed an alcoholic beverage?" The results of the survey are shown next.

0	0	1	4	1	1	1	5	1	3
0	1	0	1	0	4	0	1	0	1
0	0	0	0	2	0	0	0	0	0
1	0	2	0	0	0	1	2	1	1
2	0	1	0	1	3	1	1	0	3

(a) Is this data discrete or continuous?

(b) Draw a histogram of the data and describe its shape.

(c) Based on the shape of the histogram, do you expect the mean to be more than, equal to, or less than the median?

(d) Determine the mean and the median. What does this tell you?

(e) Determine the mode.

(f) Do you believe that Carlos's survey suffers from sampling bias? Why?

Explaining the Concepts

42. FICO Scores The Fair Isaacs Corporation has devised a model that is used to determine creditworthiness of individuals, called a FICO score. FICO scores range in value from 300 to 850, with a higher score indicating a more creditworthy individual. The distribution of FICO scores is skewed left with a median score of 723.

(a) Do you think the mean FICO score is greater than, less than, or equal to 723? Justify your response.

(b) What proportion of individuals have a FICO score above 723?

43. Why is the median resistant, but the mean is not?

44. A researcher with the Department of Energy wants to determine the mean natural gas bill of households throughout the United States. He knows the mean natural gas bill of households for each state, so he adds together these 50 values and divides by 50 to arrive at his estimate. Is this a valid approach? Why or why not?

45. Net Worth According to a Credit Suisse Survey, the mean net worth of all individuals in the United States in 2014 was $301,000, while the median net worth was $45,000.

(a) Which measure do you believe better describes the typical individual's net worth? Support your opinion.

(b) What shape would you expect the distribution of net worth to have? Why?

(c) What do you think causes the disparity in the two measures of central tendency?

46. You are negotiating a contract for the Players Association of the NBA. Which measure of central tendency will you use to support your claim that the average player's salary needs to be increased? Why? As the chief negotiator for the owners, which measure would you use to refute the claim made by the Players Association?

47. In January 2015, the mean amount of money lost per visitor to a local riverboat casino was $135. Do you think the median was more than, less than, or equal to this amount? Why?

48. For each of the following situations, determine which measure of central tendency is most appropriate and justify your reasoning.

(a) Average price of a home sold in Pittsburgh, Pennsylvania, in 2011

(b) Most popular major for students enrolled in a statistics course

(c) Average test score when the scores are distributed symmetrically

(d) Average test score when the scores are skewed right

(e) Average income of a player in the National Football League

(f) Most requested song at a radio station

(g) Typical number on the jersey of a player in the National Hockey League.

3.2   Measures of Dispersion

Objectives

① Determine the range of a variable from raw data

② Determine the standard deviation of a variable from raw data

③ Determine the variance of a variable from raw data

④ Use the Empirical Rule to describe data that are bell shaped

⑤ Use Chebyshev's Inequality to describe any set of data

In Section 3.1, we discussed measures of central tendency. These measures describe the typical *value* of a variable. We would also like to know the amount of *dispersion* in the variable. **Dispersion** is the degree to which the data are spread out. Example 1 demonstrates why measures of central tendency alone are not sufficient in describing a distribution.

EXAMPLE 1   Comparing Two Sets of Data

Problem The data in Table 7 represent the IQ scores of a random sample of 100 students from two different universities. For each university, compute the mean IQ score and draw a histogram using a lower class limit of 55 for the first class and a class width of 15. Comment on the results.

Table 7

University A										University B									
73	103	91	93	136	108	92	104	90	78	86	91	107	94	105	107	89	96	102	96
108	93	91	78	81	130	82	86	111	93	92	109	103	106	98	95	97	95	109	109
102	111	125	107	80	90	122	101	82	115	93	91	92	91	117	108	89	95	103	109
103	110	84	115	85	83	131	90	103	106	110	88	97	119	90	99	96	104	98	95
71	69	97	130	91	62	85	94	110	85	87	105	111	87	103	92	103	107	106	97
102	109	105	97	104	94	92	83	94	114	107	108	89	96	107	107	96	95	117	97
107	94	112	113	115	106	97	106	85	99	98	89	104	99	99	87	91	105	109	108
102	109	76	94	103	112	107	101	91	107	116	107	90	98	98	92	119	96	118	98
107	110	106	103	93	110	125	101	91	119	97	106	114	87	107	96	93	99	89	94
118	85	127	141	129	60	115	80	111	79	104	88	99	97	106	107	112	97	94	107

Approach We will use Minitab to compute the mean and draw a histogram for each university.

(continued)

Solution Enter the data into Minitab and determine that the mean IQ score of both universities is 100.0. Figure 10 shows the histograms.

Figure 10

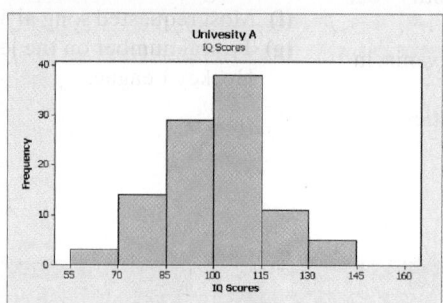
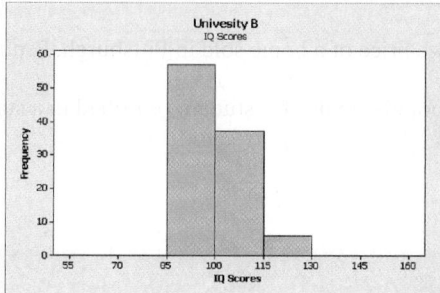

Both universities have the same mean IQ, but the histograms suggest the IQs from University A are more spread out, that is, more dispersed. While an IQ of 100.0 is typical for both universities, it appears to be a more reliable description of the typical student from University B than from University A. That is, a higher proportion of students from University B have IQ scores within, say, 15 points of the mean of 100.0 than students from University A.

The goal of this section is to discuss numerical measures of dispersion so that the spread of data may be quantified. Three numerical measures for describing the dispersion, or spread, of data will be discussed: the *range*, *standard deviation*, and *variance*. In Section 3.4, we will discuss another measure of dispersion, the *interquartile range* (IQR).

① Determine the Range of a Variable from Raw Data

The simplest measure of dispersion is the range. To compute the range, the data must be quantitative.

Definition The **range**, *R*, of a variable is the difference between the largest and the smallest data value. That is,

$$\text{Range} = R = \text{largest data value} - \text{smallest data value}$$

EXAMPLE 2 **Computing the Range of a Set of Data**

Table 8	
Student	**Score**
1. Michelle	82
2. Ryanne	77
3. Bilal	90
4. Pam	71
5. Jennifer	62
6. Dave	68
7. Joel	74
8. Sam	84
9. Justine	94
10. Juan	88

● **Now compute the range of the data in Problem 13**

Problem The data in Table 8 represent the scores on the first exam of 10 students enrolled in Introductory Statistics. Compute the range.

Approach The range is the difference between the largest and smallest data values.

Solution The highest test score is 94 and the lowest test score is 62. The range is

$$R = 94 - 62 = 32$$

All the students in the class scored between 62 and 94 on the exam. The difference between the best score and the worst score is 32 points. ●

Notice that the range is affected by extreme values in the data set, so the range is not resistant. If Jennifer scored 28, the range becomes $R = 94 - 28 = 66$. Also, the range is computed using only two values in the data set (the largest and smallest). The *standard deviation*, on the other hand, uses all the data values in the computations.

② Determine the Standard Deviation of a Variable from Raw Data

NOTE

Recall, $|a| = a$ if $a \geq 0$, and $|a| = -a$ if $a < 0$ so $|3| = 3$ and $|-3| = 3$. •

Measures of dispersion are meant to describe how spread out data are. In other words, they describe how far, on average, each observation is from the typical data value. Standard deviation is based on the **deviation about the mean**. For a population, the deviation about the mean for the ith observation is $x_i - \mu$. For a sample, the deviation about the mean for the ith observation is $x_i - \bar{x}$. The further an observation is from the mean, the larger the absolute value of the deviation.

The sum of all deviations about the mean must equal zero. That is,

$$\Sigma(x_i - \mu) = 0 \quad \text{and} \quad \Sigma(x_i - \bar{x}) = 0$$

This result follows from the fact that observations greater than the mean are offset by observations less than the mean. Because this sum is zero, we cannot use the average deviation about the mean as a measure of spread. There are two possible solutions to this "problem." We could either find the mean of the absolute values of the deviations about the mean, or we could find the mean of the squared deviations because squaring a nonzero number always results in a positive number. The first approach yields a measure of dispersion called the mean absolute deviation (MAD) (see Problem 43). The second approach leads to *variance*. The problem with variance is that squaring the deviations about the mean leads to squared units of measure, such as dollars squared. It is difficult to have a reasonable interpretation of dollars squared, so we "undo" the squaring process by taking the square root of the sum of squared deviations. We have the following definition for the *population standard deviation*.

Definition

The **population standard deviation** of a variable is the square root of the sum of squared deviations about the population mean divided by the number of observations in the population, N. That is, it is the square root of the mean of the squared deviations about the population mean.

The population standard deviation is symbolically represented by σ (lowercase Greek sigma).

$$\sigma = \sqrt{\frac{(x_1 - \mu)^2 + (x_2 - \mu)^2 + \cdots + (x_N - \mu)^2}{N}} = \sqrt{\frac{\Sigma(x_i - \mu)^2}{N}} \quad \text{(1)}$$

where $x_1, x_2, \ldots, x_N$ are the N observations in the population and μ is the population mean.

Formula (1) is sometimes referred to as the **conceptual formula** because it allows us to see how standard deviation measures spread.

A formula that is equivalent to Formula (1), called the **computational formula**, for determining the population standard deviation is

$$\sigma = \sqrt{\frac{\Sigma x_i^2 - \frac{(\Sigma x_i)^2}{N}}{N}} \quad \text{(2)}$$

where Σx_i^2 means to square each observation and then sum these squared values, and $(\Sigma x_i)^2$ means to add up all the observations and then square the sum.

Example 3 illustrates how to use both formulas.

EXAMPLE 3   Computing a Population Standard Deviation

Problem Compute the population standard deviation of the test scores in Table 8.

Approach Using Formula (1)

Step 1   Create a table with four columns. Enter the population data in Column 1. In Column 2, enter the population mean.

Step 2   Compute the deviation about the mean for each data value, $x_i - \mu$. Enter the result in Column 3.

Step 3   In Column 4, enter the squares of the values in Column 3.

Step 4   Sum the entries in Column 4, and divide this result by the size of the population, N.

Step 5   Determine the square root of the value found in Step 4.

Solution Using Formula (1)

Step 1   See Table 9. Column 1 lists the observations in the data set, and Column 2 contains the population mean.

Table 9

Score, x_i	Population Mean, μ	Deviation about the Mean, $x_i - \mu$	Squared Deviations about the Mean, $(x_i - \mu)^2$
82	79	$82 - 79 = 3$	$3^2 = 9$
77	79	$77 - 79 = -2$	$(-2)^2 = 4$
90	79	11	121
71	79	-8	64
62	79	-17	289
68	79	-11	121
74	79	-5	25
84	79	5	25
94	79	15	225
88	79	9	81
		$\Sigma(x_i - \mu) = 0$	$\Sigma(x_i - \mu)^2 = 964$

Step 2   Column 3 contains the deviations about the mean for each observation. For example, the deviation about the mean for Michelle is $82 - 79 = 3$. It is a good idea to add the entries in this column to make sure they sum to 0.

Step 3   Column 4 shows the squared deviations about the mean. Notice the farther an observation is from the mean, the larger the squared deviation.

Step 4   Sum the entries in Column 4 to obtain $\Sigma(x_i - \mu)^2$. Divide this sum by the number of students, 10:

$$\frac{\Sigma(x_i - \mu)^2}{N} = \frac{964}{10} = 96.4 \text{ points}^2$$

Step 5   The square root of the result in Step 4 is the population standard deviation.

$$\sigma = \sqrt{\frac{\Sigma(x_i - \mu)^2}{N}} = \sqrt{96.4 \text{ points}^2} \approx 9.8 \text{ points}$$

Approach Using Formula (2)

Step 1   Create a table with two columns. Enter the population data in Column 1. Square each value in Column 1 and enter the result in Column 2.

Step 2   Sum the entries in Column 1. That is, find Σx_i. Sum the entries in Column 2. That is, find Σx_i^2.

Step 3   Substitute the values found in Step 2 and the value for N into the computational formula and simplify.

Solution Using Formula (2)

Step 1   See Table 10. Column 1 lists the observations in the data set, and Column 2 contains the values in column 1 squared.

Table 10

Score, x_i	Score Squared, x_i^2
82	$82^2 = 6724$
77	$77^2 = 5929$
90	8100
71	5041
62	3844
68	4624
74	5476
84	7056
94	8836
88	7744
$\Sigma x_i = 790$	$\Sigma x_i^2 = 63{,}374$

Step 2   The last rows of Columns 1 and 2 show that $\Sigma x_i = 790$ and $\Sigma x_i^2 = 63{,}374$.

Step 3   Substitute 790 for Σx_i, 63,374 for Σx_i^2, and 10 for N into Formula (2):

$$\sigma = \sqrt{\frac{\Sigma x_i^2 - \dfrac{(\Sigma x_i)^2}{N}}{N}} = \sqrt{\frac{63{,}374 - \dfrac{(790)^2}{10}}{10}}$$

$$= \sqrt{\frac{964}{10}}$$

$$= \sqrt{96.4 \text{ points}^2}$$

$$\approx 9.8 \text{ points}$$

Look at Table 9 in Example 3. The further an observation is from the mean, 79, the larger the squared deviation. For example, because the second observation, 77, is not "far" from 79, the squared deviation, 4, is not large. However, the fifth observation, 62, is further from 79, so the squared deviation, 289, is much larger.

So, if a data set has many observations that are "far" from the mean, the sum of the squared deviations will be large, and therefore the standard deviation will be large.

Now let's look at the definition of the *sample standard deviation*.

Definition

The **sample standard deviation**, s, of a variable is the square root of the sum of squared deviations about the sample mean divided by $n - 1$, where n is the sample size.

$$s = \sqrt{\frac{(x_1 - \bar{x})^2 + (x_2 - \bar{x})^2 + \cdots + (x_n - \bar{x})^2}{n - 1}} = \sqrt{\frac{\Sigma (x_i - \bar{x})^2}{n - 1}} \quad \textbf{(3)}$$

where $x_1, x_2, \ldots, x_n$ are the n observations in the sample and $\bar{x}$ is the sample mean.

> **CAUTION!**
> When using Formula (3), be sure to use $\bar{x}$ with as many decimal places as possible to avoid round-off error.

Formula (3) is often referred to as the conceptual formula for determining the sample standard deviation.

A computational formula that is equivalent to Formula (3) for computing the sample standard deviation is

> **CAUTION!**
> When computing the sample standard deviation, be sure to divide by $n - 1$, not n.

$$s = \sqrt{\frac{\Sigma x_i^2 - \dfrac{(\Sigma x_i)^2}{n}}{n - 1}} \quad \textbf{(4)}$$

Notice that the sample standard deviation is obtained by dividing by $n - 1$. Although showing why we divide by $n - 1$ is beyond the scope of the text, the following explanation has intuitive appeal. We already know that the sum of the deviations about the mean, $\Sigma (x_i - \bar{x})$, must equal zero. Therefore, if the sample mean is known and the first $n - 1$ observations are known, then the nth observation must be the value that causes the sum of the deviations to equal zero. For example, suppose $\bar{x} = 4$ based on a sample of size 3. In addition, if $x_1 = 2$ and $x_2 = 3$, then we can determine x_3.

$$\frac{x_1 + x_2 + x_3}{3} = \bar{x}$$

$$\frac{2 + 3 + x_3}{3} = 4 \qquad x_1 = 2, x_2 = 3, \bar{x} = 4$$

$$5 + x_3 = 12$$

$$x_3 = 7$$

In Other Words
We have $n - 1$ degrees of freedom in the computation of s because an unknown parameter, μ, is estimated with $\bar{x}$. For each parameter estimated, we lose 1 degree of freedom.

We call $n - 1$ the **degrees of freedom** because the first $n - 1$ observations have freedom to be whatever value they wish, but the nth value has no freedom. It must be whatever value forces the sum of the deviations about the mean to equal zero.

Again, you should notice that typically Greek letters are used for parameters, while Roman letters are used for statistics. Do not use rounded values of the sample mean in Formula (3).

EXAMPLE 4   Computing a Sample Standard Deviation

Problem Compute the sample standard deviation of the sample obtained in Example 1(b) on page 119 from Section 3.1.

Approach We follow the same approach that we used to compute the population standard deviation, but this time using the sample data. In looking back at Example 1(b) from Section 3.1, we see that Jennifer (62), Juan (88), Ryanne (77), and Dave (68) are in the sample.

(continued)

Solution Using Formula (3)

Step 1 Create a table with four columns. Enter the sample data in Column 1. In Column 2, enter the unrounded sample mean. See Table 11.

Table 11

Score, x_i	Sample Mean, $\bar{x}$	Deviation about the Mean, $x_i - \bar{x}$	Squared Deviations about the Mean, $(x_i - \bar{x})^2$
62	73.75	$62 - 73.75 = -11.75$	$(-11.75)^2 = 138.0625$
88	73.75	14.25	203.0625
77	73.75	3.25	10.5625
68	73.75	-5.75	33.0625
		$\Sigma(x_i - \bar{x}) = 0$	$\Sigma(x_i - \bar{x})^2 = 384.75$

Step 2 Column 3 contains the deviations about the mean for each observation. For example, the deviation about the mean for Jennifer is $62 - 73.75 = -11.75$. It is a good idea to add the entries in this column to make sure they sum to 0.

Step 3 Column 4 shows the squared deviations about the mean.

Step 4 Sum the entries in Column 4 to obtain $\Sigma(x_i - \bar{x})^2$. Divide the sum of the entries in Column 4 by one fewer than the number of students, $4 - 1$:

$$\frac{\Sigma(x_i - \bar{x})^2}{n - 1} = \frac{384.75}{4 - 1} = 128.25 \text{ points}^2$$

Step 5 The square root of the result in Step 4 is the sample standard deviation.

$$s = \sqrt{\frac{\Sigma(x_i - \bar{x})^2}{n - 1}}$$
$$= \sqrt{128.25 \text{ points}^2}$$
$$\approx 11.3 \text{ points}$$

Solution Using Formula (4)

Step 1 See Table 12. Column 1 lists the observations in the data set, and Column 2 contains the values in Column 1 squared.

Table 12

Score, x_i	Score squared, x_i^2
62	$62^2 = 3844$
88	$88^2 = 7744$
77	5929
68	4624
$\Sigma x_i = 295$	$\Sigma x_i^2 = 22{,}141$

Step 2 The last rows of Columns 1 and 2 show that $\Sigma x_i = 295$ and $\Sigma x_i^2 = 22{,}141$.

Step 3 Substitute 295 for Σx_i, 22,141 for Σx_i^2, and 4 for n into the computational Formula (4).

$$s = \sqrt{\frac{\Sigma x_i^2 - \frac{(\Sigma x_i)^2}{n}}{n - 1}} = \sqrt{\frac{22{,}141 - \frac{(295)^2}{4}}{4 - 1}}$$
$$= \sqrt{\frac{384.75}{3}}$$
$$= \sqrt{128.25 \text{ points}^2}$$
$$\approx 11.3 \text{ points}$$

EXAMPLE 5   Determining the Standard Deviation Using Technology

Problem Use statistical software or a calculator to determine the population standard deviation of the data listed in Table 8 and the sample standard deviation of the sample data from Example 4.

Approach We will use a TI-84 Plus C graphing calculator. The steps for determining the standard deviation using the TI-83 or TI-84 Plus graphing calculator, Minitab, Excel, and StatCrunch are given in the Technology Step-by-Step on page 141.

Solution Figure 11(a) shows the population standard deviation, and Figure 11(b) shows the sample standard deviation. Notice that the TI graphing calculators provide both a population and sample standard deviation as output. This is because the calculator does not know whether the data entered are population data or sample data. It is up to the user of the calculator to choose the correct standard deviation. The results agree with those found in Examples 3 and 4.

Figure 11

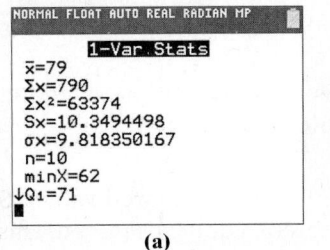
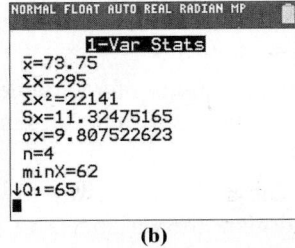

• Now Work Problem 19   (a)   (b)

Is the standard deviation resistant? To help determine the answer, suppose Jennifer's exam score (Table 8 on page 132) is 26, not 62. The population standard deviation increases from 9.8 to 18.3. Clearly, the standard deviation is *not* resistant.

Notice the sample standard deviation is larger than the population standard deviation. Why? Our sample happened to include scores that are more dispersed than the population. A different random sample of, say, Ryanne (77), Pam (71), Joel (74), and Juan (88), would result in a sample standard deviation of 7.4 points. The different sample standard deviation occurs because we have different individuals in the sample.

Interpretations of the Standard Deviation

The standard deviation is used along with the mean to describe symmetric distributions numerically. The mean measures the center of the distribution, while the standard deviation measures the spread of the distribution. So how does the value of the standard deviation relate to the spread of the distribution?

We can interpret the standard deviation as the typical deviation from the mean. So, in the data from Table 8, a typical deviation from the mean, $\mu = 79$ points, is 9.8 points.

If we are comparing two populations, then **the larger the standard deviation, the more dispersion the distribution has**, provided that the variable of interest from the two populations has the same unit of measure. The units of measure must be the same so that we are comparing apples with apples. For example, $100 is not the same as 100 Japanese yen (because recently $1 was equivalent to about 120 yen). So a standard deviation of $100 is substantially higher than a standard deviation of 100 yen.

EXAMPLE 6   Comparing the Standard Deviation of Two Data Sets

Problem Refer to the data in Example 1. Use the standard deviation to determine whether University A or University B has more dispersion in its students' IQ scores.

Approach We will use Minitab to determine the standard deviation of IQ for each university. The university with the higher standard deviation will be the university with more dispersion in IQ scores. The histograms shown in Figure 10 on page 132 imply that University A has more dispersion. Therefore, we expect University A to have a higher sample standard deviation.

Solution Enter the data into Minitab and compute the descriptive statistics. See Figure 12.

Figure 12

Descriptive statistics

Variable	N	N*	Mean	SE Mean	StDev	Minimum
Univ A	100	0	100.00	1.61	16.08	60
Univ B	100	0	100.00	0.83	8.35	86

Variable	Q1	Median	Q3	Maximum
Univ A	90	102	110	141
Univ B	94	98	107	119

The sample standard deviation is larger for University A (16.1) than for University B (8.4). Therefore, University A has IQ scores that are more dispersed.

③ **Determine the Variance of a Variable from Raw Data**

A third measure of dispersion is called the *variance*.

Definition
> The **variance** of a variable is the square of the standard deviation. The **population variance** is σ^2 and the **sample variance** is s^2.

The units of measure in variance are squared values. So, if the variable is measured in dollars, the variance is measured in dollars squared. This makes interpreting the variance difficult. However, the variance is important for conducting certain types of statistical inference, which we discuss later in the text.

EXAMPLE 7   Determining the Variance of a Variable for a Population and a Sample

Problem Use the results of Examples 3 and 4 to determine the population and sample variance of test scores on the statistics exam.

Approach The population variance is found by squaring the population standard deviation. The sample variance is found by squaring the sample standard deviation.

Solution The population standard deviation in Example 3 was found to be $\sigma = 9.8$ points, so the population variance is $\sigma^2 = (9.8 \text{ points})^2 = 96.04 \text{ points}^2$. The sample standard deviation in Example 4 was found to be $s = 11.3$ points, so the sample variance is $s^2 = (11.3 \text{ points})^2 = 127.7 \text{ points}^2$.   •

If you look carefully at Formulas (1) and (3), you will notice that the value under the radical $\left(\sqrt{} \right)$ represents the variance. So we could also find the population or sample variance while in the process of finding the population or sample standard deviation. Using this approach, we obtain a population variance of 96.4 points^2 and a sample variance of 128.3 points^2. Using a rounded value of the standard deviation to obtain the variance results in round-off error. Therefore, it is recommended that you use as many decimal places as possible when using the standard deviation to obtain the variance.

A Final Thought on Variance and Standard Deviation

The sample variance is obtained using the formula $s^2 = \frac{\Sigma (x_i - \bar{x})^2}{n - 1}$. What if we divided by n instead of $n - 1$ to obtain the sample variance, as one might expect? Then the sample variance would consistently underestimate the population variance. Whenever a statistic consistently underestimates (or overestimates) a parameter, it is said to be **biased**. To obtain an unbiased estimate of the population variance, divide the sum of squared deviations about the sample mean by $n - 1$.

Let's look at an example of a biased estimator. Suppose you work for a carnival in which you must guess a person's age. After 20 people come to your booth, you notice that you have a tendency to underestimate people's age. (You guess too low.) What would you do to correct this? You could adjust your guess higher to avoid underestimating. In other words, originally your guesses were biased. To remove the bias, you increase your guess. This is what dividing by $n - 1$ in the sample variance formula accomplishes.

Unfortunately, the sample standard deviation given in Formulas (3) and (4) is not an unbiased estimate of the population standard deviation. In fact, it is not possible to provide an unbiased estimator of the population standard deviation for all distributions. The explanation for this concept is beyond the scope of this class (it has to do with the shape of the graph of the square root function). However, for the applications presented in this text, the bias is minor and does not impact results.

④ **Use the Empirical Rule to Describe Data That Are Bell Shaped**

If data have a distribution that is bell shaped, the *Empirical Rule* can be used to determine the percentage of data that will lie within k standard deviations of the mean.

> ### The Empirical Rule
>
> If a distribution is roughly bell shaped, then
>
> * Approximately 68% of the data will lie within 1 standard deviation of the mean. That is, approximately 68% of the data lie between $\mu - 1\sigma$ and $\mu + 1\sigma$.
> * Approximately 95% of the data will lie within 2 standard deviations of the mean. That is, approximately 95% of the data lie between $\mu - 2\sigma$ and $\mu + 2\sigma$.
> * Approximately 99.7% of the data will lie within 3 standard deviations of the mean. That is, approximately 99.7% of the data lie between $\mu - 3\sigma$ and $\mu + 3\sigma$.
>
> **Note:** We can also use the Empirical Rule based on sample data with $\bar{x}$ used in place of μ and s used in place of σ.

Figure 13 illustrates the Empirical Rule.

Figure 13

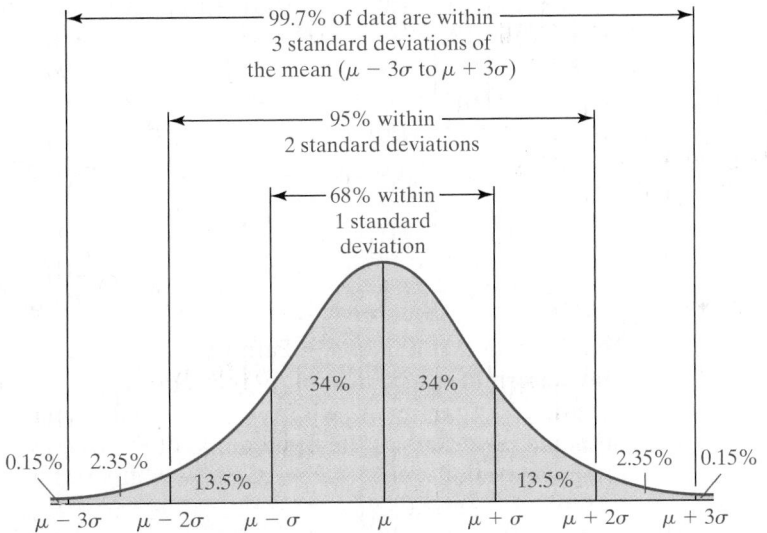

99.7% of data are within 3 standard deviations of the mean ($\mu - 3\sigma$ to $\mu + 3\sigma$)

95% within 2 standard deviations

68% within 1 standard deviation

34%   34%

0.15%   2.35%   13.5%   13.5%   2.35%   0.15%

$\mu - 3\sigma$ $\mu - 2\sigma$ $\mu - \sigma$ μ $\mu + \sigma$ $\mu + 2\sigma$ $\mu + 3\sigma$

EXAMPLE 8   Using the Empirical Rule

Figure 14

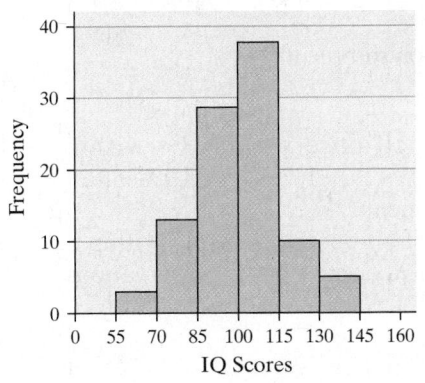

University A IQ Scores

Frequency (vertical axis: 0, 10, 20, 30, 40)

IQ Scores (horizontal axis: 0, 55, 70, 85, 100, 115, 130, 145, 160)

Problem Use the data from University A in Table 7.
(a) Determine the percentage of students who have IQ scores within 3 standard deviations of the mean according to the Empirical Rule.
(b) Determine the percentage of students who have IQ scores between 67.8 and 132.2 according to the Empirical Rule.
(c) Determine the actual percentage of students who have IQ scores between 67.8 and 132.2.
(d) According to the Empirical Rule, what percentage of students have IQ scores above 132.2?

Approach To use the Empirical Rule, a histogram of the data must be roughly bell shaped. Figure 14 shows the histogram of the data from University A.

(continued)

Solution The histogram is roughly bell shaped. From Examples 1 and 6, the mean IQ score of the students enrolled in University A is 100 and the standard deviation is 16.1. To make the analysis easier, we draw a bell-shaped curve like the one in Figure 13, with $\bar{x} = 100$ and $s = 16.1$. See Figure 15.

Figure 15

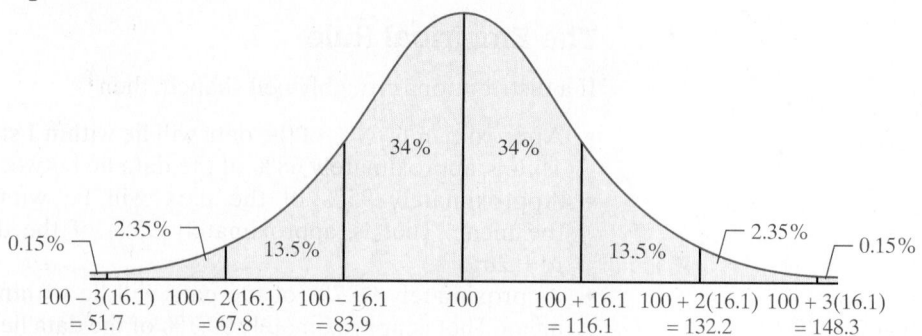

(a) According to the Empirical Rule, approximately 99.7% of the IQ scores are within 3 standard deviations of the mean [that is, greater than or equal to $100 - 3(16.1) = 51.7$ and less than or equal to $100 + 3(16.1) = 148.3$].

(b) Since 67.8 is exactly 2 standard deviations below the mean [$100 - 2(16.1) = 67.8$] and 132.2 is exactly 2 standard deviations above the mean [$100 + 2(16.1) = 132.2$], the Empirical Rule tells us that approximately 95% of the IQ scores lie between 67.8 and 132.2.

(c) Of the 100 IQ scores listed in Table 7, 96, or 96%, are between 67.8 and 132.2. This is very close to the Empirical Rule's approximation.

(d) Based on Figure 15, approximately $2.35\% + 0.15\% = 2.5\%$ of students at University A will have IQ scores above 132.2.

• Now Work Problem 29

---

⑤ **Use Chebyshev's Inequality to Describe Any Set of Data**

The Russian mathematician Pafnuty Chebyshev (1821–1894) developed an inequality that determines a minimum percentage of observations that lie within k standard deviations of the mean, where $k > 1$. What's amazing about this is that the result is obtained regardless of the basic shape of the distribution (skewed left, skewed right, or symmetric).

> **Chebyshev's Inequality**
>
> For any data set or distribution, at least $\left(1 - \frac{1}{k^2}\right) \cdot 100\%$ of the observations lie within k standard deviations of the mean, where k is any number greater than 1. That is, at least $\left(1 - \frac{1}{k^2}\right) \cdot 100\%$ of the data lie between $\mu - k\sigma$ and $\mu + k\sigma$ for $k > 1$.
>
> **Note:** We can also use Chebyshev's Inequality based on sample data.

> **CAUTION!**
> The Empirical Rule holds only if the distribution is bell shaped. Chebyshev's Inequality holds regardless of the shape of the distribution.

For example, at least $\left(1 - \frac{1}{2^2}\right) \cdot 100\% = 75\%$ of all observations lie within $k = 2$ standard deviations of the mean and at least $\left(1 - \frac{1}{3^2}\right) \cdot 100\% = 88.9\%$ of all observations lie within $k = 3$ standard deviations of the mean.

Notice the result does not state that exactly 75% of all observations lie within 2 standard deviations of the mean, but instead states that 75% or more of the observations will lie within 2 standard deviations of the mean.

EXAMPLE 9   Using Chebyshev's Inequality

Historical Note

Pafnuty Chebyshev was born on May 16, 1821, in Okatovo, Russia. In 1847, he began teaching mathematics at the University of St. Petersburg. Some of his more famous work was done on prime numbers. In particular, he discovered a way to determine the number of prime numbers less than or equal to a given number. Chebyshev also studied mechanics, including rotary motion. Chebyshev was elected a Fellow of the Royal Society in 1877. He died on November 26, 1894, in St. Petersburg.

Problem   Use the data from University A in Table 7.
(a) Determine the minimum percentage of students who have IQ scores within 3 standard deviations of the mean according to Chebyshev's Inequality.
(b) Determine the minimum percentage of students who have IQ scores between 67.8 and 132.2, according to Chebyshev's Inequality.
(c) Determine the actual percentage of students who have IQ scores between 67.8 and 132.2.

Approach
(a) Use Chebyshev's Inequality with $k = 3$.
(b) Determine the number of standard deviations 67.8 and 132.2 are from the mean of 100.0. Then substitute this value for k into Chebyshev's Inequality.
(c) Refer to Table 7 and count the number of observations between 67.8 and 132.2. Divide this result by 100, the number of observations in the data set.

Solution
(a) Using Chebyshev's Inequality with $k = 3$, at least $\left(1 - \frac{1}{3^2}\right) \cdot 100\% = 88.9\%$ of all students have IQ scores within 3 standard deviations of the mean. Since the mean of the data set is 100.0 and the standard deviation is 16.1, at least 88.9% of the students have IQ scores between $\bar{x} - ks = 100.0 - 3(16.1) = 51.7$ and $\bar{x} + ks = 100 + 3(16.1) = 148.3$.
(b) Since 67.8 is exactly 2 standard deviations below the mean $[100 - 2(16.1) = 67.8]$ and 132.2 is exactly 2 standard deviations above the mean $[100 + 2(16.1) = 132.2]$, Chebyshev's Inequality (with $k = 2$) says that at least $\left(1 - \frac{1}{2^2}\right) \cdot 100\% = 75\%$ of all IQ scores lie between 67.8 and 132.2.
(c) Of the 100 IQ scores listed, 96 or 96% are between 67.8 and 132.2. Notice that Chebyshev's Inequality provides a conservative result.   •

● **Now Work Problem 35**

Because the Empirical Rule requires that the distribution be bell shaped, while Chebyshev's Inequality applies to all distributions, the Empirical Rule gives more precise results.

Technology Step-by-Step   Determining the Range, Variance, and Standard Deviation

The same steps followed to obtain the measures of central tendency from raw data can be used to obtain the measures of dispersion.

Refer to the Technology Step-by-Step on page 126.

 ## 3.2 Assess Your Understanding

Vocabulary and Skill Building

1. The sum of the deviations about the mean always equals

_____.

2. The standard deviation is used in conjunction with the _____ to numerically describe distributions that are bell shaped. The _____ measures the center of the distribution, while the standard deviation measures the _____ of the distribution.

3. *True or False:* When comparing two populations, the larger the standard deviation, the more dispersion the distribution has, provided that the variable of interest from the two populations has the same unit of measure.

4. *True or False:* Chebyshev's Inequality applies to all distributions regardless of shape, but the Empirical Rule holds only for distributions that are bell shaped.

In Problems 5–10, by hand, find the population variance and standard deviation or the sample variance and standard deviation as indicated.

5. Sample: 20, 13, 4, 8, 10

6. Sample: 83, 65, 91, 87, 84

7. Population: 3, 6, 10, 12, 14

8. Population: 1, 19, 25, 15, 12, 16, 28, 13, 6

9. Sample: 6, 52, 13, 49, 35, 25, 31, 29, 31, 29

10. Population: 4, 10, 12, 12, 13, 21

11. Miles per Gallon The following data represent the miles per gallon for a 2013 Ford Fusion for six randomly selected vehicles. Compute the range, sample variance, and sample standard deviation miles per gallon.
Source: www.fueleconomy.gov

34.0, 33.2, 37.0, 29.4, 23.6, 25.9

12. Exam Time The following data represent the amount of time (in minutes) a random sample of eight students took to complete the online portion of an exam in Sullivan's Statistics course. Compute the range, sample variance, and sample standard deviation time.

60.5, 128.0, 84.6, 122.3, 78.9, 94.7, 85.9, 89.9

NW 13. Concrete Mix A certain type of concrete mix is designed to withstand 3000 pounds per square inch (psi) of pressure. The strength of concrete is measured by pouring the mix into casting cylinders after it is allowed to set up for 28 days. The following data represent the strength of nine randomly selected casts. Compute the range and sample standard deviation for the strength of the concrete (in psi).

3960, 4090, 3200, 3100, 2940, 3830, 4090, 4040, 3780

14. Flight Time The following data represent the flight time (in minutes) of a random sample of seven flights from Las Vegas, Nevada, to Newark, New Jersey, on United Airlines. Compute the range and sample standard deviation of flight time.

282, 270, 260, 266, 257, 260, 267

15. Which histogram depicts a higher standard deviation? Justify your answer.

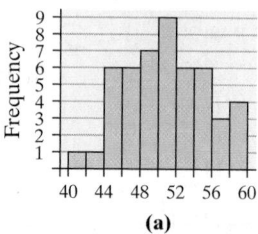
(a)

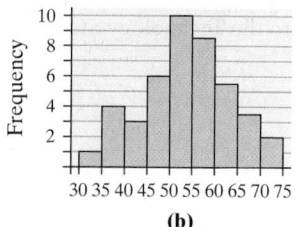
(b)

16. Match the histograms in the next column to the summary statistics given.

	Mean	Median	Standard Deviation
I	53	53	1.8
II	60	60	11
III	53	53	10
IV	53	53	22

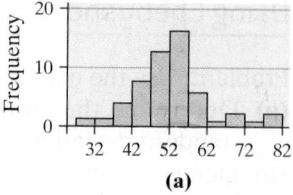
(a)

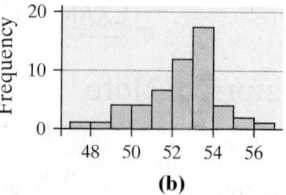
(b)

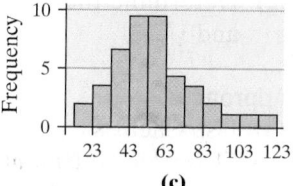
(c)

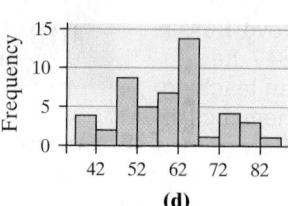
(d)

Applying the Concepts

17. Exam Scores The following data represent exam scores in a statistics class taught using traditional lecture and a class taught using a "flipped" classroom. The "flipped" classroom is one where the content is delivered via video and watched at home, while class time is used for homework and activities.

Traditional	70.8	69.1	79.4	67.6	85.3	78.2	56.2
	81.3	80.9	71.5	63.7	69.8	59.8	
Flipped	76.4	71.6	63.4	72.4	77.9	91.8	78.9
	76.8	82.1	70.2	91.5	77.8	76.5	

Source: Michael Sullivan

(a) Which course has more dispersion in exam scores using the range as the measure of dispersion?

(b) Which course has more dispersion in exam scores using the standard deviation as the measure of dispersion?

(c) Suppose the score of 59.8 in the traditional course was incorrectly recorded as 598. How does this affect the range? the standard deviation? What property does this illustrate?

18. pH in Water The acidity or alkalinity of a solution is measured using pH. A pH less than 7 is acidic, while a pH greater than 7 is alkaline. The following data represent the pH in samples of bottled water and tap water.

Tap	7.64	7.45	7.47	7.50	7.68	7.69
	7.45	7.10	7.56	7.47	7.52	7.47
Bottled	5.15	5.09	5.26	5.20	5.02	5.23
	5.28	5.26	5.13	5.26	5.21	5.24

Source: Emily McCarney, student at Joliet Junior College

(a) Which type of water has more dispersion in pH using the range as the measure of dispersion?

(b) Which type of water has more dispersion in pH using the standard deviation as the measure of dispersion?

NW 19. Pulse Rates The data on the following page represent the pulse rates (beats per minute) of nine students enrolled in a section of Sullivan's course in Introductory Statistics. Treat the nine students as a population.

Student	Pulse
Perpectual Bempah	76
Megan Brooks	60
Jeff Honeycutt	60
Clarice Jefferson	81
Crystal Kurtenbach	72
Janette Lantka	80
Kevin McCarthy	80
Tammy Ohm	68
Kathy Wojdyla	73

(a) Determine the population standard deviation.
(b) Find three simple random samples of size 3, and determine the sample standard deviation of each sample.
(c) Which samples underestimate the population standard deviation? Which overestimate the population standard deviation?

20. Travel Time The following data represent the travel time (in minutes) to school for nine students enrolled in Sullivan's College Algebra course. Treat the nine students as a population.

Student	Travel Time	Student	Travel Time
Amanda	39	Scot	45
Amber	21	Erica	11
Tim	9	Tiffany	12
Mike	32	Glenn	39
Nicole	30		

(a) Determine the population standard deviation.
(b) Find three simple random samples of size 4, and determine the sample standard deviation of each sample.
(c) Which samples underestimate the population standard deviation? Which overestimate the population standard deviation?

21. A Fish Story Ethan and Drew went on a 10-day fishing trip. The number of smallmouth bass caught and released by the two boys each day was as follows:

Ethan	9	24	8	9	5	8	9	10	8	10
Drew	15	2	3	18	20	1	17	2	19	3

(a) Find the population mean and the range for the number of smallmouth bass caught per day by each fisherman. Do these values indicate any differences between the two fishermen's catches per day? Explain.
(b) Draw a dot plot for Ethan. Draw a dot plot for Drew. Which fisherman seems more consistent?
(c) Find the population standard deviation for the number of smallmouth bass caught per day by each fisherman. Do these values present a different story about the two fishermen's catches per day? Which fisherman has the more consistent record? Explain.
(d) Discuss limitations of the range as a measure of dispersion.

22. Soybean Yield The data in the next column represent the number of pods on a sample of soybean plants for two different plot types. Which plot type do you think is superior? Why?

Plot Type	Pods								
Liberty	32	31	36	35	44	31	39	37	38
No Till	35	31	32	30	43	33	37	42	40

Source: Andrew Dieter and Brad Schmidgall, students at Joliet Junior College

23. The Empirical Rule The following data represent the weights (in grams) of a random sample of 50 M&M plain candies.

0.87	0.88	0.82	0.90	0.90	0.84	0.84
0.91	0.94	0.86	0.86	0.86	0.88	0.87
0.89	0.91	0.86	0.87	0.93	0.88	
0.83	0.95	0.87	0.93	0.91	0.85	
0.91	0.91	0.86	0.89	0.87	0.84	
0.88	0.88	0.89	0.79	0.82	0.83	
0.90	0.88	0.84	0.93	0.81	0.90	
0.88	0.92	0.85	0.84	0.84	0.86	

Source: Michael Sullivan

(a) Determine the sample standard deviation weight. Express your answer rounded to three decimal places.
(b) On the basis of the histogram drawn in Section 3.1, Problem 25, comment on the appropriateness of using the Empirical Rule to make any general statements about the weights of M&Ms.
(c) Use the Empirical Rule to determine the percentage of M&Ms with weights between 0.803 and 0.947 gram. *Hint:* $\bar{x} = 0.875$.
(d) Determine the actual percentage of M&Ms that weigh between 0.803 and 0.947 gram, inclusive.
(e) Use the Empirical Rule to determine the percentage of M&Ms with weights more than 0.911 gram.
(f) Determine the actual percentage of M&Ms that weigh more than 0.911 gram.

24. The Empirical Rule The following data represent the length of eruption for a random sample of eruptions at the Old Faithful geyser in Calistoga, California.

108	108	99	105	103	103	94
102	99	106	90	104	110	110
103	109	109	111	101	101	
110	102	105	110	106	104	
104	100	103	102	120	90	
113	116	95	105	103	101	
100	101	107	110	92	108	

Source: Ladonna Hansen, Park Curator

(a) Determine the sample standard deviation length of eruption. Express your answer rounded to the nearest whole number.
(b) On the basis of the histogram drawn in Section 3.1, Problem 26, comment on the appropriateness of using the Empirical Rule to make any general statements about the length of eruptions.
(c) Use the Empirical Rule to determine the percentage of eruptions that last between 92 and 116 seconds. *Hint:* $\bar{x} = 104$.

(d) Determine the actual percentage of eruptions that last between 92 and 116 seconds, inclusive.

(e) Use the Empirical Rule to determine the percentage of eruptions that last less than 98 seconds.

(f) Determine the actual percentage of eruptions that last less than 98 seconds.

25. Which Car Would You Buy? Suppose that you are in the market to purchase a car. You have narrowed it down to two choices and will let gas mileage be the deciding factor. You decide to conduct a little experiment in which you put 10 gallons of gas in the car and drive it on a closed track until it runs out gas. You conduct this experiment 15 times on each car and record the number of miles driven.

Car 1				
228	223	178	220	220
233	233	271	219	223
217	214	189	236	248

Car 2				
277	164	326	215	259
217	321	263	160	257
239	230	183	217	230

Describe each data set. That is, determine the shape, center, and spread. Which car would you buy and why?

26. Which Investment Is Better? You have received a year-end bonus of $5000. You decide to invest the money in the stock market and have narrowed your investment options down to two mutual funds. The following data represent the historical quarterly rates of return of each mutual fund for the past 20 quarters (5 years).

Mutual Fund A				
1.3	−0.3	0.6	6.8	5.0
5.2	4.8	2.4	3.0	1.8
7.3	8.6	3.4	3.8	−1.3
6.4	1.9	−0.5	−2.3	3.1

Mutual Fund B				
−5.4	6.7	11.9	4.3	4.3
3.5	10.5	2.9	3.8	5.9
−6.7	1.4	8.9	0.3	−2.4
−4.7	−1.1	3.4	7.7	12.9

Describe each data set. That is, determine the shape, center, and spread. Which mutual fund would you invest in and why?

27. Rates of Returns of Stocks Stocks may be categorized by sectors. Go to www.pearsonhighered.com/sullivanstats and download 3_2_27 using the file format of your choice. The data represent the one-year rate of return (in percent) for a sample of consumer cyclical stocks and industrial stocks for the period December 2013 through November 2014. Note: Consumer cyclical stocks include names such as Starbucks and Home Depot. Industrial stocks include names such as 3M and FedEx.

(a) Draw a relative frequency histogram for each sector using a lower class limit for the first class of −50 and a class width of 10. Which sector appears to have more dispersion?

(b) Determine the mean and median rate of return for each sector. Which sector has the higher mean rate of return? Which sector has the higher median rate of return?

(c) Determine the standard deviation rate of return for each sector. In finance, the standard deviation rate of return is called **risk**. Typically, an investor "pays" for a higher return by accepting more risk. Is the investor paying for higher

returns for these sectors? Do you think the higher returns are worth the cost? Explain.

28. Temperatures It is well known that San Diego has milder weather than Chicago, but which city has more deviation from normal temperatures? Use the following data, which represent the deviation from normal high temperatures for a random sample of days. In which city would you rather be a meteorologist? Why?

Deviation From Normal High Temperature, Chicago (°F)							
8	2	22	−2	0	−9	7	−15
−7	13	17	1	−5	0	7	−2
−9	11	19	−3	−4	2	1	−13
−5	15	11	6	6	8	−17	

Deviation From Normal High Temperature, San Diego (°F)							
4	−5	3	−1	−5	−3	−4	4
5	−6	7	−4	−5	−4	−3	−6
1	−1	−2	−5	−4	−5	10	−6
−4	−2	−1	−3	−6	−3	8	

Source: National Climatic Data Center

NW 29. The Empirical Rule The Stanford–Binet Intelligence Quotient (IQ) measures intelligence. IQ scores have a bell-shaped distribution with a mean of 100 and a standard deviation of 15.

(a) What percentage of people has an IQ score between 70 and 130?

(b) What percentage of people has an IQ score less than 70 or greater than 130?

(c) What percentage of people has an IQ score greater than 130?

30. The Empirical Rule SAT Math scores have a bell-shaped distribution with a mean of 515 and a standard deviation of 114.
Source: College Board

(a) What percentage of SAT scores is between 401 and 629?

(b) What percentage of SAT scores is less than 401 or greater than 629?

(c) What percentage of SAT scores is greater than 743?

31. The Empirical Rule The weight, in grams, of the pair of kidneys in adult males between the ages of 40 and 49 has a bell-shaped distribution with a mean of 325 grams and a standard deviation of 30 grams.

(a) About 95% of kidney pairs will be between what weights?

(b) What percentage of kidney pairs weighs between 235 grams and 415 grams?

(c) What percentage of kidney pairs weighs less than 235 grams or more than 415 grams?

(d) What percentage of kidney pairs weighs between 295 grams and 385 grams?

32. The Empirical Rule The distribution of the length of bolts has a bell shape with a mean of 4 inches and a standard deviation of 0.007 inch.

(a) About 68% of bolts manufactured will be between what lengths?

(b) What percentage of bolts will be between 3.986 inches and 4.014 inches?

(c) If the company discards any bolts less than 3.986 inches or greater than 4.014 inches, what percentage of bolts manufactured will be discarded?

(d) What percentage of bolts manufactured will be between 4.007 inches and 4.021 inches?

33. Which Professor? Suppose Professor Alpha and Professor Omega each teach Introductory Biology. You need to decide which professor to take the class from and have just completed your Introductory Statistics course. Records obtained from past students indicate that students in Professor Alpha's class have a mean score of 80% with a standard deviation of 5%, while past students in Professor Omega's class have a mean score of 80% with a standard deviation of 10%. Decide which instructor to take for Introductory Biology using a statistical argument.

34. Larry Summers Lawrence Summers (former Secretary of the Treasury and former president of Harvard) infamously claimed that women have a lower standard deviation IQ than men. He went on to suggest that this was a potential explanation as to why there are fewer women in top math and science positions. Suppose an IQ of 145 or higher is required to be a researcher at a top-notch research institution. Use the idea of standard deviation, the Empirical Rule, and the fact that the mean and standard deviation IQ of humans is 100 and 15, respectively, to explain Summers' argument.

NW 35. Chebyshev's Inequality In December 2014, the average price of regular unleaded gasoline excluding taxes in the United States was $3.06 per gallon, according to the Energy Information Administration. Assume that the standard deviation price per gallon is $0.06 per gallon to answer the following.

(a) What minimum percentage of gasoline stations had prices within 3 standard deviations of the mean?

(b) What minimum percentage of gasoline stations had prices within 2.5 standard deviations of the mean? What are the gasoline prices that are within 2.5 standard deviations of the mean?

(c) What is the minimum percentage of gasoline stations that had prices between $2.94 and $3.18?

36. Chebyshev's Inequality According to the U.S. Census Bureau, the mean of the commute time to work for a resident of Boston, Massachusetts, is 27.3 minutes. Assume that the standard deviation of the commute time is 8.1 minutes to answer the following:

(a) What minimum percentage of commuters in Boston has a commute time within 2 standard deviations of the mean?

(b) What minimum percentage of commuters in Boston has a commute time within 1.5 standard deviations of the mean? What are the commute times within 1.5 standard deviations of the mean?

(c) What is the minimum percentage of commuters who have commute times between 3 minutes and 51.6 minutes?

37. Comparing Standard Deviations The standard deviation of batting averages of all teams in the American League is 0.008. The standard deviation of all players in the American League is 0.02154. Why is there less variability in team batting averages?

38. Linear Transformations Benjamin owns a small Internet business. Besides himself, he employs nine other people. The salaries earned by the employees are given next in thousands of dollars (Benjamin's salary is the largest, of course):

30, 30, 45, 50, 50, 50, 55, 55, 60, 75

(a) Determine the range, population variance, and population standard deviation for the data.

(b) Business has been good! As a result, Benjamin has a total of $25,000 in bonus pay to distribute to his employees. One option for distributing bonuses is to give each employee (including himself) $2500. Add the bonuses under this plan to the original salaries to create a new data set. Recalculate the range, population variance, and population standard deviation. How do they compare to the originals?

(c) As a second option, Benjamin can give each employee a bonus of 5% of his or her original salary. Add the bonuses under this second plan to the original salaries to create a new data set. Recalculate the range, population variance, and population standard deviation. How do they compare to the originals?

(d) As a third option, Benjamin decides not to give his employees a bonus at all. Instead, he keeps the $25,000 for himself. Use this plan to create a new data set. Recalculate the range, population variance, and population standard deviation. How do they compare to the originals?

39. Resistance and Sample Size Each of the following three data sets represents the IQ scores of a random sample of adults. IQ scores are known to have a mean and median of 100. For each data set, determine the sample standard deviation. Then recompute the sample standard deviation assuming that the individual whose IQ is 106 is accidentally recorded as 160. For each sample size, state what happens to the standard deviation. Comment on the role that the number of observations plays in resistance.

Sample of Size 5				
106	92	98	103	100

Sample of Size 12					
106	92	98	103	100	102
98	124	83	70	108	121

Sample of Size 30					
106	92	98	103	100	102
98	124	83	70	108	121
102	87	121	107	97	114
140	93	130	72	81	90
103	97	89	98	88	103

40. Identical Values Compute the sample standard deviation of the following test scores: 78, 78, 78, 78. What can be said about a data set in which all the values are identical?

41. Buying a Car The following data represent the asking price, in dollars, for a random sample of 2010 coupes and a random sample of 2010 Chevy Camaros.

Coupes			Camaros		
25,991	15,900	16,900	24,949	24,948	23,061
24,948	23,791	20,990	22,150	21,855	20,990
19,900	17,900	16,888	19,950	19,593	18,995
15,995	15,891	13,991	17,849	16,900	16,440
12,900	11,995	9900	15,987	15,900	15,891

Source: autotrader.com

(continued)

(a) Find the mean and standard deviation price for each sample.

(b) Explain why the mean is higher for Camaros yet the standard deviation is less.

42. Blocking and Variability Recall that blocking refers to the idea that we can reduce the variability in a variable by segmenting the data by some other variable. The given data represent the recumbent length (in centimeters) of a sample of 10 males and 10 females who are 40 months of age.

Males		Females	
104.0	94.4	102.5	100.8
93.7	97.6	100.4	96.3
98.3	100.6	102.7	105.0
86.2	103.0	98.1	106.5
90.7	100.9	95.4	114.5

Source: National Center for Health Statistics

(a) Determine the standard deviation of recumbent length for all 20 observations.

(b) Determine the standard deviation of recumbent length for the males.

(c) Determine the standard deviation of recumbent length for the females.

(d) What effect does blocking by gender have on the standard deviation of recumbent length for each gender?

43. Mean Absolute Deviation Another measure of variation is the mean absolute deviation. It is computed using the formula

$$\text{MAD} = \frac{\Sigma |x_i - \bar{x}|}{n}$$

Compute the mean absolute deviation of the data in Problem 11 and compare the results with the sample standard deviation.

44. Coefficient of Skewness Karl Pearson developed a measure that describes the skewness of a distribution, called the **coefficient of skewness**. The formula is

$$\text{Skewness} = \frac{3(\text{mean} - \text{median})}{\text{standard deviation}}$$

The value of this measure generally lies between −3 and +3. The closer the value lies to −3, the more the distribution is skewed left. The closer the value lies to +3, the more the distribution is skewed right. A value close to 0 indicates a symmetric distribution. Find the coefficient of skewness of the following distributions and comment on the skewness.

(a) Mean = 50, median = 40, standard deviation = 10

(b) Mean = 100, median = 100, standard deviation = 15

(c) Mean = 400, median = 500, standard deviation = 120

(d) Compute the coefficient of skewness for the data in Problem 23.

(e) Compute the coefficient of skewness for the data in Problem 24.

45. Diversification A popular theory in investment states that you should invest a certain amount of money in foreign investments to reduce your risk. The risk of a portfolio is defined as the standard deviation of the rate of return. Refer to the graph in the next column, which depicts the relation between risk (standard deviation of rate of return) and reward (mean rate of return).

How Foreign Stocks Benefit a Domestic Portfolio

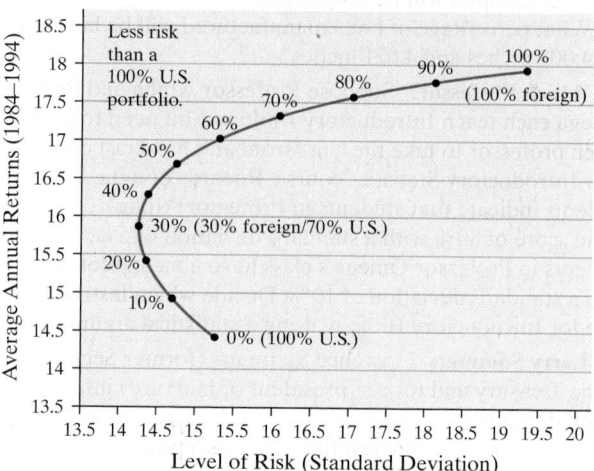

Source: T. Rowe Price

(a) Determine the average annual return and level of risk in a portfolio that is 10% foreign.

(b) Determine the percentage that should be invested in foreign stocks to best minimize risk.

(c) Why do you think risk initially decreases as the percent of foreign investments increases?

(d) A portfolio that is 30% foreign and 70% American has a mean rate of return of about 15.8%, with a standard deviation of 14.3%. According to Chebyshev's Inequality, at least 75% of returns will be between what values? According to Chebyshev's Inequality, at least 88.9% of returns will be between what two values? Should an investor be surprised if she has a negative rate of return? Why?

46. More Spread? The data set on the left represents the annual rate of return (in percent) of eight randomly sampled bond mutual funds, and the data set on the right represents the annual rate of return (in percent) of eight randomly sampled stock mutual funds.

2.0	1.9	1.8	8.4	7.2	7.6
3.2	2.4	3.4	7.4	6.9	9.4
1.6	2.7		9.1	8.1	

(a) Determine the mean and standard deviation of each data set.

(b) Based only on the standard deviation, which data set has more spread?

(c) What proportion of the observations is within one standard deviation of the mean for each data set?

(d) The **coefficient of variation**, CV, is defined as the ratio of the standard deviation to the mean of a data set, so

$$\text{CV} = \frac{\text{standard deviation}}{\text{mean}}$$

The coefficient of variation is unitless and allows for comparison in spread between two data sets by describing the amount of spread per unit mean. After all, larger numbers will likely have a larger standard deviation simply due to the size of the numbers. Compute the coefficient of variation for both data sets. Which data set do you believe has more "spread"?

(e) Let's take this idea one step further. The following data represent the height of a random sample of 8 male college students. The data set on the left has their height measured in inches, and the data set on the right has their height measured in centimeters.

74	68	71
66	72	69
69	71	

187.96	172.72	180.34
167.64	182.88	175.26
175.26	180.34	

For each data set, determine the mean and the standard deviation. Would you say that the height of the males is more dispersed based on the standard deviation of height measured in centimeters? Why? Now, determine the coefficient of variation for each data set. What did you find?

47. **Sullivan Survey** Choose two quantitative variables from the SullivanStats Survey I at www.pearsonhighered.com/sullivanstats for which a comparison is reasonable, such as number of hours of television versus number of hours of Internet. Draw a histogram for each variable. Which variable appears to have more dispersion? Determine the range and standard deviation of each variable. Based on the numerical measure, which variable has more dispersion?

48. **Sullivan Survey** Choose any quantitative variable from the SullivanStats Survey I at www.pearsonhighered.com/sullivanstats. Now choose a qualitative variable, such as gender or political philosophy. Determine the range and standard deviation by the qualitative variable chosen. For example, if you chose gender as the qualitative variable, determine the range and standard deviation by gender. Does there appear to be any difference in the measure of dispersion for each level of the qualitative variable?

Explaining the Concepts

49. Would it be appropriate to say that a distribution with a standard deviation of 10 centimeters is more dispersed than a distribution with a standard deviation of 5 inches? Support your position.

50. What is meant by the phrase *degrees of freedom* as it pertains to the computation of the sample standard deviation?

51. Are any of the measures of dispersion mentioned in this section resistant? Explain.

52. What does it mean when a statistic is biased?

53. What makes the range less desirable than the standard deviation as a measure of dispersion?

54. In one of Sullivan's statistics sections, the standard deviation of the heights of all students was 3.9 inches. The standard deviation of the heights of males was 3.4 inches and the standard deviation of females was 3.3 inches. Why is the standard deviation of the entire class more than the standard deviation of the males and females considered separately?

55. Explain how standard deviation measures spread. In your explanation include the computation of the same standard deviation for two data sets: Data set I: 3, 4, 5; Data set II: 0, 4, 8.

56. Which of the following would have a higher standard deviation? (a) IQ of students on your campus or (b) IQ of residents in your home town? Why?

57. Develop a sample of size $n = 8$ such that $\bar{x} = 15$ and $s = 0$.

58. Draw two histograms with different standard deviations and label them I and II. Which histogram has the larger standard deviation?

59. **Fast Pass** In 2000, the Walt Disney Company created the "fast pass"—a ticket issued to a rider for a popular ride and the rider is told to return at a specific time during the day. At that time, the rider is allowed to bypass the regular line, thereby reducing the wait time for that particular rider. When compared to wait times prior to creating the fast pass, overall wait times for rides at the park increased, on average, yet patrons to the park indicated they were happier with the fast-pass system. Use the concepts of central tendency and dispersion to explain why.

3.3 Measures of Central Tendency and Dispersion from Grouped Data

Preparing for This Section Before getting started, review the following:

- Organizing discrete data in tables (Section 2.2, pp. 76–77)
- Organizing continuous data in tables (Section 2.2, pp. 78–80)
- Class midpoint (Section 2.3, p. 94)

Objectives ❶ Approximate the mean of a variable from grouped data

❷ Compute the weighted mean

❸ Approximate the standard deviation of a variable from grouped data

We have discussed how to compute descriptive statistics from raw data, but often the only data available have already been summarized in frequency distributions (**grouped data**). Although we cannot find exact values of the mean or standard deviation without raw data, we can approximate these measures.

① Approximate the Mean of a Variable from Grouped Data

Because raw data cannot be retrieved from a frequency table, we assume that within each class the mean of the data values is equal to the class midpoint. We then multiply the class midpoint by the frequency. This product is expected to be close to the sum of the data that lie within the class. We repeat the process for each class and add the results. This sum approximates the sum of all the data.

Definition

> **Approximate Mean of a Variable from a Frequency Distribution**
>
> **Population Mean**
>
> $$\mu = \frac{\sum x_i f_i}{\sum f_i}$$
>
> $$= \frac{x_1 f_1 + x_2 f_2 + \cdots + x_n f_n}{f_1 + f_2 + \cdots + f_n}$$
>
> **Sample Mean**
>
> $$\bar{x} = \frac{\sum x_i f_i}{\sum f_i}$$
>
> $$= \frac{x_1 f_1 + x_2 f_2 + \cdots + x_n f_n}{f_1 + f_2 + \cdots + f_n} \qquad (1)$$
>
> where x_i is the midpoint or value of the ith class
>
> f_i is the frequency of the ith class
>
> n is the number of classes

In Formula (1), $x_1 f_1$ approximates the sum of all the data values in the first class, $x_2 f_2$ approximates the sum of all the data values in the second class, and so on. Notice that the formulas for the population mean and sample mean are essentially identical, just as they were for computing the mean from raw data.

EXAMPLE 1   Approximating the Mean for Continuous Quantitative Data from a Frequency Distribution

Table 13

Class (five-year rate of return)	Frequency
8–8.99	2
9–9.99	2
10–10.99	4
11–11.99	1
12–12.99	6
13–13.99	13
14–14.99	7
15–15.99	3
16–16.99	1
17–17.99	0
18–18.99	0
19–19.99	1

Problem The frequency distribution in Table 13 represents the five-year rate of return of a random sample of 40 large-blended mutual funds. Approximate the mean five-year rate of return.

Approach

Step 1 Determine the class midpoint of each class by adding consecutive lower class limits and dividing the result by 2.

Step 2 Compute the sum of the frequencies, Σf_i.

Step 3 Multiply the class midpoint by the frequency to obtain $x_i f_i$ for each class.

Step 4 Compute $\Sigma x_i f_i$.

Step 5 Substitute into Sample Mean Formula (1) to obtain the mean from grouped data.

Solution

Step 1 The first two lower class limits are 8 and 9. Therefore, the class midpoint of the first class is $\frac{8+9}{2} = 8.5$, so $x_1 = 8.5$. The remaining class midpoints are listed in column 2 of Table 14.

Table 14			
Class (five-year rate of return)	Class Midpoint, x_i	Frequency, f_i	$x_i f_i$
8–8.99	$\frac{8+9}{2} = 8.5$	2	$(8.5)(2) = 17$
9–9.99	9.5	2	$(9.5)(2) = 19$
10–10.99	10.5	4	42
11–11.99	11.5	1	11.5
12–12.99	12.5	6	75
13–13.99	13.5	13	175.5
14–14.99	14.5	7	101.5
15–15.99	15.5	3	46.5
16–16.99	16.5	1	16.5
17–17.99	17.5	0	0
18–18.99	18.5	0	0
19–19.99	19.5	1	19.5
		$\sum f_i = 40$	$\sum x_i f_i = 524$

Step 2 Add the frequencies in column 3 to obtain $\Sigma f_i = 2 + 2 + \cdots + 1 = 40$.

Step 3 Compute the values of $x_i f_i$ by multiplying each class midpoint by the corresponding frequency and obtain the results shown in column 4 of Table 14.

Step 4 Add the values in column 4 of Table 14 to obtain $\Sigma x_i f_i = 524$.

Step 5 Substituting into Sample Mean Formula (1), we obtain

$$\bar{x} = \frac{\sum x_i f_i}{\sum f_i} = \frac{524}{40} = 13.1$$

The approximate mean five-year rate of return is 13.1%.

> **CAUTION!**
> We computed the mean from grouped data in Example 1 even though the raw data is available. The reason for doing this was to illustrate how close the two values can be. In practice, use raw data whenever possible.

● Now compute the approximate mean of the frequency distribution in Problem 1

The mean five-year rate of return from the raw data listed in Example 3 on page 79 from Section 2.2 is 13.141%.

② Compute the Weighted Mean

When data values have different importance, or *weight*, associated with them, we compute the *weighted mean*. For example, your grade-point average is a weighted mean, with the weights equal to the number of credit hours in each course. The value of the variable is equal to the grade converted to a point value.

Definition The **weighted mean**, $\bar{x}_w$, of a variable is found by multiplying each value of the variable by its corresponding weight, adding these products, and dividing this sum by the sum of the weights. It can be expressed using the formula

$$\bar{x}_w = \frac{\sum w_i x_i}{\sum w_i} = \frac{w_1 x_1 + w_2 x_2 + \cdots + w_n x_n}{w_1 + w_2 + \cdots + w_n} \qquad (2)$$

where w_i is the weight of the ith observation

x_i is the value of the ith observation

EXAMPLE 2   Computing the Weighted Mean

Problem Marissa just completed her first semester in college. She earned an A in her four-hour statistics course, a B in her three-hour sociology course, an A in her three-hour psychology course, a C in her five-hour computer programming course, and an A in her one-hour drama course. Determine Marissa's grade-point average.

Approach Assign point values to each grade. Let an A equal 4 points, a B equal 3 points, and a C equal 2 points. The number of credit hours for each course determines its weight. So a five-hour course gets a weight of 5, a four-hour course gets a weight of 4, and so on. Multiply the weight of each course by the points earned in the course, add these products, and divide this sum by the sum of the weights, number of credit hours.

Solution

$$\text{GPA} = \bar{x}_w = \frac{\sum w_i x_i}{\sum w_i} = \frac{4(4) + 3(3) + 3(4) + 5(2) + 1(4)}{4 + 3 + 3 + 5 + 1} = \frac{51}{16} = 3.19$$

● **Now Work Problem 9**   Marissa's grade-point average for her first semester is 3.19.

③ ## Approximate the Standard Deviation of a Variable from Grouped Data

The procedure for approximating the standard deviation from grouped data is similar to that of finding the mean from grouped data. Again, because we do not have access to the original data, the standard deviation is approximate.

Definition

Approximate Standard Deviation of a Variable from a Frequency Distribution

Population Standard Deviation	Sample Standard Deviation

$$\sigma = \sqrt{\frac{\sum (x_i - \mu)^2 f_i}{\sum f_i}} \qquad\qquad s = \sqrt{\frac{\sum (x_i - \bar{x})^2 f_i}{(\sum f_i) - 1}} \qquad (3)$$

where x_i is the midpoint or value of the ith class

f_i is the frequency of the ith class

An algebraically equivalent formula for the population standard deviation is

$$\sigma = \sqrt{\frac{\sum x_i^2 f_i - \dfrac{(\sum x_i f_i)^2}{\sum f_i}}{\sum f_i}}$$

EXAMPLE 3   Approximating the Standard Deviation from a Frequency Distribution

Problem The data in Table 13 on page 148 represent the five-year rate of return of a random sample of 40 large-blended mutual funds. Approximate the standard deviation of the five-year rate of return.

Approach Use the sample standard deviation Formula (3).

Step 1   Create a table with the class in column 1, the class midpoint in column 2, the frequency in column 3, and the unrounded mean in column 4.

Step 2   Compute the deviation about the mean, $x_i - \bar{x}$, for each class, where x_i is the class midpoint of the ith class and $\bar{x}$ is the sample mean. Enter the results in column 5.

Step 3   Square the deviation about the mean and multiply this result by the frequency to obtain $(x_i - \bar{x})^2 f_i$. Enter the results in column 6.

Step 4   Add the entries in columns 3 and 6 to obtain Σf_i and $\Sigma (x_i - \bar{x})^2 f_i$.

Step 5   Substitute the values found in Step 4 into Formula (3) to obtain an approximate value for the sample standard deviation.

Solution

Step 1   Create Table 15. Column 1 contains the classes. Column 2 contains the class midpoint of each class. Column 3 contains the frequency of each class. Column 4 contains the unrounded sample mean obtained in Example 1.

Table 15

Class (five-year rate of return)	Class Midpoint, x_i	Frequency, f_i	$\bar{x}$	$x_i - \bar{x}$	$(x_i - \bar{x})^2 f_i$
8–8.99	$\frac{8+9}{2} = 8.5$	2	13.1	−4.6	42.32
9–9.99	9.5	2	13.1	−3.6	25.92
10–10.99	10.5	4	13.1	−2.6	27.04
11–11.99	11.5	1	13.1	−1.6	2.56
12–12.99	12.5	6	13.1	−0.6	2.16
13–13.99	13.5	13	13.1	0.4	2.08
14–14.99	14.5	7	13.1	1.4	13.72
15–15.99	15.5	3	13.1	2.4	17.28
16–16.99	16.5	1	13.1	3.4	11.56
17–17.99	17.5	0	13.1	4.4	0
18–18.99	18.5	0	13.1	5.4	0
19–19.99	19.5	1	13.1	6.4	40.96
		$\Sigma f_i = 40$			$\Sigma (x_i - \bar{x})^2 f_i = 185.6$

Step 2   Column 5 contains the deviation about the mean, $x_i - \bar{x}$, for each class.

Step 3   Column 6 contains the values of the squared deviation about the mean multiplied by the frequency, $(x_i - \bar{x})^2 f_i$.

Step 4   Add the entries in columns 3 and 6 to obtain $\Sigma f_i = 40$ and $\Sigma (x_i - \bar{x})^2 f_i = 185.6$.

Step 5   Substitute these values into Formula (3) to obtain an approximate value for the sample standard deviation.

$$s = \sqrt{\frac{\Sigma (x_i - \bar{x})^2 f_i}{\left(\Sigma f_i\right) - 1}} = \sqrt{\frac{185.6}{40 - 1}} \approx 2.182$$

The approximate standard deviation of the five-year rate of return is 2.182%. The standard deviation from the raw data listed in Example 3 from Section 2.2 is 2.193%. ●

EXAMPLE 4   Approximating the Mean and Standard Deviation of Grouped Data Using Technology

Problem Approximate the mean and standard deviation of the five-year rate of return data in Table 13 using a TI-84 Plus C graphing calculator.

(continued)

Approach The steps for approximating the mean and standard deviation of grouped data using the TI-83/84 Plus graphing calculator and StatCrunch are given in the Technology Step-by-Step below.

Solution Figure 16 shows the result from the TI-84 Plus C. From the output, we can see that the approximate mean is 13.1% and the approximate standard deviation is 2.18%. The results agree with our by-hand solutions.

Figure 16

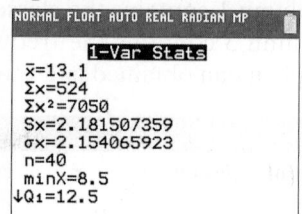

● Now compute the approximate standard deviation from the frequency distribution in Problem 1

Technology Step-by-Step Determining the Mean and Standard Deviation from Grouped Data

TI-83/84 Plus
1. Enter the class midpoint in L1 and the frequency or relative frequency in L2 by pressing STAT and then selecting 1:Edit.
2. Press STAT, highlight the CALC menu, and select 1:1-Var Stats. In the menu, select L1 for List. Select L2 for FreqList:. Highlight Calculate and press ENTER.

StatCrunch
1. If necessary, enter the summarized data into the spreadsheet. Name the columns.
2. Select **Stat**, highlight **Summary Stats**, and select **Grouped/Binned data**.
3. Choose the column that contains the class under the "Bins in:" drop-down menu. Choose the column that contains the frequencies in the "Counts in:" drop-down menu. Select the "Consecutive lower limits" radio button for defining the midpoints. Click Compute!.

 ## 3.3 Assess Your Understanding

Applying the Concepts

NW 1. Savings Recently, a random sample of 25–34 year olds was asked, "How much do you currently have in savings, not including retirement savings?" The following data represent the responses to the survey. Approximate the mean and standard deviation amount of savings.

Savings	Frequency
$0–$19,999	344
$20,000–$39,999	98
$40,000–$59,999	52
$60,000–$79,999	19
$80,000–$99,999	13
$100,000–$119,999	6
$120,000–$139,999	2

Source: Based on a poll by LearnVest

2. Square Footage of Housing The frequency distribution in the next column represents the square footage of a random sample of 500 houses that are owner occupied year round. Approximate the mean and standard deviation square footage.

Square Footage	Frequency
0–499	5
500–999	17
1000–1499	36
1500–1999	121
2000–2499	119
2500–2999	81
3000–3499	47
3500–3999	45
4000–4499	22
4500–4999	7

Source: Based on data from the U.S. Census Bureau

3. Household Winter Temperature Often, frequency distributions are reported using unequal class widths because the frequencies of some groups would otherwise be small or very large. Consider the following data, which represent the daytime household temperature the thermostat is set to when someone is home for a random sample of 750 households. Determine the class midpoint, if necessary, for each class and approximate the mean and standard deviation temperature.

Temperature (°F)	Frequency
61–64	31
65–67	67
68–69	198
70	195
71–72	120
73–76	89
77–80	50

Source: Based on data from the U.S. Department of Energy

4. Living in Poverty (See Problem 3.) The following frequency distribution represents the age of people living in poverty in 2013 based on a random sample of residents of the United States. In this frequency distribution, the class widths are not the same for each class. Approximate the mean and standard deviation age of a person living in poverty. For the open-ended class 65 and older, use 70 as the class midpoint.

Age	Frequency
0–17	14,659
18–24	5819
25–34	6694
35–44	4871
45–54	4533
55–59	2476
60–64	2036
65 and older	4231

Source: U.S. Census Bureau

5. Multiple Births The following data represent the number of live multiple-delivery births (three or more babies) in 2012 for women 15 to 54 years old.

Age	Number of Multiple Births
15–19	44
20–24	404
25–29	1204
30–34	1872
35–39	1000
40–44	332
45–49	44
50–54	19

Source: National Vital Statistics Reports

(a) Approximate the mean and standard deviation for age.
(b) Draw a frequency histogram of the data to verify that the distribution is bell shaped.
(c) According to the Empirical Rule, 95% of mothers of multiple births will be between what two ages?

6. Birth Weight The following frequency distribution represents the birth weight of all babies born in the United States in 2012.

Weight (grams)	Number of Babies
0–999	27,000
1000–1999	90,000
2000–2999	920,000
3000–3999	2,599,000
4000–4999	308,000
5000–5999	5000

Source: National Vital Statistics Report

(a) Approximate the mean and standard deviation birth weight.
(b) Draw a frequency histogram of the data to verify that the distribution is bell-shaped.
(c) According to the Empirical Rule, 68% of all babies will weigh between what two values?

7. Cigarette Tax Rates Use the frequency distribution whose class width is 0.5 obtained in Problem 29 in Section 2.2 to approximate the mean and standard deviation for cigarette tax rates. Compare these results to the actual mean and standard deviation.

8. Dividend Yield Use the frequency distribution whose class width is 0.4 obtained in Problem 30 in Section 2.2 to approximate the mean and standard deviation of the dividend yield. Compare these results to the actual mean and standard deviation.

NW **9. Grade-Point Average** Marissa has just completed her second semester in college. She earned a B in her five-hour calculus course, an A in her three-hour social work course, an A in her four-hour biology course, and a C in her three-hour American literature course. Assuming that an A equals 4 points, a B equals 3 points, and a C equals 2 points, determine Marissa's grade-point average for the semester.

10. Computing Class Average In Marissa's calculus course, attendance counts for 5% of the grade, quizzes count for 10% of the grade, exams count for 60% of the grade, and the final exam counts for 25% of the grade. Marissa had a 100% average for attendance, 93% for quizzes, 86% for exams, and 85% on the final. Determine Marissa's course average.

11. Mixed Chocolates Michael and Kevin want to buy chocolates. They can't agree on whether they want chocolate–covered almonds, chocolate-covered peanuts, or chocolate-covered raisins. They agree to create a mix. They bought 4 pounds of chocolate-covered almonds at $3.50 per pound, 3 pounds of chocolate-covered peanuts for $2.75 per pound, and 2 pounds of chocolate-covered raisins for $2.25 per pound. Determine the cost per pound of the mix.

12. Nut Mix Michael and Kevin return to the candy store, but this time they want to purchase nuts. They can't decide among peanuts, cashews, or almonds. They again agree to create a mix. They bought 2.5 pounds of peanuts for $1.30 per pound, 4 pounds of cashews for $4.50 per pound, and 2 pounds of almonds for $3.75 per pound. Determine the price per pound of the mix.

13. Population The data on the next page represent the male and female population, by age, of the United States in 2012. **Note:** Use 85 for the class midpoint of ≥80.

Age	Male Resident Pop	Female Resident Pop
0–9	20,700,000	19,826,000
10–19	21,369,000	20,475,000
20–29	21,417,000	21,355,000
30–39	19,455,000	20,011,000
40–49	20,839,000	21,532,000
50–59	20,785,000	22,058,000
60–69	14,739,000	16,362,000
70–79	7,641,000	9,474,000
≥ 80	4,230,000	6,561,000

Source: U.S. Census Bureau

(a) Approximate the population mean and standard deviation of age for males.
(b) Approximate the population mean and standard deviation of age for females.
(c) Which gender has the higher mean age?
(d) Which gender has more dispersion in age?

14. Age of Mother The following data represent the age of the mother at childbirth for 1980 and 2013.

Age	1980 Births (thousands)	2013 Births (thousands)
10–14	9.8	3.1
15–19	551.9	274.6
20–24	1226.4	902.1
25–29	1108.2	1127.6
30–34	549.9	1044.0
35–39	140.7	487.5
40–44	23.2	110.3
45–49	1.1	8.2

Source: National Vital Statistics Reports

(a) Approximate the population mean and standard deviation of age for mothers in 1980.
(b) Approximate the population mean and standard deviation of age for mothers in 2013.
(c) Which year has the higher mean age?
(d) Which year has more dispersion in age?

Problems 15 and 16 use the following steps to approximate the median from grouped data.

Approximating the Median from Grouped Data

Step 1   Construct a cumulative frequency distribution.
Step 2   Identify the class in which the median lies. Remember, the median can be obtained by determining the observation that lies in the middle.
Step 3   Interpolate the median using the formula

$$\text{Median} = M = L + \frac{\frac{n}{2} - CF}{f}(i)$$

where L is the lower class limit of the class containing the median
n is the number of data values in the frequency distribution
CF is the cumulative frequency of the class immediately preceding the class containing the median
f is the frequency of the median class
i is the class width of the class containing the median

15. Approximate the median of the frequency distribution in Problem 2.
16. Approximate the median of the frequency distribution in Problem 4.

*Problems 17 and 18 use the following definition of the modal class. The **modal class** of a variable can be obtained from data in a frequency distribution by determining the class that has the largest frequency.*

17. Determine the modal class of the frequency distribution in Problem 1.
18. Determine the modal class of the frequency distribution in Problem 2.

3.4   Measures of Position and Outliers

Objectives
1. Determine and interpret z-scores
2. Interpret percentiles
3. Determine and interpret quartiles
4. Determine and interpret the interquartile range
5. Check a set of data for outliers

In Section 3.1, we determined measures of central tendency, which describe the *typical* data value. Section 3.2 discussed measures of dispersion, which describe the amount of *spread* in a set of data. In this section, we discuss measures of position, which describe the *relative position* of a certain data value within the entire set of data.

① Determine and Interpret z-Scores

At the end of the 2014 season, the Los Angeles Angels led the American League with 773 runs scored, while the Colorado Rockies led the National League with 755 runs scored. It appears that the Angels are the better run-producing team. However, this comparison is unfair because the two teams play in different leagues. The Angels play in the American League, where the designated hitter bats for the pitcher, whereas the Rockies play in the National League, where the pitcher must bat (pitchers are typically poor hitters). To compare the two teams' scoring of runs, we need to determine their relative standings in their respective leagues. We can do this using a z-score.

Definition

The **z-score** represents the distance that a data value is from the mean in terms of the number of standard deviations. We find it by subtracting the mean from the data value and dividing this result by the standard deviation. There is both a population z-score and a sample z-score:

Population z-Score	Sample z-Score
$z = \dfrac{x - \mu}{\sigma}$	$z = \dfrac{x - \bar{x}}{s}$

(1)

The z-score is unitless. It has mean 0 and standard deviation 1.

In Other Words
The z-score provides a way to compare apples to oranges by converting variables with different centers or spreads to variables with the same center (0) and spread (1).

If a data value is larger than the mean, the z-score is positive. If a data value is smaller than the mean, the z-score is negative. If the data value equals the mean, the z-score is zero. A z-score measures the number of standard deviations an observation is above or below the mean. For example, a z-score of 1.24 means the data value is 1.24 standard deviations above the mean. A z-score of −2.31 means the data value is 2.31 standard deviations below the mean.

EXAMPLE 1 Comparing z-Scores

Problem Determine whether the Los Angeles Angels or the Colorado Rockies had a relatively better run-producing season. The Angels scored 773 runs and play in the American League, where the mean number of runs scored was $\mu = 677.4$ and the standard deviation was $\sigma = 51.7$ runs. The Rockies scored 755 runs and play in the National League, where the mean number of runs scored was $\mu = 640.0$ and the standard deviation was $\sigma = 55.9$ runs.

Approach To determine which team had the relatively better run-producing season, compute each team's z-score. The team with the higher z-score had the better season. Because we know the values of the population parameters, compute the population z-score.

Solution We compute each team's z-score, rounded to two decimal places.

$$\text{Angels:} \quad z\text{-score} = \frac{x - \mu}{\sigma} = \frac{773 - 677.4}{51.7} = 1.85$$

$$\text{Rockies:} \quad z\text{-score} = \frac{x - \mu}{\sigma} = \frac{755 - 640.0}{55.9} = 2.06$$

So the Angels had run production 1.85 standard deviations above the mean, while the Rockies had run production 2.06 standard deviations above the mean. Therefore, the Rockies had a relatively better year at scoring runs than the Angels.

● **Now Work Problem 5**

In Example 1, the team with the higher z-score was said to have a relatively better season in producing runs. With negative z-scores, we need to be careful when deciding the better outcome. For example, suppose Bob and Mary run a marathon. If Bob finished the marathon in 213 minutes, where the mean finishing time among all men was 242 minutes with a standard deviation of 57 minutes, and Mary finished the marathon in 241 minutes, where the mean finishing time among all women was 273 minutes with a standard deviation of 52 minutes, who did better in the race? Since Bob's z-score is $z_{Bob} = \dfrac{213 - 242}{57} = -0.51$ and Mary's z-score is $z_{Mary} = \dfrac{241 - 273}{52} = -0.62$, Mary did better. Even though Bob's z-score is larger, Mary did better because she is more standard deviations below the mean.

② Interpret Percentiles

Recall that the median divides the lower 50% of a set of data from the upper 50%. The median is a special case of a general concept called the *percentile*.

Definition The **kth percentile**, denoted P_k, of a set of data is a value such that k percent of the observations are less than or equal to the value.

So percentiles divide a set of data that is written in ascending order into 100 parts; thus 99 percentiles can be determined. For example, P_1 divides the bottom 1% of the observations from the top 99%, P_2 divides the bottom 2% of the observations from the top 98%, and so on. Figure 17 displays the 99 possible percentiles.

Figure 17

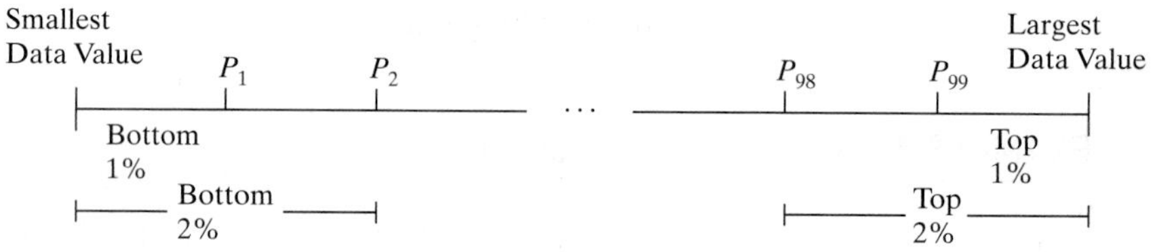

Percentiles are used to give the relative standing of an observation. Many standardized exams, such as the SAT college entrance exam, use percentiles to let students know how they scored on the exam in relation to all other students who took the exam.

EXAMPLE 2 Interpret a Percentile

Problem Jennifer just received the results of her SAT exam. Her SAT Mathematics score of 600 is in the 74th percentile. What does this mean?

Approach The kth percentile of an observation means that k percent of the observations are less than or equal to the observation.

Interpretation A percentile rank of 74% means that 74% of SAT Mathematics scores are less than or equal to 600 and 26% of the scores are greater. So 26% of the students who took the exam scored better than Jennifer.

● Now Work Problem 15

③ Determine and Interpret Quartiles

The most common percentiles are quartiles. **Quartiles** divide data sets into fourths, or four equal parts.

In Other Words
The first quartile, Q_1, is equivalent to the 25th percentile, P_{25}. The 2nd quartile, Q_2, is equivalent to the 50th percentile, P_{50}, which is equivalent to the median, M. Finally, the third quartile, Q_3, is equivalent to the 75th percentile, P_{75}.

- The first quartile, denoted Q_1, divides the bottom 25% of the data from the top 75%. Therefore, the first quartile is equivalent to the 25th percentile.
- The second quartile, Q_2, divides the bottom 50% of the data from the top 50%; it is equivalent to the 50th percentile or the median.
- The third quartile, Q_3, divides the bottom 75% of the data from the top 25%; it is equivalent to the 75th percentile.

Figure 18 illustrates the concept of quartiles.

Figure 18

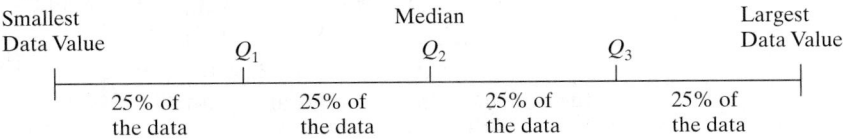

In Other Words
To find Q_2, determine the median of the data set. To find Q_1, determine the median of the "lower half" of the data set. To find Q_3, determine the median of the "upper half" of the data set.

Finding Quartiles

Step 1   Arrange the data in ascending order.

Step 2   Determine the median, M, or second quartile, Q_2.

Step 3   Divide the data set into halves: the observations below (to the left of) M and the observations above M. The first quartile, Q_1, is the median of the bottom half of the data and the third quartile, Q_3, is the median of the top half of the data.

EXAMPLE 3   Finding and Interpreting Quartiles

Problem The Highway Loss Data Institute routinely collects data on collision coverage claims. Collision coverage insures against physical damage to an insured individual's vehicle. The data in Table 16 represent a random sample of 18 collision coverage claims based on data obtained from the Highway Loss Data Institute. Find and interpret the first, second, and third quartiles for collision coverage claims.

Table 16					
$6751	$9908	$3461	$2336	$21,147	$2332
$189	$1185	$370	$1414	$4668	$1953
$10,034	$735	$802	$618	$180	$1657

Approach Follow the steps given above.

Solution

Step 1   The data written in ascending order are given as follows:

$180	$189	$370	$618	$735	$802	$1185	$1414	$1657
$1953	$2332	$2336	$3461	$4668	$6751	$9908	$10,034	$21,147

Step 2   There are $n = 18$ observations, so the median, or second quartile, Q_2, is the mean of the 9th and 10th observations. Therefore, $M = Q_2 = \frac{\$1657 + \$1953}{2} = \$1805$.

Step 3   The median of the bottom half of the data is the first quartile, Q_1. As shown next, the median of these data is the 5th observation, so $Q_1 = \$735$.

$180 $189 $370 $618 $735 $802 $1185 $1414 $1657

$$\uparrow$$
$$Q_1$$

NOTE

If the number of observations is odd, do not include the median when determining Q_1 and Q_3 by hand. •

(continued)

The median of the top half of the data is the third quartile, Q_3. As shown next, the median of these data is the 5th observation, so $Q_3 = \$4668$.

$$\$1953 \quad \$2332 \quad \$2336 \quad \$3461 \quad \$4668 \quad \$6751 \quad \$9908 \quad \$10{,}034 \quad \$21{,}147$$
$$\uparrow$$
$$Q_3$$

Interpretation Interpret the quartiles as percentiles. For example, 25% of the collision claims are less than or equal to the first quartile, $735, and 75% of the collision claims are greater than $735. Also, 50% of the collision claims are less than or equal to $1805, the second quartile, and 50% of the collision claims are greater than $1805. Finally, 75% of the collision claims are less than or equal to $4668, the third quartile, and 25% of the collision claims are greater than $4668. ●

EXAMPLE 4 Finding Quartiles Using Technology

Problem Find the quartiles of the collision coverage claims data in Table 16.

Using Technology

Statistical packages may use different formulas for obtaining the quartiles, so results may differ slightly.

Approach Use both StatCrunch and Minitab to obtain the quartiles. The steps for obtaining quartiles using a TI-83/84 Plus graphing calculator, Minitab, Excel, and StatCrunch are given in the Technology Step-by-Step on pages 160–161.

Solution The results obtained from StatCrunch [Figure 19(a)] agree with our "by hand" solution. In Figure 19(b), notice that the first quartile, 706, and the third quartile, 5189, reported by Minitab disagree with our "by hand" and StatCrunch result. This difference is due to the fact that StatCrunch and Minitab use different algorithms for obtaining quartiles.

Figure 19

Summary statistics:

Column	n	Median	Min	Max	Q1	Q3
Claim	18	1805	180	21147	735	4668

(a)

Descriptive statistics: Claim

Variable	N	N*	Mean	SE Mean	StDev	Minimum	Q1	Median	Q3	Maximum
Claim	18	0	3874	1250	5302	180	706	1805	5189	21447

(b)

● **Now Work Problem 21(b)**

4 Determine and Interpret the Interquartile Range

So far we have discussed three measures of dispersion: range, standard deviation, and variance, all of which are not resistant. Quartiles, however, are resistant. For this reason, quartiles are used to define a fourth measure of dispersion.

Definition The **interquartile range, IQR**, is the range of the middle 50% of the observations in a data set. That is, the IQR is the difference between the third and first quartiles and is found using the formula

$$\text{IQR} = Q_3 - Q_1$$

The interpretation of the interquartile range is similar to that of the range and standard deviation. That is, the more spread a set of data has, the higher the interquartile range will be.

EXAMPLE 5   Determining and Interpreting the Interquartile Range

Problem Determine and interpret the interquartile range of the collision claim data from Example 3.

Approach Use the quartiles found by hand in Example 3. The interquartile range, IQR, is found by computing the difference between the third and first quartiles. It represents the range of the middle 50% of the observations.

Solution The interquartile range is

$$IQR = Q_3 - Q_1$$
$$= \$4668 - \$735$$
$$= \$3933$$

Interpretation The IQR, that is, the range of the middle 50% of the observations, in the collision claim data is $3933.

● **Now Work Problem 21(c)**

Let's compare the measures of central tendency and dispersion discussed thus far for the collision claim data. The mean collision claim is $3874.4 and the median is $1805. The median is more representative of the "center" because the data are skewed to the right (only 5 of the 18 observations are greater than the mean). The range is $21,147 − $180 = $20,967. The standard deviation is $5301.6 and the interquartile range is $3933. The values of the range and standard deviation are affected by the extreme claim of $21,147. In fact, if this claim had been $120,000 (let's say the claim was for a totaled Mercedes S-class AMG), then the range and standard deviation would increase to $119,820 and $27,782.5, respectively. The interquartile range would not be affected. Therefore, when the distribution of data is highly skewed or contains extreme observations, it is best to use the interquartile range as the measure of dispersion because it is resistant.

Summary: Which Measures to Report		
Shape of Distribution	**Measure of Central Tendency**	**Measure of Dispersion**
Symmetric	Mean	Standard deviation
Skewed left or skewed right	Median	Interquartile range

For the remainder of this text, the direction **describe the distribution** will mean to describe its shape (skewed left, skewed right, symmetric), its center (mean or median), and its spread (standard deviation or interquartile range).

⑤ Check a Set of Data for Outliers

> **CAUTION!**
> Outliers distort both the mean and the standard deviation, because neither is resistant. Because these measures often form the basis for most statistical inference, any conclusions drawn from a set of data that contains outliers can be flawed.

When performing any type of data analysis, we should always check for extreme observations in the data set. Extreme observations are referred to as **outliers**. Outliers can occur by chance, because of error in the measurement of a variable, during data entry, or from errors in sampling. For example, in the 2000 presidential election, a precinct in New Mexico accidentally recorded 610 absentee ballots for Al Gore as 110. Workers in the Gore camp discovered the data-entry error through an analysis of vote totals.

Outliers do not always occur because of error. Sometimes extreme observations are common within a population. For example, suppose we wanted to estimate the mean price of a European car. We might take a random sample of size 5 from the population of all European automobiles. If our sample included a Ferrari F430 Spider (approximately \$175,000), it probably would be an outlier, because this car costs much more than the typical European automobile. The value of this car would be considered *unusual* because it is not a typical value from the data set.

Use the following steps to check for outliers using quartiles.

Checking for Outliers by Using Quartiles

Step 1 Determine the first and third quartiles of the data.

Step 2 Compute the interquartile range.

Step 3 Determine the fences. **Fences** serve as cutoff points for determining outliers.

$$\text{Lower fence} = Q_1 - 1.5(\text{IQR})$$
$$\text{Upper fence} = Q_3 + 1.5(\text{IQR})$$

Step 4 If a data value is less than the lower fence or greater than the upper fence, it is considered an outlier.

EXAMPLE 6   Checking for Outliers

Problem Check the collision coverage claims data in Table 16 for outliers.

Approach Follow the preceding steps. Any data value that is less than the lower fence or greater than the upper fence will be considered an outlier.

Solution

Step 1 The quartiles found in Example 3 are $Q_1 = \$735$ and $Q_3 = \$4668$.

Step 2 The interquartile range, IQR, is

$$\text{IQR} = Q_3 - Q_1$$
$$= \$4668 - \$735$$
$$= \$3933$$

Step 3 The lower fence, LF, is

$$\text{LF} = Q_1 - 1.5(\text{IQR})$$
$$= \$735 - 1.5(\$3933)$$
$$= -\$5164.5$$

The upper fence, UF, is

$$\text{UF} = Q_3 + 1.5(\text{IQR})$$
$$= \$4668 + 1.5(\$3933)$$
$$= \$10,567.5$$

Step 4 There are no observations below the lower fence. However, there is an observation above the upper fence. The claim of \$21,147 is an outlier.

● Now Work Problem 21(d)

Technology Step-by-Step   Determining Quartiles

TI-83/84 Plus
Follow the same steps given to compute the mean and median from raw data. (Section 3.1)

Minitab
Follow the same steps given to compute the mean and median from raw data. (Section 3.1)

Excel

1. Enter the raw data into column A.
2. With the data analysis Tool Pak enabled, select the Data tab and click on **Data Analysis**.
3. Select **Rank and Percentile** from the Data Analysis window. Press OK.
4. With the cursor in the **Input Range** cell, highlight the data. Press OK.

StatCrunch

Follow the same steps given to compute the mean and median from raw data. (Section 3.1)

 3.4 Assess Your Understanding

Vocabulary

1. The _____ represents the number of standard deviations an observation is from the mean.

2. The _____ _____ of a data set is a value such that k percent of the observations are less than or equal to the value.

3. _____ divide data sets into fourths.

4. The _____ _____ is the range of the middle 50% of the observations in a data set.

Applying the Concepts

NW **5. Birth Weights** Babies born after a gestation period of 32–35 weeks have a mean weight of 2600 grams and a standard deviation of 660 grams. Babies born after a gestation period of 40 weeks have a mean weight of 3500 grams and a standard deviation of 470 grams. Suppose a 34-week gestation period baby weighs 2400 grams and a 40-week gestation period baby weighs 3300 grams. What is the z-score for the 34-week gestation period baby? What is the z-score for the 40-week gestation period baby? Which baby weighs less relative to the gestation period?

6. Birth Weights Babies born after a gestation period of 32–35 weeks have a mean weight of 2600 grams and a standard deviation of 660 grams. Babies born after a gestation period of 40 weeks have a mean weight of 3500 grams and a standard deviation of 470 grams. Suppose a 34-week gestation period baby weighs 3000 grams and a 40-week gestation period baby weighs 3900 grams. What is the z-score for the 34-week gestation period baby? What is the z-score for the 40-week gestation period baby? Which baby weighs less relative to the gestation period?

7. Men versus Women The average 20- to 29-year-old man is 69.6 inches tall, with a standard deviation of 3.0 inches, while the average 20- to 29-year-old woman is 64.1 inches tall, with a standard deviation of 3.8 inches. Who is relatively taller, a 75-inch man or a 70-inch woman?
Source: CDC Vital and Health Statistics, Advance Data, Number 361, July 5, 2005

8. Men versus Women The average 20- to 29-year-old man is 69.6 inches tall, with a standard deviation of 3.0 inches, while the average 20- to 29-year-old woman is 64.1 inches tall, with a standard deviation of 3.8 inches. Who is relatively taller, a 67-inch man or a 62-inch woman?
Source: CDC Vital and Health Statistics, Advance Data, Number 361, July 5, 2005

9. ERA Champions In 2014, Clayton Kershaw of the Los Angeles Dodgers had the lowest earned-run average (ERA is the mean number of runs yielded per nine innings pitched) of any starting pitcher in the National League, with an ERA of 1.77. Also in 2014, Felix Hernandez of the Seattle Mariners had the lowest ERA of any starting pitcher in the American League with an ERA of 2.14. In the National League, the mean ERA in 2014 was 3.430 and the standard deviation was 0.721. In the American League, the mean ERA in 2014 was 3.598 and the standard deviation was 0.762. Which player had the better year relative to his peers? Why?

10. Batting Champions The highest batting average ever recorded in Major League Baseball was by Ted Williams in 1941 when he hit 0.406. That year, the mean and standard deviation for batting average were 0.2806 and 0.0328. In 2014, Jose Altuve was the American League batting champion, with a batting average of 0.341. In 2014, the mean and standard deviation for batting average were 0.2679 and 0.0282. Who had the better year relative to his peers, Williams or Altuve? Why?

11. Swim Ryan Murphy, nephew of the author, swims for the University of California at Berkeley. Ryan's best time in the 100-meter backstroke is 45.3 seconds. The mean of all NCAA swimmers in this event is 48.62 seconds with a standard deviation of 0.98 second. Ryan's best time in the 200-meter backstroke is 99.32 seconds. The mean of all NCAA swimmers in this event is 106.58 seconds with a standard deviation of 2.38 seconds. In which race is Ryan better?

12. Triathlon Roberto finishes a triathlon (750-meter swim, 5-kilometer run, and 20-kilometer bicycle) in 63.2 minutes. Among all men in the race, the mean finishing time was 69.4 minutes with a standard deviation of 8.9 minutes. Zandra finishes the same triathlon in 79.3 minutes. Among all women in the race, the mean finishing time was 84.7 minutes with a standard deviation of 7.4 minutes. Who did better in relation to their gender?

13. School Admissions A highly selective boarding school will only admit students who place at least 1.5 standard deviations above the mean on a standardized test that has a mean of 200 and a standard deviation of 26. What is the minimum score that an applicant must make on the test to be accepted?

14. Quality Control A manufacturer of bolts has a quality-control policy that requires it to destroy any bolts that are more than 2 standard deviations from the mean. The quality-control engineer knows that the bolts coming off the assembly line have a mean length of 8 cm with a standard deviation of 0.05 cm. For what lengths will a bolt be destroyed?

NW 15. You Explain It! Percentiles Explain the meaning of the following percentiles.

Source: Advance Data from Vital and Health Statistics

(a) The 15th percentile of the head circumference of males 3 to 5 months of age is 41.0 cm.

(b) The 90th percentile of the waist circumference of females 2 years of age is 52.7 cm.

(c) Anthropometry involves the measurement of the human body. One goal of these measurements is to assess how body measurements may be changing over time. The following table represents the standing height of males aged 20 years or older for various age groups. Based on the percentile measurements of the different age groups, what might you conclude?

Age	Percentile				
---	10th	25th	50th	75th	90th
20–29	166.8	171.5	176.7	181.4	186.8
30–39	166.9	171.3	176.0	181.9	186.2
40–49	167.9	172.1	176.9	182.1	186.0
50–59	166.0	170.8	176.0	181.2	185.4
60–69	165.3	170.1	175.1	179.5	183.7
70–79	163.2	167.5	172.9	178.1	181.7
80 or older	161.7	166.1	170.5	175.3	179.4

16. You Explain It! Percentiles Explain the meaning of the following percentiles.

Source: National Center for Health Statistics.

(a) The 5th percentile of the weight of males 36 months of age is 12.0 kg.

(b) The 95th percentile of the length of newborn females is 53.8 cm.

17. You Explain It! Quartiles Violent crimes include rape, robbery, assault, and homicide. The following is a summary of the violent-crime rate (violent crimes per 100,000 population) for all 50 states in the United States plus Washington, D.C., in 2012.

$$Q_1 = 252.4 \quad Q_2 = 333.8 \quad Q_3 = 454.5$$

(a) Provide an interpretation of these results.

(b) Determine and interpret the interquartile range.

(c) The violent-crime rate in Washington, D.C., in 2012 was 1243.7. Would this be an outlier?

(d) Do you believe that the distribution of violent-crime rates is skewed or symmetric? Why?

18. You Explain It! Quartiles One variable that is measured by online homework systems is the amount of time a student spends on homework for each section of the text. The following is a summary of the number of minutes a student spends for each section of the text for the fall 2014 semester in a College Algebra class at Joliet Junior College.

$$Q_1 = 42 \quad Q_2 = 51.5 \quad Q_3 = 72.5$$

(a) Provide an interpretation of these results.

(b) Determine and interpret the interquartile range.

(c) Suppose a student spent 2 hours doing homework for a section. Is this an outlier?

(d) Do you believe that the distribution of time spent doing homework is skewed or symmetric? Why?

19. Ogives and Percentiles The following graph is an ogive of IQ scores. The vertical axis in an ogive is the cumulative relative frequency and can also be interpreted as a percentile.

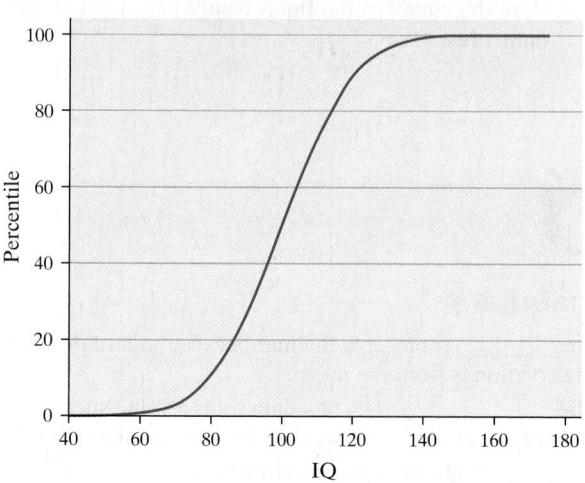

Percentile Ranks of IQ Scores

(a) Find and interpret the percentile rank of an individual whose IQ is 100.

(b) Find and interpret the percentile rank of an individual whose IQ is 120.

(c) What score corresponds to the 60th percentile for IQ?

20. Ogives and Percentiles The following graph is an ogive of the mathematics scores on the SAT. The vertical axis in an ogive is the cumulative relative frequency and can also be interpreted as a percentile.

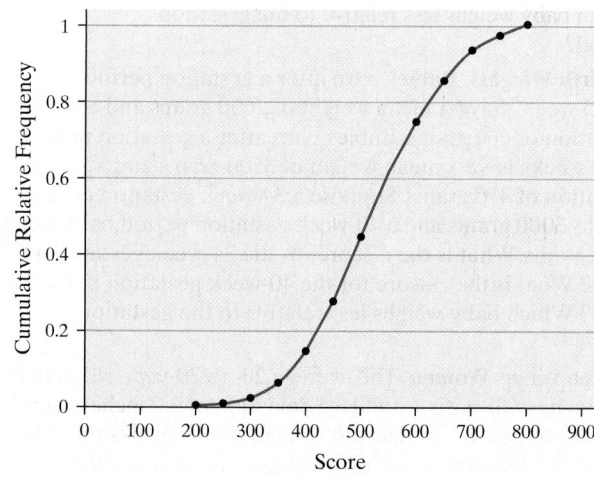

SAT Mathematics Scores

(a) Find and interpret the percentile rank of a student who scored 450 on the SAT mathematics exam.

(b) Find and interpret the percentile rank of a student who scored 750 on the SAT mathematics exam.

(c) If Jane scored at the 44th percentile, what was her score?

NW 21. SMART Car The following data represent the miles per gallon of a random sample of SMART cars with a three-cylinder, 1.0-liter engine.

31.5	36.0	37.8	38.4	40.1	42.3
34.3	36.3	37.9	38.8	40.6	42.7
34.5	37.4	38.0	39.3	41.4	43.5
35.5	37.5	38.3	39.5	41.5	47.5

Source: www.fueleconomy.gov

(a) Compute the z-score corresponding to the individual who obtained 36.3 miles per gallon. Interpret this result.
(b) Determine the quartiles.
(c) Compute and interpret the interquartile range, IQR.
(d) Determine the lower and upper fences. Are there any outliers?

22. Hemoglobin in Cats The following data represent the hemoglobin (in g/dL) for 20 randomly selected cats.

5.7	8.9	9.6	10.6	11.7
7.7	9.4	9.9	10.7	12.9
7.8	9.5	10.0	11.0	13.0
8.7	9.6	10.3	11.2	13.4

Source: Joliet Junior College Veterinarian Technology Program

(a) Compute the z-score corresponding to the hemoglobin of Blackie, 7.8 g/dL. Interpret this result.
(b) Determine the quartiles.
(c) Compute and interpret the interquartile range, IQR.
(d) Determine the lower and upper fences. Are there any outliers?

23. Rate of Return of Google The following data represent the monthly rate of return of Google common stock from its inception in January 2007 through November 2014.

−0.10	−0.02	0.00	0.02	−0.10	0.03	0.04	−0.15	−0.08
0.02	0.01	−0.18	−0.10	−0.18	0.14	0.07	−0.01	0.09
0.03	0.10	−0.17	−0.10	0.05	0.05	0.08	0.08	−0.07
0.06	0.25	−0.07	−0.02	0.10	0.01	0.09	−0.07	0.17
0.05	−0.02	0.30	−0.14	0.00	0.05	0.06	−0.08	0.17

Source: Yahoo!Finance

(a) Determine and interpret the quartiles.
(b) Check the data set for outliers.

24. CO$_2$ Emissions The following data represent the carbon dioxide emissions from the consumption of energy per capita (total carbon dioxide emissions, in tons, divided by total population) for the countries of Europe.

1.31	5.38	10.36	5.73	3.57	5.40	6.24
8.59	9.46	6.48	11.06	7.94	4.63	6.12
14.87	9.94	10.06	10.71	15.86	6.93	3.58
4.09	9.91	161.57	7.82	8.70	8.33	9.38
7.31	16.75	9.95	23.87	7.76	8.86	

Source: Carbon Dioxide Information Analysis Center

(a) Determine and interpret the quartiles.
(b) Is the observation corresponding to Albania, 1.31, an outlier?

25. Fraud Detection As part of its "Customers First" program, a cellular phone company monitors monthly phone usage. The program identifies unusual use and alerts the customer that their phone may have been used by another person. The data below represent the monthly phone use in minutes of a customer enrolled in this program for the past 20 months. The phone company decides to use the upper fence as the cutoff point for the number of minutes at which the customer should be contacted. What is the cutoff point?

346	345	489	358	471
442	466	505	466	372
442	461	515	549	437
480	490	429	470	516

26. Stolen Credit Card A credit card company has a fraud-detection service that determines if a card has any unusual activity. The company maintains a database of daily charges on a customer's credit card. Days when the card was inactive are excluded from the database. If a day's worth of charges appears unusual, the customer is contacted to make sure that the credit card has not been compromised. Use the following daily charges (rounded to the nearest dollar) to determine the amount the daily charges must exceed before the customer is contacted.

143	166	113	188	133
90	89	98	95	112
111	79	46	20	112
70	174	68	101	212

27. Student Survey of Income A survey of 50 randomly selected full-time Joliet Junior College students was conducted during the Fall 2015 semester. In the survey, the students were asked to disclose their weekly income from employment. If the student did not work, $0 was entered.

0	262	0	635	0	0	671
244	521	476	100	650	454	95
12,777	567	310	527	0	67	736
83	159	0	547	188	389	300
719	0	367	316	0	0	181
479	0	82	579	289		
375	347	331	281	628		
0	203	149	0	403		

(a) Check the data set for outliers.
(b) Draw a histogram of the data and label the outliers on the histogram.
(c) Provide an explanation for the outliers.

28. Student Survey of Entertainment Spending A survey of 40 randomly selected full-time Joliet Junior College students was conducted in the Fall 2015 semester. In the survey, the students were asked to disclose their weekly spending on entertainment. The results of the survey are as follows:

21	54	64	33	65	32	21	16
22	39	67	54	22	51	26	14
115	7	80	59	20	33	13	36
36	10	12	101	1000	26	38	8
28	28	75	50	27	35	9	48

(a) Check the data set for outliers.

(b) Draw a histogram of the data and label the outliers on the histogram.

(c) Provide an explanation for the outliers.

29. Pulse Rate Use the results of Problem 21 in Section 3.1 and Problem 19 in Section 3.2 to compute the z-scores for all the students. Compute the mean and standard deviation of these z-scores.

30. Travel Time Use the results of Problem 22 in Section 3.1 and Problem 20 in Section 3.2 to compute the z-scores for all the students. Compute the mean and standard deviation of these z-scores.

31. Fraud Detection Revisited Use the fraud-detection data from Problem 25 to do the following.

(a) Determine the standard deviation and interquartile range of the data.

(b) Suppose the month in which the customer used 346 minutes was not actually that customer's phone. That particular month, the customer did not use her phone at all, so 0 minutes were used. How does changing the observation from 346 to 0 affect the standard deviation and interquartile range? What property does this illustrate?

Explaining the Concepts

32. Write a paragraph that explains the meaning of percentiles.

33. Suppose you received the highest score on an exam. Your friend scored the second-highest score, yet you both were in the 99th percentile. How can this be?

34. Morningstar is a mutual fund rating agency. It ranks a fund's performance by using one to five stars. A one-star mutual fund is in the bottom 10% of its investment class; a five-star mutual fund is at the 90th percentile of its investment class. Interpret the meaning of a five-star mutual fund.

35. When outliers are discovered, should they always be removed from the data set before further analysis?

36. Mensa is an organization designed for people of high intelligence. One qualifies for Mensa if one's intelligence is measured at or above the 98th percentile. Explain what this means.

37. Explain the advantage of using z-scores to compare observations from two different data sets.

38. Explain the circumstances for which the interquartile range is the preferred measure of dispersion. What is an advantage that the standard deviation has over the interquartile range?

39. Explain what each quartile represents.

3.5 The Five-Number Summary and Boxplots

Objectives
1. Compute the five-number summary
2. Draw and interpret boxplots

Historical Note

John Tukey was born on July 16, 1915, in New Bedford, Massachusetts. His parents graduated numbers 1 and 2 from Bates College and were voted "the couple most likely to give birth to a genius." Tukey earned his undergraduate and master's degrees in chemistry from Brown University. In 1939, he earned his doctorate in mathematics from Princeton University. He remained at Princeton and, in 1965, became the founding chair of the Department of Statistics. Among his many accomplishments, Tukey is credited with coining the terms *software* and *bit*. In the early 1970s, he discussed the negative effects of aerosol cans on the ozone layer. In December 1976, he published *Exploratory Data Analysis*, from which the following quote appears: "Exploratory data analysis can never be the whole story, but nothing else can serve as the foundation stone—as the first step" (p. 3). Tukey also recommended that the 1990 Census be adjusted by means of statistical formulas. John Tukey died in New Brunswick, New Jersey, on July 26, 2000.

Let's consider what we have learned so far. In Chapter 2, we discussed techniques for graphically representing data. These summaries included bar graphs, pie charts, histograms, stem-and-leaf plots, and time series graphs. In Sections 3.1 to 3.4, we presented techniques for measuring the center of a distribution, spread in a distribution, and relative position of observations in a distribution of data. Why do we want these summaries? What purpose do they serve?

Well, we want these summaries to see what the data can tell us. We *explore* the data to see if they contain interesting information that may be useful in our research. The summaries make this exploration much easier. In fact, because these summaries represent an exploration, a famous statistician named John Tukey called this material **exploratory data analysis**.

Tukey defined exploratory data analysis as "detective work—numerical detective work—or graphical detective work." He believed exploration of data is best carried out the way a detective searches for evidence when investigating a crime. Our goal is only to collect and present evidence. Drawing conclusions (or inference) is like the deliberations of the jury. What we have done so far falls under the category of exploratory data analysis. We have only collected information and presented summaries, not reached any conclusions.

We have already seen one of Tukey's graphical summaries, the stem-and-leaf plot. In this section, we look at two more summaries: the five-number summary and the boxplot.

1 Compute the Five-Number Summary

Remember that the median is a measure of central tendency that divides the lower 50% of the data from the upper 50%. It is resistant to extreme values and is the preferred measure of central tendency when data are skewed right or left.

The three measures of dispersion presented in Section 3.2 (range, standard deviation, and variance) are not resistant to extreme values. However, the interquartile range, $Q_3 - Q_1$, the difference between the 75th and 25th percentiles, is resistant. It is interpreted as the range of the middle 50% of the data. However, the median, Q_1, and Q_3 do not provide information about the extremes of the data, the smallest and largest values in the data set.

The **five-number summary** of a set of data consists of the smallest data value, Q_1, the median, Q_3, and the largest data value. We organize, the five-number summary as follows:

Five-Number Summary

| MINIMUM | Q_1 | M | Q_3 | MAXIMUM |

EXAMPLE 1   Obtaining the Five-Number Summary

Problem The data shown in Table 17 show the finishing times (in minutes) of the men in the 60- to 64-year-old age group in a 5-kilometer race. Determine the five-number summary of the data.

Table 17						
19.95	23.25	23.32	25.55	25.83	26.28	42.47
28.58	28.72	30.18	30.35	30.95	32.13	49.17
33.23	33.53	36.68	37.05	37.43	41.42	54.63

Source: Laura Gillogly, student at Joliet Junior College

Approach The five-number summary requires the minimum data value, Q_1, M (the median), Q_3, and the maximum data value. Arrange the data in ascending order and then use the procedures introduced in Section 3.4 to obtain Q_1, M, and Q_3.

Solution The data in ascending order are as follows:

19.95, 23.25, 23.32, 25.55, 25.83, 26.28, 28.58, 28.72, 30.18, 30.35, 30.95,

32.13, 33.23, 33.53, 36.68, 37.05, 37.43, 41.42, 42.47, 49.17, 54.63

The smallest number (the fastest time) in the data set is 19.95. The largest number in the data set is 54.63. The first quartile, Q_1, is 26.06. The median, M, is 30.95. The third quartile, Q_3, is 37.24. The five-number summary is

19.95 26.06 30.95 37.24 54.63   ●

EXAMPLE 2   Obtaining the Five-Number Summary Using Technology

Problem Using statistical software or a graphing calculator, determine the five-number summary of the data presented in Table 17.

Approach We will use Minitab to obtain the five-number summary.

Solution Figure 20 shows the output supplied by Minitab. The five-number summary is highlighted.

Figure 20
Descriptive Statistics: Times

Variable	N	N*	Mean	SE Mean	StDev	Minimum	Q1	Median	Q3	Maximum
Times	21	0	32.89	1.90	8.68	19.95	26.06	30.95	37.24	54.63

●

2 Draw and Interpret Boxplots

The five-number summary can be used to create another graph, called the **boxplot**.

Drawing a Boxplot

Step 1 Determine the lower and upper fences:

$$\text{Lower fence} = Q_1 - 1.5(\text{IQR})$$
$$\text{Upper fence} = Q_3 + 1.5(\text{IQR})$$

where $\text{IQR} = Q_3 - Q_1$

Step 2 Draw a number line long enough to include the maximum and minimum values. Insert vertical lines at Q_1, M, and Q_3. Enclose these vertical lines in a box.

Step 3 Label the lower and upper fences.

Step 4 Draw a line from Q_1 to the smallest data value that is larger than the lower fence. Draw a line from Q_3 to the largest data value that is smaller than the upper fence. These lines are called **whiskers**.

Step 5 Any data values less than the lower fence or greater than the upper fence are outliers and are marked with an asterisk ($*$).

EXAMPLE 3 Constructing a Boxplot

Problem Use the results from Example 1 to construct a boxplot of the finishing times of the men in the 60- to 64-year-old age group.

Approach Follow the steps presented above.

Solution In Example 1, we found that $Q_1 = 26.06$, $M = 30.95$, and $Q_3 = 37.24$. Therefore, the interquartile range $= \text{IQR} = Q_3 - Q_1 = 37.24 - 26.06 = 11.18$. The difference between the 75th percentile and 25th percentile is a time of 11.18 minutes.

Step 1 Compute the lower and upper fences:

$$\text{Lower fence} = Q_1 - 1.5(\text{IQR}) = 26.06 - 1.5(11.18) = 9.29$$
$$\text{Upper fence} = Q_3 + 1.5(\text{IQR}) = 37.24 + 1.5(11.18) = 54.01$$

Step 2 Draw a horizontal number line with a scale that will accommodate our graph. Draw vertical lines at $Q_1 = 26.06$, $M = 30.95$, and $Q_3 = 37.24$. Enclose these lines in a box. See Figure 21(a).

Figure 21(a)

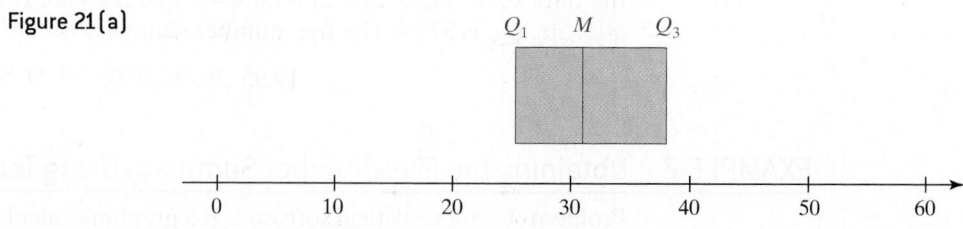

Step 3 Temporarily mark the location of the lower and upper fence with brackets ([and]). See Figure 21(b).

Figure 21(b)

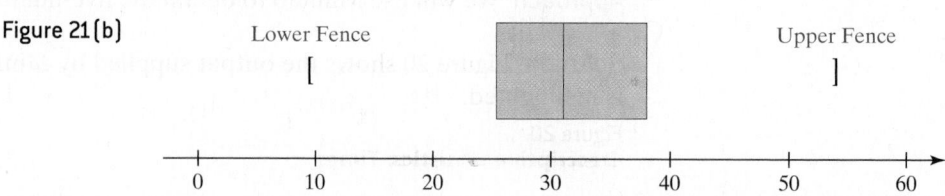

Step 4 Draw a horizontal line from Q_1 to 19.95, the smallest data value that is larger than 9.29 (the lower fence). Draw a horizontal line from Q_3 to 49.17, the largest data value that is smaller than 54.01 (the upper fence). See Figure 21(c).

Figure 21(c)

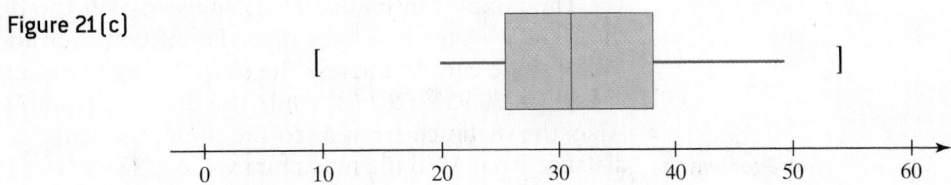

Step 5   Plot any outliers, which are values less than 9.29 (the lower fence) or greater than 54.01 (the upper fence) using an asterisk (*). So 54.63 is an outlier. Remove the temporary brackets from the graph. See Figure 21(d).

Figure 21(d)

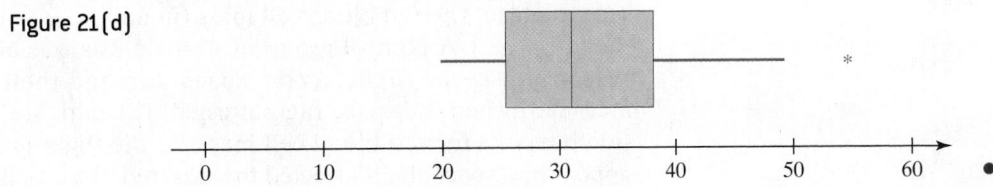

Using a Boxplot and Quartiles to Describe the Shape of a Distribution

Figure 22 shows three histograms and their corresponding boxplots with the five-number summary labeled. Notice the following from the figure.

> ┌─ **CAUTION!** ─────────────┐
> │ Identifying the shape of a │
> │ distribution from a boxplot (or from │
> │ a histogram, for that matter) is │
> │ subjective. When identifying the │
> │ shape of a distribution from a graph, │
> │ be sure to support your opinion. │
> └────────────────────────────┘

- In Figure 22(a), the histogram shows the distribution is skewed right. Notice that the median is left of center in the box, which means the distance from M to Q_1 is less than the distance from M to Q_3. In addition, the right whisker is longer than the left whisker. Finally, the distance from the median to the minimum value in the data set is less than the distance from the median to the maximum value in the data set.
- In Figure 22(b), the histogram shows the distribution is symmetric. Notice that the median is in the center of the box, so the distance from M to Q_1 is the same as the distance from M to Q_3. In addition, the left and right whiskers are roughly the same length. Finally, the distance from the median to the minimum value in the data set is the same as the distance from the median to the maximum value in the data set.
- In Figure 22(c), the histogram shows the distribution is skewed left. Notice that the median is right of center in the box, so the distance from M to Q_1 is more than the distance from M to Q_3. In addition, the left whisker is longer than the right whisker. Finally, the distance from the median to the minimum value in the data set is more than the distance from the median to the maximum value in the data set.

The guidelines given above are just that—guidelines. Judging the shape of a distribution is a subjective practice.

Figure 22

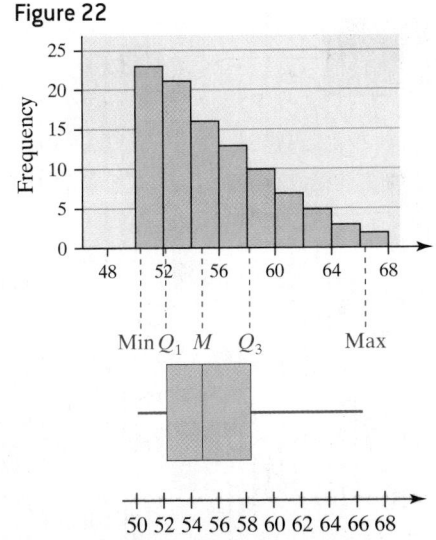

(a) Skewed right

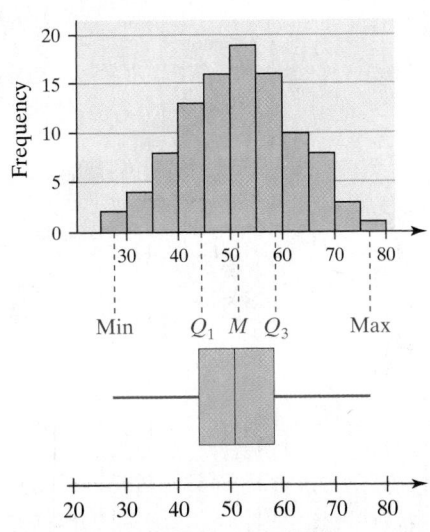

(b) Symmetric

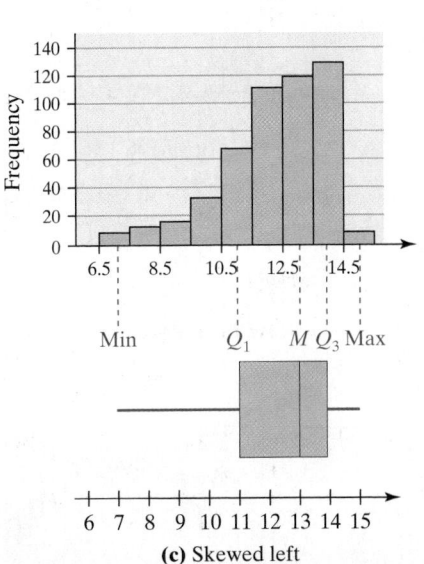

(c) Skewed left

The boxplot in Figure 21(d) suggests that the distribution is skewed right, since the right whisker is longer than the left whisker and the median is left of center in the box. We can also assess the shape using the quartiles. The distance from M to Q_1 is 4.89 ($= 30.95 - 26.06$), while the distance from M to Q_3 is 6.29 ($= 37.24 - 30.95$). Also, the distance from M to the minimum value is 11 ($= 30.95 - 19.95$), while the distance from M to the maximum value is 23.68 ($= 54.63 - 30.95$).

● Now Work Problem 11

EXAMPLE 4   Comparing Two Distributions by Using Boxplots

Problem In the Spacelab Life Sciences 2, led by Paul X. Callahan, 14 male rats were sent to space. The red blood cell mass (in milliliters) of the rats was determined when they returned. A control group of 14 male rats was held under the same conditions (except for space flight) as the space rats, and their red blood cell mass was also measured when the space rats returned. The data are in Table 18. Construct side-by-side boxplots for red blood cell mass for the flight group and control group. Does it appear that space flight affected the rats' red blood cell mass?

Table 18							
Flight				**Control**			
7.43	7.21	8.59	8.64	8.65	6.99	8.40	9.66
9.79	6.85	6.87	7.89	7.62	7.44	8.55	8.70
9.30	8.03	7.00	8.80	7.33	8.58	9.88	9.94
6.39	7.54			7.14	9.14		

Source: NASA Life Sciences Data Archive

Approach Comparing two data sets is easy if we draw side-by-side boxplots on the same horizontal number line. Graphing calculators with advanced statistical features, as well as statistical spreadsheets such as Minitab, Excel, and StatCrunch, can draw boxplots. We will use StatCrunch to draw the boxplots. The steps for drawing boxplots using a TI-83/84 Plus graphing calculator, Minitab, Excel, and StatCrunch are given in the Technology Step-by-Step on page 169.

Solution Figure 23 shows the side-by-side boxplots drawn in StatCrunch. It appears that the space flight has reduced the red blood cell mass of the rats since the median for the flight group ($M \approx 7.7$) is less than the median for the control group ($M \approx 8.6$). The spread, as measured by the interquartile range, appears to be similar for both groups.

Figure 23

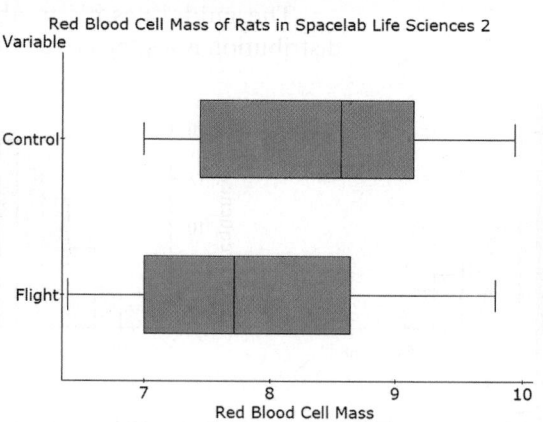

Red Blood Cell Mass of Rats in Spacelab Life Sciences 2

● Now Work Problem 15

Technology Step-by-Step Drawing Boxplots Using Technology

TI-83/84 Plus
1. Enter the raw data into L1.
2. Press 2nd Y= and select 1:Plot 1.
3. Turn the plots ON. Use the cursor to highlight the modified boxplot icon. Your screen should appear as follows:

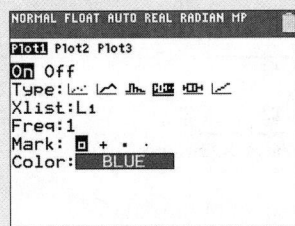

4. Press ZOOM and select 9:ZoomStat.

Minitab
1. Enter the raw data into column C1.
2. Select the **Graph** menu and then **Boxplot...**
3. For a single boxplot, select One Y, simple. For two or more boxplots, select Multiple Y's, simple.
4. Select the data to be graphed. If you want the boxplot to be horizontal rather than vertical, select the Scale button, then transpose value and category scales. Click OK.

Excel
1. Load the XLSTAT Add-in.
2. Enter the raw data into column A. If you are drawing side-by-side boxplots, enter each category of data in a separate column.
3. Select the **XLSTAT** menu and highlight **Describing data**. Then select **Descriptive statistics**.
4. In the Descriptive statistics dialogue box, place the cursor in the Quantitative data cell. Then highlight the data in column A. If you are drawing side-by-side boxplots, highlight all the data.
5. Select the Charts tab. Check Box plots under the Chart types tab. Click the Options tab. If you want the graph drawn horizontal, select the Horizontal radio button. Check the Outliers box. If you are drawing side-by-side boxplots, check the Group plots box. Click OK.

StatCrunch
1. If necessary, enter the raw data into the spreadsheet. Name the column variable.
2. Select Graph and highlight Boxplot.
3. Click on the variable whose boxplot you want to draw. Check the boxes "Use fences to identify outliers" and "Draw boxes horizontally." Enter label for the X-axis. Enter a title for the graph. Click Compute!.

 ## 3.5 Assess Your Understanding

Vocabulary and Skill Building

1. What does the five-number summary consist of?

2. In a boxplot, if the median is to the left of the center of the box and the right whisker is substantially longer than the left whisker, the distribution is skewed _____.

In Problems 3 and 4, (a) identify the shape of the distribution and (b) determine the five-number summary. Assume that each number in the five-number summary is an integer.

3.

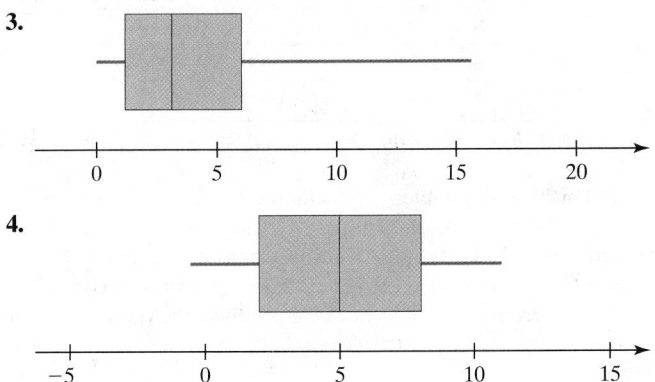

4.

5. Use the side-by-side boxplots shown to answer the questions that follow.

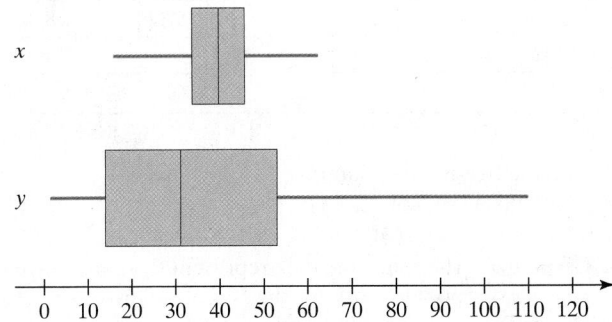

(a) To the nearest integer, what is the median of variable x?

(b) To the nearest integer, what is the third quartile of variable y?

(c) Which variable has more dispersion? Why?

(d) Describe the shape of the variable x. Support your position.

(e) Describe the shape of the variable y. Support your position.

6. Use the side-by-side boxplots shown to answer the questions that follow.

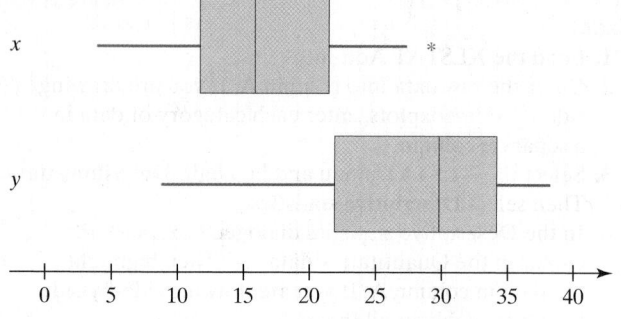

(a) To the nearest integer, what is the median of variable x?
(b) To the nearest integer, what is the first quartile of variable y?
(c) Which variable has more dispersion? Why?
(d) Does the variable x have any outliers? If so, what is the value of the outlier(s)?
(e) Describe the shape of the variable y. Support your position.

7. Exam Scores After giving a statistics exam, Professor Dang determined the following five-number summary for her class results: 60 68 77 89 98. Use this information to draw a boxplot of the exam scores.

8. Speed Reading Jessica enrolled in a course that promised to increase her reading speed. To help judge the effectiveness of the course, Jessica measured the number of words per minute she could read prior to enrolling in the course. She obtained the following five-number summary: 110 140 157 173 205. Use this information to draw a boxplot of the reading speed.

Applying the Concepts

9. Age at Inauguration The following data represent the age of U.S. presidents on their respective inauguration days (through Barack Obama).

42	47	50	52	54	55	57	61	64
43	48	51	52	54	56	57	61	65
46	49	51	54	55	56	57	61	68
46	49	51	54	55	56	58	62	69
47	50	51	54	55	57	60	64	

Source: factmonster.com

(a) Find the five-number summary.
(b) Construct a boxplot.
(c) Comment on the shape of the distribution.

10. Carpoolers The following data represent the percentage of workers who carpool to work for the 50 states plus Washington, D.C. **Note:** The minimum observation of 7.2% corresponds to Maine and the maximum observation of 16.4% corresponds to Hawaii.

7.2	8.5	9.0	9.4	10.0	10.3	11.2	11.5	13.8
7.8	8.6	9.1	9.6	10.0	10.3	11.2	11.5	14.4
7.8	8.6	9.2	9.7	10.0	10.3	11.2	11.7	16.4
7.9	8.6	9.2	9.7	10.1	10.7	11.3	12.4	
8.1	8.7	9.2	9.9	10.2	10.7	11.3	12.5	
8.3	8.8	9.4	9.9	10.3	10.9	11.3	13.6	

Source: American Community Survey by the U.S. Census Bureau

(a) Find the five-number summary.
(b) Construct a boxplot.
(c) Comment on the shape of the distribution.

NW 11. Age of Mother at Birth The data below represent the age of the mother at the time of her first birth for a random sample of 30 mothers.

21	35	33	25	22	26
21	24	16	32	25	20
30	20	20	29	21	19
18	24	33	22	23	25
17	23	25	29	25	19

Source: General Social Survey

(a) Construct a boxplot of the data.
(b) Use the boxplot and quartiles to describe the shape of the distribution.

12. Got a Headache? The following data represent the weight (in grams) of a random sample of 25 Tylenol tablets.

0.608	0.601	0.606	0.602	0.611
0.608	0.610	0.610	0.607	0.600
0.608	0.608	0.605	0.609	0.605
0.610	0.607	0.611	0.608	0.610
0.612	0.598	0.600	0.605	0.603

Source: Kelly Roe, student at Joliet Junior College

(a) Construct a boxplot.
(b) Use the boxplot and quartiles to describe the shape of the distribution.

13. M&Ms In Problem 25 from Section 3.1, we drew a histogram of the weights of M&Ms and found that the distribution is symmetric. Draw a boxplot of these data. Use the boxplot and quartiles to confirm the distribution is symmetric. For convenience, the data are displayed again.

0.87	0.88	0.82	0.90	0.90	0.84	0.84
0.91	0.94	0.86	0.86	0.86	0.88	0.87
0.89	0.91	0.86	0.87	0.93	0.88	
0.83	0.95	0.87	0.93	0.91	0.85	
0.91	0.91	0.86	0.89	0.87	0.84	
0.88	0.88	0.89	0.79	0.82	0.83	
0.90	0.88	0.84	0.93	0.81	0.90	
0.88	0.92	0.85	0.84	0.84	0.86	

Source: Michael Sullivan

14. Old Faithful In Problem 26 from Section 3.1, we drew a histogram of the length of eruption of California's Old Faithful geyser and found that the distribution is symmetric. Draw a boxplot of these data. Use the boxplot and quartiles to confirm the distribution is symmetric. For convenience, the data are displayed again on the next page.

108	108	99	105	103	103	94
102	99	106	90	104	110	110
103	109	109	111	101	101	
110	102	105	110	106	104	
104	100	103	102	120	90	
113	116	95	105	103	101	
100	101	107	110	92	108	

Source: Ladonna Hansen, Park Curator

NW 15. Dissolving Rates of Vitamins A student wanted to know whether Centrum vitamins dissolve faster than the corresponding generic brand. The student used vinegar as a proxy for stomach acid and measured the time (in minutes) it took for a vitamin to completely dissolve. The results are shown next.

Centrum					Generic Brand				
2.73	3.07	3.30	3.35	3.12	6.57	6.23	6.17	7.17	5.77
2.95	2.15	2.67	2.80	2.25	6.73	5.78	5.38	5.25	5.55
2.60	2.57	4.02	3.02	2.15	5.50	6.50	7.42	6.47	6.30
3.03	3.53	2.63	2.30	2.73	6.33	7.42	5.57	6.35	5.92
3.92	2.38	3.25	4.00	3.63	5.35	7.25	7.58	6.50	4.97
3.02	4.17	4.33	3.85	2.23	7.13	5.98	6.60	5.03	7.18

Source: Amanda A. Sindewald, student at Joliet Junior College

(a) Draw side-by-side boxplots for each vitamin type.
(b) Which vitamin type has more dispersion?
(c) Which vitamin type appears to dissolve faster?

16. Chips per Cookie Do store-brand chocolate chip cookies have fewer chips per cookie than Keebler's Chips Deluxe Chocolate Chip Cookies? To find out, a student randomly selected 21 cookies of each brand and counted the number of chips in the cookies. The results are shown next.

Keebler			Store Brand		
32	23	28	21	23	24
28	28	29	24	25	27
25	20	25	26	26	21
22	21	24	18	16	24
21	24	21	21	30	17
26	28	24	23	28	31
33	20	31	27	33	29

Source: Trina McNamara, student at Joliet Junior College

(a) Draw side-by-side boxplots for each brand of cookie.
(b) Does there appear to be a difference in the number of chips per cookie?
(c) Does one brand have a more consistent number of chips per cookie?

17. Putting It Together: Earthquakes Go to www.pearsonhighered.com/sullivanstats and download the data 3_5_17 using the file format of your choice. The data contains the depth (in kilometers) and magnitude, measured using the Richter Scale, of all earthquakes worldwide that occurred between 12/2/2014 and 12/9/2014. Depth is the distance below the surface of Earth at which the earthquake originates. A one unit increase in magnitude represents ground shaking ten times as strong (an earthquake with magnitude 4 is ten times as strong as an earthquake with magnitude 3).

(a) Find the mean, median, range, standard deviation, and quartiles for both the depth and magnitude of the earthquakes. Based on the values of the mean, median, and quartiles conjecture the shape of the distribution for depth and magnitude.
(b) Draw a relative frequency histogram for both depth and magnitude. Label the mean and median on the histogram. Describe the shape of each histogram.
(c) Draw a boxplot for both depth and magnitude. Are there outliers?
(d) Determine the lower and upper fences for identifying outliers for both depth and magnitude.

18. Putting It Together: Paternal Smoking It is well-documented that active maternal smoking during pregnancy is associated with lower-birth-weight babies. Researchers wanted to determine if there is a relationship between paternal smoking habits and birth weight. The researchers administered a questionnaire to each parent of newborn infants. One question asked whether the individual smoked regularly. Because the survey was administered within 15 days of birth, it was assumed that any regular smokers were also regular smokers during pregnancy. Birth weights for the babies (in grams) of nonsmoking mothers were obtained and divided into two groups, nonsmoking fathers and smoking fathers. The given data are representative of the data collected by the researchers. The researchers concluded that the birth weight of babies whose father smoked was less than the birth weight of babies whose father did not smoke.

Source: "The Effect of Paternal Smoking on the Birthweight of Newborns Whose Mothers Did Not Smoke," Fernando D. Martinez, Anne L. Wright, Lynn M. Taussig, *American Journal of Public Health* Vol. 84, No. 9

Nonsmokers			Smokers		
4194	3522	3454	3998	3455	3066
3062	3771	3783	3150	2986	2918
3544	3746	4019	4216	3502	3457
4054	3518	3884	3493	3255	3234
4248	3719	3668	2860	3282	2746
3128	3290	3423	3686	2851	3145
3471	4354	3544	3807	3548	4104
3994	2976	4067	3963	3892	2768
3732	3823	3302	3769	3509	3629
3436	3976	3263	4131	3129	4263

(a) Is this an observational study or a designed experiment? Why?

(b) What is the explanatory variable? What is the response variable?

(c) Can you think of any lurking variables that may affect the results of the study?

(d) In the article, the researchers stated that "birthweights were adjusted for possible confounders . . . ." What does this mean?

(e) Determine summary statistics (mean, median, standard deviation, quartiles) for each group.

(f) Interpret the first quartile for both the nonsmoker and smoker group.

(g) Draw a side-by-side boxplot of the data. Does the side-by-side boxplot confirm the conclusions of the study?

Retain Your Knowledge

19. Retain Your Knowledge: Decision Making and Hunger
Does hunger improve strategic decision making? That is, if you are hungry are you more likely to make a favorable decision when the outcome of your decision is uncertain (as in business decisions)? To test this theory, researchers randomly divided 30 normal weight individuals into two groups. All subjects were asked to refrain from eating or drinking (except water) from 11 P.M. on the day prior to their 9 A.M. meeting. At 9 A.M., the subjects were randomly assigned to one of two groups. The subjects in Group 1 were fed breakfast while the subjects in Group 2 were not fed. All subjects were administered a computerized version of an exam that assesses complex decision making under uncertain conditions. The assessment consisted of subjects choosing cards from four decks marked A, B, C, and D. Cards in decks A and B had a point value of 100 while cards in decks C and D had point values of 50. However, deck A had penalty cards that deducted points between 150 and 300; deck B had one penalty card of 1250; deck C had penalty cards between 25 and 75 points; deck D had a single penalty card of 250 points. So, decks A and B had stiffer penalties over the long haul than decks C and D and in the long haul, decks C and D resulted in more points than decks A and B. In total, the subjects would select 100 cards. However, the response variable was the number of cards selected from decks C and D out of the last 60 cards selected. The thinking here is that after 40 card selections, the subjects would be aware of the advantage of decks C and D. The researchers administered a Barret Impulsivity Scale to be sure the two groups did not differ in terms of impulsivity (e.g., "I do

things without thinking"). There was no difference in impulsivity, age, or body mass index between the two groups. Before the exam, subjects were asked to report their level of hunger and it was found that Group 2 was significantly more hungry than Group 1. After analysis of the data, it was determined that the mean number of advantageous cards (decks C and D) selected by the subjects in Group 2 was 33.36 cards while the mean was 25.86 for the subjects in Group 1. The researchers concluded that hunger improves advantageous decision making.
Source: de Ridder, D., Kroese, F., Adriaanse, M., & Evers, C., "Always Gamble on an Empty Stomach: Hunger Is Associated with Advantageous Decision Making," *PLOS One* 9(10). doi: 10.1371/journal.pone.0111081.

(a) What type of experimental design is this?

(b) Identify the experimental units.

(c) What is the response variable? Is it qualitative or quantitative? If quantitative, is it discrete or continuous?

(d) What factors that might impact the response variable are cited in the article? Which factor is manipulated? How many levels are there for the manipulated factor?

(e) What role does randomization play in the study? How do the researchers verify that randomization resulted in similar groups prior to the treatment?

(f) What are the statistics in the study?

(g) What is the conclusion of the study?

Explaining the Concepts

20. Which boxplot likely has the data with a larger standard deviation? Why?

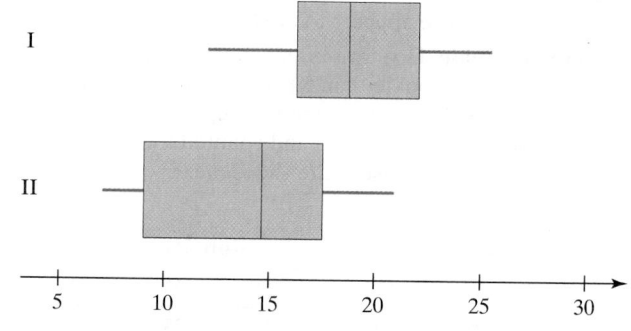

21. Explain how to determine the shape of a distribution using the boxplot and quartiles.

Chapter 3 Review

Summary

We are still in Part II of the statistical process—describing data.

This chapter concentrated on describing distributions numerically. Measures of central tendency are used to indicate the typical value in a distribution. Three measures of central tendency were discussed. The mean measures the center of gravity of the distribution. It is found by adding the observations in the data set and dividing by the number of observations. Therefore, the data must be quantitative to compute the mean. The median separates

the bottom 50% of the data from the top 50%. The data must be at least ordinal (capable of arranging the data in ascending order) to compute the median. The mode measures the most frequent observation. The data can be either quantitative or qualitative to compute the mode. Resistance refers to the idea that extreme observations in the data set do not significantly impact the value of the statistic. The median is a resistant measure of central tendency, whereas the mean is not resistant.

Measures of dispersion describe the spread in the data. The range is the difference between the highest and lowest data values. The standard deviation is based on the average squared deviation about the mean. The variance is the square of the standard deviation. The range, standard deviation, and variance, are not resistant.

The mean and standard deviation are used in many types of statistical inference.

The mean, median, mode, and standard deviation can be approximated from grouped data.

We can determine the relative position of an observation using z-scores and percentiles. A z-score denotes the number of standard deviations an observation is from the mean. Percentiles determine the percent of observations that lie below a specific observation.

Quartiles are a specific type of percentile. There is the first quartile, which is the 25th percentile; the second quartile, which is the 50th percentile, or median; the third quartile, which is the 75th percentile. The interquartile range, the difference between the third and first quartile, is a resistant measure of spread.

We use quartiles to help identify outliers in a data set. To find outliers, we must determine the lower and upper fences of the data.

The five-number summary provides an idea about the center and spread of a data set through the median and interquartile range. The range of the data can be determined from the smallest and largest data values. The five-number summary is used to construct boxplots. Boxplots can be used to describe the shape of the distribution and to visualize outliers.

Vocabulary

Arithmetic mean (p. 118)
Population arithmetic mean (p. 118)
Sample arithmetic mean (p. 118)
Mean (p. 118)
Median (p. 120)
Resistant (p. 122)
Mode (p. 124)
No mode (p. 124)
Bimodal (p. 125)
Multimodal (p. 125)
Dispersion (p. 131)
Range (p. 132)

Deviation about the mean (p. 133)
Population standard deviation (p. 133)
Conceptual formula (p. 133)
Computational formula (p. 133)
Sample standard deviation (p. 135)
Degrees of freedom (p. 135)
Population variance (p. 138)
Sample variance (p. 138)
Biased (p. 138)
The Empirical Rule (p. 139)
Chebyshev's Inequality (p. 140)
Grouped data (p. 147)

Weighted mean (p. 149)
z-score (p. 155)
kth percentile (p. 156)
Quartiles (p. 156)
Interquartile range (p. 158)
Describe the distribution (p. 159)
Outlier (p. 159)
Fences (p. 160)
Exploratory data analysis (p. 164)
Five-number summary (p. 165)
Boxplot (p. 166)
Whiskers (p. 166)

Formulas

Population Mean

$$\mu = \frac{\sum x_i}{N}$$

Sample Mean

$$\bar{x} = \frac{\sum x_i}{n}$$

Population Standard Deviation

$$\sigma = \sqrt{\frac{\sum (x_i - \mu)^2}{N}} = \sqrt{\frac{\sum x_i^2 - \frac{(\sum x_i)^2}{N}}{N}}$$

Sample Standard Deviation

$$s = \sqrt{\frac{\sum (x_i - \bar{x})^2}{n - 1}} = \sqrt{\frac{\sum x_i^2 - \frac{(\sum x_i)^2}{n}}{n - 1}}$$

Population Variance

σ^2

Sample Variance

s^2

Range = Largest Data Value − Smallest Data Value

Weighted Mean

$$\bar{x}_w = \frac{\sum w_i x_i}{\sum w_i}$$

Population Mean from Grouped Data

$$\mu = \frac{\sum x_i f_i}{\sum f_i}$$

Sample Mean from Grouped Data

$$\bar{x} = \frac{\sum x_i f_i}{\sum f_i}$$

Population Standard Deviation from Grouped Data

$$\sigma = \sqrt{\frac{\sum (x_i - \mu)^2 f_i}{\sum f_i}}$$

Sample Standard Deviation from Grouped Data

$$s = \sqrt{\frac{\sum(x_i - \bar{x})^2 f_i}{(\sum f_i) - 1}}$$

Interquartile Range

$$IQR = Q_3 - Q_1$$

Lower and Upper Fences

Lower Fence $= Q_1 - 1.5(IQR)$
Upper Fence $= Q_3 + 1.5(IQR)$

Population z-Score

$$z = \frac{x - \mu}{\sigma}$$

Sample z-Score

$$z = \frac{x - \bar{x}}{s}$$

OBJECTIVES

Section	You should be able to . . .	Example(s)	Review Exercises
3.1	**1** Determine the arithmetic mean of a variable from raw data (p. 118)	1, 4, 6	1(a), 2(a), 3(a), 4(c), 10(a)
	2 Determine the median of a variable from raw data (p. 120)	2, 3, 4, 6	1(a), 2(a), 3(a), 4(c), 10(a)
	3 Explain what it means for a statistic to be resistant (p. 122)	5	2(c), 10(h), 10(i), 12
	4 Determine the mode of a variable from raw data (p. 124)	7, 8, 9	3(a), 4(d)
3.2	**1** Determine the range of a variable from raw data (p. 132)	2	1(b), 2(b), 3(b)
	2 Determine the standard deviation of a variable from raw data (p. 133)	3–6	1(b), 2(b), 3(b), 10(d)
	3 Determine the variance of a variable from raw data (p. 138)	7	1(b)
	4 Use the Empirical Rule to describe data that are bell shaped (p. 139)	8	5(a)–(d)
	5 Use Chebyshev's Inequality to describe any set of data (p. 140)	9	5(e)–(f)
3.3	**1** Approximate the mean of a variable from grouped data (p. 148)	1, 4	6(a)
	2 Compute the weighted mean (p. 149)	2	7
	3 Approximate the standard deviation of a variable from grouped data (p. 150)	3, 4	6(b)
3.4	**1** Determine and interpret z-scores (p. 155)	1	8
	2 Interpret percentiles (p. 156)	2	11
	3 Determine and interpret quartiles (p. 156)	3, 4	10(b)
	4 Determine and interpret the interquartile range (p. 158)	5	2(b), 10(d)
	5 Check a set of data for outliers (p. 159)	6	10(e)
3.5	**1** Compute the five-number summary (p. 164)	1, 2	10(c)
	2 Draw and interpret boxplots (p. 166)	3, 4	9, 10(f)–10(g)

Review Exercises

1. Muzzle Velocity The following data represent the muzzle velocity (in meters per second) of rounds fired from a 155-mm gun.

793.8	793.1	792.4	794.0	791.4
792.4	791.7	792.3	789.6	794.4

Source: Christenson, Ronald, and Blackwood, Larry, "Tests for Precision and Accuracy of Multiple Measuring Devices," *Technometrics*, 35(4): 411–421, 1993.

(a) Compute the sample mean and median muzzle velocity.
(b) Compute the range, sample variance, and sample standard deviation.

2. Price of Chevy Cobalts The following data represent the sales price in dollars for nine 2-year-old Chevrolet Cobalts in the Los Angeles area.

14050	13999	12999	10995	9980
8998	7889	7200	5500	

Source: cars.com

(a) Determine the sample mean and median price.
(b) Determine the range, sample standard deviation, and interquartile range.
(c) Redo (a) and (b) if the data value 14,050 was incorrectly entered as 41,050. How does this change affect the mean? the median? the range? the standard deviation? the interquartile range? Which of these values is resistant?

3. Chief Justices The following data represent the ages of chief justices of the U.S. Supreme Court when they were appointed.

Justice	Age
John Jay	44
John Rutledge	56
Oliver Ellsworth	51
John Marshall	46
Roger B. Taney	59
Salmon P. Chase	56
Morrison R. Waite	58
Melville W. Fuller	55
Edward D. White	65
William H. Taft	64
Charles E. Hughes	68
Harlan F. Stone	69
Frederick M. Vinson	56
Earl Warren	62
Warren E. Burger	62
William H. Rehnquist	62
John G. Roberts	50

Source: Information Please Almanac

(a) Determine the population mean, median, and mode ages.
(b) Determine the range and population standard deviation ages.
(c) Obtain two simple random samples of size 4, and determine the sample mean and sample standard deviation ages.

4. Number of Tickets Issued As part of a statistics project, a student surveys 30 randomly selected students and asks them how many speeding tickets they have been issued in the past month. The results of the survey are as follows:

1	1	0	1	0	0
0	0	0	1	0	1
0	1	2	0	1	1
0	0	0	0	1	1
0	0	0	0	1	0

(a) Draw a frequency histogram of the data and describe its shape.
(b) Based on the shape of the histogram, do you expect the mean to be more than, equal to, or less than the median?
(c) Determine the mean and the median of the number of tickets issued.
(d) Determine the mode of the number of tickets issued.

5. Chebyshev's Inequality and the Empirical Rule Suppose that a certain brand of light bulb has a mean life of 600 hours and a standard deviation of 53 hours.

(a) A histogram of the data indicates the sample data follow a bell-shaped distribution. According to the Empirical Rule, 99.7% of light bulbs have lifetimes between _____ and _____ hours.
(b) Assuming the data are bell shaped, determine the percentage of light bulbs that will have a life between 494 and 706 hours.
(c) Assuming the data are bell shaped, what percentage of light bulbs will last between 547 and 706 hours.
(d) If the company that manufactures the light bulb guarantees to replace any bulb that does not last at least 441 hours, what

percentage of light bulbs can the firm expect to have to replace, according to the Empirical Rule?
(e) Use Chebyshev's Inequality to determine the minimum percentage of light bulbs with a life within 2.5 standard deviations of the mean.
(f) Use Chebyshev's Inequality to determine the minimum percentage of light bulbs with a life between 494 and 706 hours.

6. Travel Time to Work The frequency distribution listed in the table represents the travel time to work (in minutes) for a random sample of 895 U.S. adults.

Travel Time (minutes)	Frequency
0–9	125
10–19	271
20–29	186
30–39	121
40–49	54
50–59	62
60–69	43
70–79	20
80–89	13

Source: Based on data from the 2009 American Community Survey

(a) Approximate the mean travel time to work for U.S. adults.
(b) Approximate the standard deviation travel time to work for U.S. adults.

7. Weighted Mean Michael has just completed his first semester in college. He earned an A in his five-hour calculus course, a B in his four-hour chemistry course, an A in his three-hour speech course, and a C in his three-hour psychology course. Assuming an A equals 4 points, a B equals 3 points, and a C equals 2 points, determine Michael's grade-point average if grades are weighted by class hours.

8. Weights of Males versus Females According to the National Center for Health Statistics, the mean weight of a 20- to 29-year-old female is 156.5 pounds, with a standard deviation of 51.2 pounds. The mean weight of a 20- to 29-year-old male is 183.4 pounds, with a standard deviation of 40.0 pounds. Who is relatively heavier: a 20- to 29-year-old female who weighs 160 pounds or a 20- to 29-year-old male who weighs 185 pounds?

9. Halladay No-No On October 6, 2010, Roy Halladay of the Philadelphia Phillies threw the second post-season no-hitter in Major League history. The side-by-side boxplot shows the pitch speed (in miles per hour) for all of Halladay's pitches during the game.

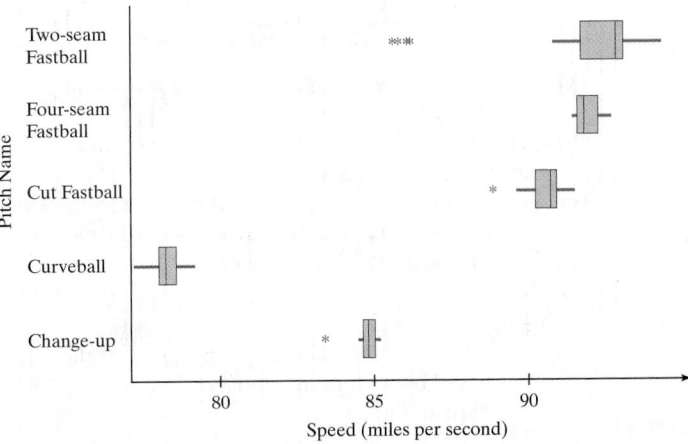

Roy Halladay's No-Hitter

(a) Which pitch is typically thrown the fastest?
(b) Which pitch is most erratic as far as pitch speed goes?
(c) Which pitch is more consistent as far as pitch speed goes, the cut fastball or the four-seam fastball?
(d) Are there any outliers for Halladay's cut fastball? If so, approximate the pitch speed of the outlier.
(e) Describe the shape of the distribution of Halladay's curveball.
(f) Describe the shape of the distribution of Halladay's four-seam fastball.

 **10. Presidential Inaugural Addresses** Ever wonder how many words are in a typical inaugural address? The following data represent the lengths of all the inaugural addresses (measured in word count) for all presidents up to Barack Obama.

1425	1125	1128	5433	2242	2283
135	1172	1337	1802	2446	1507
2308	3838	2480	1526	2449	2170
1729	8445	2978	3318	1355	1571
2158	4776	1681	4059	1437	2073
1175	996	4388	3801	2130	2406
1209	3319	2015	1883	1668	2137
3217	2821	3967	1807	1087	
4467	3634	2217	1340	2463	
2906	698	985	559	2546	

Source: infoplease.com

(a) Determine the mean and median number of words in a presidential inaugural address.
(b) Determine and interpret the quartiles for the number of words in a presidential inaugural address.
(c) Determine the five-number summary for the number of words in a presidential inaugural address.
(d) Determine the standard deviation and interquartile range for the number of words in a presidential inaugural address.
(e) Are there any outliers in the data set? If so, what is (are) the value(s)?
(f) Draw a boxplot of the data.
(g) Describe the shape of the distribution. Support your position using the boxplot and quartiles.
(h) Which measure of central tendency do you think better describes the typical number of words in an inaugural address? Why?
(i) Which measure of dispersion do you think better describes the spread of the typical number of words in an inaugural address? Why?

11. You Explain It! Percentiles According to the National Center for Health Statistics, a 19-year-old female whose height is 67.1 inches has a height that is at the 85th percentile. Explain what this means.

12. Skinfold Thickness Procedure One method of estimating body fat is through skinfold thickness measurement using from three to nine different standard anatomical sites around the body from the right side only (for consistency). The tester pinches the skin at the appropriate site to raise a double layer of skin and the underlying adipose tissue, but not the muscle. Calipers are then applied 1 centimeter below and at right angles to the pinch and a reading is taken 2 seconds later. The mean of two measurements should be taken. If the two measurements differ greatly, a third should be done and then the median value taken. Explain why a median is used as the measure of central tendency when three measures are taken, rather than the mean.

Chapter Test

1. The following data represent the amount of time (in minutes) a random sample of eight students enrolled in Sullivan's Intermediate Algebra course spent on the homework from Section 4.5, Factoring Polynomials.

48	88	57	109
111	93	71	63

Source: MyMathLab

(a) Determine the mean amount of time spent doing Section 4.5 homework.
(b) Determine the median amount of time spent doing Section 4.5 homework.
(c) Suppose the observation 109 minutes is incorrectly recorded as 1009 minutes. Recompute the mean and the median. What do you notice? What property of the median does this illustrate?

2. The Federal Bureau of Investigation classifies various larcenies. The data in the next column represent the type of larcenies based on a random sample of 15 larcenies. What is the mode type of larceny?

Pocket picking and purse snatching	Bicycles	From motor vehicles
From motor vehicles	From motor vehicles	From buildings
From buildings	Shoplifting	Motor vehicle accessories
From motor vehicles	Shoplifting	From motor vehicles
From motor vehicles	Pocket picking and purse snatching	From motor vehicles

3. Determine the range of the homework data from Problem 1.

4. (a) Determine the standard deviation of the homework data from Problem 1.
(b) By hand, determine and interpret the interquartile range of the homework data from Problem 1.
(c) Which of these two measures of dispersion is resistant? Why?

5. In a random sample of 250 toner cartridges, the mean number of pages a toner cartridge can print is 4302 and the standard deviation is 340.

(a) Suppose a histogram of the data indicates that the sample data follow a bell-shaped distribution. According to the Empirical Rule, 99.7% of toner cartridges will print between ____ and ____ pages.

(b) Assuming that the distribution of the data are bell shaped, determine the percentage of toner cartridges whose print total is between 3622 and 4982 pages.

(c) If the company that manufactures the toner cartridges guarantees to replace any cartridge that does not print at least 3622 pages, what percent of cartridges can the firm expect to be responsible for replacing, according to the Empirical Rule?

(d) Use Chebyshev's inequality to determine the minimum percentage of toner cartridges with a page count within 1.5 standard deviations of the mean.

(e) Use Chebyshev's inequality to determine the minimum percentage of toner cartridges that print between 3282 and 5322 pages.

6. The following data represent the length of time (in minutes) between eruptions of Old Faithful in Yellowstone National Park.

Time (minutes)	Frequency
40–49	8
50–59	44
60–69	23
70–79	6
80–89	107
90–99	11
100–109	1

(a) Approximate the mean length of time between eruptions.

(b) Approximate the standard deviation length of time between eruptions.

7. Yolanda wishes to develop a new type of meatloaf to sell at her restaurant. She decides to combine 2 pounds of ground sirloin (cost $2.70 per pound), 1 pound of ground turkey (cost $1.30 per pound), and $\frac{1}{2}$ pound of ground pork (cost $1.80 per pound). What is the cost per pound of the meatloaf?

8. An engineer is studying bearing failures for two different materials in aircraft gas turbine engines. The following data are failure times (in millions of cycles) for samples of the two material types.

Material A		Material B	
3.17	5.88	5.78	9.65
4.31	6.91	6.71	13.44
4.52	8.01	6.84	14.71
4.66	8.97	7.23	16.39
5.69	11.92	8.20	24.37

(a) Determine the sample mean failure time for each material.

(b) By hand, compute the median failure time for each material.

(c) Determine the sample standard deviation of the failure times for each material. Which material has its failure times more dispersed?

(d) By hand, compute the five-number summary for each material.

(e) On the same graph, draw boxplots for the two materials. Annotate the graph with some general remarks comparing the failure times.

(f) Describe the shape of the distribution of each material using the boxplot and quartiles.

9. The following data represent the weights (in grams) of 50 randomly selected quarters. Determine and interpret the quartiles. Does the data set contain any outliers?

5.49	5.58	5.60	5.62	5.68
5.52	5.58	5.60	5.63	5.70
5.53	5.58	5.60	5.63	5.71
5.53	5.58	5.60	5.63	5.71
5.53	5.58	5.60	5.65	5.72
5.56	5.58	5.60	5.66	5.73
5.57	5.59	5.60	5.66	5.73
5.57	5.59	5.61	5.66	5.73
5.57	5.59	5.62	5.67	5.74
5.57	5.59	5.62	5.67	5.84

10. Armando is filling out a college application that requires he supply either his SAT math score or his ACT math score. Armando scored 610 on the SAT math and 27 on the ACT math. Which score should Armando report, given that the mean SAT math score is 515 with a standard deviation of 114, and the mean ACT math score is 21.0 with a standard deviation of 5.1? Why?

11. According to the National Center for Health Statistics, a 10-year-old male whose height is 53.5 inches has a height that is at the 15th percentile. Explain what this means.

12. The distribution of income tends to be skewed to the right. Suppose you are running for a congressional seat and wish to portray that the average income in your district is low. Which measure of central tendency, the mean or the median, would you report? Why?

13. Answer the following based on the histograms shown on the next page.

(a) Which measure of central tendency would you recommend reporting for the data whose histogram is shown in Figure I? Why?

(b) Which one has more dispersion? Explain.

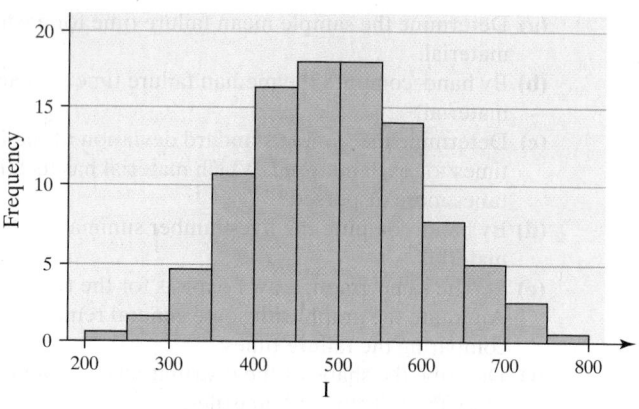

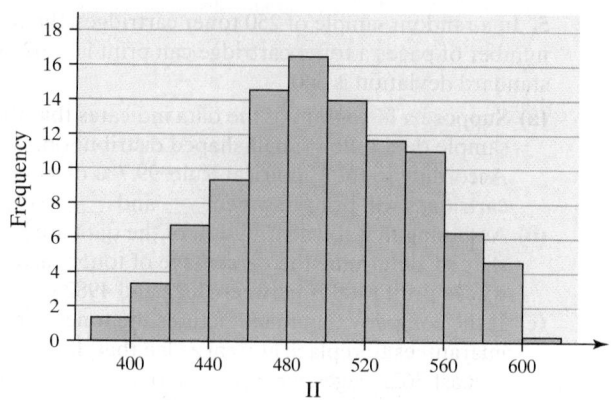

14. Explain how the standard deviation measures dispersion. In your explanation, include a discussion of deviation about the mean.

Making an Informed Decision

What Car Should I Buy?

Suppose you are in the market to purchase a used car. To make an informed decision regarding your purchase, you would like to collect as much information as possible. Among the information you might consider are the typical price of the car, the typical number of miles the car should have, its crash test results, insurance costs, and expected repair costs.

1. Make a list of at least three cars that you would consider purchasing. To be fair, the cars should be in the same class (such as compact, midsize, and so on). They should also be of the same age.

2. Collect information regarding the three cars in your list by finding at least eight cars of each type that are for sale. Obtain information such as the asking price and the number of miles the car has. Sources of data include your local newspaper, classified ads, and car websites (such as www.cars.com). Compute summary statistics for asking price, number of miles, and other variables of interest. Using the same scale, draw boxplots of each variable considered.

3. Go to the Insurance Institute for Highway Safety website (www.iihs. org). Select the Ratings link. Choose the make and model for each car you are considering. Obtain information regarding crash testing for each car under consideration. Compare cars in the same class. How does each car compare? Is one car you are considering substantially safer than the others? What about repair costs? Compute summary statistics for crash tests and repair costs.

4. Obtain information about insurance costs. Contact various insurance companies to determine the cost of insuring the cars you are considering. Compute summary statistics for insurance costs and draw boxplots.

5. Write a report supporting your conclusion regarding which car you would purchase.

Questions about authorship of unattributed documents are one of the many problems confronting historians. Statistical analyses, when combined with other historical facts, often help resolve these mysteries. One such historical conundrum is the identity of the writer known as A MOURNER. A letter appearing in the February 26, 1770, issue of *The Boston Gazette and Country Journal* is as follows:

> The general Sympathy and Concern for the Murder of the Lad by the base and infamous Richardson on the 22d Instant, will be sufficient Reason for your Notifying the Public that he will be buried from his Father's House in Frogg Lane, opposite Liberty-Tree, on Monday next, when all the Friends of Liberty may have an Opportunity of paying their last Respects to the Remains of this little Hero and first Martyr to the noble Cause—Whose manly Spirit (after this Accident happened) appear'd in his discreet Answers to his Doctor, his Thanks to the Clergymen who prayed with him, and Parents, and while he underwent the greatest Distress of bodily Pain; and with which he met the King of Terrors. These Things, together with the several heroic Pieces found in his Pocket, particularly Wolfe's Summit of human Glory, gives Reason to think he had a martial Genius, and would have made a clever Man.
> A MOURNER.

The Lad the writer refers to is Christopher Sider, an 11-year-old son of a poor German immigrant. Sider was shot and killed on February 22, 1770, in a civil disturbance involving schoolboys, patriot supporters of the nonimportation agreement, and champions of the English crown, preceding the Boston Massacre by just a couple of weeks. From a historical perspective, the identity of A

MOURNER remains a mystery. However, it seems clear from the letter's text that the author supported the patriot position, narrowing the field of possible writers.

Ordinarily, a statistical analysis of word frequencies contained in the contested document compared with frequency analyses for known authors permit an identity inference to be drawn. Unfortunately, this letter is too short for this strategy. Another possibility is based on the frequencies of word lengths. In this case, a simple letter count is generated for each word in the document. Proper names, numbers, abbreviations, and titles are removed from consideration, because these do not represent normal vocabulary use. Care must be taken in choosing comparison texts, since colonial publishers habitually incorporated their own spellings into their essays they printed. Therefore, it is desirable to get comparison texts from the same printer the contested material came from.

The following table contains the summary analysis for six passages printed in *The Boston Gazette and Country Journal* in early 1770. The Tom Sturdy (probably a pseudonym) text was included because of the use of the phrase "Friends of Liberty," which appears in A MOURNER's letter, as opposed to the more familiar "Sons of Liberty." John Hancock, James Otis, and Samuel Adams are included because they are well-known patriots who frequently wrote articles carried by the Boston papers. Samuel Adams' essay used in this analysis was signed VINDEX, one of his many pseudonyms. The table presents three summaries of work penned by Adams: The first two originate from two separate sections of the VINDEX essay and the last is a compilation of the first two.

Summary Statistics of Word Length from Sample Passages by Various Potential Authors of the Letter Signed A MOURNER						
	Tom Sturdy	John Hancock	James Otis	Samuel Adams-1	Samuel Adams-2	Samuel Adams-1 & 2
Mean	4.08	4.69	4.58	4.60	4.52	4.56
Median	4	4	4	3	4	4
Mode	2	3	2	2	2	2
Standard deviation	2.17	2.60	2.75	2.89	2.70	2.80
Sample variance	4.70	6.76	7.54	8.34	7.30	7.84
Range	13	9	14	12	16	16
Minimum	1	1	1	1	1	1
Maximum	14	10	15	13	17	17
Sum	795	568	842	810	682	1492
Count	195	121	184	176	151	27

1. Generate a data set consisting of the length of each word used in the letter signed by A MOURNER. Be sure to disregard any text that uses proper names, numbers, abbreviations, or titles.

2. Calculate the summary statistics for the letter's word lengths. Compare your findings with those of the

known authors, and speculate about the identity of A MOURNER.

3. Compare the two Adams summaries. Discuss the viability of word-length analysis as a tool for resolving disputed documents.

4. What other information would be useful to identify A MOURNER?

4

Describing the Relation between Two Variables

Outline

Making an Informed Decision

The world is a very interesting and dynamic place. How do quantitative variables relate to each other on a world scale? A website that allows us to see how the world is changing over time and, in particular, how relationships among variables in our world change over time is www.gapminder.org. See the Decisions Project on page 243.

PUTTING IT TOGETHER

In Chapters 2 and 3 we examined data in which a single variable was measured for each individual in the study (**univariate data**), such as the five-year rate of return (the variable) for various mutual funds (the individuals). We found both graphical and numerical descriptive measures for the variable.

In this chapter, we discuss graphical and numerical methods for describing **bivariate data**, data in which two variables are measured on an individual. For example, we might want to know whether the amount of cola consumed per week is related to one's bone density. The individuals would be the people in the study, and the two variables would be the amount of cola and bone density. In this study, both variables are quantitative. We present methods for describing the relation between two quantitative variables in Sections 4.1–4.3.

Suppose we want to know whether level of education is related to one's employment status (employed or unemployed). Here, both variables are qualitative. We present methods for describing the relation between two qualitative variables in Section 4.4.

Situations may also occur in which one variable is quantitative and the other is qualitative. We have already presented a technique for describing this situation. Look back at Example 4 in Section 3.5 where we considered whether space flight affected red blood cell mass. There, space flight is qualitative (rat sent to space or not), and red blood cell mass is quantitative.

4.1   Scatter Diagrams and Correlation

Preparing for This Section   Before getting started, review the following:

- Mean (Section 3.1, pp. 118–120)
- z-scores (Section 3.4, pp. 155–156)
- Standard deviation (Section 3.2, pp. 133–137)
- Confounding, Lurking Variables, and Confounding Variables (Section 1.2, pp. 15–17)

Objectives

1. Draw and interpret scatter diagrams
2. Describe the properties of the linear correlation coefficient
3. Compute and interpret the linear correlation coefficient
4. Determine whether a linear relation exists between two variables
5. Explain the difference between correlation and causation

Before we can represent bivariate data graphically, we must decide which variable will be used to predict the value of the other variable. For example, it seems reasonable to think that as the speed at which a golf club is swung increases, the distance the golf ball travels also increases. Therefore, we might use club-head speed to predict distance. We call distance the *response* (or *dependent*) *variable* and club-head speed the *explanatory* (or *predictor* or *independent*) *variable*.

Definition   The **response variable** is the variable whose value can be explained by the value of the **explanatory** or **predictor variable**.

1   Draw and Interpret Scatter Diagrams

In Other Words
We use the term *explanatory variable* because it helps to explain variability in the response variable.

The first step in identifying the type of relation that might exist between two quantitative variables is to draw a picture of the data.

Definition   A **scatter diagram** is a graph that shows the relationship between two quantitative variables measured on the same individual. Each individual in the data set is represented by a point in the scatter diagram. The explanatory variable is plotted on the horizontal axis, and the response variable is plotted on the vertical axis.

EXAMPLE 1   Drawing a Scatter Diagram

Table 1	
Club-head Speed (mph)	**Distance (yards)**
100	257
102	264
103	274
101	266
105	277
100	263
99	258
105	275

Source: Paul Stephenson, student at Joliet Junior College

Problem   A golf pro wants to investigate the relation between the club-head speed of a golf club (measured in miles per hour) and the distance (in yards) that the ball will travel. He realizes other variables besides club-head speed determine the distance a ball will travel (such as club type, ball type, golfer, and weather conditions). To eliminate the variability due to these variables, the pro uses a single model of club and ball, one golfer, and a clear, 70-degree day with no wind. The pro records the club-head speed, measures the distance the ball travels, and collects the data in Table 1. Draw a scatter diagram of the data.

Approach   Because the pro wants to use club-head speed to predict the distance the ball travels, club-head speed is the explanatory variable (horizontal axis) and distance is the response variable (vertical axis). Plot the ordered pairs (100, 257), (102, 264), and so on, in a rectangular coordinate system.

(continued)

Solution The scatter diagram is shown in Figure 1. It appears from the graph that as club-head speed increases, the distance the ball travels increases as well.

Figure 1

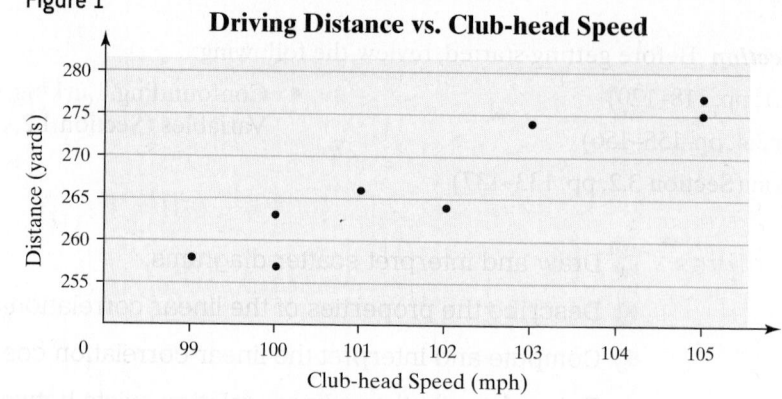

CAUTION!
Do not connect points when drawing a scatter diagram.

It is not always clear which variable should be considered the response variable and which the explanatory variable. For example, does high school GPA predict a student's SAT score or can the SAT score predict GPA? The researcher must determine which variable plays the role of explanatory variable based on the questions he or she wants answered. For example, if the researcher wants to predict SAT scores based on high school GPA, then high school GPA is the explanatory variable.

● **Now Work Problems 25(a) and 25(b)**

Scatter diagrams show the type of relation that exists between two variables. Our goal in interpreting scatter diagrams is to distinguish scatter diagrams that imply a linear relation, a nonlinear relation, or no relation. Figure 2 displays various scatter diagrams and the type of relation implied.

Figure 2

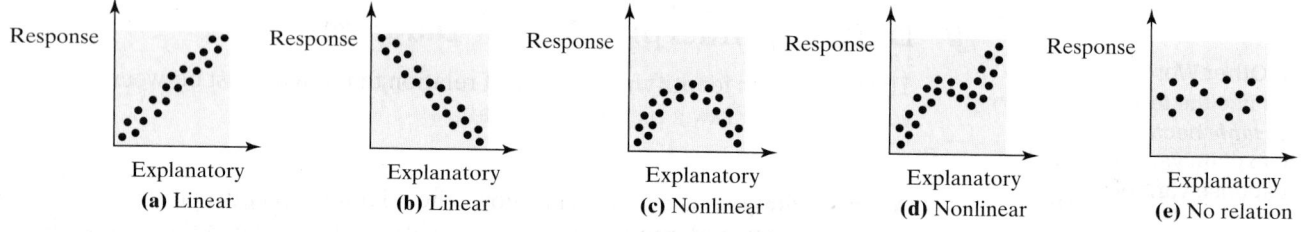

Notice the difference between Figure 2(a) and Figure 2(b). The data follow a linear pattern that slants upward to the right in Figure 2(a) and downward to the right in Figure 2(b). Figures 2(c) and 2(d) show nonlinear relations. In Figure 2(e), there is no relation between the explanatory and response variables.

Definitions

In Other Words
If two variables are positively associated, then as one goes up the other also tends to go up. If two variables are negatively associated, then as one goes up the other tends to go down.

Two variables that are linearly related are **positively associated** when above-average values of one variable are associated with above-average values of the other variable and below-average values of one variable are associated with below-average values of the other variable. That is, two variables are positively associated if, whenever the value of one variable increases, the value of the other variable also increases.

Two variables that are linearly related are **negatively associated** when above-average values of one variable are associated with below-average values of the other variable. That is, two variables are negatively associated if, whenever the value of one variable increases, the value of the other variable decreases.

● **Now Work Problem 9**

The scatter diagram in Figure 1 implies that club-head speed is positively associated with the distance a golf ball travels.

2   Describe the Properties of the Linear Correlation Coefficient

It is dangerous to use only a scatter diagram to determine if two variables are linearly related. In Figure 3, we have redrawn the scatter diagram from Figure 1 using a different vertical scale.

Figure 3

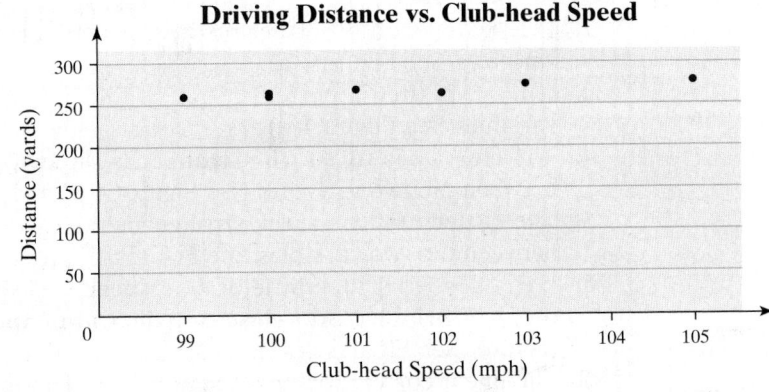

Driving Distance vs. Club-head Speed

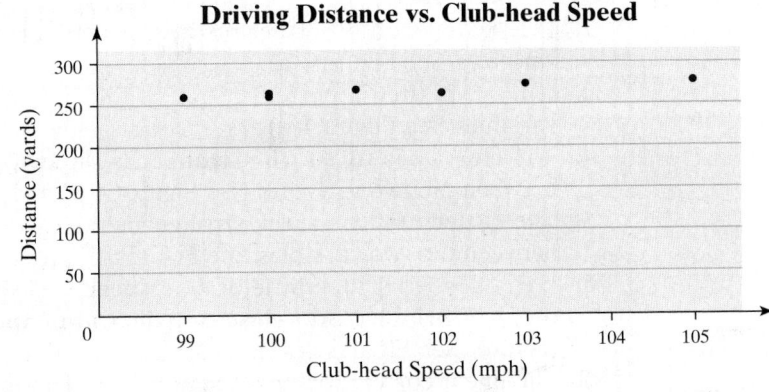

CAUTION!
The horizontal and vertical scales of a scatter diagram should be set so that the scatter diagram does not mislead a reader.

It is more difficult to conclude that the variables are related in Figure 3 than in Figure 1. The moral of the story is this: Just as we can manipulate the scale of graphs of univariate data, we can also manipulate the scale of the graphs of bivariate data, possibly resulting in incorrect conclusions. Therefore, numerical summaries of bivariate data should be used in addition to graphs to describe any relation that exists between two variables.

Definition

The **linear correlation coefficient** or **Pearson product moment correlation coefficient** is a measure of the strength and direction of the linear relation between two quantitative variables. The Greek letter ρ (rho) represents the population correlation coefficient, and r represents the sample correlation coefficient. We present only the formula for the sample correlation coefficient.

Sample Linear Correlation Coefficient*

$$r = \frac{\sum \left(\dfrac{x_i - \bar{x}}{s_x} \right)\left(\dfrac{y_i - \bar{y}}{s_y} \right)}{n - 1} \qquad (1)$$

where x_i is the ith observation of the explanatory variable
$\bar{x}$ is the sample mean of the explanatory variable
s_x is the sample standard deviation of the explanatory variable
y_i is the ith observation of the response variable
$\bar{y}$ is the sample mean of the response variable
s_y is the sample standard deviation of the response variable
n is the number of individuals in the sample

---

*An equivalent computational formula for the linear correlation coefficient is

$$r = \frac{\sum x_i y_i - \dfrac{\sum x_i \sum y_i}{n}}{\sqrt{\left(\sum x_i^2 - \dfrac{(\sum x_i)^2}{n} \right)}\sqrt{\left(\sum y_i^2 - \dfrac{(\sum y_i)^2}{n} \right)}}$$

The Pearson linear correlation coefficient is named in honor of Karl Pearson (1857–1936). See the Historical Note on page 186.

Properties of the Linear Correlation Coefficient

1. The linear correlation coefficient is always between −1 and 1, inclusive. That is, $-1 \leq r \leq 1$.
2. If $r = +1$, then a perfect positive linear relation exists between the two variables. See Figure 4(a).
3. If $r = -1$, then a perfect negative linear relation exists between the two variables. See Figure 4(d).
4. The closer r is to $+1$, the stronger is the evidence of positive association between the two variables. See Figures 4(b) and 4(c).
5. The closer r is to -1, the stronger is the evidence of negative association between the two variables. See Figures 4(e) and 4(f).
6. If r is close to 0, then little or no evidence exists of a *linear* relation between the two variables. So **r close to 0 does not imply no relation, just no *linear* relation**. See Figures 4(g) and 4(h).
7. The linear correlation coefficient is a unitless measure of association. So the unit of measure for x and y plays no role in the interpretation of r.
8. The correlation coefficient is not resistant. Therefore, an observation that does not follow the overall pattern of the data could affect the value of the linear correlation coefficient.

> **CAUTION!**
> A linear correlation coefficient close to 0 does not imply that there is no relation, just no linear relation. For example, although the scatter diagram drawn in Figure 4(h) indicates that the two variables are related, the linear correlation coefficient is close to 0.

Figure 4

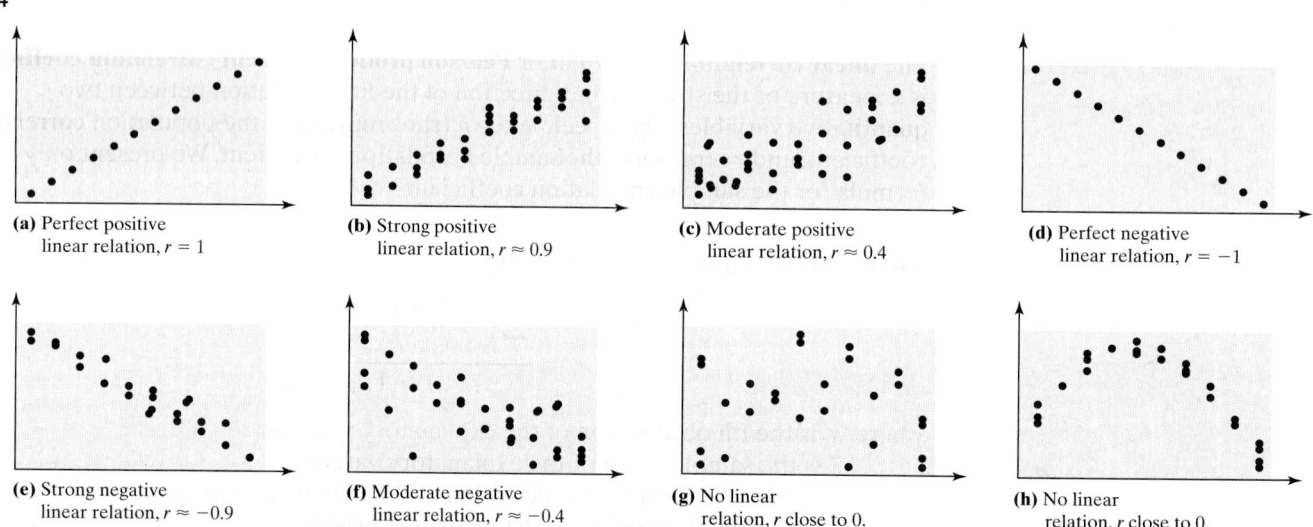

(a) Perfect positive linear relation, $r = 1$

(b) Strong positive linear relation, $r \approx 0.9$

(c) Moderate positive linear relation, $r \approx 0.4$

(d) Perfect negative linear relation, $r = -1$

(e) Strong negative linear relation, $r \approx -0.9$

(f) Moderate negative linear relation, $r \approx -0.4$

(g) No linear relation, r close to 0.

(h) No linear relation, r close to 0.

In Formula (1), notice that the numerator is the sum of the products of z-scores for the explanatory (x) and response (y) variables. A positive linear correlation coefficient means that the sum of the products of the z-scores for x and y must be positive. How does this occur? Figure 5 shows a scatter diagram with positive association between x and y. The vertical dashed line represents the value of $\bar{x}$, and the horizontal dashed line represents the value of $\bar{y}$. These two dashed lines divide our scatter diagram into four quadrants, labeled I, II, III, and IV.

Consider the data in quadrants I and III. If a certain x-value is above its mean, $\bar{x}$, then the corresponding y-value will be above its mean, $\bar{y}$. If a certain x-value is below its mean, $\bar{x}$, then the corresponding y-value will be below its mean, $\bar{y}$. Therefore, for data in quadrant I, we have $\frac{x_i - \bar{x}}{s_x}$ positive and $\frac{y_i - \bar{y}}{s_y}$ positive, so their product is positive. For data in quadrant III, we have $\frac{x_i - \bar{x}}{s_x}$ negative and $\frac{y_i - \bar{y}}{s_y}$ negative, so their product is positive. The sum of these products is positive, so the linear correlation coefficient is positive. A similar argument can be made for negative correlation.

Figure 5

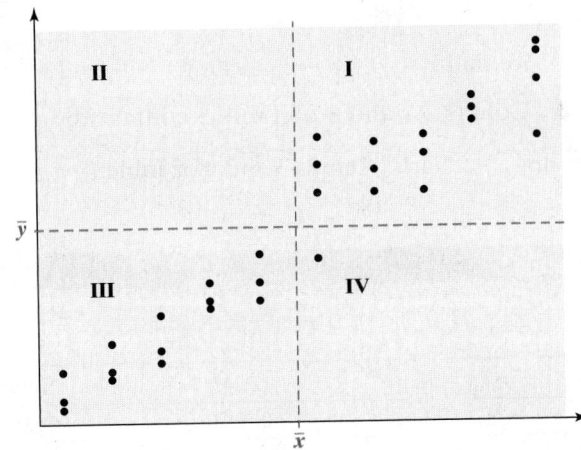

Now suppose that the data are equally dispersed in the four quadrants. Then the negative products (resulting from data in quadrants II and IV) will offset the positive products (resulting from data in quadrants I and III). The result is a linear correlation coefficient close to 0.

● Now Work Problem 13

③ Compute and Interpret the Linear Correlation Coefficient

In practice, linear correlation coefficients are found using technology. However, we present one computation by hand so that you may gain an appreciation of how the formula measures the strength of linear relation.

EXAMPLE 2   Computing the Correlation Coefficient by Hand

Problem For the data shown in Table 2, compute the linear correlation coefficient. A scatter diagram of the data is shown in Figure 6. The dashed lines on the scatter diagram represent the mean of x and the mean of y.

Table 2	
x	**y**
1	18
3	13
3	9
6	6
7	4

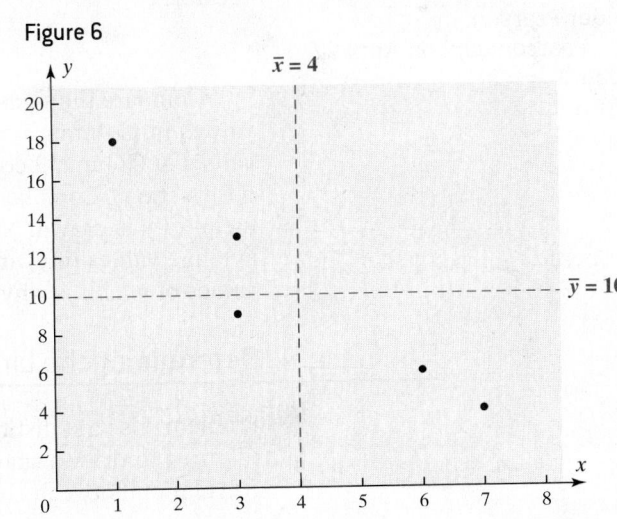

Figure 6

Approach

Step 1   Compute $\bar{x}, s_x, \bar{y},$ and s_y.

Step 2   Determine $\frac{x_i - \bar{x}}{s_x}$ and $\frac{y_i - \bar{y}}{s_y}$ for each observation.

Step 3   Compute $\left(\frac{x_i - \bar{x}}{s_x}\right)\left(\frac{y_i - \bar{y}}{s_y}\right)$ for each observation.

Step 4   Determine $\sum \left(\frac{x_i - \bar{x}}{s_x}\right)\left(\frac{y_i - \bar{y}}{s_y}\right)$ and substitute this value into Formula (1).

(continued)

Historical Note

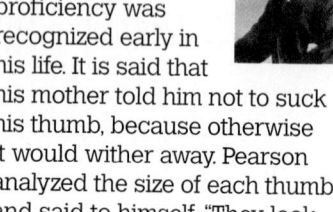

Karl Pearson was born March 27, 1857. Pearson's statistical proficiency was recognized early in his life. It is said that his mother told him not to suck his thumb, because otherwise it would wither away. Pearson analyzed the size of each thumb and said to himself, "They look alike to me. I can't see that the thumb I suck is any smaller than the other. I wonder if she could be lying to me."

Karl Pearson graduated from Cambridge University in 1879. From 1893 to 1911, he wrote 18 papers on genetics and heredity. Through this work, he developed ideas regarding correlation and the chi-square test. (See Chapter 12.) In addition, Pearson coined the term *standard deviation*.

Pearson and Ronald Fisher (see page 43) didn't get along. Their dispute was severe enough that Fisher turned down the post of chief statistician at the Galton Laboratory in 1919 because it would have meant working under Pearson.

Pearson died on April 27, 1936.

Solution

Step 1 We find $\bar{x} = 4$, $s_x = 2.44949$, $\bar{y} = 10$, and $s_y = 5.612486$.

Step 2 Columns 1 and 2 of Table 3 contain the data from Table 2. We determine $\dfrac{x_i - \bar{x}}{s_x}$ and $\dfrac{y_i - \bar{y}}{s_y}$ in Columns 3 and 4 of Table 3.

Table 3

x	y	$\dfrac{x_i - \bar{x}}{s_x}$	$\dfrac{y_i - \bar{y}}{s_y}$	$\left(\dfrac{x_i - \bar{x}}{s_x}\right)\left(\dfrac{y_i - \bar{y}}{s_y}\right)$
1	18	−1.2247	1.4254	−1.7457
3	13	−0.4083	0.5345	−0.2182
3	9	−0.4083	−0.1782	0.0727
6	6	0.8165	−0.7127	−0.5819
7	4	1.2247	−1.0690	−1.3093

$$\Sigma\left(\frac{x_i - \bar{x}}{s_x}\right)\left(\frac{y_i - \bar{y}}{s_y}\right) = -3.7824$$

Step 3 Multiply the entries in Columns 3 and 4 to obtain the entries in Column 5 of Table 3.

Step 4 Add the entries in Column 5 and substitute this value into Formula (1) to obtain the correlation coefficient.

$$r = \frac{\Sigma\left(\dfrac{x_i - \bar{x}}{s_x}\right)\left(\dfrac{y_i - \bar{y}}{s_y}\right)}{n - 1} = \frac{-3.7824}{5 - 1} = -0.946$$

We will agree to round the correlation coefficient to three decimal places. The correlation coefficient suggests a strong negative association between the two variables.

Compare the signs of the entries in Columns 3 and 4 in Table 3. Notice that negative values in Column 3 correspond with positive values in Column 4 and that positive values in Column 3 correspond with negative values in Column 4 (except for the third observation). Look back at the scatter diagram in Figure 6 where the mean of x and the mean of y is drawn. Notice that below-average values of x are associated with above-average values of y, and above-average values of x are associated with below-average values of y. This is why the linear correlation coefficient is negative.

EXAMPLE 3 Determining the Linear Correlation Coefficient Using Technology

Problem Use a statistical spreadsheet or a graphing calculator with advanced statistical features to draw a scatter diagram of the data in Table 1. Then determine the linear correlation between club-head speed and distance.

Approach We will use Excel to draw the scatter diagram and obtain the linear correlation coefficient. The steps for drawing scatter diagrams and obtaining the linear correlation coefficient using the TI-83/84 Plus graphing calculators, Minitab, Excel, and StatCrunch are given in the Technology Step-by-Step on pages 189–190.

Solution Figure 7(a) shows the scatter diagram, and Figure 7(b) shows the linear correlation coefficient of 0.939 obtained from Excel. Notice that Excel provides a

correlation matrix, which means that for every pair of columns in the spreadsheet it will compute and display the correlation in the bottom triangle of the matrix.

Because the linear correlation coefficient is positive, we know above-average values of x, club-head speed, are associated with above-average values of y, driving distance, and below-average values of x are associated with below-average values of y.

Figure 7

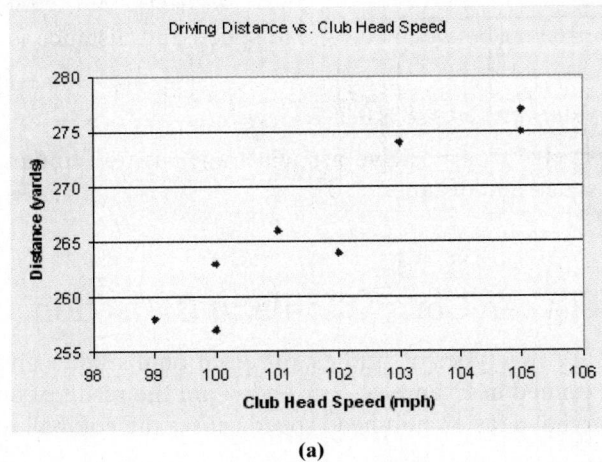

	Driving Distance	*Club Head Speed*
Driving Distance	1	
Club Head Speed	0.938695838	1

(a)   (b)

● **Now Work Problem 25(c)**

We stated in Property 8 that the linear correlation coefficient is not resistant. For example, suppose the golfer in Examples 1 and 3 hits one more golf ball. As he swings, a car driving by blows its horn, which distracts the golfer. His swing speed is 105 mph, but the mis-hit golf ball travels only 255 yards. The linear correlation coefficient with this additional observation decreases to 0.535.

④ Determine Whether a Linear Relation Exists between Two Variables

A question you may be asking yourself is "How do I know the correlation between two variables is strong enough for me to conclude that a linear relation exists between them?" Although rigorous tests can answer this question, for now we will be content with a simple comparison test.

Testing for a Linear Relation

Step 1   Determine the absolute value of the correlation coefficient.

Step 2   Find the critical value in Table II from Appendix A for the given sample size.

Step 3   If the absolute value of the correlation coefficient is greater than the critical value, we say a linear relation exists between the two variables. Otherwise, no linear relation exists.

In Other Words
We use two vertical bars to denote absolute value, as in $|5|$ or $|-4|$. Recall, $|5| = 5$, $|-4| = 4$, and $|0| = 0$.

Another way to think about the procedure is to consider Figure 8. If the correlation coefficient is positive and greater than the critical value, the variables are positively associated. If the correlation coefficient is negative and less than the opposite of the critical value, the variables are negatively associated.

Figure 8

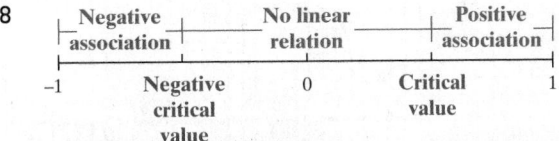

EXAMPLE 4 Does a Linear Relation Exist?

Problem Using the results from Example 3, determine whether a linear relation exists between club-head speed and distance.

Approach Follow Steps 1 through 3 given on the previous page.

Solution

Step 1 The linear correlation coefficient between club-head speed and distance is 0.939, so $|0.939| = 0.939$.

Step 2 Table II shows the critical value with $n = 8$ is 0.707.

Step 3 Because $|0.939| = 0.939 > 0.707$, a positive association (positive linear relation) exists between club-head speed and distance.

● **Now Work Problem 25(d)**

⑤ Explain the Difference between Correlation and Causation

In Chapter 1 we stated that there are two types of studies: observational studies and designed experiments. The data examined in Examples 1, 3, and 4 are the result of an experiment. Therefore, we can claim that a faster club-head speed causes the golf ball to travel a longer distance.

If data used in a study are observational, we cannot conclude the two correlated variables have a causal relationship. For example, the correlation between teenage birthrate and homicide rate since 1993 is 0.9987, but we cannot conclude that higher teenage birthrates cause a higher homicide rate because the data are observational. In fact, time-series data are often correlated because both variables happen to move in the same (or opposite) direction over time. Both teenage birthrates and homicide rates have been declining since 1993, so they have a high positive correlation.

Is there another way two variables can be correlated without there being a causal relation? Yes—through a *lurking variable*. A **lurking variable** is related to both the explanatory variable and response variable. For example, as air-conditioning bills increase, so does the crime rate. Does this mean that folks should turn off their air conditioners so that crime rates decrease? Certainly not! In this case, the lurking variable is air temperature. As air temperatures rise, both air-conditioning bills and crime rates rise.

> **CAUTION!**
> A linear correlation coefficient that implies a strong positive or negative association does not imply causation if it was computed using observational data.

EXAMPLE 5 Lurking Variables in a Bone Mineral Density Study

Problem Because colas tend to replace healthier beverages and colas contain caffeine and phosphoric acid, researchers Katherine L. Tucker and associates wanted to know whether cola consumption is associated with lower bone mineral density in women. Table 4 lists the typical number of cans of cola consumed in a week and the femoral neck bone mineral density for a sample of 15 women. The data were collected through a prospective cohort study.

Figure 9 shows the scatter diagram of the data. The correlation between number of colas per week and bone mineral density is -0.806. The critical value for correlation with $n = 15$ from Table II in Appendix A is 0.514. Because $|-0.806| > 0.514$, we conclude a negative linear relation exists between number of colas consumed and bone mineral density. Can the authors conclude that an increase in the number of colas consumed causes a decrease in bone mineral density? Identify some lurking variables in the study.

In Other Words
Confounding means that any relation that may exist between two variables may be due to some other variable not accounted for in the study.

Figure 9

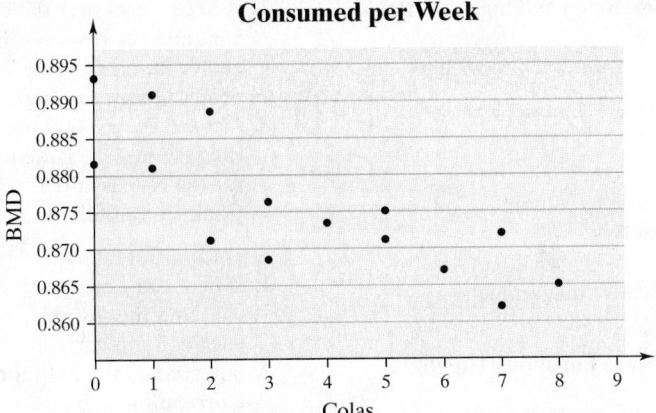

**Bone Mineral Density vs. Colas
Consumed per Week**

Table 4	
Number of Colas per Week	**Bone Mineral Density (g/cm²)**
0	0.893
0	0.882
1	0.891
1	0.881
2	0.888
2	0.871
3	0.868
3	0.876
4	0.873
5	0.875
5	0.871
6	0.867
7	0.862
7	0.872
8	0.865

Source: Based on data obtained from Katherine L. Tucker et. al., "Colas, but not other carbonated beverages, are associated with low bone mineral density in older women: The Framingham Osteoporosis Study." *American Journal of Clinical Nutrition* 2006, 84:936–942.

Approach To claim causality, we must collect the data through a designed experiment. Remember, lurking variables are related to both the explanatory and response variables in a study.

Solution In prospective cohort studies, data are collected on a group of subjects through questionnaires and surveys over time. Therefore, the data are observational. So the researchers cannot claim that increased cola consumption causes a decrease in bone mineral density.

In their article, the authors identified a number of lurking variables that could confound the results:

Variables that could potentially confound the relation between cola consumption and bone mineral density . . . included the following: age, body mass index, height, smoking, average daily intakes of alcohol, calcium, caffeine, total energy intake, physical activity, season of measurement, estrogen use, and menopause status.

The authors were careful to say that increased cola consumption is *associated* with lower bone mineral density because of potential lurking variables. They never stated that increased cola consumption *causes* lower bone mineral density.

• **Now Work Problem 41**

Technology Step-by-Step Drawing Scatter Diagrams and Determining the Correlation Coefficient

TI-83/84 Plus

Scatter Diagrams
1. Enter the explanatory variable in L1 and the response variable in L2.
2. Press 2nd Y= to bring up the StatPlot menu. Select 1: Plot1.
3. Turn Plot 1 on by highlighting the On button and pressing ENTER.
4. Highlight the scatter diagram icon and press ENTER. Be sure that Xlist is L1 and Ylist is L2.
5. Press ZOOM and select 9: ZoomStat.

Correlation Coefficient
1. Enter the explanatory variable values in L1 and the response variable values in L2.
2. Turn the diagnostics on by selecting the catalog (2nd 0). Scroll down and select DiagnosticsOn. Hit ENTER

twice to activate diagnostics (this step needs to be done only once).
3. From the HOME screen, press STAT, highlight CALC, and select 4: LinReg(ax + b). Select L1 for Xlist and L2 for Ylist. Highlight Calculate and press ENTER.

Minitab
Scatter Diagrams
1. Enter the explanatory variable in C1 and the response variable in C2. You may want to name the variables.
2. Select the **Graph** menu and then **Scatterplot** . . . .
3. Highlight the Simple icon and click OK.
4. With the cursor in the Y column, select the response variable. With the cursor in the X column, select the explanatory variable. Click OK.

Correlation Coefficient
1. With the explanatory variable in C1 and the response variable in C2, select the **Stat** menu and highlight **Basic Statistics**. Select **Correlation**.
2. Select the variables whose correlation you wish to determine and click OK.

Excel
Scatter Diagrams
1. Enter the explanatory variable in column A and the response variable in column B.
2. Highlight both sets of data.
3. Select Insert. Choose, the scatter diagram icon.

Correlation Coefficient
1. Select Formulas and then More Functions. Highlight Statistical.
2. Select CORREL.
3. With the cursor in Array 1, highlight the data containing the explanatory variable. With the cursor in Array 2, highlight the data containing the response variable. Click OK.

StatCrunch
Scatter Diagrams
1. If necessary, enter the explanatory variable in column var1 and the response variable in column var2. Name each column variable.
2. Select **Graph** and highlight **Scatter Plot**.
3. Choose the explanatory variable for the X column and the response variable for the Y column. Enter the labels for the X-axis and Y-axis. Enter a title for the graph. Click Compute!.

Correlation Coefficient
1. If necessary, enter the explanatory variable in column var1 and the response variable in column var2. Name each column variable.
2. Select **Stat**, highlight **Summary Stats**, and select **Correlation**.
3. Click on the variables whose correlation you want to determine. Click Compute!.

4.1 Assess Your Understanding

Vocabulary and Skill Building

1. What is the difference between univariate data and bivariate data?

2. The _____ variable is the variable whose value can be explained by the value of the explanatory variable.

3. A _____ _____ is a graph that shows the relation between two quantitative variables.

4. What does it mean to say two variables are positively associated? Negatively associated?

5. If $r =$ _____, then a perfect negative linear relation exists between the two quantitative variables.

6. *True or False*: If the linear correlation coefficient is close to 0, then the two variables have no relation.

7. A _____ variable is a variable that is related to both the explanatory and response variable.

8. *True or False*: Correlation implies causation.

In Problems 9–12, determine whether the scatter diagram indicates that a linear relation may exist between the two variables. If the relation is linear, determine whether it indicates a positive or negative association between the variables.

NW 9.

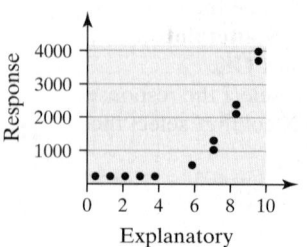

10.

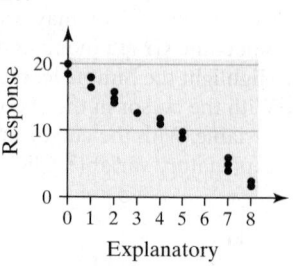

11.

(scatter diagram, Explanatory axis 10–50, Response axis 0–25)

12.

(scatter diagram, Explanatory axis 3–15, Response axis 0–25)

NW 13. Match the linear correlation coefficient to the scatter diagrams. The scales on the x- and y-axes are the same for each diagram.

(a) $r = 0.787$ **(b)** $r = 0.523$

(c) $r = 0.053$ **(d)** $r = 0.946$

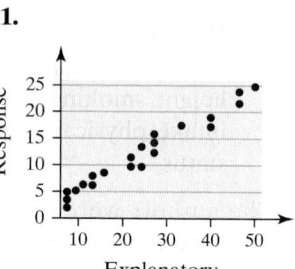

(I)

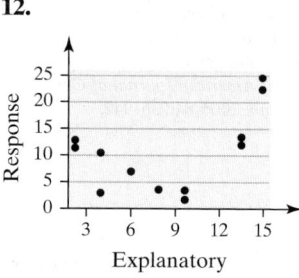

(II)

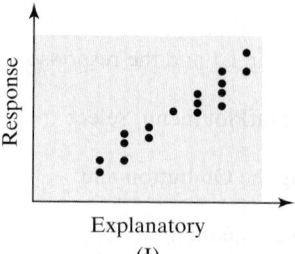

(III)

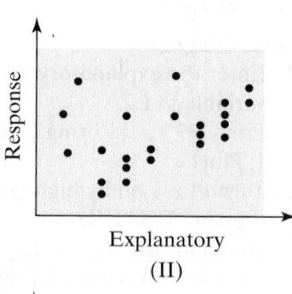

(IV)

14. Match the linear correlation coefficient to the scatter diagram. The scales on the *x*- and *y*-axes are the same for each diagram.

(a) $r = -0.969$ **(b)** $r = -0.049$
(c) $r = -1$ **(d)** $r = -0.992$

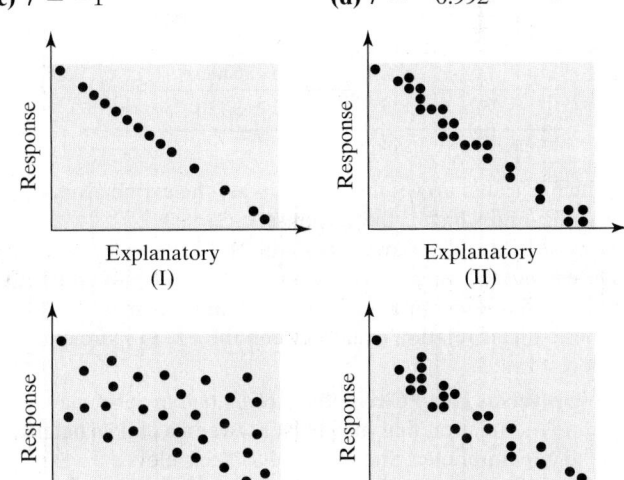

15. Does Education Pay? The scatter diagram drawn in Minitab shows the relation between the percentage of the population of a state plus Washington, DC, that has at least a bachelor's degree and the median income (in dollars) of the state for 2013.
Source: U.S. Census Bureau

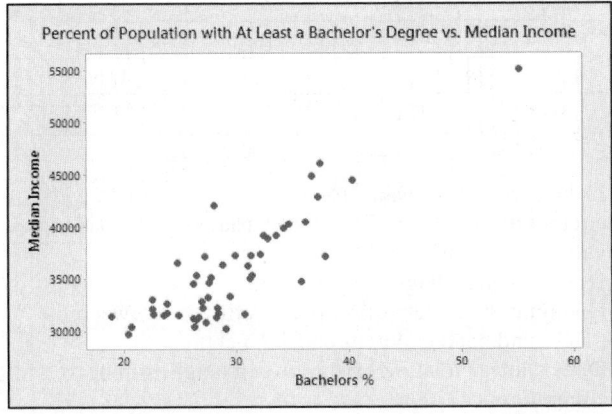

(a) Describe any relation that exists between level of education and median income.
(b) One observation appears to stick out from the rest. Which one? This particular observation is for Washington, DC. Can you think of any reasons why Washington, DC, might stick out from the rest of the states?
(c) The correlation coefficient between the percentage of the population with a bachelor's degree and median income is 0.854. Does a linear relation exist between percent of the population with at least a bachelor's degree and median income?

16. Relation between Education and Birthrate? The following scatter diagram drawn in StatCrunch shows the relation between percent of the population with at least a bachelor's degree in a state and birthrate (births per 1000 women 15 to 44 years old).

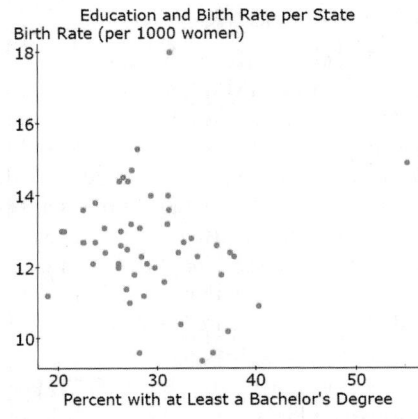

(a) Describe any relation that exists between median income and birthrate.
(b) The correlation between percent of population with at least a bachelor's degree and birth rate is −0.069. What does this imply about the relation between median income and birthrate?

In Problems 17–20, (a) draw a scatter diagram of the data, (b) by hand, compute the correlation coefficient, and (c) determine whether there is a linear relation between x and y.

17.

x	2	4	6	6	7
y	4	8	10	13	20

18.

x	2	3	5	6	6
y	10	9	7	4	2

19.

x	2	6	6	7	9
y	8	7	6	9	5

20.

x	0	5	7	8	9
y	3	8	6	9	4

21. Name the Relation, Part I For each of the following statements, explain whether you think the variables will have positive correlation, negative correlation, or no correlation. Support your opinion.

(a) Number of children in the household under the age of 3 and expenditures on diapers
(b) Interest rates on car loans and number of cars sold
(c) Number of hours per week on the treadmill and cholesterol level
(d) Price of a Big Mac and number of McDonald's French fries sold in a week
(e) Shoe size and IQ

22. Name the Relation, Part II For each of the following statements, explain whether you think the variables will have positive correlation, negative correlation, or no correlation. Support your opinion.

(a) Number of cigarettes smoked by a pregnant woman each week and birth weight of her baby
(b) Years of education and annual salary
(c) Number of doctors on staff at a hospital and number of administrators on staff
(d) Head circumference and IQ
(e) Number of movie goers and movie ticket price

Applying the Concepts

23. The TIMMS Exam The Trends in International Mathematics and Science (TIMMS) is a mathematics and science achievement exam given internationally. On each exam, students are asked to respond to a variety of background questions. For the 41 nations that participated in TIMMS, the correlation between the percentage of items answered in the background questionnaire (used as a proxy for student task persistence) and mean score on the exam was 0.79. Does this suggest there is a linear relation between student task persistence and achievement score? Write a sentence that explains what this result might mean.

24. The TIMMS Exam Part II (See Problem 23) For the 41 nations that participated in TIMMS, the correlation between the percentage of students who skipped class at least once in the past month and the mean score on the exam was −0.52. Does this suggest there is a linear relation between attendance and achievement score? Write a sentence that explains what this result might mean.

NW **25. An Unhealthy Commute** The Gallup Organization regularly surveys adult Americans regarding their commute time to work. In addition, they administer a Well-Being Survey. According to the Gallup Organization, "The Gallup-Healthways Well-Being Index Composite Score is comprised of six sub-indices: Life Evaluation, Emotional Health, Physical Health, Healthy Behavior, Work Environment and Basic Access." A complete description of the index can be found at http://www.well-beingindex.com/. The data in the following table are based on the results of the survey, which represent commute time to work (in minutes) and well-being index score.

Commute Time (in minutes)	Gallup-Healthways Well-Being Index Composite Score
5	69.2
15	68.3
25	67.5
35	67.1
50	66.4
72	66.1
105	63.9

Source: The Gallup Organization

(a) Which variable do you believe is likely the explanatory variable and which is the response variable?
(b) Draw a scatter diagram of the data.
(c) Determine the linear correlation coefficient between commute time and well-being index score.
(d) Does a linear relation exist between the commute time and well-being index score?

26. Credit Scores Your Fair Isaacs Corporation (FICO) credit score is used to determine your creditworthiness. It is used to help determine whether you qualify for a mortgage or credit and is even used to determine insurance rates. FICO scores have a range of 300 to 850, with a higher score indicating a better credit history. The given data represent the interest rate (in percent) a bank would offer on a 36-month auto loan for various FICO scores.

Credit Score	Interest Rate (percent)
545	18.982
595	17.967
640	12.218
675	8.612
705	6.680
750	5.150

Source: www.myfico.com

(a) Which variable do you believe is likely the explanatory variable and which is the response variable?
(b) Draw a scatter diagram of the data.
(c) Determine the linear correlation coefficient between FICO score and interest rate on a 36-month auto loan.
(d) Does a linear relation exist between the FICO score and interest rate?

27. Height versus Head Circumference A pediatrician wants to determine the relation that may exist between a child's height and head circumference. She randomly selects eleven 3-year-old children from her practice, measures their heights and head circumference, and obtains the data shown in the table below.

Height (inches)	Head Circumference (inches)	Height (inches)	Head Circumference (inches)
27.75	17.5	26.5	17.3
24.5	17.1	27	17.5
25.5	17.1	26.75	17.3
26	17.3	26.75	17.5
25	16.9	27.5	17.5
27.75	17.6		

Source: Denise Slucki, student at Joliet Junior College

(a) If the pediatrician wants to use height to predict head circumference, determine which variable is the explanatory variable and which is the response variable.
(b) Draw a scatter diagram of the data.
(c) Compute the linear correlation coefficient between the height and head circumference of a child.
(d) Does a linear relation exist between height and head circumference?
(e) Convert the data to centimeters (1 inch = 2.54 cm), and recompute the linear correlation coefficient. What effect did the conversion have on the linear correlation coefficient?

28. American Black Bears The American black bear (*Ursus americanus*) is one of eight bear species in the world. It is the smallest North American bear and the most common bear species on the planet. In 1969, Dr. Michael R. Pelton of the University of Tennessee initiated a long-term study of the population in the Great Smoky Mountains National Park. One aspect of the study was to develop a model that could be used to predict a bear's weight (since it is not practical to weigh bears in the field). One variable thought to be related to weight is the length of the bear. The following data represent the lengths and weights of 12 American black bears.

Total Length (cm)	Weight (kg)
139.0	110
138.0	60
139.0	90
120.5	60
149.0	85
141.0	100
141.0	95
150.0	85
166.0	155
151.5	140
129.5	105
150.0	110

Source: fieldtripearth.org

(a) Which variable is the explanatory variable based on the goals of the research?

(b) Draw a scatter diagram of the data.

(c) Determine the linear correlation coefficient between weight and length.

(d) Does a linear relation exist between the weight of the bear and its length?

29. Weight of a Car versus Miles per Gallon An engineer wanted to determine how the weight of a car affects gas mileage. The following data represent the weights of various domestic cars and their gas mileages in the city for the 2015 model year.

Car	Weight (lb)	Miles per Gallon
Buick LaCrosse	4724	17
Cadillac XTS	4006	18
Chevrolet Cruze	3097	22
Chevrolet Impala	3555	19
Chrysler 300	4029	19
Dodge Charger	3934	19
Dodge Dart	3242	24
Ford Focus	2960	26
Ford Mustang	3530	19
Lincoln MKZ	3823	18

Source: Each manufacturer's website

(a) Determine which variable is the likely explanatory variable and which is the likely response variable.

(b) Draw a scatter diagram of the data.

(c) Compute the linear correlation coefficient between the weight of a car and its miles per gallon in the city.

(d) Does a linear relation exist between the weight of a car and its miles per gallon in the city?

30. Hurricanes The data in the next column represent the maximum wind speed (in knots) and atmospheric pressure (in millibars) for a random sample of hurricanes that originated in the Atlantic Ocean.

Atmospheric Pressure (mb)	Wind Speed (knots)	Atmospheric Pressure (mb)	Wind Speed (knots)
993	50	1006	40
995	60	942	120
994	60	1002	40
997	45	986	50
1003	45	983	70
1004	40	994	65
1000	55	940	120
994	55	976	80
942	105	966	100
1006	30	982	55

Source: National Hurricane Center

(a) Draw a scatter diagram treating atmospheric pressure as the explanatory variable.

(b) Compute the linear correlation coefficient between atmospheric pressure and wind speed.

(c) Does a linear relation exist between atmospheric pressure and wind speed?

31. CEO Performance The following data represent the total compensation for 12 randomly selected chief executive officers (CEO) and the company's stock performance in 2013.

Company	Compensation (millions of dollars)	Stock Return (%)
Navistar International	14.53	75.43
Aviv REIT	4.09	64.01
Groupon	7.11	142.07
Inland Real Estate	1.05	32.72
Equity Lifestyles Properties	1.97	10.64
Tootsie Roll Industries	3.76	30.66
Catamaran	12.06	0.77
Packaging Corp of America	7.62	69.39
Brunswick	8.47	58.69
LKQ	4.04	55.93
Abbott Laboratories	20.87	24.28
TreeHouse Foods	6.63	32.21

Data from *Chicago Tribune*, June 1, 2014

(a) One would think that a higher stock return would lead to a higher compensation. Based on this, what would likely be the explanatory variable?

(b) Draw a scatter diagram of the data.

(c) Determine the linear correlation coefficient between compensation and stock return.

(d) Does a linear relation exist between compensation and stock return? Does stock performance appear to play a role in determining the compensation of a CEO?

32. Bear Markets A bear market in the stock market is defined as a condition in which the market declines by 20% or more over the course of at least two months. The following data represent the number of months and percentage change in the S&P500 (a group of 500 stocks).

Months	Percent Change	Months	Percent Change
1.9	−44.57	12.1	−20.57
8.3	−44.29	14.9	−21.63
3.3	−32.86	6.5	−27.97
3.4	−42.54	8	−22.18
6.8	−61.81	18.1	−36.06
5.8	−40.60	21	−48.20
3.1	−29.43	20.7	−27.11
13.4	−31.81	3.4	−33.51
12.9	−54.47	18.2	−36.77
5.1	−24.44	9.3	−33.75
7.6	−28.69	13.6	−51.93
17.9	−34.42	2.1	−27.62
11.8	−28.47		

Source: Gold-Eagle

(a) Treating the length of the bear market as the explanatory variable, draw a scatter diagram of the data.
(b) Determine the linear correlation coefficient between months and percent change.
(c) Does a linear relation exist between duration of the bear market and market performance?

33. Does Size Matter? Researchers wondered whether the size of a person's brain was related to the individual's mental capacity. They selected a sample of right-handed introductory psychology students who had SAT scores higher than 1350. The subjects took the Wechsler Adult Intelligence Scale-Revised to obtain their IQ scores. MRI scans were performed at the same facility for the subjects. The scans consisted of 18 horizontal MR images. The computer counted all pixels with a nonzero gray scale in each of the 18 images, and the total count served as an index for brain size.

Gender	MRI Count	IQ	Gender	MRI Count	IQ
Female	816,932	133	Male	949,395	140
Female	951,545	137	Male	1,001,121	140
Female	991,305	138	Male	1,038,437	139
Female	833,868	132	Male	965,353	133
Female	856,472	140	Male	955,466	133
Female	852,244	132	Male	1,079,549	141
Female	790,619	135	Male	924,059	135
Female	866,662	130	Male	955,003	139
Female	857,782	133	Male	935,494	141
Female	948,066	133	Male	949,589	144

Source: L. Willerman, R. Schultz, J. N. Rutledge, and E. Bigler (1991). "In Vivo Brain Size and Intelligence," *Intelligence*, 15, 223–228.

(a) Draw a scatter diagram treating MRI count as the explanatory variable and IQ as the response variable. Comment on what you see.
(b) Compute the linear correlation coefficient between MRI count and IQ. Are MRI count and IQ linearly related?

(c) A lurking variable in the analysis is gender. Draw a scatter diagram treating MRI count as the explanatory variable and IQ as the response variable, but use a different plotting symbol for each gender. For example, use a circle for males and a triangle for females. What do you notice?
(d) Compute the linear correlation coefficient between MRI count and IQ for females. Compute the linear correlation coefficient between MRI count and IQ for males. Are MRI count and IQ linearly related? What is the moral?

34. Male versus Female Drivers The following data represent the number of licensed drivers in various age groups and the number of fatal accidents within the age group by gender.

Age	Number of Male Licensed Drivers (000s)	Number of Fatal Crashes	Number of Female Licensed Drivers (000s)	Number of Fatal Crashes
<16	12	227	12	77
16–20	6,424	5,180	6,139	2,113
21–24	6,941	5,016	6,816	1,531
25–34	18,068	8,595	17,664	2,780
35–44	20,406	7,990	20,063	2,742
45–54	19,898	7,118	19,984	2,285
55–64	14,340	4,527	14,441	1,514
65–74	8,194	2,274	8,400	938
>74	4,803	2,022	5,375	980

Source: National Highway and Traffic Safety Institute

(a) On the same graph, draw a scatter diagram for both males and females. Be sure to use a different plotting symbol for each group. For example, use a square (□) for males and a plus sign (+) for females. Treat number of licensed drivers as the explanatory variable.
(b) Based on the scatter diagrams, do you think that insurance companies are justified in charging different insurance rates for males and females? Why?
(c) Compute the linear correlation coefficient between number of licensed drivers and number of fatal crashes for males.
(d) Compute the linear correlation coefficient between number of licensed drivers and number of fatal crashes for females.
(e) Which gender has the stronger linear relation between number of licensed drivers and number of fatal crashes. Why?

35. Weight of a Car versus Miles per Gallon Suppose that we add the Ford Taurus to the data in Problem 29. A Ford Taurus weighs 3917 pounds and gets 19 miles per gallon.
(a) Redraw the scatter diagram with the Taurus included.
(b) Recompute the linear correlation coefficient with the Taurus included.
(c) Compare the results of parts (a) and (b) with the results of Problem 29. Why are the results here reasonable?
(d) Now suppose that we add the Ford Fusion Hybrid to the data in Problem 29 (remove the Taurus). A Fusion Hybrid weighs 3639 pounds and gets 44 miles per gallon. Redraw the scatter diagram with the Fusion included. What do you notice?
(e) Recompute the linear correlation coefficient with the Fusion included. How did this new value affect your result?
(f) Why does this observation not follow the pattern of the data?

36. American Black Bears The website that contained the American black bear data listed in Problem 28 actually had a bear whose height is 141.0 cm and weight is 100 kg incorrectly listed as 41.0 cm tall.

(a) Redraw the scatter diagram with the incorrect entry.

(b) Recompute the linear correlation coefficient using the data with the incorrect entry.

(c) Explain how the scatter diagram can be used to identify incorrectly entered data values. Explain how the incorrectly entered data value affects the correlation coefficient.

37. Draw Your Data! Consider the four data sets shown below.

Data Set 1		Data Set 2		Data Set 3		Data Set 4	
x	y	x	y	x	y	x	y
10	8.04	10	9.14	10	7.46	8	6.58
8	6.95	8	8.14	8	6.77	8	5.76
13	7.58	13	8.74	13	12.74	8	7.71
9	8.81	9	8.77	9	7.11	8	8.84
11	8.33	11	9.26	11	7.81	8	8.47
14	9.96	14	8.10	14	8.84	8	7.04
6	7.24	6	6.13	6	6.08	8	5.25
4	4.26	4	3.10	4	5.39	8	5.56
12	10.84	12	9.13	12	8.15	8	7.91
7	4.82	7	7.26	7	6.42	8	6.89
5	5.68	5	4.47	5	5.73	19	12.50

Source: Frank Anscombe. "Graphs in Statistical Analysis," *American Statistician* 27: 17–21, 1993.

(a) Compute the linear correlation coefficient for each data set.

(b) Draw a scatter diagram for each data set. Conclude that linear correlation coefficients and scatter diagrams must be used together in any statistical analysis of bivariate data.

38. Predicting Winning Percentage The ultimate goal in any sport (besides having fun) is to win. One measure of how well a team does is winning percentage. In baseball, a lot of effort goes into figuring out the variables that help to predict a team's winning percentage. Go to www.pearsonhighered.com/sullivanstats to obtain the data file 4_1_38 using the file format of your choice for the version of the text you are using. The data represent the winning percentages of teams in both the American League (AL) and National League (NL). The difference between the AL and NL is that teams in the AL allow a designated hitter to hit for the pitcher. Pitchers are typically poor hitters, so teams in the NL tend to score fewer runs.

(a) Which variable is the best predictor of winning percentage? What does the relation between this variable and winning percentage suggest?

(b) Suppose a manager wants to score lots of runs. What variable best predicts runs scored?

(c) What does the negative correlation between runs and at-bats per home run suggest?

39. Diversification One basic theory of investing is diversification. The idea is that you want to have a basket of stocks that do not all "move in the same direction." In other words, if one investment goes down, you don't want a second investment in your portfolio that is also likely to go down. One hallmark of a good portfolio is a low correlation between investments. Go to www.pearsonhighered.com/sullivanstats to obtain the data file 4_1_39 using the file format of your choice for the version of the text you are using. The data represent the annual rates of return for various stocks. If you only wish to invest in two stocks, which two would you select if your goal is to have low correlation between the two investments? Which two would you select if your goal is to have one stock go up when the other goes down?

Source: Yahoo!Finance

40. Lyme Disease versus Drownings Lyme disease is an inflammatory disease that results in a skin rash and flulike symptoms. It is transmitted through the bite of an infected deer tick. The following data represent the number of reported cases of Lyme disease and the number of drowning deaths for a rural county in the United States.

Month	J	F	M	A	M	J	J	A	S	O	N	D
Cases of Lyme Disease	3	2	2	4	5	15	22	13	6	5	4	1
Drowning Deaths	0	1	2	1	2	9	16	5	3	3	1	0

(a) Draw a scatter diagram of the data using cases of Lyme disease as the explanatory variable.

(b) Compute the correlation coefficient for the data.

(c) Based on your results from parts (a) and (b), what type of relation exists between the number of reported cases of Lyme disease and drowning deaths? Do you believe that an increase in cases of Lyme disease causes an increase in drowning deaths? What is a likely lurking variable between cases of Lyme disease and drowning deaths?

NW 41. Television Stations and Life Expectancy Based on data obtained from the *CIA World Factbook*, the linear correlation coefficient between the number of television stations in a country and the life expectancy of residents of the country is 0.599. What does this correlation imply? Do you believe that the more television stations a country has, the longer its population can expect to live? Why or why not? What is a likely lurking variable between number of televisions and life expectancy?

42. Obesity In a study published in the *Journal of the American Medical Association*, researchers found that the length of time a mother breast-feeds is negatively associated with the likelihood a child is obese. In an interview, the head investigator stated, "It's not clear whether breast milk has obesity-preventing properties or the women who are breast-feeding are less likely to have obese kids because they are less likely to be obese themselves." Using the researcher's statement, explain what might be wrong with concluding that breast-feeding prevents obesity. Identify some lurking variables in the study.

43. Crime Rate and Cell Phones The linear correlation between violent crime rate and percentage of the population that has a cell phone is −0.918 for years since 1995. Do you believe that increasing the percentage of the population that has a cell phone will decrease the violent crime rate? What might be a lurking variable between percentage of the population with a cell phone and violent crime rate?

44. Faulty Use of Correlation On the basis of the scatter diagram on the next page, explain what is wrong with the following statement: "Because the linear correlation coefficient between age and median income is 0.012, there is no relation between age and median income."

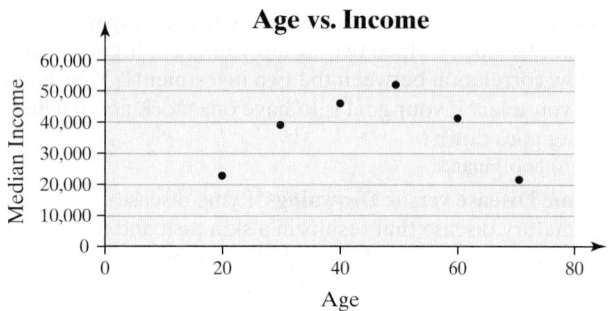

Age vs. Income

45. Influential Consider the following set of data:

x	2.2	3.7	3.9	4.1	2.6	4.1	2.9	4.7
y	3.9	4.0	1.4	2.8	1.5	3.3	3.6	4.9

(a) Draw a scatter diagram of the data and compute the linear correlation coefficient.

(b) Draw a scatter diagram of the data and compute the linear correlation coefficient with the additional data point (10.4, 9.3). Comment on the effect the additional data point has on the linear correlation coefficient. Explain why correlations should always be reported with scatter diagrams.

46. Transformations Consider the following data set:

x	5	6	7	7	8	8	8	8
y	4.2	5	5.2	5.9	6	6.2	6.1	6.9
x	9	9	10	10	11	11	12	12
y	7.2	8	8.3	7.4	8.4	7.8	8.5	9.5

(a) Draw a scatter diagram with the x-axis starting at 0 and ending at 30 and with the y-axis starting at 0 and ending at 20.

(b) Compute the linear correlation coefficient.

(c) Now multiply both x and y by 2.

(d) Draw a scatter diagram of the new data with the x-axis starting at 0 and ending at 30 and with the y-axis starting at 0 and ending at 20. Compare the scatter diagrams.

(e) Compute the linear correlation coefficient.

(f) Conclude that multiplying each value in the data set by a nonzero constant does not affect the correlation between the variables. Explain why this is the case.

47. Graduation Rates Go to www.pearsonhighered.com/ sullivanstats to obtain the data file 4_1_47 using the file format of your choice for the version of the text you are using. The variable "2013 Cost" represents the four-year cost including tuition, supplies, room and board of attending the school. The variable "Annual ROI" represents the return on investment for graduates of the school. It essentially represents how much you would earn on the investment of attending the school. The variable "Grad Rate" represents the graduation rate of the school.

(a) Draw a scatter diagram between 2013 Cost and Grad Rate treating 2013 Cost as the explanatory variable. What does the scatter diagram suggest?

(b) Find the linear correlation coefficient between 2013 Cost and Grad Rate. What does this suggest?

(c) Is there a linear association between cost and return on investment? Explain. What does your analysis suggest?

48. RateMyProfessors.com Professors Theodore Coladarci and Irv Kornfield from the University of Maine found a correlation of 0.68 between responses to questions on the RateMyProfessors.com website and typical in-class evaluations. Use this correlation to make an argument in favor of the validity of RateMyProfessors.com as a legitimate evaluation tool. RateMyProfessors.com also has a chili pepper icon, which is meant to indicate a "hotness scale" for the professor. This hotness scale serves as a proxy for the sexiness of the professor. It was found that the correlation between quality and sexiness is 0.64. In addition, it was found that the correlation between easiness of the professor and quality is 0.85 for instructors with at least 70 posts. Use this information to make an argument against RateMyProfessors.com as a legitimate evaluation tool.
Source: Theodore Coladarci and Irv Kornfield. "RateMyProfessors. com versus Formal In-class Student Evaluations of Teaching," *Practical Assessment, Research, & Evaluation,* 12:6, May, 2007.

Explaining the Concepts

49. What does it mean to say that the linear correlation coefficient between two variables equals 1? What would the scatter diagram look like?

50. What does it mean if $r = 0$?

51. Explain what is wrong with the following statement: "We have concluded that a high correlation exists between the gender of drivers and rates of automobile accidents." Suggest a better way to write the sentence.

52. Write a paragraph that explains the concept of correlation. Include a discussion of the role that $x_i - \bar{x}$ and $y_i - \bar{y}$ play in the computation.

53. Explain the difference between correlation and causation. When is it appropriate to state that the correlation implies causation?

54. Draw a scatter diagram that might represent the relation between the number of minutes spent exercising on an elliptical and calories burned. Draw a scatter diagram that might represent the relation between number of hours each week spent on Facebook and grade point average.

55. Suppose you work a part-time job and earn $15 per hour. Draw a scatter diagram that might represent the relation between your gross pay and hours worked. Is this a deterministic relation or a probabilistic relation?

56. Suppose that two variables, X and Y, are negatively associated. Does this mean that above-average values of X will always be associated with below-average values of Y? Explain.

4.2   Least-Squares Regression

Preparing for This Section Before getting started, review the following:

- Lines (Appendix B on CD, pp. B1–B5)

Objectives

① Find the least-squares regression line and use the line to make predictions

② Interpret the slope and the y-intercept of the least-squares regression line

③ Compute the sum of squared residuals

Once the scatter diagram and linear correlation coefficient show that two variables have a linear relation, we can find a linear equation that describes this relation. One way to do this is to select two points from the data that appear to provide a good fit (the line drawn appears to describe the relation between the two variables well) and to find the equation of the line through these two points.

EXAMPLE 1   Finding an Equation That Describes Linearly Related Data

Problem The data in Table 5 represent the club-head speed and the distance a golf ball travels for eight swings of the club. We found that these data are linearly related in Section 4.1.

CAUTION!

Example 1 *does not* present the least-squares method. It is used to set up the concepts of determining the line that best fits the data.

Table 5

Club-head Speed (mph) x	Distance (yd) y	(x, y)
100	257	$(100, 257)$
102	264	$(102, 264)$
103	274	$(103, 274)$
101	266	$(101, 266)$
105	277	$(105, 277)$
100	263	$(100, 263)$
99	258	$(99, 258)$
105	275	$(105, 275)$

Source: Paul Stephenson, student at Joliet Junior College

In Other Words
A *good fit* means that the line drawn appears to describe the relation between the two variables well.

(a) Find a linear equation that relates club-head speed, x (the explanatory variable), and distance, y (the response variable), by selecting two points and finding the equation of the line containing the points.
(b) Graph the line on the scatter diagram.
(c) Use the equation to predict the distance a golf ball will travel if the club-head speed is 104 miles per hour.

Approach
(a) Perform the following steps:

Step 1 Select two points so that a line drawn through the points appears to give a good fit. Call the points (x_1, y_1) and (x_2, y_2). See the scatter diagram in Figure 1 on page 182.

Step 2 Find the slope of the line containing the two points using $m = \dfrac{y_2 - y_1}{x_2 - x_1}$.

Step 3 Use the point–slope formula, $y - y_1 = m(x - x_1)$, to find the equation of the line through the points selected in Step 1. Express the equation in the form $y = mx + b$, where m is the slope and b is the y-intercept.

(continued)

(b) Draw a line through the points selected in Step 1 of part (a).
(c) Let $x = 104$ in the equation found in part (a).

Solution
(a) *Step 1* We select $(x_1, y_1) = (99, 258)$ and $(x_2, y_2) = (105, 275)$, because a line drawn through these two points seems to give a good fit based on the scatter diagram shown in Figure 10.

Step 2 $\quad m = \dfrac{y_2 - y_1}{x_2 - x_1} = \dfrac{275 - 258}{105 - 99} = \dfrac{17}{6} = 2.8333$

Step 3 Use the point–slope formula to find the equation of the line.

$$y - y_1 = m(x - x_1)$$
$$y - 258 = 2.8333(x - 99) \qquad m = 2.8333, x_1 = 99, y_1 = 258$$
$$y - 258 = 2.8333x - 280.4967$$
$$y = 2.8333x - 22.4967 \qquad\qquad\qquad\qquad \textbf{(1)}$$

> **CAUTION!**
> The line found in Step 3 of Example 1 is not the least-squares regression line.

> **CAUTION!**
> Unless otherwise noted, we will round the slope and y-intercept to four decimal places. As always, do not round until the last computation.

The slope of the line is 2.8333, and the y-intercept is -22.4967.

(b) Figure 10 shows the scatter diagram along with the line drawn through the points $(99, 258)$ and $(105, 275)$.
(c) Let $x = 104$ in equation (1) to predict the distance.

$$y = 2.8333(104) - 22.4967$$
$$= 272.2 \text{ yards}$$

We predict that a golf ball will travel 272.2 yards when it is hit with a club-head speed of 104 miles per hour.

Figure 10

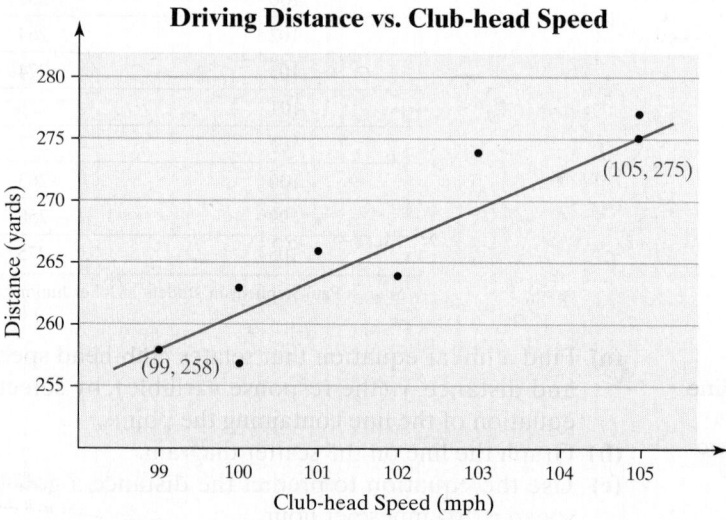

Driving Distance vs. Club-head Speed

- Now Work Problems 7(a)–(c)

---

① **Find the Least-Squares Regression Line and Use the Line to Make Predictions**

The line that we found in Example 1 appears to describe the relation between club-head speed and distance quite well. However, is there a line that fits the data better? Is there a line that fits the data *best*?

Whenever we attempt to determine the best of something, we need a criterion for determining best. For example, suppose that we are trying to identify the best domestic car. What do we mean by best? Best gas mileage? Best reliability? When describing the relation between two variables, we need a criterion for determining the best line. Consider Figure 11. Each y-coordinate on the line corresponds to a predicted distance for a given club-head speed. For example, if club-head speed

In Other Words
The residual represents how close our prediction comes to the actual observation. The smaller the residual, the better the prediction.

is 103 miles per hour, the predicted distance is $2.8333(103) - 22.4967 = 269.3$ yards. The observed distance for this club-head speed is 274 yards. The difference between the observed and predicted values of y is the error, or **residual**. For a club-head speed of 103 miles per hour, the residual is

$$\text{Residual} = \text{observed } y - \text{predicted } y$$
$$= 274 - 269.3$$
$$= 4.7 \text{ yards}$$

The residual for a club-head speed of 103 miles per hour is represented in Figure 11 by the length of the vertical line segment drawn between the points $(103, 269.3)$ and $(103, 274)$.

Figure 11

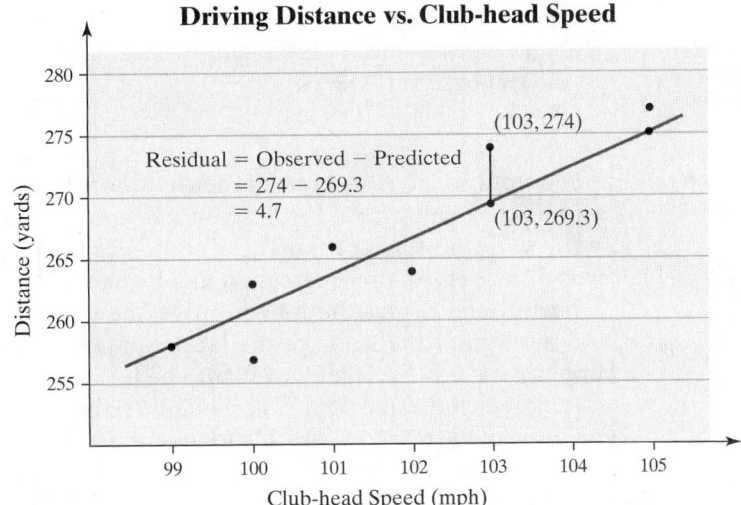

The criterion to determine the line that *best* describes the relation between two variables is based on the residuals. The most popular technique for making the residuals as small as possible is the *method of least squares*, developed by Adrien Marie Legendre.

Definition

Least-Squares Regression Criterion

The **least-squares regression line** is the line that minimizes the sum of the squared errors (or residuals). This line minimizes the sum of the squared vertical distance between the observed values of y and those predicted by the line, $\hat{y}$ (read "y-hat"). We represent this as "minimize Σ residuals²".

The advantage of the least-squares criterion is that it allows for statistical inference on the predicted value and slope (Chapter 14). Another advantage of the least-squares criterion is explained by Legendre in his text *Nouvelles méthodes pour la détermination des orbites des comètes*, published in 1806.

Of all the principles that can be proposed for this purpose, I think there is none more general, more exact, or easier to apply, than that which we have used in this work; it consists of making the sum of squares of the errors a *minimum*. By this method, a kind of equilibrium is established among the errors which, since it prevents the extremes from dominating, is appropriate for revealing the state of the system which most nearly approaches the truth.

The least-squares regression criterion leads to the following formulas for obtaining the *least-squares regression line.*

Historical Note

Adrien Marie Legendre was born on September 18, 1752, into a wealthy family and was educated in mathematics and physics at the College Mazarin in Paris. From 1775 to 1780, he taught at École Militaire. In 1783, Legendre was appointed an adjoint in the Académie des Sciences. He became a member of the committee of the Académie des Sciences in 1791 and was charged with the task of standardizing weights and measures. The committee worked to compute the length of the meter. During the French Revolution, Legendre lost his small fortune. In 1794, Legendre published *Éléments de géométrie*, which was the leading elementary text in geometry for around 100 years. In 1806, Legendre published a book on orbits, in which he developed the theory of least squares. He died on January 10, 1833.

The Least-Squares Regression Line

The equation of the least-squares regression line is given by

$$\hat{y} = b_1 x + b_0$$

where

$$b_1 = r \cdot \frac{s_y}{s_x} \text{ is the } \textbf{slope} \text{ of the least-squares regression line*} \qquad (2)$$

and

$$b_0 = \bar{y} - b_1 \bar{x} \text{ is the } \textbf{y-intercept} \text{ of the least-squares regression line} \qquad (3)$$

Note: $\bar{x}$ is the sample mean and s_x is the sample standard deviation of the explanatory variable x; $\bar{y}$ is the sample mean and s_y is the sample standard deviation of the response variable y.

The notation $\hat{y}$ is used in the least-squares regression line to remind us that it is a predicted value of y for a given value of x. The least-squares regression line, $\hat{y} = b_1 x + b_0$, always contains the point $(\bar{x}, \bar{y})$. This property can be useful when drawing the least-squares regression line by hand.

Since s_y and s_x must both be positive, the sign of the linear correlation coefficient, r, and the sign of the slope of the least-squares regression line, b_1, are the same. For example, if r is positive, then b_1 will also be positive.

The predicted value of y, $\hat{y}$, has an interesting interpretation. It is an estimate of the mean value of the response variable for any value of the explanatory variable. For example, suppose a least-squares regression equation is obtained that relates students' grade point average (GPA) to the number of hours studied each week. If the equation results in a predicted GPA of 3.14 when a student studies 20 hours each week, we would say the mean GPA of *all* students who study 20 hours each week is 3.14.

EXAMPLE 2   Finding the Least-Squares Regression Line by Hand

Problem Find the least-squares regression line for the data in Table 2 from Section 4.1.

Approach In Example 2 of Section 4.1 (page 185), we found

$$r = -0.946, \bar{x} = 4, s_x = 2.44949, \bar{y} = 10, \text{ and } s_y = 5.612486$$

Substitute the values in Formulas (2) and (3) to find the slope and intercept of the least-squares regression line.

Solution Substitute $r = -0.946$, $s_x = 2.44949$, and $s_y = 5.612486$ into Formula (2):

$$b_1 = r \cdot \frac{s_y}{s_x} = -0.946 \cdot \frac{5.612486}{2.44949} = -2.1676$$

Substitute $\bar{x} = 4, \bar{y} = 10$, and $b_1 = -2.1676$ into Formula (3):

$$b_0 = \bar{y} - b_1 \bar{x} = 10 - (-2.1676)(4) = 18.6704$$

The least-squares regression line is

$$\hat{y} = -2.1676x + 18.6704$$

•

*An equivalent formula is

$$b_1 = \frac{S_{xy}}{S_{xx}} = \frac{\sum x_i y_i - \dfrac{\left(\sum x_i\right)\left(\sum y_i\right)}{n}}{\sum x_i^2 - \dfrac{\left(\sum x_i\right)^2}{n}}$$

Rounding Rule for the Slope and Intercept

Throughout the course, we agree to round the slope and y-intercept of the least-squares regression equation to four decimal places.

In Example 2, we obtained the least-squares regression line by hand. Why? Mainly to see the role that the correlation coefficient, r, the standard deviation of y, and the standard deviation of x play in finding the slope. Plus, from the y-intercept formula we learn that all regression lines travel through the point $(\bar{x}, \bar{y})$. In practice, however, the least-squares regression line is found using technology.

EXAMPLE 3 Finding the Least-Squares Regression Line Using Technology

Problem Use the golf data in Table 5.
(a) Find the least-squares regression line.
(b) Draw the least-squares regression line on the scatter diagram of the data.
(c) Predict the distance a golf ball will travel when hit with a club-head speed of 103 miles per hour (mph).
(d) Determine the residual for the predicted value found in part (c). Is the distance in Table 5 above average or below average among all balls hit with a swing speed of 103 mph?

Approach Because technology plays a role in obtaining the least-squares regression line, we will use a TI-84 Plus C graphing calculator, Minitab, Excel, and StatCrunch to obtain the least-squares regression line. The steps for obtaining regression lines are given in the Technology Step-by-Step on page 205.

Solution
(a) Figure 12 shows the output obtained from the various technologies. The least-squares regression line is $\hat{y} = 3.1661x - 55.7966$.

Figure 12

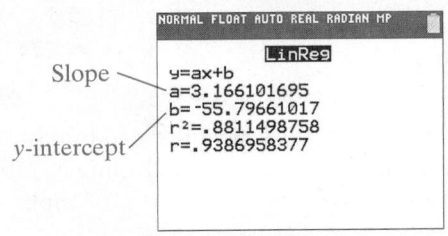

Slope
y-intercept

(a) TI-84 Plus C output

> ┌───┐
> │ CAUTION!   │
> │ Throughout the text, we will round   │
> │ the slope and y-intercept values to   │
> │ four decimal places. Predictions will   │
> │ be rounded to one more decimal   │
> │ place than the response variable.   │
> └───┘

```
The regression equation is
Distance (yards) = −55.8 + 3.17 Club Head Speed (mph)
Predictor                      Coef     SE Coef        T        P
Constant                      −55.80      48.37     −1.15    0.293
Club Head Speed (mph)         3.1661      0.4747      6.67    0.001

S = 2.88264      R − Sq = 88.1%     R − Sq(adj) = 86.1%
```

(b) Minitab output

	Coefficients	Standard Error	t Stat	P-value
Intercept	-55.79661017	48.37134953	-1.15351	0.292574312
Club-Head Speed	3.166101695	0.47470539	6.669614	0.000549825

(c) Excel output

Simple linear regression results:
Dependent Variable: Distance
Independent Variable: Club-Head Speed
Distance = -55.79661 + 3.1661017 Club-Head Speed
Sample size: 8
R (correlation coefficient) = 0.93869584
R-sq = 0.88114988
Estimate of error standard deviation: 2.8826385

(d) StatCrunch output

(b) Figure 13 on the next page shows the least-squares regression line drawn on the scatter diagram using StatCrunch.

(continued)

Historical Note

Sir Francis Galton was born on February 16, 1822. Galton came from a wealthy and well-known family. Charles Darwin was his first cousin. Galton studied medicine at Cambridge. After receiving a large inheritance, he left the medical field and traveled the world. He explored Africa from 1850 to 1852. In the 1860s, his study of meteorology led him to discover anticyclones. Influenced by Darwin, Galton always had an interest in genetics and heredity. He studied heredity through experiments with sweet peas. He noticed that the weight of the "children" of the "parent" peas reverted or *regressed* to the mean weight of all peas—hence, the term *regression analysis*. Galton died January 17, 1911.

• **Now Work Problems 17(a), (c), and (d)**

Figure 13

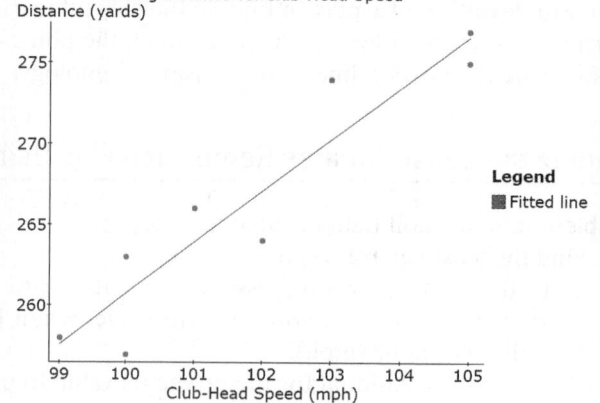

Driving Distance vs. Club-Head Speed

(c) Let $x = 103$ in the least-squares regression equation $\hat{y} = 3.1661x - 55.7966$ to predict the distance the ball will travel when hit with a club-head speed of 103 mph.

$$\hat{y} = 3.1661(103) - 55.7966$$
$$= 270.3 \text{ yards}$$

We predict the distance the golf ball will travel is 270.3 yards.

(d) From the data in Table 1 from Section 4.1, the observed distance the ball traveled when the swing speed is 103 mph was 274 yards. So,

$$\text{Residual} = y - \hat{y} \qquad \text{Residual = observed } y - \text{ predicted } y$$
$$= 274 - 270.3$$
$$= 3.7 \text{ yards}$$

Because the residual is positive (the observed value of 274 yards is greater than the predicted value of 270.3 yards), the distance of 274 yards is above average for a swing speed of 103 mph. •

2 Interpret the Slope and the y-Intercept of the Least-Squares Regression Line

Interpretation of Slope In algebra, we learned that the definition of slope is $\dfrac{\text{rise}}{\text{run}}$ or $\dfrac{\text{change in } y}{\text{change in } x}$. If a line has slope $\frac{2}{3}$, then if x increases by 3, y will increase by 2. Or if the slope of a line is $-4 = \dfrac{-4}{1}$, then if x increases by 1, y will decrease by 4.

Interpreting slope for least-squares regression lines has a minor twist. Statistical models such as a least-squares regression equation are *probabilistic*. This means that any predictions or interpretations made as a result of the model are based on uncertainty. Therefore, when we interpret the slope of a least-squares regression equation, we do not want to imply that there is 100% certainty behind the interpretation. For example, the slope of the least-squares regression line from Example 3 is 3.1661 yards per mph. In algebra, we would interpret the slope to mean "if x increases by 1 mph, then y will increase by 3.1661 yards." In statistics, this interpretation is close, but not quite accurate, because increasing the club-head speed by 1 mph does not guarantee the distance the ball travels will increase by 3.1661 yards. Instead, over the course of data we observed, an increase in club-head speed of 1 mph increased distance 3.1661 yards, *on average*—sometimes the ball might travel a shorter additional distance, sometimes a longer additional distance, but on average this is the change in distance. So two interpretations of slope are acceptable:

> If club-head speed increases by 1 mile per hour, the distance the golf ball travels increases by 3.1661 yards, on average.
>
> or
>
> If club-head speed increases by 1 mile per hour, the expected distance the golf ball will travel increases by 3.1661 yards.

Interpretation of the y-Intercept   The y-intercept of any line is the point where the graph intersects the vertical axis. It is found by letting $x = 0$ in an equation and solving for y. In general, we interpret a y-intercept as the value of the response variable when the value of the explanatory variable is 0. To interpret the y-intercept, we must first ask two questions:

1. Is 0 a reasonable value for the explanatory variable?
2. Do any observations near $x = 0$ exist in the data set?

If the answer to either of these questions is no, we do not interpret the y-intercept. In the regression equation of Example 3, a swing speed of 0 miles per hour does not make sense, so we do not interpret the y-intercept.

The second condition for interpreting the y-intercept is especially important because we should not use the regression model to make predictions **outside the scope of the model**, meaning we should not use the regression model to make predictions for values of the explanatory variable that are much larger or much smaller than those observed. This is a dangerous practice because we cannot be certain of the behavior of data for which we have no observations.

For example, we should not use the line in Example 3 to predict distance when club-head speed is 140 mph. The highest observed club-head speed is 105 mph. The linear relation between distance and club-head speed might not continue. See Figure 14.

> **CAUTION!**
> Be careful when using the least-squares regression line to make predictions for values of the explanatory variable that are much larger or much smaller than those observed.

Figure 14

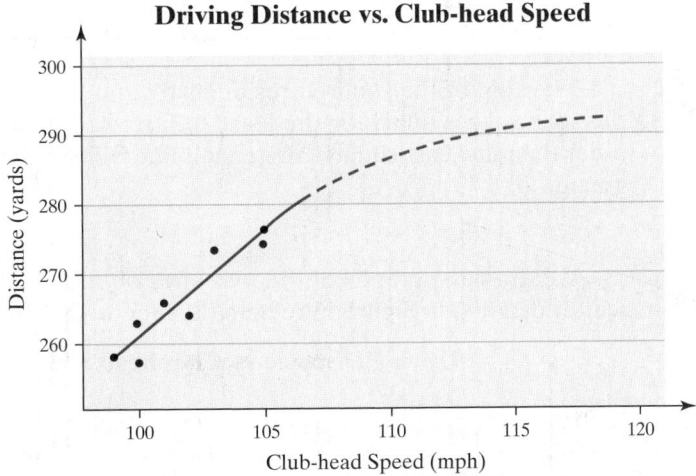

Driving Distance vs. Club-head Speed

● Now Work Problem 13

Predictions When There Is No Linear Relation
When the correlation coefficient indicates no linear relation between the explanatory and response variables, and the scatter diagram indicates no relation at all between the variables, then use the mean value of the response variable as the predicted value so that $\hat{y} = \bar{y}$.

③ Compute the Sum of Squared Residuals

Recall that the least-squares regression line is the line that minimizes the sum of the squared residuals. This means that the sum of the squared residuals, $\Sigma \text{residuals}^2$, is smaller for the least-squares line than for any other line that may describe the relation between the two variables. In particular, the sum of the squared residuals is smaller for the least-squares regression line in Example 3 than for the line obtained in Example 1. It is worthwhile to verify this result.

EXAMPLE 4 Comparing the Sum of Squared Residuals

Problem Compare the sum of squared residuals for the lines in Examples 1 and 3.

Approach Use a table to compute Σresiduals^2 using the predicted values of y, $\hat{y}$, for the equations in Examples 1 and 3.

Solution Table 6 contains the value of the explanatory variable in Column 1. Column 2 contains the corresponding response variable. Column 3 contains the predicted values using the equation found in Example 1, $\hat{y} = 2.8333x - 22.4967$. In Column 4, we compute the residuals for each observation: residual = observed y − predicted $y = y - \hat{y}$. For example, the first residual is $y - \hat{y} = 257 - 260.8 = -3.8$. Column 5 contains the squares of the residuals. Column 6 contains the predicted values using the least-squares regression equation found in Example 3: $\hat{y} = 3.1661x - 55.7966$. Column 7 represents the residuals for each observation, and Column 8 represents the squared residuals.

Table 6

Club-head Speed (mph)	Distance (yd)	Example 1 ($\hat{y} = 2.8333x - 22.4967$)	Residual $y - \hat{y}$	Residual^2 $(y - \hat{y})^2$	Example 3 ($\hat{y} = 3.1661x - 55.7966$)	Residual $y - \hat{y}$	Residual^2 $(y - \hat{y})^2$
100	257	260.8	−3.8	14.44	260.8	−3.8	14.44
102	264	266.5	−2.5	6.25	267.1	−3.1	9.61
103	274	269.3	4.7	22.09	270.3	3.7	13.69
101	266	263.7	2.3	5.29	264.0	2.0	4.00
105	277	275.0	2.0	4.00	276.6	0.4	0.16
100	263	260.8	2.2	4.84	260.8	2.2	4.84
99	258	258.0	0.0	0.00	257.6	0.4	0.16
105	275	275.0	0.0	0.00	276.6	−1.6	2.56
				Σresidual^2 $= 56.91$			Σresidual^2 $= 49.46$

The sum of the squared residuals for the line in Example 1 is 56.91; the sum of the squared residuals for the least-squares regression line is 49.46. Again, any line other than the least-squares regression line will have a sum of squared residuals that is greater than 49.46.

●

● **Now Work Problems 7(d)–(h)**

We draw the graphs of the two lines obtained in Examples 1 and 3 on the same scatter diagram in Figure 15 to help the reader visualize the difference.

Figure 15

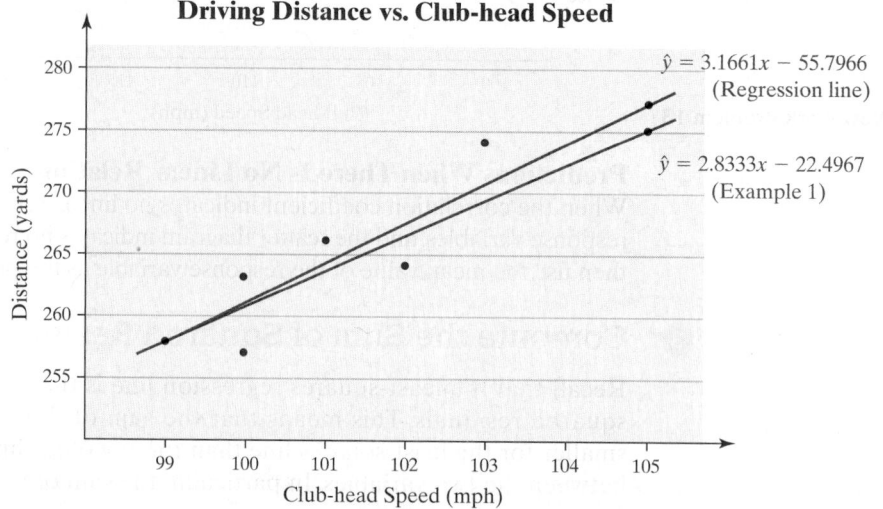

Driving Distance vs. Club-head Speed

$\hat{y} = 3.1661x - 55.7966$ (Regression line)

$\hat{y} = 2.8333x - 22.4967$ (Example 1)

Distance (yards)

Club-head Speed (mph)

Technology Step-by-Step   Determining the Least-Squares Regression Line

TI-83/84 Plus
Use the same steps that were followed to obtain the correlation coefficient. (See Section 4.1.)

Minitab
1. With the explanatory variable in C1 and the response variable in C2, select the **Stat** menu and highlight **Regression**. Select **Fitted Line Plot**...
2. Select the explanatory (predictor) variable and response variable and click OK.

Excel
1. Enter the explanatory variable in column A and the response variable in column B.
2. Select the **Data** menu and then **Data Analysis**.
3. Select the **Regression** option.
4. With the cursor in the Y-range cell, highlight the column that contains the response variable. With the cursor in

the X-range cell, highlight the column that contains the explanatory variable. Select the Output Range. Press OK.

StatCrunch
1. If necessary, enter the explanatory variable in column var1 and the response variable in column var2. Name each column variable.
2. Select Stat, highlight Regression, and select Simple Linear.
3. Choose the explanatory variable for the X variable and the response variable for the Y variable. If you want, enter a value of the explanatory variable to Predict Y for X. If you want the least-squares regression line drawn on the scatter diagram, highlight Fitted line plot under Graphs. Click Compute!.

4.2 Assess Your Understanding

Vocabulary and Skill Building

1. The difference between the observed and predicted value of y is the error, or _____.

2. If the linear correlation between two variables is negative, what can be said about the slope of the regression line?

3. *True or False*: The least-squares regression line always travels through the point $(\bar{x}, \bar{y})$.

4. If the linear correlation coefficient is 0, what is the equation of the least-squares regression line?

 5. For the data set

x	0	2	3	5	6	6
y	5.8	5.7	5.2	2.8	1.9	2.2

(a) Draw a scatter diagram. Comment on the type of relation that appears to exist between x and y.
(b) Given that $\bar{x} = 3.6667$, $s_x = 2.4221$, $\bar{y} = 3.9333$, $s_y = 1.8239$, and $r = -0.9477$, determine the least-squares regression line.
(c) Graph the least-squares regression line on the scatter diagram drawn in part (a).

6. For the data set

x	2	4	8	8	9
y	1.4	1.8	2.1	2.3	2.6

(a) Draw a scatter diagram. Comment on the type of relation that appears to exist between x and y.
(b) Given that $\bar{x} = 6.2$, $s_x = 3.03315$, $\bar{y} = 2.04$, $s_y = 0.461519$, and $r = 0.957241$, determine the least-squares regression line.
(c) Graph the least-squares regression line on the scatter diagram drawn in part (a).

In Problems 7–12:
(a) *By hand, draw a scatter diagram treating x as the explanatory variable and y as the response variable.*
(b) *Select two points from the scatter diagram and find the equation of the line containing the points selected.*
(c) *Graph the line found in part (b) on the scatter diagram.*
(d) *By hand, determine the least-squares regression line.*
(e) *Graph the least-squares regression line on the scatter diagram.*
(f) *Compute the sum of the squared residuals for the line found in part (b).*
(g) *Compute the sum of the squared residuals for the least-squares regression line found in part (d).*
(h) *Comment on the fit of the line found in part (b) versus the least-squares regression line found in part (d).*

NW 7.

x	3	4	5	7	8
y	4	6	7	12	14

8.

x	3	5	7	9	11
y	0	2	3	6	9

9.

x	−2	−1	0	1	2
y	−4	0	1	4	5

10.

x	−2	−1	0	1	2
y	7	6	3	2	0

11.

x	20	30	40	50	60
y	100	95	91	83	70

12.

x	5	10	15	20	25
y	2	4	7	11	18

NW 13. Income and Education In Problem 15 from Section 4.1, a scatter diagram and correlation coefficient suggested there is a linear relation between the percentage of individuals who have at least a bachelor's degree and median income in the states. In fact, the least-squares regression equation is $\hat{y} = 703.5x + 14{,}920$ where y is the median income and x is the percentage of individuals 25 years and older with at least a bachelor's degree in the state.

(a) Predict the median income of a state in which 25% of adults 25 years and older have at least a bachelor's degree.

(b) In North Dakota, 27.1 percent of adults 25 years and older have at least a bachelor's degree. The median income in North Dakota is $37,193. Is this income higher than what you would expect? Why?

(c) Interpret the slope.

(d) Explain why it does not make sense to interpret the intercept.

14. You Explain It! Study Time and Exam Scores After the first exam in a statistics course, Professor Katula surveyed 14 randomly selected students to determine the relation between the amount of time they spent studying for the exam and exam score. She found that a linear relation exists between the two variables. The least-squares regression line that describes this relation is $\hat{y} = 6.3333x + 53.0298$.

(a) Predict the exam score of a student who studied 2 hours.

(b) Interpret the slope.

(c) What is the mean score of students who did not study?

(d) A student who studied 5 hours for the exam scored 81 on the exam. Is this student's exam score above or below average among all students who studied 5 hours?

15. Age Gap at Marriage Is there a relation between the age difference between husband/wives and the percent of a country that is literate. Researchers found the least-squares regression between age difference (husband age minus wife age), y, and literacy rate (percent of the population that is literate), x, is $\hat{y} = -0.0527x + 7.1$. The model applied for $18 \le x \le 100$.
Source: Xu Zhang and Solomon W. Polachek, State University of New York at Binghamton "The Husband-Wife Age Gap at First Marriage: A Cross-Country Analysis"

(a) Interpret the slope.

(b) Does it make sense to interpret the y-intercept? Explain.

(c) Predict the age difference between husband/wife in a country where the literacy rate is 25 percent.

(d) Would it make sense to use this model to predict the age difference between husband/wife in a country where the literacy rate is 10 percent? Explain.

(e) The literacy rate in the United States is 99 percent and the age difference between husbands and wives is 2 years. Is this age difference above or below the average age difference among all countries whose literacy rate is 99 percent?

16. You Explain It! CO$_2$ and Energy Production The least-squares regression equation $\hat{y} = 0.7676x - 52.6841$ relates the carbon dioxide emissions (in hundred thousands of tons), y, and energy produced (hundred thousands of megawatts), x, for all countries in the world.
Source: CARMA (www.carma.org)

(a) Interpret the slope.

(b) Is the y-intercept of the model reasonable? Why? What would you expect the y-intercept of the model to equal? Why?

(c) The lowest energy-producing country is Rwanda, which produces 0.094 hundred thousand megawatts of energy. The highest energy-producing country is the United States, which produces 4190 hundred thousand megawatts of energy. Would it be reasonable to use this model to predict the CO$_2$ emissions of a country if it produces 6394 hundred thousand megawatts of energy? Why or why not?

(d) China produces 3260 hundred thousand megawatts of energy and emits 3120 hundred thousand tons of carbon dioxide. What is the residual for China? How would you interpret this residual?

Applying the Concepts

Problems 17–22 use the results from Problems 25–30 in Section 4.1.

NW 17. An Unhealthy Commute (Refer to Problem 25, Section 4.1.) The following data represent commute times (in minutes) and score on a well-being survey.

Commute Time (minutes), x	Gallup-Healthways Well-Being Index Composite Score, y
5	69.2
15	68.3
25	67.5
35	67.1
50	66.4
72	66.1
105	63.9

Source: The Gallup Organization

(a) Find the least-squares regression line treating the commute time, x, as the explanatory variable and the index score, y, as the response variable.

(b) Interpret the slope and y-intercept, if appropriate.

(c) Predict the well-being index of a person whose commute is 30 minutes.

(d) Suppose Barbara has a 20-minute commute and scores 67.3 on the survey. Is Barbara more "well-off" than the typical individual who has a 20-minute commute?

18. Credit Scores (Refer to Problem 26, Section 4.1.) An economist wants to determine the relation between one's FICO score, x, and the interest rate of a 36-month auto loan, y. The given data represent the interest rate (in percent) a bank would offer on a 36-month auto loan for various FICO scores.

Credit Score, x	Interest Rate (percent), y
545	18.982
595	17.967
640	12.218
675	8.612
705	6.680
750	5.150

Source: www.myfico.com

(a) Find the least-squares regression line treating the FICO score, x, as the explanatory variable and the interest rate, y, as the response variable.

(b) Interpret the slope and y-intercept, if appropriate.
Note: Credit scores have a range of 300 to 850.

(c) Predict the interest rate a person would pay if her FICO score were the median score of 723.

(d) Suppose Bob has a FICO score of 680 and he is offered an interest rate of 8.3%. Is this a good offer? Why?

19. Height versus Head Circumference (Refer to Problem 27, Section 4.1.) A pediatrician wants to determine the relation that exists between a child's height, x, and head circumference, y. She randomly selects 11 children from her practice, measures their heights and head circumferences, and obtains the following data.

Height (inches), x	Head Circumference (inches), y	Height (inches), x	Head Circumference (inches), y
27.75	17.5	26.5	17.3
24.5	17.1	27	17.5
25.5	17.1	26.75	17.3
26	17.3	26.75	17.5
25	16.9	27.5	17.5
27.75	17.6		

Source: Denise Slucki, student at Joliet Junior College

(a) Find the least-squares regression line treating height as the explanatory variable and head circumference as the response variable.

(b) Interpret the slope and y-intercept, if appropriate.

(c) Use the regression equation to predict the head circumference of a child who is 25 inches tall.

(d) Compute the residual based on the observed head circumference of the 25-inch-tall child in the table. Is the head circumference of this child above average or below average?

(e) Draw the least-squares regression line on the scatter diagram of the data and label the residual from part (d).

(f) Notice that two children are 26.75 inches tall. One has a head circumference of 17.3 inches; the other has a head circumference of 17.5 inches. How can this be?

(g) Would it be reasonable to use the least-squares regression line to predict the head circumference of a child who was 32 inches tall? Why?

20. American Black Bears (Refer to Problem 28, Section 4.1.) The American black bear (*Ursus americanus*) is one of eight bear species in the world. It is the smallest North American bear and the most common bear species on the planet. In 1969, Dr. Michael R. Pelton of the University of Tennessee initiated a long-term study of the population in Great Smoky Mountains National Park. One aspect of the study was to develop a model that could be used to predict a bear's weight (since it is not practical to weigh bears in the field). One variable thought to be related to weight is the length of the bear. The data in the next column represent the lengths of 12 American black bears.

Total Length (cm), x	Weight (kg), y
139.0	110
138.0	60
139.0	90
120.5	60
149.0	85
141.0	100
141.0	95
150.0	85
166.0	155
151.5	140
129.5	105
150.0	110

Source: www.fieldtripearth.org

(a) Find the least-squares regression line, treating total length as the explanatory variable and weight as the response variable.

(b) Interpret the slope and y-intercept, if appropriate.

(c) Suppose a 149.0-cm bear is captured in the field. Use the least-squares regression line to predict the weight of the bear.

(d) What is the residual of the 149.0-cm bear? Is this bear's weight above or below average for a bear of this length?

21. Weight of a Car versus Miles per Gallon (Refer to Problem 29, Section 4.1.) An engineer wants to determine how the weight of a car, x, affects gas mileage, y. The following data represent the weights of various domestic cars and their miles per gallon in the city for the 2015 model year.

Car	Weight (lb)	Miles per Gallon
Buick LaCrosse	4724	17
Cadillac XTS	4006	18
Chevrolet Cruze	3097	22
Chevrolet Impala	3555	19
Chrysler 300	4029	19
Dodge Charger	3934	19
Dodge Dart	3242	24
Ford Focus	2960	26
Ford Mustang	3530	19
Lincoln MKZ	3823	18

Source: Each manufacturer's website

(a) Find the least-squares regression line treating weight as the explanatory variable and miles per gallon as the response variable.

(b) Interpret the slope and y-intercept, if appropriate.

(c) A Cadillac CTS weighs 3649 pounds and gets 20 miles per gallon. Is the miles per gallon of a CTS above average or below average for cars of this weight?

(d) Would it be reasonable to use the least-squares regression line to predict the miles per gallon of a Toyota Prius, a hybrid gas and electric car? Why or why not?

22. Hurricanes (Refer to Problem 30, Section 4.1) The following data represent the maximum wind speed (in knots) and atmospheric pressure (in millibars) for a random sample of hurricanes that originated in the Atlantic Ocean.

Atmospheric Pressure (mb)	Wind Speed (knots)	Atmospheric Pressure (mb)	Wind Speed (knots)
993	50	1006	40
995	60	942	120
994	60	1002	40
997	45	986	50
1003	45	983	70
1004	40	994	65
1000	55	940	120
994	55	976	80
942	105	966	100
1006	30	982	55

Source: National Hurricane Center

(a) Find the least-squares regression line treating atmospheric pressure as the explanatory variable.
(b) Interpret the slope.
(c) Is it reasonable to interpret the y-intercept? Why?
(d) One hurricane had an atmospheric pressure of 997 mb. Is this hurricane's wind speed above or below average for a hurricane with this level of atmospheric pressure?

23. Cola Consumption vs. Bone Density Example 5 in Section 4.1 on page 188 discussed the effect of cola consumption on bone mineral density in the femoral neck of women.

(a) Find the least-squares regression line treating cola consumption per week as the explanatory variable.
(b) Interpret the slope.
(c) Interpret the intercept.
(d) Predict the bone mineral density of the femoral neck of a woman who consumes four colas per week.
(e) The researchers found a woman who consumed four colas per week to have a bone mineral density of 0.873 g/cm². Is this woman's bone mineral density above or below average among all women who consume four colas per week?
(f) Would you recommend using the model found in part (a) to predict the bone mineral density of a woman who consumes two cans of cola per day? Why?

24. Attending Class The following data represent the number of days absent, x, and the final grade, y, for a sample of college students in a general education course at a large state university.

No. of absences, x	0	1	2	3	4	5	6	7	8	9
Final grade, y	89.2	86.4	83.5	81.1	78.2	73.9	64.3	71.8	65.5	66.2

Source: College Teaching, 53(1), 2005.

(a) Find the least-squares regression line treating number of absences as the explanatory variable and final grade as the response variable.
(b) Interpret the slope and y-intercept, if appropriate.
(c) Predict the final grade for a student who misses five class periods and compute the residual. Is the final grade above or below average for this number of absences?

(d) Draw the least-squares regression line on the scatter diagram of the data.
(e) Would it be reasonable to use the least-squares regression line to predict the final grade for a student who has missed 15 class periods? Why or why not?

25. CEO Performance (Refer to Problem 31 in Section 4.1.) The following data represent the total compensation for 12 randomly selected chief executive officers (CEO) and the company's stock performance in 2013. Based on the analysis from Problem 31 in Section 4.1, what would be the predicted stock return for a company whose CEO made $15 million? What would be the predicted stock return for a company whose CEO made $25 million?

Company	Compensation (millions of dollars)	Stock Return (%)
Navistar International	14.53	75.43
Aviv REIT	4.09	64.01
Groupon	7.11	142.07
Inland Real Estate	1.05	32.72
Equity Lifestyles Properties	1.97	10.64
Tootsie Roll Industries	3.76	30.66
Catamaran	12.06	0.77
Packaging Corp of America	7.62	69.39
Brunswick	8.47	58.69
LKQ	4.04	55.93
Abbott Laboratories	20.87	24.28
TreeHouse Foods	6.63	32.21

Data from *Chicago Tribune,* June 1, 2014

26. Bear Markets (Refer to Problem 32, Section 4.1) A bear market in the stock market is defined as a condition in which market declines by 20% or more over the course of at least two months. The following data represent the number of months and percentage change in the S&P500 (a group of 500 stocks). Based on the analysis from Problem 32 in Section 4.1, what would be the predicted percent change in the S&P500 during a bear market that lasted 10 months? 30 months?

Months	Percent Change	Months	Percent Change
1.9	−44.57	12.1	−20.57
8.3	−44.29	14.9	−21.63
3.3	−32.86	6.5	−27.97
3.4	−42.54	8	−22.18
6.8	−61.81	18.1	−36.06
5.8	−40.60	21	−48.20
3.1	−29.43	20.7	−27.11
13.4	−31.81	3.4	−33.51
12.9	−54.47	18.2	−36.77
5.1	−24.44	9.3	−33.75
7.6	−28.69	13.6	−51.93
17.9	−34.42	2.1	−27.62
11.8	−28.47		

Source: Gold-Eagle

27. Male vs. Female Drivers (Refer to Problem 34, Section 4.1.) The following data represent the number of licensed drivers in various age groups and the number of fatal accidents within the age group by gender.

Age	Number of Male Licensed Drivers (000s)	Number of Fatal Crashes	Number of Female Licensed Drivers (000s)	Number of Fatal Crashes
<16	12	227	12	77
16–20	6,424	5,180	6,139	2,113
21–24	6,941	5,016	6,816	1,531
25–34	18,068	8,595	17,664	2,780
35–44	20,406	7,990	20,063	2,742
45–54	19,898	7,118	19,984	2,285
55–64	14,340	4,527	14,441	1,514
65–74	8,194	2,274	8,400	938
>74	4,803	2,022	5,375	980

Source: National Highway and Traffic Safety Institute

(a) Find the least-squares regression line for males treating number of licensed drivers as the explanatory variable, x, and number of fatal crashes, y, as the response variable. Repeat this procedure for females.

(b) Interpret the slope of the least-squares regression line for each gender. How might an insurance company use this information?

(c) Was the number of fatal accidents for 16- to 20-year-old males above or below average? Was the number of fatal accidents for 21- to 24-year-old-males above or below average? Was the number of fatal accidents for males greater than 74 years old above or below average? How might an insurance company use this information? Does the same relationship hold for females?

28. Graduation Rates Go to www.pearsonhighered.com/sullivanstats to obtain the data file 4_2_28 using the file format of your choice for the version of the text you are using. The variable "2013 Cost" represents the four-year cost including tuition, supplies, room and board. The variable "Annual ROI" represents the return on investment for graduates of the school. It essentially represents how much you would earn on the investment of attending the school. The variable "Grad Rate" represents the graduation rate of the school.

(a) In Problem 47 from Section 4.1, we saw that a scatter diagram between 2013 Cost and Grad Rate treating 2013 Cost as the explanatory variable suggested a positive association between the two variables. Find the least-squares regression line treating 2013 Cost as the explanatory variable. Round the slope to six decimal places.

(b) Interpret the slope.

(c) Washington University in St. Louis has a four-year cost of $234,600 and a graduation rate of 94%. Is Washington University's graduation rate above or below average among schools that cost $234,600?

(d) A scatter diagram between cost and return on investment (treating cost as the explanatory variable) suggested a negative association between the two variables. Find the least-squares regression line treating cost as the explanatory variable. Round the slope to seven decimal places.

(e) Interpret the slope.

(f) Washington University's return on investment is 6.1%. Is this above or below average among all schools that cost $234,600?

29. Putting It Together: Housing Prices One of the biggest factors in determining the value of a home is the square footage. The following data represent the square footage and selling price (in thousands of dollars) for a random sample of homes for sale in Naples, Florida in January 2015.

Square Footage, x	Selling Price ($000s), y
2204	379.9
3183	375
1128	189.9
1975	338
3101	619.9
2769	370
4113	627.7
2198	375
2609	425
1708	298.1
1786	271
3813	690.1

Source: zillow.com

(a) Which variable is the explanatory variable?

(b) Draw a scatter diagram of the data.

(c) Determine the linear correlation coefficient between square footage and asking price.

(d) Is there a linear relation between the square footage and asking price?

(e) Find the least-squares regression line treating square footage as the explanatory variable.

(f) Interpret the slope.

(g) Is it reasonable to interpret the y-intercept? Why?

(h) One home that is 1465 square feet is sold for $285,000. Is this home's price above or below average for a home of this size? What might be some reasons for this price?

30. Putting It Together: Smoking and Birth Weight It is well known that women should not smoke while pregnant, but what is the effect of smoking on a baby's birth weight? Researchers Ira M. Bernstein and associates "sought to estimate how the pattern of maternal smoking throughout pregnancy influences newborn size." To conduct this study, 160 pregnant, smoking women were enrolled in a prospective study. During the third trimester of pregnancy, the woman self-reported the number of cigarettes smoked. Urine samples were collected to measure cotinine levels (to assess nicotine levels). Birth weights (in grams) of the babies were obtained upon delivery.
Source: Ira M. Bernstein et. al. "Maternal Smoking and Its Association with Birthweight." *Obstetrics & Gynecology* 106 (Part 1) 5, 2005.

(a) The histogram on the next page, drawn in Minitab, shows the birth weight of babies whose mothers did not smoke in the third trimester of pregnancy (but did smoke prior to the third trimester). Describe the shape of the distribution. What is the class width of the histogram?

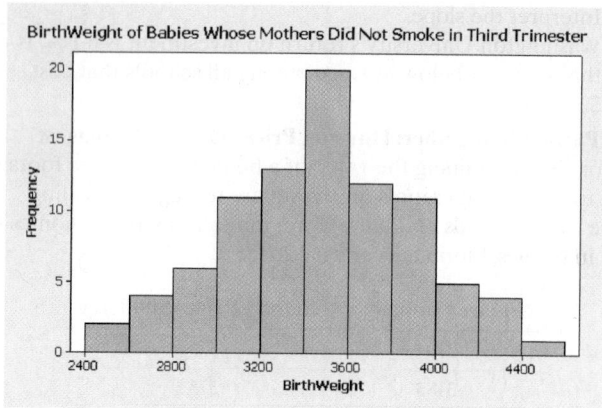

BirthWeight of Babies Whose Mothers Did Not Smoke in Third Trimester

(b) Is this an observational study or a designed experiment?

(c) What does it mean for the study to be prospective?

(d) Why would the researchers conduct a urinalysis to measure cotinine levels?

(e) What is the explanatory variable in the study? What is the response variable? Is the explanatory variable qualitative or quantitative? Is the response variable qualitative or quantitative?

(f) The scatter diagram of the data drawn using Minitab is shown below. What type of relation appears to exist between cigarette consumption in the third trimester and birth weight?

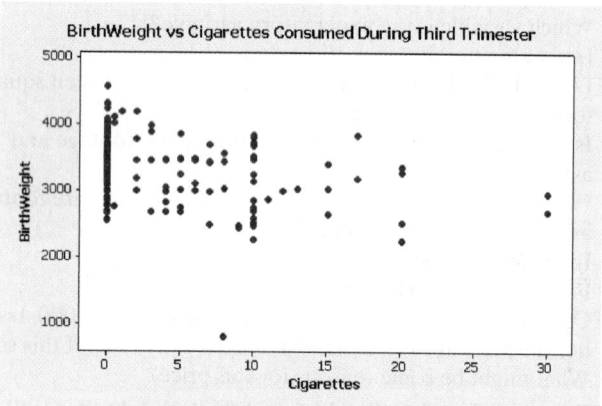

BirthWeight vs Cigarettes Consumed During Third Trimester

(g) Use the regression output from Minitab to report the least-squares regression line between cigarette consumption and birth weight.

Regression Analysis: BirthWeight versus Cigarettes

```
The regression equation is
BirthWeight = 3456 − 31.0 Cigarettes

Predictor        Coef    SE Coef       T       P
Constant      3456.04      45.44   76.07   0.000
Cigarettes    −31.014       6.498   −4.77   0.000

S = 482.603    R − Sq = 12.6%    R − Sq(adj) = 12.0%
```

(h) Interpret the slope.

(i) Interpret the y-intercept.

(j) Would you recommend using this model to predict the birth weight of a baby whose mother smoked 10 cigarettes per day during the third trimester? Why?

(k) Does this study demonstrate that smoking during the third trimester causes lower-birth-weight babies?

(l) Cite some lurking variables that may confound the results of the study.

Explaining the Concepts

31. What is a residual? What does it mean when a residual is positive?

32. Explain the phrase *outside the scope of the model*. Why is it dangerous to make predictions outside the scope of the model?

33. Explain the meaning of Legendre's quote given on page 199.

34. Explain what each point on the least-squares regression line represents.

35. Mark Twain, in his book *Life on the Mississippi* (1884), makes the following observation:

Therefore, the Mississippi between Cairo and New Orleans was twelve hundred and fifteen miles long one hundred and seventy-six years ago. It was eleven hundred and eighty after the cut-off of 1722. It was one thousand and forty after the American Bend cut-off. It has lost sixty-seven miles since. Consequently its length is only nine hundred and seventy-three miles at present.

Now, if I wanted to be one of those ponderous scientific people, and "let on" to prove what had occurred in the remote past by what had occurred in a given time in the recent past, or what will occur in the far future by what has occurred in late years, what an opportunity is here! Geology never had such a chance, nor such exact data to argue from! Nor "development of species," either! Glacial epochs are great things, but they are vague—vague. Please observe:

In the space of one hundred and seventy-six years the Lower Mississippi has shortened itself two hundred and forty-two miles. That is an average of a trifle over one mile and a third per year. Therefore, any calm person, who is not blind or idiotic, can see that in the Old Oolitic Silurian Period, just a million years ago next November, the Lower Mississippi River was upwards of one million three hundred thousand miles long, and stuck out over the Gulf of Mexico like a fishing-rod. And by the same token any person can see that seven hundred and forty-two years from now the Lower Mississippi will be only a mile and three-quarters long, and Cairo and New Orleans will have joined their streets together, and be plodding comfortably along under a single mayor and a mutual board of aldermen. There is something fascinating about Science. One gets such wholesale returns of conjecture out of such a trifling investment of fact.

Discuss how Twain's observation relates to the material presented in this section.

4.3   Diagnostics on the Least-Squares Regression Line

Preparing for This Section Before getting started, review the following:

- Outliers (Section 3.4, pp. 159–160)

Objectives ❶ Compute and interpret the coefficient of determination

❷ Perform residual analysis on a regression model

❸ Identify influential observations

In Section 4.2, we discussed how to find the least-squares regression line using formulas and technology. In this section, we present additional characteristics of the least-squares regression line, along with graphical diagnostic tests to perform on these lines.

❶ **Compute and Interpret the Coefficient of Determination**

Consider the club-head speed versus distance data introduced in Section 4.1. How could we predict the distance of a randomly selected shot? Our best guess might be the mean distance of all shots from the sample data given in Table 1, $\bar{y} = 266.75$ yards.

Now suppose we were told this particular shot resulted from a swing with a club-head speed of 103 mph. Knowing that a linear relation exists between club-head speed and distance, we can improve our estimate of the distance of the shot. In fact, we could use the least-squares regression line to adjust our guess to $\hat{y} = 3.1661(103) - 55.7966 = 270.3$ yards. In statistical terms, we say that some of the variation in distance is explained by the linear relation between club-head speed and distance.

The proportion of variation in the response variable that is explained by the least-squares regression line is called the *coefficient of determination*.

Definition | The **coefficient of determination, R^2,** measures the proportion of total variation in the response variable that is explained by the least-squares regression line.

In Other Words
The coefficient of determination is a measure of how well the least-squares regression line describes the relation between the explanatory and response variables. The closer R^2 is to 1, the better the line describes how changes in the explanatory variable affect the value of the response variable.

The coefficient of determination is a number between 0 and 1, inclusive. That is, $0 \le R^2 \le 1$. If $R^2 = 0$, the least-squares regression line has no explanatory value. If $R^2 = 1$, the least-squares regression line explains 100% of the variation in the response variable.

In Figure 16 on the following page, a horizontal line is drawn at $\bar{y} = 266.75$, the predicted distance of a shot without any knowledge of club-head speed. If we know that the club-head speed is 103 miles per hour, we increase our guess to 270.3 yards. The difference between the predicted distance of 266.75 yards and the predicted distance of 270.3 yards is due to the fact that the club-head speed is 103 miles per hour. In other words, the difference between the prediction of $\hat{y} = 270.3$ and $\bar{y} = 266.75$ is explained by the linear relation between club-head speed and distance. The observed distance when club-head speed is 103 miles per hour is 274 yards (see Table 5 on page 197). The difference between our predicted value, $\hat{y} = 270.3$, and the actual value, $y = 274$, is due to factors (variables) other than the club-head speed (wind speed, position of the ball on the club face, and so on) and also to random error. The differences just discussed are called **deviations**.

In Other Words
The word *deviations* comes from deviate. To deviate means "to stray."

The deviation between the observed and mean values of the response variable is called the **total deviation**, so total deviation = $y - \bar{y}$. The deviation between the predicted and mean values of the response variable is called the **explained deviation**, so

Figure 16

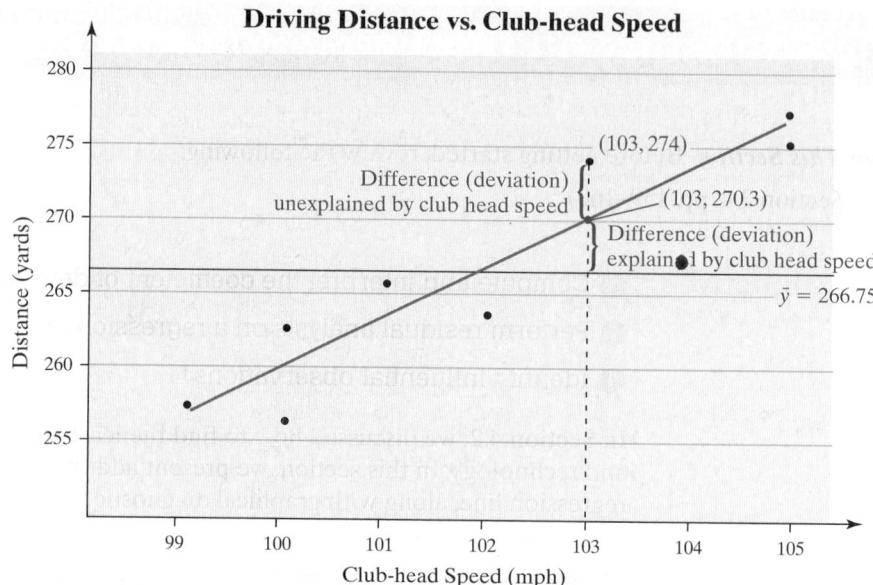

Driving Distance vs. Club-head Speed

explained deviation $= \hat{y} - \bar{y}$. Finally, the deviation between the observed and predicted values of the response variable is called the **unexplained deviation**, so unexplained deviation $= y - \hat{y}$. See Figure 17.

Figure 17

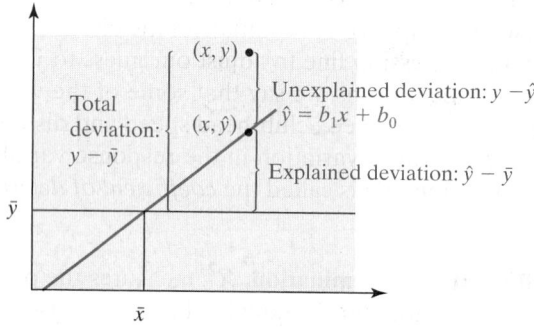

The figure illustrates that

$$\text{Total deviation} = \text{unexplained deviation} + \text{explained deviation}$$
$$y - \bar{y} \quad = \quad (y - \hat{y}) \quad + \quad (\hat{y} - \bar{y})$$

Although beyond the scope of this text, it can be shown that

$$\text{Total variation} = \text{unexplained variation} + \text{explained variation}$$
$$\sum(y - \bar{y})^2 \quad = \quad \sum(y - \hat{y})^2 \quad + \quad \sum(\hat{y} - \bar{y})^2$$

Dividing both sides by total variation, we obtain

$$1 = \frac{\text{unexplained variation}}{\text{total variation}} + \frac{\text{explained variation}}{\text{total variation}}$$

Subtracting $\dfrac{\text{unexplained variation}}{\text{total variation}}$ from both sides, we get

$$R^2 = \frac{\text{explained variation}}{\text{total variation}} = 1 - \frac{\text{unexplained variation}}{\text{total variation}}$$

Unexplained variation is found by summing the squares of the residuals, $\sum \text{residuals}^2 = \sum(y - \hat{y})^2$. So the smaller the sum of squared residuals, the smaller the unexplained variation and, therefore, the larger R^2 will be. Therefore, the closer the observed y's are to the regression line (the predicted y's), the larger R^2 will be.

CAUTION!

Squaring the linear correlation coefficient to obtain the coefficient of determination works only for the least-squares linear regression model

$$\hat{y} = b_1 x + b_0$$

The method does not work in general.

To find the coefficient of determination, R^2, for the least-squares regression model $\hat{y} = b_1 x + b_0$ (that is, a single explanatory variable to the first degree), square the linear correlation coefficient. That is, $R^2 = r^2$. This method does not work in general, however.

To reinforce the concept of the coefficient of determination, consider the three data sets in Table 7.

Table 7					
Data Set A		**Data Set B**		**Data Set C**	
x	**y**	**x**	**y**	**x**	**y**
3.6	8.9	3.1	8.9	2.8	8.9
8.3	15.0	9.4	15.0	8.1	15.0
0.5	4.8	1.2	4.8	3.0	4.8
1.4	6.0	1.0	6.0	8.3	6.0
8.2	14.9	9.0	14.9	8.2	14.9
5.9	11.9	5.0	11.9	1.4	11.9
4.3	9.8	3.4	9.8	1.0	9.8
8.3	15.0	7.4	15.0	7.9	15.0
0.3	4.7	0.1	4.7	5.9	4.7
6.8	13.0	7.5	13.0	5.0	13.0

Figures 18(a), (b), and (c) represent the scatter diagrams of data sets A, B, and C, respectively.

Figure 18

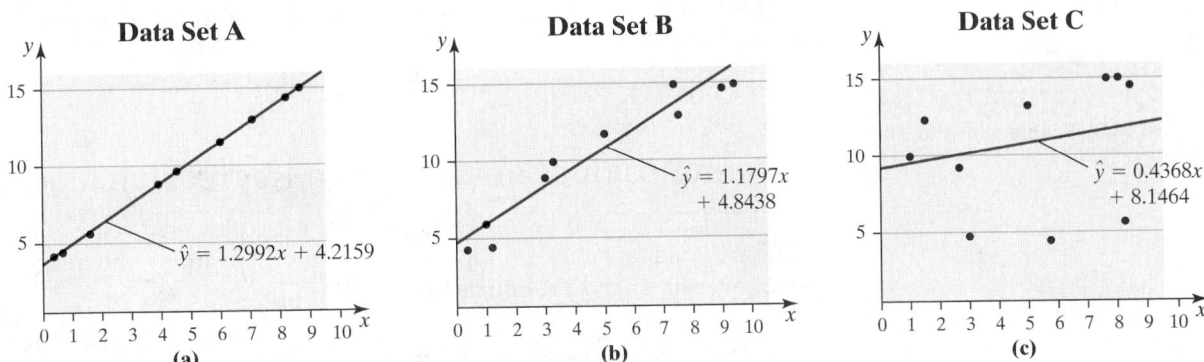

Notice that the y-values in each of the three data sets are the same. The variance of y is 17.49. In Figure 18(a), almost 100% of the variability in y can be explained by the least-squares regression line because the data lie almost perfectly on a straight line. In Figure 18(b), a high percentage of the variability in y can be explained by the least-squares regression line because the data have a strong linear relation. Finally, in Figure 18(c), a low percentage of the variability in y is explained by the least-squares regression line. If x increases, we cannot easily predict the change in y. If we compute the coefficient of determination, R^2, for the three data sets in Table 7, we obtain the following results:

Data Set	Coefficient of Determination, R^2	Interpretation
A	99.99%	99.99% of the variability in y is explained by the least-squares regression line.
B	94.7%	94.7% of the variability in y is explained by the least-squares regression line.
C	9.4%	9.4% of the variability in y is explained by the least-squares regression line.

Notice that, as the explanatory ability of the line decreases, the coefficient of determination, R^2, also decreases.

EXAMPLE 1   Determining the Coefficient of Determination

Problem Determine and interpret the coefficient of determination, R^2, for the club-head speed versus distance data shown in Table 5 on page 197.

By Hand Approach

To compute R^2, we square the linear correlation coefficient, r, found in Example 3 from Section 4.1 on page 186.

By Hand Solution

$R^2 = r^2 = 0.939^2 = 0.882 = 88.2\%$

Technology Approach

We will use Excel to determine R^2. The steps for obtaining the coefficient of determination using the TI-83/84 Plus graphing calculator, Minitab, Excel, and StatCrunch are given in the Technology Step-by-Step on pages 219–220.

Technology Solution

Figure 19 shows the results from Excel. The coefficient of determination is highlighted.

Figure 19

Summary Output

Regression Statistics	
Multiple R	0.938695838
R Square	0.881149876
Adjusted R Square	0.861341522
Standard Error	2.882638465
Observations	8

Interpretation 88.2% [Tech: 88.1%] of the variation in distance is explained by the least-squares regression line, and 11.8% of the variation in distance is explained by other factors.

●

● **Now Work Problems 9 and 19(a)**

❷ **Perform Residual Analysis on a Regression Model**

Recall that a residual is the difference between the observed value of y and the predicted value, $\hat{y}$. Residuals play an important role in determining the adequacy of the linear model. We will analyze residuals for the following purposes:

- To determine whether a linear model is appropriate to describe the relation between the explanatory and response variables
- To determine whether the variance of the residuals is constant
- To check for outliers

Is a Linear Model Appropriate?

In Section 4.1 we learned how to use the correlation coefficient to determine whether a linear relation exists between the explanatory variable and response variable. However, a correlation coefficient may indicate a linear relation exists between two variables even though the relation is not linear. To determine if a linear model is appropriate, we also need to draw a **residual plot**, which is a scatter diagram with the residuals on the vertical axis and the explanatory variable on the horizontal axis.

> If a plot of the residuals against the explanatory variable shows a discernible pattern, such as a curve, then the explanatory and response variable may not be linearly related.

The residual plot in Figure 20(a) does not show any pattern, so a linear model is appropriate. However, the residual plot in Figure 20(b) shows a U-shaped pattern, which indicates that a linear model is inappropriate.

Figure 20

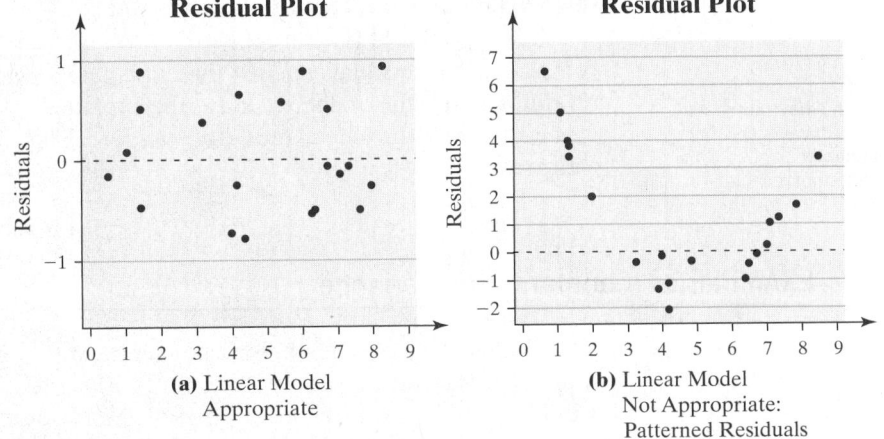

(a) Linear Model
Appropriate

(b) Linear Model
Not Appropriate:
Patterned Residuals

EXAMPLE 2   Is a Linear Model Appropriate?

Table 8	
Time (seconds)	Temperature (°F)
1	164.5
3	163.2
5	162.0
7	160.5
9	158.6
11	156.8
13	155.1

Source: Michael Sullivan

Problem The data in Table 8 were collected by placing a temperature probe in a portable heater, removing the probe, and recording the temperature (in degrees Fahrenheit) over time (in seconds). Determine whether the relation between the temperature of the probe and time is linear.

Approach Find the least-squares regression line and determine the residuals. Plot the residuals against the explanatory variable, time. If an obvious pattern results, the linear model is not appropriate. If no pattern results, a linear model is appropriate.

Solution Enter the data into statistical software or a graphing calculator with advanced statistical features. Figure 21(a) shows a scatter diagram of the data from StatCrunch. Temperature and time appear to be linearly related with a negative slope. In fact, the linear correlation coefficient between these two variables is −0.997. We use statistical software to determine the least-squares regression line and store the residuals. Figure 21(b) shows a plot of the residuals versus the explanatory variable, time, using StatCrunch. The upside-down U-shaped pattern in the plot indicates that the linear model is not appropriate. The predicted values overestimate the temperature for early and late times, whereas they underestimate the temperature for the middle times.

Figure 21

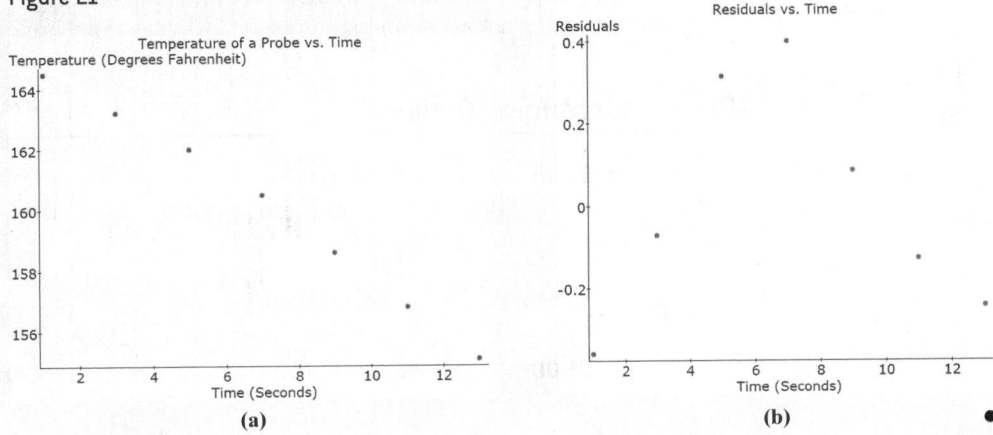

(a)

(b)

Is the Variance of the Residuals Constant?

-CAUTION!--------------------
If the model does not have constant
error variance, statistical inference
(the subject of Chapter 14) using the
regression model is not reliable.

If a plot of the residuals against the explanatory variable shows the spread of the residuals increasing or decreasing as the explanatory variable increases, then a strict requirement of the linear model is violated. This requirement is called **constant error variance**.*

EXAMPLE 3 Constant Error Variance

Figure 22 illustrates the idea behind constant error variance. In Figure 22(a) the data appear more spread out about the regression line when x is large. In the residual plot in Figure 22(b), the absolute value of the residuals is larger when x is larger and smaller when x is smaller. This means that the predictions made using the regression equation will be less reliable when x is large because there is more variability in y.

Figure 22

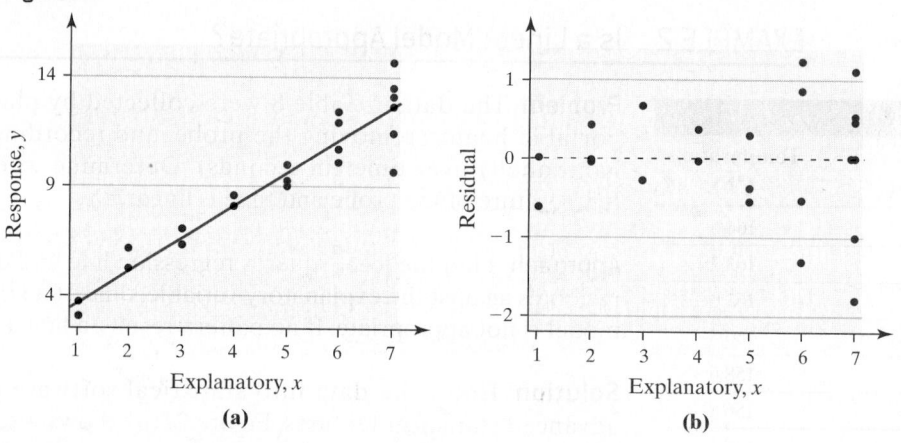

(a)   (b)

Are There Any Outliers?

In Section 3.4 we stated that outliers are extreme observations. An **outlier** can also be thought of as an observation whose response variable is inconsistent with the overall pattern of the data. To determine outliers, we can either construct a plot of residuals against the explanatory variable or draw a boxplot of the residuals.

We can find outliers using a residual plot because these residuals will lie far from the rest of the plot. Outliers can be found using a boxplot of residuals because they will appear as asterisks (*) in the plot. Statistical software typically stores the residuals upon executing a least-squares regression, so drawing a boxplot of residuals is straightforward.

EXAMPLE 4 Identifying Outliers

Problem Figure 23(a) shows a scatter diagram of a set of data. The residual plot is shown in Figure 23(b) and the boxplot of the residuals is in Figure 23(c). Do the data have any outliers?

Approach We examine the residual plot and boxplot for outliers.

Solution We can see that the data do contain an outlier. We label the outlier in Figure 23.

*The statistical term for constant error variance is **homoscedasticity**.

Figure 23

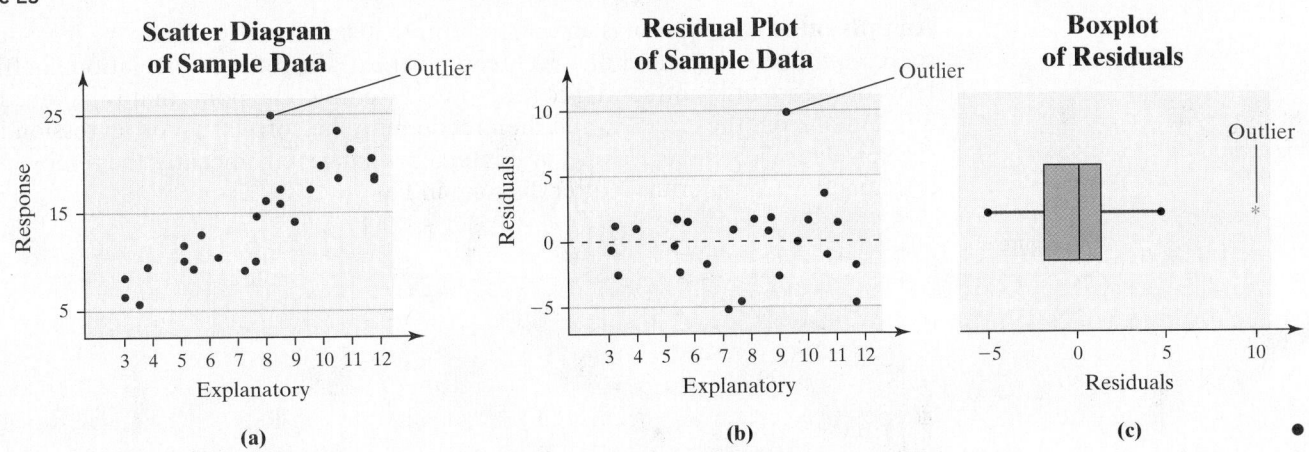

(a) **(b)** **(c)**

We will follow the practice of removing outliers only if they are the result of an error in recording, a miscalculation, or some other obvious blunder in the data-collection process (such as the observation not being from the population under study). If none of these are reasons for the outlier, it is recommended that a statistician be consulted.

EXAMPLE 5   Graphical Residual Analysis

Problem The data in Table 6 on page 204 represent the club-head speed, the distance the golf ball traveled, and the residuals (Column 7) for eight swings of a golf club. Construct a residual plot and boxplot of the residuals and comment on the appropriateness of the least-squares regression model.

Approach Plot the residuals on the vertical axis and the values of the explanatory variable on the horizontal axis. Look for any violations of the requirements of the regression model. We use a boxplot of the residuals to identify any outliers. We will use Minitab to generate the graphs.

Solution Figure 24(a) shows the residual plot and Figure 24(b) shows the boxplot.

Figure 24

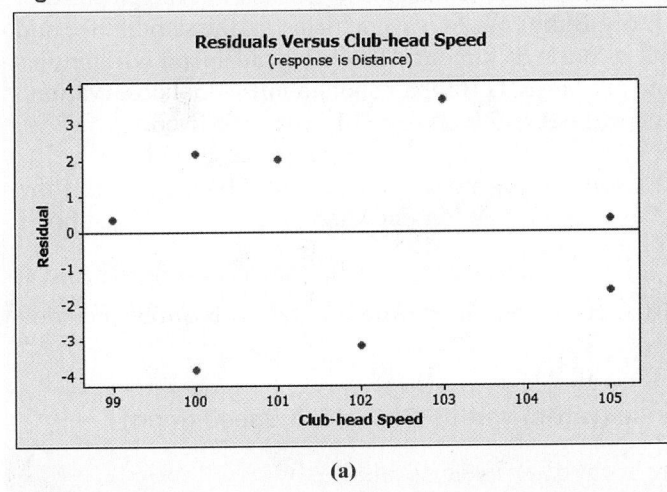

(a)

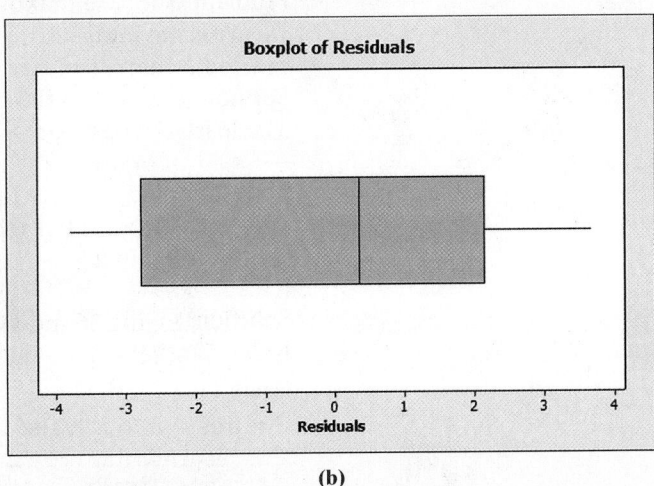

(b)

There is no discernible pattern in the residual plot, so a linear model is appropriate. The residuals display constant error variance, and no outliers appear in the boxplot of the residuals.

● **Now Work Problems 15 and 19(b)**

③ Identify Influential Observations

An **influential observation** is an observation that significantly affects the least-squares regression line's slope and/or y-intercept, or the value of the correlation coefficient. How do we identify influential observations? Remove the point that is believed to be influential from the data set, and then recompute the correlation or regression line. If the correlation coefficient, slope, or y-intercept changes significantly, the removed point is influential. Consider the scatter diagram in Figure 25.

Figure 25

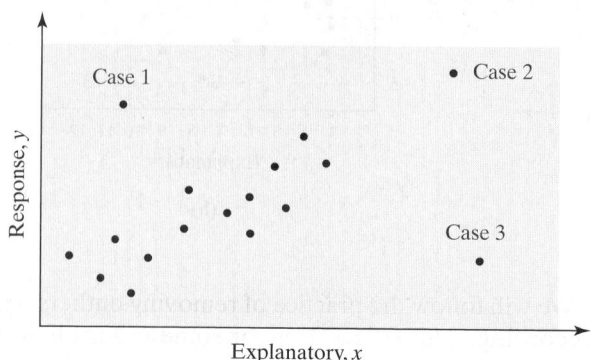

Case 1 is an outlier because its y-value is large relative to its x-value. Case 2 is large relative to both its x-value and y-value, but follows the overall pattern of the data. Finally, Case 3 is large relative to its x-value, and its y-value is not consistent with the pattern of the data.

Influence is affected by two factors:

(1) the relative vertical position of the observation (residuals) and
(2) the relative horizontal position of the observation (*leverage*).

Leverage is a measure that depends on how much the observation's value of the explanatory variable differs from the mean value of the explanatory variable. Using these terms, Case 1 has low leverage and a large residual; Case 2 has high leverage and a small residual; Case 3 has high leverage and a large residual. Observations such as Case 3 (high leverage with a large residual) tend to be influential.

In Other Words
Draw a boxplot of the explanatory variable. If an outlier appears, the observation corresponding to the outlier may be influential.

EXAMPLE 6 Identifying Influential Observations

Problem Suppose that our golf ball experiment calls for nine trials, but the player that we are using hurts his wrist. Luckily, Bubba Watson is practicing on the same range and graciously agrees to participate in our experiment. His club-head speed is 120 miles per hour and he hits the golf ball 305 yards. Is Bubba's shot an influential observation? The least-squares regression line without Bubba is $\hat{y} = 3.1661 x - 55.7966$.

Approach Recompute the least-squares regression line with Bubba included. If the slope or y-intercept of the least-squares regression line changes substantially, Bubba's shot is influential.

Solution Figure 26 shows a partial output obtained from Minitab with Bubba included in the data set.

Figure 26

Regression Analysis: Distance (yards) versus Club Head Speed (mph)

```
The regression equation is
Distance (yards) = 39.5 + 2.23 Club Head Speed (mph)

Predictor                 Coef    SE Coef      T      P
Constant                 39.46      20.15   1.96  0.091
Club Head Speed (mph)    2.2287    0.1937  11.51  0.000

S = 3.51224    R - Sq = 95.0%    R - Sq(adj) = 94.3%
```

The least-squares regression line with Bubba included in the data set is

$$\hat{y} = 2.2287x + 39.46$$

We graph the regression lines with Bubba included (red) and excluded (blue) on the same scatter diagram. See Figure 27.

Figure 27

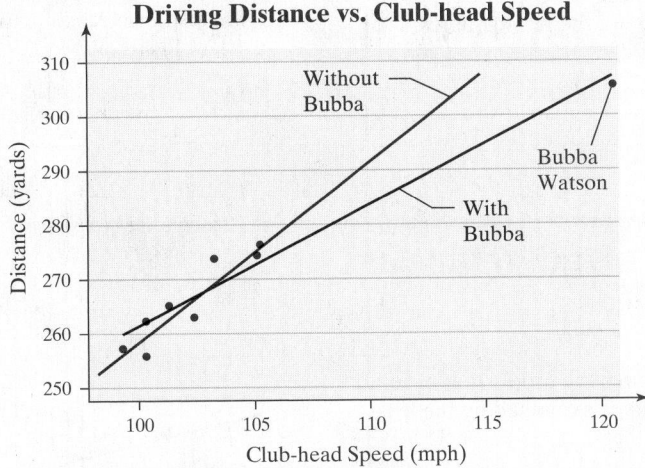

Driving Distance vs. Club-head Speed

The inclusion of Bubba Watson in the data set causes the slope of the regression line to decrease (from 3.1661 to 2.2287) and the y-intercept to increase substantially (from -55.80 to $+39.46$). In other words, the Bubba data pulls the least-squares regression line toward his observation. Look at the scatter diagram in Figure 27. Notice that Bubba's shot has high leverage and a large residual (based on predictions using the "without Bubba" line. Therefore, Bubba's data point is influential. •

Using Technology

Minitab will alert you when a data value is influential.

As with outliers, influential observations should be removed only if there is justification to do so. When an influential observation occurs in a data set and its removal is not warranted, two popular courses of action are to (1) collect more data so that additional points near the influential observation are obtained or (2) use techniques that reduce the influence of the influential observation. (These techniques are beyond the scope of this text.) In the case of Example 6, we are justified in removing Bubba Watson from the data set because our experiment called for the same player to swing the club for each trial.

• **Now Work Problem 11**

Technology Step-by-Step Determining R^2 and Residual Plots

TI-83/84 Plus

The Coefficient of Determination, R^2
Use the same steps that were followed to obtain the correlation coefficient to obtain R^2. Diagnostics must be on.

Residual Plots
1. Enter the raw data in L1 and L2. Obtain the least-squares regression line.
2. Access STAT PLOT. Select Plot1. Choose the scatter diagram icon and let XList be L1. Let YList be RESID by moving the cursor to YList, pressing 2nd STAT, and choosing the list titled RESID under the NAMES menu.
3. Press ZOOM and select 9: ZoomStat.

Minitab

The Coefficient of Determination, R^2
This is provided in the standard regression output.

Residual Plots
Follow the same steps as those used to obtain the regression output (Section 4.2). Before selecting OK, click GRAPHS. In the cell that says "Residuals versus the variables," enter the name of the explanatory variable. Click OK.

Excel

The Coefficient of Determination, R^2
This is provided in the standard regression output.

Residual Plots

Follow the same steps as those used to obtain the regression output (Section 4.2). Before selecting OK, click **Residual Plots**. Click OK.

StatCrunch

The Coefficient of Determination, R^2

Follow the same steps used to obtain the least-squares regression line. The coefficient of determination is given as part of the output (R-sq).

Residual Plots

1. If necessary, enter the explanatory variable in column var1 and the response variable in column var2. Name each column variable.
2. Select **Stat**, highlight **Regression**, and select **Simple Linear**.
3. Choose the explanatory variable for the X variable and the response variable for the Y variable. Under Graphs, select Residuals vs. X-values. Click Compute!

4.3 Assess Your Understanding

Vocabulary and Skill Building

1. The _____ _____ _____, R^2, measures the proportion of total variation in the response variable that is explained by the least-squares regression line.

2. Total deviation = _____ deviation + _____ deviation

3. A _____ _____ is a scatter diagram with the residuals on the vertical axis and the explanatory variable on the horizontal axis.

4. What is an influential observation?

In Problems 5–8, analyze the residual plots and identify which, if any, of the conditions for an adequate linear model is not met.

5.   6.

7.   8.

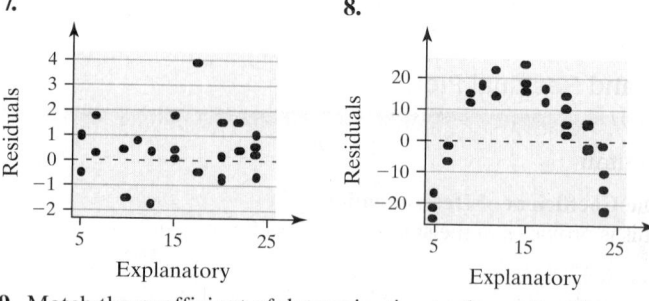

NW 9. Match the coefficient of determination to the scatter diagram in the next column. The scales on the horizontal and vertical axis are the same for each scatter diagram.

(a) $R^2 = 0.58$ (b) $R^2 = 0.90$
(c) $R^2 = 1$ (d) $R^2 = 0.12$

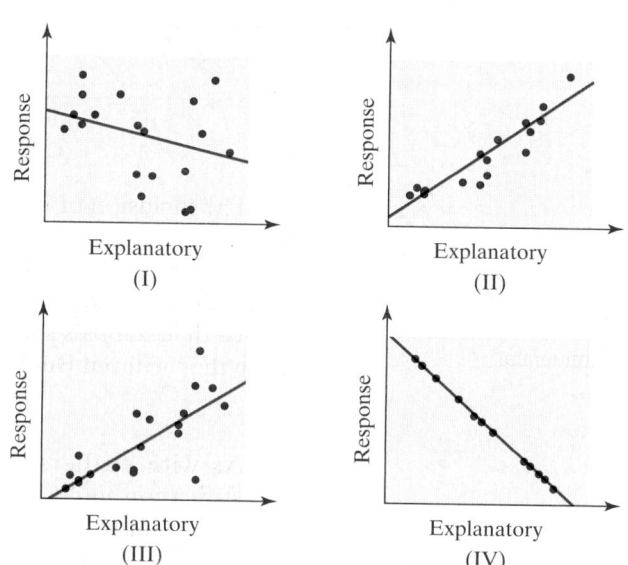

(I)   (II)

(III)   (IV)

10. Use the linear correlation coefficient given to determine the coefficient of determination, R^2. Interpret each R^2.

(a) $r = -0.32$ (b) $r = 0.13$
(c) $r = 0.40$ (d) $r = 0.93$

In Problems 11–14, a scatter diagram is given with one of the points drawn in blue. In addition, two least-squares regression lines are drawn: The line drawn in red is the least-squares regression line with the point in blue excluded. The line drawn in blue is the least-squares regression line with the point in blue included. On the basis of these graphs, do you think the point in blue is influential? Why?

NW 11.   12.

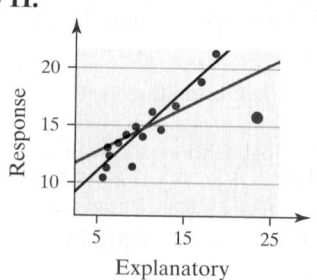   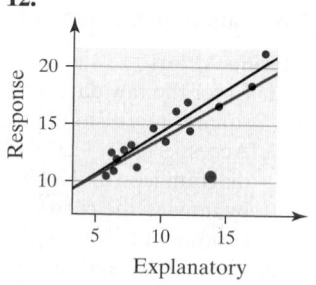

13.
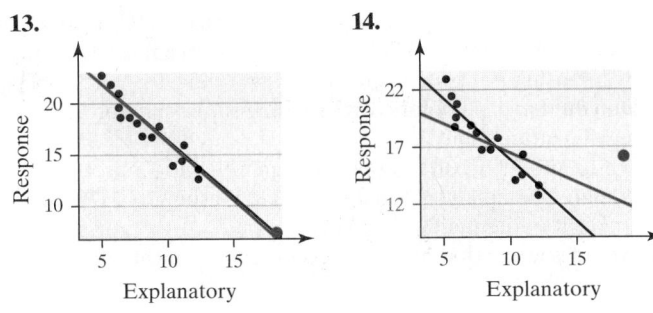

14.

Applying the Concepts

NW 15. The Other Old Faithful Perhaps you are familiar with the famous Old Faithful geyser in Yellowstone National Park. Another Old Faithful geyser is located in Calistoga in California's Napa Valley. The following data represent the time (in minutes) between eruptions and the length of eruption for 9 randomly selected eruptions.

Time between Eruptions, x	Length of Eruption, y	Time between Eruptions, x	Length of Eruption, y
12.17	1.88	11.70	1.82
11.63	1.77	12.27	1.93
12.03	1.83	11.60	1.77
12.15	1.83	11.72	1.83
11.30	1.70		

Source: Ladonna Hansen, Park Curator

(a) A scatter diagram of the data is shown next. What type of relation appears to exist between time between eruptions and length of eruption?

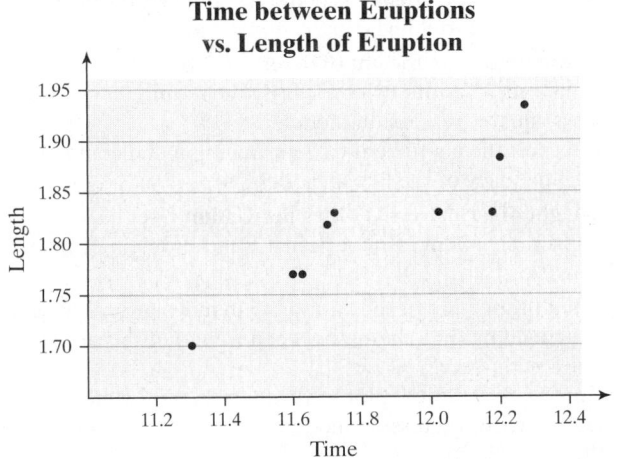

(b) The residual plot in the next column was obtained after finding the least-squares regression line. Does the residual plot confirm that the relation between time between eruptions and length of eruption is linear?

Residuals vs. Time

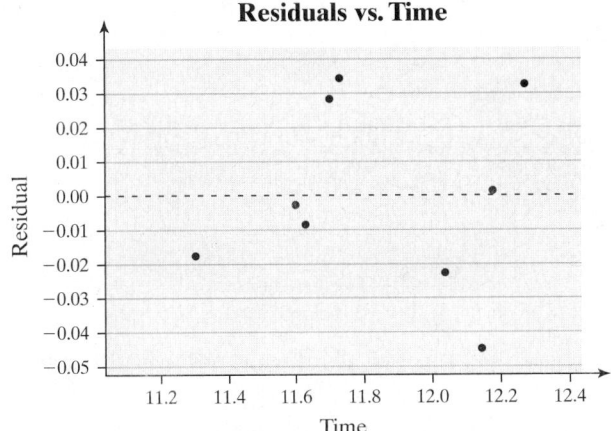

(c) The coefficient of determination is 83.0%. Provide an interpretation of this value.

16. Concrete As concrete cures, it gains strength. The following data represent the 7-day and 28-day strength (in pounds per square inch) of a certain type of concrete.

7-Day Strength, x	28-Day Strength, y	7-Day Strength, x	28-Day Strength, y
2300	4070	2480	4120
3390	5220	3380	5020
2430	4640	2660	4890
2890	4620	2620	4190
3330	4850	3340	4630

(a) A scatter diagram of the data is shown next. What type of relation appears to exist between 7-day strength and 28-day strength?

28-Day vs. 7-Day Concrete Strength

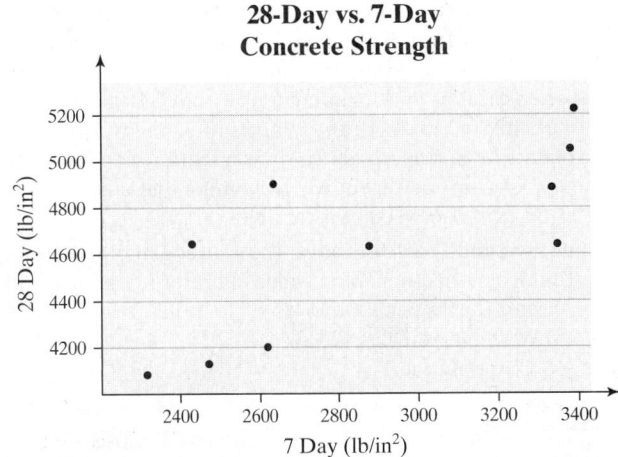

(b) The residual plot on the next page was obtained after finding the least-squares regression line. Does the residual plot confirm that the relation between 7-day and 28-day strength is linear?

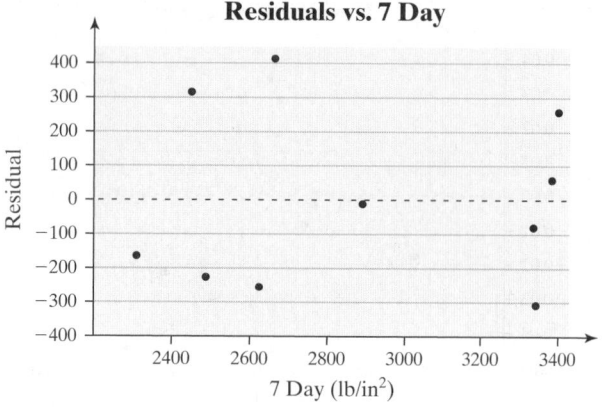

Residuals vs. 7 Day

7 Day (lb/in²)

(c) The coefficient of determination is 57.5%. Provide an interpretation of this value.

17. Calories versus Sugar The following data represent the number of calories per serving and the number of grams of sugar per serving for a random sample of high-fiber cereals.

Calories, x	Sugar, y	Calories, x	Sugar, y
200	18	210	23
210	23	210	16
170	17	210	17
190	20	190	12
200	18	190	11
180	19	200	11

Source: Consumer Reports

(a) Draw a scatter diagram of the data treating calories as the explanatory variable. What type of relation appears to exist between calories and sugar content?
(b) Determine the correlation between calories and sugar content. Is there a linear relation between calories and sugar content?
(c) Suppose that we add Kellogg's All-Bran cereal, which has 80 calories and 6 grams of sugar per serving, to the data set. Redraw the scatter diagram and recompute the correlation between calories and sugar content. Is there a linear relation between calories and sugar content with All-Bran included in the data set? Is All-Bran an influential observation?
(d) Explain why it is necessary to show a scatter diagram with the correlation coefficient when claiming that a linear relation exists between two variables.

18. Education and Teen Birthrates The scatter diagram represents the percentage of the population with a bachelor's degree, x, and teen birthrates (births per 1000 15- to 19-year-olds), y, for the 50 states and Washington, DC, in 2013.
Source: U.S. Census Bureau

Teen Birth Rate vs. Percent of State With at Least a Bachelor's Degree

Teen BirthRate

Bachelors %

(a) The least-squares regression with Washington, DC, included is $\hat{y} = -0.8276\,x + 55.7856$. The least-squares regression line with Washington, DC, excluded is $\hat{y} = -1.5470\,x + 75.8860$. Do you think the point corresponding to Washington, DC, may be influential? Why?
(b) Washington, DC, often appears as an influential observation in regression analysis. What is it about Washington, DC, that makes it influential?
(c) Would you say that higher levels of education cause lower teen birthrates?

Problems 19–24 use the results from Problems 25–30 in Section 4.1 and Problems 17–22 in Section 4.2.

NW 19. An Unhealthy Commute Use the results from Problem 25 in Section 4.1 and Problem 17 in Section 4.2 to:
(a) Determine the coefficient of determination, R^2.
(b) Construct a residual plot to verify the model requirements of the least-squares regression model.
(c) Interpret the coefficient of determination and comment on the adequacy of the linear model.

20. Credit Scores Use the results from Problem 26 in Section 4.1 and Problem 18 in Section 4.2 to:
(a) Determine the coefficient of determination, R^2.
(b) Construct a residual plot to verify the model requirements of the least-squares regression model.
(c) Interpret the coefficient of determination and comment on the adequacy of the linear model.

21. Height versus Head Circumference Use the results from Problem 27 in Section 4.1 and Problem 19 in Section 4.2 to:
(a) Compute the coefficient of determination, R^2.
(b) Construct a residual plot to verify the requirements of the least-squares regression model.
(c) Interpret the coefficient of determination and comment on the adequacy of the linear model.

22. American Black Bears Use the results from Problem 28 in Section 4.1 and Problem 20 in Section 4.2 to:
(a) Compute the coefficient of determination, R^2.
(b) Construct a residual plot to verify the requirements of the least-squares regression model.
(c) Interpret the coefficient of determination and comment on the adequacy of the linear model.

23. Weight of a Car versus Miles per Gallon Use the results from Problem 29 in Section 4.1 and Problem 21 in Section 4.2.
(a) What proportion of the variability in miles per gallon is explained by the relation between weight of the car and miles per gallon?
(b) Construct a residual plot to verify the requirements of the least-squares regression model. Are there any problems with the model?
(c) Interpret the coefficient of determination and comment on the adequacy of the linear model.

24. Hurricanes Use the results from Problem 30 in Section 4.1 and Problem 22 in Section 4.2.
(a) What proportion of the variability in the wind speed is explained by the relation between the atmospheric pressure and wind speed?
(b) Construct a residual plot to verify the requirements of the least-squares regression model.
(c) Interpret the coefficient of determination and comment on the adequacy of the linear model.

25. Kepler's Law of Planetary Motion The time it takes for a planet to complete its orbit around the sun is called the planet's sidereal year. In 1618, Johannes Kepler discovered that the sidereal year of a planet is related to the distance the planet is from the sun. The following data show the distances of the planets, and the dwarf planet Pluto, from the sun and their sidereal years.

Planet	Distance from Sun, x (millions of miles)	Sidereal Year, y
Mercury	36	0.24
Venus	67	0.62
Earth	93	1.00
Mars	142	1.88
Jupiter	483	11.9
Saturn	887	29.5
Uranus	1785	84.0
Neptune	2797	165.0
Pluto	3675	248.0

(a) Draw a scatter diagram of the data treating distance from the sun as the explanatory variable.
(b) Determine the correlation between distance and sidereal year. Does this imply a linear relation between distance and sidereal year?
(c) Compute the least-squares regression line.
(d) Plot the residuals against the distance from the sun.
(e) Do you think the least-squares regression line is a good model? Why?

26. Wind Chill Factor The wind chill factor depends on wind speed and air temperature. The following data represent the wind speed (in mph) and wind chill factor at an air temperature of 15° Fahrenheit.

Wind Speed, x (mph)	Wind Chill Factor, y	Wind Speed, x (mph)	Wind Chill Factor, y
5	12	25	−22
10	−3	30	−25
15	−11	35	−27
20	−17		

Source: Information Please Almanac

(a) Draw a scatter diagram of the data treating wind speed as the explanatory variable.
(b) Determine the correlation between wind speed and wind chill factor. Does this imply a linear relation between wind speed and wind chill factor?
(c) Compute the least-squares regression line.
(d) Plot the residuals against the wind speed.
(e) Do you think the least-squares regression line is a good model? Why?

27. Weight of a Car versus Miles per Gallon Suppose that we add the Dodge Viper to the data in Problem 21 in Section 4.2. A Dodge Viper weighs 3425 pounds and gets 11 miles per gallon.

(a) Compute the coefficient of determination of the expanded data set. What effect does the addition of the Viper to the data set have on R^2?
(b) Is the point corresponding to the Dodge Viper influential? Is it an outlier?

28. American Black Bears Suppose that we find a bear that is 205 cm tall and weighs 187 kg and add the bear to the data in Problem 20 from Section 4.2.

(a) Compute the coefficient of determination of the expanded data set. What effect does the additional bear have on R^2?
(b) Is the new bear influential? Is it an outlier?

29. CO₂ Emissions and Energy The following data represent the carbon dioxide emissions (in thousands of tons), y, and energy generated (thousands of megawatts), x, for a random sample of 14 power plants in Oregon. Is there anything unusual about the Boardman Power Plant, which produces 4618 thousand megawatts of energy and emits 4813 thousand tons of CO_2? What about the Hermiston power plant, which produces 2636 thousand megawatts of energy and emits 1311 thousand tons of CO_2?

Energy, x	Carbon, y	Energy, x	Carbon, y
4618	4813	22	15
2636	1311	13	10
535	377	64	6
357	341	13	5
370	238	33	3
15	44	24	2
25	18	17	2

Source: CARMA (www.carma.org)

30. Age versus Study Time Professor Katula feels that there is a relation between the number of hours a statistics student studies each week and the student's age. She conducts a survey in which 26 statistics students are asked their age and the number of hours they study statistics each week. She obtains the following results:

Age, x	Hours Studying, y	Age, x	Hours Studying, y	Age, x	Hours Studying, y
18	4.2	19	5.1	22	2.1
18	1.1	19	2.3	22	3.6
18	4.6	20	1.7	24	5.4
18	3.1	20	6.1	25	4.8
18	5.3	20	3.2	25	3.9
18	3.2	20	5.3	26	5.2
19	2.8	21	2.5	26	4.2
19	2.3	21	6.4	35	8.1
19	3.2	21	4.2		

(a) Draw a scatter diagram of the data. Comment on any potential influential observations.
(b) Find the least-squares regression line using all the data points.

(c) Find the least-squares regression line with the data point (35, 8.1) removed.

(d) Draw each least-squares regression line on the scatter diagram obtained in part (a).

(e) Comment on the influence that the point (35, 8.1) has on the regression line.

31. Putting It Together: A Tornado Model Is the width of a tornado related to the amount of distance for which the tornado is on the ground? Go to www.pearsonhighered.com/sullivanstats to obtain the data file 4_3_31 using the file format of your choice for the version of the text you are using. The data represent the width (yards) and length (miles) of tornadoes in the state of Oklahoma in 2013.

(a) What is the explanatory variable?

(b) Explain why this data should be analyzed as bivariate quantitative data.

(c) Draw a scatter diagram of the data. What type of relation appears to exist between the width and length of a tornado?

(d) Determine the correlation coefficient between width and length.

(e) Is there a linear relation between a tornado's width and its length on the ground?

(f) Find the least-squares regression line.

(g) Predict the length of a tornado whose width is 500 yards.

(h) Was the tornado whose width was 180 yards and length was 1.9 miles on the ground longer than would be expected?

(i) Interpret the slope.

(j) Explain why it does not make sense to interpret the intercept.

(k) What proportion of the variability in tornado length is explained by the width of the tornado?

(l) Plot residuals against the width. Does the residual plot suggest the two variables are linearly related?

(m) Draw a boxplot of the residuals. Are there any outliers?

(n) A major tornado was 4576 yards wide that had a length of 16.2 miles. Is this an influential tornado? Explain.

32. Putting It Together: Exam Scores The data in the next column represent scores earned by students in Sullivan's Elementary Algebra course for Chapter 2 (Linear Equations and Inequalities in One Variable) and Chapter 3 (Linear Equations and Inequalities in Two Variables). Completely summarize the relation between Chapter 2 and Chapter 3 exam scores, treating Chapter 2 exam scores as the explanatory variable. Write a report detailing the results of the analysis including the presence of any outliers or influential points. What does the relationship say about the role Chapter 2 plays in a student's understanding of Chapter 3?

Chapter 2 Score	Chapter 3 Score
71.4	76.1
76.2	82.6
60.3	60.9
88.1	91.3
95.2	78.3
82.5	100
100	95.7
87.3	81.5
71.4	50.7
95.2	81.5
95.2	87.0
85.7	73.9
71.4	74.6
33.3	45.7
78.0	37.3
83.3	88.0
100	100
81.0	76.1
76.2	63.0

Source: Michael Sullivan

Retain Your Knowledge

33. Sullivan Survey II Go to www.pearsonhighered.com/sullivanstats to obtain the data file SullivanStatsSurveyII using the file format of your choice for the version of the text you are using. One question asked in this survey is, "What percent of income do you believe individuals should pay in federal income tax?"

(a) Draw a relative frequency histogram of the data in the column "Tax Rate" using a lower class limit of the first class equal to 0 and a class width of 5. Comment on the shape of the distribution. What is the class with the highest relative frequency?

(b) Determine the mean and median tax rate.

(c) Determine the standard deviation and interquartile range of tax rate.

(d) Determine the lower and upper fences? Are there any outliers?

(e) In this chapter, we have focused on bivariate quantitative data. However, side-by-side boxplots may be used to see association between a qualitative and quantitative variable. Draw side-by-side boxplots of tax rate by gender. Comment on any interesting features. Which gender appears to prefer the higher tax rate?

(f) Draw side-by-side boxplots of tax rate by political philosophy. Comment on any features in the graph.

4.4   Contingency Tables and Association

Preparing for This Section Before getting started, review the following:

- Side-by-side bar graphs (Section 2.1, pp. 66–67)

Objectives   ❶ Compute the marginal distribution of a variable

❷ Use the conditional distribution to identify association among categorical data

❸ Explain Simpson's Paradox

In Sections 4.1 to 4.3, we looked at techniques for summarizing relations between two quantitative variables. We now look at techniques for summarizing relations between two qualitative variables.

Consider the data (measured in thousands) in Table 9, which represent the employment status and level of education of all U.S. residents 25 years old or older in November 2014. By definition, an individual is unemployed if he or she is actively seeking but is unable to find work. An individual is not in the labor force if he or she is not employed and not actively seeking employment.

Table 9				
	Level of Education			
Employment Status	**Did Not Finish High School**	**High School Graduate**	**Some College**	**Bachelor's Degree or Higher**
Employed	10,179	33,624	35,407	49,534
Unemployed	945	2012	1823	1615
Not in the labor force	13,271	25,806	17,089	17,415

Source: Bureau of Labor Statistics

Table 9 is referred to as a **contingency table**, or a **two-way table**, because it relates two categories of data. The **row variable** is employment status, because each row in the table describes the employment status of a group. The **column variable** is level of education. Each box inside the table is called a **cell**. For example, the cell corresponding to employed high school graduates is in the first row, second column. Each cell contains the frequency of the category: In 2014, 10,179 thousand employed individuals had not finished high school.

The data in Table 9 describe two characteristics regarding the population of U.S. residents who are 25 years or older: their employment status and their level of education. We want to investigate whether the two variables are associated. For example, are individuals who have a higher level of education more likely to be employed? Just as we did in Sections 4.1 to 4.3, we discuss both numerical and graphical methods for summarizing the data.

❶ Compute the Marginal Distribution of a Variable

The first step in summarizing data in a contingency table is to determine the distribution of each variable separately. To do so, we create *marginal distributions*.

In Other Words
The distributions are called *marginal distributions* because they appear in the right and the bottom margin of the contingency table.

Definition   A **marginal distribution** of a variable is a frequency or relative frequency distribution of either the row or column variable in the contingency table.

A marginal distribution removes the effect of either the row variable or the column variable in the contingency table.

To create a marginal distribution for a variable, calculate the row and column totals for each category of the variable. The row totals represent the distribution of the row variable. The column totals represent the distribution of the column variable.

EXAMPLE 1 Determining Frequency Marginal Distributions

Problem Find the frequency marginal distributions for employment status and level of education from the data in Table 9.

Approach Find the row total for the category "employed" by adding the number of employed individuals who did not finish high school, who finished high school, and so on. Repeat this process for each category of employment status.

Find the column total for the category "did not finish high school" by adding the number of employed individuals, unemployed individuals, and individuals not in the labor force who did not finish high school. Repeat this process for each level of education.

Solution In Table 10, the blue entries represent the marginal distribution of the row variable "employment status." For example, there were 10,179 + 33,624 + 35,407 + 49,534 = 128,744 thousand employed individuals in November 2014. The red entries represent the marginal distribution of the column variable "level of education."

The marginal distribution for employment status removes the effect of level of education; the marginal distribution for level of education removes the effect of employment status. The marginal distribution of level of education shows there were more Americans with a bachelor's degree or higher than there were Americans who were high school graduates (68,564 thousand versus 61,442 thousand) in November 2014. The marginal distribution of employment status shows that 128,744 thousand Americans were employed. The table also indicates that there were 208,720 thousand U.S. residents 25 years old or older.

Table 10

Employment Status	Level of Education				Totals
	Did Not Finish High School	High School Graduate	Some College	Bachelor's Degree or Higher	
Employed	10,179	33,624	35,407	49,534	128,744
Unemployed	945	2012	1823	1615	6395
Not in the labor force	13,271	25,806	17,089	17,415	73,581
Totals	24,395	61,442	54,319	68,564	**208,720**

● **Now Work Problem 9(a)**

We can use the row and column totals obtained in Example 1 to calculate the relative frequency marginal distribution for level of education and employment status.

EXAMPLE 2 Determining Relative Frequency Marginal Distributions

Problem Determine the relative frequency marginal distribution for level of education and employment status from the data in Table 10.

Approach The relative frequency marginal distribution for the row variable, employment status, is found by dividing the row total for each employment status by the table total, 208,720. The relative frequency marginal distribution for the column variable, level of education, is found by dividing the column total for each level of education by the table total.

> **CAUTION!**
> For relative frequency marginal distributions (such as in Table 11), row or column totals might not sum exactly to 1 due to rounding.

Solution Table 11 represents the relative frequency marginal distribution for each variable.

Table 11

Employment Status	Level of Education				
	Did Not Finish High School	**High School Graduate**	**Some College**	**Bachelor's Degree or Higher**	**Relative Frequency Marginal Distribution**
Employed	10,179	33,624	35,407	49,534	$\frac{128,744}{208,720} = 0.617$
Unemployed	945	2012	1823	1615	$\frac{6395}{208,720} = 0.031$
Not in the labor force	13,271	25,806	17,089	17,415	$\frac{73,581}{208,720} = 0.353$
Relative Frequency Marginal Distribution	$\frac{24,395}{208,720} = 0.117$	$\frac{61,442}{208,720} = 0.294$	$\frac{54,319}{208,720} = 0.260$	$\frac{68,564}{208,720} = 0.328$	1

Table 11 shows that 11.7% of U.S. residents 25 years old or older did not graduate from high school, and 61.7% of U.S. residents 25 years old or older were employed in November 2014.

● **Now Work Problems 9(b) and (c)**

② **Use the Conditional Distribution to Identify Association among Categorical Data**

As we look at the information in Tables 10 and 11, we might ask whether a higher level of education is associated with a higher likelihood of being employed. If level of education does not play any role, we would expect the relative frequencies for employment status at each level of education to be close to the relative frequency marginal distribution for employment status given in blue in Table 11. So we would expect 61.7% of individuals who did not finish high school, 61.7% of individuals who finished high school, 61.7% of individuals with some college, and 61.7% of individuals with at least a bachelor's degree to be employed. If the relative frequencies for these various levels of education are different, we might associate this difference with the level of education.

The marginal distributions in Tables 10 and 11 allow us to see the distribution of either the row variable (employment status) or the column variable (level of education) but we do not get a sense of association from these tables.

When describing any association between two categories of data, we must use relative frequencies instead of frequencies, because frequencies are difficult to compare when there are different numbers of observations for the categories of a variable.

> **CAUTION!**
> To describe the association between two categorical variables, relative frequencies must be used because there are different numbers of observations for the categories.

EXAMPLE 3 Comparing Two Categories of a Variable

Problem What proportion of the following groups of individuals is employed?
(a) Those who did not finish high school   **(b)** High school graduates
(c) Those who finished some college   **(d)** Those who have at least a bachelor's degree

Approach In part (a), we are asking, "Of the individuals who did not finish high school, what proportion is employed?" To determine this proportion, divide the number of employed individuals who did not finish high school by the number of people who did not finish high school. Repeat this process to answer parts (b)–(d).

(continued)

Solution

(a) In November 2014, 24,395 thousand individuals 25 years old or older did not finish high school. (See Table 10.) Of this number, 10,179 thousand were employed. Therefore, $\dfrac{10{,}179}{24{,}395} = 0.417$ represents the proportion of individuals who did not finish high school who are employed.

(b) In November 2014, 61,442 thousand individuals 25 years old or older were high school graduates. Of this number, 33,624 thousand were employed. Therefore, $\dfrac{33{,}624}{61{,}442} = 0.547$ represents the proportion of individuals who graduated from high school who are employed.

(c) In November 2014, 54,319 thousand individuals 25 years old or older had some college. Of this number, 35,407 thousand were employed. Therefore, $\dfrac{35{,}407}{54{,}319} = 0.652$ were employed.

(d) In November 2014, 68,564 thousand individuals 25 years old or older had at least a bachelor's degree. Of this number, 49,534 thousand were employed. Therefore, $\dfrac{49{,}534}{68{,}564} = 0.722$ were employed.

We can see from these relative frequencies that as the level of education increases, the proportion of individuals employed increases. •

In Other Words
Since we are finding the conditional distribution of employment status by level of education, the level of education is the *explanatory variable* and the employment status is the *response variable*.

The results in Example 3 are only partial. In general, we create a relative frequency distribution for each value of the explanatory variable. For our example, this means we would construct a relative frequency distribution for individuals who did not finish high school, a second relative frequency distribution for individuals who are high school graduates, and so on. These relative frequency distributions are called *conditional distributions*.

Definition

A **conditional distribution** lists the relative frequency of each category of the response variable, given a specific value of the explanatory variable in the contingency table.

An example should help to solidify our understanding of the definition.

EXAMPLE 4   Constructing a Conditional Distribution

Problem Find the conditional distribution of the response variable employment status by level of education, the explanatory variable, for the data in Table 9 on page 225. What is the association between level of education and employment status?

Approach First compute the relative frequency for each employment status, given that the individual did not finish high school. Next, compute the relative frequency for each employment status, given that the individual is a high school graduate. Continue computing the relative frequency for each employment status for the next two levels of education. This is the same approach taken in Example 3 for employed individuals.

Solution First use the number of individuals who did not finish high school (24,395 thousand) as the denominator in computing the relative frequencies for each employment status.

Then use the number of individuals who graduated from high school as the denominator in computing relative frequencies for each employment status.

Next, compute the relative frequency for each employment status, given that the individual had some college and then given that the individual had at least a bachelor's degree. We obtain Table 12. Notice that the "Employed" row in Table 12 shows the results from Example 3.

Table 12				
	Level of Education			
Employment Status	**Did Not Finish High School**	**High School Graduate**	**Some College**	**Bachelor's Degree or Higher**
Employed	$\frac{10,179}{24,395} = 0.417$	$\frac{33,624}{61,442} = 0.547$	$\frac{35,407}{54,319} = 0.652$	$\frac{49,534}{68,564} = 0.722$
Unemployed	$\frac{945}{24,395} = 0.039$	$\frac{2012}{61,442} = 0.033$	$\frac{1823}{54,319} = 0.034$	$\frac{1615}{68,564} = 0.024$
Not in the labor force	$\frac{13,271}{24,395} = 0.544$	$\frac{25,806}{61,442} = 0.420$	$\frac{17,089}{54,319} = 0.315$	$\frac{17,415}{68,564} = 0.254$
Totals	1	1	1	1

Looking at the conditional distributions of employment status by level of education, associations become apparent. Read the information in Table 12 from left to right. As the amount of schooling (the explanatory variable) increases, the proportion employed within each category also increases. As the amount of schooling increases, the proportion not in the labor force decreases. The proportion unemployed with at least a bachelor's degree is much lower than those unemployed in the other three levels of education.

Information about individuals' levels of education provides insight into their employment status. For example without any information about level of education, we might predict the number of employed individuals out of 100 to be about 62 because the employment rate for the entire United States is 61.7%. (See Table 11.) However, we would change our "guess" to 72 if we knew the 100 individuals had at least a bachelor's degree. Do you see why? The association between level of education and employment status allows us to adjust our predictions.

● Now Work Problem 9(d)

In Example 4, we were able to see how a change in level of education affected employment status. Therefore, we are treating level of education (the column variable) as the explanatory variable and employment status as the response variable.

We could also construct a conditional distribution of level of education by employment status. In this situation, we would be treating employment status (the row variable) as the explanatory variable. The procedure is the same, except the distribution uses the rows instead of the columns. In this case, the explanatory variable is employment status and the response variable is level of education.

As is usually the case, a graph can provide a powerful depiction of the data.

EXAMPLE 5   Drawing a Bar Graph of a Conditional Distribution

Problem Using the results of Example 4, draw a bar graph that represents the conditional distribution of employment status by level of education.

Approach When drawing conditional bar graphs, label the values of the explanatory variable on the horizontal axis, and use different colored bars for each value of the response variable. So, draw three bars for each level of education. Let the horizontal axis represent level of education and the vertical axis represent the relative frequency.

(continued)

Solution See Figure 28. It is clear that as level of education increases, the proportion employed also increases. As the level of education increases, the proportion not in the labor force decreases.

Figure 28

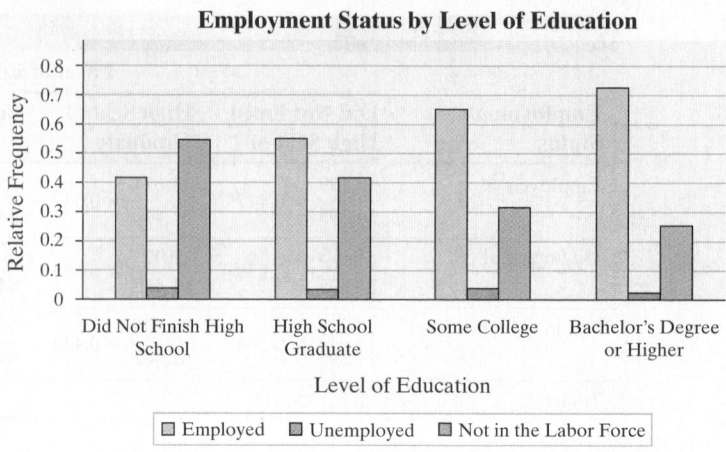

Instead of drawing a conditional bar graph as shown in Figure 28, we can draw *stacked* (or *segmented*) *bar graphs*. In these graphs, there is one bar for each value of the explanatory variable. Each bar is then divided into segments such that the height of each segment within a bar represents the proportion of observations corresponding to the response variable. Figure 29 shows the stacked bar graph for the data in Table 12. The height of the blue portion in each segment is the proportion employed for each level of education, the height of the orange portion in each segment is the proportion unemployed for each level of education, and the height of the green bar is the proportion not in the labor force for each level of education. Again, we can see as level of education increases, the proportion employed increases while the proportion not in the labor force decreases.

Figure 29

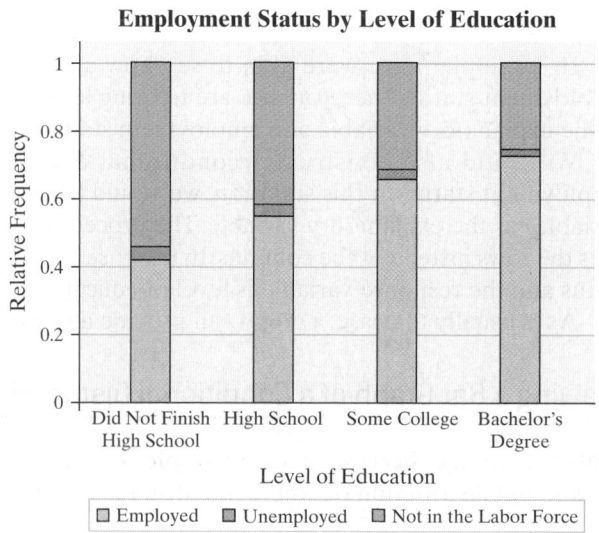

- **Now Work Problems 9(e)
and (f)**

The methods presented in this section for identifying the association between two categorical variables are different from the methods for measuring association between two quantitative variables. The measure of association in this section is based on whether there are differences in the relative frequencies of the response variable (employment status) for the different categories of the explanatory variable (level of education). If differences exist, we might attribute these differences to the explanatory variable.

In addition, because the data in Example 1 are observational, we do not make any statements regarding causation. Level of education is not said to be a cause of employment status, because a controlled experiment was not conducted.

3 Explain Simpson's Paradox

In Section 4.1, we discussed how a lurking variable can cause two quantitative variables to be correlated even though they are unrelated. This same phenomenon exists when exploring the relation between two qualitative variables.

EXAMPLE 6 Gender Bias at the University of California, Berkeley

Problem The data in Table 13 show the admission status and gender of students who applied to the University of California, Berkeley. From the data in Table 13, the proportion of accepted applications is $\frac{1748}{4425} = 0.395$. The proportion of accepted men is $\frac{1191}{2590} = 0.460$ and the proportion of accepted women is $\frac{557}{1835} = 0.304$. On the basis of these proportions, a gender bias suit was brought against the university. The university was shocked and claimed that program of study is a lurking variable that created the apparent association between admission status and gender. The university supplied Table 14 in its defense. Develop a conditional distribution by program of study to defend the university's admission policies.

Source: P. J. Bickel, E. A. Hammel, and J. W. O'Connell. "Sex Bias in Graduate Admissions: Data from Berkeley." *Science* 187(4175): 398–404, 1975.

Table 13

	Accepted (A)	Not Accepted (NA)	Total
Men	1191	1399	**2590**
Women	557	1278	**1835**
Total	**1748**	**2677**	**4425**

Table 14 Admission Status (Accepted, A, or Not Accepted, NA), for Six Programs of Study (A, B, C, D, E, F) by Gender

	A		B		C		D		E		F	
Men	A	NA	A	NA	A	NA	A	NA	A	NA	A	NA
	511	314	353	207	120	205	138	279	53	138	16	256
Women	A	NA	A	NA	A	NA	A	NA	A	NA	A	NA
	89	19	17	8	202	391	131	244	94	299	24	317

Approach Determine the proportion of accepted men for each program of study and separately determine the proportion of accepted women for each program of study. A significant difference between the proportions of men and women accepted within each program of study may be evidence of discrimination; otherwise, the university should be exonerated.

Solution The proportion of men who applied to program A and were accepted is $\frac{511}{511 + 314} = 0.619$. The proportion of women who applied to program A and were

(continued)

accepted is $\dfrac{89}{89 + 19} = 0.824$. So, within program A, a higher proportion of women were accepted. Table 15 shows the conditional distribution for the remaining programs.

Table 15 Conditional Distribution of Applicants Admitted for Men and Women by Program of Admission						
	A	**B**	**C**	**D**	**E**	**F**
Men	$\dfrac{511}{825} = 0.619$	0.630	0.369	0.331	0.277	0.059
Women	0.824	0.680	0.341	0.349	0.239	0.070

Figure 30 shows the bar graph of the conditional distribution in Table 15. The blue bars represent the proportion of men admitted for each program; the green bars represent the proportion of women admitted for each program.

Figure 30

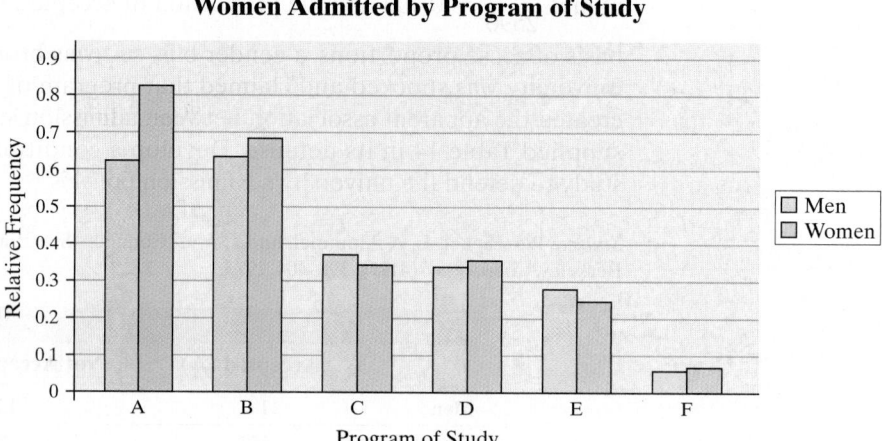

Four of the six programs actually had a higher proportion of women accepted! And the proportion of men accepted in programs C and E is not much higher than the proportion of women.

What caused the overall proportion of accepted men to be so much higher than the overall proportion of accepted women within the entire university, when, within each program, the proportions differ very little and may imply that women are accepted at a higher rate? The initial analysis did not account for the lurking variable, program of study. There were many more male applicants in programs A and B than female applicants, and these two programs happen to have higher acceptance rates. The higher acceptance rates in these programs led to the false conclusion that the University of California, Berkeley, was biased against gender in its admissions.   ●

In Example 6, the association between gender and admission reverses when the lurking variable program of study is accounted for in the analysis. This illustrates a phenomenon known as **Simpson's Paradox**, which describes a situation in which an association between two variables inverts or goes away when a third variable is introduced to the analysis. The results of Example 6 serve as a warning that an apparent association between two variables may not exist once the effect of some third variable is accounted for in the study.

● **Now Work Problem 13**

Technology Step-by-Step   Contingency Tables and Association

Minitab

1. Enter the values of the row variable in column C1 and the corresponding values of the column variable in C2. The frequency for the cell is entered in C3. For example, the data in Table 9 would be entered as follows:

↓	C1-T	C2-T	C3
	Employment Status	Level of Education	
1	Employed	No High School	10179
2	Unemployed	No High School	945
3	Not in Labor Force	No High School	13271
4	Employed	HS	33624
5	Unemployed	HS	2012
6	Not in Labor Force	HS	25806
7	Employed	Some College	35407
8	Unemployed	Some College	1823
9	Not in Labor Force	Some College	17089
10	Employed	Bachelors	49534
11	Unemployed	Bachelors	1615
12	Not in Labor Force	Bachelors	17415

2. Select the **Stat** menu and highlight **Tables**. Then select **Descriptive Statistics…**
3. In the cell "For Rows:" enter C1. In the cell "For Columns:" enter C2. In the cell "Frequencies" enter C3. Click the Options button and make sure the radio button for Display marginal statistics for Rows and columns is checked. Click OK. Click the Categorical Variables button and then select the summaries you desire. Click OK twice.

Statcrunch
Determining Marginal and Conditional Distributions
1. Enter the contingency table into the spreadsheet. The first column should be the row variable. For example, for the data in Table 9, the first column would be employment status. Each subsequent column would be

the counts of each category of the column variable. For the data in Table 9, enter the counts for each level of education. Title each column (including the first column indicating the row variable).
2. Select **Stat**, highlight **Tables**, and then **Contingency**, then select **With Summary**.
3. Select the column variables. Then select the label of the row variable. For example, the data in Table 9 has four column variables (Did Not Finish High School, and so on) and the row label is employment status. To obtain relative frequency marginal distributions and conditional distributions, choose row percent and column percent under Display. Click Compute!.

Drawing Stacked Bar Graphs of Conditional Distributions
1. Select **Applets**, highlight **Contingency Table**.
2. Under Row labels, enter the values of the variables in the rows of the contingency table. Under Column labels, enter the values of the variables in the columns of the contingency table. For the data in Table 9, enter the labels as shown below. Click Compute!

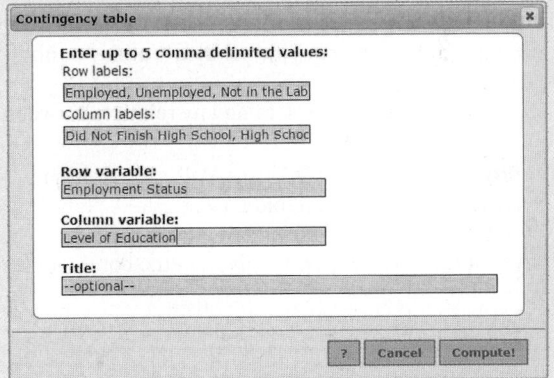

3. Enter the counts in the contingency table. If you want the conditional distribution by the row variable, select "Row distribution." If you want the conditional distribution by the column variable, select "Column distribution."

 4.4 Assess Your Understanding

Vocabulary and Skill Building

1. What is meant by a marginal distribution? What is meant by a conditional distribution?

2. Refer to Table 9. Is constructing a conditional distribution by level of education different from constructing a conditional distribution by employment status? If they are different, explain the difference.

3. Explain why we use the term *association* rather than *correlation* when describing the relation between two variables in this section.

4. Explain the idea behind Simpson's Paradox.

In Problems 5 and 6,
(a) *Construct a frequency marginal distribution.*
(b) *Construct a relative frequency marginal distribution.*
(c) *Construct a conditional distribution by x.*
(d) *Draw a bar graph of the conditional distribution found in part (c).*

5.

	x_1	x_2	x_3
y_1	20	25	30
y_2	30	25	50

6.

	x_1	x_2	x_3
y_1	35	25	20
y_2	65	75	80

Applying the Concepts

7. Made in America In a recent Harris Poll, a random sample of adult Americans (18 years and older) was asked, "When you see an ad emphasizing that a product is 'Made in America,' are you more likely to buy it, less likely to buy it, or neither more nor less likely to buy it?" The results of the survey, by age group, are presented in the contingency table below.

	18–34	35–44	45–54	55+	Total
More likely	238	329	360	402	1329
Less likely	22	6	22	16	66
Neither more nor less likely	282	201	164	118	765
Total	542	536	546	536	2160

Source: The Harris Poll

(a) How many adult Americans were surveyed? How many were 55 and older?
(b) Construct a relative frequency marginal distribution.
(c) What proportion of Americans are more likely to buy a product when the ad says "Made in America"?
(d) Construct a conditional distribution of likelihood to buy "Made in America" by age. That is, construct a conditional distribution treating age as the explanatory variable.
(e) Draw a bar graph of the conditional distribution found in part (d).
(f) Write a couple sentences explaining any relation between likelihood to buy and age.

8. Desirability Traits In a recent Harris Poll, a random sample of adult Americans (18 years and older) was asked, "Given a choice of the following, which one would you most want to be?" Results of the survey, by gender, are given in the contingency table.

	Richer	Thinner	Smarter	Younger	None of these	Total
Male	520	158	159	181	102	1120
Female	425	300	144	81	92	1042
Total	945	458	303	262	194	2162

Source: The Harris Poll

(a) How many adult Americans were surveyed? How many males were surveyed?
(b) Construct a relative frequency marginal distribution.
(c) What proportion of adult Americans want to be richer?
(d) Construct a conditional distribution of desired trait by gender. That is, construct a conditional distribution treating gender as the explanatory variable.
(e) Draw a bar graph of the conditional distribution found in part (d).
(f) Write a couple sentences explaining any relation between desired trait and gender.

NW 9. Party Affiliation Is there an association between party affiliation and gender? The data in the next column represent the gender and party affiliation of registered voters based on a random sample of 802 adults.

	Female	Male
Republican	105	115
Democrat	150	103
Independent	150	179

Source: Star Tribune Minnesota Poll

(a) Construct a frequency marginal distribution.
(b) Construct a relative frequency marginal distribution.
(c) What proportion of registered voters considers themselves to be Independent?
(d) Construct a conditional distribution of party affiliation by gender.
(e) Draw a bar graph of the conditional distribution found in part (d).
(f) Is gender associated with party affiliation? If so, how?

10. Feelings on Abortion The Pew Research Center for the People and the Press conducted a poll in which it asked individuals to disclose their level of education and feelings about the availability of abortion. The table is based on the results of the survey.

	High School or Less	Some College	College Graduate
Generally available	90	72	113
Allowed, but more limited	51	60	77
Illegal, with few exceptions	125	94	69
Never permitted	51	14	17

Source: Pew Research Center for the People and the Press

(a) Construct a frequency marginal distribution.
(b) Construct a relative frequency marginal distribution.
(c) What proportion of college graduates feel that abortion should never be permitted?
(d) Construct a conditional distribution of people's feelings about the availability of abortion by level of education.
(e) Draw a bar graph of the conditional distribution found in part (d).
(f) Is level of education associated with opinion on the availability of abortion? If so, how?

11. Health and Happiness The General Social Survey asks questions about one's happiness and health. One would think that health plays a role in one's happiness. Use the data in the table to determine whether healthier people tend to also be happier. Treat level of health as the explanatory variable.

	Poor	Fair	Good	Excellent	Total
Not too happy	696	1,386	1,629	732	4,443
Pretty happy	950	3,817	9,642	5,195	19,604
Very happy	350	1,382	4,520	5,095	11,347
Total	1,996	6,585	15,791	11,022	35,394

Source: General Social Survey

12. Happy in Your Marriage? The General Social Survey asks questions about one's happiness in marriage. Is there an association between gender and happiness in marriage? Use the data in the table to determine if gender is associated with happiness in marriage. Treat gender as the explanatory variable.

	Male	Female	Total
Very happy	7,609	7,942	15,551
Pretty happy	3,738	4,447	8,185
Not too happy	259	460	719
Total	11,606	12,849	24,455

Source: General Social Survey

NW **13. Smoking Is Healthy?** Could it be that smoking actually increases survival rates among women? The following data represent the 20-year survival status and smoking status of 1314 English women who participated in a cohort study from 1972 to 1992.

	Smoking Status		
	Smoker (S)	Nonsmoker (NS)	Total
Dead	139	230	369
Alive	443	502	945
Total	582	732	1314

Source: David R. Appleton et al. "Ignoring a Covariate: An Example of Simpson's Paradox." *American Statistician* 50(4), 1996

(a) What proportion of the smokers was dead after 20 years? What proportion of the nonsmokers was dead after 20 years? What does this imply about the health consequences of smoking?

The data in the table above do not take into account a variable that is strongly related to survival status, age. The data shown next give the survival status of women and their age at the beginning of the study. For example, 14 women who were 35 to 44 at the beginning of the study were smokers and dead after 20 years.

	Age Group													
	18–24		25–34		35–44		45–54		55–64		65–74		75 or older	
	S	NS	S	NS	S	NS	S	NS	S	NS	S	NS	S	NS
Dead	2	1	3	5	14	7	27	12	51	40	29	101	13	64
Alive	53	61	121	152	95	114	103	66	64	81	7	28	0	0

(b) Determine the proportion of 18- to 24-year-old smokers who were dead after 20 years. Determine the proportion of 18- to 24-year-old nonsmokers who were dead after 20 years.

(c) Repeat part (b) for the remaining age groups to create a conditional distribution of survival status by smoking status for each age group.

(d) Draw a bar graph of the conditional distribution from part (c).

(e) Write a short report detailing your findings.

14. Treating Kidney Stones Researchers conducted a study to determine which of two treatments, A or B, is more effective in the treatment of kidney stones. The results of their experiment are given in the table.

	Treatment A	Treatment B	Total
Effective	273	289	562
Not effective	77	61	138
Total	350	350	700

Source: C. R. Charig, D. R. Webb, S. R. Payne, and O. E. Wickham. "Comparison of Treatment Real Calculi by Operative Surgery, Percutaneous Nephrolithotomi, and Extracorporeal Shock Wave Lithotripsy." *British Medical Journal* 292(6524): 879–882.

(a) Which treatment appears to be more effective? Why?

The data in the table above do not take into account the size of the kidney stone. The data shown next indicate the effectiveness of each treatment for both large and small kidney stones.

	Small Stones		Large Stones	
Effective	A	B	A	B
	81	234	273	55
Not effective	6	36	77	25

(b) Determine the proportion of small kidney stones that were effectively dealt with using treatment A. Determine the proportion of small kidney stones that were effectively dealt with using treatment B.

(c) Repeat part (b) for the large stones to create a conditional distribution of effectiveness by treatment for each stone size.

(d) Draw a bar graph of the conditional distribution from part (c).

(e) Write a short report detailing your findings.

Retain Your Knowledge

15. Sullivan Survey II Go to www.pearsonhighered.com/sullivanstats to obtain the data file SullivanStatsSurveyII.

(a) Create a relative frequency distribution for political philosophy. What percent of the respondents are moderate?

(b) Draw a relative frequency bar graph for political philosophy.

(c) The column "GenderIncomeInequality" represents the response to the question, "Do you believe there is an income inequality discrepancy between males and females when each has the same experience and education?" Draw a pie chart for "GenderIncomeInequality." What do the results suggest?

(d) Is one's gender associated with the response to the gender income inequality question? Build a contingency table for these two variables treating gender as the row variable. Show the relative marginal distribution for both the column and row variables.

(e) Construct a conditional distribution by gender. What do you notice?

(f) Draw a bar graph of the conditional distribution from part (e).

Chapter 4 Review

Summary

This is the last chapter in Part II—Describe the Data. In Chapters 2 and 3, we described univariate data. That is, we summarized data where a single variable was measured on each individual. For example, we might consider the weight of a random sample of 30 students. In this chapter, we described bivariate data. That is, we measured two variables (such as height and weight) on each individual in the study. In particular, we looked at describing the relation between two quantitative variables (Sections 4.1–4.3) and between two qualitative variables (Section 4.4).

The first step in describing the relation between two quantitative variables is to draw a scatter diagram. The explanatory variable is plotted on the horizontal axis and the corresponding response variable on the vertical axis. The scatter diagram can be used to discover whether the relation between the explanatory and the response variables is linear. In addition, for linear relations, we can judge whether the linear relation shows positive or negative association.

A numerical measure for the strength of linear relation between two quantitative variables is the linear correlation coefficient. It is a number between −1 and 1, inclusive. Values of the correlation coefficient near −1 are indicative of a negative linear relation between the two variables. Values of the correlation coefficient near +1 indicate a positive linear relation between the two variables. If the correlation coefficient is near 0, then little *linear* relation exists between the two variables.

Be careful! Just because the correlation coefficient between two quantitative variables indicates that the variables are linearly related, it does not mean that a change in one variable *causes* a change in a second variable. It could be that the correlation is the result of a lurking variable.

Once a linear relation between the two variables has been discovered, we describe the relation by finding the least-squares regression line. This line best describes the linear relation between the explanatory and response variables. We can use the least-squares regression line to predict a value of the response variable for a given value of the explanatory variable.

The coefficient of determination, R^2, measures the percent of variation in the response variable that is explained by the least-squares regression line. It is a measure between 0 and 1, inclusive. The closer R^2 is to 1, the more explanatory value the line has. Whenever a least-squares regression line is obtained, certain diagnostics must be performed. These include verifying that the linear model is appropriate, verifying the residuals have constant variance, and checking for outliers and influential observations.

Section 4.4 introduced methods that allow us to describe any association that might exist between two qualitative variables. This is done through contingency tables. Both marginal and conditional distributions allow us to describe the effect one variable might have on the other variable in the study. We also construct bar graphs to see the association between the two variables in the study. Again, just because two qualitative variables are associated does not mean that a change in one variable *causes* a change in a second variable. We also looked at Simpson's Paradox, which represents situations in which an association between two variables inverts or goes away when a third (lurking) variable is introduced into the analysis.

Vocabulary

Univariate data (p. 180)	Residual (p. 199)	Constant error variance (p. 216)
Bivariate data (p. 180)	Least-squares regression line (p. 199)	Outlier (p. 216)
Response variable (p. 181)	Slope (p. 200)	Homoscedasticity (p. 216)
Explanatory variable (p. 181)	y-intercept (p. 200)	Influential observation (p. 218)
Predictor variable (p. 181)	Outside the scope of the model (p. 203)	Contingency (or two-way) table (p. 225)
Scatter diagram (p. 181)	Coefficient of determination (p. 211)	Row variable (p. 225)
Positively associated (p. 182)	Deviation (p. 211)	Column variable (p. 225)
Negatively associated (p. 182)	Total deviation (p. 211)	Cell (p. 225)
Linear correlation coefficient (p. 183)	Explained deviation (p. 211)	Marginal distribution (p. 225)
Correlation matrix (p. 187)	Unexplained deviation (p. 212)	Conditional distribution (p. 228)
Lurking variable (p. 188)	Residual plot (p. 214)	Simpson's Paradox (p. 232)

Formulas

Correlation Coefficient

$$r = \frac{\sum \left(\frac{x_i - \bar{x}}{s_x} \right) \left(\frac{y_i - \bar{y}}{s_y} \right)}{n - 1}$$

Equation of the Least-Squares Regression Line

$$\hat{y} = b_1 x + b_0$$

where
$\hat{y}$ is the predicted value of the response variable
$b_1 = r \cdot \frac{s_y}{s_x}$ is the slope of the least-squares regression line
$b_0 = \bar{y} - b_1 \bar{x}$ is the y-intercept of the least-squares regression line

Coefficient of Determination, R^2

$$R^2 = \frac{\text{explained variation}}{\text{total variation}}$$

$$= 1 - \frac{\text{unexplained variation}}{\text{total variation}}$$

$$= r^2 \text{ for the least-squares regression model } \hat{y} = b_1 x + b_0$$

OBJECTIVES

Section	You should be able to . . .	Example(s)	Review Exercises
4.1	1 Draw and interpret scatter diagrams (p. 181)	1, 3	2(b), 3(a), 6(a), 13(a)
	2 Describe the properties of the linear correlation coefficient (p. 183)	page 184	18
	3 Compute and interpret the linear correlation coefficient (p. 185)	2, 3	2(c), 3(b), 13(b)
	4 Determine whether a linear relation exists between two variables (p. 187)	4	2(d), 3(c)
	5 Explain the difference between correlation and causation (p. 188)	5	14, 17
4.2	1 Find the least-squares regression line and use the line to make predictions (p. 198)	2, 3	1(a), 1(b), 4(a), 4(d), 5(a), 5(c), 6(d), 12(a), 13(c), 19(c)
	2 Interpret the slope and y-intercept of the least-squares regression line (p. 202)	page 203	1(c), 1(d), 4(c), 5(b), 19(b)
	3 Compute the sum of squared residuals (p. 203)	4	6(f), 6(g)
4.3	1 Compute and interpret the coefficient of determination (p. 211)	1	1(e), 10(a), 11(a)
	2 Perform residual analysis on a regression model (p. 214)	2–5	7–9, 10(b) and (c), 11(b) and (c), 13(d) and (e), 19(d)
	3 Identify influential observations (p. 218)	6	10(d), 10(e), 11(d), 12(b), 19(e)
4.4	1 Compute the marginal distribution of a variable (p. 225)	1 and 2	15(b)
	2 Use the conditional distribution to identify association among categorical data (p. 227)	3–5	15(d), 15(e), 15(f)
	3 Explain Simpson's Paradox (p. 231)	6	16

Review Exercises

1. Basketball Spreads In sports betting, Las Vegas sports books establish winning margins for a team that is favored to win a game. An individual can place a wager on the game and will win if the team bet upon wins after accounting for the spread. For example, if Team A is favored by 5 points and wins the game by 7 points, then a bet on Team A is a winning bet. However, if Team A wins the game by only 3 points, then a bet on Team A is a losing bet. For NCAA Division I basketball games, a least-squares regression with explanatory variable home team Las Vegas spread, x, and response variable home team winning margin, y, is $\hat{y} = 1.007x - 0.012$.

Source: Justin Wolfers. "Point Shaving: Corruption in NCAA Basketball"

(a) Predict the winning margin if the home team is favored by 3 points.
(b) Predict the winning margin (of the visiting team) if the visiting team is favored by 7 points (this is equivalent to the home team being favored by −7 points).
(c) Interpret the slope.
(d) Interpret the y-intercept.
(e) The coefficient of determination is 0.39. Interpret this value.

2. Fat and Calories in Cheeseburgers A nutritionist was interested in developing a model that describes the relation between the amount of fat (in grams) in cheeseburgers at fast-food restaurants and the number of calories. She obtains the following data from the websites of the companies.

Sandwich (Restaurant)	Fat Content (g)	Calories
Quarter-pound Single with Cheese (Wendy's)	20	430
Whataburger (Whataburger)	39	750
Cheeseburger (In-n-Out)	27	480
Big Mac (McDonald's)	29	540
Quarter-pounder with cheese (McDonald's)	26	510
Whopper with cheese (Burger King)	47	760
Jumbo Jack (Jack in the Box)	35	690
Double Steakburger with cheese (Steak 'n Shake)	38	632

Source: Each company's Web site

(a) The researcher wants to use fat content to predict calories. Which is the explanatory variable?
(b) Draw a scatter diagram of the data.
(c) Compute the linear correlation coefficient between fat content and calories.
(d) Does a linear relation exist between fat content and calories in fast-food restaurant sandwiches?

3. Apartments The following data represent the square footage and rent for apartments in the borough of Queens and Nassau County, New York.

Queens (New York City)		Nassau County (Long Island)	
Square Footage, x	Rent per Month, y	Square Footage, x	Rent per Month, y
500	650	1100	1875
588	1215	588	1075
1000	2000	1250	1775
688	1655	556	1050
825	1250	825	1300
460	1805	743	1475
1259	2700	660	1315
650	1200	975	1400
560	1250	1429	1900
1073	2350	800	1650
1452	3300	1906	4625
1305	3100	1077	1395

Source: apartments.com

(a) On the same graph, draw a scatter diagram for both Queens and Nassau County apartments treating square footage as the explanatory variable. Be sure to use a different plotting symbol for each group.
(b) Compute the linear correlation coefficient between square footage and rent for each location.
(c) Does a linear relation exist between the two variables for each location?
(d) Does location appear to be a factor in rent?

4. Using the data and results from Problem 2, do the following:
(a) Find the least-squares regression line treating fat content as the explanatory variable.
(b) Draw the least-squares regression line on the scatter diagram.
(c) Interpret the slope and y-intercept, if appropriate.
(d) Predict the number of calories in a sandwich that has 30 grams of fat.
(e) A cheeseburger from Sonic has 700 calories and 42 grams of fat. Is the number of calories for this sandwich above or below average among all sandwiches with 42 grams?

5. Using the Queens data and results from Problem 3, do the following:
(a) Find the least-squares regression line, treating square footage as the explanatory variable.
(b) Interpret the slope and y-intercept, if appropriate.
(c) Is the rent on the 825-square-foot apartment in the data above or below average among 825-square-foot apartments?

6.

x	10	14	17	18	21
y	105	94	82	76	63

(a) Draw a scatter diagram treating x as the explanatory variable and y as the response variable.
(b) Select two points from the scatter diagram, and find the equation of the line containing the points selected.
(c) Graph the line found in part (b) on the scatter diagram.
(d) Determine the least-squares regression line.
(e) Graph the least-squares regression line on the scatter diagram.
(f) Compute the sum of the squared residuals for the line found in part (b).
(g) Compute the sum of the squared residuals for the least-squares regression line found in part (d).
(h) Comment on the fit of the line found in part (b) versus the least-squares regression line found in part (d).

In Problems 7–9, a residual plot is given. From the residual plot, determine whether or not a linear model is appropriate. If not, state your reason.

7. **8.**

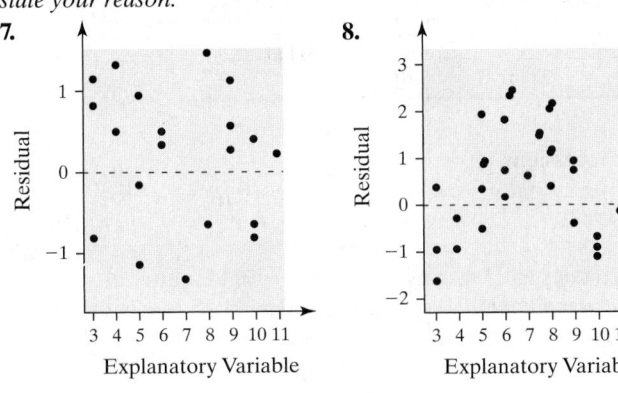

9.

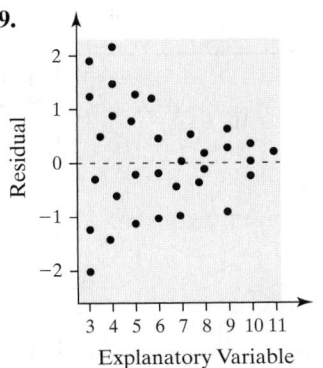

10. Use the results from Problems 2 and 4 for the following:
(a) Compute and interpret R^2.
(b) Plot the residuals against fat content.
(c) Based on the residual plot drawn in part (b), is a linear model appropriate for describing the relation between fat content and calories?
(d) Are there any outliers or influential observations?
(e) Suppose a White Castle cheeseburger (affectionately known as a slider) is added to the data set. Sliders have 9 grams of fat and 80 calories. Is this observation an outlier or influential?

11. Use Queens data and the results from Problems 3 and 5 for the following:
(a) Compute and interpret R^2.
(b) Plot the residuals against square footage.
(c) For the graph obtained in part (b), is a linear model appropriate?
(d) Are there any outliers or influential observations?

12. Apartment Rents Using the Nassau County apartment rent data and the results of Problem 3, do the following:
(a) Find the least-squares regression line treating square footage as the explanatory variable.
(b) Are there any outliers or influential observations?

13. Depreciation The following data represent the price of a random sample of used Chevy Camaros, by age.

Age, x (years)	Price, y (dollars)	Age, x (years)	Price, y (dollars)
2	15,900	1	20,365
5	10,988	2	16,463
2	16,980	6	10,824
5	9,995	1	19,995
4	11,995	1	18,650
5	10,995	4	10,488

Source: www.onlineauto.com

(a) Draw a scatter diagram treating the age of the car as the explanatory variable and price as the response variable.
(b) Find the correlation coefficient between age and price. Based on the correlation coefficient, is there a linear relation between age and price?
(c) Find the least-squares regression line treating age as the explanatory variable.
(d) Plot the residuals against the age of the car.
(e) Do you think that a linear model is appropriate for describing the relation between the age of the car and price? Why?

14. Shark Attacks The correlation between the number of visitors to the state of Florida and the number of shark attacks since 1990 is 0.946. Should the number of visitors to Florida be reduced in an attempt to reduce shark attacks? Explain your reasoning.
Source: Florida Museum of Natural History

15. New versus Used Car Satisfaction Are you more likely to be satisfied with your automobile purchase when it is new or used? The following data represent the level of satisfaction of the buyer for both new and used cars.

	New	Used	Total
Not too satisfied	11	25	36
Pretty satisfied	78	79	157
Extremely satisfied	118	85	203
Total	207	189	396

Source: General Social Survey

(a) How many were extremely satisfied with their automobile purchase?
(b) Construct a relative frequency marginal distribution.
(c) What proportion of consumers was extremely satisfied with their automobile purchase?
(d) Construct a conditional distribution of satisfaction by purchase type (new or used).
(e) Draw a bar graph of the conditional distribution found in part (d).
(f) Do you think that the purchase type (new versus used) is associated with satisfaction?

16. Unemployment Rates Recessions are an economic phenomenon that are often defined as two consecutive quarters of reduced national output. One measure to assess the severity of a recession is the rate of unemployment. The table shows the number of employed and unemployed residents of the United States at the peak of each recession (in thousands).

	Recession of 1982	Recession of 2009
Employed	98.9	130.1
Unemployed	11.3	14.5

Source: Bureau of Labor Statistics

(a) Determine the unemployment rate for each recession. Which recession appears worse as measured by unemployment rate? Note: Unemployment rate = unemployed/(employed + unemployed).

The data in the table above do not account for level of education. The following data show the unemployment rate by level of education for each recession.

	Recession of 1982			Recession of 2009		
	Less than high school	High school	Bachelor's degree or higher	Less than high school	High school	Bachelor's degree or higher
Employed	20.3	58.2	20.4	10.0	76.7	43.4
Unemployed	3.9	6.6	0.8	2.0	10.3	2.2

Source: Bureau of Labor Statistics

(b) Determine the unemployment rate for each level of education for both recessions.
(c) Draw a bar graph of the conditional distribution from part (b).
(d) Write a report that suggests the recession of 2009 is worse than that of 1982.

---

17. (a) The correlation between number of married residents and number of unemployed residents for the 50 states and Washington, DC, is 0.922. A scatter diagram of the data is shown. What type of relation appears to exist between number of marriages and number unemployed?

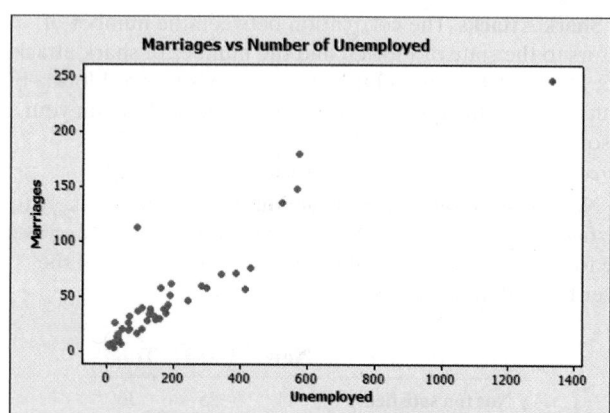

(b) A lurking variable is in the relation presented in part (a). Use the following correlation matrix to explain how population is a lurking variable.

Correlations: Unemployed, Population, Marriages

```
              Unemployed    Population
Population      0.979
Marriages       0.922        0.947

Cell Contents: Pearson correlation
```

(c) The correlation between unemployment rate (number unemployed divided by population size) and marriage rate (number married divided by population size) for the

50 states and Washington, DC, is 0.050. A scatter diagram between unemployment rate and marriage rate is shown next. What type of relation appears to exist between unemployment rate and marriage rate?

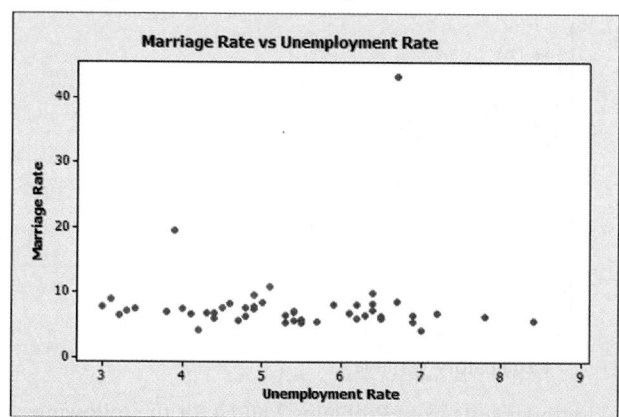

(d) Write a few sentences to explain the danger in using correlation to conclude that a relation exists between two variables without considering lurking variables.

18. List the eight properties of the linear correlation coefficient.

19. Analyzing a Newspaper Article In a newspaper article written in the *Chicago Tribune*, it was claimed that poorer school districts have shorter school days.

(a) The following scatter diagram was drawn using the data supplied in the article. In this scatter diagram, the response variable is length of the school day (in hours) and the explanatory variable is percent of the population that is low income. The correlation between length and income is −0.461. Do you think that the scatter diagram and correlation coefficient support the position of the article?

Length vs. Income

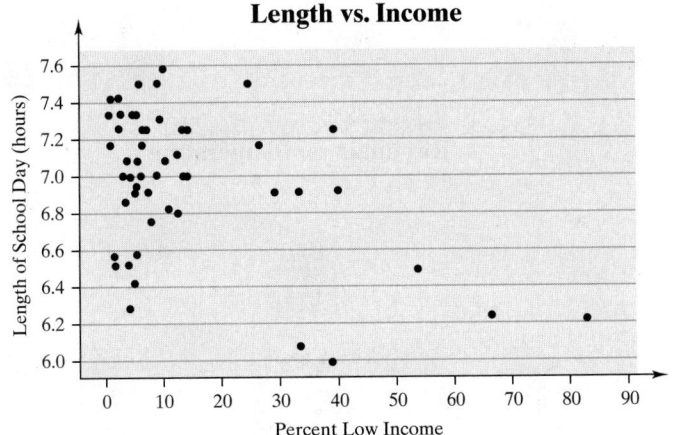

(b) The least-squares regression line between length, y, and income, x, is $\hat{y} = -0.0102\,x + 7.11$. Interpret the slope of this regression line. Does it make sense to interpret the y-intercept? If so, interpret the y-intercept.

(c) Predict the length of the school day for a district in which 20% of the population is low income by letting $x = 20$.

(d) Based on the following residual plot, do you think a linear model is appropriate for describing the relation between length of school day and income? Why?

Residuals vs. Income

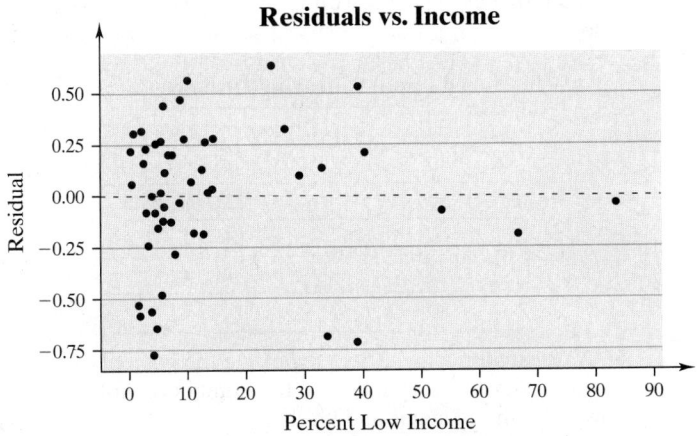

(e) Included in the output from Minitab was a notification that three observations were influential. Based on the scatter diagram in part (a), which three observations do you think might be influential?

(f) We should not remove influential observations just because they are influential. What suggestions would you make to the author of the article to improve the study and therefore give more credibility to the conclusions?

(g) This same article included average Prairie State Achievement Examination (PSAE) scores for each district. The article implied that shorter school days result in lower

PSAE scores. The correlation between PSAE score and length of school day is 0.517. A scatter diagram treating PSAE as the response variable is shown next. Do you believe that a longer school day is positively associated with a higher PSAE score?

PSAE vs. Length

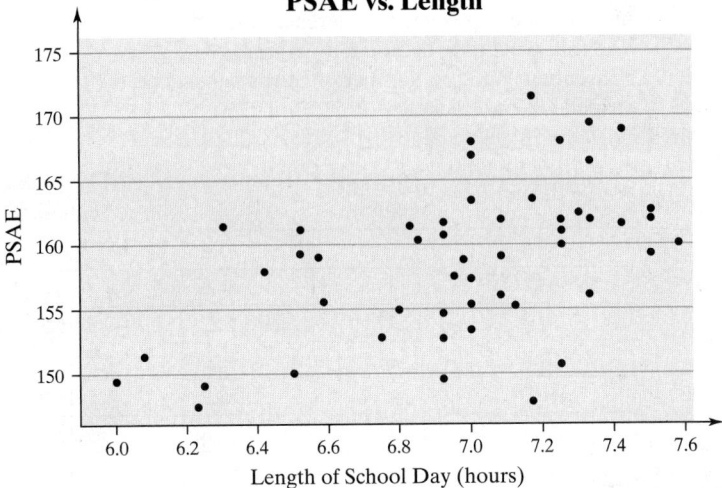

(h) The correlation between percentage of the population that is low income and PSAE score is -0.720. A scatter diagram treating PSAE score as the response variable is shown next. Do you believe that percentage of the population that is low income is negatively associated with PSAE score?

PSAE vs. Income

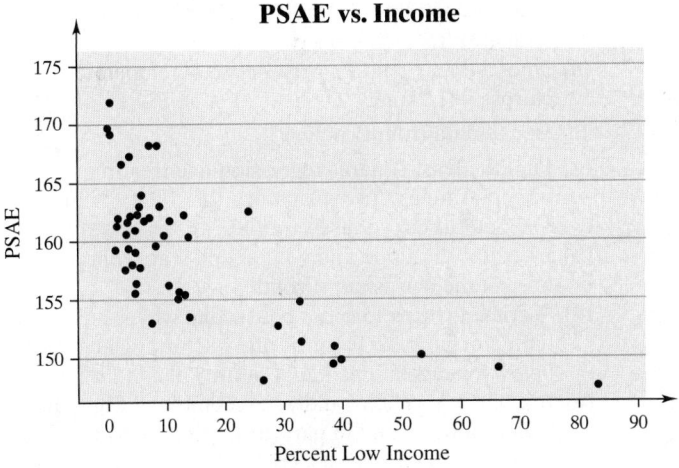

(i) Can you think of any lurking variables that are playing a role in this study?

Chapter Test

1. Crickets make a chirping noise by sliding their wings rapidly over each other. Perhaps you have noticed that the number of chirps seems to increase with the temperature. The following data list the temperature (in degrees Fahrenheit) and the number of chirps per second for the striped ground cricket.

Temperature, x	Chirps per Second, y	Temperature, x	Chirps per Second, y
88.6	20.0	71.6	16.0
93.3	19.8	84.3	18.4
80.6	17.1	75.2	15.5
69.7	14.7	82.0	17.1
69.4	15.4	83.3	16.2
79.6	15.0	82.6	17.2
80.6	16.0	83.5	17.0
76.3	14.4		

Source: George W. Pierce. *The Songs of Insects.* Cambridge, MA: Harvard University Press, 1949, pp. 12–21.

(a) What is the likely explanatory variable in these data? Why?
(b) Draw a scatter diagram of the data.
(c) Compute the linear correlation coefficient between temperature and chirps per second.
(d) Does a linear relation exist between temperature and chirps per second?

2. Use the data from Problem 1.

(a) Find the least-squares regression line treating temperature as the explanatory variable and chirps per second as the response variable.
(b) Interpret the slope and y-intercept, if appropriate.
(c) Predict the chirps per second if it is 83.3°F.
(d) A cricket chirps 15 times per second when the temperature is 82°F. Is this rate of chirping above or below average at this temperature?
(e) It is 55°F outside. Would you recommend using the linear model found in part (a) to predict the number of chirps per second of a cricket? Why or why not?

3. Use the results from Problems 1 and 2.

(a) Compute and interpret R^2.
(b) A plot of residuals against temperature is shown in the next column. Based on the residual plot, do you think a linear model is appropriate for describing the relation between temperature and chirps per second?

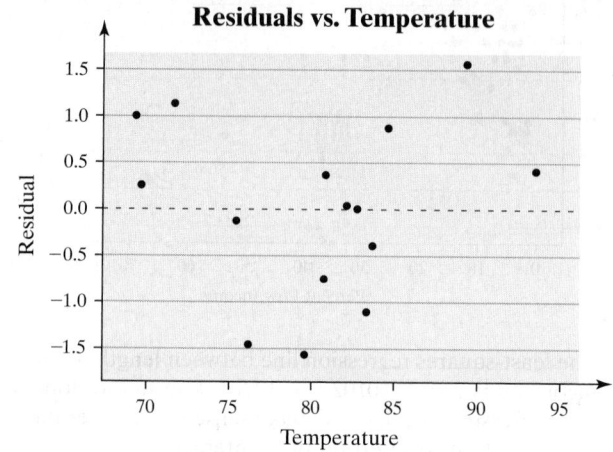

Residuals vs. Temperature

4. Use the results from Problems 1 through 3. A cricket chirps 14.5 times per second when it is 92°F outside. Is this observation influential? Why?

5. The following data represent the speed of a car (in miles per hour) and its braking distance (feet) on dry asphalt. Is the relation between speed and braking distance linear? Why?

Speed (mph)	Braking Distance (ft)
30	48
40	79
50	123
60	184
70	243
80	315

6. A researcher collects data regarding the percent of all births to unmarried women and the number of violent crimes for the 50 states and Washington, DC. The scatter diagram along with the least-squares regression line obtained from Minitab is shown. The correlation between percent of births to unmarried women and the number of violent crimes is 0.667. A politician looks over the data and claims that, for each 1% decrease in births to single mothers, violent crimes will decrease by 23. Therefore, percent of births to single mothers needs to be reduced in an effort to decrease violent crimes. Is there anything wrong with the reasoning of the politician? Explain.

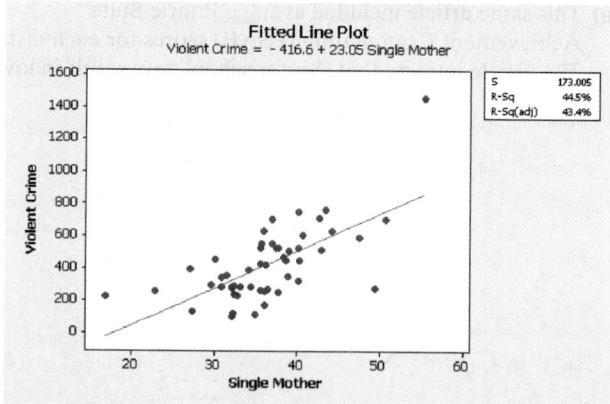

7. What is the relationship between education and belief in Heaven? The following data represent the highest level of education and belief in Heaven for a random sample of adult Americans.

	Yes, Definitely	Yes, Probably	No, Probably Not	No, Definitely Not	Total
Less than high school	316	66	21	9	**412**
High school	956	296	122	65	**1439**
Bachelor's	267	131	62	64	**524**
Total	**1539**	**493**	**205**	**138**	**2375**

Source: General Social Survey

(a) Construct a relative frequency marginal distribution.
(b) What proportion of adult Americans in the survey definitely believe in Heaven?
(c) Construct a conditional distribution of belief in Heaven by level of education.
(d) Draw a bar graph of the conditional distribution found in part (c).
(e) Is there an association between level of education and belief in Heaven?

8. Consider the following contingency table, which relates the number of applicants accepted to a college and gender.

	Accepted	Denied
Male	98	522
Female	90	200

(a) Construct a conditional distribution of acceptance status by gender.

(b) What proportion of males was accepted? What proportion of females was accepted?
(c) What might you conclude about the admittance policies of the school?
A lurking variable is the type of school applied to. This particular college has two programs of study: business and social work. The following table shows applications by type of school.

	Business School		Social Work School	
	Accepted	Denied	Accepted	Denied
Male	90	510	8	12
Female	10	60	80	140

(d) What proportion of males who applied to the business school was accepted? What proportion of females who applied to the business school was accepted?
(e) What proportion of males who applied to the social work school was accepted? What proportion of females who applied to the social work school was accepted?
(f) Explain carefully how the bias disappears when type of school is considered.

9. What would you say about a set of quantitative bivariate data whose linear correlation coefficient is −1? What would a scatter diagram of the data look like?

10. If the slope of a least-squares regression line is negative, what could be said about the correlation between the explanatory and response variable?

11. What does it mean if a linear correlation coefficient is close to zero? Draw two scatter diagrams for which the linear correlation coefficient is close to zero.

Making an Informed Decision

Relationships among Variables on a World Scale

The purpose of this Decisions Project is to decide on two quantitative variables, see how the relationship between these variables has changed over time, and determine the most current relationship between the variables.

1. Watch the video by Hans Rosling titled "Let My Dataset Change Your Mindset" at http://www.ted.com/talks/lang/eng/hans_rosling_at_state.html.
(a) Describe the "We" versus "Them" discussion in the video.
(b) In the video, there is a scatter diagram drawn between life expectancy and children per woman. Which variable is the explanatory variable in the scatter diagram? What type of relation exists between the two variables? How has the relationship changed over time when considering "We" versus "Them"?
2. Click the Data tab on the GapMinder Web site. Download the data for life expectancy and children per

woman. Draw a scatter diagram of the data. Determine the linear correlation coefficient between the two variables. Be careful; some countries may not have both variables measured, so they will need to be excluded from the analysis.

3. Find the least-squares regression line between life expectancy and children per woman. Interpret the slope of the regression line.
4. Is the life expectancy of the United States above or below average given the number of children per woman?
5. Choose any two variables in the GapMinder library and write a report detailing the relationship between the variables over time. Does the data reflect your intuition about the relationship?

In 1798, Thomas Malthus published *An Essay on the Principle of Population* in which he stated

> Population, when unchecked, increases in a geometrical ratio. Subsistence increases only in an arithmetical ratio. A slight acquaintance with numbers will show the immensity of the first power in comparison of the second.

What did he mean by this? Essentially he is claiming that population grows exponentially and food production increases linearly. The assumption is that the population grows unchecked—that is, without interference from human-caused or natural disasters such as war, famine, disease, and so on. Malthus also stated,

> By that law of our nature which makes food necessary to the life of man, the effects of these two unequal powers must be kept equal. This implies a strong and constantly operating check on population from the difficulty of subsistence. This difficulty must fall somewhere and must necessarily be severely felt by a large portion of mankind.

1. Read the article *Food, Land, Population, and the American Economy* from Carrying Capacity Network. (*Source:* www.carryingcapacity.org/resources.html)

2. Use the historic census data from the U.S. Census Bureau at www.census.gov for the United States for the years 1790 to present to find an exponential model of the form $P = a \cdot b^t$, where t is the number of years after 1790. To do this, first take the logarithm of P.

Now find the least-squares regression line between $\log P$ and t. The least-squares model is of the form $\log P = \log a + t \log b$. To find estimates of a and b, compute $10^{\log a}$, where $\log a$ is the intercept and $10^{\log b}$ is the slope of the least-squares regression line. The quantity $b - 1$ is the growth rate as a decimal. For example, if $b = 1.01$, then the annual growth rate is $1.01 - 1 = 0.01$ or 1%.

3. Multiply your model obtained in part 2 by 1.2 to obtain a function for the number of arable acres of land A required to have a diverse diet in the United States (as a function of time). [*Hint:* The model is $A = 1.2P$.]

4. Determine the number of acres of farmland A in the United States for the most recent 9 years for which data are available. Find the least-squares regression line treating year as the explanatory variable.

5. Determine when the United States will no longer be capable of providing a diverse diet for its entire population by setting the model in part 3 equal to the model in part 4 and solving for the year t.

6. Does this result seem reasonable? Discuss some reasons why this prediction might be affected or changed. For example, rather than looking at the number of arable acres of land required to have a diverse diet, consider total food production. Compare the results using this variable as a measure of subsistence.

PART

3

Probability and Probability Distributions

We now take a break from the statistical process. Why? In Chapter 1, we mentioned that inferential statistics uses methods that generalize results obtained from a sample to the population and measures their reliability. But how can we measure their reliability? It turns out that the methods we use to generalize results from a sample to a population are based on probability and probability models. Probability is a measure of the likelihood that something occurs. This part of the course will focus on methods for determining probabilities.

5

Probability

Making an Informed Decision

You are at a party and everyone is having a good time. Unfortunately, a few of the party-goers are having a little too much to drink. If they decide to drive home, what are the risks? Perhaps a view of some scary statistics about the effects of alcohol on one's driving skills will convince more people not to drink and drive. See the Decisions project on page 312.

PUTTING IT TOGETHER

In Chapter 1, we learned the methods of collecting data. In Chapters 2 through 4, we learned how to summarize raw data using tables, graphs, and numbers. As far as the statistical process goes, we have discussed the collecting, organizing, and summarizing parts of the process.

Before we begin to analyze data, we introduce probability, which forms the basis of inferential statistics. Why? Well, we can think of the probability of an outcome as the likelihood of observing that outcome. If something has a high likelihood of happening, it has a high probability (close to 1). If something has a small chance of happening, it has a low probability (close to 0). For example, it is unlikely that we would roll five straight sixes when rolling a single die, so this result has a low probability. In fact, the probability of rolling five straight sixes is 0.0001286. If we were playing a game in which a player threw five sixes in a row with a single die, we would consider the player to be lucky (or a cheater) because it is such an unusual occurrence. Statisticians use probability in the same way. If something occurs that has a low probability, we investigate to find out "what's up."

5.1 Probability Rules

Preparing for This Section Before getting started, review the following:

- Relative frequency (Section 2.1, p. 64)

Objectives
  ❶ Apply the rules of probabilities
  ❷ Compute and interpret probabilities using the empirical method
  ❸ Compute and interpret probabilities using the classical method
  ❹ Use simulation to obtain data based on probabilities
  ❺ Recognize and interpret subjective probabilities

Suppose we flip a coin 100 times and compute the proportion of heads observed after each toss of the coin. The first flip is tails, so the proportion of heads is $\frac{0}{1} = 0$; the second flip is heads, so the proportion of heads is $\frac{1}{2} = 0.5$; the third flip is heads, so the proportion of heads is $\frac{2}{3} = 0.667$, and so on. Plot the proportion of heads versus the number of flips and obtain the graph in Figure 1(a). We repeat this experiment with the results shown in Figure 1(b).

Figure 1

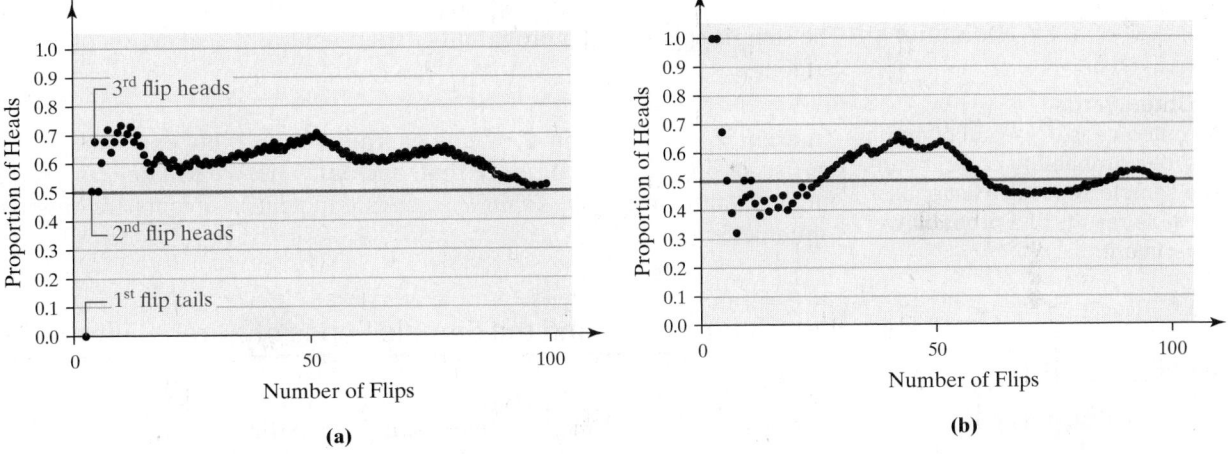

(a)   (b)

The graphs in Figures 1(a) and (b) show that in the short term (fewer flips) the observed proportion of heads is different and unpredictable for each experiment. As the number of flips increases, however, both graphs tend toward a proportion of 0.5. This is the basic premise of probability. **Probability** is the measure of the likelihood of a random phenomenon or chance behavior occurring. It deals with experiments that yield random short-term results or **outcomes** yet reveal long-term predictability. **The long-term proportion in which a certain outcome is observed is the probability of that outcome.**

So we say that the probability of observing a head is $\frac{1}{2}$ or 50% or 0.5 because, as we flip the coin more times, the proportion of heads tends toward $\frac{1}{2}$. This phenomenon, which is illustrated in Figure 1, is referred to as the *Law of Large Numbers.*

In Other Words
Probability describes how likely it is that some event will happen. If we look at the proportion of times an event has occurred over a long period of time (or over a large number of trials), we can be more certain of the likelihood of its occurrence.

CAUTION!
Probability is a measure of the likelihood of events that have yet to occur. Prior to flipping a coin, we say the probability of observing a head is $\frac{1}{2}$. However, once the coin is flipped, the probability is no longer $\frac{1}{2}$ since the outcome has been determined.

The Law of Large Numbers

As the number of repetitions of a probability experiment increases, the proportion with which a certain outcome is observed gets closer to the probability of the outcome.

Jakob Bernoulli (a major contributor to the field of probability) believed that the Law of Large Numbers was common sense. This is evident in the following quote from his text *Ars Conjectandi*: "For even the most stupid of men, by some instinct of nature, by himself and without any instruction, is convinced that the more observations have been made, the less danger there is of wandering from one's goal."

In probability, an **experiment** is any process with uncertain results that can be repeated. The result of any single trial of the experiment is not known ahead of time. However, the results of the experiment over many trials produce regular patterns that allow accurate predictions. For example, an insurance company cannot know whether a particular 16-year-old driver will have an accident over the course of a year. However, based on historical records, the company can be fairly certain that about three out of every ten 16-year-old male drivers will have a traffic accident during the course of a year. Therefore, of 825,000 male 16-year-old drivers (825,000 repetitions of the experiment), the insurance company is fairly confident that about 30%, or 247,500, will have an accident. This prediction helps to establish insurance rates for any particular 16-year-old male driver.

We now introduce some terminology that we will need to study probability.

Definitions

The **sample space**, S, of a probability experiment is the collection of all possible outcomes.

In Other Words
An outcome is the result of one trial of a probability experiment. The sample space is a list of all possible results of a probability experiment.

An **event** is any collection of outcomes from a probability experiment. An event consists of one outcome or more than one outcome. We will denote events with one outcome, sometimes called *simple events*, e_i. In general, events are denoted using capital letters such as E.

EXAMPLE 1   Identifying Events and the Sample Space of a Probability Experiment

A *fair die* is one in which each possible outcome is equally likely. For example, rolling a 2 is just as likely as rolling a 5. We contrast this with a *loaded* die, in which a certain outcome is more likely. For example, if rolling a 1 is more likely than rolling a 2, 3, 4, 5, or 6, the die is loaded.

• Now Work Problem 19

Problem A probability experiment consists of rolling a single *fair* die.
(a) Identify the outcomes of the probability experiment.
(b) Determine the sample space.
(c) Define the event $E = $ "roll an even number."

Approach The outcomes are the possible results of the experiment. The sample space is a list of all possible outcomes.

Solution
(a) The outcomes from rolling a single fair die are $e_1 = $ "rolling a one" $= \{1\}$, $e_2 = $ "rolling a two" $= \{2\}$, $e_3 = $ "rolling a three" $= \{3\}$, $e_4 = $ "rolling a four" $= \{4\}$, $e_5 = $ "rolling a five" $= \{5\}$, and $e_6 = $ "rolling a six" $= \{6\}$.
(b) The set of all possible outcomes forms the sample space, $S = \{1, 2, 3, 4, 5, 6\}$. There are 6 outcomes in the sample space.
(c) The event $E = $ "roll an even number" $= \{2, 4, 6\}$.

1   Apply the Rules of Probabilities

In the following probability rules, the notation $P(E)$ means "the probability that event E occurs."

In Other Words
Rule 1 states that probabilities less than 0 or greater than 1 are not possible. Therefore, probabilities such as −0.3 or 1.32 are not possible. Rule 2 states when the probabilities of all outcomes are added, the sum must be 1.

Rules of Probabilities

1. The probability of any event E, $P(E)$, must be greater than or equal to 0 and less than or equal to 1. That is, $0 \leq P(E) \leq 1$.
2. The sum of the probabilities of all outcomes must equal 1. That is, if the sample space $S = \{e_1, e_2, \cdots, e_n\}$, then

$$P(e_1) + P(e_2) + \cdots + P(e_n) = 1$$

A **probability model** lists the possible outcomes of a probability experiment and each outcome's probability. A probability model must satisfy Rules 1 and 2 of the rules of probabilities.

EXAMPLE 2   A Probability Model

Table 1

Color	Probability
Brown	0.13
Yellow	0.14
Red	0.13
Blue	0.24
Orange	0.20
Green	0.16

Source: M&Ms

● **Now Work Problem 7**

The color of a plain M&M milk chocolate candy can be brown, yellow, red, blue, orange, or green. Suppose a candy is randomly selected from a bag. Table 1 shows each color and the probability of drawing that color.

To verify that this is a probability model, we must show that Rules 1 and 2 are satisfied.

Each probability is greater than or equal to 0 and less than or equal to 1, so Rule 1 is satisfied.

Because

$$0.13 + 0.14 + 0.13 + 0.24 + 0.20 + 0.16 = 1$$

Rule 2 is also satisfied. The table is an example of a probability model.   ●

If an event is **impossible**, the probability of the event is 0. If an event is a **certainty**, the probability of the event is 1.

The closer a probability is to 1, the more likely the event will occur. The closer a probability is to 0, the less likely the event will occur. For example, an event with probability 0.8 is more likely to occur than an event with probability 0.75. An event with probability 0.8 will occur about 80 times out of 100 repetitions of the experiment, whereas an event with probability 0.75 will occur about 75 times out of 100.

Be careful of this interpretation. An event with a probability of 0.75 does not have to occur 75 times out of 100. Rather, we *expect* the number of occurrences to be close to 75 in 100 trials. The more repetitions of the probability experiment, the closer the proportion with which the event occurs will be to 0.75 (the Law of Large Numbers).

One goal of this course is to learn how probabilities can be used to identify *unusual events*.

Definition   An **unusual event** is an event that has a low probability of occurring.

In Other Words
An unusual event is an event that is not likely to occur.

Typically, an event with a probability less than 0.05 (or 5%) is considered unusual, but this *cutoff point* is not set in stone. The researcher and the context of the problem determine the probability that separates unusual events from *not so unusual events*.

For example, suppose that the probability of being wrongly convicted of a capital crime punishable by death is 3%. Even though 3% is below our 5% cutoff point, this probability is too high in light of the consequences (death for the wrongly convicted), so the event is not unusual (unlikely) enough. We would want this probability to be much closer to zero.

Now suppose that you are planning a picnic on a day having a 3% chance of rain. In this context, you would consider "rain" an unusual (unlikely) event and proceed with the picnic plans.

The point is this: Selecting a probability that separates unusual events from not so unusual events is subjective and depends on the situation. Statisticians typically use cutoff points of 0.01, 0.05, and 0.10.

Next, we introduce three methods for determining the probability of an event: the empirical method, the classical method, and the subjective method.

> **CAUTION!**
> A probability of 0.05 should not always be used to separate unusual events from not so unusual events.

② Compute and Interpret Probabilities Using the Empirical Method

Probabilities deal with the likelihood that a particular outcome will be observed. For this reason, we begin our discussion of determining probabilities using the idea of relative frequency. Probabilities computed in this manner rely on empirical evidence, that is, evidence based on the outcomes of a probability experiment.

Approximating Probabilities Using the Empirical Approach

The probability of an event E is approximately the number of times event E is observed divided by the number of repetitions of the experiment.

$$P(E) \approx \text{relative frequency of } E = \frac{\text{frequency of } E}{\text{number of trials of experiment}} \quad (1)$$

When we find probabilities using the empirical approach, the result is approximate because different trials of the experiment lead to different outcomes and, therefore, different estimates of $P(E)$. Consider flipping a coin 20 times and recording the number of heads. Use the results of the experiment to estimate the probability of obtaining a head. Now repeat the experiment. Because the results of the second run of the experiment do not necessarily yield the same results, we cannot say the probability *equals* the relative frequency; rather we say the probability is *approximately* the relative frequency. As we increase the number of trials of a probability experiment, our estimate becomes more accurate (again, the Law of Large Numbers).

EXAMPLE 3   Using Relative Frequencies to Approximate Probabilities

An insurance agent currently insures 182 teenage drivers (ages 16 to 19). Last year, 24 of the teenagers had to file a claim on their auto policy. Based on these results, the probability that a teenager will file a claim on his or her auto policy in a given year is

$$\frac{24}{182} \approx 0.132$$

So, for every 100 insured teenage drivers, we expect about 13 to have a claim on their auto policy. ●

Surveys are probability experiments. Why? Each time a survey is conducted, a different random sample of individuals is selected. Therefore, the results of a survey are likely to be different each time the survey is conducted because different people are included.

EXAMPLE 4   Building a Probability Model from Survey Data

Table 2	
Means of Travel	**Frequency**
Drive alone	153
Carpool	22
Public transportation	10
Walk	5
Other means	3
Work at home	7

Table 3	
Means of Travel	**Probability**
Drive alone	0.765
Carpool	0.11
Public transportation	0.05
Walk	0.025
Other means	0.015
Work at home	0.035

● **Now Work Problem 35**

Problem The data in Table 2 represent the results of a survey in which 200 people were asked their means of travel to work.
(a) Use the survey data to build a probability model for means of travel to work.
(b) Estimate the probability that a randomly selected individual carpools to work. Interpret this result.
(c) Would it be unusual to randomly select an individual who walks to work?

Approach To build a probability model, estimate the probability of each outcome by determining its relative frequency.

Solution
(a) There are $153 + 22 + \cdots + 7 = 200$ individuals in the survey. The individuals can be thought of as trials of the probability experiment. The relative frequency for "drive alone" is $\dfrac{153}{200} = 0.765$. We compute the relative frequency of the other outcomes similarly and obtain the probability model in Table 3.
(b) From Table 3, we estimate the probability to be 0.11 that a randomly selected individual carpools to work. We interpret this result by saying, "If we were to survey 1000 individuals, we would expect about 110 to carpool to work."
(c) The probability that an individual walks to work is approximately 0.025. This means if we survey 1000 individuals, we would expect about 25 to walk to work. Therefore, it is unusual to randomly choose a person who walks to work.   ●

③ Compute and Interpret Probabilities Using the Classical Method

The empirical method gives an approximate probability of an event by conducting a probability experiment.

The classical method of computing probabilities does not require that a probability experiment actually be performed. Rather, it relies on counting techniques.

The classical method of computing probabilities requires *equally likely outcomes*. An experiment has **equally likely outcomes** when each outcome has the same probability of occurring. For example, when a fair die is thrown once, each of the six outcomes in the sample space, $\{1, 2, 3, 4, 5, 6\}$, has an equal chance of occurring. Contrast this situation with a loaded die in which a five or six is twice as likely to occur as a one, two, three, or four.

Computing Probability Using the Classical Method

If an experiment has n equally likely outcomes and if the number of ways that an event E can occur is m, then the probability of E, $P(E)$, is

$$P(E) = \frac{\text{number of ways that } E \text{ can occur}}{\text{number of possible outcomes}} = \frac{m}{n} \qquad (2)$$

So, if S is the sample space of this experiment,

$$P(E) = \frac{N(E)}{N(S)} \qquad (3)$$

where $N(E)$ is the number of outcomes in E, and $N(S)$ is the number of outcomes in the sample space.

EXAMPLE 5   Computing Probabilities Using the Classical Method

Problem A pair of fair dice is rolled. Fair die are die where each outcome is equally likely.
(a) Compute the probability of rolling a seven.
(b) Compute the probability of rolling "snake eyes"; that is, compute the probability of rolling a two.
(c) Comment on the likelihood of rolling a seven versus rolling a two.

Approach To compute probabilities using the classical method, count the number of outcomes in the sample space and count the number of ways the event can occur. Then, divide the number of outcomes by the number of ways the event can occur.

Solution
(a) There are 36 equally likely outcomes in the sample space, as shown in Figure 2.

Figure 2

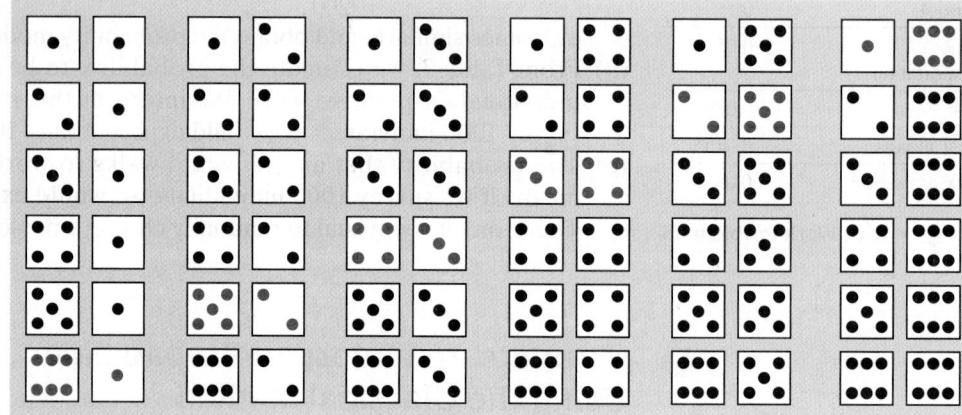

Historical Note

Girolamo Cardano (in English Jerome Cardan) was born in Pavia, Italy, on September 24, 1501. He was an illegitimate child whose father was Fazio Cardano, a lawyer in Milan. Fazio was a part-time mathematician and taught Girolamo. In 1526, Cardano earned his medical degree. Shortly thereafter, his father died. Unable to maintain a medical practice, Cardano spent his inheritance and turned to gambling to help support himself. Cardano developed an understanding of probability that helped him to win. Eventually, Cardano became a lecturer of mathematics at the Piatti Foundation. This position allowed him to practice medicine and develop a favorable reputation as a doctor. In 1545, he published his greatest work, *Ars Magna*. His booklet on probability, *Liber de Ludo Alaea*, was not printed until 1663, 87 years after his death. The booklet is a practical guide to gambling, including cards, dice, and cheating.

So, $N(S) = 36$. The event $E =$ "roll a seven" $= \{(1, 6), (2, 5), (3, 4), (4, 3), (5, 2), (6, 1)\}$ has six outcomes, so $N(E) = 6$. Using Formula (3),

$$P(E) = P(\text{roll a seven}) = \frac{N(E)}{N(S)} = \frac{6}{36} = \frac{1}{6}$$

(b) The event $F =$ "roll a two" $= \{(1, 1)\}$ has one outcome, so $N(F) = 1$.

$$P(F) = P(\text{roll a two}) = \frac{N(F)}{N(S)} = \frac{1}{36}$$

(c) Since $P(\text{roll a seven}) = \frac{6}{36}$ and $P(\text{roll a two}) = \frac{1}{36}$, rolling a seven is six times as likely as rolling a two. In other words, in 36 rolls of the dice, we *expect* to observe about 6 sevens and only 1 two. ●

We just saw that the classical probability of rolling a seven is $\frac{1}{6} \approx 0.167$. Suppose a pit boss at a casino rolls a pair of dice 100 times and obtains 15 sevens. From this empirical evidence, we would assign the probability of rolling a seven as $\frac{15}{100} = 0.15$. If the dice are fair, we would expect the relative frequency of sevens to get closer to 0.167 as the number of rolls of the dice increases. In other words, the empirical probability will get closer to the classical probability as the number of trials of the experiment increases. If the two probabilities do not get closer together, we may suspect that the dice are not fair.

In simple random sampling, each individual has the same chance of being selected. Therefore, we can use the classical method to compute the probability of obtaining a specific sample.

EXAMPLE 6   Computing Probabilities Using Equally Likely Outcomes

Problem Sophia has three tickets to a concert, but Yolanda, Michael, Kevin, and Marissa all want to go to the concert with her. To be fair, Sophia randomly selects the two people who can go with her.
(a) Determine the sample space of the experiment. In other words, list all possible simple random samples of size $n = 2$.
(b) Compute the probability of the event "Michael and Kevin attend the concert."
(c) Compute the probability of the event "Marissa attends the concert."
(d) Interpret the probability in part (c).

Approach First, determine the outcomes in the sample space by making a table. The probability of an event is the number of outcomes in the event divided by the number of outcomes in the sample space.

Solution
(a) The sample space is listed in Table 4.
(b) We have $N(S) = 6$, and there is one way the event "Michael and Kevin attend the concert" can occur. Therefore, the probability that Michael and Kevin attend the concert is $\frac{1}{6}$.
(c) We have $N(S) = 6$, and there are three ways the event "Marissa attends the concert" can occur. The probability that Marissa will attend is $\frac{3}{6} = 0.5 = 50\%$.
(d) If we conducted this experiment many times, about 50% of the experiments would result in Marissa attending the concert.

Table 4

Yolanda, Michael	Yolanda, Kevin
Yolanda, Marissa	Michael, Kevin
Michael, Marissa	Kevin, Marissa

● Now Work Problems 29 and 43

EXAMPLE 7   Comparing the Classical Method and Empirical Method

Problem Suppose that a survey asked 500 families with three children to disclose the gender of their children and found that 180 of the families had two boys and one girl.
(a) Estimate the probability of having two boys and one girl in a three-child family using the empirical method.
(b) Compute and interpret the probability of having two boys and one girl in a three-child family using the classical method, assuming boys and girls are equally likely.

Approach To answer part (a), determine the relative frequency of the event "two boys and one girl." To answer part (b), count the number of ways the event "two boys and one girl" can occur and divide this by the number of possible outcomes for this experiment.

Solution
(a) The empirical probability of the event $E =$ "two boys and one girl" is

$$P(E) \approx \text{relative frequency of } E = \frac{180}{500} = 0.36$$

The probability that a family of three children will have two boys and one girl is approximately 0.36.

(continued)

Historical Note

Pierre de Fermat was born into a wealthy family. His father was a leather merchant and second consul of Beaumont-de-Lomagne. Fermat attended the University of Toulouse. By 1631, Fermat was a lawyer and government official. He rose quickly through the ranks because of deaths from the plague. In fact, in 1653, Fermat's death was incorrectly reported. In 1654, Fermat received a correspondence from Blaise Pascal in which Pascal asked Fermat to confirm his ideas on probability. Fermat and Pascal discussed the problem of how to divide the stakes in a game that is interrupted before completion, knowing how many points each player needs to win. Their short correspondence laid the foundation for the theory of probability. They are regarded as joint founders of the subject.

Mathematics was Fermat's passionate hobby and true love. He is most famous for his Last Theorem, which states that the equation $x^n + y^n = z^n$ has no nonzero integer solutions for $n > 2$. The theorem was scribbled in the margin of a book by Diophantus, a Greek mathematician. Fermat stated, "I have discovered a truly marvelous proof of this theorem, which, however, the margin is not large enough to contain." The status of Fermat's Last Theorem baffled mathematicians until Andrew Wiles proved it to be true in 1994.

(b) To determine the sample space, construct a **tree diagram** to list the equally likely outcomes of the experiment. Draw two branches corresponding to the two possible outcomes (boy or girl) for the first repetition of the experiment (the first child). For the second child, draw four branches, and so on. See Figure 3, where B stands for boy and G stands for girl.

Figure 3

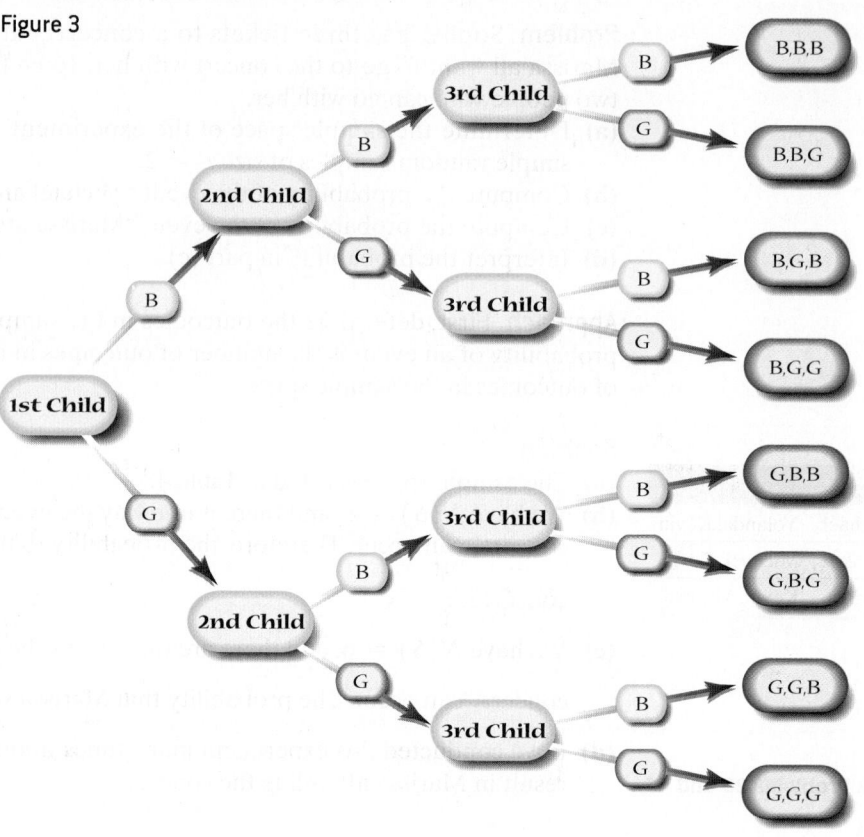

Find the sample space S of this experiment by following each branch to identify all the possible outcomes of the experiment:

$$S = \{\text{BBB, BBG, BGB, BGG, GBB, GBG, GGB, GGG}\}$$

So, $N(S) = 8$.

For the event $E = $ "two boys and a girl" $= \{\text{BBG, BGB, GBB}\}$, we have $N(E) = 3$. Since the outcomes are equally likely (for example, BBG is just as likely as BGB), we have

$$P(E) = \frac{N(E)}{N(S)} = \frac{3}{8} = 0.375$$

The probability that a family of three children will have two boys and one girl is 0.375. If we repeat this experiment 1000 times and the outcomes are equally likely (having a girl is just as likely as having a boy), we would expect about 375 of the trials to result in two boys and one girl. ●

In comparing the results of Examples 7(a) and 7(b), notice that the two probabilities are slightly different. Empirical probabilities and classical probabilities often differ in value, but, as the number of repetitions of a probability experiment increases, the empirical probability should get closer to the classical probability. However, it is possible that the two probabilities differ because having a boy or having a girl are not equally likely events. (Maybe the probability of having a boy is 0.505 and the probability of having a girl is 0.495.)

Consider the Vital Statistics data in the *Statistical Abstract of the United States*. In 2012, for example, 2,022,000 boys and 1,931,000 girls were born. Based on empirical evidence, the probability of a boy is approximately $\dfrac{2,022,000}{2,022,000 + 1,931,000} = 0.512$.

The empirical evidence, which is based on a very large number of repetitions, differs from the value of 0.50 used for classical methods (which assumes boys and girls are equally likely). This suggests having a boy is not equally likely as having a girl. Therefore, empirical probabilities of having two boys and one girl will not get closer to the classical probability as the number of repetitions of the experiment increases.

④ Use Simulation to Obtain Data Based on Probabilities

Instead of obtaining data from existing sources, we could also simulate a probability experiment using a graphing calculator or statistical software to replicate the experiment as many times as we like. Simulation is particularly helpful for estimating the probability of more complicated events. In Example 7, we used a tree diagram and the classical approach to find the probability of an event (having two boys and one girl in a three-child family). In this next example, we use simulation to estimate the same probability.

EXAMPLE 8   Simulating Probabilities

Problem

(a) Simulate the experiment of sampling 100 three-child families to estimate the probability that a three-child family has two boys.

(b) Simulate the experiment of sampling 1000 three-child families to estimate the probability that a three-child family has two boys.

Approach To simulate probabilities, use the random-number generator available in statistical software and most graphing calculators. Assume that the outcomes "have a boy" and "have a girl" are equally likely.

Solution

(a) We use Minitab to perform the simulation. Set the seed in Minitab to any value you wish, say 1970. Use the Integer Distribution* to generate random data that simulate three-child families. If we agree to let 0 represent a girl and 1 represent a boy, we can approximate the probability of having two boys by summing each row (adding up the number of boys), counting the number of 2s (the number of times we observe two boys), and dividing by 100, the number of repetitions of the experiment. See Figure 4.

Figure 4

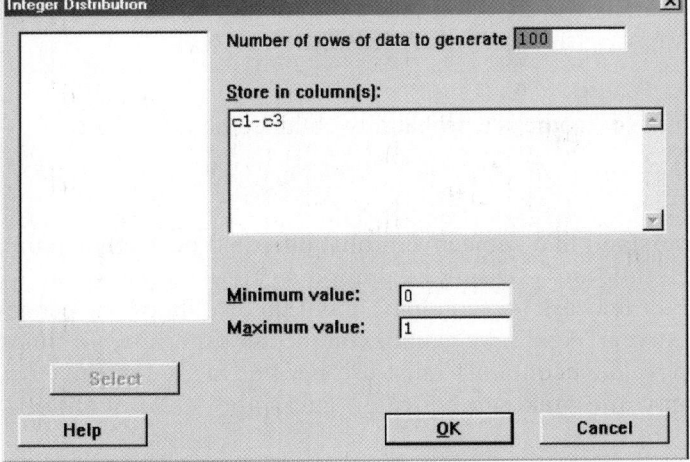
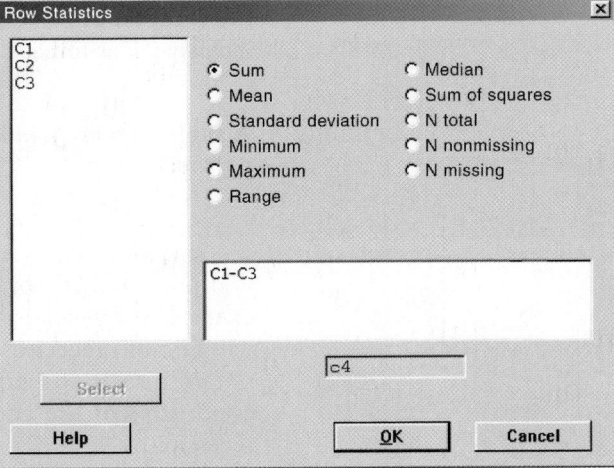

(continued)

---

*The Integer Distribution involves a mathematical formula that uses a seed number to generate a sequence of equally likely random integers. Consult the technology manuals for setting the seed and generating sequences of integers.

Historical Note

Blaise Pascal was born in 1623, in Clermont, France. Pascal's father felt that Blaise should not be taught mathematics before age 15. Pascal couldn't resist studying mathematics on his own, and at the age of 12 started to teach himself geometry. In December 1639, the Pascal family moved to Rouen, where Pascal's father had been appointed as a tax collector. Between 1642 and 1645, Pascal worked on developing a calculator to help his father collect taxes. In correspondence with Fermat, he helped develop the theory of probability. This correspondence consisted of five letters written in the summer of 1654. They considered the dice problem and the problem of points. The dice problem deals with determining the expected number of times a pair of dice must be thrown before a pair of sixes is observed. The problem of points asks how to divide the stakes if a game of dice is incomplete. They solved the problem of points for a two-player game, but did not solve it for three or more players.

Using Minitab's Tally command, we can determine the number of 2s that Minitab randomly generated. See Figure 5.

Figure 5

Tally for Discrete Variables: C4

C4	Count	Percent
0	14	14.00
1	40	40.00
2	32	32.00
3	14	14.00
N = 100		

Based on this figure, we approximate the probability that a three-child family will have two boys is 0.32.

(b) Again, set the seed to 1970. Figure 6 shows the result of simulating 1000 three-child families.

Figure 6

Tally for Discrete Variables: C4

C4	Count	Percent
0	136	13.60
1	367	36.70
2	388	38.80
3	109	10.90
N = 1000		

The approximate probability of a three-child family having two boys is 0.388. Notice that more repetitions of the experiment (100 repetitions versus 1000 repetitions) results in a probability closer to 0.375 as found in Example 7(b).

● **Now Work Problem 47**

⑤ Recognize and Interpret Subjective Probabilities

If a sports reporter is asked what he thinks the chances are for the Boston Red Sox to play in the World Series, the reporter would likely process information about the Red Sox (their pitching staff, lead-off hitter, and so on) and then make an educated guess of the likelihood. The reporter may respond that there is a 20% chance the Red Sox will play in the World Series. This forecast is a probability although it is not based on relative frequencies. We cannot, after all, repeat the experiment of playing a season under the same circumstances (same players, schedule, and so on) over and over. Nonetheless, the forecast of 20% does satisfy the criterion that a probability be between 0 and 1, inclusive. This forecast is known as a *subjective probability*.

Definition

> A **subjective probability** of an outcome is a probability obtained on the basis of personal judgment.

It is important to understand that subjective probabilities are perfectly legitimate and are often the only method of assigning likelihood to an outcome. As another example, a financial reporter may ask an economist about the likelihood the economy will fall into recession next year. Again, we cannot conduct an experiment *n* times to obtain a relative frequency. The economist must use her knowledge of the current conditions of the economy and make an educated guess about the likelihood of recession.

Technology Step-by-Step Simulation

TI-83/84 Plus
1. Set the seed by entering any number on the HOME screen. Press the STO ▶ button, press the MATH button, highlight the PROB menu, and highlight 1:rand and hit ENTER. With the cursor on the HOME screen, hit ENTER.
2. Press the MATH button and highlight the PROB menu. Highlight 5:randInt (and hit ENTER.
3. After lower: enter 1. After upper: enter the number of equally likely outcomes. After n: enter the number of repetitions of the experiment. Highlight Paste and hit ENTER. Press the STO ▶ button and then 2nd 1, and hit ENTER to store the data in L1.
4. Draw a histogram of the data using the outcomes as classes. TRACE to obtain outcomes.

Minitab
1. Set the seed by selecting the **Calc** menu and choosing **Set Base . . . .** Insert any seed you wish into the cell and click OK.
2. Select the **Calc** menu, highlight **Random Data**, and then choose **Integer**. Fill in the cells as desired.
3. Select the **Stat** menu, highlight **Tables**, and then choose **Tally Individual Variables. . . .** Enter C1 into the variables cell. Make sure that the Counts box is checked and click OK.

Excel
1. With cell A1 selected, select the Formulas menu.

2. Select Math & Trig in the Functions menu. Then choose RANDBETWEEN.
3. To simulate rolling a die 100 times, enter 1 for the Bottom and 6 for the Top. Click OK.
4. Copy the contents of cell A1 into cells A2 through A100.
5. To count the number of "1s," enter

$$=CountIf(A1:A100, 1)$$

in any cell. Repeat this to obtain counts for 2 through 6.

StatCrunch
1. Select **Data**, highlight **Simulate**, then choose **Discrete Uniform**.
2. Enter the number of random numbers you would like generated in the "Rows:" cell. For example, if we want to simulate rolling a die 100 times, enter 100. Enter 1 in the "Columns:" cell. Enter the smallest and largest integer in the "Minimum:" and "Maximum:" cell, respectively. For example, to simulate rolling a single die, enter 1 and 6, respectively.
3. Select either the dynamic seed or the fixed seed and enter a value of the seed. Click Compute!.
4. To get counts, select **Stat**, highlight **Tables**, then select **Frequency**. Select the column the simulated data are located in. Be sure Frequency and Relative Frequency are highlighted. Click Compute!.

5.1 Assess Your Understanding

Vocabulary and Skill Building

1. What is the probability of an event that is impossible? Suppose that a probability is approximated to be zero based on empirical results. Does this mean the event is impossible?

2. What does it mean for an event to be unusual? Why should the cutoff for identifying unusual events not always be 0.05?

3. *True or False:* In a probability model, the sum of the probabilities of all outcomes must equal 1.

4. *True or False:* Probability is a measure of the likelihood of a random phenomenon or chance behavior.

5. In probability, a(n) _____ is any process that can be repeated in which the results are uncertain.

6. A(n) _____ is any collection of outcomes from a probability experiment.

NW 7. Verify that the table in the next column is a probability model. What do we call the outcome "blue"?

Color	Probability
Red	0.3
Green	0.15
Blue	0
Brown	0.15
Yellow	0.2
Orange	0.2

8. Verify that the following is a probability model. If the model represents the colors of M&Ms in a bag of milk chocolate M&Ms, explain what the model implies.

Color	Probability
Red	0
Green	0
Blue	0
Brown	0
Yellow	1
Orange	0

9. Why is the following not a probability model?

Color	Probability
Red	0.3
Green	−0.3
Blue	0.2
Brown	0.4
Yellow	0.2
Orange	0.2

10. Why is the following not a probability model?

Color	Probability
Red	0.1
Green	0.1
Blue	0.1
Brown	0.4
Yellow	0.2
Orange	0.3

11. Which of the following numbers could be the probability of an event?

$$0, 0.01, 0.35, -0.4, 1, 1.4$$

12. Which of the following numbers could be the probability of an event?

$$1.5, \frac{1}{2}, \frac{3}{4}, \frac{2}{3}, 0, -\frac{1}{4}$$

13. According to Nate Silver, the probability of a senate candidate winning his/her election with a 5% lead in an average of polls with a week until the election is 0.89. Interpret this probability.

Source: fivethirtyeight.com

14. In seven-card stud poker, a player is dealt seven cards. The probability that the player is dealt two cards of the same value and five other cards of different value so that the player has a pair is 0.44. Explain what this probability means. If you play seven-card stud 100 times, will you be dealt a pair exactly 44 times? Why or why not?

15. Suppose that you toss a coin 100 times and get 95 heads and five tails. Based on these results, what is the estimated probability that the next flip results in a head?

16. Suppose that you roll a die 100 times and get six 80 times. Based on these results, what is the estimated probability that the next roll results in six?

17. Bob is asked to construct a probability model for rolling a pair of fair dice. He lists the outcomes as 2, 3, 4, 5, 6, 7, 8, 9, 10, 11, 12. Because there are 11 outcomes, he reasoned, the probability of rolling a two must be $\frac{1}{11}$. What is wrong with Bob's reasoning?

18. Blood Types A person can have one of four blood types: A, B, AB, or O. If a person is randomly selected, is the probability they have blood type A equal to $\frac{1}{4}$? Why?

NW 19. If a person rolls a six-sided die and then flips a coin, describe the sample space of possible outcomes using 1, 2, 3, 4, 5, 6 for the die outcomes and H, T for the coin outcomes.

20. If a basketball player shoots three free throws, describe the sample space of possible outcomes using S for a made free throw and F for a missed free throw.

21. According to the U.S. Department of Education, 42.8% of 3-year-olds are enrolled in day care. What is the probability that a randomly selected 3-year-old is enrolled in day care?

22. According to the American Veterinary Medical Association, the proportion of households owning a dog is 0.372. What is the probability that a randomly selected household owns a dog?

For Problems 23–26, let the sample space be $S = \{1, 2, 3, 4, 5, 6, 7, 8, 9, 10\}$. *Suppose the outcomes are equally likely.*

23. Compute the probability of the event $E = \{1, 2, 3\}$.

24. Compute the probability of the event $F = \{3, 5, 9, 10\}$.

25. Compute the probability of the event $E =$ "an even number less than 9."

26. Compute the probability of the event $F =$ "an odd number."

Applying the Concepts

27. Play Sports? A survey of 500 randomly selected high school students determined that 288 played organized sports.

(a) What is the probability that a randomly selected high school student plays organized sports?

(b) Interpret this probability.

28. Volunteer? In a survey of 1100 female adults (18 years of age or older), it was determined that 341 volunteered at least once in the past year.

(a) What is the probability that a randomly selected adult female volunteered at least once in the past year?

(b) Interpret this probability.

NW 29. Home Runs The *Wall Street Journal* regularly publishes an article entitled "The Count." In one article, The Count looked at 1000 randomly selected home runs in Major League Baseball.

Source: Wall Street Journal, September 24, 2014

(a) Of the 1000 homeruns, it was found that 85 were caught by fans. What is the probability that a randomly selected homerun is caught by a fan?

(b) Of the 1000 homeruns, it was found that 296 were dropped when a fan had a legitimate play on the ball. What is the probability that a randomly selected homerun is dropped?

(c) Of the 85 caught balls, it was determined that 34 were barehanded catches, 49 were caught with a glove, and two were caught in a hat. What is the probability a randomly selected caught ball was caught in a hat? Interpret this probability.

(d) Of the 296 dropped balls, it was determined that 234 were barehanded attempts, 54 were dropped with a glove, and eight were dropped with a failed hat attempt. What is the probability a randomly selected dropped ball was a failed hat attempt? Interpret this probability.

30. Planting Tulips A bag of 100 tulip bulbs purchased from a nursery contains 40 red tulip bulbs, 35 yellow tulip bulbs, and 25 purple tulip bulbs.

(a) What is the probability that a randomly selected tulip bulb is red?

(b) What is the probability that a randomly selected tulip bulb is purple?

(c) Interpret these two probabilities.

31. Roulette In the game of roulette, a wheel consists of 38 slots numbered 0, 00, 1, 2, . . . , 36. (See the photo.) To play the game, a metal ball is spun around the wheel and is allowed to fall into one of the numbered slots.

(a) Determine the sample space.
(b) Determine the probability that the metal ball falls into the slot marked eight. Interpret this probability.
(c) Determine the probability that the metal ball lands in an odd slot. Interpret this probability.

32. Birthdays Exclude leap years from the following calculations and assume each birthday is equally likely:

(a) Determine the probability that a randomly selected person has a birthday on the 1st day of a month. Interpret this probability.
(b) Determine the probability that a randomly selected person has a birthday on the 31st day of a month. Interpret this probability.
(c) Determine the probability that a randomly selected person was born in December. Interpret this probability.
(d) Determine the probability that a randomly selected person has a birthday on November 8. Interpret this probability.
(e) If you just met somebody and she asked you to guess her birthday, are you likely to be correct?
(f) Do you think it is appropriate to use the methods of classical probability to compute the probability that a person is born in December?

33. Genetics A gene is composed of two alleles, either dominant or recessive. Suppose that a husband and wife, who are both carriers of the sickle-cell anemia allele but do not have the disease, decide to have a child. Because both parents are carriers of the disease, each has one dominant normal-cell allele (*S*) and one recessive sickle-cell allele (*s*). Therefore, the genotype of each parent is *Ss*. Each parent contributes one allele to his or her offspring, with each allele being equally likely.

(a) List the possible genotypes of their offspring.
(b) What is the probability that the offspring will have sickle-cell anemia? In other words, what is the probability that the offspring will have genotype *ss*? Interpret this probability.
(c) What is the probability that the offspring will not have sickle-cell anemia but will be a carrier? In other words, what is the probability that the offspring will have one dominant normal-cell allele and one recessive sickle-cell allele? Interpret this probability.

34. More Genetics In Problem 33, we learned that for some diseases, such as sickle-cell anemia, an individual will get the disease only if he or she receives both recessive alleles. This is not always the case. For example, Huntington's disease only requires one dominant gene for an individual to contract the disease. Suppose that a husband and wife, who both have a dominant Huntington's disease allele (*S*) and a normal recessive allele (*s*), decide to have a child.

(a) List the possible genotypes of their offspring.
(b) What is the probability that the offspring will not have Huntington's disease? In other words, what is the probability that the offspring will have genotype *ss*? Interpret this probability.
(c) What is the probability that the offspring will have Huntington's disease?

NW 35. College Survey In a national survey conducted by the Centers for Disease Control to determine college students' health-risk behaviors, college students were asked, "How often do you wear a seatbelt when riding in a car driven by someone else?" The frequencies appear in the following table:

Response	Frequency
Never	125
Rarely	324
Sometimes	552
Most of the time	1257
Always	2518

(a) Construct a probability model for seatbelt use by a passenger.
(b) Would you consider it unusual to find a college student who never wears a seatbelt when riding in a car driven by someone else? Why?

36. College Survey In a national survey conducted by the Centers for Disease Control to determine college students' health-risk behaviors, college students were asked, "How often do you wear a seatbelt when driving a car?" The frequencies appear in the following table:

Response	Frequency
Never	118
Rarely	249
Sometimes	345
Most of the time	716
Always	3093

(a) Construct a probability model for seatbelt use by a driver.
(b) Is it unusual for a college student to never wear a seatbelt when driving a car? Why?

37. Larceny Theft A police officer randomly selected 642 police records of larceny thefts. The following data represent the number of offenses for various types of larceny thefts.

Type of Larceny Theft	Number of Offenses
Pocket picking	4
Purse snatching	6
Shoplifting	133
From motor vehicles	219
Motor vehicle accessories	90
Bicycles	42
From buildings	143
From coin-operated machines	5

Source: U.S. Federal Bureau of Investigation

(a) Construct a probability model for type of larceny theft.
(b) Are purse snatching larcenies unusual?
(c) Are bicycle larcenies unusual?

38. Multiple Births The following data represent the number of live multiple-delivery births (three or more babies) in 2012 for women 15 to 54 years old.

Age	Number of Multiple Births
15–19	44
20–24	404
25–29	1204
30–34	1872
35–39	1000
40–44	332
45–54	63

Source: National Vital Statistics Report

(a) Construct a probability model for number of multiple births.
(b) In the sample space of all multiple births, are multiple births for 15- to 19-year-old mothers unusual?
(c) In the sample space of all multiple births, are multiple births for 40- to 44-year-old mothers unusual?

In Problems 39–42, use the given table, which lists six possible assignments of probabilities for tossing a coin twice, to answer the following questions.

	Sample Space			
Assignments	HH	HT	TH	TT
A	$\frac{1}{4}$	$\frac{1}{4}$	$\frac{1}{4}$	$\frac{1}{4}$
B	0	0	0	1
C	$\frac{3}{16}$	$\frac{5}{16}$	$\frac{5}{16}$	$\frac{3}{16}$
D	$\frac{1}{2}$	$\frac{1}{2}$	$-\frac{1}{2}$	$\frac{1}{2}$
E	$\frac{1}{4}$	$\frac{1}{4}$	$\frac{1}{4}$	$\frac{1}{8}$
F	$\frac{1}{9}$	$\frac{2}{9}$	$\frac{2}{9}$	$\frac{4}{9}$

39. Which of the assignments of probabilities are consistent with the definition of a probability model?

40. Which of the assignments of probabilities should be used if the coin is known to be fair?

41. Which of the assignments of probabilities should be used if the coin is known to always come up tails?

42. Which of the assignments of probabilities should be used if tails is twice as likely to occur as heads?

NW 43. Going to Disney World John, Roberto, Clarice, Dominique, and Marco work for a publishing company. The company wants to send two employees to a statistics conference in Orlando. To be fair, the company decides that the two individuals who get to attend will have their names randomly drawn from a hat.

(a) Determine the sample space of the experiment. That is, list all possible simple random samples of size $n = 2$.

(b) What is the probability that Clarice and Dominique attend the conference?
(c) What is the probability that Clarice attends the conference?
(d) What is the probability that John stays home?

44. Six Flags In 2011, Six Flags St. Louis had 10 roller coasters: The Screamin' Eagle, The Boss, River King Mine Train, Batman the Ride, Mr. Freeze, Ninja, Tony Hawk's Big Spin, Evel Knievel, Xcalibur, and Sky Screamer. Of these, The Boss, The Screamin' Eagle, and Evel Knievel are wooden coasters. Ethan wants to ride two more roller coasters before leaving the park (not the same one twice) and decides to select them by drawing names from a hat.

(a) Determine the sample space of the experiment. That is, list all possible simple random samples of size $n = 2$.
(b) What is the probability that Ethan will ride Mr. Freeze and Evel Knievel?
(c) What is the probability that Ethan will ride the Screamin' Eagle?
(d) What is the probability that Ethan will ride two wooden roller coasters?
(e) What is the probability that Ethan will not ride any wooden roller coasters?

45. Barry Bonds On October 5, 2001, Barry Bonds broke Mark McGwire's homerun record for a single season by hitting his 71st and 72nd homeruns. Bonds went on to hit one more homerun before the season ended, for a total of 73. Of the 73 homeruns, 24 went to right field, 26 went to right center field, 11 went to center field, 10 went to left center field, and 2 went to left field.

Source: Baseball-almanac.com

(a) What is the probability that a randomly selected homerun was hit to right field?
(b) What is the probability that a randomly selected homerun was hit to left field?
(c) Was it unusual for Barry Bonds to hit a homerun to left field? Explain.

46. Rolling a Die

(a) Roll a single die 50 times, recording the result of each roll of the die. Use the results to approximate the probability of rolling a three.
(b) Roll a single die 100 times, recording the result of each roll of the die. Use the results to approximate the probability of rolling a three.
(c) Compare the results of (a) and (b) to the classical probability of rolling a three.

NW 47. Simulation Use a graphing calculator or statistical software to simulate rolling a six-sided die 100 times, using an integer distribution with numbers one through six.

(a) Use the results of the simulation to compute the probability of rolling a one.
(b) Repeat the simulation. Compute the probability of rolling a one.
(c) Simulate rolling a six-sided die 500 times. Compute the probability of rolling a one.
(d) Which simulation resulted in the closest estimate to the probability that would be obtained using the classical method?

48. Classifying Probability Determine whether the probabilities below are computed using classical methods, empirical methods, or subjective methods.

(a) The probability of having eight girls in an eight-child family is 0.00390625.

(b) On the basis of a survey of 1000 families with eight children, the probability of a family having eight girls is 0.0054.

(c) According to a sports analyst, the probability that the Chicago Bears will win their next game is about 0.30.

(d) On the basis of clinical trials, the probability of efficacy of a new drug is 0.75.

49. Checking for Loaded Dice You suspect a 6-sided die to be loaded and conduct a probability experiment by rolling the die 400 times. The outcome of the experiment is listed in the table below. Do you think the die is loaded? Why?

Value of Die	Frequency
1	105
2	47
3	44
4	49
5	51
6	104

50. Conduct a survey in your school by randomly asking 50 students whether they drive to school. Based on the results of the survey, approximate the probability that a randomly selected student drives to school.

51. In 2013, the median income of families in the United States was $52,250. What is the probability that a randomly selected family has an income greater than $52,250?

52. The middle 50% of enrolled freshmen at Washington University in St. Louis had SAT math scores in the range 700–780. What is the probability that a randomly selected freshman at Washington University has an SAT math score of 700 or higher?

53. NFL Combine Each year the National Football League (NFL) runs a combine in which players who wish to be considered for the NFL draft must participate in a variety of activities. Go to www.pearsonhighered.com/sullivanstats to obtain the data file 5_1_53 using the file format of your choice for the version of the text you are using. The data represent results of the 2015 NFL combine. Construct a probability model for the variable "POS," which represents the position of the player. For example, Nelson Ahholor plays wide receiver (WR). If a player is randomly selected from the 2015 NFL combine, what position has the highest probability of being selected? Would you be surprised if a center (C) was randomly selected? Why?

54. Putting It Together: Drug Side Effects In placebo-controlled clinical trials for the drug Viagra, 734 subjects received Viagra and 725 subjects received a placebo (subjects did not know which treatment they received). The table in the next column summarizes reports of various side effects that were reported.

Adverse Effect	Viagra ($n = 734$)	Placebo ($n = 725$)
Headache	117	29
Flushing	73	7
Dyspepsia	51	15
Nasal congestion	29	15
Urinary tract infection	22	15
Abnormal vision	22	0
Diarrhea	22	7
Dizziness	15	7
Rash	15	7

(a) Is the variable "adverse effect" qualitative or quantitative?

(b) Which type of graph would be appropriate to display the information in the table? Construct the graph.

(c) What is the estimated probability that a randomly selected subject from the Viagra group reported flushing as an adverse effect? Would this be unusual?

(d) What is the estimated probability that a subject receiving a placebo would report flushing as an adverse effect? Would this be unusual?

(e) If a subject reports flushing after receiving a treatment, what might you conclude?

(f) What type of experimental design is this?

Explaining the Concepts

55. Explain the Law of Large Numbers. How does this law apply to gambling casinos?

56. In computing classical probabilities, all outcomes must be equally likely. Explain what this means.

57. Describe what an unusual event is. Should the same cutoff always be used to identify unusual events? Why or why not?

58. You are planning a trip to a water park tomorrow and the weather forecaster says there is a 70% chance of rain. Explain what this result means.

59. Describe the difference between classical and empirical probability.

60. Ask Marilyn In a September 19, 2010, article in *Parade Magazine* written to *Ask Marilyn*, Marilyn vos Savant was asked the following: Four identical sealed envelopes are on a table, one of which contains $100. You are to select one of the envelopes. Then the game host discards two of the remaining three envelopes and informs you that they do not contain the $100. In addition, the host offers you the opportunity to switch envelopes. What should you do?

(a) Keep your envelope

(b) switch

(c) it does not matter.

61. Suppose you live in a town with two hospitals—one large and the other small. On a given day in one of the hospitals, 60% of the babies who were born were girls. Which one do you think it is? Or, is it impossible to tell? Support your decision?

5.2　The Addition Rule and Complements

Preparing for this Section　Before getting started, review the following:

- Contingency Tables (Section 4.4, p. 225)

Objectives　① Use the Addition Rule for Disjoint Events

②　Use the General Addition Rule

③　Compute the probability of an event using the Complement Rule

①　Use the Addition Rule for Disjoint Events

Before presenting more rules for computing probabilities, we must discuss *disjoint events.*

Definition　Two events are **disjoint** if they have no outcomes in common. Another name for disjoint events is **mutually exclusive** events.

In Other Words
Two events are disjoint if they cannot occur at the same time.

We can use **Venn diagrams** to represent events as circles enclosed in a rectangle. The rectangle represents the sample space, and each circle represents an event. For example, suppose we randomly select chips from a bag. Each chip is labeled 0, 1, 2, 3, 4, 5, 6, 7, 8, 9. Let E represent the event "choose a number less than or equal to 2," and let F represent the event "choose a number greater than or equal to 8." Because E and F have no outcomes in common, they are disjoint. Figure 7 shows a Venn diagram of these disjoint events.

Figure 7

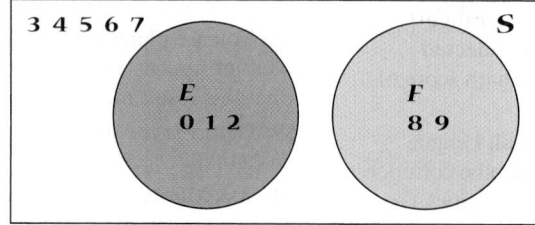

Notice that the outcomes in event E are inside circle E, and the outcomes in event F are inside circle F. All outcomes in the sample space that are not in E or F are outside the circles, but inside the rectangle. From this diagram, we know that $P(E) = \dfrac{N(E)}{N(S)} = \dfrac{3}{10} = 0.3$ and $P(F) = \dfrac{N(F)}{N(S)} = \dfrac{2}{10} = 0.2$. In addition, $P(E \text{ or } F) = \dfrac{N(E \text{ or } F)}{N(S)} = \dfrac{5}{10} = 0.5$ and $P(E \text{ or } F) = P(E) + P(F) = 0.3 + 0.2 = 0.5$. This result occurs because of the *Addition Rule for Disjoint Events.*

In Other Words
The Addition Rule for Disjoint Events states that, if you have two events that have no outcomes in common, the probability that one or the other occurs is the sum of their probabilities.

Addition Rule for Disjoint Events

If E and F are disjoint (or mutually exclusive) events, then

$$P(E \text{ or } F) = P(E) + P(F)$$

The Addition Rule for Disjoint Events can be extended to more than two disjoint events. In general, if $E, F, G, \ldots$ each have no outcomes in common (they are pairwise disjoint), then

$$P(E \text{ or } F \text{ or } G \text{ or } \ldots) = P(E) + P(F) + P(G) + \cdots$$

Let event G represent "the number is a 5 or 6." The Venn diagram in Figure 8 illustrates the Addition Rule for more than two disjoint events using the chip example. Notice that no pair of events has any outcomes in common. So, from the Venn diagram, we can see that

$$P(E) = \frac{N(E)}{N(S)} = \frac{3}{10} = 0.3, \; P(F) = \frac{N(F)}{N(S)} = \frac{2}{10} = 0.2, \text{ and } P(G) = \frac{N(G)}{N(S)} = \frac{2}{10} = 0.2.$$

In addition, $P(E \text{ or } F \text{ or } G) = P(E) + P(F) + P(G) = 0.3 + 0.2 + 0.2 = 0.7.$

Figure 8

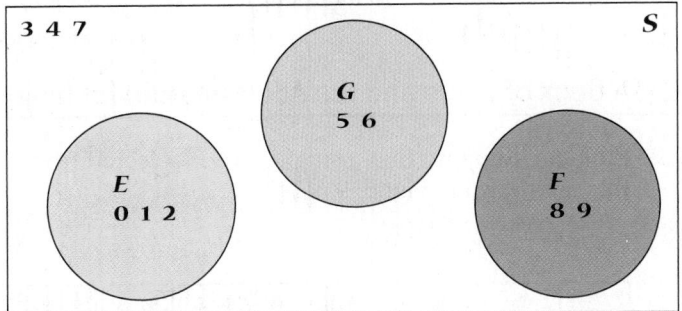

EXAMPLE 1   Benford's Law and the Addition Rule for Disjoint Events

Problem Our number system consists of the digits 0, 1, 2, 3, 4, 5, 6, 7, 8, and 9. Because we do not write numbers such as 12 as 012, the first significant digit in any number must be 1, 2, 3, 4, 5, 6, 7, 8, or 9. We may think that each digit appears with equal frequency so that each digit has a $\frac{1}{9}$ probability of being the first significant digit, but this is not true.

In 1881, Simon Newcomb discovered that digits do not occur with equal frequency. The physicist Frank Benford discovered the same result in 1938. After studying lots and lots of data, he assigned probabilities of occurrence for each of the first digits, as shown in Table 5. The probability model is now known as *Benford's Law* and plays a major role in identifying fraudulent data on tax returns and accounting books.

Table 5									
Digit	1	2	3	4	5	6	7	8	9
Probability	0.301	0.176	0.125	0.097	0.079	0.067	0.058	0.051	0.046

Source: The First Digit Phenomenon, T. P. Hill, *American Scientist*, July–August, 1998.

(a) Verify that Benford's Law is a probability model.
(b) Use Benford's Law to determine the probability that a randomly selected first digit is 1 or 2.
(c) Use Benford's Law to determine the probability that a randomly selected first digit is at least 6.

Approach For part (a), verify that each probability is between 0 and 1 and that the sum of all probabilities equals 1. For parts (b) and (c), use the Addition Rule for Disjoint Events.

Solution
(a) Each probability in Table 5 is between 0 and 1. In addition, the sum of all the probabilities, $0.301 + 0.176 + 0.125 + \cdots + 0.046$, is 1. Because Rules 1 and 2 are satisfied, Table 5 represents a probability model.

(continued)

(b)
$$P(1 \text{ or } 2) = P(1) + P(2)$$
$$= 0.301 + 0.176$$
$$= 0.477$$

If we looked at 100 numbers, we would expect about 48 to begin with 1 or 2.

(c)
$$P(\text{at least } 6) = P(6 \text{ or } 7 \text{ or } 8 \text{ or } 9)$$
$$= P(6) + P(7) + P(8) + P(9)$$
$$= 0.067 + 0.058 + 0.051 + 0.046$$
$$= 0.222$$

If we looked at 100 numbers, we would expect about 22 to begin with 6, 7, 8, or 9.   •

EXAMPLE 2   A Deck of Cards and the Addition Rule for Disjoint Events

Problem Suppose that a single card is selected from a standard 52-card deck, such as the one shown in Figure 9.

Figure 9

(a) Compute the probability of the event $E =$ "drawing a king."
(b) Compute the probability of the event $E =$ "drawing a king" or $F =$ "drawing a queen" or $G =$ "drawing a jack."

Approach Use the classical method for computing the probabilities because the outcomes are equally likely and easy to count. Use the Addition Rule for Disjoint Events to compute the probability in part (b) because the events are mutually exclusive. For example, you cannot simultaneously draw a king and a queen.

Solution The sample space consists of the 52 cards in the deck, so $N(S) = 52$.
(a) A standard deck of cards has four kings, so $N(E) = 4$. Therefore,

$$P(\text{king}) = P(E) = \frac{N(E)}{N(S)} = \frac{4}{52} = \frac{1}{13}$$

(b) A standard deck of cards has four kings, four queens, and four jacks. Because events E, F, and G are mutually exclusive, we use the Addition Rule for Disjoint Events extended to two or more disjoint events. So

$$P(\text{king or queen or jack}) = P(E \text{ or } F \text{ or } G)$$
$$= P(E) + P(F) + P(G)$$
$$= \frac{4}{52} + \frac{4}{52} + \frac{4}{52} = \frac{12}{52} = \frac{3}{13}$$

• **Now Work Problems 25(a)–(c)**

② Use the General Addition Rule

What happens when you need to compute the probability of two events that are not disjoint?

Suppose we are randomly selecting chips from a bag. Each chip is labeled 0, 1, 2, 3, 4, 5, 6, 7, 8, or 9. Let E represent the event "choose an odd number," and let F represent the event "choose a number less than or equal to 4." Because $E = \{1, 3, 5, 7, 9\}$ and $F = \{0, 1, 2, 3, 4\}$ have the outcomes 1 and 3 in common, the events are not disjoint. Figure 10 shows a Venn diagram of these events.

Figure 10

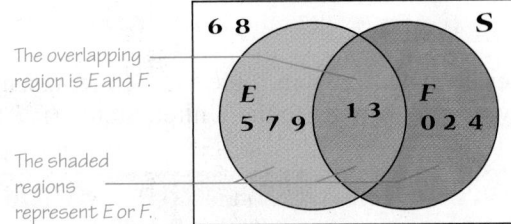

The overlapping region is E and F.

The shaded regions represent E or F.

We can compute $P(E \text{ or } F)$ directly by counting because each outcome is equally likely. There are 8 outcomes in E or F and 10 outcomes in the sample space, so

$$P(E \text{ or } F) = \frac{N(E \text{ or } F)}{N(S)} = \frac{8}{10} = \frac{4}{5}$$

Notice that using the Addition Rule for Disjoint Events to find $P(E \text{ or } F)$ would be *incorrect*:

$$P(E \text{ or } F) \neq P(E) + P(F) = \frac{5}{10} + \frac{5}{10} = 1$$

This implies that the chips labeled 6 and 8 will never be selected, which contradicts our assumption that all the outcomes are equally likely. Our result is incorrect because we counted the outcomes 1 and 3 twice: once for event E and once for event F. To avoid this double counting, we have to subtract the probability corresponding to the overlapping region, E and F. That is, we have to subtract $P(E \text{ and } F) = \frac{2}{10}$ from the result and obtain

$$P(E \text{ or } F) = P(E) + P(F) - P(E \text{ and } F)$$
$$= \frac{5}{10} + \frac{5}{10} - \frac{\overbrace{2}}{10}$$
$$= \frac{8}{10} = \frac{4}{5}$$

which agrees with the result we found by counting. The following rule generalizes these results.

The General Addition Rule

For any two events E and F,

$$P(E \text{ or } F) = P(E) + P(F) - P(E \text{ and } F)$$

EXAMPLE 3 Computing Probabilities for Events That Are Not Disjoint

Problem Suppose a single card is selected from a standard 52-card deck. Compute the probability of the event $E =$ "drawing a king" or $F =$ "drawing a diamond."

Approach The events are not disjoint because the outcome "king of diamonds" is in both events, so use the General Addition Rule.

(continued)

Solution

$$P(E \text{ or } F) = P(E) + P(F) - P(E \text{ and } F)$$

$$P(\text{king or diamond}) = P(\text{king}) + P(\text{diamond}) - P(\text{king of diamonds})$$

$$= \frac{4}{52} + \frac{13}{52} - \frac{1}{52}$$

$$= \frac{16}{52} = \frac{4}{13}$$

● Now Work Problem 31

Consider the data shown in Table 6, which represent the marital status of males and females 15 years old or older in the United States in 2013.

Table 6		
	Males (in millions)	**Females (in millions)**
Never married	41.6	36.9
Married	64.4	63.1
Widowed	3.1	11.2
Divorced	11.0	14.4
Separated	2.4	3.2

Source: U.S. Census Bureau, *Current Population Reports*

Table 6 is called a **contingency table** or **two-way table**, because it relates two categories of data. The **row variable** is marital status, because each row in the table describes the marital status of an individual. The **column variable** is gender. Each box inside the table is called a **cell**. For example, the cell corresponding to married individuals who are male is in the second row, first column. Each cell contains the frequency of the category: There were 64.4 million married males in the United States in 2013. Put another way, in the United States in 2013, there were 64.4 million individuals who were male *and* married.

EXAMPLE 4   Using the Addition Rule with Contingency Tables

Problem Using the data in Table 6,

(a) Determine the probability that a randomly selected U.S. resident 15 years old or older is male.

(b) Determine the probability that a randomly selected U.S. resident 15 years old or older is widowed.

(c) Determine the probability that a randomly selected U.S. resident 15 years old or older is widowed or divorced.

(d) Determine the probability that a randomly selected U.S. resident 15 years old or older is male or widowed.

Approach Add the entries in each row and column to get the total number of people in each category. Then determine the probabilities using either the Addition Rule for Disjoint Events or the General Addition Rule.

Solution First, add the entries in each column. For example, the "male" column shows there are 41.6 + 64.4 + 3.1 + 11.0 + 2.4 = 122.5 million males 15 years old or older in the United States. Add the entries in each row. For example, in the "never married" row we find there are 41.6 + 36.9 = 78.5 million U.S. residents 15 years old or older who have never married. Adding the row totals or column totals, we find there are 122.5 + 128.8 = 78.5 + 127.5 + 14.3 + 25.4 + 5.6 = 251.3 million U.S. residents 15 years old or older.

(a) There are 122.5 million males 15 years old or older and 251.3 million U.S. residents 15 years old or older. The probability that a randomly selected U.S. resident 15 years old or older is male is $\frac{122.5}{251.3} = 0.487$.

(b) There are 14.3 million U.S. residents 15 years old or older who are widowed. The probability that a randomly selected U.S. resident 15 years old or older is widowed is $\frac{14.3}{251.3} = 0.057$.

(c) The events widowed and divorced are disjoint. Do you see why? We use the Addition Rule for Disjoint Events.

$$P(\text{widowed or divorced}) = P(\text{widowed}) + P(\text{divorced})$$
$$= \frac{14.3}{251.3} + \frac{25.4}{251.3}$$
$$= \frac{39.7}{251.3} = 0.158$$

(d) The events male and widowed are not mutually exclusive. In fact, there are 3.1 million males who are widowed in the United States. Therefore, we use the General Addition Rule to compute $P(\text{male or widowed})$:

$$P(\text{male or widowed}) = P(\text{male}) + P(\text{widowed}) - P(\text{male and widowed})$$
$$= \frac{122.5}{251.3} + \frac{14.3}{251.3} - \frac{3.1}{251.3}$$
$$= \frac{133.7}{251.3} = 0.532$$

● **Now Work Problem 39**

③ Compute the Probability of an Event Using the Complement Rule

Suppose that the probability of an event E is known and we would like to determine the probability that E does not occur. This can easily be accomplished using the idea of *complements*.

Definition | **Complement of an Event**

Let S denote the sample space of a probability experiment and let E denote an event. The **complement of E**, denoted E^c, is all outcomes in the sample space S that are not outcomes in the event E.

Because E and E^c are mutually exclusive,

$$P(E \text{ or } E^c) = P(E) + P(E^c) = P(S) = 1$$

Subtracting $P(E)$ from both sides, we obtain the following result.

In Other Words
For any event, either the event happens or it doesn't. Use the Complement Rule when you know the probability that some event will occur and you want to know the chance it will not occur.

Complement Rule

If E represents any event and E^c represents the complement of E, then

$$P(E^c) = 1 - P(E)$$

Figure 11 illustrates the Complement Rule using a Venn diagram.

Figure 11

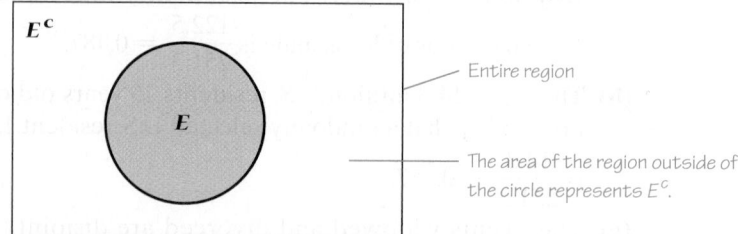

Entire region

The area of the region outside of the circle represents E^C.

EXAMPLE 5 Computing Probabilities Using Complements

Problem According to the National Gambling Impact Study Commission, 52% of Americans have played state lotteries. What is the probability that a randomly selected American has not played a state lottery?

Approach Not playing a state lottery is the complement of playing a state lottery. Compute the probability using the Complement Rule.

Solution $P(\text{not played state lottery}) = 1 - P(\text{played state lottery}) = 1 - 0.52 = 0.48$

The probability of randomly selecting an American who has not played a state lottery is 0.48.

EXAMPLE 6 Computing Probabilities Using Complements

Problem The data in Table 7 represent the income distribution of households in the United States in 2013.

Table 7			
Annual Income	**Number (in thousands)**	**Annual Income**	**Number (in thousands)**
Less than $10,000	8940	$50,000 to $74,999	21,659
$10,000 to $14,999	6693	$75,000 to $99,999	14,687
$15,000 to $24,999	13,898	$100,000 to $149,999	15,266
$25,000 to $34,999	12,756	$150,000 to $199,999	6463
$35,000 to $49,999	16,678	$200,000 or more	5913

Source: U.S. Census Bureau

Compute the probability that a randomly selected household earned the following incomes in 2013:

(a) $200,000 or more
(b) Less than $200,000
(c) At least $10,000

Approach Determine the probabilities by finding the relative frequency of each event.

Solution
(a) There were a total of $8940 + 6693 + \cdots + 5913 = 122{,}953$ thousand households in the United States in 2013 and 5913 thousand of them earned $200,000 or more, so $P(\text{earned \$200,000 or more}) = \dfrac{5913}{122{,}953} = 0.048$.

(b) We could compute the probability of randomly selecting a household that earned less than $200,000 in 2013 by adding the relative frequencies of each category less than $200,000, but it is easier to use complements. The complement of earning less than $200,000 is earning $200,000 or more. Therefore,

$$P(\text{less than } \$200,000) = 1 - P(\$200,000 \text{ or more})$$
$$= 1 - 0.048 = 0.952$$

The probability of randomly selecting a household that earned less than $200,000 in 2013 is 0.952. If we randomly selected 1000 households in 2013, we would expect 952 to have household incomes less than $200,000.

(c) The phrase *at least* means greater than or equal to. The complement of at least $10,000 is less than $10,000. In 2013, 8940 thousand households earned less than $10,000. The probability of randomly selecting a household that earned at least $10,000 is

$$P(\text{at least } \$10,000) = 1 - P(\text{less than } \$10,000)$$
$$= 1 - \frac{8940}{122,953} = 0.927$$

The probability of randomly selecting a household that earned at least $10,000 in 2013 is 0.927. If we randomly selected 1000 households in 2013, we would expect 927 to have household income of at least $10,000.

● Now Work Problems 25(d) and 29

 5.2 Assess Your Understanding

Vocabulary and Skill Building

1. What does it mean when two events are disjoint?

2. If E and F are disjoint events, then $P(E \text{ or } F) =$ _____.

3. If E and F are not disjoint events, then $P(E \text{ or } F) =$ _____.

4. What does it mean when two events are complements?

In Problems 5–12, a probability experiment is conducted in which the sample space of the experiment is $S = \{1, 2, 3, 4, 5, 6, 7, 8, 9, 10, 11, 12\}$. Let event $E = \{2, 3, 4, 5, 6, 7\}$, event $F = \{5, 6, 7, 8, 9\}$, event $G = \{9, 10, 11, 12\}$, and event $H = \{2, 3, 4\}$. Assume that each outcome is equally likely.

5. List the outcomes in E and F. Are E and F mutually exclusive?

6. List the outcomes in F and G. Are F and G mutually exclusive?

7. List the outcomes in F or G. Now find $P(F \text{ or } G)$ by counting the number of outcomes in F or G. Determine $P(F \text{ or } G)$ using the General Addition Rule.

8. List the outcomes in E or H. Now find $P(E \text{ or } H)$ by counting the number of outcomes in E or H. Determine $P(E \text{ or } H)$ using the General Addition Rule.

9. List the outcomes in E and G. Are E and G mutually exclusive?

10. List the outcomes in F and H. Are F and H mutually exclusive?

11. List the outcomes in E^c. Find $P(E^c)$.

12. List the outcomes in F^c. Find $P(F^c)$.

In Problems 13–18, find the probability of the indicated event if $P(E) = 0.25$ and $P(F) = 0.45$.

13. $P(E \text{ or } F)$ if $P(E \text{ and } F) = 0.15$

14. $P(E \text{ and } F)$ if $P(E \text{ or } F) = 0.6$

15. $P(E \text{ or } F)$ if E and F are mutually exclusive

16. $P(E \text{ and } F)$ if E and F are mutually exclusive

17. $P(E^c)$

18. $P(F^c)$

19. If $P(E) = 0.60$, $P(E \text{ or } F) = 0.85$, and $P(E \text{ and } F) = 0.05$, find $P(F)$.

20. If $P(F) = 0.30$, $P(E \text{ or } F) = 0.65$, and $P(E \text{ and } F) = 0.15$, find $P(E)$.

In Problems 21–24, a golf ball is selected at random from a golf bag. If the golf bag contains 9 Titleists, 8 Maxflis, and 3 Top-Flites, find the probability that the golf ball is:

21. A Titleist or Maxfli.

22. A Maxfli or Top-Flite.

23. Not a Titleist.

24. Not a Top-Flite.

Applying the Concepts

NW **25. Weapon of Choice** The following probability model shows the distribution of murders by type of weapon for murder cases in 2013.

Weapon	Probability
Handgun	0.472
Rifle	0.023
Shotgun	0.025
Unknown firearm	0.170
Knives	0.122
Hands, fists, etc.	0.056
Other	0.132

Source: U.S. Federal Bureau of Investigation

(a) Verify that this is a probability model.

(b) What is the probability that a randomly selected murder resulted from a rifle or shotgun? Interpret this probability.

(c) What is the probability that a randomly selected murder resulted from a handgun, rifle, or shotgun (a firearm)? Interpret this probability.

(d) What is the probability that a randomly selected murder resulted from a weapon other than a firearm? Interpret this probability.

(e) Are murders with a shotgun unusual?

26. Family Structure The following probability model shows the distribution of family structure among families with at least one child younger than 18 years of age in 2013.

Family Structure	Probability
Two married parents, first marriage	0.46
Two married parents, one or both remarried	0.15
Single parent	0.34
No parent at home	0.05

Source: Pew Research

(a) Verify that this is a probability model.

(b) What is the probability that a randomly selected family with at least one child younger than 18 years of age in 2013 had two married parents in their first marriage? Interpret this probability.

(c) What is the probability that a randomly selected family with at least one child younger than 18 years of age in 2013 had two married parents? Interpret this probability.

(d) What is the probability that a randomly selected family with at least one child younger than 18 years of age in 2013 had at least one parent at home? Interpret this probability.

27. If events *E* and *F* are disjoint and the events *F* and *G* are disjoint, must the events *E* and *G* necessarily be disjoint? Give an example to illustrate your opinion.

28. Draw a Venn diagram like that in Figure 10 that expands the general addition rule to three events. Use the diagram to write the General Addition Rule for three events.

NW **29. Medicare Fines** In an effort to reduce the number of hospital-acquired conditions (such as infection resulting from the hospital stay), Medicare officials score hospitals on a 10-point scale with a lower score representing a better patient track record. The federal government plans to reduce Medicare payments to those hospitals with the worst scores. The following data represent the scores received by Illinois hospitals.

Score	Frequency
1–1.9	3
2–2.9	12
3–3.9	16
4–4.9	23
5–5.9	23
6–6.9	21
7–7.9	17
8–8.9	5
9–10	5
Total	**125**

Source: Center for Medicare and Medicaid Services

(a) Determine the probability that a randomly selected hospital in Illinois has a score between 5 and 5.9.

(b) Determine the probability that a randomly selected hospital in Illinois has a score that is not between 5 and 5.9.

(c) Determine the probability that a randomly selected hospital in Illinois has a score less than 9.

(d) Suppose Medicare plans to reduce Medicare payments to any hospital with a score of 8 or higher. What is the probability a randomly selected hospital in Illinois will experience reduced Medicare payments? Interpret this result. Is it unusual?

30. Housing The following data represents the number of rooms in a random sample of U.S. housing units.

Rooms	Frequency
One	5
Two	11
Three	88
Four	183
Five	230
Six	204
Seven	123
Eight or more	156

Source: U.S. Census Bureau

(a) What is the probability that a randomly selected housing unit has four or more rooms? Interpret this probability.

(b) What is the probability that a randomly selected housing unit has fewer than eight rooms? Interpret this probability.

(c) What is the probability that a randomly selected housing unit has from four to six (inclusive) rooms? Interpret this probability.

(d) What is the probability that a randomly selected housing unit has at least two rooms? Interpret this probability.

NW **31. A Deck of Cards** A standard deck of cards contains 52 cards, as shown in Figure 9. One card is randomly selected from the deck.

(a) Compute the probability of randomly selecting a heart or club from a deck of cards.

(b) Compute the probability of randomly selecting a heart or club or diamond from a deck of cards.

(c) Compute the probability of randomly selecting an ace or heart from a deck of cards.

32. A Deck of Cards A standard deck of cards contains 52 cards, as shown in Figure 9. One card is randomly selected from the deck.

(a) Compute the probability of randomly selecting a two or three from a deck of cards.

(b) Compute the probability of randomly selecting a two or three or four from a deck of cards.

(c) Compute the probability of randomly selecting a two or club from a deck of cards.

33. Birthdays Exclude leap years from the following calculations:

(a) Compute the probability that a randomly selected person does not have a birthday on November 8.

(b) Compute the probability that a randomly selected person does not have a birthday on the 1st day of a month.

(c) Compute the probability that a randomly selected person does not have a birthday on the 31st day of a month.

(d) Compute the probability that a randomly selected person was not born in December.

34. Roulette In the game of roulette, a wheel consists of 38 slots numbered 0, 00, 1, 2, . . . , 36. The odd-numbered slots are red, and the even-numbered slots are black. The numbers 0 and 00 are green. To play the game, a metal ball is spun around the wheel and is allowed to fall into one of the numbered slots.

(a) What is the probability that the metal ball lands on green or red?

(b) What is the probability that the metal ball does not land on green?

35. Health Problems According to the Centers for Disease Control, the probability that a randomly selected citizen of the United States has hearing problems is 0.151. The probability that a randomly selected citizen of the United States has vision problems is 0.093. Can we compute the probability of randomly selecting a citizen of the United States who has hearing problems or vision problems by adding these probabilities? Why or why not?

36. Visits to the Doctor A National Ambulatory Medical Care Survey administered by the Centers for Disease Control found that the probability a randomly selected patient visited the doctor for a blood pressure check is 0.593. The probability a randomly selected patient visited the doctor for urinalysis is 0.064. Can we compute the probability of randomly selecting a patient who visited the doctor for a blood pressure check or urinalysis by adding these probabilities? Why or why not?

37. Language Spoken at Home According to the U.S. Census Bureau, the probability that a randomly selected household speaks only English at home is 0.81. The probability that a randomly selected household speaks only Spanish at home is 0.12.

(a) What is the probability that a randomly selected household speaks only English or only Spanish at home?

(b) What is the probability that a randomly selected household speaks a language other than only English or only Spanish at home?

(c) What is the probability that a randomly selected household speaks a language other than only English at home?

(d) Can the probability that a randomly selected household speaks only Polish at home equal 0.08? Why or why not?

38. Getting to Work According to the U.S. Census Bureau, the probability that a randomly selected worker primarily drives a car to work is 0.867. The probability that a randomly selected worker primarily takes public transportation to work is 0.048.

(a) What is the probability that a randomly selected worker primarily drives a car or takes public transportation to work?

(b) What is the probability that a randomly selected worker neither drives a car nor takes public transportation to work?

(c) What is the probability that a randomly selected worker does not drive a car to work?

(d) Can the probability that a randomly selected worker walks to work equal 0.15? Why or why not?

NW 39. Cigar Smoking The data in the following table show the results of a national study of 137,243 U.S. men that investigated the association between cigar smoking and death from cancer. **Note:** Current cigar smoker means cigar smoker at time of death.

	Died from Cancer	Did Not Die from Cancer
Never smoked cigars	782	120,747
Former cigar smoker	91	7,757
Current cigar smoker	141	7,725

Source: Shapiro, Jacobs, and Thun. "Cigar Smoking in Men and Risk of Death from Tobacco-Related Cancers," *Journal of the National Cancer Institute,* February 16, 2000.

(a) If an individual is randomly selected from this study, what is the probability that he died from cancer?

(b) If an individual is randomly selected from this study, what is the probability that he was a current cigar smoker?

(c) If an individual is randomly selected from this study, what is the probability that he died from cancer and was a current cigar smoker?

(d) If an individual is randomly selected from this study, what is the probability that he died from cancer or was a current cigar smoker?

40. Working Couples A guidance counselor at a middle school collected the following information regarding the employment status of married couples within his school's boundaries.

	Number of Children under 18 Years Old			
Worked	**0**	**1**	**2 or More**	**Total**
Husband only	172	79	174	**425**
Wife only	94	17	15	**126**
Both spouses	522	257	370	**1149**
Total	**788**	**353**	**559**	**1700**

(a) What is the probability that, for a married couple selected at random, both spouses work?

(b) What is the probability that, for a married couple selected at random, the couple has one child under the age of 18?

(c) What is the probability that, for a married couple selected at random, the couple has two or more children under the age of 18 and both spouses work?

(d) What is the probability that, for a married couple selected at random, the couple has no children or only the husband works?

(e) Would it be unusual to select a married couple at random for which only the wife works?

41. The Placebo Effect A company is testing a new medicine for migraine headaches. In the study, 150 women were given the new medicine and 100 women were given a placebo. Each participant was directed to take the medicine when the first symptoms of a migraine occurred and then to record whether the headache went away within 45 minutes or lingered. The results are recorded in the following table:

	Headache Went Away	Headache Lingered
Given medicine	132	18
Given placebo	56	44

(a) If a study participant is selected at random, what is the probability she was given the placebo?

(b) If a study participant is selected at random, what is the probability her headache went away within 45 minutes?

(c) If a study participant is selected at random, what is the probability she was given the placebo and her headache went away within 45 minutes?

(d) If a study participant is selected at random, what is the probability she was given the placebo or her headache went away within 45 minutes?

42. Social Media Harris Interactive conducted a survey in which they asked adult Americans (18 years or older) whether they used social media (Facebook, Twitter, and so on) regularly. The following table is based on the results of the survey.

	18–34	35–44	45–54	55+	Total
Use social media	117	89	83	49	**338**
Do not use social media	33	36	57	66	**192**
Total	**150**	**125**	**140**	**115**	**530**

Source: Harris Interactive

(a) If an adult American is randomly selected, what is the probability he or she uses social media?

(b) If an adult American is randomly selected, what is the probability he or she is 45 to 54 years of age?

(c) If an adult American is randomly selected, what is the probability he or she is a 35- to 44-year-old social media user?

(d) If an adult American is randomly selected, what is the probability he or she is 35 to 44 years old or uses social media?

43. Driver Fatalities The following data represent the number of drivers involved in fatal crashes in the United States in 2013 by day of the week and gender.

	Male	**Female**	**Total**
Sunday	4143	2287	**6430**
Monday	3178	1705	**4883**
Tuesday	3280	1739	**5019**
Wednesday	3197	1729	**4926**
Thursday	3389	1839	**5228**
Friday	3975	2179	**6154**
Saturday	4749	2511	**7260**
Total	**25,911**	**13,989**	**39,900**

Source: Fatality Analysis Reporting System, National Highway Traffic Safety Administration

(a) Determine the probability that a randomly selected fatal crash involved a male.

(b) Determine the probability that a randomly selected fatal crash occurred on Sunday.

(c) Determine the probability that a randomly selected fatal crash occurred on Sunday and involved a male.

(d) Determine the probability that a randomly selected fatal crash occurred on Sunday or involved a male.

(e) Would it be unusual for a fatality to occur on Wednesday and involve a female driver?

44. Putting It Together: Speeding Tickets Go to www.pearsonhighered.com/sullivanstats to obtain the data file SullivanStatsSurveyI using the file format of your choice for the version of the text you are using. The data represent the results of a survey conducted by the author. The variable "Text while Driving" represents the response to the question, "Have you ever texted while driving?" The variable "Tickets" represents the response to the question, "How many speeding tickets have you received in the past 12 months?" Treat the individuals in the survey as a random sample of all U.S. drivers.

(a) Build a contingency table treating "Text while Driving" as the row variable and "Tickets" as the column variable.

(b) Determine the marginal relative frequency distribution for both the row and column variable.

(c) What is the probability a randomly selected U.S. driver texts while driving?

(d) What is the probability a randomly selected U.S. driver received three speeding tickets in the past 12 months?

(e) What is the probability a randomly selected U.S. driver texts while driving or received three speeding tickets in the past 12 months?

(f) Interpret the marginal relative frequency distribution for the column variable, "Tickets."

45. Putting It Together: Red Light Cameras In a study of the feasibility of a red-light camera program in the city of Milwaukee, the data below summarize the projected number of crashes at 13 selected intersections over a five-year period.

Crash Type	Current System	With Red-Light Cameras
Reported injury	289	221
Reported property damage only	392	333
Unreported injury	78	60
Unreported property damage only	362	308
Total	**1121**	**922**

Source: Krig, Moran, Regan. "An Analysis of a Red-Light Camera Program in the City of Milwaukee," Spring 2006, prepared for the city of Milwaukee Budget and Management Division

(a) Identify the variables presented in the table.

(b) State whether each variable is qualitative or quantitative. If quantitative, state whether it is discrete or continuous.

(c) Construct a relative frequency distribution for each system.

(d) Construct a side-by-side relative frequency bar graph for the data.

(e) Determine the mean number of crashes per intersection in the study, if possible. If not possible, explain why.

(f) Determine the standard deviation number of crashes, if possible. If not possible, explain why.

(g) Based on the data shown, does it appear that the red-light camera program will be beneficial in reducing crashes at the intersections? Explain.

(h) For the current system, what is the probability that a crash selected at random will have reported injuries?

(i) For the camera system, what is the probability that a crash selected at random will have only property damage?

The study classified crashes further by indicating whether they were red-light running crashes or rear-end crashes. The results are as follows:

Crash Type	Rear End		Red-Light Running	
	Current	Cameras	Current	Cameras
Reported injury	67	77	222	144
Reported property damage only	157	180	235	153
Unreported injury	18	21	60	39
Unreported property damage only	145	167	217	141
Total	387	445	734	477

(j) Using Simpson's Paradox, explain how the additional classification affects your response to part (g).

(k) What recommendation would you make to the city council regarding the implementation of the red-light camera program? Would you need any additional information before making your recommendation? Explain.

Retain Your Knowledge

46. Exam Scores The following data represent the homework scores for the material on Polynomial and Rational Functions in Sullivan's College Algebra course.

37	67	76	82	89
48	70	77	83	90
54	72	77	84	90
59	73	77	84	91
61	75	77	84	92
65	75	78	85	95
65	76	80	87	95
67	76	81	88	98

Source: Michael Sullivan, MyMathLab

(a) Construct a relative frequency distribution with a lower class limit of the first class equal to 30 and a class width of 10.

(b) Draw a relative frequency histogram of the data.

(c) Determine the mean and median score.

(d) Draw a boxplot of the data. Are there any outliers?

(e) Describe the shape of the distribution based on the results from parts (b) thru (d).

(f) Determine the standard deviation and interquartile range.

(g) What is the probability a randomly selected student fails the homework (scores less than 60)?

(h) What is the probability a randomly selected student earns an A or B on the homework (scores 80 or higher)?

(i) What is the probability a randomly selected student scores less than 30 on the homework?

5.3 Independence and the Multiplication Rule

Objectives
1. Identify independent events
2. Use the Multiplication Rule for Independent Events
3. Compute at-least probabilities

1 Identify Independent Events

We use the Addition Rule for Disjoint Events to compute the probability of observing an outcome in event *E or* event *F*. We now describe a probability rule for computing the probability that *E and F* both occur.

Before we can present this rule, we must discuss the idea of *independent events*.

Definition

Two events *E* and *F* are **independent** if the occurrence of event *E* in a probability experiment does not affect the probability of event *F*. Two events are **dependent** if the occurrence of event *E* in a probability experiment affects the probability of event *F*.

Think about flipping a fair coin twice. Does the fact that you obtained a head on the first toss have any effect on the likelihood of obtaining a head on the second toss? Not unless you are a master coin flipper who can manipulate the outcome of a coin flip! For this reason, the outcome from the first flip is independent of the outcome from the second flip. Let's look at other examples.

EXAMPLE 1   Independent or Not?

In Other Words
In determining whether two events are independent, ask yourself whether the probability of one event is affected by the other event. For example, what is the probability that a 29-year-old male has high cholesterol? What is the probability that a 29-year-old male has high cholesterol, given that he eats fast food four times a week? Does the fact that the individual eats fast food four times a week change the likelihood that he has high cholesterol? If yes, the events are not independent.

(a) Suppose you flip a coin and roll a die. The events "obtain a head" and "roll a 5" are independent because the results of the coin flip do not affect the results of the die toss.

(b) Are the events "earned a bachelor's degree" and "earn more than $100,000 per year" independent? No, because knowing that an individual has a bachelor's degree affects the likelihood that the individual is earning more than $100,000 per year.

(c) Two 24-year-old male drivers who live in the United States are randomly selected. The events "male 1 gets in a car accident during the year" and "male 2 gets in a car accident during the year" are independent because the males were randomly selected. This means what happens with one of the drivers has nothing to do with what happens to the other driver. •

In Example 1(c), we are able to conclude that the events "male 1 gets in an accident" and "male 2 gets in an accident" are independent because the individuals are randomly selected. By randomly selecting the individuals, it is reasonable to conclude that the individuals are not related in any way (related in the sense that they do not live in the same town, attend the same school, and so on). If the two individuals did have a common link between them (such as they both lived on the same city block), then knowing that one male had a car accident may affect the likelihood that the other male had a car accident. After all, they could hit each other!

• **Now Work Problem 7**

Disjoint Events versus Independent Events   Disjoint events and independent events are different concepts. Recall that two events are disjoint if they have no outcomes in common, that is, if knowing that one of the events occurs, we know the other event did not occur. Independence means that one event occurring does not affect the probability of the other event occurring. Therefore, knowing two events are disjoint means that the events are not independent.

> **CAUTION!**
> Two events that are disjoint are not independent.

Consider the experiment of rolling a single die. Let E represent the event "roll an even number," and let F represent the event "roll an odd number." We can see that E and F are mutually exclusive (disjoint) because they have no outcomes in common. In addition, $P(E) = \dfrac{1}{2}$ and $P(F) = \dfrac{1}{2}$. However, if we are told that the roll of the die is going to be an even number, then what is the probability of event F? Because the outcome will be even, the probability of event F is now 0 (and the probability of event E is now 1). So knowledge of event E changes the likelihood of observing event F.

② **Use the Multiplication Rule for Independent Events**

Suppose that you flip a fair coin twice. What is the probability that you obtain a head on both flips, that is, a head on the first flip *and* you obtain a head on the second flip? If H represents the outcome "heads" and T represents the outcome "tails," the sample space of this experiment is

$$S = \{\text{HH, HT, TH, TT}\}$$

There is one outcome with both heads. Because each outcome is equally likely, we have

$$P(\text{heads on Flip 1 and heads on Flip 2}) = \frac{N(\text{heads on Flip 1 and heads on Flip 2})}{N(S)}$$

$$= \frac{1}{4}$$

We may have intuitively figured this out by recognizing $P(\text{head}) = \dfrac{1}{2}$ for each flip. So it seems reasonable that

$$P(\text{heads on Flip 2}) = P(\text{heads on Flip 1}) \cdot P(\text{heads on Flip 2})$$

$$= \frac{1}{2} \cdot \frac{1}{2}$$

$$= \frac{1}{4}$$

Because both approaches result in the same answer, $\frac{1}{4}$, we conjecture that $P(E \text{ and } F) = P(E) \cdot P(F)$, which is true.

Multiplication Rule for Independent Events

If E and F are independent events, then

$$P(E \text{ and } F) = P(E) \cdot P(F)$$

EXAMPLE 2 Computing Probabilities of Independent Events

Problem In the game of roulette, the wheel has slots numbered 0, 00, and 1 through 36. A metal ball rolls around a wheel until it falls into one of the numbered slots. What is the probability that the ball will land in the slot numbered 17 two times in a row?

Approach The sample space of the experiment has 38 outcomes. We use the classical method of computing probabilities because the outcomes are equally likely. In addition, we use the Multiplication Rule for Independent Events. The events "17 on Spin 1" and "17 on Spin 2" are independent because the ball does not remember it landed on 17 on the first spin, so this cannot affect the probability of landing on 17 on the second spin.

Solution There are 38 possible outcomes, so the probability of landing on 17 is $\frac{1}{38}$. Because the events "17 on Spin 1" and "17 on Spin 2" are independent, we have

$$P(17 \text{ on Spin 1 and } 17 \text{ on Spin 2}) = P(17 \text{ on Spin 1}) \cdot P(17 \text{ on Spin 2})$$

$$= \frac{1}{38} \cdot \frac{1}{38} = \frac{1}{1444} \approx 0.0006925$$

We expect the ball to land on 17 twice in a row about 7 times in 10,000 trials.

We can extend the Multiplication Rule for three or more independent events.

Multiplication Rule for n Independent Events

If events $E_1, E_2, E_3, \ldots, E_n$ are independent, then

$$P(E_1 \text{ and } E_2 \text{ and } E_3 \text{ and } \ldots \text{ and } E_n) = P(E_1) \cdot P(E_2) \cdot \cdots \cdot P(E_n)$$

EXAMPLE 3 Life Expectancy

Problem The probability that a randomly selected 24-year-old male will survive the year is 0.9986 according to the *National Vital Statistics Report*, Vol. 56, No. 9. (a) What is the probability that three randomly selected 24-year-old males will survive the year? (b) What is the probability that 20 randomly selected 24-year-old males will survive the year?

Approach It is safe to assume that the outcomes of the probability experiment are independent, because there is no indication that the survival of one male affects the survival of the others.

(continued)

In Other Words
In Example 3, if two of the males lived in the same house, a house fire could kill both males and we lose independence. (Knowledge that one male died in a house fire certainly affects the probability that the other died.) By randomly selecting the males, we minimize the chances that they are related in any way.

Solution

(a) $P(\text{all three males survive}) = P(\text{1st survives and 2nd survives and 3rd survives})$
$$= P(\text{1st survives}) \cdot P(\text{2nd survives}) \cdot P(\text{3rd survives})$$
$$= (0.9986)(0.9986)(0.9986)$$
$$= 0.9958$$

If we randomly selected three 24-year-old males 1000 different times, we would expect all three to survive one year in 996 of the samples.

(b) $P(\text{all 20 males survive}) = P(\text{1st survives and 2nd survives and } \ldots \text{ and 20th survives})$
$$= P(\text{1st survives}) \cdot P(\text{2nd survives}) \cdot \cdots \cdot P(\text{20th survives})$$

Multiply 0.9986 by itself 20 times.
$$= \overbrace{(0.9986) \cdot (0.9986) \cdot \cdots \cdot (0.9986)}$$
$$= (0.9986)^{20}$$
$$= 0.9724$$

● Now Work Problems 17(a)
and (b)

If we randomly selected twenty 24-year-old males 1000 different times, we would expect all twenty to survive one year in 972 of the samples. ●

③ Compute At-Least Probabilities

Usually probabilities involving the phrase *at least* use the Complement Rule.

The phrase *at least* means "greater than or equal to." For example, a person must be at least 17 years old to see an R-rated movie. This means that the person's age must be greater than or equal to 17 to watch the movie.

EXAMPLE 4   Computing At-Least Probabilities

Problem Compute the probability that at least one male out of 1000 aged 24 years will die during the course of the year if the probability that a randomly selected 24-year-old male survives the year is 0.9986.

Approach The phrase *at least* means "greater than or equal to," so we wish to know the probability that 1 or 2 or 3 or . . . or 1000 males will die during the year. These events are mutually exclusive, so

$$P(1 \text{ or } 2 \text{ or } 3 \text{ or } \ldots \text{ or } 1000 \text{ die}) = P(1 \text{ dies}) + P(2 \text{ die}) + P(3 \text{ die})$$
$$+ \cdots + P(1000 \text{ die})$$

Computing these probabilities would be very time consuming. However, notice that the complement of "at least one dying" is "none die", or all 1000 survive. Use the Complement Rule to compute the probability.

Solution

$P(\text{at least one dies}) = 1 - P(\text{none die})$
$$= 1 - P(\text{1st survives and 2nd survives and} \ldots \text{and 1000th survives})$$
$$= 1 - P(\text{1st survives}) \cdot P(\text{2nd survives}) \cdot \cdots \cdot P(\text{1000th survives})$$
$$= 1 - (0.9986)^{1000} \qquad \text{Independent events}$$
$$= 1 - 0.2464$$
$$= 0.7536$$

If we randomly selected 1000 males 24 years of age 100 different times, we would expect at least one to die in 75 of the samples. ●

● Now Work Problem 17(c)

Summary: Rules of Probability

1. The probability of any event must be between 0 and 1, inclusive. If we let E denote any event, then $0 \leq P(E) \leq 1$.
2. The sum of the probabilities of all outcomes in the sample space must equal 1. That is, if the sample space $S = \{e_1, e_2, \ldots, e_n\}$, then $P(e_1) + P(e_2) + \cdots + P(e_n) = 1$.
3. If E and F are disjoint events, then $P(E \text{ or } F) = P(E) + P(F)$. If E and F are not disjoint events, then $P(E \text{ or } F) = P(E) + P(F) - P(E \text{ and } F)$.
4. If E represents any event and E^c represents the complement of E, then $P(E^c) = 1 - P(E)$.
5. If E and F are independent events, then $P(E \text{ and } F) = P(E) \cdot P(F)$.

Notice that *or* probabilities use the Addition Rule, whereas *and* probabilities use the Multiplication Rule. Accordingly, *or* probabilities imply addition, while *and* probabilities imply multiplication.

 5.3 Assess Your Understanding

Vocabulary and Skill Building

1. Two events E and F are _____ if the occurrence of event E in a probability experiment does not affect the probability of event F.

2. The word *and* in probability implies that we use the _____ Rule.

3. The word *or* in probability implies that we use the _____ Rule.

4. *True or False:* When two events are disjoint, they are also independent.

5. If two events E and F are independent, $P(E \text{ and } F) = $ _____ .

6. Suppose events E and F are disjoint. What is $P(E \text{ and } F)$?

NW 7. Determine whether the events E and F are independent or dependent. Justify your answer.

(a) E: Speeding on the interstate.
    F: Being pulled over by a police officer.

(b) E: You gain weight.
    F: You eat fast food for dinner every night.

(c) E: You get a high score on a statistics exam.
    F: The Boston Red Sox win a baseball game.

8. Determine whether the events E and F are independent or dependent. Justify your answer.

(a) E: The battery in your cell phone is dead.
    F: The batteries in your calculator are dead.

(b) E: Your favorite color is blue.
    F: Your friend's favorite hobby is fishing.

(c) E: You are late for school.
    F: Your car runs out of gas.

9. Suppose that events E and F are independent, $P(E) = 0.3$ and $P(F) = 0.6$. What is the $P(E \text{ and } F)$?

10. Suppose that events E and F are independent, $P(E) = 0.7$ and $P(F) = 0.9$. What is the $P(E \text{ and } F)$?

Applying the Concepts

11. Flipping a Coin What is the probability of obtaining five heads in a row when flipping a fair coin? Interpret this probability.

12. Rolling a Die What is the probability of obtaining 4 ones in a row when rolling a fair, six-sided die? Interpret this probability.

13. Southpaws About 13% of the population is left-handed. If two people are randomly selected, what is the probability that both are left-handed? What is the probability that at least one is right-handed?

14. Double Jackpot Shawn lives near the border of Illinois and Missouri. One weekend he decides to play $1 in both state lotteries in hopes of hitting two jackpots. The probability of winning the Missouri Lotto is about 0.00000028357 and the probability of winning the Illinois Lotto is about 0.000000098239.

(a) Explain why the two lotteries are independent.
(b) Find the probability that Shawn will win both jackpots.

15. False Positives The ELISA is a test to determine whether the HIV antibody is present. The test is 99.5% effective, which means that the test will come back negative if the HIV antibody is not present 99.5% of the time. The probability of a test coming back positive when the antibody is not present (a false positive) is 0.005. Suppose that the ELISA is given to five randomly selected people who do not have the HIV antibody.

(a) What is the probability that the ELISA comes back negative for all five people?
(b) What is the probability that the ELISA comes back positive for at least one of the five people?

16. Christmas Lights Christmas lights are often designed with a series circuit. This means that when one light burns out the entire string of lights goes black. Suppose that the lights are designed so that the probability a bulb will last 2 years is 0.995. The success or failure of a bulb is independent of the success or failure of other bulbs.

(a) What is the probability that in a string of 100 lights all 100 will last 2 years?
(b) What is the probability that at least one bulb will burn out in 2 years?

NW 17. Life Expectancy The probability that a randomly selected 40-year-old male will live to be 41 years old is 0.99757, according to the *National Vital Statistics Report*, Vol. 56, No. 9.

(a) What is the probability that two randomly selected 40-year-old males will live to be 41 years old?

(b) What is the probability that five randomly selected 40-year-old males will live to be 41 years old?

(c) What is the probability that at least one of five randomly selected 40-year-old males will not live to be 41 years old? Would it be unusual if at least one of five randomly selected 40-year-old males did not live to be 41 years old?

18. Life Expectancy The probability that a randomly selected 40-year-old female will live to be 41 years old is 0.99855 according to the *National Vital Statistics Report*, Vol. 56, No. 9.

(a) What is the probability that two randomly selected 40-year-old females will live to be 41 years old?

(b) What is the probability that five randomly selected 40-year-old females will live to be 41 years old?

(c) What is the probability that at least one of five randomly selected 40-year-old females will not live to be 41 years old? Would it be unusual if at least one of five randomly selected 40-year-old females did not live to be 41 years old?

19. Derivatives In finance, a derivative is a financial asset whose value is determined (derived) from a bundle of various assets, such as mortgages. Suppose a randomly selected mortgage has a probability of 0.01 of default.

(a) What is the probability a randomly selected mortgage will not default (that is, pay off)?

(b) What is the probability a bundle of five randomly selected mortgages will not default assuming the likelihood any one mortgage being paid off is independent of the others? **Note:** A derivative might be an investment in which all five mortgages do not default.

(c) What is the probability the derivative becomes worthless? That is, at least one of the mortgages defaults?

(d) In part (b), we made the assumption that the likelihood of default is independent. Do you believe this is a reasonable assumption? Explain.

20. Quality Control Suppose that a company selects two people who work independently inspecting two-by-four timbers. Their job is to identify low-quality timbers. Suppose that the probability that an inspector does not identify a low-quality timber is 0.20.

(a) What is the probability that both inspectors do not identify a low-quality timber?

(b) How many inspectors should be hired to keep the probability of not identifying a low-quality timber below 1%?

(c) Interpret the probability from part (a).

21. Reliability and Redundancy In airline applications, failure of a component can result in catastrophe. As a result, many airline components utilize something called *triple modular redundancy*. This means that a critical component has two backup components that may be utilized should the initial component fail. Suppose a certain critical airline component has a probability of failure of 0.006 and the system that utilizes the component is part of a triple modular redundancy.

(a) Assuming each component's failure/success is independent of the others, what is the probability all three components fail, resulting in disaster for the flight?

(b) What is the probability at least one of the components does not fail?

22. Reliability For a parallel structure of identical components, the system can succeed if at least one of the components succeeds. Assume that components fail independently of each other and that each component has a 0.15 probability of failure.

(a) Would it be unusual to observe one component fail? Two components?

(b) What is the probability that a parallel structure with 2 identical components will succeed?

(c) How many components would be needed in the structure so that the probability the system will succeed is greater than 0.9999?

23. Reliability and Redundancy, Part II See Problem 21. Suppose a particular airline component has a probability of failure of 0.03 and is part of a triple modular redundancy system.

(a) What is the probability the system does not fail?

(b) Engineers decide the probability of failure is too high for this system. Use trial and error to determine the number of components that should be included in the system to result in a system that has greater than a 0.99999999 probability of not failing.

24. Cold Streaks Players in sports are said to have "hot streaks" and "cold streaks." For example, a batter in baseball might be considered to be in a slump, or cold streak, if he has made 10 outs in 10 consecutive at-bats. Suppose that a hitter successfully reaches base 30% of the time he comes to the plate.

(a) Find and interpret the probability that the hitter makes 10 outs in 10 consecutive at-bats, assuming that at-bats are independent events. *Hint:* The hitter makes an out 70% of the time.

(b) Are cold streaks unusual?

(c) Find the probability the hitter makes five consecutive outs and then reaches base safely.

(d) Discuss the assumption of independence in consecutive at-bats.

25. Bowling Suppose that Ralph gets a strike when bowling 30% of the time.

(a) What is the probability that Ralph gets two strikes in a row?

(b) What is the probability that Ralph gets a turkey (three strikes in a row)?

(c) When events are independent, their complements are independent as well. Use this result to determine the probability that Ralph gets a turkey, but fails to get a clover (four strikes in a row).

26. Driving under the Influence Among 21- to 25-year-olds, 29% say they have driven while under the influence of alcohol. Suppose that three 21- to 25-year-olds are selected at random.

Source: U.S. Department of Health and Human Services, reported in *USA Today*

(a) What is the probability that all three have driven while under the influence of alcohol?

(b) What is the probability that at least one has not driven while under the influence of alcohol?

(c) What is the probability that none of the three has driven while under the influence of alcohol?

(d) What is the probability that at least one has driven while under the influence of alcohol?

27. Defense System Suppose that a satellite defense system is established in which four satellites acting independently have a 0.9 probability of detecting an incoming ballistic missile. What is the probability that at least one of the four satellites detects an incoming ballistic missile? Would you feel safe with such a system?

28. Audits and Pet Ownership According to Internal Revenue Service records, 6.42% of all household tax returns are audited. According to the Humane Society, 39% of all households own a dog. Assuming dog ownership and audits are independent events, what is the probability a randomly selected household is audited and owns a dog?

29. Weight Gain and Gender According to the National Vital Statistics Report, 20.1% of all pregnancies result in weight gain in excess of 40 pounds (for singleton births). In addition, 49.5% of all pregnancies result in the birth of a baby girl. Assuming gender and weight gain are independent, what is the probability a randomly selected pregnancy results in a girl and weight gain in excess of 40 pounds?

30. Stocks Suppose your financial advisor recommends three stocks to you. He claims the likelihood that the first stock will increase in value at least 10% within the next year is 0.7, the likelihood the second stock will increase in value at least 10% within the next year is 0.55, and the likelihood the third stock will increase at least 10% within the next year is 0.20. Would it be unusual for all three stocks to increase at least 10%, assuming the stocks behave independently of each other?

31. Betting on Sports According to a Gallup Poll, about 17% of adult Americans bet on professional sports. Census data indicate that 48.4% of the adult population in the United States is male.

(a) Assuming that betting is independent of gender, compute the probability that an American adult selected at random is male and bets on professional sports.

(b) Using the result in part (a), compute the probability that an American adult selected at random is male or bets on professional sports.

(c) The Gallup poll data indicated that 10.6% of adults in the United States are males and bet on professional sports. What does this indicate about the assumption in part (a)?

(d) How will the information in part (c) affect the probability you computed in part (b)?

32. Fingerprints Fingerprints are now widely accepted as a form of identification. In fact, many computers today use fingerprint identification to link the owner to the computer. In 1892, Sir Francis Galton explored the use of fingerprints to uniquely identify an individual. A fingerprint consists of ridgelines. Based on empirical evidence, Galton estimated the probability that a square consisting of six ridgelines that covered a fingerprint could be filled in accurately by an experienced fingerprint analyst as $\frac{1}{2}$.

(a) Assuming that a full fingerprint consists of 24 of these squares, what is the probability that all 24 squares could be filled in correctly, assuming that success or failure in filling in one square is independent of success or failure in filling in any other square within the region? (This value represents the probability that two individuals would share the same ridgeline features within the 24-square region.)

(b) Galton further estimated that the likelihood of determining the fingerprint type (e.g., arch, left loop, whorl, etc.) as $\left(\frac{1}{2}\right)^4$ and the likelihood of the occurrence of the correct number of ridges entering and exiting each of the 24 regions as $\left(\frac{1}{2}\right)^8$. Assuming that all three probabilities are independent, compute Galton's estimate of the probability that a particular fingerprint configuration would occur in nature (that is, the probability that a fingerprint match occurs by chance).

33. You Explain It! Independence Suppose a mother already has three girls from three separate pregnancies. Does the fact that the mother already has three girls affect the likelihood of having a fourth girl? Explain.

34. You Explain It! Independence Ken and Dorothy like to fly to Colorado for ski vacations. Sometimes, however, they are late for their flight. On the air carrier they prefer to fly, the probability that luggage gets lost is 0.012 for luggage checked at least one hour prior to departure. However, the probability luggage gets lost is 0.043 for luggage checked within one hour of departure. Are the events "luggage check time" and "lost luggage" independent? Explain.

5.4 Conditional Probability and the General Multiplication Rule

Objectives   ① Compute conditional probabilities
② Compute probabilities using the General Multiplication Rule

① Compute Conditional Probabilities

In the last section, we learned that when two events are independent the occurrence of one event has no effect on the probability of the second event. However, we cannot always assume that two events will be independent. Will the probability of being in a car accident change depending on driving conditions? We would expect that the probability of an accident will be higher for driving on icy roads than for driving on dry roads.

According to data from the Centers for Disease Control, 33.3% of adult men in the United States are obese. So the probability is 0.333 that a randomly selected U.S. adult male is obese. However, 28% of adult men aged 20–39 are obese compared to 40% of adult men aged 40–59. The probability is 0.28 that an adult male is obese, *given* that he is aged 20–39. The probability is 0.40 that an adult male is obese, *given* that he is aged 40–59. The probability that an adult male is obese changes depending on his age group. Therefore, obesity and age are not independent. This is called *conditional probability*.

Definition

Conditional Probability

The notation $P(F|E)$ is read "the probability of event F given event E." It is the probability that the event F occurs, given that the event E has occurred.

For example, $P(\text{obese} \mid 20 \text{ to } 39) = 0.28$ and $P(\text{obese} \mid 40 \text{ to } 59) = 0.40$.

EXAMPLE 1 An Introduction to Conditional Probability

Problem Suppose a single die is rolled. What is the probability that the die comes up three? Now suppose that the die is rolled a second time, but we are told the outcome will be an odd number. What is the probability that the die comes up three?

Approach Assume that the die is fair. This means that the outcomes are equally likely, so we use the classical method of computing probabilities.

Solution In the first instance, there are six possibilities in the sample space, $S = \{1, 2, 3, 4, 5, 6\}$, so $P(3) = \dfrac{1}{6}$. In the second instance, there are three possibilities in the sample space, because the only possible outcomes are odd, so $S = \{1, 3, 5\}$. This probability is expressed symbolically as $P(3|\text{outcome is odd}) = \dfrac{1}{3}$, which is read "the probability of rolling a 3, given that the outcome is odd, is one-third." Notice that the conditional probability reduces the size of the sample space under consideration (from six outcomes to three outcomes). ●

The data in Table 8 represent the marital status of males and females 15 years old or older in the United States in 2013.

Table 8	Males (in millions)	Females (in millions)	Totals (in millions)
Never married	41.6	36.9	**78.5**
Married	64.4	63.1	**127.5**
Widowed	3.1	11.2	**14.3**
Divorced	11.0	14.4	**25.4**
Separated	2.4	3.2	**5.6**
Totals (in millions)	**122.5**	**128.8**	**251.3**

Source: U.S. Census Bureau, Current Population Reports

To find the probability that a randomly selected individual 15 years old or older is widowed, divide the number of widowed individuals by the total number of individuals who are 15 years old or older.

$$P(\text{widowed}) = \frac{14.3}{251.3}$$
$$= 0.057$$

Suppose that we know the individual is female. Does this change the probability that she is widowed? The sample space now consists only of females, so the probability that the individual is widowed, given that the individual is female, is

$$P(\text{widowed}\,|\,\text{female}) = \frac{N(\text{widowed females})}{N(\text{females})}$$
$$= \frac{11.2}{128.8} = 0.087$$

So, knowing that the individual is female increases the likelihood that the individual is widowed. This leads to the following rule.

Conditional Probability Rule

If E and F are any two events, then

$$P(F\,|\,E) = \frac{P(E \text{ and } F)}{P(E)} = \frac{N(E \text{ and } F)}{N(E)} \qquad \textbf{(1)}$$

The probability of event F occurring, given the occurrence of event E, is found by dividing the probability of E and F by the probability of E, or by dividing the number of outcomes in E and F by the number of outcomes in E.

EXAMPLE 2   Conditional Probabilities on Marital Status and Gender

Problem The data in Table 8 on the previous page represent the marital status and gender of the residents of the United States aged 15 years old or older in 2013.
(a) Compute the probability that a randomly selected individual has never married given the individual is male.
(b) Compute the probability that a randomly selected individual is male given the individual has never married.

Approach
(a) Since the randomly selected person is male, concentrate on the male column. There are 122.5 million males and 41.6 million males who never married, so $N(\text{male}) = 122.5$ million and $N(\text{male and never married}) = 41.6$ million. Compute the probability using the Conditional Probability Rule.
(b) Since the randomly selected person has never married, concentrate on the never married row. There are 78.5 million people who have never married and 41.6 million males who have never married, so $N(\text{never married}) = 78.5$ million and $N(\text{male and never married}) = 41.6$ million. Compute the probability using the Conditional Probability Rule.

Solution
(a) Substituting into Formula (1), we obtain

$$P(\text{never married}\,|\,\text{male}) = \frac{N(\text{male and never married})}{N(\text{male})} = \frac{41.6}{122.5} = 0.340$$

(continued)

The probability that the randomly selected individual has never married, given that the individual is male, is 0.340.

(b) Substituting into Formula (1), we obtain

$$P(\text{male} \mid \text{never married}) = \frac{N(\text{male and never married})}{N(\text{never married})} = \frac{41.6}{78.5} = 0.530$$

The probability that the randomly selected individual is male, given that the individual has never married, is 0.530.

● Now Work Problem 17

What is the difference between the results of Examples 2(a) and (b)? In Example 2(a), we found that 34.0% of males have never married, whereas in Example 2(b) we found that 53.0% of individuals who have never married are male.

EXAMPLE 3   Birth Weights of Preterm Babies
_____

Problem Suppose that 12.7% of all births are preterm. (The gestation period of the pregnancy is less than 37 weeks.) Also 0.22% of all births resulted in a preterm baby who weighed 8 pounds, 13 ounces or more. What is the probability that a randomly selected baby weighs 8 pounds, 13 ounces or more, given that the baby is preterm? Is this unusual? *Source:* Vital Statistics Reports

Approach We want to know the probability that the baby weighs 8 pounds, 13 ounces or more, given that the baby was preterm. Because 0.22% of all babies weigh 8 pounds, 13 ounces or more and are preterm, $P(\text{weighs 8 lb, 13 oz or more and preterm}) = 0.0022$. Since 12.7% of all births are preterm, $P(\text{preterm}) = 0.127$. The phrase "given that" suggests we use the Conditional Probability Rule to compute the probability.

Solution $P(\text{weighs 8 lb, 13 oz or more} \mid \text{preterm})$

$$= \frac{P(\text{weighs 8 lb, 13 oz or more and preterm})}{P(\text{preterm})}$$

$$= \frac{0.0022}{0.127} \approx 0.0173$$

● Now Work Problem 13

If 100 preterm babies were randomly selected, we would expect about two to weigh 8 pounds, 13 ounces or more. This is an unusual outcome.

---

❷ **Compute Probabilities Using the General Multiplication Rule**

If we solve the Conditional Probability Rule for $P(E \text{ and } F)$, we obtain the General Multiplication Rule.

> **General Multiplication Rule**
>
> The probability that two events E and F both occur is
>
> $$P(E \text{ and } F) = P(E) \cdot P(F \mid E)$$
>
> In words, the probability of E and F is the probability of event E occurring times the probability of event F occurring, given the occurrence of event E.

EXAMPLE 4   Using the General Multiplication Rule

Problem The probability that a driver who is speeding gets pulled over is 0.8. The probability that a driver gets a ticket, given that he or she is pulled over, is 0.9. What is the probability that a randomly selected driver who is speeding gets pulled over and gets a ticket?

Approach Let E represent the event "driver who is speeding gets pulled over," and let F represent the event "driver gets a ticket." Use the General Multiplication Rule to compute $P(E$ and $F)$.

Solution $P(\text{driver who is speeding gets pulled over and gets a ticket}) = P(E$ and $F) = P(E) \cdot P(F|E) = 0.8(0.9) = 0.72$. The probability that a driver who is speeding gets pulled over and gets a ticket is 0.72.

● **Now Work Problem 31**

EXAMPLE 5   Acceptance Sampling

Problem Suppose that of 100 circuits sent to a manufacturing plant, 5 are defective. The plant manager receiving the circuits randomly selects 2 and tests them. If both circuits work, she will accept the shipment. Otherwise, the shipment is rejected. What is the probability that the plant manager discovers at least 1 defective circuit and rejects the shipment?

Approach To determine the probability that at least one of the tested circuits is defective, consider four possibilities. Neither of the circuits is defective, the first is defective while the second is not, the first is not defective while the second is defective, or both circuits are defective. Note that the outcomes are not equally likely. To find the probability the manager discovers at least 1 defective circuit, we may use one of two approaches.

Approach I: Use a tree diagram to list all possible outcomes and the General Multiplication Rule to compute the probability for each outcome. Then determine the probability of at least 1 defective by adding the probability that only the first is defective, only the second is defective, or both are defective, using the Addition Rule (because they are disjoint).

Approach II: Compute the probability that both circuits are not defective and use the Complement Rule to determine the probability of at least 1 defective.
    We will illustrate both approaches.

Solution Of the 100 circuits, 5 are defective, so 95 are not defective.

Approach I: Construct a tree diagram to determine the possible outcomes for the experiment. See Figure 12 on the following page, where D stands for defective and G stands for good (not defective). Because the outcomes are not equally likely, we include the probabilities in our diagram to show how the probability of each outcome is obtained. (Multiply the individual probabilities along the corresponding path in the diagram.)

(continued)

Figure 12

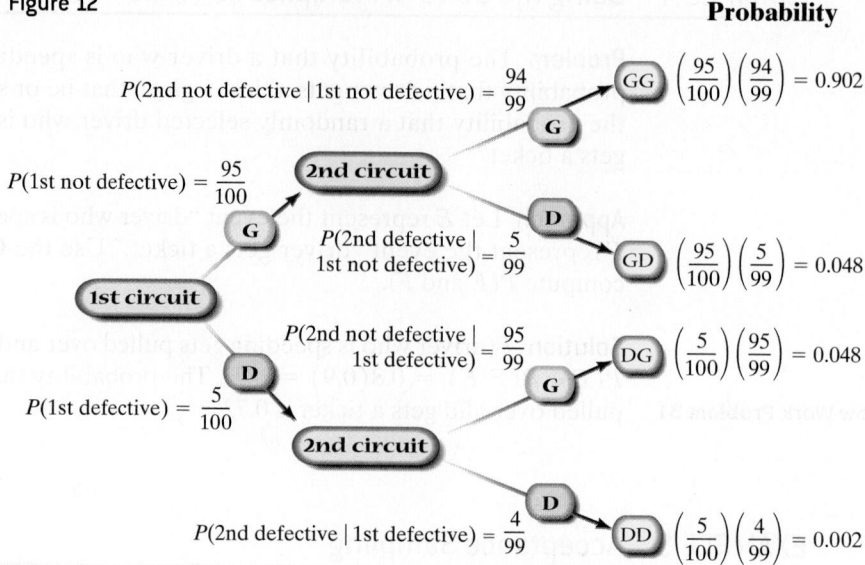

Probability

$P(\text{2nd not defective} \mid \text{1st not defective}) = \dfrac{94}{99}$ GG $\left(\dfrac{95}{100}\right)\left(\dfrac{94}{99}\right) = 0.902$

$P(\text{1st not defective}) = \dfrac{95}{100}$

$P(\text{2nd defective} \mid \text{1st not defective}) = \dfrac{5}{99}$ GD $\left(\dfrac{95}{100}\right)\left(\dfrac{5}{99}\right) = 0.048$

$P(\text{2nd not defective} \mid \text{1st defective}) = \dfrac{95}{99}$ DG $\left(\dfrac{5}{100}\right)\left(\dfrac{95}{99}\right) = 0.048$

$P(\text{1st defective}) = \dfrac{5}{100}$

$P(\text{2nd defective} \mid \text{1st defective}) = \dfrac{4}{99}$ DD $\left(\dfrac{5}{100}\right)\left(\dfrac{4}{99}\right) = 0.002$

From the tree diagram and the Addition Rule, we can write

$$P(\text{at least 1 defective}) = P(\text{GD}) + P(\text{DG}) + P(\text{DD})$$
$$= 0.048 + 0.048 + 0.002$$
$$= 0.098$$

The probability that the shipment will not be accepted is 0.098.

Approach II: Compute the probability that both circuits are not defective and use the Complement Rule to determine the probability of at least 1 defective.

$$P(\text{at least 1 defective}) = 1 - P(\text{none defective})$$
$$= 1 - P(\text{1st not defective}) \cdot P(\text{2nd not defective} \mid \text{1st not defective})$$
$$= 1 - \left(\dfrac{95}{100}\right) \cdot \left(\dfrac{94}{99}\right)$$
$$= 1 - 0.902$$
$$= 0.098$$

The probability that the shipment will not be accepted is 0.098.   ●

● **Now Work Problem 21**

Whenever a small random sample is taken from a large population, it is reasonable to compute probabilities of events assuming independence. Consider the following example.

EXAMPLE 6   Favorite Other

Problem In a study to determine whether preferences for self are more or less prevalent than preferences for others, researchers first asked individuals to identify the person who is most valuable and likeable to you, or your favorite other. Of the 1519 individuals surveyed, 42 had chosen themselves as their favorite other. *Source:* Gebauer JE, et al. Self-Love or Other-Love? Explicit Other-Preference but Implicit Self-Preference. PLoS ONE 7(7): e41789. doi:10.1371/journal.prone.0041789

(a) Suppose we randomly select 1 of the 1519 individuals surveyed. What is the probability that he or she chosen himself or herself as their favorite other?

Historical Note

Andrei Nikolaevich Kolmogorov was born on April 25, 1903, in Tambov, Russia. His parents were not married, and he was raised by his aunt. He graduated from Moscow State University in 1925. That year he published eight papers, including his first on probability. By the time he received his doctorate, in 1929, he already had 18 publications. He became a professor at Moscow State University in 1931. Kolmogorov is quoted as saying, "The theory of probability as a mathematical discipline can and should be developed from axioms in exactly the same way as Geometry and Algebra." Kolmogorov helped to educate gifted children. It did not bother him if the students did not become mathematicians; he simply wanted them to be happy. Andrei Kolmogorov died on October 20, 1987.

(b) If two individuals from this group are randomly selected, what is the probability that both chose themselves as their favorite other?

(c) Compute the probability of randomly selecting two individuals from this group who selected themselves as their favorite other assuming independence.

Approach Let event E = "themselves favorite other," so $P(E)$ = number of individuals who select themselves as favorite other divided by 1519, the number of individuals in the survey. To answer part (b), let E_1 = "first person selects themselves as favorite other" and E_2 = "second person selects themselves as favorite other." Then compute $P(E_1 \text{ and } E_2) = P(E_1) \cdot P(E_2 | E_1)$. To answer part (c), use the Multiplication Rule for Independent Events.

Solution

(a) If one individual is selected, $P(E) = \dfrac{42}{1519} = 0.02765$.

(b) Using the Multiplication Rule,

$$P(E_1 \text{ and } E_2) = P(E_1) \cdot P(E_2 | E_1) = \frac{42}{1519} \cdot \frac{41}{1518} \approx 0.0007468$$

Notice that $P(E_2 | E_1) = \dfrac{41}{1518}$ because we are sampling without replacement, so after event E_1 occurs there is one less person who considers themselves their favorite other and one less person in the sample space.

(c) The assumption of independence means that the outcome of the first trial of the experiment does not affect the probability of the second trial. (It is like sampling with replacement.) Therefore, we assume that

$$P(E_1) = P(E_2) = \frac{42}{1519}$$

Then

$$P(E_1 \text{ and } E_2) = P(E_1) \cdot P(E_2) = \frac{42}{1519} \cdot \frac{42}{1519} \approx 0.0007645 \qquad \bullet$$

The probabilities in Examples 6(b) and 6(c) are extremely close in value. Based on these results, we infer the following principle.

> If small random samples are taken from large populations without replacement, it is reasonable to assume independence of the events. As a rule of thumb, if the sample size is less than 5% of the population size, we treat the events as independent.

For example, in Example 6, we can compute the probability of randomly selecting two individuals who consider themselves their favorite other assuming independence because the sample size, 2, is only $\dfrac{2}{1519}$, or 0.13% of the population size, 1519.

We can now express independence using conditional probabilities.

● **Now Work Problem 39**

Definition Two events E and F are independent if $P(E | F) = P(E)$ or, equivalently, if $P(F | E) = P(F)$.

If either condition in our definition is true, the other is as well. In addition, for independent events,

$$P(E \text{ and } F) = P(E) \cdot P(F)$$

So the Multiplication Rule for Independent Events is a special case of the General Multiplication Rule.

Look back at Table 8 on page 280. Because $P(\text{widowed}) = 0.057$ does not equal $P(\text{widowed} \mid \text{female}) = 0.087$, the events "widowed" and "female" are not independent. In fact, knowing an individual is female increases the likelihood that the individual is also widowed.

● **Now Work Problem 41**

5.4 Assess Your Understanding

Vocabulary and Skill Building

1. The notation $P(F \mid E)$ means the probability of event _____ given event _____.

2. If $P(E) = 0.6$ and $P(E \mid F) = 0.34$, are events E and F independent?

3. Suppose that E and F are two events and that $P(E \text{ and } F) = 0.6$ and $P(E) = 0.8$. What is $P(F \mid E)$?

4. Suppose that E and F are two events and that $P(E \text{ and } F) = 0.21$ and $P(E) = 0.4$. What is $P(F \mid E)$?

5. Suppose that E and F are two events and that $N(E \text{ and } F) = 420$ and $N(E) = 740$. What is $P(F \mid E)$?

6. Suppose that E and F are two events and that $N(E \text{ and } F) = 380$ and $N(E) = 925$. What is $P(F \mid E)$?

7. Suppose that E and F are two events and that $P(E) = 0.8$ and $P(F \mid E) = 0.4$. What is $P(E \text{ and } F)$?

8. Suppose that E and F are two events and that $P(E) = 0.4$ and $P(F \mid E) = 0.6$. What is $P(E \text{ and } F)$?

9. According to the U.S. Census Bureau, the probability that a randomly selected head of household in the United States earns more than $100,000 per year is 0.202. The probability that a randomly selected head of household in the United States earns more than $100,000 per year, given that the head of household has earned a bachelor's degree, is 0.412. Are the events "earn more than $100,000 per year" and "earned a bachelor's degree" independent?

10. The probability that a randomly selected individual in the United States 25 years and older has at least a bachelor's degree is 0.094. The probability that an individual in the United States 25 years and older has at least a bachelor's degree, given that the individual lives in Washington DC, is, 0.241. Are the events "bachelor's degree" and "lives in Washington, DC," independent? *Source: American Community Survey, 2013*

Applying the Concepts

11. Drawing a Card Suppose that a single card is selected from a standard 52-card deck. What is the probability that the card drawn is a club? Now suppose that a single card is drawn from a standard 52-card deck, but we are told that the card is black. What is the probability that the card drawn is a club?

12. Drawing a Card Suppose that a single card is selected from a standard 52-card deck. What is the probability that the card drawn is a king? Now suppose that a single card is drawn from a standard 52-card deck, but we are told that the card is a heart. What is the probability that the card drawn is a king? Did the knowledge that the card is a heart change the probability that the card was a king? What term is used to describe this result?

NW 13. Rainy Days For the month of June in the city of Chicago, 37% of the days are cloudy. Also in the month of June in the city of Chicago, 21% of the days are cloudy and rainy. What is the probability that a randomly selected day in June will be rainy if it is cloudy?

14. Cause of Death According to the U.S. National Center for Health Statistics, 0.15% of deaths in the United States are 25- to 34-year-olds whose cause of death is cancer. In addition, 1.71% of all those who die are 25–34 years old. What is the probability that a randomly selected death is the result of cancer if the individual is known to have been 25–34 years old?

15. High School Dropouts According to the U.S. Census Bureau, 8.0% of 16- to 24-year-olds are high school dropouts. In addition, 2.1% of 16- to 24-year-olds are high school dropouts and unemployed. What is the probability that a randomly selected 16- to 24-year-old is unemployed, given he or she is a dropout?

16. Income by Region According to the U.S. Census Bureau, 17.9% of U.S. households are in the Northeast. In addition, 5.4% of U.S. households earn $100,000 per year or more and are located in the Northeast. Determine the probability that a randomly selected U.S. household earns more than $100,000 per year, given that the household is located in the Northeast.

NW 17. Made in America In a recent Harris Poll, a random sample of adult Americans (18 years and older) was asked, "When you see an ad emphasizing that a product is 'Made in America,' are you more likely to buy it, less likely to buy it, or neither more nor less likely to buy it?" The results of the survey, by age group, are presented in the following contingency table.

	18–34	35–44	45–54	55+	Total
More likely	238	329	360	402	1329
Less likely	22	6	22	16	66
Neither more nor less likely	282	201	164	118	765
Total	542	536	546	536	2160

Source: The Harris Poll

(a) What is the probability that a randomly selected individual is 35–44 years of age, given the individual is more likely to buy a product emphasized as "Made in America"?

(b) What is the probability that a randomly selected individual is more likely to buy a product emphasized as "Made in America," given the individual is 35–44 years of age?

(c) Are 18- to 34-year-olds more likely to buy a product emphasized as "Made in America" than individuals in general?

18. Social Media Adult Americans (18 years or older) were asked whether they used social media (Facebook, Twitter, and so on) regularly. The following table is based on the results of the survey.

	18–34	35–44	45–54	55+	Total
Use social media	117	89	83	49	**338**
Do not use social media	33	36	57	66	**192**
Total	**150**	**125**	**140**	**115**	**530**

Source: Harris Interactive

(a) What is the probability that a randomly selected adult American uses social media, given the individual is 18–34 years of age?

(b) What is the probability that a randomly selected adult American is 18–34 years of age, given the individual uses social media?

(c) Are 18- to 34-year olds more likely to use social media than individuals in general? Why?

19. Driver Fatalities The following data represent the number of drivers involved in fatal crashes in the United States in 2013 by day of the week and gender.

	Male	**Female**	**Total**
Sunday	4143	2287	**6430**
Monday	3178	1705	**4883**
Tuesday	3280	1739	**5019**
Wednesday	3197	1729	**4926**
Thursday	3389	1839	**5228**
Friday	3975	2179	**6154**
Saturday	4749	2511	**7260**
Total	**25,911**	**13,989**	**39,900**

Source: Fatality Analysis Reporting System, National Highway Traffic Safety Administration

(a) Among Sunday fatal crashes, what is the probability that a randomly selected fatality is female?

(b) Among female fatalities, what is the probability that a randomly selected fatality occurs on Sunday?

(c) Are there any days in which a fatality is more likely to be male? That is, is $P(\text{male} \mid \text{Sunday})$ much different from $P(\text{male} \mid \text{Monday})$ and so on?

20. Speeding Tickets Use the results of Problem 44 in Section 5.2 to answer the following.

(a) Among those who text while driving, what is the probability that a randomly selected individual was issued no tickets last year?

(b) Among those who were issued no tickets last year, what is the probability the individual texts while driving?

(c) Based on the results of this survey, does it appear to be the case that individuals who text while driving are less likely to be issued 0 speeding tickets than those who do not text while driving?

NW 21. Acceptance Sampling Suppose that you just received a shipment of six televisions and two are defective. If two televisions are randomly selected, compute the probability that both televisions work. What is the probability that at least one does not work?

22. Committee A committee consisting of four women and three men will randomly select two people to attend a conference in Hawaii. Find the probability that both are women.

23. Suppose that two cards are randomly selected from a standard 52-card deck.

(a) What is the probability that the first card is a king and the second card is a king if the sampling is done without replacement?

(b) What is the probability that the first card is a king and the second card is a king if the sampling is done with replacement?

24. Suppose that two cards are randomly selected from a standard 52-card deck.

(a) What is the probability that the first card is a club and the second card is a club if the sampling is done without replacement?

(b) What is the probability that the first card is a club and the second card is a club if the sampling is done with replacement?

25. Board Work This past semester, I had a small business calculus section. The students in the class were Mike, Neta, Jinita, Kristin, and Dave. Suppose that I randomly select two people to go to the board to work problems. What is the probability that Dave is the first person chosen to go to the board and Neta is the second?

26. Party My wife has organized a monthly neighborhood party. Five people are involved in the group: Yolanda (my wife), Lorrie, Laura, Kim, and Anne Marie. They decide to randomly select the first and second home that will host the party. What is the probability that my wife hosts the first party and Lorrie hosts the second? Note: Once a home has hosted, it cannot host again until all other homes have hosted.

27. Playing Music on Random Setting Suppose that a digital music player has 13 tracks. After listening to all the songs, you decide that you like 5 of them. With the random feature on your player, each of the 13 songs is played once in random order. Find the probability that among the first two songs played

(a) You like both of them. Would this be unusual?

(b) You like neither of them.

(c) You like exactly one of them.

(d) Redo (a)–(c) if a song can be replayed before all 13 songs are played (if, for example, track 2 can play twice in a row).

28. Packaging Error Because of a manufacturing error, three cans of regular soda were accidentally filled with diet soda and placed into a 12-pack. Suppose that two cans are randomly selected from the 12-pack.

(a) Determine the probability that both contain diet soda.
(b) Determine the probability that both contain regular soda. Would this be unusual?
(c) Determine the probability that exactly one is diet and one is regular?

29. Planting Tulips A bag of 30 tulip bulbs purchased from a nursery contains 12 red tulip bulbs, 10 yellow tulip bulbs, and 8 purple tulip bulbs. Use a tree diagram like the one in Example 5 to answer the following:

(a) What is the probability that two randomly selected tulip bulbs are both red?
(b) What is the probability that the first bulb selected is red and the second yellow?
(c) What is the probability that the first bulb selected is yellow and the second is red?
(d) What is the probability that one bulb is red and the other yellow?

30. Golf Balls The local golf store sells an "onion bag" that contains 35 "experienced" golf balls. Suppose that the bag contains 20 Titleists, 8 Maxflis, and 7 Top-Flites. Use a tree diagram like the one in Example 5 to answer the following:

(a) What is the probability that two randomly selected golf balls are both Titleists?
(b) What is the probability that the first ball selected is a Titleist and the second is a Maxfli?
(c) What is the probability that the first ball selected is a Maxfli and the second is a Titleist?
(d) What is the probability that one golf ball is a Titleist and the other is a Maxfli?

NW **31. Smokers** According to the National Center for Health Statistics, there is a 20.3% probability that a randomly selected resident of the United States aged 18 years or older is a smoker. In addition, there is a 44.5% probability that a randomly selected resident of the United States aged 18 years or older is female, given that he or she smokes. What is the probability that a randomly selected resident of the United States aged 18 years or older is female and smokes? Would it be unusual to randomly select a resident of the United States aged 18 years or older who is female and smokes?

32. Multiple Jobs According to the U.S. Bureau of Labor Statistics, there is a 4.9% probability that a randomly selected employed individual has more than one job (a multiple-job holder). Also, there is a 46.6% probability that a randomly selected employed individual is male, given that he has more than one job. What is the probability that a randomly selected employed individual is a multiple-job holder and male? Would it be unusual to randomly select an employed individual who is a multiple-job holder and male?

33. The Birthday Problem Determine the probability that at least 2 people in a room of 10 people share the same birthday, ignoring leap years and assuming each birthday is equally likely, by answering the following questions:

(a) Compute the probability that 10 people have 10 different birthdays. **Hint:** The first person's birthday can occur 365 ways, the second person's birthday can occur 364 ways, because he or she cannot have the same birthday as the first person, the third person's birthday can occur 363 ways, because he or she cannot have the same birthday as the first or second person, and so on.
(b) The complement of "10 people have different birthdays" is "at least 2 share a birthday." Use this information to compute the probability that at least 2 people out of 10 share the same birthday.

34. The Birthday Problem Using the procedure given in Problem 33, compute the probability that at least 2 people in a room of 23 people share the same birthday.

35. Teen Communication The following data represent the number of different communication activities (e.g., cell phone, text messaging, e-mail, Internet, and so on) used by a random sample of teenagers over the past 24 hours.

Activities	0	1–2	3–4	5+	Total
Male	21	81	60	38	200
Female	21	52	56	71	200
Total	42	133	116	109	400

(a) Are the events "male" and "0 activities" independent? Justify your answer.
(b) Are the events "female" and "5+ activities" independent? Justify your answer.
(c) Are the events "1–2 activities" and "3–4 activities" mutually exclusive? Justify your answer.
(d) Are the events "male" and "1–2 activities" mutually exclusive? Justify your answer.

36. Party Affiliation The following data represent political party by age from a random sample of registered Iowa Voters.

	17–29	30–44	45–64	65+	Total
Republican	224	340	1075	561	2200
Democrat	184	384	773	459	1800
Total	408	724	1848	1020	4000

(a) Are the events "Republican" and "30–44" independent? Justify your answer.
(b) Are the events "Democrat" and "65+" independent? Justify your answer.
(c) Are the events "17–29 and "45–64" mutually exclusive? Justify your answer.
(d) Are the events "Republican" and "45–64" mutually exclusive? Justify your answer.

37. A Flush A flush in the card game of poker occurs if a player gets five cards that are all the same suit (clubs, diamonds, hearts, or spades). Answer the following questions to obtain the probability of being dealt a flush in five cards.

(a) We initially concentrate on one suit, say clubs. There are 13 clubs in a deck. Compute $P(\text{five clubs}) = P(\text{first card is clubs and second card is clubs and third card is clubs and fourth card is clubs and fifth card is clubs})$.
(b) A flush can occur if we get five clubs or five diamonds or five hearts or five spades. Compute $P(\text{five clubs or five diamonds or five hearts or five spades})$. Note that the events are mutually exclusive.

38. A Royal Flush A royal flush in the game of poker occurs if the player gets the cards Ten, Jack, Queen, King, and Ace all in the same suit. Use the procedure given in Problem 37 to compute the probability of being dealt a royal flush.

NW **39. Independence in Small Samples from Large Populations** Suppose that a computer chip company has just shipped 10,000 computer chips to a computer company. Unfortunately, 50 of the chips are defective.

(a) Compute the probability that two randomly selected chips are defective using conditional probability.
(b) There are 50 defective chips out of 10,000 shipped. The probability that the first chip randomly selected is defective

is $\dfrac{50}{10,000} = 0.005$. Compute the probability that two randomly selected chips are defective under the assumption of independent events. Compare your results to part (a). Conclude that, when small samples are taken from large populations without replacement, the assumption of independence does not significantly affect the probability.

40. Independence in Small Samples from Large Populations
Suppose that a poll is being conducted in the village of Lemont. The pollster identifies her target population as all residents of Lemont 18 years old or older. This population has 6494 people.

(a) Compute the probability that the first resident selected to participate in the poll is Roger Cummings and the second is Rick Whittingham.

(b) The probability that any particular resident of Lemont is the first person picked is $\dfrac{1}{6494}$. Compute the probability that Roger is selected first and Rick is selected second, assuming independence. Compare your results to part (a). Conclude that, when small samples are taken from large populations without replacement, the assumption of independence does not significantly affect the probability.

NW 41. Independent? Refer to the contingency table in Problem 17 that relates age and likelihood to buy American. Determine $P(45-54 \text{ years old})$ and $P(45-54 \text{ years old} \mid \text{more likely})$. Are the events "45−54 years old" and "more likely" independent?

42. Independent? Refer to the contingency table in Problem 18 that relates social media use and age. Determine $P(\text{uses social media})$ and $P(\text{uses social media} \mid 35\text{–}44 \text{ years})$. Are the events "uses social media" and "35–44 years" independent?

43. Let's Make a Deal In 1991, columnist Marilyn Vos Savant posted her reply to a reader's question. The question posed was in reference to one of the games played on the gameshow *Let's Make a Deal* hosted by Monty Hall.

Suppose you're on a game show, and you're given the choice of three doors: Behind one door is a car; behind the others, goats. You pick a door, say No. 1, and the host, who knows what's behind the doors, opens another door, say No. 3, which has a goat.

He then says to you, "Do you want to pick door No. 2?" Is it to your advantage to take the switch?

Source: Published by Monty Hall.

Her reply generated a tremendous amount of backlash, with many highly educated individuals angrily responding that she was clearly mistaken in her reasoning.

(a) Using subjective probability, estimate the probability of winning if you switch.

(b) Load the *Let's Make a Deal* applet at www.pearsonhighered.com/sullivanstats. Simulate the probability that you will win if you switch by going through the simulation at least 100 times. How does your simulated result compare to your answer to part (a)?

(c) Research the Monty Hall Problem as well as the reply by Marilyn Vos Savant. How does the probability she gives compare to the two estimates you obtained?

(d) Write a report detailing why Marilyn was correct. One approach is to use a random variable on a wheel similar to the one shown. On the wheel, the innermost ring indicates the door where the car is located, the middle ring indicates the door you selected, and the outer ring indicates the door(s) that Monty could show you. In the outer ring, green indicates you lose if you switch while purple indicates you win if you switch.

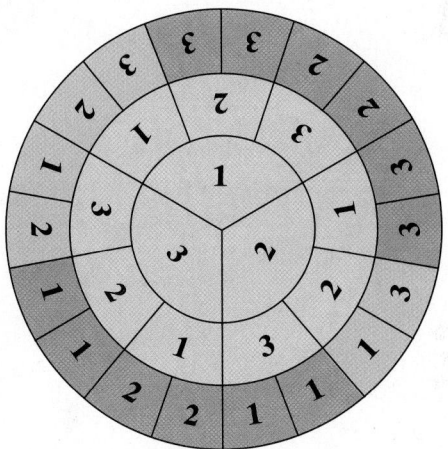

5.5   Counting Techniques

Objectives

1 Solve counting problems using the Multiplication Rule

2 Solve counting problems using permutations

3 Solve counting problems using combinations

4 Solve counting problems involving permutations with nondistinct items

5 Compute probabilities involving permutations and combinations

1   Solve Counting Problems Using the Multiplication Rule

Counting plays a major role in many diverse areas, including probability. In this section, we look at special types of counting problems and develop general techniques for solving them. We begin with an example that demonstrates a general counting principle.

EXAMPLE 1   Counting the Number of Possible Meals

Problem The fixed-price dinner at Mabenka Restaurant provides the following choices:

Appetizer: soup or salad
Entrée: baked chicken, broiled beef patty, baby beef liver, or roast beef au jus
Dessert: ice cream or cheesecake

How many different meals can be ordered?

Approach Ordering such a meal requires three separate decisions:

Choose an Appetizer	Choose an Entrée	Choose a Dessert
2 choices	4 choices	2 choices

Figure 13 is a tree diagram that lists the possible meals that can be ordered.

Figure 13

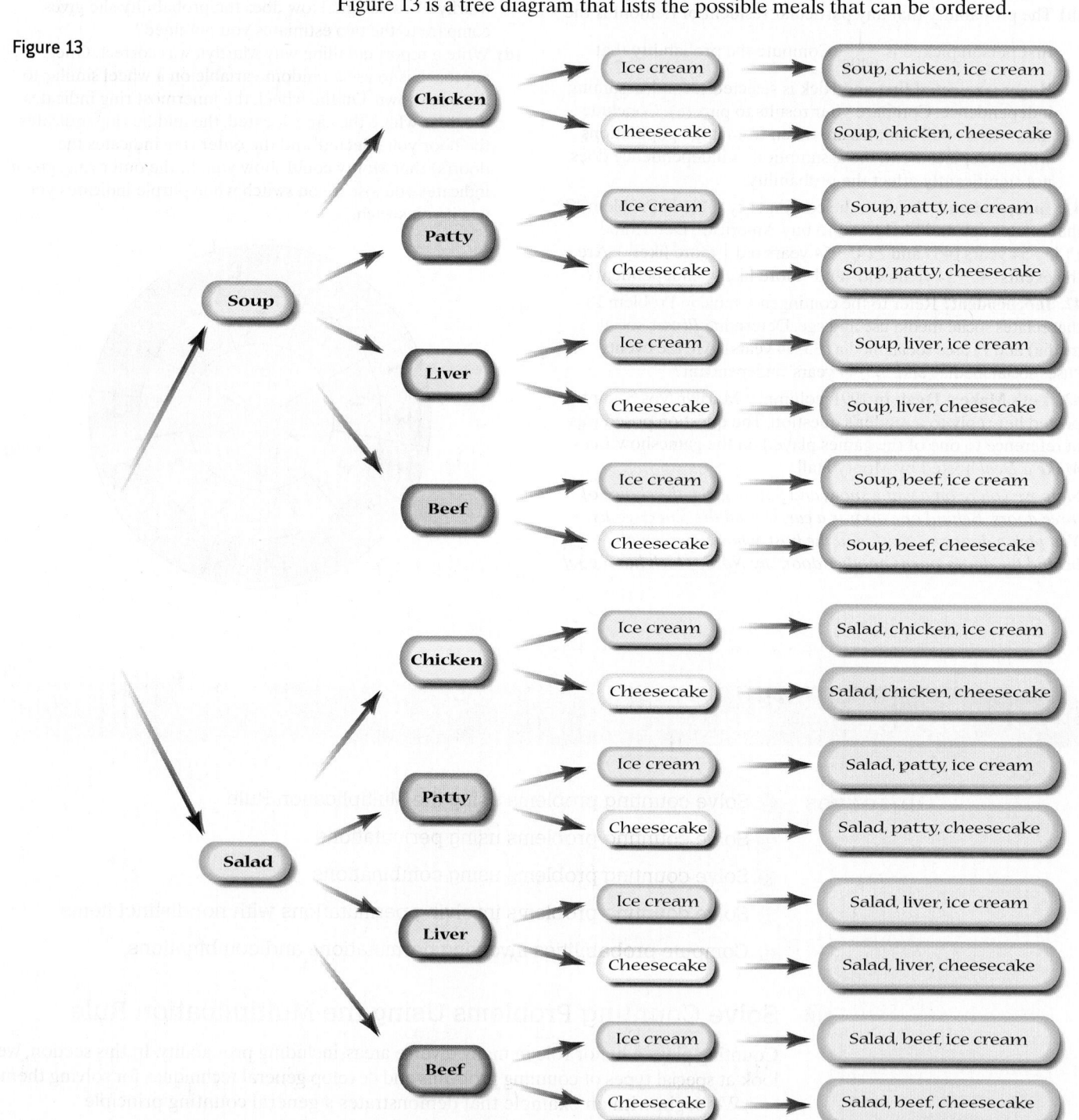

Solution Look at the tree diagram in Figure 13. For each choice of appetizer, we have 4 choices of entrée, and for each of these $2 \cdot 4 = 8$ choices, there are 2 choices for dessert. A total of $2 \cdot 4 \cdot 2 = 16$ different meals can be ordered. ●

Example 1 illustrates a general counting principle.

> ## Multiplication Rule of Counting
>
> If a task consists of a sequence of choices in which there are p selections for the first choice, q selections for the second choice, r selections for the third choice, and so on, then the task of making these selections can be done in
>
> $$p \cdot q \cdot r \cdot \cdots$$
>
> different ways.

EXAMPLE 2   Counting Airport Codes (Repetition Allowed)

Problem The International Air Transport Association (IATA) assigns three-letter codes to represent airport locations. For example, the code for Fort Lauderdale International Airport is FLL. How many different airport codes are possible?

Approach We are choosing three letters from 26 letters and arranging them in order. Notice that repetition of letters is allowed. Use the Multiplication Rule of Counting, recognizing we have 26 ways to choose the first letter, 26 ways to choose the second letter, and 26 ways to choose the third letter.

Solution By the Multiplication Rule,

$$26 \cdot 26 \cdot 26 = 26^3 = 17{,}576$$

different airport codes are possible. ●

In the following example, repetition is not allowed, unlike Example 2.

EXAMPLE 3   Counting without Repetition

Problem Three members from a 14-member committee are to be randomly selected to serve as chair, vice-chair, and secretary. The first person selected is the chair, the second is the vice-chair, and the third is the secretary. How many different committee structures are possible?

Approach The task consists of making three selections. The first selection requires choosing from 14 members. Because a member cannot serve in more than one capacity, the second selection requires choosing from the 13 remaining members. The third selection requires choosing from the 12 remaining members. (Do you see why?) We use the Multiplication Rule to determine the number of possible committee structures.

Solution By the Multiplication Rule,

$$14 \cdot 13 \cdot 12 = 2184$$

● Now Work Problem 31   different committee structures are possible. ●

The Factorial Symbol We now introduce a special symbol.

Definition

Using Technology
Your calculator has a factorial key. Use it to see how fast factorials increase in value. Find the value of 69!. What happens when you try to find 70!? In fact, 70! is larger than 10^{100} (a *googol*), the largest number most calculators can display.

If $n \geq 0$ is an integer, the **factorial symbol, $n!$**, is defined as follows:

$$n! = n(n-1) \cdot \cdots \cdot 3 \cdot 2 \cdot 1$$
$$0! = 1 \qquad 1! = 1$$

For example, $2! = 2 \cdot 1 = 2, 3! = 3 \cdot 2 \cdot 1 = 6, 4! = 4 \cdot 3 \cdot 2 \cdot 1 = 24$, and so on. Table 9 lists the values of $n!$ for $0 \leq n \leq 6$.

Table 9						
n	0	1	2	3	4	5	6
$n!$	1	1	2	6	24	120	720

EXAMPLE 4 The Traveling Salesperson

Problem You have just been hired as a book representative for Pearson Education. On your first day, you must travel to seven schools to introduce yourself. How many different routes are possible?

Approach Call the seven schools A, B, C, D, E, F, and G. School A can be visited first, second, third, fourth, fifth, sixth, or seventh. So, there are seven choices for school A. There are six choices for school B, five choices for school C, and so on. Use the Multiplication Rule and the factorial to find the solution.

● **Now Work Problems 5 and 33**

Solution $7 \cdot 6 \cdot 5 \cdot 4 \cdot 3 \cdot 2 \cdot 1 = 7! = 5040$ different routes are possible. ●

② Solve Counting Problems Using Permutations

Examples 3 and 4 illustrate a type of counting problem referred to as a *permutation*.

Definition

A **permutation** is an ordered arrangement in which r objects are chosen from n distinct (different) objects so that $r \leq n$ and repetition is not allowed. The symbol $_nP_r$, represents the number of permutations of r objects selected from n objects.

The solution in Example 3 could be represented as

$$_nP_r = {_{14}P_3} = 14 \cdot 13 \cdot 12 = 2184$$

and the solution to Example 4 as

$$_7P_7 = 7 \cdot 6 \cdot 5 \cdot 4 \cdot 3 \cdot 2 \cdot 1 = 5040$$

To arrive at a formula for $_nP_r$, note that there are n choices for the first selection, $n - 1$ choices for the second selection, $n - 2$ choices for the third selection, $\ldots$, and $n - (r - 1)$ choices for the rth selection. By the Multiplication Rule,

$$\begin{array}{cccc} \text{1st} & \text{2nd} & \text{3rd} & r\text{th} \\ _nP_r = n \cdot & (n-1) \cdot & (n-2) \cdot \cdots \cdot & [n-(r-1)] \end{array}$$
$$= n \cdot (n-1) \cdot (n-2) \cdot \cdots \cdot (n-r+1)$$

This formula for $_nP_r$ can be written in **factorial notation**:

$$_nP_r = n \cdot (n-1) \cdot (n-2) \cdots (n-r+1)$$

$$= n \cdot (n-1) \cdot (n-2) \cdots (n-r+1) \cdot \frac{(n-r) \cdots 3 \cdot 2 \cdot 1}{(n-r) \cdots 3 \cdot 2 \cdot 1}$$

$$= \frac{n!}{(n-r)!}$$

Number of Permutations of n Distinct Objects Taken r at a Time

The number of arrangements of r objects chosen from n objects, in which

1. the n objects are distinct,
2. repetition of objects is not allowed, and
3. order is important,

is given by the formula

In Other Words
"Order is important" means that
ABC is different from **BCA**.

$$_nP_r = \frac{n!}{(n-r)!} \qquad \text{(1)}$$

EXAMPLE 5   Computing Permutations

Problem Evaluate:

(a) $_7P_5$ **(b)** $_5P_5$

By Hand Approach

Use Formula (1): $_nP_r = \dfrac{n!}{(n-r)!}$

By Hand Solution

(a) $_7P_5 = \dfrac{7!}{(7-5)!} = \dfrac{7!}{2!} = \dfrac{7 \cdot 6 \cdot 5 \cdot 4 \cdot 3 \cdot 2!}{2!}$

$$= \underbrace{7 \cdot 6 \cdot 5 \cdot 4 \cdot 3}_{5 \text{ factors}}$$

$$= 2520$$

(b) $_5P_5 = \dfrac{5!}{(5-5)!} = \dfrac{5!}{0!} = 5!$

$$= 5 \cdot 4 \cdot 3 \cdot 2 \cdot 1$$

$$= 120$$

Technology Approach

We will use a TI-84 Plus C graphing calculator in part (a) and Excel in part (b) to evaluate each permutation. The steps for determining permutations using the TI-83/84 Plus graphing calculator, Excel, and StatCrunch can be found in the Technology Step-by-Step on page 299.

Technology Solution

(a) Figure 14(a) shows the results on a TI-84 Plus C calculator, so $_7P_5 = 2520$.

(b) Figure 14(b) shows the results from Excel. Enter "=permut(5,5)" in any cell, so $_5P_5 = 120$.

Figure 14

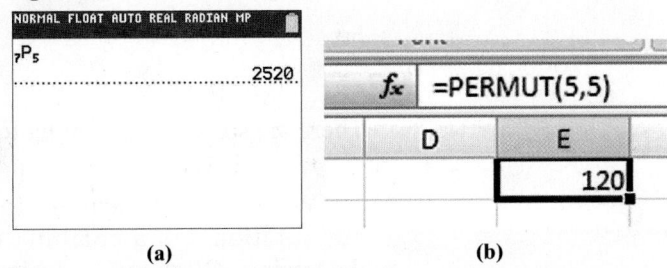

    (a)   (b)

● Now Work Problem 11

EXAMPLE 6   Betting on the Trifecta

Problem In how many ways can horses in a ten-horse race finish first, second, and third?

Approach The ten horses are distinct. Once a horse crosses the finish line, that horse will not cross the finish line again, and, in a race, finishing order is important. We have a permutation of ten objects taken three at a time.

Solution The top three horses can finish a ten-horse race in

$$_{10}P_3 = \frac{10!}{(10-3)!} = \frac{10!}{7!} = \frac{10 \cdot 9 \cdot 8 \cdot 7!}{7!}$$
$$= \underbrace{10 \cdot 9 \cdot 8}_{3 \text{ factors}}$$
$$= 720 \text{ ways}$$

● **Now Work Problem 45**

---

③ **Solve Counting Problems Using Combinations**

In a permutation, order is important. For example, the arrangements *ABC* and *BAC* are considered different arrangements of the letters *A*, *B*, and *C*. If order is unimportant, we do not distinguish *ABC* from *BAC*. In poker, the order in which the cards are received does not matter. The *combination* of the cards is what matters.

Definition   A **combination** is a collection, without regard to order, in which *r* objects are chosen from *n* distinct objects with $r \leq n$ and without repetition. The symbol $_nC_r$ represents the number of combinations of *n* distinct objects taken *r* at a time.

EXAMPLE 7   Listing Combinations

Problem Roger, Ken, Tom, and Jay are going to play golf. They will randomly select teams of two players each. List all possible team combinations. That is, list all the combinations of the four people Roger, Ken, Tom, and Jay taken two at a time. What is $_4C_2$?

Approach List the possible teams. Note that order is unimportant, so {Roger, Ken} is the same as {Ken, Roger}.

Solution The list of all such teams (combinations) is

   Roger, Ken   Roger, Tom   Roger, Jay   Ken, Tom   Ken, Jay   Tom, Jay

So,

$$_4C_2 = 6$$

There are six ways of forming teams of two from a group of four players.   ●

We can find a formula for $_nC_r$ by noting that the only difference between a permutation and a combination is that we disregard order in combinations. To determine $_nC_r$, eliminate from the formula for $_nP_r$ the number of permutations that were rearrangements of a given set of *r* objects. In Example 7, for example, selecting {Roger, Ken} was the same as selecting {Ken, Roger}, so there were 2! = 2 rearrangements of the two objects. This can be determined from the formula for $_nP_r$ by calculating $_rP_r = r!$. So, if we divide $_nP_r$ by $r!$, we will have the desired formula for $_nC_r$:

$$_nC_r = \frac{_nP_r}{r!} = \frac{n!}{r!(n-r)!}$$

Number of Combinations of n Distinct Objects Taken r at a Time

The number of different arrangements of r objects chosen from n objects, in which

1. the n objects are distinct
2. repetition of objects is not allowed, and
3. order is not important

is given by the formula

$$ {}_nC_r = \frac{n!}{r!(n-r)!} \tag{2} $$

Using Formula (2) to solve the problem presented in Example 7, we obtain

$$ {}_4C_2 = \frac{4!}{2!(4-2)!} = \frac{4!}{2!2!} = \frac{4 \cdot 3 \cdot 2!}{2 \cdot 1 \cdot 2!} = \frac{12}{2} = 6 $$

EXAMPLE 8 Computing Combinations

Problem Evaluate:
(a) ${}_4C_1$ **(b)** ${}_6C_4$ **(c)** ${}_6C_2$

By Hand Approach

Use Formula (2): ${}_nC_r = \dfrac{n!}{r!(n-r)!}$

By Hand Solution

(a) ${}_4C_1 = \dfrac{4!}{1!(4-1)!} = \dfrac{4!}{1! \cdot 3!} = \dfrac{4 \cdot 3!}{1 \cdot 3!} = 4$

(b) ${}_6C_4 = \dfrac{6!}{4!(6-4)!} = \dfrac{6!}{4! \cdot 2!} = \dfrac{6 \cdot 5 \cdot 4!}{4! \cdot 2 \cdot 1} = \dfrac{30}{2} = 15$

(c) ${}_6C_2 = \dfrac{6!}{2!(6-2)!} = \dfrac{6!}{2! \cdot 4!} = \dfrac{6 \cdot 5 \cdot 4!}{2 \cdot 1 \cdot 4!} = \dfrac{30}{2} = 15$

Technology Approach We will use a TI-84 Plus C graphing calculator in part (a) and Excel in parts (b) and (c) to evaluate each combination. The steps for determining combinations using the TI-83/84 Plus graphing calculator, Excel, and StatCrunch can be found in the Technology Step-by-Step on page 299.

Technology Solution

(a) Figure 15(a) shows the results on a TI-84 Plus C calculator. So, ${}_4C_1 = 4$.
(b) Figure 15(b) shows the results from Excel. Enter "=combin(6,4)" in any cell. So ${}_6C_4 = 15$.
(c) If we enter "=combin(6,2)" into any cell in Excel, we obtain a result of 15.

Figure 15

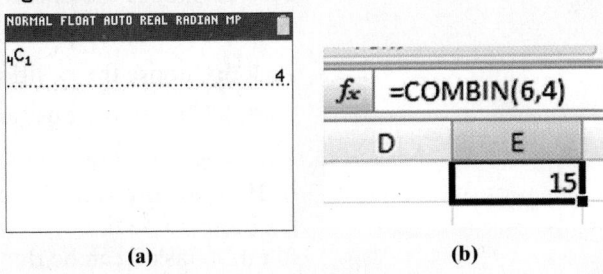

(a) (b)

● Now Work Problem 19

Notice in Examples 8(b) and (c) that ${}_6C_4 = {}_6C_2$. This result can be generalized as

$$ {}_nC_r = {}_nC_{n-r} $$

EXAMPLE 9 Simple Random Samples

Problem How many different simple random samples of size 4 can be obtained from a population whose size is 20?

Approach The 20 individuals in the population are distinct. In addition, the order in which individuals are selected is unimportant. Thus, the number of simple random samples of size 4 from a population of size 20 is a combination of $n = 20$ objects taken $r = 4$ at a time.

Solution Use Formula (2) with $n = 20$ and $r = 4$:

$$_{20}C_4 = \frac{20!}{4!(20-4)!} = \frac{20!}{4!16!} = \frac{20 \cdot 19 \cdot 18 \cdot 17 \cdot 16!}{4 \cdot 3 \cdot 2 \cdot 1 \cdot 16!} = \frac{116{,}280}{24} = 4845$$

There are 4845 different simple random samples of size 4 from a population whose size is 20.

● Now Work Problem 51

---

④ **Solve Counting Problems Involving Permutations with Nondistinct Items**

Sometimes we want to arrange objects in order, but some of the objects are not distinguishable.

EXAMPLE 10 DNA Sequence

Problem A DNA sequence consists of a series of letters representing a DNA strand that spells out the genetic code. There are four possible letters (A, C, G, and T), each representing a specific nucleotide base in the DNA strand (adenine, cytosine, guanine, and thymine, respectively). How many distinguishable sequences can be formed using two As, two Cs, three Gs, and one T?

Approach Each sequence formed will have eight letters. To construct each sequence, we need to fill in eight positions with the eight letters:

$$\overline{1} \quad \overline{2} \quad \overline{3} \quad \overline{4} \quad \overline{5} \quad \overline{6} \quad \overline{7} \quad \overline{8}$$

The process of forming a sequence consists of four tasks:

Task 1: Choose the positions for the two As.
Task 2: Choose the positions for the two Cs.
Task 3: Choose the positions for the three Gs.
Task 4: Choose the position for the one T.

Task 1 can be done in $_8C_2$ ways because we are choosing the 2 positions for A, but order does not matter (because we cannot distinguish the two As). This leaves 6 positions to be filled, so task 2 can be done in $_6C_2$ ways. This leaves 4 positions to be filled, so task 3 can be done in $_4C_3$ ways. The last position can be filled in $_1C_1$ way.

Solution By the Multiplication Rule, the number of possible sequences that can be formed is

$$_8C_2 \cdot {_6C_2} \cdot {_4C_3} \cdot {_1C_1} = \frac{8!}{2! \cdot 6!} \cdot \frac{6!}{2! \cdot 4!} \cdot \frac{4!}{3! \cdot 1!} \cdot \frac{1!}{1! \cdot 0!}$$

$$= \frac{8!}{2! \cdot 2! \cdot 3! \cdot 1! \cdot 0!}$$

$$= 1680$$

There are 1680 possible distinguishable sequences that can be formed.

Example 10 suggests a general result. Had the letters in the sequence each been different, $_8P_8 = 8!$ possible sequences would have been formed. This is the numerator of the answer. The presence of two As, two Cs, and three Gs reduces the number of different sequences, as the entries in the denominator illustrate. We are led to the following result:

Permutations with Nondistinct Items

The number of permutations of n objects of which n_1 are of one kind, n_2 are of a second kind, ..., and n_k are of a kth kind is given by

$$\frac{n!}{n_1! \cdot n_2! \cdots \cdot n_k!} \qquad \textbf{(3)}$$

where $n = n_1 + n_2 + \cdots + n_k$.

EXAMPLE 11   Arranging Flags

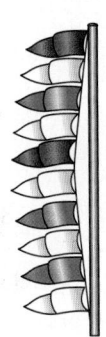

Problem How many different vertical arrangements are there of 10 flags if 5 are white, 3 are blue, and 2 are red?

Approach Because there are nondistinct items and order matters, use the formula for finding the number of permutations with nondistinct items. We seek the number of permutations of $n = 10$ objects, of which $n_1 = 5$ are of one kind (white), $n_2 = 3$ are of a second kind (blue), and $n_3 = 2$ are of a third kind (red).

Solution Using Formula (3), we find that there are

$$\frac{10!}{5! \cdot 3! \cdot 2!} = \frac{10 \cdot 9 \cdot 8 \cdot 7 \cdot 6 \cdot 5!}{5! \cdot 3! \cdot 2!}$$

$$= 2520 \text{ different vertical arrangements}$$

• Now Work Problem 55

Summary

One of the challenges in solving counting problems is selecting the appropriate formula for the given situation. Table 10 below reviews the situations in which each counting problem applies.

Table 10	Description	Formula
Combination	The selection of r objects from a set of n different objects when the order in which the objects are selected does not matter (so AB is the same as BA) and an object cannot be selected more than once (repetition is not allowed)	$_nC_r = \dfrac{n!}{r!(n-r)!}$
Permutation of Distinct Items with Replacement	The selection of r objects from a set of n different objects when the order in which the objects are selected matters (so AB is different from BA) and an object may be selected more than once (repetition is allowed)	n^r
Permutation of Distinct Items without Replacement	The selection of r objects from a set of n different objects when the order in which the objects are selected matters (so AB is different from BA) and an object cannot be selected more than once (repetition is not allowed)	$_nP_r = \dfrac{n!}{(n-r)!}$
Permutation of Nondistinct Items without Replacement	The number of ways n objects can be arranged (order matters) in which there are n_1 of one kind, n_2 of a second kind, ..., and n_k of a kth kind, where $n = n_1 + n_2 + \cdots + n_k$	$\dfrac{n!}{n_1! n_2! \cdots n_k!}$

⑤ Compute Probabilities Involving Permutations and Combinations

The counting techniques presented in this section can be used along with the classical method to compute certain probabilities. Recall that this method stated the probability of an event E is the number of ways event E can occur divided by the number of different possible outcomes of the experiment provided each outcome is equally likely.

EXAMPLE 12   Winning the Lottery

Problem In the Illinois Lottery, an urn contains balls numbered 1–52. From this urn, six balls are randomly chosen without replacement. For a $1 bet, a player chooses two sets of six numbers. To win, all six numbers must match those chosen from the urn. The order in which the balls are picked does not matter. What is the probability of winning the lottery?

Approach The probability of winning is given by the number of ways a ticket could win divided by the size of the sample space. Each ticket has two sets of six numbers and therefore two chances of winning. The size of the sample space S is the number of ways 6 objects can be selected from 52 objects without replacement and without regard to order, so $N(S) = {}_{52}C_6$.

Solution The size of the sample space is

$$N(S) = {}_{52}C_6 = \frac{52!}{6! \cdot (52-6)!} = \frac{52 \cdot 51 \cdot 50 \cdot 49 \cdot 48 \cdot 47 \cdot 46!}{6! \cdot 46!} = 20{,}358{,}520$$

Each ticket has two chances of winning. If E is the event "winning ticket," then $N(E) = 2$ and

$$P(E) = \frac{2}{20{,}358{,}520} = 0.000000098$$

There is about a 1 in 10,000,000 chance of winning the Illinois Lottery!   •

EXAMPLE 13   Acceptance Sampling

Problem A shipment of 120 fasteners that contains 4 defective fasteners was sent to a manufacturing plant. The plant's quality-control manager randomly selects and inspects 5 fasteners. What is the probability that exactly 1 of the inspected fasteners is defective?

Approach Find the probability that exactly 1 fastener is defective by calculating the number of ways of selecting exactly 1 defective fastener in 5 fasteners and dividing this result by the number of ways of selecting 5 fasteners from 120 fasteners. To choose exactly 1 defective in the 5 requires choosing 1 defective from the 4 defectives and 4 nondefectives from the 116 nondefectives. The order in which the fasteners are selected does not matter, so we use combinations.

Solution The number of ways of choosing 1 defective fastener from 4 defective fasteners is ${}_4C_1$. The number of ways of choosing 4 nondefective fasteners from 116 nondefectives is ${}_{116}C_4$. Using the Multiplication Rule, we find that the number of ways of choosing 1 defective and 4 nondefective fasteners is

$$({}_4C_1) \cdot ({}_{116}C_4) = 4 \cdot 7{,}160{,}245 = 28{,}640{,}980$$

The number of ways of selecting 5 fasteners from 120 fasteners is $_{120}C_5 = 190,578,024$. The probability of selecting exactly 1 defective fastener is

$$P(1 \text{ defective fastener}) = \frac{(_4C_1)(_{116}C_4)}{_{120}C_5} = \frac{28,640,980}{190,578,024} = 0.1503$$

The probability of randomly selecting exactly one defective fastener is 0.1503. If we selected 5 fasteners, 100 different times, we would expect about 15 of the samples to have exactly one defective fastener.

● **Now Work Problem 61**

Technology Step-by-Step Factorials, Permutations, and Combinations

TI-83/84 Plus

Factorials
1. To compute 7!, type 7 on the HOME screen.
2. Press MATH, then highlight PRB (or PROB on a TI-84 Plus C), and then select 4: ! With 7! on the HOME screen, press ENTER again.

Permutations and Combinations
1. To compute $_7P_3$, type 7 on the HOME screen.
2. Press MATH, then highlight PRB (or PROB on a TI-84 Plus C), and then select 2: $_nP_r$
3. Type 3 on the HOME screen, and press ENTER.

Note: To compute $_7C_3$, select 3: $_nC_r$ instead of 2: $_nP_r$.

Excel

Factorials In any cell, enter "=fact(n)" where n is the factorial desired.

Permutations In any cell, enter "=permut(n, r)" where n is the number of distinct objects and r is the number of objects to be selected.

Combinations In any cell, enter "=combin(n, r)" where n is the number of distinct objects and r is the number of objects to be selected.

StatCrunch
1. Place the cursor in any cell in the StatCrunch spreadsheet. Select **Data** and choose **Compute**, then **Expression**.
2. **Factorials** In the Expression box, type fact(n) to determine $n!$. For example, to find 5!, type fact(5). Click Compute!
   Permutations In the Expression box, type perm(n, r) to determine $_nP_r$. For example, to compute $_{10}P_4$, type perm(10,4). Click Compute!
   Combinations In the Expression box, type comb(n, r) to determine $_nC_r$. **Note:** To compute $_{10}C_4$, type comb(10,4). Click Compute!

 ## 5.5 Assess Your Understanding

Vocabulary and Skill Building

1. A _____ is an ordered arrangement of r objects chosen from n distinct objects without repetition.

2. A _____ is an arrangement of r objects chosen from n distinct objects without repetition and without regard to order.

3. *True or False:* In a combination problem, order is not important.

4. The factorial symbol, $n!$, is defined as $n! = $ _____ and $0! = $ ____.

In Problems 5–10, find the value of each factorial.

NW **5.** 5!   **6.** 7!
7. 10!   **8.** 12!
9. 0!   **10.** 1!

In Problems 11–18, find the value of each permutation.

NW **11.** $_6P_2$ **12.** $_7P_2$
13. $_4P_4$ **14.** $_7P_7$
15. $_5P_0$ **16.** $_4P_0$
17. $_8P_3$ **18.** $_9P_4$

In Problems 19–26, find the value of each combination.

NW **19.** $_8C_3$ **20.** $_9C_2$
21. $_{10}C_2$ **22.** $_{12}C_3$
23. $_{52}C_1$ **24.** $_{40}C_{40}$
25. $_{48}C_3$ **26.** $_{30}C_4$

27. List all the permutations of five objects $a, b, c, d,$ and e taken two at a time without repetition. What is $_5P_2$?

28. List all the permutations of four objects a, b, c, and d taken two at a time without repetition. What is $_4P_2$?

29. List all the combinations of five objects a, b, c, d, and e taken two at a time. What is $_5C_2$?

30. List all the combinations of four objects a, b, c, and d taken two at a time. What is $_4C_2$?

Applying the Concepts

NW **31. Clothing Options** A man has six shirts and four ties. Assuming that they all match, how many different shirt-and-tie combinations can he wear?

32. Clothing Options A woman has five blouses and three skirts. Assuming that they all match, how many different outfits can she wear?

NW **33. Arranging Songs** Suppose Dan is going to upload 12 songs to his digital music player. In how many ways can the 12 songs be played without repetition?

34. Arranging Students In how many ways can 15 students be lined up?

35. Traveling Salesperson A salesperson must travel to eight cities to promote a new marketing campaign. How many different trips are possible if any route between cities is possible?

36. Randomly Playing Songs A certain digital music player randomly plays each of 10 songs. Once a song is played, it is not repeated until all the songs have been played. In how many different ways can the player play the 10 songs?

37. Stocks on the NYSE Companies whose stocks are listed on the New York Stock Exchange (NYSE) have their company name represented by either one, two, or three letters (repetition of letters is allowed). What is the maximum number of companies that can be listed on the New York Stock Exchange?

38. Stocks on the NASDAQ Companies whose stocks are listed on the NASDAQ stock exchange have their company name represented by either four or five letters (repetition of letters is allowed). What is the maximum number of companies that can be listed on the NASDAQ?

39. Garage Door Code Outside a home there is a keypad that will open the garage if the correct four-digit code is entered.

(a) How many codes are possible?

(b) What is the probability of entering the correct code on the first try, assuming that the owner doesn't remember the code?

40. Social Security Numbers A Social Security number is used to identify each resident of the United States uniquely. The number is of the form xxx–xx–xxxx, where each x is a digit from 0 to 9.

(a) How many Social Security numbers can be formed?

(b) What is the probability of correctly guessing the Social Security number of the president of the United States?

41. User Names Suppose that a local area network requires eight letters for user names. Lower- and uppercase letters are considered the same. How many user names are possible for the local area network?

42. User Names How many user names are possible in Problem 41 if the last character must be a digit?

43. Combination Locks A combination lock has 50 numbers on it. To open it, you turn counterclockwise to a number, then rotate clockwise to a second number, and then counterclockwise to the third number. Repetitions are allowed.

(a) How many different lock combinations are there?

(b) What is the probability of guessing a lock combination on the first try?

44. Forming License Plate Numbers How many different license plate numbers can be made by using one letter followed by five digits selected from the digits 0 through 9?

NW **45. INDY 500** Suppose 40 cars start at the Indianapolis 500. In how many ways can the top 3 cars finish the race?

46. Betting on the Perfecta In how many ways can the top 2 horses finish in a 10-horse race?

47. Forming a Committee Four members from a 20-person committee are to be selected randomly to serve as chairperson, vice-chairperson, secretary, and treasurer. The first person selected is the chairperson; the second, the vice-chairperson; the third, the secretary; and the fourth, the treasurer. How many different leadership structures are possible?

48. Forming a Committee Four members from a 50-person committee are to be selected randomly to serve as chairperson, vice-chairperson, secretary, and treasurer. The first person selected is the chairperson; the second, the vice-chairperson; the third, the secretary; and the fourth, the treasurer. How many different leadership structures are possible?

49. Lottery A lottery exists where balls numbered 1–25 are placed in an urn. To win, you must match the four balls chosen in the correct order. How many possible outcomes are there for this game?

50. Forming a Committee In the U.S. Senate, there are 21 members on the Committee on Banking, Housing, and Urban Affairs. Nine of these 21 members are selected to be on the Subcommittee on Economic Policy. How many different committee structures are possible for this subcommittee?

NW **51. Simple Random Sample** How many different simple random samples of size 5 can be obtained from a population whose size is 50?

52. Simple Random Sample How many different simple random samples of size 7 can be obtained from a population whose size is 100?

53. Children A family has six children. If this family has exactly two boys, how many different birth and gender orders are possible?

54. Children A family has eight children. If this family has exactly three boys, how many different birth and gender orders are possible?

NW **55. DNA Sequences** (See Example 10.) How many distinguishable DNA sequences can be formed using three As, two Cs, two Gs, and three Ts?

56. DNA Sequences (See Example 10.) How many distinguishable DNA sequences can be formed using one A, four Cs, three Gs, and four Ts?

57. Landscape Design A golf-course architect has four linden trees, five white birch trees, and two bald cypress trees to plant in a row along a fairway. In how many ways can the landscaper plant the trees in a row, assuming that the trees are evenly spaced?

58. Starting Lineup A baseball team consists of three outfielders, four infielders, a pitcher, and a catcher. Assuming that the outfielders and infielders are indistinguishable, how many batting orders are possible?

59. Little Lotto In the Illinois Lottery game Little Lotto, an urn contains balls numbered 1–39. From this urn, 5 balls are chosen randomly, without replacement. For a $1 bet, a player chooses one set of five numbers. To win, all five numbers must match those chosen from the urn. The order in which the balls are selected does not matter. What is the probability of winning Little Lotto with one ticket?

60. Mega Millions In Mega Millions, an urn contains balls numbered 1–56, and a second urn contains balls numbered 1–46. From the first urn, 5 balls are chosen randomly, without replacement and without regard to order. From the second urn, 1 ball is chosen randomly. For a $1 bet, a player chooses one set of five numbers to match the balls selected from the first urn and one number to match the ball selected from the second urn. To win, all six numbers must match; that is, the player must match the first 5 balls selected from the first urn *and* the single ball selected from the second urn. What is the probability of winning the Mega Millions with a single ticket?

NW 61. Selecting a Jury The grade appeal process at a university requires that a jury be structured by selecting five individuals randomly from a pool of eight students and ten faculty.

(a) What is the probability of selecting a jury of all students?

(b) What is the probability of selecting a jury of all faculty?

(c) What is the probability of selecting a jury of two students and three faculty?

62. Selecting a Committee Suppose that there are 55 Democrats and 45 Republicans in the U.S. Senate. A committee of seven senators is to be formed by selecting members of the Senate randomly.

(a) What is the probability that the committee is composed of all Democrats?

(b) What is the probability that the committee is composed of all Republicans?

(c) What is the probability that the committee is composed of three Democrats and four Republicans?

63. Acceptance Sampling Suppose that a shipment of 120 electronic components contains 4 defective components. To determine whether the shipment should be accepted, a quality-control engineer randomly selects 4 of the components and tests them. If 1 or more of the components is defective, the shipment is rejected. What is the probability that the shipment is rejected?

64. In the Dark A box containing twelve 40-watt light bulbs and eighteen 60-watt light bulbs is stored in your basement. Unfortunately, the box is stored in the dark and you need two 60-watt bulbs. What is the probability of randomly selecting two 60-watt bulbs from the box?

65. Randomly Playing Songs Suppose a playlist you just created has 13 tracks. After listening to the playlist, you decide that you like 5 of the songs. The random feature on your music player will play each of the 13 songs once in a random order. Find the probability that among the first 4 songs played

(a) you like 2 of them;

(b) you like 3 of them;

(c) you like all 4 of them.

66. Packaging Error Through a manufacturing error, three cans marked "regular soda" were accidentally filled with diet soda and placed into a 12-pack. Suppose that three cans are randomly selected from the 12-pack.

(a) Determine the probability that exactly two contain diet soda.

(b) Determine the probability that exactly one contains diet soda.

(c) Determine the probability that all three contain diet soda.

67. Three of a Kind You are dealt 5 cards from a standard 52-card deck. Determine the probability of being dealt three of a kind (such as three aces or three kings) by answering the following questions:

(a) How many ways can 5 cards be selected from a 52-card deck?

(b) Each deck contains 4 twos, 4 threes, and so on. How many ways can three of the same card be selected from the deck?

(c) The remaining 2 cards must be different from the 3 chosen and different from each other. For example, if we drew three kings, the 4th card cannot be a king. After selecting the three of a kind, there are 12 different ranks of card remaining in the deck that can be chosen. If we have three kings, then we can choose twos, threes, and so on. Of the 12 ranks remaining, we choose 2 of them and then select one of the 4 cards in each of the two chosen ranks. How many ways can we select the remaining 2 cards?

(d) Use the General Multiplication Rule to compute the probability of obtaining three of a kind. That is, what is the probability of selecting three of a kind and two cards that are not like?

68. Two of a Kind Follow the outline presented in Problem 67 to determine the probability of being dealt exactly one pair.

69. Acceptance Sampling Suppose that you have just received a shipment of 20 modems. Although you don't know this, 3 of the modems are defective. To determine whether you will accept the shipment, you randomly select 4 modems and test them. If all 4 modems work, you accept the shipment. Otherwise, the shipment is rejected. What is the probability of accepting the shipment?

70. Acceptance Sampling Suppose that you have just received a shipment of 100 televisions. Although you don't know this, 6 are defective. To determine whether you will accept the shipment, you randomly select 5 televisions and test them. If all 5 televisions work, you accept the shipment; otherwise, the shipment is rejected. What is the probability of accepting the shipment?

71. Password Policy According to the Sefton Council Password Policy (August 2007), the United Kingdom government recommends the use of *"Environ passwords with the following format: consonant, vowel, consonant, consonant, vowel, consonant, number, number (for example, pinray45)."*

(a) Assuming passwords are not case sensitive, how many such passwords are possible (assume that there are 5 vowels and 21 consonants)?

(b) How many passwords are possible if they are case sensitive?

5.6 Putting It Together: Which Method Do I Use?

Objectives   ① Determine the appropriate probability rule to use

           ② Determine the appropriate counting technique to use

① Determine the Appropriate Probability Rule to Use

This section will help you learn when to use a particular rule. To aid you, consider the flowchart in Figure 16. While not all situations can be handled directly with the formulas provided, they can be combined and expanded to many more situations.

Figure 16

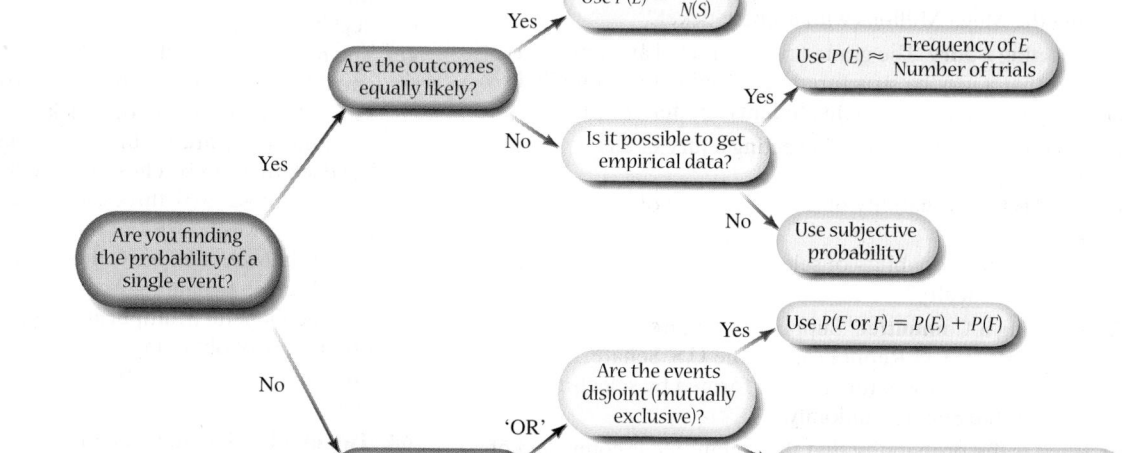

The first step is to determine whether we are finding the probability of a single event. If we are dealing with a single event, we must decide whether to use the classical method (equally likely outcomes), the empirical method (relative frequencies), or subjective probability. For experiments involving more than one event, we first decide which type of statement we have. For events involving "AND," we must know if the events are independent. For events involving "OR," we need to know if the events are disjoint (mutually exclusive).

EXAMPLE 1   Probability: Which Rule Do I Use?

Problem In the game show *Deal or No Deal?*, a contestant is presented with 26 suitcases that contain amounts ranging from $0.01 to $1,000,000. The contestant must pick an initial case that is set aside as the game progresses. The amounts are randomly distributed among the suitcases prior to the game as shown in Table 11. What is the probability that the contestant picks a case worth at least $100,000?

Table 11	
Prize	**Number of Suitcases**
$0.01–$100	8
$200–$1000	6
$5,000–$50,000	5
$100,000–$1,000,000	7

Approach Follow the flowchart in Figure 16.

Solution There is a single event, so we must decide among the empirical, classical, or subjective approaches to determine the probability. The probability experiment is selecting a suitcase. Each prize amount is randomly assigned to one of the 26 suitcases, so the outcomes are equally likely. Table 11 shows that seven cases contain at least $100,000. Letting E = "worth at least $100,000," we compute $P(E)$ using the classical approach.

$$P(E) = \frac{N(E)}{N(S)} = \frac{7}{26} = 0.269$$

The probability the contestant selects a suitcase worth at least $100,000 is 0.269. In 100 different games, we would expect about 27 games to result in a contestant choosing a suitcase worth at least $100,000. ●

EXAMPLE 2   Probability: Which Rule Do I Use?

Problem According to a Harris poll, 14% of adult Americans have one or more tattoos, 50% have pierced ears, and 65% of those with one or more tattoos also have pierced ears. What is the probability that a randomly selected adult American has one or more tattoos and pierced ears?

Approach Follow the flowchart in Figure 16.

Solution We are finding the probability of an event involving 'AND'. Letting T = "one or more tattoos" and E = "ears pierced," we must find $P(T \text{ and } E)$. We need to determine if the two events, T and E, are independent. The problem statement tells us that $P(E) = 0.50$ and $P(E|T) = 0.65$. Because $P(E) \neq P(E|T)$, the two events are not independent. We find $P(T \text{ and } E)$ using the General Multiplication Rule.

$$\begin{aligned} P(T \text{ and } E) &= P(T) \cdot P(E|T) \\ &= (0.14)(0.65) \\ &= 0.091 \end{aligned}$$

So the probability of selecting an adult American at random who has one or more tattoos and pierced ears is 0.091. ●

② Determine the Appropriate Counting Technique to Use

To determine the appropriate counting technique to use, we need the ability to distinguish between a sequence of choices and an arrangement of items. We also need to determine whether order matters in the arrangements. See Figure 17 on the following page. Keep in mind that one problem may require several counting rules.

Figure 17

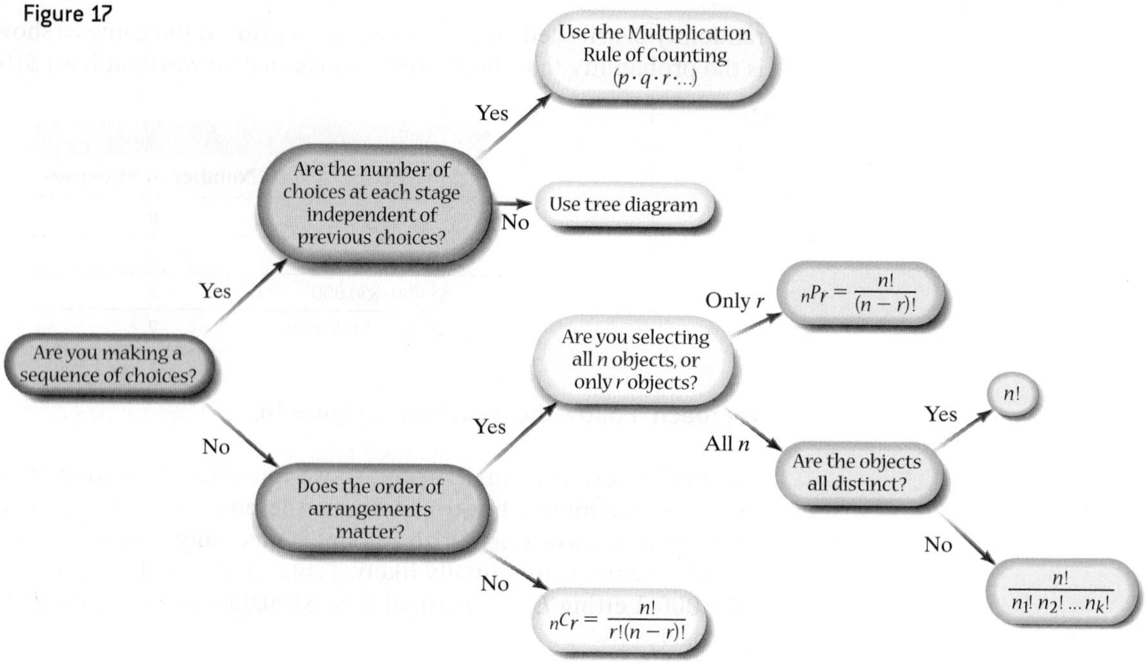

We first must decide whether we have a sequence of choices or an arrangement of items. For a sequence of choices, we use the Multiplication Rule of Counting if the number of choices at each stage is independent of previous choices. If the number of choices at each stage is not independent of previous choices, we use a tree diagram. When determining the number of arrangements of items, we want to know whether the order of selection matters. If order matters, we also want to know whether we are arranging all the items available or a subset of the items.

EXAMPLE 3   Counting: Which Technique Do I Use?

Problem The Hazelwood city council consists of five men and four women. How many different subcommittees can be formed that consist of three men and two women?

Approach Follow the flowchart in Figure 17.

Solution We need to find the number of subcommittees having three men and two women. So we consider a sequence of events: select the men, then select the women. Since the number of choices at each stage is independent of previous choices (the men chosen will not impact which women are chosen), we use the Multiplication Rule of Counting to obtain

$$N(\text{subcommittees}) = N(\text{ways to pick three men}) \cdot N(\text{ways to pick two women})$$

To select the three men, we must consider the number of arrangements of five men taken three at a time. Since the order of selection does not matter, we use the combination formula.

$$N(\text{ways to pick three men}) = {}_5C_3 = \frac{5!}{3! \cdot 2!} = 10$$

To select the two women, we must consider the number of arrangements of four women taken two at a time. Since the order of selection does not matter, we use the combination formula again.

$$N(\text{ways to pick two women}) = {}_4C_2 = \frac{4!}{2! \cdot 2!} = 6$$

Combining our results, we obtain $N(\text{subcommittees}) = 10 \cdot 6 = 60$. There are 60 possible subcommittees that contain three men and two women.

EXAMPLE 4   Counting: Which Technique Do I Use?

Problem The Daytona 500, the season opening NASCAR event, has 43 drivers in the race. In how many different ways could the top four finishers (first, second, third, and fourth place) occur?

Approach Follow the flowchart in Figure 17.

Solution We need to find the number of ways to select the top four finishers. There are two different ways to solve this problem.

1. View this as a sequence of choices, where the first choice is the first-place driver, the second choice is the second-place driver, and so on. There are 43 ways to pick the first driver, 42 ways to pick the second, 41 ways to pick the third, and 40 ways to pick the fourth. Use the Multiplication Rule of Counting. The number of ways the top four finishers can occur is

$$N(\text{top four}) = 43 \cdot 42 \cdot 41 \cdot 40 = 2{,}961{,}840$$

2. Approach this problem as an arrangement of units. Because each race position is distinguishable, order matters. We are arranging the 43 drivers taken four at a time. Using our permutation formula, we get

$$N(\text{top four}) = {}_{43}P_4 = \frac{43!}{(43-4)!} = \frac{43!}{39!} = 43 \cdot 42 \cdot 41 \cdot 40 = 2{,}961{,}840$$

Again there are 2,961,840 different ways that the top four finishers could occur.   •

5.6 Assess Your Understanding

Vocabulary and Skill Building

1. What is the difference between a permutation and a combination?

2. What method of assigning probabilities to a simple event uses relative frequencies?

3. Which type of compound event is generally associated with multiplication? Which is generally associated with addition?

4. Suppose that you roll a pair of dice 1000 times and get seven 350 times. Based on these results, what is the probability that the next roll results in seven?

For Problems 5 and 6, let the sample space be
$S = \{1, 2, 3, 4, 5, 6, 7, 8, 9, 10\}$. *Suppose that the outcomes are equally likely. Compute the probability of the event:*

5. $E = \{1, 3, 5, 10\}$.

6. $F =$ "a number divisible by three."

7. List all permutations of five objects a, b, c, d, and e taken three at a time without replacement.

In Problems 8 and 9, find the probability of the indicated event if $P(E) = 0.7$ and $P(F) = 0.2$.

8. $P(E \text{ or } F)$ if E and F are mutually exclusive

9. $P(E \text{ or } F)$ if $P(E \text{ and } F) = 0.15$

In Problems 10–12, evaluate each expression.

10. $\dfrac{6!2!}{4!}$

11. ${}_7P_3$

12. ${}_9C_4$

13. Suppose that events E and F are independent, $P(E) = 0.8$, and $P(F) = 0.5$. What is $P(E \text{ and } F)$?

14. Suppose that E and F are two events and $P(E \text{ and } F) = 0.4$ and $P(E) = 0.9$. Find $P(F|E)$.

15. Suppose that E and F are two events and $P(E) = 0.9$ and $P(F|E) = 0.3$. Find $P(E \text{ and } F)$.

16. List all combinations of five objects a, b, c, d, and e taken three at a time without replacement.

Applying the Concepts

17. Soccer? In a survey of 500 randomly selected Americans, it was determined that 22 play soccer. What is the probability that a randomly selected American plays soccer?

18. Apartment Vacancy A real estate agent conducted a survey of 200 landlords and asked how long their apartments remained vacant before a tenant was found. The results of the survey are shown in the table on the following page. The data are based on information obtained from the U.S. Census Bureau.

Duration of Vacancy	Frequency
Less than 1 month	42
1–2 months	38
2–4 months	45
4–6 months	30
6–12 months	24
1–2 years	13
2 years or more	8

(a) Construct a probability model for duration of vacancy.
(b) Is it unusual for an apartment to remain vacant for 2 years or more?
(c) Determine the probability that a randomly selected apartment is vacant for 1–4 months.
(d) Determine the probability that a randomly selected apartment is vacant for less than 2 years.

19. Seating Arrangements In how many ways can three men and three women be seated around a circular table (that seats six) assuming that women and men must alternate seats?

20. Starting Lineups Payton's futsal team consists of 10 girls, but only 5 can be on the field at any given time (four fielders and a goalie).

(a) How many starting lineups are possible if all players are selected without regard to position played?
(b) How many starting lineups are possible if either Payton or Jordyn must play goalie?

21. *Titanic* Survivors The following data represent the survival data for the ill-fated *Titanic* voyage by gender. The males are adult males and the females are adult females.

	Male	Female	Child	Total
Survived	338	316	57	**711**
Died	1352	109	52	**1513**
Total	**1690**	**425**	**109**	**2224**

Suppose a passenger is selected at random.

(a) What is the probability that the passenger survived?
(b) What is the probability that the passenger was female?
(c) What is the probability that the passenger was female or a child?
(d) What is the probability that the passenger was female and survived?
(e) What is the probability that the passenger was female or survived?
(f) If a female passenger is selected at random, what is the probability that she survived?
(g) If a child passenger is selected at random, what is the probability that the child survived?
(h) If a male passenger is selected at random, what is the probability that he survived?
(i) Do you think the adage "women and children first" was adhered to on the *Titanic*?
(j) Suppose two females are randomly selected. What is the probability both survived?

22. Marijuana Use According to the *Statistical Abstract of the United States*, about 17% of all 18- to 25-year-olds are current marijuana users.

(a) What is the probability that four randomly selected 18- to 25-year-olds are all marijuana users?

(b) What is the probability that among four randomly selected 18- to 25-year-olds at least one is a marijuana user?

23. SAT Reports Shawn is planning for college and takes the SAT test. While registering for the test, he is allowed to select three schools to which his scores will be sent at no cost. If there are 12 colleges he is considering, how many different ways could he fill out the score report form?

24. National Honor Society The distribution of National Honor Society members among the students at a local high school is shown in the table. A student's name is drawn at random.

Class	Total	National Honor Society
Senior	92	37
Junior	112	30
Sophomore	125	20
Freshman	120	0

(a) What is the probability that the student is a junior?
(b) What is the probability that the student is a senior, given that the student is in the National Honor Society?

25. Instant Winner In 2002, Valerie Wilson won $1 million in a scratch-off game (Cool Million) from the New York lottery. Four years later, she won $1 million in another scratch-off game ($3,000,000 Jubilee), becoming the first person in New York state lottery history to win $1 million or more in a scratch-off game twice. In the first game, she beat odds of 1 in 5.2 million to win. In the second, she beat odds of 1 in 705,600.

(a) What is the probability that an individual would win $1 million in both games if they bought one scratch-off ticket from each game?
(b) What is the probability that an individual would win $1 million twice in the $3,000,000 Jubilee scratch-off game?

26. Text Twist In the game Text Twist, six letters are given and the player must form words of varying lengths using the letters provided. Suppose that the letters in a particular game are ENHSIC.

(a) How many different arrangements are possible using all 6 letters?
(b) How many different arrangements are possible using only 4 letters?
(c) The solution to this game has three 6-letter words. To advance to the next round, the player needs at least one of the six-letter words. If the player simply guesses, what is the probability that he or she will get one of the six-letter words on their first guess of six letters?

27. Marriage and Education According to the U.S. Census Bureau, 20.2% of American women aged 25 years or older have a Bachelor's Degree; 16.5% of American women aged 25 years or older have never married; among American women aged 25 years or older who have never married, 22.8% have a Bachelor's Degree; and among American women aged 25 years or older who have a Bachelor's Degree, 18.6% have never married.

(a) Are the events "have a Bachelor's Degree" and "never married" independent? Explain.
(b) Suppose an American woman aged 25 years or older is randomly selected, what is the probability she has a Bachelor's Degree and has never married? Interpret this probability.

28. Weather Forecast The weather forecast says there is a 10% chance of rain on Thursday. Jim wakes up on Thursday and sees overcast skies. Since it has rained for the past three days, he believes that the chance of rain is more likely 60% or higher. What method of probability assignment did Jim use?

29. Essay Test An essay test in European History has 12 questions. Students are required to answer 8 of the 12 questions. How many different sets of questions could be answered?

30. Exercise Routines Todd is putting together an exercise routine and feels that the sequence of exercises can affect his overall performance. He has 12 exercises to select from, but only has enough time to do 9. How many different exercise routines could he put together?

31. New Cars If the 2015 Hyundai Genesis has 2 engine types, 2 vehicle styles, 3 option packages, 8 exterior color choices, and 2 interior color choices, how many different Genesis's are possible?

32. Lingo In the gameshow *Lingo*, the team that correctly guesses a mystery word gets a chance to pull two Lingo balls from a bin. Balls in the bin are labeled with numbers corresponding to the numbers remaining on their Lingo board. There are also three prize balls and three red "stopper" balls

in the bin. If a stopper ball is drawn first, the team loses their second draw. To form a Lingo, the team needs five numbers in a vertical, horizontal, or diagonal row. Consider the sample Lingo board below for a team that has just guessed a mystery word.

L	I	N	G	O
10			48	66
		34		74
		22	58	68
4	16		40	70
	26	52		64

(a) What is the probability that the first ball selected is on the Lingo board?

(b) What is the probability that the team draws a stopper ball on its first draw?

(c) What is the probability that the team makes a Lingo on their first draw?

(d) What is the probability that the team makes a Lingo on their second draw?

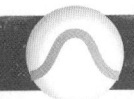

Chapter 5 Review

Summary

In this chapter, we took a break from the statistical process. The reason for this break is that probability forms the basis for statistical inference (which is the last step in the statistical process). Remember, inference takes information learned from a sample and generalizes it to a population along with a measure of reliability. Probability is used to measure the reliability in the results.

Probability is a measure of the likelihood of a random phenomenon or chance behavior. Because we are measuring a random phenomenon, there is short-term uncertainty. However, this short-term uncertainty gives rise to long-term predictability.

Probabilities are numbers between zero and one, inclusive. The closer a probability is to one, the more likely the event is to occur. If an event has probability zero, it is said to be impossible. Events with probability one are said to be certain.

We introduced three methods for computing probabilities: (1) the empirical method, (2) the classical method, and (3) subjective probabilities. Empirical probabilities rely on the relative frequency with which an event happens. Classical probabilities require the outcomes in the experiment to be equally likely. We count the number of ways an event can happen and divide this by the number of possible outcomes of the experiment. Empirical probabilities require that an experiment be performed, whereas classical probability does not. Subjective

probabilities are probabilities based on the opinion of the individual providing the probability. They are educated guesses about the likelihood of an event occurring, but still represent a legitimate way of assigning probabilities.

We are also interested in probabilities of multiple outcomes. For example, we might be interested in the probability that either event E or event F happens. The Addition Rule is used to compute the probability of E *or* F. There are two versions of the Addition Rule. There is the Addition Rule for Disjoint Events. Two events are disjoint (or mutually exclusive) if they do not have any outcomes in common. That is, mutually exclusive events cannot happen at the same time. There is also the General Addition Rule, which does not require disjoint events.

The Multiplication Rule is used to compute the probability that both E *and* F occur. There are also two versions of the Multiplication Rule. There is the Multiplication Rule for Independent Events. Two events E and F are independent if knowing that one of the events occurs does not affect the probability of the other. There is also the General Multiplication Rule, which does not require independence.

Essentially, when you see the word "or" think Addition Rule; when you see the word "and" think Multiplication Rule.

The complement of an event E, denoted E^C, is all the outcomes in the sample space that are not in E.

Finally, we introduced counting methods. The Multiplication Rule of Counting is used to count the number of ways a sequence of events can occur. Permutations are used to count the number of ways r items can be arranged from a set of n distinct items. We looked at two types of permutations—permutations without repetition (all items are distinct) and permutations with repetition (nondistinct items). Combinations are used to count the number of ways r items can be selected from a set of n distinct items without replacement and without regard to order. These counting techniques can be used to calculate probabilities using the classical method.

Vocabulary

Probability (p. 247)
Outcome (p. 247)
Law of Large Numbers (p. 248)
Experiment (p. 248)
Sample space (p. 248)
Event (p. 248)
Probability model (p. 249)
Impossible event (p. 249)
Certainty (p. 249)
Unusual event (p. 249)

Equally likely outcomes (p. 251)
Tree diagram (p. 254)
Subjective probability (p. 256)
Disjoint (p. 262)
Mutually exclusive (p. 262)
Venn diagram (p. 262)
Contingency table (p. 266)
Two-way table (p. 266)
Row variable (p. 266)
Column variable (p. 266)

Cell (p. 266)
Complement (p. 267)
Independent (p. 273)
Dependent (p. 273)
Conditional probability
   (p. 280)
Factorial symbol (p. 292)
Permutation (p. 292)
Factorial notation (p. 293)
Combination (p. 294)

Formulas

Empirical Probability

$$P(E) \approx \frac{\text{frequency of } E}{\text{number of trials of experiment}}$$

Classical Probability

$$P(E) = \frac{\text{number of ways that } E \text{ can occur}}{\text{number of possible outcomes}} = \frac{N(E)}{N(S)}$$

Addition Rule for Disjoint Events

$$P(E \text{ or } F) = P(E) + P(F)$$

General Addition Rule

$$P(E \text{ or } F) = P(E) + P(F) - P(E \text{ and } F)$$

Probabilities of Complements

$$P(E^c) = 1 - P(E)$$

Multiplication Rule for Independent Events

$$P(E \text{ and } F) = P(E) \cdot P(F)$$

Multiplication Rule for n Independent Events

$$P(E_1 \text{ and } E_2 \text{ and } E_3 \cdots \text{ and } E_n) = P(E_1) \cdot P(E_2) \cdot \cdots \cdot P(E_n)$$

Conditional Probability Rule

$$P(F \mid E) = \frac{P(E \text{ and } F)}{P(E)} = \frac{N(E \text{ and } F)}{N(E)}$$

General Multiplication Rule

$$P(E \text{ and } F) = P(E) \cdot P(F \mid E)$$

Factorial Notation

$$n! = n \cdot (n - 1) \cdot (n - 2) \cdots 3 \cdot 2 \cdot 1$$

Combination

$$_nC_r = \frac{n!}{r!(n - r)!}$$

Permutation

$$_nP_r = \frac{n!}{(n - r)!}$$

Permutations with Nondistinct Items

$$\frac{n!}{n_1! \cdot n_2! \cdots \cdots n_k!}$$

OBJECTIVES

Section	You should be able to . . .	Example(s)	Review Exercises
5.1	**1** Apply the rules of probabilities (p. 249)	2	1, 13(d), 15
	2 Compute and interpret probabilities using the empirical method (p. 250)	3, 4, 7(a)	14(a), 15, 16(a) and (b), 17(a), 30
	3 Compute and interpret probabilities using the classical method (p. 251)	5, 6, 7(b)	2–4, 13(a) and (d), 32(a) and (b)
	4 Use simulation to obtain data based on probabilities (p. 255)	8	27
	5 Recognize and interpret subjective probabilities (p. 256)		28

Section	You should be able to . . .	Example(s)	Review Exercises
5.2	**1** Use the Addition Rule for Disjoint Events (p. 262)	1 and 2	3, 4, 7, 13(b) and (c)
	2 Use the General Addition Rule (p. 265)	3 and 4	6, 16(d)
	3 Compute the probability of an event using the Complement Rule (p. 267)	5 and 6	5, 14(b), 17(b)
5.3	**1** Identify independent events (p. 273)	1	9, 16(g), 32(f)
	2 Use the Multiplication Rule for Independent Events (p. 274)	2 and 3	8, 14(c) and (d), 17(c) and (e), 18, 19
	3 Compute at-least probabilities (p. 276)	4	14(e), 17(d) and (f)
5.4	**1** Compute conditional probabilities (p. 279)	1 through 3	11, 16(f), 32(c) and (d)
	2 Compute probabilities using the General Multiplication Rule (p. 282)	4 through 6	10, 20, 29
5.5	**1** Solve counting problems using the Multiplication Rule (p. 289)	1 through 4	21
	2 Solve counting problems using permutations (p. 292)	5 and 6	12(e) and (f), 22
	3 Solve counting problems using combinations (p. 294)	7 through 9	12(c) and (d), 24
	4 Solve counting problems involving permutations with nondistinct items (p. 296)	10 and 11	23
	5 Compute probabilities involving permutations and combinations (p. 298)	12 and 13	25, 26
5.6	**1** Determine the appropriate probability rule to use (p. 302)	1 and 2	13–20, 25, 26, 29(c) and (d), 30
	2 Determine the appropriate counting technique to use (p. 303)	3 and 4	21–25

Review Exercises

1. **(a)** Which among the following numbers could be the probability of an event?

$$0, -0.01, 0.75, 0.41, 1.34$$

(b) Which among the following numbers could be the probability of an event?

$$\frac{2}{5}, \frac{1}{3}, -\frac{4}{7}, \frac{4}{3}, \frac{6}{7}$$

For Problems 2–5, let the sample space be
$S = \{\text{red, green, blue, orange, yellow}\}$. *Suppose that the outcomes are equally likely. Compute the probability of the event:*

2. $E = \{\text{yellow}\}$

3. $F = \{\text{green or orange}\}$

4. $E = \{\text{red or blue or yellow}\}$

5. Suppose that $E = \{\text{yellow}\}$. Compute the probability of E^c.

6. Suppose that $P(E) = 0.76$, $P(F) = 0.45$, and $P(E \text{ and } F) = 0.32$. What is $P(E \text{ or } F)$?

7. Suppose that $P(E) = 0.36$, $P(F) = 0.12$, and E and F are mutually exclusive. What is $P(E \text{ or } F)$?

8. Suppose that events E and F are independent. In addition, $P(E) = 0.45$ and $P(F) = 0.2$. What is $P(E \text{ and } F)$?

9. Suppose that $P(E) = 0.8$, $P(F) = 0.5$, and $P(E \text{ and } F) = 0.24$. Are events E and F independent? Why?

10. Suppose that $P(E) = 0.59$ and $P(F|E) = 0.45$. What is $P(E \text{ and } F)$?

11. Suppose that $P(E \text{ and } F) = 0.35$ and $P(F) = 0.7$. What is $P(E|F)$?

12. Determine the value of each of the following:

(a) 7! **(b)** 0!
(c) $_9C_4$ **(d)** $_{10}C_3$
(e) $_9P_2$ **(f)** $_{12}P_4$

13. **Roulette** In the game of roulette, a wheel consists of 38 slots, numbered 0, 00, 1, 2, . . . , 36. (See the photo in Problem 31 from Section 5.1.) To play the game, a metal ball is spun around the wheel and allowed to fall into one of the numbered slots. The slots numbered 0 and 00 are green, the odd numbers are red, and the even numbers are black.

(a) Determine the probability that the metal ball falls into a green slot. Interpret this probability.

(b) Determine the probability that the metal ball falls into a green or a red slot. Interpret this probability.

(c) Determine the probability that the metal ball falls into 00 or a red slot. Interpret this probability.

(d) Determine the probability that the metal ball falls into the number 31 and a black slot simultaneously. What term is used to describe this event?

14. **Traffic Fatalities** In 2013, there were 32,719 traffic fatalities in the United States. Of these, 10,076 were alcohol related.

(a) What is the probability that a randomly selected traffic fatality in 2013 was alcohol related?

(b) What is the probability that a randomly selected traffic fatality in 2013 was not alcohol related?

(c) What is the probability that two randomly selected traffic fatalities in 2013 were both alcohol related?

(d) What is the probability that neither of two randomly selected traffic fatalities in 2013 were alcohol related?

(e) What is the probability that of two randomly selected traffic fatalities in 2013 at least one was alcohol related?

15. Long Life? In a poll conducted by Genworth Financial, a random sample of adults was asked, "What age would you like to live to?" The results of the survey are given in the table.

Age	Number
18–79	126
80–89	262
90–99	263
100 or older	388

(a) Construct a probability model of the data.
(b) Is it unusual for an individual to want to live between 18 and 79 years?

16. Gestation Period versus Weight The following data represent the birth weights (in grams) of babies born in 2013, along with the period of gestation.

Birth Weight (grams)	Period of Gestation			Total
	Preterm	Term	Postterm	
Less than 1000	27,096	179	26	**27,301**
1000–1999	78,690	10,502	1063	**90,255**
2000–2999	221,778	661,729	38,665	**922,172**
3000–3999	119,161	2,317,287	160,354	**2,596,802**
4000–4999	7390	275,733	24,538	**307,661**
Over 5000	220	4019	398	**4637**
Total	**454,335**	**3,269,449**	**225,044**	**3,948,828**

Source: National Vital Statistics Report

(a) What is the probability that a randomly selected baby born in 2013 was postterm?
(b) What is the probability that a randomly selected baby born in 2013 weighed 3000–3999 grams?
(c) What is the probability that a randomly selected baby born in 2013 weighed 3000–3999 grams and was postterm?
(d) What is the probability that a randomly selected baby born in 2013 weighed 3000–3999 grams or was postterm?
(e) What is the probability that a randomly selected baby born in 2013 weighed less than 1000 grams and was postterm? Is this event impossible?
(f) What is the probability that a randomly selected baby born in 2013 weighed 3000–3999 grams, given the baby was postterm?
(g) Are the events "postterm baby" and "weighs 3000–3999 grams" independent? Why?

17. Who Do You Trust? According to the National Constitution Center, 18% of Americans trust organized religion.

(a) If an American is randomly selected, what is the probability he or she trusts organized religion?
(b) If an American is randomly selected, what is the probability he or she does not trust organized religion?
(c) In a random sample of three Americans, all three indicated that they trust organized religion. Is this result surprising?
(d) If three Americans are randomly selected, what is the probability that at least one does not trust organized religion?

(e) Would it be surprising if a random sample of five Americans resulted in none indicating they trust organized religion?
(f) If five Americans are randomly selected, what is the probability that at least one trusts organized religion?

18. Pick 3 For the Illinois Lottery's PICK 3 game, a player must match a sequence of three repeatable numbers, ranging from 0 to 9, in exact order (for example, 3–7–2). With a single ticket, what is the probability of matching the three winning numbers?

19. Pick 4 The Illinois Lottery's PICK 4 game is similar to PICK 3, except a player must match a sequence of four repeatable numbers, ranging from 0 to 9, in exact order (for example, 5–8–5–1). With a single ticket, what is the probability of matching the four winning numbers?

20. Drawing Cards Suppose that you draw 3 cards without replacement from a standard 52-card deck. What is the probability that all 3 cards are aces?

21. Forming License Plates A license plate is designed so that the first two characters are letters and the last four characters are digits (0 through 9). How many different license plates can be formed assuming that letters and numbers can be used more than once?

22. Choosing a Seat If four students enter a classroom that has 10 vacant seats, in how many ways can they be seated?

23. Arranging Flags How many different vertical arrangements are there of 10 flags if 4 are white, 3 are blue, 2 are green, and 1 is red?

24. Simple Random Sampling How many different simple random samples of size 8 can be obtained from a population whose size is 55?

25. Arizona's Pick 5 In one of Arizona's lotteries, balls are numbered 1–35. Five balls are selected randomly, without replacement. The order in which the balls are selected does not matter. To win, your numbers must match the five selected. Determine your probability of winning Arizona's Pick 5 with one ticket.

26. Packaging Error Because of a mistake in packaging, a case of 12 bottles of red wine contained five Merlot and seven Cabernet, each without labels. All the bottles look alike and have an equal probability of being chosen. Three bottles are randomly selected.

(a) What is the probability that all three are Merlot?
(b) What is the probability that exactly two are Merlot?
(c) What is the probability that none is a Merlot?

27. Simulation Use a graphing calculator or statistical software to simulate the playing of the game of roulette, using an integer distribution with numbers 1 through 38. Repeat the simulation 100 times. Let the number 37 represent 0 and the number 38 represent 00. Use the results of the simulation to answer the following questions.

(a) What is the probability that the ball lands in the slot marked 7?
(b) What is the probability that the ball lands either in the slot marked 0 or in the one marked 00?

28. Explain what is meant by a subjective probability. List some examples of subjective probabilities.

29. Playing Five-Card Stud In the game of five-card stud, one card is dealt face down to each player and the remaining four cards are dealt face up. After two cards are dealt (one down and one up), the players bet. Players continue to bet after each additional card is dealt. Suppose three cards have been dealt to each of five players at the table. You currently have three clubs in your hand, so you will attempt to get a flush (all cards in the same suit). Of the cards dealt, there are two clubs showing in other player's hands.

(a) How many clubs are in a standard 52-card deck?

(b) How many cards remain in the deck or are not known by you? Of this amount, how many are clubs?

(c) What is the probability that you get dealt a club on the next card?

(d) What is the probability that you get dealt two clubs in a row?

(e) Should you stay in the game?

30. Mark McGwire During the 1998 Major League Baseball season, Mark McGwire of the St. Louis Cardinals hit 70 home runs. Of the 70 home runs, 34 went to left field, 20 went to left center field, 13 went to center field, 3 went to right center field, and 0 went to right field. *Source:* Miklasz, B., et al. *Celebrating 70: Mark McGwire's Historic Season*, Sporting News Publishing Co., 1998, p. 179.

(a) What is the probability that a randomly selected home run was hit to left field? Interpret this probability.

(b) What is the probability that a randomly selected home run was hit to right field?

(c) Was it impossible for Mark McGwire to hit a homer to right field?

31. Lottery Luck In 1996, a New York couple won $2.5 million in the state lottery. Eleven years later, the couple won $5 million in the state lottery using the same set of numbers. The odds of winning the New York lottery twice are roughly 1 in

16 trillion, described by a lottery spokesperson as "galactically astronomical." Although it is highly unlikely that an individual will win the lottery twice, it is not "galactically astronomical" that *someone* will win a lottery twice. Explain why this is the case.

32. Coffee Sales The following data represent the number of cases of coffee or filters sold by four sales reps in a recent sales competition.

Salesperson	Gourmet	Single Cup	Filters	Total
Connor	142	325	30	**497**
Paige	42	125	40	**207**
Bryce	9	100	10	**119**
Mallory	71	75	40	**186**
Total	**264**	**625**	**120**	**1009**

(a) What is the probability that a randomly selected case was sold by Bryce? Is this unusual?

(b) What is the probability that a randomly selected case was Gourmet?

(c) What is the probability that a randomly selected Single-Cup case was sold by Mallory?

(d) What is the probability that a randomly selected Gourmet case was sold by Bryce? Is this unusual?

(e) What can be concluded from the results of parts (a) and (d)?

(f) Are the events "Mallory" and "Filters" independent? Explain.

(g) Are the events "Paige" and "Gourmet" mutually exclusive? Explain.

Chapter Test

1. Which among the following numbers could be the probability of an event? $0.23, 0, \frac{3}{2}, \frac{3}{4}, -1.32$

For Problems 2–4, let the sample space be $S = \{$Chris, Adam, Elaine, Brian, Jason$\}$. *Suppose that the outcomes are equally likely.*

2. Compute the probability of the event $E = \{$Jason$\}$.

3. Compute the probability of the event $E = \{$Chris or Elaine$\}$.

4. Suppose that $E = \{$Adam$\}$. Compute the probability of E^c.

5. Suppose that $P(E) = 0.37$ and $P(F) = 0.22$.

(a) Find $P(E \text{ or } F)$ if E and F are mutually exclusive.

(b) Find $P(E \text{ and } F)$ if E and F are independent.

6. Suppose that $P(E) = 0.15$, $P(F) = 0.45$, and $P(F|E) = 0.70$.

(a) What is $P(E \text{ and } F)$?

(b) What is $P(E \text{ or } F)$?

(c) What is $P(E|F)$?

(d) Are E and F independent?

7. Determine the value of each of the following:

(a) $8!$

(b) $_{12}C_6$

(c) $_{14}P_8$

8. Craps is a dice game in which two fair dice are cast. If the roller shoots a 7 or 11 on the first roll, he or she wins.

If the roller shoots a 2, 3, or 12 on the first roll, he or she loses.

(a) Compute the probability that the shooter wins on the first roll. Interpret this probability.

(b) Compute the probability that the shooter loses on the first roll. Interpret this probability.

9. According to Gallup, 26% of adult Americans believe their diet is very healthy.

(a) What is the probability that a randomly selected adult American believes his or her diet is very healthy? Interpret this probability.

(b) What is the probability that a randomly selected adult American does not believe his or her diet is very healthy?

10. The following probability model shows the distribution of the most-popular-selling Girl Scout Cookies®.

Cookie Type	Probability
Thin Mints	0.25
Samoas®/Caramel deLites™	0.19
Peanut Butter Patties®/Tagalongs™	0.13
Peanut Butter Sandwich/Do-si-dos™	0.11
Shortbread/Trefoils	0.09
Other varieties	0.23

Source: www.girlscouts.org

(a) Verify that this is a probability model.

(b) If a girl scout is selling cookies to people who randomly enter a shopping mall, what is the probability that the next box sold will be Peanut Butter Patties®/Tagalongs™ or Peanut Butter Sandwich/Do-si-dos™?

(c) If a girl scout is selling cookies to people who randomly enter a shopping mall, what is the probability that the next box sold will be Thin Mints, Samoas®/Caramel deLites™, or Shortbread/Trefoils?

(d) What is the probability that the next box sold will not be Thin Mints?

11. The following represent the results of a survey in which individuals were asked to disclose what they perceive to be the ideal number of children.

	0	1	2	3	4	5	6	Total
Female	5	2	87	61	28	3	2	188
Male	3	2	68	28	8	0	0	109
Total	8	4	155	89	36	3	2	297

Source: Sullivan Statistics Survey

(a) What is the probability an individual believes the ideal number of children is 2?

(b) What is the probability an individual is female and believes the ideal number of children is 2?

(c) What is the probability a randomly selected individual who took this survey is female or believes the ideal number of children is 2?

(d) Among the females, what is the probability the individual believes the ideal number of children is 2?

(e) If an individual who believes the ideal number of children is 4 is randomly selected, what is the probability the individual is male?

12. During the 2014 season, the Los Angeles Dodgers won 58% of their games. Assuming that the outcomes of the baseball games are independent and that the percentage of wins this season will be the same as in 2014, answer the following questions:

(a) What is the probability that the Dodgers will win two games in a row?

(b) What is the probability that the Dodgers will win seven games in a row?

(c) What is the probability that the Dodgers will lose at least one of their next seven games?

13. You just received a shipment of 10 DVD players. One DVD player is defective. You will accept the shipment if two randomly selected DVD players work. What is the probability that you will accept the shipment?

14. In the game of Jumble, the letters of a word are scrambled. The player must form the correct word. In a recent game in a local newspaper, the Jumble "word" was LINCEY. How many different arrangements are there of the letters in this "word"?

15. The U.S. Senate Appropriations Committee has 29 members and a subcommittee is to be formed by randomly selecting 5 of its members. How many different committees could be formed?

16. In Pennsylvania's Cash 5 lottery, balls are numbered 1 to 43. Five balls are selected randomly, without replacement. The order in which the balls are selected does not matter. To win, your numbers must match the five selected. Determine your probability of winning Pennsylvania's Cash 5 with one ticket.

17. A local area network requires eight characters for a password. The first character must be a letter, but the remaining seven characters can be either a letter or a digit (0 through 9). Lower- and uppercase letters are considered the same. How many passwords are possible for the local area network?

18. A survey distributed at the 28th Lunar and Planetary Science Conference in March 1997 asked respondents to estimate the chance that there was life on Mars. The median response was a 57% chance of life on Mars. Which method of finding probabilities was used to obtain this result? Explain why.

19. How many distinguishable DNA sequences can be formed using two As, four Cs, four Gs, and five Ts?

20. A student is taking a 40-question multiple-choice test. Each question has five possible answers. Since the student did not study for the test, he guesses on all the questions. Letting 0 or 1 indicate a correct answer, use the following line from a table of random digits to simulate the probability that the student will guess a question correctly.

73634 79304 78871 25956 59109 30573 18513 61760

Making an Informed Decision

The Effects of Drinking and Driving

Everyone knows that alcohol impairs reaction time. For example, one study indicates that reaction time at a blood alcohol concentration of 0.08% doubles from 1.5 to 3 seconds. Your job is to compile some probabilities based on existing data to help convince people that it is a really bad idea to get behind the wheel of a car after consuming alcohol. It is also a bad idea to get into a car with an individual who has had a few drinks.

Go to the Fatality Analysis Reporting System Encyclopedia, which is online at the National Highway

Traffic Safety Administration website (under Data). This encyclopedia contains records of all fatal car crashes for any given year. Click Run a Query Using the FARS Web-Based Encyclopedia. Click Query FARS data. You might consider reading the documentation provided to get a sense as to the type of data that can be obtained. Select a year you wish to analyze and click Submit.

You should see a table with variables that can be considered as it relates to fatal vehicle crashes. For example, if you want to determine the probability that a crash resulted from driver alcohol involvement, you could check the Driver Alcohol Involvement box. Then click Submit. In the drop-down menu, you can fine-tune the data obtained. Click Univariate Tabulation. The next screen allows for further refinement. Click Submit, and you will have the data for the current year. Compile a few probabilities for univariate analysis.

Next, consider relations among two variables. For example, is there a relation between gender and driver alcohol involvement? Check both variables and click Submit. Again, you can fine-tune the data obtained. Once this is complete, select Cross Tabulation. Now choose the appropriate column and row variables to create a contingency table and click Submit. Develop a few conditional probabilities to see the relation between your two variables. Are they independent? If not, what does the dependency tell us?

Write a report, complete with the probabilities showing variable dependencies, that can be used to convince people of the dangers of mixing alcohol consumption with driving.

 # CASE STUDY The Case of the Body in the Bag

While fishing on a river one spring morning, Robert Donkin slipped through the underbrush that lined the river's banks. Nearing the shore, he saw a large canvas bag floating in the water, held by foliage that leaned over the river. Upon closer inspection, Mr. Donkin observed what he believed to be hair floating through the bag's opening. Marking the spot of his discovery, the fisherman fetched the authorities.

Preliminary investigation at the scene revealed a body in the bag. Unfortunately, it was impossible to identify the corpse's sex, age, or race immediately. Forensics was assigned to identify the victim and estimate the cause and time of death. While waiting for the forensics analysis, you, the detective in charge, have gathered the information shown in the tables concerning victim–offender relationships from recent reports from the FBI.

Using the information contained in the tables, you develop a preliminary profile of the victim and offender by determining the likelihood that the:

1. Offender is at least 18.

2. Offender is white.

3. Offender is male.

4. Victim is a white female.

5. Victim is either white or female.

6. Victim and offender are from the same age category.

7. Victim and offender are from different age categories.

8. Victim and offender are of the same race.

9. Victim and offender are of different races.

10. Victim and offender are of the same sex.

11. Victim and offender are of different sexes.

12. Without knowing the contents of the forensic team's report, what is your best prediction of the age, race, and sex of the victim? Explain your reasoning.

13. What is your best prediction as to the age, race, and sex of the offender? Explain your reasoning.

Soon after you finished this analysis, the preliminary forensics report was delivered. Although no identification had been made, the autopsy suggested that the cause of death was blunt-force trauma and the body had been in the water at least 2 weeks. The victim was a white female with blonde hair, estimated as being in her mid-thirties, showed no signs of having had children, and was wearing no jewelry.

	Age of Offender		
Age of Victim	**Less Than 18**	**At Least 18**	**Unknown**
Less than 18	124	619	39
At least 18	290	5628	480
Unknown	4	80	21

	Sex of Offender		
Sex of Victim	**Male**	**Female**	**Unknown**
Male	4710	490	89
Female	1735	150	24
Unknown	50	11	26

	Race of Offender			
Race of Victim	White	Black	Other	Unknown
White	3026	573	53	57
Black	208	3034	12	49
Other	54	28	97	7
Unknown	30	27	4	26

	Race of Offender			
Sex of Victim	White	Black	Other	Unknown
Male	2162	2935	103	89
Female	1126	700	59	24
Unknown	30	27	4	26

	Sex of Offender		
Race of Victim	Male	Female	Unknown
White	3322	330	57
Black	2964	290	49
Other	159	20	7
Unknown	50	11	26

	Sex		
Race of Victim	Male	Female	Unknown
White	5067	1883	6
Black	6294	1126	1
Other	301	105	0
Unknown	131	42	34

Source: Uniform Crime Reports: Crime in the United States

Based on this new information, you develop a new offender profile by determining the likelihood that the:

14. Offender is at least 18.

15. Offender is white.

16. Offender is male.

17. Victim and offender are from the same age category.

18. Victim and offender are from different age categories.

19. Victim and offender are of the same race.

20. Victim and offender are of different races.

21. Victim and offender are of the same sex.

22. Victim and offender are of different sexes.

23. What is your best prediction of the age, race, and sex of the offender? Explain your reasoning.

24. Did your answers to the offender questions change once you knew the age, race, and sex of the victim? Explain.

Suppose that 45% of murder victims were known to be related to or acquainted with the offender, that 15% were murdered by an unrelated stranger, and that for 40% of victims relationship to their killer is unknown. Based on all the information available, complete your offender profile for this case.

6

Discrete Probability Distributions

Making an Informed Decision

A woman who was shopping in Los Angeles had her purse stolen by a young, blonde female who was wearing a ponytail. Because there were no eyewitnesses and no real evidence, the prosecution used probability to make its case against the defendant. Your job is to play the role of both the prosecution and defense attorney to make probabilistic arguments both for and against the defendant. See the Decisions project on page 353.

PUTTING IT TOGETHER

Recall, the probability of an event is the long-term proportion with which the event is observed. That is, if we conduct an experiment 1000 times and observe an outcome 300 times, we estimate that the probability of the outcome is $300/1000 = 0.3$. The more times we conduct the experiment, the more accurate this empirical probability will be. This is the Law of Large Numbers. We can use counting techniques to obtain theoretical probabilities if the outcomes in the experiment are equally likely. This is called classical probability.

A probability model lists the possible outcomes of a probability experiment and each outcome's probability. A probability model must satisfy the rules of probability. In particular, all probabilities must be between 0 and 1, inclusive, and the sum of the probabilities must equal 1.

In this chapter, we introduce probability models for *random variables*. A random variable is a numerical measure of the outcome to a probability experiment. So, rather than listing specific outcomes of a probability experiment, such as heads or tails, we might list the number of heads obtained in, say, three flips of a coin. In Section 6.1, we discuss random variables and describe the distribution of discrete random variables (shape, center, and spread). We then discuss two specific discrete probability distributions: *the binomial probability distribution* (Section 6.2) and *the Poisson probability distribution* (Section 6.3).

6.1 Discrete Random Variables

Preparing for This Section Before getting started, review the following:

- Discrete versus continuous variables (Section 1.1, pp. 7–8)
- Relative frequency histograms for discrete data (Section 2.2, pp. 77–78)
- Mean (Section 3.1, pp. 118–120)

- Standard deviation (Section 3.2, pp. 133–137)
- Mean from grouped data (Section 3.3, pp. 148–149)
- Standard deviation from grouped data (Section 3.3, pp. 150–152)

Objectives

① Distinguish between discrete and continuous random variables

② Identify discrete probability distributions

③ Graph discrete probability distributions

④ Compute and interpret the mean of a discrete random variable

⑤ Interpret the mean of a discrete random variable as an expected value

⑥ Compute the standard deviation of a discrete random variable

① **Distinguish between Discrete and Continuous Random Variables**

Consider a probability experiment in which we flip a coin two times. The outcomes of the experiment are {HH, HT, TH, TT}. Rather than being interested in a particular outcome, we might be interested in the number of heads. If the outcome of a probability experiment is a numerical result, we say the outcome is a *random variable*.

Definition   A **random variable** is a numerical measure of the outcome of a probability experiment, so its value is determined by chance. Random variables are typically denoted using capital letters such as X.

So, in our coin-flipping example, if the random variable X represents the number of heads in two flips of a coin, the possible values of X are $x = 0, 1,$ or 2. Notice that we follow the practice of using a capital letter, such as X, to identify the random variable and a lowercase letter, x, to list the possible values of the random variable, or the sample space of the experiment.

As another example, consider an experiment that measures the time between arrivals of cars at a drive-through. The random variable T describes the time between arrivals, so the sample space of the experiment is $t > 0$.

There are two types of random variables, *discrete* and *continuous*.

Definitions

In Other Words
Discrete random variables typically result from counting (0, 1, 2, 3, and so on). Continuous random variables result from measurement.

A **discrete random variable** has either a finite or countable number of values. The values of a discrete random variable can be plotted on a number line with space between each point. See Figure 1(a).

A **continuous random variable** has infinitely many values. The values of a continuous random variable can be plotted on a line in an uninterrupted fashion. See Figure 1(b).

Figure 1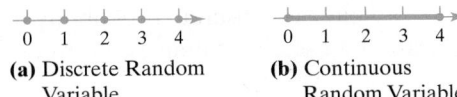

(a) Discrete Random Variable

(b) Continuous Random Variable

EXAMPLE 1   Distinguishing between Discrete and Continuous Random Variables

> **(a)** The number of As earned in a section of statistics with 15 students enrolled is a discrete random variable because its value results from counting. If the random variable X represents the number of As, then the possible values of X are $x = 0, 1, 2, \ldots, 15$.
>
> **(b)** The number of cars that travel through a McDonald's drive-through in the next hour is a discrete random variable because its value results from counting. If the random variable X represents the number of cars, the possible values of X are $x = 0, 1, 2, \ldots$.
>
> **(c)** The speed of the next car that passes a state trooper is a continuous random variable because speed is measured. If the random variable S represents the speed, the possible values of S are all positive real numbers; that is, $s > 0$. •

> **CAUTION!**
> Even though a radar gun may report the speed of a car as 37 miles per hour, it is actually any number greater than or equal to 36.5 mph and less than 37.5 mph. That is, $36.5 \le s < 37.5$.

• Now Work Problem 5

In this chapter, we will concentrate on probabilities of discrete random variables. Chapter 7 discusses how to obtain probabilities for certain continuous random variables.

② Identify Discrete Probability Distributions

Because the value of a random variable is determined by chance, we may assign probabilities to the possible values of the random variable.

Definition
> The **probability distribution** of a discrete random variable X provides the possible values of the random variable and their corresponding probabilities. A probability distribution can be in the form of a table, graph, or mathematical formula.

EXAMPLE 2   A Discrete Probability Distribution

Table 1	
x	$P(x)$
0	0.01
1	0.10
2	0.38
3	0.51

Suppose we ask a basketball player to shoot three free throws. Let the random variable X represent the number of shots made, so $x = 0, 1, 2,$ or 3. Table 1 shows a probability distribution for the random variable X.

We denote probabilities using the notation $P(x)$, where x is a specific value of the random variable. We read $P(x)$ as "the probability that the random variable X equals x." For example, $P(3) = 0.51$ is read "the probability that the random variable X equals 3 is 0.51." •

Recall from Section 5.1 that probabilities must obey certain rules. Below are the rules for a discrete probability distribution using the notation just introduced.

In Other Words
The first rule states that the sum of the probabilities must equal 1. The second rule states that each probability must be between 0 and 1, inclusive.

> ### Rules for a Discrete Probability Distribution
> Let $P(x)$ denote the probability that the random variable X equals x; then
> **1.** $\sum P(x) = 1$
> **2.** $0 \le P(x) \le 1$

Table 1 from Example 2 is a discrete probability distribution because the sum of the probabilities equals 1 and each probability is between 0 and 1, inclusive.

EXAMPLE 3 Identifying Discrete Probability Distributions

Problem Which of the following is a discrete probability distribution?

(a)

x	$P(x)$
1	0.20
2	0.35
3	0.12
4	0.40
5	−0.07

(b)

x	$P(x)$
1	0.20
2	0.25
3	0.10
4	0.14
5	0.49

(c)

x	$P(x)$
1	0.20
2	0.25
3	0.10
4	0.14
5	0.31

Approach In a discrete probability distribution, the sum of the probabilities must equal 1, and all probabilities must be between 0 and 1, inclusive.

Solution

(a) This is not a discrete probability distribution because $P(5) = -0.07$, which is less than 0.

(b) This is not a discrete probability distribution because

$$\sum P(x) = 0.20 + 0.25 + 0.10 + 0.14 + 0.49 = 1.18 \neq 1$$

(c) This is a discrete probability distribution because the sum of the probabilities equals 1, and each probability is between 0 and 1, inclusive.   ●

● **Now Work Problem 9**

Table 1 is an example of a discrete probability distribution in table form. We discuss discrete probability distributions using graphs next and using mathematical formulas in Sections 6.2 and 6.3.

③ Graph Discrete Probability Distributions

In the graph of a discrete probability distribution, the horizontal axis represents the values of the discrete random variable and the vertical axis represents the corresponding probability of the discrete random variable. When graphing a discrete probability distribution, we want to emphasize that the data are discrete. Therefore, the graph of discrete probability distributions is drawn using vertical lines above each value of the random variable to a height that is the probability of the random variable.

EXAMPLE 4 Graphing a Discrete Probability Distribution

Problem Graph the discrete probability distribution given in Table 1 from Example 2.

Approach In the graph of a discrete probability distribution, the horizontal axis represents the values of the discrete random variable and the vertical axis represents the corresponding probability of the discrete random variable. Draw the graph using vertical lines above each value of the random variable to a height that is the probability of the random variable.

Solution Figure 2 shows the graph of the distribution in Table 1.

Figure 2

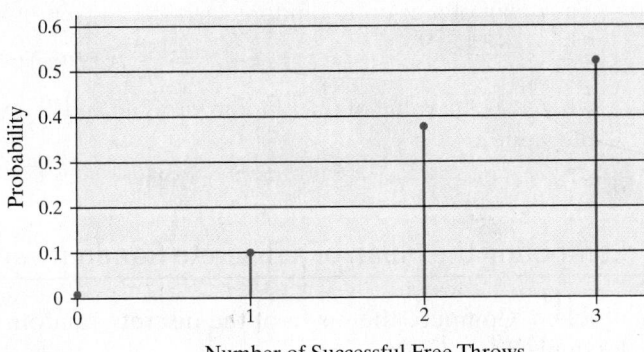

Graphs of discrete probability distributions help determine the shape of the distribution. Recall that we describe distributions as skewed left, skewed right, or symmetric. The graph in Figure 2 is skewed left.

● **Now Work Problems 17(a) and (b)**

④ **Compute and Interpret the Mean of a Discrete Random Variable**

Remember, when describing the distribution of a variable, we describe its center, spread, and shape. We will use the mean to describe the center and use the standard deviation to describe the spread.

Let's see where the formula for computing the mean of a discrete random variable comes from. One semester I asked a small statistics class of 10 students to disclose the number of people living in their households. I obtained the following data:

$$2, 4, 6, 6, 4, 4, 2, 3, 5, 5$$

What is the mean number of people in the 10 households? We could find the mean by adding the observations and dividing by 10, but we will take a different approach. Letting the random variable X represent the number of people in the household, we obtain the probability distribution in Table 2.

Now compute the mean as follows:

$$\mu = \frac{\sum x_i}{N} = \frac{2 + 4 + 6 + 6 + 4 + 4 + 2 + 3 + 5 + 5}{10}$$

$$= \frac{\overbrace{2 + 2}^{2} + \overbrace{3}^{1} + \overbrace{4 + 4 + 4}^{3} + \overbrace{5 + 5}^{2} + \overbrace{6 + 6}^{2}}{10}$$

$$= \frac{2 \cdot 2 + 3 \cdot 1 + 4 \cdot 3 + 5 \cdot 2 + 6 \cdot 2}{10}$$

$$= 2 \cdot \frac{2}{10} + 3 \cdot \frac{1}{10} + 4 \cdot \frac{3}{10} + 5 \cdot \frac{2}{10} + 6 \cdot \frac{2}{10}$$

$$= 2 \cdot P(2) + 3 \cdot P(3) + 4 \cdot P(4) + 5 \cdot P(5) + 6 \cdot P(6)$$

$$= 2(0.2) + 3(0.1) + 4(0.3) + 5(0.2) + 6(0.2)$$

$$= 4.1$$

We conclude that the mean of a discrete random variable is found by multiplying each possible value of the random variable by its corresponding probability and then adding these products.

Table 2

x	$P(x)$
2	$\frac{2}{10} = 0.2$
3	$\frac{1}{10} = 0.1$
4	$\frac{3}{10} = 0.3$
5	$\frac{2}{10} = 0.2$
6	$\frac{2}{10} = 0.2$

The Mean of a Discrete Random Variable

In Other Words
To find the mean of a discrete random variable, multiply each value of the random variable by its probability. Then add these products.

The mean of a discrete random variable is given by the formula

$$\mu_X = \sum [x \cdot P(x)] \qquad (1)$$

where x is the value of the random variable and $P(x)$ is the probability of observing the value x.

EXAMPLE 5   Computing the Mean of a Discrete Random Variable

Problem Compute the mean of the discrete random variable given in Table 1 from Example 2.

Approach Find the mean of a discrete random variable by multiplying each value of the random variable by its probability and adding these products.

Table 3		
x	$P(x)$	$x \cdot P(x)$
0	0.01	$0 \cdot 0.01 = 0$
1	0.10	$1 \cdot 0.1 = 0.1$
2	0.38	$2 \cdot 0.38 = 0.76$
3	0.51	$3 \cdot 0.51 = 1.53$

Solution Refer to Table 3. The first two columns represent the discrete probability distribution. The third column represents $x \cdot P(x)$.

Substitute into Formula (1) to find the mean number of free throws made.

$$\mu_X = \sum [x \cdot P(x)] = 0(0.01) + 1(0.10) + 2(0.38) + 3(0.51) = 2.39 \approx 2.4 \quad \bullet$$

We will follow the practice of rounding the mean and standard deviation to one more decimal place than the values of the random variable.

How to Interpret the Mean of a Discrete Random Variable

The mean of a discrete random variable can be thought of as the mean outcome of the probability experiment if we repeated the experiment many times. If we repeated the experiment in Example 5 of shooting three free throws many times, we would expect the mean number of free throws made to be around 2.4.

Interpretation of the Mean of a Discrete Random Variable

In Other Words
Recall from Chapter 5 that independent means the results of one trial of the experiment do not affect the results of other trials of the experiment.

Suppose an experiment is repeated n independent times and the value of the random variable X is recorded. As the number of repetitions of the experiment increases, the mean value of the n trials will approach μ_X, the mean of the random variable X. In other words, let x_1 be the value of the random variable X after the first experiment, x_2 be the value of the random variable X after the second experiment, and so on. Then

$$\bar{x} = \frac{x_1 + x_2 + \cdots + x_n}{n}$$

The difference between $\bar{x}$ and μ_X gets closer to 0 as n increases.

EXAMPLE 6   Interpretation of the Mean of a Discrete Random Variable

Problem The basketball player from Example 2 is asked to shoot three free throws 100 times. Compute the mean number of free throws made.

Approach The player shoots three free throws and the number made is recorded. We repeat this experiment 99 more times and then compute the mean number of free throws made.

Solution Table 4 shows the results.

First experiment → 3	2	3	3	3	3	1	2	3	2
Second experiment → 2	3	3	1	2	2	2	2	2	3
Third experiment → 3	3	2	2	3	2	3	2	2	2
3	3	2	3	2	3	3	2	3	1
3	2	2	2	2	0	2	3	1	2
3	3	2	3	2	3	2	1	3	2
2	3	3	3	1	3	3	1	3	3
3	2	2	1	3	2	2	2	3	2
3	2	2	2	3	3	2	2	3	3
2	3	2	1	2	3	3	2	3	3 ← Hundredth experiment

Table 4

In the first experiment, the player made all three free throws. In the second experiment, the player made two out of three free throws. In the hundredth experiment, the player made three free throws. The mean number of free throws made was

$$\bar{x} = \frac{3 + 2 + 3 + \cdots + 3}{100} = 2.35$$

This is close to the theoretical mean of 2.4 (from Example 5). As the number of repetitions of the experiment increases, we expect $\bar{x}$ to get even closer to 2.4. ●

Figures 3(a) and (b) further illustrate the mean of a discrete random variable. Figure 3(a) shows the mean number of free throws made versus the number of repetitions of the experiment for the data in Table 4. Figure 3(b) shows the same information when the experiment of shooting three free throws is conducted a second time for 100 repetitions. In both plots the player starts "hot," since the mean number of free throws made is above the theoretical level of 2.4. However, both graphs approach the theoretical mean of 2.4 as the number of repetitions of the experiment increases.

Figure 3

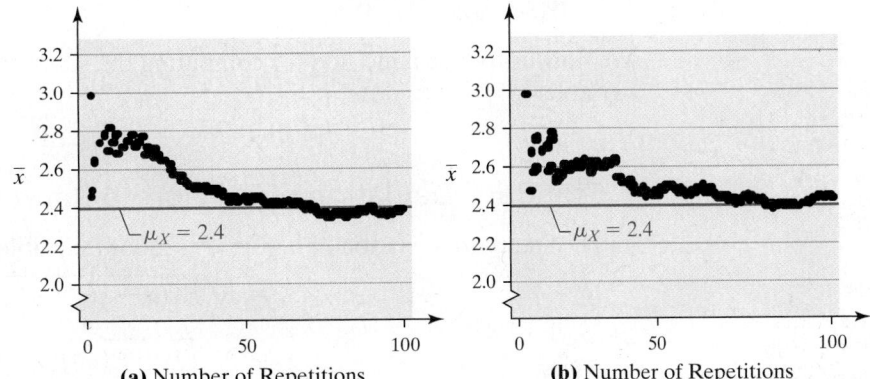

● **Now Work Problem 17(c)**

(a) Number of Repetitions

(b) Number of Repetitions

⑤ **Interpret the Mean of a Discrete Random Variable as an Expected Value**

In Other Words
The expected value of a discrete random variable is the mean of the discrete random variable.

Because the mean of a random variable represents what we would expect to happen in the long run, it is also called the **expected value**, $E(X)$. The interpretation of expected value is the same as the interpretation of the mean of a discrete random variable.

EXAMPLE 7   Finding the Expected Value

Historical Note

Christiaan Huygens
was born on April
14, 1629, into an
influential Dutch
family. He studied
Law and Mathematics
at the University of Leiden from
1645 to 1647. From 1647 to 1649,
he continued to study Law and
Mathematics at the College of
Orange at Breda. Among his
many accomplishments, Huygens
discovered the first moon of
Saturn in 1655 and the shape of
the rings of Saturn in 1656. While
in Paris sharing his discoveries,
he learned about probability
through Fermat and Pascal. In
1657, Huygens published the
first book on probability theory,
in which he introduced the idea
of expected value.

Problem A term life insurance policy will pay a beneficiary a certain sum of money upon the death of the policyholder. These policies have premiums that must be paid annually. Suppose an 18-year-old male buys a $250,000 1-year term life insurance policy for $350. According to the *National Vital Statistics Report*, Vol. 58, No. 21, the probability that the male will survive the year is 0.998937. Compute the expected value of this policy to the insurance company.

Approach The experiment has two possible outcomes: survival or death. Let the random variable X represent the *payout* (money lost or gained), depending on survival or death of the insured. Assign probabilities to each payout and substitute these values into Formula (1).

Solution

Step 1 Because $P(\text{survives}) = 0.998937$, $P(\text{dies}) = 0.001063$. If the client survives the year, the insurance company makes $350, or $x = \$350$. If the client dies during the year, the insurance company must pay $250,000 to the client's beneficiary, but still keeps the $350 premium, so $x = \$350 - \$250{,}000 = -\$249{,}650$. The value is negative because it is money paid by the insurance company. The probability distribution is listed in Table 5.

Step 2 The expected value of the policy (from the point of view of the insurance company) is

$$E(X) = \mu_X = \sum[x \cdot P(x)] = \$350(0.998937) + (-\$249{,}650)(0.001063) = \$84.25$$

Interpretation The company expects to make $84.25 for each 18-year-old male client it insures. The $84.25 profit of the insurance company is a long-term result. It does not make $84.25 on each 18-year-old male it insures, but rather the average profit per 18-year-old male insured is $84.25. Because this is a long-term result, the insurance "idea" will not work with only a few insured.

Table 5

x	$P(x)$
$350 (survives)	0.998937
−$249,650 (dies)	0.001063

• **Now Work Problem 25**

6 Compute the Standard Deviation of a Discrete Random Variable

We now introduce a method for computing the standard deviation of a discrete random variable.

In Other Words
The standard deviation of a discrete random variable is the square root of a weighted average of the squared deviations for which the weights are the probabilities.

Standard Deviation of a Discrete Random Variable

The standard deviation of a discrete random variable X is given by

$$\sigma_X = \sqrt{\sum[(x - \mu_X)^2 \cdot P(x)]} \qquad \textbf{(2a)}$$

$$= \sqrt{\sum[x^2 \cdot P(x)] - \mu_X^2} \qquad \textbf{(2b)}$$

where x is the value of the random variable, μ_X is the mean of the random variable, and $P(x)$ is the probability of observing a value of the random variable.

EXAMPLE 8   Computing the Standard Deviation of a Discrete Random Variable

Problem Find the standard deviation of the discrete random variable given in Table 1 from Example 2.

Approach We will use Formula (2a) with the unrounded mean $\mu_X = 2.39$.

Solution Refer to Table 6. Columns 1 and 2 represent the discrete probability distribution. Column 3 represents $(x - \mu_X)^2 \cdot P(x)$. Find the sum of the entries in Column 3.

Table 6		
x	$P(x)$	$(x - \mu_X)^2 \cdot P(x)$
0	0.01	$(0 - 2.39)^2 \cdot 0.01 = 0.057121$
1	0.10	$(1 - 2.39)^2 \cdot 0.10 = 0.19321$
2	0.38	$(2 - 2.39)^2 \cdot 0.38 = 0.057798$
3	0.51	$(3 - 2.39)^2 \cdot 0.51 = 0.189771$
		$\sum [(x - \mu_X)^2 \cdot P(x)] = 0.4979$

The standard deviation of the discrete random variable X is

$$\sigma_X = \sqrt{\sum [(x - \mu_X)^2 \cdot P(x)]} = \sqrt{0.4979} \approx 0.7$$

Approach We will use Formula (2b) with the unrounded mean $\mu_X = 2.39$.

Solution Refer to Table 7. Columns 1 and 2 represent the discrete probability distribution. Column 3 represents $x^2 \cdot P(x)$. Find the sum of the entries in Column 3.

Table 7		
x	$P(x)$	$x^2 \cdot P(x)$
0	0.01	$0^2 \cdot 0.01 = 0$
1	0.10	$1^2 \cdot 0.10 = 0.10$
2	0.38	$2^2 \cdot 0.38 = 1.52$
3	0.51	$3^2 \cdot 0.51 = 4.59$
		$\sum [x^2 \cdot P(x)] = 6.21$

The standard deviation of the discrete random variable X is

$$\sigma_X = \sqrt{\sum [x^2 \cdot P(x)] - \mu_X^2} = \sqrt{6.21 - 2.39^2}$$
$$= \sqrt{0.4979} \approx 0.7$$

● **Now Work Problem 17(d)**

The variance of the discrete random variable is the value under the square root in the computation of the standard deviation. The variance of the discrete random variable in Example 8 is

$$\sigma_X^2 = 0.4979 \approx 0.5$$

EXAMPLE 9 Obtaining the Mean and Standard Deviation of a Discrete Random Variable Using Technology

Problem Use statistical software or a graphing calculator to find the mean and the standard deviation of the random variable whose distribution is given in Table 1.

Figure 4

μ_X ⎯
σ_X ⎯

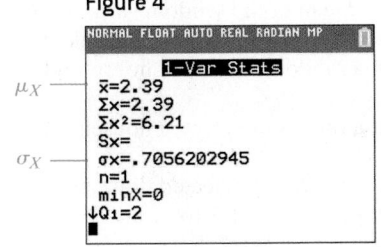

Approach We will use a TI-84 Plus C graphing calculator to obtain the mean and standard deviation. The steps for determining the mean and standard deviation using a TI-83/84 Plus graphing calculator and StatCrunch are given in the Technology Step-by-Step below.

Solution Figure 4 shows the results from a TI-84 Plus C graphing calculator.

Note: The TI-84 does not find s_X when the sum of L_2 is 1.

Technology Step-by-Step
Finding the Mean and Standard Deviation of a Discrete Random Variable Using Technology

TI-83/84 Plus
1. Enter the values of the random variable in L1 and their corresponding probabilities in L2.
2. Press STAT, highlight CALC, select
   1: 1-VAR Stats, and press ENTER.
3. Select L1 for List : ; Select L2 for FreqList:. Highlight Calculate and press ENTER.

StatCrunch
1. Enter the values of the random variable in column var1 and the corresponding probabilities in column var2. Name var1 x and var2 $P(x)$.
2. Select **Stat**, **Calculators**, **Custom**. Select x for "Values in:"; Select $P(x)$ for "Weight in:," Click Compute!.

6.1 Assess Your Understanding

Vocabulary and Skill Building

1. What is a random variable?

2. What is the difference between a discrete random variable and a continuous random variable? Provide your own examples of each.

3. What are the two requirements for a discrete probability distribution?

4. In your own words, provide an interpretation of the mean (or expected value) of a discrete random variable.

In Problems 5–8, determine whether the random variable is discrete or continuous. In each case, state the possible values of the random variable.

NW **5.** **(a)** The number of light bulbs that burn out in the next week in a room with 20 bulbs.
(b) The time it takes to fly from New York City to Los Angeles.
(c) The number of hits to a website in a day.
(d) The amount of snow in Toronto during the winter.

6. **(a)** The time it takes for a light bulb to burn out.
(b) The weight of a T-bone steak.
(c) The number of free-throw attempts before the first shot is made.
(d) In a random sample of 20 people, the number with type A blood.

7. **(a)** The amount of rain in Seattle during April.
(b) The number of fish caught during a fishing tournament.
(c) The number of customers arriving at a bank between noon and 1:00 P.M.
(d) The time required to download a file from the Internet.

8. **(a)** The number of defects in a roll of carpet.
(b) The distance a baseball travels in the air after being hit.
(c) The number of points scored during a basketball game.
(d) The square footage of a house.

In Problems 9–14, determine whether the distribution is a discrete probability distribution. If not, state why.

NW **9.**

x	P(x)
0	0.2
1	0.2
2	0.2
3	0.2
4	0.2

10.

x	P(x)
0	0.1
1	0.5
2	0.05
3	0.25
4	0.1

11.

x	P(x)
10	0.1
20	0.23
30	0.22
40	0.6
50	-0.15

12.

x	P(x)
1	0
2	0
3	0
4	0
5	1

13.

x	P(x)
100	0.1
200	0.25
300	0.2
400	0.3
500	0.1

14.

x	P(x)
100	0.25
200	0.25
300	0.25
400	0.25
500	0.25

In Problems 15 and 16, determine the required value of the missing probability to make the distribution a discrete probability distribution.

15.

x	P(x)
3	0.4
4	?
5	0.1
6	0.2

16.

x	P(x)
0	0.30
1	0.15
2	?
3	0.20
4	0.15
5	0.05

Applying the Concepts

NW **17.** **Televisions** In the Sullivan Statistics Survey I, individuals were asked to disclose the number of televisions in their household. In the following probability distribution, the random variable X represents the number of televisions in households.

Number of Televisions, x	P(x)
0	0
1	0.161
2	0.261
3	0.176
4	0.186
5	0.116
6	0.055
7	0.025
8	0.010
9	0.010

Source: Sullivan Statistics Survey I

(a) Verify this is a discrete probability distribution.
(b) Draw a graph of the probability distribution. Describe the shape of the distribution.
(c) Determine and interpret the mean of the random variable X.
(d) Determine the standard deviation of the random variable X.
(e) What is the probability that a randomly selected household has three televisions?
(f) What is the probability that a randomly selected household has three or four televisions?
(g) What is the probability that a randomly selected household has no televisions? Would you consider this to be an impossible event?

18. **Marriage** In the following probability distribution, the random variable X represents the number of marriages an individual aged 15 years or older has been involved in.

x	P(x)
0	0.272
1	0.575
2	0.121
3	0.027
4	0.004
5	0.001

Source: Based on data from the U.S. Census Bureau

(a) Verify that this is a discrete probability distribution.
(b) Draw a graph of the probability distribution. Describe the shape of the distribution.
(c) Compute and interpret the mean of the random variable X.
(d) Compute the standard deviation of the random variable X.
(e) What is the probability that a randomly selected individual 15 years or older was involved in two marriages?
(f) What is the probability that a randomly selected individual 15 years or older was involved in at least two marriages?

19. Ichiro's Hit Parade In the 2004 baseball season, Ichiro Suzuki of the Seattle Mariners set the record for the most hits in a season with a total of 262 hits. In the following probability distribution, the random variable X represents the number of hits Ichiro obtained in a game.

x	$P(x)$
0	0.1677
1	0.3354
2	0.2857
3	0.1491
4	0.0373
5	0.0248

Source: Chicago Tribune

(a) Verify that this is a discrete probability distribution.
(b) Draw a graph of the probability distribution. Describe the shape of the distribution.
(c) Compute and interpret the mean of the random variable X.
(d) Compute the standard deviation of the random variable X.
(e) What is the probability that in a randomly selected game Ichiro got 2 hits?
(f) What is the probability that in a randomly selected game Ichiro got more than 1 hit?

20. Waiting in Line A Wendy's manager performed a study to determine a probability distribution for the number of people, X, waiting in line during lunch. The results were as follows:

x	$P(x)$	x	$P(x)$
0	0.011	7	0.098
1	0.035	8	0.063
2	0.089	9	0.035
3	0.150	10	0.019
4	0.186	11	0.004
5	0.172	12	0.006
6	0.132		

(a) Verify that this is a discrete probability distribution.
(b) Draw a graph of the probability distribution. Describe the shape of the distribution.
(c) Compute and interpret the mean of the random variable X.
(d) Compute the standard deviation of the random variable X.
(e) What is the probability that eight people are waiting in line for lunch?
(f) What is the probability that 10 or more people are waiting in line for lunch? Would this be unusual?

In Problems 21 and 22, (a) construct a discrete probability distribution for the random variable X [Hint: $P(x_i) = \frac{f_i}{N}$], (b) draw a graph of the probability distribution, (c) compute and interpret the mean of the random variable X, and (d) compute the standard deviation of the random variable X.

21. The World Series The following data represent the number of games played in each World Series from 1923 to 2014.

x (games played)	Frequency
4	18
5	18
6	20
7	35

Source: Major League Baseball

22. Ideal Number of Children What is the ideal number of children to have in a family? The following data represent the ideal number of children for a random sample of 900 adult Americans.

x (number of children)	Frequency
0	10
1	30
2	520
3	250
4	70
5	17
6	3

Source: Based on data from a Gallup poll

23. Number of Births The graph of the discrete probability distribution below represents the number of live births by a mother 50–54 years old who had a live birth in 2013. *Source: National Vital Statistics Report.*

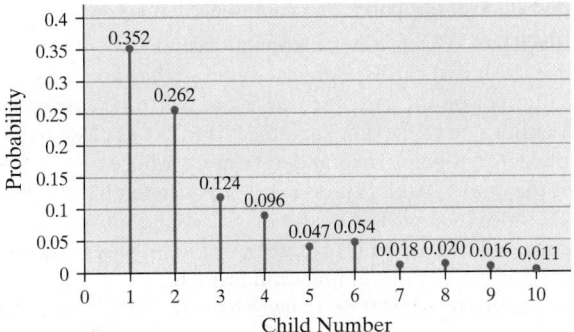

Number of Live Births, 50- to 54-Year-Old Mother

(a) What is the probability that a randomly selected 50- to 54-year-old mother who had a live birth in 2013 has had her fourth live birth?
(b) What is the probability that a randomly selected 50- to 54-year-old mother who had a live birth in 2013 has had her fourth or fifth live birth?
(c) What is the probability that a randomly selected 50- to 54-year-old mother who had a live birth in 2013 has had her sixth or more live birth?

(d) If a 50- to 54-year-old mother who had a live birth in 2013 is randomly selected, how many live births would you expect the mother to have had?

24. Rental Units The graph of the discrete probability distribution represents the number of rooms in rented housing units in 2013.
Source: U.S. Department of Commerce.

Number of Rooms in Rental Unit

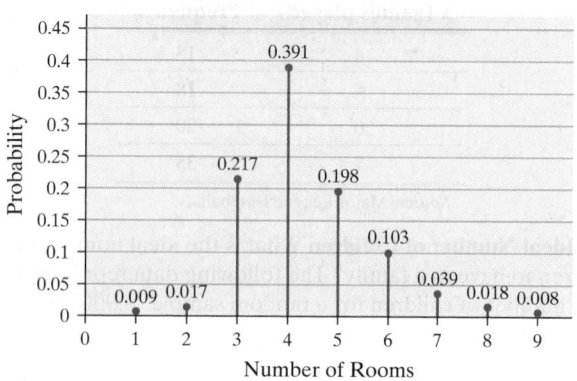

(a) What is the probability that a randomly selected rental unit has five rooms?

(b) What is the probability that a randomly selected rental unit has five or six rooms?

(c) What is the probability that a randomly selected rental unit has seven or more rooms?

(d) If a rental unit is randomly selected, how many rooms would you expect the unit to have?

NW 25. Life Insurance A life insurance company sells a $250,000 1-year term life insurance policy to a 20-year-old female for $200. According to the *National Vital Statistics Report*, 58(21), the probability that the female survives the year is 0.999544. Compute and interpret the expected value of this policy to the insurance company.

26. Life Insurance A life insurance company sells a $250,000 1-year term life insurance policy to a 20-year-old male for $350. According to the *National Vital Statistics Report*, 58(21), the probability that the male survives the year is 0.998734. Compute and interpret the expected value of this policy to the insurance company.

27. BlackJack BlackJack is a popular casino game in which a player is dealt two cards where the value of the card corresponds to the number on the card, face cards are worth ten, and aces are worth either one or eleven. The object is to get as close to 21 as possible without going over and have cards whose value exceeds that of the dealer. A blackjack is an ace and a ten in two cards. It pays 1.5 times the bet. The dealer plays last and must draw a card with sixteen and hold with seventeen or more. The following distribution shows the winnings and probability for a $20 bet. In cases where the dealer and player have the same value, there is a tie (called a "push"). *Source:* "Examining a Gambler's Claims: Probabilistic Fact-Checking and Don Johnson's Extraordinary Winning Streak" by W.J. Hurley, Jack Brimberg, and Richard Kohar. *Chance* Vol. 27.1, 2014.

Winnings	Probability
0	0.0982
$30	0.0483
$20	0.389275
−$20	0.464225

(a) Compute and interpret the expected value of the game from the player's point of view.

(b) Suppose over the course of one hour, a player can expect to be dealt about 40 hands. How much should a player expect to win or lose over the course of three hours?

28. Investment An investment counselor calls with a hot stock tip. He believes that if the economy remains strong, the investment will result in a profit of $50,000. If the economy grows at a moderate pace, the investment will result in a profit of $10,000. However, if the economy goes into recession, the investment will result in a loss of $50,000. You contact an economist who believes there is a 20% probability the economy will remain strong, a 70% probability the economy will grow at a moderate pace, and a 10% probability the economy will slip into recession. What is the expected profit from this investment?

29. Roulette In the game of roulette, a player can place a $5 bet on the number 17 and have a $\frac{1}{38}$ probability of winning. If the metal ball lands on 17, the player wins $175. Otherwise, the casino takes the player's $5. What is the expected value of the game to the player? If you played the game 1000 times, how much would you expect to lose?

30. Connecticut Lottery In the Cash Five Lottery in Connecticut, a player pays $1 for a single ticket with five numbers. Five balls numbered 1 through 35 are randomly chosen from a bin without replacement. If all five numbers on a player's ticket match the five chosen, the player wins $100,000. The probability of this occurring is $\frac{1}{324,632}$. If four numbers match, the player wins $300. This occurs with probability $\frac{1}{2164}$. If three numbers match, the player wins $10. This occurs with probability $\frac{1}{75}$. Compute and interpret the expected value of the game from the player's point of view.

31. Powerball Powerball is a multistate lottery. The following probability distribution represents the cash prizes of Powerball with their corresponding probabilities.

x (cash prize, $)	P(x)
Grand prize	0.00000000684
200,000	0.00000028
10,000	0.000001711
100	0.000153996
7	0.004778961
4	0.007881463
3	0.01450116
0	0.9726824222

Source: www.powerball.com

(a) If the grand prize is $15,000,000, find and interpret the expected cash prize. If a ticket costs $1, what is your expected profit from one ticket?

(b) To the nearest million, how much should the grand prize be so that you can expect a profit? Assume nobody else wins so that you do not have to share the grand prize.

(c) Does the size of the grand prize affect your chance of winning? Explain.

32. SAT Test Penalty Some standardized tests, such as the SAT test, incorporate a penalty for wrong answers. For example, a multiple-choice question with five possible answers will have

1 point awarded for a correct answer and $\frac{1}{4}$ point deducted for an incorrect answer. Questions left blank are worth 0 points.

(a) Find the expected number of points received for a multiple-choice question with five possible answers when a student just guesses.

(b) Explain why there is a deduction for wrong answers.

33. Simulation Use the probability distribution from Problem 19 and a DISCRETE command for some statistical software to simulate 100 repetitions of the experiment (100 games). The number of hits is recorded. Approximate the mean and standard deviation of the random variable X based on the simulation. Repeat the simulation by performing 500 repetitions of the experiment. Approximate the mean and standard deviation of the random variable. Compare your results to the theoretical mean and standard deviation. What property is being illustrated?

34. Simulation Use the probability distribution from Problem 20 and a DISCRETE command for some statistical software to simulate 100 repetitions of the experiment. Approximate the mean and standard deviation of the random variable X based on the simulation. Repeat the simulation by performing 500 repetitions of the experiment. Approximate the mean and standard deviation of the random variable. Compare your results to the theoretical mean and standard deviation. What property is being illustrated?

35. Putting It Together: Sullivan Statistics Survey I One question from the Sullivan Statistics Survey I was "How many credit cards do you currently have?" This question was asked only of those individuals who have at least one credit card. Go to www.pearsonhighered.com/sullivanstats to obtain the survey results. Answer the following questions based on the results of the survey.

(a) Determine the mean number of credit cards based on the raw data.

(b) Determine the standard deviation number of credit cards based on the raw data.

(c) Determine a probability distribution for the random variable, X, the number of credit cards issued to an individual.

(d) Draw a graph of the discrete probability distribution for the random variable X. Describe the shape of the distribution.

(e) Determine the mean and standard deviation number of credit cards from the probability distribution found in part (c).

(f) Determine the probability of randomly selecting an individual whose number of credit cards is more than two standard deviations from the mean. Is this result unusual?

(g) Determine the probability of randomly selecting two individuals who are issued exactly two credit cards. [*Hint:* Are the events independent?] Interpret this result.

Retain Your Knowledge

36. 2014 Senate Election In the 2014 election, Republicans took control of the Senate. A major theme during each senator's campaign was the popularity of President Obama (a Democrat). The scatter diagram below shows the presidential approval rating (percent of registered voters who approve of the job the president is doing) versus the percentage of votes received by the Republican senatorial candidate in the 2014 election.

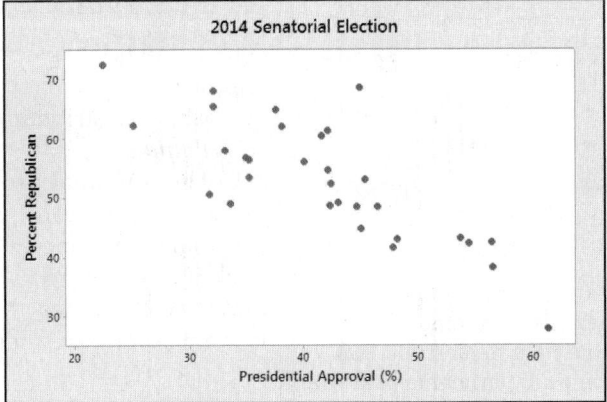

(a) What type of relation appears to exist between presidential approval ratings and percentage of Republican votes?

(b) There are 32 observations in the data set (Alabama was an uncontested election). The correlation between presidential approval and percent Republican is −0.795. Does this suggest a linear relation exists between presidential approval and percent Republican?

(c) The least-squares regression line treating presidential approval, x, as the explanatory variable and percent Republican as the response variable, y, is $\hat{y} = -0.7059x + 78.95$. Interpret the slope. What does this say about candidates of the same party as the president running in Senate elections?

(d) If President Obama's approval rating was 50%, what would you expect the percent of votes for the Republican candidate to be?

(e) In North Carolina, Pat Tillis (a Republican) won the election with 49% of the vote where President Obama's approval rating was 43%. Did Tillis get a higher percent of the votes than would be expected for this presidential approval rating?

6.2   The Binomial Probability Distribution

Preparing for This Section   Before getting started, review the following:

- Empirical Rule (Section 3.2, pp. 139–140)

- Addition Rule for Disjoint Events (Section 5.2, pp. 262–264)

- Complement Rule (Section 5.2, pp. 267–269)

- Independence (Section 5.3, pp. 273–274)

- Multiplication Rule for Independent Events (Section 5.3, pp. 274–276)

- Combinations (Section 5.5, pp. 294–296)

Objectives ❶ Determine whether a probability experiment is a binomial experiment

❷ Compute probabilities of binomial experiments

❸ Compute the mean and standard deviation of a binomial random variable

❹ Graph a binomial probability distribution

❶ Determine Whether a Probability Experiment Is a Binomial Experiment

In Section 6.1, we stated that probability distributions could be presented using tables, graphs, or mathematical formulas. In this section, we introduce a specific type of discrete probability distribution that can be presented using a formula, the *binomial probability distribution*.

The **binomial probability distribution** is a discrete probability distribution that describes probabilities for experiments in which there are two mutually exclusive (disjoint) outcomes. These two outcomes are generally referred to as *success* (such as making a free throw) and *failure* (such as missing a free throw). Experiments in which only two outcomes are possible are referred to as *binomial experiments*, provided that certain criteria are met.

In Other Words
The prefix *bi* means "two." This should help remind you that binomial experiments deal with situations in which there are only two outcomes: success or failure.

Criteria for a Binomial Probability Experiment

An experiment is said to be a **binomial experiment** if

1. The experiment is performed a fixed number of times. Each repetition of the experiment is called a **trial**.
2. The trials are independent. This means that the outcome of one trial will not affect the outcome of the other trials.
3. For each trial, there are two mutually exclusive (disjoint) outcomes: success or failure.
4. The probability of success is the same for each trial of the experiment.

Let the random variable X be the number of successes in n trials of a binomial experiment. Then X is called a **binomial random variable**. Before introducing the method for computing binomial probabilities, it is worthwhile to introduce some notation.

Notation Used in the Binomial Probability Distribution

- There are n independent trials of the experiment.
- Let p denote the probability of success for each trial so that $1 - p$ is the probability of failure for each trial.
- Let X denote the number of successes in n independent trials of the experiment. So $0 \leq x \leq n$.

EXAMPLE 1   Identifying Binomial Experiments

Problem Determine which of the following probability experiments qualify as binomial experiments. For those that are binomial experiments, identify the number of trials, probability of success, probability of failure, and possible values of the random variable X.

Historical Note

Jacob Bernoulli was born on December 27, 1654, in Basel, Switzerland. He studied philosophy and theology at the urging of his parents. (He resented this.) In 1671, he graduated from the University of Basel with a master's degree in philosophy. In 1676, he received a licentiate in theology. After earning his philosophy degree, Bernoulli traveled to Geneva to tutor. From there, he went to France to study with the great mathematicians of the time. One of Bernoulli's greatest works is *Ars Conjectandi*, published 8 years after his death. In this publication, Bernoulli proved the binomial probability formula. To this day, each trial in a binomial probability experiment is called a *Bernoulli trial*.

● **Now Work Problem 9**

● Now Work Problem 9

CAUTION!
The probability of success, p, is always associated with the random variable X, the number of successes. So if X represents the number of 18-year-olds involved in an accident, then p represents the probability of an 18-year-old being involved in an accident.

(a) An experiment in which a basketball player who historically makes 80% of his free throws is asked to shoot three free throws, and the number of free throws made is recorded.

(b) According to a recent Harris Poll, 28% of Americans state that chocolate is their favorite flavor of ice cream. Suppose a simple random sample of size 10 is obtained and the number of Americans who choose chocolate as their favorite ice cream flavor is recorded.

(c) A probability experiment in which three cards are drawn from a deck without replacement and the number of aces is recorded.

Approach Determine whether the four conditions for a binomial experiment are satisfied.
1. The experiment is performed a fixed number of times.
2. The trials are independent.
3. There are only two possible outcomes of the experiment.
4. The probability of success for each trial is constant.

Solution
(a) This is a binomial experiment because
1. There are $n = 3$ trials.
2. The trials are independent.
3. There are two possible outcomes: make or miss.
4. The probability of success (make) is 0.8 and the probability of failure (miss) is 0.2. The probabilities are the same for each trial.
The random variable X is the number of free throws made with $x = 0, 1, 2,$ or 3.
(b) This is a binomial experiment because
1. There are 10 trials (the 10 randomly selected people).
2. The trials are independent.*
3. There are two possible outcomes: finding an American who chooses chocolate as his or her favorite ice cream or not.
4. The probability of success is 0.28 and the probability of failure is $1 - 0.28 = 0.72$. The random variable X is the number of people who choose chocolate as their favorite ice cream with $x = 0, 1, 2, 3, \ldots, 10$.
(c) This is not a binomial experiment because the trials are not independent. The probability of an ace on the first trial is $\frac{4}{52} \approx 0.077$. Because we are sampling without replacement, if an ace is selected on the first trial, the probability of an ace on the second trial is $\frac{3}{51} \approx 0.059$. If an ace is not selected on the first trial, the probability of an ace on the second trial is $\frac{4}{51} \approx 0.078$.   ●

Note that the word *success* does not necessarily imply something positive. Success means that an outcome has occurred that corresponds with p, the probability of success. For example, a probability experiment might be to randomly select ten 18-year-old male drivers. If X denotes the number who have been involved in an accident within the last year, a success would mean obtaining an 18-year-old male who was involved in an accident. This outcome is not positive, but it is a success as far as the experiment goes.

② Compute Probabilities of Binomial Experiments

Now we will compute probabilities for a binomial random variable X. We present three methods for obtaining binomial probabilities: (1) the binomial probability distribution formula, (2) a table of binomial probabilities, and (3) technology. We develop the binomial probability formula in Example 2.

---

* In sampling from large populations without replacement, the trials are assumed to be independent, provided that the sample size is small in relation to the size of the population. As a rule of thumb, if the sample size is less than 5% of the population size, the trials are assumed to be independent, although they are technically dependent. See Example 6 in Section 5.4.

EXAMPLE 2 Constructing a Binomial Probability Distribution

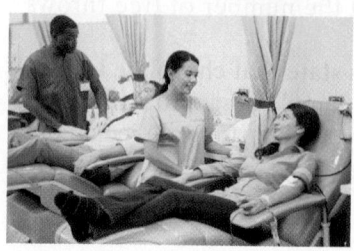

Problem According to the American Red Cross, 7% of people in the United States have blood type O-negative. A simple random sample of size 4 is obtained, and the number of people X with blood type O-negative is recorded. Construct a probability distribution for the random variable X.

Approach This is a binomial experiment with $n = 4$ trials. We define a success as selecting an individual with blood type O-negative. The probability of success, p, is 0.07, and X is the random variable representing the number of successes with $x = 0, 1, 2, 3,$ or 4.

Step 1 Construct a tree diagram listing the various outcomes of the experiment by listing each outcome as S (success) or F (failure).

Step 2 Compute the probabilities for each value of the random variable X.

Step 3 Construct the probability distribution.

Solution

Step 1 The tree diagram in Figure 5 lists the 16 possible outcomes of the experiment.

Figure 5

1st Trial	2nd Trial	3rd Trial	4th Trial	Outcome	Number of Successes, x
			S	S,S,S,S	4
		S	F	S,S,S,F	3
	S	F	S	S,S,F,S	3
			F	S,S,F,F	2
S			S	S,F,S,S	3
	F	S	F	S,F,S,F	2
		F	S	S,F,F,S	2
			F	S,F,F,F	1
			S	F,S,S,S	3
		S	F	F,S,S,F	2
	S	F	S	F,S,F,S	2
			F	F,S,F,F	1
F			S	F,F,S,S	2
	F	S	F	F,F,S,F	1
		F	S	F,F,F,S	1
			F	F,F,F,F	0

Step 2   Now compute the probability for each possible value of the random variable X.

$$P(0) = P(FFFF) = P(F) \cdot P(F) \cdot P(F) \cdot P(F)$$

Multiplication Rule for Independent Events

$$= (0.93)(0.93)(0.93)(0.93)$$
$$= (0.93)^4$$
$$= 0.74805$$

$$P(1) = P(SFFF \text{ or } FSFF \text{ or } FFSF \text{ or } FFFS)$$

Addition Rule for Disjoint Events

$$= P(SFFF) + P(FSFF) + P(FFSF) + P(FFFS)$$
$$= (0.07)^1(0.93)^3 + (0.07)^1(0.93)^3 + (0.07)^1(0.93)^3 + (0.07)^1(0.93)^3$$

Multiplication Rule for Independent Events

$$= 4(0.07)^1(0.93)^3$$
$$= 0.22522$$

$$P(2) = P(SSFF \text{ or } SFSF \text{ or } SFFS \text{ or } FSSF \text{ or } FSFS \text{ or } FFSS)$$
$$= P(SSFF) + P(SFSF) + P(SFFS) + P(FSSF) + P(FSFS) + P(FFSS)$$
$$= (0.07)^2(0.93)^2 + (0.07)^2(0.93)^2 + (0.07)^2(0.93)^2 + (0.07)^2(0.93)^2$$
$$\quad + (0.07)^2(0.93)^2 + (0.07)^2(0.93)^2$$
$$= 6(0.07)^2(0.93)^2$$
$$= 0.02543$$

Compute $P(3)$ and $P(4)$ similarly and obtain $P(3) = 0.00128$ and $P(4) = 0.00002$. You are encouraged to verify these probabilities.

Step 3   We present these results in the probability distribution in Table 8.   ●

Table 8	
x	$P(x)$
0	0.74805
1	0.22522
2	0.02543
3	0.00128
4	0.00002

Notice some interesting results in Example 2. Consider the probability of obtaining $x = 1$ success:

$$P(1) = 4(0.07)^1(0.93)^3$$

4 is the number of ways we obtain 1 success in 4 trials of the experiment. Here, it is $_4C_1$.   0.07 is the probability of success and the exponent 1 is the number of successes.   0.93 is the probability of failure and the exponent 3 is the number of failures.

The coefficient 4 is the number of ways of obtaining one success in four trials. In general, the coefficient is $_nC_x$, the number of ways of obtaining x successes in n trials. The second factor in the formula, $(0.07)^1$, is the probability of success, p, raised to the number of successes, x. The third factor in the formula, $(0.93)^3$, is the probability of failure, $1 - p$, raised to the number of failures, $n - x$. The following *binomial probability distribution function (pdf)* formula holds for all binomial experiments.

> **CAUTION!**
> Before using the binomial probability distribution function, be sure the requirements for a binomial experiment are satisfied.

Binomial Probability Distribution Function

The probability of obtaining x successes in n independent trials of a binomial experiment is given by

$$P(x) = {_nC_x}\, p^x(1 - p)^{n-x} \qquad x = 0, 1, 2, \ldots, n \qquad \textbf{(1)}$$

where p is the probability of success.

When reading probability problems, pay special attention to key phrases that translate into mathematical symbols. Table 9 lists various phrases and their corresponding mathematical equivalent.

Table 9	
Phrase	**Math Symbol**
"at least" or "no less than" or "greater than or equal to"	$\geq$
"more than" or "greater than"	$>$
"fewer than" or "less than"	$<$
"no more than" or "at most" or "less than or equal to"	$\leq$
"exactly" or "equals" or "is"	$=$

EXAMPLE 3 Using the Binomial Probability Distribution Function

Problem According to CTIA, 41% of all U.S. households are wireless-only households (no landline). In a random sample of 20 households, what is the probability that
(a) exactly 5 are wireless-only?
(b) fewer than 3 are wireless-only?
(c) at least 3 are wireless-only?
(d) the number of households that are wireless-only is between 5 and 7, inclusive?

Approach This is a binomial experiment with $n = 20$ independent trials. We define a success as selecting a household that is wireless-only. The probability of success, p, is 0.41. The possible values of the random variable X are $x = 0, 1, 2, \ldots, 20$. We use Formula (1) to compute the probabilities.

Solution
(a) $P(5) = {}_{20}C_5(0.41)^5(1 - 0.41)^{20-5}$ $n = 20, x = 5, p = 0.41$

$$= \frac{20!}{5!(20-5)!}(0.41)^5(0.59)^{15} \qquad {}_nC_x = \frac{n!}{x!(n-x)!}$$

$$= 15{,}504(0.41)^5(0.59)^{15} = 0.0656$$

Interpretation The probability of getting exactly 5 households out of 20 that are wireless-only is 0.0656. In 100 trials of this study (that is, if we surveyed 20 households 100 different times), we would expect about 7 trials to result in 5 households that are wireless-only.

(b) The phrase *fewer than* means "less than." The values of the random variable X less than 3 are $x = 0, 1$, or 2.

$$P(X < 3) = P(0 \text{ or } 1 \text{ or } 2)$$
$$= P(0) + P(1) + P(2) \qquad \text{Addition Rule for Disjoint Events}$$
$$= {}_{20}C_0(0.41)^0(1 - 0.41)^{20-0} + {}_{20}C_1(0.41)^1(1 - 0.41)^{20-1}$$
$$\quad + {}_{20}C_2(0.41)^2(1 - 0.41)^{20-2}$$
$$= 0.000026 + 0.000363 + 0.002397$$
$$= 0.0028$$

Interpretation The probability that, in a random sample of 20 households, fewer than three will be a wireless-only household is 0.0028. In 1000 trials of this study, we would expect about 3 trials to result in fewer than 3 wireless-only households.

(c) The values of the random variable X that are at least 3 are $x = 3, 4, 5, \ldots, 20$. Rather than compute $P(X \geq 3)$ directly by computing $P(3) + P(4) + \cdots + P(20)$, we can use the Complement Rule.

$$P(X \geq 3) = 1 - P(X < 3) = 1 - 0.0028 = 0.9972$$

Interpretation The probability that, in a random sample of 20 households, at least 3 will be a wireless-only household is 0.9972. In 1000 trials of this study, we would expect about 997 trials to result in at least 3 being wireless-only households.

(d) The word *inclusive* means "including," so we want to determine the probability that 5, 6, or 7 households are wireless-only.

$$
\begin{aligned}
P(5 \leq X \leq 7) &= P(5 \text{ or } 6 \text{ or } 7) \\
&= P(5) + P(6) + P(7) \qquad \text{Addition Rule for Disjoint Events} \\
&= {}_{20}C_5(0.41)^5(1 - 0.41)^{20-5} + {}_{20}C_6(0.41)^6(1 - 0.41)^{20-6} \\
&\quad + {}_{20}C_7(0.41)^7(1 - 0.41)^{20-7} \\
&= 0.0656 + 0.1140 + 0.1585 \\
&= 0.3381
\end{aligned}
$$

Interpretation The probability that the number of wireless-only households is between 5 and 7, inclusive, is 0.3381. In 100 trials of this study, we would expect about 34 trials to result in 5 to 7 households that are wireless-only. •

Obtaining Binomial Probabilities from Tables

Another method for obtaining probabilities is the binomial probability table. Table III in Appendix A gives probabilities for a binomial random variable X taking on a specific value, such as $P(10)$, for select values of n and p. Table IV in Appendix A gives cumulative probabilities of a binomial random variable X. This means that Table IV gives "less than or equal to" binomial probabilities, such as $P(X \leq 6)$. We illustrate how to use Tables III and IV in Example 4.

EXAMPLE 4   Computing Binomial Probabilities Using the Binomial Table

Problem According to the Gallup Organization, 65% of adult Americans are in favor of the death penalty for individuals convicted of murder. In a random sample of 15 adult Americans, what is the probability that
(a) exactly 10 favor the death penalty?
(b) no more than 6 favor the death penalty?

Approach We use Tables III and IV in Appendix A to obtain the probabilities.

Solution
(a) We have $n = 15, p = 0.65$, and $x = 10$. In Table III, Appendix A, go to the section that contains $n = 15$ and the column that contains $p = 0.65$. The value at which the $x = 10$ row intersects the $p = 0.65$ column is the probability we seek. See Figure 6 on the next page. So $P(10) = 0.2123$.

Interpretation In 100 trials of this study (randomly selecting 15 adult Americans), we expect about 21 trials to result in exactly 10 adult Americans who favor the death penalty for individuals convicted of murder.

(continued)

Figure 6

n	x	0.01	0.05	0.10	0.15	0.20	0.25	0.30	0.35	0.40	0.45	0.50	0.55	0.60	0.65	0.70
15	0	0.8601	0.4633	0.2059	0.0874	0.0352	0.0134	0.0047	0.0016	0.0005	0.0001	0.0000+	0.0000+	0.0000+	0.0000+	0.0000+
	1	0.1303	0.3658	0.3432	0.2312	0.1319	0.0668	0.0305	0.0126	0.0047	0.0016	0.0005	0.0001	0.0000+	0.0000+	0.0000+
	2	0.0092	0.1348	0.2669	0.2856	0.2309	0.1559	0.0916	0.0476	0.0219	0.0090	0.0032	0.0010	0.0003	0.0001	0.0000+
	3	0.0004	0.0307	0.1285	0.2184	0.2501	0.2252	0.1700	0.1110	0.0634	0.0318	0.0139	0.0052	0.0016	0.0004	0.0001
	4	0.0000+	0.0049	0.0428	0.1156	0.1876	0.2252	0.2186	0.1792	0.1268	0.0780	0.0417	0.0191	0.0074	0.0024	0.0006
	5	0.0000+	0.0006	0.0105	0.0449	0.1032	0.1651	0.2061	0.2123	0.1859	0.1404	0.0916	0.0515	0.0245	0.0096	0.0030
	6	0.0000+	0.0000+	0.0019	0.0132	0.0430	0.0917	0.1472	0.1906	0.2066	0.1914	0.1527	0.1048	0.0612	0.0298	0.0116
	7	0.0000+	0.0000+	0.0003	0.0030	0.0138	0.0393	0.0811	0.1319	0.1771	0.2013	0.1964	0.1647	0.1181	0.0710	0.0348
	8	0.0000+	0.0000+	0.0000+	0.0005	0.0035	0.0131	0.0348	0.0710	0.1181	0.1647	0.1964	0.2013	0.1771	0.1319	0.0811
	9	0.0000+	0.0000+	0.0000+	0.0001	0.0007	0.0034	0.0116	0.0298	0.0612	0.1048	0.1527	0.1914	0.2066	0.1906	0.1472
	10	0.0000+	0.0000+	0.0000+	0.0000+	0.0001	0.0007	0.0030	0.0096	0.0245	0.0515	0.0916	0.1404	0.1859	0.2123	0.2061
	11	0.0000+	0.0000+	0.0000+	0.0000+	0.0000+	0.0001	0.0006	0.0024	0.0074	0.0191	0.0417	0.0780	0.1268	0.1792	0.2186

(b) The phrase *no more than* means "less than or equal to." To compute $P(X \le 6)$, use the cumulative binomial table, Table IV in Appendix A, which lists binomial probabilities less than or equal to a specified value. We have $n = 15$, $p = 0.65$, so go to the section that contains $n = 15$ and the column that contains $p = 0.65$. The value at which the $x = 6$ row intersects the $p = 0.65$ column represents $P(X \le 6)$. See Figure 7. So $P(X \le 6) = 0.0422$.

Interpretation In 100 different trials of this study, we expect about 4 trials to result in no more than 6 adult Americans who favor the death penalty for individuals convicted of murder.

Figure 7

n	x	0.01	0.05	0.10	0.15	0.20	0.25	0.30	0.35	0.40	0.45	0.50	0.55	0.60	0.65	0.70
15	0	0.8601	0.4633	0.2059	0.0874	0.0352	0.0134	0.0047	0.0016	0.0005	0.0001	0.0000+	0.0000+	0.0000+	0.0000+	0.0000+
	1	0.9904	0.8290	0.5490	0.3186	0.1671	0.0802	0.0353	0.0142	0.0052	0.0017	0.0005	0.0001	0.0000+	0.0000+	0.0000
	2	0.9996	0.9638	0.8159	0.6042	0.3980	0.2361	0.1268	0.0617	0.0271	0.0107	0.0037	0.0011	0.0003	0.0001	0.0000+
	3	1.0000–	0.9945	0.9444	0.8227	0.6482	0.4613	0.2969	0.1727	0.0905	0.0424	0.0176	0.0063	0.0019	0.0005	0.0001
	4	1.0000–	0.9994	0.9873	0.9383	0.8358	0.6865	0.5155	0.3519	0.2173	0.1204	0.0592	0.0255	0.0093	0.0028	0.0007
	5	1.0000–	0.9999	0.9978	0.9832	0.9389	0.8518	0.7216	0.5643	0.4032	0.2608	0.1509	0.0769	0.0338	0.0124	0.0037
	6	1.0000–	1.0000–	0.9997	0.9964	0.9819	0.9434	0.8689	0.7548	0.6098	0.4522	0.3036	0.1818	0.0950	0.0422	0.0152
	7	1.0000–	1.0000–	1.0000–	0.9994	0.9958	0.9827	0.9500	0.8868	0.7869	0.6535	0.5000	0.3465	0.2131	0.1132	0.0500

Obtaining Binomial Probabilities Using Technology

Statistical software and graphing calculators also have the ability to compute binomial probabilities.

EXAMPLE 5   Obtaining Binomial Probabilities Using Technology

Problem According to the Gallup Organization, 65% of adult Americans are in favor of the death penalty for individuals convicted of murder. In a random sample of 15 adult Americans, what is the probability that

(a) exactly 10 favor the death penalty?

(b) no more than 6 favor the death penalty?

Approach Statistical software or graphing calculators with advanced statistical features have the ability to determine binomial probabilities. The steps for determining binomial probabilities using the TI-83/84 Plus graphing calculators, Minitab, Excel, and StatCrunch can be found in the Technology Step-by-Step on pages 338–339.

Solution We will use StatCrunch to determine the probability for part (a) and a TI-84 Plus C to determine the probability for part (b).

(a) Using StatCrunch's binomial calculator, we obtain the results in Figure 8(a). So
$$P(10) = 0.2123.$$

Interpretation In 100 trials of this study (randomly selecting 15 adult Americans), we expect about 21 trials to result in exactly 10 adult Americans who favor the death penalty for individuals convicted of murder.

(b) The phrase *no more than* means "less than or equal to." To compute $P(X \leq 6)$, we use the **cumulative distribution function (cdf)**, which computes probabilities less than or equal to a specified value. Using a TI-84 Plus C graphing calculator to find $P(X \leq 6)$ with $n = 15$ and $p = 0.65$, we find $P(X \leq 6) = 0.0422$. See Figure 8(b).

Interpretation In 100 trials of this study, we expect about 4 trials to result in no more than 6 adult Americans who favor the death penalty for individuals convicted of murder.

Figure 8

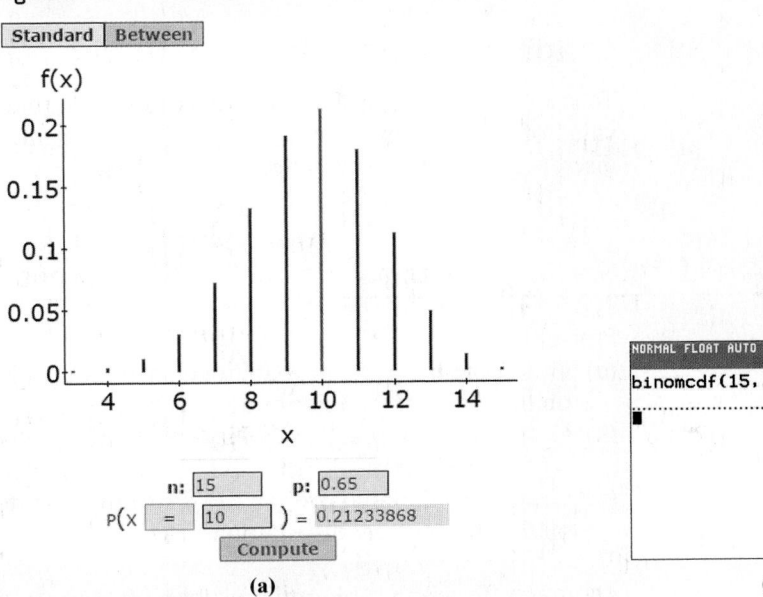

• Now Work Problem 35

(a)   **(b)**   •

③ **Compute the Mean and Standard Deviation of a Binomial Random Variable**

We discussed finding the mean (or expected value) and standard deviation of a discrete random variable in Section 6.1. These formulas can be used to find the mean and standard deviation of a binomial random variable, but a simpler method exists.

In Other Words
The mean of a binomial random variable equals the product of the number of trials of the experiment and the probability of success. It can be interpreted as the expected number of successes in n trials of the experiment.

Mean (or Expected Value) and Standard Deviation of a Binomial Random Variable

A binomial experiment with n independent trials and probability of success p has a mean and standard deviation given by the formulas

$$\mu_X = np \quad \text{and} \quad \sigma_X = \sqrt{np(1-p)} \tag{2}$$

EXAMPLE 6   Finding the Mean and Standard Deviation of a Binomial Random Variable

Problem According to CTIA, 41% of all U.S. households are wireless-only households. In a simple random sample of 300 households, determine the mean and standard deviation number of wireless-only households.

Approach This is a binomial experiment with $n = 300$ and $p = 0.41$. Use Formula (2) to find the mean and standard deviation, respectively.

Solution

$$\mu_X = np = 300(0.41) = 123$$

and

$$\sigma_X = \sqrt{np(1-p)} = \sqrt{300(0.41)(1-0.41)} = \sqrt{72.57} = 8.5$$

Interpretation We expect that, in a random sample of 300 households, 123 will be wireless-only.

● **Now Work Problems 29(a)–(c)**

④ **Graph a Binomial Probability Distribution**

To graph a binomial probability distribution, first find the probabilities for each possible value of the random variable. Then follow the same approach that was used to graph discrete probability distributions.

EXAMPLE 7   Graphing Binominal Probability Distributions

Problem
(a) Graph the binomial probability distribution with $n = 10$ and $p = 0.2$. Comment on the shape of the distribution.
(b) Graph the binomial probability distribution with $n = 10$ and $p = 0.5$. Comment on the shape of the distribution.
(c) Graph the binomial probability distribution with $n = 10$ and $p = 0.8$. Comment on the shape of the distribution.

Approach To graph a binomial probability distribution, first obtain the probability distribution and then graph the distribution using the approach in Section 6.1.

Solution
(a) Table 10 shows the probability distribution with $n = 10$ and $p = 0.2$. Although $P(9)$ is 0.000004096, it is written as 0.0000 to four significant digits. The same idea applies to $P(10)$. Figure 9 shows the graph of the distribution. The distribution is skewed right.

Figure 9

Table 10	
x	$P(x)$
0	0.1074
1	0.2684
2	0.3020
3	0.2013
4	0.0881
5	0.0264
6	0.0055
7	0.0008
8	0.0001
9	0.0000
10	0.0000

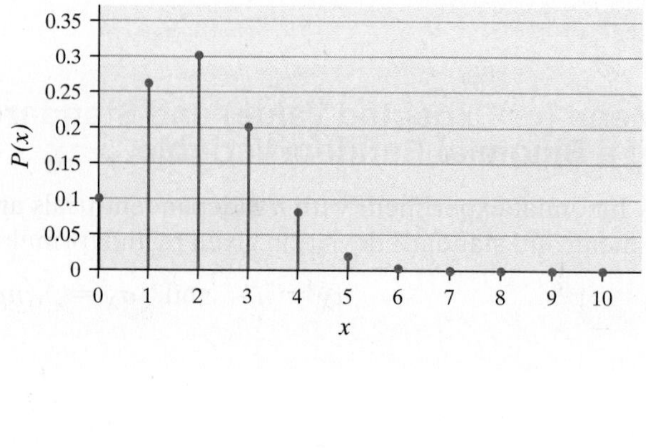

Table 11

x	P(x)
0	0.0010
1	0.0098
2	0.0439
3	0.1172
4	0.2051
5	0.2461
6	0.2051
7	0.1172
8	0.0439
9	0.0098
10	0.0010

(b) Table 11 shows the probability distribution with $n = 10$ and $p = 0.5$. Figure 10 shows the graph of the distribution. The distribution is symmetric and approximately bell shaped.

Figure 10

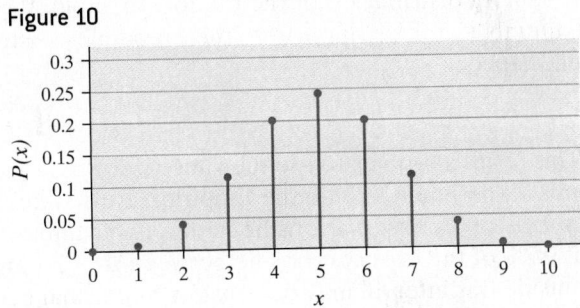

Table 12

x	P(x)
0	0.0000
1	0.0000
2	0.0001
3	0.0008
4	0.0055
5	0.0264
6	0.0881
7	0.2013
8	0.3020
9	0.2684
10	0.1074

(c) Table 12 shows the probability distribution with $n = 10$ and $p = 0.8$. Figure 11 shows the graph of the distribution. The distribution is skewed left.

Figure 11

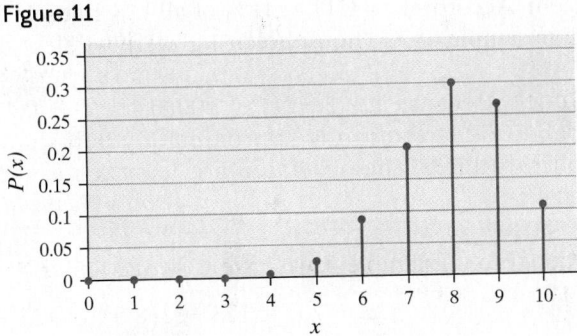

● **Now Work Problem 29(d)**

Based on the results of Example 7, we conclude that the binomial probability distribution is skewed right if $p < 0.5$, symmetric and approximately bell shaped if $p = 0.5$, and skewed left if $p > 0.5$. Notice that Figure 9 ($p = 0.2$) and Figure 11 ($p = 0.8$) are mirror images of each other.

We have seen the role that p plays in the shape of a binomial distribution, but what role does n play in its shape? Figure 12 shows the graph of three binomial probability distributions drawn in StatCrunch. In Figure 12(a), $n = 10$ and $p = 0.2$; in Figure 12(b), $n = 30$ and $p = 0.2$; in Figure 12(c), $n = 70$ and $p = 0.2$.

Figure 12

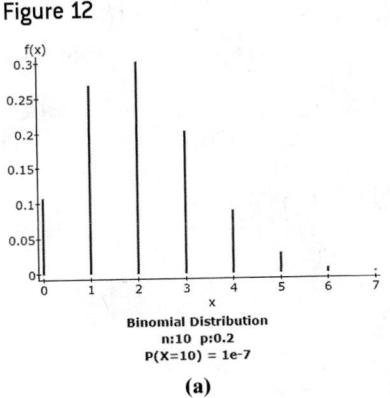

Binomial Distribution
n:10 p:0.2
P(X=10) = 1e-7

(a)

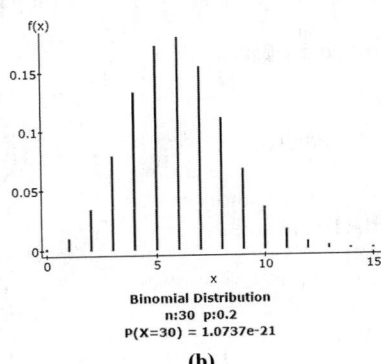

Binomial Distribution
n:30 p:0.2
P(X=30) = 1.0737e-21

(b)

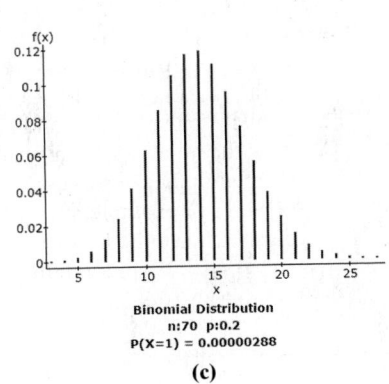

Binomial Distribution
n:70 p:0.2
P(X=1) = 0.00000288

(c)

Figure 12(a) is skewed right, Figure 12(b) is slightly skewed right, and Figure 12(c) appears bell shaped. Our conclusion follows:

> For a fixed p, as the number of trials n in a binomial experiment increases, the probability distribution of the random variable X becomes bell shaped. As a rule of thumb, if $np(1 - p) \geq 10$,* the probability distribution will be approximately bell shaped.

In Other Words
Provided that $np(1 - p) \geq 10$, the interval $\mu - 2\sigma$ to $\mu + 2\sigma$ represents the "usual" observations. Observations outside this interval may be considered unusual.

This result allows us to use the Empirical Rule to identify unusual observations in a binomial experiment. Recall the Empirical Rule states that in a bell-shaped distribution about 95% of all observations lie within two standard deviations of the mean. That is, about 95% of the observations lie between $\mu - 2\sigma$ and $\mu + 2\sigma$. Any observation that lies outside this interval may be considered unusual because the observation occurs less than 5% of the time.

EXAMPLE 8 Using the Mean, Standard Deviation, and Empirical Rule to Check for Unusual Results in a Binomial Experiment

Problem According to CTIA, 41% of all U.S. households are wireless-only. In a simple random sample of 300 households, 143 were wireless-only. Is this result unusual?

Approach Because $np(1 - p) = 300(0.41)(1 - 0.41) = 72.57 \geq 10$, the binomial probability distribution is approximately bell shaped. Therefore, we can use the Empirical Rule: If the observation is less than $\mu - 2\sigma$ or greater than $\mu + 2\sigma$, it is unusual.

Solution From Example 6, we have $\mu = 123$ and $\sigma = 8.5$.

$$\mu - 2\sigma = 123 - 2(8.5) = 123 - 17 = 106$$

and

$$\mu + 2\sigma = 123 + 2(8.5) = 123 + 17 = 140$$

Interpretation Since any value less than 106 or greater than 140 is unusual, 143 is an unusual result. We should try to identify the reason for its value. Perhaps the percentage of households that are wireless-only has increased.

● Now Work Problem 43

Technology Step-by-Step Computing Binomial Probabilities via Technology

TI-83/84 Plus
Computing $P(x)$
1. Press 2nd VARS to access the probability distribution menu.
2. Highlight `binompdf(` and hit ENTER.
3. Enter the number of trials n, the probability of success p, and the number of successes x. Highlight `Paste` and hit ENTER. For example, with $n = 15$, $p = 0.3$, and $x = 8$, you should see the following on the HOME screen
   binompdf(15,0.3,8)
Then hit ENTER.

Computing $P(X \leq x)$
1. Press 2nd VARS to access the probability distribution menu.
2. Highlight `binomcdf(` and hit ENTER.
3. Enter the number of trials n, the probability of success p, and the number of successes x. Highlight `Paste` and hit ENTER. For example, with $n = 15$, $p = 0.3$, and $x \leq 8$, you should see the following on the HOME screen
   binomcdf(15,0.3,8)
Then hit ENTER.

*P. P. Ramsey and P. H. Ramsey, "Evaluating the Normal Approximation to the Binomial Test," *Journal of Educational Statistics* 13 (1998): 173–182.

Minitab
Computing $P(x)$
1. Select the <u>CALC</u> menu, highlight **Probability Distributions**, then highlight **Binomial . . . .**
2. With $n = 15$, $p = 0.3$, and $x = 6$, fill in the window as shown. Click OK.

Binomial Distribution	✕

- ● Probability
- ○ Cumulative probability
- ○ Inverse cumulative probability

Number of trials: `15`

Event probability: `0.3`

- ○ Input column: _____
-    Optional storage: _____
- ● Input constant: `6`
-    Optional storage: _____

Select		
Help	OK	Cancel

Computing $P(X \leq x)$
Follow the same steps as those for computing $P(x)$. In the window that comes up after selecting Binomial Distribution, select **Cumulative probability** instead of **Probability**.

Excel
Computing $P(x)$
1. Click the Formulas tab. Select More Functions. Highlight Statistical in the Function category menu. Highlight BINOM.DIST in the Function name menu.
2. Fill in the window with the appropriate values. Type FALSE in the cumulative cell. Click OK.

Computing $P(X \leq x)$
Follow the same steps as those presented for computing $P(x)$. In the BINOMDIST window, type TRUE in the cumulative cell.

StatCrunch
1. Select **Stat**, highlight **Calculators**, select **Binomial**.
2. Enter the number of trials, n, and probability of success, p. If you want to compute $P(X < x)$, $P(X \leq x)$, $P(X > x)$, or $P(X \geq x)$, highlight the Standard tab. In the pull-down menu, decide if you wish to compute $P(X \leq x)$, $P(X < x)$, and so on. Finally, enter the value of x. Click Compute. If you want to compute $P(a \leq X \leq b)$, highlight the Between tab. Enter the values of a and b. Click Compute.

 6.2 Assess Your Understanding

Vocabulary and Skill Building

1. A binomial experiment is performed a fixed number of times. Each repetition of the experiment is called a _____.

2. For each trial of a binomial experiment, there are two mutually exclusive outcomes: _____ or _____.

3. *True or False*: In the binomial probability distribution function, $_nC_x$ represents the number of ways of obtaining x successes in n trials.

4. The phrase "no more than" is represented by the math symbol _____.

5. The expected number of successes in a binomial experiment with n trials and probability of success p is_____.

6. As a rule of thumb, if _____, the probability distribution of a binomial random variable X is approximately bell shaped.

In Problems 7–16, determine which of the following probability experiments represents a binomial experiment. If the probability experiment is not a binomial experiment, state why.

7. A random sample of 15 college seniors is obtained, and the individuals selected are asked to state their ages.

8. A random sample of 30 cars in a used car lot is obtained, and their mileages recorded.

NW 9. An experimental drug is administered to 100 randomly selected individuals, with the number of individuals responding favorably recorded.

10. A poll of 1200 registered voters is conducted in which the respondents are asked whether they believe Congress should reform Social Security.

11. Three cards are selected from a standard 52-card deck without replacement. The number of aces selected is recorded.

12. Three cards are selected from a standard 52-card deck with replacement. The number of kings selected is recorded.

13. A basketball player who makes 80% of her free throws is asked to shoot free throws until she misses. The number of free-throw attempts is recorded.

14. A baseball player who reaches base safely 30% of the time is allowed to bat until he reaches base safely for the third time. The number of at-bats required is recorded.

15. One hundred randomly selected U.S. parents with at least one child under the age of 18 are surveyed and asked if they have ever spanked their child. The number of parents who have spanked their child is recorded.

16. In a town with 400 citizens, 100 randomly selected citizens are asked to identify their religion. The number who identify with a Christian religion is recorded.

In Problems 17–28, a binomial probability experiment is conducted with the given parameters. Compute the probability of x successes in the n independent trials of the experiment.

17. $n = 10, p = 0.4, x = 3$

18. $n = 15, p = 0.85, x = 12$

19. $n = 40, p = 0.99, x = 38$

20. $n = 50, p = 0.02, x = 3$

21. $n = 8, p = 0.35, x = 3$

22. $n = 20, p = 0.6, x = 17$

23. $n = 9, p = 0.2, x \leq 3$

24. $n = 10, p = 0.65, x < 5$

25. $n = 7, p = 0.5, x > 3$

26. $n = 20, p = 0.7, x \geq 12$

27. $n = 12, p = 0.35, x \leq 4$

28. $n = 11, p = 0.75, x \geq 8$

In Problems 29–34, (a) construct a binomial probability distribution with the given parameters; (b) compute the mean and standard deviation of the random variable using the methods of Section 6.1; (c) compute the mean and standard deviation, using the methods of this section; and (d) draw a graph of the probability distribution and comment on its shape.

NW 29. $n = 6, p = 0.3$

30. $n = 8, p = 0.5$

31. $n = 9, p = 0.75$

32. $n = 10, p = 0.2$

33. $n = 10, p = 0.5$

34. $n = 9, p = 0.8$

Applying the Concepts

NW 35. On-Time Flights According to flightstats.com, American Airlines flights from Dallas to Chicago are on time 80% of the time. Suppose 15 flights are randomly selected, and the number of on-time flights is recorded.

(a) Explain why this is a binomial experiment.

(b) Find and interpret the probability that exactly 10 flights are on time.

(c) Find and interpret the probability that fewer than 10 flights are on time.

(d) Find and interpret the probability that at least 10 flights are on time.

(e) Find and interpret the probability that between 8 and 10 flights, inclusive, are on time.

36. Morality In a recent poll, the Gallup Organization found that 45% of adult Americans believe that the overall state of moral values in the United States is poor. Suppose a survey of a random sample of 25 adult Americans is conducted in which they are asked to disclose their feelings on the overall state of moral values in the United States.

(a) Find and interpret the probability that exactly 15 of those surveyed feel the state of morals is poor.

(b) Find and interpret the probability that no more than 10 of those surveyed feel the state of morals is poor.

(c) Find and interpret the probability that more than 16 of those surveyed feel the state of morals is poor.

(d) Find and interpret the probability that 13 or 14 believe the state of morals is poor.

(e) Would it be unusual to find 20 or more adult Americans who believe the overall state of moral values is poor in the United States? Why?

37. Toilet Flushing In the Healthy Handwashing Survey conducted by Bradley Corporation, it was found that 64% of adult Americans operate the flusher of toilets in public restrooms with their foot. Suppose a random sample of $n = 20$ adult Americans is obtained and the number x who flush public toilets with their foot is recorded.

(a) Explain why this is a binomial experiment.

(b) Find and interpret the probability that exactly 12 flush public toilets with their foot.

(c) Find and interpret the probability that at least 16 flush public toilets with their foot.

(d) Find and interpret the probability that between 9 and 11, inclusive, flush public toilets with their foot.

(e) Would it be unusual to find more than 17 who flush public toilets with their foot? Why?

38. Allergy Sufferers Clarinex-D is a medication whose purpose is to reduce the symptoms associated with a variety of allergies. In clinical trials of Clarinex-D, 5% of the patients in the study experienced insomnia as a side effect. A random sample of 20 Clarinex-D users is obtained, and the number of patients who experienced insomnia is recorded.

(a) Find the probability that exactly 3 experienced insomnia as a side effect.

(b) Find the probability that 3 or fewer experienced insomnia as a side effect.

(c) Find the probability that between 1 and 4 patients inclusive, experienced insomnia as a side effect.

(d) Would it be unusual to find 4 or more patients who experienced insomnia as a side effect? Why?

39. Sneeze According to a study done by Nick Wilson of Otago University Wellington, the probability a randomly selected individual will not cover his or her mouth when sneezing is 0.267. Suppose you sit on a bench in a mall and observe people's habits as they sneeze.

(a) What is the probability that among 10 randomly observed individuals exactly 4 do not cover their mouth when sneezing?

(b) What is the probability that among 10 randomly observed individuals fewer than 3 do not cover their mouth?

(c) Would you be surprised if, after observing 10 individuals, fewer than half covered their mouth when sneezing? Why?

40. Sneeze Revisited According to a study done by Nick Wilson of Otago University Wellington, the probability a randomly selected individual will cover his or her mouth with a tissue, handkerchief, or elbow (the method recommended by public health officials) when sneezing is 0.047. Suppose you sit on a bench in a mall and observe people's habits as they sneeze.

(a) What is the probability that among 15 randomly observed sneezing individuals exactly 2 cover their mouth with a tissue, handkerchief, or elbow?

(b) What is the probability that among 15 randomly observed sneezing individuals fewer than 3 cover their mouth with a tissue, handkerchief, or elbow?

(c) Would you be surprised if, after observing 15 sneezing individuals, more than 4 covered the mouth with a tissue, handkerchief, or elbow?

41. Jury Selection Twelve jurors are randomly selected from a population of 3 million residents. Of these 3 million residents, it is known that 45% are Hispanic. Of the 12 jurors selected, 2 are Hispanic.

(a) What proportion of the jury described is Hispanic?

(b) If 12 jurors are randomly selected from a population that is 45% Hispanic, what is the probability that 2 or fewer jurors will be Hispanic?

(c) If you were the lawyer of a Hispanic defendant, what might you argue?

42. Sullivan Survey: Car Color According to paint manufacturer DuPont, 6% of all cars worldwide are red. In the Sullivan Statistics Survey, of 175 respondents, 17, or 9.7%, indicated the color of their car is red. Determine if the results of the Sullivan Survey contradict those of DuPont by computing $P(X \geq 17)$, where X is a binomial random variable with $n = 175$ and $p = 0.06$. Explain what the probability represents.

NW 43. On-Time Flights According to flightstats.com, American Airlines flights from Dallas to Chicago are on time 80% of the time. Suppose 100 flights are randomly selected.

(a) Compute the mean and standard deviation of the random variable X, the number of on-time flights in 100 trials of the probability experiment.

(b) Interpret the mean.

(c) Would it be unusual to observe 75 on-time flights in a random sample of 100 flights from Dallas to Chicago? Why?

44. Morality In a recent poll, the Gallup Organization found that 45% of adult Americans believe that the overall state of moral values in the United States is poor.

(a) Compute the mean and standard deviation of the random variable X, the number of adults who believe that the overall state of moral values in the United States is poor based on a random sample of 500 adult Americans.

(b) Interpret the mean.

(c) Would it be unusual to identify 240 adult Americans who believe that the overall state of moral values in the United States is poor based on a random sample of 500 adult Americans? Why?

45. Toilet Flushing In the Healthy Handwashing Survey conducted by Bradley Corporation, it was found that 64% of adult Americans operate the flusher of toilets in public restrooms with their foot.

(a) If 500 adult Americans are randomly selected, how many would we expect to flush toilets in public restrooms with their foot?

(b) Would it be unusual to observe 280 adult Americans who flush toilets in public restrooms with their foot?

46. Allergy Sufferers Clarinex-D is a medication whose purpose is to reduce the symptoms associated with a variety of allergies. In clinical trials of Clarinex-D, 5% of the patients in the study experienced insomnia as a side effect.

(a) If 240 users of Clarinex-D are randomly selected, how many would we expect to experience insomnia as a side effect?

(b) Would it be unusual to observe 20 patients experiencing insomnia as a side effect in 240 trials of the probability experiment? Why?

47. Spanking In March 1995, The Harris Poll reported that 80% of parents spank their children. Suppose a recent poll of 1030 adult Americans with children finds that 781 indicated that they spank their children. If we assume parents' attitude toward spanking has not changed since 1995, how many of 1030 parents surveyed would we expect to spank? Do the results of the survey suggest that parents' attitude toward spanking may have changed since 1995? Why?

48. Government Solutions? In May, 2000, the Gallup Organization reported that 11% of adult Americans had a great deal of trust and confidence in the federal government handling domestic issues. Suppose a survey of a random sample of 1100 adult Americans finds that 84 have a great deal of trust and confidence in the federal government handling domestic issues. Would these results be considered unusual? Why?

49. Racial Profiling in New York City The following excerpt is from the *Racial Profiling Data Collection Resource Center*.

In 2006, the New York City Police Department stopped a half-million pedestrians for suspected criminal involvement. Raw statistics for these encounters suggest large racial disparities— 89 percent of the stops involved nonwhites. Do these statistics point to racial bias in police officers' decisions to stop particular pedestrians? Do they indicate that officers are particularly intrusive when stopping nonwhites?

Source: Published by Northeastern University.

Write a report that answers the questions posed using the fact that 44% of New York City residents were classified as white in 2006. In your report, cite some shortcomings in using the proportion of white residents in the city to formulate likelihoods.

50. Simulation According to the U.S. National Center for Health Statistics, there is a 98% probability that a 20-year-old male will survive to age 30.

(a) Using statistical software, simulate taking 100 random samples of size 30 from this population.

(b) Using the results of the simulation, compute the probability that exactly 29 of the 30 males survive to age 30.

(c) Compute the probability that exactly 29 of the 30 males survive to age 30, using the binomial probability distribution. Compare the results with part (b).

(d) Using the results of the simulation, compute the probability that at most 27 of the 30 males survive to age 30.

(e) Compute the probability that at most 27 of the 30 males survive to age 30 using the binomial probability distribution. Compare the results with part (d).

(f) Compute the mean number of male survivors in the 100 simulations of the probability experiment. Is it close to the expected value?

(g) Compute the standard deviation of the number of male survivors in the 100 simulations of the probability experiment. Compare the result to the theoretical standard deviation of the probability distribution.

(h) Did the simulation yield any unusual results?

51. Overbooking Flights Historically, the probability that a passenger will miss a flight is 0.0995. *Source: Passenger-Based Predictive Modeling of Airline No-show Rates* by Richard D. Lawrence, Se June Hong, and Jacques Cherrier. Airlines do not like flights with empty seats, but it is also not desirable to have overbooked flights because passengers must be "bumped" from the flight. The Lockheed L49 Constellation has a seating capacity of 54 passengers.

(a) If 56 tickets are sold, what is the probability 55 or 56 passengers show up for the flight resulting in an overbooked flight?

(b) Suppose 60 tickets are sold, what is the probability a passenger will have to be "bumped"?

(c) For a plane with seating capacity of 250 passengers, how many tickets may be sold to keep the probability of a passenger being "bumped" below 1%?

52. Athletics Participation According to the High School Athletics Participation Survey, approximately 55% of students enrolled in high schools participate in athletic programs. You are performing a study of high school students and would like at least 11 students in the study to be participating in athletics.
Source: National Federation of State High School Associations

(a) How many high school students do you expect to have to randomly select?

(b) How many high school students do you have to select to have a 99% probability that the sample contains at least 12 who participate in athletics?

53. Geometric Probability Distribution A probability distribution for the random variable X, the number of trials until a success is observed, is called the **geometric probability distribution**. It has the same criteria as the binomial distribution (see page 328), except that the number of trials is not fixed. Its probability distribution function (pdf) is

$$P(x) = p(1-p)^{x-1}, \quad x = 1, 2, 3, \ldots$$

where p is the probability of success.

(a) What is the probability that Shaquille O'Neal misses his first two free throws and makes the third? Over his career, he made 52.4% of his free throws. That is, find $P(3)$.

(b) Construct a probability distribution for the random variable X, the number of free-throw attempts of Shaquille O'Neal until he makes a free throw. Construct the distribution for $x = 1, 2, 3, \ldots, 10$. The probabilities are small for $x > 10$.

(c) Compute the mean of the distribution, using the formula presented in Section 6.1.

(d) Compare the mean obtained in part (c) with the value $\dfrac{1}{p}$. Conclude that the mean of a geometric probability distribution is $\mu_X = \dfrac{1}{p}$. How many free throws do we expect Shaq to take before we observe a made free throw?

54. Negative Binomial Probability Distribution The **negative binomial probability distribution** can be used to compute the probability of the random variable X, the number of trials necessary to observe r successes of a binomial experiment. The probability distribution function is given by

$$P(x) = (_{x-1}C_{r-1})p^r(1-p)^{x-r}$$
$$x = r, r+1, r+2, \ldots$$

Consider a roulette wheel. Remember, a roulette wheel has 2 green slots, 18 red slots, and 18 black slots.

(a) What is the probability that it will take $x = 1$ trial before observing $r = 1$ green?

(b) What is the probability that it will take $x = 20$ trials before observing $r = 2$ greens?

(c) What is the probability that it will take $x = 30$ trials before observing $r = 3$ greens?

(d) The expected number of trials before observing r successes is $\dfrac{r}{p}$. What is the expected number of trials before observing 3 greens?

55. Putting It Together: Beating the Stock Market One measure of successful investing is being able to "beat the market." To beat the market in any given year, an investor must earn a rate of return greater than the rate of return of some market basket of stocks, such as the Dow Jones Industrial Average (DJIA) or Standard and Poor's 500 (S&P 500). Suppose in any given year, there is a probability of 0.5 that a particular investment advisor beats the market for his/her clients.

(a) If there are 5000 investment advisors across the country, how many would be expected to beat the market in any given year?

(b) Assume beating the market in one year is independent of beating the market in any other year. What is the probability that a randomly selected investment advisor beats the market in two consecutive years? Based on this result, how many of 5000 investment advisors would be expected to beat the market for two consecutive years?

(c) Assume beating the market in one year is independent of beating the market in any other year. What is the probability that a randomly selected investment advisor beats the market in five consecutive years? Based on this result, how many of 5000 investment advisors would be expected to beat the market for five consecutive years?

(d) Assume beating the market in one year is independent of beating the market in any other year. What is the probability that a randomly selected investment advisor beats the market in ten consecutive years? Based on this result, how many of 5000 investment advisors would be expected to beat the market for ten consecutive years?

(e) Assume a randomly selected investment advisor can beat the market with probability 0.5 and investment results from year to year are independent. Suppose we randomly select 5000 investment advisors and determine the number x who have beaten the market the past ten years. Explain why this is a binomial experiment (assuming there are tens of thousands of investment advisors in the population) and clearly state what a success represents.

(f) Use the results of part (e) to determine the probability of identifying at least six investment advisors who will beat the market for ten consecutive years. Interpret this result. Is it unusual to identify at least six investment advisor who consistently beats the market even though his/her underlying ability to beat the market is 0.5? Explain.

Explaining the Concepts

56. State the criteria for a binomial probability experiment.

57. Explain what "success" means in a binomial probability experiment.

58. Explain how the value of n, the number of trials in a binomial experiment, affects the shape of the distribution of a binomial random variable.

59. Explain how the value of p, the probability of success, affects the shape of the distribution of a binomial random variable.

60. When can the Empirical Rule be used to identify unusual results in a binomial experiment? Why can the Empirical Rule be used to identify results in a binomial experiment?

6.3   The Poisson Probability Distribution

Objectives   ① Determine if a probability experiment follows a Poisson process

② Compute probabilities of a Poisson random variable

③ Find the mean and standard deviation of a Poisson random variable

① Determine If a Probability Experiment Follows a Poisson Process

Another discrete probability model is the *Poisson probability distribution*, named after Siméon Denis Poisson. This probability distribution can be used to compute probabilities of experiments in which the random variable X counts the number of occurrences (successes) of a particular event within a specified interval (usually time or space).

EXAMPLE 1   Illustrating a Poisson Process

A McDonald's manager knows from prior experience that cars arrive at the drive-through at an average rate of two cars per minute between the hours of 12:00 noon and 1:00 P.M. The random variable X, the number of cars that arrive between 12:20 and 12:40, follows a Poisson process.   ●

Definition   A random variable X, the number of successes in a fixed interval, follows a **Poisson process** provided the following conditions are met.

1. The probability of two or more successes in any sufficiently small subinterval* is 0.
2. The probability of success is the same for any two intervals of equal length.
3. The number of successes in any interval is independent of the number of successes in any other interval provided the intervals are not overlapping.

In the McDonald's example, if we divide the time interval into a sufficiently small length (say, 1 second), it is impossible for more than one car to arrive. This satisfies part 1 of the definition. Part 2 is satisfied because the cars arrive at an average rate of two cars per minute over the 1-hour interval. Part 3 is satisfied because the number of cars that arrive in any 1-minute interval (say between 12:23 P.M. and 12:24 P.M.) is independent of the number of cars that arrive in any other 1-minute interval (say between 12:35 P.M. and 12:36 P.M.).

② Compute Probabilities of a Poisson Random Variable

If the random variable X follows a Poisson process, we can use the following probability rule to compute Poisson probabilities.

In Other Words
Poisson probabilities are used to determine the probability of the number of successes in a fixed interval of time or space.

Poisson Probability Distribution Function

If X is the number of successes in an interval of fixed length t, then the probability of obtaining x successes in the interval is

$$P(x) = \frac{(\lambda t)^x}{x!} e^{-\lambda t} \quad x = 0, 1, 2, 3, \ldots \tag{1}$$

where λ (the Greek letter lambda) represents the average number of occurrences of the event in some interval of length 1 and $e \approx 2.71828$.

*For example, the fixed interval might be any time between 0 and 5 minutes. A subinterval could be any time between 1 and 2 seconds.

To clarify the roles of λ and t, revisit Example 1. Here, $\lambda = 2$ cars per minute, while $t = 20$ minutes (the length of time between 12:20 P.M. and 12:40 P.M.).

● Now Work Problem 3

EXAMPLE 2   Computing Probabilities of a Poisson Process

Historical Note

Siméon Denis
Poisson was
born on June 21,
1781, in Pithiviers,
France. He was
educated by
his father, who
wanted him to become a
doctor. However, one of his
first patients died within a few
hours of Poisson's treatment,
so Poisson decided never to
practice medicine again. After
this mishap, Poisson enrolled
in the Polytechnic school. While
there, he drew the attention
of Lagrange, Legendre, and
Laplace. In 1837, Poisson
published *Récherches sur la
probabilité des jugements*, in
which he first presented the
probability distribution named
after him.

Problem A McDonald's manager knows that cars arrive at the drive-through at the average rate of two cars per minute between the hours of 12 noon and 1:00 P.M. Determine and interpret the probability of the following events:
(a) Exactly six cars arrive between 12 noon and 12:05 P.M.
(b) Fewer than six cars arrive between 12 noon and 12:05 P.M.
(c) At least six cars arrive between 12 noon and 12:05 P.M.

Approach The cars arrive at a rate of two per minute over the time interval between 12 noon and 1:00 P.M. We know from Example 1 that the random variable X follows a Poisson process, where $x = 0, 1, 2, \ldots$. The Poisson probability distribution function requires a value for λ and t. Since the cars arrive at a rate of two per minute, $\lambda = 2$. The interval of time we are interested in is five minutes, so $t = 5$. Use the Poisson probability distribution function (1).

Solution
(a) The probability that exactly six cars arrive between 12 noon and 12:05 P.M. is

$$P(6) = \frac{[2(5)]^6}{6!} e^{-2(5)} = \frac{1,000,000}{720} e^{-10} = 0.0631$$

Interpretation On about 6 of every 100 days, exactly six cars will arrive between 12:00 noon and 12:05 P.M.

(b) The probability that fewer than six cars arrive between 12:00 noon and 12:05 P.M. is

$$
\begin{aligned}
P(X < 6) &= P(X \le 5) \\
&= P(0) + P(1) + P(2) + P(3) + P(4) + P(5) \\
&= \frac{[2(5)]^0}{0!} e^{-2(5)} + \frac{[2(5)]^1}{1!} e^{-2(5)} + \frac{[2(5)]^2}{2!} e^{-2(5)} \\
&\quad + \frac{[2(5)]^3}{3!} e^{-2(5)} + \frac{[2(5)]^4}{4!} e^{-2(5)} + \frac{[2(5)]^5}{5!} e^{-2(5)} \\
&= \frac{1}{1} e^{-10} + \frac{10}{1} e^{-10} + \frac{100}{2} e^{-10} + \frac{1000}{6} e^{-10} + \frac{10,000}{24} e^{-10} + \frac{100,000}{120} e^{-10} \\
&= 0.0671
\end{aligned}
$$

Interpretation On about 7 of every 100 days, fewer than six cars will arrive between 12:00 noon and 12:05 P.M.

(c) The probability that at least six cars arrive between 12 noon and 12:05 P.M. is the complement of the probability that fewer than six cars arrive during that time. That is,

$$P(X \ge 6) = 1 - P(X < 6) = 1 - 0.0671 = 0.9329$$

Interpretation On about 93 of every 100 days, at least six cars will arrive between 12:00 noon and 12:05 P.M.

● Now Work Problem 11

CAUTION!
For a Poisson process, at-least and more-than probabilities must be computed by using the Complement Rule.

An important point is that *at-least* and *more-than* probabilities for a Poisson process must be found using the Complement Rule since the random variable X can be any integer greater than or equal to 0.

③ **Find the Mean and Standard Deviation of a Poisson Random Variable**

If two cars per minute arrive at McDonald's between 12 noon and 1:00 P.M., how many cars would you expect to arrive between 12 noon and 12:05 P.M.? Considering that two cars arrive every minute (on average) and we are observing the arrival of cars for five minutes, it seems reasonable to expect $2(5) = 10$ cars to arrive. Since the expected value of a random variable is the mean of the random variable, it is reasonable that $\mu_X = \lambda t$ for interval t.

Mean and Standard Deviation of a Poisson Random Variable

A random variable X that follows a Poisson process with parameter λ has mean (or expected value) and standard deviation given by the formulas

$$\mu_X = \lambda t \quad \text{and} \quad \sigma_X = \sqrt{\lambda t} = \sqrt{\mu_X}$$

where t is the length of the interval.

Because $\mu_X = \lambda t$, we restate the Poisson probability distribution function in terms of its mean.

Poisson Probability Distribution Function

If X is the number of successes in an interval of fixed length and X follows a Poisson process with mean μ, the probability distribution function for X is

$$P(x) = \frac{\mu^x}{x!} e^{-\mu} \quad x = 0, 1, 2, 3 \ldots .$$

EXAMPLE 3   Beetles and the Poisson Distribution

Problem A biologist performs an experiment in which 2000 Asian beetles are allowed to roam in an enclosed area of 1000 square feet. The area is divided into 200 subsections of 5 square feet each.
(a) If the beetles spread evenly throughout the enclosed area, how many beetles would you expect in each subsection?
(b) What is the standard deviation of X, the number of beetles in a particular subsection?
(c) What is the probability of finding exactly eight beetles in a particular subsection?
(d) Would it be unusual to find more than sixteen beetles in a particular subsection?

Approach If the beetles spread evenly throughout the enclosed region, we can model the distribution of the beetles using Poisson probabilities.

Solution
(a) If the beetles spread evenly throughout the enclosed area, we expect

$$\mu_X = \frac{2000 \text{ beetles}}{200 \text{ subsections}} = 10 \text{ beetles per subsection}$$

(b) $\sigma_X = \sqrt{\mu_X} = \sqrt{10} \approx 3.2$

(continued)

(c) Use the expected value $\mu_X = 10$ in the Poisson probability distribution function to compute the probability of finding exactly eight beetles in a subsection.

$$P(8) = \frac{10^8}{8!} e^{-10} \qquad P(x) = \frac{\mu^x}{x!} e^{-\mu}, \mu = 10, x = 8$$
$$= 0.1126$$

Interpretation In 100 trials of this study, we expect to find eight beetles in a particular subsection about 11 times.

(d) Compute $P(X > 16)$ using the Poisson probability distribution function, the Complement Rule, and a TI-84 Plus C graphing calculator.

$$P(X > 16) = 1 - P(X \le 16)$$
$$= 1 - 0.9730 \qquad \text{See Figure 13.}$$
$$= 0.0270$$

Interpretation According to the Poisson probability model, there will be more than sixteen beetles in a subsection about 3 times in 100. To observe more than sixteen beetles in a subsection is rather unusual.

Figure 13

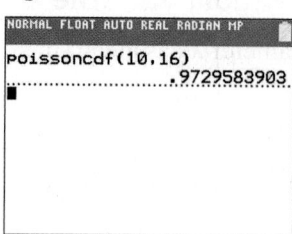

Technology Step-by-Step Computing Poisson Probabilities via Technology

TI-83/84 Plus
Computing $P(x)$
1. Press 2nd VARS to access the probability distribution menu.
2. Highlight `poissonpdf` (and hit ENTER.
3. Enter the value of λ, followed by the number of successes, x. Highlight Paste and hit ENTER. For example, with $\lambda = 10$ and $x = 4$, you should see the following on the HOME screen

    poissonpdf (10,4)

    Then hit ENTER.

Computing $P(X \le x)$
Instead of selecting `poissonpdf(`, select `poissoncdf(`. Everything else is the same.

Minitab
Computing $P(x)$
1. Select the CALC menu, highlight **Probability Distributions**, then highlight **Poisson** ....
2. Select Probability, enter the mean. Select the "Input constant:" radio button. Enter the input constant. Click OK.

Computing $P(X \le x)$
Follow the same steps as those followed for computing $P(x)$. In the window that comes up after selecting the Poisson

Distribution, select Cumulative probability instead of Probability.

Excel
Computing $P(x)$
1. With the cursor in cell where you would like the probability, select the Formulas tab. Select More Functions. Highlight Statistical, then highlight POISSON.DIST in the Function name menu.
2. In the cell labeled X, enter the value of the random variable x. In the cell labeled mean, enter the mean. In the cell labeled cumulative, type FALSE. Click OK.

Computing $P(X \le x)$
Follow the same steps as those presented for computing $P(x)$. In the POISSON.DIST window, type TRUE in the cumulative cell.

StatCrunch
1. Select **Stat**, highlight **Calculators**, select **Poisson**.
2. Enter the mean, μ. Highlight the Standard tab. In the pull-down menu, decide if you want to compute $P(X \le x), P(X < x)$, and so on. Finally, enter the value of x. If you want to compute $P(a \le X \le b)$, highlight the Between tab. Enter the values of a and b. Click Compute.

6.3 Assess Your Understanding

Vocabulary and Skill Building

1. State the conditions required for a random variable X to follow a Poisson process.

2. Explain the role of λ and t in the Poisson probability formula.

In Problems 3–6, state the values of λ and t for each Poisson process.

NW 3. The hits to a website occur at the rate of 10 per minute between 7:00 P.M. and 9:00 P.M. The random variable X is the number of hits between 7:30 P.M. and 7:35 P.M.

4. The phone calls to a computer software help desk occur at the rate of 5 per minute between 3:00 P.M. and 4:00 P.M. The random variable X is the number of phone calls between 3:10 P.M. and 3:20 P.M.

5. The flaws in a piece of timber occur at the rate of 0.07 per linear foot. The random variable X is the number of flaws in the next 20 linear feet of timber.

6. The potholes on a major highway in the city of Madison occur at the rate of 3.4 per mile. The random variable X is the number of potholes in 10 miles of randomly selected highway.

In Problems 7 and 8, the random variable X follows a Poisson process with the given mean.

7. Assuming $\mu = 5$, compute

(a) $P(6)$ **(b)** $P(X < 6)$
(c) $P(X \geq 6)$ **(d)** $P(2 \leq X \leq 4)$

8. Assuming $\mu = 7$, compute

(a) $P(10)$ **(b)** $P(X < 10)$
(c) $P(X \geq 10)$ **(d)** $P(7 \leq X \leq 9)$

In Problems 9 and 10, the random variable X follows a Poisson process with the given value of λ and t.

9. Assuming $\lambda = 0.07$ and $t = 10$, compute

(a) $P(4)$ **(b)** $P(X < 4)$
(c) $P(X \geq 4)$ **(d)** $P(4 \leq X \leq 6)$
(e) μ_X and σ_X

10. Assuming $\lambda = 0.02$ and $t = 50$, compute

(a) $P(2)$ **(b)** $P(X < 2)$
(c) $P(X \geq 2)$ **(d)** $P(1 \leq X \leq 3)$
(e) μ_X and σ_X

Applying The Concepts

NW 11. Hits to a Website The number of hits to a website follows a Poisson process; hits occur at the rate of 1.4 per minute between 7:00 P.M. and 9:00 P.M. Compute the probability that the number of hits between 7:30 P.M. and 7:35 P.M. is

(a) exactly seven. Interpret the result.
(b) fewer than seven. Interpret the result.
(c) at least seven. Interpret the result.

12. Calls to the Help Desk The phone calls to a computer software help desk occur at the rate of 2.1 per minute between 3:00 P.M. and 4:00 P.M. Compute the probability that the number of calls between 3:10 P.M. and 3:15 P.M. is

(a) exactly eight. Interpret the result.
(b) fewer than eight. Interpret the result.
(c) at least eight. Interpret the result.

13. Insect Fragments The Food and Drug Administration sets a Food Defect Action Level (FDAL) for the various foreign substances that inevitably end up in the food we eat and liquids we drink. For example, the FDAL level for insect filth in peanut butter is 0.3 insect fragment (larvae, eggs, body parts, and so on) per gram. Suppose that a supply of peanut butter contains 0.3 insect fragment per gram. Compute the probability that the number of insect fragments in a 5-gram sample of the peanut butter is

(d) exactly two. Interpret the result.
(e) fewer than two. Interpret the result.
(f) at least two. Interpret the result.
(g) at least one. Interpret the result.
(h) Would it be unusual for a 5-gram sample of this supply of peanut butter to contain four or more insect fragments?

14. Potholes The potholes on a major highway in the city of Chicago occur at the rate of 3.4 per mile. Compute the probability that the number of potholes over 3 miles of randomly selected highway is

(a) exactly seven. Interpret the result.
(b) fewer than seven. Interpret the result.
(c) at least seven. Interpret the result.
(d) Would it be unusual for a randomly selected 3-mile stretch of highway in Chicago to contain more than 15 potholes?

15. Airline Fatalities According to www.meretrix.com, airline fatalities occur at the rate of 0.05 fatal accidents per 100 million miles. Find the probability that, during the next 100 million miles of flight, there will be

(a) exactly zero fatal accidents. Interpret the result.
(b) at least one fatal accident. Interpret the result.
(c) more than one fatal accident. Interpret the result.

16. Traffic Fatalities According to www.meretrix.com, traffic fatalities occur at the rate of 1.32 fatal accidents per 100 million miles. Find the probability that, during the next 100 million vehicle miles, there will be

(a) exactly zero fatal accidents. Interpret the result.
(b) at least one fatal accident. Interpret the result.
(c) more than one fatal accident. Interpret the result.
(d) Compare these results with the results in Problem 15. Given the choice, would you rather fly or drive?

17. Florida Hurricanes From 1900 to 2014 (115 years), Florida suffered 31 direct hits from major (category 3 to 5) hurricanes. Assume that this was typical and the number of hits per year follows a Poisson distribution.
Source: National Hurricane Center

(a) What is the probability that Florida will not be hit by any major hurricanes in a single year?
(b) What is the probability that Florida will be hit by at least one major hurricane in a single year? Is this unusual?
(c) What is the probability that Florida will be hit by at least three major hurricanes in a single year, as happened in 2004? Does this indicate that the 2004 hurricane season in Florida was unusual?
(d) What is the probability that Florida will be hit by at least two major hurricanes in a single year, as happened in 2005? Does this indicate that the 2005 hurricane season in Florida was unusual?

18. Police Dispatch Officer Thompson of the Bay Ridge Police Department works the graveyard shift. He averages 4.5 calls per shift from his dispatcher. Assume the number of calls follows a Poisson distribution. Would it be unusual for Officer Thompson to get fewer than 2 calls in a shift?

19. Wendy's Drive-Through Cars arrive at Wendy's drive-through at a rate of 0.2 car per minute between the hours of 11:00 P.M. and 1:00 A.M. on Saturday evening. Wendy's begins an advertising blitz that touts its late-night service. After one week of advertising, Wendy's officials count the number of cars, X, arriving at Wendy's drive-through between the hours of 12:00 midnight and 12:30 A.M. at 200 of its restaurants. The results are shown in the following table:

x (number of cars arriving)	Frequency
1	4
2	5
3	13
4	23
5	25
6	28
7	25
8	27
9	21
10	15
11	5
12	3
13	2
14	2
15	0
16	2

(a) Construct a probability distribution for the random variable X, assuming it follows a Poisson process with $\lambda = 0.2$ and $t = 30$. This is the probability distribution of X before the advertising.

(b) Compute the expected number of restaurants that will have 0 arrivals, 1 arrival, and so on.

(c) Compare these results with the number of arrivals after the advertising. Does it appear the advertising was effective? Why?

20. Quality Control A builder ordered two hundred 8-foot grade A 2-by-4s for a construction job. To qualify as a grade A board, each 2-by-4 will have no knots and will average no more than 0.05 imperfection per linear foot. The following table lists the number of imperfections per 2-by-4 in the 200 ordered:

x (number of imperfections)	Frequency
0	124
1	51
2	20
3	5

(a) Construct a probability distribution for the random variable X, the number of imperfections per 8 feet of board, assuming that it follows a Poisson process with $\lambda = 0.05$ and $t = 8$.

(b) Compute the expected number of 2-by-4s that will have 0 imperfections, 1 imperfection, and so on.

(c) Compare these results with the number of actual imperfections. Does it appear the 2-by-4s are of grade A quality? Why?

21. Prussian Army In 1898, Ladislaus von Bortkiewicz published *The Law of Small Numbers*, in which he demonstrated the power of the Poisson probability law. Before his publication, the law was used exclusively to approximate binomial probabilities. He demonstrated the law's power, using the number of Prussian cavalry soldiers who were kicked to death by their horses. The Prussian army monitored 10 cavalry corps for 20 years and recorded the number X of annual fatalities because of horse kicks for the 200 observations. The following table shows the data:

Number of Deaths, x	Number of Times x Deaths Were Observed
0	109
1	65
2	22
3	3
4	1

Source: Adapted from *An Introduction to Mathematical Statistics* by Larsen et al., Prentice Hall, Upper Saddle River, NJ, 2001

(a) Compute the proportion of years in which there were 0 deaths, 1 death, 2 deaths, 3 deaths, and 4 deaths.

(b) From the data in the table, what was the mean number of deaths per year?

(c) Use the mean number of deaths per year found in part (b) and the Poisson probability law to determine the theoretical proportion of years that 0 deaths should occur. Repeat this for 1, 2, 3, and 4 deaths.

(d) Compare the observed proportions to the theoretical proportions. Do you think the data can be modeled by the Poisson probability law?

22. Simulation Data from the National Center for Health Statistics show that spina bifida occurs at the rate of 28 per 100,000 live births. Let the random variable X represent the number of occurrences of spina bifida in a random sample of 100,000 live births.

(a) What is the expected number of children with spina bifida per 100,000 live births in any given year?

(b) Using statistical software such as Minitab, simulate taking 200 random samples of 100,000 live births, assuming $\mu = 28$.

(c) Approximate the probability that fewer than 18 births per 100,000 result in spina bifida.

(d) In 2005, 17.96 births per 100,000 resulted in babies born with spina bifida. In light of the results of parts (b) and (c), is this an unusual occurrence? What might you conclude?

23. Simulation According to the National Center for Health Statistics, the common cold occurs at the rate of 23.6 colds per 100 people during the course of a year in the 18- to 24-year-old age group. Let the random variable X represent the number of 18- to 24-year-olds out of a sample of 500 who have had a common cold in the past year.

(a) What is the expected number of colds for every 500 18- to 24-year-olds?

(b) Using statistical software, such as Minitab, simulate taking 100 random samples of size 500 from this population.

(c) Approximate the probability that at least 150 18- to 24-year-olds in a group of 500 will have experienced a common cold in the past year.

(d) Approximate the probability that fewer than 100 18- to 24-year-olds in a group of 500 will have experienced the common cold in the past year.

(e) Given the simulation, compute the mean number of 18- to 24-year-olds out of 500 who have experienced the common cold in the past year.

(f) Given the simulation, compute the standard deviation of the number of 18- to 24-year-olds out of 500 who have experienced the common cold in the past year.

(g) Compute the 5-number summary of the data. Did the simulation result in any unusual results?

24. How Long Do I Have to Wait? The number of hits to a website follows a Poisson process and occurs at the rate of 10 hits per minute between 7:00 P.M. and 9:00 P.M. How long should you expect to wait before the probability of at least 1 hit to the site is 95%? *Hint:* $P(X \geq 1) = 1 - P(0)$

Chapter 6 Review

Summary

In this chapter, we introduced probability models for random variables. A random variable represents the numerical measurement of the outcome from a probability experiment. Discrete random variables have either a finite or a countable number of outcomes. The values of a discrete random variable can be plotted on a number line with space between each point. The term *countable* means that the values result from counting. Continuous random variables have infinitely many values and typically result from measurement. The values of a continuous random variable can be plotted on a line in an uninterrupted fashion.

We also looked at the probability distribution of discrete random variables. Discrete probability distributions must satisfy the following two criteria: (1) All probabilities must be between 0 and 1, inclusive, and (2) the sum of all probabilities must equal 1.

Discrete probability distributions can be presented by a table, graph, or mathematical formula.

The mean and standard deviation of a random variable describe the center and spread of the distribution. The mean of a random variable is also called its expected value.

We discussed two discrete probability distributions. First, we presented the binomial probability distribution function. A probability experiment is considered a binomial experiment if

1. The experiment is performed a fixed number of times, n. Each repetition of the experiment is called a trial.

2. The trials are independent. This means that the outcome of one trial will not affect the outcome of the other trials.

3. For each trial, there are two mutually exclusive (disjoint) outcomes: success or failure.

4. The probability of success is the same for each trial of the experiment. We denote the probability of success as p.

Binomial probabilities were obtained by hand, a table, or via technology using its probability distribution function.

The mean and standard deviation of a binomial random variable may be used to identify unusual results in a binomial experiment using the Empirical Rule provided $np(1 - p) \geq 10$.

We also discussed the Poisson probability distribution function. A Poisson process is one in which the following conditions are met:

1. The probability of two or more successes in any sufficiently small subinterval is 0.

2. The probability of success is the same for any two intervals of equal length.

3. The number of successes in any interval is independent of the number of successes in any other disjoint interval.

Poisson probabilities were obtained by hand or via technology using its probability distribution function. We also determined the mean and standard deviation of a Poisson random variable.

Vocabulary

Random variable (p. 316)
Discrete random variable (p. 316)
Continuous random variable (p. 316)
Probability distribution (p. 317)

Expected value (p. 321)
Binomial probability distribution (p. 328)
Binomial experiment (p. 328)
Trial (p. 328)

Binomial random variable (p. 328)
Cumulative distribution function (p. 335)
Poisson process (p. 343)

Formulas

Mean (or Expected Value) of a Discrete Random Variable

$$\mu_X = E(X) = \sum [x \cdot P(x)]$$

Standard Deviation of a Discrete Random Variable

$$\sigma_X = \sqrt{\sum [(x - \mu_X)^2 \cdot P(x)]} = \sqrt{\sum [x^2 \cdot P(x)] - \mu_X^2}$$

Binomial Probability Distribution Function

$$P(x) = {}_nC_x \, p^x (1 - p)^{n-x} \quad x = 0, 1, 2, \ldots, n$$

Mean of a Binomial Random Variable

$$\mu_X = np$$

Standard Deviation of a Binomial Random Variable

$$\sigma_X = \sqrt{np(1 - p)}$$

Poisson Probability Distribution Function

$$P(x) = \frac{(\lambda t)^x}{x!} e^{-\lambda t} = \frac{\mu^x}{x!} e^{-\mu} \quad x = 0, 1, 2, \ldots$$

Mean and Standard Deviation of a Poisson Random Variable

$$\mu_X = \lambda t \qquad \sigma_X = \sqrt{\lambda t} = \sqrt{\mu_X}$$

OBJECTIVES

Section	You should be able to ...	Example(s)	Review Exercises
6.1	1 Distinguish between discrete and continuous random variables (p. 316)	1	1
	2 Identify discrete probability distributions (p. 317)	2 and 3	2, 3(a)
	3 Graph discrete probability distributions (p. 318)	4	3(b)
	4 Compute and interpret the mean of a discrete random variable (p. 319)	5, 6, and 9	3(c)
	5 Interpret the mean of a discrete random variable as an expected value (p. 321)	7	4
	6 Compute the standard deviation of a discrete random variable (p. 322)	8 and 9	3(d)
6.2	1 Determine whether a probability experiment is a binomial experiment (p. 328)	1	5
	2 Compute probabilities of binomial experiments (p. 329)	2–5	6(a)–(e), 7(a)–(d), 8(a), 15
	3 Compute the mean and standard deviation of a binomial random variable (p. 335)	6	6(f), 7(e), 8(b)
	4 Graph a binomial probability distribution (p. 336)	7	8(c)
6.3	1 Determine if a probability experiment follows a Poisson process (p. 343)	1	11
	2 Compute probabilities of a Poisson random variable (p. 343)	2 and 3	12(a)–(d), 13, 14
	3 Find the mean and standard deviation of a Poisson random variable (p. 345)	3	12(e)

Review Exercises

1. Determine whether the random variable is discrete or continuous. In each case, state the possible values of the random variable.

(a) The number of students in a randomly selected elementary school classroom

(b) The amount of snow that falls in Minneapolis during the winter season

(c) The flight time accumulated by a randomly selected Air Force fighter pilot

(d) The number of points scored by the Miami Heat in a randomly selected basketball game

2. Determine whether the distribution is a discrete probability distribution. If not, state why.

(a)

x	$P(x)$
0	0.34
1	0.21
2	0.13
3	0.04
4	0.01

(b)

x	$P(x)$
0	0.40
1	0.31
2	0.23
3	0.04
4	0.02

3. Stanley Cup The Stanley Cup is a best-of-seven series to determine the champion of the National Hockey League. The following data represent the number of games played, X, in the Stanley Cup before a champion was determined from 1939 to 2015.

Note: There was no champion in 2005. The season was cancelled because of a labor dispute.

x	Frequency
4	20
5	18
6	22
7	16

Source: Information Please Almanac

(a) Construct a probability model for the random variable X, the number of games in the Stanley Cup.
(b) Graph the discrete probability distribution.
(c) Compute and interpret the mean of the random variable X.
(d) Compute the standard deviation of the random variable X.

4. Expected Value of Three-Card Poker A popular casino table game is three-card poker. One aspect of the game is the "pair plus" bet in which a player is paid a dollar amount for any hand of a pair or better, regardless of the hand the dealer has. The table shows the profit and probability of various hands of a player playing the $5 pair plus bet.

Outcome	Profit ($)	Probability
Straight flush	200	12/5525
Three of a kind	150	1/425
Straight	30	36/1105
Flush	20	274/5525
Pair	5	72/425
Other	−5	822/1105

Source: http://wizardofodds.com/threecardpoker

(a) What is the expected profit when playing the $5 pair plus bet in three card poker.
(b) If you play the game for 4 hours with an average of 35 hands per hour, how much would you expect to lose?

5. Determine whether the probability experiment represents a binomial experiment. If not, explain why.

(a) According to the *Chronicle of Higher Education*, the probability that a randomly selected incoming freshman will graduate from college within 6 years is 0.54. Suppose that 10 incoming freshmen are randomly selected. After 6 years, each student is asked whether he or she graduated.
(b) An experiment is conducted in which a single die is cast until a 3 comes up. The number of throws required is recorded.

6. Emergency Room Visits The probability that a randomly selected patient who visits the emergency room (ER) will die within 1 year of the visit is 0.05.

Source: SuperFreakonomics

(a) What is the probability that exactly 1 of 10 randomly selected visitors to the ER will die within 1 year? Interpret this result.
(b) What is the probability that fewer than 2 of 25 randomly selected visitors to the ER will die within 1 year? Interpret this result.
(c) What is the probability that at least 2 of 25 randomly selected visitors to the ER will die within 1 year? Interpret this result.

(d) What is the probability that at least 8 of 10 randomly selected visitors to the ER will *not* die within 1 year?
(e) Would it be unusual if more than 3 of 30 randomly selected visitors to the ER died within 1 year? Why?
(f) In a random sample of 1000 visitors to the ER, how many visitors are expected to die within the next year? What is the standard deviation number of deaths?
(g) At a particular emergency room, a researcher obtains a random sample of 800 visitors and finds that after 1 year 51 of them have died. Do you think this particular emergency room should be investigated to see if something unusual is occurring?

7. Driving Age According to a Gallup poll, 60% of U.S. women 18 years old or older stated that the minimum driving age should be 18. In a random sample of 15 U.S. women 18 years old or older, find the probability that:

(a) Exactly 10 believe that the minimum driving age should be 18.
(b) Fewer than 5 believe that the minimum driving age should be 18.
(c) At least 5 believe that the minimum driving age should be 18.
(d) Between 7 and 12, inclusive, believe that the minimum driving age should be 18.
(e) In a random sample of 200 U.S. women 18 years old or older, what is the expected number who believe that the minimum driving age should be 18? What is the standard deviation?
(f) If a random sample of 200 U.S. women 18 years old or older resulted in 110 who believe that the minimum driving age should be 18, would this be unusual? Why?

8. Consider a binomial probability distribution with parameters $n = 8$ and $p = 0.75$.

(a) Construct a binomial probability distribution with these parameters.
(b) Compute the mean and standard deviation of the distribution.
(c) Graph the discrete probability distribution. Comment on the shape of the distribution.

9. State the condition required to use the Empirical Rule to check for unusual observations in a binomial experiment.

10. In sampling from finite populations without replacement, the assumption of independence required for a binomial experiment is violated. Under what circumstances can we sample without replacement and still use the binomial probability formula to approximate probabilities?

11. State the conditions required for a Poisson process.

12. The random variable X follows a Poisson process with $\lambda = 0.05$ and $t = 30$. Find each of the following:

(a) $P(2)$
(b) $P(X < 2)$
(c) $P(X \geq 2)$
(d) $P(1 \leq X \leq 3)$
(e) What are μ_X and σ_X?

13. Carpet Flaws The mills at the Acme Carpet Company produce, on average, one flaw in every 500 yards of material produced; the carpeting is sold in 100-yard rolls. If the number of flaws in a roll follows a Poisson distribution and the quality-control department rejects any roll with two or more flaws, what percent of the rolls is rejected?

14. Copier Maintenance The student copy machine in the library requires a maintenance call an average of two times per month. Assuming that the number of required maintenance calls follows a Poisson distribution, would it be unusual for the copy machine to require more than four maintenance calls in any given month? Why or why not?

15. Self-Injury According to the article "Self-injurious Behaviors in a College Population," 17% of undergraduate or graduate students have had at least one incidence of self-injurious behavior. The researchers conducted a survey of 40 college students who reported a history of emotional abuse and found that 12 of them have had at least one incidence of self-injurious behavior. What do the results of this survey tell you about college students who report a history of emotional abuse?
Source: Janis Whitlock, John Eckenrode, and Daniel Silverman. "Self-injurious Behaviors in a College Population," *Pediatrics* 117: 1939–1948

 **Chapter Test**

1. Determine whether the random variable is discrete or continuous. In each case, state the possible values of the random variable.

 (a) The number of days with measurable rainfall in Honolulu, Hawaii, during a year
 (b) The miles per gallon of gasoline obtained by a randomly selected Toyota Prius
 (c) The number of golf balls hit into the ocean on the famous 18th hole at Pebble Beach on a randomly selected Sunday
 (d) The weight (in grams) of a randomly selected robin's egg

2. Determine whether the distribution is a discrete probability distribution. If not, state why.

(a)

x	P(x)
0	0.324
1	0.121
2	0.247
3	0.206
4	0.102

(b)

x	P(x)
0	0.34
1	0.28
2	0.26
3	0.23
4	−0.11

3. At the Wimbledon Tennis Championship, to win a match in men's singles a player must win the best of five sets. The following data represent the number of sets played, *X*, in the men's singles final match for the years 1968 to 2015.

x	Frequency
3	20
4	14
5	14

Source: www.wimbledon.org

 (a) Construct a probability model for the random variable, *X*, the number of sets played in the Wimbledon men's singles final match.
 (b) Draw a graph of the discrete probability distribution.
 (c) Compute and interpret the mean of the random variable *X*.
 (d) Compute the standard deviation of the random variable *X*.

4. A life insurance company sells a $100,000 one-year term life insurance policy to a 35-year-old male for $200. According to the *National Vital Statistics Report*, 56(9), the probability the male survives the year is 0.998725. Compute and interpret the expected value of this policy to the life insurance company.

5. State the criteria that must be met for an experiment to be a binomial experiment.

6. Determine whether the probability experiment represents a binomial experiment. If not, explain why.

 (a) An urn contains 20 colored golf balls: 8 white, 6 red, 4 blue, and 2 yellow. A child is allowed to draw balls until he gets a yellow one. The number of draws required is recorded.
 (b) According to the *Uniform Crime Report, 2013*, 16% of property crimes committed in the United States were cleared by arrest or exceptional means. Twenty-five property crimes from 2013 are randomly selected and the number that was cleared is recorded.

7. According to a study conducted by CESI Debt Solutions, 80% of married people hide purchases from their mates. In a random sample of 20 married people, find and interpret:

 (a) The probability exactly 15 hide purchases from their mates.
 (b) The probability at least 19 hide purchases from their mates.
 (c) The probability fewer than 19 hide purchases from their mates.
 (d) The probability between 15 and 17, inclusive, hide purchases from their mates.

8. Suppose the adult American population is equally split in their belief that the amount of tax (federal, state, property, sales, and so on) they pay is too high.

 (a) How many people would we expect to say they pay too much tax if we surveyed 1200 randomly selected adult Americans?
 (b) Explain why we can use the Empirical Rule with the idea of unusual events (events that occur with relative frequency less than 0.05) to identify any unusual results in a survey of 1200 adult Americans.
 (c) If a survey of 1200 adult Americans results in 640 stating they feel the amount of tax they pay is too high, would these results contradict the belief that adult Americans are equally split in their belief that the amount of tax they pay is too high? Why?

9. Consider a binomial probability distribution with parameters $n = 5$ and $p = 0.2$.

 (a) Construct a binomial probability distribution with these parameters.

(b) Compute the mean and standard deviation of the distribution.

(c) Graph the discrete probability distribution. Comment on the shape of the distribution.

10. The random variable X follows a Poisson process with $\lambda = 0.05$ and $t = 8$. Find each of the following:

(a) $P(3)$

(b) $P(X < 3)$

(c) $P(X \geq 3)$

(d) $P(3 \leq X \leq 5)$

(e) What are μ_X and σ_X?

11. The number of cars that arrive at a bank's drive-through window between 3:00 P.M. and 6:00 P.M. on Friday follows a Poisson process at the rate of 0.41 car every minute. Compute the probability that the number of cars that arrive at the bank between 4:00 P.M. and 4:10 P.M. is:

(a) Exactly four cars. Interpret this result.

(b) Fewer than four cars. Interpret this result.

(c) At least four cars. Interpret this result.

(d) What is the number of cars expected to arrive during this time?

(e) What is the standard deviation of the random variable X, the number of cars arriving between 4:00 P.M. and 4:10 P.M.?

Making an Informed Decision

Should We Convict?

A woman who was shopping in Los Angeles had her purse stolen by a young, blonde female who was wearing a ponytail. The blonde female got into a yellow car that was driven by a black male who had a mustache and a beard. The police located a blonde female named Janet Collins who wore her hair in a ponytail and had a friend who was a black male who had a mustache and beard and also drove a yellow car. The police arrested the two subjects.

Because there were no eyewitnesses and no real evidence, the prosecution used probability to make its case against the defendants. The probabilities below were presented by the prosecution for the known characteristics of the thieves.

Characteristic	Probability
Yellow car	$\frac{1}{10}$
Man with a mustache	$\frac{1}{4}$
Woman with a ponytail	$\frac{1}{10}$
Woman with blonde hair	$\frac{1}{3}$
Black man with beard	$\frac{1}{10}$
Interracial couple in car	$\frac{1}{1000}$

(a) Assuming that the characteristics listed are independent of each other, what is the probability that a randomly selected couple has all these characteristics? That is, what is P ("yellow car" and "man with a mustache" and ... and "interracial couple in a car")?

(b) Would you convict the defendants based on this probability? Why?

(c) Now let n represent the number of couples in the Los Angeles area who could have committed the crime. Let p represent the probability that a randomly selected couple has all six characteristics listed. Let the random variable X represent the number of couples who have all the characteristics listed in the table. Assuming that the random variable X follows the binomial probability function, we have

$$P(x) = {}_nC_x \cdot p^x(1-p)^{n-x} \quad x = 0, 1, 2, \ldots, n$$

Assuming that there are $n = 1{,}000{,}000$ couples in the Los Angeles area, what is the probability that more than one of them has the characteristics listed in the table? Does this result cause you to change your mind regarding the defendants' guilt?

(d) Now let's look at this case from a different point of view. We will compute the probability that more than one couple has the characteristics described, given that at least one couple has the characteristics.

$$P(X > 1 | X \geq 1) = \frac{P(X > 1 \text{ and } X \geq 1)}{P(X \geq 1)}$$
$$= \frac{P(X > 1)}{P(X \geq 1)}$$

Compute this probability, assuming that $n = 1{,}000{,}000$. Compute this probability again, but this time assume that $n = 2{,}000{,}000$. Do you think that the couple should be convicted "beyond all reasonable doubt"? Why?

CASE STUDY The Voyage of the *St. Andrew*

Throughout the picturesque valleys of mid-18th-century Germany echoed the song of the *Neuländer* (newlander), enticing journeymen who struggled to feed their families with the dream and promise of colonial America. The typical *Neuländer* sought to sign up several families from a village for immigration to a particular colony. By registering a group of neighbors, rather than isolated families, the agent increased the likelihood that his signees would not stray to the proposals of a competitor. Additionally, by signing large groups, the *Neuländer* fattened his purse, to the tune of one to two florins a head.

Generally, the Germans who chose to undertake the hardship of a trans-Atlantic voyage were poor, yet the cost of such a voyage was high. Records from a 1753 voyage indicate that the cost of an adult fare (one freight) from Rotterdam to Boston was 7.5 pistoles. Children aged 4 to 13 were assessed at half the adult rate (one-half freight) and those under 4 were not charged. To get a sense of the expense involved, an adult fare is equivalent to

approximately $2000! Many immigrants did not have the necessary funds to purchase passage and, determined to make the crossing, paid with years of indentured servitude.

As a historian studying the influence of these German immigrants on colonial America, Hans Langenscheidt is interested in describing various demographic characteristics of these people. Unfortunately, accurate records are rare. He has discovered a partially reconstructed 1752 passenger list for a ship, the *St. Andrew*, containing the names of the heads of families, a list of family members traveling, their parish of origin, and the number of freights each family purchased. Unfortunately, some data are missing for some families. Langenscheidt believes that the demographic parameters of this passenger list are likely to be similar to those of the other voyages taken from Germany to America during the mid-18th century. Assuming that he is correct, he believes that it is appropriate to create a discrete probability distribution for a number of demographic variables for this population of German immigrants, presented below.

Probability Distributions for German Immigrants on Board the 1752 Voyage of the *St. Andrew*

Families per Parish		Freights Purchased		Known Number of People in a Family	
Number of Families per Parish	**Probability**	**Number of Freights**	**Probability**	**Number in Family**	**Probability**
1	0.706	1	0.075	1	0.322
2	0.176	1.5	0.025	2	0.186
3	0.000	2	0.425	3	0.136
4	0.059	2.5	0.150	4	0.102
5	0.000	3	0.125	5	0.051
5	0.059	3.5	0.100	6	0.136
		4	0.050	7	0.034
		5	0.025	8	0.017
		6	0.025	9	0.016

1. Using the information provided, describe, through graphical summaries and numerical summaries such as the mean and standard deviation, each probability distribution.

2. Does it appear that, on average, the Neuländers were successful in signing more than one family from a parish? Does it seem likely that most of the families knew one another prior to undertaking the voyage? Explain your answers for both questions.

3. Using the mean number of freights purchased per family, estimate the average cost of the crossing for a family in pistoles and in U.S. dollars.

4. Is it appropriate to estimate the average cost of the voyage from the mean family size? Why or why not?

5. Langenscheidt came across a fragment of another ship's passenger list containing information for six families. Of these six, five families purchased more than four freights. Using the information in the appropriate probability distribution for the *St. Andrew,* calculate the probability that at least five of six German immigrant families purchased more than four freights. Does it seem likely that these families came from a population similar to that of the Germans on board the *St. Andrew*? Explain.

6. Summarize your findings in a report. Discuss any assumptions made throughout this analysis. What are the consequences to your calculations and conclusions if your assumptions are subsequently determined to be invalid?

Source: Wilford W. Whitaker and Gary T. Horlacher, *Broad Bay Pioneers* (Rockport, Maine: Picton Press, 1998), 63–68. Distributions created from the partially reconstructed 1752 passenger list of the *St. Andrew* presented by Whitaker and Horlacher.

7

The Normal Probability Distribution

Outline

Making an Informed Decision

You are interested in modeling the behavior of stocks. In particular, you want to build a model that describes the rate of return on a basket of stocks, such as large capitalization companies. To build this model, you must identify and use historical rates of return on a basket of stocks. Then your model can be used to identify high-performing companies that might be worthy of your investment. See the Decisions project on page 391.

PUTTING IT TOGETHER

In Chapter 6, we discussed discrete random variables. In particular, we learned how to find probabilities for specific distributions (binomial and Poisson) using probability distribution functions. These probability distribution functions were formulas that allowed us to determine probabilities for the discrete random variable. Recall, we mentioned that probability models can be in the form of a table, a formula, or a graph.

In this chapter, we discuss how to determine probabilities for continuous random variables. The approach used in determining probabilities for continuous random variables differs from that of discrete random variables. For discrete random variables, we used a formula to determine probabilities. For continuous random variables, we are going to use a graph that models the random variable to determine probabilities. In particular, we discuss distributions for two continuous random variables: the *uniform distribution* and the *normal distribution*. Most of the discussion will focus on the normal distribution, which has many applications.

7.1 Properties of the Normal Distribution

Preparing for This Section Before getting started, review the following:

- Continuous variable (Section 1.1, p. 7)
- The Empirical Rule (Section 3.2, pp. 139–140)
- Rules for a discrete probability distribution (Section 6.1, p. 317)

Objectives
1. Use the uniform probability distribution
2. Graph a normal curve
3. State the properties of the normal curve
4. Explain the role of area in the normal density function

1 Use the Uniform Probability Distribution

First, we discuss a uniform distribution to see the relation between area and probability.

EXAMPLE 1 The Uniform Distribution

Imagine that a friend of yours is always late. Let the random variable X represent the time from when you are supposed to meet your friend until he shows up. Suppose your friend could be on time ($x = 0$) or up to 30 minutes late ($x = 30$), with all intervals of equal time between $x = 0$ and $x = 30$ being equally likely. For example, your friend is just as likely to be 3–4 minutes late as he is to be 25–26 minutes late. The random variable X can be any value in the interval from 0 to 30, that is, $0 \le x \le 30$. Because any two intervals of equal length between 0 and 30, inclusive, are equally likely, the random variable X is said to follow a **uniform probability distribution**. ●

When computing probabilities for discrete random variables, we usually substitute the value of the random variable into a formula.

Things are not as easy for continuous random variables. Because an infinite number of outcomes are possible for continuous random variables, the probability of observing one *particular* value is zero. For example, the probability that your friend is exactly 12.9438823 minutes late is zero. This result is based on the fact that classical probability is found by dividing the number of ways an event can occur by the total number of possibilities: there is one way to observe 12.9438823, and there are an infinite number of possible values between 0 and 30. To resolve this problem, we compute probabilities of continuous random variables over an *interval* of values. For example, we might compute the probability that your friend is between 10 and 15 minutes late. To find probabilities for continuous random variables, we use *probability density functions*.

Definition A **probability density function (pdf)** is an equation used to compute probabilities of continuous random variables. It must satisfy the following two properties:

1. The total area under the graph of the equation over all possible values of the random variable must equal 1.
2. The height of the graph of the equation must be greater than or equal to 0 for all possible values of the random variable.

Property 1 is similar to the rule for discrete probability distributions that stated the sum of the probabilities must add up to 1. Property 2 is similar to the rule that stated all probabilities must be greater than or equal to 0.

Figure 1 illustrates these properties for Example 1. Since any value of the random variable between 0 and 30 is equally likely, the graph of the probability density function

Figure 1

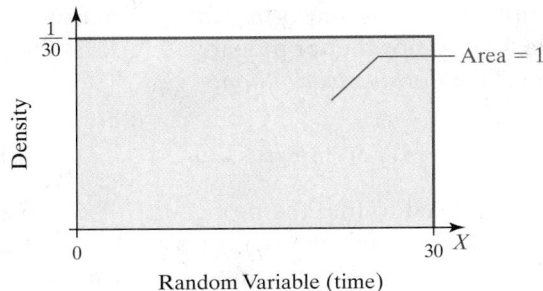

Random Variable (time)

is a rectangle. Because the random variable is any number between 0 and 30 inclusive, the width of the rectangle is 30. Since the area under the graph of the probability density function must equal 1, and the area of a rectangle equals height times width, the height of the rectangle must be $\frac{1}{30}$.

A pressing question remains: How do we use density functions to find probabilities of continuous random variables?

In Other Words
To find probabilities for continuous random variables, we do not use probability distribution functions (as we did for discrete random variables). Instead, we use probability density functions. The word *density* is used because it refers to the number of individuals per unit of area.

> The area under the graph of a density function over an interval represents the probability of observing a value of the random variable in that interval.

The following example illustrates this statement.

EXAMPLE 2   Area as a Probability

Problem Refer to the situation in Example 1.
(a) What is the probability your friend will be between 10 and 20 minutes late?
(b) It is 10 A.M. There is a 20% probability your friend will arrive within the next _____ minutes.

Approach Use the graph of the density function in Figure 1 to find the solutions.

Solution
(a) We want to find the shaded area in Figure 2(a). The width of the shaded rectangle is 10 and its height is $\frac{1}{30}$. The area between 10 and 20 is $10\left(\frac{1}{30}\right) = \frac{1}{3}$. The probability your friend is between 10 and 20 minutes late is $\frac{1}{3}$.

(b) We are given the area of the shaded region in Figure 2(b). Here we need to determine the width of the rectangle so that its area is 0.2. We solve $x_0 \cdot \frac{1}{30} = 0.2$ and find $x_0 = 30(0.2) = 6$. There is a 20% probability your friend will arrive within the next 6 minutes, or by 10:06 A.M.

Figure 2

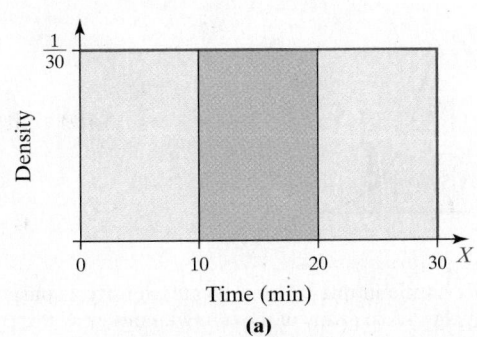

Time (min)

(a)

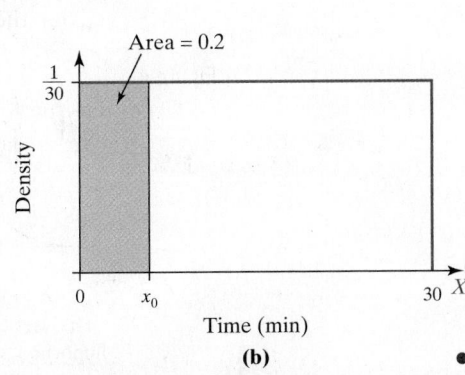

Time (min)

(b)

Now Work Problem 13

We introduced the uniform density function so we could associate probability with area. We are now better prepared to discuss the most frequently used continuous distribution, the *normal distribution*.

❷ Graph a Normal Curve

A rectangle is used to find the probability of observing an interval of numbers (such as 10–20 minutes after 10 A.M.) for a uniform random variable. However, not all continuous random variables follow a uniform distribution. For example, continuous random variables such as IQ scores and birth weights of babies have distributions that are symmetric and bell-shaped. Consider the histograms in Figure 3, which represent the IQ scores of 10,000 randomly selected adults. Notice that as the class width of the histogram decreases, the histogram becomes closely approximated by the smooth red curve. For this reason, we can use the curve to *model* the probability distribution of this continuous random variable.

Figure 3

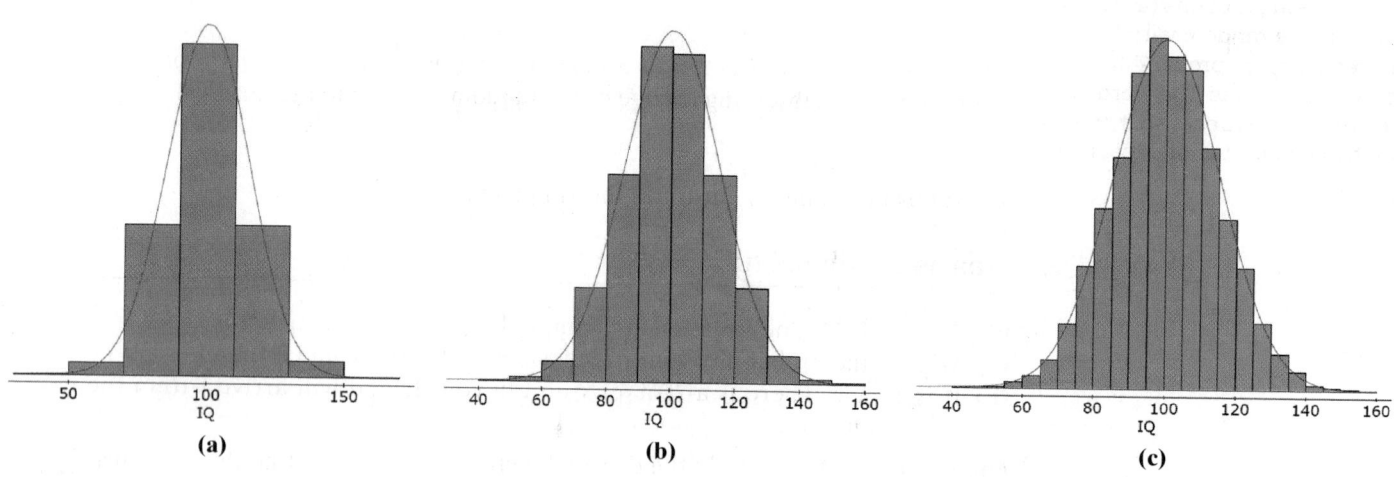

(a)   (b)   (c)

In mathematics, a **model** is an equation, table, or graph used to describe reality. The red curve in Figure 3 is a model called the **normal curve**, which is used to describe continuous random variables that are said to be *normally distributed*.

Definition
> A continuous random variable is **normally distributed**, or has a **normal probability distribution**, if its relative frequency histogram has the shape of a normal curve.

Figure 4* shows a normal curve, demonstrating the roles that the mean μ and standard deviation σ play in drawing the curve. The mode represents the "high point" of the graph of any distribution. The median represents the point where 50% of the area under the distribution is to the left and 50% is to the right. The mean represents the

Figure 4

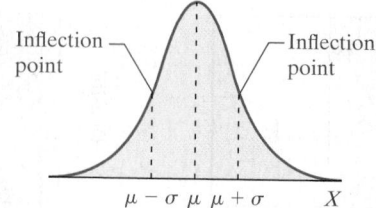

*The vertical scale on the graph, which indicates **density**, is purposely omitted. The vertical scale, while important, will not play a role in any of the computations using this curve.

balancing point of the graph of the distribution (see Figure 2 on page 120 in Section 3.1). For symmetric distributions with a single peak, such as the normal distribution, the mean = median = mode. Because of this, the mean, μ, corresponds to the high point of the graph of the distribution.

The points at $x = \mu - \sigma$ and $x = \mu + \sigma$ are the **inflection points** on the normal curve, the points on the curve where the curvature of the graph changes. To the left of $x = \mu - \sigma$ and to the right of $x = \mu + \sigma$, the curve is drawn upward (⌣ or ⌣). Between $x = \mu - \sigma$ and $x = \mu + \sigma$, the curve is drawn downward (⌒).

Figure 5 shows how changes in μ and σ change the position or shape of a normal curve. In Figure 5(a), one density curve has $\mu = 0, \sigma = 1$, and the other has $\mu = 3, \sigma = 1$. We can see that increasing the mean from 0 to 3 caused the graph to shift three units to the right but maintained its shape. In Figure 5(b), one density curve has $\mu = 0, \sigma = 1$, and the other has $\mu = 0, \sigma = 2$. We can see that increasing the standard deviation from 1 to 2 caused the graph to become flatter and more spread out but maintained its location of center.

Historical Note

Karl Pearson (of correlation fame) coined the phrase *normal curve*. He did not do this to imply that a distribution that is not normal is *abnormal*. Rather, Pearson wanted to avoid giving the name of the distribution a proper name, such as Gaussian (as in Carl Friedrich Gauss, who is incorrectly credited with the discovery of the normal curve).

Figure 5

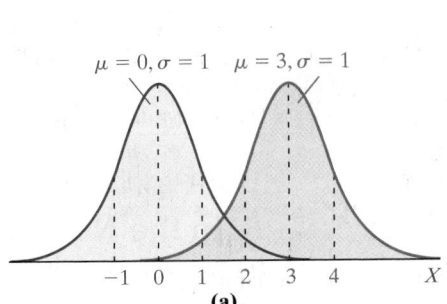
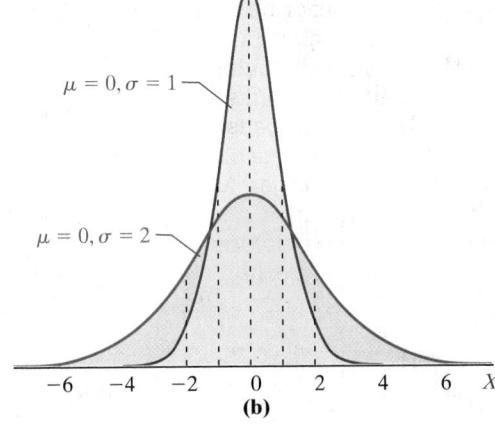

(a)   **(b)**

• **Now Work Problem 25**

③ **State the Properties of the Normal Curve**

The normal probability density function satisfies all the requirements of probability distributions. We list the properties of the normal density curve next.

Properties of the Normal Density Curve

1. The normal curve is symmetric about its mean, μ.
2. Because mean = median = mode, the normal curve has a single peak and the highest point occurs at $x = \mu$.
3. The normal curve has inflection points at $\mu - \sigma$ and $\mu + \sigma$.
4. The area under the normal curve is 1.
5. The area under the normal curve to the right of μ equals the area under the curve to the left of μ, which equals $\frac{1}{2}$.
6. As x increases without bound (gets larger and larger), the graph approaches, but never reaches, the horizontal axis. As x decreases without bound (gets more and more negative), the graph approaches, but never reaches, the horizontal axis.

(continued)

Historical Note

Abraham de Moivre was born in France on May 26, 1667. He is known as a great contributor to the areas of probability and trigonometry. In 1685, he moved to England. De Moivre was elected a fellow of the Royal Society in 1697. He was part of the commission to settle the dispute between Newton and Leibniz regarding who was the discoverer of calculus. He published *The Doctrine of Chance* in 1718. In 1733, he developed the equation that describes the normal curve. Unfortunately, de Moivre had a difficult time being accepted in English society (perhaps because of his accent) and was able to make only a meager living tutoring mathematics. An interesting piece of information regarding de Moivre: he correctly predicted the day of his death, November 27, 1754.

7. The Empirical Rule:

- Approximately 68% of the area under the normal curve is between $x = \mu - \sigma$ and $x = \mu + \sigma$;
- approximately 95% of the area is between $x = \mu - 2\sigma$ and $x = \mu + 2\sigma$;
- approximately 99.7% of the area is between $x = \mu - 3\sigma$ and $x = \mu + 3\sigma$.

See Figure 6.

Figure 6

Normal Distribution

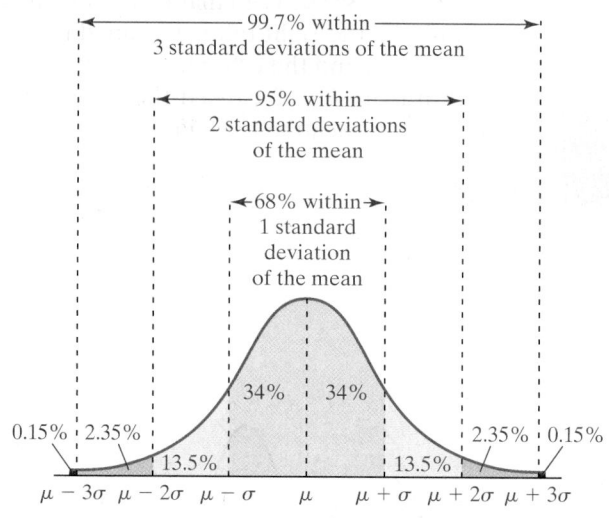

4 Explain the Role of Area in the Normal Density Function

Let's look at an example of a normally distributed random variable.

EXAMPLE 3   A Normal Random Variable

Table 1	
Height (inches)	**Relative Frequency**
29.0–29.9	0.005
30.0–30.9	0.005
31.0–31.9	0.005
32.0–32.9	0.025
33.0–33.9	0.02
34.0–34.9	0.055
35.0–35.9	0.075
36.0–36.9	0.09
37.0–37.9	0.115
38.0–38.9	0.15
39.0–39.9	0.12
40.0–40.9	0.11
41.0–41.9	0.07
42.0–42.9	0.06
43.0–43.9	0.035
44.0–44.9	0.025
45.0–45.9	0.025
46.0–46.9	0.005
47.0–47.9	0.005

Problem The relative frequency distribution given in Table 1 represents the heights of a pediatrician's three-year-old female patients. The raw data indicate that the mean height of the patients is $\mu = 38.72$ inches with standard deviation $\sigma = 3.17$ inches.

(a) Draw a relative frequency histogram of the data. Comment on the shape of the distribution.

(b) Draw a normal curve with $\mu = 38.72$ inches and $\sigma = 3.17$ inches on the relative frequency histogram. Compare the area of the rectangle for heights between 40 and 40.9 inches to the area under the normal curve for heights between 40 and 40.9 inches.

Approach

(a) Draw the relative frequency histogram. If the histogram is shaped like Figure 4, the height is approximately normal. We say "approximately normal," rather than "normal," because the normal curve is an idealized description of the data, and data rarely follow the curve exactly.

(b) Draw the normal curve on the histogram with the high point at μ and the inflection points at $\mu - \sigma$ and $\mu + \sigma$. Shade the rectangle corresponding to heights between 40 and 40.9 inches.

Solution

(a) Figure 7 shows the relative frequency histogram, which is symmetric and bell-shaped.

Figure 7

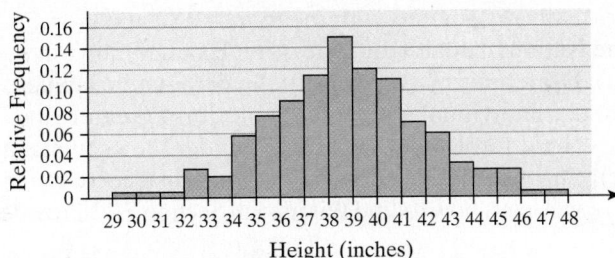

Height of Three-Year-Old Females

(b) In Figure 8, the normal curve with $\mu = 38.72$ and $\sigma = 3.17$ is superimposed on the relative frequency histogram. The normal curve describes the data fairly well. We conclude that the heights of three-year-old females are approximately normal with $\mu = 38.72$ and $\sigma = 3.17$.

Figure 8 also shows the rectangle whose area represents the proportion of three-year-old females between 40 and 40.9 inches. Notice that the area of this shaded region is very close to the area under the normal curve for the same region, so we can use the area under the normal curve to approximate the proportion of three-year-old females with heights between 40 and 40.9 inches!

Figure 8

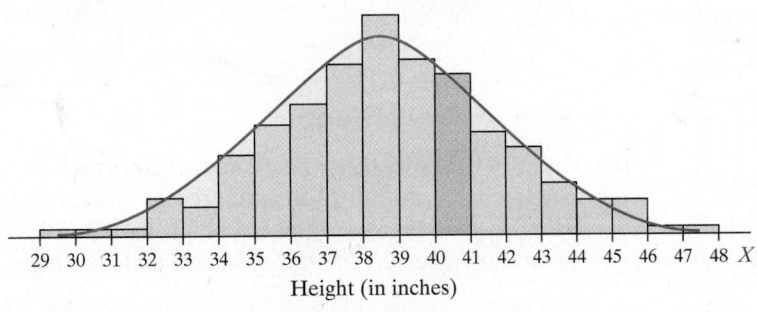

Height of Three-Year-Old Females

The equation (or model) used to determine the probability of a continuous random variable is called a **probability density function** (or **pdf**). The **normal probability density function** is given by

$$y = \frac{1}{\sigma\sqrt{2\pi}} e^{\frac{-(x-\mu)^2}{2\sigma^2}} = \frac{1}{\sigma\sqrt{2\pi}} e^{-\frac{1}{2}\left(\frac{x-\mu}{\sigma}\right)^2}$$

where μ is the mean and σ is the standard deviation of the normal random variable. Do not feel threatened by this equation, because we will not be using it in this text. Instead, we will use the normal distribution in graphical form by drawing the normal curve.

We now summarize the role area plays in the normal curve.

Area under a Normal Curve

Suppose that a random variable X is normally distributed with mean μ and standard deviation σ. The area under the normal curve for any interval of values of the random variable X represents either

- the proportion of the population with the characteristic described by the interval of values or
- the probability that a randomly selected individual from the population will have the characteristic described by the interval of values.

EXAMPLE 4 Interpreting the Area under a Normal Curve

Historical Note

The normal probability distribution is often referred to as the Gaussian distribution in honor of Carl Gauss, the individual thought to have discovered the idea. However, it was actually Abraham de Moivre who first wrote down the equation of the normal distribution. Gauss was born in Brunswick, Germany, on April 30, 1777. Mathematical prowess was evident early in Gauss's life. At age 8 he was able to instantly add the first 100 integers. In 1799, Gauss earned his doctorate. The subject of his dissertation was the Fundamental Theorem of Algebra. In 1809, Gauss published a book on the mathematics of planetary orbits. In this book, he further developed the theory of least-squares regression by analyzing the errors. The analysis of these errors led to the discovery that errors follow a normal distribution. Gauss was considered to be "glacially cold" as a person and had troubled relationships with his family. Gauss died on February 23, 1855.

● **Now Work Problems 31 and 35**

Problem The serum total cholesterol for males 20–29 years old is approximately normally distributed with mean $\mu = 180$ and $\sigma = 36.2$, based on data obtained from the National Health and Nutrition Examination Survey.

(a) Draw a normal curve with the parameters labeled.

(b) An individual with total cholesterol greater than 200 is considered to have high cholesterol. Shade the region under the normal curve to the right of $x = 200$.

(c) Suppose that the area under the normal curve to the right of $x = 200$ is 0.2903. (You will learn how to find this area in Section 7.2.) Provide two interpretations of this result.

Approach

(a) Draw the normal curve with the mean $\mu = 180$ labeled at the high point and the inflection points at $\mu - \sigma = 180 - 36.2 = 143.8$ and $\mu + \sigma = 180 + 36.2 = 216.2$.

(b) Shade the region under the normal curve to the right of $x = 200$.

(c) The two interpretations of the area under a normal curve are (1) a proportion and (2) a probability.

Solution

(a) Figure 9(a) shows the graph of the normal curve.

Figure 9

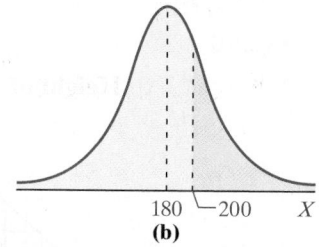

(a)   (b)

(b) Figure 9(b) shows the region under the normal curve to the right of $x = 200$ shaded.

(c) The two interpretations for the area of this shaded region are (1) the proportion of 20- to 29-year-old males that have high cholesterol is 0.2903 and (2) the probability that a randomly selected 20- to 29-year-old male has high cholesterol is 0.2903. ●

7.1 Assess Your Understanding

Vocabulary and Skill Building

1. A _____ _____ _____ is an equation used to compute probabilities of continuous random variables.

2. A _____ is an equation, table, or graph used to describe reality.

3. *True or False*: The normal curve is symmetric about its mean, μ.

4. The area under the normal curve to the right of μ equals _____.

5. The points at $x = $ _____ and $x = $ _____ are the inflection points on the normal curve.

6. The area under a normal curve can be interpreted as a _____ or _____.

For Problems 7–12, determine whether the graph can represent a normal curve. If it cannot, explain why.

7.   **8.**

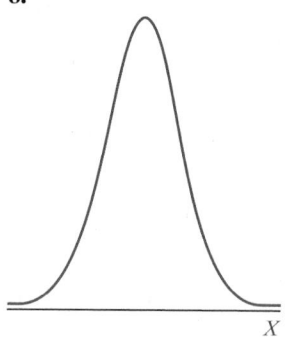

9.   **10.**

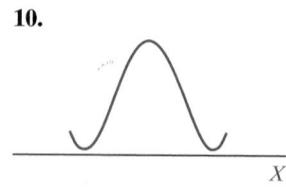

11.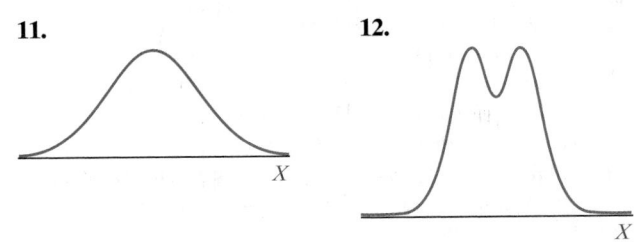

12.

Problems 13–16 use the information presented in Examples 1 and 2.

NW **13. (a)** Find the probability that your friend is between 5 and 10 minutes late.

(b) It is 10 A.M. There is a 40% probability your friend will arrive within the next _____ minutes.

14. (a) Find the probability that your friend is between 15 and 25 minutes late.

(b) It is 10 A.M. There is a 90% probability your friend will arrive within the next _____ minutes.

15. Find the probability that your friend is at least 20 minutes late.

16. Find the probability that your friend is no more than 5 minutes late.

17. Uniform Distribution The random-number generator on calculators randomly generates a number between 0 and 1. The random variable X, the number generated, follows a uniform probability distribution.

(a) Draw the graph of the uniform density function.

(b) What is the probability of generating a number between 0 and 0.2?

(c) What is the probability of generating a number between 0.25 and 0.6?

(d) What is the probability of generating a number greater than 0.95?

(e) Use your calculator or statistical software to randomly generate 200 numbers between 0 and 1. What proportion of the numbers are between 0 and 0.2? Compare the result with part (b).

18. Uniform Distribution The reaction time X (in minutes) of a certain chemical process follows a uniform probability distribution with $5 \leq X \leq 10$.

(a) Draw the graph of the density curve.

(b) What is the probability that the reaction time is between 6 and 8 minutes?

(c) What is the probability that the reaction time is between 5 and 8 minutes?

(d) What is the probability that the reaction time is less than 6 minutes?

In Problems 19–22, determine whether or not the histogram indicates that a normal distribution could be used as a model for the variable.

19. Birth Weights The relative frequency histogram represents the birth weights (in grams) of babies whose term was 36 weeks.

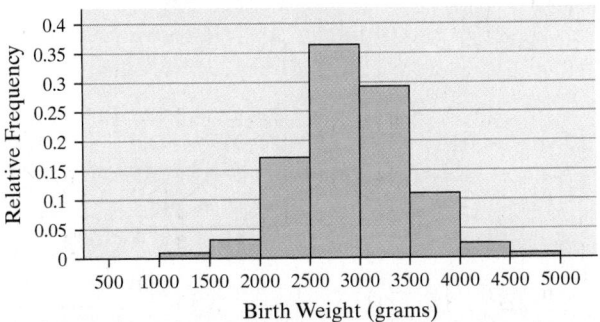

Birth Weights of Babies Whose Term Was 36 Weeks

20. Waiting in Line The relative frequency histogram represents the waiting times (in minutes) to ride the American Eagle Roller Coaster for 2000 randomly selected people on a Saturday afternoon in the summer.

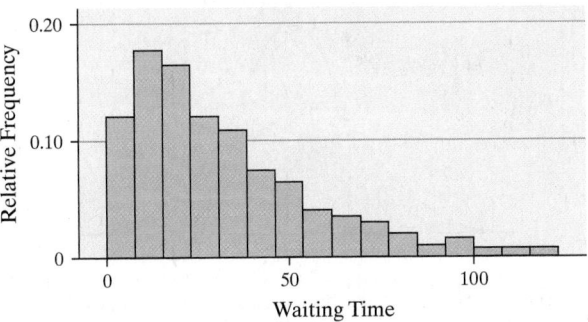

Waiting Time for the American Eagle Roller Coaster

21. Length of Phone Calls The relative frequency histogram represents the length of phone calls on my wife's cell phone during the month of September.

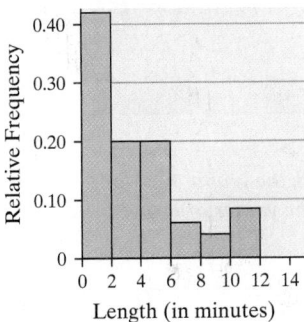

Length of Phone Calls

22. Incubation Times The relative frequency histogram represents the incubation times of a random sample of Rhode Island Red hens' eggs.

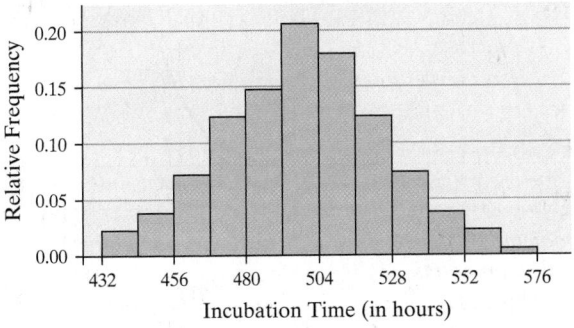

Rhode Island Red Hen Incubation Times

23. One graph in the figure on the following page represents a normal distribution with mean $\mu = 10$ and standard deviation $\sigma = 3$. The other graph represents a normal distribution with mean $\mu = 10$ and standard deviation $\sigma = 2$.

Determine which graph is which and explain how you know.

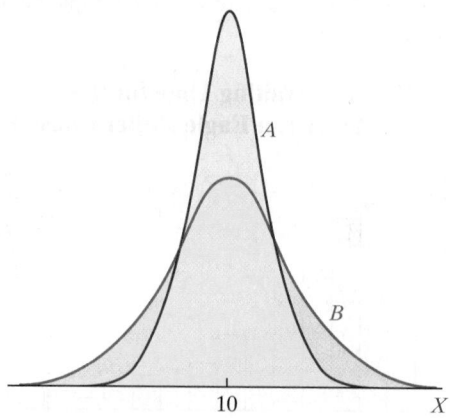

24. One graph in the figure below represents a normal distribution with mean $\mu = 8$ and standard deviation $\sigma = 2$. The other graph represents a normal distribution with mean $\mu = 14$ and standard deviation $\sigma = 2$. Determine which graph is which and explain how you know.

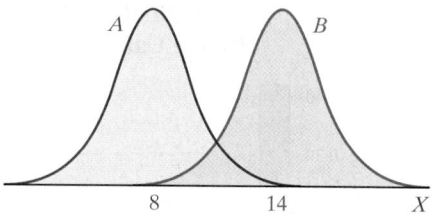

In Problems 25–28, the graph of a normal curve is given. Use the graph to identify the values of μ and σ.

NW **25.**   **26.**

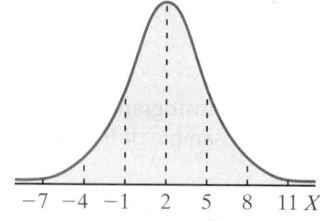   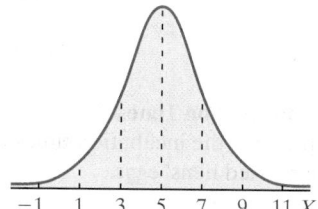

27.   **28.**

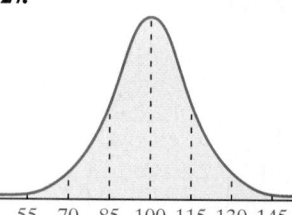   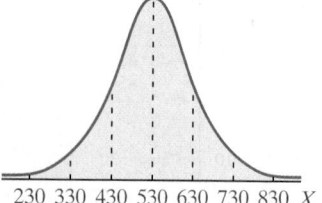

In Problems 29 and 30, draw a normal curve and label the mean and inflection points.

29. $\mu = 30$ and $\sigma = 10$ **30.** $\mu = 50$ and $\sigma = 5$

Applying the Concepts

NW 31. You Explain It! Cell Phone Rates Monthly charges for cell phone plans in the United States are normally distributed with mean $\mu = \$62$ and standard deviation $\sigma = \$18$.
Source: Based on information from *Consumer Reports*

(a) Draw a normal curve with the parameters labeled.
(b) Shade the region that represents the proportion of plans that charge less than $44.
(c) Suppose the area under the normal curve to the left of $x = \$44$ is 0.1587. Provide two interpretations of this result.

32. You Explain It! Refrigerators The lives of refrigerators are normally distributed with mean $\mu = 14$ years and standard deviation $\sigma = 2.5$ years.
Source: Based on information from *Consumer Reports*

(a) Draw a normal curve with the parameters labeled.
(b) Shade the region that represents the proportion of refrigerators that last for more than 17 years.
(c) Suppose the area under the normal curve to the right of $x = 17$ is 0.1151. Provide two interpretations of this result.

33. You Explain It! Birth Weights The birth weights of full-term babies are normally distributed with mean $\mu = 3400$ grams and $\sigma = 505$ grams.
Source: Based on data obtained from the *National Vital Statistics Report*, Vol. 48, No. 3

(a) Draw a normal curve with the parameters labeled.
(b) Shade the region that represents the proportion of full-term babies who weigh more than 4410 grams.
(c) Suppose the area under the normal curve to the right of $x = 4410$ is 0.0228. Provide two interpretations of this result.

34. You Explain It! Height of 10-Year-Old Males The heights of 10-year-old males are normally distributed with mean $\mu = 55.9$ inches and $\sigma = 5.7$ inches.

(a) Draw a normal curve with the parameters labeled.
(b) Shade the region that represents the proportion of 10-year-old males who are less than 46.5 inches tall.
(c) Suppose the area under the normal curve to the left of $x = 46.5$ is 0.0496. Provide two interpretations of this result.

NW 35. You Explain It! Gestation Period The lengths of human pregnancies are normally distributed with $\mu = 266$ days and $\sigma = 16$ days.

(a) The figure represents the normal curve with $\mu = 266$ days and $\sigma = 16$ days. The area to the right of $x = 280$ is 0.1908. Provide two interpretations of this area.

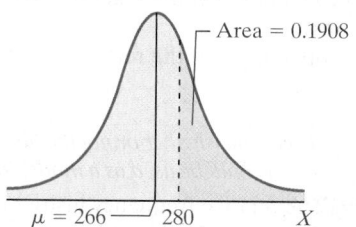

(b) The figure represents the normal curve with $\mu = 266$ days and $\sigma = 16$ days. The area between $x = 230$ and $x = 260$ is 0.3416. Provide two interpretations of this area.

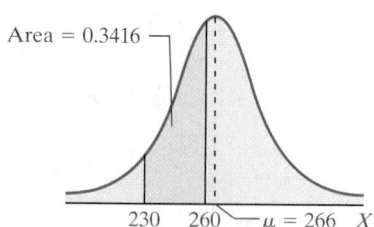

36. You Explain It! Miles per Gallon Elena conducts an experiment in which she fills up the gas tank on her Toyota Camry 40 times and records the miles per gallon for each fill-up. A histogram of the miles per gallon indicates that a normal distribution with a mean of 24.6 miles per gallon and a standard deviation of 3.2 miles per gallon could be used to model the gas mileage for her car.

(a) The figure represents the normal curve with $\mu = 24.6$ miles per gallon and $\sigma = 3.2$ miles per gallon. The area under the curve to the right of $x = 26$ is 0.3309. Provide two interpretations of this area.

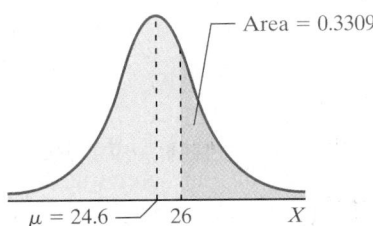

(b) The figure below represents the normal curve with $\mu = 24.6$ miles per gallon and $\sigma = 3.2$ miles per gallon. The area under the curve between $x = 18$ and $x = 21$ is 0.1107. Provide two interpretations of this area.

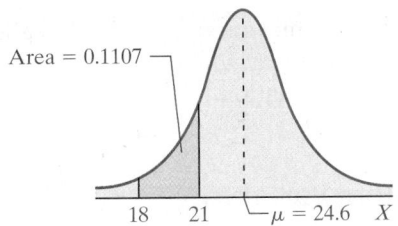

37. Hitting with a Pitching Wedge In the game of golf, distance control is just as important as how far a player hits the ball. Michael went to the driving range with his range finder and hit 75 golf balls with his pitching wedge and measured the distance each ball traveled (in yards). He obtained the following data:

100	97	101	101	103	100	99	100	100
104	100	101	98	100	99	99	97	101
104	99	101	101	101	100	96	99	99
98	94	98	107	98	100	98	103	100
98	94	104	104	98	101	99	97	103
102	101	101	100	95	104	99	102	95
99	102	103	97	101	102	96	102	99
96	108	103	100	95	101	103	105	100
94	99	95						

(a) Use statistical software to construct a relative frequency histogram. Comment on the shape of the distribution. Draw a normal density curve on the relative frequency histogram.

(b) Do you think the normal density curve accurately describes the distance Michael hits with a pitching wedge? Why?

38. Heights of Five-Year-Old Females The following data represent the heights (in inches) of 80 randomly selected five-year-old females.

44.5	42.4	42.2	46.2	45.7	44.8	43.3	39.5
45.4	43.0	43.4	44.7	38.6	41.6	50.2	46.9
39.6	44.7	36.5	42.7	40.6	47.5	48.4	37.5
45.5	43.3	41.2	40.5	44.4	42.6	42.0	40.3
42.0	42.2	38.5	43.6	40.6	45.0	40.7	36.3
44.5	37.6	42.2	40.3	48.5	41.6	41.7	38.9
39.5	43.6	41.3	38.8	41.9	40.3	42.1	41.9
42.3	44.6	40.5	37.4	44.5	40.7	38.2	42.6
44.0	35.9	43.7	48.1	38.7	46.0	43.4	44.6
37.7	34.6	42.4	42.7	47.0	42.8	39.9	42.3

(a) Use statistical software to construct a relative frequency histogram. Comment on the shape of the distribution. Draw a normal density curve on the relative frequency histogram.

(b) Do you think the normal density curve accurately describes the heights of five-year-old females? Why?

Retain Your Knowledge

39. Cardiac Arrest Researchers conducted a prospective cohort study in which male patients who had an out-of-hospital cardiac arrest were submitted to therapeutic hypothermia (intravenous infusion of cold saline followed by surface cooling with the goal of maintaining body temperature of 33 degrees Celsius for 24 hours. Note that normal body temperature is 37 degrees Celsius). The survival status, length of stay in the intensive care unit (ICU), and time spent on a ventilator were measured. Each of these variables was compared to a historical cohort of patients who were treated prior to the availability of therapeutic hypothermia. Of the 52 hypothermia patients, 37 survived; of the 74 patients in the control group, 43 survived. The median length of stay among survivors for the hypothermia patients was 14 days versus 21 days for the control group. The time on the ventilator among survivors for the hypothermia group was 219 hours versus 328 hours for the control group.
Source: Storem, Christian, et al. Mild Therapeutic Hypothermia Shortens Intensive Care Unit Stay of Survivors After Out-of-Hospital Cardiac Arrest Compared to Historical Controls. *Critical Care* 2008, 12:R78 BioMed Central

(a) What does it mean to say this is a prospective cohort study?

(b) What is the explanatory variable in the study? Is it qualitative or quantitative?

(c) What are the three response variables in the study? For each, state whether the variable is qualitative or quantitative.

(d) Is time on the ventilator a statistic or parameter? Explain.

(e) To what population does this study apply?

(f) Based on the results of this study, what is the probability a randomly selected male who has an out-of-hospital cardiac arrest and submits to therapeutic hypothermia will survive? What about those who do not submit to therapeutic hypothermia?

7.2 Applications of the Normal Distribution

Preparing for This Section Before getting started, review the following:

- z-scores (Section 3.4, pp. 155–156)
- Percentiles (Section 3.4, p. 156)
- Complement Rule (Section 5.2, p. 267)

Objectives
1. Find and interpret the area under a normal curve
2. Find the value of a normal random variable

If X is a normally distributed random variable, the area under the normal curve represents the proportion of a population with a certain characteristic, or the probability that a randomly selected individual from the population has the characteristic.

The question then is, "How do I find the area under the normal curve?" We have two options—by-hand calculations with the aid of a table or technology.

1 Find and Interpret the Area under a Normal Curve

We use z-scores to help find the area under a normal curve by hand. Recall, the z-score allows us to transform a random variable X with mean μ and standard deviation σ into a random variable Z with mean 0 and standard deviation 1.

Standardizing a Normal Random Variable

Suppose that the random variable X is normally distributed with mean μ and standard deviation σ. Then the random variable

$$Z = \frac{X - \mu}{\sigma}$$

is normally distributed with mean $\mu = 0$ and standard deviation $\sigma = 1$. The random variable Z is said to have the **standard normal distribution**.

This result is powerful! If a normal random variable X has mean different from 0 or a standard deviation different from 1, we can transform X into a **standard normal random variable Z** whose mean is 0 and standard deviation is 1. Then we can use Table V (found on the inside back cover of the text and in Appendix A) to find the area to the left of a specified z-score, z, as shown in Figure 10, which is also the area to the left of the value of x in the distribution of X. The graph in Figure 10 is called the **standard normal curve**.

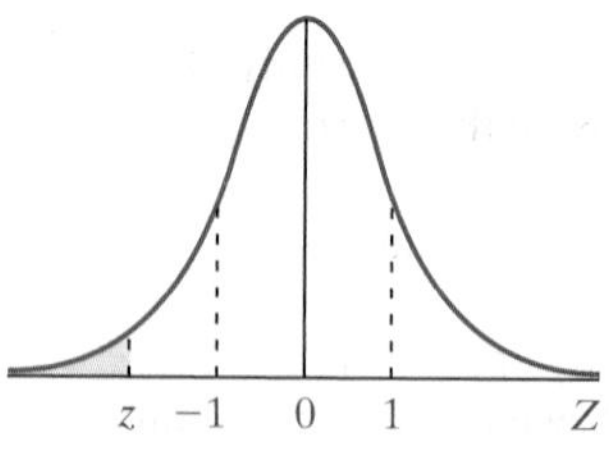

For example, IQ scores can be modeled by a normal distribution with $\mu = 100$ and $\sigma = 15$. An individual whose IQ is 120, is $z = \dfrac{x - \mu}{\sigma} = \dfrac{120 - 100}{15} = 1.33$ standard deviations above the mean (recall, we round z-scores to two decimal places). We look in Table V and find the area under the standard normal curve to the left of $z = 1.33$ is 0.9082. See Figure 11. Therefore, the area under the normal curve to the left of $x = 120$ is 0.9082 as shown in Figure 12.

To find the area to the right of the value of a random variable, use the Complement Rule and determine one minus the area to the left. For example, to find the area under the normal curve with mean $\mu = 100$ and standard deviation $\sigma = 15$ to the right of $x = 120$, compute

$$\text{Area} = 1 - 0.9082$$
$$= 0.0918$$

as shown in Figure 12.

Figure 10

Figure 11

Standard Normal Distribution							
z	.00	.01	.02	.03	.04	.05	.06
0.0	0.5000	0.5040	0.5080	0.5120	0.5160	0.5199	0.5239
0.1	0.5398	0.5438	0.5478	0.5517	0.5557	0.5596	0.5636
0.2	0.5793	0.5832	0.5871	0.5910	0.5948	0.5987	0.6026
0.3	0.6179	0.6217	0.6255	0.6293	0.6331	0.6368	0.6406
0.4	0.6554	0.6591	0.6628	0.6664	0.6700	0.6736	0.6772
0.5	0.6915	0.6950	0.6985	0.7019	0.7054	0.7088	0.7123
0.6	0.7257	0.7291	0.7324	0.7357	0.7389	0.7422	0.7454
0.7	0.7580	0.7611	0.7642	0.7673	0.7704	0.7734	0.7764
0.8	0.7881	0.7910	0.7939	0.7967	0.7995	0.8023	0.8051
0.9	0.8159	0.8186	0.8212	0.8238	0.8264	0.8289	0.8315
1.0	0.8413	0.8438	0.8461	0.8485	0.8508	0.8531	0.8554
1.1	0.8643	0.8665	0.8686	0.8708	0.8729	0.8749	0.8770
1.2	0.8849	0.8869	0.8888	0.8907	0.8925	0.8944	0.8962
1.3	0.9032	0.9049	0.9066	0.9082	0.9099	0.9115	0.9131
1.4	0.9192	0.9207	0.9222	0.9236	0.9251	0.9265	0.9279
1.5	0.9332	0.9345	0.9357	0.9370	0.9382	0.9394	0.94

Figure 12

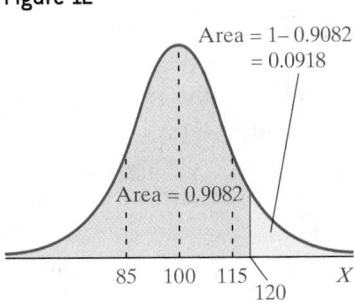

EXAMPLE 1   Finding Area under a Normal Curve

Problem A pediatrician obtains the heights of her three-year-old female patients. The heights are approximately normally distributed, with mean 38.72 inches and standard deviation 3.17 inches. Use the normal model to determine the proportion of the three-year-old females that have a height less than 35 inches.

By-Hand Approach

Step 1 Draw a normal curve and shade the desired area.

Step 2 Convert the value of x to a z-score using

$$z = \frac{x - \mu}{\sigma}$$

Step 3 Use Table V to find the area to the left of the z-score found in Step 2.

By-Hand Solution

Step 1 Figure 13 shows the normal curve with the area to the left of 35 shaded.

Figure 13

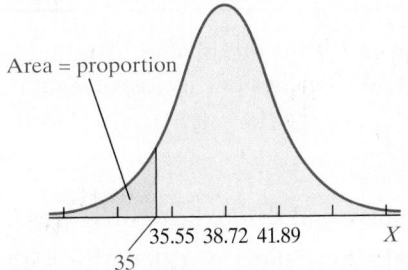

Step 2 Convert $x = 35$ to a z-score.

$$z = \frac{x - \mu}{\sigma} = \frac{35 - 38.72}{3.17} = -1.17$$

Technology Approach

Step 1 Draw a normal curve and shade the desired area.

Step 2 Use a statistical spreadsheet or calculator with advanced statistical features to find the area. The steps for determining the area under any normal curve using the TI-83/84 Plus graphing calculator, Minitab, Excel, and StatCrunch are found in the Technology Step-by-Step on pages 373–374.

Technology Solution

Step 1 Figure 14 shows the normal curve with the area to the left of 35 shaded.

Figure 14

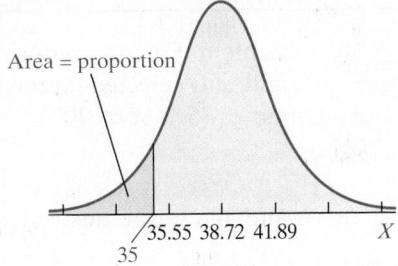

(continued)

Step 3 Look up $z = -1.17$ in Table V and find the entry. The area to the left of $z = -1.17$ is 0.1210. See Figure 15. Therefore, the area to the left of $x = 35$ is 0.1210.

Figure 15

z	.00	.01	.02	.03	.04	.05	.06	.07	.08
-3.4	0.0003	0.0003	0.0003	0.0003	0.0003	0.0003	0.0003	0.0003	0.0003
-3.3	0.0005	0.0005	0.0005	0.0004	0.0004	0.0004	0.0004	0.0004	0.0004
-3.2	0.0007	0.0007	0.0006	0.0006	0.0006	0.0006	0.0006	0.0005	0.0005
-1.4	0.0808	0.0793	0.0778	0.0764	0.0749	0.0735	0.0721	0.0708	0.0694
-1.3	0.0968	0.0951	0.0934	0.0918	0.0901	0.0885	0.0869	0.0853	0.0838
-1.2	0.1151	0.1131	0.1112	0.1093	0.1075	0.1056	0.1038	0.1020	0.1003
-1.1	0.1357	0.1335	0.1314	0.1292	0.1271	0.1251	0.1230	0.1210	0.1190
-1.0	0.1587	0.1562	0.1539	0.1515	0.1492	0.1469	0.1446	0.1423	0.1401

The normal model indicates that the proportion of the pediatrician's three-year-old females who are less than 35 inches tall is 0.1210.

Step 2 Figure 16 shows the results from Minitab. The area under the normal curve to the left of 35 is 0.1203.

Figure 16

Cumulative Distribution Function

```
Normal with mean = 38.72 and standard deviation = 3.17

  x   P( X <= x )
 35      0.120297
```

The normal model indicates that the proportion of the pediatrician's three-year-old females who are less than 35 inches tall is 0.1203.

Table 2

Height (inches)	Relative Frequency
29.0–29.9	0.005
30.0–30.9	0.005
31.0–31.9	0.005
32.0–32.9	0.025
33.0–33.9	0.02
34.0–34.9	0.055
35.0–35.9	0.075
36.0–36.9	0.09
37.0–37.9	0.115
38.0–38.9	0.15
39.0–39.9	0.12
40.0–40.9	0.11
$\vdots$	$\vdots$
47.0–47.9	0.005

CAUTION!
Notice the by-hand solution and technology solution in Example 1 differ. The difference exists because we rounded the z-score in Step 2 of the by-hand solution, which leads to rounding error.

According to the results of Example 1, the proportion of three-year-old females who are shorter than 35 inches is approximately 0.12. If the normal curve is a good model for determining proportions (or probabilities), then about 12% of the three-year-olds in Table 1 (from Section 7.1) should be shorter than 35 inches. For convenience, part of Table 1 is repeated in Table 2.

The relative frequency distribution in Table 2 shows that $0.005 + 0.005 + 0.005 + 0.025 + 0.02 + 0.055 = 0.115 = 11.5\%$ of the three-year-old females are less than 35 inches tall. The results based on the normal curve are close to the actual results. The normal curve accurately models the heights.

If we wanted to know the proportion of three-year-old females whose height is greater than 35 inches, use the Complement Rule and find the proportion is $1 - 0.1210 = 0.879$ (using the "by-hand" computation).

Because the area under the normal curve represents a proportion, we can also use the area to find percentile ranks of scores. Recall that the kth percentile divides the lower $k\%$ of a data set from the upper $(100 - k)\%$. In Example 1, 12% of the females have a height less than 35 inches, and 88% of the females have a height greater than 35 inches, so a child whose height is 35 inches is at the 12th percentile.

EXAMPLE 2   Finding the Probability of a Normal Random Variable

Problem For the pediatrician presented in Example 1, find the probability that a randomly selected three-year-old girl is between 35 and 40 inches tall, inclusive. That is, find $P(35 \leq X \leq 40)$.

By-Hand Approach

Step 1 Draw a normal curve and shade the desired area.

Step 2 Convert the values of x to z-scores using

$$z = \frac{x - \mu}{\sigma}$$

Step 3 Use Table V to find the area to the left of each z-score found in Step 2. Use this result to find the area between the z-scores.

Technology Approach

Step 1 Draw a normal curve and shade the desired area.

Step 2 Use a statistical spreadsheet or calculator with advanced statistical features to find the area. The steps for determining the area under any normal curve using the TI-83/84 Plus graphing calculator, Minitab, Excel, and StatCrunch are found in the Technology Step-by-Step on pages 373–374.

By-Hand Solution

Step 1   Figure 17 shows the normal curve with the area between 35 and 40 shaded.

Figure 17

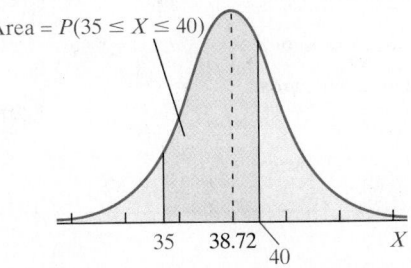

Area = $P(35 \leq X \leq 40)$

35   38.72   40 X

Step 2   Convert $x_1 = 35$ and $x_2 = 40$ to z-scores.

$$z_1 = \frac{x_1 - \mu}{\sigma} = \frac{35 - 38.72}{3.17} = -1.17$$

$$z_2 = \frac{x_2 - \mu}{\sigma} = \frac{40 - 38.72}{3.17} = 0.40$$

Step 3   Table V shows that the area to the left of $z_2 = 0.4$ (or $x_2 = 40$) is 0.6554 and the area to the left of $z_1 = -1.17$ (or $x_1 = 35$) is 0.1210, so the area between $z_1 = -1.17$ and $z_2 = 0.40$ is $0.6554 - 0.1210 = 0.5344$. The probability a randomly selected three-year-old female is between 35 and 40 inches tall is 0.5344. That is, $P(35 \leq X \leq 40) = P(-1.17 \leq Z \leq 0.40) = 0.5344$.

Technology Solution

Step 1   Figure 18 shows the normal curve with the area between 35 and 40 shaded.

Figure 18

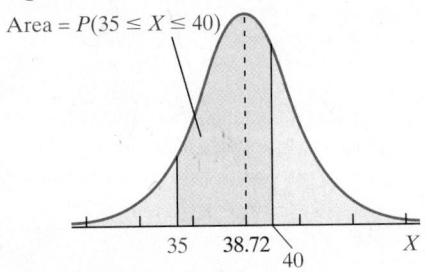

Area = $P(35 \leq X \leq 40)$

35   38.72   40 X

Step 2   Figure 19 shows the results from a TI-84 Plus C graphing calculator.

Figure 19

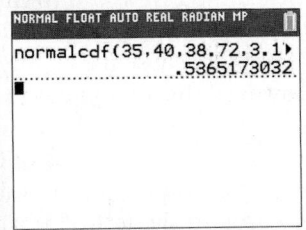

```
NORMAL FLOAT AUTO REAL RADIAN MP
normalcdf(35,40,38.72,3.1▶
            .5365173032
■
```

The area between $x = 35$ and $x = 40$ is 0.5365. The probability a randomly selected three-year-old female is between 35 and 40 inches tall is 0.5365. That is, $P(35 \leq X \leq 40) = 0.5365$.

Interpretation   If we randomly selected 100 three-year-old females, we would expect about 53 or 54 of them to be between 35 and 40 inches tall.   •

• **Now Work Problem 39**

According to the relative frequency distribution in Table 2, the proportion of three-year-old females with heights between 35 and 40 inches is $0.075 + 0.09 + 0.115 + 0.15 + 0.12 = 0.55$. This is very close to the probability found in Example 2.

We summarize the methods for obtaining the area under a normal curve in Table 3.

Table 3		
Problem	**Approach**	**Solution**
Find the area to the left of x.	Shade the area to the left of x. x X	• Convert the value of x to a z-score. Use Table V to find the row and column that correspond to z. The area to the left of x is the value where the row and column intersect. or • Use technology to find the area.
Find the area to the right of x.	Shade the area to the right of x. 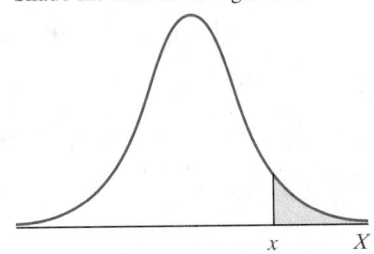 x X	• Convert the value of x to a z-score. Use Table V to find the area to the left of z (which is also the area to the left of x). The area to the right of z (also x) is 1 minus the area to the left of z. or • Use technology to find the area.

(continued)

Table 3 (Continued)

Problem	Approach	Solution
Find the area between x_1 and x_2.	Shade the area between x_1 and x_2. 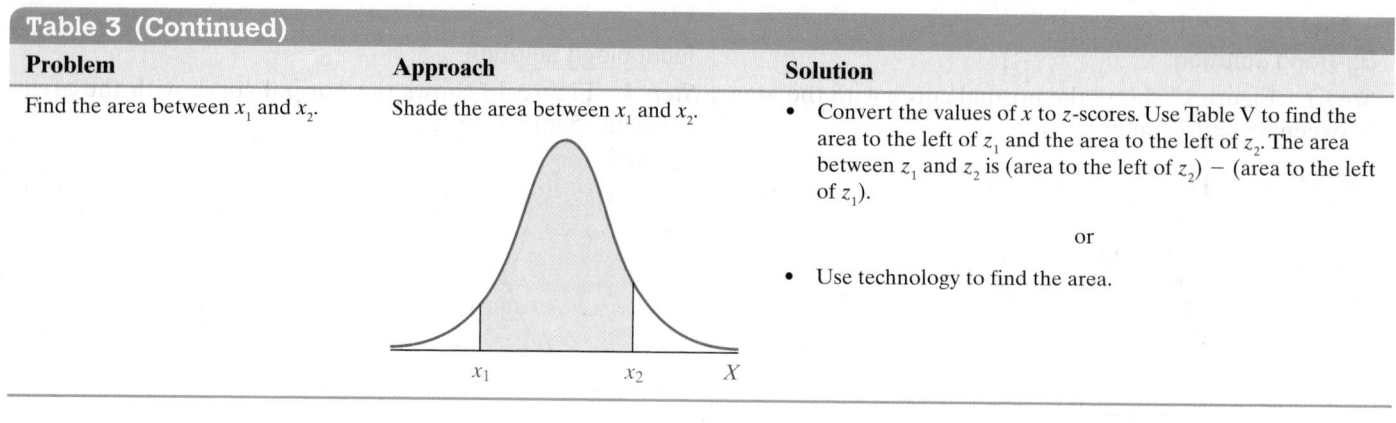	• Convert the values of x to z-scores. Use Table V to find the area to the left of z_1 and the area to the left of z_2. The area between z_1 and z_2 is (area to the left of z_2) − (area to the left of z_1). or • Use technology to find the area.

Some Cautionary Thoughts

The normal curve extends indefinitely in both directions. For this reason, there is no range of values of a normal random variable for which the area under the curve is 1. For example, if asked to find the area under a normal curve to the left of $x = 40$ with $\mu = 15$ and $\sigma = 2$, StatCrunch (as well as other software and calculators) will state the area is 1, because it can only compute a limited number of decimal places. See Figure 20. However, the area under the curve to the left of $x = 40$ is not 1; it is some value slightly less than 1. So we will follow the practice of reporting such areas as >0.9999. Similarly, if software reports an area of 0, we will report the area as <0.0001.

When finding area under the normal curve by hand using Table V, we will report any area to the left of $z = -3.49$ (the smallest value of z in the table) or to the right of $z = 3.49$ (the largest value of z in the table) as <0.0001. Any area under the normal curve to the left of $z = 3.49$ or to the right of $z = -3.49$ is stated as >0.9999.

Figure 20

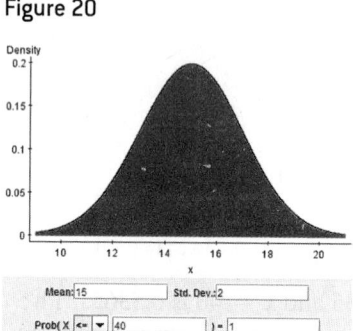

② Find the Value of a Normal Random Variable

Often, we do not want to find the proportion, probability, or percentile given a value of a normal random variable. Rather, we want to find the value of a normal random variable that corresponds to a certain proportion, probability, or percentile. For example, we might want to know the height of a three-year-old girl who is at the 20th percentile. Or we might want to know the scores on a standardized exam that separate the middle 90% of scores from the bottom and top 5%.

EXAMPLE 3   Finding the Value of a Normal Random Variable

Problem The heights of a pediatrician's three-year-old females are approximately normally distributed, with mean 38.72 inches and standard deviation 3.17 inches. Find the height of a three-year-old female at the 20th percentile.

By-Hand Approach

Step 1 Draw a normal curve and shade the desired area.

Step 2 Use Table V to find the z-score that corresponds to the shaded area.

Step 3 Obtain the normal value from the formula $x = \mu + z\sigma$.*

Technology Approach

Step 1 Draw a normal curve and shade the desired area.

Step 2 Use a statistical spreadsheet or calculator with advanced statistical features to find the score. The steps for determining the value of a normal random variable, given an area, using the TI-83/84 Plus graphing

---

*The formula provided in Step 3 of the by-hand approach is the formula for computing a z-score, solved for x.

$z = \dfrac{x - \mu}{\sigma}$ Formula for standardizing a value, x, for a random variable X

$z\sigma = x - \mu$ Multiply both sides by σ.

$x = \mu + z\sigma$ Add μ to both sides.

By-Hand Solution

Step 1   Figure 21 shows the normal curve with the unknown value of x at the 20th percentile, which separates the bottom 20% of the distribution from the top 80%.

Figure 21

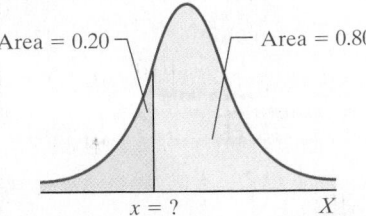

Step 2   We want to find the z-score such that the area to the left of the z-score is 0.20. Refer to Table V and look in the body of the table for the area closest to 0.20. The area closest to 0.20 is 0.2005, which corresponds to a z-score of -0.84. See Figure 22.

Figure 22

Standard Normal Distribution						
z	.00	.01	.02	.03	.04	.05
-3.4	0.0003	0.0003	0.0003	0.0003	0.0003	0.0003	0
-3.3	0.0005	0.0005	0.0005	0.0004	0.0004	0.0004	0
-0.9	0.1841	0.1814	0.1788	0.1762	0.1736	0.1711	0.
-0.8	0.2119	0.2090	0.2061	0.2033	0.2005	0.1977	0
-0.7	0.2420	0.2389	0.2358	0.2327	0.2296	0.2266	0

Step 3   The height of a three-year-old female at the 20th percentile is

$$x = \mu + z\sigma$$
$$= 38.72 + (-0.84)(3.17)$$
$$= 36.1 \text{ inches}$$

● **Now Work Problem 47(a)**

calculator, Minitab, Excel, and StatCrunch are found in the Technology Step-by-Step on pages 373–374.

Technology Solution

Step 1   Figure 23 shows the normal curve with the unknown value of x at the 20th percentile, which separates the bottom 20% of the distribution from the top 80%.

Figure 23

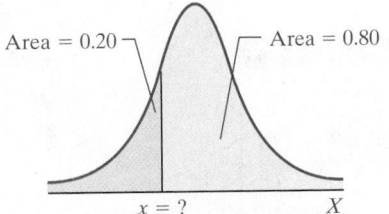

Step 2   Figure 24 shows the results obtained from StatCrunch. The height of a three-year-old female at the 20th percentile is 36.1 inches.

Figure 24

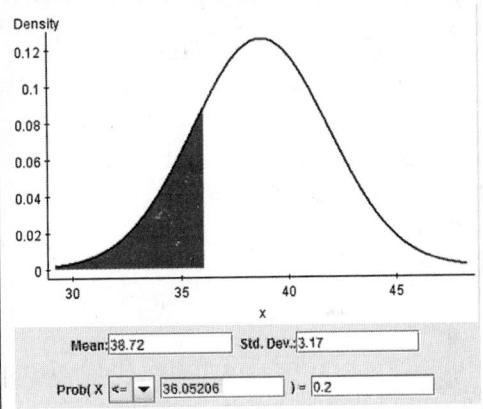

EXAMPLE 4   Finding the Value of a Normal Random Variable

Problem   The scores earned on the mathematics portion of the SAT, a college entrance exam, are approximately normally distributed with mean 516 and standard deviation 116. What scores separate the middle 90% of test takers from the bottom and top 5%? In other words, find the 5th and 95th percentiles. *Source:* The College Board

By-Hand Solution

Step 1   Figure 25 shows the normal curve with the unknown values of x separating the bottom and top 5% of the distribution from the middle 90%.

Figure 25

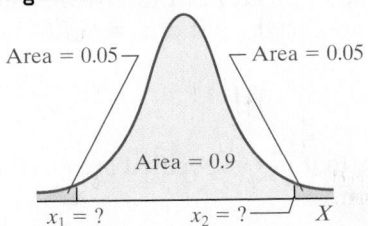

Technology Solution

Step 1   Figure 27 shows the normal curve with the unknown values of x separating the bottom and top 5% of the distribution from the middle 90%.

Figure 27

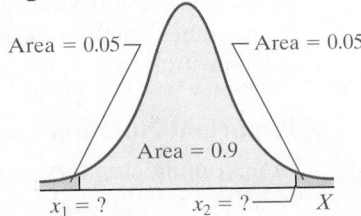

(continued)

Step 2 First, find the z-score that corresponds to an area of 0.05 to the left. In Table V, look in the body of the table and find that 0.0495 and 0.0505 are equally close to 0.05. See Figure 26. We agree to take the mean of the two z-scores corresponding to the areas. The z-score corresponding to an area of 0.0495 is -1.65, and the z-score corresponding to an area of 0.0505 is -1.64. The approximate z-score corresponding to an area of 0.05 to the left is $z_1 = -1.645$.

Figure 26

Standard Normal Distribution						
z	.00	.01	.02	.03	.04	.05
−3.4	0.0003	0.0003	0.0003	0.0003	0.0003	0.0003
−3.3	0.0005	0.0005	0.0005	0.0004	0.0004	0.0004
−1.8	0.0359	0.0351	0.0344	0.0336	0.0329	0.0322
−1.7	0.0446	0.0436	0.0427	0.0418	0.0409	0.0401
−1.6	0.0548	0.0537	0.0526	0.0516	0.0505	0.0495
−1.5	0.0668	0.0655	0.0643	0.0630	0.0618	0.0606
−1.4	0.0808	0.0793	0.0778	0.0764		

Now find the z-score corresponding to an area of 0.05 to the right, which means the area to the left is 0.95. From Table V, we find an area of 0.9495 and 0.9505, which correspond to 1.64 and 1.65. The approximate z-score, such that the area to the right is 0.05, is $Z_2 = 1.645$.

Step 3 The SAT mathematics score that separates the bottom 5% from the top 95% of scores is

$$x_1 = \mu + z_1\sigma$$
$$= 516 + (-1.645)(116)$$
$$= 325$$

The SAT mathematics score that separates the bottom 95% from the top 5% of scores is

$$x_2 = \mu + z_2\sigma$$
$$= 516 + (1.645)(116)$$
$$= 707$$

Step 2 Figure 28 shows the results obtained from Excel with the values of x_1 and x_2 highlighted.

Figure 28

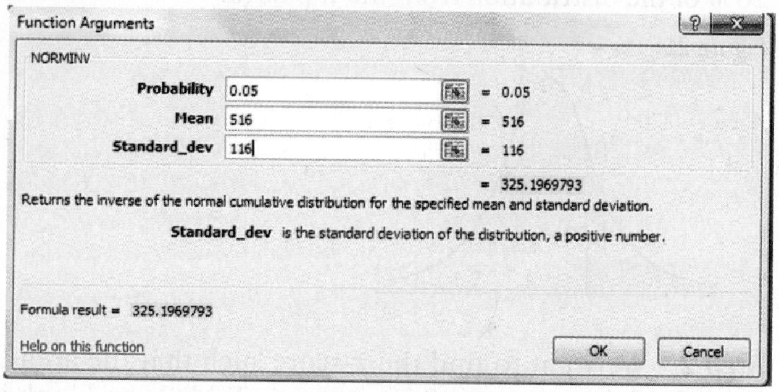

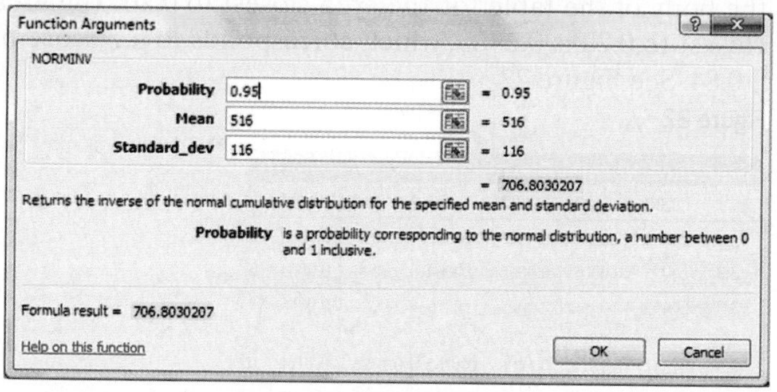

Interpretation SAT mathematics scores that separate the middle 90% of the scores from the bottom and top 5% are 325 and 707. Put another way, a student who scores 325 on the SAT math exam is at the 5th percentile. A student who scores 707 on the SAT math exam is at the 95th percentile. We might use these results to identify those scores that are unusual.

● Now Work Problem 47(b)

We could also obtain the by-hand solution to Example 4 using symmetry. Because the normal curve is symmetric about its mean, the z-score that corresponds to an area of 0.05 to the left will be the additive inverse (i.e., the opposite) of the z-score that corresponds to an area of 0.05 to the right. Since the area to the left of $z = -1.645$ is 0.05, the area to the right of $z = 1.645$ is 0.05.

Important Notation for the Future

In upcoming chapters, we will need to find the z-score that has a specified area to the right. We have special notation to represent this situation.

Figure 29

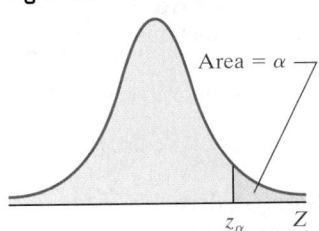

The notation z_α (pronounced "z sub alpha") is the z-score such that the area under the standard normal curve to the right of z_α is α. Figure 29 illustrates the notation.

EXAMPLE 5   Finding the Value of z_α

Problem Find the value of $z_{0.10}$.

Approach We wish to find the z-value such that the area under the standard normal curve to the right of the z-value is 0.10.

By-Hand Solution The area to the right of the unknown z-value is 0.10, so the area to the left of the z-value is $1 - 0.10 = 0.90$. We look in Table V for the area closest to 0.90. The closest area is 0.8997, which corresponds to a z-value of 1.28. Therefore, $z_{0.10} = 1.28$.

Technology Solution The area to the right of the unknown z-value is 0.10, so the area to the left is $1 - 0.10 = 0.90$. A TI-84 Plus C is used to find that the z-value such that the area to the left is 0.90 is 1.28. See Figure 30. Therefore, $z_{0.10} = 1.28$.

Figure 30

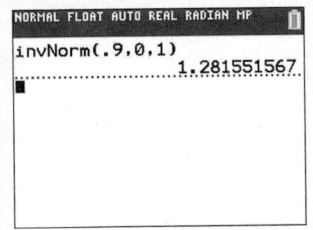

Figure 31 shows the z-value on the normal curve.

Figure 31

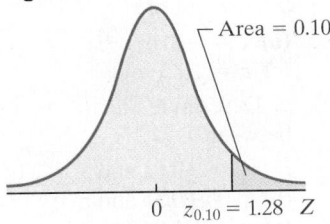

• **Now Work Problem 19**

For any continuous random variable, the probability of observing a specific value of the random variable is 0. For example, for a normal random variable, $P(a) = 0$ for any value of a, because there is no area under the normal curve associated with a single value. Therefore, the following probabilities are equivalent:

$$P(a < X < b) = P(a \le X < b) = P(a < X \le b) = P(a \le X \le b)$$

Technology Step-by-Step   The Normal Distribution

TI-83/84 Plus

Finding Area under the Normal Curve

1. From the HOME screen, press 2nd VARS to access the DISTRibution menu.
2. Select 2:normalcdf(.
3. Enter the lowerbound, upperbound, μ, and σ. Highlight Paste and hit ENTER. Hit ENTER again with the formula on the HOME screen.

Finding Normal Values Corresponding to an Area

1. From the HOME screen, press 2nd VARS to access the DISTRibution menu.
2. Select 3:invNorm(.
3. Enter the *area left*, μ, and σ. Highlight Paste and hit ENTER. Hit ENTER again with the formula on the HOME screen.

Note: When there is no lowerbound, enter −1E99. When there is no upperbound, enter 1E99. The E shown is scientific notation; it is selected by pressing 2nd then ,.

Minitab

Finding Area under the Normal Curve
1. Select the **Calc** menu, highlight **Probability Distributions**, and highlight **Normal** . . . .
2. Select **Cumulative Probability**. Enter the mean, μ, and the standard deviation, σ. Select **Input Constant**, and enter the observation. Click OK.

Finding Normal Values Corresponding to an Area
1. Select the **Calc** menu, highlight **Probability Distributions**, and highlight **Normal** . . . .
2. Select **Inverse Cumulative Probability**. Enter the mean, μ, and the standard deviation, σ. Select **Input Constant**, and enter the area to the left of the unknown normal value. Click OK.

Excel

Finding Area under the Normal Curve
1. Select the Formulas tab. Click "More Functions", then highlight "Statistical". Select "NORM.DIST".
2. Enter the specified observation, μ, and σ, and set **Cumulative** to True. Click OK.

Finding Normal Values Corresponding to an Area
1. Select the Formulas tab. Click "More Functions", then highlight "Statistical". Select "NORM.INV".
2. Enter the area left of the unknown normal value, μ, and σ. Click OK.

StatCrunch

Finding Area under the Standard Normal Curve
1. Select **Stat**, highlight **Calculators**, select **Normal**.
2. Enter the mean and the standard deviation. In the pull-down menu, decide if you wish to compute $P(X \le x)$ or $P(X \ge x)$. Finally, enter the value of x. Click Compute.

Finding Scores Corresponding to an Area
1. Select **Stat**, highlight **Calculators**, select **Normal**.
2. Enter the mean and the standard deviation. If you are given the area to the left of the unknown value, in the pull-down menu choose the $\le$ option; if given the area to the right, choose the $\ge$ option. Finally, enter the area in the right-most cell. Click Compute.

 7.2 Assess Your Understanding

Vocabulary and Skill Building

1. A random variable Z that is normally distributed with mean $\mu = 0$ and standard deviation $\sigma = 1$ is said to have the _____ _____ _____.

2. The notation z_α is the z-score such that the area under the standard normal curve to the right of z_α is _____.

3. If X is a normal random variable with mean 40 and standard deviation 10 and $P(X > 45) = 0.3085$, then $P(X < 35) = $ _____.

4. If X is normal random variable with mean 40 and standard deviation 10 and $P(X < 38) = 0.4207$, then $P(X \le 38) = $ _____.

In Problems 5–12, find the indicated areas. For each problem, be sure to draw a standard normal curve and shade the area that is to be found.

5. Determine the area under the standard normal curve that lies to the left of
(a) $z = -2.45$ (b) $z = -0.43$
(c) $z = 1.35$ (d) $z = 3.49$

6. Determine the area under the standard normal curve that lies to the left of
(a) $z = -3.49$ (b) $z = -1.99$
(c) $z = 0.92$ (d) $z = 2.90$

7. Determine the area under the standard normal curve that lies to the right of
(a) $z = -3.01$ (b) $z = -1.59$
(c) $z = 1.78$ (d) $z = 3.11$

8. Determine the area under the standard normal curve that lies to the right of
(a) $z = -3.49$ (b) $z = -0.55$
(c) $z = 2.23$ (d) $z = 3.45$

9. Determine the area under the standard normal curve that lies between
(a) $z = -2.04$ and $z = 2.04$
(b) $z = -0.55$ and $z = 0$
(c) $z = -1.04$ and $z = 2.76$

10. Determine the area under the standard normal curve that lies between
(a) $z = -2.55$ and $z = 2.55$
(b) $z = -1.67$ and $z = 0$
(c) $z = -3.03$ and $z = 1.98$

11. Determine the total area under the standard normal curve
(a) to the left of $z = -2$ or to the right of $z = 2$
(b) to the left of $z = -1.56$ or to the right of $z = 2.56$
(c) to the left of $z = -0.24$ or to the right of $z = 1.20$

12. Determine the total area under the standard normal curve
(a) to the left of $z = -2.94$ or to the right of $z = 2.94$
(b) to the left of $z = -1.68$ or to the right of $z = 3.05$
(c) to the left of $z = -0.88$ or to the right of $z = 1.23$

In Problems 13–18, find the indicated z-score. Be sure to draw a standard normal curve that depicts the solution.

13. Find the z-score such that the area under the standard normal curve to its left is 0.1.

14. Find the z-score such that the area under the standard normal curve to its left is 0.2.

15. Find the z-score such that the area under the standard normal curve to its right is 0.25.

16. Find the z-score such that the area under the standard normal curve to its right is 0.35.

17. Find z-scores that separate the middle 99% of the distribution from the area in the tails of the standard normal distribution.

18. Find the z-scores that separate the middle 94% of the distribution from the area in the tails of the standard normal distribution.

In Problems 19–22, find the value of z_α.

NW 19. $z_{0.01}$

20. $z_{0.02}$

21. $z_{0.025}$

22. $z_{0.15}$

In Problems 23–32, assume that the random variable X is normally distributed, with mean $\mu = 50$ and standard deviation $\sigma = 7$. Compute the following probabilities. Be sure to draw a normal curve with the area corresponding to the probability shaded.

23. $P(X > 35)$

24. $P(X > 65)$

25. $P(X \le 45)$

26. $P(X \le 58)$

27. $P(40 < X < 65)$

28. $P(56 < X < 68)$

29. $P(55 \le X \le 70)$

30. $P(40 \le X \le 49)$

31. $P(38 < X \le 55)$

32. $P(56 \le X < 66)$

In Problems 33–36, assume that the random variable X is normally distributed, with mean $\mu = 50$ and standard deviation $\sigma = 7$. Find each indicated percentile for X.

33. The 9th percentile

34. The 90th percentile

35. The 81st percentile

36. The 38th percentile

Applying the Concepts

37. Egg Incubation Times The mean incubation time of fertilized chicken eggs kept at 100.5°F in a still-air incubator is 21 days. Suppose that the incubation times are approximately normally distributed with a standard deviation of 1 day.
Source: University of Illinois Extension

(a) Draw a normal model that describes egg incubation times of fertilized chicken eggs.

(b) Find and interpret the probability that a randomly selected fertilized chicken egg hatches in less than 20 days.

(c) Find and interpret the probability that a randomly selected fertilized chicken egg takes over 22 days to hatch.

(d) Find and interpret the probability that a randomly selected fertilized chicken egg hatches between 19 and 21 days.

(e) Would it be unusual for an egg to hatch in less than 18 days? Why?

38. Reading Rates The reading speed of sixth-grade students is approximately normal, with a mean speed of 125 words per minute and a standard deviation of 24 words per minute.

(a) Draw a normal model that describes the reading speed of sixth-grade students.

(b) Find and interpret the probability that a randomly selected sixth-grade student reads less than 100 words per minute.

(c) Find and interpret the probability that a randomly selected sixth-grade student reads more than 140 words per minute.

(d) Find and interpret the probability that a randomly selected sixth-grade student reads between 110 and 130 words per minute. 0.3189

(e) Would it be unusual for a sixth grader to read more than 200 words per minute? Why?

NW 39. Chips Ahoy! Cookies The number of chocolate chips in an 18-ounce bag of Chips Ahoy! chocolate chip cookies is approximately normally distributed with a mean of 1262 chips and standard deviation 118 chips according to a study by cadets of the U.S. Air Force Academy.
Source: Brad Warner and Jim Rutledge, *Chance* 12(1): 10–14, 1999

(a) What is the probability that a randomly selected 18-ounce bag of Chips Ahoy! contains between 1000 and 1400 chocolate chips, inclusive?

(b) What is the probability that a randomly selected 18-ounce bag of Chips Ahoy! contains fewer than 1000 chocolate chips?

(c) What proportion of 18-ounce bags of Chips Ahoy! contains more than 1200 chocolate chips?

(d) What proportion of 18-ounce bags of Chips Ahoy! contains fewer than 1125 chocolate chips?

(e) What is the percentile rank of an 18-ounce bag of Chips Ahoy! that contains 1475 chocolate chips?

(f) What is the percentile rank of an 18-ounce bag of Chips Ahoy! that contains 1050 chocolate chips?

40. Wendy's Drive-Through Fast-food restaurants spend quite a bit of time studying the amount of time cars spend in their drive-throughs. Certainly, the faster the cars get through the drive-through, the more the opportunity for making money. *QSR Magazine* studied drive-through times for fast-food restaurants and found Wendy's had the best time, with a mean time spent in the drive-through of 138.5 seconds. Assuming drive-through times are normally distributed with a standard deviation of 29 seconds, answer the following.

(a) What is the probability that a randomly selected car will get through Wendy's drive-through in less than 100 seconds?

(b) What is the probability that a randomly selected car will spend more than 160 seconds in Wendy's drive-through?

(c) What proportion of cars spend between 2 and 3 minutes in Wendy's drive-through?

(d) Would it be unusual for a car to spend more than 3 minutes in Wendy's drive-through? Why?

41. Gestation Period The lengths of human pregnancies are approximately normally distributed, with mean $\mu = 266$ days and standard deviation $\sigma = 16$ days.

(a) What proportion of pregnancies lasts more than 270 days?

(b) What proportion of pregnancies lasts less than 250 days?

(c) What proportion of pregnancies lasts between 240 and 280 days?

(d) What is the probability that a randomly selected pregnancy lasts more than 280 days?

(e) What is the probability that a randomly selected pregnancy lasts no more than 245 days?

(f) A "very preterm" baby is one whose gestation period is less than 224 days. Are very preterm babies unusual?

42. Light Bulbs General Electric manufactures a decorative Crystal Clear 60-watt light bulb that it advertises will last 1500 hours. Suppose that the lifetimes of the light bulbs are approximately normally distributed, with a mean of 1550 hours and a standard deviation of 57 hours.

(a) What proportion of the light bulbs will last less than the advertised time?

(b) What proportion of the light bulbs will last more than 1650 hours?

(c) What is the probability that a randomly selected GE Crystal Clear 60-watt light bulb will last between 1625 and 1725 hours?

(d) What is the probability that a randomly selected GE Crystal Clear 60-watt light bulb will last longer than 1400 hours?

43. Manufacturing Steel rods are manufactured with a mean length of 25 centimeters (cm). Because of variability in the manufacturing process, the lengths of the rods are approximately normally distributed, with a standard deviation of 0.07 cm.

(a) What proportion of rods has a length less than 24.9 cm?

(b) Any rods that are shorter than 24.85 cm or longer than 25.15 cm are discarded. What proportion of rods will be discarded?

(c) Using the results of part (b), if 5000 rods are manufactured in a day, how many should the plant manager expect to discard?

(d) If an order comes in for 10,000 steel rods, how many rods should the plant manager manufacture if the order states that all rods must be between 24.9 cm and 25.1 cm?

44. Manufacturing Ball bearings are manufactured with a mean diameter of 5 millimeters (mm). Because of variability in the manufacturing process, the diameters of the ball bearings are approximately normally distributed, with a standard deviation of 0.02 mm.

(a) What proportion of ball bearings has a diameter more than 5.03 mm?

(b) Any ball bearings that have a diameter less than 4.95 mm or greater than 5.05 mm are discarded. What proportion of ball bearings will be discarded?

(c) Using the results of part (b), if 30,000 ball bearings are manufactured in a day, how many should the plant manager expect to discard?

(d) If an order comes in for 50,000 ball bearings, how many bearings should the plant manager manufacture if the order states that all ball bearings must be between 4.97 mm and 5.03 mm?

45. NCAA Basketball Point Spreads In sports betting, Las Vegas sports books establish winning margins for a team that is favored to win a game. An individual can place a wager on the game and will win if the team bet upon wins after accounting for the spread. For example, if Team A is favored by 5 points, and wins the game by 7 points, then a bet on Team A is a winning bet. However, if Team A wins the game by only 3 points, then a bet on Team A is a losing bet. In games where a team is favored by 12 or fewer points, the margin of victory for the favored team relative to the spread is approximately normally distributed with a mean of 0 points and a standard deviation of 10.9 points.
Source: Justin Wolfers, "Point Shaving: Corruption in NCAA Basketball"

(a) Explain the meaning of "the margin of victory relative to the spread has a mean of 0 points." Does this imply that the spreads are accurate for games in which a team is favored by 12 or fewer points?

(b) In games where a team is favored by 12 or fewer points, what is the probability that the favored team wins by 5 or more points relative to the spread?

(c) In games where a team is favored by 12 or fewer points, what is the probability that the favored team loses by 2 or more points relative to the spread?

46. NCAA Basketball Point Spreads Revisited See Problem 45. In games where a team is favored by more than 12 points, the margin of victory for the favored team relative to the spread is normally distributed with a mean of −1.0 point and a standard deviation of 10.9 points.
Source: Justin Wolfers, "Point Shaving: Corruption in NCAA Basketball"

(a) In games where a team is favored by more than 12 points, what is the probability that the favored team wins by 5 or more points relative to the spread?

(b) In games where a team is favored by more than 12 points, what is the probability that the favored team loses by 2 or more points relative to the spread?

(c) In games where a team is favored by more than 12 points, what is the probability that the favored team "beats the spread"? Does this imply that the possible point shaving spreads are accurate for games in which a team is favored by more than 12 points?

NW 47. Egg Incubation Times The mean incubation time of fertilized chicken eggs kept at 100.5°F in a still-air incubator is 21 days. Suppose that the incubation times are approximately normally distributed with a standard deviation of 1 day.
Source: University of Illinois Extension

(a) Determine the 17th percentile for incubation times of fertilized chicken eggs.

(b) Determine the incubation times that make up the middle 95% of fertilized chicken eggs.

48. Reading Rates The reading speed of sixth-grade students is approximately normal, with a mean speed of 125 words per minute and a standard deviation of 24 words per minute.

(a) What is the reading speed of a sixth-grader whose reading speed is at the 90th percentile?

(b) A school psychologist wants to determine reading rates for unusual students (both slow and fast). Determine the reading rates of the middle 95% of all sixth-grade students. What are the cutoff points for unusual readers?

49. Chips Ahoy! Cookies The number of chocolate chips in an 18-ounce bag of Chips Ahoy! chocolate chip cookies is approximately normally distributed, with a mean of 1262 chips and a standard deviation of 118 chips, according to a study by cadets of the U.S. Air Force Academy.
Source: Brad Warner and Jim Rutledge, *Chance* 12(1): 10–14, 1999

(a) Determine the 30th percentile for the number of chocolate chips in an 18-ounce bag of Chips Ahoy! cookies.

(b) Determine the number of chocolate chips in a bag of Chips Ahoy! that make up the middle 99% of bags.

(c) What is the interquartile range of the number of chips in Chips Ahoy! cookies?

50. Wendy's Drive-Through Fast-food restaurants spend quite a bit of time studying the amount of time cars spend in their drive-through. Certainly, the faster the cars get through the drive-through, the more the opportunity for making money. *QSR Magazine* studied drive-through times for fast-food restaurants, and found Wendy's had the best time, with a mean time a car spent in the drive-through equal to 138.5 seconds. Assume that drive-through times are normally distributed, with a standard deviation of 29 seconds. Suppose that Wendy's wants to institute a policy at its restaurants that it will not charge any patron that must wait more than a certain amount of time for an order. Management does not want to give away free meals to more than 1% of the patrons. What time would you recommend Wendy's advertise as the maximum wait time before a free meal is awarded?

51. Speedy Lube The time required for Speedy Lube to complete an oil change service on an automobile approximately follows a normal distribution, with a mean of 17 minutes and a standard deviation of 2.5 minutes.

(a) Speedy Lube guarantees customers that the service will take no longer than 20 minutes. If it does take longer, the customer will receive the service for half-price. What percent of customers receive the service for half price?

(b) If Speedy Lube does not want to give the discount to more than 3% of its customers, how long should it make the guaranteed time limit?

52. Putting It Together: Birth Weights The following data represent the distribution of birth weights (in grams) for babies in which the pregnancy went full term (37–41 weeks).

Birth Weight (g)	Number of Live Births
0–499	22
500–999	201
1000–1499	1,645
1500–1999	9,365
2000–2499	92,191
2500–2999	569,319
3000–3499	1,387,335
3500–3999	988,011
4000–4499	255,700
4500–4999	36,766
5000–5499	3,994

Source: National Vital Statistics Report

(a) Construct a relative frequency distribution for birth weight.

(b) Draw a relative frequency histogram for birth weight. Describe the shape of the distribution.

(c) Determine the mean and standard deviation birth weight.

(d) Use the normal model to determine the proportion of babies in each class.

(e) Compare the proportions predicted by the normal model to the relative frequencies found in part (a). Do you believe that the normal model is effective in describing the birth weights of babies?

Explaining the Concepts

53. Give three interpretations for the area under a normal curve.

54. Explain why $P(X < 30)$ should be reported as <0.0001 if X is a normal random variable with mean 100 and standard deviation 15.

55. Explain why $P(X \leq 220)$ should be reported as >0.9999 if X is a normal random variable with mean 100 and standard deviation 15.

56. The ACT and SAT are two college entrance exams. The composite score on the ACT is approximately normally distributed with mean 21.1 and standard deviation 5.1. The composite score on the SAT is approximately normally distributed with mean 1026 and standard deviation 210. Suppose you scored 26 on the ACT and 1240 on the SAT. Which exam did you score better on? Justify your reasoning using the normal model.

7.3 Assessing Normality

Preparing for This Section Before getting started, review the following:

- Shape of a distribution (Section 2.2, pp. 85–86)
- Correlation coefficient (Section 4.1, pp. 183–187)

Objective ① Use normal probability plots to assess normality

Up to this point, we have said that a random variable X is normally distributed, or at least approximately normal, provided the histogram of the data is symmetric and bell-shaped. This works well for large data sets, but the shape of a histogram drawn from a small sample of observations does not always accurately represent the shape of the population. For this reason, we need additional methods for assessing the normality of a random variable X when we are looking at a small set of sample data.

① **Use Normal Probability Plots to Assess Normality**

In Other Words
Normal probability plots are used to assess the normality of a data set.

A **normal probability plot** is a graph that plots observed data versus *normal scores*. A **normal score** is the expected z-score of the data value, assuming that the distribution of the random variable is normal. The expected z-score of an observed value depends on the number of observations in the data set.

Drawing a normal probability plot requires the following steps:

Drawing a Normal Probability Plot

Step 1 Arrange the data in ascending order.

Step 2 Compute $f_i = \dfrac{i - 0.375}{n + 0.25}$,* where i is the index (the position of the data value in the ordered list) and n is the number of observations. The expected proportion of observations less than or equal to the ith data value is f_i.

Step 3 Find the z-score corresponding to f_i from Table V.

Step 4 Plot the observed values on the horizontal axis and the corresponding expected z-scores on the vertical axis.

The idea behind finding the expected z-score is that, if the data come from a normally distributed population, we could predict the area to the left of each data value. The value of f_i represents the expected area to the left of the ith observation when the data come from a population that is normally distributed. For example, f_1 is the expected area to the left of the smallest data value, f_2 is the expected area to the left of the second-smallest data value, and so on. See Figure 32.

Once we determine each f_i, we find the z-scores corresponding to f_1, f_2, and so on. The smallest observation in the data set will be the smallest expected z-score, and the largest observation in the data set will be the largest expected z-score. Also, because of the symmetry of the normal curve, the expected z-scores are always paired as positive and negative values.

Values of normal random variables and their z-scores are linearly related $(x = \mu + z\sigma)$, so a plot of observations of normal variables against their expected z-scores will be linear. We conclude the following:

Figure 32

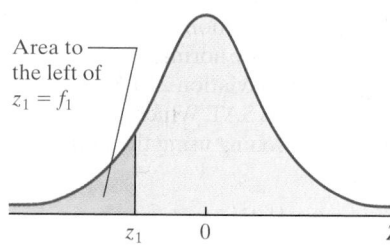

If sample data are taken from a population that is normally distributed, a normal probability plot of the observed values versus the expected z-scores will be approximately linear.

It is difficult to determine whether a normal probability plot is "linear enough." However, we can use a procedure based on the research of S. W. Looney and T. R. Gulledge in their paper "Use of the Correlation Coefficient with Normal Probability Plots," published in the *American Statistician*. Basically, if the linear correlation coefficient between the observed values and expected z-scores is greater than the critical value found in Table VI in Appendix A, then it is reasonable to conclude that the data could come from a population that is normally distributed.

Normal probability plots are typically drawn using graphing calculators or statistical software. However, it is worthwhile to go through an example that demonstrates the procedure to better understand the results supplied by technology.

EXAMPLE 1 Constructing a Normal Probability Plot

Problem The data in Table 4 represent the finishing time (in seconds) for six randomly selected races of a greyhound named Barbies Bomber in the $\dfrac{5}{16}$-mile race at Greyhound Park in Dubuque, Iowa. Is there evidence to support the belief that the variable "finishing time" is normally distributed?

Table 4

31.35	32.52
32.06	31.26
31.91	32.37

Source: Greyhound Park, Dubuque, IA

Approach Follow Steps 1 through 4.

*The derivation of this formula is beyond the scope of this text.

Solution

Step 1 Column 1 in Table 5 represents the index i. Column 2 represents the observed values in the data set, written in ascending order.

Table 5			
Index, i	**Observed Value**	**f_i**	**Expected z-score**
1	31.26	$\dfrac{1 - 0.375}{6 + 0.25} = 0.10$	-1.28
2	31.35	$\dfrac{2 - 0.375}{6 + 0.25} = 0.26$	-0.64
3	31.91	0.42	-0.20
4	32.06	0.58	0.20
5	32.37	0.74	0.64
6	32.52	0.90	1.28

Step 2 Column 3 in Table 5 represents $f_i = \dfrac{i - 0.375}{n + 0.25}$ for each observation. This value is the expected area under the normal curve to the left of the ith observation, assuming the data come from a population that is normally distributed. For example, $i = 1$ corresponds to the finishing time of 31.26, and

$$f_1 = \frac{1 - 0.375}{6 + 0.25} = 0.10$$

So the area under the normal curve to the left of 31.26 is 0.10 if the sample data come from a population that is normally distributed.

Step 3 Use Table V to find the z-scores that correspond to each f_i, then list them in Column 4 of Table 5. Look in Table V for the area closest to $f_1 = 0.1$. The expected z-score is -1.28. Notice that for each negative expected z-score there is a corresponding positive expected z-score, as a result of the symmetry of the normal curve.

Step 4 Plot the actual observations on the horizontal axis and the expected z-scores on the vertical axis. See Figure 33.

Figure 33

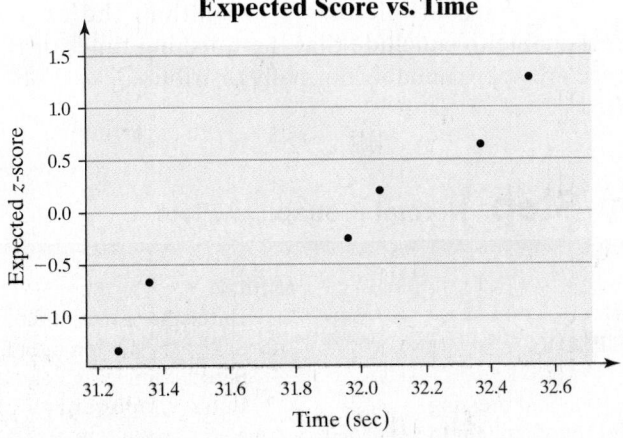

Interpretation The linear correlation between the observed values and expected z-scores from the data in Table 5 is 0.970.

The critical value in Table VI for $n = 6$ observations is 0.888. Because the correlation coefficient is greater than the critical value $(0.970 > 0.888)$, it is reasonable to conclude that the finishing times of Barbies Bomber in the 5/16-mile race are approximately normally distributed. ●

Typically, normal probability plots are drawn using either a graphing calculator with advanced statistical features or statistical software.

EXAMPLE 2   Assessing Normality Using Technology

Problem Draw a normal probability plot of the data in Table 4 using technology. Is there evidence to support the belief that the variable "finishing time" is normally distributed?

Approach We will use StatCrunch to draw the normal probability plot and find the correlation between the observed values and expected z-scores. If the correlation is greater than the critical value from Table VI, we conclude the data could come from a population that is normally distributed. The steps for constructing normal probability plots using TI-83/84 Plus graphing calculators, Minitab, Excel, and StatCrunch can be found on pages 380–381.

Solution Figure 34 shows the normal probability plot. The correlation between the observed values and expected z-scores is 0.974.

Figure 34

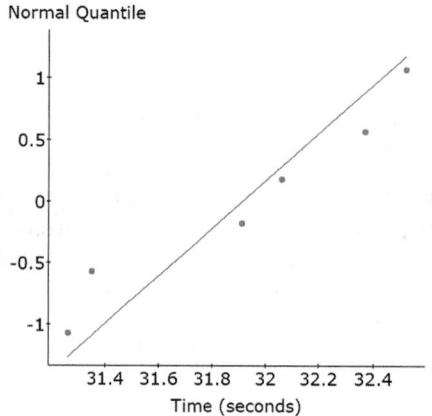

NOTE

Minitab will draw "confidence bands" in its normal probability plots. If all the data lie within these bands, then it is reasonable to conclude the data come from a population that is normally distributed.

NOTE

The correlations in Examples 1 and 2 differ due to rounding.

● **Now Work Problem 3**

The critical value in Table VI for $n = 6$ observations is 0.888. Because the correlation coefficient is greater than the critical value $(0.974 > 0.888)$, it is reasonable to conclude that the finishing times of Barbies Bomber in the 5/16-mile race are approximately normally distributed.   ●

Technology Step-by-Step   Normal Probability Plots

TI-83/84 Plus
1. Enter the raw data into L1.
2. Press 2nd Y = to access STAT PLOTS.
3. Select 1:Plot1.
4. Turn Plot1 on by highlighting On and pressing ENTER. Press the down-arrow key. Highlight the *normal probability plot* icon. Press ENTER to select this plot type. The Data List should be set at L1. The Data Axis should be the x-axis.
5. Press ZOOM, and select 9:ZoomStat. Once you have the graph, TRACE to find the values of the observations and the corresponding normal scores. Enter these observations into L1 and L2. Find the correlation coefficient for this data.

Minitab
1. Enter the raw data into C1.
2. Select the **Graph** menu. Choose **Probability Plot ....** Select "Single." Click OK.
3. In the Graph variables cell, enter the column that contains the raw data. Make sure Distribution is set to Normal. Click OK.

Excel
1. Install XLSTAT.
2. Enter the raw data into column A.
3. Select the **XLSTAT** menu. Highlight **Visualizing Data** and select **Univariate Plots**.

4. In the general tab, check Quantitative data. With the cursor in the Quantitative data box, highlight the raw data. Uncheck the box "Sample Labels." Click the charts tab and check Normal P-P plots or Normal Q-Q plots. Click OK.

StatCrunch

1. If necessary, enter the raw data into column var1. Name the column.
2. Select **Graph** and highlight **QQ Plot**.
3. Select the variable. Check the box to add the correlation statistic. Check the "Normal quantiles on y-axis" box. Click Compute!

7.3 Assess Your Understanding

Vocabulary and Skill Building

1. A _____ is a graph that plots observed data versus normal scores.

2. *True or False:* A normal score is the expected z-score of a data value, assuming the distribution of the random variable is normal.

In Problems 3–6, use the results in the table to (a) draw a normal probability plot, (b) determine the linear correlation between the observed values and expected z-scores, (c) determine the critical value in Table VI to assess the normality of the data.

NW 3.

Index, i	Observed Value	f_i	Expected z-score
1	39	0.08	−1.41
2	45	0.20	−0.84
3	48	0.32	−0.47
4	52	0.44	−0.15
5	54	0.56	0.15
6	56	0.68	0.47
7	60	0.80	0.84
8	62	0.92	1.41

4.

Index, i	Observed Value	f_i	Expected z-score
1	77	0.07	−1.48
2	80	0.18	−0.92
3	84	0.28	−0.58
4	91	0.39	−0.28
5	98	0.5	0
6	104	0.61	0.28
7	109	0.72	0.58
8	112	0.82	0.92
9	120	0.93	1.48

5.

Index, i	Observed Value	f_i	Expected z-score
1	1	0.09	−1.34
2	3	0.22	−0.77
3	6	0.36	−0.36
4	8	0.50	0
5	10	0.64	0.36
6	13	0.78	0.77
7	35	0.91	1.34

6.

Index, i	Observed Value	f_i	Expected z-score
1	1	0.10	−1.28
2	3	0.26	−0.64
3	19	0.42	−0.20
4	30	0.58	0.20
5	88	0.74	0.64
6	99	0.90	1.28

In Problems 7–10, use a normal probability plot to assess whether the sample data could have come from a population that is normally distributed.

 7. O-Ring Thickness A random sample of O-rings was obtained, and the wall thickness (in inches) of each was recorded.

0.276	0.274	0.275	0.274	0.277
0.273	0.276	0.276	0.279	0.274
0.273	0.277	0.275	0.277	0.277
0.276	0.277	0.278	0.275	0.276

8. Customer Service A random sample of weekly work logs at an automobile repair station was obtained, and the average number of customers per day was recorded.

26	24	22	25	23
24	25	23	25	22
21	26	24	23	24
25	24	25	24	25
26	21	22	24	24

9. School Loans A random sample of 20 undergraduate students receiving student loans was obtained, and the amount of their loans for the 2014–2015 school year was recorded.

2,500	1,000	2,000	14,000	1,800
3,800	10,100	2,200	29,000	16,000
5,000	2,200	6,200	9,100	2,800
2,500	1,400	13,200	750	12,000

10. Memphis Snowfall A random sample of 25 years between 1890 and 2011 was obtained, and the amount of snowfall, in inches, for Memphis was recorded.

24.0	7.9	1.5	0.0	0.3
0.4	8.1	4.3	0.0	0.5
3.6	2.9	0.4	2.6	0.1
16.6	1.4	23.8	25.1	1.6
12.2	14.8	0.4	3.7	4.2

Source: National Oceanic and Atmospheric Administration

Applying the Concepts

11. Chips per Bag In a 1998 advertising campaign, Nabisco claimed that every 18-ounce bag of Chips Ahoy! cookies contained at least 1000 chocolate chips. Brad Warner and Jim Rutledge tried to verify the claim. The following data represent the number of chips in an 18-ounce bag of Chips Ahoy! based on their study.

1087	1098	1103	1121	1132
1185	1191	1199	1200	1213
1239	1244	1247	1258	1269
1307	1325	1345	1356	1363
1135	1137	1143	1154	1166
1214	1215	1219	1219	1228
1270	1279	1293	1294	1295
1377	1402	1419	1440	1514

Source: Chance 12(1): 10–14, 1999

(a) Draw a normal probability plot to determine if the data could have come from a normal distribution.
(b) Determine the mean and standard deviation of the sample data.
(c) Using the sample mean and sample standard deviation obtained in part (b) as estimates for the population mean and population standard deviation, respectively, draw a graph of a normal model for the distribution of chips in a bag of Chips Ahoy!

(d) Using the normal model from part (c), find the probability that an 18-ounce bag of Chips Ahoy! selected at random contains at least 1000 chips.
(e) Using the normal model from part (c), determine the proportion of 18-ounce bags of Chips Ahoy! that contains between 1200 and 1400 chips, inclusive.

12. Hours of TV A random sample of college students aged 18–24 years was obtained, and the number of hours of television watched in a typical week was recorded.

36.1	30.5	2.9	17.5	21.0
23.5	25.6	16.0	28.9	29.6
7.8	20.4	33.8	36.8	0.0
9.9	25.8	19.5	19.1	18.5
22.9	9.7	39.2	19.0	8.6

(a) Draw a normal probability plot to determine if the data could have come from a normal distribution.
(b) Determine the mean and standard deviation of the sample data.
(c) Using the sample mean and sample standard deviation obtained in part (b) as estimates for the population mean and population standard deviation, respectively, draw a graph of a normal model for the distribution of weekly hours of television watched.
(d) Using the normal model from part (c), find the probability that a college student aged 18–24 years, selected at random, watches between 20 and 35 hours of television each week.
(e) Using the normal model from part (c), determine the proportion of college students aged 18–24 years who watch more than 40 hours of television per week.

13. Putting It Together: Demon Roller Coaster Retrieve the data 7_3_13 at www.pearsonhighered.com/sullivanstats using the file format of your choice for the text you are using. The data represent the time spent waiting in line (in minutes) for the Demon Roller Coaster for 100 randomly selected riders.

(a) Draw a relative frequency histogram with lower class limit of the first class equal to 0 and class width 20.
(b) Comment on the shape of the distribution.
(c) Draw a normal probability plot to assess the normality of the random variable "wait time."

7.4 The Normal Approximation to the Binomial Probability Distribution

Preparing for This Section Before getting started, review the following:

- Binomial probability distribution (Section 6.2, pp. 329–335)

Objective ① Approximate binomial probabilities using the normal distribution

① **Approximate Binomial Probabilities Using the Normal Distribution**

In Section 6.2, we discussed the binomial probability distribution. Now, we will review the criteria for a probability experiment to be a binomial experiment.

Criteria for a Binomial Probability Experiment

A probability experiment is a binomial experiment if all the following are true:

1. The experiment is performed n independent times. Each repetition of the experiment is called a **trial**. Independence means that the outcome of one trial will not affect the outcome of the other trials.
2. For each trial, there are two mutually exclusive outcomes—success or failure.
3. The probability of success, p, is the same for each trial of the experiment.

Historical Note

The normal approximation to the binomial was discovered by Abraham de Moivre in 1733. With the advance of computing technology, its importance has been diminished.

The binomial probability formula can be used to compute probabilities of events in a binomial experiment. A large number of trials of a binomial experiment, however, makes this formula difficult to use. For example, given 500 trials of a binomial experiment, to compute the probability of 400 or more successes requires that we compute the following probabilities:

$$P(X \geq 400) = P(400) + P(401) + \cdots + P(500)$$

This would be time consuming to compute by hand! Fortunately, we have an alternative means for approximating binomial probabilities, provided that certain conditions are met.

Recall, the following fact from page 338:

For a fixed p, as the number of trials n in a binomial experiment increases, the probability distribution of the random variable X becomes more nearly symmetric and bell shaped. As a rule of thumb, if $np(1 - p) \geq 10$, the probability distribution will be approximately symmetric and bell-shaped.

This result suggests that binomial probabilities can be approximated by the area under a normal curve, provided that $np(1 - p) \geq 10$.

The Normal Approximation to the Binomial Probability Distribution

If $np(1 - p) \geq 10$, the binomial random variable X is approximately normally distributed, with mean $\mu_X = np$ and standard deviation $\sigma_X = \sqrt{np(1 - p)}$.

Figure 35 shows a graph of the probability distribution for the binomial random variable X, with $n = 40$ and $p = 0.5$, drawn in StatCrunch. Because $np(1 - p) = 40(0.5)(1 - 0.5) = 10$, we can use a normal model with $\mu_X = np = 40(0.5) = 20$ and standard deviation $\sigma_X = \sqrt{np(1 - p)} = \sqrt{40(0.5)(0.5)} = \sqrt{10}$ to describe X.

Figure 35

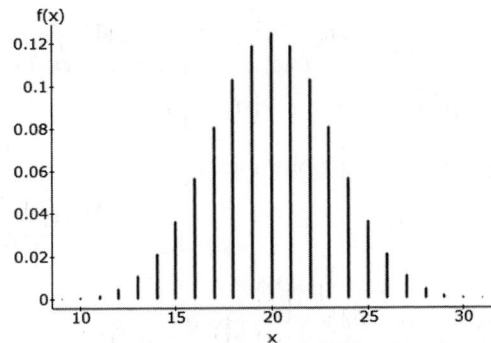

¡CAUTION!
Don't forget about the correction for continuity. It is needed because we are using a continuous density function to approximate the probability of a discrete random variable.

To approximate the probability of a specific value of the binomial random variable, such as $P(18)$, we find the area under the normal curve from $x = 17.5$ to $x = 18.5$. We add and subtract 0.5 from $x = 18$ as a **correction for continuity** because we are using a continuous density function to approximate a discrete probability.

To approximate $P(X \le 18)$, we compute the area under the normal curve for $X \le 18.5$. Do you see why?

To approximate $P(X \ge 18)$, we compute $P(X \ge 17.5)$. Do you see why? Table 6 summarizes how to use the correction for continuity.

Table 6

Exact Probability Using Binomial	Approximate Probability Using Normal	Graphical Depiction
$P(a)$	$P(a - 0.5 \le X \le a + 0.5)$	$a - 0.5$, $a + 0.5$, a
$P(X \le a)$	$P(X \le a + 0.5)$	$a + 0.5$, a
$P(X \ge a)$	$P(X \ge a - 0.5)$	$a - 0.5$, a
$P(a \le X \le b)$	$P(a - 0.5 \le X \le b + 0.5)$	$a - 0.5$, $b + 0.5$, a, b

A question remains, however. What do we do if the probability is of the form $P(X > a)$, $P(X < a)$, or $P(a < X < b)$? The solution is to rewrite the inequality in a form with $\le$ or $\ge$. For example, $P(X > 4) = P(X \ge 5)$ and $P(X < 4) = P(X \le 3)$ for binomial random variables, because the values of the random variables must be whole numbers.

EXAMPLE 1   The Normal Approximation to a Binomial Random Variable

Problem According to the American Red Cross, 7% of people in the United States have blood type O-negative. What is the probability that, in a simple random sample of 500 people in the United States, fewer than 30 have blood type O-negative?

Approach

Step 1 Verify that this is a binomial experiment.

Step 2 Computing the probability by hand would be very tedious. Verify $np(1 - p) \ge 10$. Then we may use the normal distribution to approximate the binomial probability.

Step 3 Approximate $P(X < 30) = P(X \le 29)$ by using the normal approximation to the binomial distribution.

Solution

Step 1 Each of the 500 independent trials has a probability of success equal to 0.07. This is a binomial experiment.

Step 2 Verify $np(1 - p) \geq 10$.

$$np(1 - p) = 500(0.07)(0.93) = 32.55 \geq 10$$

Use the normal distribution to approximate the binomial distribution.

Step 3 The probability that fewer than 30 people in the sample have blood type O-negative is $P(X < 30) = P(X \leq 29)$. This is approximately equal to the area under the normal curve to the left of $x = 29.5$, with $\mu_X = np = 500(0.07) = 35$ and $\sigma_X = \sqrt{np(1 - p)} = \sqrt{500(0.07)(1 - 0.07)} = \sqrt{32.55} \approx 5.71$. See Figure 36. Convert $x = 29.5$ to a z-score.

Figure 36

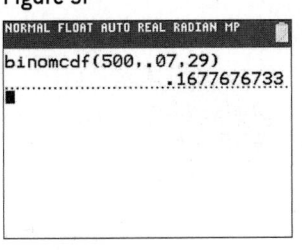
0.1685

29.5 $\mu_X = 35$ X

$$z = \frac{29.5 - 35}{\sqrt{32.55}} = -0.96$$

From Table V, we find that the area to the left of $z = -0.96$ is 0.1685. Therefore, the approximate probability that fewer than 30 people will have blood type O-negative is 0.1685.

Figure 37

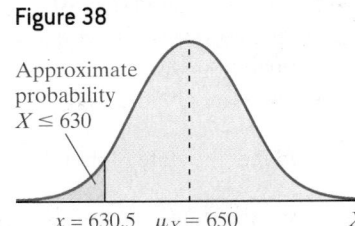
```
NORMAL FLOAT AUTO REAL RADIAN MP
binomcdf(500,.07,29)
                .1677676733
```

• **Now Work Problem 21**

Using the *binomcdf(* command on a TI-84 Plus C graphing calculator, we find that the exact probability is 0.1678. See Figure 37. The approximate result is close indeed!

EXAMPLE 2   A Normal Approximation to the Binomial

Problem According to the Gallup Organization, 65% of adult Americans are in favor of the death penalty for individuals convicted of murder. Erica selects a random sample of 1000 adult Americans in Will County, Illinois, and finds that 630 of them are in favor of the death penalty for individuals convicted of murder.

(a) Assuming that 65% of adult Americans in Will County are in favor of the death penalty, what is the probability of obtaining a random sample of no more than 630 adult Americans in favor of the death penalty from a sample of size 1000?

(b) Does the result from part (a) contradict the Gallup Organization's findings? Explain.

Approach This is a binomial experiment with $n = 1000$ and $p = 0.65$. Erica needs to determine the probability of obtaining a random sample of no more than 630 adult Americans who favor the death penalty, assuming 65% of adult Americans favor the death penalty. Computing this probability using the binomial probability formula would be difficult, so Erica will use the normal approximation to the binomial, since $np(1 - p) = 1000(0.65)(1 - 0.65) = 227.5 \geq 10$. Approximate $P(X \leq 630)$ by computing the area under the normal curve to the left of $x = 630.5$ with $\mu_X = np = 650$ and $\sigma_X = \sqrt{np(1 - p)} = \sqrt{1000(0.65)(1 - 0.65)} \approx 15.083$.

Figure 38

Approximate
probability
$X \leq 630$

$x = 630.5$ $\mu_X = 650$ X

Solution

(a) Figure 38 shows the area we wish to compute. Convert $x = 630.5$ to a z-score.

$$z = \frac{630.5 - 650}{15.083} = -1.29$$

(continued)

The area under the standard normal curve to the left of $z = -1.29$ is 0.0985. The probability of obtaining 630 or fewer adult Americans who favor the death penalty from a sample of 1000 adult Americans, assuming the proportion of adult Americans who favor the death penalty is 0.65 is 0.0985.

(b) The result of part (a) means that, if we had obtained 100 different simple random samples of size 1000, we would expect about 10 to result in 630 or fewer adult Americans favoring the death penalty if the true proportion is 0.65. Because the results obtained are not unusual under the assumption that $p = 0.65$, Erica finds that the results of her survey do not contradict those of Gallup.

● **Now Work Problem 27**

7.4 Assess Your Understanding

Vocabulary and Skill Building

1. In a binomial experiment with n trials and probability of success p, if _____, the binomial random variable X is approximately normal with $\mu_X =$ _____ and $\sigma_X =$ _____.

2. When adding or subtracting 0.5 from x, we are making a correction for _____.

3. Suppose X is a binomial random variable. To approximate $P(X < 5)$, compute _____.

4. Suppose X is a binomial random variable. To approximate $P(3 \le X \le 10)$, compute _____.

In Problems 5–14, a discrete random variable is given. Assume the probability of the random variable will be approximated using the normal distribution. Describe the area under the normal curve that will be computed. For example, if we wish to compute the probability of finding at least five defective items in a shipment, we would approximate the probability by computing the area under the normal curve to the right of $x = 4.5$.

5. The probability that at least 40 households have a gas stove

6. The probability no more than 20 people want to see *Roe v. Wade* overturned

7. The probability that exactly eight defective parts are in the shipment

8. The probability that exactly 12 students pass the course

9. The probability that the number of people with blood type O-negative is between 18 and 24, inclusive

10. The probability that the number of tornadoes that occur in the month of May is between 30 and 40, inclusive

11. The probability that more than 20 people want to see the marriage tax penalty abolished

12. The probability that fewer than 40 households have a pet

13. The probability that no more than 500 adult Americans support a bill proposing to extend daylight savings time

14. The probability that fewer than 35 people support the privatization of Social Security

In Problems 15–20, compute P(x) using the binomial probability formula. Then determine whether the normal distribution can be used as an approximation for the binomial distribution. If so, approximate P(x) and compare the result to the exact probability.

15. $n = 60, p = 0.4, x = 20$

16. $n = 80, p = 0.15, x = 18$

17. $n = 40, p = 0.25, x = 30$

18. $n = 100, p = 0.05, x = 50$

19. $n = 75, p = 0.75, x = 60$

20. $n = 85, p = 0.8, x = 70$

Applying the Concepts

NW **21. On-Time Flights** According to American Airlines, Flight 215 from Orlando to Los Angeles is on time 90% of the time. Randomly select 150 flights and use the normal approximation to the binomial to

(a) approximate the probability that exactly 130 flights are on time.

(b) approximate the probability that at least 130 flights are on time.

(c) approximate the probability that fewer than 125 flights are on time.

(d) approximate the probability that between 125 and 135 flights, inclusive, are on time.

22. Morality In a recent poll, the Gallup Organization found that 45% of adult Americans believe that the overall state of moral values in the United States is poor. Suppose a survey of a random sample of 500 adult Americans is conducted in which they are asked to disclose their feelings on the overall state of moral values in the United States. Use the normal approximation to the binomial to approximate the probability that

(a) exactly 250 of those surveyed feel the state of morals is poor.

(b) no more than 220 of those surveyed feel the state of morals is poor.

(c) more than 250 of those surveyed feel the state of morals is poor.

(d) between 220 and 250, inclusive, believe the state of morals is poor.

(e) at least 260 adult Americans believe the overall state of moral values is poor. Would you find this result unusual? Why?

23. Toilet Flushing In the Healthy Handwashing Survey conducted by Bradley Corporation, it was found that 64% of adult Americans operate the flusher of toilets in public restrooms with their foot. Suppose you survey a random sample of 740 adult American women aged 18–24 years. Use the normal approximation to the binomial to approximate the probability that

(a) exactly 490 of those surveyed flush toilets in public restrooms with their foot.

(b) no more than 490 of those surveyed flush toilets in public restrooms with their foot.

(c) at least 503 of those surveyed flush toilets in public restrooms with their foot. What does this result suggest?

24. Sneeze According to a study done by Nick Wilson of Otago University Wellington, the probability a randomly selected individual will not cover his or her mouth when sneezing is 0.267. Suppose you sit on a bench in a mall and observe 300 randomly selected individuals' habits as they sneeze. Use the normal approximation to the binomial to approximate the probability that of the 300 randomly observed individuals:

(a) exactly 100 do not cover the mouth when sneezing.

(b) fewer than 75 do not cover the mouth.

(c) Would you be surprised if, after observing 300 individuals, more than 100 did not cover the mouth when sneezing? Why?

25. Males Living at Home According to the *Current Population Survey* (Internet release date: September 15, 2004), 55% of males between the ages of 18 and 24 years lived at home in 2003. (Unmarried college students living in a dorm are counted as living at home.) Suppose a survey is administered today to 200 randomly selected males between the ages of 18 and 24 years, and 130 of them respond that they live at home.

(a) Approximate the probability that such a survey will result in at least 130 of the respondents living at home under the assumption that the true percentage is 55%.

(b) What does the result from part (a) suggest?

26. Females Living at Home According to the *Current Population Survey* (Internet release date: September 15, 2004), 46% of females between the ages of 18 and 24 years lived at home in 2003. (Unmarried college students living in a dorm are counted as living at home.) Suppose a survey is administered today to 200 randomly selected females between the ages of 18 and 24 years, and 110 of them respond that they live at home.

(a) Approximate the probability that such a survey will result in at least 110 of the respondents living at home under the assumption that the true percentage is 46%.

(b) What does the result from part (a) suggest?

NW **27. Boys Are Preferred** In a Gallup poll, 37% of survey respondents said that, if they only had one child, they would prefer the child to be a boy. You conduct a survey of 150 randomly selected students on your campus and find that 75 of them would prefer a boy.

(a) Approximate the probability that, in a random sample of 150 students, at least 75 would prefer a boy, assuming the true percentage is 37%.

(b) Does this result contradict the Gallup poll? Explain.

28. Liars According to a *USA Today* "Snapshot," 3% of Americans surveyed lie frequently. You conduct a survey of 500 college students and find that 20 of them lie frequently.

(a) Compute the probability that, in a random sample of 500 college students, at least 20 lie frequently, assuming the true percentage is 3%.

(b) Does this result contradict the *USA Today* "Snapshot"? Explain.

Chapter 7 Review

Summary

In this chapter we introduced continuous random variables and the normal probability density function. A continuous random variable is said to be approximately normally distributed if a histogram of its values is symmetric and bell-shaped. We use a normal curve to model normal random variables. The normal model is bell-shaped with the mean representing the high point of the curve. The curve also has inflection points one standard deviation on either side of the mean.

The area under a normal curve for an interval of numbers may be interpreted as a proportion, probability, or percentile. We may also reverse the process, and find the value of a random variable that corresponds to a particular proportion, probability, or percentile.

One method for determining whether a random variable might come from a population that is normally distributed is to draw a normal probability plot. If a normal probability plot is approximately linear, we say the distribution of the random variable is approximately normal. The correlation coefficient between the raw data and normal scores may be used to assess the normality of the variable.

If X is a binomial random variable with $np(1 - p) \geq 10$, we can use the area under the normal curve to approximate the probability of a particular binomial random variable. The parameters of the normal curve are $\mu_X = np$ and $\sigma_X = \sqrt{np(1 - p)}$, where n is the number of trials of the binomial experiment and p is the probability of success on a single trial.

Vocabulary

Uniform probability distribution (p. 356)

Probability density function (p. 356)

Model (p. 358)

Normal curve (p. 358)

Normally distributed (p. 358)

Normal probability distribution (p. 358)

Density (p. 358)

Inflection points (p. 359)

Probability density function (p. 361)

Normal probability density function (p. 361)

Standard normal distribution (p. 366)

Standard normal random variable Z (p. 366)

Standard normal curve (p. 366)

Normal probability plot (p. 377)

Normal score (p. 377)

Trial (p. 383)

Normal approximation to the binomial distribution (p. 383)

Correction for continuity (p. 384)

Formulas

Standardizing a Normal Random Variable

$$z = \frac{x - \mu}{\sigma}$$

Finding the Score

$$x = \mu + z\sigma$$

OBJECTIVES

Section	You should be able to . . .	Example(s)	Review Exercises
7.1	1 Use the uniform probability distribution (p. 356)	1 and 2	18
	2 Graph a normal curve (p. 358)	pages 358–359	7–9
	3 State the properties of the normal curve (p. 359)	pages 359–360	19
	4 Explain the role of area in the normal density function (p. 360)	3 and 4	1
7.2	1 Find and interpret the area under a normal curve (p. 366)	1 and 2	7–9, 10(a)–(c), 11(a)–(d), 12
	2 Find the value of a normal random variable (p. 370)	3–5	4–6, 10(d), 11(e)–(f)
7.3	1 Use normal probability plots to assess normality (p. 377)	1 and 2	14–16, 20
7.4	1 Approximate binomial probabilities using the normal distribution (p. 382)	1 and 2	13, 17

Review Exercises

1. Use the figure to answer the questions that follow.

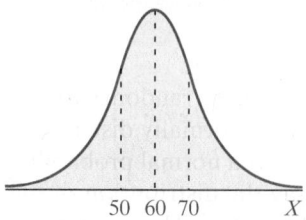

(a) What is μ?

(b) What is σ?

(c) Suppose that the area under the normal curve to the right of $x = 75$ is 0.0668. Provide two interpretations for this area.

(d) Suppose that the area under the normal curve between $x = 50$ and $x = 75$ is 0.7745. Provide two interpretations for this area.

In Problems 2 and 3, draw a standard normal curve and shade the area indicated. Then find the area of the shaded region.

2. The area to the left of $z = -1.04$.

3. The area between $z = -0.34$ and $z = 1.03$.

4. Find the z-score such that the area to the right of the z-score is 0.483.

5. Find the z-scores that separate the middle 92% of the data from the area in the tails of the standard normal distribution.

6. Find the value of $z_{0.20}$.

In Problems 7–9, draw the normal curve with the parameters indicated. Then find the probability of the random variable X. Shade the area that represents the probability.

7. $\mu = 50, \sigma = 6, P(X > 55)$

8. $\mu = 30, \sigma = 5, P(X \le 23)$

9. $\mu = 70, \sigma = 10, P(65 < X < 85)$

10. Tire Wear Suppose that Dunlop Tire manufactures a tire with a lifetime that approximately follows a normal distribution with mean 70,000 miles and standard deviation 4400 miles.

(a) What proportion of the tires will last at least 75,000 miles?
(b) Suppose that Dunlop warrants the tires for 60,000 miles. What proportion of the tires will last 60,000 miles or less?
(c) What is the probability that a randomly selected Dunlop tire lasts between 65,000 and 80,000 miles?
(d) Suppose that Dunlop wants to warrant no more than 2% of its tires. What mileage should the company advertise as its warranty mileage?

11. Wechsler Intelligence Scale The Wechsler Intelligence Scale for Children is approximately normally distributed, with mean 100 and standard deviation 15.

(a) What is the probability that a randomly selected test taker will score above 125?
(b) What is the probability that a randomly selected test taker will score below 90?
(c) What proportion of test takers will score between 110 and 140?
(d) If a child is randomly selected, what is the probability that she scores above 150?
(e) What intelligence score will place a child in the 98th percentile?
(f) If normal intelligence is defined as scoring in the middle 95% of all test takers, figure out the scores that differentiate normal intelligence from abnormal intelligence.

12. Major League Baseballs According to Major League Baseball rules, the ball must weigh between 5 and 5.25 ounces. A factory produces baseballs whose weights are approximately normally distributed, with mean 5.11 ounces and standard deviation 0.062 ounce.
Source: www.baseball-almanac.com

(a) What proportion of the baseballs produced by this factory are too heavy for use by Major League Baseball?
(b) What proportion of the baseballs produced by this factory are too light for use by Major League Baseball?
(c) What proportion of the baseballs produced by this factory can be used by Major League Baseball?
(d) If 8000 baseballs are ordered, how many baseballs should be manufactured, knowing that some will need to be discarded?

13. America Reads According to a Gallup poll, 46% of Americans 18 years old or older stated that they had read at least six books (fiction and nonfiction) within the past year. You conduct a random sample of 250 Americans 18 years old or older.

(a) Verify that the conditions for using the normal distribution to approximate the binomial distribution are met.
(b) Approximate the probability that exactly 125 read at least six books within the past year. Interpret this result.
(c) Approximate the probability that fewer than 120 read at least six books within the past year. Interpret this result.
(d) Approximate the probability that at least 140 read at least six books within the past year. Interpret this result.
(e) Approximate the probability that between 100 and 120, inclusive, read at least six books within the past year. Interpret this result.

14. Use the results in the table to (a) draw a normal plot, (b) determine the linear correlation between the observed values

and expected z-scores, (c) determine the critical value in Table VI to assess the normality of the data.

Index, i	Observed Value	f_i	Expected z-score
1	48	0.09	−1.34
2	49	0.22	−0.77
3	51	0.36	−0.36
4	52	0.50	0
5	54	0.64	0.36
6	54	0.78	0.77
7	56	0.91	1.34

15. Hector obtained a random sample of twenty recent college graduates who own cars and asked each to disclose the age of their car (in months). Is it reasonable to conclude that age of car is normally distributed? The normal probability plot is shown below and the correlation between age of car and expected z-scores is 0.914.

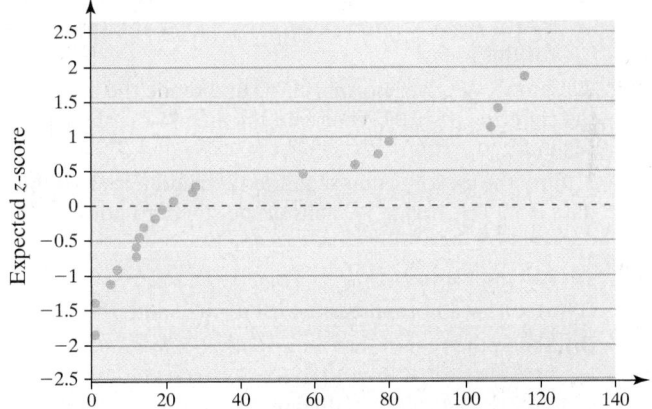

16. Density of Earth In 1798, Henry Cavendish obtained 27 measurements of the density of Earth, using a torsion balance. The following data represent his estimates, given as a multiple of the density of water. Is it reasonable to conclude that the sample data come from a population that is normally distributed?

5.50	5.34	5.57	5.30	5.42	5.36	5.63	5.27	5.55
5.47	5.10	4.88	5.86	5.62	5.58	5.44	5.53	5.29
5.29	5.65	5.34	5.39	5.26	5.61	5.79	4.07	5.85

Source: S. M. Stigler. "Do Robust Estimators Work with Real Data?" *Annals of Statistics* 5(1977), 1055–1078.

17. Creative Thinking According to a *USA Today* "Snapshot," 20% of adults surveyed do their most creative thinking while driving. You conduct a survey of 250 adults and find that 30 do their most creative thinking while driving.

(a) Compute the probability that, in a random sample of 250 adults, 30 or fewer do their most creative thinking while driving.
(b) Does this result contradict the *USA Today* "Snapshot"? Explain.

18. A continuous random variable X is uniformly distributed with $0 \le X \le 20$.

(a) Draw a graph of the uniform density function.
(b) What is $P(0 \le X \le 5)$?
(c) What is $P(10 \le X \le 18)$?

19. List the properties of the normal density curve.
20. Explain how to use a normal probability plot to assess normality.

Chapter Test

1. Use the figure to answer the questions that follow:

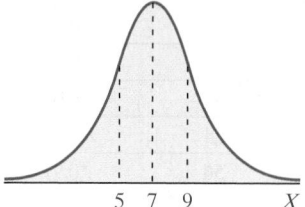

(a) What is μ?

(b) What is σ?

(c) Suppose that the area under the normal curve to the left of $x = 10$ is 0.9332. Provide two interpretations for this area.

(d) Suppose that the area under the normal curve between $x = 5$ and $x = 8$ is 0.5328. Provide two interpretations for this area.

2. Draw a standard normal curve and shade the area to the right of $z = 2.04$. Then find the area of the shaded region.

3. Find the z-scores that separate the middle 88% of the data from the area in the tails of the standard normal distribution.

4. Find the value of $z_{0.04}$.

5. (a) Draw a normal curve with $\mu = 20$ and $\sigma = 3$.

(b) Shade the region that represents $P(22 \leq X \leq 27)$ and find the probability.

6. Suppose that the talk time on the Apple iPhone is approximately normally distributed with mean 7 hours and standard deviation 0.8 hour.

(a) What proportion of the time will a fully charged iPhone last at least 6 hours?

(b) What is the probability a fully charged iPhone will last less than 5 hours?

(c) What talk time would represent the cutoff for the top 5% of all talk times?

(d) Would it be unusual for the phone to last more than 9 hours? Why?

7. The waist circumference of males 20–29 years old is approximately normally distributed, with mean 92.5 cm and standard deviation 13.7 cm.
Source: M. A. McDowell, C. D. Fryar, R. Hirsch, and C. L. Ogden. *Anthropometric Reference Data for Children and Adults: U. S. Population, 1999–2002.* Advance data from vital and health statistics: No. 361. Hyattsville, MD: National Center for Health Statistics, 2005.

(a) Use the normal model to determine the proportion of 20- to 29-year-old males whose waist circumference is less than 100 cm.

(b) What is the probability that a randomly selected 20- to 29- year-old male has a waist circumference between 80 and 100 cm?

(c) Determine the waist circumferences that represent the middle 90% of all waist circumferences.

(d) Determine the waist circumference that is at the 10th percentile.

8. Suppose the scores earned on Professor McArthur's third statistics exam are normally distributed with mean 64 and standard deviation 8. Professor McArthur wants to curve the exam scores as follows: The top 6% get an A, the next 14% get a B, the middle 60% get a C, the bottom 6% fail, and the rest earn a D. Any student who can determine these cut-offs earns five bonus points. Determine the cut-offs for Professor McArthur.

9. In a poll conducted by the Gallup organization, 16% of adult, employed Americans were dissatisfied with the amount of their vacation time. You conduct a survey of 500 adult, employed Americans.

(a) Approximate the probability that exactly 100 are dissatisfied with their amount of vacation time.

(b) Approximate the probability that less than 60 are dissatisfied with the amount of their vacation time.

10. Jane obtained a random sample of 15 college students and asked how many hours they studied last week. Is it reasonable to believe that hours studied is normally distributed? The normal probability plot is shown below and the correlation between hours studied and expected z-scores is 0.974.

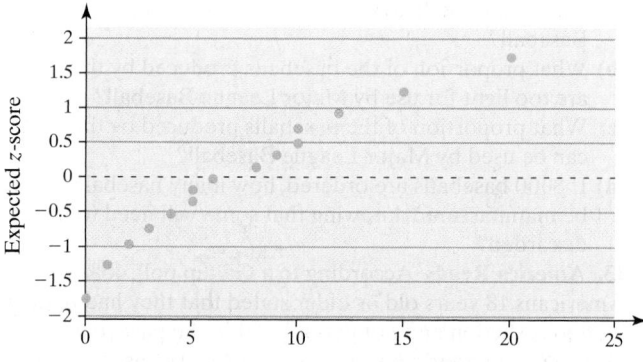

11. A continuous random variable X is uniformly distributed with $10 \leq X \leq 50$.

(a) Draw a graph of the uniform density function.

(b) What is $P(20 \leq X \leq 30)$?

(c) What is $P(X < 15)$?

Making an Informed Decision

Stock Picking

You are interested in modeling the behavior of stocks. In particular, you want to build a model that describes the rate of return on a basket of stocks, such as large capitalization companies.

(a) Go to a website that provides historical rates of return on a certain basket of stocks, such as www.morningstar.com. Decide on a certain sector of the economy you would like to model, such as consumer goods or energy. Choose the largest 100 companies in this sector and determine the time frame for which you want to build your model. For example, you might decide to build a model for the one-year rate of return on the stock.

(b) Enter your data into statistical software and construct a relative frequency histogram. Does the data appear bell-shaped? Do you think the normal model would be a good model to describe the rate of return for the sector you have chosen? Why or why not?

(c) Regardless of your answer to part (b), build a normal model by determining the mean and standard deviation for the rate of return. Draw the normal model on the relative frequency histogram from part (b).

(d) One purpose of financial models is to identify quality investments going forward. Without a crystal ball, all investment managers have is historical data. Use your historical data to determine a rate of return that falls into the top 20% of all companies within the sector. Use this rate of return as a criterion for choosing a company to invest in.

(e) Conduct further research on the stock you wish to invest in. For example, have the company's earnings been growing for the past five years? What is the company's market share? Does the company pay a dividend? If so, for how long has the company paid a dividend? Has the dividend been growing consistently?

(f) Write a report that lays out your recommendation regarding the particular stock you have researched. Perhaps include a few other models that might help to decide whether the company is a solid investment.

 **CASE STUDY**   A Tale of Blood Chemistry

Abby Tudor recently turned 40. Her knees ache and she feels shortness of breath during exercise, along with periodic dizziness and general fatigue. Despite trying various diet and exercise regimes, Abby is 20 pounds overweight. According to a drugstore machine, her blood pressure is elevated. Her family history includes cardiac disease (both parents had heart attacks), diabetes (a maternal aunt), and hypothyroidism (throughout immediate family).

Her physician advised scheduling an appointment for a full physical exam and complete blood work, hoping levels of various blood components may clarify Abby's symptoms. He is interested in her white and red blood cell counts, hemoglobin, and hematocrit figures for indications of infection or anemia. Additionally, he will test serum glucose levels for the onset of diabetes, cholesterol and triglyceride levels for potential cardiac disease, and serum TSH levels for the possibility of hypothyroidism.

Two weeks before her appointment, Abby fasted for 12 hours and immediately had blood work done at the lab. At her physical, the doctor reviewed her blood test report. He expressed concern over some results but Abby wasn't convinced she had a problem. Additional blood tests weren't an option because of the expense and amount of time they take. Abby decided to forego the offered prescriptions at that time and do some research herself.

She found that many medical measurements, such as cholesterol, are normally distributed in healthy populations. While her lab report did not provide the means and standard deviations needed to calculate the various probabilities of interest, it did provide the appropriate reference intervals. Assuming the reference intervals represent the range of values for each blood component for a healthy adult population, it is possible to estimate the various means and standard deviations for this population.

Abby estimated each mean by taking the midpoint of its reference interval. Using the Range Rule of Thumb, standard deviations were estimated by dividing the reference interval range by 4 ($\sigma \approx$ range/4). The table below shows Abby's blood test results, as well as the mean and standard deviation for a number of blood test components for the population of normal healthy adults.

For any blood component measurement that was below its population mean, Abby decided to calculate the probability that she would get a test value less than or equal to the value obtained, given that she was a member of the healthy population. For example, her HDL cholesterol reading (42 mg/dL) was below the mean of the healthy population (92.5 mg/dL), so she calculated the following probability: $P(X \leq 42 \text{ mg/dL})$ with $\mu = 92.5$ and $\sigma = 28.75$. Similarly, for any blood

component measurement reading exceeding its population mean, Abby decided to calculate the probability that she would get a test value greater than or equal to the value obtained, given that she was a member of the healthy population. For example, her LDL cholesterol value (181 mg/dL) exceeds the mean of the healthy population (64.5 mg/dL), so she calculated the following probability: $P(X \geq 181 \text{ mg/dL})$.

To help her interpret the calculated probabilities, Abby decided to only focus on blood components that had a probability of less than 0.025. By choosing this figure, she acknowledged that it is unlikely she could have such an extreme blood component reading and still be part of the healthy population.

Blood Test Components for Healthy Adults and Results for Abby Tudor*

Blood Component	Unit	Mean	Standard Deviation	Abby's Result
White blood cell count	$10^3/\mu L$	7.25	1.625	5.3
Red blood cell count	$10^6/\mu L$	4.85	0.375	4.62
Hemoglobin	g/dL	14.75	1.125	14.6
Hematocrit	%	43.0	3.5	41.7
Glucose, serum	mg/dL	87.0	11.0	95.0
Creatine, serum	mg/dL	1.00	0.25	0.8
Sodium, serum	mEq/L	141.5	3.25	143.0
Potassium, serum	mEq/L	4.5	0.5	5.1
Chloride, serum	mEq/L	102.5	3.25	100.0
Carbon dioxide, total	mEq/L	26.0	3.0	25.0
Calcium, serum	mg/dL	9.55	0.525	10.1
Total cholesterol	mg/dL	149.5	24.75	253.0
Triglycerides	mg/dL	99.5	49.75	150.0
HDL cholesterol	mg/dL	92.5	28.75	42.0
LDL cholesterol	mg/dL	64.5	32.25	181.0
LDL/HDL ratio	Ratio	1.8	0.72	4.3
TSH, high sensitivity, serum	mcIU/mL	2.925	1.2875	3.15

*Population means and standard deviations were estimated from the reference intervals derived from an actual blood test report provided by TA LabCorp, Tampa, Florida. Means were estimated by taking the midpoints of the reference intervals. Standard deviations were estimated by dividing the reference interval ranges by four. Test results attributed to Abby Tudor are actual results obtained from an anonymous patient.

1. The reference interval for HDL cholesterol is 35–150 mg/dL. Use this information to confirm the mean and standard deviation provided for this blood component.

2. Using Abby's criteria and the means and standard deviations provided in her blood test report, determine which blood components should be a cause of concern for Abby. Write up a summary report of your findings. Be sure to include a discussion concerning your assumptions and any limitations to your conclusions.

PART 4

Inference: From Samples to Population

In Chapter 1, we presented the following process of statistics:

Step 1: Identify the research objective.

Step 2: Collect the data needed to answer the question(s) posed in Step 1.

Step 3: Describe the data.

Step 4: Perform inference.

The methods for conducting Steps 1 and 2 were discussed in Chapter 1. The methods for conducting Step 3 were discussed in Chapters 2 through 4. We took a break from the statistical process in Chapters 5 through 7 so that we could develop skills that allow us to tackle Step 4.

Since it is difficult to gain access to populations, the data found in Step 2 is often from a sample. Sample data are used to make inferences about the population. For example, we might compute a sample mean from the data collected in Step 2 and use this information to draw conclusions regarding the population mean. Part 4 focuses on how sample data are used to draw conclusions about populations.

8

Sampling Distributions

Outline

Making an Informed Decision

The American Time Use Survey, conducted by the Bureau of Labor Statistics, investigates how adult Americans allocate their time during a day. As a reporter for the school newspaper, you wish to file a report that compares the typical student at your school to other Americans. See the Decisions project on page 418.

PUTTING IT TOGETHER

In Chapters 6 and 7, we learned about random variables and their probability distributions. A random variable is a numerical measure of the outcome to a probability experiment. A probability distribution provides a way to assign probabilities to the possible values of the random variable. For discrete random variables, we discussed the binomial probability distribution and the Poisson probability distribution. We assigned probabilities using a formula. For continuous random variables, we discussed the normal probability distribution. To compute probabilities for a normal random variable, we found the area under a normal density curve.

In this chapter, we continue our discussion of probability distributions where statistics, such as $\bar{x}$, will be the random variable. Statistics are random variables because the value of a statistic varies from sample to sample. For this reason, statistics have probability distributions associated with them. For example, there is a probability distribution for the sample mean, sample proportion, and so on. We use probability distributions to make probability statements regarding the statistic. So this chapter discusses the shape, center, and spread of statistics such as $\bar{x}$.

8.1   Distribution of the Sample Mean

Preparing for This Section Before getting started, review the following:

- Simple random sampling (Section 1.3, pp. 22–25)
- The mean (Section 3.1, pp. 118–120)
- The standard deviation (Section 3.2, pp. 133–137)

- Applications of the normal distribution (Section 7.2, pp. 366–372)

Objectives   ① Describe the distribution of the sample mean: normal population

② Describe the distribution of the sample mean: nonnormal population

Suppose the government wanted to determine the mean income of all U.S. households. One approach the government could take is to survey every U.S. household to determine the population mean, μ. This would be a very expensive and time-consuming survey!

A second approach the government could (and does) take is to survey a random sample of U.S. households and use the results to estimate the mean household income. The Current Population Survey is administered to approximately 250,000 randomly selected households each month. Among the many questions on the survey, respondents are asked to report the income of each individual in the household. From this information, the federal government obtains a sample mean household income for U.S. households. For example, in 2013 the mean annual household income in the United States was estimated to be $\bar{x} = \$72,641$. The government might infer from this survey that the mean annual household income of *all* U.S. households in 2013 was $\mu = \$72,641$.

The households in the Current Population Survey were determined by chance (random sampling). A second random sample of households would likely lead to a different sample mean, such as $\bar{x} = \$71,849$, and a third random sample of households would likely lead to a third sample mean, such as $\bar{x} = \$72,978$. Because the households selected will vary from sample to sample, the sample mean of household income will also vary from sample to sample. For this reason, the sample mean $\bar{x}$ is a random variable, so it has a probability distribution. Our goal in this section is to describe the distribution of the sample mean. Remember, when we describe a distribution, we do so in terms of its shape, center, and spread.

Definitions   The **sampling distribution** of a statistic is a probability distribution for all possible values of the statistic computed from a sample of size n.

The **sampling distribution of the sample mean** $\bar{x}$ is the probability distribution of all possible values of the random variable $\bar{x}$ computed from a sample of size n from a population with mean μ and standard deviation σ.

The idea behind obtaining the sampling distribution of the sample mean is as follows:

Step 1:   Obtain a simple random sample of size n.

Step 2:   Compute the sample mean.

In Other Words
If the number of individuals in a population is a positive integer, we say the population is finite. Otherwise, the population is infinite.

Step 3:   Assuming that we are sampling from a finite population, repeat Steps 1 and 2 until all distinct simple random samples of size n have been obtained.

Note: Once a particular sample is obtained, it cannot be obtained a second time.

① Describe the Distribution of the Sample Mean: Normal Population

The probability distribution of the sample mean is determined from statistical theory. We will use simulation to help justify the result that statistical theory provides. We consider two possibilities. In the first case (Examples 1, 2, and 3), we sample from a population that is normally distributed. In the second case (Examples 4 and 5), we sample from a population that is not normally distributed.

EXAMPLE 1 Sampling Distribution of the Sample Mean: Normal Population

Problem An intelligence quotient, or IQ, is a measurement of intelligence derived from a standardized test, such as the Stanford Binet IQ test. Scores on this test are approximately normally distributed with a mean score of 100 and a standard deviation of 15. What is the sampling distribution of the sample mean for a sample of size $n = 9$?

Approach The problem asks us to determine the shape, center, and spread of the distribution of the sample mean. Remember, the sampling distribution of the sample mean would be the distribution of *all* possible sample means of size $n = 9$. To get a sense of this distribution, use Minitab to simulate obtaining 1000 samples of size $n = 9$ by randomly generating 1000 rows of IQs over 9 columns. Each row represents a random sample of size 9. For each of the 1000 samples (the 1000 rows), we determine the mean IQ score. Draw a histogram to gauge the shape of the distribution of the sample mean, determine the mean of the 1000 sample means to approximate the mean of the sampling distribution, and determine the standard deviation of the 1000 sample means to approximate the standard deviation of the sampling distribution.

Solution Figure 1 shows random samples from Minitab. Row 1 contains the first sample, where the IQ scores of the nine individuals are 90, 74, 95, 88, 91, 91, 102, 91, and 96. The mean of these nine IQ scores is 90.9. Row 2 represents a second sample with nine different IQ scores; row 3 represents a third sample, and so on. Column C10 (xbar) lists the sample means for each of the different samples.

Using Technology

We are using Minitab's Random Data command under the Calc menu to generate these data. Select Normal ... from the Random Data menu.

To obtain normal random data using StatCrunch, select

Data > Simulate > Normal

Figure 1

	C1	C2	C3	C4	C5	C6	C7	C8	C9	C10 xbar	
1	90	74	95	88	91	91	102	91	96	90.9	$\bar{x} = 90.9$
2	114	86	96	82	80	68	93	136	111	96.2	$\bar{x} = 96.2$
3	89	89	86	98	96	96	89	99	107	94.3	
4	87	94	89	116	92	124	115	83	111	101.3	
5	107	103	86	86	109	104	94	82	110	97.8	
6	113	84	101	89	92	71	89	86	108	92.7	
7	75	116	107	118	112	86	104	97	106	102.4	
8	118	117	81	96	86	94	109	96	104	100.1	
9	91	91	79	80	109	89	97	81	99	90.6	
10	112	98	115	73	95	104	76	95	87	95.0	
11	103	92	100	75	95	108	105	125	62	96.1	
12	75	124	101	113	91	85	121	115	85	101.2	
13	104	103	81	134	111	108	101	88	115	105.1	
14	121	85	118	88	96	84	103	77	102	97.1	
15	119	84	119	80	99	98	88	88	89	95.9	
16	71	86	94	101	95	84	124	105	83	93.7	

Sample 1 — row 1; Sample 2 — row 2

Figure 2(a) shows the distribution of the population, and Figure 2(b) shows the distribution of the sample means from column C10 (using Minitab). The shape of the distribution of the population is normal. The histogram in Figure 2(b) shows that the shape of the distribution of the sample means is also normal. In addition, we notice that the center of the distribution of the sample means is the same as the center of the distribution of the population, but the spread of the distribution of the sample means is smaller than the spread of the distribution of the population. In fact, the mean of the 1000 sample means is 100.02, which is close to the population mean, 100; the standard deviation of the sample means is 5.01, which is less than the population standard deviation, 15.

Figure 2

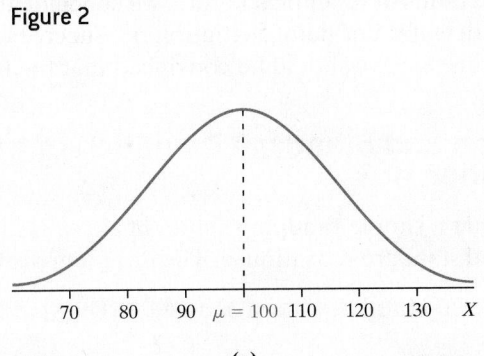

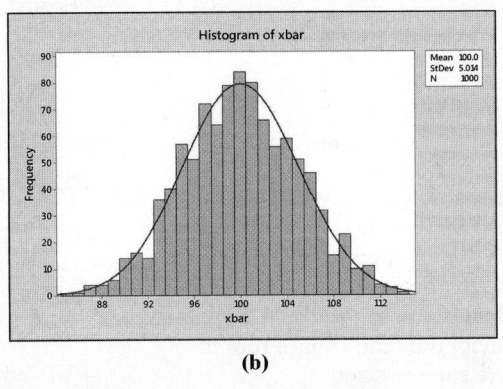

(a)   (b)

We draw the following conclusions:

- **Shape:** The shape of the distribution of the sample mean is normal.
- **Center:** The mean of the distribution of the sample mean equals the mean of the population, 100.
- **Spread:** The standard deviation of the sample mean is less than the standard deviation of the population.

Why is the standard deviation of the sample mean less than the standard deviation of the population? Consider that, if we randomly select any one individual, according to the Empirical Rule, there is about a 68% chance that the individual's IQ score is between 85 and 115 (that is, within 1 standard deviation of the mean). If we had a sample of 9 individuals, we would not expect as much spread in the sample mean as there is for a single individual, since individuals with lower IQs will offset individuals in the sample with higher IQs, resulting in a sample mean closer to the expected value of 100. Look back at Figure 1. In the first sample (row 1), the low-IQ individual ($IQ = 74$) is offset by the higher-IQ individual ($IQ = 102$), which is why the sample mean is closer to 100. In the second sample (row 2), the low-IQ individual ($IQ = 68$) is offset by the higher-IQ individual ($IQ = 136$), so the sample mean of the second sample is closer to 100. Therefore, the spread in the distribution of sample means should be less than the spread in the population from which the sample is drawn.

Based on this, what role do you think n, the sample size, plays in the standard deviation of the distribution of the sample mean?

EXAMPLE 2   The Impact of Sample Size on Sampling Variability

Problem Repeat the problem in Example 1 with a sample of size $n = 25$.

Approach Use the approach presented in Example 1, but let $n = 25$ instead of $n = 9$.

Solution Figure 3 shows the histogram of the sample means. Notice that the sample means appear to be normally distributed with the center at 100. The histogram in Figure 3 shows less dispersion than the histogram in Figure 2(b). This implies that the distribution of $\bar{x}$ with $n = 25$ has less variability than the distribution of $\bar{x}$ with $n = 9$. In fact, the mean of the 1000 sample means is 100.05, and the standard deviation is 3.08.

Figure 3

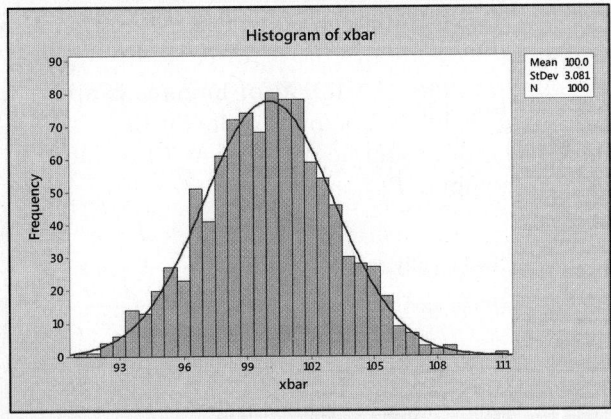

From the results of Examples 1 and 2, we conclude that, as the sample size n increases, the standard deviation of the distribution of $\bar{x}$ decreases. Although the proof is beyond the scope of this text, we should be convinced that the following result is reasonable.

In Other Words
Regardless of the distribution of the population, the sampling distribution of $\bar{x}$ will have a mean equal to the mean of the population and a standard deviation equal to the standard deviation of the population divided by the square root of the sample size!

The Mean and Standard Deviation of the Sampling Distribution of $\bar{x}$

Suppose that a simple random sample of size n is drawn from a population* with mean μ and standard deviation σ. The sampling distribution of $\bar{x}$ has mean $\mu_{\bar{x}} = \mu$ and standard deviation $\sigma_{\bar{x}} = \dfrac{\sigma}{\sqrt{n}}$. The standard deviation of the sampling distribution of $\bar{x}$, $\sigma_{\bar{x}}$, is called the **standard error of the mean**.

> **CAUTION!**
> It is important that two assumptions are satisfied with regard to sampling from a population.
> 1. The sample must be a random sample.
> 2. The sampled values must be independent. When sampling without replacement (which is the case when obtaining simple random samples), we shall verify this assumption by checking that the size is less than 5% of the population size ($n < 0.05N$).

For the population presented in Example 1, if we draw a simple random sample of size $n = 9$, the sampling distribution $\bar{x}$ will have mean $\mu_{\bar{x}} = 100$ and standard deviation

$$\sigma_{\bar{x}} = \frac{\sigma}{\sqrt{n}} = \frac{15}{\sqrt{9}} = 5$$

This standard error of the mean is close to the approximate standard error of 5.01 found in our simulation in Example 1.

In Example 2, where the simple random sample was of size $n = 25$, the sampling distribution of $\bar{x}$ will have mean $\mu_{\bar{x}} = 100$ and standard deviation

$$\sigma_{\bar{x}} = \frac{\sigma}{\sqrt{n}} = \frac{15}{\sqrt{25}} = 3$$

● **Now Work Problem 9**

This standard error of the mean is close to the approximate standard error of 3.08 found in our simulation in Example 2.

Now that we can find the mean and standard deviation for any sampling distribution of $\bar{x}$, we can concentrate on the shape of the distribution. Refer back to Figures 2(b) and 3 from Examples 1 and 2. Both histograms appear to be normal. Recall that the population from which the sample was drawn was normal. This leads us to believe that, if the population is normal, then the distribution of the sample mean is also normal.

Figure 4

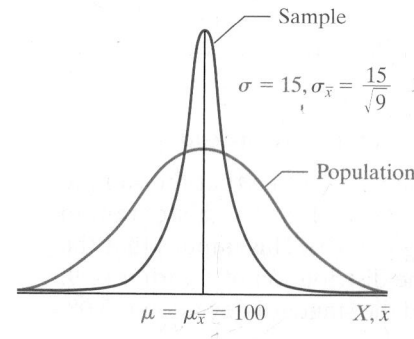

The Shape of the Sampling Distribution of $\bar{x}$ If X Is Normal

If a random variable X is normally distributed, the sampling distribution of the sample mean, $\bar{x}$, is normally distributed.

For example, the IQ scores of individuals are modeled by a normal random variable with mean $\mu = 100$ and standard deviation $\sigma = 15$. The sampling distribution of the sample mean, $\bar{x}$, the mean IQ of a simple random sample of $n = 9$ individuals, is normal, with mean $\mu_{\bar{x}} = 100$ and standard deviation $\sigma_{\bar{x}} = \dfrac{15}{\sqrt{9}}$. See Figure 4.

EXAMPLE 3   Describing the Distribution of the Sample Mean

Problem The IQ, X, of humans is approximately normally distributed with mean $\mu = 100$ and standard deviation $\sigma = 15$. Compute the probability that a simple random sample of size $n = 10$ results in a sample mean greater than 110. That is, compute $P(\bar{x} > 110)$.

---

*Technically, we assume that we are drawing a simple random sample from an infinite population. For populations of finite size N, $\sigma_{\bar{x}} = \sqrt{\dfrac{N-n}{N-1}} \cdot \dfrac{\sigma}{\sqrt{n}}$. However, if the sample size is less than 5% of the population size ($n < 0.05N$), the effect of $\sqrt{\dfrac{N-n}{N-1}}$ (the finite population correction factor) can be ignored without significantly affecting the results.

Approach The random variable X is approximately normally distributed, so the sampling distribution of $\bar{x}$ will be normally distributed. Verify the independence requirement. The mean of the sampling distribution is $\mu_{\bar{x}} = \mu$, and its standard deviation is $\sigma_{\bar{x}} = \dfrac{\sigma}{\sqrt{n}}$.

Convert the sample mean $\bar{x} = 110$ to a z-score and then find the area under the standard normal curve to the right of this z-score.

Solution The sample mean is normally distributed, with mean $\mu_{\bar{x}} = 100$ and standard deviation $\sigma_{\bar{x}} = \dfrac{\sigma}{\sqrt{n}} = \dfrac{15}{\sqrt{10}} = 4.743$. The sample size is definitely less than 5% of the population size.

Figure 5 displays the normal curve with the area we want to compute shaded. To find the area by hand, convert $\bar{x} = 110$ to a z-score and obtain

$$z = \frac{\bar{x} - \mu_{\bar{x}}}{\sigma_{\bar{x}}} = \frac{\bar{x} - \mu_{\bar{x}}}{\dfrac{\sigma}{\sqrt{n}}} = \frac{110 - 100}{\dfrac{15}{\sqrt{10}}} = 2.11$$

Figure 5

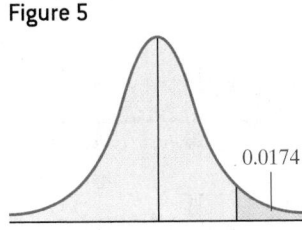

0.0174

$\mu_{\bar{x}} = 100 \quad 110 \qquad \bar{x}$

The area to the right of $z = 2.11$ is $1 - 0.9826 = 0.0174$.

Using technology, the area to the right of $\bar{x} = 110$ is 0.0175.

Interpretation The probability of obtaining a sample mean IQ greater than 110 from a population whose mean is 100 is approximately 0.02. That is, $P(\bar{x} > 110) = 0.0174$ (or 0.0175 using technology). If we take 100 simple random samples of $n = 10$ individuals from this population and if the population mean is 100, about 2 of the samples will result in a mean IQ that is greater than 110.

● **Now Work Problem 19**

② Describe the Distribution of the Sample Mean: Nonnormal Population

Now we explore the distribution of the sample mean assuming the population from which the sample is drawn is not normal. Again we use simulation.

EXAMPLE 4 Sampling from a Population That Is Not Normal

Problem The data in Table 1 represent the probability distribution of the number of people living in households in the United States. Figure 6 shows a graph of the probability distribution. From the data in Table 1, we determine the mean and standard deviation number of people living in households in the United States to be $\mu = 2.9$ and $\sigma = 1.48$.

Clearly, the distribution is not normal. In fact, the random variable is discrete! Approximate the sampling distribution of the sample mean $\bar{x}$ by obtaining, through simulation, 1000 samples of size (a) $n = 4$, (b) $n = 10$, and (c) $n = 30$ from the population.

Table 1	
Number in Household	**Proportion**
1	0.147
2	0.361
3	0.187
4	0.168
5	0.083
6	0.034
7	0.012
8	0.004
9	0.002
10	0.002

Source: General Social Survey

Figure 6

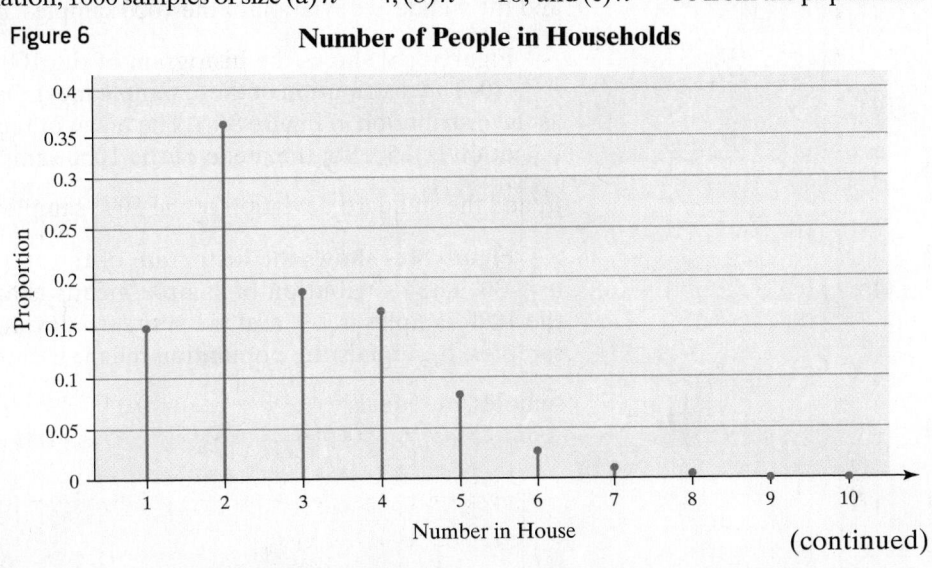

Number of People in Households

(continued)

Approach Use Minitab to obtain 1000 random samples of size $n = 4$ from the population. This simulates going to 4 households 1000 times and determining the number of people living in the household. Next, compute the mean of each of the 1000 random samples. Finally, draw a histogram, determine the mean, and determine the standard deviation of the 1000 sample means. Repeat this for samples of size $n = 10$ and $n = 30$.

Solution Figure 7 shows partial output from Minitab for random samples of size $n = 4$. Columns 1 and 2 represent the probability distribution. Each row in Columns 3 through 6 lists the number of individuals in the household for each sample. Column 7 (xbar) lists the sample mean for each sample (each row). For example, in the first sample (row 1), there are 4 individuals in the first house surveyed, and 2 individuals in the second, third, and fourth houses surveyed. The mean number of individuals in the household for the first sample is 2.5.

Figure 7

	C1 Number	C2 Proportion	C3	C4	C5	C6	C7 xbar
1	1	0.147	4	2	2	2	2.50
2	2	0.361	1	3	4	2	2.50
3	3	0.187	4	4	2	4	3.50
4	4	0.168	5	4	2	6	4.25
5	5	0.083	7	2	4	4	4.25
6	6	0.034	2	2	6	2	3.00
7	7	0.012	4	8	3	1	4.00
8	8	0.004	3	2	2	2	2.25
9	9	0.002	2	2	4	2	2.50
10	10	0.002	2	2	4	1	2.25
11			2	1	7	2	3.00
12			3	7	2	1	3.25
13			5	3	2	3	3.25
14			1	5	2	2	2.50
15			1	1	4	7	3.25

Figure 8(a) shows the histogram of the 1000 sample means for a sample of size $n = 4$. The distribution of sample means is skewed right (just like the parent population, but not as strongly). The mean of the 1000 samples is 2.9, and the standard deviation is 0.76. So, the mean of the 1000 samples, $\mu_{\bar{x}}$, equals the population mean μ, and the standard deviation of the 1000 samples, $\sigma_{\bar{x}}$, is close to $\dfrac{\sigma}{\sqrt{n}} = \dfrac{1.48}{\sqrt{4}} = 0.74$.

Figure 8(b) shows the histogram of the 1000 sample means for a sample of size $n = 10$. The distribution of these sample means is also skewed right, but not as skewed as the distribution in Figure 8(a). The mean of the 1000 samples is 2.9, and the standard deviation is 0.50. So, the mean of the 1000 samples, $\mu_{\bar{x}}$, equals the population mean, μ, and the standard deviation of the 1000 samples, $\sigma_{\bar{x}}$, is close to $\dfrac{\sigma}{\sqrt{n}} = \dfrac{1.48}{\sqrt{10}} = 0.47$.

Figure 8(c) shows the histogram of the 1000 sample means for a sample of size $n = 30$. The distribution of sample means is approximately normal! The mean of the 1000 samples is 2.9, and the standard deviation is 0.27. So, the mean of the 1000 samples, $\mu_{\bar{x}}$, equals the population mean, μ, and the standard deviation of the 1000 samples, $\sigma_{\bar{x}}$, equals $\dfrac{\sigma}{\sqrt{n}} = \dfrac{1.48}{\sqrt{30}} = 0.27$.

Figure 8

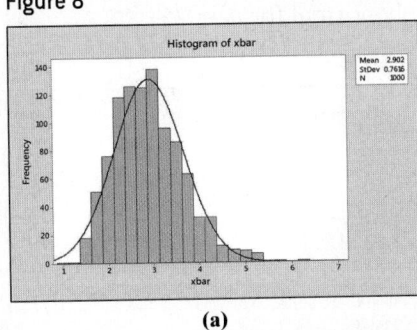

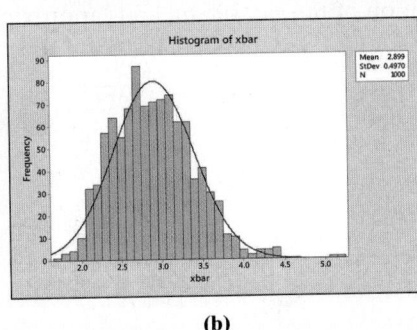

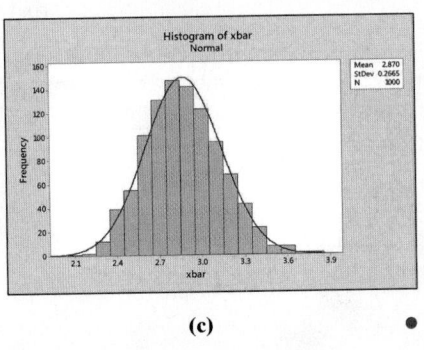

(a)   (b)   (c)

There are two key concepts to understand in Example 4.

1. The mean of the sampling distribution of the sample mean is equal to the mean of the underlying population, and the standard deviation of the sampling distribution of the sample mean is $\frac{\sigma}{\sqrt{n}}$, regardless of the size of the sample.

2. The shape of the distribution of the sample mean becomes approximately normal as the sample size n increases, regardless of the shape of the underlying population.

We formally state point 2 as the *Central Limit Theorem*.

In Other Words
For any population, regardless of its shape, as the sample size increases, the shape of the distribution of the sample mean becomes more "normal."

> **CAUTION!**
> The Central Limit Theorem only has to do with the shape of the distribution of $\bar{x}$, not the center or spread. Regardless of the size of the sample, $\mu_{\bar{x}} = \mu$ and $\sigma_{\bar{x}} = \frac{\sigma}{\sqrt{n}}$.

The Central Limit Theorem

Regardless of the shape of the underlying population, the sampling distribution of $\bar{x}$ becomes approximately normal as the sample size, n, increases.

How large does the sample size need to be before we can say that the sampling distribution of $\bar{x}$ is approximately normal? The answer depends on the shape of the distribution of the underlying population. Distributions that are highly skewed will require a larger sample size for the distribution of $\bar{x}$ to become approximately normal.

For example, the right-skewed distribution in Example 4 required a sample size of about 30 before the distribution of the sample mean became approximately normal. However, Figure 9(a) shows a uniform distribution for $0 \leq X \leq 10$. Figure 9(b) shows the distribution of the sample mean obtained via simulation using StatCrunch for $n = 4$. Even for samples as small as $n = 4$, the distribution of the sample mean is approximately normal.

Historical Note

Pierre-Simon Laplace was born on March 23, 1749, in Normandy, France. At age 16, Laplace attended Caen University, where he studied theology. While there, his mathematical talents were discovered, which led him to Paris, where he obtained a job as professor of mathematics at the École Militaire. In 1773, Laplace was elected to the Académie des Sciences. Laplace was not humble. It is reported that, in 1780, he stated that he was the best mathematician in Paris. In 1799, Laplace published the first two volumes of *Mécanique céleste*, in which he discussed methods for calculating the motion of the planets. On April 9, 1810, Laplace presented the Central Limit Theorem to the Academy.

Figure 9

Uniform Distribution

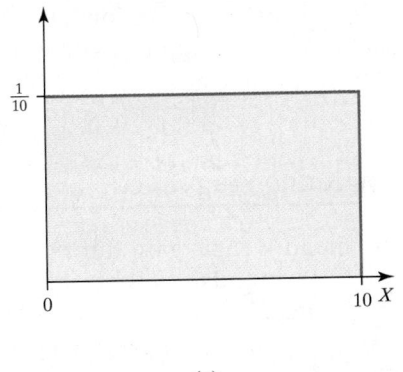

(a)

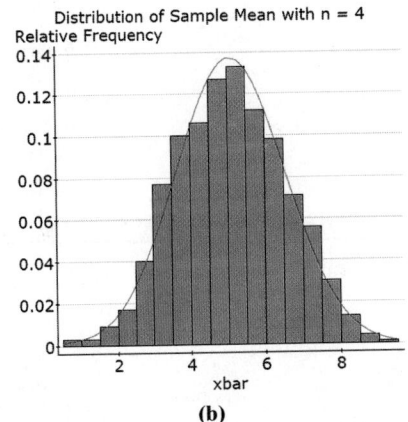

(b)

Figure 10(a) on the next page shows a distribution of household incomes for a town. Figure 10(b) shows the distribution of the sample mean for a random sample of $n = 4$ households from Minitab. Figure 10(c) shows the distribution of the sample mean for a random sample of $n = 10$ households, and Figure 10(d) shows the distribution of the sample mean for a random sample of $n = 25$ households, also from Minitab. Notice the distribution of the sample mean is approximately normal for $n = 25$.

Figure 10

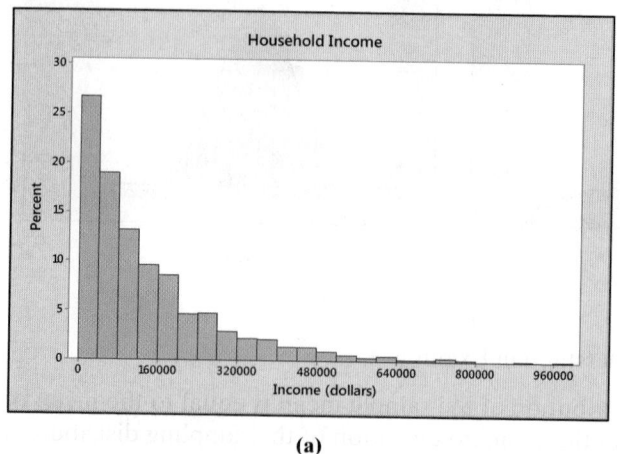

(a)

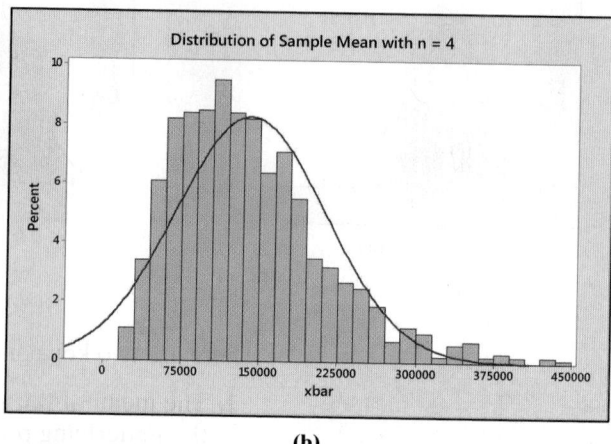

(b)

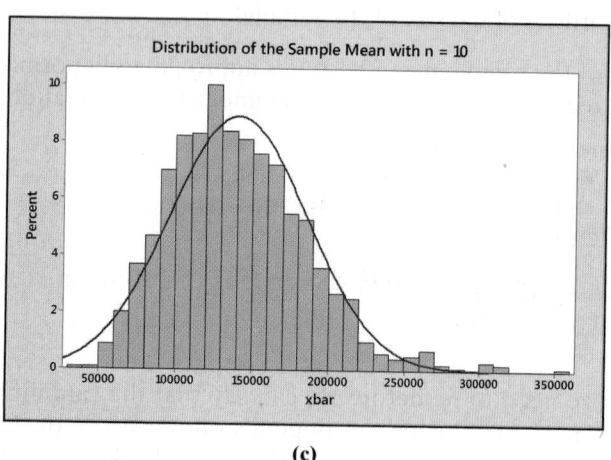

(c)

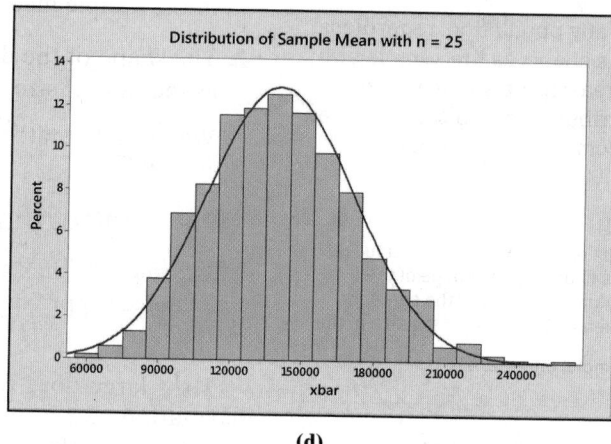

(d)

The results of Example 4 and Figures 9 and 10 confirm that the shape of the distribution of the population dictates the size of the sample required for the distribution of the sample mean to be normal. The more skewed the distribution of the population is, the larger the sample size needed to invoke the Central Limit Theorem. We will err on the side of caution and use the following rule of thumb:

If the distribution of the population is unknown or not normal, then the distribution of the sample mean is approximately normal provided that the sample size is greater than or equal to 30.

EXAMPLE 5 Weight Gain During Pregnancy

Problem The mean weight gain during pregnancy is 30 pounds, with a standard deviation of 12.9 pounds. Weight gain during pregnancy is skewed right. An obstetrician obtains a random sample of 35 low-income patients and determines their mean weight gain during pregnancy was 36.2 pounds. Does this result suggest anything unusual?

Approach We want to know whether the sample mean obtained is unusual. Therefore, determine the likelihood of obtaining a sample mean of 36.2 pounds or higher (if a 36.2-pound weight gain is unusual, certainly any weight gain above 36.2 pounds is also unusual). Assume that the patients come from the population whose mean weight gain is 30 pounds. Verify the independence assumption. Use the normal model to obtain the probability since the sample size is large enough to use the Central Limit Theorem. Determine the area under the normal curve to the right of 36.2 pounds with

$$\mu_{\bar{x}} = \mu = 30 \text{ and } \sigma_{\bar{x}} = \frac{\sigma}{\sqrt{n}} = \frac{12.9}{\sqrt{35}}.$$

Solution It seems reasonable there are at least 700 low-income pregnant women in the population. So the sample size is less than 5% of the population size. The probability is represented by the area under the normal curve to the right of 36.2. See Figure 11.
To find $P(\bar{x} \geq 36.2)$ by hand, convert the sample mean $\bar{x} = 36.2$ to a z-score.

Figure 11

Area = $P(\bar{x} \geq 36.2)$

$\mu_{\bar{x}} = 30$ $\bar{x}$

$\bar{x} = 36.2$

$$z = \frac{\bar{x} - \mu_{\bar{x}}}{\sigma_{\bar{x}}} = \frac{36.2 - 30}{\dfrac{12.9}{\sqrt{35}}} = 2.84$$

The area under the standard normal curve to the left of $z = 2.84$ is 0.9977. So the area to the right is 0.0023. Therefore, $P(\bar{x} \geq 36.2) = 0.0023$.
If we use technology to find the area to the right of $\bar{x} = 36.2$, we obtain 0.0022.

Interpretation If the population from which this sample is drawn has a mean weight gain of 30 pounds, the probability that a random sample of 35 women has a sample mean weight gain of 36.2 pounds (or more) is approximately 0.002. This means that about 2 samples in 1000 will result in a sample mean of 36.2 pounds or higher if the population mean is 30 pounds. We can conclude one of two things based on this result:

1. The mean weight gain for low-income patients is 30 pounds, and we happened to select women who, on average, gained more weight.
2. The mean weight gain for low-income patients is more than 30 pounds.

We are inclined to accept the second explanation over the first since our sample was obtained randomly. Therefore, the obstetrician should be concerned. Perhaps she should look at the diets and/or lifestyles of low-income patients while they are pregnant.

● **Now Work Problem 25**

Summary: Shape, Center, and Spread of the Sampling Distribution of $\bar{x}$

Shape, Center, and Spread of the Population	Distribution of the Sample Mean		
	Shape	**Center**	**Spread**
Population is normal with mean μ and standard deviation σ	Regardless of the sample size n, the shape of the distribution of the sample mean is normal	$\mu_{\bar{x}} = \mu$	$\sigma_{\bar{x}} = \dfrac{\sigma}{\sqrt{n}}$
Population is not normal with mean μ and standard deviation σ	As the sample size n increases, the distribution of the sample mean becomes approximately normal	$\mu_{\bar{x}} = \mu$	$\sigma_{\bar{x}} = \dfrac{\sigma}{\sqrt{n}}$

8.1 Assess Your Understanding

Vocabulary and Skill Building

1. The _____ _____ of the sample mean, $\bar{x}$, is the probability distribution of all possible values of the random variable $\bar{x}$ computed from a sample of size n from a population with mean μ and standard deviation σ.

2. Suppose a simple random sample of size n is drawn from a large population with mean μ and standard deviation σ. The sampling distribution of $\bar{x}$ has mean $\mu_{\bar{x}} = $ _____ and standard deviation $\sigma_{\bar{x}} = $ _____.

3. The standard deviation of the sampling distribution of $\bar{x}$, $\sigma_{\bar{x}}$, is called the _____ _____ of the _____.

4. *True or False:* The distribution of the sample mean, $\bar{x}$, will be normally distributed if the sample is obtained from a population that is normally distributed, regardless of the sample size.

5. *True or False:* The distribution of the sample mean, $\bar{x}$, will be normally distributed if the sample is obtained from a population that is not normally distributed, regardless of the sample size.

6. *True or False:* To cut the standard error of the mean in half, the sample size must be doubled.

7. A simple random sample of size $n = 10$ is obtained from a population that is normally distributed with $\mu = 30$ and $\sigma = 8$. What is the sampling distribution of $\bar{x}$?

8. A simple random sample of size $n = 40$ is obtained from a population with $\mu = 50$ and $\sigma = 4$. Does the population need to be normally distributed for the sampling distribution of $\bar{x}$ to be approximately normally distributed? Why? What is the sampling distribution of $\bar{x}$?

In Problems 9–12, determine $\mu_{\bar{x}}$ and $\sigma_{\bar{x}}$ from the given parameters of the population and the sample size.

NW **9.** $\mu = 80, \sigma = 14, n = 49$

10. $\mu = 64, \sigma = 18, n = 36$

11. $\mu = 52, \sigma = 10, n = 21$

12. $\mu = 27, \sigma = 6, n = 15$

13. Answer the following questions for the sampling distribution of the sample mean shown to the right.

(a) What is the value of $\mu_{\bar{x}}$?
(b) What is the value of $\sigma_{\bar{x}}$?
(c) If the sample size is $n = 16$, what is likely true about the shape of the population?
(d) If the sample size is $n = 16$, what is the standard deviation of the population from which the sample was drawn?

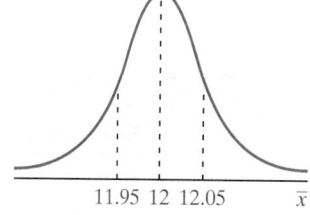

480 500 520 $\bar{x}$

14. Answer the following questions for the sampling distribution of the sample mean shown to the right.

(a) What is the value of $\mu_{\bar{x}}$?
(b) What is the value of $\sigma_{\bar{x}}$?
(c) If the sample size is $n = 9$, what is likely true about the shape of the population?
(d) If the sample size is $n = 9$, what is the standard deviation of the population from which the sample was drawn?

11.95 12 12.05 $\bar{x}$

15. A simple random sample of size $n = 49$ is obtained from a population with $\mu = 80$ and $\sigma = 14$.

(a) Describe the sampling distribution of $\bar{x}$.
(b) What is $P(\bar{x} > 83)$?
(c) What is $P(\bar{x} \le 75.8)$?
(d) What is $P(78.3 < \bar{x} < 85.1)$?

16. A simple random sample of size $n = 36$ is obtained from a population with $\mu = 64$ and $\sigma = 18$.

(a) Describe the sampling distribution of $\bar{x}$.
(b) What is $P(\bar{x} < 62.6)$?
(c) What is $P(\bar{x} \ge 68.7)$?
(d) What is $P(59.8 < \bar{x} < 65.9)$?

17. A simple random sample of size $n = 12$ is obtained from a population with $\mu = 64$ and $\sigma = 17$.

(a) What must be true regarding the distribution of the population in order to use the normal model to compute probabilities involving the sample mean? Assuming that this condition is true, describe the sampling distribution of $\bar{x}$.
(b) Assuming that the requirements described in part (a) are satisfied, determine $P(\bar{x} < 67.3)$.
(c) Assuming that the requirements described in part (a) are satisfied, determine $P(\bar{x} \ge 65.2)$.

18. A simple random sample of size $n = 20$ is obtained from a population with $\mu = 64$ and $\sigma = 17$.

(a) What must be true regarding the distribution of the population in order to use the normal model to compute probabilities involving the sample mean? Assuming that this condition is true, describe the sampling distribution of $\bar{x}$.
(b) Assuming that the requirements described in part (a) are satisfied, determine $P(\bar{x} < 67.3)$.
(c) Assuming that the requirements described in part (a) are satisfied, determine $P(\bar{x} \ge 65.2)$.
(d) Compare the results obtained in parts (b) and (c) with the results obtained in parts (b) and (c) in Problem 17. What effect does increasing the sample size have on the probabilities? Why do you think this is the case?

Applying the Concepts

NW **19. Gestation Period** The length of human pregnancies is approximately normally distributed with mean $\mu = 266$ days and standard deviation $\sigma = 16$ days.

(a) What is the probability a randomly selected pregnancy lasts less than 260 days?
(b) Suppose a random sample of 20 pregnancies is obtained. Describe the sampling distribution of the sample mean length of human pregnancies.
(c) What is the probability that a random sample of 20 pregnancies has a mean gestation period of 260 days or less?
(d) What is the probability that a random sample of 50 pregnancies has a mean gestation period of 260 days or less?
(e) What might you conclude if a random sample of 50 pregnancies resulted in a mean gestation period of 260 days or less?
(f) What is the probability a random sample of size 15 will have a mean gestation period within 10 days of the mean?

20. Upper Leg Length The upper leg length of 20- to 29-year-old males is normally distributed with a mean length of 43.7 cm and a standard deviation of 4.2 cm.

Source: "Anthropometric Reference Data for Children and Adults: U.S. Population, 1999–2002"; Volume 361, July 7, 2005.

(a) What is the probability that a randomly selected 20- to 29-year-old male has an upper leg length that is less than 40 cm?
(b) A random sample of 9 males who are 20 to 29 years old is obtained. What is the probability that the mean upper leg length is less than 40 cm?
(c) What is the probability that a random sample of 12 males who are 20–29 years old results in a mean upper leg length that is less than 40 cm?

(d) What effect does increasing the sample size have on the probability? Provide an explanation for this result.

(e) A random sample of 15 males who are 20–29 years old results in a mean upper leg length of 46 cm. Do you find this result unusual? Why?

21. Reading Rates The reading speed of second grade students is approximately normal, with a mean of 90 words per minute (wpm) and a standard deviation of 10 wpm.

(a) What is the probability a randomly selected student will read more than 95 words per minute?

(b) What is the probability that a random sample of 12 second grade students results in a mean reading rate of more than 95 words per minute?

(c) What is the probability that a random sample of 24 second grade students results in a mean reading rate of more than 95 words per minute?

(d) What effect does increasing the sample size have on the probability? Provide an explanation for this result.

(e) A teacher instituted a new reading program at school. After 10 weeks in the program, it was found that the mean reading speed of a random sample of 20 second grade students was 92.8 wpm. What might you conclude based on this result?

(f) There is a 5% chance that the mean reading speed of a random sample of 20 second grade students will exceed what value?

22. Old Faithful The most famous geyser in the world, Old Faithful in Yellowstone National Park, has a mean time between eruptions of 85 minutes. If the interval of time between eruptions is normally distributed with standard deviation 21.25 minutes, answer the following questions:

Source: www.unmuseum.org

(a) What is the probability that a randomly selected time interval between eruptions is longer than 95 minutes?

(b) What is the probability that a random sample of 20 time intervals between eruptions has a mean longer than 95 minutes?

(c) What is the probability that a random sample of 30 time intervals between eruptions has a mean longer than 95 minutes?

(d) What effect does increasing the sample size have on the probability? Provide an explanation for this result.

(e) What might you conclude if a random sample of 30 time intervals between eruptions has a mean longer than 95 minutes?

(f) On a certain day, suppose there are 22 time intervals for Old Faithful. Treating these 22 eruptions as a random sample, the likelihood the mean length of time between eruptions exceeds _____ minutes is 0.20.

23. Rates of Return in Stocks The S&P 500 is a collection of 500 stocks of publicly traded companies. Using data obtained from Yahoo! Finance, the monthly rates of return of the S&P500 since 1950 are normally distributed. The mean rate of return is 0.007233 (0.7233%), and the standard deviation for rate of return is 0.04135 (4.135%).

(a) What is the probability that a randomly selected month has a positive rate of return? That is, what is $P(x > 0)$?

(b) Treating the next 12 months as a simple random sample, what is the probability that the mean monthly rate of return will be positive? That is, with $n = 12$, what is $P(\bar{x} > 0)$?

(c) Treating the next 24 months as a simple random sample, what is the probability that the mean monthly rate of return will be positive?

(d) Treating the next 36 months as a simple random sample, what is the probability that the mean monthly rate of return will be positive?

(e) Use the results of parts (b)–(d) to describe the likelihood of earning a positive rate of return on stocks as the investment time horizon increases.

24. Winning Poker A very good poker player is expected to earn $1 per hand in $100/$200 Texas Hold'em. The standard deviation is approximately $32.

(a) What is the probability a very good poker player earns a profit (more than $0) after playing 50 hands in $100/$200 Texas Hold'em?

(b) What is the probability a very good poker player loses (earns less than $0) after playing 100 hands in $100/$200 Texas Hold'em?

(c) What proportion of the time can a very good poker player expect to earn at least $500 after playing 100 hands in $100/$200 Texas Hold'em? Hint: $500 after 100 hands is a mean of $5 per hand.

(d) Would it be unusual for a very good poker player to lose at least $1000 after playing 100 hands in $100/$200 Texas Hold'em?

(e) Suppose twenty hands are played per hour. What is the probability that a very good poker player earns a profit during a twenty-four hour marathon session?

NW 25. Oil Change The shape of the distribution of the time required to get an oil change at a 10-minute oil-change facility is unknown. However, records indicate that the mean time for an oil change is 11.4 minutes, and the standard deviation for oil-change time is 3.2 minutes.

(a) To compute probabilities regarding the sample mean using the normal model, what size sample would be required?

(b) What is the probability that a random sample of $n = 40$ oil changes results in a sample mean time of less than 10 minutes?

(c) Suppose the manager agrees to pay each employee a $50 bonus if they meet a certain goal. On a typical Saturday, the oil-change facility will perform 40 oil changes between 10 A.M. and 12 P.M. Treating this as a random sample, what mean oil-change time would there be a 10% chance of being at or below? This will be the goal established by the manager.

26. Time Spent in the Drive-Through The quality-control manager of a Long John Silver's restaurant wants to analyze the length of time that a car spends at the drive-through window waiting for an order. It is determined that the mean time spent at the window is 59.3 seconds with a standard deviation of 13.1 seconds. The distribution of time at the window is skewed right (data based on information provided by Danica Williams, student at Joliet Junior College).

(a) To obtain probabilities regarding a sample mean using the normal model, what size sample is required?

(b) The quality-control manager wishes to use a new delivery system designed to get cars through the drive-through system faster. A random sample of 40 cars results in a sample mean time spent at the window of 56.8 seconds. What is the probability of obtaining a sample mean of 56.8 seconds or less, assuming that the population mean is 59.3 seconds? Do you think that the new system is effective?

(c) Treat the next 50 cars that arrive as a simple random sample. There is a 15% chance the mean time will be at or below _____ seconds.

27. Insect Fragments The Food and Drug Administration sets Food Defect Action Levels (FDALs) for some of the various foreign substances that inevitably end up in the food we eat and liquids we drink. For example, the FDAL for insect filth in peanut butter is 3 insect fragments (larvae, eggs, body parts, and so on) per 10 grams. A random sample of 50 ten-gram portions of peanut butter is obtained and results in a sample mean of $\bar{x} = 3.6$ insect fragments per ten-gram portion.

(a) Why is the sampling distribution of $\bar{x}$ approximately normal?

(b) What is the mean and standard deviation of the sampling distribution of $\bar{x}$ assuming that $\mu = 3$ and $\sigma = \sqrt{3}$?

(c) What is the probability that a simple random sample of 50 ten-gram portions results in a mean of at least 3.6 insect fragments? Is this result unusual? What might we conclude?

28. Burger King's Drive-Through Suppose that cars arrive at Burger King's drive-through at the rate of 20 cars every hour between 12:00 noon and 1:00 P.M. A random sample of 40 one-hour time periods between 12:00 noon and 1:00 P.M. is selected and has 22.1 as the mean number of cars arriving.

(a) Why is the sampling distribution of $\bar{x}$ approximately normal?

(b) What is the mean and standard deviation of the sampling distribution of $\bar{x}$ assuming that $\mu = 20$ and $\sigma = \sqrt{20}$?

(c) What is the probability that a simple random sample of 40 one-hour time periods results in a mean of at least 22.1 cars? Is this result unusual? What might we conclude?

29. Watching Television The amount of time Americans spend watching television is closely monitored by firms such as AC Nielsen because this helps determine advertising pricing for commercials.

(a) Do you think the variable "weekly time spent watching television" would be normally distributed? If not, what shape would you expect the variable to have?

(b) According to the American Time Use Survey, adult Americans spend 2.35 hours per day watching television on a weekday. Assume that the standard deviation for "time spent watching television on a weekday" is 1.93 hours. If a random sample of 40 adult Americans is obtained, describe the sampling distribution of $\bar{x}$, the mean amount of time spent watching television on a weekday.

(c) Determine the probability that a random sample of 40 adult Americans results in a mean time watching television on a weekday of between 2 and 3 hours.

(d) One consequence of the popularity of the Internet is that it is thought to reduce television watching. Suppose that a random sample of 35 individuals who consider themselves to be avid Internet users results in a mean time of 1.89 hours watching television on a weekday. Determine the likelihood of obtaining a sample mean of 1.89 hours or less from a population whose mean is presumed to be 2.35 hours. Based on the result obtained, do you think avid Internet users watch less television?

30. ATM Withdrawals According to Crown ATM Network, the mean ATM withdrawal is $67. Assume that the standard deviation for withdrawals is $35.

(a) Do you think the variable "ATM withdrawal" is normally distributed? If not, what shape would you expect the variable to have?

(b) If a random sample of 50 ATM withdrawals is obtained, describe the sampling distribution of $\bar{x}$, the mean withdrawal amount.

(c) Determine the probability of obtaining a sample mean withdrawal amount between $70 and $75.

31. Sampling Distributions The following data represent the ages of the winners of the Academy Award for Best Actor for the years 2004–2009.

2004: Jamie Foxx	37
2005: Philip Seymour Hoffman	38
2006: Forest Whitaker	45
2007: Daniel Day-Lewis	50
2008: Sean Penn	48
2009: Jeff Bridges	60

Source: Wikipedia

(a) Compute the population mean, μ.

(b) List all possible samples with size $n = 2$. There should be $_6C_2 = 15$ samples.

(c) Construct a sampling distribution for the mean by listing the sample means and their corresponding probabilities.

(d) Compute the mean of the sampling distribution.

(e) Compute the probability that the sample mean is within 3 years of the population mean age.

(f) Repeat parts (b)–(e) using samples of size $n = 3$. Comment on the effect of increasing the sample size.

32. Sampling Distributions The following data represent the running lengths (in minutes) of the winners of the Academy Award for Best Picture for the years 2004–2009.

2004: *Million Dollar Baby*	132
2005: *Crash*	112
2006: *The Departed*	151
2007: *No Country for Old Men*	122
2008: *Slumdog Millionaire*	120
2009: *The Hurt Locker*	131

Source: The Internet Movie Database

(a) Compute the population mean, μ.

(b) List all possible samples with size $n = 2$. There should be $_6C_2 = 15$ samples.

(c) Construct a sampling distribution for the mean by listing the sample means and their corresponding probabilities.

(d) Compute the mean of the sampling distribution.

(e) Compute the probability that the sample mean is within 5 minutes of the population mean running time.

(f) Repeat parts (b)–(e) using samples of size $n = 3$. Comment on the effect of increasing the sample size.

33. Putting It Together: Playing Roulette In the game of roulette, a wheel consists of 38 slots numbered 0, 00, 1, 2, ..., 36. (See the photo.) To play the game, a metal ball is spun around the wheel and is allowed to fall into one of the numbered slots. If the number of the slot the ball falls into matches the number you selected, you win $35; otherwise you lose $1.

(a) Construct a probability distribution for the random variable X, the winnings of each spin.

(b) Determine the mean and standard deviation of the random variable X. Round your results to the nearest penny.

(c) Suppose that you play the game 100 times so that $n = 100$. Describe the sampling distribution of $\bar{x}$, the mean amount won per game.

(d) What is the probability of being ahead after playing the game 100 times? That is, what is the probability that the sample mean is greater than 0 for $n = 100$?

(e) What is the probability of being ahead after playing the game 200 times?

(f) What is the probability of being ahead after playing the game 1000 times?

(g) Compare the results of parts (d) and (e). What lesson does this teach you?

Explaining the Concepts

34. Explain what a sampling distribution is.

35. State the Central Limit Theorem.

36. We assume that we are obtaining simple random samples from infinite populations when obtaining sampling distributions. If the size of the population is finite, we technically need a finite population correction factor. However, if the sample size is small relative to the size of the population, this factor can be ignored. Explain what an "infinite population" is. What is the finite population correction factor? How small must the sample size be relative to the size of the population so that we can ignore the factor? Finally, explain why the factor can be ignored for such samples.

37. Without doing any computation, decide which has a higher probability, assuming each sample is from a population that is normally distributed with $\mu = 100$ and $\sigma = 15$. Explain your reasoning.

(a) $P(90 \le \bar{x} \le 110)$ for a random sample of size $n = 10$.
(b) $P(90 \le \bar{x} \le 110)$ for a random sample of size $n = 20$.

38. For the three probability distributions shown, rank each distribution from lowest to highest in terms of the sample size required for the distribution of the sample mean to be approximately normally distributed. Justify your choice.

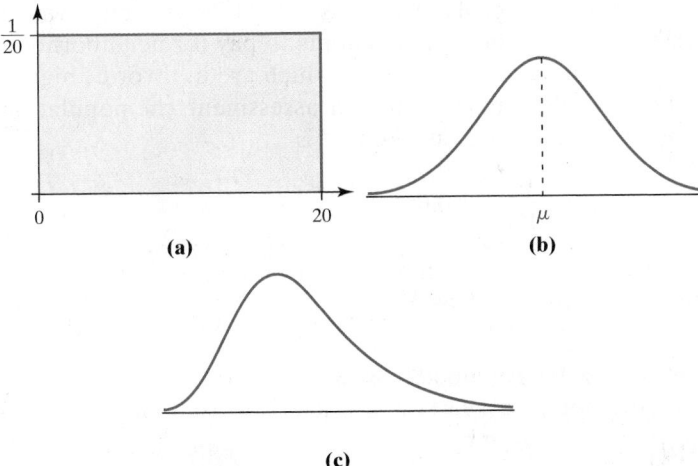
(a)

(b)

(c)

39. Suppose Jack and Diane are each attempting to use simulation to describe the sampling distribution from a population that is skewed left with mean 50 and standard deviation 10. Jack obtains 1000 random samples of size $n = 3$ from the population, finds the mean of the 1000 samples, draws a histogram of the means, finds the mean of the means, and determines the standard deviation of the means. Diane does the same simulation, but obtains 1000 random samples of size $n = 30$ from the population.

(a) Describe the shape you expect for Jack's distribution of sample means. Describe the shape you expect for Diane's distribution of sample means.

(b) What do you expect the mean of Jack's distribution to be? What do you expect the mean of Diane's distribution to be?

(c) What do you expect the standard deviation of Jack's distribution to be? What do you expect the standard deviation of Diane's distribution to be?

40. Sleepy Suppose you want to study the number of hours of sleep you get each evening. To do so, you look at the calendar and randomly select 10 days out of the next 300 days and record the number of hours you sleep.

(a) Explain why number of hours of sleep in a night by you is a random variable.

(b) Is the random variable "number of hours of sleep in a night" quantitative or qualitative?

(c) After you obtain your ten nights of data, you compute the mean number of hours of sleep. Is this a statistic or a parameter? Why?

(d) Is the mean number of hours computed in part (c) a random variable? Why? If it is a random variable, what is the source of variation?

41. Sleepy Again Suppose you want to study the number of hours of sleep full-time college students at your college get each evening. To do so, you obtain a list of full-time students at your college, obtain a simple random sample of ten students, and ask each of them to disclose how many hours of sleep they obtained the most recent Monday.

(a) What is the population of interest in this study? What is the sample?

(b) Explain why number of hours of sleep in this study is a random variable.

(c) After you obtain your ten observations, you compute the mean number of hours of sleep. Is this a statistic or a parameter? Why?

(d) Is the mean number of hours computed in part (c) a random variable? Why? If it is a random variable, what is the source of variation? How does the source of variation in this study differ from that of Problem 40?

Retain Your Knowledge

42. Bull Markets A bull market is defined as a market condition in which the price of a security rises for an extended period of time. A bull market in the stock market is often defined as a condition in which a market rises by 20% or more without a 20% decline. The data on the next page represent the number of months and percentage change in the S&P500 (a group of 500 stocks) during the 25 bull markets dating back to 1929 (the year of the famous market crash).

(a) Treating the length of the bull market as the explanatory variable, draw a scatter diagram of the data.

(b) Determine the linear correlation coefficient between months and percent change.

(c) Does a linear relation exist between duration of the bull market and market performance?
(d) Find the least-squares regression line treating length of the bull market as the explanatory variable.
(e) Interpret the slope.
(f) Did the bull market that lasted 50.4 months have a percent change above or below what would be expected? Explain.
(g) Draw a residual plot. Any outliers?
(h) Would you consider the bull market from December 4, 1987 through March 24, 2000, which lasted 149.8 months and saw a 582.15% rise in stock prices, influential? Explain. Note: After this bull market, the market entered a bear market that lasted 18.2 months and saw the stock market decline 37%. This era is often referred to as the "Tech Bubble."

Bull Months	Percent Change	Bull Months	Percent Change
4.9	46.77	86.9	267.08
2.3	25.83	50.4	86.35
0.8	25.82	44.1	79.78
1.2	30.61	26.1	48.05
2.3	111.59	32.0	73.53
4.7	120.61	74.9	125.63
3.7	37.28	61.3	228.81
24.2	131.64	149.8	582.15
7.4	62.24	3.5	21.40
6.6	26.78	60.9	101.50
5.0	20.91	1.6	24.22
49.7	157.70	48.6	127.85
13.1	23.89		

8.2 Distribution of the Sample Proportion

Preparing for This Section Before getting started, review the following:

- Applications of the normal distribution (Section 7.2, pp. 366–372)

Objectives
① Describe the sampling distribution of a sample proportion
② Compute probabilities of a sample proportion

① Describe the Sampling Distribution of a Sample Proportion

Suppose we want to determine the proportion of households in a 100-house homeowners association that favor an increase in the annual assessments to pay for neighborhood improvements. We could survey all households to learn which are in favor of higher assessments. If 65 of the 100 households favor the higher assessment, the population proportion, p, of households in favor of a higher assessment is

$$p = \frac{65}{100} = 0.65$$

Of course, gaining access to all the individuals in a population is rare, so we usually obtain estimates of population parameters such as p.

Definition
Suppose that a random sample of size n is obtained from a population in which each individual either does or does not have a certain characteristic. The **sample proportion**, denoted $\hat{p}$ (read "p-hat"), is given by

$$\hat{p} = \frac{x}{n}$$

where x is the number of individuals in the sample with the specified characteristic.*
The sample proportion, $\hat{p}$, is a statistic that estimates the population proportion, p.

*For those who studied Section 6.2 on binomial probabilities, x can be thought of as the number of successes in n trials of a binomial experiment.

EXAMPLE 1 Computing a Sample Proportion

Problem The Harris Poll conducted a survey of 1200 adult Americans who vacation during the summer and asked whether the individuals plan to work while on summer vacation. Of those surveyed, 552 indicated that they plan to work while on vacation. Find the sample proportion of individuals surveyed who plan to work while on summer vacation.

Approach Use the formula $\hat{p} = \dfrac{x}{n}$, where x is the number of individuals who plan to work on summer vacation and n is the sample size.

Solution Substituting $x = 552$ and $n = 1200$ into $\hat{p} = \dfrac{x}{n}$, we find that $\hat{p} = \dfrac{552}{1200} = 0.46$, so Harris estimates that 0.46 or 46% of adult Americans who vacation during the summer plan to work while on vacation. ●

A second survey of 1200 American adults would likely have a different estimate of the proportion of Americans who plan to work while on summer vacation because different individuals would be in the sample. Because the value of $\hat{p}$ varies from sample to sample, it is a random variable and has a probability distribution.

To get a sense of the shape, center, and spread of the sampling distribution of $\hat{p}$, we could repeat the exercise of obtaining simple random samples of 1200 adult Americans over and over. This would lead to a list of sample proportions. A histogram of the sample proportions will give us a feel for the shape of the distribution of the sample proportion. The mean of the sample proportions will give us an idea of the center of the distribution, and the standard deviation of the sample proportions will give us an idea of the spread of the distribution.

Rather than literally surveying 1200 adult Americans over and over again, we will use simulation to get an idea of the shape, center, and spread of the sampling distribution of the proportion.

EXAMPLE 2 Using Simulation to Describe the Distribution of the Sample Proportion

Problem Based on a study conducted by the Gallup organization, 76% of Americans believe that the state of moral values in the United States is getting worse. Describe the sampling distribution of the sample proportion for samples of size (a) $n = 10$, (b) $n = 25$, (c) $n = 60$.

Approach Describing a distribution means finding its shape, center, and spread. The actual sampling distribution of the sample proportion would be the distribution of *all* possible sample proportions of size $n = 10$. It is virtually impossible to find all possible samples of size $n = 10$ from the population of Americans. To get a sense as to the shape, center, and spread of the sampling distribution of the sample proportion, we will use StatCrunch's Bernoulli Random Data command with event probability 0.76 to simulate obtaining 2000 samples of size $n = 10$ by randomly generating 2000 rows of responses over 10 columns. Each row consists of a 0 or 1, with 0 representing a failure (individual does not believe the state of moral values is getting worse) and 1 representing a success (individual believes the state of moral values is getting worse). For each of the 2000 samples (the 2000 rows), determine the mean number of successes (the sample proportion). Draw a histogram of the 2000 sample proportions to gauge the shape of the distribution of the sample proportions. Determine the mean of the 2000 sample proportions to approximate the mean of the sampling distribution. Determine the standard deviation of the 2000 sample proportions to approximate the standard deviation of the sampling distribution. Repeat this process for samples of size $n = 25$ and $n = 60$.

(continued)

Solution Figure 12 shows a partial output from StatCrunch. Row 1 contains the first sample, where the results of the survey are 1 (success), 1 (success), 1(success), ... , 0 (failure), 1 (success). The mean number of successes, that is, the sample proportion, from the first sample of $n = 10$ adult Americans is 0.7.

Figure 12

Row	Bernoulli1	Bernoulli2	Bernoulli3	Bernoulli4	Bernoulli5	Bernoulli6	Bernoulli7	Bernoulli8	Bernoulli9	Bernoulli10	p-hat
Sample 1 — 1	1	1	1	1	0	1	0	1	0	1	0.7	$\hat{p} = 0.7$
Sample 2 — 2	1	1	1	1	0	1	0	1	0	1	0.7	$\hat{p} = 0.7$
3	1	0	1	1	0	1	1	1	1	1	0.8
4	1	1	0	1	1	1	1	1	0	1	0.8
5	1	1	1	1	1	1	1	1	1	1	1
6	0	1	1	0	0	0	1	1	1	1	0.6
7	1	0	1	1	1	1	1	0	1	1	0.8
8	1	1	1	1	1	1	0	0	1	1	0.8
9	1	1	1	0	1	1	1	0	0	0	0.6

Figure 13(a) shows the histogram of the 2000 sample proportions from column p-hat. Notice that the shape of the distribution is skewed left. The mean of the 2000 sample proportions is 0.76 and the standard deviation is 0.136. Notice that the mean of the sample proportions equals the population proportion.

Figure 13(b) shows the histogram for 2000 sample proportions from samples of size $n = 25$. Notice that the histogram is slightly skewed left (although not as skewed as the histogram with $n = 10$). The mean of the 2000 sample proportions for a sample of size $n = 25$ is 0.76 and the standard deviation is 0.086.

Figure 13(c) shows the histogram for 2000 sample proportions from samples of size $n = 60$. Notice that the histogram is approximately normal. The mean of the 2000 sample proportions is 0.76 and the standard deviation is 0.054.

Figure 13

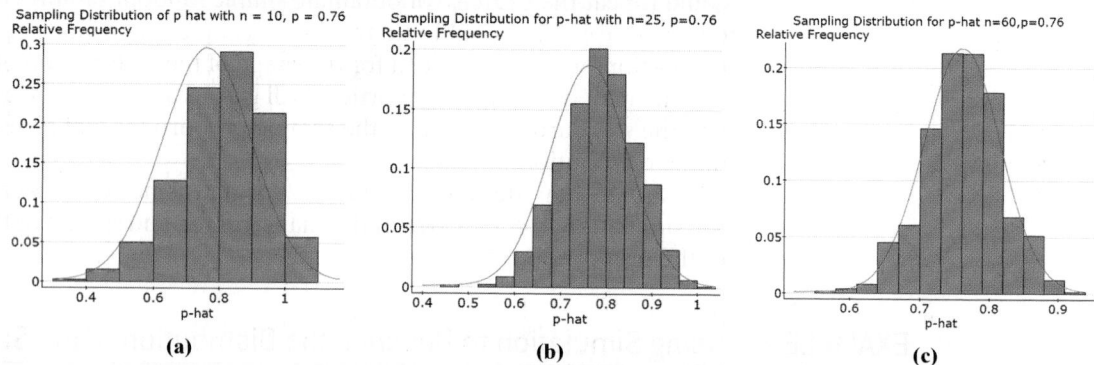

(a) (b) (c)

Notice the following points regarding the sampling distribution of the sample proportion:

- **Shape:** As the size of the sample increases, the shape of the distribution of the sample proportion becomes approximately normal.
- **Center:** The mean of the distribution of the sample proportion equals the population proportion, p.
- **Spread:** The standard deviation of the distribution of the sample proportion decreases as the sample size increases. ●

Although the proof is beyond the scope of this text, we should be convinced that the following results are reasonable.

Sampling Distribution of $\hat{p}$

For a simple random sample of size n with a population proportion p,

- The shape of the sampling distribution of $\hat{p}$ is approximately normal provided $np(1 - p) \geq 10$.
- The mean of the sampling distribution of $\hat{p}$ is $\mu_{\hat{p}} = p$.
- The standard deviation of the sampling distribution of $\hat{p}$ is $\sigma_{\hat{p}} = \sqrt{\dfrac{p(1 - p)}{n}}$.

In Other Words
The sample size cannot be more than 5% of the population size because the success or failure of identifying an individual in the population that has the specified characteristic should not be affected by earlier observations. For example, in a population of size 100 where 14 of the individuals have brown hair, the probability that a randomly selected individual has brown hair is 14/100 = 0.14. The probability that a second randomly selected individual has brown hair is 13/99 = 0.13. The probability changes because the sampling is done without replacement.

One requirement of this model is that the sampled values must be independent of each other—that is, one outcome cannot affect the success or failure of any other outcome. When sampling from finite populations, we verify the independence assumption by checking that the sample size n is no more than 5% of the population size, N ($n \leq 0.05N$).

Also, regardless of whether $np(1 - p) \geq 10$, the mean of the sampling distribution of $\hat{p}$ is p, and the standard deviation of the sampling distribution of $\hat{p}$ is $\sqrt{\dfrac{p(1 - p)}{n}}$.

EXAMPLE 3 Describing the Sampling Distribution of the Sample Proportion

Problem Based on a study conducted by the Gallup organization, 76% of Americans believe that the state of moral values in the United States is getting worse. Suppose we obtain a simple random sample of $n = 60$ Americans and determine which believe that the state of the moral values in the United States is getting worse. Describe the sampling distribution of the sample proportion for Americans with this belief.

Approach If the sample size is less than 5% of the population size and $np(1 - p)$ is at least 10, the sampling distribution of $\hat{p}$ is approximately normal, with mean $\mu_{\hat{p}} = p$ and standard deviation $\sigma_{\hat{p}} = \sqrt{\dfrac{p(1 - p)}{n}}$.

Solution The United States has over 300 million people, so the sample of $n = 60$ is less than 5% of the population size. Also, $np(1 - p) = 60(0.76)(1 - 0.76) = 10.944 \geq 10$. The distribution of $\hat{p}$ is approximately normal, with mean $\mu_{\hat{p}} = 0.76$ and standard deviation

$$\sigma_{\hat{p}} = \sqrt{\frac{p(1 - p)}{n}} = \sqrt{\frac{0.76(1 - 0.76)}{60}} = 0.055$$

Notice that the standard deviation found in Example 3 (0.055) is very close to the standard deviation of the sample proportion found using simulation with $n = 60$ in Example 2 (0.054).

● Now Work Problem 7

② Compute Probabilities of a Sample Proportion

Now that we can describe the sampling distribution of the sample proportion, we can compute probabilities involving sample proportions.

EXAMPLE 4 Compute Probabilities of a Sample Proportion

Problem According to the National Center for Health Statistics, 15% of all Americans have hearing trouble.

(a) In a random sample of 120 Americans, what is the probability at most 12% have hearing trouble?

(b) Suppose that a random sample of 120 Americans who regularly listen to music using headphones results in 26 having hearing trouble. What might you conclude?

Approach First, determine whether the sampling distribution is approximately normal by verifying that the sample size is less than 5% of the population size and that $np(1 - p) \geq 10$. Then use the normal distribution to determine the probabilities.

Solution There are over 300 million people in the United States, so the sample size of $n = 120$ is definitely less than 5% of the population size. We are told that $p = 0.15$. Because $np(1 - p) = 120(0.15)(1 - 0.15) = 15.3 \geq 10$, the shape of the distribution of the sample proportion is approximately normal. The mean of the distribution of the sample proportion $\hat{p}$ is $\mu_{\hat{p}} = 0.15$, and the standard deviation is $\sigma_{\hat{p}} = \sqrt{\dfrac{0.15(1 - 0.15)}{120}}$.

(continued)

Figure 14

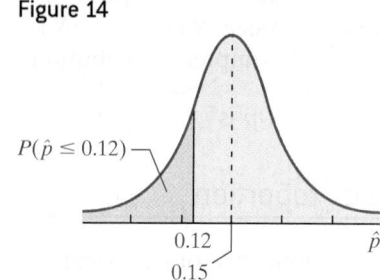

$P(\hat{p} \le 0.12)$

0.12
0.15
$\hat{p}$

(a) We want to know the probability that a random sample of 120 Americans will result in a sample proportion of at most 0.12 (or 12%). That is, we want to know $P(\hat{p} \le 0.12)$. Figure 14 shows the normal curve with the area to the left of 0.12 shaded.

To find this area by hand, convert $\hat{p} = 0.12$ to a z-score by subtracting the mean and dividing by the standard deviation. Don't forget that we round z to two decimal places.

$$z = \frac{\hat{p} - \mu_{\hat{p}}}{\sigma_{\hat{p}}} = \frac{0.12 - 0.15}{\sqrt{\dfrac{0.15(1 - 0.15)}{120}}} = -0.92$$

The area under the standard normal curve left of $z = -0.92$ is 0.1788. Remember, the area to the left of $\hat{p} = 0.12$ is the same as the area to the left of $z = -0.92$, so $P(\hat{p} \le 0.12) = 0.1788$.

If we use technology to find the area to the left of $\hat{p} = 0.12$, we obtain 0.1787, so $P(\hat{p} \le 0.12) = 0.1787$.

Interpretation The probability that a random sample of $n = 120$ Americans results in at most 12% having hearing trouble is approximately 0.18. This means that about 18 out of 100 random samples of size 120 will result in at most 12% having hearing trouble if the population proportion of Americans with hearing trouble is 0.15.

(b) A random sample of 120 Americans who regularly listen to music using headphones results in 26 having hearing trouble. The sample proportion is $\hat{p} = \dfrac{26}{120} = 0.217$.

We want to know if obtaining a sample proportion of 0.217 from a population whose proportion is assumed to be 0.15 is unusual. We compute $P(\hat{p} \ge 0.217)$, because if a sample proportion of 0.217 is unusual, then any sample proportion more than 0.217 is also unusual. Figure 15 shows the normal curve with the area to the right of 0.217 shaded.

Figure 15

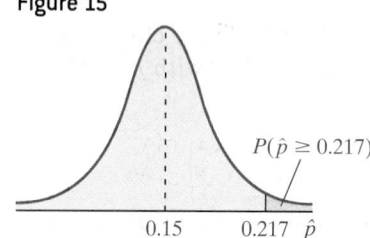

$P(\hat{p} \ge 0.217)$

0.15   0.217 $\hat{p}$

To find this area by hand, convert $\hat{p} = 0.217$ to a z-score.

$$z = \frac{\hat{p} - \mu_{\hat{p}}}{\sigma_{\hat{p}}} = \frac{0.217 - 0.15}{\sqrt{\dfrac{0.15(1 - 0.15)}{120}}} = 2.06$$

The area under the standard normal curve to the right of 2.06 is 0.0197. The area to the right of $\hat{p} = 0.217$ is the same as the area to the right of $z = 2.06$, so $P(\hat{p} \ge 0.217) = 0.0197$.

If we use technology to find the area to the right of $\hat{p} = 0.217$, we obtain 0.0199, so $P(\hat{p} \ge 0.217) = 0.0199$.

Interpretation About 2 samples in 100 will result in a sample proportion of 0.217 or more from a population whose proportion is 0.15. We obtained a result that should only happen about 2 times in 100, so the results obtained are unusual. We could make one of two conclusions:

- The population proportion of Americans with hearing trouble who regularly listen to music using headphones is 0.15, and we just happen to randomly select a higher proportion that have hearing trouble.
- The population proportion of Americans with hearing trouble who regularly listen to music using headphones is more than 0.15.

The second conclusion is more reasonable. We conclude that the proportion of Americans who regularly listen to music using headphones who have hearing trouble is higher than the general population.

• **Now Work Problem 17**

8.2 Assess Your Understanding

Vocabulary and Skill Building

1. In a town of 500 households, 220 have a dog. The population proportion of dog owners in this town (expressed as a decimal) is $p =$ _____.

2. The _____ _____, denoted $\hat{p}$, is given by the formula $\hat{p} =$ _____, where x is the number of individuals with a specified characteristic in a sample of n individuals.

3. *True or False:* The population proportion and sample proportion always have the same value.

4. *True or False:* The mean of the sampling distribution of $\hat{p}$ is p.

5. Describe the circumstances under which the shape of the sampling distribution of $\hat{p}$ is approximately normal.

6. What happens to the standard deviation of $\hat{p}$ as the sample size increases? If the sample size is increased by a factor of 4, what happens to the standard deviation of $\hat{p}$?

In Problems 7–10, describe the sampling distribution of $\hat{p}$. Assume that the size of the population is 25,000 for each problem.

NW 7. $n = 500, p = 0.4$ **8.** $n = 300, p = 0.7$

9. $n = 1000, p = 0.103$ **10.** $n = 1010, p = 0.84$

11. A simple random sample of size $n = 75$ is obtained from a population whose size is $N = 10,000$ and whose population proportion with a specified characteristic is $p = 0.8$.

(a) Describe the sampling distribution of $\hat{p}$.

(b) What is the probability of obtaining $x = 63$ or more individuals with the characteristic? That is, what is $P(\hat{p} \geq 0.84)$?

(c) What is the probability of obtaining $x = 51$ or fewer individuals with the characteristic? That is, what is $P(\hat{p} \leq 0.68)$?

12. A simple random sample of size $n = 200$ is obtained from a population whose size is $N = 25,000$ and whose population proportion with a specified characteristic is $p = 0.65$.

(a) Describe the sampling distribution of $\hat{p}$.

(b) What is the probability of obtaining $x = 136$ or more individuals with the characteristic? That is, what is $P(\hat{p} \geq 0.68)$?

(c) What is the probability of obtaining $x = 118$ or fewer individuals with the characteristic? That is, what is $P(\hat{p} \leq 0.59)$?

13. A simple random sample of size $n = 1000$ is obtained from a population whose size is $N = 1,000,000$ and whose population proportion with a specified characteristic is $p = 0.35$.

(a) Describe the sampling distribution of $\hat{p}$.

(b) What is the probability of obtaining $x = 390$ or more individuals with the characteristic?

(c) What is the probability of obtaining $x = 320$ or fewer individuals with the characteristic?

14. A simple random sample of size $n = 1460$ is obtained from a population whose size is $N = 1,500,000$ and whose population proportion with a specified characteristic is $p = 0.42$.

(a) Describe the sampling distribution of $\hat{p}$.

(b) What is the probability of obtaining $x = 657$ or more individuals with the characteristic?

(c) What is the probability of obtaining $x = 584$ or fewer individuals with the characteristic?

Applying the Concepts

15. Foreign Language According to a study done by Wakefield Research, the proportion of Americans who can order a meal in a foreign language is 0.47.

(a) Suppose a random sample of 200 Americans is asked to disclose whether they can order a meal in a foreign language. Is the response to this question qualitative or quantitative? Explain.

(b) Explain why the sample proportion, $\hat{p}$, is a random variable. What is the source of the variability?

(c) Describe the sampling distribution of $\hat{p}$, the proportion of Americans who can order a meal in a foreign language. Be sure to verify the model requirements.

(d) In the sample obtained in part (a), what is the probability the proportion of Americans who can order a meal in a foreign language is greater than 0.5?

(e) Would it be unusual that, in a survey of 200 Americans, 80 or fewer Americans can order a meal in a foreign language? Why?

16. Are You Satisfied? According to a study done by the Gallup organization, the proportion of Americans who are satisfied with the way things are going in their lives is 0.82.

(a) Suppose a random sample of 100 Americans is asked, "Are you satisfied with the way things are going in your life?" Is the response to this question qualitative or quantitative? Explain.

(b) Explain why the sample proportion, $\hat{p}$, is a random variable. What is the source of the variability?

(c) Describe the sampling distribution of $\hat{p}$, the proportion of Americans who are satisfied with the way things are going in their life. Be sure to verify the model requirements.

(d) In the sample obtained in part (a), what is the probability the proportion who are satisfied with the way things are going in their life exceeds 0.85?

(e) Would it be unusual for a survey of 100 Americans to reveal that 75 or fewer are satisfied with the way things are going in their life? Why?

NW 17. Marriage Obsolete? According to a study done by the Pew Research Center, 39% of adult Americans believe that marriage is now obsolete.

(a) Suppose a random sample of 500 adult Americans is asked whether marriage is obsolete. Describe the sampling distribution of $\hat{p}$, the proportion of adult Americans who believe marriage is obsolete.

(b) What is the probability that in a random sample of 500 adult Americans less than 38% believe that marriage is obsolete?

(c) What is the probability that in a random sample of 500 adult Americans between 40% and 45% believe that marriage is obsolete?

(d) Would it be unusual for a random sample of 500 adult Americans to result in 210 or more who believe marriage is obsolete?

18. Credit Cards According to creditcard.com, 29% of adults do not own a credit card.

(a) Suppose a random sample of 500 adults is asked, "Do you own a credit card?" Describe the sampling distribution of $\hat{p}$, the proportion of adults who do not own a credit card.

(b) What is the probability that in a random sample of 500 adults more than 30% do not own a credit card?

(c) What is the probability that in a random sample of 500 adults between 25% and 30% do not own a credit card?

(d) Would it be unusual for a random sample of 500 adults to result in 125 or fewer who do not own a credit card? Why?

19. Afraid to Fly According to a study conducted by the Gallup organization, the proportion of Americans who are afraid to fly is 0.10. A random sample of 1100 Americans results in 121 indicating that they are afraid to fly. Explain why this is not necessarily evidence that the proportion of Americans who are afraid to fly has increased since the time of the Gallup study.

20. Having Children? The Pew Research Center recently reported that 18% of women 40–44 years of age have never given birth. Suppose a random sample of 250 adult women 40–44 years of age results in 52 indicating they have never given birth. Explain why this is not necessarily evidence that the proportion of women 40–44 years of age who have not given birth has increased since the time of the Pew study.

21. Election Prediction Exit polling is a popular technique used to determine the outcome of an election prior to results being tallied. Suppose a referendum to increase funding for education is on the ballot in a large town (voting population over 100,000). An exit poll of 310 voters finds that 164 voted for the referendum. How likely are the results of your sample if the population proportion of voters in the town in favor of the referendum is 0.49? Based on your result, comment on the dangers of using exit polling to call elections. Include a discussion of the potential nonsampling error that could disrupt your findings.

22. Acceptance Sampling A shipment of 50,000 transistors arrives at a manufacturing plant. The quality control engineer at the plant obtains a random sample of 500 resistors and will reject the entire shipment if 10 or more of the resistors are defective. Suppose that 4% of the resistors in the whole shipment are defective. What is the probability the engineer accepts the shipment? Do you believe the acceptance policy of the engineer is sound?

23. Social Security Reform A researcher studying public opinion of proposed Social Security changes obtains a simple random sample of 50 adult Americans and asks them whether or not they support the proposed changes. To say that the distribution of $\hat{p}$, the sample proportion of adults who respond yes, is approximately normal, how many more adult Americans does the researcher need to sample if

(a) 10% of all adult Americans support the changes?

(b) 20% of all adult Americans support the changes?

24. ADHD A researcher studying ADHD among teenagers obtains a simple random sample of 100 teenagers aged 13–17 and asks them whether or not they have ever been prescribed medication for ADHD. To say that the distribution of $\hat{p}$, the sample proportion of teenagers who respond no, is approximately normal, how many more teenagers aged 13–17 does the researcher need to sample if

(a) 90% of all teenagers aged 13–17 have never been prescribed medication for ADHD?

(b) 95% of all teenagers aged 13–17 have never been prescribed medication for ADHD?

25. Reincarnation Suppose 21% of all American teens (age 13–17 years) believe in reincarnation.

(a) Bob and Alicia both obtain a random sample of 100 American teens and ask each participant to disclose whether they believe in reincarnation or not. Is "belief in reincarnation" qualitative or quantitative? Explain.

(b) Explain why Bob's sample of 100 randomly selected American teens might result in 18 who believe in reincarnation, while Alicia's independent sample of 100 randomly selected American teens might result in 22 who believe in reincarnation.

(c) Why is it important to randomly select American teens to estimate the population proportion who believe in reincarnation?

(d) In a survey of 100 American teens, how many would you expect to believe in reincarnation?

(e) Below is the histogram of the sample proportion of 1000 different surveys in which $n = 20$ American teens were asked to disclose whether they believed in reincarnation. Explain why the normal model should not be used to describe the distribution of the sample proportion.

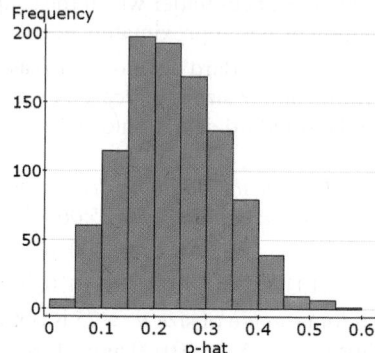

(f) What minimum sample size would you require in order for the distribution of the sample proportion to be modeled by the normal distribution?

26. Assessments Consider the homeowners association presented at the beginning of this section. A random sample of 20 households resulted in 15 indicating that they would favor an increase in assessments. Explain why the normal model could not be used to determine if a sample proportion of $\frac{15}{20} = 0.75$ or higher from a population whose proportion is 0.65 is unusual.

27. Airline Reservations In Chapter 6, we learned that the proportion of passengers who miss a flight for which they have a reservation is 0.0995.

(a) Suppose a flight has 290 reservations. What is the probability that 25 or more passengers will miss the flight?

(b) Suppose a flight has 320 reservations, but only 300 seats on the plane. What is the probability that 300 or fewer passengers show up for the flight?

(c) Suppose a popular flight has 300 seats and 300 reservations. Because this flight is popular, the proportion of passengers with a reservation who miss the flight is only 0.04. You are booked on a later flight and put yourself on the stand-by list. There are 14 passenger names ahead of you. What is the probability you get on this flight?

28. Finite Population Correction Factor In this section, we assumed that the sample size was less than 5% of the size of the population. When sampling without replacement from a finite population in which $n > 0.05N$, the standard deviation of the distribution of $\hat{p}$ is given by

$$\sigma_{\hat{p}} = \sqrt{\frac{p(1-p)}{n-1} \cdot \left(\frac{N-n}{N}\right)}$$

where N is the size of the population. A survey is conducted at a college having an enrollment of 6502 students. The student council wants to estimate the percentage of students in favor of establishing a student union. In a random sample of 500 students, it was determined that 410 were in favor of establishing a student union.

(a) Obtain the sample proportion, $\hat{p}$, of students surveyed who favor establishing a student union.

(b) Calculate the standard deviation of the sampling distribution of $\hat{p}$ using $\hat{p}$ as an estimate of p.

Retain Your Knowledge

 **29. Fumbles** The New England Patriots made headlines prior to the 2015 Super Bowl for allegedly playing with underinflated footballs. An underinflated ball is easier to grip, and therefore, less likely to be fumbled. What does the data say? The following data represent the number of plays a team has per fumble. For example, the Chicago Bears run 48 offensive plays for every fumble.

Team	Dome	Plays	Team	Dome	Plays
Atlanta Falcons	Yes	83	Saint Louis Rams	Yes	50
New Orleans Saints	Yes	80	Seattle Seahawks	No	50
New England Patriots	No	78	Detroit Lions	Yes	49
Houston Texans	Yes	61	Chicago Bears	No	48
Minnesota Vikings	Yes	59	San Francisco 49ers	No	47
Baltimore Ravens	No	58	New York Jets	No	46
Carolina Panthers	No	57	Denver Broncos	No	46
San Diego Chargers	No	57	Tampa Bay Buccaneers	No	45
Indianapolis Colts	Yes	56	Dallas Cowboys	Yes	45
Green Bay Packers	No	55	Tennessee Titans	No	45
New York Giants	No	55	Oakland Raiders	No	44
Cincinnati Bengals	No	53	Miami Dolphins	No	44
Pittsburgh Steelers	No	52	Buffalo Bills	No	43
Kansas City Chiefs	No	52	Arizona Cardinals	Yes	43
Jacksonville Jaguars	No	51	Philadelphia Eagles	No	42
Cleveland Browns	No	50	Washington Redskins	No	37

Source: Advanced Football Analytics

(a) Draw a boxplot of plays per fumble for all teams in the National Football League. Describe the shape of the distribution. Are there any outliers? If so, which team(s)?

(b) Playing in a dome (inside) removes the effect of weather (such as rain) on the game. Draw a boxplot of plays per fumble for teams who do not play in a dome. Are there any outliers? If so, which team(s)?

Chapter 8 Review

Summary

Now that we have completed our study of probability distributions, we can begin to transition to inferential statistics. Remember, inferential statistics represents taking information learned from a sample and generalizing it to a population. Often, we accompany the generalization with a measure of reliability in the results. This chapter introduces models that are used to make the generalizations. With these models in hand, we spend the rest of the course performing inference.

In Section 8.1, we discussed the sampling distribution of the sample mean. The mean of the distribution of the sample mean equals the mean of the population ($\mu_{\bar{x}} = \mu$)

and the standard deviation of the distribution of the sample mean is the standard deviation of the population divided by the square root of the sample size $\left(\sigma_{\bar{x}} = \dfrac{\sigma}{\sqrt{n}} \right)$ assuming the sample size is no more than 5% of the population size.

If a sample is obtained from a population that is known to be normally distributed, the shape of the distribution of the sample mean is also normal. If the sample is obtained from a population that is not normal, the shape of the distribution of the sample mean becomes approximately normal as the sample size increases. This result is known as the Central Limit Theorem.

In Section 8.2, we discussed the sampling distribution of the sample proportion. If a sample of size n is obtained from a population and x successes are obtained, then $\hat{p} = \dfrac{x}{n}$. If $np(1 - p) \geq 10$, then the shape of the distribution of $\hat{p}$ is approximately normal. The mean of the distribution of the sample proportion is the population proportion $(\mu_{\hat{p}} = p)$ and the standard deviation of the distribution of the sample proportion is $\sigma_{\hat{p}} = \sqrt{\dfrac{p(1 - p)}{n}}$. Because the sampled values must be independent, it must be the case that the sample size is no more than 5% of the population size.

Vocabulary

Sampling distribution (p. 395)

Sampling distribution of the sample mean (p. 395)

Standard error of the mean (p. 398)

Central Limit Theorem (p. 401)

Sample proportion (p. 408)

Sampling distribution of $\hat{p}$ (p. 410)

Formulas

Mean and Standard Deviation of the Sampling Distribution of $\bar{x}$

$$\mu_{\bar{x}} = \mu \quad \text{and} \quad \sigma_{\bar{x}} = \dfrac{\sigma}{\sqrt{n}}$$

Sample Proportion

$$\hat{p} = \dfrac{x}{n}$$

Mean and Standard Deviation of the Sampling Distribution of $\hat{p}$

$$\mu_{\hat{p}} = p \quad \text{and} \quad \sigma_{\hat{p}} = \sqrt{\dfrac{p(1 - p)}{n}}$$

Standardizing a Normal Random Variable

$$z = \dfrac{\bar{x} - \mu}{\dfrac{\sigma}{\sqrt{n}}} \quad \text{or} \quad z = \dfrac{\hat{p} - p}{\sqrt{\dfrac{p(1 - p)}{n}}}$$

OBJECTIVES

Section	You should be able to ...	Example(s)	Review Exercises
8.1	1 Describe the distribution of the sample mean: normal population (p. 395)	1–3	2, 4, 5
	2 Describe the distribution of the sample mean: nonnormal population (p. 399)	4 and 5	2, 6, 7
8.2	1 Describe the sampling distribution of a sample proportion (p. 408)	2 and 3	3, 4, 8(b), 10(a)
	2 Compute probabilities of a sample proportion (p. 411)	4	8(c), (d), 9, 10(b), (c)

Review Exercises

1. In your own words, explain what a sampling distribution is.

2. Under what conditions is the sampling distribution of $\bar{x}$ normal?

3. Under what conditions is the sampling distribution of $\hat{p}$ approximately normal?

4. What are the mean and standard deviation of the sampling distribution of $\bar{x}$? What are the mean and standard deviation of the sampling distribution of $\hat{p}$?

5. Energy Need during Pregnancy The total energy need during pregnancy is normally distributed, with mean $\mu = 2600\ \text{kcal/day}$ and standard deviation $\sigma = 50\ \text{kcal/day}$.
Source: American Dietetic Association

(a) Is total energy need during pregnancy qualitative or quantitative?

(b) What is the probability that a randomly selected pregnant woman has an energy need of more than 2625 kcal/day? Is this result unusual?

(c) Describe the sampling distribution of $\bar{x}$, the sample mean daily energy requirement for a random sample of 20 pregnant women. What might be the source of variability in the sample mean?

(d) What is the probability that a random sample of 20 pregnant women has a mean energy need of more than 2625 kcal/day? Is this result unusual?

6. Copper Tubing A machine at K&A Tube & Manufacturing Company produces a certain copper tubing component in a refrigeration unit. The tubing components produced by the manufacturer have a mean diameter of 0.75 inch with a standard deviation of 0.004 inch. The quality-control inspector takes a random sample of 30 components once each week and calculates the mean diameter of these components. If the mean is either less than 0.748 inch or greater than 0.752 inch, the inspector concludes that the machine needs an adjustment.

(a) Describe the sampling distribution of $\bar{x}$, the sample mean diameter, for a random sample of 30 such components.

(b) What is the probability that, based on a random sample of 30 such components, the inspector will conclude that the machine needs an adjustment when, in fact, the machine is correctly calibrated?

7. Number of Televisions Based on data obtained from AC Nielsen, the mean number of televisions in a household in the United States is 2.24. Assume that the population standard deviation number of television sets in the United States is 1.38.

(a) Do you believe the shape of the distribution of number of television sets follows a normal distribution? Why or why not?

(b) A random sample of 40 households results in a total of 102 television sets. What is the mean number of televisions in these 40 households?

(c) What is the probability of obtaining the sample mean obtained in part (b) if the population mean is 2.24? Does the statistic from part (b) contradict the results reported by AC Nielsen?

8. Entrepreneurship A Gallup survey indicated that 72% of 18- to 29-year-olds, if given a choice, would prefer to start their own business rather than work for someone else. A random sample of 600 18- to 29-year-olds is obtained today.

(a) Is the variable start own business versus work for someone else qualitative or quantitative?

(b) Describe the sampling distribution of $\hat{p}$, the sample proportion of 18- to 29-year-olds who would prefer to start their own business. Explain the source of variability in the sample proportion.

(c) In a random sample of 600 18- to 29-year-olds, what is the probability that no more than 70% would prefer to start their own business?

(d) Would it be unusual if a random sample of 600 18- to 29-year-olds resulted in 450 or more who would prefer to start their own business?

9. Advanced Degrees According to the U.S. Census Bureau, in 2009, 10% of adults 25 years and older in the United States had advanced degrees. A researcher with the U.S. Department of Education surveys 500 randomly selected adults 25 years of age or older and finds that 60 of them have an advanced degree. Explain why this is not necessarily evidence that the proportion of adults 25 years of age or older with advanced degrees has increased.

10. Variability in Baseball Suppose, during the course of a typical season, a batter has 500 at-bats. This means the player has the opportunity to get a hit 500 times during the course of a season. Further, suppose a batter is a career 0.280 hitter (he averages 280 hits every 1000 at-bats or he has 280 successes in 1000 trials of the experiment), so the population proportion of hits is 0.280.

(a) Assuming each at-bat is an independent event, describe the sampling distribution of $\hat{p}$, the proportion of hits in 500 at-bats over the course of a season.

(b) Would it be unusual for a player who is a career 0.280 hitter to have a season in which he hits at least 0.310?

(c) Would it be unusual for the player who hit 0.310 one season to hit below 0.255 the following season?

(d) Explain why a career 0.280 hitter could easily have a batting average between 0.260 and 0.300.

(e) Use the result of part (d) to explain that a player who hit 0.260 in a season may not be a worse player than one who hit 0.300.

 **Chapter Test**

1. State the Central Limit Theorem.

2. If a random sample of size 36 is obtained from a population with mean 50 and standard deviation 24, what is the mean and standard deviation of the sampling distribution of the sample mean?

3. The charge life of a certain lithium ion battery for camcorders is normally distributed, with mean 90 minutes and standard deviation 35 minutes.

(a) What is the probability that a randomly selected battery of this type lasts more than 100 minutes on a single charge? Is this result unusual?

(b) Describe the sampling distribution of $\bar{x}$, the sample mean charge life for a random sample of 10 such batteries.

(c) What is the probability that a random sample of 10 such batteries has a mean charge life of more than 100 minutes? Is this result unusual?

(d) What is the probability that a random sample of 25 such batteries has a mean charge life of more than 100 minutes?

(e) Explain what causes the probabilities in parts (c) and (d) to be different.

4. A machine used for filling plastic bottles with a soft drink has a known standard deviation of $\sigma = 0.05$ liter. The target mean fill volume is $\mu = 2.0$ liters.

(a) Describe the sampling distribution of $\bar{x}$, the sample mean fill volume, for a random sample of 45 such bottles.

(b) A quality-control manager obtains a random sample of 45 bottles. He will shut down the machine if the sample mean of these 45 bottles is less than 1.98 liters or greater than 2.02 liters. What is the probability that the quality-control manager will shut down the machine even though the machine is correctly calibrated?

5. According to the National Center for Health Statistics, 22.4% of adults are smokers. A random sample of 300 adults is obtained.

(a) Describe the sampling distribution of $\hat{p}$, the sample proportion of adults who smoke.

(b) In a random sample of 300 adults, what is the probability that at least 50 are smokers?

(c) Would it be unusual if a random sample of 300 adults results in 18% or less being smokers?

6. Peanut and tree nut allergies are considered to be the most serious food allergies. According to the National Institute of Allergy and Infectious Diseases, roughly 1% of Americans are allergic to peanuts or tree nuts. A random sample of 1500 Americans is obtained.

(a) Explain why a large sample is needed for the distribution of the sample proportion to be approximately normal.

(b) Would it be unusual if a random sample of 1500 Americans results in fewer than 10 with peanut or tree nut allergies?

7. Net worth is defined as total assets (value of house, cars, money, etc.) minus total liabilities (mortgage balance, credit card debt, etc.). According to a recent study by TNS Financial Services, 7% of American households had a net worth in excess of $1 million (excluding their primary residence). A random sample of 1000 American households results in 82 having a net worth in excess of $1 million. Explain why the results of this survey do not necessarily imply that the proportion of households with a net worth in excess of $1 million has increased.

Making an Informed Decision

How Much Time Do You Spend in a Day … ?

The American Time Use Survey is a survey of adult Americans conducted by the Bureau of Labor Statistics. The purpose of the survey is to learn how Americans allocate their time in a day. As a reporter for the school newspaper, you wish to file a report that compares the typical student at your school to other Americans.

1. Go to the American Time Use Survey web page. Research variables of interest to you (or that you believe would be of interest to your readers). Determine the mean amount of time Americans who are enrolled in school spend doing the activities you find interesting.

2. Use StatCrunch or an online polling site (such as surveymonkey.com) to conduct a survey of a random sample of students at your school. Write questions to learn how students at your school use their time. For example, how much time does a student at your school spend attending

class each day? Be very careful about the wording in your survey questions to avoid confusion about what is being asked.

3. For each question, describe the sampling distribution of the sample mean. Use the national norms as the population mean for each variable. Use the sample standard deviation from your survey sample as the population standard deviation.

4. Compute probabilities regarding values of the statistics obtained from the study. Are any of the results unusual?

5. Write an article for your newspaper reporting your findings. Be sure any interpretations are written so they are statistically accurate, but understandable for the statistically untrained reader.

 **CASE STUDY** Sampling Distribution of the Median

The exponential distribution is an example of a skewed distribution. It is defined by the density function

$$f(x) = \begin{cases} \lambda e^{-\lambda x} & x \geq 0 \\ 0 & x < 0 \end{cases}$$

with parameter λ. The mean of an exponential distribution is $\mu = \dfrac{1}{\lambda}$ and the median is $M = \dfrac{\ln 2}{\lambda}$.

The exponential distribution is used to model the failure rate of electrical components. For many electrical components, a plot of the failure rate over time appears to be shaped like a bathtub. The exponential distribution is used widely in reliability engineering for the constant hazard rate portion of the bathtub curve, where components spend most of their useful life. The portion of the curve prior to the constant rate is referred to as the burn-in period, and the portion after the constant rate is referred to as the wear-out period.

Another interesting property of the exponential distribution is the *memoryless property*. This says that the time to failure of a component does not depend on how long the component has been in use. For example, the probability that a component will fail in the next hour is the same whether the component has been in use for 10, 50, or 100 hours, and so on.

In our discussion of summary statistics, we have seen that the mean is used as a measure of the center for data sets whose distribution is (roughly) symmetric with no outliers. In the presence of outliers or skewness, we used the median as the measure of the center because it is resistant to extreme values. However, for skewed distributions, it is often difficult to determine which measure of center is more useful. In some cases, the mean, though not resistant, is easier to find than the median or is more intuitive to the reader within the context of the problem. Here we look at the sampling distributions of the mean and median for two different populations, one normal (symmetric) and one exponential (skewed).

1. Using statistical software, generate 250 samples (rows) of size $n = 80$ (columns) from a normal distribution with mean 50 and standard deviation 10.

2. Compute the mean for the first 10 values in each row and store the values in C81. You might label this column "n10mn" for normal distribution, sample size 10, mean.

3. Compute the median for the first 10 values in each row and store the values in C82. You might label this column "n10med" for normal distribution, sample size 10, median.

4. Repeat parts 2 and 3 for samples of size 20 (C1–C20), 40 (C1–C40), and 80 (C1–C80), storing the results in consecutive columns C83, C84, C85, and so on.

5. Compute the summary statistics for your data in columns C81–C88.

6. How do the averages of your sample means compare to the actual population mean for the different sample sizes? How do the averages of your sample medians compare to the actual population median for the different sample sizes?

7. How does the standard deviation of sample means compare to the standard deviation of sample medians for different sample sizes?

8. Based on your results in parts 6 and 7, which measure of center seems more appropriate? Explain.

9. Construct histograms for each column of summary statistics. Describe the effect, if any, of increasing the sample size on the shape of the distribution of sample means and sample medians.

10. Using statistical software, generate 250 samples of size $n = 80$ from an exponential distribution with mean 50. This could represent the distribution of failure times for a component with a failure rate of $\lambda = 0.02$.

   Note: The threshold should be set to 0. Store the results in columns C101–C180.

11. Compute the mean for the first 10 values in each row and store the values in C181. You might label this column "e10mn" for exponential distribution, sample size 10, mean.

12. Compute the median for the first 10 values in each row and store the values in C182. You might label this column "e10med" for exponential distribution, sample size 10, median.

13. Repeat parts 11 and 12 for samples of size 20 (C101–C120), 40 (C101–C140), and 80 (C101–C180), storing the results in consecutive columns C183, C184, C185, and so on.

14. Compute the summary statistics for your data in columns C181–C188.

15. How do the averages of your sample means compare to the actual population mean for the different sample sizes? How do the averages of your sample medians compare to the actual population median for the different sample sizes?

16. How does the standard deviation of sample means compare to the standard deviation of sample medians for different sample sizes?

17. Based on your results in parts 15 and 16, which measure of center seems more appropriate? Explain.

18. Construct histograms for each column of summary statistics. Describe the effect, if any, of increasing the sample size on the shape of the distribution of sample means and sample medians.

19. Can the Central Limit Theorem be used to explain any of the results in part 18? Why or why not?

9

Estimating the Value of a Parameter

Outline

Making an Informed Decision

A home purchase is one of the biggest investment decisions we make in our lifetime. Buying a home involves consideration of location, price, features, and neighborhoods. This decision could also affect your family's social life. See the Decision Project on page 470.

PUTTING IT TOGETHER

Chapters 1 through 7 laid the groundwork for the remainder of the course. These chapters dealt with data collection (Chapter 1), descriptive statistics (Chapters 2 through 4), and probability (Chapters 5 through 7). Chapter 8 formed a bridge between probability and statistical inference by giving us models we can use to make probability statements about the sample mean and sample proportion.

We know from Section 8.1 that the sample mean is a random variable and has a distribution associated with it. This distribution is called the sampling distribution of the sample mean. The mean of this distribution is equal to the mean of the population, μ, and the standard deviation is $\frac{\sigma}{\sqrt{n}}$. The shape of the distribution of the sample mean is normal if the population is normal; it is approximately normal if the sample size is large. We learned in Section 8.2 that $\hat{p}$ is also a random variable whose mean is p and whose standard deviation is $\sqrt{\frac{p\,(1-p)}{n}}$. If $np(1-p) \geq 10$, the distribution of the random variable $\hat{p}$ is approximately normal.

We now discuss inferential statistics—the process of generalizing information obtained from a sample to a population. We will study two areas of inferential statistics: (1) estimation—sample data are used to estimate the value of unknown parameters such as μ or p, and (2) hypothesis testing—statements regarding a characteristic of one or more populations are tested using sample data. In this chapter, we discuss estimation of an unknown parameter, and in the next chapter we discuss hypothesis testing.

9.1   Estimating a Population Proportion

Preparing for This Section   Before getting started, review the following:

- Parameter versus statistic (Section 1.1, p. 5)
- Simple random sampling (Section 1.3, pp. 21–25)
- Sampling error (Section 1.5, p. 39)
- z_α notation (Section 7.2, pp. 372–373)

- Finding the value of a normal random variable (Section 7.2, pp. 370–372)
- Distribution of the sample proportion (Section 8.2, pp. 408–412)

Objectives

  ❶   Obtain a point estimate for the population proportion

  ❷   Construct and interpret a confidence interval for the population proportion

  ❸   Determine the sample size necessary for estimating a population proportion within a specified margin of error

❶ Obtain a Point Estimate for the Population Proportion

Suppose we want to estimate the proportion of adult Americans who believe that the amount they pay in federal income taxes is too high. It is unreasonable to expect that we could survey every adult American. Instead, we use a sample of adult Americans to arrive at an estimate of the proportion. We call this estimate a *point estimate*.

Definition   A **point estimate** is the value of a statistic that estimates the value of a parameter.

For example, the point estimate for the population proportion is $\hat{p} = \dfrac{x}{n}$, where x is the number of individuals in the sample with a specified characteristic and n is the sample size.

EXAMPLE 1   Obtaining a Point Estimate of a Population Proportion

Problem   The Gallup Organization conducted a poll in which a simple random sample of 1016 Americans 18 and older were asked, "Do you consider the amount of federal income tax you have to pay is fair?" Of the 1016 adult Americans surveyed, 558 said yes. Obtain a point estimate for the proportion of Americans 18 and older who believe the amount of federal income tax they pay is fair.

Approach   The point estimate of the population proportion is $\hat{p} = \dfrac{x}{n}$, where $x = 558$ and $n = 1016$.

Solution   Substituting into the formula, we get $\hat{p} = \dfrac{x}{n} = \dfrac{558}{1016} = 0.549 = 54.9\%$. We

estimate that 54.9% of Americans 18 and older believe that the amount of federal income tax they have to pay is fair.

● Now Work Problem 25(a)

Note: We agree to round proportions to three decimal places.

❷ Construct and Interpret a Confidence Interval for the Population Proportion

Based on the point estimate of Example 1, can we conclude that a majority (more than 50%) of the United States adult population believes that the amount of federal income tax they have to pay is fair? Or is it possible that less than a majority of adult Americans believe that their federal income tax is fair, and we just happened to sample folks who believe they pay a fair amount in taxes? After all, statistics such as $\hat{p}$ vary from sample to sample. So a different random sample of adult Americans might result in a different point estimate of the population proportion, such as $\hat{p} = 0.498$. If the method used to select the adult Americans was done appropriately, both point estimates would be good guesses of the population proportion. Due to variability in the sample proportion, we need to report a range (or *interval*) of values, including a measure of the likelihood that the interval includes the unknown population proportion.

To understand the idea of this interval, consider the following situation. Suppose you were asked to guess the proportion of students on your campus who use Facebook. If a survey of 80 students results in 60 who use Facebook, then $\hat{p} = 0.75$. From this, you might guess that the proportion of *all* students on your campus who use Facebook is 0.75 but since you did not survey every student on campus, your estimate may be incorrect. To account for this error, you might adjust your guess by stating that the proportion of students on your campus who use Facebook is 0.75, give or take 0.05 (the *margin of error*). Mathematically, we write this as 0.75 ± 0.05. If asked how confident you are that the proportion is between 0.70 and 0.80, you might respond, "I am 90% confident that the proportion of students on my campus who use Facebook is between 0.70 and 0.80." If you want an interval for which your confidence increases to, say, 95%, what do you think will happen to the interval? Having more confidence that the interval will capture the unknown population proportion requires that your interval will need to increase to, say, 0.65 to 0.85.

In statistics, we construct an interval for a population parameter based on a guess along with a level of confidence. The guess is the point estimate of the population parameter, and the level of confidence plays a role in the width of the interval. See Figure 1.

In Other Words
The symbol $\pm$ is read "plus or minus." It means "to add and subtract the quantity following the $\pm$ symbol."

Figure 1

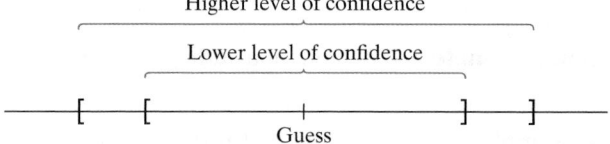

In statistics, we construct an interval for a population parameter based on a guess along with a level of confidence. The guess is the point estimate of the population parameter, and the level of confidence plays a role in the width of the interval.

Definitions

- A **confidence interval** for an unknown parameter consists of an interval of numbers based on a point estimate.
- The **level of confidence** represents the expected proportion of intervals that will contain the parameter if a large number of different samples is obtained. The level of confidence is denoted $(1 - \alpha) \cdot 100\%$.

In Other Words
A confidence interval is a range of numbers, such as 22–30. The level of confidence is the proportion of intervals that will contain the unknown parameter if repeated samples are obtained.

For example, a 95% level of confidence ($\alpha = 0.05$) implies that if 100 different confidence intervals are constructed, each based on a different sample from the same population, then we will expect 95 of the intervals to include the parameter and 5 not to include the parameter.

Confidence interval estimates for the population proportion are of the form

$$\text{point estimate} \pm \text{margin of error}$$

For example, in the poll in which Gallup was estimating the proportion of adult Americans who believe that the amount of income tax they pay is fair, the point estimate is $\hat{p} = 0.549$. From the report, Gallup indicated that the margin of error was 0.04. Gallup also reported the level of confidence to be 95%. So Gallup is 95% confident that the proportion of adult Americans who believe that the amount they pay in income tax is fair is between $0.509 (= 0.549 - 0.04)$ and $0.589 (= 0.549 + 0.04)$. Here are some unanswered questions at this point in the discussion:

- Why does the level of confidence represent the expected proportion of intervals that contain the parameter if a large number of different samples is obtained?
- How is the margin of error determined?

To help answer these questions, let's review what we know about the model that describes the sampling distribution of $\hat{p}$, the sample proportion.

- The shape of the distribution of all possible sample proportions is approximately normal provided $np(1 - p) \geq 10$ and the sample size is no more than 5% of the population size.
- The mean of the distribution of the sample proportion equals the population proportion. That is, $\mu_{\hat{p}} = p$.
- The standard deviation of the distribution of the sample proportion (the standard error) is $\sigma_{\hat{p}} = \sqrt{\dfrac{p(1 - p)}{n}}$.

Because the distribution of the sample proportion is approximately normal, we know that 95% of all sample proportions will lie within 1.96 standard deviations of the population proportion, p, and 2.5% of the sample proportions will lie in each tail. See Figure 2. The 1.96 comes from the fact that $z_{0.025}$ is the z-value such that 2.5% of the area under the standard normal curve is to its right. Recall that $z_{0.025} = 1.96$ and $-z_{0.025} = -1.96$.

Figure 2

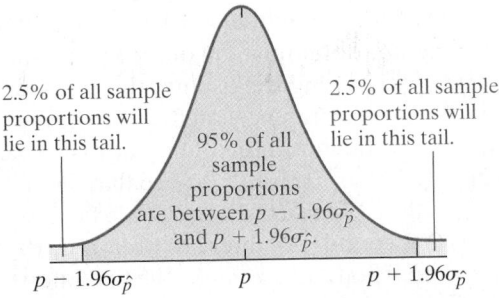

2.5% of all sample proportions will lie in this tail.

95% of all sample proportions are between $p - 1.96\sigma_{\hat{p}}$ and $p + 1.96\sigma_{\hat{p}}$.

2.5% of all sample proportions will lie in this tail.

$p - 1.96\sigma_{\hat{p}}$ p $p + 1.96\sigma_{\hat{p}}$

From Figure 2, we see that 95% of all sample proportions are in the following inequality:

$$p \quad - \quad 1.96\sigma_{\hat{p}} \quad < \quad \hat{p} \quad < \quad p \quad + \quad 1.96\sigma_{\hat{p}}$$

parameter $-$ 1.96 standard error $<$ point estimate $<$ parameter $+$ 1.96 standard error

With a little algebraic manipulation, we can rewrite this inequality with p in the middle and obtain the following:

$$\hat{p} \quad - \quad 1.96\sigma_{\hat{p}} \quad < \quad p \quad < \quad \hat{p} \quad + \quad 1.96\sigma_{\hat{p}} \quad \textbf{(1)}$$

point estimate $-$ 1.96 standard error $<$ parameter $<$ point estimate $+$ 1.96 standard error

This inequality states that 95% of *all* sample proportions will result in confidence interval estimates that contain the population proportion, whereas 5% of *all* sample proportions will result in confidence interval estimates that do not contain the population proportion. It is common to write the 95% confidence interval as

$$\hat{p} \quad \pm \quad 1.96\sigma_{\hat{p}}$$

point estimate $\pm$ 1.96 standard error

point estimate $\pm$ margin of error

So the **margin of error** for a 95% confidence interval for the population proportion is $1.96\sigma_{\hat{p}}$. This determines the width of the interval.

To visually illustrate the idea of a confidence interval, draw the sampling distribution of $\hat{p}$. Now create a "slider" that shows $\hat{p} - 1.96\sigma_{\hat{p}}$ and $\hat{p} + 1.96\sigma_{\hat{p}}$. As $\hat{p}$ leaves the blue-shaded region, the corresponding confidence interval does not capture the population proportion, p. See Figure 3 on the following page.

Figure 3 tells us that for a 95% confidence interval, 95% of all sample proportions will result in a confidence interval that includes the population proportion, while 5% of all sample proportions (those in the tails) will result in a confidence interval that does not include the population proportion.

Figure 3

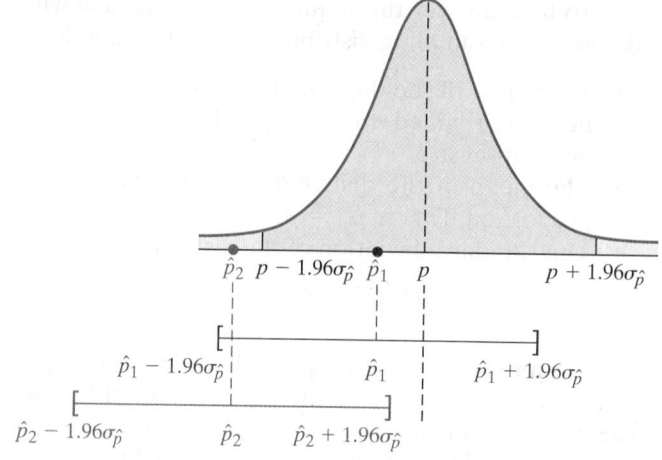

$$\hat{p}_2 \quad p - 1.96\sigma_{\hat{p}} \quad \hat{p}_1 \quad p \qquad\qquad p + 1.96\sigma_{\hat{p}}$$

$$\hat{p}_1 - 1.96\sigma_{\hat{p}} \qquad\qquad \hat{p}_1 \qquad\qquad \hat{p}_1 + 1.96\sigma_{\hat{p}}$$

$$\hat{p}_2 - 1.96\sigma_{\hat{p}} \qquad\qquad \hat{p}_2 \qquad \hat{p}_2 + 1.96\sigma_{\hat{p}}$$

EXAMPLE 2 Illustrating the Meaning of Level of Confidence Using Simulation

Let's illustrate what "95% confidence" means in a 95% confidence interval in another way. We will simulate obtaining 200 different random samples of size $n = 50$ from a population with $p = 0.7$. Figure 4 shows the confidence intervals in groups of 100. A green interval is a 95% confidence interval that includes the population proportion, 0.7. A red interval is a confidence interval that does not include the population proportion. (For now, ignore the blue intervals.) Notice that the red intervals that do not capture the population proportion 0.7 have centers that are far away (more than 1.96 standard errors) from 0.7. Of the 200 confidence intervals obtained, 10 (the red intervals) do not include the population proportion. For example, the first interval to miss has a sample proportion that is too small to result in an interval that captures 0.7. So $190/200 = 0.95$ (or 95%) of the samples have intervals that capture the population proportion.

A 95% level of confidence means that 95% of all possible samples result in confidence intervals that include the parameter (and 5% of all possible samples result in confidence intervals that do not include the parameter).

Figure 4

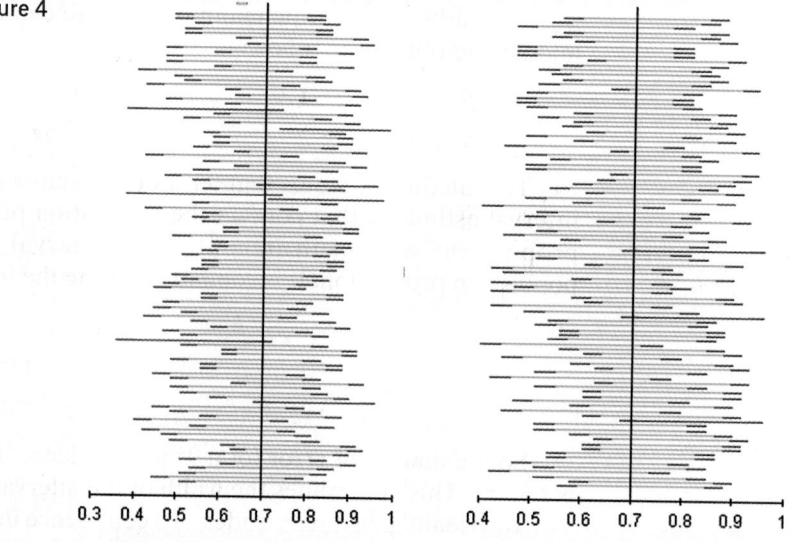

We can express the results of Example 2 in a different way. For a 95% confidence interval, any sample proportion that lies within 1.96 standard errors of the population proportion will result in a confidence interval that includes p, and any sample proportion that is more than 1.96 standard errors from the population proportion will result in a confidence interval that does not contain p.

CAUTION!

A 95% confidence interval does *not* mean that there is a 95% probability that the interval contains the parameter (such as p or μ). Remember, probability describes the likelihood of undetermined events. Therefore, it does not make sense to talk about the probability that the interval contains the parameter since the parameter is a fixed value. Think of it this way: I flip a coin and obtain a head. If I ask you to determine the probability that the flip resulted in a head, it would not be 0.5, because the outcome has already been determined. Instead, the probability is 0 or 1. Confidence intervals work the same way. Because p or μ is already determined, we do not say that there is a 95% probability that the interval contains p or μ.

Whether a confidence interval contains the population parameter depends solely on the value of the sample statistic. Any sample statistic that is in the tails of the sampling distribution will result in a confidence interval that does not include the population parameter.

In practice, we construct only one confidence interval. We do not know whether the sample results in a confidence interval that includes the parameter, but we do know that if we construct a 95% confidence interval, it will include the unknown parameter 95% of the time.

We need a method for constructing any $(1 - \alpha) \cdot 100\%$ confidence interval. [When $\alpha = 0.05$, we are constructing a $(1 - 0.05) \cdot 100\% = 95\%$ confidence interval.]

We generalize Formula (1) on page 423 by first noting that $(1 - \alpha) \cdot 100\%$ of all sample proportions are in the interval

$$p - z_{\frac{\alpha}{2}} \cdot \sqrt{\frac{p(1 - p)}{n}} < \hat{p} < p + z_{\frac{\alpha}{2}} \cdot \sqrt{\frac{p(1 - p)}{n}}$$

as shown in Figure 5.

Figure 5

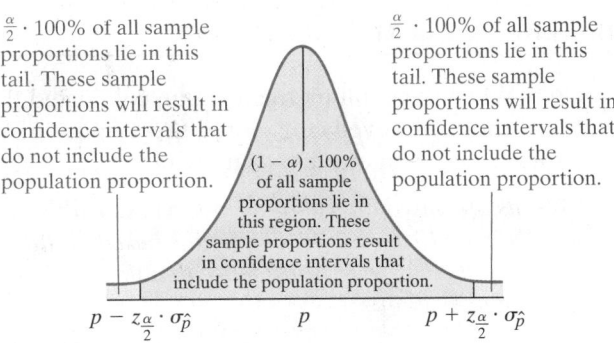

$\frac{\alpha}{2} \cdot 100\%$ of all sample proportions lie in this tail. These sample proportions will result in confidence intervals that do not include the population proportion.

$\frac{\alpha}{2} \cdot 100\%$ of all sample proportions lie in this tail. These sample proportions will result in confidence intervals that do not include the population proportion.

$(1 - \alpha) \cdot 100\%$ of all sample proportions lie in this region. These sample proportions result in confidence intervals that include the population proportion.

$p - z_{\frac{\alpha}{2}} \cdot \sigma_{\hat{p}}$ p $p + z_{\frac{\alpha}{2}} \cdot \sigma_{\hat{p}}$

Rewrite this inequality with p in the middle and obtain

$$\hat{p} - z_{\frac{\alpha}{2}} \cdot \sqrt{\frac{p(1 - p)}{n}} < p < \hat{p} + z_{\frac{\alpha}{2}} \cdot \sqrt{\frac{p(1 - p)}{n}}$$

So $(1 - \alpha) \cdot 100\%$ of all sample proportions will result in confidence intervals that contain the population proportion. The sample proportions that are in the tails of the distribution in Figure 5 will not result in confidence intervals that contain the population proportion.

The value $z_{\frac{\alpha}{2}}$ is called the **critical value** of the distribution. It represents the number of standard deviations the sample statistic can be from the parameter and still result in an interval that includes the parameter. Table 1 shows some of the common critical values used in the construction of confidence intervals. Notice that higher levels of confidence correspond to higher critical values. After all, if your level of confidence that the interval includes the unknown parameter increases, the width of your interval (using the margin of error) should increase.

NOTE

If you use technology to find the critical value for a 99% confidence interval, it will be 2.576.

Table 1		
Level of Confidence, $(1 - \alpha) \cdot 100\%$	**Area in Each Tail, $\dfrac{\alpha}{2}$**	**Critical Value, $z_{\frac{\alpha}{2}}$**
90%	0.05	1.645
95%	0.025	1.96
99%	0.005	2.575

Interpretation of a Confidence Interval

A $(1 - \alpha) \cdot 100\%$ confidence interval indicates that $(1 - \alpha) \cdot 100\%$ of all simple random samples of size n from the population whose parameter is unknown will result in an interval that contains the parameter.

In Other Words
The interpretation of a confidence interval is this: We are (*insert level of confidence*) confident that the population proportion is between (*lower bound*) and (*upper bound*). This is an abbreviated way of saying that the method is correct $(1 - \alpha) \cdot 100\%$ of the time.

For example, a 90% confidence interval for a parameter suggests that 90% of all possible samples will result in an interval that includes the unknown parameter and 10% of the samples will result in an interval that does not capture the parameter.

Look back at Figure 4 on page 424 from Example 2. The intervals, *including* the blue parts, represent 99% confidence intervals. Any interval entirely in red is an interval whose 99% confidence interval does not include the population proportion 0.7. We can see that only 2 of the 200 intervals constructed do not include 0.7, so 198/200 or 99% of the 200 intervals do include the population proportion. Therefore, a 99% confidence interval for a parameter means that 99% of all possible samples result in an interval that includes the parameter and 1% of the samples result in an interval that does not capture the parameter.

An extremely important point is that the level of confidence refers to the confidence in the *method*, not in the specific interval. A 90% confidence interval means the method "works" (that is, the interval includes the unknown parameter) for 90% of all samples. So, we do not know whether a particular sample statistic obtained is one of the 90% with an interval that includes the parameter, or one of the 10% whose interval does not include the parameter. **A 90% level of confidence *does not* tell us there is a 90% probability the parameter lies between the lower and upper bound.**

EXAMPLE 3　Interpreting a Confidence Interval

Problem When the Gallup Organization conducted the poll introduced in Example 1, 54.9% of those surveyed considered the amount of federal income tax they have to pay as fair. Gallup reported its "survey methodology" as follows:

> *Results are based on telephone interviews with a random sample of 1016 national adults, aged 18 and older. For results based on the total sample of national adults, one can say with 95% confidence that the maximum margin of sampling error is 4 percentage points.*

Determine and interpret the confidence interval for the proportion of Americans aged 18 and older who believe the amount of federal income tax they have to pay is fair.

Approach Confidence intervals for a proportion are of the form point estimate $\pm$ margin of error. So add and subtract the margin of error from the point estimate to obtain the confidence interval. Interpret the confidence interval, "We are 95% confident that the proportion of Americans aged 18 and older who believe that the amount of federal income tax they have to pay is fair is between *lower bound* and *upper bound*."

Solution The point estimate is 0.549, and the margin of error is 0.04. The confidence interval is 0.549 ± 0.04. Therefore, the lower bound of the confidence interval is $0.549 - 0.04 = 0.509$ and the upper bound of the confidence interval is $0.549 + 0.04 = 0.589$. We are 95% confident that the proportion of Americans aged 18 and older who believe that the amount of federal income tax they have to pay is fair is between 0.509 and 0.589.

● Now Work Problem 21

We are now prepared to present a method for constructing a $(1 - \alpha) \cdot 100\%$ confidence interval about the population proportion, p.

Constructing a $(1 - \alpha) \cdot 100\%$ Confidence Interval for a Population Proportion

Suppose that a simple random sample of size n is taken from a population or the data are the result of a randomized experiment. A $(1 - \alpha) \cdot 100\%$ confidence interval for p is given by the following quantities:

Lower bound: $\hat{p} - z_{\frac{\alpha}{2}} \cdot \sqrt{\dfrac{\hat{p}(1 - \hat{p})}{n}}$　　　Upper bound: $\hat{p} + z_{\frac{\alpha}{2}} \cdot \sqrt{\dfrac{\hat{p}(1 - \hat{p})}{n}}$　　**(2)**

Note: It must be the case that $n\hat{p}(1 - \hat{p}) \geq 10$ and $n \leq 0.05N$ to construct this interval.

Notice that we use $\hat{p}$ in place of p in the standard deviation. This is because p is unknown, and $\hat{p}$ is the best point estimate of p.

The width of the interval is determined by the margin of error.

> **Definition** The **margin of error, E,** in a $(1 - \alpha) \cdot 100\%$ confidence interval for a population proportion is given by
>
> $$E = z_{\frac{\alpha}{2}} \cdot \sqrt{\frac{\hat{p}(1 - \hat{p})}{n}} \qquad (3)$$

EXAMPLE 4   Constructing a Confidence Interval for a Population Proportion

Problem In the Parent–Teen Cell Phone Survey conducted by Princeton Survey Research Associates International, 800 randomly sampled 16- to 17-year-olds living in the United States were asked whether they have ever used their cell phone to text while driving. Of the 800 teenagers surveyed, 272 indicated that they text while driving. Obtain a 95% confidence interval for the proportion of 16- to 17-year-olds who text while driving.

By Hand Approach

Step 1   Compute the value of $\hat{p}$.

Step 2   Verify that $n\hat{p}(1 - \hat{p}) \geq 10$ (the normality condition) and $n \leq 0.05N$ (the sample size is no more than 5% of the population size − the independence condition).

Step 3   Determine the critical value $z_{\frac{\alpha}{2}}$.

Step 4   Use Formula (2) to determine the lower and upper bounds of the confidence interval.

Step 5   Interpret the result.

By-Hand Solution

Step 1   There are $x = 272$ successes (teens who text while driving) out of $n = 800$ individuals in the survey, so

$$\hat{p} = \frac{x}{n} = \frac{272}{800} = 0.34$$

Step 2 $n\hat{p}(1 - \hat{p}) = 800(0.34)(1 - 0.34) = 179.52 \geq 10$. There are certainly more than 1,000,000 teenagers 16 to 17 years of age in the United States, so our sample size is definitely less than 5% of the population size. The independence requirement is satisfied.

Step 3   Because we want a 95% confidence interval, we have $\alpha = 1 - 0.95 = 0.05$, so $z_{\frac{\alpha}{2}} = z_{\frac{0.05}{2}} = z_{0.025} = 1.96$.

Step 4   Substituting into Formula (2) with $n = 800$, we obtain the lower and upper bounds of the confidence interval:

Lower bound:

$$\hat{p} - z_{\frac{\alpha}{2}} \cdot \sqrt{\frac{\hat{p}(1 - \hat{p})}{n}} = 0.34 - 1.96 \cdot \sqrt{\frac{0.34(1 - 0.34)}{800}}$$
$$= 0.34 - 0.033$$
$$= 0.307$$

Technology Approach

Step 1   By hand, compute the value of $\hat{p}$.

Step 2   Verify that $n\hat{p}(1 - \hat{p}) \geq 10$ (the normality condition) and $n \leq 0.05N$ (the sample size is no more than 5% of the population size − the independence condition).

Step 3   Use a statistical spreadsheet or graphing calculator with advanced statistical features to obtain the confidence interval. We will use StatCrunch. The steps for constructing confidence intervals using StatCrunch, Minitab, Excel, and the TI-83/84 Plus graphing calculators are given in the Technology Step-by-Step on pages 430–431.

Step 4   Interpret the result.

Technology Solution

Step 1   There are $x = 272$ successes (teens who text while driving) out of $n = 800$ individuals in the survey, so

$$\hat{p} = \frac{x}{n} = \frac{272}{800} = 0.34$$

Step 2 $n\hat{p}(1 - \hat{p}) = 800(0.34)(1 - 0.34) = 179.52 \geq 10$. There are certainly more than 1,000,000 teenagers 16 to 17 years of age in the United States, so our sample size is definitely less than 5% of the population size. The independence requirement is satisfied.

Step 3   Figure 6 shows the results obtained from StatCrunch.

Figure 6

95% confidence interval results:
p : proportion of successes for population
Method: Standard-Wald

Proportion	Count	Total	Sample Prop.	Std. Err.	L. Limit	U. Limit
p	272	800	0.34	0.016748134	0.30717427	0.37282574

(continued)

Upper bound:

$$\hat{p} + z_{\frac{\alpha}{2}} \cdot \sqrt{\frac{\hat{p}(1-\hat{p})}{n}} = 0.34 + 1.96 \cdot \sqrt{\frac{0.34(1-0.34)}{800}}$$

$$= 0.34 + 0.033$$

$$= 0.373$$

The margin of error is 0.033.

Step 5 We are 95% confident that the proportion of 16- to 17-year-olds who text while driving is between 0.307 and 0.373.

The lower bound (L. Limit) is 0.307 and the upper bound (U. Limit) is 0.373. We can find the margin of error by determining the difference between the point estimate and upper bound. Here, $E = 0.373 - 0.34 = 0.033$.

Step 4 We are 95% confident that the proportion of 16- to 17-year-olds who text while driving is between 0.307 and 0.373.

Using Technology

Confidence intervals constructed by hand may differ from those using technology because of rounding.

● **Now Work Problem 25 (b), (c), and (d)**

It is important to remember the correct interpretation of a confidence interval. The statement "95% confident" means that if 1000 samples of size 800 were taken, we would expect 950 of the intervals to contain the parameter p, while 50 will not. Unfortunately, we cannot know whether the interval we constructed in Example 4 is one of the 950 intervals that contains p or one of the 50 that does not contain p.

Note: In this text we report the interval as *Lower bound:* 0.307; *Upper bound:* 0.373. Some texts will use interval notation as in (0.307, 0.373).

The Effect of Level of Confidence on the Margin of Error

We stated on page 422 that logic suggests that a higher level of confidence leads to a wider interval.

EXAMPLE 5   Role of the Level of Confidence in the Margin of Error

Problem For the problem of estimating the proportion of 16- to 17-year-old teenagers who text while driving, determine the effect on the margin of error by increasing the level of confidence from 95% to 99%.

By-Hand Approach
With a 99% level of confidence, $\alpha = 1 - 0.99 = 0.01$. So, to compute the margin of error, E, determine the value of $z_{\frac{\alpha}{2}} = z_{\frac{0.01}{2}} = z_{0.005}$. Then substitute this value into Formula (3) with $\hat{p} = 0.34$ and $n = 800$.

By-Hand Solution
After consulting Table V (Appendix A), we determine that $z_{0.005} = 2.575$. Substituting into Formula (3),

$$E = z_{\frac{\alpha}{2}} \cdot \sqrt{\frac{\hat{p}(1-\hat{p})}{n}} = 2.575 \cdot \sqrt{\frac{0.34(1-0.34)}{800}}$$

$$= 0.043$$

Technology Approach
Construct a 99% confidence interval using a statistical spreadsheet or graphing calculator with advanced statistical features. Then use the fact that a confidence interval about the population proportion is of the form $\hat{p} \pm E$. The midpoint of the upper and lower bound gives the point estimate. The difference between the point estimate and the lower bound is the margin of error.

Technology Solution
Figure 7 shows the 99% confidence interval using a TI-84 Plus C graphing calculator.

Figure 7

```
NORMAL FLOAT AUTO REAL RADIAN MP
          1-PropZInt
 (.29686,.38314)
 p̂=.34
 n=800
```

The lower bound is 0.297 and the upper bound is 0.383.

The midpoint is $\hat{p} = \dfrac{0.297 + 0.383}{2} = 0.34$. The margin of error is $E = 0.34 - 0.297 = 0.043$.

The margin of error for the 95% confidence interval found in Example 4 is 0.033, so increasing the level of confidence increases the margin of error, resulting in a wider confidence interval.

In Other Words
As the level of confidence
increases, the margin of error
also increases.

The Effect of Sample Size on the Margin of Error

We know that larger sample sizes produce more precise estimates (the Law of Large Numbers). Given that the margin of error is $z_{\frac{\alpha}{2}} \cdot \sqrt{\dfrac{\hat{p}(1-\hat{p})}{n}}$, we can see that increasing the sample size n decreases the standard error; so the margin of error decreases. This means that larger sample sizes will result in narrower confidence intervals.

To illustrate this idea, suppose the survey conducted in Example 4 resulted in $\hat{p} = 0.34$ for the proportion of 16- to 17-year-old teenagers who text while driving, but the sample size is only 200. The margin of error would be

$$E = z_{\frac{\alpha}{2}} \cdot \sqrt{\dfrac{\hat{p}(1-\hat{p})}{n}} = 1.96 \cdot \sqrt{\dfrac{0.34(1-0.34)}{200}} = 0.066$$

In Other Words
As the sample size increases,
the margin of error decreases.

So a sample size that is one-fourth the original size causes the margin of error to double. Put another way, if the sample size is quadrupled, the margin of error will be cut in half.

③ Determine the Sample Size Necessary for Estimating a Population Proportion within a Specified Margin of Error

Suppose you want to estimate a proportion with a 3% (0.03) margin of error and 95% confidence. How many individuals should be in your sample? From Formula (3), the margin of error is $E = z_{\frac{\alpha}{2}} \cdot \sqrt{\dfrac{\hat{p}(1-\hat{p})}{n}}$. If we solve this formula for n, we obtain

$$n = \hat{p}(1-\hat{p}) \cdot \left(\dfrac{z_{\frac{\alpha}{2}}}{E}\right)^2.$$

The problem with this formula is that it depends on $\hat{p}$, and $\hat{p} = \dfrac{x}{n}$ depends on the sample size, n, which is what we are trying to determine in the first place! How do we resolve this issue? There are two possibilities: (1) We could determine a preliminary value for $\hat{p}$ based on a pilot study or an earlier study, or (2) we could let $\hat{p} = 0.5$. When $\hat{p} = 0.5$, the maximum value of $\hat{p}(1-\hat{p}) = 0.25$ is obtained, as illustrated in Figure 8. Using the maximum value gives the largest possible value of n for a given level of confidence and a given margin of error.

Figure 8

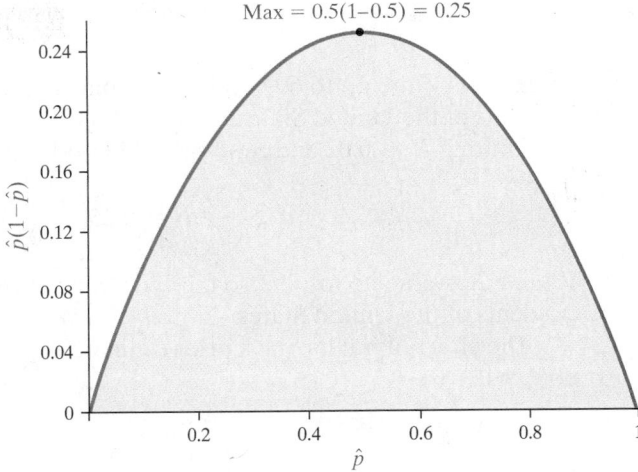

The disadvantage of the second option is that it could lead to a larger sample size than is necessary. Because of the time and expense of sampling, it is desirable to avoid too large a sample.

Sample Size Needed for Estimating the Population Proportion p

The sample size required to obtain a $(1 - \alpha) \cdot 100\%$ confidence interval for p with a margin of error E is given by

$$n = \hat{p}(1 - \hat{p})\left(\frac{z_{\frac{\alpha}{2}}}{E}\right)^2 \qquad (4)$$

rounded up to the next integer, where $\hat{p}$ is a prior estimate of p.

If a prior estimate of p is unavailable, the sample size required is

$$n = 0.25\left(\frac{z_{\frac{\alpha}{2}}}{E}\right)^2 \qquad (5)$$

rounded up to the next integer.

The margin of error should always be expressed as a decimal when using Formulas (4) and (5).

> **CAUTION!**
> Rounding *up* is different from rounding *off*. We round 12.3 up to 13; we round 12.3 off to 12.

EXAMPLE 6   Determining Sample Size

Problem An economist wants to know if the proportion of the U.S. population who commutes to work via carpooling is on the rise. What size sample should be obtained if the economist wants an estimate within 2 percentage points of the true proportion with 90% confidence if

(a) the economist uses the 2009 estimate of 10% obtained from the American Community Survey?

(b) the economist does not use any prior estimates?

Approach In both cases, $E = 0.02 (2\% = 0.02)$ and $z_{\frac{\alpha}{2}} = z_{0.1} = z_{0.05} = 1.645$. To answer part (a), let $\hat{p} = 0.10$ in Formula (4). To answer part (b), use Formula (5).

Solution

(a) Substituting $E = 0.02$, $z_{0.05} = 1.645$, and $\hat{p} = 0.10$ into Formula (4), we obtain

$$n = \hat{p}(1 - \hat{p})\left(\frac{z_{\frac{\alpha}{2}}}{E}\right)^2 = 0.10(1 - 0.10)\left(\frac{1.645}{0.02}\right)^2 = 608.9$$

Round this value up to 609, so the economist must survey 609 randomly selected residents of the United States.

> **CAUTION!**
> We always round up when determining sample size.

(b) Substituting $E = 0.02$ and $z_{0.05} = 1.645$ into Formula (5), we obtain

$$n = 0.25\left(\frac{z_{\frac{\alpha}{2}}}{E}\right)^2 = 0.25\left(\frac{1.645}{0.02}\right)^2 = 1691.3$$

Round this value up to 1692, so the economist must survey 1692 randomly selected residents of the United States.

The effect of not having a prior estimate of p is that the sample size more than doubled!

Using Technology

StatCrunch has the ability to determine sample size. See the Technology Step-by-Step on page 431.

• **Now Work Problem 35**

Technology Step-by-Step   Confidence Intervals about p

TI-83/84 Plus
1. Press STAT, highlight TESTS, and select
   A: 1-PropZInt ...
2. Enter the values of x and n.
3. Enter the confidence level following C-Level:
4. Highlight Calculate: press ENTER.

Minitab
1. If you have raw data, enter the data in column C1.
2. Select the **Stat** menu, then **Basic Statistics**, then highlight **1Proportion** ....
3. If you have raw data, select "One or more samples, each in a column" from the pull-down menu. Place the

cursor in the box, highlight the column of the data, and click Select. If you have summary statistics, select "Summarized data" from the pull-down menu. Enter the number of events (successes), x, and the number of trials, n.

4. Click the **Options** . . . button. Enter a confidence level. Select "Normal approximation" from the drop-down menu (provided that the assumptions stated are satisfied). Click OK twice.

Excel

1. Load the XLSTAT Add-in.
2. Select the XLSTAT menu. Highlight **Parametric tests**, then select **Tests for one proportion**.
3. In the Frequency cell, enter the number of successes. In the Sample size cell, enter the sample size. Be sure the Frequency radio button is checked for Data format and the z test box is checked. Click the Options tab.
4. Be sure the Sample radio button is checked under Variance and the Wald radio button is selected under Confidence Interval. For a 90% confidence interval, enter 10 for Significance Level; for a 95% confidence interval, enter 5 for Significance Level, and so on. Click OK.

StatCrunch
Confidence Intervals

1. If you have raw data, enter the data into the spreadsheet. Name the column variable.
2. Select **Stat**, then **Proportion Stats**, then **One Sample**, and then choose either **With Data** or **With Summary**.
3. If you chose **With Data**, select the column that has the observations, choose which outcome represents a success. If you chose **With Summary**, enter the number of successes and the number of trials. Choose the confidence interval radio button. Enter the level of confidence. Leave the Method as the Standard-Wald. Click Compute!.

Determining Sample Size

1. Select **Stat**, then **Proportion Stats**, then **One Sample**, and then **Power/Sample Size**.
2. Click on the "Confidence Interval Width" tab. Enter the Confidence level. For the target proportion, enter $\hat{p}$ or enter 0.5 if there is no prior estimate of p. The width is the difference between the lower bound and upper bound in the confidence interval. Therefore, the width is two times the margin of error. Clear any entry in the sample size cell. Click Compute.

 ## 9.1 Assess Your Understanding

Vocabulary and Skill Building

1. A _____ is the value of a statistic that estimates the value of a parameter.

2. The _____ represents the expected proportion of intervals that will contain the parameter if a large number of different samples is obtained. It is denoted _____.

3. *True or False:* A 95% confidence interval for a population proportion with lower bound 0.45 and upper bound 0.51 means there is a 95% probability the population proportion is between 0.45 and 0.51.

4. The value $z_{\frac{\alpha}{2}}$ represents the _____ _____ of the distribution.

5. As the level of confidence of a confidence interval increases, the margin of error _____ (increases/decreases).

6. As the sample size used to obtain a confidence interval increases, the margin of error _____ (increases/decreases).

In Problems 7–10, determine the critical value $z_{\alpha/2}$ that corresponds to the given level of confidence.

7. 90%   8. 99%

9. 98%   10. 92%

In Problems 11–14, determine the point estimate of the population proportion, the margin of error for each confidence interval, and the number of individuals in the sample with the specified characteristic, x, for the sample size provided.

11. Lower bound: 0.201, upper bound: 0.249, $n = 1200$

12. Lower bound: 0.051, upper bound: 0.074, $n = 1120$

13. Lower bound: 0.462, upper bound: 0.509, $n = 1680$

14. Lower bound: 0.853, upper bound: 0.871, $n = 10,732$

In Problems 15–20, construct a confidence interval of the population proportion at the given level of confidence.

15. $x = 30, n = 150$, 90% confidence

16. $x = 80, n = 200$, 98% confidence

17. $x = 120, n = 500$, 99% confidence

18. $x = 400, n = 1200$, 95% confidence

19. $x = 860, n = 1100$, 94% confidence

20. $x = 540, n = 900$, 96% confidence

Applying the Concepts

NW **21. You Explain It! New Deal Policies** In response to the Great Depression, Franklin D. Roosevelt enacted many New Deal policies. One such policy was the enactment of the National Recovery Administration (NRA), which required business to agree to wages and prices within their particular industry. The thought was that this would encourage higher wages among the working class, thereby spurring consumption. In a Gallup survey conducted in 1933 of 2025 adult Americans, 55% thought that wages paid to workers in industry were too low. The margin of error was 3 percentage points with 95% confidence. Which of the following represents a reasonable interpretation of the survey results? For those that are not reasonable, explain the flaw.

(a) We are 95% confident 55% of adult Americans during the Great Depression felt wages paid to workers in industry were too low.

(b) We are 92% to 98% confident 55% of adult Americans during the Great Depression felt wages paid to workers in industry were too low.

(c) We are 95% confident the proportion of adult Americans during the Great Depression who believed wages paid to workers in industry were too low was between 0.52 and 0.58.

(d) In 95% of samples of adult Americans during the Great Depression, the proportion who believed wages paid to workers in industry were too low is between 0.52 and 0.58.

22. You Explain It! Superstition A *USA Today*/Gallup poll asked l006 adult Americans how much it would bother them to stay in a room on the 13th floor of a hotel. Interestingly, 13% said it would bother them. The margin of error was 3 percentage points with 95% confidence. Which of the following represents a reasonable interpretation of the survey results? For those not reasonable, explain the flaw.

(a) We are 95% confident that the proportion of adult Americans who would be bothered to stay in a room on the 13th floor is between 0.10 and 0.16.

(b) We are between 92% and 98% confident that 13% of adult Americans would be bothered to stay in a room on the 13th floor.

(c) In 95% of samples of adult Americans, the proportion who would be bothered to stay in a room on the 13th floor is between 0.10 and 0.16.

(d) We are 95% confident that 13% of adult Americans would be bothered to stay in a room on the 13th floor.

23. You Explain It! Valentine's Day A Rasmussen Reports national survey of l000 adult Americans found that 18% dreaded Valentine's Day. The margin of error for the survey was 4.5 percentage points with 95% confidence. Explain what this means.

24. You Explain It! A Stressful Commute? A Gallup poll of 547 adult Americans employed full or part time asked, "Generally speaking, would you say your commute to work is—very stressful, somewhat stressful, not that stressful, or not stressful at all?" Gallup reported that 24% of American workers said that their commute was "very" or "somewhat" stressful. The margin of error was 4 percentage points with 95% confidence. Explain what this means.

NW 25. Giving Blood A survey of 2306 adult Americans aged 18 and older conducted by Harris Interactive found that 417 have donated blood in the past two years.

(a) Obtain a point estimate for the population proportion of adult Americans aged 18 and older who have donated blood in the past two years.

(b) Verify that the requirements for constructing a confidence interval about *p* are satisfied.

(c) Construct a 90% confidence interval for the population proportion of adult Americans who have donated blood in the past two years.

(d) Interpret the interval.

26. Saving for Retirement? A Retirement Confidence Survey of 1153 workers and retirees in the United States 25 years of age and older conducted by Employee Benefit Research Institute in January 2010 found that 496 had less than $10,000 in savings.

(a) Obtain a point estimate for the population proportion of workers and retirees in the United States 25 years of age and older who have less than $10,000 in savings.

(b) Verify that the requirements for constructing a confidence interval about *p* are satisfied.

(c) Construct a 95% confidence interval for the population proportion of workers and retirees in the United States 25 years of age and older who have less than $10,000 in savings.

(d) Interpret the interval.

27. Luxury or Necessity? A random sample of 1003 adult Americans was asked, "Do you pretty much think televisions are a necessity or a luxury you could do without?" Of the 1003 adults surveyed, 521 indicated that televisions are a luxury they could do without.

(a) Obtain a point estimate for the population proportion of adult Americans who believe that televisions are a luxury they could do without.

(b) Verify that the requirements for constructing a confidence interval about *p* are satisfied.

(c) Construct and interpret a 95% confidence interval for the population proportion of adult Americans who believe that televisions are a luxury they could do without.

(d) Is it possible that a supermajority (more than 60%) of adult Americans believe that television is a luxury they could do without? Is it likely?

(e) Use the results of part (c) to construct a 95% confidence interval for the population proportion of adult Americans who believe that televisions are a necessity.

28. Family Values In a *USA Today*/Gallup poll, 768 of 1024 randomly selected adult Americans aged 18 or older stated that a candidate's positions on the issue of family values are extremely or very important in determining their vote for president.

(a) Obtain a point estimate for the proportion of adult Americans aged 18 or older for which the issue of family values is extremely or very important in determining their vote for president.

(b) Verify that the requirements for constructing a confidence interval for *p* are satisfied.

(c) Construct a 99% confidence interval for the proportion of adult Americans aged 18 or older for which the issue of family values is extremely or very important in determining their vote for president.

(d) Is it possible that the proportion of adult Americans aged 18 or older for which the issue of family values is extremely or very important in determining their vote for president is below 70%? Is this likely?

(e) Use the results of part (c) to construct a 99% confidence interval for the proportion of adult Americans aged 18 or older for which the issue of family values is not extremely or very important in determining their vote for president.

29. AndroGel Low levels of testosterone in adult males may be treated using AndroGel 1.62%. In clinical studies of 234 adult males who were being treated with AndroGel 1.62%, it was found that 26 saw their prostate-specific antigen (PSA) elevated. The PSA is a protein produced by cells of the prostate gland. *Source:* http://www.androgelpro.com/clinical-studies/default.aspx

(a) Determine a 95% confidence interval for the proportion of adult males treated with AndroGel 1.62% who will experience elevated levels of PSA.

(b) Determine a 99% confidence interval for the proportion of adult males treated with AndroGel 1.62% who will experience elevated levels of PSA.

(c) What is the impact of increasing the level of confidence on the margin of error?

30. Reading In a survey of 700 community college students, 481 indicated that they have read a book for personal enjoyment during the school year (based on data from the Community College Survey of Student Engagement).

(a) Determine a 90% confidence interval for the proportion of community college students who have read a book for personal enjoyment during the school year.

(b) Determine a 95% confidence interval for the proportion of community college students who have read a book for personal enjoyment during the school year.

(c) What is the impact of increasing the level of confidence on the margin of error?

31. Phone in the John In a survey conducted by the marketing agency 11mark, 241 of 1000 adults 19 years of age or older confessed to bringing and using their cell phone every trip to the bathroom (confessions included texting and answering phone calls).

(a) What is the sample in this study? What is the population of interest?

(b) What is the variable of interest in this study? Is it qualitative or quantitative?

(c) Based on the results of this survey, obtain a point estimate for the proportion of adults 19 years of age or older who bring their cell phone every trip to the bathroom.

(d) Explain why the point estimate found in part (c) is a statistic. Explain why it is a random variable. What is the source of variability in the random variable?

(e) Construct and interpret a 95% confidence interval for the population proportion of adults 19 years of age or older who bring their cell phone every trip to the bathroom.

(f) What ensures that the results of this study are representative of all adults 19 years of age or older?

32. Deficit Reduction The Sullivan Statistics Survey I asks, "Would you be willing to pay higher taxes if the tax revenue went directly toward deficit reduction?" Treat the survey respondents as a random sample of adult Americans. Go to www.pearsonhighered.com/sullivanstats to obtain the data file SullivanSurveyI using the file format of your choice for the version of the text you are using. The column "Deficit" has survey responses. Construct and interpret a 90% confidence interval for the proportion of adult Americans who would be willing to pay higher taxes if the revenue went directly toward deficit reduction.

33. Random Walk Go to www.pearsonhighered.com/sullivanstats to obtain the data file 9_1_33 using the file format of your choice for the version of the text you are using. The data represent the daily (for example, Monday to Tuesday) movement of Johnson & Johnson (JNJ) stock for 200 randomly selected trading days. "Up" means the stock price increased for the time period. "Down" means the stock price decreased (or was unchanged) for the time period. Each observation is a random sample of time periods. Construct and interpret a 95% confidence interval for the proportion of days JNJ stock increases.

34. Language The Sullivan Statistics Survey I asks, "Is a language other than English the primary language spoken in your home?" Treat the survey respondents as a random sample of adult Americans. Go to www.pearsonhighered.com/sullivanstats to obtain the data file SullivanSurveyI using the file format of your choice for the version of the text you are using. The column "Language" has survey responses. Construct and interpret a 95% confidence interval for the proportion of adult Americans whose primary language spoken at home is something other than English.

NW 35. High-Speed Internet Access A researcher wishes to estimate the proportion of households that have broadband Internet access. What size sample should be obtained if she wishes the estimate to be within 0.03 with 99% confidence if

(a) she uses a 2009 estimate of 0.635 obtained from the National Telecommunications and Information Administration?

(b) she does not use any prior estimates?

36. Home Ownership An urban economist wishes to estimate the proportion of Americans who own their homes. What size sample should be obtained if he wishes the estimate to be within 0.02 with 90% confidence if

(a) he uses a 2010 estimate of 0.669 obtained from the U.S. Census Bureau?

(b) he does not use any prior estimates?

37. A Penny for Your Thoughts A researcher for the U.S. Department of the Treasury wishes to estimate the percentage of Americans who support abolishing the penny. What size sample should be obtained if he wishes the estimate to be within 2 percentage points with 98% confidence if

(a) he uses a 2006 estimate of 15% obtained from a Coinstar National Currency Poll?

(b) he does not use any prior estimate?

38. Credit-Card Debt A school administrator is concerned about the amount of credit-card debt that college students have. She wishes to conduct a poll to estimate the percentage of full-time college students who have credit-card debt of $2000 or more. What size sample should be obtained if she wishes the estimate to be within 2.5 percentage points with 94% confidence if

(a) a pilot study indicates that the percentage is 34%?

(b) no prior estimates are used?

39. Football Fans A television sports commentator wants to estimate the proportion of Americans who follow professional football. What sample size should be obtained if he wants to be within 3 percentage points with 95% confidence if

(a) he uses a 2010 estimate of 53% obtained from a Harris poll?

(b) he does not use any prior estimates?

(c) Why are the results from parts (a) and (b) so close?

40. Affirmative Action A sociologist wishes to conduct a poll to estimate the percentage of Americans who favor affirmative action programs for women and minorities for admission to colleges and universities. What sample size should be obtained if she wishes the estimate to be within 4 percentage points with 90% confidence if

(a) she uses a 2003 estimate of 55% obtained from a Gallup Youth Survey?

(b) she does not use any prior estimates?

(c) Why are the results from parts (a) and (b) so close?

41. Death Penalty In a Gallup poll, 64% of the people polled answered yes to the following question: "Are you in favor of the death penalty for a person convicted of murder?" The margin of error in the poll was 3%, and the estimate was made with 95% confidence. At least how many people were surveyed?

42. Gun Control In a Gallup Poll, 44% of the people polled answered "more strict" to the following question: "Do you feel that the laws covering the sale of firearms should be made more strict, less strict, or kept as they are now?" Suppose the margin of error in the poll was 3.5% and the estimate was made with 95% confidence. At least how many people were surveyed?

43. Senate Race Gallup polled 982 likely voters immediately preceding the 2014 North Carolina senate race. The results of the survey indicated that incumbent Kay Hagan had the support of 47% of respondents, while challenger Thom Tillis had support of 46%. The poll's margin of error was 3%. Gallup suggested the race was too close to call. Use the concept of a confidence interval to explain what this means.

44. Simulation – When Model Requirements Fail A Bernoulli random variable is a variable that is either 0 (a failure) or 1 (a success). The probability of success is denoted p.

(a) Use a statistical spreadsheet to generate 1000 Bernoulli samples of size $n = 20$ with $p = 0.15$.

(b) Estimate the population proportion for each of the 1000 Bernoulli samples.

(c) Draw a histogram of the 1000 proportions from part (b). What is the shape of the histogram?

(d) Construct a 95% confidence interval for each of the 1000 Bernoulli samples using the normal model.

(e) What proportion of the intervals do you expect to include the population proportion, p? What proportion of the intervals actually captures the population proportion? Explain any differences.

To deal with issues such as the distribution of $\hat{p}$ not following a normal distribution (Problem 44), A. Agresti and B. Coull proposed a modified approach to constructing confidence intervals for a proportion. If x is the number of successes in n trials and $n = n + z_{\alpha/2}^2$ and $\tilde{p} = \dfrac{1}{n}\left(x + \dfrac{1}{2}z_{\alpha/2}^2\right)$, then $a(1 - \alpha) \cdot 100\%$ confidence interval for p is given by

Lower bound: $\quad \tilde{p} - z_{\frac{\alpha}{2}} \cdot \sqrt{\dfrac{1}{n} \cdot \tilde{p}(1 - \tilde{p})}$

Upper bound: $\quad \tilde{p} + z_{\frac{\alpha}{2}} \cdot \sqrt{\dfrac{1}{n} \cdot \tilde{p}(1 - \tilde{p})}$

45. Cauliflower? Jane wants to estimate the proportion of students on her campus who eat cauliflower. After surveying 20 students, she finds 2 who eat cauliflower. Obtain and interpret a 95% confidence interval for the proportion of students who eat cauliflower on Jane's campus using Agresti and Coull's method.

46. Walk to Work Alan wants to estimate the proportion of adults who walk to work. In a survey of 10 adults, he finds 1 who walk to work. Explain why a 95% confidence interval using the normal model yields silly results. Then compute and interpret a 95% confidence interval for the proportion of adults who walk to work using Agresti and Coull's method.

47. Putting It Together: Hand Washing The American Society for Microbiology (ASM) and the Soap and Detergent Association (SDA) jointly commissioned two separate studies, both of which were conducted by Harris Interactive. In one of the studies, 1001 adults were interviewed by telephone and asked about their handwashing habits. In the telephone interviews, 921 of the adults said they always wash their hands in public restrooms. In the other study, the hand-washing behavior of 6076 adults was inconspicuously observed within public restrooms in four U.S. cities and 4679 of the 6076 adults were observed washing their hands.

(a) In the telephone survey, what is the variable of interest? Is it qualitative or quantitative?

(b) What is the sample in the telephone survey? What is the population to which this study applies?

(c) Verify that the requirements for constructing a confidence interval for the population proportion of adults who *say* they always wash their hands in public restrooms are satisfied.

(d) Using the results from the telephone interviews, construct a 95% confidence interval for the proportion of adults who say they always wash their hands in public restrooms.

(e) In the study where hand-washing behavior was observed, what is the variable of interest? Is it qualitative or quantitative?

(f) We are told that 6076 adults were inconspicuously observed, but were not told how these adults were selected. We know randomness is a key ingredient in statistical studies that allows us to generalize results from a sample to a population. Suggest some ways randomness might have been used to select the individuals in this study.

(g) Verify the requirements for constructing a confidence interval for the population proportion of adults who actually washed their hands while in a public restroom.

(h) Using the results from the observational study, construct a 95% confidence interval for the proportion of adults who wash their hands in public restrooms.

(i) Based on your findings in parts (a) through (h), what might you conclude about the proportion of adults who say they always wash their hands versus the proportion of adults who actually wash their hands in public restrooms?

(j) Cite some sources of variability in both studies.

Explaining the Concepts

48. Explain what "95% confidence" means in a 95% confidence interval.

49. What type of variable is required to construct a confidence interval for a population proportion?

50. Explain why quadrupling the sample size causes the margin of error to be cut in half.

51. Why do polling companies often survey 1060 individuals when they wish to estimate a population proportion with a margin of error of 3% with 95% confidence?

52. Katrina wants to estimate the proportion of adult Americans who read at least 10 books last year. To do so, she obtains a simple random sample of 100 adult Americans and constructs a 95% confidence interval. Matthew also wants to estimate the proportion of adult Americans who read at least 10 books last year. He obtains a simple random sample of 400 adult Americans and constructs a 99% confidence interval. Assuming both Katrina and Matthew obtained the same point estimate, whose estimate will have the smaller margin of error? Justify your answer.

53. Two researchers, Jaime and Mariya, are each constructing confidence intervals for the proportion of a population who is left-handed. They find the point estimate is 0.13. Each independently constructed a confidence interval based on the point estimate, but Jaime's interval has a lower bound of 0.097 and an upper bound of 0.163, while Mariya's interval has a lower bound of 0.117 and an upper bound of 0.173. Which interval is wrong? Why?

54. The 114th Congress of the United States of America has 535 members, of which 108 are women. An alien lands near the U.S. Capitol and treats members of Congress as a random sample of the human race. He reports to his superiors that a 95% confidence interval for the proportion of the human race that is female has a lower bound of 0.168 and an upper bound of 0.236. What is wrong with the alien's approach to estimating the proportion of the human race that is female?

9.2   Estimating a Population Mean

Preparing for This Section Before getting started, review the following:

- Parameter versus statistic (Section 1.1, p. 5)
- z-scores (Section 3.4, p. 155–156)
- Standard normal distribution (Section 7.2, p. 366)
- Simple random sampling (Section 1.3, pp. 21–25)

- Degrees of freedom (Section 3.2, p. 135)
- Normal probability plots (Section 7.3, pp. 377–380)
- Distribution of the sample mean (Section 8.1, pp. 395–403)

Objectives   ❶ Obtain a point estimate for the population mean

❷ State properties of Student's t-distribution

❸ Determine t-values

❹ Construct and interpret a confidence interval for a population mean

❺ Determine the sample size needed to estimate a population mean within a specified margin of error

❶ Obtain a Point Estimate for the Population Mean

Remember, the goal of statistical inference is to use information obtained from a sample and generalize the results to the population being studied. As with estimating the population proportion, the first step is to obtain a point estimate of the parameter. The point estimate of the population mean, μ, is the sample mean, $\bar{x}$.

EXAMPLE 1   Computing a Point Estimate of the Population Mean

Problem The website fueleconomy.gov allows drivers to report the miles per gallon of their vehicle. The data in Table 2 show the reported miles per gallon of 2011 Ford Focus automobiles for 16 different owners. Obtain a point estimate of the population mean miles per gallon of a 2011 Ford Focus.

Approach Treat the 16 entries as a simple random sample of all 2011 Ford Focus automobiles. To find a point estimate of the population mean, compute the sample mean miles per gallon of the 16 cars.

Solution The sample mean is

Table 2			
35.7	37.2	34.1	38.9
32.0	41.3	32.5	37.1
37.3	38.8	38.2	39.6
32.2	40.9	37.0	36.0

Source: www.fueleconomy.gov

$$\bar{x} = \frac{35.7 + 32.0 + \cdots + 36.0}{16} = \frac{588.8}{16} = 36.8 \text{ miles per hour}$$

The point estimate of μ is 36.8 miles per hour. Remember, round statistics to one more decimal point than the raw data if necessary.   ●

❷ State Properties of Student's t-distribution

In Example 1, a different random sample of 16 cars would likely result in a different point estimate of μ. For this reason, we want to construct a confidence interval for the population mean, just as we did for the population proportion.

  A confidence interval for the population mean is of the form point estimate ± margin of error (just like the confidence interval for a population proportion). To determine the margin of error, we need to know the sampling distribution of the sample mean. Recall that the distribution of $\bar{x}$ is approximately normal if the population from which the sample is drawn is normal or the sample size is sufficiently large. In addition, the distribution of $\bar{x}$ has the same mean as the parent population, $\mu_{\bar{x}} = \mu$, and a standard deviation equal to the parent population's standard deviation divided by the square root of the sample size, $\sigma_{\bar{x}} = \dfrac{\sigma}{\sqrt{n}}$. Following the same logic used in constructing a confidence interval about a population proportion, our confidence interval would be

$$\text{point estimate} \pm \text{margin of error}$$

$$\bar{x} \pm z_{\frac{\alpha}{2}} \cdot \dfrac{\sigma}{\sqrt{n}}$$

This presents a problem because we need to know the population standard deviation to construct this interval. It does not seem likely that we would know the population standard deviation but not know the population mean. So what can we do? A logical option is to use the sample standard deviation, s, as an estimate of σ. Then the standard deviation of the distribution of $\bar{x}$ would be estimated by $\dfrac{s}{\sqrt{n}}$ and our confidence interval would be

$$\bar{x} \pm z_{\frac{\alpha}{2}} \cdot \dfrac{s}{\sqrt{n}} \qquad\qquad (1)$$

Unfortunately, there is a problem with this approach. The sample standard deviation, s, is a statistic and therefore will vary from sample to sample. Using the normal model to determine the critical value, $z_{\frac{\alpha}{2}}$, in the margin of error does not take into account the additional variability introduced by using s in place of σ. This is not much of a problem for large samples because the variability in the sample standard deviation decreases as the sample size increases (Law of Large Numbers), but for small samples, we have a real problem. Put another way, the z-score of $\bar{x}$, $\dfrac{\bar{x} - \mu}{\frac{\sigma}{\sqrt{n}}}$, is normally distributed with mean 0 and standard deviation 1 (provided $\bar{x}$ is normally distributed). However, $\dfrac{\bar{x} - \mu}{\frac{s}{\sqrt{n}}}$ is *not* normally distributed with mean 0 and standard deviation 1. So a new model must be used to determine the margin of error in a confidence interval that accounts for this additional variability. This leads to the story of William Gosset.

In the early 1900s, William Gosset worked for the Guinness brewery. Gosset was in charge of conducting experiments at the brewery to identify the best barley variety. When working with beer, Gosset was limited to small data sets. At the time, the model used for constructing confidence intervals about a mean was the normal model, regardless of whether the population standard deviation was known. Gosset did not know the population standard deviation, so he simply substituted the sample standard deviation for the population standard deviation as suggested by Formula (1). While doing this, he was finding that his confidence intervals did not include the population mean at the rate expected. This led Gosset to develop a model that accounts for the additional variability introduced by using s in place of σ when determining the margin of error. Guinness would not allow Gosset to publish his results under his real name (Guinness was very secretive about its brewing practices), but did allow the results to be published under a pseudonym. Gosset chose Student. So we have Student's t-distribution.

Student's t-Distribution

Suppose that a simple random sample of size n is taken from a population. If the population from which the sample is drawn follows a normal distribution, the distribution of

$$t = \dfrac{\bar{x} - \mu}{\frac{s}{\sqrt{n}}}$$

follows Student's t-distribution with $n - 1$ degrees of freedom,* where $\bar{x}$ is the sample mean and s is the sample standard deviation.

---

* The reader may wish to review the discussion of degrees of freedom in Section 3.2 on p. 135.

The interpretation of t is the same as that of the z-score. The t-statistic represents the number of *sample* standard errors $\bar{x}$ is from the population mean, μ. It turns out that the shape of the t-distribution depends on the sample size, n.

To help see how the t-distribution differs from the standard normal (or z-) distribution and the role that the sample size n plays, we will go through the following simulation.

EXAMPLE 2 Comparing the Standard Normal Distribution to the t-Distribution Using Simulation

(a) Use statistical software such as Minitab or StatCrunch to obtain 1500 simple random samples of size $n = 5$ from a normal population with $\mu = 50$ and $\sigma = 10$. Calculate the sample mean and sample standard deviation for each sample. Compute

$$z = \frac{\bar{x} - \mu_{\bar{x}}}{\frac{\sigma}{\sqrt{n}}} \quad \text{and} \quad t = \frac{\bar{x} - \mu_{\bar{x}}}{\frac{s}{\sqrt{n}}} \quad \text{for each sample. Draw histograms for both } z \text{ and } t.$$

(b) Repeat part (a) for 1500 simple random samples of size $n = 10$.

Solution

(a) We use Minitab to obtain the 1500 simple random samples and compute the 1500 sample means and sample standard deviations. We then compute $z = \dfrac{\bar{x} - \mu_{\bar{x}}}{\frac{\sigma}{\sqrt{n}}} = $

$\dfrac{\bar{x} - 50}{\frac{10}{\sqrt{5}}}$ and $t = \dfrac{\bar{x} - \mu_{\bar{x}}}{\frac{s}{\sqrt{n}}} = \dfrac{\bar{x} - 50}{\frac{s}{\sqrt{5}}}$ for each of the 1500 samples. Figure 9(a) shows the histogram for z, and Figure 9(b) shows the histogram for t.

Figure 9

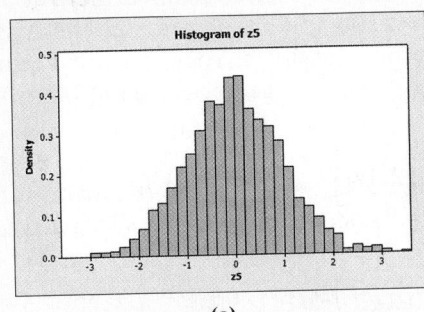
(a)

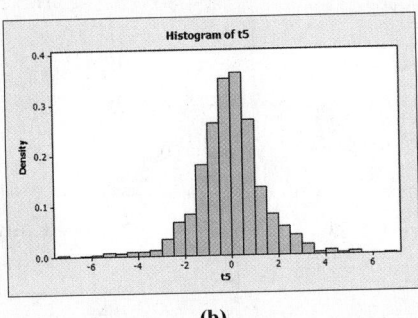
(b)

Notice that the histogram in Figure 9(a) is symmetric and bell shaped, with the histogram centered at 0, and virtually all the rectangles lying between -3 and 3. In other words, z with $n = 5$ follows a standard normal distribution. The histogram of t with $n = 5$ is also symmetric, bell shaped, and centered at 0, but the histogram of t has longer tails (that is, t is more dispersed), so it is unlikely that t follows a standard normal distribution. This additional spread is due to the fact that we divided by $\frac{s}{\sqrt{n}}$ to find t instead of by $\frac{\sigma}{\sqrt{n}}$.

(b) Repeat part (a) for samples of size $n = 10$. Figure 10(a) shows the histogram for z, and Figure 10(b) shows the histogram for t. What do you notice?

Figure 10

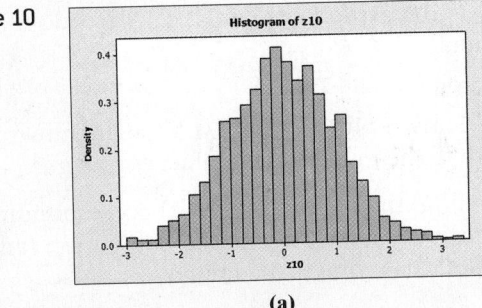
(a)

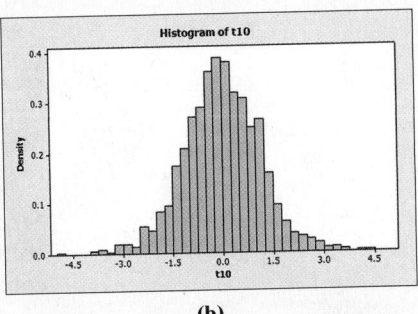
(b)

(continued)

The histogram in Figure 10(a) is symmetric, bell shaped, centered at 0, and virtually all the rectangles lie between −3 and 3. In other words, z with $n = 10$ follows a standard normal distribution. The histogram of t with $n = 10$ is also symmetric, bell shaped, and centered at 0, but the histogram has longer tails (that is, t is more dispersed) than the histogram of z with $n = 10$. So t with $n = 10$ also does not appear to follow the standard normal distribution.

One very important distinction must be made. The distribution of t with $n = 10$ (Figure 10(b)) is less dispersed than the distribution of t with $n = 5$ (Figure 9(b)).

We conclude that there are different t distributions for different sample sizes. In addition, the spread in the distribution of t decreases as the sample size n increases. In fact, it can be shown that as the sample size n increases, the distribution of t behaves more and more like the standard normal distribution.   ●

Figure 11

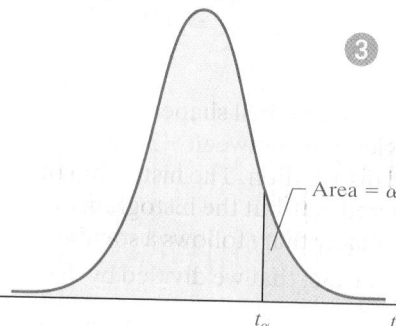

— Normal
— t with $n = 15$
— t with $n = 5$

Figure 12

> ## Properties of the *t*-Distribution
>
> 1. The t-distribution is different for different degrees of freedom.
> 2. The t-distribution is centered at 0 and is symmetric about 0.
> 3. The area under the curve is 1. The area under the curve to the right of 0 equals the area under the curve to the left of 0, which equals $\frac{1}{2}$.
> 4. As t increases or decreases without bound, the graph approaches, but never equals, zero.
> 5. The area in the tails of the t-distribution is a little greater than the area in the tails of the standard normal distribution, because we are using s as an estimate of σ, thereby introducing further variability into the t-statistic.
> 6. As the sample size n increases, the density curve of t gets closer to the standard normal density curve. This result occurs because, as the sample size increases, the values of s get closer to the value of σ, by the Law of Large Numbers.

In Figure 11, we show the t-distribution for the sample sizes $n = 5$ and $n = 15$, along with the standard normal density curve.

③ Determine *t*-Values

Recall that the notation z_α is used to represent the z-score whose area under the normal curve to the right of z_α is α. Similarly let t_α represent the t-value whose area under the t-distribution to the right of t_α is α. See Figure 12.

Area = α

The shape of the t-distribution depends on the sample size, n. Therefore, the value of t_α depends not only on α, but also on the degrees of freedom, $n - 1$. In Table VII in Appendix A, the far left column gives the degrees of freedom. The top row represents the area under the t-distribution to the right of some t-value.

t_α t

EXAMPLE 3   Finding *t*-Values

Problem Find the t-value such that the area under the t-distribution to the right of the t-value is 0.10, assuming 15 degrees of freedom (df). That is, find $t_{0.10}$ with 15 degrees of freedom.

Approach

Step 1   Draw a t-distribution with the unknown t-value labeled. Shade the area under the curve to the right of the t-value, as in Figure 12.

Step 2   Find the row in Table VII corresponding to 15 degrees of freedom and the column corresponding to an area in the right tail of 0.10. Identify where the row and column intersect. This is the unknown t-value.

Figure 13

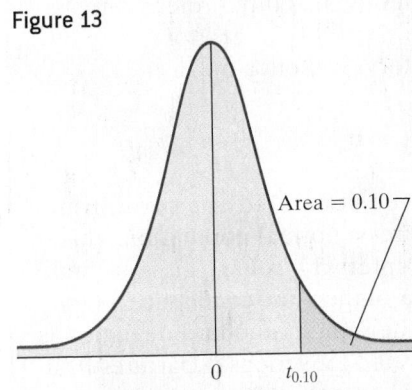

Area = 0.10

$0 \quad t_{0.10}$

Solution

Step 1 Figure 13 shows the graph of the *t*-distribution with 15 degrees of freedom. The unknown value of *t* is labeled, and the area under the curve to the right of *t* is shaded.

Step 2 A portion of Table VII is shown in Figure 14. We have enclosed the row that represents 15 degrees of freedom and the column that represents the area 0.10 in the right tail. The value where the row and column intersect is the *t*-value we are seeking. The value of $t_{0.10}$ with 15 degrees of freedom is 1.341; that is, the area under the *t*-distribution to the right of $t = 1.341$ with 15 degrees of freedom is 0.10.

Figure 14

					Area in Right Tail							
df	**0.25**	**0.20**	**0.15**	**0.10**	**0.05**	**0.025**	**0.02**	**0.01**	**0.005**	**0.0025**	**0.001**	**0.0005**
1	1.000	1.376	1.963	3.078	6.314	12.706	15.894	31.821	63.657	127.321	318.309	636.619
2	0.816	1.061	1.386	1.886	2.920	4.303	4.849	6.965	9.925	14.089	22.327	31.599
3	0.765	0.978	1.250	1.638	2.353	3.182	3.482	4.541	5.841	7.453	10.215	12.924
13	0.694	0.870	1.079	1.350	1.771	2.160	2.282	2.650	3.012	3.372	3.852	4.221
14	0.692	0.868	1.076	1.345	1.761	2.145	2.264	2.624	2.977	3.326	3.787	4.140
15	0.691	0.866	1.074	1.341	1.753	2.131	2.249	2.602	2.947	3.286	3.733	4.073
16	0.690	0.865	1.071	1.337	1.746	2.120	2.235	2.583	2.921	3.252	3.686	4.015

Notice that the critical value of *z* with an area of to the right of 0.10 is smaller— approximately 1.28. This is because the *t*-distribution has more spread than the *z*-distribution.

• **Now Work Problem 7**

⚏ Using Technology

The TI-84 Plus graphing calculator has an invT feature, which finds the value of *t* given an area to the left of the unknown *t*-value and the degrees of freedom.

If the degrees of freedom we desire are not listed in Table VII, choose the closest number in the "df" column. For example, if we have 43 degrees of freedom, we use 40 degrees of freedom from Table VII. In addition, the last row of Table VII lists the *z*-values from the standard normal distribution. Use these values when the degrees of freedom are more than 1000 because the *t*-distribution starts to behave like the standard normal distribution as *n* increases.

④ Construct and Interpret a Confidence Interval for a Population Mean

We are now ready to construct a confidence interval for a population mean.

Constructing a $(1 - \alpha) \cdot 100\%$ Confidence Interval for μ

Provided
- sample data come from a simple random sample or randomized experiment,
- sample size is small relative to the population size $(n \leq 0.05N)$, and
- the data come from a population that is normally distributed, or the sample size is large.

A $(1 - \alpha) \cdot 100\%$ confidence interval for μ is given by

$$\text{Lower bound: } \bar{x} - t_{\frac{\alpha}{2}} \cdot \frac{s}{\sqrt{n}} \qquad \qquad \text{Upper bound: } \bar{x} + t_{\frac{\alpha}{2}} \cdot \frac{s}{\sqrt{n}} \qquad \textbf{(2)}$$

where $t_{\frac{\alpha}{2}}$ is the critical value with $n - 1$ degrees of freedom.

Because this confidence interval uses the t-distribution, it is often referred to as a *t*-interval.

The **margin of error** for constructing confidence intervals about a mean is

$$E = t_{\frac{\alpha}{2}} \cdot \frac{s}{\sqrt{n}}$$

Notice that a confidence interval about μ can be computed for non-normal populations even though Student's t-distribution requires a normal population. This is because the procedure for constructing the confidence interval is **robust**—it is accurate despite minor departures from normality. If a data set has outliers, the confidence interval is not accurate because neither the sample mean nor the sample standard deviation is resistant to outliers. Sample data should always be inspected for serious departures from normality and for outliers. This is easily done with normal probability plots and boxplots.

EXAMPLE 4   Constructing a Confidence Interval about a Population Mean

Table 3

35.7	37.2	34.1	38.9
32.0	41.3	32.5	37.1
37.3	38.8	38.2	39.6
32.2	40.9	37.0	36.0

Source: www.fueleconomy.gov

Problem The website fueleconomy.gov allows drivers to report the miles per gallon of their vehicle. The data in Table 3 show the reported miles per gallon of 2011 Ford Focus automobiles for 16 different owners. Treat the sample as a simple random sample of all 2011 Ford Focus automobiles. Construct a 95% confidence interval for the mean miles per gallon of a 2011 Ford Focus. Interpret the interval.

Approach

Step 1   Verify the data are obtained randomly and the sample size is small relative to the population size. Because the sample size is small, draw a normal probability plot to verify the data come from a population that is normally distributed and a boxplot to verify that there are no outliers.

By-Hand Approach

Step 2   Compute the value of $\bar{x}$ and s.

Step 3   Determine the critical value $t_{\frac{\alpha}{2}}$ with $n - 1$ degrees of freedom.

Step 4   Use Formula (2) to determine the lower and upper bounds of the confidence interval.

Step 5   Interpret the result.

Technology Approach

Step 2   Use a statistical spreadsheet or graphing calculator with advanced statistical features to obtain the confidence interval. We will use Minitab to construct the confidence interval. The steps for constructing confidence intervals using the TI-83/84 Plus graphing calculators, Minitab, Excel, and StatCrunch are given in the Technology Step-by-Step on page 442.

Step 3   Interpret the result.

Solution

Step 1   The data are obtained from a simple random sample. In addition, there are likely thousands of 2011 Ford Focus vehicles on the road, so the sample size is small relative to the population size. Figure 15 shows a normal probability plot and boxplot for the data in Table 3. The correlation between MPG and the expected z-scores is 0.979. Because $0.979 > 0.941$ (Table VI), it is reasonable to conclude the sample data come from a population that is normally distributed. The boxplot does not reveal any outliers. The requirements for constructing the confidence interval are satisfied.

Figure 15

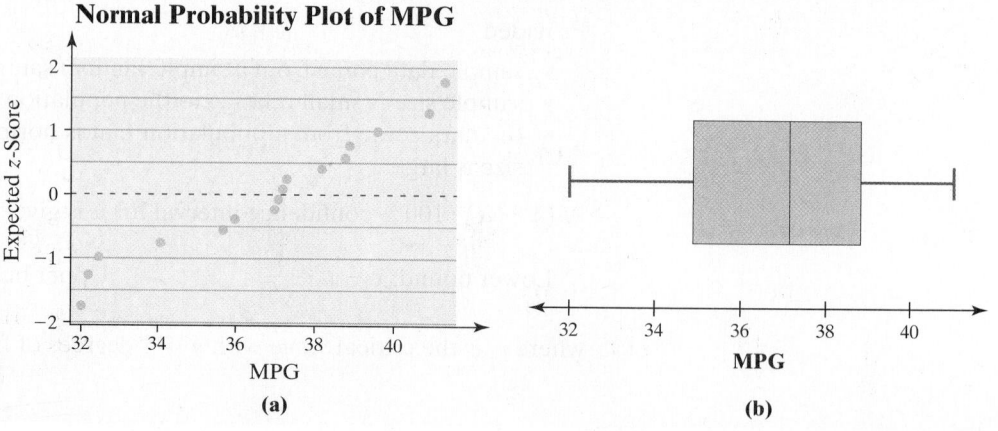

(a)   **(b)**

By-Hand Solution

Step 2   We determined the sample mean in Example 1 to be $\bar{x} = 36.8$ mpg. Using a calculator, the sample standard deviation is $s = 2.92$ mpg.

Step 3   Because we want a 95% confidence interval, we have $\alpha = 1 - 0.95 = 0.05$. The sample size is $n = 16$. So we find $t_{\frac{\alpha}{2}} = t_{\frac{0.05}{2}} = t_{0.025}$ with $16 - 1 = 15$ degrees of freedom. Table VII shows that $t_{0.025} = 2.131$.

Step 4   Substituting into Formula (2), we obtain:

Lower bound:

$$\bar{x} - t_{\frac{\alpha}{2}} \cdot \frac{s}{\sqrt{n}} = 36.8 - 2.131 \cdot \frac{2.92}{\sqrt{16}} = 35.24$$

Upper bound:

$$\bar{x} + t_{\frac{\alpha}{2}} \cdot \frac{s}{\sqrt{n}} = 36.8 + 2.131 \cdot \frac{2.92}{\sqrt{16}} = 38.36$$

Step 5   We are 95% confident that the mean miles per gallon of all 2011 Ford Focus cars is between 35.24 and 38.36 mpg.

● **Now Work Problem 31**

Technology Solution

Step 2   Figure 16 shows the results from Minitab.

Figure 16

```
One-Sample T: MPG

Variable   N    Mean  StDev  SE Mean      95% CI
MPG        16  36.800  2.917   0.729  (35.246, 38.354)
```

Minitab presents confidence intervals in the form (*lower bound, upper bound*). The lower bound is 35.25 and the upper bound is 38.35.

Step 3   We are 95% confident that the mean miles per gallon of all 2011 Ford Focus cars is between 35.25 and 38.35 mpg.

Notice that $t_{0.025} = 2.131$ for 15 degrees of freedom, while $z_{0.025} = 1.96$. The t-distribution gives a larger critical value, so the width of the interval is wider. Remember, this larger critical value is necessary to account for the increased variability due to using s as an estimate of σ.

Remember, 95% confidence refers to our confidence in the method. If we obtained 100 samples of size $n = 16$ from the population of 2011 Ford Focuses, we would expect about 95 of the samples to result in confidence intervals that include μ. We do not know whether the interval in Example 4 includes μ or does not include μ.

What should we do if the requirements to compute a t-interval are not met? We could increase the sample size beyond 30 observations, or we could try to use *nonparametric procedures*. **Nonparametric procedures** typically do not require normality, and the methods are resistant to outliers. A third option is to use resampling methods, such as bootstrapping, introduced in Section 9.5.

⑤ Determine the Sample Size Needed to Estimate a Population Mean within a Specified Margin of Error

The margin of error in constructing a confidence interval about the population mean is $E = t_{\frac{\alpha}{2}} \cdot \frac{s}{\sqrt{n}}$. Solving this for n, we obtain $n = \left(\frac{t_{\frac{\alpha}{2}} \cdot s}{E}\right)^2$. The problem with this formula is that the critical value $t_{\frac{\alpha}{2}}$ requires that we know the sample size to determine the degrees of freedom, $n - 1$. Obviously, if we do not know n we cannot know the degrees of freedom. The solution to this problem lies in the fact that the t-distribution approaches the standard normal z-distribution as the sample size increases. To convince yourself of this, look at the last few rows of Table VII and compare them with the corresponding z-scores for 95% or 99% confidence. Now if we use z in place of t and a sample standard deviation, s, from previous or pilot studies, we can write the margin of error formula as

$$E = z_{\frac{\alpha}{2}} \cdot \frac{s}{\sqrt{n}}$$ and solve it for n to obtain a formula for determining sample size.

CAUTION!
Rounding *up* is different from rounding *off*. We round 5.32 *up* to 6 and *off* to 5.

Determining the Sample Size n

The sample size required to estimate the population mean, μ, with a level of confidence $(1 - \alpha) \cdot 100\%$ within a specified margin of error, E, is given by

$$n = \left(\frac{z_{\frac{\alpha}{2}} \cdot s}{E}\right)^2 \tag{3}$$

where n is *rounded up* to the nearest whole number.

EXAMPLE 5 Determining Sample Size

Problem We again consider the problem of estimating the miles per gallon of a 2011 Ford Focus. How large a sample is required to estimate the mean miles per gallon within 0.5 mile per gallon with 95% confidence?

Using Technology

StatCrunch has the ability to determine sample size. See the Technology Step-by-Step below.

Approach Use Formula (3) with $z_{\frac{\alpha}{2}} = z_{0.025} = 1.96$, $s = 2.92$, and $E = 0.5$ to find the required sample size.

Solution Substitute the values of z, s, and E into Formula (3) and obtain

$$n = \left(\frac{z_{\frac{\alpha}{2}} \cdot s}{E}\right)^2 = \left(\frac{1.96 \cdot 2.92}{0.5}\right)^2 = 131.02$$

CAUTION!
Don't forget to round up when determining sample size.

• **Now Work Problem 41**

Round 131.02 up to 132. A sample size of $n = 132$ results in an interval estimate of the population mean miles per gallon of a 2011 Ford Focus with a margin of error of 0.5 mile per gallon with 95% confidence.

Technology Step-by-Step Confidence Intervals for μ

TI-83/84 Plus
1. If necessary, enter raw data in L1.
2. Press STAT, highlight TESTS, and select 8:TInterval.
3. If the data are raw, highlight DATA. Make sure List is set to L1 and Freq to 1. If summary statistics are known, highlight STATS and enter the summary statistics.
4. Enter the confidence level following C-Level:.
5. Highlight Calculate; press ENTER.

Minitab
1. If you have raw data, enter them in column C1.
2. Select the **Stat** menu, then **Basic Statistics**, then highlight **1-Sample** *t* . . . .
3. If you have raw data, select "One or more samples, each in a column" from the pull-down menu. Place the cursor in the box, highlight the column containing the raw data, and click "Select." If you have summarized data, select "Summarized data" from the pull-down menu and enter the summarized data. Select **Options . . .** and enter a confidence level. Click OK twice.

Excel
1. Load the XLSTAT Add-in.
2. Enter the raw data in Column A.
3. Select the **XLSTAT** menu, highlight **Parametric tests**. Select **One-sample t-test and z-test**.
4. Place the cursor in the Data cell. Highlight the raw data in the spreadsheet. Be sure the box for Student's *t* test is checked and the radio button for One sample is selected.

Click the Options tab. For a 90% confidence interval, let the Significance level (%) equal 10; for a 95% confidence interval, let the Significance level (%) equal 5, and so on. Click OK.

StatCrunch
Constructing a Confidence Interval for the Population Mean
1. If necessary, enter the raw data into column var1. Name the column.
2. Select **Stat**, highlight **T Stats**, highlight **One Sample**. Choose **With Data** if you have raw data, choose **With Summary** if you have summarized data.
3. If you chose **With Data**, highlight the column that contains the data in "Select column(s):". If you chose **With Summary**, enter the sample mean, sample standard deviation, and sample size. Choose the confidence interval radio button. Enter the level of confidence. Click Compute!.

Determining Sample Size When Estimating a Population Mean
1. Select **Stat**, highlight **Z Stats**, highlight **One Sample**, and highlight **Power/Sample Size**. Note: You may also highlight **T Stats** and follow the same steps.
2. Click on the "Confidence Interval Width" tab. Enter the Confidence level and standard deviation. The width is the difference between the lower bound and the upper bound in the confidence interval. Therefore, the width is two times the margin of error. Clear any entry in the sample size cell. Click Compute.

9.2 Assess Your Understanding

Vocabulary and Skill Building

1. As the number of degrees of freedom in the *t*-distribution increases, the spread of the distribution _____ (increases/decreases).

2. *True or False:* The *t*-distribution is centered at μ.

3. The notation t_α is the *t*-value such that the area under the *t*-distribution to the right of t_α is _____.

4. *True or False:* The value of $t_{0.10}$ with 5 degrees of freedom is greater than the value of $t_{0.10}$ with 10 degrees of freedom.

5. *True or False:* To construct a confidence interval about the mean, the population from which the sample is drawn must be approximately normal.

6. The procedure for constructing a confidence interval about a mean is _____, which means minor departures from normality do not affect the accuracy of the interval.

NW 7. (a) Find the *t*-value such that the area in the right tail is 0.10 with 25 degrees of freedom.
(b) Find the *t*-value such that the area in the right tail is 0.05 with 30 degrees of freedom.
(c) Find the *t*-value such that the area left of the *t*-value is 0.01 with 18 degrees of freedom. [*Hint:* Use symmetry.]
(d) Find the critical *t*-value that corresponds to 90% confidence. Assume 20 degrees of freedom.

8. (a) Find the *t*-value such that the area in the right tail is 0.02 with 19 degrees of freedom.
(b) Find the *t*-value such that the area in the right tail is 0.10 with 32 degrees of freedom.
(c) Find the *t*-value such that the area left of the *t*-value is 0.05 with 6 degrees of freedom. [*Hint:* Use symmetry.]
(d) Find the critical *t*-value that corresponds to 95% confidence. Assume 16 degrees of freedom.

In Problems 9–12, a simple random sample of size n < 30 for a quantitative variable has been obtained. Using the normal probability plot, the correlation between the variable and expected z-score, and the boxplot, judge whether a t-interval should be constructed.

9. $n = 12$; Correlation $= 0.987$

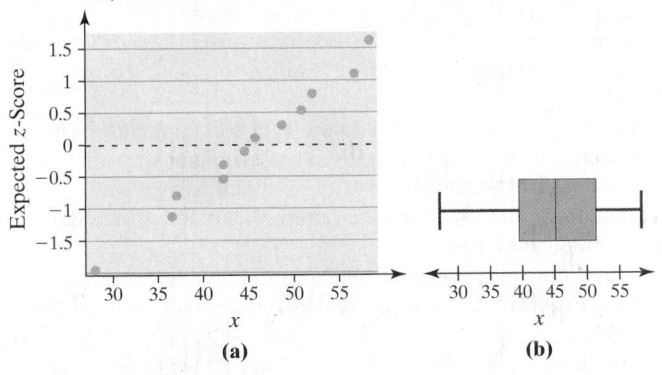

(a) (b)

10. $n = 15$; Correlation $= 0.893$

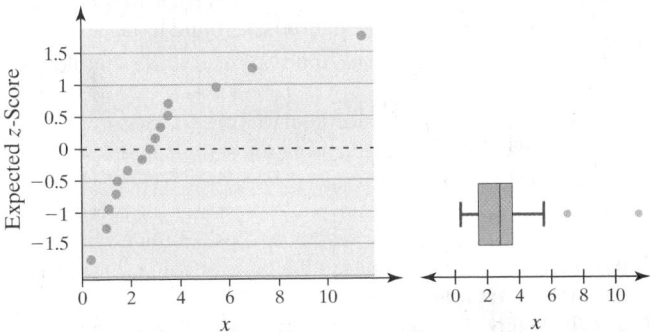

11. $n = 13$; Correlation $= 0.966$

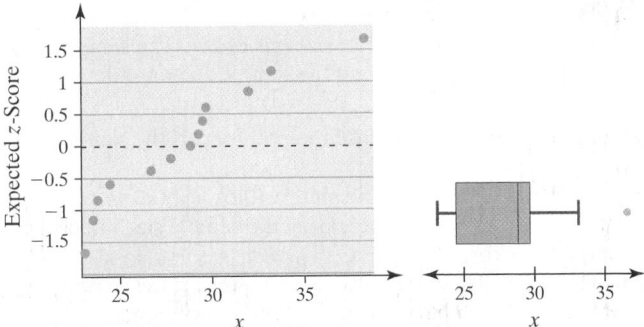

12. $n = 9$; Correlation $= 0.997$

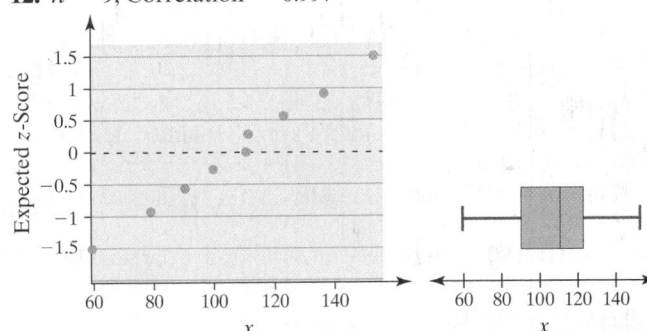

In Problems 13–16, determine the point estimate of the population mean and margin of error for each confidence interval.

13. Lower bound: 18, upper bound: 24

14. Lower bound: 20, upper bound: 30

15. Lower bound: 5, upper bound: 23

16. Lower bound: 15, upper bound: 35

17. A simple random sample of size *n* is drawn from a population that is normally distributed. The sample mean, $\bar{x}$, is found to be 108, and the sample standard deviation, *s*, is found to be 10.

(a) Construct a 96% confidence interval for μ if the sample size, *n*, is 25.

(b) Construct a 96% confidence interval for μ if the sample size, *n*, is 10. How does decreasing the sample size affect the margin of error, *E*?

(c) Construct a 90% confidence interval for μ if the sample size, *n*, is 25. Compare the results to those obtained in part (a). How does decreasing the level of confidence affect the size of the margin of error, *E*?

(d) Could we have computed the confidence intervals in parts (a)–(c) if the population had not been normally distributed? Why?

18. A simple random sample of size n is drawn from a population that is normally distributed. The sample mean, $\bar{x}$, is found to be 50, and the sample standard deviation, s, is found to be 8.

(a) Construct a 98% confidence interval for μ if the sample size, n, is 20.

(b) Construct a 98% confidence interval for μ if the sample size, n, is 15. How does decreasing the sample size affect the margin of error, E?

(c) Construct a 95% confidence interval for μ if the sample size, n, is 20. Compare the results to those obtained in part (a). How does decreasing the level of confidence affect the margin of error, E?

(d) Could we have computed the confidence intervals in parts (a)–(c) if the population had not been normally distributed? Why?

19. A simple random sample of size n is drawn. The sample mean, $\bar{x}$, is found to be 18.4, and the sample standard deviation, s, is found to be 4.5.

(a) Construct a 95% confidence interval for μ if the sample size, n, is 35.

(b) Construct a 95% confidence interval for μ if the sample size, n, is 50. How does increasing the sample size affect the margin of error, E?

(c) Construct a 99% confidence interval for μ if the sample size, n, is 35. Compare the results to those obtained in part (a). How does increasing the level of confidence affect the margin of error, E?

(d) If the sample size is $n = 15$, what conditions must be satisfied to compute the confidence interval?

20. A simple random sample of size n is drawn. The sample mean, $\bar{x}$, is found to be 35.1, and the sample standard deviation, s, is found to be 8.7.

(a) Construct a 90% confidence interval for μ if the sample size, n, is 40.

(b) Construct a 90% confidence interval for μ if the sample size, n, is 100. How does increasing the sample size affect the margin of error, E?

(c) Construct a 98% confidence interval for μ if the sample size, n, is 40. Compare the results to those obtained in part (a). How does increasing the level of confidence affect the margin of error, E?

(d) If the sample size is $n = 18$, what conditions must be satisfied to compute the confidence interval?

Applying the Concepts

21. You Explain It! Hours Worked In a survey conducted by the Gallup Organization, 1100 adult Americans were asked how many hours they worked in the previous week. Based on the results, a 95% confidence interval for mean number of hours worked was lower bound: 42.7 and upper bound: 44.5. Which of the following represents a reasonable interpretation of the result? For those that are not reasonable, explain the flaw.

(a) There is a 95% probability the mean number of hours worked by adult Americans in the previous week was between 42.7 hours and 44.5 hours.

(b) We are 95% confident that the mean number of hours worked by adult Americans in the previous week was between 42.7 hours and 44.5 hours.

(c) 95% of adult Americans worked between 42.7 hours and 44.5 hours last week.

(d) We are 95% confident that the mean number of hours worked by adults in Idaho in the previous week was between 42.7 hours and 44.5 hours.

22. You Explain It! Sleeping A 90% confidence interval for the number of hours that full-time college students sleep during a weekday is lower bound: 7.8 hours and upper bound: 8.8 hours. Which of the following represents a reasonable interpretation of the result? For those that are not reasonable, explain the flaw.

(a) 90% of full-time college students sleep between 7.8 hours and 8.8 hours.

(b) We are 90% confident that the mean number of hours of sleep that full-time college students get any day of the week is between 7.8 hours and 8.8 hours.

(c) There is a 90% probability that the mean hours of sleep that full-time college students get during a weekday is between 7.8 hours and 8.8 hours.

(d) We are 90% confident that the mean hours of sleep that full-time college students get during a weekday is between 7.8 hours and 8.8 hours.

23. You Explain It! Drive-Through Service Time The trade magazine QSR routinely checks the drive-through service times of fast-food restaurants. A 90% confidence interval that results from examining 607 customers in Taco Bell's drive-through has a lower bound of 161.5 seconds and an upper bound of 164.7 seconds. What does this mean?

24. You Explain It! MySpace.com According to Nielsen/NetRatings, the mean amount of time spent on MySpace.com per user per month in July 2014 was 171.0 minutes. A 95% confidence interval for the mean amount of time spent on MySpace.com monthly has a lower bound of 151.4 minutes and an upper bound of 190.6 minutes. What does this mean?

25. Hours Worked Revisited For the "Hours Worked" survey conducted by Gallup in Problem 21, provide two recommendations for increasing the precision of the interval.

26. Sleeping Revisited Refer to the "Sleeping" results from Problem 22. What could be done to increase the precision of the confidence interval?

27. Blood Alcohol Concentration A random sample of 51 fatal crashes in 2013 in which the driver had a positive blood alcohol concentration (BAC) from the National Highway Traffic Safety Administration results in a mean BAC of 0.167 gram per deciliter (g/dL) with a standard deviation of 0.010 g/dL.

(a) A histogram of blood alcohol concentrations in fatal accidents shows that BACs are highly skewed right. Explain why a large sample size is needed to construct a confidence interval for the mean BAC of fatal crashes with a positive BAC.

(b) In 2013, there were approximately 25,000 fatal crashes in which the driver had a positive BAC. Explain why this, along with the fact that the data were obtained using a simple random sample, satisfies the requirements for constructing a confidence interval.

(c) Determine and interpret a 90% confidence interval for the mean BAC in fatal crashes in which the driver had a positive BAC.

(d) All 50 states and the District of Columbia use a BAC of 0.08 g/dL as the legal intoxication level. Is it possible that the mean BAC of all drivers involved in fatal accidents who are found to have positive BAC values is less than the legal intoxication level? Explain.

28. Hungry or Thirsty? How much time do Americans spend eating or drinking? Suppose for a random sample of 1001 Americans age 15 or older, the mean amount of time spent eating or drinking per day is 1.22 hours with a standard deviation of 0.65 hour.
Source: American Time Use Survey conducted by the Bureau of Labor Statistics

(a) A histogram of time spent eating and drinking each day is skewed right. Use this result to explain why a large sample size is needed to construct a confidence interval for the mean time spent eating and drinking each day.

(b) There are over 200 million Americans age 15 or older. Explain why this, along with the fact that the data were obtained using a random sample, satisfies the requirements for constructing a confidence interval.

(c) Determine and interpret a 95% confidence interval for the mean amount of time Americans age 15 or older spend eating and drinking each day.

(d) Could the interval be used to estimate the mean amount of time a 9-year-old American spends eating and drinking each day? Explain.

29. Tootsie Pops A Tootsie Pop is a sucker with a candy center. A famous commercial for Tootsie Pops once asked, "How many licks to the center of a Tootsie Pop?" In an attempt to answer this question, Cory Heid of Siena Heights University asked 92 volunteers to count the number of licks required before reaching the chocolate center. The mean number of licks required was 356.1 with a standard deviation of 185.7. Find and interpret a 95% confidence interval for the number of licks required to reach the candy center of a Tootsie Pop.
Source: Heid, Cory. "Tootsie Pops: How Many Licks to the Chocolate?" *Significance*, October, 2013 Volume 10 Issue 5.

30. How Much Do You Read? A recent Gallup poll asked 1006 Americans, "During the past year, about how many books, either hardcover or paperback, did you read either all or part of the way through?" Results of the survey indicated that $\bar{x} = 13.4$ books and $s = 16.6$ books. Construct a 99% confidence interval for the mean number of books that Americans read either all or part of during the preceding year. Interpret the interval.

NW 31. pH of Rain The following data represent the pH of rain for a random sample of 12 rain dates in Tucker County, West Virginia. A normal probability plot suggests the data could come from a population that is normally distributed. A boxplot indicates there are no outliers.

4.58	5.19	5.05	4.80	4.77	4.77
5.72	4.75	5.02	4.74	4.76	4.56

Source: National Atmospheric Deposition Program

(a) Determine a point estimate for the population mean pH of rainwater in Tucker County.

(b) Construct and interpret a 95% confidence interval for the mean pH of rainwater in Tucker County, West Virginia.

(c) Construct and interpret a 99% confidence interval for the mean pH of rainwater in Tucker County, West Virginia.

(d) What happens to the interval as the level of confidence is increased? Explain why this is a logical result.

32. Travel Taxes Travelers pay taxes for flying, car rentals, and hotels. The following data represent the total travel tax for a 3-day business trip in 8 randomly selected cities. *Note:* Chicago travel taxes are the highest in the country at $101.27. A normal probability plot suggests the data could come from a population that is normally distributed. A boxplot indicates there are no outliers.

67.81	78.69	68.99	84.36
80.24	86.14	101.27	99.29

(a) Determine a point estimate for the population mean travel tax.

(b) Construct and interpret a 95% confidence interval for the mean tax paid for a three-day business trip.

(c) What would you recommend to a researcher who wants to increase the precision of the interval, but does not have access to additional data?

33. Crash Test Results The following data represent the repair cost for a low-impact collision in a simple random sample of mini- and micro-vehicles (such as the Chevrolet Aveo or Mini Cooper).

$3148	$1758	$1071	$3345	$743
$2057	$663	$2637	$773	$1370

Source: Insurance Institute for Highway Safety

(a) Draw a normal probability plot to determine if it is reasonable to conclude the data come from a population that is normally distributed.

(b) Draw boxplot to check for outliers.

(c) Construct and interpret a 95% confidence interval for the population mean cost of repair.

(d) Suppose you obtain a simple random sample of size $n = 10$ of a Mini Cooper that was in a low-impact collision and determine the cost of repair. Do you think a 95% confidence interval would be wider or narrower? Explain.

34. Crawling Babies The following data represent the age (in weeks) at which babies first crawl based on a survey of 12 mothers conducted by Essential Baby.

52	30	44	35	39	26
47	37	56	26	39	28

Source: www.essentialbaby.com

(a) Draw a normal probability plot to determine if it is reasonable to conclude the data come from a population that is normally distributed.

(b) Draw a boxplot to check for outliers.

(c) Construct and interpret a 95% confidence interval for the mean age at which a baby first crawls.

35. Housing Starts The following data represent the number of housing starts predicted for the 2nd quarter (April through June) of 2014 for a random sample of 40 economists.

984	1260	1009	992	975	993	1025	1164
1060	992	1100	942	1050	1047	1000	938
1035	1030	964	970	1061	1067	1100	1095
976	1012	1038	929	920	996	990	1095
1178	1017	980	1125	964	888	946	1004

Source: Federal Reserve Bank of Philadelphia

(a) Draw a histogram of the data. Comment on the shape of the distribution.

(b) Draw a boxplot of the data. Are there any outliers?

(c) Discuss the need for a large sample size in order to use Student's t-distribution to obtain a confidence interval for the population mean forecast of the number of housing starts in the second quarter of 2014.

(d) Construct a 95% confidence interval for the population mean forecast of the number of housing starts in the second quarter of 2014.

36. PepsiCo Stock Volume The trade volume of a stock is the number of shares traded on a given day. The following data, in millions (so that 6.16 represents 6,160,000 shares traded), represent the volume of PepsiCo stock traded for a random sample of 40 trading days in 2014.

6.16	6.39	5.05	4.41	4.16	4.00	2.37	7.71
4.98	4.02	4.95	4.97	7.54	6.22	4.84	7.29
5.55	4.35	4.42	5.07	8.88	4.64	4.13	3.94
4.28	6.69	3.25	4.80	7.56	6.96	6.67	5.04
7.28	5.32	4.92	6.92	6.10	6.71	6.23	2.42

Source: TD Ameritrade

(a) Use the data to compute a point estimate for the population mean number of shares traded per day in 2014.

(b) Construct a 95% confidence interval for the population mean number of shares traded per day in 2014. Interpret the confidence interval.

(c) A second random sample of 40 days in 2014 resulted in the data shown next. Construct another 95% confidence interval for the population mean number of shares traded per day in 2014. Interpret the confidence interval.

6.12	5.73	6.85	5.00	4.89	3.79	5.75	6.04
4.49	6.34	5.90	5.44	10.96	4.54	5.46	6.58
8.57	3.65	4.52	7.76	5.27	4.85	4.81	6.74
3.65	4.80	3.39	5.99	7.65	8.13	6.69	4.37
6.89	5.08	8.37	5.68	4.96	5.14	7.84	3.71

(d) Explain why the confidence intervals obtained in parts (b) and (c) are different.

37. Tornadoes Go to www.pearsonhighered.com/sullivanstats to obtain the data file 9_2_37 using the file format of your choice for the version of the text you are using. The data represent the width (in yards) and length (in miles) of a random sample of tornadoes that struck Oklahoma. The variable "length" represents the number of miles a particular tornado was on the ground.

(a) Draw a relative frequency histogram of the variable length using a lower class limit of the first class of 0 and a class width of 3. Comment on the shape of the distribution.

(b) Draw a boxplot of the length. Are there any outliers?

(c) Explain why a large sample is necessary to construct a confidence interval for the mean length of a tornado in Oklahoma.

(d) Use statistical software to construct and interpret a 95% confidence interval for the mean length of a tornado in Oklahoma.

38. Tax Rate The Sullivan Statistics Survey II asks, "What percent of one's income should an individual pay in federal

income taxes?" Go to www.pearsonhighered.com/sullivanstats to obtain the data file SullivanStatsSurveyII using the file format of your choice for the version of the text you are using. The data is in the column "Tax Rate."

(a) Draw a relative frequency histogram of the variable "Tax Rate" using a lower class limit of the first class of 0 and a class width of 5. Comment on the shape of the distribution.

(b) Draw a boxplot of the variable "Tax Rate." Are there any outliers?

(c) Explain why a large sample is necessary to construct a confidence interval for the mean tax rate.

(d) Treat the respondents of this survey as a simple random sample of U.S. residents. Use statistical software to construct and interpret a 90% confidence interval for the mean tax rate U.S. residents feel an individual should pay in federal income taxes.

39. Age Married The General Social Survey asked respondents, "If ever married, how old were you when you first married?" The results are summarized in the Minitab output that follows:

One-Sample T: AGEWED

```
Variable       N     Mean  StDev  SE Mean      95.0% CI
AGEWED     26540   22.150  4.885    0.030  (22.091, 22.209)
```

(a) Use the summary to determine the point estimate of the population mean and margin of error for the confidence interval.

(b) Interpret the confidence interval.

(c) Verify the results by computing a 95% confidence interval with the information provided.

(d) Why is the margin of error for this confidence interval so small?

40. Sexual Relations A question on the General Social Survey was this: "About how many times did you engage in intercourse during the month?" The question was only asked of participants who had previously indicated that they had engaged in sexual intercourse during the past month. The results are summarized in the Minitab output that follows:

One-Sample T: SEXFREQ2

```
Variable      N   Mean  StDev  SE Mean     95.0% CI
SEXFREQ2    357  7.678  6.843    0.362  (6.966, 8.390)
```

(a) Use the summary to determine the point estimate of the population mean and margin of error for the confidence interval.

(b) Interpret the confidence interval.

(c) Verify the results by computing a 95% confidence interval with the information provided.

NW 41. Sample Size Dr. Paul Oswiecmiski wants to estimate the mean serum HDL cholesterol of all 20- to 29-year-old females. How many subjects are needed to estimate the mean serum HDL cholesterol of all 20- to 29-year-old females within 2 points with 99% confidence assuming that $s = 13.4$ based on earlier studies? Suppose that Dr. Oswiecmiski would be content with 95% confidence. How does the decrease in confidence affect the sample size required?

42. Sample Size Dr. Paul Oswiecmiski wants to estimate the mean serum HDL cholesterol of all 20- to 29-year-old males. How many subjects are needed to estimate the mean serum HDL cholesterol of all 20- to 29-year-old males within

1.5 points with 90% confidence, assuming that $s = 12.5$ based on earlier studies? Suppose that Dr. Oswiecmiski would prefer 95% confidence. How does the increase in confidence affect the sample size required?

43. Reading A recent Gallup poll asked Americans to disclose the number of books they read during the previous year. Initial survey results indicate that $s = 16.6$ books.

(a) How many subjects are needed to estimate the number of books Americans read the previous year within four books with 95% confidence?

(b) How many subjects are needed to estimate the number of books Americans read the previous year within two books with 95% confidence?

(c) What effect does doubling the required accuracy have on the sample size?

(d) How many subjects are needed to estimate the number of books Americans read the previous year within four books with 99% confidence? Compare this result to part (a). How does increasing the level of confidence in the estimate affect sample size? Why is this reasonable?

44. Television A researcher wanted to determine the mean number of hours per week (Sunday through Saturday) the typical person watches television. Results from the Sullivan Statistics Survey I indicate that $s = 7.5$ hours.

(a) How many people are needed to estimate the number of hours people watch television per week within 2 hours with 95% confidence?

(b) How many people are needed to estimate the number of hours people watch television per week within 1 hour with 95% confidence?

(c) What effect does doubling the required accuracy have on the sample size?

(d) How many people are needed to estimate the number of hours people watch television per week within 2 hours with 90% confidence? Compare this result to part (a). How does decreasing the level of confidence in the estimate affect sample size? Why is this reasonable?

45. Resistance and Robustness The data sets represent simple random samples from a population whose mean is 100.

Data Set I

106	122	91	127	88
74	77	108		

Data Set II

106	122	91	127	88
74	77	108	87	88
111	86	113	115	97
122	99	86	83	102

Data Set III

106	122	91	127	88
74	77	108	87	88
111	86	113	115	97
122	99	86	83	102
88	111	118	91	102
80	86	106	91	116

(a) Compute the sample mean of each data set.

(b) For each data set, construct a 95% confidence interval about the population mean.

(c) What effect does the sample size n have on the width of the interval?

For parts (d)–(e), suppose that the data value 106 was accidentally recorded as 016.

(d) For each data set, construct a 95% confidence interval about the population mean using the incorrectly entered data.

(e) Which intervals, if any, still capture the population mean, 100? What concept does this illustrate?

46. Effect of Outliers The following small data set represents a simple random sample from a population whose mean is 50.

43	63	53	50	58	44
53	53	52	41	50	43

(a) A normal probability plot indicates that the data could come from a population that is normally distributed with no outliers. Compute a 95% confidence interval for this data set.

(b) Suppose that the observation, 41, is inadvertently entered into the computer as 14. Verify that this observation is an outlier.

(c) Construct a 95% confidence interval on the data set with the outlier. What effect does the outlier have on the confidence interval?

(d) Consider the following data set, which represents a simple random sample of size 36 from a population whose mean is 50. Verify that the sample mean for the large data set is the same as the sample mean for the small data set from part (a).

43	63	53	50	58	44
53	53	52	41	50	43
47	65	56	58	41	52
49	56	57	50	38	42
59	54	57	41	63	37
46	54	42	48	53	41

(e) Compute a 95% confidence interval for the large data set. Compare the results to part (a). What effect does increasing the sample size have on the confidence interval?

(f) Suppose that the last observation, 41, is inadvertently entered as 14. Verify that this observation is an outlier.

(g) Compute a 95% confidence interval for the large data set with the outlier. Compare the results to part (e). What effect does an outlier have on a confidence interval when the data set is large?

47. Simulation: Normal Distribution IQ scores based on the Wechsler Intelligence Scale for Children (WISC) are known to be approximately normally distributed with $\mu = 100$ and $\sigma = 15$.

(a) Use StatCrunch, Minitab, or some other statistical software to simulate obtaining 100 simple random samples of size $n = 5$ from this population.

(b) Obtain the sample mean and sample standard deviation for each of the 100 samples.

(c) Construct 95% t-intervals for each of the 100 samples.

(d) How many of the intervals do you expect to include the population mean? How many of the intervals actually include the population mean?

48. Simulation: Exponential Distribution The *exponential probability distribution* can be used to model waiting time in line or the lifetime of electronic components. Its density function is skewed right. Suppose the wait-time in a line can be modeled by the exponential distribution with $\mu = \sigma = 5$ minutes.

(a) Use StatCrunch, Minitab, or some other statistical software to generate 100 random samples of size $n = 4$ from this population.

(b) Construct 95% *t*-intervals for each of the 100 samples found in part (a).

(c) How many of the intervals do you expect to include the population mean? How many of the intervals actually contain the population mean? Explain what your results mean.

(d) Repeat parts (a)–(c) for samples of size $n = 15$ and $n = 25$. Explain what your results mean.

49. Putting It Together: Smoking Cessation Study Researchers Havar Brendryen and Pal Kraft conducted a study in which 396 subjects were randomly assigned to either an experimental smoking cessation program or control group. The experimental program consisted of the Internet and phone-based Happy Ending Intervention, which lasted 54 weeks and consisted of more than 400 contacts by e-mail, web pages, interactive voice response, and short message service (SMS) technology. The control group received a self-help booklet. Both groups were offered free nicotine replacement therapy. Abstinence was defined as "not even a puff of smoke, for the last 7 days," and assessed by means of Internet surveys or telephone interviews. The response variable was abstinence after 12 months. Of the participants in the experimental program, 22.3% reported abstinence; of the participants in the control group, 13.1% reported abstinence.

Source: "Happy Ending: A Randomized Controlled Trial of a Digital MultiMedia Smoking Cessation Intervention." Havar Brendryen and Pal Kraft. *Addiction* 103(3):478–484, 2008.

(a) What type of experimental design is this?

(b) What is the treatment? How many levels does it have?

(c) What is the response variable?

(d) What are the statistics reported by the authors?

(e) An **odds ratio** is the ratio of the odds of an event occurring in one group to the odds of it occurring in another group. These groups might be men and women, an experimental group and a control group, or any other dichotomous classification. If the probabilities of the event in each of the groups are p (first group) and q (second group), then the odds ratio is

$$\frac{\dfrac{p}{(1-p)}}{\dfrac{q}{(1-q)}} = \frac{p(1-q)}{q(1-p)}$$

An odds ratio of 1 indicates that the condition or event under study is equally likely in both groups. An odds ratio greater than 1 indicates that the condition or event is more likely in the first group. And an odds ratio less than 1 indicates that the condition or event is less likely in the first group. The odds ratio must be greater than or equal to zero. As the odds of the first group approach zero, the odds ratio

approaches zero. As the odds of the second group approach zero, the odds ratio approaches positive infinity. Verify that the odds ratio for this study is 1.90. What does this mean?

(f) The authors of the study reported a 95% confidence interval for the odds ratio to be *lower bound:* 1.12 and *upper bound* 3.26. Interpret this result.

(g) Write a conclusion that generalizes the results of this study to the population of all smokers.

Retain Your Knowledge

50. How Many Drinks? A question on the General Social Survey was, "When you drink, how many drinks do you have?" The survey was administered to a random sample of 243 adult Americans aged 21 or older. Go to www.pearsonhighered.com/sullivanstats to obtain the data file 9_2_50 using the file format of your choice for the version of the text you are using.

(a) What type of variable is "number of drinks"?

(b) Draw a histogram of the data and comment on the shape of the distribution.

(c) Determine the mean and standard deviation for number of drinks.

(d) What is the mode number of drinks?

(e) What is the probability a randomly selected individual consumes two drinks?

(f) Would it be unusual to observe an individual who consumes at least eight drinks? Why?

(g) Describe the shape of the distribution of the sample mean number of drinks. What is the source of variability in the sampling distribution?

(h) Construct a 95% confidence interval for the mean number of drinks. Interpret this interval.

Explaining the Concepts

51. Explain why the *t*-distribution has less spread as the number of degrees of freedom increases.

52. The procedure for constructing a *t*-interval is *robust*. Explain what this means.

53. Explain what is meant by *degrees of freedom*.

54. The mean age of the 43 presidents of the United States (as of 2015) on the day of inauguration is 54.6 years, with a standard deviation of 6.3 years. A researcher constructed a 95% confidence interval for the mean age of presidents on inauguration day. He wrote that he was 95% confident that the mean age of the president on inauguration day is between 52.7 and 56.5 years of age. Is there anything wrong with the researcher's analysis? Explain.

55. Suppose you have two populations: Population A—All students at Illinois State University ($N = 21,000$) and Population B—All residents of the city of Homer Glen, IL ($N = 21,000$). You want to estimate the mean age of each population using two separate samples each of size $n = 75$. If you construct a 95% confidence interval for each population mean, will the margin of error for population A be larger, the same, or smaller than the margin of error for population B? Justify your reasoning.

56. Population A has standard deviation $\sigma_A = 5$, and population B has standard deviation $\sigma_B = 10$. How many times larger than Population A's sample size does Population B's need to be to estimate μ with the same margin of error? [*Hint:* Compute n_B/n_A].

9.3   Estimating a Population Standard Deviation

Objectives   ❶ Find critical values for the chi-square distribution

❷ Construct and interpret confidence intervals for the population variance and standard deviation

In this section, we discuss methods for estimating a population variance or standard deviation. Just as we discovered the sampling distribution of $\bar{x}$ and $\hat{p}$, we must find the sampling distribution of s^2. We then construct intervals about the point estimate of σ^2 using this sampling distribution.

Why might we be interested in obtaining estimates of σ^2? Many production processes not only require accuracy on average (the mean); they also require consistency. Consider a coffee machine that consistently over- and underfills cups, but, on average, fills correctly. Customers will not be happy if the machine underfills their cups or overfills their cups and spills the coffee. They want a machine that consistently delivers the correct amount of liquid.

As another example, consider a mutual fund that claims an average rate of return of 12% per year over the past 20 years. An investor might prefer consistent returns near 12% over returns that fluctuated wildly yet resulted in a mean return of 12%. Both of these situations illustrate the importance of measuring variability.

❶ Find Critical Values for the Chi-Square Distribution

We begin by exploring the sampling distribution of s^2 through a simulation. Suppose that we obtain 2000 samples of size $n = 10$ from a population that is known to be normally distributed with mean 100 and standard deviation 15. Perform the following steps:

Step 1   Compute the sample variance of each of the 2000 samples.

Step 2   Compute $\dfrac{(n-1)s^2}{\sigma^2} = \dfrac{9s^2}{15^2}$ for each sample.

Step 3   Draw a histogram of these values as shown in Figure 17.

Figure 17

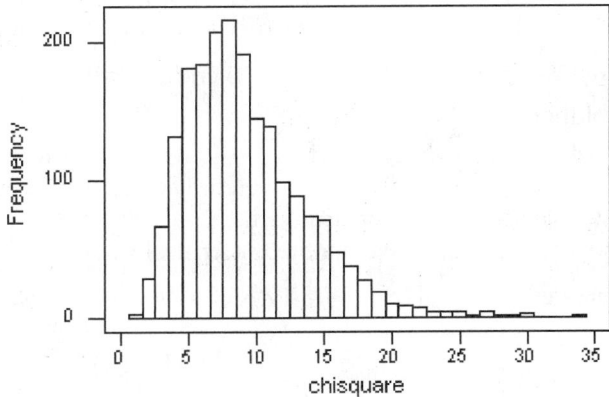

Chi-Square Distribution

The histogram suggests that the sampling distribution of $\dfrac{(n-1)s^2}{\sigma^2}$ is skewed right.

The distribution follows a *chi-square distribution*.

Chi-Square Distribution

If a simple random sample of size n is obtained from a normally distributed population with mean μ and standard deviation σ, then

$$\chi^2 = \frac{(n-1)s^2}{\sigma^2}$$

has a **chi-square distribution** with $n-1$ degrees of freedom.

The symbol χ^2, chi-square, is pronounced "kigh-square" (to rhyme with "sky-square"). We can find critical values of the chi-square distribution in Table VIII in Appendix A of the text. Before discussing how to read Table VIII, we introduce characteristics of the chi-square distribution.

Characteristics of the Chi-Square Distribution

1. It is not symmetric.
2. The shape of the chi-square distribution depends on the degrees of freedom, just like the Student's t-distribution.
3. As the number of degrees of freedom increases, the chi-square distribution becomes more nearly symmetric. See Figure 18.
4. The values of χ^2 are nonnegative (greater than or equal to 0).

Figure 18

Legend:
— χ^2 with 2 degrees of freedom
- - - χ^2 with 5 degrees of freedom
- · · χ^2 with 10 degrees of freedom
- - - χ^2 with 15 degrees of freedom
— χ^2 with 30 degrees of freedom

Because the χ^2 distribution is not symmetric, we cannot construct a confidence interval for σ^2 by computing "point estimate $\pm$ margin of error." Instead, we determine the left and right bounds by using different critical values.

Table VIII is structured similarly to Table VII for the t-distribution. The left column represents the degrees of freedom, and the top row represents the area under the chi-square distribution to the right of the critical value. We use the notation χ_α^2 to denote the critical χ^2-value such that the area under the chi-square distribution to the right of χ_α^2 is α.

EXAMPLE 1 Finding Critical Values for the Chi-Square Distribution

Problem Find the critical values that separate the middle 90% of the chi-square distribution from the 5% area in each tail, assuming 15 degrees of freedom.

Approach Perform the following steps to obtain the critical values.

Step 1 Draw a chi-square distribution with the critical values and areas labeled.

Step 2 Use Table VIII to find the critical values.

Solution

Step 1 Figure 19 shows the chi-square distribution with 15 degrees of freedom and the unknown critical values labeled. The area to the right of the right critical value is 0.05. We denote this critical value $\chi_{0.05}^2$. The area to the right of the left critical value is $1 - 0.05 = 0.95$. We denote this critical value $\chi_{0.95}^2$.

Figure 19

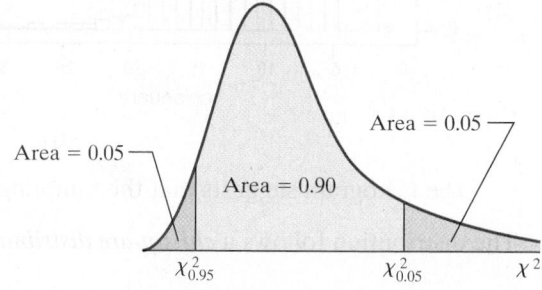

Area = 0.05
Area = 0.05
Area = 0.90
$\chi_{0.95}^2$ $\chi_{0.05}^2$ χ^2

Step 2   Figure 20 shows a partial representation of Table VIII. The row containing 15 degrees of freedom is boxed. The columns corresponding to an area to the right of 0.95 and 0.05 are also boxed. The critical values are $\chi^2_{0.95} = 7.261$ and $\chi^2_{0.05} = 24.996$.

Figure 20

Degrees of Freedom	Area to the Right of the Critical Value									
	0.995	0.99	0.975	0.95	0.90	0.10	0.05	0.025	0.01	0.005
1	—	—	0.001	0.004	0.016	2.706	3.841	5.024	6.635	7.879
2	0.010	0.020	0.051	0.103	0.211	4.605	5.991	7.378	9.210	10.597
3	0.072	0.115	0.216	0.352	0.584	6.251	7.815	9.348	11.345	12.838
12	3.074	3.571	4.404	5.226	6.304	18.549	21.026	23.337	26.217	28.299
13	3.565	4.107	5.009	5.892	7.042	19.812	22.362	24.736	27.688	29.819
14	4.075	4.660	5.629	6.571	7.790	21.064	23.685	26.119	29.141	31.319
15	4.601	5.229	6.262	7.261	8.547	22.307	24.996	27.488	30.578	32.801
16	5.142	5.812	6.908	7.962	9.312	23.542	26.296	28.845	32.000	34.267
17	5.697	6.408	7.564	8.672	10.085	24.769	27.587	30.191	33.409	35.718
18	6.365	7.215	8.231	9.288	10.265	25.200	28.601	31.595	24.265	36.456

• **Now Work Problem 5**

In studying Table VIII, notice that the degrees of freedom are numbered 1 to 30, inclusive, then 40, 50, 60, ... , 100. If the number of degrees of freedom is not in the table, choose the degrees of freedom closest to that desired. If the degrees of freedom are exactly between two values, find the mean of the values. For example, to find the critical value corresponding to 75 degrees of freedom, compute the mean of the critical values corresponding to 70 and 80 degrees of freedom.

② Construct and Interpret Confidence Intervals for the Population Variance and Standard Deviation

The sample variance, s^2, is the best point estimate of the population variance, σ^2. We use the sample standard deviation, s, as the point estimate of the population standard deviation, σ.*

We now develop a method for constructing a confidence interval for the population variance. Suppose we take a simple random sample of size n from a population that is normally distributed with mean μ and standard deviation σ; then $\chi^2 = \dfrac{(n-1)s^2}{\sigma^2}$ follows a chi-square distribution with $n - 1$ degrees of freedom. Therefore, $(1 - \alpha) \cdot 100\%$ of the values of χ^2 will lie between $\chi^2_{1-\alpha/2}$ and $\chi^2_{\alpha/2}$. Figure 21 illustrates the situation.

Figure 21

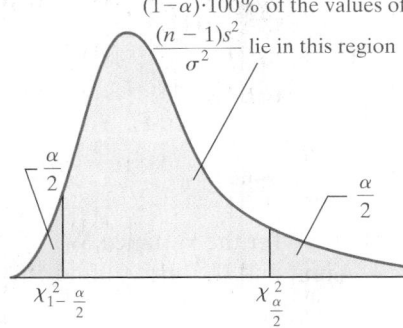

$(1-\alpha)\cdot100\%$ of the values of $\dfrac{(n-1)s^2}{\sigma^2}$ lie in this region

So $(1 - \alpha) \cdot 100\%$ of the values of $\dfrac{(n-1)s^2}{\sigma^2}$ lie within the interval defined by the inequality

$$\chi^2_{1-\alpha/2} < \frac{(n-1)s^2}{\sigma^2} < \chi^2_{\alpha/2}$$

If we rewrite this inequality with σ^2 in the center, we have the formula for a $(1 - \alpha) \cdot 100\%$ confidence interval for σ^2:

$$\frac{(n-1)s^2}{\chi^2_{\alpha/2}} < \sigma^2 < \frac{(n-1)s^2}{\chi^2_{1-\alpha/2}}$$

*It is common practice to use s as the estimator of σ, even though s is a biased estimator of σ.

CAUTION!
A confidence interval about the population variance or standard deviation is not of the form "point estimate $\pm$ margin of error" because the sampling distribution of the sample variance is not symmetric.

A $(1 - \alpha) \cdot 100\%$ Confidence Interval about σ^2

If a simple random sample of size n is taken from a normal population with mean μ and standard deviation σ, then a $(1 - \alpha) \cdot 100\%$ confidence interval about σ^2 is given by

$$\text{Lower bound: } \frac{(n - 1)s^2}{\chi_{\alpha/2}^2} \qquad \text{Upper bound: } \frac{(n - 1)s^2}{\chi_{1-\alpha/2}^2} \qquad (1)$$

where $\chi_{\alpha/2}^2$ and $\chi_{1-\alpha/2}^2$ are found using $n - 1$ degrees of freedom.

To find a $(1 - \alpha) \cdot 100\%$ confidence interval about σ, take the square root of the lower bound and upper bound.

EXAMPLE 2 Constructing a Confidence Interval for a Population Variance and Standard Deviation

Table 4

$41,844	$41,500	$39,995
$36,995	$40,990	$37,995
$41,995	$38,900	$42,995
$36,995	$43,995	$35,950

Source: cars.com

Problem Table 4 shows the sale price of 12 randomly selected 6-year-old Chevy Corvettes. Construct a 90% confidence interval for the population variance and standard deviation of the price of a 6-year-old Chevy Corvette.

Approach Step 1 Verify the data are obtained randomly and the sample size is small relative to the population size. Because the sample size is small, draw a normal probability plot to verify the data come from a population that is normally distributed and a boxplot to verify that there are no outliers.

By-Hand Approach

Step 2 Compute the value of the sample variance, s^2.

Step 3 Determine the critical values, $\chi_{1-\alpha/2}^2$ and $\chi_{\alpha/2}^2$, with $n - 1$ degrees of freedom.

Step 4 Use Formula (1) to determine the lower and upper bounds of the confidence interval for the population variance.

Step 5 Compute the square root of the lower bound and upper bound to obtain the confidence interval for the population standard deviation.

Technology Approach

Step 2 Use a statistical spreadsheet to obtain the confidence interval. We will use StatCrunch to construct the confidence interval. The steps for constructing confidence intervals using StatCrunch and Minitab are given in the Technology Step-by-Step on page 453.

Solution Step 1 A normal probability plot and boxplot indicate that the price of Corvettes could be normally distributed and no outliers are present.

By-Hand Solution

Step 2 The sample standard deviation is $2615.19, so the sample variance is ($2615.19)^2$.

Step 3 Because we want a 90% confidence interval, we have $\alpha = 1 - 0.90 = 0.10$. Using Table VIII with $n - 1 = 12 - 1 = 11$ degrees of freedom, we find the left critical value to be $\chi_{1-\alpha/2}^2 = \chi_{1-0.10/2}^2 = \chi_{0.95}^2 = 4.575$. The right critical value is $\chi_{\alpha/2}^2 = \chi_{0.10/2}^2 = \chi_{0.05}^2 = 19.675$.

Step 4 Substituting into Formula (1), we obtain:

Lower bound:

$$\frac{(n - 1)s^2}{\chi_{\alpha/2}^2} = \frac{11(2615.19)^2}{19.675} = 3,823,705.52$$

Technology Solution

Step 2 Figure 22 shows the results from StatCrunch.

Figure 22

90% confidence interval results:
σ^2 : Variance of variable

Variable	Sample Var.	DF	L. Limit	U. Limit
Price	6839205.5	11	3823671.4	16444663

StatCrunch gives confidence intervals for the variance. We determine the square root of L. Limit and U. Limit to find the confidence interval for the standard deviation.

Lower bound: $1955   Upper bound: $4055

Upper bound:

$$\frac{(n-1)s^2}{\chi^2_{1-\alpha/2}} = \frac{11(2615.19)^2}{4.575} = 16{,}444{,}023.19$$

Step 5 Taking the square root of each part,* we obtain

Lower bound: $1955   Upper bound: $4055

Interpretation We are 90% confident that the population standard deviation of the price of all 6-year-old Chevy Corvettes is between $1955 and $4055.   •

• **Now Work Problem 11**

The formula presented for constructing the confidence interval for a population variance requires that the data come from a normal distribution. Confidence intervals for the population variance (or standard deviation) are not robust. Therefore, it is vital that the requirement of normality be verified before proceeding.

Technology Step-by-Step   Confidence Intervals for σ

TI-83/84 Plus
The TI-83/84 Plus do not construct confidence intervals about σ.

Minitab
1. If necessary, enter the raw data in column C1.
2. Select the **Stat** menu, highlight **Basic Statistics**, then highlight **1 Variance**.
3. If you have raw data, select "One or more samples, each in a column" from the pull-down menu. Place the cursor in the box, highlight the column of the data, and click "Select." If you have summary statistics, select either "Sample standard deviation" or "Sample variance" from the pull-down menu.
4. Click Options. Enter the confidence level desired. Click OK twice. Note that Minitab reports two intervals,

one assuming a normal distribution and one that only assumes a continuous distribution.

Excel
Excel does not construct confidence intervals about σ.

StatCrunch
1. Enter the raw data in the first column, if necessary.
2. Select the **Stat** menu. Highlight **Variance Stats**, then highlight **One Sample.** Choose "**With Data**" or "**With Summary.**"
3. Choose the column with the raw data or enter summary statistics. Select the Confidence interval radio button. Enter the level of confidence. Click Compute!.

 9.3 Assess Your Understanding

Vocabulary and Skill Building

1. *True or False:* The chi-square distribution is symmetric.

2. *True or False:* The shape of the chi-square distribution depends on its degrees of freedom.

3. *True or False:* To construct a confidence interval about a population variance or standard deviation, either the population from which the sample is drawn must be normal, or the sample size must be large.

4. *True or False:* A confidence interval for a population standard deviation is of the form *point estimate ± margin of error.*

In Problems 5–8, find the critical values $\chi^2_{1-\alpha/2}$ and $\chi^2_{\alpha/2}$ for the given level of confidence and sample size.

NW 5. 90% confidence, $n = 20$

6. 95% confidence, $n = 25$

7. 98% confidence, $n = 23$

8. 99% confidence, $n = 14$

9. A simple random sample of size n is drawn from a population that is known to be normally distributed. The sample variance, s^2, is determined to be 12.6.

(a) Construct a 90% confidence interval for σ^2 if the sample size, n, is 20.

(b) Construct a 90% confidence interval for σ^2 if the sample size, n, is 30. How does increasing the sample size affect the width of the interval?

(c) Construct a 98% confidence interval for σ^2 if the sample size, n, is 20. Compare the results with those obtained in part (a). How does increasing the level of confidence affect the width of the confidence interval?

*Be sure to take the square root of the *unrounded* values.

10. A simple random sample of size n is drawn from a population that is known to be normally distributed. The sample variance, s^2, is determined to be 19.8.

(a) Construct a 95% confidence interval for σ^2 if the sample size, n, is 10.

(b) Construct a 95% confidence interval for σ^2 if the sample size, n, is 25. How does increasing the sample size affect the width of the interval?

(c) Construct a 99% confidence interval for σ^2 if the sample size, n, is 10. Compare the results with those obtained in part (a). How does increasing the level of confidence affect the width of the confidence interval?

Applying the Concepts

NW 11. pH of Rain The following data represent the pH of rain for a random sample of 12 rain dates in Tucker County, West Virginia. A normal probability plot suggests the data could come from a population that is normally distributed. A boxplot indicates there are no outliers. In Problem 31 from Section 9.2, we found that $s = 0.319$. Construct and interpret a 95% confidence interval for the standard deviation pH of rainwater in Tucker County, West Virginia.

4.58	5.19	5.05	4.80	4.77	4.77
5.72	4.75	5.02	4.74	4.76	4.56

Source: National Atmospheric Deposition Program

12. Travel Taxes Travelers pay taxes for flying, car rentals, and hotels. The following data represent the total travel tax for a 3-day business trip in eight randomly selected cities. *Note:* Chicago has the highest travel taxes in the country at $101.27. In Problem 32 from Section 9.2, it was verified that the data are normally distributed and that $s = 12.324$ dollars. Construct and interpret a 90% confidence interval for the standard deviation travel tax for a 3-day business trip.

67.81	78.69	68.99	84.36
80.24	86.14	101.27	99.29

13. Crash Test Results The following data represent the repair cost for a low-impact collision in a simple random sample of mini- and micro-vehicles (such as the Chevrolet Aveo or Mini Cooper). In Problem 33 from Section 9.2, it was verified that the data come from a population that is normally distributed with no outliers and $s = \$1007.4542$. Construct and interpret a 90% confidence interval for the standard deviation repair cost of a low-impact collision involving mini- and micro-vehicles.

$3148	$1758	$1071	$3345	$743
$2057	$663	$2637	$773	$1370

Source: Insurance Institute for Highway Safety

14. Crawling Babies The data in the next column represent the age (in weeks) at which babies first crawl based on a survey of 12 mothers conducted by Essential Baby. In Problem 34 from Section 9.2, it was verified that the data are normally distributed and that $s = 10.00$ weeks. Construct and interpret a 95% confidence interval for the population standard deviation of the age (in weeks) at which babies first crawl.

52	30	44	35	39	26
47	37	56	26	39	28

Source: www.essentialbaby.com

15. Peanuts A jar of peanuts is supposed to have 16 ounces of peanuts. The filling machine inevitably experiences fluctuations in filling, so a quality-control manager randomly samples 12 jars of peanuts from the storage facility and measures their contents. She obtains the following data:

15.94	15.74	16.21	15.36	15.84	15.84
15.52	16.16	15.78	15.51	16.28	16.53

(a) Verify that the data are normally distributed by constructing a normal probability plot.

(b) Determine the sample standard deviation.

(c) Construct a 90% confidence interval for the population standard deviation of the number of ounces of peanuts.

(d) The quality control manager wants the machine to have a population standard deviation below 0.20 ounce. Does the confidence interval validate this desire?

16. Investment Risk Investors not only desire a high return on their money, but they would also like the rate of return to be stable from year to year. An investment manager invests with the goal of reducing volatility (year-to-year fluctuations in the rate of return). The following data represent the rate of return (in percent) for his mutual fund for the past 12 years.

13.8	15.9	10.0	12.4	11.3	6.6
9.6	12.4	10.3	8.7	14.9	6.7

(a) Verify that the data are normally distributed by constructing a normal probability plot.

(b) Determine the sample standard deviation.

(c) Construct a 95% confidence interval for the population standard deviation of the rate of return.

(d) The investment manager wants to have a population standard deviation for the rate of return below 6%. Does the confidence interval validate this desire?

17. Critical Values Sir R. A. Fisher, a famous statistician, showed that the critical values of a chi-square distribution can be approximated by the standard normal distribution

$$\chi_k^2 = \frac{(z_k + \sqrt{2\nu - 1})^2}{2}$$

where ν is the degrees of freedom and z_k is the z-score such that the area under the standard normal curve to the right of z_k is k. Use Fisher's approximation to find $\chi_{0.975}^2$ and $\chi_{0.025}^2$ with 100 degrees of freedom. Compare the results with those found in Table VIII.

9.4   Putting It Together: Which Procedure Do I Use?

Objective   ❶ Determine the appropriate confidence interval to construct

❶ Determine the Appropriate Confidence Interval to Construct

Perhaps the most difficult aspect of constructing a confidence interval is determining which type to construct. To assist in your decision making, we present Figure 23.

Figure 23

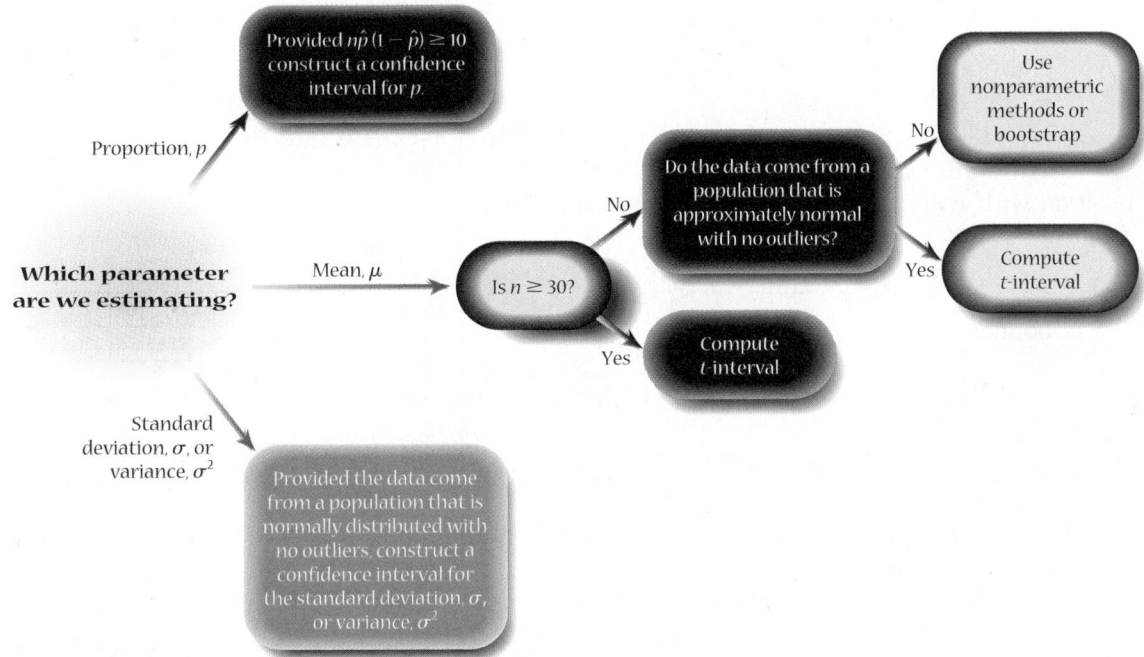

EXAMPLE 1   Constructing a Confidence Interval: Which Method Do I Use?

Table 5

323.9	326.8	370.6
450.7	368.8	423.8
398.8	417.5	
382.7	343.1	

Problem Robert wishes to estimate the mean number of miles that his Buick Lacrosse can be driven on a full tank of gas. He fills up his car with regular unleaded gasoline from the same gas station 10 times and records the number of miles that he drives until his low-tank indicator light comes on. He obtains the data shown in Table 5. Construct a 95% confidence interval for the mean number of miles driven on a full tank of gas.

Approach Follow the flow chart given in Figure 23.

Solution We are asked to construct a 95% confidence interval for the *mean* number of miles driven. We will treat the data as a simple random sample from a large population. Because the sample size is small, we verify that the data come from a population that is normally distributed with no outliers by drawing a normal probability plot and boxplot. See Figure 24 on the next page. The correlation between miles and expected z-scores in the normal probability plot is 0.987. Because $0.987 > 0.918$ (Table VI), conclude the data come from a population that is normally distributed. The boxplot shows there are no outliers. We may proceed with constructing the confidence interval for the population mean.

(continued)

Figure 24

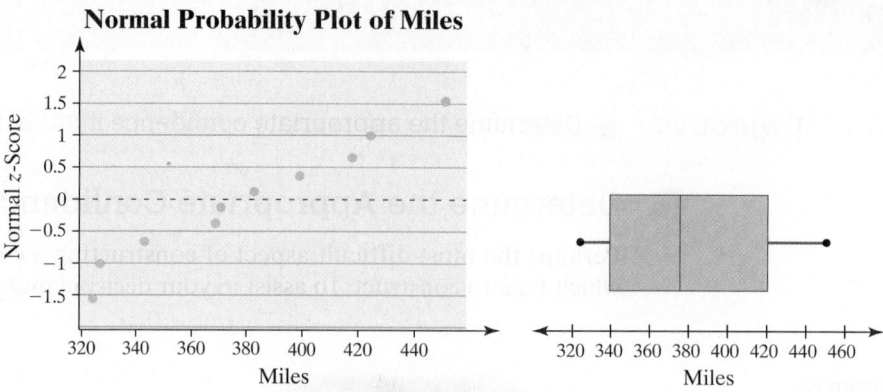

By-Hand Solution

From the sample data in Table 5, we have $n = 10$, $\bar{x} = 380.67$, and $s = 42.47$. For a 95% confidence interval with $n - 1 = 10 - 1 = 9$ degrees of freedom, we have

$$t_{\frac{\alpha}{2}} = t_{\frac{0.05}{2}} = t_{0.025} = 2.262$$

Lower bound:

$$\bar{x} - t_{\frac{\alpha}{2}} \cdot \frac{s}{\sqrt{n}} = 380.67 - 2.262 \cdot \frac{42.47}{\sqrt{10}} = 350.29$$

Upper bound:

$$\bar{x} + t_{\frac{\alpha}{2}} \cdot \frac{s}{\sqrt{n}} = 380.67 + 2.262 \cdot \frac{42.47}{\sqrt{10}} = 411.05$$

Technology Solution

Figure 25 shows the results from StatCrunch.

Figure 25

95% confidence interval results:

μ : mean of Variable

Variable	Sample Mean	Std. Err.	DF	L. Limit	U. Limit
Miles	380.67	13.429769	9	350.28976	411.05023

So

Lower bound: 350.29   Upper bound: 411.05

Interpretation   We are 95% confident that the population mean miles driven on a full tank of gas is between 350.29 and 411.05.   •

 9.4 Assess Your Understanding

Skill Building

1. For what type of variable does it make sense to construct a confidence interval about a population proportion?

2. For what type of variable does it makes sense to construct a confidence interval about a population mean?

3. What requirements must be satisfied in order to construct a confidence interval about a population proportion?

4. What requirements must be satisfied in order to construct a confidence interval about a population mean?

In Problems 5–12, construct the appropriate confidence interval.

5. A simple random sample of size $n = 300$ individuals who are currently employed is asked if they work at home at least once per week. Of the 300 employed individuals surveyed, 35 responded that they did work at home at least once per week. Construct a 99% confidence interval for the population proportion of employed individuals who work at home at least once per week.

6. A simple random sample of size $n = 785$ adults was asked if they follow college football. Of the 785 surveyed, 275 responded that they did follow college football. Construct a 95% confidence interval for the population proportion of adults who follow college football.

7. A simple random sample of size $n = 12$ is drawn from a population that is normally distributed. The sample mean is found to be $\bar{x} = 45$, and the sample standard deviation is found to be $s = 14$. Construct a 90% confidence interval for the population mean.

8. A simple random sample of size $n = 17$ is drawn from a population that is normally distributed. The sample mean is found to be $\bar{x} = 3.25$, and the sample standard deviation is found to be $s = 1.17$. Construct a 95% confidence interval for the population mean.

9. A simple random sample of size $n = 40$ is drawn from a population. The sample mean is found to be $\bar{x} = 120.5$, and the sample standard deviation is found to be $s = 12.9$. Construct a 99% confidence interval for the population mean.

10. A simple random sample of size $n = 210$ is drawn from a population. The sample mean is found to be $\bar{x} = 20.1$, and the sample standard deviation is found to be $s = 3.2$. Construct a 90% confidence interval for the population mean.

11. A simple random sample of size $n = 12$ is drawn from a population that is normally distributed. The sample variance is found to be $s^2 = 23.7$. Construct a 90% confidence interval for the population variance.

12. A simple random sample of size $n = 25$ is drawn from a population that is normally distributed. The sample variance is found to be $s^2 = 3.97$. Construct a 95% confidence interval for the population standard deviation.

Applying the Concepts

13. Aggravated Assault In a random sample of 40 felons convicted of aggravated assault, it was determined that the mean length of sentencing was 54 months, with a standard deviation of 8 months. Construct and interpret a 95% confidence interval for the mean length of sentence for an aggravated assault conviction.
Source: Based on data from the U.S. Department of Justice.

14. Click It Based on a poll conducted by the Centers for Disease Control, 862 of 1013 randomly selected adults said that they always wear seat belts. Construct and interpret a 95% confidence interval for the proportion of adults who always wear seat belts.

15. Estate Tax Returns In a random sample of 100 estate tax returns that was audited by the Internal Revenue Service, it was determined that the mean amount of additional tax owed was $3421 with a standard deviation of $2583. Construct and interpret a 90% confidence interval for the mean additional amount of tax owed for estate tax returns.

16. Muzzle Velocity Fifty rounds of a new type of ammunition were fired from a test weapon, and the muzzle velocity of the projectile was measured. The sample had a mean muzzle velocity of 863 meters per second and a standard deviation of 2.7 meters per second. Construct and interpret a 99% confidence interval for the mean muzzle velocity.

17. Worried about Retirement? In a survey of 1008 adult Americans, the Gallup organization asked, "When you retire, do you think you will have enough money to live comfortably or not?" Of the 1008 surveyed, 526 stated that they were worried about having enough money to live comfortably in retirement. Construct a 90% confidence interval for the proportion of adult Americans who are worried about having enough money to live comfortably in retirement.

18. Theme Park Spending In a random sample of 40 visitors to a certain theme park, it was determined that the mean amount of money spent per person at the park (including ticket price) was $93.43 per day with a standard deviation of $15. Construct and interpret a 99% confidence interval for the mean amount spent daily per person at the theme park.

19. Fastball Clayton Kershaw of the Los Angeles Dodgers is one of the premier pitchers in baseball. His most popular pitch is a four-seam fastball. The data in the next column represent the pitch speed (in miles per hour) for a random sample of 18 of his four-seam fastball pitches.

93.63	93.83	94.18	94.71	95.52	95.07
95.12	95.35	94.15	94.62	96.08	93.86
94.75	94.70	95.28	95.49	95.77	93.34

Source: Brooksbaseball.net

(a) Is "pitch speed" a quantitative or qualitative variable? Why is it important to know this when determining the type of confidence interval you may construct?
(b) Draw a normal probability plot to verify that "pitch speed" could come from a population that is normally distributed.
(c) Draw a boxplot to verify the data set has no outliers.
(d) Are the requirements for constructing a confidence interval for the mean pitch speed of a Clayton Kershaw four-seam fastball satisfied?
(e) Construct and interpret a 95% confidence interval for the mean pitch speed of a Clayton Kershaw four-seam fastball.
(f) Do you believe that a 95% confidence interval for the mean pitch speed of all major league pitchers' four-seam fastball would be narrower or wider? Why?

20. Annoying Behavior In March 2014, Harris Interactive conducted a poll of a random sample of 2234 adult Americans 18 years of age or older and asked, "Which is more annoying to you, tailgaters or slow drivers who stay in the passing lane?" Among those surveyed, 1184 were more annoyed by tailgaters.

(a) Explain why the variable of interest is qualitative with two possible outcomes. What are the two outcomes?
(b) Verify the requirements for constructing a 90% confidence interval for the population proportion of all adult Americans who are more annoyed by tailgaters than slow drivers in the passing lane.
(c) Construct a 90% confidence interval for the population proportion of all adult Americans who are more annoyed by tailgaters than slow drivers in the passing lane.

21. Sleep Apnea and Gum Disease Sleep apnea is a disorder in which you have one or more pauses in breathing or shallow breaths while you sleep. In a cross-sectional study of 320 individuals who suffer from sleep apnea, it was found that 192 had gum disease. *Note:* In the general population, about 17.5% of individuals have gum disease.
Source: Seo, WH et al. *The Association between Periodontitis and Obstructive Sleep Apnea: A Preliminary Study.* J Periodontal Res 2013 Aug; 48 (4)

(a) What does it mean for this study to be cross-sectional?
(b) What is the variable of interest in this study? Is it qualitative or quantitative? Explain.
(c) Estimate the proportion of individuals who suffer from sleep apnea who have gum disease with 95% confidence. Interpret your result.

22. Fastball: Revisited See Problem 19. Estimate the standard deviation of a Clayton Kershaw fastball with 95% confidence.

Retain Your Knowledge

23. Weight Gain Researchers conducted a study to see the effect of specific lifestyle and dietary changes for preventing long-term weight gain. The study involved the consolidation of three cohorts from the Nurses' Health Study (NHS): (1) cohort of 121,701 female registered nurses, who enrolled in 1976; (2) the Nurses' Health Study II (NSH II), a cohort of 116,686 younger female nurses, who enrolled in 1989; and (3) the Health Professionals Follow-up Study (HPFS), a cohort of 51,529 male health professionals, who enrolled in 1986. Participants were followed with biennial questionnaires concerning medical history, lifestyle, and health practices. The baseline year of the study was used to determine diet, physical activity, and smoking habits of the participants. The final analysis included 50,422 from the NHS, 49,898 from the NHS II, and 22,557 from the HPFS (those with health issues or obesity were excluded from the study). Individuals were categorized based on various lifestyle choices. For example, those who disclosed an increase in consumption of french fries had a mean weight gain of 3.35 pounds over a 4-year period with a 95% confidence interval of 2.28 pounds to 4.42 pounds. Among those who disclosed a change in smoking status from former smoker to current smoker the mean weight gain was −2.47 pounds over a 4-year period with a 95% confidence interval of −3.82 pounds to −1.12 pounds. *Source:* Dariush Mozaffarian, M.D. et al. "Changes in Diet and Lifestyle and Long-Term Weight Gain in Women and Men," *The New England Journal of Medicine* 364;25

(a) Explain what it means for this study to be a cohort study.

(b) Is the variable weight gain quantitative or qualitative? How was the variable measured?

(c) Within the cohort "individuals who increased their consumption of french fries during the observation period," there was a mean weight gain of 3.35 pounds over the 4-year period. Is the variable "increase in consumption of french fries" quantitative or qualitative? Explain.

(d) What is the margin of error for the 95% confidence interval for increase in weight gain among those who increased their consumption of french fries?

(e) What does a mean weight gain of −2.47 pounds among those whose smoking status changed from former to current smoker suggest?

(f) Interpret the 95% confidence interval for the mean weight gain among those whose smoking status changed from former to current smoker.

(g) Describe the population to which the results of this study apply.

In Problems 24 through 27 indicate whether a confidence interval for a proportion or mean should be constructed to estimate the variable of interest. Justify your response.

24. A developmental mathematics instructor wishes to estimate the typical amount of time students dedicate to studying mathematics in a week. She asks a random sample of 50 students enrolled in developmental mathematics at her school to report the amount of time spent studying mathematics in the past week.

25. Researchers at the Gallup Organization asked a random sample 1016 adult Americans aged 21 years or older, "Right now, do you think the state of moral values in the country as a whole is getting better, or getting worse?"

26. A researcher wanted to know whether consumption of green tea on a daily basis reduces LDL (bad) cholesterol. She obtains a random sample of 500 subjects. Each subject consumes at least 1 cup of green tea daily for 1 year. After 1 year, the researcher determines whether the subjects LDL cholesterol decreased, or not.

27. Does chewing your food for a longer period of time reduce one's caloric intake of food at dinner? A researcher requires a sample of 75 healthy males to chew their food twice as long as they normally do. The researcher then records the calorie consumption at dinner.

9.5 Estimating with Bootstrapping

Objective ❶ Estimate a parameter using the bootstrap method

❶ Estimate a Parameter Using the Bootstrap Method

The methods for estimating parameters (such as μ or p) by constructing confidence intervals introduced in Sections 9.1 to 9.3 relied on parametric methods. **Parametric statistics** use estimates of parameters along with probability distributions to make inferences about the parameter. For example, we use the normal probability distribution to make inferences about the population proportion by constructing a confidence interval. To make inferences about a parameter using parametric statistics certain conditions (such as a normality requirement) must be satisfied. What if the conditions for using parametric statistics are not satisfied? In this case, **nonparametric statistics**, which do not require sample data follow a specified distribution, may be used.

One nonparametric method for estimation that is gaining in popularity due to increased computational power is *bootstrapping*.

Definition

Bootstrapping is a computer-intensive approach to statistical inference whereby parameters are estimated by treating a set of sample data as a population. A computer is used to resample *with replacement* n observations from the sample data. This process is repeated many (say, 1000) times. For each resample, the statistic (such as the sample mean) is obtained.

In Other Words
To "bootstrap" is to rely on oneself without the aid of others.

The bootstrap method was invented by Bradley Efron, from Stanford University, in 1979.

To understand the logic behind the bootstrap, let's consider an example. The birth weight of full term babies is a normal random variable whose mean is $\mu = 3375$ g and standard deviation is $\sigma = 500$ g. Figure 26(a) shows the distribution of the population. Suppose we obtained a simple random sample of $n = 10$ babies from this population. Statistical theory says the sampling distribution of the sample mean for this sample is also going to be normal with mean $\mu_{\bar{x}} = 3375$ g and standard deviation $\sigma_{\bar{x}} = \dfrac{500}{\sqrt{10}}$.

The red curve in Figure 26(b) represents the theoretical distribution of sample means (based on the results discussed in Section 8.1). The histogram in Figure 26(b) represents the distribution of sample means for 1000 different simple random samples of size $n = 10$ from this population.

Now suppose that we have 10 randomly selected full term baby's weights as shown in Table 6. Treat the n observations in the sample as though they were a population. To build a proxy for the sampling distribution of the sample mean, obtain $B = 1000$ samples of size $n = 10$ *with replacement* from this sample data. For each sample, determine the sample mean. Construct a histogram of the 1000 sample means. This distribution is a close approximation to the theoretical sampling distribution. See Figure 26(c). The histograms in Figures 26(b) and 26(c) are remarkably close! Plus, the standard deviation of the 1000 bootstrap sample means from Figure 26(c) is 159.5, which is close to the theoretical standard error, $\sigma_{\bar{x}} = \dfrac{500}{\sqrt{10}} \approx 158.1$.

Table 6

3673	3212	3643	3751	2962
4087	3371	2859	2970	3480

Figure 26

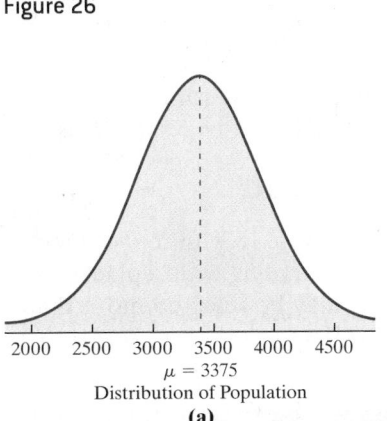

$\mu = 3375$
Distribution of Population
(a)

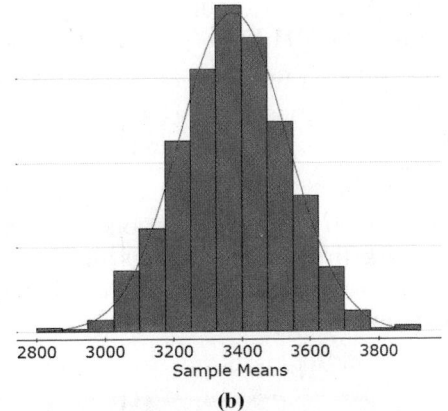

Sample Means
(b)

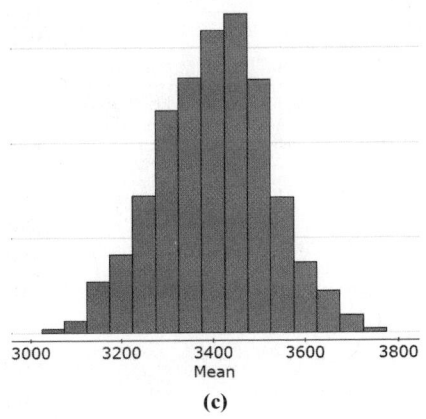

Mean
(c)

So, bootstrapping a set of sample data allows us to create a sampling distribution that is remarkably close to the theoretical distribution of the sample mean based on the population! Amazing!

To use a bootstrap, two basic requirements must be satisfied:

1. The "center" of the bootstrap distribution must be close to the "center" of the original sample data. For example, the mean of all bootstrap means must be close to the mean of the original data.
2. The distribution of the bootstrap sample statistic must be symmetric.

If using the bootstrap method to obtain a confidence interval for a population mean, we can use the distribution of the sample means obtained from the B resamples as an approximation for the actual sampling distribution. A $(1 - \alpha) \cdot 100\%$ confidence interval may be obtained using the **percentile method**. With this method, determine the $\frac{\alpha}{2} \cdot 100$th percentile (the lower bound) and the $\left(1 - \frac{\alpha}{2}\right) \cdot 100$th percentile (the upper bound) of the distribution. For example, the lower bound of a 95% confidence interval would be the 2.5th percentile and the upper bound would be the 97.5th percentile.

The bootstrap method may be summarized in the following algorithm.

The Bootstrap Algorithm

Step 1 Select B independent bootstrap samples of size n with replacement. Note that n is the number of observations in the original sample. So, if the original sample has 10 observations, each bootstrap sample will have 10 observations. Sampling with replacement means that once an observation is selected to be in the sample, its value is "put back into the hat" so it may be sampled again. The number of resamples B required is typically 1000.

Step 2 Determine the value of the statistic of interest, such as the mean, for each of the B samples.

Step 3 Use the distribution of the B statistics to make a judgment about the value of the parameter. For example, find the 2.5th and 97.5th percentiles to determine the lower and upper bound of a 95% confidence interval.

EXAMPLE 1   Using the Bootstrap Method to Construct a 95% Confidence Interval

Problem The website *fueleconomy.gov* allows drivers to report the miles per gallon of their vehicle. The data in Table 7 shows the reported miles per gallon of 2011 Ford Focus automobiles for 16 different owners. Treat the sample as a simple random sample of all 2011 Ford Focus automobiles. Construct a 95% confidence interval for the mean miles per gallon of a 2011 Ford Focus using a bootstrap sample. Interpret the interval.

Approach We will use Minitab to obtain the 95% confidence interval using a bootstrap sample. The steps for constructing confidence intervals using a bootstrap sample using Minitab and StatCrunch are given in the Technology Step-by-Step on pages 462–463.

Solution

Step 1 Obtain 1000 independent bootstrap samples of size $n = 16$ with replacement. Table 8 shows the first few bootstrap samples. Notice that sampling with replacement means that some observations from the original sample may be selected more than once. *Note:* Your results will differ.

Table 7			
35.7	37.2	34.1	38.9
32.0	41.3	32.5	37.1
37.3	38.8	38.2	39.6
32.2	40.9	37.0	36.0

Source: www.fueleconomy.gov

Table 8	
Bootstrap Sample	**Observations From Each Sample**
1	37.3 32.5 35.7 41.3 38.8 37.0 34.1 38.9 41.3 32.2 38.8 35.7 34.1 40.9 37.0 40.9
2	37.2 35.7 41.3 32.5 40.9 37.2 41.3 37.3 37.3 37.0 35.7 37.0 32.5 38.9 32.2 37.2
3	36.0 32.2 32.5 37.0 32.2 36.0 37.1 40.9 32.2 38.9 36.0 40.9 37.2 38.2 32.0 32.2

Step 2   Determine the sample mean for each of the 1000 bootstrap samples.

Step 3   Identify the 2.5th (the 25th observation with the data written in ascending order) and the 97.5th percentile (the 975th observation with the data written in ascending order). The lower bound is the 2.5th percentile, 35.35 mpg, and the upper bound is the 97.5th percentile, 38.09 mpg. Figure 27 shows a histogram of the 1000 sample means with the lower and upper bounds labeled. This histogram represents an approximation of the sampling distribution of the sample mean based on the bootstrap resamples. Notice that the distribution is symmetric and centered around the mean of the original data in Table 7, $\bar{x} = 36.8$ mpg.

Figure 27

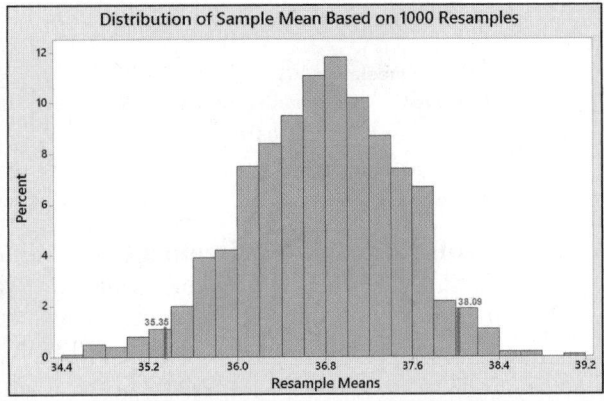

Interpretation We are 95% confident the mean miles per gallon of a 2011 Ford Focus is between 35.35 and 38.09 miles per gallon.   ●

The confidence interval constructed using the bootstrap method is narrower than the confidence interval using Student's *t*-distribution (lower bound: 35.24 mpg, upper bound: 38.36 mpg), but the two intervals are similar.

Because bootstrapping relies on randomization, results will vary each time the method is used. So the results you obtain by following Example 1 may differ from the results we obtained, but your results should be close.

The approach used in Example 1 illustrates the bootstrap method and logic, but many statistical spreadsheets (such as StatCrunch) have built-in algorithms that do bootstrapping.

EXAMPLE 2   Using StatCrunch's Resample Command to Obtain a Bootstrap Confidence Interval

Problem Use StatCrunch to estimate a 95% confidence interval for the mean and median miles per gallon of a 2011 Ford Focus based on the sample data in Table 7.

Approach The steps for constructing confidence intervals using a bootstrap sample using StatCrunch's Resample command are given in the Technology Step-by-Step on page 463.

Solution Figure 28 on the next page shows the results for the mean obtained from StatCrunch based on $B = 1000$ resamples. For a 95% confidence interval, the lower bound is the 2.5th percentile, 35.46 miles per gallon, and the upper bound is the 97.5th percentile, 38.14 miles per gallon. These bounds are close to those obtained in Example 1. They are also close to the 95% *t*-interval constructed in Example 4 from Section 9.2 (lower bound: 35.24 mpg; upper bound: 38.36 mpg).

CAUTION!
Watch out for the following situations
in which a bootstrap will not work.

- The bootstrap must be symmetric.
- The mean of the B bootstrap statistics must be close to the mean of the sample data.
- If the distribution of the bootstrap means is highly skewed or has major gaps, consult a professional statistician. The methods required to obtain a bootstrap confidence interval for these circumstances are beyond the scope of this course.

● Now Work Problem 7

Figure 28

Statistic: mean(MPG)

Observed	n	Mean	Std. dev.	2.5th per.	5th per.	50th per.	95th per.	97.5th per.
36.8	1000	36.811381	0.70530151	35.459375	35.61875	36.8125	38.00625	38.140625

Figure 29 shows the results for the median obtained from StatCrunch based on $B = 1000$ resamples. For a 95% confidence interval about the median, the lower bound is the 2.5th percentile, 35.7 mpg, and the upper bound is the 97.5th percentile, 38.85 mpg.

Figure 29

Statistic: median(MPG)

Observed	n	Mean	Std. dev.	2.5th per.	5th per.	50th per.	95th per.	97.5th per.
37.15	1000	37.16505	0.77255928	35.7	35.85	37.15	38.55	38.85

Bootstrapping to Estimate a Population Proportion

To construct a bootstrap confidence interval for a proportion, we need raw data using 0 for a failure and 1 for a success. The sample proportion is then estimated using the mean of the 0s and 1s. See Problems 13 and 14.

Technology Step-by-Step Bootstrapping for Confidence Intervals

Minitab

1. Enter the raw data into column C1.
2. Select the **Calc** menu, highlight **Random Data**, then highlight **Sample from Columns** . . . . Multiply the number of observations in the data set by B, the number of resamples (usually 1000) and enter this result in the cell marked "Number of rows to sample:" For example, if there are 12 observations in the data set, multiply 12 by 1000 and enter 12,000 in the cell. In the "From columns:" cell, enter C1, the column containing the raw data. Enter C2 in the "Store samples in:" cell. Check "Sample with replacement" and hit OK.
3. Select the **Calc** menu, highlight **Make Patterned Data**, then highlight **Simple Set of Numbers** . . . . Enter C3 in the cell labeled "Store patterned data in:" Enter 1 in the cell "From first value:", enter 1000 in the cell "To last value:", enter 1 in the cell "In steps of:." In the cell "Number of times to list each value:" enter the number of observations in the original data set. If the original data set has 12 observations, enter 12. Enter 1 in the cell "Number of times to list the sequence:" Press OK.
4. Select the **Stat** menu, highlight **Basic Statistics**, then highlight **Store Descriptive Statistics** . . . . In the cell "Variables:" enter C2, in the cell "By variables:" enter C3. Press the Statistics . . . button and check the statistic whose value you are estimating. Click OK twice.
5. Select the **Data** menu, highlight **Sort** . . . . In the "Sort column(s):" cell and "By column:" cell, enter the column that has the statistics (such as Mean1). Click OK.
6. Now identify the percentiles that represent the middle $(1 - \alpha) \cdot 100\%$ values of the statistic. For example, to find

a 95% confidence interval, identify the 2.5th percentile (the 25th observation) and the 97.5th percentile (the 975th observation). These values represent the lower and upper bound of the confidence interval.

StatCrunch

Confidence Intervals using Simulation

1. Enter the raw data into column var1. Name the column.
2. Select the **Data** menu, highlight **Sample**. Select the variable from column var1. In the "Sample Size:" cell, enter the number of observations, n. In the "Number of samples:" cell, enter the number of resamples (usually 1000). Check the "Sample with replacement" box. Check the radio button "Stacked with a sample id". Click Compute!.
3. Select the **Stat** menu, highlight **Summary Stats**, and select **Columns**. Select the Sample(var1) under "Select Columns:" For example, if the variable name in column 1 is MPG, select Sample(MPG). In the dropdown menu "Group by:" select "Sample". Select only the Mean (or statistic of interest) in the Statistics menu. Check the box "Store output in data table". Click Compute!. A popup window will appear that says, "Whoa! Lots of numeric. . ." Click Cancel.
4. Select the **Stat** menu, highlight **Summary Stats**, and select **Columns**. Select the column "Mean" (or statistic of interest). Highlight "Mean" (or statistic of interest) in the "Statistics:" cell. Enter the percentiles corresponding to the lower and upper bound in the "Percentiles" cell. For example, enter 2.5, 97.5 for a 95% confidence interval. Click Compute!.

Confidence Intervals Using Resample Command
1. Select the **Stat** menu, highlight **Resample**, and select **Statistic**. Under "Columns to resample:" select the variable you wish to estimate. Enter the statistic you wish to estimate in the "Statistic:" cell. For example, enter "mean(*variable name*)" for the mean,

"median(*variable name*)" for the median. Select the "Bootstrap" radio button. Select the "Univariate" radio button. Enter *B*, the number of resamples, in the "Number of resamples:" cell. Enter the percentiles corresponding to the confidence interval you wish to determine. Check any boxes you wish. Click Compute!.

 9.5 Assess Your Understanding

Vocabulary

1. What is bootstrapping? Be sure to include a discussion of the general bootstrap algorithm.

2. If we wish to obtain a 95% confidence interval of a parameter using the bootstrap percentile method, we determine the _____ percentile and the _____ percentile of the resampled distribution.

Skill Building

3. Suppose the following data represent the heights (in inches) of a random sample of males: 68, 72, 73, 70, 75, 71. Which of the following could be a possible bootstrap sample?

(a) 68, 72, 72, 68, 70, 71
(b) 75, 72, 73, 73, 68
(c) 70, 71, 68, 73, 73, 71, 68
(d) 72, 73, 75, 71, 73, 63
(e) 68, 72, 73, 71, 72, 73

4. Suppose the following data represent the amount of time (in hours) a random sample of students enrolled in College Algebra spent working on a homework assignment: 3.2, 4.1, 1.2, 0.6, and 2.5. Which of the following could be a possible bootstrap sample?

(a) 4.1, 1.2, 1.2, 2.5
(b) 2.5, 0.6, 3.2, 0.6, 2.5
(c) 1.2, 1.2, 3.2, 2.5, 0.6, 4.1
(d) 4.1, 1.2, 1.2, 2.5, 0.3
(e) 1.2, 3.2, 0.6, 0.6, 0.6

5. Suppose the following data represent the heights (in inches) of a random sample of males: 68, 72, 73, 70, 75, 71. Below are three bootstraps samples. For each bootstrap sample, determine the bootstrap sample mean.

Bootstrap Sample 1: 68, 72, 68, 73, 72, 75
Bootstrap Sample 2: 71, 72, 73, 72, 72, 71
Bootstrap Sample 3: 72, 73, 68, 68, 73, 71

6. Suppose the following data represent the amount of time (in hours) a random sample of students enrolled in College Algebra spent working on a homework assignment: 3.2, 4.1, 1.2, 0.6, and 2.5. Below are three bootstraps samples. For each bootstrap sample, determine the bootstrap sample mean.

Bootstrap Sample 1: 1.2, 0.6, 3.2, 3.2, 1.2
Bootstrap Sample 2: 0.6, 4.1, 4.1, 0.6, 4.1
Bootstrap Sample 3: 4.1, 3.2, 3.2, 0.6, 1.2

Applying the Concepts

NW 7. pH of Rain The following data represent the pH of rain for a random sample of rain dates in Tucker County, West Virginia. A normal probability plot suggests the data could come from a population that is normally distributed. A boxplot indicates there are no outliers. In Problem 31 from Section 9.2, we found a 95% confidence interval for the mean pH of rainwater in Tucker County, West Virginia.

4.58	5.19	5.05	4.80	4.77	4.77
5.72	4.75	5.02	4.74	4.76	4.56

Source: National Atmospheric Deposition Program

(a) Construct a 95% confidence interval for the mean pH of rainwater in Tucker County, West Virginia using a bootstrap sample with 1000 resamples. Compare the bootstrap confidence interval to the *t*-interval.
(b) Draw a histogram of the 1000 sample means. What is the shape of the histogram?

8. Travel Taxes Travelers pay taxes for flying, car rentals, and hotels. The following data represent the total travel tax for a 3-day business trip in eight randomly selected cities. *Note:* Chicago has the highest travel taxes in the country at $101.27. In Problem 32 from Section 9.2, we obtained a 95% confidence interval for the mean travel tax.

67.81	78.69	68.99	84.36
80.24	86.14	101.27	99.29

(a) Construct a 95% confidence interval for the mean travel tax using a bootstrap sample with 1000 resamples.
(b) Draw a histogram of the 1000 sample means. What is the shape of the histogram?
(c) Compare the bootstrap confidence interval to the *t*-interval.

9. Crash Test Results The following data represent the repair cost for a low-impact collision in a simple random sample of mini- and micro-vehicles (such as the Chevrolet Aveo or Mini Cooper). In Problem 33 from Section 9.2, we obtained a 95% confidence interval for the mean repair cost of a low-impact collision involving mini- and micro-vehicles. Construct a 95% confidence interval for the mean repair cost of a low-impact collision involving mini- and micro-vehicles using a bootstrap sample with 1000 resamples. Compare the bootstrap confidence interval to the *t*-interval.

$3148	$1758	$1071	$3345	$743
$2057	$663	$2637	$773	$1370

Source: Insurance Institute for Highway Safety

10. Crawling Babies The following data represent the age (in weeks) at which babies first crawl based on a survey of 12 mothers conducted by Essential Baby. In Problem 34 from Section 9.2, we constructed a 95% confidence interval for the mean age at which a baby first crawls. Construct a 95% confidence interval for the mean age at which a baby first crawls using a bootstrap sample with 1000 resamples. Compare the bootstrap interval to the *t*-interval constructed in Problem 34 of Section 9.2.

52	30	44	35	39	26
47	37	56	26	39	28

Source: www.essentialbaby.com

11. Home Prices The following data represent the selling price (in thousands of dollars) of oceanfront condominiums in Daytona Beach Shores, Florida.

476	525	410	145	1250	360
205	425	200	205	344	1080

Source: Zillow.com

(a) Draw a boxplot of the data. Explain why a *t*-interval should not be constructed.
(b) Obtain 1000 bootstrap samples and determine the mean of each sample. Draw a histogram of the 1000 means. Comment on the shape of the distribution.
(c) Construct a 95% confidence interval for the mean selling price of oceanfront condominiums in Daytona Beach Shores, Florida using the 1000 bootstrap samples from part (b).

12. Baseball Salaries The following data represent the annual 2015 salary (in thousands of dollars) of a random sample of professional baseball players.

680	500	1250	21000	20000
414	500	1500	6250	1550
2000	750	1000	12000	425

Source: SportsCity

(a) Draw a boxplot of the data. Explain why a *t*-interval should not be constructed.
(b) Obtain 1000 bootstrap samples and determine the mean of each sample. Draw a histogram of the 1000 means. Comment on the shape of the distribution.
(c) Construct a 95% confidence interval for the mean salary of a professional baseball player in 2015 using the 1000 bootstrap samples from part (b).

13. Bootstrap Proportions To estimate proportions using bootstrapping methods, report successes as 1 and failures as 0. Then follow the same procedures that we used to estimate a mean using bootstrapping. A random sample of 85 adult Americans was asked, "Have you ever ended a budding relationship because a kiss did not go well?" Go to www.pearsonhighered.com/sullivanstats to obtain the data file 9_5_13 using the file format of your choice for the version of the text you are using. In the dataset, 0 represents "No" and 1 represents "Yes".

(a) What is the sample proportion of adult Americans who have ended a relationship because a kiss did not go well?
(b) Construct a 95% confidence interval for the proportion of adult Americans who have ended a budding relationship because a kiss did not go well by finding 1000 bootstrap samples.
(c) Explain why a 95% confidence interval could be constructed using the normal model. Then find the interval. Compare your results to part (b).

14. Bootstrap Proportions To estimate proportions using bootstrapping methods, report successes as 1 and failures as 0. Then follow the same procedures that we used to estimate a mean using bootstrapping. Suppose a random sample of 25 50- to 64-year-old Americans were asked if they feel younger than their actual age. In the data, a 1 indicates the individual feels younger and a 0 indicates the individual does not feel younger.

1	1	0	0	1
1	0	1	1	0
1	1	0	1	1
0	1	1	1	0
0	0	1	0	1

(a) What is the sample proportion of 50- to 64-year-olds who feel younger than their actual age?
(b) Explain why the normal model cannot be used to construct a confidence interval about the population proportion.
(c) Construct a 95% confidence interval for the proportion of 50- to 64-year-olds who feel younger than their actual age by obtaining 1000 bootstrap samples.

A second method for finding a bootstrap confidence interval is called the **Bootstrap *t*-Method**. This method requires estimating the standard error of the estimate (such as the standard error of the mean) from the bootstrap sample. For any given set of estimates of a parameter, the standard error of the estimate is found by determining the sample standard deviation of the *B* bootstrap estimates. For example, in Example 1, we found 1000 estimates of the sample mean mpg. The standard deviation of these 1000 sample means is found to be 0.692. The standard deviation of the 16 observations is 2.917, so $\dfrac{s}{\sqrt{n}} = \dfrac{2.917}{\sqrt{n}} = 0.729$, is close to the standard error of the estimate found from the bootstrap samples. The estimate of the standard error from Figure 28 in Example 2 is 0.705. We can use an estimate of the standard error (SE_{est}) along with critical values from Student's *t*-distribution (Table VII) to construct confidence intervals as follows:

$$statistic \pm t_{\frac{\alpha}{2}} \cdot SE_{est}$$

For example, in Example 1 we know $\bar{x} = 36.8$ and $n = 16$, so for a 95% confidence interval, $t_{0.025} = 2.131$ using 15 degrees of freedom. Using the standard error estimate from Example 1, we find the lower bound of the confidence interval to be $36.8 - 2.131(0.692) = 35.33$ mpg and the upper bound to be $36.8 + 2.131(0.692) = 38.27$.

15. pH of Rain Revisited See Problem 7. Use the Bootstrap *t*-Method to find a 95% confidence for the mean pH of rainwater in Tucker County, West Virginia.

16. Travel Taxes Revisited See Problem 8. Use the Bootstrap *t*-Method to find a 95% confidence interval for the mean travel tax.

17. Crash Test Results Revisited See Problem 9. Use the Bootstrap *t*-Method to find a 95% confidence interval for the mean repair cost of a low-impact collision involving mini- and micro-vehicles.

18. Crawling Babies Revisited See Problem 10. Use the Bootstrap *t*-Method to find a 95% confidence interval for the mean age at which a baby first crawls.

19. Simulation IQ scores are known to be approximately normally distributed with mean 100 and standard deviation 15.

(a) Simulate obtaining a random sample of 12 IQ scores from this population.

(b) Use the data from part (a) to construct a 95% confidence interval for the mean IQ using Student's *t*-distribution.

(c) Use the data from part (a) to obtain 1000 bootstrap samples. For each sample, find the mean.

(d) Determine an estimate of the standard error of the mean from the 1000 bootstrap means found in part (c). Compare this result to the theoretical standard error of the mean, $\frac{\sigma}{\sqrt{n}}$. Compare this result to the estimate of the standard error of the mean based on the sample data, $\frac{s}{\sqrt{n}}$.

(e) Construct a 95% confidence interval for the mean IQ using the bootstrap sample from part (c).

20. Simulation The exponential probability distribution can be used to model waiting time in line or the lifetime of electronic components. Its density function is skewed right. Suppose the wait-time in a line can be modeled by the exponential distribution with $\mu = \sigma = 5$ minutes.

(a) Simulate obtaining a random sample of 15 wait-times.

(b) Explain why constructing a 95% confidence interval using Student's *t*-distribution is a bad idea. Nonetheless, construct a 95% confidence interval using Student's *t*-distribution.

(c) Use the data to construct a 95% confidence interval for the mean using 1000 resamples.

21. Simulation: What about the Median? Certain statistics are difficult to bootstrap. One such statistic is the median. Consider the following to see why.

(a) Simulate obtaining a random sample of 12 IQ scores. Recall IQ scores are approximately normally distributed with mean 100 and standard deviation 15.

(b) Given that IQ scores are normally distributed, what is the median IQ score?

(c) Obtain 1000 bootstrap samples from the data in part (a). Find the median of each bootstrap sample.

(d) Draw a histogram of the bootstrap medians from part (c). What do you notice about the distribution? Find the 95% confidence interval based on the 1000 bootstrap medians using the percentile method.

(e) Repeat parts (a) through (d) using a random sample of 13 IQ scores.

(f) Conclude that finding confidence intervals for medians is best if done where the sample size is even.

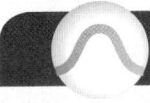

Chapter 9 Review

Summary

In this chapter, we discussed estimation methods. We estimated the values of the parameters $p, \mu,$ and σ. We started by estimating the value of a population proportion, p. A confidence interval is an interval of numbers for an unknown parameter that is reported with a level of confidence. The level of confidence represents the proportion of intervals that will contain the parameter if a large number of different samples is obtained and is denoted $(1 - \alpha) \cdot 100\%$. For example, when constructing a 95% confidence interval, we would expect 95 of 100 different random samples from the same population to include the unknown parameter.

In Section 9.1, we learned how to estimate a population proportion. A $(1 - \alpha) \cdot 100\%$ confidence interval for the population proportion p is given by $\hat{p} \pm z_{\alpha/2} \cdot \sqrt{\frac{\hat{p}(1 - \hat{p})}{n}}$, provided the data are obtained using simple random sampling or through a randomized experiment, $n\hat{p}(1 - \hat{p}) \geq 10$, and the sample size is no more than 5% of the population size (independence requirement).

In Section 9.2 we learned how to estimate the population mean, μ. To construct this interval, we required that the sample come from a population that is normally distributed or the sample size is large (so that the distribution of $\bar{x}$ is approximately normal). We also required that the sample size be no more than 5% of the

population size (independence requirement) and that the data be obtained using simple random sampling or through a randomized experiment. Because the population standard deviation is likely unknown (since we do not know μ, how could we expect to know σ), we use Student's *t*-distribution with $n - 1$ degrees of freedom to construct a confidence interval for the population mean. A $(1 - \alpha) \cdot 100\%$ confidence interval for the population mean, μ, is given by $\bar{x} \pm t_{\alpha/2} \cdot \frac{s}{\sqrt{n}}$ where $t_{\alpha/2}$ has $n - 1$ degrees of freedom.

Next we introduced a method for estimating the population standard deviation. To perform this estimation, the population from which the sample is drawn must be normal, and the sampling method must be simple random sampling. If these requirements are satisfied, then $\chi^2 = \frac{(n-1)s^2}{\sigma^2}$ follows the chi-square distribution with $n - 1$ degrees of freedom. The $(1 - \alpha) \cdot 100\%$ confidence interval about σ^2 is $\frac{(n-1)s^2}{\chi^2_{\alpha/2}} < \sigma^2 < \frac{(n-1)s^2}{\chi^2_{1-\alpha/2}}$. To construct the $(1 - \alpha) \cdot 100\%$ confidence interval for σ, take the square root of each part of the inequality and obtain $\sqrt{\frac{(n-1)s^2}{\chi^2_{\alpha/2}}} < \sigma < \sqrt{\frac{(n-1)s^2}{\chi^2_{1-\alpha/2}}}$. We are $(1 - \alpha) \cdot 100\%$ confident that the unknown value of σ lies within the interval.

Finally, we discussed the bootstrap approach to estimating a population parameter. This method has the advantage that it uses a random sample to estimate the distribution of the sample statistic by obtaining many, many random samples with replacement from the sample data. Determine the statistic corresponding to the parameter of interest and then find the percentiles corresponding to the cutoffs for the interval desired. For example, for a 95% confidence interval, find the 2.5th and 97.5th percentile of the estimated sample distribution.

Vocabulary

Point estimate (p. 421)
Confidence interval (p. 422)
Level of confidence (p. 422)
Margin of error (pp. 423, 440)
Critical value (p. 425)

Student's t-distribution (p. 436)
t-Interval (p. 440)
Robust (p. 440)
Nonparametric procedures (p. 441)
Chi-square distribution (p. 450)

Parametric statistics (p. 458)
Nonparametric statistics (p. 458)
Bootstrapping (p. 459)
Percentile method (p. 460)
Bootstrap t-Method (p. 464)

Formulas

Confidence Intervals

- A $(1 - \alpha) \cdot 100\%$ confidence interval for p is
$\hat{p} \pm z_{\frac{\alpha}{2}} \cdot \sqrt{\dfrac{\hat{p}(1 - \hat{p})}{n}}$, provided that $n\hat{p}(1 - \hat{p}) \geq 10$
and $n \leq 0.05N$.

- A $(1 - \alpha) \cdot 100\%$ confidence interval for μ is
$\bar{x} \pm t_{\frac{\alpha}{2}} \cdot \dfrac{s}{\sqrt{n}}$, where $t_{\alpha/2}$ has $n - 1$ degrees of freedom,
provided that the population from which the sample was drawn is normal or that the sample size is large $(n \geq 30)$ and $n \leq 0.05N$.

- A $(1 - \alpha) \cdot 100\%$ confidence interval for σ^2 is
$\dfrac{(n - 1)s^2}{\chi^2_{\alpha/2}} < \sigma^2 < \dfrac{(n - 1)s^2}{\chi^2_{1-\alpha/2}}$, where χ^2 has $n - 1$
degrees of freedom, provided that the population from which the sample was drawn is normal.

Sample Size

- To estimate the population proportion within a margin of error E at a $(1 - \alpha) \cdot 100\%$ level of confidence requires a sample of size $n = \hat{p}(1 - \hat{p})\left(\dfrac{z_{\frac{\alpha}{2}}}{E}\right)^2$ (rounded up to the next integer), where $\hat{p}$ is a prior estimate of the population proportion.

- To estimate the population proportion within a margin of error E at a $(1 - \alpha) \cdot 100\%$ level of confidence requires a sample of size $n = 0.25\left(\dfrac{z_{\frac{\alpha}{2}}}{E}\right)^2$ (rounded up to the next integer) when no prior estimate is available.

- To estimate the population mean within a margin of error E at a $(1 - \alpha) \cdot 100\%$ level of confidence requires a sample of size $n = \left(\dfrac{z_{\frac{\alpha}{2}} \cdot s}{E}\right)^2$ (rounded up to the next integer).

OBJECTIVES

Section	You should be able to . . .	Example(s)	Review Exercises
9.1	**1** Obtain a point estimate for the population proportion (p. 421)	1	16(a)
	2 Construct and interpret a confidence interval for the population proportion (p. 421)	2–5	16(b)
	3 Determine the sample size necessary for estimating a population proportion within a specified margin of error (p. 429)	6	16(c), 16(d)
9.2	**1** Obtain a point estimate for the population mean (p. 435)	1	15(a)
	2 State properties of Student's t-distribution (p. 435)	2	6, 7, 8
	3 Determine t-values (p. 438)	3	1
	4 Construct and interpret a confidence interval for a population mean (p. 439)	4	9, 10(a)–(c), 11(b), 12(b), 13(b), 13(d), 14(b), 14(c), 15(c)
	5 Determine the sample size needed to estimate a population mean within a specified margin of error (p. 441)	5	11(c)
9.3	**1** Find critical values for the chi-square distribution (p. 449)	1	2
	2 Construct and interpret confidence intervals for the population variance and standard deviation (p. 451)	2	10(d), 15(d)

Section	You should be able to . . .	Example(s)	Review Exercises
9.4	1 Determine the appropriate confidence interval to construct (p. 455)	1	9–16
9.5	1 Estimate a parameter using the bootstrap method (p. 458)	1, 2	17

Review Exercises

1. Find the critical *t*-value for constructing a confidence interval for a population mean at the given level of confidence for the given sample size, *n*.
(a) 99% confidence; $n = 18$
(b) 90% confidence; $n = 27$

2. Find the critical values $\chi^2_{1-\alpha/2}$ and $\chi^2_{\alpha/2}$ required to construct a confidence interval for σ for the given level of confidence and sample size.
(a) 95% confidence; $n = 22$
(b) 99% confidence; $n = 12$

3. IQ Scores Many of the examples and exercises in the text have dealt with IQ scores. We now know that IQ scores based on the Stanford–Binet IQ test are approximately normally distributed with a mean of 100 and standard deviation 15. If you were to obtain 100 different simple random samples of size 20 from the population of all adult humans and determine 95% confidence intervals for each of them, how many of the intervals would you expect to include 100? What causes a particular interval to not include 100?

4. What does the 95% represent in a 95% confidence interval?

5. For what proportion of samples will a 90% confidence interval for a population mean not capture the true population mean?

6. The area under the *t*-distribution with 18 degrees of freedom to the right of $t = 1.56$ is 0.0681. What is the area under the *t*-distribution with 18 degrees of freedom to the left of $t = -1.56$? Why?

7. Which is larger, the area under the *t*-distribution with 10 degrees of freedom to the right of $t = 2.32$ or the area under the standard normal distribution to the right of $z = 2.32$? Why?

8. State the properties of Student's *t*-distribution.

9. A simple random sample of size *n* is drawn from a population. The sample mean, $\bar{x}$, is 54.8 and the sample standard deviation is 10.5.
(a) Construct the 90% confidence interval for the population mean if the sample size, *n*, is 30.
(b) Construct the 90% confidence interval for the population mean if the sample size, *n*, is 51. How does increasing the sample size affect the width of the interval?
(c) Construct the 99% confidence interval for the population mean if the sample size, *n*, is 30. Compare the results to those obtained in part (a). How does increasing the level of confidence affect the confidence interval?

10. A simple random sample of size *n* is drawn from a population that is known to be normally distributed. The sample mean, $\bar{x}$, is determined to be 104.3 and the sample standard deviation, *s*, is determined to be 15.9.
(a) Construct the 90% confidence interval for the population mean if the sample size, *n*, is 15.
(b) Construct the 90% confidence interval for the population mean if the sample size, *n*, is 25. How does increasing the sample size affect the width of the interval?
(c) Construct the 95% confidence interval for the population mean if the sample size, *n*, is 15. Compare the results to those obtained in part (a). How does increasing the level of confidence affect the confidence interval?
(d) Construct the 90% confidence interval for the population standard deviation if the sample size, *n*, is 15.

11. Long Life? In a survey of 35 adult Americans, it was found that the mean age (in years) that people would like to live to is 87.9 with a standard deviation of 15.5. An analysis of the raw data indicates the distribution is skewed left.
(a) Explain why a large sample size is necessary to construct a confidence interval for the mean age that people would like to live to.
(b) Construct and interpret a 95% confidence interval for the mean.
(c) How many adult Americans would need to be surveyed to estimate the mean age that people would like to live to within 2 years with 95% confidence?

12. E-mail The General Social Survey asked: "How many e-mails do you send in a day?" The results of 928 respondents indicate that the mean number of e-mails sent in a day is 10.4, with a standard deviation of 28.5.
(a) Given the fact that 1 standard deviation to the left of the mean results in a negative number of e-mails being sent, what shape would you expect the distribution of e-mails sent to have?
(b) Construct and interpret a 90% confidence interval for the mean number of e-mails sent per day.

13. Caffeinated Sports Drinks Researchers conducted an experiment to determine the effectiveness of a commercial caffeinated carbohydrate–electrolyte sports drink compared with a placebo. Sixteen highly trained cyclists each completed two trials of prolonged cycling in a warm environment, one while receiving the sports drink and another while receiving a placebo. For a given trial, one beverage treatment was administered throughout a 2-hour variable-intensity cycling bout followed by a 15-minute performance ride. Total work (in kilojoules) performed during the final 15 minutes was used to measure performance. The beverage order for individual subjects was randomly assigned with a period of at least five days separating the trials. Assume that the researchers verified the normality

of the population of total work performed for each treatment.
Source: Kirk J. Cureton, Gordon L. Warren, et al., "Caffeinated Sports Drink: Ergogenic Effects and Possible Mechanisms." *International Journal of Sport Nutrition and Exercise Metabolism* 17(1):35–55, 2007

(a) Why do you think the sample size was small ($n = 16$) for this experiment?

(b) For the sports-drink treatment, the mean total work performed during the performance ride for the $n = 16$ riders was 218 kilojoules, with standard deviation 31 kilojoules. Construct and interpret a 95% confidence interval for the population mean total work performed.

(c) Is it possible for the population mean total work performed for the sports-drink treatment to be less than 198 kilojoules? Do you think this is likely?

(d) For the placebo treatment, the mean total work performed during the performance ride for the $n = 16$ riders was 178 kilojoules, with standard deviation 31 kilojoules. Construct and interpret a 95% confidence interval for the population mean total work performed.

(e) Is it possible for the population mean total work performed for the placebo treatment to be more than 198 kilojoules? Do you think this is likely?

(f) The researchers concluded that the caffeinated carbohydrate–electrolyte sports drink substantially enhanced physical performance during prolonged exercise compared with the placebo. Do your findings in parts (b) and (d) support the researchers' conclusion? Explain.

14. Family Size A random sample of 60 married couples who have been married 7 years was asked the number of children they have. The results of the survey are as follows:

0	0	0	3	3	3	1	3	2	2	3	1
3	2	4	0	3	3	3	1	0	2	3	3
1	4	2	3	1	3	3	5	0	2	3	0
4	4	2	2	3	2	2	2	2	3	4	3
2	2	1	4	3	2	4	2	1	2	3	2

Note: $\bar{x} = 2.27$, $s = 1.22$.

(a) What is the shape of the distribution of the sample mean? Why?

(b) Compute a 95% confidence interval for the mean number of children of all couples who have been married 7 years. Interpret this interval.

(c) Compute a 99% confidence interval for the mean number of children of all couples who have been married 7 years. Interpret this interval.

15. Diameter of Douglas Fir Trees The diameter of the Douglas fir tree is measured at a height of 1.37 meters. The following data represent the diameter in centimeters of a random sample of 12 Douglas firs in the western Washington Cascades.

156	190	147	173	159	181
162	130	101	147	113	109

Source: L. Winter. "Live Tree and Tree-Ring Records to Reconstruct the Structural Development of an Old-Growth Douglas Fir/Western Hemlock Stand in the Western Washington Cascades." Corvallis, OR: Forest Science Data Bank, 2005.

(a) Obtain a point estimate for the mean and standard deviation diameter of a Douglas fir tree in the western Washington Cascades.

(b) Because the sample size is small, we must verify that the data come from a population that is normally distributed and

that the data do not contain any outliers. The figures show the normal probability plot and boxplot. The correlation between the tree diameters and expected z-scores is 0.982. Are the conditions for constructing a confidence interval for the population mean diameter satisfied?

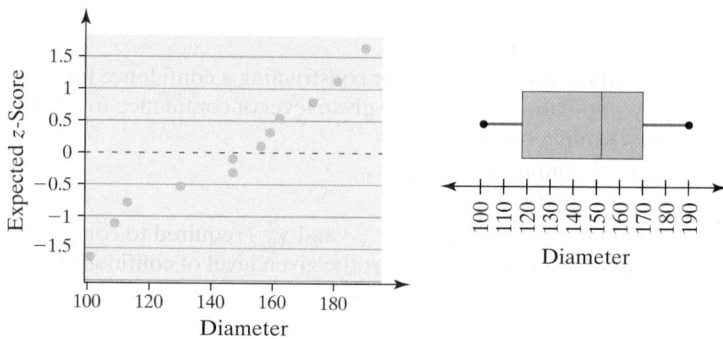

Normal Probability Plot of Diameter of Douglas Fir Trees

(c) Construct a 95% confidence interval for the mean diameter of a Douglas fir tree in the western Washington Cascades.

(d) Construct a 95% confidence interval for the standard deviation diameter of a Douglas fir tree in the western Washington Cascades.

16. Hypertension In a random sample of 678 adult males 20 to 34 years of age, it was determined that 58 of them have hypertension (high blood pressure).
Source: The Centers for Disease Control.

(a) Obtain a point estimate for the proportion of adult males 20 to 34 years of age who have hypertension.

(b) Construct a 95% confidence interval for the proportion of adult males 20 to 34 years of age who have hypertension. Interpret the confidence interval.

(c) You wish to conduct your own study to determine the proportion of adult males 20 to 34 years old who have hypertension. What sample size would be needed for the estimate to be within 3 percentage points with 95% confidence if you use the point estimate obtained in part (a)?

(d) You wish to conduct your own study to determine the proportion of adult males 20 to 34 years old who have hypertension. What sample size would be needed for the estimate to be within 3 percentage points with 95% confidence if you don't have a prior estimate?

17. Resampling Douglas Fir Trees Use the data from Problem 15 on the diameter of Douglas fir trees.

(a) Construct and interpret a 95% confidence interval for the mean diameter of a Douglas fir tree using a bootstrap sample with 1000 resamples. Compare the bootstrap confidence interval to the *t*-interval found in Problem 15(c).

(b) Draw a histogram of the 1000 means. What is the shape of the histogram? Why does the result not surprise you?

(c) Construct and interpret a 95% confidence interval for the median diameter of a Douglas fir tree using a bootstrap sample with 1000 resamples.

Chapter Test

1. Find the critical t-value for constructing a confidence interval about a population mean at the given level of confidence for the given sample size, n.

(a) 96% confidence; $n = 26$
(b) 98% confidence; $n = 18$

2. Find the critical values $\chi^2_{1-\alpha/2}$ and $\chi^2_{\alpha/2}$ required to construct a confidence interval about σ for the given level of confidence and sample size.

(a) 98% confidence; $n = 16$
(b) 90% confidence; $n = 30$

3. Determine the point estimate of the population mean and margin of error if the confidence interval has lower bound: 125.8 and upper bound: 152.6.

4. A question on the General Social Survey was this: "How many family members do you know that are in prison?" The results of 499 respondents indicate that the mean number of family members in jail is 1.22, with a standard deviation of 0.59.

(a) What shape would you expect the distribution of this variable to have? Why?
(b) Construct and interpret a 99% confidence interval for the mean number of family members in jail.

5. A random sample of 50 recent college graduates results in a mean time to graduate of 4.58 years, with a standard deviation of 1.10 years.
Source: Based on data from *The Toolbox Revisited* by Clifford Adelman, U.S. Department of Education

(a) Compute and interpret a 90% confidence interval for time to graduate with a bachelor's degree.
(b) Does this evidence contradict the widely held belief that it takes 4 years to complete a bachelor's degree? Why?

6. The campus at Joliet Junior College has a lake. A student used a Secchi disk to measure the clarity of the lake's water by lowering the disk into the water and measuring the distance below the water surface at which the disk is no longer visible. The following measurements (in inches) were taken on the lake at various points in time over the course of a year.

82	64	62	66	68	43
38	26	68	56	54	66

Source: Virginia Piekarski, Joliet Junior College

(a) Use the data to compute a point estimate for the population mean and population standard deviation.
(b) Because the sample size is small, we must verify that the data are normally distributed and do not contain any outliers. The figures in the next column show the normal probability plot and boxplot. The correlation between depth and expected z-scores is 0.960. Are the conditions for constructing a confidence interval about μ satisfied?

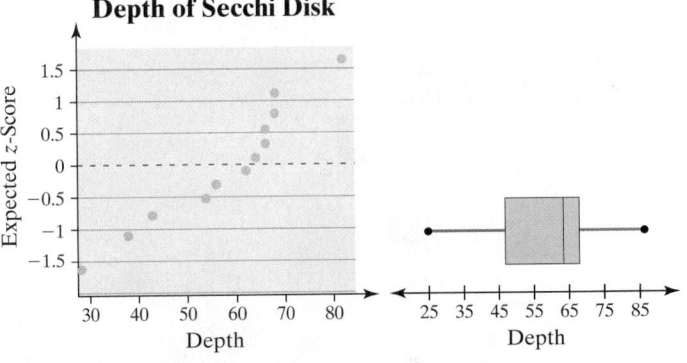

Normal Probability Plot of Depth of Secchi Disk

(c) Construct a 95% confidence interval for the mean Secchi disk measurement. Interpret this interval.
(d) Construct a 99% confidence interval for the mean Secchi disk measurement. Interpret this interval.
(e) Construct a 95% confidence interval for the population standard deviation Secchi disk measurement. Interpret this interval.

7. From a random sample of 1201 Americans, it was discovered that 1139 of them lived in neighborhoods with acceptable levels of carbon monoxide.
Source: Environmental Protection Agency.

(a) Obtain a point estimate for the proportion of Americans who live in neighborhoods with acceptable levels of carbon monoxide.
(b) Construct a 99% confidence interval for the proportion of Americans who live in neighborhoods with acceptable levels of carbon monoxide.
(c) You wish to conduct your own study to determine the proportion of Americans who live in neighborhoods with acceptable levels of carbon monoxide. What sample size would be needed for the estimate to be within 1.5 percentage points with 90% confidence if you use the estimate obtained in part (a)?
(d) You wish to conduct your own study to determine the proportion of Americans who live in neighborhoods with acceptable levels of carbon monoxide. What sample size would be needed for the estimate to be within 1.5 percentage points with 90% confidence if you do not have a prior estimate?

8. Wimbledon Match Lengths A tennis enthusiast wants to estimate the mean length of men's singles matches held during the Wimbledon tennis tournament. From the Wimbledon history archives, he randomly selects 40 matches played during the tournament since the year 1968 (when professional players were first allowed to participate). The lengths of the 40 selected matches, in minutes, follow:

110	76	84	231	122	115	87	137
101	119	138	132	136	111	194	92
198	153	256	146	149	103	116	163
182	132	123	178	140	151	115	107
202	128	60	89	94	95	89	182

Source: www.wimbledon.org

(a) Obtain a point estimate of the population mean length of men's singles matches during Wimbledon.

(b) A frequency histogram of the data is shown next. Explain why a large sample is necessary to construct a confidence interval for the population mean length of men's singles matches during Wimbledon.

Wimbledon Match Lengths

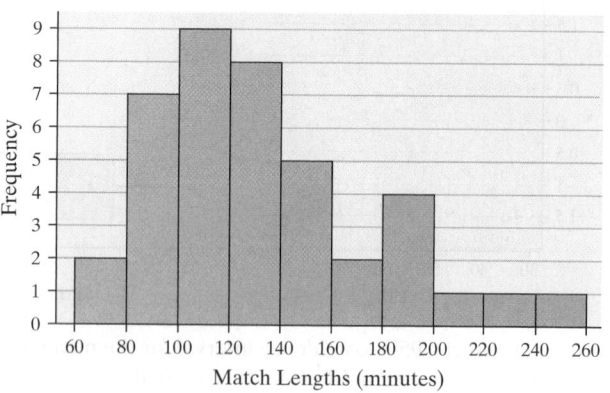

(c) Construct and interpret a 99% confidence interval for the population mean length of men's singles matches during Wimbledon.

(d) Construct and interpret a 95% confidence interval for the population mean length of men's singles matches during Wimbledon.

(e) What effect does increasing the level of confidence have on the interval?

(f) Do the confidence intervals computed in parts (c) and (d) represent an estimate for the population mean length of men's singles matches during all professional tennis tournaments? Why?

9. The given data represent the wait time (in seconds) for a random sample of 14 individuals before connecting with a customer service agent.

66	37	4	150	47	12	16
62	12	21	124	25	49	3

(a) A boxplot of the data is shown next. The quality control manager of the customer service department wants to estimate the mean wait time. Explain why you would not recommend the construction of a t-interval to estimate the mean.

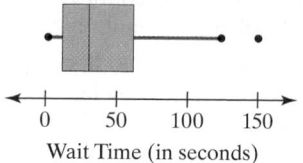

Wait Time (in seconds)

(b) Explain how the bootstrapping method can be used to obtain a 95% confidence interval for the mean using 1000 resamples.

(c) Obtain a 95% confidence interval for the mean wait time using 1000 resamples.

Making an Informed Decision

How Much Should I Spend for this House?

One of the biggest purchases we make in our lifetimes is for a home. Questions that we all ask are these:

- How much should I spend for a particular home?
- How many bathrooms are there?
- How long should I expect a home to be on the market?
- What is the cost per square foot?

The purpose of this project is to help you make an informed decision about housing values. This will help to ensure you receive a good deal when purchasing a home.

(a) Go to a real estate Web site such as www.realtor.com or www.zillow.com and enter the particular zip code you are interested in moving to. Randomly select at least 30 homes for sale and record the following information:

- Asking price
- Square footage
- Number of days on the market
- Cost per square foot (asking price divided by square footage)

(b) For each of the variables identified, determine a 95% confidence interval. Interpret the interval.

(c) Now randomly select 30 recently sold homes and determine the percentage discount from the asking price. This is determined by computing

$$\frac{\text{asking price} - \text{closing price}}{\text{asking price}}$$

(d) Determine a 95% confidence interval for percentage discount. Interpret the interval.

(e) For the type of house you are considering (such as a 2400 square foot 3-bedroom/2-bath home), identify at least 20 homes that are for sale in the neighborhood you are considering. Compute a 95% confidence interval for the asking price of this type of home.

(f) Write a report that details how much you should expect to pay for the type of house you are considering.

CASE STUDY Fire-Safe Cigarettes

In 2011, 90,000 fires in the United States were started by a lighted tobacco product. The dollar value of the property lost in these fires is staggering. In that same year, fires caused by dropped or discarded cigarettes resulted in 540 deaths. Twenty-five percent of the victims of cigarette induced fires were not the smokers, and about 40% of cigarette-induced fires started in the bedroom. (*Source:* www.firesafecigarettes.org)

Certainly, it makes sense to develop a cigarette that extinguishes itself when left unattended (as would happen if someone falls asleep while smoking). The state of New York decided to require cigarettes sold in its stores to be designed so they self-extinguished when left unattended. Two types of cigarettes were tested to determine their propensity to self-extinguish. For each type (A and B), 40 cigarettes were lit and allowed to burn unattended until the cigarette either extinguished or did not. Cigarette brand A was designed to have ultra thin concentric paper bands affixed to the traditional cigarette paper. These bands are referred to as "speed bumps" and cause extinguishing of the cigarette by restricting the flow of oxygen to the burning ember. Cigarette brand B was a traditionally designed model. The results of the experiment are in the following table, where E represents an extinguished cigarette and F represents a full-burn cigarette.

BRAND A

E	E	E	E	E	E	F	F	F	F
E	E	F	E	E	E	E	E	E	E
E	E	E	E	E	F	E	F	F	E
F	E	E	F	F	F	E	F	E	E

BRAND B

F	F	F	F	F	F	F	F	F	F
F	F	F	F	F	F	E	F	F	F
F	F	F	F	F	F	F	F	F	F
F	F	F	F	F	F	F	F	F	F

Naturally, cigarette manufacturers were concerned about the additional cost of manufacturing such a cigarette. Plus, the manufacturers were concerned that consumers might prefer the cigarettes without "speed bumps" to those with design changes. One particular measure the manufacturers were concerned about was the amount of nicotine in each cigarette. The following data represent the amount of nicotine in each brand of cigarette for a random sample of 15 cigarettes.

BRAND A

1.40	1.36	1.42	1.30	1.40
1.09	1.06	1.06	1.12	1.12
1.14	1.14	1.13	1.21	1.15

BRAND B

1.24	1.36	1.32	1.32	1.36
1.23	1.13	1.20	1.28	1.10
1.18	1.19	1.20	1.20	1.20

Write a report detailing the propensity of each brand to self-extinguish and the level of nicotine in each brand. Include any relevant confidence intervals. If the normal model or Student's *t*-distribution cannot be used to construct a confidence interval, use alternative models (such as the Agresti–Coull model for estimating proportions—see Problems 45 and 46 from Section 9.1 or the bootstrap methods of Section 9.5). Would you support legislation that reduced the risk of fire from unattended cigarettes? How much extra would you be willing to pay for such a cigarette if you were a smoker?

10

Hypothesis Tests Regarding a Parameter

Making an Informed Decision

Suppose you have just received a $1000 bonus at your job. Rather than waste the money on frivolous items, you decide to invest the money so you can use it later to buy a home. You have many investment options. Your family and friends who have some experience investing recommend mutual funds. Which fund should you choose? See the Decision project on page 522.

PUTTING IT TOGETHER

In Chapter 9, we mentioned there are two types of inferential statistics: (1) estimation and (2) hypothesis testing. We have already discussed the procedures for estimating the population proportion, the population mean, and the population standard deviation.

We now focus our attention on hypothesis testing. Hypothesis testing is used to test statements regarding a characteristic of one or more populations. In this chapter, we will test various hypotheses regarding a single population parameter. The hypotheses that we test concern the population proportion, the population mean, and the population standard deviation.

10.1   The Language of Hypothesis Testing

Preparing for This Section   Before getting started, review the following:

- Parameter versus statistic (Section 1.1, p. 5)
- Simple random sampling (Section 1.3, pp. 22–25)
- Table 9 (Section 6.2, p. 332)
- Sampling distribution of $\bar{x}$ (Section 8.1, pp. 395–403)

Objectives   ❶ Determine the null and alternative hypotheses

❷ Explain Type I and Type II errors

❸ State conclusions to hypothesis tests

We begin with an example.

EXAMPLE 1   Is Your Friend Cheating?

Problem   A friend of yours wants to play a simple coin-flipping game. If the coin comes up heads, you win; if it comes up tails, your friend wins. Suppose the outcome of five plays of the game is T, T, T, T, T. Is your friend cheating?

Approach   To decide whether your friend is cheating, determine the likelihood of obtaining five tails in a row. Assume that the coin is fair so $P(\text{tail}) = P(\text{head}) = \dfrac{1}{2}$ and the flips of the coin are independent. Next ask, "Is it unusual to obtain five tails in a row with a fair coin?"

Solution   We will determine the probability of getting five tails in a row, assuming that the coin is fair. The flips of the coin are independent, so

$$
\begin{aligned}
P(\text{five tails in a row}) &= P(\text{T and T and T and T and T}) \\
&= P(\text{T}) \cdot P(\text{T}) \cdot P(\text{T}) \cdot P(\text{T}) \cdot P(\text{T}) \\
&= \frac{1}{2} \cdot \frac{1}{2} \cdot \frac{1}{2} \cdot \frac{1}{2} \cdot \frac{1}{2} \\
&= \left(\frac{1}{2}\right)^5 \\
&= 0.03125
\end{aligned}
$$

If we flipped a *fair* coin 5 times, 100 different times, we would expect about 3 of the 100 experiments to result in all tails. So what we observed is possible but not likely. You can make one of two conclusions:

1. Your friend is not cheating and happens to be lucky.
2. Your friend is not using a fair coin (that is, the probability of obtaining a tail on one flip is greater than $\dfrac{1}{2}$) and is cheating.

Is your friend cheating, or did you just happen to experience an unusual result from a fair coin?

This is at the heart of *hypothesis testing*. We make an assumption about reality (in this case, the probability of obtaining a tail is $\dfrac{1}{2}$). We then look at (or gather) sample evidence to determine whether it contradicts our assumption.   ●

❶ Determine the Null and Alternative Hypotheses

According to dictionary.com, a **hypothesis** is a proposition assumed as a premise in an argument. The word *hypothesis* comes from the Greek word *hypotithenai*, which means "to suppose." The definition of *hypothesis* in statistics is given next.

Definition   A **hypothesis** is a statement regarding a characteristic of one or more populations.

In this chapter, we look at hypotheses regarding a single population parameter. Consider the following:

(A) According to a Gallup poll conducted in 2008, 80% of Americans felt satisfied with the way things were going in their personal lives. A researcher wonders if the percentage of satisfied Americans is different today (a statement regarding a population proportion).

(B) The packaging on a light bulb states that the bulb will last 500 hours under normal use. A consumer advocate would like to know if the mean lifetime of a bulb is less than 500 hours (a statement regarding the population mean).

(C) The standard deviation of the rate of return for a certain class of mutual funds is 0.08 percent. A mutual fund manager believes the standard deviation of the rate of return for his fund is less than 0.08 percent (a statement regarding the population standard deviation).

We test these types of statements using sample data because it is usually impossible or impractical to gain access to the entire population. The procedure (or process) we use to test such statements is called *hypothesis testing*.

> **CAUTION!**
> If population data are available, there is no need for inferential statistics.

Definition   **Hypothesis testing** is a procedure, based on sample evidence and probability, used to test statements regarding a characteristic of one or more populations.

The basic steps in conducting a hypothesis test are these:

Steps in Hypothesis Testing

1. Make a statement regarding the nature of the population.
2. Collect evidence (sample data) to test the statement.
3. Analyze the data to assess the plausibility of the statement.

Because we use sample data to test hypotheses, we cannot state with 100% certainty that the statement is true; we can only determine whether the sample data support the statement or not. In fact, because the statement can be either true or false, hypothesis testing is based on two types of hypotheses. In this chapter, the hypotheses will be statements regarding the value of a population parameter.

Definitions   The **null hypothesis**, denoted H_0 (read "H-naught"), is a statement to be tested. The null hypothesis is a statement of no change, no effect, or no difference and is assumed true until evidence indicates otherwise.

The **alternative hypothesis**, denoted H_1 (read "H-one"), is a statement that we are trying to find evidence to support.

In Other Words
The null hypothesis is a statement of *status quo* or *no difference* and always contains a statement of equality. The null hypothesis is assumed to be true until we have evidence to the contrary. We seek evidence that supports the statement in the alternative hypothesis.

In this chapter, there are three ways to set up the null and alternative hypotheses.

1. Equal hypothesis versus not equal hypothesis **(two-tailed test)**
$$H_0: \text{parameter} = \text{some value}$$
$$H_1: \text{parameter} \neq \text{some value}$$

2. Equal versus less than **(left-tailed test)**
$$H_0: \text{parameter} = \text{some value}$$
$$H_1: \text{parameter} < \text{some value}$$

3. Equal versus greater than **(right-tailed test)**
$$H_0: \text{parameter} = \text{some value}$$
$$H_1: \text{parameter} > \text{some value}$$

Left- and right-tailed tests are referred to as **one-tailed tests**. Notice that in the left-tailed test the direction of the inequality sign in the alternative hypothesis points to the left ($<$), while in the right-tailed test the direction of the inequality sign in the alternative hypothesis points to the right ($>$). In all three tests, the null hypothesis contains a statement of equality.

Refer to the three hypotheses made on the previous page. In Situation A, the null hypothesis is expressed using the notation H_0: $p = 0.80$. This is a statement of *status quo* or no difference. The Latin phrase *status quo* means "the existing state or condition." So, the statement in the null hypothesis means that American opinions have not changed from 2008. We are trying to show that the proportion is different today, so the alternative hypothesis is H_1: $p \neq 0.80$. In Situation B, the null hypothesis is H_0: $\mu = 500$ hours. This is a statement of no difference between the population mean and the lifetime stated on the label. We are trying to show that the mean lifetime is less than 500 hours, so the alternative hypothesis is H_1: $\mu < 500$ hours. In Situation C, the null hypothesis is H_0: $\sigma = 0.08$ percent. This is a statement of no difference between the population standard deviation rate of return of the manager's mutual fund and all mutual funds. The alternative hypothesis is H_1: $\sigma < 0.08$ percent. Do you see why?

The statement we are trying to gather evidence for, which is dictated by the researcher before any data are collected, determines the structure of the alternative hypothesis (two-tailed, left-tailed, or right-tailed). For example, the label on a can of soda states that the can contains 12 ounces of liquid. A consumer advocate would be concerned only if the mean contents are less than 12 ounces, so the alternative hypothesis is H_1: $\mu < 12$ ounces. However, a quality-control engineer for the soda manufacturer would be concerned if there is too little or too much soda in the can, so the alternative hypothesis would be H_1: $\mu \neq 12$ ounces. In both cases, however, the null hypothesis is a statement of no difference between the manufacturer's assertion on the label and the actual mean contents of the can, so the null hypothesis is H_0: $\mu = 12$ ounces.

EXAMPLE 2   Forming Hypotheses

Problem Determine the null and alternative hypotheses. State whether the test is two-tailed, left-tailed, or right-tailed.

(a) The Medco pharmaceutical company has just developed a new antibiotic for children. Two percent of children taking competing antibiotics experience headaches as a side effect. A researcher for the Food and Drug Administration wishes to know if the percentage of children taking the new antibiotic who experience headaches as a side effect is more than 2%.

(b) The *Blue Book* price of a used three-year-old Chevy Corvette ZR1 is \$86,012. Grant wonders if the mean price of a used three-year-old Chevy Corvette ZR1 in the Miami metropolitan area is different from \$86,012.

(c) The standard deviation of the contents in a 64-ounce bottle of detergent using an old filling machine is 0.23 ounce. The manufacturer wants to know if a new filling machine has less variability.

Approach In each case, we must determine the parameter to be tested, the statement of no change or no difference (status quo), and the statement we are attempting to gather evidence for.

In Other Words
Structuring the null and alternative hypotheses:
1. Identify the parameter to be tested.
2. Determine the status quo value of the parameter. This gives the null hypothesis.
3. Determine the statement that reflects what we are trying to gather evidence for. This gives the alternative hypothesis.

Solution
(a) The hypothesis deals with a population proportion, p. If the new drug is no different from competing drugs, the proportion of individuals taking it who experience a headache will be 0.02; so the null hypothesis is H_0: $p = 0.02$. We want to determine if the proportion of individuals who experience a headache is more than 0.02, so the alternative hypothesis is H_1: $p > 0.02$. This is a right-tailed test because the alternative hypothesis contains a $>$ symbol.

(b) The hypothesis deals with a population mean, μ. If the mean price of a three-year-old Corvette ZR1 in Miami is no different from the *Blue Book* price, then the population mean in Miami will be \$86,012, so the null hypothesis is

(continued)

In Other Words
Look for key phrases when forming the alternative hypothesis. For example, *more than* means >; *different from* means ≠ and *less than* means <. See Table 9 on page 332 for a list of key phrases and the symbols they translate into.

● **Now Work Problem 17(a)**

H_0: μ = \$86,012. Grant wants to know if the mean price is different from \$86,012, so the alternative hypothesis is H_1: $\mu \neq$ \$86,012. This is a two-tailed test because the alternative hypothesis contains a $\neq$ symbol.

(c) The hypothesis deals with a population standard deviation, σ. If the new machine is no different from the old one, the standard deviation of the amount in the bottles filled by the new machine will be 0.23 ounce, so the null hypothesis is H_0: σ = 0.23 ounce. The company wants to know if the new machine has *less* variability than the old machine, so the alternative hypothesis is H_1: σ < 0.23 ounce. This is a left-tailed test because the alternative hypothesis contains a < symbol.   ●

② Explain Type I and Type II Errors

In Other Words
When you are testing a hypothesis, there is always the possibility that your conclusion will be wrong. To make matters worse, you won't know whether you are wrong or not! Don't fret, however; we have tools to help manage these incorrect conclusions.

Sample data is used to decide whether or not to reject the statement in the null hypothesis. Because this decision is based on incomplete (sample) information, there is the possibility of making an incorrect decision. In fact, there are four possible outcomes from hypothesis testing.

Four Outcomes from Hypothesis Testing

1. Reject the null hypothesis when the alternative hypothesis is true. This decision would be correct.
2. Do not reject the null hypothesis when the null hypothesis is true. This decision would be correct.
3. Reject the null hypothesis when the null hypothesis is true. This decision would be incorrect. This type of error is called a **Type I error**.
4. Do not reject the null hypothesis when the alternative hypothesis is true. This decision would be incorrect. This type of error is called a **Type II error**.

Figure 1 illustrates the two types of errors that can be made in hypothesis testing.

Figure 1

		Reality	
		H_0 Is True	H_1 Is True
Conclusion	Do Not Reject H_0	Correct Conclusion	Type II Error
	Reject H_0	Type I Error	Correct Conclusion

We illustrate the idea of Type I and Type II errors by looking at hypothesis testing from the point of view of a criminal trial. In any trial, the defendant is assumed to be innocent. (We give the defendant the benefit of the doubt.) The district attorney must collect and present evidence proving that the defendant is guilty beyond all reasonable doubt.

Because we are seeking evidence for guilt, it becomes the alternative hypothesis. Innocence is assumed, so it is the null hypothesis.

H_0: the defendant is innocent

H_1: the defendant is guilty

In a trial, the jury obtains information (sample data). It then deliberates about the evidence (the data analysis). Finally, it either convicts the defendant (rejects the null hypothesis) or declares the defendant not guilty (fails to reject the null hypothesis).

Note that the defendant is never declared innocent. That is, the null hypothesis is never declared true. The two correct decisions are to declare an innocent person not guilty or declare a guilty person to be guilty. The two incorrect decisions are to convict

In Other Words
A Type I error is like putting an innocent person in jail. A Type II error is like letting a guilty person go free.

an innocent person (a Type I error) or to let a guilty person go free (a Type II error). It is helpful to think in this way when trying to remember the difference between a Type I and a Type II error.

EXAMPLE 3   Type I and Type II Errors

Problem The Medco pharmaceutical company has just developed a new antibiotic. Two percent of children taking competing antibiotics experience headaches as a side effect. A researcher for the Food and Drug Administration wishes to know if the percentage of children taking the new antibiotic who experience a headache as a side effect is more than 2%. The researcher conducts a hypothesis test with $H_0: p = 0.02$ and $H_1: p > 0.02$. Explain what it would mean to make a (a) Type I error and (b) Type II error.

Approach A Type I error occurs if we reject the null hypothesis when it is true. A Type II error occurs if we do not reject the null hypothesis when the alternative hypothesis is true.

Solution
(a) A Type I error is made if the sample evidence leads the researcher to believe that $p > 0.02$ (that is, we reject the null hypothesis) when, in fact, the proportion of children who experience a headache is not greater than 0.02.
(b) A Type II error is made if the researcher does not reject the null hypothesis that the proportion of children experiencing a headache is equal to 0.02 when, in fact, the proportion of children who experience a headache is more than 0.02. In other words, the sample evidence led the researcher to believe $p = 0.02$ when in fact the true proportion is some value larger than 0.02.

● **Now Work Problems 17(b) and (c)**

The Probability of Making a Type I or Type II Error

When we studied confidence intervals, we learned that we never know whether a confidence interval contains the unknown parameter. We only know the likelihood that a confidence interval captures the parameter. Similarly, we never know whether the conclusion of a hypothesis test is correct. However, just as we place a level of confidence in the construction of a confidence interval, we can assign probabilities to making Type I or Type II errors when testing hypotheses. The following notation is commonplace:

$$\alpha = P(\text{Type I error}) = P(\text{rejecting } H_0 \text{ when } H_0 \text{ is true})$$
$$\beta = P(\text{Type II error}) = P(\text{not rejecting } H_0 \text{ when } H_1 \text{ is true})$$

The symbol β is the Greek letter beta (pronounced "BAY tah"). The probability of making a Type I error, α, is chosen by the researcher *before* the sample data are collected. This probability is referred to as the *level of significance*.

Definition   The **level of significance**, α, is the probability of making a Type I error.

The choice of the level of significance depends on the consequences of making a Type I error. If the consequences are severe, the level of significance should be small (say, $\alpha = 0.01$). However, if the consequences are not severe, a higher level of significance can be chosen (say $\alpha = 0.05$ or $\alpha = 0.10$).

Why is the level of significance not always set at $\alpha = 0.01$? Reducing the probability of making a Type I error increases the probability of making a Type II error, β. Using our court analogy, a jury is instructed that the prosecution must provide proof of guilt "beyond all reasonable doubt." This implies that we are choosing to make α small so that the probability of convicting an innocent person is very small. The consequence of the small α, however, is a large β, which means many guilty defendants will go free. For now, we are content to recognize the inverse relation between α and β (as one goes up the other goes down).

In Other Words
As the probability of a Type I error increases, the probability of a Type II error decreases, and vice versa.

③ **State Conclusions to Hypothesis Tests**

--CAUTION!------------------
We never accept the null hypothesis,
because, without having access to
the entire population, we don't know
the exact value of the parameter
stated in the null hypothesis. Rather,
we say that we do not reject the
null hypothesis. This is just like the
court system. We never declare a
defendant innocent, but rather say
the defendant is not guilty.

Once the decision whether or not to reject the null hypothesis is made, the researcher must state his or her conclusion. It is important to recognize that we never *accept* the null hypothesis. Again, the court system analogy helps to illustrate the idea. The null hypothesis is H_0: innocent. When the evidence presented to the jury is not enough to convict beyond all reasonable doubt, the jury's verdict is "not guilty."

Notice that the verdict does not state that the null hypothesis of innocence is true; it states that there is not enough evidence to conclude guilt. This is a huge difference. Being told that you are not guilty is very different from being told that you are innocent!

So, sample evidence can never prove the null hypothesis to be true. By not rejecting the null hypothesis, we are saying that the evidence indicates the null hypothesis *could* be true. That is, there is not enough evidence to reject our assumption that the null hypothesis is true.

EXAMPLE 4   Stating the Conclusion

Problem The Medco pharmaceutical company has just developed a new antibiotic. Two percent of children taking competing antibiotics experience a headache as a side effect. A researcher for the Food and Drug Administration believes that the proportion of children taking the new antibiotic who experience a headache as a side effect is more than 0.02. From Example 2(a), we know the null hypothesis is H_0: $p = 0.02$ and the alternative hypothesis is H_1: $p > 0.02$.

Suppose that the sample evidence indicates that

(a) the null hypothesis is rejected. State the conclusion.
(b) the null hypothesis is not rejected. State the conclusion.

In Other Words
The conclusion to a hypothesis test is always as follows: There (*is/is not*) sufficient evidence to conclude that *insert statement in alternative hypothesis*.

Approach When the null hypothesis is rejected, we say that there is sufficient evidence to support the statement in the alternative hypothesis. When the null hypothesis is not rejected, we say that there is not sufficient evidence to support the statement in the alternative hypothesis. We never say that the null hypothesis is true!

Solution
(a) The statement in the alternative hypothesis is that the proportion of children taking the new antibiotic who experience a headache as a side effect is more than 0.02. Because the null hypothesis ($p = 0.02$) is rejected, there is sufficient evidence to conclude that the proportion of children who experience a headache as a side effect is more than 0.02.
(b) Because the null hypothesis is not rejected, there is not sufficient evidence to say that the proportion of children who experience a headache as a side effect is more than 0.02.

● Now Work Problem 25

10.1 Assess Your Understanding

Vocabulary and Skill Building

1. A _____ is a statement regarding a characteristic of one or more populations.

2. _____ _____ is a procedure, based on sample evidence and probability, used to test statements regarding a characteristic of one or more populations.

3. The _____ is a statement of no change, no effect, or no difference.

4. The _____ _____ is a statement we are trying to find evidence to support.

5. If we reject the null hypothesis when the statement in the null hypothesis is true, we have made a Type ____ error.

6. If we do not reject the null hypothesis when the statement in the alternative hypothesis is true, we have made a Type _____ error.

7. The _____ — _____ is the probability of making a Type I error.

8. *True or False*: Sample evidence can prove a null hypothesis is true.

In Problems 9–14, the null and alternative hypotheses are given. Determine whether the hypothesis test is left-tailed, right-tailed, or two-tailed. What parameter is being tested?

9. H_0: $\mu = 5$
    H_1: $\mu > 5$

10. H_0: $p = 0.2$
    H_1: $p < 0.2$

11. $H_0: \sigma = 4.2$
$H_1: \sigma \neq 4.2$

12. $H_0: p = 0.76$
$H_1: p > 0.76$

13. $H_0: \mu = 120$
$H_1: \mu < 120$

14. $H_0: \sigma = 7.8$
$H_1: \sigma \neq 7.8$

In Problems 15–22, (a) determine the null and alternative hypotheses, (b) explain what it would mean to make a Type I error, and (c) explain what it would mean to make a Type II error.

15. Complete College For students who first enrolled in two-year public institutions in fall 2007, the proportion who earned a bachelor's degree within six years was 0.399. The president of Joliet Junior College believes that the proportion of students who enroll in her institution have a higher completion rate.

16. Pizza Historically, the time to order and deliver a pizza at Jimbo's pizza was 48 minutes. Jim, the owner, implements a new system for ordering and delivering pizzas that he believes will reduce the time required to get a pizza to his customers.

NW **17. Single-Family Home Price** According to the National Association of Home Builders, the mean price of an existing single-family home in 2013 was $245,700. A real estate broker believes that existing home prices in her neighborhood are lower.

18. Fair Packaging and Labeling Federal law requires that a jar of peanut butter that is labeled as containing 32 ounces must contain at least 32 ounces. A consumer advocate feels that a certain peanut butter manufacturer is shorting customers by underfilling the jars.

19. Valve Pressure The standard deviation in the pressure required to open a certain valve is known to be $\sigma = 0.7$ psi. Due to changes in the manufacturing process, the quality-control manager feels that the pressure variability has been reduced.

20. Overweight According to the Centers for Disease Control and Prevention, 19.6% of children aged 6 to 11 years are overweight. A school nurse thinks that the percentage of 6- to 11-year-olds who are overweight is different in her school district.

21. Cell Phone Service According to the *CTIA–The Wireless Association*, the mean monthly revenue per cell phone was $48.79 in 2014. A researcher suspects the mean monthly revenue per cell phone is different today.

22. SAT Reading Scores In 2014, the standard deviation of SAT score on the Critical Reading Test for all students taking the exam was 112. A teacher believes that, due to changes to high school curricula, the standard deviation of SAT math scores has decreased.

In Problems 23–34, state the conclusion based on the results of the test.

23. For the hypotheses in Problem 15, the null hypothesis is rejected.

24. For the hypotheses in Problem 16, the null hypothesis is not rejected.

NW **25.** For the hypotheses in Problem 17, the null hypothesis is not rejected.

26. For the hypotheses in Problem 18, the null hypothesis is rejected.

27. For the hypotheses in Problem 19, the null hypothesis is not rejected.

28. For the hypotheses in Problem 20, the null hypothesis is not rejected.

29. For the hypotheses in Problem 21, the null hypothesis is rejected.

30. For the hypotheses in Problem 22, the null hypothesis is not rejected.

31. For the hypotheses in Problem 15, the null hypothesis is not rejected.

32. For the hypotheses in Problem 16, the null hypothesis is rejected.

33. For the hypotheses in Problem 17, the null hypothesis is rejected.

34. For the hypotheses in Problem 18, the null hypothesis is not rejected.

Applying the Concepts

35. Quality Control A can of soda is labeled as containing 12 fluid ounces. The quality control manager wants to verify that the filling machine is neither over-filling nor under-filling the cans.

(a) Determine the null and alternative hypotheses that would be used to determine if the filling machine is calibrated correctly.

(b) The quality control manager obtains a sample of 75 cans and measures the contents. The sample evidence leads the manager to reject the null hypothesis. Write a conclusion for this hypothesis test.

(c) Suppose, in fact, the machine is not out of calibration. Has a Type I or Type II error been made?

(d) Management has informed the quality control department that it does not want to shut down the filling machine unless the evidence is overwhelming that the machine is out of calibration. What level of significance would you recommend the quality control manager use? Explain.

36. Popcorn Consumption According to popcorn.org, the mean consumption of popcorn annually by Americans is 54 quarts. The marketing division of popcorn.org unleashes an aggressive campaign designed to get Americans to consume even more popcorn.

(a) Determine the null and alternative hypotheses that would be used to test the effectiveness of the marketing campaign.

(b) A sample of 800 Americans provides enough evidence to conclude that the marketing campaign was effective. Provide a statement that should be put out by the marketing department.

(c) Suppose, in fact, that the mean annual consumption of popcorn after the marketing campaign is 53.4 quarts. Has a Type I or Type II error been made by the marketing department? If they tested the hypothesis at the $\alpha = 0.05$ level of significance, what is the probability of making a Type I error?

37. E-Cigs According to the Centers for Disease Control and Prevention, 2.8% of high school students currently use electronic cigarettes. A high school counselor is concerned the use of e-cigs at her school is higher.

(a) Determine the null and alternative hypotheses.

(b) If the sample data indicate that the null hypothesis should not be rejected, state the conclusion of the school counselor.

(c) Suppose, in fact, that the proportion of students at the counselor's high school who use electronic cigarettes is 0.034. Was a Type I or Type II error committed?

38. Migraines According to the Centers for Disease Control, 15.2% of American adults experience migraine headaches. Stress is a major contributor to the frequency and intensity of headaches. A massage therapist feels that she has a technique that can reduce the frequency and intensity of migraine headaches.

(a) Determine the null and alternative hypotheses that would be used to test the effectiveness of the massage therapist's techniques.

(b) A sample of 500 American adults who participated in the massage therapist's program results in data that indicate that the null hypothesis should be rejected. Provide a statement that supports the massage therapist's program.

(c) Suppose, in fact, that the percentage of patients in the program who experience migraine headaches is 15.3%. Was a Type I or Type II error committed?

39. Prolong Engine Treatment The manufacturer of Prolong Engine Treatment claims that if you add one 12-ounce bottle of its $20 product, your engine will be protected from excessive wear. An infomercial claims that a woman drove 4 hours without oil, thanks to Prolong. *Consumer Reports* magazine tested engines in which they added Prolong to the motor oil, ran the engines, drained the oil, and then determined the time until the engines seized.

(a) Determine the null and alternative hypotheses that *Consumer Reports* will test.

(b) Both engines took exactly 13 minutes to seize. What conclusion might *Consumer Reports* draw based on this evidence?

40. Refer to Problem 18. Researchers must choose the level of significance based on the consequences of making a Type I error. In your opinion, is a Type I error or Type II error more serious? Why? On the basis of your answer, decide on a level of significance, α. Be sure to support your opinion.

Retain Your Knowledge

41. Retirement Savings Designed by Bill Bengen, the 4 percent rule says that a retiree may withdraw 4% of savings during the first year of retirement, and then each year after that withdraw the same amount plus an adjustment for inflation. Under this rule, your retirement savings should be expected to last 30 years, which is longer than most retirements.

(a) If your retirement savings is $750,000, how much may you withdraw in your first year of retirement if you want the retirement savings to last 30 years?

(b) According to the American College of Financial Services, the proportion of people 60 to 75 years of age who believe it would be safe to withdraw 6 to 8 percent of their retirement savings annually is 0.16. Suppose you conduct a survey of twenty 60 to 75 year olds and ask them if it is safe to withdraw 6 to 8 percent of retirement savings annually if they wish their retirement savings to last 30 years. Explain why this is a binomial experiment. What are the values of n and p?

(c) In a random sample of twenty 60 to 75 year olds, what is the probability exactly 8 individuals will believe it is safe to withdraw 6 to 8 percent of retirement savings annually if they wish their retirement savings to last 30 years.

(d) In a random sample of twenty 60 to 75 year olds, what is the probability fewer than 8 individuals will believe it is safe to

withdraw 6 to 8 percent of retirement savings annually if they wish their retirement savings to last 30 years.

(e) Suppose you obtain a random sample of five hundred 60 to 75 year olds. Explain why the normal model may be used to describe the sampling distribution of $\hat{p}$ the sample proportion of 60 to 75 year olds who believe it is safe to withdraw 6 to 8 percent of their retirement savings annually. Describe this sampling distribution. That is, find the shape, center, and spread of the sampling distribution of the sample proportion.

(f) Use the normal model from part (e) to approximate the probability of obtaining a random sample of at least one hundred 60 to 75 years olds who believe it would be safe to withdraw 6 to 8 percent of their retirement savings annually assuming the true proportion is 0.16. Is this result unusual? Explain.

Explaining the Concepts

42. If the consequences of making a Type I error are severe, would you choose the level of significance, α, to equal 0.01, 0.05, or 0.10? Why?

43. What happens to the probability of making a Type II error, β, as the level of significance, α, decreases? Why?

44. The following is a quotation from Sir Ronald A. Fisher, a famous statistician.

For the logical fallacy of believing that a hypothesis has been proved true, merely because it is not contradicted by the available facts, has no more right to insinuate itself in statistics than in other kinds of scientific reasoning. . . . It would, therefore, add greatly to the clarity with which the tests of significance are regarded if it were generally understood that tests of significance, when used accurately, are capable of rejecting or invalidating hypotheses, in so far as they are contradicted by the data; but that they are never capable of establishing them as certainly true. . . .

Source: Letter by Ronald A Fisher in Nature. Copyright © by Nature Publishing Group.

In your own words, explain what this quotation means.

45. In your own words, explain the difference between "beyond all reasonable doubt" and "beyond all doubt." Use these phrases to explain why we never "accept" the statement in the null hypothesis.

10.2 Hypothesis Tests for a Population Proportion

Preparing for This Section Before getting started, review the following:

- Using probabilities to identify unusual events (Section 5.1, p. 249)
- z_α notation (Section 7.2, pp. 372–373)
- Sampling distribution of the sample proportion (Section 8.2, pp. 408–412)
- Computing normal probabilities (Section 7.2, pp. 366–370)
- Binomial probability distribution (Section 6.2, pp. 328–338)

Objectives   ❶ Explain the logic of hypothesis testing

❷ Test hypotheses about a population proportion

❸ Test hypotheses about a population proportion using the binomial probability distribution

① Explain the Logic of Hypothesis Testing

Recall that the best point estimate of p, the proportion of the population with a certain characteristic, is given by

$$\hat{p} = \frac{x}{n}$$

where x is the number of individuals in the sample with the specified characteristic and n is the sample size. We learned in Section 8.2 that the sampling distribution of $\hat{p}$ is approximately normal, with mean $\mu_{\hat{p}} = p$ and standard deviation $\sigma_{\hat{p}} = \sqrt{\dfrac{p(1-p)}{n}}$, provided that the following requirements are satisfied.

1. The sample is a simple random sample.
2. $np(1-p) \geq 10$.
3. The sampled values are independent of each other ($n \leq 0.05N$).

We will present three methods for testing hypotheses. The first method is called the classical (traditional) approach, the second method is the P-value approach, and the third method uses confidence intervals. Your instructor may choose to cover one, two, or all three approaches to hypothesis testing.

First, we lay out a scenario that will be used to understand both the classical and P-value approaches. Suppose a politician wants to know if a majority (more than 50%) of her constituents are in favor of a certain policy. We are therefore testing the following hypotheses:

$$H_0: p = 0.5 \quad \text{versus} \quad H_1: p > 0.5$$

The politician hires a polling firm to obtain a random sample of 1000 registered voters in her district and finds that 534 are in favor of the policy, so $\hat{p} = \dfrac{534}{1000} = 0.534$. Do these results suggest that among *all* registered voters more than 50% favor the policy? Or is it possible that the true proportion of registered voters who favor the policy is 0.5 and we just happened to survey a majority in favor of the policy? In other words, would it be unusual to obtain a sample proportion of 0.534 or higher from a population whose proportion is 0.5? What is convincing, or *statistically significant*, evidence?

Definition	When observed results are unlikely under the assumption that the null hypothesis is true, we say the result is **statistically significant** and we reject the null hypothesis.

To determine if a sample proportion of 0.534 is statistically significant, we build a probability model. After all, a second random sample of 1000 registered voters will likely result in a different sample proportion, and we want to describe this variability so we can determine if the results we obtained are unusual. Since $np(1-p) = 1000(0.5)(1-0.5) = 250 \geq 10$ and the sample size ($n = 1000$) is sufficiently smaller than the population size (provided there are at least $N = 20{,}000$ registered voters in the politician's district), we can use a normal model to describe the variability in $\hat{p}$. The mean of the distribution of $\hat{p}$ is $\mu_{\hat{p}} = 0.5$ (since we assume the statement in the null hypothesis is true) and the standard deviation of the distribution of $\hat{p}$ is $\sigma_{\hat{p}} = \sqrt{\dfrac{p(1-p)}{n}} = \sqrt{\dfrac{0.5(1-0.5)}{1000}} \approx 0.016$. Figure 2 shows the sampling distribution of the sample proportion for the "politician" example.

Figure 2

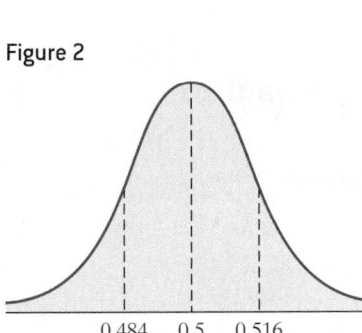

0.484   0.5   0.516 $\hat{p}$

Now that we have a model that describes the distribution of the sample proportion, we can use it to look at the logic of the classical and P-value approaches to test if a majority of the politician's constituents are in favor of the policy.

The Logic of the Classical Approach

We may consider the sample evidence to be statistically significant (or sufficient) if the sample proportion is too many standard deviations, say 2, above the assumed population proportion of 0.5:

Recall that $z = \dfrac{\hat{p} - \mu_{\hat{p}}}{\sigma_{\hat{p}}}$ represents the number of standard deviations that $\hat{p}$ is from the population proportion, p. Our simple random sample of 1000 registered voters results in a sample proportion of 0.534, so under the assumption that the null hypothesis is true we have

$$z = \frac{\hat{p} - \mu_{\hat{p}}}{\sigma_{\hat{p}}} = \frac{0.534 - 0.5}{\sqrt{\dfrac{0.5(1 - 0.5)}{1000}}} = 2.15$$

The sample proportion is 2.15 standard deviations above the hypothesized population proportion of 0.5, which is more than 2 standard deviations (that is, "too far") above the hypothesized population proportion. So we will reject the null hypothesis. There is statistically significant (sufficient) evidence to conclude that a majority of registered voters are in favor of the policy.

Why does it make sense to reject the null hypothesis if the sample proportion is more than 2 standard deviations away from the hypothesized proportion? The area under the standard normal curve to the right of $z = 2$ is 0.0228, as shown in Figure 3.

Figure 3

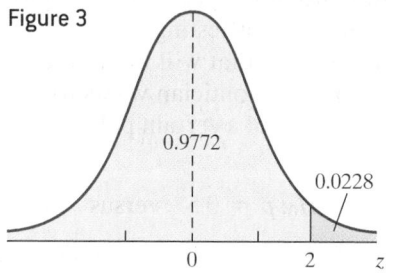

Figure 4

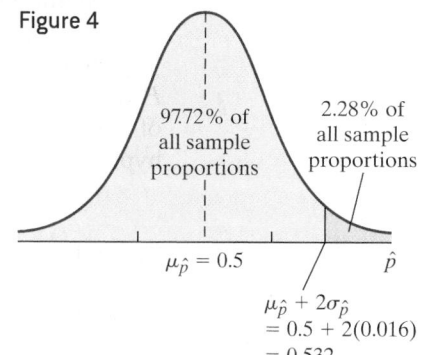

Figure 4 shows that if the null hypothesis is true (that is, if the population proportion is 0.5), then 97.72% of all sample proportions will be 0.532 or less, and only 2.28% of the sample proportions will be more than 0.532 (0.532 is 2 standard deviations above the hypothesized proportion of 0.5). If a sample proportion lies in the blue region, we are inclined to believe it came from a population whose proportion is greater than 0.5, rather than believe that the population proportion equals 0.5 and our sample just happened to result in a proportion of registered voters much higher than 0.5.

Notice that our criterion for rejecting the null hypothesis will lead to making a Type I error (rejecting a true null hypothesis) 2.28% of the time. This is because 2.28% of all sample proportions are more than 0.532, even though the population proportion is 0.5.

This discussion illustrates the following point.

Hypothesis Testing Using the Classical Approach

If the sample statistic is too many standard deviations from the population parameter stated in the null hypothesis, we reject the null hypothesis.

The Logic of the *P*-Value Approach

A second criterion we may use for testing hypotheses is to determine how likely it is to obtain a sample proportion of 0.534 or higher from a population whose proportion is 0.5. If a sample proportion of 0.534 or higher is unlikely (or unusual), we have evidence against the statement in the null hypothesis. Otherwise, we do not have sufficient evidence against the statement in the null hypothesis.

We can compute the probability of obtaining a sample proportion of 0.534 or higher from a population whose proportion is 0.5 using the normal model. See Figure 5.

Figure 5

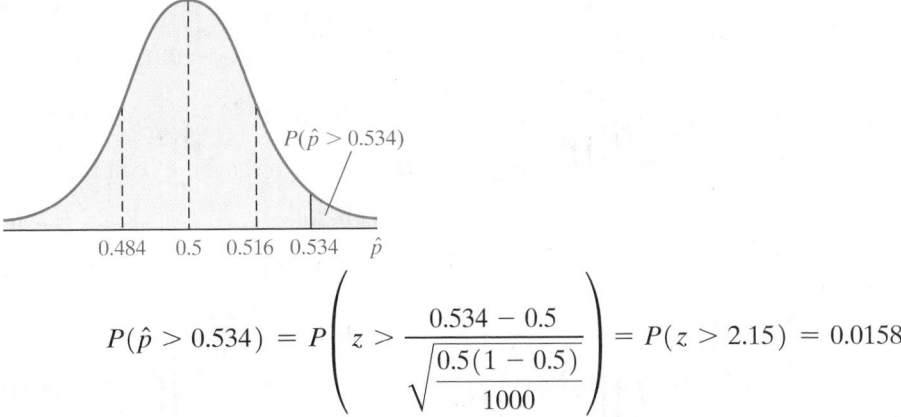

$$P(\hat{p} > 0.534) = P\left(z > \dfrac{0.534 - 0.5}{\sqrt{\dfrac{0.5(1 - 0.5)}{1000}}} \right) = P(z > 2.15) = 0.0158$$

The value 0.0158 is called the *P-value*, which means about 2 samples in 100 will give a sample proportion as high or higher than the one we obtained *if* the population proportion really is 0.5. Because these results are unusual, we take this as evidence against the statement in the null hypothesis.

Definition   A **P-value** is the probability of observing a sample statistic as extreme or more extreme than one observed under the assumption that the statement in the null hypothesis is true. Put another way, the *P*-value is the likelihood or probability that a sample will result in a statistic such as the one obtained if the null hypothesis is true.

This discussion illustrates the idea behind hypothesis testing using the *P*-value approach.

> ## Hypothesis Testing Using the *P*-Value Approach
>
> If the probability of getting a sample statistic as extreme or more extreme than the one obtained is small under the assumption the statement in the null hypothesis is true, reject the null hypothesis.

Figure 6 illustrates the situation for both the classical and *P*-value approaches. The distribution in blue shows the distribution of the sample proportion assuming the statement in the null hypothesis is true. The sample proportion of 0.534 is too far from the assumed population proportion of 0.5. Therefore, we reject the null hypothesis that $p = 0.5$ and conclude that $p > 0.5$, as indicated by the distribution in red. We do not know what the population proportion of registered voters who are in favor of the policy is, but we have evidence to say it is greater than 0.5 (a majority).

Figure 6   Distribution of $\hat{p}$ if H_0 is true

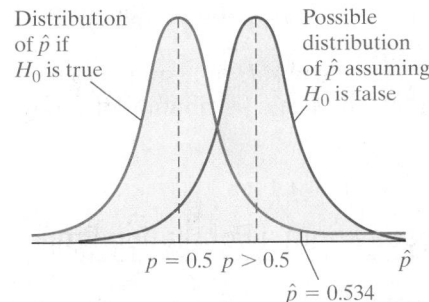

Possible distribution of $\hat{p}$ assuming H_0 is false

② Test Hypotheses about a Population Proportion

We now formalize the procedure for testing hypotheses regarding a population proportion.

Testing Hypotheses Regarding a Population Proportion, p

Use Steps 1 through 5, provided that

- the sample is obtained by simple random sampling or the data result from a randomized experiment.
- $np_0(1 - p_0) \geq 10$.
- the sampled values are independent of each other.

Step 1 Determine the null and alternative hypotheses. The hypotheses can be structured in one of three ways:

Two-Tailed	Left-Tailed	Right-Tailed
$H_0: p = p_0$	$H_0: p = p_0$	$H_0: p = p_0$
$H_1: p \neq p_0$	$H_1: p < p_0$	$H_1: p > p_0$

Note: p_0 is the assumed value of the population proportion.

Step 2 Select a level of significance α, depending on the seriousness of making a Type I error.

Classical Approach

Step 3 Compute the **test statistic**

$$z_0 = \frac{\hat{p} - p_0}{\sqrt{\dfrac{p_0(1 - p_0)}{n}}}$$

Use Table V to determine the critical value.

	Two-Tailed	Left-Tailed	Right-Tailed
Critical value	$-z_{\frac{\alpha}{2}}$ and $z_{\frac{\alpha}{2}}$	$-z_\alpha$	z_α

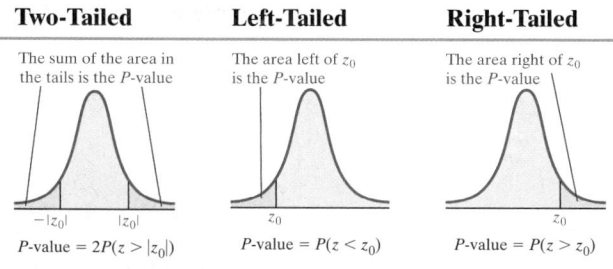

Step 4 Compare the critical value with the test statistic.

Two-Tailed	Left-Tailed	Right-Tailed
If $z_0 < -z_{\frac{\alpha}{2}}$ or $z_0 > z_{\frac{\alpha}{2}}$, reject the null hypothesis.	If $z_0 < -z_\alpha$, reject the null hypothesis.	If $z_0 > z_\alpha$, reject the null hypothesis.

P-Value Approach

By Hand Step 3 Compute the **test statistic**

$$z_0 = \frac{\hat{p} - p_0}{\sqrt{\dfrac{p_0(1 - p_0)}{n}}}$$

Use Table V to determine the P-value.

Two-Tailed	Left-Tailed	Right-Tailed
The sum of the area in the tails is the P-value	The area left of z_0 is the P-value	The area right of z_0 is the P-value
P-value $= 2P(z > |z_0|)$	P-value $= P(z < z_0)$	P-value $= P(z > z_0)$

Technology Step 3 Use a statistical spreadsheet or calculator with statistical capabilities to obtain the P-value. The directions for obtaining the P-value using the TI-83/84 Plus graphing calculators, Minitab, Excel, and StatCrunch are in the Technology Step-by-Step on pages 489–490.

Step 4 If P-value $< \alpha$, reject the null hypothesis.

Step 5 State the conclusion.

Notice in Step 3 that we are using p_0 (the proportion stated in the null hypothesis) in computing the standard error rather than $\hat{p}$, as we did in constructing confidence intervals about p. This is because H_0 is assumed to be true when performing a hypothesis test, so the assumed mean of the distribution of $\hat{p}$ is $\mu_{\hat{p}} = p_0$ and the assumed standard error is $\sigma_{\hat{p}} = \sqrt{\dfrac{p_0(1 - p_0)}{n}}$.

EXAMPLE 1 Testing Hypotheses about a Population Proportion: Left-Tailed Test

Problem The two major college entrance exams that a majority of colleges accept for admission are the SAT and ACT. ACT looked at historical records and established 22 as the minimum ACT math score for a student to be considered prepared for college mathematics. [**Note:** "Being prepared" means there is a 75% probability of successfully completing College Algebra in college.] An official with the Illinois State Department of Education wonders whether less than half of the students in her state are prepared for College Algebra. She obtains a simple random sample of 500 records of students who have taken the ACT and finds that 219 are prepared for college mathematics (that is, scored at least 22 on the ACT math test). Does this represent significant evidence that less than half of Illinois students who have taken the ACT are prepared for college mathematics upon graduation from a high school? Use the $\alpha = 0.05$ level of significance. *Source:* ACT High School Profile Report.

Approach This problem deals with a hypothesis test of a proportion. We want to determine if the sample evidence shows that less than half of the students are prepared for college mathematics. Symbolically, we represent this as $p < \frac{1}{2}$ or $p < 0.5$.

Verify the three requirements to perform the hypothesis test: the sample must be a simple random sample, $np_0(1 - p_0) \geq 10$, and the sample size cannot be more than 5% of the population size (for independence). Then follow Steps 1 through 5.

Solution Assume that $p = 0.5$. The sample is a simple random sample. Also, $np_0(1 - p_0) = 500(0.5)(1 - 0.5) = 125 > 10$. Provided that there are over 10,000 students in the state, the sample size is less than 5% of the population size. Assuming that this is the case, the requirements are satisfied. Now proceed with Steps 1 through 5.

Step 1 The burden of proof lies in showing $p < 0.5$. We assume there is no difference between the proportion of students *ready* for college math and the proportion of students *not ready* for college math. Therefore, the statement in the null hypothesis is that $p = 0.5$. So we have

$$H_0: p = 0.5 \qquad \text{versus} \qquad H_1: p < 0.5$$

Step 2 The level of significance is $\alpha = 0.05$.

Classical Approach

Step 3 The assumed value of the population proportion is $p_0 = 0.5$. The sample proportion is $\hat{p} = \frac{x}{n} = \frac{219}{500} = 0.438$. We want to know if it is unusual to obtain a sample proportion of 0.438 or less from a population whose proportion is assumed to be 0.5.

The test statistic is

$$z_0 = \frac{\hat{p} - p_0}{\sqrt{\dfrac{p_0(1 - p_0)}{n}}} = \frac{0.438 - 0.5}{\sqrt{\dfrac{0.5(1 - 0.5)}{500}}} = -2.77$$

Because this is a left-tailed test, we determine the critical value at the $\alpha = 0.05$ level of significance to be $-z_{0.05} = -1.645$. The critical region is shown in Figure 7.

Figure 7

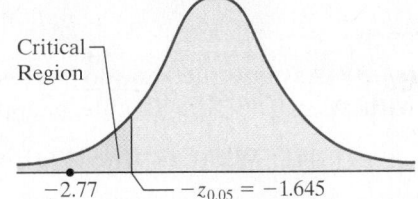

Critical Region

-2.77 $-z_{0.05} = -1.645$

Step 4 The test statistic, $z_0 = -2.77$, is labeled in Figure 7. Because the test statistic is less than the critical value $(-2.77 < -1.645)$, we reject the null hypothesis.

P-Value Approach

By Hand Step 3 The assumed value of the population proportion is $p_0 = 0.5$. The sample proportion is $\hat{p} = \frac{x}{n} = \frac{219}{500} = 0.438$. We want to know how likely it is to obtain a sample proportion of 0.438 or less from a population whose proportion is assumed to be 0.5.

The test statistic is

$$z_0 = \frac{\hat{p} - p_0}{\sqrt{\dfrac{p_0(1 - p_0)}{n}}} = \frac{0.438 - 0.5}{\sqrt{\dfrac{0.5(1 - 0.5)}{500}}} = -2.77$$

Because this is a left-tailed test, the P-value is the area under the standard normal distribution to the left of the test statistic, $z_0 = -2.77$, as shown in Figure 8. So, P-value $= P(z < z_0) = P(z < -2.77) = 0.0028$.

Figure 8

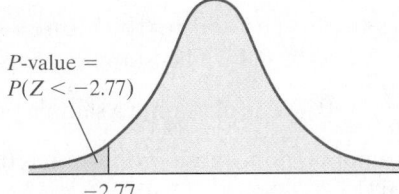

P-value = $P(Z < -2.77)$

-2.77

Technology Step 3 Using Minitab, we find the P-value is 0.003. See Figure 9 on the next page.

(continued)

Figure 9

```
Test and CI for One Proportion

Test of p = 0.5 vs p < 0.5


Sample    X    N   Sample p   95% Upper Bound   Z-Value   P-Value
1        219  500  0.438000          0.474496     -2.77     0.003
```

Step 4 The *P*-value of 0.003 means that *if* the null hypothesis that $p = 0.5$ is true, we expect 219 or fewer successes in 500 trials in 3 out of 1000 repetitions of this study! The observed results are unusual, indeed. Because the *P*-value is less than the level of significance, $\alpha = 0.05$ $(0.003 < 0.05)$, we reject the null hypothesis.

Step 5 There is sufficient evidence at the $\alpha = 0.05$ level of significance to conclude that fewer than half of the Illinois students are prepared for college mathematics. In other words, the data suggest less than a majority of the students in the state of Illinois who take the ACT are prepared for college mathematics. •

• **Now Work Problem 17**

EXAMPLE 2 Testing Hypotheses about a Population Proportion: Two-Tailed Test

---

Problem What do you think is more important—to protect the right of Americans to own guns or to control gun ownership? When asked this question, 46% of all Americans said protecting the right to own guns is more important. The Pew Research Center surveyed 1267 randomly selected Americans with at least a bachelor's degree and found that 559 believed that protecting the right to own guns is more important. Does this result suggest the proportion of Americans with at least a bachelor's degree feel differently than the general American population when it comes to gun control? Use the $\alpha = 0.1$ level of significance.

Approach This problem deals with a hypothesis test of a proportion. Verify the three requirements to perform the hypothesis test. Then follow Steps 1 through 5.

Solution We want to know if the proportion of Americans with at least a bachelor's degree who believe protecting the right of Americans to own guns is more important is *different from* 0.46. To conduct the test, assume the sample comes from a population with $p = 0.46$. The sample is a simple random sample. Also, $np_0(1 - p_0) = 1267(0.46)(1 - 0.46) = 314.7 \geq 10$. Because there are well over 10 million Americans with at least a bachelor's degree, the sample size is less than 5% of the population. We now proceed with Steps 1 through 5.

Step 1 We want to know whether the proportion is *different from* 0.46, which can be written $p \neq 0.46$, so this is a two-tailed test.

$$H_0: p = 0.46 \quad \text{versus} \quad H_1: p \neq 0.46$$

Step 2 The level of significance is $\alpha = 0.1$.

Classical Approach

Step 3 Assume the sample comes from a population with $p_0 = 0.46$. The sample proportion is $\hat{p} = \dfrac{x}{n} = \dfrac{559}{1267} = 0.441$. Is obtaining a sample proportion of 0.441 from a population whose proportion is 0.46 likely, or is it unusual?

The test statistic is

$$z_0 = \frac{\hat{p} - p_0}{\sqrt{\dfrac{p_0(1 - p_0)}{n}}} = \frac{0.441 - 0.46}{\sqrt{\dfrac{0.46(1 - 0.46)}{1267}}} = -1.36$$

P-Value Approach

By Hand Step 3 Assume the sample comes from a population with $p_0 = 0.46$. The sample proportion is $\hat{p} = \dfrac{x}{n} = \dfrac{559}{1267} = 0.441$. What is the likelihood of obtaining a sample proportion of 0.441 from a population whose proportion is 0.46?

The test statistic is

$$z_0 = \frac{\hat{p} - p_0}{\sqrt{\dfrac{p_0(1 - p_0)}{n}}} = \frac{0.441 - 0.46}{\sqrt{\dfrac{0.46(1 - 0.46)}{1267}}} = -1.36$$

Because this is a two-tailed test, we determine the critical values at the $\alpha = 0.10$ level of significance to be $-z_{0.1/2} = -z_{0.05} = -1.645$ and $z_{0.1/2} = z_{0.05} = 1.645$. The critical regions are shown in Figure 10.

Figure 10

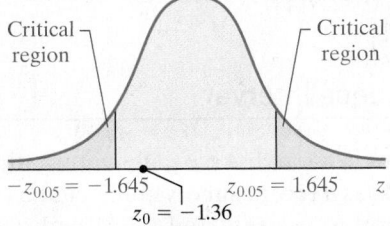

Step 4 The test statistic, $z_0 = -1.36$, is labeled in Figure 10. Because the test statistic does not lie in the critical region, do not reject the null hypothesis.

Because this is a two-tailed test, the *P*-value is the area under the standard normal distribution to the left of $-|z_0| = -1.36$ and to the right of $|z_0| = 1.36$, as shown in Figure 11.

Figure 11

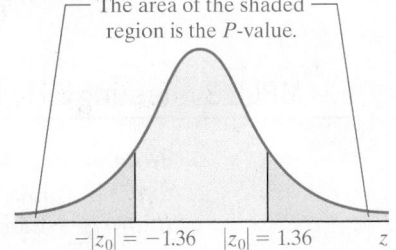

The area of the shaded region is the *P*-value.

$$P\text{-value} = P(Z < -|z_0|) + P(Z > |z_0|)$$
$$= 2P(Z < -1.36) \quad \text{Use symmetry}$$
$$= 2(0.0869)$$
$$= 0.1738$$

Technology Step 3 Using StatCrunch, we find the *P*-value is 0.1794. See Figure 12.

Figure 12

Hypothesis test results:
p : proportion of successes for population
$H_0 : p = 0.46$
$H_A : p \neq 0.46$

Proportion	Count	Total	Sample Prop.	Std. Err.	Z-Stat	P-value
p	559	1267	0.4411997	0.014001917	-1.3426958	0.1794

Step 4 The *P*-value of 0.1738 [Technology: 0.1794] means that *if* the null hypothesis that $p = 0.46$ is true, we expect the type of results we observed (or more extreme results) in about 17 or 18 out of 100 samples. The observed results are not unusual. Because the *P*-value is greater than the level of significance, $\alpha = 0.1$ $(0.1738 > 0.1)$, we do not reject the null hypothesis.

Step 5 There is not sufficient evidence at the $\alpha = 0.1$ level of significance to conclude that Americans with at least a bachelor's degree feel differently than the general American population when it comes to gun control. ●

> **CAUTION!**
> In Example 2, we do not have enough evidence to reject the statement in the null hypothesis. In other words, it is not unusual to obtain a sample proportion of 0.441 from a population whose proportion is 0.46. However, this does not imply that we are accepting the statement in the null hypothesis (that is, we are not saying that the proportion equals 0.46). We are only saying we do not have enough evidence to conclude that the proportion is different from 0.46. Be sure that you understand the difference between "accepting" and "not rejecting." It is similar to the difference between being declared "innocent" versus "not guilty."
> Also, be sure you understand that the *P*-value is the probability of obtaining a sample statistic as extreme or more extreme than the one observed *if* the statement in the null hypothesis is true. The *P*-value does not represent the probability that the null hypothesis is true. The statement in the null hypothesis is either true or false, we just don't know which.

In practice, the level of significance is not reported using the *P*-value approach. Instead, only the *P*-value is given, and the reader of the study must interpret its value and judge its significance.

● **Now Work Problem 21**

Test a Hypothesis Using a Confidence Interval

Recall, the level of confidence, $(1 - \alpha) \cdot 100\%$, in a confidence interval represents the percentage of intervals that will contain the unknown parameter if repeated samples are obtained.

Two-Tailed Hypothesis Testing Using Confidence Intervals

When testing $H_0: p = p_0$ versus $H_1: p \neq p_0$, if a $(1 - \alpha) \cdot 100\%$ confidence interval contains p_0, do not reject the null hypothesis. However, if the confidence interval does not contain p_0, conclude that $p \neq p_0$ at the level of significance, α.

EXAMPLE 3   Testing a Hypothesis Using a Confidence Interval

Problem A 2009 study by Princeton Survey Research Associates International found that 34% of teenagers text while driving. Does a recent survey conducted by *Consumer Reports*, which found that 353 of 1200 randomly selected teens had texted while driving, suggest that the proportion of teens who text while driving has changed since 2009? Use a 95% confidence interval to answer the question.

Approach Construct a 95% confidence about the proportion of teens who text while driving based on the *Consumer Reports* survey. If the interval does not include 0.34, reject the null hypothesis $H_0: p = 0.34$ in favor of $H_1: p \neq 0.34$.

Solution The 95% confidence interval for p based on the *Consumer Reports* survey has a lower bound of 0.268 and an upper bound of 0.320. Because 0.34 is not within the bounds of the confidence interval, there is sufficient evidence to conclude that the proportion of teens who text while driving has changed since 2009.

● Now Work Problem 23

Note: The same *Consumer Reports* article cited in Example 3 states that 75% of teens have friends who text while driving. What does this say about the difficulty in finding truthful responses to questions while conducting a survey?

③ **Test Hypotheses about a Population Proportion Using the Binomial Probability Distribution**

For the sampling distribution of $\hat{p}$ to be approximately normal, we require that $np(1 - p)$ be at least 10. What if this requirement is not satisfied? In Section 6.2, we used the binomial probability formula to identify unusual events. We stated that an event was unusual if the probability of observing the event was less than 0.05. This criterion is based on the *P*-value approach to testing hypotheses; the probability that we computed was the *P*-value. We use this same approach to test hypotheses regarding a population proportion for small samples.

EXAMPLE 4   Hypothesis Test for a Population Proportion: Small Sample Size

Problem According to the U.S. Department of Agriculture (USDA), 48.9% of males aged 20 to 39 years consume the recommended daily requirement of calcium. After an aggressive "Got Milk" advertising campaign, the USDA conducts a survey of 35 randomly selected males aged 20 to 39 and finds that 21 of them consume the recommended daily allowance (RDA) of calcium. At the $\alpha = 0.10$ level of significance, is there evidence to conclude that the percentage of males aged 20 to 39 who consume the RDA of calcium has increased?

Approach We use the following steps:

Step 1   Determine the null and alternative hypotheses.

Step 2   Check whether $np_0(1 - p_0)$ is greater than or equal to 10, where p_0 is the proportion stated in the null hypothesis. If it is, then the sampling distribution of $\hat{p}$ is approximately normal and we can use the steps on page 484. Otherwise, we use Steps 3 and 4, presented next.

Step 3   Compute the *P*-value. For right-tailed tests, the *P*-value is the probability of obtaining *x* or more successes. For left-tailed tests, the *P*-value is the probability of obtaining *x* or fewer successes.* The *P*-value is always computed with the proportion given in the null hypothesis. Remember, assume that the null is true until we have evidence to the contrary.

Step 4   If the *P*-value is less than the level of significance, α, reject the null hypothesis.

Solution

Step 1   The status quo or no change proportion of 20- to 39-year-old males who consume the recommended daily requirement of calcium is 0.489. We wish to know whether the advertising campaign increased this proportion. Therefore,

$$H_0: p = 0.489 \quad \text{and} \quad H_1: p > 0.489$$

Step 2   From the null hypothesis, we have $p_0 = 0.489$. There were $n = 35$ individuals surveyed, so $np_0(1 - p_0) = 35(0.489)(1 - 0.489) = 8.75$. Because $np_0(1 - p_0) < 10$, the sampling distribution of $\hat{p}$ is not approximately normal.

Step 3   Let the random variable *X* represent the number of individuals who consume the daily requirement of calcium. We have $x = 21$ successes in $n = 35$ trials, so $\hat{p} = \dfrac{21}{35} = 0.6$. We want to judge whether the larger proportion is due to an increase in the population proportion or to sampling error. We obtained $x = 21$ successes in the survey and this is a right-tailed test, so the *P*-value is $P(X \geq 21)$.

$$P\text{-value} = P(X \geq 21) = 1 - P(X < 21) = 1 - P(X \leq 20)$$

Figure 13

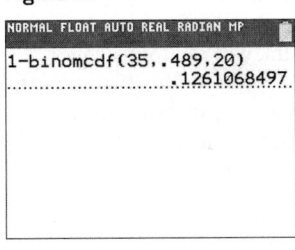

```
NORMAL FLOAT AUTO REAL RADIAN MP
1-binomcdf(35,.489,20)
                    .1261068497
```

We will compute this *P*-value using a TI-84 Plus C graphing calculator, with $n = 35$ and $p = 0.489$. Figure 13 shows the results.

The *P*-value is 0.1261. Minitab, StatCrunch, and Excel will compute exact *P*-values using this approach as well.

Step 4   The *P*-value is greater than the level of significance ($0.1261 > 0.10$), so we do not reject H_0. There is not sufficient evidence (at the $\alpha = 0.1$ level of significance) to conclude that the proportion of 20- to 39-year-old males who consume the recommended daily allowance of calcium has increased.

● **Now Work Problem 27**

Technology Step-by-Step   Hypothesis Tests Regarding a Population Proportion

TI-83/84 Plus

1. Press STAT, highlight TESTS, and select 5:1-PropZTest.
2. For the value of p_0, enter the value of the population proportion stated in the null hypothesis.
3. Enter the number of successes, *x*, and the sample size, *n*.
4. Select the direction of the alternative hypothesis.
5. Highlight **Calculate** or **Draw**, and press ENTER.

Minitab

1. If you have raw data, enter them in C1, using 0 for failure and 1 for success.
2. Select the **Stat** menu, highlight **Basic Statistics**, then highlight **1-Proportion**.
3. If you have raw data, select "One or more samples, each in a column" from the drop-down menu. Place the

cursor in the box, highlight the column containing the raw data, and click "Select." If you have summarized data, select "Summarized data" from the drop-down menu. Enter the number of successes in the "Number of events" box and enter the number of trials. Check the "Perform hypothesis test" box and enter the value of the population proportion stated in the null hypothesis.

4. Click Options. Enter the direction of the alternative hypothesis. Assuming $np_0(1 - p_0) \geq 10$, select "Normal approximation" from the drop-down menu. Click OK twice.

Excel

1. Load the XLSTAT Add-In.
2. Select the XLSTAT menu and select **Parametric Tests**. From the drop-down menu, select **Tests for one proportion**.

---

*We will not address *P*-values for two-tailed hypothesis tests. For those who are interested, the *P*-value is two times the probability of obtaining *x* or more successes if $\hat{p} > p$ and two times the probability of obtaining *x* or fewer successes if $\hat{p} < p$.

3. In the cell marked Frequency, enter the number of successes. In the cell marked Sample size, enter the number of trials. In the cell marked Test proportion, enter the proportion stated in the null hypothesis. Check the Frequency radio button. Click Options. Choose the appropriate direction for the alternative hypothesis. Be sure Hypothesize difference (D): is set to zero. Enter the level of significance. Click OK.

StatCrunch

1. If you have raw data, enter them into the spreadsheet. Name the column variable.

2. Select **Stat**, highlight **Proportion Stats**, select **One Sample**, and then choose either **With Data** or **With Summary**.

3. If you chose **With Data**, select the column that has the observations, choose which outcome represents a success. If you chose **With Summary**, enter the number of successes and the number of trials. Choose the hypothesis test radio button. Enter the value of the proportion stated in the null hypothesis and choose the direction of the alternative hypothesis from the pull-down menu. Click Compute!.

10.2 Assess Your Understanding

Vocabulary and Skill Building

1. When observed results are unlikely under the assumption that the null hypothesis is true, we say the result is _____ _____ and we reject the null hypothesis.

2. *True or False:* When testing a hypothesis using the Classical Approach, if the sample proportion is too many standard deviations from the proportion stated in the null hypothesis, we reject the null hypothesis.

3. *True or False:* When testing a hypothesis using the *P*-value Approach, if the *P*-value is large, we reject the null hypothesis.

4. Determine the critical value for a right-tailed test regarding a population proportion at the $\alpha = 0.01$ level of significance.

5. Determine the critical value for a left-tailed test regarding a population proportion at the $\alpha = 0.1$ level of significance.

6. Determine the critical value for a two-tailed test regarding a population proportion at the $\alpha = 0.05$ level of significance.

In Problems 7–12, test the hypothesis using (a) the classical approach and (b) the P-value approach. Be sure to verify the requirements of the test.

7. $H_0: p = 0.3$ versus $H_1: p > 0.3$
$n = 200; x = 75; \alpha = 0.05$

8. $H_0: p = 0.6$ versus $H_1: p < 0.6$
$n = 250; x = 124; \alpha = 0.01$

9. $H_0: p = 0.55$ versus $H_1: p < 0.55$
$n = 150; x = 78; \alpha = 0.1$

10. $H_0: p = 0.25$ versus $H_1: p < 0.25$
$n = 400; x = 96; \alpha = 0.1$

11. $H_0: p = 0.9$ versus $H_1: p \neq 0.9$
$n = 500; x = 440; \alpha = 0.05$

12. $H_0: p = 0.4$ versus $H_1: p \neq 0.4$
$n = 1000; x = 420; \alpha = 0.01$

13. You Explain It! Stock Analyst Throwing darts at the stock pages to decide which companies to invest in could be a successful stock-picking strategy. Suppose a researcher decides to test this theory and randomly chooses 100 companies to invest in. After 1 year, 53 of the companies were considered winners; that is, they outperformed other companies in the same investment class. To assess whether the dart-picking strategy resulted in a majority

of winners, the researcher tested $H_0: p = 0.5$ versus $H_1: p > 0.5$ and obtained a *P*-value of 0.2743. Explain what this *P*-value means and write a conclusion for the researcher.

14. You Explain It! ESP Suppose an acquaintance claims to have the ability to determine the birth month of randomly selected individuals. To test such a claim, you randomly select 80 individuals and ask the acquaintance to state the birth month of the individual. If the individual has the ability to determine birth month, then the proportion of correct birth months should exceed $\dfrac{1}{12}$, the rate one would expect from simply guessing.

(a) State the null and alternative hypotheses for this experiment.

(b) Suppose the individual was able to guess nine correct birth months. The *P*-value for such results is 0.1726. Explain what this *P*-value means and write a conclusion for the test.

Applying the Concepts

15. Cramer Correct Less Than Half the Time? The website pundittracker.com keeps track of predictions made by individuals in finance, politics, sports, and entertainment. Jim Cramer is a famous TV financial personality and author. Pundittracker monitored 678 of his stock predictions (such as a recommendation to buy the stock) and found that 320 were correct predictions. Treat these 678 predictions as a random sample of all of Cramer's predictions.

(a) Determine the sample proportion of predictions Cramer got correct.

(b) Suppose that we want to know whether the evidence suggests Cramer is correct less than half the time. State the null and alternative hypotheses.

(c) Verify the normal model may be used to determine the *P*-value for this hypothesis test.

(d) Draw a normal model with area representing the *P*-value shaded for this hypothesis test.

(e) Determine the *P*-value based on the model from part (d).

(f) Interpret the *P*-value.

(g) Based on the *P*-value, what does the sample evidence suggest? That is, what is the conclusion of the hypothesis test? Assume an $\alpha = 0.05$ level of significance.

16. Political Pundits In his book, "The Signal and the Noise," Nate Silver analyzed 733 predictions made by experts regarding political events. Of the 733 predictions, 338 were mostly true.

(a) Determine the sample proportion of political predictions that were mostly true.

(b) Suppose that we want to know whether the evidence suggests the political predictions were mostly true less than half the time. State the null and alternative hypotheses.

(c) Verify the normal model may be used to determine the *P*-value for this hypothesis test.

(d) Draw a normal model with the area representing the *P*-value shaded for this hypothesis test.

(e) Determine the *P*-value based on the model from part (d).

(f) Interpret the *P*-value.

(g) Based on the *P*-value, what does the sample evidence suggest? That is, what is the conclusion of the hypothesis test? Assume an $\alpha = 0.1$ level of significance.

NW **17. Lipitor** The drug Lipitor is meant to reduce cholesterol and LDL cholesterol. In clinical trials, 19 out of 863 patients taking 10 mg of Lipitor daily complained of flulike symptoms. Suppose that it is known that 1.9% of patients taking competing drugs complain of flulike symptoms. Is there evidence to conclude that more than 1.9% of Lipitor users experience flulike symptoms as a side effect at the $\alpha = 0.01$ level of significance?

18. Nexium Nexium is a drug that can be used to reduce the acid produced by the body and heal damage to the esophagus due to acid reflux. The manufacturer of Nexium claims that more than 94% of patients taking Nexium are healed within 8 weeks. In clinical trials, 213 of 224 patients suffering from acid reflux disease were healed after 8 weeks. Test the manufacturer's claim at the $\alpha = 0.01$ level of significance.

19. Fatal Traffic Accidents According to the National Highway and Traffic Safety Administration, the proportion of fatal traffic accidents in the United States in which the driver had a positive blood alcohol concentration (BAC) is 0.36. Suppose a random sample of 105 traffic fatalities in the state of Hawaii results in 51 that involved a positive BAC. Does the sample evidence suggest that Hawaii has a higher proportion of traffic fatalities involving a positive BAC than the United States at the $\alpha = 0.05$ level of significance?

20. Eating Together In December 2001, 38% of adults with children under the age of 18 reported that their family ate dinner together seven nights a week. In a recent poll, 403 of 1122 adults with children under the age of 18 reported that their family ate dinner together seven nights a week. Has the proportion of families with children under the age of 18 who eat dinner together seven nights a week decreased? Use the $\alpha = 0.05$ significance level.

NW **21. Taught Enough Math?** In 1994, 52% of parents with children in high school felt it was a serious problem that high school students were not being taught enough math and science. A recent survey found that 256 of 800 parents with children in high school felt it was a serious problem that high school students were not being taught enough math and science. Do parents feel differently today than they did in 1994? Use the $\alpha = 0.05$ level of significance?
Source: Based on "Reality Check: Are Parents and Students Ready for More Math and Science?" *Public Agenda*, 2006.

22. Living Alone? In 2000, 58% of females aged 15 and older lived alone, according to the U.S. Census Bureau. A sociologist tests whether this percentage is different today by conducting a random sample of 500 females aged 15 and older and finds that 285 are living alone. Is there sufficient evidence at the $\alpha = 0.1$ level of significance to conclude the proportion has changed since 2000?

NW **23. Quality of Education** In August 2002, 47% of parents with children in grades K–12 were satisfied with the quality of education the students receive. A recent Gallup poll found that 437 of 1013 parents with children in grades K–12 were satisfied with the quality of education the students receive. Construct a 95% confidence interval to assess whether this represents evidence that parents' attitudes toward the quality of education in the United States has changed since August 2002.

24. Infidelity According to menstuff.org, 22% of married men have "strayed" at least once during their married lives.

(a) Describe how you might go about administering a survey to assess the accuracy of this statement.

(b) A survey of 500 married men indicated that 122 have "strayed" at least once during their married life. Construct a 95% confidence interval for the population proportion of married men who have strayed. Use this interval to assess the accuracy of the statement made by menstuff.org.

25. Accuracy of the Drive Thru According to QSR Magazine, Chick-fil-A has the best accuracy of drive thru orders with 96.4% of all its drive thru orders filled correctly. The manager of a competing fast food restaurant wants to advertise that her drive thru is more accurate that Chick-fil-A. In a random sample of 350 drive thru orders, how many accurate orders would the manager need out of 350 to be able to claim her drive thru has a statistically significantly better accuracy record than Chick-fil-A at the 0.1 level of significance?

26. Talk to the Animals In an American Animal Hospital Association survey, 37% of respondents stated that they talk to their pets on the telephone. A veterinarian found this result hard to believe, so he randomly selected 150 pet owners and discovered that 54 of them spoke to their pet on the telephone. Does the veterinarian have the right to be skeptical? Use a 0.05 level of significance.

NW **27. Small-Sample Hypothesis Test** Professors Honey Kirk and Diane Lerma of Palo Alto College developed a "learning community curriculum that blended the developmental mathematics and the reading curriculum with a structured emphasis on study skills." In a typical developmental mathematics course at Palo Alto College, 50% of the students complete the course with a letter grade of A, B, or C. In the experimental course, of the 16 students enrolled, 11 completed the course with a letter grade of A, B, or C. Do you believe the experimental course was effective at the $\alpha = 0.05$ level of significance?

(a) State the appropriate null and alternative hypotheses.

(b) Verify that the normal model may not be used to estimate the *P*-value.

(c) Explain why this is a binomial experiment.

(d) Determine the *P*-value using the binomial probability distribution. State your conclusion to the hypothesis test.

(e) Suppose the course is taught with 48 students and 33 complete the course with a letter grade of A, B, or C. Verify the normal model may now be used to estimate the *P*-value.

(f) Use the normal model to obtain and interpret the *P*-value. State your conclusion to the hypothesis test.

(g) Explain the role that sample size plays in the ability to reject statements in the null hypothesis.
Source: Kirk, Honey & Lerma, Diane, "Reading Your Way to Success in Mathematics: A Paired Course of Developmental Mathematics and Reading." *MathAMATYC Educator*, Vol. 1. No. 2, 2010.

28. Small-Sample Hypothesis Test In 1997, 4% of mothers smoked more than 21 cigarettes during their pregnancy. An obstetrician believes that the percentage of mothers who smoke 21 cigarettes or more is less than 4% today. She randomly selects

120 pregnant mothers and finds that 3 of them smoked 21 or more cigarettes during pregnancy. Does the sample data support the obstetrician's belief? Use the $\alpha = 0.05$ level of significance.

29. Small Sample Hypothesis Test: Super Bowl Investing From Super Bowl I (1967) through Super Bowl XXXI (1997), the stock market increased if an NFL team won the Super Bowl and decreased if an AFL team won. This condition held 28 out of 31 years.

(a) Suppose the likelihood of predicting the direction of the stock market (increasing or decreasing) in any given year is 0.50. Decide on the appropriate null and alternative hypotheses to test whether the outcome of the Super Bowl can be used to predict the direction of the stock market.

(b) Use the binomial probability distribution to determine the P-value for the hypothesis test from part (a).

(c) Comment on the dangers of using the outcome of the hypothesis test to judge investments. Be sure your comment includes a discussion of circumstances in which associations have a causal relationship.

30. Statistics in the Media A headline read, "More Than Half of Americans Say Federal Taxes Too High." The headline was based on a random sample of 1026 adult Americans in which 534 stated the amount of federal tax they have to pay is too high. Is this an accurate headline?
Source: Gallup Organization, April 14, 2014

31. Gender Income Inequality The Sullivan Statistics Survey II asked, "Do you believe there is an income inequality discrepancy between males and females when each has the same experience and education?" Go to www.pearsonhighered.com/sullivanstats to obtain the data file SullivanStatsSurveyII using the file format of your choice for the version of the text you are using. The data may be found under the column "GenderIncomeInequality." Treat the sample as a random sample of adult Americans. Do the survey results suggest a supermajority (more than 60%) of adult Americans believe there is income inequality among males and females with the same experience and education? Use an $\alpha = 0.05$ level of significance.

32. Political Philosophy According to Gallup, 21% of adult Americans consider themselves to be liberal. Respondents of the Sullivan Statistics Survey I were asked to disclose their political philosophy: Conservative, Liberal, Moderate. Go to www.pearsonhighered.com/sullivanstats to obtain the data file SullivanStatsSurveyI using the file format of your choice for the version of the text you are using. The data may be found under the column "Political philosophy." Treat the results of the survey as a random sample of adult Americans. Do the survey results suggest the proportion is higher than that reported by Gallup? Use an $\alpha = 0.05$ level of significance.

33. Are Spreads Accurate? For every NFL game, there is a team that is expected to win by a certain number of points. In betting parlance, this is called the spread. For example, if the Chicago Bears are expected to beat the Green Bay Packers by three points, a sports bettor would say, "Chicago is minus three." So, if the Bears lose to Green Bay, or do not win by more than three points, a bet on Chicago would be a loser. If point spreads are accurate, we would expect about half of all games played to result in the favored team winning (beating the spread) and about half of all games to result in the team favored to not beat the spread. The following data represent the results of 45 randomly selected games where a 0 indicates the favored team did not beat the spread and a 1 indicates the favored team beat the spread. Do the data suggest that sport books establish accurate spreads?

0	0	0	0	0
1	0	0	1	0
0	0	0	1	0
1	1	0	0	1
0	0	1	1	1
0	0	1	1	0
1	0	0	1	0
0	1	1	0	1
0	0	1	1	1

Source: http://www.vegasinsider.com

34. Accept versus Do Not Reject In the United States, historically, 40% of registered voters are Republican. Suppose you obtain a simple random sample of 320 registered voters and find 142 registered Republicans.

(a) Consider the hypotheses $H_0: p = 0.4$ versus $H_1: p > 0.4$. Explain what the researcher would be testing. Perform the test at the $\alpha = 0.05$ level of significance. Write a conclusion for the test.

(b) Consider the hypotheses $H_0: p = 0.41$ versus $H_1: p > 0.41$. Explain what the researcher would be testing. Perform the test at the $\alpha = 0.05$ level of significance. Write a conclusion for the test.

(c) Consider the hypotheses $H_0: p = 0.42$ versus $H_1: p > 0.42$. Explain what the researcher would be testing. Perform the test at the $\alpha = 0.05$ level of significance. Write a conclusion for the test.

(d) Based on the results of parts (a)–(c), write a few sentences that explain the difference between "accepting" the statement in the null hypothesis versus "not rejecting" the statement in the null hypothesis.

35. Interesting Results Suppose you wish to find out the answer to the age-old question, "Do Americans prefer Coke or Pepsi?" You conduct a blind taste test in which individuals are randomly asked to drink one of the colas first, followed by the other cola, and then asked to disclose which drink they prefer. Results of your taste test indicate that 53 of 100 individuals prefer Pepsi.

(a) Conduct a hypothesis test (preferably using technology) $H_0: p = p_0$ versus $H_1: p \neq p_0$ for $p_0 = 0.42, 0.43, 0.44, \ldots, 0.64$ at the $\alpha = 0.05$ level of significance. For which values of p_0 do you not reject the null hypothesis? What do each of the values of p_0 represent?

(b) Construct a 95% confidence interval for the proportion of individuals who prefer Pepsi.

(c) Suppose you changed the level of significance in conducting the hypothesis test to $\alpha = 0.01$? What would happen to the range of values of p_0 for which the null hypothesis is not rejected? Why does this make sense?

36. Simulation Simulate drawing 100 simple random samples of size $n = 40$ from a population whose proportion is 0.3.

(a) Test the null hypothesis $H_0: p = 0.3$ versus $H_1: p \neq 0.3$ for each simulated sample.

(b) If we test the hypothesis at the $\alpha = 0.1$ level of significance, how many of the 100 samples would you expect to result in a Type I error?

(c) Count the number of samples that lead to a rejection of the null hypothesis. Is it close to the expected value determined in part (b)?

(d) How do we know that a rejection of the null hypothesis results in making a Type I error in this situation?

37. Simulation: Predicting the Future Parapsychology (psi) is a field of study that deals with clairvoyance or precognition. Psi made its way back into the news when a professional, refereed journal published an article by Cornell psychologist Daryl Bem, in which he claimed to demonstrate that psi is a real phenomenon. In the article Bem stated that certain individuals behave today as if they already know what is going to happen in the future. That is, individuals adjust current behavior in anticipation of events that are going to happen in the future. Here, we will present a simplified version of Bem's research.

(a) Suppose an individual claims to have the ability to predict the color (red or black) of a card from a standard 52-card deck. Of course, simply by guessing we would expect the individual to get half the predictions correct, and half incorrect. What is the statement of no change or no effect in this type of experiment? What statement would we be looking to demonstrate? Based on this, what would be the null and alternative hypotheses?

(b) Suppose you ask the individual to guess the correct color of a card 40 times, and the alleged savant (wise person) guesses the correct color 24 times. Would you consider this to be convincing evidence that that individual can guess the color of the card at better than a 50/50 rate? To answer this question, we want to determine the likelihood of getting 24 or more colors correct even if the individual is simply guessing. To do this, we assume the individual is guessing so that the probability of a successful guess is 0.5. Explain how 40 coins flipped independently with heads representing a successful guess can be used to model the card-guessing experiment.

(c) Now, use a random number generator, or applet such as the Coin-Flip applet in StatCrunch to flip 40 fair coins, 1000 different times. What proportion of time did you observe 24 or more heads due to chance alone? What does this tell you? Do you believe the individual has the ability to guess card color based on the results of the simulation, or could the results simply have occurred due to chance?

(d) Explain why guessing card color (or flipping coins) 40 times and recording the number of correct guesses (or heads) is a binomial experiment.

(e) Use the binomial probability function to find the probability of at least 24 correct guesses in 40 trials assuming the probability of success is 0.5.

(f) Look at the graph of the outcomes of the simulation from part (c). Explain why the normal model might be used to estimate the probability of obtaining at least 24 correct guesses in 40 trials assuming the probability of success is 0.5. Use the model to estimate the P-value.

(g) Based on the probabilities found in parts (c), (e), and (f), what might you conclude about the alleged savants ability to predict card color?

38. Putting It Together: Lupus Based on historical birthing records, the proportion of males born worldwide is 0.51. In other words, the commonly held belief that boys are just as likely as girls is false. Systematic lupus erythematosus (SLE), or lupus for short, is a disease in which one's immune system attacks healthy cells and tissue by mistake. It is well known that lupus tends to exist more in females than in males. Researchers wondered, however, if families with a child who had lupus had a lower ratio of males to females than the general population. If this were true, it would suggest that something happens during conception that causes males to be conceived at a lower rate when the SLE gene is present. To determine if this hypothesis is true, the researchers obtained records of families with a child who had SLE. A total of 23 males and 79 females were found to have SLE. The 23 males with SLE had a total of 23 male siblings and 22 female siblings. The 79 females with SLE had a total of 69 male siblings and 80 female siblings.
Source: L.N. Moorthy, M.G.E. Peterson, K.B. Onel, and T.J.A. Lehman. "Do Children with Lupus Have Fewer Male Siblings" *Lupus* 2008 17:128–131, 2008.

(a) Explain why this is an observational study.

(b) Is the study retrospective or prospective? Why?

(c) There are a total of $23 + 69 = 92$ male siblings in the study. How many female siblings are in the study?

(d) Draw a relative frequency bar graph of gender of the siblings.

(e) Find a point estimate for the proportion of male siblings in families where one of the children has SLE.

(f) Does the sample evidence suggest that the proportion of male siblings in families where one of the children has SLE is less than 0.51, the accepted proportion of males born in the general population? Use the $\alpha = 0.05$ level of significance.

(g) Construct a 95% confidence interval for the proportion of male siblings in a family where one of the children has SLE.

39. Putting It Together: Naughty or Nice? Yale University graduate student J. Kiley Hamlin conducted an experiment in which 16 ten-month-old babies were asked to watch a climber character attempt to ascend a hill. On two occasions, the baby witnesses the character fail to make the climb. On the third attempt, the baby witnesses either a helper toy push the character up the hill or a hinderer toy prevent the character from making the ascent. The helper and hinderer toys were shown to each baby in a random fashion for a fixed amount of time. The baby was then placed in front of each toy and allowed to choose which toy he or she wished to play with. In 14 of the 16 cases, the baby chose the helper toy.
Source: J. Kiley Hamlin et al., "Social Evaluation by Preverbal Infants." *Nature,* Nov. 2007.

(a) Why is it important to randomly expose the baby to the helper or hinderer toy first?

(b) What would be the appropriate null and alternative hypotheses if the researcher is attempting to show that babies prefer helpers over hinderers?

(c) Use the binomial probability formula to determine the P-value for this test.

(d) In testing 12 six-month-old babies, all 12 preferred the helper toy. The P-value was reported as 0.0002. Interpret this result.

Explaining the Concepts

40. Explain what a P-value is. What is the criterion for rejecting the null hypothesis using the P-value approach?

41. Suppose we are testing the hypothesis $H_0: p = 0.3$ versus $H_1: p > 0.3$ and we find the P-value to be 0.23. Explain what this means. Would you reject the null hypothesis? Why?

42. Suppose we are testing the hypothesis $H_0: p = 0.65$ versus $H_1: p \neq 0.65$ and we find the P-value to be 0.02. Explain what this means. Would you reject the null hypothesis? Why?

43. Discuss the advantages and disadvantages of using the Classical Approach to hypothesis testing. Discuss the advantages and disadvantages of using the P-value approach to hypothesis testing.

44. The headline reporting the results of a poll conducted by the Gallup organization stated "Majority of Americans at Personal Best in the Morning." The results indicated that a survey of 1100 Americans resulted in 55% stating they were at their personal best in the morning. The poll's results were reported with a margin of error of 3%. Explain why the Gallup organization's headline is accurate.

45. Explain what "statistical significance" means.

10.3 Hypothesis Tests for a Population Mean

Preparing for This Section Before getting started, review the following:

- Sampling distribution of $\bar{x}$ (Section 8.1, pp. 395–403)
- The t-distribution (Section 9.2, pp. 435–439)
- Using probabilities to identify unusual results (Section 5.1, p. 249)
- Confidence intervals for a mean (Section 9.2, pp. 439–441)

Objectives   ① Test hypotheses about a mean

② Understand the difference between statistical significance and practical significance

① Test Hypotheses about a Mean

In Section 8.1, we learned that the distribution of $\bar{x}$ is approximately normal with mean $\mu_{\bar{x}} = \mu$ and standard deviation $\sigma_{\bar{x}} = \dfrac{\sigma}{\sqrt{n}}$ provided the population from which the sample was drawn is normally distributed or the sample size is sufficiently large (because of the Central Limit Theorem). So $z = \dfrac{\bar{x} - \mu}{\dfrac{\sigma}{\sqrt{n}}}$ follows a standard normal distribution.

However, it is unreasonable to expect to know σ without knowing μ. This problem was resolved by William Gosset, who determined that $t = \dfrac{\bar{x} - \mu}{\dfrac{s}{\sqrt{n}}}$ follows Student's t-distribution with $n - 1$ degrees of freedom. We use this distribution to perform hypothesis tests on a mean.

Testing hypotheses about a mean follows the same logic as testing a hypothesis about a population proportion. The only difference is that we use Student's t-distribution, rather than the normal distribution.

Testing Hypotheses Regarding a Population Mean

To test hypotheses regarding the population mean, use the following steps, provided that

- the sample is obtained using simple random sampling or from a randomized experiment.
- the sample has no outliers and the population from which the sample is drawn is normally distributed, or the sample size, n, is large ($n \geq 30$).
- the sampled values are independent of each other.

Step 1 Determine the null and alternative hypotheses. The hypotheses can be structured in one of three ways:

Two-Tailed	Left-Tailed	Right-Tailed
$H_0: \mu = \mu_0$	$H_0: \mu = \mu_0$	$H_0: \mu = \mu_0$
$H_1: \mu \neq \mu_0$	$H_1: \mu < \mu_0$	$H_1: \mu > \mu_0$

Note: μ_0 is the assumed value of the population mean.

Step 2 Select a level of significance, α, depending on the seriousness of making a Type I error.

Classical Approach

Step 3 Compute the **test statistic**

$$t_0 = \frac{\bar{x} - \mu_0}{\frac{s}{\sqrt{n}}}$$

which follows Student's *t*-distribution with $n - 1$ degrees of freedom.

Use Table VII to determine the critical value.

	Two-Tailed	Left-Tailed	Right-Tailed
Critical value(s)	$-t_{\frac{\alpha}{2}}$ and $t_{\frac{\alpha}{2}}$	$-t_\alpha$	t_α

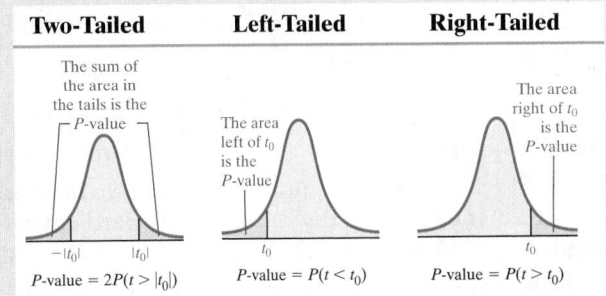

Step 4 Compare the critical value to the test statistic.

Two-Tailed	Left-Tailed	Right-Tailed
If $t_0 < -t_{\frac{\alpha}{2}}$ or $t_0 > t_{\frac{\alpha}{2}}$, reject the null hypothesis.	If $t_0 < -t_\alpha$, reject the null hypothesis.	If $t_0 > t_\alpha$, reject the null hypothesis.

P-Value Approach

By Hand Step 3 Compute the **test statistic**

$$t_0 = \frac{\bar{x} - \mu_0}{\frac{s}{\sqrt{n}}}$$

which follows Student's *t*-distribution with $n - 1$ degrees of freedom.

Use Table VII to approximate the *P*-value.

Two-Tailed	Left-Tailed	Right-Tailed
The sum of the area in the tails is the *P*-value	The area left of t_0 is the *P*-value	The area right of t_0 is the *P*-value
P-value $= 2P(t > |t_0|)$	P-value $= P(t < t_0)$	P-value $= P(t > t_0)$

Technology Step 3 Use a statistical spreadsheet or calculator with statistical capabilities to obtain the *P*-value. The directions for obtaining the *P*-value using the TI-83/84 Plus graphing calculators, Minitab, Excel, and StatCrunch are in the Technology Step-by-Step on page 500.

Step 4 If the *P*-value $< \alpha$, reject the null hypothesis.

Step 5 State the conclusion.

Notice that the procedure just presented requires either that the population from which the sample was drawn be normal or that the sample size be large $(n \geq 30)$. The procedure is robust, so minor departures from normality will not adversely affect the results of the test. However, if the data include outliers, the procedure should not be used.

We will verify these assumptions by constructing normal probability plots (to assess normality) and boxplots (to discover whether there are outliers). If the normal probability plot indicates that the data do not come from a normal population or if the boxplot reveals outliers, nonparametric tests should be performed (Chapter 15).

Before we look at a couple of examples, it is important to understand that we cannot find exact *P*-values using the *t*-distribution table (Table VII) because the table provides *t*-values only for certain areas. However, we can use the table to calculate lower and upper bounds on the *P*-value. To find exact *P*-values, use statistical software or a graphing calculator with advanced statistical features.

EXAMPLE 1   Testing a Hypothesis about a Population Mean: Large Sample

Problem The mean height of American males is 69.5 inches. The heights of the 43 male U.S. presidents* (Washington through Obama) have a mean 70.78 inches and a standard deviation of 2.77 inches. Treating the 43 presidents as a simple random sample, determine if there is evidence to suggest that U.S. presidents are taller than the average American male. Use the $\alpha = 0.05$ level of significance.

Approach Assume that all U.S. presidents come from a population whose height is 69.5 inches (that is, there is no difference between heights of U.S. presidents and the general American male population). Then determine the likelihood of obtaining a sample mean of 70.78 inches or higher from a population whose mean is 69.5 inches.

(continued)

*Grover Cleveland was elected to two non-consecutive terms, so there have technically been 44 presidents of the United States.

If the result is unlikely, reject the assumption stated in the null hypothesis in favor of the more likely notion that the mean height of U.S. presidents is greater than 69.5 inches. However, if obtaining a sample mean of 70.78 inches from a population whose mean is assumed to be 69.5 inches is not unusual, do not reject the null hypothesis (and attribute the difference to sampling error). Assume the population of potential U.S. presidents is large (for independence). Because the sample size is large, the distribution of $\bar{x}$ is approximately normal. Follow Steps 1 through 5.

Solution

Step 1 We want to know if U.S. presidents are taller than the typical American male who is 69.5 inches. We assume there is no difference between the height of a typical American male and U.S. presidents, so

$$H_0\colon \mu = 69.5 \text{ inches} \quad \text{versus} \quad H_1\colon \mu > 69.5 \text{ inches}$$

Step 2 The level of significance is $\alpha = 0.05$.

Classical Approach

Step 3 The sample mean is $\bar{x} = 70.78$ inches and the sample standard deviation is $s = 2.77$ inches.

The test statistic is

$$t_0 = \frac{\bar{x} - \mu_0}{\frac{s}{\sqrt{n}}} = \frac{70.78 - 69.5}{\frac{2.77}{\sqrt{43}}} = 3.030$$

Because this is a right-tailed test, determine the critical value at the $\alpha = 0.05$ level of significance with $43 - 1 = 42$ degrees of freedom to be $t_{0.05} = 1.684$ (using 40 degrees of freedom since this is closest to 42). The critical region is shown in Figure 14.

Figure 14

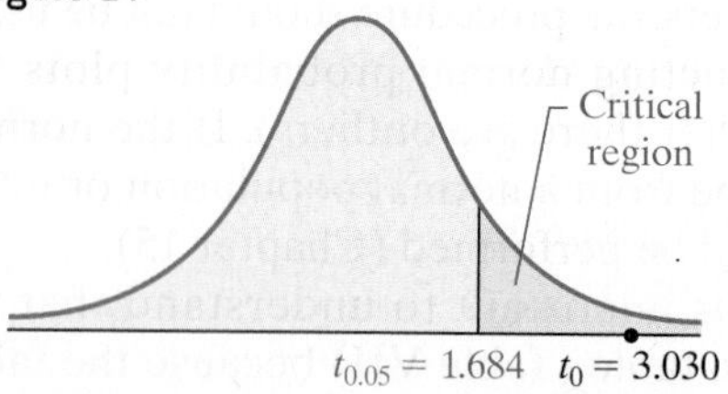

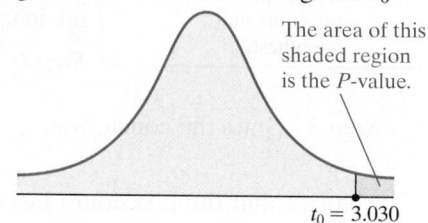

$t_{0.05} = 1.684 \quad t_0 = 3.030$

Step 4 The test statistic, $t_0 = 3.030$, is labeled in Figure 14. Because the test statistic lies in the critical region, reject the null hypothesis.

P-Value Approach

By Hand Step 3 The sample mean is $\bar{x} = 70.78$ inches and the sample standard deviation is $s = 2.77$ inches.

The test statistic is

$$t_0 = \frac{\bar{x} - \mu_0}{\frac{s}{\sqrt{n}}} = \frac{70.78 - 69.5}{\frac{2.77}{\sqrt{43}}} = 3.030$$

Because this is a right-tailed test, the P-value is the area under the t-distribution with 42 degrees of freedom to the right of $t_0 = 3.030$ as shown in Figure 15.

Figure 15

The area of this shaded region is the P-value.

$t_0 = 3.030$

Using Table VII, find the row that corresponds to 40 degrees of freedom (we use 40 degrees of freedom because it is closest to the actual degrees of freedom, $43 - 1 = 42$). The value 3.030 lies between 2.971 and 3.307. The value of 2.971 has an area of 0.0025 to the right under the t-distribution with 40 degrees of freedom. The area under the t-distribution with 40 degrees of freedom to the right of 3.307 is 0.001.

Because 3.030 is between 2.971 and 3.307, the P-value is between 0.001 and 0.0025. So

$$0.001 < P\text{-value} < 0.0025$$

There is an alternate form of Table VII (see Table XVII in Appendix A) that is useful for finding more accurate P-values. The table is set up similarly to Table V (the standard normal table). To use this alternate version, find the column that corresponds to the degrees of freedom, and the row that corresponds to the test statistic (rounded to the nearest tenth). The intersection of the row and column represents the area under the t-distribution to the right of the test statistic. Using 40 degrees of freedom (because 42 df is not in the table) with a test statistic of 3.0, the P-value is 0.002.

Technology Step 3 Using a TI-84 Plus C graphing calculator, the P-value is 0.0021. See Figure 16.

Figure 16

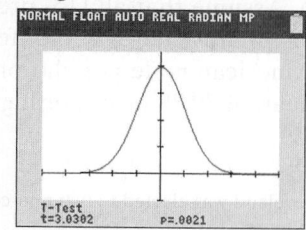

Note

If you are using Table XVII to find *P*-values for a left-tailed test, use the symmetry of the *t*-distribution. That is,

$$P(t < -t_0) = P(t > t_0) \bullet$$

• **Now Work Problem 13**

Step 4 The *P*-value of 0.0021 [by hand: $0.001 < P\text{-value} < 0.0025$] means that, *if* the null hypothesis that $\mu = 69.5$ inches is true, we expect a sample mean of 70.78 inches or higher in about 2 out of 1000 samples. The results we obtained do not seem to be consistent with the assumption that the mean height of this population is 69.5 inches. Put another way, because the *P*-value is less than the level of significance, $\alpha = 0.05$ $(0.0021 < 0.05)$, we reject the null hypothesis.

Step 5 There is sufficient evidence at the $\alpha = 0.05$ level of significance to conclude that U.S. presidents are taller than the typical American male. •

EXAMPLE 2 Testing a Hypothesis about a Population Mean: Small Sample

Table 1

19.68	20.66	19.56
19.98	20.65	19.61
20.55	20.36	21.02
21.50	19.74	

Source: Michael Carlisle, student at Joliet Junior College

Problem The "fun size" of a Snickers bar is supposed to weigh 20 grams. Because the penalty for selling candy bars under their advertised weight is severe, the manufacturer calibrates the machine so the mean weight is 20.1 grams. The quality-control engineer at M&M–Mars, the Snickers manufacturer, is concerned about the calibration. He obtains a random sample of 11 candy bars, weighs them, and obtains the data shown in Table 1. Should the machine be shut down and calibrated? Because shutting down the plant is very expensive, he decides to conduct the test at the $\alpha = 0.01$ level of significance.

Approach Assume that the machine is calibrated correctly. So there is no difference between the actual mean weight and the calibrated weight of the candy. We want to know whether the machine is incorrectly calibrated, which would result in a mean weight that is too high or too low. Therefore, this is a two-tailed test.

Before performing the hypothesis test, verify that the data come from a population that is normally distributed with no outliers by constructing a normal probability plot and boxplot. Then proceed to follow Steps 1 through 5.

Solution Figure 17 displays the normal probability plot and boxplot. The correlation between the weights and expected *z*-scores is 0.967 [Tech: 0.970]. Because $0.967 > 0.923$ (Table VI), the normal probability plot indicates that the data could come from a population that is approximately normal. The boxplot has no outliers. We can proceed with the hypothesis test.

Figure 17

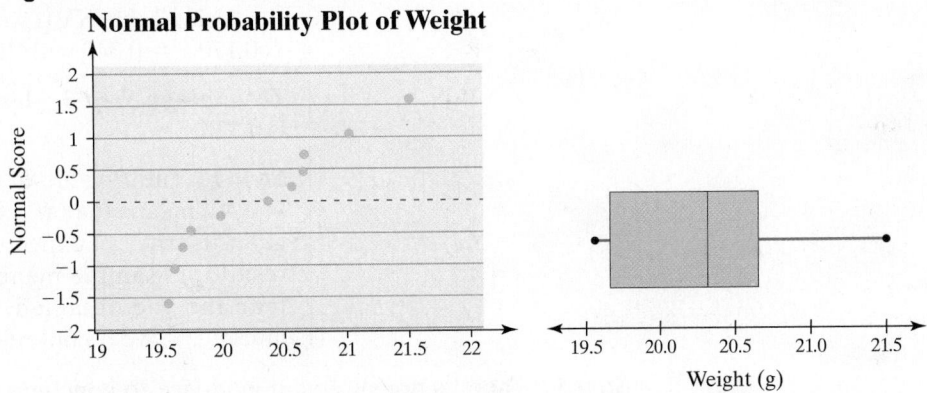

Normal Probability Plot of Weight

Step 1 The engineer wishes to determine whether the Snickers have a mean weight of 20.1 grams or not. The hypotheses can be written

$$H_0: \mu = 20.1 \text{ grams} \quad \text{versus} \quad H_1: \mu \neq 20.1 \text{ grams}$$

This is a two-tailed test.

Step 2 The level of significance is $\alpha = 0.01$.

(continued)

Classical Approach

Step 3 From the data in Table 1, the sample mean is $\bar{x} = 20.301$ grams and the sample standard deviation is $s = 0.64$ gram.

The test statistic is

$$t_0 = \frac{\bar{x} - \mu_0}{\frac{s}{\sqrt{n}}} = \frac{20.301 - 20.1}{\frac{0.64}{\sqrt{11}}} = 1.042$$

Because this is a two-tailed test, determine the critical values at the $\alpha = 0.01$ level of significance with $11 - 1 = 10$ degrees of freedom to be $-t_{0.01/2} = -t_{0.005} = -3.169$ and $t_{0.01/2} = t_{0.005} = 3.169$. The critical regions are shown in Figure 18.

Figure 18

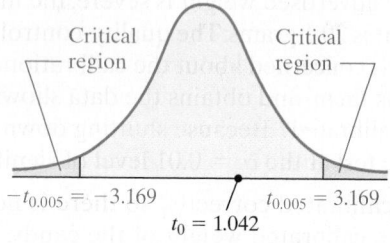

Critical region   Critical region

$-t_{0.005} = -3.169$ $t_{0.005} = 3.169$
$t_0 = 1.042$

Step 4 Because the test statistic, $t_0 = 1.042$, does not lie in the critical region, do not reject the null hypothesis.

P-Value Approach

By Hand Step 3 From the data in Table 1, the sample mean is $\bar{x} = 20.301$ grams and the sample standard deviation is $s = 0.64$ gram.

The test statistic is

$$t_0 = \frac{\bar{x} - \mu_0}{\frac{s}{\sqrt{n}}} = \frac{20.301 - 20.1}{\frac{0.64}{\sqrt{11}}} = 1.042$$

Because this is a two-tailed test, the P-value is the area under the t-distribution with $n - 1 = 11 - 1 = 10$ degrees of freedom to the left of $-t_0 = -1.042$ and to the right of $t_0 = 1.042$, as shown in Figure 19. That is, P-value $= P(t < -1.042) + P(t > 1.042) = 2P(t > 1.042)$, with 10 degrees of freedom.

Figure 19 The sum of these two areas is the P-value

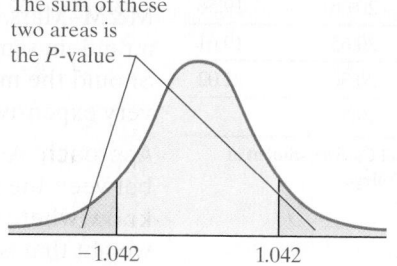

-1.042 1.042

Using Table VII, we find the row that corresponds to 10 degrees of freedom. The value 1.042 lies between 0.879 and 1.093. The value of 0.879 has an area of 0.20 to the right under the t-distribution. The area under the t-distribution to the right of 1.093 is 0.15.

Because 1.042 is between 0.879 and 1.093, the P-value is between 2(0.15) and 2(0.20). So

$$0.30 < P\text{-value} < 0.40$$

Using Table XVII we find P-value $= 2P(t > 1.0) = 2(0.170) = 0.340$ with 10 degrees of freedom.

Technology Step 3 Using Minitab, the exact P-value is 0.323.

Step 4 The P-value of 0.323 [by-hand: $0.30 < P$-value < 0.40] means that, *if* the null hypothesis that $\mu = 20.1$ grams is true, we expect about 32 out of 100 samples to result in a sample mean as extreme or more extreme than the one obtained. The result we obtained is not unusual, so we do not reject the null hypothesis.

Step 5 There is not sufficient evidence to conclude that the Snickers have a mean weight different from 20.1 grams at the $\alpha = 0.01$ level of significance. The machine should not be shut down.

● **Now Work Problem 21**

In Other Words
Results are statistically significant if the difference between the observed result and the statement made in the null hypothesis is unlikely to occur due to chance alone.

② ## Understand the Difference between Statistical Significance and Practical Significance

When a large sample size is used in a hypothesis test, the results could be statistically significant even though the difference between the sample statistic and mean stated in the null hypothesis may have no *practical significance*.

Definition | **Practical significance** refers to the idea that, while small differences between the statistic and parameter stated in the null hypothesis are statistically significant, the difference may not be large enough to cause concern or be considered important.

EXAMPLE 3   Statistical versus Practical Significance

Problem According to the American Community Survey, the mean travel time to work in Collin County, Texas, in 2013 was 27.5 minutes. The Department of Transportation reprogrammed all the traffic lights in Collin County in an attempt to reduce travel time. To determine if there is evidence that travel time has decreased as a result of the reprogramming, the Department of Transportation obtains a random sample of 2500 commuters, records their travel time to work, and finds a sample mean of 27.2 minutes with a standard deviation of 8.5 minutes. Does this result suggest that travel time has decreased at the $\alpha = 0.05$ level of significance?

Approach We will use both the classical and P-value approach to test the hypothesis.

Solution

Step 1 The Department of Transportation wants to know if the mean travel time to work has decreased from 27.5 minutes. From this, we have

$$H_0: \mu = 27.5 \text{ minutes} \quad \text{versus} \quad H_1: \mu < 27.5 \text{ minutes}$$

Step 2 The level of significance is $\alpha = 0.05$.

Step 3 The test statistic is

$$t_0 = \frac{\bar{x} - \mu_0}{\frac{s}{\sqrt{n}}} = \frac{27.2 - 27.5}{\frac{8.5}{\sqrt{2500}}} = -1.765$$

Classical Approach

Because this is a left-tailed test, the critical value with $\alpha = 0.05$ and $2500 - 1 = 2499$ degrees of freedom is $-t_{0.05} \approx -1.645$ (use the last row of Table VII when the degrees of freedom is greater than 1000).

Step 4 Because the test statistic is less than the critical value (the test statistic falls in the critical region), we reject the null hypothesis.

P-Value Approach

Because this is a left-tailed test, P-value $= P(t_0 < -1.765)$. From Table VII, we find the approximate P-value is $0.025 < P$-value < 0.05 [Technology: P-value $= 0.0389$].

Step 4 Because the P-value is less than the level of significance, $\alpha = 0.05$, we reject the null hypothesis.

Step 5 There is sufficient evidence at the $\alpha = 0.05$ level of significance to conclude the mean travel time to work has decreased.

While the difference between 27.2 minutes and 27.5 minutes is statistically significant, it has no practical meaning. After all, is 0.3 minute (18 seconds) really going to make anyone feel better about his or her commute to work?   •

The reason that the results from Example 3 were statistically significant had to do with the large sample size. The moral of the story is this:

> **CAUTION!**
> Beware of studies with large sample sizes that claim statistical significance because the differences may not have any practical meaning.

Large sample sizes can lead to results that are statistically significant, while the difference between the statistic and parameter in the null hypothesis is not enough to be considered practically significant.

Technology Step-by-Step Hypothesis Tests Regarding μ

TI-83/84 Plus
1. If necessary, enter raw data in L1.
2. Press STAT, highlight TESTS, and select 2:T-Test.
3. If the data are raw, highlight DATA; make sure that List is set to L1 and Freq is set to 1. If summary statistics are known, highlight STATS and enter the summary statistics. For the value of μ_0, enter the value of the mean stated in the null hypothesis.
4. Select the direction of the alternative hypothesis.
5. Highlight **Calculate** or **Draw** and press ENTER. The TI-83/84 gives the *P*-value.

Minitab
1. Enter raw data in column C1.
2. Select the **Stat** menu, highlight **Basic Statistics**, then highlight **1-Sample t . . . .**
3. If you have raw data, select "One or more samples, each in a column" from the drop-down menu. Place the cursor in the box, highlight the column containing the raw data, and click "Select." If you have summarized data, select "Summarized data" from the drop-down menu. Enter the sample size, sample mean, and sample standard deviation. Check the "Perform hypothesis test" box and enter the value of the population mean stated in the null hypothesis.
4. Click Options. Enter the direction of the alternative hypothesis. Click OK twice.

Excel
1. Enter raw data into Column A.
2. Load the XLSTAT Add-in, if necessary.
3. Select the XLSTAT menu and highlight Parametric tests. Select One-sample *t*-test and *z*-test.
4. Place the cursor in the Data: cell and then highlight the data in the spreadsheet. Check Student's *t*-test.
5. Click the Options tab. Choose the appropriate direction of the alternative hypothesis. Enter the mean stated in the null hypothesis in the Theoretical mean: cell. Enter the level of significance required for a confidence interval. For example, enter 10 for a 90% confidence interval. Click OK.

StatCrunch
1. If you have raw data, enter them into the spreadsheet. Name the column variable.
2. Select **Stat**, highlight **T Stats**, select **One Sample**, and then choose either **With Data** or **With Summary**.
3. If you chose **With Data**, select the column that has the observations. If you chose **With Summary**, enter the mean, standard deviation, and sample size. Choose the hypothesis test radio button. Enter the value of the mean stated in the null hypothesis and choose the direction of the alternative hypothesis from the pull-down menu. Click Compute!.

10.3 Assess Your Understanding

Skill Building

1. (a) Determine the critical value for a right-tailed test of a population mean at the $\alpha = 0.01$ level of significance with 15 degrees of freedom.
 (b) Determine the critical value for a left-tailed test of a population mean at the $\alpha = 0.05$ level of significance based on a sample size of $n = 20$.
 (c) Determine the critical values for a two-tailed test of a population mean at the $\alpha = 0.05$ level of significance based on a sample size of $n = 13$.

2. (a) Determine the critical value for a right-tailed test of a population mean at the $\alpha = 0.1$ level of significance with 22 degrees of freedom.
 (b) Determine the critical value for a left-tailed test of a population mean at the $\alpha = 0.01$ level of significance based on a sample size of $n = 40$.
 (c) Determine the critical values for a two-tailed test of a population mean at the $\alpha = 0.01$ level of significance based on a sample size of $n = 33$.

3. To test $H_0: \mu = 50$ versus $H_1: \mu < 50$, a simple random sample of size $n = 24$ is obtained from a population that is known to be normally distributed.
 (a) If $\bar{x} = 47.1$ and $s = 10.3$, compute the test statistic.
 (b) If the researcher decides to test this hypothesis at the $\alpha = 0.05$ level of significance, determine the critical value.

 (c) Draw a *t*-distribution that depicts the critical region.
 (d) Will the researcher reject the null hypothesis? Why?

4. To test $H_0: \mu = 40$ versus $H_1: \mu > 40$, a simple random sample of size $n = 25$ is obtained from a population that is known to be normally distributed.
 (a) If $\bar{x} = 42.3$ and $s = 4.3$, compute the test statistic.
 (b) If the researcher decides to test this hypothesis at the $\alpha = 0.1$ level of significance, determine the critical value.
 (c) Draw a *t*-distribution that depicts the critical region.
 (d) Will the researcher reject the null hypothesis? Why?

5. To test $H_0: \mu = 100$ versus $H_1: \mu \neq 100$, a simple random sample of size $n = 23$ is obtained from a population that is known to be normally distributed.
 (a) If $\bar{x} = 104.8$ and $s = 9.2$, compute the test statistic.
 (b) If the researcher decides to test this hypothesis at the $\alpha = 0.01$ level of significance, determine the critical values.
 (c) Draw a *t*-distribution that depicts the critical region.
 (d) Will the researcher reject the null hypothesis? Why?
 (e) Construct a 99% confidence interval to test the hypothesis.

6. To test $H_0: \mu = 80$ versus $H_1: \mu < 80$, a simple random sample of size $n = 22$ is obtained from a population that is known to be normally distributed.
 (a) If $\bar{x} = 76.9$ and $s = 8.5$, compute the test statistic.
 (b) If the researcher decides to test this hypothesis at the $\alpha = 0.02$ level of significance, determine the critical value.

(c) Draw a t-distribution that depicts the critical region.

(d) Will the researcher reject the null hypothesis? Why?

7. To test $H_0: \mu = 20$ versus $H_1: \mu < 20$, a simple random sample of size $n = 18$ is obtained from a population that is known to be normally distributed.

(a) If $\bar{x} = 18.3$ and $s = 4.3$, compute the test statistic.

(b) Draw a t-distribution with the area that represents the P-value shaded.

(c) Approximate and interpret the P-value.

(d) If the researcher decides to test this hypothesis at the $\alpha = 0.05$ level of significance, will the researcher reject the null hypothesis? Why?

8. To test $H_0: \mu = 4.5$ versus $H_1: \mu > 4.5$, a simple random sample of size $n = 13$ is obtained from a population that is known to be normally distributed.

(a) If $\bar{x} = 4.9$ and $s = 1.3$, compute the test statistic.

(b) Draw a t-distribution with the area that represents the P-value shaded.

(c) Approximate and interpret the P-value.

(d) If the researcher decides to test this hypothesis at the $\alpha = 0.1$ level of significance, will the researcher reject the null hypothesis? Why?

9. To test $H_0: \mu = 105$ versus $H_1: \mu \neq 105$, a simple random sample of size $n = 35$ is obtained.

(a) Does the population have to be normally distributed to test this hypothesis by using the methods presented in this section? Why?

(b) If $\bar{x} = 101.9$ and $s = 5.9$, compute the test statistic.

(c) Draw a t-distribution with the area that represents the P-value shaded.

(d) Approximate and interpret the P-value.

(e) If the researcher decides to test this hypothesis at the $\alpha = 0.01$ level of significance, will the researcher reject the null hypothesis? Why?

10. To test $H_0: \mu = 45$ versus $H_1: \mu \neq 45$, a simple random sample of size $n = 40$ is obtained.

(a) Does the population have to be normally distributed to test this hypothesis by using the methods presented in this section? Why?

(b) If $\bar{x} = 48.3$ and $s = 8.5$, compute the test statistic.

(c) Draw a t-distribution with the area that represents the P-value shaded.

(d) Approximate and interpret the P-value.

(e) If the researcher decides to test this hypothesis at the $\alpha = 0.01$ level of significance, will the researcher reject the null hypothesis? Why?

(f) Construct a 99% confidence interval to test the hypothesis.

Applying the Concepts

11. You Explain It! ATM Withdrawals According to the Crown ATM Network, the mean ATM withdrawal is $67. PayEase, Inc., manufactures an ATM that allows one to pay bills (electric, water, parking tickets, and so on), as well as withdraw money. A review of 40 withdrawals shows the mean withdrawal is $73 from a PayEase ATM machine. Do people withdraw more money from a PayEase ATM machine?

(a) Determine the appropriate null and alternative hypotheses to answer the question.

(b) Suppose the P-value for this test is 0.02. Explain what this value represents.

(c) Write a conclusion for this hypothesis test assuming an $\alpha = 0.05$ level of significance.

12. You Explain It! Are Women Getting Taller? In 1990, the mean height of women 20 years of age or older was 63.7 inches based on data obtained from the Centers for Disease Control and Prevention's *Advance Data Report*, No. 347. Suppose that a random sample of 45 women who are 20 years of age or older today results in a mean height of 63.9 inches.

(a) State the appropriate null and alternative hypotheses to assess whether women are taller today.

(b) Suppose the P-value for this test is 0.35. Explain what this value represents.

(c) Write a conclusion for this hypothesis test assuming an $\alpha = 0.10$ level of significance.

NW 13. Ready for College? The ACT is a college entrance exam. ACT has determined that a score of 22 on the mathematics portion of the ACT suggests that a student is ready for college-level mathematics. To achieve this goal, ACT recommends that students take a core curriculum of math courses: Algebra I, Algebra II, and Geometry. Suppose a random sample of 200 students who completed this core set of courses results in a mean ACT math score of 22.6 with a standard deviation of 3.9. Do these results suggest that students who complete the core curriculum are ready for college-level mathematics? That is, are they scoring above 22 on the math portion of the ACT?

(a) State the appropriate null and alternative hypotheses.

(b) Verify that the requirements to perform the test using the t-distribution are satisfied.

(c) Use the classical or P-value approach at the $\alpha = 0.05$ level of significance to test the hypotheses in part (a).

(d) Write a conclusion based on your results to part (c).

14. SAT Verbal Scores Do students who learned English and another language simultaneously score worse on the SAT Critical Reading exam than the general population of test takers? The mean score among all test takers on the SAT Critical Reading exam is 501. A random sample of 100 test takers who learned English and another language simultaneously had a mean SAT Critical Reading score of 485 with a standard deviation of 116. Do these results suggest that students who learn English as well as another language simultaneously score worse on the SAT Critical Reading exam?

(a) State the appropriate null and alternative hypotheses.

(b) Verify that the requirements to perform the test using the t-distribution are satisfied.

(c) Use the classical or P-value approach at the $\alpha = 0.1$ level of significance to test the hypotheses in part (a).

(d) Write a conclusion based on your results to part (c).

15. Effects of Alcohol on the Brain In a study published in the *American Journal of Psychiatry* (157:737–744, May 2000), researchers wanted to measure the effect of alcohol on the hippocampal region, the portion of the brain responsible for long-term memory storage, in adolescents. The researchers randomly selected 12 adolescents with alcohol use disorders to determine whether the hippocampal volumes in the alcoholic adolescents were less than the normal volume of 9.02 cubic centimeters (cm^3). An analysis of the sample data revealed that the hippocampal volume is approximately normal with $\bar{x} = 8.10$ cm^3 and $s = 0.7$ cm^3. Conduct the appropriate test at the $\alpha = 0.01$ level of significance.

16. Effects of Plastic Resin Para-nonylphenol is found in polyvinyl chloride (PVC) used in the food processing and packaging industries. Researchers wanted to determine the effect this substance had on the organ weight of first-generation mice when both parents were exposed to 50 micrograms per liter $(\mu\text{g}/\text{L})$ of para-nonylphenol in drinking water for 4 weeks.

After 4 weeks, the mice were bred. After 100 days, the offspring of the exposed parents were sacrificed and the kidney weights were determined. The mean kidney weight of the 12 offspring was found to be 396.9 milligrams (mg), with a standard deviation of 45.4 mg. Is there significant evidence to conclude that the kidney weight of the offspring whose parents were exposed to 50 μg/L of para-nonylphenol in drinking water for 4 weeks is greater than 355.7 mg, the mean weight of kidneys in normal 100-day-old mice at the $\alpha = 0.05$ level of significance?
Source: Vendula Kyselova et al., "Effects of p-nonylphenol and resveratrol on body and organ weight and in vivo fertility of outbred CD-1 mice," *Reproductive Biology and Endocrinology*, 2003.

17. Credit Scores A Fair Isaac Corporation (FICO) score is used by credit agencies (such as mortgage companies and banks) to assess the creditworthiness of individuals. Values range from 300 to 850, with a FICO score over 700 considered to be a quality credit risk. According to Fair Isaac Corporation, the mean FICO score is 703.5. A credit analyst wondered whether high-income individuals (incomes in excess of $100,000 per year) had higher credit scores. He obtained a random sample of 40 high-income individuals and found the sample mean credit score to be 714.2 with a standard deviation of 83.2. Conduct the appropriate test to determine if high-income individuals have higher FICO scores at the $\alpha = 0.05$ level of significance.

18. TVaholics According to the American Time Use Survey, the typical American spends 154.8 minutes (2.58 hours) per day watching television. A survey of 50 Internet users results in a mean time watching television per day of 128.7 minutes, with a standard deviation of 46.5 minutes. Conduct the appropriate test to determine if Internet users spend less time watching television at the $\alpha = 0.05$ level of significance.
Source: Norman H. Nie and D. Sunshine Hillygus. "Where Does Internet Time Come From? A Reconnaissance." *IT & Society*, 1(2).

19. Age of Death-Row Inmates In 2002, the mean age of an inmate on death row was 40.7 years, according to data from the U.S. Department of Justice. A sociologist wondered whether the mean age of a death-row inmate has changed since then. She randomly selects 32 death-row inmates and finds that their mean age is 38.9, with a standard deviation of 9.6. Construct a 95% confidence interval about the mean age. What does the interval imply?

20. Energy Consumption In 2001, the mean household expenditure for energy was $1493, according to data from the U.S. Energy Information Administration. An economist wanted to know whether this amount has changed significantly from its 2001 level. In a random sample of 35 households, he found the mean expenditure (in 2001 dollars) for energy during the most recent year to be $1618, with a standard deviation $321. Construct a 95% confidence interval about the mean energy expenditure. What does the interval imply?

NW **21. Waiting in Line** The mean waiting time at the drive-through of a fast-food restaurant from the time an order is placed to the time the order is received is 84.3 seconds. A manager devises a new drive-through system that he believes will decrease wait time. He initiates the new system at his restaurant and measures the wait time for 10 randomly selected orders. The wait times are provided in the table.

108.5	67.4	58.0	75.9	65.1
80.4	95.5	86.3	70.9	72.0

(a) Because the sample size is small, the manager must verify that wait time is normally distributed and the sample does not contain any outliers. The normal probability plot and boxplot are shown. The correlation between waiting time

and expected z-scores is 0.971. Are the conditions for testing the hypothesis satisfied?

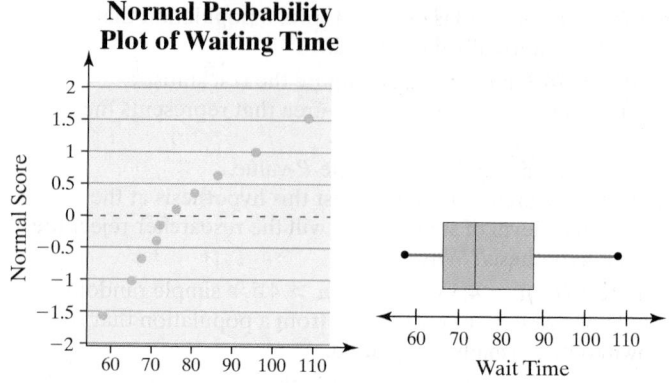

Normal Probability Plot of Waiting Time

(b) Is the new system effective? Use the $\alpha = 0.1$ level of significance.

22. Reading Rates Michael Sullivan, son of the author, decided to enroll in a reading course that allegedly increases reading speed and comprehension. Prior to enrolling in the class, Michael read 198 words per minute (wpm). The following data represent the words per minute read for 10 different passages read after the course.

206	217	197	199	210
210	197	212	227	209

(a) Because the sample size is small, we must verify that reading speed is normally distributed and the sample does not contain any outliers. The normal probability plot and boxplot are shown. The correlation between reading rate and expected z-scores is 0.964. Are the conditions for testing the hypothesis satisfied?

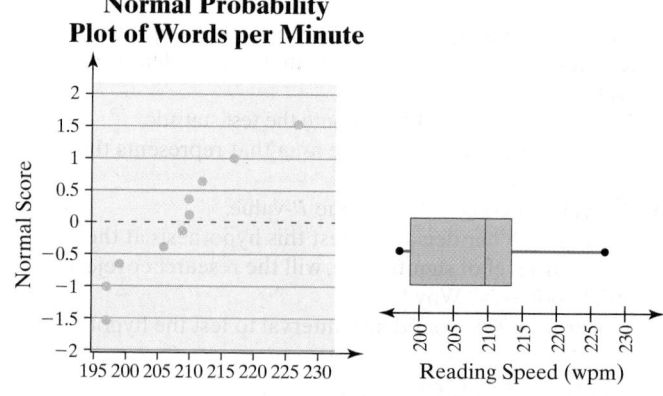

Normal Probability Plot of Words per Minute

(b) Was the class effective? Use the $\alpha = 0.10$ level of significance.

23. Calcium in Rainwater Calcium is essential to tree growth. In 1990, the concentration of calcium in precipitation in Chautauqua, New York, was 0.11 milligram per liter (mg/L). A random sample of 10 precipitation dates in 2014 results in the following data:

0.065	0.087	0.070	0.262	0.126
0.183	0.120	0.234	0.313	0.108

Source: National Atmospheric Deposition Program

A normal probability plot suggests the data could come from a population that is normally distributed. A boxplot does not show any outliers. Does the sample evidence suggest that calcium concentrations have changed since 1990? Use the $\alpha = 0.05$ level of significance.

24. Filling Bottles A certain brand of apple juice is supposed to have 64 ounces of juice. Because the penalty for underfilling bottles is severe, the target mean amount of juice is 64.05 ounces. However, the filling machine is not precise, and the exact amount of juice varies from bottle to bottle. The quality-control manager wishes to verify that the mean amount of juice in each bottle is 64.05 ounces so that she can be sure that the machine is not over- or underfilling. She randomly samples 22 bottles of juice, measures the content, and obtains the following data:

64.05	64.05	64.03	63.97	63.95	64.02
64.01	63.99	64.00	64.01	64.06	63.94
63.98	64.05	63.95	64.01	64.08	64.01
63.95	63.97	64.10	63.98		

A normal probability plot suggests the data could come from a population that is normally distributed. A boxplot does not show any outliers.

(a) Should the assembly line be shut down so that the machine can be recalibrated? Use a 0.01 level of significance.

(b) Explain why a level of significance of $\alpha = 0.01$ is more reasonable than $\alpha = 0.1$. [*Hint:* Consider the consequences of incorrectly rejecting the null hypothesis.]

25. Starbucks Stock The volume of a stock is the number of shares traded for a given day. In 2011, Starbucks stock had a mean daily volume of 7.52 million shares according to Yahoo!Finance. A random sample of 40 trading days in 2014 was obtained and the volume of shares traded on those days was recorded. Go to www.pearsonhighered.com/sullivanstats to obtain the data file 10_3_25 using the file format of your choice for the version of the text you are using.

(a) Draw a histogram of the data. Describe the shape of the distribution.

(b) Draw a boxplot of the data. Are there any outliers?

(c) Based on the shape of the histogram and boxplot, explain why a large sample size is necessary to perform inference on the mean using the normal model.

(d) Does the evidence suggest that the volume of Starbucks stock has changed since 2014? Use an $\alpha = 0.05$ level of significance.

26. Study Time Go to www.pearsonhighered.com/sullivanstats to obtain the data file 10_3_26 using the file format of your choice for the version of the text you are using. The data represent the amount of time students in Sullivan's online statistics course spent studying for Section 4.1—Scatter Diagrams and Correlation.

(a) Draw a histogram of the data. Describe the shape of the distribution.

(b) Draw a boxplot of the data. Are there any outliers?

(c) Based on the shape of the histogram and boxplot, explain why a large sample size is necessary to perform inference on the mean using the normal model.

(d) According to MyStatLab, the mean time students would spend on this assignment nationwide is 95 minutes. Treat the data as a random sample of all Sullivan online statistics students. Do the sample data suggest that Sullivan's students are any different from the country as far as time spent on Section 4.1 goes? Use an $\alpha = 0.05$ level of significance.

Test the hypothesis in the problem given by constructing a 95% confidence interval.

27. Problem 23   **28.** Problem 24

29. Problem 25   **30.** Problem 26

31. Statistical Significance versus Practical Significance A math teacher claims that she has developed a review course that increases the scores of students on the math portion of the SAT exam. Based on data from the College Board, SAT scores are normally distributed with $\mu = 515$. The teacher obtains a random sample of 1800 students, puts them through the review class, and finds that the mean SAT math score of the 1800 students is 519 with a standard deviation of 111.

(a) State the null and alternative hypotheses.

(b) Test the hypothesis at the $\alpha = 0.10$ level of significance. Is a mean SAT math score of 519 significantly higher than 515?

(c) Do you think that a mean SAT math score of 519 versus 515 will affect the decision of a school admissions administrator? In other words, does the increase in the score have any practical significance?

(d) Test the hypothesis at the $\alpha = 0.10$ level of significance with $n = 400$ students. Assume the same sample statistics. Is a sample mean of 519 significantly more than 515? What do you conclude about the impact of large samples on the hypothesis test?

32. Statistical Significance versus Practical Significance The manufacturer of a daily dietary supplement claims that its product will help people lose weight. The company obtains a random sample of 950 adult males aged 20 to 74 who take the supplement and finds their mean weight loss after 8 weeks to be 0.9 pound with standard deviation weight loss of 7.2 pounds.

(a) State the null and alternative hypotheses.

(b) Test the hypothesis at the $\alpha = 0.1$ level of significance. Is a mean weight loss of 0.9 pound significant?

(c) Do you think that a mean weight loss of 0.9 pound is worth the expense and commitment of a daily dietary supplement? In other words, does the weight loss have any practical significance?

(d) Test the hypothesis at the $\alpha = 0.1$ level of significance with $n = 40$ subjects. Assume the same sample statistics. Is a sample mean weight loss of 0.9 pound significantly more than 0 pound? What do you conclude about the impact of large samples on the hypothesis test?

33. Accept versus Do Not Reject The mean IQ score of humans is 100. Suppose the director of Institutional Research at Joliet Junior College (JJC) obtains a simple random sample of 40 JJC students and finds the mean IQ is 103.4 with a standard deviation of 13.2.

(a) Consider the hypotheses $H_0: \mu = 100$ versus $H_1: \mu > 100$. Explain what the director of Institutional Research is testing. Perform the test at the $\alpha = 0.05$ level of significance. Write a conclusion for the test.

(b) Consider the hypotheses $H_0: \mu = 101$ versus $H_1: \mu > 101$. Explain what the director of Institutional Research is testing. Perform the test at the $\alpha = 0.05$ level of significance. Write a conclusion for the test.

(c) Consider the hypotheses $H_0: \mu = 102$ versus $H_1: \mu > 102$. Explain what the director of Institutional Research is testing. Perform the test at the $\alpha = 0.05$ level of significance. Write a conclusion for the test.

(d) Based on the results of parts (a)–(c), write a few sentences that explain the difference between "accepting" the statement in the null hypothesis versus "not rejecting" the statement in the null hypothesis.

34. Simulation Simulate drawing 100 simple random samples of size $n = 15$ from a population that is normally distributed with mean 100 and standard deviation 15.

(a) Test the null hypothesis $H_0: \mu = 100$ versus $H_1: \mu \neq 100$ for each of the 100 simple random samples.

(b) If we test this hypothesis at the $\alpha = 0.05$ level of significance, how many of the 100 samples would you expect to result in a Type I error?

(c) Count the number of samples that lead to a rejection of the null hypothesis. Is it close to the expected value determined in part (b)?

(d) Describe how we know that a rejection of the null hypothesis results in making a Type I error in this situation.

35. Simulation The *exponential probability distribution* can be used to model waiting time in line or the lifetime of electronic components. Its density function is skewed right. Suppose the wait time in a line can be modeled by the exponential distribution with $\mu = \sigma = 5$ minutes.

(a) Simulate obtaining 100 simple random samples of size $n = 10$ from the population described. That is, simulate obtaining a simple random sample of 10 individuals waiting in a line where the wait time is expected to be 5 minutes.

(b) Test the null hypothesis $H_0: \mu = 5$ versus the alternative $H_1: \mu \neq 5$ for each of the 100 simulated simple random samples.

(c) If we test this hypothesis at the $\alpha = 0.05$ level of significance, how many of the 100 samples would you expect to result in a Type I error?

(d) Count the number of samples that lead to a rejection of the null hypothesis. Is it close to the expected value determined in part (c)? What might account for any discrepancies?

Retain Your Knowledge

36. Reading at Bedtime It is well-documented that watching TV, working on a computer, or any other activity involving artificial light can be harmful to sleep patterns. Researchers wanted to determine if the artificial light from e-Readers also disrupted sleep. In the study, 12 young adults were given either an iPad or printed book for four hours before bedtime. Then, they switched reading

devices. Whether the individual received the iPad or book first was determined randomly. Bedtime was 10 P.M. and the time to fall asleep was measured each evening. It was found that participants took an average of 10 minutes longer to fall asleep after reading on an iPad. The P-value for the test was 0.009.
Source: Anne-Marie Chang, et. al. "Evening Use of Light-Emitting eReaders Negatively Affects Sleep, Circarian Timing, and Next-Morning Alertness" *PNAS* 2015 112(4) 1232-1277. doi:10.1073/pnas.1418490112

(a) What is the research objective?

(b) What is the response variable? It is quantitative or qualitative?

(c) What is the treatment?

(d) Is this a designed experiment or observational study? What type?

(e) The null hypothesis for this test would be that there is no difference in time to fall asleep with an e-Reader and printed book. The alternative is that there is a difference. Interpret the P-value.

Explaining the Concepts

37. What's the Problem? The head of institutional research at a university believes that the mean age of full-time students is declining. In 1995, the mean age of a full-time student was known to be 27.4 years. After looking at the enrollment records of all 4934 full-time students in the current semester, he found that the mean age was 27.1 years, with a standard deviation of 7.3 years. He conducted a hypothesis of $H_0: \mu = 27.4$ years versus $H_1: \mu < 27.4$ years and obtained a P-value of 0.0019. He concluded that the mean age of full-time students did decline. Is there anything wrong with his research?

38. The procedures for testing a hypothesis regarding a population mean are robust. What does this mean?

39. Explain the difference between *statistical significance* and *practical significance*.

40. Wanna Live Longer? Become a Chief Justice The life expectancy of a male during the course of the past 100 years is approximately 27,725 days. Go to Wikipedia.com and download the data that represent the life span of chief justices of Canada for those who have died. Conduct a test to determine whether the evidence suggests that chief justices of Canada live longer than the general population of males. Suggest a reason why the conclusion drawn may be flawed.

10.4 Hypothesis Tests for a Population Standard Deviation

Preparing for This Section Before getting started, review the following:

- Confidence intervals for a population standard deviation (Section 9.3, pp. 451–453)

Objective   ❶ Test hypotheses about a population standard deviation

In this section, we discuss methods for testing hypotheses regarding a population variance or standard deviation.

Why might we be interested in testing hypotheses regarding σ^2 or σ? Many production processes require not only accuracy on average (the mean) but also consistency. Consider a filling machine (such as a coffee machine) that over- and underfills cups, but, on average, fills correctly. Customers are not happy about underfilled cups, and they are dissatisfied with overfilled cups because of spilling. They want a machine that *consistently* delivers the correct amount. As another example, consider a mutual fund that invests in the stock market. An investor would prefer consistent year-to-year returns near 12%, rather than returns that fluctuate wildly, yet result in a mean return of 12%. Both of these situations illustrate the importance of measuring variability. The main idea is that the standard deviation and the variance are measures of consistency. The less consistent the values of a variable are, the higher the standard deviation of the variable.

We begin by reviewing the chi-square distribution.

Chi-Square Distribution

If a simple random sample of size n is obtained from a normally distributed population with mean μ and standard deviation σ, then

$$\chi^2 = \frac{(n-1)s^2}{\sigma^2}$$

has a chi-square distribution with $n - 1$ degrees of freedom, where s^2 is a sample variance.

Remember, the symbol χ^2 is pronounced "kigh-square." Critical values of the chi-square distribution are in Table VIII in Appendix A.

Figure 20
Chi-Square Distributions

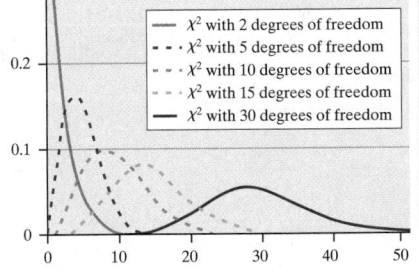

Characteristics of the Chi-Square Distribution

1. It is not symmetric.
2. The shape of the chi-square distribution depends on the degrees of freedom, just as with Student's t-distribution.
3. As the number of degrees of freedom increases, the chi-square distribution becomes more nearly symmetric, as is illustrated in Figure 20.
4. The values of χ^2 are nonnegative (greater than or equal to 0).

Recall that Table VIII is structured similarly to Table VII (for the t-distribution). The left column represents the degrees of freedom, and the top row represents the area under the chi-square distribution to the right of the critical value. We use the notation χ^2_α to denote the critical χ^2-value for which the area under the chi-square distribution to the right of χ^2_α is α. For example, the area under the chi-square distribution to the right of $\chi^2_{0.10}$ is 0.10.

❶ Test Hypotheses about a Population Standard Deviation

Testing Hypotheses about a Population Variance or Standard Deviation

To test hypotheses about the population variance or standard deviation, use the following steps, provided that

- the sample is obtained using simple random sampling or from a randomized experiment.
- the population is normally distributed.

Step 1   Determine the null and alternative hypotheses. The hypotheses can be structured in one of three ways:

Two-Tailed	Left-Tailed	Right-Tailed
$H_0: \sigma = \sigma_0$	$H_0: \sigma = \sigma_0$	$H_0: \sigma = \sigma_0$
$H_1: \sigma \neq \sigma_0$	$H_1: \sigma < \sigma_0$	$H_1: \sigma > \sigma_0$

Note: σ_0 is the assumed value of the population standard deviation.

Step 2   Select a level of significance, α, based on the seriousness of making a Type I error.

Classical Approach

Step 3   Compute the **test statistic**

$$\chi_0^2 = \frac{(n-1)s^2}{\sigma_0^2}$$

Use Table VIII to determine the critical value using $n - 1$ degrees of freedom.

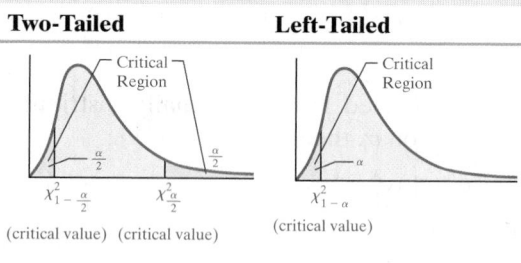

Two-Tailed	Left-Tailed

Critical Region $\chi_{1-\frac{\alpha}{2}}^2$ (critical value) $\chi_{\frac{\alpha}{2}}^2$ (critical value)

Critical Region $\chi_{1-\alpha}^2$ (critical value)

Right-Tailed

Critical Region χ_α^2 (critical value)

Step 4   Compare the critical value to the test statistic.

Two-Tailed	Left-Tailed	Right-Tailed
If $\chi_0^2 < \chi_{1-\alpha/2}^2$ or $\chi_0^2 > \chi_{\alpha/2}^2$, reject the null hypothesis.	If $\chi_0^2 < \chi_{1-\alpha}^2$, reject the null hypothesis.	If $\chi_0^2 > \chi_\alpha^2$, reject the null hypothesis.

Step 5   State the conclusion.

P-Value Approach

By Hand Step 3   Compute the **test statistic**

$$\chi_0^2 = \frac{(n-1)s^2}{\sigma_0^2}$$

Use Table VIII to approximate the P-value for a left- or right-tailed test by determining the area under the chi-square distribution with $n - 1$ degrees of freedom to the left (for a left-tailed test) or right (for a right-tailed test) of the test statistic. For two-tailed tests, it is recommended that technology be used to find the P-value or obtain a confidence interval.

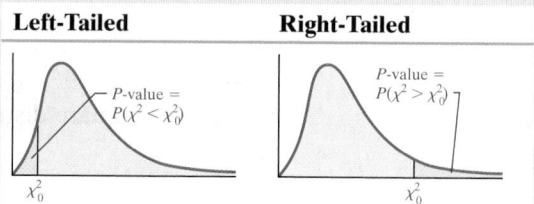

Left-Tailed	Right-Tailed

P-value = $P(\chi^2 < \chi_0^2)$ χ_0^2

P-value = $P(\chi^2 > \chi_0^2)$ χ_0^2

Technology Step 3   Use a statistical spreadsheet or calculator with statistical capabilities to obtain the P-value. The directions for obtaining the P-value using Minitab and StatCrunch are in the Technology Step-by-Step on pages 507–508.

Step 4   If P-value $< \alpha$, reject the null hypothesis.

Step 5   State the conclusion.

> **CAUTION!**
> The procedures in this section are not robust.

The methods presented for testing a hypothesis about a population variance or standard deviation are not robust. Therefore, if analysis of the data indicates that the variable does not come from a population that is normally distributed, the procedures presented in this section are not valid. Again, always be sure to verify that the requirements are satisfied before proceeding with the test!

EXAMPLE 1   Testing a Hypothesis about a Population Standard Deviation, Left-Tailed Test

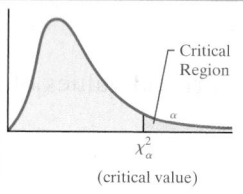

Table 2		
19.68	20.66	19.56
19.98	20.65	19.61
20.55	20.36	21.02
21.50	19.74	

Source: Michael Carlisle, student at Joliet Junior College

Problem   In Example 2 from Section 10.3, the quality-control engineer for M&M–Mars tested whether the mean weight of fun-size Snickers was 20.1 grams. Suppose that the standard deviation of the weight of the candy was 0.75 gram before a machine recalibration. The engineer wants to know if the recalibration resulted in more consistent weights. Conduct the appropriate test at the $\alpha = 0.05$ level of significance. The data are reproduced in Table 2.

Approach   If the recalibration results in more consistent weights, then the standard deviation weight should be less than it was prior to recalibration. We want to test if $\sigma < 0.75$ gram. Before we can perform the hypothesis test, verify that the data come from a population that is normally distributed by constructing a normal probability plot. Then follow Steps 1 through 5.

Solution   A normal probability plot was shown for these data in Figure 17. (See page 497.) The plot indicates that the weights of the candy could come from a population that is normally distributed.

> **CAUTION!**
> The test statistic requires s^2 and σ^2, so be careful if the problem gives s or σ.

Step 1   We need evidence to conclude that the population standard deviation is less than 0.75 gram. This can be written $\sigma < 0.75$. We have

$$H_0\text{: } \sigma = 0.75 \text{ gram} \quad \text{versus} \quad H_1\text{: } \sigma < 0.75 \text{ gram}$$

This is a left-tailed test.

Step 2   The level of significance is $\alpha = 0.05$.

Classical Approach

Step 3   The standard deviation of the data in Table 2 is 0.6404 gram. The test statistic is

$$\chi_0^2 = \frac{(n-1)s^2}{\sigma_0^2} = \frac{(11-1)(0.6404)^2}{0.75^2} = 7.291$$

Because this is a left-tailed test, determine the critical χ^2-value at the $\alpha = 0.05$ level of significance with $n - 1 = 11 - 1 = 10$ degrees of freedom to be $\chi_{1-0.05}^2 = \chi_{0.95}^2 = 3.940$. The critical region is displayed in Figure 21.

Figure 21

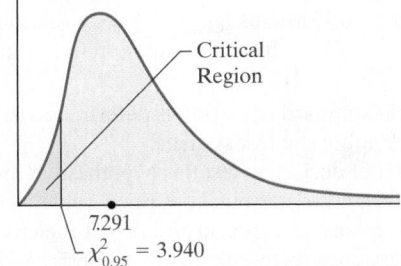

7.291
$\chi_{0.95}^2 = 3.940$

Step 4   Because the test statistic $\chi_0^2 = 7.291$ is greater than the critical value $\chi_{0.95}^2 = 3.940$, the quality-control engineer does not reject the null hypothesis.

P-Value Approach

By Hand Step 3   The standard deviation of the data in Table 2 is 0.6404 gram. The test statistic is

$$\chi_0^2 = \frac{(n-1)s^2}{\sigma_0^2} = \frac{(11-1)(0.6404)^2}{0.75^2} = 7.291$$

Because this is a left-tailed test, the P-value is the area under the χ^2-distribution with $11 - 1 = 10$ degrees of freedom to the left of the test statistic, $\chi_0^2 = 7.291$, as shown in Figure 22. That is, P-value $= P(\chi^2 < \chi_0^2) = P(\chi^2 < 7.291)$, with 10 degrees of freedom.

Using Table VIII, find the row that corresponds to 10 degrees of freedom. The value 7.291 is greater than 4.865. The value of 4.865 has an area under the χ^2-distribution of 0.10 to the left. So the P-value is greater than 0.10.

Figure 22

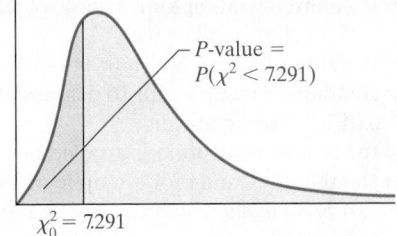

P-value = $P(\chi^2 < 7.291)$

$\chi_0^2 = 7.291$

Technology Step 3   Figure 23 shows the results from StatCrunch. The P-value is reported as 0.3022.

Figure 23

Hypothesis test results:
σ^2 : variance of Variable
$H_0 : \sigma^2 = 0.5625$
$H_A : \sigma^2 < 0.5625$

Variable	Sample Var.	DF	Chi-Square Stat	P-value
Weight	0.41006908	10	7.2901173	0.3022

Step 4   The P-value is greater than the level of significance $(0.3022 > 0.05)$, so we do not reject the null hypothesis.

Step 5   There is not sufficient evidence at the $\alpha = 0.05$ level of significance to conclude the standard deviation of the weight of fun-size Snickers is less than 0.75 gram. It appears that the recalibration did not result in more consistent weights.

● **Now Work Problem 9**

Technology Step-by-Step   Hypothesis Tests Regarding a Population Standard Deviation

Minitab

1. Enter the raw data into column C1, if necessary.
2. Select the **Stat** menu, highlight **Basic Statistics**, then choose **1 Variance**.
3. If you have raw data, select "One or more samples, each in column" from the drop-down menu. Place the cursor in the box, highlight the column containing the raw data, and

click "Select." If you have summarized data, select either "Sample standard deviation" or "Sample variance" from the drop-down menu. Enter the sample size and sample standard deviation or sample variance. Check the "Perform hypothesis test" box and enter the value of the population standard deviation or variance stated in the null hypothesis using the drop-down menu. Click Options and choose the direction of the alternative hypothesis. Click OK twice.

StatCrunch

1. If you have raw data, enter them into the spreadsheet. Name the column variable.
2. Select **Stat**, highlight **Variance Stats**, highlight **One Sample**, and then chose either **With Data** or **With Summary**.

3. If you chose **With Data**, select the column that has the observations. If you chose **With Summary**, enter the sample variance and sample size. Choose the hypothesis test radio button. Enter the value of the variance stated in the null hypothesis and choose the direction of the alternative hypothesis from the pull-down menu. Click Compute!.

 ## 10.4 Assess Your Understanding

Skill Building

1. (a) Determine the critical value for a right-tailed test of a population standard deviation with 18 degrees of freedom at the $\alpha = 0.05$ level of significance.
(b) Determine the critical value for a left-tailed test of a population standard deviation for a sample of size $n = 23$ at the $\alpha = 0.1$ level of significance.
(c) Determine the critical values for a two-tailed test of a population standard deviation for a sample of size $n = 30$ at the $\alpha = 0.05$ level of significance.

2. (a) Determine the critical value for a right-tailed test of a population standard deviation with 16 degrees of freedom at the $\alpha = 0.01$ level of significance.
(b) Determine the critical value for a left-tailed test of a population standard deviation for a sample of size $n = 14$ at the $\alpha = 0.01$ level of significance.
(c) Determine the critical values for a two-tailed test of a population standard deviation for a sample of size $n = 61$ at the $\alpha = 0.05$ level of significance.

3. To test $H_0: \sigma = 50$ versus $H_1: \sigma < 50$, a random sample of size $n = 24$ is obtained from a population that is known to be normally distributed.

(a) If the sample standard deviation is determined to be $s = 47.2$, compute the test statistic.
(b) If the researcher decides to test this hypothesis at the $\alpha = 0.05$ level of significance, determine the critical value.
(c) Draw a chi-square distribution and depict the critical region.
(d) Will the researcher reject the null hypothesis? Why?

4. To test $H_0: \sigma = 35$ versus $H_1: \sigma > 35$, a random sample of size $n = 15$ is obtained from a population that is known to be normally distributed.

(a) If the sample standard deviation is determined to be $s = 37.4$, compute the test statistic.
(b) If the researcher decides to test this hypothesis at the $\alpha = 0.01$ level of significance, determine the critical value.
(c) Draw a chi-square distribution and depict the critical region.
(d) Will the researcher reject the null hypothesis? Why?

5. To test $H_0: \sigma = 1.8$ versus $H_1: \sigma > 1.8$, a random sample of size $n = 18$ is obtained from a population that is known to be normally distributed.

(a) If the sample standard deviation is determined to be $s = 2.4$, compute the test statistic.

(b) If the researcher decides to test this hypothesis at the $\alpha = 0.10$ level of significance, determine the critical value.
(c) Draw a chi-square distribution and depict the critical region.
(d) Will the researcher reject the null hypothesis? Why?

6. To test $H_0: \sigma = 0.35$ versus $H_1: \sigma < 0.35$, a random sample of size $n = 41$ is obtained from a population that is known to be normally distributed.

(a) If the sample standard deviation is determined to be $s = 0.23$, compute the test statistic.
(b) If the researcher decides to test this hypothesis at the $\alpha = 0.01$ level of significance, determine the critical value.
(c) Draw a chi-square distribution and depict the critical region.
(d) Will the researcher reject the null hypothesis? Why?

7. To test $H_0: \sigma = 4.3$ versus $H_1: \sigma \neq 4.3$, a random sample of size $n = 12$ is obtained from a population that is known to be normally distributed.

(a) If the sample standard deviation is determined to be $s = 4.8$, compute the test statistic.
(b) If the researcher decides to test this hypothesis at the $\alpha = 0.05$ level of significance, determine the critical values.
(c) Draw a chi-square distribution and depict the critical regions.
(d) Will the researcher reject the null hypothesis? Why?

8. To test $H_0: \sigma = 1.2$ versus $H_1: \sigma \neq 1.2$, a random sample of size $n = 22$ is obtained from a population that is known to be normally distributed.

(a) If the sample standard deviation is determined to be $s = 0.8$, compute the test statistic.
(b) If the researcher decides to test this hypothesis at the $\alpha = 0.10$ level of significance, determine the critical values.
(c) Draw a chi-square distribution and depict the critical regions.
(d) Will the researcher reject the null hypothesis? Why?

Applying the Concepts

NW 9. Mutual Fund Risk One measure of the risk of a mutual fund is the standard deviation of its rate of return. Suppose a mutual fund qualifies as having moderate risk if the standard deviation of its monthly rate of return is less than 4%. A mutual-fund manager claims that his fund has moderate risk. A mutual-fund rating agency does not believe this claim and randomly selects 25 months and determines the rate of return for the fund. The standard deviation of the rate of return is computed to be 3.01%. Is there sufficient evidence to conclude that the fund has moderate risk at the $\alpha = 0.05$ level of significance? A normal probability plot indicates that the monthly rates of return are normally distributed.

10. Filling Machine A machine fills bottles with 64 fluid ounces of liquid. The quality-control manager determines that the fill levels are normally distributed with a mean of 64 ounces and a standard deviation of 0.42 ounce. He has an engineer recalibrate the machine in an attempt to lower the standard deviation. After the recalibration, the quality-control manager randomly selects 19 bottles from the line and determines that the standard deviation is 0.38 ounce. Is there less variability in the filling machine? Use the $\alpha = 0.01$ level of significance.

11. Pump Design The piston diameter of a certain hand pump is 0.5 inch. The quality-control manager determines that the diameters are normally distributed, with a mean of 0.5 inch and a standard deviation of 0.004 inch. The machine that controls the piston diameter is recalibrated in an attempt to lower the standard deviation. After recalibration, the quality-control manager randomly selects 25 pistons from the production line and determines that the standard deviation is 0.0025 inch. Was the recalibration effective? Use the $\alpha = 0.01$ level of significance.

12. Counting Carbs The manufacturer of processed deli meats reports that the standard deviation of the number of carbohydrates in its smoked turkey breast is 0.5 gram per 2-ounce serving. A dietitian does not believe the manufacturer and randomly selects eighteen 2-ounce servings of the smoked turkey breast and determines the number of carbohydrates per serving. The standard deviation of the number of carbs is computed to be 0.62 gram per serving. Is there sufficient evidence to indicate that the standard deviation is not 0.5 gram per serving at the $\alpha = 0.05$ level of significance? A normal probability plot indicates that the number of carbohydrates per serving is normally distributed.

13. Waiting in Line In Problem 21 from Section 10.3, we considered the mean waiting time at the drive-through of a fast-food restaurant. The manager is also worried about the variability in wait times. Prior to the new drive-through system, the standard deviation of wait time was 18.0 seconds. Use the data in the following table to decide whether the evidence suggests the standard deviation of wait time is less than 18.0 seconds. In Problem 21 from Section 10.3, we verified the data could come from a population that is normally distributed. Use the $\alpha = 0.05$ level of significance.

108.5	67.4	58.0	75.9	65.1
80.4	95.5	86.3	70.9	72.0

14. Filling Bottles In Problem 24 from Section 10.3, we considered whether a filling machine was calibrated correctly. Another aspect of the calibration is the variability in the amount of juice in the bottle. Suppose the machine is calibrated so that it fills the bottles with a standard deviation of 0.04 ounce. Do the sample data suggest the machine may be "out of control," that is, have too much variability? In Problem 24 from Section 10.3, we were told that the data could come from a population that is normally distributed. Use the $\alpha = 0.05$ level of significance.

64.05	64.05	64.03	63.97	63.95	64.02
64.01	63.99	64.00	64.01	64.06	63.94
63.98	64.05	63.95	64.01	64.08	64.01
63.95	63.97	64.10	63.98		

15. An Inconsistent Player Sports announcers often say, "I wonder which player will show up to play today." This is the announcer's way of saying that the player's performance varies dramatically from game to game. Suppose that the standard deviation of the number of points scored by shooting guards in the NBA is 8.3. A random sample of 25 games played by Derrick Rose results in a sample standard deviation of 6.7 points. Assume that a normal probability plot indicates that the points scored are approximately normally distributed. Is Derrick Rose more consistent than other shooting guards in the NBA at the $\alpha = 0.10$ level of significance?

16. Waiting in Line One aspect of queuing theory is to consider waiting time in lines. A fast-food chain is trying to determine whether it should switch from having four cash registers with four separate lines to four cash registers with a single line. It has been determined that the mean wait-time in both lines is equal. However, the chain is uncertain about which line has less variability in wait time. From experience, the chain knows that the wait times in the four separate lines are normally distributed with $\sigma = 1.2$ minutes. In a study, the chain reconfigured five restaurants to have a single line and measured the wait times for 50 randomly selected customers. The sample standard deviation was determined to be $s = 0.84$ minute. Is the variability in wait time less for a single line than for multiple lines at the $\alpha = 0.05$ level of significance?

17. Heights of Baseball Players Data obtained from the National Center for Health Statistics show that men between the ages of 20 and 29 have a mean height of 69.3 inches, with a standard deviation of 2.9 inches. A baseball analyst wonders whether the standard deviation of heights of major-league baseball players is less than 2.9 inches. The heights (in inches) of 20 randomly selected players are shown in the table.

72	74	71	72	76
70	77	75	72	72
77	72	75	70	73
73	75	73	74	74

Source: espn.com

(a) Verify that the data are normally distributed by drawing a normal probability plot.
(b) Compute the sample standard deviation.
(c) Test the notion at the $\alpha = 0.01$ level of significance.

18. NCAA Softball NCAA rules require the circumference of a softball to be 12 ± 0.125 inches. Suppose that the NCAA also requires that the standard deviation of the softball circumferences not exceed 0.05 inch. A representative from the NCAA believes the manufacturer does not meet this requirement. She collects a random sample of 20 softballs from the production line and finds that $s = 0.09$ inch. Is there enough evidence to support the representative's belief at the $\alpha = 0.05$ level of significance?

10.5 Putting It Together: Which Method Do I Use?

Objective ❶ Determine the appropriate hypothesis test to perform

❶ **Determine the Appropriate Hypothesis Test to Perform**

Perhaps the most difficult aspect of testing hypotheses is determining which hypothesis test to conduct. To assist in the decision making, we present Figure 24, which shows which approach to take in testing hypotheses for the three parameters discussed in this chapter.

Figure 24

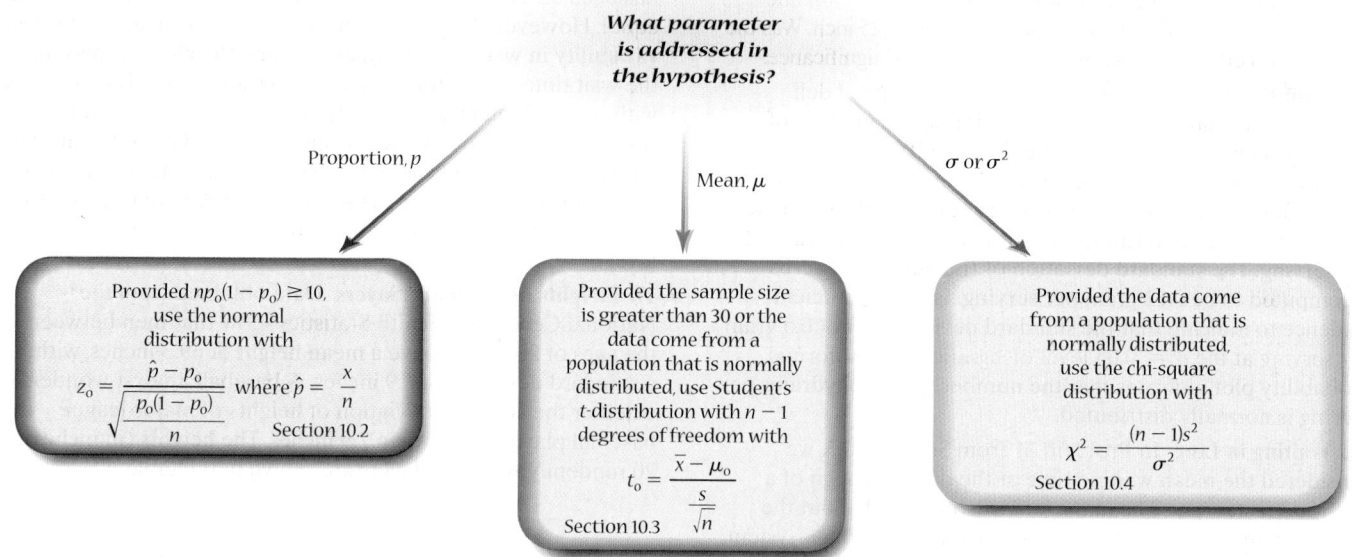

10.5 Assess Your Understanding

Skill Building

1. A simple random sample of size $n = 19$ is drawn from a population that is normally distributed. The sample mean is found to be 0.8, and the sample standard deviation is found to be 0.4. Test whether the population mean is less than 1.0 at the $\alpha = 0.01$ level of significance.

2. A simple random sample of size $n = 200$ individuals with a valid driver's license is asked if they drive an American-made automobile. Of the 200 individuals surveyed, 115 responded that they drive an American-made automobile. Determine if a majority of those with a valid driver's license drive an American-made automobile at the $\alpha = 0.05$ level of significance.

3. A simple random sample of size $n = 20$ is drawn from a population that is normally distributed. The sample variance is found to be 49.3. Determine whether the population variance is less than 95 at the $\alpha = 0.1$ level of significance.

4. A simple random sample of size $n = 15$ is drawn from a population that is normally distributed. The sample mean is found to be 23.8, and the sample standard deviation is found to be 6.3. Is the population mean different from 25 at the $\alpha = 0.01$ level of significance?

5. A simple random sample of size $n = 16$ is drawn from a population that is normally distributed. The sample variance is found to be 13.7. Test whether the population variance is greater than 10 at the $\alpha = 0.05$ level of significance.

6. A simple random sample of size $n = 65$ is drawn from a population. The sample mean is found to be 583.1, and the sample standard deviation is found to be 114.9. Is the population mean different from 600 at the $\alpha = 0.1$ level of significance?

7. A simple random sample of size $n = 40$ is drawn from a population. The sample mean is found to be 108.5, and the sample standard deviation is found to be 17.9. Is the population mean greater than 100 at the $\alpha = 0.05$ level of significance?

8. A simple random sample of size $n = 320$ adults was asked their favorite ice cream flavor. Of the 320 individuals surveyed, 58 responded that they preferred mint chocolate chip. Do less than 25% of adults prefer mint chocolate chip ice cream? Use the $\alpha = 0.01$ level of significance.

Applying the Concepts

9. Smarter Kids? A psychologist obtains a random sample of 20 mothers in the first trimester of their pregnancy. The mothers are asked to play Mozart in the house at least 30 minutes each day until they give birth. After 5 years, the child is administered an IQ test. We know that IQs are normally distributed with a mean of 100. If the IQs of the 20 children in the study result in a sample mean of 104.2, and a sample standard deviation of 14.7, is there evidence that the children have higher IQs? Use the $\alpha = 0.05$ level of significance.

10. The Atomic Bomb In October 1945, the Gallup organization asked 1487 randomly sampled Americans, "Do you think we can develop a way to protect ourselves from atomic bombs in case other countries tried to use them against us?" with 788 responding yes. Did a majority of Americans feel the United States could develop a way to protect itself from atomic bombs in 1945? Use the $\alpha = 0.05$ level of significance.

11. Yield Strength A manufacturer of high-strength, low-alloy steel beams requires that the standard deviation of yield strength not exceed 7000 pounds per square inch (psi). The quality-control manager selected a sample of 20 steel beams and measured their yield strength. The standard deviation of the sample was 7500 psi. Assume that yield strengths are normally distributed. Does the evidence suggest that the standard deviation of yield strength exceeds 7000 psi at the $\alpha = 0.01$ level of significance?

12. Pharmaceuticals A pharmaceutical company manufactures a 200-milligram (mg) pain reliever. Company specifications require that the standard deviation of the amount of the active ingredient must not exceed 5 mg. The quality-control manager selects a random sample of 30 tablets from a certain batch and finds that the sample standard deviation is 7.3 mg. Assume that the amount of the active ingredient is normally distributed. Determine whether the standard deviation of the amount of the active ingredient is greater than 5 mg at the $\alpha = 0.05$ level of significance.

13. Course Redesign Pass rates for Intermediate Algebra at a community college are 52.6%. In an effort to improve pass rates in the course, faculty of a community college develop a mastery-based learning model where course content is delivered in a lab through a computer program. The instructor serves as a learning mentor for the students. Of the 480 students who enroll in the mastery-based course, 267 pass.

(a) What is the variable of interest in this study? What type of variable is it?

(b) At the 0.01 level of significance, decide whether the sample evidence suggests the mastery-based learning model improved pass rates.

(c) Explain why a 0.01 level of significance might be used to test this hypothesis.

14. Number of Credit Cards According to the Federal Reserve Bank of Boston, among individuals who had credit cards in 2014, the mean number of cards was 3.5. Treat the individuals who have credit cards in the SullivanStatsSurveyI as a random sample of credit card holders. Go to www.pearsonhighered.com/sullivanstats to obtain the data file SullivanStatsSurveyI using the file format of your choice for the version of the text you are using. The results of the survey are under the column "Number of cards."

(a) What is the variable of interest in this study? Is it qualitative or quantitative?

(b) Do the results of the survey imply that the mean number of cards per individual is less than 3.5? Use the $\alpha = 0.05$ level of significance.

15. Gas Mileage The Environmental Protection Agency (EPA) states that a 2013 Kia Optima should get 28 miles per gallon, on average. The website www.fueleconomy.gov allows users to report the miles per gallon that they get on their vehicle. Treat the following data as a random sample of ten 2013 Kia Optima owners. The data represent the miles per gallon on their vehicle. Is there reason to believe that individuals are getting different gas mileage than the EPA states should be attained? Be sure to verify all conditions necessary to conduct the appropriate test.

30.2	30.3	19.4	26.7	17.6
19.9	23.3	20.2	16.7	27.5

Source: www.fueleconomy.gov

16. Sleepy? According to the National Sleep Foundation, children between the ages of 6 and 11 years should get 10 hours of sleep each night. In a survey of 56 parents of 6 to 11 year olds, it was found that the mean number of hours the children slept was 8.9 with a standard deviation of 3.2. Does the sample data suggest that 6 to 11 year olds are sleeping less than the required amount of time each night? Use the 0.01 level of significance.

17. Text while Driving According to the research firm Toluna, the proportion of individuals who text while driving is 0.26. Suppose a random sample of 60 individuals are asked to disclose if they text while driving. Results of the survey are shown next, where 0 indicates no and 1 indicates yes. Do the data contradict the results of Toluna? Use the $\alpha = 0.05$ level of significance.

0	0	0	0	0	0	1	0	0	1
0	0	1	1	1	0	1	1	1	0
1	0	0	0	0	0	0	1	0	1
1	0	1	0	1	0	0	0	0	0
0	1	1	0	0	0	0	0	0	1
0	1	1	0	1	0	0	0	0	1

18. Student Loan Balances Student loan debt has reached record levels in the United States. In a random sample of 100 individuals who have student loan debt, it was found the mean debt was $23,979 with a standard deviation of $31,400. Data based on results from the Federal Reserve Bank of New York.

(a) What do you believe is the shape of the distribution of student loan debt? Explain.

(b) Use this information to estimate the mean student loan debt among all with such debt at the 95% level of confidence. Interpret this result.

(c) What could be done to increase the precision of the estimate?

19. Political Decision Politicians often form their positions on various policies through polling. Suppose the U.S. Congress is considering passage of a tax increase to pay down the national debt and national polls suggest the general population is equally split on the matter. A congresswoman wants to poll her constituency on this controversial tax increase. To get a good sense as to how the citizens of her very populous district feel (well over 1 million registered voters), she decides to poll 8250 individuals within the district. Of those surveyed, 4205 indicated they are in favor of the tax increase. Given that politicians are generally leery of voting for tax increases, what level of significance would you recommend the congresswoman use in conducting this hypothesis test? Do the results of the survey represent statistically significant evidence a majority of the district favor the tax increase? What would you recommend to the congresswoman?

20. Quality Control Suppose the mean wait-time for a telephone reservation agent at a large airline is 43 seconds. A manager with the airline is concerned that business may be lost due to customers having to wait too long for an agent. To address this concern, the manager develops new airline reservation policies that are intended to reduce the amount of time an agent needs to spend with each customer. A random sample of 250 customers results in a sample mean wait-time of 42.3 seconds with a standard deviation of 4.2 seconds. Using an $\alpha = 0.05$ level of significance, do you believe the new policies were effective? Do you think the results have any practical significance?

Retain Your Knowledge

21. Ideal Number of Children A survey from the Gallup organization asked, "What do you think is the ideal number of children for a family to have?" Go to www.pearsonhighered.com/sullivanstats to obtain the data file 10_5_21 using the file format of your choice for the version of the text you are using to get the survey results.

(a) Draw a dot plot of the data. Comment on the shape of the distribution.

(b) What is the mode ideal number of children?

(c) Determine the mean, median, standard deviation, and interquartile range ideal number of children. Round your answers to the nearest thousandth.

(d) Explain why a large sample size is needed to perform any inference regarding this population.

(e) In May 1997, the ideal number of children was considered to be 2.64. Do the results of this poll indicate that people's beliefs as to the ideal number of children have changed? Use the 0.05 level of significance.

22. Confidence Intervals Suppose you wish to determine if the mean IQ of students on your campus is different from the mean IQ in the general population, 100. To conduct this study, you obtain a simple random sample of 50 students on your campus, administer an IQ test, and record the results. The mean IQ of the sample of 50 students is found to be 107.3 with a standard deviation of 13.6.

(a) Conduct a hypothesis test (preferably using technology) $H_0: \mu = \mu_0$ versus $H_1: \mu \neq \mu_0$ for

$\mu_0 = 103, 104, 105, 106, 107, 108, 109, 110, 111, 112$ at the $\alpha = 0.05$ level of significance. For which values of μ_0 do you not reject the null hypothesis?

(b) Construct a 95% confidence interval for the mean IQ of students on your campus. What might you conclude about how the lower and upper bounds of a confidence interval relate to the values for which the null hypothesis is rejected?

(c) Suppose you changed the level of significance in conducting the hypothesis test to $\alpha = 0.01$. What would happen to the range of values of μ_0 for which the null hypothesis is not rejected? Why does this make sense?

In Problems 23–29, decide whether the problem requires a confidence interval or hypothesis test, and determine the variable of interest. For any problem requiring a confidence interval, state whether the confidence interval will be for a population proportion or population mean. For any problem requiring a hypothesis test, write the null and alternative hypothesis.

23. An investigator with the Food and Drug Administration wanted to determine whether a typical bag of potato chips contained less than the 16 ounces claimed by the manufacturer.

24. A researcher wanted to estimate the average length of time mothers who gave birth via Caesarean section spent in a hospital after delivery of the baby.

25. An official with the Internal Revenue Service wished to estimate the proportion of high-income (greater than $100,000 annually) earners who under-reported their net income (and, therefore, their tax liability).

26. According to the Pew Research Center, 55% of adult Americans support the death penalty for those convicted of murder. A social scientist wondered whether a higher proportion of adult Americans with at least a bachelor's degree support the death penalty for those convicted of murder.

27. In 2014, of the 37 million borrowers who have outstanding student loan balances, 14% have at least one past due student loan account. A researcher with the United States Department of Education believes this proportion has increased since then. *Source:* American Student Assistance

28. Researchers measured regular testosterone levels in a random sample of athletes and then measured testosterone levels prior to an athletic event. They wanted to know whether testosterone levels increase prior to athletic events.

29. Is the dispersion of IQ scores among undergraduate students lower than that of the general population, 15?

10.6 The Probability of a Type II Error and the Power of the Test

Preparing for This Section Before getting started, review the following:

- Computing normal probabilities (Section 7.2, pp. 366–370)

Objectives
① Determine the probability of making a Type II error
② Compute the power of the test

① Determine the Probability of Making a Type II Error

The probability of making a Type I error, rejecting a true null hypothesis, is denoted α, and is also called the level of significance. So, if the null hypothesis is true and $\alpha = 0.05$, the probability of rejecting H_0 is 0.05.

If the null hypothesis is false, but we fail to reject it, we have made a Type II error. We denote the probability of making a Type II error β. We know there is an inverse relation between α and β, but how can we determine the probability of β? Because the alternative hypothesis contains a range of values (such as $p > 0.3$), the probability of making a Type II error has multiple values, each corresponding to a specific value of the parameter from the alternative hypothesis.

EXAMPLE 1   Computing the Probability of a Type II Error

Problem In Example 1 from Section 10.2 on page 484, we tested $H_0: p = 0.5$ versus $H_1: p < 0.5$, where p is the proportion of Illinois high school students who have taken the ACT and are prepared for college-level mathematics. To conduct this test, a random sample of $n = 500$ high school students who took the ACT was obtained and the number of students who were prepared for college mathematics (ACT score at least 22) was determined. The test was conducted with $\alpha = 0.05$. If the true population proportion of Illinois high school students with a score of 22 or higher on the ACT is 0.48, which means the alternative hypothesis is true, what is the probability of making a Type II error, β? That is, what is the probability of failing to reject the null hypothesis when, in fact, $p < 0.5$?

Approach To compute β, we need to determine the likelihood of obtaining a sample proportion that would lead to failing to reject the null hypothesis assuming the true population proportion is 0.48. This can be done in two steps.

Step 1   Determine the values of the sample proportion that lead to rejecting H_0.

Step 2   Determine the probability that we do not reject H_0 assuming the true population proportion is 0.48.

Solution

Step 1   Because we are testing $H_0: p = 0.5$ and $n = 500$, the distribution for the sample proportion is approximately normally distributed with $\mu_{\hat{p}} = 0.5$ and standard deviation $\sigma_{\hat{p}} = \sqrt{\dfrac{0.5(1 - 0.5)}{500}}$. Any test statistic less than $z = -1.645$ will lead to rejecting the statement in the null hypothesis (since this results in a P-value < 0.05). Put another way, if $\hat{p}$ is more than 1.645 standard errors below $p_0 = 0.5$, the null hypothesis will be rejected. So we reject H_0 if

$$\hat{p} < 0.5 - 1.645 \cdot \sqrt{\frac{0.5(1 - 0.5)}{500}} = 0.463$$

Figure 25

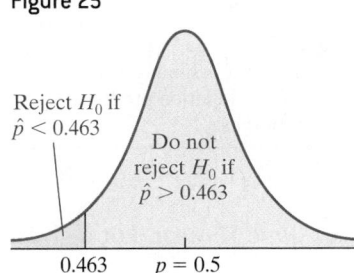

Reject H_0 if
$\hat{p} < 0.463$

Do not
reject H_0 if
$\hat{p} > 0.463$

0.463 $p = 0.5$

Figure 25 shows the reject and do not reject regions.

Note: We agree to round the sample proportion that makes the "cut-off" point to three decimal places.

Step 2 If the true population proportion is 0.48, then the mean of the sampling distribution is $\mu_{\hat{p}} = 0.48$ and standard deviation $\sigma_{\hat{p}} = \sqrt{\dfrac{0.48(1 - 0.48)}{500}}$. Remember, the probability of a Type II error is the probability of not rejecting H_0 when H_1 is true. So we need to determine the probability of obtaining a sample proportion of $\hat{p} = 0.463$ or higher from a population whose proportion is $p = 0.48$.

$$
\begin{aligned}
\beta &= P(\text{Type II error}) \\
&= P(\text{do not reject } H_0 \text{ when } H_1 \text{ is true}) \\
&= P(\hat{p} > 0.463 \text{ given that } p = 0.48) \\
&= P\left(z > \frac{0.463 - 0.48}{\sqrt{\dfrac{0.48(1 - 0.48)}{500}}}\right) = P(z > -0.76) = 0.7764
\end{aligned}
$$

The probability of not rejecting $H_0: p = 0.50$ when, in fact, $H_1: p < 0.5$ is true is 0.7764. ●

Figure 26 shows the distributions of $\hat{p}$ for both the assumption that the null hypothesis is true, $p = 0.5$ (red), and if a specific alternative hypothesis is true, $p = 0.48$ (blue).

Figure 26

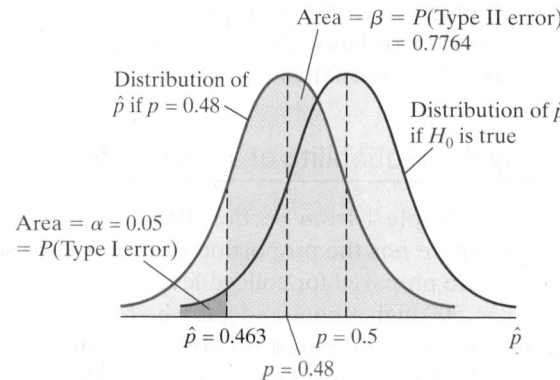

Area = β = P(Type II error)
= 0.7764

Distribution of
$\hat{p}$ if $p = 0.48$

Distribution of $\hat{p}$
if H_0 is true

Area = α = 0.05
= P(Type I error)

$\hat{p} = 0.463$ $p = 0.5$ $\hat{p}$
$p = 0.48$

For each possible value of p less than 0.5, there is a unique value of β. For example, if the true population proportion for the scenario in Example 1 was $p = 0.45$, then $\beta = 0.2810$. You should confirm this for yourself. Notice the probability of making a Type II error decreases as the specific value in the alternative moves farther away from the statement in the null ($p = 0.5$).

The procedure for computing the probability of making a Type II error is shown next.

Determining the Probability of a Type II Error

Step 1 Determine the sample proportion that separates the rejection region from the non-rejection region.

Left-Tailed Test (H_1: $p < p_0$)	Right-Tailed Test (H_1: $p > p_0$)	Two-Tailed Test (H_1: $p \neq p_0$)
$\hat{p} = p_0 - z_\alpha \cdot \sqrt{\dfrac{p_0(1-p_0)}{n}}$	$\hat{p} = p_0 + z_\alpha \cdot \sqrt{\dfrac{p_0(1-p_0)}{n}}$	$\hat{p}_L = p_0 - z_{\alpha/2} \cdot \sqrt{\dfrac{p_0(1-p_0)}{n}}$
		$\hat{p}_U = p_0 + z_{\alpha/2} \cdot \sqrt{\dfrac{p_0(1-p_0)}{n}}$

Note: z_α is the critical z-value corresponding to the level of significance, α.

Step 2

Left-Tailed Test (H_1: $p < p_0$)	Right-Tailed Test (H_1: $p > p_0$)	Two-Tailed Test (H_1: $p \neq p_0$)
Find the area under the normal curve to the right of the sample proportion found in Step 1 assuming a particular value of the population proportion in the alternative hypothesis.	Find the area under the normal curve to the left of the sample proportion found in Step 1 assuming a particular value of the population proportion in the alternative hypothesis.	Find the area under the normal curve between $\hat{p}_L$ and $\hat{p}_U$ found in Step 1 assuming a particular value of the population proportion in the alternative hypothesis.

Note two important points about the probability of making a Type II error. For a given value of the level of significance, α:

- As the value of the parameter moves farther away from the value stated in the null hypothesis, the likelihood of making a Type II error decreases.
- As the sample size n increases, the probability of making a Type II error decreases.

The second point should not shock you. We know as the sample size increases the standard error decreases, thereby reducing the chance of making an error. The second point also emphasizes the need to determine a sample size appropriate for your level of comfort in making an error. Whenever designing an experiment or survey, it is good practice to determine β for the level of α chosen. The value of the parameter that should be used to represent a possible value in the alternative hypothesis should be one that represents the *effect size* you are looking for. For example, if the researchers from Example 1 were really concerned if p actually was 0.45, this is a good choice for determining the probability of making a Type II error for the sample size chosen. If this value of β is too large, the researchers should increase the sample size.

② Compute the Power of the Test

In Other Words
The power of the test is the probability that you will correctly reject H_0.

The probability of rejecting the null hypothesis when the alternative hypothesis is true is $1 - \beta$. The value of $1 - \beta$ is called the **power of the test**. The higher the power of the test, the more likely the test will reject the null when the alternative hypothesis is true.

EXAMPLE 2   Computing the Power of the Test

Problem Compute the power of the test for the situation in Example 1.

Approach The power of the test is $1 - \beta$. In Example 1, we found that β is 0.7764 when the true population proportion is 0.48.

Solution The power of the test is $1 - 0.7764 = 0.2236$. There is a 0.2236 probability of rejecting the null hypothesis if the true population proportion is 0.48.   ●

In Other Words
As the true proportion gets closer to the value of the proportion stated in the null hypothesis, it becomes more difficult to correctly reject the null hypothesis.

Table 3 shows both the probability of making a Type II error and the power of the test for a number of different values of the population proportion that are less than 0.5. Notice the closer the population proportion gets to 0.5, the larger the probability of a Type II error becomes. This is because when the true population proportion is close to the population proportion stated in the null hypothesis, there is a high probability that we will not reject the null hypothesis and will make a Type II error.

We can use the values of p and their corresponding powers to construct a *power curve*. A **power curve** is a graph that shows the power of the test against values of the parameter that make the null hypothesis false. Figure 27 shows the power curve for the values in Table 3. The power curve shows that as the true population proportion approaches the proportion stated in the null hypothesis, $p = 0.5$, the power of the test decreases.

Table 3		
Value of p	**P (Type II error), β**	**Power, $1 - \beta$**
0.40	0.0020	0.9980
0.42	0.0256	0.9744
0.44	0.1492	0.8508
0.46	0.4483	0.5517
0.47	0.6217	0.3783
0.48	0.7764	0.2236
0.49	0.8869	0.1131

Figure 27

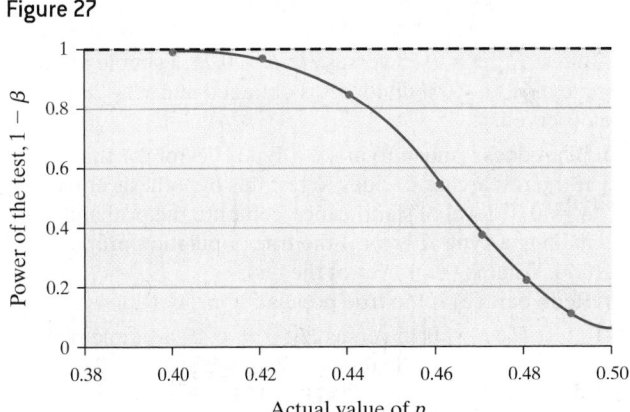

Technology Step-by-Step   Probability of a Type II Error and Power

StatCrunch

1. Select **Stat**, highlight **T Stats** or **Proportion Stats** (depending on whether finding power for a mean or proportion), select **One Sample**, then **Power/Sample Size**.

2. For means: Enter Alpha (the level of significance), the standard deviation, whether the alternative hypothesis is one-sided or two-sided, the difference between the mean stated in the null hypothesis and the true mean, and the sample size. Delete any entry in the Power cell. Leave the cursor in the Power cell and click Compute.

For proportions: Enter Alpha (the level of significance), the target proportion (proportion stated in the null hypothesis), whether the alternative hypothesis is one-sided or two-sided, the true population proportion, and the sample size. Delete any entry in the Power cell. Leave the cursor in the Power cell and click Compute.

10.6 Assess Your Understanding

Vocabulary and Skill Building

1. Explain what it means to make a Type II error.

2. Explain the term *power of the test*.

3. To test H_0: $p = 0.30$ versus H_1: $p < 0.30$, a simple random sample of $n = 300$ individuals is obtained and $x = 86$ successes are observed.
(a) What does it mean to make a Type II error for this test?
(b) If the researcher decides to test this hypothesis at the $\alpha = 0.05$ level of significance, compute the probability of making a Type II error if the true population proportion is 0.28. What is the power of the test?
(c) Redo part (b) if the true population proportion is 0.25.

4. To test H_0: $p = 0.40$ versus H_1: $p > 0.40$, a simple random sample of $n = 200$ individuals is obtained and $x = 84$ successes are observed.
(a) What does it mean to make a Type II error for this test?
(b) If the researcher decides to test this hypothesis at the $\alpha = 0.05$ level of significance, compute the probability of making a Type II error if the true population proportion is 0.44. What is the power of the test?
(c) Redo part (b) if the true population proportion is 0.47.

5. To test H_0: $p = 0.65$ versus H_1: $p > 0.65$, a simple random sample of $n = 100$ individuals is obtained and $x = 69$ successes are observed.
(a) What does it mean to make a Type II error for this test?
(b) If the researcher decides to test this hypothesis at the $\alpha = 0.01$ level of significance, compute the probability of making a Type II error if the true population proportion is 0.70. What is the power of the test?
(c) Redo part (b) if the true population proportion is 0.72.

6. To test H_0: $p = 0.75$ versus H_1: $p < 0.75$, a simple random sample of $n = 400$ individuals is obtained and $x = 280$ successes are observed.
(a) What does it mean to make a Type II error for this test?
(b) If the researcher decides to test this hypothesis at the $\alpha = 0.01$ level of significance, compute the probability of making a Type II error if the true population proportion is 0.71. What is the power of the test?
(c) Redo part (b) if the true population proportion is 0.68.

7. To test H_0: $p = 0.45$ versus H_1: $p \neq 0.45$, a simple random sample of $n = 500$ individuals is obtained and $x = 245$ successes are observed.
(a) What does it mean to make a Type II error for this test?
(b) If the researcher decides to test this hypothesis at the $\alpha = 0.1$ level of significance, compute the probability of making a Type II error if the true population proportion is 0.49. What is the power of the test?
(c) Redo part (b) if the true population proportion is 0.47.

8. To test H_0: $p = 0.25$ versus H_1: $p \neq 0.25$, a simple random sample of $n = 350$ individuals is obtained and $x = 74$ successes are observed.
(a) What does it mean to make a Type II error for this test?
(b) If the researcher decides to test this hypothesis at the $\alpha = 0.05$ level of significance, compute the probability of making a Type II error if the true population proportion is 0.23. What is the power of the test?
(c) Redo part (b) if the true population proportion is 0.28.

Applying the Concepts

9. Worried about Retirement? In April 2009, the Gallup organization surveyed 676 adults aged 18 and older and found that 352 believed they would not have enough money to live comfortably in retirement. The folks at Gallup want to know if this represents sufficient evidence to conclude a majority (more than 50%) of adults in the United States believe they will not have enough money in retirement.
(a) What does it mean to make a Type II error for this test?
(b) If the researcher decides to test this hypothesis at the $\alpha = 0.05$ level of significance, determine the probability of making a Type II error if the true population proportion is 0.53. What is the power of the test?
(c) Redo part (b) if the true proportion is 0.55.

10. Ready for College? ACT, a college entrance exam used for admission, looked at historical records and established 21 as the minimum score on the ACT reading portion of the exam for a student to be considered prepared for social science in college. (**Note:** "Being prepared" means there is a 75% probability of successfully completing a social science course in college.) An official with the Illinois State Department of Education wonders whether a majority of the students in her state who took the ACT are prepared to take social science. She obtains a simple random sample of 500 records of students who have taken the ACT and finds that 269 are prepared. Does this represent significant evidence that a majority (more than 50%) of the students in the state of Illinois are prepared for social science in college upon graduation?
(a) What does it mean to make a Type II error for this test?
(b) If the researcher decides to test this hypothesis at the $\alpha = 0.05$ level of significance, determine the probability of making a Type II error if the true population proportion is 0.52. What is the power of the test?
(c) Redo part (b) if the true proportion is 0.55.

11. Quality of Education In August 2002, 47% of parents who had children in grades K–12 were satisfied with the quality of education the students receive. A recent Gallup poll found that 437 of 1013 parents of children in grades K–12 were satisfied with the quality of education the students receive. Does this suggest the proportion of parents satisfied with the quality of education has decreased?
(a) What does it mean to make a Type II error for this test?
(b) If the researcher decides to test this hypothesis at the $\alpha = 0.10$ level of significance, determine the probability of making a Type II error if the true population proportion is 0.42. What is the power of the test?
(c) Redo part (b) if the true proportion is 0.46.

12. Eating Together In December 2001, 38% of adults with children under the age of 18 reported that their family ate dinner together seven nights a week. In a recent poll, 403 of 1122 adults with children under the age of 18 reported that their family ate dinner together seven nights a week. Has the proportion of families with children under the age of 18 who eat dinner together seven nights a week decreased?
(a) What does it mean to make a Type II error for this test?
(b) If the researcher decides to test this hypothesis at the $\alpha = 0.10$ level of significance, determine the probability of making a Type II error if the true population proportion is 0.35. What is the power of the test?
(c) Redo part (b) if the true proportion is 0.32.

13. Taught Enough Math In 1994, 52% of parents of children in high school felt it was a serious problem that high school students were not being taught enough math and science. A recent survey found that 256 of 800 parents of children in high school felt it was a serious problem that high school students were not being taught enough math and science. Do parents feel differently today than they did in 1994?

(a) What does it mean to make a Type II error for this test?

(b) If the researcher decides to test this hypothesis at the $\alpha = 0.05$ level of significance, determine the probability of making a Type II error if the true population proportion is 0.50. What is the power of the test?

(c) Redo part (b) if the true proportion is 0.48.

14. Infidelity According to menstuff.org, 22% of married men have "strayed" at least once during their married lives. A survey of 500 married men indicated that 122 have strayed at least once during their married life. Does this survey result contradict the results of menstuff.org?

(a) What does it mean to make a Type II error for this test?

(b) If the researcher decides to test this hypothesis at the $\alpha = 0.05$ level of significance, determine the probability of making a Type II error if the true population proportion is 0.25. What is the power of the test?

(c) Redo part (b) if the true proportion is 0.20.

15. Effect of α Redo Problem 3(b) with $\alpha = 0.01$. What effect does lowering the level of significance have on the power of the test? Why does this make sense?

16. Effect of α Redo Problem 4(b) with $\alpha = 0.01$. What effect does lowering the level of significance have on the power of the test? Why does this make sense?

17. Power Curve Draw a power curve for the scenario in Problem 3 by finding the power of the test for $p = 0.2, 0.22, 0.24, 0.26, 0.28$ [done in part (b)], and 0.29.

18. Power Curve Draw a power curve for the scenario in Problem 4 by finding the power of the test for $p = 0.41, 0.42, 0.44$ [done in part (b)], 0.46, 0.48, and 0.50.

19. Power in Tests on Means To test $H_0: \mu = 50$ versus $H_1: \mu < 50$, a simple random sample of size $n = 24$ is obtained from a population that is known to be normally distributed, and the sample standard deviation is found to be 6.

(a) A researcher decides to test the hypothesis at the $\alpha = 0.05$ level of significance. Determine the sample mean that separates the rejection region from the nonrejection region. [*Hint:* Follow the same approach as that laid out on page 514, but use Student's t-distribution to find the critical value.]

(b) Suppose the true population mean is $\mu = 48.9$. Use technology to find the area under the t-distribution to the right of the sample mean found in part (a) assuming $\mu = 48.9$. [*Hint:* This can be accomplished by performing a one-sample t-test.] This represents the probability of making a Type II error, β. What is the power of the test?

Explaining the Concepts

20. What happens to the power of the test as the true value of the parameter gets closer to the value of the parameter stated in the null hypothesis? Why is this result reasonable?

21. What effect does increasing the sample size have on the power of the test, assuming all else remains unchanged?

22. How do the probability of making a Type II error and effect size play a role in determining an appropriate sample size when performing a hypothesis test?

Chapter 10 Review

Summary

In this chapter, we discussed hypothesis testing. Hypothesis testing is the second type of inferential statistics (recall the other type of inference is estimation via confidence intervals).

In hypothesis testing, a statement is made regarding a population parameter, which leads to a null, H_0, and alternative hypothesis, H_1. The null hypothesis is a statement of "no change" or "no difference". We build a probability model under the assumption the statement in the null hypothesis is true, and we use sample data to decide whether to reject or not reject the statement in the null hypothesis. There are three options for structuring the null and alternative hypotheses on a single parameter.

Determining the Null and Alternative Hypotheses

1. Equal hypothesis versus not equal hypothesis **(two-tailed test)**

$$H_0: \text{parameter} = \text{some value}$$
$$H_1: \text{parameter} \neq \text{some value}$$

2. Equal hypothesis versus less than hypothesis **(left-tailed test)**

$$H_0: \text{parameter} = \text{some value}$$
$$H_1: \text{parameter} < \text{some value}$$

3. Equal hypothesis versus greater than hypothesis **(right-tailed test)**

$$H_0: \text{parameter} = \text{some value}$$
$$H_1: \text{parameter} > \text{some value}$$

In performing a hypothesis test, there is always the possibility of making a Type I error (rejecting the null hypothesis when it is true) or making a Type II error (not rejecting the null hypothesis when it is false). The probability of making a Type I error is equal to the level of significance, α, of the test.

The first test introduced was hypothesis tests about a population proportion p. Provided certain requirements were satisfied (simple random sample or randomized experiment, independence, and a large sample size), we were able to use the normal model to assess statements made in the null hypothesis.

Then we discussed hypothesis testing on the mean. Here we also required a simple random sample (or randomized experiment) and independence, but we also required that either the sample come from a population that is normally distributed with no outliers or a large ($n \geq 30$) sample size. When dealing with small sample sizes, we tested the normality requirement with a normal probability plot and the outlier requirement with a boxplot. These requirements allow us to use Student's t-distribution to test the hypothesis.

Finally, we performed tests regarding a population standard deviation (or variance). This test requires that the population from which the sample is drawn be normally distributed. This test is not robust, so deviation from the normality requirement is not allowed.

All three hypothesis tests were performed using both the classical method and the P-value approach. The P-value approach to testing hypotheses has appeal because the rejection rule is always to reject the null hypothesis if the P-value is less than the level of significance, α.

In testing hypotheses, it is important to remember that we never *accept* the null hypothesis, because, without having access to the entire population, we don't know the exact value of the parameter stated in the null hypothesis. Rather, we say that we do not reject the null hypothesis.

Remember, statistical significance refers to the idea that the observed results are unlikely to occur if the statement in the null hypothesis is true. Practical significance, on the other hand, refers to the idea that, while small differences between the statistic and parameter stated in the null hypothesis are statistically significant, the difference may not be large enough to cause concern or be considered important.

We concluded the section by learning how to compute the probability of making a Type II error and the power of the test. The power of the test is the probability of rejecting the null hypothesis when the alternative hypothesis is true.

Vocabulary

Formulas

Test Statistics

- $z_0 = \dfrac{\hat{p} - p_0}{\sqrt{\dfrac{p_0(1 - p_0)}{n}}}$ follows the standard normal

distribution if $np_0(1 - p_0) \geq 10$ and $n \leq 0.05N$

- $t_0 = \dfrac{\bar{x} - \mu_0}{\dfrac{s}{\sqrt{n}}}$ follows Student's t-distribution with $n - 1$

degrees of freedom if the population from which the sample was drawn is normal or if the sample size is large ($n \geq 30$).

- $\chi_0^2 = \dfrac{(n - 1)s^2}{\sigma_0^2}$ follows the χ^2-distribution with $n - 1$

degrees of freedom if the population from which the sample was drawn is normal.

Type I and Type II Errors

- $\alpha = P(\text{Type I error})$
  $= P(\text{rejecting } H_0 \text{ when } H_0 \text{ is true})$
- $\beta = P(\text{Type II error})$
  $= P(\text{not rejecting } H_0 \text{ when } H_1 \text{ is true})$

OBJECTIVES

Section	You should be able to . . .	Example(s)	Review Exercises
10.1	1 Determine the null and alternative hypotheses (p. 473)	2	1(a), 2(a)
	2 Explain Type I and Type II errors (p. 476)	3	1(b), 1(c), 2(b), 2(c), 3, 4, 13(e), 18(a)
	3 State conclusions to hypothesis tests (p. 478)	4	1(d), 1(e), 2(d), 2(e)
10.2	1 Explain the logic of hypothesis testing (p. 481)	pp. 481–483	23, 25, 26
	2 Test hypotheses about a population proportion (p. 483)	1, 2, 3	7, 8, 11, 12, 17, 20
	3 Test hypotheses about a population proportion using the binomial probability distribution (p. 488)	4	21

Section	You should be able to . . .	Example(s)	Review Exercises
10.3	**1** Test hypotheses about a mean (p. 494)	1, 2	5, 6, 13, 14, 15, 16, 22
	2 Understand the difference between statistical significance and practical significance (p. 498)	3	20, 22
10.4	**1** Test hypotheses about a population standard deviation (p. 505)	1	9, 10, 19
10.5	**1** Determine the appropriate hypothesis test to perform (p. 510)		5–17, 19–22
10.6	**1** Determine the probability of making a Type II error (p. 512)	1	18(b)
	2 Compute the power of the test (p. 515)	2	4, 18(b)

Review Exercises

For Problems 1 and 2, (a) determine the null and alternative hypotheses, (b) explain what it would mean to make a Type I error, (c) explain what it would mean to make a Type II error, (d) state the conclusion that would be reached if the null hypothesis is not rejected, and (e) state the conclusion that would be reached if the null hypothesis is rejected.

1. Credit-Card Debt According to creditcard.com, the mean outstanding credit-card debt of college undergraduates was $3173 in 2010. A researcher believes that this amount has decreased since then.

2. More Credit-Card Debt Among all credit cards issued, the proportion of cards that result in default was 0.13 in 2010. A credit analyst with Visa believes this proportion is different today.

3. A test is conducted at the $\alpha = 0.05$ level of significance. What is the probability of a Type I error?

4. β is computed to be 0.113. What is the probability of a Type II error? What is the power of the test? How would you interpret the power of the test?

5. To test H_0: $\mu = 100$ versus H_1: $\mu > 100$, a simple random sample of size $n = 35$ is obtained from an unknown distribution. The sample mean is 104.3 and the sample standard deviation is 12.4.

(a) To use the t-distribution, why must the sample size be large?

(b) Use the classical or P-value approach to decide whether to reject the statement in the null hypothesis at the $\alpha = 0.05$ level of significance.

6. To test H_0: $\mu = 50$ versus H_1: $\mu \neq 50$, a simple random sample of size $n = 15$ is obtained from a population that is normally distributed. The sample mean is 48.1 and the sample standard deviation is 4.1.

(a) Why must it be the case that the population from which the sample was drawn is normally distributed?

(b) Use the classical or P-value approach to decide whether to reject the statement in the null hypothesis at the $\alpha = 0.05$ level of significance.

In Problems 7 and 8, test the hypothesis at the $\alpha = 0.05$ level of significance, using (a) the classical approach and (b) the P-value approach. Be sure to verify the requirements of the test.

7. H_0: $p = 0.6$ versus H_1: $p > 0.6$
$n = 250$; $x = 165$

8. H_0: $p = 0.35$ versus H_1: $p \neq 0.35$
$n = 420$; $x = 138$

9. To test H_0: $\sigma = 5.2$ versus H_1: $\sigma \neq 5.2$, a simple random sample of size $n = 18$ is obtained from a population that is known to be normally distributed.

(a) If the sample standard deviation is determined to be $s = 4.9$, compute the test statistic.

(b) Test this hypothesis at the $\alpha = 0.05$ level of significance.

10. To test H_0: $\sigma = 15.7$ versus H_1: $\sigma > 15.7$, a simple random sample of size $n = 25$ is obtained from a population that is known to be normally distributed.

(a) If the sample standard deviation is determined to be $s = 16.5$, compute the test statistic.

(b) Test this hypothesis at the $\alpha = 0.1$ level of significance.

11. Sneeze According to work done by Nick Wilson of Otago University Wellington, the proportion of individuals who cover their mouth when sneezing is 0.733. As part of a school project, Mary decides to confirm this by observing 100 randomly selected individuals sneeze and finds that 78 covered their mouth when sneezing.

(a) What are the null and alternative hypotheses for Mary's project?

(b) Verify the requirements that allow use of the normal model to test the hypothesis are satisfied.

(c) Does the sample evidence contradict Professor Wilson's findings?

12. Emergency Room The proportion of patients who visit the emergency room (ER) and die within the year is 0.05. *Source:* SuperFreakonomics. Suppose a hospital administrator is concerned that his ER has a higher proportion of patients who die within the year. In a random sample of 250 patients who have visited the ER in the past year, 17 have died. Should the administrator be concerned?

13. Linear Rotary Bearing A linear rotary bearing is designed so that the distance between the retaining rings is 0.875 inch. The quality-control manager suspects that the manufacturing process needs to be recalibrated because the mean distance between the retaining rings is greater than 0.875 inch. In a random sample of 36 bearings, he finds the sample mean distance between the retaining rings is 0.876 inch with standard deviation 0.005 inch.

(a) Are the requirements for conducting a hypothesis test satisfied?

(b) State the null and alternative hypotheses.

(c) The quality-control manager decides to use an $\alpha = 0.01$ level of significance. Why do you think this level of significance was chosen?

(d) Does the evidence suggest the machine be recalibrated?

(e) What does it mean for the quality-control engineer to make a Type I error? A Type II error?

14. Normal Temperature Carl Reinhold August Wunderlich said that the mean temperature of humans is 98.6°F. Researchers Philip Mackowiak, Steven Wasserman, and Myron Levine [*JAMA*, Sept. 23–30 1992; 268(12):1578–80] thought that the mean temperature of humans is less than 98.6°F. They measured the temperature of 148 healthy adults one to four times daily for 3 days, obtaining 700 measurements. The sample data resulted in a sample mean of 98.2°F and a sample standard deviation of 0.7°F. Test whether the mean temperature of humans is less than 98.6°F at the $\alpha = 0.01$ level of significance.

15. Conforming Golf Balls The U.S. Golf Association (USGA) requires that golf balls have a diameter that is 1.68 inches. To determine if Maxfli XS golf balls conform to USGA standards, a random sample of Maxfli XS golf balls was selected. Their diameters are shown in the table.

1.683	1.684	1.677	1.684	1.681	1.673
1.685	1.685	1.678	1.682	1.686	1.674

Source: Michael McCraith, Joliet Junior College

(a) Because the sample size is small, the engineer must verify that the diameter is normally distributed and the sample does not contain any outliers. The normal probability plot and boxplot are shown. The correlation between ball diameter and expected z-scores is 0.951. Are the conditions for testing the hypothesis satisfied?

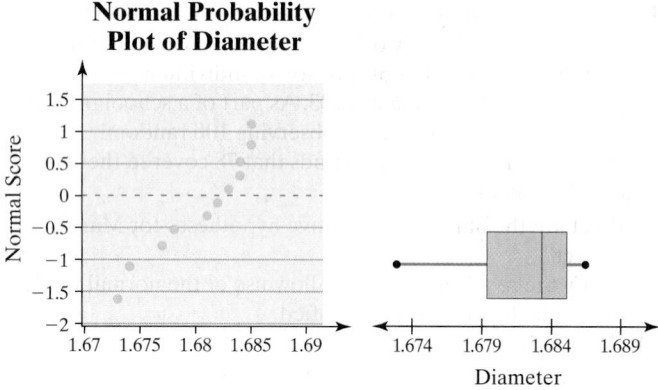

Normal Probability Plot of Diameter

(b) Construct a 95% confidence interval to judge whether the golf balls conform to USGA standards. Be sure to state the null and alternative hypotheses and write a conclusion.

16. Studying Enough? College mathematics instructors suggest that students spend 2 hours outside class studying for every hour in class. So, for a 4-credit-hour math class, students should spend at least 8 hours (480 minutes) studying each week. The given data, from Michael Sullivan's College Algebra class, represent the time spent on task recorded in MyMathLab (in minutes) for randomly selected students during the third week of the semester. Determine if the evidence suggests students may not, in fact, be following the advice. That is, does the evidence suggest students are studying less than 480 minutes each week? Use $\alpha = 0.05$ level of significance. **Note:** A normal probability plot and boxplot indicate that the data come from a population that is normally distributed with no outliers.

504	267	220	322	538	542
428	481	413	302	602	

Source: MyMathLab

17. Sleeping Patterns of Pregnant Women A random sample of 150 pregnant women indicated that 81 napped at least twice per week. Do a majority of pregnant women nap at least twice a week? Use the $\alpha = 0.05$ level of significance.
Source: National Sleep Foundation

18. Refer to Problem 17.

(a) Explain what it would mean to make a Type II error for this test.

(b) Determine the probability of making a Type II error if the true population proportion of pregnant women who nap is 0.53. What is the power of the test?

19. Birth Weights An obstetrician maintains that preterm babies (gestation period less than 37 weeks) have a higher variability in birth weight than do full-term babies (gestation period 37 to 41 weeks). According to the *National Vital Statistics Report*, the birth weights of full-term babies are normally distributed, with standard deviation 505.6 grams. A random sample of 41 preterm babies results in a standard deviation equal to 840 grams. Test the researcher's hypothesis that the variability in the birth weight of preterm babies is more than the variability in birth weight of full-term babies, at the $\alpha = 0.01$ level of significance.

20. Grim Report Throughout the country, the proportion of first-time, first-year community college students who return for their second year of studies is 0.52 according to the Community College Survey of Student Engagement. Suppose a community college institutes new policies geared toward increasing student retention. The first year of this policy there were 2843 first-time, first-year students of which 1516 returned for their second year of studies. Treat these students as a random sample of all first-time, first-year students. Do a statistically significant higher proportion of students return for their second year under this policy? Use the $\alpha = 0.1$ level of significance. Would you say the results are practically significant? In other words, do you believe the results are significant enough that other community colleges might consider emulating these same policies?

21. Teen Prayer In 1995, 40% of adolescents stated they prayed daily. A researcher wants to know whether this percentage has risen since then. He surveys 40 adolescents and finds that 18 pray on a daily basis. Is this evidence that the proportion of adolescents who pray daily has increased at the $\alpha = 0.05$ level of significance?

22. A New Teaching Method A large university has a college algebra enrollment of 5000 students each semester. Because of space limitations, the university decides to offer its college algebra courses in a self-study format in which students learn independently, but have access to tutors and other help in a lab setting. Historically, students in traditional college algebra scored 73.2 points on the final exam, and the coordinator of this course is concerned that test scores are going to decrease in the new format. At the end of the first semester using the new delivery system, 3851 students took the final exams and had a mean score of 72.8 and a standard deviation of 12.3. Treating these students as a simple random sample of all students, determine whether or not the scores decreased significantly at the $\alpha = 0.05$ level of significance. Do you think that the decrease in scores has any practical significance?

23. Explain the difference between "accepting" and "not rejecting" a null hypothesis.

24. According to the American Time Use Survey, the mean number of hours each day Americans, aged 15 and older, spend eating and drinking is 1.22. A researcher wanted to know if Americans, aged 15 to 19, spent less time eating and drinking. After surveying 50 Americans, aged 15 to 19, and running the appropriate hypotheses test, she obtained a *P*-value of 0.0329.

State the null and alternative hypothesis this researcher used and interpret the *P*-value.

25. Explain the procedure for testing a hypothesis using the Classical Approach. What is the criterion for judging whether to reject the null hypothesis?

26. Explain the procedure for testing a hypothesis using the *P*-value Approach. What is the criterion for judging whether to reject the null hypothesis?

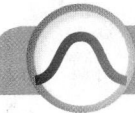 **Chapter Test**

1. According to the American Time Use Survey, adult Americans spent 42.6 minutes per day on phone calls and answering or writing email in 2006.

(a) Suppose that we want to judge whether the amount of daily time spent on phone calls and answering or writing email has increased. Write the appropriate null and alternative hypotheses.

(b) The sample data indicated that the null hypothesis should be rejected. Write a conclusion.

(c) Explain what it would mean if we made a Type I error in conducting the test from part (a).

(d) Explain what it would mean if we made a Type II error in conducting the test from part (a).

2. The trade magazine *QSR* routinely examines fast-food drive-through service times. Their recent research indicates that the mean time a car spends in a McDonald's drive-through is 167.1 seconds. A McDonald's manager in Salt Lake City feels that she has instituted a drive-through policy that results in lower drive-through service times. A random sample of 70 cars results in a mean service time of 163.9 seconds, with a standard deviation of 15.3 seconds. Determine whether the policy is effective in reducing drive-through service times.

(a) State the null and alternative hypotheses.

(b) Because the cost of instituting the policy is quite high, the quality-control researcher at McDonald's chooses to test the hypothesis using an $\alpha = 0.01$ level of significance. Why is this a good idea?

(c) Conduct the appropriate test to determine if the policy is effective.

3. "Did you get your 8 hours of sleep last night?" is a common question. In a recent survey of 151 postpartum women, the folks at the National Sleep Foundation found that the mean sleep time was 7.8 hours, with a standard deviation of 1.4 hours. Does the evidence suggest that postpartum women do not get enough sleep? Use $\alpha = 0.05$ level of significance.

4. The outside diameter of a manufactured part must be 1.3825 inches, according to customer specifications. The data shown represent a random sample of ten parts. Use a 95% confidence interval to judge whether the part has been manufactured to specifications.

Note: A normal probability plot and boxplot indicate that the data come from a population that is normally distributed with no outliers.

1.3821	1.3830	1.3823	1.3829	1.3830
1.3829	1.3826	1.3825	1.3823	1.3824

Source: Dennis Johnson, student at Joliet Junior College

5. In many parliamentary procedures, a supermajority is defined as an excess of 60% of voting members. In a poll conducted by the Gallup organization on May 10, 1939, 1561 adult Americans were asked, "Do you think the United States will have to fight Japan within your lifetime?" Of the 1561 respondents, 954 said no. Does this constitute sufficient evidence that a supermajority of Americans did not feel the United States would have to fight Japan within their lifetimes at the $\alpha = 0.05$ level of significance?

6. A Zone diet is one with a 40%–30%–30% distribution of carbohydrate, protein, and fat, respectively, and is based on the book *Enter the Zone*. In a study conducted by researchers Christopher Gardner and associates, 79 subjects were administered the Zone diet. After 12 months, the mean weight loss was 1.6 kg, with a standard deviation of 5.4 kg. Do these results suggest that the weight loss was statistically significantly greater than zero at the 0.05 level of significance? Do you believe the weight loss has any practical significance? Why?

Source: Christopher D. Gardner, Alexandre Kiazand, and Sofiya Alhassan, et al. "Comparison of the Atkins, Zone, Ornish, and LEARN Diets for Change in Weight and Related Risk Factors among Overweight Premenopausal Women. The A to Z Weight Loss Study: A Randomized Trial." *Journal of the American Medical Association,* March 2007.

7. According to the Pew Research Center, the proportion of the American population who use only a cellular telephone (no landline) is 0.37. Jason conducts a survey of thirty 20- to 24-year-olds who live on their own and finds that 16 do not have a landline to their home. Does this provide sufficient evidence to conclude that the proportion of 20- to 24-year-olds who live on their own and don't have a landline is greater than 0.37? Use an $\alpha = 0.10$ level of significance.

8. Standard deviation rate of return is a measure of risk in the stock market. An investment manager claims that the standard deviation rate of return for his portfolio is less than the general market, which is known to be 18%. Treating the past 10 years as a random sample, you find the standard deviation rate of return of the manager's portfolio is 16%. Does this represent sufficient evidence to conclude that the investment manager's portfolio has less risk than the general market? Use $\alpha = 0.05$.

9. Refer to Problem 5. Determine the probability of making a Type II error if the true population proportion is 0.63. Assume the test is conducted at the $\alpha = 0.05$ level of significance. What is the power of the test?

Making an Informed Decision

Selecting a Mutual Fund

Suppose you have just received a $1000 bonus at your job. Rather than waste the money on frivolous items, you decide to invest the money so that you can apply it toward the purchase of a home one day. Many investment options are available. Your family and friends, who have some experience in investing, recommend mutual funds. Which mutual funds should you choose?

Thousands of mutual funds are available to invest in. According to Wikipedia, "a **mutual fund** is a professionally managed type of collective investment scheme that pools money from many investors and invests typically in investment securities (stocks, bonds, and so on)."

(a) Research the various classifications (Growth, Value, and so on) of mutual funds. Decide on a particular mutual fund classification.

(b) Go to www.morningstar.com. Morningstar is a mutual fund rating agency that ranks mutual funds according to a star rating. The 5-star rating system divides the mutual funds into percentile classes. A

mutual fund with a five-star rating is in the top 20% of all mutual funds in the category. Choose a mutual fund that has at least a four-star rating and has been around for at least 5 years.

(c) Research the historical monthly rates of return of the mutual fund you selected in part (b). You will use two criteria in selecting a mutual fund. First, the mean rate of return of the fund over the past 48 months must exceed 7%. Second, the proportion of the past 48 months where the rate of return is positive must exceed 0.7. Treat the selected months as a random sample of rates of return. Conduct the appropriate tests to see if the mutual fund you selected meets your criteria.

 **CASE STUDY** How Old Is Stonehenge?

Approximately eight miles north of Salisbury, Wiltshire, England, stands a large prehistoric circular stone monument surrounded by an earthwork, called Stonehenge. The monument consists of an outer ring of sarsen stones, surrounding two inner circles of bluestones. The entire structure is surrounded by a ditch and bank with 56 pits named the Aubrey Holes, after their discoverer, that appear to have been filled shortly after their excavation.

Corinn Dillion is interested in dating the construction of the structure. Excavations at the site uncovered a number of unshed antlers, antler tines, and animal bones. Carbon-14 dating methods were used to estimate the ages of the Stonehenge artifacts. The ratio of carbon-14 to carbon-12 remains constant in living organisms. Once the organism dies, the amount of carbon-14 in the remains begins to decline, because it is radioactive, with a half-life of 5730 years (the "Cambridge half-life"). The decay of carbon-14 into ordinary nitrogen makes possible a reliable estimate about the time of death of the organism. The counted carbon-14 decay events can be modeled by the normal distribution.

Stonehenge's main ditch was dug in a series of segments. Excavations at the base of the ditch uncovered

antlers bearing signs of heavy use, possibly used by the builders as picks or rakes and buried in the ditch shortly after its completion. Another researcher, Phillip Corbin had previously claimed that the mean date for the construction of the ditch was 2950 B.C. A sample of nine age estimates from unshed antlers excavated from the ditch produced a mean of 3033.1 B.C., with standard deviation 66.9 years. Assume that the ages are normally distributed with no outliers. At an $\alpha = 0.05$ significance level, is there any reason to dispute Corbin's claim?

Four animal bone samples were discovered in the ditch terminals, bearing signs of attempts at artificial preservation and possibility of use for a substantial period of time before being placed at Stonehenge. When dated, these bones had a mean age of 3187.5 B.C. and standard deviation of 67.4 years. Assume that the ages are normally distributed with no outliers. Use an $\alpha = 0.05$ significance level to test the hypothesis that the population mean age of the site is different from 2950 B.C.

In the center of the monument are two concentric circles of igneous rock pillars. Excavation here revealed an antler, an antler tine, and an animal bone. Each artifact was submitted for dating and this sample of three artifacts had a mean age of 2193.3 B.C., with a standard deviation of 104.1 years. Assume that the ages are normally distributed with no outliers. Use an $\alpha = 0.05$ significance level to test the hypothesis that the population mean age of the formations is different from Corbin's declared mean age of the ditch, that is, 2950 B.C.

Finally, three additional antler samples were uncovered at the Y and Z holes, part of a formation of concentric circles. The sample mean age of these antlers is 1671.7 B.C. with a standard deviation of 99.7 years. Assume that the ages are normally distributed with no outliers. Use an $\alpha = 0.05$ significance level to test whether the population mean age of the Y and Z holes is different from Corbin's stated mean age of the ditch, that is, 2950 B.C.

From your analysis, does it appear that the mean ages of the artifacts from the ditch, the ditch terminals, the Bluestones, and the Y and Z holes dated by Dillion are consistent with Corbin's claimed mean age of 2950 B.C. for construction of the ditch? Can you use the results from your hypothesis tests to infer the likely construction order of the various Stonehenge structures? Explain.

Using Dillion's data, construct a 95% confidence interval for the population mean ages of the various sites. Do these confidence intervals support Corbin's claim? Can you use these confidence intervals to infer the likely construction order of the various Stonehenge structures? Explain.

Which statistical technique, hypothesis testing or confidence intervals, is more useful in assessing the age and likely construction order of the Stonehenge structures? Explain.

Discuss the limitations and assumptions of your analysis. Is there any additional information that you would like to have before publishing your findings? Would another statistical procedure be more useful in analyzing these data? If so, which one? Explain. Write a report to Corinn Dillion detailing your analysis.

Source: This fictional account is based on information obtained from *Archaeometry and Stonehenge* (www.eng_h.gov.uk/stoneh). The means and standard deviations used throughout this case study were constructed by calculating the statistics from the midpoint of the calibrated date range supplied for each artifact.

Inferences on Two Samples

Making an Informed Decision

You have decided to purchase a car. A couple areas of concern for you are:

1. What kind of gas mileage does the car get?
2. Will the car hold its value?

See the Decisions Project on page 577.

PUTTING IT TOGETHER

In Chapters 9 and 10, we discussed inferences regarding a single population parameter. The inferential methods presented in those chapters will be modified slightly in this chapter so that we can compare two population parameters.

The first section presents inferential methods for comparing two population proportions. That is, inference when the response variable is qualitative with two possible outcomes (success or failure). The first order of business is to decide whether the data are obtained from an independent or dependent sample—simply put, we determine if the observations in one sample are somehow related to the observations in the other. We then discuss methods for comparing two proportions from independent samples. Methods for comparing proportions from dependent samples are covered in Section 12.3.

Section 11.2 presents inferential methods used to handle dependent samples when the response variable is quantitative. For example, we might want to know whether the reaction time in an individual's dominant hand is different from the reaction time in the nondominant hand.

Section 11.3 presents inferential methods used to handle independent samples when there are two levels of treatment and the response variable is quantitative. For example, we might randomly divide 100 volunteers who have a common cold into two groups. The control group would receive a placebo and the experimental group would receive a specific amount of some experimental drug. The response variable might be time until the cold symptoms go away.

Section 11.4 is a discussion for comparing two population standard deviations.

We wrap up the chapter with a "Putting It Together" section. One of the more difficult aspects of inference is determining which inferential method to use. This section helps develop this skill.

11.1   Inference about Two Population Proportions

Preparing for This Section Before getting started, review the following:

- Completely randomized design (Section 1.6, pp. 44–46)
- Matched-pairs design (Section 1.6, pp. 46–47)
- Estimating a population proportion (Section 9.1, pp. 421–429)
- Hypothesis tests about a population proportion (Section 10.2, pp. 483–488)
- Statistical versus practical significance (Section 10.3, pp. 498–499)

Objectives

1. Distinguish between independent and dependent sampling
2. Test hypotheses regarding two proportions from independent samples
3. Construct and interpret confidence intervals for the difference between two population proportions
4. Determine the sample size necessary for estimating the difference between two population proportions

1 Distinguish between Independent and Dependent Sampling

Let's consider two scenarios:

Scenario 1: Among competing acne medications, does one perform better than the other? To answer this question, researchers applied Medication A to one part of the subject's face and Medication B to a different part of the subject's face to determine the proportion of subjects whose acne cleared up for each medication. The part of the face that received Medication A was randomly determined.

Scenario 2: Do individuals who make fast-food purchases with a credit card tend to spend more than those who pay with cash? To answer this question, a marketing manager randomly selects 30 credit-card receipts and 30 cash receipts to determine if the credit-card receipts have a significantly higher dollar amount, on average.

Is there a difference in the approach taken to select the individuals in each study? Yes! In scenario 1, once an individual is selected, one part of his or her face is "matched-up" with a second part of the face. In scenario 2, the receipts selected from the credit-card group have nothing at all to do with the receipts selected from the cash group.

Definitions

A sampling method is **independent** when an individual selected for one sample does not dictate which individual is to be in a second sample. A sampling method is **dependent** when an individual selected to be in one sample is used to determine the individual in the second sample. Dependent samples are often referred to as **matched-pairs** samples. It is possible for an individual to be matched against him- or herself.

So, the sampling method in scenario 1 is dependent (or a matched-pairs sample), while the sampling method in scenario 2 is independent.

EXAMPLE 1   Distinguishing between Independent and Dependent Sampling

Problem Decide whether the sampling method is independent or dependent. Then determine whether the response variable is qualitative or quantitative.

(a) Joliet Junior College decided to implement a course redesign of its developmental mathematics program. Students either enrolled in a traditional lecture format

(continued)

course or a lab-based format in which lectures and homework are done using video and the course management system MyMathLab. There were 1200 students enrolled in the traditional lecture format and 300 enrolled in the lab-based format. Once the course ended, the researchers determined whether the student passed the course with an A, B, or C, or not. The goal of the study was to determine whether the proportion of students who passed the lab-based format exceeded that of the lecture format.

(b) Do women tend to select a spouse who has an IQ higher than their own? To answer this question, researchers randomly selected 20 women and their husbands. They measured the IQ of each husband–wife team to determine if there is a significant difference in IQ.

Approach Determine whether the individuals in one group were used to determine the individuals in the other group. If so, the sampling method is dependent. If not, the sampling method is independent. Finally, consider the response variable in the study. Is it qualitative with two outcomes? If so, inferential methods based on proportions are appropriate. Is it quantitative? If so, inferential methods based on means may be appropriate.

Solution

(a) The sampling method is independent because the individuals in the lecture format are not related to the individuals in the lab-based format. The response variable is whether the student passed the course or not. Because there are two outcomes, pass or do not pass, the researchers can compare the proportion of students passing the lecture course to those passing the lab-based course.

(b) The sampling method is dependent because once a wife is selected, her husband is automatically enrolled in the study. That is, the wife–husband team is "matched up." The response variable, IQ, is quantitative.

● **Now Work Problem 3**

② Test Hypotheses Regarding Two Proportions from Independent Samples

In Sections 9.1 and 10.2, we discussed inference regarding a single population proportion. We now discuss inference for comparing two proportions. We begin with inference comparing two proportions from independent samples. We will discuss inference comparing two proportions from dependent samples in Chapter 12.

For example, in clinical trials of the drug Nasonex, a drug that is meant to relieve allergy symptoms, 26% of patients receiving 200 micrograms (μg) of Nasonex reported a headache as a side effect, while 22% of patients receiving a placebo reported a headache as a side effect. Researchers want to determine whether the proportion of patients receiving Nasonex and complaining of headaches is significantly higher than the proportion of patients receiving the placebo and complaining of headaches.

To conduct inference about two population proportions from independent samples, we must first determine the sampling distribution of the difference of two proportions. Recall that the point estimate of a population proportion, p, is given by $\hat{p} = \dfrac{x}{n}$, where x is the number of the n individuals in the sample that have a specific characteristic. In addition, recall that the sampling distribution of $\hat{p}$ is approximately normal with mean $\mu_{\hat{p}} = p$ and standard deviation $\sigma_{\hat{p}} = \sqrt{\dfrac{p(1-p)}{n}}$, provided that $np(1-p) \geq 10$ and $n \leq 0.05N$, so $Z = \dfrac{\hat{p} - p}{\sqrt{\dfrac{p(1-p)}{n}}}$ is approximately normal with mean 0 and standard deviation 1. Using this information along with the idea of independent sampling from two populations, we obtain the sampling distribution of the difference between two proportions.

Sampling Distribution of the Difference between Two Proportions (Independent Sample)

Suppose a simple random sample of size n_1 is taken from a population where x_1 of the individuals have a specified characteristic, and a simple random sample of size n_2 is independently taken from a different population where x_2 of the individuals have a specified characteristic. The sampling distribution of $\hat{p}_1 - \hat{p}_2$, where $\hat{p}_1 = \dfrac{x_1}{n_1}$ and $\hat{p}_2 = \dfrac{x_2}{n_2}$, is approximately normal, with mean $\mu_{\hat{p}_1 - \hat{p}_2} = p_1 - p_2$ and standard deviation $\sigma_{\hat{p}_1 - \hat{p}_2} = \sqrt{\dfrac{p_1(1 - p_1)}{n_1} + \dfrac{p_2(1 - p_2)}{n_2}}$, provided that $n_1 \hat{p}_1(1 - \hat{p}_1) \geq 10$ and $n_2 \hat{p}_2(1 - \hat{p}_2) \geq 10$ and each sample size is no more than 5% of the population size. The standardized version of $\hat{p}_1 - \hat{p}_2$ is then written as

$$Z = \frac{(\hat{p}_1 - \hat{p}_2) - (p_1 - p_2)}{\sqrt{\dfrac{p_1(1 - p_1)}{n_1} + \dfrac{p_2(1 - p_2)}{n_2}}}$$

which has an approximate standard normal distribution.

Now that we know the approximate sampling distribution of the difference of two sample proportions, we can introduce a procedure that can be used to test hypotheses regarding two population proportions. We first consider the test statistic. Following the discussion for testing hypotheses on a single proportion, it seems reasonable that the test statistic for the difference of two population proportions would be

$$z_0 = \frac{(\hat{p}_1 - \hat{p}_2) - (p_1 - p_2)}{\sqrt{\dfrac{p_1(1 - p_1)}{n_1} + \dfrac{p_2(1 - p_2)}{n_2}}} \qquad \textbf{(1)}$$

When comparing two population proportions, the null hypothesis is a statement of "no difference" (as always), so H_0: $p_1 = p_2$. Because the null hypothesis is assumed to be true, the test assumes that $p_1 = p_2$, or $p_1 - p_2 = 0$. Because we assume that both p_1 and p_2 equal p, where p is the common population proportion, substitute p into Equation (1), and obtain

$$z_0 = \frac{(\hat{p}_1 - \hat{p}_2) - (p_1 - p_2)}{\sqrt{\dfrac{p_1(1 - p_1)}{n_1} + \dfrac{p_2(1 - p_2)}{n_2}}} = \frac{\hat{p}_1 - \hat{p}_2 - 0}{\sqrt{\dfrac{p(1 - p)}{n_1} + \dfrac{p(1 - p)}{n_2}}} = \frac{\hat{p}_1 - \hat{p}_2}{\sqrt{p(1 - p)}\sqrt{\dfrac{1}{n_1} + \dfrac{1}{n_2}}} \qquad \textbf{(2)}$$

In Other Words
The pooled estimate of p is obtained by summing the number of individuals in the two samples that have a certain characteristic and dividing this result by the sum of the two sample sizes.

We need a point estimate of p because it is unknown. The best point estimate of p is called the **pooled estimate of p**, denoted $\hat{p}$, where

$$\hat{p} = \frac{x_1 + x_2}{n_1 + n_2}$$

Substituting the pooled estimate of p into Equation (2), we obtain

$$z_0 = \frac{\hat{p}_1 - \hat{p}_2}{\sigma_{\hat{p}_1 - \hat{p}_2}} = \frac{\hat{p}_1 - \hat{p}_2}{\sqrt{\hat{p}(1 - \hat{p})}\sqrt{\dfrac{1}{n_1} + \dfrac{1}{n_2}}}$$

This test statistic will be used to test hypotheses regarding two population proportions.

Hypothesis Test Regarding the Difference between Two Population Proportions

To test hypotheses regarding two population proportions, p_1 and p_2, use the steps that follow, provided that

- the samples are independently obtained using simple random sampling or through a completely randomized experiment with two levels of treatment,
- $n_1 \hat{p}_1(1 - \hat{p}_1) \geq 10$ and $n_2 \hat{p}_2(1 - \hat{p}_2) \geq 10$, and
- $n_1 \leq 0.05N_1$ and $n_2 \leq 0.05N_2$ (the sample size is no more than 5% of the population size); this requirement ensures the independence necessary for a binomial experiment.

(continued)

Step 1 Determine the null and alternative hypotheses. The hypotheses can be structured in one of three ways:

Two-Tailed	Left-Tailed	Right-Tailed
$H_0: p_1 = p_2$	$H_0: p_1 = p_2$	$H_0: p_1 = p_2$
$H_1: p_1 \neq p_2$	$H_1: p_1 < p_2$	$H_1: p_1 > p_2$

Note: p_1 is the population proportion for population 1, and p_2 is the population proportion for population 2.

Step 2 Select a level of significance, α, depending on the seriousness of making a Type I error.

Classical Approach

Step 3 Compute the **test statistic**

$$z_0 = \frac{\hat{p}_1 - \hat{p}_2}{\sqrt{\hat{p}(1 - \hat{p})}\sqrt{\dfrac{1}{n_1} + \dfrac{1}{n_2}}}, \text{ where } \hat{p} = \frac{x_1 + x_2}{n_1 + n_2}$$

Use Table V to determine the critical value.

	Two-Tailed	Left-Tailed	Right-Tailed
Critical value	$-z_{\frac{\alpha}{2}}$ and $z_{\frac{\alpha}{2}}$	$-z_{\alpha}$	z_{α}
Critical region	Critical Region Critical Region	Critical Region	Critical Region

Step 4 Compare the critical value to the test statistic.

Two-Tailed	Left-Tailed	Right-Tailed
If $z_0 < -z_{\frac{\alpha}{2}}$ or $z_0 > z_{\frac{\alpha}{2}}$, reject the null hypothesis.	If $z_0 < -z_{\alpha}$, reject the null hypothesis.	If $z_0 > z_{\alpha}$, reject the null hypothesis.

P-Value Approach

By Hand Step 3 Compute the **test statistic**

$$z_0 = \frac{\hat{p}_1 - \hat{p}_2}{\sqrt{\hat{p}(1 - \hat{p})}\sqrt{\dfrac{1}{n_1} + \dfrac{1}{n_2}}}, \text{ where } \hat{p} = \frac{x_1 + x_2}{n_1 + n_2}$$

Use Table V to determine the P-value.

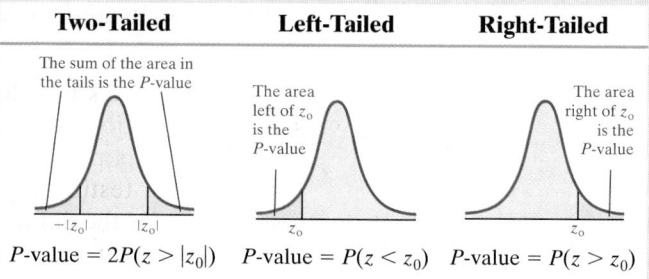

Two-Tailed	Left-Tailed	Right-Tailed
The sum of the area in the tails is the P-value	The area left of z_0 is the P-value	The area right of z_0 is the P-value
P-value $= 2P(z > \lvert z_0 \rvert)$	P-value $= P(z < z_0)$	P-value $= P(z > z_0)$

Technology Step 3 Use a statistical spreadsheet or calculator with statistical capabilities to obtain the P-value. The directions for obtaining the P-value using the TI-83/84 Plus graphing calculators, Excel, Minitab, and StatCrunch are in the Technology Step-by-Step on page 533.

Step 4 If P-value $< \alpha$, reject the null hypothesis.

Step 5 State the conclusion.

EXAMPLE 2 Testing a Hypothesis Regarding Two Population Proportions

Problem In clinical trials of Nasonex, 3774 adult and adolescent allergy patients (patients 12 years and older) were randomly divided into two groups. The patients in group 1 (experimental group) received 200 μg of Nasonex, while the patients in group 2 (control group) received a placebo. Of the 2103 patients in the experimental group, 547 reported headaches as a side effect. Of the 1671 patients in the control group, 368 reported headaches as a side effect. Is there significant evidence to conclude that the proportion of Nasonex users who experienced headaches as a side effect is greater than the proportion in the control group at the $\alpha = 0.05$ level of significance?

Approach Note that this is a completely randomized design. The response variable is whether or not the patient reports a headache. The treatments are the drugs: Nasonex or a placebo. The subjects are the 3774 adult and adolescent allergy patients.

We are attempting to determine if the evidence suggests that the proportion of patients who report a headache and are receiving Nasonex is greater than the proportion of patients who report a headache and are receiving a placebo. We will call the group that received Nasonex sample 1 and the group that received a placebo sample 2.

Verify the requirements to perform the hypothesis test. That is, the sample must be a simple random sample or the result of a randomized experiment and $n_1 \hat{p}_1(1 - \hat{p}_1) \geq 10$ and $n_2 \hat{p}_2(1 - \hat{p}_2) \geq 10$. In addition, the sample size cannot be more than 5% of the population size. Then follow the preceding Steps 1 through 5.

Solution First we verify that the requirements are satisfied.

1. The samples are independent because the subjects were randomly assigned to the treatment.
2. We have $x_1 = 547$, $n_1 = 2103$, $x_2 = 368$, and $n_2 = 1671$, so

$$\hat{p}_1 = \frac{x_1}{n_1} = \frac{547}{2103} = 0.260 \text{ and } \hat{p}_2 = \frac{x_2}{n_2} = \frac{368}{1671} = 0.220. \text{ Therefore,}$$

$$n_1 \hat{p}_1(1 - \hat{p}_1) = 2103(0.260)(1 - 0.260) = 404.6172 \geq 10$$
$$n_2 \hat{p}_2(1 - \hat{p}_2) = 1671(0.220)(1 - 0.220) = 286.7436 \geq 10$$

3. More than 10 million Americans 12 years old or older are allergy sufferers, so both sample sizes are less than 5% of the population size.

All three requirements are satisfied, so we proceed to follow Steps 1 through 5.

Step 1 Is the proportion of patients taking Nasonex who experience a headache greater than the proportion of patients taking the placebo who experience a headache? Letting p_1 represent the population proportion of patients taking Nasonex who experience a headache and p_2 represent the population proportion of patients taking the placebo who experience a headache, we want to know if $p_1 > p_2$. This is a right-tailed hypothesis with

$$H_0: p_1 = p_2 \quad \text{versus} \quad H_1: p_1 > p_2$$

or, equivalently,

$$H_0: p_1 - p_2 = 0 \quad \text{versus} \quad H_1: p_1 - p_2 > 0$$

Step 2 The level of significance is $\alpha = 0.05$.

Classical Approach

Step 3 From verifying requirement 2, we have $\hat{p}_1 = 0.260$ and $\hat{p}_2 = 0.220$. To find the test statistic, first compute the pooled estimate of p:

$$\hat{p} = \frac{x_1 + x_2}{n_1 + n_2} = \frac{547 + 368}{2103 + 1671} = 0.242$$

The test statistic is

$$z_0 = \frac{\hat{p}_1 - \hat{p}_2}{\sqrt{\hat{p}(1 - \hat{p})}\sqrt{\frac{1}{n_1} + \frac{1}{n_2}}} = \frac{0.260 - 0.220}{\sqrt{0.242(1 - 0.242)}\sqrt{\frac{1}{2103} + \frac{1}{1671}}} = 2.85$$

Because this is a right-tailed test, we determine the critical value at the $\alpha = 0.05$ level of significance to be $z_{0.05} = 1.645$. The critical region is shown in Figure 1.

Figure 1

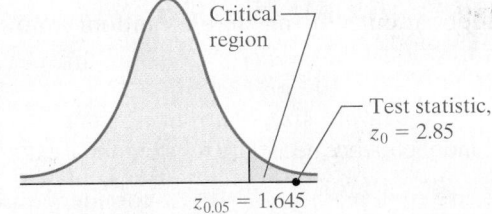

Critical region

Test statistic, $z_0 = 2.85$

$z_{0.05} = 1.645$

Step 4 The test statistic is $z_0 = 2.85$. We label this point in Figure 1. Because the test statistic is greater than the critical value (2.85 > 1.645), we reject the null hypothesis.

P-Value Approach

By Hand Step 3 From verifying requirement 2, we have $\hat{p}_1 = 0.260$ and $\hat{p}_2 = 0.220$. To find the test statistic, first compute the pooled estimate of p:

$$\hat{p} = \frac{x_1 + x_2}{n_1 + n_2} = \frac{547 + 368}{2103 + 1671} = 0.242$$

The test statistic is

$$z_0 = \frac{\hat{p}_1 - \hat{p}_2}{\sqrt{\hat{p}(1 - \hat{p})}\sqrt{\frac{1}{n_1} + \frac{1}{n_2}}} = \frac{0.260 - 0.220}{\sqrt{0.242(1 - 0.242)}\sqrt{\frac{1}{2103} + \frac{1}{1671}}} = 2.85$$

Because this is a right-tailed test, the P-value is the area under the standard normal distribution to the right of the test statistic, $z_0 = 2.85$, as shown in Figure 2.

Figure 2

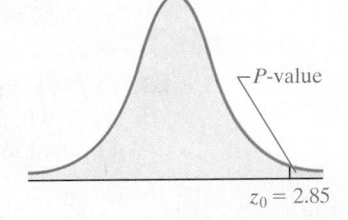

P-value

$z_0 = 2.85$

$$P\text{-value} = P(z_0 > 2.85)$$
$$= 0.0022$$

Technology Step 3 Using Minitab, we find the P-value is 0.002. See Figure 3 on the next page.

(continued)

Figure 3

```
Test and CI for Two Proportions

Sample    X      N    Sample p
1        547   2103   0.260105
2        368   1671   0.220227


Difference = p (1) - p (2)
Estimate for difference:  0.0398772
95% lower bound for difference:  0.0169504
Test for difference = 0 (vs > 0):   Z = 2.84   P-Value = 0.002
```

Step 4 The P-value of 0.002 means that *if* the null hypothesis that $p_1 - p_2 = 0$ (or $p_1 = p_2$) is true, we expect 2 samples in 1000 repetitions of this experiment to yield the results we obtained! The observed results are unusual. Because the P-value is less than the level of significance, $\alpha = 0.05$ (0.002 < 0.05), we reject the null hypothesis.

Step 5 There is sufficient evidence at the $\alpha = 0.05$ level of significance to conclude the proportion of individuals 12 years and older taking 200 μg of Nasonex who experience headaches is greater than the proportion of individuals 12 years and older taking a placebo who experience headaches. ●

> **CAUTION!**
> In any statistical study, be sure to consider practical significance. Many statistically significant results can be produced simply by increasing the sample size.

Looking back at the results of Example 1, we notice that the proportion of individuals taking 200 μg of Nasonex who experience headaches is *statistically significantly* greater than the proportion of individuals 12 years and older taking a placebo who experience headaches. However, we need to ask ourselves a pressing question. Would you not take an allergy medication because 26% of patients experienced a headache taking the medication versus 22% who experienced a headache taking a placebo? Most people would be willing to accept the additional risk of a headache to relieve their allergy symptoms. While the difference of 4% is statistically significant, it does not have any *practical significance*.

● **Now Work Problem 17**

③ Construct and Interpret Confidence Intervals for the Difference between Two Population Proportions

The sampling distribution of the difference of two proportions, $\hat{p}_1 - \hat{p}_2$, from independent samples can also be used to construct confidence intervals for the difference of two proportions.

> ### Constructing a $(1 - \alpha) \cdot 100\%$ Confidence Interval for the Difference between Two Population Proportions (Independent Samples)
>
> To construct a $(1 - \alpha) \cdot 100\%$ confidence interval for the difference between two population proportions from independent samples, the following requirements must be satisfied:
>
> **1.** The samples are obtained independently, using simple random sampling or from a randomized experiment.
> **2.** $n_1 \hat{p}_1 (1 - \hat{p}_1) \geq 10$ and $n_2 \hat{p}_2 (1 - \hat{p}_2) \geq 10$.
> **3.** $n_1 \leq 0.05 N_1$ and $n_2 \leq 0.05 N_2$ (the sample size is no more than 5% of the population size); this ensures the independence necessary for a binomial experiment.
>
> Provided that these requirements are met, a $(1 - \alpha) \cdot 100\%$ confidence interval for $p_1 - p_2$ is given by
>
> Lower bound: $(\hat{p}_1 - \hat{p}_2) - z_{\frac{\alpha}{2}} \cdot \sqrt{\dfrac{\hat{p}_1(1 - \hat{p}_1)}{n_1} + \dfrac{\hat{p}_2(1 - \hat{p}_2)}{n_2}}$
>
> Upper bound: $(\hat{p}_1 - \hat{p}_2) + z_{\frac{\alpha}{2}} \cdot \sqrt{\dfrac{\hat{p}_1(1 - \hat{p}_1)}{n_1} + \dfrac{\hat{p}_2(1 - \hat{p}_2)}{n_2}}$
>
> (3)

Notice that we do not pool the sample proportions. This is because we are not making any assumptions regarding their equality, as we did in hypothesis testing.

EXAMPLE 3   Constructing a Confidence Interval for the Difference between Two Population Proportions

Problem The Gallup organization surveyed 1100 adult Americans on May 6–9, 2002, and conducted an independent survey of 1100 adult Americans on May 8–11, 2014. In both surveys they asked the following: "Right now, do you think the state of moral values in the country as a whole is getting better or getting worse?" On May 8–11, 2014, 816 of the 1100 surveyed responded that the state of moral values is getting worse; on May 6–9, 2002, 737 of the 1100 surveyed responded that the state of moral values is getting worse. Construct and interpret a 90% confidence interval for the difference between the two population proportions.

Approach We can compute a 90% confidence interval for the two population proportions provided that the stated requirements are satisfied. We then construct the interval by hand using Formula (3) or using technology.

Solution We have to verify the requirements for constructing a confidence interval for the difference between two population proportions.

1. The samples were obtained independently through a random sample.
2. For the May 8–11, 2014, survey (sample 1), we have $n_1 = 1100$ and $x_1 = 816$, so
$$\hat{p}_1 = \frac{x_1}{n_1} = \frac{816}{1100} = 0.742.$$ For the May 6–9, 2002, survey (sample 2), we have
$$n_2 = 1100 \text{ and } x_2 = 737, \text{ so } \hat{p}_2 = \frac{x_2}{n_2} = \frac{737}{1100} = 0.67. \text{ Therefore,}$$
$$n_1\hat{p}_1(1 - \hat{p}_1) = 1100\,(0.742)(1 - 0.742) = 210.58 \geq 10$$
$$n_2\hat{p}_2(1 - \hat{p}_2) = 1100\,(0.67)(1 - 0.67) = 243.21 \geq 10$$
3. The population of adult Americans exceeded 100 million in 2002 and 2014, so the sample size is definitely less than 5% of the population size.

By-Hand Solution

Substituting into Formula (3) with $\hat{p}_1 = 0.742$, $n_1 = 1100$, $\hat{p}_2 = 0.67$, and $n_2 = 1100$, we obtain the lower and upper bounds on the confidence interval:

Lower bound: $(\hat{p}_1 - \hat{p}_2) - z_{\frac{\alpha}{2}} \cdot \sqrt{\dfrac{\hat{p}_1(1 - \hat{p}_1)}{n_1} + \dfrac{\hat{p}_2(1 - \hat{p}_2)}{n_2}}$

$= (0.742 - 0.67) - 1.645 \cdot \sqrt{\dfrac{0.742(1 - 0.742)}{1100} + \dfrac{0.67(1 - 0.67)}{1100}}$

$= 0.072 - 0.032$

$= 0.04$

Upper bound: $(\hat{p}_1 - \hat{p}_2) + z_{\frac{\alpha}{2}} \cdot \sqrt{\dfrac{\hat{p}_1(1 - \hat{p}_1)}{n_1} + \dfrac{\hat{p}_2(1 - \hat{p}_2)}{n_2}}$

$= (0.742 - 0.67) + 1.645 \cdot \sqrt{\dfrac{0.742(1 - 0.742)}{1100} + \dfrac{0.67(1 - 0.67)}{1100}}$

$= 0.072 + 0.032$

$= 0.104$

Technology Solution

Figure 4 shows the 90% confidence interval using a TI-84 Plus C graphing calculator.

Figure 4

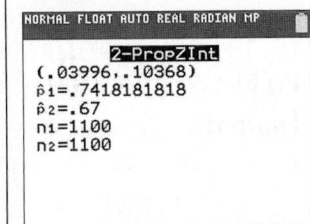

In Figure 4, the lower bound is 0.040. The upper bound is 0.104.

Interpretation We are 90% confident that the difference between the proportion of adult Americans who believed that the state of moral values in the country as a whole was getting worse from 2002 to 2014 is between 0.040 and 0.104. To put this statement into everyday language, we might say that we are 90% confident that the percentage of adult Americans who believe that the state of moral values in the country as a whole was getting worse increased between 4% and 10.4% from 2002 to 2014. Because this interval does not contain 0, we might conclude that a higher proportion of the country believed that the state of moral values was getting worse in the United States in 2014 than in 2002.

• **Now Work Problem 21**

4 Determine the Sample Size Necessary for Estimating the Difference between Two Population Proportions

In Section 9.1, we introduced a method for determining the sample size, n, required to estimate a single population proportion within a specified margin of error, E, with a specified level of confidence. This formula was obtained by solving the margin of error,

$$E = z_{\frac{\alpha}{2}} \cdot \sqrt{\frac{\hat{p}(1 - \hat{p})}{n}},$$ for n. We follow the same approach to determine the sample size

when estimating two population proportions. The margin of error, E, in constructing a confidence interval for the difference between two population proportions is

$$E = z_{\frac{\alpha}{2}} \cdot \sqrt{\frac{\hat{p}_1(1 - \hat{p}_1)}{n_1} + \frac{\hat{p}_2(1 - \hat{p}_2)}{n_2}}.$$ Assuming that $n_1 = n_2 = n$, we can solve this

expression for $n = n_1 = n_2$ and obtain the following result:

> **Sample Size for Estimating $p_1 - p_2$**
>
> The sample size required to obtain a $(1 - \alpha) \cdot 100\%$ confidence interval with a margin of error, E, is given by
>
> $$n = n_1 = n_2 = [\hat{p}_1(1 - \hat{p}_1) + \hat{p}_2(1 - \hat{p}_2)]\left(\frac{z_{\frac{\alpha}{2}}}{E}\right)^2 \qquad (4)$$
>
> rounded up to the next integer, if prior estimates of p_1 and p_2, $\hat{p}_1$ and $\hat{p}_2$, are available. If prior estimates of p_1 and p_2 are unavailable, the sample size is
>
> $$n = n_1 = n_2 = 0.5\left(\frac{z_{\frac{\alpha}{2}}}{E}\right)^2 \qquad (5)$$
>
> rounded up to the next integer.
>
> The margin of error should always be expressed as a decimal when using Formulas (4) and (5).

CAUTION!
When doing sample size calculations, always round up.

EXAMPLE 4 Determining Sample Size

Problem A nutritionist wants to estimate the difference between the proportion of males and females who consume the USDA's recommended daily intake of calcium. What sample size should be obtained if she wishes the estimate to be within 3 percentage points with 95% confidence, assuming that
(a) she uses the results of the USDA's 1994–1996 Diet and Health Knowledge Survey, according to which 51.1% of males and 75.2% of females consume the USDA's recommended daily intake of calcium, and
(b) she does not use any prior estimates?

Approach We have $E = 0.03$ and $z_{\frac{\alpha}{2}} = z_{\frac{0.05}{2}} = z_{0.025} = 1.96$. To answer part (a), let $\hat{p}_1 = 0.511$ (for males) and $\hat{p}_2 = 0.752$ (for females) in Formula (4). To answer part (b), use Formula (5).

Solution

(a)
$$n_1 = n_2 = [\hat{p}_1(1 - \hat{p}_1) + \hat{p}_2(1 - \hat{p}_2)]\left(\frac{z_{\frac{\alpha}{2}}}{E}\right)^2$$

$$= [0.511(1 - 0.511) + 0.752(1 - 0.752)]\left(\frac{1.96}{0.03}\right)^2$$

$$= 1862.6$$

Round this value up to 1863. The nutritionist must survey 1863 randomly selected males and 1863 randomly selected females.

(b)
$$n_1 = n_2 = 0.5\left(\frac{z_{\frac{\alpha}{2}}}{E}\right)^2 = 0.5\left(\frac{1.96}{0.03}\right)^2 = 2134.2$$

Round this value up to 2135. The nutritionist must survey 2135 randomly selected

● Now Work Problem 33 males and 2135 randomly selected females.

In Other Words
If possible, obtain a prior estimate of $\hat{p}$ when doing sample size computations.

Notice that having prior estimates of the population proportions reduces the number of individuals that need to be surveyed from 2135 to 1863. Because fewer individuals need to be surveyed, the costs of the study (in terms of time and money) are less when prior estimates are available.

Technology Step-by-Step Inference for Two Population Proportions

TI-83/84 Plus

Hypothesis Tests

1. Press STAT, highlight TESTS, and select
   6:2-PropZTest....
2. Enter the values of x_1, n_1, x_2, and n_2.
3. Highlight the appropriate relation between p_1 and p_2 in the alternative hypothesis.
4. Highlight Calculate or Draw and press ENTER. Calculate gives the test statistic and P-value. Draw will draw the Z-distribution with the P-value shaded.

Confidence Intervals

Follow the same steps given for hypothesis tests, except select B:2-PropZInt.... Also, select a confidence level (such as 95% = 0.95).

Minitab

1. Enter the raw data into columns C1 and C2, if necessary.
2. Select the **Stat** menu, highlight **Basic Statistics**, then highlight **2 Proportions** ....
3. If you have raw data in different columns, select "Each sample in its own column" and enter C1 for first sample and C2 for second sample. If you have raw data in a single column, select "Both samples are in one column" and enter C1 for the sample IDs and C2 for the Samples. If you have summary statistics, select "Summarized data." Enter the number of successes in the "events" cell and the sample size in the "trials" cell for each sample.
4. Click Options . . . and enter the level of confidence desired, the "test difference" (usually 0), and the direction of the alternative hypothesis. For confidence intervals, select "Estimate the proportions separately"; for hypothesis tests, select "Use the pooled estimate of the proportion" under Test method. Click OK twice.

Excel

Hypothesis Tests or Confidence Intervals

1. Load the XLSTAT Add-in. Select the XLSTAT menu, highlight Parametric tests. From the pull-down menu, select Tests for two proportions.
2. Enter the number of successes for sample 1 in the Frequency 1: cell; enter the sample size for sample 1 in the Sample size 1: cell; and so on. Make sure Data format: is set to Frequencies. Click the Options tab. Select the correct alternative hypothesis. Enter the Hypothesized difference (D) (usually 0), and select the Significance level (%). For a 95% confidence interval, enter 5. For confidence intervals, select the first radio button under Variance; for hypothesis tests, select the second radio button. Click OK.

StatCrunch

Hypothesis Tests or Confidence Intervals

1. If you have raw data, enter them into the spreadsheet. Name each column variable.
2. Select **Stat**, highlight **Proportion Stats**, highlight **Two Sample**, and then choose either **With Data** or **With Summary**.
3. If you chose **With Data**, select the column that has the observations, choose which outcome represents a success for each sample. If you chose **With Summary**, enter the number of successes and the number of trials for each sample. If you choose the hypothesis test radio button, enter the value of the proportion stated in the null hypothesis and choose the direction of the alternative hypothesis from the pull-down menu. If you choose the confidence interval radio button, enter the level of confidence. Click Calculate.

11.1 Assess Your Understanding

Vocabulary and Skill Building

1. A sampling method is _____ when the individuals selected for one sample do not dictate which individuals are selected to be in a second sample.

2. A sampling method is _____ when the individuals selected for one sample are used to determine the individuals in the second sample.

In Problems 3–8, determine whether the sampling is dependent or independent. Indicate whether the response variable is qualitative or quantitative.

NW **3.** A sociologist wishes to compare the annual salaries of married couples in which both spouses work and determines each spouse's annual salary.

4. A researcher wishes to determine the effects of alcohol on people's reaction time to a stimulus. She randomly divides 100 people aged 21 or older into two groups. Group 1 is asked to drink 3 ounces of alcohol, while group 2 drinks a placebo. Both drinks taste the same, so the individuals in the study do not know which group they belong to. Thirty minutes after consuming the drink, the subjects in each group perform a series of tests meant to measure reaction time.

5. The Gallup Organization asked 1050 randomly selected adult Americans age 18 or older who consider themselves to be religious, "Do you believe it is morally acceptable or morally wrong [rotated] to conduct medical research using stem cells obtained from human embryos?" The same question was asked to 1050 randomly selected adult Americans age 18 or older who do not consider themselves to be religious. The goal of the study was to determine whether the proportion of religious adult Americans who believe it is morally wrong to conduct medical research using stem cells obtained from human embryos differed from the proportion of non-religious adult Americans who believe it is morally wrong to conduct medical research using stem cells obtained from human embryos.

6. A political scientist wants to know how a random sample of 18- to 25-year-olds feel about Democrats and Republicans in Congress. She obtains a random sample of 1030 registered voters 18 to 25 years of age and asks, "Do you have favorable/unfavorable [rotated] opinion of the Democratic/Republican [rotated] party?" Each individual was asked to disclose his or her opinion about each party.

7. An educator wants to determine whether a new curriculum significantly improves standardized test scores for third grade students. She randomly divides 80 third-graders into two groups. Group 1 is taught using the new curriculum, while group 2 is taught using the traditional curriculum. At the end of the school year, both groups are given the standardized test and the mean scores are compared.

8. A psychologist wants to know whether subjects respond faster to a go/no go stimulus or a choice stimulus. With the go/no go stimulus, subjects must respond to a particular stimulus by pressing a button and disregard other stimuli. In the choice stimulus, the subjects respond differently depending on the stimulus. The psychologist randomly selects 20 subjects, and each subject is presented a series of go/no go stimuli and choice stimuli. The mean reaction time to each stimulus is compared.

In Problems 9–12, conduct each test at the $\alpha = 0.05$ level of significance by determining (a) the null and alternative hypotheses, (b) the test statistic, (c) the critical value, and (d) the P-value. Assume that the samples were obtained independently using simple random sampling.

9. Test whether $p_1 > p_2$. Sample data: $x_1 = 368, n_1 = 541, x_2 = 351, n_2 = 593$

10. Test whether $p_1 < p_2$. Sample data: $x_1 = 109, n_1 = 475, x_2 = 78, n_2 = 325$

11. Test whether $p_1 \neq p_2$. Sample data: $x_1 = 28, n_1 = 254, x_2 = 36, n_2 = 301$

12. Test whether $p_1 \neq p_2$. Sample data: $x_1 = 804, n_1 = 874, x_2 = 902, n_2 = 954$

In Problems 13–16, construct a confidence interval for $p_1 - p_2$ at the given level of confidence.

13. $x_1 = 368, n_1 = 541, x_2 = 421, n_2 = 593$, 90% confidence

14. $x_1 = 109, n_1 = 475, x_2 = 78, n_2 = 325$, 99% confidence

15. $x_1 = 28, n_1 = 254, x_2 = 36, n_2 = 301$, 95% confidence

16. $x_1 = 804, n_1 = 874, x_2 = 892, n_2 = 954$, 95% confidence

Applying the Concepts

NW 17. Prevnar The drug Prevnar is a vaccine meant to prevent certain types of bacterial meningitis, typically administered to infants around 2 months. In randomized, double-blind clinical trials of Prevnar, infants were randomly divided into two groups. Subjects in group 1 received Prevnar, while subjects in group 2 received a control vaccine. After the first dose, 107 of 710 subjects in the experimental group (group 1) experienced fever as a side effect. After the first dose, 67 of 611 of the subjects in the control group (group 2) experienced fever as a side effect. Does the evidence suggest that a higher proportion of subjects in group 1 experienced fever as a side effect than subjects in group 2 at the $\alpha = 0.05$ level of significance?

18. Prevnar Part 2 In randomized, double-blind clinical trials of Prevnar, infants were randomly divided into two groups. Subjects in group 1 received Prevnar, while subjects in group 2 received a control vaccine. After the second dose, 137 of 452 subjects in the experimental group (group 1) experienced drowsiness as a side effect. After the second dose, 31 of 99 subjects in the control group (group 2) experienced drowsiness as a side effect. Does the evidence suggest that a lower proportion of subjects in group 1 experienced drowsiness as a side effect than subjects in group 2 at the $\alpha = 0.05$ level of significance?

19. Abstain from Alcohol In October 1947, the Gallup organization surveyed 1100 adult Americans and asked, "Are you a total abstainer from, or do you on occasion consume, alcoholic beverages?" Of the 1100 adults surveyed, 407 indicated that they were total abstainers. In a recent survey, the same question was asked of 1100 adult Americans and 333 indicated that they were total abstainers. Has the proportion of adult Americans who totally abstain from alcohol changed? Use the $\alpha = 0.05$ level of significance.

20. Views on the Death Penalty The Pew Research Group conducted a poll in which they asked, "Are you in favor of, or opposed to, executing persons as a general policy when the crime was committed while under the age of 18?" Of the 580 Catholics surveyed, 180 indicated they favored capital punishment; of the 600 seculars (those who do not associate with a religion) surveyed, 238 favored capital punishment. Is there a significant difference in the proportion of individuals in these groups in favor of capital punishment for persons under the age of 18? Use the $\alpha = 0.01$ level of significance.

NW 21. Tattoos The Harris Poll conducted a survey in which they asked, "How many tattoos do you currently have on your body?" Of the 1205 males surveyed, 181 responded that they had at least one tattoo. Of the 1097 females surveyed, 143 responded that they had at least one tattoo. Construct a 95% confidence interval to judge whether the proportion of males that have at least one tattoo differs significantly from the proportion of females that have at least one tattoo. Interpret the interval.

22. Body Mass Index The body mass index (BMI) of an individual is a measure used to judge whether an individual is overweight or not. A BMI between 20 and 25 indicates a normal weight. In a survey of 750 men and 750 women, the Gallup organization found that 203 men and 270 women were normal weight. Construct a 90% confidence interval to gauge whether there is a difference in the proportion of men and women who are normal weight. Interpret the interval.

23. Public Cell Phone Conversations Researchers at Harris Interactive wondered if there was a difference between males and females in regard to some common annoyances. They asked a random sample of males and females, the following question: "Are you annoyed by people who repeatedly check their mobile phones while having an in-person conversation?" Among the 540 males surveyed, 178 responded "Yes"; among the 560 females surveyed, 206 responded "Yes." Does the evidence suggest a higher proportion of females are annoyed by this behavior?

(a) Explain why this study can be analyzed using the methods for conducting a hypothesis test regarding two independent proportions.

(b) What are the null and alternative hypotheses?

(c) Describe the sampling distribution of $\hat{p}_{female} - \hat{p}_{male}$. Draw a normal model with the area representing the P-value shaded for this hypothesis test.

(d) Determine the P-value based on the model from part (c).

(e) Interpret the P-value.

(f) Based on the P-value, what does the sample evidence suggest? That is, what is the conclusion of the hypothesis test? Assume an $\alpha = 0.05$ level of significance.

24. Name Brand Researchers at Harris Interactive wondered if there was a difference between males and females in regard to whether they typically buy name-brand or store-brand products. They asked a random sample of males and females the following question: "For each of the following types of products, please indicate whether you typically buy name-brand products or store-brand products?" Among the 1104 males surveyed, 343 indicated they buy name-brand over-the-counter drugs; among the 1172 females surveyed, 295 indicated they buy name-brand over-the-counter drugs. Does the evidence suggest a lower proportion of females buy name-brand over-the-counter drugs?

(a) Explain why this study can be analyzed using the methods for conducting a hypothesis test regarding two independent proportions.

(b) What are the null and alternative hypotheses?

(c) Describe the sampling distribution of $\hat{p}_{female} - \hat{p}_{male}$. Draw a normal model with the area representing the P-value shaded for this hypothesis test.

(d) Determine the P-value based on the model from part (c).

(e) Interpret the P-value.

(f) Based on the P-value, what does the sample evidence suggest? That is, what is the conclusion of the hypothesis test? Assume an $\alpha = 0.05$ level of significance.

Retain Your Knowledge

25. Side Effects In clinical trials of the allergy medicine Clarinex (5 mg), 3307 allergy sufferers were randomly assigned to either a Clarinex group or a placebo group. It was reported that 50 out of 1655 individuals in the Clarinex group and 31 out of 1652 individuals in the placebo group experienced dry mouth as a side effect of their respective treatments. *Source:* www.clarinex.com

(a) What type of experimental design is this?

(b) What is the response variable? Is it qualitative or quantitative?

(c) What is the explanatory variable? How many levels does the treatment have?

(d) The clinical trial was double-blind. What does this mean?

(e) Why is it important to have a placebo group?

(f) Does the sample evidence suggest that the proportion of individuals experiencing dry mouth is greater for those taking Clarinex than for those taking a placebo at the $\alpha = 0.05$ level of significance?

(g) Do you believe the difference between the groups is practically significant?

26. Practical versus Statistical Significance In clinical trials for treatment of a skin disorder, 642 of 2105 patients receiving the current standard treatment were cured of the disorder and 697 of 2115 patients receiving a new proposed treatment were cured of the disorder.

(a) Does the new procedure cure a higher proportion of patients at the $\alpha = 0.05$ level of significance?

(b) Do you think that the difference in success rates is practically significant? What factors might influence your decision?

27. Iraq War In March 2003, the Pew Research Group surveyed 1508 adult Americans and asked, "Do you believe the United States made the right or wrong decision to use military force in Iraq?" Of the 1508 adult Americans surveyed, 1086 stated the United States made the right decision. In August 2010, the Pew Research Group asked the same question of 1508 adult Americans and found that 618 believed the United States made the right decision.

(a) In the survey question, the choices "right" and "wrong" were randomly rotated. Why?

(b) Construct and interpret a 90% confidence interval for the difference between the two population proportions, $p_{2003} - p_{2010}$.

28. Course Redesign In many colleges and universities around the country, educators are changing their approach to instruction from a "teacher/lecture-centered model" to a "student-centered model" where students learn in a laboratory environment in which lecture is deemphasized and students can proceed at a pace suitable to their learning needs. In one school where this model was being introduced, of the 743 students who enrolled in the traditional lecture model, 364 passed; of the 567 in the student-centered model, 335 passed.

(a) What is the response variable in this study? What is the explanatory variable?

(b) Does the evidence suggest that the student-centered model results in a higher pass rate than the traditional model? Use the $\alpha = 0.05$ level of significance.

29. Deficit Reduction In the Sullivan Statistics Survey I, respondents were asked, "Would you be willing to pay higher taxes if the tax revenue went directly toward deficit reduction?" Treat the respondents as a simple random sample of adult Americans. The results of the survey may be downloaded at www.pearsonhighered.com/sullivanstats.

(a) What proportion of the males who took the survey is willing to pay higher taxes to reduce the deficit? What proportion of the females who took the survey is willing to pay higher taxes to reduce the deficit?

(b) Is there significant evidence to suggest the proportions of males and females who are willing to pay higher taxes to reduce the deficit differs at the $\alpha = 0.05$ level of significance?

30. Web Page Design John has an online company that sells custom rims for cars and designed two different web pages that he wants to use to sell his rims online. However, he cannot decide which page to go with, so he decides to collect data to see which site results in a higher proportion of sales. He hires a firm that has the ability to randomly assign one of his two web page designs to potential customers. With web page design I, John secures a sale from 54 out of 523 hits to the page. With web page design II, John secures a sale from 62 out of 512 hits to the page.

(a) What is the response variable in this study? What is the explanatory variable?

(b) Based on these results, which web page, if any, should John go with? Why? **Note:** This problem is based on the type of research done by Adobe Test & Target.

31. Political Grammar Psychologists asked students to read two sentences about hypothetical politicians. Ninety-eight students read, "Last year, Mark was having an affair with his assistant and was taking hush money from a prominent constituent." Let's call this sentence A. Ninety-eight other students read, "Last year, Mark had an affair with his assistant and took hush money from a prominent constituent." We will call this sentence B. *Source:* Fausey, C.M., and Matlock, T. Can Grammar Win Elections? *Political Psychology,* no. doi: 10.1111/j. 1467-9221.2010.00802.x

(a) What are the specific differences in the way the two sentences are phrased?

(b) In the study, 71 of the 98 students who read sentence A and 49 of the 98 students who read sentence B felt the politician would not be re-elected. Do these results suggest that the sentence structure makes a difference in deciding whether the politician would be re-elected?

(c) Research "imperfect aspect" and "perfect aspect." Does this help explain any differences in the results of the survey?

32. Confidence in Public Schools In June 2014, the Gallup organization surveyed 1134 American adults and found that 298 had confidence in public schools. In June 2013, the Gallup organization had surveyed 1134 American adults and found that 360 had confidence in public schools. Suppose that a newspaper article has a headline that reads, "Confidence in Public Schools Deteriorates." Is this an accurate headline? Why?

NW **33. Determining Sample Size** A physical therapist wants to determine the difference in the proportion of men and women who participate in regular, sustained physical activity. What sample size should be obtained if she wishes the estimate to be within 3 percentage points with 95% confidence, assuming that

(a) she uses the 1998 estimates of 21.9% male and 19.7% female from the U.S. National Center for Chronic Disease Prevention and Health Promotion?

(b) she does not use any prior estimates?

34. Determining Sample Size An educator wants to determine the difference between the proportion of males and females who have completed four or more years of college. What sample size should be obtained if she wishes the estimate to be within 2 percentage points with 90% confidence, assuming that

(a) she uses the 1999 estimates of 27.5% male and 23.1% female from the U.S. Census Bureau?

(b) she does not use any prior estimates?

35. Designing a Study Stock fund managers are investment professionals who decide which stocks should be part of a portfolio. In a recent article in the Wall Street Journal (*Not a Stock-Picker's Market, WSJ January 25, 2014*), the performance of stock fund managers was considered based on dispersion in the market. In the stock market, risk is measured by the standard deviation rate of return of stock (dispersion). When dispersion is low, then the rate of return of the stocks that make up the market are not as spread out. That is, the return on Company X is close to that of Y is close to that of Z, and so on. When dispersion is high, then the rate of return of stocks is more spread out; meaning some stocks outperform others by

a substantial amount. Since 1991, the dispersion of stocks has been about 7.1%. In some years, the dispersion is higher (such as 2001 when dispersion was 10%), and in some years it is lower (such as 2013 when dispersion was 5%). So, in 2001, stock fund managers would argue, one needed to have more investment advice in order to identify the stock market winners, whereas in 2013, since dispersion was low, virtually all stocks ended up with returns near the mean, so investment advice was not as valuable.

(a) Suppose you want to design a study to determine whether the proportion of fund managers who outperform the market in low-dispersion years is less than the proportion of fund managers who outperform the market in high-dispersion years. What would be the response variable in this study? What is the explanatory variable in this study?

(b) What or who are the individuals in this study?

(c) To what population does this study apply?

(d) What would be the null and alternative hypothesis?

(e) Suppose this study was conducted and the data yielded a *P*-value of 0.083. Explain what this result suggests.

36. Putting It Together: Salk Vaccine On April 12, 1955, Dr. Jonas Salk released the results of clinical trials for his vaccine to prevent polio. In these clinical trials, 400,000 children were randomly divided in two groups. The subjects in group 1 (the experimental group) were given the vaccine, while the subjects in group 2 (the control group) were given a placebo. Of the 200,000 children in the experimental group, 33 developed polio. Of the 200,000 children in the control group, 115 developed polio.

(a) What type of experimental design is this?

(b) What is the response variable?

(c) What are the treatments?

(d) What is a placebo?

(e) Why is such a large number of subjects needed for this study?

(f) Does it appear to be the case that the vaccine was effective?

Explaining the Concepts

37. Why do we use a pooled estimate of the population proportion when testing a hypothesis about two proportions? Why do we not use a pooled estimate of the population proportion when constructing a confidence interval for the difference of two proportions?

38. Explain the difference between an independent and dependent sample.

11.2 Inference about Two Means: Dependent Samples

Preparing for This Section Before getting started, review the following:

- Matched-pairs design (Section 1.6, pp. 46–47)
- Confidence intervals about μ (Section 9.2, pp. 439–441)

- Hypothesis tests about μ (Section 10.3, pp. 494–498)
- Type I and Type II errors (Section 10.1, pp. 476–477)

Objectives ① Test hypotheses for a population mean from matched-pairs data

② Construct and interpret confidence intervals about the population mean difference of matched-pairs data

In this section, we discuss inference on the difference of two means for dependent sampling. We will address inference when the sampling is independent in Section 11.3.

❶ Test Hypotheses for a Population Mean from Matched-Pairs Data

Inference on matched-pairs data is similar to inference regarding a population mean. Recall that if the population from which the sample was drawn is normally distributed or the sample size is large $(n \geq 30)$, we said that

$$t = \frac{\overline{x} - \mu}{\frac{s}{\sqrt{n}}}$$

In Other Words
Statistical inference methods on matched-pairs data use the same methods as inference on a single population mean, except that the *differences* are analyzed.

follows Student's t-distribution with $n - 1$ degrees of freedom.

When analyzing matched-pairs data, we compute the difference in each matched pair and then perform inference on the differenced data using the methods of Section 9.2 or 10.3.

Testing Hypotheses Regarding the Difference of Two Means Using a Matched-Pairs Design

To test hypotheses regarding the mean difference of data obtained from a dependent sample (matched-pairs data), use the following steps, provided that

- the sample is obtained by simple random sampling or the data result from a matched-pairs design experiment.
- the sample data are dependent (matched pairs).
- the differences are normally distributed with no outliers or the sample size, n, is large $(n \geq 30)$.
- the sampled values are independent (sample size is no more than 5% of population size).

Step 1   Determine the null and alternative hypotheses. The hypotheses can be structured in one of three ways, where μ_d is the population mean difference of the matched-pairs data.

Two-Tailed	Left-Tailed	Right-Tailed
$H_0: \mu_d = 0$	$H_0: \mu_d = 0$	$H_0: \mu_d = 0$
$H_1: \mu_d \neq 0$	$H_1: \mu_d < 0$	$H_1: \mu_d > 0$

Step 2   Select a level of significance, α, depending on the seriousness of making a Type I error.

Classical Approach	P-Value Approach

Classical Approach

Step 3   Compute the **test statistic**

$$t_0 = \frac{\overline{d} - 0}{\frac{s_d}{\sqrt{n}}} = \frac{\overline{d}}{\frac{s_d}{\sqrt{n}}}$$

which follows Student's t-distribution with $n - 1$ degrees of freedom. The values of $\overline{d}$ and s_d are the mean and standard deviation of the differenced data. Use Table VII to determine the critical value.

P-Value Approach

By Hand Step 3   Compute the **test statistic**

$$t_0 = \frac{\overline{d} - 0}{\frac{s_d}{\sqrt{n}}} = \frac{\overline{d}}{\frac{s_d}{\sqrt{n}}}$$

which follows Student's t-distribution with $n - 1$ degrees of freedom. The values of $\overline{d}$ and s_d are the mean and standard deviation of the differenced data. Use Table VII to approximate the P-value.

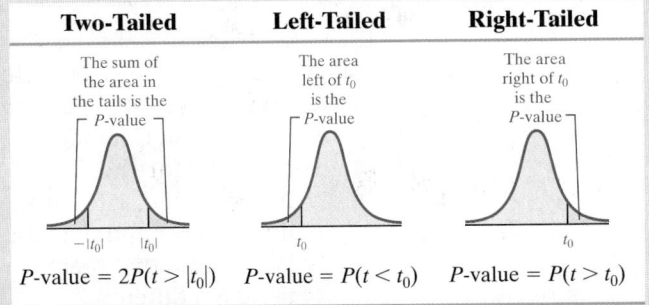

(continued)

Step 4 Compare the critical value to the test statistic.

Two-Tailed	Left-Tailed	Right-Tailed
If $t_0 < -t_{\frac{\alpha}{2}}$ or $t_0 > t_{\frac{\alpha}{2}}$, reject the null hypothesis.	If $t_0 < -t_\alpha$, reject the null hypothesis.	If $t_0 > t_\alpha$, reject the null hypothesis.

Technology Step 3 Use a statistical spreadsheet or calculator with statistical capabilities to obtain the *P*-value. The directions for obtaining the *P*-value using the TI-83/84 Plus graphing calculators, Minitab, Excel, and StatCrunch are given in the Technology Step-by-Step on page 542.

Step 4 If *P*-value $< \alpha$, reject the null hypothesis.

Step 5 State the conclusion.

The procedures just presented are **robust**, which means that minor departures from normality will not adversely affect the results of the test. If the data have outliers, however, the procedure should not be used.

Verify the assumption that the differenced data come from a population that is normally distributed by constructing a normal probability plot. Use a boxplot to determine whether there are outliers. If the normal probability plot indicates that the differenced data are not normally distributed or the boxplot reveals outliers, nonparametric tests should be performed, as discussed in Section 15.4.

EXAMPLE 1 Testing Hypotheses Regarding Matched-Pairs Data

Problem Professor Andy Neill measured the time (in seconds) required to catch a falling meter stick for 12 randomly selected students' dominant hand and nondominant hand. Professor Neill wants to know if the reaction time in an individual's dominant hand is less than the reaction time in his or her nondominant hand. A coin flip is used to determine whether reaction time is measured using the dominant or nondominant hand first. Conduct the test at the $\alpha = 0.05$ level of significance. The data obtained are presented in Table 1.

Table 1		
Student	**Dominant Hand, X_i**	**Nondominant Hand, Y_i**
1	0.177	0.179
2	0.210	0.202
3	0.186	0.208
4	0.189	0.184
5	0.198	0.215
6	0.194	0.193
7	0.160	0.194
8	0.163	0.160
9	0.166	0.209
10	0.152	0.164
11	0.190	0.210
12	0.172	0.197

Source: Professor Andy Neill, Joliet Junior College

Approach This is a matched-pairs design because the variable is measured on the same subject for both the dominant and nondominant hand, the treatment in this experiment. Compute the difference between the dominant time and the nondominant time. So, for the first student compute $X_1 - Y_1$, for the second student compute $X_2 - Y_2$, and so on. If the reaction time in the dominant hand is less than the reaction time in the nondominant hand, we would expect the values of $X_i - Y_i$ to be negative. We assume that there is no difference and seek evidence that leads us to believe that there is a difference.

Before performing the hypothesis test, verify that the differences come from a population that is approximately normally distributed with no outliers because the sample size is small. Construct a normal probability plot and boxplot of the differenced data to verify these requirements. Then proceed to follow Steps 1 through 5.

Solution We compute the differences as $d_i = X_i - Y_i =$ time of dominant hand for ith student minus time of nondominant hand for ith student. We expect these differences to be negative, so we want to determine if $\mu_d < 0$. Table 2 displays the differences.

Table 2			
Student	**Dominant Hand, X_i**	**Nondominant Hand, Y_i**	**Difference, d_i**
1	0.177	0.179	$0.177 - 0.179 = -0.002$
2	0.210	0.202	$0.210 - 0.202 = 0.008$
3	0.186	0.208	-0.022
4	0.189	0.184	0.005
5	0.198	0.215	-0.017
6	0.194	0.193	0.001
7	0.160	0.194	-0.034
8	0.163	0.160	0.003
9	0.166	0.209	-0.043
10	0.152	0.164	-0.012
11	0.190	0.210	-0.020
12	0.172	0.197	-0.025
			$\sum d_i = -0.158$

> **CAUTION!**
> The way that we define the difference determines the direction of the alternative hypothesis in one-tailed tests. In Example 1, we expect $X_i < Y_i$, so the difference $X_i - Y_i$ is expected to be negative. Therefore, the alternative hypothesis is $H_1: \mu_d < 0$, and we have a left-tailed test. However, if we computed the differences as $Y_i - X_i$, we would expect the differences to be positive, and we have a right-tailed test!

We compute the mean and standard deviation of the differences and obtain $\bar{d} = -0.0132$ and $s_d = 0.0164$, each rounded to four decimal places. We must verify that the data come from a population that is approximately normal with no outliers. Figure 5 shows the normal probability plot and boxplot of the differenced data.

Figure 5

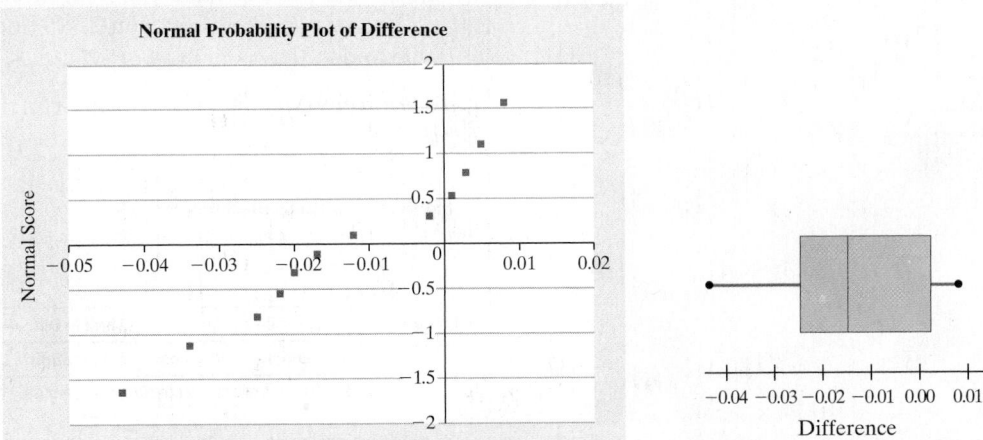

The correlation between the differenced data and expected z-scores is 0.978. Because $0.978 > 0.928$ (Table VI), it is reasonable to conclude the data come from a population that is normally distributed.

Step 1 Professor Neill wants to know if the reaction time in the dominant hand is less than the reaction time in the nondominant hand. We express this as $\mu_d < 0$. We have

$$H_0: \mu_d = 0 \text{ second} \quad \text{versus} \quad H_1: \mu_d < 0 \text{ second}$$

This test is left-tailed.

Step 2 The level of significance is $\alpha = 0.05$.

(continued)

Classical Approach

Step 3 The sample mean is $\bar{d} = -0.0132$ second and the sample standard deviation is $s_d = 0.0164$ second.

The test statistic is

$$t_0 = \frac{\bar{d}_0}{\frac{s_d}{\sqrt{n}}} = \frac{-0.0132}{\frac{0.0164}{\sqrt{12}}} = -2.788$$

Because this is a left-tailed test, we determine the critical value at the $\alpha = 0.05$ level of significance with $12 - 1 = 11$ degrees of freedom to be $-t_{0.05} = -1.796$. The critical region is shown in Figure 6.

Figure 6

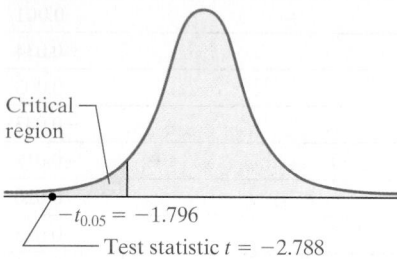

Critical region

$-t_{0.05} = -1.796$

Test statistic $t = -2.788$

Step 4 The test statistic, $t_0 = -2.788$, is labeled in Figure 6. Because the test statistic lies in the critical region, we reject the null hypothesis.

P-Value Approach

By Hand Step 3 The sample mean is $\bar{d} = -0.0132$ second and the sample standard deviation is $s_d = 0.0164$ second.

The test statistic is

$$t_0 = \frac{\bar{d}_0}{\frac{s_d}{\sqrt{n}}} = \frac{-0.0132}{\frac{0.0164}{\sqrt{12}}} = -2.788$$

Because this is a left-tailed test, the P-value is the area under the t-distribution with $12 - 1 = 11$ degrees of freedom to the left of the test statistic, $t_0 = -2.788$, as shown in Figure 7(a). That is, P-value $= P(t < t_0) = P(t < -2.788)$ with 11 degrees of freedom.

Because of the symmetry of the t-distribution, the area under the t-distribution to the left of -2.788 equals the area under the t-distribution to the right of 2.788. So the P-value $= P(t_0 < -2.788) = P(t_0 > 2.788)$. See Figure 7(b).

Figure 7

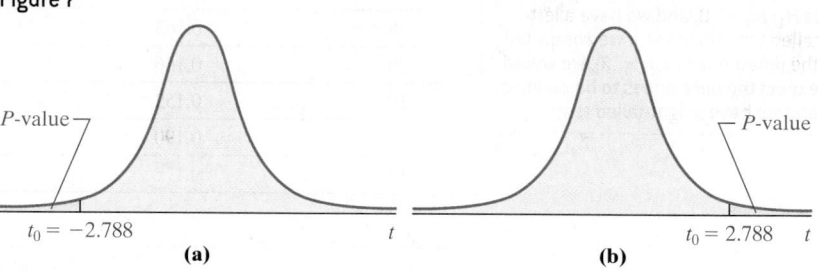

P-value

$t_0 = -2.788$ t

(a)

P-value

$t_0 = 2.788$ t

(b)

Using Table VII, find the row that corresponds to 11 degrees of freedom. The value 2.788 lies between 2.718 and 3.106. The value of 2.718 has an area of 0.01 to the right under the t-distribution with 11 degrees of freedom. The value of 3.106 has an area of 0.005 to the right under the t-distribution with 11 degrees of freedom.

Because 2.788 is between 2.718 and 3.106, the P-value is between 0.005 and 0.01. So $0.005 < P$-value < 0.01.

Technology Step 3 Using StatCrunch, we find the P-value is 0.009. See Figure 8.

Figure 8 **Hypothesis test results:**

$\mu_1 - \mu_2$: mean of the paired difference between Dominant and Nondominant
$H_0 : \mu_1 - \mu_2 = 0$
$H_A : \mu_1 - \mu_2 < 0$

Difference	Sample Diff.	Std. Err.	DF	T-Stat	P-value
Dominant - Nondominant	-0.013166667	0.0047431504	11	-2.7759328	0.009

Differences stored in column, Differences.

Step 4 The P-value of 0.009 [by hand: $0.005 < P$-value < 0.01] means that *if* the null hypothesis that the mean difference is zero is true, we expect a sample mean of -0.0132 second or lower in about 9 out of 1000 repetitions of this experiment. The results we obtained are not consistent with the assumption the mean difference in reaction time between the dominant and nondominant hand is 0 second. Simply put, because the P-value is less than the level of significance, $\alpha = 0.05$ $(0.009 < 0.05)$, Professor Neill rejects the null hypothesis.

Step 5 The sample data provide sufficient evidence at the $\alpha = 0.05$ level of significance to conclude that the reaction time in the dominant hand is less than the reaction time in the nondominant hand.

● **Now Work Problem 7 (a) and (b)**

2   Construct and Interpret Confidence Intervals about the Population Mean Difference of Matched-Pairs Data

We can also obtain a confidence interval for the population mean difference, μ_d, using the sample mean difference, $\overline{d}$, the sample standard deviation difference, s_d, the sample size, and $t_{\frac{\alpha}{2}}$. Remember, a confidence interval about a population mean is given in the following form:

$$\text{Point estimate} \pm \text{margin of error}$$

Based on this, we compute the confidence interval for μ_d as follows:

Confidence Interval for Matched-Pairs Data

A $(1 - \alpha) \cdot 100\%$ confidence interval for μ_d is given by

$$\text{Lower bound: } \overline{d} - t_{\frac{\alpha}{2}} \cdot \frac{s_d}{\sqrt{n}} \qquad \text{Upper bound: } \overline{d} + t_{\frac{\alpha}{2}} \cdot \frac{s_d}{\sqrt{n}} \qquad \textbf{(1)}$$

The critical value $t_{\alpha/2}$ is determined using $n - 1$ degrees of freedom. The values of $\overline{d}$ and s_d are the mean and standard deviation of the differenced data.

Note: The interval is exact when the population is normally distributed and approximately correct for nonnormal populations, provided that n is large.

EXAMPLE 2   Constructing a Confidence Interval for Matched-Pairs Data

Problem Using the data from Table 2 on page 539, construct a 95% confidence interval estimate of the mean difference, μ_d.

By Hand Approach

Step 1 Compute the differenced data. Because the sample size is small, we must verify that the differenced data come from a population that is approximately normal with no outliers.

Step 2 Compute the sample mean difference, $\overline{d}$, and the sample standard deviation difference, s_d.

Step 3 Determine the critical value, $t_{\frac{\alpha}{2}}$, with $\alpha = 0.05$ and $n - 1$ degrees of freedom.

Step 4 Use Formula (1) to determine the lower and upper bounds.

Step 5 Interpret the results.

By-Hand Solution

Step 1 We computed the differenced data and verified that they come from a population that is approximately normally distributed with no outliers in Example 1.

Step 2 We computed the sample mean difference, $\overline{d}$, to be -0.0132 and the sample standard deviation of the difference, s_d, to be 0.0164 in Example 1.

Step 3 Using Table VII with $\alpha = 0.05$ and $12 - 1 = 11$ degrees of freedom, we find $t_{\frac{\alpha}{2}} = t_{0.025} = 2.201$.

Technology Approach

Step 1 Compute the differenced data. Because the sample size is small, verify the differenced data come from a population that is approximately normal with no outliers.

Step 2 Use a statistical spreadsheet or graphing calculator with advanced statistical features to obtain the confidence interval. We will use a TI-84 Plus C to construct the confidence interval. The steps for constructing confidence intervals using StatCrunch, Minitab, Excel, and the TI-83/84 graphing calculators are given in the Technology Step-by-Step on page 542.

Step 3 Interpret the result.

Technology Solution

Step 1 This was done in Example 1 on page 538.

Step 2 Figure 9 shows the results from a TI-84 Plus C graphing calculator.

Figure 9

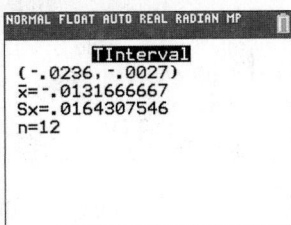

```
NORMAL FLOAT AUTO REAL RADIAN MP
         TInterval
( -.0236, -.0027)
x̄=-.0131666667
Sx=.0164307546
n=12
```

(continued)

Step 4 Substituting into Formula (1), we find

Lower bound: $\bar{d} - t_{\frac{\alpha}{2}} \cdot \dfrac{s_d}{\sqrt{n}} = -0.0132 - 2.201 \cdot \dfrac{0.0164}{\sqrt{12}} = -0.0236$

Upper bound: $\bar{d} + t_{\frac{\alpha}{2}} \cdot \dfrac{s_d}{\sqrt{n}} = -0.0132 + 2.201 \cdot \dfrac{0.0164}{\sqrt{12}} = -0.0028$

Step 5 We are 95% confident the mean difference between the dominant hand's reaction time and the nondominant hand's reaction time is between −0.0236 and −0.0028 second. In other words, we are 95% confident the dominant hand has a mean reaction time that is somewhere between 0.0028 and 0.0236 second faster than the nondominant hand. Because the confidence interval does not contain zero, the evidence suggests the reaction time of a person's dominant hand is different from the reaction time of the nondominant hand.

Step 3 We are 95% confident the mean difference between the dominant hand's reaction time and the nondominant hand's reaction time is between −0.0236 and −0.0027 second. In other words, we are 95% confident the dominant hand has a mean reaction time that is somewhere between 0.0027 and 0.0236 second faster than the nondominant hand. Because the confidence interval does not contain zero, the evidence suggests the reaction time of the dominant hand is different from the reaction time of the nondominant hand.

• Now Work Problem 7 (c)

Technology Step-by-Step Two-Sample *t*-Tests, Dependent Sampling

TI-83/84 Plus

Hypothesis Tests

1. If necessary, enter raw data in L1 and L2. Let L3 = L1 − L2 (or L2 − L1), depending on how the alternative hypothesis is defined.
2. Press STAT, highlight TESTS, and select 2:T-Test.
3. If the data are raw, highlight DATA, making sure that List is set to L3 with frequency set to 1. If summary statistics are known, highlight STATS and enter the summary statistics.
4. Highlight the appropriate relation in the alternative hypothesis.
5. Highlight Calculate or Draw and press ENTER. Calculate gives the test statistic and *P*-value. Draw will draw the *t*-distribution with the *P*-value shaded.

Confidence Intervals

Follow the same steps given for hypothesis tests, except select 8:TInterval. Also, select a confidence level (such as 95% = 0.95).

Minitab

1. If necessary, enter raw data in columns C1 and C2.
2. Select the **Stat** menu, highlight **Basic Statistics**, and then select **Paired-t . . . .**
3. If you have raw data, select "Each sample is in a column" from the drop-down menu. Enter C1 in the cell marked "Sample 1" and enter C2 in the cell marked "Sample 2." If you have summarized data, select "Summarized data (differences)" from the drop-down menu and enter the summary statistics. Click Options . . . , select the

direction of the alternative hypothesis and select a confidence level. Click OK twice.

Excel

1. Enter the raw data into Columns A and B.
2. Select the Formulas menu. Select More Functions. Highlight Statistical, and select T.Test from the drop-down menu.
3. Place the cursor in Array1. Highlight the data in Column A. Place the cursor in Array2. Highlight the data in Column B. Place the cursor in the Tails cell. Enter the number corresponding to the test you desire (1 for a one-tailed distribution; 2 for a two-tailed distribution). Place the cursor in the Type cell. Enter 1 for a paired *t*-test. Click OK.

StatCrunch

1. If necessary, enter the raw data into the first two columns of the spreadsheet. Name each column variable.
2. Select **Stat**, highlight **T Stats**, select **Paired**.
3. Select the column that contains the data for Sample 1. Select the column that contains the data for Sample 2. Note that the differences are computed Sample 1 – Sample 2. If you choose the hypothesis test radio button, enter the value of the mean stated in the null hypothesis and choose the direction of the alternative hypothesis from the pull-down menu. If you choose the confidence interval radio button, enter the level of confidence. Click Compute!.

Note: If you have summarized data, then follow the steps for performing a single sample *t* test.

11.2 Assess Your Understanding

Skill Building

1. A researcher wants to show the mean from population 1 is less than the mean from population 2 in matched-pairs data. If the observations from sample 1 are X_i and the observations from sample 2 are Y_i, and $d_i = X_i - Y_i$, then the null hypothesis is $H_0: \mu_d = 0$ and the alternative hypothesis is $H_1: \mu_d$ _____ 0.

2. A researcher wants to show the mean from population 1 is less than the mean from population 2 in matched-pairs data. If the observations from sample 1 are X_i and the observations from sample 2 are Y_i and $d_i = Y_i - X_i$, then the null hypothesis is $H_0: \mu_d = 0$ and the alternative hypothesis is $H_1: \mu_d$ _____ 0.

In Problems 3 and 4, assume that the differences are normally distributed.

3.

Observation	1	2	3	4	5	6	7
X_i	7.6	7.6	7.4	5.7	8.3	6.6	5.6
Y_i	8.1	6.6	10.7	9.4	7.8	9.0	8.5

(a) Determine $d_i = X_i - Y_i$ for each pair of data.
(b) Compute $\overline{d}$ and s_d.
(c) Test if $\mu_d < 0$ at the $\alpha = 0.05$ level of significance.
(d) Compute a 95% confidence interval about the population mean difference μ_d.

4.

Observation	1	2	3	4	5	6	7	8
X_i	19.4	18.3	22.1	20.7	19.2	11.8	20.1	18.6
Y_i	19.8	16.8	21.1	22.0	21.5	18.7	15.0	23.9

(a) Determine $d_i = X_i - Y_i$ for each pair of data.
(b) Compute $\overline{d}$ and s_d.
(c) Test if $\mu_d \neq 0$ at the $\alpha = 0.01$ level of significance.
(d) Compute a 99% confidence interval about the population mean difference μ_d.

Applying the Concepts

5. Naughty or Nice? An experiment was conducted in which 16 ten-month-old babies were asked to watch a climber character attempt to ascend a hill. On two occasions, the baby witnesses the character fail to make the climb. On the third attempt, the baby witnesses either a helper toy push the character up the hill, or a hinderer toy preventing the character from making the ascent. The helper and hinderer toys were shown to each baby in a random fashion for a fixed amount of time. In Problem 39 from Section 10.2, we learned that, after watching both the helper and hinderer toy in action, 14 of 16 ten-month-old babies preferred to play with the helper toy when given a choice as to which toy to play with. A second part of this experiment showed the climber approach the helper toy, which is not a surprising action, and then alternatively the climber approached the hinderer toy, which is a surprising action. The amount of time the ten-month-old watched the event was recorded. The mean difference in time spent watching the climber approach the hinderer toy versus watching the climber approach the helper toy was 1.14 seconds with a standard deviation of 1.75 second. *Source:* J. Kiley Hamlin et al., "Social Evaluation by Preverbal Infants," *Nature*, Nov. 2007.

(a) State the null and alternative hypothesis to determine if babies tend to look at the hinderer toy longer than the helper toy.
(b) Assuming the differences are normally distributed with no outliers, test if the difference in the amount of time the baby will watch the hinderer toy versus the helper toy is greater than 0 at the 0.05 level of significance.
(c) What do you think the results of this experiment imply about 10-month-olds' ability to assess surprising behavior?

6. Caffeine-Enhanced Workout? Since its removal from the banned substances list in 2004 by the World Anti-Doping Agency, caffeine has been used by athletes with the expectancy that it enhances their workout and performance. However, few studies look at the role caffeine plays in sedentary females. Researchers at the University of Western Australia conducted a test in which they determined the rate of energy expenditure (kilojoules) on 10 healthy, sedentary females who were nonregular caffeine users. Each female was randomly assigned either a placebo or caffeine pill (6 mg/kg) 60 minutes prior to exercise. The subject rode an exercise bicycle for 15 minutes at 65% of their maximum heart rate, and the energy expenditure was measured. The process was repeated on a separate day for the remaining treatment. The mean difference in energy expenditure (caffeine – placebo) was 18 kJ with a standard deviation of 19 kJ. *Source:* Wallman, Karen E. "Effect of Caffeine on Exercise Performance in Sedentary Females," *Journal of Sports Science and Medicine* (2010) 9, 183–189

(a) State the null and alternative hypotheses to determine if caffeine increases energy expenditure.
(b) Assuming the differences are normally distributed, determine if caffeine appears to increase energy expenditure at the $\alpha = 0.05$ level of significance.

NW 7. Muzzle Velocity The following data represent the muzzle velocity (in feet per second) of rounds fired from a 155-mm gun. For each round, two measurements of the velocity were recorded using two different measuring devices, with the following data obtained:

Observation	1	2	3	4	5	6
A	793.8	793.1	792.4	794.0	791.4	792.4
B	793.2	793.3	792.6	793.8	791.6	791.6

Observation	7	8	9	10	11	12
A	791.7	792.3	789.6	794.4	790.9	793.5
B	791.6	792.4	788.5	794.7	791.3	793.5

Source: Ronald Christenson and Larry Blackwood. "Tests for Precision and Accuracy of Multiple Measuring Devices." *Technometrics*, 35(4):411–421, 1993.

(a) Why are these matched-pairs data?
(b) Is there a difference in the measurement of the muzzle velocity between device A and device B at the $\alpha = 0.01$ level of significance? **Note:** A normal probability plot and boxplot of the data indicate that the differences are approximately normally distributed with no outliers.
(c) Construct a 99% confidence interval about the population mean difference. Interpret your results.
(d) Draw a boxplot of the differenced data. Does this visual evidence support the results obtained in part (b)?

8. Reaction Time In an experiment conducted online at the University of Mississippi, study participants are asked to react to a stimulus. In one experiment, the participant must press a key on seeing a blue screen and reaction time (in seconds) to press the key is measured. The same person is then asked to press a key on seeing a red screen, again with reaction time measured. The results for six randomly sampled study participants are as follows:

Participant	1	2	3	4	5	6
Blue	0.582	0.481	0.841	0.267	0.685	0.450
Red	0.408	0.407	0.542	0.402	0.456	0.533

Source: PsychExperiments at the University of Mississippi

(a) Why are these matched-pairs data?

(b) In this study, the color that the participant was first asked to react to was randomly selected. Why is this a good idea in this experiment?

(c) Is the reaction time to the blue stimulus different from the reaction time to the red stimulus at the $\alpha = 0.01$ level of significance? **Note:** A normal probability plot and boxplot of the data indicate that the differences are approximately normally distributed with no outliers.

(d) Construct a 99% confidence interval about the population mean difference. Interpret your results.

(e) Draw a boxplot of the differenced data. Does this visual evidence support the results obtained in part (c)?

9. SUV versus Car It is a commonly held belief that SUVs are safer than cars. If an SUV and car are in a collision, does the SUV sustain less damage (as suggested by the cost of repair)? The Insurance Institute for Highway Safety crashed SUVs into cars, with the SUV moving 10 miles per hour and the front of the SUV crashing into the rear of the car.

SUV into Car	SUV Damage	Car Damage
Honda CR-V into Honda Civic	1721	1274
Toyota RAV4 into Toyota Corolla	1434	2327
Hyundai Tucson into Kia Forte	850	3223
Volkswagen Tiguan into VW Golf	2329	2058
Jeep Patriot into Dodge Caliber	1415	3095
Ford Escape into Ford Focus	1470	3386
Nissan Rogue into Nissan Sentra	2884	4560

Source: Insurance Institute for Highway Safety

(a) Why are these matched-pairs data?

(b) Draw a boxplot of the differenced data. Does the visual evidence support the belief that SUVs have a lower repair cost?

(c) Do the data suggest the repair cost for the car is higher? Use an $\alpha = 0.05$ level of significance.

Note: A normal probability plot indicates the differenced data are approximately normal with no outliers.

10. Secchi Disk A Secchi disk is an 8-inch-diameter weighted disk that is painted black and white and attached to a rope. The disk is lowered into water and the depth (in inches) at which it is no longer visible is recorded. The measurement is an indication of water clarity. An environmental biologist interested in determining whether the water clarity of the lake at Joliet Junior College is improving takes measurements at the same location on eight dates during the course of a year and repeats the measurements on the same dates five years later. She obtains the following results:

Observation	1	2	3	4	5	6	7	8
Date	5/11	6/7	6/24	7/8	7/27	8/31	9/30	10/12
Initial depth, X_i	38	58	65	74	56	36	56	52
Depth five years later, Y_i	52	60	72	72	54	48	58	60

Source: Virginia Piekarski, Joliet Junior College

(a) Why is it important to take the measurements on the same date?

(b) Does the evidence suggest that the clarity of the lake is improving at the $\alpha = 0.05$ level of significance? **Note:** A normal probability plot and boxplot of the data indicate that the differences are approximately normally distributed with no outliers.

(c) Draw a boxplot of the differenced data. Does this visual evidence support the results obtained in part (b)?

11. Getting Taller? To test the belief that sons are taller than their fathers, a student randomly selects 13 fathers who have adult male children. She records the height of both the father and son in inches and obtains the following data. Are sons taller than their fathers? Use the $\alpha = 0.1$ level of significance. **Note:** A normal probability plot and boxplot of the data indicate that the differences are approximately normally distributed with no outliers.

	1	2	3	4	5	6	7
Height of father, X_i	70.3	67.1	70.9	66.8	72.8	70.4	71.8
Height of son, Y_i	74.1	69.2	66.9	69.2	68.9	70.2	70.4

	8	9	10	11	12	13
Height of father, X_i	70.1	69.9	70.8	70.2	70.4	72.4
Height of son, Y_i	69.3	75.8	72.3	69.2	68.6	73.9

Source: Anna Behounek, student at Joliet Junior College

12. Waiting in Line A quality-control manager at an amusement park feels that the amount of time that people spend waiting in line for the American Eagle roller coaster is too long. To determine if a new loading/unloading procedure is effective in reducing wait time in line, he measures the amount of time (in minutes) people are waiting in line on 7 days. After implementing the new procedure, he again measures the amount of time (in minutes) people are waiting in line on 7 days and obtains the following data. To make a reasonable comparison, he chooses days when the weather conditions are similar. Is the new loading/unloading procedure effective in reducing wait time at the $\alpha = 0.05$ level of significance? **Note:** A normal probability plot and boxplot of the data indicate that the differences are approximately normally distributed with no outliers.

Day	Mon (2 P.M.)	Tues (2 P.M.)	Wed (2 P.M.)	Thurs (2 P.M.)	Fri (2 P.M.)
Wait time before, X_i	11.6	25.9	20.0	38.2	57.3
Wait time after, Y_i	10.7	28.3	19.2	35.9	59.2

Day	Sat (11 A.M.)	Sat (4 P.M.)	Sun (12 noon)	Sun (4 P.M.)
Wait time before, X_i	32.1	81.8	57.1	62.8
Wait time after, Y_i	31.8	75.3	54.9	62.0

13. Hardness Testing The manufacturer of hardness testing equipment uses steel-ball indenters to penetrate a metal that is being tested. However, the manufacturer thinks it would be better to use a diamond indenter so that all types of metal can be tested. Because of differences between the two types of indenters, it is suspected that the two methods will produce different hardness readings. The metal specimens to be tested are large enough so that two indentions can be made. Therefore, the manufacturer uses both indenters on each specimen and compares the hardness readings. Construct a 95% confidence interval to judge whether the two indenters result in different measurements.

Specimen	1	2	3	4	5	6	7	8	9
Steel ball	50	57	61	71	68	54	65	51	53
Diamond	52	56	61	74	69	55	68	51	56

Note: A normal probability plot and boxplot of the data indicate that the differences are approximately normally distributed with no outliers.

14. Car Rentals The following data represent the daily rental for a compact automobile charged by two car rental companies, Thrifty and Hertz, in ten locations. Test whether Thrifty is less expensive than Hertz at the $\alpha = 0.1$ level of significance. **Note:** A normal probability plot and boxplot of the data indicate that the differences are approximately normally distributed with no outliers.

City	Thrifty	Hertz
Chicago	21.81	18.99
Los Angeles	29.89	37.99
Houston	17.90	19.99
Orlando	27.98	35.99
Boston	24.61	25.60
Seattle	21.96	22.99
Pittsburgh	20.90	19.99
Phoenix	37.75	36.99
New Orleans	33.81	26.99
Minneapolis	33.49	30.99

Source: Yahoo!Travel

15. DUI Simulator To illustrate the effects of driving under the influence (DUI) of alcohol, a police officer brought a DUI simulator to a local high school. Student reaction time in an emergency was measured with unimpaired vision and also while wearing a pair of special goggles to simulate the effects of alcohol on vision. For a random sample of nine teenagers, the time (in seconds) required to bring the vehicle to a stop from a speed of 60 miles per hour was recorded.

Subject	1	2	3	4	5	6	7	8	9
Normal, X_i	4.47	4.24	4.58	4.65	4.31	4.80	4.55	5.00	4.79
Impaired, Y_i	5.77	5.67	5.51	5.32	5.83	5.49	5.23	5.61	5.63

(a) Whether the student had unimpaired vision or wore goggles first was randomly selected. Why is this a good idea in designing the experiment?

(b) Use a 95% confidence interval to test if there is a difference in braking time with impaired vision and normal vision where the differences are computed as "impaired minus normal." **Note:** A normal probability plot and boxplot of the data indicate that the differences are approximately normally distributed with no outliers.

16. Braking Distance An automotive researcher wanted to estimate the difference in distance required to come to a complete stop while traveling 40 miles per hour on wet versus dry pavement. Because car type plays a role, the researcher used eight different cars with the same driver and tires. The braking distance (in feet) on both wet and dry pavement is shown in the table below. Construct a 95% confidence interval for the mean difference in braking distance on wet versus dry pavement where the differences are computed as "wet minus dry." Interpret the interval. **Note:** A normal probability plot and boxplot of the data indicate that the differences are approximately normally distributed with no outliers.

Car	1	2	3	4	5	6	7	8
Wet	106.9	100.9	108.8	111.8	105.0	105.6	110.6	107.9
Dry	71.8	68.8	74.1	73.4	75.9	75.2	75.7	81.0

17. Does Octane Matter? Octane is a measure of how much fuel can be compressed before it spontaneously ignites. Some people believe that higher-octane fuels result in better gas mileage for their cars. To test this claim, a researcher randomly selected 11 individuals (and their cars) to participate in the study. Each participant received 10 gallons of gas and drove his or her car on a closed course that simulated both city and highway driving. The number of miles driven until the car ran out of gas was recorded. A coin flip was used to determine whether the car was filled up with 87-octane or 92-octane fuel first, and the driver did not know which type of fuel was in the tank. The results are in the following table:

Driver	1	2	3	4	5	6
Miles on 87 octane	234	257	243	215	114	287
Miles on 92 octane	237	238	229	224	119	297

Driver	7	8	9	10	11
Miles on 87 octane	315	229	192	204	547
Miles on 92 octane	351	241	186	209	562

(a) Why is it important that the matching be done by driver and car?

(b) Why is it important to conduct the study on a closed track?

(c) The normal probability plots for miles on 87 octane and miles on 92 octane are shown. The correlation between 87 octane and the expected z-scores is 0.877. The correlation between 92 octane and the expected z-scores is 0.879. Are either of these variables normally distributed?

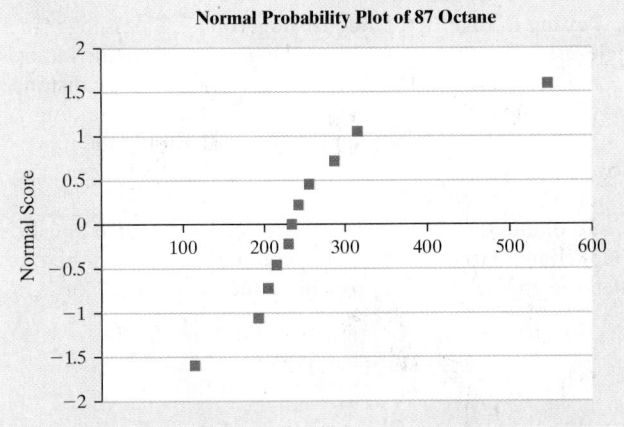

Normal Probability Plot of 92 Octane

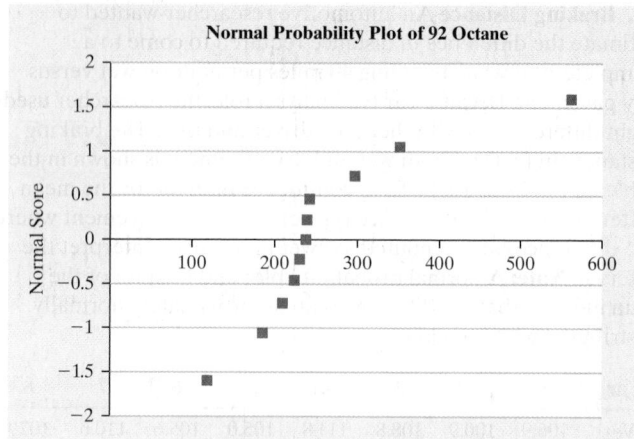

(d) The differences are computed as 92 octane minus 87 octane. The normal probability plot of the differences is shown. The correlation between the differenced data and the expected z-scores is 0.966. Is there reason to believe that the differences are normally distributed? Conclude that the differences can be normally distributed even though the original data are not.

Normal Probability Plot of Difference

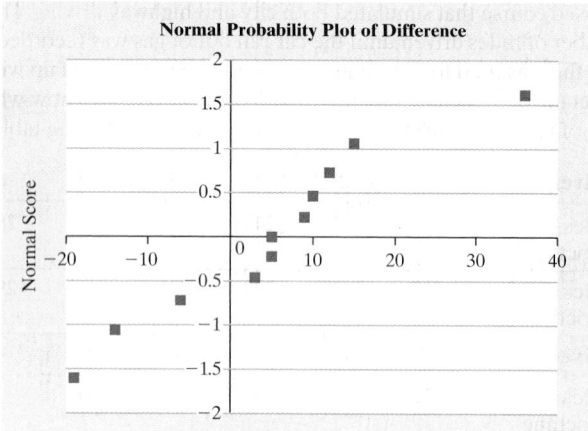

(e) The researchers used Minitab to determine whether the mileage from 92 octane is greater than the mileage from 87 octane. The results are as follows:

Paired T-Test and CI: 92 Octane, 87 Octane
```
Paired T for 92 Octane - 87 Octane

                N     Mean     StDev   SE Mean
92 Octane      11   263.000   115.041   34.686
87 Octane      11   257.909   109.138   32.906
Difference     11     5.09091   14.87585    4.48524

95% lower bound for mean difference: -3.03841
T-test of mean difference = 0 (vs > 0): T-value = 1.14 P-value = 0.141
```

What do you conclude? Why?

18. Putting It Together: Glide Testing You are a passenger in a single-propeller-driven aircraft that experiences engine failure in the middle of a flight. The pilot wants to maximize the distance that the plane can glide to increase the likelihood of finding a safe place to land. To accomplish this goal, should the pilot allow the propeller to "windmill" or should the pilot force the propeller to stop?

To obtain the data needed to answer the research question, a pilot climbed to 8000 feet at a speed of 60 knots and then killed the engine with the propeller either windmilling or

stopped. Because the time to descend is directly proportional to glide distance, the time to descend to 7200 feet was recorded in seconds and used as a proxy for glide distance. The design called for randomly choosing the order in which the propeller would windmill or be stopped. The data in the table represent the time to descend 800 feet for each of 27 trials. **Note:** Visit www.aceaerobaticschool.com to see footage of this scenario.

Trial	Windmilling	Stopped	Trial	Windmilling	Stopped
1	73.4	82.3	15	64.2	82.5
2	68.9	75.8	16	67.5	81.1
3	74.1	75.7	17	71.2	72.3
4	71.7	71.7	18	75.6	77.7
5	74.2	68.8	19	73.1	82.6
6	63.5	74.2	20	77.4	79.5
7	64.4	78.0	21	77.0	82.3
8	60.9	68.5	22	77.8	79.5
9	79.5	90.6	23	77.0	79.7
10	74.5	81.9	24	72.3	73.4
11	76.5	72.9	25	69.2	76.0
12	70.3	75.7	26	63.9	74.2
13	71.3	77.6	27	70.3	79.0
14	72.7	174.3			

Source: Catherine Elizabeth Cavagnaro. "Glide Testing: A Paired Samples Experiment." *Stats* 46, Fall 2006.

(a) The trials took place over the course of a few days. However, for each trial, the pilot conducted both windmilling and stopped propeller one right after the other to minimize any impact of a change in weather conditions. Knowing this, explain why this is matched-pair data.

(b) Why does the researcher randomly determine whether to windmill or stop the propeller first for each trial?

(c) Explain why blinding is not possible for this experiment.

(d) What is the response variable in the study? What are the treatments?

(e) Compute the difference as "difference = stopped − windmilling." Draw a boxplot of the differenced data. What do you notice?

(f) From part (e), you should notice that trial 14 results in an outlier. Because our sample size is small, this outlier will have a major effect on any results. The author of the article indicated that it was possibly a situation in which there was an updraft of wind, causing the plane to take quite a bit longer than normal to fall 800 feet. Explain why this explanation makes it reasonable to eliminate trial 14 from the analysis.

(g) Redraw a boxplot of the data with trial 14 eliminated. Based on the shape of the boxplot, do you believe it is reasonable to proceed with a matched-pair t-test?

(h) The researchers wanted to determine if stopping the propeller resulted in a longer glide distance. Based on this goal, determine the null and alternative hypotheses.

(i) Conduct the appropriate test to answer the researcher's question.

(j) Write a few sentences outlining your recommendations to pilots who experience engine failure.

11.3   Inference about Two Means: Independent Samples

Preparing for This Section Before getting started, review the following:

- The Completely Randomized Design (Section 1.6, pp. 44–46)
- Confidence intervals about μ (Section 9.2, pp. 439–441)
- Hypothesis tests about μ (Section 10.3, pp. 494–498)
- Type I and Type II errors (Section 10.1, pp. 476–477)

Objectives

① Test hypotheses regarding the difference of two independent means

② Construct and interpret confidence intervals regarding the difference of two independent means

We now turn our attention to inferential methods for comparing means from two independent samples. For example, suppose we want to know whether a new experimental drug relieves symptoms attributable to the common cold. The response variable might be the time until the cold symptoms go away—a quantitative variable. If the drug is effective, the mean time until the cold symptoms go away should be less for individuals taking the drug than for those not taking the drug. If we let μ_1 represent the mean time until cold symptoms go away for the individuals taking the drug and μ_2 represent the mean time until cold symptoms go away for individuals taking a placebo, the null and alternative hypotheses will be

$$H_0: \mu_1 = \mu_2 \quad \text{versus} \quad H_1: \mu_1 < \mu_2$$

or, equivalently,

$$H_0: \mu_1 - \mu_2 = 0 \quad \text{versus} \quad H_1: \mu_1 - \mu_2 < 0$$

To conduct this test, we might randomly divide 500 volunteers who have a common cold into two groups: an experimental group (group 1) and a control group (group 2). The control group will receive a placebo and the experimental group will receive a predetermined amount of the experimental drug. Next, determine the time until the cold symptoms go away. Compute $\bar{x}_1$, the sample mean time until cold symptoms go away in the experimental group, and $\bar{x}_2$, the sample mean time until cold symptoms go away in the control group. Now we determine whether the difference in the sample means, $\bar{x}_1 - \bar{x}_2$, is significantly less than 0, the assumed difference stated in the null hypothesis. To do this, we need to know the sampling distribution of $\bar{x}_1 - \bar{x}_2$.

It is unreasonable to expect to know information regarding σ_1 and σ_2 without knowing information regarding the population means. Therefore, we must develop a sampling distribution for the difference of two means when the population standard deviations are unknown.

The comparison of two means with unequal (and unknown) population variances is called the Behrens–Fisher problem. While an exact method for performing inference on the equality of two means with unequal population standard deviations does not exist, an approximate solution is available. The approach that we use is known as **Welch's approximate *t*,** in honor of English statistician Bernard Lewis Welch (1911–1989).

Sampling Distribution of the Difference of Two Means: Independent Samples with Population Standard Deviations Unknown (Welch's *t*)

Suppose that a simple random sample of size n_1 is taken from a population with unknown mean μ_1 and unknown standard deviation σ_1. In addition, a simple random sample of size n_2 is taken from a second population with unknown mean μ_2 and

(continued)

unknown standard deviation σ_2. If the two populations are normally distributed or the sample sizes are sufficiently large ($n_1 \geq 30$ and $n_2 \geq 30$), then

$$t = \frac{(\bar{x}_1 - \bar{x}_2) - (\mu_1 - \mu_2)}{\sqrt{\dfrac{s_1^2}{n_1} + \dfrac{s_2^2}{n_2}}} \qquad \text{(1)}$$

approximately follows Student's t-distribution with the smaller of $n_1 - 1$ or $n_2 - 1$ degrees of freedom, where $\bar{x}_1$ is the sample mean and s_1 is the sample standard deviation from population 1, and $\bar{x}_2$ is the sample mean and s_2 is the sample standard deviation from population 2.

❶ Test Hypotheses Regarding the Difference of Two Independent Means

Now that we know the approximate sampling distribution of $\bar{x}_1 - \bar{x}_2$, we can introduce a procedure that can be used to test hypotheses regarding two population means.

Testing Hypotheses Regarding the Difference of Two Means

To test hypotheses regarding two population means, μ_1 and μ_2, with unknown population standard deviations, we can use the following steps, provided that

- the samples are obtained using simple random sampling or through a completely randomized experiment with two levels of treatment.
- the samples are independent.
- the populations from which the samples are drawn are normally distributed or the sample sizes are large ($n_1 \geq 30$ and $n_2 \geq 30$).
- for each sample, the sample size is no more than 5% of the population size.

Step 1   Determine the null and alternative hypotheses. The hypotheses are structured in one of three ways:

Two-Tailed	Left-Tailed	Right-Tailed
$H_0: \mu_1 = \mu_2$	$H_0: \mu_1 = \mu_2$	$H_0: \mu_1 = \mu_2$
$H_1: \mu_1 \neq \mu_2$	$H_1: \mu_1 < \mu_2$	$H_1: \mu_1 > \mu_2$

Note: μ_1 is the population mean for population 1, and μ_2 is the population mean for population 2.

Step 2   Select a level of significance α, depending on the seriousness of making a Type I error.

Classical Approach

Step 3   Compute the **test statistic**

$$t_0 = \frac{(\bar{x}_1 - \bar{x}_2) - (\mu_1 - \mu_2)}{\sqrt{\dfrac{s_1^2}{n_1} + \dfrac{s_2^2}{n_2}}}$$

which approximately follows Student's t-distribution. Use Table VII to determine the critical value using the smaller of $n_1 - 1$ or $n_2 - 1$ degrees of freedom.

P-Value Approach

By Hand Step 3   Compute the **test statistic**

$$t_0 = \frac{(\bar{x}_1 - \bar{x}_2) - (\mu_1 - \mu_2)}{\sqrt{\dfrac{s_1^2}{n_1} + \dfrac{s_2^2}{n_2}}}$$

which approximately follows Student's t-distribution. Use Table VII to approximate the P-value using the smaller of $n_1 - 1$ or $n_2 - 1$ degrees of freedom.

	Two-Tailed	Left-Tailed	Right-Tailed
Critical value(s)	$-t_{\frac{\alpha}{2}}$ and $t_{\frac{\alpha}{2}}$	$-t_\alpha$	t_α

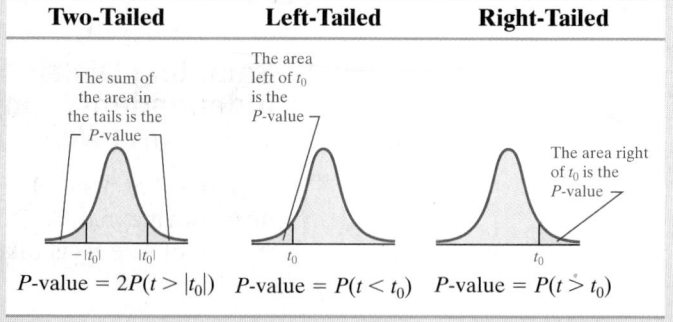

$$P\text{-value} = 2P(t > |t_0|) \qquad P\text{-value} = P(t < t_0) \qquad P\text{-value} = P(t > t_0)$$

Step 4 Compare the critical value to the test statistic.

Two-Tailed	Left-Tailed	Right-Tailed
If $t_0 < -t_{\frac{\alpha}{2}}$ or $t_0 > t_{\frac{\alpha}{2}}$, reject the null hypothesis.	If $t_0 < -t_\alpha$, reject the null hypothesis.	If $t_0 > t_\alpha$, reject the null hypothesis.

Technology Step 3 Use a statistical spreadsheet or calculator with statistical capabilities to obtain the P-value. The directions for obtaining the P-value using the TI-83/84 Plus graphing calculators, Excel, Minitab, and StatCrunch are in the Technology Step-by-Step on pages 553–554.

Step 4 If P-value $< \alpha$, reject the null hypothesis.

Step 5 State the conclusion.

The procedure just presented is robust, so minor departures from normality will not adversely affect the results of the test. If the data have outliers, however, the procedure should not be used.

We verify these requirements by constructing normal probability plots (to assess normality) and boxplots (to determine whether there are outliers). If the normal probability plots indicate that the data come from populations that are not normally distributed or the boxplots reveals outliers, then nonparametric tests should be performed, as discussed in Section 15.4.

EXAMPLE 1   Testing Hypotheses Regarding Two Means

Problem In the Spacelab Life Sciences 2 payload, 14 male rats were sent to space. Upon their return, the red blood cell mass (in milliliters) of the rats was determined. A control group of 14 male rats was held under the same conditions (except for space flight) as the space rats, and their red blood cell mass was also determined when the space rats returned. The project resulted in the data listed in Table 3. Does the evidence suggest that the flight animals have a different red blood cell mass from the control animals at the $\alpha = 0.05$ level of significance?

Table 3									
Flight					**Control**				
8.59	8.64	7.43	7.21	6.39	8.65	6.99	8.40	9.66	7.14
6.87	7.89	9.79	6.85	7.54	7.62	7.44	8.55	8.70	9.14
7.00	8.80	9.30	8.03		7.33	8.58	9.88	9.94	

Source: NASA Life Sciences Data Archive

Approach This experiment is a completely randomized design with response variable red blood cell mass. The treatment is space flight, which is set at two levels: space flight or no space flight. The experimental units are the 28 rats. The expectation is that all other variables that may affect red blood cell mass are accounted for by holding both groups of rats in the same condition (other than space flight) and through random assignment.

We are attempting to determine if the evidence suggests that space flight affects red blood cell mass. We assume no difference, so we assume the mean red blood cell mass of the flight group equals that of the control group. We want to show that the mean of the flight group is different from the mean of the control group.

Verify that each sample comes from a population that is approximately normal with no outliers by drawing normal probability plots and boxplots. The boxplots will be drawn on the same graph so that we can visually compare the two samples. Then follow Steps 1 through 5, listed on pages 548–549.

Solution Figure 10 on the next page shows the normal probability plots of the data. The correlation between the red blood cell mass in the flight group and expected z-scores is 0.984. The correlation between the red blood cell mass in the control group

(continued)

and expected z-scores is 0.973. Because $0.984 > 0.935$ (Table VI) and $0.973 > 0.935$, it is reasonable to conclude the data come from populations that are normally distributed. On the basis of the boxplots, it seems that there is not much difference in the red blood cell mass of the two samples, although the flight group might have a slightly lower red blood cell mass. We have to determine if this difference is significant or due to random sampling error (chance).

Figure 10

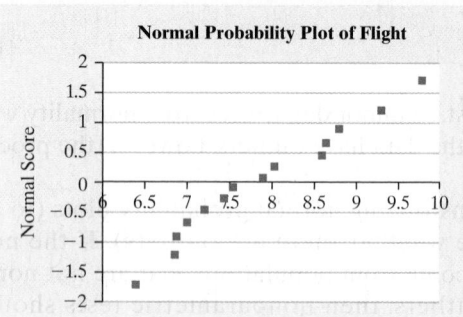
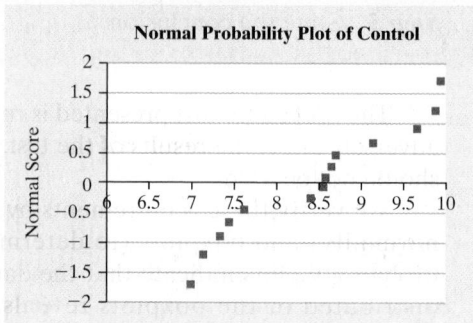
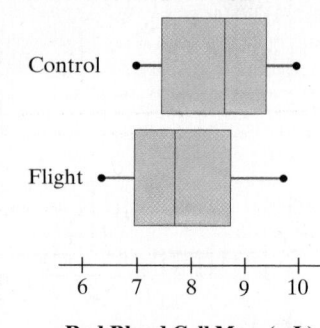

Step 1 We want to know whether the flight animals have a different red blood cell mass from the control animals. Let μ_1 represent the mean red blood cell mass of the flight animals and μ_2 represent the mean red blood cell mass of the control animals. We are attempting to gather evidence that shows $\mu_1 \neq \mu_2$, and we have the hypotheses

$$H_0: \mu_1 = \mu_2 \qquad H_0: \mu_1 - \mu_2 = 0$$
$$\text{versus} \quad \text{or} \quad \text{versus}$$
$$H_1: \mu_1 \neq \mu_2 \qquad H_1: \mu_1 - \mu_2 \neq 0$$

Step 2 The level of significance $\alpha = 0.05$.

Classical Approach

Step 3 The statistics for sample 1 (the flight rats) are $n_1 = 14$, $\bar{x}_1 = 7.881$, and $s_1 = 1.017$.

The statistics for sample 2 (the control rats) are $n_2 = 14$, $\bar{x}_2 = 8.430$, and $s_2 = 1.005$.

The test statistic is

$$t_0 = \frac{(\bar{x}_1 - \bar{x}_2) - (\mu_1 - \mu_2)}{\sqrt{\frac{s_1^2}{n_1} + \frac{s_2^2}{n_2}}}$$

$$= \frac{(7.881 - 8.430) - 0}{\sqrt{\frac{1.017^2}{14} + \frac{1.005^2}{14}}}$$

$$= \frac{-0.549}{0.3821288115}$$

$$= -1.437$$

Step 4 The test statistic is $t_0 = -1.437$. We label this point in Figure 11.

This is a two-tailed test with $\alpha = 0.05$. Since the sample sizes of the experimental group and control group are both 14, we have $n_1 - 1 = 14 - 1 = 13$ degrees of freedom. The critical values are $t_{\frac{\alpha}{2}} = t_{\frac{0.05}{2}} = t_{0.025} = 2.160$ and $-t_{0.025} = -2.160$. The critical region is displayed in Figure 11.

P-Value Approach

By Hand Step 3 The statistics for sample 1 (the flight rats) are $n_1 = 14$, $\bar{x}_1 = 7.881$, and $s_1 = 1.017$. The statistics for sample 2 (the control rats) are $n_2 = 14$, $\bar{x}_2 = 8.430$, and $s_2 = 1.005$.

The test statistic is

$$t_0 = \frac{(\bar{x}_1 - \bar{x}_2) - (\mu_1 - \mu_2)}{\sqrt{\frac{s_1^2}{n_1} + \frac{s_2^2}{n_2}}} = \frac{(7.881 - 8.430) - 0}{\sqrt{\frac{1.017^2}{14} + \frac{1.005^2}{14}}}$$

$$= \frac{-0.549}{0.3821288115}$$

$$= -1.437$$

Because this is a two-tailed test, the P-value is the area under the t-distribution to the left of $t_0 = -1.437$ plus the area under the t-distribution to the right of $t_0 = 1.437$. See Figure 12.

Figure 12

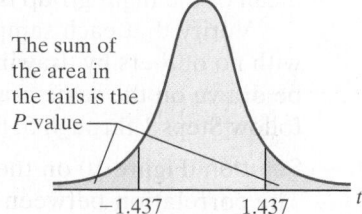

The sum of the area in the tails is the P-value

Since the sample size of the experimental group and control group are both 14, we have $n_1 - 1 = 14 - 1 = 13$ degrees

Figure 11

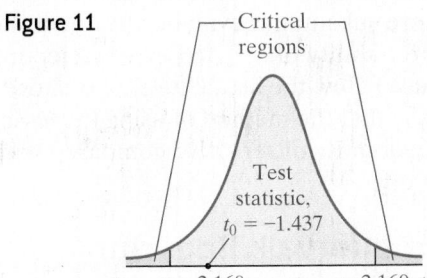

Because the test statistic does not lie within a critical region, we do not reject the null hypothesis.

of freedom. Because of symmetry, use Table VII to estimate the area under the t-distribution to the right of $t_0 = 1.437$ and double it.

$$P\text{-value} = P(t_0 < -1.437 \text{ or } t_0 > 1.437) = 2P(t_0 > 1.437)$$

Using Table VII, find the row that corresponds to 13 degrees of freedom. The value 1.437 lies between 1.350 and 1.771. The area under the t-distribution with 13 degrees of freedom to the right of 1.350 is 0.10. The area under the t-distribution with 13 degrees of freedom to the right of 1.771 is 0.05. After doubling these values, we have

$$0.10 < P\text{-value} < 0.20$$

Technology Step 3   Using StatCrunch, we find the P-value is 0.1627. See Figure 13.

Figure 13 **Hypothesis test results:**
μ_1 : Mean of Flight
μ_2 : Mean of Control
$\mu_1 - \mu_2$: Difference between two means
$H_0 : \mu_1 - \mu_2 = 0$
$H_A : \mu_1 - \mu_2 \neq 0$
(without pooled variances)

Difference	Sample Diff.	Std. Err.	DF	T-Stat	P-value
$\mu_1 - \mu_2$	-0.54928571	0.38230283	25.996352	-1.4367817	0.1627

Step 4   The P-value of 0.1627 [by hand: $0.10 < P\text{-value} < 0.20$] means that *if* the null hypothesis (the mean difference of red blood cell mass in the flight group and control group is zero) is true, we expect a sample difference at least as extreme as the one obtained in about 16 out of 100 repetitions of this experiment. The results obtained are not unusual if the null hypothesis is true. Simply put, because the P-value is greater than the level of significance ($0.1627 > 0.05$), we do not reject the null hypothesis.

Step 5   The sample data suggest that there is not sufficient evidence to conclude that the flight animals have a different red blood cell mass from the control animals at the $\alpha = 0.05$ level of significance.   •

The degrees of freedom used to determine the critical value(s) in the Classical Approach and by-hand P-value presented in Example 1 are conservative (this means we would require more evidence against the statement in the null hypothesis than if we use the formula for degrees of freedom given below). Results that are more accurate can be obtained by using the following formula for degrees of freedom:

$$\text{df} = \frac{\left(\dfrac{s_1^2}{n_1} + \dfrac{s_2^2}{n_2}\right)^2}{\dfrac{\left(\dfrac{s_1^2}{n_1}\right)^2}{n_1 - 1} + \dfrac{\left(\dfrac{s_2^2}{n_2}\right)^2}{n_2 - 1}} \tag{2}$$

When using Formula (2) to compute degrees of freedom, round down to the nearest integer to use Table VII. For by-hand inference, it is recommended that you use the smaller of $n_1 - 1$ or $n_2 - 1$ as the degrees of freedom to ease computation. However, for increased precision in determining the P-value, computer software will use Formula (2) when computing the degrees of freedom.

Notice that the degrees of freedom in the technology solution are 25.996352 versus 13 in the conservative solution done by hand in Example 1. With the lower degrees of freedom, the critical t is larger (2.160 with 13 degrees of freedom versus 2.056 with approximately 26 degrees of freedom). The larger critical value increases the number of standard deviations the difference in the sample means must be from the hypothesized

CAUTION!
The degrees of freedom in by-hand solutions will not equal the degrees of freedom in technology solutions unless you use Formula (2) to compute degrees of freedom.

• Now Work Problem 13

mean difference before the null hypothesis is rejected. Therefore, in using the smaller of $n_1 - 1$ or $n_2 - 1$ degrees of freedom, we need more substantial evidence to reject the null hypothesis. This requirement decreases the probability of a Type I error (rejecting the null hypothesis when the null hypothesis is true) below the actual level of α chosen by the researcher. This is what we mean when we say that the method of using the lesser of $n_1 - 1$ and $n_2 - 1$ as a proxy for degrees of freedom is conservative compared with using Formula (2).

❷ Construct and Interpret Confidence Intervals Regarding the Difference of Two Independent Means

Constructing a confidence interval for the difference of two means is an extension of the results presented in Section 9.2.

> ### Constructing a $(1 - \alpha) \cdot 100\%$ Confidence Interval for the Difference of Two Means
>
> A simple random sample of size n_1 is taken from a population with unknown mean μ_1 and unknown standard deviation σ_1. Also, a simple random sample of size n_2 is taken from a population with unknown mean μ_2 and unknown standard deviation σ_2. If the two populations are normally distributed or the sample sizes are sufficiently large ($n_1 \geq 30$ and $n_2 \geq 30$), a $(1 - \alpha) \cdot 100\%$ confidence interval about $\mu_1 - \mu_2$ is given by
>
> $$\text{Lower bound:} \quad (\bar{x}_1 - \bar{x}_2) - t_{\frac{\alpha}{2}} \cdot \sqrt{\frac{s_1^2}{n_1} + \frac{s_2^2}{n_2}}$$
>
> and $\qquad\qquad\qquad\qquad\qquad\qquad\qquad\qquad\qquad\qquad\qquad\qquad$ **(3)**
>
> $$\text{Upper bound:} \quad (\bar{x}_1 - \bar{x}_2) + t_{\frac{\alpha}{2}} \cdot \sqrt{\frac{s_1^2}{n_1} + \frac{s_2^2}{n_2}}$$
>
> where $t_{\frac{\alpha}{2}}$ is computed using the smaller of $n_1 - 1$ or $n_2 - 1$ degrees of freedom or Formula (2).

EXAMPLE 2 Constructing a Confidence Interval for the Difference of Two Means

Problem Construct a 95% confidence interval for $\mu_1 - \mu_2$ using the data presented in Table 3 on page 549.

Approach The normal probability plots and boxplot (Figure 10) indicate that the data are approximately normal with no outliers. Construct the confidence interval with $\alpha = 0.05$ using Formula (3) or using technology.

By-Hand Solution
We have already found the sample statistics in Example 1. In addition, we found $t_{\frac{\alpha}{2}} = t_{0.025}$ with 13 degrees of freedom to be 2.160. Substituting into Formula (3), we obtain the following results:

Lower bound: $(\bar{x}_1 - \bar{x}_2) - t_{\frac{\alpha}{2}} \cdot \sqrt{\dfrac{s_1^2}{n_1} + \dfrac{s_2^2}{n_2}}$

$= (7.881 - 8.430) - 2.160 \cdot \sqrt{\dfrac{1.017^2}{14} + \dfrac{1.005^2}{14}}$

$= -0.549 - 0.825$

$= -1.374$

Technology Solution Figure 14 shows the results from Minitab.

Figure 14

```
Two-Sample T-Test and CI: Flight, Control

Two-sample T for Flight vs Control

          N   Mean  StDev  SE Mean
Flight   14   7.88   1.02     0.27
Control  14   8.43   1.01     0.27

Difference = μ (Flight) - μ (Control)
Estimate for difference:  -0.549
95% CI for difference:  (-1.337, 0.238)
T-Test of difference = 0 (vs ≠): T-Value = -1.44  P-Value = 0.163  DF = 25
```

Upper bound: $(\bar{x}_1 - \bar{x}_2) + t_{\frac{\alpha}{2}} \cdot \sqrt{\dfrac{s_1^2}{n_1} + \dfrac{s_2^2}{n_2}}$

$= (7.881 - 8.430) + 2.160 \cdot \sqrt{\dfrac{1.017^2}{14} + \dfrac{1.005^2}{14}}$

$= -0.549 + 0.825$

$= 0.276$

Interpretation We are 95% confident the mean difference between the red blood cell mass of the flight animals and control animals is between -1.374 [Tech: -1.337] mL and 0.276 [Tech: 0.238] mL. The confidence interval contains zero, so there is not sufficient evidence to conclude there is a difference in the red blood cell mass of the flight group and the control group. That is, space flight does not seem to affect red blood cell mass significantly.

● Now Work Problem 19

> **CAUTION!**
> We would use the pooled two-sample t-test when the two samples come from populations that have the same variance. *Pooling* refers to finding a weighted average of the two sample variances from the independent samples. It is difficult to verify that two population variances might be equal based on sample data, so we will always use Welch's t when comparing two means.

What about the Pooled Two-Sample t-Tests?

Perhaps you have noticed that statistical software and graphing calculators with advanced statistical features provide an option for two types of two-sample t-tests: one that assumes equal population variances (pooling) and one that does not assume equal population variances. Welch's t-statistic does not assume that the population variances are equal and can be used whether the population variances are equal or not. The test that assumes equal population variances is referred to as the *pooled t-statistic*.

The **pooled t-statistic** is computed by finding a weighted average of the sample variances and uses this average in the computation of the test statistic. The advantage of this test statistic is that it exactly follows Student's t-distribution with $n_1 + n_2 - 2$ degrees of freedom.

The disadvantage of the test statistic is that it requires that the population variances be equal. How is this requirement to be verified? While a test for determining the equality of variances does exist (F-test, Section 11.4), the test *requires* that each population be normally distributed. However, the F-test is not robust. Any minor departures from normality will make the results of the F-test unreliable. It has been recommended by many statisticians[*] that a preliminary F-test to check the requirement of equality of variance not be performed. In fact, George Box once said, "To make preliminary tests on variances is rather like putting to sea in a rowing boat to find out whether conditions are sufficiently calm for an ocean liner to leave port!"

Because the formal F-test for testing the equality of variances is so volatile, we are content to use Welch's t. Welch's t-test is more conservative than the pooled t. The price that must be paid for the conservative approach is that the probability of a Type II error is higher with Welch's t than with the pooled t when the population variances are equal. However, the two tests typically provide the same conclusion, even if the assumption of equal population standard deviations seems reasonable.

Technology Step-by-Step   Two-Sample t-Tests, Independent Sampling

TI-83/84 Plus

Hypothesis Tests

1. If necessary, enter raw data in L1 and L2.
2. Press STAT, highlight TESTS, and select
   `4:2-SampTTest ....`
3. If the data are raw, highlight Data, making sure that List1 is set to L1 and List2 is set to L2, with frequencies set to 1. If summary statistics are known, highlight STATS and enter the summary statistics.

4. Highlight the appropriate relation between μ_1 and μ_2 in the alternative hypothesis. Set Pooled to No.
5. Highlight Calculate or Draw and press ENTER. Calculate gives the test statistic and P-value. Draw will draw the t-distribution with the P-value shaded.

Confidence Intervals

Follow the steps given for hypothesis tests, except select Q: `2-SampTInt`. Also, select a confidence level (such as 95% = 0.95).

----

*Moser and Stevens, "Homogeneity of Variance in the Two-Sample Means Test." *American Statistician* 46(1).

Minitab

1. If necessary, enter raw data in columns C1 and C2.
2. Select the **Stat** menu, highlight **Basic Statistics**, then highlight **2-Sample t . . . .**
3. If you have raw data, select "Each sample is in its own column." Enter C1 in the cell marked "Sample 1" and enter C2 in the cell marked "Sample 2:." If you have summarized data, select "Summarized data." Enter the summary statistics in the appropriate cell. Click Options . . . . Select the direction of the alternative hypothesis, enter the "test difference" (usually zero), and select a confidence level. Click OK twice.

Excel

1. Enter the raw data into Columns A and B.
2. Select the Formulas menu. Select More Functions. Highlight Statistical, and select T.TEST from the drop-down menu.
3. Place the cursor in Array1. Highlight the data in Column A. Place the cursor in Array2. Highlight the data in

Column B. Place the cursor in the Tails cell. Enter the number corresponding to the test you desire (1 for a one-tailed distribution; 2 for a two-tailed distribution). Place the cursor in the Type cell. Enter 3 for a two-sample unequal variance test.

StatCrunch

1. If necessary, enter the raw data into the first two columns of the spreadsheet. Name the column variables.
2. Select **Stat**, highlight **T Stats**, select **Two Sample**, and then choose either **With Data** or **With Summary**.
3. Select the column that contains the data for Sample 1. Select the column that contains the data for Sample 2. Note that the differences are computed Sample 1− Sample 2. Uncheck the box "Pool variances." If you choose the hypothesis test radio button, enter the value of the mean stated in the null hypothesis and choose the direction of the alternative hypothesis from the pull-down menu. If you choose the confidence interval radio button, enter the level of confidence. Click Compute!.

 11.3 Assess Your Understanding

Skill Building*

In Problems 1–6, assume that the populations are normally distributed.

1. (a) Test whether $\mu_1 \neq \mu_2$ at the $\alpha = 0.05$ level of significance for the given sample data.
(b) Construct a 95% confidence interval about $\mu_1 - \mu_2$.

	Sample 1	Sample 2
n	15	15
$\bar{x}$	15.3	14.2
s	3.2	3.5

2. (a) Test whether $\mu_1 \neq \mu_2$ at the $\alpha = 0.05$ level of significance for the given sample data.
(b) Construct a 95% confidence interval about $\mu_1 - \mu_2$.

	Sample 1	Sample 2
n	20	20
$\bar{x}$	111	104
s	8.6	9.2

3. (a) Test whether $\mu_1 > \mu_2$ at the $\alpha = 0.1$ level of significance for the given sample data.
(b) Construct a 90% confidence interval about $\mu_1 - \mu_2$.

	Sample 1	Sample 2
n	25	18
$\bar{x}$	50.2	42.0
s	6.4	9.9

4. (a) Test whether $\mu_1 < \mu_2$ at the $\alpha = 0.05$ level of significance for the given sample data.
(b) Construct a 95% confidence interval about $\mu_1 - \mu_2$.

	Sample 1	Sample 2
n	40	32
$\bar{x}$	94.2	115.2
s	15.9	23.0

5. Test whether $\mu_1 < \mu_2$ at the $\alpha = 0.02$ level of significance for the given sample data.

	Sample 1	Sample 2
n	32	25
$\bar{x}$	103.4	114.2
s	12.3	13.2

6. Test whether $\mu_1 > \mu_2$ at the $\alpha = 0.05$ level of significance for the given sample data.

	Sample 1	Sample 2
n	23	13
$\bar{x}$	43.1	41.0
s	4.5	5.1

Applying the Concepts

7. Elapsed Time to Earn a Bachelor's Degree A researcher with the Department of Education followed a cohort of students who graduated from high school in 1992, monitoring the progress the students made toward completing a bachelor's degree. One aspect of his research was to determine whether students who first attended community college took longer to attain a bachelor's degree than those who immediately attended and remained at a 4-year institution. The data in the table summarize the results of his study.

*The by-hand confidence intervals in the back of the text were computed using the smaller of $n_1 - 1$ or $n_2 - 1$ degrees of freedom.

	Community College to Four-Year Transfer	No Transfer
n	268	1145
Sample mean time to graduate, in years	5.43	4.43
Sample standard deviation time to graduate, in years	1.162	1.015

Source: Clifford Adelman, *The Toolbox Revisited.* United States Department of Education, 2006.

(a) What is the response variable in this study? What is the explanatory variable?

(b) Explain why this study can be analyzed using the methods of this section.

(c) Does the evidence suggest that community college transfer students take longer to attain a bachelor's degree? Use an $\alpha = 0.01$ level of significance.

(d) Construct a 95% confidence interval for $\mu_{\text{community college}} - \mu_{\text{no transfer}}$ to approximate the mean additional time it takes to complete a bachelor's degree if you begin in community college.

(e) Do the results of parts (c) and (d) imply that community college causes you to take extra time to earn a bachelor's degree? Cite some reasons that you think might contribute to the extra time to graduate.

8. Sugary Beverages It has been reported that consumption of sodas and other sugar-sweetened beverages cause excessive weight gain. Researchers conducted a randomized study in which 224 overweight and obese adolescents who regularly consumed sugar-sweetened beverages were randomly assigned to experimental and control groups. The experimental groups received a one-year intervention designed to decrease consumption of sugar-sweetened beverages, with follow-up for an additional year without intervention. The response variable in the study was body mass index (BMI—the weight in kilograms divided by the square of the height in meters). Results of the study appear in the following table. *Source:* Cara B. Ebbeling, PhD and associates, "A Randomized Trial of Sugar-Sweetened Beverages and Adolescent Body Weight" *N Engl J Med* 2012;367:1407–16. DOI: 10.1056/NEJMoal2Q3388

	Experimental Group ($n = 110$)	Control Group ($n = 114$)
Start of Study	Mean BMI = 30.4 Standard Deviation BMI = 5.2	Mean BMI = 30.1 Standard Deviation BMI = 4.7
After One Year	Mean Change in BMI = 0.06 Standard Deviation Change in BMI = 0.20	Mean Change in BMI = 0.63 Standard Deviation Change in BMI = 0.20
After Two Years	Mean Change in BMI = 0.71 Standard Deviation Change in BMI = 0.28	Mean Change in BMI = 1.00 Standard Deviation Change in BMI = 0.28

(a) What type of experimental design is this?

(b) What is the response variable? What is the explanatory variable?

(c) One aspect of statistical studies is to verify that the subjects in the various treatment groups are similar. Does the sample evidence support the belief that the BMIs of the subjects in the experimental group is not different from the BMIs in the control group at the start of the study? Use an $\alpha = 0.05$ level of significance.

(d) One goal of the research was to determine if the change in BMI for the experimental group was less than that for the control group after one year. Conduct the appropriate test to see if the evidence suggests this goal was met. Use an $\alpha = 0.05$ level of significance. What does this result suggest?

(e) Does the sample evidence suggest the change in BMI is less for the experimental group than the control group after two years? Use an $\alpha = 0.05$ level of significance. What does this result suggest?

(f) To what population do the results of this study apply?

9. Walking in the Airport, Part I Do people walk faster in the airport when they are departing (getting on a plane) or when they are arriving (getting off a plane)? Researcher Seth B. Young measured the walking speed of travelers in San Francisco International Airport and Cleveland Hopkins International Airport. His findings are summarized in the table.

Direction of Travel	Departure	Arrival
Mean speed (feet per minute)	260	269
Standard deviation (feet per minute)	53	34
Sample size	35	35

Source: Seth B. Young. "Evaluation of Pedestrian Walking Speeds in Airport Terminals." *Transportation Research Record.* Paper 99-0824.

(a) Is this an observational study or a designed experiment? Why?

(b) Explain why it is reasonable to use Welch's *t*-test.

(c) Do individuals walk at different speeds depending on whether they are departing or arriving at the $\alpha = 0.05$ level of significance?

10. Walking in the Airport, Part II Do business travelers walk at a different pace than leisure travelers? Researcher Seth B. Young measured the walking speed of business and leisure travelers in San Francisco International Airport and Cleveland Hopkins International Airport. His findings are summarized in the table.

Type of Traveler	Business	Leisure
Mean speed (feet per minute)	272	261
Standard deviation (feet per minute)	43	47
Sample size	20	20

Source: Seth B. Young. "Evaluation of Pedestrian Walking Speeds in Airport Terminals." *Transportation Research Record*, Paper 99-0824.

(a) Is this an observational study or a designed experiment? Why?

(b) What must be true regarding the populations to use Welch's *t*-test to compare the means?

(c) Assuming that the requirements listed in part (b) are satisfied, determine whether business travelers walk at a different speed from leisure travelers at the $\alpha = 0.05$ level of significance.

11. Priming Two Dutch researchers conducted a study in which two groups of students were asked to answer 42 questions from Trivial Pursuit. The students in group 1 were asked to spend 5 minutes thinking about what it would mean to be a professor, while the students in group 2 were asked to think about soccer hooligans. The 200 students in group 1 had a mean score of 23.4 with a standard deviation of 4.1, while the 200 students in group 2

had a mean score of 17.9 with a standard deviation of 3.9.
Source: Based on the research of Dijksterhuis, Ap, and Ad van Knippenberg. "The relation between perception and behavior, or how to win a game of Trivial Pursuit." *Journal of Personality and Social Psychology* 74.4 (1998): 865+. *Academic OneFile.* Web. 5 July 2010.

(a) Determine the 95% confidence interval for the difference in scores, $\mu_1 - \mu_2$. Interpret the interval.

(b) What does this say about priming?

12. Government Waste In a Gallup poll 513 national adults aged 18 years or older who consider themselves to be Republican were asked, "Of every tax dollar that goes to the federal government in Washington, D.C., how many cents of each dollar would you say are wasted?" The mean wasted was found to be 54 cents with a standard deviation of 2.9 cents. The same question was asked of 513 national adults aged 18 years or older who consider themselves to be Democrat. The mean wasted was found to be 41 cents with a standard deviation of 2.6 cents. Construct a 95% confidence interval for the mean difference in government waste, $\mu_R - \mu_D$. Interpret the interval.

NW **13. Ramp Metering** Ramp metering is a traffic engineering idea that requires cars entering a freeway to stop for a certain period of time before joining the traffic flow. The theory is that ramp metering controls the number of cars on the freeway and the number of cars accessing the freeway, resulting in a freer flow of cars, which ultimately results in faster travel times. To test whether ramp metering is effective in reducing travel times, engineers in Minneapolis, Minnesota, conducted an experiment in which a section of freeway had ramp meters installed on the on-ramps. The response variable for the study was speed of the vehicles. A random sample of 15 cars on the highway for a Monday at 6 P.M. with the ramp meters on and a second random sample of 15 cars on a different Monday at 6 P.M. with the meters off resulted in the following speeds (in miles per hour).

Ramp Meters On			Ramp Meters Off		
28	48	56	24	26	42
38	31	25	34	37	31
43	46	50	47	38	17
35	55	40	29	23	40
42	26	47	37	52	41

(a) Draw side-by-side boxplots of each data set. Does there appear to be a difference in the speeds? Are there any outliers?

(b) Are the ramp meters effective in maintaining a higher speed on the freeway? Use the $\alpha = 0.10$ level of significance.

Note: Normal probability plots indicate the data could come from a population that is normally distributed.

14. Measuring Reaction Time Researchers wanted to determine whether the reaction time (in seconds) of males differed from that of females to a go/no go stimulus. The researchers randomly selected 20 females and 15 males to participate in the study. The go/no go stimulus required the student to respond to a particular stimulus and not to respond to other stimuli. The results are as follows:

Female Students				Male Students		
0.588	0.652	0.442	0.293	0.375	0.256	0.427
0.340	0.636	0.391	0.367	0.654	0.563	0.405
0.377	0.646	0.403	0.377	0.374	0.465	0.402
0.380	0.403	0.617	0.434	0.373	0.488	0.337
0.443	0.481	0.613	0.274	0.224	0.477	0.655

Source: PsychExperiments at the University of Mississippi

(a) Is it reasonable to use Welch's *t*-test? Why?
Note: Normal probability plots indicate that the data are approximately normal and boxplots indicate that there are no outliers.

(b) Test whether there is a difference in the reaction times of males and females at the $\alpha = 0.05$ level of significance.

(c) Draw boxplots of each data set using the same scale. Does this visual evidence support the results obtained in part (b)?

15. Bacteria in Hospital Carpeting Researchers wanted to determine if carpeted rooms contained more bacteria than uncarpeted rooms. To determine the amount of bacteria in a room, researchers pumped the air from the room over a Petri dish at the rate of 1 cubic foot per minute for eight carpeted rooms and eight uncarpeted rooms. Colonies of bacteria were allowed to form in the 16 Petri dishes. The results are presented in the table. A normal probability plot and boxplot indicate that the data are approximately normally distributed with no outliers. Do carpeted rooms have more bacteria than uncarpeted rooms at the $\alpha = 0.05$ level of significance?

Carpeted Rooms (bacteria/cubic foot)				Uncarpeted Rooms (bacteria/cubic foot)			
11.8	10.8	7.1	14.6	12.1	12.0	3.8	10.1
8.2	10.1	13.0	14.0	8.3	11.1	7.2	13.7

Source: William G. Walter and Angie Stober. "Microbial Air Sampling in a Carpeted Hospital." *Journal of Environmental Health,* 30 (1968), p. 405.

16. Visual versus Textual Learners Researchers wanted to know whether there was a difference in comprehension among students learning a computer program based on the style of the text. They randomly divided 36 students of similar educational level, age, and so on, into two groups of 18 each. Group 1 individuals learned the software using a visual manual (*multimodal instruction*), while group 2 individuals learned the software using a textual manual (*unimodal instruction*). The following data represent scores that the students received on an exam given to them after they studied from the manuals.

Visual Manual		Textual Manual	
51.08	60.35	64.55	56.54
57.03	76.60	57.60	39.91
44.85	70.77	68.59	65.31
75.21	70.15	50.75	51.95
56.87	47.60	49.63	49.07
75.28	46.59	43.58	48.83
57.07	81.23	57.40	72.40
80.30	67.30	49.48	42.01
52.20	60.82	49.57	61.16

Source: Mark Gellevij et al. "Multimodal versus Unimodal Instruction in a Complex Learning Context." *Journal of Experimental Education* 70(3):215–239, 2002.

(a) What type of experimental design is this?

(b) What are the treatments?

(c) A normal probability plot and boxplot indicate it is reasonable to use Welch's *t*-test. Is there a difference in test scores at the $\alpha = 0.05$ level of significance?

17. Rates of Returns of Stocks Stocks may be categorized by sectors. Go to www.pearsonhighered.com/sullivanstats to obtain the data file 11_3_17 using the file format of your choice for the version of the text you are using. The data represent the one-year rate of return (in percent) for a sample of consumer cyclical stocks and industrial stocks for the period December, 2013, through November, 2014. **Note:** Consumer cyclical stocks include names such as Starbucks and Home Depot. Industrial stocks include names such as 3M and FedEx.

(a) Draw side-by-side boxplots of one-year rate of return by sector. Does there appear to be a difference in the one-year rate of return for these two sectors?

(b) Explain why the methods of this section may be used to test whether the mean rate of return for the two sectors differ.

(c) Test whether the mean one-year rate of return for consumer cyclical stocks is different from that of industrial stocks at the $\alpha = 0.05$ level of significance.

(d) Construct a 95% confidence interval for the mean difference in rate of return between industrial stocks and consumer cyclical stocks. Interpret the interval.

18. Tax Rates Do women feel differently from men when it comes to federal tax rates? One question on the Sullivan Statistics Survey II was, "What percent of income do you believe individuals should pay in federal income tax?" Results of the survey may be found at www.pearsonhighered.com/sullivanstats. Select the data file SullivanSurveyII using the file format of your choice for the version of the text you are using. The Tax Rate column contains the response.

(a) Draw side-by-side boxplots of tax rates by gender. Does there appear to be a difference in the tax rates for the genders?

(b) Explain why the methods of this section may be used to test whether the mean tax rates for the two genders differ.

(c) Test whether the mean tax rate for females differs from that of males at the $\alpha = 0.05$ level of significance.

NW 19. Kids and Leisure Young children require a lot of time and this time commitment cuts into a parent's leisure time. A sociologist wanted to estimate the difference in the amount of daily leisure time (in hours) of adults who do not have children under the age of 18 years and adults who have children under the age of 18 years. A random sample of 40 adults with no children under the age of 18 years results in a mean daily leisure time of 5.62 hours, with a standard deviation of 2.43 hours. A random sample of 40 adults with children under the age of 18 years results in a mean daily leisure time of 4.10 hours, with a standard deviation of 1.82 hours. Construct and interpret a 90% confidence interval for the mean difference in leisure time between adults with no children and adults with children. *Source: American Time Use Survey*

20. Aluminum Bottles The aluminum bottle, first introduced in 1991 by CCL Container for mainly personal and household items such as lotions, has become popular with beverage manufacturers. Besides being lightweight and requiring less packaging, the aluminum bottle is reported to cool faster and stay cold longer than typical glass bottles. A small brewery tests this claim and obtains the following information regarding the time (in minutes) required to chill a bottle of beer from room temperature (75°F) to serving temperature (45°F). Construct and interpret a 90% confidence interval for the mean difference in cooling time for clear glass versus aluminum.

	Clear Glass	Aluminum
Sample size	42	35
Mean time to chill (minutes)	133.8	92.4
Sample standard deviation (minutes)	9.9	7.3

21. Comparing Step Pulses A physical therapist wanted to know whether the mean step pulse of men was less than the mean step pulse of women. She randomly selected 51 men and 70 women to participate in the study. Each subject was required to step up and down onto a 6-inch platform for 3 minutes. The pulse of each subject (in beats per minute) was then recorded. After the data were entered into Minitab, the following results were obtained

Two Sample T-Test and Confidence Interval

```
Two sample T for Men vs Women

         N     Mean    StDev    SE Mean
Men     51    112.3    11.3     1.6
Women   70    118.3    14.2     1.7

95% CI for mu Men − mu Women: (−10.7, −1.5)
T-Test mu Men = mu Women (vs <):T = −2.61 P = 0.0051 DF = 118
```

(a) State the null and alternative hypotheses.

(b) Identify the *P*-value and state the researcher's conclusion if the level of significance was $\alpha = 0.01$.

(c) What is the 95% confidence interval for the mean difference in pulse rates of men versus women? Interpret this interval.

22. Comparing Flexibility A physical therapist believes that women are more flexible than men. She measures the flexibility of 31 randomly selected women and 45 randomly selected men by determining the number of inches subjects could reach while sitting on the floor with their legs straight out and back perpendicular to the ground. The more flexible an individual is, the higher the measured flexibility will be. After entering the data into Minitab, she obtained the following results:

Two Sample T-Test and Confidence Interval

```
Two sample T for Men vs Women

         N     Mean    StDev    SE Mean
Men     45    18.64    3.29     0.49
Women   31    20.99    2.07     0.37

95% CI for mu Men − mu Women: (−3.58, −1.12)
T-Test mu Men = mu Women (vs <):T = −3.82 P = 0.0001 DF = 73
```

(a) State the null and alternative hypotheses.

(b) Identify the *P*-value and state the researcher's conclusion if the level of significance was $\alpha = 0.01$.

(c) What is the 95% confidence interval for the mean difference in flexibility of men versus women? Interpret this interval.

23. Putting It Together: Online Homework Professor Stephen Zuro of Joliet Junior College wanted to determine whether an online homework system (meaning students did homework online and received instant feedback with helpful guidance about their answers) improved scores on a final exam. In the fall semester, he taught a precalculus class using an online homework system. In the spring semester, he taught a precalculus class without the homework system (which meant students were responsible for doing their homework the old-fashioned way—paper and pencil). Professor Zuro made sure to teach the two courses identically (same text, syllabus, tests, meeting time, meeting location, and so on).

The table summarizes the results of the two classes on their final exam.

	Fall Semester	**Spring Semester**
Number of students	27	25
Mean final exam score	73.6	67.9
Standard deviation final exam score	10.3	12.4

(a) What type of experimental design is this?

(b) What is the response variable? What are the treatments in the study?

(c) What factors are controlled in the experiment?

(d) In many experiments, the researcher will recruit volunteers and randomly assign the individuals to a treatment group. In what regard was this done for this experiment?

(e) Did the students perform better on the final exam in the fall semester? Use an $\alpha = 0.05$ level of significance.

(f) Can you think of any factors that may confound the results? Could Professor Zuro have done anything about these confounding factors?

Retain Your Knowledge

24. Graduation Rates PayScale is a company that reports statistics on colleges and universities. Go to www.pearsonhighered.com/sullivanstats to obtain the data file 11_3_24 using the file format of your choice for the version of the text you are using. The data contain the four-year cost and graduation rate for over 1300 colleges and universities. Do schools that charge more have higher graduation rates? The variable "2013 Cost" represents the four-year cost of attending the college or university. The variable "Grad Rate" represents the percentage of incoming freshman who graduate within six years.

(a) Draw a scatter diagram treating "2013 Cost" as the explanatory variable and "Grad Rate" as the response variable.

(b) Determine the correlation coefficient between "2013 Cost" and "Grad Rate."

(c) Is there a linear relation between "2013 Cost" and "Grad Rate"?

(d) Find the least-squares regression line.

(e) Is the graduation rate of Harvey Mudd College higher than would be expected among all schools that charge $229,500? Explain,

(f) What proportion of the variability in graduation rates is explain by the 2013 cost of attending?

Explaining the Concepts

25. College Skills The Collegiate Learning Assessment Plus (CLA+) is an exam that is meant to assess the intellectual gains made between one's freshman and senior year of college. The exam, graded on a scale of 400 to 1600, assesses critical thinking, analytical reasoning, document literacy, writing, and communication. The exam was administered to 135 freshman in Fall 2012 at California State University Long Beach (CSULB). The mean score on the exam was 1191 with a standard deviation of 187. The exam was also administered to graduating seniors of CSULB in Spring 2013. The mean score was 1252 with a standard deviation of 182 . Explain the type of analysis that could be applied to these data to assess whether CLA+ scores increase while at CSULB. Explain the shortcomings in the data available and provide a better data collection technique.

26. MythBusters In a MythBusters episode, the question was asked, "Which is better? A four-way stop or a roundabout?" "Better" was determined based on determining the number of vehicles that travel through the four-way stop over a 5-minute interval of time. Suppose the folks at MythBusters conducted this experiment 15 times for each intersection design.

(a) What is the variable of interest in this study? Is it qualitative or quantitative?

(b) Explain why the data might be analyzed by comparing two independent means. Include in this explanation what the null and alternative hypotheses would be and what each mean represents.

(c) A potential improvement on the experimental design might be to identify 15 groups of different drivers and ask each group to drive through each intersection design for the 5- minute time interval. Explain why this is a better design and explain the role randomization would play. What would be the null and alternative hypotheses here and how would the mean be computed?

27. Explain why using the smaller of $n_1 - 1$ or $n_2 - 1$ degrees of freedom to determine the critical t instead of Formula (2) is conservative.

11.4 Inference about Two Population Standard Deviations

Objectives ❶ Find critical values of the F-distribution

❷ Test hypotheses regarding two population standard deviations

❶ Find Critical Values of the F-distribution

In this section, we discuss methods for comparing two population standard deviations (or variances). For example, we might be interested in determining whether the rate of return for Cisco Systems stock is more volatile than General Electric (GE) stock. If Cisco Systems stock is more volatile than GE, the standard deviation rate of return on Cisco Systems would be higher than the standard deviation rate of return on GE. To test hypotheses regarding population standard deviations, we use Fisher's **F-distribution**, named in honor of Sir Ronald A. Fisher. Certain requirements must be satisfied to test hypotheses regarding two population standard deviations.

Requirements for Testing Hypotheses Regarding Two Population Standard Deviations

1. The samples are independent simple random samples or from a completely randomized design.
2. The populations from which the samples are drawn are normally distributed.

The second requirement is critical. The procedures introduced in this section are **not robust**, so any departures from normality will adversely affect the results of the test and make them unreliable. One reason for this is that the standard deviation is not resistant to extreme values. Populations that are skewed will have extreme values that inflate the value of the standard deviation. Therefore, whenever performing inference regarding two population standard deviations, we must verify the requirement of the normality of the population using normal probability plots.

> **CAUTION!**
> If the populations from which the samples are drawn are not normal, do not use the inferential procedures discussed in this section.

We use the following notation when describing the two populations.

Notation Used When Comparing Two Population Standard Deviations

σ_1^2: Variance for population 1 σ_2^2: Variance for population 2

s_1^2: Sample variance for population 1 s_2^2: Sample variance for population 2

n_1: Sample size for population 1 n_2: Sample size for population 2

Before we can perform statistical inference regarding two population standard deviations, we need to know the sampling distribution of the test statistic. In comparing two population standard deviations, the test statistic follows Fisher's F-distribution.

Fisher's F-distribution

If $\sigma_1^2 = \sigma_2^2$ and s_1^2 and s_2^2 are sample variances from independent simple random samples of size n_1 and n_2, respectively, drawn from normal populations, then

$$F = \frac{s_1^2}{s_2^2}$$

follows the F-distribution with $n_1 - 1$ degrees of freedom in the numerator and $n_2 - 1$ degrees of freedom in the denominator.

We can find critical values of the F-distribution using Table IX in Appendix A. Before discussing how to read Table IX, we present characteristics of the F-distribution.

Figure 15
F-distributions

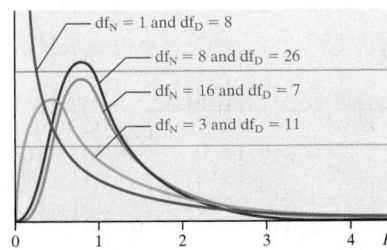

$df_N = 1$ and $df_D = 8$
$df_N = 8$ and $df_D = 26$
$df_N = 16$ and $df_D = 7$
$df_N = 3$ and $df_D = 11$

Characteristics of the F-distribution

1. The F-distribution is skewed right.
2. The shape of the F-distribution depends on the degrees of freedom in the numerator and denominator. See Figure 15. This is similar to the χ^2 distribution and Student's t-distribution, whose shapes depend on their degrees of freedom.
3. The total area under the curve is 1.
4. The values of F are always greater than or equal to zero.

Table IX is structured differently from those for the t- or χ^2-distributions. The row across the top provides the degrees of freedom in the numerator, while the column on the left provides the degrees of freedom in the denominator. Corresponding to the degrees of freedom in the numerator and denominator, we have various areas (0.1, 0.05, 0.025, 0.01, 0.001) in the right tail of the F-distribution. The body of the table provides the critical F values. We use the notation

$$F_{\alpha, n_1 - 1, n_2 - 1}$$

where $n_1 - 1$ is the degrees of freedom in the numerator, $n_2 - 1$ is the degrees of freedom in the denominator, and α is the area to the right of $F_{\alpha, n_1 - 1, n_2 - 1}$. To determine the critical value that has an area of α to the left, use the following property:

$$F_{1 - \alpha, n_1 - 1, n_2 - 1} = \frac{1}{F_{\alpha, n_2 - 1, n_1 - 1}}$$

So, to find the critical F that has an area of $\alpha = 0.05$ to the left with 12 degrees of freedom in the numerator and 20 degrees of freedom in the denominator, find the critical F that has an area of $\alpha = 0.05$ to the right with 20 degrees of freedom in the numerator and 12 degrees of freedom in the denominator and compute its reciprocal.

EXAMPLE 1 Finding Critical Values for the F-distribution

Problem Find the critical F-values
(a) for a right-tailed test with $\alpha = 0.05$, degrees of freedom in the numerator $= 10$, and degrees of freedom in the denominator $= 7$.
(b) for a two-tailed test with $\alpha = 0.05$, degrees of freedom in the numerator $= 15$, and degrees of freedom in the denominator $= 20$.

Approach Perform the following steps to obtain the critical values.
Step 1 Draw an F-distribution with the critical value(s) and area labeled.
Step 2 Use Table IX to find the critical value(s).

Solution
(a) **Step 1** Figure 16 shows the F-distribution with 10 degrees of freedom in the numerator and 7 degrees of freedom in the denominator. The area to the right of the unknown critical value is 0.05. We denote this critical value $F_{0.05, 10, 7}$.

Step 2 Figure 17 shows a partial representation of Table IX. We box the column corresponding to 10 degrees of freedom (the numerator degrees of freedom) and the row corresponding to 7 degrees of freedom (the denominator degrees of freedom) with $\alpha = 0.05$. The critical value is $F_{0.05, 10, 7} = 3.64$.

Figure 16

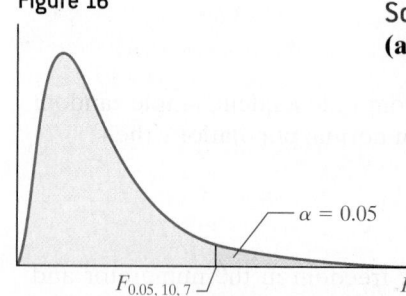

$\alpha = 0.05$

$F_{0.05, 10, 7}$ F

Figure 17

Area to the Right of Critical Value	Degrees of Freedom in the Numerator							
	9	10	15	20	30	60	120	1000
0.100	59.86	60.19	61.22	61.74	62.26	62.79	63.06	63.30
0.050	240.54	241.88	245.95	248.01	250.10	252.20	253.25	254.19
1 0.025	963.28	968.63	984.87	993.10	1001.4	1009.8	1014	1017.7
0.010	6022.5	6055.8	6157.3	6208.7	6260.6	6313	6339.4	6362.7
0.001	602284	605621	615764	620908	626099	631337	633972	636301
0.100	2.72	2.70	2.63	2.59	2.56	2.51	2.49	2.47
0.050	3.68	3.64	3.51	3.44	3.38	3.30	3.27	3.23
7 0.025	4.82	4.76	4.57	4.47	4.36	4.25	4.20	4.15
0.010	6.72	6.62	6.31	6.16	5.99	5.82	5.74	5.66
0.001	14.33	14.08	13.32	12.93	12.53	12.12	11.91	11.72

(b) *Step 1*   Figure 18 shows the F-distribution with 15 degrees of freedom in the numerator and 20 degrees of freedom in the denominator. The area to the right of the right critical value is $\frac{\alpha}{2} = 0.025$. We denote this critical value $F_{0.025,15,20}$. The area to the right of the left critical value is $1 - \frac{\alpha}{2} = 0.975$. We denote this critical value $F_{0.975,\,15,\,20}$.

Figure 18

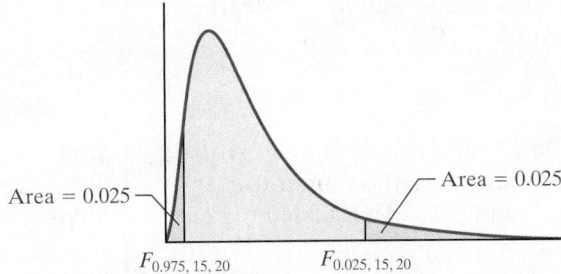

Area = 0.025

Area = 0.025

$F_{0.975,\,15,\,20}$ $F_{0.025,\,15,\,20}$

Step 2   Refer to Table IX. To find the right critical value, identify the column that represents 15 degrees of freedom (the numerator degrees of freedom) and the row that represents 20 degrees of freedom (the denominator degrees of freedom) with $\frac{\alpha}{2} = 0.025$. The right critical value is $F_{0.025,\,15,\,20} = 2.57$. To find the left critical value, use the fact that $F_{1-\alpha,\,n_1-1,\,n_2-1} = \dfrac{1}{F_{\alpha,\,n_2-1,\,n_1-1}}$. Therefore, use Table IX and find that $F_{0.025,20,15} = 2.76$. So the left critical value is $F_{0.975,15,20} = \dfrac{1}{F_{0.025,20,15}} = \dfrac{1}{2.76} = 0.36$.

In studying Table IX, notice that some values of the degrees of freedom for either the numerator or denominator are not in the table. If the number of degrees of freedom is not found in the table, follow the practice of choosing the degrees of freedom closest to that desired. If the degrees of freedom is exactly between two values, find the mean of the values. For example, to find the critical value corresponding to 35 degrees of freedom in the numerator, compute the mean of the critical values corresponding to 30 and 40 degrees of freedom in the numerator.

● Now Work Problem 1

② **Test Hypotheses Regarding Two Population Standard Deviations**

Now that we know the approximate sampling distribution of $\dfrac{s_1^2}{s_2^2}$ and how to find critical values in the F-distribution, we can introduce a procedure that can be used to test hypotheses regarding two population standard deviations (or variances).

Test Hypotheses Regarding Two Population Standard Deviations

To test hypotheses regarding two population standard deviations, σ_1 and σ_2, use the following steps, provided that

1. the samples are obtained using simple random sampling or through a randomized experiment.
2. the sample data are independent.
3. the populations from which the samples are drawn are normally distributed.

Step 1   Determine the null and alternative hypotheses. The hypotheses can be structured in one of three ways:

Two-Tailed	Left-Tailed	Right-Tailed
$H_0\colon \sigma_1 = \sigma_2$	$H_0\colon \sigma_1 = \sigma_2$	$H_0\colon \sigma_1 = \sigma_2$
$H_1\colon \sigma_1 \neq \sigma_2$	$H_1\colon \sigma_1 < \sigma_2$	$H_1\colon \sigma_1 > \sigma_2$

Note: σ_1 is the population standard deviation for population 1; σ_2 is the population standard deviation for population 2.

Step 2 Select a level of significance, α, depending on the seriousness of making a Type I error.

Step 3 Compute the test statistic

$$F_0 = \frac{s_1^2}{s_2^2}$$

which follows Fisher's F-distribution with $n_1 - 1$ degrees of freedom in the numerator and $n_2 - 1$ degrees of freedom in the denominator.

Classical Approach

Step 3 (continued) Use Table IX to determine the critical value(s) using $n_1 - 1$ degrees of freedom in the numerator and $n_2 - 1$ degrees of freedom in the denominator. The shaded regions represent the critical region.

	Two-Tailed	Left-Tailed	Right-Tailed
Critical value(s)	$F_{1-\alpha/2, n_1-1, n_2-1}$ and $F_{\alpha/2, n_1-1, n_2-1}$	$F_{1-\alpha, n_1-1, n_2-1}$	F_{α, n_1-1, n_2-1}
Critical region(s)			

Step 4 Compare the critical value with the test statistic.

Two-Tailed	Left-Tailed	Right-Tailed
If $F_0 < F_{1-\alpha/2, n_1-1, n_2-1}$ or $F_0 > F_{\alpha/2, n_1-1, n_2-1}$, reject the null hypothesis.	If $F_0 < F_{1-\alpha, n_1-1, n_2-1}$, reject the null hypothesis.	If $F_0 > F_{\alpha, n_1-1, n_2-1}$, reject the null hypothesis.

P-Value Approach

Step 3 (continued) Use technology to determine the P-value. See the Technology Step-by-Step on page 565.

Step 4 If P-value $< \alpha$, reject the null hypothesis.

Step 5 State the conclusion.

> **CAUTION!**
> The test for equality of population standard deviations is not robust. Thus, any departures from normality make the results of the inference useless.

Because the procedures just presented are **not robust**, minor departures from normality will adversely affect the results of the test. Therefore, the test should be used only when the requirement of normality has been verified. We will verify this requirement by constructing normal probability plots.

EXAMPLE 2   Testing Hypotheses Regarding Two Population Standard Deviations

Problem An investor believes that Cisco Systems is a more volatile stock than General Electric. The volatility of a stock is measured by the standard deviation rate of return on the stock. The data in Table 4 represent the monthly rate of return between 1990 and 2014 for 10 randomly selected months for Cisco Systems stock and 14 randomly selected months for General Electric stock. Does the evidence suggest that Cisco Systems stock is more volatile than General Electric stock at the $\alpha = 0.05$ level of significance?

Table 4							
Monthly Rate of Return for Cisco Systems Stock (%)				**Monthly Rate of Return for General Electric Stock (%)**			
1.93	31.11	−7.60	9.72	1.15	−1.92	−0.55	5.29
11.64	4.87	5.18	8.91	4.68	−2.32	−5.88	9.00
23.13	−5.56			3.23	−5.19	−3.26	−0.60
				0.98	4.80		

Source: Yahoo!Finance

Approach We want to know if the evidence suggests that Cisco Systems is more volatile than General Electric. Symbolically, this is represented as $\sigma_1 > \sigma_2$, where σ_1 is the standard deviation rate of return of Cisco Systems stock and σ_2 is the standard deviation rate of return of General Electric stock. We assume that there is no difference, or $\sigma_1 = \sigma_2$. Verify that both variables are normally distributed by constructing normal probability plots. Then follow Steps 1 through 5.

Solution Figure 19(a) shows the normal probability plot for Cisco Systems, while Figure 19(b) shows the normal probability plot for General Electric. The correlation between return on Cisco stock and expected z-scores is 0.971. Because $0.971 > 0.918$ (Table VI), we conclude the data come from a population that is normally distributed. The correlation between return on GE stock and expected z-scores is 0.989. Because $0.989 > 0.935$ (Table VI), we conclude the data come from a population that is normally distributed.

Figure 19

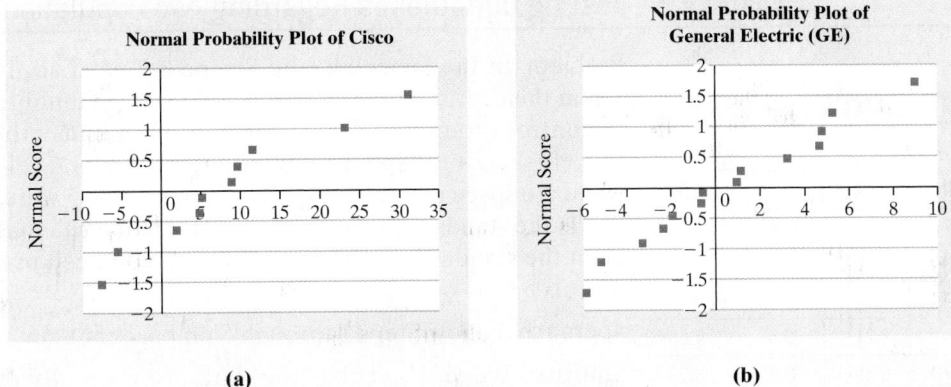

(a)   (b)

Step 1 The investor wants to know if Cisco Systems is a more volatile stock than General Electric. This is written $\sigma_1 > \sigma_2$. The null hypothesis is a statement of no difference, so we have

$$H_0: \sigma_1 = \sigma_2 \quad \text{versus} \quad H_1: \sigma_1 > \sigma_2$$

This is a right-tailed test.

Step 2 The level of significance is $\alpha = 0.05$.

Step 3 Compute the sample standard deviation of the rate of return for Cisco Systems, s_1, to be 11.84% and the sample standard deviation of the rate of return for General Electric, s_2, to be 4.32%. Because the data are normally distributed, the test statistic is

$$F_0 = \frac{s_1^2}{s_2^2} = \frac{11.84^2}{4.32^2} = 7.51$$

Classical Approach

Step 3 (continued) Because this is a right-tailed test, we determine the critical value at the $\alpha = 0.05$ level of significance with $n_1 - 1 = 10 - 1 = 9$ degrees of freedom in the numerator and $n_2 - 1 = 14 - 1 = 13$ degrees of freedom in the denominator and find it to be $F_{0.05, 9, 13} \approx 2.80$ (using 12 degrees of freedom in the denominator, because that is closest). The critical region is displayed in Figure 20.

Figure 20

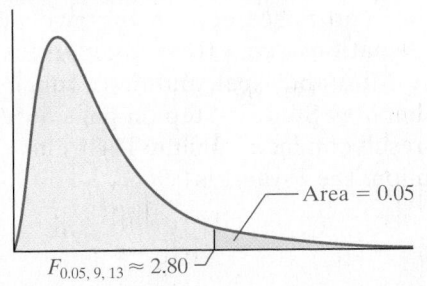

Area = 0.05

$F_{0.05, 9, 13} \approx 2.80$

P-Value Approach Using Technology

Step 3 (continued) We will use StatCrunch to obtain the P-value. The steps for testing hypotheses comparing two population standard deviations using the TI–83/84 Plus graphing calculators, Minitab, Excel, and StatCrunch are given in the Technology Step-by-Step on page 565. Figure 21 shows the results obtained from StatCrunch.

Figure 21

Hypothesis test results:
σ_1^2 : Variance of Cisco
σ_2^2 : Variance of GE
σ_1^2/σ_2^2 : Ratio of two variances
$H_0 : \sigma_1^2/\sigma_2^2 = 1$
$H_A : \sigma_1^2/\sigma_2^2 > 1$

Ratio	Num. DF	Den. DF	Sample Ratio	F-Stat	P-value
σ_1^2/σ_2^2	9	13	7.5192967	7.5192967	0.0007

(continued)

Step 4 Because the test statistic $F_0 = 7.51$ is greater than the critical value $F_{0.05,9,13} \approx 2.80$, we reject the null hypothesis.

Step 4 The P-value is 0.0007. If the statement in the null hypothesis were true, we would expect to get results at least as extreme as the results obtained in about 7 samples out of 10,000. Very unusual results, indeed! Because the P-value is less than the level of significance $\alpha = 0.05$, we reject the null hypothesis.

Step 5 There is sufficient evidence to conclude that Cisco Systems stock is more volatile than General Electric stock at the $\alpha = 0.05$ level of significance. Investors would use this result to demand a higher rate of return on investments in Cisco Systems.

● Now Work Problem 17

EXAMPLE 3 Testing Hypotheses Regarding Two Population Standard Deviations

Table 5			
Flight		**Control**	
8.59	8.64	8.65	6.99
6.87	7.89	7.62	7.44
7.00	8.80	7.33	8.58
6.39	7.54	7.14	9.14
7.43	7.21	8.40	9.66
9.79	6.85	8.55	8.70
9.30	8.03	9.88	9.94

Source: NASA, Life Sciences Data Archive

Problem In the Spacelab Life Sciences 2 payload, 14 male rats were sent to space. Upon their return, the red blood cell mass (in milliliters) of the rats was determined. A control group of 14 male rats was held under the same conditions (except for spaceflight) as the space rats, and their red blood cell mass was likewise determined when the space rats returned. The data in Table 5 were obtained.

Is the standard deviation of the red blood cell mass in the flight animals different from the standard deviation of the red blood cell mass in the control animals at the $\alpha = 0.05$ level of significance?

Approach Follow Steps 1 through 5 on pages 561–562.

Solution We verified that the data are normally distributed in Example 1 from Section 11.3.

Step 1 We want to know whether the standard deviation of the red blood cell mass in the flight animals is different from the standard deviation of the red blood cell mass in the control animals. Letting σ_1 represent the standard deviation of the red blood cell mass for the flight animals and σ_2 represent the standard deviation of the red blood cell mass for the control animals, this can be written $\sigma_1 \neq \sigma_2$. We have

$$H_0: \sigma_1 = \sigma_2 \quad \text{versus} \quad H_1: \sigma_1 \neq \sigma_2$$

This is a two-tailed test.

Step 2 The level of significance is $\alpha = 0.05$.

Step 3 In Example 1 from Section 11.3, we computed the sample standard deviation of the red blood cell mass for the experimental group, s_1, to be 1.017 and the sample standard deviation of the red blood cell mass for the control group, s_2, to be 1.005. Because the data are normally distributed, the test statistic is

$$F_0 = \frac{s_1^2}{s_2^2} = \frac{1.017^2}{1.005^2} = 1.024$$

Classical Approach

Step 3 (continued) Because this is a two-tailed test, we determine the critical value at the $\alpha = 0.05$ level of significance with $n_1 - 1 = 14 - 1 = 13$ degrees of freedom in the numerator and $n_2 - 1 = 14 - 1 = 13$ degrees of freedom in the denominator and find them to be $F_{0.025,13,13} \approx 3.28$ (using 12 degrees of freedom in the numerator and denominator, because that is closest) and

$$F_{0.975,13,13} = \frac{1}{F_{0.025,13,13}} = \frac{1}{3.28} = 0.30$$

P-Value Approach Using Technology

Step 3 (continued) We will use a TI-84 Plus C graphing calculator to obtain the P-value. The steps for testing hypotheses comparing two population standard deviations using the TI-83/84 Plus graphing calculators, Minitab, Excel, and StatCrunch are given in the Technology Step-by-Step on page 565. Figure 23 shows the result obtained from the TI-84 Plus C using the DRAW option. The P-value is 0.9666.

The critical regions are displayed in Figure 22.

Figure 22

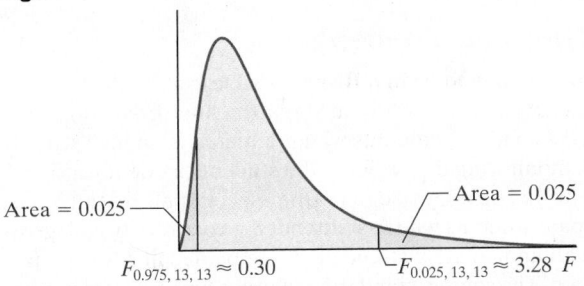

Area = 0.025

$F_{0.975, 13, 13} \approx 0.30$

Area = 0.025

$F_{0.025, 13, 13} \approx 3.28$ F

Step 4 Because the test statistic $F_0 = 1.024$ lies between the critical values 0.30 and 3.28, we do not reject the null hypothesis.

Figure 23

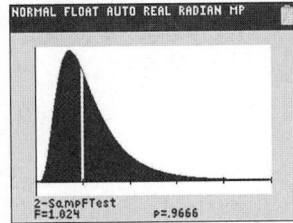

Step 4 Because the P-value is greater than the level of significance $\alpha = 0.05$, we do not reject the null hypothesis.

Step 5 There is not sufficient evidence to conclude that the standard deviation of the red blood cell mass in the experimental group is significantly different from the red blood cell mass in the control group at the $\alpha = 0.05$ level of significance. •

Technology Step-by-Step Comparing Two Population Standard Deviations

TI–83/84 Plus

1. If necessary, enter raw data in L1 and L2.
2. Press STAT, highlight TESTS, and select D:2-SampleFTest.
3. If the data are raw, highlight DATA and make sure List1 is set to L1 and List2 is set to L2, with frequencies set to 1. If summary statistics are known, highlight STATS and enter the summary statistics.
4. Highlight the appropriate relation between σ_1 and σ_2 in the alternative hypothesis.
5. Highlight Calculate or Draw and press ENTER. Calculate gives the test statistic and P-value. Draw draws the F-distribution with the P-value shaded.

Minitab

1. If necessary, enter the raw data in columns C1 and C2.
2. Select the **Stat** menu, highlight **Basic Statistics**, then highlight **2 Variances** . . . .
3. If you have raw data, click "Each sample is in its own column," and enter C1 in the cell marked "Sample 1:" and enter C2 in the cell marked "Sample 2:" If you have summarized data, select either "Sample standard deviations" or "Sample variances." Enter the values in the cells, as indicated.
4. Click Options . . . . Select the correct ratio from the drop-down menu, the confidence level, the hypothesized ratio (usually 1), and the direction of the alternative hypothesis. Click OK twice.

Excel

1. Enter raw data in columns A and B.
2. Select the Formulas menu; highlight more Functions and select Statistical.
3. Select F.TEST from the pull-down menu. With the cursor in the "Array 1" cell, highlight the data in column A. With the cursor in the "Array 2" cell, highlight the data in column B. Click OK.

StatCrunch

1. If necessary, enter the raw data into the first two columns of the spreadsheet. Name the column variables.
2. Select **Stat**, highlight **Variance Stats**, select **Two Sample**, and then choose either **With Data** or **With Summary**.
3. If you chose **With Data**, select each column that has the observations. If you chose **With Summary**, enter the variance and sample size for each sample. If you chose the hypothesis test radio button, enter the value of the ratio of the variances stated in the null hypothesis and choose the direction of the alternative hypothesis from the pull-down menu. Click Compute!.

11.4 Assess Your Understanding

Skill Building

For Problems 1–8, find the critical value(s) for α:

NW 1. right-tailed test with $\alpha = 0.05$, degrees of freedom in the numerator = 9, degrees of freedom in the denominator = 10.

2. right-tailed test with $\alpha = 0.01$, degrees of freedom in the numerator = 20, degrees of freedom in the denominator = 25.

3. two-tailed test with $\alpha = 0.05$, degrees of freedom in the numerator = 6, degrees of freedom in the denominator = 8.

4. two-tailed test with $\alpha = 0.02$, degrees of freedom in the numerator = 5, degrees of freedom in the denominator = 7.

5. left-tailed test with $\alpha = 0.10$, degrees of freedom in the numerator = 25, degrees of freedom in the denominator = 20.

6. left-tailed test with $\alpha = 0.01$, degrees of freedom in the numerator = 15, degrees of freedom in the denominator = 20.

7. right-tailed test with $\alpha = 0.05$, degrees of freedom in the numerator = 45, degrees of freedom in the denominator = 15.

8. right-tailed test with $\alpha = 0.01$, degrees of freedom in the numerator = 55, degrees of freedom in the denominator = 50.

In Problems 9–14, assume that the populations are normally distributed. Test the given hypothesis.

9. $\sigma_1 \neq \sigma_2$ at the $\alpha = 0.05$ level of significance

	Sample 1	Sample 2
n	16	16
s	3.2	3.5

10. $\sigma_1 \neq \sigma_2$ at the $\alpha = 0.1$ level of significance

	Sample 1	Sample 2
n	21	21
s	8.6	9.2

11. $\sigma_1 > \sigma_2$ at the $\alpha = 0.01$ level of significance

	Sample 1	Sample 2
n	26	19
s	9.9	6.4

12. $\sigma_1 < \sigma_2$ at the $\alpha = 0.05$ level of significance

	Sample 1	Sample 2
n	21	26
s	15.9	23.0

13. $\sigma_1 < \sigma_2$ at the $\alpha = 0.1$ level of significance

	Sample 1	Sample 2
n	51	26
s	8.3	13.2

14. $\sigma_1 > \sigma_2$ at the $\alpha = 0.05$ level of significance

	Sample 1	Sample 2
n	23	13
s	7.5	5.1

Applying the Concepts

15. Elapsed Time to Earn a Bachelor's Degree Clifford Adelman, a researcher with the Department of Education, followed a cohort of students who graduated from high school in 1992, monitoring the progress the students made toward completing a bachelor's degree. One aspect of his research was to compare students who first attended a community college to those who immediately attended and remained at a four-year institution. The sample standard deviation time to complete a bachelor's degree of the 268 students who transferred to a four-year school after attending community college was 1.162. The sample standard deviation time to complete a bachelor's degree of the 1145 students who immediately attended and remained at a four-year institution was 1.015. Assuming the time to earn a bachelor's degree is normally distributed, does the evidence suggest the standard deviation time to earn a bachelor's degree is different between the two groups? Use the $\alpha = 0.05$ level of significance.

16. Walking in the Airport Researcher Seth B. Young measured the walking speed of travelers in San Francisco International Airport and Cleveland Hopkins International Airport. The standard deviation speed of the 35 travelers who were departing was 53 feet per minute. The standard deviation speed of the 35 travelers who were arriving was 34 feet per minute. Assuming walking speed is normally distributed, does the evidence suggest the standard deviation walking speed is different between the two groups? Use the $\alpha = 0.05$ level of significance.

NW 17. Vitamin A Supplements in Low-Birth-Weight Babies Low-birth-weight babies are at increased risk of respiratory infections in the first few months of life and have low liver stores of vitamin A. In a randomized, double-blind experiment, 130 low-birth-weight babies were randomly divided into two groups. Subjects in group 1 (the treatment group, $n_1 = 65$) were given 25,000 IU of vitamin A on study days 1, 4, and 8 where study day 1 was between 36 and 60 hours after delivery. Subjects in group 2 (the control group, $n_2 = 65$) were given a placebo. The treatment group had a mean serum retinol concentration of 45.77 micrograms per deciliter ($\mu g/dL$), with a standard deviation of 17.07 $\mu g/dL$. The control group had a mean serum retinol concentration of 12.88 $\mu g/dL$, with a standard deviation of 6.48 $\mu g/dL$. Does the treatment group have a higher standard deviation for serum retinol concentration than the control group at the $\alpha = 0.01$ level of significance? It is known that serum retinol concentration is normally distributed.

18. SAT Test Scores A researcher wants to know if students who do not plan to apply for financial aid had more variability on the SAT I math test than those who plan to do so. She obtains a random sample of 35 students who do not plan to apply for financial aid and a random sample of 38 students who do plan to apply for financial aid and obtains the following results:

Do Not Plan to Apply for Financial Aid	Plan to Apply for Financial Aid
$n_1 = 35$	$n_2 = 38$
$s_1 = 123.1$	$s_2 = 119.4$

Do students who do not plan to apply for financial aid have a higher standard deviation on the SAT I math exam than students who plan to apply for financial aid at the $\alpha = 0.01$ level of significance? SAT I math exam scores are known to be normally distributed.

19. Waiting Time in Line McDonald's executives want to experiment with redesigning its restaurants so that the customers form one line leading to four registers to place orders, rather than four lines leading to four separate registers. They redesign 30 randomly selected restaurants with the single line. In addition, they randomly select 30 restaurants with the four-line configuration to participate in the study. At each restaurant, an employee monitors the wait time (in minutes) of randomly selected patrons. The following data are collected.

Single Line				Multiple Lines			
1.2	2.1	1.7	2.8	1.1	1.6	2.9	4.6
1.9	2.1	2.4	3.0	3.8	2.9	2.8	2.9
2.1	2.3	1.9	2.9	4.3	2.3	2.0	2.6
2.7	2.0	2.9	2.3	1.3	1.3	2.0	2.7
2.8	1.1	3.1	1.8	2.0	3.2	2.3	0.9

(a) Is the variability in wait time in the single line less than that for the multiple lines at the $\alpha = 0.05$ level of significance? **Note:** Normal probability plots indicate that the data are normally distributed.
(b) Draw boxplots of each data set to confirm the results of part (a) visually.

20. Filling Machines A quality-control engineer wants to find out whether or not a new machine that fills bottles with liquid has less variability than the machine currently in use. The engineer calibrates each machine to fill bottles with 16 ounces of a liquid. After running each machine for 5 hours, she randomly selects 15 filled bottles from each machine and measures their contents. She obtains the following results:

Old Machine			New Machine		
16.01	16.04	15.96	16.02	15.96	16.05
16.00	16.07	15.89	15.95	15.99	16.02
16.04	16.05	15.91	16.00	15.97	16.03
16.10	16.01	16.00	16.06	16.05	15.94
15.92	16.16	15.92	16.08	15.96	15.95

(a) Is the variability in the new machine less than that of the old machine at the $\alpha = 0.05$ level of significance? **Note:** Normal probability plots indicate that the data are normally distributed.
(b) Draw boxplots of each data set to confirm the results of part (a) visually.

21. Rates of Returns of Stocks Stocks may be categorized by sectors. Go to www.pearsonhighered.com/sullivanstats to obtain the data file 11_4_21 using the file format of your choice for the version of the text you are using. The data represent the one-year rate of return (in percent) for a sample of consumer cyclical stocks and industrial stocks for the period December 2013 through November 2014. **Note:** Consumer cyclical stocks include names such as Starbucks and Home Depot. Industrial stocks include names such as 3M and FedEx. In Section 11.3, we learned that industrial stocks appear to have a higher mean rate of return. Does this suggest that industrial stocks are riskier as measured by the standard deviation rate of return?

22. Measuring Reaction Time Researchers at the University of Mississippi wanted to discover whether variability for reaction time to a go/no go stimulus of males was different from that of females. The researchers randomly selected 20 females and 15 males to participate in the study. The go/no go stimulus required the student to respond to a particular stimulus and not to respond to other stimuli. The data are as follows:

Female Students						
0.588	0.403	0.293	0.377	0.613	0.377	0.391
0.367	0.442	0.274	0.434	0.403	0.636	0.481
0.652	0.443	0.380	0.646	0.340	0.617	

Male Students						
0.375	0.477	0.374	0.465	0.402	0.337	0.655
0.488	0.427	0.373	0.224	0.654	0.563	0.405
0.256						

Source: PsychExperiments at the University of Mississippi (www.olemiss.edu/psychexps)

Normal probability plots indicate that the requirement of normality is satisfied. The output shown is from a TI-84 Plus C.

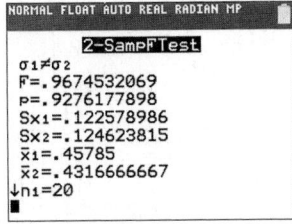

(a) Use the results to determine whether there is a difference in variability of reaction time in males and females at the $\alpha = 0.05$ level of significance.
(b) Draw boxplots of each data set, using the same scale. Does this visual evidence support the results obtained in part (a)?

11.5 Putting It Together: Which Method Do I Use?

Objective   ① Determine the appropriate hypothesis test to perform

① Determine the Appropriate Hypothesis Test to Perform

Again, the ability to recognize the type of test to perform is one of the most important aspects of statistical analysis. To help decide which test to use when conducting inference on two samples, we provide the flow chart in Figure 24 on the next page. Use it to assist you in the problems that follow.

Figure 24

Dependent → See Section 12.3.

Dependent or independent sampling?

Independent → Provided $n\hat{p}(1 - \hat{p}) \geq 10$ for each sample and the sample size is no more than 5% of the population size, use the normal distribution with

$$z_0 = \frac{\hat{p}_1 - \hat{p}_2}{\sqrt{\hat{p}(1 - \hat{p})}\sqrt{\dfrac{1}{n_1} + \dfrac{1}{n_2}}} \quad \text{where } \hat{p} = \frac{x_1 + x_2}{n_1 + n_2}$$

Proportion, p

What parameter is addressed in the hypothesis?

σ or σ^2 → Provided the data are normally distributed, use the F-distribution with

$$F_0 = \frac{s_1^2}{s_2^2}$$

Mean, μ

Independent → Provided each sample size is greater than 30 or each population is normally distributed, use Student's t-distribution

$$t_0 = \frac{(\overline{x}_1 - \overline{x}_2) - (\mu_1 - \mu_2)}{\sqrt{\dfrac{s_1^2}{n_1} + \dfrac{s_2^2}{n_2}}}$$

Dependent or independent sampling?

Dependent → Provided each sample size is greater than 30 or the differences come from a population that is normally distributed, use Student's t-distribution with $n - 1$ degrees of freedom with

$$t_0 = \frac{\overline{d} - \mu_d}{\dfrac{s_d}{\sqrt{n}}}$$

11.5 Assess Your Understanding

Skill Building

In Problems 1–8, perform the appropriate hypothesis test.

1. A random sample of $n_1 = 120$ individuals results in $x_1 = 43$ successes. An independent sample of $n_2 = 130$ individuals results in $x_2 = 56$ successes. Does this represent sufficient evidence to conclude that $p_1 \neq p_2$ at the $\alpha = 0.01$ level of significance?

2. If $n_1 = 31, s_1 = 12, n_2 = 51$, and $s_2 = 10$, test whether $\sigma_1 > \sigma_2$ at the $\alpha = 0.05$ level of significance.

3. A random sample of size $n = 13$ obtained from a population that is normally distributed results in a sample mean of 45.3 and sample standard deviation of 12.4. An independent sample of size $n = 18$ obtained from a population that is normally distributed results in a sample mean of 52.1 and sample standard deviation of 14.7. Does this constitute sufficient evidence to

conclude that the population means differ at the $\alpha = 0.05$ level of significance?

4. A random sample of $n_1 = 135$ individuals results in $x_1 = 40$ successes. An independent sample of $n_2 = 150$ individuals results in $x_2 = 60$ successes. Does this represent sufficient evidence to conclude that $p_1 < p_2$ at the $\alpha = 0.05$ level of significance?

5. If $n_1 = 61, s_1 = 18.3, n_2 = 57$, and $s_2 = 13.5$, test whether the population standard deviations differ at the $\alpha = 0.05$ level of significance.

6. A random sample of size $n = 41$ results in a sample mean of 125.3 and a sample standard deviation of 8.5. An independent sample of size $n = 50$ results in a sample mean of 130.8 and sample standard deviation of 7.3. Does this constitute sufficient evidence to conclude that the population means differ at the $\alpha = 0.01$ level of significance?

7. The following data represent the measure of a variable before and after a treatment.

Individual	1	2	3	4	5
Before, X_i	93	102	90	112	107
After, Y_i	95	100	95	115	107

Does the sample evidence suggest that the treatment is effective in increasing the value of the response variable? Use the $\alpha = 0.05$ level of significance.
Note: Assume that the differenced data come from a population that is normally distributed with no outliers.

8. The following data represent the measure of a variable before and after a treatment.

Individual	1	2	3	4	5
Before, X_i	40	32	53	48	38
After, Y_i	38	33	49	48	33

Does the sample evidence suggest that the treatment is effective in decreasing the value of the response variable? Use the $\alpha = 0.10$ level of significance.
Note: Assume that the differenced data come from a population that is normally distributed with no outliers.

Applying the Concepts

9. Collision Claims Automobile collision insurance is used to pay for any claims made against the driver in the event of an accident. This type of insurance will typically pay to repair any assets that your vehicle damages.
(a) Collision claims tend to be skewed right. Why do you think this is the case?
(b) A random sample of 40 collision claims of 30- to 59-year-old drivers results in a mean claim of $3669 with a standard deviation of $2029. An independent random sample of 40 collision claims of 20- to 24-year-old drivers results in a mean claim of $4586 with a standard deviation of $2302. Using the concept of hypothesis testing, provide an argument that justifies charging a higher insurance premium to 20- to 24-year-old drivers. *Source:* Based on data obtained from the Insurance Institute for Highway Safety.

10. TIMS Report and Kumon TIMS is an acronym for the Third International Mathematics and Science Study. Kumon promotes a method of studying mathematics that it claims develops mathematical ability. Do data support this claim? In one particular question on the TIMS exam, a random sample of 400 non-Kumon students resulted in 73 getting the correct answer. For the same question, a random sample of 400 Kumon students resulted in 130 getting the correct answer. Perform the appropriate test to substantiate Kumon's claims. Are there any confounding factors regarding the claim? *Source:* TIMS and PROM/SE Assessment of Kumon Students: What the Results Indicate, Kumon North America.

11. Age Difference in Married Couples What is the typical age difference between husband and wife? The following data represent the ages of husbands and wives, based on results from the Current Population Survey.

	Couple 1	Couple 2	Couple 3	Couple 4
Husband	47	53	45	50
Wife	43	43	45	48

	Couple 5	Couple 6	Couple 7	Couple 8
Husband	28	65	25	56
Wife	29	61	27	51

(a) What is the response variable in this study?
(b) Is the sampling method dependent or independent? Explain.
(c) Use the data to estimate the mean difference in age of husband and wives with 95% confidence. Explain the technique that you used.

12. Cash or Credit? Do people tend to spend more money on fast-food when they use a credit card? The following data represent a random sample of credit-card and cash purchases.

Credit				
16.78	23.89	13.89	15.54	10.35
12.76	18.32	20.67	18.36	19.16

Cash				
10.76	6.26	18.98	11.36	6.78
21.76	8.90	15.64	13.78	9.21

Source: Brian Ortiz, student at Joliet Junior College

(a) Draw boxplots of each data set using the same scale. What do the boxplots imply for cash versus credit?
(b) Test whether the sample evidence suggests that people spend more when using a credit card. Use the $\alpha = 0.01$ level of significance. **Note:** Normal probability plots indicate that each sample could come from a population that is normally distributed.
(c) Suppose that you were looking to gather evidence to convince your manager that an effort needs to be made to boost credit-card sales. Would it be legitimate to change the level to $\alpha = 0.05$ after seeing the results from part (b)? Why?

13. Major League Fastball The average major league fastball is 92.0 miles per hour (mph). While there are many other factors other than velocity that are used to judge the quality of this pitch (location and movement, for example), velocity is a major factor in deciding whether a pitcher has "big league stuff." Suppose a scout is judging a college player and records the velocity of a random sample of 14 fastballs. Does the evidence imply that this pitcher has better than average major league "stuff" in regards to velocity?

91.3	93.5	93.4	91.9	91.8	90.3	92.3
92.8	91.6	93.0	91.9	92.6	92.6	92.5

14. Wet Suits Do wet suits allow a swimmer to swim faster? Researchers measured the speed (in meters per second) of swimmers both with and without a wetsuit. The results of the study are shown in the table. Conduct the appropriate test to determine whether the data suggest that wet suits allow a swimmer to swim faster. Be sure to check all model requirements. Use an $\alpha = 0.05$ level of significance. If you reject the null hypothesis, estimate the average difference in speed of the swimmer with the wet suit with 95% confidence.

Swimmer	1	2	3	4	5	6
Without	1.49	1.37	1.35	1.27	1.12	1.64
With	1.57	1.47	1.42	1.35	1.22	1.75

Swimmer	7	8	9	10	11	12
Without	1.59	1.52	1.50	1.45	1.44	1.41
With	1.64	1.57	1.56	1.53	1.49	1.51

Source: Data from de Lucas, R.D., Balildan, P., Neiva, C.M., Greco, C.C., & Denadai, B.S. (2000). The effects of wet suits on physiological and biomechanical indices during swimming. *Journal of Science and Medicine in Sport 3* (1): 1–8.

15. Predicting Election Outcomes Researchers conducted an experiment in which 695 individuals were shown black and white photographs of individuals running for Congress (either the U.S. Senate or House of Representatives). In each instance, the individuals were exposed to the photograph of both the winner and runner-up (in random order) for 1 second. The individuals were then asked to decide who they believed was more competent (and, therefore, more likely to receive their vote). Of the 695 individuals exposed to the photos, 469 correctly predicted the winner of the race. Do the results suggest that a quick 1-second view of a black and white photo represents enough information to judge the winner of an election (based on perceived level of competence of the individual) more often than not? Use the $\alpha = 0.05$ level of significance. *Source:* Todorov, Mandisodza, Goren, Hall, "Inferences of Competence from Faces Predict Election Outcomes." *Science* Vol. 308.

16. Bribe 'em with Chocolate In a study published in the journal *Teaching of Psychology*, the article "Fudging the Numbers: Distributing Chocolate Influences Student Evaluations of an Undergraduate Course" states that distributing chocolate to students prior to teacher evaluations increases results. The authors randomly divided three sections of a course taught by the same instructor into two groups. Fifty of the students were given chocolate by an individual not associated with the course and 50 of the students were not given chocolate. The mean score from students who received chocolate was 4.2, while the mean score for the nonchocolate groups was 3.9. Suppose that the sample standard deviation of both the chocolate and nonchocolate groups was 0.8. Does chocolate appear to improve teacher evaluations? Use the $\alpha = 0.1$ level of significance.

17. Unwed Women Having Children The Pew Research Group asked the following question of individuals who earned in excess of $100,000 per year and those who earned less than $100,000 per year: "Do you believe that it is morally wrong for unwed women to have children?" Of the 1205 individuals who earned in excess of $100,000 per year, 710 said yes; of the 1310 individuals who earned less than $100,000 per year, 695 said yes. Construct a 95% confidence interval to determine if there is a difference in the proportion of individuals who believe it is morally wrong for unwed women to have children.

18. Extramarital Affairs Is there a difference in the attitude toward extramarital affairs in the United States versus Canada? Pew Research surveyed a random sample of adults from each country and asked, "Do you personally believe that married people having an affair is morally acceptable?" Results of the surveys are below:

	Number Surveyed	Number Responding "Morally Acceptable"
Canada	701	533
United States	1002	841

(a) What is the response variable in this study?
(b) Is the sampling method dependent or independent? Explain.
(c) What approach should be used to determine whether the attitudes in Canada differ from the United States when it comes to extramarital affairs?
(d) Conduct the appropriate test to determine whether the attitudes in Canada differ from the United States when it comes to extramarital affairs. Use the $\alpha = 0.05$ level of significance.
(e) Why do you think fewer individuals were surveyed in Canada?

19. Multitasking in School In today's "wired" society, students believe that they can multitask. Research suggests that 5% of individuals truly have the ability to multitask. Can students multitask, or do they perform worse while multitasking? In an introductory accounting class, students were randomly given an instruction sheet at the beginning of class. Half the instruction sheets required students to send three text messages to the instructor during the lecture. The other half of class was told to turn off their cell phones. At the end of class, a quiz was administered based on information shared during class. The results of the quiz are shown in the table below.

	n	Sample Mean	Sample Standard Deviation
Texting Group	31	42.81	9.91
Cell-phone Off Group	31	58.67	10.42

Source: Ellis, Y., Daniels, W. and Jauregui, A. (2010) "The Effect of Multitasking on the Grade Performance of Business Students." *Research in Higher Education Journal*, 8.

(a) What is the response variable? What is the explanatory variable?
(b) How is randomization used in this study?
(c) Is the sampling method dependent or independent? Explain.
(d) Conduct the appropriate test to determine whether the data suggest students score worse in the texting group. Use an $\alpha = 0.05$ level of significance.
(e) Estimate the mean difference in scores between the texting group and cell-phone off group with 90% confidence.
(f) Does the sample evidence suggest there is a difference in the dispersion of the scores between the two groups?

20. Comparing Stock Sectors The following data represent the 5-year rate of return (in percent) for a random sample of stocks in financial services and an independent random sample of stocks in health care. *Source:* Monrnigstar.com

Financial Services

17.34	16.13	10.26	15.62	15.91
10.16	21.25	22.32	28.30	26.31
23.90	17.51	16.93	24.48	21.76
27.53	21.63	15.41	20.86	17.19

Health Care

22.08	21.81	31.07	15.96	24.30
24.42	12.96	10.35	30.71	9.18
11.07	21.66	23.08	14.70	15.08
17.76	9.83	33.67	28.28	

(a) Has there been a difference in the rate of return of companies in the Financial Services sector versus those in the Health Care sector over the past 5 years? Use an $\alpha = 0.05$ level of significance.

(b) Is there a difference in the standard deviation rate of return between the two sectors? Remember, standard deviation is a measure of risk in equities.

21. Hospital Readmission As of October 1, 2012, hospitals in the United States with excessive numbers of readmissions based on the Centers for Medicare and Medicaid Services (CMS) data were penalized. Therefore, it is important for hospitals to identify the risk factors associated with readmission. The following data represent statistics on patients readmitted within 30 days and those not readmitted within 30 days of being discharged at a community hospital. Conduct the appropriate test for each variable that might be associated with readmission (age, length of stay, and so on). Researchers wonder if:

- the readmits tend to be older
- is length of stay longer for readmits
- were a higher proportion of readmits admitted the previous calendar year
- were a higher proportion of readmits discharged in winter
- were a higher proportion of readmits on the cardiac floor

Write a short report detailing your findings.
Source: Lee Park, Danielle Andrade, Andrew Mastey, James Sun, LeRoi Hicks "Institution Specific Risk Factors for 30 Day Readmission at a Community Hospital: A Retrospective Observational Study" *BioMedCentral* 2014 14:40

	Readmit	Non-Readmit
n	637	3137
Age	Mean: 78.0 Standard deviation: 13.7	Mean: 76.0 Standard deviation: 15.8
Length of stay on previous visit	Mean: 4.8 Standard deviation: 4.3	Mean: 3.9 Standard deviation: 3.1
Admission in previous calendar year	139 out of 637	445 out of 3137
Season	199 out of 637 readmits were discharged in winter	765 out of 3137 non-readmits were discharged in winter
Floor	332 out of 637 readmits were on the cardiac floor	1617 out of 3137 non-readmits were on the cardiac floor

22. Behavior and Gender at Work Researchers wanted to determine whether individuals considered certain employee work behaviors to be required or optional. For each work behavior cited, participants rated the behavior on a scale from 1 (definitely required) to 7 (completely optional). There

were a total of 41 individuals who rated the behavior, with 20 individuals rating a female employee and 21 individuals rating a male employee. The following table shows the work behavior, mean, and standard deviation for female and male employees. For example, of the 20 ratings on the female employee, the mean score for resolving conflict between coworkers was 2.50, while the mean of the 21 ratings on male employees for this behavior was 3.52. Although the sample size is small, assume the distribution of the sample mean is approximately normal for all four work behavior ratings. Conduct the appropriate test for all four work behaviors to determine if there is a difference in ratings for female versus male employees. Write a report that presents your conclusions. *Source:* Heilman, Madeline and Chen, Julie, "Same Behavior, Different Consequences: Reactions to Men's and Women's Altruistic Citizenship Behavior" *Journal of Applied Psychology* 2005, Vol. 90, No. 3, 431–441

Work Behavior	Female Employee	Male Employee
Resolving conflict between coworkers	Mean: 2.50 Standard deviation: 0.76	Mean: 3.52 Standard deviation: 0.68
Alerting management to potentially troublesome issues	Mean: 2.40 Standard deviation: 1.00	Mean: 1.57 Standard deviation: 0.60
Working extra hours during busy times	Mean: 3.00 Standard deviation: 0.73	Mean: 2.29 Standard deviation: 1.06
Attending special training sessions	Mean: 4.00 Standard deviation: 0.86	Mean: 4.19 Standard deviation: 1.29

23. Gender Wage Gap It has long been a concern that there is a wage gap between men and women in the United States with some reports suggesting that women only make $0.77 for every dollar earned by a man. Design a study that would allow you to confirm whether a wage gap does actually exist.

Explaining the Concepts

In Problems 24–33, for each study, explain which statistical procedure (estimating a single proportion; estimating a single mean; hypothesis test for a single proportion; hypothesis test for a single mean; hypothesis test or estimation of two proportions, hypothesis test or estimation of two means, dependent or independent) would most likely be used for the research objective given. Assume all model requirements for conducting the appropriate procedure have been satisfied.

24. Is the mean IQ of the students in Professor Dang's statistics class higher than that of the general population, 100?

25. Do adult males who take a single aspirin daily experience a lower rate of heart attacks than adult males who do not take aspirin daily?

26. Does Marriott Courtyard charge more than Holiday Inn Express for a one-night stay?

27. What is the typical amount of time 20- to 24-year-old males spend brushing their teeth (each time they brush)?

28. What proportion of registered voters is in favor of a tax increase to reduce the federal debt?

29. Does drinking two cups of water before a meal assist with weight loss?

30. Does turmeric (an antioxidant that can be added to foods) help with depression? Researchers randomly assigned 200 adult women who were clinically depressed to two groups. Group 1 had turmeric added to their regular diet for one week; group 2 had no additives in their diet. At the end of one week, the change in their scores on the Beck Depression Inventory was compared.

31. While exercising by climbing stairs, is it better to take one stair, or two stairs, at a time? Researchers identified 30 volunteers who were asked to climb stairs for two different 15-minute intervals taking both one stair and two stairs at a

time. Whether the volunteer did one stair or two stairs first was determined randomly. The goal of the research was to determine if energy expenditure for each exercise routine was different.

32. By how much does adiposity (a measure of body fat) differ between adult women who maintain a regular sleep schedule versus women whose sleep schedule fluctuates by 90 minutes or more?

33. Do recent graduates from college who have no debt start their own business at a higher rate than recent graduates who have debt between $20,000 and $40,000?

Chapter 11 Review

Summary

This chapter continues the presentation of inferential statistics. Here, we discussed performing statistical inference by comparing two population parameters. In particular, we compared two population proportions and two population means. The first step in performing this inference is to decide whether the sampling method is independent or dependent. A sampling method is independent when the choice of individuals for one sample does not dictate which individuals will be in a second sample. Data obtained from a completely randomized design with two treatments can be analyzed using the methods based on independent sampling presented in this chapter. A sampling method is dependent when the individuals selected for one sample are used to determine the individuals in the second sample. Data obtained from a matched-pairs experimental design can be analyzed using the dependent sampling inferential techniques presented in this chapter.

Section 11.1 presented statistical inference for comparing two population proportions from independent samples. To perform these tests, the samples must be obtained independently using simple random sampling or the data result from a completely randomized experiment with two levels of treatment. The response variable is qualitative with two outcomes.

In addition, $n\hat{p}(1 - \hat{p})$ must be greater than or equal to 10 for each sample, and each sample size can be no more than 5% of the population size. The distribution of $\hat{p}_1 - \hat{p}_2$ is approximately normal, with mean $p_1 - p_2$ and standard deviation $\sqrt{\dfrac{p_1(1 - p_1)}{n_1} + \dfrac{p_2(1 - p_2)}{n_2}}$. We use this distribution to perform hypothesis tests using either the classical approach or P-value approach. Follow the same five steps for conducting a hypothesis test that we introduced in Chapter 10. We also use the model to construct confidence intervals about the difference of two independent proportions. We will present inference on two dependent proportions in Section 12.3.

Section 11.2 presented inference on the difference of two means from dependent sampling. We use Student's t-distribution to perform hypothesis testing or construct confidence intervals. However, the following conditions must be satisfied:

1. Data must be obtained through a simple random sample or result from a randomized experiment. Data must be matched-pairs (dependent). Response variable is quantitative.
2. Differenced data come from a population that is normally distributed or the sample size is large.
3. Sample size may be no more than 5% of the population size (independence requirement).

Again, use the same five steps for conducting a hypothesis test introduced in Section 10.3.

In Section 11.3, we presented inference on the difference of two means from independent sampling. Here, we use Welch's approximate t provided:

1. The samples are independently obtained using simple random sampling or the data result from a completely randomized experiment with two levels of treatment. Response variable is quantitative.
2. The sample data come from a population that is normally distributed with no outliers, or the sample sizes are large.
3. For each sample, the sample size is no more than 5% of the population size.

The steps for conducting a hypothesis test are the same as they have been throughout the chapters.

In Section 11.4, we discussed inference for comparing two population standard deviations. We use the F-test, but require that both populations be normally distributed. This test is not robust, so if there are any departures from normality, the test should not be used.

To help determine which test to use, we included the flow chart in Figure 24 within Section 11.5.

Vocabulary

Independent (p. 525)
Dependent (p. 525)
Matched pairs (p. 525)

Pooled estimate of p (p. 527)
Robust (p. 538)
Welch's approximate t (p. 547)

Pooled t-statistic (p. 553)
F-distribution (p. 558)

Formulas

- Test statistic comparing two population proportions (independent sampling):

$$z_0 = \frac{(\hat{p}_1 - \hat{p}_2) - (p_1 - p_2)}{\sqrt{\hat{p}(1 - \hat{p})}\sqrt{\frac{1}{n_1} + \frac{1}{n_2}}}$$

where $\hat{p} = \dfrac{x_1 + x_2}{n_1 + n_2}$

- Confidence interval for the difference of two proportions (independent sampling):

Lower bound: $(\hat{p}_1 - \hat{p}_2) - z_{\frac{\alpha}{2}} \cdot \sqrt{\dfrac{\hat{p}_1(1 - \hat{p}_1)}{n_1} + \dfrac{\hat{p}_2(1 - \hat{p}_2)}{n_2}}$

Upper bound: $(\hat{p}_1 - \hat{p}_2) + z_{\frac{\alpha}{2}} \cdot \sqrt{\dfrac{\hat{p}_1(1 - \hat{p}_1)}{n_1} + \dfrac{\hat{p}_2(1 - \hat{p}_2)}{n_2}}$

- Sample size for estimating $p_1 - p_2$:

$$n = n_1 = n_2 = [\hat{p}_1(1 - \hat{p}_1) + \hat{p}_2(1 - \hat{p}_2)]\left(\frac{z_{\alpha/2}}{E}\right)^2$$

$$n = n_1 = n_2 = 0.5\left(\frac{z_{\alpha/2}}{E}\right)^2$$

- Test statistic for matched-pairs data:

$$t_0 = \frac{\overline{d} - \mu_d}{\frac{s_d}{\sqrt{n}}}$$

where $\overline{d}$ is the mean and s_d is the standard deviation of the differenced data

- Confidence interval for matched-pairs data:

Lower bound: $\overline{d} - t_{\frac{\alpha}{2}} \cdot \dfrac{s_d}{\sqrt{n}}$

Upper bound: $\overline{d} + t_{\frac{\alpha}{2}} \cdot \dfrac{s_d}{\sqrt{n}}$

- Test statistic comparing two means (independent sampling):

$$t_0 = \frac{(\overline{x}_1 - \overline{x}_2) - (\mu_1 - \mu_2)}{\sqrt{\frac{s_1^2}{n_1} + \frac{s_2^2}{n_2}}}$$

- Confidence interval for the difference of two means (independent samples):

Lower bound: $(\overline{x}_1 - \overline{x}_2) - t_{\frac{\alpha}{2}} \cdot \sqrt{\dfrac{s_1^2}{n_1} + \dfrac{s_2^2}{n_2}}$

Upper bound: $(\overline{x}_1 - \overline{x}_2) + t_{\frac{\alpha}{2}} \cdot \sqrt{\dfrac{s_1^2}{n_1} + \dfrac{s_2^2}{n_2}}$

- Test statistic for comparing two population standard deviations:

$$F_0 = \frac{s_1^2}{s_2^2}$$

- Finding a critical F for the left tail:

$$F_{1-\alpha,\, n_1-1,\, n_2-1} = \frac{1}{F_{\alpha,\, n_2-1,\, n_1-1}}$$

OBJECTIVES

Section	You should be able to ...	Example(s)	Review Exercises
11.1	1 Distinguish between independent and dependent sampling (p. 525)	1	1, 2
	2 Test hypotheses regarding two proportions from independent samples (p. 526)	2	7, 10(c)
	3 Construct and interpret confidence intervals for the difference between two population proportions (p. 530)	3	10(d)
	4 Determine the sample size necessary for estimating the difference between two population proportions (p. 532)	4	11
11.2	1 Test hypotheses for a population mean from matched-pairs data (p. 537)	1	4(c), 8(b)
	2 Construct and interpret confidence intervals about the population mean difference of matched-pairs data (p. 541)	2	4(d), 13
11.3	1 Test hypotheses regarding the difference of two independent means (p. 548)	1	5(a), 6(a), 9(c)
	2 Construct and interpret confidence intervals regarding the difference of two independent means (p. 552)	2	5(b), 14

Section	You should be able to ...	Example(s)	Review Exercises
11.4	**1** Find critical values of the *F*-distribution (p. 558)	1	3
	2 Test hypotheses regarding two population standard deviations (p. 561)	2, 3	5(c), 6(b), 12
11.5	**1** Determine the appropriate hypothesis test to perform (p. 567)		4–10, 12–14

Review Exercises

In Problems 1 and 2, determine if the sampling is dependent or independent. Indicate whether the response variable is qualitative or quantitative.

1. A researcher wants to know if the mean length of stay in for-profit hospitals is different from that in not-for-profit hospitals. He randomly selected 20 individuals in the for-profit hospital and matched them with 20 individuals in the not-for-profit hospital by diagnosis.

2. An urban economist believes that commute times to work in the South are less than commute times to work in the Midwest. He randomly selects 40 employed individuals in the South and 40 employed individuals in the Midwest and determines their commute times.

3. (a) Find the critical *F*-value for a right-tailed test with $\alpha = 0.05$, degrees of freedom in the numerator = 8, and degrees of freedom in the denominator = 9.
   (b) Find the critical *F*-value for a two-tailed test with $\alpha = 0.05$, degrees of freedom in the numerator = 10, and degrees of freedom in the denominator = 5.

In Problem 4, assume that the paired differences come from a population that is normally distributed.

4.

Observation	1	2	3	4	5	6
X_i	34.2	32.1	39.5	41.8	45.1	38.4
Y_i	34.9	31.5	39.5	41.9	45.5	38.8

(a) Compute $d_i = X_i - Y_i$ for each pair of data.
(b) Compute $\bar{d}$ and s_d.
(c) Test the hypothesis that $\mu_d < 0$ at the $\alpha = 0.05$ level of significance.
(d) Compute a 98% confidence interval for the population mean difference μ_d.

In Problems 5 and 6, assume that the populations are normally distributed and that independent sampling occurred.

5.

	Sample 1	Sample 2
n	13	8
$\bar{x}$	32.4	28.2
s	4.5	3.8

(a) Test the hypothesis that $\mu_1 \neq \mu_2$ at the $\alpha = 0.1$ level of significance for the given sample data.
(b) Construct a 90% confidence interval for $\mu_1 - \mu_2$.
(c) Test the hypothesis that $\sigma_1 \neq \sigma_2$ at the $\alpha = 0.05$ level of significance for the given sample data.

6.

	Sample 1	Sample 2
n	45	41
$\bar{x}$	48.2	45.2
s	8.4	10.3

(a) Test the hypothesis that $\mu_1 > \mu_2$ at the $\alpha = 0.01$ level of significance for the given sample data.
(b) Test the hypothesis that $\sigma_1 < \sigma_2$ at the $\alpha = 0.01$ level of significance for the given sample data.

7. A random sample of $n_1 = 555$ individuals results in $x_1 = 451$ successes. An independent sample of $n_2 = 600$ individuals results in $x_2 = 510$ successes. Does this represent sufficient evidence to conclude that $p_1 \neq p_2$ at the $\alpha = 0.05$ level of significance?

8. Height versus Arm Span A statistics student heard that an individual's arm span is equal to the individual's height. To test this hypothesis, the student used a random sample of ten students and obtained the following data.

Student:	1	2	3	4	5
Height (inches)	59.5	69	77	59.5	74.5
Arm span (inches)	62	65.5	76	63	74

Student:	6	7	8	9	10
Height (inches)	63	61.5	67.5	73	69
Arm span (inches)	66	61	69	70	71

Source: John Climent, Cecil Community College

(a) Is the sampling method dependent or independent? Why?
(b) Does the sample evidence contradict the belief that an individual's height and arm span are the same at the $\alpha = 0.05$ level of significance?

Note: A normal probability plot indicates that the data and differenced data are normally distributed. A boxplot indicates that the data and differenced data have no outliers.

9. McDonald's versus Wendy's A student wanted to determine whether the wait time in the drive-through at McDonald's differed from that at Wendy's. She used a random sample of 30 cars at McDonald's and 27 cars at Wendy's and obtained these results:

Wait Time at McDonald's Drive-Through (seconds)

151.09	227.38	111.84	131.21	128.75
191.60	126.91	137.90	195.44	246.59
141.78	127.35	121.21	101.03	95.09
122.06	122.62	100.04	71.37	153.34
140.44	126.62	116.72	131.69	100.94
115.66	147.28	81.43	86.31	156.34

Wait Time at Wendy's Drive-Through (seconds)

281.90	71.02	204.29	128.59	133.56
187.53	199.86	190.91	110.55	110.64
196.84	233.65	171.01	182.54	183.79
284.48	363.34	270.82	390.50	471.62
123.66	174.43	385.90	386.71	155.53
203.62	119.61			

Source: Catherine M. Simmons, student at Joliet Junior College

Note: The sample size for Wendy's is less than 30. However, the data do not contain any outliers, so the Central Limit Theorem can be used.

(a) Is the sampling method dependent or independent?

(b) What is the variable of interest? Is it qualitative or quantitative?

(c) Is there a difference in wait times at each restaurant's drive-through? Use the $\alpha = 0.1$ level of significance.

(d) Draw boxplots of each data set using the same scale. Does this visual evidence support the results obtained in part (b)?

10. Treatment for Osteoporosis Osteoporosis is a condition in which people experience decreased bone mass and an increase in the risk of bone fracture. Actonel is a drug that helps combat osteoporosis in postmenopausal women. In clinical trials, 1374 postmenopausal women were randomly divided into experimental and control groups. The subjects in the experimental group were administered 5 milligrams (mg) of Actonel, while the subjects in the control group were administered a placebo. The number of women who experienced a bone fracture over the course of one year was recorded. Of the 696 women in the experimental group, 27 experienced a fracture during the course of the year. Of the 678 women in the control group, 49 experienced a fracture during the course of the year.

(a) What type of experimental design is this? What is the response variable? Is it qualitative or quantitative? What are the treatments?

(b) The experiment was double-blind. What does this mean?

(c) Does the sample evidence suggest the drug is effective in preventing bone fractures? Use the $\alpha = 0.01$ level of significance.

(d) Construct a 95% confidence interval for the difference between the two population proportions, $p_{exp} - p_{control}$.

11. Determining Sample Size A nutritionist wants to estimate the difference between the percentage of men and women who have high cholesterol. What sample size should be obtained if she wishes the estimate to be within 2 percentage points with 90% confidence, assuming that

(a) she uses the 1994 estimates of 18.8% male and 20.5% female from the National Center for Health Statistics?

(b) she does not use any prior estimates?

12. Wait Time Using the data from Problem 9, test whether the standard deviation of wait time at Wendy's is more than that at McDonald's at the $\alpha = 0.05$ level of significance.

13. Height versus Arm Span Construct and interpret a 95% confidence interval for the population mean difference between height and arm span using the data from Problem 8. What does the interval lead us to conclude regarding any differences between height and arm span?

14. McDonald's versus Wendy's Construct and interpret a 95% confidence interval about $\mu_M - \mu_W$ using the data from Problem 9. How might a marketing executive with McDonald's use this information?

Chapter Test

In Problems 1 and 2, determine whether the sampling method is independent or dependent.

1. A stock analyst wants to know if there is a difference between the mean rate of return from energy stocks and that from financial stocks. He randomly selects 13 energy stocks and computes the rate of return for the past year. He randomly selects 13 financial stocks and computes the rate of return for the past year.

2. A prison warden wants to know if men receive longer sentences for crimes than women. He randomly samples 30 men and matches them with 30 women by type of crime committed and records their lengths of sentence.

In Problem 3, assume that the paired differences come from a population that is normally distributed.

3.

Observation	1	2	3	4	5	6	7
X_i	18.5	21.8	19.4	22.9	18.3	20.2	23.1
Y_i	18.3	22.3	19.2	22.3	18.9	20.7	23.9

(a) Compute $d_i = X_i - Y_i$ for each pair of data.

(b) Compute $\bar{d}$ and s_d.

(c) Test the hypothesis that $\mu_d \neq 0$ at the $\alpha = 0.01$ level of significance.

(d) Compute a 95% confidence interval for the population mean difference μ_d.

In Problems 4 and 5, assume that the populations are normally distributed and that independent sampling occurred.

4.

	Sample 1	Sample 2
n	24	27
$\bar{x}$	104.2	110.4
s	12.3	8.7

(a) Test the hypothesis that $\mu_1 \neq \mu_2$ at the $\alpha = 0.1$ level of significance for the given sample data.

(b) Construct a 95% confidence interval for $\mu_1 - \mu_2$.

(c) Test the hypothesis that $\sigma_1 > \sigma_2$ at the $\alpha = 0.1$ level of significance for the given sample data.

5.

	Sample 1	Sample 2
n	13	8
$\bar{x}$	96.6	98.3
s	3.2	2.5

(a) Test the hypothesis that $\mu_1 < \mu_2$ at the $\alpha = 0.05$ level of significance for the given sample data.

(b) Test the hypothesis that $\sigma_1 \neq \sigma_2$ at the $\alpha = 0.05$ level of significance for the given sample data.

6. A random sample of $n_1 = 650$ individuals results in $x_1 = 156$ successes. An independent sample of $n_2 = 550$ individuals results in $x_2 = 143$ successes. Does this represent sufficient evidence to conclude that $p_1 < p_2$ at the $\alpha = 0.05$ level of significance?

7. A researcher wants to know whether the acidity of rain (pH) near Houston, Texas, is significantly different from that near Chicago, Illinois. He randomly selects 12 rain dates in Texas and 14 rain dates in Illinois and obtains the following data:

Texas					
4.69	5.10	5.22	4.46	4.93	4.65
5.22	4.76	4.25	5.14	4.11	4.71

Illinois						
4.40	4.69	4.22	4.64	4.54	4.35	4.69
4.40	4.75	4.63	4.45	4.49	4.36	4.52

Source: National Atmospheric Deposition Program

(a) Is the sampling method dependent or independent? Why?

(b) Because the sample sizes are small, what must be true regarding the populations from which the samples were drawn?

(c) Draw side-by-side boxplots of the data. What does the visual evidence imply about the pH of rain in the two states?

(d) Does the evidence suggest that there is a difference in the pH of rain in Chicago and Houston? Use the $\alpha = 0.05$ level of significance.

8. In a study conducted to determine the role that sleep disorders play in academic performance, researcher Jane Gaultney conducted a survey of 1845 college students to determine if they had a sleep disorder (such as narcolepsy, insomnia, or restless leg syndrome). Of the 503 students with a sleep disorder, the mean grade point average was 2.65 with a standard deviation of 0.87. Of the 1342 students without a sleep disorder, the mean grade point average was 2.82 with a standard deviation of 0.83. *Source:* SLEEP 2010: Associated Professional Sleep Societies 24th Annual Meeting

(a) What is the response variable in this study? What is the explanatory variable?

(b) Is there evidence to suggest sleep disorders adversely affect one's GPA at the $\alpha = 0.05$ level of significance?

9. Researchers had a car traveling 10 miles per hour collide into the rear bumper of an SUV and recorded the amount of damage, in dollars. The data are below. Do the given data suggest the repair cost of the car is higher? Use the $\alpha = 0.1$ level of significance.

Car into SUV	SUV Damage	Car Damage
Kia Forte into Hyundai Tucson	2091	1510
Dodge Caliber into Jeep Patriot	1338	2559
Honda Civic into Honda CR-V	1053	4921
Volkswagen Golf into VW Tiguan	1872	4555
Nissan Sentra into Nissan Rogue	1428	5114
Ford Focus into Ford Escape	2208	5203
Toyota Corolla into Toyota RAV4	6015	3852

Source: Insurance Institute for Highway Safety

10. Zoloft is a drug used to treat obsessive-compulsive disorder (OCD). In randomized, double-blind clinical trials, 926 patients diagnosed with OCD were randomly divided into two groups. Subjects in group 1 (experimental group) received 200 milligrams per day (mg/day) of Zoloft, while subjects in group 2 (control group) received a placebo. Of the 553 subjects in the experimental group, 77 experienced dry mouth as a side effect. Of the 373 subjects in the control group, 34 experienced dry mouth as a side effect.

(a) What type of experimental design is this?

(b) What is the response variable?

(c) Do a higher proportion of subjects experience dry mouth who are taking Zoloft versus the proportion taking the placebo? Use the $\alpha = 0.05$ level of significance.

11. Does hypnotism result in a different success rate for men and women who are trying to quit smoking? Researchers at *Science* magazine analyzed studies involving 5600 male and female smokers. Of the 2800 females, 644 quit smoking; of the 2800 males, 840 quit smoking. Construct a 90% confidence interval for the difference in proportion of males and females, $p_M - p_F$, and use the interval to judge whether there is a difference in the proportions.

12. A researcher wants to estimate the difference between the percentage of individuals without a high school diploma who smoke and the percentage of individuals with bachelor's degrees who smoke. What sample size should be obtained if she wishes the estimate to be within 4 percentage points with 95% confidence, assuming that

(a) she uses the 1999 estimates of 32.2% of those without a high school diploma and 11.1% of those with a bachelor's degree from the National Center for Health Statistics?

(b) she does not use any prior estimates?

13. It is commonplace to gain weight after quitting smoking. To determine the effectiveness of the drug Naltrexone in limiting weight gain after quitting smoking, 147 subjects who smoked 20 or more cigarettes daily were randomly divided into two groups. All 147 subjects received a 21-milligram (mg) transdermal nicotine patch (to curb nicotine cravings while trying to quit smoking). However, seventy-two subjects received a placebo, while 75 received 25 mg of Naltrexone. After 6 weeks, the placebo subjects had a mean weight gain of 1.9 kg, with a standard deviation of 0.22 kg; the Naltrexone subjects had a mean weight gain of 0.8 kg with a standard deviation of 0.21 kg. Construct a 95% confidence interval for the mean difference in weight gain for the two groups. Based on this interval, do you believe that Naltrexone is effective in controlling weight gain following smoking cessation? Do the results have any practical significance? *Source:* Stephanie O'Malley et al. "A Controlled Trial of Naltrexone Augmentation of Nicotine Replacement Therapy for Smoking Cessation." *Archives of Internal Medicine*, Vol. 166, 667–674

14. Using the data from Problem 7, test whether the standard deviation of acidity in the rain near Houston, Texas, is different from that near Chicago, Illinois, at the $\alpha = 0.05$ level of significance.

Making an Informed Decision

Which Car Should I Buy?

You have decided to purchase a car and have narrowed your choice down to two cars. However, you have two areas of concern. First, you want to purchase the car that gets the better gas mileage. Second, you want to purchase the car that holds its value better. To help make an informed decision, you decide to collect some data and run some tests.

(a) Decide on two cars that are similar that you would consider purchasing.

(b) Go to www.fueleconomy.gov and obtain data on the fuel economy of each car you are considering. Treat the data as a random sample of all cars.

(c) Draw side-by-side boxplots of the fuel economy to verify there are no outliers and to verify it is reasonable to conclude the data come from a population that is normally distributed (if your sample size is small).

(d) Conduct the appropriate test to determine if there is a significant difference in the gas mileage of the two cars.

(e) Go to an online website that lists used cars for sale. Obtain a matched-pairs random sample of cars where each car is paired with the second car based on age of the car and  mileage. For example, if you are considering a Camry or Accord, then match a two-year-old Camry with 18,000 miles with a two-year-old Accord with 18,000 miles. However, to determine how well the car holds its value, subtract the asking price of the car from the price when the car is new.

(f) Conduct the appropriate test to determine if one car holds its value better than the other car.

(g) Write a report detailing which car you would purchase.

 CASE STUDY   Control in the Design of an Experiment

Dr. Penelope Nicholls is interested in exploring a possible connection between high plasma homocysteine (a toxic amino acid created by the body as it metabolizes protein) levels and cardiac hypertrophy (enlargement of the heart) in humans. Because there are many complex relationships among human characteristics, it will be difficult to answer her research question due to a significant risk that confounding factors will cloud her inferences. She wants to be sure that any differences in cardiac hypertrophy are due to high plasma homocysteine levels and not to other factors. Consequently, she needs to design her experiment carefully so that she controls lurking variables to the extent possible. Therefore, she decides to design a two-sample experiment with independent sampling: one group will be the experimental group, the other a control group. Knowing that many factors can affect the degree of cardiac hypertrophy (the response variable),

Dr. Nicholls controls these factors by randomly assigning the experimental units to the experimental or control group. She hopes the randomization will result in both groups having similar characteristics.

In her preliminary literature review, Dr. Nicholls uncovered an article in which the authors hypothesized that there might be a relationship between high plasma homocysteine levels in patients with end-stage renal disease (ESRD) and cardiac hypertrophy. She has asked you, as her assistant, to review this article.

Upon reading the article, you discover that the authors used a nonrandom process to select a control group and an ESRD group. The researchers enlisted 75 stable ESRD patients into their study, all on hemodialysis for between 6 and 312 months. The control group subjects were chosen so as to eliminate any intergroup differences in terms of mean blood pressure (BP) and gender. In an effort to minimize situational contaminants, all physical and biochemical measurements were made after an overnight fast. The results for the control and ESRD groups are reproduced in the following tables:

Clinical Characteristics (Mean ± Standard Deviation)		
Parameters	Controls (n = 57)	ESRD Subjects (n = 75)
Age (years)	49.2 ± 14.7	57.3 ± 15.1
Sex (M/F ratio)	1.4 ± 0.50	1.4 ± 0.50
Body surface area (m²)	1.85 ± 0.25	1.67 ± 0.20
Body mass index (kg/m²)	26.0 ± 4.70	23.7 ± 3.90
Systolic BP (mmHg)	145.0 ± 15.5	148.8 ± 29.7
Diastolic BP (mmHg)	85.0 ± 14.8	80.2 ± 14.3
Mean BP (mmHg)	104.5 ± 14.2	103.6 ± 17.4
Pulse pressure (mmHg)	59.4 ± 15.5	68.6 ± 24.3
Heart rate (beats/min)	63.0 ± 8.0	70.0 ± 9.0

Biological Findings (Mean ± Standard Deviation)		
Parameters	Controls (n = 57)	ESRD Subjects (n = 75)
Total cholesterol (mmol/L)	5.28 ± 1.04	4.91 ± 1.06
HDL cholesterol (mmol/L)	1.38 ± 0.39	1.07 ± 0.38
Triglycerides (mmol/L)	1.39 ± 0.63	1.90 ± 1.02
Serum albumin (g/L)	44.7 ± 2.60	39.9 ± 3.00
Plasma fibrinogen (g/L)	3.21 ± 0.78	4.75 ± 1.04
Plasma creatinine (mmol/L)	0.10 ± 0.01	0.90 ± 0.13
Blood urea (mmol/L)	6.10 ± 1.20	24.3 ± 2.00
Calcium (mmol/L)	2.46 ± 0.08	2.45 ± 0.12
Phosphates (mmol/L)	1.03 ± 0.21	1.88 ± 0.38

Source: Jacques Blacher et al., "Association between Plasma Homocysteine Concentrations and Cardiac Hypertrophy in End-Stage Renal Disease." *Journal of Nephrology* 12(4): 248–255, 1999. Article available at www.sin_italia.org/jnonline/vol12n4/blacher/blacher.htm.

1. Which type of sampling method, independent or dependent, was used in this experiment? Explain.

2. Using the appropriate hypothesis-testing procedure, determine whether the control and ESRD groups have equivalent population means for each of the various clinical and biochemical parameters. Dr. Nicholls requires that you indicate those parameters that have P-values less than 0.05 and those less than 0.01.

3. Detail any assumptions and the rationale behind making them that you made while carrying out your analysis. Is there any additional information that you would like to have? Explain. Are there any additional statistical procedures that you think might be useful for analyzing these data? Explain.

4. Based on your findings, does it appear that the control and ESRD groups have similar initial clinical characteristics and biochemical findings? Does it appear that the authors of this article were successful in reducing the likelihood that a confounding effect would obscure their results?

5. Even though Dr. Nicholls does not wish to restrict her research to patients with end-stage renal disease, how might the information presented for this research assist her in designing her own experiment?

6. Submit a written report to Dr. Nicholls that outlines all your findings and recommendations.

12

Inference on Categorical Data

Making an Informed Decision

Are there benefits to attending college? If so, what are they? See the Decision project on page 615.

PUTTING IT TOGETHER

In Chapters 9 through 11, we introduced statistical methods that can be used to test hypotheses regarding a population parameter such as p or μ.

Often, however, rather than being interested in testing a hypothesis regarding a parameter of a probability distribution, we want to test a hypothesis regarding the entire probability distribution. For example, we might want to test whether the distribution of colors in a bag of plain M&M candies is 13% brown, 14% yellow, 13% red, 20% orange, 24% blue, and 16% green. Methods for testing such hypotheses are covered in Section 12.1.

In Section 12.2, we discuss a method for determining whether two qualitative variables are independent based on a sample. If they are not independent, the value of one variable affects the value of the other variable, so the variables are somehow related. We conclude Section 12.2 by introducing a test for homogeneity of proportions, which compare proportions from two or more populations. This test is an extension of the two-sample z-test for proportions from independent samples discussed in Section 11.1.

We end the chapter with a discussion for comparing two proportions from dependent samples in Section 12.3.

12.1 Goodness-of-Fit Test

Preparing for This Section Before getting started, review the following:

- Mutually exclusive (Section 5.2, p. 262)
- Expected value (Section 6.1, pp. 321–322)
- Mean of a binomial random variable (Section 6.2, pp. 335–336)
- Chi-square distribution (Section 9.3, pp. 449–451)

Objective   ① Perform a goodness-of-fit test

① Perform a Goodness-of-Fit Test

In this section, we present a procedure that can be used to test hypotheses regarding a probability distribution. For example, we might want to test whether the distribution of plain M&M candies in a bag is 13% brown, 14% yellow, 13% red, 20% orange, 24% blue, and 16% green. Or we might want to test whether the number of hits a player gets in his next four at bats follows a binomial distribution with $n = 4$ and $p = 0.298$.

The methods for testing these types of hypotheses use the chi-square distribution, introduced in Section 9.3. Recall the following properties of the chi-square distribution.

> ### Characteristics of the Chi-Square Distribution
>
> 1. It is not symmetric.
> 2. Its shape depends on the degrees of freedom, just like Student's t-distribution.
> 3. As the number of degrees of freedom increases, it becomes more symmetric, as illustrated in Figure 1.
> 4. The values of χ^2 are nonnegative. That is, the values of χ^2 are greater than or equal to 0.

The critical values of the chi-square distribution can be found in Table VIII.

Definition   A **goodness-of-fit test** is an inferential procedure used to determine whether a frequency distribution follows a specific distribution.

Figure 1
Chi-square distributions

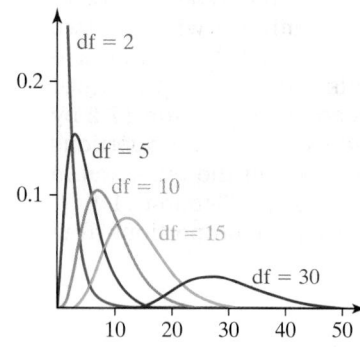

As an example, we might want to test whether a die is fair. This means that the probability of each outcome is $\dfrac{1}{6}$ when a die is cast. Because we give the die the benefit of the doubt (that is, assume that the die is fair), the null hypothesis is:

$$H_0: p_1 = p_2 = p_3 = p_4 = p_5 = p_6 = \frac{1}{6}$$

As another example, according to the Census Bureau in 2000, 19.0% of the population of the United States resided in the Northeast, 22.9% in the Midwest, 35.6% in the South, and 22.5% in the West. We might want to test whether the distribution of residents is the same today as it was in 2000. Since the null hypothesis is a statement of "no change," we have

H_0: The distribution of residents in the United States is the same today as it was in 2000.

The idea behind testing these types of hypotheses is to compare the actual number of observations for each category of data with the number of observations we would expect if the null hypothesis were true. If a significant difference exists between the observed counts and expected counts, we have evidence against the null hypothesis.

The method for obtaining the **expected counts** is an extension of the expected value of a binomial random variable. Recall that the mean (and therefore expected value) of a binomial random variable with n independent trials and probability of success, p, is given by $E = \mu = np$.

Expected Counts

In Other Words
The expected count for each category is the number of trials of the experiment times the probability of success in the category.

Suppose there are n independent trials of an experiment with $k \geq 3$ mutually exclusive possible outcomes. Let p_1 represent the probability of observing the first outcome and E_1 represent the expected count of the first outcome, p_2 represent the probability of observing the second outcome and E_2 represent the expected count of the second outcome, and so on. The expected counts for each possible outcome are given by

$$E_i = \mu_i = np_i \quad \text{for} \quad i = 1, 2, \ldots, k$$

EXAMPLE 1   Finding Expected Counts

Table 1 Distribution of Household Income, 2000	
Income	**Percent**
Under $15,000	7.0
$15,000 to $24,999	8.6
$25,000 to $34,999	9.3
$35,000 to $49,999	14.3
$50,000 to $74,999	19.7
$75,000 to $99,999	15.0
At least $100,000	26.1

Source: U.S. Census Bureau

Problem One growing concern regarding the U.S. economy is the inequality in the distribution of income. The data in Table 1 represent the distribution of household income for various levels of income in 2000. An economist wants to know if the distribution of income is changing, so she randomly selects 1500 households and obtains the household income. Find the expected number of households at each income level assuming that the distribution of income has not changed since 2000.
Note: The income data has been adjusted for inflation.

Approach Any quantitative variable (such as income) can become qualitative when categories of the variable are created. For example, in this data, we have seven categories of income. There are two steps to follow to determine the expected counts for each category.

Step 1 Determine the probabilities for each outcome.

Step 2 There are $n = 1500$ trials (the 1500 households surveyed) of the experiment. We expect $np_{\text{under } \$15,000}$ of the households surveyed to have an income under $15,000.

Solution

Step 1 The probabilities are the relative frequencies from the 2000 distribution:

$p_{\text{under } 15,000} = 0.070, p_{15,000-24,999} = 0.086, p_{25,000-34,999} = 0.093, p_{35,000-49,999} = 0.143,$
$p_{50,000-74,999} = 0.197, p_{75,000-99,999} = 0.150, p_{\text{at least } 100,000} = 0.261$

Step 2 The expected counts for each income are as follows:

Expected count of under $15,000:	$np_{\text{under } 15,000} = 1500(0.070) = 105$
Expected count of $15,000 to $24,999:	$np_{15,000 \text{ to } 24,999} = 1500(0.086) = 129$
Expected count of $25,000 to $34,999:	$np_{25,000 \text{ to } 34,999} = 1500(0.093) = 139.5$
Expected count of $35,000 to $49,999:	$np_{35,000 \text{ to } 49,999} = 1500(0.143) = 214.5$
Expected count of $50,000 to $74,999:	$np_{50,000 \text{ to } 74,999} = 1500(0.197) = 295.5$
Expected count of $75,000 to $99,999:	$np_{75,000 \text{ to } 99,999} = 1500(0.150) = 225$
Expected count of at least $100,000:	$np_{\text{At least } \$100,000} = 1500(0.261) = 391.5$

Of the 1500 households surveyed, the economist expects to have 105 households that earned under $15,000, 129 that earned $15,000 to $24,999, and so on, assuming the distribution of income has not changed since 2000.

● Now Work Problem 5

To conduct a hypothesis test, we compare the observed counts to the expected counts for each category. If the observed counts are significantly different from the expected counts, we have evidence against the null hypothesis. To perform this test, we need a test statistic and sampling distribution.

Test Statistic for Goodness-of-Fit Tests

Let O_i represent the observed counts of category i, E_i represent the expected counts of category i, k represent the number of categories, and n represent the number of independent trials of an experiment. Then

$$\chi^2 = \sum \frac{(O_i - E_i)^2}{E_i} \qquad i = 1, 2, \ldots, k$$

approximately follows the chi-square distribution with $k - 1$ degrees of freedom, provided that

1. all expected frequencies are greater than or equal to 1 (all $E_i \geq 1$) and
2. no more than 20% of the expected frequencies are less than 5.

Note: $E_i = np_i$ for $i = 1, 2, \ldots, k$.

> **CAUTION!**
> Goodness-of-fit tests are used to test hypotheses regarding the distribution of a variable based on a single population. If you wish to compare two or more populations, you must use the tests for homogeneity presented in Section 12.2.

In Example 1, there were $k = 7$ categories (seven levels of income).

Now that we know the distribution of goodness-of-fit tests, we can present a method for testing hypotheses regarding the distribution of a random variable.

The Goodness-of-Fit Test

To test hypotheses regarding a distribution, use the steps that follow.

Step 1 Determine the null and alternative hypotheses:

H_0: The random variable follows a certain distribution.

H_1: The random variable does not follow the distribution in the null hypothesis.

> **CAUTION!**
> If the requirements in Step 3(b) are not satisfied, one option is to combine two or more low-frequency categories into a single category.

Step 2 Decide on a level of significance, α, depending on the seriousness of making a Type I error.

Step 3

(a) Calculate the expected counts, E_i, for each of the k categories: $E_i = np_i$ for $i = 1, 2, \ldots, k$, where n is the number of trials and p_i is the probability of the ith category, assuming that the null hypothesis is true.

(b) Verify that the requirements for the goodness-of-fit test are satisfied.

1. All expected counts are greater than or equal to 1 (all $E_i \geq 1$).
2. No more than 20% of the expected counts are less than 5.

Classical Approach

Step 3 (continued)

(c) Compute the **test statistic**

$$\chi_0^2 = \sum \frac{(O_i - E_i)^2}{E_i}$$

Note: O_i is the observed count for the ith category.

Step 4 Determine the critical value using Table VIII. All goodness-of-fit tests are right-tailed tests, so the critical value is χ_α^2 with $k - 1$ degrees of freedom. See Figure 2.

P-Value Approach

By Hand Step 3 (continued)

(c) Compute the **test statistic**

$$\chi_0^2 = \sum \frac{(O_i - E_i)^2}{E_i}$$

Note: O_i is the observed count for the ith category.

(d) Use Table VIII to approximate the P-value by determining the area under the chi-square distribution with $k - 1$ degrees of freedom to the right of the test statistic. See Figure 3.

Figure 2

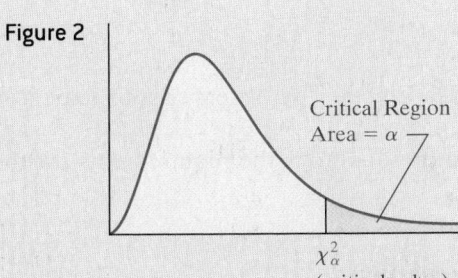

Critical Region
Area = α

χ_{α}^2
(critical value)

Compare the critical value to the statistic. If $\chi_0^2 > \chi_{\alpha}^2$, reject the null hypothesis.

Figure 3

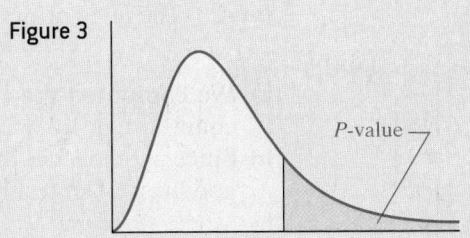

P-value

χ_0^2

Technology Step 3 (continued)

(c) Use a statistical spreadsheet or calculator with statistical capabilities to obtain the P-value. The directions for obtaining the P-value using the TI-84 Plus graphing calculators, Minitab, Excel, and StatCrunch are given in the Technology Step-by-Step on pages 586–587.

Step 4   If P-value $< \alpha$, reject the null hypothesis.

Step 5   State the conclusion.

EXAMPLE 2   Conducting a Goodness-of-Fit Test

Problem One growing concern regarding the U.S. economy is the inequality in the distribution of income. An economist wants to know if the distribution of income is changing, so she randomly selects 1500 households and obtains the household income shown in Table 2. Table 2 also contains the expected counts under the assumption the distribution has not changed since 2000 (obtained in Example 1). Does the evidence suggest that the distribution of income has changed since 2000 at the $\alpha = 0.05$ level of significance? **Note:** The data in Table 2 is based on the 2013 Current Population Survey and has been adjusted for inflation.

Table 2		
Income	**Observed Counts**	**Expected Counts**
Under $15,000	130	105
$15,000 to $24,999	137	129
$25,000 to $34,999	150	139.5
$35,000 to $49,999	207	214.5
$50,000 to $74,999	291	295.5
$75,000 to $99,999	202	225
At least $100,000	383	391.5

Approach Follow Steps 1 through 5 just listed.

Solution

Step 1   The null hypothesis is always a statement of "no difference." Here, that means there is no difference in distribution of income between 2000 and today. We are looking for the sample evidence (sample data) to show that the distribution is different today.

H_0: The distribution of household income in the United States is the same today as it was in 2000.

H_1: The distribution of household income in the United States is different today from what it was in 2000.

(continued)

Step 2   The level of significance is $\alpha = 0.05$.

Step 3

(a) We computed the expected counts in Example 1. The observed and expected counts are in Table 2.

(b) Since all expected counts are greater than or equal to 5, the requirements for the goodness-of-fit test are satisfied.

Classical Approach

Step 3 (continued)

(c) The test statistic is

$$\chi_0^2 = \sum \frac{(O_i - E_i)^2}{E_i}$$
$$= \frac{(130 - 105)^2}{105} + \frac{(137 - 129)^2}{129} + \frac{(150 - 139.5)^2}{139.5}$$
$$+ \frac{(207 - 214.5)^2}{214.5} + \frac{(291 - 295.5)^2}{295.5}$$
$$+ \frac{(202 - 225)^2}{225} + \frac{(383 - 391.5)^2}{391.5} = 10.105$$

Step 4   There are $k = 7$ categories, so we find the critical value using $7 - 1 = 6$ degrees of freedom. The critical value is $\chi_{0.05}^2 = 12.592$. See Figure 4.

Figure 4

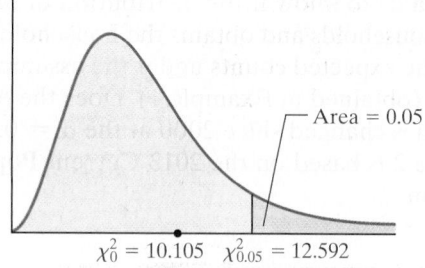

Area = 0.05

$\chi_0^2 = 10.105$ $\chi_{0.05}^2 = 12.592$

Because the test statistic, 10.105, is less than the critical value, 12.592, we do not reject the null hypothesis.

P-Value Approach

By Hand Step 3 (continued)

(c) The test statistic is

$$\chi_0^2 = \sum \frac{(O_i - E_i)^2}{E_i}$$
$$= \frac{(130 - 105)^2}{105} + \frac{(137 - 129)^2}{129} + \frac{(150 - 139.5)^2}{139.5}$$
$$+ \frac{(207 - 214.5)^2}{214.5} + \frac{(291 - 295.5)^2}{295.5}$$
$$+ \frac{(202 - 225)^2}{225} + \frac{(383 - 391.5)^2}{391.5} = 10.105$$

(d) There are $k = 7$ categories. The P-value is the area under the chi-square distribution with $7 - 1 = 6$ degrees of freedom to the right of $\chi_0^2 = 10.105$, as shown in Figure 5.

Figure 5

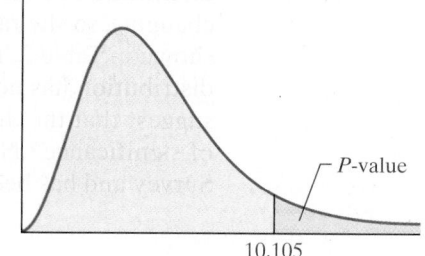

P-value

10.105

Using Table VIII, find the row that corresponds to 6 degrees of freedom. The value of 10.105 lies between 2.204, which corresponds to an area to the right of 0.9, and 10.645, which corresponds to an area to the right of 0.10. Therefore, the P-value is greater than 0.10. So, P-value > 0.10.

Technology Step 3 (continued)

(c) Figure 6 shows the result of the Goodness-of-Fit test from StatCrunch. The P-value is reported as 0.1203.

Figure 6   **Chi-Square goodness-of-fit results:**
Observed: Observed
Expected: Expected

N	DF	Chi-Square	P-value
1500	6	10.105251	0.1203

Step 4   The P-value of 0.1203 means that *if* the null hypothesis "the distribution of income is the same today as in 2000" is true, we would expect about 12 samples in 100 to yield the results at least as extreme as those we obtained. The observed results are not unusual under

Historical Note

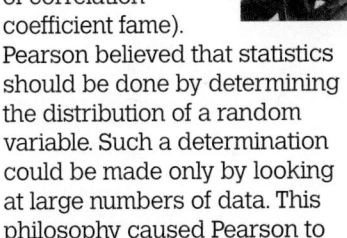

The goodness-of-fit test was invented by Karl Pearson (the Pearson of correlation coefficient fame). Pearson believed that statistics should be done by determining the distribution of a random variable. Such a determination could be made only by looking at large numbers of data. This philosophy caused Pearson to "butt heads" with Ronald Fisher, because Fisher believed in analyzing small samples.

the assumption the null hypothesis is true. Because the P-value is greater than the level of significance (Tech: $0.1203 > 0.05$), we do not reject the null hypothesis.

Step 5 The sample evidence suggests there is not sufficient evidence at the $\alpha = 0.05$ level of significance to conclude the distribution of income in the United States is different today than it was in 2000.

If we compare the observed and expected counts, we see that a shift in income is occurring (even though it is not a statistically significant shift). Lower levels of income have observed counts above what is expected under the assumption of no change in the distribution of household income. For example, we would expect 105 households in the survey to have incomes under $15,000 if there was no change in the distribution of income, but we observed 130 households in this income class. In addition, higher levels of income have lower than expected counts.

EXAMPLE 3 Conducting a Goodness-of-Fit Test

Table 3

Day of Week	Frequency
Sunday	46
Monday	76
Tuesday	83
Wednesday	81
Thursday	81
Friday	80
Saturday	53

Problem An obstetrician wants to know whether the proportion of children born on each day of the week is the same. She randomly selects 500 birth records and obtains the data shown in Table 3 (based on data obtained from *Vital Statistics of the United States*, 2012).

Is there reason to believe that the day on which a child is born does not occur with equal frequency at the $\alpha = 0.01$ level of significance?

Approach Follow Steps 1 through 5 presented on pages 582–583.

Solution

Step 1 The null hypothesis is a statement of "no difference," so we assume that the day on which a child is born occurs with equal frequency. If 1 represents Sunday, 2 represents Monday, and so on, we have

$$H_0: p_1 = p_2 = p_3 = p_4 = p_5 = p_6 = p_7 = \frac{1}{7}$$

H_1: At least one of the proportions is different from the others.

Step 2 The level of significance is $\alpha = 0.01$.

Step 3
(a) The expected count for each category (day of the week), assuming the null hypothesis is true, is

$$500\left(\frac{1}{7}\right) \approx 71.4$$

(b) Since all expected counts are greater than or equal to 5, the requirements for the goodness-of-fit test are satisfied.
(c) The test statistic is

$$\chi_0^2 = \frac{(46 - 500/7)^2}{500/7} + \frac{(76 - 500/7)^2}{500/7} + \frac{(83 - 500/7)^2}{500/7} + \frac{(81 - 500/7)^2}{500/7}$$
$$+ \frac{(81 - 500/7)^2}{500/7} + \frac{(80 - 500/7)^2}{500/7} + \frac{(53 - 500/7)^2}{500/7} = 19.568$$

(continued)

Classical Approach

Step 4 There are $k = 7$ categories, so we find the critical value using $7 - 1 = 6$ degrees of freedom. The critical value is $\chi^2_{0.01} = 16.812$. See Figure 7.

Figure 7

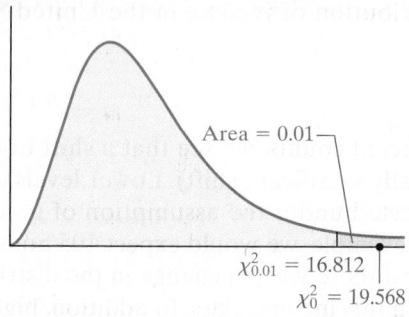

Area = 0.01

$\chi^2_{0.01} = 16.812$
$\chi^2_0 = 19.568$

Because the test statistic, 19.568, is greater than the critical value, 16.812, we reject the null hypothesis.

P-Value Approach

Step 3 (continued)

(d) There are $k = 7$ categories. The P-value is the area under the chi-square distribution with $7 - 1 = 6$ degrees of freedom to the right of $\chi^2_0 = 19.568$, as shown in Figure 8.

Figure 8

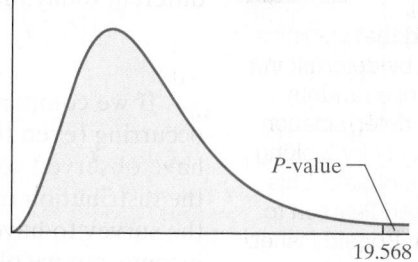

P-value

19.568

Using Table VIII, we find the row that corresponds to 6 degrees of freedom. The value of 19.568 is greater than 18.548, which has an area under the chi-square distribution of 0.005 to the right, so the P-value is less than 0.005 (P-value < 0.005). Using technology, the exact P-value is 0.0033.

Step 4 Because the P-value is less than the level of significance, $\alpha = 0.01$, we reject the null hypothesis.

Step 5 There is sufficient evidence at the $\alpha = 0.01$ level of significance to reject the belief that the day of the week on which a child is born occurs with equal frequency. ●

● **Now Work Problem 19**

Whenever we are testing a hypothesis, the evidence can never prove the null hypothesis to be true. For example, in performing the goodness-of-fit test from Example 2, we tested

H_0: the distribution of household income is the same today as in 2000

Although we failed to reject the null hypothesis, we are not saying that the distribution today is the same as it was in 2000. We are saying that we don't have enough evidence to conclude that the distribution has changed significantly. Unfortunately, goodness-of-fit tests cannot be used to test whether sample data follow a specific distribution. We can only say the data are consistent with a distribution stated in the null hypothesis.

Technology Step-by-Step Goodness-of-Fit Test

TI-84 Plus*

1. Enter the observed counts in L1 and enter the expected counts in L2.
2. Press STAT, highlight TESTS, and select
   D:χ^2-GOF-Test....
3. Enter L1 after Observed:, and enter L2 after Expected:. Enter the appropriate degrees of freedom following df:. Highlight either Calculate or Draw and press ENTER.

*The TI-83 and older TI-84 Plus graphing calculators do not have this capability.

Minitab

1. Enter the observed counts in column C1. Enter expected proportions or counts, assuming that the null hypothesis is true, in column C2, if necessary.
2. Select the **Stat** menu, highlight **Tables**, then choose **Chi-square Goodness-of-Fit Test (One Variable)** ....
3. Select the Observed Counts radio button and enter C1 in the cell. If you are testing equal proportions, select this radio button. If you entered expected proportions in C2, select the "Specific proportions" radio button and enter C2. If you entered expected counts in C2, select the "Proportions specified by historical counts" radio button and enter C2. Click OK.

Excel
1. Load the XLSTAT Add-in.
2. Enter the category names in column A. Enter the proportions used to formulate expected counts in column B. Enter the observed counts in column C.
3. Select **Parametric tests**, highlight Multinomial goodness of fit.
4. Place the cursor in the Frequencies cell. Highlight the data in column C. Place the cursor in the Expected proportions cell. Highlight the data in column B. Be sure the proportions radio button is selected. Check the Column labels box. Check the Chi-square test box. Enter a level of significance. Click OK.

StatCrunch
1. Enter the observed counts in the first column. Enter the expected counts in the second column. Name the columns observed and expected.
2. Select **Stat**, highlight **Goodness-of-fit**, then highlight **Chi-Square Test**.
3. Select the column that contains the observed counts and select the column that contains the expected counts. Click Compute!

12.1 Assess Your Understanding

Vocabulary and Skill Building

1. *True or False*: The shape of the chi-square distribution depends on the degrees of freedom.

2. A _____ test is an inferential procedure used to determine whether a frequency distribution follows a specific distribution.

3. Suppose there are n independent trials of an experiment with $k > 3$ mutually exclusive outcomes, where p_i represents the probability of observing the ith outcome. The _____ _____ for each possible outcome are given by $E_i = $ ___.

4. What are the two requirements that must be satisfied to perform a goodness-of-fit test?

In Problems 5 and 6, determine the expected counts for each outcome.

NW 5.

$n = 500$				
p_i	0.2	0.1	0.45	0.25
Expected counts				

6.

$n = 700$				
p_i	0.15	0.3	0.35	0.20
Expected counts				

In Problems 7–10, determine (a) the χ^2 test statistic, (b) the degrees of freedom, (c) the critical value using $\alpha = 0.05$, and (d) test the hypothesis at the $\alpha = 0.05$ level of significance.

7. $H_0: p_A = p_B = p_C = p_D = \dfrac{1}{4}$

H_1: At least one of the proportions is different from the others.

Outcome	A	B	C	D
Observed	30	20	28	22
Expected	25	25	25	25

8. $H_0: p_A = p_B = p_C = p_D = p_E = \dfrac{1}{5}$

H_1: At least one of the proportions is different from the others.

Outcome	A	B	C	D	E
Observed	38	45	41	33	43
Expected	40	40	40	40	40

9. H_0: The random variable X is binomial with $n = 4, p = 0.8$
H_1: The random variable X is not binomial with $n = 4$, $p = 0.8$

X	0	1	2	3	4
Observed	1	38	132	440	389
Expected	1.6	25.6	153.6	409.6	409.6

10. H_0: The random variable X is binomial with $n = 4, p = 0.3$
H_1: The random variable X is not binomial with $n = 4$, $p = 0.3$

X	0	1	2	3	4
Observed	260	400	280	50	10
Expected	240.1	411.6	264.6	75.6	8.1

Applying the Concepts

NW 11. Plain M&Ms According to the manufacturer of M&Ms, 13% of the plain M&Ms in a bag should be brown, 14% yellow, 13% red, 24% blue, 20% orange, and 16% green. A student randomly selected a bag of plain M&Ms. He counted the number of M&Ms that were each color and obtained the results shown in the table. Test whether plain M&Ms follow the distribution stated by M&M/Mars at the $\alpha = 0.05$ level of significance.

Color	Frequency
Brown	57
Yellow	64
Red	54
Blue	75
Orange	86
Green	64

12. Peanut M&Ms According to the manufacturer of M&Ms, 12% of the peanut M&Ms in a bag should be brown, 15% yellow, 12% red, 23% blue, 23% orange, and 15% green. A student randomly selected a bag of peanut M&Ms. He counted the number of M&Ms that were each color and obtained the results shown in the table. Test whether peanut M&Ms follow the distribution stated by M&M/Mars at the $\alpha = 0.05$ level of significance.

Color	Frequency
Brown	53
Yellow	66
Red	38
Blue	96
Orange	88
Green	59

13. Benford's Law, Part I Our number system consists of the digits 0, 1, 2, 3, 4, 5, 6, 7, 8, and 9. The first significant digit in any number must be 1, 2, 3, 4, 5, 6, 7, 8, or 9 because we do not write numbers such as 12 as 012. Although we may think that each first digit appears with equal frequency so that each digit has a $\frac{1}{9}$ probability of being the first significant digit, this is not true. In 1881, Simon Newcomb discovered that first digits do not occur with equal frequency. This same result was discovered again in 1938 by physicist Frank Benford. After studying much data, he was able to assign probabilities of occurrence to the first digit in a number as shown.

Digit	1	2	3	4	5
Probability	0.301	0.176	0.125	0.097	0.079

Digit	6	7	8	9
Probability	0.067	0.058	0.051	0.046

Source: T. P. Hill, "The First Digit Phenomenon," *American Scientist,* July–August, 1998.

The probability distribution is now known as Benford's Law and plays a major role in identifying fraudulent data on tax returns and accounting books. For example, the following distribution represents the first digits in 200 allegedly fraudulent checks written to a bogus company by an employee attempting to embezzle funds from his employer.

First digit	1	2	3	4	5	6	7	8	9
Frequency	36	32	28	26	23	17	15	16	7

Source: State of Arizona vs. Wayne James Nelson

(a) Because these data are meant to prove that someone is guilty of fraud, what would be an appropriate level of significance when performing a goodness-of-fit test?
(b) Using the level of significance chosen in part (a), test whether the first digits in the allegedly fraudulent checks obey Benford's Law.
(c) Based on the results of part (b), do you think that the employee is guilty of embezzlement?

14. Benford's Law, Part II Refer to Problem 13. The distribution in the next column lists the first digit of the surface area (in square miles) of 335 rivers. Is there evidence at the $\alpha = 0.05$ level of significance to support the belief that the distribution follows Benford's Law?

First digit	1	2	3	4	5	6	7	8	9
Frequency	104	55	36	38	24	29	18	14	17

Source: Eric W. Weisstein, Benford's Law, from *MathWorld*—A Wolfram Web Resource.

15. Always Wear a Helmet The National Highway Traffic Safety Administration publishes reports about motorcycle fatalities and helmet use. The distribution shows the proportion of fatalities by location of injury for motorcycle accidents.

Location of injury	Multiple Locations	Head	Neck	Thorax	Abdomen/Lumbar/Spine
Proportion	0.57	0.31	0.03	0.06	0.03

The following data show the location of injury and number of fatalities for 2068 riders not wearing a helmet.

Location of injury	Multiple Locations	Head	Neck	Thorax	Abdomen/Lumbar/Spine
Number	1036	864	38	83	47

(a) Does the distribution of fatal injuries for riders not wearing a helmet follow the distribution for all riders? Use the $\alpha = 0.05$ level of significance.
(b) Compare the observed and expected counts for each category. What does this information tell you?

16. Religion in Congress Is the religious make-up of the United States Congress reflective of that in the general population? The following table shows the religious affiliation of the 535 members of the 114th Congress along with the religious affiliation of a random sample of 1200 adult Americans.

Religion	Number of Members	Sample of Residents
Protestant	306	616
Catholic	164	287
Mormon	16	20
Orthodox Christian	5	7
Jewish	28	20
Buddhist/Muslim/Hindu/Other	6	57
Unaffiliated/Don't Know/Refused	10	193

Source: Congressional Quarterly Roll Call and Pew Research

(a) Determine the probability distribution for the religious affiliation of the members of the 114th Congress.
(b) Assuming the distribution of the religious affiliation of the adult American population is the same as that of the Congress, determine the number of adult Americans we would expect for each religion from a random sample of 1200 individuals.
(c) The data in the third column represent the declared religion of a random sample of 1200 adult Americans (based on data obtained from Pew Research). Do the sample data suggest that the American population has the same distribution of religious affiliation as the 114th Congress?
(d) Explain what the results of your analysis suggest.

17. Does It Matter Where I Sit? Does the location of your seat in a classroom play a role in attendance or grade? To answer this question, professors randomly assigned 400 students* in a general education physics course to one of four groups.

*The number of students was increased so that goodness-of-fit procedures could be used.

Source: Perkins, Katherine K. and Wieman, Carl E, "The Surprising Impact of Seat Location on Student Performance" *The Physics Teacher*, Vol. 43, Jan. 2005.

The 100 students in group 1 sat 0 to 4 meters from the front of the class, the 100 students in group 2 sat 4 to 6.5 meters from the front, the 100 students in group 3 sat 6.5 to 9 meters from the front, and the 100 students in group 4 sat 9 to 12 meters from the front.

(a) For the first half of the semester, the attendance for the whole class averaged 83%. So, if there is no effect due to seat location, we would expect 83% of students in each group to attend. The data show the attendance history for each group. How many students in each group attended, on average? Is there a significant difference among the groups in attendance patterns? Use the $\alpha = 0.05$ level of significance.

Group	1	2	3	4
Attendance	0.84	0.84	0.84	0.81

(b) For the second half of the semester, the groups were rotated so that group 1 students moved to the back of class and group 4 students moved to the front. The same switch took place between groups 2 and 3. The attendance for the second half of the semester averaged 80%. The data show the attendance records for the original groups (group 1 is now in back, group 2 is 6.5 to 9 meters from the front, and so on). How many students in each group attended, on average? Is there a significant difference in attendance patterns? Use the $\alpha = 0.05$ level of significance. Do you find anything curious about these data?

Group	1	2	3	4
Attendance	0.84	0.81	0.78	0.76

(c) At the end of the semester, the proportion of students in the top 20% of the class was determined. Of the students in group 1, 25% were in the top 20%; of the students in group 2, 21% were in the top 20%; of the students in group 3, 15% were in the top 20%; of the students in group 4, 19% were in the top 20%. How many students would we expect to be in the top 20% of the class if seat location plays no role in grades? Is there a significant difference in the number of students in the top 20% of the class by group?

(d) In earlier sections, we discussed results that were statistically significant, but did not have any practical significance. Discuss the practical significance of these results. In other words, given the choice, would you prefer sitting in the front or back?

18. Racial Profiling On January 1, 2004, it became mandatory for all police departments in Illinois to record data pertaining to race from every traffic stop. Mundelein, Illinois, has been collecting data since 2000. Rather than using census data to determine the racial distribution of the village, they thought it better to use data based on who is using the roads in Mundelein. So they collected data on at-fault drivers involved in car accidents in the village (the implicit assumption here is that race is independent of fault in a crash) and obtained the following distribution of race for users of roads in Mundelein.

Race	White	African American	Hispanic	Asian
Proportion	0.719	0.028	0.207	0.046

Source: Village of Mundelein, Illinois

The following data represent the races of all 9868 drivers who were stopped for a moving violation in the village of Mundelein in a recent year.

Race	White	African American	Hispanic	Asian
Proportion	7079	273	2025	491

Source: Village of Mundelein, Illinois

(a) Does the distribution of race in traffic stops reflect the distribution of drivers in Mundelein? In other words, is there any evidence of racial profiling in Mundelein? Use the $\alpha = 0.05$ level of significance.

(b) Compare observed and expected counts for each category. What does this information tell you?

NW 19. The NHL In his book *Outliers*, Malcolm Gladwell claims that more hockey players are born in January through March than in October through December. The following data show the number of players in the National Hockey League in the 2014–2015 season according to their birth month. Is there evidence to suggest that professional hockey players' birth dates are not uniformly distributed throughout the year at the $\alpha = 0.05$ level of significance?

Birth Month	Frequency
January–March	278
April–June	246
July–September	163
October–December	143

Source: http://freezethepuck.wordpress.com

20. Bicycle Deaths A researcher wanted to determine whether bicycle deaths were uniformly distributed over the days of the week. She randomly selected 200 deaths that involved a bicycle, recorded the day of the week on which the death occurred, and obtained the following results (the data are based on information obtained from the Insurance Institute for Highway Safety).

Day of the Week	Frequency	Day of the Week	Frequency
Sunday	16	Thursday	34
Monday	35	Friday	41
Tuesday	16	Saturday	30
Wednesday	28		

Is there reason to believe that bicycle fatalities occur with equal frequency with respect to day of the week at the $\alpha = 0.05$ level of significance?

21. Pedestrian Deaths A researcher wanted to determine whether pedestrian deaths were uniformly distributed over the days of the week. She randomly selected 300 pedestrian deaths, recorded the day of the week on which the death occurred, and obtained the following results (the data are based on information obtained from the Insurance Institute for Highway Safety).

Day of the Week	Frequency	Day of the Week	Frequency
Sunday	39	Thursday	41
Monday	40	Friday	49
Tuesday	30	Saturday	61
Wednesday	40		

Test the belief that the day of the week on which a fatality happens involving a pedestrian occurs with equal frequency at the $\alpha = 0.05$ level of significance.

22. Is the Die Loaded? A player in a craps game suspects that one of the dice is loaded. A loaded die is one in which all the possibilities (1, 2, 3, 4, 5, and 6) are not equally likely. The player throws the die 400 times, records the outcome after each throw, and obtains the following results:

Outcome	Frequency	Outcome	Frequency
1	62	4	62
2	76	5	57
3	76	6	67

(a) Do you think that the die is loaded? Use the $\alpha = 0.01$ level of significance.

(b) Why do you think the player might conduct the test at the $\alpha = 0.01$ level of significance rather than, say, the $\alpha = 0.1$ level of significance?

23. Grade Distributions At Joliet Junior College, the mathematics department decided to offer a redesigned course in Intermediate Algebra, called the Math Redesign Program (MRP). Laura Egner, the coordinator of the program, wanted to determine if the grade distribution in the course differed from that of traditional courses. The following shows the grade distribution of traditional courses based on historical records and the observed grades in three pilot classes in which the MRP program was utilized.

	A	B	C	D	F	W
Traditional Distribution	0.133	0.191	0.246	0.104	0.114	0.212
Observed Counts in MRP Program	7	16	10	13	6	12

(a) How many students were enrolled in the MRP program for the three pilot courses? Based on this result, determine the expected number of students for each grade assuming there is no difference in the distribution of MRP student grades and traditional grades.

(b) Does the sample evidence suggest that the distribution of grades is different from the traditional classes at the $\alpha = 0.01$ level of significance?

(c) Explain why it makes sense to use 0.01 as the level of significance.

(d) Suppose the MRP pilot program continues in three more classes with the grades earned for all six pilot courses shown below. Notice that the sample size was simply doubled with the grade distribution remaining unchanged. Does this sample evidence suggest that the distribution of grades is different from the traditional classes at the $\alpha = 0.01$ level of significance? What does this result suggest about the role of sample size in the ability to reject a statement in the null hypothesis?

	A	B	C	D	F	W
Observed Counts in MRP Program	14	32	20	26	12	24

24. Population Shift An urban economist wonders if the distribution of U.S. residents in the United States is different today than it was in 2000. The table shows the distribution of residents in 2000 along with the observed counts of residents today based on a random sample of 1500 US residents.

Region	Distribution in 2000	Observed Counts Today
Northeast	0.190	269
Midwest	0.229	327
South	0.356	554
West	0.225	350

(a) Does the sample evidence suggest the distribution of U.S. residents has changed since 2000? Use the $\alpha = 0.05$ level of significance.

(b) Suppose the survey was instead conducted on 5000 U.S. residents with the results shown in the following table. Compare the proportion of U.S. residents in each region based on the sample of 1500 U.S. residents versus 5000 U.S. residents. What do you notice?

Region	Observed Counts Today
Northeast	900
Midwest	1089
South	1846
West	1165

(c) Does the sample evidence for the survey of 5000 U.S. residents suggest that the distribution of residents in the United States has changed since 2000?

(d) Discuss the role sample size can play in determining whether the statement in the null hypothesis is rejected.

In Section 10.2, we tested hypotheses regarding a population proportion using a z-test. However, we can also use the chi-square goodness-of-fit test to test hypotheses with $k = 2$ possible outcomes. In Problems 25 and 26, we test hypotheses with the use of both methods.

25. Low Birth Weight According to the U.S. Census Bureau, 7.1% of all babies born are of low birth weight (<5 lb, 8 oz). An obstetrician wanted to know whether mothers between the ages of 35 and 39 years give birth to a higher percentage of low-birth-weight babies. She randomly selected 240 births for which the mother was 35 to 39 years old and found 22 low-birth-weight babies.

(a) If the proportion of low-birth-weight babies for mothers in this age group is 0.071, compute the expected number of low-birth-weight births to 35- to 39-year-old mothers. What is the expected number of births to mothers 35 to 39 years old that are not low birth weight?

(b) Answer the obstetrician's question at the $\alpha = 0.05$ level of significance using the chi-square goodness-of-fit test.

(c) Answer the question by using the approach presented in Section 10.2.

26. Living Alone? In 2000, 25.8% of Americans 15 years of age or older lived alone, according to the Census Bureau. A sociologist, who believes that this percentage is greater today, conducts a random sample of 400 Americans 15 years of age or older and finds that 164 are living alone.

(a) If the proportion of Americans aged 15 years or older living alone is 0.258, compute the following expected numbers: Americans 15 years of age or older who live alone; Americans 15 years of age or older who do not live alone.

(b) Test the sociologist's belief at the $\alpha = 0.05$ level of significance using the goodness-of-fit test.

(c) Test the belief by using the approach presented in Section 10.2.

27. Putting It Together: The V-2 Rocket in London In Thomas Pynchon's book *Gravity Rainbow*, the characters discuss whether the Poisson probabilistic model can be used to describe the locations that Germany's feared V-2 rocket would land in. They divided London into 0.25-km² regions. They then counted the number of rockets that landed in each region, with the following results:

Number of rocket hits	0	1	2	3	4	5	6	7
Observed number of regions	229	211	93	35	7	0	0	1

Source: Lawrence Lesser. "Even More Fun Learning Statistics." *Stats: The Magazine for Students of Statistics*, Issue 49.

(a) Estimate the mean number of rocket hits in a region by computing $\mu = \sum xP(x)$. Round your answer to four decimal places.

(b) Explain why the requirements for conducting a goodness-of-fit test are not satisfied.

(c) After consolidating the table, we obtain the following distribution for rocket hits. Using the Poisson probability model, $P(x) = \dfrac{\mu^x}{x!}e^{-\mu}$, where μ is the mean from part (a), we can obtain the probability distribution for the number of rocket hits. Find the probability of 0 hits in a region. Then find the probability of 1 hit, 2 hits, 3 hits, and 4 or more hits.

Number of rocket hits	0	1	2	3	4 or more
Observed number of regions	229	211	93	35	8

(d) A total of $n = 576$ rockets was fired. Determine the expected number of rocket hits by computing "expected number of rockets" $= np$, where p is the probability of observing that particular number of hits in the region.

(e) Conduct a goodness-of-fit test for the distribution using the $\alpha = 0.05$ level of significance. Do the rocket hits appear to be modeled by a Poisson random variable?

28. Putting It Together: Weldon's Dice On February 2, 1894, Frank Raphael Weldon wrote a letter to Francis Galton that included the results of 26,306 rolls of 12 dice. Weldon recorded the results such that a roll of a 5 or 6 resulted in a success, while a roll of 1, 2, 3, or 4 was a failure. The number of successes in each roll of the 12 dice are recorded in the table.

Number of Successes	Frequency	Number of Successes	Frequency
0	185	7	1331
1	1149	8	403
2	3265	9	105
3	5475	10	14
4	6114	11	4
5	5194	12	0
6	3067		

Source: Chance Magazine, Vol. 22, No. 4, 2009

(a) What is the probability of rolling a 5 or 6 when throwing a six-sided fair die?

(b) Treating the probability determined in part (a) as the probability of success, compute the theoretical probability of $0, 1, 2, \ldots, 12$ successes in throwing 12 dice.

(c) Use the probabilities found in part (b) to determine the expected frequency in observing $0, 1, 2, \ldots, 12$ successes after throwing the 12 dice 26,306 times.

(d) Conduct a goodness-of-fit test to determine if the number of successes follows a binomial probability distribution. **Note:** Combine 11 and 12 into a single bin.

Retain Your Knowledge

29. Buying a New Car How much does the typical person pay for a new 2015 Buick Regal? The following data represent the selling price of a random sample of new Regals (in dollars).

41,215	41,303	41,453	41,898	40,988
40,078	41,215	39,623	42,352	41,898
40,533	42,580	40,306	41,670	39,851

Source: TrueCar.com

(a) Is this data quantitative or qualitative?

(b) Find the mean and median price of a new 2015 Regal.

(c) Find the standard deviation and interquartile range.

(d) Verify it is reasonable to conclude that this data come from a population that is normally distributed.

(e) Draw a boxplot of the data.

(f) Estimate the typical price paid for a new 2015 Buick Regal with 90% confidence.

(g) Would a 90% confidence interval for all new 2015 domestic vehicles be wider or narrower? Explain.

Explaining the Concepts

30. Why is goodness of fit a good choice for the title of the procedures used in this section?

31. Explain why chi-square goodness-of-fit tests are always right tailed.

32. If the expected count of a category is less than 1, what can be done to the categories so that a goodness-of-fit test can still be performed?

12.2 Tests for Independence and the Homogeneity of Proportions

Preparing for This Section Before getting started, review the following:

- The language of hypothesis tests (Section 10.1, pp. 473–478)
- Independent events (Section 5.3, pp. 273–274)
- Contingency tables and association (Section 4.4, pp. 225–231)

- Mean of a binomial random variable (Section 6.2, pp. 335–336)
- Testing a hypothesis about two population proportions from independent samples (Section 11.1, pp. 526–530)

Objectives ❶ Perform a test for independence

❷ Perform a test for homogeneity of proportions

In Section 4.4, we discussed the association between two variables by examining the conditional distribution in a contingency table. The data presented in a contingency table represent the measurement of two categorical variables on each individual in the study and allow us to describe the association between the two variables.

We now introduce an inferential method to determine whether an apparent association between two categorical variables is statistically significant.

❶ Perform a Test for Independence

Is there a relationship between marital status and happiness? The data in Table 4 show the marital status and happiness of individuals who participated in the General Social Survey.

Table 4			Marital Status		
		Married	**Widowed**	**Divorced/ Separated**	**Never Married**
Happiness	**Very Happy**	600	63	112	144
	Pretty Happy	720	142	355	459
	Not Too Happy	93	51	119	127

For the data in this contingency table, we are looking at counts of a single group of individuals categorized based on the row variable "Happiness" and column variable "Marital Status." It seems reasonable to wonder whether one's marital status is associated with one's level of happiness. To determine whether there is a statistically significant association between two categorical variables, we perform a *chi-square test of independence*.

Definition The **chi-square test for independence** is used to determine whether there is an association between a row variable and column variable in a contingency table constructed from sample data.

- The null hypothesis is that the variables are not associated, or independent.
- The alternative hypothesis is that the variables are associated, or dependent.

In Other Words
In a chi-square independence test, the null hypothesis is always

H_0: The variables are independent

The alternative hypothesis is always

H_1: The variables are not independent

The idea behind testing these types of hypotheses is to compare actual counts to the counts we would expect assuming that the variables are independent (that is, assuming that the null hypothesis were true). If a significant difference between the actual counts and expected counts exists, we have evidence against the null hypothesis.

To obtain the **expected counts**, compute the number of observations expected within each cell under the assumption of independence (the null hypothesis is true). Recall that if two events E and F are independent, then $P(E \text{ and } F) = P(E) \cdot P(F)$. We can use the Multiplication Rule for Independent Events to obtain the expected proportion of observations within each cell under the assumption of independence. Then multiply each proportion by n, the sample size, to obtain the expected count within each cell.*

EXAMPLE 1 Determining the Expected Counts in a Test for Independence

Problem Is there a relationship between marital status and happiness? The data in Table 4 show the marital status and happiness of individuals who participated in the General Social Survey. Compute the expected counts within each cell, assuming that marital status and happiness are independent.

Approach

Step 1 Compute the row and column totals.

Step 2 Compute the relative marginal frequencies for the row variable and column variable.

Step 3 Use the Multiplication Rule for Independent Events to compute the expected proportion of observations within each cell, assuming independence.

Step 4 Multiply the proportions by 2985, the sample size, to obtain the expected counts within each cell.

Solution

Step 1 The row totals (blue) and column totals (red) are presented in Table 5.

Table 5

		Marital Status				
		Married	**Widowed**	**Divorced/Separated**	**Never Married**	**Row Totals**
Happiness	**Very Happy**	600	63	112	144	919
	Pretty Happy	720	142	355	459	1676
	Not Too Happy	93	51	119	127	390
	Column Totals	1413	256	586	730	2985

Step 2 The relative marginal frequencies for the row variable (happiness) and column variable (marital status) are presented in Table 6.

Table 6

		Marital Status				
		Married	**Widowed**	**Divorced/Separated**	**Never Married**	**Relative Frequency**
Happiness	**Very Happy**	600	63	112	144	$\frac{919}{2985} \approx 0.308$
	Pretty Happy	720	142	355	459	$\frac{1676}{2985} \approx 0.561$
	Not Too Happy	93	51	119	127	$\frac{390}{2985} \approx 0.131$
	Relative Frequency	$\frac{1413}{2985} \approx 0.473$	$\frac{256}{2985} \approx 0.086$	$\frac{586}{2985} \approx 0.196$	$\frac{730}{2985} \approx 0.245$	1

(continued)

---

*Recall that the expected value of a binomial random variable for n independent trials of a binomial experiment with probability of success p is given by $E(X) = \mu = np$.

Step 3 Assume the variables are independent and use the Multiplication Rule for Independent Events to compute the expected proportions for each cell. For example, the proportion of individuals who are "very happy" and "married" would be

$$\begin{pmatrix} \text{Proportion "very happy"} \\ \text{and "married"} \end{pmatrix} = (\text{proportion "very happy"}) \cdot (\text{proportion "married"})$$

$$= \left(\frac{919}{2985}\right)\left(\frac{1413}{2985}\right)$$

$$= 0.145737$$

Table 7 shows the expected proportion in each cell, assuming independence.

Table 7

		Marital Status			
		Married	**Widowed**	**Divorced/Separated**	**Never Married**
	Very Happy	0.145737	0.026404	0.060440	0.075292
Happiness	**Pretty Happy**	0.265783	0.048153	0.110226	0.137312
	Not Too Happy	0.061847	0.011205	0.025649	0.031952

Step 4 Multiply the expected proportions in Table 7 by 2985, the sample size, to obtain the expected counts. See Table 8.

Table 8

		Marital Status			
		Married	**Widowed**	**Divorced/Separated**	**Never Married**
	Very Happy	2985(0.145737)	2985(0.026404)	2985(0.060440)	2985(0.075292)
		= 435.025	= 78.816	= 180.413	= 224.747
Happiness	**Pretty Happy**	793.362	143.737	329.025	409.876
	Not Too Happy	184.613	33.447	76.562	95.377

If happiness and marital status are independent, we would expect a random sample of 2985 individuals to contain about 435 who are "very happy" and "married."   ●

The technique used in Example 1 to find the expected counts might seem rather tedious. It certainly would be more pleasant if we could determine a shortcut formula to obtain the expected counts. Consider the expected count for "very happy" and "married." This expected count was obtained by multiplying the proportion of individuals who are "very happy," the proportion of individuals who are "married," and the number of individuals in the sample. That is,

$$\text{Expected count} = (\text{proportion "very happy"})(\text{proportion "married"})(\text{sample size})$$

$$= \frac{919}{2985} \cdot \frac{1413}{2985} \cdot 2985$$

$$= \frac{919 \cdot 1413}{2985} \quad \text{Cancel the 2985s}$$

$$= \frac{(\text{row total for happiness})(\text{column total for marital status})}{\text{table total}}$$

This leads to the following general result:

Expected Frequencies in a Chi-Square Test for Independence

To find the expected frequency in a cell when performing a chi-square independence test, multiply the cell's row total by its column total and divide this result by the table total. That is,

$$\text{Expected frequency} = \frac{(\text{row total})(\text{column total})}{\text{table total}} \tag{1}$$

For example, the expected frequency for "very happy and married" is

$$\text{Expected frequency} = \frac{(\text{row total})(\text{column total})}{\text{table total}} = \frac{(919)(1413)}{2985} = 435.024$$

● Now Work Problem 7(a)

This result is close to that obtained in Table 8 (the difference exists because of rounding error; in fact, 435.024 is more accurate).

We need a test statistic and sampling distribution to see whether the expected and observed counts are significantly different.

Test Statistic for the Test of Independence

Let O_i represent the observed number of counts in the ith cell and E_i represent the expected number of counts in the ith cell. Then

$$\chi^2 = \sum \frac{(O_i - E_i)^2}{E_i}$$

approximately follows the chi-square distribution with $(r - 1)(c - 1)$ degrees of freedom, where r is the number of rows and c is the number of columns in the contingency table, provided that (1) all expected frequencies are greater than or equal to 1 and (2) no more than 20% of the expected frequencies are less than 5.

In Example 1, there were $r = 3$ rows and $c = 4$ columns.

We now present a method for testing hypotheses regarding the association between two variables in a contingency table.

Chi-Square Test for Independence

To test hypotheses regarding the association between (or independence of) two variables in a contingency table, use the steps that follow:

Step 1 Determine the null and alternative hypotheses.

H_0: The row variable and column variable are independent.
H_1: The row variable and column variable are dependent.

Step 2 Choose a level of significance, α, depending on the seriousness of making a Type I error.

(continued)

Step 3
(a) Calculate the expected frequencies (counts) for each cell in the contingency table using Formula (1).
(b) Verify that the requirements for the chi-square test for independence are satisfied:

   1. All expected frequencies are greater than or equal to 1 (all $E_i \geq 1$).
   2. No more than 20% of the expected frequencies are less than 5.

Classical Approach	P-Value Approach
Step 3 (continued)	**By Hand Step 3 (continued)**
(c) Compute the **test statistic**	**(c)** Compute the **test statistic**

Classical Approach:

$$\chi_0^2 = \sum \frac{(O_i - E_i)^2}{E_i}$$

Note: O_i is the observed frequency for the *i*th cell.

Step 4 Determine the critical value using Table VIII. All chi-square tests for independence are right-tailed tests, so the critical value is χ_α^2 with $(r - 1)(c - 1)$ degrees of freedom, where r is the number of rows and c is the number of columns in the contingency table. See Figure 9.

Figure 9

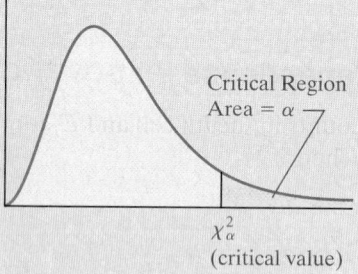

Critical Region
Area = α

χ_α^2
(critical value)

Compare the critical value to the test statistic. If $\chi_0^2 > \chi_\alpha^2$, reject the null hypothesis.

P-Value Approach:

$$\chi_0^2 = \sum \frac{(O_i - E_i)^2}{E_i}$$

Note: O_i is the observed frequency for the *i*th cell.
(d) Use Table VIII to determine an approximate P-value by determining the area under the chi-square distribution with $(r - 1)(c - 1)$ degrees of freedom to the right of the test statistic, where r is the number of rows and c is the number of columns in the contingency table. See Figure 10.

Figure 10

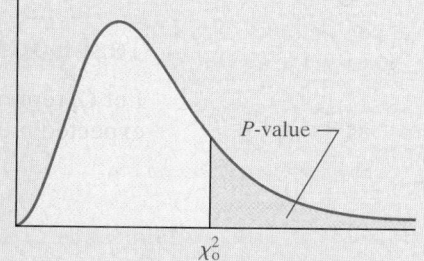

P-value

χ_0^2

Technology Step 3 (continued)
(c) Use a statistical spreadsheet or calculator with statistical capabilities to obtain the P-value. The directions for obtaining the P-value using the TI-83/84 Plus graphing calculators, Minitab, Excel, and StatCrunch are in the Technology Step-by-Step on pages 602–603.

Step 4 If P-value $< \alpha$, reject the null hypothesis.

Step 5 State the conclusion.

EXAMPLE 2 Performing a Chi-Square Test for Independence

Problem Does one's happiness depend on one's marital status? We present the data from Table 4 in Example 1 again in Table 9 to answer this question. Use the $\alpha = 0.05$ level of significance.

Table 9					
		Marital Status			
		Married	**Widowed**	**Divorced/Separated**	**Never Married**
	Very Happy	600	63	112	144
Happiness	**Pretty Happy**	720	142	355	459
	Not Too Happy	93	51	119	127

Approach Follow Steps 1 through 5 just given.

Solution

Step 1 We want to determine whether happiness and marital status are dependent or independent. The null hypothesis is a statement of "no effect," so we state the hypotheses as follows:

H_0: Happiness and marital status are independent (not related)

H_1: Happiness and marital status are dependent (related)

Step 2 The level of significance is $\alpha = 0.05$.

Step 3

(a) The expected frequencies were determined in Example 1. Table 10 shows the observed frequencies, with the expected frequencies in parentheses.

Table 10

		Marital Status			
		Married	**Widowed**	**Divorced/Separated**	**Never Married**
Happiness	**Very Happy**	600 (435.025)	63 (78.816)	112 (180.413)	144 (224.747)
	Pretty Happy	720 (793.362)	142 (143.737)	355 (329.025)	459 (409.876)
	Not Too Happy	93 (184.613)	51 (33.447)	119 (76.562)	127 (95.377)

(b) Since none of the expected frequencies are less than 5, the requirements for the Goodness-of-Fit Test are satisfied.

Classical Approach
Step 3 (continued)

(c) $\chi_0^2 = \dfrac{(600 - 435.025)^2}{435.025} + \dfrac{(63 - 78.816)^2}{78.816} + \dfrac{(112 - 180.413)^2}{180.413} +$

$\cdots + \dfrac{(119 - 76.562)^2}{76.562} + \dfrac{(127 - 95.377)^2}{95.377}$

$= 224.116$

Step 4 There are $r = 3$ rows and $c = 4$ columns. The critical value is $\chi_{0.05}^2 = 12.592$ using $(r - 1)(c - 1) = (3 - 1)(4 - 1) = 6$ degrees of freedom. See Figure 11.

Figure 11

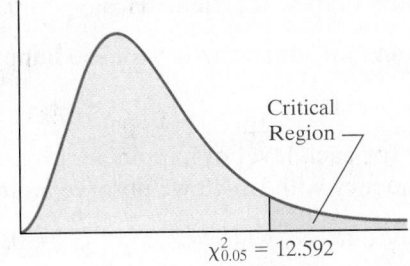

$\chi_{0.05}^2 = 12.592$

Critical Region

Because the test statistic, 224.116, is greater than the critical value, $\chi_{0.05}^2 = 12.592$, we reject the null hypothesis.

P-Value Approach
By Hand Step 3 (continued)

(c) $\chi_0^2 = \dfrac{(600 - 435.025)^2}{435.025} + \dfrac{(63 - 78.816)^2}{78.816} + \dfrac{(112 - 180.413)^2}{180.413} +$

$\cdots + \dfrac{(119 - 76.562)^2}{76.562} + \dfrac{(127 - 95.377)^2}{95.377}$

$= 224.116$

(d) There are $r = 3$ rows and $c = 4$ columns, so we find the P-value using $(r - 1)(c - 1) = (3 - 1)(4 - 1) = 6$ degrees of freedom. The P-value is the area under the chi-square distribution with 6 degrees of freedom to the right of $\chi_0^2 = 224.116$.

Using Table VIII, we find the row that corresponds to 6 degrees of freedom. The value of 224.116 is greater than 18.548. The area under the chi-square distribution to the right of 18.548 is 0.005. Because 224.116 is greater than 18.548, the P-value is less than 0.005, so we have P-value < 0.005.

Technology Step 3 (continued)

(c) Figure 12 on the next page shows the P-value obtained from a TI-84 Plus C graphing calculator using the Calculate feature. The P-value is reported as 1.38×10^{-45}, which is very close to 0.

(continued)

Figure 12

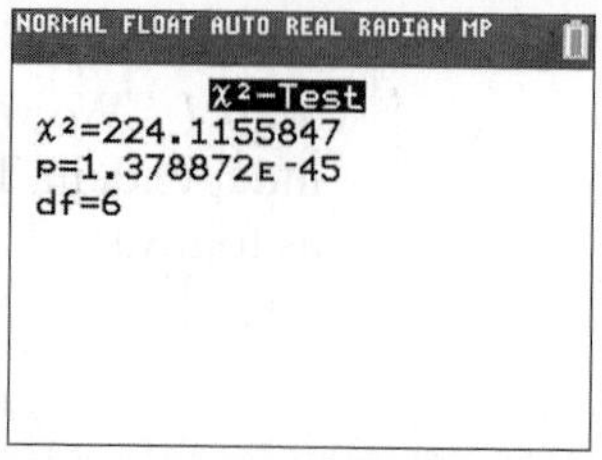

```
NORMAL FLOAT AUTO REAL RADIAN MP
              X²-Test
X²=224.1155847
P=1.378872ᴇ-45
df=6
```

Step 4 Because the P-value is less than the level of significance, we reject the null hypothesis.

Step 5 The data suggest there is sufficient evidence, at the $\alpha = 0.05$ level of significance, to conclude that happiness and marital status are dependent. We conclude that happiness and marital status are associated with each other.

● **Now Work Problems 7(b)–(e)**

To see the association between happiness and marital status, draw bar graphs of the conditional distributions of happiness by marital status. Recall that a conditional distribution lists the relative frequency of each category of a variable, given a specific value of the other variable in a contingency table. For example, we can calculate the relative frequency of "very happy," given that an individual is "married." We repeat this for each remaining category of marital status.

EXAMPLE 3   Constructing a Conditional Distribution and Bar Graph

Problem Find the conditional distribution of happiness by marital status for the data in Table 9. Then draw a bar graph that represents the conditional distribution of happiness by marital status.

Approach First, compute the relative frequency for happiness, given that the individual is "married." Then compute the relative frequency for happiness, given that the individual is "widowed," and so on. For each marital status, draw three bars side by side. The horizontal axis represents marital status, and the vertical axis represents the relative frequency of the level of happiness.

Solution Start with the individuals who are "married." The relative frequency with which we observe an individual who is "very happy," given that the individual is "married," is $\frac{600}{1413} = 0.425$. The relative frequency with which we observe an individual who is "pretty happy," given that the individual is "married," is $\frac{720}{1413} = 0.510$. The relative frequency with which we observe an individual who is "not too happy," given that the individual is "married," is $\frac{93}{1413} = 0.066$.

Now compute the relative frequency for each level of happiness, given that the individual is "widowed." The relative frequency with which we observe an individual who is "very happy," given that the individual is "widowed," is $\frac{63}{256} = 0.246$. The relative frequency with which we observe an individual who is "pretty happy," given that the individual is "widowed," is $\frac{142}{256} = 0.555$. The relative frequency with which we observe an individual who is "not too happy," given that the individual is "widowed," is $\frac{51}{256} = 0.199$.

We repeat the process for "divorced/separated" and "never married" and obtain Table 11.

Table 11		Marital Status			
		Married	**Widowed**	**Divorced/Separated**	**Never Married**
Happiness	**Very Happy**	$\dfrac{600}{1413} = 0.425$	$\dfrac{63}{256} = 0.246$	$\dfrac{112}{586} = 0.191$	$\dfrac{144}{730} = 0.197$
	Pretty Happy	$\dfrac{720}{1413} = 0.510$	$\dfrac{142}{256} = 0.555$	$\dfrac{355}{586} = 0.606$	$\dfrac{459}{730} = 0.629$
	Not Too Happy	$\dfrac{93}{1413} = 0.066$	$\dfrac{51}{256} = 0.199$	$\dfrac{119}{586} = 0.203$	$\dfrac{127}{730} = 0.174$

From the conditional distribution by marital status, the association between happiness and marital status should be apparent. The proportion of individuals who are "very happy" is highest in the "married" category. In addition, the proportion of individuals who are "not too happy" is lowest in the "married" category.

Figure 13 shows the graph of the conditional distribution. The blue bars represent the proportion of individuals who are "Very Happy" for each marital status, the red bars represent the proportion of individuals who are "Pretty Happy" for each marital status, and the green bars represent the proportion of individuals who are "Not Too Happy" for each marital status. From the bar graph, it is clear that a higher proportion of married individuals are "very happy" compared with the other categories of marital status.

Figure 13

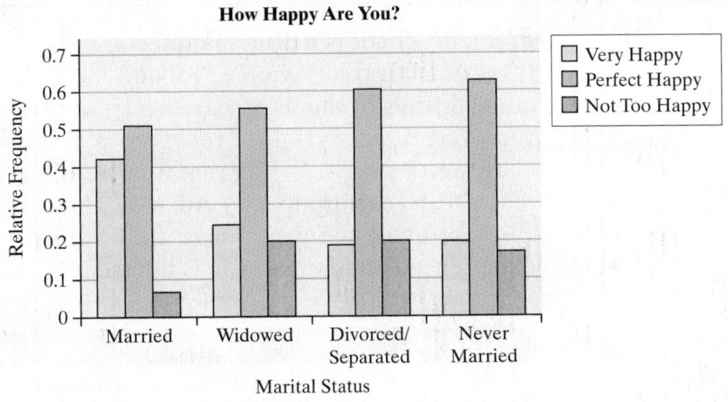

How Happy Are You?

Legend: Very Happy / Perfect Happy / Not Too Happy

X-axis: Married, Widowed, Divorced/Separated, Never Married — Marital Status
Y-axis: Relative Frequency

• Now Work Problem 7(f)

Perform a Test for Homogeneity of Proportions

A second type of chi-square test can be used to compare the population proportions from two or more independent samples. The test is an extension of the two-sample Z-test introduced in Section 11.1, where we compared two population proportions from independent samples.

Definition

In a **chi-square test for homogeneity of proportions**, we test whether different populations have the same proportion of individuals with some characteristic.

In Other Words
The chi-square test for homogeneity of proportions is used to compare proportions from two or more populations.

For example, we might look at the proportion of individuals who experience headaches as a side effect for a placebo group (group 1) and for one experimental group that receives 50 milligrams (mg) per day of a medication (group 2) and another that receives 100 mg per day (group 3). We assume that the proportion of individuals who experience a headache as a side effect is the same in each population (because the null hypothesis is a statement of "no difference"). So our hypotheses are

$$H_0: p_1 = p_2 = p_3$$
H_1: At least one of the population proportions is different from the others.

> The procedures for performing a test of homogeneity are identical to those for a test of independence.

While the procedures for the test for independence and the test of homogeneity of proportions are the same, the data differ. How? In the test for independence, we are measuring two variables (such as marital status and level of happiness) on each individual. In other words, a single population is segmented based on the value of two variables. In the test of homogeneity of proportions, we consider whether the proportion of individuals among different populations have the same value.

So, if you have a single population in which two variables are measured on each individual to assess whether one variable might be associated with another, conduct a test of independence. If you have two or more populations in which you want to determine equality of proportions among the populations, conduct a test of homogeneity of proportions.

EXAMPLE 4   A Test for Homogeneity of Proportions

Problem Zocor is a drug manufactured by Merck and Co. that is meant to reduce the level of LDL (bad) cholesterol and increase the level of HDL (good) cholesterol. In clinical trials of the drug, patients were randomly divided into three groups. Group 1 received Zocor; group 2 received a placebo; group 3 received cholestyramine, a cholesterol-lowering drug that is currently available. Table 12 contains the number of patients in each group who did and did not experience abdominal pain as a side effect. Is there evidence to indicate that the proportion of subjects in each group who experienced abdominal pain is different at the $\alpha = 0.01$ level of significance?

Table 12

	Group 1 (Zocor)	Group 2 (placebo)	Group 3 (cholestyramine)
Number of people who experienced abdominal pain	51	5	16
Number of people who did not experience abdominal pain	1532	152	163

Source: Merck and Co.

Approach Follow Steps 1 through 5 on pages 595–596.

Solution

Step 1   The null hypothesis is a statement of "no difference," so the proportion of subjects in each group who experienced abdominal pain are equal. We state the hypotheses as follows:

$$H_0: p_1 = p_2 = p_3$$
H_1: At least one of the proportions is different from the others.

Here, p_1, p_2, and p_3 are the proportions in groups 1, 2, and 3, respectively.

Step 2   The level of significance is $\alpha = 0.01$.

Step 3

(a) The expected frequency of subjects who experienced abdominal pain in group 1 is the product of the row total of individuals who experienced abdominal pain and the column total of individuals in group 1, divided by the total number of subjects in the study. So

$$E = \frac{72 \cdot 1583}{1919} = 59.393$$

Table 13 contains the row and column totals, the observed frequencies, and the expected frequencies (in parentheses).

Table 13

	Observed (and Expected) Frequencies			
	Group 1 (Zocor)	Group 2 (placebo)	Group 3 (cholestyramine)	Row Totals
Number of people who experienced abdominal pain	51 (59.393)	5 (5.891)	16 (6.716)	72
Number of people who did not experience abdominal pain	1532 (1523.607)	152 (151.109)	163 (172.284)	1847
Column totals	1583	157	179	1919

(b) Since none of the expected frequencies are less than 5, the requirements for the test of homogeneity of proportions are satisfied.

Classical Approach

Step 3 (continued)

(c) The test statistic is

$$\chi_0^2 = \frac{(51 - 59.393)^2}{59.393} + \frac{(5 - 5.891)^2}{5.891} + \frac{(16 - 6.716)^2}{6.716}$$
$$+ \frac{(1532 - 1523.607)^2}{1523.607} + \frac{(152 - 151.109)^2}{151.109} + \frac{(163 - 172.284)^2}{172.284}$$
$$= 14.707$$

Step 4 There are $r = 2$ rows and $c = 3$ columns, so we find the critical value using $(2 - 1)(3 - 1) = 2$ degrees of freedom. The critical value is $\chi_{0.01}^2 = 9.210$. See Figure 14.

Figure 14

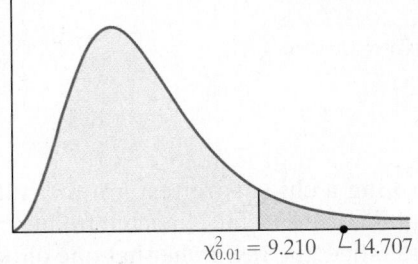

$\chi_{0.01}^2 = 9.210$ 14.707

Because the test statistic, 14.707, is greater than the critical value, 9.210, we reject the null hypothesis.

P-Value Approach

By Hand Step 3 (continued)

(c) The test statistic is

$$\chi_0^2 = \frac{(51 - 59.393)^2}{59.393} + \frac{(5 - 5.891)^2}{5.891} + \frac{(16 - 6.716)^2}{6.716}$$
$$+ \frac{(1532 - 1523.607)^2}{1523.607} + \frac{(152 - 151.109)^2}{151.109} + \frac{(163 - 172.284)^2}{172.284}$$
$$= 14.707$$

(d) There are $r = 2$ rows and $c = 3$ columns, so we find the P-value using $(2 - 1)(3 - 1) = 2$ degrees of freedom. The P-value is the area under the chi-square distribution with 2 degrees of freedom to the right of $\chi_0^2 = 14.707$, as shown in Figure 15.

Figure 15

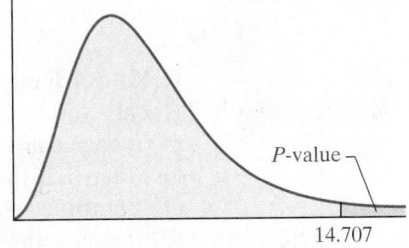

P-value

14.707

Using Table VIII, in the row corresponding to 2 degrees of freedom, the area under the chi-square distribution with 2 degrees of freedom, to the right of 10.597, is 0.005. Because 14.707 is to the right of 10.597, the P-value is less than 0.005. So P-value < 0.005.

(continued)

Technology Step 3 (continued)

(c) Figure 16 shows the *P*-value obtained from StatCrunch. The *P*-value is reported as 0.0006.

Figure 16 **Contingency table results:**
Rows: var1
Columns: None

	Group 1	Group 2	Group 3	Total
Pain	51	5	16	72
No pain	1532	152	163	1847
Total	1583	157	179	1919

Chi-Square test:

Statistic	df	Value	*P*-value
Chi-square	2	14.706513	0.0006

Step 4 Because the *P*-value is less than the level of significance, $\alpha = 0.01$, we reject the null hypothesis.

Step 5 The data suggest that there is sufficient evidence at the $\alpha = 0.01$ level of significance to reject the null hypothesis that the proportion of subjects in each group who experience abdominal pain are equal. We conclude that at least one of the three groups experiences abdominal pain at a rate that is different from the other two groups.

> **CAUTION!**
> If we reject the null hypothesis in a chi-square test for homogeneity, we are saying there is sufficient evidence to believe that at least one proportion is different from the others. However, it does not tell us which proportions differ.

Figure 17 shows the conditional bar graph. From the graph, it is apparent that a higher proportion of patients taking cholestyramine experience abdominal pain as a side effect.

Figure 17

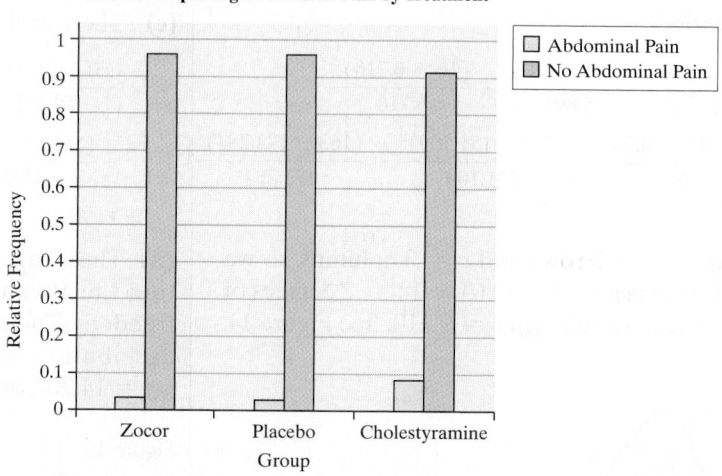

Patients Reporting Abdominal Plan by Treatment

• **Now Work Problem 15**

If Model Requirements Fail

Recall that the requirements for performing a chi-square test are that all expected frequencies are greater than 1 and that at most 20% of the expected frequencies can be less than 5. If these requirements are not satisfied, the researcher has one of two options: (1) combine two or more columns (or rows) to increase the expected frequencies or (2) increase the sample size.

Technology Step-by-Step Chi-Square Tests

TI-83/84 Plus
1. Access the MATRIX menu. Highlight the EDIT menu, and select `1:[A]`.
2. Enter the number of rows and columns of the contingency table (matrix).

3. Enter the cell entries for the observed matrix, and press `2nd QUIT`.
4. Press `STAT`, highlight the TESTS menu, and select `C:χ² - Test`.

5. With the cursor after Observed:, enter matrix [A] by accessing the MATRIX menu, highlighting NAMES, and selecting 1 : [A].

6. With the cursor after Expected:, enter matrix [B] by accessing the MATRIX menu, highlighting NAMES, and selecting 2 : [B].

7. Highlight Calculate or Draw, and press ENTER.

Minitab

1. Enter the data into the MINITAB spreadsheet.

2. Select the **Stat** menu, highlight **Tables**, and select **Chi-Square Test for Association ….**

3. From the pull-down menu, select "Summarized data in a two-way table." Select the columns that contain the data, and click OK.

Excel

1. Load the XLSTAT Add-in.

2. Enter the data in a contingency table. Column A should be the row variable. For example, for the data in Table 4, column A is level of happiness. Each subsequent column is the counts of each category of the column variable. For the data in Table 4, enter the counts for each marital status. Title each column (including the first column indicating the row variable).

3. Select **Describing data:**, highlight **Contingency table.**

4. Place the cursor in the Contingency table cell. Highlight the table in the spreadsheet. Check the box Labels included. Click the Options tab.

5. Check the box Chi-square test. Choose a level of significance. Click OK.

StatCrunch

1. If the data are already in a contingency table, enter them into the spreadsheet. The first column is the row variable. For example, for the data in Table 4, the first column is level of happiness. Each subsequent column is the counts of each category of the column variable. For the data in Table 4, enter the counts for each marital status. Title each column (including the first column indicating the row variable). If the data are not in a contingency table, enter each variable in a column and name the column variable.

2. Select **Stat**, highlight **Tables**, select **Contingency**, then highlight **With Data** or **With Summary.**

3. Select the column variable(s). Then select the row variable. For example, the data in Table 4 has four column variables ("Married," and so on) and the row label is "Happiness." Decide what values you want displayed. Highlight "Chi-Square test for independence" under Hypothesis Tests:. Click Calculate!.

12.2 Assess Your Understanding

Vocabulary and Skill Building

1. *True or False*: The expected frequencies in a chi-square test for independence are found using the formula

$$\text{Expected frequency} = \frac{(\text{row total})(\text{column total})}{\text{table total}}$$

2. In a chi-square test for _____ of proportions, we test whether different populations have the same proportion of individuals with some characteristic.

3. The following table contains observed values and expected values in parentheses for two categorical variables, X and Y, where variable X has three categories and variable Y has two categories:

	X_1	X_2	X_3
Y_1	34 (36.26)	43 (44.63)	52 (48.11)
Y_2	18 (15.74)	21 (19.37)	17 (20.89)

(a) Compute the value of the chi-square test statistic.

(b) Test the hypothesis that X and Y are independent at the $\alpha = 0.05$ level of significance.

4. The table in the next column contains observed values and expected values in parentheses for two categorical variables, X and Y, where variable X has three categories and variable Y has two categories.

	X_1	X_2	X_3
Y_1	87 (75.12)	74 (80.43)	34 (39.46)
Y_2	12 (23.88)	32 (25.57)	18 (12.54)

(a) Compute the value of the chi-square test statistic.

(b) Test the hypothesis that X and Y are independent at the $\alpha = 0.05$ level of significance.

5. The following table contains the number of successes and failures for three categories of a variable.

	Category 1	Category 2	Category 3
Success	76	84	69
Failure	44	41	49

Test whether the proportions are equal for each category at the $\alpha = 0.01$ level of significance.

6. The following table contains the number of successes and failures for three categories of a variable.

	Category 1	Category 2	Category 3
Success	204	199	214
Failure	96	121	98

Test whether the proportions are equal for each category at the $\alpha = 0.01$ level of significance.

Applying the Concepts

NW **7. Family Structure and Sexual Activity** A sociologist wants to discover whether the sexual activity of females between the ages of 15 and 19 years and family structure are associated. She randomly selects 380 females between the ages of 15 and 19 years and asks each to disclose her family structure at age 14 and whether she has had sexual intercourse. The results are shown in the table. Data are based on information obtained from the National Center for Health Statistics.

	Family Structure			
Had Sexual Intercourse	**Both Biological or Adoptive Parents**	**Single Parent**	**Parent and Stepparent**	**Nonparental Guardian**
Yes	64	59	44	32
No	86	41	36	18

(a) Compute the expected values of each cell under the assumption of independence.

(b) Verify that the requirements for performing a chi-square test of independence are satisfied.

(c) Compute the chi-square test statistic.

(d) Test whether family structure and sexual activity of 15- to 19-year-old females are independent at the $\alpha = 0.05$ level of significance.

(e) Compare the observed frequencies with the expected frequencies. Which cell contributed most to the test statistic? Was the expected frequency greater than or less than the observed frequency? What does this information tell you?

(f) Construct a conditional distribution by family structure and draw a bar graph. Does this evidence support your conclusion in part (d)?

8. Prenatal Care An obstetrician wants to learn whether the amount of prenatal care and the wantedness of the pregnancy are associated. He randomly selects 939 women who had recently given birth and asks them to disclose whether their pregnancy was intended, unintended, or mistimed. In addition, they were to disclose when they started receiving prenatal care, if ever. The results of the survey are as follows:

	Months Pregnant before Prenatal Care Began		
Wantedness of Pregnancy	**Less Than 3 Months**	**3 to 5 Months**	**More Than 5 Months (or never)**
Intended	593	26	33
Unintended	64	8	11
Mistimed	169	19	16

(a) Compute the expected values of each cell under the assumption of independence.

(b) Verify that the requirements for performing a chi-square test of independence are satisfied.

(c) Compute the chi-square test statistic.

(d) Test whether prenatal care and the wantedness of pregnancy are independent at the $\alpha = 0.05$ level of significance.

(e) Compare the observed frequencies with the expected frequencies. Which cell contributed most to the test statistic? Was the expected frequency greater or less than the observed frequency? What does this information tell you?

(f) Construct a conditional distribution by wantedness of the pregnancy and draw a bar graph. Does this evidence support your conclusion in part (d)?

9. Health and Happiness Are health and happiness related? The following data represent the level of happiness and level of health for a random sample of individuals from the General Social Survey.

		Health			
		Excellent	**Good**	**Fair**	**Poor**
Happiness	**Very Happy**	271	261	82	20
	Pretty Happy	247	567	231	53
	Not Too Happy	33	103	92	36

Source: General Social Survey

(a) Does the evidence suggest that health and happiness are related? Use the $\alpha = 0.05$ level of significance.

(b) Construct a conditional distribution of happiness by level of health and draw a bar graph.

(c) Write a few sentences that explain the relation, if any, between health and happiness.

10. Health and Education Does amount of education play a role in the healthiness of an individual? The following data represent the level of health and the highest degree earned for a random sample of individuals from the General Social Survey.

		Health			
		Excellent	**Good**	**Fair**	**Poor**
Education	**Less than High School**	72	202	199	62
	High School	465	877	358	108
	Junior College	80	138	49	11
	Bachelor	229	276	64	12
	Graduate	130	147	32	2

Source: General Social Survey

(a) Does the evidence suggest that health and education are related? Use the $\alpha = 0.05$ level of significance.

(b) Construct a conditional distribution of health by level of education and draw a bar graph.

(c) Write a few sentences that explain the relation, if any, between health and education. Can you think of any lurking variables that help to explain the relation between these two variables?

11. Social Well-Being and Obesity The Gallup Organization conducted a survey in 2014 asking individuals questions pertaining to social well-being such as strength of relationship with spouse, partner, or closest friend, making time for trips or vacations, and having someone who encourages them to be healthy. Social well-being scores were determined based on answers to these questions and used to categorize individuals as thriving, struggling, or suffering in their social well-being. In addition, body mass index (BMI) was determined based on height and weight of the individual. This allowed for classification as obese, overweight, normal weight, or underweight. The data in the following contingency table are based on the results of this survey.

	Thriving	**Struggling**	**Suffering**
Obese	202	250	102
Overweight	294	302	110
Normal Weight	300	295	103
Underweight	17	17	8

(a) Researchers wanted to determine whether the sample data suggest there is an association between weight classification and social well-being. Explain why this data should be analyzed using a chi-square test for independence.

(b) Do the sample data suggest that weight classification and social well-being are related?

(c) Draw a conditional bar graph of the data by weight classification.

(d) Write some general conclusions based on the results from parts (b) and (c).

12. Profile of Smokers The following data represent the smoking status from a random sample of 1054 U.S. residents 18 years or older by level of education.

Number of Years of Education	Smoking Status		
	Current	**Former**	**Never**
<12	178	88	208
12	137	69	143
13–15	44	25	44
16 or more	34	33	51

Source: National Health Interview Survey

(a) Test whether smoking status and level of education are independent at the $\alpha = 0.05$ level of significance.

(b) Construct a conditional distribution of smoking status by number of years of education, and draw a bar graph. Does this evidence support your conclusion in part (a)?

13. Efficacy of e-Cigs Do electronic cigarettes assist in helping individuals quit smoking? Researchers found 300 current smokers to volunteer for a study in which each was randomly assigned to one of three treatment groups. Group 1 received an electronic cigarette (e-cig) in which each cartridge contained 7.2 mg of nicotine, Group 2 received an e-cig that contained 5.4 mg of nicotine, and Group 3 received an e-cig that contained no nicotine. The subjects did not know which group they were assigned. During the course of the 52-week intervention, subjects dropped out of the study. At the end of the study 65 subjects remained in Group 1, 63 in Group 2, and 55 in Group 3. After 52 weeks, it was determined via questionnaire whether the subject quit smoking entirely. Results of the study are presented in the following table.

	Group 1	**Group 2**	**Group 3**
Did Not Quit	52	54	51
Quit	13	9	4

Source: Caponnetto P, Campagna D, Cibella F, Morjaria JB, Caruso M, et al. (2013) EffiCiency and Safety of an eLectronic cigAreTte (ECLAT) as Tobacco Cigarettes Substitute: A Prospective 12-Month Randomized Control Design Study. *PLoS ONE* 8(6): e66317.

doi:10.1371/journal.pone.0066317

(a) What type of experimental design was used in this study?

(b) Researchers wanted to know whether electronic cigarettes may be used to help individuals abstain from cigarette smoking. What is the response variable? Is it qualitative or quantitative?

(c) To what population do the results of this study apply?

(d) State the null and alternative hypotheses.

(e) Does the evidence suggest e-cigs are effective in helping individuals abstain from cigarette smoking?

(f) Draw a conditional bar graph by group. Explain what the graph suggests.

(g) Write a conclusion for this hypothesis test.

14. Celebrex Celebrex, a drug manufactured by Pfizer, Inc., is used to relieve symptoms associated with osteoarthritis and rheumatoid arthritis in adults. In clinical trials of the medication, some subjects reported dizziness as a side effect. The researchers wanted to discover whether the proportion of subjects taking Celebrex who reported dizziness as a side effect differed significantly from that for other treatment groups. The following data were collected.

Side Effect	Drug				
	Celebrex	**Placebo**	**Naproxen**	**Diclofenac**	**Ibuprofen**
Dizziness	83	32	36	5	8
No dizziness	4063	1832	1330	382	337

Source: Pfizer, Inc.

(a) Test whether the proportion of subjects within each treatment group who experienced dizziness are the same at the $\alpha = 0.01$ level of significance.

(b) Construct a conditional distribution of side effect by treatment and draw a bar graph. Does this evidence support your conclusion in part (a)?

NW 15. What's in a Word? In a recent survey conducted by the Pew Research Center, a random sample of adults 18 years of age or older living in the continental United States was asked their reaction to the word *socialism*. In addition, the individuals were asked to disclose which political party they most associate with. Results of the survey are given in the table.

	Democrat	**Independent**	**Republican**
Positive	220	144	62
Negative	279	410	351

Source: Pew Research

(a) Explain why this data should be analyzed by homogeneity of proportions.

(b) Does the evidence suggest individuals within each political affiliation react differently to the word *socialism*? Use the $\alpha = 0.05$ level of significance.

(c) Construct a conditional distribution of reaction by political party.

(d) Write a summary about the "partisan divide" regarding reaction to the word *socialism*.

16. What's in a Word? Part II In a recent survey conducted by the Pew Research Center, a random sample of adults 18 years of age or older living in the continental United States was asked their reaction to the word *capitalism*. In addition, the individuals were asked to disclose which political party they most associate with. Results of the survey are given in the table on the next page.

	Democrat	Independent	Republican
Positive	235	288	256
Negative	264	266	157

Source: Pew Research

(a) Does the evidence suggest individuals within each political affiliation react differently to the word *capitalism*? Use the $\alpha = 0.05$ level of significance.

(b) Construct a conditional distribution of reaction by political party.

(c) Write a summary about the "partisan divide" regarding reaction to the word *capitalism*.

(d) Compare the results of this problem with that of Problem 15. Write a short report detailing the findings of the two survey questions.

17. Dropping a Course A survey was conducted at a community college of 50 randomly selected students who dropped a course in the current semester to learn why students drop courses. "Personal" drop reasons include financial, transportation, family issues, health issues, and lack of child care. "Course" drop reasons include reducing one's load, being unprepared for the course, the course was not what was expected, dissatisfaction with teaching, and not getting the desired grade. "Work" drop reasons include an increase in hours, a change in shift, and obtaining fulltime employment. Go to www.pearsonhighered.com/sullivanstats to obtain the data file 12_2_17 using the file format of your choice for the version of the text you are using.

(a) Construct a contingency table for the two variables.

(b) Test whether gender is independent of drop reason at the $\alpha = 0.1$ level of significance.

(c) Construct a conditional distribution of drop reason by gender and draw a bar graph. Does this evidence support your conclusion in part (b)?

18. Political Affiliation In the Sullivan Statistics Survey, respondents were asked to disclose their political affiliation (Democrat, Independent, Republican) and also answer the question: "Would you be willing to pay higher taxes if the tax revenue went directly toward deficit reduction?" Go to www.pearsonhighered.com/sullivanstats to obtain the data file SullivanSurveyI using the file format of your choice for the version of the text you are using. Create a contingency table and determine whether the results suggest there is an association between political affiliation and willingness to pay higher taxes to directly reduce the federal debt. Use the $\alpha = 0.05$ level of significance.

19. Credit Risk Traditional underwriting to determine the risks associated with lending include credit scores, income, and employment history. The online lender ZestFinance used data analysis to find that people who fill out loan applications using all capital letters default more often than those who use all lower case letters. In addition, people who fill out the application using upper and lowercase letters accurately default at the lowest rate. Explain how to obtain and analyze data to determine whether the method used to fill out loan applications results in different default rates.

20. Insurance and Credit Scores A study by InsuranceQuotes.com found that homeowners with poor credit pay 91% more for home insurance than people with excellent credit.

(a) A quote in the article stated, "Insurers have found a direct correlation between a consumer's credit and the likelihood that he or she will make a home (or auto) claim." Explain what is wrong with this quote.

(b) Credit scores may be classified as Excellent, Good, Fair, and Poor. Explain how you might go about deciding whether credit scores might be used to determine whether an individual files a claim on his or her homeowner's insurance policy or not. Include an explanation of the type of inferential procedure you would use.

21. Putting It Together: Women, Aspirin, and Heart Attacks
In a famous study by the Physicians Health Study Group from Harvard University from the late 1980s, 22,000 healthy male physicians were randomly divided into two groups; half the physicians took aspirin every other day, and the others were given a placebo. Of the physicians in the aspirin group, 104 heart attacks occurred; of the physicians in the placebo group, 189 heart attacks occurred. The results were statistically significant, which led to the advice that males should take an aspirin every other day in the interest of reducing the chance of having a heart attack. Does the same advice apply to women?

In a randomized, placebo-controlled study, 39,876 healthy women 45 years of age or older were randomly divided into two groups. The women in group 1 received 100 mg of aspirin every other day; the women in group 2 received a placebo every other day. The women were monitored for 10 years to determine if they experienced a cardiovascular event (such as heart attack or stroke). Of the 19,934 in the aspirin group, 477 experienced a heart attack. Of the 19,942 women in the placebo group, 522 experienced a heart attack.
Source: Paul M. Ridker et al. "A Randomized Trial of Low-Dose Aspirin in the Primary Prevention of Cardiovascular Disease in Women." *New England Journal of Medicine* 352:1293–1304.

(a) What is the population being studied? What is the sample?

(b) What is the response variable? Is it qualitative or quantitative?

(c) What are the treatments?

(d) What type of experimental design is this?

(e) How does randomization deal with the explanatory variables that were not controlled in the study?

(f) Determine whether the proportion of cardiovascular events in each treatment group is different using a two-sample Z-test for comparing two proportions. Use the $\alpha = 0.05$ level of significance. What is the test statistic?

(g) Determine whether the proportion of cardiovascular events in each treatment group is different using a chi-square test for homogeneity of proportions. Use the $\alpha = 0.05$ level of significance. What is the test statistic?

(h) Square the test statistic from part (f) and compare it to the test statistic from part (g). What do you conclude?

Retain Your Knowledge

22. Homeruns Go to www.pearsonhighered.com/sullivanstats to obtain the data file 12_2_22 using the file format of your choice for the version of the text you are using. The variable "TrueDist" represents the distance, in feet, that the homerun traveled for all homeruns hit in the 2014 season.

(a) Draw a relative frequency histogram of the distance a homerun traveled in 2014 using a lower class limit of the first class of 300 and a class width of 10. Describe the shape of the distribution.

(b) Find the population mean and population standard deviation distance.

(c) Find the quartiles of distance.

(d) Draw a boxplot of distance. Are there any outliers?

(e) Use a normal model to determine the proportion of homeruns that exceeded 450 feet. Compare this to the actual proportion of homeruns that exceeded 450 feet.

(f) Use a normal model to determine the first and third quartiles. Compare this result to the quartiles found in part (c).

Explaining the Concepts

23. Explain the differences between the chi-square test for independence and the chi-square test for homogeneity. What are the similarities?

24. Why does the test for homogeneity follow the same procedures as the test for independence?

12.3 Inference about Two Population Proportions: Dependent Samples

Preparing for This Section Before getting started, review the following:

- Independent versus dependent sampling (Section 11.1, pages 525–526)

- The chi-square distribution (Section 9.3, pages 449–451)

Objective ① Test Hypotheses Regarding Two Proportions from Dependent Samples

① Test Hypotheses Regarding Two Proportions from Dependent Samples

In Section 11.1, we compared two proportions obtained from independent samples. But what if we want to compare two proportions with matched-pairs data (i.e., dependent samples)? In this case, we can use **McNemar's Test**.

Suppose we want to determine whether there is a difference between two ointments meant to treat poison ivy. Rather than randomly divide a group of individuals with poison ivy into two groups, where group 1 gets ointment A and group 2 gets ointment B and compare the proportion of individuals who heal in each group, we might apply ointment A on one arm and ointment B on the other arm of each individual and record whether the poison ivy cleared up, or not. The individuals in each group are not independent because each ointment is applied to the same individual. How might we determine whether the proportion of healed poison ivy sufferers differs between the two treatments? To answer this question, let's say the results of the experiment are as given in Table 14.

Table 14		Ointment A	
		Healed (success)	Did not heal (failure)
Ointment B	Healed (success)	293	43
	Did not heal (failure)	31	103

Just as we did when comparing two population proportions from independent samples, we assume that the proportion of subjects healed with ointment A equals the proportion of subjects healed with ointment B (there is no healing difference in the ointments). So the null hypothesis is $H_0: p_A = p_B$. There are a total of $293 + 43 + 31 + 103 = 470$ subjects in the study. In this experiment, the sample proportion of subjects who healed with ointment A is $\hat{p}_A = \dfrac{293 + 31}{470} = 0.689$, and the sample proportion of subjects who healed with ointment B is $\hat{p}_B = \dfrac{293 + 43}{470} = 0.715$.

McNemar's Test will be used to determine whether the difference in sample proportions is due to sampling error or whether the differences are significant enough to conclude that the proportions are different. So the alternative hypothesis is $H_0: p_A \neq p_B$.

Testing a Hypothesis Regarding the Difference of Two Proportions: Dependent Samples

To test hypotheses regarding two population proportions, p_1 and p_2, where the samples are dependent, arrange the data in a contingency table as follows:

		Treatment A	
		Success	**Failure**
Treatment B	**Success**	f_{11}	f_{12}
	Failure	f_{21}	f_{22}

We can use the steps that follow provided that

1. the samples are dependent and are obtained randomly and
2. the total number of observations where the outcomes differ are greater than or equal to 10. That is, $f_{12} + f_{21} \geq 10$.

> **Step 1** Determine the null and alternative hypotheses.
>
> H_0: the proportions between the two populations are equal $(p_1 = p_2)$
> H_1: the proportions between the two populations differ $(p_1 \neq p_2)$
>
> **Step 2** Select a level of significance, α, depending on the seriousness of making a Type I error.
>
> **Step 3** Compute the test statistic*
>
> $$\chi_0^2 = \frac{(f_{12} - f_{21})^2}{f_{12} + f_{21}}$$

Classical Approach

Step 3 (continued) Use Table VIII to determine the critical value with 1 degree of freedom.

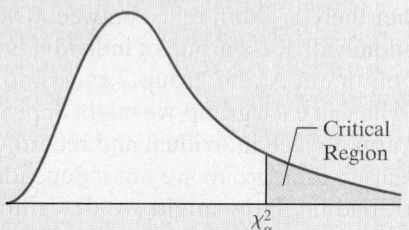

Step 4 If $\chi_0^2 > \chi_\alpha^2$, reject the null hypothesis.

P-Value Approach

By Hand Step 3 (continued) Use Table VIII to determine the P-value by determining the area under the chi-square distribution with 1 degree of freedom to the right of the test statistic.

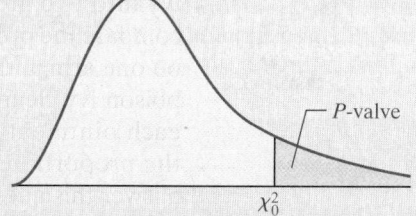

Technology Step 3 (continued) Use a statistical spreadsheet or calculator with statistical capabilities to obtain the P-value. The directions for obtaining the P-value using the TI-84 plus graphing calculators, Minitab, or StatCrunch are given in the Technology Step-by-Step on page 610.

Step 4 If the P-value $< \alpha$, reject the null hypothesis.

Step 5 State the conclusion.

EXAMPLE 1   Analyzing the Difference of Two Proportions from Matched-Pairs Data

Problem A recent General Social Survey asked the following two questions of a random sample of 1492 adult Americans under the hypothetical scenario that the government suspected that a terrorist act was about to happen:

*   Do you believe that the authorities should have the right to tap people's telephone conversations?
*   Do you believe that the authorities should have the right to stop and search people on the street at random?

*We will not discuss the underlying theory behind the test statistic as it is beyond the scope of this course.

The results of the survey are shown in Table 15.

Table 15		Random Stop	
		Agree	**Disagree**
Tap Phone	**Agree**	494	335
	Disagree	126	537

Do the proportion of people who agree with each scenario differ significantly? Use the $\alpha = 0.05$ level of significance.

Approach The sample proportion of individuals who believe that the authorities should be able to tap phones is $\hat{p}_T = \dfrac{494 + 335}{1492} = 0.556$. The sample proportion of individuals who believe that the authorities should be able to randomly stop and search an individual on the street is $\hat{p}_R = \dfrac{494 + 126}{1492} = 0.416$. We want to determine whether the difference in sample proportions is due to sampling error or to the fact that the population proportions differ.

The samples are dependent and were obtained randomly. The total number of individuals who agree with one scenario, but disagree with the other, is $335 + 126 = 461$, which is greater than 10. We can proceed with McNemar's test.

Solution

Step 1 The hypotheses are as follows:

H_0: the proportions between the two populations are equal ($p_T = p_R$)

H_1: the proportions between the two populations differ ($p_T \neq p_R$)

Step 2 The level of significance is $\alpha = 0.05$.

Step 3 The test statistic is

$$\chi_0^2 = \frac{(f_{12} - f_{21})^2}{f_{12} + f_{21}}$$

$$= \frac{(335 - 126)^2}{335 + 126} = 94.753$$

Classical Approach

Step 3 (continued) The critical value with an $\alpha = 0.05$ level of significance is $\chi_{0.05}^2 = 3.841$. The critical region is displayed in Figure 18.

Figure 18

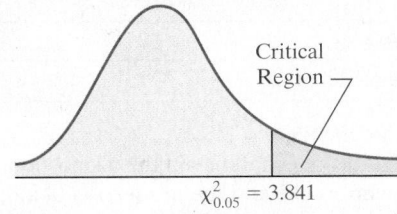

$\chi_{0.05}^2 = 3.841$

Step 4 Because $\chi_0^2 > \chi_{0.05}^2$ (the test statistic lies in the critical region), we reject the null hypothesis.

P-Value Approach

By Hand Step 3 (continued) The P-value is the area under the chi-square distribution to the right of the test statistic, $\chi_0^2 = 94.753$. The test statistic is so large that the P-value < 0.0005.

Technology Step 3 (continued) Figure 19 shows the P-value obtained from StatCrunch. The P-value is reported as < 0.0001.

Figure 19

Contingency table results:
Rows: Tap Phone
Columns: None

	Stop Agree	Stop Disagree	Total
Agree	494	335	829
Disagree	126	537	663
Total	620	872	1492

McNemar's test for marginal homogeneity

Measure	DF	Value	P-value
Chi-Square	1	94.752711	<0.0001

Step 4 Because P-value < α, we reject the null hypothesis.

(continued)

Step 5 There is sufficient evidence to conclude that there is a difference in the proportion of adult Americans who believe it is okay to phone tap versus stopping and searching people on the street in the event that the government believed a terrorist plot was about to happen. In fact, looking at the sample proportions, it seems clear that more people are willing to allow phone tapping than they are to allow stopping and searching on the street.

● Now Work Problem 3

Technology Step-by-Step McNemar's Test

TI-83/84 Plus
1. By hand, compute the value of the test statistic.
2. Find the area under the chi-square distribution with 1 degree of freedom to the right of the test statistic. To do this, press 2nd VARS to access the DISTRibution menu. Select 8: χ^2-cdf.
3. Enter the lower bound (the value of the test statistic), the upper bound (1E99), and 1 for df (degrees of freedom). Highlight Paste and hit ENTER. From the HOME screen, press ENTER a second time.

Minitab
1. Enter the contingency table into the spreadsheet. The first column is the row variable. Each subsequent column is the counts of each category of the column variable.
2. Select **Stat**, highlight **Tables**, highlight **Cross Tabulation and Chi-Square...** Click **Other Stats...** and check the McNemar's test box. Click OK.

3. From the drop-down menu, select "Summarized data in a two-way table." With the cursor in the "Columns containing the table" box, highlight the columns containing the column variables and click Select. Click OK.

StatCrunch
1. Enter the contingency table into the spreadsheet. The first column is the row variable. Each subsequent column is the counts of each category of the column variable.
2. Select **Stat**, highlight **Tables**, highlight **Contingency**, and then highlight **With Summary**.
3. Highlight the column variables. Then select the column that contains the row labels. Under Hypothesis tests, select McNemar's test for marginal homogeneity (2 × 2 only). Click Compute!.

12.3 Assess Your Understanding

Skill Building

In Problems 1 and 2, test whether the population proportions differ at the $\alpha = 0.05$ level of significance by determining (a) the null and alternative hypotheses, (b) the test statistic, (c) the critical value, and (d) the P-value. Assume that the samples are dependent and were obtained randomly.

1.

		Treatment A	
		Success	Failure
Treatment B	Success	45	19
	Failure	14	23

2.

		Treatment A	
		Success	Failure
Treatment B	Success	84	21
	Failure	11	37

Applying the Concepts

NW **3. Hazardous Activities** In a survey of 3029 adult Americans, the Harris Poll asked people whether they smoked cigarettes

and whether they always wear a seat belt in a car. The table shows the results of the survey. For each activity, we define a success as finding an individual who participates in the hazardous activity.

	No Seat Belt (success)	Seat Belt (failure)
Smoke (success)	67	448
Do not smoke (failure)	327	2187

(a) Why is this a dependent sample?
(b) Is there a significant difference in the proportion of individuals who smoke and the proportion of individuals who do not wear a seat belt? In other words, is there a significant difference between the proportion of individuals who engage in hazardous activities? Use the $\alpha = 0.05$ level of significance.

4. Income Taxes A Gallup organization survey asked 1041 adult Americans whether they felt low-income people paid their fair share of taxes. Then they asked whether they felt high-income people paid their fair share of taxes. For each income group, a success is identifying an individual who feels the person pays his or her fair share of taxes. Is there a significant difference in the proportion of adult Americans who feel high-income people pay their fair share of taxes and the proportion who feel low-income people pay their fair share of taxes? Use the $\alpha = 0.05$ level of significance.

		Low-Income	
		Fair Share (success)	**Not Fair Share (failure)**
High-Income	**Fair Share (success)**	73	146
	Not Fair Share (failure)	274	548

5. Voice-Recognition Systems Have you ever been frustrated by computer telephone systems that do not understand your voice commands? Quite a bit of effort goes into designing these systems to minimize voice-recognition errors. Researchers at the Oregon Graduate Institute of Science and Technology developed a new method of voice recognition (called a remapped network) that was thought to be an improvement over an existing neural network. The data shown are based on results of their research. Does the evidence suggest that the remapped network has a different proportion of errors than the neural network? Use the $\alpha = 0.05$ level of significance.

		Remapped Network	
		Recognized Word (success)	**Did Not Recognize Word (failure)**
Neural Network	**Recognized Word (success)**	9326	385
	Did Not Recognize Word (failure)	456	29

Source: Wei, Wei, Leen, Todd, Barnard, and Etienne, "A Fast Histogram-Based Postprocessor That Improves Posterior Probability Estimates." *Neural Computation* 11:1235–1248.

6. Poison Ivy Look back at the data in Table 14 on page 607, which compared the effectiveness of two ointments. Conduct the appropriate test to determine if one ointment has a different effectiveness than the other. Use the $\alpha = 0.05$ level of significance.

7. Gun Laws and Capital Punishment The General Social Survey asked a random sample of adult Americans two questions: (1) Would you favor or oppose a law which would require a person to obtain a police permit before he or she could buy a gun? (2) Do you favor or oppose the death penalty for persons convicted of murder? Results of the survey are below. Does the sample evidence suggest the proportion of adult Americans who favor permits for guns is different from the proportion of adult Americans who favor the death penalty for murder convictions? Use the $\alpha = 0.05$ level of significance.

		Death Penalty	
		Favor	**Oppose**
Gun Laws	**Favor**	176	69
	Oppose	58	17

8. Children and Childcare The General Social Survey asked a random sample of adult Americans with children two questions: (1) Do you believe there should be paid leave for childcare? (2) Do you believe children are a financial burden on parents? Results of the survey are in the following table. Does the sample evidence suggest there is a difference in the proportion of adult Americans with children who feel there should be paid leave for childcare and the proportion who feel children are a financial burden on parents? Use an $\alpha = 0.05$ level of significance.

		Financial Burden	
		Agree	**Disagree**
Paid Leave	**Yes**	259	616
	No	64	101

Chapter 12 Review

Summary

In this chapter, we continued our discussion of inferential statistics. Here, we introduced chi-square methods. Chi-square methods revolve around analyzing qualitative data.

The first chi-square method involved tests for goodness-of-fit. The null hypothesis is always a statement of no change/no effect/no difference. So the null hypothesis in a goodness-of-fit test is that the random variable follows some specified distribution. That is, there is no difference in the distribution of the sample data and that stated in the null hypothesis. The alternative hypothesis (or statement we are gathering evidence to demonstrate) is that the random variable does not follow the specified distribution. We used the chi-square distribution to test these hypotheses. This is done by comparing the expected values based on the distribution

of the random variable to the observed values. If the observed values differ significantly from those expected, we have evidence against the statement in the null hypothesis.

Next, we introduced the chi-square test for independence. In such a test, the researcher obtains random data for two variables and tests whether the variables are associated. The null hypothesis in these tests is always that the variables are not associated (independent). The test statistic compares the expected values if the variables were independent to those observed. If the expected and observed values differ significantly, we reject the null hypothesis and conclude that there is evidence to support the belief that the variables are not independent (they are associated). We draw bar graphs of the conditional distributions to help us see the association, if any.

The chi-square test for homogeneity of proportions is an extension of the test for comparing two independent proportions. Here, we are testing the null hypothesis that three or more proportions are equal versus the alternative that at least one proportion differs from the others. This test uses the exact same approach as that for independence. While the procedures for the test for independence and the test of homogeneity of proportions are the same, the data differ.

In the test for independence, we are measuring two variables (such as marital status and level of happiness) on each individual. In other words, a *single* population is segmented based on the value of two variables. In the

test of homogeneity of proportions, we consider whether the proportion of individuals among *different* populations have the same value. So, if you have a single population in which two variables are measured on each individual to assess whether one variable might be associated with another, conduct a test of independence. If you have two or more populations in which you want to determine equality of proportions among the choices, conduct a test of homogeneity of proportions.

We wrapped up the chapter by looking at McNemar's test. McNemar's test is used to compare two proportions from dependent samples. The hypothesis testing procedure is once again identical to the other hypothesis test procedures.

Vocabulary

Goodness-of-fit test (p. 580)
Expected counts (pp. 580, 593)

Test statistic (pp. 582, 596)
Chi-square test for independence (p. 592)

Chi-square test for homogeneity of proportions (p. 599)
McNemar's Test (p. 607)

Formulas

Expected Counts in a Goodness-of-Fit Test

$E_i = \mu_i = np_i$ for $i = 1, 2, \ldots, k$

Chi-Square Test Statistic

$\chi_0^2 = \sum \frac{(O_i - E_i)^2}{E_i}$ $i = 1, 2, \ldots, k$

Expected Frequencies in a Test for Independence or Homogeneity of Proportions

$$\text{Expected frequency} = \frac{(\text{row total})(\text{column total})}{\text{table total}}$$

Test Statistic for McNemar's Test

$$\chi_0^2 = \frac{(f_{12} - f_{21})^2}{f_{12} + f_{21}}$$

OBJECTIVES

Section	You should be able to ...	Example(s)	Review Exercises
12.1	**1** Perform a goodness-of-fit test (p. 580)	1–3	1 and 2
12.2	**1** Perform a test for independence (p. 592)	1–3	3 and 4
	2 Perform a test for homogeneity of proportions (p. 599)	4	5
12.3	**1** Test hypotheses regarding two proportions from dependent samples (p. 607)	1	6

Review Exercises

1. Roulette Wheel A pit boss suspects that a roulette wheel is out of balance. A roulette wheel has 18 black slots, 18 red slots, and 2 green slots. The pit boss spins the wheel 500 times and records the following frequencies:

Outcome	Frequency
Black	233
Red	237
Green	30

Is the wheel out of balance? Use the $\alpha = 0.05$ level of significance.

2. World Series Are the teams that play in the World Series evenly matched? To win a World Series, a team must win four games. If the teams are evenly matched, we would expect the number of games played in the World Series to follow the distribution shown in the first two columns of the following table. The third column represents the actual number of games played in each World Series from 1930 to 2014. Do the data support the distribution that would exist if the teams are evenly matched and the outcome of each game is independent? Use the $\alpha = 0.05$ level of significance.

Number of Games	Probability	Observed Frequency
4	0.125	16
5	0.25	17
6	0.3125	19
7	0.3125	32

Source: Major League Baseball

3. *Titanic* With 20% of men, 74% of women, and 52% of children surviving the infamous *Titanic* disaster, it is clear that the saying "women and children first" was followed. But what, if any, role did the class of service play in the survival of passengers? The data shown represent the survival status of passengers by class of service.

		Class		
		First	**Second**	**Third**
Survival Status	**Survived**	203	118	178
	Did Not Survive	122	167	528

(a) Is class of service independent of survival rate? Use the $\alpha = 0.05$ level of significance.
(b) Construct a conditional distribution of survival status by class of service and draw a bar graph. What does this summary tell you?

4. Premature Birth and Education Does the length of term of pregnancy play a role in the level of education of the baby? Researchers in Norway followed over 1 million births between 1967 and 1988 and looked at the educational attainment of the children. The following data are based on the results of their research. Note that a full-term pregnancy is 38 weeks. Is gestational period independent of completing a high school diploma? Use the $\alpha = 0.05$ level of significance.

		Gestational Period (weeks)				
		22–27	**28–32**	**33–36**	**37–42**	**43+**
Less Than High School Degree	**Yes**	14	34	140	1010	81
	No	26	65	343	3032	208

Source: Chicago Tribune, March 26, 2008.

5. Roosevelt versus Landon One of the most famous presidential elections (from a statistician's point of view) is the 1936 contest between incumbent Franklin D. Roosevelt (FDR) and Republican challenger Alf Landon. The notoriety of the election comes from the fact that polling done by *Literary Digest* suggested that Landon would win by a 3 to 2 margin. However, Roosevelt actually crushed Landon in the general election.

After the election, author and editor David Lawrence studied the vote. Lawrence asked the following question: "To what extent was the campaign a reflection of a new trend in American politics, a trend in which the federal government's paternalistic interest in the citizen brought an amazing reward to the party in power?"

As part of FDR's New Deal policies to help get the United States out of the Great Depression, the federal government created the Agricultural Adjustment Administration (AAA). The AAA tried to force higher prices for commodities by paying farmers not to farm and not to bring cattle to market.

To determine whether farm subsidies may have played a role in the election results, Lawrence looked at election results in 2052 counties. He segmented the counties based on AAA funding between 1934 and 1935 as follows:

- High-funded AAA counties received $500,000 or more in AAA money
- Medium-funded AAA counties received between $100,000 and $500,000 in AAA money
- Low-funded AAA counties received some AAA money, but less than $100,000
- Counties with no AAA-fund did not receive any funds.

Treat the following data as a random sample of voters within each county type. The results are based on the election results in the various counties.

	High-Funded	Medium-Funded	Low-Funded	No Funding
Roosevelt	745	641	513	411
Landon	498	484	437	465

Source: Folsom, Burton, "New Deal or Raw Deal?" Simon & Schuster, 2008.

(a) Do the data suggest that the level of funding received by counties through the AAA is associated with the candidate? Use the $\alpha = 0.05$ level of significance.
(b) Construct a conditional distribution of candidate by level of AAA funding and draw a bar graph.

6. Obligations to Vote and Serve In the General Social Survey, individuals were asked whether civic duty included voting and whether it included serving on a jury. The results of the survey are shown in the table. Is there a difference in the proportion of individuals who feel jury duty is a civic duty and the proportion of individuals who feel voting is a civic duty? Use the $\alpha = 0.05$ level of significance.

		Jury	
		Duty (success)	**Not Duty (failure)**
Voting	**Duty (success)**	1322	65
	Not duty (failure)	45	17

Chapter Test

1. A pit boss is concerned that a pair of dice being used in a craps game is not fair. The distribution of the expected sum of two fair dice is as follows:

Sum of Two Dice	Probability	Sum of Two Dice	Probability
2	$\frac{1}{36}$	8	$\frac{5}{36}$
3	$\frac{2}{36}$	9	$\frac{4}{36}$
4	$\frac{3}{36}$	10	$\frac{3}{36}$
5	$\frac{4}{36}$	11	$\frac{2}{36}$
6	$\frac{5}{36}$	12	$\frac{1}{36}$
7	$\frac{6}{36}$		

The pit boss rolls the dice 400 times and records the sum of the dice. The table shows the results. Do you think the dice are fair? Use the $\alpha = 0.01$ level of significance.

Sum of Two Dice	Frequency	Sum of Two Dice	Frequency
2	16	8	59
3	23	9	45
4	31	10	34
5	41	11	19
6	62	12	11
7	59		

2. A researcher wanted to determine if the distribution of educational attainment of Americans today is different from the distribution in 2000. The distribution of educational attainment in 2000 was as follows:

Education	Relative Frequency
Not a high school graduate	0.158
High school graduate	0.331
Some college	0.176
Associate's degree	0.078
Bachelor's degree	0.170
Advanced degree	0.087

Source: Statistical Abstract of the United States.

The researcher randomly selects 500 Americans, learns their levels of education, and obtains the data shown in the table. Do the data suggest that the distribution of educational attainment has changed since 2000? Use the $\alpha = 0.1$ level of significance.

Education	Frequency
Not a high school graduate	72
High school graduate	159
Some college	85
Associate's degree	44
Bachelor's degree	92
Advanced degree	48

3. The Harris Poll asked a random sample of adult Americans, "How important are moral values when deciding how to vote?" The results of the survey by disclosed political affiliation are shown in the table.

		Political Affiliation		
		Republican	Independent	Democrat
Morality	**Important**	644	662	670
	Not important	56	155	147

(a) Do the sample data suggest that the proportion of adults who feel morality is important differ based on political affiliation? Use the $\alpha = 0.05$ level of significance.

(b) Construct a conditional distribution of morality by party affiliation and draw a bar graph. What does this summary tell you?

4. The General Social Survey regularly asks individuals to disclose their religious affiliation. The following data represent the religious affiliation of young adults, aged 18 to 29, in the 1970s, 1980s, 1990s, and 2000s. Do the data suggest different proportions of 18- to 29-year-olds have been affiliated with religion in the past four decades? Use the $\alpha = 0.05$ level of significance.

	1970s	1980s	1990s	2000s
Affiliated	2395	3022	2121	624
Unaffiliated (no religion)	327	412	404	2087

Source: General Social Survey

5. Many municipalities are passing legislation that forbids smoking in restaurants and bars. Bar owners claim that these laws hurt their business. Are their concerns legitimate? The following data represent the smoking status and frequency of visits to bars from the General Social Survey. Do smokers tend to spend more time in bars? Use the $\alpha = 0.05$ level of significance.

	Almost Daily	Several Times a Week	Several Times a Month	Once a Month	Several Times a Year	Once a Year	Never
Smoker	80	409	294	362	433	336	1265
Nonsmoker	57	350	379	471	573	568	3297

6. The General Social Survey asked the following two questions of randomly selected individuals.

- Do you favor or oppose the death penalty for persons convicted of murder?
- Should it be possible for a woman to obtain a legal abortion?

The results of the survey are given in the table. Is there a difference in the proportion who favor the death penalty and favor abortion? Use $\alpha = 0.05$ level of significance.

	Favor Death Penalty	Oppose Death Penalty
Favor abortion	7956	2549
Oppose abortion	11685	4272

Making an Informed Decision

Benefits of College

Are there benefits to attending college? If so, what are they? In this project, we will identify some of the perks that a college education provides. Obtain a random sample of at least 50 people aged 21 years or older and administer the following survey.

Please answer the following questions:

1. What is the highest level of education you have attained?

_____ Have not completed high school

_____ High school graduate

_____ College graduate

2. What is your employment status?

_____ Employed

_____ Unemployed, but actively seeking work

_____ Unemployed, but not actively seeking work

3. If you are employed, what is your annual income?

_____ Less than $20,000

_____ $20,000–$39,999

_____ $40,000–$60,000

_____ More than $60,000

4. If you are employed, which statement best describes the level of satisfaction you have with your career? Answer this question only if you are employed.

_____ Satisfied—I enjoy my job and am happy with my career.

_____ Somewhat satisfied—Work is work, but I am not unhappy with my career.

_____ Somewhat dissatisfied—I do not enjoy my work, but I also have no intention of leaving.

_____ Dissatisfied—Going to work is painful. I would quit tomorrow if I could.

(a) Use the results of the survey to create a contingency table for each of the following categories:
- Level of education/employment status
- Level of education/annual income
- Level of education/job satisfaction
- Annual income/job satisfaction

(b) Perform a chi-square test for independence on each contingency table from part (a).

(c) Draw bar graphs for each contingency table from part (a).

(d) Write a report that details your findings.

 **CASE STUDY** Feeling Lucky? Well, Are You?

In fiscal year (FY) 2013–2014 (July 2013–June 2014), the Florida Lottery generated $5.37 billion in total sales. Over that period, the state spent $37.5 million on advertising to promote its various games. Rand Advertising is interested in gaining access to this lucrative market. You are assigned the task of preparing a report on the lottery sales structure for three of Florida's online (nonscratch-off ticket) games: Fantasy 5, Lucky Money, and Lotto. Your findings will become part of a proposal by Rand to the Florida Lottery.

In Fantasy 5, a player picks five numbers from 1 to 36, at $1 per play. Drawings are held seven days a week. If there is no jackpot winner for a drawing, the money allocated for the top prize rolls down to the next prize tier (4 of 5).

The Lucky Money game costs $1 per ticket. Players pick four numbers from 1 to 47 and one Lucky Money number from 1 to 22. Drawings are held on Tuesdays and Fridays. If there is no jackpot winner for a drawing, the money allocated for the top prize rolls over to the next drawing, adding to the total of the next jackpot. The prize structures for both games is as shown.

Prize Structure for Florida's Fantasy 5 Game		
Match	**Estimated Prize Amount per Winner**	**Probability**
5 of 5	$200,000	$\frac{1}{376,992}$
4 of 5	$100	$\frac{1}{2432}$
3 of 5	$10	$\frac{1}{81}$
2 of 5	Free ticket	$\frac{1}{8}$

Prize Structure for Florida's Lucky Money Game		
Match	**Estimated Prize Amount per Winner**	**Probability**
4 of 4 + Lucky Ball	$500,000	$\frac{1}{3,032,205}$
4 of 4	$1044	$\frac{1}{189,513}$
3 of 4 + Lucky Ball	$292	$\frac{1}{17,629}$
3 of 4	$71.50	$\frac{1}{1102}$
2 of 4 + Lucky Ball	$20	$\frac{1}{560}$
1 of 4 + Lucky Ball	$2.50	$\frac{1}{61}$
2 of 4	$2	$\frac{1}{35}$
0 of 4 + Lucky Ball	Free ticket	$\frac{1}{25}$

In Lotto, players pick six numbers from 1 to 53, at $1 per play. Drawings are held on Wednesdays and Saturdays. If there is no jackpot winner, the top prize is rolled over to the next drawing. It is difficult to win the Lotto jackpot, so there are numerous jackpot rollovers. Rollovers make it difficult to determine an estimated prize payout per winner. However, the odds structure is shown in the table at the top of the next column.

Match	Probability
6 of 6	$\frac{1}{22,957,480}$
5 of 6	$\frac{1}{81,410}$
4 of 6	$\frac{1}{1416}$
3 of 6	$\frac{1}{71}$

To conduct your study, you have obtained the sales figures for each of the three games by district sales office for one week. These data are as follows:

	Florida Lottery Game		
District	**Fantasy 5**	**Lucky Money**	**Lotto**
1	87,030	27,221	270,256
3	186,780	56,822	934,451
4	267,955	94,123	1,004,651
5	232,451	61,019	707,934
6	727,390	262,840	2,746,438
8	462,874	135,321	1,563,491
9	353,532	139,421	1,492,549
10	377,627	140,285	1,433,077
11	470,993	153,591	1,743,370
12	588,154	185,946	1,852,986
13	1,283,042	474,067	2,883,453

Are the numbers of tickets sold for each lottery game and sales district independent? Construct a bar graph that represents the conditional distribution of game by sales district. Does this graphical evidence support your conclusion regarding the relationship between the type of game and the sales district? Explain.

Additionally, you are interested in the daily sales structure for the various districts. The following are the numbers of tickets sold each day for a randomly selected week for the three games in District 1.

	Florida Lottery Game		
Day	**Fantasy 5**	**Lucky Money**	**Lotto**
Monday	12,794	2,082	13,093
Tuesday	11,564	9,983	18,200
Wednesday	12,562	1,040	60,746
Thursday	11,408	1,677	18,054
Friday	15,299	10,439	38,513
Saturday	15,684	1,502	115,686
Sunday	7,728	498	5,964

Does the evidence suggest that the proportion of Fantasy 5 sales is the same for each day of the week? Perform a similar test for Lucky Money and Lotto.

Write a report detailing your assumptions, analyses, findings, and conclusions.

13

Comparing Three
or More Means

Outline

Making an Informed Decision

Remember the bonus you received in Chapter 10? Suppose you have decided that you now want to try investing in individual stocks rather than mutual funds. Which type of stocks should you invest in to earn the highest rate of return? See the Decision project on page 671.

PUTTING IT TOGETHER

Do you remember the progression for comparing proportions? Chapters 9 and 10 discussed inference for a single proportion, Section 11.1 discussed inference for two proportions from independent samples, and Section 12.2 presented a discussion of inference for three or more proportions (homogeneity of proportions) from independent samples.

We have this same progression of topics for inference on means. In Chapters 9 and 10, we discussed inferential techniques for a single population mean. In Section 11.3, we discussed inferential techniques for comparing two means from independent samples. In this chapter, we learn inferential techniques for comparing three or more means. Just as we used a different distribution to compare multiple proportions (the chi-square distribution), we use a different distribution for comparing three or more means. Although the choice of distribution initially may seem strange, once the logic of the procedure is understood, the choice of distribution makes sense.

13.1 Comparing Three or More Means (One-Way Analysis of Variance)

Preparing for This Section Before getting started, review the following:

- Completely randomized design (Section 1.6, pp. 42–46)
- The language of hypothesis testing (Section 10.1, pp. 473–478)
- Comparing two population means (Section 11.3, pp. 547–552)
- Normal probability plots (Section 7.3, pp. 377–380)
- *F*-distribution (Section 11.4, pp. 558–561)
- Boxplots (Section 3.5, pp. 166–168)

Objectives ➊ Verify the requirements to perform a one-way ANOVA

➋ Test a hypothesis regarding three or more means using one-way ANOVA

We now extend the concept of comparing two population means (Section 11.3) to comparing three or more population means. The procedure for doing this is called *analysis of variance*, or *ANOVA* for short.

Definition **Analysis of Variance (ANOVA)** is an inferential method used to test the equality of three or more population means.

For example, a marketing firm might wonder whether web page design plays a role in the average sale per visit to the website. To answer the question, the firm develops three different web pages that are randomly assigned to customers as they visit the site and determine the mean sale per visit for each design. In any hypothesis test, the null hypothesis is always a statement of "no difference," so we assume that the mean sale per visit is the same for all three designs. If we call the web page designs Design1, Design2, and Design3, our null hypothesis would be

$$H_0: \mu_1 = \mu_2 = \mu_3$$

versus the alternative hypothesis

H_1: at least one population mean is different from the others.

In Other Words
In ANOVA, the null hypothesis is always that the means of the different populations are equal. The alternative hypothesis is always that at least one population mean is different from the others.

As another example, a medical researcher might compare the effect that different levels of an experimental drug have on hair growth. The researcher might randomly divide a group of subjects into three different treatment groups. Once a day, group 1 receives a placebo, group 2 receives 50 milligrams (mg) of the experimental drug, and group 3 receives 100 mg of the experimental drug. The researcher then compares the mean number of new hair follicles for each of the three treatment groups.

Suppose we are testing a hypothesis regarding three population means, we have

$$H_0: \mu_1 = \mu_2 = \mu_3$$

H_1: at least one population mean is different from the others.

Figure 1(a) shows the distribution of each population if the null hypothesis is true, and Figure 1(b) shows one example of what the distributions of the populations might look like if the alternative hypothesis is true.

Figure 1

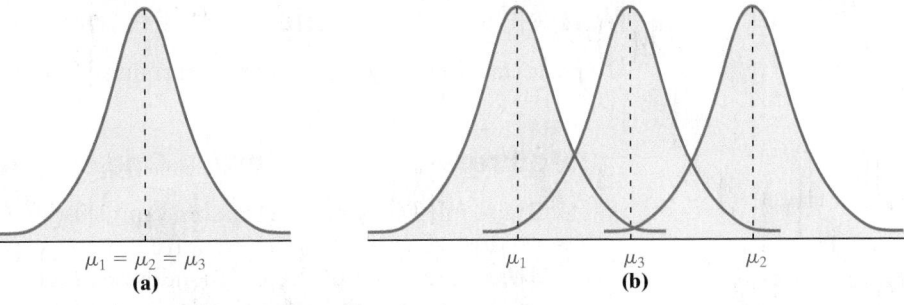

$$\mu_1 = \mu_2 = \mu_3$$
(a)

$$\mu_1 \qquad \mu_3 \qquad \mu_2$$
(b)

It may be tempting to test the null hypothesis by comparing the population means two at a time using the techniques introduced in Section 11.3. If we proceeded this way, we would need to test three different hypotheses:*

$$\begin{array}{c} H_0: \mu_1 = \mu_2 \\ H_1: \mu_1 \neq \mu_2 \end{array} \quad \text{and} \quad \begin{array}{c} H_0: \mu_1 = \mu_3 \\ H_1: \mu_1 \neq \mu_3 \end{array} \quad \text{and} \quad \begin{array}{c} H_0: \mu_2 = \mu_3 \\ H_1: \mu_2 \neq \mu_3 \end{array}$$

Each test would have a probability of a Type I error (rejecting the null hypothesis when it is true) of α. Using an $\alpha = 0.05$ level of significance, each test would have a 95% probability of correctly not rejecting the null hypothesis. The probability that all three tests correctly do not reject the null hypothesis is $0.95^3 = 0.86$ (assuming that the tests are independent). There is a $1 - 0.95^3 = 0.14$ probability that at least one test will lead to an incorrect rejection of H_0, which is higher than the desired 0.05 probability of making a Type I error. As the number of populations to be compared increases, the probability of making a Type I error using multiple t-tests for a given value of α also increases.

One possible solution to this problem is to adjust the individual α-values so the overall probability of a Type I error is the desired value. For example, we might test three pairs of means using $\alpha = 0.017$. The probability that all three tests correctly reject the null if the alternative hypothesis is true is $0.983^3 \approx 0.95$ (again assuming that the tests are independent). Thus, the overall probability of a Type I error is the desired 0.05. However, decreasing the probability of making a Type I error in each test increases the probability of making a Type II error for those tests. So conducting multiple t tests ultimately increases the chance of making a mistake.

To address this problem, Sir Ronald A. Fisher (1890–1962) introduced the method called analysis of variance (ANOVA). This term may seem odd because we are conducting a test on means, not variances. However, the name refers to the *approach* we are using, which will involve a comparison of two estimates of the same population variance. The justification for the name will become clear as we develop the test statistic.

The procedure used in this section is called *one-way* ANOVA because only one factor distinguishes the various populations in the study. For example, in comparing the sales from the web page, only page design distinguishes the three groups. In the hair growth drug, only the amount of the experimental drug received (placebo, 50 mg, or 100 mg) distinguishes the three groups.

We use a one-way ANOVA when analyzing data from a completely randomized design with three or more levels of treatment where the response variable is quantitative. In this design, the subjects must be similar in all characteristics, except the level of the treatment. Fisher stated that this is easiest to accomplish through randomization, which neutralizes the effect of uncontrolled variables. In the hair growth example, the subjects should be similar in terms of eating habits, age, and other factors that may affect hair growth. By randomly assigning the subjects to the three groups, any factors not controlled "even out."

Note that the methods of one-way ANOVA also apply when comparing only two population means. This, however, does not suggest that we should abandon our previous method for comparing two means. First, the ANOVA alternative hypothesis is always nondirectional. That is, we could use ANOVA to test whether two means are different, not whether one mean is larger than the other. Second, ANOVA requires that the population variances be equal. If they are not, we must use Welch's t-test, which does not assume equal variances. For these reasons, we typically utilize ANOVA only when comparing three or more means.

> **CAUTION!**
> Do not test $H_0: \mu_1 = \mu_2 = \mu_3$ by conducting three separate hypothesis tests, because the probability of making a Type I error will be much higher than α.

> **CAUTION!**
> It is vital that individuals be randomly assigned to treatments.

* In general, there would be $\binom{k}{2} = {}_kC_2$ pairs to test, where k equals the number of population means to be compared.

① Verify the Requirements to Perform a One-Way ANOVA

To perform a one-way ANOVA test, certain requirements must be satisfied.

Requirements to Perform a One-Way ANOVA Test

1. There must be k simple random samples, one from each of k populations or a randomized experiment with k treatments.
2. The k samples must be independent of each other; that is, the subjects in one group cannot be related in any way to subjects in a second group.
3. The populations must be normally distributed.
4. The populations must have the same variance; that is, each treatment group has population variance σ^2.

In Other Words
Try to design experiments or samples that use ANOVA so that each treatment group has the same number of experimental units.

The methods of one-way ANOVA are **robust**, so small departures from the normality requirement will not significantly affect the results of the procedure. In addition, the requirement of equal population variances does not need to be strictly adhered to, *especially if the sample size for each treatment group is the same*. Therefore, it is worthwhile to design an experiment in which the samples from the populations are roughly equal in size.

We can verify the requirement of normality by constructing normal probability plots. The requirement of equal population variances is more difficult to verify. In Section 11.4, we warned about testing the equality of population variances because the test is not robust. Also, since multiple paired comparisons increase the chance of making a mistake, pairwise comparison of population variances is not recommended. However, a general rule of thumb is as follows:

Verifying the Requirement of Equal Population Variances

The one-way ANOVA procedures may be used if the largest sample standard deviation is no more than twice the smallest sample standard deviation.

EXAMPLE 1 Testing the Requirements of One-Way ANOVA

Problem Prosthodontists specialize in the restoration of oral function, including the use of dental implants, veneers, dentures, and crowns. Since repairing chipped veneer is less costly and time-consuming than complete restoration, a researcher wanted to compare the shear bond strength of different repair kits for repairs of chipped porcelain veneer in fixed prosthodontics. He randomly divided 20 porcelain specimens into four treatment groups. Group 1 specimens used the Cojet system, group 2 used the Silistor system, group 3 used the Cimara system, and group 4 specimens used the Ceramic Repair system. At the conclusion of the study, shear bond strength (in megapascals, MPa) was measured according to ISO 10477. The data in Table 1 are based on the results of the study. Verify that the requirements to perform one-way ANOVA are satisfied.

Table 1

Cojet	Silistor	Cimara	Ceramic Repair
15.4	17.2	5.5	11.0
12.9	14.3	7.7	12.4
17.2	17.6	12.2	13.5
16.6	21.6	11.4	8.9
19.3	17.5	16.4	8.1

Source: P. Schmage et al. "Shear Bond Strengths of Five Intraoral Porcelain Repair Systems," *Journal of Adhesion Science & Technology* 21(5–6):409–422, 2007

Approach Verify each of the requirements listed on page 620.

Solution

1. The data are obtained from a completely randomized design with four treatments.
2. The specimens were assigned randomly to each group, so the samples are independent.
3. Figure 2 shows the normal probability plots for each group drawn in StatCrunch along with the correlation between the score and expected z-score. Each correlation is greater than 0.880, the critical value for $n = 5$ observations from Table VI. Therefore, it is reasonable to conclude each data set comes from a population that is normally distributed.

Figure 2

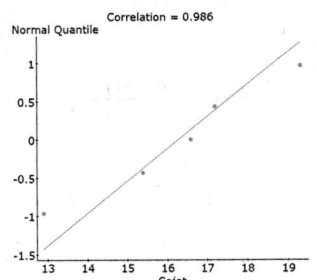 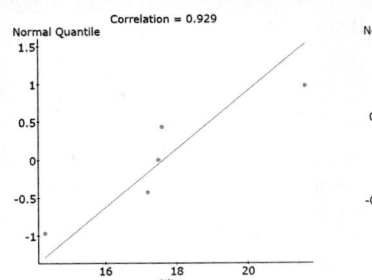 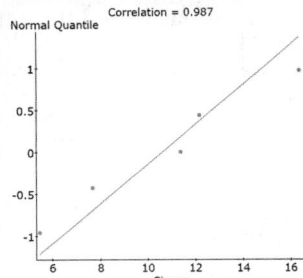 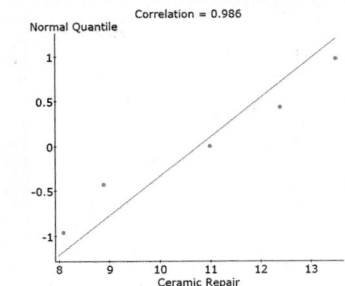

4. The sample standard deviations for each sample are computed using Minitab and presented as part of Figure 3. The largest standard deviation is 4.22 MPa; the smallest is 2.28 MPa. Because the largest standard deviation is not more than twice the smallest $(2 \cdot 2.28 = 4.56 > 4.22)$, the requirement of equal population variances is satisfied.

Figure 3

Descriptive Statistics: Cojet, Silistor, Cimara, Ceramic Repair

Variable	Mean	StDev	Variance	Minimum	Q1	Median	Q3	Maximum
Cojet	16.28	2.36	5.57	12.90	14.15	16.60	18.25	19.30
Silistor	17.64	2.60	6.76	14.30	15.75	17.50	19.60	21.60
Cimara	10.64	4.22	17.81	5.50	6.60	11.40	14.30	16.40
Ceramic Repair	10.78	2.28	5.20	8.10	8.50	11.00	12.95	13.50

Since all four requirements are satisfied, we can perform a one-way ANOVA.   ●

② Test a Hypothesis Regarding Three or More Means Using One-Way ANOVA

A Conceptual Understanding of One-Way ANOVA

The basic idea in one-way ANOVA is to determine if the sample data could come from populations with the same mean, μ, or suggests that at least one sample comes from a population whose mean is different from the others.

    To help understand the idea behind the ANOVA procedure, consider the data sets in Tables 2(a) and 2(b) on the following page. Each set of data represents samples from three populations. Computing the sample means for each set of data gives $\bar{x}_1 = \bar{y}_1 = 5$, $\bar{x}_2 = \bar{y}_2 = 8$, and $\bar{x}_3 = \bar{y}_3 = 11$. Figures 4(a) and 4(b) on the next page show dot plots for the data in Tables 2(a) and 2(b), respectively. The blue dots represent sample 1, red squares represent sample 2, and green diamonds represent sample 3.

    The dot plot in Figure 4(a) of the data in Table 2(a) reveals that the sample data come from populations with different means, because the variability *within* each sample is small relative to the amount of variability between the three sample means (in fact, there is not even any overlap among the samples). However, the difference in the means is not as apparent from the dot plot in Figure 4(b) for the data in Table 2(b). So, even though the sample means of the data in Table 2(b) vary by the same amount as they

Table 2					
x_1	x_2	x_3	y_1	y_2	y_3
4	7	10	14	10	6
5	8	10	2	3	7
6	9	11	2	15	12
6	7	11	7	7	15
4	9	13	0	5	15
(a)			**(b)**		

do in Table 2(a), the variability within each sample in Table 2(b) makes it difficult to determine whether the means differ enough to conclude that the samples come from populations with different means.

Figure 4

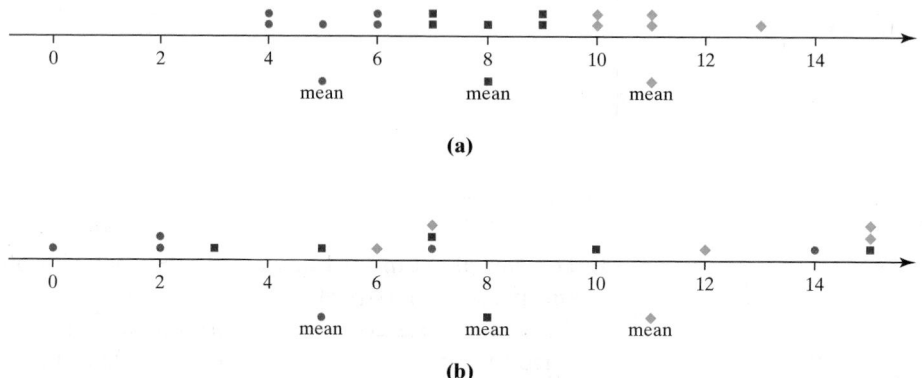

In general, we call the variability among the sample means the **between-sample variability** and the variability of each sample the **within-sample variability**. The between-sample variability for the data in Tables 2(a) and 2(b) is the same (since both data sets have the same sample means, overall mean, and sample size), but the within-sample variability for the data in Table 2(b) is much larger than it is for the data in Table 2(a).

So, if the between-sample variability is large relative to the within-sample variability as in Table 2(a), we have evidence to suggest that the samples come from populations with different means. This is the logic behind *analysis of variance* to test equality of *means*. The test statistic that we use to judge whether the population means might be equal is based on this idea.

ANOVA *F*-Test Statistic

The analysis of variance **F-test statistic** is given by

$$F_0 = \frac{\text{between-sample variability}}{\text{within-sample variability}}$$

How to Compute the *F*-Test Statistic

Remember, in testing any hypothesis, the null hypothesis is assumed to be true until the evidence indicates otherwise. In testing hypotheses regarding k population means, we assume that $\mu_1 = \mu_2 = \cdots = \mu_k = \mu$. That is, we assume that all k samples come from the same normal population whose mean is μ and variance is σ^2. Table 3 shows the statistics that result by sampling from each of the k populations.

Table 3			
Sample	**Sample Size**	**Sample Mean**	**Sample Standard Deviation**
1	n_1	$\bar{x}_1$	s_1
2	n_2	$\bar{x}_2$	s_2
3	n_3	$\bar{x}_3$	s_3
$\vdots$	$\vdots$	$\vdots$	$\vdots$
k	n_k	$\bar{x}_k$	s_k

The computation of the *F*-test statistic is based on *mean squares*. A **mean square** is an average (mean) of squared values. For example, any variance is a mean square. In Section 3.2, we saw that the population variance, σ^2, is estimated by the sample variance, $s^2 = \dfrac{\sum (x_i - \bar{x})^2}{n - 1}$. The numerator of s^2 is called the **sum of squares**. It gives a measure of the variability of the data about the sample mean. The denominator of s^2, $n - 1$, represents the *degrees of freedom*. So, in general, a mean square is a sum of squares divided by the corresponding degrees of freedom. The *F*-test statistic is the ratio of two mean squares, or two estimates of the variance: the between-sample variability and the within-sample variability.

The quantity $\sum (x_i - \bar{x})^2$ represents the total variability about the mean of the entire set of data, $\bar{x}$, and is called the **total sum of squares** or **SS (Total)**. Within any given set of data, there are two types of variability: variability due to differences in the treatment groups (or samples) and variability within each treatment group (or sample).

The variability within each treatment group, or within-sample variability estimate of σ^2 (or denominator of the *F*-test statistic), is a weighted average of the sample variances from each treatment (or sample), where the weights are based on the size of each sample. We call this estimate of σ^2 the **mean square due to error**, or **MSE**. To find the MSE, first compute the **sum of squares due to error**, or **SSE**. The SSE is the sum of all sample variances weighted by the degrees of freedom. So

$$\text{SSE} = (n_1 - 1)s_1^2 + (n_2 - 1)s_2^2 + \cdots + (n_k - 1)s_k^2$$

For each sample variance in the sum of squares due to error, the degrees of freedom is $n_i - 1$. The total degrees of freedom for the SSE is the sum of the individual degrees of freedom, or

$$(n_1 - 1) + (n_2 - 1) + \cdots + (n_k - 1) = (n_1 + n_2 + \cdots + n_k) - k = n - k$$

where n is the total number of observations. That is, $n = n_1 + n_2 + \cdots + n_k$ and k is the number of treatments (or samples).

The mean square due to error is an unbiased estimator of σ^2 whether the null hypothesis of equal means is true or not. It is computed as follows:

$$\text{MSE} = \frac{\text{SSE}}{\text{degrees of freedom}} = \frac{(n_1 - 1)s_1^2 + (n_2 - 1)s_2^2 + \cdots + (n_k - 1)s_k^2}{n - k}$$

The between-sample variability estimate of σ^2 (or numerator of the *F*-test statistic) is called the **mean square due to treatment**, or **MST**, since any differences in the sample means could be attributed to the different levels of the treatment. To find the MST, first compute the **sum of squares due to treatment** or **SST**. The SST is the sum of the squared differences between each treatment (or sample) mean and the overall mean, where each squared difference is weighted by its sample size. That is,

$$\text{SST} = n_1(\bar{x}_1 - \bar{x})^2 + n_2(\bar{x}_2 - \bar{x})^2 + \cdots + n_k(\bar{x}_k - \bar{x})^2$$

Since there are k sample means and SST measures how these vary from the overall mean, this sum of squares has $k - 1$ degrees of freedom. The MST is computed as follows:

$$\text{MST} = \frac{\text{SST}}{\text{degrees of freedom}} = \frac{n_1(\bar{x}_1 - \bar{x})^2 + n_2(\bar{x}_2 - \bar{x})^2 + \cdots + n_k(\bar{x}_k - \bar{x})^2}{k - 1}$$

The mean square due to treatment is an unbiased estimate of σ^2 only if the null hypothesis is true. So, if H_0 is true (that is, if the k populations means are equal), the mean square due to treatment should be close in value to the mean square due to error. However, if H_0 is false (that is, at least one population mean differs from the others), the mean square due to treatment is *not* a good estimate of σ^2. In fact, if the statement in the null hypothesis is false, the MST will overestimate the value of σ^2, because at least one of the sample means in a treatment (or sample) will be "far away" from the overall mean.

The F-test statistic is based on the ratio of the mean squares.

F-Test Statistic

$$F_0 = \frac{\text{mean square due to treatment}}{\text{mean square due to error}} = \frac{\text{MST}}{\text{MSE}}$$

If the F-test statistic is large, then the MST is greater than the MSE. This would be evidence against the statement in the null hypothesis.

We now present the steps used to compute the F-test statistic.

Computing the F-Test Statistic by Hand

Step 1 Compute the sample mean of the combined data set by adding up all the observations and dividing by the number of observations. Call this value $\bar{x}$.

Step 2 Find the sample mean for each sample (or treatment). Let $\bar{x}_1$ represent the sample mean of sample 1, $\bar{x}_2$ represent the sample mean of sample 2, and so on.

Step 3 Find the sample variance for each sample (or treatment). Let s_1^2 represent the sample variance for sample 1, s_2^2 represent the sample variance for sample 2, and so on.

Step 4 Compute the sum of squares due to treatments, SST, and the sum of squares due to error, SSE. (See page 623.)

Step 5 Divide each sum of squares by its corresponding degrees of freedom $(k - 1$ and $n - k$, respectively) to obtain the mean squares MST and MSE.

Step 6 Compute the F-test statistic:

$$F_0 = \frac{\text{mean square due to treatment}}{\text{mean square due to error}} = \frac{\text{MST}}{\text{MSE}}$$

EXAMPLE 2   Computing the F-Test Statistic by Hand

x_1	x_2	x_3
4	7	10
5	8	10
6	9	11
6	7	11
4	9	13

Problem Compute the F-test statistic for the data in Table 2(a). The data are presented again for convenience.

Approach Follow Steps 1 through 6.

Solution

Step 1 Compute the mean of the entire set of data, or the overall mean.

$$\bar{x} = \frac{4 + 5 + 6 + \cdots + 11 + 13}{15} = 8$$

Step 2 Find the sample mean for each treatment (or sample).

$$\bar{x}_1 = 5 \qquad \bar{x}_2 = 8 \qquad \bar{x}_3 = 11$$

Step 3   Find the sample variance for each treatment (or sample).

$$s_1^2 = \frac{(4-5)^2 + (5-5)^2 + (6-5)^2 + (6-5)^2 + (4-5)^2}{5-1} = 1$$

$$s_2^2 = 1 \qquad s_3^2 = 1.5$$

Step 4   Compute the sum of squares due to treatment, SST, and the sum of squares due to error, SSE.

$$\text{SST} = 5(5-8)^2 + 5(8-8)^2 + 5(11-8)^2 = 90$$
$$\text{SSE} = (5-1)1 + (5-1)1 + (5-1)1.5 = 14$$

Step 5   Compute the mean square due to treatment, MST, and the mean square due to error, MSE.

$$\text{MST} = \frac{\text{SST}}{k-1} = \frac{90}{3-1} = 45 \qquad \text{MSE} = \frac{\text{SSE}}{n-k} = \frac{14}{15-3} = 1.1667$$

Step 6   Compute the F-test statistic.

$$F_0 = \frac{\text{MST}}{\text{MSE}} = \frac{45}{1.1667} = 38.57$$

- **Now Work Problem 11**

The large value of F_0 suggests that the sample means for each treatment differ.   •

For the data in Table 2(b), MST = 45, MSE = 24.1667, and $F_0 = 1.86$. You should verify these results. The small F-test statistic is evidence that the sample means for each treatment do not differ. In addition, notice that MST = 45 for the data in both tables. Why?

The results of the computations for the data in Table 2(a) that led to the F-test statistic are presented in Table 4, which is called an **ANOVA table**.

Table 4				
Source of Variation	**Sum of Squares**	**Degrees of Freedom**	**Mean Squares**	**F-Test Statistic**
Treatment	90	2	45	38.57
Error	14	12	1.1667	
Total	104	14		

Remember, the mean square due to error is an unbiased estimate of σ^2 whether the null hypothesis is true or not. However, the mean square due to treatment is an unbiased estimate of σ^2 only if the null hypothesis is true (that is, only if the population means from each treatment, or sample, are the same). If at least one mean for a treatment is significantly different from the others, the mean square due to treatment will substantially overestimate the value of σ^2. Because the F-test statistic is the ratio of MST to MSE, ANOVA tests are always right-tailed tests.

As you may have realized, the computations of an ANOVA test are tedious, so researchers use statistical software to conduct the test. When using software, it is easiest to use the P-value approach. The nice thing about P-values is that the decision rule is always the same regardless of the type of hypothesis being tested.

In Other Words
Remember, "if the P-value is too low, the null must go."

Decision Rule in the One-Way ANOVA Test

If the P-value is less than the level of significance, α, reject the null hypothesis.

EXAMPLE 3 Performing One-Way ANOVA Using Technology

Problem The researcher in Example 1 wants to determine if there is a difference in the mean shear bond strength among the four treatment groups at the $\alpha = 0.05$ level of significance.

Approach We will use StatCrunch, Minitab, Excel, and a TI-84 Plus C graphing calculator to conduct the test. If the P-value is less than the level of significance, reject the null hypothesis. The steps for performing one-way ANOVA using the TI-83/84 Plus graphing calculators, Minitab, Excel, and StatCrunch are given in the Technology Step-by-Step on pages 628–629.

Solution The null hypothesis is always a statement of "no difference." So the null hypothesis is that the mean shear bond strength among the four treatment groups is the same, or

$$H_0: \mu_{\text{Cojet}} = \mu_{\text{Silistor}} = \mu_{\text{Cimara}} = \mu_{\text{Ceramic Repair}}$$

versus the alternative hypothesis

$$H_1: \text{at least one population mean is different from the others}$$

Figure 5 shows (a) the output from StatCrunch, (b) the output from Minitab, (c) the output from Excel, and (d) the output from a TI-84 Plus C graphing calculator.

Figure 5 **Analysis of Variance results:**
Data stored in separate columns.

Column statistics

Column	n	Mean	Std. Dev.	Std. Error
Cojet	5	16.28	2.3594491	1.0551777
Silistor	5	17.64	2.6005769	1.1630133
Cimara	5	10.64	4.220545	1.8874851
Ceramic Repair	5	10.78	2.279693	1.0195097

ANOVA table

Source	DF	SS	MS	F-Stat	P-value
Columns	3	199.9855	66.661833	7.545199	0.0023
Error	16	141.36	8.835		
Total	19	341.3455			

(a) StatCrunch Output

```
One-way ANOVA: Cojet, Silistor, Cimara, Ceramic Repair

Analysis of Variance

Source  DF  Adj SS  Adj MS  F-Value  P-Value
Factor   3   200.0  66.662     7.55    0.002
Error   16   141.4   8.835
Total   19   341.3

Model Summary

     S   R-sq  R-sq(adj)  R-sq(pred)
2.97237  58.59%    50.82%      35.29%
```

(b) Minitab Output

Anova: Single Factor

SUMMARY

Groups	Count	Sum	Average	Variance
Cojet	5	81.4	16.28	5.567
Silistor	5	88.2	17.64	6.763
Cimara	5	53.2	10.64	17.813
Ceramic Repair	5	53.9	10.78	5.197

ANOVA

Source of Variation	SS	df	MS	F	P-value	F crit
Between Groups	199.9855	3	66.66183	7.545199	0.002295	3.238872
Within Groups	141.36	16	8.835			
Total	341.3455	19				

(c) Excel Output

P-value —

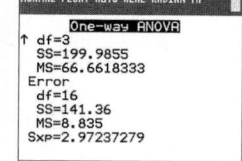

(d) TI-84 Plus C Output

Because the P-value, 0.002, is less than the level of significance, α, we reject the null hypothesis. There is sufficient evidence to indicate that at least one population mean shear bond strength is different from the others. •

Whenever performing ANOVA, it is a good idea to present visual evidence that supports the conclusions of the test. Side-by-side boxplots are a great way to help visually reinforce the results of the ANOVA procedure. Figure 6 shows the side-by-side boxplots of the data presented in Table 1. The boxplots support the ANOVA results from Example 3.

• **Now Work Problem 13(a)–(e)**

Figure 6

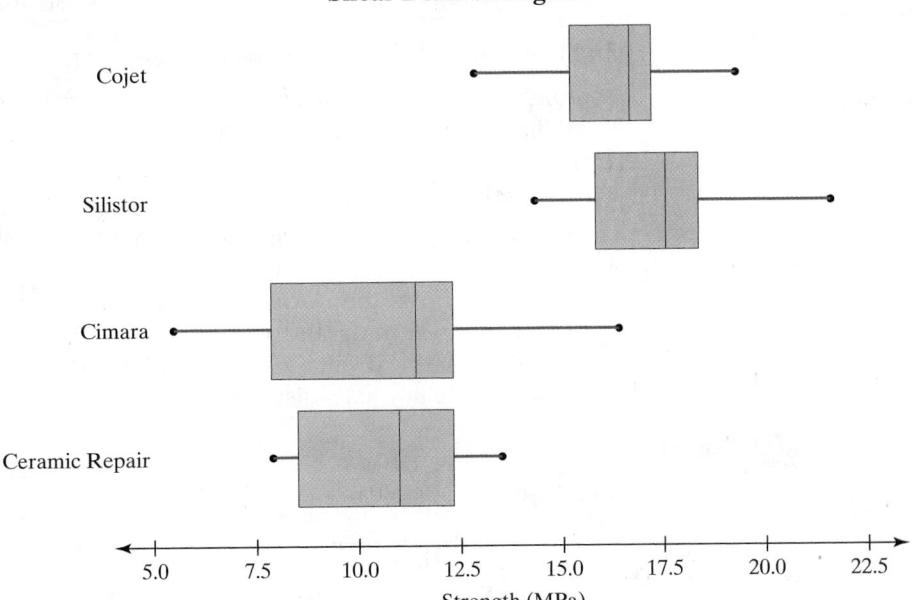

Shear Bond Strengths

Figure 7

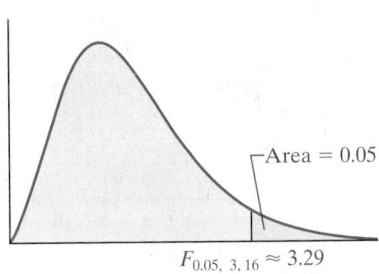

Area = 0.05

$F_{0.05,\,3,\,16} \approx 3.29$

> **CAUTION!**
> If we reject the null hypothesis when using ANOVA, we are rejecting that the population means are all equal. However, the test doesn't tell us which means differ.

If we were conducting this ANOVA test by hand using the Classical Method, we would compare the F-test statistic with a critical F-value. The ANOVA test is a right-tailed test, so the **critical F-value** is the F-value whose area in the right tail is α with $k - 1$ degrees of freedom in the numerator and $n - k$ degrees of freedom in the denominator. The critical F-value for the study in Example 1 at the $\alpha = 0.05$ level of significance is $F_{0.05,3,16} \approx 3.29$ from Table IX. Note that, because the table is limited, we must use the value for $F_{0.05,3,15}$ as our critical value. In Figure 5(c), we see that the technology gives us a more accurate critical value, $F_{0.05,3,16} \approx 3.239$. Since the F-test statistic, 7.55, is greater than the critical F, we reject the null hypothesis (see Figure 7).

When we reject the null hypothesis of equal population means, as in Example 3, we know that at least one population mean differs from the others. However, we do not know which means differ. Side-by-side boxplots or confidence intervals, like those in Figure 6 and Figure 5(b), can give us some idea, but we can more formally answer this question using the Tukey test, which will be discussed in the next section.

Verifying the Normality Requirement in a One-Way ANOVA

One requirement that must be satisfied to conduct a one-way ANOVA is that the k populations must be normally distributed. We verified this requirement in the shear bond strength example by drawing normal probability plots for each of the $k = 4$ samples to see if the plots suggested that the data could come from a population that is normally distributed.

A potential problem with this approach is that if the sample sizes are small, then drawing normal probability plots for each sample does not shed enough light on whether the normality requirement is satisfied. To address this issue, we can take another route. In a one-way ANOVA, a single factor is varied. We will call T_j the effect of the jth level of the factor. Then the ANOVA model is

$$Y_{ij} = \mu + T_j + \varepsilon_{ij}$$

where Y_{ij} represents the ith observation on the jth treatment,
          T_j represents the effect of the jth treatment, and
          ε_{ij} represents the random error in the ith observation of the jth treatment.

The error term ε_{ij} is a normally distributed variable with a mean of 0 and a variance that is the same for each treatment. In a one-way ANOVA, the residuals are $Y_{ij} - \hat{Y}_{ij}$, where $\hat{Y}_{ij}$ is the predicted value of the ith observation from the jth level of the factor (or treatment). The value of $\hat{Y}_{ij}$ can be estimated by computing the mean value of the response variable for each level of the factor, $\overline{Y}_{\cdot j}$. Then $\overline{Y}_{\cdot j}$ is subtracted from each observation to obtain the residuals, $Y_{ij} - \overline{Y}_{\cdot j}$. Finally, a normal probability plot of the residuals is drawn to verify the normality requirement.

EXAMPLE 4 Verifying the Normality Requirement by Analyzing Residuals

Problem Verify the normality requirement for the data analyzed in Example 1.

Approach Compute the mean value for each level of the factor and subtract the result from each observation to obtain the residuals. Then draw a normal probability plot of the residuals.

Solution The first two columns of Table 5 show the level of the factor and the value of the response variable, shear bond strength. Column 3 of Table 5 shows the sample mean for each level of the factor. Column 4 shows the residuals.

Figure 8 shows the normal probability plot of the residuals. The correlation between the residuals and expected z-scores is 0.993 [Tech: 0.991]. Because $0.993 > 0.951$ (critical value from Table VI with $n = 20$), it is reasonable to conclude the residuals are normally distributed.

Table 5

Level of Factor	Shear Bond Strength	Sample Mean of Factor	Residual
Cojet	15.4	16.28	$15.4 - 16.28 = -0.88$
Cojet	12.9	16.28	$12.9 - 16.28 = -3.38$
Cojet	17.2	16.28	0.92
Cojet	16.6	16.28	0.32
Cojet	19.3	16.28	3.02
Silistor	17.2	17.64	-0.44
Silistor	14.3	17.64	-3.34
Silistor	17.6	17.64	-0.04
Silistor	21.6	17.64	3.96
Silistor	17.5	17.64	-0.14
Cimara	5.5	10.64	-5.14
Cimara	7.7	10.64	-2.94
Cimara	12.2	10.64	1.56
Cimara	11.4	10.64	0.76
Cimara	16.4	10.64	5.76
Ceramic Repair	11.0	10.78	0.22
Ceramic Repair	12.4	10.78	1.62
Ceramic Repair	13.5	10.78	2.72
Ceramic Repair	8.9	10.78	-1.88
Ceramic Repair	8.1	10.78	-2.68

Figure 8

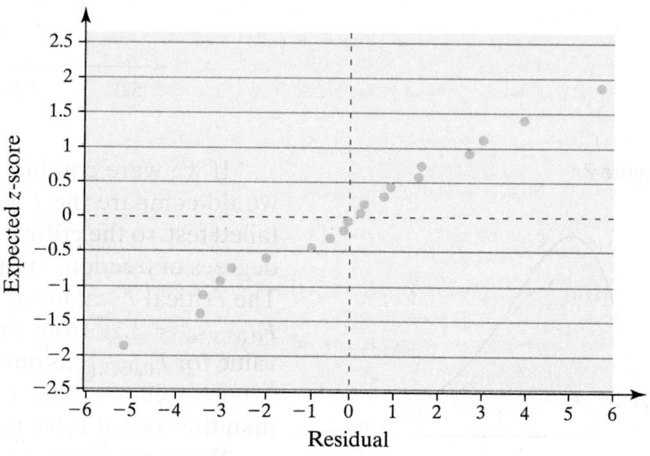

Technology Step-by-Step ANOVA

TI-83/84 Plus

1. Enter the raw data into L1, L2, L3, and so on, for each sample or treatment.
2. Press STAT, highlight TESTS, and select ANOVA (.
3. Enter the list names for each sample or treatment after ANOVA (. For example, if there are three treatments in L1, L2, and L3, enter

    ANOVA (L1, L2, L3)

    Press ENTER.

Minitab

1. Enter the raw data into C1, C2, C3, and so on, for each sample or treatment.
2. Select **Stat**; then highlight **ANOVA** and select **One-way**.
3. In the pull-down menu, select "Response data are in a separate column for each factor level." Enter the column names in the cell marked "Responses." Click OK.

Excel

1. Enter the raw data in columns A, B, C, and so on, for each sample or treatment.
2. Be sure the Data Analysis ToolPak is activated. This is done by selecting File and selecting Options. Now select Add-ins, select Excel Add-ins under "Manage," click Go, and choose Analysis ToolPak.
3. Select Data. Then choose Data Analysis. Select ANOVA: Single Factor and click OK.
4. With the cursor in the "Input Range:" cell, highlight the data. Click OK.

StatCrunch

1. Either enter the raw data in separate columns for each sample or treatment, or enter the value of the variable in a single column with indicator variables for each sample or treatment in a second column.
2. Select **Stat**, highlight **ANOVA**, and select **One Way**.
3. If the raw data are in separate columns, select "Compare selected columns" and then click the columns you want to compare. If the raw data are in a single column, select "Compare values in a single column" and then choose the column that contains the value of the variables and the column that indicates the treatment or sample. Click Compute!.

13.1 Assess Your Understanding

Vocabulary and Skill Building

1. The acronym ANOVA stands for _____ __ _____.

2. *True or False:* To perform a one-way ANOVA, the populations do not need to be normally distributed.

3. *True or False:* To perform a one-way ANOVA, the populations must have the same variance.

4. The variability among the sample means is called _____ sample variability, and the variability of each sample is the _____ sample variability.

5. The variability within each treatment group, which is a weighted average of the sample variances from each treatment where the weights are based on the size of each sample, is called the mean square due to _____ and is denoted _____.

6. *True or False:* The F-test statistic is $F_0 = \dfrac{\text{MST}}{\text{MSE}}$.

In Problems 7 and 8, fill in the ANOVA table.

7.

Source of Variation	Sum of Squares	Degrees of Freedom	Mean Squares	F-Test Statistic
Treatment	387	2		
Error	8042	27		
Total				

8.

Source of Variation	Sum of Squares	Degrees of Freedom	Mean Squares	F-Test Statistic
Treatment	2814	3		
Error	4915	36		
Total				

In Problems 9 and 10, determine the F-test statistic based on the given summary statistics. [Hint: $\bar{x} = \dfrac{\sum n_i \bar{x}_i}{\sum n_i}$.]

9.

Population	Sample Size	Sample Mean	Sample Variance
1	10	40	48
2	10	42	31
3	10	44	25

10.

Population	Sample Size	Sample Mean	Sample Variance
1	15	105	34
2	15	110	40
3	15	108	30
4	15	90	38

NW 11. The following data represent a simple random sample of $n = 4$ from three populations that are known to be normally distributed. Verify that the F-test statistic is 2.04.

Sample 1	Sample 2	Sample 3
28	22	25
23	25	24
30	17	19
27	23	30

12. The following data represent a simple random sample of $n = 5$ from three populations that are known to be normally distributed. Verify that the F-test statistic is 2.599.

Sample 1	Sample 2	Sample 3
73	67	72
82	77	80
82	66	87
81	67	77
97	83	96

Applying the Concepts

NW 13. Corn Production The data in the table represent the number of corn plants in randomly sampled rows (a 17-foot by 5-inch strip) for various types of plot. An agricultural researcher wants to know whether the mean number of plants for each plot type are equal.

Plot Type	Number of Plants					
Sludge plot	25	27	33	30	28	27
Spring disc	32	30	33	35	34	34
No till	30	26	29	32	25	29

Source: Andrew Dieter and Brad Schmidgall, Joliet Junior College

(a) Write the null and alternative hypotheses.
(b) State the requirements that must be satisfied to use the one-way ANOVA procedure.
(c) Use the following partial Minitab output to test the hypothesis of equal means at the $\alpha = 0.05$ level of significance.

One-way ANOVA: Sludge Plot, Spring Disc, No Till

Source	df	SS	MS	F	P
Factor	2	84.11	42.06	7.10	0.007
Error	15	88.83	5.92		
Total	17	172.94			

(d) Shown are side-by-side boxplots of each type of plot. Do these boxplots support the results obtained in part (c)?

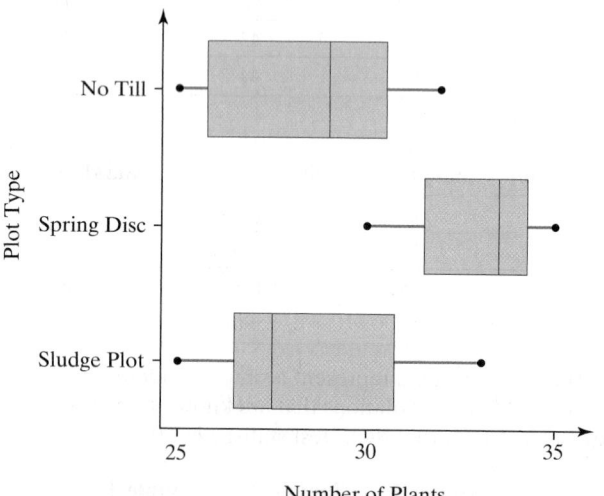

(e) Verify that the F-test statistic is 7.10.
(f) Verify the residuals are normally distributed.

14. Soybean Yield The data in the table represent the number of pods on a random sample of soybean plants for various plot types. An agricultural researcher wants to determine if the mean numbers of pods for each plot type are equal.

Plot Type	Pods								
Liberty	32	31	36	35	41	34	39	37	38
No till	34	30	31	27	40	33	37	42	39
Chisel plowed	34	37	24	23	32	33	27	34	30

Source: Andrew Dieter and Brad Schmidgall, Joliet Junior College

(a) Write the null and alternative hypotheses.
(b) State the requirements that must be satisfied to use the one-way ANOVA procedure.
(c) Use the following Minitab output to test the hypothesis of equal means at the $\alpha = 0.05$ level of significance.

One-way ANOVA: Liberty, No Till Chisel Plowed

Source	df	SS	MS	F	P
Factor	2	149.0	74.5	3.77	0.038
Error	24	474.7	19.8		
Total	26	623.6			

(d) Shown are side-by-side boxplots of each type of plot. Do these boxplots support the results obtained in part (c)?

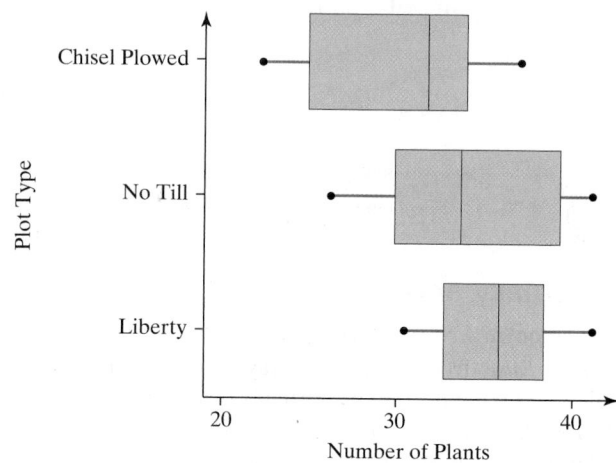

(e) Verify that the F-test statistic is 3.77.
(f) Verify the residuals are normally distributed.

15. Which Delivery Method Is Best? At a community college, the mathematics department has been experimenting with four different delivery mechanisms for content in their Intermediate Algebra courses. One method is the traditional lecture (method I), the second is a hybrid format in which half the class time is online and the other half is face-to-face (method II), the third is online (method III), and the fourth is an emporium model from which students obtain their lectures and do their work in a lab with an instructor available for assistance (method IV). To assess the effectiveness of the four methods, students in each approach are given a final exam with the results shown next. Do the data suggest that any method has a different mean score from the others?

Method I	81	81	85	67	88	72	80	63	62
	92	82	49	69	66	74	80		
Method II	85	53	80	75	64	39	60	61	83
	66	75	66	90	93				
Method III	81	59	70	70	64	78	75	80	52
	45	87	85	79					
Method IV	86	90	81	61	84	72	56	68	82
	98	79	74	82					

(a) What is the response variable in this study? Is it quantitative or qualitative?
(b) What is the factor in this study? How many levels does it have?
(c) Write the null and alternative hypotheses.

(d) State the requirements that must be satisfied to use the one-way ANOVA procedure.

(e) Assuming the requirements stated in part (d) are satisfied, use the following StatCrunch output to test the hypothesis of equal means at the $\alpha = 0.05$ level of significance.

Analysis of Variance results:
Data stored in separate columns.

Column statistics

Column◆	n◆	Mean◆	Std. Dev.◆	Std. Error◆
Method I	16	74.4375	11.266285	2.8165711
Method II	14	70.714286	15.101852	4.0361397
Method III	13	71.153846	12.88957	3.5749236
Method IV	13	77.923077	11.586354	3.2134763

ANOVA table

Source	DF	SS	MS	F-Stat	P-value
Columns	3	448.42926	149.47642	0.91731355	0.439
Error	52	8473.41	162.95019		
Total	55	8921.8393			

(f) Shown are side-by-side boxplots drawn in StatCrunch of each delivery method. Do these support the results obtained in part (e)?

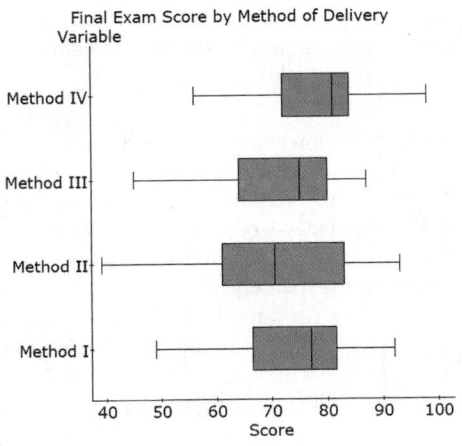

Final Exam Score by Method of Delivery

(g) Interpret the *P*-value.

(h) Verify the residuals are normally distributed.

16. Births by Day of Week An obstetrician knew that there were more live births during the week than on weekends. She wanted to determine whether the mean number of births was the same for each of the five days of the week. She randomly selected eight dates for each of the five days of the week and obtained the following data:

Monday	Tuesday	Wednesday	Thursday	Friday
10,456	11,621	11,084	11,171	11,545
10,023	11,944	11,570	11,745	12,321
10,691	11,045	11,346	12,023	11,749
10,283	12,927	11,875	12,433	12,192
10,265	12,577	12,193	12,132	12,422
11,189	11,753	11,593	11,903	11,627
11,198	12,509	11,216	11,233	11,624
11,465	13,521	11,818	12,543	12,543

Source: National Center for Health Statistics

(a) Write the null and alternative hypotheses.

(b) State the requirements that must be satisfied to use the one-way ANOVA procedure.

(c) Use the following Minitab output to test the hypothesis of equal means at the $\alpha = 0.01$ level of significance.

One-way ANOVA: Mon, Tues, Wed, Thurs, Fri

Source	df	SS	MS	F	P
Factor	4	11507633	2876908	9.80	0.000
Error	35	10270781	293451		
Total	39	21778414			

(d) Shown are side-by-side boxplots for each weekday. Do these boxplots support the results obtained in part (c)?

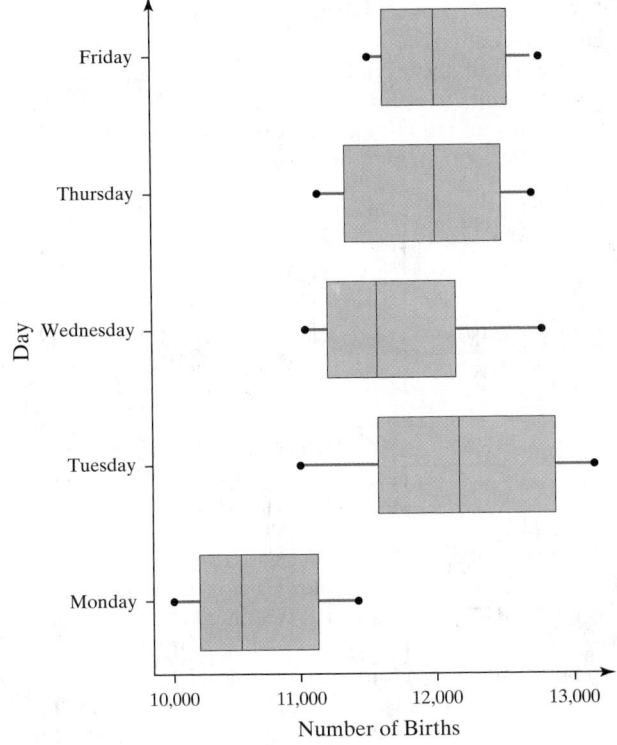

17. Rates of Return A stock analyst wondered whether the mean rate of return of financial, energy, and utility stocks differed over the past five years. He obtained a simple random sample of eight companies from each of the three sectors and obtained the five-year rates of return shown in the following table (in percent):

Financial	Energy	Utilities
10.76	12.72	11.88
15.05	13.91	5.86
17.01	6.43	13.46
5.07	11.19	9.90
19.50	18.79	3.95
8.16	20.73	3.44
10.38	9.60	7.11
6.75	17.40	15.70

Source: Morningstar.com

(a) State the null and alternative hypotheses.

(b) Verify that the requirements to use the one-way ANOVA procedure are satisfied. Normal probability plots indicate that the sample data come from normal populations.

(c) Are the mean rates of return different at the $\alpha = 0.05$ level of significance?

(d) Draw boxplots of the three sectors to support the results obtained in part (c).

18. Reaction Time In an online psychology experiment sponsored by the University of Mississippi, researchers asked study participants to respond to various stimuli. Participants were randomly assigned to one of three groups: Group 1, the simple group, required to respond as quickly as possible after a stimulus was presented; Group 2, the go/no-go group, required to respond to a particular stimulus while disregarding other stimuli; and Group 3, the choice group, required to respond differently depending on the type of whistle sound, the subject must press a certain button. The researcher wants to determine if the mean reaction times for each stimulus are equal. The reaction time (in seconds) for each stimulus is presented in the table.

Simple	Go/No-Go	Choice
0.430	0.588	0.561
0.498	0.375	0.498
0.480	0.409	0.519
0.376	0.613	0.538
0.402	0.481	0.464
0.329	0.355	0.625

Source: PsychExperiments; The University of Mississippi; www.olemiss.edu/psychexps/

(a) What type of experimental design is this?

(b) What is the response variable? What is the explanatory variable? How many levels of treatment are there in this experiment?

(c) State the null and alternative hypotheses.

(d) Verify that the requirements to use the one-way ANOVA procedure are satisfied. Normal probability plots indicate that the sample data come from a normal population.

(e) Test the hypothesis that the mean reaction times for the three stimuli are the same at the $\alpha = 0.05$ level of significance.

(f) Draw boxplots of the three stimuli to support the analytic results obtained in part (c).

19. Car-Buying Discrimination To determine if there is gender and/or race discrimination in car buying, Ian Ayres put together a team of fifteen white males, five white females, four black males, and seven black females who were each asked to obtain an initial offer price from the dealer on a certain model car. The 31 individuals were made to appear as similar as possible to account for other variables that may play a role in the offer price of a car. The following data are based on the results in the article and represent the profit on the initial price offered by the dealer. Ayres wanted to determine if the profit based on the initial offer differed among the four groups.

White Male		Black Male	White Female	Black Female
1300	853	1241	951	1899
646	727	1824	954	2053
951	559	1616	754	1943
794	429	1537	706	2168
661	1181		596	2325
824	853			1982
1038	877			1780
754				

Source: Ian Ayres. "Fair Driving: Gender and Race Discrimination in Retail Car Negotiations." *Harvard Law Review.* Vol. 104, No. 4, Feb. 1991.

(a) What is the response variable in this study? Is it qualitative or quantitative?

(b) State the null and alternative hypotheses.

(c) A normal probability plot of each group suggests the data come from a population that is normally distributed. Verify the requirement of equal variances is satisfied.

(d) Test the hypothesis stated in part (b).

(e) Draw side-by-side boxplots of the four groups to support the analytic results of part (d).

(f) What do the results of the analysis suggest?

(g) Because the group of black males has a small sample size, the normality requirement is best verified by assessing the normality of the residuals. Verify the normality requirement by drawing a normal probability plot of the residuals.

20. Crash Data The Insurance Institute for Highway Safety conducts experiments in which cars are crashed into a fixed barrier at 40 mph. The barrier's deformable face is made of aluminum honeycomb, which makes the forces in the test similar to those involved in a frontal offset crash between two vehicles of the same weight, each going just less than 40 mph. Suppose you want to know if the mean head injury resulting from this offset crash is the same for large family cars, passenger vans, and midsize utility vehicles. The researcher wants to determine if the means for head injury for each class of vehicle are different. The following data were collected from the institute's study.

Large Family Cars	Head Injury (hic)
Hyundai XG300	264
Ford Taurus	134
Buick LeSabre	409
Chevrolet Impala	530
Chrysler 300	149
Pontiac Grand Prix	627
Toyota Avalon	166

Passenger Vans	Head Injury (hic)
Toyota Sienna	148
Honda Odyssey	238
Ford Freestar	340
Mazda MPV	693
Chevrolet Uplander	550
Nissan Quest	470
Kia Sedona	322

Midsize Utility Vehicles	Head Injury (hic)
Honda Pilot	225
Toyota 4Runner	216
Mitsubishi Endeavor	186
Nissan Murano	307
Ford Explorer	353
Kia Sorento	552
Chevy Trailblazer	397

Source: Insurance Institute for Highway Safety

(a) State the null and alternative hypotheses.
(b) Verify that the requirements to use the one-way ANOVA procedure are satisfied. Normal probability plots indicate that the sample data come from normal populations.
(c) Test the hypothesis that the mean head injury for each vehicle type is the same at the $\alpha = 0.01$ level of significance.
(d) Draw boxplots of the three vehicle types to support the analytic results obtained in part (c).

21. Lower Your Cholesterol Researcher Francisco Fuentes and his colleagues wanted to determine the most effective diet for reducing LDL cholesterol, the so-called "bad" cholesterol, among three diets: (1) a saturated-fat diet: 15% protein, 47% carbohydrates, and 38% fat (20% saturated fat, 12% monounsaturated fat, and 6% polyunsaturated fat); (2) the Mediterranean diet: 47% carbohydrates, and 38% fat (10% saturated fat, 22% monounsaturated fat, and 6% polyunsaturated fat); and (3) the US National Cholesterol Education Program or NCEP-1 Diet: 10% saturated fat, 12% monounsaturated fat, and 6% polyunsaturated fat. Participants in the study were shown to have the same levels of LDL cholesterol before the study and were randomly assigned to one of the three diets, or treatment groups. After 28 days, their LDL cholesterol levels were recorded. The data in the following table are based on this study.

Saturated Fat		Mediterranean		NCEP-1	
245	218	56	131	125	184
123	173	78	125	100	116
166	223	101	160	140	144
104	177	158	130	151	101
196	193	145	83	138	135
300	224	118	243	208	144
140	149	145	150	75	130
240		211		71	

(a) Why is this study a completely randomized design?
(b) What is the response variable? What is the explanatory variable that is controlled and set at three levels?
(c) The participants were randomly assigned to one of three treatment groups. What is the purpose of randomization in this study?
(d) State the null and alternative hypotheses.
(e) Verify that the requirements to use the one-way ANOVA procedure are satisfied. Normal probability plots indicate that the sample data come from normal populations.
(f) Are the mean LDL cholesterol levels different at the $\alpha = 0.05$ level of significance?
(g) Draw boxplots of the LDL cholesterol levels for the three groups to support the analytic results obtained in part (f).

22. pH in Rain An environmentalist wanted to determine if the mean acidity of rain differed among Alaska, Florida, and Texas. He randomly selected six rain dates at each of the three locations and obtained the following data:

Alaska	Florida	Texas
5.41	4.87	5.46
5.39	5.18	5.89
4.90	4.52	5.57
5.14	5.12	5.15
4.80	4.89	5.45
5.24	5.06	5.30

Source: National Atmospheric Deposition Program

(a) State the null and alternative hypotheses.
(b) Verify that the requirements to use the one-way ANOVA procedure are satisfied. Normal probability plots indicate that the sample data come from a normal population.
(c) Test the hypothesis that the mean pHs in the rainwater are the same at the $\alpha = 0.05$ level of significance.
(d) Draw boxplots of the pH in rain for the three states to support the results obtained in part (c).

23. NFL Combine The National Football League (NFL) combine is a forum in which players who want to play in the NFL demonstrate various skills. One skill that is measured is the 40-yard dash. Go to www.pearsonhighered.com/sullivanstats to obtain the data file 13_1_23 using the file format of your choice for the version of the text you are using.

(a) What type of variable is 40-yard dash time?
(b) Excluding quarterback (QB), the skill positions are cornerback (CB), safety (S), running back (RB), and wide receiver (WR). Suppose we want to know whether there is a difference between the 40-yard dash times for these positions. What would be the null and alternative hypotheses?
(c) Assuming we treat the players as a random sample of all players in each position, explain why we may use a one-way ANOVA to determine if there is a significant difference between 40-yard dash times for these positions. How many means are being compared?
(d) Does the evidence suggest a difference exists for the 40-yard dash times for the skill positions (except quarterback)? Use the $\alpha = 0.05$ level of significance.
(e) Draw boxplots of the 40-yard dash times for the skill positions (except quarterback) to support the results from part (d).

24. Age and Politics Do people's political philosophy tend to change with age? One technique we may use to answer this question is to see if the mean age of conservatives, moderates, and liberals differ. Go to www.pearsonhighered.com/sullivanstats to obtain the data file SullivanStatsSurveyI using the file format of your choice for the version of the text you are using.

(a) State the null and alternative hypotheses.
(b) Verify that the requirements to use the one-way ANOVA procedure are satisfied. Normal probability plots indicate that the sample data come from normal populations.
(c) Are the mean ages for the three political philosophies different at the $\alpha = 0.1$ level of significance?
(d) Draw boxplots of the ages of the three political philosophies to support the analytic results obtained in part (c).

25. Concrete Strength An engineer wants to know if the mean strengths of three different concrete mix designs differ significantly. He randomly selects nine cylinders that measure

6 inches in diameter and 12 inches in height in which mixture 67-0-301 is poured, nine cylinders of mixture 67-0-400, and nine cylinders of mixture 67-0-353. After 28 days, he measures the strength (in pounds per square inch) of the cylinders. The results are presented in the following table:

Mixture 67-0-301		Mixture 67-0-400		Mixture 67-0-353	
3960	4090	4070	4120	4150	3820
4040	3830	4330	4640	3820	3750
3780	3940	4620	4190	4010	3990
3890	4080	3730	3850	4150	4320
3990		4890		4190	

(a) State the null and alternative hypotheses.

(b) Explain why we cannot use one-way ANOVA to test these hypotheses.

26. Analyzing Journal Article Results Researchers (Brian G. Feagan et al. "Erythropoietin with Iron Supplementation to Prevent Allogeneic Blood Transfusion in Total Hip Joint Arthroplasty," *Annals of Internal Medicine*, Vol. 133, No. 11) wanted to determine whether epoetin alfa was effective in increasing the hemoglobin concentration in patients undergoing hip arthroplasty. A complete medical history and physical of the patients was performed for screening purposes and eligible patients were identified. The researchers used a computer-generated schedule to assign the patients to the high-dose epoetin group, low-dose epoetin group, or placebo group. The study was double-blind. Based on ANOVA, it was determined that there were significant differences in the increase in hemoglobin concentration in the three groups with a *P*-value less than 0.001. The mean increase in hemoglobin in the high-dose epoetin group was 19.5 grams per liter (g/L), the mean increase in hemoglobin in the low-dose epoetin group was 17.2 g/L, and mean increase in hemoglobin in the placebo group was 1.2 g/L.

(a) Why do you think it was necessary to screen patients for eligibility?

(b) Why was a computer-generated schedule used to assign patients to the various treatment groups?

(c) What does it mean for a study to be double-blind? Why do you think the researchers desired a double-blind study?

(d) Interpret the reported *P*-value.

27. Putting It Together: Psychological Profiles Researchers wanted to determine if the psychological profile of healthy children was different than for children suffering from recurrent abdominal pain (RAP) or recurring headaches. A total of 210 children and adolescents were studied and their psychological profiles were graded according to the Child Behavior Checklist 4–18 (CBCL). Children were stratified in two age groups: 4 to 11 years and 12 to 18 years. The results of the study are summarized in the following table:

	n	Sample Mean	Sample Variance
Control group	70	11.7	21.6
RAP	70	9.0	13.0
Headache	70	12.4	8.4

Source: Galli, et al. "Headache and Recurrent Abdominal Pain: A Controlled Study by Means of the Child Behaviour Checklist (CBCL)." *Cephalalgia* 27, 211–219, 2007

(a) Compute the sample standard deviations for each group.

(b) What sampling method was used for each treatment group? Why?

(c) Use a two sample *t*-test for independent samples to determine if there is a significant difference in mean CBCL scores between the control group and the RAP group (assume that both samples are simple random samples).

(d) Is it necessary to check the normality assumption to answer part (c)? Explain.

(e) Use the one-way ANOVA procedure with $\alpha = 0.05$ to determine if the mean CBCL scores are different for the three treatment groups.

(f) Based on your results from parts (c) and (e), can you determine if there is a significant difference between the mean scores of the RAP group and the headache group? Explain.

Retain Your Knowledge

28. Diversity and pH The following data represent the number of fish species living in various Andirondack Lakes and the pH of the lakes. From chemistry, we know pH is a measure of the acidity or basicity of a solution. Solutions with pH less than 7 are said to be acidic. As pH increases, the solution is said to be less acidic.

pH	Species	pH	Species
4.6	0	5.8	8
4.7	0	6	3
4.8	0	6.1	4
5	0	6.2	9
5	2	6.25	9
5.2	2	6.3	2
5.2	1	6.3	4
5.25	0	6.3	9
5.3	1	6.4	5
5.35	1	6.7	6
5.5	5	6.7	8
5.7	4	6.7	8
5.75	3	6.8	10

Data based on "Taking the Sting Out of Acid Rain" by David Malakoff *Science* 12 November 2010 pp. 910–911

(a) Draw a scatter diagram of the data treating pH as the explanatory variable.

(b) Determine the linear correlation coefficient between pH and number of fish species.

(c) Does a linear relation exist between pH and number of fish species?

(d) Find the least-squares regression line treating pH as the explanatory variable.

(e) Interpret the slope.

(f) Is it reasonable to interpret the intercept? Explain.

(g) What proportion of the variability in number of fish species is explained by pH?

(h) Is the number of fish species in the lake whose pH is 5.5 above or below average? Explain.

(i) In part (g), you found the proportion of variability in number of fish species that is explained by the variability in pH. Can you think of other variables that might also explain the variability in the number of fish species?

29. Another E-Cig Study In a study of adult smokers who wanted to quit, 657 subjects were randomly assigned to one of three treatment groups: subjects in Group 1 were given nicotine e-cigarettes, subjects in Group 2 were given nicotine patches, and subjects in Group 3 were given placebo e-cigarettes. The response variable was whether the subject maintained abstinence from cigarettes for six months, verified using exhaled breath carbon monoxide measurement.

(a) What type of experimental design is this?

(b) What is the response variable?

(c) What are the treatments? How many levels does the treatment have?

(d) Explain why it is not appropriate to use a one-way ANOVA to analyze the effectiveness of e-cigs with nicotine. Suggest an appropriate analysis.

(e) The results of the study are shown to the right. Using the procedure suggested in part (d), perform the analysis regarding the effectiveness of the treatment. Be sure to state the null and alternative hypotheses.

	e-Cig with Nicotine	Nicotine Patch	e-Cig with Placebo
Abstained after six months	21	17	3
Did not abstain after six months	268	278	70

Source: Bullen, Christopher, *The Lancet* Volume 382, Issue 9905, 16 Nov 2013 doi:10.1016/S0140-6736(13)61842-5

Explaining the Concepts

30. What are the requirements to perform a one-way ANOVA? Is the test robust?

31. What is the mean square due to treatment estimate of σ^2? What is the mean square due to error estimate of σ^2?

32. Why does a large value of the F statistic provide evidence against the null hypothesis $H_0: \mu_1 = \mu_2 = \cdots = \mu_k$?

33. In a one-way ANOVA, explain what it means to reject the statement in the null hypothesis if three treatment groups are being compared.

13.2 Post Hoc Tests on One-Way Analysis of Variance

Preparing for This Section Before getting started, review the following:

- Using Confidence Intervals to Test Hypotheses (Section 10.2, pp. 487–488)

Objective ❶ Perform the Tukey Test

Suppose the results of a one-way ANOVA show that at least one population mean is different from the others. To determine which means differ significantly, we make additional comparisons between means, using procedures called **multiple comparison methods**.

Suppose a one-way ANOVA led us to reject the null hypothesis $H_0: \mu_1 = \mu_2 = \mu_3 = \mu_4$. We probably would want to know which means differ. For example, if we suspected that μ_1 differed from μ_2, we would test $H_0: \mu_1 = \mu_2$ versus $H_1: \mu_1 \neq \mu_2$. Or if we suspected that μ_1 and μ_2 were equal, but different from μ_3, we would test $H_0: \mu_1 + \mu_2 = 2\mu_3$ versus $H_1: \mu_1 + \mu_2 \neq 2\mu_3$.

The researcher typically has some idea as to the type of comparisons to make. In most situations, and in this text, we will compare only pairs of means, as in $H_0: \mu_1 = \mu_2$ versus $H_1: \mu_1 \neq \mu_2$. Many different procedures can be used to compare two means. In this text, we present the method introduced by John Tukey (the same Tukey of boxplot fame).

In Other Words

Post hoc is Latin for *after this.* Post hoc tests are performed after rejecting the null hypothesis that three or more population means are equal.

❶ **Perform the Tukey Test**

The Tukey test, also known as the *Honestly Significant Difference Test* and the *Wholly Significant Difference Test*, compares pairs of means after the null hypothesis of equal means has been rejected. That is, it tests $H_0: \mu_i = \mu_j$ versus $H_1: \mu_i \neq \mu_j$ for all means where $i \neq j$. The goal of the test is to determine which population means differ significantly.

The computation of the test statistic for Tukey's test follows the same logic as the test for comparing two means from independent samples. Suppose that we want to test $H_0: \mu_1 = \mu_2$ versus $H_1: \mu_1 \neq \mu_2$. This is equivalent to testing $H_0: \mu_2 - \mu_1 = 0$ versus $H_1: \mu_2 - \mu_1 \neq 0$. The test statistic is based on the sample mean difference, $\bar{x}_2 - \bar{x}_1$, and the standard error of the sample mean difference. The standard error is not the same as the standard error used in comparing two means from independent samples. Instead, the standard error is

$$SE = \sqrt{\frac{s^2}{2} \cdot \left(\frac{1}{n_1} + \frac{1}{n_2}\right)}$$

where s^2 is the mean square error estimate (MSE) of σ^2 from the one-way ANOVA, n_1 is the sample size from population 1, and n_2 is the sample size from population 2.

Test Statistic for Tukey's Test

The test statistic for Tukey's test when testing $H_0: \mu_1 = \mu_2$ versus $H_1: \mu_1 \neq \mu_2$ is given by

$$q_0 = \frac{(\bar{x}_2 - \bar{x}_1) - (\mu_2 - \mu_1)}{\sqrt{\frac{s^2}{2} \cdot \left(\frac{1}{n_1} + \frac{1}{n_2}\right)}} = \frac{\bar{x}_2 - \bar{x}_1}{\sqrt{\frac{s^2}{2} \cdot \left(\frac{1}{n_1} + \frac{1}{n_2}\right)}}$$

where $\bar{x}_2 > \bar{x}_1$

s^2 is the mean square error estimate of σ^2 (MSE) from ANOVA
n_1 is the sample size from population 1
n_2 is the sample size from population 2

Because we assume that the null hypothesis is true, until we have evidence to the contrary, we let $\mu_2 - \mu_1 = 0$ in the computation of the test statistic.

The question now becomes "What distribution does this test statistic follow?" The q-test statistic follows the **Studentized range distribution**. The shape of the distribution depends on the error degrees of freedom, ν, and the total number of means being compared, k. We will compare the test statistic, q_0, to a critical value from this distribution, $q_{\alpha,\nu,k}$, where α is the level of significance of the test and therefore the probability of making a Type I error. The level of significance α is called the **experimentwise error rate** or the **familywise error rate**.

In Other Words
Remember, a Type I error means rejecting a true null hypothesis.

Critical Value for Tukey's Test

The critical value for Tukey's test using a familywise error rate α is given by

$$q_{\alpha,\nu,k}$$

where
ν is the degrees of freedom due to error, which is the total number of subjects sampled minus the number of means being compared, or $n - k$
k is the total number of means being compared

We determine the critical value from the Studentized range distribution by referring to Table X. We have provided tables for $\alpha = 0.01$ and $\alpha = 0.05$.

EXAMPLE 1 Finding the Critical Value from the Studentized Range Distribution

Problem Find the critical value from the Studentized range distribution with $\nu = 7$ degrees of freedom and $k = 3$ degrees of freedom, with a familywise error rate $\alpha = 0.05$.

Approach Look in Table X with $\alpha = 0.05$ where the row corresponding to $\nu = 7$ intersects with the column corresponding to $k = 3$. The value in the cell is the critical value.

Solution See Figure 9. The critical value is $q_{0.05,7,3} = 4.165$.

Figure 9

$\alpha = 0.05$									
ν k(or p):2	3	4	5	6	7	8	9	10
1 17.97	26.98	32.82	37.08	40.41	43.12	45.40	47.36	49.07
2 6.085	8.331	9.798	10.88	11.74	12.44	13.03	13.54	13.99
3 4.501	5.910	6.825	7.502	8.037	8.478	8.853	9.177	9.462
4 3.927	5.040	5.757	6.287	6.707	7.053	7.347	7.602	7.826
5 3.635	4.602	5.218	5.673	6.033	6.330	6.582	6.802	6.995
6 3.461	4.339	4.896	5.305	5.628	5.895	6.122	6.319	6.493
7 3.344	4.165	4.681	5.060	5.359	5.606	5.815	5.998	6.158
8 3.261	4.041	4.529	4.886	5.167	5.399	5.597	5.767	5.918
9 3.199	3.949	4.415	4.756	5.024	5.244	5.432	5.595	5.739
10 3.151	3.877	4.327	4.654	4.912	5.124	5.305	5.461	5.599

● Now Work Problem 1

Now that we know the distribution of the test statistic q and how to find the critical value, we can use Tukey's test to make multiple comparisons.

Tukey's Test

After rejecting the null hypothesis $H_0: \mu_1 = \mu_2 = \cdots = \mu_k$, use Steps 1 through 6 to compare pairs of means for significant differences.

Step 1 Arrange the sample means in ascending order.

Step 2 Compute the pairwise differences, $\bar{x}_i - \bar{x}_j$, where $\bar{x}_i > \bar{x}_j$.

Step 3 Compute the test statistic, $q_0 = \dfrac{\bar{x}_i - \bar{x}_j}{\sqrt{\dfrac{s^2}{2} \cdot \left(\dfrac{1}{n_i} + \dfrac{1}{n_j} \right)}}$, for each pairwise difference.

Step 4 Determine the critical value, $q_{\alpha,\nu,k}$, where α is the level of significance (the familywise error rate).

Step 5 If $q_0 \geq q_{\alpha,\nu,k}$, reject the null hypothesis $H_0: \mu_i = \mu_j$ and conclude that the means are significantly different.

Step 6 Compare all pairwise differences to identify which means differ.

EXAMPLE 2   Performing Tukey's Test by Hand

Problem In Example 3 from Section 13.1, we rejected the null hypothesis $H_0: \mu_{\text{Cojet}} = \mu_{\text{Silistor}} = \mu_{\text{Cimara}} = \mu_{\text{Ceramic}}$. Use Tukey's test to determine which pairwise means differ using a familywise error rate of $\alpha = 0.05$.

Approach Follow Steps 1 through 6 to determine which pairwise means differ.

Solution

Step 1 The means for each category are given in Table 6. The means, in ascending order, are $\bar{x}_{\text{Ci}} = 10.64$, $\bar{x}_{\text{Ce}} = 10.78$, $\bar{x}_{\text{Co}} = 16.28$, and $\bar{x}_{\text{Si}} = 17.64$.

Step 2 Now compute the pairwise differences, $\bar{x}_i - \bar{x}_j$, where $\bar{x}_i > \bar{x}_j$. That is, subtract the smaller sample mean from the larger sample mean, as shown in Column 2 of Table 7 on the next page. It is helpful to write the differences in descending order.

(continued)

Table 6

Cojet	Silistor	Cimara	Ceramic Repair
16.28	17.64	10.64	10.78

Step 3 Compute the test statistic q_0 for each pairwise difference. From the ANOVA table in Figure 5, on page 626, we have the mean square error = 8.835. This is the value of s^2 in the computation of the test statistic q_0. The test statistic for each pairwise difference is shown in Column 3 of Table 7.

Step 4 Find the critical value using an $\alpha = 0.05$ familywise error rate with $\nu = n - k = 20 - 4 = 16$ and $k = 4$. From Table X, the critical value is $q_{0.05,16,4} = 4.046$. (If the critical value were not in Table X because of the value for ν, use the entry corresponding to the value closest to ν.) The critical value is in Column 4 of Table 7.

Step 5 If $q_0 > q_{0.05,16,4}$, reject the null hypothesis and state the conclusion in Column 5 of Table 7. By writing the differences in descending order, once we "do not reject H_0," we can stop because we will not reject smaller differences either!

Table 7

Comparison	Difference, $\bar{x}_i - \bar{x}_j$	Test Statistic, q_0	Critical Value	Decision
Silistor vs. Cimara (Silistor − Cimara)	$17.64 - 10.64 = 7.00$	$q_0 = \dfrac{\bar{x}_{Si} - \bar{x}_{Ci}}{\sqrt{\dfrac{s^2}{2}\cdot\left(\dfrac{1}{n_{Si}}+\dfrac{1}{n_{Ci}}\right)}} = \dfrac{17.64 - 10.64}{\sqrt{\dfrac{8.835}{2}\cdot\left(\dfrac{1}{5}+\dfrac{1}{5}\right)}} = 5.266$	4.046	Reject H_0: $\mu_{Si} = \mu_{Ci}$ since $q_0 > q_{0.05,16,4}$
Silistor vs. Ceramic (Silistor − Ceramic)	$17.64 - 10.78 = 6.86$	$q_0 = \dfrac{\bar{x}_{Si} - \bar{x}_{Ce}}{\sqrt{\dfrac{s^2}{2}\cdot\left(\dfrac{1}{n_{Si}}+\dfrac{1}{n_{Ce}}\right)}} = \dfrac{17.64 - 10.78}{\sqrt{\dfrac{8.835}{2}\cdot\left(\dfrac{1}{5}+\dfrac{1}{5}\right)}} = 5.161$	4.046	Reject H_0: $\mu_{Si} = \mu_{Ce}$ since $q_0 > q_{0.05,16,4}$
Cojet vs. Cimara (Cojet − Cimara)	$16.28 - 10.64 = 5.64$	$q_0 = \dfrac{\bar{x}_{Co} - \bar{x}_{Ci}}{\sqrt{\dfrac{s^2}{2}\cdot\left(\dfrac{1}{n_{Co}}+\dfrac{1}{n_{Ci}}\right)}} = \dfrac{16.28 - 10.64}{\sqrt{\dfrac{8.835}{2}\cdot\left(\dfrac{1}{5}+\dfrac{1}{5}\right)}} = 4.243$	4.046	Reject H_0: $\mu_{Co} = \mu_{Ci}$ since $q_0 > q_{0.05,16,4}$
Cojet vs. Ceramic (Cojet − Ceramic)	$16.28 - 10.78 = 5.50$	$q_0 = \dfrac{\bar{x}_{Co} - \bar{x}_{Ce}}{\sqrt{\dfrac{s^2}{2}\cdot\left(\dfrac{1}{n_{Co}}+\dfrac{1}{n_{Ce}}\right)}} = \dfrac{16.28 - 10.78}{\sqrt{\dfrac{8.835}{2}\cdot\left(\dfrac{1}{5}+\dfrac{1}{5}\right)}} = 4.138$	4.046	Reject H_0: $\mu_{Co} = \mu_{Ce}$ since $q_0 > q_{0.05,16,4}$
Silistor vs. Cojet (Silistor − Cojet)	$17.64 - 16.28 = 1.36$	$q_0 = \dfrac{\bar{x}_{Si} - \bar{x}_{Co}}{\sqrt{\dfrac{s^2}{2}\cdot\left(\dfrac{1}{n_{Si}}+\dfrac{1}{n_{Co}}\right)}} = \dfrac{17.64 - 16.28}{\sqrt{\dfrac{8.835}{2}\cdot\left(\dfrac{1}{5}+\dfrac{1}{5}\right)}} = 1.023$	4.046	Do Not Reject H_0: $\mu_{Si} = \mu_{Co}$ since $q_0 < q_{0.05,16,4}$
Ceramic vs. Cimara (Ceramic − Cimara)	$10.78 - 10.64 = 0.14$	$q_0 = \dfrac{\bar{x}_{Ce} - \bar{x}_{Ci}}{\sqrt{\dfrac{s^2}{2}\cdot\left(\dfrac{1}{n_{Ce}}+\dfrac{1}{n_{Ci}}\right)}} = \dfrac{10.78 - 10.64}{\sqrt{\dfrac{8.835}{2}\cdot\left(\dfrac{1}{5}+\dfrac{1}{5}\right)}} = 0.105$	4.046	Do Not Reject H_0: $\mu_{Ce} = \mu_{Ci}$ since $q_0 < q_{0.05,16,4}$

Step 6 The conclusions of Tukey's test are

$$\mu_{Co} = \mu_{Si} \neq \mu_{Ci} = \mu_{Ce}$$

Use lines to indicate which sample means are not significantly different:

$$\underline{\mu_{Co} \ \mu_{Si}} \quad \underline{\mu_{Ci} \ \mu_{Ce}}$$

The means listed over a common line are equal. So the evidence indicates that veneer repairs using the Cojet or Silistor treatment have the highest shear bond strength. The Cimara and Ceramic Repair treatments were not significantly different, but were weaker in terms of shear bond strength than the other treatments. •

EXAMPLE 3   Tukey's Test Using Technology

Problem Use Minitab to obtain the results of Tukey's test for the shear bond strength data from Example 1 in Section 13.1.

Approach The steps for obtaining the results of Tukey's test using Minitab, Excel, and StatCrunch are given in the Technology Step-by-Step on page 640.

Solution Figure 10 shows the results of Tukey's test using Minitab. Instead of reporting P-values to test each hypothesis that the pairwise means are equal, Minitab reports confidence intervals. Recall, we can use confidence intervals to test $H_0: \mu_1 = \mu_2$ versus $H_1: \mu_1 \neq \mu_2$ (or, equivalently, $H_0: \mu_1 - \mu_2 = 0$ versus $H_1: \mu_1 - \mu_2 \neq 0$) by determining whether the interval contains 0. If the interval contains 0, we do not reject the null hypothesis, but if the interval does not contain 0, we reject the null hypothesis at the α level of significance.

Figure 10   Tukey 95% Simultaneous Confidence Intervals All Pairwise Comparisons

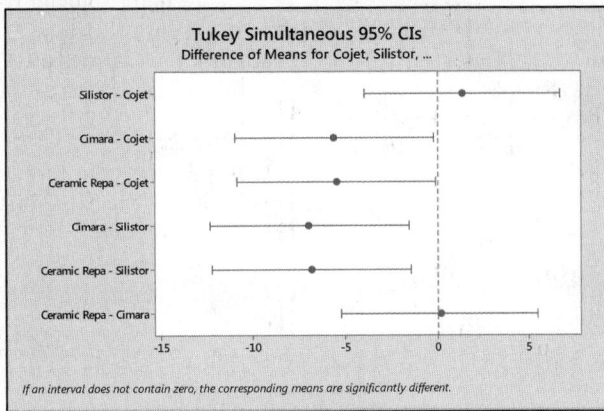

If we reject, $H_0: \mu_1 = \mu_2$, we can look at the bounds of the confidence interval to determine which mean (μ_1 or μ_2) is greater. For example, if the interval contains only positive numbers, then $\bar{x}_1 - \bar{x}_2$ is positive, so we have evidence that $\mu_1 > \mu_2$. Similarly, if the confidence interval contains only negative numbers, we have evidence that $\mu_1 < \mu_2$.

If we look at the first group of confidence intervals, we are comparing means from the Cojet treatment and the other three treatments. The lower bound on the confidence interval $\mu_{\text{Silistor}} - \mu_{\text{Cojet}}$ is -4.024 MPa and the upper bound is 6.744 MPa. Because the interval contains 0, do not reject the null hypothesis that the mean shear strength of the Silistor treatment is the same as the mean shear strength of the Cojet treatment. Next, compare the Cojet treatment to the Cimara treatment. The lower bound is -11.024 MPa and the upper bound is -0.256 MPa. Because the interval does not contain 0, reject the null hypothesis that the mean shear strength of the Cojet treatment is the same as the mean shear strength of the Cimara treatment. In addition, because $\mu_{\text{Cimara}} - \mu_{\text{Cojet}}$ is negative, we know that $\mu_{\text{Ci}} < \mu_{\text{Co}}$. Reading the remainder of the output in a similar manner, we have $\mu_{\text{Co}} > \mu_{\text{Ce}}$, $\mu_{\text{Si}} > \mu_{\text{Ci}}$, $\mu_{\text{Si}} > \mu_{\text{Ce}}$, and $\mu_{\text{Ce}} = \mu_{\text{Ci}}$. Summarizing all six statements, we have $\mu_{\text{Co}} = \mu_{\text{Si}} > \mu_{\text{Ce}} = \mu_{\text{Ci}}$ which agrees with our results from Example 2.

● **Now Work Problem 9**

Some Cautions Regarding Tukey's Test

Sometimes the results of Tukey's test are ambiguous. Suppose the null hypothesis $H_0: \mu_1 = \mu_2 = \mu_3 = \mu_4$ is rejected and the results of Tukey's test indicate the following:

$$\underline{\mu_1 \quad \mu_2 \quad \underline{\mu_3} \quad \mu_4}$$

It appears that populations 1, 2, and 3 have a population mean that is different from that of populations 3 and 4. Clearly, this is impossible. Do you see why? The population mean from treatment 3 cannot equal the population mean from treatments 1 and 2 while at the same time equal the population mean from treatment 4, whose mean is different from that of populations 1 and 2! This means that at least one Type II error has been committed by Tukey's test.

We conclude from this result that $\mu_1 = \mu_2 \neq \mu_4$, but we cannot tell how μ_3 is related to μ_1, μ_2, or μ_4. A solution to this problem is to increase the sample size so that the test is more powerful.

It can also happen that the one-way ANOVA rejects $H_0: \mu_1 = \mu_2 = \cdots = \mu_k$ but the Tukey test does not detect any pairwise differences. This result occurs because one-way ANOVA is more powerful than Tukey's test. Again, the solution is to increase the sample size.

Technology Step-by-Step Tukey's Test

Minitab

1. Enter the raw data in C1, C2, and so on, for each population or treatment.
2. Select **Stat**; then highlight **ANOVA** and then select **One-Way**.
3. In the pull-down menu, select "Response data are in a separate column for each factor level". Enter the column names in the cell marked "Responses". Click "Comparisons . . . ". Check the box labeled "Tukey". Enter the familywise error rate in the box. For $\alpha = 0.05$, enter 5. Click OK twice.

Excel

1. Load the XLSTAT Add-in.
2. Enter the values of the response variable in Column A. Enter the corresponding values of the qualitative explanatory variable in Column B.
3. Select the XLSTAT menu. Now select Modeling data, and choose the ANOVA option.
4. Place the cursor in the Quantitative Y/Dependent variables: cell. Highlight the data in Column A. Place the cursor in the Qualitative X/Explanatory variables: cell. Highlight the data in Column B.
5. Click the Outputs tab. Check the Pairwise comparisons: tab and select Tukey (HSD). Click OK.

StatCrunch

1. Repeat the steps for conducting a one-way ANOVA. In Step 3, check the box "Compute Tukey HSD:". Click Compute!.

13.2 Assess Your Understanding

Skill Building

For Problems 1 and 2, find the critical value from the Studentized range distribution for:

NW **1. (a)** $\alpha = 0.05, \nu = 10, k = 3$
   (b) $\alpha = 0.05, \nu = 24, k = 5$
   (c) $\alpha = 0.05, \nu = 32, k = 8$
   (d) $H_0: \mu_1 = \mu_2 = \mu_3 = \mu_4 = \mu_5$, with $n = 65$ at $\alpha = 0.05$.

2. (a) $\alpha = 0.05, \nu = 12, k = 3$
   (b) $\alpha = 0.05, \nu = 20, k = 4$
   (c) $\alpha = 0.05, \nu = 42, k = 5$
   (d) $H_0: \mu_1 = \mu_2 = \mu_3 = \mu_4$, with $n = 34$ at $\alpha = 0.05$.

3. Suppose that there is sufficient evidence to reject $H_0: \mu_1 = \mu_2 = \mu_3$ using a one-way ANOVA. The mean square error from ANOVA is determined to be 26.2. The sample means are $\bar{x}_1 = 9.5, \bar{x}_2 = 9.1, \bar{x}_3 = 18.1$, with $n_1 = n_2 = n_3 = 5$. Use Tukey's test to determine which pairwise means are significantly different using a familywise error rate of $\alpha = 0.05$.

4. Suppose there is sufficient evidence to reject $H_0: \mu_1 = \mu_2 = \mu_3 = \mu_4$ using a one-way ANOVA. The mean square error from ANOVA is determined to be 26.2. The sample means are $\bar{x}_1 = 42.6, \bar{x}_2 = 49.1, \bar{x}_3 = 46.8, \bar{x}_4 = 63.7$, with $n_1 = n_2 = n_3 = n_4 = 6$. Use Tukey's test to determine which pairwise means are significantly different using a familywise error rate of $\alpha = 0.05$.

 5. The following data are taken from three different populations known to be normally distributed, with equal population variances based on independent simple random samples.

Sample 1	Sample 2	Sample 3
35.4	42.0	43.3
35.0	39.4	48.6
39.2	33.4	42.0
44.8	35.1	53.9
36.9	32.4	46.8
28.9	22.0	51.7

(a) Test the hypothesis that each sample comes from a population with the same mean at the $\alpha = 0.05$ level of significance. That is, test $H_0: \mu_1 = \mu_2 = \mu_3$.
(b) If you rejected the null hypothesis in part (a), use Tukey's test to determine which pairwise means differ using a familywise error rate of $\alpha = 0.05$.
(c) Draw boxplots of each set of sample data to support your results from parts (a) and (b).

6. The following data are taken from four different populations that are known to be normally distributed, with equal population variances based on independent simple random samples.

Sample 1	Sample 2	Sample 3	Sample 4
110	138	98	130
85	140	100	116
83	130	94	157
95	115	110	137
103	101	104	144
105	130	118	124
107	123	102	139

(a) Test the hypothesis that each sample comes from a population with the same mean at the $\alpha = 0.05$ level of significance. That is, test $H_0: \mu_1 = \mu_2 = \mu_3 = \mu_4$.

(b) If you rejected the null hypothesis in part (a), use Tukey's test to determine which pairwise means differ using a familywise error rate of $\alpha = 0.05$.

(c) Draw boxplots of each set of sample data to support your results from parts (a) and (b).

7. Corn Production (*See Problem 13 from Section 13.1*) An agricultural researcher wanted to know whether the mean number of plants for each plot type differed. A one-way ANOVA was performed to test $H_0: \mu_{SP} = \mu_{SD} = \mu_{NT}$. The null hypothesis was rejected with a P-value of 0.007. The researcher proceeded to conduct Tukey's test using StatCrunch to determine which means differed. The results are presented next. Which pairwise means differ? Which plot type would you recommend?

Tukey 95% Simultaneous Confidence Intervals:

Sludge Plot subtracted from

	Difference	Lower	Upper	P-value
Spring Disk	4.6666667	1.0171756	8.3161577	0.0122
No Till	0.16666667	-3.4828244	3.8161577	0.9923

Spring Disk subtracted from

	Difference	Lower	Upper	P-value
No Till	-4.5	-8.149491	-0.85050897	0.0154

8. Soybean Yield (*See Problem 14 from Section 13.1*) An agricultural researcher wanted to determine if the mean soybean yield differed for different plot types. A one-way ANOVA was performed to test $H_0: \mu_L = \mu_{NT} = \mu_{CP}$. The null hypothesis was rejected with a P-value of 0.038. The researcher proceeded to conduct Tukey's test to determine which pairwise means differed using Minitab. The results are presented next. Which pairwise means differ? Which plot type would you recommend?

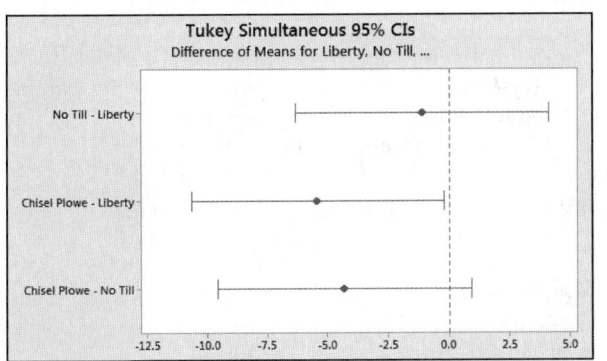

Applying the Concepts

NW 9. Car-Buying Discrimination In Problem 19 from Section 13.1, we rejected $H_0: \mu_{WM} = \mu_{BM} = \mu_{WF} = \mu_{BF}$ and concluded that the mean profit for the four race–gender groups differ. Use Tukey's test to determine which pairwise means differ using a familywise error rate of $\alpha = 0.05$.

10. Lower Your Cholesterol In Problem 21 from Section 13.1, we rejected $H_0: \mu_{SF} = \mu_{MED} = \mu_{NCEP}$ and concluded that there is at least one treatment that results in a mean LDL cholesterol that differs from the rest. Use Tukey's test to determine which pairwise means differ using a familywise error rate of $\alpha = 0.05$.

11. Attention Deficit Hyperactivity Disorder Researchers studied the effect that varying levels of Ritalin and Adderall had on a child's ability to follow rules when the child is diagnosed with attention deficit hyperactivity disorder (ADHD). They randomly assigned children to one of five treatment groups: placebo, 10 mg of Ritalin, 17.5 mg of Ritalin, 7.5 mg of Adderall, or 12.5 mg of Adderall twice a day. They recorded a score that indicated the child's ability to follow rules, with a higher score indicating a higher ability to follow rules. The following data are based on their study, "A Comparison of Ritalin and Adderall: Efficacy and Time-Course in Children with Attention Deficit/Hyperactivity Disorder," *Pediatrics*, Vol. 103, No. 4.

Placebo	Ritalin 10 mg	Ritalin 17.5 mg	Adderall 7.5 mg	Adderall 12.5 mg
47	96	37	57	84
20	75	83	71	105
34	57	76	85	83
60	44	43	92	92
28	83	87	75	74
72	69	62	55	92
25	57	99	73	64

(a) The data was collected using a completely randomized design. Explain what this means.

(b) What is the response variable in this study? What explanatory variable was controlled and fixed at five levels?

(c) The children were randomly assigned to one of five treatment groups. What role does randomization play in this study?

(d) What are the null and alternative hypotheses in this study?

(e) Test the null hypothesis that the mean score for each treatment is the same at the $\alpha = 0.05$ level of significance. **Note:** The requirements for a one-way ANOVA are satisfied.

(f) If the null hypothesis is rejected in part (e), use Tukey's test to determine which pairwise means differ using a familywise error rate of $\alpha = 0.05$.

(g) Draw boxplots of the five treatment levels to support the analytic results obtained in parts (e) and (f).

12. Nutrition Researchers Sharon Peterson and Madeleine Sigman-Grant wanted to compare the overall nutrient intake of American children (ages 2 to 19) who exclusively use lean meats, mixed meats, or higher-fat meats. The data given on the next page represent the daily consumption of calcium (in mg) for a random sample of eight children in each category and are based on the results presented in their article "Impact of Adopting Lower-Fat Food Choices on Nutrient Intake of American Children," *Pediatrics*, Vol. 100, No. 3.

Lean Meats	Mixed Meats	Higher-Fat Meats
844.2	897.7	843. 4
745.0	908.1	862.2
773.1	948.8	790.5
823.6	836.6	876.5
812.0	871.6	790.8
758.9	945.9	847.2
810.7	859.4	772.0
790.6	920.2	851.3

(a) The data was collected using a cohort observational study. Explain what this means.
(b) What is the response variable in this study?
(c) What are the null and alternative hypotheses in this study?
(d) Test the null hypothesis that the mean calcium for each category is the same at the $\alpha = 0.05$ level of significance. **Note:** The requirements for a one-way ANOVA are satisfied.
(e) If the null hypothesis is rejected in part (d), use Tukey's test to determine which pairwise means differ using a familywise error rate of $\alpha = 0.05$.
(f) Draw boxplots of the three treatment levels to support the analytic results obtained in parts (d) and (e).
(g) Can any statements of causality between meat consumption and consumption of calcium be made based on the results of this study? Explain.

13. Heart Rate of Smokers A researcher wants to determine the effect that smoking has on resting heart rate. She randomly selects seven individuals from three categories: (1) nonsmokers, (2) light smokers (fewer than 10 cigarettes per day), and (3) heavy smokers (10 or more cigarettes per day) and obtains the following heart rate data (beats per minute):

Nonsmokers	Light Smokers	Heavy Smokers
70	67	79
58	75	80
51	65	77
56	78	77
53	62	86
53	70	68
65	73	83

(a) Test the null hypothesis that the mean resting heart rate for each category is the same at the $\alpha = 0.05$ level of significance.
Note: The requirements for a one-way ANOVA are satisfied.
(b) If the null hypothesis is rejected in part (a), use Tukey's test to determine which pairwise means differ using a familywise error rate of $\alpha = 0.05$.
(c) Draw boxplots of the three treatment levels to support the results obtained in parts (a) and (b).

14. Price to Earnings Ratios One measure of the value of a stock is its price to earnings ratio (or P/E ratio): the ratio of the price of a stock per share to the earnings per share and can be thought of as the price an investor is willing to pay for $1 of earnings in a company. A stock analyst wants to know whether the P/E ratios for three industry categories differ significantly. The following data represent simple random samples of companies from three categories: (1) financial, (2) food, and (3) leisure goods.

Financial	Food	Leisure Goods
8.83	19.75	14.10
12.75	17.87	10.12
13.48	15.18	15.57
14.42	22.84	13.48
10.06	15.60	11.27

Source: Yahoo!Finance

(a) Test the null hypothesis that the mean P/E ratio for each category is the same at the $\alpha = 0.05$ level of significance. **Note:** The requirements for a one-way ANOVA are satisfied.
(b) If the null hypothesis is rejected in part (a), use Tukey's test to determine which pairwise means differ using a familywise error rate of $\alpha = 0.05$.
(c) Draw boxplots of the three categories to support the analytic results obtained in parts (a) and (b).

15. Got Milk? Researchers Sharon Peterson and Madeleine Sigman-Grant wanted to compare the overall nutrient intake of American children (ages 2 to 19) who exclusively use skim milk instead of 1%, 2%, or whole milk. The researchers combined children who consumed 1% or 2% milk into a "mixed milk" category. The following data represent the daily calcium intake (in mg) for a random sample of eight children in each category and are based on the results presented in their article "Impact of Adopting Lower-Fat Food Choices on Nutrient Intake of American Children," *Pediatrics,* Vol. 100, No. 3.

Skim Milk	Mixed Milk	Whole Milk
916	1024	870
886	1013	874
854	1065	881
856	1002	836
857	1006	879
853	991	938
865	1015	841
904	1035	818

(a) Is there sufficient evidence to support the belief that at least one of the means is different from the others at the $\alpha = 0.05$ level of significance?
Note: The requirements for a one-way ANOVA are satisfied.
(b) If the null hypothesis is rejected in part (a), use Tukey's test to determine which pairwise means differ using a familywise error rate of $\alpha = 0.05$.
(c) Draw boxplots of the three categories to support the analytic results obtained in parts (a) and (b).

16. The **comparisonwise error rate,** denoted α_c, is the probability of making a Type I error when comparing two means. It is related to the familywise error rate, α, through the formula $1 - \alpha = (1 - \alpha_c)^k$, where k is the number of means being compared.

(a) If the familywise error rate is $\alpha = 0.05$ and $k = 3$ means are being compared, what is the comparisonwise error rate?
(b) If the familywise error rate is $\alpha = 0.05$ and $k = 5$ means are being compared, what is the comparisonwise error rate?

(c) Based on the results of parts (a) and (b), what happens to the comparisonwise error rate as the number of means compared increases?

17. Putting It Together: Time to Complete a Degree A researcher wanted to determine if the mean time to complete a bachelor's degree was different depending on the selectivity of the first institution of higher education that was attended. The following data represent a random sample of 12th-graders who earned their degree within eight years. Probability plots indicate that the data for each treatment level are normally distributed.

Highly Selective	Selective	Nonselective	Open-door	Not Rated
2.5	4.6	4.7	5.9	5.1
5.1	4.3	2.3	4.2	4.3
4.3	4.4	4.3	6.4	4.4
4.8	4.0	4.2	5.6	3.8
5.5	4.1	5.1	6.0	4.7
2.6	3.1	4.7	5.0	4.5
4.1	3.8	4.2	7.3	5.5
3.4	5.2	5.7	5.6	5.5
4.3	4.5	2.3	6.9	4.6
3.9	4.0	4.5	5.1	4.1

Source: C. Adelman. "The Toolbox Revisited: Paths to Degree Completion from High School Through College," U.S. Department of Education, 2006.

(a) What type of observational study was conducted? What is the response variable?

(b) Find the sample mean for each treatment level.

(c) Find the sample standard deviation for each treatment level. Using the general rule presented in this chapter, does it appear that the population variances are the same?

(d) Use the time to degree completion for students first attending highly selective institutions to construct a 95% confidence interval estimate for the population mean.

(e) How many pairwise comparisons are possible among the treatment levels?

(f) Consider the null hypothesis $H_0: \mu_1 = \mu_2 = \mu_3 = \mu_4 = \mu_5$. If we test this hypothesis using t-tests for each pair of treatments, use your answer from part (e) to compute the probability of making a Type I error, assuming that each test uses an $\alpha = 0.05$ level of significance.

(g) Use the one-way ANOVA procedure to determine if there is a difference in the mean time to degree completion for the different types of initial institutions. If the null hypothesis is rejected, use Tukey's test to determine which pairwise differences are significant using a familywise error rate of $\alpha = 0.05$.

13.3   The Randomized Complete Block Design

Preparing for This Section Before getting started, review the following:

• Randomized Block Design (Section 1.6, pp. 47–48)

• Inference on Two Means: Dependent Samples (Section 11.2, pp. 536–542)

Objectives ❶ Conduct analysis of variance on the randomized complete block design

❷ Perform the Tukey test

In the completely randomized design, a single factor is manipulated and fixed at different levels. The experimental units are then randomly assigned to one of the factor levels, or treatments. If there are only two levels of the factor (or two treatments), we analyze the data using methods from Section 11.3. For three or more treatments, we use the methods of one-way ANOVA.

Remember, in designing an experiment, the researcher identifies the various factors that may affect the value of the response variable. The researcher can then deal with the factors in one of two ways:

1. Control their levels so that they remain fixed throughout the experiment or allow them to vary with predetermined fixed levels.
2. Randomize so any effects not identified or uncontrollable are minimized.

In the completely randomized design, the researcher manipulates a single factor and fixes it at two or more levels and then randomly assigns experimental units to a treatment. The researcher may have the ability to control certain factors in this design setting so that they remain fixed. The remaining factors are dealt with by randomly assigning the experimental units to treatment groups.

In Other Words
A block is a method for controlling experimental error. Blocks should form a homogeneous group. For example, if gender is thought to explain some of the variability in the response variable, remove gender from the experimental error by forming blocks of experimental units with the same gender.

The completely randomized design is not always sufficient, because the researcher may be aware of additional factors that cannot be fixed at a single level throughout the experiment.

One experimental design that captures more information (and therefore reduces experimental error) is the randomized block design, first introduced in Section 1.6, where we discussed the effect fertilizer might have on crop yield. We designed an experiment in which three levels of fertilizer (the three treatments) were each analyzed on two different plant types (the blocks).

The following example illustrates the purpose of the completely randomized block design.

EXAMPLE 1   Randomized Complete Block Design

A researcher wants to determine the tread wear on four different tires after 5000 miles of highway driving. The researcher has four of each brand of tire, four cars, and four drivers. For simplicity, we call the tire types A, B, C, and D. The researcher wants to test $H_0: \mu_A = \mu_B = \mu_C = \mu_D$ versus the alternative that at least one of the means differs, where μ_i represents the mean tread loss for each tire type.

The researcher must first determine the variables that affect tire wear: the type of car, the driver, tire inflation, and road conditions are a few. To save time, the researcher decides to use all four cars in the experiment rather than one. He decides to set the inflation of each tire at 32 pounds, drive the cars on the same oval track, and assign one driver to each car.

One possible experimental design would be to randomly assign the tires to the cars. First, name each tire: name the first A-tire, tire 1A, the second A-tire, tire 2A, and so on. Then number car and location of the tires in the cells as shown in Table 8(a). Now we use a random-number generator to assign the tires to the cars: the first number generated is 15, so we place tire 1A on car 4's back-left position. If the next random number is 12, tire 2A is placed on car 3's back-right position. We continue until all the tires are assigned to the cars. Let X_{1A} represent the tread loss for the first tire A, let X_{2A} represent the tread loss for the second tire A, and so on. Table 8(b) shows the design we could end up with.

Table 8

	Front Left	Front Right	Back Left	Back Right		Front Left	Front Right	Back Left	Back Right
Car 1	1	2	3	4	**Car 1**	X_{4C}	X_{4B}	X_{3D}	X_{4D}
Car 2	5	6	7	8	**Car 2**	X_{2B}	X_{3A}	X_{3B}	X_{1B}
Car 3	9	10	11	12	**Car 3**	X_{4A}	X_{1D}	X_{2C}	X_{2A}
Car 4	13	14	15	16	**Car 4**	X_{2D}	X_{3C}	X_{1A}	X_{1C}
	(a)					**(b)**			

A quick look at Table 8(b) indicates that tire A is not on car 1, tires C and D are not on car 2, and tire B is not on cars 3 or 4. Therefore, any effect that the car might have on tire wear is not completely accounted for. That is, tire wear variability might be confounded with car variability.

To resolve this issue, we might require that each car must have all four tire types on it. This would result in a **randomized complete block design**. We say *complete* because each block (car) gets every treatment (tire type). We say *randomized* because the order in which the treatment is applied within each block is random. To obtain the randomness, randomly assign one treatment (tire brand) to each block (car). For example, we have four treatments (tire brand), so randomly select an integer between 1 and 4 for the tire number and a second random integer between 1 and 4 for the tire brand. For example, if a random-number generator gives 2 then 1, place tire 2 of brand A on car 1. Continuing to place tires on car 1, we repeat the process, but omitting brand A because each brand must be represented on car 1. By randomly assigning each treatment to each block, we might end up with the design in Table 9.*

Table 9				
	Front Left	**Front Right**	**Back Left**	**Back Right**
Car 1	X_{2A}	X_{3C}	X_{2D}	X_{3B}
Car 2	X_{1A}	X_{2B}	X_{1D}	X_{2C}
Car 3	X_{4D}	X_{4A}	X_{1B}	X_{4C}
Car 4	X_{1C}	X_{4B}	X_{3A}	X_{3D}

In Table 9, notice that horizontal lines help the reader see how the blocks are formed. Using this randomized complete block design, the null hypothesis is

$$H_0: \mu_A = \mu_B = \mu_C = \mu_D$$

versus

$$H_1: \text{at least one of the means is different}$$

<table>
<tr><td>

CAUTION!

When we block, we are not interested in determining whether the block is significant. We only want to remove experimental error to reduce the mean square error.

</td></tr>
</table>

One type of randomized complete block design is the matched-pairs design introduced in Section 1.6. This type of experimental design requires dependent sampling in which the experimental units are somehow related, such as in husband/wife, the same experimental unit, or some other characteristic that relates the individuals. The related experimental units form the blocks in this experimental design. We learned how to analyze data obtained from this experimental design in Section 11.2 using the paired *t*-test. For example, in Section 11.2, Example 1, we analyzed matched-pairs data where the blocks were the students.

The methods used in this section can also be used if an observational study has three or more categories whose means we want to compare (the categories act like the treatment in an experiment) along with blocks. For example, we obtain a simple random sample of individuals and ask them to disclose their number of years of education. Suppose we wonder if the mean number of years of education differs by the following age groups: 25 to 34, 35 to 49, and 50 years and older. Since gender or ethnicity might be factors that contribute to variability in the number of years of education, we might use gender or ethnicity as a block.

① Conduct Analysis of Variance on the Randomized Complete Block Design

The following requirements must be satisfied to analyze data from a randomized complete block design.

> ### Requirements to Perform a Randomized Complete Block Design
>
> 1. The response variable for each of the k populations must be normally distributed.
> 2. The response variable for each of the k populations must have the same variance; that is, each treatment group must have population variance σ^2.[†]

---

* Table 9 reveals another flaw. Tire B is never in the front-left location. The position of the tire on the car also affects tire wear. We could account for this by rotating the tires every 1250 miles, or we could require that each treatment (tire) must be placed in each position on the car, so one car must have tire A in the front left, a second car must have tire A in the front right, and so on. If we impose this second requirement, we would have a **Latin square design**.

† There are many tests for comparing variances. Minitab compares variances using two different tests: Bartlett's test and Levene's test. Levene's test will be conducted provided there are no missing observations. (We do not cover analysis when there are missing observations because it is beyond the scope of the text.) Levene's test is more robust than Bartlett's test, but neither is very reliable. The best approach is to design your experiment with equal sample sizes for each factor because ANOVA is not affected too much when we have equal sample sizes, even if the variances are unequal.

As in one-way ANOVA, we will verify the requirement of equal population variances by comparing the sample standard deviations. The procedures can be used provided the largest sample standard deviation is no more than twice the smallest sample standard deviation.

Because the computations in performing the analysis of the randomized complete block design are tedious, we will use statistical software with the P-value approach to test the hypothesis. Again, the decision rule will be to reject the null hypothesis if the P-value is less than the level of significance, α.

EXAMPLE 2 Analyzing the Randomized Complete Block Design

Problem A researcher is interested in the effect of four diets on the weight gain of newborn rats. The researcher randomly assigns rats from the same mother to each treatment so that a set of four rats constitutes a block. In blocking by mother, we are removing any variability in weight gain due to genetic differences. The researcher identifies five mother rats that each had four offspring, for a total of 20 rats. For each block (sibling rats), the researcher randomly assigns the experimental units (rats) to a treatment using randomly generated integers from 1 to 4. After six months, she records the weight gain of the 20 rats (in grams) and obtains the data in Table 10. Is there sufficient evidence to conclude that the weight gains of the rats on the four diets differ at the $\alpha = 0.05$ level of significance?

Table 10				
	Diet 1	**Diet 2**	**Diet 3**	**Diet 4**
Block 1	15.0	14.6	17.7	10.4
Block 2	17.9	17.4	16.0	12.2
Block 3	17.5	14.8	14.2	14.8
Block 4	16.3	17.3	14.4	12.0
Block 5	15.4	19.3	18.8	14.3

Approach Verify the requirements for analyzing data from a randomized complete block design are satisfied and then analyze the results of the experiment using Minitab. The steps for performing a randomized complete block design using Minitab and StatCrunch are given in the Technology Step-by-Step on page 649.

Solution We want to test

$$H_0: \mu_1 = \mu_2 = \mu_3 = \mu_4 \quad \text{vs.} \quad H_1: \text{at least one of the means is different}$$

A normal probability plot for the data for each of the four diets indicates that the requirement of normality is satisfied. The highest standard deviation, diet 3 with $s_3 = 2.018$, is not more than twice the lowest standard deviation, diet 1 with $s_1 = 1.268$, so we can proceed with the analysis.

Enter the data into Minitab in three columns. Column 1 represents the block, Column 2 the diet, and Column 3 the response variable, weight gain. So, for the block 1, diet 1 rat, we enter a 1 in column C1, a 1 in C2, and 15 in C3. For the block 1, diet 2 rat, we enter 1 in C1, 2 in C2, and 14.6 in C3, and so on.

The results from Minitab are in Figure 11.

Figure 11

Two-way ANOVA: Weight Gain versus Block, Treatment

```
Analysis of Variance for Weight G
Source       df      SS      MS       F       P
Block         4   14.71    3.68    1.21   0.359  ———— P-value for treatment
Treatment     3   51.87   17.29    5.67   0.012  ─╯
Error        12   36.62    3.05
Total        19  103.21
```

In Other Words
Remember, the smaller the
P-value is, the stronger the
evidence against the null
hypothesis.

● Now Work Problem 1(a) and (b)

We are not interested in whether the weight gains among the blocks are equal, so we will ignore the *P*-value for blocks. We are interested in whether the treatment appears to result in different weight gains. The *P*-value of 0.012 due to treatment is evidence against the null hypothesis of equal weight gain among treatments. Therefore, we conclude that the treatment, diet, does result in significantly different weight gain. However, we do not know which diets result in different mean weight gains. To determine this, we again rely on the pairwise comparisons of Tukey's test. ●

Verifying the Normality Requirement in a Randomized Complete Block Design

Just as we required that the response variable for each of the *k* populations be normally distributed in a one-way ANOVA, we require the response variable to be normally distributed for each of the *k* populations to perform ANOVA on the randomized complete block design.

The model for the randomized complete block design is

$$Y_{ij} = \mu + \beta_i + T_j + \varepsilon_{ij}$$

where Y_{ij} represents the *i*th observation on the *j*th treatment
β_i represents the effect of the *i*th block
T_j represents the effect of the *j*th treatment
ε_{ij} represents the random error in the *i*th observation of the *j*th treatment

The error term ε_{ij} is a normally distributed variable with a mean of 0 and a variance that is the same for each treatment.

The impact of block *i* is $\beta_i = \mu_i - \mu$ (the difference between the mean of the values from the *i*th block and the overall mean). The impact of treatment *j* is $T_j = \mu_j - \mu$ (the difference between the mean of the values from the *j*th treatment and the overall mean). So the model becomes

$$Y_{ij} = \mu + (\mu_i - \mu) + (\mu_j - \mu) + (Y_{ij} - \mu_i - \mu_j + \mu)$$

The last term in the model represents the error term (residual). To estimate the residual for the *i*th block and *j*th treatment compute

$$Y_{ij} - \overline{Y}_i - \overline{Y}_j + \overline{Y}$$

where Y_{ij} represents the *i*th observation on the *j*th treatment
$\overline{Y}_i$ represents the sample mean of the *i*th block
$\overline{Y}_j$ represents the sample mean of the *j*th treatment
$\overline{Y}$ represents the overall mean

Draw a normal probability plot of the residuals to verify the normality requirement.

EXAMPLE 3   Verifying the Normality Requirement in the Randomized Complete Block Design

Problem Verify the normality requirement for the data analyzed in Example 2.

Approach Obtain the residuals and draw the normal probability plot.

Solution Figure 12 on the next page shows the normal probability plot of the residuals drawn in StatCrunch. The correlation between the residuals and expected *z*-scores is 0.99. Because $0.99 > 0.959$ (critical value from Table VI with $n = 25$). It is reasonable to conclude the residuals are normally distributed.

(continued)

Figure 12

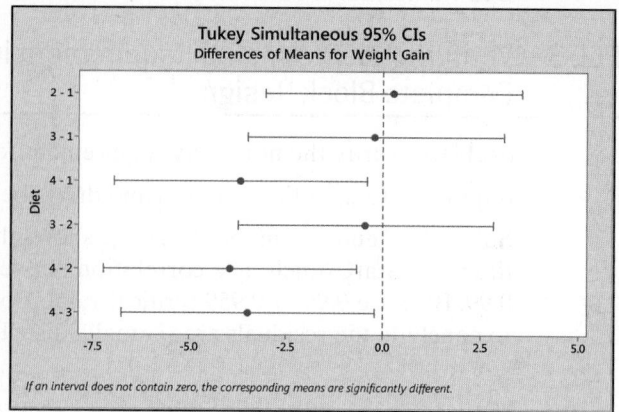

Normal Quantile

Correlation= 0.99

2 Perform the Tukey Test

After rejecting the null hypothesis of equal population means, we determine which means differ significantly. The *post hoc* (after this) test we use to conduct the analysis is Tukey's test. The steps are identical to those in Section 13.2. The critical value is $q_{\alpha,\nu,k}$, using a familywise error rate of α with $\nu = (r-1)(c-1) =$ the error degrees of freedom (r is the number of blocks and c is the number of treatments) and k is the number of means being tested.

EXAMPLE 4   Multiple Comparisons Using Tukey's Test

Problem Use Tukey's test to determine which pairwise means differ for the data in Example 2, with a familywise error rate of $\alpha = 0.05$, using Minitab.

Approach The steps for conducting Tukey's test using Minitab and StatCrunch are given in the Technology Step-by-Step on page 649.

Solution The results of Tukey's test using Minitab are presented in Figure 13. Minitab performs pairwise comparisons using confidence intervals. Figure 13 shows the pairwise comparisons using 95% confidence intervals. The first sets of confidence intervals compare diet 1 to diets 2, 3, and 4. We can see that the interval $\mu_{\text{Diet 4}} - \mu_{\text{Diet 1}}$ does not include 0. Therefore, we reject the null hypothesis that the mean weight gain from diets 1 and 4 are equal. In addition, the interval contains all negative numbers, so the mean weight gain from diet 1 is greater than that from diet 4. Continuing the comparisons leads us to conclude that diet 4 results in significantly less weight gain than the other diets.

Figure 13

Tukey Simultaneous 95% CIs
Differences of Means for Weight Gain

If an interval does not contain zero, the corresponding means are significantly different.

• **Now Work Problem 1(c)**

Technology Step-by-Step Randomized Complete Block Design

Minitab
Conducting the Analysis of Variance
1. In C1, enter the block of the response variable; in C2, enter the treatment of the response variable; and in C3, enter the value of the response variable.
2. Select the **Stat** menu. Highlight **ANOVA** and select **General Linear Model**. Then select **Fit General Linear Model . . . .**
3. Enter C3 in the box marked "Responses," enter C1 and C2 in the box marked "Factors," and click OK.

Tukey's Test
1. With the data entered into the Minitab spreadsheet as specified in Step 1 above, select the **Stat** menu. Highlight **ANOVA**, then **General Linear Model . . . .** Select **Comparisons**.
2. In the pull-down menu labeled Response, select the column containing the values of the response variable.

Check the Tukey box. In the box labeled "Choose terms for comparisons:" select the column containing the treatment. Click on the C = Compare levels for this item. Click OK.

StatCrunch
1. In column var1, enter the block of the response variable; in column var2, enter the treatment of the response variable; and in column var3, enter the value of the response variable. Name the columns.
2. Select **Stat**, highlight **ANOVA**, and select **Repeated Measures**.
3. Select the column containing the values of response variable from the pull-down menu "Responses in:". Select the column containing the treatments in the pull-down menu "Treatments in:". Select the column containing the blocks in the pull-down menu "Blocks in:". Check the box "Compute Tukey HSD." Click Compute!

13.3 Assess Your Understanding

Skill Building

NW 1. Given the following ANOVA output, answer the questions that follow:

Source	df	SS	MS	F	P
Block	4	768.27	192.067	10.96	0.002
Treatment	2	278.53	139.267	7.95	0.013
Error	8	140.13	17.517		
Total	14	1186.93			

(a) The researcher wants to test $H_0: \mu_1 = \mu_2 = \mu_3$ against H_1: at least one of the means is different. Based on the ANOVA table, what should the researcher conclude?
(b) What is the mean square due to error?
(c) The following output represents the results of Tukey's test. What should the researcher conclude?

```
Tukey Simultaneous Tests
Response Variable Response
All Pairwise Comparisons among Levels
of Treatment
Treatment = 1 subtracted from:

Treat-  Difference   SE of              Adjusted
ment    of Means   Difference  T-Value   P-Value
2        -1.000      2.647     -0.3778    0.9251
3         8.600      2.647      3.2489    0.0282

Treatment = 2 subtracted from:

Treat-  Difference   SE of              Adjusted
ment    of Means   Difference  T-Value   P-Value
3         9.600      2.647      3.627     0.0165
```

2. Given the following ANOVA output, answer the questions that follow:

Source	df	SS	MS	F	P
Block	6	1637.81	272.968	7.91	0.001
Treatment	2	707.43	353.714	10.25	0.003
Error	12	413.90	34.492		
Total	20	2759.14			

(a) The researcher wants to test $H_0: \mu_1 = \mu_2 = \mu_3$ against H_1: at least one of the means is different. Based on the ANOVA table, what should the researcher conclude?
(b) What is the mean square due to error?
(c) The following output represents the results of Tukey's test. What should the researcher conclude?

```
Tukey 95% Simultaneous Confidence Intervals
Response Variable Response
All Pairwise Comparisons among Levels
of Treatment
Treatment = 1 subtracted from:

Treatment  Lower  Center  Upper   -----+---------+---------+---------+-
2         -3.511   4.857  13.23   (----------*-----------)
3          5.631  14.000  22.37             (-----------*-----------)
                                   -----+---------+---------+---------+-
                                     0.0      7.0     14.0     21.0

Treatment = 2 subtracted from:

Treatment  Lower  Center  Upper   -----+---------+---------+---------+-
3          0.7743  9.143  17.51        (-----------*-----------)
                                   -----+---------+---------+---------+-
                                     0.0      7.0     14.0     21.0
```

3. Given the following ANOVA output, answer the questions that follow:

```
Analysis of Variance for Response
```

Source	df	SS	MS	F	P
Block	8	2105.436	263.179	278.66	0.000
Treatment	3	6.393	2.131	2.26	0.108
Error	24	22.667	0.944		
Total	35	2134.496			

(a) The researcher wants to test $H_0: \mu_1 = \mu_2 = \mu_3 = \mu_4$ against H_1: at least one of the means is different. Based on the ANOVA table, what should the researcher conclude?
(b) What is the mean square due to error?
(c) Explain why it is not necessary to use Tukey's test on these data.

4. Given the following ANOVA output, answer the questions that follow:

```
Analysis of Variance for Response
Source       df        SS        MS        F        P
Block         6    1712.37    285.39    134.20    0.000
Treatment     3       2.27      0.76      0.36    0.786
Error        18      38.28      2.13
Total        27    1752.91
```

(a) The researcher wants to test $H_0: \mu_1 = \mu_2 = \mu_3 = \mu_4$ against H_1: at least one of the means is different. Based on the ANOVA table, what should the researcher conclude?

(b) What is the mean square due to error?

(c) Explain why it is not necessary to use Tukey's test on these data.

In Problems 5 and 6, assume that the data come from populations that are normally distributed with the same variance.

5.

Block	Treatment 1	Treatment 2	Treatment 3
1	9.7	8.4	8.8
2	10.4	8.9	8.5
3	10.5	9.3	9.0
4	10.7	10.5	9.3
5	11.1	10.7	10.3

(a) Test $H_0: \mu_1 = \mu_2 = \mu_3$ against H_1: at least one of the means is different, where μ_1 is the mean for treatment 1, and so on, at the $\alpha = 0.05$ level of significance.

(b) If the null hypothesis from part (a) was rejected, use Tukey's test to determine which pairwise means differ using a familywise error rate of $\alpha = 0.05$.

(c) Draw boxplots of the data for each treatment using the same scale to support the analytical results obtained in parts (a) and (b).

6.

Block	Treatment 1	Treatment 2	Treatment 3
1	15.8	15.0	15.3
2	16.0	15.8	17.2
3	21.6	18.3	21.5
4	21.6	20.8	21.3
5	22.5	21.5	23.5
6	17.5	16.2	16.8

(a) Test $H_0: \mu_1 = \mu_2 = \mu_3$ against H_1: at least one of the means is different, where μ_1 is the mean for treatment 1, and so on, at the $\alpha = 0.05$ level of significance.

(b) If the null hypothesis from part (a) was rejected, use Tukey's test to determine which pairwise means differ using a familywise error rate of $\alpha = 0.05$.

(c) Draw boxplots of the data for each treatment using the same scale to support the analytical results obtained in parts (a) and (b).

Applying the Concepts

7. Octane An automotive engineer wanted to determine whether the octane of gasoline used in a car increases gas mileage. Recognizing that car and driver are variables that affect gas mileage, he selected six different brands of car and assigned a driver to each car, so he blocked by car type and driver. For each car (and driver), the researcher randomly selected a number from 1 to 3, with 1 representing 87-octane gasoline, 2 representing 89-octane gasoline, and 3 representing 92-octane gasoline. Then 5 gallons of the gasoline selected was placed in the car. The car was driven around a closed track at 40 miles per hour until the car ran out of gas. The number of miles driven was recorded and then divided by 5 to obtain the miles per gallon. He obtained the following results:

	87 Octane	89 Octane	92 Octane
Chevrolet Impala	28.3	28.4	28.7
Chrysler 300M	27.1	26.9	27.2
Ford Taurus	26.4	26.1	26.8
Lincoln LS	26.1	26.4	27.3
Toyota Camry	28.4	28.9	29.1
Volvo S60	25.3	25.1	25.8

(a) Normal probability plots for each treatment indicate that the requirement of normality is satisfied. Verify that the requirement of equal population variances for each treatment is satisfied.

(b) Explain the role blocking plays in reducing the variability of gas mileage.

(c) Is there sufficient evidence that the mean miles per gallon are different among the three octane levels at the $\alpha = 0.05$ level of significance?

(d) If the null hypothesis from part (c) was rejected, use Tukey's test to determine which pairwise means differ using a familywise error rate of $\alpha = 0.05$.

(e) Based on your results for part (d), what do you conclude?

8. Healing Rate A medical researcher wanted to determine the effectiveness of coagulants on the healing rate of a razor cut on lab mice. Because healing rates of mice vary from mouse to mouse, the researcher decided to block by mouse. First, the researcher gave each mouse a local anesthesia and then made a 5-mm incision that was 2 mm deep on each mouse. He randomly selected one of the three treatments and recorded the time it took for the wound to stop bleeding (in minutes). He repeated this process two more times on each mouse and obtained the results shown.

Mouse	No Drug	Experimental Drug 1	Experimental Drug 2
1	3.2	3.4	3.4
2	4.8	4.4	3.4
3	6.6	5.9	5.4
4	6.5	6.3	5.2
5	6.4	6.3	6.1

(a) Explain how each mouse forms a block. Explain how blocking might reduce variability of time to heal.

(b) Normal probability plots for each treatment indicate that the requirement of normality is satisfied. Verify that the requirement of equal population variances for each treatment is satisfied.

(c) Is there sufficient evidence that the mean healing time is different among the three treatments at the $\alpha = 0.05$ level of significance?

(d) If the null hypothesis from part (c) was rejected, use Tukey's test to determine which pairwise means differ using a familywise error rate of $\alpha = 0.05$.

(e) Based on your results for part (d), what do you conclude?

9. Crash Tests The Insurance Institute for Highway Safety regularly tests cars for various safety factors. In one such test, the institute tests the bumpers in 5-mile per hour (mph) crashes. The following data represent the cost of repairs (in dollars) after four different 5-mph crashes on small utility vehicles. The institute blocks by location of crash, and the treatment is car model.

	Jeep Cherokee	Saturn VUE	Toyota RAV4	Hyundai Santa Fe
Front into flat barrier	652	416	489	539
Rear into flat barrier	824	556	1897	1504
Front into angle barrier	1448	1179	1151	1578
Rear into pole	1553	1335	2377	1988

Source: Insurance Institute for Highway Safety

(a) Normal probability plots for each treatment indicate that the requirement of normality is satisfied. Verify that the requirement of equal population variances for each treatment is satisfied.

(b) Is there sufficient evidence that the mean cost of repairs is different among the four SUVs at the $\alpha = 0.05$ level of significance?

(c) If the null hypothesis from part (b) was rejected, use Tukey's test to determine which pairwise means differ using a familywise error rate of $\alpha = 0.05$.

10. Lodging A travel agent wanted to know whether the price (in dollars) of Marriott, Hyatt, and Sheraton Hotels differed significantly. She knew that location of the hotel is a factor in determining price, so she blocked each hotel by location. After randomly selecting six cities and obtaining the room rate for each hotel, she obtained the following data:

	Marriott	Hyatt	Sheraton
Chicago	179	139.40	150
Los Angeles	169	161.50	161
Houston	163	187	189
Boston	189	179.10	169
Denver	179	168	112
Orlando	147	159	147

Source: Expedia.com

(a) Normal probability plots for each treatment indicate that the requirement of normality is satisfied. Verify that the requirement of equal population variances for each treatment is satisfied.

(b) Is there sufficient evidence that the mean cost of the room is different among the three hotel chains at the $\alpha = 0.05$ level of significance?

(c) If the null hypothesis from part (b) was rejected, use Tukey's test to determine which pairwise means differ using a familywise error rate of $\alpha = 0.05$.

11. Rats in Space Researchers at NASA wanted to determine the effects of space flight on a rat's daily consumption of water. The following data represent the water consumption (in milliliters per day) at lift-off minus 1, return plus 1, and 1 month after return for six rats sent to space on the Spacelab Sciences 1 flight.

Rat	Lift-Off Minus 1	Return Plus 1	Return Plus 1 Month
1	18.5	32	30
2	17.5	18	34
3	28.0	31	39
4	28.5	29	44
5	31.0	48	54
6	22.5	25	32

Source: NASA Life Sciences Data Archive

(a) What is the response variable in this study? What is the treatment? How many levels does it have?

(b) Normal probability plots for each treatment indicate that the requirement of normality is satisfied. Verify that the requirement of equal population variances for each treatment is satisfied.

(c) Is there sufficient evidence that the water consumption is different for the three days at the $\alpha = 0.05$ level of significance?

(d) If the null hypothesis from part (c) was rejected, use Tukey's test to determine which pairwise means differ using a familywise error rate of $\alpha = 0.05$.

12. Concrete Strength Researchers Olivia Carrillo-Gamboa and Richard Gunst presented the following data in their article, "Measurement-Error-Model Collinearities." The data represent the compressive strength (in pounds per square inch) of a random sample of concrete 2 days, 7 days, and 28 days after pouring.

Concrete	2 Days	7 Days	28 Days
1	2830	3505	4470
2	3295	3430	4740
3	2710	3670	5115
4	2855	3355	4880
5	2980	3985	4445
6	3065	3630	4080
7	3765	4570	5390

Source: "Measurement-Error-Model Collinearities," *Technometrics* 34(454–464).

(a) Normal probability plots for each treatment indicate that the requirement of normality is satisfied. Verify that the requirement of equal population variances for each treatment is satisfied.

(b) Is there sufficient evidence that the mean strength is different among the three days at the $\alpha = 0.05$ level of significance?

(c) If the null hypothesis from part (b) was rejected, use Tukey's test to determine which pairwise means differ using a familywise error rate of $\alpha = 0.05$.

13. Waiting in Line A quality-control manager at an amusement park feels that the amount of time that people spend waiting in line for the American Eagle roller coaster is too long. To determine if a new loading/unloading procedure is effective in reducing wait time in line, he measures the amount of time (in minutes) people are waiting in line for seven days. After implementing the new procedure, he again measures the amount of time (in minutes) people are waiting in line for seven days and obtains the data on the next page. To make a reasonable comparison, he chooses days when weather conditions are alike.

Treat each day as a block and the wait times before and after the procedure as the treatment.

Day	Mon	Tues	Wed	Thurs	Fri	Sat	Sun
Wait time before new procedure	11.6	25.9	20.0	38.2	57.3	32.1	81.8
Wait time after new procedure	10.7	28.3	19.2	35.9	59.2	31.8	75.3

(a) Using the methods introduced in this section, determine whether there is sufficient evidence to conclude that the two loading procedures are resulting in different measurements of the wait time at the $\alpha = 0.05$ level of significance.

(b) Using the methods introduced in Section 11.2, determine whether there is sufficient evidence to conclude that the two loading procedures are resulting in different measurements of the wait time at the $\alpha = 0.05$ level of significance.

(c) Compare the P-values of both approaches. Can you conclude that the method presented in this section is a generalization of the matched-pairs t-test?

14. **Design a Study** David Elkington, the Chief Executive Officer (CEO) of InsideSales.com, wondered whether the cycles of the moon affect sales. He got the idea from his father-in-law, who said that emergency rooms are crowded when the moon is full. There are eight phases of the moon: New Moon, Waning Crescent, Third Quarter, Waning Gibbous, Full, Waxing Gibbous, First Quarter, and Waxing Crescent. Explain how to obtain and analyze data to determine whether moon cycles affect sales.

Explaining the Concepts

15. How does the completely randomized design differ from a randomized complete block design?

16. What is blocking? Why might a researcher want to block?

17. What does *randomized* mean in the randomized complete block design? What does *complete* mean in the randomized complete block design?

18. What requirements must be satisfied to analyze a randomized complete block design?

19. Name the two ways that a researcher can deal with explanatory variables in a designed experiment.

20. How is a matched-pairs design related to the randomized complete block design?

13.4 Two-Way Analysis of Variance

Objectives
1. Analyze a two-way ANOVA design
2. Draw interaction plots
3. Perform the Tukey test

One-way ANOVA is used to compare k population means. In this design, one factor is set at k levels. The other factors are fixed at a single level or dealt with through random assignment of the experimental units to the treatment groups. For example, we randomly assign 40 males between the ages of 20 and 29 to $k = 3$ different treatment groups. Group 1 receives 5 mg per day of an experimental drug, group 2 receives 7.5 mg per day of an experimental drug, and group 3 receives a placebo. The subjects all have the same diet and exercise routine. Other factors not controlled are expected to "even out" by randomly assigning the subjects to the three treatment groups. The experimental design that allows for this type of analysis is the completely randomized design.

The methods of one-way ANOVA can also be used in analyzing observational data. For example, a researcher wonders whether the mean of hemoglobin levels varies with age. Here, the factor is age, such as 20 to 29, 30 to 39, and 40 to 49 years old. In this case, the factor has $k = 3$ levels. Of course, in observational studies, there is never any talk of causation when it comes to the relation between the factor (age) and response variable (hemoglobin). In both the experimental design and the observational study, we should notice that the factor is a categorical (qualitative) variable that classifies an individual into one of k groups.

An improvement on analysis involving one-way ANOVA is the randomized complete block design in which experimental units are organized through a technique called *blocking*. For example, if the researcher attempting to determine the effectiveness of a drug feels that genetics may explain some of the variation in the response variable, he may select families with three male siblings and block by family. He would randomly assign one sibling to group 1, one sibling to group 2, and the third sibling to group 3. It is important to recognize that we are still only interested in determining whether the factor, drug, affects the response variable. We block to reduce experimental error.

> **CAUTION!**
> The randomized complete block design is a design that varies a single factor. We are not interested in determining whether the block is significant! If we want to analyze the effect of fixing two different factors at different levels, we use the methods of two-way ANOVA.

① Analyze a Two-Way ANOVA Design

We now present analysis in which *two* factors can explain variability in the response variable. We deal with the two factors by fixing them at different levels. Remember that we can deal with factors through control by fixing them at one level or at different levels, and randomizing so that the effect of uncontrolled variables on the response variable is minimized. In both the completely randomized design and the randomized complete block design, we manipulated one factor to see how varying it affected the response variable. In this section, we manipulate two factors.

Throughout the section, we will call the two factors factor A and factor B. If factor A has two levels and factor B has two levels, we have a **2 × 2 factorial design** (read "two by two factorial design"). For example, suppose we want to determine the effect an experimental drug has on hemoglobin levels in humans. We might have two levels in the first factor (placebo or 5 mg per day of the drug) and two levels in the second factor (male or female). In this way, our design will look like Figure 14.

Figure 14

		Factor B	
		Placebo	**5 mg per day**
Factor A	**Male**	Hemoglobin of males receiving the placebo	Hemoglobin of males receiving 5 mg per day
	Female	Hemoglobin of females receiving the placebo	Hemoglobin of females receiving 5 mg per day

In general, the two-factor design looks as presented in Figure 15, where there are *n* replications for each combination of factor A and factor B. Factor A has *a* levels and factor B has *b* levels. Each combination of factor levels is a treatment.

Figure 15

		Factor B			
		1	**2**	**· · ·**	**b**
Factor A	**1**	*n* observations of the response variable with level 1 of factor A and level 1 of factor B	*n* observations of the response variable with level 1 of factor A and level 2 of factor B	· · ·	*n* observations of the response variable with level 1 of factor A and level *b* of factor B
	2	*n* observations of the response variable with level 2 of factor A and level 1 of factor B	*n* observations of the response variable with level 2 of factor A and level 2 of factor B	· · ·	*n* observations of the response variable with level 2 of factor A and level *b* of factor B
	⋮	⋮	⋮		⋮
	a	*n* observations of the response variable with level *a* of factor A and level 1 of factor B	*n* observations of the response variable with level *a* of factor A and level 2 of factor B	· · ·	*n* observations of the response variable with level *a* of factor A and level *b* of factor B

So, if there are three levels of factor A and two levels of factor B, we have a 3 × 2 factorial design as shown in Figure 16(a). If there are three levels of factor A and three levels of factor B, we have a 3 × 3 factorial design as shown in Figure 16(b) on the next page.

Figure 16

		Factor B	
		1	**2**
Factor A	**1**	*n* observations of the response variable with level 1 of factor A and level 1 of factor B	*n* observations of the response variable with level 1 of factor A and level 2 of factor B
	2	*n* observations of the response variable with level 2 of factor A and level 1 of factor B	*n* observations of the response variable with level 2 of factor A and level 2 of factor B
	3	*n* observations of the response variable with level 3 of factor A and level 1 of factor B	*n* observations of the response variable with level 3 of factor A and level 2 of factor B

(a)

Figure 16
(continued)

		Factor B		
		1	**2**	**3**
Factor A	**1**	n observations of the response variable with level 1 of factor A and level 1 of factor B	n observations of the response variable with level 1 of factor A and level 2 of factor B	n observations of the response variable with level 1 of factor A and level 3 of factor B
	2	n observations of the response variable with level 2 of factor A and level 1 of factor B	n observations of the response variable with level 2 of factor A and level 2 of factor B	n observations of the response variable with level 2 of factor A and level 3 of factor B
	3	n observations of the response variable with level 3 of factor A and level 1 of factor B	n observations of the response variable with level 3 of factor A and level 2 of factor B	n observations of the response variable with level 3 of factor A and level 3 of factor B

(b)

Each of the n observations of the response variable for the different levels of the factors exists within a **cell**. In a 2×3 factorial design, there are six cells with n observations within each cell. Each cell corresponds to a specific treatment.

EXAMPLE 1 A 2 × 3 Factorial Design

High-density lipoprotein (HDL) cholesterol is called good cholesterol because it helps to reduce the amount of bad cholesterol in your system. A pharmaceutical company developed a drug that is meant to increase HDL levels in patients. They obtained volunteers whose HDL cholesterol was roughly the same and randomly assigned them to one of three groups. Group 1, a placebo group, received a sugar tablet. Group 2 received 5 mg of the drug, and group 3 received 10 mg of the drug. In addition to drug dosage, the researchers considered age to be a factor in the analysis; so they divided patients into an 18- to 39-year-old category and a 40 and older category. The design results in a 2×3 factorial design. Table 11 shows the increase in HDL cholesterol in milligrams per deciliter (mg/dL) for each patient after 10 weeks. Each cell has $n = 3$ replications.

Table 11									
				Drug Dosage					
Age		**Placebo**			**5 mg**			**10 mg**	
18 to 39 years	4	3	−1	9	5	6	14	12	10
40 or older	3	2	0	3	6	7	10	8	7

● **Now Work Problem 17(a)**

In the factorial design, all levels of factors A and B are combined, so we say that the factors are **crossed**. The effect of factor A is the change in the response variable that results from changing the level of factor A. The effect of factor B is the change in the response variable that results from changing the level of factor B. Together these two effects are called the **main effects**. If changes in the level of factor A result in different changes in the value of the response variable for the different levels of factor B, we say that there is an **interaction effect** between the factors. An example will help clarify this idea.

In Other Words
An example of interaction is sleeping pills and alcohol. They are usually not fatal when taken alone, but can be fatal when combined.

EXAMPLE 2 Main Effects and Interaction Effect

Suppose we have two factors, A and B, each fixed at two levels, high and low. The value of the response variable at each level is presented in Table 12. For example, the response variables when both factor A and B are set to low are 5 and 3.

Table 12

Factor A	Factor B			
	Low		High	
Low	5	3	6	8
High	8	4	18	17

The main effect of factor A is the difference between the mean values of the response variable when factor A is set at high and at low:

$$\text{Main effect of factor A} = \overbrace{\frac{8 + 4 + 18 + 17}{4}}^{\text{Factor A High}} - \overbrace{\frac{5 + 3 + 6 + 8}{4}}^{\text{Factor A Low}} = 6.25 \text{ units}$$

Increasing factor A from low level to high level increases the value of the response variable by 6.25 units, on average.

$$\text{Main effect of factor B} = \overbrace{\frac{6 + 8 + 18 + 17}{4}}^{\text{Factor B High}} - \overbrace{\frac{5 + 3 + 8 + 4}{4}}^{\text{Factor B Low}} = 7.25 \text{ units}$$

Increasing factor B from low level to high level increases the value of the response variable by 7.25 units, on average.

To see the interaction effect between the factors, we first look at one level of one factor and see how changing the other factor affects the value of the response variable. For example, if we look only at the low level of factor B, we can see that the effect of changing factor A from low to high is

$$\overbrace{\frac{8 + 4}{2}}^{\substack{\text{Factor A} \\ \text{High}}} - \overbrace{\frac{5 + 3}{2}}^{\substack{\text{Factor A} \\ \text{Low}}} = 2 \text{ units}$$

So, if we are at the low level of factor B, increasing factor A from low to high increases the response variable by 2 units, on average. If we look only at the high level of factor B, we see the effect of changing factor A from low to high is

$$\frac{18 + 17}{2} - \frac{6 + 8}{2} = 10.5 \text{ units}$$

So, if we are at the high level of factor B, increasing factor A from low to high increases the response variable by 10.5, on average.

> **CAUTION!**
> If interaction between two factors exists, interpret the main effects with extreme caution.

The main effect of factor A is 6.25 units, but this is misleading because, at the low level of factor B, the mean increase in the response variable is only 2 units; but at the high level of factor B, the mean increase in the response variable is 10.5 units. So the increase in the value of the response variable depends on the level of factor B. For this reason, we say there is an interaction effect between factors A and B.

The moral: If interaction between two factors exists, looking at main effects is misleading.

● **Now Work Problem 9(a)**

Now let's learn to test hypotheses that involve a factorial design.

Requirements to Perform the Two-Way Analysis of Variance

1. The populations from which the samples are drawn must be normal.
2. The samples must be independent.
3. The populations must have the same variance.

We can check the normality requirement through a normal probability plot of the data within each cell. The requirement for equal variances can be checked by verifying that the largest sample standard deviation is no more than twice the smallest sample standard deviation.

Although not a requirement to conduct two-way ANOVA, we also assume that there are an equal number of observations for each combination of the factors because this increases the power of the test. That is, it reduces the likelihood of making a Type II error.

In a two-way ANOVA, we will test three separate hypotheses. The three hypotheses go with the three types of effects presented earlier. The first hypothesis deals with the significance of any interaction effect. So we have

Hypotheses Regarding Interaction Effect

H_0: there is no interaction between the factors
H_1: there is interaction between the factors

Hypotheses Regarding Main Effects

H_0: there is no effect of factor A on the response variable
H_1: there is an effect of factor A on the response variable

H_0: there is no effect of factor B on the response variable
H_1: there is an effect of factor B on the response variable

When we conduct a two-way ANOVA, we always first test the hypothesis regarding interaction effect. If the null hypothesis of no interaction is rejected, we do not interpret the results of the hypotheses involving the main effects. This is because the interaction clouds the interpretation of the main effects.

The decision rule for each of these hypotheses will be the same as always: if the P-value is less than the level of significance, we reject the null hypothesis in favor of the alternative. We will obtain the P-value from the output of statistical software. Let's look at an example.

EXAMPLE 3 Examining a Two-Way ANOVA

Problem In Example 1, we presented data for an experimental drug that was meant to increase HDL cholesterol. The data are represented in Table 13 for convenience.

Table 13									
	Drug Dosage								
Age	**Placebo**			**5 mg**			**10 mg**		
18 to 39 years	4	3	−1	9	5	6	14	12	10
40 or older	3	2	0	3	6	7	10	8	7

(a) HDL cholesterol levels are known to have a distribution that is approximately normal. Verify that the largest sample standard deviation of any cell is no more than twice the smallest sample standard deviation of any cell.
(b) Use Minitab to test whether there is an interaction effect between the drug dosage and age.
(c) If the null hypothesis of no interaction is not rejected, determine whether there is sufficient evidence to conclude that the mean increase in HDL cholesterol is different (i) among each drug dosage group, (ii) for each age group.

Approach We use Minitab to obtain the two-way ANOVA. The steps for performing two-way ANOVA using Minitab and StatCrunch are given in the Technology Step-by-Step on page 662. If the P-value corresponding to each hypothesis is small (say, less than $\alpha = 0.05$), reject the null hypothesis in favor of the alternative hypothesis.

Solution

(a) We obtain the descriptive statistics from Minitab shown in Figure 17 (a partial printout). The largest sample standard deviation, 2.65 mg/dL, is not twice as large as the smallest sample standard deviation, 1.528 mg/dL, so the requirement of equal population variances is satisfied.

Figure 17

Descriptive Statistics: P/18–39, 5/18–39, 10/18–39, P/40, 5/40, 10/40

Variable	N	Mean	Median	TrMean	StDev	SE Mean
P/18–39	3	2.00	3.00	2.00	2.65	1.53
5/18–39	3	6.67	6.00	6.67	2.08	1.20
10/18–39	3	12.00	12.00	12.00	2.00	1.15
P/40	3	1.667	2.000	1.667	1.528	0.882
5/40	3	5.33	6.00	5.33	2.08	1.20
10/40	3	8.333	8.000	8.333	1.528	0.882

(b) First test the hypotheses

H_0: there is no interaction between drug dosage and age

H_1: there is an interaction between drug dosage and age

Enter the data into Minitab. Let column 1 represent the drug dosage; so enter 1 if the observation is from the placebo group, 2 if the observation is from 5 mg, and so on. Let column 2 represent the age and enter 1 for 18- to 39-year-olds and 2 for 40 or older. Column 3 gets the value of the response variable, increase in HDL cholesterol. The results of the two-way ANOVA appear in Figure 18.

Figure 18

Two-way ANOVA: HDL versus Drug, Age

Analysis of Variance for HDL

Source	df	SS	MS	F	P
Drug	2	208.33	104.17	25.68	0.000
Age	1	14.22	14.22	3.51	0.086
Interaction	2	8.78	4.39	1.08	0.370
Error	12	48.67	4.06		
Total	17	280.00			

The P-value for the interaction effect is 0.370. The P-value is large, so we do not reject the null hypothesis and conclude that there is no interaction effect.

(c) Now test the hypothesis regarding the main effect, drug dosage, which is

$$H_0: \mu_{\text{placebo}} = \mu_{5\,\text{mg}} = \mu_{10\,\text{mg}}$$
$$H_1: \text{at least one of the means differs}$$

and the hypothesis regarding the main effect, age, which is

$$H_0: \mu_{18\text{–}39} = \mu_{40\,\text{or older}} \quad \text{versus} \quad H_1: \text{The means differ}$$

> **CAUTION!**
> Although we do not reject the null hypothesis that the mean increase in HDL is the same at the different age levels, there is some evidence that age plays a role in HDL levels.

The P-value for drug dosage is 0.000. The small P-value is taken as evidence against the null hypothesis. We conclude that at least one of the mean increases in HDL is different for the different levels of drug dosage. Therefore, the level of the drug dosage is a significant contributor to explaining the increase in HDL cholesterol.

The P-value for age is given as 0.086. Because this is greater than the level of significance for the test ($\alpha = 0.05$), we do not reject the null hypothesis and conclude that the mean increase in HDL cholesterol does not change with age. •

● **Now Work Problem 17(b) and (c)**

Verifying the Normality Requirement in a Two-Way ANOVA

Again, to conduct any ANOVA, the response variable must be normally distributed. This requirement can be verified by drawing a normal probability plot of the residuals in the ANOVA model.

EXAMPLE 4 Verifying the Normality Requirement in a Two-Way Analysis of Variance

Problem Verify the normality requirement for the HDL data analyzed in Example 3.

Approach We will use Minitab to obtain the residuals and draw the normal probability plot.

Solution Figure 19 shows the normal probability plot of the residuals. Because the residuals all lie within the confidence bands, the normality requirement is satisfied.

If we use StatCrunch to assess normality of the residuals, we find the correlation between the residuals and expected z-scores is 0.981. Because $0.981 > 0946$ (critical value from Table VI with $n = 18$), it is reasonable to conclude the residuals are normally distributed.

Figure 19

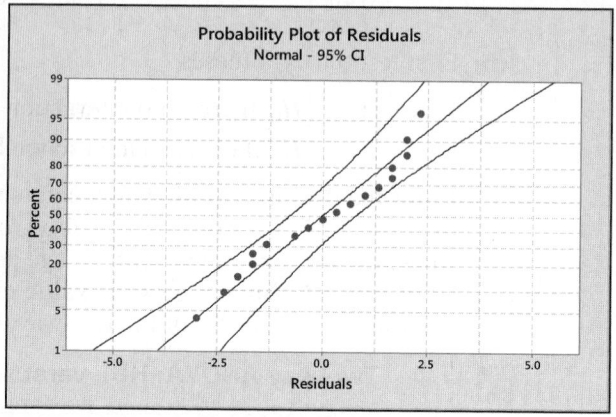

• Now Work Problem 17(f)

② Draw Interaction Plots

As usual in statistics, we like to support results graphically. To graphically represent the role interaction plays in any factorial design, we use **interaction plots**.

Constructing Interaction Plots by Hand

Step 1 Compute the mean value of the response variable within each cell.

Step 2 In a Cartesian plane, label the horizontal axis for each level of factor A. Let the vertical axis represent the mean value of the response variable. For each level of factor A, plot the mean value of the response variable for each level of factor B. Draw straight lines connecting the points for the common level of factor B. You should have as many lines as there are levels of factor B. The more difference there is in the slopes of the lines, the stronger the evidence of interaction.

EXAMPLE 5 Drawing an Interaction Plot

Problem Draw an interaction plot for the data from Example 3.

By-Hand Approach
Follow Steps 1 and 2.

By Hand Solution
Step 1 Table 14 shows the mean value of the response variable for each cell.

Technology Approach
Use Minitab or StatCrunch to draw the interaction plot. The steps to follow are given in the Technology Step-by-Step on page 662.

Technology Solution
Figure 21 shows the interaction plot from Minitab using the data in Example 3.

Table 14

| Age | Drug Dosage | | |
	Placebo	5 mg	10 mg
18 to 39 years	$\dfrac{4 + 3 + (-1)}{3} = 2$	$\dfrac{9 + 5 + 6}{3} = 6.7$	$\dfrac{14 + 12 + 10}{3} = 12$
40 or older	$\dfrac{3 + 2 + 0}{3} = 1.7$	$\dfrac{3 + 6 + 7}{3} = 5.3$	$\dfrac{10 + 8 + 7}{3} = 8.3$

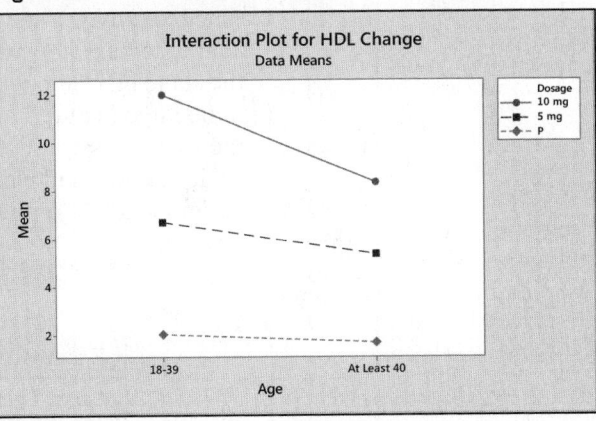

Figure 21

Step 2 Draw a Cartesian plane, label the horizontal axis Age, and indicate the levels 18 to 39 and 40 or older. The vertical axis is labeled as the response. For each age group, plot the mean values of the response variable for drug. Connect the points and label each line with its drug dosage. See Figure 20.

Figure 20

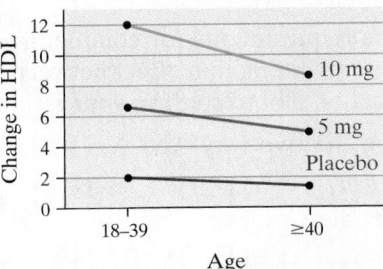

Interpretation When interpreting interaction plots, look at the level of parallelism among the lines. The more parallel the lines seem to be, the stronger the visual evidence is of no interaction. From Figures 20 and 21, the roughly parallel lines verify our conclusion of no interaction. Another bit of information that can be learned from the interaction plot is the effect age has on the response variable, change in HDL. Notice that change in HDL decreases as age increases. The increase in HDL cholesterol is largest for the 10-mg dose and smallest for the placebo.

● **Now Work Problem 17(e)**

Some interaction plots are shown in Figure 22, where A_1, A_2, and A_3 represent three levels of factor A, and B_1, B_2, and B_3 represent three levels of factor B. Remember, although the interaction plots are meant to help us visualize the interaction (just as boxplots allow us to visualize differences among treatments in one-way ANOVA), they are not meant to be used to test for interaction. We can test for interaction by looking at the F-test statistic and its P-value for interaction in the ANOVA.

Figure 22

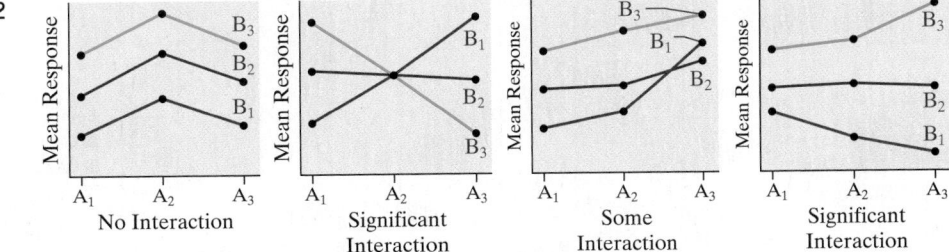

EXAMPLE 6   Analyzing a 2 × 2 Factorial Design

Problem An educational psychologist conducts an experiment to determine whether varying the conditions in which learning and testing take place affects test results. She randomly selects 20 students who are unfamiliar with the Battle at Gettysburg from

(continued)

the Civil War. After presenting the material to the students and allowing them ample time to study, the students are given an exam. Ten of the students receive the lecture in a large lecture hall and 10 receive the lecture in a small classroom (both lectures have the same instructor). Of the students in the large lecture hall, five take the exam in the same large lecture hall, and five take the exam in a small classroom. Of the students in the small classroom, five take the exam in a large lecture hall and five take the exam in the same small classroom. The experimental design is, therefore, a 2×2 factorial design with five replications in each cell. Table 15 gives the scores for each student.

Table 15

	Lecture Administered	
Exam Given	**Large Lecture Hall**	**Classroom**
Large Lecture Hall	70 63 95 84 72	63 46 47 71 66
Classroom	51 73 65 72 43	76 67 80 90 79

Preliminary analysis indicates that the requirements for conducting a two-way ANOVA are satisfied. Test whether there is an interaction effect between the lecture location and exam location. Draw an interaction plot to confirm your results.

Approach We use StatCrunch to perform the two-way ANOVA and to draw the interaction plot.

Solution Figure 23 shows the ANOVA from StatCrunch.

Figure 23

Analysis of Variance results:
Responses: Score
Row factor: Exam
Column factor: Lecture

ANOVA table

Source	DF	SS	MS	F-Stat	P-value
Exam	1	18.05	18.05	0.13482726	0.7183
Lecture	1	0.45	0.45	0.0033613445	0.9545
Interaction	1	1602.05	1602.05	11.96676	0.0032
Error	16	2142	133.875		
Total	19	3762.55			

The P-value for the interaction effect is 0.0032, indicating that there is a significant interaction effect between lecture location and exam location. Therefore, we will not test the main effects. Figure 24 shows the interaction plot from StatCrunch, which confirms the analytic results from the ANOVA.

Figure 24

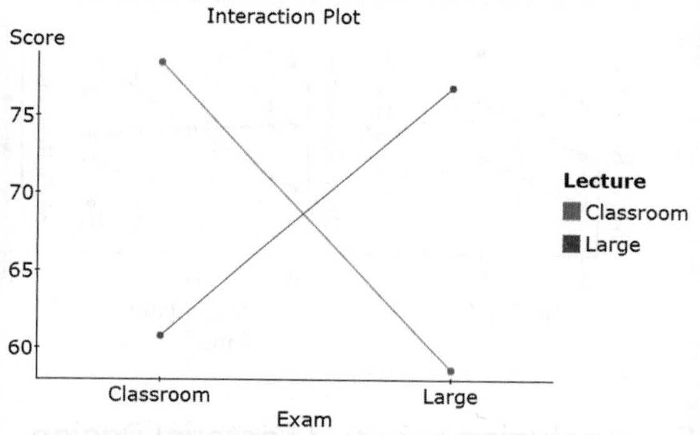

Interaction Plot

The interaction plot suggests that students perform best if they take their exams in the same classroom where their lecture was given. The two high means are the means for (large lecture, large exam) and (classroom lecture, classroom exam).

③ Perform the Tukey Test

Once the null hypothesis of equal population means for either factor is rejected, we determine which means differ significantly using Tukey's test. The steps are identical to those presented for one-way ANOVA. However, the critical value and the estimate of the standard error are a little different. The critical value is $q_{\alpha,\nu,k}$, using a familywise error rate of α with $\nu = N - ab$, where N is the total number of observations (or replications), a is the number of levels for factor A, and b is the number of levels for factor B; and k is the number of means being tested for the factor. The standard error is

$$SE = \sqrt{\frac{MSE}{m}}$$

where m is the product of the number of levels for the factor and the number of observations within each cell.

EXAMPLE 7   Multiple Comparisons Using Tukey's Test

Problem Use statistical software to perform Tukey's test to determine which pairwise means differ for the data presented in Example 3 using a familywise error rate of $\alpha = 0.05$.

Approach We will use Minitab to perform Tukey's test. The steps for performing this test using either Minitab or StatCrunch can be found in the Technology Step-by-Step on page 662.

Solution Figure 25 shows the results of Tukey's test using Minitab. To find the upper and lower bounds on the intervals, scroll over the graphic. For example, the confidence interval for $\mu_p - \mu_{5\,mg}$ does not contain 0 and is negative (lower bound: -7.2 mg/dL; upper bound: -1.1 mg/dL). Therefore, the mean increase in HDL cholesterol resulting from 5 mg of the drug is significantly greater than the mean increase due to the placebo. The confidence interval for $\mu_p - \mu_{10\,mg}$ does not contain 0 and is negative. This indicates that the mean increase in HDL cholesterol resulting from 10 mg of the drug is also greater than the mean increase due to the placebo. Finally, $\mu_{5\,mg} - \mu_{10\,mg}$ shows that the interval also does not contain 0, so the mean increase in HDL is higher with 10 mg than with 5 mg of the drug. Notice that the pairwise comparisons for age contains zero because we did not reject the null hypothesis $H_0: \mu_{18-39} = \mu_{40}$ or older for age.

Figure 25

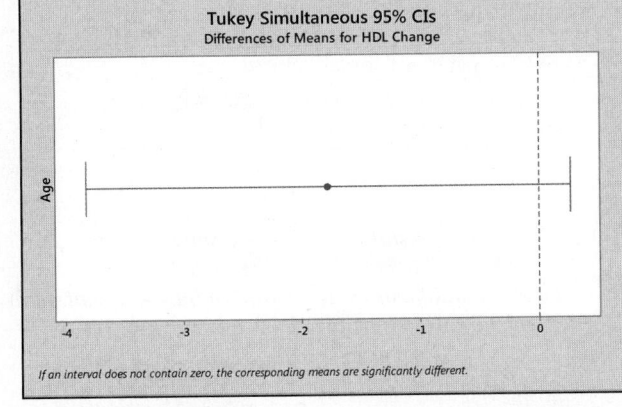

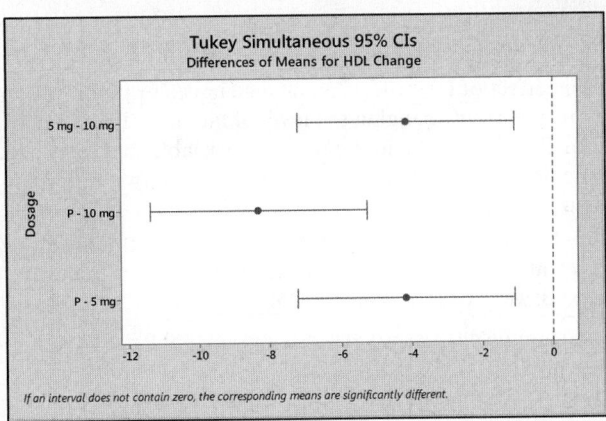

● **Now Work Problem 17(g)**

Technology Step-by-Step Two-Way ANOVA

Minitab
Obtaining Two-Way ANOVA
1. In C1, enter the level of factor A (that is for level 1 enter a 1, and so on); in C2; enter the level of factor B, and in C3 enter the value of the response variable.
2. Select the **Stat** menu. Highlight **ANOVA** and select **General Linear Model**. Then select **Fit General Linear Model** . . . .
3. Enter C3 in the box marked "Responses:". Enter C1 and C2 in the box marked "Factors:".
4. Select Model. Highlight C1 and C2 in the box "Factors and covariates:". Be sure "2" is selected under the drop-down menu "Interactions through order:". Click Add. Click OK. Click Storage and check the Residual box if you want to save the residuals. Click OK twice.

Interaction Plots
1. Enter the data as described in Step 1 above. Select the **Stat** menu. Highlight **ANOVA** and select **Interaction Plot** . . . .
2. Enter C3 in the box marked "Response Variable." Enter C1 and C2 in the box marked "Factors." Click OK.
Note: The factor entered second will be plotted on the horizontal axis.

Tukey's Test
1. With the data entered into the Minitab spreadsheet as specified in Step 1 above, select the **Stat** menu. Highlight **ANOVA** select **General Linear Model...**, then select **Comparisons** . . . .
2. In the drop-down menu under "Response:", select the column containing the response variable. In the

drop-down menu under "Type of comparison:", select Pairwise. Check the Tukey box under Method. Highlight the columns containing the factors in the box "Choose terms for comparisons:". Click on the box "C = Compare levels for this item". Click OK.

Main Effects Plots
1. Enter the data as described in Step 1. Select the **Stat** menu. Highlight **ANOVA** and select **Main Effects Plot** . . . .
2. Enter C3 in the box marked "Response Variable." Enter C1 and C2 in the box marked "Factors." Click OK.

StatCrunch
Obtaining Two-Way ANOVA
1. In column var1, enter the level of factor A; in column var2, enter the level of factor B; and in column var3, enter the value of the response variable. Name the columns.
2. Select **Stat**, highlight **ANOVA**, and select **Two Way**.
3. Select the column containing the values of the response variable from the pull-down menu "Responses in:". Select the column containing the row factor in the pull-down menu "Row factor in:". Select the column containing the column factor in the pull-down menu "Column factor in:". Select the "Compute Tukey HSD" box to conduct Tukey's test. To save residuals to draw a normal probability plot, highlight Residuals under Save:. Click Compute!

Interaction Plots
In the same screen where you checked "Compute Tukey HSD," also check "Plot interactions."

13.4 Assess Your Understanding

Vocabulary and Skill Building

1. If factor A has 2 levels and factor B has 3 levels in a two-way ANOVA, we have a _ × _ factorial design.

2. The effect of factor A is the change in the response variable that results from changing the level of factor A. The effect of factor B is the change in the response variable that results from changing the level of factor B. Together, these two effects are called ____ _____.

3. If changes in the level of factor A result in different changes in the value of the response variable for different levels of factor B, we say there is an _____ ____.

4. To graphically display the role interaction plays in a factorial design, we draw _____ ____.

In Problems 5–8, determine whether the interaction plot suggests that significant interaction exists among the factors.

5.

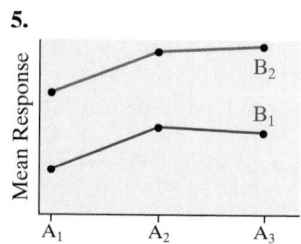

6.

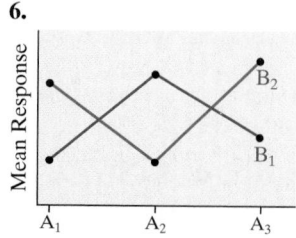

7.

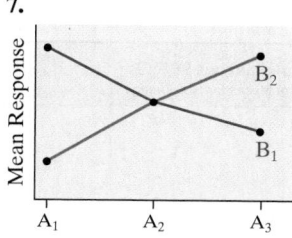

8.

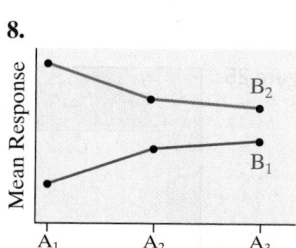

In Problems 9 and 10, (a) use the same analysis as that presented in Example 2 to conjecture whether interaction exists between factor A and factor B. (b) Draw an interaction plot to verify your conjecture from part (a).

NW 9.

		Factor B	
		Low	**High**
Factor A	**Low**	12	6
		8	7
	High	20	1
		14	3

10.

		Factor B	
		Low	High
	Low	104	143
		96	121
Factor A	High	84	55
		76	43

11. Given the following ANOVA output, answer the questions that follow.

Analysis of Variance for Response

Source	df	SS	MS	F	P
Factor A	1	531.2	531.2	11.73	0.003
Factor B	2	3018.0	1509.0	33.33	0.000
Interaction	2	16.3	8.2	0.18	0.836
Error	18	814.9	45.3		

(a) Is there evidence of an interaction effect? Why or why not?
(b) Based on the P-value, is there evidence of a difference in the means from factor A? Based on the P-value, is there evidence of a difference in the means from factor B?
(c) What is the mean square error?

12. Given the following ANOVA output, answer the questions that follow.

Analysis of Variance for Response

Source	df	SS	MS	F	P
Factor A	2	156	78	0.39	0.679
Factor B	2	132	66	0.33	0.720
Interaction	4	311	78	0.39	0.813
Error	27	5354	198		
Total	35	5952			

(a) Is there evidence of an interaction effect? Why or why not?
(b) Based on the P-value, is there evidence of a difference in the means from factor A? Based on the P-value, is there evidence of a difference in the means from factor B?
(c) What is the mean square error?

13. Given the following ANOVA output, answer the questions that follow.

Analysis of Variance for Response

Source	df	SS	MS	F	P
Factor A	2	2269.8	1134.9	35.63	0.000
Factor B	2	115.2	57.6	1.81	0.183
Interaction	4	1694.8	423.7	13.30	0.000
Error	27	860.0	31.9		
Total	35	4939.8			

(a) Is there evidence of an interaction effect? Why or why not?
(b) Based on the P-value, is there evidence of a difference in the means from factor A? Based on the P-value, is there evidence of a difference in the means from factor B?
(c) What is the mean square error?

14. Given the following ANOVA output, answer the questions that follow.

Analysis of Variance for Response

Source	df	SS	MS	F	P
Factor A	2	1209.7	604.8	8.32	0.002
Factor B	2	577.6	288.8	3.97	0.031
Interaction	4	1474.9	368.7	5.07	0.004
Error	27	1962.2	72.7		
Total	35	5224.3			

(a) Is there evidence of an interaction effect? Why or why not?
(b) Based on the P-value, is there evidence of a difference in the means from factor A? Based on the P-value, is there evidence of a difference in the means from factor B?
(c) What is the mean square error?

In Problems 15 and 16, assume that the data come from populations that are normally distributed with the same variance.

15.

		Factor B		
		Level 1	Level 2	Level 3
		64.9	53.3	51.9
		59.3	44.2	61.8
	Level 1	32.9	46.2	68.5
Factor A		59.4	43.3	65.8
		50.7	38.4	53.1
	Level 2	59.3	50.7	57.9
		58.1	40.2	55.7
		33.5	52.6	74.4

(a) Determine whether there is significant interaction between factor A and factor B.
(b) If there is no significant interaction, determine if there is a significant difference in the means for the two levels of factor A. If there is no significant interaction, determine if there is a significant difference in the means for the three levels of factor B.
(c) Draw an interaction plot of the data to support the results of parts (a) and (b).
(d) If there is a significant difference in the means for the two levels of factor A, use Tukey's test to determine which pairwise means differ using a familywise error rate of $\alpha = 0.05$. If there is a significant difference in the means for the three levels of factor B, use Tukey's test to determine which pairwise means differ using a familywise error rate of $\alpha = 0.05$.

16.

		Factor B		
		Level 1	Level 2	Level 3
		104	111	95
	Level 1	82	104	84
		81	74	104
		111	109	106
Factor A	Level 2	112	115	110
		82	97	99
		112	108	92
	Level 3	108	99	129
		117	104	120

(a) Determine whether there is significant interaction between factor A and factor B.
(b) If there is no significant interaction, determine whether there is a significant difference in the means for the three levels of factor A. If there is no significant interaction, determine whether there is a significant difference in the means for the three levels of factor B.
(c) Draw an interaction plot of the data to support the results of parts (a) and (b).

(a) Is there evidence of an interaction effect? Why or why not?
(b) Based on the P-value, is there evidence of a difference in the means from factor A? Based on the P-value, is there evidence of a difference in the means from factor B?
(c) What is the mean square error?

(d) If there is a significant difference in the means for the three levels of factor A, use Tukey's test to determine which pairwise means differ using a familywise error rate of $\alpha = 0.05$. If there is a significant difference in the means for the three levels of factor B, use Tukey's test to determine which pairwise means differ using a familywise error rate of $\alpha = 0.05$.

Applying the Concepts

NW 17. Cholesterol Levels A family physician wanted to know if age and gender were factors that explained levels of serum cholesterol (in mg/dL) in her adult patients. She randomly selects two patients for each category of data and obtains the following results:

Gender	Age (years)		
	18–34	**35–54**	**55 and older**
Female	180, 192	205, 226	218, 231
Male	175, 193	213, 222	203, 185

Source: National Center for Health Statistics

Serum cholesterols are known to be approximately normally distributed and the population variances are equal.

(a) What type of factorial design is this? How many replications are there within each cell?

(b) What is the response variable? What are the two factors?

(c) Determine if there is significant interaction between age and gender.

(d) If there is no significant interaction, determine whether there is significant difference in the means for the three age groups. If there is no significant interaction, determine whether there is significant difference in the means for the genders.

(e) Draw an interaction plot of the data to support the results of parts (c) and (d).

(f) The residuals are normally distributed. Verify this.

(g) If there is significant difference in the means for the three age groups, use Tukey's test to determine which pairwise means differ using a familywise error rate of $\alpha = 0.05$. If there is significant difference in the means for gender, use Tukey's test to determine which pairwise means differ using a familywise error rate of $\alpha = 0.05$.

18. Reaction Time In an online psychology experiment sponsored by the University of Mississippi, researchers asked study participants to respond to various stimuli. Participants were randomly assigned to one of three treatment groups: Group 1, the simple group, required to respond as quickly as possible after a stimulus was presented; Group 2, the go/no-go group, required to respond to a particular stimulus while disregarding other stimuli; and Group 3, the choice group, required to respond differently depending on the stimuli presented. The researcher felt that age may be a factor in determining reaction time, so she organized the experimental units by age and obtained the following data:

Age (years)	Stimulus		
	Simple	**Go/No-Go**	**Choice**
18–24	0.248	0.338	0.586
	0.428	0.631	0.364
	0.191	0.485	0.626
25–34	0.303	0.389	0.858
	0.467	0.629	0.529
	0.494	0.585	0.520
35 and older	0.384	0.782	0.854
	0.567	0.529	0.509
	0.302	0.495	0.700

Source: PsychExperiments; University of Mississippi; psychexps.olemiss.edu/

(a) What type of factorial design is this? How many replications are there within each cell?

(b) Normal probability plots indicate that it is reasonable to believe that the data come from populations that are normally distributed. Verify the requirement of equal population variances.

(c) Determine if there is significant interaction between stimulus and age.

(d) If there is no significant interaction, determine whether there is significant difference in the means for the three types of stimulus. If there is no significant interaction, determine whether there is significant difference in the means for the three categories of age.

(e) The residuals are normally distributed. Verify this.

(f) Draw an interaction plot of the data to support the results of parts (c) and (d).

19. Concrete Strength An engineer wants to know if the mean strengths of three different concrete mix designs differ significantly. He also suspects that slump may be a predictor of concrete strength. Slump is a measure of the uniformity of the concrete, with a higher slump indicating a less uniform mixture. The following data represent the 28-day strength (in pounds per square inch) of three different mixtures with three different slumps.

Slump	Mixture 67-0-301	Mixture 67-0-400	Mixture 67-0-353
3.75	3960	4815	4595
	4005	4595	4145
	3445	4185	4585
4	4010	4070	3855
	3415	4545	3675
	3710	4175	4010
5	3290	4020	3875
	3390	4355	3700
	3740	3935	3350

(a) Normal probability plots indicate that it is reasonable to believe that the data come from populations that are normally distributed. Verify the requirement of equal population variances.

(b) Determine whether there is significant interaction between mixture type and slump.

(c) If there is no significant interaction, determine whether there is significant difference in the means for the three types of mixture. If there is no significant interaction, determine whether there is significant difference in the means for the slumps.

(d) Draw an interaction plot of the data to support the results of parts (b) and (c).

(e) The residuals are normally distributed. Verify this.

(f) If there is significant difference in the means for the three mixture types, use Tukey's test to determine which pairwise means differ using a familywise error rate of $\alpha = 0.05$. If there is significant difference in the means for the slumps, use Tukey's test to determine which pairwise means differ using a familywise error rate of $\alpha = 0.05$.

20. Diet and Birth Weight An obstetrician wanted to determine the impact that three experimental diets had on the birth weights of pregnant mothers. She randomly selected 27 pregnant mothers in the first trimester of whom 9 were 20 to 29 years old, 9 were 30 to 39 years old, and 9 were 40 or older. For each age group, she randomly assigned the mothers to one of the three diets. After delivery, she measured the birth weight (in grams) of the babies and obtained the following data:

Age (years)	Diet		
	Diet 1	**Diet 2**	**Diet 3**
	4473	3961	3667
20–29	3878	3557	3139
	3936	3321	3356
	3886	3330	2762
30–39	4147	3644	3551
	3693	2811	3272
	3878	2937	2781
40 or older	4002	3228	3138
	3382	2732	3435

(a) Birth weights are known to be approximately normally distributed. Verify the requirement of equal population variances.

(b) Determine whether there is significant interaction between age and diet.

(c) If there is no significant interaction, determine whether there is significant difference in the means for the three

age groups. If there is no significant interaction, determine whether there is significant difference in the means for the diets.

(d) Draw an interaction plot of the data to support the results of parts (b) and (c).

(e) If there is significant difference in the means for the three age groups, use Tukey's test to determine which pairwise means differ using a familywise error rate of $\alpha = 0.05$. If there is significant difference in the means for the diets, use Tukey's test to determine which pairwise means differ using a familywise error rate of $\alpha = 0.05$.

21. Oil Changes The following data represent the cost of an oil change (in dollars) in three different geographic regions for two types of service centers. A specialty chain is an oil change facility that specializes in oil changes, while a general service station provides a wide array of services in addition to oil changes.

Service Center	Location		
	Chicago	**Bolingbrook**	**Peoria**
	19.95	23.99	24.99
Specialty chain	27.95	29.95	26.99
	23.99	28.99	19.95
	21.99	22.45	22.99
General service	26.95	29.95	24.95
	24.95	28.13	27.99

Source: Anna Paris, student at Joliet Junior College

(a) The prices of oil changes are approximately normally distributed. Verify the requirement of equal population variances.

(b) Determine if there is significant interaction between location and service center type.

(c) If there is no significant interaction, determine whether there is significant difference in the means for the three locations. If there is no significant interaction, determine whether there is significant difference in the means for the two service center types.

(d) Draw an interaction plot of the data to support the results of parts (b) and (c).

(e) If there is significant difference in the means for the three locations, use Tukey's test to determine which pairwise means differ using a familywise error rate of $\alpha = 0.05$. If there is significant difference in the means for the service center type, use Tukey's test to determine which pairwise means differ using a familywise error rate of $\alpha = 0.05$.

22. Putting It Together: Students at Ease Do gender and seating arrangement in college classrooms affect student attitude? In a study at a large public university in the United States, researchers surveyed students to measure their level of feeling at ease in the classroom. Participants were shown different classroom layouts and asked questions regarding their attitude toward each layout. The following data represent feeling-at-ease scores for a random sample of 32 students (four students for each possible treatment).

Gender		Classroom Layout							
		Tablet-Arm Chairs		**U-Shaped**		**Clusters**		**Tables with Chairs**	
	Female	19.8	18.4	19.2	19.2	18.1	17.5	17.3	17.1
		18.1	18.5	18.6	18.7	17.8	18.3	17.7	17.6
	Male	18.8	18.2	20.6	19.2	18.4	17.7	17.7	16.9
		18.9	18.9	19.8	19.7	17.1	18.2	17.8	17.5

Source: Brigitte Burgess and Naz Kaya. "Gender Differences in Student Attitude for Seating Layout in College Classrooms," *College Student Journal* 41(4), December 2007.

(a) What is the population of interest?

(b) Is this study an experiment or an observational study? Which type?

(c) What are the response and explanatory variables? Identify each as qualitative or quantitative.

(d) Compute the mean and standard deviation for the scores in the male/U-shaped cell.

(e) Assuming that feeling-at-ease scores for males on the U-shaped layout are normally distributed with $\mu = 19.1$ and $\sigma = 0.8$, what is the probability that you would observe a sample mean as large or larger than actually observed? Would this be unusual?

(f) Determine whether the mean feeling-at-ease score is different for males than females using a two-sample t-test for independent samples. Use the $\alpha = 0.05$ level of significance.

(g) Determine whether the mean feeling-at-ease scores for the classroom layouts are different using one-way ANOVA. Use the $\alpha = 0.05$ level of significance.

(h) Determine if there is an interaction effect between the two factors. If not, determine if either main effect is significant.

(i) Draw an interaction plot of the data. Does the plot support your conclusions in part (h)?

(j) In the original study, the researchers sent out e-mails to a random sample of 100 professors at the university asking permission to survey students in their class. Only 32 respondents agreed to allow their students to be surveyed. What type of nonsampling error is this? How might this affect the results of the study?

Explaining the Concepts

23. Explain the differences among the completely randomized design, randomized complete block design, and factorial design.

24. Explain what an interaction effect is. Why is it dangerous to analyze main effects if there is an interaction effect?

25. What is an interaction plot? Why are they useful?

Chapter 13 Review

Summary

We began the chapter with a discussion of one-way analysis of variance (ANOVA). One-way ANOVA is used to compare k means for equality when there is a single factor that is manipulated at k levels. To perform any ANOVA test, the response variable must come from a population that is normally distributed. The samples must be obtained independently, and the largest sample standard deviation can be no more than two times the smallest standard deviation. The ANOVA procedures are robust, so minor departures from the normality requirement do not seriously affect the results.

If the null hypothesis, $H_0: \mu_1 = \mu_2 = \cdots = \mu_k$, is rejected in a one-way ANOVA, at least one of the population means is different from the others. To determine which means differ, we use Tukey's test, which compares each pair of sample means for significant differences. That is, it tests $H_0: \mu_i = \mu_j$ versus $H_1: \mu_i \neq \mu_j$ for $i \neq j$.

A different ANOVA procedure is used for analyzing experiments designed as randomized complete block designs. Blocking describes a way of organizing experimental units according to some common characteristic for which the response variable is expected

to be similar, such as gender. By blocking, we reduce experimental error.

When data are analyzed from a randomized complete block design, we are not interested in whether the block is significant or not. Once again, if the null hypothesis of equal population means is rejected, we can use Tukey's test to compare each pair of the means.

Finally, we analyzed data in which there are two factors. Factor A can be set at a levels and factor B can be set at b levels. When there are two factors, use a two-way ANOVA procedure to test $H_0: \mu_1 = \mu_2 = \cdots = \mu_k$ versus H_1: at least one of the means differs. When performing a two-way ANOVA, first look for a significant interaction effect. This means the changes in factor A result in different changes in the value of the response variable for different levels of factor B. If interaction exists, do not consider main effects.

If no interaction exists, look for significant main effects. An example of a main effect would be to see how changes in the levels of factor B affect changes in the response variable. If the main effects indicate that at least one of the means differs, use Tukey's test to determine which means differ.

Vocabulary

Analysis of variance (p. 618)
Robust (p. 620)
Between-sample variability (p. 622)
Within-sample variability (p. 622)
F-test statistic (p. 622)
Mean square (p. 623)
Sum of squares (p. 623)
Total sum of squares (p. 623)
Mean square due to error (p. 623)
Sum of squares due to error (p. 623)

Mean square due to treatment (p. 623)
Sum of squares due to treatment (p. 623)
ANOVA table (p. 625)
Critical F-value (p. 627)
Multiple comparison methods (p. 635)
Studentized range distribution (p. 636)
Experimentwise error rate (p. 636)
Familywise error rate (p. 636)

Comparison error rate (p. 642)
Randomized complete block design (p. 644)
Latin square design (p. 645)
2×2 factorial design (p. 653)
Cell (p. 654)
Crossed (p. 654)
Main effects (p. 654)
Interaction effect (p. 654)
Interaction plots (p. 658)

Formulas

Test Statistic for One-Way ANOVA

$$F_0 = \frac{\text{mean square due to treatment}}{\text{mean square due to error}} = \frac{\text{MST}}{\text{MSE}}, \text{ where}$$

$$\text{MST} = \frac{\text{SST}}{k-1} = \frac{n_1(\bar{x}_1 - \bar{x})^2 + n_2(\bar{x}_2 - \bar{x})^2 + \cdots + n_k(\bar{x}_k - \bar{x})^2}{k-1}$$

$$\text{MSE} = \frac{\text{SSE}}{n-k} = \frac{(n_1 - 1)s_1^2 + (n_2 - 1)s_2^2 + \cdots + (n_k - 1)s_k^2}{n-k}$$

Test Statistic for Tukey's Test after One-Way ANOVA

$$q_0 = \frac{(\bar{x}_2 - \bar{x}_1) - (\mu_2 - \mu_1)}{\sqrt{\frac{s^2}{2} \cdot \left(\frac{1}{n_1} + \frac{1}{n_2}\right)}} = \frac{\bar{x}_2 - \bar{x}_1}{\sqrt{\frac{s^2}{2} \cdot \left(\frac{1}{n_1} + \frac{1}{n_2}\right)}}$$

OBJECTIVES

Section	You should be able to . . .	Example(s)	Review Exercises
13.1	1 Verify the requirements to perform a one-way ANOVA (p. 620)	1	3(d)
	2 Test a hypothesis regarding three or more means using one-way ANOVA (p. 621)	2 and 3	3(e), 5(b)
13.2	1 Perform the Tukey test (p. 635)	1–3	2, 3(h), 4, 5(c)
13.3	1 Conduct analysis of variance on the randomized complete block design (p. 645)	2	6(c)
	2 Perform the Tukey test (p. 648)	4	6(d)
13.4	1 Analyze a two-way ANOVA design (p. 653)	1–3 and 6	7(b), (c)
	2 Draw interaction plots (p. 658)	5 and 6	7(d)
	3 Perform the Tukey test (p. 661)	7	7(e)

Review Exercises

1. Find the critical value from the Studentized range distribution for:

   (a) $\alpha = 0.05$, $\nu = 16$, $k = 7$

   (b) $\alpha = 0.05$, $\nu = 30$, $k = 6$

   (c) $\alpha = 0.01$, $\nu = 42$, $k = 4$

   Note: Use $q_{0.01,40,4}$ since $q_{0.01,42,4}$ is not in the table.

   (d) Find the critical value from the Studentized range distribution for $H_0: \mu_1 = \mu_2 = \mu_3 = \mu_4 = \mu_5 = \mu_6$, with $n = 46$, $\alpha = 0.05$.

2. Suppose that there is sufficient evidence to reject $H_0: \mu_1 = \mu_2 = \mu_3 = \mu_4$ using a one-way ANOVA. The mean square error from ANOVA is determined to be 373. The sample means are $\bar{x}_1 = 78$, $\bar{x}_2 = 64$, $\bar{x}_3 = 70$, $\bar{x}_4 = 47$, with $n_1 = n_2 = n_3 = n_4 = 31$. Use Tukey's test to determine which pairwise means are significantly different using a familywise error rate of $\alpha = 0.05$.

3. Soil Testing A researcher took water samples from a stream running through a forest and groundwater in the forest. The samples were collected over the course of a year. Each sample was analyzed for the concentration of dissolved organic carbon (mg/L) in it and categorized according to the type of water collected. Water was collected from streams (surface water), groundwater was collected from organic soil, and groundwater was collected from mineral soil. The researcher wanted to determine if the mean concentration of dissolved organic carbon was the same for each collection area. The results are presented in the following table.

Organic				Mineral					Surface				
22.74	14.90	17.90	11.40	8.50	5.50	3.02	7.31	4.85	10.83	11.94	15.76	15.03	15.77
29.80	14.86	18.30	5.30	3.91	4.71	7.45	16.92	11.97	8.74	12.40	10.96	14.53	6.40
27.10	15.91	5.20	15.72	9.29	7.66	11.33	4.60	7.85	9.20	10.30	19.78	12.04	14.19
16.51	15.35	11.90	20.46	21.00	11.72	7.11	8.50	9.11	8.12	10.48	20.56	16.82	13.89
6.51	9.72	14.00	16.87	10.89	11.80	17.99	4.80	8.79	7.60	12.88	17.93	10.70	8.69
8.81	19.80	7.40	15.42	10.30	8.05	21.40	4.90	9.60	6.30	19.01	14.28	16.00	10.46
5.29	14.86	17.50	22.49	11.56	10.72	8.37	9.10	12.57	6.68	19.19	13.11	20.70	16.23
20.46	8.09	10.30		7.00	21.82	7.92	7.90	12.89	7.34	13.14	12.27	13.70	15.43
				3.99	22.62	17.90	11.72	9.81	9.52	12.51	16.47	16.12	
				3.79	10.74								

Source: Lisa Emili, PhD Candidate, Department of Geography and Wetlands Research Centre, University of Waterloo

(a) What is the response variable in this study?
(b) What is the factor? How many levels of treatment does the factor have?
(c) State the null and alternative hypotheses.
(d) A normal probability plot of each sample (organic, mineral, and surface) indicates the data come from a normally distributed population. Use the following summary statistics from StatCrunch to verify the equal population standard deviation requirement.

Summary statistics:

Column	n	Mean	Variance	Std. Dev.	Std. Err.	Median	Range	Min	Max	Q1	Q3
organic	31	14.866775	39.01497	6.2461963	1.12185	15.35	24.6	5.2	29.8	9.72	18.3
mineral	47	10.026596	24.789036	4.978859	0.72624123	9.1	19.6	3.02	22.62	7.11	11.72
surface	44	13.045455	15.418835	3.9266825	0.59196967	12.995	14.4	6.3	20.7	10.38	15.885

(e) Use the following StatCrunch output to test if the means are the same at the $\alpha = 0.01$ level of significance.

ANOVA table

Source	df	SS	MS	F-Stat	P-value
Treatments	2	471.35263	235.67632	9.431001	0.0002
Error	119	2973.7546	24.989534		
Total	121	3445.1072			

(f) Interpret the P-value.
(g) Shown are side-by-side boxplots of each type of collection area drawn in StatCrunch. Do the boxplots support the analytic results? Which collection area type appears to differ from the rest?

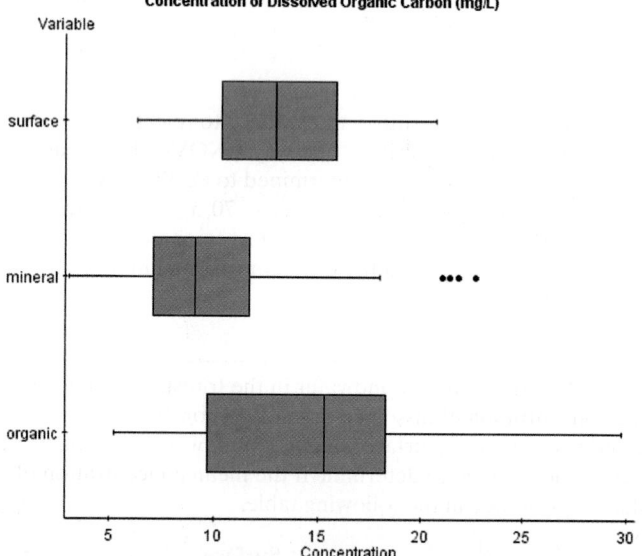

Concentration of Dissolved Organic Carbon (mg/L)

(h) The following display represents the results of Tukey's test obtained from StatCrunch. Which pairwise means differ?

Tukey 95% Simultaneous Confidence Intervals

organic subtracted from

	Lower	Upper
mineral	-7.5853386	-2.0950184
surface	-4.603425	0.9607859

mineral subtracted from

	Lower	Upper
surface	0.5300268	5.507691

4. Grading Timber Desiring to grade timber mechanically, engineers studied the modulus of rupture in lb/in.², for a sample of untreated green, 7 inch by 9 inch, mixed oak timbers, sorted by grade: select structural, No. 1, No. 2, and below grade. A one-way ANOVA was performed to test $H_0: \mu_{select} = \mu_{No.\,1} = \mu_{No.\,2} = \mu_{below\;grade}$. The null hypothesis was rejected with a P-value of 0.018. The engineers proceeded to conduct Tukey's test to determine which pairwise means differed using Minitab. The results are presented next. Which pairwise means differ? What do you conclude from the test?

Note: 1 = select structural, 2 = No. 1, 3 = No. 2, and 4 = below grade.

Source: "Mechanical Grading of Oak Timbers," *Journal of Materials in Civil Engineering*, 11(2).

```
Tukey's Pairwise Comparison

Family error rate = 0.0500
Individual error rate = 0.0108

Critical value = 3.86

Intervals for (column level mean) - (row level mean)

                     1              2              3

   2            -6.914
                13.064

   3             1.936         -1.139
                21.914         18.839

   4            -2.989         -6.064        -14.914
                16.989         13.914          5.064
```

5. Seating Choice versus GPA In a study of students at Lewis & Clark Community College, a professor compared the chosen seat location of students in mathematics courses to their overall grade point average (GPA). He randomly selected 10 students from each of three seat locations: (1) front, (2) middle, and (3) back. The GPAs follow:

Front	Middle	Back
2.690	3.080	2.598
3.523	2.937	2.879
3.332	3.091	2.926
3.885	2.655	3.221
3.559	2.526	2.646
3.062	2.859	2.583
3.894	2.639	2.653
2.966	3.634	3.090
3.575	3.564	3.060
4.000	2.115	2.463

(a) Explain why this data may be analyzed using a one-way ANOVA.

(b) Test the null hypothesis that the mean score for each location is the same at the $\alpha = 0.05$ level of significance.

(c) If the null hypothesis is rejected in part (a), use Tukey's test to determine which pairwise means differ using a familywise error rate of $\alpha = 0.05$.

(d) Draw boxplots of the three treatment levels to support the analytic results obtained in parts (b) and (c).

6. Catapults Four different catapult designs are being tested for their ability to launch water balloons. The experimenter decides to block by person firing. The distances (in feet) achieved for each catapult are given.

	Mark IV V2	Balloon Ballista	Hydrolaunch	Waterworks II
Peter	92	94	97	100
Shawn	94	93	98	99
Mark	92	94	95	98
Jake	97	96	101	103

(a) What is the response variable in this study?

(b) Explain why it makes sense to block by person firing the catapult.

(c) Test H_0: $\mu_1 = \mu_2 = \mu_3 = \mu_4$ against H_1: at least one of the means is different, where μ_1 is the mean for the Mark IV V2, μ_2 is the mean for the Balloon Ballista, and so on, at the $\alpha = 0.05$ level of significance.

(d) If the null hypothesis from part (c) was rejected, use Tukey's test to determine which pairwise means differ using a familywise error rate of $\alpha = 0.05$.

(e) Draw boxplots of the data for each catapult design using the same scale to support the analytical results obtained in parts (c) and (d).

7. Defensive Driving An investigator for the state police wants to determine the effectiveness of three different defensive driving training programs to see if there are gender differences. Five subjects of each gender who recently received speeding tickets are assigned to each program. At the end of the program, each is given a written test on his or her knowledge of defensive driving. The scores (out of 100) are given next.

Gender	Program		
	One 8-h Session	Two 4-h Sessions	Two 2-h Sessions
Male	89	95	77
	96	87	78
	95	90	83
	90	91	78
	96	92	78
Female	88	87	80
	92	92	82
	98	91	79
	99	94	86
	91	93	88

(a) The scores are approximately normally distributed. Verify the requirement of equal population variances.

(b) Determine whether there is significant interaction between program and gender.

(c) If there is no significant interaction, determine whether there is significant difference in the means for the three programs.

(d) Draw an interaction plot of the data to support the results of parts (b) and (c).

(e) If there is significant difference in the means for the three programs, use Tukey's test to determine which pairwise means differ using a familywise error rate of $\alpha = 0.05$.

Chapter Test

1. Many fast-food restaurants use automatic soft-drink dispensing machines that fill cups once an order is placed at a drive-through window. A regional manager for a fast-food chain wants to determine if the machines in his region are dispensing the same amount of product. He samples four different machines and measures a random sample of 16-ounce drinks from each machine.

Machine I	Machine II	Machine III	Machine IV
16.5	15.0	16.0	16.6
16.6	15.4	16.3	15.9
16.5	15.3	16.5	15.5
15.8	15.7	16.4	16.2
15.6	15.2	17.3	17.0
16.4	16.0	16.7	15.5
16.1	15.6	15.7	16.3

(a) Do the data indicate that the mean amount dispensed from the machines is not the same? Use an $\alpha = 0.05$ level of significance.

(b) Use Tukey's test, if necessary, to determine which pairwise means differ using a familywise error rate of $\alpha = 0.05$.

(c) Draw boxplots of the four treatment levels to support your conclusions in parts (a) and (b).

2. Lighting Effect on Plant Growth Which type of light results in the tallest rosemary plants after 38 days of growth? To answer this question, Joliet Junior College student Michael Bueno planted 90 rosemary seeds in plant pods. He randomly assigned each seed to a green, blue, or red light bulb. All the seeds were planted in the same type of soil, received the same amount of water and same amount of light each day, and were all located in a room maintained at 70 degrees Fahrenheit.

The following data represent the heights (in inches) of the plants after 38 days.

Blue			Red			Green		
0.3	1.0	1.6	4.0	2.6	2.2	0.3	0.8	0.9
1.1	3.0	1.5	3.1	2.8	4.2	1.3	0.7	0.2
1.2	1.3	2.1	2.4	2.0	1.7	0.2	1.0	0.5
0.6	1.0	0.4	3.3	3.9	2.3	0.2	0.9	1.6
1.8	0.7	1.2	1.0	3.6	2.7	1.9	0.4	0.5
2.0	1.3	0.4	4.2	3.1	3.9	0.2	0.8	1.1
2.2	0.7	1.9	2.5	3.0	2.7	0.4	0.7	1.3
2.6	1.2	0.3	3.2	3.7	4.0	0.9	0.6	1.3
1.0	3.0	2.1	4.4	1.3	2.8	1.6	1.2	1.8
1.4	1.1	0.8	1.6	4.6	2.3	0.3	0.7	0.5

Source: Michael Bueno, student at Joliet Junior College

(a) What type of experimental design did Michael perform?

(b) What is the response variable in his experiment? Is it quantitative or qualitative?

(c) What is the explanatory variable in his experiment? Is it quantitative or qualitative?

(d) What other factors did Michael identify that may affect plant growth? How did he handle these factors?

(e) What technique was used to address the fact that some seeds may grow at different rates? In other words, just as not all humans grow to the same height, all rosemary seeds will not grow to the same height (due to genetics). What did Michael do to address this concern?

(f) Michael wanted to know whether the color of the light played a role in determining plant height after 38 days. State the null and alternative hypotheses.

(g) Following are the descriptive statistics for each treatment group. Is the requirement of equal population variances satisfied? Why?

Descriptive Statistics: Blue, Red, Green

```
Variable  N  N*  Mean   SE Mean  StDev  Minimum   Q1     Median  Q3     Maximum
Blue      30  0  1.360  0.136    0.746  0.300   0.775   1.200  1.925  3.000
Red       30  0  2.970  0.173    0.950  1.000   2.300   2.900  3.900  4.600
Green     30  0  0.860  0.109    0.595  0.200   0.400   0.750  1.225  2.900
```

(h) Use the results of a one-way ANOVA from Minitab given next to test the hypotheses stated in part (f).

One-way ANOVA: Blue, Red, Green

```
Source  DF    SS       MS      F      P
Factor   2   74.704   37.352  65.68  0.000
Error   87   49.474    0.569
Total   89  124.178
```

(i) In the next column are side-by-side boxplots of each plant's growth for each light color. Based on the boxplots and the results of part (h), do you believe a certain light color promotes plant growth more than the others?

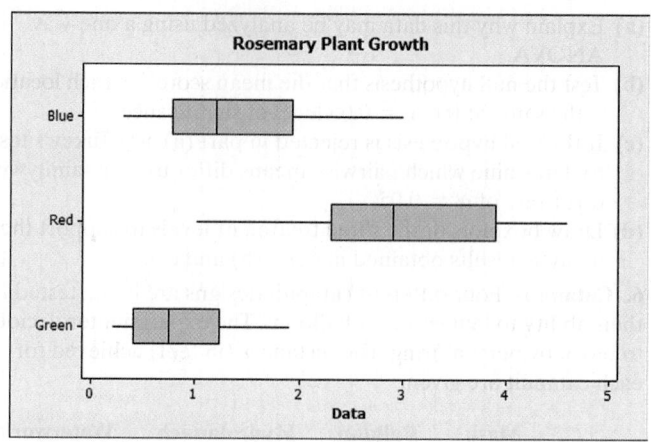

(j) Following are the results of Tukey's test. Determine which pairwise means may differ using a familywise error rate of $\alpha = 0.05$.

```
Tukey 95% Simultaneous Confidence Intervals
All Pairwise Comparisons

Individual confidence level = 98.06%


Blue subtracted from:

         Lower    Center   Upper    -------+---------+---------+---------+--
Red     1.1460   1.6100   2.0740                                (--*--)
Green  -0.9973  -0.5333  -0.0694                   (--*---)
                                    -------+---------+---------+---------+--
                                        -1.5       0.0       1.5       3.0


Red subtracted from:

         Lower    Center   Upper    -------+---------+---------+---------+--
Green  -2.6073  -2.1433  -1.6794    (--*--)
                                    -------+---------+---------+---------+--
                                        -1.5       0.0       1.5       3.0
```

3. A rock climber wants to test the tensile strength of three different types of braided ropes. She knows that the rope thickness can make a difference, so she decides to block by thickness. The tensile strengths for each rope type are given next.

	Nylon Braided	Polyester Braided	Polypropylene Braided
$\frac{1}{4}$ inch	2500	1550	1325
$\frac{3}{8}$ inch	4900	3420	2575
$\frac{1}{2}$ inch	8400	5940	4170
$\frac{5}{8}$ inch	12300	9500	6315

(a) Test $H_0: \mu_1 = \mu_2 = \mu_3$ against H_1: at least one of the means is different, where μ_1 is the mean for the nylon braided rope, μ_2 is the mean for the polyester braided rope, and μ_3 is the mean for the polypropylene braided rope, at the $\alpha = 0.05$ level of significance.

(b) If the null hypothesis from part (a) was rejected, use Tukey's test to determine which pairwise means differ using a familywise error rate of $\alpha = 0.05$.

(c) Draw boxplots of the data for each treatment using the same scale to support the analytical results obtained in parts (a) and (b).

4. A manufacturing researcher wants to determine if age or gender significantly affects the time required to learn an assembly line task. He randomly selected 24 adults aged 20 to 64 years old, of whom 8 are 20 to 34 years old (4 male, 4 female), 8 are 35 to 49 years old (4 male, 4 female), and 8 are 50 to 64 years old (4 male, 4 female). He then measured the time (in minutes) required to complete a certain task. The data obtained are shown next.

Gender	Age (years)		
	20–34	**35–49**	**50–64**
Male	5.2, 5.1	4.8, 5.8	5.2, 4.3
	5.7, 6.1	5.0, 4.8	5.5, 4.7
Female	5.3, 5.5	5.0, 5.4	4.9, 5.5
	4.9, 5.6	5.6, 5.1	5.5, 5.0

(a) The learning times are approximately normally distributed. Verify the requirement of equal population variances.

(b) Determine whether there is significant interaction between age and gender.

(c) If there is no significant interaction, determine whether there is significant difference in the means for the three age groups.

(d) Draw an interaction plot of the data to support the results of parts (b) and (c).

(e) If there is significant difference in the means for the three age groups, use Tukey's test to determine which pairwise means differ using a familywise error rate of $\alpha = 0.05$.

5. The following table summarizes the sample mean earnings, in thousands of dollars, of a random sample of year-round, full-time workers by gender and age.

	25–34	**35–44**	**45–54**
Male	42.9	59.5	67
Female	35.2	41.9	43.1

(a) Construct an interaction plot for the data.

(b) Does the interaction plot indicate a significant interaction effect? Explain?

(c) What, if anything, can you conclude about the significance of the main effects?

Making an Informed Decision

Where Should I Invest?

Remember that bonus you received in Chapter 10? You have now decided that you want to try investing in individual stocks rather than mutual funds. You are told by a friend who is a finance major that there are three broad stock types: large-capitalization stocks, midcapitalization stocks, and small-capitalization stocks. Large-capitalization stocks are companies that have a market value in excess of $10 billion. Small-capitalization stocks are companies with a market value below $2 billion. Midcaps fall in between. Your finance friend also tells you that there are two investment styles: value and growth. Value stocks are companies thought to be undervalued relative to their peers, while growth stocks are companies thought to have a higher growth rate than their peers.

(a) Obtain a random sample of five stocks within each category. Determine the 1- and 5-year rates of return for each stock.

(b) Verify that the rates of return are normally distributed. Also verify that the largest standard deviation is less

than twice the smallest standard deviation. If any requirements are not satisfied, increase the sample size until the requirements become satisfied.

(c) Perform a one-way ANOVA with stock type as the factor. Is there any difference in the mean rate of return among the three stock types? If so, perform a Tukey test to determine which mean rates of return differ.

(d) Perform a two-way ANOVA with stock type and investment style as the factors. Is there any interaction between stock type and investment style? If not, do the means of either factor differ? If the means do differ, perform a Tukey test to determine which mean rates of return differ.

(e) Write a report that details which investment category seems to be best.

In the mid-19th century, Paul Broca, a professor of clinical surgery, effectively argued that the degree of a person's intelligence was directly related to the size of the brain.

Arguments regarding brain size diminished in the 20th century as scientists turned to intelligence tests—a more direct method to measure and compare mental capacities, they claimed. However, in a 1978 book designed to acquaint educators with modern brain research, H. T. Epstein declared, "First we shall ask if there is any indication of a linkage of any kind between brain and intelligence. It is generally stated that there is no such linkage. . . . But the one set of data I have found seems to show clearly that there is a substantial connection. Hooton studied the head circumferences of white Bostonians as part of his massive study of criminals. The following table shows that the ordering of people according to head size yields an entirely plausible ordering according to vocational status. It is not at all clear how the impression has been spread that there is no such correlation." *Source:* Excerpt from The Mismeasure of Man by Stephen Jay Gould. Published by W. W. Norton & Company, © 1981, p. 109. Epstein's table is reproduced here:

Head Circumference by Vocational Status			
Vocational Status	***n***	**Mean (mm)**	**SD (mm)**
Professional	25	569.9	1.9
Semiprofessional	61	566.5	1.5
Clerical	107	566.2	1.1
Trades	194	565.7	0.8
Public service	25	564.1	2.5
Skilled trades	351	562.9	0.6
Personal services	262	562.7	0.7
Laborers	647	560.7	0.3

Source: Stephen Jay Gould, *The Mismeasure of Man.* W.W. Norton & Company. New York, 1981, p. 109

Initially, this table seems to support the assertion that those with more prestigious vocations have a larger head circumference. However, we can examine the alleged relationship more closely through a one-way ANOVA. Specifically, we test the claim that all the head-circumference population means are equal. We use an $\alpha = 0.05$ significance level and obtain the following ANOVA table, derived from Epstein's statistics:

Source of Variation	Sum of Squares	Degrees of Freedom	Mean Squares	*F*-Test Statistic
Treatment	7865.359	7	1123.6284	1998.6275
Error	935.450	1664	0.5622	
Total	8800.809	1671		

From these results, does it appear that at least one of the head-circumference means is different from the others? Explain.

Examine Epstein's table closely. Does it appear that the ANOVA procedure's requirement of equal variances is satisfied? Explain.

Subsequently, there was an error in Epstein's table. The column labeled SD (standard deviation) should have been labeled *standard error*. For each vocational category, calculate the correct standard deviation by multiplying the tabled value by the square root of its sample size, *n*. If the ANOVA table was recalculated with the correct standard deviations, predict the effect on the *F*-test statistic. Would you be more or less likely to reject the null hypothesis? Explain.

Using the formula for MSE, calculate a revised sum of squared errors, mean square due to error, and *F*-test statistic. Retest the claim that all the head-circumference population means are equal. Use an $\alpha = 0.05$ significance level. From the corrected values for the standard deviation, does it appear that the ANOVA procedure's requirement of equal variances is satisfied? Did any of your conclusions change in light of the new information? Explain.

If the null hypothesis of equal population means was rejected, use Tukey's test to determine which of the pairwise means differ using a familywise error rate of $\alpha = 0.05$. Which of these categories appear to have significantly different means?

Further research revealed that results for three vocational status groups were not presented in Epstein's table: factory workers (rank 7 out of 11 in status), transportation employees (rank 8), and extractive trades (farming and mining, rank 11). These mean head-circumference sizes were 564.7, 564.9, and 564.7 mm, respectively. Also, Epstein's table did not list each group in the order of Hooton's prestige ranking but, rather, in order of mean head circumference, suggesting a perfect correlation where one does not exist. (*Source:* Stephen Jay Gould, *The Mismeasure of Man.* W. W. Norton & Company, New York, 1981, p. 110) Using this new information and your previous analyses, write a brief report summarizing your findings and conclusions.

14

Inference on the Least-Squares Regression Model and Multiple Regression

Making an Informed Decision

The website zillow.com provides information regarding homes for sale and homes that have recently sold for a given region. One piece of information supplied by Zillow is the Zestimate of a home. The Zestimate represents the predicted selling price of a home according to the folks at Zillow. Least-squares regression may be used to develop a model that uses Zillow's Zestimate to predict the actual selling price of a home. See the Decision project on page 741.

PUTTING IT TOGETHER

In Chapter 4, we learned methods for describing the relation between bivariate quantitative data. We also learned to perform diagnostic tests, such as determining whether a linear model is appropriate, identifying outliers, and identifying influential observations.

In this chapter, we begin by extending hypothesis testing and confidence intervals to least-squares regression models. In Section 14.1, we test whether a linear relation exists between two quantitative variables using methods based on those presented in Chapter 10. In Section 14.2, we construct confidence intervals about the predicted value of the response variable.

Finally, in Sections 14.3 through 14.6, we introduce *multiple regression* in which more than one explanatory variable is used to predict the value of a response variable. We perform the same diagnostic tests for multiple regression that we used for *simple linear regression* (one explanatory variable). We also test whether the relation between the explanatory variables and response variable is significant, and we construct confidence intervals about the predicted value of the response variable. Throughout this section, we let statistical software do the "heavy lifting" so we are not bogged down by computations. This approach will allow us to concentrate on analyzing our results rather than calculating.

14.1 Testing the Significance of the Least-Squares Regression Model

Preparing for This Section Before getting started, review the following:

- Scatter diagrams; correlation (Section 4.1, pp. 181–189)
- Least-squares regression (Section 4.2, pp. 197–204)
- Diagnostics on the least-squares regression line (Section 4.3, pp. 211–219)

- Sampling distribution of the sample mean $\bar{x}$ (Section 8.1, pp. 395–403)
- Testing a hypothesis about μ (Section 10.3, pp. 494–498)
- Confidence intervals about a mean (Section 9.2, pp. 435–441)

Objectives

1. State the requirements of the least-squares regression model
2. Compute the standard error of the estimate
3. Verify that the residuals are normally distributed
4. Conduct inference on the slope of the least-squares regression model
5. Construct a confidence interval about the slope of the least-squares regression model

As a quick review of the topics discussed in Chapter 4, we present the following example:

EXAMPLE 1   Least-Squares Regression

Problem A family doctor is interested in examining the relationship between a patient's age and total cholesterol (in mg/dL). He randomly selects 14 of his female patients and obtains the data presented in Table 1. The data are based on results obtained from the National Center for Health Statistics. Draw a scatter diagram, compute the correlation coefficient, find the least-squares regression equation, and determine the coefficient of determination.

Figure 1

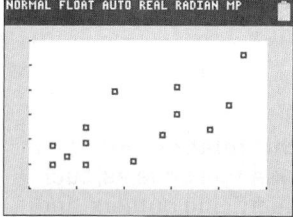

Figure 2

Table 1			
Age, x	**Total Cholesterol, y**	**Age, x**	**Total Cholesterol, y**
25	180	42	183
25	195	48	204
28	186	51	221
32	180	51	243
32	210	58	208
32	197	62	228
38	239	65	269

Approach We will use a TI-84 Plus C graphing calculator.

Solution Figure 1 displays the scatter diagram. Figure 2 displays the output. The linear correlation coefficient is 0.718. The least-squares regression equation for these data is $\hat{y} = 1.3991x + 151.3537$, where $\hat{y}$ represents the predicted total cholesterol for a female whose age is x.

The coefficient of determination, R^2, is 0.515, so 51.5% of the variation in total cholesterol is explained by the regression line. Figure 3 shows a graph of the least-squares regression equation on the scatter diagram.   •

Figure 3

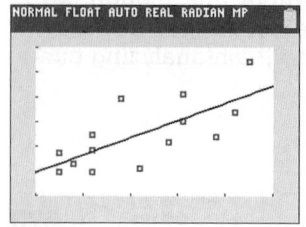

The information obtained in Example 1 is descriptive in nature. Notice that the descriptions are both graphical (as in the scatter diagram) and numerical (as in the correlation coefficient, the least-squares regression equation, and the coefficient of determination).

① State the Requirements of the Least-Squares Regression Model

In the least-squares regression equation $\hat{y} = b_1 x + b_0$, the values for the slope, b_1, and intercept, b_0, are statistics, just as the sample mean, $\bar{x}$, and sample standard deviation, s, are statistics. The statistics b_0 and b_1 are estimates for the population intercept, β_0, and the population slope, β_1. The true linear relation between the explanatory variable, x, and the response variable, y, is given by $y = \beta_1 x + \beta_0$.

In Other Words
Because b_0 and b_1 are statistics, they have sampling distributions.

Because b_0 and b_1 are statistics, their values vary from sample to sample, so a sampling distribution is associated with each. We use this sampling distribution to perform inference on b_0 and b_1. For example, we might want to test whether β_1 is different from 0. If we have sufficient evidence to this effect, we conclude that there is a linear relation between the explanatory variable, x, and response variable, y.

To find the sampling distributions of b_0 and b_1, we have some requirements of the population from which the bivariate data (x_i, y_i) were sampled. Just as in Section 8.1 when we discussed the sampling distribution of $\bar{x}$, we start by asking what would happen if we took many samples for a given value of the explanatory variable, x. For example, in Table 1 notice that our sample included three women aged 32 years with different corresponding values of y: 180 mg/dL, 210 mg/dL, and 197 mg/dL. This means that y varies for a given value of x, so there is a distribution of total cholesterol levels for $x = 32$ years. If we looked at *all* women aged 32 we could find the population mean total cholesterol for *all* 32-year-old women, denoted $\mu_{y|32}$. The notation $\mu_{y|32}$ is read "the mean value of the response variable y given that the value of the explanatory variable is 32." We could repeat this process for any other age. In general, different ages have a different population mean total cholesterol. This brings us to our first requirement regarding inference on the least-squares regression model.

Requirement 1 for Inference on the Least-Squares Regression Model

For any particular value of the explanatory variable x (such as 32 in Example 1), the mean of the corresponding responses in the population depends linearly on x. That is,

$$\mu_{y|x} = \beta_1 x + \beta_0$$

for some numbers β_0 and β_1, where $\mu_{y|x}$ represents the population mean response when the value of the explanatory variable is x.

In Other Words
When doing inference on the least-squares regression model, we require (1) for any explanatory variable, x, the mean of the response variable, y, depends on the value of x through a linear equation, and (2) the response variable, y, is normally distributed with a constant standard deviation, σ. The mean increases/decreases at a constant rate depending on the slope, while the standard deviation remains constant.

We learned how to check this requirement in Section 4.3. If a plot of the residuals against the explanatory variable shows any discernible pattern (such as a U-shape), then the linear model is not appropriate and requirement 1 is violated.

We also have a requirement regarding the distribution of the response variable for any particular value of the explanatory variable.

Requirement 2 for Inference on the Least-Squares Regression Model

The response variable is normally distributed with mean $\mu_{y|x} = \beta_1 x + \beta_0$ and standard deviation σ.

The second requirement states that the mean of the response variable changes linearly, but the standard deviation remains constant, and the distribution of the response variable is normal. For example, a sample of the total cholesterol of many 32-year-old females would be normal with mean $\mu_{y|32} = \beta_1(32) + \beta_0$ and standard deviation σ. For 43-year-old females, the distribution would be normal with mean $\mu_{y|43} = \beta_1(43) + \beta_0$ and standard deviation σ. See Figure 4.

Figure 4

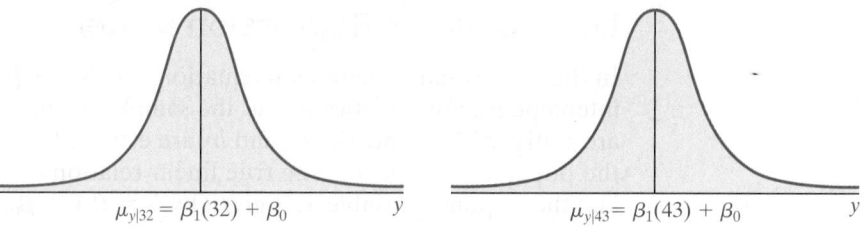

$$\mu_{y|32} = \beta_1(32) + \beta_0 \qquad y \qquad\qquad \mu_{y|43} = \beta_1(43) + \beta_0 \qquad y$$

In Other Words

The larger σ is, the more spread out the data are around the regression line.

A large value of σ, the population standard deviation, indicates that the data are widely dispersed about the regression line, and a small value of σ indicates that the data lie fairly close to the regression line. Figure 5 illustrates these ideas. The regression line represents the mean value of each normal distribution at a specified value of x. The standard deviation of each distribution is σ.

Figure 5

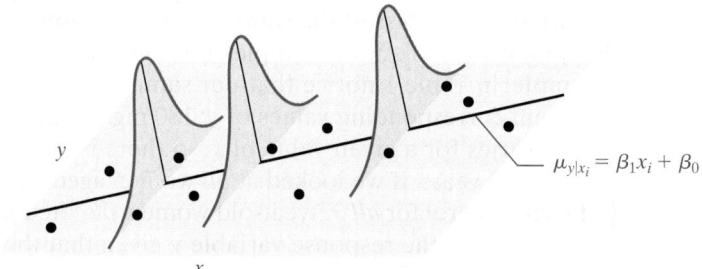

$$\mu_{y|x_i} = \beta_1 x_i + \beta_0$$

Of course, not all the observed values of the response variable lie on the true regression line $\mu_{y|x} = \beta_1 x + \beta_0$. The difference between the observed and predicted value of the response variable is an error term or residual, ε_i. We now present the least-squares regression model.

Definition

The **least-squares regression model** is given by

$$y_i = \beta_1 x_i + \beta_0 + \varepsilon_i \tag{1}$$

where

y_i is the value of the response variable for the ith individual
x_i is the value of the explanatory variable for the ith individual
β_0 and β_1 are the parameters to be estimated based on sample data
ε_i is a random error term with mean 0 and standard deviation $\sigma_{\varepsilon_i} = \sigma$, the error terms are independent
$i = 1, \ldots, n$, where n is the sample size (number of ordered pairs in the data set)

Because the expected value, or mean, of y_i is $\beta_1 x_i + \beta_0$ and the expression on the left side of Equation (1) equals the expression on the right side, the expected value, or mean, of the error term, ε_i, is 0.

• **Now Work Problem 13(a)**

② Compute the Standard Error of the Estimate

In Section 4.2, we learned how to estimate β_0 and β_1. We now present the method for obtaining the estimate of σ, the standard deviation of the response variable y for any given value of x. The unbiased estimator of σ is called the *standard error of the estimate*.

Recall the formula for the sample standard deviation from Section 3.2:

$$s = \sqrt{\frac{\sum(x_i - \bar{x})^2}{n-1}}$$

We compute the deviations about the mean, square them, add the squared deviations, divide by $n - 1$, and take the square root of the result. We divide by $n - 1$ because we lose 1 degree of freedom since one parameter, $\bar{x}$, is estimated. The same logic is used to compute the standard error of the estimate.

As we mentioned, the predicted values of y, denoted $\hat{y}_i$, represent the mean value of the response variable for any given value of the explanatory variable, x_i. So $y_i - \hat{y}_i$ = residual represents the difference between the observed value, y_i, and the mean value, $\hat{y}_i$. This calculation is used to compute the standard error of the estimate.

Definition   The **standard error of the estimate**, s_e, is found using the formula

$$s_e = \sqrt{\frac{\sum (y_i - \hat{y}_i)^2}{n - 2}} = \sqrt{\frac{\sum \text{residuals}^2}{n - 2}} \tag{2}$$

We divide by $n - 2$ because in a least-squares regression we have estimated two parameters, β_0 and β_1. That is, we lose 2 degrees of freedom.

EXAMPLE 2   Computing the Standard Error by Hand

Problem Compute the standard error of the estimate for the data in Table 1 on page 674.

Approach Use the following steps to compute the standard error of the estimate.

Step 1 Find the least-squares regression line.

Step 2 Obtain predicted values for each observation in the data set.

Step 3 Compute the residuals for each observation in the data set.

Step 4 Compute $\sum \text{residuals}^2$.

Step 5 Compute the standard error of the estimate using Formula (2).

Solution

Step 1 In Example 1, the least-squares regression was $\hat{y} = 1.3991x + 151.3537$.

Step 2 Column 3 of Table 2 shows the predicted values for the $n = 14$ observations.

Step 3 Column 4 of Table 2 shows the residuals for the 14 observations.

Table 2

Age, x	Total Cholesterol, y	$\hat{y} = 1.3991\,x + 151.3537$	Residuals, $y - \hat{y}$	Residuals^2, $(y - \hat{y})^2$
25	180	186.33	−6.33	40.0689
25	195	186.33	8.67	75.1689
28	186	190.53	−4.53	20.5209
32	180	196.12	−16.12	259.8544
32	210	196.12	13.88	192.6544
32	197	196.12	0.88	0.7744
38	239	204.52	34.48	1188.8704
42	183	210.12	−27.12	735.4944
48	204	218.51	−14.51	210.5401
51	221	222.71	−1.71	2.9241
51	243	222.71	20.29	411.6841
58	208	232.50	−24.50	600.2500
62	228	238.10	−10.10	102.0100
65	269	242.30	26.70	712.8900

$\sum \text{residuals}^2 = 4553.705$

(continued)

Step 4 Sum the squared residuals in column 5 to find the sum of squared errors:

$$\Sigma \text{ residuals}^2 = 4553.705$$

> **CAUTION!**
> Be sure to divide by $n - 2$ when computing the standard error of the estimate.

Step 5 Use Formula (2) to compute the standard error of the estimate.

$$s_e = \sqrt{\frac{\Sigma \text{ residuals}^2}{n - 2}} = \sqrt{\frac{4553.705}{14 - 2}} = 19.48$$

EXAMPLE 3 Obtaining the Standard Error of the Estimate Using Technology

Figure 6

Regression Statistics	
Multiple R	0.7178106
R square	0.5152521
Adjusted R square	0.4748564
Standard error	19.480535
Observations	14

Problem Obtain the standard error of the estimate for the data in Table 1 using statistical software.

Approach Use Excel to obtain the standard error. The steps for obtaining the standard error of the estimate using TI-83/84 Plus graphing calculators, Minitab, Excel, and StatCrunch are given in the Technology Step-by-Step on pages 683–684.

Solution Figure 6 shows the partial output from Excel. The results agree with the by-hand computation.

 Now Work Problem 13(b)

③ Verify That the Residuals Are Normally Distributed

> **CAUTION!**
> The residuals must be normally distributed to perform inference on the least-squares regression line.

The least-squares regression model $y_i = \beta_1 x_i + \beta_0 + \varepsilon_i$ requires the response variable, y_i, to be normally distributed. Because $\beta_1 x_i + \beta_0$ is constant for any x_i, if y_i is normal, then the residuals, ε_i, must be normal. To perform statistical inference on the regression line, we verify that the residuals are normally distributed by examining a normal probability plot.

EXAMPLE 4 Verifying That the Residuals Are Normally Distributed

Problem Verify that the residuals obtained in Table 2 from Example 2 are normally distributed.

Approach Construct a normal probability plot to assess normality. If the correlation between the residuals and expected z-scores is greater than the critical value in Table VI, the residuals are said to be normal.

Solution Figure 7 contains the normal probability plot. The correlation between the residuals and expected z-scores is 0.988. Because $0.988 > 0.935$ (Table VI), it is reasonable to conclude the residuals are normally distributed. Therefore, we can perform inference on the least-squares regression equation.

Figure 7

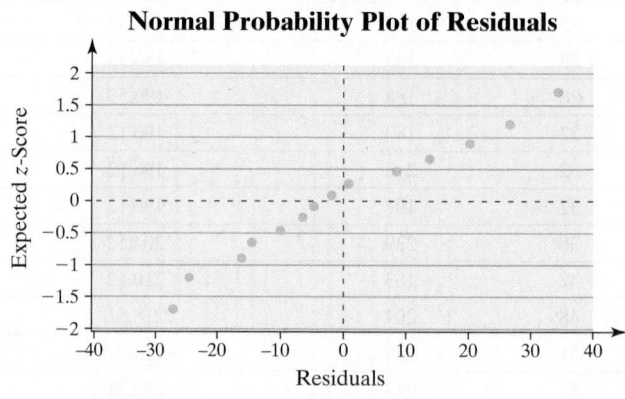

Normal Probability Plot of Residuals

● **Now Work Problem 13(c)**

④ Conduct Inference on the Slope of the Least-Squares Regression Model

At this point, we know how to estimate the intercept and slope of the least-squares regression model. We can also compute the standard error of the estimate, s_e, which is an estimate of σ, the standard deviation of the response variable about the true least-squares regression model, and we know how to assess the normality of the residuals. We will now use this information to test whether a linear relation exists between the explanatory and the response variables.

We want to answer the following question: Do the sample data provide sufficient evidence to conclude that a linear relation exists between the two variables? If there is no linear relation between the response and explanatory variables, the slope of the true regression line will be zero. Do you know why? A slope of zero means that information about the explanatory variable, x, does not change our estimate of the value of the response variable, y.

Using the notation of hypothesis testing, we can perform one of three tests:

In Other Words
Remember, the null hypothesis is a statement of "no effect." So the null hypothesis $H_0: \beta_1 = 0$ means there is no linear relation between the explanatory and response variables.

Two-Tailed	Left-Tailed	Right-Tailed
$H_0: \beta_1 = 0$	$H_0: \beta_1 = 0$	$H_0: \beta_1 = 0$
$H_1: \beta_1 \neq 0$	$H_1: \beta_1 < 0$	$H_1: \beta_1 > 0$

The null hypothesis is $\beta_1 = 0$, the statement of "no effect." We want to find evidence of a relation in the alternative hypothesis. The two-tailed test determines whether a linear relation exists between two variables without regard to the sign of the slope. The left-tailed test determines whether the slope of the true regression line is negative. The right-tailed test determines whether the slope of the true regression line is positive.

To test any one of these hypotheses, we need to know the sampling distribution of b_1. It turns out that when certain conditions are met,

$$t = \frac{b_1 - \beta_1}{\dfrac{s_e}{\sqrt{\Sigma(x_i - \bar{x})^2}}} = \frac{b_1 - \beta_1}{s_{b_1}}$$

CAUTION!
Before testing $H_0: \beta_1 = 0$, be sure to draw a residual plot to verify that a linear model is appropriate.

NOTE

Because $s_x = \sqrt{\dfrac{\Sigma(x_i - \bar{x})^2}{n-1}}$, the standard deviation of b_1 can be calculated using

$$s_{b_1} = \frac{s_e}{\sqrt{n-1}\,s_x}$$

follows Student's t-distribution with $n - 2$ degrees of freedom, where n is the number of observations, b_1 is the estimate of the slope of the regression line β_1, and s_{b_1} is the sample standard deviation of b_1.

Note that

$$s_{b_1} = \frac{s_e}{\sqrt{\Sigma(x_i - \bar{x})^2}}$$

Hypothesis Test Regarding the Slope Coefficient, β_1

To test whether two quantitative variables are linearly related, use the following steps provided that

- The sample is obtained using random sampling or from a randomized experiment.
- The residuals are normally distributed with constant error variance.

Step 1 Determine the null and alternative hypotheses. The hypotheses can be structured in one of three ways:

Two-Tailed	Left-Tailed	Right-Tailed
$H_0: \beta_1 = 0$	$H_0: \beta_1 = 0$	$H_0: \beta_1 = 0$
$H_1: \beta_1 \neq 0$	$H_1: \beta_1 < 0$	$H_1: \beta_1 > 0$

Step 2 Select a level of significance, α, depending on the seriousness of making a Type I error.

Classical Approach

Step 3 Compute the **test statistic**

$$t_0 = \frac{b_1 - \beta_1}{s_{b_1}} = \frac{b_1}{s_{b_1}}$$

P-Value Approach

By-Hand Step 3 Compute the **test statistic**

$$t_0 = \frac{b_1 - \beta_1}{s_{b_1}} = \frac{b_1}{s_{b_1}}$$

(continued)

which follows Student's t-distribution with $n - 2$ degrees of freedom. Use Table VII to determine the critical value.

	Two-Tailed	Left-Tailed	Right-Tailed
Critical value	$-t_{\frac{\alpha}{2}}$ and $t_{\frac{\alpha}{2}}$	$-t_\alpha$	t_α

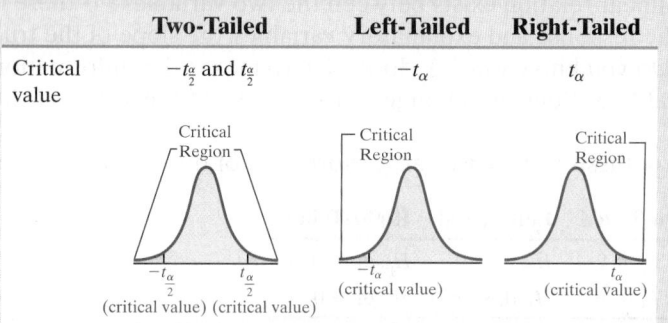

Step 4 Compare the critical value to the test statistic.

Two-Tailed	Left-Tailed	Right-Tailed
If $t_0 < -t_{\frac{\alpha}{2}}$ or $t_0 > t_{\frac{\alpha}{2}}$, reject the null hypothesis.	If $t_0 < -t_\alpha$, reject the null hypothesis.	If $t_0 > t_\alpha$, reject the null hypothesis.

which follows Student's t-distribution with $n - 2$ degrees of freedom. Use Table VII to approximate the P-value.

Two-Tailed	Left-Tailed	Right-Tailed

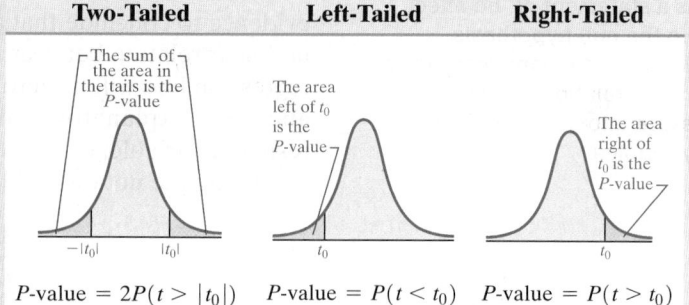

$P\text{-value} = 2P(t > |t_0|)$ $P\text{-value} = P(t < t_0)$ $P\text{-value} = P(t > t_0)$

Technology Step 3 Use a statistical spreadsheet or calculator with statistical capabilities to obtain the P-value. The directions for obtaining the P-value using the TI-83/84 Plus graphing calculators, Minitab, Excel, and StatCrunch are in the Technology Step-by-Step on pages 683–684.

Step 4 If P-value $< \alpha$, reject the null hypothesis.

Step 5 State the conclusion.

Because these procedures are **robust**, minor departures from normality will not adversely affect the results of the test. In fact, for large samples ($n \geq 30$), inferential procedures regarding b_1 can be used even with significant departures from normality.

EXAMPLE 5   Testing for a Linear Relation

Problem Test whether a linear relation exists between age and total cholesterol at the $\alpha = 0.05$ level of significance using the data in Table 1 from Example 1.

Approach Verify that the requirements to perform the inference are satisfied. Then follow Steps 1 through 5.

Solution In Example 1, we were told that the individuals were randomly selected. In Example 4, we confirmed that the residuals are normally distributed.

Verify the requirement of constant error variance by plotting the residuals against the values of the explanatory variable as shown in Figure 8. The errors are evenly spread around a horizontal line drawn at 0, so the requirement of constant error variance is satisfied. Now follow Steps 1 through 5.

Step 1 We want to know if there is a linear relation between age and total cholesterol without regard to the sign of the slope. This is a two-tailed test and we have

$$H_0: \beta_1 = 0 \quad \text{versus} \quad H_1: \beta_1 \neq 0$$

Step 2 The level of significance is $\alpha = 0.05$.

Figure 8

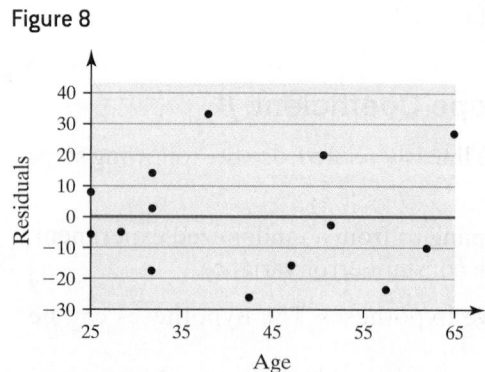

CAUTION!
In Step 3, use unrounded values of the sample mean in the computation of $\Sigma(x_i - \bar{x})^2$ to avoid round-off error.

Classical Approach

Step 3 We obtained an estimate of β_1 in Example 1 to be $b_1 = 1.3991$, and we computed the standard error, $s_e = 19.48$, in Example 2. To determine the standard deviation of b_1, compute $\Sigma(x_i - \bar{x})^2$, where the x_i are the values of the explanatory variable, age, and $\bar{x}$ is the sample mean. We compute this value in Table 3.

P-Value Approach

By-Hand Step 3 From Step 3 of the Classical Approach, we have that the test statistic is $t_0 = 3.572$.

Because this is a two-tailed test, the P-value is the sum of the area under the t-distribution with $14 - 2 = 12$ degrees of freedom to the left of $-t_0 = -3.572$, and to the right of $t_0 = 3.572$, as shown in Figure 10. That is,

Table 3

Age, x	$\bar{x}$	$x_i - \bar{x}$	$(x_i - \bar{x})^2$
25	42.07143	−17.07143	291.4337
25	42.07143	−17.07143	291.4337
28	42.07143	−14.07143	198.0051
32	42.07143	−10.07143	101.4337
32	42.07143	−10.07143	101.4337
32	42.07143	−10.07143	101.4337
38	42.07143	−4.07143	16.5765
42	42.07143	−0.07143	0.0051
48	42.07143	5.92857	35.1479
51	42.07143	8.92857	79.7194
51	42.07143	8.92857	79.7194
58	42.07143	15.92857	253.7193
62	42.07143	19.92857	397.1479
65	42.07143	22.92857	525.7193

$$\sum (x_i - \bar{x})^2 = 2472.9284$$

We have

$$s_{b_1} = \frac{s_e}{\sqrt{\sum (x_i - \bar{x})^2}} = \frac{19.48}{\sqrt{2472.9284}} = 0.3917$$

The test statistic is

$$t_0 = \frac{b_1}{s_{b_1}} = \frac{1.3991}{0.3917} = 3.572$$

Because this is a two-tailed test, we determine the critical t-values at the $\alpha = 0.05$ level of significance with $n - 2 = 14 - 2 = 12$ degrees of freedom to be $-t_{0.05/2} = -t_{0.025} = -2.179$ and $t_{0.05/2} = t_{0.025} = 2.179$. The critical regions are displayed in Figure 9.

Figure 9

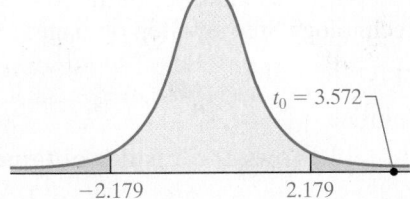

$t_0 = 3.572$

-2.179 2.179

Step 4 The test statistic is $t_0 = 3.572$. We label this point in Figure 9. Because the test statistic is greater than the critical value $t_{0.025} = 2.179$, we reject the null hypothesis.

P-value $= P(t < -3.572) + P(t > 3.572)$
$= 2P(t > 3.572)$, with 12 degrees of freedom.

Figure 10

The sum of these two areas is the P-value

-3.572 3.572

Using Table VII, we find the row that corresponds to 12 degrees of freedom. The value 3.572 lies between 3.428 and 3.930. The area under the t-distribution with 12 degrees of freedom to the right of 3.428 is 0.0025. The area under the t-distribution with 12 degrees of freedom to the right of 3.930 is 0.001.

Because 3.572 is between 3.428 and 3.930, the P-value is between 2(0.001) and 2(0.0025). So,

$$0.002 < P\text{-value} < 0.005.$$

Using Technology Step 3 Using Minitab, we find the P-value is 0.004. See Figure 11.

Figure 11

```
Regression Analysis: Cholesterol versus Age

Analysis of Variance

Source        DF  Adj SS  Adj MS  F-Value  P-Value
Regression     1  4840.5  4840.5    12.76    0.004
  Age          1  4840.5  4840.5    12.76    0.004
Error         12  4553.9   379.5
  Lack-of-Fit  8  3746.7   468.3     2.32    0.217
  Pure Error   4   807.2   201.8
Total         13  9394.4


Model Summary

      S   R-sq  R-sq(adj)  R-sq(pred)
19.4805  51.53%    47.49%      33.55%


Coefficients

Term       Coef  SE Coef  T-Value  P-Value   VIF
Constant  151.4     17.3     8.76    0.000
Age       1.399    0.392     3.57    0.004  1.00


Regression Equation

Cholesterol = 151.4 + 1.399 Age
```

Step 4 The P-value of 0.004 suggests that about 4 samples in 1000 would yield a slope estimate that is as extreme as or more extreme than the one obtained if the null hypothesis of no linear relation were true. Because the P-value is less than the level of significance, $\alpha = 0.05$, we reject the null hypothesis.

Step 5 There is sufficient evidence at the $\alpha = 0.05$ level of significance to conclude that a linear relation exists between age and total cholesterol. •

CAUTION!
If we do not reject H_0, then we use the sample mean of y to predict the value of the response for any value of the explanatory variable.

• Now Work Problems 13(d)
and 13(e)

⑤ **Construct a Confidence Interval about the Slope of the Least-Squares Regression Model**

We can also obtain confidence intervals for the slope of the least-squares regression line. The procedure is identical to that for obtaining confidence intervals for a mean. As was the case with confidence intervals for a population mean, the confidence interval for the slope of the least-squares regression line is of the form

$$\text{Point estimate} \pm \text{margin of error}$$

> **Confidence Intervals for the Slope of the Regression Line**
>
> A $(1 - \alpha) \cdot 100\%$ confidence interval for the slope of the true regression line, β_1, is given by the following formulas:
>
> Lower bound: $b_1 - t_{\alpha/2} \cdot \dfrac{s_e}{\sqrt{\sum (x_i - \bar{x})^2}}$
>
> Upper bound: $b_1 + t_{\alpha/2} \cdot \dfrac{s_e}{\sqrt{\sum (x_i - \bar{x})^2}}$ (3)
>
> Here, $t_{\alpha/2}$ is computed with $n - 2$ degrees of freedom.
>
> **Note:** This interval can be computed only if the data are randomly obtained, the residuals are normally distributed, and there is constant error variance.

EXAMPLE 6   Constructing a Confidence Interval for the Slope of the True Regression Line

Problem Determine a 95% confidence interval for the slope of the true regression line for the data presented in Table 1 in Example 1.

By-Hand Approach

Step 1   Determine the least-squares regression line.

Step 2   Verify that the requirements for inference on the regression line are satisfied.

Step 3   Compute s_e.

Step 4   Determine the critical value $t_{\alpha/2}$ with $n - 2$ degrees of freedom.

Step 5   Compute the bounds on the $(1 - \alpha) \cdot 100\%$ confidence interval for β_1 using Formula (3).

Step 6   Interpret the result by stating, "We are 95% confident that β_1 is between *lower bound* and *upper bound*."

By-Hand Solution

Step 1   The least-squares regression line was determined in Example 1 and is $\hat{y} = 1.3991x + 151.3537$.

Step 2   The requirements were verified in Examples 1–5.

Step 3   We computed s_e in Example 2, obtaining $s_e = 19.48$.

Step 4   Because we wish to determine a 95% confidence interval, we have $\alpha = 0.05$. Therefore, we need to find $t_{0.05/2} = t_{0.025}$ with $14 - 2 = 12$ degrees of freedom. Referring to Table VII, we find that $t_{0.025} = 2.179$.

Technology Approach

Step 1   Use a statistical spreadsheet or graphing calculator with advanced statistical features to obtain the confidence interval. We will use StatCrunch. The steps for constructing confidence intervals using StatCrunch, the TI-83/84 Plus graphing calculators, Minitab, and Excel are given in the Technology Step-by-Step on pages 683–684.

Step 2   Interpret the result.

Technology Solution

Step 1   Figure 12 shows the results obtained from StatCrunch.

Figure 12

Simple linear regression results:
Dependent Variable: Cholesterol
Independent Variable: Age
Cholesterol = 151.35365 + 1.3990642 Age
Sample size: 14
R (correlation coefficient) = 0.7178
R-sq = 0.5152521
Estimate of error standard deviation: 19.480536

Parameter estimates:

Parameter	Estimate	Std. Err.	DF	95% L. Limit	95% U. Limit
Intercept	151.35365	17.28376	12	113.69558	189.01173
Slope	1.3990642	0.39173746	12	0.5455416	2.2525868

Step 5   We use Formula (3) to find the lower and upper bounds.

Lower bound: $b_1 - t_{\alpha/2} \cdot \dfrac{s_e}{\sqrt{\sum (x_i - \bar{x})^2}} = 1.3991 - 2.179 \cdot \dfrac{19.48}{\sqrt{2472.9284}}$

$= 1.3991 - 0.8536 = 0.5455$

Upper bound: $b_1 + t_{\alpha/2} \cdot \dfrac{s_e}{\sqrt{\sum (x_i - \bar{x})^2}} = 1.3991 + 2.179 \cdot \dfrac{19.48}{\sqrt{2472.9284}}$

$= 1.3991 + 0.8536 = 2.2527$

Step 6   We are 95% confident that the mean increase in cholesterol for each additional year of life is somewhere between 0.5455 and 2.2527. Because the 95% confidence interval does not include 0, we reject $H_0: \beta_1 = 0$.

The lower bound (95% L. Limit) is 0.5455 and the upper bound (95% U. Limit) is 2.2526.

Step 2   We are 95% confident that the mean increase in cholesterol for each additional year of life is between 0.5455 and 2.2526. Because the confidence interval does not contain 0, we reject $H_0: \beta_1 = 0$.

● Now Work Problem 13(f)

CAUTION!

It is best that the values of the explanatory variable be spread out when doing regression analysis.

In looking carefully at the formula for the standard deviation of b_1, we should notice that the larger the value of $\sum (x_i - \bar{x})^2$, the smaller the value of s_{b_1}. This result implies that whenever we are finding a least-squares regression line, we should attempt to make the values of the explanatory variable, x, as evenly spread out as possible so that b_1, our estimate of β_1, is as precise as possible.

Inference on the Linear Correlation Coefficient

Perhaps you are wondering why we have not presented hypothesis tests regarding the linear correlation coefficient in this section. Recall in Chapter 4 that we introduced a quick method for testing the significance of the correlation coefficient, even though we did not have a full appreciation of statistical inference. At this point, we intentionally avoid discussion of inference on the correlation coefficient for two reasons: (1) the hypothesis test on the slope and a hypothesis test on the linear correlation coefficient will yield the same conclusion, and (2) inferential methods on the linear correlation coefficient, ρ, require that the y's at any given x be normally distributed and that the x's at any given y be normally distributed. That is, testing a hypothesis such as $H_0: \rho = 0$ versus $H_1: \rho \neq 0$ requires that the two variables follow a **bivariate normal distribution** or be **jointly normally distributed**. Verifying this requirement is a difficult task. Although a normal probability plot of the x_i's and a separate normal probability plot of the y_i's generally mean that the joint distribution is normal, it is not guaranteed. For these two reasons, we will be content in verifying the linearity of the data by performing inference on the slope coefficient only.

Technology Step-by-Step   Testing the Least-Squares Regression Model

TI-83/84 Plus
Hypothesis Test on the Slope
1. Enter the explanatory variable in L1 and the response variable in L2.
2. Press STAT, highlight TESTS, and select `F:LinRegTTest....`
3. Be sure that Xlist is L1 and Ylist is L2. Make sure that `Freq:` is set to 1. Select the direction of the alternative hypothesis. Place the cursor on Calculate and press ENTER.

Confidence Interval for the Slope
1. Enter the explanatory variable in L1 and the response variable in L2.
2. Press STAT, highlight TESTS, and select `G: LinRegTInt . . . .`

3. Be sure that Xlist is L1 and Ylist is L2. Make sure the `Freq:` is set to 1. Select the confidence level. Highlight Calculate. Press ENTER.

Minitab
1. With the explanatory variable in C1 and the response variable in C2, select the **Stat** menu and highlight **Regression**. Highlight **Regression**, then select **Fit Regression Model....**
2. Place the cursor in the "Responses:" box. Highlight the column containing the response variable. Click Select. Place the cursor in the "Continuous predictors:" box. Highlight the column containing the explanatory variable. Click Select. Click OK.

Excel
1. Make sure the Data Analysis Tool Pack is activated by selecting File. Click Options, then click Add-Ins. From the drop-down menu, select "Excel Add-ins." Click Go. Check the boxes "Analysis ToolPak" and "Analysis ToolPak-VBA." Click OK.
2. Enter the explanatory variable in column A and the response variable in column B.
3. Select the Data menu and then select **Data Analysis . . . .**
4. Select the **Regression** option.
5. With the cursor in the Y-range cell, highlight the range of cells that contains the response variable. With the cursor in the X-range cell, highlight the range of cells that contains the explanatory variable. Click OK.

StatCrunch
Hypothesis Test on the Slope
1. Enter the explanatory variable in column var1 and the response variable in column var2.
2. Select **Stat**, highlight **Regression**, choose **Simple Linear**. Choose var1 for the X-variable, choose var2 for the Y-variable. Select the Hypothesis tests radio button. Choose the appropriate values in the null hypothesis for both the intercept and slope. Choose the direction of the alternative hypothesis. Click Compute!

Confidence Interval for the Slope
1. Enter the explanatory variable in column var1 and the response variable in column var2.
2. Select **Stat**, highlight **Regression**, choose **Simple Linear**. Choose var1 for the X-variable, choose var2 for the Y-variable. Select the Confidence intervals radio button. Choose the confidence level. Click Compute!

14.1 Assess Your Understanding

Vocabulary and Skill Building

1. Suppose a least-squares regression line is given by $\hat{y} = 4.302x - 3.293$. What is the mean value of the response variable if $x = 20$?

2. *True or False*: In a least-squares regression, the response variable is normally distributed with mean $\mu_{y|x}$ and standard deviation σ.

3. In the least-squares regression model, $y_i = \beta_1 x_i + \beta_0 + \varepsilon_i$, ε_i is a random error term with mean _____ and standard deviation $\sigma_{\varepsilon_i} =$ _____.

4. If $H_0: \beta_1 = 0$ is not rejected, what is the best estimate for the value of the response variable for any value of the explanatory variable?

In Problems 5–10, use the results of Problems 7–12, respectively, from Section 4.2 to answer the following questions:

(a) What are the estimates of β_0 and β_1?
(b) Compute the standard error, the point estimate for σ.
(c) Determine s_{b_1}.
(d) Assuming the residuals are normally distributed, test $H_0: \beta_1 = 0$ versus $H_1: \beta_1 \neq 0$ at the $\alpha = 0.05$ level of significance.

5.

x	3	4	5	7	8
y	4	6	7	12	14

6.

x	3	5	7	9	11
y	0	2	3	6	9

7.

x	-2	-1	0	1	2
y	-4	0	1	4	5

8.

x	-2	-1	0	1	2
y	7	6	3	2	0

9.

x	20	30	40	50	60
y	100	95	91	83	70

10.

x	5	10	15	20	25
y	2	4	7	11	18

Applying the Concepts

 11. An Unhealthy Commute The following data represent commute times (in minutes) and a score on a well-being survey.

Commute Time (minutes), x	Gallup-Healthways Well-Being Index Composite Score, y
5	69.2
15	68.3
25	67.5
35	67.1
50	66.4
72	66.1
105	63.9

Source: The Gallup Organization

Use the results from Problem 17 in Section 4.2 to answer the following questions:

(a) Treating commute time as the explanatory variable, x, determine the estimates of β_0 and β_1.
(b) Compute the standard error of the estimate, s_e.
(c) Determine s_{b_1}.
(d) A normal probability plot of the residuals indicates it is reasonable to conclude the residuals are normally distributed. Test whether a linear relation exists between commute time and well-being index composite score at the $\alpha = 0.05$ level of significance.
(e) Construct a 95% confidence interval about the slope of the true least-squares regression line.

12. Credit Scores An economist wants to determine the relation between one's FICO score, x, and the interest rate of a 36-month auto loan, y. The data represent the interest rate (in percent) a bank might offer on a 36-month auto loan for various FICO scores.

Credit Score, x	Interest Rate (percent), y
545	18.982
595	17.967
640	12.218
675	8.612
705	6.680
750	5.150

Source: www.myfico.com

Use the results from Problem 18 in Section 4.2 to answer the following questions:

(a) Treating credit score as the explanatory variable, x, determine the estimates of β_0 and β_1.

(b) Compute the standard error of the estimate, s_e.

(c) Determine s_{b_1}.

(d) A normal probability plot of the residuals indicates it is reasonable to conclude the residuals are normally distributed. Test whether a linear relation exists between credit score and interest rate at the $\alpha = 0.05$ level of significance.

(e) Construct a 95% confidence interval about the slope of the true least-squares regression line.

NW 13. Height versus Head Circumference A pediatrician wants to determine the relation that may exist between a child's height and head circumference. She randomly selects 11 children from her practice, measures their heights and head circumferences, and obtains the following data:

Height (inches), x	Head Circumference (inches), y	Height (inches), x	Head Circumference (inches), y
27.75	17.5	26.5	17.3
24.5	17.1	27	17.5
25.5	17.1	26.75	17.3
26	17.3	26.75	17.5
25	16.9	27.5	17.5
27.75	17.6		

Source: Denise Slucki, student at Joliet Junior College

Use the results from Problem 19 in Section 4.2 to answer the following questions:

(a) Treating height as the explanatory variable, x, determine the estimates of β_0 and β_1.

(b) Compute the standard error of the estimate, s_e.

(c) Determine whether the residuals are normally distributed.

(d) Determine s_{b_1}.

(e) If the residuals are normally distributed, test whether a linear relation exists between height and head circumference at the $\alpha = 0.01$ level of significance.

(f) If the residuals are normally distributed, construct a 95% confidence interval about the slope of the true least-squares regression line.

(g) A child comes in for a physical, and the nurse determines his height to be 26.5 inches. However, the child is being rather uncooperative, so the nurse is unable to measure the head circumference of the child. What would be a good estimate of this child's head circumference? Why is this a good estimate?

14. Hurricanes The following data represent the maximum wind speed (in knots) and atmospheric pressure (in millibars) for a random sample of hurricanes that originated in the Atlantic Ocean. Does atmospheric pressure play a role in the wind speed of a hurricane?

Atmospheric Pressure (mb), x	Wind Speed (knots), y	Atmospheric Pressure (mb), x	Wind Speed (knots), y
993	50	1006	40
995	60	942	120
994	60	1002	40
997	45	986	50
1003	45	983	70
1004	40	994	65
1000	55	940	120
994	55	976	80
942	105	966	100
1006	30	982	55

Source: National Hurricane Center

Use the results from Problem 22 in Section 4.2 to answer the following questions:

(a) Treating atmospheric pressure as the explanatory variable, x, determine the estimates of β_0 and β_1.

(b) Compute the standard error of the estimate.

(c) Determine whether the residuals are normally distributed.

(d) Determine s_{b_1}.

(e) If the residuals are normally distributed, test whether a linear relation exists between the atmospheric pressure and wind speed at the $\alpha = 0.01$ level of significance.

(f) If the residuals are normally distributed, construct a 99% confidence interval for the slope of the true least-squares regression line.

(g) What is the mean wind speed of a hurricane whose atmospheric pressure is 995 mb?

15. Concrete As concrete cures, it gains strength. The following data represent the 7-day and 28-day strength (in pounds per square inch) of a certain type of concrete:

7-Day Strength, x	28-Day Strength, y	7-Day Strength, x	28-Day Strength, y
2300	4070	2480	4120
3390	5220	3380	5020
2430	4640	2660	4890
2890	4620	2620	4190
3330	4850	3340	4630

(a) Treating the 7-day strength as the explanatory variable, x, determine the estimates of β_0 and β_1.

(b) Compute the standard error of the estimate.

(c) Determine s_{b_1}.

(d) Assuming the residuals are normally distributed, test whether a linear relation exists between 7-day strength and 28-day strength at the $\alpha = 0.05$ level of significance.

(e) Assuming the residuals are normally distributed, construct a 95% confidence interval for the slope of the true least-squares regression line.

(f) What is the estimated mean 28-day strength of this concrete if the 7-day strength is 3000 psi?

16. Tar and Nicotine Every year the Federal Trade Commission (FTC) must report tar and nicotine levels in cigarettes to Congress. Tar and nicotine levels of over 1200 brands of cigarettes are given to Congress and a random sample of those appear in the following table:

Brand	Tar (mg), x	Nicotine (mg), y
Barclay 100	5	0.4
Benson and Hedges King	16	1.1
Camel Regular	24	1.7
Chesterfield King	24	1.4
Doral	8	0.5
Kent Golden Lights	9	0.8
Kool Menthol	9	0.8
Lucky Strike	24	1.5
Marlboro Gold	15	1.2
Newport Menthol	18	1.3
Salem Menthol	17	1.3
Virginia Slims Ultra Light	5	0.5
Winston Light	10	0.8

Source: Federal Trade Commission

(a) Treating the amount of tar as the explanatory variable, x, determine the estimates of β_0 and β_1.
(b) Compute the standard error of the estimate.
(c) Determine s_{b_1}.
(d) Assuming the residuals are normally distributed, test whether a linear relation exists between the amount of tar, x, and the amount of nicotine, y, at the $\alpha = 0.1$ level of significance.
(e) Assuming the residuals are normally distributed, construct a 90% confidence interval for the slope of the true least-squares regression line.
(f) What is the mean amount of nicotine in a cigarette that has 12 milligrams of tar?

17. Invest in Education Go to www.pearsonhighered.com/sullivanstats to obtain the data file 14_1_17. The variable "2013 Cost" represents the four-year cost including tuition, supplies, room and board, the variable "Annual ROI" represents the return on investment for graduates of the school—essentially how much you would earn on the investment of attending the school—the variable "Grad Rate" represents the graduation rate of the school.

(a) In Problem 47 from Section 4.1, we saw that a scatter diagram between "2013 Cost" and "Grad Rate" treating "2013 Cost" as the explanatory variable suggested a positive association between the two variables. Treating "2013 Cost" as the explanatory variable, x, test whether a negative association exists between the 2013 cost and annual ROI for graduates of four-year schools at the $\alpha = 0.01$ level of significance. Normal probability plots suggest the residuals are normally distributed.
(b) Construct a 90% confidence interval for the slope of the true least-squares regression line.
(c) What is the mean annual ROI for a four-year school whose 2013 cost is $180,000?

18. American Black Bears In 1969, Dr. Michael R. Pelton of the University of Tennessee initiated a long-term study of the American black bear (*Ursus americanus*) population in Great Smoky Mountains National Park. One aspect of the study was

to develop a model that could be used to predict a bear's weight (since it is not practical to weigh bears in the field). One variable that is thought to be related to weight is the length of the bear. The following data represent the lengths and weights of 12 American black bears.

Total Length (cm), x	Weight (kg), y
139.0	110
138.0	60
139.0	90
120.5	60
149.0	85
141.0	100
141.0	95
150.0	85
166.0	155
151.5	140
129.5	105
150.0	110

Source: fieldtripearth.org

Use the results from Problem 20 in Section 4.2 to answer the following questions:

(a) Treating total length as the explanatory variable, x, determine the estimates of β_0 and β_1.
(b) Assuming the residuals are normally distributed, test whether a linear relation exists between total length and weight at the $\alpha = 0.05$ level of significance.
(c) Assuming the residuals are normally distributed, construct a 95% confidence interval for the slope of the true least-squares regression line.
(d) What is the mean weight of American black bears of length 146.0 cm?

19. CEO Performance (Refer to Problem 31 in Section 4.1) The following data represent the total compensation for 12 randomly selected chief executive officers (CEOs) and the company's stock performance in 2013.

Company	Compensation (millions of dollars)	Stock Return (%)
Navistar International	14.53	75.43
Aviv REIT	4.09	64.01
Groupon	7.11	142.07
Inland Real Estate	1.05	32.72
Equity Lifestyles Properties	1.97	10.64
Tootsie Roll Industries	3.76	30.66
Catamaran	12.06	0.77
Packaging Corp of America	7.62	69.39
Brunswick	8.47	58.69
LKQ	4.04	55.93
Abbott Laboratories	20.87	24.28
TreeHouse Foods	6.63	32.21

Data from *Chicago Tribune*, June 1, 2014

(a) Treating compensation as the explanatory variable, x, determine the estimates of β_0 and β_1.

(b) Assuming the residuals are normally distributed, test whether a linear relation exists between compensation and stock return at the $\alpha = 0.05$ level of significance.

(c) Assuming the residuals are normally distributed, construct a 95% confidence interval for the slope of the true least-squares regression line.

(d) Based on your results to parts (b) and (c), would you recommend using the least-squares regression line to predict the stock return of a company based on the CEO's compensation? Why? What would be a good estimate of the stock return based on the data in the table?

20. Bear Markets (Refer to Problem 32, Section 4.1) A bear market is a market condition in which the price of the security falls. A bear market in the stock market is defined as a condition in which market declines by 20% or more over the course of at least two months. The following data represent the number of months and percentage change in the S&P500 (a group of 500 stocks).

Months	Percent Change	Months	Percent Change
1.9	−44.57	12.1	−20.57
8.3	−44.29	14.9	−21.63
3.3	−32.86	6.5	−27.97
3.4	−42.54	8.0	−22.18
6.8	−61.81	18.1	−36.06
5.8	−40.60	21.0	−48.20
3.1	−29.43	20.7	−27.11
13.4	−31.81	3.4	−33.51
12.9	−54.47	18.2	−36.77
5.1	−24.44	9.3	−33.75
7.6	−28.69	13.6	−51.93
17.9	−34.42	2.1	−27.62
11.8	−28.47		

Source: Gold-Eagle

(a) Treating months as the explanatory variable, x, determine the estimates for β_0 and β_1.

(b) Assuming the residuals are normally distributed, test whether a linear relation exists between the number of months of a bear market and percent change at the $\alpha = 0.05$ level of significance.

(c) Assuming the residuals are normally distributed, construct a 95% confidence interval for the slope of the true least-squares regression line.

(d) Based on your results to parts (b) and (c), would you recommend using the least-squares regression line to predict the percent change in the S&P500 during a bear market? Why? What would be a good estimate of the percent change in the S&P500 during a bear market?

21. Age versus HDL Cholesterol A doctor wanted to determine whether there is a relation between a male's age and his HDL (so-called good) cholesterol. He randomly selected 17 of his patients and determined their HDL levels. He obtained the following data:

Age, x	HDL Cholesterol, y	Age, x	HDL Cholesterol, y
38	57	38	44
42	54	66	62
46	34	30	53
32	56	51	36
55	35	27	45
52	40	52	38
61	42	49	55
61	38	39	28
26	47		

Source: Data based on information obtained from the National Center for Health Statistics

(a) Draw a scatter diagram of the data, treating age as the explanatory variable. What type of relation, if any, appears to exist between age and HDL cholesterol?

(b) Determine the least-squares regression equation from the sample data.

(c) Plot the residuals against the explanatory variable, age. Does a linear model seem appropriate on the basis of the residual plot? (*Hint:* See Section 4.3.)

(d) Are there any outliers or influential observations?

(e) Assuming the residuals are normally distributed, test whether a linear relation exists between age and HDL cholesterol levels at the $\alpha = 0.01$ level of significance.

(f) Assuming the residuals are normally distributed, construct a 95% confidence interval for the slope of the true least-squares regression line.

(g) For a 42-year-old male patient who visits the doctor's office, would you recommend using the least-squares regression line obtained in part (b) to predict the HDL cholesterol of this patient? Why? What would be a good estimate for the HDL cholesterol of this patient?

22. The U.S. Population The following data represent the population of the United States for the years 1900–2010:

Year, x	Population, y	Year, x	Population, y
1900	76,212,168	1960	179,323,175
1910	92,228,496	1970	203,302,031
1920	106,021,537	1980	226,542,203
1930	123,202,624	1990	248,709,873
1940	132,164,569	2000	281,421,906
1950	151,325,798	2010	308,745,538

Source: U.S. Census Bureau

An ecologist is interested in finding an equation that describes the population of the United States over time.

(a) Determine the least-squares regression equation, treating year as the explanatory variable.

(b) A normal probability plot of the residuals indicates that the residuals are approximately normally distributed. Test whether a linear relation exists between year and population.

(c) Draw a scatter diagram, treating year as the explanatory variable.

(d) Plot the residuals against the explanatory variable, year.

(e) Does a linear model seem appropriate based on the scatter diagram and residual plot? (*Hint:* See Section 4.3.)

(f) What is the moral?

23. Kepler's Law of Planetary Motion The time it takes for a planet to complete its orbit around the sun is called the planet's *sidereal year*. Johann Kepler studied the relation between the sidereal year of a planet and its distance from the sun in 1618. The following data show the distances that the planets are from the sun and their sidereal years.

Planet	Distance from Sun, x (millions of miles)	Sidereal Year, y
Mercury	36	0.24
Venus	67	0.62
Earth	93	1.00
Mars	142	1.88
Jupiter	483	11.9
Saturn	887	29.5
Uranus	1785	84.0
Neptune	2797	165.0
Pluto*	3675	248.0

*Pluto's status was reduced to a dwarf planet in September 2006.

(a) Determine the least-squares regression equation, treating distance from the sun as the explanatory variable.

(b) A normal probability plot of the residuals indicates that the residuals are approximately normally distributed. Test whether a linear relation exists between distance from the sun and sidereal year.

(c) Draw a scatter diagram, treating distance from the sun as the explanatory variable.

(d) Plot the residuals against the explanatory variable, distance from the sun.

(e) Does a linear model seem appropriate based on the scatter diagram and residual plot? (*Hint:* See Section 4.3.)

(f) What is the moral?

24. The output shown was obtained from Minitab.

```
The regression equation is
y = 12.4 + 1.40 x

Predictor    Coef     StDev      T       P
Constant    12.396    1.381     8.97   0.000
x           1.3962    0.1245   11.21   0.000

S = 2.167    R-Sq = 91.3%   R-Sq(adj) = 90.6%
```

(a) The least-squares regression equation is $\hat{y} = 1.3962x + 12.396$. What is the predicted value of y at $x = 10$?

(b) What is the mean of y at $x = 10$?

(c) The standard error, s_e, is 2.167. What is an estimate of the standard deviation of y at $x = 10$?

(d) If the requirements for inference on the least-squares regression model are satisfied, what is the distribution of y at $x = 10$?

25. Influential Observations Zillow.com is a site that can be used to assess the value of homes in your neighborhood. The organization provides a list of homes for sale as well as a Zestimate, which is the price Zillow believes the home will sell for. The following data represent the Zestimate and sale price (in thousands of dollars) of a random sample of recently sold homes in Charleston, South Carolina.

Zestimate	Sale Price
362	370
309	315
365.5	371.9
215	218
184	186.5
252.5	260
247.5	250.8
244	251

Source: zillow.com

(a) Draw a scatter diagram of the data, treating the Zestimate as the explanatory variable and sale price as the response variable.

(b) Determine the least-squares regression line. Test whether there is a relation between the Zestimate and sale price at the $\alpha = 0.05$ level of significance.

(c) A home with a Zestimate of $370,000 recently sold for $150,000. Determine the least-squares regression line with this home included. Test whether there is a relation between the Zestimate and sale price at the $\alpha = 0.05$ level of significance. Do you think this observation is influential?

Explaining the Concepts

26. Why is it important to perform graphical as well as analytical analyses when analyzing relations between two quantitative variables?

27. What do the y-coordinates on the least-squares regression line represent?

28. Why is it desirable to have the explanatory variables spread out to test a hypothesis regarding β_1 or to construct confidence intervals about β_1?

29. Why don't we conduct inference on the linear correlation coefficient?

14.2 Confidence and Prediction Intervals

Preparing for This Section Before getting started, review the following:

- Confidence intervals about μ (Section 9.2, pp. 435–441)

Objectives

1 Construct confidence intervals for a mean response

2 Construct prediction intervals for an individual response

We know how to obtain the least-squares regression equation of best fit from data. We also know how to use the least-squares regression equation to obtain a predicted value. For example, the least-squares regression equation for the cholesterol data introduced in Example 1 from Section 14.1 is

$$\hat{y} = 1.3991x + 151.3537$$

where $\hat{y}$ represents the predicted total cholesterol for a female whose age is x. The predicted value of total cholesterol for a given age x actually has two interpretations:

1. It represents the mean total cholesterol for all females whose age is x.
2. It represents the predicted total cholesterol for a randomly selected female whose age is x.

So if we let $x = 42$ in the least-squares regression equation $\hat{y} = 1.3991x + 151.3537$, we obtain $\hat{y} = 1.3991(42) + 151.3537 = 210.1$ mg/dL. We can interpret this result in one of two ways:

1. The mean total cholesterol for all 42-year-old females is 210.1 mg/dL.
2. The predicted total cholesterol for a randomly selected 42-year-old female is 210.1 mg/dL.

Of course, there is a margin of error in making predictions, so we construct intervals about any predicted value to describe its accuracy. The type of interval constructed will depend on whether we are predicting a mean total cholesterol for all 42-year-old females or the total cholesterol for an individual 42-year-old female. In other words, the margin of error is going to be different for predicting the mean total cholesterol for all females who are 42 years old versus the total cholesterol for one individual. Which prediction (the mean or the individual) do you think will have a wider confidence interval? It seems logical that the distribution of means should have less variability (and therefore a lower margin of error) than the distribution of individuals. After all, in the distribution of means, high total cholesterols can be offset by low total cholesterols.

Definitions

Confidence intervals for a mean response are intervals constructed about the predicted value of y, at a given level of x, that are used to measure the accuracy of the mean response of all the individuals in the population.

Prediction intervals for an individual response are intervals constructed about the predicted value of y that are used to measure the accuracy of a single individual's predicted value.

In Other Words
Confidence intervals are intervals for the mean of the population. Prediction intervals are intervals for an individual from the population.

If we use the least-squares regression equation to predict the mean total cholesterol for all 42-year-old females, we construct a confidence interval for a mean response. If we use the least-squares regression equation to predict the total cholesterol for a single 42-year-old female, we construct a prediction interval for an individual response.

① **Construct Confidence Intervals for a Mean Response**

The structure of a confidence interval is the same as it was in Section 9.1. The interval is of the form

$$\text{Point estimate} \pm \text{margin of error}$$

> **Confidence Interval for the Mean Response of y, $\hat{y}$**
>
> A $(1 - \alpha) \cdot 100\%$ confidence interval for $\hat{y}$, the mean response of y for a specified value of x, is given by
>
> $$\text{Lower bound:} \quad \hat{y} - t_{\alpha/2} \cdot s_e \sqrt{\frac{1}{n} + \frac{(x^* - \bar{x})^2}{\sum(x_i - \bar{x})^2}}$$
>
> $$\text{Upper bound:} \quad \hat{y} + t_{\alpha/2} \cdot s_e \sqrt{\frac{1}{n} + \frac{(x^* - \bar{x})^2}{\sum(x_i - \bar{x})^2}}$$
>
> (1)
>
> where x^* is the given value of the explanatory variable, n is the number of observations, and $t_{\alpha/2}$ is the critical value with $n - 2$ degrees of freedom.

NOTE

The interval may be constructed provided the residuals are normally distributed or the sample size is large.

EXAMPLE 1 **Constructing a Confidence Interval for a Mean Response by Hand**

Problem Construct a 95% confidence interval for the predicted mean total cholesterol of all 42-year-old females using the data in Table 1 on page 674.

Approach Determine a confidence interval for the predicted mean total cholesterol at $x^* = 42$ using Formula (1), since our estimate is for the mean cholesterol of all 42-year-old females.

Solution The least-squares regression equation is $\hat{y} = 1.3991x + 151.3537$. To find the predicted mean total cholesterol of all 42-year-olds, let $x^* = 42$ in the regression equation and obtain $\hat{y} = 1.3991(42) + 151.3537 = 210.1$. From Example 2 in Section 14.1, we found that $s_e = 19.48$, and from Example 5 in Section 14.1, we found that $\sum(x_i - \bar{x})^2 = 2472.9284$ and $\bar{x} = 42.07143$. The critical t-value, $t_{\alpha/2} = t_{0.025}$, with $n - 2 = 14 - 2 = 12$ degrees of freedom is 2.179. The 95% confidence interval for the predicted mean total cholesterol for all 42-year-old females is therefore

$$\text{Lower bound:} \quad \hat{y} - t_{\alpha/2} \cdot s_e \cdot \sqrt{\frac{1}{n} + \frac{(x^* - \bar{x})^2}{\sum(x_i - \bar{x})^2}} = 210.1 - 2.179 \cdot 19.48 \cdot \sqrt{\frac{1}{14} + \frac{(42 - 42.07143)^2}{2472.9284}}$$

$$= 198.8$$

$$\text{Upper bound:} \quad \hat{y} + t_{\alpha/2} \cdot s_e \cdot \sqrt{\frac{1}{n} + \frac{(x^* - \bar{x})^2}{\sum(x_i - \bar{x})^2}} = 210.1 + 2.179 \cdot 19.48 \cdot \sqrt{\frac{1}{14} + \frac{(42 - 42.07143)^2}{2472.9284}}$$

$$= 221.4$$

● **Now Work Problems 3(a) and (b)**

We are 95% confident that the mean total cholesterol of all 42-year-old females is between 198.8 mg/dL and 221.4 mg/dL.

●

---

② **Construct Prediction Intervals for an Individual Response**

In Other Words

Prediction intervals are wider than confidence intervals because it is tougher to guess the value for an individual than the mean of a population.

The procedure for obtaining a prediction interval for an individual response is similar to that for finding a confidence interval for a mean response. The only difference is the standard error. More variability is associated with individuals than with means. Therefore, the computation of the interval must account for this increased variability. Again, the form of the interval is

$$\text{Point estimate} \pm \text{margin of error}$$

Prediction Interval for an Individual Response about $\hat{y}$

A $(1 - \alpha) \cdot 100\%$ prediction interval for $\hat{y}$, the individual response of y, is given by

$$\text{Lower bound:} \quad \hat{y} - t_{\alpha/2} \cdot s_e \sqrt{1 + \frac{1}{n} + \frac{(x^* - \bar{x})^2}{\sum (x_i - \bar{x})^2}}$$

$$\text{Upper bound:} \quad \hat{y} + t_{\alpha/2} \cdot s_e \sqrt{1 + \frac{1}{n} + \frac{(x^* - \bar{x})^2}{\sum (x_i - \bar{x})^2}}$$

(2)

where x^* is the given value of the explanatory variable, n is the number of observations, and $t_{\alpha/2}$ is the critical value with $n - 2$ degrees of freedom.

NOTE
The interval may be constructed provided the residuals are normally distributed or the sample size is large.

Notice that the only difference between Formulas (1) and (2) is the "$1 +$" under the radical in Formula (2).

EXAMPLE 2 Constructing a Prediction Interval for an Individual Response by Hand

Problem Construct a 95% prediction interval for the predicted total cholesterol of a 42-year-old female.

Approach Determine the predicted total cholesterol at $x^* = 42$ and use Formula (2), since our estimate is for a particular 42-year-old female.

Solution The least-squares regression equation is $\hat{y} = 1.3991x + 151.3537$. To find the predicted total cholesterol of a 42-year-old, let $x^* = 42$ in the regression equation and obtain $\hat{y} = 1.3991(42) + 151.3537 = 210.1$. From Example 2 in Section 14.1, we found that $s_e = 19.48$; from Example 5 in Section 14.1, we found that $\sum (x_i - \bar{x})^2 = 2472.9284$ and $\bar{x} = 42.07143$. We find $t_{\alpha/2} = t_{0.025}$ with $n - 2 = 14 - 2 = 12$ degrees of freedom to be 2.179.

The 95% prediction interval for the predicted total cholesterol for a 42-year-old female is

$$\text{Lower bound:} \quad \hat{y} - t_{\alpha/2} \cdot s_e \sqrt{1 + \frac{1}{n} + \frac{(x^* - \bar{x})^2}{\sum (x_i - \bar{x})^2}} = 210.1 - 2.179 \cdot 19.48 \cdot \sqrt{1 + \frac{1}{14} + \frac{(42 - 42.07143)^2}{2472.9284}} = 166.2$$

$$\text{Upper bound:} \quad \hat{y} + t_{\alpha/2} \cdot s_e \sqrt{1 + \frac{1}{n} + \frac{(x^* - \bar{x})^2}{\sum (x_i - \bar{x})^2}} = 210.1 + 2.179 \cdot 19.48 \cdot \sqrt{1 + \frac{1}{14} + \frac{(42 - 42.07143)^2}{2472.9284}} = 254.0$$

● **Now Work Problems 3(c) and (d)**

We are 95% confident that the total cholesterol of a randomly selected 42-year-old female is between 166.2 mg/dL and 254.0 mg/dL.

●

Notice that the interval about the individual (prediction interval for an individual response) is wider than the interval about the mean (confidence interval for a mean response). The reason for this should be clear: More variability is associated with individuals than with groups of individuals. That is, it is more difficult to predict a single 42-year-old female's total cholesterol than it is to predict the mean total cholesterol for all 42-year-old females.

EXAMPLE 3 Confidence and Prediction Intervals Using Technology

Problem Construct a 95% confidence interval for the predicted mean total cholesterol of all 42-year-old females using statistical software. Construct a 95% prediction interval for the predicted total cholesterol for a 42-year-old female using statistical software.

Approach We will use Minitab to obtain the intervals. The steps for obtaining confidence and prediction intervals using Minitab, Excel, and StatCrunch are given in the Technology Step-by-Step on page 692.

Using Technology
The bounds for confidence and prediction intervals obtained using statistical software may differ from bounds computed by hand due to rounding error.

(continued)

Solution Figure 13 shows the results obtained from Minitab.

Figure 13

Predicted Values

```
        Fit    StDev Fit         95.0% CI              95.0% PI
     210.11         5.21   ( 198.77, 221.46)    ( 166.18, 254.05)
```

Technology Step-by-Step   Confidence and Prediction Intervals

TI-83/84 Plus
The TI-83/84 Plus graphing calculators do not compute confidence or prediction intervals.

Minitab
1. With the model entered as in Section 14.1, select the **Stat** menu, highlight **Regression**, highlight **Regression** again, and select **Predict....**
2. Select the response variable. From the pull-down menu select "Enter individual values." Enter the values of the explanatory variable in the table.
3. Click the Options . . . button. Enter the desired level of confidence. Click OK twice.

Excel
1. Enter the values of the explanatory variable in Column A and the corresponding values of the response variable in Column B. Enter the value of the explanatory variable for which you wish to make predictions in a cell.

2. Select the XLSTAT menu. Choose the Modeling Data menu and select Linear Regression.
3. With the cursor in the Y/Dependent variables quantitative cell, highlight the data in Column B. With the cursor in the X/Explanatory variables quantitative cell, highlight the data in Column A.
4. Select the Prediction menu. Check the Prediction box. With the cursor in the X/Explanatory variable quantitative cell, highlight the cell that contains the value of the explanatory variable. Click OK.

StatCrunch
Follow the steps given in Section 14.1 for testing the significance of the least-squares regression model. Enter the value of the explanatory variable for which you wish to make a prediction in the box following "X value(s)" under "Prediction of Y:". Enter a level of confidence. Click Compute!

14.2 Assess Your Understanding

Vocabulary and Skill Building

1. Intervals constructed about the predicted value of y, at a given level of x, that are used to measure the accuracy of the mean response of all individuals in the population are called _____ intervals for a(n) _____ response.

2. Intervals constructed about the predicted value of y, at a given level of x, that are used to measure the accuracy of a single individual's prediction are called _____ intervals for a(n) _____ response.

In Problems 3–6, use the results of Problems 5–8 in Section 14.1.

NW 3. Using the sample data from Problem 5 in Section 14.1,
(a) Predict the mean value of y if $x = 7$.
(b) Construct a 95% confidence interval for the mean value of y if $x = 7$.
(c) Predict the value of y if $x = 7$.
(d) Construct a 95% prediction interval for the value of y if $x = 7$.
(e) Explain the difference between the prediction in parts (a) and (c).

4. Using the sample data from Problem 6 in Section 14.1,
(a) Predict the mean value of y if $x = 8$.
(b) Construct a 95% confidence interval for the mean value of y if $x = 8$.
(c) Predict the value of y if $x = 8$.
(d) Construct a 95% prediction interval for the value of y if $x = 8$.
(e) Explain the difference between the prediction in parts (a) and (c).

5. Using the sample data from Problem 7 in Section 14.1,
(a) Predict the mean value of y if $x = 1.4$.
(b) Construct a 95% confidence interval for the mean value of y if $x = 1.4$.
(c) Predict the value of y if $x = 1.4$.
(d) Construct a 95% prediction interval for the value of y if $x = 1.4$.

6. Using the sample data from Problem 8 in Section 14.1,
(a) Predict the mean value of y if $x = 1.8$.
(b) Construct a 95% confidence interval for the mean value of y if $x = 1.8$.
(c) Predict the value of y if $x = 1.8$.
(d) Construct a 95% prediction interval for the value of y if $x = 1.8$.

Applying the Concepts

7. An Unhealthy Commute Use the results of Problem 11 from Section 14.1 to answer the following questions:

(a) Predict the mean well-being index composite score of all individuals whose commute time is 20 minutes.

(b) Construct a 90% confidence interval for the mean well-being index composite score of all individuals whose commute time is 20 minutes.

(c) Predict the well-being index composite score of Jane, whose commute time is 20 minutes.

(d) Construct a 90% prediction interval for the well-being index composite score of Jane, whose commute time is 20 minutes.

(e) Explain the difference between the predictions made in parts (a) and (c).

8. Credit Scores Use the results of Problem 12 from Section 14.1 to answer the following questions:

(a) Predict the mean interest rate of all individuals whose credit score is 730.

(b) Construct a 90% confidence interval for the mean interest rate of all individuals whose credit score is 730.

(c) Predict the interest rate of Kaleigh, whose credit score is 730.

(d) Construct a 90% prediction interval for the interest rate of Kaleigh, whose credit score is 730.

(e) Explain the difference between the prediction made in parts (a) and (c).

9. Height versus Head Circumference Use the results of Problem 13 from Section 14.1 to answer the following questions:

(a) Predict the mean head circumference of children who are 25.75 inches tall.

(b) Construct a 95% confidence interval for the mean head circumference of children who are 25.75 inches tall.

(c) Predict the head circumference of a randomly selected child who is 25.75 inches tall.

(d) Construct a 95% prediction interval for the head circumference of a child who is 25.75 inches tall.

(e) Explain the difference between the predictions in parts (a) and (c).

10. Hurricanes Use the results of Problem 14 in Section 14.1 to answer the following questions:

(a) Predict the mean wind speed of all hurricanes whose atmospheric pressure is 950 mb.

(b) Construct a 95% confidence interval for the mean wind speed found in part (a).

(c) Predict the wind speed of a randomly selected hurricane whose atmospheric pressure is 950 mb.

(d) Construct a 95% prediction interval for the wind speed found in part (c).

(e) Explain why the predicted lengths in parts (a) and (c) are the same, yet the intervals constructed in parts (b) and (d) are different.

11. Concrete Use the results of Problem 15 from Section 14.1 to answer the following questions:

(a) Predict the mean 28-day strength of concrete whose 7-day strength is 2550 psi.

(b) Construct a 95% confidence interval for the mean 28-day strength of concrete whose 7-day strength is 2550 psi.

(c) Predict the 28-day strength of concrete whose 7-day strength is 2550 psi.

(d) Construct a 95% prediction interval for the 28-day strength of concrete whose 7-day strength is 2550 psi.

(e) Explain the difference between the predictions in parts (a) and (c).

12. Tar and Nicotine Use the results of Problem 16 in Section 14.1 to answer the following questions:

(a) Predict the mean nicotine content of all cigarettes whose tar content is 12 mg.

(b) Construct a 95% confidence interval for the tar content found in part (a).

(c) Predict the nicotine content of a randomly selected cigarette whose tar content is 12 mg.

(d) Construct a 95% prediction interval for the nicotine content found in part (c).

(e) Explain why the predicted nicotine contents found in parts (a) and (c) are the same, yet the intervals constructed in parts (b) and (d) are different.

13. Invest in Education Use the results of Problem 17 in Section 14.1 to answer the following questions:

(a) Predict the mean annual ROI of all four-year schools whose 2013 cost is $180,000.

(b) Construct a 95% confidence interval for the mean annual ROI found in part (a).

(c) Predict the annual ROI of a randomly selected four-year school whose 2013 cost is $180,000.

(d) Construct a 95% prediction interval for the annual ROI found in part (c).

(e) Explain why the predicted lengths in parts (a) and (c) are the same, yet the intervals constructed in parts (b) and (d) are different.

14. American Black Bears Use the results of Problem 18 from Section 14.1 to answer the following questions:

(a) Predict the mean weight of American black bears with a total length of 154.5 cm.

(b) Construct a 95% confidence interval for the mean weight of American black bears with a total length of 154.5 cm.

(c) Predict the weight of a randomly selected American black bear that is 154.5 cm long.

(d) Construct a 95% prediction interval for the weight of an American black bear that is 154.5 cm long.

(e) Explain why the predicted weights in parts (a) and (c) are the same, yet the intervals constructed in parts (b) and (d) are different.

15. CEO Performance Use the results of Problem 19 from Section 14.1 to answer the following:

(a) Explain why it does not make sense to construct confidence or prediction intervals based on the least-squares regression equation.

(b) Construct a 95% confidence interval for the mean stock return.

16. Bear Markets Use the results of Problem 20 from Section 14.1 to answer the following:

(a) Explain why it does not make sense to construct confidence or prediction intervals based on the least-squares regression equation.

(b) Construct a 95% confidence interval for the mean percent change during a bear market.

17. Putting It Together: Predicting Intelligence Can a photograph of an individual be used to predict their intelligence? Researchers at Charles University in Prague, Czech Republic, had 160 raters analyze the photos of 80 students and asked each rater to rate the intelligence and attractiveness of the individual in the photo on a scale from one to seven. To eliminate individual bias in ratings, each rater's scores were converted to z-scores using each individual's mean rating. The perceived intelligence and attractiveness of each photo was calculated as the mean z-score. Go to www.pearsonhighered.com/sullivanstats to obtain the data file 14_2_17 using the file format of your choice. The following explains the variables in the data:

> sex: Gender of the individual in the photo
> age: Age of the individual in the photo
> perceived intelligence (ALL): Mean z-score of the
>   perceived intelligence of all 160 raters
> perceived intelligence (WOMEN): Mean z-score of the
>   perceived intelligence of the female raters
> perceived intelligence (MEN): Mean z-score of the
>   perceived intelligence of the male raters
> attractiveness (ALL): Mean z-score of the attractiveness
>   rating of all 160 raters
> attractiveness (MEN): Mean z-score of the attractiveness
>   rating of the male raters
> attractiveness (WOMEN): Mean z-score of the
>   attractiveness rating of the female raters
> IQ: Intelligence quotient based on the Czech version of
>   Intelligence Structure Test

Source: Kleisner K, Chvátalová V, Flegr J (2014) Perceived Intelligence Is Associated with Measured Intelligence in Men but Not Women. *PLoS One* 9(3): e81237. doi:10.1371/journal.pone.0081237

(a) Are attractive people perceived as more intelligent? Draw a scatter diagram between attractiveness (ALL) and perceived intelligence (ALL) for all 160 raters treating perceived intelligence as the response variable.

(b) What is the linear correlation coefficient between attractiveness and perceived intelligence for all 160 raters? Based on the linear correlation coefficient, does a linear relation exist between attractiveness and perceived intelligence?

(c) Treating perceived intelligence (ALL) as the response variable and attractiveness (ALL) as the explanatory variable, find the least-squares regression equation between these two variables.

(d) Provide an interpretation of the intercept.

(e) A normal probability plot confirms the residuals are normally distributed. Test whether a positive linear relation exists between perceived intelligence and attractiveness.

(f) Are higher IQs associated with higher perceived intelligence? Draw a scatter diagram between IQ and perceived intelligence for all 160 raters treating IQ as the response variable. What is the linear correlation coefficient between IQ and perceived intelligence (ALL)? Is this linear correlation coefficient suggestive of a linear relation between the two variables? Explain.

(g) Treating IQ as the response variable, find the least-squares regression between IQ and perceived intelligence (ALL) for females only (sex = F). Test whether a positive linear relation exists between perceived intelligence for females only and IQ. Use an $\alpha = 0.1$ level of significance.

(h) Treating IQ as the response variable, find the least-squares regression between IQ and perceived intelligence (ALL) for males only (sex = M). Test whether a positive linear relation exists between perceived intelligence for males only and IQ. Use an $\alpha = 0.1$ level of significance.

(i) Construct a 95% confidence interval for the mean IQ of males who have perceived intelligence of 1.28.

(j) Construct a 95% prediction interval for the IQ of a male whose perceived intelligence is 1.28.

14.3 Introduction to Multiple Regression

Preparing for This Section Before getting started, review the following:

- Correlation (Section 4.1, pp. 183–187)
- Least-Squares Regression (Section 4.2, pp. 197–204)
- Diagnostics on the Least-Squares Regression Line (Section 4.3, pp. 211–219)

Objectives
① Obtain the correlation matrix
② Use technology to find a multiple regression equation
③ Interpret the coefficients of a multiple regression equation
④ Determine R^2 and adjusted R^2
⑤ Perform an F-test for lack of fit
⑥ Test individual regression coefficients for significance
⑦ Construct confidence and prediction intervals

① Obtain the Correlation Matrix

The least-squares regression model $y_i = \beta_0 + \beta_1 x_i + \varepsilon_i$ introduced in Section 14.1 is a model for describing the linear relation between a single explanatory variable, x, and a response variable, y. However, it is often the case that more than one explanatory variable can be used to predict the value of a response variable. In these circumstances, we use a *multiple regression model*.

Definition

A **multiple regression model** is given by

$$y_i = \beta_0 + \beta_1 x_{1i} + \beta_2 x_{2i} + \cdots + \beta_k x_{ki} + \varepsilon_i \qquad \textbf{(1)}$$

where

y_i is the value of the response variable for the ith individual

x_{1i} is the value of the first explanatory variable for the ith observation, x_{2i} is the value of the second explanatory variable for the ith observation, and so on

$\beta_0, \beta_1, \beta_2, \ldots, \beta_k$ are the parameters to be estimated based on sample data

ε_i is a random error term that is normally distributed with mean 0 and standard deviation $\sigma_{\varepsilon_i} = \sigma$

The error terms are independent, and $i = 1, 2, \ldots, n$, where n is the sample size.

If there are two explanatory variables, model (1) reduces to

$$y_i = \beta_0 + \beta_1 x_{1i} + \beta_2 x_{2i} + \varepsilon_i \qquad \textbf{(2)}$$

If there are three explanatory variables, model (1) reduces to

$$y_i = \beta_0 + \beta_1 x_{1i} + \beta_2 x_{2i} + \beta_3 x_{3i} + \varepsilon_i \qquad \textbf{(3)}$$

Because we typically do not have access to population data, the parameters, $\beta_0, \beta_1, \beta_2$, and so on, must be estimated using sample data. Let b_0 represent the estimate of β_0, b_1 represent the estimate of β_1, and so on. Although formulas do exist for determining the values of $b_0, b_1, \ldots, b_k$, they are beyond the scope of this text. Therefore, we will be content to obtain estimates of the parameters using statistical software, such as Minitab, Excel, or StatCrunch.

In obtaining a multiple regression model, we must decide which variables to include in the model and which to exclude. One tool that will aid us in this decision is the *correlation matrix*.

Definition

A **correlation matrix** shows the linear correlation between each pair of variables under consideration in a multiple regression model.

When determining which explanatory variables to include in a multiple regression model, choose the ones that have a high linear correlation with the response variable. However, some caution is in order. We must guard against including explanatory variables that are highly correlated among themselves. That is, we must guard against *multicollinearity*.

Definition

Multicollinearity exists between two explanatory variables if they have a high linear correlation.

To help understand the idea of multicollinearity, consider the following scenario. We are trying to predict sales of lemonade at a corner store. Certainly, one variable that might help to explain lemonade sales is the outside temperature. Another variable that might help explain lemonade sales is air-conditioning bills. If the researcher includes both explanatory variables in the model, he may get results that are a little strange, because the two explanatory variables, temperature and air-conditioning bills, are themselves highly correlated. As temperatures increase, so do air-conditioning bills. It would be silly to include both variables in the model because they are both doing the same job when it comes to explaining lemonade sales. Of course, identifying multicollinearity is not always this obvious.

CAUTION!
If two explanatory variables in the regression model are highly correlated with each other, watch out for strange results in the regression output.

Multicollinearity does not necessarily cause problems for us, but it can. Some of the problems that result from multicollinearity include getting estimates of slope coefficients that are the opposite sign of what we would expect or obtaining estimates of slope coefficients that are not as large (or small) as we would expect.

We will look at a correlation matrix to determine which, if any, explanatory variables have a high correlation between them. For the sake of having a general rule, we will say that a linear correlation between two explanatory variables less than -0.7 or greater than 0.7 may be cause for concern. If there are two explanatory variables in the regression model that are highly correlated, we will keep an eye out for unexpected results. We will have more to say about this at the end of the section.

EXAMPLE 1   Constructing a Correlation Matrix

Problem A family doctor wishes to further examine the variables that affect his female patients' total cholesterol. He randomly selects 14 of his female patients and asks them to determine their average daily consumption of saturated fat. He then measures their total cholesterol and obtains the data in Table 4.

Table 4

Age, x_1	Saturated Fat (g), x_2	Total Cholesterol, y
25	19	180
25	28	195
28	19	186
32	16	180
32	24	210
32	20	197
38	31	239
42	20	183
48	26	204
51	24	221
51	32	243
58	21	208
62	21	228
65	30	269

Source: Based on results obtained from the National Center for Health Statistics

Approach Enter the data into Minitab and create the correlation matrix. The steps for creating a correlation matrix using Minitab, Excel, and StatCrunch are given in the Technology Step-by-Step on page 705.

Solution Figure 14 shows the correlation matrix from Minitab.

Figure 14   **Correlations: Age, Fat, Cholesterol**

```
                    Age        Fat
Fat               0.324
Cholesterol       0.718      0.778

Cell Contents: Pearson correlation
```

The linear correlation between total cholesterol and age is 0.718. The linear correlation between total cholesterol and average daily consumption of saturated fat is 0.778. Because the linear correlation between average daily consumption and age, the two explanatory variables, is only 0.324, we are not concerned with multicollinearity.

● **Now Work Problem 17(a)**

② Use Technology to Find a Multiple Regression Equation

When obtaining a multiple regression equation, perform the same diagnostic tests that we performed for the least-squares regression model with one explanatory variable. Always draw residual plots to verify that the model is appropriate. The residual plots that should be drawn are (1) a plot of residuals against fitted values, $\hat{y}$, to see if the linear model is appropriate and if the errors have constant variance, (2) a plot of residuals against each of the explanatory variables to make sure that the relation between the explanatory variable and response variable is linear, and (3) a boxplot of the residuals to check for outliers.

EXAMPLE 2   Multiple Regression

Problem Use the data in Table 4 presented in Example 1.

(a) Find the least-squares regression equation $\hat{y} = b_0 + b_1x_1 + b_2x_2$, where x_1 represents the patient's age, x_2 represents the patient's daily consumption of saturated fat, and y represents the patient's total cholesterol.

(b) Draw residual plots and a boxplot of the residuals to assess the adequacy of the model.

Approach Enter the data into Minitab to obtain the least-squares regression equation and to draw the residual plots and boxplot of the residuals. The steps for determining the multiple regression equation and residual plots using Minitab, Excel, and StatCrunch are given in the Technology Step-by-Step on page 705.

Solution

(a) Figure 15 shows the output from Minitab. The least-squares regression equation for these data is $\hat{y} = 90.8 + 1.014x_1 + 3.244x_2$.

Figure 15

```
Regression Analysis: Cholesterol versus Age, Fat

Analysis of Variance

Source       DF  Adj SS  Adj MS  F-Value  P-Value
Regression    2    7960  3980.1    30.53    0.000
  Age         1    2276  2276.4    17.46    0.002
  Fat         1    3120  3119.8    23.93    0.000
Error        11    1434   130.4
Total        13    9394


Model Summary

      S   R-sq  R-sq(adj)  R-sq(pred)
11.4179  84.73%    81.96%      75.32%


Coefficients

Term      Coef  SE Coef  T-Value  P-Value   VIF
Constant  90.8     16.0     5.68    0.000
Age      1.014    0.243     4.18    0.002  1.12
Fat      3.244    0.663     4.89    0.000  1.12


Regression Equation

Cholesterol = 90.8 + 1.014 Age + 3.244 Fat
```

(b) Figure 16(a) shows a plot of residuals against the predicted values, $\hat{y}$. To obtain this plot by hand, first find all the fitted values. For example, the predicted (or fitted) value for the first observation in Table 4 (age = 25, fat consumption = 19) is $\hat{y} = 90.8 + 1.014(25) + 3.244(19) = 177.8$ mg/dL. Therefore, residual = observed − predicted = $180 - 177.8 = 2.2$ mg/dL. Now plot the point $(177.8, 2.2)$ on the graph. Repeat this for the remaining observations. Figure 16(b) shows a plot of residuals against the explanatory variable, age. Figure 16(c) shows a plot of residuals against the explanatory variable saturated fat. Figure 16(d) shows a boxplot of the residuals. None of the residual plots show any discernible pattern, and the boxplot does not show any outliers. Therefore, the linear model is appropriate.

(continued)

Figure 16

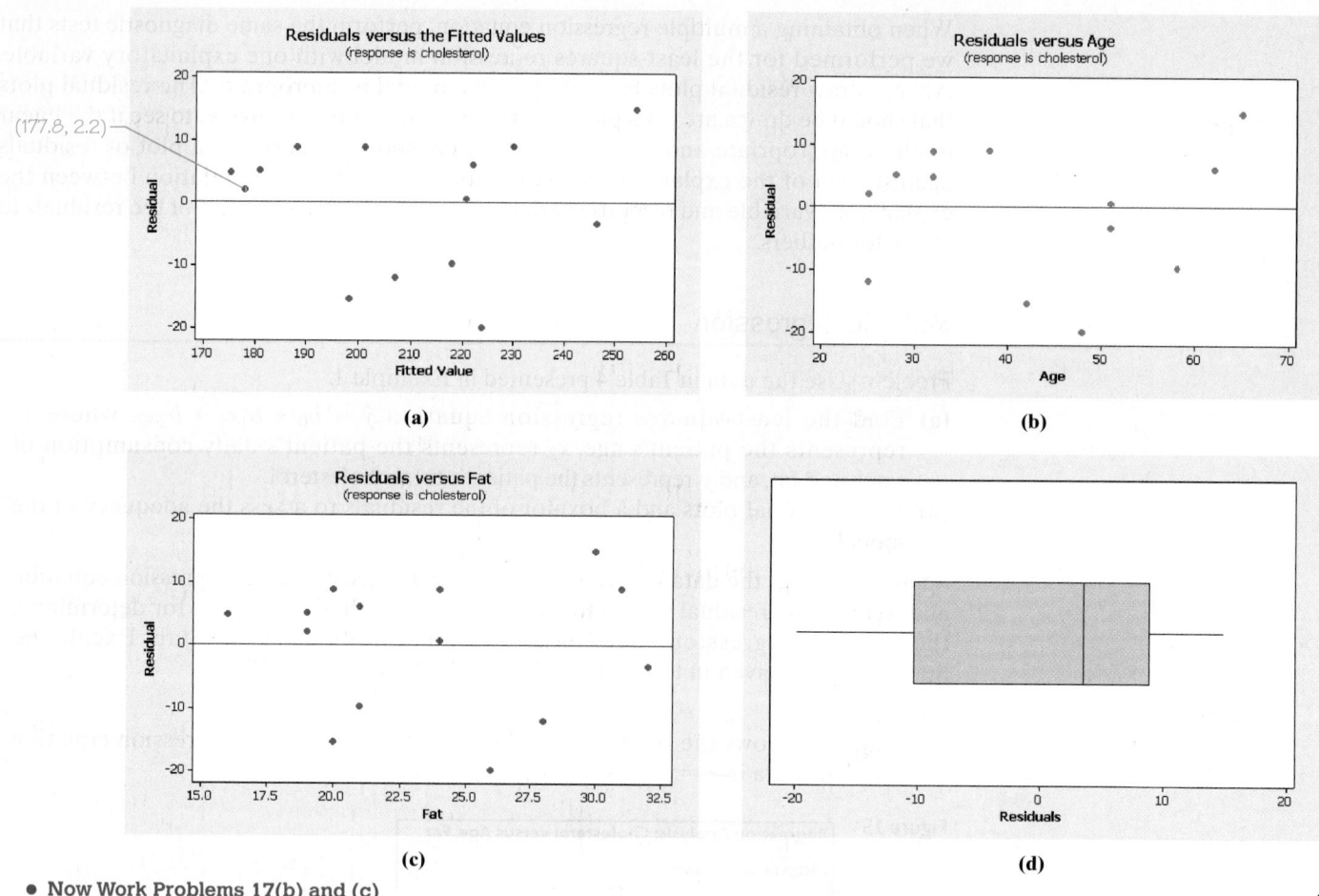

(a)

(b)

(c)

(d)

● **Now Work Problems 17(b) and (c)**

③ Interpret the Coefficients of a Multiple Regression Equation

For the simple linear regression model the slope, β_1, represents the *average* or *expected* change in the value of the response variable for a 1-unit change in the value of the explanatory variable. For example, if income, y, and years of education, x, are related by the equation $\hat{y} = 17{,}203 + 3423x$, then the slope, 3423, can be interpreted as follows: If the number of years of education increases by one year, income increases by \$3423, on average.

In the multiple regression model, we interpret the regression coefficients similarly. However, because of other variables in the model, we assume that the remaining variables stay constant when interpreting the slope coefficient of a particular variable. If we have three explanatory variables, the multiple regression is $\hat{y} = b_0 + b_1x_1 + b_2x_2 + b_3x_3$. So the interpretation of the regression coefficient b_1 is that "If x_1 increases by 1 unit, then the response variable y will increase by b_1 units, on average, while x_2 and x_3 are held constant." The interpretation of the regression coefficients b_2 and b_3 is similar.

EXAMPLE 3   Interpreting Regression Coefficients

Problem Interpret the regression coefficients for the least-squares regression equation found in Example 2.

Approach Interpret the coefficients the same way as we did for a regression with one explanatory variable. However, assume that the remaining explanatory variables are constant.

Solution The least-squares regression equation found in Example 2 was $\hat{y} = 90.8 + 1.014x_1 + 3.244x_2$, where x_1 represents the patient's age and x_2 represents the patient's average daily consumption of saturated fat (in grams). The regression coefficient, $b_1 = 1.014$, means that, if the patient's age increases by one year then total cholesterol increases by 1.014 mg/dL, on average, assuming that the average daily consumption of saturated fat remains unchanged. The regression coefficient, $b_2 = 3.244$, means that, if the patient's average daily consumption of saturated fat increases by 1 gram, total cholesterol increases by 3.244 mg/dL, on average, assuming that age remains unchanged.

● Now Work Problem 17(d)

Interaction Effects

In Example 3, we were able to interpret the regression coefficients by holding the value of one explanatory variable fixed and determining how a 1-unit change in the other explanatory variable affects the response variable. We were able to do this because the explanatory variables had an **additive effect**, or did not interact. One hint that the explanatory variables do not interact comes from the low correlation between them. A positive correlation between two explanatory variables, x_1 and x_2, implies that, if x_1 changes then x_2 is likely to change. For example, if we wish to describe the percentage of body fat in an individual, we probably do not want our explanatory variables to be skinfold thickness on the tricep and skinfold thickness on the thigh. Because these two explanatory variables are highly correlated, an increase in skinfold thickness of the tricep is accompanied by an increase in skinfold thickness of the thigh. We will have more to say about interaction between two explanatory variables in the next section.

④ ## Determine R^2 and Adjusted R^2

Recall from Section 4.3 that the coefficient of determination, R^2, measures the percentage of total variation in the response variable that is explained by the least-squares regression line. We also said that the computation of R^2 for the least-squares regression model with one explanatory variable was as easy as computing the square of the linear correlation coefficient, r, between the explanatory and response variable. That is, $R^2 = r^2$. While the interpretation of R^2 is the same for the least-squares regression model with two or more explanatory variables, its computation is by no means that straightforward. Once again, we will be content in using software to supply the value of R^2. Although we do not provide detail as to the computation of R^2, it is important to recognize that R^2 is the ratio of variation in the response variable that is explained by the regression equation to the total variation in the response variable. That is,

$$R^2 = \frac{\text{explained variation}}{\text{total variation}} = 1 - \frac{\text{unexplained variation}}{\text{total variation}}$$

Remember, the unexplained variation is found by summing the squares of the residuals. The sum of the squared residuals will never increase if more explanatory variables are added. The total variation is constant regardless of the number of explanatory variables. Therefore, the coefficient of determination, R^2, will never decrease with the addition of more explanatory variables; instead, it generally increases. So R^2 can often be made as large as we like simply by adding more explanatory variables. To compensate for the ability to artificially inflate R^2 by adding more explanatory variables, it is recommended that the *adjusted* R^2 be used when working with least-squares regression models with two or more explanatory variables.

In Other Words
The value of R^2 always increases by adding one more explanatory variable.

Definition | The **adjusted R^2**, denoted R^2_{adj} or $\overline{R}^2$, is the adjusted coefficient of determination. It modifies the value of R^2 based on the sample size, n, and the number of explanatory variables, k. The formula for the adjusted R^2 is

$$R^2_{adj} = 1 - \left(\frac{n-1}{n-k-1} \right)(1 - R^2)$$

The adjusted R^2 will decrease if an explanatory variable is added to the model that does little to explain the variation in the response variable. This is because the decrease in the unexplained variation is offset by the increase in the number of explanatory variables k.

EXAMPLE 4   Coefficient of Determination

Problem For the model obtained in Example 2, determine the coefficient of determination and the adjusted R^2. Compare the R^2 with the two explanatory variables age and daily saturated fat to the R^2 with the single explanatory variable age. Comment on the effect the additional explanatory variable has on the value of the model.

Approach Use Minitab to determine the values of the coefficient of determination and the adjusted coefficient of determination.

Solution Looking back to the output in Figure 15, we see that $R^2 = 0.847 = 84.7\%$. This means that 84.7% of the variation in the response variable is explained by the least-squares regression model. In addition, we find that $R^2_{adj} = 0.820 = 82.0\%$. If we look back to Figure 2 on page 674, we see that $R^2 = 51.5\%$ with age as the only explanatory variable. Based on the adjusted R^2 with the two explanatory variables age and daily saturated fat consumption, we conclude that the additional variable increases the proportion of explained variation.

> **CAUTION!**
> Never use R^2 to compare regression models with a different number of explanatory variables. Rather, use the adjusted R^2.

● Now Work Problem 17(e)

⑤   Perform an F-Test for Lack of Fit

We now wish to determine whether there is a significant linear relation between the explanatory variables and response variable. That is, we want to test

$$H_0: \beta_1 = \beta_2 = \cdots = \beta_k = 0 \quad \text{versus} \quad H_1: \text{at least one } \beta_i \neq 0$$

In Other Words
The null hypothesis states that there is no linear relation between the explanatory variables and the response variable. The alternative hypothesis states that there is a linear relation between at least one explanatory variable and the response variable.

Just as in analysis of variance, the computations in multiple regression can be cumbersome. Therefore, we are content to let statistical software perform the computations.

We use ANOVA to test if there is a significant linear relation between the explanatory variables and response variable. Recall from Section 13.1 that $F_0 = \dfrac{\text{MST}}{\text{MSE}}$ follows Fisher's F-distribution, where MST is the mean square due to treatment and MSE is the mean square error. In the context of regression analysis, the treatment is the regression line itself, and the residuals make up the error, which leads to the F-test statistic.

Test Statistic for Multiple Regression

$$F_0 = \frac{\text{mean square due to regression}}{\text{mean square error}} = \frac{\text{MSR}}{\text{MSE}}$$

with $k - 1$ degrees of freedom in the numerator and $n - k$ degrees of freedom in the denominator, where k is the number of explanatory variables and n is the sample size.

Because the coefficient of determination, R^2, represents the proportion of explained variation in the response variable and $1 - R^2$ represents the proportion of unexplained variation, we can express the F-test statistic using the coefficient of determination.

F-Test Statistic for Multiple Regression Using R^2

$$F_0 = \frac{R^2}{1 - R^2} \cdot \frac{n - (k + 1)}{k}$$

where R^2 is the coefficient of determination
k is the number of explanatory variables
n is the sample size

A large F-test statistic provides evidence against the null hypothesis that all the slope coefficients are zero because a large F-test statistic implies that the ratio of explained variation to unexplained variation is large. Care must be taken in using this as the sole judge in rejecting the null hypothesis, because a large number of explanatory variables (that is, a large value of k) will cause the ratio $\dfrac{n - (k + 1)}{k}$ to decrease and therefore cause the F-test statistic to decrease.

Rather than comparing the F-test statistic to a critical value, we continue to use the P-value approach to testing hypotheses. We will use the P-value reported in the output of the statistical software package.

Decision Rule for Testing H_0: $\beta_1 = \beta_2 = \cdots = \beta_k = 0$

If the P-value is less than the level of significance, α, reject the null hypothesis. Otherwise, do not reject the null hypothesis.

EXAMPLE 5   Inference on the Regression Model

Problem Test H_0: $\beta_1 = \beta_2 = 0$ versus H_1: at least one $\beta_i \neq 0$ for the regression model found in Example 2.

Approach To perform this inference, we need to determine whether it is reasonable to believe that the residuals are normally distributed with no outliers. To verify this requirement, we draw a normal probability plot. Once this requirement has been verified, we look at the P-value associated with the F-test statistic. If the P-value is less than the level of significance, α, we reject the null hypothesis.

Solution Figure 17 shows the normal probability plot of the residuals drawn in Minitab.

Figure 17

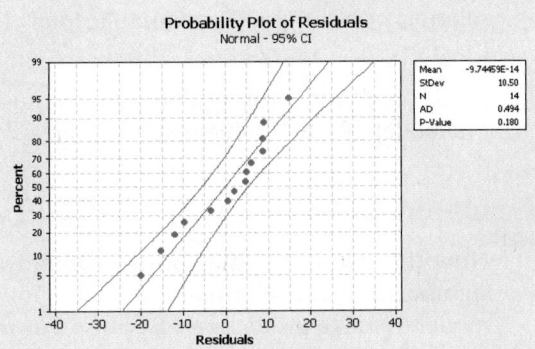

Because the points are roughly linear (or lie within the confidence bounds drawn by Minitab), it is reasonable to conclude that the residuals are normally distributed. Alternatively, the correlation between the residuals and expected z-scores is 0.967. Because $0.967 > 0.935$ (Table VI), the residuals appear to be normally distributed.

(continued)

In addition, the boxplot of the residuals drawn in Figure 16(d) on page 698 indicates that there are no outliers. Therefore, we can continue with the inference.

If we look back at the output in Figure 15 on page 697, we see that the F-test statistic is 30.53 with a P-value reported as 0.000. This small P-value is sufficient evidence against the null hypothesis. So we reject the null hypothesis and conclude that at least one of the regression coefficients is different from zero. There is a linear relation between at least one of the explanatory variables and the response variable.

● **Now Work Problem 17(f)**

--- **CAUTION!** ---------------
If we reject the null hypothesis that all the slope coefficients are zero, then we are saying that at least one of the slopes is different from zero, *not* that they *all* are different from zero.

By rejecting the null hypothesis in Example 5, we know that *at least* one explanatory variable has a coefficient that is different from zero. Of course, this conclusion leads to this question: Which explanatory variables have a significant linear relation with the response variable?

6 **Test Individual Regression Coefficients for Significance**

We can use the t-distribution, along with P-values, to determine whether a specific explanatory variable has a significant linear relation with the response variable. Again, if the P-value is sufficiently small (say, less than $\alpha = 0.05$), we conclude that the explanatory variable has a significant linear relation with the response variable.

EXAMPLE 6 Testing the Significance of Individual Predictor Variables

Problem For our model presented in Example 2, test the hypotheses

$$H_0: \beta_1 = 0 \qquad H_0: \beta_2 = 0$$
$$\text{vs.} \qquad \text{vs.}$$
$$H_1: \beta_1 \neq 0 \qquad H_1: \beta_2 \neq 0$$

where β_1 is the slope coefficient for the explanatory variable age, and β_2 is the slope coefficient for the explanatory variable saturated fat.

Approach We will use Minitab to determine the P-values for each explanatory variable. If the P-value is less than the level of significance, we reject the null hypothesis and conclude that the slope coefficient is different from zero.

Solution Looking again to the output in Figure 15 on page 697, we see that the test statistic for the explanatory variable age is $t_0 = 4.18$ with a P-value = 0.002. We also see that the test statistic for the explanatory variable saturated fat is $t_0 = 4.89$ with a P-value = 0.000. Since both P-values are sufficiently small, we reject each null hypothesis. We conclude that both explanatory variables have a significant linear relation with the response variable, total cholesterol.

● **Now Work Problem 17(g)**

If the P-value for a slope coefficient is large, remove the explanatory variable from the model. If the P-value is small, keep the explanatory variable in the model.

7 **Construct Confidence and Prediction Intervals**

Now that we know that both explanatory variables age and saturated fat have a significant linear relation with the response variable total cholesterol, we can use the model to make predictions. Because our model is based on sample data, we know that variability will be associated with any predictions. Therefore, we will construct intervals about our predicted value and describe the accuracy of our predictions.

Recall from Section 14.2 that there are two interpretations to a predicted value:

1. It represents a mean value of the response variable.
2. It represents a predicted value for a specific individual.

Because of this, we again require two types of intervals. One type, the confidence interval for a mean response, corresponds to an interval about the mean value; the other type, the prediction interval for an individual response, corresponds to an interval for a specific individual. Once again, we will leave the heavy work to Minitab and let it construct the intervals. Our job will be to interpret the results.

EXAMPLE 7   Confidence and Prediction Intervals

Problem Construct a 95% confidence interval for a mean response and a 95% prediction interval for an individual response for a 32-year-old female who consumes 23 grams of saturated fat daily using the model determined in Example 2. Be sure to interpret the results.

Approach We will use Minitab to construct the intervals. The steps for obtaining confidence and prediction intervals using Minitab and StatCrunch are given in the Technology Step-by-Step on page 705.

Solution Figure 18 shows the partial output from Minitab.

Figure 18

```
Prediction for Cholesterol

Regression Equation

Cholesterol = 90.8 + 1.014 Age + 3.244 Fat


Variable  Setting
Age          32
Fat          23


    Fit    SE Fit       95% CI            95% PI
197.914   3.84623  (189.449, 206.380)  (171.396, 224.432)
```

Based on the Minitab output, we are 95% confident that the mean total cholesterol of all 32-year-old females who consume 23 grams of saturated fat daily is between 189.4 mg/dL and 206.4 mg/dL. We are 95% confident that the total cholesterol of a particular 32-year-old female who consumes 23 grams of saturated fat daily is between 171.4 mg/dL and 224.4 mg/dL.

• **Now Work Problems 17(h), (i), and (j)**

Multicollinearity Revisited

Suppose that two explanatory variables that have a strong linear association with the response variable are highly correlated with each other. One consequence of this correlation can be that their t-test statistics are small and their P-values are large. The reason for this is that the t-test statistics are *marginal*. The word *marginal* means incremental. For example, if we had a regression model with three explanatory variables x_1, x_2, and x_3, the t-statistic for x_1 is a measure for how much additional explanation the variable x_1 adds to the model when x_2 and x_3 are already present. Likewise, the t-statistic for x_2 is a measure for how much additional explanation the variable x_2 adds to the model when x_1 and x_3 are already present.

So a small t-test statistic for an explanatory variable means that a variable does not yield much additional explanation. Suppose that x_1 and x_2 are highly correlated to each other and to the response variable. If we include both explanatory variables in the model, the high correlation may lead to small t-test statistics for x_1 and x_2. This leads us to believe that neither variable is important.

In fact, the model may be "confused." This confusion results because the model believes that not much additional information is learned by adding x_1 to the model when x_2 is already in the model. In addition, x_2 does not provide much additional information when x_1 is already in the model. The model concludes that neither variable is important.

EXAMPLE 8   Effects of Multicollinearity

Problem Using the data presented in Table 5:

(a) Find the correlation matrix among all three variables.

(b) Find the least-squares regression model using both x_1 and x_2 as explanatory variables.

(c) Comment on the effect that including both x_1 and x_2 has on the t-test statistics.

Approach We will use StatCrunch to conduct the analysis.

Solution

(a) Figure 19 shows the correlation matrix. An extremely high correlation exists between x_1 and x_2, so multicollinearity exists between the two variables.

Table 5

x_1	x_2	y
11.4	−9.7	14.7
12.5	−11.5	38.8
16.4	−15.9	42.9
14.4	−13.9	45.7
15.3	−14.2	52.3
18	−18.5	55.9
19.5	−21.2	60.1
25.2	−27.2	72.6

Figure 19   **Correlation matrix:**

	x1♦	x2♦
x2	-0.99607612	
y	0.89091722	-0.89445738

(b) Figure 20 shows the regression output from StatCrunch. First, notice that the P-value for the F-test statistic is 0.0179, indicating that at least one of the slope coefficients is different from zero. However, if we look at each individual t-test statistic, we see that each has a very high P-value (P-value for x_1: 0.9987; P-value for x_2: 0.7074), indicating that neither coefficient is different from zero.

Figure 20   **Multiple linear regression results:**
Dependent Variable: y
Independent Variable(s): x1, x2

Parameter estimates:

Parameter♦	Estimate♦	Std. Err.♦	Alternative♦	DF♦	T-Stat♦	P-value♦
Intercept	3.1966894	35.467992	≠ 0	5	0.090128851	0.9317
x1	-0.015176864	8.8287255	≠ 0	5	-0.0017190323	0.9987
x2	-2.7209724	6.8440747	≠ 0	5	-0.39756615	0.7074

Analysis of variance table for multiple regression model:

Source	DF	SS	MS	F-stat	P-value
Model	2	1645.8514	822.92568	10.003384	0.0179
Error	5	411.32364	82.264729		
Total	7	2057.175			

Summary of fit:
Root MSE: 9.0699906
R-squared: 0.8001
R-squared (adjusted): 0.7201

(c) The contradictory results of the regression output occur because both x_1 and x_2 are related to the response variable y, as indicated by the correlation matrix. However, x_1 and x_2 are also related to each other. So, with x_1 in the model, x_2 adds little explanation. Likewise, with x_2 in the model, x_1 adds little explanation. The solution is to use only one explanatory variable. Which explanatory variable we choose is up to us. We can choose either the explanatory variable with the lower P-value or the explanatory variable that has the higher correlation with the response.

● Now Work Problem 19

Technology Step-by-Step   Multiple Regression

TI-83/84 Plus
The TI-83/84 Plus graphing calculators do not compute multiple regression equations.

Minitab
Correlation Matrix
1. Enter the explanatory variables and response variable into the spreadsheet.
2. Select the **Stat** menu and highlight **Basic Statistics**. Now select **Correlation . . . .**
3. Highlight all the variables in the list of variables and click Select. Click OK.

Determining the Multiple Regression Equation and Residual Plots
1. Select the **Stat** menu and highlight **Regression**. Highlight **Regression**. Select **Fit Regression Model . . . .**
2. Place the cursor in the "Responses:" cell. Highlight the response variable and click Select. Place the cursor in the "Continuous predictors:" cell. Highlight the explanatory variables and click Select.
3. Click **Graphs . . . .** In the cell that says "Residuals versus the variables" enter the names of the explanatory variables. Select the box that says "Residuals versus fits." Click OK.
4. Click **Storage . . . .** Select the box that says "Residuals." Select OK twice. The residuals are stored in the spreadsheet. Draw a normal probability plot of the residuals as indicated in Section 7.3 and a boxplot of the residuals as indicated in Section 3.5.

Prediction and Confidence Intervals
See the steps given on page 692 in Section 14.2.

Excel
Correlation Matrix
1. Enter the values for the explanatory variables and response variable into the spreadsheet.
2. Be sure the Data Analysis ToolPak is activated. Select the Data menu; then select **Data Analysis**. Select Correlation and click OK.
3. With the cursor in the "Input Range:" cell, highlight the data. Click OK.

Determining the Multiple Regression Equation and Residual Plots
1. Select the Data menu; then select **Data Analysis**. Select Regression and click OK.
2. With the cursor in the "Input Y Range:" cell, highlight the data for the response variable. With the cursor in the "Input X Range:" cell, highlight the data for the explanatory variables. Check the boxes for <u>R</u>esiduals, Residual Plots, and <u>N</u>ormal Probability Plots. Click OK. The residuals are stored in the spreadsheet. Draw a boxplot of the residuals as indicated in Section 3.5.

   Note: We can also use the XLSTAT add-in to compute the multiple regression equation and to construct some of the related graphs.

Prediction and Confidence Intervals
Excel does not compute prediction and confidence intervals for predictions from multiple regression equations.

StatCrunch
Correlation Matrix
1. Enter the explanatory variables and response variable into the spreadsheet.
2. Select **Stat**, highlight **Summary Stats**, and select **Correlation**.
3. Click on the variables whose correlation you wish to determine. Click Calculate.

Determining the Multiple Regression Equation and Residual Plots
1. Enter the explanatory variables and the response variable into the spreadsheet. Name each column variable.
2. Select **Stat**, highlight **Regression**, and select **Multiple Linear**.
3. Choose the response variable for the Y variable, the explanatory variables for the X variables. Choose None for variable selection. Check any of the options you want. Click Compute!

Prediction and Confidence Intervals
Enter the values of each explanatory variable in the spreadsheet. Leave the value of the response variable blank. Highlight "95% confidence interval for the mean response" or "95% confidence interval for individual prediction". Click Compute!

 14.3 Assess Your Understanding

Vocabulary and Skill Building

1. A _____ _____ shows the linear correlation between each pair of variables under consideration in a multiple regression model.
2. If the correlation between two explanatory variables is high, the least-squares regression model may suffer from _____.
3. Suppose a multiple regression model is given by $\hat{y} = 4.39x_1 - 8.75x_2 + 34.09$. An interpretation of the coefficient of x_1 would be, "if x_1 increases by 1 unit, then the response variable will increase by _____ units, on average, while holding x_2 constant."
4. *True or False*: The value of R^2 never decreases as more explanatory variables are added to a regression model.
5. You obtain the multiple regression equation $\hat{y} = 5 + 3x_1 - 4x_2$ from a set of sample data.

   (a) Interpret the slope coefficients for x_1 and x_2.
   (b) Determine the regression equation with $x_1 = 10$. Graph the regression equation with $x_1 = 10$.
   (c) Determine the regression equation with $x_1 = 15$. Graph the regression equation with $x_1 = 15$.

(d) Determine the regression equation with $x_1 = 20$. Graph the regression equation with $x_1 = 20$.

(e) What is the effect of changing the value x_1 on the graph of the regression equation?

6. You obtain the multiple regression equation $\hat{y} = -5 - 9x_1 + 4x_2$ from a set of sample data.

(a) Interpret the slope coefficients for x_1 and x_2.

(b) Determine the regression equation with $x_1 = 10$. Graph the regression equation with $x_1 = 10$.

(c) Determine the regression equation with $x_1 = 15$. Graph the regression equation with $x_1 = 15$.

(d) Determine the regression equation with $x_1 = 20$. Graph the regression equation with $x_1 = 20$.

(e) What is the effect of changing the value x_1 on the graph of the regression equation?

7. A multiple regression model has $k = 3$ explanatory variables. The coefficient of determination, R^2, is found to be 0.653 based on a sample of $n = 25$ observations.

(a) Compute the adjusted R^2.

(b) Compute the F-test statistic.

(c) If one additional explanatory variable is added to the model and R^2 increases to 0.665, compute the adjusted R^2. Would you recommend adding the additional explanatory variable to the model? Why or why not?

8. A multiple regression model has $k = 4$ explanatory variables. The coefficient of determination, R^2, is found to be 0.542 based on a sample of $n = 40$ observations.

(a) Compute the adjusted R^2.

(b) Compute the F-test statistic.

(c) If one additional explanatory variable is added to the model and R^2 increases to 0.579, compute the adjusted R^2. Would you recommend adding the additional explanatory variable to the model? Why or why not?

9. For the data set

x_1	x_2	x_3	y
0.8	2.8	2.5	11.0
3.9	2.6	5.7	10.8
1.8	2.4	7.8	10.6
5.1	2.3	7.1	10.3
4.9	2.5	5.9	10.3
8.4	2.1	8.6	10.3
12.9	2.3	9.2	10.0
6.0	2.0	1.2	9.4
14.6	2.2	3.7	8.7
9.3	1.1	5.5	8.7

(a) Construct a correlation matrix between x_1, x_2, x_3, and y. Is there any evidence that multicollinearity exists? Why?

(b) Determine the multiple regression line with x_1, x_2, and x_3 as the explanatory variables.

(c) Assuming that the requirements of the model are satisfied, test $H_0: \beta_1 = \beta_2 = \beta_3 = 0$ versus H_1: at least one of the β_i is different from zero at the $\alpha = 0.05$ level of significance.

(d) Assuming that the requirements of the model are satisfied, test $H_0: \beta_i = 0$ versus $H_1: \beta_i \neq 0$ for $i = 1, 2, 3$ at the $\alpha = 0.05$ level of significance.

10. For the data set

x_1	x_2	x_3	y
24.9	13.5	3.7	59.8
26.7	15.7	11.4	66.3
30.6	13.8	15.7	76.5
39.6	8.8	8.8	77.1
33.1	10.6	18.3	81.9
41.1	9.7	21.8	84.6
25.4	9.8	16.4	87.3
33.8	6.8	25.9	88.5
23.5	7.5	15.5	90.7
39.8	6.8	30.8	93.4

(a) Construct a correlation matrix between x_1, x_2, x_3, and y. Is there any evidence that multicollinearity exists? Why?

(b) Determine the multiple regression line with x_1, x_2, and x_3 as the explanatory variables.

(c) Assuming that the requirements of the model are satisfied, test $H_0: \beta_1 = \beta_2 = \beta_3 = 0$ versus H_1: at least one of the β_i is different from zero at the $\alpha = 0.05$ level of significance.

(d) Assuming that the requirements of the model are satisfied, test $H_0: \beta_i = 0$ versus $H_1: \beta_i \neq 0$ for $i = 1, 2, 3$ at the $\alpha = 0.05$ level of significance. Should a variable be removed from the model? Why?

(e) Remove the variable identified in part (d) and recompute the regression model. Test whether at least one regression coefficient is different from zero. Then test whether each individual regression coefficient is significantly different from zero.

11. For the data set

x_1	x_2	x_3	x_4	y
43	19.6	7.1	32	200
44	13.1	58.5	37	204
40	24.7	2.1	32	215
35	30.4	41.4	39	229
38	28.2	7.7	30	231
39	24.9	25.0	26	243
39	45.7	28.5	25	266
40	38.4	27.7	24	278
47	36.9	26.2	17	287
35	66.3	4.2	23	298
36	112.8	26.2	21	339
44	108.4	22.3	24	359

(a) Construct a correlation matrix between x_1, x_2, x_3, x_4, and y. Is there any evidence that multicollinearity may be a problem?

(b) Determine the multiple regression line using all the explanatory variables listed. Does the F-test indicate that we should reject $H_0: \beta_1 = \beta_2 = \beta_3 = \beta_4 = 0$? Which explanatory variables have slope coefficients that are not significantly different from zero?

(c) Remove the explanatory variable with the highest P-value from the model and recompute the regression model. Does the F-test still indicate that the model is significant? Remove any additional explanatory variables on the basis of the

P-value of the slope coefficient. Then compute the model with the variable removed.

(d) Draw residual plots and a boxplot of the residuals to assess the adequacy of the model.

(e) Use the model constructed in part (c) to predict the value of y if $x_1 = 34$, $x_2 = 35.6$, $x_3 = 12.4$, and $x_4 = 29$.

(f) Draw a normal probability plot of the residuals. Is it reasonable to construct confidence and prediction intervals?

(g) Construct 95% confidence and prediction intervals if $x_1 = 34$, $x_2 = 35.6$, $x_3 = 12.4$, and $x_4 = 29$.

12. For the data set

x_1	x_2	x_3	x_4	y
47.3	0.9	4	76	105.5
53.1	0.8	6	55	113.8
56.7	0.8	4	65	115.2
48.8	0.5	7	67	118.9
42.7	1.1	7	74	148.9
44.3	1.1	6	76	120.2
44.5	0.7	8	68	121.6
37.7	0.7	7	79	140.0
36.9	1.0	5	73	141.5
28.1	1.8	6	68	141.9
32.0	0.8	8	81	152.8
34.7	0.8	10	68	156.5

(a) Construct a correlation matrix between x_1, x_2, x_3, x_4, and y. Is there any evidence that multicollinearity may be a problem?

(b) Determine the multiple regression line using all the explanatory variables listed. Does the F-test indicate that we should reject $H_0: \beta_1 = \beta_2 = \beta_3 = \beta_4 = 0$? Which explanatory variables have slope coefficients that are not significantly different from zero?

(c) Remove the explanatory variable with the highest P-value from the model and recompute the regression model. Does the F-test still indicate that the model is significant? Remove any additional explanatory variables on the basis of the P-value of the slope coefficient. Then compute the model with the variable removed.

(d) Draw residual plots and a boxplot of the residuals to assess the adequacy of the model.

(e) Use the final model constructed in part (c) to predict the value of y if $x_1 = 44.3$, $x_2 = 1.1$, $x_3 = 7$, and $x_4 = 69$.

(f) Draw a normal probability plot of the residuals. Is it reasonable to construct confidence and prediction intervals?

(g) Construct 95% confidence and prediction intervals if $x_1 = 44.3$, $x_2 = 1.1$, $x_3 = 7$, and $x_4 = 69$.

Applying the Concepts

13. Resisting the Computer Researchers Alfred P. Rovai and Marcus D. Childress asked the following question: "How can resistance to reduction of computer anxiety among teacher education students be explained and predicted?" (*Journal of Research on Technology in Education*, 35(2)). To answer this research question, they identified 86 undergraduate teacher education students enrolled in a computer literacy course and administered a series of questionnaires that quantified various predictors of the response variable, computer anxiety, y. For

example, computer anxiety was measured by administering each student the Computer Anxiety Scale. This score ranges from 20 to 100, with higher scores indicating higher levels of computer anxiety. The explanatory variables were:

x_1: Computer confidence on a scale from 10 to 40, with higher scores indicating higher confidence

x_2: Computer knowledge on a scale from 0 to 33, with 0 indicating no computer knowledge and 33 indicating superior computer knowledge

x_3: Computer liking on a scale from 10 to 40, with higher scores indicating a greater like of computers

x_4: Trait anxiety on a scale from 20 to 80, with higher scores indicating a higher level of overall anxiety

The multiple regression model was

$$\hat{y} = 84.04 - 0.87x_1 - 0.51x_2 - 0.45x_3 + 0.33x_4$$

(a) The reported P-value of the regression model was less than 0.0001. Would you reject the null hypothesis $H_0: \beta_1 = \beta_2 = \beta_3 = \beta_4 = 0$?

(b) Interpret the slope coefficients of the model in part (a). Are they all reasonable?

(c) Predict the computer anxiety score of an individual whose computer confidence score was 25, computer knowledge score was 19, computer liking score was 20, and trait anxiety score was 43.

(d) The coefficient of determination for this model is 0.69. Interpret this value.

(e) The article states that "regression assumptions were tested and found to be tenable." Explain what this means.

14. Pistol Shooting Researchers at Victoria University wanted to determine the factors that affect precision in shooting air pistols. "Inter- and Intra-Individual Analysis in Elite Sport: Pistol Shooting," *Journal of Applied Biomechanics, 28–38*, 2003. The explanatory variables were

x_1: Percent of the time the shooter's aim was on target (a measure of accuracy)

x_2: Percent of the time the shooter's aim was within a certain region (a measure of consistency or steadiness)

x_3: Distance (mm) the barrel of the pistol moves horizontally while aiming

x_4: Distance (mm) the barrel of the pistol moves vertically while aiming

(a) One response variable in the study was the score that the individual received on the shot, with a higher score indicating a better shooter. The regression model presented was $\hat{y} = 10.6 + 0.02x_1 - 0.03x_3$. The reported P-value of the regression model was 0.05. Would you reject the null hypothesis $H_0: \beta_1 = \beta_3 = 0$?

(b) Interpret the slope coefficients of the model in part (a).

(c) Predict the score of an individual whose aim was on target $x_1 = 20\%$ of the time with a distance the pistol barrel moves horizontally of $x_3 = 12$ mm using the model from part (a).

(d) A second response variable in the study was the vertical distance that the bullet hole was from the target. The regression model for this response variable was $\hat{y} = -24.6 - 0.13x_1 + 0.21x_2 + 0.13x_3 + 0.22x_4$. The reported P-value of the regression model was 0.04. Would you reject the null hypothesis $H_0: \beta_1 = \beta_2 = \beta_3 = \beta_4 = 0$?

(e) Interpret the slope coefficients of the model in part (d).

(f) Based on your answer to part (e), do you think that the model is useful in predicting vertical distance from the target? Why?

15. Wind Chill Temperature Wanting to know if there is a linear relation among wind chill temperature, air temperature (in degrees Fahrenheit), and wind speed (in miles per hour), a researcher collected the following data for various days.

Air Temp.	Wind Speed	Wind Chill
15	10	3
15	15	0
15	25	−4
0	5	−11
0	20	−22
−5	10	−22
−5	25	−31
−10	15	−32
−10	20	−35
−15	25	−44
−15	35	−48
−15	50	−52
5	40	−22
10	45	−16

(a) Find the least-squares regression equation $\hat{y} = b_0 + b_1x_1 + b_2x_2$, where x_1 is air temperature, x_2 is wind speed, and y is the response variable "wind chill."

(b) Draw residual plots to assess the adequacy of the model. What might you conclude based on the plot of residuals against wind speed?

16. Heat Index Wanting to know if there is a linear relation among heat index, air temperature (in degrees Fahrenheit), and dew point, a researched collected the following data for various days.

Air Temp.	Dew Point	Heat Index
90	64	93
90	68	95
94	66	99
94	70	102
96	70	105
96	76	111
99	68	107
99	72	111
100	74	114
100	80	123
93	72	103
93	78	109
97	80	118
92	82	114
95	66	100
95	82	118

(a) Find the least-squares regression equation $\hat{y} = b_0 + b_1x_1 + b_2x_2$, where x_1 is air temperature, x_2 is dew point, and y is the response variable, heat index.

(b) Draw residual plots to assess the adequacy of the model. What might you conclude based on the residual plots?

NW 17. Concrete A researcher wants to determine a model that can be used to predict the 28-day strength of a concrete mixture. The following data represent the 28-day and 7-day strength (in pounds per square inch) of a certain type of concrete along with the concrete's slump. Slump is a measure of the uniformity of the concrete, with a higher slump indicating a less uniform mixture.

Slump (inches)	7-Day psi	28-Day psi
4.5	2330	4025
4.25	2640	4535
3	3360	4985
4	1770	3890
3.75	2590	3810
2.5	3080	4685
4	2050	3765
5	2220	3350
4.5	2240	3610
5	2510	3875
2.5	2250	4475

(a) Construct a correlation matrix between slump, 7-day psi, and 28-day psi. Is there any reason to be concerned with multicollinearity based on the correlation matrix?

(b) Find the least-squares regression equation $\hat{y} = b_0 + b_1x_1 + b_2x_2$, where x_1 is slump, x_2 is 7-day strength, and y is the response variable, 28-day strength.

(c) Draw residual plots and a boxplot of the residuals to assess the adequacy of the model.

(d) Interpret the regression coefficients for the least-squares regression equation.

(e) Determine and interpret R^2 and the adjusted R^2.

(f) Test $H_0: \beta_1 = \beta_2 = 0$ versus H_1: at least one of the $\beta_i \neq 0$ at the $\alpha = 0.05$ level of significance.

(g) Test the hypotheses $H_0: \beta_1 = 0$ versus $H_1: \beta_1 \neq 0$ and $H_0: \beta_2 = 0$ versus $H_1: \beta_2 \neq 0$ at the $\alpha = 0.05$ level of significance.

(h) Predict the mean 28-day strength of all concrete for which slump is 3.5 inches and 7-day strength is 2450 psi.

(i) Predict the 28-day strength of a specific sample of concrete for which slump is 3.5 inches and 7-day strength is 2450 psi.

(j) Construct 95% confidence and prediction intervals for concrete for which slump is 3.5 inches and 7-day strength is 2450 psi. Interpret the results.

18. Income An economist was interested in modeling the relation among annual income, level of education, and work experience. The level of education is the number of years of education beyond eighth grade, so 1 represents completing 1 year of high school, 8 means completing 4 years of college, and so on. Work experience is the number of years employed in the current profession. From a random sample of 12 individuals, he obtained the following data:

Work Experience (years)	Level of Education	Annual Income ($ thousands)
21	6	34.7
14	3	17.9
4	8	22.7
16	8	63.1
12	4	33.0
20	4	41.4
25	1	20.7
8	3	14.6
24	12	97.3
28	9	72.1
4	11	49.1
15	4	52.0

(a) Construct a correlation matrix between work experience, level of education, and annual income. Is there any reason to be concerned with multicollinearity based on the correlation matrix?

(b) Find the least-squares regression equation $\hat{y} = b_0 + b_1x_1 + b_2x_2$, where x_1 is work experience, x_2 is level of education, and y is the response variable, annual income.

(c) Draw residual plots and a boxplot of the residuals to assess the adequacy of the model.

(d) Interpret the regression coefficients for the least-squares regression equation.

(e) Determine and interpret R^2 and the adjusted R^2.

(f) Test $H_0: \beta_1 = \beta_2 = 0$ versus H_1: at least one of the $\beta_i \neq 0$ at the $\alpha = 0.05$ level of significance.

(g) Test the hypotheses $H_0: \beta_1 = 0$ versus $H_1: \beta_1 \neq 0$ and $H_0: \beta_2 = 0$ versus $H_1: \beta_2 \neq 0$ at the $\alpha = 0.05$ level of significance.

(h) Predict the mean income of all individuals whose experience is 12 years and level of education is 4.

(i) Predict the income of a single individual whose experience is 12 years and level of education is 4.

(j) Construct 95% confidence and prediction intervals for income when experience is 12 years and level of education is 4.

NW 19. SAT Scores and Salary Go to www.pearsonhighered.com/ sullivanstats to obtain the data file 14_3_19 using the file format of your choice. The data represent the SAT Math scores for students who took the exam in 2012, along with the percent of students who took the exam in the state, and the average and starting salaries of the teachers (in thousands of dollars).

(a) Find the least-squares regression equation $\hat{y} = b_0 + b_1x_1 + b_2x_2 + b_3x_3$, where x_1 is the percent of students who take the exam in the state, x_2 is the average salary, x_3 is the starting salary, and y is the SAT Math score.

(b) Does the F-test indicate that we should reject $H_0: \beta_1 = \beta_2 = \beta_3 = 0$? Do any of the explanatory variables have slope coefficients that are not significantly different from zero?

(c) Interpret each of the slope coefficients of the model in part (a). Are they all reasonable? Explain. Hint: Obtain a correlation matrix.

(d) Find the least-squares regression equation $\hat{y} = b_0 + b_1x_1 + b_2x_2$. Does the F-test indicate that we should reject $H_0: \beta_1 = \beta_2 = 0$? Do any of the explanatory variables have slope coefficients that are not significantly different from zero? Do the slope coefficients have signs that are consistent with what you would expect?

Explaining the Concepts

20. When testing whether there is a linear relation between the response variable and the explanatory variables, we use an F-test. If the P-value indicates that we reject the null hypothesis, $H_0: \beta_1 = \beta_2 = \cdots = \beta_k = 0$, what conclusion should we come to? Is it possible that one of the β_i is zero if we reject the null hypothesis?

21. What does it mean when we say that the explanatory variables have an additive effect or do not interact?

22. Explain the difference between the coefficient of determination, R^2, and the adjusted coefficient of determination, R^2_{adj}. Which is better for determining whether an additional explanatory variable should be added to the regression model?

23. What is multicollinearity? How can we check for it? What are the consequences of multicollinearity?

24. What does it mean when we say that t-test statistics are marginal?

Objectives   ❶ Work with multiple regression models with interaction

❷ Build multiple regression models with indicator variables

❶ Work with Multiple Regression Models with Interaction

Suppose we want to build a regression model with two explanatory variables, x_1 and x_2. However, a 1-unit change in x_1 does not cause the response variable y to change by a fixed amount regardless of the value of x_2; instead, the change in the response variable for a 1-unit change in x_1 is different for different values of x_2. In this case, there is **interaction** between x_1 and x_2 and we build a model of the form

$$y_i = \beta_0 + \beta_1x_{1i} + \beta_2x_{2i} + \beta_3x_{1i}x_{2i} + \varepsilon_i \qquad \text{(1)}$$

The meaning of the coefficients β_1 and β_2 in this model is not the same as it is for a model of the form $y_i = \beta_0 + \beta_1 x_{1i} + \beta_2 x_{2i} + \varepsilon_i$ because of the **interaction term** $\beta_3 x_{1i} x_{2i}$. In a model in the form of equation (1), if we hold x_2 fixed, then the expected change in the response variable, y, for a 1-unit change in the explanatory variable x_1 is

$$\beta_1 + \beta_3 x_2$$

So the expected change in the response variable for a 1-unit change in the explanatory variable x_1 depends not only on the coefficient of x_1 but also on the coefficient of the interaction term, β_3, and the value of the other explanatory variable, x_2.

Similarly, if we hold x_1 fixed, then the expected change in the response variable, y, for a 1-unit change in the explanatory variable x_2 is

$$\beta_2 + \beta_3 x_1$$

To help visualize this interaction, suppose we are attempting to explain the relation between high-density cholesterol (HDL), y, with two explanatory variables, body mass index (BMI), x_1, and total cholesterol, x_2. High-density cholesterol is the so-called good cholesterol because the higher the level of HDL cholesterol, the greater protection one has against heart disease. Initially, the researcher fits an additive model of the form $y_i = \beta_0 + \beta_1 x_{1i} + \beta_2 x_{2i} + \varepsilon_i$. The researcher obtains the following regression equation:

$$\hat{y} = 65 - 2x_1 + 0.1x_2$$

This model gives the expected (or mean) value of HDL cholesterol for a specific BMI and total cholesterol for BMIs between 16 and 40 and total cholesterol levels between 130 mg/dL and 300 mg/dL. From this model we interpret the coefficients as follows: For a 1-unit increase in body mass index, high-density cholesterol is expected to decrease by 2 mg/dL. For a 1-unit increase in total cholesterol, high-density cholesterol is expected to increase 0.1 mg/dL.

The following represent the expected HDL cholesterol of an individual with three different levels of BMI:

$$x_1 = 20: \hat{y} = 65 - 2(20) + 0.1x_2 = 25 + 0.1x_2$$
$$x_1 = 25: \hat{y} = 65 - 2(25) + 0.1x_2 = 15 + 0.1x_2$$
$$x_1 = 30: \hat{y} = 65 - 2(30) + 0.1x_2 = 5 + 0.1x_2$$

Notice that each of these lines has the same slope, 0.1, but a different intercept. Therefore, the lines are parallel, as indicated in Figure 21. We can interpret the slope as follows: if the total cholesterol of an individual increases by 1 mg/dL, then HDL cholesterol will increase by 0.1 mg/dL, on average, regardless of the level of body mass index or total cholesterol.

Figure 21

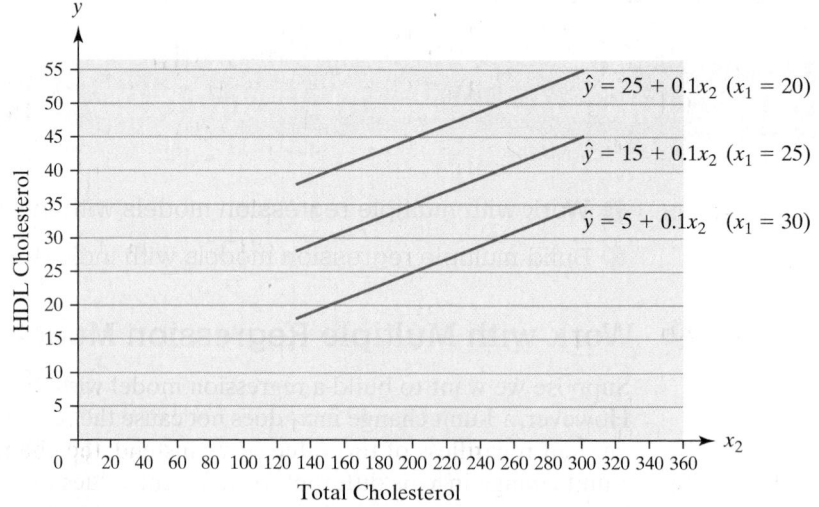

Now suppose the researcher believes that the increase in HDL cholesterol when total cholesterol (x_2) increases is larger for lower levels of BMI because higher levels of BMI typically coincide with higher levels of total cholesterol while adversely affecting HDL cholesterol. Put another way, the researcher expects that as total cholesterol increases, lower levels of body mass index will yield larger changes in HDL than will higher levels of body mass index. This means that the lines for the different levels of body mass index will not be parallel. Lines with a higher level of BMI will be "flatter" than the lines with a lower level of BMI. To account for this interaction between BMI and total cholesterol, the researcher adds a term for x_1x_2 and obtains the following regression equation:

$$\hat{y} = -54 + 2.4x_1 + 0.65x_2 - 0.02x_1x_2$$

This model also gives the expected (or mean) value of HDL cholesterol for a specific BMI and total cholesterol, except now there is an interaction term. For example, the following represent expected HDL cholesterol of an individual with three different levels of BMI:

$$x_1 = 20: \hat{y} = -54 + 2.4(20) + 0.65x_2 - 0.02(20)x_2 = -6 + 0.25x_2$$
$$x_1 = 25: \hat{y} = -54 + 2.4(25) + 0.65x_2 - 0.02(25)x_2 = 6 + 0.15x_2$$
$$x_1 = 30: \hat{y} = -54 + 2.4(30) + 0.65x_2 - 0.02(30)x_2 = 18 + 0.05x_2$$

Notice now that each of the three regression lines has a different slope coefficient on x_2, total cholesterol. We can see that as the BMI increases (from 20 to 25 to 30), the expected change in HDL cholesterol decreases from 0.25 to 0.15 to 0.05 mg/dL for a 1-unit change in total cholesterol. So higher levels of BMI (higher values of x_1) result in a smaller increase in the predicted change in HDL cholesterol for a 1-unit change in total cholesterol. See Figure 22.

Figure 22

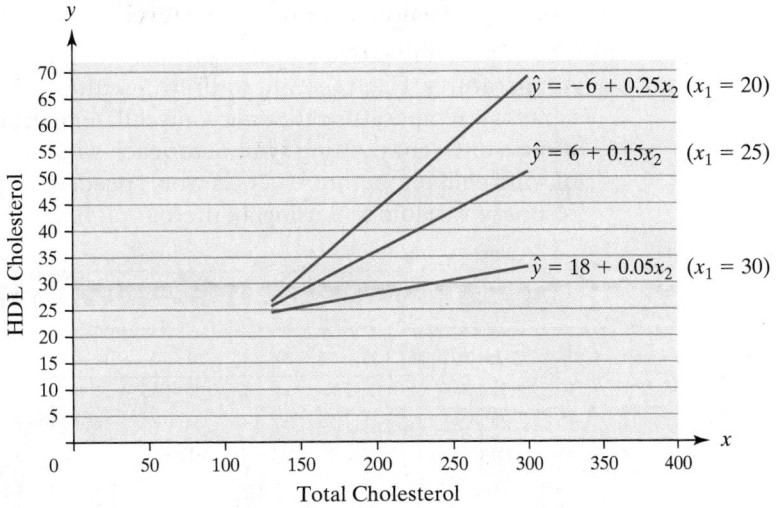

Definition   If the mean value of the response variable y in a least-squares regression associated with a 1-unit change in an explanatory variable depends on a second explanatory variable, there is **interaction** between the two explanatory variables. When interaction exists between two explanatory variables, x_1 and x_2, we introduce a term with the variable x_1x_2 in the regression model as an explanatory variable.

When building models with more than one explanatory variable, it is important to consider the possibility of interaction between two (or more) of the explanatory variables, because these models may better explain the variability in the response variable, y.

If there are three explanatory variables, $x_1, x_2,$ and x_3, a potential model is

$$y_i = \beta_0 + \beta_1 x_{1i} + \beta_2 x_{2i} + \beta_3 x_{3i} + \beta_4 x_{1i} x_{2i} + \beta_5 x_{1i} x_{3i} + \beta_6 x_{2i} x_{3i} + \varepsilon_i$$

It is even possible to consider three-way interaction, $x_1 x_2 x_3$, but this is rarely done in practice. In addition, a regression model can include quadratic terms such as x_1^2 or x_2^2 when a scatter diagram of the response variable and an explanatory variable follow a U-shaped pattern. The **complete second-order model** with two explanatory variables x_1 and x_2 is

$$y_i = \beta_0 + \beta_1 x_{1i} + \beta_2 x_{2i} + \beta_3 x_{1i} x_{2i} + \beta_4 x_{1i}^2 + \beta_5 x_{2i}^2 + \varepsilon_i$$

These type of models are discussed in Section 14.5. We recommend that scatter diagrams be drawn against each explanatory variable to determine the type of relationship that may exist between the two variables.

EXAMPLE 1   A Multiple Regression Model with Interaction

Problem The wind chill index represents what the air feels like on exposed skin while factoring in wind speed. The wind chill index is determined by considering the wind speed at a height of 5 feet, the typical height of an adult human face. A model for determining wind chill based on the air temperature and wind speed developed by the National Weather Service is

$$W = 35.74 + 0.6215T - 35.75V + 0.4275TV$$

where W is the wind chill (in °F)

T is the air temperature (in °F)

V is the velocity of the wind (in miles per hour) raised to the 0.16 power (so $V = v^{0.16}$)

The variable velocity is transformed by raising the actual wind speed to the 0.16 power to account for the nonlinear relation between wind chill and wind speed.

(a) The data in Table 6 show the wind chill temperature for various wind speeds and temperatures. Use the data to draw a scatter diagram of wind chill temperature versus air temperature, treating wind chill temperature as the response variable and using a different plotting symbol for each wind speed. Then draw a scatter diagram of wind chill temperature versus wind speed, treating wind chill temperature as the response variable and using a different plotting symbol for each air temperature.

Table 6

Temperature (°F)

Wind (mph) \ Calm	40	35	30	25	20	15	10	5	0	−5	−10	−15	−20	−25	−30	−35	−40	−45
5	36	31	25	19	13	7	1	−5	−11	−16	−22	−28	−34	−40	−46	−52	−57	−63
10	34	27	21	15	9	3	−4	−10	−16	−22	−28	−35	−41	−47	−53	−59	−66	−72
15	32	25	19	13	6	0	−7	−13	−19	−26	−32	−39	−45	−51	−58	−64	−71	−77
20	30	24	17	11	4	−2	−9	−15	−22	−29	−35	−42	−48	−55	−61	−68	−74	−81
25	29	23	16	9	3	−4	−11	−17	−24	−31	−37	−44	−51	−58	−64	−71	−78	−84
30	28	22	15	8	1	−5	−12	−19	−26	−33	−39	−46	−53	−60	−67	−73	−80	−87
35	28	21	14	7	0	−7	−14	−21	−27	−34	−41	−48	−55	−62	−69	−76	−82	−89
40	27	20	13	6	−1	−8	−15	−22	−29	−36	−43	−50	−57	−64	−71	−78	−84	−91
45	26	19	12	5	−2	−9	−16	−23	−30	−37	−44	−51	−58	−65	−72	−79	−86	−93
50	26	19	12	4	−3	−10	−17	−24	−31	−38	−45	−52	−60	−67	−74	−81	−88	−95
55	25	18	11	4	−3	−11	−18	−25	−32	−39	−46	−54	−61	−68	−75	−82	−89	−97
60	25	17	10	3	−4	−11	−19	−26	−33	−40	−48	−55	−62	−69	−76	−84	−91	−98

(b) Use the model to predict the wind chill temperature if the air temperature is 3°F and the wind speed is 13 miles per hour (mph).

(c) Interpret the model for a wind speed of 20 mph. Interpret the model for a wind speed of 30 mph.

Solution

(a) Figure 23 shows the scatter diagram drawn in Minitab of wind chill temperature versus air temperature using a different plotting symbol for each wind speed. Notice that the points for each wind speed are not parallel. The differences in wind chill temperature at very low air temperature are more significant for different wind speeds than the differences in wind chill temperature at higher temperatures for different wind speeds. This implies an interaction effect between air temperature and wind speed.

Figure 23

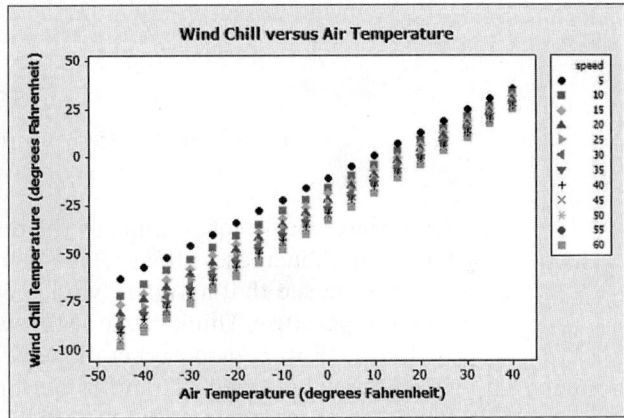

Figure 24 below shows the scatter diagram of wind chill temperature versus wind velocity using a different plotting symbol for each air temperature. Not only are the points for each air temperature not parallel, but the data do not follow a linear pattern. For example, at an air temperature of $-45°F$ (the • symbol), it is clear that the data do not follow a linear pattern. This is the reason that we needed to transform the variable wind speed by raising the variable to the 0.16 power. Popular transformations include the square root transformation and the logarithmic transformation in order to linearize data [see Section 4.5 on the companion website (www.pearsonhighered.com/sullivanstats) that accompanies this text].

Figure 24

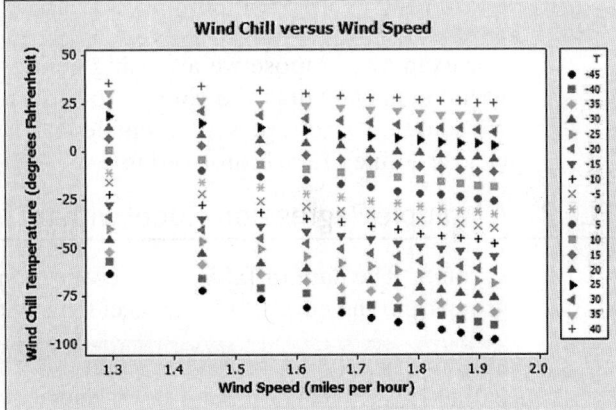

(b) Let $T = 3°F$ and $V = 13^{0.16}$ in the model to predict the wind chill temperature at 3°F with a wind speed of 13 mph.

$$W = 35.74 + 0.6215T - 35.75V + 0.4275TV$$
$$= 35.74 + 0.6215(3) - 35.75(13^{0.16}) + 0.4275(3)(13^{0.16})$$
$$= -14$$

Rounded to the nearest degree, the wind chill temperature at an air temperature of 3°F with a wind speed of 13 mph is $-14°F$.

(continued)

(c) At a wind velocity of 20 mph, the expected wind chill is

$$W = 35.74 + 0.6215T - 35.75V + 0.4275TV$$
$$= 35.74 + 0.6215T - 35.75(20^{0.16}) + 0.4275T(20^{0.16})$$
$$= 35.74 + 0.6215T - 57.7352 + 0.6904T$$
$$= -21.9952 + 1.3119T$$

At a wind velocity of 20 mph, as the temperature increases by 1°F, the wind chill temperature increases about 1.3119°F, on average.

At a wind velocity of 30 mph, the expected wind chill is

$$W = 35.74 + 0.6215T - 35.75V + 0.4275TV$$
$$= 35.74 + 0.6215T - 35.75(30^{0.16}) + 0.4275T(30^{0.16})$$
$$= 35.74 + 0.6215T - 61.6049 + 0.7367T$$
$$= -25.8649 + 1.3582T$$

At a wind velocity of 30 mph, as the temperature increases by 1°F, the wind chill temperature increases about 1.3582°F, on average.

We can see that a higher wind speed results in a higher impact on the wind chill temperature. Think about it this way, if the air temperature decreases from 20°F to 19°F, we would expect the wind chill temperature to decrease by 1.3119°F when the wind speed is 20 miles per hour, while we would expect the wind chill temperature to decrease by 1.3582°F when the wind speed is 30 miles per hour.

• **Now Work Problem 7**

② Build Multiple Regression Models with Indicator Variables

Thus far, we have only considered quantitative explanatory variables in our regression models. However, suppose we are conducting a regression analysis in which we believe a qualitative variable, such as gender, can explain some of the variability in the response variable. Can we handle this in regression analysis? Yes! We use an *indicator* or *dummy variable*.

Definition | An **indicator** (or **dummy**) **variable** is a qualitative explanatory variable in a multiple regression model that takes on the value 0 or 1.

For example, suppose we are using gender as an explanatory variable. The indicator variable may assume the value 0 to represent a male and 1 to represent a female. Or we could use 0 to represent a female and 1 to represent a male. It is important that we clearly define the indicator variable.

EXAMPLE 2 | **A Multiple Regression Model with an Indicator Variable**

Problem The data in Table 7 represent the number of licensed drivers in various age groups and the number of fatal accidents within the age group by gender.

Table 7				
Age	**Number of Male Licensed Drivers (000s)**	**Number of Fatal Crashes**	**Number of Female Licensed Drivers (000s)**	**Number of Fatal Crashes**
<16	12	227	12	77
16–20	6424	5180	6139	2113
21–24	6941	5016	6816	1531
25–34	18,068	8595	17,664	2780
35–44	20,406	7990	20,063	2742
45–54	19,898	7118	19,984	2285
55–64	14,430	4527	14,441	1514
65–74	8194	2274	8400	938
>74	4803	2022	5375	980

Source: National Highway and Traffic Safety Institute

(a) Draw a scatter diagram of the data treating number of licensed drivers as the explanatory variable and number of fatal crashes as the response variable using a different plotting symbol for each gender.

(b) Find the least-squares equation $\hat{y} = b_0 + b_1 x_1 + b_2 x_2$, where x_1 is the number of licensed drivers, x_2 is an indicator variable where $0 =$ male, $1 =$ female, and y is the number of fatal crashes.

(c) Interpret the coefficient of each explanatory variable.

Solution

(a) Figure 25 below shows the scatter diagram of the data drawn in Minitab. For males (0) we use the symbol ♦ and for females (1) we use the symbol ■. For each gender, there is a positive association between the variables. In addition, the slope for the males appears to be much steeper than that of the females.

Figure 25

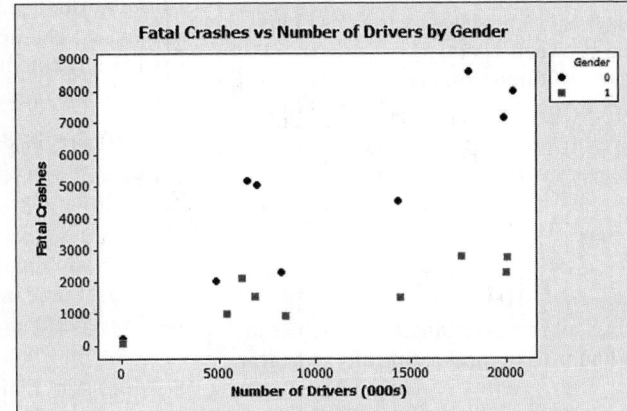

(b) Using Minitab, we find that the least-squares regression equation is

$$\hat{y} = 2289 + 0.2254x_1 - 3103x_2$$

(c) Because the model is additive, we interpret the coefficients the same way as we did for a regression with one explanatory variable. However, we assume the other variable is held constant. So, for the variable $x_1 =$ number of licensed drivers, which is measured in thousands, the coefficient is 0.2254. Therefore, if the number of licensed drivers increases by 1 thousand, we expect the number of fatal crashes to increase by 0.2254. For the variable $x_2 =$ gender, the coefficient is -3103. Therefore, we expect 3103 fewer fatal crashes for females than males at any level of number of drivers. This explains why females have lower insurance premiums than males!

● **Now Work Problem 9**

NOTE

If you look carefully at the scatter diagram in Figure 25, it appears there may be interaction between number of licensed drivers and gender. Using StatCrunch, a better model for the data in Table 7 is

$\hat{y} = 760 + 0.3579x_1 - 0.2695x_1x_2$

This occurs because there is a bigger difference in the number of fatalities as the difference in the number of male versus female drivers for any age group increases.

What if the qualitative variable has more than two possible categories? For example, what if we believe political affiliation (Democrat, Republican, or Independent) helps explain some of the variability in the response variable? We might think to simply designate 0 as Democrat, 1 as Republican, and 2 as Independent. Unfortunately, this is incorrect and will lead to errors. Instead, we must use two indicator variables as follows:

$$x_1 = \begin{cases} 1 & \text{if Democrat} \\ 0 & \text{otherwise} \end{cases}$$

$$x_2 = \begin{cases} 1 & \text{if Republican} \\ 0 & \text{otherwise} \end{cases}$$

Notice that if the individual is Independent both x_1 and x_2 assume a value of 0. In addition, notice it is impossible to have both x_1 and x_2 assume a value of 1 (one cannot simultaneously be a Democrat and Republican).

In general, if there are c categories for a qualitative explanatory variable, the regression model will require $c - 1$ indicator variables, each taking on a value of 0 or 1.

14.4 Assess Your Understanding

Vocabulary and Skill Building

1. If there is _____ between x_1 and x_2 in a least-squares regression model, we would build a model of the form $\hat{y}_i = \beta_0 + \beta_1 x_{1i} + \beta_2 x_{2i} + \beta_3 x_{1i} x_{2i} + \varepsilon_i$.

2. A(n) _____ or _____ variable is a qualitative explanatory variable in a multiple regression model that takes on the value 0 or 1.

3. Suppose we wish to develop a model with three explanatory variables, x_1, x_2, and x_3.

(a) Write a model that utilizes all three explanatory variables with no interaction.

(b) Write a model that utilizes the explanatory variables x_1 and x_2 along with interaction between x_1 and x_2.

(c) Write a model that utilizes all three explanatory variables with interaction between x_2 and x_3.

4. Suppose you want to develop a model that predicts the gas mileage of a car. The explanatory variables you are going to utilize are

x_1: city or highway driving
x_2: weight of the car
x_3: tire pressure

(a) Write a model that utilizes all three explanatory variables in an additive model with linear terms and define any indicator variables.

(b) Suppose you suspect there is interaction between weight and tire pressure. Write a model that incorporates this interaction term into the model from part (a).

5. Suppose that the response variable y is related to the explanatory variables x_1 and x_2 by the regression equation.

$$\hat{y} = 4 + 0.4x_1 - 1.3x_2$$

(a) Construct a graph similar to Figure 21 showing the relationship between the expected value of y and x_1 for $x_2 = 10, 20$, and 30.

(b) Construct a graph showing the relationship between the expected value of y and x_2 for $x_1 = 40, 50$, and 60.

(c) How can we tell from the graphs alone that there is no interaction between x_1 and x_2?

(d) Redo parts (a) and (b) with the interaction term $0.05x_1x_2$ added to the regression equation. How do the graphs differ?

6. Suppose that the response variable y is related to the explanatory variables x_1 and x_2 by the regression equation

$$\hat{y} = 6 - 0.3x_1 + 1.7x_2$$

(a) Construct a graph similar to Figure 21 showing the relationship between the expected value of y and x_1 for $x_2 = 10, 20$, and 30.

(b) Construct a graph showing the relationship between the expected value of y and x_2 for $x_1 = 40, 50$, and 60.

(c) How can we tell from the graphs alone that there is no interaction between x_1 and x_2?

(d) Redo parts (a) and (b) with the interaction term $0.04x_1x_2$ added to the regression equation. How do the graphs differ?

Applying the Concepts

NW 7. Estimating Age In the European Union, it has become important to be able to determine an individual's age when legal documentation of the birth date of an individual is unavailable. In the article "Age Estimation in Children by Measurement of Open Apices in Teeth: a European Formula" (*International Journal of Legal Medicine* [2007]:121: 449–453), researchers developed a model to predict the age, y, of an individual based on the gender of the individual, x_1 (0 = female, 1 = male), the height of the second premolar, x_2, the number of teeth with root development, x_3, and the sum of the normalized heights of seven teeth on the left side of the mouth, x_4. The normalized height of the seven teeth was found by dividing the distance between teeth by the height of the tooth. Their model is

$$\hat{y} = 9.063 + 0.386x_1 + 1.268x_2 + 0.676x_3 - 0.913x_4 - 0.175x_3x_4$$

(a) Based on this model, what is the expected age of a female with $x_2 = 28$ mm, $x_3 = 8$, and $x_4 = 18$ mm?

(b) Based on this model, what is the expected age of a male with $x_2 = 28$ mm, $x_3 = 8$, and $x_4 = 18$ mm?

(c) What is the interaction term? What variables interact?

(d) The coefficient of determination for this model is 86.3%. Explain what this means.

8. More Age Estimation In the article "Bigger Teeth for Longer Life? Longevity and Molar Height in Two Roe Deer Populations" (*Biology Letters* [June, 2007] vol. 3 no. 3 268–270), researchers developed a model to predict the tooth height (in mm), y, of roe deer based on their age, x_1, gender, x_2 (0 = female, 1 = male), and location, x_3 (Trois Fontaines deer, which have a shorter life expectancy, and Chizé, which have a longer life expectancy, $x_3 = 0$ for Trois Fontaines, $x_3 = 1$ for Chizé). The model is

$$\hat{y} = 7.790 - 0.382x_1 - 0.587x_2 - 0.925x_3 + 0.091x_2x_3$$

(a) What is the expected tooth length of a female roe deer who is 12 years old and lives in Trois Fontaines?

(b) What is the expected tooth length of a male roe deer who is 8 years old and lives in Chizé?

(c) What is the interaction term? What does the coefficient of the interaction term imply about tooth length?

NW 9. Does Size Matter? Researchers wondered whether the size of a person's brain was related to the individual's mental capacity. They selected a sample of right-handed Anglo introductory psychology students who had Scholastic Aptitude Test scores higher than 1350. The subjects were administered the Wechsler Adult Intelligence Scale–Revised to obtain their IQ scores. The MRI scans, performed at the same facility, consisted of 18 horizontal MR images. The computer counted all pixels with nonzero gray scale in each of the 18 images, and the total count served as an index for brain size. The resulting data are presented in the following table:

Gender	MRI Count	IQ	Gender	MRI Count	IQ
Female	816,932	133	Male	949,395	140
Female	951,545	137	Male	1,001,121	140
Female	991,305	138	Male	1,038,437	139
Female	833,868	132	Male	965,353	133
Female	856,472	140	Male	955,466	133
Female	852,244	132	Male	1,079,549	141
Female	790,619	135	Male	924,059	135
Female	866,662	130	Male	955,003	139
Female	857,782	133	Male	935,494	141
Female	948,066	133	Male	949,589	144

Source: L. Willerman, R. Schultz, I. N. Rutledge, and E. Bigler. "In Vivo Brain Size and Intelligence." *Intelligence*, 15(223–228), 1991.

(a) Find the least-squares regression equation $\hat{y} = b_0 + b_1x_1$, where x_1 is MRI count and y is the response variable IQ.
(b) Test the hypotheses $H_0: \beta_1 = 0$ versus $H_1: \beta_1 \neq 0$. What do you conclude?
(c) Draw a scatter diagram, treating MRI count as the explanatory variable and IQ as the response variable, but use a different plotting symbol for males and females. For example, use a circle for males and a square for females.
(d) Find the least-squares regression equation $\hat{y} = b_0 + b_1x_1 + b_2x_2$, where x_1 is MRI count and $x_2 = 0$ for males and $x_2 = 1$ for females.
(e) Test the hypotheses $H_0: \beta_1 = 0$ versus $H_1: \beta_1 \neq 0$ and $H_0: \beta_2 = 0$ versus $H_1: \beta_2 \neq 0$.
(f) What do you conclude from this analysis?

10. Drill Time The following data represent the time (minutes) it takes to drill an additional 5 feet from the depth indicated for both wet drilling and dry drilling conditions.

Depth	Conditions	Time	Depth	Conditions	Time
5	Wet	8.68	60	Wet	9.34
5	Wet	8.61	60	Dry	7.58
5	Dry	7.07	80	Wet	8.96
10	Wet	7.71	80	Dry	8.13
15	Dry	7.43	105	Wet	8.60
20	Wet	8.26	120	Dry	7.47
20	Dry	6.65	130	Wet	9.18
30	Wet	8.27	130	Dry	8.37
30	Dry	7.95	150	Wet	9.54
40	Wet	9.04	160	Dry	8.20
40	Dry	7.80			

Source: American Statistician, 45(1)

(a) Draw a scatter diagram, treating depth as the explanatory variable and time as the response variable, but use a different plotting symbol for wet and dry. For example, use a circle for wet and a square for dry.
(b) Find the least-squares regression equation $\hat{y} = b_0 + b_1x_1 + b_2x_2$, where x_1 is depth and $x_2 = 0$ for wet and $x_2 = 1$ for dry.
(c) Test the hypotheses $H_0: \beta_1 = 0$ versus $H_1: \beta_1 \neq 0$ and $H_0: \beta_2 = 0$ versus $H_1: \beta_2 \neq 0$.
(d) Construct 95% confidence and prediction intervals for time to drill an additional 5 feet in dry conditions where drilling starts at 100 feet. Interpret the results.

11. Suppose we wish to develop a regression equation that models the selling price of a home. The researcher wishes to include the variable garage in the model. She has identified three possibilities for a garage: (1) attached, (2) detached, (3) no garage. Define the indicator variables necessary to incorporate the variable "garage" into the model.

14.5 Polynomial Regression

Objective ① Find a Quadratic Regression Equation

① Find a Quadratic Regression Equation

We mentioned that a regression model can include quadratic terms. For example, a complete second-order model with two explanatory variables x_1 and x_2 is

$$y_i = \beta_0 + \beta_1x_{1i} + \beta_2x_{2i} + \beta_3x_{1i}x_{2i} + \beta_4x_{1i}^2 + \beta_5x_{2i}^2 + \varepsilon_i$$

Scatter diagrams of the response variable drawn against each explanatory variable are used to determine the type of relation that may exist between the two variables. In this section, we consider quadratic models of the form

$$y_i = \beta_0 + \beta_1x_{1i} + \beta_2x_{1i}^2 + \varepsilon_i$$

Once a quadratic model is obtained, we go through the usual analysis of (1) drawing residual plots to verify the model is appropriate, (2) performing an F-test for lack of fit, and (3) testing individual regression coefficients for significance. We may also construct confidence and prediction intervals using the quadratic model. As always, we use technology to obtain the results.

EXAMPLE 1   Building a Quadratic Model

Photosynthetic efficiency is the percentage of light energy that is converted into chemical energy by a plant. The chemical energy is used by a plant to maintain the plant's activities. A sample of grass was used to determine the efficiency of the grass species. Table 8 shows the results of the study at various temperatures, measured in degrees Celsius. The efficiency is measured as a percent.

Table 8			
Temperature, x	Efficiency, y	Temperature, x	Efficiency, y
−1.5	33	15	91
0	46	17	89
2.5	55	20	77
5	80	22	72
7	87	25	54
10	93	27	46
12	95		

Source: D. Brown and P. Rothery, *Models in Biology: Mathematics, Statistics, and Computing*

(a) Draw a scatter diagram of the data. What type of relation appears to exist between temperature and efficiency?
(b) Find the quadratic regression equation $\hat{y} = b_0 + b_1 x + b_2 x^2$.
(c) Draw a residual plot against the fitted values, x, and x^2. Also, draw a boxplot of the residuals. Are there any problems with the model?
(d) Interpret the coefficient of determination.
(e) Does the F-test indicate that we should reject $H_0: \beta_1 = \beta_2 = 0$? Is either coefficient not significantly different from zero?
(f) Construct and interpret 95% confidence and prediction intervals for $x = 18$ degrees Celsius.

Approach We will follow the same procedures that we used in multiple regression (Section 14.3). Enter the data into StatCrunch to obtain the quadratic regression. The steps for obtaining polynomial regression equation using Minitab, Excel, and StatCrunch are given in the Technology Step-by-Step on page 720.

Solution
(a) Figure 26 shows a scatter diagram of the data. Clearly, there is a quadratic relation between temperature and efficiency.

Figure 26

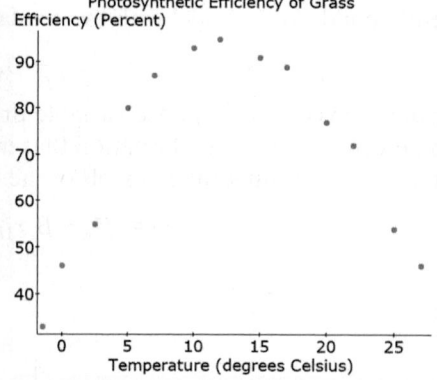

Photosynthetic Efficiency of Grass

(b) Figure 27 shows the output from StatCrunch. In anticipation of answering parts (c) and (f), we save the residuals and fitted values in the spreadsheet, and ask StatCrunch to make a prediction for $x = 18$ degrees Celsius.

Figure 27

Polynomial Regression Results:
Dependent Variable: Efficiency
Independent Variable: Temperature

Parameter	Estimate	Std. Err.	Alternative	DF	T-Stat	P-value
Intercept	45.325324	2.0707437	≠ 0	10	21.888427	<0.0001
X	7.3327911	0.38411809	≠ 0	10	19.08994	<0.0001
X^2	-0.27669598	0.01455614	≠ 0	10	-19.008884	<0.0001

Analysis of variance table for polynomial regression model:

Source	DF	SS	MS	F-stat	P-value
Model	2	5253.6032	2626.8016	185.67421	<0.0001
Error	10	141.4737	14.14737		
Total	12	5395.0769			

Summary of fit:
Root MSE: 3.7612989
R-squared: 0.9738
R-squared (adjusted): 0.9685

The coefficients for the quadratic regression equation are in the first table under the heading "Estimate." The quadratic regression equation for these data is $\hat{y} = -0.2767x^2 + 7.3328x + 45.3253$.

(c) Figure 28(a)–(c) shows residual plots of the fitted values, x, and x^2. Figure 28(d) shows a boxplot of the residuals. None of the residual plots show any obvious pattern, and the boxplot does not show any outliers. Therefore, the quadratic model is appropriate.

Figure 28

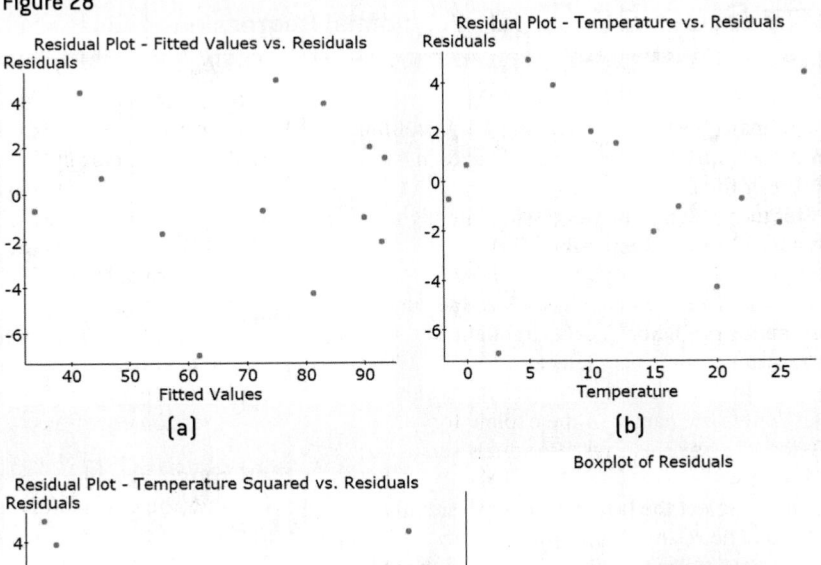

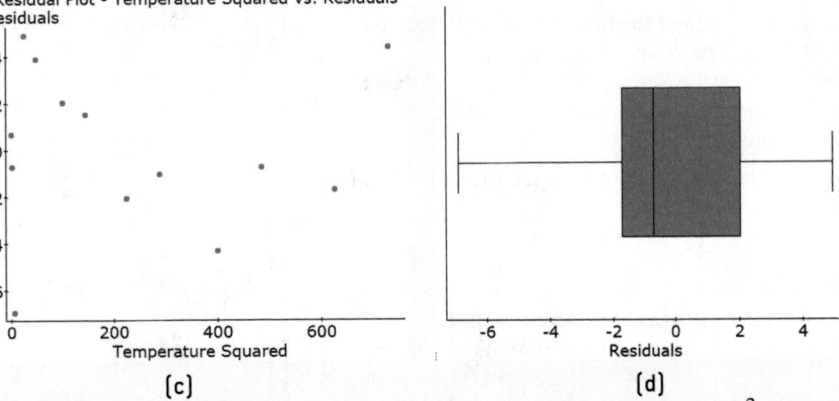

(d) From the output in Figure 27, the coefficient of determination, R^2, is $0.974 = 97.4\%$. Therefore, 97.4% of the variation in efficiency is explained by the quadratic model.

(continued)

(e) To test H_0: $\beta_1 = \beta_2 = 0$ versus H_1: at least one of the regression coefficients is different from 0, we look at the F-test statistic from the analysis of variance output and see that $F = 185.67$ with a P-value reported as <0.0001. This small P-value is sufficient evidence against the statement in the null hypothesis. Therefore, we reject the null hypothesis and conclude that at least one of the regression coefficients is different from zero. Now, we test each regression coefficient separately by testing

$$H_0: \beta_1 = 0 \quad \text{vs.} \quad H_1: \beta_1 \neq 0$$

and

$$H_0: \beta_2 = 0 \quad \text{vs.} \quad H_1: \beta_2 \neq 0$$

Looking again at the output in Figure 27, we see that the test statistic for temperature, x, is $t_0 = 19.09$ with a P-value <0.0001. We also see that the test statistic for temperature^2, x^2, is $t_0 = -19.01$ with a P-value <0.0001. Since both P-values are sufficiently small, we reject each null hypothesis. We conclude that both temperature and temperature^2 are significantly different from zero.

(f) Figure 29 shows partial output from StatCrunch.

Figure 29 **Predicted values:**

X value	Pred. Y	s.e.(Pred. y)	95% C.I. for mean	95% P.I. for new
18	87.666066	1.457411	(84.418752, 90.91338)	(78.678234, 96.653897)

Based on the output, we are 95% confident that the mean efficiency at a temperature of 18 degrees Celsius is between 84.4% and 90.9%. We are 95% confident that the efficiency for a particular grass plant at a temperature of 18 degrees Celsius is between 78.7% and 96.7%.

• Now Work Problem 3

Technology Step-by-Step Polynomial Regression

Minitab

1. Enter the explanatory variable in column C1. In column C2, determine the squared values of the entries in column C1. Enter the response variable in column C3.
2. Select the **Stat** menu, highlight **Regression,** highlight **Regression** and select **Fit Regression Model. . . .**
3. Place the cursor in the "Responses:" cell. Highlight the response variable and click Select. Place the cursor in the "Continuous predictors:" cell. Highlight the explanatory variables and click Select.
4. Click **Graphs. . .** In the cell that says "Residuals versus the variables" enter the names of the explanatory variables. Select the box that says "Residuals versus fits." Click OK.
5. Click **Storage. . . .** Select the box that says "Residuals." Select OK twice. The residuals are stored in the spreadsheet. Draw a normal probability plot of the residuals as indicated in Section 7.3 and a boxplot of the residuals as indicated in Section 3.5.

For prediction and confidence intervals, see the steps given on page 692 in Section 14.2.

Excel

1. Enter the explanatory variable in column A. In column B, determine the squared values of the entries in column A. Enter the response variable in column C.
2. Select the Data menu; then select **Data Analysis.** Select Regression and click OK.
3. With the cursor in the "Input Y Range:" cell, highlight the data for the response variable. With the cursor in the "Input X Range:" cell, highlight the data in columns A and B. Check the boxes for Residuals, Residual Plots, and Normal Probability Plots. Click OK. The residuals are stored in the spreadsheet. Draw a boxplot of the residuals as indicated in Section 3.5. **Note:** We can also use the XLSTAT add-in. Excel does not compute prediction and confidence intervals from multiple regression equations.

StatCrunch

1. Enter the explanatory and response variables into the spreadsheet.
2. Select **Stat**, highlight **Regression,** and select **Polynomial**.
3. Choose 2 for Poly. Order; choose the explanatory variable for the X-Variable; choose the response variable for the Y-Variable.
4. Choose between Hypothesis Tests (and direction of alternative) or Confidence Intervals (and level of confidence).
5. Choose any other options you want. Click Compute!

14.5 Assess Your Understanding

Skill Building

In Problems 1 and 2,

(a) Draw a scatter diagram of the data. What type of relation appears to exist between x and y?
(b) Find the quadratic regression equation $\hat{y} = b_0 + b_1x + b_2x^2$.
(c) Draw a residual plot against the fitted values, x, and x^2. Also, draw a boxplot of the residuals. Are there any problems with the model?
(d) Interpret the coefficient of determination.
(e) Does the F-test indicate that we should reject $H_0: \beta_1 = \beta_2 = 0$? Is either coefficient not significantly different from zero?
(f) Construct and interpret 95% confidence and prediction intervals for $x = 4$.

1.

x	y
1	25.1
1.5	19.3
2.1	13.9
3.6	8.5
4.3	7.7
5.9	12.9
6.2	14.3

2.

x	y
2.3	19.3
2.7	14.8
3.2	10.2
4.1	4.8
4.9	2.9
5.6	3.9
6.4	7.9

Applying the Concepts

3. Life Cycle Hypothesis In the 1950s, Franco Modigliani
NW developed the *Life Cycle Hypothesis*. One tenet of this hypothesis is that income varies with age. The following data represent the annual income and age of a random sample of 15 adult Americans.

Age, x	Income, y	Age, x	Income, y
25	25,490	47	41,398
27	26,910	52	36,474
32	32,141	54	38,934
37	35,893	57	35,775
42	36,451	62	30,629
42	38,093	67	22,708
47	36,266	72	20,506

Source: Based on data obtained from the U.S. Census Bureau

(a) Draw a scatter diagram of the data. What type of relation appears to exist between x and y?
(b) Find the quadratic regression equation $\hat{y} = b_0 + b_1x + b_2x^2$.
(c) Draw a residual plot against the fitted values, x, and x^2. Also, draw a boxplot of the residuals. Are there any problems with the model?
(d) Interpret the coefficient of determination.
(e) Does the F-test indicate that we should reject $H_0: \beta_1 = \beta_2 = 0$? Is either coefficient not significantly different from zero?
(f) Construct and interpret 95% confidence and prediction intervals incomes for an age of 45 years.

4. Divorce Rates The given data represent the percentage, y, of the population that is divorced for various ages, x, in the United States in 2010 based on sample data obtained from the *United States Statistical Abstract* in 2012.

Age, x	Percentage Divorced, y
22	0.9
27	3.6
32	7.4
37	10.4
42	12.7
50	15.7
60	16.2
70	13.1
80	6.5

Source: United States Statistical Abstract, 2012

(a) Draw a scatter diagram of the data. What type of relation appears to exist between x and y?
(b) Find the quadratic regression equation $\hat{y} = b_0 + b_1x + b_2x^2$.
(c) Draw a residual plot against the fitted values, x, and x^2. Also, draw a boxplot of the residuals. Are there any problems with the model?
(d) Interpret the coefficient of determination.
(e) Does the F-test indicate that we should reject $H_0: \beta_1 = \beta_2 = 0$? Is either coefficient not significantly different from zero?
(f) Construct and interpret a 95% confidence interval for percent divorced among all 30 years olds.

5. Miles per Gallon An engineer collects the following data showing the speed and miles per gallon of a Toyota Camry.

Speed, y	Miles per Gallon, x	Speed, x	Miles per Gallon, y
30	18	55	30
35	20	60	29
40	23	65	26
40	25	65	25
45	25	70	25
50	28		

(a) Draw a scatter diagram of the data. What type of relation appears to exist between x and y?
(b) Find the quadratic regression equation $\hat{y} = b_0 + b_1x + b_2x^2$.
(c) Draw a residual plot against the fitted values, x, and x^2. Also, draw a boxplot of the residuals. Are there any problems with the model?

(d) Interpret the coefficient of determination.

(e) Does the F-test indicate that we should reject $H_0: \beta_1 = \beta_2 = 0$? Is either coefficient not significantly different from zero?

(f) Construct and interpret 95% confidence and prediction intervals for miles per gallon at a speed of 48 miles per hour.

6. Effect of Wind Direction on a Race The data to the right represent the time (in seconds) to complete a 400-meter race while a wind is blowing 2 meters per second in different directions. The wind direction is measured from the finishing straight (so 180 degrees is a head on wind). See the figure.
Note: The wind speed of 2 meters per second is the strongest wind speed allowed for a world record to be recorded without the notation "wind assisted."

Wind Direction (degrees), x	Time (seconds), y	Wind Direction (degrees), x	Time (seconds), y
0	43.03	180	42.85
30	43.04	210	42.82
60	43.01	240	42.82
90	42.97	270	42.86
120	42.92	300	42.92
150	42.88	330	42.99

Source: "The Effects of Wind and Altitude in the 400 m Sprint with Various IAAF Track Geometries" by Vanessa Alday and Michael Frantz. *Mathematics and Sports*, MAA.

(a) Draw a scatter diagram of the data. What type of relation appears to exist between x and y?

(b) Find the quadratic regression equation $\hat{y} = b_0 + b_1 x + b_2 x^2$.

(c) Draw a residual plot against the fitted values, x, and x^2. Also, draw a boxplot of the residuals. Are there any problems with the model?

7. Age Difference at Marriage Researchers developed a model to explain the age gap between husbands and wives at first marriage. The model is below:

$$\hat{y} = 0.0321x_1 + 0.9848x_2 + 0.5391x_3 - 0.000145x_4^2 + 3.8483$$

where

y: Age gap at first marriage (male − female)
x_1: Percent of children aged 10 to 14 involved in child labor
x_2: Indicator variable where 1 is an African country, 0 otherwise
x_3: Percent of the population that is Muslim
x_4: Percent of the population that is literate

Source: Xu Zhang and Solomon W. Polachek, State University of New York at Binghamton "The Husband-Wife Age Gap at First Marriage: A Cross-Country Analysis"

(a) Use the model to predict the age gap at first marriage for an African country where the percent of children aged 10 to 14 who are involved in child labor is 12, the percent of the population that is Muslim is 30, and the percent of the population that is literate is 75.

(b) What would be the mean difference in age gap between an African country and a non-African country?

(c) Interpret the coefficient of "percent of children aged 10 to 14 involved in child labor."

(d) The coefficient of determination for this model is 0.593. Interpret this result.

(e) The P-value for the test $H_1: \beta_1 \neq 0$ versus $H_1: \beta_1 \neq 0$ is 0.008. What would you conclude about this test?

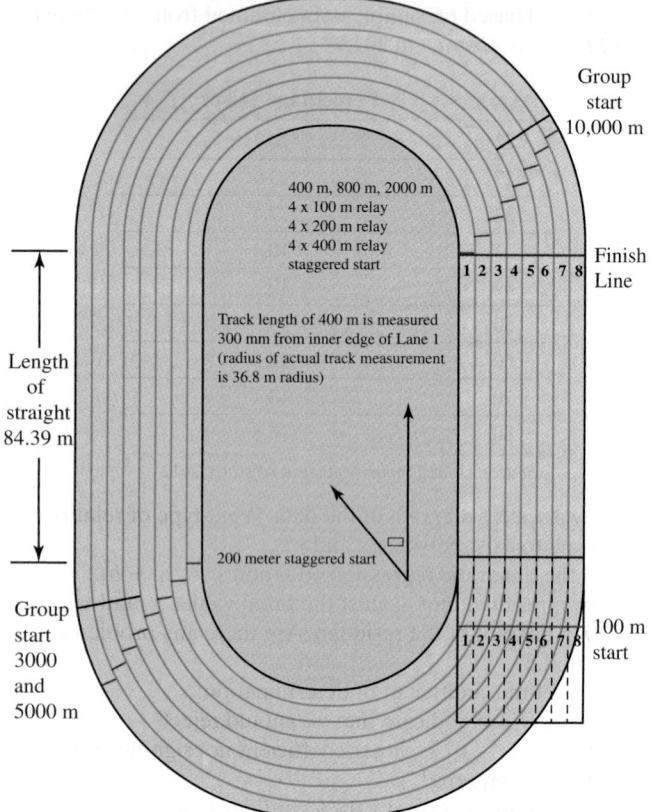

14.6 Building a Regression Model

Objectives ❶ Perform a partial F-test

❷ Build a regression model

❶ Perform a Partial F-test

Up to this point, we have conducted two tests on the coefficients of a multiple regression model. The first test is the F-test, which is used to determine if at least one of the coefficients in a least-squares regression model is different from 0. The second test is a *t*-test to determine if a specific coefficient in a least-squares regression model is different from zero. We now introduce a third test, called the **partial *F*-test**, which is used to determine if at least one of a subset of regression coefficients in a multiple regression model is different from zero.

For example, suppose we believe that five explanatory variables x_1, x_2, x_3, x_4, and x_5 may be used to predict the value of a response variable y. The multiple regression model would be

$$y_i = \beta_0 + \beta_1 x_{1i} + \beta_2 x_{2i} + \beta_3 x_{3i} + \beta_4 x_{4i} + \beta_5 x_{5i} + \varepsilon_i \tag{1}$$

Suppose we suspect that the explanatory variables x_4 and x_5 do not contribute significantly to the knowledge of y. That is, we believe a better regression model to predict the value of y is

$$y_i = \beta_0 + \beta_1 x_{1i} + \beta_2 x_{2i} + \beta_3 x_{3i} + \varepsilon_i \tag{2}$$

To determine if (2) is a better model than (1), we would test

$$H_0: \beta_4 = \beta_5 = 0$$

versus

H_1: at least one of the coefficients, β_4 or β_5, is different from zero

In general, a partial *F*-test is used to compare two regression models.

Model 1 (the reduced model): $y_i = \beta_0 + \beta_1 x_{1i} + \beta_2 x_{2i} + \cdots + \beta_r x_{ri} + \varepsilon_i$

versus

Model 2 (the full model):
$$y_i = \underbrace{\beta_0 + \beta_1 x_{1i} + \beta_2 x_{2i} + \cdots + \beta_r x_{ri}}_{\text{Model 1}} + \underbrace{\beta_{r+1} x_{(r+1)i} + \beta_{r+2} x_{(r+2)i} + \cdots + \beta_k x_{ki}}_{\text{Additional Terms in Model 2}} + \varepsilon_i$$

We wish to test whether any of the additional coefficients in Model 2 are significantly different from 0 (contribute significantly to the ability to predict y). That is, we wish to test

$$H_0: \beta_{r+1} = \beta_{r+2} = \cdots = \beta_k = 0$$

versus

H_1: at least one of the coefficients, $\beta_{r+1}, \beta_{r+2}, \ldots, \beta_k$ is different from zero

To test this hypothesis, find the least-squares regression of both Model 1 and Model 2. If Model 2 contributes significantly to explaining the value of y, then we would expect the sum of squared errors (SSE) in Model 2 to be significantly less than the SSE in Model 1. So, the greater the difference in sum of squared errors between the two models, the stronger the evidence against H_0. The test statistic for such a hypothesis test is

$$F_0 = \dfrac{\dfrac{\text{SSE}(1) - \text{SSE}(2)}{k - r}}{\dfrac{\text{SSE}(2)}{n - (k + 1)}} = \dfrac{\dfrac{\text{SSE}(1) - \text{SSE}(2)}{k - r}}{\text{MSE}(2)}$$

where   SSE(1) is the sum of squared errors from Model 1 (the reduced model)

SSE(2) is the sum of squared errors from Model 2 (the full model)

k is the number of explanatory variables in Model 2

r is the number of explanatory variables in Model 1

The test statistic, F_0, is based on $k - r$ degrees of freedom in the numerator and $n - (k + 1)$ degrees of freedom in the denominator. So, the critical value, F_α, is found using Table IX with $k - r$ degrees of freedom in the numerator and $n - (k + 1)$ degrees of freedom in the denominator. If $F_0 > F_\alpha$, reject the statement in H_0 and conclude that at least one of the coefficients $\beta_{r+1}, \beta_{r+2}, \ldots, \beta_k$ is different from zero. That is, conclude at least one of the additional explanatory variables contributes to predicting the value of y.

EXAMPLE 1   A Partial F-test

Problem The data in Table 9 represent the miles per gallon (MPG) along with potential explanatory variables weight (in thousands of pounds), number of cylinders, and horsepower. Use a partial F-test to determine whether both cylinders and horsepower do not contribute significantly to the ability to predict miles per gallon. Use the $\alpha = 0.05$ level of significance.

Table 9

MPG, y	Weight, x_1	Cylinders, x_2	Horsepower, x_3	MPG, y	Weight, x_1	Cylinders, x_2	Horsepower, x_3
28.4	2.67	4	90	18.5	3.94	8	150
30.9	2.23	4	75	27.2	2.3	4	97
20.8	3.07	6	85	18.1	3.41	6	120
37.3	2.13	4	69	20.6	3.38	6	105
16.2	3.41	6	133	21.6	2.795	4	115
31.9	1.925	4	71	31.8	2.02	4	65
34.2	2.2	4	70	17.6	3.725	8	129
34.1	1.975	4	65	17	3.14	6	125
16.9	4.36	8	155	18.2	3.83	8	135
20.3	2.83	5	103	27.5	2.56	4	95
19.2	3.605	8	125	30	2.155	4	68
18.6	3.62	6	110	21.9	2.91	6	109
30.5	2.19	4	78	27.4	2.67	4	80
26.5	2.585	4	88	15.5	4.054	8	142
35.1	1.915	4	80	21.5	2.6	4	110
22	2.815	6	97	33.5	2.556	4	90
31.5	1.99	4	71	29.5	2.135	4	68
28.8	2.595	6	115	16.5	3.955	8	138
26.8	2.7	6	115	17	3.84	8	130

Source: Professor Butler, University of Toronto-Scarborough via StatCrunch

Approach To perform a partial F-test, obtain the sum of squared errors from the full regression model with all three explanatory variables (weight, number of cylinders, and horsepower). Also obtain the sum of squared errors from the reduced model with only the explanatory variable weight. Then, compute the F-test statistic and compare it to the critical value to see whether the two explanatory variables, number of cylinders and horsepower, contribute significantly to explain the variability in miles per gallon.

Solution The hypotheses to be tested are

$$H_0: \beta_2 = \beta_3 = 0$$

versus

$$H_1: \text{at least one of } \beta_2 \text{ or } \beta_3 \text{ is different from zero}$$

Figure 30(a) shows the least-squares regression output for the full model and Figure 30(b) shows the least-squares regression output for the reduced model both obtained from StatCrunch.

Figure 30

Multiple linear regression results:
Dependent Variable: MPG
Independent Variable(s): Weight, Cylinders, Horsepower

Parameter estimates:

Parameter	Estimate	Std. Err.	Alternative	DF	T-Stat	P-value
Intercept	49.380231	1.9690473	≠ 0	34	25.078236	<0.0001
Weight	-7.3898246	2.0797337	≠ 0	34	-3.5532553	0.0011
Cylinders	0.74591885	0.72524617	≠ 0	34	1.0285044	0.311
Horsepower	-0.073596259	0.044100334	≠ 0	34	-1.6688368	0.1043

Analysis of variance table for multiple regression model:

Source	DF	SS	MS	F-stat	P-value
Model	3	1320.2373	440.07909	56.281704	<0.0001
Error	34	265.85352	7.8192212		
Total	37	1586.0908			

(a)

Simple linear regression results:
Dependent Variable: MPG
Independent Variable: Weight
MPG = 48.707495 - 8.3645999 Weight
Sample size: 38
R (correlation coefficient) = -0.90307083
R-sq = 0.81553692
Estimate of error standard deviation: 2.8508049

Parameter estimates:

Parameter	Estimate	Std. Err.	Alternative	DF	T-Stat	P-value
Intercept	48.707495	1.9536817	≠ 0	36	24.931131	<0.0001
Slope	-8.3645999	0.66302032	≠ 0	36	-12.615903	<0.0001

Analysis of variance table for regression model:

Source	DF	SS	MS	F-stat	P-value
Model	1	1293.5156	1293.5156	159.161	<0.0001
Error	36	292.57519	8.1270887		
Total	37	1586.0908			

(b)

From Figure 30(a), the sum of squared error for the full model (Model 2) is $SSE(2) = 265.85352$ and the mean squared error for the full model is $MSE(2) = 7.819221$. From Figure 30(b), the sum of squared error for the reduced model (Model 1) is $SSE(1) = 292.5752$. The test statistic is

$$F_0 = \frac{\dfrac{SSE(1) - SSE(2)}{k - r}}{MSE(2)} = \frac{\dfrac{292.5752 - 265.85352}{3 - 1}}{7.819221}$$
$$= 1.71$$

The critical value with numerator degrees of freedom $= 2$ and denominator degrees of freedom $= 34$ is $F_{0.05, 2, 34} \approx 3.39$ (using 25 degrees of freedom in the denominator). Because $F_0 < F_\alpha$, do not reject the null hypothesis. There is not sufficient evidence to suggest either number of cylinders or horsepower explains variability in the response variable miles per gallon. Both variables should be removed from the model.   ●

② Build a Regression Model

Now we discuss an approach to determine the "best" regression model. We put the word *best* in quotes because there is really no best model. It is up to the researcher to decide which model is best.

When building a least-squares regression model, there are often a large number of explanatory variables to choose from. The ultimate question is, "Which explanatory variables should be included in the model?" There are three accepted approaches to building a regression model.

- Forward selection
- Backward elimination
- Stepwise regression

When utilizing these three approaches, the regression model will be such that the explanatory variable is first-order with no interaction. That is, the model is of the form

$$y_i = \beta_0 + \beta_1 x_{1i} + \beta_2 x_{2i} + \cdots + \beta_k x_{ki} + \varepsilon_i$$

The **forward selection** method proceeds in steps. At first the model only contains the intercept. At each step at most one variable can be added to the model. For an explanatory variable to be added it must have the smallest P-value when it is the only variable added to the model and must be significant (P-value less than a specified level of significance). The process ends when any remaining explanatory variables not in the model have P-values greater than the level of significance when added to the model.

The **backward elimination** method also works in steps. The process begins with all the explanatory variables in the model. At each step one explanatory variable is eliminated from the model. For an explanatory variable to be eliminated, it must have the largest P-value that is greater than the specified level of significance. The process ends when any explanatory variables that remain in the model have P-values less than the level of significance.

Next, we present the steps utilized for **stepwise regression**.

Stepwise Regression Procedure

1. Find a simple linear regression equation for each of the explanatory variables. Choose the explanatory variable that has the lowest P-value from the F-test. If no explanatory variable has a P-value considered to be statistically significant, then none of the explanatory variables is useful in predicting the value of the response variable and the procedure ends.

2. Fit a regression model with the explanatory variable from Step 1 and each of the remaining explanatory variables. For each regression model, determine the partial F-test statistic to test whether the additional explanatory variable is different from zero when the explanatory variable from Step 1 is already in the model. Test the hypothesis at the $\alpha = 0.05$ level of significance. Choose the model with the largest F-test statistic that is significant. If no explanatory variable yields a partial F-test statistic that is significant, the procedure stops.

3. The procedure now verifies whether any of the explanatory variables already in the model should be dropped (other than the explanatory variable just added in Step 2) based on the predetermined rejection criterion.

4. The procedure then determines whether another explanatory variable should be added to the model, and then determines whether any explanatory variables already in the model should be dropped. This process continues until no explanatory variables are added or dropped from the model, at which point the procedure terminates.

It is important to recognize that the stepwise regression procedure allows for an explanatory variable to enter the model, but then be dropped in later iterations if it is deemed to be a variable that does not significantly contribute to the prediction of the response variable (given that other variables are in the model).

EXAMPLE 2   Building a Regression Model Using Forward Selection

Problem An engineer wants to develop a model to describe the gas mileage of a sport utility vehicle. He collects the data presented in Table 10. The final drive ratio is the ratio of the gear set that is farthest from the engine. A higher ratio generally means better acceleration and pulling power. Use the forward selection procedure to find the best model to describe miles per gallon. Use a level of significance of $\alpha = 0.05$ as the criteria for an explanatory variable to enter the model.

Table 10

Sport Utility Vehicle	Curb Weight (pounds)	Engine Size (liters)	Cylinders	Final Drive Ratio (x : 1)	Ground Clearance (inches)	Miles per Gallon
Mercedes-Benz G500	5510	5.0	8	4.38	8.3	13
Jeep Wrangler X	3760	3.8	6	3.21	8.8	16
Mitsubishi Endeavor LS 2WD	3902	3.8	6	4.01	8.3	18
Toyota Land Cruiser	5690	5.7	8	3.91	8.9	15
Kia Sorento 4 × 2	4068	3.3	6	3.33	8.2	18
Jeep Commander Sport 4 × 2	4649	4.7	8	3.73	8.6	15
Dodge Durango SLT 4 × 2	4709	4.7	8	3.55	8.7	15
Lincoln Navigator 4 × 2	5555	5.4	8	3.73	9.2	15
Chevrolet Tahoe LS 2WD	5265	4.8	8	3.23	9.1	16
Ford Escape FWD	3387	3.0	6	2.93	8.5	20
Ford Expedition XLT	5578	5.4	8	3.31	8.7	14
Buick Enclave CX FWD	4780	3.6	6	3.16	8.4	19
Cadillac Escalade 2WD	5459	6.2	8	3.42	9.0	14
Hummer H3 SUV	4689	3.7	5	4.56	9.1	15
Saab 9-7X	4720	4.2	6	3.73	7.7	16

Source: www.newcars.org

Approach We will use Minitab to find the best model.

Solution Use Minitab to find a simple linear regression for each of the explanatory variables listed in Table 10. Figure 31 shows partial output for all the linear regressions for each explanatory variable.

Figure 31

```
Coefficients

Term               Coef    SE Coef   T-Value   P-Value   VIF
Constant          25.57      2.39     10.69     0.000
Curb Weight   -0.002015   0.000495   -4.07     0.001    1.00


Regression Equation

MPG = 25.57 - 0.002015 Curb Weight
```
(a)

```
Coefficients

Term             Coef    SE Coef   T-Value   P-Value   VIF
Constant        23.39      1.61     14.55     0.000
Engine Size    -1.663     0.351     -4.74     0.000    1.00


Regression Equation

MPG = 23.39 - 1.663 Engine Size
```
(b)

```
Coefficients

Term            Coef    SE Coef   T-Value   P-Value   VIF
Constant       23.71      2.65      8.95     0.000
Cylinders     -1.111     0.374     -2.97     0.011    1.00


Regression Equation

MPG = 23.71 - 1.111 Cylinders
```
(c)

```
Coefficients

Term                 Coef    SE Coef   T-Value   P-Value   VIF
Constant            23.96      3.72      6.44     0.000
Final Drive Ratio   -2.22      1.02     -2.17     0.049    1.00


Regression Equation

MPG = 23.96 - 2.22 Final Drive Ratio
```
(d)

```
Coefficients

Term                 Coef    SE Coef   T-Value   P-Value   VIF
Constant            30.5      10.9      2.81      0.015
Ground Clearance   -1.69      1.26     -1.34      0.203    1.00


Regression Equation

MPG = 30.5 - 1.69 Ground Clearance
```
(e)

(continued)

The explanatory variable with the lowest significant P-value, engine size (or largest, in absolute value, t test statistic), is added to the model. In Figure 31(b) we see the P-value for engine size is 0.000 (that is, <0.0005), which suggests engine size contributes significantly to the ability to predict gas mileage. **Note:** We choose engine size over weight, cylinders, or drive ratio because the test statistic for engine size is largest. Now, repeat the process by finding least-squares regressions with engine size and each of the remaining explanatory variables. The additional explanatory variable with the lowest P-value is added to the model (provided the P-value is less than 0.05). We will not show each possible regression. Figure 32 shows that the explanatory variable drive ratio should be added to the model.

Figure 32   **Regression Analysis: MPG versus Engine Size, Final Drive Ratio**

```
Analysis of Variance

Source              DF  Adj SS  Adj MS  F-Value  P-Value
Regression           2  42.546  21.273    20.61    0.000
  Engine Size        1  27.915  27.915    27.04    0.000
  Final Drive Ratio  1   7.772   7.772     7.53    0.018
Error               12  12.387   1.032
Total               14  54.933


Model Summary

      S    R-sq  R-sq(adj)  R-sq(pred)
1.01601  77.45%     73.69%      64.65%


Coefficients

Term                  Coef  SE Coef  T-Value  P-Value   VIF
Constant             28.68     2.33    12.30    0.000
Engine Size         -1.516    0.291    -5.20    0.000  1.04
Final Drive Ratio   -1.647    0.600    -2.74    0.018  1.04


Regression Equation

MPG = 28.68 - 1.516 Engine Size - 1.647 Final Drive Ratio
```

Note: The regression model including both weight and engine size results in a model with an F-test statistic that is not significant. This result likely occurs due to the high correlation between engine size and weight (0.870). Put simply, the model became confused when two highly correlated variables are included in the model. That is, the model was confused due to multicollinearity.

Now, we repeat the process by finding least-squares regressions with engine size and drive ratio and each of the remaining explanatory variables. None of the remaining explanatory variables have P-values less than the level of significance, 0.05. Therefore, the regression model from the forward selection method is

$$\hat{y} = 28.68 - 1.516x_1 - 1.647x_2$$

where x_1 is engine size and x_2 is final drive ratio.

EXAMPLE 3   Building a Regression Model Using Backward Elimination

Problem Use the backward elimination procedure to find the best model to describe miles per gallon for the data in Table 10. Use a level of significance of $\alpha = 0.05$ as the criteria for an explanatory variable to be removed from the model.

Approach Again, we will use Minitab to find the least-squares regression models.

Solution First, find the least-squares regression model using all the explanatory variables. See Figure 33.

Figure 33

Regression Analysis: MPG versus Curb Weight, Engine Size, Cylinders, Final Drive , ...

```
Analysis of Variance

Source              DF    Adj SS    Adj MS   F-Value   P-Value
Regression           5   42.7881   8.55762      6.34     0.009
  Curb Weight        1    0.0011   0.00106      0.00     0.978
  Engine Size        1    2.6350   2.63498      1.95     0.196
  Cylinders          1    0.2295   0.22953      0.17     0.690
  Final Drive Ratio  1    7.0180   7.01800      5.20     0.049
  Ground Clearance   1    0.0191   0.01910      0.01     0.908
Error                9   12.1453   1.34947
Total               14   54.9333


Model Summary

      S   R-sq  R-sq(adj)  R-sq(pred)
1.16167  77.89%     65.61%      13.69%


Coefficients

Term                   Coef    SE Coef   T-Value   P-Value   VIF
Constant              30.38       7.75      3.92     0.004
Curb Weight        0.000025   0.000900      0.03     0.978  4.55
Engine Size          -1.267      0.907     -1.40     0.196  7.66
Cylinders            -0.231      0.561     -0.41     0.690  4.19
Final Drive Ratio    -1.769      0.776     -2.28     0.049  1.32
Ground Clearance     -0.101      0.851     -0.12     0.908  1.26


Regression Equation

MPG = 30.38 + 0.000025 Curb Weight - 1.267 Engine Size - 0.231 Cylinders
      - 1.769 Final Drive Ratio - 0.101 Ground Clearance
```

The *P*-value for the *F*-test statistic is very low at 0.009. This indicates that our model is reasonable. However, looking at the individual *t*-test statistics for slope coefficients, we see that some have *P*-values that are very large. We notice that the *P*-value for the slope coefficient of the explanatory variable curb weight is the largest (0.978), so we will remove weight from the model.

Notice that this slope coefficient is positive, which indicates that SUVs with more weight will get better gas mileage. This contradicts common sense. The correlation between curb weight and cylinders (0.720) explains the strange results (multicollinearity).

(continued)

Figure 34 shows the new regression model.

Figure 34

Regression Analysis: MPG versus Engine Size, Cylinders, Final Drive Ratio, Ground Clearance

```
Analysis of Variance

Source              DF   Adj SS   Adj MS   F-Value   P-Value
Regression           4  42.7870  10.6968      8.81     0.003
  Engine Size        1   4.1869   4.1869      3.45     0.093
  Cylinders          1   0.2286   0.2286      0.19     0.674
  Final Drive Ratio  1   7.6732   7.6732      6.32     0.031
  Ground Clearance   1   0.0192   0.0192      0.02     0.903
Error               10  12.1463   1.2146
Total               14  54.9333
```

```
Model Summary

      S   R-sq  R-sq(adj)  R-sq(pred)
1.10210  77.89%    69.04%      36.50%
```

```
Coefficients

Term                 Coef  SE Coef  T-Value  P-Value   VIF
Constant            30.40     7.31     4.16    0.002
Engine Size         -1.251   0.674    -1.86    0.093  4.70
Cylinders           -0.231   0.531    -0.43    0.674  4.19
Final Drive Ratio   -1.763   0.701    -2.51    0.031  1.20
Ground Clearance    -0.101   0.807    -0.13    0.903  1.26
```

```
Regression Equation

MPG = 30.40 - 1.251 Engine Size - 0.231 Cylinders - 1.763 Final Drive Ratio
        - 0.101 Ground Clearance
```

The P-value for the F-test decreased to 0.003, which indicates that our model has improved. However, we still have some explanatory variables whose P-values are quite large. We will next remove the explanatory variable "ground clearance" from the model because it has the largest insignificant P-value. Notice also that the adjusted R^2 increased from 65.6% to 69.0% when we removed curb weight from the model. This illustrates that the adjusted R^2 punishes a model that is unnecessarily complex.

Figure 35 shows the new regression model.

Figure 35

Regression Analysis: MPG versus Engine Size, Cylinders, Final Drive Ratio

```
Analysis of Variance

Source              DF   Adj SS   Adj MS   F-Value   P-Value
Regression           3  42.7679  14.2560     12.89     0.001
  Engine Size        1   4.7421   4.7421      4.29     0.063
  Cylinders          1   0.2218   0.2218      0.20     0.663
  Final Drive Ratio  1   7.6623   7.6623      6.93     0.023
Error               11  12.1655   1.1060
Total               14  54.9333
```

```
Model Summary

      S   R-sq  R-sq(adj)  R-sq(pred)
1.05164  77.85%    71.81%      59.50%
```

```
Coefficients

Term                 Coef  SE Coef  T-Value  P-Value   VIF
Constant            29.58     3.14     9.42    0.000
Engine Size         -1.275   0.616    -2.07    0.063  4.31
Cylinders           -0.227   0.506    -0.45    0.663  4.17
Final Drive Ratio   -1.756   0.667    -2.63    0.023  1.19
```

```
Regression Equation

MPG = 29.58 - 1.275 Engine Size - 0.227 Cylinders - 1.756 Final Drive Ratio
```

The *P*-value for the *F*-test continues to decline as we eliminate variables! However, we still have explanatory variables with large *P*-values. Next we remove the explanatory variable cylinders because it has the largest insignificant *P*-value.

Figure 36 shows the new regression model.

Figure 36   **Regression Analysis: MPG versus Engine Size, Final Drive Ratio**

```
Analysis of Variance

Source             DF  Adj SS  Adj MS  F-Value  P-Value
Regression          2  42.546  21.273    20.61    0.000
  Engine Size       1  27.915  27.915    27.04    0.000
  Final Drive Ratio 1   7.772   7.772     7.53    0.018
Error              12  12.387   1.032
Total              14  54.933


Model Summary

      S    R-sq  R-sq(adj)  R-sq(pred)
1.01601  77.45%     73.69%      64.65%


Coefficients

Term               Coef  SE Coef  T-Value  P-Value   VIF
Constant          28.68     2.33    12.30    0.000
Engine Size      -1.516    0.291    -5.20    0.000  1.04
Final Drive Ratio -1.647    0.600    -2.74    0.018  1.04


Regression Equation

MPG = 28.68 - 1.516 Engine Size - 1.647 Final Drive Ratio
```

Now the *P*-value for the *F*-test is very low at 0.000 (so *P*-value < 0.0005), and the *P*-values for the slope coefficients are also low. In addition, the estimates of the slope coefficients are reasonable. We expect both to be negative because (1) a larger engine should get worse gas mileage and (2) a higher final drive ratio (which indicates better initial acceleration) is likely to get worse gas mileage. We believe that the "best" model is

$$\hat{y} = 28.68 - 1.516x_1 - 1.647x_2$$

where x_1 is engine size and x_2 is final drive ratio.   •

EXAMPLE 4   Building a Regression Model Using Stepwise Regression

Problem Use the stepwise regression procedure to find the best model to describe miles per gallon for the data in Table 10. Use a level of significance of $\alpha = 0.15$ as the criteria for an explanatory variables to be removed from, or included in, the model.

Approach We will use the stepwise regression feature in StatCrunch. A similar feature exists in Minitab.

(continued)

Solution Figure 37 shows the results obtained from StatCrunch.

Figure 37

Multiple linear regression results:
Dependent Variable: MPG
Independent Variable(s): Weight, Engine Size, Cylinders, Final Drive Ratio, Ground Clearance

Stepwise results:
P-value to enter: 0.15
P-value to leave: 0.25

Step	Variable	Action	P-value	RMSE	R-squared	R-squared (adj)
1	Engine Size	Entered	0.0004	1.2452701	0.633	0.6048
2	Final Drive Ratio	Entered	0.0178	1.0160078	0.7745	0.7369

Parameter estimates:

Parameter ◆	Estimate ◆	Std. Err. ◆	Alternative ◆	DF ◆	T-Stat ◆	P-value ◆
Intercept	28.6832	2.3317894	≠ 0	12	12.300939	<0.0001
Engine Size	-1.5156649	0.29145949	≠ 0	12	-5.2002593	0.0002
Final Drive Ratio	-1.6468675	0.60019824	≠ 0	12	-2.7438726	0.0178

Analysis of variance table for multiple regression model:

Source	DF	SS	MS	F-stat	P-value
Model	2	42.546071	21.273035	20.607977	0.0001
Error	12	12.387263	1.0322719		
Total	14	54.933333			

Summary of fit:
Root MSE: 1.0160078
R-squared: 0.7745
R-squared (adjusted): 0.7369

The stepwise regression procedure selects

$$\hat{y} = 28.6832 - 1.5157x_1 - 1.6469x_2$$

where x_1 is engine size and x_2 is final drive ratio as the best model.   ●

Regardless of whether you use forward selection, backward elimination, or stepwise regression, the regression model should always be verified by drawing residual plots. Figure 38 shows the residual plots as well as the normal probability plots of the residuals.

Figure 38

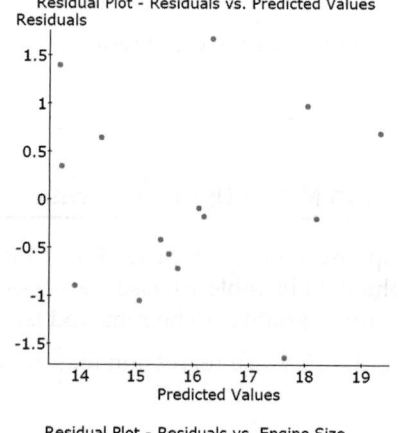

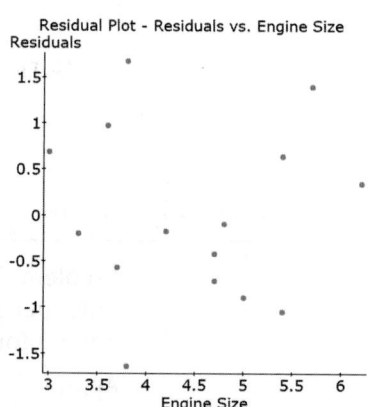

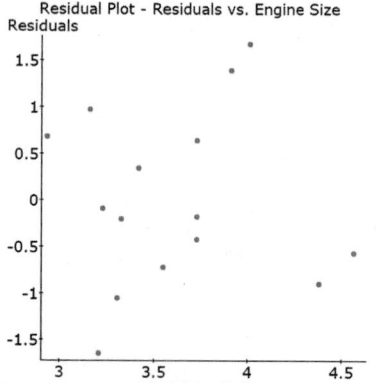

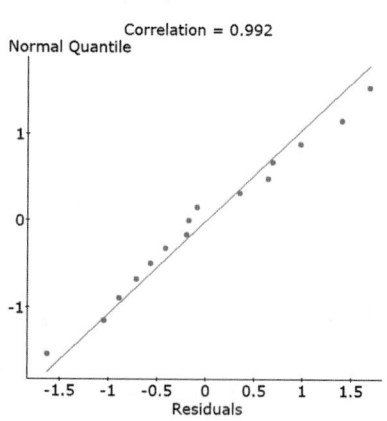

The residual plots do not indicate that there are any problems in the model, and the normal probability plot indicates that the residuals appear to be normally distributed because 0.992 > 0.939 (Table VI) and a boxplot of the residuals (not shown) indicates there are no outliers, so we accept the model.

Technology Step-by-Step Forward Selection, Backward Elimination, and Stepwise Regression

Minitab
1. With the data in the spreadsheet, select **Stat**, highlight **Regression**, highlight **Regression**, and select **Fit Regression Model**.
2. Select the Response variable. Select all the quantitative explanatory variables and enter them into the "Continuous predictors" cell. Select all the qualitative explanatory variables and enter them into the "Categorical predictors" cell.
3. Click **Stepwise**. From the drop-down menu under "Method," select Stepwise, Forward selection, or Backward elimination. Click OK twice.

StatCrunch
1. With the data in the spreadsheet, select **Stat**, highlight **Regression**, then select **Multiple Linear**.
2. Select the Y Variable (response variable) from the drop-down menu. Choose all explanatory variables as the X Variables.
3. Choose Stepwise, Forward selection, or Backward selection and enter the P-value to enter or P-value to leave (level of significance). Choose any additional output to save. Click Compute!

14.6 Assess Your Understanding

Skill Building

 1. For the data set below, use a partial F-test to determine whether the variables x_4 and x_5 do not significantly help to predict the response variable, y. Use the $\alpha = 0.05$ level of significance.

x_1	x_2	x_3	x_4	x_5	y
0.8	2.8	2.5	10.6	15.7	11.0
3.9	2.6	5.7	9.2	4.2	10.8
1.8	2.4	7.8	10.1	1.5	10.6
5.1	2.3	7.1	9.2	1.9	10.3
4.9	2.5	5.9	11.2	5.6	10.3
8.4	2.1	8.6	10.4	4.9	10.3
12.9	2.3	9.2	11.1	1.9	10.0
6.0	2.0	1.2	8.6	22.3	9.4
14.6	2.2	3.7	10.5	11.5	8.7
9.3	1.1	5.5	8.8	6.1	8.7

2. For the data set below, use a partial F-test to determine whether the variables x_1 and x_2 do not significantly help to predict the response variable, y. Use the $\alpha = 0.05$ level of significance.

x_1	x_2	x_3	x_4	y	x_1	x_2	x_3	x_4	y
24.9	66.3	13.5	3.7	59.8	41.1	83.5	9.7	21.8	84.6
26.7	100.6	15.7	11.4	66.3	25.4	112.7	9.8	16.4	87.3
30.6	77.8	13.8	15.7	76.5	33.8	68.8	6.8	25.9	88.5
39.6	83.4	8.8	8.8	77.1	23.5	69.5	7.5	15.5	90.7
33.1	69.4	10.6	18.3	81.9	39.8	63.0	6.8	30.8	93.4

3. Housing Prices A realtor wanted to find a model that describes the asking price of houses in Greenville, South Carolina. She obtains a random sample of homes from the area.

Square Footage	Bedrooms	Baths	Asking Price ($ thousands)
3800	4	3.5	498
2600	4	3	449
2600	5	3.5	435
2250	4	4	400
3300	4	3	379
2750	3	2.5	375
2200	3	2.5	356
3000	4	2.5	350
2300	3	2	340
2600	4	2.5	332
2300	4	2	298
2000	4	3	280
2200	3	2.5	260

Source: remax.com

(a) Find the least-squares regression equation $\hat{y} = b_0 + b_1x_1 + b_2x_2 + b_3x_3$, where x_1 is square footage, x_2 is number of bedrooms, x_3 is number of baths, and y is the response variable, asking price.

(b) Use a partial F-test to determine whether number of bedrooms and number of baths do not significantly help to predict the response variable, asking price.

(c) Use either forward selection, backward elimination, or stepwise regression to identify the best model in predicting asking price.

(d) Draw residual plots, a boxplot of residuals, and a normal probability plot of residuals to assess the adequacy of the model found in part (c).

(e) Interpret the regression coefficients for the least-squares regression equation found in part (c).

(f) Construct 95% confidence and prediction intervals for the asking price of a 2600-square foot house in Greenville, South Carolina, with four bedrooms and three baths. Interpret the results.

4. Another Mileage Model A researcher is interested in developing a model that describes the gas mileage, measured in miles per gallon (mpg), of automobiles. Based on input from an engineer, she decides that the explanatory variables might be engine size (liters), curb weight (pounds), and horsepower. From a random sample of 13 automobiles, she obtains the following data:

Engine Size	Curb Weight	Horsepower	Miles per Gallon
2.4	3289	177	24
2.4	3263	158	25
2.5	3230	170	24
3.5	3580	272	22
2.8	3175	255	18
3.5	3643	263	22
3.5	3497	306	20
3.0	3340	230	21
3.6	3861	263	19
2.4	3287	173	24
3.3	3629	234	21
2.5	3270	170	22
3.5	3292	270	22

Source: www.newcars.org

(a) Find the least-squares regression equation $\hat{y} = b_0 + b_1x_1 + b_2x_2 + b_3x_3$, where x_1 is engine size, x_2 is curb weight, x_3 is horsepower, and y is the response variable, miles per gallon.

(b) Use a partial F-test to determine whether engine size and curb weight do not significantly help to predict the response variable, miles per gallon.

(c) Use forward selection, backward elimination, or stepwise regression to identify the best model in predicting asking price.

(d) Draw residual plots, a boxplot of residuals, and a normal probability plot of residuals to assess the adequacy of the model found in part (c).

(e) Interpret the regression coefficients for the least-squares regression equation found in part (c).

(f) Construct 95% confidence and prediction intervals for the gas mileage of an automobile that weighs 3100 pounds, has a 2.5-liter engine, and 200 horsepower. Interpret the results.

5. Go to www.pearsonhighered.com/sullivanstats to obtain the data file 14_6_5 using the file format of your choice. The data represent a variety of variables that may explain a pitcher's ERA (earned run average), or the mean number of earned runs (runs given up excluding those due to fielding errors) yielded by a pitcher over the course of nine innings pitched. The potential explanatory variables are strikeout to walk ratio (K/BB), (strikeouts per nine innings (K/9), pitchers per batter faced (P/PA), pitches per inning pitched (P/IP), percentage of games won when pitching (W%), ground balls allowed (GB), fly balls allowed (FB), ratio of ground balls to fly balls (G/F), runs scored by pitchers team or run support (RS), and walks and hits per inning pitched (WHIP). Use forward selection, backward elimination, or stepwise regression to build a model that can be used to predict the earned run average of a pitcher.

6. Go to www.pearsonhighered.com/sullivanstats to obtain the data file 14_6_6 using the file format of your choice. Suppose a nutritionist wants to develop a model that describes the amount of calories in a single serving of cereal. The potential explanatory variables for the model are protein (grams), fat (grams), sodium (milligrams), fiber (grams), carbs (grams of carbohydrates), sugars (grams), potass (milligrams of potassium), and shelf the cereal is sold on in the store. Use forward selection, backward elimination, or stepwise regression to build a model that can be used to predict the calories in cereal.

7. Tires The following data represent the cost of tires (in dollars) along with a variety of potential explanatory variables. Slalom time is the amount of time it took for a 3-series BMW to get through a slalom track, lap time is the amount of time it took the same car to complete a 1/3-mile lap, and stopping distance is the distance it took the BMW to stop on wet pavement traveling 60 miles per hour. Find the best regression model using each of the three techniques presented in the section. What do you notice?

TIRE	Cost (dollars)	MPG	Slalom Time (seconds)	Lap Time (seconds)	Stopping Distance (feet)	Cornering g-Force
BFGoodrich g-Force Sport COMP-2	114	30.5	5.13	30.24	80.0	0.90
Bridgestone Potenza RE760 Sport	126	30.2	5.08	30.14	79.4	0.91
Firestone Firehawk Wide Oval Indy 500	111	30.4	5.16	30.58	83.3	0.88
Yokohama S.drive	119	31.0	5.20	30.61	82.2	0.90
Bridgestone Turanza Serenity Plus	154	32.2	5.10	31.13	90.4	0.84
Continental PureContact	134	32.7	5.15	31.18	91.2	0.85
Michelin Primacy MXV4	135	32.3	5.15	31.18	90.2	0.85
Yokohama AVID Ascend	134	32.3	5.17	31.11	91.4	0.86

8. Putting It Together: Purchasing Diamonds The value of a diamond is determined by the four C's: carat weight, color, clarity, and cut. *Carat weight* is the standard measure for the size of a diamond. Generally, the more a diamond weighs, the more valuable it will be. The Gemological Institute of America (GIA) determines the *color* of diamonds using a 22-grade scale from D (almost clear white) to Z (light yellow). Colorless diamonds are generally considered the most desirable. The clarity of a diamond refers to how "free" the diamond is of imperfections and is determined using an 11-grade scale: flawless (FL), internally flawless (IF), very, very slightly imperfect (VVS1, VVS2), very slightly imperfect (VS1, VS2), slightly imperfect (SI1, SI2), and imperfect (I1, I2, I3). The *cut* of a diamond refers to the diamond's proportions and finish. Put simply, the better the diamond's cut is, the better it reflects and refracts light, which makes it more beautiful and thus more valuable. The cut of a diamond is rated using a five-grade scale: Excellent, Very Good, Good, Fair, and Poor. Finally, the *shape* of a diamond (which is not one of the four C's) refers to its basic form: round, oval, pear-shaped, marquis, and so on. A novice might confuse shape with cut, so be careful not to confuse the two. Go to www.pearsonhighered.com/sullivanstats to obtain the data file 14_6_8 using the file format of your choice for the version of the text you are using.

The data represent a random sample of 40 unmounted, round-shaped diamonds. Use the data to answer the questions that follow:

(a) Determine the level of measurement for each variable.

    (i) Carat weight   (iv) Cut
    (ii) Color   (v) Price
    (iii) Clarity   (vi) Shape

(b) Construct a correlation matrix. To do so, first convert the variables color, clarity, and cut to numeric values as follows:

Color: D = 1, E = 2, F = 3, G = 4, H = 5, I = 6, J = 7
Clarity: FL = 1, IF = 2, VVS1 = 3, VVS2 = 4, VS1 = 5, VS2 = 6, SI1 = 7, SI2 = 8
Cut: Excellent = 1, Very Good = 2, Good = 3

If price is to be the response variable in our model, is there reason to be concerned about multicollinearity? Explain.

(c) Find the "best" model for predicting the price of a diamond.
(d) Draw residual plots, a boxplot of the residuals, and a normal probability plot of the residuals to assess the adequacy of the "best" model.
(e) For the "best" model, interpret each regression coefficient.
(f) Determine and interpret R^2 and the adjusted R^2.
(g) Predict the mean price of a round-shaped diamond with the following characteristics: 0.85 carat, E, VVS1, Excellent.
(h) Construct a 95% confidence interval for the mean price found in part (g).
(i) Predict the price of an individual round-shaped diamond with the following characteristics: 0.85 carat, E, VVS1 Excellent.
(j) Construct a 95% prediction interval for the price found in part (i).
(k) Explain why the predictions in parts (g) and (i) are the same, yet the intervals in parts (h) and (j) are different.

Chapter 14 Review

Summary

The first two sections of this chapter dealt with inferential techniques that can be used on the least-squares regression model $y_i = \beta_0 + \beta_1 x_i + \varepsilon_i$.

In Section 14.1, we used sample data to obtain estimates of an intercept and slope. The residuals are required to be normally distributed, with mean 0 and constant standard deviation σ. The residuals for each observation should be independent. We verified these requirements through residual plots and a normal probability plot of the residuals. Provided that these requirements are satisfied, we can test hypotheses regarding the slope to determine whether the relation between the explanatory and response variables is linear.

In Section 14.2, we learned how to construct confidence and prediction intervals for a predicted value. We constructed confidence intervals for a mean response and prediction intervals for an individual response.

Sections 14.3 through 14.6 discussed how to build a multiple linear regression model $y_i = \beta_0 + \beta_1 x_{1i} + \beta_2 x_{2i} + \cdots + \beta_k x_{ki} + \varepsilon_i$. We used a correlation matrix to identify the explanatory variables that are linearly related to the response variable, y. We used technology to obtain estimates for the coefficients of each explanatory variable and the intercept.

The multiple linear regression model has the same requirements as the least-squares regression model with one explanatory variable. We used an F-test to test whether at least one coefficient is different from zero. If the null hypothesis that all coefficients are zero is rejected, we then use t-tests on each coefficient to determine which is different from zero.

When building a model, we use either forward selection, backward elimination, or stepwise regression.

Watch out for multicollinearity among explanatory variables. It can easily distort results by giving coefficients that are opposite in sign to what we would expect or have all coefficients appear to equal zero, even though the F-test leads us to reject the null hypothesis that all coefficients are zero.

Vocabulary

Formulas

Standard Error of the Estimate

$$s_e = \sqrt{\frac{\sum (y_i - \hat{y}_i)^2}{n - 2}} = \sqrt{\frac{\sum \text{residuals}^2}{n - 2}}$$

Standard Error of b_1

$$s_{b_1} = \frac{s_e}{\sqrt{\sum (x_i - \bar{x})^2}}$$

Confidence Intervals for the Slope of the Regression Line

A $(1 - \alpha) \cdot 100\%$ confidence interval for the slope of the true regression line, β_1, is given by the following formulas:

Lower bound: $b_1 - t_{\alpha/2} \cdot \dfrac{s_e}{\sqrt{\sum (x_i - \bar{x})^2}} = b_1 - t_{\alpha/2} \cdot s_{b_1}$

Upper bound: $b_1 + t_{\alpha/2} \cdot \dfrac{s_e}{\sqrt{\sum (x_i - \bar{x})^2}} = b_1 + t_{\alpha/2} \cdot s_{b_1}$

Here, $t_{\alpha/2}$ is computed with $n - 2$ degrees of freedom.

Confidence Interval about the Mean Response of $\hat{y}$

A $(1 - \alpha) \cdot 100\%$ confidence interval about the mean response of y, $\hat{y}$, is given by the following formulas:

Lower bound: $\hat{y} - t_{\alpha/2} \cdot s_e \sqrt{\dfrac{1}{n} + \dfrac{(x^* - \bar{x})^2}{\sum (x_i - \bar{x})^2}}$

Upper bound: $\hat{y} + t_{\alpha/2} \cdot s_e \sqrt{\dfrac{1}{n} + \dfrac{(x^* - \bar{x})^2}{\sum (x_i - \bar{x})^2}}$

Here, x^* is the given value of the explanatory variable, and $t_{\alpha/2}$ is the critical value with $n - 2$ degrees of freedom.

Prediction Interval about an Individual Response, $\hat{y}$

A $(1 - \alpha) \cdot 100\%$ prediction interval for the individual response of y, $\hat{y}$, is given by

Lower bound: $\hat{y} - t_{\alpha/2} \cdot s_e \sqrt{1 + \dfrac{1}{n} + \dfrac{(x^* - \bar{x})^2}{\sum (x_i - \bar{x})^2}}$

Upper bound: $\hat{y} + t_{\alpha/2} \cdot s_e \sqrt{1 + \dfrac{1}{n} + \dfrac{(x^* - \bar{x})^2}{\sum (x_i - \bar{x})^2}}$

where x^* is the given value of the explanatory variable and $t_{\alpha/2}$ is the critical value with $n - 2$ degrees of freedom.

OBJECTIVES

Section	You should be able to ...	Example(s)	Review Exercises
14.1	1 State the requirements of the least-squares regression model (p. 675)	p. 675	1
	2 Compute the standard error of the estimate (p. 676)	2 and 3	2(b), 3(b), 4(b)
	3 Verify that the residuals are normally distributed (p. 678)	4	2(c), 3(c), 4(c)
	4 Conduct inference on the slope of the least-squares regression model (p. 678)	5	2(e), 3(e), 4(f), 5(b)
	5 Construct a confidence interval about the slope of the least-squares regression model (p. 682)	6	2(f), 3(f), 4(g)
14.2	1 Construct confidence intervals for a mean response (p. 690)	1 and 3	2(g), 3(g)
	2 Construct prediction intervals for an individual response (p. 690)	2 and 3	2(i), 3(i)
14.3	1 Obtain the correlation matrix (p. 694)	1	7(a)
	2 Use technology to find a multiple regression equation (p. 697)	2	7(b), 10(a)
	3 Interpret the coefficients of a multiple regression equation (p. 698)	3	7(d)
	4 Determine R^2 and adjusted R^2 (p. 699)	4	7(e)
	5 Perform an F-test for lack of fit (p. 700)	5	7(f), 8(c)
	6 Test individual regression coefficients for significance (p. 702)	6	7(g), 8(d)
	7 Construct confidence and prediction intervals (p. 702)	7	7(i), 8(e)
14.4	1 Work with multiple regression models with interaction (p. 709)	1	8(c)
	2 Build multiple regression models with indicator variables (p. 714)	2	9
14.5	1 Find a quadratic regression equation (p. 717)	1	8(b)
14.6	1 Perform a partial F-test (p. 722)	1	10(b)
	2 Build a regression model (p. 725)	2–4	10(c)

Review Exercises

1. What is the simple least-squares regression model? What are the requirements to perform inference on a simple least-squares regression line? How do we verify that these requirements are met?

2. Seat Choice and GPA A biology professor wants to investigate the relation between the seat location chosen by a student on the first day of class and their cumulative grade point average (GPA). He randomly selected an introductory biology class and obtained the following information for the 38 students in the class.

Row Chosen, x	GPA, y	Row Chosen, x	GPA, y
1	4.00	6	2.63
2	3.35	6	3.15
2	3.50	6	3.69
2	3.67	6	3.71
2	3.75	7	2.88
3	3.37	7	2.93
3	3.62	7	3.00
4	2.35	7	3.21
4	2.71	7	3.53
4	3.75	7	3.74
5	3.10	7	3.75
5	3.22	7	3.90
5	3.36	8	2.30
5	3.58	8	2.54
5	3.67	8	2.61
5	3.69	9	2.71
5	3.72	9	3.74
5	3.84	9	3.75
6	2.35	11	1.71

Source: S. Kalinowski and Taper M. "The Effect of Seat Location on Exam Grades and Student Perceptions in an Introductory Biology Class." *Journal of College Science Teaching,* 36(4):54–57, 2007.

(a) Treating row as the explanatory variable, determine the estimates of β_0 and β_1. What is the mean GPA of students who choose a seat in the fifth row?

(b) Compute the standard error of the estimate, s_e.

(c) Determine whether the residuals are normally distributed.

(d) Determine s_{b_1}.

(e) If the residuals are normally distributed, test whether a linear relation exists between the explanatory variable, row choice, and response variable, GPA, at the $\alpha = 0.05$ level of significance.

(f) If the residuals are normally distributed, construct a 95% confidence interval for the slope of the true least-squares regression line.

(g) Construct a 95% confidence interval for the mean GPA of students who choose a seat in the fifth row.

(h) Predict the GPA of a randomly selected student who chooses a seat in the fifth row.

(i) Construct a 95% prediction interval for the GPA found in part (h).

(j) Explain why the predicted GPAs found in parts (a) and (h) are the same, yet the intervals are different.

3. Apartments The following data represent the square footage and rents for apartments in Queens, New York and Nassau County, New York. For this problem, only consider the Queens data.

Queens (New York City)		Nassau County (Long Island)	
Square Footage, x	Rent per Month, y	Square Footage, x	Rent per Month, y
500	650	1100	1875
588	1215	588	1075
1000	2000	1250	1775
688	1655	556	1050
825	1250	825	1300
1259	2700	743	1475
650	1200	660	1315
560	1250	975	1400
1073	2350	1429	1900
1452	3300	800	1650
1305	3100	1077	1395

Source: apartments.com

(a) What are the estimates of β_0 and β_1? What is the mean rent of a 900-square-foot apartment in Queens?

(b) Compute the standard error of the estimate, s_e.

(c) Determine whether the residuals are normally distributed.

(d) Determine s_{b_1}.

(e) If the residuals are normally distributed, test whether a linear relation exists between the explanatory variable, x, and response variable, y, at the $\alpha = 0.05$ level of significance.

(f) If the residuals are normally distributed, construct a 95% confidence interval for the slope of the true least-squares regression line.

(g) Construct a 90% confidence interval for the mean rent of all 900-square-foot apartments in Queens.

(h) Predict the rent of a particular 900-square-foot apartment in Queens.

(i) Construct a 90% prediction interval for the rent of a particular 900-square-foot apartment in Queens.

(j) Explain why the predicted rents found in parts (a) and (h) are the same, yet the intervals are different.

4. Calories versus Sugar The following data represent the number of calories per serving and the number of grams of sugar per serving for a random sample of high-protein and moderate-protein energy bars.

Calories, x	Sugar, y	Calories, x	Sugar, y
180	10	270	20
200	18	320	2
210	14	110	10
220	20	180	12
220	0	200	22
230	28	220	24
240	2	230	24

Source: Consumer Reports

(a) Draw a scatter diagram of the data, treating calories as the explanatory variable. What type of relation, if any, appears to exist between calories and sugar?

(b) Determine the least-squares regression equation from the sample data.

(c) Compute the standard error of the estimate.

(d) Determine whether the residuals are normally distributed.

(e) Determine s_{b_1}.

(f) If the residuals are normally distributed, test whether a linear relation exists between calories and sugar content at the $\alpha = 0.01$ level of significance.

(g) If the residuals are normally distributed, construct a 95% confidence interval about the slope of the true least-squares regression line.

(h) For a randomly selected energy bar, would you recommend using the least-squares regression line obtained in part (b) to predict the sugar content of the energy bar? Why? What would be a good estimate for the sugar content of the energy bar?

5. Depreciation The following data represent the price of a random sample of used Chevy Camaros by age.

Age (years), x	Price ($), y	Age (years), x	Price ($), y
2	15,900	1	20,365
5	10,988	2	16,463
2	16,980	6	10,824
5	9,995	1	19,995
4	11,995	1	18,650
5	10,995	4	10,488

Source: www.onlineauto.com

(a) Determine the least-squares regression equation, treating age as the explanatory variable.

(b) A normal probability plot of the residuals indicates that the residuals are approximately normally distributed. Test whether a linear relation exists between age and price at the $\alpha = 0.05$ level of significance.

(c) Plot the residuals against the explanatory variable, age.

(d) Does a linear model seem appropriate based on the scatter diagram and residual plot? (*Hint:* See Section 4.3.) What is the moral?

6. Wine Quality In the book *Super Crunchers*, author Ian Ayres presents the model $\hat{y} = 12.15 + 0.00117x_1 + 0.0614x_2 - 0.00386x_3$, where y represents the quality of a wine, x_1 represents total rainfall in the winter prior to harvest (in inches), x_2 represents the average temperature during the growing season (in degrees Fahrenheit), and x_3 represents total rainfall during the harvest season (in inches).

(a) Interpret the slope coefficients of the model.

(b) Predict the wine quality rating of a wine where the total rainfall the winter prior to harvest was 8 inches, average temperature during the growing season was 58°F, and total rainfall during the harvest season was 2.9 inches.

7. Course Grade A statistics instructor wishes to investigate the relation between a student's final course grade and grades on a midterm exam and a major project. She selects a random sample of 10 statistics students and obtains the following information:

Midterm	Project	Course Grade
88	90	83
95	80	83
91	90	92
94	93	94
95	90	89
93	58	77
91	90	91
87	74	73
82	92	70
87	74	75

(a) Construct a correlation matrix between course grade, midterm grade, and project grade. Is there any reason to be concerned with multicollinearity based on the correlation matrix?

(b) Find the least-squares regression equation $\hat{y} = b_0 + b_1x_1 + b_2x_2$, where x_1 is the midterm exam score, x_2 is the project score, and y is the final course grade.

(c) Draw residual plots, a boxplot of residuals, and a normal probability plot of the residuals to assess the adequacy of the model.

(d) Interpret the regression coefficients for the least-squares regression equation.

(e) Determine and interpret R^2 and the adjusted R^2.

(f) Test $H_0: \beta_1 = \beta_2 = 0$ versus H_1: at least one $\beta_i \neq 0$ at the $\alpha = 0.05$ level of significance.

(g) Test the hypotheses $H_0: \beta_1 = 0$ versus $H_1: \beta_1 \neq 0$ and $H_0: \beta_2 = 0$ versus $H_1: \beta_2 \neq 0$ at the $\alpha = 0.05$ level of significance. Should any of the explanatory variables be removed from the model?

(h) Predict the mean final course grade of all statistics students who have an 85 on their midterm and a 75 on their project.

(i) Construct and interpret 95% confidence and prediction intervals for statistics students who score an 83 on their midterm and a 92 on their project. Interpret the results.

8. Suppose you want to develop a model that predicts the revenue per visit to a company's web site. The explanatory variables you are going to use are

x_1: household income

x_2: time spent on the web site

x_3: gender

(a) Write a model that utilizes all three explanatory variables in an additive model with linear terms. Be sure to define any indicator variables.

(b) Write a quadratic model that only utilizes time spent on the website.

(c) Suppose you suspect there is interaction between household income and time spent on the web site. Write a model that incorporates this interaction term into the model from part (a).

9. Rents in New York The data in Problem 3 represent the square footage and rent for apartments in the borough of Queens and Nassau County, New York.

(a) On the same graph, draw a scatter diagram for both Queens and Nassau County apartments treating square footage as the explanatory variable, but use a different plotting symbol for Queens and Nassau County.

(b) Find the least-squares regression equation $\hat{y} = b_0 + b_1x_1 + b_2x_2$, where x_1 is square footage and $x_2 = 0$ for Queens and $x_2 = 1$ for Nassau County.

(c) Test the hypotheses $H_0: \beta_1 = \beta_2 = 0$ versus H_1: at least one of the $\beta_i \neq 0$ at the $\alpha = 0.05$ level of significance.

(d) Test the hypothesis $H_0: \beta_2 = 0$ versus $H_1: \beta_2 \neq 0$ at the $\alpha = 0.05$ level of significance. What does this say about the role of location in rents in New York?

(e) Construct a 95% confidence interval for a 950-square-foot apartment in Queens.

10. NFL Combine Download the data file 14_cr_10.txt at www.pearsonhighered.com/sullivanstats. The data represent a variety of variables measured on players in the 2015 National Football League (NFL) combine. Suppose you want to build a model to predict a player's weight. The following variables were measured:

y: Weight—weight of the player in pounds
x_1: 40 YD—time (in seconds) to complete the 40-yard dash
x_2: Vertical—standing vertical jump (in inches)
x_3: Bench—number of times player can bench press 225 pounds
x_4: Shuttle—time (in seconds) of the shuttle run
x_5: Cone—time (in seconds) of the 3-cone run
x_6: 20 YD—time (in seconds) of the 20-yard shuttle drill

(a) Find the least-squares regression equation $\hat{y} = b_0 + b_1x_1 + b_2x_2 + b_3x_3 + b_4x_4 + b_5x_5 + b_6x_6$.

(b) Use a partial F test to test whether 3-cone run and 20-yard shuttle drill do not significantly help to explain variability in weight.

(c) Use forward selection, backward elimination, or stepwise regression to identify the best model in predicting weight.

Chapter Test

1. State the requirements to perform inference on a simple least-squares regression line.

 **2.** Crickets make a chirping noise by sliding their wings rapidly over each other. Perhaps you have noticed that the number of chirps seems to increase with the temperature. The following table lists the temperature (in degrees Fahrenheit, °F) and the number of chirps per second for the striped ground cricket.

Temperature, x	Chirps per Second, y	Temperature, x	Chirps per Second, y
88.6	20.0	71.6	16.0
93.3	19.8	84.3	18.4
80.6	17.1	75.2	15.5
69.7	14.7	82.0	17.1
69.4	15.4	83.3	16.2
79.6	15.0	82.6	17.2
80.6	16.0	83.5	17.0
76.3	14.4		

Source: George W. Pierce. *The Songs of Insects*, Cambridge, MA: Harvard University Press, 1949, pp. 12–21.

(a) What are the estimates of β_0 and β_1? What is the mean number of chirps when the temperature is 80.2°F?

(b) Compute the standard error of the estimate, s_e.

(c) Determine whether the residuals are normally distributed.

(d) Determine s_{b_1}.

(e) If the residuals are normally distributed, test whether a linear relation exists between the explanatory variable, x, and response variable, y, at the $\alpha = 0.05$ level of significance.

(f) If the residuals are normally distributed, construct a 95% confidence interval for the slope of the true least-squares regression line.

(g) Construct a 90% confidence interval for the mean number of chirps found in part (a).

(h) Predict the number of chirps on a day when the temperature is 80.2°F.

(i) Construct a 90% prediction interval for the number of chirps found in part (h).

(j) Explain why the predicted number of chirps found in parts (a) and (h) are the same, yet the intervals are different.

 **3.** The following data represent the height (inches) of boys between the ages of 2 and 10 years.

Age, x	Boy Height, y	Age, x	Boy Height, y	Age, x	Boy Height, y
2	36.1	5	45.6	8	48.3
2	34.2	5	44.8	8	50.9
2	31.1	5	44.6	9	52.2
3	36.3	6	49.8	9	51.3
3	39.5	7	43.2	10	55.6
4	41.5	7	47.9	10	59.5
4	38.6	8	51.4		

Source: National Center for Health Statistics

(a) Treating age as the explanatory variable, determine the estimates of β_0 and β_1. What is the mean height of a 7-year-old boy?

(b) Compute the standard error of the estimate, s_e.

(c) Assuming the residuals are normally distributed, test whether a linear relation exists between the explanatory variable, age, and response variable, height, at the $\alpha = 0.05$ level of significance.

(d) Assuming the residuals are normally distributed, construct a 95% confidence interval for the slope of the true least-squares regression line.

(e) Construct a 90% confidence interval for the mean height found in part (a).

(f) Predict the height of a 7-year-old boy.

(g) Construct a 90% prediction interval for the height found in part (f).

(h) Explain why the predicted heights found in parts (a) and (f) are the same, yet the intervals are different.

4. A researcher believes that as age increases, the grip strength (pounds per square inch, psi) of an individual's dominant hand decreases. From a random sample of 17 females, he obtains the following data.

Age, x	Grip Strength, y	Age, x	Grip Strength, y
15	65	34	45
16	60	37	58
28	58	41	70
61	60	43	73
53	46	49	45
43	66	53	60
16	56	61	56
25	75	68	30
28	46		

Source: Kevin McCarthy, student at Joliet Junior College

(a) Treating age as the explanatory variable, determine the estimates of β_0 and β_1.

(b) Assuming the residuals are normally distributed, test whether a linear relation exists between the explanatory

variable, age, and response variable, grip strength, at the $\alpha = 0.05$ level of significance.

(c) Based on your answer to (b), what would be a good estimate of the grip strength of a randomly selected 42-year-old female?

5. First-Year College GPA Researchers at the College Board wanted to build a model that describes one's first-year college GPA. The researchers obtained the following model:

$$\hat{y} = 0.06x_1 + 0.07x_2 + 0.18x_3 + 0.29x_4$$

where y represents the z-score for first-year college grade point average (GPA)

x_1 represents the z-score on the math portion of the SAT

x_2 represents the z-score on the critical reading portion of the SAT

x_3 represents the z-score on the writing portion of the SAT

x_4 represents the z-score of the student's high school grade point average (GPA)

Source: Kobrin, J., Sinharay, S., Haberman, S.J., & Chajewski, M. *An Investigation of the Fit of Linear Regression Models to Data from an SAT Validity Study.* College Board Research Report 2011–13.

(a) Suppose a student has a z-score of 1.52 on the math portion of the SAT. Explain what this result represents.

(b) What is the impact of a z-score of -1 for the student's high school GPA?

(c) Interpret the slope coefficient for high school GPA

(d) The coefficient of determination for this model is 0.24. Interpret this value.

(e) The correlation coefficient between x_2 and x_3 is 0.71. What might this suggest?

(f) Predict the z-score for first-year college GPA if $x_1 = -0.54, x_2 = 1.32, x_3 = 0.98$, and $x_4 = 0.36$. Would we expect a student with these credentials to have an above or below average first-year college GPA?

---

6. A nutritionist wants to develop a model that describes the relation between the calories, total fat content, protein, sugar, and carbohydrates in cheeseburgers at fast-food restaurants. She obtains the following data from the websites of the companies. She will use calories as the response variable and the others as explanatory variables.

Restaurant	Fat (g)	Protein (g)	Sugar (g)	Carbs (g)	Calories
1/4-pound single with cheese (Wendy's)	20	25	9	39	430
Whataburger (Whataburger)	32	30	10	61	640
Cheeseburger (In-n-Out)	27	22	10	39	480
Big Mac (McDonald's)	29	25	9	45	540
Whopper with cheese (Burger King)	47	33	11	52	760
Jumbo Jack (Jack in the Box)	42	25	12	54	690
1/4 Pounder with Cheese (McDonald's)	26	29	9	40	510
Cheeseburger (Sonic)	31	29	15	59	630

Source: Each company's Web site

(a) Construct a correlation matrix. Is there any reason to be concerned about multicollinearity?

(b) Find the least-squares regression equation $\hat{y} = b_0 + b_1x_1 + b_2x_2 + b_3x_3 + b_4x_4$, where x_1 is fat content, x_2 is protein, x_3 is sugar, x_4 is carbohydrates, and y is the response variable, calories.

(c) Test $H_0: \beta_1 = \beta_2 = \beta_3 = \beta_4 = 0$ versus H_1: at least one of the $\beta_i \neq 0$ at the $\alpha = 0.05$ level of significance.

(d) Test the hypotheses $H_0: \beta_i = 0$ versus $H_1: \beta_i \neq 0$ for $i = 1, 2, 3, 4$ at the $\alpha = 0.05$ level of significance. Should any

of the explanatory variables be removed from the model? If so, which one?

(e) Determine the regression model with the explanatory variable identified in part (d) removed. Are the remaining slope coefficients significantly different from zero? If not, remove the appropriate explanatory variable and compute the least-squares regression equation.

(f) Draw residual plots, a boxplot of the residuals, and a normal probability plot of the residuals to assess the adequacy of the model found in part (e).

(g) Interpret the regression coefficients for the least-squares regression equation found in part (e).

(h) Determine and interpret R^2 and the adjusted R^2.

(i) Construct 95% confidence and prediction intervals for the calories in a fast-food cheeseburger that has 38 g of fat, 29 g of protein, 11 g of sugar, and 52 g of carbohydrates. Interpret the results.

Making an Informed Decision

Buying a Home

You are in the market to buy a home. Buying a home can be a difficult and trying experience. How much is the home you want worth? What is a fair selling price for the home? The abundance of data on the Internet can be helpful in determining a fair selling price. One website, in particular, is www.zillow.com. The site provides a "Zestimate," which is an estimate of the true value of a home based on their models.

(a) Decide on a location in which you would like to buy a home. Go to www.zillow.com and check "Recently Sold." Randomly select at least 15 homes that have recently sold and record the Zestimate and selling price.

(b) Draw a scatter diagram of the data from part (a), treating the Zestimate as the explanatory variable. Comment on the association between the Zestimate and selling price.

(c) Find the least-squares regression line, treating the Zestimate as the explanatory variable.

(d) Interpret the slope of the least-squares regression line.

(e) Draw a residual plot to verify a linear model is appropriate.

(f) Draw a boxplot of the residuals. Are there any outliers?

(g) Assuming the residuals are normally distributed, test whether a linear relation exists between the Zestimate and selling price. Use the $\alpha = 0.05$ level of significance.

(h) Choose a Zestimate price that is not outside the scope of your model. Use the regression model from part (c) to predict the mean sale price of all homes with the Zestimate price you selected.

(i) Construct a 95% confidence interval for the mean sale price of all homes whose Zestimate is equal to the value chosen in part (h).

(j) Construct a 95% prediction interval for the sale price of a particular home whose Zestimate is equal to the value chosen in part (h).

 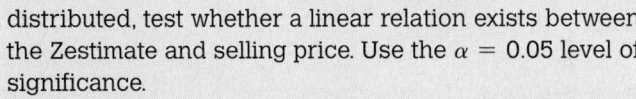 **CASE STUDY** Housing Boom

During the early 2000s, the United States experienced a boom in the housing industry, in large part due to efforts by the government to boost consumer spending. For many, the lure of low interest rates put them in the market for a house. When house shopping, a natural question is "How much is the house worth?" This is difficult to answer because it depends on what the market will bear—the house is worth what someone else is willing to pay for it.

A real estate agent wishes to examine several recent house sales in his territory and develop a model that could be used to give a rough idea of a house's fair market value.

Articles on how to determine the value of a house often suggest comparing square footage, number of bedrooms, number of bathrooms, and size of the lot. The agent decided to examine these four variables, along with the age of the house and number of rooms, in an effort to predict the house's selling price. The data on the next page summarize his findings.

Price ($1000s)	Acres	Bedrooms	Bathrooms	Sq. ft.	Age (yr)	Rooms
104.9	0.19	3	1.0	900	44	8
109.0	0.15	3	2.0	1431	34	6
94.9	0.20	3	1.5	1064	49	6
96.5	0.18	2	1.0	780	52	4
127.9	0.17	3	2.5	1140	47	6
129.9	0.18	3	1.5	1140	41	7
145.0	0.18	3	1.5	1845	45	6
199.9	0.17	4	3.5	1974	17	8
255.9	0.24	5	3.0	2460	22	8
310.0	0.23	4	3.5	2490	14	9
169.0	0.20	4	2.0	1896	37	8
344.5	0.24	4	4.5	2709	11	8
123.0	0.13	3	1.0	828	63	5
139.9	0.16	2	2.0	1131	75	5
169.9	0.15	2	2.0	1002	96	7
194.9	0.19	3	1.0	1024	55	6
210.0	0.23	3	1.0	1694	53	9
275.0	0.17	4	2.0	2380	10	8
299.5	0.17	4	2.0	1936	97	8
319.9	0.27	3	2.0	1648	77	8
397.5	0.30	4	2.5	2500	106	10
189.9	0.18	2	1.0	1016	71	6
349.9	0.40	4	2.5	1816	70	8
454.9	0.96	3	3.0	2160	37	7
499.9	1.00	5	3.0	3104	48	10
615.0	0.66	4	3.5	3205	26	10
635.0	0.44	4	3.5	3084	27	10
929.0	0.90	5	4.5	4470	16	14

(a) Construct a correlation matrix with price, acres, bedrooms, bathrooms, square feet, age, and rooms. Is there any reason to be concerned with multicollinearity based on the correlation matrix?

(b) Find the least-squares regression equation $\hat{y} = b_0 + b_1 x_1 + b_2 x_2 + b_3 x_3 + b_4 x_4 + b_5 x_5 + b_6 x_6$, where x_1 is acres, x_2 is bedrooms, and so on.

(c) Test H_0: $\beta_i = 0$ versus H_1: at least one of the $\beta_i \neq 0$ at the $\alpha = 0.05$ level of significance.

(d) Test the hypotheses H_0: $\beta_i = 0$ versus H_1: $\beta_i \neq 0$ for $i = 1, 2, \ldots, 6$, at the $\alpha = 0.05$ level of significance.

(e) Examine your regression results and remove any explanatory variable whose coefficient is not significantly different from 0 to obtain the model of best fit.

(f) Once you have obtained your model of best fit, draw residual plots and a boxplot of the residuals to assess the adequacy of the model.

(g) Determine and interpret R^2 and adjusted R^2. How well does your model appear to fit the data?

(h) Use your model to predict the selling price for another house from the agent's territory that has the following characteristics: 0.18 acre, 3 bedrooms, 1 bath, 1176 square feet, 47 years old, and 6 total rooms. Compare your prediction to the actual selling price: $99,900.

Location, location, location! The location of the house can have a large effect on its selling price. The first 12 houses listed are from the same zip code, the next 10 are from a second zip code, and the last 6 are from a third zip code.

(i) Construct side-by-side boxplots of selling price for the three zip codes. Is there any reason to believe that selling prices vary from one zip code to the next within the agent's territory?

(j) Introduce dummy explanatory variables to represent zip code and repeat parts (b)–(g) to find the model of best fit for price.

(k) Repeat part (h) assuming the house comes from the first zip code. Which model did a better job predicting the selling price?

(l) Use the model you obtained in part (j) to predict the selling price of a house in your area. How well did the model work?

(m) Explain the limitations of this model. Which, if any, can be dealt with, and how would you do so?

(n) What other variables might affect the cost of a house? Support your choices; then collect data from your area for the subset of variables you think best predicts house price. Find the model of best fit and test it with additional houses in your area, as well as houses in other areas. How well does your model work?

15

Nonparametric Statistics

Making an Informed Decision

You have just graduated from college and must determine where you would like to live. To do this, you must judge which characteristics of a neighborhood are most important to you, such as spending on education, access to health care, and typical commute times. See the Decisions project on page 794.

PUTTING IT TOGETHER

Chapters 9 through 11, 13, and 14 introduced inferential methods regarding population parameters. These methods required that certain conditions be satisfied before proceeding with the *parametric* test. We have not yet addressed how to handle the situations in which these requirements are not satisfied. For example, suppose we wish to test a hypothesis about a population mean for which we have a small sample size, but the population is not normal.

To deal with this circumstance, we could increase the sample size and utilize the Central Limit Theorem, but we could also use nonparametric statistics. Nonparametric methods use techniques to test hypotheses that are *distribution-free*. In other words, we do not have a requirement that a sample comes from a population that fits any particular distribution (such as a normal distribution). Because of this, the tests can be used more generally than parametric tests.

15.1 An Overview of Nonparametric Statistics

Objective   ① Distinguish between parametric and nonparametric statistical procedures

① Distinguish between Parametric and Nonparametric Statistical Procedures

Up to this point, the inferential statistics that we have performed have been based on *parametric statistical procedures*. To use these procedures requirements about the underlying distribution of the random variable must be met. For example, to test a hypothesis regarding a population mean, μ, the population had to be normally distributed if the sample size was small.

Definition   **Parametric statistical procedures** are inferential procedures conducted under the assumption that the underlying distribution of the data belongs to some parametric family of distributions (such as the normal distribution).

Distributions belong to a parametric family of distributions when they are defined by certain parameters. For example, we saw in Chapter 7 that the normal distribution is defined by specifying the mean, μ, and standard deviation, σ.

The inferential procedures presented in Chapters 9 through 11, 13, and 14 were all parametric statistical procedures. But what if the requirements to conduct parametric statistical procedures are not satisfied? Then we use *nonparametric statistical procedures*.

Definitions   **Nonparametric statistical procedures** are inferential procedures that make no assumptions about the underlying distribution of the data. Since they do not require that the population belong to any particular parametric family of distributions, they are often referred to as **distribution-free procedures**.

The term *nonparametric* may seem to imply that such procedures do not involve parameters, but this is not the case. Just like parametric procedures, some nonparametric procedures can involve parameters (such as the median). However, nonparametric procedures do not require any particular underlying distribution, which is why they are often referred to as *distribution-free* procedures. This does not mean that nonparametric procedures have no requirements. In general, there will be some requirements, but not as many as with parametric procedures, and they are usually easier to satisfy.

You might wonder why we would ever use parametric procedures when nonparametric procedures exist? Well, there are advantages and disadvantages to using nonparametric statistical procedures.

Advantages of Nonparametric Statistical Procedures

- Most nonparametric tests have very few requirements, so it is unlikely that these tests will be used improperly.
- For some nonparametric procedures, the computations are fairly easy.
- The procedures can be used for count data or rank data, so nonparametric methods can be used on ordinal data, such as the rankings of a movie as excellent, good, fair, or poor.

Disadvantages of Nonparametric Statistical Procedures

- Nonparametric procedures are less efficient than parametric procedures. This means that a larger sample size is required when conducting a nonparametric procedure to have the same probability of a Type I error as the equivalent parametric procedure.

- Nonparametric procedures often discard useful information. For example, the sign test uses only the sign of the data and rank tests merely preserve order, so the magnitudes of the actual data values are lost. As a result, nonparametric procedures are typically less powerful. Recall that the **power of a test** is the probability that the null hypothesis is rejected when the alternative hypothesis is true.

> **CAUTION!**
> Do not use nonparametric procedures if parametric procedures can be used.

- Because fewer requirements must be satisfied to conduct these tests, researchers sometimes incorrectly use these procedures when parametric procedures can be used.

Typically, a test with more requirements has stronger results. So, if the requirements to perform parametric statistical procedures are satisfied, these tests should be used because the results will be more powerful and efficient. Nonparametric statistical procedures should only be used if the requirements are not satisfied.

In Other Words
The lower the efficiency is, the larger the sample size must be for a nonparametric test to have the probability of a Type I error the same as it would be for its equivalent parametric test.

Let's explore the idea of **efficiency** a little more. If a nonparametric statistical test has an efficiency of 0.85, a sample size of 100 would be required in the nonparametric test to achieve the same results that a sample of 85 would produce in the equivalent parametric test. The cost of fewer requirements is that additional individuals must be sampled. Table 1 shows some of the nonparametric tests that we will study in this chapter, along with their efficiencies and their corresponding parametric tests.

Table 1

Nonparametric Test	Parametric Test	Efficiency of Nonparametric Test
Runs test for randomness	No corresponding test	—
Sign test	Single-sample *t*-test	0.955 (for small samples that come from a normal population) 0.637 (for large samples if data are normal)
Wilcoxon matched-pairs signed-ranks test	Inference about the difference of two means—dependent samples	0.955 (if the differences are normal)
Mann–Whitney test	Inference about the difference of two means—independent samples	0.955 (if data are normal)
Spearman's rank-correlation test	Linear correlation	0.912 (if the data are bivariate normal)
Kruskal–Wallis test	One-way ANOVA	0.955 (if the data are normal)

15.1 Assess Your Understanding

Explaining the Concepts

1. Describe the difference between parametric statistical procedures and nonparametric statistical procedures.

2. Explain the idea of efficiency.

3. Explain the concept of the power of a test.

4. List the advantages of using nonparametric statistical procedures.

5. List the disadvantages of using nonparametric statistical procedures.

6. Why is it appropriate to call nonparametric statistical procedures distribution-free procedures?

15.2 Runs Test for Randomness

Preparing for This Section Before getting started, review the following:

- Mutually exclusive (Section 5.2, p. 262)
- The language of hypothesis testing (Section 10.1, pp. 473–478)

Objective ❶ Perform a runs test for randomness

❶ Perform a Runs Test for Randomness

In many situations, we would like to know whether a set of data is random. By this we mean that we would like to know whether a series of observations appears to follow a particular pattern, even if the data were obtained in a systematic fashion. For example, we might record the gender of the next 20 people who leave a store. Certainly, our method for obtaining the individuals is not random, but we would like to test whether the observations occurred randomly, as if we pulled them from a hat.

If a researcher is not certain that data are random, a *runs test for randomness* can be conducted.

Definitions
> A **runs test for randomness** is used to test whether data have been obtained, or occur, randomly. A **run** is a sequence of similar events, items, or symbols that is followed by an event, item, or symbol that is mutually exclusive from the first event, item, or symbol. The number of events, items, or symbols in a run is called its **length**.

Suppose we record the gender of the 15 students enrolled in an introductory statistics course as they enter the classroom. The males are denoted by a blue M and the females are denoted by a red F.

M M F M M F M M F F M M M M F

The first two males constitute a run of length 2 because they are the same gender. The third individual is a run of length 1. The longest run is length 4. Do you see it? The number of runs is 8.

The goal of this section is to discover whether a sequence of observations is random. The randomness of a sequence would be called into question if there are too few or too many runs. For example, the sequence

M M M M M M M M F F F F F F F

appears to be nonrandom, because it contains too few runs (only 2 runs). The sequence

M F M F M F M F M F M F M F M

is also nonrandom, because it contains too many runs (15 runs).

The following notation is used in testing hypotheses regarding randomness:

> **CAUTION!**
> Runs tests are used to test whether it is reasonable to conclude that data occur randomly, not whether the data are collected randomly. For example, we might wonder whether defective parts come off an assembly line randomly or systematically. If broken parts occur systematically (such as every fourth part), we might be led to believe that we have a broken machine. We don't collect the data randomly; instead, we select 100 consecutive parts. We want to know whether the defective parts in the 100 selected occur randomly.

Notation Used in Conducting a Runs Test for Randomness

Let n represent the sample size of which there are two mutually exclusive types.
Let n_1 represent the number of observations of the first type.
Let n_2 represent the number of observations of the second type.
Let r represent the number of runs.

EXAMPLE 1 Notation in a Runs Test for Randomness

Problem Suppose we record the gender of the 15 students enrolled in an introductory statistics course as they enter the classroom. The males are denoted by a blue M and the females are denoted by a red F.

M M F M M F M M F F M M M M F

Identify the values of n, n_1, n_2, and r.

Approach Let n represent the number of students sampled. Let n_1 represent the number of male students and n_2 represent the number of female students. Finally, let r represent the number of runs.

Solution We have $n = 15$ students in the sample, $n_1 = 10$ males, $n_2 = 5$ females, and $r = 8$ runs.

● Now Work Problem 3(a)

To conduct a runs test for randomness, we need a test statistic and a critical value.

Test Statistic for a Runs Test for Randomness

Small-Sample Case: If $n_1 \leq 20$ and $n_2 \leq 20$, the test statistic in the runs test for randomness is r, the number of runs.

Large-Sample Case: If $n_1 > 20$ or $n_2 > 20$, the test statistic in the runs test for randomness is

$$z_0 = \frac{r - \mu_r}{\sigma_r}$$

where

$$\mu_r = \frac{2n_1 n_2}{n} + 1 \quad \text{and} \quad \sigma_r = \sqrt{\frac{2n_1 n_2 (2n_1 n_2 - n)}{n^2(n-1)}}$$

Critical Values for a Runs Test for Randomness

Small-Sample Case: To find the critical value at the $\alpha = 0.05$ level of significance for a runs test, we use Table XI if $n_1 \leq 20$ and $n_2 \leq 20$.

Large-Sample Case: If $n_1 > 20$ or $n_2 > 20$, the critical value is found from Table V, the standard normal table.

EXAMPLE 2   Obtaining Critical Values from Table XI

Problem Find the upper and lower critical values at the $\alpha = 0.05$ level of significance from Table XI if $n_1 = 10$ and $n_2 = 5$.

Approach Determine the intersection of the row corresponding to $n_1 = 10$ and the column corresponding to $n_2 = 5$ to identify the lower and upper critical values.

Solution The lower critical value is 3 and the upper critical value is 12. Figure 1 shows a partial display of Table XI.

Figure 1

Critical Values for the Number of Runs

Value of n_1		Value of n_2												
		2	**3**	**4**	**5**	**6**	**7**	**8**	**9**	**10**	**11**	**12**	**13**	**14**
	2	1 / 6	1 / 6	1 / 6	1 / 6	1 / 6	1 / 6	1 / 6	1 / 6	1 / 6	1 / 6	2 / 6	2 / 6	2 / 6
	9	1 / 6	2 / 8	3 / 10	3 / 12	4 / 13	4 / 14	5 / 14	5 / 15	5 / 16	6 / 16	6 / 16	6 / 17	7 / 17
	10	1 / 6	2 / 8	3 / 10	3 / 12	4 / 13	5 / 14	5 / 15	5 / 16	6 / 16	6 / 17	7 / 17	7 / 18	7 / 18
	11	1 / 6	2 / 8	3 / 10	4 / 12	4 / 13	5 / 14	5 / 15	6 / 16	6 / 17	7 / 17	7 / 18	7 / 19	8 / 19
	12	2 / 6	2 / 8	3 / 10	4 / 12	4 / 13	5 / 14	6 / 16	6 / 16	7 / 17	7 / 18	7 / 19	8 / 19	8 / 20

● Now Work Problem 3(b)

We now present the steps required to conduct a runs test for randomness.

Runs Test for Randomness

To test the randomness of data, use the following steps, provided that

1. the sample is a sequence of observations recorded in the order of their occurrence,
2. the observations can be categorized into two mutually exclusive categories.

Step 1 Assume the data are random. This forms the basis of the null and alternative hypotheses, which are structured as follows:

$$H_0: \text{The sequence of data is random.}$$

$$H_1: \text{The sequence of data is not random.}$$

Step 2 Determine a level of significance, α, based on the seriousness of making a Type I error. **Note:** For the small-sample case, we must use the level of significance $\alpha = 0.05$.

By-Hand Approach

Step 3 Use the number of runs, r, to compute the **test statistic**.

Small-Sample Case	Large-Sample Case
r	$z_0 = \dfrac{r - \mu_r}{\sigma_r}$

Step 4 Determine the critical value from Table XI (small sample) or Table V (large sample). Compare the critical value to the test statistic.

Small-Sample Case	Large-Sample Case
If $r \leq$ lower critical value or $r \geq$ upper critical value, reject the null hypothesis.	If $z_0 < -z_{\alpha/2}$ or $z_0 > z_{\alpha/2}$, reject the null hypothesis.

Technology Approach

Step 3 Use a statistical spreadsheet to obtain the P-value. The directions for obtaining the P-value using Minitab are in the Technology Step-by-Step on page 750.

Step 4 If P-value $< \alpha$, reject the null hypothesis.

Step 5 State the conclusion.

EXAMPLE 3   Testing for Randomness (Small-Sample Case)

Problem We record the gender of the 15 students enrolled in an introductory statistics course as they enter the classroom. The males are denoted by a blue M and the females are denoted by a red F:

M M F M M F M M F F M M M M F

Is there sufficient evidence to conclude that the individuals enter the room in a nonrandom way as it pertains to gender at the $\alpha = 0.05$ level of significance?

Approach We want to see if the data indicate that students enter the class randomly (independent of gender). After verifying the requirements that the data were collected in the order they occurred and that there are two mutually exclusive categories of data, we follow Steps 1 through 5.

Solution The sample is a sequence of observations (the first person who walked in was male, second person male, third person female, and so on) recorded in the order of occurrence. The observations are in two mutually exclusive categories: male or female. The requirements are satisfied.

Step 1 We are testing the hypothesis that the sequence of observations is random. Thus,

H_0: the sequence of data is random.

H_1: the sequence of data is not random.

Step 2 The level of significance is $\alpha = 0.05$.

By-Hand Approach

Step 3 The test statistic is $r = 8$ (Example 1).

Step 4 The lower critical value is 3, and the upper critical value is 12 (Example 2). Because the test statistic, $r = 8$, is not less than or equal to the lower critical value, 3, and is not greater than or equal to the upper critical value, 12, we do not reject the null hypothesis.

Technology Approach

Step 3 Figure 2 shows the results obtained from Minitab. The P-value is 0.839.

Figure 2

Runs Test: Student

```
Runs test for Student

Runs above and below K = 0.5

The observed number of runs = 8
The expected number of runs = 7.66667
5 observations above K, 10 below
* N is small, so the following approximation may be invalid.
P-value = 0.839
```

Step 4 Because the P-value is greater than the level of significance ($0.839 > 0.05$), we do not reject the null hypothesis.

Step 5 There is not sufficient evidence to conclude that the students enter the room in a nonrandom way as it pertains to gender. We have reason to believe that students enter the classroom, as it pertains to gender, randomly. •

● Now Work Problem 3(c)

We only show the by-hand approach for the large-sample case.

EXAMPLE 4 ## Testing for Randomness (Large-Sample Case)

Historical Note

The idea behind this problem is based on the book *A Random Walk Down Wall Street* by W.W. Burton Malkiel, Norton & Co., New York, 1996.

Problem The data in Table 2 represent the monthly rates of return of the Standard and Poor's Index of 500 Stocks from January 2012 through March 2015. Test the randomness of positive monthly rates of return at the $\alpha = 0.05$ level of significance.

Approach Let P represent a positive monthly rate of return and N represent a negative or zero monthly rate of return. We list the sequence of Ps and Ns in chronological order and determine n (the sample size), n_1 (the number of Ps), n_2 (the number of Ns), and r (the number of runs). We then verify that the requirements are satisfied and follow Steps 1 through 5.

Table 2							
Date	**Return (%)**	**Date**	**Return (%)**	**Date**	**Return (%)**	**Date**	**Return (%)**
1/2012	4.06	1/2013	1.11	1/2014	4.31	1/2015	5.49
2/2012	3.13	2/2013	3.60	2/2014	0.69	2/2015	−1.74
3/2012	−0.75	3/2013	1.81	3/2014	0.62	3/2015	2.27
4/2012	−6.27	4/2013	2.08	4/2014	2.10		
5/2012	3.96	5/2013	−1.50	5/2014	1.91		
6/2012	1.26	6/2013	4.95	6/2014	−1.51		
7/2012	1.98	7/2013	−3.13	7/2014	3.77		
8/2012	2.42	8/2013	2.97	8/2014	−1.55		
9/2012	−1.98	9/2013	4.46	9/2014	2.32		
10/2012	0.28	10/2013	2.80	10/2014	2.45		
11/2012	0.71	11/2013	2.36	11/2014	−0.42		
12/2012	5.04	12/2013	−3.56	12/2014	−3.10		

Source: Yahoo!Finance

(continued)

Solution The rates of return are represented by the following sequence:

P P N N P P P P P N P P P P P P P P N P N P P P P P N P P P P P N P N P P N N P N P

There are $n = 39$ months of observations, with $n_1 = 28$ positive months and $n_2 = 11$ negative months. There are $r = 19$ runs. Because (1) the sample is a sequence of observations recorded in the order of their occurrence and (2) the observations can be categorized into two mutually exclusive categories (P and N), the requirements are satisfied. Now follow Steps 1 through 5 to test the randomness of the data.

Step 1 The null and alternative hypotheses are as follows:

H_0: the sequence of data is random.

H_1: the sequence of data is not random.

Step 2 The level of significance is $\alpha = 0.05$.

Step 3 Because $n_1 > 20$, this is a large-sample case. First, we calculate μ_r and σ_r:

$$\mu_r = \frac{2n_1n_2}{n} + 1 = \frac{2(28)(11)}{39} + 1 = 16.795$$

$$\sigma_r = \sqrt{\frac{2n_1n_2(2n_1n_2 - n)}{n^2(n-1)}} = \sqrt{\frac{2(28)(11)[2(28)(11) - 39]}{39^2(39-1)}} = 2.480$$

Now we can compute the test statistic:

$$z_0 = \frac{r - \mu_r}{\sigma_r} = \frac{19 - 16.795}{2.480} = 0.89$$

Step 4 The critical values at the $\alpha = 0.05$ level of significance are $-z_{0.025} = -1.96$ and $z_{0.025} = 1.96$. The test statistic, 0.89, does not lie within the critical region, so we do not reject the null hypothesis.

Step 5 There is not sufficient evidence to reject the null hypothesis that monthly rates of return are random. It is possible that monthly rates of return occur in a random fashion.

● Now Work Problem 13

Technology Step-by-Step Conducting a Runs Test

TI-83/84 Plus
The TI-83/84 Plus calculators do not have this feature.

Excel
Excel does not have this feature.

Minitab
1. Enter raw data in column C1. For qualitative data, enter 0 for one category and 1 for the other category.
2. Select the **Stat** menu, highlight **Nonparametrics**, then highlight **Runs Test . . . .**
3. Enter C1 in the cell marked "Variables." If the data are qualitative, select "Above and Below," and enter 0.5. If the data are quantitative, select "Above and Below the mean," and enter the appropriate value. Click OK.

StatCrunch
StatCrunch does not have this feature.

 15.2 Assess Your Understanding

Vocabulary and Skill Building

1. A _____ is a sequence of similar events, items, or symbols that is followed by an event, item, or symbol that is mutually exclusive from the first event, item, or symbol.

2. In a runs test for randomness, we let _____ represent the sample size of which there are two mutually exclusive types; _____ represent the number of observations of the first type; _____ represent the number of observations of the second type; and _____ represent the number of runs.

In Problems 3–6, a sequence is given: (a) Identify the values of n, n_1, n_2, and r. (b) Determine the critical values at the $\alpha = 0.05$ level of significance. (c) Conduct a hypothesis test on the randomness of the data.

NW 3. M M M F F F M M F M F F F F

4. M M M F F F F M M F F F F F F

5. Y Y N Y Y Y Y Y N N N N N N N N Y Y

6. Y Y Y N N Y N N Y Y Y N N N Y Y N Y Y N N

Applying the Concepts

7. Baseball Madison Bumgarner is a pitcher for the San Francisco Giants. He throws a four-seam fastball pitch (F) to complement his cutter (C). One inning, Bumgarner threw 20 pitches. The following sequence represents the pitches he threw, in order.

F F C F C C F F F C F C C F C C C C F C

Conduct a runs test for randomness on the data at the $\alpha = 0.05$ significance level. Is there enough evidence to support the hypothesis that Bumgarner's pitches are not random?

8. Football A certain NFL coach likes to select the first 15 offensive plays before the game begins. The coach can select either a running play (R) or a pass play (P). The following sequence represents the first 15 plays, in order:

R P P P R P R R P R R R P R R

Test the randomness of the coach's play selection at the $\alpha = 0.05$ level of significance.

9. On-Time Flights The following data represent the arrival status of Southwest Flight 0426 from Dallas' Love Field to San Antonio International Airport for 14 consecutive flights.
Source: U.S. Department of Transportation

L O O L L L O O L O L O O O O L

Status is indicated as on time (O) or late (L), defined as more than 10 minutes after the scheduled arrival. What can you conclude about the randomness of the arrival status at the $\alpha = 0.05$ level of significance?

10. On-Time Flights The following data represent the arrival status of American Airlines Flight 0715 from Washington DC's Ronald Reagan Airport to Chicago's O'Hare International Airport for 15 consecutive flights.
Source: U.S. Department of Transportation

L O O O O O O L L L O L O L O O

Status is indicated as on time (O) or late (L), defined as more than 10 minutes after the scheduled arrival. What can you conclude about the randomness of the arrival status at the $\alpha = 0.05$ level of significance?

11. Quality Control The quality-control manager of a bottling company wants to discover whether a filling machine over- or underfills 16-ounce bottles randomly. The following data represent the filling status of 20 consecutive bottles:

A A A A A A A R R R R A A A A A A A R R

A bottle is rejected (R) if it is either overfilled or underfilled and accepted (A) if it is filled according to specification. Test the randomness of the filling machine in the way that it over- or underfills at the $\alpha = 0.05$ level of significance.

12. Quality Control The quality-control manager of a candy company wants to discover whether a filling machine over- or underfills 16-ounce bags randomly. The following data represent the filling status of 18 consecutive bags:

A A A A A R R R A A A A A A A A A A

A bag is rejected (R) if it is either overfilled or underfilled and accepted (A) if it is filled according to specification. Test the randomness of the filling machine in the way that it over- or underfills at the $\alpha = 0.05$ level of significance.

NW 13. Random Walk Down Wall Street? Does a stock price fluctuate randomly from day to day? A stock analyst selected 45 consecutive days in 2015 in which the stock of Boeing Corporation was traded and computed the daily percentage change in the stock. He indicated positive percentage changes with a P and negative percentage changes with an N. The results are in the following table. What conclusion will the analyst reach if he tests the hypothesis that the stock price fluctuates randomly from day to day at the $\alpha = 0.05$ level of significance?

Date	Return	Date	Return	Date	Return
2/24	N	3/17	P	4/8	P
2/25	N	3/18	P	4/9	P
2/26	N	3/19	N	4/10	P
2/27	N	3/20	P	4/13	N
3/2	P	3/23	N	4/14	N
3/3	P	3/24	N	4/15	N
3/4	N	3/25	N	4/16	N
3/5	P	3/26	N	4/17	N
3/6	N	3/27	P	4/20	P
3/9	P	3/30	P	4/21	P
3/10	N	3/31	N	4/22	N
3/11	N	4/1	N	4/23	N
3/12	P	4/2	P	4/24	N
3/13	N	4/6	P	4/27	N
3/16	P	4/7	P	4/28	N

14. Random Walk Down Wall Street? Does the Dow Jones Industrial Average (DJIA) fluctuate randomly from quarter to quarter? (A quarter refers to a quarter of the year. For example, January–March is the first quarter.) A stock analyst selected 40 consecutive quarters and computed the percentage change in the DJIA. He indicated positive percentage changes with a P and negative percentage changes with an N. The following are the results. Conduct a runs test for randomness on the sequence of quarterly market fluctuations at the $\alpha = 0.05$ level of significance.

Quarter	Return	Quarter	Return	Quarter	Return
First, 2005	N	Third, 2008	N	First, 2012	P
Second, 2005	N	Fourth, 2008	N	Second, 2012	N
Third, 2005	P	First, 2009	N	Third, 2012	P
Fourth, 2005	P	Second, 2009	P	Fourth, 2012	P
First, 2006	P	Third, 2009	P	First, 2013	P
Second, 2006	P	Fourth, 2009	P	Second, 2013	P
Third, 2006	P	First, 2010	P	Third, 2013	P
Fourth, 2006	P	Second, 2010	N	Fourth, 2013	P
First, 2007	N	Third, 2010	P	First, 2014	P
Second, 2007	P	Fourth, 2010	P	Second, 2014	P
Third, 2007	P	First, 2011	P	Third, 2014	P
Fourth, 2007	N	Second, 2011	N	Fourth, 2014	P
First, 2008	N	Third, 2011	N		
Second, 2008	N	Fourth, 2011	P		

Source: The Vanguard Group

15. Random Residuals In a least-squares regression model, the residuals are assumed to be random. The following data represent the life expectancy of a male born in the given year.

Year, x	Life Expectancy, y	Year, x	Life Expectancy, y
1999	73.9	2005	74.9
2000	74.1	2006	75.1
2001	74.2	2007	75.4
2002	74.3	2008	75.5
2003	74.5	2009	76.0
2004	74.9	2010	76.2

Source: Statistical Abstract of the United States

The least-squares regression equation, treating year as the independent variable, is $\hat{y} = 0.2056x - 337.1973$. The residuals from left to right are

0.114	0.109	0.003	−0.103	−0.108	0.086
−0.119	−0.125	−0.031	−0.136	0.158	0.153

(a) Denote residuals above zero with an A and those below zero with a B to form a sequence.

(b) Test the assumption that the residuals are random at the $\alpha = 0.05$ level of significance.

16. Random Residuals In a least-squares regression model, the residuals are assumed to be random. The following data represent the life expectancy of a female born in the given year.

Year, x	Life Expectancy, y	Year, x	Life Expectancy, y
1999	79.4	2005	79.9
2000	79.3	2006	80.2
2001	79.4	2007	80.4
2002	79.5	2008	80.6
2003	79.6	2009	80.9
2004	79.9	2010	81.0

The least-squares regression equation treating the year as the independent variable is $\hat{y} = 0.1633x - 247.2999$. The residuals from left to right are

0.290	0.026	−0.037	−0.100	−0.163	−0.027
−0.190	−0.053	−0.017	0.020	0.157	0.094

(a) Denote residuals above 0 with an A and those below 0 with a B to form a sequence.

(b) Test the assumption that the residuals are random at the $\alpha = 0.05$ level of significance.

17. Quality Control A quality-control inspector tracks the compressive strength of parts created by an injection molding process. If the process is in control, the strengths should fluctuate randomly around the target value of 75 psi. The inspector measures the strength of 20 parts as they come off the production line and obtains the following strengths, in order:

82.0, 78.3, 73.5, 74.4, 72.6, 79.8, 77.0, 83.4, 76.2, 75.2, 81.5, 69.8, 71.3, 69.4, 82.1, 77.6, 76.9, 77.1, 72.7, 73.6

Conduct a runs test for randomness of the compression strengths at the $\alpha = 0.05$ level of significance. Should the quality-control inspector be concerned about the process?

18. Random-Number Generators Using statistical software or a graphing calculator with advanced statistical features, randomly generate a sequence of 20 integer values, each either 0 or 1. Conduct a runs test at the $\alpha = 0.05$ level of significance to verify that the sequence of integers is random.

Explaining the Concepts

19. What is meant by *random*? Explain what a run is.

20. We have evidence against the null hypothesis that the data are random if there are either too few or too many runs. Explain the logic behind this criterion.

15.3 Inference about Measures of Central Tendency

Preparing for This Section Before getting started, review the following:

- Median (Section 3.1, pp. 120–121)
- Binomial probability distribution (Section 6.2, pp. 327–335)
- The language of hypothesis testing (Section 10.1, pp. 473–478)
- Normal approximation to the binomial (Section 7.4, pp. 382–386)
- Testing hypotheses regarding a population mean (Section 10.3, pp. 494–498)

Objective ❶ Conduct a one-sample sign test

In Section 10.3, we introduced a parametric technique that is used to test hypotheses regarding a population mean, μ. To test these hypotheses, we required that either the population be normal (if $n < 30$) or the sample size be large ($n \geq 30$). If these requirements are not satisfied, we can use nonparametric procedures to test hypotheses regarding the central tendency of the population. These nonparametric procedures make inferences regarding the median, rather than the mean. In this section, we will discuss the *one-sample sign test*.

Remember, in hypothesis testing, the null hypothesis must always contain a statement of equality, and the null hypothesis is assumed to be true. In the context of testing a hypothesis regarding the median, there are three ways to structure the hypothesis test:

Two-Tailed	Left-Tailed	Right-Tailed
$H_0: M = M_0$	$H_0: M = M_0$	$H_0: M = M_0$
$H_1: M \neq M_0$	$H_1: M < M_0$	$H_1: M > M_0$

Note: M_0 is the assumed value of the median.

❶ Conduct a One-Sample Sign Test

The sign test is one of the oldest nonparametric procedures. It is sometimes attributed to John Arbuthnot, who reported its use in 1710.

Definition   The **One-Sample Sign Test** is a nonparametric test that uses data, converted to plus and minus signs, to test a hypothesis regarding the median of a population. Data values equal to the assumed value of the median are ignored during the test.

In Other Words
The one-sample sign test is the nonparametric equivalent of tests regarding a single population mean.

For example, suppose we have the following set of data:

$$3, 3, 5, 6, 6, 7, 8, 9, 9, 9, 10, 12$$

We might want to test the hypotheses

$$H_0: M = 7$$
$$H_1: M \neq 7$$

To use the sign test, convert all observations less than 7 to minus $(-)$ signs and all observations greater than 7 to plus $(+)$ signs. We have five minus signs and six plus signs in this set of data. The sample size, n, is the number of minus and plus signs, so $n = 11$. We ignore data values equal to the value of the median stated in the null hypothesis.

Remember, the median lies in the middle of the ordered data. Ignoring values equal to the median, we would expect half of the remaining values to be above the median and half to be below. Therefore, if the null hypothesis were true, we would expect an equal number of minus signs and plus signs. If there are significantly different numbers of minus and plus signs, we have evidence against the null hypothesis.

Test Statistic for a One-Sample Sign Test

The test statistic will depend on the structure of the hypothesis test and on the sample size.

Small-Sample Case ($n \leq 25$)

Two-Tailed	Left-Tailed	Right-Tailed
$H_0: M = M_0$	$H_0: M = M_0$	$H_0: M = M_0$
$H_1: M \neq M_0$	$H_1: M < M_0$	$H_1: M > M_0$
The test statistic, k, will be the smaller of the number of minus signs or plus signs.	The test statistic, k, will be the number of plus signs.	The test statistic, k, will be the number of minus signs.

(continued)

Historical Note

John Arbuthnot was born in April 1667 in Inverbervie, Kincardine, Scotland. He earned a degree in medicine in 1696 from the University of St. Andrews. Around 1700, he went to London to tutor mathematics. In 1704, he was elected a Fellow of the Royal Society. After curing Prince George when he was ill, he was appointed the physician to Queen Anne in 1705. In 1710, he published "An Argument of Divine Providence Taken from the Constant Regularity Observed in the Birth of Both Sexes," in which he developed the sign test. This is thought to be the first paper that applied probability to social statistics. He is also famous for his wit. One of his famous quotes is "All political parties die at last of swallowing their own lies." Arbuthnot died on February 27, 1735.

Large-Sample Case ($n > 25$)

The test statistic, z_0, is

$$z_0 = \frac{(k + 0.5) - \frac{n}{2}}{\frac{\sqrt{n}}{2}}$$

where n is the number of minus and plus signs and k is obtained as described in the small-sample case.

In the previous data set, we had five minus signs and six plus signs, so the test statistic would be $k = 5$ with $n = 11$.

Critical Values for a One-Sample Sign Test

Small-Sample Case: To find the critical value for a one-sample sign test, we use Table XII if $n \le 25$.

Large-Sample Case: If $n > 25$, the critical value is found from Table V, the standard normal table. The critical value is always located in the left tail of the standard normal distribution. For a two-tailed test, the critical value is $-z_{\alpha/2}$. For a left-tailed or right-tailed test, the critical value is $-z_\alpha$.

Let's take a closer look at the test statistic for the left-tailed test. The alternative hypothesis is $H_1: M < M_0$. So, if the true median is less than M_0, there should not be many observations that are above the hypothesized median, M_0, so the value of k will be small. If k is *sufficiently* small, we reject the null hypothesis. A similar argument could be made for the right-tailed and the two-tailed tests. (See Problems 22 and 23.)

Note that for a left-tailed test we would not conduct the test if the number of plus signs was larger than the number of minus signs. If the alternative hypothesis, $H_1: M < M_0$, were true, we would expect there to be more minus signs than plus signs. If more values fall above the hypothesized median than below (more plus signs than minus signs), we will not be able to reject the null hypothesis at any reasonable significance level. Similarly, for a right-tailed test if there are more minus signs than plus signs, we would not reject the null hypothesis. The end result is that in an appropriately conducted sign test the value of k will represent the least-occurring sign. This is why we always reject the null hypothesis if the test statistic is less than the critical value.

One-Sample Sign Test

To test hypotheses regarding the median of a population, use the following steps, provided that the sample is a random sample.

Step 1 Determine the null and alternative hypotheses. The hypotheses can be structured in one of three ways:

Two-Tailed	Left-Tailed	Right-Tailed
$H_0: M = M_0$	$H_0: M = M_0$	$H_0: M = M_0$
$H_1: M \ne M_0$	$H_1: M < M_0$	$H_1: M > M_0$

Note: M_0 is the assumed value of the median.

Step 2 Select a level of significance, α, depending on the seriousness of making a Type I error.

By-Hand Approach

Step 3   Count the number of observations below M_0, and assign them minus $(-)$ signs. Count the number of observations above M_0, and assign them plus $(+)$ signs. Ignore the observations that are equal to M_0.

Determine the test statistic.

Small-Sample Case	Large-Sample Case
k	$z_0 = \dfrac{(k + 0.5) - \dfrac{n}{2}}{\dfrac{\sqrt{n}}{2}}$

Note that k is the smaller of the number of minus signs and plus signs in the two-tailed test, k is the number of plus signs in the left-tailed test, and k is the number of minus signs in the right-tailed test. In addition, n is the total number of plus and minus signs.

Step 4   The level of significance is used to determine the critical value. The critical value for small samples $(n \le 25)$ is found from Table XII. The critical value for large samples $(n > 25)$ is found from Table V.

Compare the critical value to the test statistic.

Small-Sample Case	Large-Sample Case	
If $k \le$ critical value, reject the null hypothesis.	**Two-tailed:** If $z_0 < -z_{\alpha/2}$, reject the null hypothesis.	**Left-tailed or right-tailed:** If $z_0 < -z_\alpha$, reject the null hypothesis.

Step 5   State the conclusion.

Technology (*P*-Value) Approach

Step 3   Use a statistical spreadsheet to obtain the *P*-value. The directions for obtaining the *P*-value using Minitab and StatCrunch are in the Technology Step-by-Step on page 757.

Step 4   If *P*-value $< \alpha$, reject the null hypothesis.

EXAMPLE 1   Conducting a One-Sample Sign Test (Small-Sample Case)

Table 3

6000	870	1530	1660
1060	1790	1630	3180
2180	2370	1800	2170
1210	410	1720	1270
570	1050	2320	1120

Problem   An article at Mobilize.org reported that the median credit-card balance for undergraduate students was \$1770 for those who carried a balance from month to month. A professor at a community college believes that the median credit-card balance of students at his college is different than \$1770. To test this hypothesis, he obtains a random sample of 20 students enrolled at the college who carry a credit-card balance from month to month and asks them to disclose their credit-card debt. The results of the survey are presented in Table 3 in dollars. Do the data indicate that the median credit-card debt of students at the professor's college differs from \$1770 at the $\alpha = 0.05$ level of significance?

Approach   A normal probability plot indicates that the sample data do not come from a population that is normal, and the sample size is small, so parametric procedures for testing the mean cannot be used. However, the data were randomly selected, so we can use the nonparametric sign test for the median.

Solution

Step 1   We want to know if the median credit-card debit is different from \$1770, so this is a two-tailed test.

$$H_0: M = \$1770 \quad \text{versus} \quad H_1: M \ne \$1770$$

Step 2   We use the $\alpha = 0.05$ level of significance.

(continued)

By-Hand Solution

Step 3 There are 12 observation less than \$1770 and 8 observations greater than \$1770. Therefore, we have 12 minus signs and 8 plus signs, so $n = 20$.

The test statistic is the smaller of the number of plus signs and minus signs. The test statistic is $k = 8$, the number of plus signs.

Step 4 Because we are performing a two-tailed test and $n \leq 25$, we find the critical value from Table XII for the two-tailed test at the $\alpha = 0.05$ level of significance with $n = 20$. The critical value is 5. See Figure 3. Because the test statistic, $k = 8$, is greater than the critical value, 5, we do not reject the null hypothesis.

Figure 3

	Critical Values for the Sign Test			
		α		
n	0.005 (one tail) 0.01 (two tails)	0.01 (one tail) 0.02 (two tails)	0.025 (one tail) 0.05 (two tails)	0.05 (one tail) 0.10 (two tails)
1	*	*	*	*
2	*	*	*	*
3	*	*	*	*
17	2	3	4	4
18	3	3	4	5
19	3	4	4	5
20	3	4	5	5
21	4	4	5	6
22	4	5	5	6

* Indicates that it is not possible to get a value in the critical region.

Technology Solution

Step 3 Figure 4 shows the results obtained from StatCrunch. The P-value is 0.5034.

Figure 4

Hypothesis test results:
Parameter : median of Variable
H_0 : Parameter = 1770
H_A : Parameter ≠ 1770

Variable	n	n for test	Sample Median	Below	Equal	Above	P-value
var1	20	20	1645	12	0	8	0.5034

Step 4 Because the P-value is greater than the level of significance $(0.5034 > 0.05)$, we do not reject the null hypothesis.

Step 5 There is not sufficient evidence to support the hypothesis that the median credit-card debt for the professor's students is different from \$1770. ●

● **Now Work Problem 11**

Rationale for the Large-Sample Test Statistic When determining critical values for the sign test, we only use Table XII for values of n up to 25. For $n > 25$, we use the test statistic, z_0, by means of the normal approximation to the binomial. Under our null hypothesis, H_0: $M = M_0$, we would expect an equal number of plus and minus signs. That is, we expect 50% of the signs to be positive and 50% to be negative, which gives $p = 1 - p = 0.5$. In Section 7.4, we saw that a binomial random variable X, as a rule of thumb, is approximately normal provided $np(1 - p) \geq 10$. An alternative that produces acceptable results is to require both $np \geq 5$ and $n(1 - p) \geq 5$. The former requirement is met when $n \geq 40$, while the latter is met when $n \geq 10$. Therefore, using $n > 25$ offers a reasonable compromise between the two.

In Section 6.2, we found the mean and standard deviation for binomial distributions to be $\mu = np$ and $\sigma = \sqrt{np(1 - p)}$. Under the null hypothesis, these become $\mu = \dfrac{n}{2}$ and $\sigma = \sqrt{n\left(\dfrac{1}{2}\right)\left(1 - \dfrac{1}{2}\right)} = \sqrt{\dfrac{n}{4}} = \dfrac{\sqrt{n}}{2}$ $\left(\text{since } p = 0.5 = \dfrac{1}{2}\right)$.

Standardizing k gives

$$z_0 = \frac{k - \dfrac{n}{2}}{\dfrac{\sqrt{n}}{2}}$$

Since we are approximating a discrete distribution with a continuous distribution, we implement the *correction for continuity* introduced in Section 7.4. We noted earlier that, in an appropriately conducted sign test, the value of k always represents the least-occurring sign and that the critical region always lies to the left of the critical value. Therefore, to correct for continuity, we replace k by $k + 0.5$ to obtain our test statistic:

$$z_0 = \frac{(k + 0.5) - \dfrac{n}{2}}{\dfrac{\sqrt{n}}{2}}$$

We only show the by-hand approach for the large-sample case.

EXAMPLE 2   Conducting a One-Sample Sign Test (Large-Sample Case)

Table 4		
285	310	300
300	320	308
310	293	329
293	326	310
297	301	315
332	305	340
242	310	312
329	320	300
311	286	309
292	287	305

Source: espn.com

Problem A sports reporter thinks that the median weight of offensive linemen in the NFL is greater than 300 pounds. He obtains a random sample of 30 offensive linemen and obtains the data shown in Table 4. Do the data support the reporter's belief at the $\alpha = 0.1$ level of significance?

Approach Because the hypothesis is in regard to a median, use the sign test. Verify that the data were obtained from a random sample and then follow Steps 1 through 5.

Solution The data were obtained through a random sample.

Step 1 We want to know if the median weight of NFL offensive linemen is greater than 300 pounds. We have

$$H_0: M = 300 \quad \text{versus} \quad H_1: M > 300$$

This is a right-tailed test.

Step 2 We use the $\alpha = 0.1$ level of significance.

Step 3 There are 8 observations below 300, 19 observations above 300, and 3 observations equal to 300. Because this is a right-tailed test, k is the number of minus signs, so $k = 8$. There are 8 minus signs and 19 plus signs, so $n = 27$.
The test statistic is

$$z_0 = \frac{(k + 0.5) - \dfrac{n}{2}}{\dfrac{\sqrt{n}}{2}} = \frac{(8 + 0.5) - \dfrac{27}{2}}{\dfrac{\sqrt{27}}{2}} = -1.92$$

Figure 5

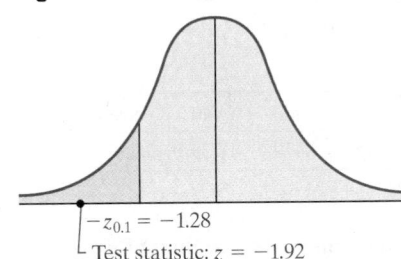

$-z_{0.1} = -1.28$
Test statistic: $z = -1.92$

Step 4 Because this is a right-tailed test and $n > 25$, we calculate the critical value for the right-tailed test at the $\alpha = 0.1$ level of significance as $-z_{0.1} = -1.28$. The test statistic is less than the critical value of -1.28, so we reject the null hypothesis. See Figure 5.

Step 5 There is sufficient evidence at the $\alpha = 0.1$ level of significance to support the hypothesis that the median weight of offensive linemen in the NFL is greater than 300 pounds.

● Now Work Problem 3

Technology Step-by-Step   One-Sample Sign Test

Minitab
1. Enter raw data in column C1.
2. Select the **Stat** menu, highlight **Nonparametrics**, then highlight **1-Sample Sign** . . . .
3. Enter C1 in the cell marked "Variables." Select "Test Median," and enter the assumed value of the median, M_0. Determine the direction of the alternative. Click OK.

StatCrunch
1. Enter the raw data into column var1. Name the column.
2. Select **Stat**. Highlight **Nonparametrics**. Then select **Sign Test**.
3. Select the variable from the Select Columns: list. Enter the appropriate value of the median in the null hypothesis. Choose the correct alternative. Click Compute!

15.3 Assess Your Understanding

Skill Building

In Problems 1–6, use the sign test to test the given alternative hypothesis at the $\alpha = 0.05$ level of significance.

1. The median is less than 8. An analysis of the data reveals that there are 13 minus signs and 8 plus signs.

2. The median is more than 50. An analysis of the data reveals that there are 7 minus signs and 12 plus signs.

NW 3. The median is different from 100. An analysis of the data reveals that there are 21 minus signs and 28 plus signs.

4. The median is different from 68. An analysis of the data reveals that there are 45 plus signs and 27 minus signs.

5. The median is more than 12. The following data were obtained from a random sample:

18	16	15	13	9
15	13	11	15	18
10	14	20	14	12

6. The median is less than 15. The following data were obtained from a random sample:

18	18	33	2	34	18	6	12
36	6	21	14	12	9	13	

Applying the Concepts

7. Acid Rain In 2002, the median pH level of the rain in Glacier National Park, Montana, was 5.25. A biologist thinks that the acidity of rain has decreased since then, which would suggest the pH has increased. She obtains a random sample of 15 rain dates in 2015 and obtains the following data:

5.31	5.19	5.55	5.38	5.37
5.19	5.26	5.29	5.27	5.19
5.27	5.36	5.22	5.28	5.24

Source: National Atmospheric Deposition Program

Test the hypothesis that the median pH level has increased from 5.25 at the $\alpha = 0.05$ level of significance.

8. Age of the Bride A sociologist feels that the median age at which women marry in Cook County, Illinois, is less than the median age of 26.9 throughout the United States (obtained from the U.S. Census Bureau). Based on a random sample of 20 marriage certificates from the county, she obtains the ages shown in the following table:

31	27	24	30	24
27	32	24	23	22
25	23	28	22	26
21	30	25	24	27

Do the data support the sociologist's feelings at the $\alpha = 0.05$ level of significance?

9. Teacher Salaries According to TeacherSalaryInfo.com, the median salary of a teacher in Illinois is $57,283. A politician

obtains the following data based on a random sample of 12 teachers in a district currently experiencing financial difficulty. Do the data support the belief that the teachers in the district are paid less than the average teacher in the state? Use the $\alpha = 0.05$ level of significance.

Note: The data are skewed right with an outlier.

53,620	40,836	48,152	92,006
45,470	49,893	56,838	47,508
73,316	48,876	69,899	113,924

Source: www.familytaxpayers.org

10. Housing Prices on Zillow Zillow.com allows people to research real estate listings. One feature is the "Zestimate," which represents the price Zillow believes the property should sell for. If the Zestimate is accurate, the median difference between the actual selling price and the Zestimate should be zero. The following data represent the difference between the actual selling price and Zestimate for a random sample of 10 real estate transactions. Do the data suggest the Zestimate is inaccurate? Use the $\alpha = 0.05$ level of significance.

Note: A normal probability plot indicates the data are not normal.

2500	−500	−79,500	−13,500	−500
−1000	14,500	3,500	−12,500	−6000

Source: zillow.com

NW 11. Baseball Salaries The following data represent the salaries of 14 randomly selected baseball players in the 2014 season (data are in thousands of dollars, so 3250 means $3,250,000).

445	2,800	19,000	3,125	400	1,440	890
3,250	3,400	550	1,625	2,470	1,320	5,050

Source: sportscity.com

Test the hypothesis that the median salary is more than $1000 thousand ($1 million) at the $\alpha = 0.05$ level of significance.

Note: A normal probability plot indicates that the data are not normal.

12. Miles on an Impala A random sample of 18 three-year-old Chevrolet Impalas was obtained in the St. Louis, Missouri area, and the number of miles on each car was recorded as follows:

27,647	30,858	67,476	35,874	39,943	55,702
31,739	31,832	38,194	29,949	36,231	67,097
32,707	25,672	45,109	26,199	60,798	66,592

Source: carsoup.com

Test the hypothesis that the median number of miles on three-year-old Impalas is less than 40,428 miles, the median miles driven over three years nationwide, at the $\alpha = 0.05$ level of significance.

Note: A normal probability plot indicates that the data are not normal.

Problems 13–16 discuss the by-hand P-value approach to testing hypotheses regarding the median by using the sign test.

We can use the *P*-value approach when determining whether to reject the null hypothesis regarding a median by using the sign test. Recall that the *P*-value is the probability of observing a test statistic as extreme or more extreme than what was actually observed, under the assumption that the null hypothesis is true. In the sign test, we assume that the median is M_0, so 50% of the data should be less than M_0 and 50% of the data greater than M_0. So we *expect* half of the data to result in minus signs and half of the data to result in plus signs. We can think of the data as a bunch of plus and minus signs that follow a binomial probability distribution with $p = \dfrac{1}{2}$ if the null hypothesis is true. So the *P*-value is computed from the binomial probability formula, with $X = k$ and n equal to the number of plus and minus signs:

$$P\text{-value} = P(X \le k) = {}_nC_k 0.5^k (1 - 0.5)^{n-k}$$
$$+ {}_nC_{k-1} 0.5^{k-1}(1 - 0.5)^{n-(k-1)} + \cdots + {}_nC_0(1 - 0.5)^n$$

For Example 1 in this section, the *P*-value is

$$P\text{-value} = P(X \le 8) = {}_{20}C_8 \cdot 0.5^8 \cdot (1 - 0.5)^{20-8}$$
$$+ {}_{20}C_7 \cdot 0.5^7 \cdot (1 - 0.5)^{13} + \cdots + {}_{20}C_0(1 - 0.5)^{20} = 0.2517$$

Because the *P*-value is greater than the level of significance, $\alpha = 0.05$, we do not reject the null hypothesis. These binomial probabilities are easiest to compute with statistical software or a graphing calculator with advanced statistical features.

In Problems 13–16, determine the P-value of the hypothesis test performed in:

13. Problem 7

14. Problem 8

15. Problem 9

16. Problem 10

Problems 17 and 18 illustrate the use of the sign test to test hypotheses regarding a population proportion.

The only requirement for the sign test is that our sample be obtained randomly. When dealing with nominal data, we can identify a characteristic of interest and then determine whether each individual in the sample possesses this characteristic. Under the null hypothesis in the sign test, we expect that half of the data will result in minus signs and half in plus signs. If we let a plus sign indicate the presence of the characteristic (and a minus sign indicate the absence), we expect half of our sample to possess the characteristic while the other half will not. Letting p represent the proportion of the population that possesses the characteristic, our null hypothesis will be $H_0: p = 0.5$. Use the sign test for Problems 17 and 18, following the sign convention indicated previously.

17. Women Gamers A study of 100 randomly selected computer and video game players found that 38 were women (based on data from the Entertainment Software Association). Using an $\alpha = 0.01$ level of significance, does this indicate that less than 50% of gamers are women?

18. Trusting the Press In a study of 2302 U.S. adults surveyed online by Harris Interactive 1243 respondents indicated that they tend to not trust the press. Using an $\alpha = 0.05$ level of significance, does this indicate that more than half of U.S. adults tend to not trust the press?

19. You Explain It! Respondents for the survey in Problem 18 were selected from those who agreed to participate in Harris Interactive Surveys. The survey's methodology includes the following statement: "Because the sample is based on those who agreed to participate in the Harris Interactive panel, no estimates of theoretical sampling error can be calculated." Explain why this could cause someone to question the results of the hypothesis test that was conducted.

Explaining the Concepts

20. Locate the efficiency of the sign test in Section 15.1 for both large samples and small samples. What do these values indicate?

21. In the large-sample case, we compare

$$z_0 = \dfrac{(k + 0.5) - \dfrac{n}{2}}{\dfrac{\sqrt{n}}{2}}$$

to $-z_\alpha$, regardless of whether we are conducting a left-tailed or right-tailed test. Provide a justification for this.

22. Write a paragraph that describes the logic of the test statistic in a right-tailed sign test.

23. Write a paragraph that describes the logic of the test statistic in a two-tailed sign test.

15.4 Inference about the Difference between Two Medians: Dependent Samples

Preparing for This Section Before getting started, review the following:

- Median (Section 3.1, pp. 120–121)
- Boxplots (Section 3.5, pp. 164–168)

- Dependent versus independent sampling (Section 11.1, pp. 525–526)
- Inference about two means: dependent samples (Section 11.2, pp. 536–540)

Objective ① Test a hypothesis about the difference between the medians of two dependent samples

In this section, we compare two population medians using dependent sampling. Recall that dependent sampling occurs when the observations in each sample are somehow related, such as the measurement of a person's pulse rate before and after a workout regimen to discover the effect that the workout has on pulse rate.

We introduced inferential techniques on dependent samples in Section 11.2, where we first computed the differences between the matched observations. The inference was then performed on these differences by comparing the sample mean difference to the assumed difference (usually, the assumed difference is 0). If the sample size was small ($n < 30$), the differences had to be normally distributed. In this section, we introduce a nonparametric procedure that can be used to compare population medians even if the normality requirement is not satisfied. Again, we first compute differences in the matched observations.

① Test a Hypothesis about the Difference between the Medians of Two Dependent Samples

In 1945, Frank Wilcoxon wrote an article entitled "Individual Comparisons by Ranking Methods" that was published in *Biometrics*. In this article, Wilcoxon introduced the *Wilcoxon matched-pairs signed-ranks test.*

Definition   The **Wilcoxon Matched-Pairs Signed-Ranks Test** is a nonparametric procedure used to test the equality of two population medians by dependent sampling.

The idea behind the Wilcoxon matched-pairs signed-ranks test is to first compute the differences in the matched observations and then rank the absolute value of the differences from smallest to largest. If we have matched-pairs data, we compute the differences as follows:

Historical Note

Frank Wilcoxon was a chemist at American Cyanamid. He was conducting *t*-tests on experiments for which he expected a significant difference in the treatment effect, but Student's *t*-test would not always support these beliefs. It turned out that the data sets he was analyzing were small and contained outliers, therefore affecting the *t*-test statistic. To address this problem, Wilcoxon wrote an article in 1945 entitled "Individual Comparisons by Ranking Methods" that was published in *Biometrics*. In this article, Wilcoxon introduced the Wilcoxon matched-pairs signed-ranks test.

X	Y	D = Y − X
5	8	8 − 5 = 3
3	4	4 − 3 = 1
7	6	6 − 7 = −1
9	9	9 − 9 = 0
8	6	6 − 8 = −2

Any difference that equals 0 is eliminated from analysis, so the sample size is reduced to $n = 4$. Now rank the absolute values of the differences from smallest to largest. For positive differences, the rank will be positive; for negative differences, the rank will be negative. In the case of ties, assign the mean value of the ranks that would have been assigned. For example, notice that the absolute value difference 1 occurs twice. To find the ranks, note that the 1s occupy the first and second ranking positions, so the mean rank is $\dfrac{1 + 2}{2} = 1.5$.

X	Y	D = Y − X	|D|	Rank
5	8	8 − 5 = 3	|3| = 3	+4
3	4	4 − 3 = 1	|1| = 1	+1.5
7	6	6 − 7 = −1	|−1| = 1	−1.5
8	6	6 − 8 = −2	|−2| = 2	−3

Once the ranks have been identified, sum the ranks of the *positive differences* and denote this sum as T_+. Then sum the ranks of the *negative differences* and denote this sum as T_-. For example, $T_+ = 4 + 1.5 = +5.5$, and $T_- = −1.5 + (−3) = −4.5$. The value of T_+ or $|T_-|$ will be the test statistic, depending on the structure of the alternative hypothesis.

If we conduct a two-tailed test and the true median difference is 0 (that is, if the null hypothesis is true), we would expect T_+ to be close to $|T_-|$. If the median for population X is greater than the median for population Y (the left-tailed test), what would we expect for the values of T_+ and T_-? Remember, we decided to compute the differences by computing (sample population Y) $-$ (sample population X). So if the median for population X is greater than the median for population Y, we would expect a majority of the differences to be negative. This will lead to large values of $|T_-|$ and small values of T_+. If T_+ is sufficiently small, reject the null hypothesis. A similar argument can be made for the right-tailed test. The test statistic is based on this rationale.

In Other Words
Just as in Section 11.2, the direction of the alternative hypothesis will depend on how the differences are computed.

Test Statistic for the Wilcoxon Matched-Pairs Signed-Ranks Test

The test statistic will depend on the size of the sample and on the alternative hypothesis. Let n represent the number of nonzero differences.

Small-Sample Case ($n \leq 30$)

Two-Tailed	Left-Tailed	Right-Tailed
H_0: $M_D = 0$	H_0: $M_D = 0$	H_0: $M_D = 0$
H_1: $M_D \neq 0$	H_1: $M_D < 0$	H_1: $M_D > 0$
Test Statistic: T is the smaller of T_+ or $|T_-|$	**Test Statistic:** $T = T_+$	**Test Statistic:** $T = |T_-|$

Large-Sample Case ($n > 30$)

The test statistic is given by

$$z_0 = \frac{T - \dfrac{n(n + 1)}{4}}{\sqrt{\dfrac{n(n + 1)(2n + 1)}{24}}}$$

where T is the test statistic from the small-sample case.

For a left-tailed test, reject H_0 if T_+ is small. Do you see why? If T_+ is small, the sum of the ranks for the positive differences is small. This happens if the values in the sample from population 1 (the x-values) are greater than the corresponding values in the sample from population 2 (the y-values), which implies that $M_D < 0$. For a right-tailed test, reject H_0 if $|T_-|$ is small. Can you explain why?

Critical Value for Wilcoxon Matched-Pairs Signed-Ranks Test

Small-Sample Case ($n \leq 30$)

Using α as the level of significance, the critical value(s) is obtained from Table XIII.

Two-Tailed	Left-Tailed	Right-Tailed
$T_{\alpha/2}$	T_α	T_α

Large-Sample Case ($n > 30$)

Using α as the level of significance, the critical value(s) is obtained from Table V in Appendix A. The critical value is always in the left tail of the standard normal distribution.

Two-Tailed	Left-Tailed	Right-Tailed
$-z_{\alpha/2}$	$-z_\alpha$	$-z_\alpha$

We now present the steps required to conduct a Wilcoxon matched-pairs signed-ranks test.

Wilcoxon Matched-Pairs Signed-Ranks Test

If a hypothesis is made regarding the medians of two populations, use the following steps to test the hypothesis, provided that

1. the samples are dependent random samples, and
2. the distribution of the differences is symmetric.

Although tests for verifying the symmetry of data exist, we do not present them in this text. All the data given satisfy the second requirement.

Step 1 Determine the null and alternative hypotheses. The hypotheses can be structured in one of three ways:

Two-Tailed	Left-Tailed	Right-Tailed
$H_0: M_D = 0$	$H_0: M_D = 0$	$H_0: M_D = 0$
$H_1: M_D \neq 0$	$H_1: M_D < 0$	$H_1: M_D > 0$

Note: M_D is the median of the differences of matched pairs.

Step 2 Select a level of significance, α, depending on the seriousness of making a Type I error.

By-Hand Approach

Step 3 Compute the differences in the matched-pairs observations. Rank the absolute value of all sample differences from smallest to largest after discarding those differences that equal 0. Handle ties by finding the mean of the ranks for tied values. Assign negative values to the ranks where the differences are negative and positive values to the ranks where the differences are positive. Find the sum of the positive ranks, T_+, and the sum of the negative ranks, T_-.

Compute the test statistic.

Small-Sample Case ($n \leq 30$)

Two-Tailed	Left-Tailed	Right-Tailed
$H_0: M_D = 0$	$H_0: M_D = 0$	$H_0: M_D = 0$
$H_1: M_D \neq 0$	$H_1: M_D < 0$	$H_1: M_D > 0$
Test Statistic: T is the smaller of T_+ and $\lvert T_- \rvert$.	**Test Statistic:** $T = T_+$	**Test Statistic:** $T = \lvert T_- \rvert$

Large-Sample Case ($n > 30$)

$$z_0 = \frac{T - \dfrac{n(n+1)}{4}}{\sqrt{\dfrac{n(n+1)(2n+1)}{24}}}$$

where T is the test statistic from the small-sample case.

Step 4 The critical value is found from Table XIII for small samples ($n \leq 30$). The critical value is found from Table V for large samples ($n > 30$).

Compare the critical value to the test statistic.

Small-Sample Case	Large-Sample Case
Two-tailed: If $T < T_{\alpha/2}$, reject H_0.	**Two-tailed:** If $z_0 < -z_{\alpha/2}$, reject H_0.
Left-tailed: If $T < T_\alpha$, reject H_0.	**Left-tailed:** If $z_0 < -z_\alpha$, reject H_0.
Right-tailed: If $T < T_\alpha$, reject H_0.	**Right-tailed:** If $z_0 < -z_\alpha$, reject H_0.

Technology (P-value) Approach

Step 3 Use a statistical spreadsheet to obtain the P-value. The directions for obtaining the P-value using Minitab, Excel, and StatCrunch are in the Technology Step-by-Step on page 766.

Step 4 If P-value $< \alpha$, reject the null hypothesis.

Step 5 State the conclusion.

EXAMPLE 1   Wilcoxon Matched-Pairs Signed-Ranks Test: Small-Sample Case

Problem One important variable to consider in trading stock is the daily volume. Volume is measured in number of shares traded in the stock. Stocks with lower volume tend to have more variability in the stock price. A stock analyst believes the median number of shares traded in Walgreens Boots Alliance (WBA) stock is greater than that in McDonalds (MCD). Because national news can play a role in volume of stock traded, the analyst records the volume (in millions of shares) for each of the two stocks on the same day for 14 randomly selected trading days. The results are in Table 5. Test the analyst's belief at the $\alpha = 0.05$ level of significance. **Note:** A normal probability plot indicates the differenced data do not come from a population that is normally distributed.

Table 5					
Day	**WBA**	**MCD**	**Day**	**WBA**	**MCD**
1	8.9	8.5	8	6.0	4.7
2	6.3	7.6	9	15.6	9.0
3	6.2	8.3	10	5.2	6.0
4	7.2	10.4	11	6.3	5.6
5	2.8	2.5	12	10.1	5.0
6	3.3	2.6	13	4.0	4.4
7	23.6	3.5	14	8.4	5.6

Source: Yahoo! Finance

Approach This is matched-pairs data since the volume is measured on the same day for each stock. The differenced data are not normal, so we cannot use Student's t-distribution as a model. Therefore, we use Wilcoxon's matched-pairs signed-ranks test.

Solution The data were obtained randomly. Throughout the section, we assume the requirement of symmetry is satisfied.

Step 1 The analyst is looking for evidence that suggests the median volume of Walgreens is greater than that of McDonald's. We will find the differenced data by computing WBA – MCD. If Walgreens stock has greater volume, we would expect this difference to be positive so that $M_D > 0$. We have

$$H_0: M_D = 0 \quad \text{versus} \quad H_1: M_D > 0$$

This is a right-tailed test.

Step 2 The level of significance is $\alpha = 0.05$.

By-Hand Solution

Step 3 See Table 6. The differences in the matched-pairs observations are in column 4. The absolute values of the differences are in column 5. The ranks (from smallest to largest) of the absolute value of the differences are in column 6. Remember, ties are handled by averaging the ranks of tied differences. This is a right-tailed test, so we determine T_- by adding the negative signed ranks. So,

$$T_- = -7.5 + (-9) + (-11) + (-6) + (-2.5) = -36$$

The test statistic is $T = |T_-| = |-36| = 36$.

Technology (*P*-value) Solution

Step 3 Figure 6 shows the results obtained from Minitab.

Figure 6

Wilcoxon Signed Rank Test: Diff

```
Test of median = 0.000000 versus median > 0.000000

          N for  Wilcoxon          Estimated
     N    Test   Statistic    P      Median
Diff 14   14       69.0    0.158    0.7500
```

(continued)

Table 6

Day	WBA, X	MCD, Y	$D = X - Y$	$|D|$	Signed Ranks
1	8.9	8.5	0.4	0.4	+2.5
2	6.3	7.6	−1.3	1.3	−7.5
3	6.2	8.3	−2.1	2.1	−9
4	7.2	10.4	−3.2	3.2	−11
5	2.8	2.5	0.3	0.3	+1
6	3.3	2.6	0.7	0.7	+4.5
7	23.6	3.5	20.1	20.1	+14
8	6.0	4.7	1.3	1.3	+7.5
9	15.6	9.0	6.6	6.6	+13
10	5.2	6.0	−0.8	0.8	−6
11	6.3	5.6	0.7	0.7	+4.5
12	10.1	5.0	5.1	5.1	+12
13	4.0	4.4	−0.4	0.4	−2.5
14	8.4	5.6	2.8	2.8	+10

Step 4 Because we are performing a right-tailed test and the sample size is less than 30, we find the critical value with $n = 14$ at the $\alpha = 0.05$ level of significance by using Table XIII and obtain $T_{0.05} = 25$. See Figure 7.

Figure 7

	Critical Values for the Wilcoxon Signed-Rank Test			
	Level of Significance, α			
n	0.005	0.01	0.025	0.05
5	*	*	*	0
6	*	*	0	2
12	7	9	13	17
13	9	12	17	21
14	12	15	21	25
15	15	19	25	30

* Indicates that it is not possible to get a value in the critical region.

Because the test statistic, $T = 36$, is greater than the critical value, $T_{0.05} = 25$, we do not reject the null hypothesis.

Step 4 Because the P-value is greater than the level of significance $(0.158 > 0.05)$, we do not reject the null hypothesis.

Step 5 There is not sufficient evidence to conclude the median daily volume of Walgreens stock is greater than that of McDonald's at the $\alpha = 0.05$ level of significance.

● **Now Work Problem 11**

Using the Wilcoxon Signed-Ranks Test on a Single Sample

The Wilcoxon matched-pairs signed-ranks test can also be used as a single sample test for the median of a population, provided that the population is symmetric. To begin the test, the required differences are found by subtracting the median stated in the null hypothesis from each data value. That is, we let $D_i = x_i - M_0$. Once these differences are computed, the test is conducted in the same manner as for two-sample dependent data.

Note that if the symmetry assumption is not met, the null hypothesis, H_0: $M = M_0$, may be rejected even if $M = M_0$. This is possible because the lack of symmetry could cause T_+ and $|T_-|$ to no longer be approximately equal. Since the Wilcoxon signed-ranks test takes both the sign and magnitude of the differences into account, we are really testing that the distribution is symmetric *and* that the distribution has median M_0.

We only show the by-hand approach.

EXAMPLE 2   Wilcoxon One-Sample Ranked-Sums Test

Table 7

$79.01	$81.58	$107.60	$167.38
$134.22	$92.79	$39.65	$147.18
$68.76	$126.22	$187.56	
$178.34	$95.47		

Problem According to Colliers International, the median monthly parking rate in the United States for 2012 was $164.80. The data in Table 7 represent a random sample of monthly parking rates in Milwaukee, Wisconsin.

Using the Wilcoxon signed-ranks test at the $\alpha = 0.05$ level of significance, do the data indicate that the median monthly parking rate in Milwaukee is below normal? Note that a normal probability plot indicates that the data are not normally distributed.

Approach As indicated, we use the Wilcoxon signed-ranks test. Because this is not a matched-pairs sample, we obtain the required differences by subtracting the median stated in the null hypothesis from each observation.

Solution The data were obtained randomly and we are assuming that the symmetry requirement is satisfied.

Step 1 We want to gather evidence to show that the median monthly parking rate in Milwaukee is less than the U.S. median rate of $164.80. We subtract 164.80 from the monthly rate for each observation. If the alternative hypothesis is true, the median differences should be negative so that $M_D < 0$. The null hypothesis is a statement of "no difference." So the null hypothesis is that there is no difference in the parking rate in Milwaukee from those in the United States as a whole. We have

$$H_0\text{: } M = \$164.80 \quad \text{versus} \quad H_1\text{: } M < \$164.80$$

This is a left-tailed test.

Step 2 The level of significance is $\alpha = 0.05$.

Step 3 First, compute the required differences (column 2 of Table 8). Next, calculate the absolute values of the differences (column 3 of the table) and then rank these values from smallest to largest. Assign negative values to the ranks whose differences are negative and positive values to the ranks whose differences are positive (column 4 of the table). The sample size is small, and we are conducting a left-tailed test, so the test statistic is $T = T_+ = 1 + 2 + 4 = 7$.

Table 8

Monthly Parking Rate, X_i	$D_i = X_i - \$164.80$	$|D_i|$	Signed Ranks
79.01	−85.79	85.79	−11
81.58	−83.22	83.22	−10
107.60	−57.20	57.20	−7
167.38	2.58	2.58	1
134.22	−30.58	30.58	−5
92.79	−72.01	72.01	−9
39.65	−125.15	125.15	−13
147.18	−17.62	17.62	−3
68.76	−96.04	96.04	−12
126.22	−38.58	38.58	−6
187.56	22.76	22.76	4
178.34	13.54	13.54	2
95.47	−69.33	69.33	−8

(continued)

Step 4 We are testing the hypothesis at the $\alpha = 0.05$ level of significance. Because we are performing a left-tailed test and the sample size is less than 30, we find the critical value with $n = 13$ at the $\alpha = 0.05$ level of significance by using Table XIII and obtain $T_{0.05} = 21$. The test statistic is less than the critical value $(7 < 21)$, so we reject the null hypothesis.

Step 5 There is sufficient evidence at the 0.05 significance level to conclude that the median monthly parking rate in Milwaukee, Wisconsin, is less than the U.S. median monthly parking rate.

• **Now Work Problem 21**

Technology Step-by-Step Wilcoxon Signed-Ranks Test of Matched-Pairs Data

Minitab
1. Enter the raw paired data in the first two columns. Name the variables. Determine the differenced data in the next column.
2. Select the **Stat** menu, highlight **Nonparametrics**, then highlight **1-Sample Wilcoxon . . . .**
3. Enter the column containing the differenced data in the cell marked "Variables." Click "Test Median," and enter "0" in the cell. Click OK.

Excel
1. Load the XLSTAT Add-in.
2. Enter the raw paired data in Columns A and B.
3. Select the XLSTAT menu. Click Nonparametric tests. Then select Comparison of Two Samples.
4. Place the cursor in the Sample 1: cell and highlight the data in Column A. Place the cursor in the Sample 2:

cell and highlight the data in Column B. Make sure the Paired samples radio button is selected. Make sure Column labels is unchecked (unless you entered column names). Check the box "Wilcoxon signed-rank test." Click the Options tab.
5. Choose the appropriate Alternative hypothesis. Enter 0 as the Hypothesized difference (D). Click OK.

StatCrunch
1. Enter the raw paired data in the first two columns. Name the variables.
2. Select **Stat**. Highlight **Nonparametrics**. Select **Wilcoxon Signed Ranks**.
3. Choose the Paired radio button. Select Sample 1 and Sample 2 from the drop-down menu. Choose the Hypothesis test radio button. Enter 0 for the Null Median. Choose the appropriate alternative hypothesis. Click Compute!

15.4 Assess Your Understanding

Skill Building

1. A researcher believes the median from population 1 is less than the median from population 2 in matched-pairs data. How would you define M_D? How would you compute the differences?

2. When conducting a left-tailed Wilcoxon matched-pairs signed-ranks test, the test statistic is $T =$ _____.

In Problems 3–10, use the Wilcoxon matched-pairs signed-ranks test to test the given hypotheses at the $\alpha = 0.05$ level of significance. The dependent samples were obtained randomly.

3. Hypotheses: H_0: $M_D = 0$ versus H_1: $M_D > 0$ with $n = 12$ and $T_- = -16$.

4. Hypotheses: H_0: $M_D = 0$ versus H_1: $M_D > 0$ with $n = 20$ and $T_- = -65$.

5. Hypotheses: H_0: $M_D = 0$ versus H_1: $M_D < 0$ with $n = 15$ and $T_+ = 33$.

6. Hypotheses: H_0: $M_D = 0$ versus H_1: $M_D < 0$ with $n = 25$ and $T_+ = 95$.

7. Hypotheses: H_0: $M_D = 0$ versus H_1: $M_D \neq 0$ with $n = 18$, $T_- = -121$, and $T_+ = 50$.

8. Hypotheses: H_0: $M_D = 0$ versus H_1: $M_D \neq 0$ with $n = 14$, $T_- = -45$, and $T_+ = 60$.

9. Hypotheses: H_0: $M_D = 0$ versus H_1: $M_D > 0$ with $n = 40$ and $T_- = -300$.

10. Hypotheses: H_0: $M_D = 0$ versus H_1: $M_D < 0$ with $n = 35$ and $T_+ = 210$.

Applying the Concepts

NW 11. Effects of Exercise To determine the effectiveness of an exercise regimen, a physical therapist randomly selects 10 women to participate in a study. She measures their waistlines (in inches) both before and after a rigorous 8-week exercise program and obtains the data shown. Is the median waistline before the exercise program more than the median waistline after the exercise program? Use the $\alpha = 0.05$ level of significance.

Woman	Before	After	Woman	Before	After
1	23.5	19.75	6	19.75	19.5
2	18.5	19.25	7	35	34.25
3	21.5	21.75	8	36.5	35
4	24	22.5	9	52	51.5
5	25	25	10	30	31

12. Effects of Exercise A physical therapist wishes to learn whether an exercise program increases flexibility. She measures the flexibility (in inches) of 12 randomly selected subjects both before and after an intensive 8-week training program and obtains the data shown. Is the median flexibility before the exercise program less than the median flexibility after the exercise program? Use the $\alpha = 0.05$ level of significance.

Subject	Before	After	Subject	Before	After
1	18.5	19.25	7	19.75	19.5
2	21.5	21.75	8	15.75	17
3	16.5	16.5	9	18	19.25
4	21	20.25	10	22	19.5
5	20	22.25	11	15	16.5
6	15	16	12	20.5	20

13. Reaction Time In an experiment conducted online at the University of Mississippi, participants are asked to react to a stimulus such as pressing a key upon seeing a blue screen. The time to press the key (in seconds) is measured and the same person is then asked to press a key upon seeing a red screen, again with the time to react measured. The results for six study participants are shown. Does the evidence suggest that the median reaction time to the blue stimulus is different from the median reaction time to the red stimulus? Use the $\alpha = 0.05$ level of significance.

Participant	Reaction Time Blue	Reaction Time Red
1	0.582	0.408
2	0.481	0.407
3	0.841	0.542
4	0.267	0.402
5	0.685	0.456
6	0.450	0.533

Source: PsychExperiments at the University of Mississippi

14. Rat Hemoglobin Hemoglobin helps the red blood cells transport oxygen and remove carbon dioxide. Researchers at NASA wanted to discover the effects of space flight on a rat's hemoglobin. The following data represent the hemoglobin (in grams per deciliter) at lift-off minus 3 days (H-L3) and immediately upon return (H-R0) for 12 randomly selected rats sent to space on the Spacelab Sciences 1 flight. Is the median hemoglobin level at lift-off minus 3 days less than the median hemoglobin level upon return? Use the $\alpha = 0.05$ level of significance.

Rat	H-L3	H-R0	Rat	H-L3	H-R0
1	15.2	15.8	7	14.3	16.4
2	16.1	16.5	8	14.5	16.5
3	15.3	16.7	9	15.2	16.0
4	16.4	15.7	10	16.1	16.8
5	15.7	16.9	11	15.1	17.6
6	14.7	13.1	12	15.8	16.9

Source: NASA Life Sciences Data Archive

15. Secchi Disk A Secchi disk is an 8-inch-diameter weighted disk painted black and white and attached to a rope. The disk is lowered into water, and the depth (in inches) at which it is no longer visible is recorded. The measurement is an indication of water clarity. An environmental biologist interested in discovering whether the water clarity of the lake at Joliet Junior College is improving takes measurements at the same location during the course of a year and repeats the measurements on the same dates five years later. She obtains the results shown.

Do you believe that the clarity of the lake is improving? Use the $\alpha = 0.05$ level of significance.

Observation	Date	Initial Depth	Depth 5 Years Later
1	5/11	38	52
2	6/7	58	60
3	6/24	65	72
4	7/8	74	72
5	7/27	56	54
6	8/31	36	48
7	9/30	56	58
8	10/12	52	60

Source: Virginia Piekarski, Joliet Junior College

16. Effect of Aspirin on Blood Clotting Blood clotting is due to a sequence of chemical reactions. The protein thrombin initiates blood clotting by working with another protein, prothrombin. It is common to measure an individual's blood clotting time as *prothrombin time*, the time between the start of the thrombin–prothrombin reaction and the formation of the clot. Researchers wanted to study the effect of aspirin on prothrombin time. They randomly selected 12 subjects and measured the prothrombin time (in seconds) without taking aspirin and 3 hours after taking two aspirin tablets. They obtained the following data. Does the evidence suggest that aspirin affects the median time it takes for a clot to form? Use the $\alpha = 0.05$ level of significance.

Subject	Before Aspirin	After Aspirin	Subject	Before Aspirin	After Aspirin
1	12.3	12.0	7	11.3	10.3
2	12.0	12.3	8	11.8	11.3
3	12.0	12.5	9	11.5	11.5
4	13.0	12.0	10	11.0	11.5
5	13.0	13.0	11	11.0	11.0
6	12.5	12.5	12	11.3	11.5

Source: Donald Yochem and Darrell Roach. "Aspirin: Effect of Thrombus Formation Time and Prothrombin Time of Human Subjects." *Angiology*, 22:70–76, 1971

17. Car Rentals The following data represent the weekday rental rate for a compact car charged by two car-rental companies, Avis and Hertz, in 11 locations. Is Avis cheaper than Hertz? Use the $\alpha = 0.05$ level of significance.

City	Avis	Hertz
Chicago	67.54	59.14
Los Angeles	97.16	97.15
Houston	103.43	103.14
Orlando	38.79	36.63
Boston	87.90	87.80
Seattle	109.86	110.09
Pittsburgh	109.35	109.35
Phoenix	82.86	83.03
New Orleans	112.65	113.32
Minneapolis	116.46	114.53
St. Louis	105.49	107.04

Source: Yahoo! Travel

18. Does Octane Affect Miles per Gallon? A researcher wants to know if the octane level of gasoline affects the gas mileage of a car. She randomly selects 10 cars and puts 5 gallons of 87-octane gasoline in the tank. On a closed track, each car is driven at 50 miles per hour until it runs out of gas. The experiment is repeated, with each car getting 5 gallons of 92-octane gasoline. The miles per gallon for each car are then computed. The results are shown. Would you recommend purchasing 92 octane? Why or why not?

87 Octane	92 Octane	87 Octane	92 Octane
18.0	18.5	23.4	22.8
23.2	23.1	23.1	23.5
31.5	31.9	19.0	19.5
24.9	26.7	26.8	26.2
24.1	25.1	31.8	30.7

19. Rats' Red Blood Cell Counts Researchers at NASA wanted to learn the effects of spaceflight on a rat's red blood cell count (RBC). The following data represent the RBC at lift-off minus 3 days (RBC-L3) and immediately upon return (RBC-R0) of 27 rats sent to space on the Spacelab Sciences 1 flight.

Rat	RBC-L3	RBC-R0	Rat	RBC-L3	RBC-R0
112	7.53	8.16	79	7.79	8.52
145	7.79	9.15	47	8.24	8.59
15	7.84	9.09	109	7.69	7.29
142	6.86	8.42	13	7.23	8.93
45	7.93	8.96	74	7.83	8.64
150	7.48	9.25	126	8.09	9.65
162	7.94	7.40	157	8.27	8.76
136	7.95	9.55	156	7.27	9.11
127	8.44	9.38	128	8.23	9.81
153	6.70	9.66	99	14.50	9.86
97	6.95	8.83	90	7.73	9.90
94	6.73	8.46	82	7.31	7.64
124	7.21	9.04	55	7.84	9.54
117	6.95	9.48			

Source: NASA Life Sciences Data Archive

They used Minitab to test whether the RBC three days prior to lift-off was different from the RBC upon return. The results are as follows:

Wilcoxon Signed-Rank Test
```
Test of median = 0.000000 versus median not = 0.000000

                  N for   Wilcoxon                Estimated
            N    Test    Statistic        P        Median
differen    27    27       343.0       0.000       1.255
```

(a) State the null and alternative hypotheses.
(b) Does it appear that the flight to space affected the RBC of the rats? Use the $\alpha = 0.05$ level of significance. Support your answer.

20. Reaction-Time Experiment Researchers at the University of Mississippi wanted to learn the reaction times of students to different stimuli. In the following data, the reaction times for subjects were measured after they received a simple stimulus and a go/no-go stimulus. The simple stimulus was an auditory cue, and the time from when the cue was given to when the student reacted was measured. The go/no-go stimulus required the student to respond to a particular stimulus and not respond to other stimuli. Again, the reaction time was measured. The following data were obtained:

Subject Number	Simple	Go/No-Go	Subject Number	Simple	Go/No-Go
1	0.220	0.375	16	0.498	0.565
2	0.430	1.759	17	0.262	0.402
3	0.338	0.652	18	0.620	0.643
4	0.266	0.467	19	0.300	0.351
5	0.381	0.651	20	0.424	0.380
6	0.738	0.442	21	0.478	0.434
7	0.885	1.246	22	0.305	0.452
8	0.683	0.224	23	0.281	0.745
9	0.250	0.654	24	0.291	0.290
10	0.255	0.442	25	0.453	0.790
11	0.198	0.347	26	0.376	0.792
12	0.352	0.698	27	0.328	0.613
13	0.285	0.803	28	0.952	1.179
14	0.259	0.488	29	0.355	0.636
15	0.200	0.281	30	0.368	0.391

Source: PsychExperiments at The University of Mississippi

The researchers used Minitab to test whether the simple stimulus had a lower reaction time than the go/no-go stimulus. The results of the analysis are as follows:

Wilcoxon Signed-Rank Test
```
Test of median = 0.000000 versus median    >    0.000000

                      N for   Wilcoxon                Estimated
                N    Test    Statistic        P        Median
difference     30    30       408.0        0.000       1.1920
```

(a) State the null and alternative hypotheses.
(b) Is the median reaction time for the go/no-go stimulus higher than the median reaction time for the simple stimulus? Use the $\alpha = 0.05$ level of significance. Why?

NW 21. Outpatient Treatment The median length of stay for substance-abuse outpatient treatment completers is 107 days for those referred by the criminal justice system. The following data represent the length of stays for a random sample of substance-abuse outpatient treatment completers who were referred by an employer.

80	108	95	107	89	100
85	102	115	109	99	94

Source: Department of Health and Human Services

Using the Wilcoxon signed-ranks test at the $\alpha = 0.05$ level of significance, does the median length of stay seem different for employer referrals than for those referred by the criminal justice system?

22. Outpatient Treatment The median length of stay for male substance-abuse outpatient treatment completers is 105 days. The following data represent the length of stays for a random sample of female substance-abuse outpatient treatment completers.

117	111	81	110	103	94
92	91	116	108	98	84

Source: Department of Health and Human Services

Using the Wilcoxon signed-ranks test at the $\alpha = 0.05$ level of significance, does the median length of stay seem different for males and females?

23. You Explain It! Dietary Habits A student in an exercise science program wishes to study dietary habits of married couples. She surveys married couples from her local gym and asks them (individually) what percent of their daily calories are from fat. She analyzes the results using the Wilcoxon signed-ranks test. Explain why her results are questionable.

15.5 Inference about the Difference between Two Medians: Independent Samples

Preparing for This Section Before getting started, review the following:

- Median (Section 3.1, pp. 120–121)
- Boxplots (Section 3.5, pp. 166–168)
- Independent versus dependent sampling (Section 11.1, pp. 525–526)
- Inference about two means: independent samples (Section 11.3, pp. 547–552)

Objective ❶ Test a hypothesis about the difference between the medians of two independent samples

In Section 11.3, we learned how to test for the equality of two population means with independent sampling. With independent sampling, the individuals in the sample from population 1 are not used to determine the individuals in the sample from population 2. The methods used to test for the difference between two population means required that each population be normally distributed when the sample size was small ($n_1 < 30, n_2 < 30$). In this section, we introduce nonparametric procedures that can test hypotheses regarding the equality of two measures of central tendency even if this requirement is not satisfied.

❶ Test a Hypothesis about the Difference between the Medians of Two Independent Samples

In 1947, H. B. Mann and D. R. Whitney introduced a nonparametric technique that can be used to test the equality of two population medians in the case of independent sampling.

Definition The **Mann–Whitney Test** is a nonparametric procedure that is used to test the equality of two population medians from independent samples.

The idea behind the Mann–Whitney test is to combine the two samples and rank *all* the observations from smallest to largest. We handle ties by finding the mean of the ranks for tied values. For example, if the data for the sample corresponding to population *X* are

$$7, 8, 10, 13$$

and the data for the sample corresponding to population *Y* are

$$6, 8, 9$$

Historical Note

Henry B. Mann and his graduate student D. Ransom Whitney were comparing wage data from 1940 with wage data from 1944. Their goal was to show that the distribution of wages in 1940 was less than the distribution of wages in 1944. To conduct this analysis, they devised the Mann–Whitney test. They published the technique used to analyze the data in 1947, in an article entitled "On a Test of Whether One of Two Random Variables Is Stochastically Larger Than the Other" in the *Annals of Mathematical Statistics*.

we combine and rank the data as follows:

Combined Data	Population	Rank
6	Y	1
7	X	2
8	X	3.5
8	Y	3.5
9	Y	5
10	X	6
13	X	7

Notice that we observed an 8 twice. To find the rank of 8, we recognize that the 8s occupy the third- and fourth-ranking positions. So the mean rank is $\dfrac{3+4}{2} = 3.5$.

Once the ranks have been identified, we sum the ranks of the sample observations from population X *only*. The sum of the ranks from the sample observations from population X is $2 + 3.5 + 6 + 7 = 18.5$. If the two populations have the same median (that is, if the null hypothesis is true), we would expect the sum of the ranks for the sample observations from population X to be close to the sum of the ranks for the sample observations from population Y. If the median for population X is less than the median for population Y (left-tailed test), we would expect the sum of the ranks for the sample observations from population X to be less than the sum of the ranks for the sample observations from population Y.

Suppose we conduct a right-tailed test and the alternative hypothesis that $M_X > M_Y$ is true. We should expect the sum of the ranks for the sample observations from population X to be greater than the sum of the ranks for the sample observations from population Y. The test statistic is based on this rationale.

Test Statistic for the Mann–Whitney Test

The test statistic will depend on the size of the samples from each population. Let n_1 represent the sample size for population X and n_2 represent the sample size for population Y.

Small-Sample Case ($n_1 \leq 20$ and $n_2 \leq 20$)
If S is the sum of the ranks corresponding to the sample from population X, then the test statistic, T, is given by

$$T = S - \frac{n_1(n_1 + 1)}{2}$$

Note: The value of S is always obtained by summing the ranks of the sample data that correspond to M_X, the median of population X, in the hypothesis.

Large-Sample Case ($n_1 > 20$ or $n_2 > 20$)
From the Central Limit Theorem, the test statistic is given by

$$z_0 = \frac{T - \dfrac{n_1 n_2}{2}}{\sqrt{\dfrac{n_1 n_2(n_1 + n_2 + 1)}{12}}}$$

where T is the test statistic from the small-sample case.

For a two-tailed test, we reject $H_0: M_X = M_Y$ if T is sufficiently large or sufficiently small. To test $H_1: M_X < M_Y$ (a left-tailed test), we reject $H_0: M_X = M_Y$ if T is sufficiently small. Do you see why? If T is small, the sum of the ranks for population X is small. This happens if the sample observations from population X are less than the sample observations from population Y, which implies that $M_X < M_Y$. For a right-tailed test, we reject H_0 if T is sufficiently large. Do you know why?

Critical Value for the Mann–Whitney Test

Small-Sample Case ($n_1 \leq 20$ and $n_2 \leq 20$)

Using α as the level of significance, the critical value(s) is(are) obtained from Table XIII.

Two-Tailed	Left-Tailed	Right-Tailed
$w_{\alpha/2}$	w_α	$w_{1-\alpha} = n_1 n_2 - w_\alpha$
$w_{1-\alpha/2} = n_1 n_2 - w_{\alpha/2}$		

Large-Sample Case ($n_1 > 20$ or $n_2 > 20$)

Using α as the level of significance, the critical value(s) is(are) obtained from Table V in Appendix A.

Two-Tailed	Left-Tailed	Right-Tailed
$z_{\alpha/2}$ and $-z_{\alpha/2}$	$-z_\alpha$	z_α

We now present the steps required to conduct a Mann–Whitney test.

Mann–Whitney Test

To test hypotheses regarding the medians of two populations, we can use the following steps, provided that

1. the samples are independent random samples, and
2. the shapes of the distributions are the same.

Throughout this section, we will assume that the condition that the shapes of the distributions be the same is satisfied.

Step 1 Draw a side-by-side boxplot to compare the sample data from the two populations. This helps to visualize the difference in the medians.

Determine the null and alternative hypotheses. The hypotheses are structured as follows:

Two-Tailed	Left-Tailed	Right-Tailed
$H_0: M_X = M_Y$	$H_0: M_X = M_Y$	$H_0: M_X = M_Y$
$H_1: M_X \neq M_Y$	$H_1: M_X < M_Y$	$H_1: M_X > M_Y$

Note: M_X is the median of population X and M_Y is the median of population Y.

Step 2 Select a level of significance, α, depending on the seriousness of making a Type I error.

By-Hand Approach

Step 3 Rank all sample observations from smallest to largest. Handle ties by finding the mean of the ranks for tied values. Find the sum of the ranks for the sample from population X.

Compute the test statistic. Note that S is the sum of the ranks obtained from the sample observations from population X. In addition, n_1 is the size of the sample from population X, and n_2 is the size of the sample from population Y.

Small-Sample Case	Large-Sample Case
$T = S - \dfrac{n_1(n_1 + 1)}{2}$	$z_0 = \dfrac{T - \dfrac{n_1 n_2}{2}}{\sqrt{\dfrac{n_1 n_2 (n_1 + n_2 + 1)}{12}}}$

Technology (P-Value) Approach

Step 3 Use a statistical spreadsheet to obtain the P-value. The directions for obtaining the P-value using Minitab, Excel, and StatCrunch are in the Technology Step-by-Step on page 775.

Step 4 If P-value $< \alpha$, reject the null hypothesis.

Step 4 The level of significance is used to determine the critical value. The critical value is found from Table XIV for small samples ($n_1 \leq 20$ and $n_2 \leq 20$) and from Table V for large samples ($n_1 > 20$ or $n_2 > 20$).

Compare the critical value to the test statistic.

Small-Sample Case	Large-Sample Case
Two-tailed: If $T < w_{\alpha/2}$ or $T > w_{1-\alpha/2}$, reject H_0. *Note:* $w_{1-\alpha/2} = n_1 n_2 - w_{\alpha/2}$	**Two-tailed:** If $z_0 < -z_{\alpha/2}$ or $z_0 > z_{\alpha/2}$, reject H_0.
Left-tailed: If $T < w_\alpha$, reject H_0.	**Left-tailed:** If $z_0 < -z_\alpha$, reject H_0.
Right-tailed: If $T > w_{1-\alpha}$, reject H_0. *Note:* $w_{1-\alpha} = n_1 n_2 - w_\alpha$	**Right-tailed:** If $z_0 > z_\alpha$, reject H_0.

Step 5   State the conclusion.

EXAMPLE 1   Mann–Whitney Test (Small-Sample Case)

Problem When exposed to an infection, a person typically develops antibodies. The extent to which the antibodies respond can be measured by looking at a person's titer, which is a measure of the number of antibodies present. The higher the titer is, the more antibodies that are present. The data in Table 9 represent the titers of 11 ill people and 11 healthy people exposed to the tularemia virus in Vermont. Is the level of titer in the ill group greater than the level of titer in the healthy group? Use the $\alpha = 0.1$ level of significance.

Table 9							
Ill				**Healthy**			
640	160	1280	320	10	320	160	160
80	640	640	160	320	320	10	320
1280	640	160		320	80	640	

Source: Adapted from *An Introduction to Mathematical Statistics and Its Applications*, by R. Larsen and M. Marx. Prentice Hall, 2001.

Approach Normal probability plots indicate that the sample data do not come from a population that is normal. Therefore, we will use the Mann–Whitney test. We then proceed to follow Steps 1 through 5.

Solution The titer levels were obtained from independent random samples.

Step 1   Figure 8 shows the boxplots of the two samples. The median for the ill group equals 640, the same as Q_3. The median for the healthy group equals 320, the same as Q_3.

Figure 8

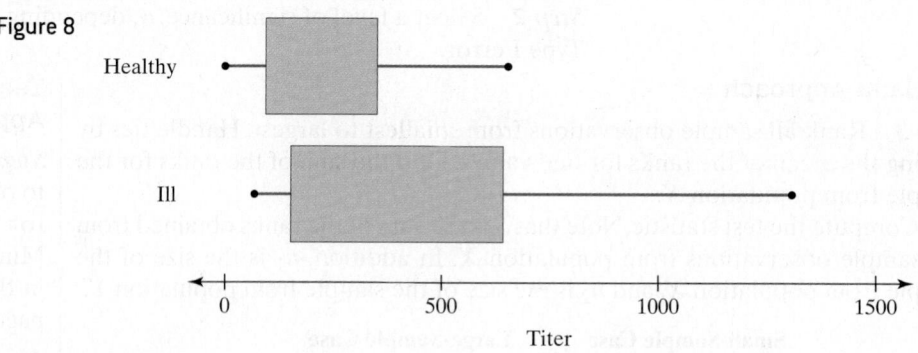

The sample median for the ill individuals is more than the sample median for the healthy individuals. Is this difference due to differences in the population medians or to sampling error?

We want to gather evidence that demonstrates that the median titer for the ill group is greater than the median titer for the healthy group. This can be written $M_{\text{ILL}} > M_{\text{HEALTHY}}$. We assume there is no difference between the two groups.

Therefore, we have

$$H_0: M_{\text{ILL}} = M_{\text{HEALTHY}} \quad \text{versus} \quad H_1: M_{\text{ILL}} > M_{\text{HEALTHY}}$$

This is a right-tailed test.

Step 2 The level of significance is $\alpha = 0.1$.

By-Hand Approach

Step 3 We combine the two sample data sets into one data set and arrange the data in ascending order. Be sure to keep track of the sample from which the titer was drawn. Then assign ranks to each observation. See Table 10. We now add up the ranks corresponding to ill individuals (because this group corresponds to the median on the left in the hypothesis) and obtain

$$S = 18 + 3.5 + 21.5 + 7 + 18 + 18 + 21.5 + 18 + 7 + 12.5 + 7 = 152$$

Table 10					
Titer	**Sample**	**Rank**	**Titer**	**Sample**	**Rank**
10	Healthy	1.5	320	Healthy	12.5
10	Healthy	1.5	320	Healthy	12.5
80	Healthy	3.5	320	Healthy	12.5
80	Ill	3.5	320	Ill	12.5
160	Healthy	7	640	Healthy	18
160	Healthy	7	640	Ill	18
160	Ill	7	640	Ill	18
160	Ill	7	640	Ill	18
160	Ill	7	640	Ill	18
320	Healthy	12.5	1280	Ill	21.5
320	Healthy	12.5	1280	Ill	21.5

The test statistic is

$$T = S - \frac{n_1(n_1 + 1)}{2} = 152 - \frac{11(11 + 1)}{2} = 86$$

Step 4 Because we are performing a right-tailed test and both sample sizes are less than 20, we determine the right critical value with $n_1 = 11$ and $n_2 = 11$ at the $\alpha = 0.10$ level of significance from Table XIV and obtain $w_{0.9} = n_1 n_2 - w_{0.10} = (11)(11) - 41 = 80$. See Figure 9.

Figure 9

Critical Values of the Mann-Whitney Test Statistic															
n_1	p	$n_2 = 2$	3	4	5	6	7	8	9	10	11	12	13	14	15
	0.001	0	0	1	2	4	6	7	9	11	13	15	18	20	22
	0.005	0	1	3	5	7	10	12	14	17	19	22	25	27	30
10	0.01	0	2	4	7	9	12	14	17	20	23	25	28	31	34
	0.005	0	1	3	6	8	11	14	17	19	22	25	28	31	34
11	0.01	0	2	5	8	10	13	16	19	23	26	29	32	35	38
	0.025	1	4	7	10	14	17	20	24	27	31	34	38	41	45
	0.05	2	6	9	13	17	20	24	28	32	35	39	43	47	51
	0.10	4	8	12	16	20	24	28	32	37	41	45	49	53	58
	0.001	0	0	1	2	5	8	10	13	15	18	21	24	26	29

Because the test statistic ($T = 86$) is greater than the critical value, 80, we reject the null hypothesis.

Technology Approach

Step 3 Figure 10 shows the results obtained from StatCrunch. The P-value is 0.0466.

Figure 10

Hypothesis test results:
m1 = median of Ill
m2 = median of Healthy
Parameter : m1 - m2
H_0 : Parameter = 0
H_A : Parameter > 0

Difference	n1	n2	Diff. Est.	Test Stat.	P-value	Method
m1 - m2	11	11	320	152	0.0466	Norm. Approx.

Step 4 Because the P-value is less than the level of significance ($0.0466 < 0.1$), we reject the null hypothesis.

Step 5 There is sufficient evidence to conclude that the median titer of the ill individuals is greater than the median titer of the healthy individuals at the $\alpha = 0.1$ level of significance. ●

● **Now Work Problem 9**

We now present an example of the Mann–Whitney test with large samples. We only show the by-hand approach.

EXAMPLE 2   Mann–Whitney Test (Large-Sample Case)

Problem An engineer is comparing the time to failure (in flight hours) of two different air conditioners for airplanes and wants to determine if the median time to failure for model Y is longer than the median time to failure for model X. She obtains a random sample of 22 failure times for model X and an independent random sample of 17 failure times for model Y. Do the data in Table 11 suggest that the time to failure for model Y is longer? Use the 0.05 level of significance. Normal probability plots indicate that neither data set is normally distributed.

Table 11						
	Model X			**Model Y**		
7	7	33	4	115	412	200
20	4	59	91	55	62	253
5	76	287	472	219	225	122
52	19	128	28	245	129	168
103	25	68		239	71	118
17	109	3		130	12	

Approach Verify that the requirement of independent random samples is satisfied and then proceed to follow Steps 1 through 5. Continue to assume that the shapes of the distributions are the same. This can be roughly confirmed by considering the boxplots in Figure 11.

Solution The samples are independent random samples.

Step 1 To visualize the difference in the medians, we construct side-by-side boxplots of the two data sets. See Figure 11. The median of the sample for model Y is greater than the median of the sample for model X. Is this difference due to a difference in the population medians or to sampling error?

Figure 11

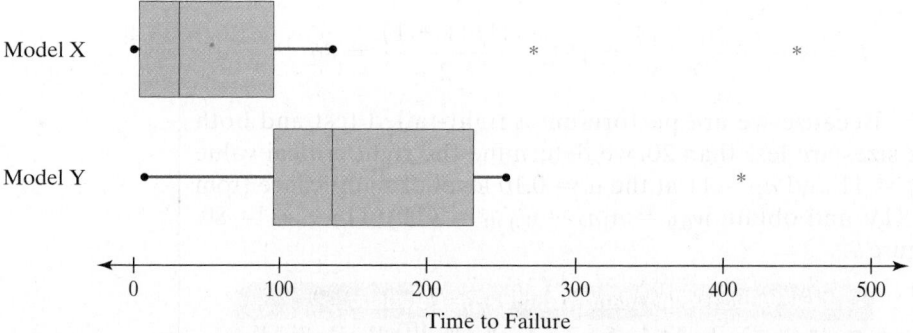

We want to determine if the median time to failure for model Y is larger than the median time to failure for model X. Therefore, our hypotheses are

$$H_0: M_X = M_Y \quad \text{versus} \quad H_1: M_X < M_Y$$

This is a left-tailed test.

Step 2 The level of significance is $\alpha = 0.05$.

Step 3 Combine the two sample data sets into one data set and arrange the data in ascending order. Remember to keep track of the population from which each sample was drawn. Next assign ranks to each observation (see Table 12) and add the ranks corresponding to model X (because this group corresponds to the median on the left in the hypothesis). We obtain

$$S = 1 + 2.5 + 2.5 + \cdots + 27 + 37 + 39 = 321$$

Table 12

Model	Time to Failure	Rank	Model	Time to Failure	Rank	Model	Time to Failure	Rank
X	3	1	X	52	14	X	128	27
X	4	2.5	Y	55	15	Y	129	28
X	4	2.5	X	59	16	Y	130	29
X	5	4	Y	62	17	Y	168	30
X	7	5.5	X	68	18	Y	200	31
X	7	5.5	Y	71	19	Y	219	32
Y	12	7	X	76	20	Y	225	33
X	17	8	X	91	21	Y	239	34
X	19	9	X	103	22	Y	245	35
X	20	10	X	109	23	Y	253	36
X	25	11	Y	115	24	X	287	37
X	28	12	Y	118	25	Y	412	38
X	33	13	Y	122	26	X	472	39

Compute the value of T.

$$T = S - \frac{n_1(n_1 + 1)}{2} = 321 - \frac{22(22 + 1)}{2} = 68$$

The test statistic is

$$z_0 = \frac{T - \frac{n_1 n_2}{2}}{\sqrt{\frac{n_1 n_2(n_1 + n_2 + 1)}{12}}} = \frac{68 - \frac{22(17)}{2}}{\sqrt{\frac{22(17)(22 + 17 + 1)}{12}}} = -3.37$$

Step 4 Because we are performing a left-tailed test and one of the sample sizes is greater than 20, we locate the critical value in Table V. For the $\alpha = 0.05$ level of significance, we obtain $-z_{0.05} = -1.645$.

Because the test statistic, $z_0 = -3.37$, is less than the critical value, $-z_{0.05} = -1.645$, we reject the null hypothesis.

Step 5 There is sufficient evidence to conclude that the median time to failure for model Y is longer than the median time to failure for model X.

• Now Work Problem 13

Technology Step-by-Step Mann–Whitney Test

Minitab

1. Enter the first-sample data in column C1. Enter the second-sample data in column C2.
2. Select the **Stat** menu and highlight **Nonparametrics**; then highlight **Mann-Whitney . . . .**
3. Enter C1 in the cell marked "First Sample." Enter C2 in the cell marked "Second Sample." Select the appropriate alternative hypothesis. Click OK.

Excel

1. Load the XLSTAT Add-in.
2. Enter the raw data into Columns A and B.
3. Select the XLSTAT menu; click Nonparametric tests. Select Comparison of two samples.
4. Place the cursor in the Sample 1 cell. Highlight the data in Column A. Place the cursor in the Sample 2 cell. Highlight the data in Column B. Select the "One column

per sample" radio button. Be sure the "Column labels" box is not checked unless you have named Columns A and B. Check the "Mann–Whitney test" box.
5. Click the Options tab. Choose the appropriate Alternative hypothesis. Set the Hypothesized difference to 0. Check the "Exact P-value" box. Click OK.

StatCrunch

1. Enter the raw data in the first two columns. Name the variables. Select **Stat**. Highlight **Nonparametrics**. Select **Mann–Whitney**.
2. Choose the column that contains the data for Sample 1 from the drop-down menu. Choose the column that contains the data for Sample 2 from the drop-down menu. Enter 0 for the Null Median. Choose the appropriate alternative hypothesis. Click Compute!

15.5 Assess Your Understanding

Skill Building

In Problems 1–8, use the Mann–Whitney test to test the given hypotheses at the $\alpha = 0.05$ level of significance. The independent samples were obtained randomly.

1. Hypotheses: H_0: $M_x = M_y$ versus H_1: $M_x \neq M_y$ with $n_1 = 12, n_2 = 15$, and $S = 170$

2. Hypotheses: H_0: $M_x = M_y$ versus H_1: $M_x \neq M_y$ with $n_1 = 10, n_2 = 8$, and $S = 100$

3. Hypotheses: H_0: $M_x = M_y$ versus H_1: $M_x < M_y$ with $n_1 = 18, n_2 = 16$, and $S = 210$

4. Hypotheses: H_0: $M_x = M_y$ versus H_1: $M_x < M_y$ with $n_1 = 13, n_2 = 17$, and $S = 180$

5. Hypotheses: H_0: $M_x = M_y$ versus H_1: $M_x > M_y$ with $n_1 = 15, n_2 = 15$, and $S = 250$

6. Hypotheses: H_0: $M_x = M_y$ versus H_1: $M_x > M_y$ with $n_1 = 12, n_2 = 15$, and $S = 220$

7. Hypotheses: H_0: $M_x = M_y$ versus H_1: $M_x \neq M_y$ with $n_1 = 22, n_2 = 25$, and $S = 590$

8. Hypotheses: H_0: $M_x = M_y$ versus H_1: $M_x > M_y$ with $n_1 = 34, n_2 = 30$, and $S = 1310$

Applying the Concepts

NW 9. Emotions The EAS Temperament Survey can be used to measure the activity, emotionality, and sociability of adults. In a study on the effects of emotions on time perception, a researcher asked subjects to complete the EAS Temperament Survey and recorded the following results for sociability subscale temperament.

Women					Men				
7	15	9	7	18	12	8	9	15	13
17	16	15	10	12	14	17	18	13	15
14	6	11	13	3	10	20	13		

Do the data indicate that the median sociability score for women is different from the median sociability score for men at the $\alpha = 0.05$ level of significance? **Note:** Since scores are discrete, the data do not follow a normal distribution.

10. Weights of Linemen A researcher wants to know if the median weight of NFL offensive tackles is higher than the median weight of NFL defensive tackles. He randomly selected 10 offensive tackles and 8 defensive tackles and obtained the following data:

Offensive Linemen				
323	295	305	380	309
320	328	313	318	305

Defensive Linemen			
289	250	305	310
295	278	300	339

Source: nfl.com

Do the data indicate that offensive tackles are heavier? Use the $\alpha = 0.05$ level of significance.

Note: A normal probability plot indicates that the weight of offensive linemen is not normally distributed.

11. Bacteria in Hospital Carpeting Researchers wanted to discover whether the median amount of bacteria in carpeted rooms was greater than that in uncarpeted rooms. To determine the amount of bacteria in a room, researchers pumped the air from the room over a Petri dish at the rate of 1 cubic foot per minute for 8 carpeted rooms and 8 uncarpeted rooms. Colonies of bacteria were allowed to form in the 16 Petri dishes. The results (bacteria/cubic foot) are presented in the following table:

Carpeted Rooms		Uncarpeted Rooms	
11.8	10.8	12.1	12.0
8.2	10.1	8.3	11.1
7.1	14.6	3.8	10.1
13.0	14.0	7.2	13.7

Source: William G. Walter and Angie Stober. "Microbial Air Sampling in a Carpeted Hospital." *Journal of Environmental Health* **30**:405, 1968.

Is the median amount of bacteria in carpeted rooms greater than the median amount of bacteria in uncarpeted rooms? Use the $\alpha = 0.05$ level of significance.

12. Rats in Space Researchers at NASA wanted to judge the effects of spaceflight on a rat's tibia. The following data represent the length (in millimeters) of the right tibia upon return from the Spacelab Life Sciences 2 flight for a flight (experimental) group and ground (control) group of six rats each.

Experimental Group		Control Group	
36.05	34.58	34.93	34.41
35.57	34.20	34.68	34.42
35.57	34.73	34.20	34.93

Source: NASA Life Sciences Data Archive

Does the evidence suggest that the median tibia length in the experimental group is greater than the median tibia length in the control group at the $\alpha = 0.1$ level of significance?

NW 13. Calcium in Rainwater An environmentalist wants to determine if the median amount of calcium (mg/L) in rainwater in Lincoln County, Nebraska, is different from that in the rainwater in Clarendon County, South Carolina. She randomly selects 22 weeks in the Nebraska location in which it rained at least once during the week and 20 weeks in the South Carolina location in which it rained at least once during the week and determines the calcium levels in the rainwater.

Lincoln County Calcium Level (mg/L)				Clarendon County Calcium Level (mg/L)			
0.11	0.41	0.19	0.33	0.06	0.12	0.14	0.10
0.09	0.33	0.67	0.20	0.09	0.29	0.14	0.21
0.21	0.20	0.75	0.42	0.14	0.10	0.12	0.16
0.09	0.22	0.19	0.25	0.16	0.41	0.08	0.13
0.07	0.34	0.30	0.47	0.03	0.08	0.09	0.12
0.30	0.46						

Source: National Atmospheric Deposition Program

Do the data indicate a difference in calcium levels at the $\alpha = 0.05$ level of significance?

14. Potassium in Rainwater An environmentalist wants to determine if the median amount of potassium (mg/L) in rainwater in Lincoln County, Nebraska, is different from that in the rainwater in Clarendon County, South Carolina. She randomly selects 22 weeks in the Nebraska location in which it rained at least once during the week and 20 weeks in the South Carolina location in which it rained at least once during the week and determines the potassium levels in the rainwater. Her data are shown in the following table:

Lincoln County Potassium Level (mg/L)				Clarendon County Potassium Level (mg/L)			
0.013	0.062	0.078	0.064	0.016	0.059	0.037	0.019
0.014	0.044	0.096	0.028	0.023	0.033	0.031	0.019
0.016	0.018	0.101	0.065	0.057	0.064	0.058	0.022
0.085	0.100	0.035	0.030	0.118	0.025	0.035	0.051
0.011	0.116	0.043	0.054	0.024	0.069	0.027	0.053
0.005	0.080						

Source: National Atmospheric Deposition Program

Do the data indicate a difference in potassium levels at the $\alpha = 0.05$ level of significance?

15. Mann–Whitney Using Ordinal Data The Mann–Whitney test can be performed on ordinal data. For example, a letter grade received in a class is ordinal data because it can be ranked: an A ranks higher than a B. A department chair wants to discover whether the grades of students in two different teachers' statistics courses are different. The chair randomly selects 15 students from professor A's class and 15 students from professor B's class and obtains the following data.

Professor A					Professor B				
C	D	F	C	C	B	B	A	A	C
B	B	D	C	A	B	B	C	D	A
A	C	B	D	C	B	C	F	C	B

Do the following to test the belief that the grades administered in each class are equivalent.

(a) Rank the grades in descending order (A's first, then B's, and so on.)

(b) Perform a two-tailed test on the hypotheses

H_0: the grades administered in each class are equivalent
H_1: the grades administered in each class are different

by following Steps 1 through 5 (boxplots cannot be drawn).

16. Mann–Whitney Using Ordinal Data See Problem 15. A restaurant critic ranks restaurants as excellent, good, fair, or poor. We want to judge whether two restaurant critics have the same distribution of opinions. The opinions of critic A regarding 15 randomly selected restaurants are presented together with the opinions of critic B regarding 15 randomly selected restaurants (the samples are independent).

Critic A				
Excellent	Good	Fair	Fair	Poor
Excellent	Good	Good	Fair	Fair
Fair	Good	Excellent	Poor	Fair

Critic B				
Fair	Fair	Good	Good	Fair
Fair	Poor	Fair	Good	Good
Excellent	Fair	Good	Fair	Poor

Do the following to test the belief that the opinions of each critic are equivalent.

(a) Rank the opinions in ascending order (excellent first, then good, and so on).

(b) Perform a two-tailed test on the hypotheses

H_0: the opinions given are equivalent
H_1: the opinions given are different

by following Steps 1 through 5 (boxplots cannot be drawn).

Explaining the Concepts

17. Explain the rationale behind the test statistic for the Mann–Whitney test.

18. Explain the rationale behind the decision rule for the Mann–Whitney test.

19. Explain the primary difference between the Wilcoxon signed-ranks test and the Mann–Whitney test.

20. For the large-sample case, the test statistic for the Mann–Whitney test follows a normal distribution. Explain why this test is still considered a nonparametric test.

15.6 Spearman's Rank-Correlation Test

Preparing for This Section Before getting started, review the following:

- Linear correlation coefficient (Section 4.1, pp. 183–188)
- Tests for a linear relation (Section 14.1, pp. 678–681)

Objective ➊ Perform Spearman's rank-correlation test

In Section 4.1, we introduced Pearson's linear correlation coefficient. Recall that the correlation coefficient measures the strength of the linear relation between two quantitative variables that are collected as ordered pairs [that is, $(x_1, y_1), (x_2, y_2), \ldots, (x_n, y_n)$].

In Section 14.1, we learned that the data had to be bivariate normal to test hypotheses regarding Pearson's correlation coefficient. This requirement is difficult to verify. We therefore tested the linear relation between the two variables by determining whether the slope coefficient, β_1, was significantly different from 0. But this test required that the residuals be normally distributed. In this section, we introduce the nonparametric *Spearman's rank-correlation test*, which can be used to test the hypothesis that two variables are associated.

① Perform Spearman's Rank-Correlation Test

Definition   The **Spearman's Rank-Correlation Test** is a nonparametric procedure used to test hypotheses regarding the association between two variables.

When performing the rank-correlation test, we test the hypothesis that an association exists between the two variables. There are three ways to structure the hypothesis test. Assume that the data are represented as ordered pairs (X, Y).

Two-Tailed	One-Tailed	One-Tailed
H_0: X and Y are not associated	H_0: X and Y are not associated	H_0: X and Y are not associated
H_1: X and Y are associated	H_1: X and Y are positively associated	H_1: X and Y are negatively associated

X	Y	Rank of X-values	Rank of Y-values
5	8	2	4
3	4	1	1
7	6	3	2.5
9	9	5	5
8	6	4	2.5

The idea behind the rank correlation test is to rank the X-values and then separately rank the Y-values, from smallest to largest. Handle ties by finding the mean of the ranks for tied values. For example, if the bivariate data are $(5, 8), (3, 4), (7, 6), (9, 9),$ and $(8, 6)$, we form a table of X-values and their corresponding Y-values. Rank the X-values and then rank the Y-values. The smallest X-value is 3, so we assign it rank 1; the second smallest X-value is 5, so we assign it rank 2; and so on. The smallest Y-value is 4, so we assign it rank 1. The next smallest Y-value is 6, but it occurs twice, so we assign the value 6 the rank $\dfrac{2 + 3}{2} = 2.5$; and so on.

Once the ranks have been identified, compute the differences between them by subtracting the rank value of Y from the rank value of X. If there is a positive association between X and Y, the small values of X will tend to be paired with small values of Y; so their ranks are both small, resulting in a small difference. Large values of X will tend to be paired with large values of Y, so their ranks are both large. Again, the difference between the ranks will be close to zero for each ordered pair. If there is a negative association between X and Y, the small values of X will tend to be paired with the large values of Y (and vice versa); so when one rank is small, the other will be large. Thus, we expect to see large differences in ranks for data in the extremes and smaller differences for data in the middle. The test statistic is based on this rationale.

Test Statistic for Spearman's Rank-Correlation Test

The test statistic depends on the size of the sample, n, and on the sum of the squared differences and is given by

$$r_s = 1 - \frac{6\sum d_i^2}{n(n^2 - 1)}$$

where d_i is the difference in the ranks of the two observations in the ith ordered pair.

CAUTION!
$\sum d_i^2$ means to square the differences first and then add up the squared differences.

Historical Note

Charles Spearman was born in London in 1863. After earning a degree at Leamington College, he joined the army in 1883. After a very good career, he resigned with the rank of captain in 1897 to pursue additional studies. He earned his Ph.D. in experimental psychology in 1906 from the University of Leipzig. He is most famous for his *two-factor theory of intelligence*. This theory basically states that the ability to perform an intellectual task is based on an individual's general intelligence, *g*, and specific intelligence, *s*, for the task. The rank-correlation coefficient is named in honor of Spearman, who first published the idea in 1904 in an article entitled "The Proof and Measurement of Association between Two Things" in the *American Journal of Psychiatry*. Spearman died in 1945 in London.

The test statistic, r_s, is also called **Spearman's rank-correlation coefficient**.

Critical Value for Spearman's Rank-Correlation Test

Using α as the level of significance, the critical value(s) is (are) obtained from Table XV. For a two-tailed test, be sure to divide the level of significance, α, by 2.

We now present the steps required to conduct a Spearman's rank-correlation test.

Spearman's Rank-Correlation Test

To test hypotheses regarding the association between two variables X and Y, use the following steps, provided that

1. the data are a random sample of n ordered pairs, and
2. each pair of observations is two measurements taken on the same individual.

Notice that there is no requirement about the form of the distribution of the data.

Step 1 Determine the null and alternative hypotheses, which are structured as follows:

Two-Tailed	One-Tailed	One-Tailed
H_0: X and Y are not associated	H_0: X and Y are not associated	H_0: X and Y are not associated
H_1: X and Y are associated	H_1: X and Y are positively associated	H_1: X and Y are negatively associated

Step 2 Choose the level of significance, α, based on the seriousness of making a Type I error.

Step 3 Rank the X-values and rank the Y-values. Compute the differences between ranks and then square these differences. Compute the sum of the squared differences. Compute the test statistic

$$r_s = 1 - \frac{6\sum d_i^2}{n(n^2 - 1)}$$

where n is the sample size and d_i is the difference in the ranks of the two observations in the ith ordered pair.

Step 4 Find the critical value in Table XV. Compare the critical value with the test statistic.

Hypothesis	Decision Rule
H_0: X and Y are not associated H_1: X and Y are associated	Reject H_0 if r_s is greater than the critical value in Table XV or less than the negative of the critical value in Table XV.
H_0: X and Y are not associated H_1: X and Y are positively associated	Reject H_0 if r_s is greater than the critical value in Table XV.
H_0: X and Y are not associated H_1: X and Y are negatively associated	Reject H_0 if r_s is less than the negative of the critical value in Table XV.

Step 5 State the conclusion.

EXAMPLE 1   Spearman's Rank-Correlation Test

Table 13	
Club-Head Speed (mph)	**Distance (yards)**
100	257
102	264
103	274
101	266
105	277
100	263
99	258
105	275

Problem A golf pro wanted to learn the relation between the club-head speed of a golf-club (measured in miles per hour) and the distance the ball travels. He realized that there are other variables besides club-head speed that determine the distance a ball travels (such as club type, ball type, golfer, and weather conditions). To eliminate the variability due to these variables, the pro used a single model of club and ball. One golfer was chosen to swing the club on a clear, 70° day with no wind. The pro recorded the club-head speed and measured the distance that the ball traveled and collected the data given in Table 13. The scatter diagram is in Figure 1 on page 182. Are club-head speed and distance associated at the $\alpha = 0.05$ level of significance?

Approach Verify the requirements and follow Steps 1 through 5.

Solution Treat each swing as a simple random sample of all swings by the individual. Each pair of observations is taken for the same swing.

Step 1 We are looking for evidence that club-head speed and distance are associated. Let X represent club-head speed and Y represent distance. The null and alternative hypotheses are as follows:

$$H_0: X \text{ and } Y \text{ are not associated}$$
$$H_1: X \text{ and } Y \text{ are associated}$$

Step 2 The level of significance is $\alpha = 0.05$.

Step 3 Rank the X-values and rank the Y-values (columns 3 and 4 of Table 14). Compute the differences in ranks (column 5), and then square the differences (column 6). Compute the sum of the squared differences.

Table 14					
Club-Head Speed (mph)	**Distance (yards)**	**Rank of X**	**Rank of Y**	$d_i = X - Y$	d_i^2
100	257	2.5	1	1.5	2.25
102	264	5	4	1	1
103	274	6	6	0	0
101	266	4	5	−1	1
105	277	7.5	8	−0.5	0.25
100	263	2.5	3	−0.5	0.25
99	258	1	2	−1	1
105	275	7.5	7	0.5	0.25

Adding up the entries in column 6, we find that

$$\Sigma d_i^2 = 2.25 + 1 + 0 + 1 + 0.25 + 0.25 + 1 + 0.25 = 6$$

Compute the test statistic.

$$r_s = 1 - \frac{6\Sigma d_i^2}{n(n^2 - 1)} = 1 - \frac{6(6)}{8(8^2 - 1)} = 0.929$$

Step 4 The level of significance is given as $\alpha = 0.05$. In Table XV, we look for the row that corresponds to $n = 8$ (the 8 swings) and the column with $\alpha = 0.05$ for a two-tailed test. The critical value is 0.738. See Figure 12.

Figure 12

Critical Values of Spearman's Rank Correlation Coefficient								
α(2): 0.50	0.20	0.10	0.05	0.02	0.01	0.005	0.002	0.001
α(1): 0.25	0.10	0.05	0.025	0.01	0.005	0.0025	0.001	0.0005
n								
4 0.600	1.000	1.000						
5 0.500	0.800	0.900	1.000	1.000				
6 0.371	0.657	0.829	0.886	0.943	1.000	1.000		
7 0.321	0.571	0.714	0.786	0.893	0.929	0.964	1.000	1.000
8 0.310	0.524	0.643	0.738	0.833	0.881	0.905	0.952	0.976
9 0.267	0.483	0.600	0.700	0.783	0.833	0.867	0.917	0.933
10 0.248	0.455	0.564	0.648	0.745	0.794	0.830	0.879	0.903
11 0.236	0.427	0.536	0.618	0.709	0.755	0.800	0.845	0.873

Compare the critical value with the test statistic. Because the test statistic, $r_s = 0.929$, is greater than the critical value, 0.738, we reject the null hypothesis.

Step 5 There is sufficient evidence to conclude that club-head speed and distance are associated. ●

Figure 13 shows the results of Spearman's Rank Correlation Test using Statcrunch. The *P*-value 0.0009 suggests there is an association between club-head speed and distance.

Using Technology

Although Minitab and the TI-83/84 Plus graphing calculator do not have the ability to compute Spearman's rank correlation directly, it can be done indirectly by listing the ranks of the *X*- and *Y*-values and then finding the linear correlation of the ranks.

● **Now Work Problem 5**

Figure 13   Spearman's correlation between Club-Head Speed and Distance is:
0.92777818(0.0009)

Notice that critical values do not exist for $n > 100$ in Table XV. In this case, we can utilize the *large-sample approximation*.

Large-Sample ($n > 100$) Approximation

If $n > 100$, the test statistic for Spearman's rank-correlation test is

$$z_0 = r_s \sqrt{n - 1}$$

Compare this test statistic with the critical value obtained from the standard normal table, Table V, in Appendix A. For a two-tailed test, the critical values are $\pm z_{\alpha/2}$. When testing for positive association, the critical value is z_α. When testing for negative association, the critical value is $-z_\alpha$.

Technology Step-by-Step Spearman's Rank Correlation Test

StatCrunch

Spearman's Rank Correlation Test
1. Enter the raw data into the first two columns. Name the columns.

2. Select **Stat**, highlight **Nonparametrics**, select **Spearman's Correlation**.
3. Select the variables under consideration under "Select column(s)." Check the box "Two-sided *P*-value." Click Compute!

15.6 Assess Your Understanding

Skill Building

In Problems 1–4, (a) draw a scatter diagram, (b) compute r_s, and (c) determine if X and Y are associated at the α = 0.05 level of significance.

1.

X	2	4	8	8	9
Y	1.4	1.8	2.1	2.3	2.6

2.

X	0	1.2	1.8	2.3	3.5	4.1
Y	2	6.4	7	7.3	4.5	2.4

3.

X	0	2	3	5	6	6
Y	5.8	5.7	5.2	2.8	1.9	2.2

4.

X	0	0.5	1.4	1.4	3.9	4.6
Y	0.8	2.3	1.9	2.5	5.0	6.8

Applying the Concepts

NW **5. Income versus Education** Does the notion that the more education an individual receives, the higher the person's income is likely to be translate to states? The following data represent the percentage of the population that has at least a bachelor's degree and the per capita personal income for a random sample of states.

State	Percentage of Population with at Least a Bachelor's Degree	Per Capita Personal Income (Dollars)
Arkansas	18.8	31,946
California	29.6	42,325
Iowa	24.3	36,751
Louisiana	20.3	35,507
Maine	25.4	36,745
Nebraska	27.1	38,081
Ohio	24.1	35,381
Oregon	28.1	35,667
Texas	25.3	36,484

Source: Statistical Abstract of the United States

(a) Do states with higher percentages of population with at least a bachelor's degree have higher per capita personal incomes at the $\alpha = 0.05$ level of significance?

(b) Draw a scatter diagram to support your conclusion.

6. Crime Rate versus Population Density Is a state with a higher population density likely to have a higher violent crime rate? The following data represent the population density (people per square mile) and violent crime rate (crimes per 100,000 population) for a random sample of states.

State	Population Density	Crime Rate
Alabama	92.8	448
Colorado	48.4	348
Illinois	232.3	533
Kansas	34.5	453
Minnesota	66.2	289
North Carolina	192.6	466
South Dakota	10.7	169
Virginia	199.1	270

Source: Statistical Abstract of the United States

(a) Do states with higher population densities have higher crime rates at the $\alpha = 0.05$ level of significance?

(b) Draw a scatter diagram to support your conclusion.

7. Per Capita Personal Income versus Birthrate A sociologist believes that as the per capita personal incomes of states increase the birthrates decrease. She randomly selects eight states and the District of Columbia and obtains the following data:

State	Per Capita Personal Income ($)	Birthrate
Delaware	39,817	14.1
Maryland	48,285	13.9
District of Columbia	66,000	15.1
Virginia	43,874	14.1
West Virginia	32,219	12.1
North Carolina	34,453	14.5
South Carolina	31,799	14.3
Georgia	33,786	15.8
Florida	37,780	13.1

Source: Statistical Abstract of the United States

(a) Do the data support the sociologist's belief? Use the $\alpha = 0.05$ level of significance.

(b) Draw a scatter diagram to support your conclusion.

8. Does Defense Win? "Defense wins championships" is a common phrase used in the National Football League. Is defense associated with winning? The following data represent the winning percentage and the yards per game allowed during the 2014–2015 season for a random sample of teams.

Team	Winning Percentage	Total Yards
Baltimore Ravens	0.625	336.9
Cleveland Browns	0.438	366.1
Denver Broncos	0.750	305.2
Jacksonville Jaguars	0.188	370.8
New England Patriots	0.750	344.1
Oakland Raiders	0.188	357.6
Pittsburgh Steelers	0.688	353.4

(a) Test the belief that defense wins championships by determining whether a higher winning percentage is associated with a lower number of total yards given up at the $\alpha = 0.10$ level of significance.

(b) Draw a scatter diagram to support your conclusion.

9. Top Cities Every year *Money* magazine publishes its list of top places to live. The following data represent a list of top places to live for a recent year, along with the median family income and median commute time.

City	Family Income ($1000s)	Commute Time (minutes)
Woodridge, Illinois	83	27.1
Urbandale, Idaho	82	17.0
La Palma, California	86	26.9
Friendswood, Texas	90	26.0
Suwanee, Georgia	101	32.1
Somers, Connecticut	83	22.6

Source: Money magazine

(a) Does a positive association exist between income and commute time at the $\alpha = 0.10$ level of significance?

(b) Draw a scatter diagram to support your conclusion.

10. College Football Polls A sports reporter wants to determine if the preseason Associated Press (AP) poll is positively related to the final AP poll. The following data represent the preseason and final AP rankings for a random sample of teams for the 2014 college football season.

Team	Preseason AP Rank	Final AP Rank
Auburn	6	22
Wisconsin	14	13
Florida	27	41
Georgia	12	9
Missouri	24	11
Michigan	37	81
Texas A&M	21	34
University of Southern California	15	21
Ohio State	5	1

Source: espn.com

(a) Does a positive relationship exist between preseason AP rank and final AP rank? Use the $\alpha = 0.05$ level of significance.

(b) Draw a scatter diagram to support your conclusion.

For Problems 11 and 12, Compute r_s for the following data. What type of relation exists between X and Y? Draw a scatter diagram to confirm your results.

11.

X	Y	X	Y
2	5	5	14
3	8	6	17
4	11		

12.

X	Y	X	Y
2	23	5	11
3	19	6	7
4	15		

Explaining the Concepts

13. The Pearson correlation coefficient requires that the data be quantitative. Does the Spearman rank correlation require that data be quantitative? Explain.

14. Provide an intuitive explanation of how the Spearman rank correlation measures association.

15.7 Kruskal–Wallis Test

Preparing for This Section Before getting started, review the following:

- Median (Section 3.1, pp. 120–121)
- Boxplots (Section 3.5, pp. 166–168)
- One-way analysis of variance (Section 13.1, pp. 618–628)

Objective   ① Test a hypothesis using the Kruskal–Wallis test

In Section 13.1, we introduced one-way analysis of variance (ANOVA). This is used to test the null hypothesis that the means of different populations are equal. To perform the one-way ANOVA test, we require that the populations have equal population variances and that each population be normally distributed. If either of these requirements is not satisfied, we cannot perform the test.

① Test a Hypothesis Using the Kruskal–Wallis Test

Definition   The **Kruskal–Wallis Test** is a nonparametric procedure that is used to test whether k independent samples come from populations with the same distribution.

Recall that the null hypothesis in one-way ANOVA is stated as

$$H_0: \mu_1 = \mu_2 = \cdots = \mu_k$$

versus the alternative hypothesis

H_1: at least one of the means is not equal to the others

The null hypothesis in the Kruskal–Wallis test is stated as

H_0: the distributions of the populations are identical

versus

H_1: the distributions of the populations are not identical

Historical Note

In 1952, W. H. Kruskal and W. A. Wallis published "Use of Ranks in One-Criterion Variance Analysis" in the *Journal of the American Statistical Association*. This article presented a nonparametric version of one-way ANOVA.

The idea behind the Kruskal–Wallis test is to combine the samples from the different populations and rank *all* the observations from smallest to largest. We handle ties by finding the mean of the ranks for tied values. For example, if the data for samples from three populations are

Sample 1	Sample 2	Sample 3
5.1	7.6	7.2
6.5	5.2	8.7
7.2	4.7	6.2

we combine and rank the data as

Sample	Combined Data	Rank
2	4.7	1
1	5.1	2
2	5.2	3
3	6.2	4
1	6.5	5
1	7.2	6.5
3	7.2	6.5
2	7.6	8
3	8.7	9

Notice that we observed 7.2 twice. To find the rank of 7.2, we recognize that the two occurrences occupy the sixth- and seventh-ranking positions. So the mean rank is $\dfrac{6+7}{2} = 6.5$.

Once the ranks have been identified, sum the ranks of the observations from each sample. The sum of the ranks from sample 1 is $2 + 5 + 6.5 = 13.5$. If the populations have the same distribution (if the null hypothesis is true), we expect the sums of the ranks for all the samples to be close to each other. The test statistic is based on this rationale.

Test Statistic for the Kruskal–Wallis Test

The test statistic for the Kruskal–Wallis test is

$$H = \frac{12}{N(N+1)}\sum \frac{1}{n_i}\left[R_i - \frac{n_i(N+1)}{2}\right]^2 \tag{1}$$

A computational formula for the test statistic is

$$H = \frac{12}{N(N+1)}\left[\frac{R_1^2}{n_1} + \frac{R_2^2}{n_2} + \cdots + \frac{R_k^2}{n_k}\right] - 3(N+1) \tag{2}$$

where

R_i is the sum of the ranks of the *i*th sample

R_1^2 is the sum of the ranks squared for the first sample

R_2^2 is the sum of the ranks squared for the second sample, and so on

n_1 is the number of observations in the first sample

n_2 is the number of observations in the second sample, and so on

N is the total number of observations ($N = n_1 + n_2 + \cdots + n_k$)

k is the number of populations being compared.

The Kruskal–Wallis test is always a right-tailed test, so we reject the null hypothesis if H is sufficiently large. We use Formula (1) to justify the rejection of the null hypothesis if H is sufficiently large. In Formula (1), the expression $\dfrac{n_i(N+1)}{2}$ is the expected sum of ranks for the ith population if the null hypothesis is true. To compute H, square the deviation between the actual sum of ranks, R_i, and the expected sum of ranks. If the sum of the squared deviations is large, we have evidence against the null hypothesis.

Now that we know the procedure for obtaining the test statistic, we want to obtain a critical value.

Critical Value for the Kruskal–Wallis Test

Small-Sample Case
When three populations are being compared and when the sample size from each population is 5 or less, the critical value is obtained from Table XVI.

Large-Sample Case
When four or more populations are being compared or the sample size from one population is more than 5, the critical value is χ^2_α with $k-1$ degrees of freedom, where k is the number of populations being compared and α is the level of significance.

We now present the steps required to conduct a Kruskal–Wallis test.

Kruskal–Wallis Test

To test hypotheses regarding the distribution of three or more populations, use the following steps, provided that two requirements are satisfied:

1. The samples are independent random samples.
2. The data can be ranked.

Step 1 Draw side-by-side boxplots to compare the sample data from the populations. Doing so helps to visualize the differences, if any, between the medians.
State the null and alternative hypotheses, which are structured as follows:

H_0: the distributions of the populations are the same

H_1: the distributions of the populations are not the same

Step 2 Choose a level of significance, α, depending on the seriousness of making a Type I error.

By-Hand Approach

Step 3 Rank all sample observations from smallest to largest. Handle ties by finding the mean of the ranks for tied values. Find the sum of the ranks for each sample.
Compute the **test statistic**.

$$H = \frac{12}{N(N+1)}\left[\frac{R_1^2}{n_1} + \frac{R_2^2}{n_2} + \cdots + \frac{R_k^2}{n_k}\right] - 3(N+1)$$

Step 4 The level of significance is used to determine the critical value. The critical value is found from Table XVI for small samples. The critical value is χ^2_α with $k-1$ degrees of freedom (found in Table VIII) for large samples.
Compare the critical value to the test statistic. Reject the null hypothesis if the test statistic is greater than the critical value.

Technology (*P*-Value) Approach

Step 3 Use a statistical spreadsheet to obtain the P-value. The directions for obtaining the P-value using Minitab, Excel, and StatCrunch are in the Technology Step-by-Step on page 787.

Step 4 If P-value $< \alpha$, reject the null hypothesis.

Step 5 State the conclusion.

EXAMPLE 1 Kruskal–Wallis Test

Problem A family doctor wants to determine if the distributions of HDL cholesterol in males for the age groups 20 to 29 years, 40 to 49 years, and 60 to 69 years old are different. He obtains a simple random sample of 12 individuals from each age group and determines their HDL cholesterol. The results are presented in Table 15. Do the data indicate the distributions vary depending on age? Use the $\alpha = 0.05$ level of significance.

Table 15			
Subject Number	**Age (yr)**		
	20 to 29	**40 to 49**	**60 to 69**
1	54 (29)	61 (31.5)	44 (18)
2	43 (16)	41 (14)	65 (34.5)
3	38 (11.5)	44 (18)	62 (33)
4	30 (2)	47 (21)	53 (27.5)
5	61 (31.5)	33 (3)	51 (26)
6	53 (27.5)	29 (1)	49 (22.5)
7	35 (7.5)	59 (30)	49 (22.5)
8	34 (4.5)	35 (7.5)	42 (15)
9	39 (13)	34 (4.5)	35 (7.5)
10	46 (20)	74 (36)	44 (18)
11	50 (24.5)	50 (24.5)	37 (10)
12	35 (7.5)	65 (34.5)	38 (11.5)

Approach Verify that the requirements needed to perform the Kruskal–Wallis test are satisfied. We then proceed to follow Steps 1 through 5.

Solution The samples obtained are independent random samples.

Step 1 Figure 14 shows side-by-side boxplots of the sample data. Notice that the sample median for the 20- to 29-year-olds is less than the sample medians of the other two groups. Is this difference statistically significant or attributable to random chance?

Figure 14

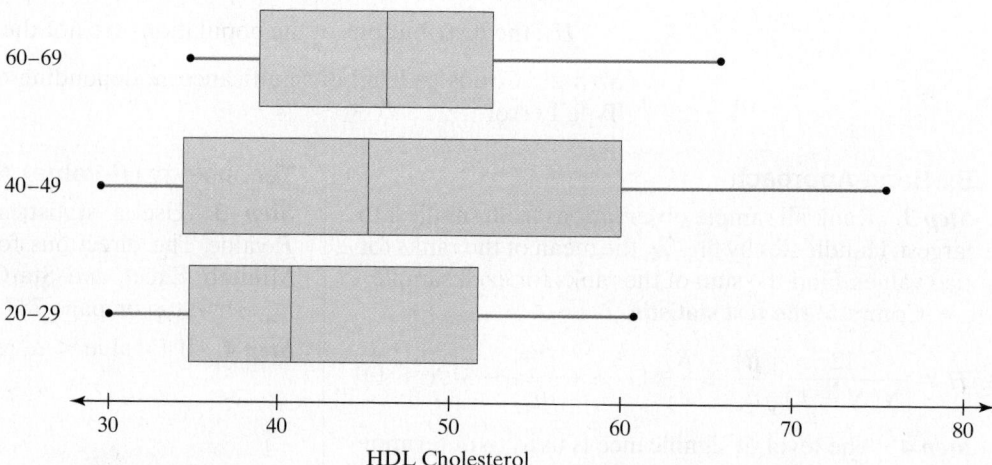

The doctor is looking for evidence that the distributions of HDL cholesterol in males for the age groups 20 to 29, 40 to 49, and 60 to 69 years old are not the same, so

H_0: the distributions of HDL cholesterol for the three age groups are the same

H_1: the distributions of HDL cholesterol for the three age groups are not the same

Step 2 The level of significance is $\alpha = 0.05$.

By-Hand Approach

Step 3   The ranks of the data are presented in parentheses in Table 15. The smallest data value is 29, and we assign this observation the rank 1. Sum the ranks in the first sample (20- to 29-year-olds) to obtain R_1, so $R_1 = 29 + 16 + \cdots + 7.5 = 194.5$. Similar computations are performed to obtain $R_2 = 225.5$ and $R_3 = 246$. The results are summarized as follows:

	Age (yr)		
	20 to 29	**40 to 49**	**60 to 69**
Sample size	$n_1 = 12$	$n_2 = 12$	$n_3 = 12$
Sum of ranks	$R_1 = 194.5$	$R_2 = 225.5$	$R_3 = 246$

Note that $N = n_1 + n_2 + n_3 = 36$. The test statistic is computed as follows:

$$H = \frac{12}{N(N+1)}\left[\frac{R_1^2}{n_1} + \frac{R_2^2}{n_2} + \frac{R_3^2}{n_3}\right] - 3(N+1)$$

$$= \frac{12}{36(36+1)}\left[\frac{194.5^2}{12} + \frac{225.5^2}{12} + \frac{246^2}{12}\right] - 3(36+1)$$

$$= 1.009$$

Step 4   Because the sample size for each population is greater than 5, we find the critical value from the chi-square distribution with $k - 1 = 3 - 1 = 2$ degrees of freedom, with $\alpha = 0.05$. Thus, the critical value is $\chi_{0.05}^2 = 5.991$.

Because the test statistic, 1.009, is not greater than the critical value, 5.991, we do not reject the null hypothesis.

Technology Approach

Step 3   Figure 15 shows the results from Minitab. The P-value is 0.604.

Figure 15

Kruskal-Wallis Test

```
C4            N     Median    Ave Rank          Z
1            12      41.00        16.2      -0.92
2            12      45.50        18.8       0.12
3            12      46.50        20.5       0.81
Overall      36                   18.5

H = 1.01    df = 2    P = 0.604
H = 1.01    df = 2    P = 0.603   (adjusted for ties)
```

Step 4   Since the P-value is greater than the level of significance ($0.604 > 0.05$), we do not reject the null hypothesis.

Step 5   There is not sufficient evidence to indicate that the distributions of HDL cholesterol for the three age groups are different.

● **Now Work Problem 3**

Technology Step-by-Step   Kruskal–Wallis Test

Minitab
1. Enter the level of the factor in column C1. Enter the value of the response variable in column C2.
2. Select the **Stat** menu, highlight **Nonparametrics**, then highlight **Kruskal–Wallis** . . . . Place the cursor in the Response cell and select the column containing the response variable. Place the cursor in the Factor cell and select the column containing the factor. Click OK.

Excel
1. Load the XLSTAT Add-in.
2. Enter the level of the factor in Column A. Enter the value of the response in Column B.
3. Select the XLSTAT menu. Highlight Nonparametric tests. Select Comparison of k samples.
4. Place the cursor in the Data cell and highlight the data in Column B. Place the cursor in the Sample identifiers cell and highlight the data in Column A. Be sure the One column per variable radio button is selected. Check the Kruskal–Wallis test box. Click OK.

StatCrunch
1. Enter the level of the factor in the first column. Enter the value of the response in the second column.
2. Select **Stat**. Highlight **Nonparametrics**. Select **Kruskal–Wallis**.
3. Check the Values in a single column radio button under Compare. Choose the appropriate columns for the response and factor. Click Compute!

15.7 Assess Your Understanding

Skill Building

In Problems 1 and 2, (a) determine the test statistic, H, (b) determine the critical value at the $\alpha = 0.05$ level of significance, and (c) test whether the distributions of the populations are different.

1.

X	Y	Z
13	16	12
9	18	14
17	11	9
12	13	15

2.

X	Y	Z
7	9	4
4	4	5
3	8	10
3		7

Applying the Concepts

NW **3. Births by Day of Week** An obstetrician knew that there were more live births during the week than on weekends and wanted to discover whether the distribution for number of births was the same for each of the 5 days of the week. She randomly selected eight dates for each of the 5 days of the week and obtained the data shown.

Monday	Tuesday	Wednesday	Thursday	Friday
10,456	11,621	11,084	11,171	11,545
10,023	11,944	11,570	11,745	12,321
10,691	11,045	11,346	12,023	11,749
10,283	12,927	11,875	12,433	12,192
10,265	12,577	12,193	12,132	12,422
11,189	11,753	11,593	11,903	11,627
11,198	12,509	11,216	11,233	11,624
11,465	13,521	11,818	12,543	12,543

Source: National Center for Health Statistics

(a) State the null and alternative hypotheses.
(b) The sums of the ranks in each category are as follows:

	Monday	Tuesday	Wednesday	Thursday	Friday
Sample size	8	8	8	8	8
Sum of ranks	48	226	144	194.5	207.5

Use this information to compute the test statistic.
(c) What is the critical value at the $\alpha = 0.05$ level of significance?
(d) State your conclusion.
(e) Draw side-by-side boxplots to support your conclusion.

4. Births by Season An obstetrician wants to determine if the distribution of births is the same for the periods January to March, April to June, July to September, and October to December. She randomly selected 10 days within each time frame and obtained the results shown in the table in the next column.

January–March	April–June	July–September	October–December
10,456	10,799	11,465	11,261
11,574	11,743	12,475	11,406
12,321	11,657	12,454	11,627
11,661	11,608	12,193	8,706
8,521	10,982	11,244	12,022
11,621	9,184	12,081	11,930
11,321	12,357	11,387	11,054
7,706	12,251	9,055	11,431
10,872	11,916	12,319	12,618
10,837	11,212	12,927	10,824

Source: National Center for Health Statistics

(a) State the null and alternative hypotheses.
(b) The sums of the ranks in each category are as follows:

	January–March	April–June	July–September	October–December
Sample size	10	10	10	10
Sum of ranks	149	207	267	197

Use this information to compute the test statistic.
(c) What is the critical value at the $\alpha = 0.05$ level of significance?
(d) State your conclusion.
(e) Draw side-by-side boxplots to support your conclusion.

5. Corn Production The following data represent the number of corn plants in randomly sampled rows (a 17-foot by 5-inch strip) for various types of plot.

Plot Type	Number of Plants					
Sludge plot	25	27	33	30	28	27
Spring disk	32	30	33	35	34	34
No till	30	26	29	32	25	29
Spring chisel	30	32	26	28	31	29
Great lakes bt	28	32	27	30	29	27

Source: Andrew Dieter and Brad Schmidgall, Joliet Junior College

Do the data indicate that the distribution for each type of plot is the same at the $\alpha = 0.05$ level of significance?

6. Soybean Yield The following data represent the number of pods on a random sample of soybean plants for various types of plot.

Plot Type	Pods								
Liberty	32	31	36	35	41	34	39	37	38
Fall plowed	29	31	33	22	19	30	36	30	
No till	34	30	31	27	40	33	37	42	39
Chisel plowed	34	37	24	23	32	33	27	34	30
Round-up ready	27	34	36	32	35	29	35		

Source: Andrew Dieter and Brad Schmidgall, Joliet Junior College

Do the data indicate that the distribution for each type of plot is the same at the $\alpha = 0.05$ level of significance?

7. Reaction Time In an online psychology experiment sponsored by the University of Mississippi, researchers asked study participants to respond to various stimuli. Participants were randomly assigned to one of three groups. Subjects in the simple group were required to respond as quickly as possible after a stimulus was presented. Subjects in the go/no-go group were required to respond to a particular stimulus while disregarding other stimuli. Finally, subjects in the choice group needed to respond differently to different stimuli: depending on the type of whistle sound, the subject had to press a certain button. The reaction time (in seconds) for each stimulus is presented in the table below. Is the distribution for each stimulus the same at the $\alpha = 0.01$ level of significance?

Simple	Go/No-Go	Choice
0.430	0.588	0.561
0.498	0.375	0.498
0.480	0.409	0.519
0.376	0.613	0.538
0.402	0.481	0.464
0.329	0.355	0.725

Source: PsychExperiments; University of Mississippi, www.olemiss.edu/psychexps/

8. Math Scores Researchers wanted to compare math test scores of students at the end of secondary school from various countries. Eight randomly selected students from Canada, Denmark, and the United States each were administered the same exam; the results are presented in the following table. Can the researchers conclude that the distribution of exam scores is the same for each country at the $\alpha = 0.01$ level of significance?

Canada		Denmark		United States	
578	548	568	563	506	458
548	530	530	535	518	456
521	502	571	561	485	513
555	492	569	513	480	491

Source: Based on data obtained from the International Association for the Evaluation of Educational Achievement

For Problems 9 and 10, use the following scenario. The Insurance Institute for Highway Safety conducts experiments in which cars are crashed into a fixed barrier at 40 mph. In a 40-mph offset test, 40% of the total width of each vehicle strikes a barrier on the driver side. The barrier's deformable face is made of aluminum honeycomb, which makes the forces in the test similar to those involved in a frontal offset crash between two vehicles of the same weight, each going just less than 40 mph.

9. Crash Data You are in the market to buy a new family car and want to know whether the distribution of chest compression (mm) resulting from this crash is the same for each vehicle category at the $\alpha = 0.01$ level of significance. The following data were collected from the study.

Large Family Cars	Chest Compression	Passenger Vans	Chest Compression	Midsize Utility Vehicles	Chest Compression
Hyundai XG300	33	Toyota Sienna	29	Honda Pilot	29
Ford Taurus	40	Honda Odyssey	28	Toyota 4Runner	36
Buick LeSabre	28	Ford Freestar	27	Mitsubishi Endeavor	35
Chevrolet Impala	30	Mazda MPV	30	Nissan Murano	32
Chrysler 300	34	Chevrolet Uplander	26	Ford Explorer	34
Pontiac Grand Prix	34	Nissan Quest	33	Jeep Liberty	42
Toyota Avalon	31	Kia Sedona	36	Buick Rendezvous	29

Source: Insurance Institute for Highway Safety

10. Crash Data You are in the market to buy a new family car and want to know whether the distribution of head injury resulting from this offset crash is the same for large family cars, passenger vans, and midsize utility vehicles at the $\alpha = 0.01$ level of significance. The following data were collected from the study.

Large Family Cars	Head Injury (hic)	Passenger Vans	Head Injury (hic)	Midsize Utility Vehicles	Head Injury (hic)
Hyundai XG300	264	Toyota Sienna	148	Honda Pilot	225
Ford Taurus	134	Honda Odyssey	238	Toyota 4Runner	216
Buick LeSabre	409	Ford Freestar	340	Mitsubishi Endeavor	186
Chevrolet Impala	530	Mazda MPV	693	Nissan Murano	307
Chrysler 300	149	Chevrolet Uplander	550	Ford Explorer	353
Pontiac Grand Prix	627	Nissan Quest	470	Kia Sorento	411
Toyota Avalon	166	Kia Sedona	322	Chevy Trailblazer	397

Source: Insurance Institute for Highway Safety

Chapter 15 Review

Summary

In this chapter, we studied nonparametric procedures. These procedures paralleled parametric procedures introduced throughout the text.

Nonparametric procedures allow researchers to test hypotheses regarding measures of central tendency without requiring that certain assumptions about the underlying probability distribution, such as normality, be satisfied. If the populations from which samples are drawn are normal, nonparametric tests are less efficient than the corresponding parametric procedures.

We summarize the procedures presented in this chapter in the following table, along with the corresponding parametric procedures.

Nonparametric Test	Parametric Test	Purpose
Runs Test (Section 15.2)	No equivalent procedure	To test for randomness
Sign Test (Section 15.3) or Wilcoxon Signed-Ranks Test (Section 15.4)	t-test (Section 10.3)	To test a hypothesis regarding a measure of central tendency
Wilcoxon Matched-Pairs Signed-Ranks Test (Section 15.4)	Test for the difference of means of dependent samples (Section 11.2)	To test a hypothesis regarding the difference between two measures of central tendency when the sampling is dependent
Mann–Whitney Test (Section 15.5)	Test for the difference of means of independent samples (Section 11.3)	To test a hypothesis regarding the difference between two measures of central tendency when the sampling is independent
Spearman Rank-Correlation Test (Section 15.6)	Test for linear relation (Section 14.1)	To test whether two variables are associated
Kruskal–Wallis Test (Section 15.7)	One-way analysis of variance (Section 13.1)	To test whether three or more populations have the same distribution

Vocabulary

Parametric statistical procedures (p. 744)
Nonparametric (or distribution-free) statistical procedures (p. 744)
Power of a test (p. 745)
Efficiency (p. 745)

Runs test for randomness (p. 746)
Run (p. 746)
Length (p. 746)
One-sample sign test (p. 753)
Wilcoxon matched-pairs signed-ranks test (p. 760)

Mann–Whitney test (p. 769)
Spearman's rank-correlation test (p. 778)
Kruskal–Wallis test (p. 783)

Formulas

- **Test Statistic for a Runs Test for Randomness**
  Let n represent the sample size, containing two mutually exclusive types. Let n_1 represent the number of observations of the first type. Let n_2 represent the number of observations of the second type. Let r represent the number of runs.

  Small-Sample Case:
  If $n_1 \leq 20$ and $n_2 \leq 20$, the test statistic in the runs test for randomness is r, the number of runs.

  Large-Sample Case:
  If $n_1 > 20$ or $n_2 > 20$, the test statistic in the runs test for randomness is

  $$z_0 = \frac{r - \mu_r}{\sigma_r}$$

  where

  $$\mu_r = \frac{2n_1 n_2}{n} + 1 \quad \text{and} \quad \sigma_r = \sqrt{\frac{2n_1 n_2 (2n_1 n_2 - n)}{n^2(n-1)}}$$

- **Test Statistic for a One-Sample Sign Test**
  Small-Sample Case ($n \leq 25$)

Two-Tailed	Left-Tailed	Right-Tailed
$H_0: M = M_0$	$H_0: M = M_0$	$H_0: M = M_0$
$H_1: M \neq M_0$	$H_1: M < M_0$	$H_1: M > M_0$
The test statistic, k, is the smaller of the number of minus signs or plus signs.	The test statistic, k, is the number of plus signs.	The test statistic, k, is the number of minus signs.

  Large-Sample Case ($n > 25$)
  The test statistic is

  $$z_0 = \frac{(k + 0.5) - \dfrac{n}{2}}{\dfrac{\sqrt{n}}{2}}$$

  where n is the number of minus and plus signs and k is obtained as described in the small-sample case.

- **Test Statistic for the Wilcoxon Matched-Pairs Signed-Ranks Test**
  The test statistic depends on the size of the sample and the alternative hypothesis. Let n represent the number of nonzero differences.

Small-Sample Case ($n \leq 30$)

Two-Tailed	Left-Tailed	Right-Tailed
$H_0: M_D = 0$	$H_0: M_D = 0$	$H_0: M_D = 0$
$H_1: M_D \neq 0$	$H_1: M_D < 0$	$H_1: M_D > 0$
Test Statistic	**Test Statistic**	**Test Statistic**
T is the smaller of T_+ or $\|T_-\|$	$T = T_+$	$T = \|T_-\|$

Large-Sample Case ($n > 30$)
The test statistic is given by

$$z_0 = \frac{T - \frac{n(n+1)}{4}}{\sqrt{\frac{n(n+1)(2n+1)}{24}}}$$

where T is the test statistic from the small-sample case.

- **Test Statistic for the Mann–Whitney Test**
  The test statistic depends on the size of the samples from each population. Let n_1 represent the sample size for population X and n_2 represent the sample size for population Y.

Small-Sample Case ($n_1 \leq 20$ and $n_2 \leq 20$)
If S is the sum of the ranks corresponding to the sample from population X, the test statistic, T, is given by

$$T = S - \frac{n_1(n_1+1)}{2}$$

Note: The value of S is always obtained by summing the ranks of the sample data that correspond to M_X in the hypothesis.

Large-Sample Case ($n_1 > 20$ or $n_2 > 20$)
The test statistic is given by

$$z_0 = \frac{T - \frac{n_1 n_2}{2}}{\sqrt{\frac{n_1 n_2 (n_1 + n_2 + 1)}{12}}}$$

where T is the test statistic from the small-sample case.

- **Test Statistic for Spearman's Rank-Correlation Test**
  The test statistic depends on the size of the sample, n, and on the sum of the squared differences and is given by

$$r_s = 1 - \frac{6 \sum d_i^2}{n(n^2 - 1)}$$

where d_i is the difference in the ranks of the two observations in the ith ordered pair.

- **Test Statistic for the Kruskal–Wallis Test**
  The test statistic for the Kruskal–Wallis Test is

$$H = \frac{12}{N(N+1)} \sum \frac{1}{n_i} \left[R_i - \frac{n_i(N+1)}{2} \right]^2$$
$$= \frac{12}{N(N+1)} \left[\frac{R_1^2}{n_1} + \frac{R_2^2}{n_2} + \cdots + \frac{R_k^2}{n_k} \right] - 3(N+1)$$

where

R_i is the sum of the ranks of the ith sample
R_1^2 is the sum of the ranks squared for the first sample
R_2^2 is the sum of the ranks squared for the second sample, and so on
n_1 is the number of observations in the first sample
n_2 is the number of observations in the second sample, and so on
N is the total number of observations ($N = n_1 + n_2 + \cdots + n_k$)
k is the number of populations being compared

OBJECTIVES

Section	You should be able to ...	Example(s)	Review Exercises
15.1	1 Distinguish between parametric and nonparametric statistical procedures (p. 744)	pp. 744–745	7
15.2	1 Perform a runs test for randomness (p. 745)	1–4	1
15.3	1 Conduct a one-sample sign test (p. 753)	1 and 2	2
15.4	1 Test a hypothesis about the difference between the medians of two dependent samples (p. 760)	1 and 2	3
15.5	1 Test a hypothesis about the difference between the medians of two independent samples (p. 769)	1 and 2	4
15.6	1 Perform Spearman's rank-correlation test (p. 778)	1	5
15.7	1 Test a hypothesis using the Kruskal–Wallis test (p. 783)	1	6

Review Exercises

1. **The Stanley Cup** The division from which the winner of the Stanley Cup came for the years 1990 to 2015 is given in the following list, where the Western Division is represented by a W and the Eastern Division is represented by an E.

W E E E E E W W W W E W W E E E W W E W E W E W W W

Is there sufficient evidence to conclude that the winning division occurs randomly at the $\alpha = 0.05$ level of significance?

2. On the Phone An introductory statistics instructor wants to know if the median number of hours that students talk on the phone per week is more than 15. He randomly sampled 20 students, asked them to record the number of hours they talked on the phone in one week, and obtained the following data:

30	30	60	70	30
5	60	100	60	20
20	5	60	1	20
10	20	2	10	5

Source: Michael McCraith, Joliet Junior College

Do the data indicate that the median number of hours that students talk on the phone per week is more than 15? Use the $\alpha = 0.05$ level of significance.

3. Height versus Arm Span A statistics student thinks that an individual's arm span is equal to the individual's height. To test this belief, the student obtained the following data from a random sample of 10 students. Is there enough evidence to indicate that an individual's height and arm span are different at the $\alpha = 0.05$ level of significance?

Student	1	2	3	4	5	6	7	8	9	10
Height (inches)	59.5	69	77	59.5	74.5	63	61.5	67.5	73	69
Arm Span (inches)	62	65.5	76	63	74	66	61	69	70	71

Source: John Climent, Cecil Community College

4. Measuring Reaction Time Researchers at the University of Mississippi wanted to determine if the median reaction time (in seconds) of males differed from the median reaction time of females to a choice stimulus. The researchers randomly selected 16 females and 12 males to participate in the study. The choice stimulus required the student to respond differently to different stimuli. The results are as below. Are the reaction times different at the $\alpha = 0.01$ level of significance?

Female Students

0.474	0.436	0.398	0.633	0.831	0.887	0.711
0.743	0.480	0.561	0.596	0.725	0.905	0.338
0.538	0.531					

Male Students

0.541	1.050	0.577	0.849	0.464	0.626
0.659	0.880	0.752	0.544	0.675	0.393

Source: PsychExperiments at the University of Mississippi

5. Engine Displacement versus Fuel Economy The following data represent the size of a car's engine (in liters) versus its miles per gallon in the city for various 2008 domestic automobiles. Are engine displacement and fuel economy negatively associated at the $\alpha = 0.05$ level of significance?

Car	Engine Displacement (in liters), x	City Miles per Gallon, y
Acura RL	3.5	16
Buick LaCrosse	3.8	17
Cadillac DTS	4.6	15
Chevrolet Cobalt	2.2	24
Chevrolet Impala	3.5	18
Chevrolet Malibu	2.4	22
Chrysler Sebring	2.4	21
Dodge Magnum	2.7	18
Ford Taurus	3.5	18
Ford Focus	2.0	24
Ford Mustang	4.0	17
Lexus LS 460	4.6	16
Pontiac Grand Prix	3.8	18
Pontiac G5	2.2	24
Saturn Aura	2.4	22

Source: Road & Track magazine

6. Water Samples Over the course of a year, a researcher took water samples from three areas in a forest with a stream running through it. Each sample was analyzed for the concentration of dissolved organic carbon (mg/L), with the results presented in the following table.

Organic							
22.74	6.51	14.90	9.72	17.90	14.00	11.40	16.87
29.80	8.81	14.86	19.80	18.30	7.40	5.30	15.42
27.10	5.29	15.91	14.86	5.20	17.50	15.72	22.49
16.51	20.46	15.35	8.09	11.90	10.30	20.46	

Mineral									
8.50	10.30	5.50	8.05	3.02	21.40	16.92	9.10	7.85	12.89
3.91	11.56	4.71	10.72	7.45	8.37	4.60	7.90	9.11	9.81
9.29	7.00	7.66	21.82	11.33	7.92	8.50	11.72	8.79	
21.00	3.99	11.72	22.62	7.11	17.90	4.80	4.85	9.60	
10.89	3.79	11.80	10.74	17.99	7.31	4.90	11.97	12.57	

Surface								
10.83	6.30	12.40	19.19	19.78	12.27	16.82	16.12	8.69
8.74	6.68	10.30	13.14	20.56	16.47	10.70	15.77	10.46
9.20	7.34	10.48	12.51	17.93	15.03	16.00	6.40	16.23
8.12	9.52	12.88	15.76	14.28	14.53	20.70	14.19	15.43
7.60	11.94	19.01	10.96	13.11	12.04	13.70	13.89	

Source: Lisa Emili, PhD candidate, Department of Geography and Wetlands Research Centre, University of Waterloo

Each sample was categorized according to the water type collected. Water was collected from streams (surface water), groundwater was collected from organic soil, and groundwater was collected from mineral soil. The researcher wanted to determine if the distributions of concentration of dissolved organic carbon were the same for each collection area.

(a) State the null and alternative hypotheses.

(b) The sums of the ranks in each category are as follows:

	Organic	Mineral	Surface
Sample size	31	47	44
Sum of ranks	2355.5	2119.5	3028

Use this information to compute the test statistic.

(c) What is the critical value at the $\alpha = 0.05$ level of significance?

(d) State your conclusion.

(e) Draw side-by-side boxplots to support your conclusion.

7. In general, how do parametric tests differ from nonparametric tests? For each nonparametric test listed in this chapter, identify the corresponding parametric test and explain the difference between the two.

Chapter Test

1. The division from which the winner of the National Basketball Association's champion came for the years 1991 to 2015 is given in the following list, where the Western Division is represented by a W and the Eastern Division is represented by an E. Is there sufficient evidence to support the claim that the winning division occurs randomly at the $\alpha = 0.05$ level of significance?

E E E W W E E E W W W W W W E W E W E W W W E E W W

2. An introductory statistics instructor believes that the median number of hours that students study each week is less than 15. He randomly sampled 20 students, asked them to disclose the number of hours they study each week, and obtained the following data:

8	17	8	10	5
12	8	8	12	5
22	10	20	20	2
10	4	8	5	15

Source: Michael McCraith, Joliet Junior College

Do the data indicate that the median number of hours that students study each week is less than 15? Use the $\alpha = 0.05$ level of significance.

3. A physical therapist wants to investigate whether a new exercise program reduces the pulse rate of subjects. She randomly selects 10 women to participate in the study and asks each subject to step up and down on a 6-inch step for 3 minutes. Her pulse (in beats per minute) is then recorded. After a 10-week training program, the pulse is again measured with the same technique. The results are presented in the following table:

Woman	Before	After	Woman	Before	After
1	136	128	6	113	112
2	120	111	7	89	98
3	129	129	8	122	103
4	143	148	9	102	103
5	115	110	10	122	103

Is the exercise program effective at the $\alpha = 0.05$ level of significance?

4. A researcher wants to know whether the median pH of rain near Houston, Texas, is significantly greater than that near Chicago, Illinois. He randomly selects 12 rain dates in Texas and obtains the following data:

Houston, Texas					
4.69	5.10	5.22	4.46	4.93	4.65
5.22	4.76	4.25	5.14	4.11	4.71

Source: National Atmospheric Deposition Program

Independently, he randomly selects 14 rain dates in Illinois and obtains the following data:

Chicago, Illinois						
4.40	4.69	4.22	4.64	4.54	4.35	4.69
4.40	4.75	4.63	4.45	4.49	4.36	4.52

Source: National Atmospheric Deposition Program

Is the median pH of rain in Houston greater than the median pH of rain near Chicago at the $\alpha = 0.05$ level of significance?

5. Crickets make a chirping noise by sliding their wings rapidly over each other. Perhaps you have noticed that the number of chirps seems to increase with the temperature. The given data list the temperature (in degrees Fahrenheit) and the number of chirps per second for the striped ground cricket. Are temperature and number of chirps per second associated at the $\alpha = 0.05$ level of significance?

Temperature	Chirps	Temperature	Chirps
88.6	20.0	71.6	16.0
93.3	19.8	84.3	18.4
80.6	17.1	75.2	15.5
69.7	14.7	82.0	17.1
69.4	15.4	83.3	16.2
79.6	15.0	82.6	17.2
80.6	16.0	83.5	17.0
76.3	14.4		

Source: George W. Pierce. *The Songs of Insects.* Cambridge, MA: Harvard University Press, 1949, pp. 12–21

6. The Insurance Institute for Highway Safety conducts experiments in which cars are crashed into a fixed barrier at 40 mph. In a 40-mph offset test, 40% of the total width of each vehicle strikes a barrier on the driver side. The barrier's deformable face is made of aluminum honeycomb, which makes the forces in the test similar to those involved in a frontal offset crash between two vehicles of the same weight, each going just less than 40 mph. Determine if the distribution of femur force in kilonewtons (kN) on the left leg resulting from this offset crash is the same for large family cars, passenger vans, and midsize utility vehicles at the $\alpha = 0.05$ level of significance. The following data were collected from the study.

Large Family Cars	Femur Force (kN)	Passenger Vans	Femur Force (kN)	Midsize Utility Vehicles	Femur Force (kN)
Chevrolet Lumina	5.8	Toyota Sienna	4.7	Mazda CX-7	1.0
Ford Taurus	1.9	Honda Odyssey	4.3	Toyota 4Runner	0.2
Buick LeSabre	2.8	Ford Windstar	2.1	Mitsubishi Montero	3.6
Chevrolet Impala	0.9	Mazda MPV	6.3	Nissan Xterra	1.8
Chrysler LHS	4.2	Chevrolet Astro	8.2	Ford Explorer	0.2
Pontiac Grand Prix	2.4	Nissan Quest	2.6	Jeep Grand Cherokee	2.3
Dodge Intrepid	4.7	Pontiac Trans Sport	6.7	Nissan Pathfinder	1.9

Source: Insurance Institute for Highway Safety

7. Explain what it means if a nonparametric test has an efficiency of 0.713.

Making an Informed Decision

Where Should I Live?

When you graduate from college, you must choose in what part of the country you would like to live. To do this, you must judge which characteristics of a neighborhood are most important to you, such as spending on education, access to health care, and typical commute times.

To determine the best place to live, go to money.cnn.com/best/bplive/. (This is *Money* magazine's "Best Places to Live" site.) Select the detailed search option. Fill in the survey questions presented on the basis of your personal desires for a community. Choose two regions of the country. (You might want to select one region at a time to avoid confusion.) Use the Mann–Whitney test to learn whether there are differences in certain characteristics of one area of the country versus the other area that you have chosen. For example, is the average spending per pupil different for cities in the South versus those in the Midwest?

Use the results of this type of analysis to help you make a decision about the city in which you would like to live.

 **CASE STUDY** Evaluating Alabama's 1891 House Bill 504

Before the Civil War, Confederate State constitutions paid little attention to public education. Nonetheless, blacks were legally prohibited formal schooling and black literacy was discouraged. This situation would soon change. In 1868, the pre–Civil War general educational clauses of Alabama's constitution were replaced by new ones that provided for sufficient funds for a state-wide public school system and discouraged donations from private sources. It promised equal educational opportunity regardless of race or economic status, but did not mandate racial segregation of public schools—a decision left to the state board of education.

While equal educational opportunity for blacks was not a reality, blacks made significant educational gains during the next two decades. However, white Democrats regained power and the white-authored Alabama state constitution of 1875 severely restricted public funding for education. While segregated schools became the law of the state, black education continued to flourish while conditions for whites eroded. In 1891 the Alabama House Bill 504, preserving the constitutionally guaranteed per capita education funding mechanism, passed both legislative houses but was circumvented when the bill turned state education revenues over to white county officials who dispersed the money at their discretion.

As a historian, you are interested in evaluating the impact of House Bill 504. You have uncovered the average length of the school year and average monthly teacher salary by race for various counties in Alabama in 1887 (before the passage of the bill) and 1915 (after its passage). The data are presented in the following tables:

Average Length of Public School Year in Days				
	1887		**1915**	
County	**Black**	**White**	**Black**	**White**
Autauga	89	61	93	140
Barbour	85	71	91	148
Bullock	106	70	86	163
Butler	82	65	84	122
Chambers	131	120	91	156
Choctaw	65	65	56	120
Clarke	82	72	72	110
Dallas	75	60	108	172
Greene	106	74	94	158
Hale	80	64	102	115
Lee	95	71	89	156
Lowndes	83	60	91	142
Macon	70	60	101	158
Marengo	95	70	93	126
Monroe	59	63	65	120
Montgomery	100	97	121	174
Perry	96	72	109	152
Pickens	60	60	80	108
Russell	83	71	80	152
Sumter	79	79	86	152
Wilcox	78	60	81	151

Average Monthly Pay (in dollars) of Teachers				
	1887		**1915**	
County	**Black**	**White**	**Black**	**White**
Autauga	22.08	26.50	24.78	47.93
Barbour	29.30	23.33	27.54	55.72
Bullock	27.50	20.00	25.57	60.48
Butler	26.04	25.56	25.54	55.30
Chambers	31.57	38.00	30.87	49.76
Choctaw	25.93	22.95	22.12	52.58
Clarke	22.33	24.75	30.65	64.23
Dallas	34.33	17.00	22.93	71.73
Greene	29.22	19.30	24.29	55.89
Hale	36.86	19.53	25.68	63.55
Lee	20.75	18.00	32.93	62.28
Lowndes	29.47	26.44	27.84	74.58
Macon	20.00	20.00	28.87	56.65
Marengo	34.50	25.18	22.48	68.66
Monroe	19.93	21.77	30.52	49.27
Montgomery	22.00	30.00	31.52	78.96
Perry	32.39	20.57	26.08	53.65
Pickens	14.00	16.00	20.75	50.22
Russell	31.00	36.95	29.26	70.71
Sumter	27.75	27.07	26.46	66.68
Wilcox	25.60	16.30	19.88	62.63

Source: From *School and Society: Historical and Contemporary Perspectives*, by Steven Tozer, Paul C. Violas, Guy B. Senese. Copyright © 1998 by McGraw-Hill Companies.

Use the Wilcoxon matched-pairs signed-ranks test procedure with an $\alpha = 0.05$ significance level to answer the following questions:

1. In 1887, was the median school-year length different for black and white children? How about in 1915?

2. In 1887, was the median salary different for black and white teachers? How about in 1915?

Are there any other analyses that you would like to conduct using these data? Explain. Conduct these analyses.

Write a report detailing your assumptions, analyses, findings, and conclusions regarding the impact of Alabama House Bill 504 on the education provided to black and white children.

Photo Credits

COVER   Lucky Dragon USA/Fotolia

Chapter 1   Page 2, Jocic/Shutterstock; Page 8, Ministr-84/Shutterstock; Page 37, Oleksii Sergieiev/Fotolia; Page 43, Pearson Education, Inc.; Page 46, Tomasz Trojanowski/Shutterstock; Page 49, Emde71/Fotolia; Page 51, Matka_Wariatka/Shutterstock; Page 59, Jocic/Shutterstock.

Chapter 2   Page 62, Serg64/Shutterstock; Page 76, Tony Freeman/PhotoEdit; Page 80, Pearson Education, Inc.; Page 115, Serg64/Shutterstock.

Chapter 3   Page 117, GreenPimp/iStock/Getty Images; Page 141, Pearson Education, Inc.; Page 164, Pearson Education, Inc.; Page 178, GreenPimp/iStock/Getty Images.

Chapter 4   Page 180, GSFC/Reto Stöckli/Nazmi El Saleous/Marit Jentoft-Nilsen/NASA; Page 186, SPL/Science Source; Page 200, Classic Image/Alamy; Page 202, Pearson Education, Inc.; Page 243, GSFC/Reto Stöckli/Nazmi El Saleous/Marit Jentoft-Nilsen/NASA.

Chapter 5   Page 246, Maxsol7/Fotolia; Page 252, Mary Evans Picture Library/The Image Works; Page 254, Pearson Education, Inc.; Page 256, Pearson Education, Inc.; Page 259, Victor de Schwanberg/Alamy; Page 275, Victor de Schwanberg/Alamy; Page 285, Ria Novosti/Alamy; Page 294, Churchill Downs/AP Images; Page 298, Andy Marlin/NHL/Pool BB/dk/VP/Cck/REUTERS; Page 312, Maxsol7/Fotolia.

Chapter 6   Page 315, Jim Barber/Shutterstock; Page 322, Juulijs/Fotolia; Page 329, Ivan Vdovin/Alamy; Page 330, Monkey Business/Fotolia; Page 344, Pictorial Press Ltd/Alamy; Page 353, Jim Barber/Shutterstock.

Chapter 7   Page 355, Alan Schein Photography/Fuse/Getty Images; Page 359, SPL/Science Source; Page 360, Pearson Education, Inc.; Page 362, Pearson Education, Inc.; Page 383, Pearson Education, Inc.; Page 391, Alan Schein Photography/Fuse/Getty Images.

Chapter 8   Page 394, David Vernon/E+/Getty Images; Page 401, Pearson Education, Inc.; Page 406, Victor de Schwanberg/Alamy; Page 418, David Vernon/E+/Getty Images.

Chapter 9   Page 420, Rafael Ramirez Lee/123RF; Page 470, Rafael Ramirez Lee/123RF.

Chapter 10   Page 472, Blackwaterimages/E+/Getty Images; 476, Ingram Publishing/Getty Images; Page 522, Blackwaterimages/E+/Getty Images.

Chapter 11   Page 524, CandyBox Images/Shutterstock; Page 577, CandyBox Images/Shutterstock.

Chapter 12   Page 579, Joe Gough/Shutterstock; Page 580, Eric Carr/Alamy; Page 615, Joe Gough/Shutterstock.

Chapter 13   Page 617, Art2002/Fotolia; Page 671, Art2002/Fotolia.

Chapter 14   Page 673, Andy Dean/Fotolia; Page 741T, Andy Dean/Fotolia; Page 741B, Andy Dean/Fotolia.

Chapter 15   Page 743, Donald R. Swartz/Shutterstock; Page 794, Donald R. Swartz/Shutterstock.

FM   Page vt, Jocic/Shutterstock; Page vc, Serg64/Shutterstock; Page vb, GreenPimp/iStock/Getty Images; Page via, NASA GSFC/Reto Stöckli/Nazmi El Saleous/ Marit Jentoft-Nilsen; Page vib, Maxsol7/Fotolia; Page vic, Jim Barber/Shutterstock; Page vid, Alan Schein Photography/Fuse/Getty Images; Page viia, David Vernon/E+/Getty Images; Page viib, Rafael Ramirez Lee/123RF; Page viic, Blackwaterimages/E+/Getty Images; Page viid, CandyBox Images/Shutterstock; Page viiia, Joe Gough/Shutterstock; Page viiib, Art2002/Fotolia; Page viiic, Andy Dean/Fotolia; Page viiid, Donald R. Swartz/Shutterstock; Page xiv, Michael Sullivan.

ONLINE 4   4-2, Pearson Education, Inc.

ONLINE 5   5-2, Pearson Education, Inc.

ONLINE 6   6-2, Pearson Education, Inc.

Screenshots from Minitab. Courtesy of Minitab Corporation.

Screenshot from Texas Instruments. Courtesy of Texas Instruments.

Screenshots from StaCrunch. Used by permission of StatCrunch.

Screenshots from Microsoft® Excel®. Used by permission of Microsoft Corporation.

Appendix

A Tables

Table I

Random Numbers
Column Number

Row Number	01–05	06–10	11–15	16–20	21–25	26–30	31–35	36–40	41–45	46–50
01	89392	23212	74483	36590	25956	36544	68518	40805	09980	00467
02	61458	17639	96252	95649	73727	33912	72896	66218	52341	97141
03	11452	74197	81962	48443	90360	26480	73231	37740	26628	44690
04	27575	04429	31308	02241	01698	19191	18948	78871	36030	23980
05	36829	59109	88976	46845	28329	47460	88944	08264	00843	84592
06	81902	93458	42161	26099	09419	89073	82849	09160	61845	40906
07	59761	55212	33360	68751	86737	79743	85262	31887	37879	17525
08	46827	25906	64708	20307	78423	15910	86548	08763	47050	18513
09	24040	66449	32353	83668	13874	86741	81312	54185	78824	00718
10	98144	96372	50277	15571	82261	66628	31457	00377	63423	55141
11	14228	17930	30118	00438	49666	65189	62869	31304	17117	71489
12	55366	51057	90065	14791	62426	02957	85518	28822	30588	32798
13	96101	30646	35526	90389	73634	79304	96635	06626	94683	16696
14	38152	55474	30153	26525	83647	31988	82182	98377	33802	80471
15	85007	18416	24661	95581	45868	15662	28906	36392	07617	50248
16	85544	15890	80011	18160	33468	84106	40603	01315	74664	20553
17	10446	20699	98370	17684	16932	80449	92654	02084	19985	59321
18	67237	45509	17638	65115	29757	80705	82686	48565	72612	61760
19	23026	89817	05403	82209	30573	47501	00135	33955	50250	72592
20	67411	58542	18678	46491	13219	84084	27783	34508	55158	78742

Table II

Critical Values for Correlation Coefficient

n	
3	0.997
4	0.950
5	0.878
6	0.811
7	0.754
8	0.707
9	0.666
10	0.632
11	0.602
12	0.576
13	0.553
14	0.532
15	0.514
16	0.497
17	0.482
18	0.468
19	0.456
20	0.444
21	0.433
22	0.423
23	0.413
24	0.404
25	0.396
26	0.388
27	0.381
28	0.374
29	0.367
30	0.361

Table III

Binomial Probability Distribution

n	x										p										
		0.01	0.05	0.10	0.15	0.20	0.25	0.30	0.35	0.40	0.45	0.50	0.55	0.60	0.65	0.70	0.75	0.80	0.85	0.90	0.95
2	0	0.9801	0.9025	0.8100	0.7225	0.6400	0.5625	0.4900	0.4225	0.3600	0.3025	0.2500	0.2025	0.1600	0.1225	0.0900	0.0625	0.0400	0.0225	0.0100	0.0025
	1	0.0198	0.0950	0.1800	0.2550	0.3200	0.3750	0.4200	0.4550	0.4800	0.4950	0.5000	0.4950	0.4800	0.4550	0.4200	0.3750	0.3200	0.2550	0.1800	0.0950
	2	0.0001	0.0025	0.0100	0.0225	0.0400	0.0625	0.0900	0.1225	0.1600	0.2025	0.2500	0.3025	0.3600	0.4225	0.4900	0.5625	0.6400	0.7225	0.8100	0.9025
3	0	0.9703	0.8574	0.7290	0.6141	0.5120	0.4219	0.3430	0.2746	0.2160	0.1664	0.1250	0.0911	0.0640	0.0429	0.0270	0.0156	0.0080	0.0034	0.0010	0.0001
	1	0.0294	0.1354	0.2430	0.3251	0.3840	0.4219	0.4410	0.4436	0.4320	0.4084	0.3750	0.3341	0.2880	0.2389	0.1890	0.1406	0.0960	0.0574	0.0270	0.0071
	2	0.0003	0.0071	0.0270	0.0574	0.0960	0.1406	0.1890	0.2389	0.2880	0.3341	0.3750	0.4084	0.4320	0.4436	0.4410	0.4219	0.3840	0.3251	0.2430	0.1354
	3	0.0000+	0.0001	0.0010	0.0034	0.0080	0.0156	0.0270	0.0429	0.0640	0.0911	0.1250	0.1664	0.2160	0.2746	0.3430	0.4219	0.5120	0.6141	0.7290	0.8574
4	0	0.9606	0.8145	0.6561	0.5220	0.4096	0.3164	0.2401	0.1785	0.1296	0.0915	0.0625	0.0410	0.0256	0.0150	0.0081	0.0039	0.0016	0.0005	0.0001	0.0000+
	1	0.0388	0.1715	0.2916	0.3685	0.4096	0.4219	0.4116	0.3845	0.3456	0.2995	0.2500	0.2005	0.1536	0.1115	0.0756	0.0469	0.0256	0.0115	0.0036	0.0005
	2	0.0006	0.0135	0.0486	0.0975	0.1536	0.2109	0.2646	0.3105	0.3456	0.3675	0.3750	0.3675	0.3456	0.3105	0.2646	0.2109	0.1536	0.0975	0.0486	0.0135
	3	0.0000+	0.0005	0.0036	0.0115	0.0256	0.0469	0.0756	0.1115	0.1536	0.2005	0.2500	0.2995	0.3456	0.3845	0.4116	0.4219	0.4096	0.3685	0.2916	0.1715
	4	0.0000+	0.0000+	0.0001	0.0005	0.0016	0.0039	0.0081	0.0150	0.0256	0.0410	0.0625	0.0915	0.1296	0.1785	0.2401	0.3164	0.4096	0.5220	0.6561	0.8145
5	0	0.9510	0.7738	0.5905	0.4437	0.3277	0.2373	0.1681	0.1160	0.0778	0.0503	0.0313	0.0185	0.0102	0.0053	0.0024	0.0010	0.0003	0.0001	0.0000+	0.0000+
	1	0.0480	0.2036	0.3281	0.3915	0.4096	0.3955	0.3602	0.3124	0.2592	0.2059	0.1563	0.1128	0.0768	0.0488	0.0284	0.0146	0.0064	0.0022	0.0005	0.0000+
	2	0.0010	0.0214	0.0729	0.1382	0.2048	0.2637	0.3087	0.3364	0.3456	0.3369	0.3125	0.2757	0.2304	0.1811	0.1323	0.0879	0.0512	0.0244	0.0081	0.0011
	3	0.0000+	0.0011	0.0081	0.0244	0.0512	0.0879	0.1323	0.1811	0.2304	0.2757	0.3125	0.3369	0.3456	0.3364	0.3087	0.2637	0.2048	0.1382	0.0729	0.0214
	4	0.0000+	0.0000+	0.0005	0.0022	0.0064	0.0146	0.0284	0.0488	0.0768	0.1128	0.1563	0.2059	0.2592	0.3124	0.3602	0.3955	0.4096	0.3915	0.3281	0.2036
	5	0.0000+	0.0000+	0.0000+	0.0001	0.0003	0.0010	0.0024	0.0053	0.0102	0.0185	0.0313	0.0503	0.0778	0.1160	0.1681	0.2373	0.3277	0.4437	0.5905	0.7738
6	0	0.9415	0.7351	0.5314	0.3771	0.2621	0.1780	0.1176	0.0754	0.0467	0.0277	0.0156	0.0083	0.0041	0.0018	0.0007	0.0002	0.0001	0.0000+	0.0000+	0.0000+
	1	0.0571	0.2321	0.3543	0.3993	0.3932	0.3560	0.3025	0.2437	0.1866	0.1359	0.0938	0.0609	0.0369	0.0205	0.0102	0.0044	0.0015	0.0004	0.0001	0.0000+
	2	0.0014	0.0305	0.0984	0.1762	0.2458	0.2966	0.3241	0.3280	0.3110	0.2780	0.2344	0.1861	0.1382	0.0951	0.0595	0.0330	0.0154	0.0055	0.0012	0.0001
	3	0.0000+	0.0021	0.0146	0.0415	0.0819	0.1318	0.1852	0.2355	0.2765	0.3032	0.3125	0.3032	0.2765	0.2355	0.1852	0.1318	0.0819	0.0415	0.0146	0.0021
	4	0.0000+	0.0001	0.0012	0.0055	0.0154	0.0330	0.0595	0.0951	0.1382	0.1861	0.2344	0.2780	0.3110	0.3280	0.3241	0.2966	0.2458	0.1762	0.0984	0.0305
	5	0.0000+	0.0000+	0.0001	0.0004	0.0015	0.0044	0.0102	0.0205	0.0369	0.0609	0.0938	0.1359	0.1866	0.2437	0.3025	0.3560	0.3932	0.3993	0.3543	0.2321
	6	0.0000+	0.0000+	0.0000+	0.0000+	0.0001	0.0002	0.0007	0.0018	0.0041	0.0083	0.0156	0.0277	0.0467	0.0754	0.1176	0.1780	0.2621	0.3771	0.5314	0.7351
7	0	0.9321	0.6983	0.4783	0.3206	0.2097	0.1335	0.0824	0.0490	0.0280	0.0152	0.0078	0.0037	0.0016	0.0006	0.0002	0.0001	0.0000+	0.0000+	0.0000+	0.0000+
	1	0.0659	0.2573	0.3720	0.3960	0.3670	0.3115	0.2471	0.1848	0.1306	0.0872	0.0547	0.0320	0.0172	0.0084	0.0036	0.0013	0.0004	0.0001	0.0000+	0.0000+
	2	0.0020	0.0406	0.1240	0.2097	0.2753	0.3115	0.3177	0.2985	0.2613	0.2140	0.1641	0.1172	0.0774	0.0466	0.0250	0.0115	0.0043	0.0012	0.0002	0.0000+
	3	0.0000+	0.0036	0.0230	0.0617	0.1147	0.1730	0.2269	0.2679	0.2903	0.2918	0.2734	0.2388	0.1935	0.1442	0.0972	0.0577	0.0287	0.0109	0.0026	0.0002
	4	0.0000+	0.0002	0.0026	0.0109	0.0287	0.0577	0.0972	0.1442	0.1935	0.2388	0.2734	0.2918	0.2903	0.2679	0.2269	0.1730	0.1147	0.0617	0.0230	0.0036
	5	0.0000+	0.0000+	0.0002	0.0012	0.0043	0.0115	0.0250	0.0466	0.0774	0.1172	0.1641	0.2140	0.2613	0.2985	0.3177	0.3115	0.2753	0.2097	0.1240	0.0406
	6	0.0000+	0.0000+	0.0000+	0.0001	0.0004	0.0013	0.0036	0.0084	0.0172	0.0320	0.0547	0.0872	0.1306	0.1848	0.2471	0.3115	0.3670	0.3960	0.3720	0.2573
	7	0.0000+	0.0000+	0.0000+	0.0000+	0.0001	0.0001	0.0002	0.0006	0.0016	0.0037	0.0078	0.0152	0.0280	0.0490	0.0824	0.1335	0.2097	0.3206	0.4783	0.6983

Note: 0.0000+ means the probability is 0.0000 rounded to four decimal places. However, the probability is *not* zero.

This table computes the probability of obtaining x successes in n trials of a binomial experiment with probability of success p.

Table III (continued)

n	x											p									
		0.01	0.05	0.10	0.15	0.20	0.25	0.30	0.35	0.40	0.45	0.50	0.55	0.60	0.65	0.70	0.75	0.80	0.85	0.90	0.95
8	0	0.9227	0.6634	0.4305	0.2725	0.1678	0.1001	0.0576	0.0319	0.0168	0.0084	0.0039	0.0017	0.0007	0.0002	0.0001	0.0000+	0.0000+	0.0000+	0.0000+	0.0000+
	1	0.0746	0.2793	0.3826	0.3847	0.3355	0.2670	0.1977	0.1373	0.0896	0.0548	0.0313	0.0164	0.0079	0.0033	0.0012	0.0004	0.0001	0.0000+	0.0000+	0.0000+
	2	0.0026	0.0515	0.1488	0.2376	0.2936	0.3115	0.2965	0.2587	0.2090	0.1569	0.1094	0.0703	0.0413	0.0217	0.0100	0.0038	0.0011	0.0002	0.0000+	0.0000+
	3	0.0001	0.0054	0.0331	0.0839	0.1468	0.2076	0.2541	0.2786	0.2787	0.2568	0.2188	0.1719	0.1239	0.0808	0.0467	0.0231	0.0092	0.0026	0.0004	0.0000+
	4	0.0000+	0.0004	0.0046	0.0185	0.0459	0.0865	0.1361	0.1875	0.2322	0.2627	0.2734	0.2627	0.2322	0.1875	0.1361	0.0865	0.0459	0.0185	0.0046	0.0004
	5	0.0000+	0.0000+	0.0004	0.0026	0.0092	0.0231	0.0467	0.0808	0.1239	0.1719	0.2188	0.2568	0.2787	0.2786	0.2541	0.2076	0.1468	0.0839	0.0331	0.0054
	6	0.0000+	0.0000+	0.0000+	0.0002	0.0011	0.0038	0.0100	0.0217	0.0413	0.0703	0.1094	0.1569	0.2090	0.2587	0.2965	0.3115	0.2936	0.2376	0.1488	0.0515
	7	0.0000+	0.0000+	0.0000+	0.0000+	0.0001	0.0004	0.0012	0.0033	0.0079	0.0164	0.0313	0.0548	0.0896	0.1373	0.1977	0.2670	0.3355	0.3847	0.3826	0.2793
	8	0.0000+	0.0000+	0.0000+	0.0000+	0.0000+	0.0000+	0.0001	0.0002	0.0007	0.0017	0.0039	0.0084	0.0168	0.0319	0.0576	0.1001	0.1678	0.2725	0.4305	0.6634
9	0	0.9135	0.6302	0.3874	0.2316	0.1342	0.0751	0.0404	0.0207	0.0101	0.0046	0.0020	0.0008	0.0003	0.0001	0.0000+	0.0000+	0.0000+	0.0000+	0.0000+	0.0000+
	1	0.0830	0.2985	0.3874	0.3679	0.3020	0.2253	0.1556	0.1004	0.0605	0.0339	0.0176	0.0083	0.0035	0.0013	0.0004	0.0001	0.0000+	0.0000+	0.0000+	0.0000+
	2	0.0034	0.0629	0.1722	0.2597	0.3020	0.3003	0.2668	0.2162	0.1612	0.1110	0.0703	0.0407	0.0212	0.0098	0.0039	0.0012	0.0003	0.0000+	0.0000+	0.0000+
	3	0.0001	0.0077	0.0446	0.1069	0.1762	0.2336	0.2668	0.2716	0.2508	0.2119	0.1641	0.1160	0.0743	0.0424	0.0210	0.0087	0.0028	0.0006	0.0001	0.0000+
	4	0.0000+	0.0006	0.0074	0.0283	0.0661	0.1168	0.1715	0.2194	0.2508	0.2600	0.2461	0.2128	0.1672	0.1181	0.0735	0.0389	0.0165	0.0050	0.0008	0.0000+
	5	0.0000+	0.0000+	0.0008	0.0050	0.0165	0.0389	0.0735	0.1181	0.1672	0.2128	0.2461	0.2600	0.2508	0.2194	0.1715	0.1168	0.0661	0.0283	0.0074	0.0006
	6	0.0000+	0.0000+	0.0001	0.0006	0.0028	0.0087	0.0210	0.0424	0.0743	0.1160	0.1641	0.2119	0.2508	0.2716	0.2668	0.2336	0.1762	0.1069	0.0446	0.0077
	7	0.0000+	0.0000+	0.0000+	0.0000+	0.0003	0.0012	0.0039	0.0098	0.0212	0.0407	0.0703	0.1110	0.1612	0.2162	0.2668	0.3003	0.3020	0.2597	0.1722	0.0629
	8	0.0000+	0.0000+	0.0000+	0.0000+	0.0000+	0.0001	0.0004	0.0013	0.0035	0.0083	0.0176	0.0339	0.0605	0.1004	0.1556	0.2253	0.3020	0.3679	0.3874	0.2985
	9	0.0000+	0.0000+	0.0000+	0.0000+	0.0000+	0.0000+	0.0000+	0.0001	0.0003	0.0008	0.0020	0.0046	0.0101	0.0207	0.0404	0.0751	0.1342	0.2316	0.3874	0.6302
10	0	0.9044	0.5987	0.3487	0.1969	0.1074	0.0563	0.0282	0.0135	0.0060	0.0025	0.0010	0.0003	0.0001	0.0000+	0.0000+	0.0000+	0.0000+	0.0000+	0.0000+	0.0000+
	1	0.0914	0.3151	0.3874	0.3474	0.2684	0.1877	0.1211	0.0725	0.0403	0.0207	0.0098	0.0042	0.0016	0.0005	0.0001	0.0000+	0.0000+	0.0000+	0.0000+	0.0000+
	2	0.0042	0.0746	0.1937	0.2759	0.3020	0.2816	0.2335	0.1757	0.1209	0.0763	0.0439	0.0229	0.0106	0.0043	0.0014	0.0004	0.0001	0.0000+	0.0000+	0.0000+
	3	0.0001	0.0105	0.0574	0.1298	0.2013	0.2503	0.2668	0.2522	0.2150	0.1665	0.1172	0.0746	0.0425	0.0212	0.0090	0.0031	0.0008	0.0001	0.0000+	0.0000+
	4	0.0000+	0.0010	0.0112	0.0401	0.0881	0.1460	0.2001	0.2377	0.2508	0.2384	0.2051	0.1596	0.1115	0.0689	0.0368	0.0162	0.0055	0.0012	0.0001	0.0000+
	5	0.0000+	0.0001	0.0015	0.0085	0.0264	0.0584	0.1029	0.1536	0.2007	0.2340	0.2461	0.2340	0.2007	0.1536	0.1029	0.0584	0.0264	0.0085	0.0015	0.0001
	6	0.0000+	0.0000+	0.0001	0.0012	0.0055	0.0162	0.0368	0.0689	0.1115	0.1596	0.2051	0.2384	0.2508	0.2377	0.2001	0.1460	0.0881	0.0401	0.0112	0.0010
	7	0.0000+	0.0000+	0.0000+	0.0001	0.0008	0.0031	0.0090	0.0212	0.0425	0.0746	0.1172	0.1665	0.2150	0.2522	0.2668	0.2503	0.2013	0.1298	0.0574	0.0105
	8	0.0000+	0.0000+	0.0000+	0.0000+	0.0001	0.0004	0.0014	0.0043	0.0106	0.0229	0.0439	0.0763	0.1209	0.1757	0.2335	0.2816	0.3020	0.2759	0.1937	0.0746
	9	0.0000+	0.0000+	0.0000+	0.0000+	0.0000+	0.0000+	0.0001	0.0005	0.0016	0.0042	0.0098	0.0207	0.0403	0.0725	0.1211	0.1877	0.2684	0.3474	0.3874	0.3151
	10	0.0000+	0.0000+	0.0000+	0.0000+	0.0000+	0.0000+	0.0000+	0.0000+	0.0001	0.0003	0.0010	0.0025	0.0060	0.0135	0.0282	0.0563	0.1074	0.1969	0.3487	0.5987

Note: 0.0000+ means the probability is 0.0000 rounded to four decimal places. However, the probability is *not* zero.

Table III (continued)

n	x	0.01	0.05	0.10	0.15	0.20	0.25	0.30	0.35	0.40	0.45	0.50	0.55	0.60	0.65	0.70	0.75	0.80	0.85	0.90	0.95
												p									
11	0	0.8953	0.5688	0.3138	0.1673	0.0859	0.0422	0.0198	0.0088	0.0036	0.0014	0.0005	0.0002	0.0000+	0.0000+	0.0000+	0.0000+	0.0000+	0.0000+	0.0000+	0.0000+
	1	0.0995	0.3293	0.3835	0.3248	0.2362	0.1549	0.0932	0.0518	0.0266	0.0125	0.0054	0.0021	0.0007	0.0002	0.0000+	0.0000+	0.0000+	0.0000+	0.0000+	0.0000+
	2	0.0050	0.0867	0.2131	0.2866	0.2953	0.2581	0.1998	0.1395	0.0887	0.0513	0.0269	0.0126	0.0052	0.0018	0.0005	0.0001	0.0000+	0.0000+	0.0000+	0.0000+
	3	0.0002	0.0137	0.0710	0.1517	0.2215	0.2581	0.2568	0.2254	0.1774	0.1259	0.0806	0.0462	0.0234	0.0102	0.0037	0.0011	0.0002	0.0000+	0.0000+	0.0000+
	4	0.0000+	0.0014	0.0158	0.0536	0.1107	0.1721	0.2201	0.2428	0.2365	0.2060	0.1611	0.1128	0.0701	0.0379	0.0173	0.0064	0.0017	0.0003	0.0000+	0.0000+
	5	0.0000+	0.0001	0.0025	0.0132	0.0388	0.0803	0.1321	0.1830	0.2207	0.2360	0.2256	0.1931	0.1471	0.0985	0.0566	0.0268	0.0097	0.0023	0.0003	0.0001
	6	0.0000+	0.0000+	0.0003	0.0023	0.0097	0.0268	0.0566	0.0985	0.1471	0.1931	0.2256	0.2360	0.2207	0.1830	0.1321	0.0803	0.0388	0.0132	0.0025	0.0001
	7	0.0000+	0.0000+	0.0000+	0.0003	0.0017	0.0064	0.0173	0.0379	0.0701	0.1128	0.1611	0.2060	0.2365	0.2428	0.2201	0.1721	0.1107	0.0536	0.0158	0.0014
	8	0.0000+	0.0000+	0.0000+	0.0000+	0.0002	0.0011	0.0037	0.0102	0.0234	0.0462	0.0806	0.1259	0.1774	0.2254	0.2568	0.2581	0.2215	0.1517	0.0710	0.0137
	9	0.0000+	0.0000+	0.0000+	0.0000+	0.0000+	0.0001	0.0005	0.0018	0.0052	0.0126	0.0269	0.0513	0.0887	0.1395	0.1998	0.2581	0.2953	0.2866	0.2131	0.0867
	10	0.0000+	0.0000+	0.0000+	0.0000+	0.0000+	0.0000+	0.0000+	0.0002	0.0007	0.0021	0.0054	0.0125	0.0266	0.0518	0.0932	0.1549	0.2362	0.3248	0.3835	0.3293
	11	0.0000+	0.0000+	0.0000+	0.0000+	0.0000+	0.0000+	0.0000+	0.0000+	0.0000+	0.0002	0.0005	0.0014	0.0036	0.0088	0.0198	0.0422	0.0859	0.1673	0.3138	0.5688
12	0	0.8864	0.5404	0.2824	0.1422	0.0687	0.0317	0.0138	0.0057	0.0022	0.0008	0.0002	0.0001	0.0000+	0.0000+	0.0000+	0.0000+	0.0000+	0.0000+	0.0000+	0.0000+
	1	0.1074	0.3413	0.3766	0.3012	0.2062	0.1267	0.0712	0.0368	0.0174	0.0075	0.0029	0.0010	0.0003	0.0001	0.0000+	0.0000+	0.0000+	0.0000+	0.0000+	0.0000+
	2	0.0060	0.0988	0.2301	0.2924	0.2835	0.2323	0.1678	0.1088	0.0639	0.0339	0.0161	0.0068	0.0025	0.0008	0.0002	0.0000+	0.0000+	0.0000+	0.0000+	0.0000+
	3	0.0002	0.0173	0.0852	0.1720	0.2362	0.2581	0.2397	0.1954	0.1419	0.0923	0.0537	0.0277	0.0125	0.0048	0.0015	0.0004	0.0001	0.0000+	0.0000+	0.0000+
	4	0.0000+	0.0021	0.0213	0.0683	0.1329	0.1936	0.2311	0.2367	0.2128	0.1700	0.1208	0.0762	0.0420	0.0199	0.0078	0.0024	0.0005	0.0001	0.0000+	0.0000+
	5	0.0000+	0.0002	0.0038	0.0193	0.0532	0.1032	0.1585	0.2039	0.2270	0.2225	0.1934	0.1489	0.1009	0.0591	0.0291	0.0115	0.0033	0.0006	0.0000+	0.0000+
	6	0.0000+	0.0000+	0.0005	0.0040	0.0155	0.0401	0.0792	0.1281	0.1766	0.2124	0.2256	0.2124	0.1766	0.1281	0.0792	0.0401	0.0155	0.0040	0.0005	0.0000+
	7	0.0000+	0.0000+	0.0000+	0.0006	0.0033	0.0115	0.0291	0.0591	0.1009	0.1489	0.1934	0.2225	0.2270	0.2039	0.1585	0.1032	0.0532	0.0193	0.0038	0.0002
	8	0.0000+	0.0000+	0.0000+	0.0001	0.0005	0.0024	0.0078	0.0199	0.0420	0.0762	0.1208	0.1700	0.2128	0.2367	0.2311	0.1936	0.1329	0.0683	0.0213	0.0021
	9	0.0000+	0.0000+	0.0000+	0.0000+	0.0001	0.0004	0.0015	0.0048	0.0125	0.0277	0.0537	0.0923	0.1419	0.1954	0.2397	0.2581	0.2362	0.1720	0.0852	0.0173
	10	0.0000+	0.0000+	0.0000+	0.0000+	0.0000+	0.0000+	0.0002	0.0008	0.0025	0.0068	0.0161	0.0339	0.0639	0.1088	0.1678	0.2323	0.2835	0.2924	0.2301	0.0988
	11	0.0000+	0.0000+	0.0000+	0.0000+	0.0000+	0.0000+	0.0000+	0.0001	0.0003	0.0010	0.0029	0.0075	0.0174	0.0368	0.0712	0.1267	0.2062	0.3012	0.3766	0.3413
	12	0.0000+	0.0000+	0.0000+	0.0000+	0.0000+	0.0000+	0.0000+	0.0000+	0.0000+	0.0001	0.0002	0.0008	0.0022	0.0057	0.0138	0.0317	0.0687	0.1422	0.2824	0.5404

Note: 0.0000+ means the probability is 0.0000 rounded to four decimal places. However, the probability is *not* zero.

Table III (continued)

n	x	0.01	0.05	0.10	0.15	0.20	0.25	0.30	0.35	0.40	0.45	0.50	0.55	0.60	0.65	0.70	0.75	0.80	0.85	0.90	0.95
											p										
15	0	0.8601	0.4633	0.2059	0.0874	0.0352	0.0134	0.0047	0.0016	0.0005	0.0001	0.0000+	0.0000+	0.0000+	0.0000+	0.0000+	0.0000+	0.0000+	0.0000+	0.0000+	0.0000+
	1	0.1303	0.3658	0.3432	0.2312	0.1319	0.0668	0.0305	0.0126	0.0047	0.0016	0.0005	0.0001	0.0000+	0.0000+	0.0000+	0.0000+	0.0000+	0.0000+	0.0000+	0.0000+
	2	0.0092	0.1348	0.2669	0.2856	0.2309	0.1559	0.0916	0.0476	0.0219	0.0090	0.0032	0.0010	0.0003	0.0001	0.0000+	0.0000+	0.0000+	0.0000+	0.0000+	0.0000+
	3	0.0004	0.0307	0.1285	0.2184	0.2501	0.2252	0.1700	0.1110	0.0634	0.0318	0.0139	0.0052	0.0016	0.0004	0.0001	0.0000+	0.0000+	0.0000+	0.0000+	0.0000+
	4	0.0000+	0.0049	0.0428	0.1156	0.1876	0.2252	0.2186	0.1792	0.1268	0.0780	0.0417	0.0191	0.0074	0.0024	0.0006	0.0001	0.0000+	0.0000+	0.0000+	0.0000+
	5	0.0000+	0.0006	0.0105	0.0449	0.1032	0.1651	0.2061	0.2123	0.1859	0.1404	0.0916	0.0515	0.0245	0.0096	0.0030	0.0007	0.0001	0.0000+	0.0000+	0.0000+
	6	0.0000+	0.0000+	0.0019	0.0132	0.0430	0.0917	0.1472	0.1906	0.2066	0.1914	0.1527	0.1048	0.0612	0.0298	0.0116	0.0034	0.0007	0.0001	0.0000+	0.0000+
	7	0.0000+	0.0000+	0.0003	0.0030	0.0138	0.0393	0.0811	0.1319	0.1771	0.2013	0.1964	0.1647	0.1181	0.0710	0.0348	0.0131	0.0035	0.0005	0.0000+	0.0000+
	8	0.0000+	0.0000+	0.0000+	0.0005	0.0035	0.0131	0.0348	0.0710	0.1181	0.1647	0.1964	0.2013	0.1771	0.1319	0.0811	0.0393	0.0138	0.0030	0.0003	0.0000+
	9	0.0000+	0.0000+	0.0000+	0.0001	0.0007	0.0034	0.0116	0.0298	0.0612	0.1048	0.1527	0.1914	0.2066	0.1906	0.1472	0.0917	0.0430	0.0132	0.0019	0.0000+
	10	0.0000+	0.0000+	0.0000+	0.0000+	0.0001	0.0007	0.0030	0.0096	0.0245	0.0515	0.0916	0.1404	0.1859	0.2123	0.2061	0.1651	0.1032	0.0449	0.0105	0.0006
	11	0.0000+	0.0000+	0.0000+	0.0000+	0.0000+	0.0001	0.0006	0.0024	0.0074	0.0191	0.0417	0.0780	0.1268	0.1792	0.2186	0.2252	0.1876	0.1156	0.0428	0.0049
	12	0.0000+	0.0000+	0.0000+	0.0000+	0.0000+	0.0000+	0.0001	0.0004	0.0016	0.0052	0.0139	0.0318	0.0634	0.1110	0.1700	0.2252	0.2501	0.2184	0.1285	0.0307
	13	0.0000+	0.0000+	0.0000+	0.0000+	0.0000+	0.0000+	0.0000+	0.0001	0.0003	0.0010	0.0032	0.0090	0.0219	0.0476	0.0916	0.1559	0.2309	0.2856	0.2669	0.1348
	14	0.0000+	0.0000+	0.0000+	0.0000+	0.0000+	0.0000+	0.0000+	0.0000+	0.0000+	0.0001	0.0005	0.0016	0.0047	0.0126	0.0305	0.0668	0.1319	0.2312	0.3432	0.3658
	15	0.0000+	0.0000+	0.0000+	0.0000+	0.0000+	0.0000+	0.0000+	0.0000+	0.0000+	0.0000+	0.0000+	0.0001	0.0005	0.0016	0.0047	0.0134	0.0352	0.0874	0.2059	0.4633
20	0	0.8179	0.3585	0.1216	0.0388	0.0115	0.0032	0.0008	0.0002	0.0000+	0.0000+	0.0000+	0.0000+	0.0000+	0.0000+	0.0000+	0.0000+	0.0000+	0.0000+	0.0000+	0.0000+
	1	0.1652	0.3774	0.2702	0.1368	0.0576	0.0211	0.0068	0.0020	0.0005	0.0001	0.0000+	0.0000+	0.0000+	0.0000+	0.0000+	0.0000+	0.0000+	0.0000+	0.0000+	0.0000+
	2	0.0159	0.1887	0.2852	0.2293	0.1369	0.0669	0.0278	0.0100	0.0031	0.0008	0.0002	0.0000+	0.0000+	0.0000+	0.0000+	0.0000+	0.0000+	0.0000+	0.0000+	0.0000+
	3	0.0010	0.0596	0.1901	0.2428	0.2054	0.1339	0.0716	0.0323	0.0123	0.0040	0.0011	0.0002	0.0000+	0.0000+	0.0000+	0.0000+	0.0000+	0.0000+	0.0000+	0.0000+
	4	0.0000+	0.0133	0.0898	0.1821	0.2182	0.1897	0.1304	0.0738	0.0350	0.0139	0.0046	0.0013	0.0003	0.0000+	0.0000+	0.0000+	0.0000+	0.0000+	0.0000+	0.0000+
	5	0.0000+	0.0022	0.0319	0.1028	0.1746	0.2023	0.1789	0.1272	0.0746	0.0365	0.0148	0.0049	0.0013	0.0003	0.0000+	0.0000+	0.0000+	0.0000+	0.0000+	0.0000+
	6	0.0000+	0.0003	0.0089	0.0454	0.1091	0.1686	0.1916	0.1712	0.1244	0.0746	0.0370	0.0150	0.0049	0.0012	0.0002	0.0000+	0.0000+	0.0000+	0.0000+	0.0000+
	7	0.0000+	0.0000+	0.0020	0.0160	0.0545	0.1124	0.1643	0.1844	0.1659	0.1221	0.0739	0.0366	0.0146	0.0045	0.0010	0.0002	0.0000+	0.0000+	0.0000+	0.0000+
	8	0.0000+	0.0000+	0.0004	0.0046	0.0222	0.0609	0.1144	0.1614	0.1797	0.1623	0.1201	0.0727	0.0355	0.0136	0.0039	0.0008	0.0001	0.0000+	0.0000+	0.0000+
	9	0.0000+	0.0000+	0.0001	0.0011	0.0074	0.0271	0.0654	0.1158	0.1597	0.1771	0.1602	0.1185	0.0710	0.0336	0.0120	0.0030	0.0005	0.0000+	0.0000+	0.0000+
	10	0.0000+	0.0000+	0.0000+	0.0002	0.0020	0.0099	0.0308	0.0686	0.1171	0.1593	0.1762	0.1593	0.1171	0.0686	0.0308	0.0099	0.0020	0.0002	0.0000+	0.0000+
	11	0.0000+	0.0000+	0.0000+	0.0000+	0.0005	0.0030	0.0120	0.0336	0.0710	0.1185	0.1602	0.1771	0.1597	0.1158	0.0654	0.0271	0.0074	0.0011	0.0001	0.0000+
	12	0.0000+	0.0000+	0.0000+	0.0000+	0.0001	0.0008	0.0039	0.0136	0.0355	0.0727	0.1201	0.1623	0.1797	0.1614	0.1144	0.0609	0.0222	0.0046	0.0004	0.0000+
	13	0.0000+	0.0000+	0.0000+	0.0000+	0.0000+	0.0002	0.0010	0.0045	0.0146	0.0366	0.0739	0.1221	0.1659	0.1844	0.1643	0.1124	0.0545	0.0160	0.0020	0.0000+
	14	0.0000+	0.0000+	0.0000+	0.0000+	0.0000+	0.0000+	0.0002	0.0012	0.0049	0.0150	0.0370	0.0746	0.1244	0.1712	0.1916	0.1686	0.1091	0.0454	0.0089	0.0003
	15	0.0000+	0.0000+	0.0000+	0.0000+	0.0000+	0.0000+	0.0000+	0.0003	0.0013	0.0049	0.0148	0.0365	0.0746	0.1272	0.1789	0.2023	0.1746	0.1028	0.0319	0.0022
	16	0.0000+	0.0000+	0.0000+	0.0000+	0.0000+	0.0000+	0.0000+	0.0000+	0.0003	0.0013	0.0046	0.0139	0.0350	0.0738	0.1304	0.1897	0.2182	0.1821	0.0898	0.0133
	17	0.0000+	0.0000+	0.0000+	0.0000+	0.0000+	0.0000+	0.0000+	0.0000+	0.0000+	0.0002	0.0011	0.0040	0.0123	0.0323	0.0716	0.1339	0.2054	0.2428	0.1901	0.0596
	18	0.0000+	0.0000+	0.0000+	0.0000+	0.0000+	0.0000+	0.0000+	0.0000+	0.0000+	0.0000+	0.0002	0.0008	0.0031	0.0100	0.0278	0.0669	0.1369	0.2293	0.2852	0.1887
	19	0.0000+	0.0000+	0.0000+	0.0000+	0.0000+	0.0000+	0.0000+	0.0000+	0.0000+	0.0000+	0.0000+	0.0001	0.0005	0.0020	0.0068	0.0211	0.0576	0.1368	0.2702	0.3774
	20	0.0000+	0.0000+	0.0000+	0.0000+	0.0000+	0.0000+	0.0000+	0.0000+	0.0000+	0.0000+	0.0000+	0.0000+	0.0000+	0.0002	0.0008	0.0032	0.0115	0.0388	0.1216	0.3585

Note: 0.0000+ means the probability is 0.0000 rounded to four decimal places. However, the probability is *not* zero.

Table IV

Cumulative Binomial Probability Distribution

n	x	0.01	0.05	0.10	0.15	0.20	0.25	0.30	0.35	0.40	0.45	0.50	0.55	0.60	0.65	0.70	0.75	0.80	0.85	0.90	0.95
2	0	0.9801	0.9025	0.8100	0.7225	0.6400	0.5625	0.4900	0.4225	0.3600	0.3025	0.2500	0.2025	0.1600	0.1225	0.0900	0.0625	0.0400	0.0225	0.0100	0.0025
	1	0.9999	0.9975	0.9900	0.9775	0.9600	0.9375	0.9100	0.8775	0.8400	0.7975	0.7500	0.6975	0.6400	0.5775	0.5100	0.4375	0.3600	0.2775	0.1900	0.0975
	2	1	1	1	1	1	1	1	1	1	1	1	1	1	1	1	1	1	1	1	1
3	0	0.9703	0.8574	0.7290	0.6141	0.5120	0.4219	0.3430	0.2746	0.2160	0.1664	0.1250	0.0911	0.0640	0.0429	0.0270	0.0156	0.0080	0.0034	0.0010	0.0001
	1	0.9997	0.9928	0.9720	0.9393	0.8960	0.8438	0.7840	0.7183	0.6480	0.5748	0.5000	0.4253	0.3520	0.2818	0.2160	0.1563	0.1040	0.0608	0.0280	0.0073
	2	1.0000−	0.9999	0.9990	0.9966	0.9920	0.9844	0.9730	0.9571	0.9360	0.9089	0.8750	0.8336	0.7840	0.7254	0.6570	0.5781	0.4880	0.3859	0.2710	0.1426
	3	1	1	1	1	1	1	1	1	1	1	1	1	1	1	1	1	1	1	1	1
4	0	0.9606	0.8145	0.6561	0.5220	0.4096	0.3164	0.2401	0.1785	0.1296	0.0915	0.0625	0.0410	0.0256	0.0150	0.0081	0.0039	0.0016	0.0005	0.0001	0.0000+
	1	0.9994	0.9860	0.9477	0.8905	0.8192	0.7383	0.6517	0.5630	0.4752	0.3910	0.3125	0.2415	0.1792	0.1265	0.0837	0.0508	0.0272	0.0120	0.0037	0.0005
	2	1.0000−	0.9995	0.9963	0.9880	0.9728	0.9492	0.9163	0.8735	0.8208	0.7585	0.6875	0.6090	0.5248	0.4370	0.3483	0.2617	0.1808	0.1095	0.0523	0.0140
	3	1.0000−	1.0000−	0.9999	0.9995	0.9984	0.9961	0.9919	0.9850	0.9744	0.9590	0.9375	0.9085	0.8704	0.8215	0.7599	0.6836	0.5904	0.4780	0.3439	0.1855
	4	1	1	1	1	1	1	1	1	1	1	1	1	1	1	1	1	1	1	1	1
5	0	0.9510	0.7738	0.5905	0.4437	0.3277	0.2373	0.1681	0.1160	0.0778	0.0503	0.0313	0.0185	0.0102	0.0053	0.0024	0.0010	0.0003	0.0001	0.0000+	0.0000+
	1	0.9990	0.9774	0.9185	0.8352	0.7373	0.6328	0.5282	0.4284	0.3370	0.2562	0.1875	0.1312	0.0870	0.0540	0.0308	0.0156	0.0067	0.0022	0.0005	0.0000+
	2	1.0000−	0.9988	0.9914	0.9734	0.9421	0.8965	0.8369	0.7648	0.6826	0.5931	0.5000	0.4069	0.3174	0.2352	0.1631	0.1035	0.0579	0.0266	0.0086	0.0012
	3	1.0000−	1.0000−	0.9995	0.9978	0.9933	0.9844	0.9692	0.9460	0.9130	0.8688	0.8125	0.7438	0.6630	0.5716	0.4718	0.3672	0.2627	0.1648	0.0815	0.0226
	4	1.0000−	1.0000−	0.9999	0.9999	0.9997	0.9990	0.9976	0.9947	0.9898	0.9815	0.9688	0.9497	0.9222	0.8840	0.8319	0.7627	0.6723	0.5563	0.4095	0.2262
	5	1	1	1	1	1	1	1	1	1	1	1	1	1	1	1	1	1	1	1	1
6	0	0.9415	0.7351	0.5314	0.3771	0.2621	0.1780	0.1176	0.0754	0.0467	0.0277	0.0156	0.0083	0.0041	0.0018	0.0007	0.0002	0.0001	0.0000+	0.0000+	0.0000+
	1	0.9985	0.9672	0.8857	0.7765	0.6554	0.5339	0.4202	0.3191	0.2333	0.1636	0.1094	0.0692	0.0410	0.0223	0.0109	0.0046	0.0016	0.0004	0.0001	0.0000+
	2	1.0000−	0.9978	0.9842	0.9527	0.9011	0.8306	0.7443	0.6471	0.5443	0.4415	0.3438	0.2553	0.1792	0.1174	0.0705	0.0376	0.0170	0.0059	0.0013	0.0001
	3	1.0000−	0.9999	0.9987	0.9941	0.9830	0.9624	0.9295	0.8826	0.8208	0.7447	0.6563	0.5585	0.4557	0.3529	0.2557	0.1694	0.0989	0.0473	0.0159	0.0022
	4	1.0000−	1.0000−	0.9999	0.9996	0.9984	0.9954	0.9891	0.9777	0.9590	0.9308	0.8906	0.8364	0.7667	0.6809	0.5798	0.4661	0.3446	0.2235	0.1143	0.0328
	5	1.0000−	1.0000−	1.0000−	1.0000−	0.9999	0.9998	0.9993	0.9982	0.9959	0.9917	0.9844	0.9723	0.9533	0.9246	0.8824	0.8220	0.7379	0.6229	0.4686	0.2649
	6	1	1	1	1	1	1	1	1	1	1	1	1	1	1	1	1	1	1	1	1
7	0	0.9321	0.6983	0.4783	0.3206	0.2097	0.1335	0.0824	0.0490	0.0280	0.0152	0.0078	0.0037	0.0016	0.0006	0.0002	0.0001	0.0000+	0.0000+	0.0000+	0.0000+
	1	0.9980	0.9556	0.8503	0.7166	0.5767	0.4449	0.3294	0.2338	0.1586	0.1024	0.0625	0.0357	0.0188	0.0090	0.0038	0.0013	0.0004	0.0001	0.0000+	0.0000+
	2	1.0000−	0.9962	0.9743	0.9262	0.8520	0.7564	0.6471	0.5323	0.4199	0.3164	0.2266	0.1529	0.0963	0.0556	0.0288	0.0129	0.0047	0.0012	0.0002	0.0000+
	3	1.0000−	0.9998	0.9973	0.9879	0.9667	0.9294	0.8740	0.8002	0.7102	0.6083	0.5000	0.3917	0.2898	0.1998	0.1260	0.0706	0.0333	0.0121	0.0027	0.0002
	4	1.0000−	1.0000−	0.9998	0.9988	0.9953	0.9871	0.9712	0.9444	0.9037	0.8471	0.7734	0.6836	0.5801	0.4677	0.3529	0.2436	0.1480	0.0738	0.0257	0.0038
	5	1.0000−	1.0000−	1.0000−	0.9999	0.9996	0.9987	0.9962	0.9910	0.9812	0.9643	0.9375	0.8976	0.8414	0.7662	0.6706	0.5551	0.4233	0.2834	0.1497	0.0444
	6	1.0000−	1.0000−	1.0000−	1.0000−	1.0000−	0.9999	0.9998	0.9994	0.9984	0.9963	0.9922	0.9848	0.9720	0.9510	0.9176	0.8665	0.7903	0.6794	0.5217	0.3017
	7	1	1	1	1	1	1	1	1	1	1	1	1	1	1	1	1	1	1	1	1

Note: 0.0000+ means the probability is 0.0000 rounded to four decimal places. However, the probability is *not* zero.
1.0000− means the probability is 1.0000 rounded to four decimal places. However, the probability is *not* one.

This table computes the cumulative probability of obtaining x successes in n trials of a binomial experiment with probability of success p.

Table IV (continued)

n	x	0.01	0.05	0.10	0.15	0.20	0.25	0.30	0.35	0.40	0.45	0.50	0.55	0.60	0.65	0.70	0.75	0.80	0.85	0.90	0.95
8	0	0.9227	0.6634	0.4305	0.2725	0.1678	0.1001	0.0576	0.0319	0.0168	0.0084	0.0039	0.0017	0.0007	0.0002	0.0001	0.0000+	0.0000+	0.0000+	0.0000+	0.0000+
	1	0.9973	0.9428	0.8131	0.6572	0.5033	0.3671	0.2553	0.1691	0.1064	0.0632	0.0352	0.0181	0.0085	0.0036	0.0013	0.0004	0.0001	0.0000+	0.0000+	0.0000+
	2	0.9999	0.9942	0.9619	0.8948	0.7969	0.6785	0.5518	0.4278	0.3154	0.2201	0.1445	0.0885	0.0498	0.0253	0.0113	0.0042	0.0012	0.0002	0.0000+	0.0000+
	3	1.0000−	0.9996	0.9950	0.9786	0.9437	0.8862	0.8059	0.7064	0.5941	0.4770	0.3633	0.2604	0.1737	0.1061	0.0580	0.0273	0.0104	0.0029	0.0004	0.0000+
	4	1.0000−	1.0000−	0.9996	0.9971	0.9896	0.9727	0.9420	0.8389	0.8263	0.7396	0.6367	0.5230	0.4059	0.2936	0.1941	0.1138	0.0563	0.0214	0.0050	0.0004
	5	1.0000−	1.0000−	1.0000−	0.9998	0.9988	0.9958	0.9887	0.9747	0.9502	0.9115	0.8555	0.7799	0.6846	0.5722	0.4482	0.3215	0.2031	0.1052	0.0381	0.0058
	6	1.0000−	1.0000−	1.0000−	1.0000−	0.9999	0.9996	0.9987	0.9964	0.9915	0.9819	0.9648	0.9368	0.8936	0.8309	0.7447	0.6329	0.4967	0.3428	0.1869	0.0572
	7	1.0000−	1.0000−	1.0000−	1.0000−	1.0000−	1.0000−	0.9999	0.9998	0.9993	0.9983	0.9961	0.9916	0.9832	0.9681	0.9424	0.8999	0.8322	0.7275	0.5695	0.3366
	8	1	1	1	1	1	1	1	1	1	1	1	1	1	1	1	1	1	1	1	1
9	0	0.9135	0.6302	0.3874	0.2316	0.1342	0.0751	0.0404	0.0207	0.0101	0.0046	0.0020	0.0008	0.0003	0.0001	0.0000+	0.0000+	0.0000+	0.0000+	0.0000+	0.0000+
	1	0.9966	0.9288	0.7748	0.5995	0.4362	0.3003	0.1960	0.1211	0.0705	0.0385	0.0195	0.0091	0.0038	0.0014	0.0004	0.0001	0.0000+	0.0000+	0.0000+	0.0000+
	2	0.9999	0.9916	0.9470	0.8591	0.7382	0.6007	0.4628	0.3373	0.2318	0.1495	0.0898	0.0498	0.0250	0.0112	0.0043	0.0013	0.0003	0.0000+	0.0000+	0.0000+
	3	1.0000−	0.9994	0.9917	0.9661	0.9144	0.8343	0.7297	0.6089	0.4826	0.3614	0.2539	0.1658	0.0994	0.0536	0.0253	0.0100	0.0031	0.0006	0.0001	0.0000+
	4	1.0000−	1.0000−	0.9991	0.9944	0.9804	0.9511	0.9012	0.8283	0.7334	0.6214	0.5000	0.3786	0.2666	0.1717	0.0988	0.0489	0.0196	0.0056	0.0009	0.0000+
	5	1.0000−	1.0000−	0.9999	0.9994	0.9969	0.9900	0.9747	0.9464	0.9006	0.8342	0.7461	0.6386	0.5174	0.3911	0.2703	0.1657	0.0856	0.0339	0.0083	0.0006
	6	1.0000−	1.0000−	1.0000−	1.0000−	0.9997	0.9987	0.9957	0.9888	0.9750	0.9502	0.9102	0.8505	0.7682	0.6627	0.5372	0.3993	0.2618	0.1409	0.0530	0.0084
	7	1.0000−	1.0000−	1.0000−	1.0000−	1.0000−	0.9999	0.9996	0.9986	0.9962	0.9909	0.9805	0.9615	0.9295	0.8789	0.8040	0.6997	0.5638	0.4005	0.2252	0.0712
	8	1.0000−	1.0000−	1.0000−	1.0000−	1.0000−	1.0000−	1.0000−	0.9999	0.9997	0.9992	0.9980	0.9954	0.9899	0.9793	0.9596	0.9249	0.8658	0.7684	0.6126	0.3698
	9	1	1	1	1	1	1	1	1	1	1	1	1	1	1	1	1	1	1	1	1
10	0	0.9044	0.5987	0.3487	0.1969	0.1074	0.0563	0.0282	0.0135	0.0060	0.0025	0.0010	0.0003	0.0001	0.0000+	0.0000+	0.0000+	0.0000+	0.0000+	0.0000+	0.0000+
	1	0.9957	0.9139	0.7361	0.5443	0.3758	0.2440	0.1493	0.0860	0.0464	0.0233	0.0107	0.0045	0.0017	0.0005	0.0001	0.0000+	0.0000+	0.0000+	0.0000+	0.0000+
	2	0.9999	0.9885	0.9298	0.8202	0.6778	0.5256	0.3828	0.2616	0.1673	0.0996	0.0547	0.0274	0.0123	0.0048	0.0016	0.0004	0.0001	0.0000+	0.0000+	0.0000+
	3	1.0000−	0.9990	0.9872	0.9500	0.8791	0.7759	0.6496	0.5138	0.3823	0.2660	0.1719	0.1020	0.0548	0.0260	0.0106	0.0035	0.0009	0.0001	0.0000+	0.0000+
	4	1.0000−	0.9999	0.9984	0.9901	0.9672	0.9219	0.8497	0.7515	0.6331	0.5044	0.3770	0.2616	0.1662	0.0949	0.0473	0.0197	0.0064	0.0014	0.0001	0.0000+
	5	1.0000−	1.0000−	0.9999	0.9986	0.9936	0.9803	0.9527	0.9051	0.8338	0.7384	0.6230	0.4956	0.3669	0.2485	0.1503	0.0781	0.0328	0.0099	0.0016	0.0001
	6	1.0000−	1.0000−	1.0000−	0.9999	0.9991	0.9965	0.9894	0.9740	0.9452	0.8980	0.8281	0.7340	0.6177	0.4862	0.3504	0.2241	0.1209	0.0500	0.0128	0.0010
	7	1.0000−	1.0000−	1.0000−	1.0000−	0.9999	0.9996	0.9984	0.9952	0.9877	0.9726	0.9453	0.9004	0.8327	0.7384	0.6172	0.4744	0.3222	0.1798	0.0702	0.0115
	8	1.0000−	1.0000−	1.0000−	1.0000−	1.0000−	1.0000−	0.9999	0.9995	0.9983	0.9955	0.9893	0.9767	0.9536	0.9140	0.8507	0.7560	0.6242	0.4557	0.2639	0.0861
	9	1.0000−	1.0000−	1.0000−	1.0000−	1.0000−	1.0000−	1.0000−	1.0000−	0.9999	0.9997	0.9990	0.9975	0.9940	0.9865	0.9718	0.9437	0.8926	0.8031	0.6513	0.4013
	10	1	1	1	1	1	1	1	1	1	1	1	1	1	1	1	1	1	1	1	1

Note: 0.0000+ means the probability is 0.0000 rounded to four decimal places. However, the probability is *not* zero.
1.0000− means the probability is 1.0000 rounded to four decimal places. However, the probability is *not* one.

Table IV (continued)

n	x												p									
		0.01	0.05	0.10	0.15	0.20	0.25	0.30	0.35	0.40	0.45	0.50	0.55	0.60	0.65	0.70	0.75	0.80	0.85	0.90	0.95
11	0	0.8953	0.5688	0.3138	0.1673	0.0859	0.0422	0.0198	0.0088	0.0036	0.0014	0.0005	0.0002	0.0000+	0.0000+	0.0000+	0.0000+	0.0000+	0.0000+	0.0000+	0.0000+
	1	0.9948	0.8981	0.6974	0.4922	0.3221	0.1971	0.1130	0.0606	0.0302	0.0139	0.0059	0.0022	0.0007	0.0002	0.0000+	0.0000+	0.0000+	0.0000+	0.0000+	0.0000+
	2	0.9998	0.9848	0.9104	0.7788	0.6174	0.4552	0.3127	0.2001	0.1189	0.0652	0.0327	0.0148	0.0059	0.0020	0.0006	0.0001	0.0000+	0.0000+	0.0000+	0.0000+
	3	1.0000−	0.9984	0.9815	0.9306	0.8389	0.7133	0.5696	0.4256	0.2963	0.1911	0.1133	0.0610	0.0293	0.0122	0.0043	0.0012	0.0002	0.0000+	0.0000+	0.0000+
	4	1.0000−	0.9999	0.9972	0.9841	0.9496	0.8854	0.7897	0.6683	0.5326	0.3971	0.2744	0.1738	0.0994	0.0501	0.0216	0.0076	0.0020	0.0003	0.0000+	0.0000+
	5	1.0000−	1.0000−	0.9997	0.9973	0.9883	0.9657	0.9218	0.8513	0.7535	0.6331	0.5000	0.3669	0.2465	0.1487	0.0782	0.0343	0.0117	0.0027	0.0003	0.0000+
	6	1.0000−	1.0000−	1.0000−	0.9997	0.9980	0.9924	0.9784	0.9499	0.9006	0.8262	0.7256	0.6029	0.4672	0.3317	0.2103	0.1146	0.0504	0.0159	0.0028	0.0001
	7	1.0000−	1.0000−	1.0000−	1.0000−	0.9998	0.9988	0.9957	0.9878	0.9707	0.9390	0.8867	0.8089	0.7037	0.5744	0.4304	0.2867	0.1611	0.0694	0.0185	0.0016
	8	1.0000−	1.0000−	1.0000−	1.0000−	1.0000−	0.9999	0.9994	0.9980	0.9941	0.9852	0.9673	0.9348	0.8811	0.7999	0.6873	0.5448	0.3826	0.2212	0.0896	0.0152
	9	1.0000−	1.0000−	1.0000−	1.0000−	1.0000−	1.0000−	0.9994	0.9998	0.9993	0.9976	0.9941	0.9861	0.9698	0.9394	0.8870	0.8029	0.6779	0.5078	0.3026	0.1019
	10	1.0000−	1.0000−	1.0000−	1.0000−	1.0000−	1.0000−	1.0000−	1.0000−	1.0000−	0.9996	0.9995	0.9986	0.9964	0.9912	0.9802	0.9578	0.9141	0.8327	0.6862	0.4312
	11	1	1	1	1	1	1	1	1	1	1	1	1	1	1	1	1	1	1	1	1
12	0	0.8864	0.5404	0.2824	0.1422	0.0687	0.0317	0.0138	0.0057	0.0022	0.0008	0.0002	0.0001	0.0000+	0.0000+	0.0000+	0.0000+	0.0000+	0.0000+	0.0000+	0.0000+
	1	0.9938	0.8816	0.6590	0.4435	0.2749	0.1584	0.0850	0.0424	0.0196	0.0083	0.0032	0.0011	0.0003	0.0001	0.0000+	0.0000+	0.0000+	0.0000+	0.0000+	0.0000+
	2	0.9998	0.9804	0.8891	0.7358	0.5583	0.3907	0.2528	0.1513	0.0834	0.0421	0.0193	0.0079	0.0028	0.0008	0.0002	0.0000+	0.0000+	0.0000+	0.0000+	0.0000+
	3	1.0000−	0.9978	0.9744	0.9078	0.7946	0.6488	0.4925	0.3467	0.2253	0.1345	0.0730	0.0356	0.0153	0.0056	0.0017	0.0004	0.0001	0.0000+	0.0000+	0.0000+
	4	1.0000−	0.9998	0.9957	0.9761	0.9274	0.8424	0.7237	0.5833	0.4382	0.3044	0.1938	0.1117	0.0573	0.0255	0.0095	0.0028	0.0006	0.0001	0.0000+	0.0000+
	5	1.0000−	1.0000−	0.9995	0.9954	0.9806	0.9456	0.8822	0.7873	0.6652	0.5269	0.3872	0.2607	0.1582	0.0846	0.0386	0.0143	0.0039	0.0007	0.0001	0.0000+
	6	1.0000−	1.0000−	0.9999	0.9993	0.9961	0.9857	0.9614	0.9154	0.8418	0.7393	0.6128	0.4731	0.3348	0.2127	0.1178	0.0544	0.0194	0.0046	0.0005	0.0000+
	7	1.0000−	1.0000−	1.0000−	0.9999	0.9994	0.9972	0.9905	0.9745	0.9427	0.8883	0.8062	0.6956	0.5618	0.4167	0.2763	0.1576	0.0726	0.0239	0.0043	0.0002
	8	1.0000−	1.0000−	1.0000−	1.0000−	0.9999	0.9996	0.9983	0.9944	0.9847	0.9644	0.9270	0.8655	0.7747	0.6533	0.5075	0.3512	0.2054	0.0922	0.0256	0.0022
	9	1.0000−	1.0000−	1.0000−	1.0000−	1.0000−	1.0000−	0.9998	0.9992	0.9972	0.9921	0.9807	0.9579	0.9166	0.8487	0.7472	0.6093	0.4417	0.2642	0.1109	0.0196
	10	1.0000−	1.0000−	1.0000−	1.0000−	1.0000−	1.0000−	1.0000−	0.9999	0.9997	0.9989	0.9968	0.9917	0.9804	0.9576	0.9150	0.8416	0.7251	0.5565	0.3410	0.1184
	11	1.0000−	1.0000−	1.0000−	1.0000−	1.0000−	1.0000−	1.0000−	1.0000−	1.0000−	0.9999	0.9998	0.9992	0.9978	0.9943	0.9862	0.9683	0.9313	0.8578	0.7176	0.4596
	12	1	1	1	1	1	1	1	1	1	1	1	1	1	1	1	1	1	1	1	1

Note: 0.0000+ means the probability is 0.0000 rounded to four decimal places. However, the probability is *not zero*.
1.0000− means the probability is 1.0000 rounded to four decimal places. However, the probability is *not one*.

Table IV (continued)

Values shown are cumulative binomial probabilities under p.

n	x	0.01	0.05	0.10	0.15	0.20	0.25	0.30	0.35	0.40	0.45	0.50	0.55	0.60	0.65	0.70	0.75	0.80	0.85	0.90	0.95
15	0	0.8601	0.4633	0.2059	0.0874	0.0352	0.0134	0.0047	0.0016	0.0005	0.0001	0.0000+	0.0000+	0.0000+	0.0000+	0.0000+	0.0000+	0.0000+	0.0000+	0.0000+	0.0000+
	1	0.9904	0.8290	0.5490	0.3186	0.1671	0.0802	0.0353	0.0142	0.0052	0.0017	0.0005	0.0001	0.0000+	0.0000+	0.0000+	0.0000+	0.0000+	0.0000+	0.0000+	0.0000+
	2	0.9996	0.9638	0.8159	0.6042	0.3980	0.2361	0.1268	0.0617	0.0271	0.0107	0.0037	0.0011	0.0003	0.0001	0.0000+	0.0000+	0.0000+	0.0000+	0.0000+	0.0000+
	3	1.0000−	0.9945	0.9444	0.8227	0.6482	0.4613	0.2969	0.1727	0.0905	0.0424	0.0176	0.0063	0.0019	0.0005	0.0001	0.0000+	0.0000+	0.0000+	0.0000+	0.0000+
	4	1.0000−	0.9994	0.9873	0.9383	0.8358	0.6865	0.5155	0.3519	0.2173	0.1204	0.0592	0.0255	0.0093	0.0028	0.0007	0.0001	0.0000+	0.0000+	0.0000+	0.0000+
	5	1.0000−	0.9999	0.9978	0.9832	0.9389	0.8516	0.7216	0.5643	0.4032	0.2608	0.1509	0.0769	0.0338	0.0124	0.0037	0.0008	0.0001	0.0000+	0.0000+	0.0000+
	6	1.0000−	1.0000−	0.9997	0.9964	0.9819	0.9434	0.8689	0.7548	0.6098	0.4522	0.3036	0.1818	0.0950	0.0422	0.0152	0.0042	0.0008	0.0001	0.0000+	0.0000+
	7	1.0000−	1.0000−	1.0000−	0.9994	0.9958	0.9827	0.9500	0.8868	0.7869	0.6535	0.5000	0.3465	0.2131	0.1132	0.0500	0.0173	0.0042	0.0006	0.0000+	0.0000+
	8	1.0000−	1.0000−	1.0000−	0.9999	0.9992	0.9958	0.9848	0.9578	0.9050	0.8182	0.6964	0.5478	0.3902	0.2452	0.1311	0.0566	0.0181	0.0036	0.0003	0.0000+
	9	1.0000−	1.0000−	1.0000−	1.0000−	0.9999	0.9992	0.9963	0.9876	0.9662	0.9231	0.8491	0.7392	0.5968	0.4357	0.2784	0.1484	0.0611	0.0168	0.0022	0.0001
	10	1.0000−	1.0000−	1.0000−	1.0000−	1.0000−	0.9999	0.9993	0.9972	0.9907	0.9745	0.9408	0.8796	0.7827	0.6481	0.4845	0.3135	0.1642	0.0617	0.0127	0.0006
	11	1.0000−	1.0000−	1.0000−	1.0000−	1.0000−	1.0000−	0.9999	0.9995	0.9981	0.9937	0.9824	0.9576	0.9095	0.8273	0.7031	0.5387	0.3518	0.1773	0.0556	0.0055
	12	1.0000−	1.0000−	1.0000−	1.0000−	1.0000−	1.0000−	1.0000−	0.9999	0.9997	0.9989	0.9963	0.9893	0.9729	0.9383	0.8732	0.7639	0.6020	0.3958	0.1841	0.0362
	13	1.0000−	1.0000−	1.0000−	1.0000−	1.0000−	1.0000−	1.0000−	1.0000−	1.0000−	0.9999	0.9995	0.9983	0.9948	0.9858	0.9647	0.9198	0.8329	0.6814	0.4510	0.1710
	14	1.0000−	1.0000−	1.0000−	1.0000−	1.0000−	1.0000−	1.0000−	1.0000−	1.0000−	1.0000−	1.0000−	0.9999	0.9995	0.9984	0.9953	0.9866	0.9648	0.9126	0.7941	0.5367
	15	1	1	1	1	1	1	1	1	1	1	1	1	1	1	1	1	1	1	1	1
20	0	0.8179	0.3585	0.1216	0.0388	0.0115	0.0032	0.0008	0.0002	0.0000+	0.0000+	0.0000+	0.0000+	0.0000+	0.0000+	0.0000+	0.0000+	0.0000+	0.0000+	0.0000+	0.0000+
	1	0.9831	0.7358	0.3917	0.1756	0.0692	0.0243	0.0076	0.0021	0.0005	0.0001	0.0000+	0.0000+	0.0000+	0.0000+	0.0000+	0.0000+	0.0000+	0.0000+	0.0000+	0.0000+
	2	0.9990	0.9245	0.6769	0.4049	0.2061	0.0913	0.0355	0.0121	0.0036	0.0009	0.0002	0.0000+	0.0000+	0.0000+	0.0000+	0.0000+	0.0000+	0.0000+	0.0000+	0.0000+
	3	1.0000−	0.9841	0.8670	0.6477	0.4114	0.2252	0.1071	0.0444	0.0160	0.0049	0.0013	0.0003	0.0000+	0.0000+	0.0000+	0.0000+	0.0000+	0.0000+	0.0000+	0.0000+
	4	1.0000−	0.9974	0.9568	0.8298	0.6296	0.4148	0.2375	0.1182	0.0510	0.0189	0.0059	0.0015	0.0003	0.0000+	0.0000+	0.0000+	0.0000+	0.0000+	0.0000+	0.0000+
	5	1.0000−	0.9997	0.9887	0.9327	0.8042	0.6172	0.4164	0.2454	0.1256	0.0553	0.0207	0.0064	0.0016	0.0003	0.0000+	0.0000+	0.0000+	0.0000+	0.0000+	0.0000+
	6	1.0000−	1.0000−	0.9976	0.9781	0.9133	0.7858	0.6080	0.4166	0.2500	0.1299	0.0577	0.0214	0.0065	0.0015	0.0003	0.0000+	0.0000+	0.0000+	0.0000+	0.0000+
	7	1.0000−	1.0000−	0.9996	0.9941	0.9679	0.8982	0.7723	0.6010	0.4159	0.2520	0.1316	0.0580	0.0210	0.0060	0.0013	0.0002	0.0000+	0.0000+	0.0000+	0.0000+
	8	1.0000−	1.0000−	0.9999	0.9987	0.9900	0.9591	0.8867	0.7624	0.5956	0.4143	0.2517	0.1308	0.0565	0.0196	0.0051	0.0009	0.0001	0.0000+	0.0000+	0.0000+
	9	1.0000−	1.0000−	1.0000−	0.9998	0.9974	0.9861	0.9520	0.8782	0.7553	0.5914	0.4119	0.2493	0.1275	0.0532	0.0171	0.0039	0.0006	0.0000+	0.0000+	0.0000+
	10	1.0000−	1.0000−	1.0000−	1.0000−	0.9994	0.9961	0.9829	0.9468	0.8725	0.7507	0.5881	0.4086	0.2447	0.1218	0.0480	0.0139	0.0026	0.0002	0.0000+	0.0000+
	11	1.0000−	1.0000−	1.0000−	1.0000−	0.9999	0.9991	0.9949	0.9804	0.9435	0.8692	0.7483	0.5857	0.4044	0.2376	0.1133	0.0409	0.0100	0.0013	0.0001	0.0000+
	12	1.0000−	1.0000−	1.0000−	1.0000−	1.0000−	0.9998	0.9987	0.9940	0.9790	0.9420	0.8684	0.7480	0.5841	0.3990	0.2277	0.1018	0.0321	0.0059	0.0004	0.0000+
	13	1.0000−	1.0000−	1.0000−	1.0000−	1.0000−	1.0000−	0.9997	0.9985	0.9935	0.9786	0.9423	0.8701	0.7500	0.5834	0.3920	0.2142	0.0867	0.0219	0.0024	0.0000+
	14	1.0000−	1.0000−	1.0000−	1.0000−	1.0000−	1.0000−	1.0000−	0.9997	0.9984	0.9936	0.9793	0.9447	0.8744	0.7546	0.5836	0.3828	0.1958	0.0673	0.0113	0.0003
	15	1.0000−	1.0000−	1.0000−	1.0000−	1.0000−	1.0000−	1.0000−	1.0000−	0.9997	0.9985	0.9941	0.9811	0.9490	0.8818	0.7625	0.5852	0.3704	0.1702	0.0432	0.0026
	16	1.0000−	1.0000−	1.0000−	1.0000−	1.0000−	1.0000−	1.0000−	1.0000−	1.0000−	0.9997	0.9987	0.9951	0.9840	0.9556	0.8929	0.7748	0.5886	0.3523	0.1330	0.0159
	17	1.0000−	1.0000−	1.0000−	1.0000−	1.0000−	1.0000−	1.0000−	1.0000−	1.0000−	1.0000−	0.9998	0.9991	0.9964	0.9879	0.9645	0.9087	0.7939	0.5951	0.3231	0.0755
	18	1.0000−	1.0000−	1.0000−	1.0000−	1.0000−	1.0000−	1.0000−	1.0000−	1.0000−	1.0000−	1.0000−	0.9999	0.9995	0.9979	0.9924	0.9757	0.9308	0.8244	0.6083	0.2642
	19	1.0000−	1.0000−	1.0000−	1.0000−	1.0000−	1.0000−	1.0000−	1.0000−	1.0000−	1.0000−	1.0000−	1.0000−	1.0000−	0.9998	0.9992	0.9968	0.9885	0.9612	0.8784	0.6415
	20	1	1	1	1	1	1	1	1	1	1	1	1	1	1	1	1	1	1	1	1

Note: 0.0000+ means the probability is 0.0000 rounded to four decimal places. However, the probability is *not zero*.
1.0000− means the probability is 1.0000 rounded to four decimal places. However, the probability is *not one*.

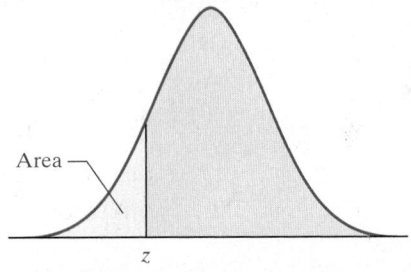

Area

z

Table V

Standard Normal Distribution

z	0.00	0.01	0.02	0.03	0.04	0.05	0.06	0.07	0.08	0.09
−3.4	0.0003	0.0003	0.0003	0.0003	0.0003	0.0003	0.0003	0.0003	0.0003	0.0002
−3.3	0.0005	0.0005	0.0005	0.0004	0.0004	0.0004	0.0004	0.0004	0.0004	0.0003
−3.2	0.0007	0.0007	0.0006	0.0006	0.0006	0.0006	0.0006	0.0005	0.0005	0.0005
−3.1	0.0010	0.0009	0.0009	0.0009	0.0008	0.0008	0.0008	0.0008	0.0007	0.0007
−3.0	0.0013	0.0013	0.0013	0.0012	0.0012	0.0011	0.0011	0.0011	0.0010	0.0010
−2.9	0.0019	0.0018	0.0018	0.0017	0.0016	0.0016	0.0015	0.0015	0.0014	0.0014
−2.8	0.0026	0.0025	0.0024	0.0023	0.0023	0.0022	0.0021	0.0021	0.0020	0.0019
−2.7	0.0035	0.0034	0.0033	0.0032	0.0031	0.0030	0.0029	0.0028	0.0027	0.0026
−2.6	0.0047	0.0045	0.0044	0.0043	0.0041	0.0040	0.0039	0.0038	0.0037	0.0036
−2.5	0.0062	0.0060	0.0059	0.0057	0.0055	0.0054	0.0052	0.0051	0.0049	0.0048
−2.4	0.0082	0.0080	0.0078	0.0075	0.0073	0.0071	0.0069	0.0068	0.0066	0.0064
−2.3	0.0107	0.0104	0.0102	0.0099	0.0096	0.0094	0.0091	0.0089	0.0087	0.0084
−2.2	0.0139	0.0136	0.0132	0.0129	0.0125	0.0122	0.0119	0.0116	0.0113	0.0110
−2.1	0.0179	0.0174	0.0170	0.0166	0.0162	0.0158	0.0154	0.0150	0.0146	0.0143
−2.0	0.0228	0.0222	0.0217	0.0212	0.0207	0.0202	0.0197	0.0192	0.0188	0.0183
−1.9	0.0287	0.0281	0.0274	0.0268	0.0262	0.0256	0.0250	0.0244	0.0239	0.0233
−1.8	0.0359	0.0351	0.0344	0.0336	0.0329	0.0322	0.0314	0.0307	0.0301	0.0294
−1.7	0.0446	0.0436	0.0427	0.0418	0.0409	0.0401	0.0392	0.0384	0.0375	0.0367
−1.6	0.0548	0.0537	0.0526	0.0516	0.0505	0.0495	0.0485	0.0475	0.0465	0.0455
−1.5	0.0668	0.0655	0.0643	0.0630	0.0618	0.0606	0.0594	0.0582	0.0571	0.0559
−1.4	0.0808	0.0793	0.0778	0.0764	0.0749	0.0735	0.0721	0.0708	0.0694	0.0681
−1.3	0.0968	0.0951	0.0934	0.0918	0.0901	0.0885	0.0869	0.0853	0.0838	0.0823
−1.2	0.1151	0.1131	0.1112	0.1093	0.1075	0.1056	0.1038	0.1020	0.1003	0.0985
−1.1	0.1357	0.1335	0.1314	0.1292	0.1271	0.1251	0.1230	0.1210	0.1190	0.1170
−1.0	0.1587	0.1562	0.1539	0.1515	0.1492	0.1469	0.1446	0.1423	0.1401	0.1379
−0.9	0.1841	0.1814	0.1788	0.1762	0.1736	0.1711	0.1685	0.1660	0.1635	0.1611
−0.8	0.2119	0.2090	0.2061	0.2033	0.2005	0.1977	0.1949	0.1922	0.1894	0.1867
−0.7	0.2420	0.2389	0.2358	0.2327	0.2296	0.2266	0.2236	0.2206	0.2177	0.2148
−0.6	0.2743	0.2709	0.2676	0.2643	0.2611	0.2578	0.2546	0.2514	0.2483	0.2451
−0.5	0.3085	0.3050	0.3015	0.2981	0.2946	0.2912	0.2877	0.2843	0.2810	0.2776
−0.4	0.3446	0.3409	0.3372	0.3336	0.3300	0.3264	0.3228	0.3192	0.3156	0.3121
−0.3	0.3821	0.3783	0.3745	0.3707	0.3669	0.3632	0.3594	0.3557	0.3520	0.3483
−0.2	0.4207	0.4168	0.4129	0.4090	0.4052	0.4013	0.3974	0.3936	0.3897	0.3859
−0.1	0.4602	0.4562	0.4522	0.4483	0.4443	0.4404	0.4364	0.4325	0.4286	0.4247
−0.0	0.5000	0.4960	0.4920	0.4880	0.4840	0.4801	0.4761	0.4721	0.4681	0.4641

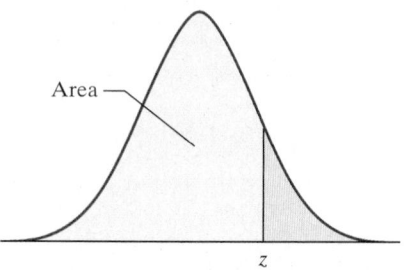

Area

z

Table V (continued)

Standard Normal Distribution

z	0.00	0.01	0.02	0.03	0.04	0.05	0.06	0.07	0.08	0.09
0.0	0.5000	0.5040	0.5080	0.5120	0.5160	0.5199	0.5239	0.5279	0.5319	0.5359
0.1	0.5398	0.5438	0.5478	0.5517	0.5557	0.5596	0.5636	0.5675	0.5714	0.5753
0.2	0.5793	0.5832	0.5871	0.5910	0.5948	0.5987	0.6026	0.6064	0.6103	0.6141
0.3	0.6179	0.6217	0.6255	0.6293	0.6331	0.6368	0.6406	0.6443	0.6480	0.6517
0.4	0.6554	0.6591	0.6628	0.6664	0.6700	0.6736	0.6772	0.6808	0.6844	0.6879
0.5	0.6915	0.6950	0.6985	0.7019	0.7054	0.7088	0.7123	0.7157	0.7190	0.7224
0.6	0.7257	0.7291	0.7324	0.7357	0.7389	0.7422	0.7454	0.7486	0.7517	0.7549
0.7	0.7580	0.7611	0.7642	0.7673	0.7704	0.7734	0.7764	0.7794	0.7823	0.7852
0.8	0.7881	0.7910	0.7939	0.7967	0.7995	0.8023	0.8051	0.8078	0.8106	0.8133
0.9	0.8159	0.8186	0.8212	0.8238	0.8264	0.8289	0.8315	0.8340	0.8365	0.8389
1.0	0.8413	0.8438	0.8461	0.8485	0.8508	0.8531	0.8554	0.8577	0.8599	0.8621
1.1	0.8643	0.8665	0.8686	0.8708	0.8729	0.8749	0.8770	0.8790	0.8810	0.8830
1.2	0.8849	0.8869	0.8888	0.8907	0.8925	0.8944	0.8962	0.8980	0.8997	0.9015
1.3	0.9032	0.9049	0.9066	0.9082	0.9099	0.9115	0.9131	0.9147	0.9162	0.9177
1.4	0.9192	0.9207	0.9222	0.9236	0.9251	0.9265	0.9279	0.9292	0.9306	0.9319
1.5	0.9332	0.9345	0.9357	0.9370	0.9382	0.9394	0.9406	0.9418	0.9429	0.9441
1.6	0.9452	0.9463	0.9474	0.9484	0.9495	0.9505	0.9515	0.9525	0.9535	0.9545
1.7	0.9554	0.9564	0.9573	0.9582	0.9591	0.9599	0.9608	0.9616	0.9625	0.9633
1.8	0.9641	0.9649	0.9656	0.9664	0.9671	0.9678	0.9686	0.9693	0.9699	0.9706
1.9	0.9713	0.9719	0.9726	0.9732	0.9738	0.9744	0.9750	0.9756	0.9761	0.9767
2.0	0.9772	0.9778	0.9783	0.9788	0.9793	0.9798	0.9803	0.9808	0.9812	0.9817
2.1	0.9821	0.9826	0.9830	0.9834	0.9838	0.9842	0.9846	0.9850	0.9854	0.9857
2.2	0.9861	0.9864	0.9868	0.9871	0.9875	0.9878	0.9881	0.9884	0.9887	0.9890
2.3	0.9893	0.9896	0.9898	0.9901	0.9904	0.9906	0.9909	0.9911	0.9913	0.9916
2.4	0.9918	0.9920	0.9922	0.9925	0.9927	0.9929	0.9931	0.9932	0.9934	0.9936
2.5	0.9938	0.9940	0.9941	0.9943	0.9945	0.9946	0.9948	0.9949	0.9951	0.9952
2.6	0.9953	0.9955	0.9956	0.9957	0.9959	0.9960	0.9961	0.9962	0.9963	0.9964
2.7	0.9965	0.9966	0.9967	0.9968	0.9969	0.9970	0.9971	0.9972	0.9973	0.9974
2.8	0.9974	0.9975	0.9976	0.9977	0.9977	0.9978	0.9979	0.9979	0.9980	0.9981
2.9	0.9981	0.9982	0.9982	0.9983	0.9984	0.9984	0.9985	0.9985	0.9986	0.9986
3.0	0.9987	0.9987	0.9987	0.9988	0.9988	0.9989	0.9989	0.9989	0.9990	0.9990
3.1	0.9990	0.9991	0.9991	0.9991	0.9992	0.9992	0.9992	0.9992	0.9993	0.9993
3.2	0.9993	0.9993	0.9994	0.9994	0.9994	0.9994	0.9994	0.9995	0.9995	0.9995
3.3	0.9995	0.9995	0.9995	0.9996	0.9996	0.9996	0.9996	0.9996	0.9996	0.9997
3.4	0.9997	0.9997	0.9997	0.9997	0.9997	0.9997	0.9997	0.9997	0.9997	0.9998

(b) Confidence Interval Critical Values, $z_{\alpha/2}$

Level of Confidence	Critical Value, $z_{\alpha/2}$
0.90 or 90%	1.645
0.95 or 95%	1.96
0.98 or 98%	2.33
0.99 or 99%	2.575

(c) Hypothesis Testing Critical Values

Level of Significance, α	Left-Tailed	Right-Tailed	Two-Tailed
0.10	−1.28	1.28	±1.645
0.05	−1.645	1.645	±1.96
0.01	−2.33	2.33	±2.575

Table VI	
Sample Size, n	**Critical Value**
5	0.880
6	0.888
7	0.898
8	0.906
9	0.912
10	0.918
11	0.923
12	0.928
13	0.932
14	0.935
15	0.939
16	0.941
17	0.944
18	0.946
19	0.949
20	0.951
21	0.952
22	0.954
23	0.956
24	0.957
25	0.959
30	0.960

Source: S. W. Looney and T. R. Gulledge, Jr. "Use of the Correlation Coefficient with Normal Probability Plots," *American Statistician* 39(Feb. 1985): 75–79.

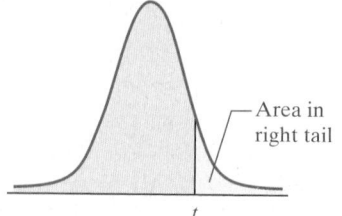
Area in right tail

Table VII

t-Distribution
Area in Right Tail

Degrees of Freedom	0.25	0.20	0.15	0.10	0.05	0.025	0.02	0.01	0.005	0.0025	0.001	0.0005
1	1.000	1.376	1.963	3.078	6.314	12.706	15.894	31.821	63.657	127.321	318.309	636.619
2	0.816	1.061	1.386	1.886	2.920	4.303	4.849	6.965	9.925	14.089	22.327	31.599
3	0.765	0.978	1.250	1.638	2.353	3.182	3.482	4.541	5.841	7.453	10.215	12.924
4	0.741	0.941	1.190	1.533	2.132	2.776	2.999	3.747	4.604	5.598	7.173	8.610
5	0.727	0.920	1.156	1.476	2.015	2.571	2.757	3.365	4.032	4.773	5.893	6.869
6	0.718	0.906	1.134	1.440	1.943	2.447	2.612	3.143	3.707	4.317	5.208	5.959
7	0.711	0.896	1.119	1.415	1.895	2.365	2.517	2.998	3.499	4.029	4.785	5.408
8	0.706	0.889	1.108	1.397	1.860	2.306	2.449	2.896	3.355	3.833	4.501	5.041
9	0.703	0.883	1.100	1.383	1.833	2.262	2.398	2.821	3.250	3.690	4.297	4.781
10	0.700	0.879	1.093	1.372	1.812	2.228	2.359	2.764	3.169	3.581	4.144	4.587
11	0.697	0.876	1.088	1.363	1.796	2.201	2.328	2.718	3.106	3.497	4.025	4.437
12	0.695	0.873	1.083	1.356	1.782	2.179	2.303	2.681	3.055	3.428	3.930	4.318
13	0.694	0.870	1.079	1.350	1.771	2.160	2.282	2.650	3.012	3.372	3.852	4.221
14	0.692	0.868	1.076	1.345	1.761	2.145	2.264	2.624	2.977	3.326	3.787	4.140
15	0.691	0.866	1.074	1.341	1.753	2.131	2.249	2.602	2.947	3.286	3.733	4.073
16	0.690	0.865	1.071	1.337	1.746	2.120	2.235	2.583	2.921	3.252	3.686	4.015
17	0.689	0.863	1.069	1.333	1.740	2.110	2.224	2.567	2.898	3.222	3.646	3.965
18	0.688	0.862	1.067	1.330	1.734	2.101	2.214	2.552	2.878	3.197	3.610	3.922
19	0.688	0.861	1.066	1.328	1.729	2.093	2.205	2.539	2.861	3.174	3.579	3.883
20	0.687	0.860	1.064	1.325	1.725	2.086	2.197	2.528	2.845	3.153	3.552	3.850
21	0.686	0.859	1.063	1.323	1.721	2.080	2.189	2.518	2.831	3.135	3.527	3.819
22	0.686	0.858	1.061	1.321	1.717	2.074	2.183	2.508	2.819	3.119	3.505	3.792
23	0.685	0.858	1.060	1.319	1.714	2.069	2.177	2.500	2.807	3.104	3.485	3.768
24	0.685	0.857	1.059	1.318	1.711	2.064	2.172	2.492	2.797	3.091	3.467	3.745
25	0.684	0.856	1.058	1.316	1.708	2.060	2.167	2.485	2.787	3.078	3.450	3.725
26	0.684	0.856	1.058	1.315	1.706	2.056	2.162	2.479	2.779	3.067	3.435	3.707
27	0.684	0.855	1.057	1.314	1.703	2.052	2.158	2.473	2.771	3.057	3.421	3.690
28	0.683	0.855	1.056	1.313	1.701	2.048	2.154	2.467	2.763	3.047	3.408	3.674
29	0.683	0.854	1.055	1.311	1.699	2.045	2.150	2.462	2.756	3.038	3.396	3.659
30	0.683	0.854	1.055	1.310	1.697	2.042	2.147	2.457	2.750	3.030	3.385	3.646
31	0.682	0.853	1.054	1.309	1.696	2.040	2.144	2.453	2.744	3.022	3.375	3.633
32	0.682	0.853	1.054	1.309	1.694	2.037	2.141	2.449	2.738	3.015	3.365	3.622
33	0.682	0.853	1.053	1.308	1.692	2.035	2.138	2.445	2.733	3.008	3.356	3.611
34	0.682	0.852	1.052	1.307	1.691	2.032	2.136	2.441	2.728	3.002	3.348	3.601
35	0.682	0.852	1.052	1.306	1.690	2.030	2.133	2.438	2.724	2.996	3.340	3.591
36	0.681	0.852	1.052	1.306	1.688	2.028	2.131	2.434	2.719	2.990	3.333	3.582
37	0.681	0.851	1.051	1.305	1.687	2.026	2.129	2.431	2.715	2.985	3.326	3.574
38	0.681	0.851	1.051	1.304	1.686	2.024	2.127	2.429	2.712	2.980	3.319	3.566
39	0.681	0.851	1.050	1.304	1.685	2.023	2.125	2.426	2.708	2.976	3.313	3.558
40	0.681	0.851	1.050	1.303	1.684	2.021	2.123	2.423	2.704	2.971	3.307	3.551
50	0.679	0.849	1.047	1.299	1.676	2.009	2.109	2.403	2.678	2.937	3.261	3.496
60	0.679	0.848	1.045	1.296	1.671	2.000	2.099	2.390	2.660	2.915	3.232	3.460
70	0.678	0.847	1.044	1.294	1.667	1.994	2.093	2.381	2.648	2.899	3.211	3.435
80	0.678	0.846	1.043	1.292	1.664	1.990	2.088	2.374	2.639	2.887	3.195	3.416
90	0.677	0.846	1.042	1.291	1.662	1.987	2.084	2.368	2.632	2.878	3.183	3.402
100	0.677	0.845	1.042	1.290	1.660	1.984	2.081	2.364	2.626	2.871	3.174	3.390
1000	0.675	0.842	1.037	1.282	1.646	1.962	2.056	2.330	2.581	2.813	3.098	3.300
z	0.674	0.842	1.036	1.282	1.645	1.960	2.054	2.326	2.576	2.807	3.090	3.291

Table VIII

Chi-Square (χ^2) Distribution
Area to the Right of Critical Value

Degrees of Freedom	0.995	0.99	0.975	0.95	0.90	0.10	0.05	0.025	0.01	0.005
1	—	—	0.001	0.004	0.016	2.706	3.841	5.024	6.635	7.879
2	0.010	0.020	0.051	0.103	0.211	4.605	5.991	7.378	9.210	10.597
3	0.072	0.115	0.216	0.352	0.584	6.251	7.815	9.348	11.345	12.838
4	0.207	0.297	0.484	0.711	1.064	7.779	9.488	11.143	13.277	14.860
5	0.412	0.554	0.831	1.145	1.610	9.236	11.070	12.833	15.086	16.750
6	0.676	0.872	1.237	1.635	2.204	10.645	12.592	14.449	16.812	18.548
7	0.989	1.239	1.690	2.167	2.833	12.017	14.067	16.013	18.475	20.278
8	1.344	1.646	2.180	2.733	3.490	13.362	15.507	17.535	20.090	21.955
9	1.735	2.088	2.700	3.325	4.168	14.684	16.919	19.023	21.666	23.589
10	2.156	2.558	3.247	3.940	4.865	15.987	18.307	20.483	23.209	25.188
11	2.603	3.053	3.816	4.575	5.578	17.275	19.675	21.920	24.725	26.757
12	3.074	3.571	4.404	5.226	6.304	18.549	21.026	23.337	26.217	28.300
13	3.565	4.107	5.009	5.892	7.042	19.812	22.362	24.736	27.688	29.819
14	4.075	4.660	5.629	6.571	7.790	21.064	23.685	26.119	29.141	31.319
15	4.601	5.229	6.262	7.261	8.547	22.307	24.996	27.488	30.578	32.801
16	5.142	5.812	6.908	7.962	9.312	23.542	26.296	28.845	32.000	34.267
17	5.697	6.408	7.564	8.672	10.085	24.769	27.587	30.191	33.409	35.718
18	6.265	7.015	8.231	9.390	10.865	25.989	28.869	31.526	34.805	37.156
19	6.844	7.633	8.907	10.117	11.651	27.204	30.144	32.852	36.191	38.582
20	7.434	8.260	9.591	10.851	12.443	28.412	31.410	34.170	37.566	39.997
21	8.034	8.897	10.283	11.591	13.240	29.615	32.671	35.479	38.932	41.401
22	8.643	9.542	10.982	12.338	14.041	30.813	33.924	36.781	40.289	42.796
23	9.260	10.196	11.689	13.091	14.848	32.007	35.172	38.076	41.638	44.181
24	9.886	10.856	12.401	13.848	15.659	33.196	36.415	39.364	42.980	45.559
25	10.520	11.524	13.120	14.611	16.473	34.382	37.652	40.646	44.314	46.928
26	11.160	12.198	13.844	15.379	17.292	35.563	38.885	41.923	45.642	48.290
27	11.808	12.879	14.573	16.151	18.114	36.741	40.113	43.195	46.963	49.645
28	12.461	13.565	15.308	16.928	18.939	37.916	41.337	44.461	48.278	50.993
29	13.121	14.256	16.047	17.708	19.768	39.087	42.557	45.722	49.588	52.336
30	13.787	14.953	16.791	18.493	20.599	40.256	43.773	46.979	50.892	53.672
40	20.707	22.164	24.433	26.509	29.051	51.805	55.758	59.342	63.691	66.766
50	27.991	29.707	32.357	34.764	37.689	63.167	67.505	71.420	76.154	79.490
60	35.534	37.485	40.482	43.188	46.459	74.397	79.082	83.298	88.379	91.952
70	43.275	45.442	48.758	51.739	55.329	85.527	90.531	95.023	100.425	104.215
80	51.172	53.540	57.153	60.391	64.278	96.578	101.879	106.629	112.329	116.321
90	59.196	61.754	65.647	69.126	73.291	107.565	113.145	118.136	124.116	128.299
100	67.328	70.065	74.222	77.929	82.358	118.498	124.342	129.561	135.807	140.169

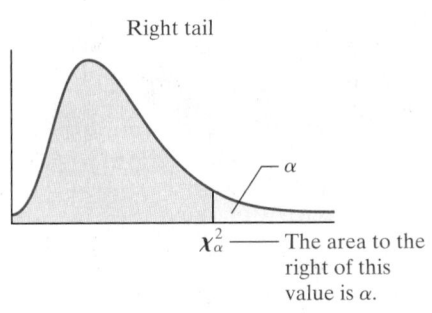

Right tail

χ^2_α —— The area to the right of this value is α.

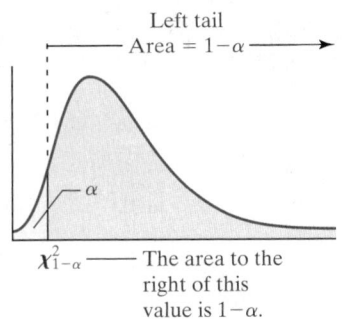

Left tail
Area = $1-\alpha$

$\chi^2_{1-\alpha}$ —— The area to the right of this value is $1-\alpha$.

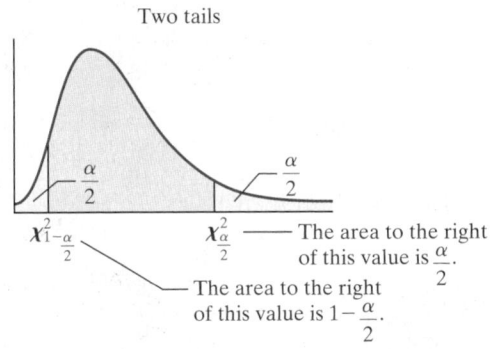

Two tails

$\chi^2_{1-\frac{\alpha}{2}}$ $\chi^2_{\frac{\alpha}{2}}$ —— The area to the right of this value is $\frac{\alpha}{2}$.

—— The area to the right of this value is $1-\frac{\alpha}{2}$.

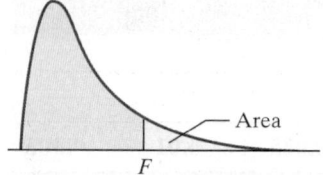
Area

F

Table IX

F-Distribution Critical Values

	Area in Right Tail	Degrees of Freedom in the Numerator							
		1	**2**	**3**	**4**	**5**	**6**	**7**	**8**
1	0.100	39.86	49.59	53.59	55.83	57.24	58.20	58.91	59.44
	0.050	161.45	199.50	215.71	224.58	230.16	233.99	236.77	238.88
	0.025	647.79	799.50	864.16	899.58	921.85	937.11	948.22	956.66
	0.010	4052.20	4999.50	5403.35	5624.58	5763.65	5858.99	5928.36	5981.07
	0.001	405284.07	499999.50	540379.20	562499.58	576404.56	585937.11	592873.29	598144.16
2	0.100	8.53	9.00	9.16	9.24	9.29	9.33	9.35	9.37
	0.050	18.51	19.00	19.16	19.25	19.30	19.33	19.35	19.37
	0.025	38.51	39.00	39.17	39.25	39.30	39.33	39.36	39.37
	0.010	98.50	99.00	99.17	99.25	99.30	99.33	99.36	99.37
	0.001	998.50	999.00	999.17	999.25	999.30	999.33	999.36	999.37
3	0.100	5.54	5.46	5.39	5.34	5.31	5.28	5.27	5.25
	0.050	10.13	9.55	9.28	9.12	9.01	8.94	8.89	8.85
	0.025	17.44	16.04	15.44	15.10	14.88	14.73	14.62	14.54
	0.010	34.12	30.82	29.46	28.71	28.24	27.91	27.67	27.49
	0.001	167.03	148.50	141.11	137.10	134.58	132.85	131.58	130.62
4	0.100	4.54	4.32	4.19	4.11	4.05	4.01	3.98	3.95
	0.050	7.71	6.94	6.59	6.39	6.26	6.16	6.09	6.04
	0.025	12.22	10.65	9.98	9.60	9.36	9.20	9.07	8.98
	0.010	21.20	18.00	16.69	15.98	15.52	15.21	14.98	14.80
	0.001	74.14	61.25	56.18	53.44	51.71	50.53	49.66	49.00
5	0.100	4.06	3.78	3.62	3.52	3.45	3.40	3.37	3.34
	0.050	6.61	5.79	5.41	5.19	5.05	4.95	4.88	4.82
	0.025	10.01	8.43	7.76	7.39	7.15	6.98	6.85	6.76
	0.010	16.26	13.27	12.06	11.39	10.97	10.67	10.46	10.29
	0.001	47.18	37.12	33.20	31.09	29.75	28.83	28.16	27.65
6	0.100	3.78	3.46	3.29	3.18	3.11	3.05	3.01	2.98
	0.050	5.99	5.14	4.76	4.53	4.39	4.28	4.21	4.15
	0.025	8.81	7.26	6.60	6.23	5.99	5.82	5.70	5.60
	0.010	13.75	10.92	9.78	9.15	8.75	8.47	8.26	8.10
	0.001	35.51	27.00	23.70	21.92	20.80	20.03	19.46	19.03
7	0.100	3.59	3.26	3.07	2.96	2.88	2.83	2.78	2.75
	0.050	5.59	4.74	4.35	4.12	3.97	3.87	3.79	3.73
	0.025	8.07	6.54	5.89	5.52	5.29	5.12	4.99	4.90
	0.010	12.25	9.55	8.45	7.85	7.46	7.19	6.99	6.84
	0.001	29.25	21.69	18.77	17.20	16.21	15.52	15.02	14.63
8	0.100	3.46	3.11	2.92	2.81	2.73	2.67	2.62	2.59
	0.050	5.32	4.46	4.07	3.84	3.69	3.58	3.50	3.44
	0.025	7.57	6.06	5.42	5.05	4.82	4.65	4.53	4.43
	0.010	11.26	8.65	7.59	7.01	6.63	6.37	6.18	6.03
	0.001	25.41	18.49	15.83	14.39	13.48	12.86	12.40	12.05

Degrees of Freedom in the Denominator

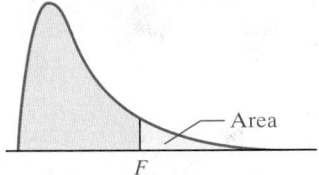
Area

F

Table IX (*continued*)

F-Distribution Critical Values

	Area in Right Tail	Degrees of Freedom in the Numerator							
		9	**10**	**15**	**20**	**30**	**60**	**120**	**1000**
1	0.100	59.86	60.19	61.22	61.74	62.26	62.79	63.06	63.30
	0.050	240.54	241.88	245.95	248.01	250.10	252.20	253.25	254.19
	0.025	963.28	968.63	984.87	993.10	1001.41	1009.80	1014.02	1017.75
	0.010	6022.47	6055.85	6157.28	6208.73	6260.65	6313.03	6339.39	6362.68
	0.001	602283.99	605620.97	615763.66	620907.67	626098.96	631336.56	633972.40	636301.21
2	0.100	9.38	9.39	9.42	9.44	9.46	9.47	9.48	9.49
	0.050	19.38	19.40	19.43	19.45	19.46	19.48	19.49	19.49
	0.025	39.39	39.40	39.43	39.45	39.46	39.48	39.49	39.50
	0.010	99.39	99.40	99.43	99.45	99.47	99.48	99.49	99.50
	0.001	999.39	999.40	999.43	999.45	999.47	999.48	999.49	999.50
3	0.100	5.24	5.23	5.20	5.18	5.17	5.15	5.14	5.13
	0.050	8.81	8.79	8.70	8.66	8.62	8.57	8.55	8.53
	0.025	14.47	14.42	14.25	14.17	14.08	13.99	13.95	13.91
	0.010	27.35	27.23	26.87	26.69	26.50	26.32	26.22	26.14
	0.001	129.86	129.25	127.37	126.42	125.45	124.47	123.97	123.53
4	0.100	3.94	3.92	3.87	3.84	3.82	3.79	3.78	3.76
	0.050	6.00	5.96	5.86	5.80	5.75	5.69	5.66	5.63
	0.025	8.90	8.84	8.66	8.56	8.46	8.36	8.31	8.26
	0.010	14.66	14.55	14.20	14.02	13.84	13.65	13.56	13.47
	0.001	48.47	48.05	46.76	46.10	45.43	44.75	44.40	44.09
5	0.100	3.32	3.30	3.24	3.21	3.17	3.14	3.12	3.11
	0.050	4.77	4.74	4.62	4.56	4.50	4.43	4.40	4.37
	0.025	6.68	6.62	6.43	6.33	6.23	6.12	6.07	6.02
	0.010	10.16	10.05	9.72	9.55	9.38	9.20	9.11	9.03
	0.001	27.24	26.92	25.91	25.39	24.87	24.33	24.06	23.82
6	0.100	2.96	2.94	2.87	2.84	2.80	2.76	2.74	2.72
	0.050	4.10	4.06	3.94	3.87	3.81	3.74	3.70	3.67
	0.025	5.52	5.46	5.27	5.17	5.07	4.96	4.90	4.86
	0.010	7.98	7.87	7.56	7.40	7.23	7.06	6.97	6.89
	0.001	18.69	18.41	17.56	17.12	16.67	16.21	15.98	15.77
7	0.100	2.72	2.70	2.63	2.59	2.56	2.51	2.49	2.47
	0.050	3.68	3.64	3.51	3.44	3.38	3.30	3.27	3.23
	0.025	4.82	4.76	4.57	4.47	4.36	4.25	4.20	4.15
	0.010	6.72	6.62	6.31	6.16	5.99	5.82	5.74	5.66
	0.001	14.33	14.08	13.32	12.93	12.53	12.12	11.91	11.72
8	0.100	2.56	2.54	2.46	2.42	2.38	2.34	2.32	2.30
	0.050	3.39	3.35	3.22	3.15	3.08	3.01	2.97	2.93
	0.025	4.36	4.30	4.10	4.00	3.89	3.78	3.73	3.68
	0.010	5.91	5.81	5.52	5.36	5.20	5.03	4.95	4.87
	0.001	11.77	11.54	10.84	10.48	10.11	9.73	9.53	9.36

Degrees of Freedom in the Denominator

Table IX (*continued*)

F-Distribution Critical Values

	Area in Right Tail	Degrees of Freedom in the Numerator									
		1	2	3	4	5	6	7	8	9	10
9	0.100	3.36	3.01	2.81	2.69	2.61	2.55	2.51	2.47	2.44	2.42
	0.050	5.12	4.26	3.86	3.63	3.48	3.37	3.29	3.23	3.18	3.14
	0.025	7.21	5.71	5.08	4.72	4.48	4.32	4.20	4.10	4.03	3.96
	0.010	10.56	8.02	6.99	6.42	6.06	5.80	5.61	5.47	5.35	5.26
	0.001	22.86	16.39	13.90	12.56	11.71	11.13	10.70	10.37	10.11	9.89
10	0.100	3.29	2.92	2.73	2.61	2.52	2.46	2.41	2.38	2.35	2.32
	0.050	4.96	4.10	3.71	3.48	3.33	3.22	3.14	3.07	3.02	2.98
	0.025	6.94	5.46	4.83	4.47	4.24	4.07	3.95	3.85	3.78	3.72
	0.010	10.04	7.56	6.55	5.99	5.64	5.39	5.20	5.06	4.94	4.85
	0.001	21.04	14.91	12.55	11.28	10.48	9.93	9.52	9.20	8.96	8.75
12	0.100	3.18	2.81	2.61	2.48	2.39	2.33	2.28	2.24	2.21	2.19
	0.050	4.75	3.89	3.49	3.26	3.11	3.00	2.91	2.85	2.80	2.75
	0.025	6.55	5.10	4.47	4.12	3.89	3.73	3.61	3.51	3.44	3.37
	0.010	9.33	6.93	5.95	5.41	5.06	4.82	4.64	4.50	4.39	4.30
	0.001	18.64	12.97	10.80	9.63	8.89	8.38	8.00	7.71	7.48	7.29
15	0.100	3.07	2.70	2.49	2.36	2.27	2.21	2.16	2.12	2.09	2.06
	0.050	4.54	3.68	3.29	3.06	2.90	2.79	2.71	2.64	2.59	2.54
	0.025	6.20	4.77	4.15	3.80	3.58	3.41	3.29	3.20	3.12	3.06
	0.010	8.68	6.36	5.42	4.89	4.56	4.32	4.14	4.00	3.89	3.80
	0.001	16.59	11.34	9.34	8.25	7.57	7.09	6.74	6.47	6.26	6.08
20	0.100	2.97	2.59	2.38	2.25	2.16	2.09	2.04	2.00	1.96	1.94
	0.050	4.35	3.49	3.10	2.87	2.71	2.60	2.51	2.45	2.39	2.35
	0.025	5.87	4.46	3.86	3.51	3.29	3.13	3.01	2.91	2.84	2.77
	0.010	8.10	5.85	4.94	4.43	4.10	3.87	3.70	3.56	3.46	3.37
	0.001	14.82	9.95	8.10	7.10	6.46	6.02	5.69	5.44	5.24	5.08
25	0.100	2.92	2.53	2.32	2.18	2.09	2.02	1.97	1.93	1.89	1.87
	0.050	4.24	3.39	2.99	2.76	2.60	2.49	2.40	2.34	2.28	2.24
	0.025	5.69	4.29	3.69	3.35	3.13	2.97	2.85	2.75	2.68	2.61
	0.010	7.77	5.57	4.68	4.18	3.85	3.63	3.46	3.32	3.22	3.13
	0.001	13.88	9.22	7.45	6.49	5.89	5.46	5.15	4.91	4.71	4.56
50	0.100	2.81	2.41	2.20	2.06	1.97	1.90	1.84	1.80	1.76	1.73
	0.050	4.03	3.18	2.79	2.56	2.40	2.29	2.20	2.13	2.07	2.03
	0.025	5.34	3.97	3.39	3.05	2.83	2.67	2.55	2.46	2.38	2.32
	0.010	7.17	5.06	4.20	3.72	3.41	3.19	3.02	2.89	2.78	2.70
	0.001	12.22	7.96	6.34	5.46	4.90	4.51	4.22	4.00	3.82	3.67
100	0.100	2.76	2.36	2.14	2.00	1.91	1.83	1.78	1.73	1.69	1.66
	0.050	3.94	3.09	2.70	2.46	2.31	2.19	2.10	2.03	1.97	1.93
	0.025	5.18	3.83	3.25	2.92	2.70	2.54	2.42	2.32	2.24	2.18
	0.010	6.90	4.82	3.98	3.51	3.21	2.99	2.82	2.69	2.59	2.50
	0.001	11.50	7.41	5.86	5.02	4.48	4.11	3.83	3.61	3.44	3.30
200	0.100	2.73	2.33	2.11	1.97	1.88	1.80	1.75	1.70	1.66	1.63
	0.050	3.89	3.04	2.65	2.42	2.26	2.14	2.06	1.98	1.93	1.88
	0.025	5.10	3.76	3.18	2.85	2.63	2.47	2.35	2.26	2.18	2.11
	0.010	6.76	4.71	3.88	3.41	3.11	2.89	2.73	2.60	2.50	2.41
	0.001	11.15	7.15	5.63	4.81	4.29	3.92	3.65	3.43	3.26	3.12
1000	0.100	2.71	2.31	2.09	1.95	1.85	1.78	1.72	1.68	1.64	1.61
	0.050	3.85	3.00	2.61	2.38	2.22	2.11	2.02	1.95	1.89	1.84
	0.025	5.04	3.70	3.13	2.80	2.58	2.42	2.30	2.20	2.13	2.06
	0.010	6.66	4.63	3.80	3.34	3.04	2.82	2.66	2.53	2.43	2.34
	0.001	10.89	6.96	5.46	4.65	4.14	3.78	3.51	3.30	3.13	2.99

Degrees of Freedom in the Denominator

Table IX (*continued*)

F-Distribution Critical Values

Degrees of Freedom in the Numerator

Degrees of Freedom in the Denominator	Area in Right Tail	12	15	20	25	30	40	50	60	120	1000
9	0.100	2.38	2.34	2.30	2.27	2.25	2.23	2.22	2.21	2.18	2.16
	0.050	3.07	3.01	2.94	2.89	2.86	2.83	2.80	2.79	2.75	2.71
	0.025	3.87	3.77	3.67	3.60	3.56	3.51	3.47	3.45	3.39	3.34
	0.010	5.11	4.96	4.81	4.71	4.65	4.57	4.52	4.48	4.40	4.32
	0.001	9.57	9.24	8.90	8.69	8.55	8.37	8.26	8.19	8.00	7.84
10	0.100	2.28	2.24	2.20	2.17	2.16	2.13	2.12	2.11	2.08	2.06
	0.050	2.91	2.85	2.77	2.73	2.70	2.66	2.64	2.62	2.58	2.54
	0.025	3.62	3.52	3.42	3.35	3.31	3.26	3.22	3.20	3.14	3.09
	0.010	4.71	4.56	4.41	4.31	4.25	4.17	4.12	4.08	4.00	3.92
	0.001	8.45	8.13	7.80	7.60	7.47	7.30	7.19	7.12	6.94	6.78
12	0.100	2.15	2.10	2.06	2.03	2.01	1.99	1.97	1.96	1.93	1.91
	0.050	2.69	2.62	2.54	2.50	2.47	2.43	2.40	2.38	2.34	2.30
	0.025	3.28	3.18	3.07	3.01	2.96	2.91	2.87	2.85	2.79	2.73
	0.010	4.16	4.01	3.86	3.76	3.70	3.62	3.57	3.54	3.45	3.37
	0.001	7.00	6.71	6.40	6.22	6.09	5.93	5.83	5.76	5.59	5.44
15	0.100	2.02	1.97	1.92	1.89	1.87	1.85	1.83	1.82	1.79	1.76
	0.050	2.48	2.40	2.33	2.28	2.25	2.20	2.18	2.16	2.11	2.07
	0.025	2.96	2.86	2.76	2.69	2.64	2.59	2.55	2.52	2.46	2.40
	0.010	3.67	3.52	3.37	3.28	3.21	3.13	3.08	3.05	2.96	2.88
	0.001	5.81	5.54	5.25	5.07	4.95	4.80	4.70	4.64	4.47	4.33
20	0.100	1.89	1.84	1.79	1.76	1.74	1.71	1.69	1.68	1.64	1.61
	0.050	2.28	2.20	2.12	2.07	2.04	1.99	1.97	1.95	1.90	1.85
	0.025	2.68	2.57	2.46	2.40	2.35	2.29	2.25	2.22	2.16	2.09
	0.010	3.23	3.09	2.94	2.84	2.78	2.69	2.64	2.61	2.52	2.43
	0.001	4.82	4.56	4.29	4.12	4.00	3.86	3.77	3.70	3.54	3.40
25	0.100	1.82	1.77	1.72	1.68	1.66	1.63	1.61	1.59	1.56	1.52
	0.050	2.16	2.09	2.01	1.96	1.92	1.87	1.84	1.82	1.77	1.72
	0.025	2.51	2.41	2.30	2.23	2.18	2.12	2.08	2.05	1.98	1.91
	0.010	2.99	2.85	2.70	2.60	2.54	2.45	2.40	2.36	2.27	2.18
	0.001	4.31	4.06	3.79	3.63	3.52	3.37	3.28	3.22	3.06	2.91
50	0.100	1.68	1.63	1.57	1.53	1.50	1.46	1.44	1.42	1.38	1.33
	0.050	1.95	1.87	1.78	1.73	1.69	1.63	1.60	1.58	1.51	1.45
	0.025	2.22	2.11	1.99	1.92	1.87	1.80	1.75	1.72	1.64	1.56
	0.010	2.56	2.42	2.27	2.17	2.10	2.01	1.95	1.91	1.80	1.70
	0.001	3.44	3.20	2.95	2.79	2.68	2.53	2.44	2.38	2.21	2.05
100	0.100	1.61	1.56	1.49	1.45	1.42	1.38	1.35	1.34	1.28	1.22
	0.050	1.85	1.77	1.68	1.62	1.57	1.52	1.48	1.45	1.38	1.30
	0.025	2.08	1.97	1.85	1.77	1.71	1.64	1.59	1.56	1.46	1.36
	0.010	2.37	2.22	2.07	1.97	1.89	1.80	1.74	1.69	1.57	1.45
	0.001	3.07	2.84	2.59	2.43	2.32	2.17	2.08	2.01	1.83	1.64
200	0.100	1.58	1.52	1.46	1.41	1.38	1.34	1.31	1.29	1.23	1.16
	0.050	1.80	1.72	1.62	1.56	1.52	1.46	1.41	1.39	1.30	1.21
	0.025	2.01	1.90	1.78	1.70	1.64	1.56	1.51	1.47	1.37	1.25
	0.010	2.27	2.13	1.97	1.87	1.79	1.69	1.63	1.58	1.45	1.30
	0.001	2.90	2.67	2.42	2.26	2.15	2.00	1.90	1.83	1.64	1.43
1000	0.100	1.55	1.49	1.43	1.38	1.35	1.30	1.27	1.25	1.18	1.08
	0.050	1.76	1.68	1.58	1.52	1.47	1.41	1.36	1.33	1.24	1.11
	0.025	1.96	1.85	1.72	1.64	1.58	1.50	1.45	1.41	1.29	1.13
	0.010	2.20	2.06	1.90	1.79	1.72	1.61	1.54	1.50	1.35	1.16
	0.001	2.77	2.54	2.30	2.14	2.02	1.87	1.77	1.69	1.49	1.22

Table X

$\alpha = 0.05$

Critical Values for Tukey's Test

v	k (or p): 2	3	4	5	6	7	8	9	10	11	12	13	14	15	16	17	18	19
1	17.97	26.98	32.82	37.08	40.41	43.12	45.40	47.36	49.07	50.59	51.96	53.20	54.33	55.36	56.32	57.22	58.04	58.83
2	6.085	8.331	9.798	10.88	11.74	12.44	13.03	13.54	13.99	14.39	14.75	15.08	15.38	15.65	15.91	16.14	16.37	16.57
3	4.501	5.910	6.825	7.502	8.037	8.478	8.853	9.177	9.462	9.717	9.946	10.15	10.35	10.53	10.69	10.84	10.98	11.11
4	3.927	5.040	5.757	6.287	6.707	7.053	7.347	7.602	7.826	8.027	8.208	8.373	8.525	8.664	8.794	8.914	9.028	9.134
5	3.635	4.602	5.218	5.673	6.033	6.330	6.582	6.802	6.995	7.168	7.324	7.466	7.596	7.717	7.828	7.932	8.030	8.122
6	3.461	4.339	4.896	5.305	5.628	5.895	6.122	6.319	6.493	6.649	6.789	6.917	7.034	7.143	7.244	7.338	7.426	7.508
7	3.344	4.165	4.681	5.060	5.359	5.606	5.815	5.998	6.158	6.302	6.431	6.550	6.658	6.759	6.852	6.939	7.020	7.097
8	3.261	4.041	4.529	4.886	5.167	5.399	5.597	5.767	5.918	6.054	6.175	6.287	6.389	6.483	6.571	6.653	6.510	6.802
9	3.199	3.949	4.415	4.756	5.024	5.244	5.432	5.595	5.739	5.867	5.983	6.089	6.186	6.276	6.359	6.437	6.339	6.579
10	3.151	3.877	4.327	4.654	4.912	5.124	5.305	5.461	5.599	5.722	5.833	5.935	6.028	6.114	6.194	6.269	6.202	6.405
11	3.113	3.820	4.256	4.574	4.823	5.028	5.202	5.353	5.487	5.605	5.713	5.811	5.901	5.984	6.062	6.134	6.089	6.265
12	3.082	3.773	4.199	4.508	4.751	4.950	5.119	5.265	5.395	5.511	5.615	5.710	5.798	5.878	5.953	6.023	5.995	6.151
13	3.055	3.735	4.151	4.453	4.690	4.885	5.049	5.192	5.318	5.431	5.533	5.625	5.711	5.789	5.862	5.931	5.915	6.055
14	3.033	3.702	4.111	4.407	4.639	4.829	4.990	5.131	5.254	5.364	5.463	5.554	5.637	5.714	5.786	5.852	5.846	5.974
15	3.014	3.674	4.076	4.367	4.595	4.782	4.940	5.077	5.198	5.306	5.404	5.493	5.574	5.649	5.720	5.785	5.786	5.904
16	2.998	3.649	4.046	4.333	4.557	4.741	4.897	5.031	5.150	5.256	5.352	5.439	5.520	5.593	5.662	5.727	5.734	5.843
17	2.984	3.628	4.020	4.303	4.524	4.705	4.858	4.991	5.108	5.212	5.307	5.392	5.471	5.544	5.612	5.675	5.688	5.790
18	2.971	3.609	3.997	4.277	4.495	4.673	4.824	4.956	5.071	5.174	5.267	5.352	5.429	5.501	5.568	5.630	5.647	5.743
19	2.960	3.593	3.977	4.253	4.469	4.645	4.794	4.924	5.038	5.140	5.231	5.315	5.391	5.462	5.528	5.589	5.610	5.701
20	2.950	3.578	3.958	4.232	4.445	4.620	4.768	4.896	5.008	5.108	5.199	5.282	5.357	5.427	5.493	5.553	5.494	5.663
24	2.919	3.532	3.901	4.166	4.373	4.541	4.684	4.807	4.915	5.012	5.099	5.179	5.251	5.319	5.381	5.439	5.379	5.545
30	2.888	3.486	3.845	4.102	4.302	4.464	4.602	4.720	4.824	4.917	5.001	5.077	5.147	5.211	5.271	5.327	5.266	5.429
40	2.858	3.442	3.791	4.039	4.232	4.389	4.521	4.635	4.735	4.824	4.904	4.977	5.044	5.106	5.163	5.216	5.154	5.313
60	2.829	3.399	3.737	3.977	4.163	4.314	4.441	4.550	4.646	4.732	4.808	4.878	4.942	5.001	5.056	5.107	5.044	5.199
120	2.800	3.356	3.685	3.917	4.096	4.241	4.363	4.468	4.560	4.641	4.714	4.781	4.842	4.898	4.950	4.998	4.934	5.086
∞	2.772	3.314	3.633	3.858	4.030	4.170	4.286	4.387	4.474	4.552	4.622	4.685	4.743	4.796	4.845	4.891	4.891	4.974

Table X (continued)

Critical Values for Tukey's Test

$\alpha = 0.05$

ν	k (or p): 38	40	50	60	70	80	90	100
1	68.26	68.92	71.73	73.97	75.82	77.40	78.77	79.98
2	19.11	19.28	20.05	20.66	21.16	21.59	21.96	22.29
3	12.75	12.87	13.36	13.76	14.08	14.36	14.61	14.82
4	10.44	10.53	10.93	11.24	11.51	11.73	11.92	12.09
5	9.250	9.330	9.674	9.949	10.18	10.38	10.54	10.69
6	8.529	8.601	8.913	9.163	9.370	9.548	9.702	9.839
7	8.043	8.110	8.400	8.632	8.824	8.989	9.133	9.261
8	7.693	7.756	8.029	8.248	8.430	8.586	8.722	8.843
9	7.428	7.488	7.749	7.958	8.132	8.281	8.410	8.526
10	7.220	7.279	7.529	7.730	7.897	8.041	8.166	8.276
11	7.053	7.110	7.352	7.546	7.708	7.847	7.968	8.075
12	6.916	6.970	7.205	7.394	7.552	7.687	7.804	7.909
13	6.800	6.854	7.083	7.267	7.421	7.552	7.667	7.769
14	6.702	6.754	6.979	7.159	7.309	7.438	7.550	7.650
15	6.618	6.669	6.888	7.065	7.212	7.339	7.449	7.546
16	6.544	6.594	6.810	6.984	7.128	7.252	7.360	7.457
17	6.479	6.529	6.741	6.912	7.054	7.176	7.283	7.377
18	6.422	6.471	6.680	6.848	6.989	7.109	7.213	7.307
19	6.371	6.419	6.626	6.792	6.930	7.048	7.152	7.244
20	6.325	6.373	6.576	6.740	6.877	6.994	7.097	7.187
24	6.181	6.226	6.421	6.579	6.710	6.822	6.920	7.008
30	6.037	6.080	6.267	6.417	6.543	6.650	6.744	6.827
40	5.893	5.934	6.112	6.255	6.375	6.477	6.566	6.645
60	5.750	5.789	5.958	6.093	6.206	6.303	6.387	6.462
120	5.607	5.644	5.802	5.929	6.035	6.126	6.205	6.275
∞	5.463	5.498	5.646	5.764	5.863	5.947	6.020	6.085

ν	k (or p): 20	22	24	26	28	30	32	34	36
1	59.56	60.91	62.12	63.22	64.23	65.15	66.01	66.81	67.56
2	16.77	17.13	17.45	17.75	18.02	18.27	18.50	18.72	18.92
3	11.24	11.47	11.68	11.87	12.05	12.21	12.36	12.50	12.63
4	9.233	9.418	9.584	9.736	9.875	10.00	10.12	10.23	10.34
5	8.208	8.368	8.512	8.643	8.764	8.875	8.979	9.075	9.165
6	7.587	7.730	7.861	7.979	8.088	8.189	8.283	8.370	8.452
7	7.170	7.303	7.423	7.533	7.634	7.728	7.814	7.895	7.972
8	6.870	6.995	7.109	7.212	7.307	7.395	7.477	7.554	7.625
9	6.644	6.763	6.871	6.970	7.061	7.145	7.222	7.295	7.363
10	6.467	6.582	6.686	6.781	6.868	6.948	7.023	7.093	7.159
11	6.326	6.436	6.536	6.628	6.712	6.790	6.863	6.930	6.994
12	6.209	6.317	6.414	6.503	6.585	6.660	6.731	6.796	6.858
13	6.112	6.217	6.312	6.398	6.478	6.551	6.620	6.684	6.744
14	6.029	6.132	6.224	6.309	6.387	6.459	6.526	6.588	6.647
15	5.958	6.059	6.149	6.233	6.309	6.379	6.445	6.506	6.564
16	5.897	5.995	6.084	6.166	6.241	6.310	6.374	6.434	6.491
17	5.842	5.940	6.027	6.107	6.181	6.249	6.313	6.372	6.427
18	5.794	5.890	5.977	6.055	6.128	6.195	6.258	6.316	6.371
19	5.752	5.846	5.932	6.009	6.081	6.147	6.209	6.267	6.321
20	5.714	5.807	5.891	5.968	6.039	6.104	6.165	6.222	6.275
24	5.594	5.683	5.764	5.838	5.906	5.968	6.027	6.081	6.132
30	5.475	5.561	5.638	5.709	5.774	5.833	5.889	5.941	5.990
40	5.358	5.439	5.513	5.581	5.642	5.700	5.753	5.803	5.849
60	5.241	5.319	5.389	5.453	5.512	5.566	5.617	5.664	5.708
120	5.126	5.200	5.266	5.327	5.382	5.434	5.481	5.526	5.568
∞	5.012	5.081	5.144	5.201	5.253	5.301	5.346	5.388	5.427

Table X (continued)

Critical Values for Tukey's Test

$\alpha = 0.01$

v	k (or p): 2	3	4	5	6	7	8	9	10	11	12	13	14	15	16	17	18	19
1	90.03	135.0	164.3	185.6	202.2	215.8	227.2	237.0	245.6	253.2	260.0	266.2	271.8	277.0	281.8	286.3	290.4	294.3
2	14.04	19.02	22.29	24.72	26.63	28.20	29.53	30.68	31.69	32.59	33.40	34.13	34.81	35.43	36.00	36.53	37.03	37.50
3	8.261	10.62	12.17	13.33	14.24	15.00	15.64	16.20	16.69	17.13	17.53	17.89	18.22	18.52	18.81	19.07	19.32	19.55
4	6.512	8.120	9.173	9.958	10.58	11.10	11.55	11.93	12.27	12.57	12.84	13.09	13.32	13.53	13.73	13.91	14.08	14.24
5	5.702	6.976	7.804	8.421	8.913	9.321	9.669	9.972	10.24	10.48	10.70	10.89	11.08	11.24	11.40	11.55	11.68	11.81
6	5.243	6.331	7.033	7.556	7.973	8.318	8.613	8.869	9.097	9.301	9.485	9.653	9.808	9.951	10.08	10.21	10.32	10.43
7	4.949	5.919	6.543	7.005	7.373	7.679	7.939	8.166	8.368	8.548	8.711	8.860	8.997	9.124	9.242	9.353	9.456	9.554
8	4.746	5.635	6.204	6.625	6.960	7.237	7.474	7.681	7.863	8.027	8.176	8.312	8.436	8.552	8.659	8.760	8.854	8.943
9	4.596	5.428	5.957	6.348	6.658	6.915	7.134	7.325	7.495	7.647	7.784	7.910	8.025	8.132	8.232	8.325	8.412	8.495
10	4.482	5.270	5.769	6.136	6.428	6.669	6.875	7.055	7.213	7.356	7.485	7.603	7.712	7.812	7.906	7.993	8.076	8.153
11	4.392	5.146	5.621	5.970	6.247	6.476	6.672	6.842	6.992	7.128	7.250	7.362	7.465	7.560	7.649	7.732	7.809	7.883
12	4.320	5.046	5.502	5.836	6.101	6.321	6.507	6.670	6.814	6.943	7.060	7.167	7.265	7.356	7.441	7.520	7.594	7.665
13	4.260	4.964	5.404	5.727	5.981	6.192	6.372	6.528	6.667	6.791	6.903	7.006	7.101	7.188	7.269	7.345	7.417	7.485
14	4.210	4.895	5.322	5.634	5.881	6.085	6.258	6.409	6.543	6.664	6.772	6.871	6.962	7.047	7.126	7.199	7.268	7.333
15	4.168	4.836	5.252	5.556	5.796	5.994	6.162	6.309	6.439	6.555	6.660	6.757	6.845	6.927	7.003	7.074	7.142	7.204
16	4.131	4.786	5.192	5.489	5.722	5.915	6.079	6.222	6.349	6.462	6.564	6.658	6.744	6.823	6.898	6.967	7.032	7.093
17	4.099	4.742	5.140	5.430	5.659	5.847	6.007	6.147	6.270	6.381	6.480	6.572	6.656	6.734	6.806	6.873	6.937	6.997
18	4.071	4.703	5.094	5.379	5.603	5.788	5.944	6.081	6.201	6.310	6.407	6.497	6.579	6.655	6.725	6.792	6.854	6.912
19	4.046	4.670	5.054	5.334	5.554	5.735	5.889	6.022	6.141	6.247	6.342	6.430	6.510	6.585	6.654	6.719	6.780	6.837
20	4.024	4.639	5.018	5.294	5.510	5.688	5.839	5.970	6.087	6.191	6.285	6.371	6.450	6.523	6.591	6.654	6.714	6.771
24	3.956	4.546	4.907	5.168	5.374	5.542	5.685	5.809	5.919	6.017	6.106	6.186	6.261	6.330	6.394	6.453	6.510	6.563
30	3.889	4.455	4.799	5.048	5.242	5.401	5.536	5.653	5.756	5.849	5.932	6.008	6.078	6.143	6.203	6.259	6.311	6.361
40	3.825	4.367	4.696	4.931	5.114	5.265	5.392	5.502	5.559	5.686	5.764	5.835	5.900	5.961	6.017	6.069	6.119	6.165
60	3.762	4.282	4.595	4.818	4.991	5.133	5.253	5.356	5.447	5.528	5.601	5.667	5.728	5.785	5.837	5.886	5.931	5.974
120	3.702	4.200	4.497	4.709	4.872	5.005	5.118	5.214	5.299	5.375	5.443	5.505	5.562	5.614	5.662	5.708	5.750	5.790
∞	3.643	4.120	4.403	4.603	4.757	4.882	4.987	5.078	5.157	5.227	5.290	5.348	5.400	5.448	5.493	5.535	5.574	5.611

Table X (continued)

Critical Values for Tukey's Test

α = 0.01

v	k (or p): 20	22	24	26	28	30	32	34	36
1	298.0	304.7	310.8	316.3	321.3	326.0	330.3	334.3	338.0
2	37.95	38.76	39.49	40.15	40.76	41.32	41.84	42.33	42.78
3	19.77	20.17	20.53	20.86	21.16	21.44	21.70	21.95	22.17
4	14.40	14.68	14.93	15.16	15.37	15.57	15.75	15.92	16.08
5	11.93	12.16	12.36	12.54	12.71	12.87	13.02	13.15	13.28
6	10.54	10.73	10.91	11.06	11.21	11.34	11.47	11.58	11.69
7	9.646	9.815	9.970	10.11	10.24	10.36	10.47	10.58	10.67
8	9.027	9.182	9.322	9.450	9.569	9.678	9.779	9.874	9.964
9	8.573	8.717	8.847	8.966	9.075	9.177	9.271	9.360	9.443
10	8.226	8.361	8.483	8.595	8.698	8.794	8.883	8.966	9.044
11	7.952	8.080	8.196	8.303	8.400	8.491	8.575	8.654	8.728
12	7.731	7.853	7.964	8.066	8.159	8.246	8.327	8.402	8.473
13	7.548	7.665	7.772	7.870	7.960	8.043	8.121	8.193	8.262
14	7.395	7.508	7.611	7.705	7.792	7.873	7.948	8.018	8.084
15	7.264	7.374	7.474	7.566	7.650	7.728	7.800	7.869	7.932
16	7.152	7.258	7.356	7.445	7.527	7.602	7.673	7.739	7.802
17	7.053	7.158	7.253	7.340	7.420	7.493	7.563	7.627	7.687
18	6.968	7.070	7.163	7.247	7.325	7.398	7.465	7.528	7.587
19	6.891	6.992	7.082	7.166	7.242	7.313	7.379	7.440	7.498
20	6.823	6.922	7.011	7.092	7.168	7.237	7.302	7.362	7.419
24	6.612	6.705	6.789	6.865	6.936	7.001	7.062	7.119	7.173
30	6.407	6.494	6.572	6.644	6.710	6.772	6.828	6.881	6.932
40	6.209	6.289	6.362	6.429	6.490	6.547	6.600	6.650	6.697
60	6.015	6.090	6.158	6.220	6.277	6.330	6.378	6.424	6.467
120	5.827	5.897	5.959	6.016	6.069	6.117	6.162	6.204	6.244
∞	5.645	5.709	5.766	5.818	5.866	5.911	5.952	5.990	6.026

v	k (or p): 38	40	50	60	70	80	90	100
1	341.5	344.8	358.9	370.1	379.4	387.3	394.1	400.1
2	43.21	43.61	45.33	46.70	47.83	48.80	49.64	50.38
3	22.39	22.59	23.45	24.13	24.71	25.19	25.62	25.99
4	16.23	16.37	16.98	17.46	17.86	18.02	18.50	18.77
5	13.40	13.52	14.00	14.39	14.72	14.99	15.23	15.45
6	11.80	11.90	12.31	12.65	12.92	13.16	13.37	13.55
7	10.77	10.85	11.23	11.52	11.77	11.99	12.17	12.34
8	10.05	10.13	10.47	10.75	10.97	11.17	11.34	11.49
9	9.521	9.594	9.912	10.17	10.38	10.57	10.73	10.87
10	9.117	9.187	9.486	9.726	9.927	10.10	10.25	10.39
11	8.798	8.864	9.148	9.377	9.568	9.732	9.875	10.00
12	8.539	8.603	8.875	9.094	9.277	9.434	9.571	9.693
13	8.326	8.387	8.648	8.859	9.035	9.187	9.318	9.436
14	8.146	8.204	8.457	8.661	8.832	8.978	9.106	9.219
15	7.992	8.049	8.295	8.492	8.658	8.800	8.924	9.035
16	7.860	7.916	8.154	8.347	8.507	8.646	8.767	8.874
17	7.745	7.799	8.031	8.219	8.377	8.511	8.630	8.735
18	7.643	7.696	7.924	8.107	8.261	8.393	8.508	8.611
19	7.553	7.605	7.828	8.008	8.159	8.288	8.401	8.502
20	7.473	7.523	7.742	7.919	8.067	8.194	8.305	8.404
24	7.223	7.270	7.476	7.642	7.780	7.900	8.004	8.097
30	6.978	7.023	7.215	7.370	7.500	7.611	7.709	7.796
40	6.740	6.782	6.960	7.104	7.225	7.328	7.419	7.500
60	6.507	6.546	6.710	6.843	6.954	7.050	7.133	7.207
120	6.281	6.316	6.467	6.588	6.689	6.776	6.852	6.919
∞	6.060	6.092	6.228	6.338	6.429	6.507	6.575	6.636

Source: Zar, Jerrold H., Biostatistical Analysis, 4th Edition, © 1999. Reprinted by permission of Pearson Education, Inc., Upper Saddle River, NJ.

Table XI

Critical Values for the Number of Runs

Value of n_2

Value of n_1

	2	3	4	5	6	7	8	9	10	11	12	13	14	15	16	17	18	19	20
2	1/6	1/6	1/6	1/6	1/6	1/6	1/6	1/6	1/6	1/6	2/6	2/6	2/6	2/6	2/6	2/6	2/6	2/6	2/6
3	1/6	1/8	1/8	1/8	2/8	2/8	2/8	2/8	2/8	2/8	2/8	2/8	2/8	3/8	3/8	3/8	3/8	3/8	3/8
4	1/6	1/8	1/9	2/9	2/9	2/10	3/10	3/10	3/10	3/10	3/10	3/10	3/10	3/10	4/10	4/10	4/10	4/10	4/10
5	1/6	1/8	2/9	2/10	3/10	3/11	3/11	3/12	3/12	4/12	4/12	4/12	4/12	4/12	4/12	4/12	5/12	5/12	5/12
6	1/6	2/8	2/9	3/10	3/11	3/12	3/12	4/13	4/13	4/13	4/13	5/14	5/14	5/14	5/14	5/14	5/14	6/14	6/14
7	1/6	2/8	2/10	3/11	3/12	3/13	4/13	4/14	5/14	5/14	5/14	5/15	5/15	6/15	6/16	6/16	6/16	6/16	6/16
8	1/6	2/8	3/10	3/11	3/12	4/13	4/14	5/14	5/15	5/15	6/16	6/16	6/16	6/16	6/17	7/17	7/17	7/17	7/17
9	1/6	2/8	3/10	3/12	4/13	4/14	5/14	5/15	5/16	6/16	6/16	6/17	7/17	7/18	7/18	7/18	8/18	8/18	8/18
10	1/6	2/8	3/10	3/12	4/13	5/14	5/15	5/16	6/16	6/17	7/17	7/18	7/18	7/18	8/19	8/19	8/19	8/20	9/20
11	1/6	2/8	3/10	4/12	4/13	5/14	5/15	6/16	6/17	7/17	7/18	7/19	8/19	8/19	8/20	9/20	9/20	9/21	9/21
12	2/6	2/8	3/10	4/12	4/13	5/14	6/16	6/16	7/17	7/18	7/19	8/19	8/20	8/20	9/21	9/21	9/21	10/22	10/22
13	2/6	2/8	3/10	4/12	5/14	5/15	6/16	6/17	7/18	7/19	8/19	8/20	9/20	9/21	9/21	10/22	10/22	10/23	10/23
14	2/6	2/8	3/10	4/12	5/14	5/15	6/16	7/17	7/18	8/19	8/20	9/20	9/21	9/22	10/22	10/23	10/23	11/23	11/24
15	2/6	3/8	3/10	4/12	5/14	6/15	6/16	7/18	7/18	8/19	8/20	9/21	9/22	10/22	10/23	11/23	11/24	11/24	12/25
16	2/6	3/8	4/10	4/12	5/14	6/16	6/17	7/18	8/19	8/20	9/21	9/21	10/22	10/23	11/23	11/24	11/25	12/25	12/25
17	2/6	3/8	4/10	4/12	5/14	6/16	7/17	7/18	8/19	9/20	9/21	10/22	10/23	11/23	11/24	11/25	12/25	12/26	13/26
18	2/6	3/8	4/10	5/12	5/14	6/16	7/17	8/18	8/19	9/20	9/21	10/22	10/23	11/24	11/25	12/25	12/26	13/26	13/27
19	2/6	3/8	4/10	5/12	6/14	6/16	7/17	8/18	8/20	9/21	10/22	10/23	11/23	11/24	12/25	12/26	13/26	13/27	13/27
20	2/6	3/8	4/10	5/12	6/14	6/16	7/17	8/18	9/20	9/21	10/22	10/23	11/24	12/25	12/25	13/26	13/27	13/27	14/28

Source: Triola, Mario F., *Elementary Statistics*, 10th Edition © 2007. Reprinted by permission of Pearson Education, Inc., Upper Saddle River, NJ.

Table XII

Critical Values for the Sign Test

n	0.005 (one tail) 0.01 (two tails)	0.01 (one tail) 0.02 (two tails)	0.025 (one tail) 0.05 (two tails)	0.05 (one tail) 0.10 (two tails)
1	*	*	*	*
2	*	*	*	*
3	*	*	*	*
4	*	*	*	*
5	*	*	*	0
6	*	*	0	0
7	*	0	0	0
8	0	0	0	1
9	0	0	1	1
10	0	0	1	1
11	0	1	1	2
12	1	1	2	2
13	1	1	2	3
14	1	2	2	3
15	2	2	3	3
16	2	2	3	4
17	2	3	4	4
18	3	3	4	5
19	3	4	4	5
20	3	4	5	5
21	4	4	5	6
22	4	5	5	6
23	4	5	6	7
24	5	5	6	7
25	5	6	7	7

* Indicates that it is not possible to get a value in the critical region.

Source: Zar, Jerrold H., *Biostatistical Analysis,* 4th Edition, © 1999. Reprinted by permission of Pearson Education, Inc., Upper Saddle River, NJ.

Table XIII

Critical Values for the Wilcoxon Signed-Rank Test

	α			
n	0.005	0.01	0.025	0.05
5	*	*	*	0
6	*	*	0	2
7	*	0	2	3
8	0	1	3	5
9	1	3	5	8
10	3	5	8	10
11	5	7	10	13
12	7	9	13	17
13	9	12	17	21
14	12	15	21	25
15	15	19	25	30
16	19	23	29	35
17	23	27	34	41
18	27	32	40	47
19	32	37	46	53
20	37	43	52	60
21	42	49	58	67
22	48	55	65	75
23	54	62	73	83
24	61	69	81	91
25	68	76	89	100
26	75	84	98	110
27	83	92	107	119
28	91	101	116	130
29	100	110	126	140
30	109	120	137	151

* Indicates that it is not possible to get a value in the critical region.
Source: Zar, Jerrold H., *Biostatistical Analysis,* 4th Edition, © 1999.
Reprinted by permission of Pearson Education, Inc., Upper Saddle River, NJ.

Table XIV

Critical Values of the Mann–Whitney Test Statistic

n_1	α	2	3	4	5	6	7	8	9	10	11	12	13	14	15	16	17	18	19	20
2	0.001	0	0	0	0	0	0	0	0	0	0	0	0	0	0	0	0	0	0	0
	0.005	0	0	0	0	0	0	0	0	0	0	0	0	0	0	0	0	0	1	1
	0.01	0	0	0	0	0	0	0	0	0	0	0	1	1	1	1	1	1	2	2
	0.025	0	0	0	0	0	0	1	1	1	1	2	2	2	2	2	3	3	3	3
	0.05	0	0	0	1	1	1	2	2	2	2	3	3	4	4	4	4	5	5	5
	0.10	0	1	1	2	2	2	3	3	4	4	5	5	5	6	6	7	7	8	8
3	0.001	0	0	0	0	0	0	0	0	0	0	0	0	0	0	0	1	1	1	1
	0.005	0	0	0	0	0	0	0	1	1	1	2	2	2	3	3	3	3	4	4
	0.01	0	0	0	0	0	1	1	2	2	2	3	3	3	4	4	5	5	5	6
	0.025	0	0	0	1	2	2	3	3	4	4	5	5	6	6	7	7	8	8	9
	0.05	0	1	1	2	3	3	4	5	5	6	6	7	8	8	9	10	10	11	12
	0.10	1	2	2	3	4	5	6	6	7	8	9	10	11	11	12	13	14	15	16
4	0.001	0	0	0	0	0	0	0	0	1	1	1	2	2	2	3	3	4	4	4
	0.005	0	0	0	0	1	1	2	2	3	3	4	4	5	6	6	7	7	8	9
	0.01	0	0	0	1	2	2	3	4	4	5	6	6	7	8	8	9	10	10	11
	0.025	0	0	1	2	3	4	5	5	6	7	8	9	10	11	12	12	13	14	15
	0.05	0	1	2	3	4	5	6	7	8	9	10	11	12	13	15	16	17	18	19
	0.10	1	2	4	5	6	7	8	10	11	12	13	14	16	17	18	19	21	22	23
5	0.001	0	0	0	0	0	0	1	2	2	3	3	4	4	5	6	6	7	8	8
	0.005	0	0	0	1	2	2	3	4	5	6	7	8	8	9	10	11	12	13	14
	0.01	0	0	1	2	3	4	5	6	7	8	9	10	11	12	13	14	15	16	17
	0.025	0	1	2	3	4	6	7	8	9	10	12	13	14	15	16	18	19	20	21
	0.05	1	2	3	5	6	7	9	10	12	13	14	16	17	19	20	21	23	24	26
	0.10	2	3	5	6	8	9	11	13	14	16	18	19	21	23	24	26	28	29	31
6	0.001	0	0	0	0	0	0	2	3	4	5	5	6	7	8	9	10	11	12	13
	0.005	0	0	1	2	3	4	5	6	7	8	10	11	12	13	14	16	17	18	19
	0.01	0	0	2	3	4	5	7	8	9	10	12	13	14	16	17	19	20	21	23
	0.025	0	2	3	4	6	7	9	11	12	14	15	17	18	20	22	23	25	28	28
	0.05	1	3	4	6	8	9	11	13	15	17	18	20	22	24	26	27	29	31	33
	0.10	2	4	6	8	10	12	14	16	18	20	22	24	26	28	30	32	35	37	39
7	0.001	0	0	0	0	1	2	3	4	6	7	8	9	10	11	12	14	15	16	17
	0.005	0	0	1	2	4	5	7	8	10	11	13	14	16	17	19	20	22	23	25
	0.01	0	1	2	4	5	7	8	10	12	13	15	17	18	20	22	24	25	27	29
	0.025	0	2	4	6	7	9	11	13	15	17	19	21	23	25	27	29	31	33	35
	0.05	1	3	5	7	9	12	14	16	18	20	22	25	27	29	31	34	36	38	40
	0.10	2	5	7	9	12	14	17	19	22	24	27	29	32	34	37	39	42	44	47
8	0.001	0	0	0	1	2	3	5	6	7	9	10	12	13	15	16	18	19	21	22
	0.005	0	0	2	3	5	7	8	10	12	14	16	18	19	21	23	25	27	29	31
	0.01	0	1	3	5	7	8	10	12	14	16	18	21	23	25	27	29	31	33	35
	0.025	1	3	5	7	9	11	14	16	18	20	23	25	27	30	32	35	37	39	42
	0.05	2	4	6	9	11	14	16	19	21	24	27	29	32	34	37	40	42	45	48
	0.10	3	6	8	11	14	17	20	23	25	28	31	34	37	40	43	46	49	52	55
9	0.001	0	0	0	2	3	4	6	8	9	11	13	15	16	18	20	22	24	26	27
	0.005	0	1	2	4	6	8	10	12	14	17	19	21	23	25	28	30	32	34	37
	0.01	0	2	4	6	8	10	12	15	17	19	22	24	27	29	32	34	37	39	41
	0.025	1	3	5	8	11	13	16	18	21	24	27	29	32	35	38	40	43	46	49
	0.05	2	5	7	10	13	16	19	22	25	28	31	34	37	40	43	46	49	52	55
	0.10	3	6	10	13	16	19	23	26	29	32	36	39	42	46	49	53	56	59	63
10	0.001	0	0	1	2	4	6	7	9	11	13	15	18	20	22	24	26	28	30	33
	0.005	0	1	3	5	7	10	12	14	17	19	22	25	27	30	32	35	38	40	43
	0.01	0	2	4	7	9	12	14	17	20	23	25	28	31	34	37	39	42	45	48
	0.025	1	4	6	9	12	15	18	21	24	27	30	34	37	40	43	46	49	53	56
	0.05	2	5	8	12	15	18	21	25	28	32	35	38	42	45	49	52	56	59	63
	0.10	4	7	11	14	18	22	25	29	33	37	40	44	48	52	55	59	63	67	71
11	0.001	0	0	1	3	5	7	9	11	13	16	18	21	23	25	28	30	33	35	38
	0.005	0	1	3	6	8	11	14	17	19	22	25	28	31	34	37	40	43	46	49
	0.01	0	2	5	8	10	13	16	19	23	26	29	32	35	38	42	45	48	51	54
	0.025	1	4	7	10	14	17	20	24	27	31	34	38	41	45	48	52	56	59	63
	0.05	2	6	9	13	17	20	24	28	32	35	39	43	47	51	55	58	62	66	70
	0.10	4	8	12	16	20	24	28	32	37	41	45	49	53	58	62	66	70	74	79

Table XIV (*continued*)

Critical Values of the Mann–Whitney Test Statistic

n_1	α	2	3	4	5	6	7	8	9	10	11	12	13	14	15	16	17	18	19	20
12	0.001	0	0	1	3	5	8	10	13	15	18	21	24	26	29	32	35	38	41	43
	0.005	0	2	4	7	10	13	16	19	22	25	28	32	35	38	42	45	48	52	55
	0.01	0	3	6	9	12	15	18	22	25	29	32	36	39	43	47	50	54	57	61
	0.025	2	5	8	12	15	19	23	27	30	34	36	42	46	50	54	58	62	66	70
	0.05	3	6	10	14	18	22	27	31	35	39	43	48	52	56	61	65	69	73	78
	0.10	5	9	13	18	22	27	31	36	40	45	50	54	59	64	68	73	78	82	87
13	0.001	0	0	2	4	6	9	12	15	18	21	24	27	30	33	36	39	43	46	49
	0.005	0	2	4	8	11	14	18	21	25	28	32	35	39	43	46	50	54	58	61
	0.01	1	3	6	10	13	17	21	24	28	32	36	40	44	48	52	56	60	64	68
	0.025	2	5	9	13	17	21	25	29	34	38	42	46	51	55	60	64	68	73	77
	0.05	3	7	11	16	20	25	29	34	38	43	48	52	57	62	66	71	76	81	85
	0.10	5	10	14	19	24	29	34	39	44	49	54	59	64	69	75	80	85	90	95
14	0.001	0	0	2	4	7	10	13	16	20	23	26	30	33	37	40	44	47	51	55
	0.005	0	2	5	8	12	16	19	23	27	31	35	39	43	47	51	55	59	64	68
	0.01	1	3	7	11	14	18	23	27	31	35	39	44	48	52	57	61	66	70	74
	0.025	2	6	10	14	18	23	27	32	37	41	46	51	56	60	65	70	75	79	84
	0.05	4	8	12	17	22	27	32	37	42	47	52	57	62	67	72	78	83	88	93
	0.10	5	11	16	21	26	32	37	42	48	53	59	64	70	75	81	86	92	98	103
15	0.001	0	0	2	5	8	11	15	18	22	25	29	33	37	41	44	48	52	56	60
	0.005	0	3	6	9	13	17	21	25	30	34	38	43	47	52	56	61	65	70	74
	0.01	1	4	8	12	16	20	25	29	34	38	43	48	52	57	62	67	71	76	81
	0.025	2	6	11	15	20	25	30	35	40	45	50	55	60	65	71	76	81	86	91
	0.05	4	8	13	19	24	29	34	40	45	51	56	62	67	73	78	84	89	95	101
	0.10	6	11	17	23	28	34	40	46	52	58	64	69	75	81	87	93	99	105	111
16	0.001	0	0	3	6	9	12	16	20	24	28	32	36	40	44	49	53	57	61	66
	0.005	0	3	6	10	14	19	23	28	32	37	42	46	51	56	61	66	71	75	80
	0.01	1	4	8	13	17	22	27	32	37	42	47	52	57	62	67	72	77	83	88
	0.025	2	7	12	16	22	27	32	38	43	48	54	60	65	71	76	82	87	93	99
	0.05	4	9	15	20	26	31	37	43	49	55	61	66	72	78	84	90	96	102	108
	0.10	6	12	18	24	30	37	43	49	55	62	68	75	81	87	94	100	107	113	120
17	0.001	0	1	3	6	10	14	18	22	26	30	35	39	44	48	53	58	62	67	71
	0.005	0	3	7	11	16	20	25	30	35	40	45	50	55	61	66	71	76	82	87
	0.01	1	5	9	14	19	24	29	34	39	45	50	56	61	67	72	78	83	89	94
	0.025	3	7	12	18	23	29	35	40	46	52	58	64	70	76	82	88	94	100	106
	0.05	4	10	16	21	27	34	40	46	52	58	65	71	78	84	90	97	103	110	118
	0.10	7	13	19	26	32	39	46	53	59	66	73	80	86	93	100	107	114	121	128
18	0.001	0	1	4	7	11	15	19	24	28	33	38	43	47	52	57	62	67	72	77
	0.005	0	3	7	12	17	22	27	32	38	43	48	54	59	65	71	76	82	88	93
	0.01	1	5	10	15	20	25	31	37	42	48	54	60	66	71	77	83	89	95	101
	0.025	3	8	13	19	25	31	37	43	49	56	62	68	75	81	87	94	100	107	113
	0.05	5	10	17	23	29	36	42	49	56	62	69	76	83	89	96	103	110	117	124
	0.10	7	14	21	28	35	42	49	56	63	70	78	85	92	99	107	114	121	129	136
19	0.001	0	1	4	8	12	16	21	26	30	35	41	46	51	56	61	67	72	78	83
	0.005	1	4	8	13	18	23	29	34	40	46	52	58	64	70	75	82	88	94	100
	0.01	2	5	10	16	21	27	33	39	45	51	57	64	70	76	83	89	95	102	108
	0.025	3	8	14	20	26	33	39	46	53	59	66	73	79	86	93	100	107	114	120
	0.05	5	11	18	24	31	38	45	52	59	66	73	81	88	95	102	110	117	124	131
	0.10	8	15	22	29	37	44	52	59	67	74	82	90	98	105	113	121	129	136	144
20	0.001	0	1	4	8	13	17	22	27	33	38	43	49	55	60	66	71	77	83	89
	0.005	1	4	9	14	19	25	31	37	43	49	55	61	68	74	80	87	93	100	106
	0.01	2	6	11	17	23	29	35	41	48	54	61	68	74	81	88	94	101	108	115
	0.025	3	9	15	21	28	35	42	49	56	63	70	77	84	91	99	106	113	120	128
	0.05	5	12	19	26	33	40	48	55	63	70	78	85	93	101	108	116	124	131	139
	0.10	8	16	23	31	39	47	55	63	71	79	87	95	103	111	120	128	136	144	152

Table XV

Critical Values of Spearman's Rank Correlation Coefficient

$\alpha(2)$: $\alpha(1)$:	0.50 0.25	0.20 0.10	0.10 0.05	0.05 0.025	0.02 0.01	0.01 0.005	0.005 0.0025	0.002 0.001	0.001 0.0005
4	0.600	1.000	1.000						
5	0.500	0.800	0.900	1.000	1.000				
6	0.371	0.657	0.829	0.886	0.943	1.000	1.000		
7	0.321	0.571	0.714	0.786	0.893	0.929	0.964	1.000	1.000
8	0.310	0.524	0.643	0.738	0.833	0.881	0.905	0.952	0.976
9	0.267	0.483	0.600	0.700	0.783	0.833	0.867	0.917	0.933
10	0.248	0.455	0.564	0.648	0.745	0.794	0.830	0.879	0.903
11	0.236	0.427	0.536	0.618	0.709	0.755	0.800	0.845	0.873
12	0.217	0.406	0.503	0.587	0.678	0.727	0.769	0.818	0.846
13	0.209	0.385	0.484	0.560	0.648	0.703	0.747	0.791	0.824
14	0.200	0.367	0.464	0.538	0.626	0.679	0.723	0.771	0.802
15	0.189	0.354	0.446	0.521	0.604	0.654	0.700	0.750	0.779
16	0.182	0.341	0.429	0.503	0.582	0.635	0.679	0.729	0.762
17	0.176	0.328	0.414	0.485	0.566	0.615	0.662	0.713	0.748
18	0.170	0.317	0.401	0.472	0.550	0.600	0.643	0.695	0.728
19	0.165	0.309	0.391	0.460	0.535	0.584	0.628	0.677	0.712
20	0.161	0.299	0.380	0.447	0.520	0.570	0.612	0.662	0.696
21	0.156	0.292	0.370	0.435	0.508	0.556	0.599	0.648	0.681
22	0.152	0.284	0.361	0.425	0.496	0.544	0.586	0.634	0.667
23	0.148	0.278	0.353	0.415	0.486	0.532	0.573	0.622	0.654
24	0.144	0.271	0.344	0.406	0.476	0.521	0.562	0.610	0.642
25	0.142	0.265	0.337	0.398	0.466	0.511	0.551	0.598	0.630
26	0.138	0.259	0.331	0.390	0.457	0.501	0.541	0.587	0.619
27	0.136	0.255	0.324	0.382	0.448	0.491	0.531	0.577	0.608
28	0.133	0.250	0.317	0.375	0.440	0.483	0.522	0.567	0.598
29	0.130	0.245	0.312	0.368	0.433	0.475	0.513	0.558	0.589
30	0.128	0.240	0.306	0.362	0.425	0.467	0.504	0.549	0.580
31	0.126	0.236	0.301	0.356	0.418	0.459	0.496	0.541	0.571
32	0.124	0.232	0.296	0.350	0.412	0.452	0.489	0.533	0.563
33	0.121	0.229	0.291	0.345	0.405	0.446	0.482	0.525	0.554
34	0.120	0.225	0.287	0.340	0.399	0.439	0.475	0.517	0.547
35	0.118	0.222	0.283	0.335	0.394	0.433	0.488	0.510	0.539
36	0.116	0.219	0.279	0.330	0.388	0.427	0.462	0.504	0.533
37	0.114	0.216	0.275	0.325	0.383	0.421	0.456	0.497	0.526
38	0.113	0.212	0.271	0.321	0.378	0.415	0.450	0.491	0.519
39	0.111	0.210	0.267	0.317	0.373	0.410	0.444	0.485	0.513
40	0.110	0.207	0.264	0.313	0.368	0.405	0.439	0.479	0.507
41	0.108	0.204	0.261	0.309	0.384	0.400	0.433	0.473	0.501
42	0.107	0.202	0.257	0.305	0.359	0.395	0.428	0.468	0.495
43	0.105	0.199	0.254	0.301	0.355	0.391	0.423	0.463	0.490
44	0.104	0.197	0.251	0.298	0.351	0.386	0.419	0.458	0.484
45	0.103	0.194	0.248	0.294	0.347	0.382	0.414	0.453	0.479
46	0.102	0.192	0.246	0.291	0.343	0.378	0.410	0.448	0.474
47	0.101	0.190	0.243	0.288	0.340	0.374	0.406	0.443	0.469
48	0.100	0.188	0.240	0.285	0.336	0.370	0.401	0.439	0.465
49	0.098	0.186	0.238	0.282	0.333	0.366	0.397	0.434	0.460
50	0.097	0.184	0.235	0.279	0.329	0.363	0.393	0.430	0.456

n

Table XV (*continued*)

Critical Values of Spearman's Rank Correlation Coefficient

$\alpha(2)$: $\alpha(1)$:	0.50 0.25	0.20 0.10	0.10 0.05	0.05 0.025	0.02 0.01	0.01 0.005	0.005 0.0025	0.002 0.001	0.001 0.0005
51	0.096	0.182	0.233	0.276	0.326	0.359	0.390	0.426	0.451
52	0.095	0.180	0.231	0.274	0.323	0.356	0.386	0.422	0.447
53	0.095	0.179	0.228	0.271	0.320	0.352	0.382	0.418	0.443
54	0.094	0.177	0.226	0.268	0.317	0.349	0.379	0.414	0.439
55	0.093	0.175	0.224	0.266	0.314	0.346	0.375	0.411	0.435
56	0.092	0.174	0.222	0.264	0.311	0.343	0.372	0.407	0.432
57	0.091	0.172	0.220	0.261	0.308	0.340	0.369	0.404	0.428
58	0.090	0.171	0.218	0.259	0.306	0.337	0.366	0.400	0.424
59	0.089	0.169	0.216	0.257	0.303	0.334	0.363	0.397	0.421
60	0.089	0.168	0.214	0.255	0.300	0.331	0.360	0.394	0.418
61	0.088	0.166	0.213	0.252	0.298	0.329	0.357	0.391	0.414
62	0.087	0.165	0.211	0.250	0.296	0.326	0.354	0.388	0.411
63	0.086	0.163	0.209	0.248	0.293	0.323	0.351	0.385	0.408
64	0.086	0.162	0.207	0.246	0.291	0.321	0.348	0.382	0.405
65	0.085	0.161	0.206	0.244	0.289	0.318	0.346	0.379	0.402
66	0.084	0.160	0.204	0.243	0.287	0.316	0.343	0.376	0.399
67	0.084	0.158	0.203	0.241	0.284	0.314	0.341	0.373	0.396
68	0.083	0.157	0.201	0.239	0.282	0.311	0.338	0.370	0.393
69	0.082	0.156	0.200	0.237	0.280	0.309	0.336	0.368	0.390
70	0.082	0.155	0.198	0.235	0.278	0.307	0.333	0.365	0.388
71	0.081	0.154	0.197	0.234	0.276	0.305	0.331	0.363	0.385
72	0.081	0.153	0.195	0.232	0.274	0.303	0.329	0.360	0.382
73	0.080	0.152	0.194	0.230	0.272	0.301	0.327	0.358	0.380
74	0.080	0.151	0.193	0.229	0.271	0.299	0.324	0.355	0.377
75	0.079	0.150	0.191	0.227	0.269	0.297	0.322	0.353	0.375
76	0.078	0.149	0.190	0.226	0.267	0.295	0.320	0.351	0.372
77	0.078	0.148	0.189	0.224	0.265	0.293	0.318	0.349	0.370
78	0.077	0.147	0.188	0.223	0.264	0.291	0.316	0.346	0.368
79	0.077	0.146	0.186	0.221	0.262	0.289	0.314	0.344	0.365
80	0.076	0.145	0.185	0.220	0.260	0.287	0.312	0.342	0.363
81	0.076	0.144	0.184	0.219	0.259	0.285	0.310	0.340	0.361
82	0.075	0.143	0.183	0.217	0.257	0.284	0.308	0.338	0.359
83	0.075	0.142	0.182	0.216	0.255	0.282	0.306	0.336	0.357
84	0.074	0.141	0.181	0.215	0.254	0.280	0.305	0.334	0.355
85	0.074	0.140	0.180	0.213	0.252	0.279	0.303	0.332	0.353
86	0.074	0.139	0.179	0.212	0.251	0.277	0.301	0.330	0.351
87	0.073	0.139	0.177	0.211	0.250	0.276	0.299	0.328	0.349
88	0.073	0.138	0.176	0.210	0.248	0.274	0.298	0.327	0.347
89	0.072	0.137	0.175	0.209	0.247	0.272	0.296	0.325	0.345
90	0.072	0.136	0.174	0.207	0.245	0.271	0.294	0.323	0.343
91	0.072	0.135	0.173	0.206	0.244	0.269	0.293	0.321	0.341
92	0.071	0.135	0.173	0.205	0.243	0.268	0.291	0.319	0.339
93	0.071	0.134	0.172	0.204	0.241	0.267	0.290	0.318	0.338
94	0.070	0.133	0.171	0.203	0.240	0.265	0.288	0.316	0.336
95	0.070	0.133	0.170	0.202	0.239	0.264	0.287	0.314	0.334
96	0.070	0.132	0.169	0.201	0.238	0.262	0.285	0.313	0.332
97	0.069	0.131	0.168	0.200	0.236	0.261	0.284	0.311	0.331
98	0.069	0.130	0.167	0.199	0.235	0.260	0.282	0.310	0.329
99	0.068	0.130	0.166	0.198	0.234	0.258	0.281	0.308	0.327
100	0.068	0.129	0.165	0.197	0.233	0.257	0.279	0.307	0.326

Source: Zar, Jerrold H., *Biostatistical Analysis,* 4th Edition, © 1999. Reprinted by permission of Pearson Education, Inc., Upper Saddle River, NJ.

Table XVI

Critical Values of the Kruskal–Wallis Test Statistic

n_1	n_2	n_3	Critical Value	α
2	1	1	2.7000	0.500
2	2	1	3.6000	0.200
2	2	2	4.5714	0.067
			3.7143	0.200
3	1	1	3.2000	0.300
3	2	1	4.2857	0.100
			3.8571	0.133
3	2	2	5.3572	0.029
			4.7143	0.048
			4.5000	0.067
			4.4643	0.105
3	3	1	5.1429	0.043
			4.5714	0.100
			4.0000	0.129
3	3	2	6.2500	0.011
			5.3611	0.032
			5.1389	0.061
			4.5556	0.100
			4.2500	0.121
3	3	3	7.2000	0.004
			6.4889	0.011
			5.6889	0.029
			5.6000	0.050
			5.0667	0.086
			4.6222	0.100
4	1	1	3.5714	0.200
4	2	1	4.8214	0.057
			4.5000	0.076
			4.0179	0.114
4	2	2	6.0000	0.014
			5.3333	0.033
			5.1250	0.052
			4.4583	0.100
			4.1667	0.105
4	3	1	5.8333	0.021
			5.2083	0.050
			5.0000	0.057
			4.0556	0.093
			3.8889	0.129
4	3	2	6.4444	0.008
			6.3000	0.011
			5.4444	0.046
			5.4000	0.051
			4.5111	0.098
			4.4444	0.102
4	3	3	6.7455	0.010
			6.7091	0.013
			5.7909	0.046
			5.7273	0.050
			4.7091	0.092

n_1	n_2	n_3	Critical Value	α
			4.7000	0.101
4	4	1	6.6667	0.010
			6.1667	0.022
			4.9667	0.048
			4.8667	0.054
			4.1667	0.082
			4.0667	0.102
4	4	2	7.0364	0.006
			6.8727	0.011
			5.4545	0.046
			5.2364	0.052
			4.5545	0.098
			4.4455	0.103
4	4	3	7.1439	0.010
			7.1364	0.011
			5.5985	0.049
			5.5758	0.051
			4.5455	0.099
			4.4773	0.102
4	4	4	7.6538	0.008
			7.5385	0.011
			5.6923	0.049
			5.6538	0.054
			4.6539	0.097
			4.5001	0.104
5	1	1	3.8571	0.143
5	2	1	5.2500	0.036
			5.0000	0.048
			4.4500	0.071
			4.2000	0.095
			4.0500	0.119
5	2	2	6.5333	0.008
			6.1333	0.013
			5.1600	0.034
			5.0400	0.056
			4.3733	0.090
			4.2933	0.122
5	3	1	6.4000	0.012
			4.9600	0.048
			4.8711	0.052
			4.0178	0.095
			3.8400	0.123
5	3	2	6.9091	0.009
			6.8218	0.010
			5.2509	0.049
			5.1055	0.052
			4.6509	0.091
			4.4945	0.101
5	3	3	7.0788	0.009
			6.9818	0.011

Table XVI (*continued*)

Critical Values of the Kruskal–Wallis Test Statistic

n_1	n_2	n_3	Critical Value	α	n_1	n_2	n_3	Critical Value	α
5	3	3	5.6485	0.049	5	5	1	6.8364	0.011
			5.5152	0.051				5.1273	0.046
			4.5333	0.097				4.9091	0.053
			4.4121	0.109				4.1091	0.086
5	4	1	6.9545	0.008				4.0364	0.105
			6.8400	0.011	5	5	2	7.3385	0.010
			4.9855	0.044				7.2692	0.010
			4.8600	0.056				5.3385	0.047
			3.9873	0.098				5.2462	0.051
			3.9600	0.102				4.6231	0.097
5	4	2	7.2045	0.009				4.5077	0.100
			7.1182	0.010	5	5	3	7.5780	0.010
			5.2727	0.049				7.5429	0.010
			5.2682	0.050				5.7055	0.046
			4.5409	0.098				5.6264	0.051
			4.5182	0.101				4.5451	0.100
5	4	3	7.4449	0.010				4.5363	0.102
			7.3949	0.011	5	5	4	7.8229	0.010
			5.6564	0.049				7.7914	0.010
			5.6308	0.050				5.6657	0.049
			4.5487	0.099				5.6429	0.050
			4.5231	0.103				4.5229	0.099
5	4	4	7.7604	0.009				4.5200	0.101
			7.7440	0.011	5	5	5	8.0000	0.009
			5.6571	0.049				7.9600	0.010
			5.6176	0.050				5.7800	0.049
			4.6187	0.100				5.6600	0.051
			4.5527	0.102				4.5600	0.100
5	5	1	7.3091	0.009				4.5000	0.102

Source: Based on W. H. Kruskal and W. A. Wallis, Use of Ranks in One-Criterion Analysis of Variance, *J. Amer. Statist. Assoc.*, **47** (1952), 583–621, Addendum, *Ibid.*, **48** (1953), 907–911.

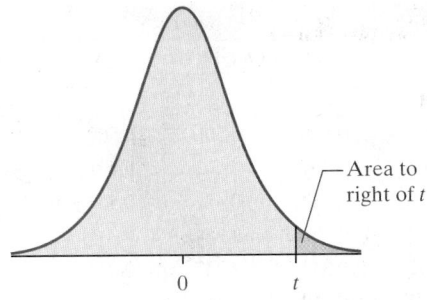

Area to right of t

0   t

Table XVII

Student's t-Distribution

t \ df	1	2	3	4	5	6	7	8	9	10	11	12
0.0	0.500	0.500	0.500	0.500	0.500	0.500	0.500	0.500	0.500	0.500	0.500	0.500
0.1	0.468	0.465	0.463	0.463	0.462	0.462	0.462	0.461	0.461	0.461	0.461	0.461
0.2	0.437	0.430	0.427	0.426	0.425	0.424	0.424	0.423	0.423	0.423	0.423	0.422
0.3	0.407	0.396	0.392	0.390	0.388	0.387	0.386	0.386	0.385	0.385	0.385	0.385
0.4	0.379	0.364	0.358	0.355	0.353	0.352	0.351	0.350	0.349	0.349	0.348	0.348
0.5	0.352	0.333	0.326	0.322	0.319	0.317	0.316	0.315	0.315	0.314	0.313	0.313
0.6	0.328	0.305	0.295	0.290	0.287	0.285	0.284	0.283	0.282	0.281	0.280	0.280
0.7	0.306	0.278	0.267	0.261	0.258	0.255	0.253	0.252	0.251	0.250	0.249	0.249
0.8	0.285	0.254	0.241	0.234	0.230	0.227	0.225	0.223	0.222	0.221	0.220	0.220
0.9	0.267	0.232	0.217	0.210	0.205	0.201	0.199	0.197	0.196	0.195	0.194	0.193
1.0	0.250	0.211	0.196	0.187	0.182	0.178	0.175	0.173	0.172	0.170	0.169	0.169
1.1	0.235	0.193	0.176	0.167	0.161	0.157	0.154	0.152	0.150	0.149	0.147	0.146
1.2	0.221	0.177	0.158	0.148	0.142	0.138	0.135	0.132	0.130	0.129	0.128	0.127
1.3	0.209	0.162	0.142	0.132	0.125	0.121	0.117	0.115	0.113	0.111	0.110	0.109
1.4	0.197	0.148	0.128	0.117	0.110	0.106	0.102	0.100	0.098	0.096	0.095	0.093
1.5	0.187	0.136	0.115	0.104	0.097	0.092	0.089	0.086	0.084	0.082	0.081	0.080
1.6	0.178	0.125	0.104	0.092	0.085	0.080	0.077	0.074	0.072	0.070	0.069	0.068
1.7	0.169	0.116	0.094	0.082	0.075	0.070	0.066	0.064	0.062	0.060	0.059	0.057
1.8	0.161	0.107	0.085	0.073	0.066	0.061	0.057	0.055	0.053	0.051	0.050	0.049
1.9	0.154	0.099	0.077	0.065	0.058	0.053	0.050	0.047	0.045	0.043	0.042	0.041
2.0	0.148	0.092	0.070	0.058	0.051	0.046	0.043	0.040	0.038	0.037	0.035	0.034
2.1	0.141	0.085	0.063	0.052	0.045	0.040	0.037	0.034	0.033	0.031	0.030	0.029
2.2	0.136	0.079	0.058	0.046	0.040	0.035	0.032	0.029	0.028	0.026	0.025	0.024
2.3	0.131	0.074	0.052	0.041	0.035	0.031	0.027	0.025	0.023	0.022	0.021	0.020
2.4	0.126	0.069	0.048	0.037	0.031	0.027	0.024	0.022	0.020	0.019	0.018	0.017
2.5	0.121	0.065	0.044	0.033	0.027	0.023	0.020	0.018	0.017	0.016	0.015	0.014
2.6	0.117	0.061	0.040	0.030	0.024	0.020	0.018	0.016	0.014	0.013	0.012	0.012
2.7	0.113	0.057	0.037	0.027	0.021	0.018	0.015	0.014	0.012	0.011	0.010	0.010
2.8	0.109	0.054	0.034	0.024	0.019	0.016	0.013	0.012	0.010	0.009	0.009	0.008
2.9	0.106	0.051	0.031	0.022	0.017	0.014	0.011	0.010	0.009	0.008	0.007	0.007
3.0	0.102	0.048	0.029	0.020	0.015	0.012	0.010	0.009	0.007	0.007	0.006	0.006
3.1	0.099	0.045	0.027	0.018	0.013	0.011	0.009	0.007	0.006	0.006	0.005	0.005
3.2	0.096	0.043	0.025	0.016	0.012	0.009	0.008	0.006	0.005	0.005	0.004	0.004
3.3	0.094	0.040	0.023	0.015	0.011	0.008	0.007	0.005	0.005	0.004	0.004	0.003
3.4	0.091	0.038	0.021	0.014	0.010	0.007	0.006	0.005	0.004	0.003	0.003	0.003
3.5	0.089	0.036	0.020	0.012	0.009	0.006	0.005	0.004	0.003	0.003	0.002	0.002
3.6	0.086	0.035	0.018	0.011	0.008	0.006	0.004	0.003	0.003	0.002	0.002	0.002
3.7	0.084	0.033	0.017	0.010	0.007	0.005	0.004	0.003	0.002	0.002	0.002	0.002
3.8	0.082	0.031	0.016	0.010	0.006	0.004	0.003	0.003	0.002	0.002	0.001	0.001
3.9	0.080	0.030	0.015	0.009	0.006	0.004	0.003	0.002	0.002	0.001	0.001	0.001
4.0	0.078	0.029	0.014	0.008	0.005	0.004	0.003	0.002	0.002	0.001	0.001	0.001

Table XVII (continued)

Student's *t*-Distribution

t \ *df*	13	14	15	16	17	18	19	20	21	22	23	24
0.0	0.500	0.500	0.500	0.500	0.500	0.500	0.500	0.500	0.500	0.500	0.500	0.500
0.1	0.461	0.461	0.461	0.461	0.461	0.461	0.461	0.461	0.461	0.461	0.461	0.461
0.2	0.422	0.422	0.422	0.422	0.422	0.422	0.422	0.422	0.422	0.422	0.422	0.422
0.3	0.384	0.384	0.384	0.384	0.384	0.384	0.384	0.384	0.384	0.383	0.383	0.383
0.4	0.348	0.348	0.347	0.347	0.347	0.347	0.347	0.347	0.347	0.347	0.346	0.346
0.5	0.313	0.312	0.312	0.312	0.312	0.312	0.311	0.311	0.311	0.311	0.311	0.311
0.6	0.279	0.279	0.279	0.278	0.278	0.278	0.278	0.278	0.277	0.277	0.277	0.277
0.7	0.248	0.248	0.247	0.247	0.247	0.246	0.246	0.246	0.246	0.246	0.245	0.245
0.8	0.219	0.219	0.218	0.218	0.217	0.217	0.217	0.217	0.216	0.216	0.216	0.216
0.9	0.192	0.192	0.191	0.191	0.190	0.190	0.190	0.189	0.189	0.189	0.189	0.189
1.0	0.168	0.167	0.167	0.166	0.166	0.165	0.165	0.165	0.164	0.164	0.164	0.164
1.1	0.146	0.145	0.144	0.144	0.143	0.143	0.143	0.142	0.142	0.142	0.141	0.141
1.2	0.126	0.125	0.124	0.124	0.123	0.123	0.122	0.122	0.122	0.121	0.121	0.121
1.3	0.108	0.107	0.107	0.106	0.105	0.105	0.105	0.104	0.104	0.104	0.103	0.103
1.4	0.092	0.092	0.091	0.090	0.090	0.089	0.089	0.088	0.088	0.088	0.087	0.087
1.5	0.079	0.078	0.077	0.077	0.076	0.075	0.075	0.075	0.074	0.074	0.074	0.073
1.6	0.067	0.066	0.065	0.065	0.064	0.064	0.063	0.063	0.062	0.062	0.062	0.061
1.7	0.056	0.056	0.055	0.054	0.054	0.053	0.053	0.052	0.052	0.052	0.051	0.051
1.8	0.048	0.047	0.046	0.045	0.045	0.044	0.044	0.043	0.043	0.043	0.042	0.042
1.9	0.040	0.039	0.038	0.038	0.037	0.037	0.036	0.036	0.036	0.035	0.035	0.035
2.0	0.033	0.033	0.032	0.031	0.031	0.030	0.030	0.030	0.029	0.029	0.029	0.028
2.1	0.028	0.027	0.027	0.026	0.025	0.025	0.025	0.024	0.024	0.024	0.023	0.023
2.2	0.023	0.023	0.022	0.021	0.021	0.021	0.020	0.020	0.020	0.019	0.019	0.019
2.3	0.019	0.019	0.018	0.018	0.017	0.017	0.016	0.016	0.016	0.016	0.015	0.015
2.4	0.016	0.015	0.015	0.014	0.014	0.014	0.013	0.013	0.013	0.013	0.012	0.012
2.5	0.013	0.013	0.012	0.012	0.011	0.011	0.011	0.011	0.010	0.010	0.010	0.010
2.6	0.011	0.010	0.010	0.010	0.009	0.009	0.009	0.009	0.008	0.008	0.008	0.008
2.7	0.009	0.009	0.008	0.008	0.008	0.007	0.007	0.007	0.007	0.007	0.006	0.006
2.8	0.008	0.007	0.007	0.006	0.006	0.006	0.006	0.006	0.005	0.005	0.005	0.005
2.9	0.006	0.006	0.005	0.005	0.005	0.005	0.005	0.004	0.004	0.004	0.004	0.004
3.0	0.005	0.005	0.004	0.004	0.004	0.004	0.004	0.004	0.003	0.003	0.003	0.003
3.1	0.004	0.004	0.004	0.003	0.003	0.003	0.003	0.003	0.003	0.003	0.003	0.002
3.2	0.003	0.003	0.003	0.003	0.003	0.002	0.002	0.002	0.002	0.002	0.002	0.002
3.3	0.003	0.003	0.002	0.002	0.002	0.002	0.002	0.002	0.002	0.002	0.002	0.002
3.4	0.002	0.002	0.002	0.002	0.002	0.002	0.002	0.001	0.001	0.001	0.001	0.001
3.5	0.002	0.002	0.002	0.001	0.001	0.001	0.001	0.001	0.001	0.001	0.001	0.001
3.6	0.002	0.001	0.001	0.001	0.001	0.001	0.001	0.001	0.001	0.001	0.001	0.001
3.7	0.001	0.001	0.001	0.001	0.001	0.001	0.001	0.001	0.001	0.001	0.001	0.001
3.8	0.001	0.001	0.001	0.001	0.001	0.001	0.001	0.001	0.001	0.000	0.000	0.000
3.9	0.001	0.001	0.001	0.001	0.001	0.001	0.000	0.000	0.000	0.000	0.000	0.000
4.0	0.001	0.001	0.001	0.001	0.000	0.000	0.000	0.000	0.000	0.000	0.000	0.000

Table XVII (*continued*)

Student's *t*-Distribution

t \ *df*	25	26	27	28	29	30	35	40	60	120	∞ (= *z*)
0.0	0.500	0.500	0.500	0.500	0.500	0.500	0.500	0.500	0.500	0.500	0.500
0.1	0.461	0.461	0.461	0.461	0.461	0.461	0.460	0.460	0.460	0.460	0.460
0.2	0.422	0.422	0.421	0.421	0.421	0.421	0.421	0.421	0.421	0.421	0.421
0.3	0.383	0.383	0.383	0.383	0.383	0.383	0.383	0.383	0.383	0.382	0.382
0.4	0.346	0.346	0.346	0.346	0.346	0.346	0.346	0.346	0.345	0.345	0.345
0.5	0.311	0.311	0.311	0.310	0.310	0.310	0.310	0.310	0.309	0.309	0.309
0.6	0.277	0.277	0.277	0.277	0.277	0.277	0.276	0.276	0.275	0.275	0.274
0.7	0.245	0.245	0.245	0.245	0.245	0.245	0.244	0.244	0.243	0.243	0.242
0.8	0.216	0.215	0.215	0.215	0.215	0.215	0.215	0.214	0.213	0.213	0.212
0.9	0.188	0.188	0.188	0.188	0.188	0.188	0.187	0.187	0.186	0.185	0.184
1.0	0.163	0.163	0.163	0.163	0.163	0.163	0.162	0.162	0.161	0.160	0.159
1.1	0.141	0.141	0.141	0.140	0.140	0.140	0.139	0.139	0.138	0.137	0.136
1.2	0.121	0.120	0.120	0.120	0.120	0.120	0.119	0.119	0.117	0.116	0.115
1.3	0.103	0.103	0.102	0.102	0.102	0.102	0.101	0.101	0.099	0.098	0.097
1.4	0.087	0.087	0.086	0.086	0.086	0.086	0.085	0.085	0.083	0.082	0.081
1.5	0.073	0.073	0.073	0.072	0.072	0.072	0.071	0.071	0.069	0.068	0.067
1.6	0.061	0.061	0.061	0.060	0.060	0.060	0.059	0.059	0.057	0.056	0.055
1.7	0.051	0.051	0.050	0.050	0.050	0.050	0.049	0.048	0.047	0.046	0.045
1.8	0.042	0.042	0.042	0.041	0.041	0.041	0.040	0.040	0.038	0.037	0.036
1.9	0.035	0.034	0.034	0.034	0.034	0.034	0.033	0.032	0.031	0.030	0.029
2.0	0.028	0.028	0.028	0.028	0.027	0.027	0.027	0.026	0.025	0.024	0.023
2.1	0.023	0.023	0.023	0.022	0.022	0.022	0.022	0.021	0.020	0.019	0.018
2.2	0.019	0.018	0.018	0.018	0.018	0.018	0.017	0.017	0.016	0.015	0.014
2.3	0.015	0.015	0.015	0.015	0.014	0.014	0.014	0.013	0.012	0.012	0.011
2.4	0.012	0.012	0.012	0.012	0.012	0.011	0.011	0.011	0.010	0.009	0.008
2.5	0.010	0.010	0.009	0.009	0.009	0.009	0.009	0.008	0.008	0.007	0.006
2.6	0.008	0.008	0.007	0.007	0.007	0.007	0.007	0.006	0.006	0.005	0.005
2.7	0.006	0.006	0.006	0.006	0.006	0.006	0.005	0.005	0.004	0.004	0.003
2.8	0.005	0.005	0.005	0.005	0.004	0.004	0.004	0.004	0.003	0.003	0.003
2.9	0.004	0.004	0.004	0.004	0.004	0.003	0.003	0.003	0.003	0.002	0.002
3.0	0.003	0.003	0.003	0.003	0.003	0.003	0.002	0.002	0.002	0.002	0.001
3.1	0.002	0.002	0.002	0.002	0.002	0.002	0.002	0.002	0.001	0.001	0.001
3.2	0.002	0.002	0.002	0.002	0.002	0.002	0.001	0.001	0.001	0.001	0.001
3.3	0.001	0.001	0.001	0.001	0.001	0.001	0.001	0.001	0.001	0.001	0.000
3.4	0.001	0.001	0.001	0.001	0.001	0.001	0.001	0.001	0.001	0.000	0.000
3.5	0.001	0.001	0.001	0.001	0.001	0.001	0.001	0.001	0.000	0.000	0.000
3.6	0.001	0.001	0.001	0.001	0.001	0.001	0.000	0.000	0.000	0.000	0.000
3.7	0.001	0.001	0.000	0.000	0.000	0.000	0.000	0.000	0.000	0.000	0.000
3.8	0.000	0.000	0.000	0.000	0.000	0.000	0.000	0.000	0.000	0.000	0.000
3.9	0.000	0.000	0.000	0.000	0.000	0.000	0.000	0.000	0.000	0.000	0.000
4.0	0.000	0.000	0.000	0.000	0.000	0.000	0.000	0.000	0.000	0.000	0.000

Answers

CHAPTER 1 Data Collection

1.1 Assess Your Understanding (page 10)

1. Statistics is the science of collecting, organizing, summarizing, and analyzing information to draw a conclusion and answer questions. In addition, statistics is about providing a measure of confidence in any conclusions.

3. individual **5.** statistic; parameter

7. Parameter **9.** Statistic **11.** Parameter

13. Statistic **15.** Qualitative **17.** Quantitative

19. Quantitative **21.** Qualitative **23.** Discrete

25. Continuous **27.** Continuous **29.** Discrete

31. Nominal **33.** Ratio **35.** Ordinal

37. Ratio

39. Population: teenagers 13 to 17 years of age who live in the United States. Sample: 1028 teenagers 13 to 17 years of age who live in the United States

41. Population: entire soybean crop. Sample: 100 plants selected

43. Population: women 27 to 44 years of age with hypertension. Sample: 7373 women 27 to 44 years of age with hypertension

45. Individuals: Alabama, Colorado, Indiana, North Carolina, Wisconsin
Variables: minimum age for driver's license (unrestricted), mandatory belt use seating positions, maximum allowable speed limit on rural interstates in 2011
Data for minimum age for driver's license: 17, 17, 18, 16, 18
Data for mandatory belt use seating positions: front, front, all, all, all
Data for maximum allowable speed limit on rural interstates in 2011: 70, 75, 70, 70, 65 (mph)
The variable *minimum age for driver's license* is continuous; the variable *mandatory belt use seating positions* is qualitative; the variable *maximum allowable speed limit on rural interstates in 2011* is continuous.

47. **(a)** To determine if adolescents 18–21 who smoke have a lower IQ than nonsmokers
(b) All adolescents 18–21; the 20,211 18-year-old Israeli male military recruits.
(c) The average IQ of smokers was 94; the average IQ of nonsmokers was 101.
(d) Lower IQ individuals are more likely to choose to smoke.

49. **(a)** To determine the proportion of adult Americans who believe the federal government wastes 51 cents or more of every dollar
(b) American adults aged 18 years or older
(c) 1017 American adults aged 18 years or older
(d) Of the 1017 individuals surveyed, 35% indicated that 51 cents or more is wasted.
(e) Gallup is 95% certain that the percentage of all adult Americans who believe the federal government wastes 51 cents or more of every dollar received is between 31% and 39%.

51. Nominal: ordinal, the level of measurement changes because the goal of the research has changed. Rather than the number being used to identify the player, it is now also used to explain the caliber of the player, with a higher number implying a lower caliber of player.

53. **(a)** Does season of birth affect mood?
(b) 400 people **(c)** Qualitative
(d) Those born in summer are prone to mood swings; those born in winter are less likely to be irritable.
(e) Season of birth plays a role in one's temperament.

55. A discrete variable is a quantitative variable that has a finite or countable number of possible values. A discrete variable cannot take on every possible value between any two possible values. Continuous variables are also quantitative variables, but there are an infinite number of possible values that are not countable. A continuous variable may take on every possible value between any two values.

57. This means that the values of the variable change from individual to individual. In addition, certain variables can change over time for certain individuals. Because data vary, two different statistical analyses of the same variable can lead to different results.

59. No. We measure age to as much accuracy as we wish.

1.2 Assess Your Understanding (page 19)

1. The explanatory variable is the variable that affects the value of the response variable. In research, we want to see how changes in the value of the explanatory variable affect the value of the response variable.

3. Confounding exists in a study when the effects of two or more explanatory variables are not separated. So any relation that appears to exist between a certain explanatory variable and the response variable may be due to some other variable or variables not accounted for in the study. A lurking variable is an explanatory variable that was not considered in a study, but that affects the value of the response variable (and, typically, the explanatory variable) in the study. A confounding variable is an explanatory variable that was considered in a study whose effect cannot be distinguished from a second explanatory variable in the study. Lurking variables tend to exist in observational studies, while confounding variables tend to exist in designed experiments.

5. Cross-sectional studies collect information at a specific point in time (or over a very short period of time). Case-control studies are retrospective (they look back in time). Also, individuals that have a certain characteristic (such as cancer) in a case-control study are matched with those that do not have the characteristic. Case-control studies are typically superior to cross-sectional studies.

7. There is a perceived benefit to obtaining a flu shot, so there are ethical issues in intentionally denying certain seniors access.

9. Observational study **11.** Experiment

13. Observational study **15.** Experiment

17. **(a)** Cohort since people were followed for 10 years.
(b) Whether the individual has heart disease or not; whether the individual is happy or not
(c) Confounding due to lurking variables

19. **(a)** The researchers administered a questionnaire to obtain their results, so there is no control of the explanatory variable. This is a cross-sectional study.
(b) Body mass index; whether a TV is in the bedroom or not
(c) Answers will vary. Some lurking variables might be the amount of exercise per week and eating habits.
(d) The researchers made an effort to avoid confounding by accounting for potential lurking variables.
(e) No. This is an observational study, so all we can say is that a television in the bedroom is associated with a higher body mass index.

21. **(a)** The data was collected at a specific point in time.
(b) Delivery scenario
(c) Method of delivery (qualitative); Cost (quantitative)

23. Answers will vary. This is a prospective, cohort observational study. Some possible lurking variables include smoking habits, eating habits, exercise, and family history of cancer. The study concluded that there might be an increased risk of certain blood cancers in people with prolonged exposure to electromagnetic fields. The author of the article reminds us that this is an observational study, so there is no control of variables that may affect the likelihood of getting certain cancers. In addition, the author states that we should do things in our lives that promote health.

1.3 Assess Your Understanding (page 26)

1. The frame is a list of the individuals in the population we are studying.

3. Once an individual is selected, he or she cannot be selected again.

5. Answers will vary. One option would be to write each name on a sheet of paper and choose the names out of a hat. A second option would be to number the books from 1 to 9 and randomly select three numbers.

7. (a) 616, 630; 616, 631; 616, 632; 616, 645; 616, 649; 616, 650; 630, 631; 630, 632; 630, 645; 630, 649; 630, 650; 631, 632; 631, 645; 631, 649; 631, 650; 632, 645; 632, 649; 632, 650; 645, 649; 645, 650; 649, 650
(b) There is a 1 in 21 chance the pair of courses will be EPR 630 and EPR 645.

9. (a) 83, 67, 84, 38, 22, 24, 36, 58, 34
(b) Answers may vary depending on the type of technology used.

11. (a) Answers will vary.   **(b)** Answers will vary.

13. (a) The list provided by the administration serves as the frame. Number the students on the list from 1 through 19,935. Using a random-number generator, set a seed (or select a starting point if using Table I in Appendix A). Generate 25 different numbers randomly. The students corresponding to these numbers will be the 25 students in the sample.
(b) Answers will vary.

15. Answers will vary. However, the members should be numbered from 1 to 32. Then, using the table of random numbers or a random-number generator, four different numbers should be selected. The names corresponding to these numbers will represent the simple random sample.

1.4 Assess Your Understanding (page 34)

1. If the population can be divided into homogeneous, nonoverlapping groups

3. Convenience samples are not random samples. Individuals who participate are often self-selected, so the results are not likely to be representative of the population.

5. stratified   **7.** False   **9.** True

11. Systematic   **13.** Cluster   **15.** Simple random

17. Cluster   **19.** Convenience   **21.** Stratified

23. 16, 41, 66, 91, 116, 141, 166, 191, 216, 241, 266, 291, 316, 341, 366, 391, 416, 441, 466, 491

25. Answers will vary. To obtain the sample, number the Democrats 1 to 16 and obtain a simple random sample of size 2. Then number the Republicans 1 to 16 and obtain a simple random sample of size 2. Be sure to use a different starting point in Table I or a different seed for each stratum.

27. (a) 90
(b) Randomly select a number between 1 and 90. Suppose that we randomly select 15; then the individuals in the survey will be 15, 105, 195, 285, . . ., 4425.

29. SRS: number from 1 to 1280. Randomly select 128 students to survey
Stratified: Randomly select four students from each section to survey.
Cluster: Randomly select four sections; survey all students in these four sections.
Answers will vary.

31. Answers will vary. A good choice might be stratified sampling with the strata being commuters and noncommuters.

33. Answers will vary. One option would be cluster sampling. The clusters could be the city blocks. Randomly select clusters and then survey all the households on the selected city blocks.

35. Answers will vary. Simple random sampling will work fine here, especially because a list of 6600 individuals who meet the needs of our study already exists (the frame).

37. (a) Registered voters who have voted in the past few elections.
(b) Because each individual has an equally likely chance of being selected, there is a chance that one group may be over- or underrepresented.
(c) By using a stratified sample, the strategist can obtain a simple random sample within each strata so that the number of individuals in the sample is proportionate to the number of individuals in the population.

1.5 Assess Your Understanding (page 40)

1. A closed question has fixed choices for answers, whereas an open question is a free-response question. Closed questions are easier to analyze, but limit the responses. Open questions allow respondents to state exactly how they feel, but are harder to analyze due to the variety of answers and possible misinterpretation of answers.

3. Bias means that the results of the sample are not representative of the population. Three types of bias: sampling bias, nonresponse bias, response bias. Sampling bias could mean that the sampling was not done using random sampling, but rather through a convenience sample. Nonresponse bias could be the result of an inability to contact an individual selected to be in a sample. Response bias could be the result of misunderstanding a question on a survey. A census could have bias through misinterpreted questions on a survey.

5. (a) Sampling bias due to undercoverage since the first 60 customers may not be representative of the customer population.
(b) Since a complete frame is not possible, systematic random sampling could be used to make the sample more representative of the customer population.

7. (a) Response bias due to a poorly worded question.
(b) The survey should begin by stating the current penalty for selling a gun illegally. The question might be rewritten as "Do you approve or disapprove of harsher penalties for individuals who sell guns illegally?" The words *approve* and *disapprove* should be rotated from individual to individual.

9. (a) Nonresponse bias; if the survey is only written in English, non-English speaking homes will not complete the survey, resulting in undercoverage
(b) The survey can be improved through face-to-face or telephone interviews.

11. (a) Sampling bias due to undercoverage, since the readers of the magazine may not be representative of all Australian women, and interviewer error, since advertisements and images in the magazine could affect the women's view of themselves.
(b) A well-designed sampling plan (not in a magazine), such as a cluster sample, could make the sample more representative of the intended population.

13. (a) Response bias due to a poorly worded question
(b) The question should be reworded so that it doesn't imply the opinion of the editors. One possibility might be "Do you believe that a marriage can be maintained after an extramarital relation?"

15. (a) Response bias because the students are not likely to be truthful when the teacher is administering the survey
(b) The survey should be administered by an impartial party so that the students will be more likely to respond truthfully.

17. No; the survey still suffers from undercoverage (sampling bias), nonresponse bias, and potentially response bias.

19. The ordering of the questions is likely to affect the survey results. Perhaps question B should be asked first. Another possibility is to rotate the questions randomly.

21. The company is using a reward in the form of the $5.00 payment and an incentive by telling the readers that his or her input will make a difference.

23. The frame is anyone with a landline phone. Any household without a landline phone, households on the do-not-call registry, and homeless individuals are excluded. This could result in sampling bias due to undercoverage.

25. Definitely, especially if households that are on the do-not-call registry have a trait that is not part of those households that are not on the do-not-call registry.

27. Poorly trained interviewers; interviewer bias, surveyed too many female voters.

29. The words used in a survey question can have a significant impact on the response. Whenever you read survey results, be mindful of the way the survey question was written. In this particular situation, the words "along with allies in Europe" had a large impact on survey results.

31. Answers will vary.

33. Sampling bias: using an incorrect frame led to undercoverage. Nonresponse bias: the low response rate.

35. (a) Answers may vary. However, you should stratify by political affiliation (Democrat, Republican, Independent, for example).

(b) Answers may vary, but historically Democratic turnout is higher in presidential election cycles.

(c) Answers may vary. A higher percentage of Democrats in polls versus turnout will lead to overstating the predicted percentage of Democratic votes.

37. Callbacks, incentives

39. The researcher can learn common answers.

41. Better survey results. A low response rate may mean that some segments of the population are underrepresented or that only individuals with strong opinions have participated.

43. The question is ambiguous since CD could mean compact disc or certificate of deposit. The question could be improved by not using the abbreviation.

1.6 Assess Your Understanding (page 49)

1. (a) A person, object, or some other well-defined item upon which a treatment is applied

(b) Any combination of the values of the factors (explanatory variables)

(c) A quantitative or qualitative variable that represents the variable of interest

(d) The variable whose effect on the response variable is to be assessed by the experimenter

(e) An innocuous medication, such as a sugar tablet, that looks, tastes, and smells like the experimental medication

(f) The effect of two factors (explanatory variables) on the response variable cannot be distinguished.

3. In a single-blind experiment, the subject does not know which treatment is received. In a double-blind experiment, neither the subject nor the researcher in contact with the subject knows which treatment is received.

5. blocking

7. (a) To determine the association between number of times one chews food and food consumption.

(b) The response variable is food consumption; quantitative

(c) The explanatory variable is chew level (100%, 150%, 200%); qualitative

(d) The 45 individuals 18 to 45 years of age.

(e) Determine a baseline number of chews before swallowing; same type of food used in baseline as in experiment; same time of day (lunch)

(f) Randomization reduces the effect of the order in which number of chews required plays. For example, perhaps the first time through subjects are more diligent about their chewing than the last time through the study.

9. (a) Score on achievement test

(b) Method of teaching, grade level, intelligence, school district, teacher. Fixed: grade level, school district, teacher. Set at a predetermined level: method of teaching

(c) New teaching method and traditional method; 2

(d) Random assignment

(e) Group 2

(f) Completely randomized design

(g) 500 first-grade students in District 203

(h)

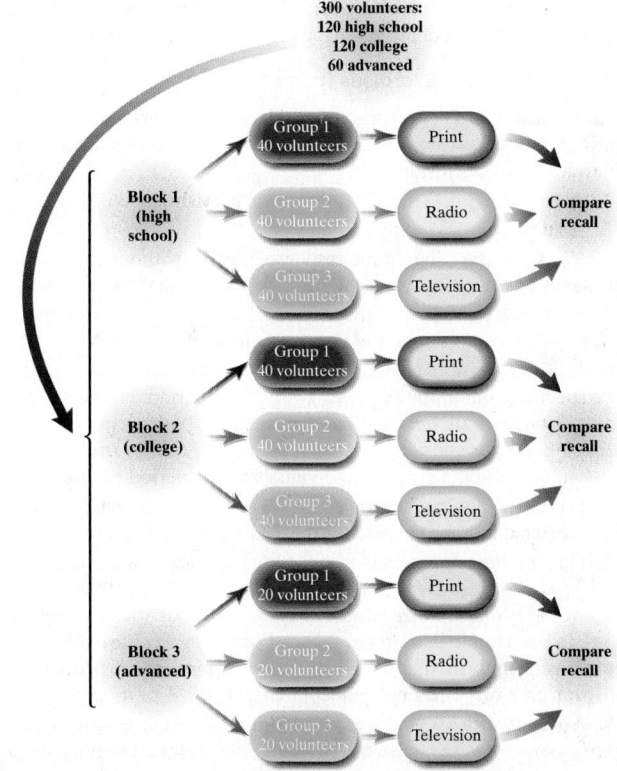

11. (a) Matched pair

(b) Whiteness level

(c) Crest Whitestrips Premium with brushing and flossing versus brushing and flossing alone

(d) Answers will vary. One possible variable would be diet. Certain foods and tobacco products are more likely to stain teeth. This could affect the whiteness level.

(e) Answers will vary. One possible answer is that using twins helps control for genetic factors (like weak teeth) that may affect the results of the study.

13. (a) Completely randomized design

(b) Adults with insomnia

(c) Wake time after sleep onset (WASO)

(d) CBT, RT, placebo

(e) 75 adults with insomnia

(f)

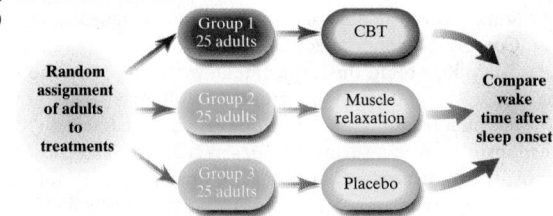

15. (a) Completely randomized design

(b) Adults older than 60 years in good health

(c) Score on a standardized test of learning and memory

(d) Drug: 40 mg of ginkgo 3 times per day or a matching placebo

(e) 98 men and 132 women older than 60 in good health

(f) The placebo group

(g)

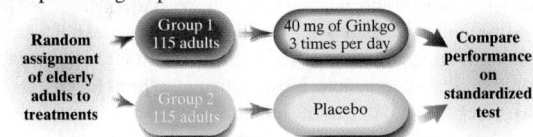

17. (a) Matched-pairs design

(b) Distance yardstick falls

(c) Dominant versus nondominant hand

(d) 15 students

(e) To eliminate bias due to starting on dominant or nondominant first for each trial

(f)

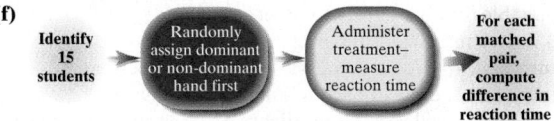

19. (a) Randomized block design

(b) Number of correct answers on recall exam

(c) Print, radio, or television advertising

(d) Level of education

(e)

21. Answers will vary.

23. (a) This is an observational study because there is no intent to manipulate an explanatory variable. The explanatory variable is whether the individual is a green tea drinker or not, which is a qualitative variable.
(b) Some lurking variables include diet, exercise, genetics, gender, age
(c) Completely randomized design
(d) To double-blind this experiment, we would need the placebo to look, taste, and smell like green tea. Subjects would not know which treatment is being delivered. Plus, the individual responsible for measuring changes in LDL cholesterol would not know the treatment either (odds are this blinding is not necessary the way the experiment is set up since the subjects have completed the experiment by the time the LDL is measured).
(e) The factor that is manipulated is the tea, which is set at three levels. It is qualitative.
(f) Exercise, diet
(g) Number the subjects 1 to 120. Randomly select 40 subjects and assign them to the placebo group. Then, randomly select 40 from the remaining 80 and assign them to the one cup of green tea group. The remaining subjects will be in the two cups of green tea group. By randomly assigning subjects the expectation is that those variables not controlled (such as genetic history) are neutralized (even out).
(h) Exercise is a confounding variable because any difference in the change in LDL cholesterol cannot be attributed to the tea. It may be the exercise that caused the change in LDL cholesterol.

25. Answers will vary. Completely randomized design is likely best.

27. Answers will vary. Randomized block design is likely best. Block by type of car.

29. (a) Blood pressure
(b) Daily consumption of salt, daily consumption of fruits and vegetables, body's ability to process salt
(c) Salt: controlled; fruits/vegetables: controlled, body: cannot be controlled. To deal with variability in body types, randomly assign experimental units to each treatment group.
(d) Answers will vary. Three might be a good choice; one level below RDA, one equal to RDA, one above RDA.

31. Answers will vary.

33. Control groups are needed in a designed experiment to serve as a baseline against which other treatments can be compared.

35. The blocks are like the homogeneous groups in stratified sampling. Blocking is done to reduce the variability in the response variable and to remove potential confounding resulting from the block variable.

Chapter 1 Review Exercises (page 55)

1. The science of collecting, organizing, summarizing, and analyzing data to draw conclusions or answer questions; provides a measure of confidence to conclusions.

2. A group of individuals that a researcher wishes to study

3. A subset of the population

4. Measures the value of the response variable without attempting to influence the value of either the response or explanatory variables.

5. Applies a treatment to the individuals in the study to isolate the effect of the treatment on the response variable.

6. (1) *Cross-sectional studies*: collect information about individuals at a specific point in time or over a very short period of time. (2) *Case-control studies*: look back in time and match individuals possessing certain characteristic with those that do not. (3) *Cohort studies*: collect information about a group of individuals over a period of time.

7. (1) Identify the research objective, (2) collect the data needed to answer the research questions, (3) describe the data, (4) perform inference

8. (1) *Sampling bias* occurs when the techniques used to select individuals to be in the sample favor one part of the population over another. (2) *Nonresponse bias* occurs when the individuals selected to be in the sample that do not respond to the survey have different opinions from those that do respond. (3) *Response bias* exists when the answers on a survey do not reflect the true feelings of the respondent.

9. *Nonsampling errors* are errors that result from undercoverage, nonresponse bias, response bias, or data-entry errors. *Sampling errors* are errors that result from using a sample to estimate information about a population. They result because samples contain incomplete information regarding a population.

10. (1) Identify the problem to be solved. (2) Determine the factors that affect the response variable. (3) Determine the number of experimental units. (4) Determine the level of each factor. (5) Conduct the experiment. (6) Test the claim.

11. Quantitative; discrete   **12.** Quantitative; continuous
13. Qualitative   **14.** Statistic
15. Parameter   **16.** Interval
17. Nominal   **18.** Ordinal
19. Ratio   **20.** Observational study
21. Experiment   **22.** Cohort study
23. Convenience sample   **24.** Cluster sample
25. Stratified sample   **26.** Systematic sample

27. (a) Sampling bias; undercoverage or nonrepresentative sample due to poor sampling frame
(b) Response bias; interviewer error
(c) Data-entry error

28. Answers will vary.

29. Answers will vary.

30. Answers will vary. Label each goggle with pairs of digits from 00 to 99. Using row 12, column 1 of Table I in Appendix A and reading down, the selected labels would be 55, 96, 38, 85, 10, 67, 23, 39, 45, 57, 82, 90, and 76. The goggles with these labels would be inspected for defects.

31. (a) To determine the ability of chewing gum to remove stains from teeth
(b) Designed experiment, because treatments were intentionally imposed on the experimental units (teeth) to determine their effect on a response variable (percentage of stain removed)
(c) Completely randomized design
(d) Percentage of stain removed
(e) Type of stain remover (gum or saliva); qualitative
(f) The 64 stained teeth (bovine incisors)
(g) The amount of time chewing (fixed at 120 minutes); method of chewing (fixed by simulator)
(h) Gums A and B remove significantly more stain than gum C or saliva. Gum C removes more stain than saliva.

32. (a) Matched-pair design
(b) Reaction time; quantitative
(c) Alcohol (two drinks: placebo and 40% vodka mixed with orange juice)
(d) Food consumption; caffeine intake
(e) Weight; gender
(f) To act as a placebo to control for psychosomatic effects of alcohol
(g) Alcohol delays reaction time significantly in seniors for low levels of alcohol consumption. The study applies to healthy seniors who are not regular drinkers.

33. (a) Randomized block design   **(b)** Exam grade
(c) Notecard use: with notecard or without notecard
(d) The instructor's statistics students
(e)

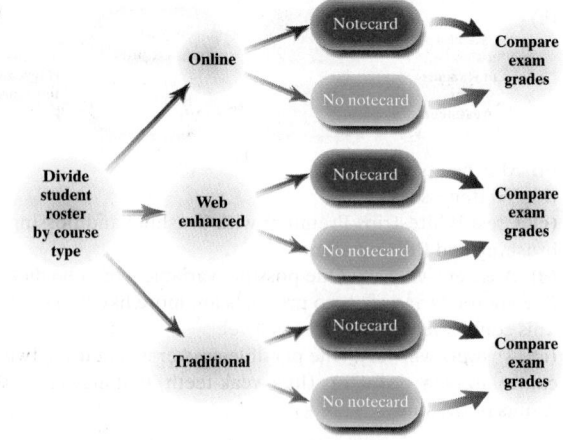

34. Answers will vary.

35. (a) Answers will vary. **(b)** Answers will vary.

36. In a completely randomized design, the experimental units are randomly assigned to one of the treatments. The value of the response variable is compared for each treatment. In a matched-pairs design, experimental units are matched up on the basis of some common characteristic (such as husband–wife or twins). The difference in the matched-up experimental units is analyzed.

37. Answers will vary. **38.** Answers will vary.

39. Randomization is meant to even out the affect of those variables that are not controlled for in a designed experiment. Answers to the randomization question may vary; however, each experimental unit must be assigned randomly. For example, a researcher might randomly select 25 experimental units from the 100 using the methods of simple random sampling and assign them to treatment 1. Then the researcher could randomly select another 25 experimental units from the remaining 75 and assign them to treatment 2 and so on.

Chapter 1 Test (page 57)

1. Collect information, organize and summarize the information, analyze the information to draw conclusions, provide a measure of confidence in the conclusions drawn from the information.

2. (1) Identify the research objective, (2) collect the data needed to answer the research questions, (3) describe the data, (4) perform inference

3. Quantitative, continuous, ratio **4.** Qualitative, ordinal

5. Quantitative, discrete, ratio **6.** Experiment, battery life

7. Observational study; fan opinion on presence of asterisk on record ball

8. A *cross-sectional study* collects data at a specific point in time or over a short period of time; a *cohort study* collects data over a period of time, sometimes over a long period of time; a *case-controlled study* is retrospective, looking back in time to collect data.

9. An experiment involves the researcher actively imposing treatments on experimental units in order to observe any difference between the treatments in terms of effect on the response variable. In an observational study, the researcher observes the individuals in the study without attempting to influence the response variable in any way. Only an experiment will allow a researcher to establish causality.

10. A control group is necessary for a baseline comparison. Comparing other treatments to the control group allows the researcher to identify which, if any, of the other treatments are superior to the current treatment (or no treatment at all). Blinding is important to eliminate bias due to the individual or experimenter knowing which treatment is being applied.

11. (1) Identify the problem to be solved, (2) determine the factors that affect the response variable, (3) determine the number of experimental units, (4) determine the level of each factor, (5) conduct the experiment, (6) test the claim.

12. Number the franchise locations 1 to 15. Use Table I in Appendix A or a random-number generator to determine four unique numbers. The franchise locations corresponding to these numbers are the franchises in the sample. Results will vary.

13. Obtain a simple random sample for each stratum. Be sure to use a different starting point in Table I in Appendix A or a different seed for each stratum. Results will vary.

14. Number the blocks from 1 to 2500 and obtain a simple random sample of size 10. The blocks corresponding to these numbers represent the blocks analyzed. Analyze all trees on each of the selected blocks. Results will vary.

15. Ideally, k would be $600/14 = 42$ (rounded down). Randomly select a number between 1 and 42. This represents the first slot machine inspected. Then inspect every 42nd machine thereafter. Results will vary.

16. In a completely randomized design, the experimental units are randomly assigned to one of the treatments. The value of the response variable is compared for each treatment. In a randomized block design,

the experimental units are first divided according to some common characteristic (such as gender). Then each experimental unit within each block is randomly assigned to one treatment. The value of the response variable is compared for each treatment. By blocking, we prevent the effect of the blocked variable from confounding with the treatment.

17. (a) Sampling bias (voluntary response)
 (b) Nonresponse bias
 (c) Response bias (poorly worded question)
 (d) Sampling bias (undercoverage)

18. (a) Matched-pairs design
 (b) 159 social drinkers
 (c) Type of beer glass
 (d) Time to complete the drink; quantitative
 (e) The type of glass used in the first week is determined randomly. This is to neutralize the effect of drinking out of a specific glass first.
 (f)

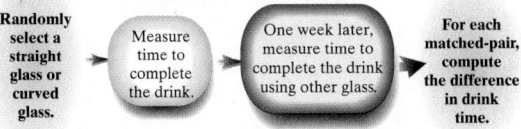

19. (a) Completely randomized design
 (b) Topical cream; 0.5%, 1.0%, 0%
 (c) Neither the subjects nor the person applying the treatments were aware of which treatment was being given.
 (d) Placebo group (0% cream)
 (e) 225 patients with skin irritation
 (f)

20. (a) Subjects were observed over a long period of time and certain characteristics were recorded. The response variable was recorded at the end of the study.
 (b) Bone mineral density; weekly cola consumption
 (c) Quantitative
 (d) The researchers observed values of variables that could potentially impact bone mineral density (besides cola consumption) so their effect could be isolated from the variable of interest.
 (e) Smoking status; alcohol consumption, physical activity, calcium intake.
 (f) Women who consumed at least one cola per day (on average) had a bone mineral density that was significantly lower at the femoral neck than those who consumed less than one cola per day. No, they are associated.

21. Lurking variables tend to occur in observational studies. In addition, lurking variables are related to both an explanatory variable and the response variable. Confounding variables tend to occur in experiments. The effect of a confounding variable on the response variable cannot be distinguished from a second explanatory variable.

Chapter 2 Organizing and Summarizing Data

2.1 Assess Your Understanding (page 70)

1. Raw data are data that are not organized.

3. One (although rounding may cause the result to vary slightly)

5. (a) Washing your hands; 61%
 (b) Drinking orange juice; 2%
 (c) 25%

7. (a) OF **(b)** 15 **(c)** 15
(d) Each of these positions should be reported as MVPs, rather than treating the three positions as a single position.
9. (a) 69% **(b)** 55.2 million **(c)** Inferential
11. (a) 0.42; 0.61 **(b)** 55+
(c) 18–34 **(d)** As age increases, so does likelihood to buy American

13. (a)

Response	Relative Frequency
Never	0.0262
Rarely	0.0678
Sometimes	0.1156
Most of the time	0.2632
Always	0.5272

(b) 52.7% **(c)** 9.4%
(d)

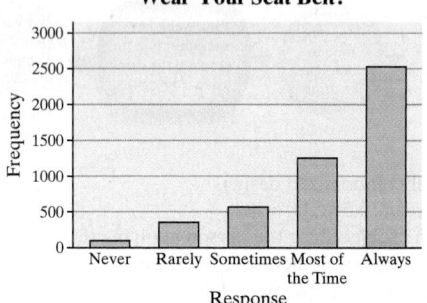

"How Often Do You Wear Your Seat Belt?"

(e)

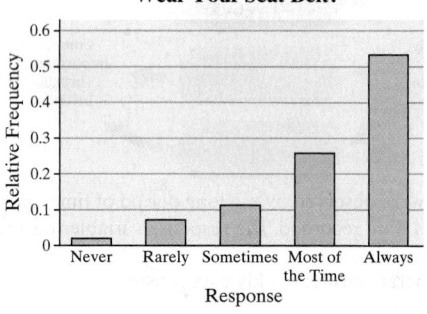

"How Often Do You Wear Your Seat Belt?"

(f)

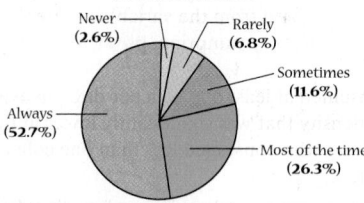

"How Often Do You Wear Your Seat Belt?"

(g) This is a descriptive statement because it is reporting a result of the sample.

15. (a)

Response	Relative Frequency
More than 1 hour a day	0.3678
Up to 1 hour a day	0.1873
A few times a week	0.1288
A few times a month or less	0.0790
Never	0.2371

(b) 0.2371 (about 24%)

(c)

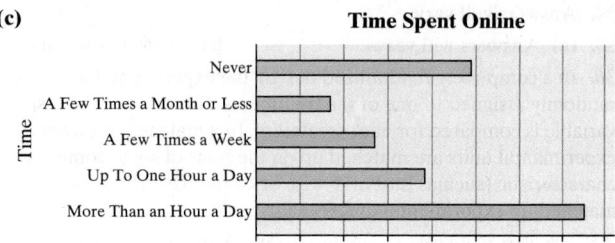

Time Spent Online

(d)

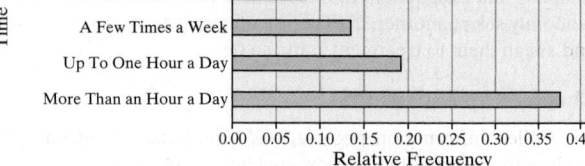

Time Spent Online

(e)

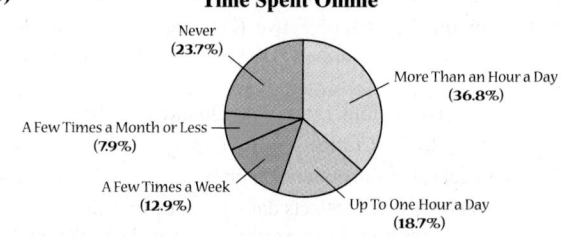

Time Spent Online

(f) No level of confidence is provided along with the estimate.

17. (a), (b)

Number of Texts	Adults	Teens
None	0.0894	0.0207
1 to 10	0.5052	0.2201
11 to 20	0.1286	0.1100
21 to 50	0.1286	0.1802
51 to 100	0.0692	0.1802
101+	0.0790	0.2887

(c)

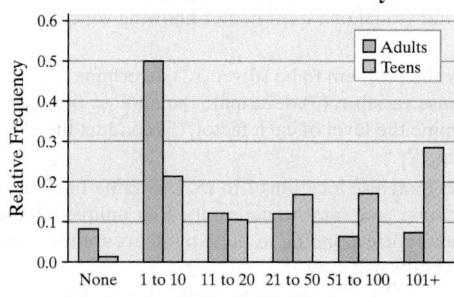

Number of Texts Each Day

(d) Adults are much more likely to send fewer texts per day, while teens are much more likely to do more texting per day.

19. (a)

Dream Job	Males	Females
Professional Athlete	0.4040	0.18
Actor/Actress	0.2626	0.37
President of the United States	0.1313	0.13
Rock Star	0.1313	0.13
Not Sure	0.0707	0.19

(b)

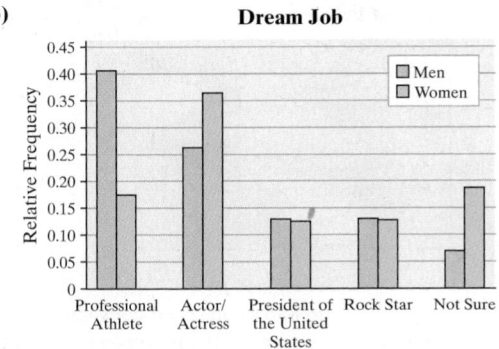

Dream Job

(c) Answers will vary.

21. (a), (b)

Price Change	Frequency	Relative Frequency
Down	15	0.5
No Change	2	0.067
Up	13	0.433

(c)

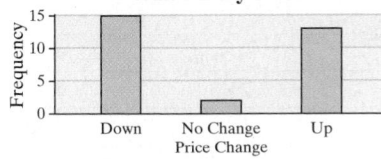

Daily Price Change of Walt Disney Stock

(d)

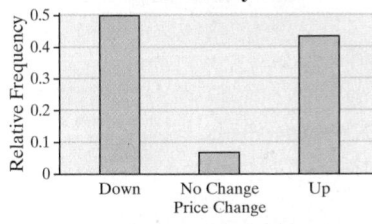

Daily Price Change of Walt Disney Stock

(e)

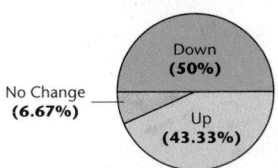

Daily Price Change in Walt Disney Stock

23. (a), (b)

Day	Frequency	Relative Frequency
Sunday	3	0.075
Monday	2	0.05
Tuesday	5	0.125
Wednesday	6	0.15
Thursday	2	0.05
Friday	14	0.35
Saturday	8	0.2

(c) Answers will vary. Monday would likely be the worst bet, and Thursday may be the best choice.

(d)

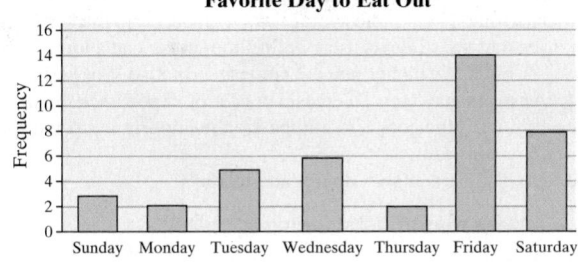

Favorite Day to Eat Out

(e)

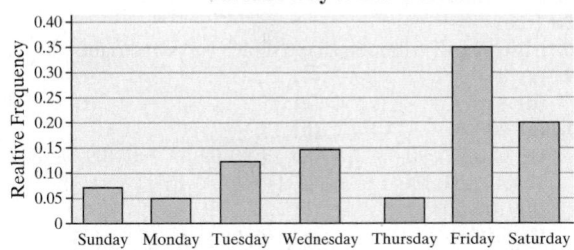

Favorite Day to Eat Out

(f)

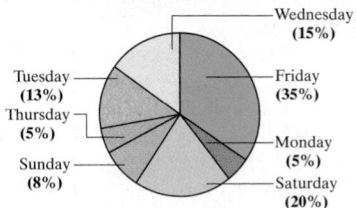

Favorite Day to Eat Out

25. (a)

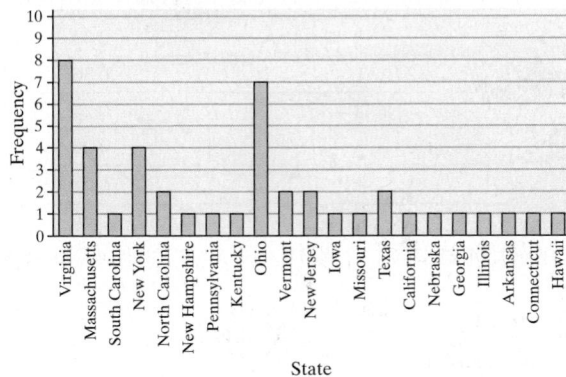

The U.S. Presidents' Birthplaces

(b) More presidents were born in Virginia than in any other state.
(c) Virginia is one of the original 13 colonies.
27. Answers will vary.
29. (a) To determine if online homework improves student learning over traditional pencil-and-paper homework.
(b) Experiment
(c) Answers will vary. Some examples are same teacher, same semester, same course.
(d) There could be differences between the classes. The instructor may give more instruction to one class than the other. The instructor is not blinded, so he or she may treat one group differently from the other.
(e) *Number of students*: quantitative, discrete; *average age*: quantitative, continuous; *average exam score*: quantitative, continuous; *type of homework*: qualitative; *college experience*: qualitative
(f) Qualitative; ordinal. Answers will vary. Likely yes because it gives more "weight" to the higher grade and the researcher is attempting to convey that a higher percent of students passed using online homework.
(g) Side-by-side relative frequency bar graph
(h) Yes; the "whole" is the set of students who received a grade for the course for each class type.

(i) The table shows the two groups with no prior college experience had roughly the same average exam grade. From the bar graph, we see that the students using online homework had a lower percent for As, but had a higher percent who passed with a C or better.
31. Answers may vary. Decreasing order of importance seems to be a good choice if the goal is to emphasize the importance. A pie chart does not allow for order.
33. No. The percentages do not add to 100%.

2.2 Assess Your Understanding (page 88)

1. classes
3. class width
5. True
7. False. The distribution shape shown is skewed right.
9. (a) 8   **(b)** 2   **(c)** 15
    (d) 4   **(e)** 15%   **(f)** Bell shaped
11. (a) 200   **(b)** 10
    (c) 60–69, 2; 70–79, 3; 80–89, 13; 90–99, 42; 100–109, 58; 110–119, 40; 120–129, 31; 130–139, 8; 140–149, 2; 150–159, 1
    (d) 100–109   **(e)** 150–159
    (f) 5.5%   **(g)** No
13. (a) Likely skewed right. Most household incomes will be to the left (perhaps in the $50,000 to $150,000 range), with fewer higher incomes to the right (in the millions).
    (b) Likely bell-shaped. Most scores will occur near the middle range, with scores tapering off equally in both directions.
    (c) Likely skewed right. Most households will have, say, 1 to 4 occupants, with fewer households having a higher number of occupants.
    (d) Likely skewed left. Most Alzheimer's patients will fall in older-aged categories, with fewer patients being younger.
15. (a)

Number of Children under 5	Relative Frequency
0	0.32
1	0.36
2	0.24
3	0.06
4	0.02

    (b) 24%   **(c)** 60%
17. 10, 11, 14, 21, 24, 24, 27, 29, 33, 35, 35, 35, 37, 37, 38, 40, 40, 41, 42, 46, 46, 48, 49, 49, 53, 53, 55, 58, 61, 62
19. 1.2, 1.4, 1.6, 2.1, 2.4, 2.7, 2.7, 2.9, 3.3, 3.3, 3.3, 3.5, 3.7, 3.7, 3.8, 4.0, 4.1, 4.1, 4.3, 4.6, 4.6, 4.8, 4.8, 4.9, 5.3, 5.4, 5.5, 5.8, 6.2, 6.4
21. (a) 6 classes
    (b) Lower class limits: 10, 14, 18, 22, 26, 30; upper class limits: 13.9, 17.9, 21.9, 25.9, 29.9, 33.9
    (c) Class width: 4
23. (a)

Speed (km/hr)	Relative Frequency
10–13.9	0.0066
14–17.9	0.0115
18–21.9	0.0280
22–25.9	0.1499
26–29.9	0.4646
30–33.9	0.3394

(b)

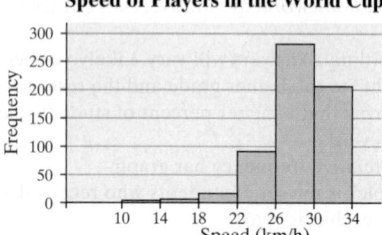

Speed of Players in the World Cup

(c)

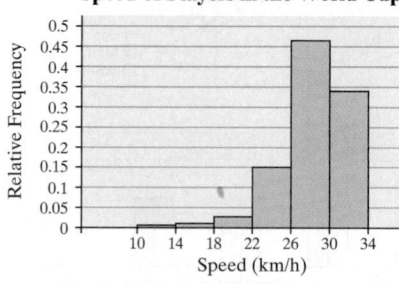

Speed of Players in the World Cup

The percent of players who had a top speed between 30 and 33.9 km/h is 33.94%. The percent of the players who had a top speed less than 13.9 km/h is 0.66%.

25. (a) Discrete. The possible values for the number of televisions are countable.
(b), (c)

Number of Televisions	Frequency	Relative Frequency
0	1	0.025
1	14	0.35
2	14	0.35
3	8	0.2
4	2	0.05
5	1	0.025

(d) 20%   **(e)** 7.5%
(f)

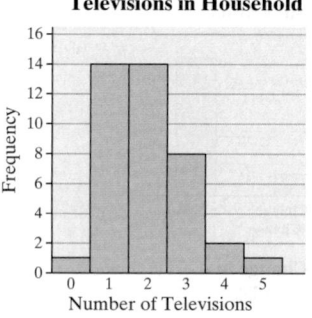

Televisions in Household

(g)

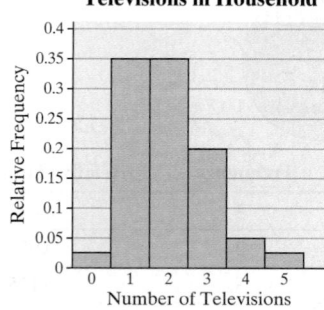

Televisions in Household

(h) Skewed right
27. (a), (b)

Gini Index	Frequency	Relative Frequency
20–24.9	5	0.037
25–29.9	16	0.118
30–34.9	28	0.206
35–39.9	27	0.199
40–44.9	20	0.147
45–49.9	17	0.125
50–54.9	13	0.096
55–59.9	5	0.037
60–64.9	5	0.037

(c)

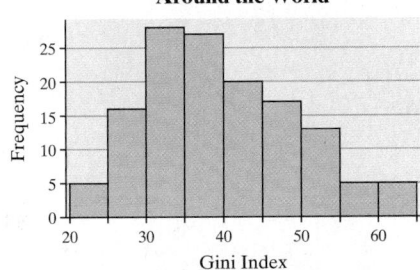

Gini Index for Countries Around the World

(d)

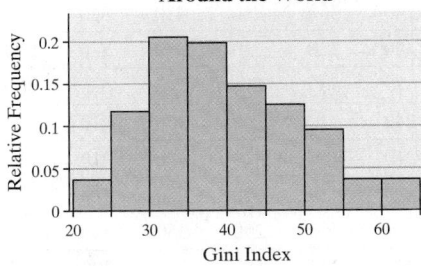

Gini Index for Countries Around the World

(e) Skewed right

(f)

Gini Index	Frequency	Relative Frequency
20–29.9	21	0.154
30–39.9	55	0.404
40–49.9	37	0.272
50–59.9	18	0.132
60–69.9	5	0.037

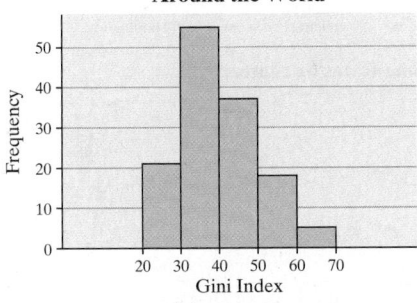

Gini Index for Countries Around the World

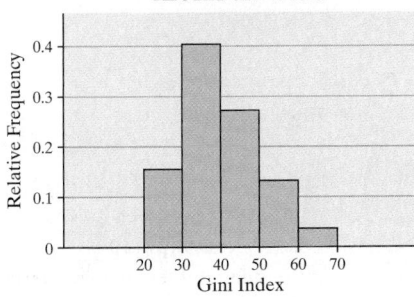

Gini Index for Countries Around the World

Skewed right

(g) Answers may vary. However, the summary with a class width of 5 seems superior.

29. (a), (b)

Tax	Frequency	Relative Frequency
0.0–0.499	7	0.1373
0.5–0.999	13	0.2549
1.0–1.499	7	0.1373
1.5–1.999	8	0.1569
2.0–2.499	5	0.0980
2.5–2.999	5	0.0980
3.0–3.499	3	0.0588
3.5–3.999	2	0.0392
4.0–4.499	1	0.0196

(c)

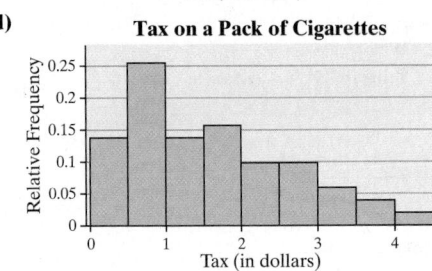

Tax on a Pack of Cigarettes

(d)

Tax on a Pack of Cigarettes

(e) Skewed right

(f)

Tax	Frequency	Relative Frequency
0.0–0.999	20	0.3922
1.0–1.999	15	0.2941
2.0–2.999	10	0.1961
3.0–3.999	5	0.0980
4.0–4.999	1	0.0196

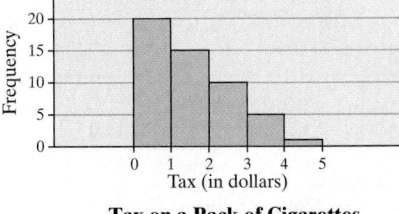

Tax on a Pack of Cigarettes

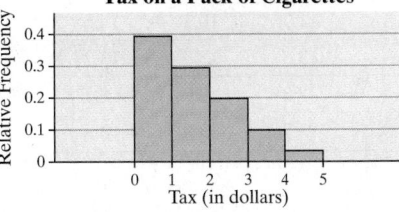

Tax on a Pack of Cigarettes

The distribution is skewed right.

(g) Answers will vary. Both do a nice job of summarizing the data.

31. Answers will vary. One possible answer follows.

(a) Choose a lower class limit for the first class of 0 with a class width of 200.

(b), (c)

Violent Crime Rate	Frequency	Relative Frequency
0–199.9	4	0.0784
200–399.9	26	0.5098
400–599.9	17	0.3333
600–799.9	3	0.0588
800–999.9	0	0
1000–1199.9	0	0
1200–1399.9	1	0.0196

(d)

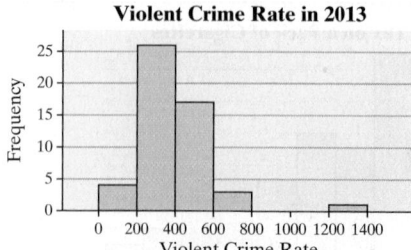

Violent Crime Rate in 2013

(e)

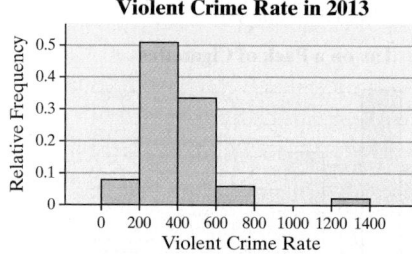

Violent Crime Rate in 2013

(f) Skewed right

33. (a)   **President Ages at Inauguration**

```
4 | 23
4 | 6677899
5 | 0011112244444
5 | 555566677778
6 | 0111244
6 | 589
```
Legend: 4|2 represents 42 years.

(b) Bell shaped

35. (a)   **Fat in McDonald's Breakfast (b)** Bell shaped

```
0 | 39
1 | 1266
2 | 1224577
3 | 0012267
4 | 6
5 | 159
```
Legend: 5|1 represents 51 grams of fat.

37. (a)

10.9	14.2	12.4	13.6	13.0
10.5	10.3	13.1	15.7	14.9
14.1	12.8	13.3	9.9	15.6
12.3	13.9	13.4	19.4	13.4
12.2	14.8	11.9	10.1	13.6
14.6	14.8	13.5	13.9	13.2
14.0	15.2	8.3	9.0	8.7
14.9	16.0	13.7	13.9	12.8

(b) Five-Year Rate of Return

```
 8 | 37
 9 | 09
10 | 1359
11 | 9
12 | 23488
13 | 0123445667999
14 | 01268899
15 | 267
16 | 0
17 |
18 |
19 | 4
```
Legend: 8|3 represents 8.3%

(c) Bell-shaped

39. (a)

450	1240	350	260	410	560	320
600	490	220	450	350	320	280
430	380	500	270	240	640	200
470	240	120	260	300	410	
420	210	480	610	470	210	
310	410	410	190	250	140	
280	350	450	290	350	190	
550	260	230	560	250	300	

(b) Violent Crime Rates by State, 2013

```
 1 | 2499
 2 | 0112344556667889
 3 | 0012255558
 4 | 1111235557789
 5 | 0566
 6 | 014
 7 |
 8 |
 9 |
10 |
11 |
12 | 4
```
Legend: 1|2 represents 120 violent crimes per 100,000 population

(c) Violent Crime Rates by State, 2013

```
 1 | 24
 1 | 99
 2 | 0112344
 2 | 556667889
 3 | 00122
 3 | 55558
 4 | 111123
 4 | 5557789
 5 | 0
 5 | 566
 6 | 014
 6 |
 7 |
 7 |
 8 |
 8 |
 9 |
 9 |
10 |
10 |
11 |
11 |
12 | 4
```
Legend: 1|2 represents 120 violent crimes per 100,000 population

(d) Answers will vary, but the stem-and-leaf in part (b) seems like a better graph.

41. (a)

Home Run Distances

McGwire   Bonds

```
              32 | 0 0
              33 |
          1 0  34 | 7
          0 0  35 | 0
       9 0 0 0  36 | 0 0 0 1 5
      7 0 0 0 0  37 | 0 0 5 5 5 5
  8 5 5 0 0 0 0  38 | 0 0 0 0 0 5
      8 0 0 0 0  39 | 0 0 1 4 6
        9 0 0  40 | 0 0 0 0 4 5
    0 0 0 0 0  41 | 0 0 0 0 0 0 0 0 0 0 1 5 5 6 7 7
  5 3 0 0 0 0  42 | 0 0 0 0 0 0 0 0 9
0 0 0 0 0 0 0  43 | 0 0 0 0 0 5 5 6
      0 0 0  44 | 0 0 0 0 2
  8 2 0 0 0 0  45 | 0 4
        1 0 0  46 |
      8 0 0 0  47 |
            0  48 | 8
              49 |
            0  50 |
          0 0  51 |
            7  52 |
              53 |
              54 |
            0  55 |
```

Legend: 0 | 34 | 7 represents 340 feet
for McGwire and 347 feet for Bonds.

(b) Answers will vary.
43. It is disconcerting to note that some schools have a negative ROI.

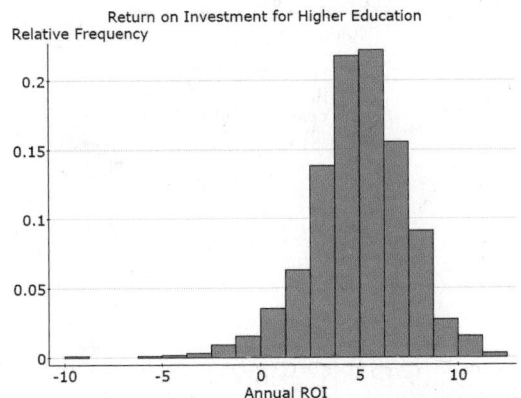

45.

Televisions in Household

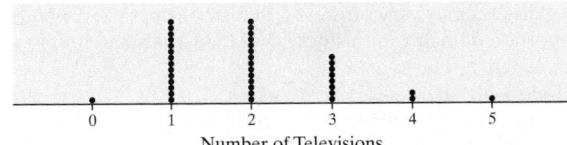

47. Answers will vary. However, because the data is continuous, it makes sense to draw a frequency or relative frequency histogram. A stem-and-leaf plot would work too.
49. Classes shouldn't overlap so there is no confusion as to which class an observation belongs.
51. There is no such thing as a correct class width, however some choices are better than others.
53. Yes. This exercise illustrates the fact that there is no such thing as the "correct" histogram. However, some histograms are better than others and class width can affect the shape of the graph.

2.3 Assess Your Understanding (page 98)

1. An ogive is a graph that represents the cumulative frequency or cumulative relative frequency of the class.
3. True
5. (a) 0.5; 6 **(b)** 1.25; 1.00–1.49 **(c)** 3.75; 3.50–3.99 **(d)** 2.50–2.99
(e) 1.00–1.49
7. (a) 5 **(b)** 34.9% **(c)** 60% **(d)** 90%
9. (a) 9% **(b)** 2010 **(c)** 2008 **(d)** 2001; 2009 **(e)** 2014
(f) Declining due to the decreases in unemployment

11. (a), (b)

Speed (km/hr)	Cumulative Frequency	Cumulative Relative Frequency
10–13.9	4	0.0066
14–17.9	11	0.0181
18–21.9	28	0.0461
22–25.9	119	0.1960
26–29.9	401	0.6606
30–33.9	607	1

(c)

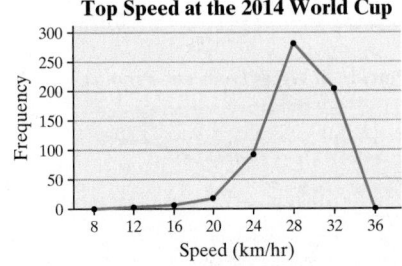

(d)

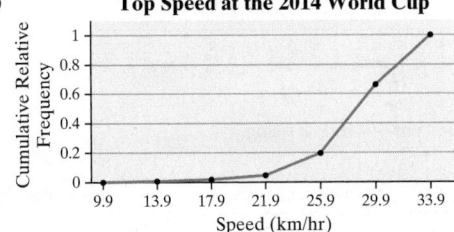

13. (a), (b)

Gini Index	Cumulative Frequency	Cumulative Relative Frequency
20–24.9	5	0.0368
25–29.9	21	0.1544
30–34.9	49	0.3603
35–39.9	76	0.5588
40–44.9	96	0.7059
45–49.9	113	0.8309
50–54.9	126	0.9265
55–59.9	131	0.9632
60–64.9	136	1

(c)

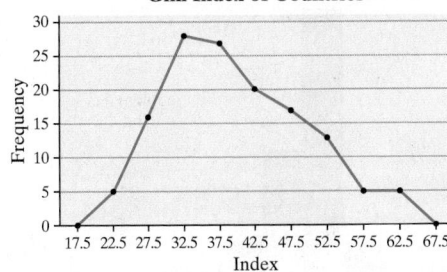

(d)

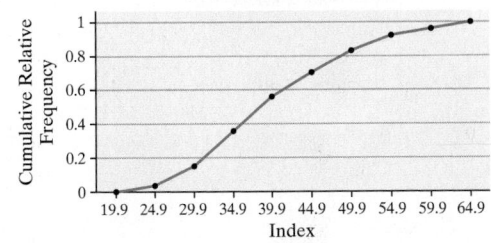

15. (a)

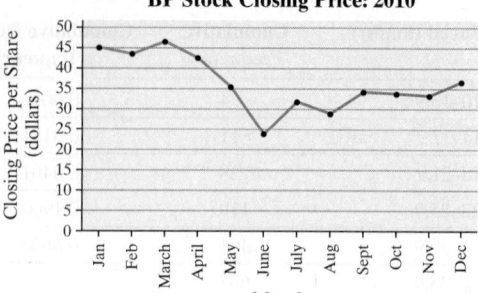

BP Stock Closing Price: 2010

(b) Percentage change $= \dfrac{24.02 - 35.72}{35.72} = -0.328 = -32.8\%$

17.

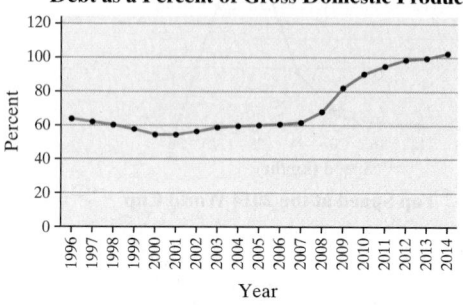

Debt as a Percent of Gross Domestic Product

Debt as a percent of GDP is rising.

19. (a)

Return (Percent)	Consumer Defensive	Industrial
−20−−10.01	0.0435	0
−10−−0.01	0.0217	0.0235
0–9.99	0.0543	0.1882
10–19.99	0.5217	0.2471
20–29.99	0.2391	0.3059
30–39.99	0.0543	0.1647
40–49.99	0.0326	0.0235
50–59.99	0.0109	0.0235
60–69.99	0	0
70–79.99	0.0217	0.0118
80–89.99	0	0.0118

(b)

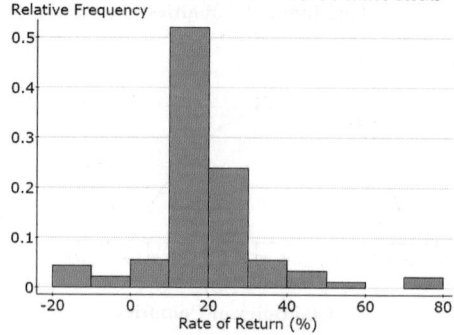

Three-Year Rate of Return in Consumer Defensive Stocks

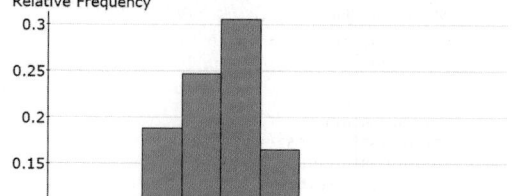

Three-Year Return on Industrial Stocks

(c)

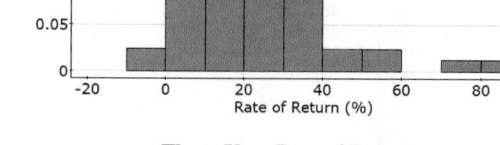

Three-Year Rate of Return

(d)

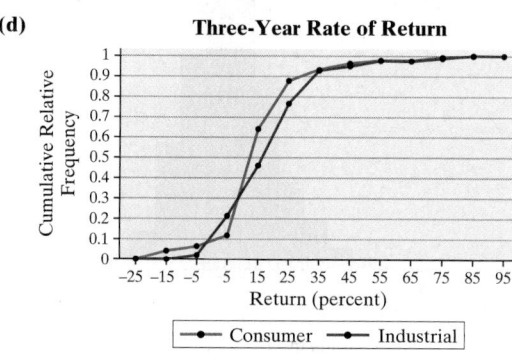

Three-Year Rate of Return

(e) Answers may vary. However, industrial stocks had a higher proportion of high-return stocks, and industrials had fewer stocks with negative returns.

21. Answers will vary.

23. Time-series plots are drawn with quantitative variables. They are drawn to see trends in the data.

2.4 Assess Your Understanding (page 106)

1. The lengths of the bars are not proportional. For example, the bar representing the cost of Clinton's inauguration should be slightly more than 9 times the one for Carter's cost and twice as long as the bar representing Reagan's cost.

3. (a) The vertical axis starts at $21,500 instead of $0. Doing this suggests that income for females declined much more than they actually did.

(b) This graph indicates that the median income for females has declined, but not as significantly as suggested by the graph in part (a).

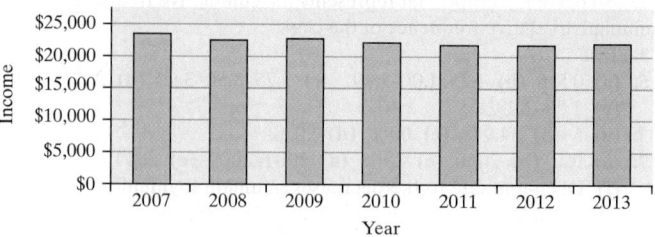

Median Income for Females in Constant 2013 Dollars

5. The bar for 12p–6p covers twice as many hours as the other bars. By combining two 3-hour periods, this bar looks larger compared to the others, making afternoon hours look more dangerous. When the bar is split into two periods, the graph may give a different impression.

7. This graph is misleading because it does not take into account the size of the population of each state. Certainly, Vermont is going to pay less in total taxes than California simply because its population is so much lower. The moral of the story here is that many variables should be considered on per capita (per person) basis. For example, this graph should be drawn to represent taxes paid per capita (per person).

9. (a) The bar for housing should be a little more than twice the length of the bar for transportation, but it is not.
(b) Adjust the graph so that the lengths of the bars are proportional.

11. (a) The politician's view:

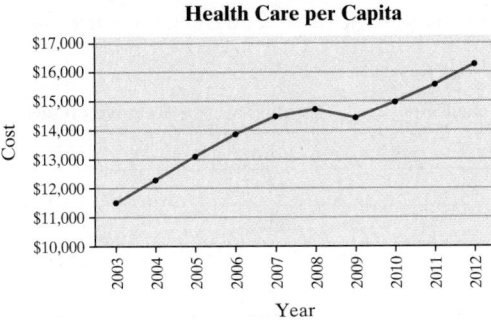

Health Care per Capita

(b) The health care industry's view:

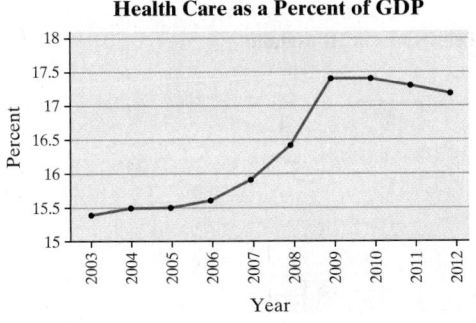

Health Care as a Percent of GDP

(c) Answers may vary. Not only does the scale affect the message, so does the variable used to measure the argument.

13. (a) Graphic that is not misleading:

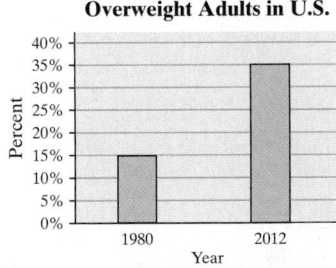

Overweight Adults in U.S.

(b) Graphic that is misleading (graphics may vary):

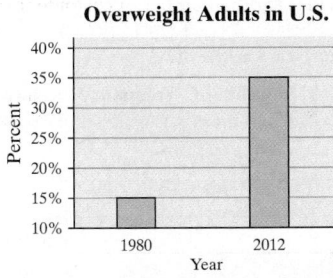

Overweight Adults in U.S.

15. Three-dimensional graphs are deceptive. The area for P (pitcher) looks substantially larger than the area for 3B (third base) even though both are the same percentage. Graphs should not be drawn using three dimensions. Instead, use a two-dimensional graph.

Chapter 2 Review Exercises (page 110)

1. (a) 2216 participants
(b) 0.653
(c)

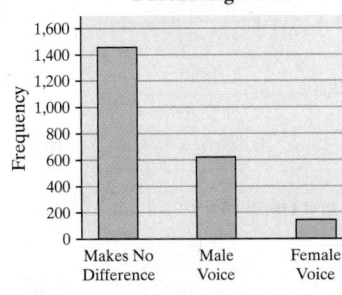

Convincing Voice in Purchasing a Car

2. (a)

Type of Weapon	Relative Frequency
Firearms	0.6899
Knives or cutting instruments	0.1215
Personal weapons	0.0560
Other weapon	0.1325

(b) 68.99% of homicides were committed using a firearm.
(c)

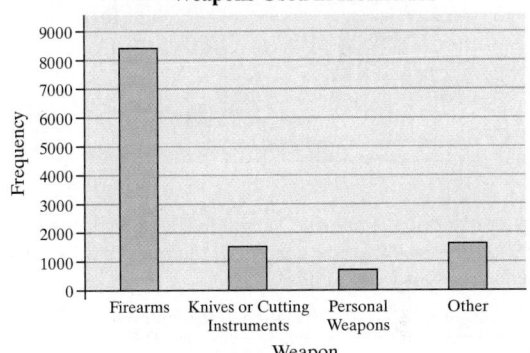

Weapons Used in Homicides

(d)

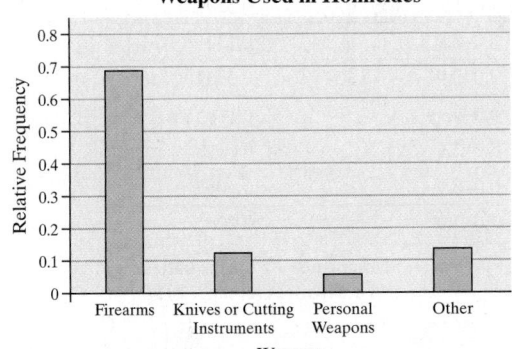

Weapons Used in Homicides

(e)

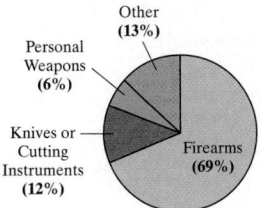

Weapons Used in Homicides

22. (a) Eight classes
(b) Lower class limits: 0, 1.0, 2.0, 3.0, 4.0, 5.0, 6.0, 7.0;
upper class limits: 0.9, 1.9, 2.9, 3.9, 4.9, 5.9, 6.9, 7.9
(c) Class width: 1.0

23. (a)

Speed (km/hr)	Relative Frequency
10–13.9	0.0066
14–17.9	0.0115
18–21.9	0.0280
22–25.9	0.1499
26–29.9	0.4646
30–33.9	0.3394

(b)

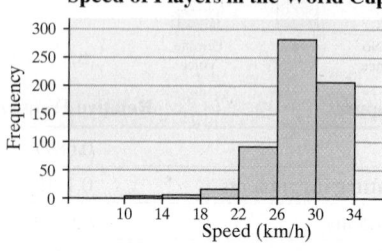

Speed of Players in the World Cup

(c)

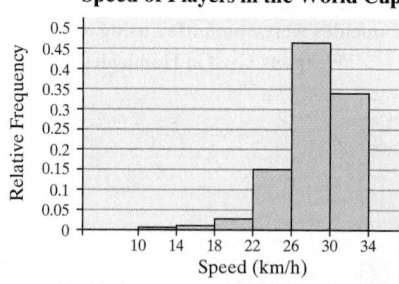

Speed of Players in the World Cup

The percent of players who had a top speed between 30 and 33.9 km/h is 33.94%. The percent of the players who had a top speed less than 13.9 km/h is 0.66%.

24. (a)

Magnitude	Relative Frequency
0–0.9	0.3560
1.0–1.9	0.3591
2.0–2.9	0.1306
3.0–3.9	0.0302
4.0–4.9	0.1104
5.0–5.9	0.0128
6.0–6.9	0.0007
7.0–7.9	0.0002

(b)

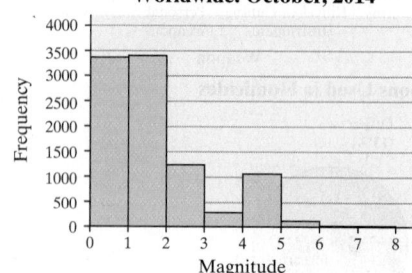

Magnitude of Earthquakes Worldwide: October, 2014

(c)

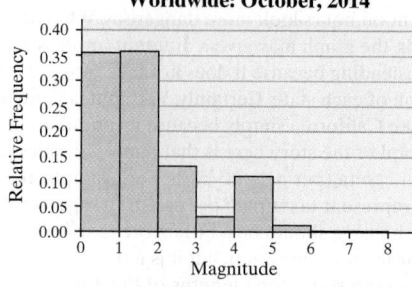

Magnitude of Earthquakes Worldwide: October, 2014

The percentage of earthquakes that registered 4.0 to 4.9 is 11.04%. The percentage of earthquakes that registered 4.9 or less is 98.63%.

25. (a) Discrete. The possible values for the number of televisions are countable.
(b), (c)

Number of Televisions	Frequency	Relative Frequency
0	1	0.025
1	14	0.35
2	14	0.35
3	8	0.2
4	2	0.05
5	1	0.025

(d) 20%   **(e)** 7.5%
(f)

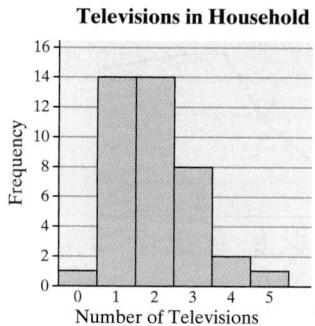

Televisions in Household

(g)

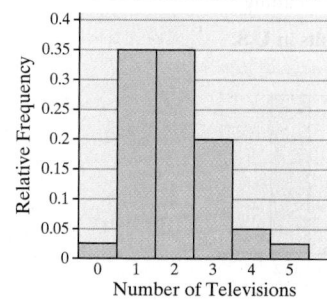

Televisions in Household

(h) Skewed right
26. (a) Discrete. The possible values for the number of customers are countable.
(b), (c)

Number of Customers	Frequency	Relative Frequency	Number of Customers	Frequency	Relative Frequency
3	2	0.05	9	4	0.1
4	3	0.075	10	4	0.1
5	3	0.075	11	4	0.1
6	5	0.125	12	0	0
7	4	0.1	13	2	0.05
8	8	0.2	14	1	0.025

(e)

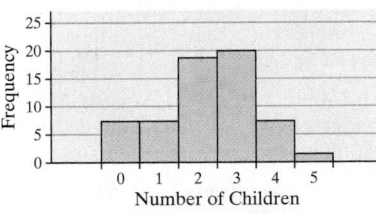

Number of Children for Couples Married 7 Years

The distribution is symmetric.

(f)

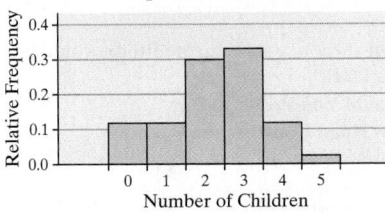

Number of Children for Couples Married 7 Years

(g) 30% of the couples has two children.
(h) 76.7% of the couples has at least two children.
(i)

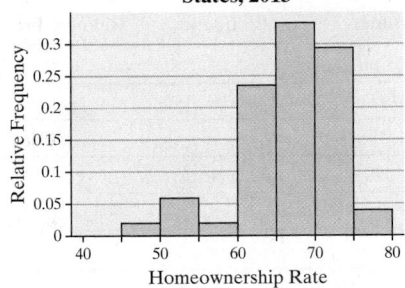

6. (a), (b)

Homeownership Rate	Frequency	Relative Frequency
45–49.9	1	0.0196
50–54.9	3	0.0588
55–59.9	1	0.0196
60–64.9	12	0.2353
65–69.9	17	0.3333
70–74.9	15	0.2941
75–79.9	2	0.0392

(c)

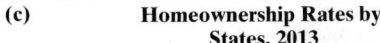

Homeownership Rates by States, 2013

(d)

Homeownership Rates by States, 2013

(e) The distribution is slightly skewed to the left.

(f)

Homeownership Rate	Frequency	Relative Frequency
40–49.9	1	0.0196
50–59.9	4	0.0784
60–69.9	29	0.5686
70–79.9	17	0.3333

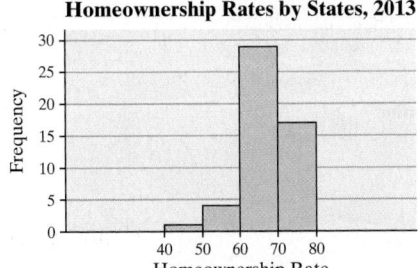

Homeownership Rates by States, 2013

Homeownership Rates by States, 2013

(g) Answers may vary. However, the class width of 5 appears to be the better summary.

7. (a), (b), (c), (d) Answers will vary. Using 2.2000 as the lower class limit of the first class and 0.0200 as the class width, we obtain the following:

Class	Frequency	Relative Frequency	Cumulative Frequency	Cumulative Relative Frequency
2.2000–2.2199	2	0.0588	2	0.0588
2.2200–2.2399	3	0.0882	5	0.1470
2.2400–2.2599	5	0.1471	10	0.2941
2.2600–2.2799	6	0.1765	16	0.4706
2.2800–2.2999	4	0.1176	20	0.5882
2.3000–2.3199	7	0.2059	27	0.7941
2.3200–2.3399	5	0.1471	32	0.9412
2.3400–2.3599	1	0.0294	33	0.9706
2.3600–2.3799	1	0.0294	34	1

(e)

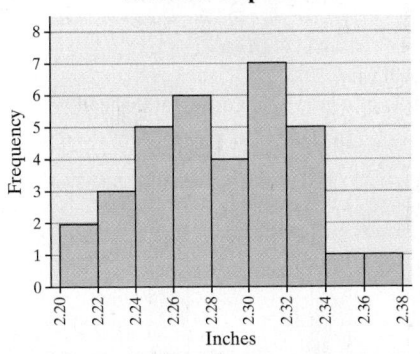

Diameter of Chocolate Chip Cookies

The distribution is roughly symmetric.

(f)

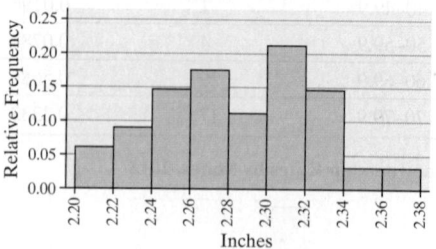

Diameter of Chocolate Chip Cookies

8.   Hours Spent Online

```
13 | 467
14 | 05578
15 | 1236
16 | 456
17 | 113449
18 | 066889
19 | 2
20 | 168
21 | 119
22 | 29
23 | 48
24 | 4
25 | 7
```

Legend: 13 | 4 = average 13.4 hours per week.

The distribution is skewed right.

9. (a) Yes. Grade point averages have increased every time period for all schools.
(b) GPAs increased 5.6% for public schools. GPAs increased 6.8% for private schools. Private schools have higher inflation both because the GPAs are higher, and they are increasing faster.
(c) The graph is misleading because it starts at 2.6 on the vertical axis.

10. (a)

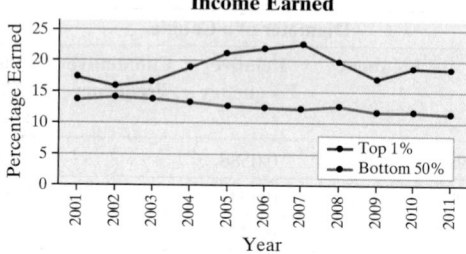

Percentage of Adjusted Gross Income Earned

(b)

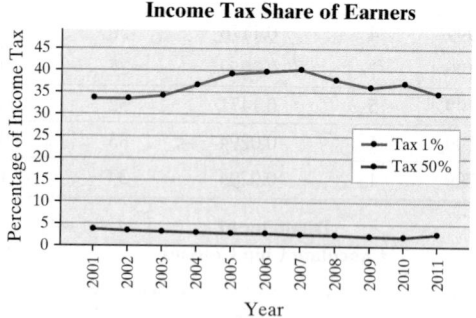

Income Tax Share of Earners

11. (a) Answers will vary.
(b) An example of a graph that does not mislead:

2013 Average Earnings

12. (a) Flats are preferred most; extra-high heels are preferred least.
(b) The bar heights and areas are not proportional.

Chapter 2 Test (page 113)

1. (a) 5 Stars   **(b)** 2747   **(c)** 951   **(d)** 61%
(e) No. This is a bar graph. We only talk about distribution shape for graphs of quantitative data, such as histograms.

2. (a)

Response	Frequency	Relative Frequency
New tolls	412	0.4100
Increase gas tax	181	0.1801
No new roads	412	0.4100

(b) About 18% of respondents indicated they would like to see an increase in gas taxes.

(c)

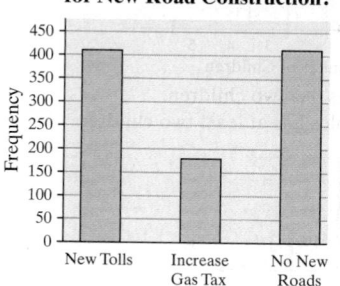

How Would You Prefer to Pay for New Road Construction?

(d)

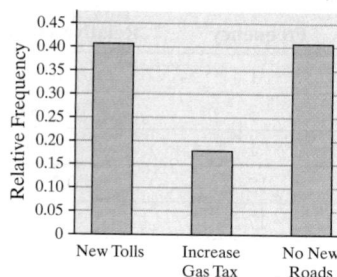

How Would You Prefer to Pay for New Road Construction?

(e)

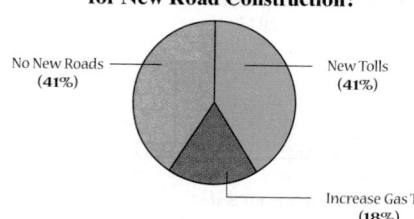

How Would You Prefer to Pay for New Road Construction?

3. (a), (b)

Educational Attainment	Frequency	Relative Frequency
No high school diploma	9	0.18
High school graduate	16	0.32
Some college	9	0.18
Associate's degree	4	0.08
Bachelor's degree	8	0.16
Advanced degree	4	0.08

(c)

Educational Attainment of Commuters

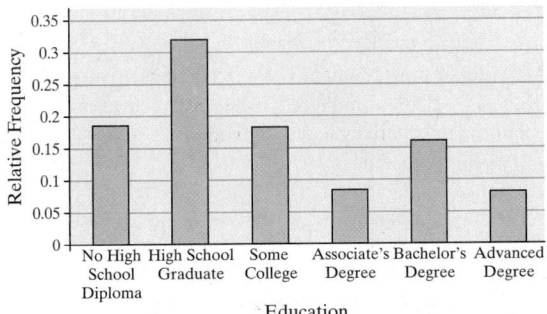

(d) Educational Attainment of Commuters

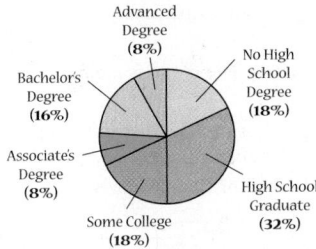

(e) High school graduate is the most common educational level.

4. (a), (b), (c), (d)

Number of Customers	Frequency	Relative Frequency	Cumulative Frequency	Cumulative Relative Frequency
1	5	0.10	5	0.10
2	7	0.14	12	0.24
3	12	0.24	24	0.48
4	6	0.12	30	0.60
5	8	0.16	38	0.76
6	5	0.10	43	0.86
7	2	0.04	45	0.90
8	4	0.08	49	0.98
9	1	0.02	50	1

(e)

Number of Cars Arriving at McDonald's

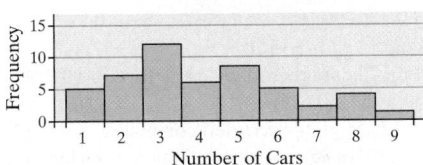

The distribution is skewed right.

(f)

Number of Cars Arriving at McDonald's

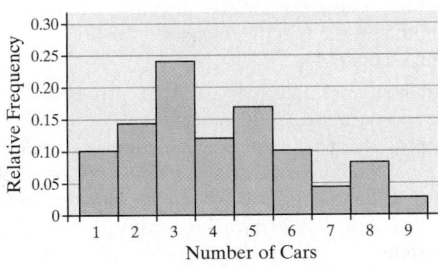

(g) 24% of weeks, three cars arrived between 11:50 A.M. and noon.

(h) 76% of weeks, at least three cars arrived between 11:50 A.M. and noon.

(i)

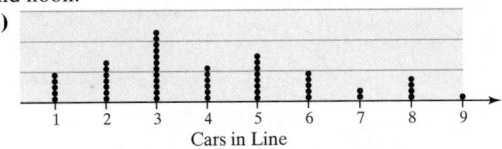

5. Answers may vary. One possibility follows:

(a), (b) Using a lower class limit of the first class of 20 and a class width of 10:

Class	Frequency	Relative Frequency
20–29	1	0.025
30–39	6	0.15
40–49	10	0.25
50–59	14	0.35
60–69	6	0.15
70–79	3	0.075

(c) **Serum HDL of 20–29 Year Olds**

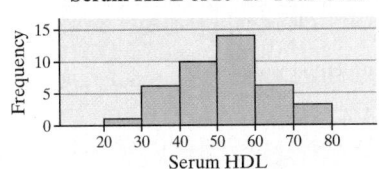

(d) **Serum HDL of 20–29 Year Olds**

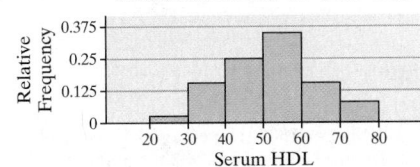

(e) Bell shaped

6. (a), (b)

Closing Price	Cumulative Frequency	Cumulative Relative Frequency
50,000–99,999	4	0.08
100,000–149,999	17	0.34
150,000–199,999	36	0.72
200,000–249,999	43	0.86
250,000–299,999	46	0.92
300,000–349,999	48	0.96
350,000–399,999	49	0.98
400,000–449,999	49	0.98
450,000–499,999	50	1

(c) 72% of the homes sold for less than $200,000.

(d)   **Closing Price of Homes Sold**

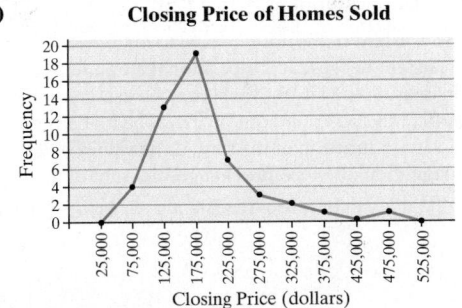

The distribution is skewed right.

(e)

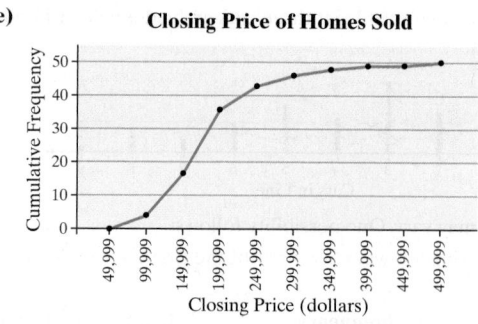

Closing Price of Homes Sold

(f)

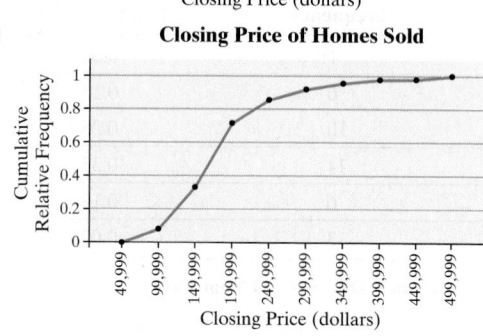

Closing Price of Homes Sold

7. (a) Time Spent on Homework

```
 4 | 0567
 5 | 26
 6 | 13
 7 | 01338
 8 | 59
 9 | 1369
10 | 3899
11 | 0018
12 | 556
```
Legend: 4 | 0 represents 40 minutes.

The distribution is symmetric (uniform).

8.   **Birth Rates (births per 1000 women aged 15–44 and per Capita Income (thousands of 2011 dollars)**

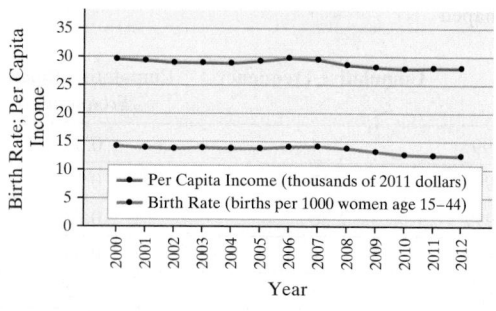

The graphs appear to have the same trend—as per capita income increases, so does birth rate.
9. Answers will vary.

Chapter 3 Numerically Summarizing Data

3.1 Assess Your Understanding (page 126)

1. A statistic is resistant if it is not sensitive to extreme values. The median is resistant because it is a positional measure of central tendency, and increasing the largest value or decreasing the smallest value does not affect the position of the center. The mean is not resistant because it is a function of the sum of the values of data. Changing the magnitude of one value changes the sum of the values.

3. HUD uses the median because the data are skewed. Explanations will vary.

5. The median is between the 5000th and the 5001st ordered values.
7. $\bar{x} = 11$ **9.** $\mu = 9$
11. Mean price per ticket was $2646.
13. Mean: 30.52 mpg; Median: 31.3 mpg; there is no mode.
15. The mean, median, and mode strengths are 3670, 3830, and 4090 pounds per square inch, respectively.
17. (a) mean > median
     (b) mean = median
     (c) mean < median
     Justification will vary.
19. (a) Traditional: $\bar{x} = 71.82$; $M = 70.8$;
     Flipped: $\bar{x} = 77.48$; $M = 76.8$
     (b) $\bar{x} = 113.22$; $M = 71.5$; resistance
21. (a) The mean pulse rate is 72.2 beats per minute.
     (b) Samples and sample means will vary.
     (c) Answers will vary.
23. The distribution is symmetric. The mean is the better measure of central tendency.
25. $\bar{x} = 0.875$ gram; $M = 0.875$ gram. The distribution is symmetric, so the mean is the better measure of central tendency.

Weight of Plain M&Ms

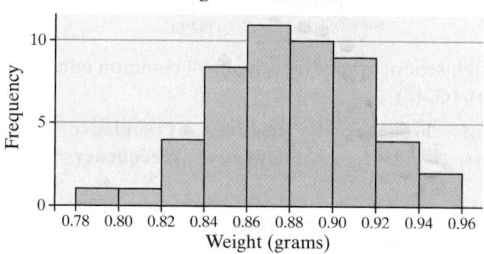

27. The distribution is skewed left; $\bar{x} = 22$ hours; $M = 25$ hours. The median is the better measure of central tendency.

Hours Worked per Week

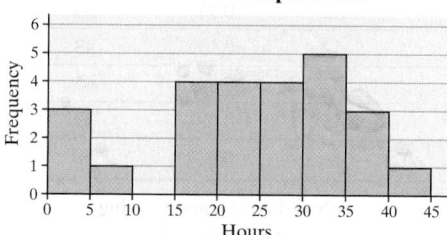

29. (a) Moderate
     (b) Yes, to avoid response bias
31. Sample of size 5: All data recorded correctly: $\bar{x} = 99.8$; $M = 100$; 106 recorded as 160: $\bar{x} = 110.6$; $M = 100$
     Sample of size 12: All data recorded correctly:
     $\bar{x} = 100.4$; $M = 101$; 106 recorded as 160: $\bar{x} = 104.9$; $M = 101$
     Sample of size 30: All data recorded correctly:
     $\bar{x} = 100.6$; $M = 99$; 106 recorded as 160: $\bar{x} = 102.4$; $M = 99$
     For each sample size, the mean becomes larger, but the median remains constant. As the sample size increases, the effect of the misrecorded data on the mean decreases.
33. The unreadable score is 44.
35. The histogram is skewed right. $\bar{x} = 0.061$; $M = 0$. Because the distribution is skewed right, we would expect to report the median as the measure of central tendency. However, a median of 0 does not tell the whole story of the fatal accidents, so it could be argued that both the mean and median be reported. With this data, because the median is 0, at least half of all fatal accidents involved drivers with no alcohol in their system.

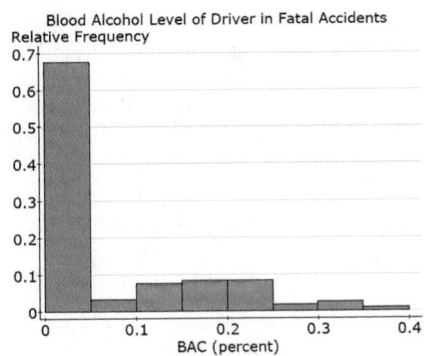

Blood Alcohol Level of Driver in Fatal Accidents

37. (a) Mean = $50,000; median = $50,000; mode = $50,000

(b) New data set: 32.5, 32.5, 47.5, 52.5, 52.5, 52.5, 57.5, 57.5, 62.5, 77.5; mean = $52,500; median = $52,500; mode = $52,500. All three measures increased by $2500.

(c) New data set: 31.5, 31.5, 47.25, 52.5, 52.5, 52.5, 57.75, 57.75, 63, 78.75; mean = $52,500; median = $52,500; mode = $52,500. All three measures increased by 5%.

(d) New data set: 30, 30, 45, 50, 50, 50, 55, 55, 60, 100; mean = $52,500; median = $50,000; mode = $50,000. The mean increased by $2500, but the median and the mode remained at $50,000.

39. The trimmed mean is 0.875. The trimmed mean is resistant only if the extreme observation is the highest or lowest value.

41. (a) Discrete   **(b)** Skewed right

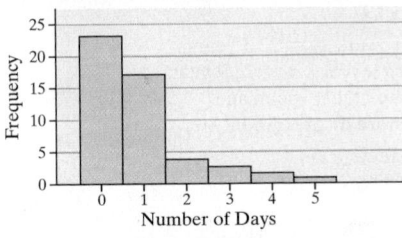

Number of Days High School Students Consumed Alcohol in the Past Week

(c) Since the data are skewed right, we would expect the mean to be greater than the median.

(d) Mean: 0.94; median: 1; the mean can be less than the median in skewed-right data. Therefore, using the rule *mean greater than median implies the data are skewed right* does not always work.

(e) 0

(f) Yes. It is difficult to get truthful responses to this type of question. Carlos would need to ensure that the identity of the respondents is anonymous.

43. The median is resistant because it is a positional measure of central tendency, and increasing the largest value or decreasing the smallest value does not affect the position of the center. The mean is not resistant because it is a function of the sum of the values of data. Changing the magnitude of one value changes the sum of the values.

45. (a) Median

(b) Skewed right because the mean is much larger than the median.

(c) There are a few households with very high net worth.

47. The distribution is skewed right, so the median amount of money lost is less than the mean amount lost.

3.2 Assess Your Understanding (page 141)

1. zero   **3.** True

5. $s^2 = 36; s = 6$ **7.** $\sigma^2 = 16; \sigma = 4$ **9.** $s^2 = 196; s = 14$

11. $R = 13.4$ mpg; $s^2 = 26.35$ mpg^2; $s = 5.13$ mpg

13. $R = 1150$ psi; $s = 459.6$ psi

15. (b) has the higher standard deviation because the data go from 30 to 75; in **(a)** they are clustered between 40 and 60.

17. (a) Traditional (29.1 vs. 28.4)

(b) Traditional (8.84 vs. 7.85)

(c) $R = 541.8; s = 145.88$; neither measure is resistant.

19. (a) $\sigma = 7.7$ beats per minute   **(b), (c)** Answers will vary.

21. (a) Ethan: $\mu = 10$ fish, $R = 19$ fish; Drew: $\mu = 10$ fish, $R = 19$ fish. There is no difference between Ethan and Drew based on these measures.

(b)

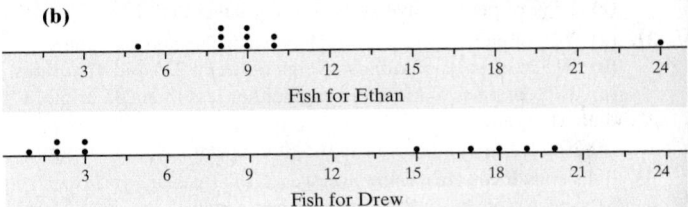

Ethan appears to be more consistent.

(c) Ethan: $\sigma = 4.9$ fish, Drew: $\sigma = 7.9$ fish; Ethan has the more consistent record.

(d) Answers will vary.

23. (a) $s = 0.036$ gram

(b) The histogram is approximately symmetric, so the Empirical Rule is applicable.

(c) 95% of the M&Ms should weigh between 0.803 and 0.947 gram.

(d) 96% of the M&Ms actually weigh between 0.803 and 0.947 gram.

(e) 16% of the M&Ms should weigh more than 0.911 gram.

(f) 12% of the M&Ms actually weigh more than 0.911 gram.

25.

	Car 1	Car 2
Sample mean	$\bar{x} = 223.5$ mi	$\bar{x} = 237.2$ mi
Median	$M = 223$ mi	$M = 230$ mi
Mode	None	None
Range	$R = 93$ mi	$R = 166$ mi
Sample variance	$s^2 = 475.1$ mi^2	$s^2 = 2406.9$ mi^2
Sample standard deviation	$s = 21.8$ mi	$s = 49.1$ mi

Answers will vary.

27. (a)

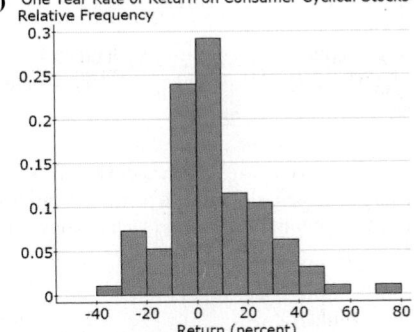

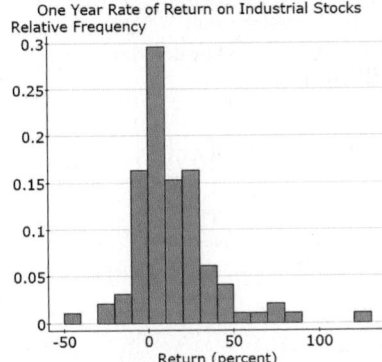

It appears that industrial stocks have more dispersion.

(b) Consumer Cyclical: $\bar{x} = 6.595$; $M = 3.915\%$; Industrial: $\bar{x} = 14.425\%$; $M = 9.595\%$. Industrial stocks have the higher mean and median rate of return.

(c) Consumer Cyclical: $s = 19.078\%$; Industrial: $s = 23.851\%$. The sector with the higher rate of return also has the higher risk. Answers may vary regarding the determination as to whether it is worth the cost.

29. (a) 95% of people have an IQ score between 70 and 130.
(b) 5% of people have an IQ score either less than 70 or greater than 130.
(c) 2.5% of people have an IQ score greater than 130.

31. (a) 95% of pairs of kidneys weigh between 265 and 385 grams.
(b) 99.7% of pairs of kidneys weigh between 235 and 415 grams.
(c) 0.3% of pairs of kidneys weigh either less than 235 or more than 415 grams.
(d) 81.5% of pairs of kidneys weigh between 295 and 385 grams.

33. It depends. If you are a below average student (meaning you expect to have a mean score below 80%), you are better off with Professor Alpha, since about 97.5% of students will score 70% or better in the class. If you are an above average student, you would go with Professor Omega, since you likely want an A, and about 16% of the class will score 90% or higher.

35. (a) 88.9% of gas stations have prices within 3 standard deviations of the mean.
(b) 84% of gas stations have prices within 2.5 standard deviations of the mean. Gasoline priced from $2.91 to $3.21 is within 2.5 standard deviations of the mean.
(c) At least 75% of gas stations have prices between $2.94 and $3.18.

37. There is more variation among individuals than among means.

39. Sample of size 5: correct $s = 5.3$, incorrect $s = 27.9$
Sample of size 12: correct $s = 14.7$, incorrect $s = 22.7$
Sample of size 30: correct $s = 15.9$, incorrect $s = 19.2$
As the sample size increases, the effect of the misrecorded observation on standard deviation decreases.

41. (a) Coupe: $\bar{x} = \$17,592$; $s = \$4742.0$
Camaro: $\bar{x} = \$19,697.2$; $s = \$3206.0$
(b) Camaro's tend to cost more than the typical 2-door vehicle. The standard deviation is lower because there is less variability in prices of a specific vehicle.

43. MAD $= 4.22$ mpg, $s = 5.13$ mpg

45. (a) Average rate of return $= 14.9\%$, risk level $= 14.7\%$
(b) To minimize risk, invest 30% in foreign stock.
(c) Answers will vary.
(d) At least 75% of returns are between -12.8% and 44.4%. At least 88.9% of returns are between -27.1% and 58.7%. Negative returns are not unusual.

47. Answers will vary.

49. No. It is not appropriate to compare standard deviations for two units of measure. Assuming the same variable is being measured, each should be reported in the same unit of measure (such as converting the inches to centimeters).

51. No. The range, standard deviation, and variance are all sensitive to extreme observations.

53. The range only uses two observations from a data set, whereas the standard deviation uses all the observations.

55. Standard deviation measures spread by determining the mean distance to the typical observation.

Data Set 1:

x	$\bar{x}$	$x_i - \bar{x}$	$(x_i - \bar{x})^2$	
3	4	-1	1	$s = \sqrt{\dfrac{2}{3-1}}$
4	4	0	0	
5	4	1	1	$= 1$
			$\Sigma(x_i - \bar{x})^2 = 2$	

Data Set 2:

x	$\bar{x}$	$x_i - \bar{x}$	$(x_i - \bar{x})^2$	
0	4	-4	16	$s = \sqrt{\dfrac{32}{3-1}}$
4	4	0	0	
8	4	4	16	$= 4$
			$\Sigma(x_i - \bar{x})^2 = 32$	

Data set 2 has the higher standard deviation because the observations are farther from the mean, on average.

57. 15, 15, 15, 15, 15, 15, 15, 15

59. Even though the mean wait time increased, patrons were happy because there was less variability in wait time (so fastpass decreases standard deviation wait times). This suggests that people get less upset at long lines than they do at not having their expectations (as to how long they will need to wait in line) satisfied.

3.3 Assess Your Understanding (page 152)

1. $\bar{x} = \$23,221.0$; $s = \$22,484.5$

3. $\bar{x} = 70.6°F$, $s = 3.5°F$

5. (a) $\mu = 32.2$ years, $\sigma = 5.6$ years
(b)

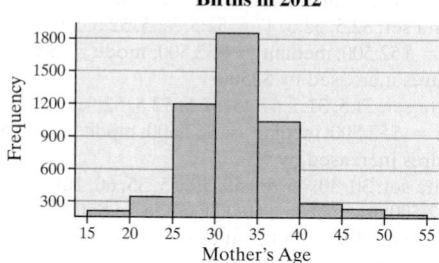

Number of Multiple Births in 2012

(c) 95% of mothers of multiple births are between 21.0 and 43.4 years of age.

7. Grouped data: $\mu = \$1.583$, $\sigma = \$1.051$; raw data: $\mu = \$1.531$, $\sigma = \$1.003$

9. GPA $= 3.27$

11. Cost per pound $= \$2.97$

13. (a) Males: $\mu = 37.1$ years, $\sigma = 22.3$ years
(b) Females: $\mu = 39.0$ years, $\sigma = 22.9$ years
(c) Females have the higher mean age.
(d) Females have more dispersion in age.

15. $M = 2298.3$ square feet

17. Modal class: $0–$19,999

3.4 Assess Your Understanding (page 161)

1. z-score   **3.** Quartiles

5. -0.30; -0.43; the 40-week gestation baby weighs less relative to the gestation period.

7. The man is relatively taller.

9. Clayton Kershaw had the better year because his ERA was 2.30 standard deviations below the National League mean ERA, while Hernandez's ERA was only 1.91 standard deviations below the American League's mean ERA.

11. Ryan is 3.39 standard deviations below the mean in the 100 m back and 3.05 standard deviations below the mean in the 200 m back. So, Ryan is better at the 100 m race.

13. 239

15. (a) 15% of 3- to 5-month-old males have a head circumference that is 41.0 cm or less, and 85% of 3- to 5-month-old males have a head circumference that is greater than 41.0 cm.
(b) 90% of 2-year-old females have a waist circumference that is 52.7 cm or less, and 10% of 2-year-old females have a waist circumference that is more than 52.7 cm.
(c) The heights at each percentile decrease as the age increases. This implies that adult males are getting taller.

17. (a) 25% of the states have a violent crime rate that is 252.4 crimes per 100,000 population or less, and 75% of the states have a violent crime rate more than 252.4. 50% of the states have a violent crime rate that is 333.8 crimes per 100,000 population or less, while 50% of the states have a violent crime rate more than 333.8. 75% of the states have a violent crime rate that is 454.5 crimes per 100,000 population or less, and 25% of the states have a violent crime rate more than 454.5.

(b) 202.1 crimes per 100,000 population; the middle 50% of all observations have a range of 202.1 crimes per 100,000 population.
(c) Yes
(d) Skewed right. The difference between Q_1 and Q_2 is quite a bit less than the difference between Q_2 and Q_3. In addition, the outlier in the right tail of the distribution implies that the distribution is skewed right.

19. (a) 50th percentile
(b) 90th percentile
(c) 105

21. (a) $z = -0.73$. The individual whose car got 36.3 miles per gallon was 0.73 standard deviations below the mean.
(b) By hand, TI-83 or 84, StatCrunch: $Q_1 = 36.85$ mpg, $Q_2 = 38.35$ mpg, $Q_3 = 41.0$ mpg; MINITAB: $Q_1 = 36.575$ mpg, $Q_2 = 38.35$ mpg, $Q_3 = 41.2$ mpg
(c) By hand, TI-83 or 84, StatCrunch: IQR = 4.15 mpg; MINITAB: IQR = 4.625 mpg
(d) By hand, TI-83 or 84, StatCrunch: lower fence = 30.625 mpg, upper fence = 47.225 mpg. Yes, 47.5 mpg is an outlier, MINITAB: lower fence = 29.6375 mpg, upper fence = 48.1375 mpg. There are no outliers using MINITAB's quartiles.

23. (a) By hand, TI-83/84: Q_1: −0.075, Q_2: 0.02, Q_3: 0.075; StatCrunch: Q_1: −0.07, Q_2: 0.02, Q_3: 0.07; MINITAB: Q_1: −0.075, Q_2: 0.02, Q_3: 0.075. Using the by-hand quartiles, 25% of the monthly returns are less than or equal to the first quartile, −0.075, and about 75% of the monthly returns are greater than −0.075; 50% of the monthly returns are less than or equal to the second quartile, 0.02, and about 50% of the monthly returns are greater than 0.02; about 75% of the monthly returns are less than or equal to the third quartile, 0.075, and about 25% of the monthly returns are greater than 0.075.
(b) 0.30 is an outlier.

25. By hand, TI-83/84, StatCrunch: The cutoff point is 574 minutes. MINITAB: The cutoff point is 578 minutes. If more minutes are used, the customer is contacted.

27. (a) $12,777 is an outlier.
(b)

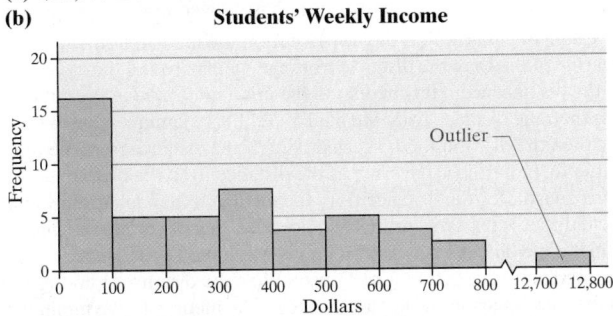

Students' Weekly Income

(c) Answers will vary.
29. Mean of the z-scores is 0.0; standard deviation of the z-scores is 1.0.

Student	z-Score
Perpectual Bempah	0.49
Megan Brooks	−1.58
Jeff Honeycutt	−1.58
Clarice Jefferson	1.14
Crystal Kurtenbach	−0.03
Janette Lantka	1.01
Kevin McCarthy	1.01
Tommy Ohm	−0.55
Kathy Wojdyla	0.10

31. (a) $s = 58.0$ minutes; by hand, TI-83/84: IQR = 56.5 minutes; MINITAB: IQR = 58.8 minutes
(b) $s = 115.0$ minutes; by hand, TI-83/84: IQR = 56.5 minutes; MINITAB: IQR = 58.8 minutes. The standard deviation almost doubles in value, while the interquartile range is not affected. The

standard deviation is not resistant to extreme observations, but the interquartile range is resistant.
33. Since the percentile of a score is rounded to the nearest integer, it is possible for two different scores to have the same percentile.
35. No; when an outlier is discovered, it should be investigated to find its cause.
37. Comparing z-scores allows a unitless comparison of the number of standard deviations an observation is from the mean.
39. The first quartile is the 25th percentile, which means that 25% of the observations are less than or equal to the value and 75% of the observations are greater than the value. The second quartile is the 50th percentile, which means that 50% of the observations are less than or equal to the value and 50% of the observations are greater than the value. The third quartile is the 75th percentile, which means that 75% of the observations are less than or equal to the value and 25% of the observations are greater than the value.

3.5 Assess Your Understanding (page 169)

1. The five-number summary consists of the minimum value in the data set, the first quartile, median, third quartile, and the maximum value in the data set.
3. (a) Skewed right   **(b)** 0, 1, 3, 6, 16
5. (a) 40   **(b)** 52
(c) y   **(d)** Symmetric
(e) Skewed right
7.

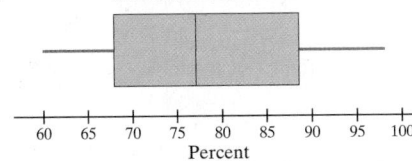

Statistics Exams Scores

9. (a) By hand, TI-83 or 84, StatCrunch: 42, 50.5, 54.5, 57.5, 69; MINITAB: 42, 50.25, 54.5, 57.75, 69
(b)

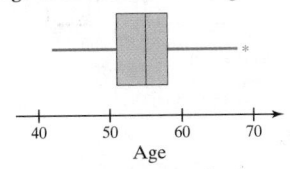

Age of Presidents at Inauguration

(c) Symmetric with an outlier.
11. (a)

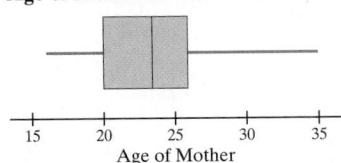

Age of Mother at Time of First Birth

(b) Slightly skewed right
13.

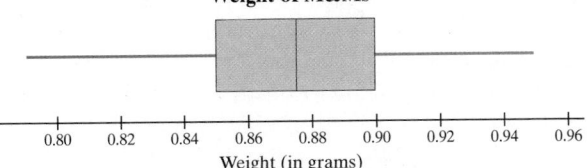

Weight of M&Ms

Since the range of the data between the minimum value and the median is roughly the same as the range between the median and the maximum value, and because the range of the data between the first quartile and median is the same as the range of the data between the median and third quartile, the distribution is symmetric.

15. (a)

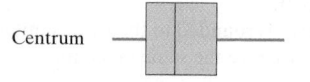

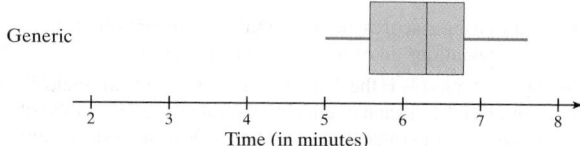

Dissolving Time of Vitamins

(b) Generic
(c) Centrum

17. (a) Depth: $\mu = 19.483$ km; $M = 7.26$ km; Range $= 534.05$ km; $\sigma = 41.498$ km; $Q_1 = 2.99$ km; $Q_3 = 15.38$ km; Magnitude: $\mu = 1.6967$; $M = 1.38$; Range $= 6.8$; $\sigma = 1.2061$; $Q_1 = 0.9$; $Q_3 = 2.01$

For depth, the mean is much larger than the median, so the distribution of depth is likely skewed right. For magnitude, the mean is larger than the median, and the distance from Q_1 to M is less than the distance from M to Q_3, which suggests the distribution of magnitude is skewed right.

(b)

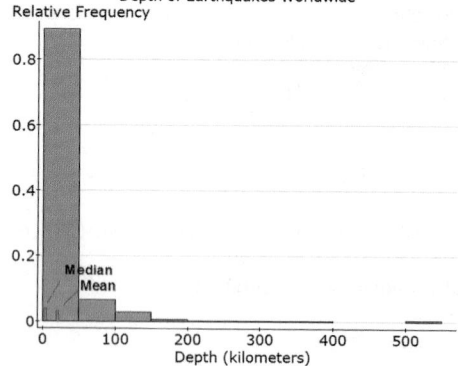

Depth of Earthquakes Worldwide

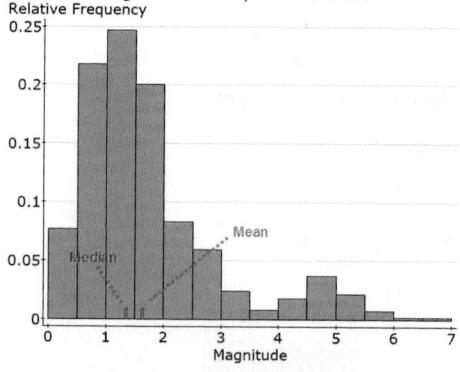

Magnitude of Earthquakes Worldwide

(c) Both data sets have many outliers.

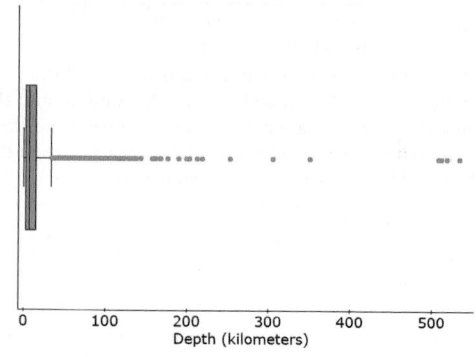

Depth of Earthquakes Worldwide

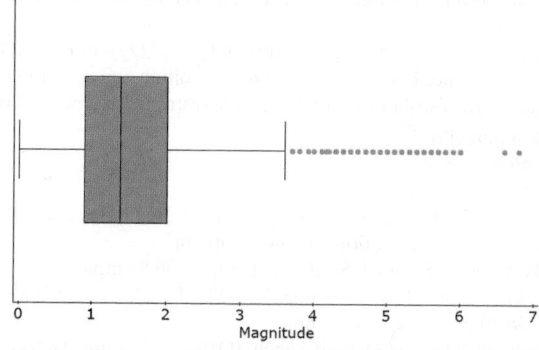

Magnitude of Earthquakes Worldwide

(d) Depth: Lower Fence $= -15.595$; Upper Fence $= 33.965$; Magnitude: Lower Fence $= -0.765$; Upper Fence $= 3.675$

19. (a) Completely randomized design
(b) 30 subjects
(c) Number of advantageous cards; quantitative; discrete
(d) Hunger, impulsivity, BMI, age; Hunger; 2
(e) In an attempt to create groups that were similar in terms of hunger, impulsivity, body mass index, and age, the subjects were randomly assigned to one of two treatment groups. In addition, the expectation is that randomization "evens out" the effect of any other explanatory variables not considered. The researchers verified that the impulsivity, ages, and body mass index of each group were not significantly different.
(f) $\bar{x}_1 = 33.36$; $\bar{x}_2 = 25.86$
(g) Hunger improves advantageous decision making.

21. Using the boxplot: If the median is left of center in the box, and the right whisker is longer than the left whisker, the distribution is skewed right. If the median is in the center of the box, and the left and right whiskers are roughly the same length, the distribution is symmetric. If the median is right of center in the box, and the left whisker is longer than the right whisker, the distribution is skewed left.

Using the quartiles: If the distance from the median to the first quartile is less than the distance from the median to the third quartile, or the distance from the median to the minimum value in the data set is less than the distance from the median to the maximum value in the data set, then the distribution is skewed right. If the distance from the median to the first quartile is the same as the distance from the median to the third quartile, or the distance from the median to the minimum value in the data set is the same as the distance from the median to the maximum value in the data set, the distribution is symmetric. If the distance from the median to the first quartile is more than the distance from the median to the third quartile, or the distance from the median to the minimum value in the data set is more than the distance from the median to the maximum value in the data set, the distribution is skewed left.

Chapter 3 Review Exercises (page 174)

1. (a) Mean $= 792.51$ m/sec, median $= 792.40$ m/sec
(b) Range $= 4.8$ m/sec, $s^2 = 2.03$, $s = 1.42$ m/sec
2. (a) $\bar{x} = \$10,178.9$, $M = \$9980$
(b) Range: $R = \$8550$, $s = \$3074.9$, IQR $= \$5954.5$; using MINITAB: IQR $= \$5954$; using StatCrunch: IQR $= \$5110$
(c) $\bar{x} = \$13,178.9$, $M = \$9980$; range: $R = \$35,550$, $s = \$10,797.5$, IQR $= \$5954.5$ [using MINITAB: $\$5954$]. The median and interquartile range are resistant.
3. (a) $\mu = 57.8$ years, $M = 58.0$ years, bimodal: 56 and 62
(b) Range $= 25$ years, $\sigma = 7.0$ years
(c) Answers will vary.

4. (a) Skewed right

**Number of Tickets
Issued in a Month**

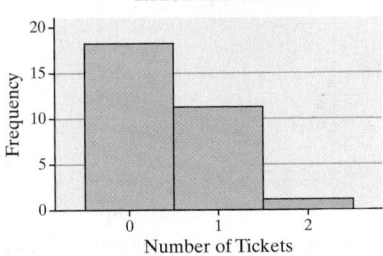

(b) Mean greater than median
(c) $\bar{x} = 0.4, M = 0$
(d) 0

5. (a) 441; 759
(b) 95% of the light bulbs have a life between 494 and 706 hours.
(c) 81.5% of the light bulbs have a life between 547 and 706 hours.
(d) The firm can expect to replace 0.15% of the light bulbs.
(e) At least 84% of the light bulbs have a life within 2.5 standard deviations of the mean.
(f) At least 75% of the light bulbs have a life between 494 and 706 hours.

6. (a) $\bar{x} = 27.7$ minutes   **(b)** $s = 19.1$ minutes

7. 3.33

8. Female, since her weight is more standard deviations above the mean

9. (a) Two-seam fastball
(b) Two-seam fastball
(c) Four-seam fastball
(d) 88 mph
(e) Symmetric
(f) Skewed right

10. (a) $\mu = 2354.9$ words; $M = 2137$ words
(b) $Q_1 = 1425$ words; $Q_2 = M = 2137$ words; $Q_3 = 2906$ words; MINITAB: $Q_1 = 1390$ words; $Q_2 = M = 2137$ words; $Q_3 = 2942$ words
(c) 135, 1425, 2137, 2906, 8445; MINITAB: 135, 1390, 2137, 2942, 8445
(d) $\sigma = 1381.2$ words; IQR = 1481 words [MINITAB: 1552 words]
(e) Lower Fence $= -796.5$ words [MINITAB: -938]; Upper Fence $= 5127.5$ words [MINITAB: 5270]; Outliers: 5433 and 8445
(f)

Number of Words in an Inaugural Speech

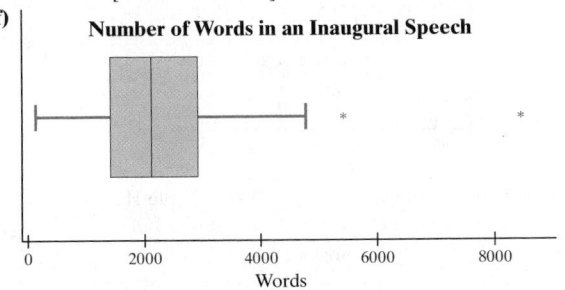

(g) Slightly skewed right since the right whisker is longer than the left whisker and considering the outliers.
(h) The median is the better measure since the outliers inflate the value of the mean.
(i) The interquartile range is the better measure since the outliers inflate the value of the standard deviation.

11. 85% of 19-year-old females have a height that is 67.1 inches or less, and 15% of 19-year-old females have a height that is more than 67.1 inches.

12. The median is used for three measures since it is likely the case that one of the three measures is extreme relative to the other two, thus substantially affecting the value of the mean. Since the median is resistant to extreme values, it is the better measure of central tendency.

Chapter 3 Test (page 176)

1. (a) $\bar{x} = 80$ min
(b) $M = 79.5$ min
(c) $\bar{x} = 192.5$ min, $M = 79.5$ min; the median is resistant.

2. From motor vehicles   **3.** 63 min

4. (a) 23.8 min
(b) 41 min; the middle 50% of all study times has a range of 41 min.
(c) The interquartile range is resistant; the standard deviation is not resistant.

5. (a) 3282; 5322
(b) 95% of the cartridges print between 3622 and 4982 pages.
(c) The firm can expect to replace 2.5% of the cartridges.
(d) At least 55.6% of the cartridges have a page count within 1.5 standard deviations of the mean.
(e) At least 88.9% of the cartridges print between 3282 and 5322 pages.

6. (a) 74.9 minutes   **(b)** 14.7 minutes

7. $2.17

8. (a) Material A: $\bar{x} = 6.404$ million cycles
Material B: $\bar{x} = 11.332$ million cycles
(b) Material A: $M = 5.785$ million cycles
Material B: $M = 8.925$ million cycles
(c) Material A: $s = 2.626$ million cycles
Material B: $s = 5.900$ million cycles
Material B is more dispersed.
(d) Material A (in million cycles): 3.17, 4.52, 5.785, 8.01, 11.92
Material B (in million cycles): 5.78, 6.84, 8.925, 14.71, 24.37
(e)   **Bearing Failures**

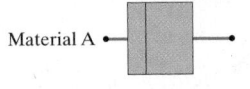
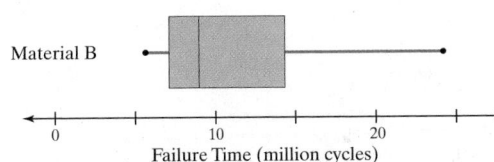

Answers will vary.
(f) The distributions are skewed right.

9. Quartiles by hand, TI-83/84 (in grams): $Q_1 = 5.58$, $Q_2 = 5.60$, $Q_3 = 5.66$; quartiles by MINITAB (in grams): $Q_1 = 5.58$, $Q_2 = 5.60$, $Q_3 = 5.6625$. Using the by-hand quartiles: 25% of quarters have a weight that is 5.58 grams or less, 75% of the quarters have a weight more than 5.58 grams; 50% of the quarters have a weight that is 5.60 grams or less, 50% of the quarters have a weight more than 5.60 grams; 75% of the quarters have a weight that is 5.66 grams or less, 25% of the quarters have a weight that is more than 5.66 grams. The quarter whose weight is 5.84 grams is an outlier.

10. Armando should report his ACT math score since it is more standard deviations above the mean.

11. 15% of 10-year-old males have a height that is 53.5 inches or less, and 85% of 10-year-old males have a height that is more than 53.5 inches.

12. The median will be less than the mean for income data, so you should report the median.

13. (a) Report the mean since the distribution is symmetric.
(b) Histogram I has more dispersion. The range of classes is larger.

14. The standard deviation can be thought of as a mean of the deviations from the mean. We find the standard deviation by squaring the deviations about the mean because, otherwise, the deviations below the mean would offset deviations above the mean. The further a particular observation is from the mean, the higher the squared deviation—this is what causes data sets with more dispersion to have a higher standard deviation. Of course, to "undo" the squaring of the deviations, we need to take the square root after we add the squared deviations and divide by N (for population standard deviation) or $n - 1$ (for sample standard deviations).

Chapter 4 Describing the Relation between Two Variables

4.1 Assess Your Understanding (page 190)

1. Univariate data measure the value of a single variable for each individual in the study. Bivariate data measure values of two variables for each individual.

3. scatter diagram

5. -1

7. lurking

9. Nonlinear

11. Linear, positive

13. (a) III   **(b)** IV
(c) II   **(d)** I

15. (a) Linear; positive association
(b) The point $(55, 55000)$ appears to stick out. Reasons may vary. One explanation is the high concentration of government jobs that require a bachelor's degree, and the high proportion of lobbyists and attorneys who live in D.C.
(c) Yes, because $|0.854| > 0.361$ (critical value from Table II with $n = 30$).

17. (a)

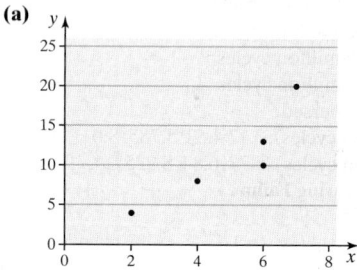

(b) $r = 0.896$ **(c)** Linear relation

19. (a)

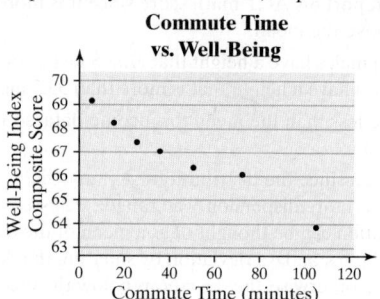

(b) $r = -0.496$ **(c)** No linear relation

21. (a) Positive   **(b)** Negative
(c) Negative   **(d)** Negative
(e) No correlation

23. Yes; $0.79 > 0.361$ (critical r for $n = 30$), so countries in which students answered a greater percentage of items in the background questionnaire tended to have higher mean scores on the TIMMS exam.

25. (a) Explanatory: commute time; response: well-being score
(b)

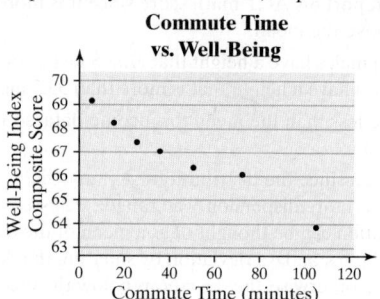

(c) -0.981
(d) $|-0.981| = 0.981 > 0.754$ (critical r from Appendix Table II), so a negative association exists between commute time and well-being index score.

27. (a) Explanatory: height; response: head circumference
(b)

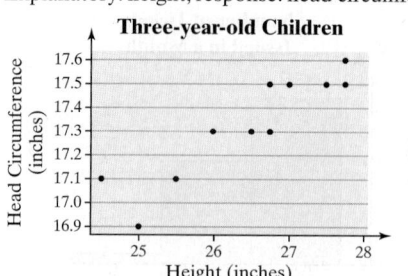

(c) $r = 0.911$

(d) Yes, positive association because $0.911 > 0.602$ (critical r from Table II).
(e) Converting to cm has no effect on the linear correlation coefficient.

29. (a) Explanatory: weight; response: miles per gallon
(b)

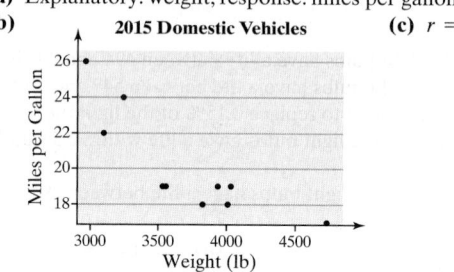

(c) $r = -0.842$

(d) Yes; $|-0.842| > 0.632$ (Table II). There is a negative association between weight of a car and miles per gallon.

31. (a) Stock return
(b)

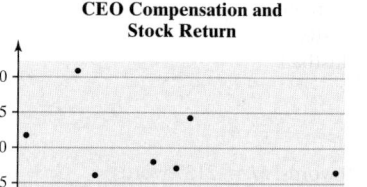

(c) -0.028

(d) No, because $|-0.028| < 0.576$ (Table II). Stock performance does not play a role in determining CEO compensation.

33. (a)

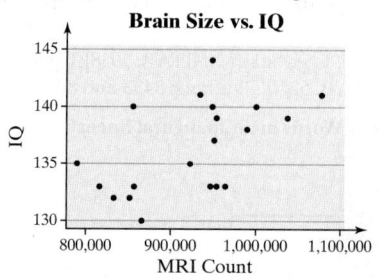

(b) $r = 0.548$. Because $0.548 > 0.444$ (Table II), a positive association exists between MRI count and IQ.
(c)

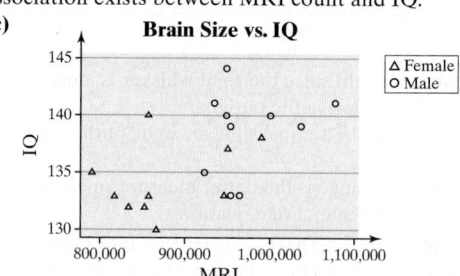

The females have lower MRI counts. Looking at each gender's plot, we see that the relation that appeared to exist between brain size and IQ disappears.

(d) Females: $r = 0.359$; males: $r = 0.236$. No linear relation exists between brain size and IQ.

35. (a)

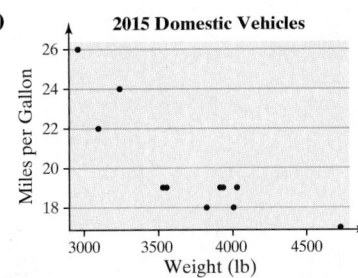

2015 Domestic Vehicles

(b) Correlation coefficient (with Taurus included): $r = -0.844$

(c) The results are reasonable because the Taurus follows the overall pattern of the data.

(d)

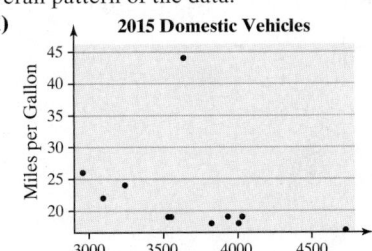

2015 Domestic Vehicles

(e) Correlation coefficient (with Fusion included): $r = -0.331$

(f) The Fusion is a hybrid car; the other cars are not hybrids.

37. (a) 1: 0.816; 2: 0.817; 3: 0.816; 4: 0.817

(b)

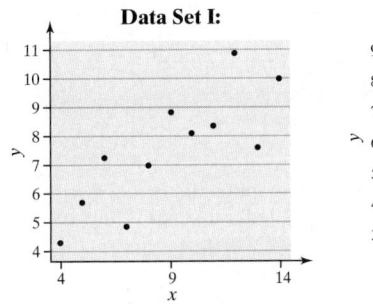

Data Set I:

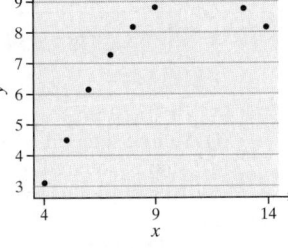

Data Set II:

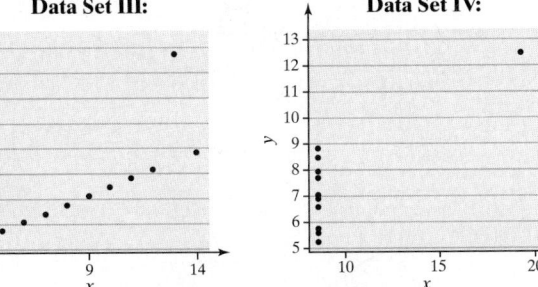

Data Set III: Data Set IV:

39. First Energy and General Electric have the lowest correlation (-0.025). However, you could also invest in Johnson & Johnson and Cisco Systems (0.058). First Energy and Cisco Systems are negatively correlated (-0.248).

41. $r = 0.599$ implies that a positive linear relation exists between the number of television stations and life expectancy, but this is correlation, not causation. The more television stations a country has, the more affluent it is. The more affluent, the better the health care, so wealth is a likely lurking variable.

43. No; a likely lurking variable is the economy. In a strong economy, crime rates tend to decrease, and consumers are better able to afford cell phones.

45. (a)

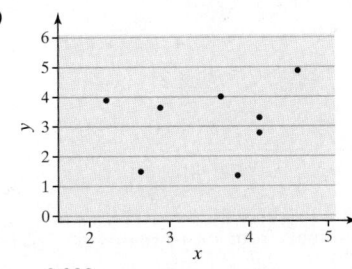

$r = 0.228$

(b)

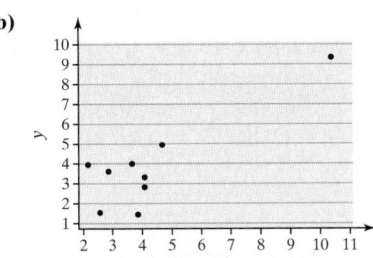

$r = 0.860$; A single point may affect the value of the correlation coefficient making it seem as if a strong linear relation exists between two variables, when it does not.

47. (a)

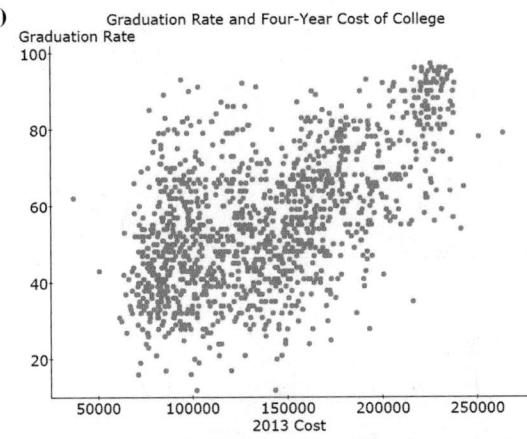

Graduation Rate and Four-Year Cost of College

There appears to be slight positive association between cost and graduation rates.

(b) 0.556; this correlation is high enough to suggest that cost and graduation rates are positively associated.

(c)

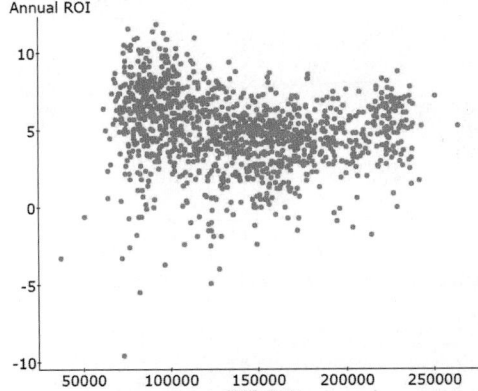

Return on Investment and Four-Year Cost of College

The correlation between cost and return on investment (ROI) is -0.201. The scatter diagram shows some negative association and the correlation coefficient confirms this. The more expensive the school, the lower the ROI.

49. If the correlation coefficient equals 1, then a perfect positive linear relation exists between the variables, and the points of the scatter diagram lie exactly on a straight line with a positive slope.

51. The linear correlation coefficient can only be calculated from bivariate quantitative data, and the gender of a driver is a qualitative variable. A possible better way to write the sentence: Gender is associated with the rate of automobile accidents.

53. Correlation describes a relation between two variables in an observational study. Causation describes a conditional (if–then) relation in an experimental study.

55. Deterministic

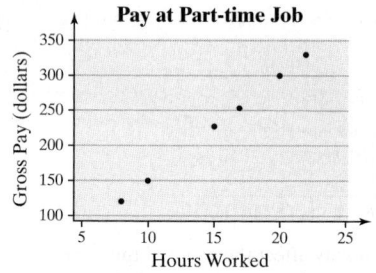

4.2 Assess Your Understanding (page 205)

1. residual   **3.** True

5. (a)

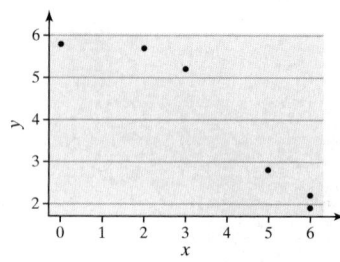

(b) $\hat{y} = -0.7136x + 6.5499$

(c)

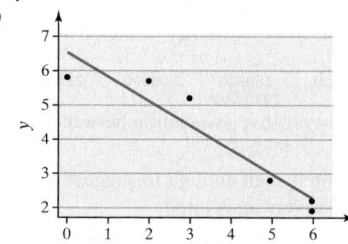

7. (a)

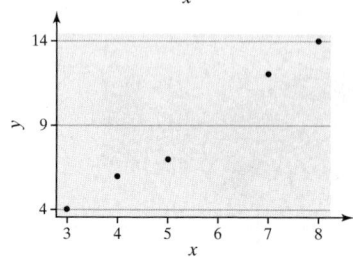

(b) Using points $(3, 4)$ and $(8, 14)$: $\hat{y} = 2x - 2$

(c)

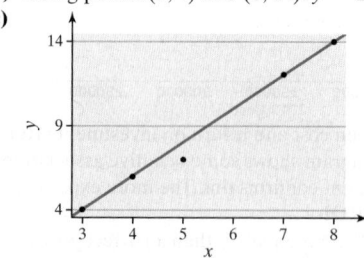

(d) $\hat{y} = 2.0233x - 2.3256$

(e)

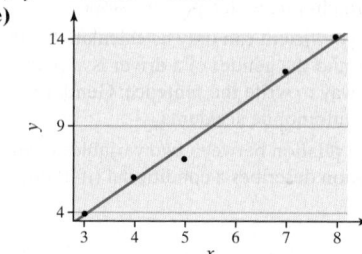

(f) Sum of squared residuals (computed line): 1
(g) Sum of squared residuals (least-squares line): 0.7907
(h) Answers will vary.

9. (a)

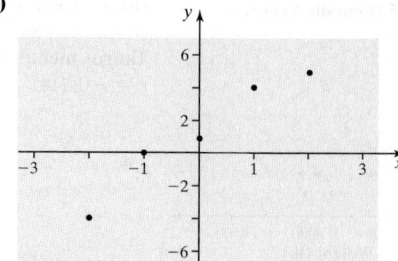

(b) Using points $(-2, -4)$ and $(2, 5)$: $\hat{y} = \frac{9}{4}x + \frac{1}{2} = 2.25x + 0.5$
(c)   **(d)** $\hat{y} = 2.2x + 1.2$

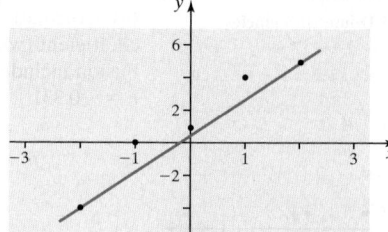

(e)

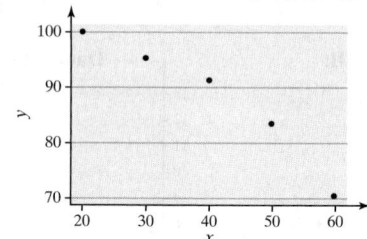

(f) Sum of squared residuals (computed line): 4.875
(g) Sum of squared residuals from least-squares line: 2.4
(h) Answers will vary.

11. (a)

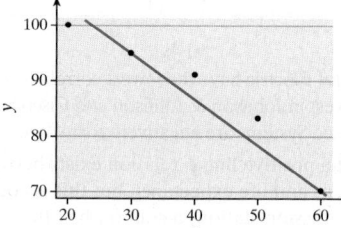

(b) Using points $(30, 95)$ and $(60, 70)$: $\hat{y} = -\frac{5}{6}x + 120$
(c)

(d) $\hat{y} = -0.72x + 116.6$
(e)

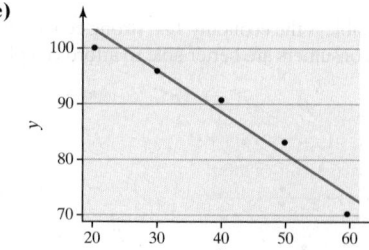

(f) Sum of squared residuals (computed line): 51.6667
(g) Sum of squared residuals from least-squares line: 32.4
(h) Answers will vary.

13. (a) $32,507.5

(b) Higher because the predicted income for a state with 27.1% of adults with at least a bachelor's degree is $33,984.85 and North Dakota's median income is $39,193.

(c) If the percentage of the population with at least a bachelor's degree increases 1 percent, median income increases $703.5, on average.

(d) There are no observations near 0%, so this is outside the scope of the model.

15. (a) If literacy rate increases 1 percent, the age difference decreases 0.0527 years, on average.

(b) No. A 0% literacy rate is outside the scope of the model.

(c) 5.8 years

(d) No. Outside the scope of the model.

(e) The average age difference for a literacy rate of 99 percent is 1.88 years, so the United States has an age difference slightly above average.

17. (a) $\hat{y} = -0.0479x + 69.0296$

(b) Slope: for each 1-minute increase in commute time, the index score decreases by 0.0479, on average; y-intercept: The average index score for a person with a 0-minute commute is 69.0296. A person who works from home will have a 0-minute commute, so the y-intercept has a meaningful interpretation.

(c) 67.6

(d) Barbara's score is less than the mean score for people with a 20-minute commute, 68.1, so she is less well off.

19. (a) $\hat{y} = 0.1827x + 12.4932$

(b) If height increases by 1 inch, head circumference increases by about 0.1827 inch, on average. It is not appropriate to interpret the y-intercept. It is outside the scope of the model.

(c) $\hat{y} = 17.06$ inches

(d) Residual = −0.16 inch; below

(e)

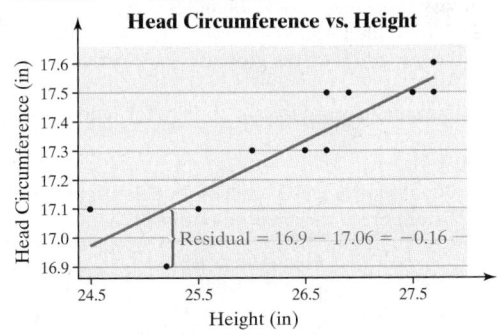
Head Circumference vs. Height
Residual = 16.9 − 17.06 = −0.16

(f) For children who are 26.75 inches tall, head circumference varies.

(g) No; 32 inches is outside the scope of the model.

21. (a) $\hat{y} = -0.0047x + 37.3569$

(b) For every pound added to the weight of the car, gas mileage in the city will decrease by 0.0047 mile per gallon, on average. It is not appropriate to interpret the y-intercept.

(c) $\hat{y} = 20.2$ [Tech: 20.3] when $x = 3649$, so the CTS mileage is only slightly below average.

(d) It is not reasonable to use this least-squares regression to predict the miles per gallon of a Toyota Prius because a Prius is a different type of car (hybrid).

23. (a) $\hat{y} = -0.0029x + 0.8861$

(b) For each additional cola consumed per week, bone mineral density will decrease by 0.0029 g/cm², on average.

(c) For a woman who does not drink cola, the mean bone mineral density will be 0.8861 g/cm².

(d) The predicted bone mineral density is 0.8745 g/cm².

(e) This bone mineral density is below average for women who consume 4 cans of cola per week.

(f) No; 2 cans of cola per day equates to 14 cans per week, which is outside the scope of the model.

25. In both cases, $\hat{y} = \bar{y} = 49.7\%$.

27. (a) Males: $\hat{y} = 0.3428x + 998.4488$
Females: $\hat{y} = 0.1045x + 514.1520$

(b) Males: If the number of licensed drivers increases by 1 (thousand), then the number of fatal crashes increases by 0.3428, on average. Females: If the number of licensed drivers increases by 1 (thousand), then the number of fatal crashes increases by 0.1045, on average. Since females tend to be involved in fewer fatal crashes, an insurance company may use this information to argue for higher rates for male customers.

(c) Above average; above average; below average. An insurance company may use this information to argue for higher rates for younger drivers and lower rates for older drivers. The same relationship holds for females.

29. (a) Square footage   **(b)**

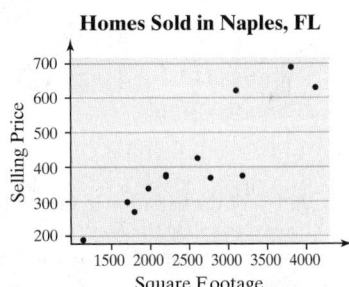
Homes Sold in Naples, FL

(c) $r = 0.903$ **(d)** Yes, because $0.903 > 0.576$ (Table II).

(e) $\hat{y} = 0.1563x + 14.8740$

(f) For each square foot added to the area, the selling price of the house will increase by $156.3 (that is, 0.1563 thousand dollars), on average.

(g) It is not reasonable to interpret the intercept.

(h) Above average; factors that could affect the price include location or number of bedrooms.

31. A residual is the difference between the observed value of y and the predicted value of y. If the residual is positive, the observed value is above average for the given value of the explanatory variable.

33. Answers will vary.

35. Answers may vary. Discussions should involve the scope of the model and the idea of being "outside the scope."

4.3 Assess Your Understanding (page 220)

1. coefficient of determination   **3.** residual plot

5. The variance of the residuals is not constant.

7. There is an outlier.

9. (a) III   **(b)** II
(c) IV   **(d)** I

11. Influential   **13.** Not influential

15. (a) Linear; positive association

(b) Yes; the residual plot does not show any violations of the model.

(c) 83.0% of the variation in the length of eruption is explained by the least-squares regression equation.

17. (a)

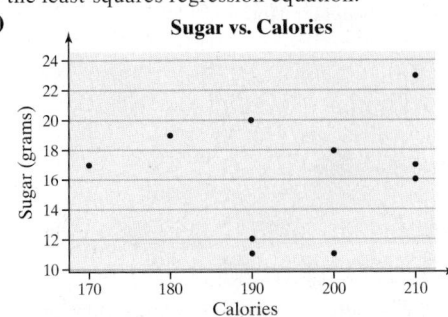
Sugar vs. Calories

(b) $r = 0.261$; no, there is no significant linear relation.

(c)

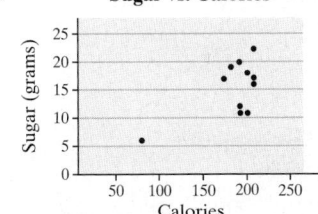
Sugar vs. Calories

$r = 0.649$; yes, there is significant linear relation; All-Bran is an influential observation.

(d) Influential observations can cause the correlation coefficient to increase substantially, thereby increasing the apparent strength of the linear relation between two variables.

19. (a) $R^2 = 96.1\%$ **(b)**

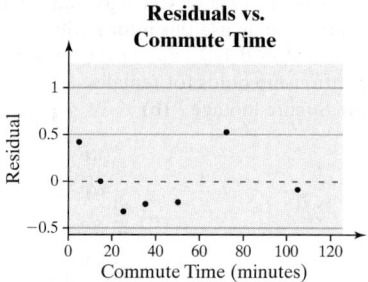

Residuals vs. Commute Time

(c) 96.1% of the variation in the well-being index composite score can be explained by the least-squares regression equation. The linear model appears to be appropriate, based on the residual plot.

21. (a) $R^2 = 83.0\%$ **(b)**

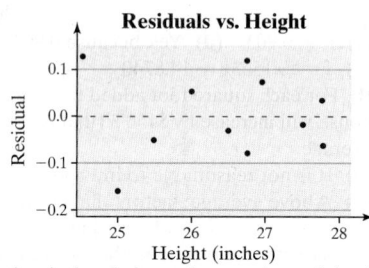

Residuals vs. Height

(c) 83.0% of the variation in head circumference is explained by the least-squares regression equation. The linear model appears to be appropriate, based on the residual plot.

23. (a) $R^2 = 70.8\%$

(b) The residuals do not show a pattern and there are no outliers. The linear model is appropriate.

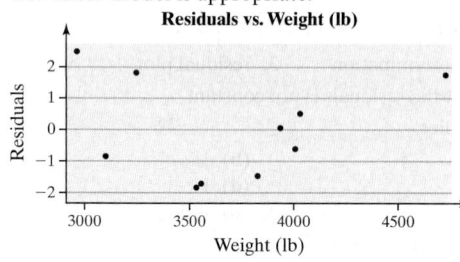

Residuals vs. Weight (lb)

(c) 70.8% of the variation in gas mileage is explained by the linear model. The least-squares regression model appears to be appropriate.

25. (a)

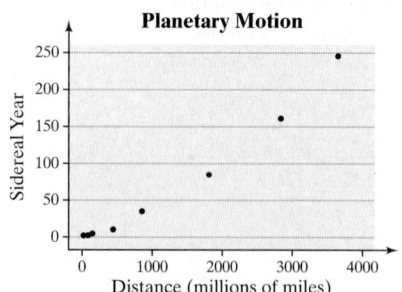

Planetary Motion

(b) $r = 0.989$; yes, there appears to be a linear relation.
(c) $\hat{y} = 0.0657x - 12.4967$
(d)

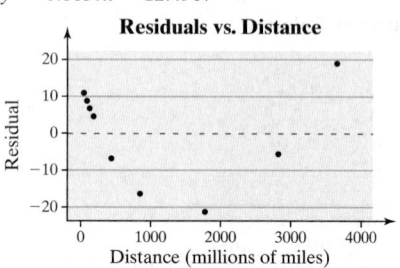

Residuals vs. Distance

(e) No; the residuals follow a U-shaped pattern.

27. (a) The coefficient of determination with the Viper included is $R^2 = 23.0\%$. Adding the Viper reduces the amount of variability explained by the model by approximately 47.8%.
(b) The Viper is influential; the slope and the y-intercept are significantly altered; the Viper is an outlier.

29. The Boardman Power Plant is unusual in that it emits more carbon than expected from the overall trend. The observation is an outlier and influential. The Hermiston power plant is not unusual; it follows the overall trend of the data. Although it has a large residual initially, this effect goes away if the Boardman observation is removed from the data.

31. (a) Width
(b) For each tornado, two variables are measured: width and length. Both of these variables are quantitative.
(c)

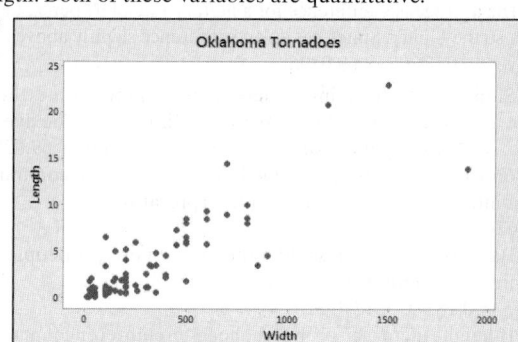

Oklahoma Tornadoes

(d) 0.851
(e) Yes, because $0.851 > 0.361$ (for $n = 30$ in Table II).
(f) $\hat{y} = 0.0111x + 0.0575$ **(g)** 5.6 miles
(h) No, it was expected to last 2.06 miles.
(i) If the width of the tornado increases by 1 yard, the distance for which the tornado is on the ground is expected to increase by 0.0111 mile.
(j) It does not make sense to talk about a tornado whose width is 0 yards.
(k) 72.3%
(l) The residual plot suggests a linear model is appropriate (no pattern, constant error variance, no outliers).

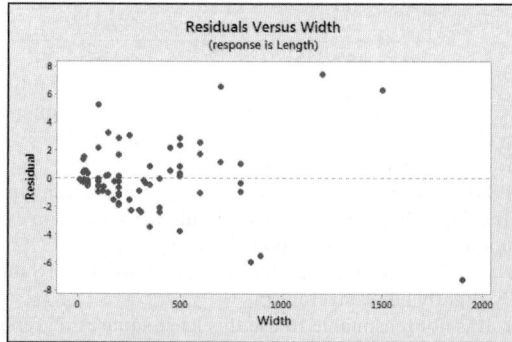

Residuals Versus Width
(response is Length)

(m) Yes, there are many outliers.

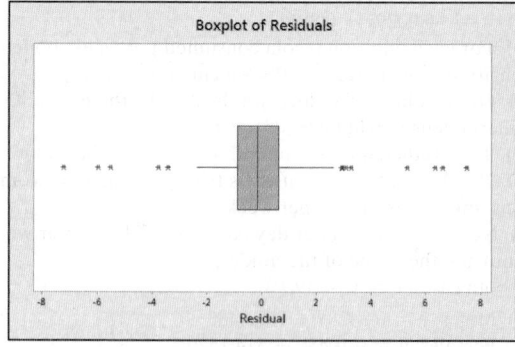

Boxplot of Residuals

(n) Yes. With the point (4576, 16.2) included, the slope decreases from 0.0111 to 0.0058 and the intercept increases from 0.0575 to 1.491.

33. (a) The graph is skewed right. Two classes have a relative frequency of 0.23: 10 − 14.9; 15 − 19.9.

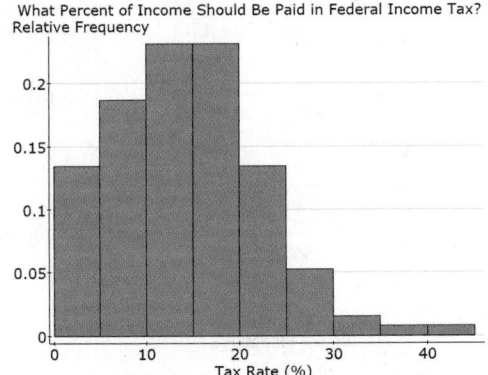

What Percent of Income Should Be Paid in Federal Income Tax?
Relative Frequency

(b) $\bar{x} = 12.3\%$; $M = 10\%$ **(c)** $s = 7.6\%$; $IQR = 12\%$
(d) $LF = -13\%$; $UF = 35\%$; 40% is the only outlier. (Interestingly, the top marginal federal tax rate at the time of this printing was 39.6%).
(e) It appears males are willing to pay more in federal income tax. The median for males appears to be 15%, while the median for females is 10%. However, the range of responses for females is greater than that of males, although both genders have similar interquartile ranges. Both genders have an outlier.

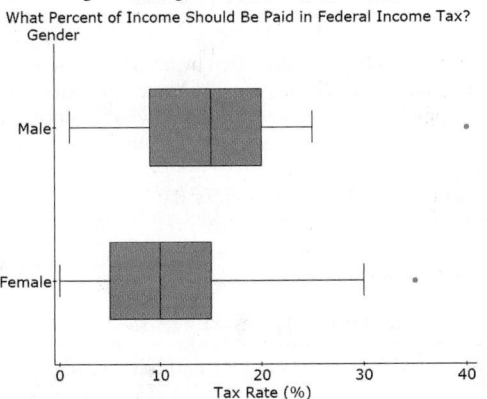

What Percent of Income Should Be Paid in Federal Income Tax?
Gender

(f) None of the boxplots have outliers. Moderates have the highest median (around 12%). Liberals have the highest tax rate (40%) and the most dispersion (as measured by both the interquartile range and range). Conservatives have the least dispersion (as measured by both the interquartile range and range). Interestingly, liberals and conservatives have the same median tax rate.

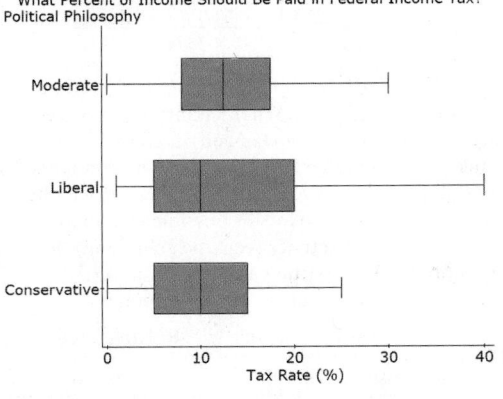

What Percent of Income Should Be Paid in Federal Income Tax?
Political Philosophy

4.4 Assess Your Understanding (page 233)

1. A marginal distribution is a frequency or relative frequency distribution of either the row or column variable in a contingency table. A conditional distribution is the relative frequency of each category of one variable, given a specific value of the other variable in a contingency table.

3. Correlation is used with quantitative variables; association is used with categorical variables.

5. (a)

	x_1	x_2	x_3	**Marginal Distribution**
y_1	20	25	30	75
y_2	30	25	50	105
Marginal Distribution	50	50	80	180

(b)

	x_1	x_2	x_3	**Relative Frequency Marginal Distribution**
y_1	20	25	30	0.417
y_2	30	25	50	0.583
Relative Frequency Marginal Distribution	0.278	0.278	0.444	1

(c)

	x_1	x_2	x_3
y_1	0.400	0.500	0.375
y_2	0.600	0.500	0.625
Total	1	1	1

(d)

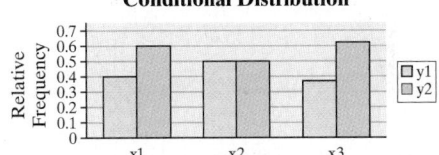

Conditional Distribution

7. (a) 2160 adult Americans were surveyed; 536 were 55 and older.
(b)

Likely to Buy	Age 18–34	35–44	45–54	55+	Relative Frequency Marginal Distribution
More likely	238	329	360	402	0.615
Less likely	22	6	22	16	0.031
Neither	282	201	164	118	0.354
Relative Frequency Marginal Distribution	0.251	0.248	0.253	0.248	1

(c) 0.615
(d)

Likely to Buy	Age 18–34	35–44	45–54	55+
More likely	0.439	0.614	0.659	0.750
Less likely	0.041	0.011	0.040	0.030
Neither	0.520	0.375	0.300	0.220
Total	1	1	1	1

(e)

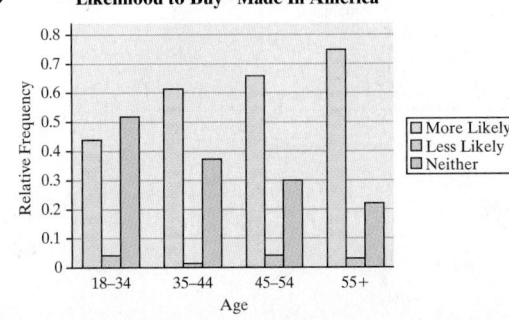

Likelihood to Buy "Made In America"

Likelihood to Buy "Made In America"

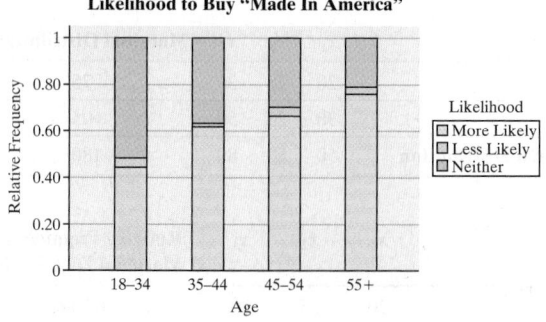

(f) The number of people more likely to buy a product because it is made in America increases with age. On the other hand, age does not seem to be a significant factor in whether a person is less likely to buy a product because it is made in America.

9. (a)

Political Party	Gender		Frequency Marginal Distribution
	Female	**Male**	
Republican	105	115	220
Democrat	150	103	253
Independent	150	179	329
Frequency Marginal Distribution	405	397	802

(b)

Political Party	Gender		Relative Frequency Marginal Distribution
	Female	**Male**	
Republican	105	115	0.274
Democrat	150	103	0.315
Independent	150	179	0.410
Relative Frequency Marginal Distribution	0.505	0.495	1

(c) 0.410

(d)

Political Party	Gender	
	Female	**Male**
Republican	0.259	0.290
Democrat	0.370	0.259
Independent	0.370	0.451
Total	1	1

(e)

Party Affiliation

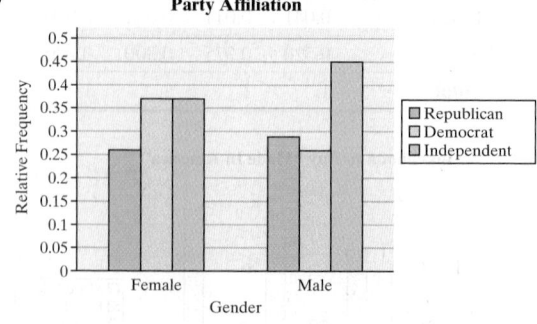

(f) Yes; males are more likely to be Independents and less likely to be Democrats.

Party Affiliation

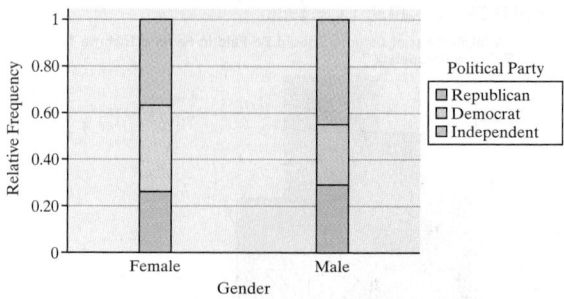

11.

Health and Happiness

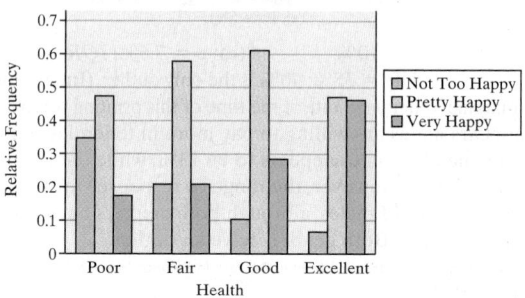

Based on the conditional distribution by health, we can see that healthier people tend to be happier. As health increases, the percent who are very happy increases, while the percent who are not happy decreases.

13. (a) Smokers: 0.239; nonsmokers: 0.314; this implies that it is healthier to smoke.

(b) Smokers: 0.036; Nonsmokers: 0.016

(c) Proportion deceased after 20 years by smoking status and age

	Age						
	18–24	**25–34**	**35–44**	**45–54**	**55–64**	**65–74**	**75 or Older**
Smoker	0.036	0.024	0.128	0.208	0.443	0.806	1
Nonsmoker	0.016	0.032	0.058	0.154	0.331	0.783	1

(d)

Death Rate After 20 Years

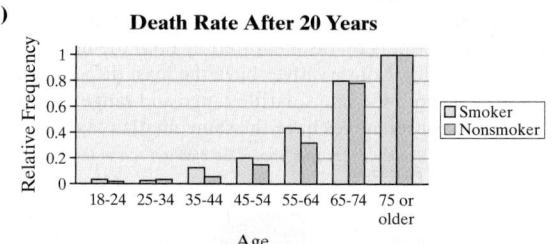

(e) Answers may vary. When taking age into account, the direction of association changed. In almost all age groups, smokers had a higher death rate than nonsmokers. The most notable exception is for the 25 to 34 age group, the largest age group for the nonsmokers. A possible explanation could be rigorous physical activity (e.g., rock climbing) that nonsmokers are more likely to participate in than smokers.

15. (a) Moderates make up 50.75% of respondents.

Political Philosophy	Relative Frequency
Conservative	0.2761
Liberal	0.2164
Moderate	0.5075

(b)

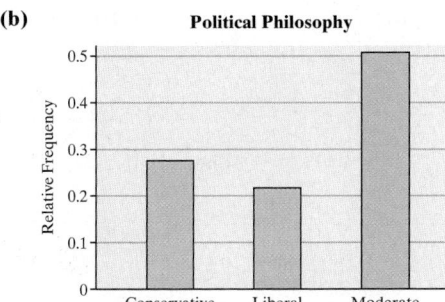

Political Philosophy

(c) It appears that a majority of the respondents believe there is gender income inequality.

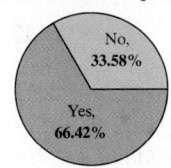

Do You Believe There is Gender Income Inequality?

No, 33.58%

Yes, 66.42%

(d)

Gender	Yes	No	Relative Frequency Marginal Distribution
Male	24	22	0.343
Female	65	23	0.657
Relative Frequency Marginal Distribution	0.664	0.336	

(e) The conditional distribution by gender shows there is a gender gap when it comes to the belief there is gender income inequality.

Gender	Yes	No
Male	0.522	0.478
Female	0.739	0.261

(f)

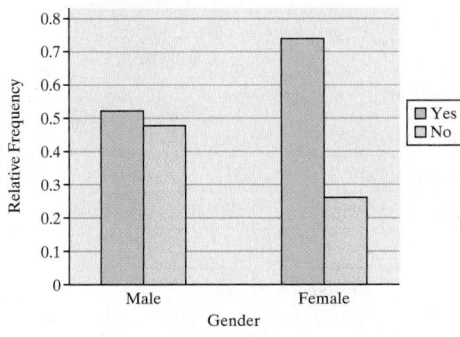

Do You Believe There is an Income Inequality Discrepancy between Male and Females?

Chapter 4 Review Exercises (page 238)

1. (a) If the home team is favored by 3 points, the winning margin predicted by the regression equation is 3.009 points.

(b) If the visiting team is favored by 7 points, the winning margin (for the visiting team) predicted by the regression equation is 7.061 points.
(c) For each 1-point increase in the spread, the winning margin increases by 1.007 points, on average.
(d) If the spread is 0, the home team is expected to lose by 0.012 point, on average.
(e) 39% of the variation in winning margins can be explained by the least-squares regression equation.

2. (a) Fat content

(b)

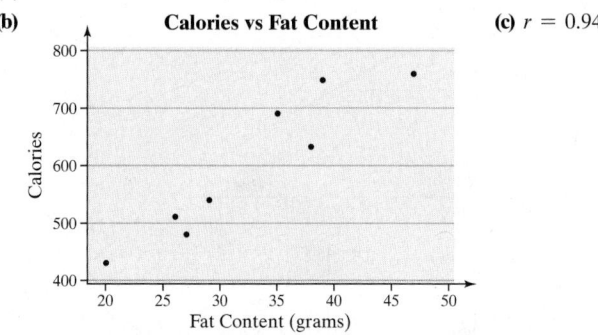

Calories vs Fat Content

(c) $r = 0.944$

(d) Yes; a strong linear relation exists between fat content and calories in fast-food restaurant sandwiches because $0.944 > 0.707$ (Table II).

3. (a)

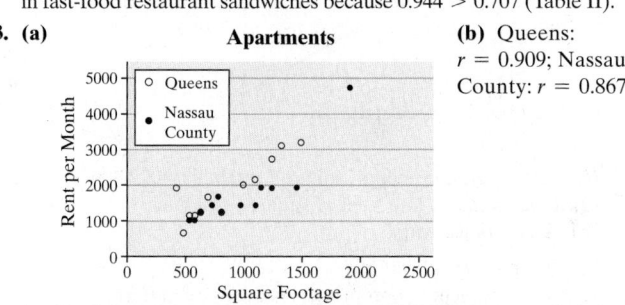

Apartments

(b) Queens: $r = 0.909$; Nassau County: $r = 0.867$

(c) Both locations appear to have a positive linear association between square footage and monthly rent because $r > 0.576$ (Table II) for each location.
(d) For small apartments (those less than 1000 square feet in area), there seems to be no difference in rent between Queens and Nassau County. In larger apartments, Queens seems to have higher rents than Nassau County.

4. (a) $\hat{y} = 13.7334x + 150.9469$

(b)

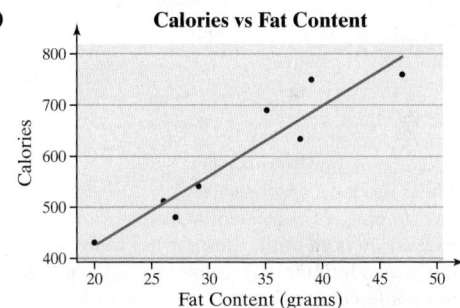

Calories vs Fat Content

(c) The slope indicates that each additional gram of fat in a sandwich adds approximately 13.73 calories, on average. The y-intercept indicates that a sandwich with no fat will contain about 151 calories.
(d) 562.9 calories
(e) Below average (the average value is 727. 7 [Tech: 727.8] calories)

5. (a) $\hat{y} = 2.2091x - 34.3148$
(b) The slope of the least-squares regression line indicates that, for each additional square foot of floor area, the rent increases by \$2.21, on average. It is not appropriate to interpret the y-intercept since it is not possible to have an apartment with 0 square footage.
(c) This apartment's rent is below average.

6. (a)

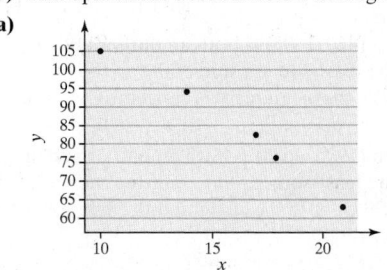

(b) Using points (10, 105) and (18, 76), $\hat{y} = -\frac{29}{8}x + \frac{565}{4}$.

(c)

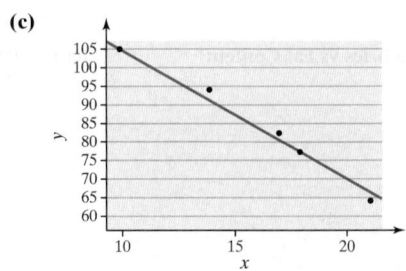

(d) $\hat{y} = -3.8429x + 145.4857$

(e)

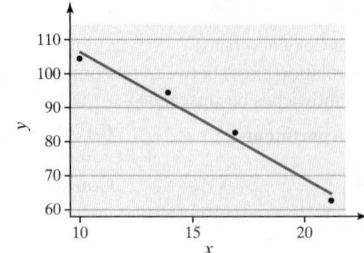

(f) Computed line: sum of squared residuals = 22.406
(g) Least-squares line: sum of squared residuals = 16.271
(h) Answers will vary.

7. Linear model is appropriate.

8. Linear model is not appropriate; residuals are patterned.

9. Linear model is not appropriate; nonconstant error variance.

10. (a) $R^2 = 89.1\%$; 89.1% of the variation in calories is explained by the least-squares regression line.

(b)

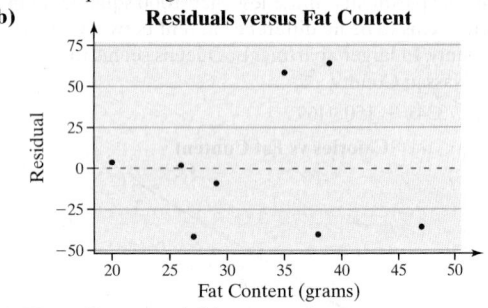

(c) Yes, a linear model is appropriate.
(d) There are no outliers or influential observations.
(e) This observation is an outlier since it lies away from the rest of the data. It is also influential because it significantly changes the slope and y-intercept of the regression line.

11. (a) $R^2 = 82.7\%$; 82.7% of the variance in rent is explained by the least-squares regression model.

(b)

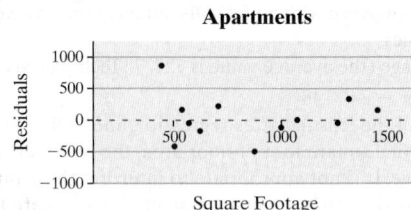

(c) The plot of residuals shows no discernible pattern, so a linear model is appropriate.
(d) The observation (460, 1805) is both an outlier and an influential point. The least-squares line omitting the outlier is $\hat{y} = 2.5315x - 399.25$; the y-intercept is substantially different.

12. (a) $\hat{y} = 2.0995x - 347.3641$
(b) The observation (1906, 4625) is both an outlier and an influential point. The least-squares line omitting the outlier is $\hat{y} = 0.8779x + 675.2945$; both the slope and the $\hat{y}$-intercept are substantially different.

13. (a)

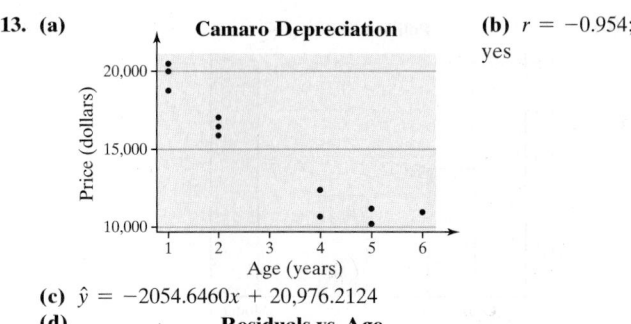

(b) $r = -0.954$; yes

(c) $\hat{y} = -2054.6460x + 20{,}976.2124$
(d)

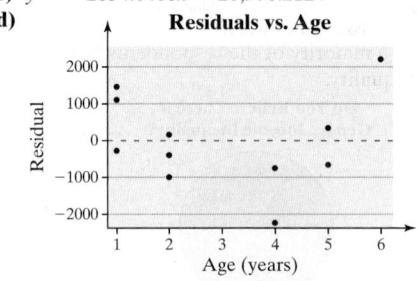

(e) No; the residuals follow a U-shaped pattern.

14. No; correlation does not imply causation. Florida has a large number of tourists in warmer months, times when more people will be in the water to cool off. The larger number of people in the water splashing around leads to a larger number of shark attacks.

15. (a) 203
(b)

Satisfaction	Car Type		Relative Frequency Marginal Distribution
	New	Used	
Not too satisfied	11	25	0.091
Pretty satisfied	78	79	0.396
Extremely satisfied	118	85	0.513
Relative Frequency Marginal Distribution	0.523	0.477	1

(c) 0.513
(d)

Satisfaction	Car Type	
	New	Used
Not too satisfied	0.053	0.132
Pretty satisfied	0.377	0.418
Extremely satisfied	0.570	0.450
Total	1	1

(e)

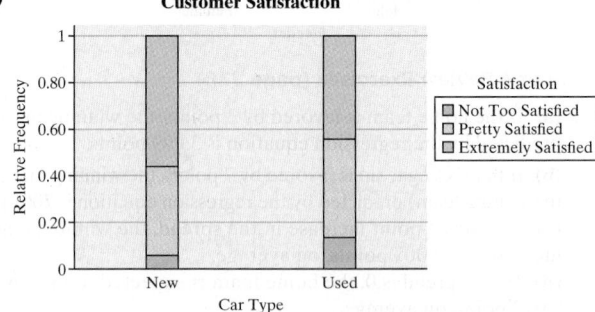

(f) There appears to be some association. Buyers of used cars are more likely to be dissatisfied and are less likely to be extremely satisfied than buyers of new cars.

16. (a) In 1982, the unemployment rate was approximately 10.3%. In 2009, the unemployment rate was approximately 10.0%. The 1982 recession appears to be worse.

(b) The unemployment rates are displayed in the table.

Recession	Level of Education		
	Less than High School	**High School**	**Bachelor's Degree or Higher**
Recession of 1982	16.1%	10.2%	3.8%
Recession of 2009	16.7%	11.8%	4.8%

(c)

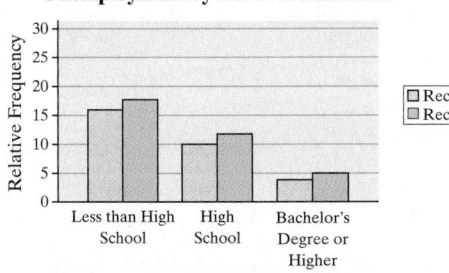

Unemployment by Level of Education

(d) Answer will vary. The discussion should include the observation that, although the overall unemployment rate was higher in 1982, the unemployment rate within each level of education was higher in 2009.

17. (a) A positive linear relation appears to exist between number of marriages and number unemployed.
(b) Population is highly correlated with both the number of marriages and the number unemployed. The size of the population affects both variables.
(c) No association exists between the two variables.
(d) Answers may vary. A strong correlation between two variables may be due to a third variable that is highly correlated with the two original variables.

18. The eight properties of a linear correlation coefficient are the following:
1. The linear correlation coefficient is always between -1 and 1, inclusive. That is, $-1 \leq r \leq 1$.
2. If $r = +1$, there is a perfect positive linear relation between the two variables. See Figure 4(a) on page 184.
3. If $r = -1$, there is a perfect negative linear relation between the two variables. See Figure 4(d).
4. The closer r is to $+1$, the stronger is the evidence of positive association between the two variables. See Figures 4(b) and 4(c).
5. The closer r is to -1, the stronger is the evidence of negative association between the two variables. See Figures 4(e) and 4(f).
6. If r is close to 0, there is no evidence of a *linear* relation between the two variables. Because the linear correlation coefficient is a measure of the strength of the linear relation, r close to 0 does not imply no relation, just no linear relation. See Figures 4(g) and 4(h).
7. The linear correlation coefficient is a unitless measure of association. So the unit of measure for x and y plays no role in the interpretation of r.
8. The correlation coefficient is not resistant.

19. (a) Answers will vary.
(b) The slope can be interpreted as "the school day decreases by 0.01 hour for each 1% increase in percent low income, on average." The y-intercept can be interpreted as the length of the school day when 0% of the population is low income.
(c) $\hat{y} = 6.91$ hours   **(d) – (i)** Answers will vary.

Chapter 4 Test (page 242)

1. (a) Temperature
(b)

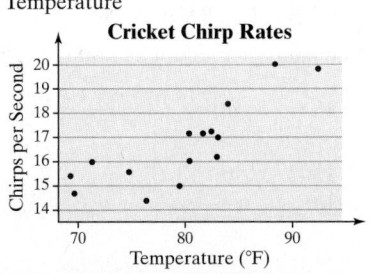

Cricket Chirp Rates

(c) $r = 0.835$
(d) Yes, a positive linear relation exists between temperature and chirps per second because $0.835 > 0.514$ (Table II).

2. (a) $\hat{y} = 0.2119x - 0.3091$
(b) If the temperature increases $1°F$, the number of chirps per second increases by 0.2119, on average. Since there are no observations near $0°F$, it does not make sense to interpret the y-intercept.
(c) At $83.3°F$, $\hat{y} = 17.3$ chirps per second.
(d) Below average
(e) No; outside the scope of the model

3. (a) $R^2 = 69.7\%$; 69.7% of the variation in number of chirps per second is explained by the least-squares regression line.
(b) Yes; the residual plot does not indicate any patterns.

4. Yes; when added, both the slope and y-intercept change substantially.

5. No; although there is a strong correlation, the residual plot is U-shaped, so a linear model is not appropriate.

6. Correlation does not imply causation. It is possible that a lurking variable, such as income level or educational level, is affecting both the explanatory and response variables.

7. (a)

Education Level	Belief				Relative Frequency Marginal Distribution
	Yes, Definitely	**Yes, Probably**	**No, Probably Not**	**No, Definitely Not**	
Less than high school	316	66	21	9	0.173
High school	956	296	122	65	0.606
Bachelor's	267	131	62	64	0.221
Relative Frequency Marginal Distribution	0.648	0.208	0.086	0.058	1

(b) 0.648
(c)

Education Level	Belief				Total
	Yes, Definitely	**Yes, Probably**	**No, Probably Not**	**No, Definitely Not**	
Less than high school	0.767	0.160	0.051	0.022	1
High school	0.664	0.206	0.085	0.045	1
Bachelor's	0.510	0.250	0.118	0.122	1

(d)

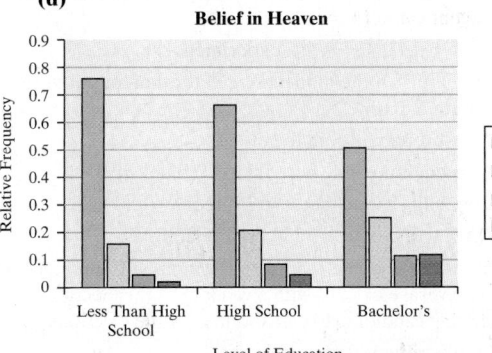

Belief in Heaven

(e) Yes; as education level increases, the percent who definitely believe in Heaven decreases (i.e., doubt in Heaven increases).

8. (a)

College Applicants			
	Accepted	**Denied**	**Total**
Male	0.158	0.842	1
Female	0.310	0.690	1

(b) 0.158 of the males who applied was accepted. 0.310 of the females who applied was accepted.
(c) Conclusion: a higher proportion of females was accepted.
(d) 0.150 of the males applying to the business school was accepted. 0.143 of the females applying to the business school was accepted.
(e) 0.40 of the males applying to the social work school was accepted. 0.364 of the females applying to the social work school was accepted.
(f) Answers will vary. A larger number of males applied to the business school, which has an overall lower acceptance rate than the social work school, so more male applicants were declined.
9. The data would have a perfect negative linear relation. A scatter diagram would show all the observations being collinear (falling on the same line) with a negative slope.
10. If the slope of the least-squares regression line is negative, the correlation between the explanatory and response variables is also negative.
11. If a linear correlation coefficient is close to zero, this means that there is no linear relation between the explanatory and response variable. This does not mean there is no relation, just no *linear* relation.

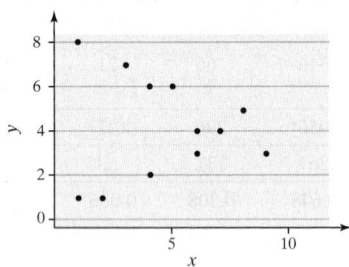

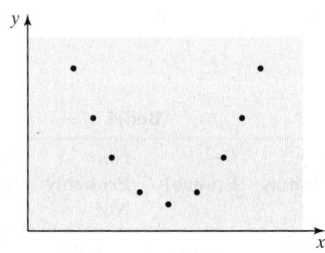

Chapter 5 Probability

5.1 Assess Your Understanding (page 257)

1. The probability of an impossible event is zero. No; when a probability is based on an empirical experiment, a probability of zero does not mean the event cannot occur.
3. True   **5.** experiment
7. Rule 1: All probabilities in the model are greater than or equal to zero and less than or equal to one.
Rule 2: The sum of the probabilities in the model is 1.
The outcome "blue" is an impossible event since $P(\text{blue}) = 0$.
9. This is not a probability model because $P(\text{green}) < 0$.
11. The numbers $0, 0.01, 0.35$, and 1 could be probabilities.
13. In 100 elections, where a senate candidate is winning his/her election with a 5% lead in the average of polls with a week until the election, we would expect the leading candidate to win about 89 of the elections.
15. The empirical probability is 0.95.
17. $P(2) \neq \dfrac{1}{11}$ since the outcomes are not equally likely.
19. $S = \{1H, 1T, 2H, 2T, 3H, 3T, 4H, 4T, 5H, 5T, 6H, 6T\}$

21. 0.428   **23.** $P(E) = \dfrac{3}{10} = 0.3$ **25.** $P(E) = \dfrac{2}{5} = 0.4$

27. (a) $P(\text{plays sports}) = \dfrac{288}{500} = 0.576$
(b) If we sampled 1000 high school students, we would expect that about 576 of the students play organized sports.

29. (a) $P(\text{caught}) = \dfrac{85}{1000} = 0.085$
(b) $P(\text{dropped}) = \dfrac{296}{1000} = 0.296$
(c) $P(\text{hat}) = \dfrac{2}{85} \approx 0.024$; If 1000 caught homeruns are randomly selected, we would expect about 24 to be caught in a hat.
(d) $P(\text{failed hat}) = \dfrac{8}{296} \approx 0.027$; If 1000 dropped homeruns are randomly selected, we would expect about 27 to be a failed hat attempt.

31. (a) $S = \{0, 00, 1, 2, 3, 4, \ldots, 35, 36\}$
(b) $P(8) = \dfrac{1}{38} = 0.0263$, if we spun the wheel 1000 times, we would expect about 26 of those times to result in the ball landing in slot 8.
(c) $P(\text{odd}) = 9/19 = 0.4737$, if we spun the wheel 100 times, we would expect about 47 of the spins to result in an odd number.

33. (a) $\{SS, Ss, sS, ss\}$
(b) $P(ss) = \dfrac{1}{4}$; the probability that a randomly selected offspring will have sickle-cell anemia is 0.25.
(c) $P(Ss \text{ or } sS) = \dfrac{1}{2} = 0.5$; the probability that a randomly selected offspring will be a carrier of sickle-cell anemia is 0.5.

35. (a)

Response	Probability
Never	0.026
Rarely	0.068
Sometimes	0.116
Most of the time	0.263
Always	0.527

(b) It would be unusual to randomly find a college student who never wears a seatbelt when riding in a car driven by someone else, because $P(\text{never}) < 0.05$.

37. (a)

Type of Larceny Theft	Probability
Pocket picking	0.006
Purse snatching	0.009
Shoplifting	0.207
From motor vehicles	0.341
Motor vehicle accessories	0.140
Bicycles	0.065
From buildings	0.223
From coin-operated machines	0.008

(b) Purse snatching larcenies are unusual.
(c) Bicycle larcenies are not unusual.
39. A, B, C, and F are consistent with the definition of a probability model.
41. Use model B if the coin is known to always come up tails.
43. (a) $S = \{JR, JC, JD, JM, RC, RD, RM, CD, CM, DM\}$
(b) $P(\text{Clarice and Dominique attend}) = \dfrac{1}{10} = 0.10$
(c) $P(\text{Clarice attends}) = \dfrac{4}{10} = 0.40$
(d) $P(\text{John stays home}) = \dfrac{6}{10} = 0.60$

45. **(a)** $P(\text{right field}) = \dfrac{24}{73} = 0.329$

(b) $P(\text{left field}) = \dfrac{2}{73} = 0.027$

(c) It was unusual for Barry Bonds to hit a home run to left field; the probability is less than 5%.

47. Answers will vary.

49. The dice appear to be loaded since the outcomes are not equally likely.

51. $P(\text{income is greater than } \$52,250) = \dfrac{1}{2} = 0.5$

53.

Position	Relative Frequency
C	0.015
CB	0.120
DE	0.081
DT	0.089
FB	0.015
ILB	0.046
LS	0.004
OG	0.062
OLB	0.081
OT	0.081
P	0.004
QB	0.046
RB	0.097
S	0.062
TE	0.042
WR	0.154

Wide receiver has the highest probability of being selected since this position has the highest relative frequency. It would be surprising if a center was randomly selected since this should occur with probability 0.015 (or, 15 out of every 1000 selections).

55. The Law of Large Numbers states that, as the number of repetitions of a probability experiment increases (long term), the proportion with which a certain outcome is observed (relative frequency) gets closer to the probability of the outcome. The games at a gambling casino are designed to benefit the casino in the long run; the risk to the casino is minimal because of the large number of gamblers.

57. An event is unusual if it has a low probability of occurring. The same cutoff should not always be used to identify unusual events. Selecting a cutoff is subjective and should take into account the consequences of incorrectly identifying an event as unusual.

59. Empirical probability is based on the outcomes of a probability experiment and is the relative frequency of the event. Classical probability is based on counting techniques and is equal to the ratio of the number of ways an event can occur to the number of possible outcomes of the experiment. Classical probability requires equally likely outcomes.

61. The smaller hospital due to the Law of Large Numbers. The larger hospital likely has more births so it is less likely that the births would deviate from the expected proportion of girls.

5.2 Assess Your Understanding (page 269)

1. Two events are disjoint (mutually exclusive) if they have no outcomes in common.

3. $P(E) + P(F) - P(E \text{ and } F)$

5. E and $F = \{5, 6, 7\}$; E and F are not mutually exclusive.

7. F or $G = \{5, 6, 7, 8, 9, 10, 11, 12\}$; $P(F \text{ or } G) = \dfrac{2}{3}$

9. There are no outcomes in event "E and G." Events E and G are mutually exclusive.

11. $E^c = \{1, 8, 9, 10, 11, 12\}$; $P(E^c) = \dfrac{1}{2}$

13. $P(E \text{ or } F) = 0.55$

15. $P(E \text{ or } F) = 0.70$

17. $P(E^c) = 0.75$

19. $P(F) = 0.30$

21. $P(\text{Titleist or Maxfli}) = \dfrac{17}{20} = 0.85$

23. $P(\text{not a Titleist}) = \dfrac{11}{20} = 0.55$

25. **(a)** All the probabilities are nonnegative; the sum of the probabilities equals 1.

(b) $P(\text{rifle or shotgun}) = 0.048$. If 1000 murders in 2013 were randomly selected, we would expect 48 of them to be the result of a rifle or shotgun.

(c) $P(\text{handgun, rifle, or shotgun}) = 0.52$. If 100 murders in 2013 were randomly selected, we would expect 52 of them to be the result of a handgun, rifle, or shotgun.

(d) $P(\text{not a gun}) = 0.31$. If 100 murders in 2013 were randomly selected, we would expect 31 of them to be the result of a weapon other than a firearm.

(e) A murder with a shotgun is unusual.

27. No; for example, on one draw of a card from a standard deck, let $E = $ diamond, $F = $ club, and $G = $ red card.

29. **(a)** $P(5 - 5.9) = \dfrac{23}{125} = 0.184$

(b) $P(\text{not between } 5 - 5.9) = 1 - 0.184 = 0.816$

(c) $P(<9) = 0.96$

(d) $P(\text{reduced payments}) = 0.08$. If 100 hospitals in Illinois are randomly selected, we would expect eight to received reduced Medicare payments. Given the fact that reduced Medicare payments result due to a poor patient track record, we would like the proportion of hospitals receiving lower payments from Medicare to be close to zero. Therefore, this result is not unusual enough (unfortunately).

31. **(a)** $P(\text{heart or club}) = \dfrac{1}{2} = 0.5$

(b) $P(\text{heart or club or diamond}) = \dfrac{3}{4} = 0.75$

(c) $P(\text{ace or heart}) = \dfrac{4}{13} = 0.308$

33. **(a)** $P(\text{birthday is not November 8}) = \dfrac{364}{365} = 0.997$

(b) $P(\text{birthday is not 1st of month}) = \dfrac{353}{365} = 0.967$

(c) $P(\text{birthday is not 31st of month}) = \dfrac{358}{365} = 0.981$

(d) $P(\text{birthday is not in December}) = \dfrac{334}{365} = 0.915$

35. No; some people have both vision and hearing problems, but we do not know the proportion.

37. **(a)** $P(\text{only English or only Spanish is spoken}) = 0.93$

(b) $P(\text{language other than only English or only Spanish}) = 0.07$

(c) $P(\text{not only English is spoken}) = 0.19$

(d) No; the sum of the probabilities would be greater than 1, and there would be no probability model.

39. **(a)** $P(\text{died from cancer}) = 0.007$

(b) $P(\text{current smoker}) = 0.057$

(c) $P(\text{died from cancer and was current cigar smoker}) = 0.001$

(d) $P(\text{died from cancer or was current cigar smoker}) = 0.063$

41. **(a)** $P(\text{placebo}) = \dfrac{2}{5} = 0.40$

(b) $P(\text{headache went away}) = \dfrac{94}{125} = 0.752$

(c) $P(\text{placebo and headache went away}) = \dfrac{28}{125} = 0.224$

(d) $P(\text{placebo or headache went away}) = \dfrac{116}{125} = 0.928$

43. **(a)** $P(\text{male}) = 0.649$ **(b)** $P(\text{Sunday}) = 0.161$

(c) $P(\text{male and Sunday}) = 0.104$

(d) $P(\text{male or Sunday}) = 0.707$

(e) Yes, a fatality on Wednesday involving a female driver is unusual because the probability of this event is 0.043.

45. (a) Crash type, system, and number of crashes
(b) Crash type: qualitative; system: qualitative; number of crashes: quantitative, discrete
(c)

Crash Type	Current System	Red-Light Cameras
Reported injury	0.26	0.24
Reported property damage only	0.35	0.36
Unreported injury	0.07	0.07
Unreported property damage only	0.32	0.33
Total	**1**	**1**

(d)

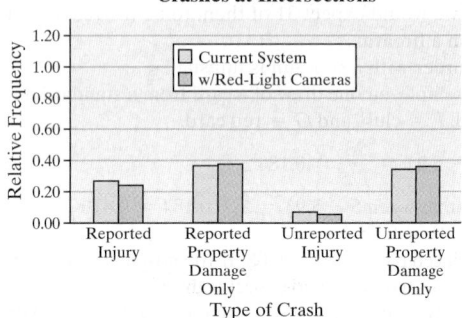

Crashes at Intersections

(e) Current system: $\dfrac{1121}{13} \approx 86.2$ crashes

With cameras: $\dfrac{922}{13} \approx 70.9$ crashes

(f) Not possible; we would need to know the number of crashes at each intersection.
(g) Since the mean number of crashes is less with the cameras, it appears that the program will be beneficial.
(h) $P(\text{reported injuries}) = \dfrac{289}{1121} = 0.258$
(i) $P(\text{only property damage}) = \dfrac{641}{922} = 0.695$
(j) When accounting for the cause of the crash (rear end vs. red-light running), the camera system does not reduce all types of accidents. Under the camera system, red-light running crashes decreased, but rear-end crashes increased.
(k) Recommendations may vary. The benefits of the decrease in red-light running crashes must be weighed against the negative of increased rear-end crashes. Seriousness of injuries and amount of property damage may need to be considered.

5.3 Assess Your Understanding (page 277)

1. independent   **3.** Addition   **5.** $P(E) \cdot P(F)$
7. (a) Dependent   **(b)** Dependent   **(c)** Independent
9. $P(E \text{ and } F) = 0.18$
11. $P(5 \text{ heads}) = \dfrac{1}{32} = 0.03125$; if we flipped a coin 5 times, one hundred different times, we would expect to observe 5 heads in a row about 3 times.
13. $P(\text{two left-handed people are chosen}) = 0.0169$
$P(\text{at least one person chosen is right-handed}) = 0.9831$
15. (a) $P(\text{all five are negative}) = 0.9752$
(b) $P(\text{at least one is positive}) = 0.0248$
17. (a) $P(\text{two will live to 41 years}) = 0.99515$
(b) $P(\text{five will live to 41 years}) = 0.98791$
(c) $P(\text{at least one of five dies}) = 0.01209$; it would be unusual that at least one of the five dies before age 41.
19. (a) $P(\text{not default}) = 0.99$
(b) $P(\text{five not default}) = 0.951$
(c) $P(\text{worthless}) = 0.049$

(d) No. National economic conditions (such as recession) will impact all mortgages. So, if one mortgage defaults, the likelihood of a second mortgage defaulting increases.
21. (a) $P(\text{all three fail}) = 2.16 \times 10^{-7} = 0.000000216$
(b) $P(\text{at least one does not fail}) = 0.999999784$
23. (a) $P(\text{system does not fail}) = 0.999973$
(b) Six components
25. (a) $P(\text{two strikes in a row}) = 0.09$
(b) $P(\text{turkey}) = 0.027$
(c) $P(3 \text{ strikes followed by nonstrike}) = 0.0189$
27. $P(\text{at least 1}) = 0.9999$; answers will vary.
29. $P(\text{girl and in excess of 40 pounds}) = 0.099$
31. (a) $P(\text{male and bets on professional sports}) = 0.0823$
(b) $P(\text{male or bets on professional sports}) = 0.5717$
(c) The independence assumption is not correct.
(d) $P(\text{male or bets on professional sports}) = 0.548$
33. No, assuming gender of children for different births are independent events.

5.4 Assess Your Understanding (page 286)

1. F occurring; E has occurred   **3.** $P(F|E) = 0.75$
5. $P(F|E) = 0.568$ **7.** $P(E \text{ and } F) = 0.32$
9. The events are not independent.
11. $P(\text{club}) = \dfrac{1}{4}$; $P(\text{club}|\text{black card}) = \dfrac{1}{2}$
13. $P(\text{rainy}|\text{cloudy}) = \dfrac{0.21}{0.37} = 0.568$
15. $P(\text{unemployed}|\text{dropout}) = \dfrac{0.021}{0.080} = 0.263$
17. (a) $P(35\text{–}44|\text{more likely}) = \dfrac{329}{1329} = 0.248$
(b) $P(\text{more likely}|35\text{–}44) = \dfrac{329}{536} = 0.614$
(c) No; they are less likely to buy American; 0.439 for 18- to 34-year-olds compared to 0.615 in general.
19. (a) $P(\text{female}|\text{Sunday}) = \dfrac{2287}{6430} = 0.356$
(b) $P(\text{Sunday}|\text{female}) = \dfrac{2287}{13{,}989} = 0.163$
(c) No; for any given day of the week, $P(\text{male}|\text{day})$ is between 0.64 and 0.66, so the probability that a fatality is male is essentially the same for every day of the week.
21. $P(\text{both TVs work}) = 0.4$; $P(\text{at least one TV does not work}) = 0.6$
23. (a) $P(\text{first card is a king and the second card is a king}) = \dfrac{1}{221} = 0.005$
(b) $P(\text{first card is a king and the second card is a king}) = \dfrac{1}{169} = 0.006$
25. $P(\text{Dave and then Neta are chosen}) = \dfrac{1}{20} = 0.05$
27. (a) $P(\text{like both songs}) = \dfrac{5}{39} = 0.128$; it is not unusual to like both songs.
(b) $P(\text{like neither song}) = \dfrac{14}{39} = 0.359$
(c) $P(\text{like exactly one song}) = \dfrac{20}{39} = 0.513$
(d) $P(\text{like both songs}) = \dfrac{25}{169} = 0.148$;
$P(\text{like neither song}) = \dfrac{64}{169} = 0.379$;
$P(\text{like exactly one song}) = \dfrac{80}{169} = 0.473$

29.

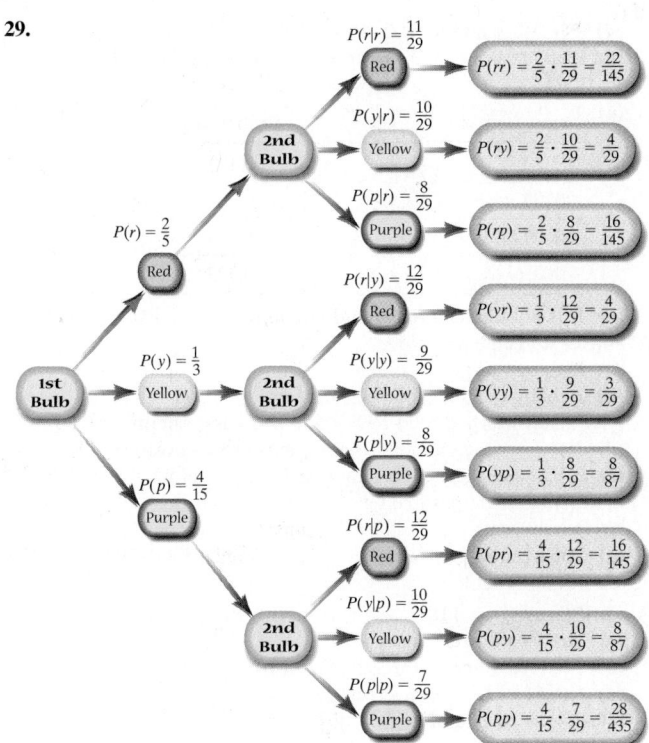

(a) $P(\text{two red bulbs}) = \dfrac{22}{145} = 0.152$

(b) $P(\text{first a red and then a yellow bulb}) = \dfrac{4}{29} = 0.138$

(c) $P(\text{first a yellow and then a red bulb}) = \dfrac{4}{29} = 0.138$

(d) $P(\text{a red bulb and a yellow bulb}) = \dfrac{8}{29} = 0.276$

31. $P(\text{female and smokes}) = 0.090$; it would not be unusual to randomly select a female who smokes.

33. (a) $P(10 \text{ people each have a different birthday}) = 0.883$

(b) $P(\text{at least two of the ten people have the same birthday}) = 0.117$

35. (a) Yes; $P(\text{male}) = \dfrac{1}{2} = P(\text{male} \mid 0 \text{ activities}) = \dfrac{1}{2}$

(b) No; $P(\text{female}) = \dfrac{1}{2} \neq P(\text{female} \mid 5+ \text{ activities}) = \dfrac{71}{109} = 0.651$

(c) Yes; $P(1\text{–}2 \text{ activities and } 3\text{–}4 \text{ activities}) = 0$

(d) No; $P(\text{male and } 1\text{–}2 \text{ activities}) = 0.2025 \neq 0$

37. (a) $P(\text{being dealt five clubs}) = 0.000495$

(b) $P(\text{being dealt a flush}) = 0.002$

39. (a) $P(\text{two defective chips}) = 0.0000245$

(b) Assuming independence: $P(\text{two defective chips}) = 0.000025$

41. $P(45\text{–}54 \text{ years old}) = \dfrac{546}{2160} = 0.253$; $P(45\text{–}54 \text{ years old} \mid \text{more}$

likely$) = \dfrac{360}{1329} = 0.271$; the events are not independent.

43. Answers will vary.

5.5 Assess Your Understanding (page 299)

1. permutation   **3.** True   **5.** $5! = 120$

7. $10! = 3,628,800$ **9.** $0! = 1$ **11.** $_6P_2 = 30$

13. $_4P_4 = 24$ **15.** $_5P_0 = 1$ **17.** $_8P_3 = 336$

19. $_8C_3 = 56$ **21.** $_{10}C_2 = 45$ **23.** $_{52}C_1 = 52$

25. $_{48}C_3 = 17{,}296$

27. $ab, ac, ad, ae, ba, bc, bd, be, ca, cb, cd, ce, da, db, dc, de, ea, eb, ec, ed$; $_5P_2 = 20$

29. $ab, ac, ad, ae, bc, bd, be, cd, ce, de$; $_5C_2 = 10$

31. He can wear 24 different shirt-and-tie combinations.

33. Dan can arrange the songs $12! = 479{,}001{,}600$ ways.

35. The salesperson can take $8! = 40{,}320$ different routes.

37. At most 18,278 companies can be listed on the NYSE.

39. (a) 10,000 different codes are possible.

(b) $P(\text{correct code is guessed}) = \dfrac{1}{10{,}000}$

41. $26^8 \approx 2.09 \times 10^{11}$ different user names are possible.

43. (a) There are $50^3 = 125{,}000$ lock combinations.

(b) $P(\text{guessing the correct combination}) = \dfrac{1}{50^3} = \dfrac{1}{125{,}000}$

45. The top three cars can finish in $_{40}P_3 = 59{,}280$ ways.

47. The officers can be chosen in $_{20}P_4 = 116{,}280$ ways.

49. There are $_{25}P_4 = 303{,}600$ possible outcomes.

51. There are $_{50}C_5 = 2{,}118{,}760$ possible simple random samples of size 5.

53. $_6C_2 = 15$ different birth and gender orders are possible.

55. $\dfrac{10!}{3! \cdot 2! \cdot 2! \cdot 3!} = 25{,}200$ different sequences are possible.

57. The trees can be planted $\dfrac{11!}{4! \cdot 5! \cdot 2!} = 6930$ different ways.

59. $P(\text{winning}) = \dfrac{1}{_{39}C_5} = \dfrac{1}{575{,}757}$

61. (a) $P(\text{jury has only students}) = \dfrac{1}{153} = 0.0065$

(b) $P(\text{jury has only faculty}) = \dfrac{1}{34} = 0.0294$

(c) $P(\text{jury has two students and three faculty}) = \dfrac{20}{51} = 0.3922$

63. $P(\text{shipment is rejected}) = 0.1283$

65. (a) $P(\text{you like 2 of 4 songs}) = 0.3916$

(b) $P(\text{you like 3 of 4 songs}) = 0.1119$

(c) $P(\text{you like all 4 songs}) = 0.0070$

67. (a) Five cards can be selected from a deck $_{52}C_5 = 2{,}598{,}960$ ways.

(b) Three of the same card can be chosen $13 \cdot {_4C_3} = 52$ ways.

(c) The remaining two cards can be chosen $_{12}C_2 \cdot {_4C_1} \cdot {_4C_1} = 1056$ ways.

(d) $P(\text{three of a kind}) = \dfrac{52 \cdot 1056}{2{,}598{,}960} = 0.0211$

69. $P(\text{all 4 modems tested work}) = 0.4912$

71. (a) $21 \cdot 5 \cdot 21 \cdot 21 \cdot 5 \cdot 21 \cdot 10 \cdot 10 = 486{,}202{,}500$ passwords are possible.

(b) $42 \cdot 10 \cdot 42 \cdot 42 \cdot 10 \cdot 42 \cdot 10 \cdot 10 = 420^4 \approx 3.11 \times 10^{10}$ passwords are possible.

5.6 Assess Your Understanding (page 305)

1. A permutation is an arrangement in which order matters; a combination is an arrangement in which order does not matter.

3. AND is generally associated with multiplication; OR is generally associated with addition.

5. $P(\text{E}) = \dfrac{4}{10} = 0.4$

7. *abc, acb, abd, adb, abe, aeb, acd, adc, ace, aec, ade, aed, bac, bca, bad, bda, bae, bea, bcd, bdc, bce, bec, bde, bed, cbe, ceb, cab, cba, cad, cda, cae, cea, cbd, cdb, cde, ced, dab, dba, dac, dca, dae, dea, dbc, dcb, dbe, deb, dce, dec, eab, eba, eac, eca, ead, eda, ebc, ecb, ebd, edb, ecd, edc*

9. $P(E \text{ or } F) = 0.7 + 0.2 - 0.15 = 0.75$

11. $_7P_3 = 210$

13. $P(E \text{ and } F) = P(E) \cdot P(F) = 0.4$

15. $P(E \text{ and } F) = P(E) \cdot P(F \mid E) = 0.27$

17. $P(\text{plays soccer}) = \dfrac{22}{500} = 0.044$

19. There are $3 \cdot 3 \cdot 2 \cdot 2 \cdot 1 \cdot 1 = 36$ ways to arrange the three men and three women.

21. (a) $P(\text{survived}) = \dfrac{711}{2224} = 0.320$

(b) $P(\text{female}) = \dfrac{425}{2224} = 0.191$

(c) $P(\text{female or child}) = \dfrac{534}{2224} = 0.240$

(d) $P(\text{female and survived}) = \dfrac{316}{2224} = 0.142$

(e) $P(\text{female or survived}) = \dfrac{820}{2224} = 0.369$

(f) $P(\text{survived}\,|\,\text{female}) = \dfrac{316}{425} = 0.744$

(g) $P(\text{survived}\,|\,\text{child}) = \dfrac{57}{109} = 0.523$

(h) $P(\text{survived}\,|\,\text{male}) = \dfrac{338}{1690} = 0.2$

(i) Yes, the survival rate was much higher for women and children.

(j) $P(\text{both females survived}) = \dfrac{316}{425} \cdot \dfrac{315}{424} = 0.552$

23. Shawn can fill out the report in $_{12}C_3 = 220$ different ways.

25. (a) $P(\text{win both}) = \dfrac{1}{5.2 \times 10^6} \cdot \dfrac{1}{705,600} = 0.00000000000027$

(b) $P(\text{win Jubilee twice}) = \left(\dfrac{1}{705,600}\right)^2 = 0.000000000002$

27. (a) No $P(\text{Bachelor's}) \neq P(\text{Bachelor's}\,|\,\text{never married})$

(b) $P(\text{Bachelor's and Never Married}) =$
$P(\text{Bachelor's}) \cdot P(\text{Never Married}\,|\,\text{Bachelor's})$
$= (0.202)(0.186) = 0.038$

29. $_{12}C_8 = 495$ different sets of questions can be answered.

31. $2 \cdot 2 \cdot 3 \cdot 8 \cdot 2 = 192$ different cars are possible.

Chapter 5 Review Exercises (page 309)

1. (a) Possible probabilities are $0, 0.75, 0.41$.

(b) Possible probabilities are $\dfrac{2}{5}, \dfrac{1}{3}, \dfrac{6}{7}$.

2. $P(E) = \dfrac{1}{5} = 0.2$ **3.** $P(F) = \dfrac{2}{5} = 0.4$

4. $P(E) = \dfrac{3}{5} = 0.6$ **5.** $P(E^c) = \dfrac{4}{5} = 0.8$

6. $P(E \text{ or } F) = 0.89$ **7.** $P(E \text{ or } F) = 0.48$ **8.** $P(E \text{ and } F) = 0.09$

9. Events E and F are not independent because
$P(E \text{ and } F) \neq P(E) \cdot P(F)$.

10. $P(E \text{ and } F) = 0.2655$ **11.** $P(E|F) = 0.5$

12. (a) $7! = 5040$ **(b)** $0! = 1$
(c) $_9C_4 = 126$ **(d)** $_{10}C_3 = 120$
(e) $_9P_2 = 72$ **(f)** $_{12}P_4 = 11,880$

13. (a) $P(\text{green}) = \dfrac{1}{19} = 0.0526$; If the wheel is spun 100 times, we would expect about 5 spins to end with the ball in a green slot.

(b) $P(\text{green or red}) = \dfrac{10}{19} = 0.5263$; If the wheel is spun 100 times, we would expect about 53 spins to end with the ball in either a green slot or a red slot.

(c) $P(00 \text{ or red}) = \dfrac{19}{38} = 0.5$; If the wheel is spun 100 times, we would expect about 50 spins to end with the ball either in 00 or in a red slot.

(d) $P(31 \text{ and black}) = 0$; this is called an impossible event.

14. (a) $P(\text{fatality was alcohol related}) = \dfrac{10,076}{32,719} = 0.308$

(b) $P(\text{fatality was not alcohol related}) = 1 - 0.308 = 0.692$

(c) $P(\text{both fatalities were alcohol related}) = 0.095$

(d) $P(\text{neither fatality was alcohol related}) = 0.479$

(e) $P(\text{at least one fatality was alcohol related}) = 1 - 0.479 = 0.521$

15. (a)

Age	Probability
18–79	0.121
80–89	0.252
90–99	0.253
100 or older	0.373

(b) It is not unusual for an individual to want to live between 18 and 79 years.

16. (a) $P(\text{baby was postterm}) = 0.057$
(b) $P(\text{baby weighed } 3000\text{–}3999 \text{ grams}) = 0.658$
(c) $P(\text{baby weighed } 3000\text{–}3999 \text{ grams and was postterm}) = 0.041$
(d) $P(\text{baby weighed } 3000\text{–}3999 \text{ grams or was postterm}) = 0.674$
(e) $P(\text{baby weighed } {<}1000 \text{ grams and was postterm}) = 0.000007$; this event is not impossible.
(f) $P(\text{baby weighed } 3000\text{–}3999 \text{ grams}\,|\,\text{baby was postterm}) = 0.713$
(g) The events "postterm baby" and "weighs 3000–3999 grams" are not independent.

17. (a) $P(\text{trusts}) = 0.18$ **(b)** $P(\text{does not trust}) = 0.82$
(c) Yes, $P(\text{all three trust}) = 0.006$
(d) $P(\text{at least one of three does not trust}) = 0.994$
(e) No, $P(\text{all five do not trust}) = 0.371$
(f) $P(\text{at least one of five trusts}) = 0.629$

18. $P(\text{matching the three winning numbers}) = 0.001$

19. $P(\text{matching the four winning numbers}) = 0.0001$

20. $P(\text{drawing three aces}) = 0.00018$

21. $26^2 \cdot 10^4 = 6,760,000$ different license plates can be formed.

22. The students can be seated in $_{10}P_4 = 5040$ different arrangements.

23. There are $\dfrac{10!}{4! \cdot 3! \cdot 2!} = 12,600$ different vertical arrangements of the flags.

24. There are $_{55}C_8 = 1,217,566,350$ possible simple random samples.

25. $P(\text{winning Pick 5}) = \dfrac{1}{_{35}C_5} = \dfrac{1}{324,632} = 0.000003$

26. (a) $P(\text{three Merlot}) = 0.0455$
(b) $P(\text{two Merlot}) = 0.3182$
(c) $P(\text{no Merlot}) = 0.1591$

27. Answers will vary, but should be reasonably close to $\dfrac{1}{38}$ for part (a) and $\dfrac{1}{19}$ for part (b).

28. Answers will vary. Subjective probability is based on personal experience or intuition (e.g., "There is a 70% chance that the Packers will make it to the NFL playoffs next season.")

29. (a) There are 13 clubs in the deck.
(b) Thirty-seven cards remain in the deck. Forty-one cards are not known by you. Eight of the unknown cards are clubs.
(c) $P(\text{next card dealt is a club}) = \dfrac{8}{41}$
(d) $P(\text{two clubs in a row}) = 0.0341$ **(e)** No

30. (a) $P(\text{home run to left field}) = \dfrac{34}{70} = 0.486$; the probability that a randomly selected home run by McGwire went to left field is 0.486.
(b) $P(\text{home run to right field}) = 0$
(c) No; it was not impossible for McGwire to hit a home run to right field. He just never did it.

31. Someone winning a lottery twice is not that unlikely considering millions of people play lotteries who have already won a lottery (sometimes more than one) each week, and many lotteries have multiple drawings each week.

32. (a) $P(\text{Bryce}) = \dfrac{119}{1009} = 0.118$; not unusual

(b) $P(\text{Gourmet}) = \dfrac{264}{1009} = 0.262$

(c) $P(\text{Mallory} | \text{Single Cup}) = \dfrac{3}{25} = 0.12$

(d) $P(\text{Bryce} | \text{Gourmet}) = \dfrac{3}{88} = 0.034$; yes, this is unusual.

(e) While it is not unusual for Bryce to sell a case, it is unusual for him to sell a Gourmet case.

(f) No; $P(\text{Mallory}) = \dfrac{186}{1009} \neq P(\text{Mallory} | \text{Filter}) = \dfrac{1}{3}$.

(g) No; $P(\text{Paige and Gourmet}) = \dfrac{42}{1009} \neq 0$.

Chapter 5   Test (page 311)

1. Possible probabilities are $0.23, 0, \dfrac{3}{4}$.

2. $P(\text{Jason}) = \dfrac{1}{5}$ **3.** $P(\text{Chris or Elaine}) = \dfrac{2}{5}$

4. $P(E^c) = \dfrac{4}{5}$

5. (a) $P(E \text{ or } F) = P(E) + P(F) = 0.59$
(b) $P(E \text{ and } F) = P(E) \cdot P(F) = 0.0814$

6. (a) $P(E \text{ and } F) = P(E) \cdot P(F | E) = 0.105$
(b) $P(E \text{ or } F) = P(E) + P(F) - P(E \text{ and } F) = 0.495$

(c) $P(E | F) = \dfrac{P(E \text{ and } F)}{P(F)} = \dfrac{7}{30} = 0.233$

(d) No; $P(E) = 0.15 \neq P(E | F) = 0.233$

7. (a) $8! = 40,320$ **(b)** $_{12}C_6 = 924$ **(c)** $_{14}P_8 = 121,080,960$

8. (a) $P(7 \text{ or } 11) = \dfrac{8}{36} = \dfrac{2}{9} = 0.222$; If the die are thrown 100 times, we would expect the player will win on the first roll about 22 times.

(b) $P(2, 3, \text{ or } 12) = \dfrac{4}{36} = \dfrac{1}{9} = 0.111$; If the die are thrown 100 times, we would expect that the player will lose on the first roll about 11 times.

9. (a) $P(\text{healthy}) = 0.26$; If we randomly selected 100 adult Americans, we would expect 26 to believe his/her diet is healthy.
(b) $P(\text{not healthy}) = 0.74$

10. (a) All the probabilities are greater than or equal to zero and less than or equal to one, and the sum of the probabilities is 1.
(b) $P(\text{a peanut butter cookie}) = 0.24$
(c) $P(\text{next box sold is mint, caramel, or shortbread}) = 0.53$
(d) $P(\text{next box sold is not mint}) = 0.75$

11. (a) $P(\text{ideal is 2}) = \dfrac{155}{297} = 0.522$

(b) $P(\text{female and ideal is 2}) = \dfrac{87}{297} = 0.293$

(c) $P(\text{female or ideal is 2}) = \dfrac{256}{297} = 0.862$

(d) $P(\text{ideal is 2} | \text{female}) = \dfrac{87}{188} = 0.463$

(e) $P(\text{male} | \text{ideal is 4}) = \dfrac{8}{36} = 0.222$

12. (a) $P(\text{win two in a row}) = 0.336$
(b) $P(\text{win seven in a row}) = 0.022$
(c) $P(\text{lose at least one of next seven}) =$
$1 - P(\text{win seven in a row}) = 0.978$

13. $P(\text{shipment accepted}) = 0.8$

14. There are $6! = 720$ different arrangements of the letters LINCEY.

15. $_{29}C_5 = 118,755$ different subcommittees can be formed.

16. $P(\text{winning Cash 5}) = \dfrac{1}{_{43}C_5} = \dfrac{1}{962,598} = 0.00000104$

17. $26 \cdot 36^7 \approx 2.04 \times 10^{12}$ different passwords are possible.

18. Subjective probability was used. It is not possible to repeat probability experiments to estimate the probability of life on Mars.

19. $\dfrac{15!}{2! \cdot 4! \cdot 4! \cdot 5!} = 9,459,450$ different sequences are possible.

20. $P(\text{guessing correctly}) = \dfrac{9}{40} = 0.225$

Chapter 6   Discrete Probability Distributions

6.1 Assess Your Understanding (page 324)

1. A random variable is a numerical measure of the outcome of a probability experiment.

3. Each probability must be between 0 and 1, inclusive, and the sum of the probabilities must equal 1.

5. (a) Discrete, $x = 0, 1, 2, \ldots, 20$ **(b)** Continuous, $t > 0$
(c) Discrete, $x = 0, 1, 2, \ldots$ **(d)** Continuous, $s \geq 0$

7. (a) Continuous, $r \geq 0$ **(b)** Discrete, $x = 0, 1, 2, 3, \ldots$
(c) Discrete, $x = 0, 1, 2, 3, \ldots$ **(d)** Continuous, $t > 0$

9. Yes, it is a probability distribution.

11. No, $P(50) < 0$.   **13.** No, $\Sigma P(x) \neq 1$ **15.** $P(4) = 0.3$

17. (a) Each probability is between 0 and 1, inclusive, and the sum of the probabilities equals 1.

(b)

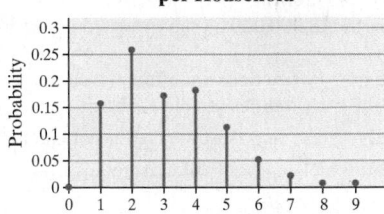

Number of Televisions per Household

The distribution is skewed right.

(c) $\mu_X = 3.2$ televisions; if we surveyed many households, we would expect the mean number of televisions per household to be 3.2.
(d) $\sigma_X = 1.7$ televisions   **(e)** $P(3) = 0.176$
(f) $P(3 \text{ or } 4) = 0.362$
(g) $P(0) = 0$; no, this is not an impossible event.

19. (a) Each probability is between 0 and 1, inclusive, and the sum of the probabilities equals 1.

(b)

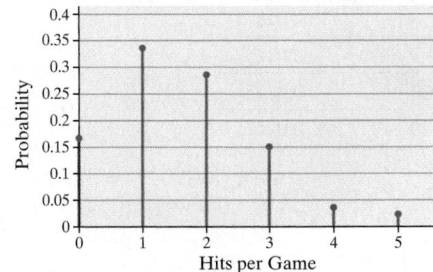

Ichiro's Hit Parade

The distribution is skewed right.

(c) $\mu_X = 1.6$ hits; over many games, Ichiro is expected to average about 1.6 hits per game.
(d) $\sigma_X = 1.2$ hits   **(e)** $P(2) = 0.2857$
(f) $P(X > 1) = 0.4969$

21. (a)

x (games played)	P(x)
4	0.1978
5	0.1978
6	0.2198
7	0.3846

(b)

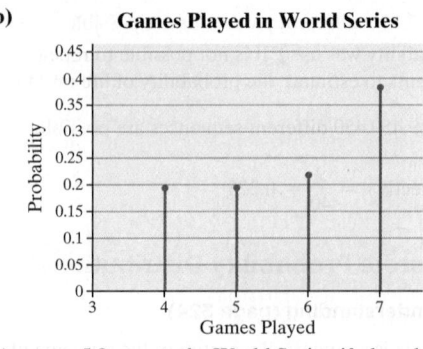

Games Played in World Series

(c) $\mu_X = 5.8$ games; the World Series, if played many times, would be expected to last about 5.8 games, on average.
(d) $\sigma_X = 1.2$ games

23. (a) $P(4) = 0.096$ **(b)** $P(4 \text{ or } 5) = 0.143$
(c) $P(X \geq 6) = 0.119$
(d) $\mu_X = 2.7$ children; we would expect the mother to have had 2.7 children, on average.

25. $E(X) = \$86.00$; the insurance company expects to make an average profit of $86.00 on every 20-year-old female it insures for one year.

27. (a) $-\$0.05$ **(b)** Lose $6

29. $E(X) = -\$0.26$; if you played 1000 times, you would expect to lose about $260.

31. (a) The expected cash prize is $0.30. After paying $1.00 to play, your expected profit is $-\$0.70$.
(b) A grand prize of $118,000,000 has an expected profit greater than zero.
(c) The size of the grand prize does not affect the chance of winning provided the probabilities remain constant.

33. Answers will vary. The simulations illustrate the Law of Large Numbers.

35. (a) 3.3 credit cards   **(b)** 2.3 credit cards
(c)

x (credit cards)	P(x)
1	0.14375
2	0.275
3	0.26875
4	0.1125
5	0.08125
6	0.04375
7	0.025
8	0.01875
9	0.0125
10	0.0125
20	0.00625

(d)

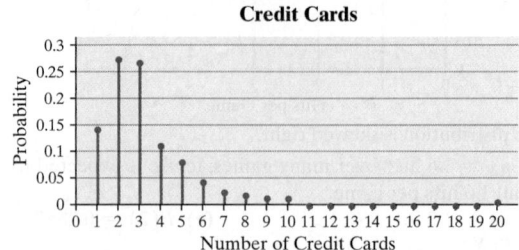

Credit Cards

The distribution is skewed to the right.
(e) $\mu_X = 3.3$; $\sigma_X = 2.3$
(f) 0.05; this result is a little unusual.
(g) $P(\text{two with exactly two credit cards}) = 0.0744$. If we surveyed two individuals 100 different times, we would expect about seven of the surveys to result in two people with two credit cards.

6.2 Assess Your Understanding (page 339)

1. trial   **3.** True   **5.** np

7. Not binomial, because the random variable is continuous
9. Binomial
11. Not binomial, because the trials are not independent
13. Not binomial, because the number of trials is not fixed
15. Binomial
17. $P(3) = 0.2150$ **19.** $P(38) = 0.0532$ **21.** $P(3) = 0.2786$
23. $P(X \leq 3) = 0.9144$ **25.** $P(X > 3) = 0.5$
27. $P(X \leq 4) = 0.5833$
29. (a)

x	P(x)	x	P(x)
0	0.1176	4	0.0595
1	0.3025	5	0.0102
2	0.3241	6	0.0007
3	0.1852		

(b) $\mu_X = 1.8$; $\sigma_X = 1.1$ **(c)** $\mu_X = 1.8$; $\sigma_X = 1.1$
(d)

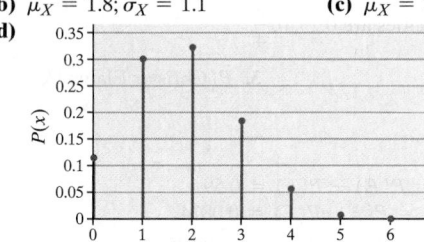

The distribution is skewed right.

31. (a)

x	P(x)	x	P(x)
0	0.0000	5	0.1168
1	0.0001	6	0.2336
2	0.0012	7	0.3003
3	0.0087	8	0.2253
4	0.0389	9	0.0751

(b) $\mu_X = 6.75$; $\sigma_X = 1.3$ **(c)** $\mu_X = 6.75$; $\sigma_X = 1.3$
(d)

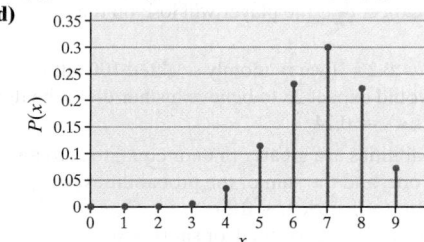

The distribution is skewed left.

33. (a)

x	P(x)	x	P(x)
0	0.0010	6	0.2051
1	0.0098	7	0.1172
2	0.0439	8	0.0439
3	0.1172	9	0.0098
4	0.2051	10	0.0010
5	0.2461		

(b) $\mu_X = 5$; $\sigma_X = 1.6$ **(c)** $\mu_X = 5$; $\sigma_X = 1.6$
(d)

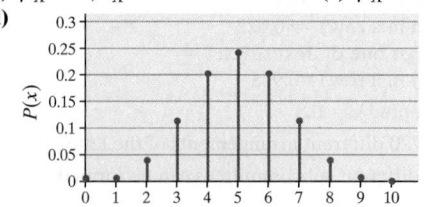

The distribution is symmetric.

35. (a) This is a binomial experiment because:
  1. It is performed a fixed number of times, $n = 15$.
  2. The trials are independent.
  3. For each trial, there are two possible mutually exclusive outcomes: on time and not on time.
  4. The probability of "on time" is fixed at $p = 0.80$.
(b) $P(10) = 0.1032$; in 100 trials of this study, we expect about 10 trials to result in exactly 10 flights being on time.
(c) $P(X < 10) = 0.0611$; in 100 trials of this study, we expect about 6 trials to result in fewer than 10 flights being on time.
(d) $P(X \geq 10) = 0.9389$; in 100 trials of this study, we expect about 94 trials to result in at least 10 flights being on time.
(e) $P(8 \leq X \leq 10) = 0.16$; in 100 trials of this study, we expect about 16 trials to result in between 8 and 10 flights, inclusive, being on time.

37. (a) This is a binomial experiment because:
  1. It is performed a fixed number of times, $n = 20$.
  2. The trials are independent.
  3. For each trial there are two mutually exclusive outcomes: either the individual flushes a public toilet with their foot, or not.
  4. The probability an individual flushes a public toilet with their foot is fixed at $p = 0.64$.
(b) $P(12) = 0.1678$. In 100 different surveys of 20 adult Americans, we would expect 17 of the surveys to result in exactly 12 who flush a public toilet with their foot.
(c) $P(X \geq 16) = 0.1011$. In 100 different surveys of 20 adult Americans, we would expect 10 of the surveys to result in at least 16 who flush a public toilet with their foot.
(d) $P(9 \leq X \leq 11) = 0.2436$. In 100 different surveys of 20 adult Americans, we would expect 24 of the surveys to result in 9 to 11, inclusive, who flush a public toilet with their foot.
(e) Yes, $P(X > 17) = 0.0096$ under the assumption the proportion of adult Americans who flush a public toilet with their foot is 0.64. We would expect these results in only 1 of every 100 surveys.

39. (a) $P(4) = 0.1655$ **(b)** $P(X < 3) = 0.4752$
(c) Yes; the probability is 0.0272.

41. (a) 0.1667 **(b)** $P(X \leq 2) = 0.0421$
(c) The number of Hispanics on the jury is unusually low, given the composition of the population from which it came.

43. (a) $\mu_X = 80$ flights; $\sigma_X = 4$ flights
(b) We expect that, in a random sample of 100 flights from Dallas to Chicago, 80 will be on time.
(c) It would not be unusual to have 75 on-time flights because 75 is within 2 standard deviations of the mean.

45. (a) $E(X) = \mu_X = 320$
(b) Yes, because 280 is more than 2 standard deviations below what we would expect.

47. We would expect 824 out of 1030 parents to spank their children. The results suggest that parents' attitudes have changed since $781 < \mu_X - 2\sigma_X = 824 - 2(12.8) = 798.4$.

49. We would expect $500,000(0.56) = 280,000$ of the stops to be pedestrians who are nonwhite. Because $500,000(0.56)(1 - 0.56) = 123,200 \geq 10$, we can use the Empirical Rule to identify cutoff points for unusual results. The standard deviation number of stops is 351. If the number of stops of nonwhite pedestrians exceeds $280,000 + 2(351) = 280,702$, we would say the result is unusual. The actual number of stops is $500,000(0.89) = 445,000$, which is definitely unusual. A potential criticism of this analysis is the use of 0.44 as the proportion of whites, since the actual proportion of whites may be different due to individuals commuting back and forth to the city.

51. (a) $P(55 \text{ or } 56) = P(55) + P(56) = 0.0203$
(b) $P(\text{bumped}) = P(X \geq 55) = 0.4423$
(c) 266 tickets

53. (a) $P(3) = 0.1187$
(b)

x	$P(x)$	x	$P(x)$
1	0.5240	6	0.0128
2	0.2494	7	0.0061
3	0.1187	8	0.0029
4	0.0565	9	0.0014
5	0.0269	10	0.0007

(c) $\mu_X = 1.9$ free throws
(d) $\mu_X = 1.9$ free throws; 1.9 free throws

55. (a) 2500 **(b)** 0.25; 1250
(c) 0.03125; 156.25 **(d)** 0.00098; 4.9
(e) 1. The experiment is performed a fixed number of times, $n = 5000$.
  2. The trials are independent because the sample size is small relative to the size of the population (assuming there are tens of thousands of investment advisors in the population).
  3. For each trial there are two mutually exclusive outcomes: either the investment advisor beats the market, or does not.
  4. The probability an investment advisor beats the market is $p = 0.5$.
(f) $P(\text{at least 6 beat the market}) = 0.3635$; If we randomly selected five thousand investment advisors 100 different times and the probability any individual advisor beats the market in a single year is 0.5, we would expect 36 of the time for at least 6 advisors to beat the market 10 years in a row. This is not an unusual result and suggests that those advisors who beat the market are not necessarily better than those who do not. That is, we would expect these types of results simply due to randomness.

57. Success means the outcome you are measuring.

59. When n is small, the shape of the binomial distribution is determined by p. If p is close to zero, the distribution is skewed right; if p is close to 0.5, the distribution is approximately symmetric; and if p is close to one, the distribution is skewed left.

6.3 Assess Your Understanding (page 347)

1. To follow a Poisson process, a random variable X must meet the following:
  1. The probability of two or more successes in a sufficiently small subinterval is 0.
  2. The probability of success is the same for any two intervals of equal length.
  3. The number of successes in any interval is independent of the number of successes in any other disjoint interval.

3. $\lambda = 10$ per minute; $t = 5$ minutes

5. $\lambda = 0.07$ per linear foot; $t = 20$ linear feet

7. (a) $P(6) = 0.1462$ **(b)** $P(X < 6) = 0.6160$
(c) $P(X \geq 6) = 0.3840$ **(d)** $P(2 \leq X \leq 4) = 0.4001$

9. (a) $P(4) = 0.0050$ **(b)** $P(X < 4) = 0.9942$
(c) $P(X \geq 4) = 0.0058$ **(d)** $P(4 \leq X \leq 6) = 0.0057$
(e) $\mu_X = 0.7$; $\sigma_X = 0.8$

11. (a) $P(7) = 0.1490$; on about 15 of every 100 days, there will be exactly 7 hits to the website between 7:30 and 7:35 P.M.
(b) $P(X < 7) = 0.4497$; on about 45 of every 100 days, there will be fewer than 7 hits to the website between 7:30 and 7:35 P.M.
(c) $P(X \geq 7) = 0.5503$; on about 55 of every 100 days, there will be at least 7 hits to the website between 7:30 and 7:35 P.M.

13. (a) $P(2) = 0.2510$; in about 25 of every 100 five-gram samples of this supply of peanut butter, we would expect to find 2 insect fragments.
(b) $P(X < 2) = 0.5578$; in about 56 of every 100 five-gram samples of this supply of peanut butter, we would expect to find fewer than 2 insect fragments.
(c) $P(X \geq 2) = 0.4422$; in about 44 of every 100 five-gram samples of this supply of peanut butter, we would expect to find at least 2 insect fragments.

(d) $P(X \geq 1) = 0.7769$; in about 78 of every 100 five-gram samples of this supply of peanut butter, we would expect to find at least 1 insect fragment.
(e) It would not be unusual. About 7 of every 100 five-gram samples of this supply will contain at least 4 insect fragments.

15. (a) $P(0) = 0.9512$; there will be no fatal accidents in approximately 95 of 100 randomly selected 100 million miles of flight.
(b) $P(X \geq 1) = 0.0488$; there will be at least 1 fatal accident in approximately 5 of 100 randomly selected 100 million miles of flight.
(c) $P(X > 1) = 0.0012$; we expect more than 1 fatal accident in approximately 12 of 10,000 randomly selected 100 million miles of flight.

17. (a) $P(0) = 0.7637$
(b) $P(X \geq 1) = 0.2363$; it is not unusual because the probability is greater than 0.05.
(c) $P(X \geq 3) = 0.0027$; this is unusual because the probability is less than 0.05.
(d) $P(X \geq 2) = 0.0304$; this is unusual because the probability is less than 0.05.

19. (a), (b)

Number of Cars, x	$P(x)$	Expected Number of Restaurants	Number of Cars, x	$P(x)$	Expected Number of Restaurants
0	0.0025	0.50	9	0.0688	13.77
1	0.0149	2.97	10	0.0413	8.26
2	0.0446	8.92	11	0.0225	4.51
3	0.0892	17.85	12	0.0113	2.25
4	0.1339	26.77	13	0.0052	1.04
5	0.1606	32.12	14	0.0022	0.45
6	0.1606	32.12	15	0.0009	0.18
7	0.1377	27.54	16	0.0003	0.07
8	0.1033	20.65			

(c) The observed frequencies are lower for fewer cars (x small) and higher for more cars (x large), so the advertising campaign appears to be effective.

21. (a)

Number of Deaths, x	Proportion of Years
0	0.545
1	0.325
2	0.110
3	0.015
4	0.005

(b) $\mu_X = 0.61$
(c)

Number of Deaths, x	$P(x)$
0	0.5434
1	0.3314
2	0.1011
3	0.0206
4	0.0031

(d) Answers will vary.
23. (a) $E(X) = 118$ colds   **(b)–(g)** Answers will vary.

Chapter 6   Review Exercises (page 350)

1. (a) Discrete, $s = 0, 1, 2, \ldots$ **(b)** Continuous, $s \geq 0$
(c) Continuous, $h \geq 0$ **(d)** Discrete, $x = 0, 1, 2, \ldots$
2. (a) It is not a probability distribution because $\sum P(x) \neq 1$.
(b) It is a probability distribution.

3. (a)

x	$P(x)$
4	0.2632
5	0.2368
6	0.2895
7	0.2105

(b)

Stanley Cup Finals

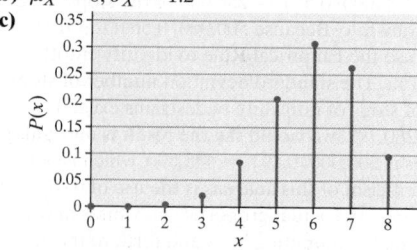

Number of Games

(c) $\mu_X = 5.4$ games; Over the course of many Stanley Cup Finals, the mean number of games required is 5.4.
(d) $\sigma_X = 1.1$ games

4. (a) $E(X) = -\$0.12$ **(b)** $\$16.80$
5. (a) Binomial experiment
(b) Not a binomial experiment because the number of trials is not fixed.

6. (a) $P(1) = 0.3151$; If we randomly selected 10 visitors to the ER 100 different times, we would expect about 32 of those times to result in exactly 1 of 10 visitors dieing within one year.
(b) $P(X < 2) = 0.6424$ **(c)** $P(X \geq 2) = 0.3576$
(d) $P(X \leq 2) = 0.9885$ **(e)** No; $P(X > 3) = 0.0608$
(f) $\mu_X = 50$; $\sigma_X = 6.9$
(g) When $n = 800$, $\mu_X = 40$ and $\sigma_X = 6.2$, so 51 lies within two standard deviations of the mean. The hospital should not be investigated.

7. (a) $P(10) = 0.1859$ **(b)** $P(X < 5) = 0.0093$
(c) $P(X \geq 5) = 0.9907$
(d) $P(7 \leq X \leq 12) = 0.8779$ (tech: 0.8778)
(e) $\mu_X = 120$ women; $\sigma_X = 6.9$ women
(f) It would not be unusual to have 110 females in a sample of 200 believe that the driving age should be 18, because 110 is within 2 standard deviations of the mean.

8. (a)

x	$P(x)$	x	$P(x)$
0	0.00002	5	0.20764
1	0.00037	6	0.31146
2	0.00385	7	0.26697
3	0.02307	8	0.10011
4	0.08652		

(b) $\mu_X = 6$; $\sigma_X = 1.2$
(c)

The distribution is skewed left.

9. As a rule of thumb, if X is binomially distributed, the Empirical Rule can be used when $np(1 - p) \geq 10$.

10. We can sample without replacement and use the binomial probability distribution to approximate probabilities when the sample size is small in relation to the population size. As a rule of thumb, if the sample size is less than 5% of the population size, the trials can be considered nearly independent.

11. To follow a Poisson process, a random variable X must meet the following:

    1. The probability of two or more successes in any sufficiently small subinterval is zero.

    2. The probability of success is the same for any two intervals of equal length.

    3. The number of successes in any interval is independent of the number of successes in any other disjoint interval.

12. **(a)** $P(2) = 0.2510$ **(b)** $P(X < 2) = 0.5578$

    (c) $P(X \geq 2) = 0.4422$ **(d)** $P(1 \leq X \leq 3) = 0.7112$

    (e) $\mu_X = 1.5; \sigma_X = 1.2$

13. 1.75% of the rolls are rejected.

14. It would not be unusual to have more than four maintenance calls; $P(X > 4) = 0.0527$.

15. $P(X \geq 12) = 0.03$, so the result of the survey is unusual. This suggests that emotional abuse may be a factor that increases the likelihood of self-injurious behavior.

Chapter 6 Test (page 352)

1. **(a)** Discrete; $r = 0, 1, 2, \ldots, 365$

    (b) Continuous; $m > 0$ **(c)** Discrete; $x = 0, 1, 2, \ldots$

    (d) Continuous; $w > 0$

2. **(a)** It is a probability distribution.

    (b) It is not a probability distribution because $P(4) = -0.11$, which is negative.

3. **(a)**

x	$P(x)$
3	0.4167
4	0.2917
5	0.2917

    (b)

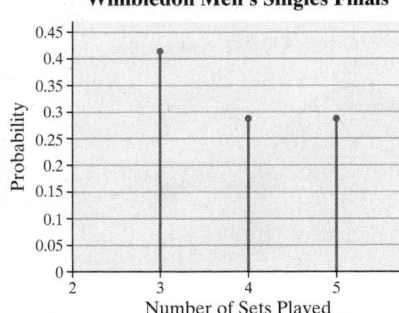

Wimbledon Men's Singles Finals

    (c) $\mu_X = 3.9$ sets. On average, we expect the Wimbledon men's singles final match for the championship to require 3.9 sets of play.

    (d) $\sigma_X = 0.8$ sets

4. $E(X) = 72.50$; the insurance company expects to earn an average profit of $72.50 for each 35-year-old male it insures for one year for $100,000.

5. An experiment is binomial provided:

    1. The experiment consists of a fixed number, n, of trials.

    2. The trials are independent.

    3. Each trial has two possible mutually exclusive outcomes: success and failure.

    4. The probability of success, p, remains constant for each trial of the experiment.

6. **(a)** Not a binomial experiment because the trials are not independent and the number of trials is not fixed

    (b) Binomial experiment.

7. **(a)** $P(15) = 0.1746$; in 100 trials of this experiment, we expect about 17 trials to result in exactly 15 married people who hide purchases from their mates.

    (b) $P(X \geq 19) = 0.0692$; in 100 trials of this experiment, we expect about 7 trials to result in at least 19 married people who hide purchases from their mates.

    (c) $P(X < 19) = 0.9308$; in 100 trials of this experiment, we expect about 93 trials to result in fewer than 19 married people who hide purchases from their mates.

    (d) $P(15 \leq X \leq 17) = 0.5981$; in 100 trials of this experiment, we expect about 60 trials to result in between 15 and 17 married people, inclusive, who hide purchases from their mates.

8. **(a)** We would expect 600 out of 1200 adult Americans to say they pay too much tax.

    (b) We can use the Empirical Rule to identify unusual events since $np(1-p) = 300 \geq 10$.

    (c) Yes; these results do contradict the belief since 640 is more than two standard deviations above the mean.

9. **(a)**

x	$P(x)$
0	0.3277
1	0.4096
2	0.2048
3	0.0512
4	0.0064
5	0.0003

    (b) $\mu_X = 1; \sigma_X = 0.9$

    (c)

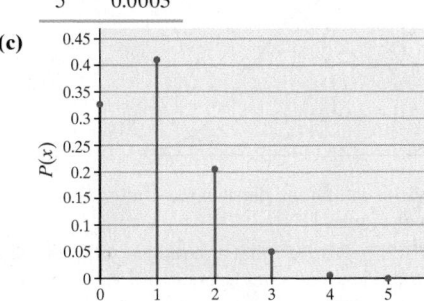

The distribution is skewed right.

10. **(a)** $P(3) = 0.0072$ **(b)** $P(X < 3) = 0.9921$

    (c) $P(X \geq 3) = 0.0079$ **(d)** $P(3 \leq X \leq 5) = 0.0079$

    (e) $\mu_X = 0.4; \sigma_X = 0.6$

11. **(a)** $P(4) = 0.1951$; on about 20 of every 100 randomly chosen days, exactly four cars will arrive at the bank's drive-through window between 4:00 P.M. and 4:10 P.M.

    (b) $P(X < 4) = 0.4142$; on about 41 of every 100 randomly chosen days, fewer than four cars will arrive at the bank's drive-through window between 4:00 P.M. and 4:10 P.M.

    (c) $P(X \geq 4) = 0.5858$; on about 59 of every 100 randomly chosen days, at least four cars will arrive at the bank's drive-through window between 4:00 P.M. and 4:10 P.M.

    (d) $\mu_X = 4.1$ cars   **(e)** $\sigma_X = 2.0$ cars

Chapter 7 The Normal Probability Distribution

7.1 Assess Your Understanding (page 362)

1. probability density function   **3.** True

5. $\mu - \sigma; \mu + \sigma$

7. This graph is not symmetric; it cannot represent a normal density function.

9. This graph is not always greater than zero; it cannot represent a normal density function.

11. The graph can represent a normal density function.

13. **(a)** $P(5 \leq X \leq 10) = \dfrac{1}{6}$ **15.** $P(X \geq 20) = \dfrac{1}{3}$

    (b) 12

17. **(a)**

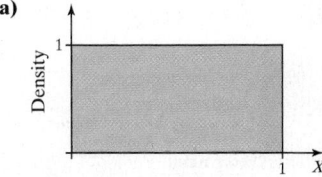

    (b) $P(0 \leq X \leq 0.2) = 0.2$ **(c)** $P(0.25 \leq X \leq 0.6) = 0.35$

    (d) $P(X > 0.95) = 0.05$ **(e)** Answers will vary.

19. Normal   **21.** Not normal

23. Graph A: $\mu = 10$, $\sigma = 2$; graph B: $\mu = 10$, $\sigma = 3$. A larger standard deviation makes the graph lower and more spread out.

25. $\mu = 2$, $\sigma = 3$ **27.** $\mu = 100$, $\sigma = 15$

29.

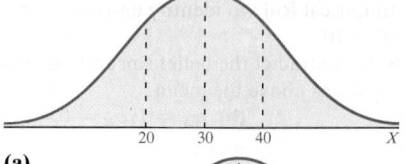

31. (a)

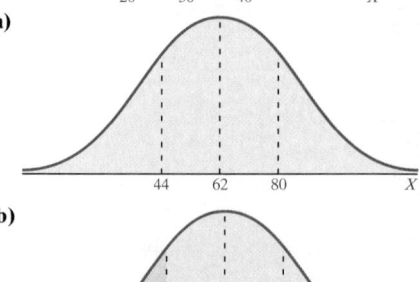

(b)

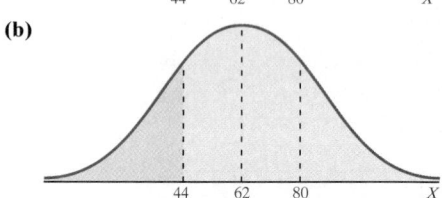

(c) (i) 15.87% of the cell phone plans in the United States are less than $44.00 per month.

(ii) The probability is 0.1587 that a randomly selected cell phone plan in the United States is less than $44.00 per month.

33. (a)

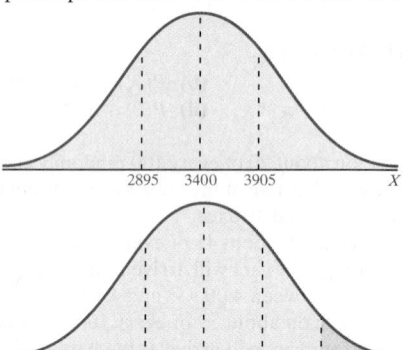

(b)

(c) (i) 2.28% of all full-term babies have a birth weight of more than 4410 grams.
(ii) The probability is 0.0228 that the birth weight of a randomly chosen full-term baby is more than 4410 grams.

35. (a) (i) The proportion of human pregnancies that last more than 280 days is 0.1908.
(ii) The probability that a randomly selected human pregnancy lasts more than 280 days is 0.1908.
(b) (i) The proportion of human pregnancies that last between 230 and 260 days is 0.3416.
(ii) The probability that a randomly selected human pregnancy lasts between 230 and 260 days is 0.3416.

37. (a)

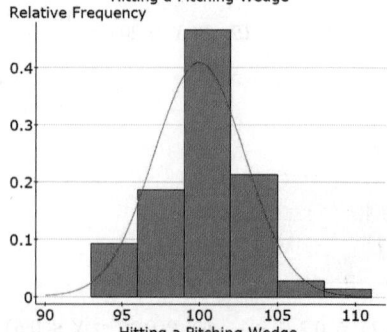

(b) Answers will vary.

39. (a) This is an observational study in which the data are collected over time.
(b) Explanatory variable: hypothermia or not. It is qualitative.
(c) Survival status (qualitative); ICU stay (quantitative); time on ventilator (quantitative)
(d) Statistic
(e) Male patients who have an out-of-hospital cardiac arrest
(f) $P(\text{survive} \mid \text{hypothermia}) = 37/52 = 0.711$; $P(\text{survive} \mid \text{no hypothermia}) = 43/74 = 0.581$

7.2 Assess Your Understanding (page 374)

1. standard normal distribution   **3.** 0.3085

5. (a) Area = 0.0071   **(b)** Area = 0.3336
(c) Area = 0.9115   **(d)** Area = 0.9998

7. (a) Area = 0.9987   **(b)** Area = 0.9441
(c) Area = 0.0375   **(d)** Area = 0.0009

9. (a) Area = 0.9586   **(b)** Area = 0.2088
(c) Area = 0.8479

11. (a) Area = 0.0456 [Tech: 0.0455]   **(b)** Area = 0.0646
(c) Area = 0.5203 [Tech: 0.5202]

13. $z = -1.28$ **15.** $z = 0.67$

17. $z_1 = -2.575$ [Tech: -2.576];
$z_2 = 2.575$ [Tech: 2.576]

19. $z_{0.01} = 2.33$

21. $z_{0.025} = 1.96$

23.

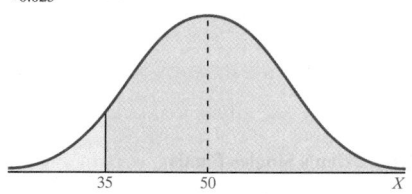

$P(X > 35) = 0.9838$
Tech: $P(X > 35) = 0.9839$

25.

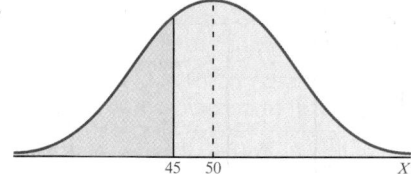

$P(X \leq 45) = 0.2389$
Tech: $P(X \leq 45) = 0.2375$

27.

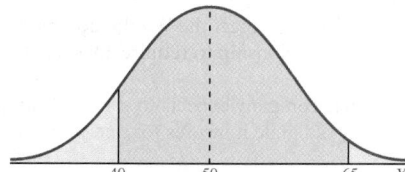

$P(40 < X < 65) = 0.9074$

29.

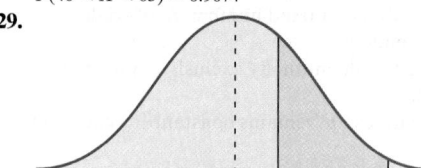

$P(55 \leq X \leq 70) = 0.2368$
Tech: $P(55 \leq X \leq 70) = 0.2354$

31.

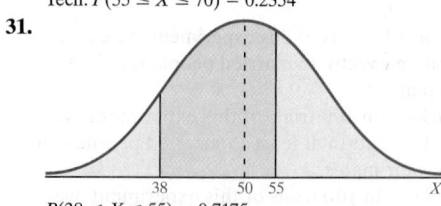

$P(38 < X \leq 55) = 0.7175$
Tech: $P(38 < X \leq 55) = 0.7192$

33. $x = 40.62$ [Tech: 40.61] is at the 9th percentile.

35. $x = 56.16$ [Tech: 56.15] is at the 81st percentile.

37. (a)

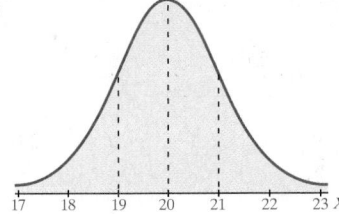

(b) $P(X < 20) = 0.1587$. If 100 eggs are randomly selected, we would expect 16 to incubate in less than 20 days.

(c) $P(X > 22) = 0.1587$. If 100 eggs are randomly selected, we would expect 16 to incubate in more than 22 days.

(d) $P(19 \le X \le 21) = 0.4772$. If 100 eggs are randomly selected, we would expect 48 to incubate in between 19 and 21 days.

(e) Yes; $P(X < 18) = 0.0013$. The model suggests that about 1 egg in 1000 incubates in less than 18 days.

39. (a) $P(1000 \le X \le 1400) = 0.8658$ [Tech: 0.8657]

(b) $P(X < 1000) = 0.0132$

(c) 0.7019 [Tech: 0.7004] of the bags have more than 1200 chocolate chips.

(d) 0.1230 [Tech: 0.1228] of the bags have fewer than 1125 chocolate chips.

(e) A bag that contains 1475 chocolate chips is at the 96th percentile.

(f) A bag that contains 1050 chocolate chips is at the 4th percentile.

41. (a) 0.4013 of pregnancies last more than 270 days.

(b) 0.1587 of pregnancies last fewer than 250 days.

(c) 0.7590 [Tech: 0.7571] of pregnancies last between 240 and 280 days.

(d) $P(X > 280) = 0.1894$ [Tech: 0.1908]

(e) $P(X \le 245) = 0.0951$ [Tech: 0.0947]

(f) Yes; 0.0043 of births are very preterm. So about 4 births in 1000 births are very preterm.

43. (a) 0.0764 [Tech: 0.0766] of the rods have a length of less than 24.9 cm.

(b) 0.0324 [Tech: 0.0321] of the rods will be discarded.

(c) The plant manager expects to discard 162 [Tech: 161] of the 5000 rods manufactured.

(d) To meet the order, the plant manager should manufacture 11,804 [Tech: 11,808] rods.

45. (a) The favored team is equally likely to win or lose relative to the spread. Yes; a mean of 0 implies the spreads are accurate.

(b) $P(X \ge 5) = 0.3228$ [Tech: 0.3232]

(c) $P(X \le -2) = 0.4286$ [Tech: 0.4272]

47. (a) The 17th percentile for incubation time is 20 days.

(b) From 19 to 23 days make up the middle 95% of incubation times of the eggs.

49. (a) The 30th percentile for the number of chips in an 18-ounce bag is 1201 [Tech: 1200] chips.

(b) The middle 99% of the bags contain between 958 and 1566 chocolate chips.

(c) $Q_1 = 1183$ [Tech: 1182]; $Q_3 = 1341$ [Tech: 1342]; $IQR = 158$ [Tech: 160]

51. (a) 11.51% of customers receive the service for half-price.

(b) So that no more than 3% receives the discount, the guaranteed time limit should be 22 minutes.

53. The area under a normal curve can be interpreted as probability, proportion, or percentile.

55. Reporting the probability as > 0.9999 accurately describes the event as highly likely, but not certain. Reporting the probability as 1.0000 might be incorrectly interpreted to mean that the event is certain.

7.3 Assess Your Understanding (page 381)

1. normal probability plot

3. (a)

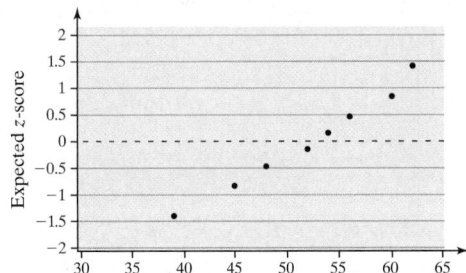

(b) 0.991

(c) Critical value: 0.906; the sample data come from a population that is normally distributed because $0.991 > 0.906$.

5. (a)

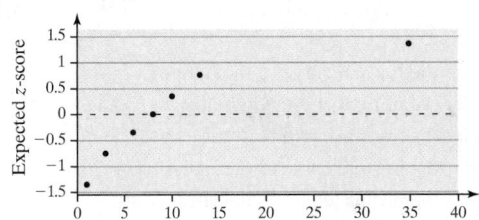

(b) 0.873

(c) Critical value: 0.898; the sample data do not come from a population that is normally distributed because $0.873 < 0.898$.

7. Because $0.979 > 0.951$ the sample data could come from a normally distributed population.

9. Because $0.88 < 0.951$ the sample data do not come from a normally distributed population.

11. (a) The sample data come from a normally distributed population.

(b) $\bar{x} = 1247.4$ chips, $s = 101.0$ chips

(c)

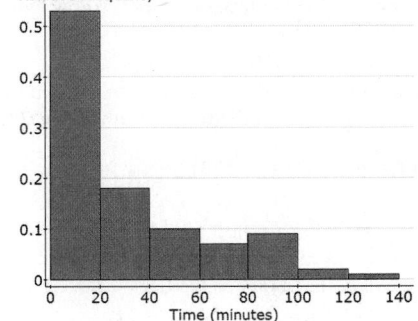

(d) $P(X \ge 1000) = 0.9929$ [Tech: 0.9928]

(e) $P(1200 \le X \le 1400) = 0.6153$ [Tech: 0.6152]

13. (a)

Wait Times for Demon Roller Coaster

(b) Skewed right

(c) Because $0.922 < 0.960$ (Table VI with $n = 30$), the data do not come from a population that is normally distributed.

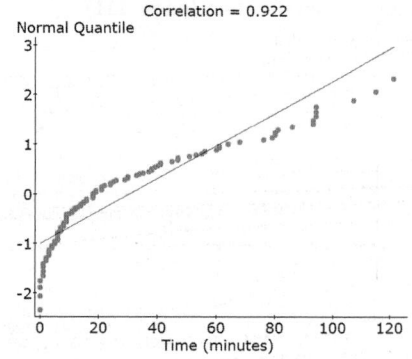

Correlation = 0.922

7.4 Assess Your Understanding (page 386)

1. $np(1 - p) \geq 10; np; \sqrt{np(1 - p)}$

3. $P(X \leq 4.5)$

5. Area under the normal curve to the right of $x = 39.5$

7. Area under the normal curve between $x = 7.5$ and $x = 8.5$

9. Area under the normal curve between $x = 17.5$ and $x = 24.5$

11. Area under the normal curve to the right of $x = 20.5$

13. Area under the normal curve to the left of $x = 500.5$

15. Using the binomial formula, $P(20) = 0.0616; np(1 - p) = 14.4$, the normal distribution can be used. Approximate probability is 0.0618. [Tech: 0.0603]

17. Using the binomial formula, $P(30) < 0.0001; np(1 - p) = 7.5$, the normal distribution cannot be used.

19. Using the binomial formula, $P(60) = 0.0677; np(1 - p) = 14.1$, the normal distribution can be used. Approximate probability is 0.0630. [Tech: 0.0645]

21. (a) $P(130) \approx 0.0444$ [Tech: 0.0431]
(b) $P(X \geq 130) \approx 0.9332$ [Tech: 0.9328]
(c) $P(X < 125) \approx 0.0021$
(d) $P(125 \leq X \leq 135) \approx 0.5536$ [Tech: 0.5520]

23. (a) $P(490) \approx 0.0127$ [Tech: 0.0139]
(b) $P(X \leq 490) \approx 0.9015$ [Tech: 0.9022]
(c) $P(X \geq 503) \approx 0.0136$ [Tech: 0.0134]; the result suggests the proportion of adult women 18–24 years of age is higher than 0.64.

25. (a) $P(X \geq 130) \approx 0.0028$
(b) The result is unusual. Less than 3 samples in 1000 will result in 130 or more living at home if the true percentage is 55%. Perhaps a higher percentage of males are now living at home.

27. (a) $P(X \geq 75) \approx 0.0007$
(b) Yes; less than 1 sample in 1000 will result in 75 or more respondents preferring a male if the true percentage is 37%.

Chapter 7 Review Exercises (page 388)

1. (a) $\mu = 60$ (b) $\sigma = 10$
(c) (i) The proportion of values for the random variable to the right of $x = 75$ is 0.0668.
(ii) The probability that a randomly selected value is greater than $x = 75$ is 0.0668.
(d) (i) The proportion of values for the random variable between $x = 50$ and $x = 75$ is 0.7745.
(ii) The probability that a randomly selected value is between $x = 50$ and $x = 75$ is 0.7745.

2.

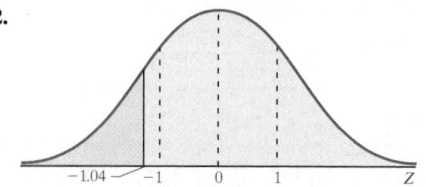

Area under the normal curve to the left of -1.04 is 0.1492.

3.

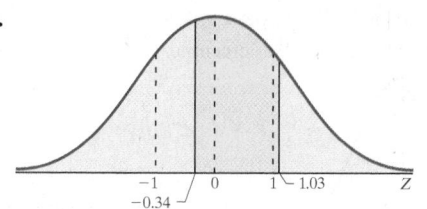

Area under the normal curve between -0.34 and 1.03 is 0.4816.

4. $z = 0.04$ **5.** $z_1 = -1.75$ and $z_2 = 1.75$

6. $z_{0.20} = 0.84$

7.

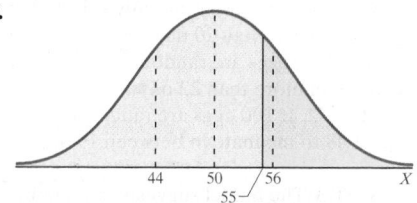

$P(X > 55) = 0.2033$ [Tech: 0.2023]

8.

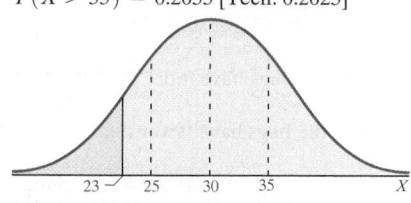

$P(X \leq 23) = 0.0808$

9.

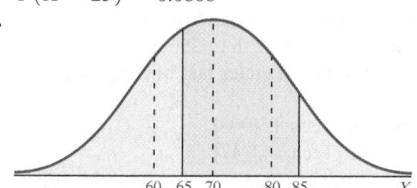

$P(65 < X < 85) = 0.6247$

10. (a) 0.1271 [Tech: 0.1279] of the tires last at least 75,000 miles.
(b) 0.0116 [Tech: 0.0115] of the tires last at most 60,000 miles.
(c) $P(65,000 \leq X \leq 80,000) = 0.8613$ [Tech: 0.8606]
(d) The company should advertise 60,980 [Tech: 60,964] miles as its mileage warranty.

11. (a) The probability is 0.0475 [Tech: 0.0478] that the test taker scores above 125.
(b) The probability is 0.2514 [Tech: 0.2525] that the test taker scores below 90.
(c) 0.2476 [Tech: 0.2487] of the test takers score between 110 and 140.
(d) $P(X > 150) = 0.0004$
(e) A score of 131 places a child at the 98th percentile.
(f) Normal children score between 71 and 129 points.

12. (a) 0.0119 [Tech: 0.0120] of the baseballs produced are too heavy for use.
(b) 0.0384 [Tech: 0.0380] of the baseballs produced are too light to use.
(c) 0.9497 [Tech: 0.9500] of the baseballs produced are within acceptable weight limits.
(d) 8424 [Tech: 8421] should be manufactured.

13. (a) $np(1 - p) = 62.1 > 10$, so the normal distribution can be used to approximate the binomial probabilities.
(b) $P(125) \approx 0.0213$ [Tech: 0.0226] Interpretation: Approximately 2 of every 100 random samples of 250 adult Americans will result in exactly 125 who state that they have read at least 6 books within the past year.
(c) $P(X < 120) \approx 0.7157$ [Tech: 0.7160] Interpretation: Approximately 72 of every 100 random samples of 250 adult Americans will result in fewer than 120 who state that they have read at least 6 books within the past year.

(d) $P(X \geq 140) \approx 0.0009$ Interpretation: Approximately 1 of every 1000 random samples of 250 adult Americans will result in 140 or more who state that they have read at least 6 books within the past year.

(e) $P(100 \leq X \leq 120) \approx 0.7336$ [Tech: 0.7328] Interpretation: Approximately 73 of every 100 random samples of 250 adult Americans will result in between 100 and 120, inclusive, who state that they have read at least six books within the past year.

14. (a)

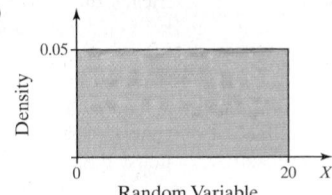

(b) 0.986

(c) Because $0.986 > 0.898$ (Table VI), the sample data could come from a population that is normally distributed.

15. Because $0.914 < 0.951$ (Table VI), the sample data do not come from a population that is normally distributed.

16. Because $0.873 < 0.959$ (Table VI with $n = 25$), the sample data do not come from a population that is normally distributed.

17. (a) $P(X \leq 30) = 0.0010$

(b) Yes; this contradicts the *USA Today* "Snapshot." About 1 sample in 1000 will result in 30 or fewer who do their most creative thinking while driving, if the true percentage is 20%.

18. (a)

(b) $P(0 \leq X \leq 5) = 0.25$

(c) $P(10 \leq X \leq 18) = 0.4$

19. A normal curve has the following properties:

1. It is symmetric about its mean μ.

2. Its highest point occurs at μ.

3. It has inflection points at $\mu - \sigma$ and $\mu + \sigma$.

4. The area under the curve is 1.

5. The area under the curve to the right of μ equals the area under the curve to the left of μ. Both equal $\frac{1}{2}$.

6. As the value of X increases, the graph approaches but never equals zero. When the value of X decreases, the graph approaches but never equals zero.

7. The Empirical Rule: Approximately 68% of the area under the standard normal curve is between $\mu - \sigma$ and $\mu + \sigma$. Approximately 95% of the area under the standard normal curve is between $\mu - 2\sigma$ and $\mu + 2\sigma$. Approximately 99.7% of the area under the standard normal curve is between $\mu - 3\sigma$ and $\mu + 3\sigma$.

20. The graph plots actual observations against expected z-scores, assuming that the data are normal. If the plot is not linear, then we have evidence that the data are not normal.

Chapter 7 Test (page 390)

1. (a) $\mu = 7$ **(b)** $\sigma = 2$

(c) (i) The proportion of values for the random variable to the left of $x = 10$ is 0.9332.

(ii) The probability that a randomly selected value is less than $x = 10$ is 0.9332.

(d) (i) The proportion of values for the random variable between $x = 5$ and $x = 8$ is 0.5328.

(ii) The probability that a randomly selected value is between $x = 5$ and $x = 8$ is 0.5328.

2.

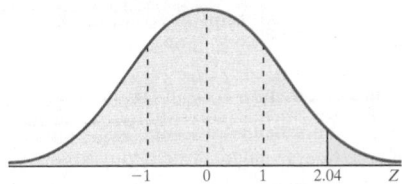

Area under the normal curve to the right of 2.04 is 0.0207.

3. $z_1 = -1.555; z_2 = 1.555$ **4.** $z_{0.04} = 1.75$

5. (a)

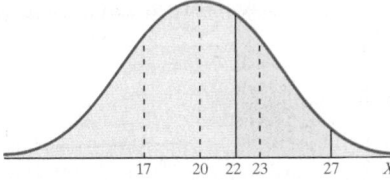

(b) $P(22 \leq X \leq 27) = 0.2415$ [Tech: 0.2427]

6. (a) 0.8944 of the time, the iPhone will last at least 6 hours.

(b) There is a 0.0062 probability that the iPhone will last less than 5 hours. This is an unusual result.

(c) About 8.3 hours would be the cutoff for the top 5% of all talk times.

(d) Yes; $P(X > 9) = 0.0062$. About 6 out of every 1000 full charges will result in the iPhone lasting more than 9 hours.

7. (a) The proportion of 20- to 29-year-old males whose waist circumference is less than 100 cm is 0.7088 [Tech: 0.7080].

(b) $P(80 \leq X \leq 100) = 0.5274$ [Tech: 0.5272]

(c) Waist circumferences between 70 and 115 cm make up the middle 90% of all waist circumferences.

(d) A waist circumference of 75 cm is at the 10th percentile.

8. A: ≥ 76.4; B: 70.7–76.3; C: 57.3–70.6; D: 51.6–57.2; F: <51.6

9. (a) $P(100) \approx 0.0025$ **(b)** $P(X < 60) \approx 0.0062$

10. The data likely come from a population that is normally distributed because $0.974 > 0.939$ (Table VI).

11. (a)

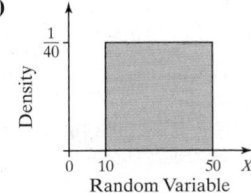

(b) $P(20 \leq X \leq 30) = 0.25$

(c) $P(X < 15) = 0.125$

Chapter 8 Sampling Distributions

8.1 Assess Your Understanding (page 403)

1. sampling distribution   **3.** standard error; mean

5. False

7. The sampling distribution of $\bar{x}$ is normal with $\mu_{\bar{x}} = 30$ and $\sigma_{\bar{x}} = \dfrac{8}{\sqrt{10}} \approx 2.530$.

9. $\mu_{\bar{x}} = 80, \sigma_{\bar{x}} = 2$

11. $\mu_{\bar{x}} = 52, \sigma_{\bar{x}} = \dfrac{10}{\sqrt{21}} \approx 2.182$

13. (a) $\mu_{\bar{x}} = 500$ **(b)** $\sigma_{\bar{x}} = 20$

(c) The population must be normally distributed.

(d) $\sigma = 80$

15. (a) $\bar{x}$ is approximately normal with $\mu_{\bar{x}} = 80, \sigma_{\bar{x}} = 2$.

(b) $P(\bar{x} > 83) = 0.0668$. If we take 100 simple random samples of size $n = 49$ from a population with $\mu = 80$ and $\sigma = 14$, then about 7 of the samples will result in a mean that is greater than 83.

(c) $P(\bar{x} \leq 75.8) = 0.0179$. If we take 100 simple random samples of size $n = 49$ from a population with $\mu = 80$ and $\sigma = 14$, then about 2 of the samples will result in a mean that is less than or equal to 75.8.

(d) $P(78.3 < \bar{x} < 85.1) = 0.7969$ [Tech: 0.7970]. If we take 100 simple random samples of size $n = 49$ from a population with $\mu = 80$ and $\sigma = 14$, then about 80 of the samples will result in a mean that is between 78.3 and 85.1.

17. (a) The population must be normally distributed to compute probabilities involving the sample mean. If the population is normally distributed, then the sampling distribution of $\bar{x}$ is also normally distributed with $\mu_{\bar{x}} = 64$ and $\sigma_{\bar{x}} = \dfrac{17}{\sqrt{12}} \approx 4.907$.

(b) $P(\bar{x} < 67.3) = 0.7486$ [Tech: 0.7493]. If we take 100 simple random samples of size $n = 12$ from a population that is normally distributed with $\mu = 64$ and $\sigma = 17$, then about 75 of the samples will result in a mean that is less than 67.3.

(c) $P(\bar{x} \geq 65.2) = 0.4052$ [Tech: 0.4034]. If we take 100 simple random samples of size $n = 12$ from a population that is normally distributed with $\mu = 64$ and $\sigma = 17$, then about 40 or 41 of the samples will result in a mean that is greater than or equal to 65.2.

19. (a) $P(X < 260) = 0.3520$ [Tech: 0.3538]. If we randomly select 100 human pregnancies, then about 35 of the pregnancies will last less than 260 days.

(b) The sampling distribution of $\bar{x}$ is normal with $\mu_{\bar{x}} = 266$ and $\sigma_{\bar{x}} = \dfrac{16}{\sqrt{20}} \approx 3.578$.

(c) $P(\bar{x} \leq 260) = 0.0465$ [Tech: 0.0468]. If we take 100 simple random samples of size $n = 20$ human pregnancies, then about 5 of the samples will result in a mean gestation period of 260 days or less.

(d) $P(\bar{x} \leq 260) = 0.0040$. If we take 1000 simple random samples of size $n = 50$ human pregnancies, then about 4 of the samples will result in a mean gestation period of 260 days or less.

(e) This result would be unusual, so the sample likely came from a population whose mean gestation period is less than 266 days.

(f) $P(256 \leq \bar{x} \leq 276) = 0.9844$ [Tech: 0.9845]. If we take 100 simple random samples of size $n = 15$ human pregnancies, then about 98 of the samples will result in a mean gestation period between 256 and 276 days, inclusive.

21. (a) $P(X > 95) = 0.3085$. If we select a simple random sample of $n = 100$ second grade students, then about 31 of the students will read more than 95 words per minute.

(b) $P(\bar{x} > 95) = 0.0418$ [Tech: 0.0416]. If we take 100 simple random samples of size $n = 12$ second grade students, then about 4 of the samples will result in a mean reading rate that is more than 95 words per minute.

(c) $P(\bar{x} > 95) = 0.0071$ [Tech: 0.0072]. If we take 1000 simple random samples of size $n = 24$ second grade students, then about 7 of the samples will result in a mean reading rate that is more than 95 words per minute.

(d) Increasing the sample size decreases $P(\bar{x} > 95)$. This happens because $\sigma_{\bar{x}}$ decreases as n increases.

(e) A mean reading rate of 92.8 wpm is not unusual since $P(\bar{x} \geq 92.8) = 0.1056$ [Tech: 0.1052]. This means that the new reading program is not abundantly more effective than the old program.

(f) There is a 5% chance that the mean reading speed of a random sample of 20 second-grade students will exceed 93.7 words per minute.

23. (a) $P(X > 0) = 0.5675$ [Tech: 0.5694]. If we select a simple random sample of $n = 100$ months, then about 57 of the months will have positive rates of return.

(b) $P(\bar{x} > 0) = 0.7291$ [Tech: 0.7277]. If we take 100 simple random samples of size $n = 12$ months, then about 73 of the samples will result in a mean monthly rate of return that is positive.

(c) $P(\bar{x} > 0) = 0.8051$ [Tech: 0.8043]. If we take 100 simple random samples of size $n = 24$ months, then about 81 of the samples will result in a mean monthly rate of return that is positive.

(d) $P(\bar{x} > 0) = 0.8531$ [Tech: 0.8530]. If we take 100 simple random samples of size $n = 36$ months, then about 85 of the samples will result in a mean monthly rate of return that is positive.

(e) The likelihood of earning a positive rate of return increases as the investment time horizon increases.

25. (a) A sample size of at least 30 is needed to compute the probabilities.

(b) $P(\bar{x} < 10) = 0.0028$. If we take 1000 simple random samples of size $n = 40$ oil changes, then about 3 of the samples will result in a mean time of less than 10 minutes.

(c) There is a 10% chance of being at or below a mean oil-change time of 10.8 minutes.

27. (a) Since we have a large sample ($n = 50 \geq 30$) the Central Limit Theorem allows us to say that the sampling distribution of the mean is approximately normal.

(b) $\mu_{\bar{x}} = 3$, $\sigma_{\bar{x}} = \dfrac{\sqrt{3}}{\sqrt{50}} = \sqrt{\dfrac{3}{50}} \approx 0.245$

(c) $P(\bar{x} \geq 3.6) = 0.0071$ [Tech: 0.0072]. If we take 1000 simple random samples of size $n = 50$ ten-gram portions of peanut butter, then about 7 of the samples will result in a mean of at least 3.6 insect fragments. This result is unusual. We might conclude that the sample comes from a population with a mean higher than 3 insect fragments per ten-gram portion.

29. (a) No; the variable "weekly time spent watching television" is likely skewed right.

(b) $\bar{x}$ is approximately normal with $\mu_{\bar{x}} = 2.35$ and $\sigma_{\bar{x}} = \dfrac{1.93}{\sqrt{40}} \approx 0.305$.

(c) $P(2 \leq \bar{x} \leq 3) = 0.8583$ [Tech: 0.8577]. If we take 100 simple random samples of size $n = 40$ adult Americans, then about 86 of the samples will result in a mean time between 2 and 3 hours watching television on a weekday.

(d) $P(\bar{x} \leq 1.89) = 0.0793$. If we take 100 simple random samples of size $n = 35$ adult Americans, then about 8 of the samples will result in a mean time of 1.89 hours or less watching television on a weekday. This result is not unusual, so this evidence is insufficient to conclude that avid Internet users watch less television.

31. (a) $\mu = 46.3$ years old

(b) Samples: $(37, 38)$; $(37, 45)$; $(37, 50)$; $(37, 48)$; $(37, 60)$; $(38, 45)$; $(38, 50)$; $(38, 48)$; $(38, 60)$; $(45, 50)$; $(45, 48)$; $(45, 60)$; $(50, 48)$; $(50, 60)$; $(48, 60)$

(c)

$\bar{x}$	Probability	$\bar{x}$	Probability
37.5	$\dfrac{1}{15}$	46.5	$\dfrac{1}{15}$
41	$\dfrac{1}{15}$	47.5	$\dfrac{1}{15}$
41.5	$\dfrac{1}{15}$	48.5	$\dfrac{1}{15}$
42.5	$\dfrac{1}{15}$	49	$\dfrac{2}{15}$
43	$\dfrac{1}{15}$	52.5	$\dfrac{1}{15}$
43.5	$\dfrac{1}{15}$	54	$\dfrac{1}{15}$
44	$\dfrac{1}{15}$	55	$\dfrac{1}{15}$

(d) $\mu_{\bar{x}} = 46.3$ years old

(e) $P(43.3 \leq \bar{x} \leq 49.3) = \dfrac{7}{15} = 0.467$

(f)

Samples		Sampling Distribution	
		$\bar{x}$	Probability
37, 38, 45	38, 45, 50	40	0.05
37, 38, 50	38, 45, 48	41	0.05
37, 38, 48	38, 45, 60	41.7	0.05
37, 38, 60	38, 50, 48	43.3	0.05
37, 45, 50	38, 50, 60	43.7	0.05
37, 45, 48	38, 48, 60	44	0.05
37, 45, 60	45, 50, 48	44.3	0.05
37, 50, 48	45, 50, 60	45	0.10
37, 50, 60	45, 48, 60	45.3	0.05
37, 48, 60	50, 48, 60	47.3	0.05
		47.7	0.10
		48.3	0.05
		48.7	0.05
		49	0.05
		49.3	0.05
		51	0.05
		51.7	0.05
		52.7	0.05

$\mu_{\bar{x}} = 46.3$ years old;
$P(43.3 \leq \bar{x} \leq 49.3) = 0.7$

As the sample size increases, the probability of obtaining a sample mean age within three years of the population mean age increases.

33. (a)

x	$P(x)$
35	0.0263
−1	0.9737

(b) $\mu = -\$0.05, \sigma = \$5.76$
(c) $\bar{x}$ is approximately normal with $\mu_{\bar{x}} = -\$0.05$ and

$\sigma_{\bar{x}} = \dfrac{5.76}{\sqrt{100}} = \$0.576.$

(d) $P(\bar{x} > 0) = 0.4641$ [Tech: 0.4654]
(e) $P(\bar{x} > 0) = 0.4522$ [Tech: 0.4511]
(f) $P(\bar{x} > 0) = 0.3936$ [Tech: 0.3918]
(g) The probability of being ahead decreases as the number of games played increases.

35. The Central Limit Theorem states that, regardless of the distribution of the population, the sampling distribution of the sample means becomes approximately normal as the sample size, n, increases.

37. (b); The distribution for the larger sample size has the smaller standard deviation.

39. (a) We would expect that Jack's distribution would be skewed left but not as much as the original distribution. Diane's distribution should be bell shaped and symmetric, that is, approximately normal.
(b) We would expect both distributions to have a mean of 50.
(c) We expect Jack's distribution to have standard deviation $\dfrac{10}{\sqrt{3}} \approx 5.8$. We expect Diane's distribution to have standard

deviation $\dfrac{10}{\sqrt{30}} \approx 1.8$.

41. (a) The population of interest is full-time students at your college. The sample is the ten students in the simple random sample.
(b) The number of hours of sleep is a random variable because different people sleep different amounts.
(c) Because the mean is based on a sample of ten randomly selected students, the mean is a sample mean, and therefore, a statistic.

(d) The sample mean from part (c) is ... its value will change depending on the ... sample. In this study, there are two sour... is person-to-person variability since diff... sleeping habits. Second, there is variabili... individual day-to-day (this variability was ... to Problem 40(d)).

8.2 Assess Your Understanding (page 413)

1. 0.44

3. False

5. The sampling distribution of $\hat{p}$ is approximately normal when $n \leq 0.05N$ and $np(1-p) \geq 10$.

7. The sampling distribution of $\hat{p}$ is approximately normal with

$\mu_{\hat{p}} = 0.4$ and $\sigma_{\hat{p}} = \sqrt{\dfrac{0.4(0.6)}{500}} \approx 0.022.$

9. The sampling distribution of $\hat{p}$ is approximately normal with

$\mu_{\hat{p}} = 0.103$ and $\sigma_{\hat{p}} = \sqrt{\dfrac{0.103(0.897)}{1000}} \approx 0.010.$

11. (a) The sampling distribution of $\hat{p}$ is approximately normal with

$\mu_{\hat{p}} = 0.8$ and $\sigma_{\hat{p}} = \sqrt{\dfrac{0.8(0.2)}{75}} \approx 0.046.$

(b) $P(\hat{p} \geq 0.84) = 0.1922$ [Tech: 0.1932]. About 19 out of 100 random samples of size $n = 75$ will result in 63 or more individuals (that is, 84% or more) with the characteristic.
(c) $P(\hat{p} \leq 0.68) = 0.0047$. About 5 out of 1000 random samples of size $n = 75$ will result in 51 or fewer individuals (that is, 68% or less) with the characteristic.

13. (a) The sampling distribution of $\hat{p}$ is approximately normal with

$\mu_{\hat{p}} = 0.35$ and $\sigma_{\hat{p}} = \sqrt{\dfrac{0.35(0.65)}{1000}} \approx 0.015.$

(b) $P(\hat{p} \geq 0.39) = 0.0040$. About 4 out of 1000 random samples of size $n = 1000$ will result in 390 or more individuals (that is, 39% or more) with the characteristic.
(c) $P(\hat{p} \leq 0.32) = 0.0233$ [Tech: 0.0234]. About 2 out of 100 random samples of size $n = 1000$ will result in 320 or fewer individuals (that is, 32% or less) with the characteristic.

15. (a) Qualitative with two possible outcomes-order a meal in a foreign language, or not.
(b) The source of the variability is the individuals in the survey and their ability to order a meal in a foreign language.
(c) The sampling distribution of $\hat{p}$ is approximately normal with

$\mu_{\hat{p}} = 0.47$ and $\sigma_{\hat{p}} = \sqrt{\dfrac{0.47(0.53)}{200}} \approx 0.035.$

(d) $P(\hat{p} > 0.5) = 0.1977$ [Tech: 0.1976]. About 20 out of 100 random samples of size $n = 200$ Americans will result in more than 100 individuals (that is, more than 50%) who can order a meal in a foreign language.
(e) $P(\hat{p} \leq 0.4) = 0.0239$ [Tech: 0.0237]. About 2 out of 100 random samples of size $n = 200$ Americans will result in 80 or fewer individuals (that is, 40% or less) who can order a meal in a foreign language. This result is unusual.

17. (a) The sampling distribution of $\hat{p}$ is approximately normal with

$\mu_{\hat{p}} = 0.39$ and $\sigma_{\hat{p}} = \sqrt{\dfrac{0.39(0.61)}{500}} \approx 0.022.$

(b) $P(\hat{p} < 0.38) = 0.3228$ [Tech: 0.3233]. About 32 out of 100 random samples of size $n = 500$ adult Americans will result in fewer than 190 individuals (that is, less than 38%) who believe that marriage is obsolete.
(c) $P(0.40 < \hat{p} < 0.45) = 0.3198$ [Tech: 0.3203]. About 32 out of 100 random samples of size $n = 500$ adult Americans will result in between 200 and 225 individuals (that is, between 40% and 45%) who believe that marriage is obsolete.
(d) $P(\hat{p} \geq 0.42) = 0.0838$ [Tech: 0.0845]. About 8 out of 100 random samples of size $n = 500$ adult Americans will result in 210 or more individuals (that is, 42% or more) who believe that marriage is obsolete. This result is not unusual.

$1) = P(\hat{p} \geq 0.11) = 0.1335$ [Tech: 0.1345]. This result ~~ual, so this evidence is insufficient to conclude that the ~~ion of Americans who are afraid to fly has increased above 0.10. ~~other words, the results obtained could be due to sampling error and the true proportion is still 0.10.

21. $P(X \geq 164) = P(\hat{p} \geq 0.529) = 0.0853$ [Tech: 0.0846]. When the final election results will show 49% of voters supporting an increase in funding for education, approximately 9 out of 100 random samples of 310 voters will result in 52.9% or more who support the increase. This result is not unusual, so it would not be unusual for a wrong call to be made in an election if exit polling alone was considered. The exit polling could be biased in favor of an increase in funding for education, since a voter who voted against it in the privacy of the voting booth might not want to admit it to a pollster.

23. (a) 62 more adult Americans must be polled to make $np(1-p) \geq 10$.
(b) 13 more adult Americans must be polled if $p = 0.2$.

25. (a) Qualitative with two outcomes—believe in reincarnation, or not.
(b) There is variability in the sample proportion due to sampling variability. Bob likely has different individuals from Alicia in his sample.
(c) Randomly selecting individuals is important so that the results of the survey are representative of the population. This allows the researcher to generalize the results from the sample to the population.
(d) $E(X) = np = 100(0.21) = 21$
(e) The sample size is not large enough for the distribution of the sample proportion to be approximately normal. In fact, $np(1-p) = 20(0.21)(1-0.21) = 3.318 < 10$.
(f) Solve $n(0.21)(1-0.21) > 10$ and find that $n > 60.28$. Therefore, the sample size must be at least 61 to use the normal model to describe the distribution of the sample proportion.

27. (a) $P(X \geq 25) = P\left(\hat{p} \geq \dfrac{25}{290}\right) = P(\hat{p} \geq 0.0862) =$
0.7764 [Tech: 0.7753]
(b) Let X represent the number who show up so that $p = 0.9005$.
$P(X \leq 300) = P\left(\hat{p} \leq \dfrac{300}{320}\right) = P(\hat{p} \leq 0.9375)$
$= 0.9864$ [Tech: 0.9865]
(c) Let X represent the number who do not show up. Here, $p = 0.04$. We need at least 15 to not show up.
$P(X \geq 15) = P\left(\hat{p} \geq \dfrac{15}{300}\right) = P(\hat{p} \geq 0.05)$
$= 0.1894$ [Tech: 0.1884]

29. (a) The distribution is slightly skewed right. There are three outliers (Atlanta Falcons, New Orleans Saints, New England Patriots).

Plays per Fumble

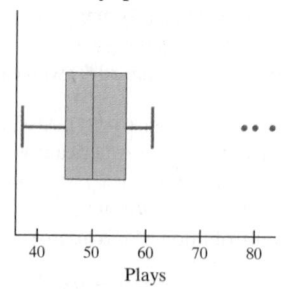

(b) The only outlier is the New England Patriots.

Plays per Fumble

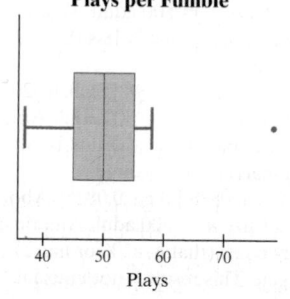

Chapter 8 Review Exercises (page 416)

1. A sampling distribution is a probability distribution for all possible values of a statistic computed from a sample of size n.

2. The sampling distribution of $\bar{x}$ is exactly normal when the underlying population distribution is normal. The sampling distribution of $\bar{x}$ is approximately normal when the sample size is large, usually greater than 30, regardless of how the population is distributed.

3. The sampling distribution of $\hat{p}$ is approximately normal when $np(1-p) \geq 10$, provided that $n \leq 0.05N$.

4. $\mu_{\bar{x}} = \mu$, $\sigma_{\bar{x}} = \dfrac{\sigma}{\sqrt{n}}$, and $\mu_{\hat{p}} = p$, $\sigma_{\hat{p}} = \sqrt{\dfrac{p(1-p)}{n}}$

5. (a) Quantitative
(b) $P(x > 2625) = 0.3085$. If we select a simple random sample of $n = 100$ pregnant women, then about 31 will have energy needs of more than 2625 kcal/day. This result is not unusual.
(c) $\bar{x}$ is normal with $\mu_{\bar{x}} = 2600$ kcal and $\sigma_{\bar{x}} = \dfrac{50}{\sqrt{20}} = \dfrac{25}{\sqrt{5}} \approx 11.180$ kcal. Mother-to-mother variation.
(d) $P(\bar{x} > 2625) = 0.0125$ [Tech: 0.0127]. If we take 100 simple random samples of size $n = 20$ pregnant women, then about 1 of the samples will result in a mean energy need of more than 2625 kcal/day. This result is unusual.

6. (a) $\bar{x}$ is approximately normal with $\mu_{\bar{x}} = 0.75$ inch and $\sigma_{\bar{x}} = \dfrac{0.004}{\sqrt{30}} \approx 0.001$ inch.
(b) $P(\bar{x} < 0.748) + P(\bar{x} > 0.752) = 0.0062$. There is a 0.0062 probability that the inspector will conclude that the machine needs adjustment when, in fact, it does not need adjustment.

7. (a) No, the variable "number of televisions" is likely skewed right.
(b) $\bar{x} = 2.55$ televisions
(c) $P(\bar{x} \geq 2.55) = 0.0778$ [Tech: 0.0777]. If we take 100 simple random samples of size $n = 40$ households, then about 8 of the samples will result in a mean of 2.55 televisions or more. This result is not unusual, so it does not contradict the results reported by A. C. Nielsen.

8. (a) Qualitative with two possible outcomes.
(b) The sampling distribution of $\hat{p}$ is approximately normal with $\mu_{\hat{p}} = 0.72$ and $\sigma_{\hat{p}} = \sqrt{\dfrac{0.72(0.28)}{600}} \approx 0.018$. Person-to-person variability.
(c) $P(\hat{p} \leq 0.70) = 0.1379$ [Tech: 0.1376]. About 14 out of 100 random samples of size $n = 600$ 18- to 29-year-olds will result in no more than 70% who would prefer to start their own business.
(d) Yes; it would be a little unusual for 450 of the 600 randomly selected 18- to 29-year-olds to prefer to start their own business. $P(X \geq 450) = P(\hat{p} \geq 0.75) = 0.0505$ [Tech: 0.0509].

9. $P(X \geq 60) = P(\hat{p} \geq 0.12) = 0.0681$ [Tech: 0.0680]. This result is not that unusual. There is some evidence to suggest that the proportion of adults 25 years of age or older with advanced degrees has increased above 10%.

10. (a) The sampling distribution of $\hat{p}$ is approximately normal with $\mu_{\hat{p}} = 0.280$ and $\sigma_{\hat{p}} = \sqrt{\dfrac{0.28(0.72)}{500}} \approx 0.020$.
(b) $P(\hat{p} \geq 0.310) = 0.0681$ [Tech: 0.0676]. It would not be unusual for a career 0.280 hitter to have a season in which he hits 0.310.
(c) $P(\hat{p} \leq 0.255) = 0.1056$ [Tech: 0.1066]. It would not be unusual for a career 0.280 hitter to have a season in which he hits 0.255.
(d) Batting averages between 0.260 and 0.300 lie within 1 standard deviation of the mean of the sampling distribution.
(e) It is unlikely that a career 0.280 hitter hits 0.280 each season, so he will have seasons where he bats below 0.280 and seasons where he bats above 0.280, and batting averages as low as 0.260 and as high as 0.300 are not unusual $[P(\hat{p} \leq 0.260) \approx 0.16$ and $P(\hat{p} \geq 0.300) \approx 0.16]$. Based on a single season, we cannot conclude that a player who hit 0.260 is worse than a player who hit 0.300 because neither result would be unusual for players who had identical career batting averages of 0.280.

Chapter 8 Test (page 417)

1. Regardless of the shape of the population, the sampling distribution of $\bar{x}$ becomes approximately normal as the sample size n increases.

2. $\mu_{\bar{x}} = 50, \sigma_{\bar{x}} = \dfrac{24}{\sqrt{36}} = 4$

3. **(a)** $P(X > 100) = 0.3859$ [Tech: 0.3875]. If we select a simple random sample of $n = 100$ batteries of this type, then about 39 batteries would last more than 100 minutes. This result is not unusual.
(b) $\bar{x}$ is normally distributed with $\mu_{\bar{x}} = 90$ minutes and $\sigma_{\bar{x}} = \dfrac{35}{\sqrt{10}} \approx 11.068$ minutes.
(c) $P(\bar{x} > 100) = 0.1841$ [Tech: 0.1831]. If we take 100 simple random samples of size $n = 10$ batteries of this type, then about 18 of the samples will result in a mean charge life of more than 100 minutes. This result is not unusual.
(d) $P(\bar{x} > 100) = 0.0764$ [Tech: 0.0766]. If we take 100 simple random samples of size $n = 25$ batteries of this type, then about 8 of the samples will result in a mean charge life of more than 100 minutes.
(e) The probabilities are different because a change in n causes a change in $\sigma_{\bar{x}}$.

4. **(a)** $\bar{x}$ is approximately normally distributed with $\mu_{\bar{x}} = 2.0$ liters and $\sigma_{\bar{x}} = \dfrac{0.05}{\sqrt{45}} \approx 0.007$ liter.
(b) $P(\bar{x} < 1.98) + P(\bar{x} > 2.02) = 0.0074$ [Tech: 0.0073]. There is a 0.0074 probability that the quality-control manager will shut down the machine even though it is correctly calibrated.

5. **(a)** The sampling distribution of $\hat{p}$ is approximately normal with $\mu_{\hat{p}} = 0.224$ and $\sigma_{\hat{p}} = \sqrt{\dfrac{0.224(0.776)}{300}} \approx 0.024$.
(b) $P(X \geq 50) = P(\hat{p} \geq 0.1667) = 0.9913$ [Tech: 0.9914]. About 99 out of 100 random samples of size $n = 300$ adults will result in at least 50 adults (that is, at least 16.7%) who are smokers.
(c) Yes; it would be unusual for 18% or less of 300 randomly selected adults to be smokers. $P(\hat{p} \leq 0.18) = 0.0336$ [Tech: 0.0338].

6. **(a)** For the sample proportion to be normal, the sample size must be large enough to meet the condition $np(1 - p) \geq 10$. Since $p = 0.01$, we have
$$n(0.01)(1 - 0.01) \geq 10$$
$$0.0099n \geq 10$$
$$n \geq \frac{10}{0.0099} \approx 1010.1$$
Thus, the sample size must be at least 1011 to satisfy the condition.
(b) No; it would not be unusual for a random sample of 1500 Americans to result in fewer than 10 with peanut or tree nut allergies. $P(X \leq 9) = P(\hat{p} \leq 0.006) = 0.0594$ [Tech: 0.0597].

7. $P(X \geq 82) = P(\hat{p} \geq 0.082) = 0.0681$ [Tech: 0.0685]. This result is not unusual, so this evidence is insufficient to conclude that the proportion of households with a net worth in excess of $1 million has increased above 7%.

Chapter 9 Estimating the Value of a Parameter

9.1 Assess Your Understanding (page 431)

1. point estimate   **3.** False   **5.** increases
7. $z_{\alpha/2} = z_{0.05} = 1.645$ **9.** $z_{\alpha/2} = z_{0.01} = 2.33$
11. $\hat{p} = 0.225, E = 0.024, x = 270$
13. $\hat{p} = 0.4855, E = 0.0235, x = 816$
15. Lower bound: 0.146, upper bound: 0.254
17. Lower bound: 0.191, upper bound: 0.289
19. Lower bound: 0.759 [Tech: 0.758], upper bound: 0.805
21. **(a)** Flawed; no interval has been provided about the population proportion.
(b) Flawed; this interpretation indicates that the level of confidence is varying.

(c) Correct
(d) Flawed; this interpretation suggests that this interval sets the standard for all the other intervals, which is not true.

23. We are 95% confident that the population proportion of adult Americans who dread Valentine's Day is between 0.135 and 0.225.

25. **(a)** $\hat{p} = 0.181$
(b) $n\hat{p}(1 - \hat{p}) = 341.8 \geq 10$, and the sample is less than 5% of the population.
(c) Lower bound: 0.168, upper bound: 0.194
(d) We are 90% confident that the proportion of adult Americans 18 years and older who have donated blood in the past two years is between 0.168 and 0.194.

27. **(a)** $\hat{p} = 0.519$
(b) $n\hat{p}(1 - \hat{p}) = 250.39 \geq 10$, and the sample is less than 5% of the population.
(c) Lower bound: 0.488, upper bound: 0.550 [Tech: (0.489, 0.550)]
(d) Yes; it is possible that the population proportion is more than 60%, because it is possible that the true proportion is not captured in the confidence interval. It is not likely.
(e) Lower bound: 0.450, upper bound: 0.512 [Tech: (0.450, 0.511)]

29. **(a)** Lower bound: 0.071; upper bound: 0.151
(b) Lower bound: 0.058; upper bound: 0.164
(c) As the level of confidence increases, the margin of error increases.

31. **(a)** The sample is the 1000 adults 19 years of age or older; the population is all adults 19 years of age or older.
(b) The variable of interest is whether the individual brings and uses his or her cell phone every trip to the bathroom. It is qualitative with two possible outcomes—either the individual brings and uses his or her cell phone every trip to the bathroom, or not.
(c) 0.241
(d) The point estimate in part (c) is a statistic because its value is based on a sample. The point estimate is a random variable because its value may change depending on the individuals in the survey. The main source of variability are the individuals selected to be in the study. That is, there is person-to-person variability.
(e) Lower bound: 0.214; upper bound: 0.268. We are 95% confident that the proportion of adults 19 years of age or older who bring and use their cell phone every trip to the bathroom is between 0.214 and 0.268.
(f) To generalize these results to the population, the sample must be representative of the population. Random sampling is required to ensure the individuals in the sample are representative of the population.

33. Lower bound: 0.451; upper bound: 0.589; We are 95% confident that the proportion of days JNJ stock will increase is between 0.451 and 0.589.

35. **(a)** Using $\hat{p} = 0.635, n = 1708$. [Tech: 1726]
(b) Using $\hat{p} = 0.5, n = 1842$. [Tech: 1844]

37. **(a)** Using $\hat{p} = 0.15, n = 1731$. [Tech: 1726]
(b) Using $\hat{p} = 0.5, n = 3394$. [Tech: 3383]

39. **(a)** Using $\hat{p} = 0.53, n = 1064$.   **(b)** Using $\hat{p} = 0.5, n = 1068$.
(c) The results are close because $0.53(1 - 0.53) = 0.2491$ is very close to 0.25.

41. At least $n = 984$ people were surveyed.

43. The difference between the two point estimates is within the margin of error.

45. Lower bound: 0.013, upper bound: 0.313

47. **(a)** The variable of interest in the observational study is whether the individual says they wash their hands in a public rest room, or not. It is qualitative with two possible outcomes.
(b) The sample is the 1001 adults interviewed. The population is all adults.
(c) $n\hat{p}(1 - \hat{p}) = 73.7 \geq 10$ and the sample size is less than 5% of the population size.
(d) Lower bound: 0.903; upper bound: 0.937
(e) The variable of interest in the observational study is whether the individual is observed washing their hands in a public rest room, or not. It is qualitative with two possible outcomes.
(f) Randomness could be achieved by using systematic sampling. For example, select every 10th individual who enters the

bathroom. Also, to be able to generalize the results, we would want about half the individuals to be female, and half male.

(g) $n\hat{p}(1 - \hat{p}) = 1076.1 \geq 10$ and the sample size is less than 5% of the population size.

(h) Lower bound: 0.759; upper bound: 0.781

(i) The proportion who say that they wash their hands is greater than the proportion who actually do. Explanations as to why may vary. One possibility is that people lie about their hand-washing habits out of embarrassment.

(j) In the telephone survey, there is certainly person-to-person variability. That is, different samples will result in different individuals surveyed, and therefore, different results. In the observational study, there is also person-to-person variability. There may also be individual variability. Perhaps an individual who does not always wash his or her hands is in the study, but during this particular observation does wash his or her hands. Another source of variability might be the type of public restroom the observation takes place in. For example, do people wash their hands more often in an airport rest room than they do at a sporting event?

49. Data must be qualitative with two possible outcomes to construct confidence intervals for a proportion.

51. They use the margin of error formula with $\hat{p} = 0.5$.

53. Mariya's interval is wrong. The upper and lower bounds are not the same distance from the point estimate.

9.2 Assess Your Understanding (page 443)

1. decreases   **3.** α

5. False; the population does not need to be normally distributed if the sample size is sufficiently large.

7. (a) $t_{0.10} = 1.316$ **(b)** $t_{0.05} = 1.697$
   (c) $t_{0.99} = -2.552$ **(d)** $t_{0.05} = 1.725$

9. Yes, $0.987 > 0.928$ (Table VI) and there are no outliers.

11. No, there are outliers.

13. $\bar{x} = 21, E = 3$ **15.** $\bar{x} = 14, E = 9$

17. (a) Lower bound: 103.7, upper bound: 112.3
   (b) Lower bound: 100.4, upper bound: 115.6; decreasing the sample size increases the margin of error.
   (c) Lower bound: 104.6, upper bound: 111.4; decreasing the level of confidence decreases the margin of error.
   (d) No; the sample sizes were too small.

19. (a) Lower bound: 16.85, upper bound: 19.95
   (b) Lower bound: 17.12, upper bound: 19.68; increasing the sample size decreases the margin of error.
   (c) Lower bound: 16.32, upper bound: 20.48 [Tech: (16.33, 20.48)]; increasing the level of confidence increases the margin of error.
   (d) If $n = 15$, the population must be normal.

21. (a) Flawed; this interpretation implies that the population mean varies rather than the interval.
   (b) Correct
   (c) Flawed; this interpretation makes an implication about individuals rather than the mean.
   (d) Flawed; the interpretation should be about the mean number of hours worked by adult Americans, not about adults in Idaho.

23. We are 90% confident that the mean drive-through service time of Taco Bell restaurants is between 161.5 and 164.7 seconds.

25. (1) Increase the sample size, and (2) decrease the level of confidence to narrow the confidence interval.

27. (a) Since the distribution of blood alcohol concentrations is not normally distributed (highly skewed right), the sample must be large so that the distribution of the sample mean will be approximately normal.
   (b) The sample size is less than 5% of the population.
   (c) Lower bound: 0.1647, upper bound: 0.1693; we are 90% confident that the mean BAC in fatal crashes where the driver had a positive BAC is between 0.1647 and 0.1693 g/dL.
   (d) Yes; it is possible that the mean BAC is less than 0.08 g/dL, because it is possible that the true mean is not captured in the confidence interval, but it is not likely.

29. Lower bound: 317.63 licks [Tech: 317.64]; upper bound: 394.57 licks [Tech: 394.56]; We are 95% confident the mean number of licks to the center of a Tootsie pop is between 317.63 and 394.57.

31. (a) 4.893
   (b) Lower bound: 4.690, upper bound: 5.096 [Tech: 5.095]; We are 95% confident the mean pH of rain water in Tucker County, West Virginia, is between 4.690 and 5.096.
   (c) Lower bound: 4.607 [Tech: 4.606], upper bound: 5.179; We are 99% confident the mean pH of rain water in Tucker County, West Virginia, is between 4.607 and 5.179.
   (d) As the level of confidence increases, the margin of error also increases.

33. (a) 0.961 [Tech: 0.966] > 0.918 (Table VI), so it is reasonable to conclude the data come from a normal population.

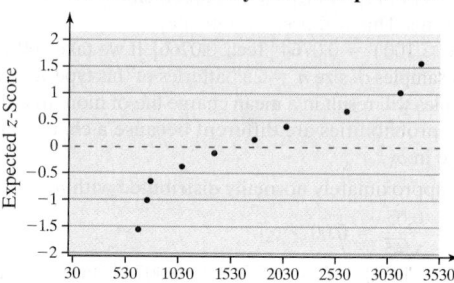
Normal Probability Plot of Repair Cost

(b) There are no outliers.

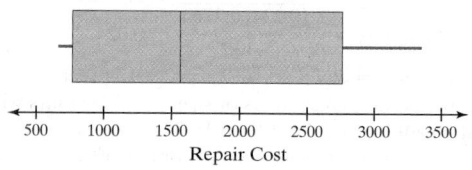
Repair Cost for Low-Impact Collision

(c) Lower bound: $1035.8, upper bound: $2477.2; We are 95% confident the mean repair cost for a low-impact collision involving mini- or micro-vehicles is between $1035.8 and $2477.2.
(d) The 95% confidence interval would likely be narrower because there is less variability in the data because variability associated with make of the vehicle has been removed.

35. (a) Histogram is slightly skewed to the right.

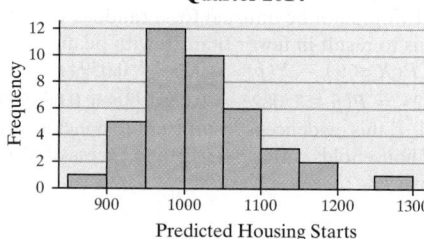
Predicted Housing Starts, Second Quarter 2014

(b) The prediction of 1260 housing starts is an outlier.

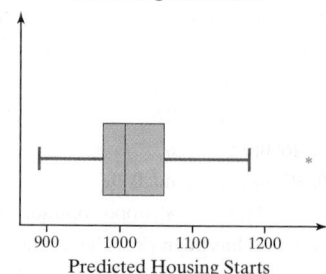
Predicted Housing Starts, Second Quarter 2014

(c) Because of the outlier and slightly skewed distribution of the sample data, it is necessary to have a large sample size to invoke the Central Limit Theorem so that the sampling distribution of the sample mean is approximately normal.

(d) Lower bound: 998.8 housing starts [Tech: 998.7]; upper bound: 1046.8 housing starts. We are 95% confident the mean number of predicted housing starts is between 998.8 and 1046.8 in the second quarter of 2014.

37. (a)   skewed right

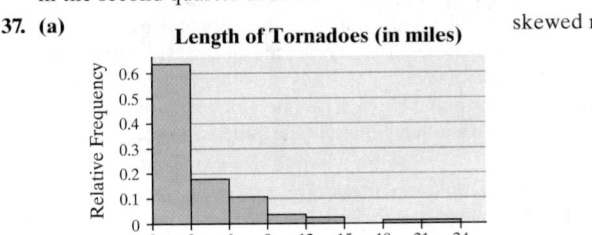

(b)   Yes, there are outliers.

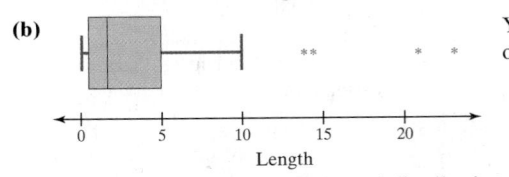

(c) Because of the outliers and skewed distribution of the sample data, it is necessary to have a large sample size to invoke the Central Limit Theorem so that the sampling distribution of the sample mean is approximately normal.

(d) Lower bound: 2.400 miles; upper bound: 4.280 miles; We are 95% confident that the mean length of a tornado in Oklahoma is between 2.400 miles and 4.280 miles.

39. (a) $\bar{x} = 22.150$ years, $E = 0.059$ year

(b) We are 95% confident that the mean age of people when first married is between 22.091 and 22.209 years.

(c) Lower bound: $22.150 - 1.962 \cdot \dfrac{4.885}{\sqrt{26,540}} \approx 22.091$

Upper bound: $22.150 + 1.962 \cdot \dfrac{4.885}{\sqrt{26,540}} \approx 22.209$

41. For a 99% confidence level, $n = 298$ patients; for a 95% confidence level, $n = 173$ patients. Decreasing the confidence level decreases the sample size needed.

43. (a) To estimate within four books with 95% confidence, $n = 67$ subjects are needed.

(b) To estimate within two books with 95% confidence, $n = 265$ subjects are needed.

(c) Doubling the required accuracy quadruples the sample size.

(d) To estimate within four books with 99% confidence, $n = 115$ subjects are needed. Increasing the level of confidence increases the sample size. For a fixed margin of error, greater confidence can be achieved with a larger sample size.

45. (a) Set I: $\bar{x} \approx 99.1$; set II: $\bar{x} \approx 99.1$; set III: $\bar{x} \approx 99.0$

(b) Set I: Lower bound: 82.6, upper bound: 115.6 [Tech: 115.7]

Set II: Lower bound: 91.7, upper bound: 106.5

Set III: Lower bound: 93.5, upper bound: 104.5 [Tech: 104.6]

(c) As the size of the sample increases, the width of the confidence interval decreases.

(d) Set I: Lower bound: 58.6, upper bound: 117.2

Set II: Lower bound: 83.2, upper bound: 106.0

Set III: Lower bound: 88.1, upper bound: 103.9

(e) Each interval contains the population mean. The procedure for constructing the confidence interval is robust. This also illustrates the Law of Large Numbers, and Central Limit Theorem.

47. (a) Answers will vary.   **(b)** Answers will vary.

(c) Answers will vary.

(d) Answers will vary. Expect 95% of the intervals to contain the population mean.

49. (a) Completely randomized design

(b) The treatment is the smoking cessation program. There are 2 levels.

(c) The response variable is whether or not the smoker had 'even a puff' from a cigarette in the past 7 days.

(d) The statistics reported are 22.3% of participants in the experimental group reported abstinence and 13.1% of participants in the control group reported abstinence.

(e) $\dfrac{p(1-q)}{q(1-p)} = \dfrac{0.223(1-0.131)}{0.131(1-0.223)} \approx 1.90$; this means that reported abstinence is almost twice as likely in the experimental group than in the control group.

(f) The authors are 95% confident that the population odds ratio is between 1.12 and 3.26.

(g) Answers will vary. One possibility: smoking cessation is more likely when the Happy Ending Intervention program is used rather than the control method.

51. The t-distribution has less spread as the degrees of freedom increase because as n increases s becomes closer to σ by the Law of Large Numbers.

53. The degrees of freedom are the number of data values that are free to vary.

55. We expect that the margin of error for population A will be smaller since there should be less variation in the ages of college students than in the ages of the residents of a town, resulting in a smaller standard deviation for population A.

9.3 Assess Your Understanding (page 453)

1. False; the chi-square distribution is skewed to the right.

3. False; to construct a confidence interval about a population variance or standard deviation, the sample must come from a normally distributed population.

5. $\chi^2_{0.95} = 10.117$, $\chi^2_{0.05} = 30.144$ **7.** $\chi^2_{0.99} = 9.542$, $\chi^2_{0.01} = 40.289$

9. (a) Lower bound: 7.94, upper bound: 23.66

(b) Lower bound: 8.59, upper bound: 20.63; the width of the interval decreases.

(c) Lower bound: 6.61, upper bound: 31.36; the width of the interval increases.

11. Lower bound: 0.226, upper bound: 0.542; we can be 95% confident that the population standard deviation of the pH of rainwater in Tucker County, WV is between 0.226 and 0.542.

13. Lower bound: $734.8, upper bound: $1657.5; we can be 90% confident that the population standard deviation of repair costs of a low-impact bumper crash on a mini- or micro-car is between $734.8 and $1657.5.

15. (a) Correlation between number of ounces and expected z-scores is $0.987 > 0.928$ (Table VI). It is reasonable to conclude the sample data come from a population that is normally distributed.

(b) $s = 0.349$ ounce

(c) Lower bound: 0.261, upper bound: 0.541; the quality-control manager can be 90% confident that the population standard deviation of the number of ounces of peanuts is between 0.261 and 0.541 ounce.

(d) No; 0.20 oz is not in the confidence interval.

17. Fisher's approximation: $\chi^2_{0.975} \approx 73.772$, $\chi^2_{0.025} \approx 129.070$
Actual values: $\chi^2_{0.975} = 74.222$, $\chi^2_{0.025} = 129.561$

9.4 Assess Your Understanding (page 456)

1. Confidence intervals for a population proportion are constructed on qualitative variables for which there are two possible outcomes.

3. (1) data from a simple random sample or the result of a randomized experiment; (2) $n\hat{p}(1-\hat{p}) \geq 10$; (3) sample size no more than 5% of the population size.

5. Lower bound: 0.069, upper bound: 0.165 [Tech: 0.164]

7. Lower bound: 37.74, upper bound: 52.26

9. Lower bound: 114.98, upper bound: 126.02

11. Lower bound: 13.25, upper bound: 56.98.

13. Lower bound: 51.4, upper bound: 56.6; we can be 95% confident that felons convicted of aggravated assault serve a mean sentence between 51.4 and 56.6 months.

15. Lower bound: 2992.2, [Tech: 2992.1] upper bound: 3849.9; the Internal Revenue Service can be 90% confident that the mean additional tax owed is between $2992.2 and $3849.9.

17. Lower bound: 0.496, upper bound: 0.548; the Gallup organization can be 90% confident that the proportion of adult Americans who are worried about having enough money for retirement is between 0.496 and 0.548.

19. (a) Pitch speed is quantitative. This is important to know because confidence intervals for a mean are constructed on quantitative data, while confidence intervals for a proportion are constructed on qualitative data with two possible outcomes.
(b) Correlation between raw data and normal score is 0.990 [Tech: 0.992], which is greater than the critical value from Table VI for $n = 18$ (0.946). Therefore, it is reasonable to conclude pitch speed comes from a population that is normally distributed.

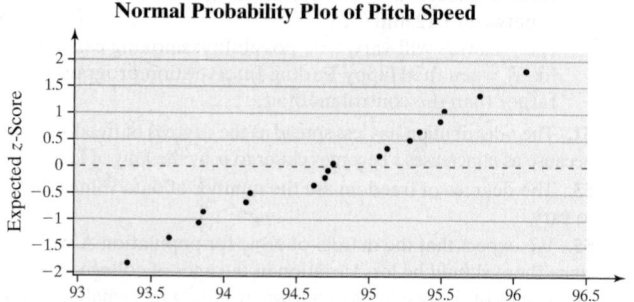

Normal Probability Plot of Pitch Speed

(c) There are no outliers.

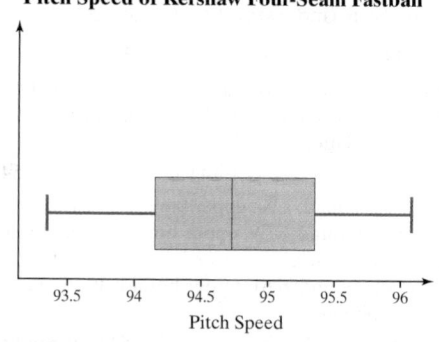

Pitch Speed of Kershaw Four-Seam Fastball

Pitch Speed

(d) Because it is reasonable to conclude the data come from a population that is normally distributed, there are no outliers, and the data is the result of a random sample of pitches, it is reasonable to construct a confidence interval for the mean pitch speed.
(e) Lower bound: 94.358 mph; upper bound: 95.136 mph. We are 95% confident the mean pitch speed of a Clayton Kershaw four-seam fastball is between 94.358 mph and 95.136 mph.
(f) A 95% confidence interval for the mean pitch speed of all major league pitchers' four-seam fastballs would be wider because of the pitcher-to-pitcher variability in pitch speed is now part of the analysis.

21. (a) The study is cross-sectional because the data was obtained at a specific point in time (or over a very short period of time). It also means the study is an observational study.
(b) The variable of interest is whether an individual with sleep apnea has gum disease, or not. This is a qualitative variable because it categorizes the individual.
(c) Lower bound: 0.546; upper bound: 0.654. We are 95% confident the proportion of individuals with sleep apnea who have gum disease is between 0.546 and 0.654.

23. (a) A cohort study is an observational study in which a group of individuals are followed over a period of time (prospective) and information about the individual is obtained over this time period.
(b) Weight gain is quantitative. The variable is measured by determining the change in weight of the individual from a baseline measure to a second point in time (4 years later).
(c) Increase in consumption of french fries is qualitative. Either an individual is categorized as increasing french fry consumption, or not.
(d) The margin of error is $4.42 - 3.35 = 1.07$ pounds.
(e) Individuals whose smoking status changed from former to current smoker actually lost 2.47 pounds over the 4-year period, on average.
(f) We are 95% confident the mean weight loss over a 4-year period of individuals whose smoking status changed from former to current smoker was between 1.12 pounds and 3.82 pounds.
(g) The population applies to males and females who are healthy and not obese.

25. Confidence interval for a proportion since the variable of interest is qualitative with two possible outcomes—morals getting better, or not.

27. Confidence interval for a mean since the variable of interest is quantitative and we would like to estimate the typical reduction in caloric intake.

9.5 Assess Your Understanding (page 463)

1. Bootstrapping is a method of statistical inference that uses a set of sample data as the population. Many samples of size equal to the original sample size are chosen. The samples are chosen with replacement so that the samples are not necessarily the original sample. The statistic of interest is determined for each sample. The distribution of these statistics is used to make a judgment about the population parameter.

3. (a) Yes   **(b)** No, sample size too small.
(c) No, sample size too large.
(d) No, 63 is not in original sample.
(e) Yes

5. Sample 1: 71.3; Sample 2: 71.8; Sample 3: 70.8

7. (a), (b) Answers will vary. The results below are from StatCrunch. The 95% confidence interval using Student's t-distribution is (4.690, 5.076). The bootstrap interval is narrower.

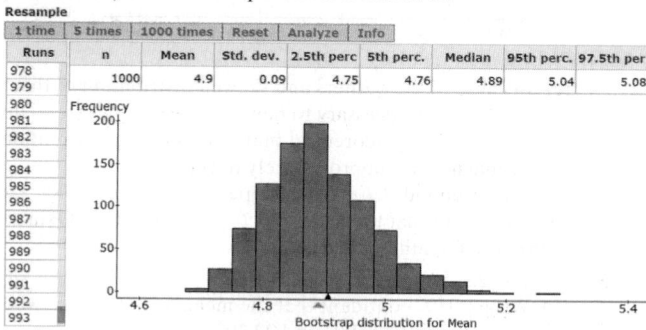

Bootstrap distribution for Mean

9. (a), (b) Answers will vary. The results below are from StatCrunch. The 95% confidence interval using Student's t-distribution is (1035.8, 2477.2). The bootstrap interval is narrower.

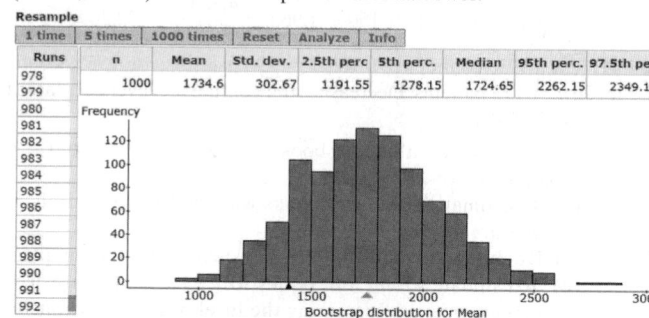

Bootstrap distribution for Mean

11. (a)   **Home Prices**

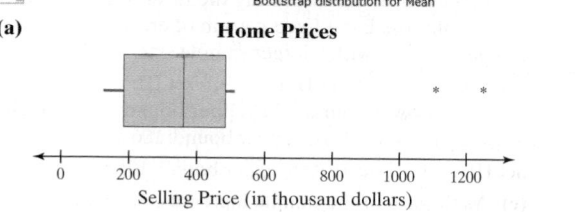

Selling Price (in thousand dollars)

A t-interval cannot be constructed because there are outliers.
(b) Answers will vary. A sample answer generated by StatCrunch is given.

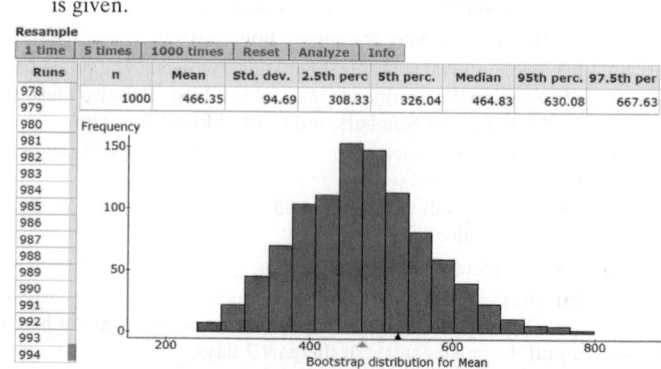

Bootstrap distribution for Mean

(c) We are 95% confident the mean selling price of a condominium in Daytona Beach Shores is between $308.3 thousand and $667.6 thousand.

13. (a) 0.271 **(b)** Answers will vary.
(c) Lower bound: 0.176; upper bound: 0.365

15. Answers will vary. **17.** Answers will vary.
19. Answers will vary. **21.** Answers will vary.

Chapter 9 Review Exercises (page 467)

1. (a) $t_{0.005} = 2.898$ **(b)** $t_{0.05} = 1.706$
2. (a) $\chi^2_{0.975} = 10.283, \chi^2_{0.025} = 35.479$
(b) $\chi^2_{0.995} = 2.603, \chi^2_{0.005} = 26.757$

3. We would expect 95 of the 100 intervals to include the mean 100. Random chance in sampling causes a particular interval to not include the mean 100. Any sample statistic in the tails of the distribution will result in a confidence interval that does not include the parameter.

4. In a 95% confidence interval, the 95% represents the proportion of intervals that would contain the parameter (e.g., the population mean, population proportion, or population standard deviation) if a large number of different samples is obtained.

5. If a large number of different samples is obtained, a 90% confidence interval for a population mean will not capture the true population mean 10% of the time.

6. Area to the left of $t = -1.56$ is 0.0681 because the t-distribution is symmetric about zero.

7. There is more area under the t-distribution to the right of $t = 2.32$ than under the standard normal distribution to the right of $z = 2.32$, because the t-distribution uses s to approximate σ, making it more dispersed than the z-distribution.

8. The properties of Student's t-distribution:
1. It is symmetric around $t = 0$.
2. It is different for different sample sizes.
3. The area under the curve is 1; half the area is to the right of 0 and half the area is to the left of 0.
4. As t gets extremely large, the graph approaches, but never equals, zero. Similarly, as t gets extremely small (negative), the graph approaches, but never equals, zero.
5. The area in the tails of the t-distribution is greater than the area in the tails of the standard normal distribution.
6. As the sample size n increases, the distribution (and the density curve) of the t-distribution becomes more like the standard normal distribution.

9. (a) Lower bound: 51.54, upper bound: 58.06
(b) Lower bound: 52.34, upper bound: 57.26; increasing the sample size decreases the width of the interval.
(c) Lower bound: 49.52, upper bound: 60.08; increasing the level of confidence increases the width of the interval.

10. (a) Lower bound: 97.07, upper bound: 111.53
(b) Lower bound: 98.86, upper bound: 109.74; increasing the sample size decreases the width of the interval.
(c) Lower bound: 95.49, upper bound: 113.11; increasing the level of confidence increases the width of the interval.
(d) Lower bound: 12.22, upper bound: 23.21

11. (a) The distribution is skewed left, but according to the Central Limit Theorem, the sampling distribution of $\bar{x}$ is approximately normal when the sample size is large.
(b) Lower bound: 82.58, upper bound: 93.22; we are 95% confident that the mean age adult Americans would like to live is between 82.58 and 93.22 years.
(c) A sample size of $n = 231$ is required to estimate the mean age adult Americans would like to live within 2 years with 95% confidence.

12. (a) The distribution is skewed right.
(b) Lower bound: 8.86, upper bound: 11.94; we are 90% confident that the population mean number of e-mails sent per day is between 8.86 and 11.94 e-mails.

13. (a) The sample is probably small because of the difficulty and expense of gathering data.
(b) Lower bound: 201.5, upper bound: 234.5; the researchers can be 95% confident that the population "mean total work

performed" for the sports-drink treatment is between 201.5 and 234.5 kilojoules.
(c) Yes; it is possible that the population "mean total work performed" for the sports-drink treatment is less than 198 kilojoules, since it is possible that the true mean is not captured in the confidence interval. It is not likely.
(d) Lower bound: 161.5, upper bound: 194.5; the researchers can be 95% confident that the population "mean total work performed" for the placebo treatment is between 161.5 and 194.5 kilojoules.
(e) Yes; it is possible that the population "mean total work performed" for the placebo treatment is more than 198 kilojoules, since it is possible that the true mean is not captured in the confidence interval. It is not likely.
(f) Yes; our findings support the researchers' conclusion. The confidence intervals do not overlap, so we are confident that the mean for the sports-drink treatment is greater than the mean for the placebo treatment.

14. (a) From the Central Limit Theorem, when the sample is large, $\bar{x}$ is approximately normally distributed.
(b) Lower bound: 1.95, upper bound: 2.59 [Tech: 2.58]; we can be 95% confident that couples who have been married for 7 years have a mean number of children between 1.95 and 2.59.
(c) Lower bound: 1.85, upper bound: 2.69; we can be 99% confident that couples who have been married for 7 years have a mean number of children between 1.85 and 2.69.

15. (a) $\bar{x} = 147.3$ cm, $s = 28.8$ cm
(b) Because 0.982 [Tech: 0.985] > 0.928 (Table VI) and there are no outliers, the conditions are met.
(c) Lower bound: 129.0, upper bound: 165.6; we are 95% confident that the population mean diameter of a Douglas fir tree in the western Washington Cascades is between 129.0 and 165.6 cm.
(d) Lower bound: 20.4, upper bound: 48.9; we are 95% confident that the population standard deviation diameter of a Douglas fir tree in the western Washington Cascades is between 20.4 and 48.9 cm.

16. (a) $\hat{p} = 0.086$
(b) Lower bound: 0.065, upper bound: 0.107 [Tech: (0.064, 0.107)]; the Centers for Disease Control is 95% confident that the proportion of adult males 20 to 34 years old who have hypertension is between 0.065 and 0.107.
(c) 336 subjects would be needed if we use the point estimate of the proportion found in part (a).
(d) 1068 subjects would be needed if no prior estimate is available.

17. Answers will vary. A sample answer generated by StatCrunch is given.
(a) Summary Statistics

Column	Mean	2.5th Per.	97.5th Per
Mean	147.25333	132.625	162.29167

Lower bound: 132.625, upper bound: 162.292; the confidence interval constructed using the bootstrapping method is narrower.
(b)

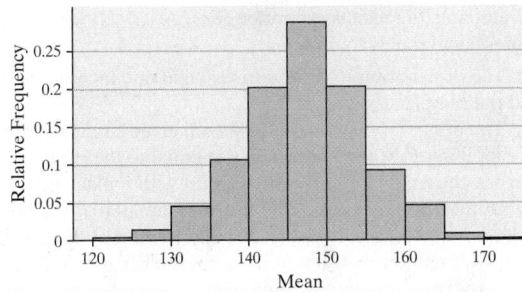

Histogram of Means

The histogram resembles that of a normal distribution, which is not surprising since the distribution of the original data was approximately normal.
(c) Summary Statistics

Column	Median	2.5th Per.	97.5th Per.
Median	151.5	121.5	167.5

Lower bound: 121.5, upper bound: 167.5

Chapter 9 Test (page 469)

1. (a) $t_{0.02} = 2.167$ (b) $t_{0.01} = 2.567$

2. (a) $\chi^2_{0.99} = 5.229$, $\chi^2_{0.01} = 30.578$
   (b) $\chi^2_{0.95} = 17.708$, $\chi^2_{0.05} = 42.557$

3. $\bar{x} = 139.2$, $E = 13.4$

4. (a) The distribution is skewed right.
   (b) Lower bound: 1.151, upper bound: 1.289 [Tech: (1.152, 1.288)];
   we are 99% confident that the population "mean number of family
   members in jail" is between 1.151 and 1.289 members.

5. (a) Lower bound: 4.319, upper bound: 4.841; we are 90% confident
   that the population "mean time to graduate" is between 4.319 and
   4.841 years.
   (b) Yes; our result from part (a) indicates that the mean time to
   graduate is more than 4 years. The entire interval is above 4.

6. (a) $\bar{x} = 57.8$ inches; $s = 15.4$ inches
   (b) Yes; the conditions are met. The distribution is approximately
   normal (since 0.960 [Tech: 0.957] > 0.928–Table VI), and there are
   no outliers.
   (c) Lower bound: 48.0 [Tech: 47.9], upper bound: 67.6; the student
   is 95% confident that the mean depth of visibility of the Secchi disk
   is between 48.0 and 67.6 inches.
   (d) Lower bound: 44.0 [Tech: 43.9], upper bound: 71.6; the student
   is 99% confident that the mean depth of visibility of the Secchi disk
   is between 44.0 and 71.6 inches.
   (e) Lower bound: 10.9, upper bound: 26.2; the student is 95%
   confident that the standard deviation of the depth of visibility of
   the Secchi disk is between 10.9 and 26.2 inches.

7. (a) $\hat{p} = 0.948$
   (b) Lower bound: 0.932, upper bound: 0.964 [Tech: 0.965]; the
   EPA is 99% confident that the proportion of Americans who live
   in neighborhoods with acceptable levels of carbon monoxide is
   between 0.932 and 0.964.
   (c) 593 Americans must be sampled if the prior estimate of p is
   used.
   (d) 3007 Americans must be sampled for the estimate to be within
   1.5 percentage points with 90% confidence if no prior estimate is
   available.

8. (a) $\bar{x} = 133.4$ minutes
   (b) Since the distribution of the lengths of matches is not normally
   distributed, the sample must be large so that the distribution of the
   sample mean will be approximately normal.
   (c) Lower bound: 114.9, upper bound: 151.9; the tennis enthusiast
   is 99% confident that the population "mean length of matches at
   Wimbledon" is between 114.9 and 151.9 minutes.
   (d) Lower bound: 119.6, upper bound: 147.2; the tennis enthusiast
   is 95% confident that the population "mean length of matches at
   Wimbledon" is between 119.6 and 147.2 minutes.
   (e) Increasing the level of confidence increases the width of the
   interval.
   (f) No; because the tennis enthusiast only sampled matches at
   Wimbledon, the results cannot be generalized to include other
   professional tennis tournaments.

9. (a) The distribution is skewed to the right and includes an outlier,
   and the sample size is small.
   (b) The bootstrapping method uses the given set of sample data as
   the population. Many samples of size equal to the original sample
   size are chosen. The samples are chosen with replacement so that
   the samples are not necessarily the original sample. The mean
   is determined for each sample. The distribution of these sample
   means is used to construct the 95% confidence interval.
   (c) Answers will vary. A sample answer generated by StatCrunch
   is given.

Summary Statistics

Column	Mean	2.5th Per.	97.5th Per.
Mean	45.220856	25.857143	67.67857

Lower bound: 25.857, upper bound: 67.679

Chapter 10 Hypothesis Tests Regarding a Parameter

10.1 Assess Your Understanding (page 478)

1. hypothesis   3. null hypothesis   5. I

7. level of significance   9. Right-tailed, μ

11. Two-tailed, σ 13. Left-tailed, μ

15. (a) $H_0: p = 0.399$; $H_1: p > 0.399$
   (b) The sample evidence leads the president to conclude the
   proportion of students who earn a bachelor's degree within six
   years is greater than 0.399, when, in fact, the proportion is 0.399.
   (c) The sample evidence leads the president to conclude the proportion
   of students who earn a bachelor's degree within six years is not greater
   than 0.399, when, in fact, the proportion is greater than 0.399.

17. (a) $H_0: \mu = \$245,700$; $H_1: \mu < \$245,700$
   (b) The sample evidence led the real estate broker to conclude that
   the mean price of an existing single-family home is lower in her
   neighborhood, when in fact the mean price is not lower.
   (c) The sample evidence led the real estate broker to conclude that
   the mean price of an existing single-family home is not lower in her
   neighborhood, when in fact the mean price is lower.

19. (a) $H_0: \sigma = 0.7$ psi, $H_1: \sigma < 0.7$ psi
   (b) The quality-control manager rejects the hypothesis that
   the variability in the pressure required is 0.7 psi, when the true
   variability is 0.7 psi.
   (c) The quality-control manager fails to reject that the variability in the
   pressure required is 0.7 psi, when the variability is less than 0.7 psi.

21. (a) $H_0: \mu = \$48.79$, $H_1: \mu \neq \$48.79$
   (b) The sample evidence led the researcher to believe the mean
   monthly cell phone bill is different from $48.79, when in fact the
   mean bill is $48.79.
   (c) The sample evidence did not lead the researcher to believe the
   mean monthly cell phone bill is different from $48.79, when in fact
   the mean bill is different from $48.79.

23. There is sufficient evidence to conclude the proportion of students
who enroll at Joliet Junior College and earn a bachelor's degree within
six years exceeds 0.399.

25. There is not sufficient evidence to conclude that the mean price of
an existing single-family home in the realtor's neighborhood is less than
$245,700.

27. There is not sufficient evidence to conclude that the variability in
pressure has been reduced.

29. There is sufficient evidence to conclude that the mean monthly
revenue per cell phone is different from $48.79.

31. There is not sufficient evidence to conclude the proportion of
students who enroll at Joliet Junior College and earn a bachelor's
degree within six years exceeds 0.399.

33. There is sufficient evidence to conclude that the mean price of an
existing single-family home in the realtor's neighborhood is less than
$245,700.

35. (a) $H_0: \mu = 12$ oz.; $H_1: \mu \neq 12$ oz.
   (b) Based on the sample evidence, there is sufficient evidence to
   conclude that the machine is out of calibration.
   (c) A Type I error has been made since the sample evidence led
   the quality-control manager to reject the null hypothesis, when the
   null hypothesis (not out of calibration) is true.
   (d) The level of significance should be 0.01 because this makes the
   probability of a Type I error small.

37. (a) $H_0: p = 0.028$; $H_1: p > 0.028$
   (b) There is not sufficient evidence to conclude the proportion of
   high school students exceeds 0.028 at this counselor's high school.
   (c) A Type II error was committed because the sample evidence
   led the counselor to conclude the proportion of e-cig users at her
   school was 0.028, when, in fact, the proportion is higher.

39. (a) $H_0: \mu = 4$ hours, $H_1: \mu < 4$ hours
   (b) *Consumer Reports* might reject the null hypothesis and
   conclude that a car with Prolong will not run 4 hours without oil.

41. (a) $30,000

(b) This is a binomial experiment because (a) there are a fixed number of trials, (b) there are two mutually exclusive outcomes, (c) the trials are independent because the sample size is small relative to the population size, and (d) the probability of success is fixed for all trials. Here, $n = 20$, $p = 0.16$.

(c) $P(8) = 0.0067$ **(d)** $P(X < 8) = 0.9912$

(e) Assume that $p = 0.16$, so $np(1 - p) = 500(0.16)(1 - 0.16) = 67.2 > 10$ and the sample size is less than 5% of the population size assuming there are more than 10,000 60- to 75-year-olds in the population (which is reasonable). Therefore, the sample proportion is approximately normal with mean $500(0.16) = 80$ and standard deviation $\sqrt{500(0.16)(1 - 0.16)/500} = 0.016$.

(f) $P(X \geq 100) = 0.0073$. This result is unusual. If the true proportion of 60- to 75-year-olds who believe it is safe to withdraw 6 to 8 percent of their savings annually is 0.16, we would expect about 7 of every 1000 samples of size 500 to result in at least 100 who believe this. We might conclude from this that the true proportion of 60- to 75-year-olds who believe it is safe to withdraw 6% to 8% of their savings annually is greater than 0.16.

43. As the level of significance, α, decreases, the probability of making a Type II error, β, increases. As we decrease the probability of rejecting a true null hypothesis, we increase the probability of not rejecting the null hypothesis when the alternative hypothesis is true.

45. Answers will vary.

10.2 Assess Your Understanding (page 490)

1. statistically significant   **3.** False   **5.** -1.28

7. $np_0(1 - p_0) = 42 > 10$

(a) Classical approach: $z_0 = 2.31 > z_{0.05} = 1.645$; reject the null hypothesis.

(b) P-value approach: P-value $= 0.0104$ [Tech: 0.0103] $< \alpha = 0.05$; reject the null hypothesis. There is sufficient evidence at the $\alpha = 0.05$ level of significance to reject the null hypothesis.

9. $np_0(1 - p_0) = 37.1 > 10$

(a) Classical approach: $z_0 = -0.74 > -z_{0.10} = -1.28$; do not reject the null hypothesis.

(b) P-value approach:
P-value $= 0.2296$ [Tech: 0.2301] $> \alpha = 0.10$; do not reject the null hypothesis. There is not sufficient evidence at the $\alpha = 0.10$ level of significance to reject the null hypothesis.

11. $np_0(1 - p_0) = 45 > 10$

(a) Classical approach: $z_0 = -1.49$ is between $-z_{0.025} = -1.96$ and $z_{0.025} = 1.96$; do not reject the null hypothesis.

(b) P-value approach: P-value $= 0.1362$ [Tech: 0.1360] $> \alpha = 0.05$; do not reject the null hypothesis. There is not sufficient evidence at the $\alpha = 0.05$ level of significance to reject the null hypothesis.

13. About 27 in 100 samples will give a sample proportion as high or higher than the one obtained if the population proportion really is 0.5. Because this probability is not small, we do not reject the null hypothesis. There is not sufficient evidence to conclude that the dart-picking strategy resulted in a majority of winners.

15. (a) $\hat{p} = 0.472$ **(b)** H_0: $p = 0.5$ versus H_1: $p < 0.5$

(c) $np_0(1 - p_0) = 169.5 \geq 10$. The sample size is less than 5% of the population size, provided Cramer has made more than 13,560 predictions (which is likely).

(d)

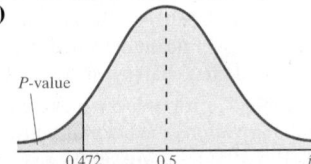

(e) P-value $= P(\hat{p} \leq 0.472) = 0.0721$ [Tech: 0.0722]

(f) If we obtained 100 different samples of size 678 from the population of Cramer predictions and the true proportion of correct predictions was 0.5, we would expect about seven of the samples to result in a sample proportion of correct predictions of 0.472 or less.

(g) Because the P-value is greater than the level of significance, do not reject the null hypothesis. The sample data does not provide

sufficient evidence to conclude that Cramer's predictions are correct less than half the time.

17. $np_0(1 - p_0) = 16.1 > 10$ and $n \leq 0.05\,N$. Hypotheses: H_0: $p = 0.019$, H_1: $p > 0.019$

(a) Classical approach: $z_0 = 0.65 < z_{0.01} = 2.33$; do not reject the null hypothesis.

(b) P-value approach: P-value $= 0.2578$ [Tech: 0.2582] $> \alpha = 0.01$; do not reject the null hypothesis. There is not sufficient evidence at the $\alpha = 0.01$ level of significance to conclude that more than 1.9% of Lipitor users experience flulike symptoms as a side effect.

19. $np_0(1 - p_0) = 24.192 > 10$ and $n \leq 0.05N$. Hypotheses: H_0: $p = 0.36$, H_1: $p > 0.36$. Classical approach: $z_0 = 2.69 > z_{0.05} = 1.645$; reject the null hypothesis. P-value approach: P-value $= 0.0036 < \alpha = 0.05$; reject the null hypothesis. There is sufficient evidence at the $\alpha = 0.05$ level of significance to conclude Hawaii has a higher proportion of traffic fatalities in which the driver has a positive BAC than in the United States.

21. H_0: $p = 0.52$, H_1: $p \neq 0.52$; $np_0(1 - p_0) = 199.68 > 10$ and $n \leq 0.05N$. Classical approach: $z_0 = -11.32 < -z_{0.05/2} = -1.96$; reject the null hypothesis. P-value approach: P-value $< 0.0001 < \alpha = 0.05$; reject the null hypothesis. There is sufficient evidence to conclude that the proportion of parents with children in high school who feel it was a serious problem that high school students were not being taught enough math and science has changed since 1994.

23. H_0: $p = 0.47$, H_1: $p \neq 0.47$; $\hat{p} = 0.431$; $n\hat{p}(1 - \hat{p}) = 248.427 > 10$ and $n \leq 0.05N$. Lower bound: 0.401, upper bound: 0.461 [Tech: 0.462]. Since 0.47 is not contained in the interval, we reject the null hypothesis. There is sufficient evidence to conclude that parents' attitude toward the quality of education in the United States has changed since August 2002.

25. 342

27. (a) H_0: $p = 0.5$ versus H_1: $p > 0.5$.

(b) $np(1 - p) = 4 < 10$. The normal model may not be used to describe the distribution of the sample proportion.

(c) There are a fixed number of trials with two mutually exclusive outcomes (pass or not). The trials are independent and the probability of success is fixed at 0.5 for each trial.

(d) P-value $= P(X \geq 11) = 0.1051$. If we taught the class 100 times with 16 students enrolled, we would expect 11 or more to pass in about 10 or 11 of the classes, assuming the probability of passing is 0.5.

(e) $np(1 - p) = 12 > 10$. Assuming there are over 960 students in the population (very likely), the trials are independent. The normal model may be used.

(f) P-value $= 0.0045$ [Tech: 0.0047]. The P-value is less than the level of significance. There is sufficient evidence to support the belief that the blended course has a higher proportion who pass.

(g) When there are small sample sizes, the evidence against the statement in the null hypothesis must be substantial. The moral is that you should be beware of studies that do not reject the null hypothesis when the test was conducted with a small sample size.

29. (a) H_0: $p = 0.5$ versus H_1: $p > 0.5$.

(b) P-value $= P(X \geq 28) = 0.000002$

(c) Answers may vary. However, always be careful about drawing conclusions of causation from apparent associations.

31. H_0: $p = 0.6$ vs. H_1: $p > 0.6$. Classical: $z = 1.51 < 1.645$, do not reject the null hypothesis. P-value $= 0.0655$ [Tech: 0.0647]. Do not reject the null hypothesis. There is not sufficient evidence to support the belief that a supermajority of adult Americans believe there is gender inequality between males and females when each has the same experience and education.

33. Hypotheses: H_0: $p = 0.5$, H_1: $p \neq 0.5$. $np_0(1 - p_0) = 11.25 > 10$ and $n \leq 0.05N$. P-value $= 0.2938$ [Tech: 0.2967]; do not reject the null hypothesis. Yes, the data suggest that the spreads are accurate.

35. (a) We do not reject the null hypothesis for values of p_0 between 0.44 and 0.62, inclusive. Each of these values of p_0 represents a possible value of the population proportion at the $\alpha = 0.05$ level of significance.

(b) Lower bound: 0.432, upper bound: 0.628

(c) At $\alpha = 0.01$, we do not reject the null hypothesis for any of the values of p_0 given in part (a), so that the range of values of p_0 for which we do not reject the null hypothesis increases. The lower value of α means we need more convincing evidence to reject the null hypothesis, so we would expect a larger range of possible values for the population proportion.

37. (a) $H_0: p = 0.5$ vs. $H_1: p > 0.5$
(b) Answers will vary. **(c)** Answers will vary.
(d) There are a fixed number of trials with two mutually exclusive outcomes. The trials are independent and the probability of success is fixed at 0.5.
(e) $P(X \geq 24) = 0.1341$ **(f)** Answers will vary.
(g) The sample evidence suggests that savants do not have the ability to predict card color as their results could easily be obtained by chance.

39. (a) The randomness in the order in which the baby is exposed to the toys is important to avoid bias.
(b) $H_0: p = 0.5$, $H_1: p > 0.5$
(c) P-value $= 0.0021$; there is sufficient evidence to suggest the proportion of babies who choose the "helper" toy is greater than 0.5.
(d) If the population proportion of babies who choose the helper toy is 0.5, a sample where all 12 babies choose the helper toy will occur in about 2 out of 10,000 samples of 12 babies.

41. If the P-value for a particular test statistic is 0.23, we expect results at least as extreme as the test statistic in about 23 of 100 samples if the null hypothesis is true. Since this event is not unusual, we do not reject the null hypothesis.

43. For the Classical Approach, the calculation of the test statistic, is simple, but you need to be very careful when determining the rejection region to interpret the result. For the P-value method, the P-value is harder to calculate than the test statistic, but the decision to reject or not is the same, regardless of the test. Plus, evidence as to the strength of evidence against the null hypothesis is reported. Most software will calculate both the test statistic and the P-value, which eliminates the disadvantages of the P-value method. Since the P-value is easier to interpret, the P-value method is easier to use with technology.

45. Statistical significance means that the result observed in a sample is unusual when the null hypothesis is assumed to be true.

10.3 Assess Your Understanding (page 500)

1. (a) $t_{0.01} = 2.602$ **(b)** $-t_{0.05} = -1.729$
(c) $\pm t_{0.025} = \pm 2.179$

3. (a) $t_0 = -1.379$ **(b)** $-t_{0.05} = -1.714$
(c)

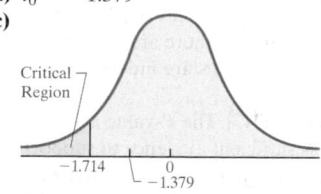

(d) There is not enough evidence for the researcher to reject the null hypothesis because it is a left-tailed test and the test statistic is greater than the critical value $(-1.379 > -1.714)$.

5. (a) $t_0 = 2.502$ **(b)** $-t_{0.005} = -2.819$; $t_{0.005} = 2.819$
(c)

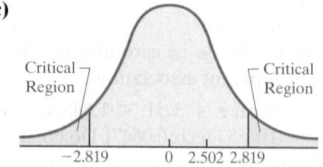

(d) There is not sufficient evidence for the researcher to reject the null hypothesis, since the test statistic is between the critical values $(-2.819 < 2.502 < 2.819)$.
(e) Lower bound: 99.39, upper bound: 110.21. Because the 99% confidence interval includes 100, we do not reject the statement in the null hypothesis.

7. (a) $t_0 = -1.677$
(b)

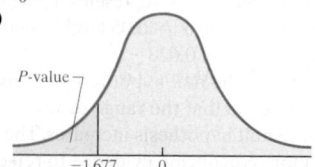

(c) $0.05 < P$-value < 0.10 [Tech: $P = 0.0559$]. If we take 100 random samples of size 18, we would expect about six of the samples to result in a sample mean of 18.3 or less if $\mu = 20$.
(d) The researcher will not reject the null hypothesis at the $\alpha = 0.05$ level of significance because the P-value is greater than the level of significance.

9. (a) No; $n \geq 30$ **(b)** $t_0 = -3.108$
(c)

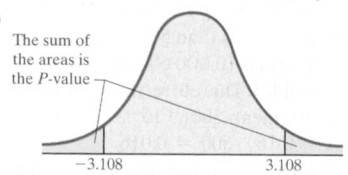

(d) $0.002 < P$-value < 0.005 [Tech: P-value $= 0.0038$]. If we obtain 1000 random samples of size $n = 35$, we would expect about four samples to result in a mean as extreme or more extreme than the one observed if $\mu = 105$.
(e) The researcher will reject the null hypothesis at the $\alpha = 0.01$ level of significance because the P-value is less than the level of significance $(0.0038 < 0.01)$.

11. (a) $H_0: \mu = \$67$, $H_1: \mu > \$67$
(b) There is a 0.02 probability of obtaining a sample mean of \$73 or higher from a population whose mean is \$67. So, if we obtained 100 simple random samples of size $n = 40$ from a population whose mean is \$67, we would expect about two of these samples to result in sample means of \$73 or higher.
(c) Because the P-value is low (P-value $= 0.02 < \alpha = 0.05$), we reject the statement in the null hypothesis. There is sufficient evidence to conclude that the mean dollar amount withdrawn from a PayEase ATM is more than the mean amount from a standard ATM (that is, more than \$67).

13. (a) $H_0: \mu = 22$; $H_1: \mu > 22$
(b) The sample is random. The sample size is large, $n = 200 \geq 30$. We can reasonably assume that the sample is small relative to the population, so the scores are independent.
(c) Classical approach: $t_0 = 2.176 > t_{0.05} = 1.660$; reject the null hypothesis. P-value approach: $0.01 < P$-value < 0.02 [Tech: $P = 0.0154$]; reject the null hypothesis.
(d) There is sufficient evidence to conclude that students who complete the core curriculum are scoring above 22 on the math portion of the ACT.

15. Hypotheses: $H_0: \mu = 9.02$ cm^3, $H_1: \mu < 9.02$ cm^3. Classical approach: $t_0 = -4.553 < -t_{0.01} = -2.718$ cm^2; reject the null hypothesis. P-value approach: P-value < 0.0005 [Tech: P-value $= 0.0004$] $< \alpha = 0.01$; reject the null hypothesis. There is sufficient evidence to conclude that the mean hippocampal volume in alcoholic adolescents is less than the normal mean volume of 9.02 cm^3.

17. $H_0: \mu = 703.5$, $H_1: \mu > 703.5$. Classical approach: $t_0 = 0.813 < t_{0.05} = 1.685$ (39 degrees of freedom); do not reject the null hypothesis. P-value approach: $0.25 > P$-value > 0.20 [Tech: P-value $= 0.2105] > \alpha = 0.05$; do not reject the null hypothesis. There is not sufficient evidence to conclude that the mean FICO score of high-income individuals is greater than that of the general population. In other words, it is not unlikely to obtain a mean credit score of 714.2 or higher even though the true population mean credit score is 703.5.

19. $H_0: \mu = 40.7$ years, $H_1: \mu \neq 40.7$ years; 95% confidence interval: Lower bound: 35.44 years, upper bound: 42.36 years. Because the interval includes 40.7 years, there is not significant evidence to conclude that the mean age of a death-row inmate has changed since 2002.

21. (a) Because $0.971 > 0.918$ (Table VI), it is reasonable to conclude the data come from a normal population. The boxplot does not show any outliers. Therefore, the conditions are satisfied.
(b) Hypotheses: $H_0: \mu = 84.3$ seconds, $H_1: \mu < 84.3$ seconds. Classical approach: $t_0 = -1.310 > -t_{0.10} = -1.383$ with 9 degrees of freedom; do not reject the null hypothesis. P-value approach: $0.15 > P$-value > 0.10 [Tech: P-value $= 0.1113] > \alpha = 0.10$; do not reject the null hypothesis. There is not sufficient evidence to conclude that the new system is effective.

23. H_0: $\mu = 0.11$ mg/L, H_1: $\mu \neq 0.11$ mg/L. Classical approach: $t_0 = 1.707$ is between $-t_{0.025} = -2.262$ and $t_{0.025} = 2.262$; do not reject the null hypothesis. P-value approach: $0.1 < P$-value < 0.2 [Tech: $P = 0.122$]; do not reject the null hypothesis. Conclusion: There is not sufficient evidence to indicate that the calcium concentration in rainwater in Chautauqua, New York, has changed since 1990.

25. (a)   **(b)**

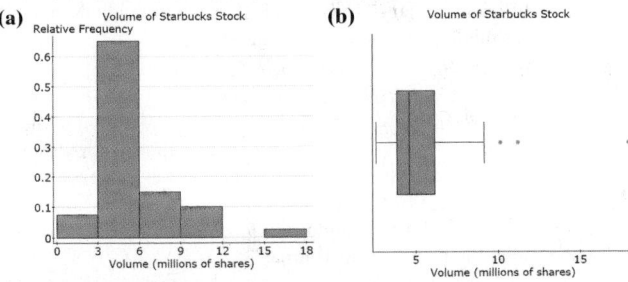

(c) The histogram indicates that the data is skewed right and the boxplot shows outliers.
(d) H_0: $\mu = 7.52$ million shares versus H_1: $\mu \neq 7.52$ million shares. Classical Approach: $t_0 = -4.202$, $-t_{0.025} = -2.023$; since $t_0 < -t_{\alpha/2}$, reject H_0. P-value approach: P-value < 0.001 [Tech: P-value $= 0.0001$], reject H_0. There is sufficient evidence to conclude that the volume of Starbucks stock has changed since 2011.

27. H_0: $\mu = 0.11$ mg/L, H_1: $\mu \neq 0.11$ mg/L. Lower bound: 0.0948, upper bound: 0.2188. Because 0.11 is in the 95% confidence interval, we do not reject the statement in the null hypothesis. There is not sufficient evidence to indicate that the calcium concentration in rainwater in Chautauqua, New York, has changed since 1990.

29. H_0: $\mu = 7.52$, H_1: $\mu \neq 7.52$. Lower bound: 4.668, upper bound: 6.521. Because 7.52 is not in the 95% confidence interval, we reject the statement in the null hypothesis. The evidence suggests that the volume of Starbucks stock has changed since 2011.

31. (a) H_0: $\mu = 515$, H_1: $\mu > 515$
(b) Classical approach: $t_0 = 1.529 > t_{0.10} \approx 1.282$; reject the null hypothesis. P-value approach: $0.05 < P$-value < 0.10 [Tech: 0.0632] $< \alpha = 0.10$; reject the null hypothesis.
(c) Answers will vary.
(d) With $n = 400$ students: Classical approach: $t_0 = 0.721 <$ $t_{0.10} \approx 1.29$; do not reject the null hypothesis. P-value approach: $0.20 < P$-value < 0.25 [Tech: 0.2358] $> \alpha = 0.10$; do not reject the null hypothesis.

33. (a) $0.05 < P$-value < 0.10 [Tech: $P = 0.0557$]; do not reject the null hypothesis.
(b) $0.10 < P$-value < 0.15 [Tech: $P = 0.1286$]; do not reject the null hypothesis.
(c) P-value > 0.25 [Tech: $P = 0.2532$]; do not reject the null hypothesis.
(d) If we "accept" rather than "not reject" the null hypothesis, we are saying that the population mean is a specific value, such as 100, 101, or 102, and so we have used the same data to conclude that the population mean is three different values. However, if we do not reject the null hypothesis, we are saying that the population mean could be 100, 101, or 102 or even some other value; we are simply not ruling them out as the value of the population mean. "Accepting" the null hypothesis can lead to contradictory conclusions, whereas "not rejecting" does not.

35. (a) Answers will vary.   **(b)** Answers will vary.
(c) We would expect five to result in a Type I error.
(d) Answers will vary. The distribution is skewed right and the sample size is small.

37. Yes; because the head of institutional research has access to the entire population, inference is unnecessary. He can say with 100% confidence that the mean age decreased because the mean age in the current semester is less than the mean age in 1995.

39. Statistical significance means that the sample statistic likely does not come from the population whose parameter is stated in the null hypothesis. Practical significance refers to whether the difference between the sample statistic and the parameter stated in the null hypothesis is large enough to be considered important in an application. A statistically significant result may be of no practical significance.

10.4 Assess Your Understanding (page 508)

1. (a) $\chi^2_{0.05} = 28.869$ **(b)** $\chi^2_{0.9} = 14.041$
(c) $\chi^2_{0.975} = 16.047$; $\chi^2_{0.025} = 45.722$
3. (a) $\chi^2_0 = 20.496$ **(b)** $\chi^2_{0.95} = 13.091$
(c)

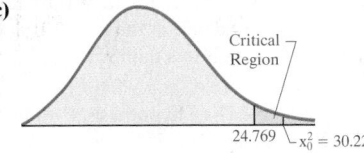

(d) Do not reject H_0, because $\chi^2_0 > \chi^2_{0.95}$.
5. (a) $\chi^2_0 = 30.222$
(b) $\chi^2_{0.1} = 24.769$
(c)

(d) Reject H_0, because $\chi^2_0 > \chi^2_{0.1}$.
7. (a) $\chi^2_0 = 13.707$ **(b)** $\chi^2_{0.975} = 3.816$, $\chi^2_{0.025} = 21.920$
(c)

(d) Do not reject H_0, because $\chi^2_{0.975} < \chi^2_0 < \chi^2_{0.025}$.
9. Hypotheses: H_0: $\sigma = 0.04$, H_1: $\sigma < 0.04$. $\chi^2_0 = 13.590 < \chi^2_{0.95} = 13.848$; $0.025 < P$-value < 0.05 [Tech: P-value $= 0.045$] $< \alpha = 0.05$; reject the null hypothesis at the $\alpha = 0.05$ level of significance. There is sufficient evidence to conclude that the mutual fund has moderate risk.
11. Hypotheses: H_0: $\sigma = 0.004$, H_1: $\sigma < 0.004$. $\chi^2_0 = 9.375 < \chi^2_{0.99} = 10.856$; P-value < 0.005 [Tech: P-value $= 0.003$] $< \alpha = 0.01$; reject the null hypothesis at the $\alpha = 0.01$ level of significance. There is sufficient evidence at the $\alpha = 0.01$ level of significance for the manager to conclude that the standard deviation has decreased. The recalibration was effective.
13. Hypotheses: H_0: $\sigma = 18.0$, H_1: $\sigma < 18.0$. $\chi^2_0 = 6.422 > \chi^2_{0.95} = 3.325$; $0.10 < P$-value < 0.90 [Tech: 0.303]; do not reject the null hypothesis at the $\alpha = 0.05$ level of significance. There is not sufficient evidence at the $\alpha = 0.05$ level of significance to conclude that the standard deviation wait-time is less than 18.0 seconds.
15. Hypotheses: H_0: $\sigma = 8.3$, H_1: $\sigma < 8.3$. $\chi^2_0 = 15.639 < \chi^2_{0.90} = 15.659$; $0.05 < P$-value < 0.10 [Tech: 0.0993] $< \alpha = 0.10$; reject the null hypothesis at the $\alpha = 0.10$ level of significance. There is sufficient evidence at the $\alpha = 0.10$ level of significance to conclude that Rose is a more consistent player than other shooting guards in the NBA.

17. (a)

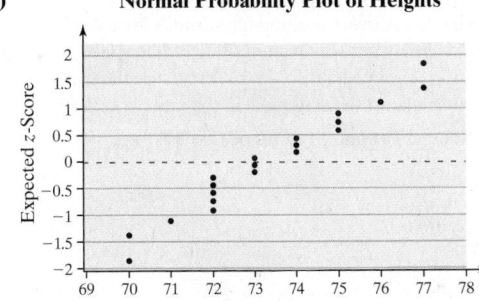

The correlation between the heights and expected z-scores is 0.982. Because 0.982 [Tech: 0.984] > 0.951 (Table VI), we conclude the data could come from a population that is normally distributed.
(b) $s = 2.059$ inches

(c) Hypotheses: H_0: $\sigma = 2.9$, H_1: $\sigma < 2.9$. $\chi_0^2 = 9.5779 > \chi_{0.99}^2 = 7.633$; $0.05 > P$-value > 0.025 [Tech: 0.0374] $> \alpha = 0.01$; do not reject the null hypothesis at the $\alpha = 0.01$ level of significance. There is not sufficient evidence at the $\alpha = 0.01$ level of significance to support the claim that the standard deviation of the heights of baseball players is less than 2.9 inches.

10.5 Assess Your Understanding (page 510)

1. Hypotheses: H_0: $\mu = 1$, H_1: $\mu < 1$. Classical approach: $t_0 = -2.179 > -t_{0.01} = -2.552$ with 18 degrees of freedom; do not reject the null hypothesis. P-value approach: $0.025 > P$-value > 0.02 [Tech: P-value $= 0.0214$] $> \alpha = 0.01$; do not reject the null hypothesis. There is not sufficient evidence at the $\alpha = 0.01$ level of significance to conclude that the mean is less than 1.

3. Hypotheses: H_0: $\sigma^2 = 95$, H_1: $\sigma^2 < 95$. Classical approach: $\chi_0^2 = 9.86 < \chi_{0.90}^2 = 11.651$; reject the null hypothesis. P-value approach: $0.025 < P$-value < 0.05 [Tech: P-value $= 0.0437$] $< \alpha = 0.10$; reject the null hypothesis. There is sufficient evidence at the $\alpha = 0.10$ level of significance to conclude that the variance is less than 95.

5. Hypotheses: H_0: $\sigma^2 = 10$, H_1: $\sigma^2 > 10$. Classical approach: $\chi_0^2 = 20.55 < \chi_{0.05}^2 = 24.996$; do not reject the null hypothesis. P-value approach: P-value > 0.10 [Tech: P-value $= 0.1518$] $> \alpha = 0.05$; do not reject the null hypothesis. There is not sufficient evidence at the $\alpha = 0.05$ level of significance to conclude that the population variance is greater than 10.

7. Hypotheses: H_0: $\mu = 100$, H_1: $\mu > 100$. Classical approach: $t_0 = 3.003 > t_{0.05} = 1.685$; reject the null hypothesis. P-value approach: $0.001 < P$-value < 0.0025 [Tech: P-value $= 0.0023$] $< \alpha = 0.05$; reject the null hypothesis. There is sufficient evidence at the $\alpha = 0.05$ level of significance to conclude that the population mean is greater than 100.

9. H_0: $\mu = 100$, H_1: $\mu > 100$. Classical approach: $t_0 = 1.278 < t_{0.05} = 1.729$; do not reject the null hypothesis. P-value approach: $0.10 < P$-value < 0.15 [Tech: 0.1084] $> \alpha = 0.05$; do not reject the null hypothesis. There is not sufficient evidence to conclude that mothers who listen to Mozart have children with higher IQs.

11. Hypotheses: H_0: $\sigma = 7000$, H_1: $\sigma > 7000$. Classical approach: $\chi_0^2 = 21.81 < \chi_{0.01}^2 = 36.191$; do not reject the null hypothesis. P-value approach: P-value > 0.10 [Tech: P-value $= 0.2938$] $> \alpha = 0.01$; do not reject the null hypothesis. There is not sufficient evidence at the $\alpha = 0.01$ level of significance to conclude that the standard deviation of yield strength is not acceptable. The evidence suggests that the standard deviation yield strength does not exceed 7000 psi.

13. (a) The variable of interest is whether the student passes, or not. It is qualitative with two outcomes.
(b) H_0: $p = 0.526$, H_1: $p > 0.526$; $np_0(1 - p_0) = 119.7 > 10$. Assuming there are over 9600 students eligible for this course (reasonable for large schools), $n \leq 0.05N$. Classical approach: $z_0 = 1.32 < z_{0.01} = 2.33$; do not reject the null hypothesis. P-value approach: P-value $= 0.0934$ [Tech: 0.0922] $> \alpha = 0.01$; do not reject the null hypothesis. There is not sufficient evidence to conclude that the proportion of students who pass is greater than 0.526.
(c) Changing the delivery method for a course for the entire campus would be extremely time-consuming. Plus, building labs would be expensive. The researchers would want to be sure the evidence is overwhelming against the statement in the null hypothesis.

15.

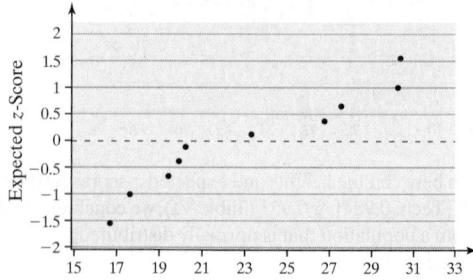

Normal Probability Plot of Gas Mileage

The correlation between gas mileage and expected z-scores is 0.962 [Tech: 0.968] > 0.918 (Table VI). It is reasonable to conclude that the data come from a population that is normally distributed. A boxplot indicates that there are no outliers. H_0: $\mu = 28$ mpg, H_1: $\mu \neq 28$ mpg; Classical approach: $t_0 = -2.963 < -t_{0.05} = -1.833$, reject H_0. P-value approach: $0.01 < P$-value < 0.02 [Tech: 0.0159] $< \alpha = 0.05$, reject H_0. There is sufficient evidence to conclude that the individuals are getting different gas mileage than the EPA suggests.

17. Hypotheses: H_0: $p = 0.26$, H_1: $p \neq 0.26$; $np_0(1 - p_0) = 11.544 > 10$ and $n \leq 0.05N$. Classical approach: $z_0 = 1.589$ is between $-z_{0.025} = -1.96$ and $z_{0.025} = 1.96$; do not reject the null hypothesis. P-value approach: P-value $= 0.1118$ [Tech: 0.112] $> \alpha = 0.05$; do not reject the null hypothesis. There is not sufficient evidence at the $\alpha = 0.05$ level of significance to conclude that the proportion of individuals who text while driving is different from 0.26. The data do not contradict the results from Toluna.

19. Assuming the congresswoman wants to avoid voting for the tax increase unless she is confident that the majority of her constituents are in favor of it, the value of α should be small, such as $\alpha = 0.01$ or $\alpha = 0.05$, to avoid a Type I error. Let p be the population proportion in favor of the tax increase. Hypotheses: H_0: $p = 0.5$, H_1: $p > 0.5$; $np_0(1 - p_0) = 2062.5 > 10$ and $n \leq 0.05 N$. P-value $= 0.0392$ [Tech: 0.0391]. The P-value is greater than $\alpha = 0.01$ but less than $\alpha = 0.05$, so the conclusion depends on the value of α. Recommendations will vary.

21. (a) The distribution is skewed right.

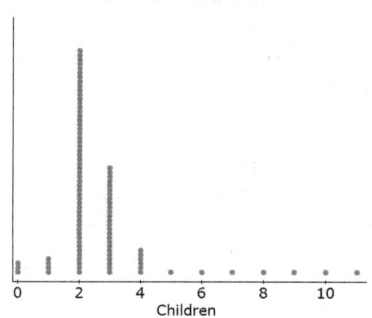

Ideal Number of Children
Up to 11 values per dot

(b) 2
(c) $\bar{x} = 2.418$, $M = 2$, $s = 1.064$, IQR $= 1$
(d) A large sample is needed because the distribution is skewed right.
(e) H_0: $\mu = 2.64$, H_1: $\mu \neq 2.64$. Classical approach: $t_0 = -6.597 < -t_{0.05} = -1.646$; reject H_0. P-value approach: P-value < 0.0005 [Tech: P-value < 0.0001]; reject H_0. There is sufficient evidence to conclude that the mean ideal number of children has changed since May 1997.

23. Hypothesis test; mean; H_0: $\mu = 16$ ounces vs. H_1: $\mu < 16$ ounces
25. Confidence interval; proportion
27. Hypothesis test; proportion; H_0: $p = 0.14$ vs. H_1: $p > 0.14$
29. Hypothesis test; standard deviation; ; H_0: $\sigma = 15$ vs. H_1: $\sigma < 15$

10.6 Assess Your Understanding (page 516)

1. A Type II error is made when we fail to reject the null hypothesis and the alternative is true.

3. (a) If the null hypothesis is not rejected and the true proportion is less than 0.30, a Type II error has been made.
(b) $\beta = 0.8238$ [Tech: 0.8179]; power of the test: $1 - \beta = 0.1762$ [Tech: 0.1821]
(c) $\beta = 0.4052$; power of the test: $1 - \beta = 0.5948$.

5. (a) If the null hypothesis is not rejected and the true proportion is greater than 0.65, a Type II error has been made.
(b) $\beta = 0.9082$ [Tech: 0.9083]; power of the test: $1 - \beta = 0.0918$ [Tech: 0.0917]
(c) $\beta = 0.8186$ [Tech: 0.8192]; power of the test: $1 - \beta = 0.1814$ [Tech: 0.1808]

7. (a) If the null hypothesis is not rejected and the true proportion is not 0.45, a Type II error has been made. **(b)** $\beta = 0.4480$ [Tech: 0.4392]; power of the test: $1 - \beta = 0.5520$ [Tech: 0.5608] **(c)** $\beta = 0.7710$ [Tech: 0.7658]; power of the test: $1 - \beta = 0.2290$ [Tech: 0.2342]

9. (a) If the null hypothesis is not rejected and the true proportion of adults in the United States who believe they will not have enough money in retirement is greater than 0.5, a Type II error has been made. **(b)** $\beta = 0.5398$ [Tech: 0.5339]; power of the test: $1 - \beta = 0.4602$ [Tech: 0.4661] **(c)** $\beta = 0.1736$ [Tech: 0.1685]; power of the test: $1 - \beta = 0.8264$ [Tech: 0.8315]

11. (a) If the null hypothesis is not rejected and the true proportion of parents who are satisfied with the quality of education is less than 0.47, a Type II error has been made. **(b)** $\beta = 0.0268$ [Tech: 0.0269]; power of the test: $1 - \beta = 0.9732$ [Tech: 0.9731] **(c)** $\beta = 0.7389$ [Tech: 0.7405], power of the test: $1 - \beta = 0.2611$ [Tech: 0.2595]

13. (a) If the null hypothesis is not rejected and the true proportion of parents who feel students are not being taught enough math and science is not 0.52, a Type II error has been made. **(b)** $\beta = 0.8014$ [Tech: 0.7949]; power of the test: $1 - \beta = 0.1986$ [Tech: 0.2051] **(c)** $\beta = 0.3897$ [Tech: 0.3803]; power of the test: $1 - \beta = 0.6103$ [Tech: 0.6197]

15. $\beta = 0.9474$ [Tech: 0.9455]; power of the test: $1 - \beta = 0.0526$; [Tech: 0.0545]; when the level of significance was lowered, the power of the test decreased. As we lower the probability of rejecting a true null hypothesis, the probability of not rejecting a false null hypothesis increases, lowering the power of the test.

17.

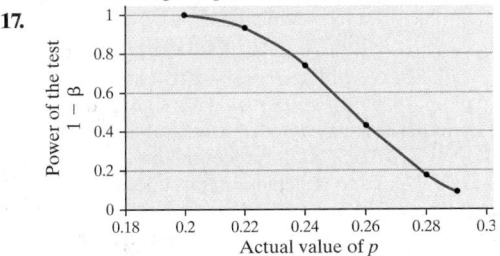

19. (a) The sample mean that separates the rejection region from the nonrejection region is 47.9. **(b)** Area $= \beta = 0.7887$; power of the test: $1 - \beta = 0.2113$

21. Increasing the sample size increases the power of the test.

Chapter 10 Review Exercises (page 519)

1. (a) H_0: $\mu = 3173$; H_1: $\mu < 3173$ **(b)** If we reject the null hypothesis and the mean credit card debt of college undergraduates has not decreased, then we make a Type I error. **(c)** If we do not reject the null hypothesis and the mean credit card debt of college undergraduates has decreased, then we make a Type II error. **(d)** There is not sufficient evidence at the α level to support the researcher's belief that the mean credit card debt of college undergraduates has decreased. **(e)** There is sufficient evidence at the α level to support the researcher's belief that the mean credit card debt of college undergraduates has decreased.

2. (a) H_0: $\mu = 0.13$; H_1: $\mu \neq 0.13$ **(b)** If we reject the null hypothesis and the proportion of cards that result in default is not different today, we make a Type I error. **(c)** If we do not reject the null hypothesis and the proportion of cards that result in default is different today, we make a Type II error.

(d) There is not sufficient evidence at the α level to support the credit analyst's belief that the proportion of cards that result in default is different today. **(e)** There is sufficient evidence at the α level to support the credit analyst's belief that the proportion of cards that result in default is different today.

3. The probability of a Type I error is 0.05.

4. The probability of a Type II error is 0.113. The power of the test is $1 - \beta = 0.887$. The power of the test is the probability of correctly rejecting the null hypothesis.

5. (a) The sample size must be large to use the Central Limit Theorem because the distribution of the population may not be normal. **(b)** Classical approach: $t_0 = 2.0515 > t_{0.05} = 1.691$ for 34 degrees of freedom; reject the null hypothesis. P-value approach: $0.02 < P$-value < 0.025 [Tech: $P = 0.0240$]; reject the null hypothesis.

6. (a) The distribution must be normal because the sample size is small. **(b)** Classical approach: $t_0 = -1.795$ is between $-t_{0.025} = -2.145$ and $t_{0.025} = 2.145$ for 14 degrees of freedom; do not reject the null hypothesis. P-value approach: $0.05 < P$-value < 0.10 [Tech: $P = 0.0943$]; do not reject the null hypothesis.

7. Hypotheses: H_0: $p = 0.6$, H_1: $p > 0.6$; $np_0(1 - p_0) = 60 > 10$ and $n \leq 0.05N$. Classical approach: $z_0 = 1.94 > z_{0.05} = 1.645$; reject the null hypothesis. P-value approach: P-value $= 0.0262$ [Tech: 0.0264] $< \alpha = 0.05$; reject the null hypothesis. There is sufficient evidence at the $\alpha = 0.05$ level of significance to conclude that $p > 0.6$.

8. Hypotheses: H_0: $p = 0.35$, H_1: $p \neq 0.35$; $np_0(1 - p_0) = 95.6 > 10$ and $n \leq 0.05N$. Classical approach: $z_0 = -0.92$ is between $-z_{0.025} = -1.96$ and $z_{0.025} = 1.96$; do not reject the null hypothesis. P-value approach: P-value $= 0.3576$ [Tech: 0.3572] $> \alpha = 0.05$; do not reject the null hypothesis. There is not sufficient evidence at the $\alpha = 0.05$ level of significance to conclude that $p \neq 0.35$.

9. (a) $\chi_0^2 = 15.095$ **(b)** $\chi_{0.975}^2 = 7.564$; $\chi_{0.025}^2 = 30.191$. Do not reject H_0, because $\chi_{0.975}^2 < \chi_0^2 < \chi_{0.025}^2$.

10. (a) $\chi_0^2 = 26.508$ **(b)** $\chi_{0.1}^2 = 33.196$. Do not reject H_0 because $\chi_0^2 < \chi_{0.1}^2$.

11. (a) H_0: $p = 0.733$, H_1: $p \neq 0.733$. **(b)** The sample is random. $np_0(1 - p_0) = 19.57 \geq 10$. We can reasonably assume that the sample is less than 5% of the population. **(c)** The P-value is 0.2892 [Tech: 0.2881], so a result this far from the proportion stated in the null hypothesis will occur in about 29 of 100 samples of this size when the null hypothesis is true. We do not reject the null hypothesis at any commonly used level of significance α. Mary's sample evidence does not contradict Professor Wilson's findings.

12. Hypotheses: H_0: $p = 0.05$, H_1: $p > 0.05$. $np_0(1 - p_0) = 11.875 \geq 10$. The P-value is 0.0951 [Tech: 0.0958]. We reject the null hypothesis at $\alpha = 0.10$, but we do not reject it for $\alpha = 0.01$ or $\alpha = 0.05$. The administrator should be a little concerned.

13. (a) Yes, we can reasonably assume that the distances between retaining rings is normally distributed. In addition, the sample size is large. **(b)** H_0: $\mu = 0.875$, H_1: $\mu > 0.875$ **(c)** Recalibrating the machine could be very costly, so he wants to avoid the consequences of making a Type I error. **(d)** Classical approach: $t_0 = 1.2 < t_{0.01} = 2.438$; do not reject the null hypothesis. P-value approach: $0.10 < P$-value < 0.15 [Tech: P-value $= 0.1191$]; do not reject the null hypothesis. No, the evidence does not suggest that the machine be recalibrated. **(e)** The quality-control engineer would make a Type I error if he recalibrated the machine when it did not need to be recalibrated. He would make a Type II error if he did not recalibrate the machine when it actually did need to be recalibrated.

14. Hypotheses: H_0: $\mu = 98.6$, H_1: $\mu < 98.6$. Classical approach: $t_0 = -15.119 < -t_{0.01} = -2.364$, using 100 degrees of freedom; reject the null hypothesis. P-value approach: P-value < 0.0005 [Tech: P-value < 0.0001]; reject the null hypothesis. The evidence suggests that the mean temperature of humans is less than 98.6°F.

15. (a) Yes, because $0.951 > 0.928$ (Table VI), it is reasonable to conclude the data come from a population that is normally distributed and the boxplot shows no outliers.

(b) Hypotheses: $H_0: \mu = 1.68$, $H_1: \mu \neq 1.68$. Lower bound: 1.6781 [Tech: 1.6782], upper bound: 1.6839 [Tech: 1.6838]. The mean stated in the null hypothesis is included in the 95% confidence interval, so do not reject the null hypothesis. Conclusion: There is not sufficient evidence to conclude that the golf balls do not conform to USGA standards.

16. Hypotheses: $H_0: \mu = 480$, $H_1: \mu < 480$. Classical approach: $t_0 = -1.577 > -t_{0.05} = -1.812$; do not reject the null hypothesis. P-value approach: $0.05 < P$-value < 0.10 [Tech: P-value < 0.0730]; do not reject the null hypothesis. There is not sufficient evidence to suggest that the students are studying less than 480 minutes per week.

17. $H_0: p = 0.5$, $H_1: p > 0.5$; $np_0(1 - p_0) = 150(0.5)(1 - 0.5) = 37.5 > 10$ and $n \leq 0.05N$. Classical approach: $z_0 = 0.98 < z_{0.05} = 1.645$; do not reject the null hypothesis; P-value approach: P-value $= 0.1635$ [Tech: 0.1636] $> \alpha = 0.05$; do not reject the null hypothesis. There is not sufficient evidence to suggest that a majority of pregnant women nap at least twice each week.

18. (a) If we do not reject the null hypothesis and the majority of pregnant women do nap at least twice per week, we make a Type II error.

(b) $\beta = 0.8186$ [Tech: 0.8190]; power of the test: $1 - \beta = 0.1814$ [Tech: 0.1810]

19. Hypotheses: $H_0: \sigma = 505.6$, $H_1: \sigma > 505.6$. Classical approach: $\chi_0^2 = 110.409 > \chi_{0.01}^2 = 63.691$; reject the null hypothesis. P-value approach: P-value < 0.005 [Tech: <0.0001] $< \alpha = 0.01$; reject the null hypothesis. There is sufficient evidence at the $\alpha = 0.01$ level of significance to support the hypothesis that the variance of the birth weight of preterm babies is greater than that of full-term babies.

20. Hypotheses: $H_0: p = 0.52$, $H_1: p > 0.52$. The P-value is 0.0793 [Tech: 0.0788]. Reject the null hypothesis at the $\alpha = 0.10$ level of significance. The results are statistically significant. The new retention rate was about 53.3%, not a great increase over 52%, so the results might not be considered practically significant. The cost of the new policies would have to be weighed against such a small increase in the retention rate when considering whether other community colleges should implement the policies.

21. Hypotheses: $H_0: p = 0.4$, $H_1: p > 0.4$; $np_0(1 - p_0) = 9.6 < 10$. P-value $= 0.3115 > \alpha = 0.05$; do not reject the null hypothesis. There is not sufficient evidence at the $\alpha = 0.05$ level of significance to support the researcher's notion that the proportion of adolescents who prays daily has increased.

22. Hypotheses: $H_0: \mu = 73.2$, $H_1: \mu < 73.2$. Classical approach: $t_0 = -2.018 < -t_{0.05} = -1.645$; reject the null hypothesis. P-value approach: P-value $= 0.0218 < \alpha = 0.05$; reject the null hypothesis. There is sufficient evidence at the $\alpha = 0.05$ level of significance to support the coordinator's concern that the mean test scores decreased. The results do not illustrate any practical significance. The average score only declined 0.4 point.

23. If we accept the null hypothesis, we are saying it is true. If we do not reject the null hypothesis, we are saying that we do not have enough evidence to conclude that it is not true. It is the difference between saying H_0 is true and saying we are not convinced that it is false.

24. $H_0: \mu = 1.22$, $H_1: \mu < 1.22$. The P-value indicates that if the population mean is 1.22 hours as stated in the null hypothesis, then results as low as or lower than those obtained by the researcher would occur in about 3 of 100 similar samples.

25. In the Classical Approach, a test statistic is calculated and a rejection region based on the level of significance α is determined within the appropriate distribution for the population parameter stated in the null hypothesis. The null hypothesis is rejected if the test statistic is in the rejection region.

26. In the P-value approach, the probability of getting a result as extreme as the one obtained in the sample is calculated assuming the null hypothesis to be true. If the P-value is less than a predetermined level of significance α, we reject the null hypothesis.

Chapter 10 Test (page 521)

1. (a) $H_0: \mu = 42.6$ minutes, $H_1: \mu > 42.6$ minutes

(b) There is sufficient evidence to conclude that the mean amount of daily time spent on phone calls and answering or writing emails has increased since 2006.

(c) We would reject the null hypothesis that the mean is 42.6 minutes, when, in fact, the mean amount of daily time spent on phone calls and answering or writing emails is 42.6 minutes.

(d) We would not reject the null hypothesis that the mean is 42.6 minutes, when, in fact, the mean amount of time spent on phone calls and answering or writing emails is greater than 42.6 minutes.

2. (a) $H_0: \mu = 167.1$ seconds, $H_1: \mu < 167.1$ seconds

(b) By choosing a level of significance of 0.01, the probability of rejecting the null hypothesis that the mean is 167.1 seconds in favor of the alternative hypothesis that the mean is less than 167.1 seconds, when, in fact, the mean is 167.1 seconds, is small.

(c) Classical approach: $t_0 = -1.75 > -t_{0.01} = -2.381$ (using 70 degrees of freedom); do not reject the null hypothesis. P-value approach: $0.025 < P$-value < 0.05 [Tech: P-value $= 0.0423$] is greater than $\alpha = 0.01$; do not reject the null hypothesis. There is not sufficient evidence to conclude that the drive-through service time has decreased.

3. $H_0: \mu = 8$, $H_1: \mu < 8$. Classical approach: $t_0 = -1.755 < -t_{0.05} = -1.660$ (using 100 degrees of freedom); reject the null hypothesis; P-value approach: $0.025 < P$-value < 0.05 [Tech: P-value $= 0.0406$] $< \alpha = 0.05$; reject the null hypothesis. There is sufficient evidence to conclude that postpartum women get less than 8 hours of sleep each night.

4. $H_0: \mu = 1.3825$ inches, $H_1: \mu \neq 1.3825$ inches. Lower bound: 1.3824 inches; upper bound: 1.3828 inches. Because the interval includes the mean stated in the null hypothesis, we do not reject the null hypothesis. There is not sufficient evidence to conclude that the mean is different from 1.3825 inches. Therefore, we presume that the part has been manufactured to specifications.

5. $H_0: p = 0.6$, $H_1: p > 0.6$; $np_0(1 - p_0) = 1.561(0.6)(0.4) = 374.64 > 10$ and $n \leq 0.05N$. We will use an $\alpha = 0.05$ level of significance. Classical approach: $z_0 = 0.89 < z_{0.05} = 1.645$; do not reject the null hypothesis. P-value approach: P-value $= 0.1867$ [Tech: 0.1843] $> \alpha = 0.05$; do not reject the null hypothesis. There is not sufficient evidence to conclude that a supermajority of Americans did not feel that the United States would need to fight Japan in their lifetimes. It is interesting, however, that significantly fewer than half of all Americans felt the United States would not have to fight Japan in their lifetimes. About 2.5 years after this survey, the Japanese attack on Pearl Harbor thrust the United States into World War II.

6. $H_0: \mu = 0$, $H_1: \mu > 0$. Classical approach: $t_0 = 2.634 > t_{0.05} = 1.664$ (using 80 df); reject the null hypothesis. P-value approach: $0.0025 < P$-value < 0.005 [Tech: P-value $= 0.0051$] $< \alpha = 0.05$; reject the null hypothesis. There is sufficient evidence to suggest that the diet is effective. However, losing 1.6 kg of weight over the course of a year does not seem to have much practical significance.

7. $H_0: p = 0.37$, $H_1: p > 0.37$, $np_0(1 - p_0) = 6.993 < 10$. P-value $= P(\hat{p} \geq 0.37) = P(X \geq 16) = 0.0501 < \alpha = 0.10$; reject the null hypothesis. There is sufficient evidence to conclude that the proportion of 20- to 24-year-olds who live on their own and do not have a landline is greater than 0.37.

8. $H_0: \sigma = 18$, $H_1: \sigma < 18$; $\chi_0^2 = 7.111 > \chi_{0.95}^2 = 3.325$; P-value $= 0.3744$; do not reject the null hypothesis. There is not sufficient evidence to conclude that the investment manager's portfolio is less risky than the market.

9. $\beta = 0.2061$ [Tech: 0.2159]; power of the test: $1 - \beta = 0.7939$ [Tech: 0.7841].

Chapter 11 Inferences on Two Samples

11.1 Assess Your Understanding (page 533)

1. Independent

3. Dependent; quantitative

5. Independent; qualitative

7. Independent; quantitative

9. (a) $H_0: p_1 = p_2$ versus $H_1: p_1 > p_2$
(b) $z_0 = 3.07$
(c) $z_{0.05} = 1.645$
(d) P-value $= 0.0011$ [Tech: 0.0010]. Because $z_0 > z_{0.05}$ (or P-value $< \alpha$), we reject the null hypothesis. There is sufficient evidence to support the claim that $p_1 > p_2$.

11. (a) $H_0: p_1 = p_2$ versus $H_1: p_1 \neq p_2$
(b) $z_0 = -0.34$
(c) $-z_{0.025} = -1.96$; $z_{0.025} = 1.96$
(d) P-value $= 0.7264$ [Tech: 0.7307]. Because $-z_{0.025} < z_0 < z_{0.025}$ (or P-value $> \alpha$), we do not reject the null hypothesis. There is not sufficient evidence to support the claim that $p_1 \neq p_2$.

13. Lower bound: -0.075, upper bound: 0.015

15. Lower bound: -0.063, upper bound: 0.043 [Tech: 0.044]

17. Each sample is the result of a completely randomized design experiment; the response variable is qualitative with two outcomes. $n_1\hat{p}_1(1 - \hat{p}_1) = 91 \geq 10$ and $n_2\hat{p}_2(1 - \hat{p}_2) = 60 \geq 10$; and each sample is less than 5% of the population size. $H_0: p_1 = p_2; H_1: p_1 > p_2$. Classical approach: $z_0 = 2.20 > z_{0.05} = 1.645$; reject H_0. P-value approach: P-value $= 0.0139 < \alpha = 0.05$; reject H_0. There is sufficient evidence at the $\alpha = 0.05$ level of significance to conclude that a higher proportion of subjects in the treatment group (taking Prevnar) experienced fever as a side effect than in the control (placebo) group.

19. $\hat{p}_{1947} = 0.37$, $\hat{p}_{recent} = 0.303$. Each independent sample is a simple random sample; the variable of interest is qualitative with two outcomes. $n_{1947}\hat{p}_{1947}(1 - \hat{p}_{1947}) \geq 10$ and $n_{recent}\hat{p}_{recent}(1 - \hat{p}_{recent}) \geq 10$. The sample size is less than 5% of the population for each sample. $H_0: p_{1947} = p_{recent}$ vs. $H_1: p_{1947} \neq p_{recent}$. Classical approach: $z_0 = 3.33$ is greater than $z_{0.025} = 1.96$; reject H_0. P-value approach: P-value $= 0.0008 < \alpha = 0.05$; reject H_0. There is sufficient evidence at the $\alpha = 0.05$ level of significance to conclude that the proportion of adult Americans who were abstainers in 1947 is different from the recent proportion of abstainers.

21. $H_0: p_m = p_f$ vs. $H_1: p_m \neq p_f$. Lower bound: -0.008 [Tech: -0.009], upper bound: 0.048. We are 95% confident that the difference in the proportion of males and females that have at least one tattoo is between -0.008 and 0.048. Because the interval includes zero, we do not reject the null hypothesis. There is no significant difference in the proportion of males and females that have tattoos.

23. (a) The data were obtained from two independent simple random samples, $n\hat{p}(1 - \hat{p}) \geq 10$ for each independent sample, and the sample size is less than 5% of the population size for each sample.
(b) $H_0: p_f = p_m$ vs. $H_1: p_f > p_m$
(c) $\hat{p}_f - \hat{p}_m$ is approximately normal with $\mu_{\hat{p}_f - \hat{p}_m} = 0$ and
$$\sigma_{\hat{p}_f - \hat{p}_m} = \sqrt{0.349(1 - 0.349)\left(\frac{1}{540} + \frac{1}{560}\right)} \approx 0.029.$$

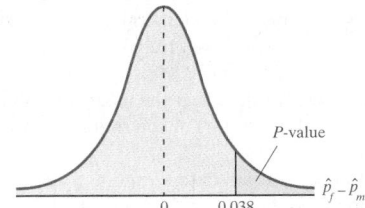

(d) P-value $= p(\hat{p}_f - \hat{p}_m > 0.038) = 0.0934$ [Tech: 0.0918]
(e) If we obtain 1000 different simple random samples as described in the original problem, we would expect 93 to have a difference in sample proportion of 0.038 or higher if the population proportion difference was 0.
(f) Since P-value $> \alpha$, do not reject H_0. There is not sufficient evidence to conclude the proportion of females annoyed by people who repeatedly check their mobile phones while having an in-person conversation is greater than the proportion of males.

25. (a) Completely randomized design.
(b) The response variable is whether the subject experiences dry mouth, or not. It is qualitative with two possible outcomes.
(c) The explanatory variable is type of drug. It has two levels— Clarinex or placebo.
(d) Double-blind means that neither the subject nor the individual monitoring the subject knows which treatment the subject is receiving (Clarinex or placebo).
(e) It is important to have a placebo group for two reasons. (1) So there is a baseline group against which to judge the Clarinex group, and (2) to eliminate any effect due to psychosomatic behavior.
(f) The conditions for using the normal model to conduct the hypothesis test are satisfied. Classical approach: $z_0 = 2.07 > z_{0.05} = 1.645$, so reject H_0. P-value approach: P-value $= 0.0192$ [Tech: 0.0166] $< \alpha = 0.05$, reject H_0.
There is sufficient evidence at the 0.05 level of significance to conclude that the proportion of individuals taking Clarinex who experience dry mouth is greater than the proportion of individuals taking a placebo who experience dry mouth.
(g) The difference in sample proportions is 0.011 (0.030 vs. 0.019). This is a small difference (about 1 in 100) and no reason to avoid Clarinex (and suffer from allergies) simply to avoid a potential dry mouth. The statistical significance is partially due to the large sample size (which was necessary to be able to use the normal model).

27. (a) To remove any potential nonsampling error due to the respondent hearing the word "right" or "wrong" first.
(b) Lower bound: 0.282, upper bound: 0.338 [Tech: 0.339]. We are 90% confident that the difference in the proportion of adult Americans who believe the United States made the right decision to use military force in Iraq from 2003 to 2010 is between 0.282 and 0.338. The attitude regarding the decision to go to war changed substantially.

29. (a) $\hat{p}_{male} = 0.317$, $\hat{p}_{female} = 0.214$
(b) $H_0: p_{male} = p_{female}$ vs. $H_1: p_{male} \neq p_{female}$. Classical approach: $z_0 = 1.64$ is between $-z_{0.025} = -1.96$ and $z_{0.025} = 1.96$; do not reject H_0. P-value approach: P-value $= 0.1010$ [Tech: 0.1001] $> \alpha = 0.05$; do not reject H_0. There is not sufficient evidence at the $\alpha = 0.05$ level of significance to conclude that the proportion of males and females willing to pay higher taxes to reduce the deficit differs.

31. (a) In sentence A, the verbs are "was having" and "was taking." In sentence B, the verbs are "had" and "took."
(b) $H_0: p_A = p_B$ vs. $H_1: p_A \neq p_B$, P-value $= 0.0013$. A level of significance is not specified, but since the P-value is very small we reject H_0. There is sufficient evidence to conclude that the sentence structure makes a difference.
(c) Answers will vary. The wording in sentence A suggests that the actions were taking place over a period of time and habitual behavior may be continuing in the present or might be expected to continue at some time in the future. The wording in sentence B suggests events that are concluded and were possibly brief in duration or one-time occurrences.

33. (a) $n = n_1 = n_2 = 1406$
(b) $n = n_1 = n_2 = 2135$

35. (a) The response variable is whether the fund manager outperforms the market, or not. The explanatory variable is whether it is a high-dispersion year or a low-dispersion year.
(b) The individuals are the fund managers.
(c) This study applies to the population of fund managers.
(d) $H_0: p_{high} = p_{low}$ vs. $H_1: p_{high} > p_{low}$
(e) If the null hypothesis were true and we conducted the study 100 times, we would expect to observe the results as extreme or more extreme than the results observed in about eight of the studies. There is some evidence to suggest that the proportion of fund managers who outperform the market in high-dispersion years is greater than the proportion of fund managers who outperform the market in low-dispersion years.

37. A pooled estimate of p is the best point estimate of the common population proportion p. However, when finding a confidence interval, the sample proportions are not pooled because no assumption about their equality is made.

11.2 Assess Your Understanding (page 543)

1. $<$

3. (a)

Observation	1	2	3	4	5	6	7
d_i	-0.5	1	-3.3	-3.7	0.5	-2.4	-2.9

(b) $\bar{d} = -1.614$; $s_d = 1.915$

(c) Classical approach: $t_0 = -2.230 < -t_{0.05} = -1.943$; reject the null hypothesis. P-value approach: $0.025 < P\text{-value} < 0.05$ [Tech: P-value $= 0.0336] < \alpha = 0.05$; reject the null hypothesis. There is sufficient evidence at the $\alpha = 0.05$ level of significance to reject the null hypothesis that $\mu_d = 0$.

(d) We can be 95% confident that the mean difference is between -3.39 and 0.16.

5. (a) $H_0: \mu_d = 0$ vs. $H_1: \mu_d > 0$

(b) Classical approach: $t_0 = 2.606 > t_{0.05} = 1.753$; reject the null hypothesis. P-value approach: $0.005 < P\text{-value} < 0.01$ [Tech: P-value $= 0.0099] < \alpha = 0.05$; reject the null hypothesis. There is sufficient evidence at the $\alpha = 0.05$ level of significance that a baby will watch the climber approach the hinderer toy for a longer time, on average, than the baby will watch the climber approach the helper toy.

(c) Answers will vary. The fact that the babies watch the surprising behavior for a longer period of time suggests that they are curious about it.

7. (a) This is matched-pairs data because two measurements (A and B) are taken on the same round.

(b) Hypotheses: $H_0: \mu_d = 0$, $H_1: \mu_d \neq 0$; $d_i = A_i - B_i$. Classical approach: $t_0 = 0.852$; do not reject the null hypothesis since $-t_{0.005} = -3.106 < 0.852 < t_{0.005} = 3.106$. P-value approach: $0.50 > P\text{-value} > 0.40$ [Tech: P-value $= 0.4125] > \alpha = 0.01$; do not reject the null hypothesis. There is not sufficient evidence at the $\alpha = 0.01$ level of significance to conclude that there is a difference in the measurements of velocity between device A and device B.

(c) Lower bound: -0.309, upper bound: 0.542. We are 99% confident that the mean difference in measurement is between -0.31 and 0.54 feet per second.

(d)

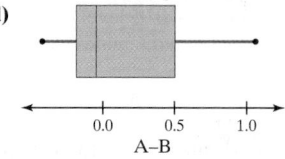

Yes, the boxplot supports that there is no difference in measurements.

9. (a) These are matched pairs because the car and the SUV were involved in the same collision.

(b)

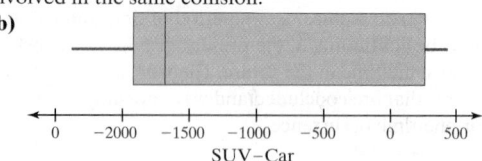

The median of the differences is well to the left of 0 suggesting that SUVs do have a lower repair cost.

(c) $H_0: \mu_d = 0$ vs. $H_1: \mu_d < 0$. Classical approach: $t_0 = -2.685 < -t_{0.05} = -1.943$; reject the null hypothesis. P-value approach: $0.01 < P\text{-value} < 0.02$ [Tech: P-value $= 0.0181] < \alpha = 0.05$; reject the null hypothesis. There is sufficient evidence at the $\alpha = 0.05$ level of significance to suggest that the mean repair cost for the car is higher.

11. Hypotheses: $H_0: \mu_d = 0$, $H_1: \mu_d > 0$; $d_i = Y_i - X_i$. Classical approach: $t_0 = 0.392 < t_{0.10} = 1.356$; do not reject the null hypothesis. P-value approach: P-value > 0.25 [Tech: P-value $= 0.3508] > \alpha = 0.10$; do not reject the null hypothesis. No; there is not sufficient evidence at the $\alpha = 0.10$ level of significance to conclude that sons are taller than their fathers.

13. $H_0: \mu_d = 0$; $H_1: \mu_d \neq 0$; $d_i = \text{diamond} - \text{steel}$, $\bar{d} = 1.333$; $s_d = 1.5$, $t_{0.025} = 2.306$. Lower bound: 0.2; upper bound: 2.5. We are 95% confident that the difference in hardness reading is between 0.2 and 2.5. Because the interval does not include 0, we

reject the null hypothesis. There is sufficient evidence to conclude that the two indenters produce different hardness readings.

15. (a) To control for any "learning" that may occur in using the simulator.

(b) Hypotheses: $H_0: \mu_d = 0$, $H_1: \mu_d \neq 0$. Lower bound: 0.688, upper bound: 1.238. $d_i = Y_i - X_i$. We can be 95% confident that the mean difference in reaction time when teenagers are driving impaired from when driving normally is between 0.688 second and 1.238 seconds. Because the interval does not contain zero, we reject the null hypothesis. There is sufficient evidence to conclude there is a difference in braking time with impaired vision and normal vision.

17. (a) Drivers and cars behave differently, so this reduces variability in mpg attributable to the driver's driving style.

(b) Driving conditions also affect mpg. By conducting the experiment on a closed track, driving conditions are constant.

(c) Neither variable is normally distributed.

(d) The difference in mileage appears to be approximately normal.

(e) $H_0: \mu_d = 0$; $H_1: \mu_d > 0$; $d_i = 92 \text{ Oct}_i - 87 \text{ Oct}_i$. P-value $= 0.141$. We would expect to get the results we obtained or a more extreme difference in about 14 samples out of 100 samples if the statement in the null hypothesis were true. Our results are not unusual; therefore, do not reject H_0.

11.3 Assess Your Understanding (page 554)

1. (a) $H_0: \mu_1 = \mu_2$, $H_1: \mu_1 \neq \mu_2$. Classical approach: $t_0 = 0.898$ is between $-t_{0.025} = -2.145$ and $t_{0.025} = 2.145$; do not reject H_0. P-value approach: $0.40 > P\text{-value} > 0.30$ [Tech: P-value $= 0.3767] > \alpha = 0.05$; do not reject H_0. There is not sufficient evidence at the $\alpha = 0.05$ level of significance to conclude that the population means are different.

(b) Lower bound: -1.53 [Tech: -1.41], upper bound: 3.73 [Tech: 3.61]

3. (a) $H_0: \mu_1 = \mu_2$, $H_1: \mu_1 > \mu_2$. Classical approach: $t_0 = 3.081 > t_{0.10} = 1.333$; reject H_0. P-value approach: $0.0025 < P\text{-value} < 0.005$ [Tech: P-value $= 0.0024] < \alpha = 0.10$; reject H_0. There is sufficient evidence at the $\alpha = 0.10$ level of significance to conclude that $\mu_1 > \mu_2$.

(b) Lower bound: 3.57 [Tech: 3.67], upper bound: 12.83 [Tech: 12.73]

5. $H_0: \mu_1 = \mu_2$; $H_1: \mu_1 < \mu_2$. Classical approach: $t_0 = -3.158 < -t_{0.02} = -2.172$; reject H_0. P-value approach: $0.001 < P\text{-value} < 0.0025$ [Tech: P-value $= 0.0013] < \alpha = 0.02$; reject H_0. There is sufficient evidence at the $\alpha = 0.02$ level of significance to conclude that $\mu_1 < \mu_2$.

7. (a) The response variable is time to graduate. The explanatory variable is community college first, or not.

(b) There are two independent groups: those who enroll directly in four-year institutions, and those who first enroll in community college and the response variable is quantitative. Treat each sample as a simple random sample. Each sample size is large, so each sample mean is approximately normal. Each population is small relative to its population size.

(c) $H_0: \mu_{CC} = \mu_{NT}$ vs. $H_1: \mu_{CC} > \mu_{NT}$. Classical approach: $t_0 = 12.977 > t_{0.01} \approx 2.364$ (using 100 df); reject H_0. P-value approach: P-value < 0.0005 [Tech: P-value < 0.0001]; reject H_0. The evidence suggests that the mean time to graduate for students who first start in community college is longer than the mean time to graduate for those who do not transfer.

(d) Lower bound: 0.847 [Tech: 0.848], upper bound: 1.153 [Tech: 1.152]. We are 95% confident that the mean additional time to graduate for students who start in community college is between 0.847 year and 1.153 years.

(e) No; this is observational data. Community college students may be working more hours, which does not allow them to take additional classes.

9. (a) This is an observational study. The researcher did not influence the data.

(b) (1) Treat the data as a simple random sample, (2) the samples are obtained independently, (3) the sample sizes are large, and (4) each sample is small relative to the population size.

(c) $H_0: \mu_A = \mu_D$, $H_1: \mu_A \neq \mu_D$. Classical approach: $t_0 = 0.846$ is between $-t_{0.025} = -2.032$ and $t_{0.025} = 2.032$; do not reject H_0. P-value approach: $0.50 > P\text{-value} > 0.40$ [Tech: P-value $= 0.4013] > \alpha = 0.05$; do not reject H_0. There is not sufficient evidence at the $\alpha = 0.05$ level of significance to say that travelers walk at different speeds depending on whether they are arriving or departing an airport.

11. (a) Lower bound: 4.71, upper bound: 6.29, using 100 degrees of freedom. We are 95% confident that the mean difference in scores between students who think about being a professor and students who think about soccer hooligans is between 4.71 and 6.29.
(b) Since the 95% confidence interval does not contain 0, the results suggest that priming does have an effect on scores.

13. (a)

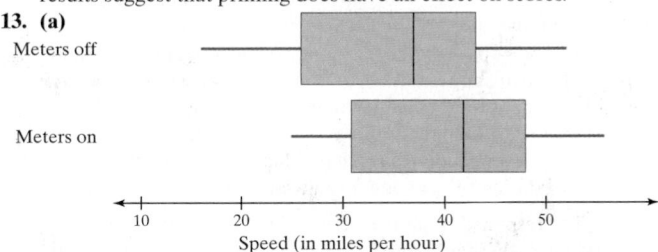

Speed (in miles per hour)

There are no outliers. The speed with meters on appears to be higher than with meters off.
(b) $H_0: \mu_{on} = \mu_{off}, H_1: \mu_{on} > \mu_{off}$. Classical approach: $t_0 = 1.713 > t_{0.10} = 1.345$; reject the null hypothesis. P-value approach: $0.05 < P$-value < 0.10 [Tech: P-value $= 0.0489$]; reject the null hypothesis. There is sufficient evidence at the $\alpha = 0.10$ level of significance that the ramp meters are effective in maintaining higher speed on the freeway.

15. $H_0: \mu_{carpet} = \mu_{no\ carpet}$, $H_1: \mu_{carpet} > \mu_{no\ carpet}$. Classical approach: $t_0 = 0.956 < t_{0.05} = 1.895$; do not reject H_0. P-value approach: $0.20 > P$-value > 0.15 [Tech: P-value $= 0.1780$] $> \alpha = 0.05$; do not reject H_0. There is not sufficient evidence at the $\alpha = 0.05$ level of significance to conclude that carpeted rooms have more bacteria than uncarpeted rooms.

17. (a)

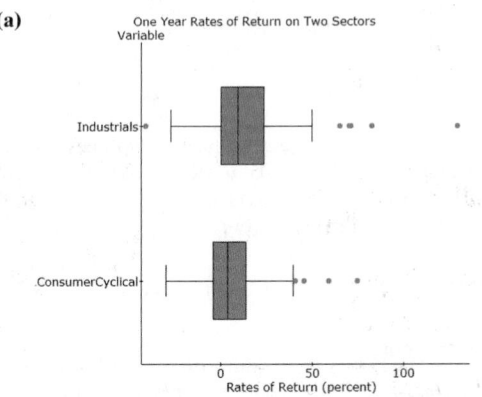

Rates of Return (percent)

Industrial stocks appear to have a higher median rate of return.
(b) (1) Treat each sample as a simple random sample, (2) each sample is obtained independently of the other, (3) each sample size is large, and (4) each sample size is small relative to the size of its population. The response variable is quantitative and there are two groups to compare.
(c) $H_0: \mu_{cc} = \mu_I$ vs. $H_0: \mu_{cc} \neq \mu_I$. Classical approach: $t_0 = -2.528 < -t_{0.025} = -1.984$ (100 df). Reject H_0. P-value approach: $0.01 < P$-value < 0.02 [Tech: P-value $= 0.0123$]. Reject H_0. There is sufficient evidence to conclude that the mean rate of return on consumer cyclical stocks differs from the mean rate of return of industrial stocks.
(d) Lower bound: 1.684 [Tech: 1.719], Upper bound : 13.976 [Tech: 13.942]
We are 95% confident the mean difference in rate of return of industrial stocks versus consumer cyclical stocks is between 1.684% and 13.976%. This suggests that the one-year rate of return on industrial stocks was higher than consumer cyclical stocks by somewhere between 1.684% and 13.976% for this time period.

19. Lower bound: 0.71 [Tech: 0.72], upper bound: 2.33 [Tech: 2.32]. We can be 90% confident that the mean difference in daily leisure time between adults without children and those with children is between 0.71 and 2.33 hours. Since the confidence interval does not include zero, we can conclude that there is a significant difference in the leisure time of adults without children and those with children.

21. (a) $H_0: \mu_{men} = \mu_{women}$ versus $H_1: \mu_{men} < \mu_{women}$
(b) P-value $= 0.0051$. Because P-value $< \alpha$, we reject the null hypothesis. There is sufficient evidence at the $\alpha = 0.01$ level of significance to conclude that the mean step pulse of men is lower than the mean step pulse of women.
(c) Lower bound: -10.7, upper bound: -1.5. We are 95% confident that the mean step pulse of men is between 1.5 and 10.7 beats per minute lower than the mean step pulse of women.

23. (a) Completely randomized design
(b) Final exam score; online versus traditional homework
(c) Teacher; location; time; text; syllabus; tests
(d) The assumption is that the students "randomly" enrolled in the course.
(e) $H_0: \mu_F = \mu_S$; $H_1: \mu_F > \mu_S$. Classical: $t_0 = 1.795 > t_{0.05} = 1.711$; reject H_0. $0.025 < P$-value < 0.05 [Tech: P-value $= 0.0395$] $< \alpha = 0.05$; reject H_0. There is sufficient evidence at the $\alpha = 0.05$ level of significance to conclude that the final exam scores in the fall semester were higher than the final exam scores in the spring semester. It would appear to be the case that the online homework system helps in raising final exam scores.
(f) One factor is the fact that the weather is pretty lousy at the end of the fall semester, but pretty nice at the end of the spring semester. If "spring fever" kicked in for the spring semester students, then they probably studied less for the final exam.

25. The sampling method is independent (the freshman cannot be matched to the corresponding seniors). Therefore, the inferential method that may be applied is a two-sample t-test. This comparison, however, has major shortcomings. The goal of the CLA+ is to measure gains in critical thinking, analytical reasoning, and so on, as a result of four years of college. The logical design to measure this as a matched-pairs design where the exam is administered before and after college to the same student.

27. The degrees of freedom obtained from Formula (2) are larger than the smaller of $n_1 - 1$ or $n_2 - 1$, and t_α decreases as the degrees of freedom increase. The larger the critical value is, the harder it is to reject the null hypothesis.

11.4 Assess Your Understanding (page 566)

1. $F_{0.05,9,10} = 3.02$
3. $F_{0.975,6,8} = 0.18$, $F_{0.025,6,8} = 4.65$
5. $F_{0.90,25,20} = 0.58$ **7.** $F_{0.05,45,15} = 2.19$
9. Classical approach: $F_0 = 0.84$ is between $F_{0.975,15,15} = 0.35$ and $F_{0.025,15,15} = 2.86$; do not reject the null hypothesis. P-value approach: P-value $= 0.7730 > \alpha = 0.05$; do not reject the null hypothesis. There is not sufficient evidence at the $\alpha = 0.05$ level of significance to conclude that $\sigma_1 \neq \sigma_2$.
11. Classical approach: $F_0 = 2.39 < F_{0.01,25,18} = 2.84$; do not reject the null hypothesis. P-value approach: P-value $= 0.0303 > \alpha = 0.01$; do not reject the null hypothesis. There is not sufficient evidence at the $\alpha = 0.01$ level of significance to conclude that $\sigma_1 > \sigma_2$.
13. Classical approach: $F_0 = 0.40 < F_{0.90,50,25} = 0.65$; reject the null hypothesis. P-value approach: P-value $= 0.0026 < \alpha = 0.10$; reject the null hypothesis. There is sufficient evidence at the $\alpha = 0.10$ level of significance to conclude that $\sigma_1 < \sigma_2$.
15. Hypotheses: $H_0: \sigma_1 = \sigma_2, H_1: \sigma_1 \neq \sigma_2$. Classical approach: $F_0 = 1.311 > F_{0.025,267,1144} = 1.29$; reject the null hypothesis. P-value approach: P-value $= 0.0036 < \alpha = 0.05$; reject the null hypothesis. There is sufficient evidence at the $\alpha = 0.05$ level of significance to suggest that the standard deviation time to earn a bachelor's degree is different for students who attend a community college than for students who immediately attend a four-year institution.
17. Hypotheses: $H_0: \sigma_1 = \sigma_2, H_1: \sigma_1 > \sigma_2$. Classical approach: $F_0 = 6.94 > F_{0.01,64,64} = 1.91$; reject the null hypothesis. P-value approach: P-value $< 0.0001 < \alpha = 0.01$; reject the null hypothesis. There is sufficient evidence at the $\alpha = 0.01$ level of significance to conclude that the treatment group had a higher standard deviation for serum retinal concentration than did the control group.
19. (a) Hypotheses: $H_0: \sigma_1 = \sigma_2, H_1: \sigma_1 < \sigma_2$. Classical approach: $F_0 = 0.32 < F_{0.95,19,19} \approx 0.47$; reject the null hypothesis. P-value

approach: P-value $= 0.0087 < \alpha = 0.05$; reject the null hypothesis. There is sufficient evidence at the $\alpha = 0.05$ level of significance to conclude that the standard deviation for wait time in the single line is less than the standard deviation for wait time in the multiple lines. The variability in the single line is less than that of the multiple line.

(b)

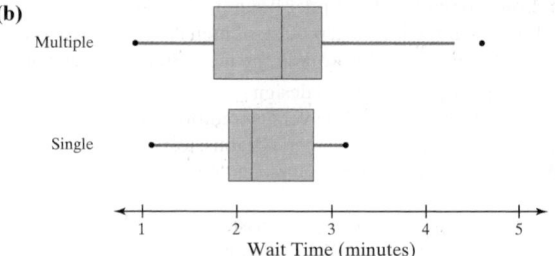

21. The normal probability plots suggest that the data are approximately normally distributed. Classical approach: $F_0 = 1.56 > F_{0.05,60,100} = 1.45$. Reject H_0; P-value approach: P-value $= 0.015 < \alpha = 0.05$. Reject H_0. There is sufficient evidence to suggest that industrial stocks are riskier than consumer cyclical stocks.

11.5 Assess Your Understanding (page 568)

1. $H_0: p_1 = p_2$ vs. $H_1: p_1 \neq p_2$. All requirements to conduct the test are satisfied. Classical: $z_0 = -1.18$ is between $-z_{0.005} = -2.575$ and $z_{0.005} = 2.575$; do not reject the null hypothesis. P-value $= 0.238$ [Tech: 0.242] $> \alpha = 0.01$; do not reject the null hypothesis. There is not sufficient evidence at the $\alpha = 0.05$ level of significance to conclude that there is a difference in the proportions.

3. $H_0: \mu_1 = \mu_2$ vs. $H_1: \mu_1 \neq \mu_2$. All requirements to conduct the test are satisfied. Classical: $t_0 = -1.39$ is between $-t_{0.025} = -2.179$ and $t_{0.025} = 2.179$; do not reject the null hypothesis. $0.20 > P$-value > 0.10 [Tech: P-value $= 0.1745$] $> \alpha = 0.05$; do not reject the null hypothesis. There is not sufficient evidence at the $\alpha = 0.05$ level of significance to conclude that there is a difference in the means.

5. $H_0: \sigma_1 = \sigma_2$ vs. $H_1: \sigma_1 \neq \sigma_2$. Classical: $F = 1.84 > F_{0.025,60,56} = 1.72$; reject the null hypothesis. P-value $= 0.0228 < \alpha = 0.05$; reject the null hypothesis. There is sufficient evidence at the $\alpha = 0.05$ level of significance to conclude that $\sigma_1 \neq \sigma_2$.

7. $H_0: \mu_d = 0$ vs. $H_1: \mu_d > 0$; $d_i = Y_i - X_i$. Classical: $t_0 = 1.324 < t_{0.05} = 2.132$; do not reject the null hypothesis. $0.15 > P$-value > 0.10 [Tech: P-value $= 0.128$] $> \alpha = 0.05$; do not reject the null hypothesis. There is not sufficient evidence at the $\alpha = 0.05$ level of significance to conclude that the treatment is effective.

9. (a) There are a few very large collision claims relative to the majority of claims.
(b) $H_0: \mu_{30-59} = \mu_{20-24}$ vs. $H_1: \mu_{30-59} < \mu_{20-24}$. All requirements to conduct the test are satisfied; use $\alpha = 0.05$. Classical: $t_0 = -1.890 < -t_{0.05} = -1.685$; reject the null hypothesis. $0.025 < P$-value < 0.05 [Tech: P-value $= 0.0313$] $< \alpha = 0.05$; reject the null hypothesis. There is sufficient evidence at the $\alpha = 0.05$ level of significance to conclude that the mean collision claim of a 30- to 59-year-old is less than the mean claim of a 20- to 24-year-old. Given that 20- to 24-year-olds tend to claim more for each accident, it makes sense to charge them more for coverage.

11. (a) The response variable is age.
(b) The sampling method is dependent because each husband is matched with the wife.
(c) To estimate the mean difference, construct a 95% confidence interval for the mean difference of the data. First, verify the differences are approximately normal by drawing a normal probability plot and verify the differences contain no outliers by drawing a boxplot. Compute the differences as "Husband minus Wife." Lower bound: –0.5; upper bound: 6.0. We are 95% confident the mean difference in age between a husband and wife is between −0.5 year and 6.0 years.

13. This requires a single sample t-test for a population mean. $H_0: \mu = 92.0$ mph versus $H_1: \mu > 92.0$ mph. The sample size is small. A normal probability plot indicates the data could come from a population that is normally distributed. A boxplot indicates there are no outliers. Classical approach: $t_0 = 1.086 < t_{0.05} = 1.771$. Do not reject

the null hypothesis. P-value approach: $0.10 < P$-value < 0.15 [Tech: P-value $= 0.1485$]. Do not reject the null hypothesis. There is not sufficient evidence to conclude the pitcher has a fastball that exceeds 92.0 mph.

15. $H_0: p = 0.5$ vs. $H_1: p > 0.5$. All requirements to conduct the test are satisfied. Classical approach: $z_0 = 9.217 > z_{0.05} = 1.645$; reject the null hypothesis. P-value approach: P-value < 0.0002 [Tech: P-value < 0.0001] $< \alpha = 0.05$; reject the null hypothesis. There is sufficient evidence at the $\alpha = 0.05$ level of significance to suggest that a quick 1-second view of a black and white photo represents enough information to judge the winner of an election.

17. $H_0: p_{>100K} = p_{<100K}$ vs. $H_1: p_{>100K} \neq p_{<100K}$. Lower bound: 0.019 [Tech: 0.020], upper bound: 0.097. Because the confidence interval does not include 0, there is sufficient evidence at the $\alpha = 0.05$ level of significance to conclude that there is a difference in the proportions. It seems that a higher proportion of individuals who earn over \$100,000 per year feel it is morally wrong for unwed women to have children.

19. (a) The response variable is score on the quiz. The explanatory variable is whether texting was required or the cell phone was turned off (no texting allowed).
(b) Students were randomly given instructions at the beginning of the class. Both groups received the same lecture.
(c) The sampling method is independent.
(d) $H_0: \mu_{text} = \mu_{cell\,off}$ versus $H_1: \mu_{text} < \mu_{cell\,off}$; the sample sizes are large enough. Assume this study applies to all students, so the sample size is less than 5% of the population size. Classical approach: $t_0 = -6.141 < -t_{0.05} = -1.697$. Reject the null hypothesis. P-value approach: P-value < 0.0001. There is sufficient evidence to suggest the mean score in the texting group is less than the mean score in the cellphone off group. Apparently, students cannot multitask.
(e) Lower bound: -20.243 [Tech: -20.17], upper bound: -11.477 [Tech: -11.55]. We are 90% confident the texting group scored between 11.477 points and 20.243 points worse, on average, than the cell-phone off group.
(f) $H_0: \sigma^2_{text} = \sigma^2_{cell\,off}$ versus $H_0: \sigma^2_{text} \neq \sigma^2_{cell\,off}$. Assume the data are normally distributed. Classical approach: $F_0 = 1.106$ (using cell-phone off variance in the numerator). Because $F_0 < F_{0.025,\,30,\,25} = 2.18$, do not reject the null hypothesis. P-value approach: P-value $= 0.7852 > 0.05$, so we do not reject the null hypothesis. There is not sufficient evidence to suggest the variability in scores differs between the two groups.

21.

Variable	Hypothesis	P-value (Using Technology)	Conclusion
Age	$H_0: \mu_{Readmit} = \mu_{Non\text{-}readmit}$ $H_1: \mu_{Readmit} > \mu_{Non\text{-}readmit}$	0.0006	Reject the null hypothesis
Length of stay	$H_0: \mu_{Readmit} = \mu_{Non\text{-}readmit}$ $H_1: \mu_{Readmit} > \mu_{Non\text{-}readmit}$	< 0.0001	Reject the null hypothesis
Admission in previous calendar year	$H_0: p_{Readmit} = p_{Non\text{-}readmit}$ $H_1: p_{Readmit} > p_{Non\text{-}readmit}$	< 0.0001	Reject the null hypothesis
Season	$H_0: p_{Readmit} = p_{Non\text{-}readmit}$ $H_1: p_{Readmit} > p_{Non\text{-}readmit}$	0.0001	Reject the null hypothesis
Floor	$H_0: p_{Readmit} = p_{Non\text{-}readmit}$ $H_1: p_{Readmit} > p_{Non\text{-}readmit}$	0.3959	Do not reject the null hypothesis

From the analysis, the readmits were older and had a longer length of stay. A higher proportion of readmits were admitted in the previous calendar year and a higher proportion of readmits were discharged in the winter. The proportion of readmits who were on the cardiac floor is not significantly different from the proportion of non-readmits who were on the cardiac floor.

23. This study is best done using a matched-pairs design. Researchers should match males and females based on the characteristics that determine salary (career, experience, level of education, geographic location, and so on). Determine the difference in each pair and see if the mean difference (men salary minus women salary) is significantly greater than 0.

25. Hypothesis test for two proportions, independent sample (using a completely randomized design)

27. Confidence interval for a single mean

29. Two sample t-test of independent means

31. Matched-pairs design. Analyze using t-test for a dependent sample

33. Hypothesis test of two independent proportions

Chapter 11 Review Exercises (page 574)

1. Dependent; quantitative
2. Independent; quantitative
3. (a) $F_{0.05,8,9} = 3.23$
(b) $F_{0.975,10,5} = 0.24$, $F_{0.025,10,5} = 6.62$
4. (a)

Observation	1	2	3	4	5	6
$d_i = X_i - Y_i$	−0.7	0.6	0	−0.1	−0.4	−0.4

(b) $\bar{d} = -0.167$; $s_d = 0.450$
(c) Hypotheses: $H_0: \mu_d = 0$, $H_1: \mu_d < 0$. Classical approach: $t_0 = -0.909 > -t_{0.05} = -2.015$; do not reject the null hypothesis. P-value approach: $0.25 > P$-value > 0.20 [Tech: P-value $= 0.2030$] $> \alpha = 0.05$; do not reject the null hypothesis. There is not sufficient evidence to conclude that the mean difference is less than zero.
(d) Lower bound: −0.79, upper bound: 0.45

5. (a) Hypotheses: $H_0: \mu_1 = \mu_2$, $H_1: \mu_1 \neq \mu_2$. Classical approach: $t_0 = 2.290 > t_{0.05} = 1.895$; reject the null hypothesis. P-value approach: $0.05 < P$-value < 0.10 [Tech: P-value $= 0.0351$] $< \alpha = 0.10$; reject the null hypothesis. There is sufficient evidence at the $\alpha = 0.1$ level of significance to conclude that $\mu_1 \neq \mu_2$.
(b) Lower bound: 0.73 [Tech: 1.01], upper bound: 7.67 [Tech: 7.39]
(c) Hypotheses: $H_0: \sigma_1 = \sigma_2$, $H_1: \sigma_1 \neq \sigma_2$. Classical approach: $F_0 = 1.40$ is between $F_{0.975,12,7} = 0.28$ and $F_{0.025,12,7} \approx 4.76$; do not reject the null hypothesis. P-value approach: P-value $= 0.6734 > \alpha = 0.05$; do not reject the null hypothesis. There is not sufficient evidence at the $\alpha = 0.05$ level of significance to conclude that the standard deviation in population 1 is different from the standard deviation in population 2.

6. (a) Hypotheses: $H_0: \mu_1 = \mu_2$, $H_1: \mu_1 > \mu_2$. Classical approach: $t_0 = 1.472 < t_{0.01} = 2.423$; do not reject the null hypothesis. P-value approach: $0.10 > P$-value > 0.05 [Tech: P-value $= 0.0726$] $> \alpha = 0.01$; do not reject the null hypothesis. There is not sufficient evidence at the $\alpha = 0.01$ level of significance to conclude that the mean of population 1 is larger than the mean of population 2.
(b) Hypotheses: $H_0: \sigma_1 = \sigma_2$, $H_1: \sigma_1 < \sigma_2$. Classical approach: $F_0 = 0.67 > F_{0.99,44,40} \approx 0.50$; do not reject the null hypothesis. P-value approach: P-value $= 0.0940 > \alpha = 0.01$; do not reject the null hypothesis. There is not sufficient evidence at the $\alpha = 0.01$ level of significance to conclude that the standard deviation in population 1 is less than the standard deviation in population 2.

7. $H_0: p_1 = p_2$, $H_1: p_1 \neq p_2$. Classical: $z_0 = -1.68$ is between $-z_{0.025} = -1.96$ and $z_{0.025} = 1.96$; do not reject the null hypothesis. P-value $= 0.0930$ [Tech: 0.0895] $> \alpha = 0.05$; do not reject the null hypothesis. There is not sufficient evidence at the $\alpha = 0.05$ level of significance to conclude that the proportion in population 1 is different from the proportion in population 2.

8. (a) The sampling method is dependent because the same individual is used for both measurements.
(b) Hypotheses: $H_0: \mu_d = 0$, $H_1: \mu_d \neq 0$. Classical approach: $t_0 = 0.512$ is between $-t_{0.025} = -2.262$ and $t_{0.025} = 2.262$; do not reject the null hypothesis. P-value approach: P-value > 0.50 [Tech: P-value $= 0.6209$] $> \alpha = 0.05$; do not reject the null hypothesis. There is not sufficient evidence at the $\alpha = 0.05$ level of significance to conclude that arm span is different from height. The sample evidence does not contradict the belief that arm span and height are the same.

9. (a) The sampling method is independent since the cars selected for the McDonald's sample had no bearing on the cars chosen for the Wendy's sample.

(b) Time in drive-through, which is quantitative.
(c) Hypotheses: $H_0: \mu_{McD} = \mu_W$, $H_1: \mu_{McD} \neq \mu_W$. Classical approach: $t_0 = -4.059 < -t_{0.05} = -1.706$; reject the null hypothesis. P-value approach: P-value < 0.0005 [Tech: P-value $= 0.0003$] $< \alpha = 0.10$; reject the null hypothesis. There is sufficient evidence at the $\alpha = 0.10$ level of significance to conclude that the wait times in the drive-throughs of the two restaurants differ.
(d)

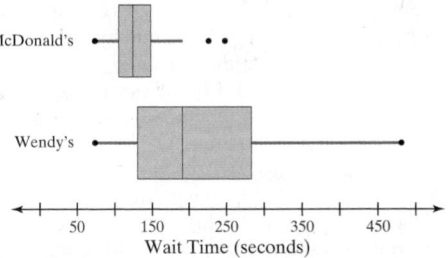

Based on the boxplots, it would appear to be the case that the wait time at McDonalds is less than the wait time at Wendy's.

10. (a) This is a completely randomized design with two treatments: placebo and 5 mg of Actonel. The response variable is whether the subject has a bone fracture over the course of one year, or not. It is qualitative with two outcomes.
(b) A double-blind experiment is one in which neither the subject nor the individual administering the treatment knows which group (experimental or control) the subject is in.
(c) Each sample is the result of a randomized experiment; $n_1\hat{p}_1(1 - \hat{p}_1) = 26 \geq 10$ and $n_2\hat{p}_2(1 - \hat{p}_2) = 45 \geq 10$; and each sample is less than 5% of the population size. Hypotheses: $H_0: p_{exp} = p_{control}$, $H_1: p_{exp} < p_{control}$. Classical approach: $z_0 = -2.68 < -z_{0.01} = -2.33$; reject the null hypothesis. P-value approach: P-value $= 0.0037$ [Tech: 0.0033] $< \alpha = 0.01$; reject the null hypothesis. There is sufficient evidence at the $\alpha = 0.01$ level of significance to conclude that a lower proportion of women in the experimental group experienced a bone fracture than in the control group.
(d) Lower bound: −0.06, upper bound: −0.01. We are 95% confident that the difference in the proportion of women who experienced a bone fracture between the experimental and control group is between −0.06 and −0.01.

11. (a) $n = n_1 = n_2 = 2136$ **(b)** $n = n_1 = n_2 = 3383$

12. Hypotheses: $H_0: \sigma_{McD} = \sigma_W$, $H_1: \sigma_{McD} < \sigma_W$. Classical approach: $F_0 = 0.15 < F_{0.95,29,26} \approx 0.45$; reject the null hypothesis. P-value $< 0.0001 < \alpha = 0.05$; reject the null hypothesis. There is sufficient evidence at the $\alpha = 0.05$ level of significance to conclude that the standard deviation in wait time at Wendy's is more than the standard deviation in wait time at McDonald's.

13. Lower bound: −1.37, upper bound: 2.17. Since the interval includes zero, we conclude that there is not sufficient evidence at the $\alpha = 0.05$ level of significance to reject the claim that arm span and height are equal.

14. Lower bound: −128.868 [Tech: −128.42]; upper bound: −42.218 [Tech: −42.67]. A marketing campaign could be initiated by McDonald's touting the fact that wait times are up to 2 minutes less at McDonald's.

Chapter 11 Test (page 575)

1. Independent
2. Dependent
3. (a)

Observation	1	2	3	4	5	6	7
$d_i = X_i - Y_i$	0.2	−0.5	0.2	0.6	−0.6	−0.5	−0.8

(b) $\bar{d} = -0.2$; $s_d = 0.526$
(c) Hypotheses: $H_0: \mu_d = 0$, $H_1: \mu_d \neq 0$. Classical approach: $t_0 = -1.006$ is between $-t_{0.005} = -3.707$ and $t_{0.005} = 3.707$; do not reject the null hypothesis. $0.40 > P$-value > 0.30 [Tech: P-value $= 0.3532$] $> \alpha = 0.01$; do not reject the null hypothesis. There is not sufficient evidence to conclude that the mean difference is different from zero.
(d) Lower bound: −0.69, upper bound: 0.29

4. (a) Hypotheses: $H_0: \mu_1 = \mu_2$, $H_1: \mu_1 \neq \mu_2$. Classical approach: $t_0 = -2.054 < -t_{0.05} = -1.714$. Reject the null hypothesis. $0.05 < P\text{-value} < 0.10$ [Tech: $P\text{-value} = 0.0464$] $< \alpha = 0.1$; reject the null hypothesis. There is sufficient evidence at the $\alpha = 0.1$ level of significance to conclude that the means are different.
(b) Lower bound: -12.44 [Tech: -12.3], upper bound: 0.04 [Tech: -0.10]
(c) Hypotheses: $H_0: \sigma_1 = \sigma_2$, $H_1: \sigma_1 > \sigma_2$. Classical approach: $F_0 = 2.0 > F_{0.10,23,26} \approx 1.72$; reject the null hypothesis. $P\text{-value} = 0.0448 < \alpha = 0.10$; reject the null hypothesis. There is sufficient evidence at the $\alpha = 0.10$ level of significance to conclude that the standard deviation in population 1 is greater than the standard deviation in population 2.

5. (a) Hypotheses: $H_0: \mu_1 = \mu_2$, $H_1: \mu_1 < \mu_2$. Classical approach: $t_0 = -1.357 > -t_{0.05} = -1.895$; do not reject the null hypothesis. $0.15 > P\text{-value} > 0.10$ [Tech: $P\text{-value} = 0.0959$] $> \alpha = 0.05$; do not reject the null hypothesis. There is not sufficient evidence at the $\alpha = 0.05$ level of significance to conclude that the mean of population 1 is less than the mean of population 2.
(b) Hypotheses: $H_0: \sigma_1 = \sigma_2$, $H_1: \sigma_1 \neq \sigma_2$. Classical approach: $F_0 = 1.64$ is between $F_{0.975,12,7} = 0.28$ and $F_{0.025,12,7} \approx 4.76$; do not reject the null hypothesis. $P\text{-value} = 0.5243 > \alpha = 0.05$; do not reject the null hypothesis. There is not sufficient evidence at the $\alpha = 0.05$ level of significance to conclude that the standard deviation in population 1 is different from the standard deviation in population 2.

6. $H_0: p_1 = p_2$ vs. $H_1: p_1 < p_2$ Classical: $z_0 = -0.80 > -z_{0.05} = -1.645$; do not reject the null hypothesis. $P\text{-value} = 0.2119$ [Tech: 0.2124] $> \alpha = 0.05$; do not reject the null hypothesis. There is not sufficient evidence at the $\alpha = 0.05$ level of significance to conclude that the proportion in population 1 is less than the proportion in population 2.

7. (a) The sampling method is independent since the dates selected for the Texas sample have no bearing on the dates chosen for the Illinois sample.
(b) Both samples must come from populations that are normally distributed.
(c)

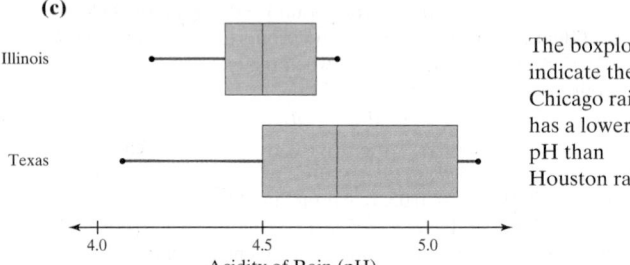

The boxplots indicate the Chicago rain has a lower pH than Houston rain.

(d) Hypotheses: $H_0: \mu_{\text{Texas}} = \mu_{\text{Illinois}}$, $H_1: \mu_{\text{Texas}} \neq \mu_{\text{Illinois}}$. Classical approach: $t_0 = 2.276 > t_{0.025} = 2.201$; reject the null hypothesis. $0.04 < P\text{-value} < 0.05$ [Tech: $P\text{-value} = 0.0387$] $< \alpha = 0.05$; reject the null hypothesis. There is sufficient evidence at the $\alpha = 0.05$ level of significance to conclude that the acidity of the rain near Houston is different from its acidity near Chicago.

8. (a) The response variable is the student's GPA. The explanatory variable is whether the student has a sleep disorder or not.
(b) Hypotheses: $H_0: \mu_{\text{sleep disorder}} = \mu_{\text{no sleep disorder}}$, $H_1: \mu_{\text{sleep disorder}} < \mu_{\text{no sleep disorder}}$. Classical approach: $t_0 = -3.784$ is less than $-t_{0.05} = -1.660$; reject H_0. $P\text{-value}$ approach: $P\text{-value} < 0.0005$ [Tech: $P\text{-value} < 0.0001$] $< \alpha = 0.05$; reject H_0. There is sufficient evidence at the $\alpha = 0.05$ level of significance to suggest that sleep disorders adversely affect a student's GPA.

9. Hypotheses: $H_0: \mu_d = 0$ vs. $H_1: \mu_d < 0$; Classical approach: $t_0 = -1.927 < -t_{0.10} = -1.440$; reject the null hypothesis. $P\text{-value}$ approach: $0.05 < P\text{-value} < 0.10$ [Tech: $P\text{-value} = 0.0511$] $< \alpha = 0.1$; reject the null hypothesis. There is sufficient evidence at the $\alpha = 0.1$ level of significance to suggest that the repair cost for the car is higher.

10. (a) Completely randomized design.
(b) Whether the subject gets dry mouth or not.
(c) Each sample is the result of a randomized experiment. $n_1\hat{p}_1(1 - \hat{p}_1) = 66 \geq 10$ and $n_2\hat{p}_2(1 - \hat{p}_2) = 31 \geq 10$; and each sample is less than 5% of the population size. Hypotheses: $H_0: p_{\text{exp}} = p_{\text{control}}$ $H_1: p_{\text{exp}} > p_{\text{control}}$. Classical

approach: $z_0 = 2.20 > z_{0.05} = 1.645$; reject the null hypothesis. $P\text{-value} = 0.0139$ [Tech: 0.0136] $< \alpha = 0.05$; reject the null hypothesis. There is sufficient evidence at the $\alpha = 0.05$ level of significance to conclude that a higher proportion of subjects in the experimental group experienced a dry mouth than in the control group.

11. $H_0: p_M = p_F$, $H_1: p_M \neq p_F$. Lower bound: 0.051, upper bound: 0.089. Because the confidence interval does not include 0, we reject the null hypothesis. There is sufficient evidence at the $\alpha = 0.1$ level of significance to conclude that the proportion of males and females for which hypnotism led to quitting smoking is different.

12. (a) $n = n_1 = n_2 = 762$
(b) $n = n_1 = n_2 = 1201$

13. $H_0: \mu_p = \mu_n$; $H_1: \mu_p \neq \mu_n$. Lower bound: 1.03 kg, upper bound: 1.17 kg. Because the confidence interval does not contain 0, we reject the null hypothesis. There is sufficient evidence to conclude that Naltrexone is effective in preventing weight gain among individuals who quit smoking. Answers will vary regarding practical significance, but one must ask, "Do I want to take a drug so that I can keep about 1 kg off?" Probably not.

14. Hypotheses: $H_0: \sigma_{\text{Tex}} = \sigma_{\text{Ill}}$, $H_1: \sigma_{\text{Tex}} \neq \sigma_{\text{Ill}}$. Classical approach: $F_0 = 5.64 > F_{0.025,11,13} \approx 3.325$; reject the null hypothesis. $P\text{-value} = 0.0044 < \alpha = 0.05$; reject the null hypothesis. There is sufficient evidence at the $\alpha = 0.05$ level of significance to conclude that the standard deviation in the acidity of rain near Houston is different from the standard deviation in the acidity of rain near Chicago.

Chapter 12 Inference on Categorical Data

12.1 Assess Your Understanding (page 587)

1. True   **3.** expected counts; np_i
5.

p_i	0.2	0.1	0.45	0.25
Expected counts	100	50	225	125

7. (a) $\chi_0^2 = 2.72$ **(b)** df $= 3$ **(c)** $\chi_{0.05}^2 = 7.815$
(d) Do not reject H_0, since $\chi_0^2 < \chi_{0.05}^2$. There is not sufficient evidence at the $\alpha = 0.05$ level of significance to conclude that any one of the proportions is different from the others.

9. (a) $\chi_0^2 = 12.56$ **(b)** df $= 4$ **(c)** $\chi_{0.05}^2 = 9.488$
(d) Reject H_0, since $\chi_0^2 > \chi_{0.05}^2$. There is sufficient evidence at the $\alpha = 0.05$ level of significance to conclude that X is not binomial with $n = 4$, $p = 0.8$.

11. H_0: the distribution of colors is as stated by M&M; H_1: the distribution is different from that stated by M&M. Classical approach: $\chi_0^2 = 6.744 < \chi_{0.05}^2 = 11.071$; do not reject the null hypothesis. $P\text{-value}$ approach: $P\text{-value} > 0.10 > \alpha = 0.05$ [Tech: $P\text{-value} = 0.2404$]; do not reject the null hypothesis. There is not sufficient evidence at the $\alpha = 0.05$ level of significance to conclude that the distribution of candies in a bag of M&Ms is different from 13% brown, 14% yellow, 13% red, 20% orange, 24% blue, and 16% green.

13. (a) Answers will vary depending on the desired probability of making a Type I error. One possible choice: $\alpha = 0.01$.
(b) H_0: the digits follow Benford's Law; H_1: the digits do not follow Benford's Law. Classical approach: $\chi_0^2 = 21.693$; compare χ_0^2 to your chosen χ_α^2. $P\text{-value}$ approach: $P\text{-value}$ is between 0.01 and 0.005 [Tech: $P\text{-value} = 0.0055$].
(c) Answers will vary. One possible answer: Since $P\text{-value} < 0.01$, there is sufficient evidence to conclude that Benford's Law is not followed and the employee is guilty of embezzlement.

15. (a) Classical approach: $\chi_0^2 = 121.367 > \chi_{0.05}^2 = 9.488$; reject the null hypothesis. $P\text{-value}$ approach: $P\text{-value} < 0.005 < \alpha = 0.05$ [Tech: $P\text{-value} < 0.0001$]; reject the null hypothesis. There is sufficient evidence at the $\alpha = 0.05$ level of significance to conclude that the distribution of fatal injuries for riders not wearing a helmet does not follow the distribution for all riders.

(b)

Location of Injury	Multiple Locations	Head	Neck	Thorax	Abdomen/Lumbar/Spine
Observed	1036	864	38	83	47
Expected	1178.76	641.08	62.04	124.08	62.04

We notice that the observed count for head injuries is much higher than expected, while the observed counts for all the other categories are lower. We might conclude that motorcycle fatalities from head injuries occur more frequently for riders not wearing a helmet.

17. **(a)** Group 1: 84; Group 2: 84; Group 3: 84; Group 4: 81. Classical approach: $\chi_0^2 = 0.084$ [Tech: 0.081] $< \chi_{0.05}^2 = 7.815$; do not reject the null hypothesis. P-value approach: P-value $> 0.99 > \alpha = 0.05$ [Tech: P-value $= 0.994$]; do not reject the null hypothesis. There is not sufficient evidence at the $\alpha = 0.05$ level of significance to conclude that there are differences among the groups in attendance patterns.
(b) Group 1: 84; Group 2: 81; Group 3: 78; Group 4: 76. Classical approach: $\chi_0^2 = 0.463$ [Tech: 0.461] $< \chi_{0.05}^2 = 7.815$; do not reject the null hypothesis. P-value approach: P-value $> 0.90 > \alpha = 0.05$ [Tech: P-value $= 0.9274$]; do not reject the null hypothesis. There is not sufficient evidence at the $\alpha = 0.05$ level of significance to conclude that there are differences among the groups in attendance patterns. It is curious that the farther a group's original position is located from the front of the room, the more the attendance rate for the group decreases.
(c) Group 1: 20; Group 2: 20; Group 3: 20; Group 4: 20. Classical approach: $\chi_0^2 = 2.60 < \chi_{0.05}^2 = 7.815$; do not reject the null hypothesis. P-value approach: P-value $> 0.10 > \alpha = 0.05$ [Tech: P-value $= 0.4575$]; do not reject the null hypothesis. There is not sufficient evidence at the $\alpha = 0.05$ level of significance to conclude that there is a significant difference in the number of students in the top 20% of the class by group.
(d) Though not statistically significant, the group located in the front had both better attendance and a larger number of students in the top 20%. Choose the front.

19. Classical approach: $\chi_0^2 = 60.689 > \chi_{0.01}^2 = 11.345$; reject the null hypothesis. P-value approach: P-value < 0.005 [Tech: P-value < 0.0001]; reject the null hypothesis. There is sufficient evidence to suggest that hockey players' birthdates are not evenly distributed throughout the year. Many more than expected are born in the early part of the year.

21. Classical approach: $\chi_0^2 = 13.227 > \chi_{0.05}^2 = 12.592$; reject the null hypothesis. P-value approach: P-value $< 0.05 = \alpha$ [Tech: P-value $= 0.0396$]; reject the null hypothesis. There is sufficient evidence at the $\alpha = 0.05$ level of significance to reject the belief that pedestrian deaths occur with equal frequency over the days of the week. We might conclude that fewer pedestrian deaths occur on Tuesdays and more occur on Saturdays.

23. **(a)** 64 students

Grade	Expected Number of Students
A	8.512
B	12.224
C	15.744
D	6.656
F	7.296
W	13.568

(b) Classical approach: $\chi_0^2 = 9.989 < \chi_{0.01}^2 = 15.086$; Do not reject H_0. P-value approach: $0.05 < P$-value < 0.10 [Tech: P-value $= 0.0756$]; Do not reject H_0. There is not sufficient evidence to conclude the grade distribution in the MRP program differs from traditional classes at the $\alpha = 0.01$ level of significance.
(c) It is a big change to adjust an entire curriculum, so we don't want to make a Type I error. That is, we don't want to conclude the grade distribution in the MRP program differs from traditional, when, in fact, it does not differ.
(d) Classical approach: $\chi_0^2 = 19.977 > \chi_{0.01}^2 = 15.086$; reject H_0. P-value approach: P-value < 0.005 [Tech: P-value $= 0.0013$] $< \alpha = 0.01$; reject H_0. There is sufficient evidence to conclude the distribution of grades in the MRP program differs from that of the traditional classes. Looking at the observed versus expected values, there is a higher number of Bs than expected, a lower number of Cs, and a higher number of Ds in the MRP program. With small sample sizes, the evidence against the null hypothesis must be overwhelming to be able to reject the statement in the null hypothesis. So, watch out for studies that suggest there is not significant evidence when the sample size is small.

25. **(a)** Expected number with low birth weight: 17.04; expected number without low birth weight: 222.96.
(b) $H_0: p = 0.071$ vs. $H_1: p > 0.071$. Classical approach: $\chi_0^2 = 1.554 < \chi_{0.05}^2 = 3.841$; do not reject the null hypothesis. P-value approach: P-value $> 0.10 > \alpha = 0.05$ Tech: P-value $= 0.2125$]; do not reject the null hypothesis. There is not sufficient evidence at the $\alpha = 0.05$ level of significance to conclude that mothers between the ages of 35 and 39 have a higher percentage of low-birth-weight babies.
(c) $H_0: p = 0.71$ vs. $H_1: p > 0.71$; $np_0(1 - p_0) = 15.83 > 10$. Classical approach: $z_0 = 1.25 < z_{0.05} = 1.645$; do not reject the null hypothesis. P-value approach: P-value $= 0.1056 > \alpha = 0.05$ [Tech: P-value $= 0.1063$]; do not reject the null hypothesis. There is not sufficient evidence at the $\alpha = 0.05$ level of significance to conclude that mothers between the ages of 35 and 39 have a higher percentage of low-birth-weight babies.

27. **(a)** $\mu = 0.9323$ hits
(b) All expected frequencies will not be greater than or equal to 1. Also, more than 20% (37.5%) of the expected frequencies are less than 5.
(c), (d)

x	$P(x)$	Expected Number of Regions
0	0.3936	226.741
1	0.3670	211.390
2	0.1711	98.540
3	0.0532	30.623
4 or more	0.0151	8.698

(e) Classical approach: $\chi_0^2 = 1.016 < \chi_{0.05}^2 = 9.488$; do not reject the null hypothesis. P-value approach: P-value $> 0.90 > \alpha = 0.05$ [Tech: P-value $= 0.9073$]; do not reject the null hypothesis. There is not sufficient evidence at the $\alpha = 0.05$ level of significance to conclude that the distribution of rocket hits is different from the Poisson distribution. That is, the rocket hits do appear to be modeled by a Poisson random variable.

29. **(a)** Quantitative   **(b)** $\bar{x} = \$41,130.9$, $M = \$41,215$
(c) $s = \$897.8$, IQR $= \$1592$
(d) The correlation between the raw data and normal scores is $0.991 > 0.939$ (Table VI), so it is reasonable to conclude that the data come from a population is normally distributed.
(e)

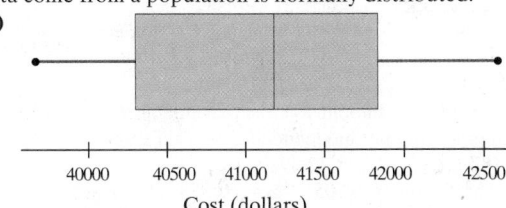

Cost (dollars)

(f) Lower bound: $\$40,722.7$ [Tech: $\$40,772.6$], upper bound: $\$41,539.1$ [Tech: $\$41,539.2$]; We are 90% confident the mean price of a new 2015 Buick Regal is between $\$40,722.7$ and $\$41,539.1$.
(g) Wider because there is more variability in the data. By removing the variability due to car type, the interval becomes more precise.

31. The χ^2 goodness-of-fit tests are always right tailed because the numerator in the test statistic is squared, making every test statistic other than a perfect fit positive. So we are measuring if $\chi_0^2 > \chi_\alpha^2$.

12.2 Assess Your Understanding (page 603)

1. True

3. **(a)** $\chi_0^2 = 1.698$
(b) Classical approach: $\chi_0^2 = 1.698 < \chi_{0.05}^2 = 5.991$; do not reject the null hypothesis. P-value approach: P-value $> 0.10 > \alpha = 0.05$ [Tech: P-value $= 0.4272$]; do not reject the null hypothesis. There is evidence at the $\alpha = 0.05$ level of significance to conclude that X and Y are independent. We conclude that X and Y are not related.

5. Classical approach: $\chi_0^2 = 1.989 < \chi_{0.01}^2 = 9.210$; do not reject the null hypothesis. P-value approach: P-value $> 0.10 > \alpha = 0.01$ [Tech: P-value $= 0.3699$]; do not reject the null hypothesis. There is not sufficient evidence at the $\alpha = 0.01$ level of significance to conclude that at least one of the proportions is different from the others.

7. (a)

Had Sexual Intercourse?	Both Biological/ Adoptive Parents	Single Parent	Parent and Stepparent	Nonparental Guardian
Yes	78.553	52.368	41.895	26.184
No	71.447	47.632	38.105	23.816

(b) (1) All expected frequencies are greater than or equal to 1, and (2) no more than 20% of the expected frequencies are less than 5.
(c) $\chi_0^2 = 10.357$
(d) H_0: family structure and sexual activity are independent; H_1: family structure and sexual activity are not independent. Classical approach: $\chi_0^2 = 10.357 > \chi_{0.05}^2 = 7.815$; reject the null hypothesis. P-value approach: P-value $< 0.025 < \alpha = 0.05$ [Tech: P-value $= 0.0158$]; reject the null hypothesis. There is sufficient evidence at the $\alpha = 0.05$ level of significance to conclude that sexual activity and family structure are associated.
(e) The biggest difference between observed and expected occurs under the family structure in which both parents are present. Fewer females were sexually active than was expected when both parents were present. This means that having both parents present seems to have an effect on whether the child is sexually active.
(f)

Had Sexual Intercourse?	Both Biological/ Adoptive Parents	Single Parent	Parent and Stepparent	Nonparental Guardian
Yes	0.427	0.59	0.55	0.64
No	0.573	0.41	0.45	0.36

Family Structure and Sexual Activity

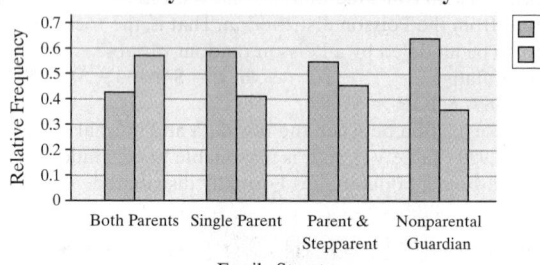

9. (a) H_0: health and happiness are independent; H_1: health and happiness are not independent. Classical approach: $\chi_0^2 = 182.174 > \chi_{0.05}^2 = 12.592$; reject the null hypothesis. P-value approach: P-value $< 0.005 < \alpha = 0.05$ [Tech: P-value < 0.0001]; reject the null hypothesis. There is sufficient evidence at the $\alpha = 0.05$ level of significance to conclude that happiness and health are dependent. That is, happiness and health are related to each other.
(b)

Happiness	Health			
	Excellent	Good	Fair	Poor
Very Happy	0.492	0.280	0.202	0.183
Pretty Happy	0.448	0.609	0.570	0.486
Not Too Happy	0.060	0.111	0.227	0.330

Health and Happiness

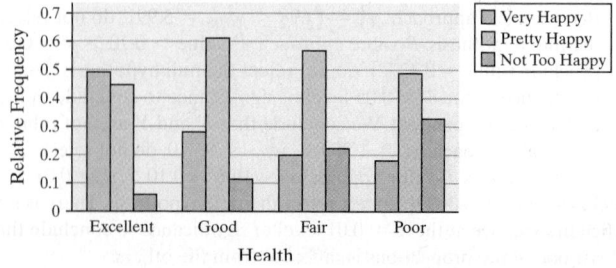

(c) The proportion of individuals who are "very happy" is much higher for individuals in "excellent" health than any other health category. Further, the proportion of individuals who are "not too happy" is much lower for individuals in "excellent" health compared to the other health categories. Put simply, the level of happiness seems to decline as health status declines.

11. (a) The data represent the measurement of two variables (weight classification and social well-being) on each individual in the study. Because two variables are measured on a single individual, use a chi-square test for independence.
(b) Classical: $\chi_0^2 = 7.167 < \chi_{0.05}^2 = 12.592$; do not reject H_0. P-value approach: P-value > 0.10 [Tech: P-value $= 0.306$]; do not reject H_0.
There is not sufficient evidence at the $\alpha = 0.05$ level of significance to conclude there is an association between social well-being and weight classification.
(c)

Social Well-Being and Weight

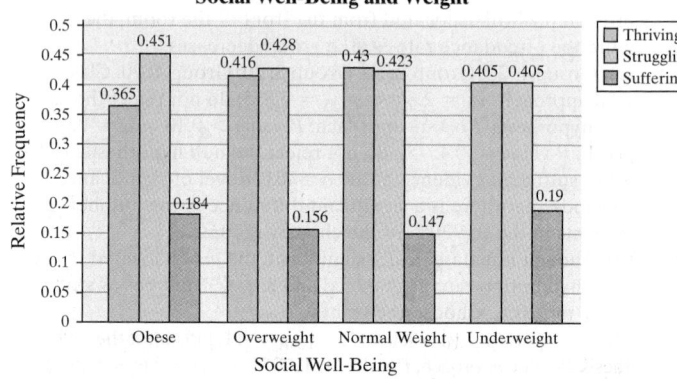

(d) Answers may vary. There is not enough sample evidence to suggest that one's social well-being is associated with one's weight classification. However, it is worth noting some differences in the relative frequencies. For example, both obese and underweight individuals have a higher relative frequency for "suffering."

13. (a) A completely randomized design with three levels of treatment.
(b) The response variable is whether the subject abstained from cigarette smoking, or not. It is qualitative with two possible outcomes.
(c) Current smokers
(d) $H_0: p_1 = p_2 = p_3$
H_1: At least one proportion differs from the others
(e) Classical approach: $\chi_0^2 = 3.960 < \chi_{0.05}^2 = 5.991$; do not reject H_0. P-value approach: P-value > 0.1 [Tech: P-value $= 0.1381$] $> \alpha = 0.05$; do not reject H_0.
(f)

Efficacy of e-Cigs

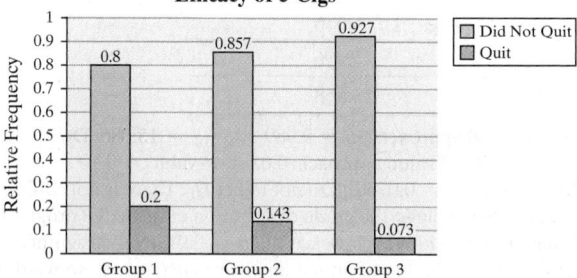

(g) There is not sufficient evidence at the 0.05 level of significance to suggest that the proportion of individuals who abstain from cigarette smoking in the three groups differs.

15. (a) Because there are three distinct populations that are being surveyed (Democrat, Republican, Independent) and the response variable is qualitative with two outcomes (positive/negative), we analyze the data using homogeneity of proportions.
(b) $H_0: p_D = p_I = p_R$; H_1: at least one proportion differs. Classical approach: $\chi_0^2 = 96.733 > \chi_{0.05}^2 = 5.991$; reject the null hypothesis. P-value approach: P-value < 0.005 [Tech: P-value < 0.0001]; reject the null hypothesis. There is sufficient evidence at the $\alpha = 0.05$ level of significance that a different proportion of individuals within each political affiliation reacts positively to the word socialism.

(c)

	Democrat	Independent	Republican
Positive	0.4409	0.2599	0.1501
Negative	0.5591	0.7401	0.8499
Total	1	1	1

(d) Independents and Republicans are far more likely to react negatively to the word *socialism* than Democrats are. However, it is important to note that a majority of Democrats in the sample did have a negative reaction, so the word *socialism* has a negative connotation among all groups.

17. (a)

	Course	Personal	Work
Female	13	5	3
Male	10	6	13

(b) Classical approach: $\chi_0^2 = 5.595 > \chi_{0.10}^2 = 4.605$; reject the null hypothesis. P-value approach: P-value $< 0.10 = \alpha$ [Tech: P-value $= 0.0609$]; reject the null hypothesis. The evidence suggests a relation between gender and drop reason. Females are more likely to drop because of the course, while males are more likely to drop because of work.

(c)

	Reason		
	Course	Personal	Work
Female	0.619	0.238	0.143
Male	0.345	0.207	0.448

Why Did You Drop?

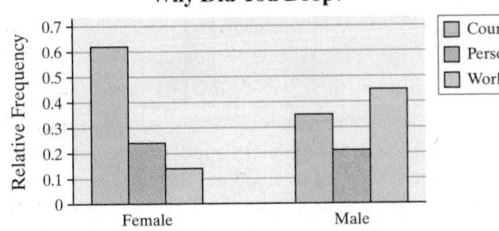

19. There are three different populations to study: (1) all capital letters, (2) all lower case letters, and (3) upper and lower case accurately. Obtain a random sample of applications from each population. Determine whether the loan corresponding to the application resulted in default. Conduct a hypothesis test for homogeneity of proportions to determine if the sample data suggest at least one population differs from the others.

21. (a) All healthy women 45 years of age or older
(b) Whether or not a cardiovascular event (such as heart attack or stroke) occurs: qualitative
(c) 100 mg aspirin, placebo   **(d)** Completely randomized
(e) The women were randomly assigned into two groups (aspirin vs. placebo). This randomization controls for any other explanatory variables, since individuals affected by these lurking variables should be equally dispersed between the two groups.
(f) Classical approach: $z_0 = -1.44 > z_{0.025} = -1.96$; do not reject the null hypothesis. P-value approach: P-value $= 0.1498$ [Tech: 0.1511]; do not reject the null hypothesis. There is not sufficient evidence to conclude that a difference exists between the proportions of cardiovascular events in the aspirin group versus the placebo group. The test statistic, rounded to three places, is $z_0 = -1.435$.
(g) Classical approach: $\chi_0^2 = 2.061 < \chi_{0.05}^2 = 3.841$; do not reject the null hypothesis. P-value approach: P-value $> 0.10 > \alpha = 0.05$ [Tech: P-value $= 0.1511$]; do not reject the null hypothesis. There is not sufficient evidence to conclude that a difference exists between the proportions of cardiovascular events in the aspirin group versus the placebo group.
(h) $(z_0)^2 = (-1.435)^2 = 2.059 \approx \chi_0^2 = 2.061$. Conclusion: $(z_0)^2 = \chi_0^2$.

23. Answers may vary, but differences should include chi-square test for independence compares two characteristics from a single population, whereas the chi-square test for homogeneity compares a single characteristic

from two (or more) populations. Similarities should include the procedures of the two tests and the assumptions of the two tests are the same.

12.3 Assess Your Understanding (page 610)

1. (a) $H_0: p_A = p_B$ vs. $H_1: p_A \neq p_B$
(b) $\chi_0^2 = 0.758$ **(c)** $\chi_{0.05}^2 = 3.841$
(d) P-value > 0.10 [Tech: 0.3841]

3. (a) This is a dependent sample because two variables (seat belt or not and smoke or not) are measured on the same individual.
(b) $H_0: p_{smoke} = p_{no\ seat\ belt}$ vs. $H_1: p_{smoke} \neq p_{no\ seat\ belt}$. Classical: $\chi_0^2 = 18.892 > \chi_{0.05}^2 = 3.841$; reject H_0. P-value: P-value < 0.005 [Tech: P-value < 0.0001] $< \alpha = 0.05$; reject H_0. There is sufficient evidence to suggest that there is a difference in the proportion who do not use a seat belt and the proportion who smoke. The sample proportion of smokers is 0.17, while the sample proportion of those who do not wear a seat belt is 0.13. Smoking appears to be the more popular hazardous activity.

5. $H_0: p_N = p_R$ vs. $H_1: p_N \neq p_R$. Classical: $\chi_0^2 = 5.994 > \chi_{0.05}^2 = 3.841$; reject H_0. P-value: $0.01 < P$-value < 0.025 [Tech: 0.0144] $< \alpha = 0.05$; reject H_0. There is sufficient evidence to conclude there is a difference in the proportion of words recognized in the two systems.

7. $H_0: p_g = p_d$ vs. $H_1: p_g \neq p_d$. Classical: $\chi_0^2 = 0.953 < \chi_{0.05}^2 = 3.841$; do not reject H_0. P-value: P-value > 0.10 [Tech: 0.329] $> \alpha = 0.05$; do not reject H_0. There is not sufficient to suggest the proportion of adult Americans who favor a law which would require a person to obtain a permit prior to purchasing a gun is different from the proportion who favor the death penalty for persons convicted of murder.

Chapter 12 Review Exercises (page 612)

1. H_0: the wheel is in balance; H_1: the wheel is out of balance. Classical approach: $\chi_0^2 = 0.578 < \chi_{0.05}^2 = 5.991$; do not reject the null hypothesis. P-value approach: P-value $> 0.10 > \alpha = 0.05$ [Tech: P-value $= 0.7489$]; do not reject the null hypothesis. There is not sufficient evidence at the $\alpha = 0.05$ level of significance to conclude that the wheel is out of balance. That is, the evidence suggests that the wheel is in balance.

2. H_0: the teams are evenly matched; H_1: the teams are not evenly matched. Classical approach: $\chi_0^2 = 6.905 < \chi_{0.05}^2 = 7.815$; do not reject the null hypothesis. P-value approach: $0.05 < P$-value $< 0.10 = \alpha$ [Tech: P-value $= 0.075$]; do not reject the null hypothesis. There is not sufficient evidence at the $\alpha = 0.05$ level of significance to conclude that the teams playing in the World Series have not been evenly matched. That is, the evidence suggests that the teams have been evenly matched. On the other hand, the P-value is suggestive of an issue. In particular, there are fewer six-game series than we would expect. Perhaps the team that is down "goes all out" in game 6, trying to force game 7.

3. (a) H_0: class is independent of survival status; H_1: class is associated with survival status. Classical approach: $\chi_0^2 = 133.052 > \chi_{0.05}^2 = 5.991$; reject H_0. P-value approach: P-value $< 0.005 < \alpha = 0.05$ [Tech: P-value < 0.0001]; reject H_0. There is sufficient evidence at the $\alpha = 0.05$ level of significance to conclude that the class of service and survival rates are dependent.

(b)

Titanic	Class		
	First	Second	Third
Survived	0.625	0.414	0.252
Did not survive	0.375	0.586	0.748

Titanic Survival

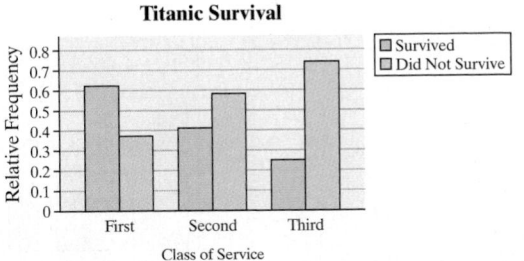

This summary supports the existence of a relationship between class and survival rate. Individuals with higher-class tickets survived in greater proportions than individuals with lower-class tickets.

4. H_0: gestational period is independent of degree; H_1: gestational period is not independent of degree. Classical approach: $\chi_0^2 = 10.239 > \chi_{0.05}^2 = 9.488$; reject the null hypothesis. P-value approach: P-value < 0.05 [Tech: P-value $= 0.0366$]; reject the null hypothesis. There is sufficient evidence at the $\alpha = 0.05$ level of significance to conclude that length of the gestational period and completion of a high school diploma are dependent.

5. (a) Classical approach: $\chi_0^2 = 37.518 > \chi_{0.05}^2 = 7.815$; reject the null hypothesis. P-value approach: P-value < 0.005 [Tech: P-value < 0.0001]; reject the null hypothesis. There is sufficient evidence at the $\alpha = 0.05$ level of significance that the level of funding received by the counties is associated with the candidate.

(b)

	High-Funded	Medium-Funded	Low-Funded	No Funding
Roosevelt	0.5994	0.5698	0.5400	0.4692
Landon	0.4006	0.4302	0.4600	0.5308
Total	1	1	1	1

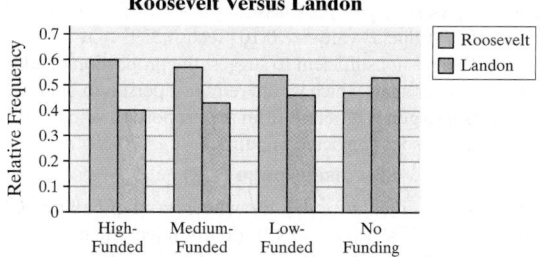

6. $H_0: p_v = p_j$ vs. $H_1: p_v \neq p_j$. Classical: $\chi_0^2 = 3.636 < \chi_{0.05}^2 = 3.841$; do not reject H_0. P-value: $0.05 < P$-value < 0.10 [Tech: 0.0565] $> \alpha = 0.05$; do not reject H_0. There is not sufficient evidence at the 0.05 level of significance to suggest the proportion of individuals who believe voting is a civic duty differs from the proportion who believe jury duty is a civic duty.

Chapter 12 Test (page 614)

1. H_0: the dice are fair; H_1: the dice are not fair. Classical approach: $\chi_0^2 = 4.940 < \chi_{0.01}^2 = 23.209$; do not reject the null hypothesis. P-value approach: P-value $> 0.10 > \alpha = 0.01$ [Tech: P-value $= 0.8951$]; do not reject the null hypothesis. There is not sufficient evidence at the $\alpha = 0.01$ level of significance to conclude that the dice are loaded. That is, the evidence suggests that the dice are fair.

2. H_0: educational attainment is the same today as 2000; H_1: educational attainment has changed since 2000. Classical approach: $\chi_0^2 = 2.661 < \chi_{0.10}^2 = 9.236$; do not reject the null hypothesis. P-value approach: P-value $> 0.10 = \alpha$ [Tech: P-value $= 0.7521$]; do not reject the null hypothesis. There is not sufficient evidence at the $\alpha = 0.10$ level of significance to conclude that the distribution of educational attainment of Americans today is different from the distribution of education attainment in 2000. That is, the evidence suggests that educational attainment has not changed.

3. (a) $H_0: p_R = p_I = p_D$; H_1: at least one proportion differs. Classical approach: $\chi_0^2 = 41.767 > \chi_{0.05}^2 = 5.991$; reject the null hypothesis. P-value approach: P-value $< 0.005 < \alpha = 0.05$ [Tech: P-value < 0.0001]; reject the null hypothesis. There is sufficient evidence at the $\alpha = 0.05$ level of significance to conclude that at least one proportion is different from the others. That is, the evidence suggests that the proportion of adults who feel morality is important when deciding how to vote is different for at least one political affiliation.

(b)

	Republican	Independent	Democrat
Important	0.920	0.810	0.820
Not Important	0.080	0.190	0.180

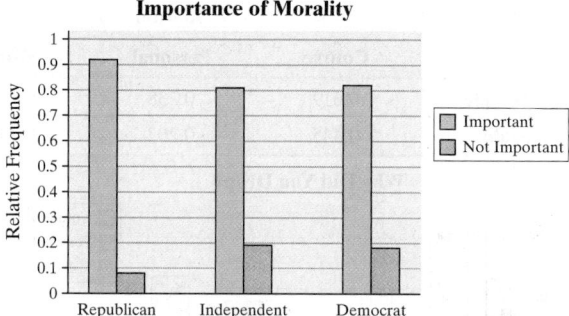

A higher proportion of Republicans appears to feel that morality is important when deciding how to vote than do Democrats or Independents.

4. Classical approach: $\chi_0^2 = 4155.584 > \chi_{0.05}^2 = 7.815$; reject the null hypothesis. P-value approach: P-value < 0.005 [Tech: P-value < 0.0001]; reject the null hypothesis. There is sufficient evidence at the $\alpha = 0.05$ level of significance that different proportions of 18- to 29-year-olds have been affiliated with religion in the past four decades.

5. Classical approach: $\chi_0^2 = 330.803 > \chi_{0.05}^2 = 12.592$; reject the null hypothesis. P-value approach: P-value $< 0.005 < \alpha = 0.05$ [Tech: P-value < 0.0001]; reject the null hypothesis. There is sufficient evidence at the $\alpha = 0.05$ level of significance to conclude that the proportions for the time categories spent in bars differ between smokers and nonsmokers. From the conditional distribution and bar graph, it appears that a higher proportion of smokers spends more time in bars than nonsmokers.

	Almost Daily	Several Times a Week	Several Times a Month	Once a Month	Several Times a Year	Once a Year	Never
Smoker	0.584	0.539	0.437	0.435	0.430	0.372	0.277
Nonsmoker	0.416	0.461	0.563	0.565	0.570	0.628	0.723

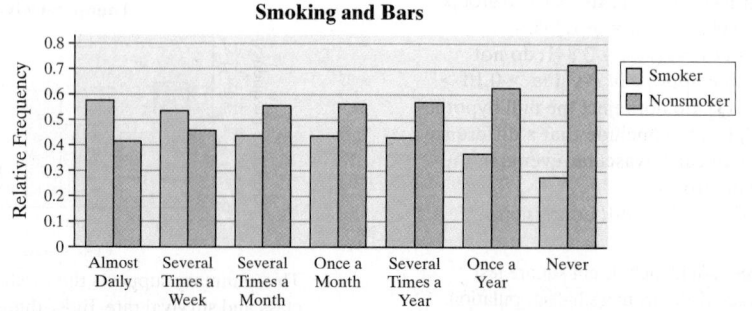

6. $H_0: p_a = p_d$ vs. $H_1: p_a \neq p_d$. Classical:
$\chi_0^2 = 5863.882 > \chi_{0.05}^2 = 3.841$; reject H_0. P-value: P-value < 0.005 [Tech: P-value < 0.0001]; reject H_0. There is sufficient evidence to suggest the proportion of individuals who favor the death penalty for persons convicted of murder differs from the proportion of individuals who favor legal abortion.

Chapter 13 Comparing Three or More Means

13.1 Assess Your Understanding (page 629)

1. analysis of variance   **3.** True   **5.** error; MSE

7.

Source of Variation	Sum of Squares	Degrees of Freedom	Mean Squares	F–Test Statistic
Treatment	387	2	193.5	0.650
Error	8042	27	297.852	
Total	8429	29		

9. $F_0 = 1.154$

11. $F_0 = \dfrac{MST}{MSE} = \dfrac{27.5833}{13.5278} = 2.039$

13. (a) $H_0: \mu_{\text{sludge plot}} = \mu_{\text{spring disc}} = \mu_{\text{no till}}$ versus H_1: at least one of the plot types has a different mean number of plants
(b) 1. Each sample is a simple random sample.
2. The three samples are independent of each other.
3. The samples are from normally distributed populations.
4. The populations have equal variances.
(c) Since the P-value $= 0.007 < \alpha = 0.05$, we reject the null hypothesis and conclude that at least one of the means is different.
(d) Yes, the boxplots support the result obtained in part (c). The boxplots suggest that significantly more plants are growing in the spring disk plot.
(e) $F_0 = \dfrac{MST}{MSE} = \dfrac{42.06}{5.92} = 7.10$
(f) The correlation between the residuals and expected z-scores is 0.988 [Tech: 0.989]. Because $0.988 > 0.946$ (critical value from Table VI with $n = 18$), it is reasonable to conclude the residuals are normally distributed.

15. (a) Final exam score; quantitative
(b) Method of instruction; 4
(c) $H_0: \mu_{\text{I}} = \mu_{\text{II}} = \mu_{\text{III}} = \mu_{\text{IV}}$ versus H_1: at least one of the delivery methods has a different mean score
(d) 1. Each sample is a simple random sample.
2. The four samples are independent of each other.
3. The samples are from normally distributed populations.
4. The populations have equal variances.
(e) Since the P-value $= 0.439 > \alpha = 0.05$, we do not reject the null hypothesis and conclude that there is not enough evidence to support the hypothesis that at least one of the means is different.
(f) Yes, the boxplots support the results obtained in part (e).
(g) Assuming the null hypothesis is true, we expect this much or more variability in the four sample means about 44 times out of every 100 times the experiment is repeated.
(h) The correlation between the residuals and expected z-scores is 0.989. Because $0.989 > 0.960$ (Table VI with $n = 30$), it is reasonable to conclude the residuals are normally distributed.

17. (a) $H_0: \mu_{\text{financial}} = \mu_{\text{energy}} = \mu_{\text{utilities}}$ versus H_1: at least one of the means is different
(b) 1. Each sample is a simple random sample.
2. The three samples are independent of each other.
3. The samples are from normally distributed populations.
4. The largest sample standard deviation is less than twice the smallest sample standard deviation ($5.12 < 2 \cdot 4.53$), so the requirement that the populations have equal variances is satisfied.
(c) $F = 2.077$, P-value $= 0.1502$. Since the P-value is greater than $\alpha = 0.05$, we do not reject the null hypothesis and conclude that there is not enough evidence to support the hypothesis that at least one of the means is different.

(d)

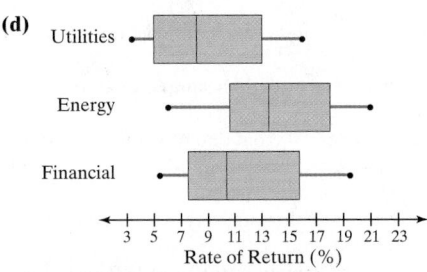

19. (a) Profit on initial offer; Quantitative
(b) $H_0: \mu_{\text{WM}} = \mu_{\text{BM}} = \mu_{\text{WF}} = \mu_{\text{BF}}$ versus H_1: at least one of the means is different
(c) The largest sample standard deviation, 241.5, is less than twice the smallest sample standard deviation, 157.1, so the requirement of equal variances is satisfied.
(d) Since the P-value < 0.0001, we reject the null hypothesis and conclude that there is enough evidence to support the hypothesis that at least one of the means is different.
(e)

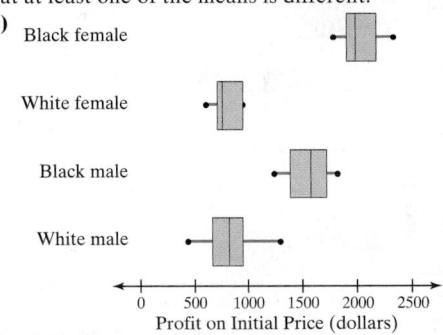

(f) The results suggest that the mean initial price offered to blacks and especially black females is higher than that offered to whites, so there does seem to be race discrimination. While gender discrimination against females is apparent among blacks, it is not among whites. In fact, white females in the sample were offered the lowest mean price with the least variability of the four groups. A separate study might be needed to test for gender discrimination specifically.
(g) The correlation between the residuals and expected z-score is 0.994 [Tech: 0.993]. Because $0.994 > 0.960$ (Table VI with $n = 30$), it is reasonable to conclude the residuals are normally distributed.

21. (a) The subjects were randomly assigned to one of three treatment groups.
(b) LDL cholesterol; diet
(c) Randomization is used to neutralize the effect of factors not controlled in the study. The thinking is that the individuals within each treatment group will be similar prior to manipulating the diet.
(d) $H_0: \mu_{\text{SatFat}} = \mu_{\text{Med}} = \mu_{\text{NCEP-1}}$ versus H_1: at least one of the means is different
(e) 1. Each sample is based on a randomized experiment.
2. The three samples are independent of each other.
3. The samples are from normally distributed populations.
4. The largest sample standard deviation is less than twice the smallest sample standard deviation ($51.94 < 2 \cdot 47.14$), so the requirement that the populations have equal variances is satisfied.
(f) $F = 8.029 > F_{0.05,2,42} \approx 3.18$; P-value $= 0.0011 < \alpha = 0.05$. We reject the null hypothesis and conclude that there is sufficient evidence to support the hypothesis that at least one of the means is different.
(g)

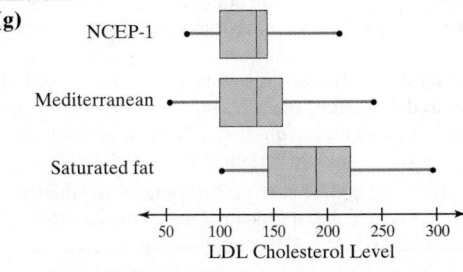

23. (a) Quantitative

(b) $H_0: \mu_{CB} = \mu_S = \mu_{RB} = \mu_{WR}$ vs. H_1: at least one mean differs from the others

(c) 1. Treat each position as a random sample of all players from that position.
    2. The individuals selected from one position are not used to determine the individuals from other positions, so the different samples are independent.
    3. A normal probability plot suggests each position's times are normally distributed.
    4. The largest standard deviation (0.105 for WR) is not more than two times the smallest standard deviation (0.082 for S). There are four means being compared.

(d) $F = 8.433$; P-value < 0.0001. The small P-value leads us to reject the statement in the null hypothesis. There is sufficient evidence to suggest the mean 40-yard dash times differs among the skill positions (excluding QB).

(e)

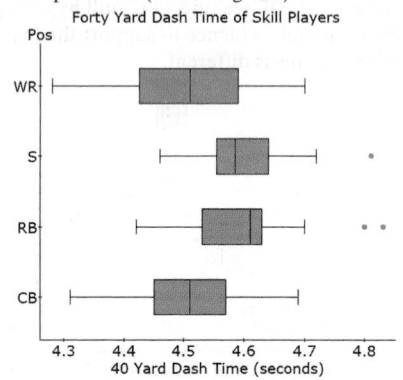

Forty Yard Dash Time of Skill Players

25. (a) $H_0: \mu_{301} = \mu_{400} = \mu_{353}$ versus H_1: at least one of the means is different

(b) The standard deviation for mixture 67-0-400 is more than two times larger than the standard deviation for mixture 67-0-301 ($384.75 > 2(107.8)$).

27. (a) $s_C = 4.6$, $s_R = 3.6$, $s_H = 2.9$

(b) Stratified sampling; the researchers wanted to be sure that both children and adolescents were represented in the samples.

(c) $t_0 = 3.87$, P-value $= 0.00017$, reject H_0; there is enough evidence to indicate that the mean CBCL scores are different for the two groups.

(d) No; since we have large samples, we can assume that the distributions of the sample means are approximately normal.

(e) $F = 15.74 > F_{0.05,2,207} \approx 3.04$; reject H_0; there is enough evidence to indicate that at least one of the mean scores is different.

(f) No; the result from part (e) only indicates that *at least one* mean is different. It does not indicate which pairs are different nor even how many. The result from part (c) indicates that the mean scores for the control group and the RAP group are different, but additional pairs could also be significantly different.

29. (a) Completely randomized design

(b) Smoke after six months, or not

(c) Treatments: e-Cig with nicotine, nicotine patch, e-Cig with placebo; three levels

(d) The response variable is qualitative with two possible outcomes. Should be analyzed using a homogeneity of proportions.

(e) $H_0: p_{e\text{-Cig/N}} = p_{patch} = p_{e\text{-Cig/placebo}}$ vs. H_1: at least one proportion differs from the others; Classical $\chi_0^2 = 1.202 < \chi_{0.05}^2 = 5.991$ (with 2 df). Do not reject H_0. P-value: P-value > 0.10 [Tech: P-value $= 0.5484$] $> \alpha = 0.05$; Do not reject H_0. There is not sufficient evidence to conclude the proportion of abstainers for each group differs.

31. The mean square due to treatment estimate of σ^2 is a weighted average of the squared deviations of each sample mean from the grand mean of all the samples. The mean square due to error estimate of σ^2 is the weighted average of the sample variances.

33. Rejecting the statement in the null hypothesis means that there is sufficient evidence to conclude that the mean of one of the populations is different than the other two or that all three population means are different.

13.2 Assess Your Understanding (page 640)

1. (a) $q_{0.05,10,3} = 3.877$
  (b) $q_{0.05,24,5} = 4.166$
  (c) $q_{0.05,32,8} \approx q_{0.05,30,8} = 4.602$
  (d) $q_{0.05,65,5} \approx q_{0.05,60,5} = 3.977$

3. $q_{0.05,12,3} = 3.773$;
$q_{\mu_3,\mu_2} = 3.932 > q_{0.05,12,3} = 3.773$, so reject $H_0: \mu_2 = \mu_3$
$q_{\mu_3,\mu_1} = 3.757 < q_{0.05,12,3} = 3.773$, so do not reject $H_0: \mu_1 = \mu_3$
$q_{\mu_1,\mu_2} = 0.175 < q_{0.05,12,3} = 3.773$, so do not reject $H_0: \mu_1 = \mu_2$
Conclusion: Results are ambiguous, $\underline{\mu_2 \ \mu_1 \mu_3}$

5. (a) $H_0: \mu_1 = \mu_2 = \mu_3$; H_1: at least one of the means is different. $F = 9.72$; P-value $= 0.002 < \alpha = 0.05$. We reject the null hypothesis and conclude that at least one of the means is different.

(b) The result of Tukey's test indicates that $\mu_1 = \mu_2 \neq \mu_3$ $(\underline{\mu_1 \ \mu_2} \ \mu_3)$.

(c)

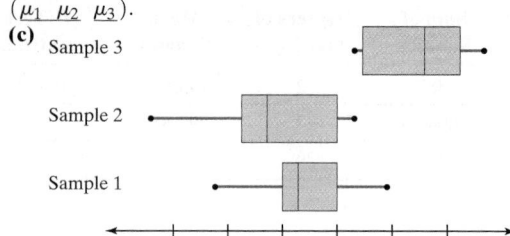

7. The result of Tukey's test indicates that $\mu_{SP} = \mu_{NT} \neq \mu_{SD}$, $(\underline{\mu_{SP} \ \mu_{NT}} \ \mu_{SD})$. Recommend using the spring-disc method of planting.

9. The result of Tukey's test indicates that $\mu_{WM} = \mu_{WF} \neq \mu_{BM} \neq \mu_{BF}$ $(\underline{\mu_{WM} \mu_{WF}} \ \underline{\mu_{BM}} \ \mu_{BF})$.

11. (a) The individuals in the study were randomly assigned to one of five treatment groups. The factor drug was intentionally manipulated while other factors that may affect ADHD were either controlled by fixing them or dealt with through randomization.

(b) Score on ability to follow rules; the drug was controlled at five levels.

(c) Randomization is used to "even out" those variables that may affect the symptoms associated with ADHD that were not controlled in the study.

(d) $H_0: \mu_P = \mu_{R10} = \mu_{R17.5} = \mu_{A7.5} = \mu_{A12.5}$; H_1: at least one of the means is different.

(e) $F = 5.75$; P-value $= 0.0015 < \alpha = 0.05$. We reject the null hypothesis and conclude that at least one of the mean abilities to follow rules is different.

(f) The result of Tukey's test indicates that $\mu_P \neq \mu_{R10} = \mu_{R17.5} = \mu_{A7.5} = \mu_{A12.5}$; $(\mu_P \ \underline{\mu_{R10} \ \mu_{R17.5} \ \mu_{A7.5} \ \mu_{A12.5}})$; that is, the mean ability to follow rules of a child taking the placebo is less than the mean ability to follow rules of a child taking any of the four drug treatments.

(g)

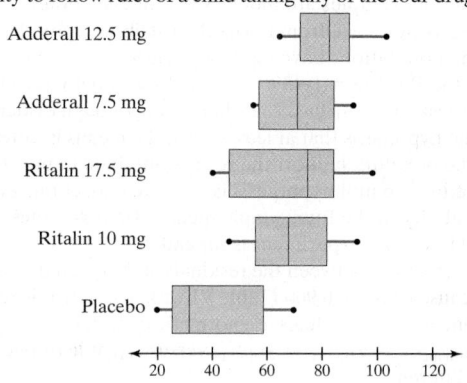

13. (a) $H_0: \mu_{nonsmoker} = \mu_{light} = \mu_{heavy}$ versus H_1: at least one of the means is different. $F = 19.62$; P-value $< 0.0001 < \alpha = 0.05$. We reject the null hypothesis and conclude that at least one of the mean heart rates is different.

(b) The result of Tukey's test indicates that $\mu_{nonsmoker} \neq \mu_{light} \neq \mu_{heavy}$; $\underline{\mu_{nonsmoker}} \ \underline{\mu_{light}} \ \underline{\mu_{heavy}}$. We conclude that the mean heart rate of nonsmokers is lower than the mean heart rate of light smokers and that the mean heart rate of light smokers is lower than the mean heart rate of heavy smokers.

(c)

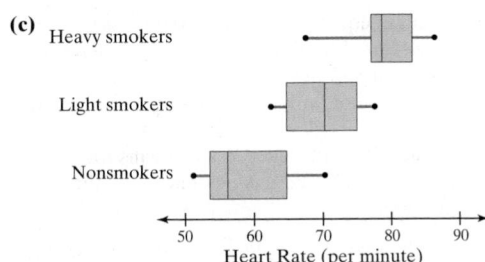

Heavy smokers

Light smokers

Nonsmokers

50   60   70   80   90
Heart Rate (per minute)

15. (a) H_0: $\mu_{\text{skim milk}} = \mu_{\text{mixed milk}} = \mu_{\text{whole milk}}$ versus H_1: at least one of the means is different. $F = 70.51$; P-value $< 0.0001 < \alpha = 0.05$. We reject the null hypothesis and conclude that there is sufficient evidence to support the hypothesis that at least one of the mean calcium intake levels is different.

(b) The result of Tukey's test indicates that $\mu_{\text{skim milk}} = \mu_{\text{whole milk}} \neq \mu_{\text{mixed milk}}$; $\underline{\mu_{\text{skim milk}} \ \mu_{\text{whole milk}}} \ \mu_{\text{mixed milk}}$. We conclude that the mean calcium intake of children who drank mixed milk is different from the mean calcium intakes of children who drank skim or whole milk.

(c)

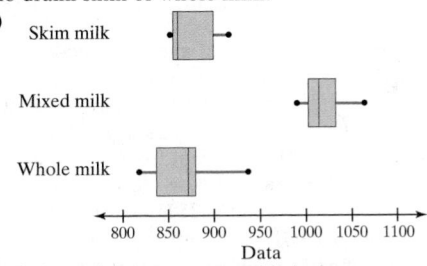

Skim milk

Mixed milk

Whole milk

800 850 900 950 1000 1050 1100
Data

17. (a) Cohort; the response variable is time to complete a degree
(b) $\bar{x}_{\text{HS}} = 4.05$, $\bar{x}_{\text{S}} = 4.20$, $\bar{x}_{\text{NS}} = 4.20$, $\bar{x}_{\text{OD}} = 5.80$, $\bar{x}_{\text{NR}} = 4.65$; the means do not appear to be equal.
(c) $1.099 < 2(0.5538) = 1.108$; the population variances can be assumed to be equal.
(d) Lower bound: 3.341; Upper bound: 4.759
(e) 10 **(f)** 0.4013
(g) $F = 7.03$, P-value $= 0.002$, reject H_0; there is enough evidence to indicate that at least one mean time to degree completion is different; the mean time to degree completion appears to be the same for all initial institution types except for those with an open-door policy. It appears that students starting at an institution with an open-door policy take longer to complete their bachelor's degree.

13.3 Assess Your Understanding (page 649)

1. (a) Since the P-value $= 0.013$, the researcher rejects the null hypothesis that the population means are equal.
(b) The mean square due to error is MSE $= 17.517$.
(c) The Tukey test indicates that means 1 and 2 are the same, but mean 3 is different ($\mu_1 = \mu_2 \neq \mu_3$ or $\underline{\mu_1 \ \mu_2} \ \mu_3$).
3. (a) Since the treatment P-value $= 0.108$, the researcher does not reject the null hypothesis that the population means are equal.
(b) The mean square due to error is MSE $= 0.944$.
(c) The Tukey test is only used if the null hypothesis is rejected.
5. (a) Since the treatment P-value $= 0.0018$, we reject the null hypothesis that the population means are equal and conclude at least one of the means is different.
(b) The Tukey test indicates that means of treatment 2 and treatment 3 are equal, but the mean of treatment 1 is different. ($\mu_1 \neq \mu_2 = \mu_3$ or $\mu_1 \ \underline{\mu_2 \ \mu_3}$)
(c)

Treatment 3

Treatment 2

Treatment 1

8   9   10   11

7. (a) The requirement for equal population variances is satisfied since the largest standard deviation is less than twice the smallest standard deviation, $1.439 < 2(1.225)$.

(b) Blocking is used to remove the effect car type has on gas mileage.
(c) Since the treatment P-value $= 0.0034$, there is evidence that the mean miles per gallon is different among the three octane levels at the $\alpha = 0.05$ level of significance.
(d) The Tukey test indicates that $\mu_{87} = \mu_{89} < \mu_{92}$; $\underline{\mu_{87} \ \mu_{89}} \ \mu_{92}$.
(e) Conclude that the mean gas mileage is equal if 87- or 89-octane gasoline is used, but the mean miles per gallon increases when 92-octane gasoline is used.

9. (a) The requirement for equal population variances is satisfied since the largest standard deviation is less than twice the smallest standard deviation, $830 < 2(448)$.
(b) Since the treatment P-value $= 0.105$, there is not sufficient evidence to conclude that the mean cost of repair is different among the car brands at the $\alpha = 0.05$ level of significance.
(c) Since the null hypothesis was not rejected, there is no need to do post hoc analysis.

11. (a) The response variable is water consumption (in ml/day). The treatment is space flight. The treatment has three levels.
(b) The requirement for equal population variances is satisfied since the largest standard deviation is less than twice the smallest standard deviation, $9.97 < 2(5.65)$.
(c) Since the treatment P-value $= 0.0008$, there is sufficient evidence to indicate that the mean water consumption of mice is different among the three experiment levels at the $\alpha = 0.05$ level of significance.
(d) The Tukey test indicates that $\mu_{\text{LO}-1} = \mu_{\text{R}+1} \neq \mu_{\text{R}+1\text{ month}}$ ($\underline{\mu_{\text{LO}-1} \ \mu_{\text{R}+1}} \ \mu_{\text{R}+1\text{ month}}$).

13. (a) Since the treatment P-value is 0.438, there is not sufficient evidence to conclude that there is a difference in mean wait time between the two loading procedures.
(b) Hypotheses: H_0: $\mu_d = 0$ versus H_1: $\mu_d \neq 0$. Classical approach: $t_0 = 0.83 < t_{0.025,6} = 2.447$, so we do not reject the null hypothesis. P-value approach: P-value $= 0.438 > \alpha = 0.05$, so we do not reject the null hypothesis. There is not sufficient evidence at the $\alpha = 0.05$ level of significance to conclude that the new loading procedure changed the mean wait time.
(c) The P-value for both procedures is 0.438, which supports that the method presented in this section is a generalization of the matched-pairs t-test.

15. In a completely randomized design, the researcher examines a single factor fixed at k levels and randomly distributes the experimental units among the levels. In a randomized complete block design, there is a factor whose level cannot be fixed. So the researcher partitions the experimental units according to this factor, forming blocks, which may form homogeneous groups. Within each block, the experimental units are assigned to one of the treatments.

17. Randomized means that the assignment of the experimental unit to the treatment within each block is random. Complete refers to the fact that each block gets every treatment.

19. A researcher can deal with explanatory variables by:
1. Controlling their levels so that they remain fixed throughout the experiment or setting them at fixed levels.
2. Randomizing so that any effects that cannot be identified or controlled are minimized.

13.4 Assess Your Understanding (page 662)

1. 2; 3 **3.** interaction effect

5. The plots are relatively parallel, so there is no significant interaction.

7. The plots cross, so there is significant interaction.

9. (a) Main effect of factor A $= 1.25$; main effect of factor B $= -9.25$.
Interaction effects: If factor B is fixed at low and factor A changes from low to high, the response variable increases by 3.5 units. On the other hand, if factor B is fixed at high and factor A changes from low to high, the response variable increases by -4.5 (decreases by 4.5) units. So the change in the response variable for factor A depends on the level of factor B, meaning there is an interaction effect between factors A and B.

If factor A is fixed at low and factor B changes from low to high, the response variable increases by −3.5 (decreases by 3.5) units. On the other hand, if factor A is fixed at high and factor B changes from low to high, the response variable increases by −15 (decreases by 15) units. So the change in the response variable for factor B depends on the level of factor A, meaning that there is an interaction effect between factors A and B.

(b)

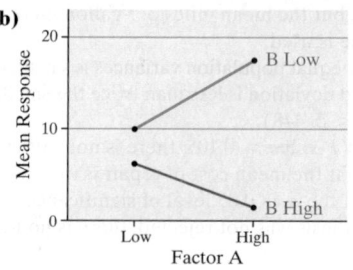

11. (a) There is no evidence of an interaction effect since the interaction P-value is 0.836.
(b) Since the factor A P-value $= 0.003$, there is evidence of difference in means from factor A. The factor B P-value < 0.001, giving evidence that there is also a difference in means from factor B.
(c) MSE $= 45.3$

13. (a) There is evidence of an interaction effect since the interaction P-value < 0.001.
(b) Whenever there is evidence of an interaction effect, we do not consider main effects because they could be misleading.
(c) MSE $= 31.9$

15. (a) There is no evidence of an interaction effect since interaction P-value is 0.965.
(b) Since the factor A P-value $= 0.579$, there is no evidence of difference in means from factor A. That the factor B P-value $= 0.021$ is evidence that there is a difference in means from factor B.
(c)

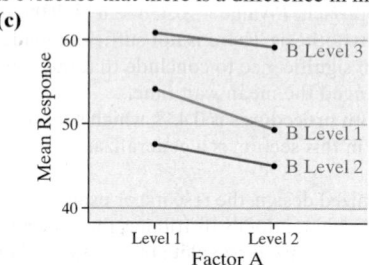

(d) Tukey's test for factor A is unnecessary. Tukey's test for factor B is ambiguous. It shows $B_1 = B_2$ and $B_1 = B_3$, but $B_2 \neq B_3$; $\underline{B_2 \ B_1} \ \underline{B_3}$.

17. (a) This is a 2×3 factorial design with two replications per cell.
(b) The response variable is serum cholesterol; the factors are gender and age.
(c) There is no evidence of a significant interaction effect since the interaction P-value is $= 0.159$, which is large.
(d) There is a significant difference in the means for the three age groups; age P-value $= 0.016$. There is no significant difference in the means for the two genders; gender P-value $= 0.164$.
(e)

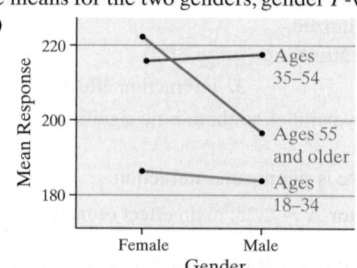

(f) The correlation between residuals and expected z-scores is 0.945. Because $0.945 > 0.928$ (Table VI with $n = 12$), it is reasonable to conclude the residuals are normally distributed.
(g) The mean of age group 18 to 34 is significantly different. Tukey's test shows $\underline{B_1} \ \underline{B_2 B_3}$.

19. (a) The requirement of equal population variances is satisfied since the largest standard deviation is less than twice the smallest standard deviation $[320 < 2(168) = 336]$.
(b) There is no evidence of an interaction effect. Interaction P-value $= 0.469$.
(c) There is evidence of a difference in the means for the three mixtures (P-value < 0.001). There is evidence of a difference in the means for the three types of slump (P-value $= 0.002$).
(d)

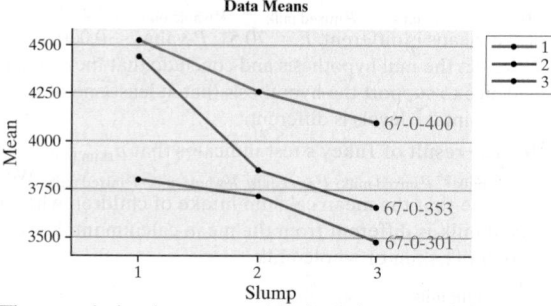

(e) The correlation between residuals and expected Z-scores is 0.968. Because $0.968 > 0.960$ [Table VI with $n = 30$], it is reasonable to conclude the residuals are normally distributed.
(f) Tukey's test indicates that the mean 28-day strength of mixture 67-0-400 is significantly different from the mean 28-day strength of the other two mixtures. Tukey's test indicates that the mean 28-day strength for slump 3.75 is significantly different from the mean 28-day strength for slumps 4 and 5.

21. (a) The requirement of equal population variances is satisfied since the largest standard deviation is less than twice the smallest standard deviation $[4.00 < 2(2.5) = 5.0]$.
(b) There is no evidence of an interaction effect. Interaction P-value $= 0.854$.
(c) There is no evidence of a difference in the means for three locations. Location P-value $= 0.287$. There is no evidence of a difference in the means for type of service station; type P-value $= 0.804$.
(d)

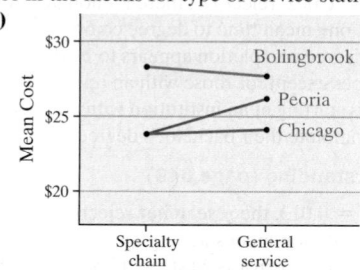

(e) Since there were no significant main effects, there is no need to do Tukey's tests.

23. In a completely randomized design, the researcher examines a single factor fixed at k levels and randomly distributes the experimental units among the levels. In a randomized complete block design, there is a factor whose level cannot be fixed. So the researcher partitions the experimental units according to this factor, forming blocks. The experimental units are then randomly assigned to the blocks, being sure that each block has each treatment. In a factorial design, the researcher has two variables A and B, called factor A and factor B. The factors are fixed, so factor A has a levels and factor B has b levels. The experimental units are then uniformly distributed among the cells formed by the factors, and the response variable is measured at each of the ab cells.

25. An interaction plot graphically displays the mean response for each treatment. These plots are useful for illustrating potential interaction between factors.

Chapter 13 Review Exercises (page 667)

1. (a) $q_{0.05,16,7} = 4.741$ **(b)** $q_{0.05,30,6} = 4.302$
(c) $q_{0.01,42,4} \approx q_{0.01,40,4} = 4.696$
(d) $q_{0.05,46,6} \approx q_{0.05,40,6} = 4.232$

2. $q_{0.05,27,4} \approx 3.873$; $q_{x1,x4} = 8.937$; $q_{x1,x2} = 4.036$; $q_{x1,x3} = 2.306$; $q_{x3,x4} = 6.631$; $q_{x3,x2} = 1.730$; $q_{x2,x4} = 4.901$. Conclude that Tukey's test is ambiguous. $\mu_1 \neq \mu_2$, but $\mu_1 = \mu_3 \neq \mu_4$ and $\mu_2 = \mu_3 \neq \mu_4$.

3. **(a)** Concentration of dissolved organic carbon

(b) Type of water; 3

(c) H_0: $\mu_{water-organic\ soil} = \mu_{water-mineral\ soil} = \mu_{surface\ water}$ versus H_1: at least one of the means is different

(d) The largest sample standard deviation, 6.246, is less than twice the smallest sample standard deviation, 3.927, so the requirement of equal variances is satisfied.

(e) Since P-value $= 0.0002 < \alpha = 0.01$, the researcher will reject the null hypothesis and conclude that the mean concentration of dissolved carbon is different in at least one of the collection areas.

(f) Results at least as extreme as these only occur in about 2 out of 10,000 similar samples if the null hypothesis is true.

(g) Yes, the boxplots support the conclusion that at least one of the means is different. The plot for water collected from mineral soil appears to differ from the other two.

(h) Tukey's test indicates that the mean dissolved carbon is equal in the surface water and in the groundwater taken from organic soil, but it differs from the mean dissolved carbon concentration in groundwater taken from mineral soil $\left(\underline{\mu_{surface\ water}\quad \mu_{water-organic\ soil}}\quad \mu_{water-mineral\ soil}\right)$.

4. Tukey's test is ambiguous. It indicates that $\mu_{select\ structural} \neq \mu_{no.\ 2}$, but that the $\mu_{below\ grade}$ is equal to all other means $\left(\underline{\mu_{select}\quad \mu_{No.1}\quad \mu_{below}}\quad \mu_{No.2}\right)$.

5. **(a)** There are three independent samples corresponding to Front/Middle/Back seat position. The response variable is quantitative (GPA).

(b) The P-value $= 0.0026$, so we reject the null hypothesis and conclude that at least one of the mean GPAs is different.

(c) Tukey's test indicates that $\mu_{middle} = \mu_{back} \neq \mu_{front}$. There is evidence that students who choose seats in the front of the mathematics classroom have a different mean GPA from that of students choosing seats in the middle or back of the classroom.

(d)
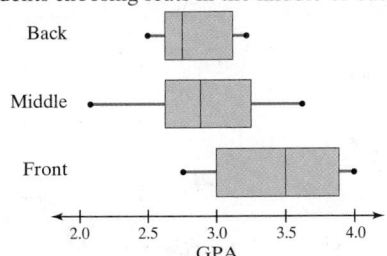

6. **(a)** Distance

(b) There is variability due to person firing.

(c) The P-value < 0.001, so we reject the null hypothesis and conclude at least one of the means is different.

(d) Tukey's test indicates that $\mu_1 = \mu_2 \neq \mu_3 \neq \mu_4$: $\underline{\mu_1\quad \mu_2}\quad \mu_3\quad \mu_4$.

(e)
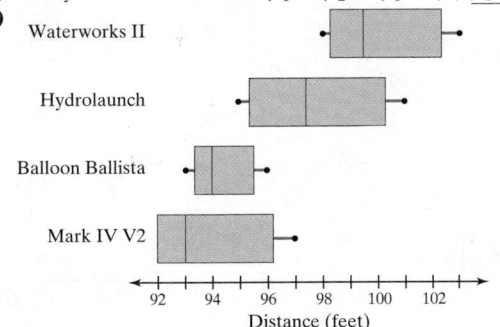

7. **(a)** Since the largest standard deviation is less than twice the smallest deviation, the requirement of equal population variances is satisfied $[4.72 < 2(2.39) = 4.78]$.

(b) There is no evidence of an interaction effect. Interaction P-value $= 0.374$.

(c) There is no evidence of a difference in the means for gender; gender P-value $= 0.196$. There is evidence of a difference in the means for type of driving program; program P-value < 0.001. So the null hypothesis H_0: $\mu_{8\ hour} = \mu_{4\ hour} = \mu_{2\ hour}$ is rejected.

(d)
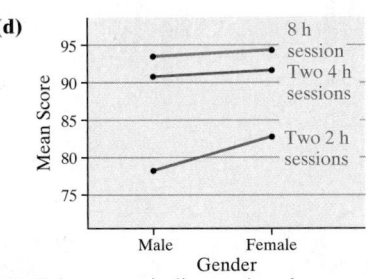

(e) Tukey's test indicates that the mean score after taking one 8-hour defensive driving class is equal to the mean score after taking two 4-hour classes, but that the mean score after taking two 2-hour classes is different $\underline{\mu_{8\ hour}\quad \mu_{4\ hour}}\quad \mu_{2\ hour}$.

Chapter 13 Test (page 669)

1. **(a)** Reject H_0 since P-value $= 0.0039$. Conclude at least one of the means is different.

(b) $\mu_1 = \mu_3 = \mu_4 \neq \mu_2$; it appears that machine II is dispensing less product than the other machines.

(c)
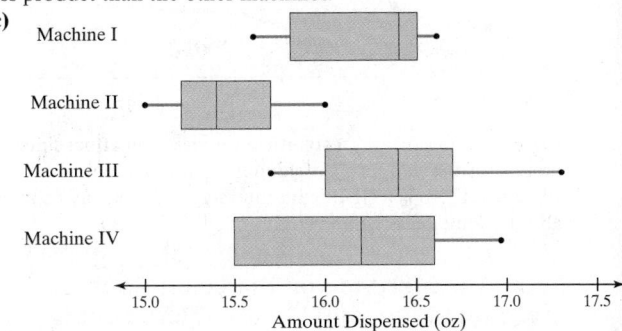

2. **(a)** Michael used a completely randomized design for his experiment.

(b) The response variable is the height after 38 days; this is a quantitative variable.

(c) The explanatory variable is the color of the light bulb; this is a qualitative variable.

(d) Other factors identified that might affect plant growth include type of soil, amount of water, amount of light, and temperature. Michael kept these consistent for all three treatments.

(e) Michael randomly assigned the seeds to each treatment group.

(f) H_0: $\mu_B = \mu_R = \mu_G$ versus H_1: at least one of the means is different

(g) The largest sample standard deviation, 0.950, is less than twice the smallest sample standard deviation, 0.595, so the requirement of equal variances is satisfied.

(h) Since P-value < 0.001, we reject the null hypothesis and conclude that the mean height is different in at least one type of light.

(i) Yes, it appears that red light promotes plant growth more than the others.

(j) Tukey's test indicates that all three means are different $\left(\underline{\mu_B}\quad \underline{\mu_R}\quad \underline{\mu_G}\right)$.

3. **(a)** The P-value $= 0.011$, so we reject the null hypothesis and conclude at least one of the means is different.

(b) Tukey's test is ambiguous. It indicates that $\mu_3 \neq \mu_1 = \mu_2$, but $\mu_2 = \mu_3$; $\underline{\mu_1\quad \mu_2}\quad \mu_3$.

(c)
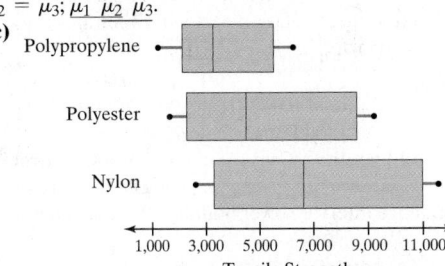

4. **(a)** Since the largest standard deviation is less than twice the smallest standard deviation, the requirement of equal population variances is satisfied $[0.53 < 2(0.28) = 0.56]$.

(b) There is no evidence of significant interaction. Interaction P-value $= 0.459$.

(c) There is no evidence of a difference in means for gender; gender P-value $= 0.589$. There is no evidence of a difference in the means for age: age P-value $= 0.243$. So the null hypothesis H_0: $\mu_{20-34} = \mu_{35-39} = \mu_{50-64}$ is not rejected.

(d)

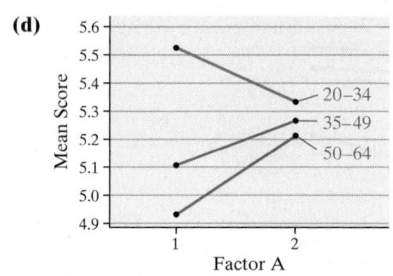

(e) No significant difference in means

5. (a)

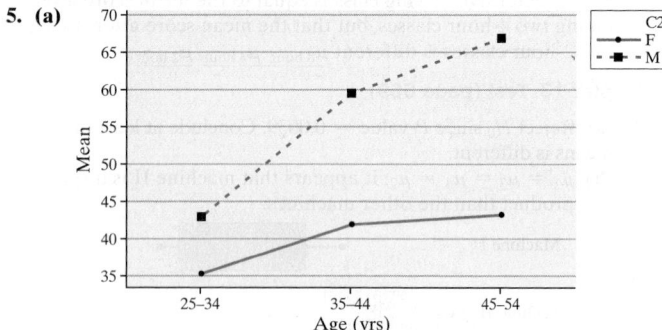

(b) The plot indicates a significant interaction effect since the graphs for the two levels of gender are not parallel.
(c) Since there is a significant interaction effect, any conclusions about main effects will be confounded.

Chapter 14 Inference on the Least-Squares Regression Model and Multiple Regression

14.1 Assess Your Understanding (page 684)

1. 82.7   **3.** $0; \sigma$

5. (a) $\beta_0 \approx b_0 = -2.3256, \beta_1 \approx b_1 = 2.0233$
(b) $s_e = 0.5134$ is the point estimate for σ.
(c) $s_{b_1} = 0.1238$
(d) Because the P-value $< 0.001 < \alpha = 0.05$ (or $t_0 = 16.343$ [Tech: 16.344] $> t_{0.025} = 3.182$), we reject the null hypothesis and conclude that a linear relation exists between x and y.

7. (a) $\beta_0 \approx b_0 = 1.200, \beta_1 \approx b_1 = 2.200$
(b) $s_e = 0.8944$ is the point estimate for σ.
(c) $s_{b_1} = 0.2828$
(d) Because the P-value $= 0.0044 < \alpha = 0.05$ (or $t_0 = 7.778$ [Tech: 7.779] $> t_{0.025} = 3.182$), we reject the null hypothesis and conclude that a linear relation exists between x and y.

9. (a) $\beta_0 \approx b_0 = 116.600, \beta_1 \approx b_1 = -0.7200$
(b) $s_e = 3.2863$ is the point estimate for σ.
(c) $s_{b_1} = 0.1039$
(d) Because the P-value $= 0.0062 < \alpha = 0.05$ (or $t_0 = -6.929$ [Tech: -6.928] $< -t_{0.025} = -3.182$), we reject the null hypothesis and conclude that a linear relation exists between x and y.

11. (a) $\beta_0 \approx b_0 = 69.0296; \beta_1 \approx b_1 = -0.0479$
(b) $s_e = 0.3680$ **(c)** $s_{b_1} = 0.0043$
(d) Because the P-value < 0.001 [Tech: P-value $= 0.0001$] $< \alpha = 0.05$ (or $t_0 = -11.140$ [Tech: -11.157] $< -t_{0.025} = -2.571$), we reject the null hypothesis and conclude that a linear relation exists between commute time and score on a well-being survey.
(e) 95% confidence interval: lower bound: -0.0589, upper bound: -0.0369

13. (a) $\beta_0 \approx b_0 = 12.4932, \beta_1 \approx b_1 = 0.1827$
(b) $s_e = 0.0954$
(c) The correlation between the residuals and expected z-scores is 0.986. Because 0.986 [Tech: 0.987] > 0.923 (Table VI), conclude the residuals are approximately normally distributed.
(d) $s_{b_1} = 0.0276$
(e) Because the P-value $< 0.001 < \alpha = 0.01$ (or $t_0 = 6.62$ [Tech: 6.63] $> t_{0.005} = 3.250$], we reject the null

hypothesis and conclude that a linear relation exists between a child's height and head circumference.
(f) 95% confidence interval: lower bound: 0.1204; upper bound: 0.2451
(g) A good estimate of the child's head circumference would be 17.33 inches.

15. (a) $\beta_0 \approx b_0 = 2675.6, \beta_1 \approx b_1 = 0.6764$
(b) $s_e = 271.04$ **(c)** $s_{b_1} = 0.2055$
(d) Because the P-value $= 0.011 < \alpha = 0.05$ (or $t_0 = 3.291 > t_{0.025} = 2.306$), we reject the null hypothesis and conclude that a linear relation exists between 7-day strength and 28-day strength.
(e) 95% confidence interval: lower bound: 0.2025; upper bound: 1.1503
(f) The mean 28-day strength of this concrete if the 7-day strength is 3000 psi is 4704.8 psi.

17. (a) $H_0: \beta_1 = 0$ vs. $H_1: \beta_1 < 0$. $t_0 = -7.38 < -t_{0.01} = -2.330$, therefore reject H_0. P-value $< 0.0001 < \alpha = 0.01$, therefore reject H_0. Conclude that a linear relation exists between 2013 cost and ROI.
(c) 90% confidence interval: lower bound: -0.000013; upper bound: -0.0000084
(d) When the 2013 cost is $180,000, the mean return on investment is 4.51%.

19. (a) $\beta_0 \approx b_0 = 51.1310; \beta_1 \approx b_1 = -0.1819$
(b) Because the P-value > 0.5 [Tech: P-value $= 0.9317$] $> \alpha = 0.05$ (or $t_0 = -0.088 > -t_{0.025} = -2.228$), we do not reject the null hypothesis. There is not sufficient evidence to conclude that a linear relation exists between compensation and stock return.
(c) 95% confidence interval: lower bound: -4.7916, upper bound: 4.4278
(d) No, the results do not indicate that a linear relation exists. We could use $\bar{y} = 49.733\%$ as an estimate of the stock return.

21. (a)

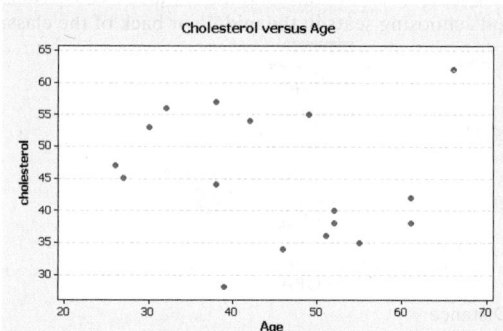

No linear relation appears to exist.
(b) $\hat{y} = 50.7841 - 0.1298x$
(c)

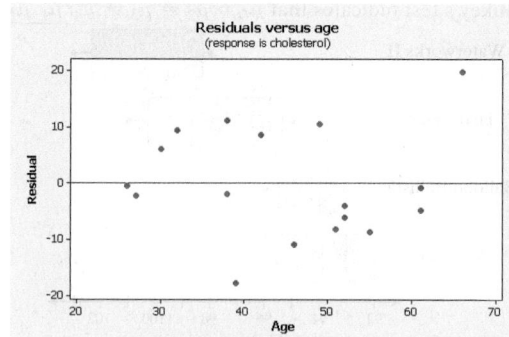

Because the residuals are evenly spread around the horizontal line drawn at 0, the requirement of constant error variance is satisfied, so the linear model seems appropriate.
(d) The lower fence is -28.08, the upper fence is 30.72; so there are no outliers or influential observations.
(e) Because the P-value $= 0.530 > \alpha = 0.01$ (or $t_0 = -0.642$ [Tech: -0.643] $> -t_{0.005} = -2.947$), we do not reject the null hypothesis and conclude that a linear relation does not exist between the age and HDL levels.

(f) 95% confidence interval: lower bound: -0.5605; upper bound: 0.3009

(g) Do not recommend using the least-squares regression line to predict the HDL cholesterol levels since we did not reject the null hypothesis. A good estimate for the HDL cholesterol level would be $\bar{y} = 44.9$.

23. (a) $\hat{y} = -12.4967 + 0.0657x$

(b) $t_0 = 17.61$; P-value $< 0.001 < \alpha = 0.05$; reject H_0: $\beta_1 = 0$.

(c)

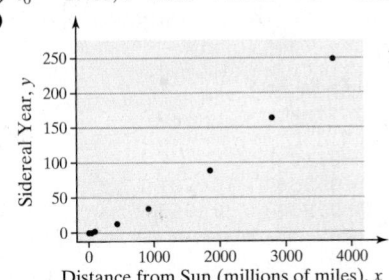

(d)

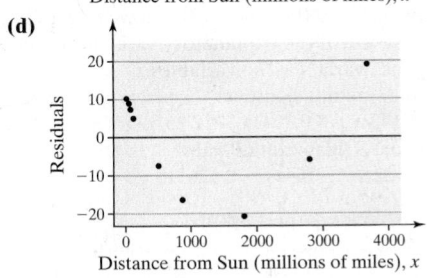

(e) Linear model is not appropriate because there is a pattern in the residual plot.

(f) The moral is that the inferential procedures may lead us to believe that a linear relation between the two variables exists even though diagnostic tools (such as residual plots) indicate that a linear model is inappropriate.

25. (a)

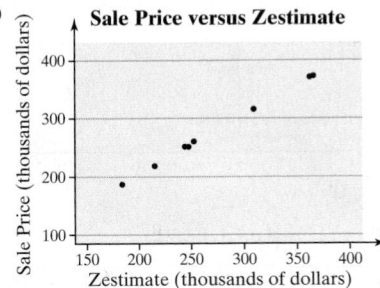

(b) $\hat{y} = 1.0228x - 0.7590$; because the P-value $< 0.001 < \alpha = 0.05$ (or $t_0 = 105.426 > t_{0.025} = 2.447$), we reject the null hypothesis and conclude that there is a linear relation between the Zestimate and the sale price.

(c) $\hat{y} = 0.5220x + 115.8094$; because $0.20 > P$-value > 0.10 [Tech: P-value $= 0.1927$] $> \alpha = 0.05$ (or $t_0 = 1.441 < t_{0.025} = 2.365$), we do not reject the null hypothesis. There is not sufficient evidence to conclude that a linear relation exists between the Zestimate and the sale price. Yes, this observation is influential.

27. The y-coordinates on the least-squares regression line represent the mean value of the response variable for any given value of the explanatory variable.

29. We do not conduct inference on the linear correlation coefficient because a hypothesis test on the slope and a hypothesis test on the linear correlation coefficient yield the same conclusion. Moreover, the requirements for conducting inference on the linear correlation coefficient are very hard to verify.

14.2 Assess Your Understanding (page 692)

1. confidence; mean

3. (a) $\hat{y} = 11.8$

(b) Lower bound: 10.8 [Tech: 10.9]; upper bound: 12.8

(c) $\hat{y} = 11.8$

(d) Lower bound: 9.9; upper bound: 13.7

(e) The confidence interval is an interval estimate for the mean value of y at $x = 7$, whereas the prediction interval is an interval estimate for a single value of y at $x = 7$.

5. (a) $\hat{y} = 4.3$

(b) Lower bound: 2.5; upper bound: 6.1   **(c)** $\hat{y} = 4.3$

(d) Lower bound: 0.9; upper bound: 7.7

7. (a) $\hat{y} = 68.07$

(b) Lower bound: 67.723; upper bound: 68.420   **(c)** $\hat{y} = 68.07$

(d) Lower bound: 67.252; upper bound: 68.891

(e) The prediction made in part (a) is an estimate of the mean well-being index composite score for all individuals whose commute time is 20 minutes. The prediction made in part (c) is an estimate of the well-being index composite score of one individual, Jane, whose commute time is 20 minutes.

9. (a) $\hat{y} = 17.20$ inches

(b) 95% confidence interval: lower bound: 17.12 inches; upper bound: 17.28 inches

(c) $\hat{y} = 17.20$ inches

(d) 95% prediction interval: lower bound: 16.97 inches; upper bound: 17.43 inches

(e) The confidence interval is an interval estimate for the mean head circumference of all children who are 25.75 inches tall. The prediction interval is an interval estimate for the head circumference of a single child who is 25.75 inches tall.

11. (a) $\hat{y} = 4400.4$ psi

(b) 95% confidence interval: lower bound: 4147.8 psi; upper bound: 4653.1 psi

(c) $\hat{y} = 4400.4$ psi

(d) 95% prediction interval: lower bound: 3726.3 psi; upper bound: 5074.6 psi

(e) The confidence interval is an interval estimate for the mean 28-day strength of all concrete cylinders that have a 7-day strength of 2550 psi. The prediction interval is an interval estimate for the 28-day strength of a single cylinder whose 7-day strength is 2550 psi.

13. (a) $\hat{y} = 4.51\%$

(b) 95% confidence interval: lower bound: 4.33%; upper bound: 4.69%

(c) $\hat{y} = 4.51\%$

(d) 95% prediction interval: lower bound: -0.18%; upper bound: 9.21%

(e) Although the predicted return on investment in parts (a) and (c) are the same, the intervals are different because the distribution of the mean rate of return, part (a), has less variability than the distribution of the individual rate of return of a particular S&P 500 rate of return, part (c).

15. (a) It does not make sense to construct either a confidence interval or prediction interval based on the least-squares regression equation because the evidence indicated that there is no linear relation between CEO compensation and stock return.

(b) Since there is no linear relation between x and y, we use the techniques of Section 9.2 on the y-data to construct the confidence interval. 95% confidence interval for the mean stock return: lower bound: 25.938%, upper bound: 73.528% [Tech: 73.529%].

17. (a)

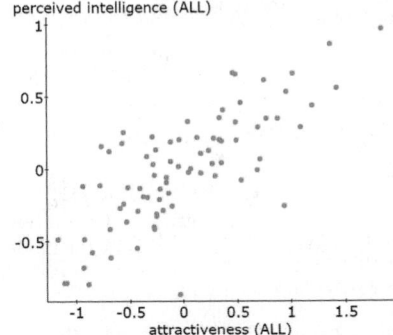

(b) $r = 0.762$; because $0.762 > 0.361$ (Table II with $n = 30$), we conclude there is a linear relation between attractiveness and perceived intelligence.

(c) $\hat{y} = 4.6008\,x + 0.0000$
(d) A person of average attractiveness is perceived to be of average intelligence.
(e) $H_0: \beta_1 = 0$ vs. $H_1: \beta_1 > 0$, P-value < 0.0001. Reject H_0.
(f)

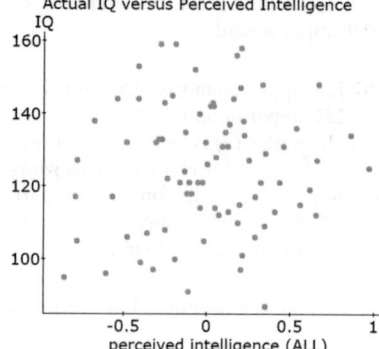

Actual IQ versus Perceived Intelligence

$r = 0.086 < 0.361$, No linear relation between perceived intelligence and IQ.
(g) $\hat{y} = -1.8893x + 122.6038$. $H_0: \beta_1 = 0$ vs. $H_1: \beta_1 > 0$. P-value $= 0.6329 > \alpha = 0.1$. Do not reject H_0.
(h) $\hat{y} = 18.1100x + 128.8345$, $H_0: \beta_1 = 0$ vs. $H_1: \beta_1 > 0$. P-value $= 0.0287 < \alpha = 0.1$. Reject H_0.
(i) Lower bound: 126.1; upper bound: 177.9
(j) Lower bound: 106.3; upper bound: 197.7

14.3 Assess Your Understanding (page 705)

1. correlation matrix   **3.** 4.39

5. (a) The slope coefficient of x_1 is 3. This indicates that $\hat{y}$ will increase 3 units, on average, for every 1-unit increase in x_1, provided that x_2 remains constant. The slope coefficient of x_2 is -4. This indicates that $\hat{y}$ will decrease 4 units, on average, for every 1-unit increase in x_2, provided that x_1 remains constant.
(b) $\hat{y} = 35 - 4x_2$ **(c)** $\hat{y} = 50 - 4x_2$

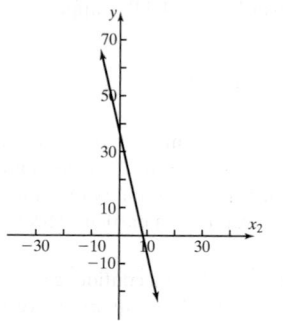

(d) $\hat{y} = 65 - 4x_2$

(e) Changing the value of x_1 has the effect of changing the y-intercept of the graph of the regression line.

7. (a) $R^2_{\text{adj}} = 0.603$ **(b)** $F_0 = 13.173$
(c) $R^2_{\text{adj}} = 0.598$; do not add the new variable. Its addition reduces the proportion of variance explained by the model.

9. (a) Correlations: x_1, x_2, x_3, y

	x_1	x_2	x_3
x_2	-0.460		
x_3	0.183	-0.012	
y	-0.748	0.788	0.258

There is no evidence that multicollinearity exists.

(b) $\hat{y} = 7.9647 - 0.1035x_1 + 0.9359x_2 + 0.1152x_3$
(c) $F_0 = 31.33$. Since the P-value $< 0.001 < \alpha = 0.05$, we reject the null hypothesis and conclude that at least one of the explanatory variables is linearly related to the response variable.
(d) Reject $H_0: \beta_1 = 0$ since $t_{\beta1} = -4.98$ and the P-value $= 0.003 < \alpha = 0.05$
Reject $H_0: \beta_2 = 0$ since $t_{\beta2} = 4.68$ and the P-value $= 0.003 < \alpha = 0.05$.
Reject $H_0: \beta_3 = 0$ since $t_{\beta3} = 3.61$ and the P-value $= 0.011 < \alpha = 0.05$.

11. (a) Correlations: x_1, x_2, x_3, x_4, y

	x_1	x_2	x_3	x_4
x_2	-0.184			
x_3	0.182	-0.106		
x_4	-0.166	-0.558	0.291	
y	-0.046	0.932	-0.063	-0.750

There is no evidence that multicollinearity will be a problem.

(b) $\hat{y} = 279.02 + 0.068x_1 + 1.146x_2 + 0.3897x_3 - 2.9378x_4$
$F_0 = 40.90$ and P-value $< 0.001 < \alpha = 0.05$, so we reject H_0 and conclude that at least one of the explanatory variables is linearly related to the response variable. The variables x_1 and x_3 have slope coefficients that are not significantly different from zero.
(c) $\hat{y} = 282.33 + 1.1426x_2 + 0.3941x_3 - 2.9570x_4$: the F-test still indicates that the model is significant (P-value $< 0.001 < \alpha = 0.05$). Remove x_3: $\hat{y} = 281.62 + 1.1584x_2 - 2.6267x_4$
(d)

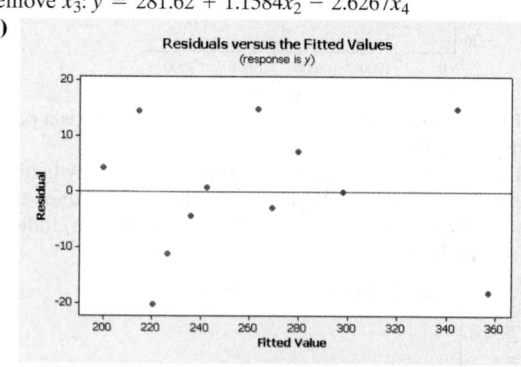

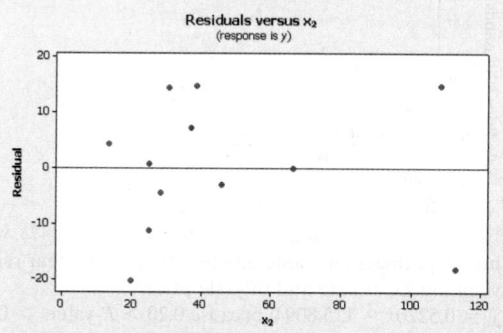

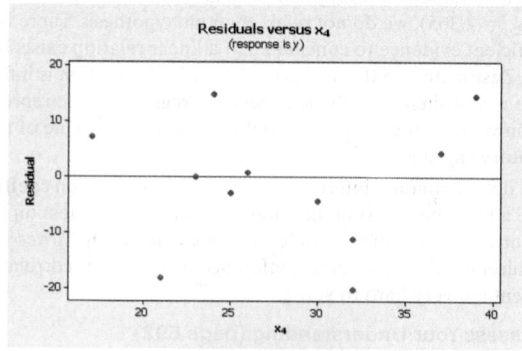

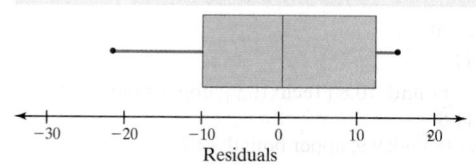

(e) $\hat{y} = 246.7$

(f)

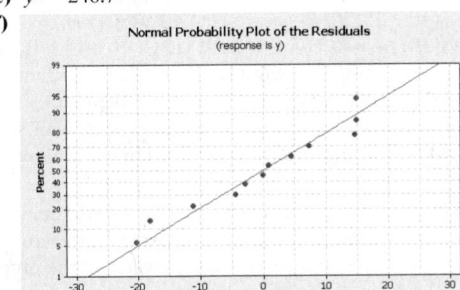

Normal Probability Plot of the Residuals
(response is y)

The residuals are normally distributed, so it is reasonable to construct confidence and prediction intervals.

(g) 95% confidence interval: lower bound: 237.45; upper bound: 255.91; 95% prediction interval: lower bound: 214.93; upper bound: 278.43.

13. (a) Yes; conclude that at least one of the slope coefficients is linearly related to the response variable, computer anxiety.

(b) $b_1 = -0.87$: A 1-point increase in the computer confidence variable reduces the computer anxiety measurement by 0.87 point, on average, provided that all other explanatory variables remain constant. $b_2 = -0.51$: A 1-point increase in the computer knowledge variable reduces the computer anxiety measurement by 0.51 point, on average, provided that all other explanatory variables remain constant. $b_3 = -0.45$: A 1-point increase in the computer liking scale reduces the computer anxiety measurement by 0.45 point, on average, provided that all other explanatory variables remain constant. $b_4 = 0.33$: A 1-point increase in the trait anxiety scale increases the computer anxiety measurement by 0.33 point, on average, provided that all other explanatory variables remain constant. They are all reasonable.

(c) $\hat{y} = 57.79$

(d) $R^2 = 0.69$ indicates that 69% of the variance in computer anxiety is explained by this model.

(e) The statement means that the requirements of a multiple linear regression model were checked and were met.

15. (a) $\hat{y} = -12.5091 + 1.3341x_1 - 0.4143x_2$

(b)

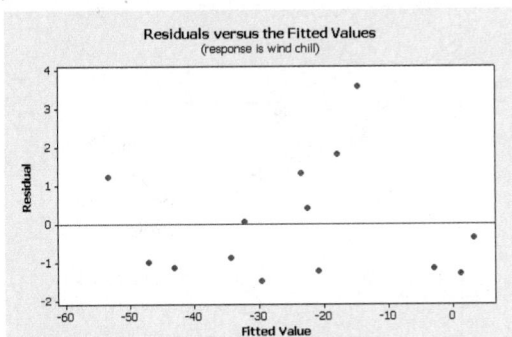

Residuals versus the Fitted Values
(response is wind chill)

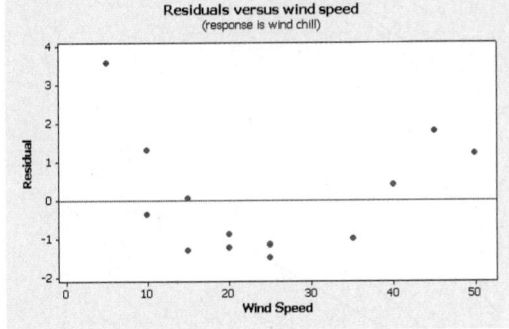

Residuals versus wind speed
(response is wind chill)

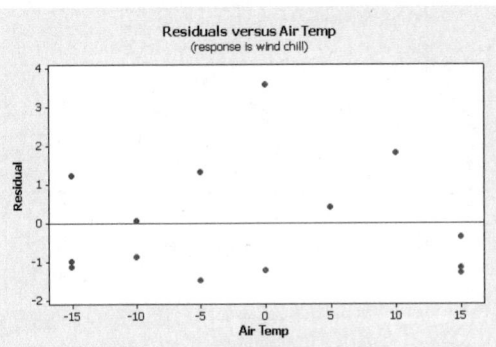

Residuals versus Air Temp
(response is wind chill)

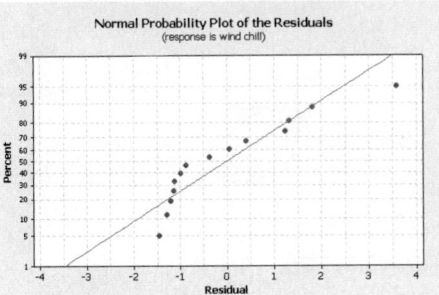

Normal Probability Plot of the Residuals
(response is wind chill)

Conclude that the linear model is not appropriate because of the patterned residuals.

17. (a) Correlations: SLUMP, 7 DAY, 28 DAY

```
                SLUMP       7 DAY
  7 DAY     -0.460
 28 DAY     -0.753        0.737
```

There is no reason to be concerned about multicollinearity based on the model.

(b) $\hat{y} = 3890.5 - 295.9x_1 + 0.552x_2$

(c)

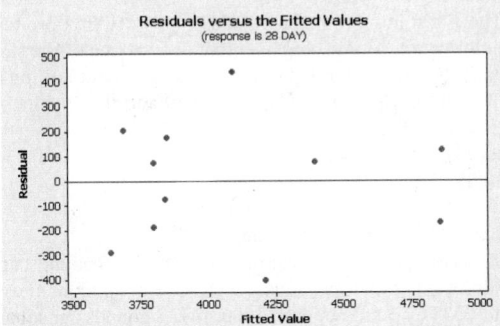

Residuals versus the Fitted Values
(response is 28 DAY)

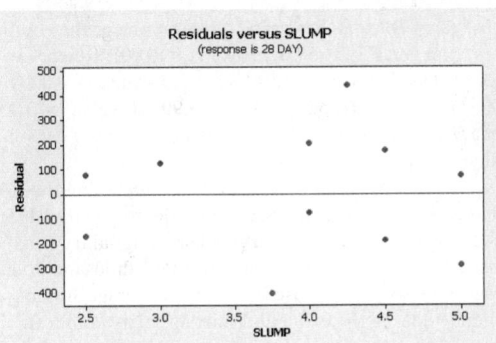

Residuals versus SLUMP
(response is 28 DAY)

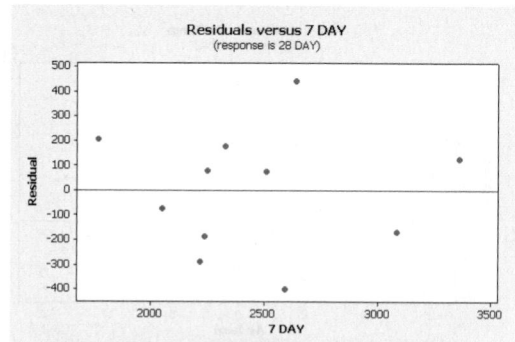

Residuals versus 7 DAY
(response is 28 DAY)

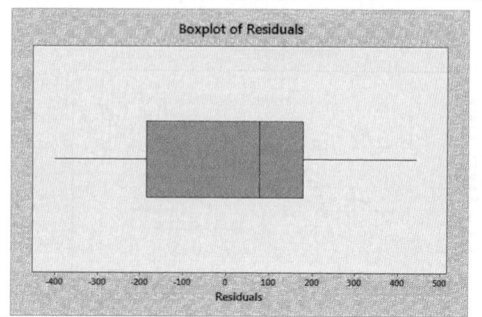

Boxplot of Residuals

(d) $b_1 = -295.9$: A 1-unit increase in slump of the concrete decreases the 28-day strength of the concrete by 295.9 pounds per square inch, on average, provided that the other variable remains constant.
$b_2 = 0.552$: A 1-pound per square inch increase in 7-day strength increases the 28-day strength of the concrete by 0.552 pound per square inch, on average, provided that slump remains constant.
(e) $R^2 = 76.0\%$ represents the variance in 28-day strength explained by the model. $R^2_{adj} = 70.1\%$ modifies the value of R^2 based on the sample size and the number of explanatory variables in the model.
(f) The F-test has a P-value $= 0.003 < \alpha = 0.05$, so we reject the null hypothesis and conclude that at least one of the slope coefficients is linearly related to the 28-day strength of the concrete.
(g) Reject H_0: $\beta_1 = 0$ since $t_{\beta 1} = -2.69$ and the P-value $= 0.027 < \alpha = 0.05$.
Reject H_0: $\beta_2 = 0$ since $t_{\beta 2} = 2.54$ and the P-value $= 0.034 < \alpha = 0.05$.
(h) $\hat{y} = 4207.3$ pounds per square inch.
(i) $\hat{y} = 4207.3$ pounds per square inch.
(j) 95% confidence interval: lower bound: 3988.6 pounds; upper bound: 4427.2 pounds per square inch; 95% prediction interval: lower bound: 3533.6 pounds; upper bound: 4882.2 pounds per square inch.

19. (a) $\hat{y} = -1.1752x_1 + 2.3463x_2 - 2.2489x_3 + 539.4308$
(b) H_0: $\beta_1 = \beta_2 = \beta_3 = 0$ vs. H_1: at least one of the coefficients differs from zero. $F_0 = 57.07$. P-value < 0.001. Reject H_0.
H_0: $\beta_1 = 0$ vs. H_1: $\beta_1 \neq 0$. $t_0 = -11.97$. P-value < 0.0001. Reject H_0. $H_0 : \beta_2 = 0$ vs. $H_1 : \beta_2 \neq 0$. $t_0 = 3.907$. P-value $= 0.0003$, Reject H_0. H_0: $\beta_3 = 0$ vs. H_1: $\beta_3 \neq 0$. $t_0 = -2.096$. P-value $= 0.0416$. Reject H_0.
(c) $b_1 = -1.175$: If the percent of students taking the SAT increases by 1, the average SAT score decreases by 1.175, on average, holding average salary and starting salary constant.
$b_2 = 2.346$: If average salary increases by 1 thousand dollars, average SAT scores increase by 2.346, on average, holding percent of students taking the exam and starting salary constant.
$b_3 = -2.249$: If starting salary increases by 1 thousand dollars, average SAT scores decrease by 2.249, on average, holding percent of students taking the exam and average salary constant.
The coefficient for starting salary is the opposite sign we would expect. One would think higher starting salaries result in better teachers hired, which would translate into higher SAT scores. However, the correlation between average salary and starting salary is 0.811, which suggests the model suffers from multicollinearity.

(d) $\hat{y} = -1.2122x_1 + 1.4007x_2 + 511.6738$. H_0: $\beta_1 = 0$ vs. H_1: $\beta_1 \neq 0$. $t_0 = -12.120$. P-value < 0.0001. Reject H_0.
H_0: $\beta_2 = 0$ vs. H_1: $\beta_2 \neq 0$. $t_0 = 3.412$. P-value $= 0.0013$. Reject H_0.
Each coefficient has a sign we would expect. The coefficient on percent taking the exam is negative. The higher the percentage of students who take the exam, the lower the average SAT score. The coefficient for average salary is positive. Higher salaries for teachers correspond to higher SAT scores.

21. When the explanatory variables have an additive effect, or do not interact, it means that the effect of any one explanatory variable on the response variable does not depend on the value of the other explanatory variables.

23. Multicollinearity exists when two explanatory variables are highly correlated. We use a correlation matrix to check if multicollinearity exists. A consequence of multicollinearity is that t-test statistics might be small and their P-values large, falsely indicating that neither explanatory variables is significant.

14.4 Assess Your Understanding (page 716)

1. interaction

3. (a) $y_i = \beta_0 + \beta_1 x_1 + \beta_2 x_2 + \beta_3 x_3 + \varepsilon_i$
(b) $y_i = \beta_0 + \beta_1 x_1 + \beta_2 x_2 + \beta_3 x_1 x_2 + \varepsilon_i$
(c) $y_i = \beta_0 + \beta_1 x_1 + \beta_2 x_2 + \beta_3 x_3 + \beta_4 x_2 x_3 + \varepsilon_i$

5. (a)

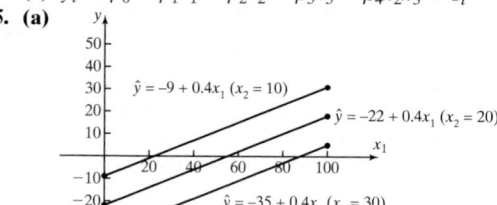

(b)

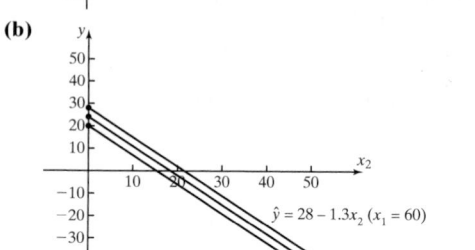

(c) The lines are parallel.
(d)

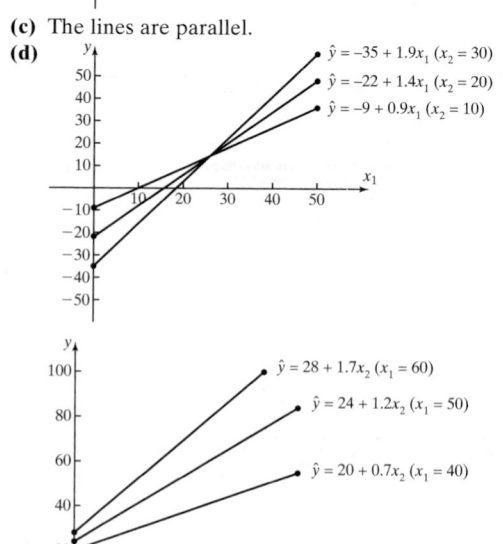

7. (a) 8.3 years   **(b)** 8.7 years

(c) $-0.175x_3x_4$: the number of teeth with root development, x_3, and the sum of the normalized heights of seven teeth on the left side of the mouth, x_4, interact.

(d) 86.3% of the variation in the response variable is explained by the least-squares regression model.

9. (a) $\hat{y} = 109.894 + 0.0000286x_1$

(b) We reject the null hypothesis that $\beta_1 = 0$ since the P-value $= 0.012$ and conclude that there is a linear relation between MRI count and IQ.

(c)

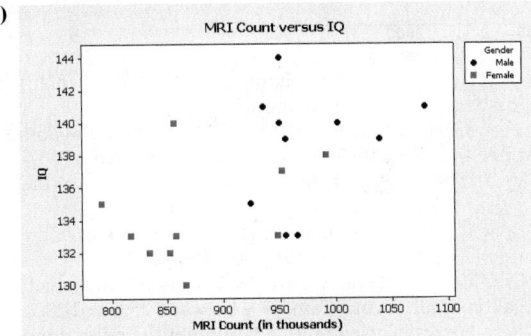

(d) $\hat{y} = 121.58 + 0.0000174x_1 - 2.486x_2$

(e) For the explanatory variable gender, $t_0 = -1.23$ and the P-value $= 0.234$. So we do not reject the null hypothesis and conclude that there is not significant linear relation between gender and intelligence.

For the explanatory MRI count. $t_0 = 1.27$ and the P-value $= 0.221$. So we do not reject the null hypothesis and conclude that there is no significant linear relation between MRI count and intelligence.

(f) The variable x_1, MRI count, originally has a slope coefficient that was significantly different from 0. Adding the variable x_2. gender, caused this relationship to change so that the slope coefficient of both variables are not significantly different from 0. This might lead us to conclude that gender is a lurking variable.

11. We use two indicator variables.

$$x_1 = \begin{cases} 1 & \text{if attached} \\ 0 & \text{otherwise} \end{cases} \qquad x_2 = \begin{cases} 1 & \text{if detached} \\ 0 & \text{otherwise} \end{cases}$$

14.5 Assess Your Understanding (page 721)

1. (a)

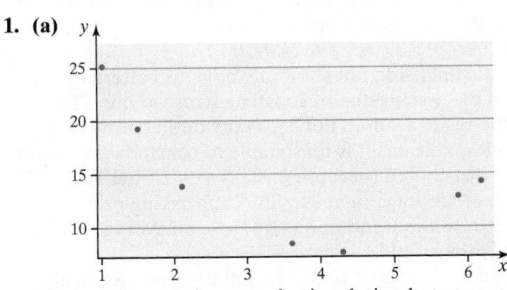

There appears to be a quadratic relation between x and y.

(b) $\hat{y} = 1.7046x^2 - 14.1746x + 36.9806$

(c)

Residuals Plot – Fitted Values vs. Residuals

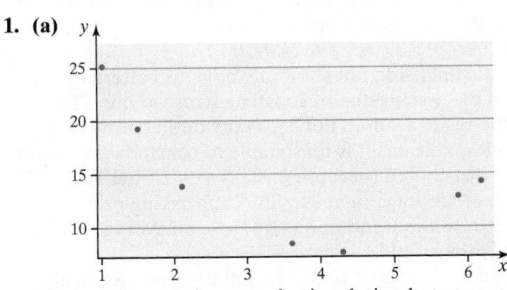

Residuals Plot – x vs. Residuals

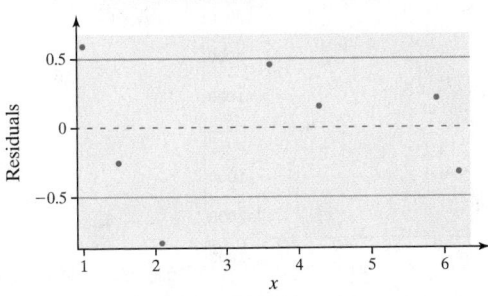

Residuals Plot – x Squared vs. Residuals

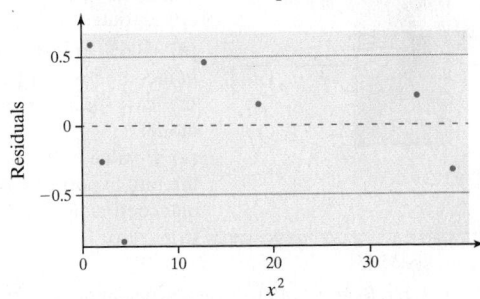

Boxplot of Residuals

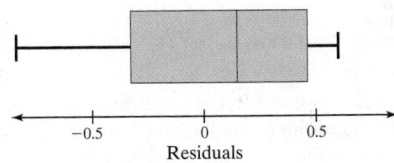

The quadratic model appears to be appropriate.

(d) 99.3% of the variation in y is explained by the least-squares regression model.

(e) $H_0: \beta_1 = \beta_2 = 0$ vs. H_1: at least one coefficient is different from 0. $F_0 = 295.4$ P-value < 0.0001. Reject H_0.

$H_0: \beta_1 = 0$ vs. $H_1: \beta_1 \neq 0$. $t_0 = 19.33$. P-value < 0.0001. Reject H_0

$H_0: \beta_2 = 0$ vs. $H_1: \beta_2 \neq 0$. $t_0 = -21.7$. P-value < 0.0001. Reject H_0.

(f) Confidence internal: lower bound: 6.46, upper bound: 8.65; prediction interval: lower bound: 5.54, upper bound: 9.57

3. (a)

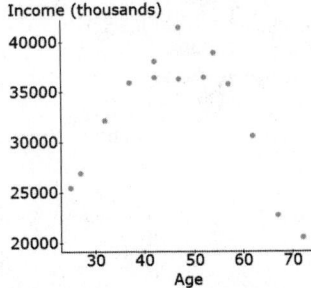

The relation appears to be a quadratic relation that opens down.

(b) $\hat{y} = -29.7706x^2 + 2756.9751x - 25403.904$

(c)

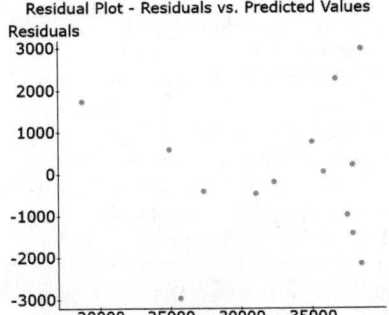

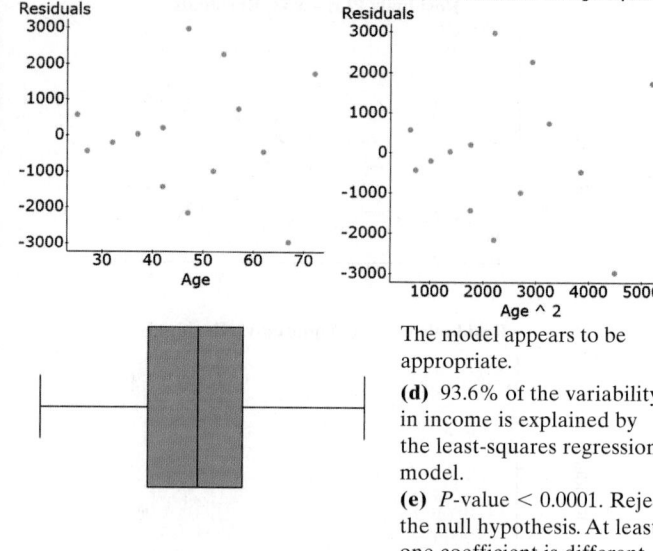

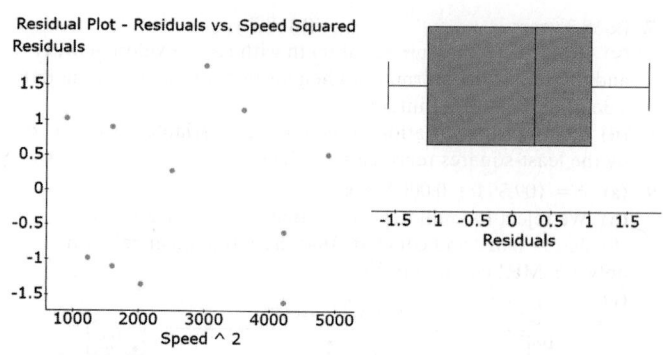

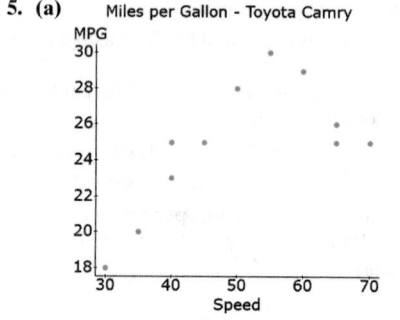

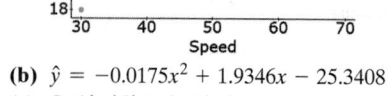

The model appears to be appropriate.

(d) 93.6% of the variability in income is explained by the least-squares regression model.

(e) P-value < 0.0001. Reject the null hypothesis. At least, one coefficient is different from zero.

$H_0: \beta_1 = 0$ vs. $H_1: \beta_1 \neq 0$, where β_1 is coefficient of linear term.
$t_0 = 11.876$. P-value < 0.0001. Reject H_0.

$H_0: \beta_2 = 0$ vs. $H_1: \beta_2 \neq 0$, where β_2 is coefficient of squared term.
$t_0 = -12.394$. P-value < 0.0001. Reject H_0.

(f) Confidence interval: we are 95% confident the mean income of all 45-year-olds is between $36,934.2 and $39,814.8. Prediction interval: we are 95% confident the income of a particular 45-year-old is between $34,199.1 and $42,550.0.

5. (a)

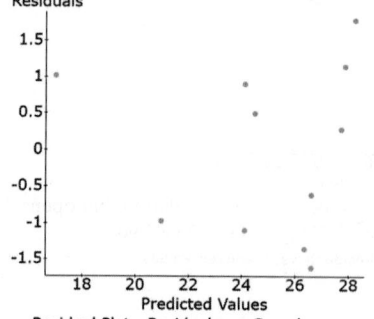

The relation appears quadratic that opens downward.

(b) $\hat{y} = -0.0175x^2 + 1.9346x - 25.3408$

(c)

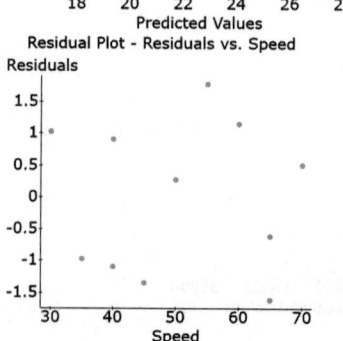

(d) 89.5% of the variability in miles per gallon is explained by the least-squares regression model.

(e) $F = 33.949$; P-value $= 0.0001$. Reject the null hypothesis. At least one of the coefficients is different from zero.
$H_0: \beta_1 = 0$ vs. $H_1: \beta_1 \neq 0$, where β_1 is the coefficient of the linear term. $t_0 = 6.532$. P-value $= 0.0002$. Reject H_0.
$H_0: \beta_2 = 0$ vs. $H_1: \beta_2 \neq 0$, where β_2 is the coefficient of the squared term. $t_0 = -5.979$. P-value $= 0.0003$. Reject H_0.

(f) Confidence interval: we are 95% confident the mean miles per gallon of all Toyota Camrys is between 25.9 and 28.7. Prediction interval: we are 95% confident the miles per gallon of a particular Toyota Camry is between 24.0 and 30.6.

7. (a) 20.6 years   **(b)** 0.9848 year

(c) If the percent of children aged 10 to 14 years involved in child labor increases by 1, the age gap at first marriage increases by 0.0321 year, on average.

(d) 59.3% of the variation in age gap at first marriage is explained by the regression model.

(e) Because the P-value is small, reject the null hypothesis. Percent of children aged 10 to 14 years involved in child labor is significant in explaining age gap at first marriage.

14.6 Assess Your Understanding (page 733)

1. $H_0: \beta_4 = \beta_5 = 0$ vs. H_1: at least one of β_4 and β_5 is different from zero. $F_0 = 0.71 < F_{0.05,2,4} = 6.94$. Do not reject H_0.
There is not sufficient evidence to suggest that either x_4 or x_5 explain the variability in the response variable y.

3. (a) $\hat{y} = 0.0719x_1 - 0.8099x_2 + 55.3491x_3 + 25.6281$

(b) $H_0: \beta_2 = \beta_3 = 0$ vs. H_1: at least one of β_2 and β_3 is different from zero.
$F_0 = 2.79 < F_{0.05,2,9} = 4.26$. Do not reject H_0. There is not sufficient evidence to suggest number of bedrooms or number of baths explain the variability in asking price.

(c) $\hat{y} = 0.0718x_1 + 54.9163x_3 + 24.0920$

(d) The residual plots do not show any obvious patterns nor a violation of the assumption of constant error variance. The boxplot of the residuals does not show any outliers. The normal probability plot indicates it is reasonable to conclude the residuals are normally distributed (because $0.991 > 0.932$ from Table VI).

(e) If the square footage increases by 1, the asking price will increase by 0.0718 thousand dollars ($71.8), on average, assuming the number baths is held constant.
If the number of baths increases by 1, the asking price will increase by 54.9163 thousand dollars ($54,916), on average, assuming the square footage is held constant.

(f) Confidence interval: lower bound: $346.8 thousand, upper bound: $404.3 thousand; prediction interval: lower bound: $273.6 thousand, upper bound: $477.4 thousand.

5. Answers may vary. Using $\alpha = 0.05$ Cu: $\hat{y} = 2.9176$ WHIP $- 0.5504$ G/F $- 3.0253$W% $+ 0.3577$RS $+ 0.6176$

7. Using Stepwise and forward selection: $\hat{y} = 558.2815$ Cornering -325.2538 Slalom $+ 64.2139$ Lap Time -662.7487
Using backward selection:
$\hat{y} = -223.8081$ Slalom Time $+ 28.4755$ Lap Time $+ 403.0802$

Chapter 14 Review Exercises (page 737)

1. The least-squares regression model is $y_i = \beta_0 + \beta_1 x_i + \varepsilon_i$. The requirements to perform inference on the least-squares regression line are (1) for any particular values of the explanatory variable x, the mean of the corresponding responses in the population depends linearly on x, and (2) the response variables, y_i, are normally distributed with mean $\mu_{y|x} = \beta_0 + \beta_1 x$ and standard deviation σ. We verify these requirements by checking to see that the residuals are normally distributed, with mean 0 and constant variance σ^2 and that the residuals are independent. We do this by constructing residual plots and a normal probability plot of the residuals.

2. (a) $\beta_0 \approx b_0 = 3.8589$; $\beta_1 \approx b_1 = -0.1049$; $\hat{y} = 3.334$
(b) $s_e = 0.5102$
(c)

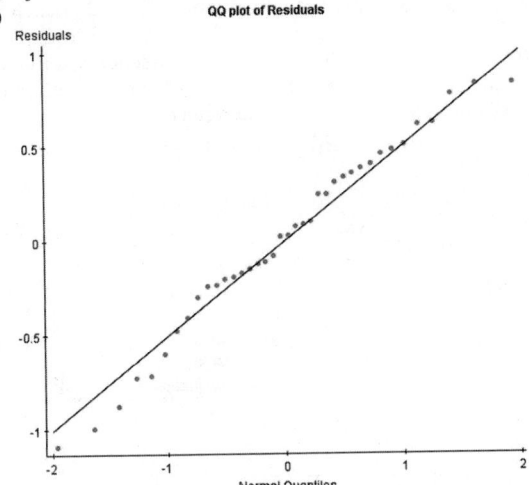

QQ plot of Residuals

The residuals are normally distributed.

(d) $s_{b_1} = 0.0366$
(e) Because the P-value $= 0.007 < \alpha = 0.05$ (or $t_0 = -2.866$ [Tech: -2.868] $< t_{0.025} = -2.028$), we reject the null hypothesis and conclude that a linear relation exists between the row chosen by students on the first day of class and their cumulative GPAs.
(f) 95% confidence interval: lower bound: -0.1791; upper bound: -0.0307
(g) 95% confidence interval: lower bound: 3.159; upper bound: 3.509
(h) $\hat{y} = 3.334$
(i) 95% prediction interval: lower bound: 2.285; upper bound: 4.383 [Tech: 4.384]
(j) Although the predicted GPAs in parts (a) and (h) are the same, the intervals are different because the distribution of the mean GPAs, part (a), has less variability than the distribution of individual GPAs, part (h).

3. (a) $\beta_0 \approx b_0 = -399.2496$; $\beta_1 \approx b_1 = 2.5315$; $\hat{y} = 1879.1$, so the mean rent of a 900-square-foot apartment in Queens is $1879.10.
(b) $s_e = 229.547$
(c)

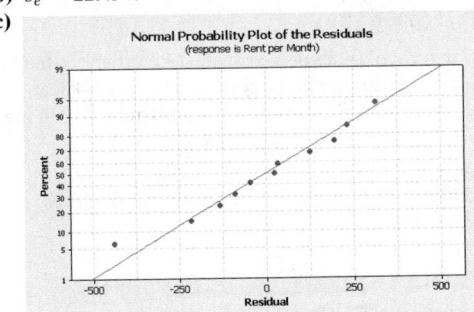

Normal Probability Plot of the Residuals
(response is Rent per Month)

The correlation between the residuals and expected z-scores is 0.985. Because $0.985 > 0.923$, the residuals are normally distributed.
(d) $s_{b_1} = 0.2166$
(e) There is evidence that a linear relation exists between the square footage of an apartment in Queens, New York, and the monthly rent.
(f) 95% confidence interval about the slope of the true least-squares regression line: lower bound: 2.0416, upper bound: 3.0214.
(g) 90% confidence interval about the mean rent of 900-square-foot apartments: lower bound: $1752.20 [Tech: $1752.20]; upper bound: $2006.00 [Tech: $2006.96].

(h) When an apartment has 900 square feet, $\hat{y} = \$1879.1$ [Tech: 1879.1].
(i) 90% prediction interval for the rent of a particular 900-square-foot apartment: lower bound: $1439.60 [Tech: $1439.59]; upper bound: $2318.60 [Tech: $2318.59].
(j) Although the predicted rents in parts (a) and (h) are the same, the intervals are different because the distribution of the means, part (a), has less variability than the distribution of the individuals, part (h).

4. (a)

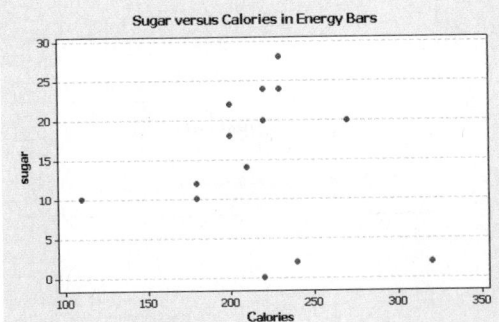

Sugar versus Calories in Energy Bars

No linear relation appears to exist.
(b) $\hat{y} = 17.8675 - 0.0146x$ **(c)** $s_e = 9.3749$
(d) A normal probability plot of the residuals shows that they are approximately normally distributed.
(e) $s_{b_1} = 0.0549$
(f) Because the P-value $= 0.795 > \alpha = 0.01$ (or $t_0 = -0.266$ [Tech: -0.265] $> -t_{0.005} = -3.055$), we do not reject the null hypothesis. We conclude that a linear relation does not exist between the number of calories per serving and the number of grams per serving in high-protein and moderate-protein energy bars.
(g) 95% confidence interval: lower bound: -0.1342 [Tech: -0.1343]; upper bound: 0.1050 [Tech: 0.1051]
(h) Do not recommend using the least-squares regression line to predict the sugar content of the energy bars because we did not reject the null hypothesis. A good estimate for the sugar content is $\bar{y} = 14.7$ grams.

5. (a) $\hat{y} = 20{,}976.2 - 2054.6x$
(b) Reject H_0. There is sufficient evidence at the $\alpha = 0.05$ level of significance that a linear relation exists between the age of used Chevy Camaros and their selling price (P-value < 0.001).
(c)

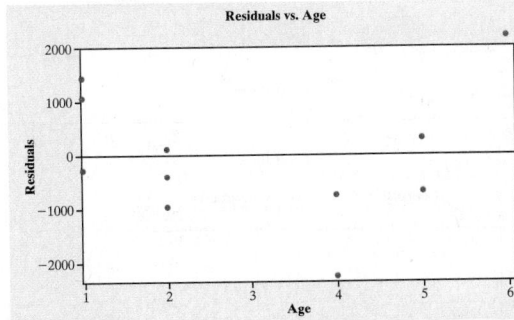

Residuals vs. Age

(d) The residual plot shows a pattern indicating that a linear model is not appropriate. The moral is to perform graphical diagnostic tests along with the inferential procedures before drawing any conclusions regarding the model.

6. (a) Coefficient of x_1: if rainfall during the winter increases by 1 inch while temperature and rainfall during the harvest remain constant, then the wine quality rating increases by 0.00117. Coefficient of x_2: if temperature increases by 1°F while rainfall during the winter and rainfall during the harvest remain constant, then the wine quality rating increases by 0.0614. Coefficient of x_3: if rainfall during the harvest increases by 1 inch while rainfall during the winter and temperature remain constant, then the wine quality rating decreases by 0.00386.
(b) The wine quality rating is predicted to be 15.7.

7. (a) **Correlations: Midterm, Project, Course Grade**

	Midterm	Project
Project	-0.076	
Course Grade	0.704	0.526

There is no problem with multicollinearity between the explanatory variables.

(b) $\hat{y} = -91.29 + 1.5211x_1 + 0.4407x_2$

(c)

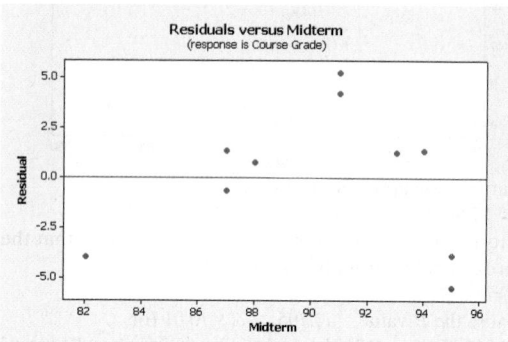

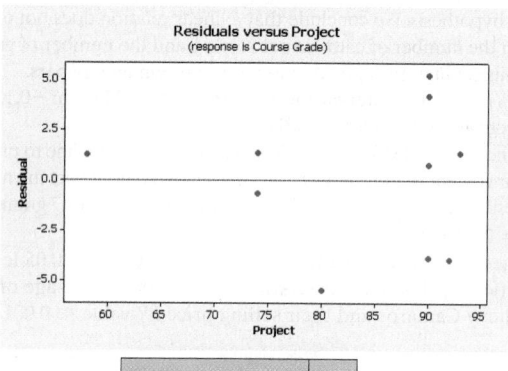

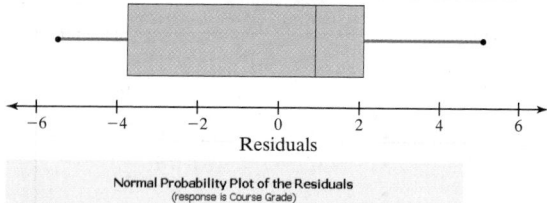

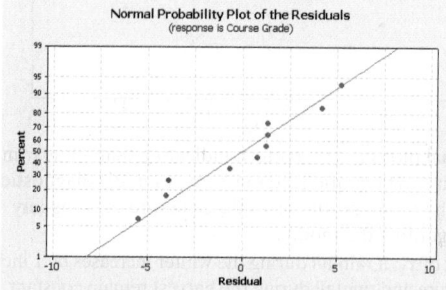

(d) The regression coefficient 1.5211 indicates that when midterm grade increases by 1 point (while holding project grade constant), the course grade increases by 1.5211 points, on average. The regression coefficient 0.4407 indicates that when project grade increases by 1 point (while holding midterm grade constant), the course grade increases by 0.4407 points, on average.

(e) $R^2 = 0.833$; $R^2_{adj} = 0.785$ are measures of the proportion of the variance in the course grade that is explained by the linear regression model.

(f) There is sufficient evidence at the $\alpha = 0.05$ level of significance to reject the null hypothesis and conclude that at least one $\beta_i \ne 0$ ($F_0 = 17.47$; P-value = 0.002).

(g) For the explanatory variable midterm, $t_0 = 4.83$ and the P-value = 0.002. For the explanatory variable project, $t_0 = 3.77$ and the P-value = 0.007. We reject both null hypotheses and conclude that both the explanatory variable midterm and the explanatory variable project are linearly related to course grade.

(h) $\hat{y} = 71.1$

(i) 95% confidence interval: lower bound: 69.00; upper bound: 82.04. We are 95% confident that the mean course grade for students who have an 83 on their midterm and a 92 on their project will be between 69 and 82. 95% prediction interval: lower bound: 64.04; upper bound: 87.00. We are 95% confident that a randomly selected student who has an 83 on the midterm and a 92 on the project will have a course grade between 64 and 87.

8. (a) $y = \beta_1 x_1 + \beta_2 x_2 + \beta_3 x_3 + \beta_0$; x_3: $0 =$ male, $1 =$ female

(b) $y = \beta_0 + \beta_1 x_2 + \beta_2 x_2^2$

(c) $y = \beta_1 x_1 + \beta_2 x_2 + \beta_3 x_3 + \beta_4 x_1 x_2 + \beta_0$

9. (a)

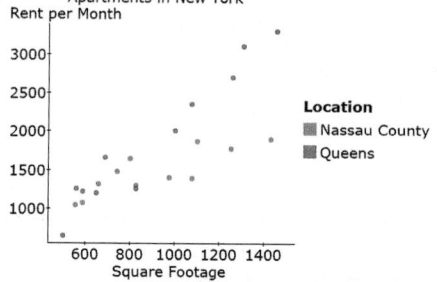

(b) $\hat{y} = 216.0137 + 1.8479x_1 - 422.7573x_2$

(c) $F_0 = 35.29$. P-value < 0.0001.

Reject the null hypothesis. There is sufficient evidence to conclude that at least one coefficient is different from 0.

(d) H_0: $\beta_2 = 0$ vs. H_1: $\beta_2 \ne 0$. $t_0 = -3.052$. P-value = 0.0066. Reject H_0.

There is sufficient evidence to conclude that the coefficient is different from zero. The coefficient suggests that rents in Nassau County are $422.76 cheaper per month.

(e) Lower bound: $1765.1, upper bound: $2240.3. We are 95% confident the mean monthly rent for 950-square-foot apartment in Queens is between $1765.1 and $2177.9.

10. (a) $\hat{y} = -493.2621 + 80.5515x_1 + 0.7237x_2 + 1.9861x_3 + 49.1947x_4 + 7.4312x_5 + 7.8615x_6$

(b) $F_0 = 0.666 < F_{0.05,2,146} \approx 3.09$. Do not reject H_0.

There is not sufficient evidence to conclude that at least one of the coefficients is different from zero.

(c) $\hat{y} = -422.0052 + 83.3777x_1 + 2.0282x_3 + 52.5186x_4$

Chapter 14 Test (page 739)

1. (1) For any particular values of the explanatory variable x, the mean of the corresponding responses in the population depends linearly on x.

(2) The response variable, y_i, is normally distributed with mean $\mu_{y|x_i} = \beta_0 + \beta_1 x$ and standard deviation σ.

2. (a) $\beta_0 \approx b_0 = -0.3091$; $\beta_1 \approx b_1 = 0.2119$; $\hat{y} = 16.685$ [Tech: 16.687]; so the mean number of chirps when the temperature is 80.2°F is 16.69.

(b) $s_e = 0.9715$

(c)

The residuals are normally distributed.

(d) $s_{b_1} = 0.0387$

(e) There is sufficient evidence that a linear relation exists between temperature and the cricket's chirps (P-value = 0.0001).
(f) 95% confidence interval about the slope of the true least-squares regression line: lower bound: 0.1283; upper bound: 0.2955.
(g) 90% confidence interval about the mean number of chirps at 80.2°F: lower bound: 16.25 [Tech: 16.24]; upper bound: 17.13.
(h) When the temperature is 80.2°F, $\hat{y}$ = 16.69 chirps.
(i) 90% prediction interval for the number of chirps of a particular cricket at 80.2°F: lower bound: 14.91; upper bound: 18.47 [Tech: 18.46].
(j) Although the predicted numbers of chirps in parts (a) and (h) are the same, the intervals are different because the distribution of the means, part (a), has less variability than the distribution of the individuals, part (h).

3. **(a)** $\beta_0 \approx b_0$ = 29.7048; $\beta_1 \approx b_1$ = 2.6351; $\hat{y}$ = 48.15; so the mean height of a 7-year-old boy is 48.15 inches.
(b) s_e = 2.45
(c) There is sufficient evidence that a linear relation exists between boys' ages and heights (P-value < 0.0001).
(d) 95% confidence interval about the slope of the true least-squares regression line: lower bound: 2.2009; upper bound: 3.0692.
(e) 90% confidence interval about the mean height of 7-year-old boys: lower bound: 47.11; upper bound: 49.19 inches.
(f) The predicted height of a randomly chosen 7-year-old boy is 48.15 inches, $\hat{y}$ = 48.15.
(g) 90% prediction interval for the height of a particular 7-year-old boy: lower bound: 43.78; upper bound: 52.52 inches.
(h) Although the predicted heights in parts (a) and (f) are the same, the intervals are different because the distribution of the means, part (a), has less variability than the distribution of the individuals, part (f).

4. **(a)** $\beta_0 \approx b_0$ = 67.388; $\beta_1 \approx b_1$ = −0.2632
(b) There is not sufficient evidence at the α = 0.05 level of significance to support that a linear relation exists between a woman's age and grip strength (P-value = 0.138).
(c) Based on the answer to (b), a good estimate of the grip strength of a 42-year-old female would be the mean strength of the population, 57 psi.

5. **(a)** The student scored 1.52 standard deviations above the mean on the Math portion of the SAT.
(b) The student's expected first-year college GPA will be 0.29 z-scores lower.
(c) If high school GPA z-score increases by 1, the first-year college GPA z-score will increase 0.29, on average.
(d) 24% of the variability in first-year college GPA is explained by the least-squares regression model.
(e) The model may suffer from multicollinearity.
(f) 0.34; the expected first-year college GPA is 0.34 standard deviations above the mean.

6. **(a)** **Correlations: Fat, Protein, Sugar, Carbs, Calories**

	Fat	Protein	Sugar	Carbs
Protein	0.507			
Sugar	0.412	0.247		
Carbs	0.576	0.550	0.671	
Calories	0.950	0.661	0.542	0.784

There is no reason to be concerned about multicollinearity.
(b) $\hat{y}$ = −11.9 + 9.0765x_1 + 4.4676x_2 + 1.488x_3 + 3.5202x_4
(c) Since the P-value < 0.001 (F_0 = 1157.97), we reject the null hypothesis and conclude that at least one of the explanatory variables is linearly associated with calories.
(d) For the explanatory variable fat content, t_0 = 37.49 and the P-value < 0.001. For the explanatory variable protein, t_0 = 7.53 and the P-value is 0.005. For the explanatory variable sugar, t_0 = 1.34 and the P-value is 0.273. For the explanatory variable carbohydrates, t_0 = 11.60 and the P-value is 0.001. We reject the null hypotheses for fat, protein, and carbohydrates and conclude that they are linearly related to calories. We do not reject the null hypothesis for sugar and conclude that sugar is not linearly related to calories.
(e) Remove x_3 = sugar; $\hat{y}$ = −4.42 + 9.1099x_1 + 4.2937x_2 + 3.7672x_4. The remaining slope coefficients are significantly different from zero.

(f)

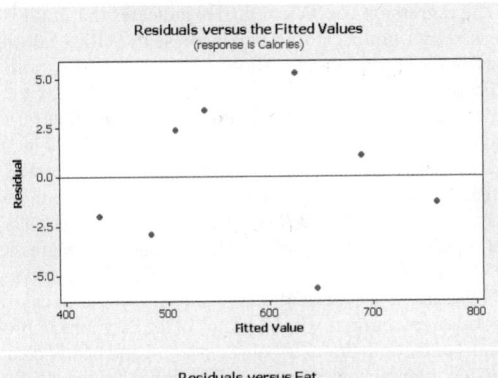

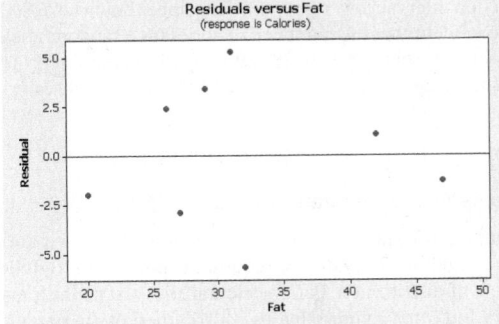

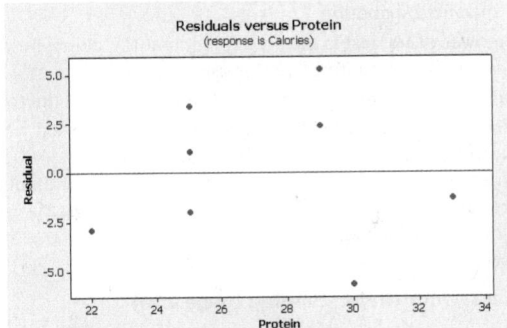

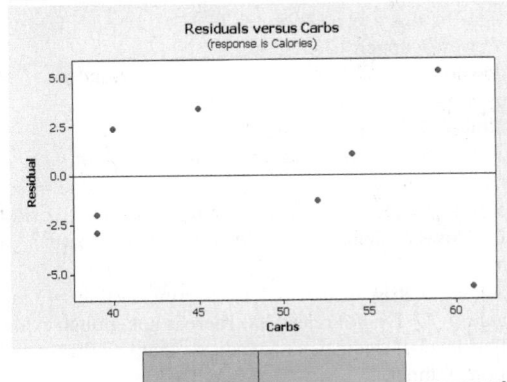

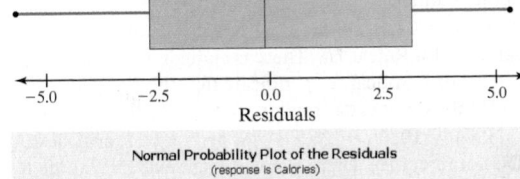

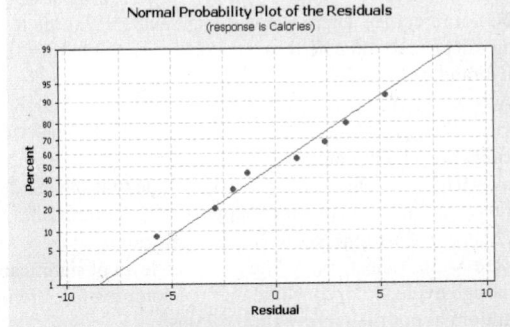

(g) The regression coefficient 9.1099 indicates that if fat is increased by 1 gram, calories will increase by 9.1099 calories, assuming that protein and carbohydrates remain constant. The coefficient 4.2937 indicates that if protein is increased by 1 gram, calories will increase by 4.2937 calories, assuming that fat and carbohydrates remain constant. The coefficient 3.7672 indicates that if carbohydrates are increased by 1 gram, calories will increase by 3.7672 calories, assuming that fat and protein remain constant. **(h)** $R^2 = 99.9\%$; adjusted $R^2 = 99.8\%$; virtually all (99.9%) of the variation in calories is explained by the least-squares regression model. **(i)** 95% confidence interval: lower bound: 656.16, upper bound: 668.17. We are 95% confident that the mean number of calories for all fast-food cheeseburgers with 38 grams of fat, 29 grams of protein, and 52 grams of carbohydrates is between 656.16 and 668.17 calories. 95% prediction interval: lower bound: 647.53, upper bound: 676.80. We are 95% confident that the number of calories for a randomly selected fast-food cheeseburger with 38 grams of fat, 39 grams of protein, and 52 grams of carbohydrates is between 647.53 and 676.80 calories.

Chapter 15 Nonparametric Statistics

15.1 Assess Your Understanding (page 745)

1. Nonparametric statistics are inferential procedures that are not based on parameters. They do not require the population to follow a specific type of distribution. Parametric statistics test claims regarding parameters and often require that the distribution of the population follow a specific distribution.
3. The power of the test is the probability that the null hypothesis is rejected when the alternative hypothesis is true.
5. Disadvantages of nonparametric procedures include the following:
1. They are often misused. That is, they are often used when more powerful parametric methods are appropriate.
2. Nonparametric methods are less powerful than parametric procedures.
3. Nonparametric methods are less efficient than parametric procedures.

15.2 Assess Your Understanding (page 750)

1. run
3. **(a)** $n = 14; n_M = 6; n_F = 8; r = 6$
(b) Lower: 3; upper: 12
(c) Do not reject H_0. The sequence of data is random.
5. **(a)** $n = 19; n_Y = 9; n_N = 10; r = 5$
(b) Lower: 5; upper: 16
(c) Reject H_0. The sequence of data is not random.
7. $n = 20, n_F = 9, n_C = 11, r = 12$; lower critical value $= 6$, upper critical value $= 16$. Do not reject H_0. There is not enough evidence at the $\alpha = 0.05$ level of significance to indicate that Bumgarner's pitches are not random.
9. $n = 14, n_O = 8, n_L = 6, r = 7$; lower critical value $= 3$, upper critical value $= 12$. Do not reject H_0. There is not enough evidence at the $\alpha = 0.05$ level of significance to indicate that late flights occur in a nonrandom manner.
11. $n = 20, n_A = 14, n_R = 6, r = 4$; lower critical value $= 5$, upper critical value $= 14$. Reject H_0. There is enough evidence at the $\alpha = 0.05$ level of significance to indicate that the machine's overfilling and underfilling do not occur randomly.
13. $n = 45, n_P = 19, n_N = 26, r = 19, z_0 = -1.22$; critical values $= \pm 1.96$. Do not reject H_0. There is not enough evidence at the $\alpha = 0.05$ level of significance to indicate that the stock price fluctuations do not behave in a random manner.
15. **(a)** A A A B B A B B B B A A
(b) $n = 12, n_A = 6, n_B = 6, r = 5$; lower critical value: 3, upper critical value: 11. Do not reject H_0. There is not enough evidence at the $\alpha = 0.05$ level of significance to conclude that the residuals are not random. The data appear to support the assumption.
17. $n = 20, n_A = 12, n_B = 8, r = 6$; lower critical value $= 6$; upper critical value $= 16$. Reject H_0. At the $\alpha = 0.05$ level of significance, there is enough evidence to conclude that the compression strengths are not random about the target value of 75 psi.

19. Random means that the sequence of outcomes follows no particular pattern. A run is a sequence of similar events, items, or symbols that is followed by an event, item, or symbol that is mutually exclusive from the first event, item, or symbol.

15.3 Assess Your Understanding (page 758)

1. $H_0: M = 8$ versus $H_1: M < 8; k = 8$; critical value: 6. Do not reject H_0.
3. $H_0: M = 100$ versus $H_1: M \neq 100; z_0 = -0.86; z_{0.025} = -1.96$. Do not reject H_0.
5. $H_0: M = 12$ versus $H_1: M > 12; k = 3$; critical value: 3. Reject H_0.
7. $H_0: M = 5.25$ versus $H_1: M > 5.25; k = 5$; critical value: 3. Do not reject H_0. There is not enough evidence to conclude that the median pH level is higher than 5.25.
9. $H_0: M = 57,283$ versus $H_1: M < 57,283; k = 4$; critical value: 2. Do not reject H_0. There is not enough evidence to support the belief that teachers in the district are paid less than the median (average) teacher in the state.
11. $H_0: M = 1000$ versus $H_1: M > 1000; k = 4$; critical value: 3. Do not reject H_0. There is not sufficient evidence to indicate that the median baseball salary is greater than $1,000,000.
13. P-value $= 0.1509$ **15.** P-value $= 0.1938$
17. $H_0: p = 0.5$ versus $H_1: p < 0.5; z_0 = -2.30$; critical value: -2.33. Do not reject H_0. There is not enough evidence at the $\alpha = 0.01$ level of significance to indicate that less than 50% of gamers are women.
19. Since participants had to agree to be part of the survey, the sample may not be truly random, which is required to use the sign test.
21. Regardless of whether the test is left-tailed or right-tailed, we are testing whether k is sufficiently small. In a left-tailed test, we are testing whether the number of positive signs is sufficiently small to reject the null hypothesis. In a right-tailed test, we are testing whether the number of negative signs is sufficiently small to reject the null hypothesis.
23. Answers will vary.

15.4 Assess Your Understanding (page 766)

1. M_D is the median of the differences between the observations in the matched-pairs data. The differences are computed by subtracting the first observation in the pair from the second observation in the pair.
3. Critical value: 17; reject H_0.
5. Critical value: 30; do not reject H_0.
7. Critical value: 40; do not reject H_0.
9. $z_0 = -1.48, -z_{0.05} = -1.645$; do not reject H_0.
11. $H_0: M_D = 0$ versus $H_1: M_D > 0; T = 12$; critical value: 8. Do not reject H_0. There is not sufficient evidence to indicate that the exercise program is effective in reducing waistline.
13. $H_0: M_D = 0$ versus $H_1: M_D \neq 0; T = 5$; critical value: 0. Do not reject H_0. There is not sufficient evidence to conclude that there is a difference in reaction time to blue versus to red.
15. $H_0: M_D = 0$ versus $H_1: M_D > 0; T = 5$; critical value: 5. Do not reject H_0. There is not sufficient evidence to determine that the clarity of the lake is improving.
17. $H_0: M_D = 0$ versus $H_1: M_D > 0; T = 35$; critical value: 10. Do not reject H_0. There is not sufficient evidence to suggest that Avis is less expensive than Hertz.
19. **(a)** $H_0: M_D = 0$ versus $H_1: M_D \neq 0$
(b) Yes; because the P-value < 0.001
21. $H_0: M_D = 0$ versus $H_1: M_D \neq 0; T = 8.5$; critical value: 10. Reject H_0. There is enough evidence at the $\alpha = 0.05$ level of significance to conclude that the median length of stay is different for employer referrals than those referred by the criminal justice system.
23. Her results are suspect because the data were not randomly selected.

15.5 Assess Your Understanding (page 776)

1. $T = 92$, critical values $= 50, 130$; do not reject H_0.
3. $T = 39$, critical value $= 96$; reject H_0.

5. $T = 130$, critical value $= 152$; do not reject H_0.

7. $T = 337$, $z_0 = 1.32$, $-z_{0.025} = -1.96$, $z_{0.025} = 1.96$; do not reject H_0.

9. H_0: $M_W = M_M$ versus H_1: $M_W \neq M_M$; $w_{0.025} = 55$, $w_{0.975} = 140$, $T = 73.5$; do not reject H_0. There is not sufficient evidence at the $\alpha = 0.05$ significance level to indicate that the median sociability score is different for women and men.

11. H_0: $M_{\text{Carpet}} = M_{\text{No Carpet}}$ versus H_1: $M_{\text{Carpet}} > M_{\text{No Carpet}}$; $T = 38.5$, critical value $= 48$; do not reject H_0. There is not sufficient evidence to conclude that the median amount of bacteria in carpeted rooms is more than that in uncarpeted rooms.

13. H_0: $M_{\text{Lincoln}} = M_{\text{Clarendon}}$ versus H_1: $M_{\text{Lincoln}} \neq M_{\text{Clarendon}}$; $z_0 = 3.43$, critical values: $-z_{0.025} = -1.96$, $z_{0.025} = 1.96$; reject H_0. There is sufficient evidence to conclude that the median calcium level of rainwater in Lincoln County is different from that in Clarendon County.

15. (a)

Professor A		Professor B	
Grade	Numerical Equivalent	Grade	Numerical Equivalent
A	4	A	4
A	4	A	4
B	3	A	4
B	3	B	3
B	3	B	3
C	2	B	3
C	2	B	3
C	2	B	3
C	2	B	3
C	2	C	2
C	2	C	2
D	1	C	2
D	1	C	2
D	1	D	1
F	0	F	0

(b) H_0: $M_A = M_B$ versus H_1: $M_A \neq M_B$; $T = 83$, critical values: $65, 160$; do not reject H_0. There is not sufficient evidence to conclude that there is a difference in the grades earned in the two classes.

17. The rationale behind the test statistic T is that, if the null hypothesis is true and the two medians are equal, the sum of the ranks of the sample from population X should be the same as the sum of the ranks of sample from population Y. If the true median of population X is less than that of population Y, more of the X sample ranks will be small, and so the sum of the ranks of the sample from population X should be less than the sum of the ranks of sample from population Y. Finally, if the true median of population X is greater than that of population Y, more of the X sample ranks will be large, making the sum of the ranks of the sample from population X larger than the sum of the ranks of sample from population Y.

19. The Wilcoxon matched-pairs signed-ranks test is used for dependent samples, while the Mann–Whitney test is used for independent samples.

15.6 Assess Your Understanding (page 781)

1. (a)

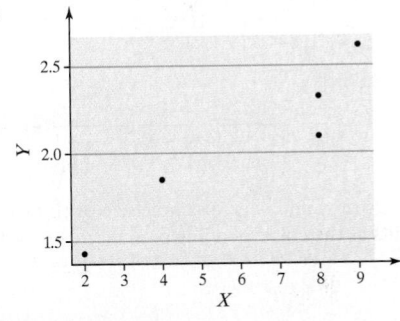

(b) $r_s = 0.975$

(c) H_0: X and Y are not associated versus H_1: X and Y are associated. Critical values: -1 and 1; do not reject H_0.

3. (a)

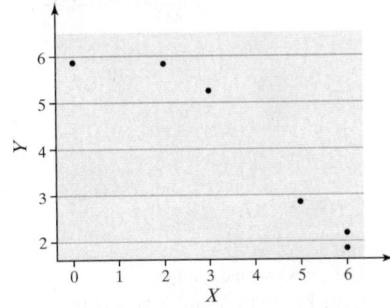

(b) $r_s = -0.957$

(c) H_0: X and Y are not associated versus H_1: X and Y are associated. Critical values: -0.886 and 0.886; reject H_0.

5. (a) H_0: education and income are not associated versus H_1: education and income are positively associated. $r_s = 0.767$. Critical value: 0.600; reject H_0. There is sufficient evidence at the $\alpha = 0.05$ level of significance to conclude that a positive association exists between having at least a bachelor's degree and personal income.

(b)

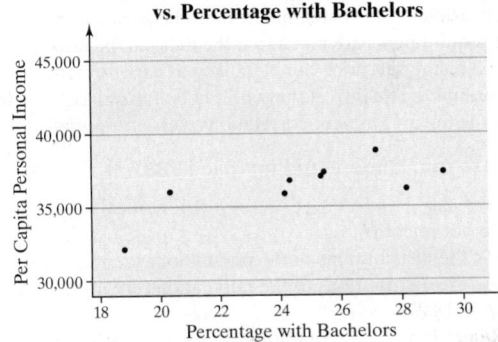

Scatterplot of per Capita Personal Income vs. Percentage with Bachelors

7. (a) H_0: per capita income and birthrate are not associated versus H_1: per capita income and birthrate are negatively associated. $r_s = 0.046$. Critical value: -0.600; do not reject H_0. There is not sufficient evidence at the $\alpha = 0.05$ level of significance to conclude that a negative association exists between per capita income and birth rate.

(b)

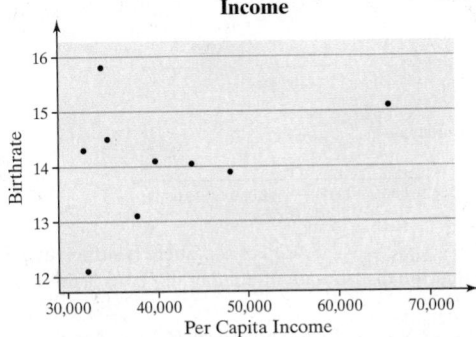

Scatterplot of Birthrate vs. per Capita Income

9. (a) H_0: family income and commute time are not associated versus H_1: family income and commute time are positively associated. $r_s = 0.700$. Critical value: 0.657; reject H_0. There is sufficient evidence at the $\alpha = 0.10$ level of significance to conclude that a positive association exists between family income and average commute time.

(b)

Scatterplot of Average Commute Time vs. Family Income

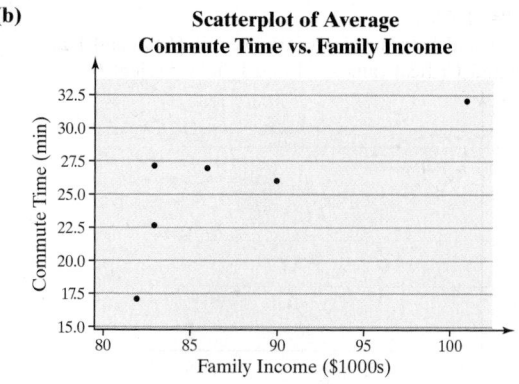

11. A positive relation exists between X and Y, $r_s = 1.0$.

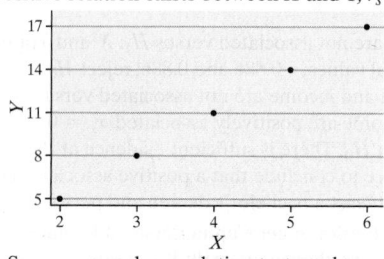

13. The Spearman rank-correlation test can be performed on qualitative data if the requirements are met and if the data can be ranked. For example, if a restaurant price can be ranked as expensive, moderate, or inexpensive, and the quality of the food can be ranked as good, fair, or poor, then the Spearman rank-correlation test is appropriate.

15.7 Assess Your Understanding (page 788)

1. (a) $H = 0.875$ **(b)** Critical value: 5.6923
(c) Do not reject H_0.
3. (a) H_0: the distributions of the populations are the same.
    H_1: the distributions of the populations are not the same.
(b) $H = 18.77$ **(c)** $\chi^2_{0.05} = 9.488$
(d) Reject H_0. The distributions of births by day of the week are not the same.
(e)

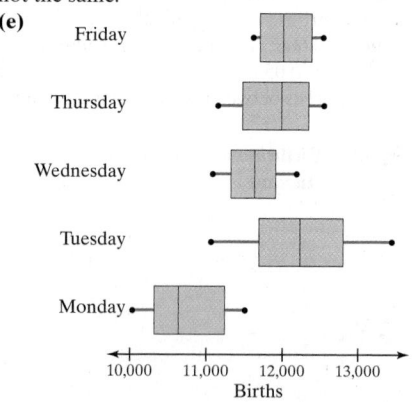

Births

5. $H = 10.83$, $\chi^2_{0.05} = 9.488$. Reject H_0. There is sufficient evidence to conclude that the distribution of number of plants is not the same for each plot type.
7. $H = 5.24$, $\chi^2_{0.01} = 9.210$. Do not reject H_0. There is not sufficient evidence to conclude that the distribution of reaction time is not the same for all stimuli.
9. $H = 3.66$, $H = 3.69$ adjusted for ties; $\chi^2_{0.01} = 9.210$. Do not reject H_0. There is not sufficient evidence to conclude that the distribution of chest compression for each car type is different.

Chapter 15 Review Exercises (page 791)

1. $n = 25$, $n_E = 11$, $n_W = 14$, $r = 11$; lower critical value: 8, upper critical value: 19. Do not reject H_0. There is evidence to conclude that the winning division occurs randomly.
2. H_0: $M = 15$ versus H_1: $M > 15$, $k = 7$, critical value: 5. Do not reject H_0. There is not sufficient evidence to indicate that the median number of hours students talk on the phone each week is more than 15.

3. H_0: $M_D = 0$ versus H_1: $M_D \neq 0$; $T = |T_-| = 23$, critical value = 8. Do not reject H_0. There is not sufficient evidence to conclude that an individual's height and arm span are different.
4. H_0: $M_F = M_M$ versus H_1: $M_F \neq M_M$; $T = 80$, critical values: 42 and 150. Do not reject H_0. There is not sufficient evidence to indicate that there is a difference in the median reaction time of males and females to these stimuli.
5. H_0: engine displacement and city miles per gallon are not associated. H_1: engine displacement and city miles per gallon are negatively associated. $r_s = -0.88$, critical value: -0.446; reject H_0. There is sufficient evidence at the $\alpha = 0.05$ level of significance to conclude that a negative association exists between engine displacement and city miles per gallon.
6. (a) H_0: the distributions of the populations are the same.
        H_1: the distributions of the populations are not the same.
(b) $H = 17.20$ **(c)** $\chi^2_{0.05} = 5.991$
(d) Reject H_0. There is evidence to conclude that the distributions of concentration of dissolved organic carbon are not the same.
(e)

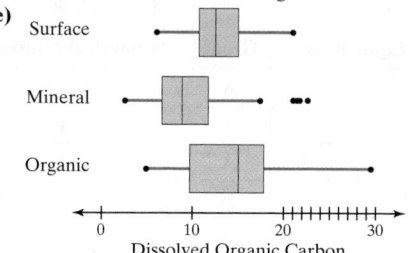

Dissolved Organic Carbon

7. See Table 1.

Chapter 15 Test (page 793)

1. $n = 25$, $n_E = 11$, $n_W = 14$, $r = 12$; lower critical value: 8, upper critical value: 19. Do not reject H_0. There is evidence to indicate that the winning division occurs randomly.
2. H_0: $M = 15$ versus H_1: $M < 15$; $k = 4$, critical value: 5. Reject H_0. There is sufficient evidence to conclude that the median number of hours students spend studying is less than 15.
3. H_0: $M_D = 0$ versus H_1: $M_D < 0$; $T = T_+ = 11.5$; critical value: 8. Do not reject H_0. There is not sufficient evidence to conclude that participation in the exercise program reduces pulse rate.
4. H_0: $M_H = M_C$ versus H_1: $M_H > M_C$; $T = 127$, critical value: 116. Reject H_0. There is sufficient evidence to indicate that the median pH level of rain in Houston is greater than the median pH of rain near Chicago.
5. H_0: temperature and the number of chirps per second are not associated. H_1: temperature and the number of chirps per second are associated. $r_s = 0.851$; critical values: -0.521 and 0.521. Reject H_0. There is sufficient evidence at the $\alpha = 0.05$ level of significance to conclude that an association exists between temperature and the number of cricket chirps per second.
6. H_0: the distributions of the populations are the same. H_1: the distributions of the populations are not the same. $H = 9.14$, $H = 9.16$ adjusted for ties; $\chi^2_{0.05} = 5.991$. Reject H_0. There is enough evidence to indicate that the distributions of femur force are different for the three types of cars.

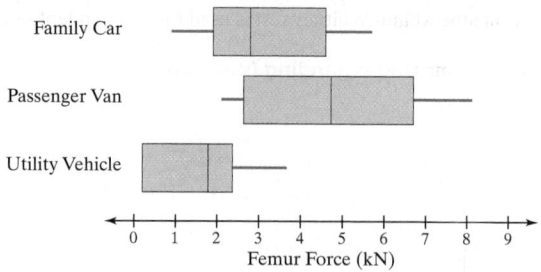

Femur Force (kN)

7. The nonparametric test would need a sample size of 1000 to achieve the same results as the corresponding parametric test with a sample size of 713.

Index

Table VIII

Chi-Square (χ^2) Distribution
Area to the Right of Critical Value

Degrees of Freedom	0.995	0.99	0.975	0.95	0.90	0.10	0.05	0.025	0.01	0.005
1	—	—	0.001	0.004	0.016	2.706	3.841	5.024	6.635	7.879
2	0.010	0.020	0.051	0.103	0.211	4.605	5.991	7.378	9.210	10.597
3	0.072	0.115	0.216	0.352	0.584	6.251	7.815	9.348	11.345	12.838
4	0.207	0.297	0.484	0.711	1.064	7.779	9.488	11.143	13.277	14.860
5	0.412	0.554	0.831	1.145	1.610	9.236	11.070	12.833	15.086	16.750
6	0.676	0.872	1.237	1.635	2.204	10.645	12.592	14.449	16.812	18.548
7	0.989	1.239	1.690	2.167	2.833	12.017	14.067	16.013	18.475	20.278
8	1.344	1.646	2.180	2.733	3.490	13.362	15.507	17.535	20.090	21.955
9	1.735	2.088	2.700	3.325	4.168	14.684	16.919	19.023	21.666	23.589
10	2.156	2.558	3.247	3.940	4.865	15.987	18.307	20.483	23.209	25.188
11	2.603	3.053	3.816	4.575	5.578	17.275	19.675	21.920	24.725	26.757
12	3.074	3.571	4.404	5.226	6.304	18.549	21.026	23.337	26.217	28.300
13	3.565	4.107	5.009	5.892	7.042	19.812	22.362	24.736	27.688	29.819
14	4.075	4.660	5.629	6.571	7.790	21.064	23.685	26.119	29.141	31.319
15	4.601	5.229	6.262	7.261	8.547	22.307	24.996	27.488	30.578	32.801
16	5.142	5.812	6.908	7.962	9.312	23.542	26.296	28.845	32.000	34.267
17	5.697	6.408	7.564	8.672	10.085	24.769	27.587	30.191	33.409	35.718
18	6.265	7.015	8.231	9.390	10.865	25.989	28.869	31.526	34.805	37.156
19	6.844	7.633	8.907	10.117	11.651	27.204	30.144	32.852	36.191	38.582
20	7.434	8.260	9.591	10.851	12.443	28.412	31.410	34.170	37.566	39.997
21	8.034	8.897	10.283	11.591	13.240	29.615	32.671	35.479	38.932	41.401
22	8.643	9.542	10.982	12.338	14.041	30.813	33.924	36.781	40.289	42.796
23	9.260	10.196	11.689	13.091	14.848	32.007	35.172	38.076	41.638	44.181
24	9.886	10.856	12.401	13.848	15.659	33.196	36.415	39.364	42.980	45.559
25	10.520	11.524	13.120	14.611	16.473	34.382	37.652	40.646	44.314	46.928
26	11.160	12.198	13.844	15.379	17.292	35.563	38.885	41.923	45.642	48.290
27	11.808	12.879	14.573	16.151	18.114	36.741	40.113	43.195	46.963	49.645
28	12.461	13.565	15.308	16.928	18.939	37.916	41.337	44.461	48.278	50.993
29	13.121	14.256	16.047	17.708	19.768	39.087	42.557	45.722	49.588	52.336
30	13.787	14.953	16.791	18.493	20.599	40.256	43.773	46.979	50.892	53.672
40	20.707	22.164	24.433	26.509	29.051	51.805	55.758	59.342	63.691	66.766
50	27.991	29.707	32.357	34.764	37.689	63.167	67.505	71.420	76.154	79.490
60	35.534	37.485	40.482	43.188	46.459	74.397	79.082	83.298	88.379	91.952
70	43.275	45.442	48.758	51.739	55.329	85.527	90.531	95.023	100.425	104.215
80	51.172	53.540	57.153	60.391	64.278	96.578	101.879	106.629	112.329	116.321
90	59.196	61.754	65.647	69.126	73.291	107.565	113.145	118.136	124.116	128.299
100	67.328	70.065	74.222	77.929	82.358	118.498	124.342	129.561	135.807	140.169

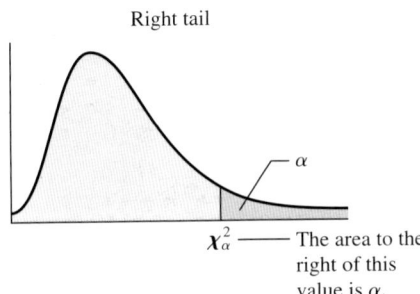

Right tail

χ^2_α — The area to the right of this value is α.

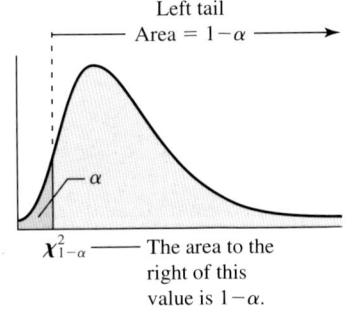

Left tail

Area = $1-\alpha$

$\chi^2_{1-\alpha}$ — The area to the right of this value is $1-\alpha$.

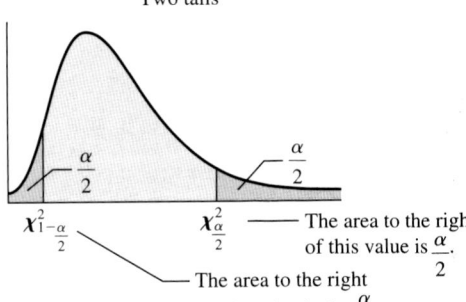

Two tails

$\chi^2_{1-\frac{\alpha}{2}}$

$\chi^2_{\frac{\alpha}{2}}$ — The area to the right of this value is $\frac{\alpha}{2}$.

The area to the right of this value is $1-\frac{\alpha}{2}$.

Chapter 2   Organizing and Summarizing Data

- Relative frequency $= \dfrac{\text{frequency}}{\text{sum of all frequencies}}$

- Class midpoint: The sum of consecutive lower class limits divided by 2.

Chapter 3   Numerically Summarizing Data

- Population Mean: $\mu = \dfrac{\sum x_i}{N}$

- Sample Mean: $\bar{x} = \dfrac{\sum x_i}{n}$

- Range $=$ Largest Data Value $-$ Smallest Data Value

- Population Standard Deviation:

$$\sigma = \sqrt{\dfrac{\sum(x_i - \mu)^2}{N}} = \sqrt{\dfrac{\sum x_i^2 - \dfrac{(\sum x_i)^2}{N}}{N}}$$

- Sample Standard Deviation

$$s = \sqrt{\dfrac{\sum(x_i - \bar{x})^2}{n-1}} = \sqrt{\dfrac{\sum x_i^2 - \dfrac{(\sum x_i)^2}{n}}{n-1}}$$

- Population Variance: σ^2

- Sample Variance: s^2

- **Empirical Rule:** If the shape of the distribution is bell-shaped, then
  - Approximately 68% of the data lie within 1 standard deviation of the mean
  - Approximately 95% of the data lie within 2 standard deviations of the mean
  - Approximately 99.7% of the data lie within 3 standard deviations of the mean

- Population Mean from Grouped Data: $\mu = \dfrac{\sum x_i f_i}{\sum f_i}$

- Sample Mean from Grouped Data: $\bar{x} = \dfrac{\sum x_i f_i}{\sum f_i}$

- Weighted Mean: $\bar{x}_w = \dfrac{\sum w_i x_i}{\sum w_i}$

- Population Standard Deviation from Grouped Data:

$$\sigma = \sqrt{\dfrac{\sum(x_i - \mu)^2 f_i}{\sum f_i}} = \sqrt{\dfrac{\sum x_i^2 f_i - \dfrac{(\sum x_i f_i)^2}{\sum f_i}}{\sum f_i}}$$

- Sample Standard Deviation from Grouped Data:

$$s = \sqrt{\dfrac{\sum(x_i - \mu)^2 f_i}{(\sum f_i) - 1}} = \sqrt{\dfrac{\sum x_i^2 f_i - \dfrac{(\sum x_i f_i)^2}{\sum f_i}}{\sum f_i - 1}}$$

- Population z-score: $z = \dfrac{x - \mu}{\sigma}$

- Sample z-score: $z = \dfrac{x - \bar{x}}{s}$

- Interquartile Range: $\text{IQR} = Q_3 - Q_1$

- Lower and Upper Fences: Lower fence $= Q_1 - 1.5(\text{IQR})$
  Upper fence $= Q_3 + 1.5(\text{IQR})$

- Five-Number Summary

$$\text{Minimum, } Q_1, M, Q_3, \text{Maximum}$$

Chapter 4   Describing the Relation between Two Variables

- Correlation Coefficient: $r = \dfrac{\sum\left(\dfrac{x_i - \bar{x}}{s_x}\right)\left(\dfrac{y_i - \bar{y}}{s_y}\right)}{n-1}$

- The equation of the least-squares regression line is

$\hat{y} = b_1 x + b_0$, where $\hat{y}$ is the predicted value, $b_1 = r \cdot \dfrac{s_y}{s_x}$

is the slope, and $b_0 = \bar{y} - b_1\bar{x}$ is the intercept.

- Residual $=$ observed $y -$ predicted $y = y - \hat{y}$

- $R^2 = r^2$ for the least-squares regression model $\hat{y} = b_1 x + b_0$

- The coefficient of determination, R^2, measures the proportion of total variation in the response variable that is explained by the least-squares regression line.

Chapter 5   Probability

- Empirical Probability

$$P(E) \approx \dfrac{\text{frequency of } E}{\text{number of trials of experiment}}$$

- Classical Probability

$$P(E) = \dfrac{\text{number of ways that } E \text{ can occur}}{\text{number of possible outcomes}} = \dfrac{N(E)}{N(S)}$$

- Addition Rule for Disjoint Events

$$P(E \text{ or } F) = P(E) + P(F)$$

- Addition Rule for n Disjoint Events

$$P(E \text{ or } F \text{ or } G \text{ or } \cdots) = P(E) + P(F) + P(G) + \cdots$$

- General Addition Rule

$$P(E \text{ or } F) = P(E) + P(F) - P(E \text{ and } F)$$

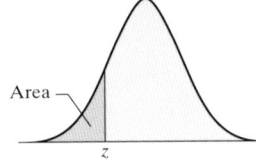

Area

z

Table V

Standard Normal Distribution

z	.00	.01	.02	.03	.04	.05	.06	.07	.08	.09
−3.4	0.0003	0.0003	0.0003	0.0003	0.0003	0.0003	0.0003	0.0003	0.0003	0.0002
−3.3	0.0005	0.0005	0.0005	0.0004	0.0004	0.0004	0.0004	0.0004	0.0004	0.0003
−3.2	0.0007	0.0007	0.0006	0.0006	0.0006	0.0006	0.0006	0.0005	0.0005	0.0005
−3.1	0.0010	0.0009	0.0009	0.0009	0.0008	0.0008	0.0008	0.0008	0.0007	0.0007
−3.0	0.0013	0.0013	0.0013	0.0012	0.0012	0.0011	0.0011	0.0011	0.0010	0.0010
−2.9	0.0019	0.0018	0.0018	0.0017	0.0016	0.0016	0.0015	0.0015	0.0014	0.0014
−2.8	0.0026	0.0025	0.0024	0.0023	0.0023	0.0022	0.0021	0.0021	0.0020	0.0019
−2.7	0.0035	0.0034	0.0033	0.0032	0.0031	0.0030	0.0029	0.0028	0.0027	0.0026
−2.6	0.0047	0.0045	0.0044	0.0043	0.0041	0.0040	0.0039	0.0038	0.0037	0.0036
−2.5	0.0062	0.0060	0.0059	0.0057	0.0055	0.0054	0.0052	0.0051	0.0049	0.0048
−2.4	0.0082	0.0080	0.0078	0.0075	0.0073	0.0071	0.0069	0.0068	0.0066	0.0064
−2.3	0.0107	0.0104	0.0102	0.0099	0.0096	0.0094	0.0091	0.0089	0.0087	0.0084
−2.2	0.0139	0.0136	0.0132	0.0129	0.0125	0.0122	0.0119	0.0116	0.0113	0.0110
−2.1	0.0179	0.0174	0.0170	0.0166	0.0162	0.0158	0.0154	0.0150	0.0146	0.0143
−2.0	0.0228	0.0222	0.0217	0.0212	0.0207	0.0202	0.0197	0.0192	0.0188	0.0183
−1.9	0.0287	0.0281	0.0274	0.0268	0.0262	0.0256	0.0250	0.0244	0.0239	0.0233
−1.8	0.0359	0.0351	0.0344	0.0336	0.0329	0.0322	0.0314	0.0307	0.0301	0.0294
−1.7	0.0446	0.0436	0.0427	0.0418	0.0409	0.0401	0.0392	0.0384	0.0375	0.0367
−1.6	0.0548	0.0537	0.0526	0.0516	0.0505	0.0495	0.0485	0.0475	0.0465	0.0455
−1.5	0.0668	0.0655	0.0643	0.0630	0.0618	0.0606	0.0594	0.0582	0.0571	0.0559
−1.4	0.0808	0.0793	0.0778	0.0764	0.0749	0.0735	0.0721	0.0708	0.0694	0.0681
−1.3	0.0968	0.0951	0.0934	0.0918	0.0901	0.0885	0.0869	0.0853	0.0838	0.0823
−1.2	0.1151	0.1131	0.1112	0.1093	0.1075	0.1056	0.1038	0.1020	0.1003	0.0985
−1.1	0.1357	0.1335	0.1314	0.1292	0.1271	0.1251	0.1230	0.1210	0.1190	0.1170
−1.0	0.1587	0.1562	0.1539	0.1515	0.1492	0.1469	0.1446	0.1423	0.1401	0.1379
−0.9	0.1841	0.1814	0.1788	0.1762	0.1736	0.1711	0.1685	0.1660	0.1635	0.1611
−0.8	0.2119	0.2090	0.2061	0.2033	0.2005	0.1977	0.1949	0.1922	0.1894	0.1867
−0.7	0.2420	0.2389	0.2358	0.2327	0.2296	0.2266	0.2236	0.2206	0.2177	0.2148
−0.6	0.2743	0.2709	0.2676	0.2643	0.2611	0.2578	0.2546	0.2514	0.2483	0.2451
−0.5	0.3085	0.3050	0.3015	0.2981	0.2946	0.2912	0.2877	0.2843	0.2810	0.2776
−0.4	0.3446	0.3409	0.3372	0.3336	0.3300	0.3264	0.3228	0.3192	0.3156	0.3121
−0.3	0.3821	0.3783	0.3745	0.3707	0.3669	0.3632	0.3594	0.3557	0.3520	0.3483
−0.2	0.4207	0.4168	0.4129	0.4090	0.4052	0.4013	0.3974	0.3936	0.3897	0.3859
−0.1	0.4602	0.4562	0.4522	0.4483	0.4443	0.4404	0.4364	0.4325	0.4286	0.4247
−0.0	0.5000	0.4960	0.4920	0.4880	0.4840	0.4801	0.4761	0.4721	0.4681	0.4641
0.0	0.5000	0.5040	0.5080	0.5120	0.5160	0.5199	0.5239	0.5279	0.5319	0.5359
0.1	0.5398	0.5438	0.5478	0.5517	0.5557	0.5596	0.5636	0.5675	0.5714	0.5753
0.2	0.5793	0.5832	0.5871	0.5910	0.5948	0.5987	0.6026	0.6064	0.6103	0.6141
0.3	0.6179	0.6217	0.6255	0.6293	0.6331	0.6368	0.6406	0.6443	0.6480	0.6517
0.4	0.6554	0.6591	0.6628	0.6664	0.6700	0.6736	0.6772	0.6808	0.6844	0.6879
0.5	0.6915	0.6950	0.6985	0.7019	0.7054	0.7088	0.7123	0.7157	0.7190	0.7224
0.6	0.7257	0.7291	0.7324	0.7357	0.7389	0.7422	0.7454	0.7486	0.7517	0.7549
0.7	0.7580	0.7611	0.7642	0.7673	0.7704	0.7734	0.7764	0.7794	0.7823	0.7852
0.8	0.7881	0.7910	0.7939	0.7967	0.7995	0.8023	0.8051	0.8078	0.8106	0.8133
0.9	0.8159	0.8186	0.8212	0.8238	0.8264	0.8289	0.8315	0.8340	0.8365	0.8389
1.0	0.8413	0.8438	0.8461	0.8485	0.8508	0.8531	0.8554	0.8577	0.8599	0.8621
1.1	0.8643	0.8665	0.8686	0.8708	0.8729	0.8749	0.8770	0.8790	0.8810	0.8830
1.2	0.8849	0.8869	0.8888	0.8907	0.8925	0.8944	0.8962	0.8980	0.8997	0.9015
1.3	0.9032	0.9049	0.9066	0.9082	0.9099	0.9115	0.9131	0.9147	0.9162	0.9177
1.4	0.9192	0.9207	0.9222	0.9236	0.9251	0.9265	0.9279	0.9292	0.9306	0.9319
1.5	0.9332	0.9345	0.9357	0.9370	0.9382	0.9394	0.9406	0.9418	0.9429	0.9441
1.6	0.9452	0.9463	0.9474	0.9484	0.9495	0.9505	0.9515	0.9525	0.9535	0.9545
1.7	0.9554	0.9564	0.9573	0.9582	0.9591	0.9599	0.9608	0.9616	0.9625	0.9633
1.8	0.9641	0.9649	0.9656	0.9664	0.9671	0.9678	0.9686	0.9693	0.9699	0.9706
1.9	0.9713	0.9719	0.9726	0.9732	0.9738	0.9744	0.9750	0.9756	0.9761	0.9767
2.0	0.9772	0.9778	0.9783	0.9788	0.9793	0.9798	0.9803	0.9808	0.9812	0.9817
2.1	0.9821	0.9826	0.9830	0.9834	0.9838	0.9842	0.9846	0.9850	0.9854	0.9857
2.2	0.9861	0.9864	0.9868	0.9871	0.9875	0.9878	0.9881	0.9884	0.9887	0.9890
2.3	0.9893	0.9896	0.9898	0.9901	0.9904	0.9906	0.9909	0.9911	0.9913	0.9916
2.4	0.9918	0.9920	0.9922	0.9925	0.9927	0.9929	0.9931	0.9932	0.9934	0.9936
2.5	0.9938	0.9940	0.9941	0.9943	0.9945	0.9946	0.9948	0.9949	0.9951	0.9952
2.6	0.9953	0.9955	0.9956	0.9957	0.9959	0.9960	0.9961	0.9962	0.9963	0.9964
2.7	0.9965	0.9966	0.9967	0.9968	0.9969	0.9970	0.9971	0.9972	0.9973	0.9974
2.8	0.9974	0.9975	0.9976	0.9977	0.9977	0.9978	0.9979	0.9979	0.9980	0.9981
2.9	0.9981	0.9982	0.9982	0.9983	0.9984	0.9984	0.9985	0.9985	0.9986	0.9986
3.0	0.9987	0.9987	0.9987	0.9988	0.9988	0.9989	0.9989	0.9989	0.9990	0.9990
3.1	0.9990	0.9991	0.9991	0.9991	0.9992	0.9992	0.9992	0.9992	0.9993	0.9993
3.2	0.9993	0.9993	0.9994	0.9994	0.9994	0.9994	0.9994	0.9995	0.9995	0.9995
3.3	0.9995	0.9995	0.9995	0.9996	0.9996	0.9996	0.9996	0.9996	0.9996	0.9997
3.4	0.9997	0.9997	0.9997	0.9997	0.9997	0.9997	0.9997	0.9997	0.9997	0.9998

Confidence Interval Critical Values, $z_{\alpha/2}$

Level of Confidence	Critical Value, $z_{\alpha/2}$
0.90 or 90%	1.645
0.95 or 95%	1.96
0.98 or 98%	2.33
0.99 or 99%	2.575

Hypothesis Testing Critical Values

Level of Significance, α	Left-Tailed	Right-Tailed	Two-Tailed
0.10	−1.28	1.28	±1.645
0.05	−1.645	1.645	±1.96
0.01	−2.33	2.33	±2.575

Chapter 10   Hypothesis Tests Regarding a Parameter

Test Statistics

- $z_0 = \dfrac{\hat{p} - p_0}{\sqrt{\dfrac{p_0(1 - p_0)}{n}}}$

- $t_0 = \dfrac{\bar{x} - \mu_0}{s / \sqrt{n}}$

- $\chi_0^2 = \dfrac{(n - 1)s^2}{\sigma_0^2}$

Chapter 11   Inferences on Two Samples

- Test Statistic Comparing Two Population Proportions (Independent Samples)

$$z_0 = \dfrac{\hat{p}_1 - \hat{p}_2 - (p_1 - p_2)}{\sqrt{\hat{p}(1 - \hat{p})}\sqrt{\dfrac{1}{n_1} + \dfrac{1}{n_2}}} \quad \text{where} \ \ \hat{p} = \dfrac{x_1 + x_2}{n_1 + n_2}$$

- Confidence Interval for the Difference of Two Proportions (Independent Samples)

$$(\hat{p}_1 - \hat{p}_2) \pm z_{\alpha/2}\sqrt{\dfrac{\hat{p}_1(1 - \hat{p}_1)}{n_1} + \dfrac{\hat{p}_2(1 - \hat{p}_2)}{n_2}}$$

- Test Statistic for Matched-Pairs Data

$$t_0 = \dfrac{\bar{d} - \mu_d}{s_d / \sqrt{n}}$$

where $\bar{d}$ is the mean and s_d is the standard deviation of the differenced data.

- Confidence Interval for Matched-Pairs Data

$$\bar{d} \pm t_{\alpha/2} \cdot \dfrac{s_d}{\sqrt{n}}$$

Note: $t_{\alpha/2}$ is found using $n - 1$ degrees of freedom.

- Test Statistic Comparing Two Means (Independent Sampling)

$$t_0 = \dfrac{(\bar{x}_1 - \bar{x}_2) - (\mu_1 - \mu_2)}{\sqrt{\dfrac{s_1^2}{n_1} + \dfrac{s_2^2}{n_2}}}$$

- Confidence Interval for the Difference of Two Means (Independent Samples)

$$(\bar{x}_1 - \bar{x}_2) \pm t_{\alpha/2}\sqrt{\dfrac{s_1^2}{n_1} + \dfrac{s_2^2}{n_2}}$$

Note: $t_{\alpha/2}$ is found using the smaller of $n_1 - 1$ or $n_2 - 1$ degrees of freedom.

- Test Statistic for Comparing Two Population Standard Deviations

$$F_0 = \dfrac{s_1^2}{s_2^2}$$

- Finding a Critical F for the Left Tail

$$F_{1-\alpha, n_1-1, n_2-1} = \dfrac{1}{F_{\alpha, n_2-1, n_1-1}}$$

Chapter 12   Inference on Categorical Data

- Expected Counts (when testing for goodness of fit)

$$E_i = \mu_i = np_i \quad \text{for} \quad i = 1, 2, \ldots, k$$

- Expected Frequencies (when testing for independence or homogeneity of proportions)

$$\text{Expected frequency} = \dfrac{(\text{row total})(\text{column total})}{\text{table total}}$$

- Chi-Square Test Statistic

$$\chi_0^2 = \sum \dfrac{(\text{observed} - \text{expected})^2}{\text{expected}} = \sum \dfrac{(O_i - E_i)^2}{E_i}$$

$$i = 1, 2, \ldots, k$$

All $E_i \geq 1$ and no more than 20% less than 5.

- Test Statistic for Comparing Two Proportions (Dependent Samples)

$$\chi_0^2 = \dfrac{(f_{12} - f_{21})^2}{f_{12} + f_{21}}$$

Chapter 13   Comparing Three or More Means

- Test Statistic for One-Way ANOVA

$$F = \dfrac{\text{Mean square due to treatment}}{\text{Mean square due to error}} = \dfrac{\text{MST}}{\text{MSE}}$$

where

$$\text{MST} = \dfrac{n_1(\bar{x}_1 - \bar{x})^2 + n_2(\bar{x}_2 - \bar{x})^2 + \cdots + n_k(\bar{x}_k - \bar{x})^2}{k - 1}$$

$$\text{MSE} = \dfrac{(n_1 - 1)s_1^2 + (n_2 - 1)s_2^2 + \cdots + (n_k - 1)s_k^2}{n - k}$$

- Test Statistic for Tukey's Test after One-Way ANOVA

$$q = \dfrac{(\bar{x}_2 - \bar{x}_1) - (\mu_2 - \mu_1)}{\sqrt{\dfrac{s^2}{2} \cdot \left(\dfrac{1}{n_1} + \dfrac{1}{n_2}\right)}} = \dfrac{\bar{x}_2 - \bar{x}_1}{\sqrt{\dfrac{s^2}{2} \cdot \left(\dfrac{1}{n_1} + \dfrac{1}{n_2}\right)}}$$

Chapter 14   Inference on the Least-Squares Regression Model and Multiple Regression

- Standard Error of the Estimate

$$s_e = \sqrt{\frac{\sum (y_i - \hat{y}_i)^2}{n - 2}} = \sqrt{\frac{\sum \text{residuals}^2}{n - 2}}$$

- Standard error of b_1

$$s_{b_1} = \frac{s_e}{\sqrt{\sum (x_i - \bar{x})^2}}$$

- Test Statistic for the Slope of the Least-Squares Regression Line

$$t_0 = \frac{b_1 - \beta_1}{s_e / \sqrt{\sum (x_i - \bar{x})^2}} = \frac{b_1 - \beta_1}{s_{b_1}}$$

- Confidence Interval for the Slope of the Regression Line

$$b_1 \pm t_{\alpha/2} \cdot \frac{s_e}{\sqrt{\sum (x_i - \bar{x})^2}}$$

where $t_{\alpha/2}$ is computed with $n - 2$ degrees of freedom.

- Confidence Interval about the Mean Response of y, $\hat{y}$

$$\hat{y} \pm t_{\alpha/2} \cdot s_e \sqrt{\frac{1}{n} + \frac{(x^* - \bar{x})^2}{\sum (x_i - \bar{x})^2}}$$

where x^* is the given value of the explanatory variable and $t_{\alpha/2}$ is the critical value with $n - 2$ degrees of freedom.

- Prediction Interval about an Individual Response, $\hat{y}$

$$\hat{y} \pm t_{\alpha/2} \cdot s_e \sqrt{1 + \frac{1}{n} + \frac{(x^* - \bar{x})^2}{\sum (x_i - \bar{x})^2}}$$

where x^* is the given value of the explanatory variable and $t_{\alpha/2}$ is the critical value with $n - 2$ degrees of freedom.

Chapter 15   Nonparametric Statistics

- Test Statistic for a Runs Test for Randomness

Small-Sample Case If $n_1 \leq 20$ and $n_2 \leq 20$, the test statistic in the runs test for randomness is r, the number of runs.

Large-Sample Case If $n_1 > 20$ or $n_2 > 20$, the test statistic is

$$z_0 = \frac{r - \mu_r}{\sigma_r} \text{ where}$$

$$\mu_r = \frac{2n_1 n_2}{n} + 1 \quad \text{and} \quad \sigma_r = \sqrt{\frac{2n_1 n_2 (2n_1 n_2 - n)}{n^2 (n - 1)}}$$

- Test Statistic for a One-Sample Sign Test

Small-Sample Case ($n \leq 25$)

Two-Tailed	Left-Tailed	Right-Tailed
$H_0: M = M_0$	$H_0: M = M_0$	$H_0: M = M_0$
$H_1: M \neq M_0$	$H_1: M < M_0$	$H_1: M > M_0$
The test statistic, k, is the smaller of the number of minus signs or plus signs.	The test statistic, k, is the number of plus signs.	The test statistic, k, is the number of minus signs.

Large-Sample Case ($n > 25$) The test statistic, z_0, is

$$z_0 = \frac{(k + 0.5) - \frac{n}{2}}{\frac{\sqrt{n}}{2}}$$

where n is the number of minus and plus signs and k is obtained as described in the small sample case.

- Test Statistic for the Wilcoxon Matched-Pairs Signed-Ranks Test

Small-Sample Case ($n \leq 30$)

Two-Tailed	Left-Tailed	Right-Tailed
$H_0: M_D = 0$	$H_0: M_D = 0$	$H_0: M_D = 0$
$H_1: M_D \neq 0$	$H_1: M_D < 0$	$H_0: M_D > 0$
Test Statistic: T is the smaller of T_+ or T_-	**Test Statistic:** $T = T_+$	**Test Statistic:** $T = |T_-|$

Large-Sample Case ($n > 30$)

$$z_0 = \frac{T - \frac{n(n + 1)}{4}}{\sqrt{\frac{n(n + 1)(2n + 1)}{24}}}$$

where T is the test statistic from the small-sample case.

- Test Statistic for the Mann–Whitney Test

Small-Sample Case ($n_1 \leq 20$ and $n_2 \leq 20$)
If S is the sum of the ranks corresponding to the sample from population X, then the test statistic, T, is given by

$$T = S - \frac{n_1(n_1 + 1)}{2}$$

Note: The value of S is always obtained by summing the ranks of the sample data that correspond to M_X in the hypothesis.
Large-Sample Case ($n_1 > 20$) or ($n_2 > 20$)

$$z_0 = \frac{T - \frac{n_1 n_2}{2}}{\sqrt{\frac{n_1 n_2 (n_1 + n_2 + 1)}{12}}}$$

- Test Statistic for Spearman's Rank Correlation Test

$$r_s = 1 - \frac{6 \sum d_i^2}{n(n^2 - 1)}$$

where $d_i =$ the difference in the ranks of the two observations in the ith ordered pair.

- Test Statistic for the Kruskal–Wallis Test

$$H = \frac{12}{N(N + 1)} \sum \frac{1}{n_i} \left[R_i - \frac{n_i(N + 1)}{2} \right]^2$$

$$= \frac{12}{N(N + 1)} \left[\frac{R_1^2}{n_1} + \frac{R_2^2}{n_2} + \cdots + \frac{R_k^2}{n_k} \right] - 3(N + 1)$$

where R_i is the sum of the ranks in the ith sample.

Table I

Random Numbers
Column Number

Row Number	01–05	06–10	11–15	16–20	21–25	26–30	31–35	36–40	41–45	46–50
01	89392	23212	74483	36590	25956	36544	68518	40805	09980	00467
02	61458	17639	96252	95649	73727	33912	72896	66218	52341	97141
03	11452	74197	81962	48443	90360	26480	73231	37740	26628	44690
04	27575	04429	31308	02241	01698	19191	18948	78871	36030	23980
05	36829	59109	88976	46845	28329	47460	88944	08264	00843	84592
06	81902	93458	42161	26099	09419	89073	82849	09160	61845	40906
07	59761	55212	33360	68751	86737	79743	85262	31887	37879	17525
08	46827	25906	64708	20307	78423	15910	86548	08763	47050	18513
09	24040	66449	32353	83668	13874	86741	81312	54185	78824	00718
10	98144	96372	50277	15571	82261	66628	31457	00377	63423	55141
11	14228	17930	30118	00438	49666	65189	62869	31304	17117	71489
12	55366	51057	90065	14791	62426	02957	85518	28822	30588	32798
13	96101	30646	35526	90389	73634	79304	96635	06626	94683	16696
14	38152	55474	30153	26525	83647	31988	82182	98377	33802	80471
15	85007	18416	24661	95581	45868	15662	28906	36392	07617	50248
16	85544	15890	80011	18160	33468	84106	40603	01315	74664	20553
17	10446	20699	98370	17684	16932	80449	92654	02084	19985	59321
18	67237	45509	17638	65115	29757	80705	82686	48565	72612	61760
19	23026	89817	05403	82209	30573	47501	00135	33955	50250	72592
20	67411	58542	18678	46491	13219	84084	27783	34508	55158	78742

Table II

Critical Values (CV) for Correlation Coefficient

n	CV	n	CV	n	CV	n	CV
3	0.997	10	0.632	17	0.482	24	0.404
4	0.950	11	0.602	18	0.468	25	0.396
5	0.878	12	0.576	19	0.456	26	0.388
6	0.811	13	0.553	20	0.444	27	0.381
7	0.754	14	0.532	21	0.433	28	0.374
8	0.707	15	0.514	22	0.423	29	0.367
9	0.666	16	0.497	23	0.413	30	0.361

Table VI

Critical Values for Normal Probability Plots

Sample Size, n	Critical Value	Sample Size, n	Critical Value	Sample Size, n	Critical Value
5	0.880	13	0.932	21	0.952
6	0.888	14	0.935	22	0.954
7	0.898	15	0.939	23	0.956
8	0.906	16	0.941	24	0.957
9	0.912	17	0.944	25	0.959
10	0.918	18	0.946	30	0.960
11	0.923	19	0.949		
12	0.928	20	0.951		

Source: S. W. Looney and T. R. Gulledge, Jr. "Use of the Correlation Coefficient with Normal Probability Plots," *American Statistician* 39(Feb. 1985): 75–79.

- Complement Rule

$$P(E^c) = 1 - P(E)$$

- Multiplication Rule for Independent Events

$$P(E \text{ and } F) = P(E) \cdot P(F)$$

- Multiplication Rule for n Independent Events

$$P(E \text{ and } F \text{ and } G \cdots) = P(E) \cdot P(F) \cdot P(G) \cdots$$

- Conditional Probability Rule

$$P(F|E) = \frac{P(E \text{ and } F)}{P(E)} = \frac{N(E \text{ and } F)}{N(E)}$$

- General Multiplication Rule

$$P(E \text{ and } F) = P(E) \cdot P(F|E)$$

- Factorial

$$n! = n \cdot (n-1) \cdot (n-2) \cdots \cdot 3 \cdot 2 \cdot 1$$

- Permutation of n objects taken r at a time:

$$_nP_r = \frac{n!}{(n-r)!}$$

- Combination of n objects taken r at a time:

$$_nC_r = \frac{n!}{r!(n-r)!}$$

- Permutations with Repetition:

$$\frac{n!}{n_1! \cdot n_2! \cdot \cdots \cdot n_k!}$$

Chapter 6   Discrete Probability Distributions

- Mean (Expected Value) of a Discrete Random Variable

$$\mu_X = \Sigma x \cdot P(x)$$

- Standard Deviation of a Discrete Random Variable

$$\sigma_X = \sqrt{\Sigma(x-\mu)^2 \cdot P(x)} = \sqrt{\Sigma[x^2 P(x)] - \mu_X^2}$$

- Binomial Probability Distribution Function

$$P(x) = {}_nC_x p^x (1-p)^{n-x}$$

- Mean and Standard Deviation of a Binomial Random Variable

$$\mu_X = np \qquad \sigma_X = \sqrt{np(1-p)}$$

- Poisson Probability Distribution Function

$$P(x) = \frac{(\lambda t)^x}{x!} e^{-\lambda t} \quad x = 0, 1, 2, \ldots$$

- Mean and Standard Deviation of a Poisson Random Variable

$$\mu_X = \lambda t \quad \sigma_X = \sqrt{\lambda t}$$

Chapter 7   The Normal Distribution

- Standardizing a Normal Random Variable

$$z = \frac{x-\mu}{\sigma}$$

- Finding the Score: $x = \mu + z\sigma$

Chapter 8   Sampling Distributions

- Mean and Standard Deviation of the Sampling Distribution of $\bar{x}$

$$\mu_{\bar{x}} = \mu \quad \text{and} \quad \sigma_{\bar{x}} = \frac{\sigma}{\sqrt{n}}$$

- Sample Proportion: $\hat{p} = \dfrac{x}{n}$

- Mean and Standard Deviation of the Sampling Distribution of $\hat{p}$

$$\mu_{\hat{p}} = p \text{ and } \sigma_{\hat{p}} = \sqrt{\frac{p(1-p)}{n}}$$

Chapter 9   Estimating the Value of a Parameter

Confidence Intervals
- A $(1-\alpha) \cdot 100\%$ confidence interval about p is

$$\hat{p} \pm z_{\alpha/2} \cdot \sqrt{\frac{\hat{p}(1-\hat{p})}{n}}$$

- A $(1-\alpha) \cdot 100\%$ confidence interval about μ is

$$\bar{x} \pm t_{\alpha/2} \cdot \frac{s}{\sqrt{n}}$$

Note: $t_{\alpha/2}$ is computed using $n-1$ degrees of freedom.

- A $(1-\alpha) \cdot 100\%$ confidence interval about σ is

$$\sqrt{\frac{(n-1)s^2}{\chi_{\alpha/2}^2}} < \sigma < \sqrt{\frac{(n-1)s^2}{\chi_{1-\alpha/2}^2}}$$

Sample Size
- To estimate the population proportion with a margin of error E at a $(1-\alpha) \cdot 100\%$ level of confidence:

$n = \hat{p}(1-\hat{p})\left(\dfrac{z_{\alpha/2}}{E}\right)^2$ rounded up to the next integer,

where $\hat{p}$ is a prior estimate of the population proportion,

or $n = 0.25\left(\dfrac{z_{\alpha/2}}{E}\right)^2$ rounded up to the next integer when no prior estimate of p is available.

- To estimate the population mean with a margin of error E at a $(1-\alpha) \cdot 100\%$ level of confidence: $n = \left(\dfrac{z_{\alpha/2} \cdot s}{E}\right)^2$ rounded up to the next integer.

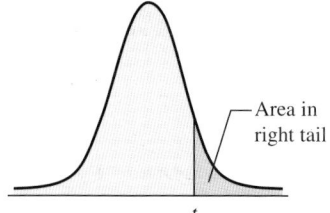
Area in right tail

Table VII

t-Distribution
Area in Right Tail

df	0.25	0.20	0.15	0.10	0.05	0.025	0.02	0.01	0.005	0.0025	0.001	0.0005
1	1.000	1.376	1.963	3.078	6.314	12.706	15.894	31.821	63.657	127.321	318.309	636.619
2	0.816	1.061	1.386	1.886	2.920	4.303	4.849	6.965	9.925	14.089	22.327	31.599
3	0.765	0.978	1.250	1.638	2.353	3.182	3.482	4.541	5.841	7.453	10.215	12.924
4	0.741	0.941	1.190	1.533	2.132	2.776	2.999	3.747	4.604	5.598	7.173	8.610
5	0.727	0.920	1.156	1.476	2.015	2.571	2.757	3.365	4.032	4.773	5.893	6.869
6	0.718	0.906	1.134	1.440	1.943	2.447	2.612	3.143	3.707	4.317	5.208	5.959
7	0.711	0.896	1.119	1.415	1.895	2.365	2.517	2.998	3.499	4.029	4.785	5.408
8	0.706	0.889	1.108	1.397	1.860	2.306	2.449	2.896	3.355	3.833	4.501	5.041
9	0.703	0.883	1.100	1.383	1.833	2.262	2.398	2.821	3.250	3.690	4.297	4.781
10	0.700	0.879	1.093	1.372	1.812	2.228	2.359	2.764	3.169	3.581	4.144	4.587
11	0.697	0.876	1.088	1.363	1.796	2.201	2.328	2.718	3.106	3.497	4.025	4.437
12	0.695	0.873	1.083	1.356	1.782	2.179	2.303	2.681	3.055	3.428	3.930	4.318
13	0.694	0.870	1.079	1.350	1.771	2.160	2.282	2.650	3.012	3.372	3.852	4.221
14	0.692	0.868	1.076	1.345	1.761	2.145	2.264	2.624	2.977	3.326	3.787	4.140
15	0.691	0.866	1.074	1.341	1.753	2.131	2.249	2.602	2.947	3.286	3.733	4.073
16	0.690	0.865	1.071	1.337	1.746	2.120	2.235	2.583	2.921	3.252	3.686	4.015
17	0.689	0.863	1.069	1.333	1.740	2.110	2.224	2.567	2.898	3.222	3.646	3.965
18	0.688	0.862	1.067	1.330	1.734	2.101	2.214	2.552	2.878	3.197	3.610	3.922
19	0.688	0.861	1.066	1.328	1.729	2.093	2.205	2.539	2.861	3.174	3.579	3.883
20	0.687	0.860	1.064	1.325	1.725	2.086	2.197	2.528	2.845	3.153	3.552	3.850
21	0.686	0.859	1.063	1.323	1.721	2.080	2.189	2.518	2.831	3.135	3.527	3.819
22	0.686	0.858	1.061	1.321	1.717	2.074	2.183	2.508	2.819	3.119	3.505	3.792
23	0.685	0.858	1.060	1.319	1.714	2.069	2.177	2.500	2.807	3.104	3.485	3.768
24	0.685	0.857	1.059	1.318	1.711	2.064	2.172	2.492	2.797	3.091	3.467	3.745
25	0.684	0.856	1.058	1.316	1.708	2.060	2.167	2.485	2.787	3.078	3.450	3.725
26	0.684	0.856	1.058	1.315	1.706	2.056	2.162	2.479	2.779	3.067	3.435	3.707
27	0.684	0.855	1.057	1.314	1.703	2.052	2.158	2.473	2.771	3.057	3.421	3.690
28	0.683	0.855	1.056	1.313	1.701	2.048	2.154	2.467	2.763	3.047	3.408	3.674
29	0.683	0.854	1.055	1.311	1.699	2.045	2.150	2.462	2.756	3.038	3.396	3.659
30	0.683	0.854	1.055	1.310	1.697	2.042	2.147	2.457	2.750	3.030	3.385	3.646
31	0.682	0.853	1.054	1.309	1.696	2.040	2.144	2.453	2.744	3.022	3.375	3.633
32	0.682	0.853	1.054	1.309	1.694	2.037	2.141	2.449	2.738	3.015	3.365	3.622
33	0.682	0.853	1.053	1.308	1.692	2.035	2.138	2.445	2.733	3.008	3.356	3.611
34	0.682	0.852	1.052	1.307	1.691	2.032	2.136	2.441	2.728	3.002	3.348	3.601
35	0.682	0.852	1.052	1.306	1.690	2.030	2.133	2.438	2.724	2.996	3.340	3.591
36	0.681	0.852	1.052	1.306	1.688	2.028	2.131	2.434	2.719	2.990	3.333	3.582
37	0.681	0.851	1.051	1.305	1.687	2.026	2.129	2.431	2.715	2.985	3.326	3.574
38	0.681	0.851	1.051	1.304	1.686	2.024	2.127	2.429	2.712	2.980	3.319	3.566
39	0.681	0.851	1.050	1.304	1.685	2.023	2.125	2.426	2.708	2.976	3.313	3.558
40	0.681	0.851	1.050	1.303	1.684	2.021	2.123	2.423	2.704	2.971	3.307	3.551
50	0.679	0.849	1.047	1.299	1.676	2.009	2.109	2.403	2.678	2.937	3.261	3.496
60	0.679	0.848	1.045	1.296	1.671	2.000	2.099	2.390	2.660	2.915	3.232	3.460
70	0.678	0.847	1.044	1.294	1.667	1.994	2.093	2.381	2.648	2.899	3.211	3.435
80	0.678	0.846	1.043	1.292	1.664	1.990	2.088	2.374	2.639	2.887	3.195	3.416
90	0.677	0.846	1.042	1.291	1.662	1.987	2.084	2.368	2.632	2.878	3.183	3.402
100	0.677	0.845	1.042	1.290	1.660	1.984	2.081	2.364	2.626	2.871	3.174	3.390
1000	0.675	0.842	1.037	1.282	1.646	1.962	2.056	2.330	2.581	2.813	3.098	3.300
z	0.674	0.842	1.036	1.282	1.645	1.960	2.054	2.326	2.576	2.807	3.090	3.291